Tolley's
Health and Safety
at Work

While every care has been taken to ensure the accuracy of this work, no responsibility for loss or damage occasioned to any person acting or refraining from action as a result of any statement in it can be accepted by the authors, editors or publishers.

Tolley's
Health and Safety
at Work

Tolley's Health and Safety at Work

LexisNexis® UK & Worldwide

United Kingdom RELX (UK) Limited trading as LexisNexis®, 1-3 Strand, London WC2N 5JR and 9-10 St Andrew Square, Edinburgh EH2 2AF

LNUK Global Partners LexisNexis® encompasses authoritative legal publishing brands dating back to the 19th century including: Butterworths® in the United Kingdom, Canada and the Asia-Pacific region; Les Editions du Juris Classeur in France; and Matthew Bender® worldwide. Details of LexisNexis® locations worldwide can be found at www.lexisnexis.com

First published in 1996

© 2019 RELX (UK) Limited.

Published by LexisNexis

This is a Tolley title

All rights reserved. No part of this publication may be reproduced in any material form (including photocopying or storing it in any medium by electronic means and whether or not transiently or incidentally to some other use of this publication) without the written permission of the copyright owner except in accordance with the provisions of the Copyright, Designs and Patents Act 1988 or under the terms of a licence issued by the Copyright Licensing Agency Ltd, 5th Floor, Shackleton House, 4 Battlebridge Lane, London, SE1 2HX. Applications for the copyright owner's written permission to reproduce any part of this publication should be addressed to the publisher.
Warning: The doing of an unauthorised act in relation to a copyright work may result in both a civil claim for damages and criminal prosecution.

Crown copyright material is reproduced with the permission of the Controller of HMSO and the Queen's Printer for Scotland. Parliamentary copyright material is reproduced with the permission of the Controller of Her Majesty's Stationery Office on behalf of Parliament. Any European material in this work which has been reproduced from EUR-lex, the official European Union legislation website, is European Union copyright.
A CIP Catalogue record for this book is available from the British Library.

ISBN for this volume: 9781474311298

Printed and bound by CPI Group (UK) Ltd, Croydon, CR0 4YY

Visit LexisNexis UK at www.lexisnexis.co.uk

Acknowledgements

Crown Copyright material is reproduced with the permission of the Controller of Her Majesty's Stationery Office.

Extracts from British Standards are reproduced with the kind permission of the British Standards Institution. Complete copies of the documents can be obtained from the British Standards Institution (BSI), 389 Chiswick High Road, London W4 4AL. Telephone: (020) 8996 9000.

Health and Safety at Work Editorial Board

The following are members of the appointed Health and Safety Editorial Board for the handbook. Each member brings with them a wealth of experience and knowledge of health and safety matters. The benefits to be gained from introducing this board are to ensure that this publication continues to provide practical, authoritative coverage of health and safety law in all aspects of industrial and office workplaces.

Lawrence Bamber is currently Managing Director of his own OSH/risk management consultancy, Risk Solutions International, which has been operational since 2001. He has worked in the OSH/ risk management field for over 40 years with General Accident, Stenhouse (Aon), as well as Norwich Union (Aviva), and is well known for his training, consultancy and presentation skills. Lawrence is a past president of the Institution of Occupational Safety and Health (IOSH) and is the author of many papers and publications on OSH risk management.

Nicola Coote is a founder member of PHSC plc and continues to work both operationally and strategically within the group. She was the first female Fellow of the Institution of Occupational Safety and Health (IOSH) in the South East, and worked on the Executive Committee of the South East Branch of IOSH for several years. She also sits on the peer review panel for safety practitioners applying for both Chartered and Fellowship status. Previously she has been an examiner for the National Examination Board in Occupational Safety and Health (NEBOSH). As a contributor and managing editor of many well-known safety publications, Nicola is a popular and authoritative speaker on a wide range of subjects.

Stephen Granger, CFIOSH, is a Consultant and Company Director of The Granger Partnership Ltd providing OSH consultancy and training management. He a former IOSH President and is currently a Director of The Center for Safety and Health Sustainability (CSHS), who have the objective of protecting people at work, as part of a sustainable world. Prior to starting his own business Stephen worked in health and safety for private and public employers, in business continuity, facilities management and lecturing in Higher and Further Education. He was the chairman of a UK working party developing standardised health and safety qualifications and is involved in developing a European competency framework. As President of IOSH Stephen represented the world's largest professional body for OSH in both national and international forums with business and regulatory leaders and minister level government officials.

Andrea Oates qualified and worked as an Environmental Health Officer (EHO) before taking up a career as a journalist in the trade union movement. She was a researcher at the independent research organisation Labour Research Department (LRD) for several years, writing on health and safety, environmental issues and other workplace issues from a trade union perspective. Since 2007 she has worked as a freelance journalist and researcher, writing feature articles on health and safety topics in health and safety, trade union and business-to-business publications. In addition to writing several LRD Booklets on health and safety issues, she is also the author of Tolley's Corporate Manslaughter and Homicide: A guide to compliance.

Mark Tyler MA, LL.M, CMIOSH is a Solicitor of the Senior Courts of England & Wales and a Chartered Safety and Health Practitioner. He is member of the IOSH Professional Ethics Committee, and is nominated as an independent panel member to the Health and Safety Executive's Fee for Intervention Dispute Panel. He has acted in a number of major cases which include the Southall and Ladbroke Grove Rail Inquiries, the Organophosphate group litigation, and the South Kensington Legionnaires' Disease outbreak. He has also defended numerous serious HSE and other regulatory prosecutions. Mark is also a co-author of the books Product Safety, Safer by Design and Tolley's Workplace Accident Handbook, and a consultant editor of Halsbury's Laws of England.

List of contributors or authors

Lawrence Bamber, BSc, DIS, CFIOSH, FIRM MASSE Managing Director, Risk Solutions International — refer to the Editorial Board for details

Kevin Chicken, Kevin gained executive engineering experience as a marine engineering officer when he served as Chief Engineer on both oil tankers and container ships were he had responsibility for all engineering aspects of the vessels he sailed on. He subsequently he worked as an engineer surveyor for a specialist engineering insurance inspection and certification company when he undertook statutory compliance thorough examinations for both new build and in service pressure plant. Other work-streams have included the delivery of client services through performing insurance risk surveys of their facilities and the preparation of detailed written reports for the UK engineering construction insurance market. Since 1996 Kevin provided health and safety consultancy, initially as a health and safety consultant working for Aviva Risk Management Solutions (formerly known as Norwich Union Risk Services). He was subsequently promoted to the role of Training and Consultancy Manager with responsibility for managing the delivery team as well as continuing to provide health and safety training and consultancy to a selection of key clients until 2012. He now heads up his own consultancy business KC Safety Solutions Ltd where he provides risk management solutions to a wide range of clients.

Nicola Coote, CFIOSH, MIIRSM, MCIPD, Sp Dip Env Man Executive Director, PHSC plc — refer to the Editorial Board for details

Neville Craddock, MA(Cantab), CSci, FIFST, Neville Craddock Associates — Neville Craddock spent his whole career in the food industry, including over 30 years practical experience of EU and UK food law development and compliance, and the management of food and food law-related consumer issues. Formerly Head of Regulatory Affairs for a major international food company, he now operates his own Consultancy and has undertaken projects and workshops for national and international governmental organisations, food businesses and trade associations. He is a former member of the UK Food Advisory Committee and UK Advisory Committee on Novel Foods and Processes (advising the UK government on food law, labelling and safety matters and Novel and GM foods, respectively) and is currently Vice-President of the Institute of Food Science and Technology (IFST) and a Council member of the European Food Law Association (UK).

Phil Grace, BSc (Hons), MRSC Chartered Chemist, ACII Chartered Insurer, CMIOSH, Casualty Risk Manager, Norwich Union Insurance — Phil Grace graduated with a degree in industrial chemistry in 1972 and then worked in the primary aluminium smelting sector for ten years. It was during this time that he developed his interest in health, safety and occupational hygiene. In 1985 Phil moved to the insurance industry to work as a surveyor/consultant, advising policyholders on how to improve their risk management, achieve legal compliance and reduce the risk of civil claims. After some years he moved into a more strategic post working alongside underwriters, working on risk assessment and embedding risk management thinking into the underwriting process. He is currently employed by a leading UK insurer and is both a Chartered Chemist and a Chartered Insurer, a Chartered Member of IOSH and a member of the British Occupational Hygiene Society

Steve Granger, CFIOSH Immediate Past President of the Institution of Occupational Safety and Health, Director and Consultant — refer to the Editorial Board for details

Alexander Green, M.Theol (Hons), LL.B, LL.M, M.Litt, FSAScot — Alexander Green is the founder and Managing Director of the Law Agency, a boutique firm of solicitors specialising in health & safety law and employment law. He is an Affiliate Member of IOSH. Prior to setting up the Law Agency in 2006, Alexander was a partner with a major international law firm. He is a solicitor of the Senior Courts of England & Wales and an enrolled solicitor in Scotland. He has over 20 years' of experience and has acted in high profile industrial accidents in the offshore oil and gas industry and the aviation industry.

Robert Greenfield, DipOSH Grad IOSH MBIFM — Robert has over 25 years experience in FM with the last 14 specialising in Safety Health Environmental & Quality (SHEQ) having started his career as an engineering apprentice within the Atomic Weapons Research Establishment. Robert has worked in a number of building managing agent companies as well as within the hard FM industry where he has a wealth of experience in the development of Safety, Health, Environmental and Quality (SHEQ) operational strategy for large multi site portfolios. Robert is the past Deputy Chairman of the British Institute of Facilities Management (BIFM), and the current Chairman of the BIFM Health & Safety Specialist Interest Group.

Nick Humphreys, LLB, LLM, Solicitor, Barrister (Non-practising), Partner, Bircham Dyson Bell — Nick Humphreys is the London Head of Employment at Hill Dickinson LLP. Nick's advises a number of major global corporations in relation to their employment law and industrial relations concerns.

Gam Jhutti, CMIOSH, Senior Health, Safety and Environment Adviser — Gam Jhutti is a chartered member of IOSH and has expertise in creating and implementing management systems and in-depth experience of ISO 9001, 18001, 14001 and 45001. Gam advises on wellbeing, health, safety, environment and quality and her areas of practice are wellbeing, health, safety, environment, quality, auditing, culture and behaviour and management systems.

Subash Ludhra, BSc (Hons), CFIOSH, OSHCR, EnvDipNEBOSH, JP Managing Director, Anntara Management Ltd — Subash Ludra is a qualified Occupational Hygienist, now specialising in the subject of Risk Management and Loss Control. He has a wealth of experience gained from industry and commerce, both in the UK and overseas. In 2002 Subash Established Anntara Management Ltd, a Risk management and Loss control consultancy and has since provided services to several blue chip companies (both in the UK and overseas) as well as many SMEs. Subash is a chartered safety and health practitioner and experienced trainer and is a contributor to several health and safety publications and has been involved with the production of several Health and Safety documents for Trade Associations and the Health and Safety Executive.

Lynda Macdonald, MA, FCIPD, LLM Employment Law and Management Training Consultant — Lynda Macdonald has worked for over 20 years as a freelance management trainer, advisor and writer specialising in employment law, and she regularly designs and conducts training courses on all aspects of employment law. She also sits as a panel member for the Employment Tribunals service in Aberdeen, where she lives. Previously, Lynda worked as an HR Manager in the oil industry for over ten years. She is a university graduate in language, a Chartered Fellow of the Chartered Institute of Personnel and Development and has a masters degree in employment law and practice. Lynda has written sixteen books on various aspects of employment law, and co-authored several others, and she currently contributes extensively to various HR and employment law hard-copy and on-line products.

Alison Newstead, LLB, Partner, Shook, Hardy & Bacon International LLP — Alison practices in the defence of health and safety investigations and prosecutions. She has represented clients at inquests and in criminal proceedings resulting from Police, HSE/and EHO investigations. She also regularly advises on the defence of civil proceedings arising from such action. Alison is a member of the Law Society of England and Wales, the British Insurance Law Association, the Defence Research Institute (DRI), the Institution of Occupational Safety and Health (IOSH), the Health and Safety Lawyers Association (HSLA), and sits on the CBI Health and Safety Panel.

Andrea Oates, BSc Freelance writer on health and safety issues — refer to the Editorial Board for details

Mark Rutter, BSc, PhD, Environment Science, Law and Policy Writer and Consultant — Mark Rutter currently writes and advises on the implications of new environmental policy and laws. He has previously worked for the law firm CMS Cameron McKenna, monitoring and analyzing new developments in this area, both for in-house lawyers and external clients. With a PhD in Environmental Science and experience working as a research scientist, Mark also has an excellent understanding of the scientific and technical issues frequently encountered in the field of environmental law.

Mark Tyler — refer to the Editorial Board for details

Ian Wallace — BSc (HONS), Dip SM, CFIOSH, FLLA, HSE Consultant

Contents

List of Illustrations	xxxvii
List of Abbreviations	xli
Introduction	**INT-1**
Lawrence Bamber and Andrea Oates	
The Handbook	IN01
Drivers for health and safety performance	IN03
The changing world of work	IN06
Legal framework for health and safety at work	IN07
The structure of UK health and safety law	IN17
Management of health and safety at work	IN23
Access, Traffic Routes and Vehicles	**A10**
Nicola Coote	
Introduction to access, traffic routes and vehicles	A1001
Statutory duties concerning workplace access and egress – Workplace (Health, Safety and Welfare) Regulations 1992 (SI 1992 No 3004)	A1002
Vehicles	A1008
Common law duty of care	A1021
Confined spaces	A1022
Accident Reporting and Investigation	**A30**
Gam Jhutti	
Introduction to accident reporting and investigation	A3001
Reporting of Injuries, Diseases and Dangerous Occurrences Regulations 2013 (RIDDOR) – (SI 2013 No 1471)	A3002
Persons responsible for notification and reporting	A3003
What is covered?	A3004
Fatalities	A3005
Specified injuries to workers	A3006
Injuries incapacitating a worker for more than seven consecutive days	A3007
Injuries to non-workers	A3008
Road accidents	A3009
Duty to report gas incidents	A3010
Some problem areas	A3012

Contents

Dangerous occurrences	A3013
Occupational diseases	A3014
Reporting procedures	A3015
Records and record-keeping	A3017
Action to be taken by employers and others when accidents occur at work	A3019
Exemptions from RIDDOR reporting	A3020
Penalties	A3021
Obligations on employees	A3023
Objectives of accident reporting	A3025
Duty of disclosure of accident data	A3026
Accident Investigation	A3029
Further guidance on accident reporting and investigation	A3034
List of Reportable Dangerous Occurrences	A3035
Asbestos	**A50**
Andrea Oates	
Introduction to asbestos	A5001
Historical exposures to asbestos – manufacturing	A5003
Asbestos prohibitions	A5004
Asbestos removal	A5005
Exposure of building maintenance workers	A5006
The Control of Asbestos Regulations 2012	A5007
Duties under the Control of Asbestos Regulations 2012	A5007.1
Asbestos building surveys	A5023
Business Continuity	**B80**
Lawrence Bamber	
What is business continuity planning?	B8001
Business continuity planning in context	B8002
Why is business continuity planning of importance?	B8003
Business continuity drivers	B8004
Convincing the board – the business case	B8011
The business continuity model	B8012
Crisis communication and public relations	B8039
Civil Contingencies Act, 2004	B8044

Compensation for Work Injuries/Diseases — C60

Andrea Oates

Introduction to compensation for work injuries/diseases	C6001
Industrial injuries scheme	C6002
Accident and personal injury provisions	C6003
Industrial injuries disablement benefit	C6011
Constant attendance allowance and exceptionally severe disablement allowance	C6014
Other sickness/disability benefits	C6015
Disability Element of Working Tax Credit	C6016
Damages for occupational injuries and diseases	C6021
Awards of damages and recovery of state benefits	C6044
Complications	C6046
Appeals against certificates of recoverable benefits	C6050

Community Health and Safety — C65

Mike Bateman and Andrea Oates

Introduction to community health and safety	C6501
Relevant legal requirements relating to community health and safety	C6502
Neighbours and Passers-by	C6509
Trespassers on Work Premises	C6512
Customers and Users of Facilities and Services	C6515
Visiting Individuals and Groups	C6518
Volunteers	C6521
Major Public Events	C6524
Use of Contractors	C6527
Further information on community health and safety	C6528

Confined Spaces — C70

Angus Withington

Introduction to confined spaces	C7001
Relevant legal requirements relating to work in confined spaces	C7002
Confined Spaces Regulations 1997	C7003
Other risks in confined spaces	C7008
Safe systems of work	C7009
Precautions for work in confined spaces	C7014
Emergency arrangements	C7029

Contents

Construction and Building Operations — C80
Andrea Oates

Introduction to construction and building operations	C8001
Part 4, Construction (Design And Management) Regulations 2015	C8002
Typical issues occurring in building and construction projects	C8018

Construction, Design and Management — C85
Andrea Oates

Introduction to construction, design and management	C8501
Information in the Schedules	C8518
The Corporate Manslaughter and Corporate Homicide Act 2007	C8520

Corporate Manslaughter — C90
Andrea Oates

Background to the Corporate Manslaughter and Corporate Homicide Act 2007	C9001
Corporate manslaughter: the offence	C9002
The meaning of 'relevant duty of care'	C9003
Public policy decisions, exclusively public functions and statutory inspections	C9004
Military activities	C9005
Policing and law enforcement	C9006
Emergencies and the emergency services	C9007
Child protection and probation functions	C9008
Factors for the jury to consider when deciding if there has been a 'gross' breach	C9009
Penalties and sentencing	C9010
Fines	C9011
Remedial Orders	C9012
Publicity Orders	C9013
Application to Crown bodies	C9014
Application to the armed forces	C9015
Application to police forces	C9016
Application to partnerships	C9017
Procedure, evidence and sentencing	C9018
Transfer of functions	C9019
Director of Public Prosecutions (DPP) consent for proceedings	C9020
No individual liability for corporate manslaughter or corporate homicide	C9021

Contents

Convictions under the Act and under health and safety legislation	C9022
Abolition of liability of corporations for manslaughter at common law	C9023
Extent and territorial application	C9024
Schedule 1: List of government departments etc	C9025
Dangerous Goods – Carriage	**D04**
John Wintle	
Introduction to dangerous goods – carriage	D0401
General requirements	D0405
Exemptions, revocations and radioactive materials	D0439
Acknowledgement	D0448
Further Information	D0449
Dangerous Substances and Explosive Atmospheres	**D09**
John Wintle and William Delamare	
Introduction to dangerous substances and explosive atmospheres	D0901
The Regulations	D0904
Risk assessment	D0912
Management of risks	D0915
Classification of places with explosive atmospheres	D0920
Equipment and protection systems	D0924
Specific products and areas	D0930
Arrangements to deal with incidents and emergencies	D0934
Provision of information, instruction and training	D0935
Identification of hazardous contents of pipes and containers	D0936
Exemption	D0937
Amendments to modernise petroleum and other legislation	D0938
Repeals and revocations of existing legislation	D0939
Enforcement	D0940
Regulatory impact assessment	D0941
Approved codes of practice and guidance etc.	D0942
Other information	D0949
Case histories	D0951
Sources and acknowledgement	D0952
Disaster and Emergency Management Systems (DEMS)	**D60**
Roger Bentley	
Historical development of disaster and emergency management	D6001

Contents

Origin of disaster and emergency management as a modern discipline	**D6002**
Definitions relating to disasters and emergencies	**D6003**
The need for effective disaster and emergency management	**D6004**
Disaster and emergency management systems (DEMS)	**D6034**
External and internal factors	**D6035**
Establish a disaster and emergency policy	**D6038**
Organise for disasters and emergencies	**D6043**
Disaster and emergency planning	**D6048**
Monitor the disaster and emergency plan(s)	**D6066**
Audit and review	**D6067**
Conclusion to disaster and emergency management systems (DEMS)	**D6070**
Sources of information on disaster and emergency management systems (DEMS)	**D6071**
Display Screen Equipment	**D82**
Andrea Oates	
Introduction to display screen equipment	**D8201**
The Regulations summarised	**D8202**
DSE workstation assessments	**D8210**
Guidance on completing the DSE workstation self-assessment	**D8217**
Special situations concerning display screen equipment	**D8218**
After the assessment	**D8222**
Electricity	**E30**
Chris Buck and Andrea Oates	
Introduction to electricity	**E3001**
Understanding electricity	**E3002**
Legal background and standards	**E3003**
Dangers of electricity	**E3004**
Fundamentals of controlling electrical risks	**E3010**
Competence	**E3011**
Design and construction of electrical equipment and systems	**E3012**
Maintenance	**E3021**
Safe systems of work	**E3022**
Special situations concerning electricity	**E3025**
Other relevant legislation on electricity	**E3029**

Emissions into the Atmosphere — E50
Andrea Oates

Introduction to emissions into the atmosphere	**E5001**
Regulatory drivers	**E5002**
Regional and global issues	**E5021**
International and European agreements	**E5026**
UK regulations relating to emissions into the atmosphere	**E5041**

Employers' Liability Insurance — E130
Phil Grace

Introduction to employers' liability insurance	**E13001**
Purpose of compulsory employers' liability insurance	**E13002**
General law relating to insurance contracts	**E13003**
Subrogation	**E13007**
Duty of employer to take out and maintain insurance	**E13008**
Issue, display and retention of certificates of insurance	**E13013**
Penalties	**E13014**
Cover provided by a typical policy	**E13016**
'Prohibition' of certain conditions	**E13020**
Trade endorsements for certain types of work	**E13021**
Insolvent employers' liability insurers	**E13024**
Limitation Act	**E13025**
Long Tail Disease	**E13026**
Vicarious Liability	**E13027**
Making the Market Work	**E13028**
Woolf Reforms	**E13029**
Benchmarking and performance indicators	**E13030**
NHS Recoveries	**E13031**
Corporate Manslaughter	**E13032**
Compensation Act 2006	**E13033**
Employers' Liability Tracing Office	**E13034**
Lord Young Report	**E13035**
Lofstedt Review	**E13036**
Fees for Intervention (FFI)	**E13037**

Contents

Employment Protection — E140
Andrea Oates

Introduction to employment protection — **E14001**
Sources of contractual terms — **E14002**
Employee employment protection rights — **E14009**
Enforcement of safety rules by the employer — **E14012**
Employment protection: dismissal — **E14017**
Health and safety duties in relation to women at work — **E14023**

Enforcement — E150
Malcolm Galloway and Alice Jarratt

Introduction to enforcement — **E15001**
The role of the Health and Safety Executive — **E15005**
Fee for Intervention Cost Recovery Scheme — **E15005.1**
Enforcing authorities — **E15007**
'Relevant statutory provisions' — **E15012**
National Enforcement Code — **E15012.1**
Part A: Enforcement Powers of Inspectors — **E15013**
Part B: Offences and Penalties — **E15030**

Environmental Management — E160
Mark Rutter

Introduction to environmental management — **E16001**
Internal drivers — **E16002**
External drivers — **E16005**
Environmental management guidelines — **E16013**
Implementing environmental management — **E16017**
Benefits of environmental management — **E16026**
The future — **E16027**

Equality Act 2010 — E165
Alexander M S Green

Introduction to the Equality Act 2010 — **E16501**
The HSE stance on equality and diversity — **E16502**
Prohibited conduct — **E16503**
Health and Safety for Disabled People — **E16509**
Health and Safety in Relation to Age — **E16513**
Health and Safety in Relation to Race — **E16517**

Health and Safety in Relation to Gender	E16520
Remedies and Enforcement in Relation to Unlawful Discrimination	E16522
Ergonomics	**E170**
Margaret Hanson, Jill Cleaver and Andrea Oates	
Introduction to ergonomics	E17001
The ergonomic approach	E17004
Designing for people	E17005
Designing the task	E17010
Workstation design	E17018
Physical work environment	E17030
Job design and work organisation	E17036
Musculoskeletal disorders	E17044
Ergonomics tools	E17051
Reducing error and influencing behaviour	E17059A
Relevant legislation and guidance for ergonomics	E17060
Europe – Health and Safety	**E180**
Steve Granger	
Introduction to Europe — health and safety	E18001
Working in the European Economic Area	E18002
Law and Europe	E18003
The global influence of European legal systems	E18010
Providing employers with OSH advice in the EEA	E18011
Corporate accountability – who is listening to the OSH practitioner?	E18012
Worker participation	E18013
What is a regulated profession?	E18014
Identifying countries where the OSH profession is regulated	E18015
State regulated OSH professionals	E18017
Development of a European Vocational Occupational Standard	E18018
Monitoring OSH performance standards	E18020
European organisations and resources	E18022
Working essentials	E18023
Facilities Management – An Overview	**F30**
Robert Greenfield	
Introduction to facilities management	F3001
The definition of facilities management	F3002

Contents

Facilities management	**F3003**
Hard and soft facilities management	**F3004**
Managing	**F3005**
Company approach to facilities management	**F3006**
In-house or outsourced facilities management?	**F3008**
Advantages to the company	**F3009**
Responsibilities of facilities management	**F3011**
Four most common health and safety issues for facilities management	**F3024**
How facilities management and building design can help a company	**F3028**
Recognition of facilities management	**F3029**
Innovations within the facilities management industry	**F3030**
The effects of safety, health and environmental legislation on facilities management	**F3031**
Summary of facilities management	**F3035**
Fire Prevention and Control	**F50**
Adair Lewis (updated by Andrea Oates)	
Introduction	**F5001**
Elements of fire	**F5001.1**
Fire statistics	**F5002**
Fire classification	**F5003**
Electrical fires	**F5004**
Fire extinction – active fire protection measures	**F5005**
Passive and active fire protection	**F5007**
Fire procedures and portable equipment	**F5008**
Good 'housekeeping'	**F5018**
Pre-planning of fire prevention	**F5019**
Fire drills	**F5022**
Fire and fire precautions – legislation	**F5030**
Building regulations	**F5030.1**
The independent review of building regulations and fire safety	**F5030.2**
Fire Safety Legislation	**F5031**
Fire certification	**F5032**
The Regulatory Reform (Fire Safety) Order 2005	**F5033**
Fire risk assessment	**F5037**
Enforcement	**F5056**

Offences and appeals	**F5060**
Appeals against enforcement notices	**F5060.1**
Application to the Crown	**F5061**
Further guidance on specific types of premises	**F5061.1**
The dangerous substances and explosive atmospheres regulations (DSEAR) 2002	**F5062**
Liability of occupier	**F5073**
Fire insurance	**F5074**
First-Aid	**F70**

Subash Ludhra

Introduction to first-aid	**F7001**
The employer's duty to make provision for first-aid	**F7002**
The assessment of first-aid needs	**F7003**
Duty of the employer to inform employees of first-aid arrangements	**F7014**
First-aid and the self-employed	**F7015**
Number of first-aiders	**F7016**
Appointed persons	**F7019**
Records and record keeping	**F7020**
First-aid resources	**F7021**
Rooms designated as first-aid areas	**F7025**
Food Safety and Standards	**F90**

Neville Craddock and Andrea Oates

Introduction	**F9001**
Food Safety Act 1990	**F9006**
Food premises regulations	**F9026**
Food hygiene regulations	**F9034**
Food Safety Manual – food safety policy and procedures	**F9070**
Gas Safety	**G10**

Andrea Oates

Introduction to gas safety	**G1001**
Gas Supply	**G1002**
Gas supply management – the Gas Safety (Management) Regulations 1996 (SI 1996 No 551)	**G1003**
Rights of entry – the Gas Safety (Rights of Entry) Regulations 1996 (SI 1996 No 2535)	**G1006**
Pipelines – the Pipelines Safety Regulations 1996 (SI 1996 No 825)	**G1009**

Contents

Gas systems and appliances	**G1010**
HSE Guidance on gas safety	**G1025C**
Interface with other legislation	**G1026**
Recent Significant Gas Safety Prosecutions	**G1026A**
Harassment in the Workplace	**H17**

Nick Humphreys (Partner, Penningtons Manches Cooper LLP)

Introduction to harassment in the workplace	**H1701**
Elements of a harassment claim	**H1703**
Limitations to the definition of harassment	**H1709**
Employer's liability for harassment	**H1710**
Legal action and remedies for harassment	**H1716**
Preventing and handling claims of harassment	**H1734**
Hazardous Substances in the Workplace	**H21**

Andrea Oates

Introduction to hazardous substances in the workplace	**H2101**
How hazardous substances harm the body	**H2102**
The COSHH Regulations	**H2104**
Carrying out COSHH risk assessments	**H2106**
Practical aspects of COSHH assessments	**H2114**
COSHH assessment records	**H2122**
Further requirements of the COSHH Regulations	**H2126**
Other important regulations on hazardous substances in the workplace	**H2136**
International Health and Safety	**I20**

Steve Granger

Introduction to international health and safety	**I2001**
Health and safety in the USA	**I2003**
Health and safety in Canada	**I2010**
Joint Consultation in Safety – Safety Representatives, Safety Committees, Collective Agreements and Works Councils	**J30**

Nick Humphreys, Partner, Penningtons Manches Cooper LLP

Regulatory framework for joint consultation in safety	**J3001**
Regulatory framework for joint consultation in safety	**J3002**
Consultation obligations for unionised employers	**J3003**
Safety committees	**J3017**
Non-unionised workforce – consultation obligations	**J3020**

Recourse for safety representatives	**J3030**
European developments in health and safety	**J3031**
Lifting Operations	**L30**
Kevin Chicken	
Introduction to lifting operations	**L3001**
Statutory requirements relating to the manufacture of lifting equipment	**L3002**
Lifts Regulations 1997	**L3003**
Lifting operations and equipment failure	**L3028**
Hoists and lifts	**L3032**
Fork lift trucks	**L3034**
Patient/bath hoist	**L3036**
Vehicle lifting table	**L3037**
Lifting accessories	**L3038**
Shackles	**L3048**
Once only use accessories	**L3050**
Mobile lifting equipment	**L3051**
Further information regarding lifting operations	**L3052**
Lighting	**L50**
Nicola Coote	
Introduction to lighting	**L5001**
Statutory lighting requirements	**L5002**
Sources of light	**L5007**
Standards of lighting or illuminance	**L5010**
Qualitative aspects of lighting and lighting design	**L5015**
Machinery Safety	**M10**
Andrea Oates	
Introduction to machinery safety	**M1001**
Legal requirements relating to machinery	**M1002**
Machinery safety – the risk based – approach	**M1003**
The Provision and Use of Work Equipment Regulations 1998 (PUWER'98)	**M1004**
The Supply of Machinery (Safety) Regulations 2008 (as amended)	**M1009**
The Transposed Harmonised European Machinery Safety Standards	**M1019**
Machinery risk assessment	**M1022**
Machinery hazards identification	**M1025**

Contents

Options for machinery risk reduction	**M1026**
Types of safeguards and safety devices	**M1027**
Fixed guards	**M1029**
Safety devices	**M1036**
Major Accident Hazards	**M11**
Andrea Oates	
Background to major accident hazards	**M1101**
Land-use planning	**M1135**
Sources of information on major accident hazards	**M1141**
Managing Absence	**M15**
Lynda Macdonald	
Introduction	**M1501**
Managing long-term sickness absence	**M1502**
Managing short-term absences	**M1518**
Managing Health and Safety	**M20**
Abigail Cohen	
Introduction to managing health and safety	**M2001**
Legal requirements	**M2002**
The Management Regulations	**M2005**
Construction (Design and Management) Regulations 2007	**M2022**
Health and safety management systems	**M2029**
Policy	**M2031**
Organisation	**M2033**
Planning	**M2038**
Implementation	**M2040**
Monitoring	**M2041**
Audit	**M2042**
Review	**M2043**
Continual improvement	**M2044**
POPIMAR in action	**M2045**
References	**M2046**
Managing Work-related Road Safety	**M21**
Roger Bibbings and Andrea Oates	
Overview of managing work-related road safety	**M2101**
Managing work-related road safety – employers' duties	**M2102**

Legal responsibilities	M2103
The business case for action	M2104
Extending health and safety management systems to Cover Work-Related Road Safety (WRRS) hse management systems to cover WRRS	M2105
Assessing risks on the road	M2111
Road risk control measures	M2112
Monitoring and evaluation	M2117
Where is the organisation now?	M2123
Useful publications	M2125
Useful websites	M2126
Manual Handling	**M30**
Andrea Oates	
Introduction to manual handling	M3001
What the Regulations require	M3002
Assessing the risk of injury from manual handling	M3003
Avoiding or reducing risks	M3008
Planning and preparation	M3016
Consulting and involving workers	M3020
After the assessment	M3024
Noise at Work	**N30**
Andrea Oates	
Introduction to noise at work	N3001
Scale of the problem	N3002
Overview of the Control of Noise at Work Regulations 2005	N3002A
Hearing damage	N3003
The perception of sound	N3004
Noise indices	N3005
General legal requirements relating to noise at work	N3012
Specific legal requirements regarding noise at work	N3015
Compensation for occupational deafness	N3018
Action against employer at common law	N3022
General guidance on noise at work	N3023
Reducing noise in specific working environments	N3024
Workplace noise assessments	N3029
Noise reduction	N3035

Contents

Hearing protection	**N3036**
Occupational Health and Diseases – An Overview	**O10**
Leslie Hawkins and Andrea Oates	
What is occupational health?	01001
Why is it important?	01001.1
Relationships between occupational health and safety	01006
Prevention is better than cure	01007
The occupational health team	01010
Closure of the fit for work service	01015.1
GP fit notes	01015.2
Health records	01016
Buying services and priorities	01019
Occupational diseases and disorders	01020
Musculoskeletal disorders	01020.1
Psychological disorders (stress, anxiety and depression)	01020.6
Hand arm vibration syndrome (HAVS) and whole body vibration	01027
Diseases and disorders of the eye	01032
Diseases of the skin	01037
Occupational respiratory diseases	01042
Other occupational health concerns	01056
Occupiers' Liability	**O30**
Alison Newstead	
Introduction to occupiers' liability	03001
Duties owed under the Occupiers' Liability Act 1957	03002
Duty owed to trespassers, at common law and under the Occupiers' Liability Act 1984	03008
Dangers to guard against	03012
Waiver of duty and the unfair contract terms act 1977 (ucta 1977)	03016
Risks willingly accepted	03017
Actions against factory occupiers	03018
Occupier's duties under HSWA 1974	03019
Offshore Operations	**O70**
Fred Osliff, Ian Wallace and Andrea Oates	
Introduction to offshore operations	07001
Offshore Installations (Offshore Safety Directive (Safety Case Etc)	

Regulations 2005 (SI 2005 No 3117)	**07009.1**
Offshore Installations (Safety Case) Regulations 2005 (SI 2005 No 3117)	**07010**
Offshore Installations and Wells (Design and Construction, etc.) Regulations 1996 (SI 1996 No 913)	**07024**
Offshore Installations (Prevention of Fire and Explosion, and Emergency Response) Regulations 1995 (SI 1995 No 743)	**07028**
Offshore Installations and Pipeline Works (Management and Administration) Regulations 1995 (SI 1995 No 738)	**07029**
Construction and use of submarine pipelines	**07034**
Pipelines Safety Regulations 1996 (SI 1996 No 825)	**07035**
The Diving at Work Regulations 1997	**07036**
Other offshore specific legislation	**07039**
Offshore Installations (Safety Representatives and Safety Committees) Regulations 1989	**07046**
First-aid – the Offshore Installations and Pipeline Works (First-Aid) Regulations 1989	**07056.1**
The Offshore Electricity and Noise Regulations 1997	**07063**
Legislation for other hazardous offshore activities	**07064**
Written procedures and training programmes for dealing with hazards	**07069**
Hazards in offshore activities	**07070**
Personal Protective Equipment	**P30**
Nicola Coote	
Introduction to personal protective equipment	**P3001**
The Personal Protective Equipment at Work Regulations 1992	**P3002**
Work activities/processes requiring personal protective equipment	**P3003**
Statutory requirements in connection with personal protective equipment	**P3004**
Increased importance of uniform European standards	**P3009**
Main types of personal protection	**P3019**
Common law requirements	**P3026**
Pressure Systems	**P70**
Kevin Chicken	
Introduction to pressure systems	**P7001**
Simple pressure vessels	**P7002**
The Simple Pressure Vessels (Safety) Regulations 2016 (SI 2016 No 1092)	**P7003**
EC Certificate of Adequacy	**P7004**

Contents

EC Type-Examination Certificate	**P7005**
The Pressure Equipment (Safety) Regulations 2016 (SI 2016 No 1105)	**P7006**
Essential safety requirements	**P7007**
Conformity assessment procedure	**P7008**
Notified bodies	**P7009**
Pressure Systems Safety Regulations 2000	**P7011**
Systems and equipment covered by the regulations	**P7012**
Systems and equipment excepted from some or all the regulations	**P7013**
Approved code of practice and guidance	**P7014**
Interpretation	**P7015**
Design and construction	**P7016**
The provision of information and marking	**P7017**
Competent persons	**P7018**
Safe operating limits	**P7018A**
Written scheme of examination	**P7019**
Operations, maintenance, modification and repair	**P7026**
Record keeping	**P7030**
Related health and safety legislation	**P7031**
Product Safety	**P90**
Alison Newstead	
Introduction to product safety	**P9001**
Consumer products	**P9002**
Industrial products	**P9009**
Regulations made under the Health and Safety at Work etc Act 1974	**P9010**
Criminal liability for breach of statutory duties	**P9011**
Civil liability for unsafe products – historical background	**P9024**
Consumer Protection Act 1987	**P9027**
Contractual liability for sub-standard products	**P9047**
Unfair Terms	**P9058**
Public & Products Liability Insurance	**P95**
Phil Grace	
Introduction to public & products liability insurance	**P9501**
General Law and Practice Relating to Liability Insurance Policies	**P9502**
Public Liability Insurance	**P9514**
Products Liability Insurance	**P9530**

General matters relating to public and products insurance	P9546
Radiation	**R10**
Andrea Oates	
Non-ionising and ionising radiation	R1001
Rehabilitation	**R20**
Andrea Oates	
What is Rehabilitation?	R2001
Why should we be concerned?	R2002
Managing sickness absence	R2007
Long-term absence and rehabilitation	R2009
Using professional advice	R2010
REACH – Registration, Evaluation and Authorisation of Chemicals in Europe	**R25**
John Wintle and Andrea Oates	
Introduction to REACH	R2501
What does reach aim to achieve?	R2502
The need for REACH	R2503
Chemicals covered and exempted	R2504
Duty Holders within REACH	R2505
Preparation for REACH registration 2018	R2506
The Registration process	R2507
Timescale for Registration	R2508
Information required for Registration	R2509
Safety Data Sheets and downstream use	R2510
European Chemicals Agency	R2511
Evaluation	R2512
Authorisation	R2513
Restrictions	R2514
UK Competent Authority and Helpdesk	R2515
Enforcement	R2516
Further information	R2517
Acknowledgement	R2518
Risk Assessment	**R30**
Andrea Oates	
Introduction to risk assessment	R3001
HSWA 1974 requirements	R3002

Contents

The Management of Health and Safety at Work Regulations 1999	**R3005**
Common regulations requiring risk assessment	**R3006**
Specialist regulations requiring risk assessment	**R3016**
Related health and safety concepts	**R3017**
Management Regulations requirements	**R3018**
Children and young people	**R3025**
New or expectant mothers	**R3026**
Other vulnerable persons	**R3028**
Who should carry out the risk assessment?	**R3029**
Assessment units	**R3030**
Relevant sources of information	**R3031**
Consider who might be at risk	**R3032**
Identify the issues to be addressed	**R3033**
Variations in work practices	**R3034**
Checklist of possible risks	**R3035**
Carrying out the risk assessment	**R3036**
Assessment records	**R3042**
After the assessment	**R3044**
Content of assessment records	**R3048**
Sensible risk management	**R3052**
References	**R3053**
Safe Systems of Work	**S30**
Lawrence Bamber	
Introduction to safe systems of work	**S3001**
Components of a safe system of work	**S3002**
Which type of safe system of work is appropriate for the level of risk?	**S3007**
Development of safe systems of work	**S3008**
Permit to work systems	**S3011**
Isolation procedures	**S3023**
Further reading	**S3024**
Statements of Health and Safety Policy	**S70**
Andrea Oates	
Introduction to statements of health and safety policy	**S7001**
The legal requirements of statements of health and safety policy	**S7002**
Content of the policy statement	**S7003**

Stress at Work	**S110**
Nicola Coote	
Introduction to stress at work	S11001
Definition of stress	S11002
Understanding the stress response	S11003
Why manage stress?	S11004
Occupational stress risk factors	S11010
The work	S11016
Company structure and job organisation	S11025
Personal factors	S11032
Stress management techniques	S11036
Training and Competence in Occupational Safety and Health	**T70**
Steve Granger	
Introduction to training and competence in occupational safety and health	T7001
What is competence?	T7002
The education framework	T7007
Assessment	T7014
Achieving and maintaining competence in safety practice	T7016
Continuing Professional development (CPD)	T7021
Working in the global market	T7022
Conclusions	T7023
Ventilation	**V30**
Andrea Oates	
Introduction to ventilation	V3001
Sources of contamination	V3002
Local exhaust ventilation (LEV)	V3003
Legal requirements relating to ventilation	V3015
Ventilation: summary	V3016
Vibration	**V50**
Andrea Oates	
Vibration: the scope of this chapter	V5001
An introduction to the effects of vibration on people	V5002
Vibration measurement	V5007
Causes and effects of hand-arm vibration exposure	V5010

xxxiii

Contents

CTS and HAV as reportable occupational diseases	V5011
The Control of Vibration at Work Regulations, 2005 (SI 2005 No 1093)	V5015
Advice and obligations in respect of hand-arm vibration	V5017
Personal protection against hand-arm vibration	V5019
Advice and obligations in respect of whole-body vibration	V5023
HSE guidance on hand-arm and whole-body vibration	V5029
Violence in the Workplace	**V80**
Andrea Oates	
Introduction to violence in the workplace	V8001
The nature and extent of the problem of workplace violence	V8002
Work-related violence – the legislation	V8004
Developing and implementing policy	V8016
Risk assessment for workplace violence	V8044
Reporting, recording and monitoring system for workplace violence	V8057
Risk reduction measures for workplace violence	V8072
Training	V8096
Incident management	V8116
Post-incident management	V8129
Sources of further information	V8142
Vulnerable Persons	**V120**
Andrea Oates	
Introduction to vulnerable persons	V12001
Relevant legislation	V12002
Children and young persons	V12006
New or expectant mothers	V12017
Lone workers	V12027
Disabled people	V12032
Inexperienced workers	V12038
Work at Height	**W90**
Andrea Oates	
Introduction to work at height	W9001
Commencement and scope	W9003
Who do the regulations apply to?	W9004
Organisation and planning, competence and avoidance of risks	W9005
Work equipment	W9009

Fragile surfaces	**W9010**
Falling objects and danger areas	**W9011**
Inspection for work at height	**W9012**
Duties of persons at work	**W9013**
Protection from falls	**W9014**
APPG report on working at height	**W9014.1**
Further information concerning work at height	**W9015**
Working Time	**W100**

Mark Greaves, Pupil Barrister at Old Square Chambers and Nicola Coote, Personnel Health & Safety Consultants Ltd

Introduction to working time	**W10001**
Definitions	**W10002**
Maximum weekly working time	**W10010**
Rest periods and breaks	**W10014**
Night work	**W10019**
Annual leave	**W10022**
Records	**W10025**
Excluded sectors	**W10026**
Enforcement	**W10038**
Workplaces – Health, Safety and Welfare	**W110**

Andrea Oates

Introduction to workplaces – health, safety and welfare	**W11001**
Workplace (Health, Safety and Welfare) Regulations 1992	**W11002**
Safety signs at work – Health and Safety (Safety Signs and Signals) Regulations 1996 (SI 1996 No 341)	**W11030**
Equality Act 2010	**W11043**
Building Regulations	**W11044**
Factories Act 1961	**W11045**
Table of Cases	**TC-1**
Table of Statutes	**TS-1**
Table of Statutory Instruments	**TSI-1**
Index	**I_{ND}-1**

List of illustrations

Plan-Do-Check-Act	Int22
Action to be taken by employers and others when accidents occur at work	A3019
Internal incident notification form	A3025
Elements of the investigation process	A3030
Asbestos Management Plan	A5012
Confined spaces – identification of risks	C7010
Permit to work – entry into confined spaces	C7010
Confined spaces – control measures and other precautions	C7010
UN classification system	D0407
ADR Approved List	D0411
Example of UN mark	D0415
Proper Shipping Name (PSN) and UN Number	D0420
Hazard labels	D0420
Truckmarking	D0421
Example of ADR vehicle labelling	D0422
Limited quality package exemptions	D0441
Classification of major incident types	D6003
Disaster and emergency management systems (DEMS)	D6034
Major Incident Matrix	D6050
Decisional Matrix	D6051
Monitoring Matrix	D6066
Distance Scales	E5001
Ergonomic Approach	E17004
The 'S' Shaped Curve of the Spine	E17006
Zone of Convenient Reach	E17022
Viewing Angle	E17026
Body Map	E17057
Detention of food notice	F9014
Withdrawal of detention of food notice	F9014
Food condemnation warning notice	F9014
Improvement notice	F9015
Prohibition order	F9016
Emergency prohibition notice	F9018

List of illustrations

Form of application for registration of food premises	F9027
Employee safety representatives	J3005
Relationship between the manufacturers and the owner/users	L3002
Example of an inspection record sheet	M1007
The Risk Matrix	M1023
Risk assessment framework	M1024
Examples of mechanical hazards	M1025
Design/selection of safeguards	M1028
Example of a fixed guard	M1029
Fixed guard with adjustable element	M1030
Main types of electrical interlocking switches	M1032
Trapped key system for power interlocking	M1034
Example of a mechanical interlocking system	M1035
Risk contours and zones around a hazardous installation	M1140
The costs of accidents	M2001
Systematic management cycle	M2008
Continual improvement loop	M2029
ILO model of continual improvement	M2044
The Plan, Do, Check, Act Cycle	M2105
Suggested risk assessment framework	M2111
Lifting and lowering	M3003
Handling while seated	M3003
Manual handling assessment checklist	M3020
Examples of manual handling assessment records	M3024
Sign for informing that ear protectors must be worn (white on a circular blue background)	N3013
Normal audiogram with the responses close to 0dB across the frequency range	O1023
CE mark of conformity for personal protective equipment	P3013
Relationship between Manufacturers' Duties	P7001
Pressure Systems Safety Regulations 2000 – Duties Decision Flow Chart	P7012
Stress Response Curve	S11003
Components of a local exhaust ventilation system	V3003
Duct pressures	V3013
Model for effective policy development group	V8024

List of illustrations

Model for monitoring and evaluating policy	V8042
Risk assessment model	V8050
Likelihood of harm	V8053
Severity of harm	V8053
A model for reporting and monitoring incidents of workplace violence	V8057
General training and development model	V8097
Identifying the need	V8099
A typical delegate feedback process combining individual and organisational needs	V8110
Incident management – planning and practice	V8116
Timescale of reactions to workplace violence	V8130
Supporting the victim	V8135
The process of prosecution	V8140
Maybo Risk Management Model	V8144
Examples of prohibitory signs	W11037
Examples of warning signs	W11038
Examples of mandatory signs	W11039
Examples of emergency escape or first-aid signs	W11040
Examples of fire-fighting signs	W11041
Hand signals	W11042

List of abbreviations

Many abbreviations occur only in one section of the work and are set out in full there. The following is a list of abbreviations used more frequently or throughout the work.

Legislation

ACM	Asbestos-containing Material
BRM	Business Risk Management
BS	British Standard
BSE	Bovine Spongiform Encephalepathy
CCTV	Closed Circuit Television
CHAN	Chemical Hazards Alert Notice
CRI	Corporate Responsibility Index
CSR	Corporate Social Responsibility
DSE	Display Screen Equipment
EHO	Environmental Health Officer
ELCI (ELI)	Employer's Liability (Compulsory) Insurance
EMF	Electromagnetic Field
EMS	Environmental Management System
FM	Facilities Management
FMD	Foot and Mouth Disease
FOPS	Falling Object Protective Structure
HAVS	Hand-arm Vibration Syndrome
IR	Infra-red (radiation) or Ionising Radiation
LEV	Local Exhaust Ventilation
EL	Lower Explosive Limit
LOCAE	List of Classified and Authorised Explosives
LPG	Liquefied Petroleum Gas
MDF	Medium Density Fibreboard

List of abbreviations

MDHS	Methods for the Determination of Hazardous Substances
MEL	Maximum Exposure Limit
MEWP	Mobile Elevating Work Platform
MPE	Maximum Permissible Exposure
MSDS	Material Safety Data Sheet
NOEM	New or Expectant Mother
OEL	Occupational Exposure Limit
OES	Occupational Exposure Standard
OHSAS	Occupational Health and Safety Assessment Series
OHSMS	Occupational Health and Safety Management Systems
PAH	Polycyclic Aromatic Hydrocarbons
PAT	Portable Appliance Testing
PCB	Polychlorinated Biphenyls
PLI	Public Liability Insurance
PPE	Personal Protective Equipment
ppm	parts per million
PTW	Permit to Work
RHS	Revitalising Health and Safety
ROPS	Roll-over Protective Structure
RPE	Respiratory Protective System
RSI	Repetitive Strain Injury
RSP	Registered Safety Practitioner
SHT	Securing Health Together
SWL	Safe Working Load
UEL	Upper Explosive Limit
ULD	Upper Limb Disorder
UV	Ultra-violet (radiation)
VCM	Vinyl Chloride Monomer
VDU	Visual Display Unit
WRULD	Work-related Upper Limb Disorder
WSA	Workers' Safety Advisors
YP	Young Person

Introduction

Lawrence Bamber and Andrea Oates

The Handbook

[IN01] In the early days of this publication, health and safety was still a matter for specialists – often professional safety advisers, HR directors and others who were required to 'deal with' health and safety on behalf of their organisations. Now, whether it is a debate on work-related travel, stress or the hazards associated with power generation, everyone is engaged in discussing 'risk'. In the boardroom this is often linked to corporate governance and corporate responsibility. On the shop floor, it is associated with a radical rejection of a 'compensation culture' in favour of a desire to be able to return home from work healthy and uninjured. The development of these arguments over more recent years has led to a greater focus on the 'health' part of health and safety.

Also over recent years, particularly under the 2010–2015 Conservative-led Coalition government, Britain's health and safety system has seen the introduction of a raft of changes.

These have included the following:

- New rules on Health and Safety Executive (HSE) and local authority inspections exempting hundreds of thousands of businesses from health and safety inspections. A National Enforcement Code, introduced in May 2013, targets proactive inspections on 'higher risk' activities in specified sectors or when there is intelligence of workplaces putting employees or the public at risk. Businesses are now only inspected if they are operating in 'high risk' areas, such as construction, or if they have a poor safety record. Research by academics and safety campaigners shows that enforcement activity has also declined in some high-risk areas.
- The repeal of several sets of health and safety regulations and the withdrawal of a number of approved codes of practice (ACoPs). The Department for Work and Pensions' *A final progress report on implementation of health and safety reforms* showed that the HSE had reduced the overall stock of legislation by 50% by early 2015.
- A change brought in by the *Enterprise and Regulatory Reform Act 2013 (ERRA 2013)* means that employers are now only liable for civil damages in health and safety cases if they can be shown to have acted negligently. 'Strict liability' was established by a few workplace regulations made under the *Health and Safety at Work etc Act 1974*. If the employer breached those regulations, they would be liable in a civil claim even if they could argue they had taken reasonably practical steps

to try and prevent the incident. Where a breach of duty occurs on or after 1 October 2013, when the ERRA 2013 came into force, workers can no longer rely on an employer's breach of health and safety law to win a personal injury claim, but must provide proof of negligence.
- A change brought in by the *Deregulation Act 2015* means that only 'relevant' self-employed people are now covered by health and safety law. HSE guidance explains that if a self-employed person's work activity is specifically mentioned in the *Health and Safety at Work etc Act 1974 (General Duties of Self-Employed Persons) (Prescribed Undertakings) Regulations 2015*, or if their work activity poses a risk to the health and safety of others, then health and safety law applies. The regulations refer to agriculture (including forestry), work with asbestos, work on a construction site, an activity to which the *Gas Safety (Installation and Use) Regulations 1998* applies, work with genetically modified organisms and work on the railways.

At the time of updating this chapter (December 2018), there was still considerable uncertainty about the UK's exit from the European Union (EU). How Brexit will affect UK health and safety law, if indeed it goes ahead at all, will depend on the terms of the UK withdrawal. The UK remains a member of the EU and is covered by all its provisions, including health and safety legislation, until the exit negotiations are complete and a new relationship is defined, or the UK leaves the EU without a deal on 29 March 2019.

The Grenfell Tower fire in west London in June 2017, which killed 72 people, has resulted in intense scrutiny of building control and fire safety legislation and enforcement of this law. There is an ongoing public inquiry (see www.grenfelltowerinquiry.org.uk), the Metropolitan Police are leading a criminal investigation, and former HSE chair Dame Judith Hackitt has carried out an independent review of building regulations and fire safety. She published her final report in May 2018 and made 50 recommendations for the government on delivering a more robust regulatory system for high-rise homes. In November 2018, the government made its first change to fire safety laws following the fire, laying new regulations to ban combustible materials on new high-rise homes, hospitals and some schools.

This Introduction is in five main parts

[IN02] They are:
- drivers for health and safety performance
- the changing world of work
- recent trends in UK health and safety law and its enforcement
- the structure of UK health and safety law
- the management of health and safety at work

Drivers for health and safety performance

Why do organisations seek to improve their performance?

[IN03] Research in the UK and around the world provides evidence of distinct differences between the perception of small businesses and larger companies.

The usual definition of small and medium-sized enterprises (SMEs) is any business with fewer than 250 employees. Microbusinesses have 0–9 employees. In small companies, the attitudes and beliefs of the owners/managers are a dominant force, while in larger corporations and public bodies the social norms are subject to the broader forces of corporate governance. A combination of the following appears to be of greatest influence, but each overlap with the other and may be present in different proportions and combinations:

- beliefs and attitudes of owners/managers in very small organisations;
- risk management and reputation risk;
- corporate governance; and
- legal compliance.

A European Agency for Safety and Health at Work survey asked employers about the drivers for managing health and safety risks and found that 90% pointed to health and safety legislation impelling them to act.

Beliefs and attitudes of owner/managers in very small organisations

[IN04] There were around 5.7 million SMEs in the UK in 2018 accounting for over 99% of all businesses. Around 5.4 million microbusinesses account for 96% of all businesses, 33% of employment and 21% of turnover (https://researchbriefings.files.parliament.uk/documents/SN06152/SN06152.pdf). The modern economy increasingly comprises services which are often provided by new organisations and by the self-employed, including in the so-called 'gig economy' where people use apps to sell their labour. Identifying what motivates the leaders of these organisations is also important to governments and health and safety enforcement bodies.

It appears that the level of awareness of specific health and safety requirements embodied in regulations is often poor among small businesses and almost wholly dependent on suppliers, customers and peers. The suppliers of goods and services are regarded as a reliable source of information. For example, the labels on containers of materials and the accompanying safety data sheets are used by small businesses to establish working practices without, typically, any reference to the requirements of the *Control of Substances Hazardous to Health Regulations 2002*. For some small businesses, which supply major clients, larger organisations increasingly impose operating conditions and provide support which incorporates health and safety as part of the management of a secure supply chain. The last significant element of advice and guidance is from informal contacts with other owners/managers.

Whatever their source of information as to what they should be doing and how to do it, the owners/managers of small businesses who do act in this area seem to have a clear set of reasons for getting health and safety right. These motivators are:

- the *focus on employees* who are so well known because of daily contact that injury or ill health is treated almost as if a family member were affected (and in many cases family members do work in the small organisation, and children may attend the workplace in the school holidays, for example);
- the *focus on satisfying customers* and responding to guidance from suppliers, a general approach to 'doing the right thing';
- a conviction among a minority that looking after worker health and safety and similar matters contributes to a *productive, healthier, happier workforce* and may have benefits such as controlling insurance premiums.

The European Agency for Safety and Health at Work factsheet, *The business benefits of good occupational safety and health*, highlights the following motivators for good occupational safety and health (OSH) performance for small and medium businesses:

- meeting the OSH requirements of business clients in order to win and retain contracts;
- avoiding business disruption and loss of key staff;
- motivating staff and retaining their commitment; and
- the availability and affordability of insurance.

Why do larger organisations seek to improve their performance?

[IN05] Most large organisations have been on a learning curve since the post-war period. At first there was an emphasis on managing the pure risk of getting health and safety 'wrong', and the impact that this could have on people (injuries), products (damage) and property (losses such as by fire). By the 1970s this effort had broadened to encompass environmental issues, and also the concept of business recovery – and as a result events which could lead to business disruption were being investigated to both prevent their occurrence but also to respond effectively if they did occur. By the 1980s, risk management had matured into a discipline which also encompassed such matters as brand management and corporate reputation. This does not mean that large companies have eliminated such risks. The impact on companies involved in high-profile health and safety failures is a testament to the correct evaluation that these are significant matters for corporate value – but most have implemented systems which are designed to identify and control such risks. Thus, protecting the reputation of a company as an efficient and well-managed organisation is one of the most significant drivers to improving health and safety performance.

A series of collapses of high-profile companies and press reports of 'fat cats' drawing huge salaries and bonuses for incompetent management, not surprisingly, resulted in considerable public criticism and governments set up several committees to report on remedial steps – Cadbury (1992), Greenbury (1995) and Hampel (1998). The latter called for a new approach to corporate governance which included a component of risk management at the highest level. The Turnbull Report (1999, updated 2005), which required organisations to establish a risk management strategy, was adopted by the London Stock Exchange and obliged listed companies (PLCs) to publish an annual risk management report as part of their annual reporting arrangements. This prompted many large businesses to adopt formal plan-do-check-act management systems to achieve the level of assurance required for directors to sign such reports.

The 2001 Myners Report considered the role of institutional investors and fund managers. In 2003, the Higgs Report looked at the role of non-executive directors (NEDs) and in 2009 the Walker Review made a number of recommendations for changes to corporate governance in banks and financial institutions. In 2010, the Financial Reporting Council consolidated the previous Combined Code for listed companies into the UK Corporate Governance Code and published a stewardship code for the first time.

The Modern Slavery Act 2015 requires large organisations to develop a slavery and human trafficking statement each year, setting out what steps they have taken to ensure modern slavery is not taking place in their business or supply chains.

In November 2016, the government published a green paper on corporate governance, focussing on executive pay, private companies, and workers on boards and in 2017 the Business, Energy and Industrial Strategy Committee reported on its inquiry into corporate governance. This focussed on executive pay, directors' duties and the composition of boardrooms, including worker representation and gender balance in executive positions. The inquiry followed corporate governance failings highlighted by the committee's inquiries into the collapse of BHS and working practices at Sports Direct, and government commitments to overhaul corporate governance. HSE guidance on managing for health and safety has also moved away from using the POPMAR (Policy, Organising, Planning, Measuring performance, Auditing and Review) model to a 'Plan, Do, Check, Act' approach. It says that the move 'achieves a better balance between the systems and behavioural aspects of management' and 'treats health and safety management as an integral part of good management generally, rather than as a stand-alone system'.

The approach to risk management also dovetailed with an increasing interest in corporate social responsibility and the pressures on companies to be seen to look after their staff, the environment and their neighbouring communities.

In the public and voluntary sectors, similar developments have been encouraged by the Audit Commission (which was replaced by Public Sector Audit Appointments Ltd, National Audit Office, Financial Reporting Council and Cabinet Office in April 2015) and Charities Commission (see '*Worth the risk: An introduction to risk management and its benefits*' (2001), Audit Commission; Charity Reporting and Accounting: The essentials (March 2015), Charity Commission (www.gov.uk/government/publications/charity-reporting-and-accounting-the-essentials-march-2015-cc15c).

Across the public and private sectors, therefore, the development of corporate governance is linked to and integrated with risk management – all organisations are asked to identify what could blow them off course, resulting in a diminution of services, a loss of profitability, and the development of an unsustainable approach to their work. Health and safety is an important part of the matrix of risks identified in any such exercise, and the management and control of the health and safety risks has become part of what the board of directors or trustees have to achieve and assure. It was against this background that the guide on the duties of directors was developed and published by the Health and Safety Executive (HSE) (INDG 417 *Leading health and safety at work*) (www.hse.gov.uk/leadership/index.htm).

Drawn up by the HSE and the Institute of Directors (IOD), with input from employer, safety and trade union organisations, this sets out an agenda for the effective leadership of health and safety. It is designed for use by all directors, governors, trustees, officers and their equivalents in the private, public and third sectors. It applies to organisations of all sizes:

- Strong and active leadership from the top
  - visible, active commitment from the board
  - establishing effective downward communication systems and management structures
  - integration of good health and safety management with business decisions
- Worker involvement
  - engaging the workforce in the promotion and achievement of safe and healthy conditions
  - effective upward communication
  - providing high quality training
- Assessment and review
  - identifying and managing health and safety risks
  - accessing and following competent advice
  - monitoring, reporting and reviewing performance

The third pressure for health and safety performance is the need to achieve legal compliance. Most people, including those who lead organisations, are social beings who accept that the rules they need to live and work by represent the framework for a civilised life. Even if an individual driver occasionally breaks the speed limit, he or she is unlikely to wish for unlimited speeds to be agreed for the street on which they live. For small organisations, there is an often-expressed view that they are very unlikely to be visited by an inspector, and there may be a low awareness of both the statutory requirements and often a misperception of the level of risks associated with their work. It is in the large organisation with a nervous board of directors and a management commitment to legal compliance that represents a significant stimulus. Identifying what regulations need to be complied with, those which are specifically applicable to the mix of work being undertaken, and then assuring that compliance, is a further push towards formal management systems.

Greater incentives for effective management of health and safety have been provided by:

- the *Corporate Manslaughter and Corporate Homicide Act 2007 (CMCHA 2007)* (see the chapter dealing with **CORPORATE MANSLAUGHTER**);

- increased penalties introduced by the *Health and Safety (Offences) Act 2008 and the Legal Aid, Sentencing and Punishment of Offenders Act 2012* (which came into force in March 2015);
- the Sentencing Council guidelines for health and safety and corporate manslaughter offences which came into force on 1 February 2016; and
- a new Sentencing Council guideline for gross negligence manslaughter offences which came into effect on 1 November 2018 and recommends prison sentences of up to 18 years. Gross negligence manslaughter 'occurs when the offender is in breach of a duty of care towards the victim and amounts to a criminal act or omission'. In a work setting, the Sentencing Council explains 'it could cover employers who completely disregard the safety of employees'. The guideline is expected to increase sentences in gross negligence manslaughter cases where, for example, 'an employer's long-standing and serious disregard for the safety of employees, motivated by cost-cutting, has led to someone being killed'.

Examples of the increased penalties introduced by the *Health and Safety (Offences) Act 2008* and the *Legal Aid, Sentencing and Punishment of Offenders Act 2012* are as follows:

	Offence	Date offence committed		
		Prior to 16th January, 2009	After 16th January, 2009 and before 12th March 2015	After 12th March 2015
Magistrates' Court	Breach of General Duties under HSWA Act	Fine up to £20,000	Fine up to £20,000 and/or six months' imprisonment	Unlimited fine and/or six months' imprisonment*
	Breaches of Safety Regulations	Fine up to £5,000	Fine up to £20,000 and/or six months' imprisonment	Unlimited fine and/or six months' imprisonment*
Crown Court	Breaches of Safety Regulations	Unlimited fine	Unlimited fine and/or two years' imprisonment	Unlimited fine and/or two years' imprisonment

* When s 154(1) of the *Criminal Justice Act 2003* is brought into force the maximum term will be increased to 12 months.

The *2008 Act* gave magistrates and sheriffs greater powers to send an offender to prison. In the past custodial sentences were reserved for specific cases, but offenders can now be sent to prison for almost all health and safety offences. In addition, certain offences that in the past could only be tried in the lower courts, such as the failure to comply with an improvement order, were made triable in either court, meaning the offender could face a much tougher sentence if their case was referred to the Crown Court.

Section 85 of the *Legal Aid, Sentencing and Punishment of Offenders Act 2012* came into force on 12 March 2015 and increased the level of most fines available for magistrates' courts to an unlimited fine (previously £20,000 for most health and safety offences) as the table above shows.

Sentencing Council guidelines, which came into force on 1 February 2016, are aimed at increasing penalties for serious offending. They aim to ensure that fines are fair and proportionate to the seriousness of the offence and the means of offenders. Under the guidelines, organisations with a turnover of £50 million and over which commit serious health and safety offences, where there is a very high level of culpability and a high risk of harm, can expect to face fines of up to £10 million. Large organisations found guilty of corporate manslaughter offences can expect to face fines of up to £20 million. Where their turnover greatly exceeds this threshold, the guidelines say that a higher fine may be necessary in order to achieve a proportionate sentence.

The 2016 guidelines had a huge impact on the level of fines handed down to large organisations found guilty of health and safety offences. The HSE report, *Enforcement in Great Britain 2017*, shows that between 1 April 2016 and 31 March 2017 there were 38 fines at or above £500,000. The maximum fine was £5 million, handed down to Merlin Attractions Operations in September 2016 following an incident on a ride at Alton Towers Theme Park in June 2015 which left several people with life-changing injuries. In 2014/15, the last full year before the guidelines were introduced, the largest single fine was £750,000 and just five cases were at or above £500,000.

There was also a large increase in the total amount of fines, from £38.8 million in 2015/16 to £69.9 million in 2016/17, the first full year for which the sentencing guidelines were in effect. The average level of fines also sharply increased. In 2014/15, it was £29,000 per conviction, in 2015/16 (which included two months when the guidelines were in force), it increased to £58,000, and in 2016/17, it reached an average of £126,000 per conviction.

The latest (2018) HSE enforcement report includes provisional 2017/18 data on prosecutions showing a levelling of the total amount of fines handed down, with little change from £71.8 million in 2016/17 to £72.6 million in 2017/18.

The guidelines are examined in more detail in **CORPORATE MANSLAUGHTER**.

The changing world of work

[IN06] There is a wide range of evidence about changes in the world of work. In the UK, we now have many more people working in call centres than in mining. This move towards service industries, as traditional manufacturing continues to decline in terms of the numbers of people employed, is matched by the increasing dominance of small businesses and the self-employed. A third (33%) of the UK workforce is employed in businesses with fewer than ten people, compared with four in ten who work in the UK's 8,000 large businesses with more than 250 employees. Businesses with more than 250 employees account for 0.1% of businesses but 40% of employment and 48% of turnover (https://researchbriefings.files.parliament.uk/documents/SN06152/SN06152.pdf).

A 2017 report on modern work practices by two parliamentary select committees put the number of self-employed people in the UK workforce at five million. The joint Work and Pensions and Business, Energy and Industrial Strategy committees report, *A framework for modern employment*, pointed to Office for National Statistics (ONS) figures showing the level of self-employment in the UK increased from 3.8 million in 2008 to 4.6 million in 2015. It found that there were more than 900,000 people on zero-hour contracts, 1.6 million temporary and agency workers, and an estimated 1.3 million people working in the gig economy.

Flexible work patterns and the temporary nature of much employment, particularly for many young workers, have also altered the balance of health and safety issues. Psychosocial risks give rise to stress-related sickness, which is now most common cause of work-related illness. Stress, depression or anxiety and musculoskeletal disorders accounted for the majority of days lost due to work-related ill health, 15.4 million and 6.6 million respectively, according to the latest HSE statistics (www.hse.gov.uk/statistics/dayslost.htm). The big increase in female employment and the pressure on an ageing workforce to stay at work longer is also changing the nature of the 'at risk' groups.

The changes being experienced at work suggests that the law, which often moves slowly, the enforcement authorities, employers, employees and professional safety advisers are all under pressure to adapt and change. Many of these changes are reflected in individual chapters and some are developed further in the remainder of this introduction.

To combat the growing threat of health (as opposed to safety) issues within UK plc, the first ever review into the health of the health of the working age population, *Working for a healthier tomorrow* was published in March 2008. Sickness absence costs the UK economy some £12 billion a year, according to the HSE, with over 130 million days lost to sickness absence each year.

A subsequent 2011 government-commissioned review of sickness absence, *Health at work*, reported that while much sickness absence ended in a swift return to work, a significant number of absences last longer than they need to. Over 300,000 people fall out of work and onto health-related state benefits each year.

In response, the government introduced tax relief on expenditure up to £500 on medical treatment to help an employee return to work and set up a new national occupational health service, Fit for Work, to support employees to return to work after a period of sick absence. The Fit for Work assessment services closed in spring 2018 as a result of low referral rates.

In November 2017, the government published a ten-year strategy, *Improving lives: the future of work, health and disability*. This includes extending fit note

[IN06] Introduction

certification beyond GPs to a wider group of healthcare professionals, including physiotherapists, psychiatrists and senior nurses; responding to the 40 recommendations of the recent Stevenson/Farmer review of mental health and employers, *Thriving at work*; and appointing an expert working group on occupational health to champion,shape and drive a programme of work to take an in-depth look at the sector.

More information about the Improving lives strategy can be found at: www.gov.uk/government/publications/improving-lives-the-fut ure-of-work-health-and-disability.

Legal framework for health and safety at work

The role of the European Union and European Directives and Brexit

[IN07]–[IN08] As set out above, how the so-called Brexit vote to leave the European Union (EU) will affect UK health and safety law depends on the terms of the UK withdrawal. The UK remains a member of the EU and is covered by all its provisions, including health and safety legislation, until the exit negotiations are complete and a new relationship is defined, or the UK leaves the EU without agreement (a no deal Brexit). At the time of writing (December 2018) the situation was very uncertain, with a parliamentary vote on the deal negotiated by Prime Minister Theresa May delayed until January 2019.

The European Union (EU) has been an important source of legislation improving health and safety protection for UK workers since 1978, when the Commission adopted its first action programme. Article 118A in the Treaty of Rome 1957 gives health and safety prominence in the objectives of the EU. The Social Charter also contains a declaration on health and safety, although this has no legal force. While the EU can (and does) issue its own directly-acting regulations, it mainly operates through directives requiring Member States to pass their own legislation. The 'Framework Directive' was adopted in 1989 as part of the creation of the Single Market. It contained many broad duties, including the requirement to assess risks and introduce appropriate control measures. There have been 23 further EU health and safety directives issued since the 1989 Framework Directive.

Developments have slowed down in recent years, and the European Commission is no longer the driving force for improving health and safety protection it once was. However, more recent UK regulations emanating from European directives include: the *Control of Major Accident Hazards (COMAH) Regulations 2015* which repealed and replaced the *1999 COMAH Regulations* and implemented provisions of the Seveso III Directive; the *Control of Electromagnetic Fields at Work Regulations 2016*, which implemented the provisions of the electromagnetic fields directive (2013/35/EU); and the *Ionising Radiations Regulations 2017*, which came into force in January 2018 and repealed and replaced the *Ionising Radiations Regulations 1999* and implemented the requirements of the Basic Safety Standards Directive (BSSD).

Implementation and enforcement

[IN09] The introduction and implementation of health and safety legislation in the UK is mainly overseen by the Health and Safety Executive (HSE).

For regulatory initiatives, the HSE generally circulates consultative documents incorporating the proposed regulations together with any related Approved Code of Practice (ACoP) and/or guidance for interested parties (such as trades unions, employers' organisations, professional bodies and local authorities) to comment on. Following the consultation process, the HSE submits a final version to the Secretary of State who then lays the regulations before Parliament. It is at this stage that a date for their coming into force is determined – usually on common commencement dates, 6 April and 1 October, unless the regulations are implementing an EU directive stipulating a different date.

Enforcement of health and safety legislation in most 'high-risk' work sectors is the province of the HSE although the Office of Road and Rail (ORR) enforces safety on the railways and other guided transport systems, for example. In most other work activities, health and safety enforcement is normally the responsibility of the local authority in whose area the activity is carried out. There are other enforcement authorities, including the police, that may have a role – see the chapters on CORPORATE MANSLAUGHTER AND ENFORCEMENT for more details.

Health and Safety Executive – health and work strategy

[IN10] The HSE's Health and Work strategy increases its focus on occupational ill health. The strategy is accompanied by three health priority plans on work-related stress, musculoskeletal disorders (MSDs) and occupational lung disease – conditions with widespread prevalence, the largest lost-time and economic cost consequences, and life-limiting or life-altering impacts.

These plans include:

- stress pilots in priority sectors and developing a suite of leading indicators to measure stress risk management;
- continued MSD intervention programmes in food manufacturing and construction and regulatory interventions in other sectors; and
- a new leadership body to provide direction and coordinate activity on occupational respiratory disease, and regulatory activity in sectors/activities where lung cancer, occupational asthma and legionella pose the highest risks.

More information about the strategy and plans can be found on the HSE website at: www.hse.gov.uk/aboutus/strategiesandplans/health-and-work-strategy/index.htm.

'Fee for Intervention (FFI)'

[IN11] Since October 2012, a 'fee for intervention' (FFI) cost recovery scheme has been in place (following Parliamentary approval of the *Health and Safety*

[IN11] Introduction

(Fees) Regulations 2012). The Regulations require the HSE to recover its costs for carrying out its regulatory functions from employers and other duty holders found to be 'in material breach' of health and safety law. The current fees are based on an hourly rate of £129. HSE data on FFI shows that the number of invoices issued, the number of companies invoiced and the average cost of a single invoice have all significantly increased since the scheme was introduced. The first FFI invoices were issued in January 2013 and covered the period from October to November 2012. The HSE issued 1,418 invoices to 1,399 companies raising £727,645. The average cost of a single invoice was £513.15. The latest data published on the HSE website (www.hse.gov.uk/fee-for-intervention/assets/docs/ffi-nvoice-information.pdf) shows that FFI income was around £15 million in 2017/18.

The HSE brought in a change to the process for considering FFI disputes. From 1 September 2017, disputed invoices raised under FFI are considered by a fully independent panel.

More information about the scheme is available on the HSE website at www.hse.gov.uk/fee-for-intervention/index.htm.

HSE Business Plan

[IN12] HSE Business Plans set out the main focus of HSE work over the year. The 2018/19 plan set out that the year ahead would be significant in terms of supporting the government in its preparations for the UK's exit from the European Union and any changes that may follow the Grenfell fire. 2019 will see a new HSE chief executive appointed to succeed Dr Richard Judge who resigned in August 2018. Former HSE head of regulation Dr David Snowball has been acting chief executive in the interim.

Business plans can be downloaded from the HSE website at: www.hse.gov.uk/aboutus/strategiesandplans/businessplans/.

Penalties and prosecutions

[IN13] The penalties for breaches of health and safety legislation have increased steadily over recent years, although the record fine for a single offence remains the £15 million imposed in 2005 on gas transportation company Transco for a breach of *Section 3(1)* of *HSWA*. This related to a large gas explosion in a residential area of Lanarkshire in 1999 which killed a family of four. Transco were heavily criticised for a lack of effective inspection and maintenance arrangements and inaccurate records of gas pipes. The ductile iron gas main involved in the explosion was badly corroded.

The HSE report *Enforcement statistics in Great Britain, 2018* provides information about enforcement action taken by the HSE, local authorities and, in Scotland, the Crown Office and Procurator Fiscal Service (COPFS).

In 2017/18, 517 HSE and COPFS prosecution cases reached a verdict, a decrease of 16% from 2016/17. A conviction was secured (for at least one offence) in 493 of the 517 cases where a verdict was reached in 2017/18; a conviction rate of 95%. Those found guilty of health and safety offences in

2017/18 received fines totalling £72.6 million, an average penalty of around £147,000 per case resulting in conviction. This is a slight increase compared to the average fine of just under £126,000 per case resulting in conviction in 2016/17.

Last year saw a fall in the number of cases where a verdict was reached, continuing the fall seen in 2016/17. The 517 figure is the lowest in the last five years. The proportion of cases resulting in a conviction (for at least one offence) has been between 93–95% for the last five years.

The 2017/18 data show a levelling of the total amount of fines handed down, with little change from £71.8 million in 2016/17 to £72.6 million in 2017/18. The 2017/18 figure follows two years in which there were large increases in the amount of fines resulting from convictions for health and safety offences following the introduction of new sentencing guidelines in 2016.

Because these guidelines link fines to turnover, large organisations convicted of offences are receiving larger fines than seen prior to their introduction. In 2017/18, the single largest fine was £3 million and a total of 45 cases received fines over £500,000. This contrasts with the 2014/15 period, the last full year without these guidelines, where the single largest fine was £750,000 and five cases were at or above £500,000. The average level of fine has also increased since the sentencing guidelines came into effect, from £29,000 per conviction in 2014/15 to £58,000 in 2015/16, the last two months of which were under the guidelines. Fines reached an average of £147,000 per conviction in 2017/18.

Fines are the most common penalty handed down by the courts for health and safety offences. The use of immediate and suspended custodial sentences has remained comparable with 2016/17 with 7% of offences resulting in an immediate custodial sentence compared to 6% in 2016/17. Nine percent of offences resulted in a suspended custodial sentence compared to 12% in 2016/17.

Large fines and custodial sentences have also resulted from manslaughter charges brought against individuals in respect of work-related deaths. The *Corporate Manslaughter and Corporate Homicide Act 2007 (CMCHA 2007)* provides a further opportunity for unlimited fines to be imposed on organisations.

The first company to stand trial under the *CMCHA 2007*, Cotswold Geotechnical Holdings Ltd (CGH), was fined £385,000 after being found guilty by a jury at Winchester Crown Court on 17 February 2011. By December 2017, 26 companies had been convicted of corporate manslaughter charges.

The highest fines to date were handed down to Martinisation London in July 2017. It received two concurrent £1.2 million fines for the deaths of two workers who fell from a balcony while attempting to hoist a sofa into a first floor flat.

In November 2017, Birmingham-based waste company Master Construction Products (Skips) Ltd (MCP) became the 26th company to be convicted of corporate manslaughter under the *Corporate Manslaughter and Corporate*

Homicide Act 2007 following the death of factory worker in January 2015. He died after becoming entangled in a trommel machine used to sort waste material at the company's recycling yard. MCP was handed down a £255,000 fine and its director was sentenced to 12 months' imprisonment, suspended for two years, and given 300 hours of community service. He was also disqualified from being a company director for eight years and ordered to pay £11,500 in prosecution costs.

Further information on the Sentencing Council guidelines and corporate manslaughter convictions is contained in **CORPORATE MANSLAUGHTER**.

Companies and individuals convicted of health and safety offences are 'named and shamed' on the HSE website – www.hse.gov.uk/enforce/prosecutions.htm.

Risk assessments

[IN14] Much recent UK legislation has included a requirement for some type of risk assessment. The concept was introduced in the early 1980s in regulations applying to asbestos and lead but it came to wider prominence as a core requirement of the *Control of Substances Hazardous to Health Regulations 1988* ('COSHH') *(SI 1988 No 1657)* (since repealed and replaced by the *Control of Substances Hazardous to Health Regulations 2002 (SI 2002 No 2677)*). The group of Regulations which came into force on 1 January 1993 and derived from the EU Framework and Daughter directives continued this trend with four of the six requiring an assessment of one kind or another. This 'six pack' of UK regulations deal with the management of health and safety at work, work equipment, display screen equipment (DSE), manual handling, personal protective equipment (PPE), and health, safety and welfare in workplaces. The most important is the general requirement for risk assessment contained in the *Management of Health and Safety at Work Regulations 1999 (SI 1999 No 3242)* ('the *Management Regulations*'), Reg 3. Other important regulations requiring more specific types of risk assessment relate to noise, manual handling operations, PPE, DSE and vibration.

In practice, a less formal type of risk assessment was already required by health and safety legislation – an assessment of the level of risk, the adequacy of existing precautions and the costs of additional precautions was necessary in order to determine what was 'reasonably practicable'. Risk assessment lies at the heart of the original concept of self-regulation enshrined in the 1972 Robens Report – which led to the enactment of the *Health and Safety at Work etc Act 1974*. Employers are required to demonstrate that they have identified relevant risks together with appropriate precautions and, in most cases, must have records available to prove this.

The HSE has provided numerous guides and templates to help SMEs and other organisations undertake pragmatic risk assessments with minimal effort.

Example risk assessments can be found on the HSE website at www.hse.gov.uk/risk/casestudies.

Supply-chain pressure

[IN15] Since the introduction of *HSWA 1974*, many court decisions, most notably those involving *Mitchell v Vickers Armstrong Ltd and Swan Hunter Shipbuiders Ltd* ([1984] IRLR 93–116) and *R v Associated Octel Co Ltd* [1996] 4 All ER 846, [1996] 1 WLR 1543, HL, have emphasised that employers often have responsibilities for the activities of employees of other organisations. The structure of HSWA 1974 and much subsidiary legislation is such that responsibilities overlap between employers rather than being neatly apportioned between them. Employers must do more than simply not turn a blind eye to the obvious health and safety failings of those with who they come into contact. They must often take a pro-active interest in the health and safety standards of others.

In recent years, a growing number of larger companies, local authorities and other public bodies have implemented increasingly formalised procedures for checking the health and safety standards of contractors wishing to work for them. This process was accelerated by the demands of the original *Construction (Design and Management) Regulations 1994* ('CDM') (SI 1994 No 3140). The current *Construction (Design and Management) Regulations 2015* (SI 2015 No 51) specifically require clients to satisfy themselves that contractors are capable of dealing with the health and safety issues associated with construction work of any type or size. The Regulations also place responsibilities on principal contractors in respect of sub-contractors. Consequently, contractors are frequently required to provide details of their health and safety policies and generic risk assessments together with risk assessments and/or method statements for specific projects or activities.

Some companies take an extremely hands-on approach in policing the activities of contractors working on their premises or on projects where they are the principal contractor. There are also several schemes for vetting the health and safety standards of contractors, and the single national Safety Schemes in Procurement (SSIP) is recognised by the majority of organisations providing contractor vetting schemes. SSIP is an umbrella organisation set up to address the spread of OSH pre-qualification schemes. The scheme is voluntary and aims to reduce bureaucracy and to facilitate mutual recognition of core competence criteria within the construction and allied industries.

Further information is available at: www.ssip.org.uk.

Recent changes in legislation

[IN16] Changes in health and safety legislation normally come into force on two dates each year – 6 April and 1 October.

As set out above, over recent years there have been numerous changes to health and safety legislation. These have resulted from the recommendations of two government-commissioned reviews of health and safety legislation, Lord Young's 2009/2010 '*Common Sense: Common Safety*' and Professor Ragnar Löfstedt's 2011 '*Reclaiming health and safety for all: An independent review of health and safety legislation*', together with the Coalition government's '*Good health and safety, good for everyone*' initiative to support its growth agenda and ease the regulatory 'burdens on business'. European directives and regulations have also resulted in changes to UK health and safety law.

For example, the *Reporting of Injuries, Diseases and Dangerous Occurrences Regulations (RIDDOR) 2013* replaced the *1995 RIDDOR Regulations*. The regulations require employers to report 'over-seven-day' injures (rather than the previous requirement to report over-three-day injuries) and replaced the list of 'major injuries' to workers with a shorter list of 'specified injuries'. They also replaced the previous schedule detailing 47 types of industrial disease with six categories of reportable occupational diseases and require fewer types of 'dangerous occurrence' to be reported.

New regulations controlling explosives, acetylene and genetically modified organisms (GMOs) came into force in October 2014; new *Construction, Design and Management (CDM) Regulations* and *Control of Major Accident Hazards (COMAH) Regulations* came into force in 2015, along with the *Health and Safety at Work etc Act 1974 (General Duties of Self-Employed Persons) (Prescribed Undertakings) Regulations* (see **IN01** above). In July 2016, new regulations controlling Electromagnetic Fields (EMFs) came into force and in January 2018 the *Ionising Radiations Regulations (IRR) 2017* came into force, repealing and replacing the previous 1999 IRR regulations.

The structure of UK health and safety law

[IN17] Health and safety in the workplace involves two different branches of the law – criminal law and civil law (detailed below).

Criminal law

[IN18] Criminal law is the process by which society, through the courts, punishes organisations and individuals for breaches of its rules. These rules, known as 'statutory duties', are comprised in Acts passed by Parliament (eg *HSWA 1974*) or regulations which are made by government ministers using powers given to them by virtue of Acts (eg the *Manual Handling Operations Regulations 1992 (SI 1992 No 2793)* as amended).

Cases involving breaches of criminal law may be brought before the courts by the enforcement authorities which, in the case of most health and safety laws, are the HSE, the ORR and local authorities via their environmental health departments. Magistrates' courts hear the vast majority of health and safety prosecutions although more serious cases can be heard by the Crown Courts. As in all criminal prosecutions the case must be proved 'beyond all reasonable doubt'. Appeals from the Crown Court will go to the High Court, and potentially to the Court of Appeal or even the Supreme Court, although in practice very few health and safety cases go to appeal.

Unlimited fines and prison sentences of up to six months (if the case is heard in the magistrates' courts) or up to two years (if the case is heard in the Crown Court) are now an option for most health and safety offences (see **IN05** above). Deaths involving work activities can result in manslaughter charges which can lead to more severe custodial sentences.

A new Sentencing Council guideline for gross negligence manslaughter offences came into effect on 1 November 2018, regardless of when the offence took place, and recommends prison sentences of up to 18 years. Gross negligence manslaughter 'occurs when the offender is in breach of a duty of care towards the victim and amounts to a criminal act or omission'.

In a work setting, the Sentencing Council says 'it could cover employers who completely disregard the safety of employees'. The guideline is expected to increase sentences in gross negligence manslaughter cases where, for example, 'an employer's long-standing and serious disregard for the safety of employees, motivated by cost-cutting, has led to someone being killed'.

The new guideline specifies the range of sentences appropriate for the type of offence. The offence range is split into category ranges – sentences appropriate for each level of seriousness – and starting points define the position within a category range from which the court should start calculating the provisional sentence. The guideline also sets out aggravating and mitigating factors for the court to consider.

For offences with a 'very high' level of culpability where, for example, the offender 'showed a blatant disregard for a very high risk of death resulting from the negligent conduct' and 'the negligent conduct was motivated by financial gain', the starting point is 12 years' custody and the category range 10 to 18 years.

Where culpability is judged as 'high', the starting point is eight years' custody and the range between 6 and 12 years; for 'medium' culpability the starting point is four years' custody and the category range three to seven years; and for 'lower' culpability the starting point is two years' custody and the category range one to four years.

Aggravating factors include previous convictions and ignoring previous warnings. Mitigating factors include a lack of previous convictions, remorse, and where for reasons beyond their control, the offender lacked the necessary expertise, equipment, support or training which contributed to the negligent conduct.

After determining the offence category, and the starting point and category range, the court should consider factors which indicate a reduction for assistance to the prosecution, such as assistance given to the prosecutor or investigator. It will also take account of any potential reduction for a guilty plea.

The new guideline, *Manslaughter definitive guideline*, can be found on the Sentencing Council website at: www.sentencingcouncil.org.uk/wp-content/uploads/Manslaughter_Definitive-Guideline_WEB.pdf.

The Health and Safety at Work etc Act 1974 (HSWA 1974)

[IN19] *HSWA 1974* is the most important Act of Parliament relating to health and safety. It applies to the vast majority of people 'at work' – employers, 'relevant' self-employed people (see the changes to the *HSWA 1974* brought in by the *Deregulation Act 2015* at **IN01** above) and employees (with the exception of domestic servants in private households). It also protects the general public who may be affected by work activities. Some of the key sections of the Act are listed below.

Section 2 – Duties of employers

HSWA 1974, s 2(1) is the catch-all provision: 'It shall be the duty of every employer to ensure, so far as is reasonably practicable, the health, safety and welfare at work of all his employees.' See below for further discussion of the term 'reasonably practicable'.

HSWA 1974, s 2(2) goes on to detail more specific requirements relating to:

- the provision and maintenance of plant and systems of work;
- the use, handling, storage and transport of articles and substances;
- the provision of information, instruction, training and supervision;
- places of work and means of access and egress; and
- the working environment, facilities and welfare arrangements.

These are also qualified by the term 'reasonably practicable'.

HSWA 1974, s 2(3) requires an employer with five or more employees to prepare a written health and safety policy statement, together with the organisation and arrangements for carrying it out, and bring this to the notice of employees.

Section 3 – Duties to others

HSWA 1974, s 3(1) provides: 'It shall be the duty of every employer to conduct his undertaking in such a way as to ensure, so far as is reasonably practicable, that persons not in his employment who may be affected thereby are not exposed to risks to their health or safety.'

Employers therefore have duties to contractors (and their employees), visitors, customers, members of the emergency services, neighbours, passers-by and the public at large. This may extend to include trespassers, particularly if it is 'reasonably foreseeable' that they could be endangered, for example where high-risk workplaces are left unfenced.

'Relevant' self-employed people are placed under a similar duty and must also take care of themselves. If they have employees, they must comply with section 2 (see **IN01** above).

Section 4 – Duties relating to premises

Under *HSWA 1974, s 4* persons in total or partial control of work premises (and plant or substances within them) must take 'reasonable' measures to ensure the health and safety of those who are not their employees. These responsibilities might be held by landlords or managing agents etc, even if they have no presence on the premises.

Section 6 – Duties of manufacturers, suppliers etc

Those who design, manufacture, import, supply, erect or install any article, plant, machinery, equipment or appliances for use at work, or who manufacture, import or supply any substance for use at work, have duties under *HSWA 1974*, s 6.

Section 7 – Duties of employees

It shall be the duty of every employee while at work:

(a) to take reasonable care for the health and safety of himself and of other persons who may be affected by his acts or omissions at work; and
(b) as regards any duty or requirement imposed on his employer or any other person by or under any of the relevant statutory provisions, to co-operate with him so far as is necessary to enable that duty or requirement to be complied with.

[*HSWA 1974*, s 7]

Consequently, employees must not do, or fail to do, anything which could endanger themselves or others. It should be noted that managers and supervisors also hold these duties as employees.

Section 8 – Interference and misuse

No person shall intentionally or recklessly interfere with or misuse anything provided in the interests of health, safety or welfare in pursuance of any of the relevant statutory provisions.

[*HSWA 1974*, s 8]

Section 9 – Duty not to charge

No employer shall levy or permit to be levied on any employee of his any charge in respect of anything done or provided in pursuance of any specific requirement of the relevant statutory provisions.

[*HSWA 1974*, s 9]

Levels of duty

[IN20] Health and safety law contains different levels of duty:

Absolute

Absolute requirements must be complied with whatever the practicalities of the situation or the economic burden.

Practicable

The term 'practicable' means that measures must be possible in the light of current knowledge and invention.

Reasonably practicable

This term is contained in the main sections of *HSWA 1974* and many important regulations. It requires the risk to be weighed against the costs necessary to avert it (including time and trouble as well as financial cost). If, compared with the costs involved, the risk is very small (the HSE gives the example: to spend £1m to prevent five staff suffering bruised knees is obviously grossly disproportionate) then the precautions need not be taken – it should be noted that such a comparison should be made before any incident has occurred. The burden of proof, however, rests on the person with the duty (usually the employer) – they must prove why something was not reasonably practicable at a particular point in time. The duty holder's ability to meet the cost is not a factor to be taken into account. The HSE also points out that to spend £1 million to prevent a major explosion capable of killing 150 people is obviously proportionate.

In effect, considering what is 'reasonably practicable' requires that a risk assessment be carried out. The existence of a well-documented and carefully considered risk assessment would go a long way towards supporting a case on what was or was not reasonably practicable. Neither risks nor costs remain the same forever and what is practicable or reasonably practicable will change with time – hence the need to keep risk assessments up to date.

Key health and safety regulations

[IN20.1] Some of the most important health and safety regulations include the following (in alphabetical order rather than in order of importance):

Carriage of Dangerous Goods and Use of Transportable Pressure Equipment Regulations 2009 (SI 2009 No 1348)

[IN21] These Regulations impose requirements and prohibitions in relation to the carriage of dangerous goods by road and by rail and, in so far as they relate to safety advisers, by inland waterway.

See the chapter entitled DANGEROUS GOODS - CARRIAGE.

European Regulation on Classification, Labelling and Packaging of Substances and Mixtures (CLP Regulation) (Council Regulation (EC) 1272/2008)

The directly-acting European CLP Regulation aims to protect human health and the environment and allow the free movement of substances, mixtures and articles by harmonising the criteria for the classification of, and the rules on labelling and packaging for hazardous substances and mixtures. See the chapter entitled HAZARDOUS SUBSTANCES IN THE WORKPLACE.

Construction (Design and Management) Regulations 2015 (CDM) (SI 2015 No 51)

The Construction (Design and Management) Regulations 2015, SI 2015/51 cover the management of health, safety and welfare when carrying out construction projects and set out requirements for clients, domestic clients, designers, principal designers, principle contractors, contractors and workers. Organisations or individuals can take on the role of one or more dutyholders for a project, provided they have the skills, knowledge, experience and (in the case of an organisation) the organisational capacity to carry out that role in a way that secures health and safety.

The Regulations aim to ensure that those responsible for health and safety on construction sites: manage the risks by applying the general principles of prevention; appoint the right people and organisations at the right time; make sure that everyone has the information, instruction, training and supervision they need to carry out their jobs in a way that secures health and safety; cooperate and communicate with one another and coordinate their work; and consult workers and engage with them to promote and develop effective measures to secure health, safety and welfare.

See the chapters entitled CONSTRUCTION, DESIGN AND MANAGEMENT and CONSTRUCTION AND BUILDING OPERATIONS.

Control of Asbestos Regulations 2012 (SI 2012 No 632)

The regulations contain many quite detailed requirements relating to work with asbestos-containing materials (ACMs). Some types of asbestos work may only be carried out by organisations holding an appropriate licence, but even where this is not the case strict precautions must be taken. Many employers will be affected by the duty to manage asbestos in non-domestic premises contained in regulation 4. An assessment must be carried out of the possible presence of ACMs in the premises and also their condition. Suitable measures must then be put in place to manage the asbestos risk. This may involve the removal of ACMs but is more likely to require the establishment of an ACM inspection program and arrangements to prevent ACMs being disturbed or damaged.

See the chapter dealing with ASBESTOS.

Control of Major Accident Hazards (COMAH) Regulations 2015 (SI 2015 No 483)

The *Control of Major Accident Hazards (COMAH) Regulations 2015* lay down a comprehensive set of control systems with the aim of preventing major accidents, or mitigating the consequences should one occur. Both upper tier and lower tier operators must provide public information about their site and its hazards electronically and keep it up to date. See the chapter entitled MAJOR ACCIDENT HAZARDS.

Control of Noise at Work Regulations 2005 (SI 2005 No 1643)

These Regulations impose duties on employers and 'relevant' self-employed persons (see IN01 above) to protect both employees who may be exposed to risk from exposure to noise at work and others who might be affected by that work.

See the chapter NOISE AT WORK.

Control of Substances Hazardous to Health Regulations 2002 (COSHH) (SI 2002 No 2677)

These Regulations require an assessment to be made of all substances hazardous to health in order to identify means of preventing or controlling exposure. There are also requirements for the proper use and maintenance of control measures and for workplace monitoring and health surveillance in certain circumstances.

More detail is available in the chapter HAZARDOUS SUBSTANCES IN THE WORKPLACE.

[IN21] Introduction

Control of Vibration at Work Regulations 2005 (SI 2005 No 1093)

Employers are required to carry out assessments of risks to their employees from vibration, whether hand-arm vibration (HAV) or whole-body vibration. If exposure action values are exceeded employers must introduce technical and organisational measures to reduce exposure. The regulations set down exposure limit values which must not be exceeded.

See the chapter dealing with **VIBRATION**.

Dangerous Substances and Explosive Atmospheres Regulations 2002 (DSEAR) (SI 2002 No 2776)

These regulations apply to substances which have the potential to create risks from fire, explosion and exothermic reactions, including flammable gases and liquids and explosive dusts. Employers are required to carry out risk assessments of activities involving dangerous substances and to implement appropriate control measures.

See the chapter dealing with **DANGEROUS SUBSTANCES AND EXPLOSIVE ATMOSPHERES**.

Electricity at Work Regulations 1989 (SI 1989 No 635)

These Regulations contain requirements relating to the construction and maintenance of all electrical systems and work activities on or near such systems. They apply to all electrical equipment, from a battery-operated torch to a high-voltage transmission line.

See the chapter entitled **ELECTRICITY**.

Employers' Liability (Compulsory Insurance) Regulations 1998 (SI 1998 No 2573)

The Employers' Liability (Compulsory Insurance) Regulations 1998 are relevant to all employers in Great Britain holding employers' liability compulsory insurance. Most employers are legally required to insure against liability for injury or disease to their employees arising out of their employment.

See the chapter entitled **EMPLOYERS' LIABILITY INSURANCE**.

Health and Safety (Consultation with Employees) Regulations 1996 (SI 1996 No 1513)

These Regulations extend the previous requirements (contained in the *Safety Representatives and Safety Committees Regulations 1977 (SI 1977 No 500)* so that employers must now also consult workers not covered by trade union safety representatives.

See the chapter dealing with **JOINT CONSULTATION IN SAFETY – SAFETY REPRESENTATIVES, SAFETY COMMITTEES, COLLECTIVE AGREEMENTS AND WORKS COUNCILS**.

Health and Safety (Display Screen Equipment) Regulations 1992 (SI 1992 No 2792)

Where there is significant use of display screen equipment (DSE), employers must assess DSE workstations and offer 'users' eye and eyesight tests (which may necessitate provision of spectacles for DSE work).

See the chapter entitled **DISPLAY SCREEN EQUIPMENT**.

Health and Safety (First-Aid) Regulations 1981 (SI 1981 No 917)

Basic first-aid equipment controlled by an 'appointed person' must be provided for all workplaces. Higher risk activities or larger numbers of employees may require additional equipment and fully trained first-aiders.

See the chapter entitled **FIRST-AID**.

Health and Safety (Safety Signs and Signals) Regulations 1996 (SI 1996 No 341)

These Regulations require safety signs to be provided, where appropriate, for risks which cannot adequately be controlled by other means. Signs must be of the prescribed design and colours.

This subject is dealt with in some detail in the chapter on WORKPLACES – HEALTH, SAFETY AND WELFARE.

Health and Safety (Training for Employment) Regulations 1990 (SI 1990 No 1380)

Those receiving 'relevant training' (through training for employment schemes or work experience programmes) are treated as being 'at work' for the purposes of health and safety law. The provider of the 'relevant training' is deemed to be their employer – youth trainees and students on work experience placements therefore have the status of employees and must be protected accordingly. Matters relating to children and young persons are covered in the chapters **RISK ASSESSMENT** and **VULNERABLE PERSONS**.

Ionising Radiations Regulations 2017 (SI 2017 No 1075)

These regulations came into force on 1 January 2018 replacing the earlier *Ionising Radiations Regulation 1999*. The main changes introduced are a reduction in the dose limit for exposure to the lens of the eye, from 150mSv to 20mSv in a year; and changes to the requirement to notify, register or gain consent from the HSE to work with ionising radiation.

Lifting Operations and Lifting Equipment Regulations 1998 (LOLER) (SI 1998 No 2307)

Lifting equipment is defined in these regulations as 'work equipment for lifting and lowering loads'. As well as cranes, chain-blocks, passenger lifts and mobile elevating work platforms, this definition includes vehicle tail lifts, bath hoists and many other types of equipment. All lifting equipment must be of adequate strength and stability and meet other specific design and installation criteria. It must be subjected to periodic thorough examinations and inspections and all lifting operations must be properly planned, appropriately supervised and carried out in a safe manner. See the chapter dealing with **LIFTING OPERATIONS**.

Management of Health and Safety at Work Regulations 1999 (SI 1999 No 3242)

Employers and 'relevant' self-employed people (see **IN01**) are required to manage the health and safety aspects of their activities in a systematic and responsible way. The Regulations include requirements for risk assessment, the availability of competent health and safety advice and emergency procedures. Several of these management issues are dealt with later in this Introduction (see **IN25**) and the Regulations are covered in some detail in the chapter entitled **MANAGING HEALTH AND SAFETY**.

Manual Handling Operations Regulations 1992 (SI 1992 No 2793)

Manual handling operations involving risk of injury must either be avoided or be assessed by the employer, with steps taken to reduce the risk to the lowest level reasonably practicable. They require employers to provide general indications, and where reasonably practicable, precise information on the weight of the load and the heaviest side (where the weight of the load is uneven) to employees who are undertaking such operations. See the chapter entitled **MANUAL HANDLING**.

Personal Protective Equipment at Work Regulations 1992 (SI 1992 No 2966)

Employers must assess the personal protective equipment (PPE) needs created by their work activities, provide the necessary PPE, and take reasonable steps to ensure its use.

See the chapter entitled **PERSONAL PROTECTIVE EQUIPMENT**.

Provision and Use of Work Equipment Regulations 1998 (PUWER 1998) (SI 1998 No 2306)

These Regulations cover equipment safety, including the guarding of machinery. The definition of 'work equipment' also includes hand tools, vehicles, laboratory apparatus, lifting equipment, access equipment etc. The 1998 Regulations introduced additional requirements in respect of mobile work equipment and also replaced previous specific regulations relating to power presses, woodworking machines and abrasive wheels.

See the chapter entitled **MACHINERY SAFETY**.

Regulatory Reform (Fire Safety) Order 2005 (SI 2005 No 1541)

This Order replaced a vast range of previous legislation dealing with fire safety matters (including requirements for Fire Certificates). Enforcement of the Order is normally carried out by the local fire and rescue authority. It places a duty on the 'responsible person', who is either the employer in respect of a workplace under his control to any extent, or the person who has control of the premises, or the owner, to take general fire precautions to ensure the safety of employees and others. They must assess the fire risks and the adequacy of existing fire precautions, taking into account fire prevention measures, fire detection and alarm equipment, fire escape routes and evacuation procedures, firefighting equipment and fire-related signs. Fire safety regulation, including the Order, is under scrutiny following the Grenfell Tower fire (see **IN01**).

See the chapter dealing with **FIRE PREVENTION AND CONTROL**.

Reporting of Injuries, Diseases and Dangerous Occurrences Regulations 2013 (RIDDOR) (SI 2013 No 1471)

Fatal accidents, specified injuries, occupational diseases and dangerous occurrences (as defined in the Regulations) must be reported immediately to the enforcing authority. Accidents involving seven or more days' incapacity must be reported within 15 days.

See the chapter on **ACCIDENT REPORTING AND INVESTIGATION**.

Safety Representatives and Safety Committees Regulations 1977 (SI 1977 No 500)

Members of recognised trade unions may appoint safety representatives to represent them formally in consultations with their employer in respect of health and safety issues. The functions and rights of safety representatives are detailed in the Regulations. The employer must establish a safety committee if at least two representatives request this in writing.

The *Trade Union Act 2016* gave ministers the power to make new regulations requiring public authorities, that is employers that are mainly publicly funded, with at least one union representative to record and publish details about the time they spend on their union role (known as facility time), and any facilities provided.

The *Trade Union (Facility Time Publication Requirements) Regulations 2017* came into force on 1 April 2017 and require public authorities with 49 or more full-time equivalent staff for the 12-month period monitored to provide information on: the number of employees who were union officials in that period; the number who spent set proportions of their working hours on facility time (0%, 1–50%, 51–99% and 100%); the percentage of the organisation's pay bill spent on facility time; and the proportion of paid facility time that was spent on trade union 'activities' (which do not require paid time off by law).

The Act also contains reserve powers allowing ministers to make regulations capping the total amount of paid facility time in public authorities, including facility time for safety reps carrying out their statutory functions under these regulations. They cannot exercise this power until at least three years' worth of data on facility time has been collected and analysed, and only after consultation, followed by 12 months' notice to the employer.

See the chapter entitled **JOINT CONSULTATION IN SAFETY – SAFETY REPRESENTATIVES, SAFETY COMMITTEES, COLLECTIVE AGREEMENTS AND WORKS COUNCILS**.

Work at Height Regulations 2005 (SI 2005 No 735)

The Regulations apply to all work at height where there is a risk of a fall liable to cause personal injury. They require risk assessments to be carried out for all work at height activities to make sure all work at height is planned, organised and carried out by a competent person. They set out a hierarchy for managing the risks from work at height.

See the chapter on **WORK AT HEIGHT**.

Working Time Regulations 1998 (SI 1998/1833)

The majority of workers are legally entitled to a maximum working week, paid holidays, and to breaks and rest periods, as a result of the *Working Time Regulations 1998*.

See the chapter entitled **WORKING TIME**.

Workplace (Health, Safety and Welfare) Regulations 1992 (SI 1992 No 3004)

Physical working conditions in the workplace together with safe access for pedestrians and vehicles and welfare provisions are covered by these Regulations.

See the chapters entitled **WORKPLACES – HEALTH, SAFETY AND WELFARE** and **ACCESS, TRAFFIC ROUTES AND VEHICLES**.

Civil law

[IN22] A civil action can be initiated by an employee who has suffered injury or damage to health caused by their work. This may be based upon the law of negligence, ie where the employer has been in breach of the duty of care which he owes to the employee. Being part of the common law, the law of negligence has evolved, and continues to evolve, by virtue of decisions in the courts – Parliament has had virtually no role to play in its development.

As a result of an amendment to the *Health and Safety at Work etc Act 1974*, introduced by the Enterprise and Regulatory Reform Act 2013, employers are now only liable for civil damages in health and safety cases if they can be shown to have acted negligently. This ended the previous situation where businesses could be automatically liable for damages if they breached health and safety law, even if they were not actually negligent (see **IN01** above).

Duty of care

Every member of society is under a 'duty of care', ie to take reasonable care to avoid acts or omissions which they can reasonably foresee are likely to injure their neighbour (anyone who ought reasonably to have been kept in mind). What is 'reasonable' will depend upon the circumstances.

Employers owe a duty of care not only to employees but also to such people as contractors, visitors, customers, and people on neighbouring property. In the case of the duty of care owed by employers to employees, it includes the duty to provide:

- safe premises;
- a safe system of work;
- safe plant, equipment and tools; and
- safe fellow workers.

Occupiers of premises are under statutory duties comprised in the *Occupiers' Liability Acts* of 1957 and 1984 (see OCCUPIERS' LIABILITY). Those suffering an injury caused by a defect in a product may sue the producer or importer under the *Consumer Protection Act 1987* (see PRODUCT SAFETY).

Vicarious liability

Employers are liable to people injured by the wrongful acts of their employees, if such acts are committed in the course of their employment. If an employee's careless driving of a forklift truck injures another employee (or a contractor or customer), the employer is likely to be liable. There is no vicarious liability if the act is not committed in the course of employment. The employer is not likely to be held liable if one employee assaults another.

(See VIOLENCE IN THE WORKPLACE for recent case law on interpreting 'in the course of employment'.)

Civil procedure

Civil actions for negligence must commence within three years from the time of knowledge of the cause of action, the date on which the plaintiff knew or should have known that there was a significant injury and that it was caused by the employer's negligence. The plaintiff must be prepared to prove their case in the courts, but in practice most cases are settled out of court following negotiations between the plaintiff's legal representatives and the employer's insurers or their representatives. The *Employers' Liability (Compulsory Insurance) Act 1969* requires employers to be insured against such actions (see **EMPLOYERS' LIABILITY INSURANCE**), although some public bodies, for example local authorities, are exempt from the provisions of this Act.

Damages

Damages are assessed under a number of headings including:

- *loss of earnings (prior to trial);*
- *damage to property etc;*
- *pain and suffering (before and after trial);*
- *future loss of earnings;*
- *disfigurement;*
- *medical or nursing expenses; and*
- *inability to pursue personal or social interests or activities.*

Defences

The plaintiff must prove breach of a statutory duty or of the duty of care on a balance of probabilities. However, a number of defences are available to the employer, including:

Contributory negligence
- The employer may claim that the injured person was careless or reckless, for example, that they ignored clear safety rules or disobeyed instructions. Accidental errors are distinguished from a failure to take reasonable care. Damages will be reduced by the percentage of contributory negligence established, which will vary with the facts of each case.

Injuries not reasonably foreseeable
- The employer may claim that the injuries were beyond normal expectation or control (an act of God). In cases of noise-induced hearing damage, mesothelioma (an asbestos-related cancer) or vibration-induced white finger, the courts have established dates after which a reasonable employer should have been aware of the relevant risks and taken precautions.

Voluntary assumption of risk
- If an employee consents to take risks as part of the job, the employer may escape liability. However, this defence (*volenti non fit injuria*) cannot be used for cases involving breach of statutory duty – no one can contract out of their statutory obligations or be deprived of statutory protection.

Other civil actions

Other health and safety related situations may result in civil actions by employees. Employment protection legislation relates to dismissals or redundancies resulting from health and safety activities (including refusal to work in situations of serious and imminent danger). Suspension or dismissal on maternity or medical grounds may also give a right of action. See EMPLOYMENT PROTECTION.

Management of health and safety at work

[IN23] The Institution of Occupational Safety and Health (IOSH) guidance to health and safety management systems, states in its introduction that:

> 'IOSH recognises that work-related accidents and ill-health can be prevented and well-being at work can be improved if organisations manage health and safety competently and apply the same or better standards as they do to other core business activities. We believe that the formal [occupational safety and health management systems] OSHMSs mentioned in this guidance, and others based on similar principles, provide a useful approach to achieving these goals.
> ['*Systems in focus - guidance on occupational safety and health management systems*' IOSH, Wigston.]'

The continuing development and adoption of formal OHSMSs is a reflection of this conviction – such systems include those developed by HSE 'Managing for Health and Safety', (HSG65), by the ILO 'Guidelines on occupational health and safety management systems', (ILO-OSH 2001), ILO Geneva and the new BS 45002-0:2018 Occupational health and safety management systems. General guidelines for the application of ISO 45001, published in March 2018 (BSI, London). Each system exhibits differences when compared to the others, but there is a similarity to the core elements of Plan-Do-Check-Act (see Figure 1).

This theme is developed considerably in the chapter entitled MANAGING HEALTH AND SAFETY.

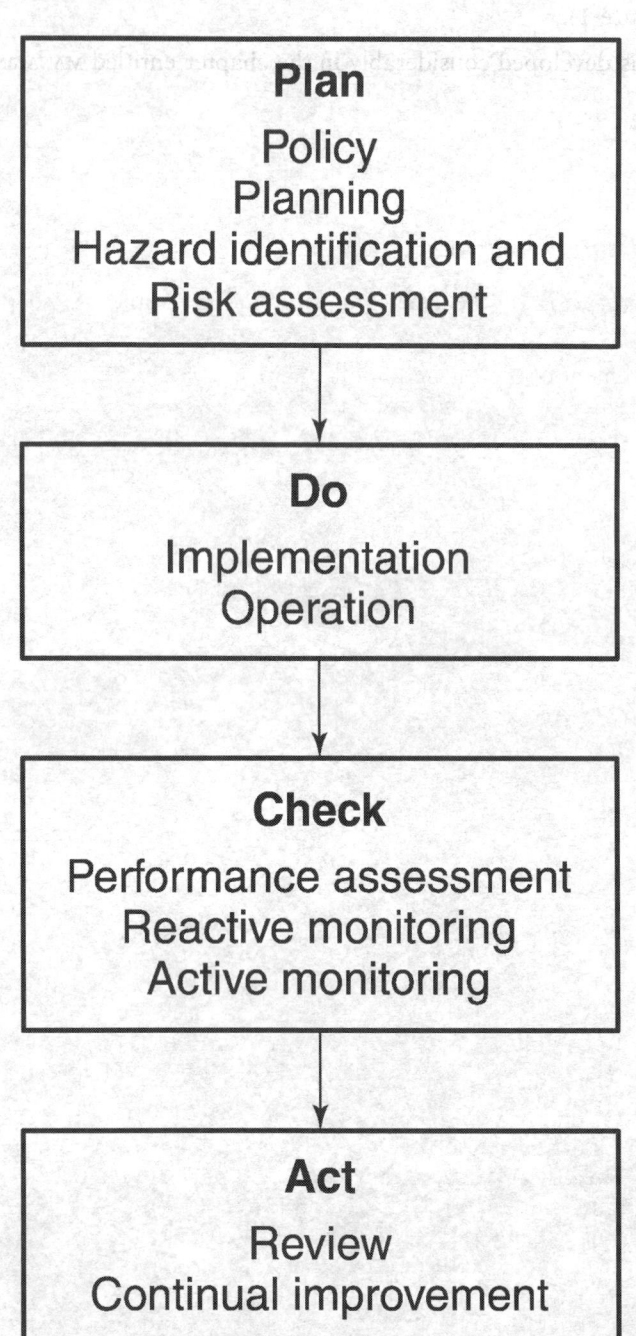

Figure 1: Plan-Do-Check-Act

The costs of accidents and ill health

[IN24] The latest HSE statistics show that 30.7 million working days were lost due to work-related ill health and non-fatal workplace injuries in 2017/18. The total annual cost of work-related injuries and ill health is estimated to be around £15 billion.

The majority of costs fall on individuals, while employers, government and taxpayers bear a similar proportion of the costs of workplace injury and ill health.

Health and safety policies

[IN25] *Section 2(3)* of the *HSWA 1974* requires employers to prepare in writing:

- a statement of their general policy with respect to the health and safety at work of their employees; and
- the organisation and arrangements for carrying out the policy.

It also requires the statement to be brought to the notice of all employees – employers with fewer than five employees are exempt from this requirement.

Policies are normally divided into three sections, to meet the three separate demands of *HSWA 1974*:

(i) The statement of intent

This involves a general statement of good intent, usually linked to a commitment to comply with relevant legislation. Many employers extend their policies so as to relate also to the health and safety of others affected by their activities. In order to demonstrate clearly that there is commitment at a high level, the statement should preferably be signed by the chair, chief executive or someone in a similar position of seniority.

(ii) Organisational responsibilities

It is vitally important that the responsibilities for putting the good intentions into practice are clearly identified. In a small organisation, this may be relatively simple but larger employers should identify the responsibilities held by those at different levels in the management structure. While reference to employees' responsibilities may be included, it should be emphasised that the law requires the employer's organisation to be detailed in writing. Types of responsibilities to be covered in the policy might include:

- making adequate resources available to implement the policy;
- setting health and safety objectives;
- developing suitable procedures and safe systems;
- delegating specific responsibilities to others;
- monitoring the effectiveness of others in carrying out their responsibilities;
- monitoring standards within the workplace; and
- feeding concerns up through the organisation.

(iii) Arrangements

The policy need not contain all of the organisation's arrangements relating to health and safety but should contain information as to where they might be found, for example risk assessment records, in a separate health and safety manual or within various procedural documents. Topics which may require detailed arrangements to be specified are:

- operational procedures relating to health and safety;
- training;
- personal protective equipment;
- health and safety inspection programmes;
- accident and incident investigation arrangements;
- fire and other emergency procedures;
- first aid;
- occupational health;
- control of contractors and visitors;
- consultation with employees; and
- audits of health and safety arrangements.

Employees must be aware of the policy and, in particular, must understand the arrangements which affect them and what their own responsibilities might be. They may be given their own copy (for example, within an employee handbook) or the policy might be displayed around the workplace. With regard to some arrangements detailed briefings may be necessary, for example as part of induction training.

Employers must revise their policies as often 'as may be appropriate'. Larger employers are likely to need to arrange for formal review and, where necessary, for revision to take place on a regular basis. Dating of the policy document is an important part of this process.

More detail is provided in the chapter entitled STATEMENTS OF HEALTH AND SAFETY POLICY.

Sources of health and safety advice

[IN26] In *Managing for health and safety*, the HSE emphasises the importance of establishing a positive health and safety culture within an organisation as a prerequisite of effective health and safety management. It refers to the 'four Cs' as key components in establishing such a culture: control, competence, communication and co-operation.

HSE guidance on managing for health and safety has moved from using the POPMAR (Policy, Organising, Planning, Measuring performance, Auditing and Review) model to a 'Plan, Do, Check, Act' approach in order to achieve a better balance between the systems and behavioural aspects of management. This approach also treats health and safety management as an integral part of good management generally, rather than as a stand-alone system.

While competence in health and safety matters is relevant throughout any workforce, it is particularly important at management levels. The *Management of Health and Safety at Work Regulations 1999 (SI 1999 No 3242)* have taken this concept further by requiring (in *Reg* 7) every employer to appoint one or more competent persons to assist them in complying with the law. Full-time or part-time specialists may be appointed, or use may be made of external consultants, although the 1999 Management Regulations state a preference for employees. Smaller employers may appoint themselves, provided that they are competent. Competence involves having the necessary skills, experience and knowledge to manage health and safety. An awareness of the limits of one's own knowledge and capabilities is also important.

Health and safety training is available from many different sources. The following organisations either provide training themselves or oversee training through accredited training centres.

- National Examination Board in Occupational Safety and Health (NEBOSH)
  tel: 0116 263 4700/www.nebosh.org.uk
- Institution of Occupational Safety and Health (IOSH)
  tel: 0116 257 3100/www.iosh.co.uk
- Chartered Institute of Environmental Health (CIEH)
  tel: 020 7827 5800/www.cieh.org
- Royal Society for the Prevention of Accidents (RoSPA)
  tel: 0121 248 2000/www.rospa.com
- British Safety Council (BSC)
  tel: 020 3510 8355/www.britsafe.org

The Occupational Safety and Health Consultants Register (OSHCR) can be found at www.oshcr.org. In order to join the register, consultants must have achieved at least one of the following:

- Chartered status with IOSH; Chartered Member of the British Psychological Society; CIEH; or Royal Environmental Health Institute of Scotland with health and safety qualifications;
- Fellow status with the International Institute of Risk and Safety Management with degree level qualifications;
- Chartered Member or Chartered Fellow status with the British Occupational Hygiene Society Faculty of Occupational Hygiene; or
- Registered Member or Fellow status with the Chartered Institute of Ergonomics and Human Factors.

All consultants wishing to join the register are also asked to declare that they will:

- demonstrate adequate continuing professional development;
- abide by their professional body's code of conduct;
- provide sensible and proportionate advice; and
- have professional indemnity insurance or equivalent to cover the nature of their duties.

Consultants who have joined the register are searchable via industry, topic, location and keyword; many also list the specific services which they are competent to provide.

The HSE is also a valuable source of information and advice:

- HSE website www.hse.gov.uk

- HSE Books https://books.hse.gov.uk; hseorders@tso.co.uk; 0333 202 5070
    The HSE's huge range of publications is freely available as pdf downloads from its website. Priced publications are also available from HSE Books, as are a number of free-issue leaflets. The online guidance, *The health and safety toolbox: How to control risks at work*, is an excellent starting point for SMEs and can be found at: www.hse.gov.uk/toolbox.

Access, Traffic Routes and Vehicles

Nicola Coote

Introduction to access, traffic routes and vehicles

[A1001] With the regular daily flow of labour to and from the workplace and vehicles making deliveries and collecting items, access and egress points cause many hazardous situations. For this reason there is a duty on employers and occupiers to 'provide and maintain' safe access to and egress from a place of work both under statute and at common law. As part and parcel of compliance with the *Building Regulations 2010 (SI 2010/2214)*, this includes access facilities for disabled workers and visitors (that is, persons who have difficulty walking or are wheelchair users, or, have a hearing problem or impaired vision) (see further **W11043** WORKPLACES – HEALTH, SAFETY AND WELFARE). Requirements to ensure safe access to those with physical, sensory or mental impairments is further stipulated in the *Equality Act 2010 (2010, c 15)*.

It also includes proper precautions to effect entry or exit to or from confined spaces, where a build-up of gas/combustible substances can be a real though not obvious danger (see **A1022** below). Statutory requirements consist of general duties under *s 2(2)(d)* of the *Health and Safety at Work etc Act 1974 (HSWA)*, which apply to all employers, and the more specific duties of the *Workplace (Health, Safety and Welfare) Regulations 1992 (SI 1992 No 3004)*.

The term 'access' is a comprehensive one and refers to just about anything that can reasonably be regarded as means of entrance/exit to a workplace or working area, even if it is not the usual method of access/egress. Unreasonable means of access/egress would not be included, such as a dangerous short-cut, particularly if management has drawn a worker's attention to the danger, though the fact that a worker is a trespasser has not prevented recovery of damages. This is demonstrated in *Westwood v The Post Office* [1973] 3 All ER 184 where an employee claimed damages for injuries sustained after walking through an access door which had a notice stating not to enter. The proper procedure is for the employer/factory occupier to designate points of access/egress for workers and see that they are safe, well-lit, maintained and (if necessary) manned and de-iced. In the case of *Fildes v International Computers (1984), unreported*, an employee slipped on a patch of ice and injured his back whilst walking from his car in the car park to the factory gates. It was an agreed means of access.

Moreover, the statutory duties apply to access/egress points to any place where any employees have to work, and not merely their normal workplace. This includes getting in and out of the workplace, and around the workplace. It includes inside and outside of buildings, and general walkways through working areas, walkways in car parks and other external areas of the work site and access into non-routine areas such as basements.

[A1001] Access, Traffic Routes and Vehicles

The *Occupiers' Liability Act 1984* (1984, c 3) imposes a duty on occupiers to take reasonable care for the safety of trespassers in respect of any risk of their suffering injury by reason of any danger due to the state of the premises or to things done or omitted to be done on them. This naturally includes access routes. The threshold test in *s 1(3)* of the Act requires that a duty is owed to trespassers in respect of any such risk if:

(a) the occupier is aware of the danger or has reasonable grounds to believe that it exists, eg access into an area where the trespasser could be seriously injured;

(b) the occupier knows or has reasonable grounds to believe that the trespasser is in the vicinity of the danger or that he may come into the vicinity of the danger; and

(c) the risk is one against which, in all the circumstances of the case, the occupier may reasonably be expected to offer the trespasser some protection, ie that the access route not controlled to deter or prevent anyone from getting through.

This chapter summarises key statutory and common law duties in connection with:

(a) access and egress;
(b) vehicular traffic routes for internal traffic and deliveries;
(c) work vehicles and delivery vehicles – with particular emphasis on hazardous activities involving such vehicles; and
(d) work in confined spaces, of necessity involving access and egress points.

In addition, where access routes lead to routes above ground level, requirements for safe work at heights may need to be implemented (see further **W9001** – WORK AT HEIGHTS).

Access for those with disabilities

[A1001.1] The *Equalities Act 2010* requires employers to consider the needs of people with physical, sensory or mental impairments when designating and designing access/egress of workers, and BS8300:2010 is a very useful guidance document for employers to refer when deciding what should be considered for a range of access needs.

The *Equality Act 2010*, defines a disabled person as someone with:

> 'a physical or mental impairment which has a substantial and long-term adverse effect on his ability to carry out normal day-to-day activities.'

Since 1 October 2004 duties under *Disability Discrimination Act 1995* affect all employers with fewer than 15 employees and anyone who provides a service to the public. Whilst this has been replaced by the *Equality Act 2010* the general requirements remain the same with regards to provision for access and egress at the workplace.

The provisions, include requiring employers to consider making changes to the physical features of premises that they occupy. One way in which an employer might unlawfully discriminate against a disabled employee is by not making reasonable adjustments (without justification) and this includes adjusting the work to enable access and safe traffic routes. A Code of Practice - 'Elimination of discrimination in the field of employment against disabled persons or persons who have had a disability' gives general guidance on the main employment provisions of the Act. Further practical information on standards that should be achieved to cater for the needs of those with physical or sensory impairment can be found in BS 8300:2009+A1:2010 Design of buildings and their approaches to meet the needs of disabled people.

Statutory duties concerning workplace access and egress – Workplace (Health, Safety and Welfare) Regulations 1992 (SI 1992 No 3004)

[A1002] Statutory duties centre around:

(a) the organisation of safe workplace transport systems;
(b) the suitability of traffic routes for vehicles and pedestrians; and
(c) the need to keep vehicles and pedestrians separate.

Organisation of safe workplace transport systems

[A1003] Every workplace must be so organised that pedestrians and vehicles can circulate in a safe manner. [*Reg 17(1)*].

Suitability of traffic routes

[A1004] Traffic routes in a workplace must be suitable for the persons or vehicles using them, sufficient in number, in suitable positions and of sufficient size. [*Reg 17(2)*].

More particularly:

(a) pedestrians or vehicles must be able to use traffic routes without endangering those at work;
(b) there must be sufficient separation of traffic routes from doors, gates and pedestrian traffic routes, in the case of vehicles;
(c) where vehicles and pedestrians use the same traffic routes, there must be sufficient space between them; and
(d) where necessary, all traffic routes must be suitably indicated.

[Reg 17(3), (4)].

Compliance with these statutory duties involves provision of safe access for:

(i) vehicles, with attention being paid to design and layout of road systems, loading bays and parking spaces for employees and visitors; and
(ii) pedestrians, so as to avoid their coming into contact with vehicles.

(Traffic routes in existence before 1 January 1993 shall comply with regulation 17(2) and (3) only to the extent that it is reasonably practicable).

Traffic routes for vehicles

[A1005] There should be sufficient traffic routes to allow pedestrians and vehicles to circulate safely and without difficulty. The onus of protection is upon the pedestrians to ensure that they cannot come into contact/be injured by moving traffic. This can be a challenge for pre-existing workplace layouts where there is limited room for vehicles to circulate and where historically no room has been made available for pedestrian walkways. Workplaces being designed or redeveloped should ensure that where there are vehicles and pedestrians likely to be in the same working area, that suitable walkways, barriers and other protective measures are included in the design specification.

As for internal traffic, lines marked on roads/access routes in and between buildings should clearly indicate where vehicles are to pass eg fork lift trucks. Obstructions, such as limited headroom, are acceptable if clearly indicated. Temporary obstacles should be brought to the attention of drivers by warning signs or hazard cones; alternatively access should be prevented or restricted. Both internal and delivery traffic should be subject to sensible speed limits (eg 10 mph), which should be clearly displayed. Speed ramps (sleeping policemen), preceded by a warning sign or mark, are necessary on workplace approaches, except where fork lift trucks are used. The traffic route should be wide enough to allow vehicles to pass and repass oncoming or parked traffic, and it may be advisable to introduce one way systems or parking restrictions. It may also be necessary to make allowance for pedestrian routes, where these interact with the vehicle traffic route. Traffic signs on roads, for example speed limit signs, must conform with those on public roads, whether or not the road is subject to the *Road Traffic Regulation Act 1984* (1984, c 7).

Checklist – safe traffic routes

[A1006] Safe traffic routes should:

(a) provide the safest route possible between places where vehicles have to call or deliver;

(b) be wide enough for the safe movement of the largest vehicle, including visiting vehicles (eg articulated lorries, ambulances etc) and should allow vehicles to pass oncoming or parked vehicles safely. One way systems or parking restrictions are desirable;

(c) avoid vulnerable areas/items, such as fuel or chemical tanks or pipes, open or unprotected edges, and structures likely to collapse;

(d) incorporate safe areas for loading/unloading;

(e) avoid sharp or blind bends; if this is not possible, hazards should be indicated (eg blind corner) or facilities to help reduce risk such as mirrors to see round the bend, should be used;

(f) ensure that road/rail crossings are kept to a minimum and are clearly signed and controlled to prevent access when there is danger, so far as is reasonably practicable;

(g) ensure that entrances/gateways are wide enough; if necessary, to accommodate a second vehicle that may have stopped, without causing obstruction;

(h) set sensible speed limits, which are clearly signposted. Where necessary, ramps should be used to reduce speed, and road humps or bollards to restrict the width of the road. These should be preceded by a warning sign or mark on the road;
(i) ensure that fork lift trucks should not have to pass over road humps, unless of a type capable of doing so;
(j) give prominent warning of limited headroom, both in advance and at an obstruction. Overhead electric cables or pipes containing flammable/hazardous chemicals should be shielded, ie using goal posts, height gauge posts or barriers;
(k) ensure that routes on open manoeuvring areas/yards are marked and signposted, and banksmen are employed to supervise the safe movement of vehicles, particularly when reversing manoeuvres are undertaken;
(l) ensure that people at risk from exhaust fumes or material falling from vehicles are screened or protected;
(m) restrict vehicle access where high-risk substances are stored (eg LPG) and where refuelling takes place;
(n) avoid the need to reverse or have use of banksmen and instead enable one way systems;
(o) consider installation of refuge points (safe havens) where one-way systems are not feasible and vehicles need to reverse into delivery areas or dead ends, or position barriers to prevent vehicles reversing from colliding into people; and
(p) ensure safety of users from falling if the access route is above ground level (see also section W9001 Work at Height and W9004 Application to Work at Height Regulations).

Traffic routes for pedestrians

Checklist – safe traffic routes

[A1007] In the case of pedestrians, the main object of the traffic route is to prevent their coming into contact with vehicles. Safe traffic routes should:

(a) provide separate routes/pavements for pedestrians, to keep them away from vehicles;
(b) where necessary, provide suitable barriers/guard rails at entrances/exits and at the corners of buildings;
(c) where traffic routes are used by both pedestrians and vehicles, be wide enough to allow vehicles to pass pedestrians safely;
(d) where pedestrian and vehicle routes cross, provide appropriate crossing points. These should be clearly marked and signposted. If necessary, barriers or rails should be provided to prevent pedestrians crossing at dangerous points and to direct them to designated crossing points;
(e) where traffic volume is high, traffic lights, bridges or subways should be used to control movement and ensure a smooth, safe flow;
(f) where crowds use or are likely to use roadways, eg at the end of a shift, stop vehicles from using them at such times;

(g) provide separate vehicle and pedestrian doors in premises, with vision panels on all doors;
(h) provide high visibility clothing for people permitted in delivery areas (eg bright jackets/overalls);
(i) where the public has access (eg at a farm or factory shop), public access points should be as near as possible to shops and separate from work activities;
(j) ensure the walkways next to traffic routes are constructed from non-slip materials and are free from obstacles that could cause someone to slip or trip into the road.

(See also W11014 WORKPLACES.)

Vehicles

[A1008] Vehicles account for a high percentage of deaths and injuries at work. According to the latest statistics from the Health and Safety Executive (HSE) being struck by a moving vehicle accounted for 31 fatal injuries to workers in 2016/17 compared with 28 in 2015/16 and an annual average of 25 over the period 2012/13–2016/17 (17% of the total fatalities recorded) and 7% of serious injuries were caused by moving objects.

Many of these casualties occur whilst vehicles are reversing, though activities such as loading and unloading, sheeting and unsheeting, as well as cleaning, can similarly lead to injuries, especially where employees are struck by a falling load or a fall from a height on, say, a tanker or HGV.

So, too, climbing and descending ladders on tankers during delivery and 'dipping' at petrol forecourts can be hazardous, access onto vehicles and egress being as important as design and construction. Tipping also has its dangers, with tipping vehicles, tipping trailers and tankers overturning in considerable numbers. Sheeting and unsheeting operations have led to sheeters slipping or losing their grip or falling whilst walking on top of loads or in consequence of ropes breaking; absence of, or inadequate, training being an additional factor in injuries involving work vehicles. This part of the chapter considers the general statutory requirements relating to vehicles at work, precautions in connection with potentially hazardous operations involving vehicles, as well as providing a checklist for vehicle safety. Specific construction and use requirements are not considered. When considering the risk of sheeting operations etc when workers are gaining access to the top of loads, employers must also consider risks associated with Work at Heights (see section W9001).

General statutory requirements

[A1009] Both work and private vehicles come within the parameters of health and safety at work. Regarding work vehicles, employers have the direct responsibilities of provision and maintenance generally under *HSWA, s 2*, and, more specifically, under the *Provision and Use of Work Equipment Regulations 1998 (SI 1998 No 2306) (PUWER)*, for vehicles qualifying as 'work equipment'. Regarding private vehicles being driven on workplace access routes, employers have much less control – at least, as far as design, construction and use of the vehicle is concerned. However, they do have control of the access route itself and should ensure regulated use via:

(a) restricted routes and access;
(b) provision of clearly signposted parking areas away from hazardous activities and operations;
(c) enforcement of speed limits; and
(d) clearly marked areas allocated for parking/areas where parking is not permitted.

Work vehicles

[A1010] Generally, work vehicles should be as safe, stable, efficient and roadworthy as private vehicles on public roads. As work equipment, they are subject to the controls of *PUWER*, which specifies provision, maintenance, access and safety provisions in the event of rolling or falling over. Employers must also ensure that drivers are suitably trained in conformity with the requirements of the *Management of Health and Safety at Work Regulations 1999 (SI 1999 No 3242)* and *PUWER*.

Provision

[A1011] All employers must ensure that vehicles:

(a) are constructed and adapted as to be suitable for its purpose;
(b) when selected, caters for risks to the health and safety of persons where the vehicles are to be used; and
(c) are only used for operations specified and under suitable conditions.

[PUWER, Reg 5].

Compliance with these requirements on the part of operators of HGVs, fork lift trucks, dump trucks and mobile cranes presupposes conformity with the following checklist, namely:

(i) a high level of stability;
(ii) safe means of access and egress to and from the cab;
(iii) suitable and effective service and parking brakes;
(iv) windscreens with wipers and external mirrors giving optimum all-round visibility;
(v) a horn, vehicle lights, reflectors, reversing lights, reversing alarms;
(vi) suitable painting/markings so as to be conspicuous;
(vii) provision of a seat and seat belts;
(viii) guards on dangerous parts (eg power take-offs);
(ix) driver protection to prevent injury from overturning, and from falling objects or materials;
(x) driver protection from adverse weather;
(xi) training and competence in line with current industry guidance for the specific vehicle; and
(xii) medicals for drivers of HGVs (everyone wishing to drive an HGV or PCV professionally must undergo a comprehensive medical exam prior to earning a first licence, and at regular intervals afterwards). Existing

[A1011] Access, Traffic Routes and Vehicles

lorry drivers are required by the DVLA to have a medical at the age of 45 and then every five years until the age of 65. After 65, the medical has to be done every year). The medical will include: eyesight; general health; cardiovascular health; mental health; nervous system problems; diabetes mellitus.

Maintenance

[A1012] All employers must ensure that work equipment (including vehicles) is maintained in an efficient state, in efficient working order and in good repair. [*PUWER, Reg 6*]. This combines the need for basic daily safety checks by the driver before using the vehicle, as well as preventive inspections and services carried out at regular intervals of time and/or mileage, in accordance with manufacturers recommendations. As regards basic daily safety checks, employers should provide drivers with a log book in which to record visual inspections undertaken and the findings of the following:

– brakes, tyres, steering, mirrors, windscreen washers and wipers, warning signals and specific safety systems (eg control interlocks)

Employers should see that drivers carry out the checks.

Training of drivers

[A1013] All employers must:

(a) in entrusting tasks to employees, take into account their capabilities as regards health and safety; and

(b) ensure that employees are provided with adequate health and safety training on recruitment and exposure to new or increased risks.

[Management of Health and Safety at Work Regulations 1999 (SI 1999 No 3242)].

In order to conform with these requirements, employers should ensure that, for general purposes, drivers of work vehicles are over 17 and have passed their driving test or, in the case of drivers of HGVs, that they are over 21 and have passed the HGV test, and associated medical (see **A1010**). Moreover, to ensure continued competence, or to accommodate new risks at work or a changing work environment, employers should provide safety updates on an on-going basis as well as refresher training. They should further require approved drivers to report any conviction for a driving offence, whether or not involving a company vehicle. One method for ensuring this is to require all drivers should undergo at least an annual licence check to ensure it remains valid. This now requires application to the DVLA as the paper part of the driving license is no longer issued.

Drivers of company vehicles may not be subject to a medical but should be subject to the other forms of monitoring discussed above. This includes at least an annual driving licence check to ensure it remains valid, and consideration of additional training. This often has additional benefits of reduced insurance premia as well as reduced risk of accidents whilst driving, and damage to the vehicle.

Vehicles [A1015]

Contractors and subcontractors

[A1014] Similar assurances (see A1013 above) should be obtained from drivers of contractors and subcontractors visiting an employer's workplace. If they are not forthcoming, permission to work on site or in-house should be refused until either the contractor's vehicles comply with statutory requirement and/or his drivers are adequately trained. Training of contractor's drivers would normally be undertaken by contractors themselves, though site or in-house hazards, routes to be used etc should be communicated to contractors by employers or occupiers. Contractors should be left in no doubt of the penalties involved for failure to conform with safe working practices – a useful way of ensuring enforcement on the part of contractors and subcontractors is to issue a licence.

Increasingly, contractors are multinational and many have limited or no ability to speak, read or write English. Therefore employers should take this into account when vehicles from other countries arrive, and provide them with means so they can understand the access and traffic route arrangements for the site. This is often best achieved either via translation into the main language(s) of delivery drivers or via pictures and diagrams.

Access to vehicles

[A1015] In addition to *Regulation 5* of *PUWER* (see A1011 above), employers (and others having control, to any extent, of workplaces (see OCCUPIERS' LIABILITY)), who operate/use vehicles, are subject to:

(a) the fall prevention requirements of *Regulation 13* of the *Workplace (Health, Safety and Welfare) Regulations 1992 (SI 1992 No 3004)* (see W9003 WORK AT HEIGHTS). As far as possible, compliance with this regulation would obviate the need for climbing on top of vehicles (by bottom-filling) and also require vehicle operators to ensure that loads are evenly distributed, packaged properly and secured in the interests of drivers going down slopes and up steep hills (see further A1018 below).

(b) the co-operation requirements of *Regulation 11* of the *Management of Health and Safety at Work Regulations 1999 (SI 1999 No 3242)*, specifying that where activities of different employers interact, different employers may need to co-operate with each other and co-ordinate preventive and protective measures. This regulation is particularly relevant for example, to tanker deliveries and 'dipping' at petrol forecourts as well as loading and/or unloading operations. For 'dipping' purposes or gaining top access to tankers, access should be by a ladder at the front or rear, such ladders being properly constructed, maintained and securely fixed; ideally, they should incline inwards towards the top. There should be a means of preventing people from falling whilst on top of the tanker. Failing this, employers of tanker drivers and forecourt owners should liaise on potential risks involved in tanker deliveries, eg the provision of suitable step-ladders and a safe system of work on the part of the latter. As for carriage of goods and loading/unloading operations, consignors should ensure that goods are evenly distributed, properly packaged and secured.

Potentially hazardous operations

[A1016] The following activities and operations are potentially hazardous in connection with vehicles.

Reversing

[A1017] Approximately a quarter of all deaths involving vehicles at work are caused by reversing vehicles; in addition, negligent reversing can result in costly damage to premises, plant and goods. Where possible, workplace design should aspire to obviate the need for reversing by the incorporation of one-way traffic systems. Failing this, reversing areas should be clearly identified and marked, and non-essential personnel excluded from the area. Ideally, banksmen wearing high-visibility clothing should be in attendance to guide drivers through, and keep non-essential personnel and pedestrians away from the reversing area. Refuge points, also known as safe havens, should be constructed where possible to enable an escape route or safe place for people to stay in the event of a vehicle reversing in an unsafe manner. Vehicles should be fitted with external side-mounted and rear-view mirrors – as, indeed, many now are. Mirrors are advisable for enabling drivers to see round 'blind spots' and corners, and closed-circuit television systems enable remote monitoring of traffic circulation to ensure that procedures are being followed.

Loading and unloading

[A1018] Because employees can be seriously injured by falling loads or overturning vehicles, loading/unloading should not be carried out:

(a) near passing traffic, pedestrians and other employees;
(b) where there is a possibility of contact with overhead electric cables;
(c) on steep gradients;
(d) unless the load is spread evenly (racking will assist load stability);
(e) on soft/uneven ground;
(f) unless the vehicle has its brakes applied or is stabilised (similarly with trailers); or
(g) with the driver in the cab.

(See also OFFICES AND SHOPS.)

Tipping

[A1019] Overturning of lorries and trailers is the main hazard associated with tipping. In order to minimise the potential for injuries, tipping operations should only occur:

(a) after drivers have consulted with site operators and checked that loads are evenly distributed;
(b) when non-essential personnel are not present;
(c) on level and stable ground away from power lines and pipework; and
(d) with the driver in the cab and the cab door closed.

Moreover, after discharge, drivers should ensure that the body of the vehicle is completely empty and should not drive the vehicle in an endeavour to free a stuck load.

Giving unauthorised lifts

[A1020] Giving unauthorised lifts in work vehicles is both a criminal offence and can lead to employers being involved in civil liability. Thus, 'every employer shall ensure that work equipment is used only for operations for which, and under conditions for which, it is suitable'. [*PUWER, Reg 5*].

Where a driver of a work vehicle gives employees and/or others unauthorised lifts, his employer could find himself prosecuted for breach of the above regulation, whilst the driver himself may be similarly prosecuted for breach of *HSWA, s 7*, as endangering co-employees and members of the public. In addition, although acting in an unauthorised manner and contrary to instructions, the employee may well involve his employer in vicarious liability for any subsequent injury to a co-employee and/or member of the public (see further *Rose v Plenty* [1976] 1 All ER 97).

Where instructions to employees not to give unauthorised lifts are clearly displayed in a work vehicle, but an employee nevertheless gives an unauthorised lift and a co-employee or member of the public is injured or killed as a result of the employee's negligent driving, it can be argued that the employee has exceeded the scope of his employment and so the employer is absolved from liability. (In *Twine v Bean's Express Ltd* [1946] 1 All ER 202 a driver gave a lift to a third party who was killed in consequence of his negligent driving. There was a notice in the van prohibiting drivers from giving lifts. It was held that the employer was not liable, as the driver was acting outside the parameters of his employment when giving a lift. The injured passenger knew that the driver should not give lifts.)

Conversely, courts have taken the view that such conduct, on the part of drivers, does not circumscribe the scope of employment but rather constitutes performance of work in an unauthorised manner, so leaving the employer liable as the employee is doing what he is employed to do but doing it wrongly (see further *Rose v Plenty and Century Insurance Co v Northern Ireland Road Transport Board* [1942] AC 509). Yet again, if the passenger, having seen the prohibition on unauthorised lifts in the vehicle, nevertheless accepted a lift and was injured, arguably, if an adult rather than a minor, he has agreed to run the risk of negligent injury and so will forfeit the right to compensation.

Authorisation to Drive

[A1020.1] Employees driving vehicles for which they are not authorised by the employer to drive may be liable to prosecution under *HSWA 1974, s 7(1)* due to their acts and omissions. In addition, an employer may remain responsible if an unauthorised worker gains access to a vehicle and subsequently causes injury. In the civil case of *Ilkiw v Samuels* [1963] 2 All ER 879, [1963] 1 WLR 991, 107 Sol Jo 680, CA a lorry driver who had strict instructions from his employers not to allow the lorry to be driven by anybody else, allowed a contractor to move the lorry, showing him how to start the engine and remaining in the back of the lorry while the contractor drove it. The lorry driver did not undertake any checks of competence, and the contractor had never driven a lorry before; he had no driving licence and, though he thought himself competent to move the lorry, he was quite incompetent to manoeuvre a heavily loaded vehicle in the confined space. He drove the lorry a few yards forward but could not stop it and it ran into the base of two conveyor belts, between which the claimant was standing, injuring him.

Common law duty of care

[A1021] At common law every employer owes all his employees a duty to provide and maintain safe means of access to and egress from places of work. Moreover, this duty extends to the workforce of another employer/contractor who happens to be working temporarily on the premises. Hence the common law duty covers all workplaces, out of doors as well as indoors, above ground or below, and extends to factories, mines, schools, universities, aircraft, ships, buses and even fire engines and appliances (*Cox v Angus* [1981] ICR 683, where a fireman injured in a cab was entitled to damages at common law).

Confined spaces

[A1022] Accidents and fatalities such as drowning, poisoning by fumes or gassing, have happened as a result of working in confined spaces. See, for instance, the case of *Baker v T E Hopkins & Son Ltd* [1959] 3 All ER 225 where a doctor was overcome by carbon monoxide fumes while going to rescue two workmen down a well – the defence of *volenti non fit injuria* failed). Normal safe practice is a formalised permit to work system or checklist tailored to a particular task and requiring appropriate and sufficient personal protective equipment. Hazards typical of this sort of operation are:

(a) atmospheric hazards – oxygen deficiency, enrichment toxic gases (eg carbon monoxide), explosive atmospheres (eg methane in sewers);
(b) physical hazards – low entry headroom or low working headroom, protruding pipes, wet surfaces underfoot as well as any electrical or mechanical hazards;
(c) chemical hazards – concentration of toxic gas can quickly build up, where there is a combination of chemical cleaning substances and restricted air flow or movement.

In order to combat this variety of hazards peculiar to work in confined spaces, use of both gas detection equipment and suitable personal protective equipment are a prerequisite, since entry/exit paths are necessarily restricted.

Prior to entry, gas checks should test for (*a*) oxygen deficiency/enrichment, then (*b*) combustible gas and (*c*) toxic gas, by detection equipment being lowered into the space. This will determine the nature of personal protective equipment necessary. If gas is present in any quantity, the offending space should then be either naturally or mechanically ventilated. Where gas is present, entry should only take place in emergencies, subject to the correct respiratory protective equipment being worn. Assuming gas checks establish that there is no gaseous atmosphere, entry can then be made without use of respiratory equipment. Gas detection equipment should continue to be used

Confined spaces [A1023]

whilst people are in the confined space so that any atmospheric change can subsequently be registered on the gas detection equipment. It is essential that an emergency plan is devised when gaining access to confined spaces, which includes effective two way communication between people inside and immediately outside the confined space, contact with emergency services and first aid, and fire prevention personnel on hand.

Statutory requirements

[A1023] It should be noted that the *Confined Spaces Regulations 1997 (SI 1997 No 1713)* came into force on 28 January 1998. These repeal *Factories Act 1961, s 30*, and impose requirements and prohibitions with respect to the health and safety of persons carrying out work in confined spaces.

A 'confined space' is defined in *Reg 1(2)* as 'any place, including any chamber, tank, vat, silo, pit, trench, pipe, sewer, flue, well or other similar space in which, by virtue of its enclosed nature, there arises a reasonably foreseeable specified risk'.

A 'specified risk' means a risk of:

(a) serious injury to any person at work arising from a fire or explosion;
(b) without prejudice to paragraph (a) –
    (i) the loss of consciousness of any person at work arising from an increase in body temperature;
    (ii) the loss of consciousness or asphyxiation of any person at work arising from gas, fume, vapour or the lack of oxygen;
(c) the drowning of any person at work arising from an increase in the level of a liquid; or
(d) the asphyxiation of any person at work arising from a free flowing solid or the inability to reach a respirable environment due to entrapment by a free flowing solid.

Regulation 4 prohibits a person from entering a confined space to carry out work for any purpose where it is reasonably practicable to carry out the work by other means.

If, however, a person is required to work in a confined space, a risk assessment must be undertaken to comply with the requirements of the *Management of Health and Safety at Work Regulations 1999 (SI 1999 No 3242), Reg 3*. The risk assessment must be undertaken by a competent person and the outcome of the risk assessment process will then provide the basis for the development of a safe system of work (*Confined Spaces Regulations 1997, Reg 3*).

The risk assessment process should make use of all available information such as engineering drawings, working plans, soil or geological information and take into consideration factors such as the general condition of the confined space, work to be undertaken in the space to minimise hazards produced in the area, need for isolation of the space and the requirements for emergency rescue. In particular, information should be collected and assessed on the previous contents of the confined space, residues that still may be present, contamination that may arise from adjacent plant, processes, gas mains, surrounding soil, land or strata; oxygen level and physical dimensions of the space that may limit safe access and/or egress. The work to be undertaken should be assessed to determine if additional risks will be produced as a result of this work and systems developed to control these risks. All information collected should be recorded and a safe system of work developed for safe entry.

The main elements to consider when designing a safe system of work include the following:

(a) supervision,
(b) competence levels for personnel working in confined spaces,
(c) two-way communications,
(d) testing/monitoring the atmosphere prior to and during access,
(e) gas purging,
(f) ventilation,
(g) removal of residues,
(h) isolation from gases, liquids and other flowing materials,
(i) isolation from mechanical and electrical equipment,
(j) selection and use of suitable equipment,
(k) personal protective equipment (PPE) and respiratory protective equipment (RPE),
(l) portable gas cylinders and internal combustion engines,
(m) gas supplied by pipes and hoses,
(n) access and egress,
(o) fire prevention,
(p) lighting,
(q) static electricity,
(r) smoking,
(s) emergencies and rescue,
(t) limited working time.

[Confined Spaces Regulations 1997, Reg 4].

Regulation 6 provides for circumstances allowing the Health and Safety Executive to grant exemption certificates.

Accident Reporting and Investigation

Gam Jhutti

Introduction to accident reporting and investigation

[A3001] Employers and other 'responsible persons' (see A3003 below) who have control over employees and work premises are required to notify and report to the relevant enforcing authority a range of events occurring at work where they give rise to death or certain specified injuries; where one of eight occupational diseases has been diagnosed; or where there has been an event which is designated as one of eighty-seven types of dangerous occurrence.

The duty to report applies not only in the case of incidents involving employees, but also to visitors, customers and members of the public killed or (in some circumstances) injured by work activities [*Reporting of Injuries, Diseases and Dangerous Occurrences Regulations 2013 (RIDDOR) (SI 2013 No 1471), Reg 3(1)*].

Employees also have certain obligations to report accidents to their employers.

It has been estimated by the HSE that over half of non-fatal injuries to employees (approximately 70,000) are not duly reported, with even higher under-reporting by the self-employed. Nevertheless enforcement over violations has historically been at low levels as well. In the period 2009-2013 only around 20 convictions were made for RIDDOR offences in proceedings brought by the HSE.

RIDDOR data is often used by organisations and the authorities for monitoring and benchmarking purposes. The HSE annually measures numbers of fatal and other significant day injuries as well as the actual numbers of such accidents and other reportable matters in order to understand trends nationally. It also analyses the patterns of specific types of injuries, accidents or other occurrences.

As well as providing an invaluable database the *RIDDOR* system also alerts inspectors to events which they may need to investigate for other purposes, in particular to ensure that workplaces have been made safe or to make enquiries with a view to possible enforcement action if the incident is deemed significantly serious. By no means do most matters reported under *RIDDOR* result in any formal action, and often the report is simply logged. The Health & Safety Executive has published its formal criteria for selection for investigation of RIDDOR notifications ('*Revised Incident Selection Criteria (2014)*').

RIDDOR data consists of only very basic information about injuries and other occurrences. It does not comprise a sufficient analysis for the purposes of the responsible employer or other organisation properly investigating accidents. Later in the chapter the rationale for a fuller investigation is explained, and the processes involved are outlined.

Ministers are now required to carry out a review of RIDDOR at least every five years and to publish reports on the effectiveness of the Regulations, whether their objectives remain appropriate, and the extent to which the objectives could be met with more streamlined Regulations.

Reporting of Injuries, Diseases and Dangerous Occurrences Regulations 2013 (RIDDOR) – (SI 2013 No 1471)

[A3002] The *Reporting of Injuries, Diseases and Dangerous Occurrences Regulations 2013 (RIDDOR)* came into effect on 1 October 2013 and replaces (with significant changes) previous Regulations with the same title dated 2005. These Regulations cover:

(a) reportable worker fatalities and certain other injuries (see **A3005** below);

(b) fatalities and hospitalisation of members of the public due to work-related accidents (see **A3008** below);

(c) reportable occupational diseases (see **A3014** below);

(d) reportable dangerous occurrences (see **A3013** and **A3037** below);

(e) (to a limited extent) road accidents involving work (see **A3009** below); and

(f) gas incidents (see **A3010** and **A3011** below).

Reporting is made to the 'relevant enforcing authority' (see **E15007**) which is to say the enforcing authority applicable to the activities of 'responsible person' for the purposes of the *Health and Safety Enforcing Authority Regulations 1998 (SI 1998 No 494)*. The reporting procedures are described at **A3015** below.

Records must be kept by employers and other 'responsible persons' of such injuries, diseases and dangerous occurrences for a minimum of three years from the date they were made (see **A3017** below). In addition, employers must also keep an Accident Book (Form BI510).

Persons responsible for notification and reporting

[A3003] The person generally responsible for reporting injury-causing accidents, deaths or diseases is the employer. Failing that, the person having control of the work premises or activity will be the responsible person. [*RIDDOR (SI 2013 No 1471) Reg 3(1)*]. In certain cases these normal rules are displaced and there are specifically designated 'responsible persons', eg in the case of mines, quarries, offshore installations, vehicles, diving operations and pipelines (see Table 1 below).

Table 1
Persons generally responsible for reporting accidents

Death, specified injury, over-seven-day injury or specified occupational disease or dangerous occurrence	involving an employee at work	that person's employer
	involving a self-employed person	the person for the time being having control of the premises in connection with the carrying on by him of any trade, business or undertaking
Specified injury or condition, or over-seven-day injury:	involving a self-employed person	the person who by means of their carrying on any undertaking was in control of the premises, at the time it happened
Death, or injury:	of a person who is not at work eg a member of the public	the person who by means of their carrying on any undertaking was in control of the premises, at the time it happened

[Reg 2(1) (b), (c)].

Persons responsible for reporting accidents in specific locations	
A mine	the mine manager
A closed tip	the owner of the mine with which that tip is associated
A quarry	the quarry operator
An offshore installation (except in the case of reportable diseases)	the duty holder (see O7029)
A dangerous occurrence at a pipeline	the pipeline operator
A dangerous occurrence at a well	the person appointed to organise and supervise drilling and operation or, failing that, the licensee
A diving operation (except in the case of reportable diseases)	the diving contractor

[*RIDDOR Reg 3(2)*].

In situations where the responsible person is difficult to identify because of the shared control of a site or operations, arrangements should be made to determine who will deal with *RIDDOR* reporting in line with the duty to co-operate and co-ordinate under the *Management of Health and Safety at Work Regulations 1999 (SI 1999 No 3242), Reg 11*.

[A3004] Accident Reporting and Investigation

What is covered?

[A3004] In spite of much rationalisation of the reporting obligations in the 2013 version of RIDDOR their scope is still very wide. Whenever there is any kind of accident or near miss, or an occupational disease is diagnosed, the employers and other organisations responsible for the work area need to carefully check and determine whether or not the event is one which is reportable. This inevitably requires scrutiny of the reporting criteria set out in the Regulations themselves.

Certain situations are exempted from the RIDDOR reporting obligations, in particular medical accidents, and deaths, injuries and diseases suffered by members of the armed forces while on duty. (See **A3012** below). Also where other specified legislation contains reporting obligations (eg the *Ionising Radiations Regulations 2017*) there does not need to be parallel reporting under RIDDOR. [*RIDDOR (SI 2013 No 1471), Regs 14 (1), (5) and (6)*].

As will be seen below, many of the dangerous occurrences designated as reportable by RIDDOR are qualified by reference to particular areas of industry so that particular rules apply to operations offshore, rail-ways, mines and quarries.

Fatalities

[A3005] Where any person (including not only a worker but a member of the public or other person not at work at the time) dies as a result of a work related accident (or occupational exposure to a biological agent), the incident must be notified by the responsible person by the quickest practicable means (in practice, by telephone) and a formal report form submitted to the enforcing authority within ten days.

There is an additional rule that where an employee, as a result of an accident at work, has suffered a reportable injury which is the cause of his death within one year of the date of the accident, the employer must inform the enforcing authority in writing of the death without delay, whether or not the accident has been already reported previously [*RIDDOR (SI 2013 No 1471), Reg 6(3)*].

Specified injuries to workers

[A3006] The following 'specified injuries' suffered by a person who was at work must also be reported by the quickest practicable means by the responsible person (in practice, by telephone):

(a)   any bone fracture diagnosed by a registered medical practitioner, other than to a finger, thumb or toe;
(b)   amputation of an arm, hand, finger, thumb, leg, foot or toe;
(c)   any injury diagnosed by a registered medical practitioner as being likely to cause permanent blinding or reduction in sight in one or both eyes;
(d)   any crush injury to the head or torso causing damage to the brain or internal organs in the chest or abdomen;

(e) any burn injury (including scalding) which—
 – covers more than 10% of the whole body's total surface area; or
 – causes significant damage to the eyes, respiratory system or other vital organs;
(f) any degree of scalping requiring hospital treatment;
(g) loss of consciousness caused by head injury or asphyxia; or
(h) any other injury arising from working in an enclosed space which—
 (i) leads to hypothermia, heat-induced illness or unconsciousness,
 (ii) requires resuscitation, or admittance to hospital for more than 24 hours;

[*RIDDOR (SI 2013 No 1471), Reg 4(1), Sch 1*]

Injuries incapacitating a worker for more than seven consecutive days

[A3007] Where a person at work is incapacitated for more than seven consecutive days from their normal contractual work (excluding the day of the accident but including any days which would not have been working days) owing to an injury resulting from an accident at work (other than an injury reportable as a Specified Injury listed in **A3006** above), a report of the accident must be made by the quickest practicable means (effectively this means online to the ICC) and in any event within 15 days of the accident. [*RIDDOR (SI 2013 No 1471), Reg 4(2), Sch 1*].

It does not matter what the nature of the injury is in such situations. It is simply the over-seven day absence period which triggers the reporting obligation for the responsible person.

The seven day threshold was introduced in 2012, replacing the previous rule which applied to over-three day injuries. There still remains a duty to keep records of over-three-day injuries — see **A3017**.

Injuries to non-workers

[A3008] These rules apply to any injuries suffered by persons who are not themselves involved in the work activities in question. Specifically they require that where a non-worker, as a result of a work-related accident, sufferers—(a) an injury, and that person is taken from the site of the accident to a hospital for treatment in respect of that injury; or (b) a Specified Injury on hospital premises, the responsible person must make a report of the accident as soon as practicable, (which will be online to the ICC). [*RIDDOR (SI 2013 No 1471), Reg 5, Sch 1*].

For example, reporting requirements would apply to a shopper who falls and is injured on an escalator, so long as the injury was connected with the escalator (*Woking Borough Council v BHS plc* (1994) 93 LGR 396, 159 JP 427); or a member of the public is overcome by fumes on a visit to a factory; or a patient in a nursing home who falls over an electrical cable lying across the floor and was injured; or a pupil or student killed or injured in the course of their curricular work which was supervised by a lecturer or teacher. (There is a specific exemption so that there is no requirement to notify or report the injury or death of a patient undergoing treatment in a hospital, a doctor's or dentist's surgery [RIDDOR (SI 2013 No 1471), Reg 14(1)].)

Road accidents

[A3009] Deaths and injuries involving vehicles on roads are not reportable unless they are caused by one of the following events:

(a) exposure to any substance conveyed by the vehicle;
(b) loading or unloading vehicles;
(c) construction, demolition, alteration or maintenance activities alongside a road; or
(d) an accident involving a train.

[RIDDOR (SI 2013 No 1471), Reg 14(3) and (4)]

These provisions should be treated as having potentially wide scope. For example, where a refuse vehicle was involved in a fatal accident when it was driving to empty a bin (ie not yet in the act of unloading or loading) this was reportable (*R (on the application of Aineto) v Brighton and Hove District Coroner* [2003] EWHC 1896 (Admin), [2003] All ER (D) 353 (Jul)).

The person injured, whether fatally or not, may or may not be engaged in the above-mentioned activities. Thus, an employee struck by a passing vehicle or a motorist injured by falling scaffolding, are covered. In addition, certain dangerous occurrences on public highways and private roads are covered (see A3035 below).

Duty to report gas incidents

[A3010] Where a conveyor of flammable gas through a fixed pipe distribution system, or a filler, importer or supplier (not by way of the retail trade) of a refillable container containing liquefied petroleum gas receives notification of any death, loss of consciousness or taking to hospital of a person because of an injury arsing in connection with that gas, that person must immediately notify the Health and Safety Executive and then send a report on the approved form within 14 days of the incident. [RIDDOR (SI 2013 No 1471), Reg 11(1)].

Gas fitters

[A3011] Where an approved person has sufficient information to decide that the design, construction, manner of installation, modification of servicing of a gas fitting (including any related flue or ventilation) is, or could have been likely to cause, death, loss of consciousness or taking to hospital of a person because of accidental leakage, incomplete combustion of gas or inadequate removal of the products of gas combustion, they are required to report that information to the Health and Safety Executive on the prescribed form within 14 days. [*RIDDOR (SI 2013 No 1471), Reg 11(2)*].

Some problem areas

[**A3012**] 'Accident' and 'injury' for these purposes are not defined, and are both words which have given rise to difficulty of interpretation in other legal contexts such as insurance policy wording and claims for asbestos diseases and mental conditions. They are not defined in the Regulations, except that they state an accident is deemed to include 'non-consensual violence' (so that for example professional sports injuries are usually not reportable).

Deaths and injuries are reportable when they result from an accident 'arising out of or in connection with work'. This phrase is only partly explained by the Regulations as including reference to accidents or dangerous occurrences 'attributable to the manner of conducting an undertaking, or the plant or substances used for the purposes of an undertaking, or the condition of premises used for the purposes of an undertaking or any part of them'. [*RIDDOR (SI 2013 No 1471), Reg 2(2)*]. Clearly there needs to be some causal connection with the work that is going on. The mere fact that someone suffers an injury while in a workplace at the time is not by itself enough to engage these Regulations.

Particular problems can arise with the reporting acts of violence. Generally, physical harm done to a member of staff in an assault would be reportable but not shock and distress as this is not classed as a direct injury of the incident.

Dangerous occurrences

[**A3013**] In all 87 types of reportable dangerous occurrences specified in *RIDDOR (SI 2013 No 1471), Sch 2*, many are limited to specific industrial sectors and only 27 of them apply generically to all (onshore) workplaces.

Common types of specified occurrences include for example the collapse of lifting equipment; structural collapses; and outbreaks of fire which result in stoppages of plant or normal operations for more than 24 hours.

An especially wide category of dangerous occurrence is hazardous 'escapes of substances'. This is defined as 'the unintentional release or escape of any substance which could cause personal injury to any person other than through the combustion of flammable liquids or gases'. (There is another category for releases of flammable liquids and gases). These situations can range from large scale process failures to plant failures and spills from containers. In the past HSE guidance has suggested that releases which are sufficiently well controlled to ensure that no person is put at risk would not be reportable. Determining whether or not there is a duty to report in such cases can be problematic. It will usually involve a careful assessment of the all the possible harmful properties of the substances in question, the extent and duration of their release, and whether anyone was actually put in danger. See **A3035** below for a complete list of the reportable dangerous occurrences.

Occupational diseases

[A3014] Probably the most significant changes made to the RIDDOR regime by the *2013 Regulations* has been the reduction in the number of reportable types of occupational disease in most work situations, from 47 to the following:

(1) Carpal Tunnel Syndrome, where the person's work involves regular use of percussive or vibrating tools;
(2) cramp in the hand or forearm, where the person's work involves prolonged periods of repetitive movement of the fingers, hand or arm;
(3) occupational dermatitis, where the person's work involves significant or regular exposure to a known skin sensitizer or irritant;
(4) Hand Arm Vibration Syndrome, where the person's work involves regular use of percussive or vibrating tools, or the holding of materials which are subject to percussive processes, or processes causing vibration;
(5) occupational asthma, where the person's work involves significant or regular exposure to a known respiratory sensitizer;
(6) tendonitis or tenosynovitis in the hand or forearm, where the person's work is physically demanding and involves frequent, repetitive movements,
(7) any cancer attributed to an occupational exposure to a known human carcinogen or mutagen (including ionising radiation);
(8) any disease attributed to an occupational exposure to a biological agent.

[*RIDDOR (SI 2013 No 1471), Regs 8 and 9*]

An important qualifying provision is that to be reportable each of these conditions must first be formally 'diagnosed'. For these purposes this means the identification by a registered medical practitioner of new symptoms or symptoms which have significantly worsened. Where the person affected is an employee (as opposed to self-employed) this condition needs to be confirmed in writing by the diagnosing doctor.

Separate requirements apply to offshore workplaces. There is a longer list of twenty-five diseases which are reportable in relation to persons working offshore. These are set out in Schedule 3 of RIDDOR, and include chickenpox, measles, mumps and food poisoning. [*RIDDOR (SI 2013 No 1471), Reg 10*].

Where a worker is diagnosed with one of the occupational diseases listed in the Regulations, a report must be sent to the enforcing authority 'without delay'. (No specific time limit is set for reporting diagnoses of cancers and diseases attributable to exposure to biological agents.) [*RIDDOR (SI 2013 No 1471), Sch 1*].

The *Industrial Diseases (Notification) Act 1981* and the *Registration of Births and Deaths Regulations 1987 (SI 1987 No 2088), Sch 2*, Form 14, require that particulars are to be included on the death certificate as to whether death might have been due to, or contributed to by, the deceased's employment. Such particulars are to be supplied by the doctor who attended the deceased during the last illness.

Reporting procedures

[A3015] The submission of reports on paper forms has now been largely eliminated. After various changes in recent years there are now two main reporting routes available:

(a) Injuries, diseases and dangerous occurrences can be reported to the relevant enforcing authority online at www.hse.gov.uk/riddor/report.htm using the electronic forms provided; or

(b) Reports can be made by telephone to the national Incident Contact Centre (ICC) on 0345 300 9923, but this is for reporting fatal/specified and major incidents only. The ICC has opening hours of Monday to Friday 8.30 am to 5 pm. There is an emergency out of hours number as well, 0151 922 9235. For contact details of local authorities (where these are the enforcing authority), see www.gov.uk.

The main reporting forms and their contents fields can be seen online at www.hse.gov.uk/riddor/index.htm.

There are some additional reporting procedures for incidents at mines and quarries [*RIDDOR (SI 2013 No 1471), Sch 1, para 4*].

Incidents in Northern Ireland need to be reported direct to HSE IN (www.hseni.gov.uk). Separate arrangements exist for reporting of accidents etc on railways (see www.rail-reg.gov.uk) and in marine cases under the Merchant Shipping (Accident Reporting and Investigation) Regulations 2012, (SI 2012 No 1743). (See www.maib.gov.uk/report_an_accident/index.cfm.)

As for notification of major accident hazards (see **M1112 CONTROL OF MAJOR ACCIDENT HAZARDS**), notification and reporting under RIDDOR is sufficient for the purposes of the *Control of Major Accident Hazards Regulations 1999 (SI 1999 No 743), (as amended in 2015) Reg 15(4)*.

[A3016] In taking advantage of these arrangements, the employer or other reporting person may not have a formal copy record of a statutory RIDDOR form, so the ICC will send out confirmation copies of reports which should be checked for accuracy and retained (see **A3017**).

Records and record-keeping

[A3017] The prescribed Records of injury-causing accidents, dangerous occurrences and specified diseases must be kept by responsible persons for at least three years from the date on which they are made. These records are to be kept either where the relevant work was carried on or at the responsible person's usual place of business. [*RIDDOR (SI 2013 No 1471), Reg 12(2)*].

These records cover information which relates to all reportable matters. In addition, records also have to be kept of injuries to persons at work resulting from accidents at work which cause 'over-three day injuries' (even though these are no longer reportable to the enforcing authorities). These are injuries which incapacitate a person from routine work for more than three consecutive days (not counting the day of the accident). [*RIDDOR (SI 2013 No 1471), Reg 12(1)(c)*].

[A3018] The prescribed records are as follows:

Particulars to be kept in records of any reportable death, injury or dangerous occurrence reportable under Regulations 4 to 7

(1) The date and time of the accident or dangerous occurrence.
(2) In respect of an accident injuring a person at work, that person's —
   (a) full name;
   (b) occupation;
   (c) injury.
(3) In respect of an accident injuring a person not at work, that person's —
   (a) full name;
   (b) status (for example 'passenger', 'customer', 'visitor', or 'bystander');
   (c) injury.
   unless these are not known and it is not reasonably practicable to ascertain them.
(4) The place where the accident or dangerous occurrence happened.
(5) A brief description of the circumstances in which the accident or dangerous occurrence happened.
(6) The date on which the accident or dangerous occurrence was first notified or reported to the relevant enforcing authority.
(7) The method by which the accident or dangerous occurrence was first notified or reported.

Particulars to be kept in records of any injury resulting from an accident incapacitating a worker for routine work for more than three consecutive days under Regulation 12 (1)(c)

(8) The date and time of the accident.
(9) The following particulars of the injured person —
   (a) occupation;
   (b) occupation;
   (c) injury.
(10) The place where the accident happened.
(11) A brief description of the circumstances in which the accident happened.

Particulars to be kept in records of any diagnosis reportable under Regulations 8 to 10

(12) The date of diagnosis of the disease.
(13) The name of the person affected.
(14) The occupation of the person affected.
(15) The name or nature of the disease.
(16) The date on which the disease was first reported to the relevant enforcing authority.
(17) The method by which the disease was reported.

[RIDDOR (SI 2013 No 1471),Sch 1, Part 2]

Action to be taken by employers and others when accidents occur at work

[A3019] The following should be completed —

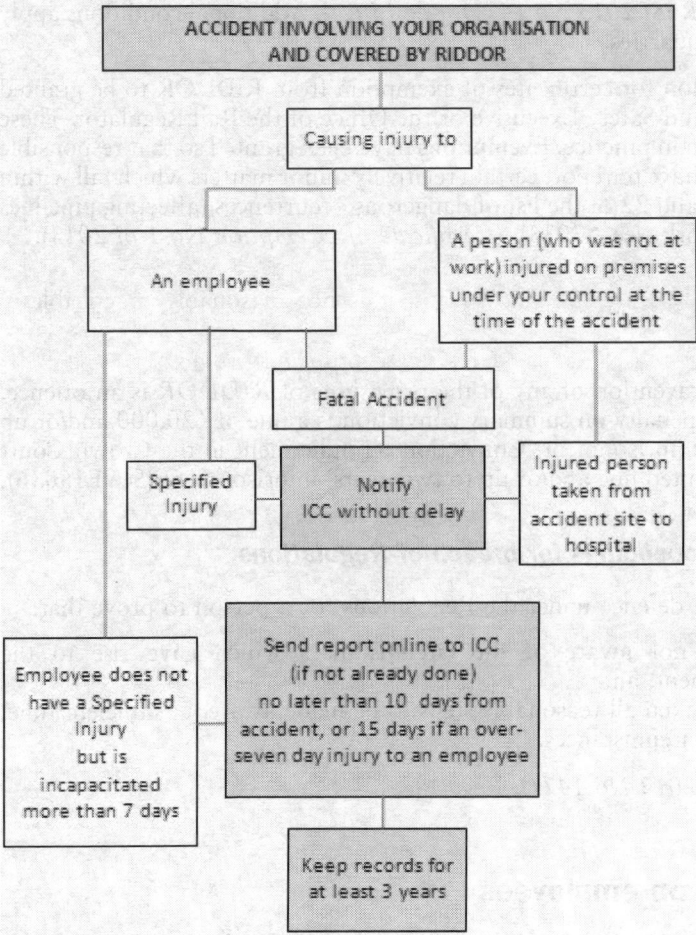

Exemptions from RIDDOR reporting

[A3020] Much of RIDDOR does not apply to deaths, injuries and diseases to members of the armed forces when on duty at the time. [*RIDDOR (SI 2013 No 1471), Reg 14(5)*]. Deaths and injuries arising from operations and other medical or dental treatment are not reportable. [*RIDDOR (SI 2013 No 1471), Reg 14(1)*].

Another provision makes it unnecessary for a responsible person to report the same details twice or more where the same circumstances would trigger two or

[A3020] Accident Reporting and Investigation

more RIDDOR reporting requirements. [*RIDDOR (SI 2013 No 1471), Reg 15 (1)*]. However care is needed when relying on this exemption because the conditions must be that the facts giving rise to each reporting requirement are identical; all relevant reportable information is contained by all the requirements is provided; and the shortest applicable time reporting limit is complied with. [*RIDDOR (SI 2013 No 1471), Reg 15 (2)*]. Additional conditions apply for mines and quarries.

There is provision for certificates of exemption from RIDDOR to be granted by the Health and Safety Executive or the Office of the Rail Regulator. These are not common in practice. Exemptions have been granted so that responsible persons do not have to report certain relatively minor matters which fall within paragraphs 21 and 22 of the list of dangerous occurrences (affecting pipelines and pipeline work, see **A3035**). (*Certificate of Exemption No 1 of 2013*).

Penalties

[A3021] Contravention of any of the provisions of *RIDDOR* is an offence. The maximum penalty on summary conviction is a fine of £20,000 and/or up to 12 months' imprisonment. Conviction on indictment in the Crown Court carries an unlimited fine and/or up to two years' imprisonment (See **E15036**).

Defence in proceedings for breach of Regulations

[A3022] It is a defence under the Regulations for a person to prove that:

(a) he was not aware of the circumstances which gave rise to the requirement; and
(b) he had taken all reasonable steps to be made aware, in sufficient time, of such circumstances.

[*RIDDOR (SI 2013 No 1471), Reg 16*].

Obligations on employees

[A3023] All employees have duties to inform their employers of serious and immediate dangers and recognisable short comings in safety arrangements (*Management of Health and Safety at Work Regulations 1999* – see **M2017**).

[A3024] There are separate reporting requirements on employees by the *Social Security (Claims and Payments) Regulations 1979 (SI 1979 No 628), Reg 24*. Although not actually bound in with the separate scheme of health and safety legislation, these reporting requirements complement those which employers have under *RIDDOR*, and there is an overlap in the record keeping requirements for both sets of provisions – see further below.

An accident to which the employees' reporting requirements apply is one 'in respect of which benefit may be payable . . . '. [*Social Security (Claims and Payments) Regulations 1979 (SI 1979 No 628), Reg 24(1)*]. Under *Regulation*

24, every 'employed earner' who suffers personal injury by an accident for the purposes of the Regulations is required by to give either *oral or written* notice to the employer in one of number of different ways. These ways include notice given to any supervisor or direct to a person designated by the employer. The method that has probably become the norm through custom and practice is by way of an entry of the appropriate particulars in a 'book kept specifically for these purposes' under the Regulations. These books are available from HSE Books as form BI 510. As well as containing sections for completing details of accidents etc form BI 150 includes pages of instructions to employees and employers about their responsibilities. It is important only to use the latest (2012) version which has been updated so that it is compliant with the *Data Protection Act 1988*.

The entry in an Accident Book is to be made as soon as practicable after the happening of an accident by the employed earner or by some other person acting on his behalf. [*Social Security (Claims and Payments) Regulations 1979 (SI 1979 No 628), Reg 24(3)*]. Failure to do so is an offence punishable by a fine. [*Social Security (Claims and Payments) Regulations 1979 (SI 1979 No 628), Reg 31*].

Under *Reg 25* of the 1979 Regulations, once an accident is reported by the employee, the employer must take reasonable steps to investigate the circumstances and, if there appear to be any discrepancies between the circumstances reported and the findings in these investigations, these should be recorded.

Where the accident in question is one to which *RIDDOR* applies there is, as well as the usual reporting requirements a requirement to keep a record of the accident (see **A3017**. The 2012 edition HSE guide to *RIDDOR* (see **A3034**) provides that an employer may choose to utilise the form BI 150 Accident Book as this record, and the format of the form includes space for the employer to initial the report as being one reportable under *RIDDOR*.

Objectives of accident reporting

[A3025] There should be an effective accident reporting and investigation system in all organisations. Accident reporting procedures should be clearly established in writing with individual reporting responsibilities specified. Staff should be trained in the system and disciplinary action may have to be taken where there is a failure to comply with it. Moreover, there is a case for all incidents, no matter how trivial they may seem and of the absence of any resulting injury being reported through the internal reporting procedures.

The capture of information about accidents and other incidents can be enhanced by the provision of a simple form used throughout the organisation.

Figure 2: Internal Incident Notification Form

INTERNAL INCIDENT NOTIFICATION FORM

Time/date: 00.00 hrs   dd/mm/yy

Location:

Description of incident:

Person(s) injured:

Nature of injury:

Other persons involved:

First aid administered:

Accident Book completed: Y/N

Form completed by:

Date: dd/mm/yy

Accident Investigation [A3029]

Duty of disclosure of accident data

[A3026] Employers are under a duty to disclose accident data to works safety representatives (see J3015) and, in the course of litigation, to legal representatives of persons claiming damages for death or personal injury (see E13029).

Safety representatives

[A3027] An employer must make available to safety representatives of both unionised and non-unionised workforces the information within the employer's knowledge necessary to enable them to fulfil their functions. [*Safety Representatives and Safety Committees Regulations 1977 (SI 1977 No 500), Reg 7(2); Health and Safety (Consultation with Employees) Regulations 1996 (SI 1996 No 1513), Reg 5(1)*]. The Approved Code of Practice in association with these Regulations states that such information should include information which the employer keeps relating to the occurrence of any accident, dangerous occurrence or notifiable industrial disease and any associated statistical records. (Code of Practice: Safety Representatives and Safety Committees 1976, para 6(c)). See J3015.

Legal representatives

[A3028] In the course of litigation (and sometimes before proceedings have actually begun under the Pre-Action Protocols and civil procedure rules) obligations may arise to give disclosure of relevant documents concerning an accident – even if they are confidential. Legal representatives of an injured party would expect reasonable access to *RIDDOR* information. See generally E13029. The decision in *Waugh v British Railways Board* [1979] 2 All ER 1169 established that where an employer seeks to withhold on grounds of privilege a report made following an accident, he can only do so if its dominant purpose is related to actual or potential hostile legal proceedings. In *Waugh* a report was commissioned for two purposes following the death of an employee: (*a*) to recommend improvements in safety measures, and (*b*) to gather material for the employer's defence. It was held that the report was not privileged.

Accident Investigation

Rationale for undertaking investigations

[A3029] Unlike the highly prescriptive legal requirements for notification under *RIDDOR (SI 2013 No 1471)*, there are no specific legal obligations requiring the investigation of accidents or the production of accident reports. The HSE consulted on proposals for a statutory duty to undertake investigations into accidents, diseases and dangerous occurrences in 2001, but the outcome was a decision to develop new investigations guidelines with a view to encouraging the implementation of best practice voluntarily (see A3035).

Nevertheless a number of statutory requirements demand some level of investigation is carried out. In particular, one needs to determine whether it is necessary to undertake a review of the adequacy of existing risk assessments for the relevant activity in order to comply to with *Regulation 3(3)* of the *Management of Health and Safety at Work Regulations 1999 (SI 1999 No 3242)* (see **R3022**).

Accident investigation does however have wider rationales. The main drivers for it are:

(a) *learning lessons*: identifying and understanding the immediate and the underlying causes of accidents and identifying ways in which to prevent the recurrence of similar accidents.

(b) *reassurance and explanation*: particularly the victim of an accident, but also others connected with it often need to be reassured that appropriate action is being taken or that a satisfactory explanation has been given for what has happened.

(c) *monitoring of performance*: investigations produce data for the reactive monitoring of health and safety performance, measurement of results against an organisation's targets or for benchmarking, recording of essential information in databases ('corporate memory'), and the provision of management reports – ultimately to directors – on the management of operational risks.

(d) *providing information to other interested parties*: the information is likely to be needed with which to brief insurers, safety representatives, also those who are responsible for dealing with media. There is also an expectation on the part of health and safety inspectors as well that an accident investigation will be produced, and disclosed to them once completed, and an absence of a formal report may be taken as an indicator of management failure.

(e) *disciplinary procedures*: where there is evidence of deliberate reckless failure to follow procedures an investigation may form part of the disciplinary process and evidence upon which the fairness of action perhaps ultimately resulting in dismissal might be judged.

(f) *allocation of blame/liability*: this will be part and parcel of any investigation undertaken by enforcement agencies and insurers, any information may need to be obtained and analysed by the organisation involved in the accident in preparation for what are inevitably adversarial proceedings.

Accident investigation literature often tends to be focussed on the risk management and accident prevention benefits that can accrue from the process, while treating disciplinary and liability issues as subsidiary to other objectives. This can result in a number of difficulties and potential conflicts. It is difficult in reality for accidents to be investigated in an entirely blame-free context. Accident investigations can – if there is not careful consideration given to legal and disciplinary implications – prejudice not just the organisation but also the position of individual employees and managers who may be subject to actual or implied criticism or even legal action. It should also be borne in mind that the HSE's internal work instructions for inspectors have been known to steer inspectors towards directing or using an organisation's investigation process in order to achieve enforcement-related goals, which include providing *'an early insight into the duty holder's thoughts regarding cause and blame enabling any potential defence or mitigation to any subsequent proceedings to be identified'.*

It is not possible to produce a single system of investigation which ultimately resolves all the potential conflicts between the different rationales and needs for the investigation. However, the processes do need to be operated in a way which takes into account these conflicts and seeks to minimise them by appropriately involving different parts of the organisation, insurers and legal representatives. Particular attention needs to be given to identifying the information which is obtained for advice where legal privilege may apply and where the wider dissemination of that material might cause prejudice in future legal proceedings.

The main elements of investigations

[A3030] The following sections outline the main elements of an accident investigation process, and each stage of this process provides a framework within which more or less details an elaborate examination of relevant issues which can be carried out depending on what is considered necessary or proportionate (see FIGURE 3). The process should not however be inflexible, and it may need to be adapted if for example there are parallel investigations taking place by enforcement authorities with a view to future prosecution.

Figure 3: Elements of the investigation process

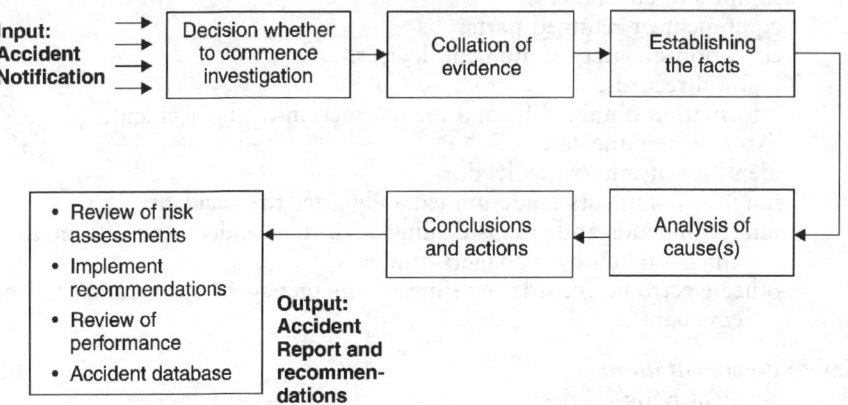

It is possible to devise ad hoc arrangements for each investigation as it arises, but ideally (and essentially in large organisations where there is a steady occurrence of accidents) there should be defined procedures. Responsibilities should be designated, and there should be basic but clear criteria as to the types of occurrence which should be investigated more thoroughly than is necessary merely for the purposes of carrying out RIDDOR notifications. Thought needs to be given to how an investigation team is to be constituted, setting terms of reference, and determining what the reporting lines are during the process and not just for the delivery of the accident report. An agreed format for the contents of accident reports is desirable so that consistent and comprehensive results are produced. Those with responsibility for the investigation process may require training in some areas such as interviewing skills, accident causation principles and report writing.

A HSE investigation or claim for personal injury will raise liability issues and opportunities for internal or external legal advice should be built into the process at each stage.

It is important that the process does not exist in isolation of what is happening elsewhere in the organisation. There may need to be liaison not just with legal advisors, but with others who may be dealing with necessary actions following an accident such as HR managers, insurance managers and of course line managers responsible for the activity in which the accident has occurred who may need immediate guidance on whether (or how) to recommence or continue the activities.

Collating evidence

[A3031] Every situation will be different, but typically, the evidence will comprise of some or all of the following:

Conditions at the accident scene:
- photographs and videos;
- sketches and plans;
- measurements;
- records of weather/environmental conditions;
- samples of substances;
- equipment or retained parts;
- condition of safety equipment/devices;
- first-aid records;
- information obtained from a reconstruction of the accident;
- lists of eye-witnesses;
- identities of others involved;
- notes of comments made immediately after the accident;
- internal memos and emails dealing with the accident and aftermath;
- statements taken by managers/others;
- other electronic records (eg time of emergency calls, or computerised process data).

Safety documentation:
- accident book entries;
- RIDDOR records;
- relevant risk assessments;
- method statements/safe systems of work;
- training materials;
- training records;
- equipment operating instructions;
- equipment logs/maintenance records;
- personnel records of those involved in the accident;
- reports on relevant previous incidents;

- relevant health and safety audits, management reports, consultants' advice;
- relevant emergency procedures.

The most difficult part of this stage of the process is likely to be obtaining statements promptly from witnesses, either because they are distressed or because they are unco-operative (which may be for a variety of reasons including concern that they or their workmates might be criticised). It is nevertheless important to obtain witnesses' evidence as soon as possible after an accident. The accuracy of their information is reduced the more time elapses, and recall can be affected by *post-event enhancement* – the unconscious incorporation of ideas and perceptions based on discussing events with others in group setting and by *false confidence* – a hardening of a version of events from recalling and re-telling others or committing evidence to paper.

If witnesses are unco-operative or hostile, it may serve no useful purpose to persevere in efforts to interview them. A short period of delay can sometimes assist, in which it may be possible to resolve the underlying problems. Where the witness is an employee, ultimately the failure to co-operate with the employer may be treated as a disciplinary matter.

Another difficulty with interviewing witnesses can be that the HSE, police or other authorities may wish to restrict access to them – so that they can interview them first and take formal statements. There is no legal power to control witnesses in this way, but inspectors sometimes threaten employers with prosecution for obstructing them if they do not agree to postpone their interviews. It is therefore advisable to be open with the authorities about the interview process, and sometimes to offer inspectors or other officers an opportunity to attend interviews with witnesses about whom they have concerns.

Planning the interview
- Identify what information is sought from this interviewee and whether the interviewee is willing or reluctant.
- Draft a list of principal topics and documents to ask about.
- Key interviewees should be interviewed as soon as possible after the event.
- Will the interviewee be accompanied, and if so, by whom?
- Arrange for the venue to be neutral and under the interviewer's control if possible.
- Confirm date, time and venue with interviewee.
- Arrange the furniture appropriately, for example no glaring spotlights, relatively informal seating arrangements, removal of telephone.
- Availability of refreshments, tissues and props such as plans, models, etc.

Introduce the interview:
- Introductions – interviewer(s), interviewee and accompanying observer.
- Summarise purpose of interview, include note-taking, recording, confidentiality.

Planning the interview
- Outline structure of interview – chronological, beginning with interviewee's own account, then followed by questions on points of interest.
- Turn off mobile phones, OK to ask for comfort break or pause if needed.

Body of interview:
- Use broad questions to initiate free account if interviewee is willing.
- Use simple probing questions if the interviewee is reluctant.
- Closed questions used sparingly, avoiding counterproductive questions.
- Use verbal and non-verbal prompts.
- Actively listen to the answers provided, and evaluate and respond appropriately.
- Use a range of questions to obtain more detailed information.
- Keep asking 'why' to find out about critical acts and decisions (human factors).
- Use cognitive interviewing techniques if appropriate.
- Use hunches and normalisation where appropriate.
- Use summary/trailer questions for inconsistencies with other accounts/evidence.
- Pay attention to drop-it cues, returning to issue later if important.
- Deal with interviewee's emotional reactions without becoming involved.
- Ensure your verbal and non-verbal communication is consistent.
- Maintain a neutral, open and non-judgemental manner throughout.

Closing the interview:
- Summarise key points and ask for confirmation of account and summary.
- Explain what will happen next (further interviews, testing and analysis, provision of statement and/or transcript).
- Thank interviewee for contribution, mention possibility of follow-up.
- Provide your contact details in case of query or further information.
- Be prepared for any off-the-record statement as interview closes.

(Source: *Tolley's Workplace Accident Handbook* (2nd edition, 2007) ISBN 0 7506 8151 9)

Analysis of causes

[A3032] This stage of the process is concerned with two inter-related questions: 'what happened?' and 'why did it happen?' These apparently simple questions can be subjected to highly sophisticated analytical techniques. The following paragraphs summarise a basic approach, but those charged with accident investigations should have a fuller understanding of accident causation theory and knowledge of the background to *human factors* (see **E17001**).

Having collated the information about nature and circumstances of the accident the investigation team needs to decide what issues it needs to analyse. A useful starting point is often the organisation's standard template for accident investigation reports (see A3034 below) but not all the headings will necessarily be relevant in each case, and a degree of flexibility is needed so that the investigation does not become diverted by issues of low or no relevance.

Examples of the questions, which are typically explored, are:

- What was the chain of events leading up to the accident?
- What planning of the activity had taken place?
- Were the risks known?
- Was the planning deficient?
- Were defective premises, equipment or maintenance a factor?
- Were the correct materials being used?
- Were those involved properly trained?
- Was the training adequate?
- Did people depart from instructions/procedures?
- Was supervision properly undertaken?
- Were the instructions/procedures adequate?
- Did organisational factors of the work play a role eg excessive hours, rushing to meet a deadline or target?
- Was there a failure to learn lessons from previous incidents or to follow-up previous warnings or complaints?
- Did the emergency procedures and provision of first-aid operate effectively?

It is not unusual for these or other questions to require additional evidence or for further queries to be raised with witnesses, and that the collating analysis stages may have to overlap.

Answers to these questions are used to inform the next stage of analysis which involves reaching an understanding as to the cause or causes of the accident, and determining whether the risk control measures that were in place are in need of improvement.

Causes of accidents can be complex and multi-factorial. It is often the unusual combination of a series of events which leads to injury or damage in what were hitherto routine workplace activities. Each event may be traced back to factors that preceded it, and so it has become common to analysis accidents in terms of three types of 'causes'.

(a) 'Immediate' cause: an unsafe act (or omission) or an unsafe condition which leads directly to the injury or damage. (An example of an unsafe act is improper stacking of a load; an example of an unsafe condition is a machine with a missing guard). Occasionally the immediate cause of an accident may be a 'violation': a deliberate or reckless departure from correct procedures by an individual or group.

(b) 'Underlying' cause: these precede the time of the accident and consist of management control factors (eg inadequate risk assessment or the absence of supervision), job factors (eg lack of competence or insufficient staffing), or environmental factors (such as working in harsh conditions).

(c) 'Root' causes: a failing from which all the other causes have grown – usually a deficiency in the planning and organisation of management of health and safety.

Caution is needed in using this approach for a number of reasons. First, it may be difficult to separate and distinguish between several different causes as neatly in practice as it is in theory. Second, this approach does not give weight to the *gravity* of causal factors, which may be relevant in the wider employment or liability context. Third, and most importantly, there is a large element of subjectivity in determining what are the relevant underlying, and particular root causes. The causal analysis therefore needs to be firmly based on the factual analysis, and the reasoning process explained at each stage by reference to the facts as they have found during the first stage of the analysis.

Preparing the report

[A3033] There are no hard and fast rules about the structure or content of accident investigation reports. The essential features however are:

- description of circumstances of the accident;
- consideration of control measures;
- conclusions as to causes;
- recommendations for remedial action.

A basic template which can be used for reports is as follows.

OUTLINE FOR ACCIDENT INVESTIGATION REPORT

Title:	Including details of the time, location and outcome(s) of the accident.
Summary:	Providing background information, which will include factual and historical information. Conclusions and recommendations can be copied and pasted to finish the summary, which will then provide an overview of the investigation.
History of accident:	Including detailed factual and historical information, the sequence of events, and any actions taken in immediate response to mitigate or make the site safe, including emergency response and attendance of first aid or medical personnel.
Investigation and analysis	Including evidence from interviews, engineering and forensic testing and/or analyses, other sources of information.
Conclusions:	Including the most likely or probable immediate and underlying causes, and an indication of the extent to which the conclusions were unanimous. Where there are discrepancies in the evidence, this should be pointed out.
Recommendations:	These should address causes and should have a clear link with the investigation and analysis.

(Source: *Tolley's Workplace Accident Handbook* (2nd edition, 2007) ISBN 0 7506 8151 9)

Before finalising the report it is sometimes desirable to seek comments or corrections from those involved in the accident or connected with the background. Partly this is out of courtesy where individuals might perceive the conclusions as implied criticism, but it also provides a final opportunity to eliminate factual or analytical errors. It is also sensible that proposed recommendations for action are at least canvassed with the relevant managers who may become responsible for implementing them to ensure that they are not impractical or unrealistic. A draft may also be provided to the organisations' legal advisors to consider how it may affect liability issues and whether the conclusions are consistent with the findings of privileged investigations that have been, or are still being carried out.

When the report is finalised there needs to be a clear and timely management plan of action to implement the recommendations.

The document itself will be disclosable in any subsequent legal proceedings and a copy may be requested by enforcing authorities. Delay or incompleteness in giving effect to the recommendations may expose the organisation to subsequent criticism.

Further guidance on accident reporting and investigation

[A3034] Further guidance can be found at —

- *Guide to the Reporting of Accidents, Diseases and Dangerous Occurrences Regulations 1995* (Fourth edition (2012) HSE Books, L73, ISBN 9780717664597) (Due for further new edition in late 2013)
- *Reporting Accidents and Incidents at Work – A brief guide* (HSE Books, INDG 453 Rev 1, 2013)
- *Incident reporting in schools (accidents, diseases and dangerous occurrences): Guidance for employers* (HSE Books EDIS1 (rev3) 2013).
- *Reporting injuries, diseases and dangerous occurrences in health and social care* (HSIS1 (rev3) 2013).
- *Investigating Accidents and Incidents – A Workbook for Employers, Unions, Safety Representatives and Safety Professionals* (HSE Books, HSG245, ISBN 0717628272)
- *Reducing Error and Influencing Behaviour* (HSE Books, HSG48, ISBN 0717624528)
- *Tolley's Workplace Accident Handbook*, 2nd Edition (Butterworths-Heinemann, ISBN 9780750681513)

Appendix A

List of Reportable Dangerous Occurrences

[A3035] —

2013 No 1471

REPORTING OF INJURIES, DISEASES AND DANGEROUS OCCURRENCES REGULATIONS 2013

SCHEDULE 2
DANGEROUS OCCURRENCES

Regulation 7

PART I

GENERAL

1 Lifting equipment
The collapse, overturning or failure of any load-bearing part of any lifting equipment, other than an accessory for lifting.

2 Pressure systems
The failure of any closed vessel or of any associated pipework (other than a pipeline) forming part of a pressure system as defined by regulation 2(1) of the Pressure Systems Safety Regulations 2000, where that failure could cause the death of any person.

3 Overhead electric lines
Any plant or equipment unintentionally coming into—
(a) contact with an uninsulated overhead electric line in which the voltage exceeds 200 volts; or
(b) close proximity with such an electric line, such that it causes an electrical discharge.

4 Electrical incidents causing explosion or fire
Any explosion or fire caused by an electrical short circuit or overload (including those resulting from accidental damage to the electrical plant) which either—
(a) results in the stoppage of the plant involved for more than 24 hours; or
(b) causes a significant risk of death.

5 Explosives
Any unintentional—
(a) fire, explosion or ignition at a site where the manufacture or storage of explosives requires a licence or registration, as the case may be, under regulation 9, 10 or 11 of the Manufacture and Storage of Explosives Regulations 2005; or
(b) explosion or ignition of explosives (unless caused by the unintentional discharge of a weapon, where, apart from that unintentional discharge, the weapon and explosives functioned as they were designed to), except where a fail-safe device or safe

system of work prevented any person being endangered as a result of the fire, explosion or ignition.

6
The misfire of explosives (other than at a mine or quarry, inside a well or involving a weapon) except where a fail-safe device or safe system of work prevented any person being endangered as a result of the misfire.

7
Any explosion, discharge or intentional fire or ignition which causes any injury to a person requiring first-aid or medical treatment, other than at a mine or quarry.

8
(1) The projection of material beyond the boundary of the site on which the explosives are being used, or beyond the danger zone of the site, which caused or might have caused injury, except at a quarry.
(2) In this paragraph, "danger zone" means the area from which persons have been excluded or forbidden to enter to avoid being endangered by any explosion or ignition of explosives.

9
The failure of shots to cause the intended extent of collapse or direction of fall of a structure in any demolition operation.

10 Biological agents
Any accident or incident which results or could have resulted in the release or escape of a biological agent likely to cause severe human infection or illness.

11 Radiation generators and radiography
(1) The malfunction of—
    (a) a radiation generator or its ancillary equipment used in fixed or mobile industrial radiography, the irradiation of food or the processing of products by irradiation, which causes it to fail to de-energise at the end of the intended exposure period; or
    (b) equipment used in fixed or mobile industrial radiography or gamma irradiation, which causes a radioactive source to fail to return to its safe position by the normal means at the end of the intended exposure period.
(2) In this paragraph, "radiation generator" means any electrical equipment emitting ionising radiation and containing components operating at a potential difference of more than 5kV.

12 Breathing apparatus
The malfunction of breathing apparatus—
    (a) where the malfunction causes a significant risk of personal injury to the user; or
    (b) during testing immediately prior to use, where the malfunction would have caused a significant risk to the health and safety of the user had it occurred during use, other than at a mine.

13 Diving operations
The failure, damaging or endangering of—
    (a) any life support equipment, including control panels, hoses and breathing apparatus; or
    (b) the dive platform, or any failure of the dive platform to remain on station, which causes a significant risk of personal injury to a diver.

14

The failure or endangering of any lifting equipment associated with a diving operation.

15

The trapping of a diver.

16

Any explosion in the vicinity of a diver.

17

Any uncontrolled ascent or any omitted decompression which causes a significant risk of personal injury to a diver.

18 Collapse of scaffolding

The complete or partial collapse (including falling, buckling or overturning) of—
- (a) a substantial part of any scaffold more than 5 metres in height;
- (b) any supporting part of any slung or suspended scaffold which causes a working platform to fall (whether or not in use); or
- (c) any part of any scaffold in circumstances such that there would be a significant risk of drowning to a person falling from the scaffold.

19 Train collisions

The collision of a train with any other train or vehicle, other than a collision reportable under Part 5 of this Schedule, which could have caused the death, or specified injury, of any person.

20 Wells

In relation to a well (other than a well sunk for the purpose of the abstraction of water)—
- (a) a blow-out (which includes any uncontrolled flow of well-fluids from a well);
- (b) the coming into operation of a blow-out prevention or diversion system to control flow of well-fluids where normal control procedures fail;
- (c) the detection of hydrogen sulphide at a well or in samples of well-fluids where the responsible person did not anticipate its presence in the reservoir drawn on by the well;
- (d) the taking of precautionary measures additional to any contained in the original drilling programme where a planned minimum separation distance between adjacent wells was not maintained; or
- (e) the mechanical failure of any part of a well whose purpose is to prevent or limit the effect of the unintentional release of fluids from a well or a reservoir being drawn on by a well, or whose failure would cause or contribute to such a release.

21 Pipelines or pipeline works

In relation to a pipeline or pipeline works—
- (a) any damage to, accidental or uncontrolled release from or inrush of anything into a pipeline;
- (b) the failure of any pipeline isolation device, associated equipment or system; or
- (c) the failure of equipment involved with pipeline works,

which could cause personal injury to any person, or which results in the pipeline being shut down for more than 24 hours.

22

The unintentional change in position of a pipeline, or in the subsoil or seabed in the vicinity, which requires immediate attention to safeguard the pipeline's integrity or safety.

PART II

DANGEROUS OCCURRENCES WHICH ARE REPORTABLE EXCEPT IN RELATION TO OFFSHORE WORKPLACES

23 Structural collapse

The unintentional collapse or partial collapse of—
- (a) any structure, which involves a fall of more than 5 tonnes of material; or
- (b) any floor or wall of any place of work, arising from, or in connection with, ongoing construction work (including demolition, refurbishment and maintenance), whether above or below ground.

24

The unintentional collapse or partial collapse of any falsework.

25 Explosion or fire

Any unintentional explosion or fire in any plant or premises which results in the stoppage of that plant, or the suspension of normal work in those premises, for more than 24 hours.

26 Release of flammable liquids and gases

The sudden, unintentional and uncontrolled release—
- (a) inside a building—
    - (i) of 100 kilograms or more of a flammable liquid;
    - (ii) of 10 kilograms or more of a flammable liquid at a temperature above its normal boiling point;
    - (iii) of 10 kilograms or more of a flammable gas; or
- (b) in the open air, of 500 kilograms or more of a flammable liquid or gas.

27 Hazardous escapes of substances

The unintentional release or escape of any substance which could cause personal injury to any person other than through the combustion of flammable liquids or gases.

PART III

DANGEROUS OCCURRENCES REPORTABLE IN RELATION TO A MINE

28 Fires or ignition of gas

Any outbreak of fire below ground.

29

Any person being caused to leave any place pursuant to regulation 11(1) of the Coal and Other Mines (Fire and Rescue) Regulations 1956 or section 79 of the 1954 Act, as a result of smoke or other indication that a fire may have broken out below ground.

30

Any fire on the surface which endangers the operation of any winding or haulage apparatus installed at a shaft or unwalkable outlet or of any mechanically operated apparatus for producing ventilation below ground.

31

The ignition of any gas (other than in a safety lamp) or dust below ground.

32
The unintentional ignition of any gas in part of a firedamp drainage system on the surface or in an exhauster house.

33 Escapes of gas with solid matter
The violent unintentional escape of gas together with coal or other solid matter into the mine workings.

34 Failures of plant or equipment
The breakage or unintentional uncoupling of any belt, rope, chain, coupling, balance rope, guide rope, rope tensioning system, suspension gear or other gear used for or in connection with—
 (a) carrying persons through any shaft or staple shaft;
 (b) transporting persons below ground; or
 (c) a belt conveyor designated by the mine manager as a man-riding conveyor.

35
The overwinding of—
 (a) any conveyance being used for the carriage of persons; or
 (b) any other conveyance, which becomes detached from its winding rope.

36
The bringing to rest of any conveyance operated using the friction of a rope on a winding sheave by the apparatus provided—
 (a) in the headframe of the shaft; or
 (b) in the part of the shaft below the lowest landing for the time being in use, for the purpose of bringing the conveyance to rest in the event of it being overwound.

37
The stoppage of any ventilating apparatus (other than an auxiliary fan) for over 30 minutes, except for planned maintenance, which causes a reduction in mine ventilation resulting in dangerous levels of noxious or flammable gases.

38
The collapse of any headframe, winding engine house, fan house or storage bunker.

39 Breathing apparatus
The malfunction of, or development of a defect in, breathing apparatus or a smoke helmet or other apparatus serving the same purpose or a self-rescuer where—
 (a) the malfunction or defect causes, or is likely to cause, a significant risk of personal injury to the user; or
 (b) immediately after use and as a result of its use any person receives first-aid or medical treatment because of that person's unfitness or suspected unfitness.

40 Emergency escape apparatus
The use of any apparatus—
 (a) provided at a mine in accordance with regulation 4 of the Mines (Safety of Exit) Regulations 1988; or
 (b) used to leave a mine when apparatus and equipment normally so used is unavailable, other than for the purpose of training and practice.

41 Inrushes of gas or flowing material
The inrush of noxious or flammable gas from old workings.

42

The inrush of water or material which flows when wet from any source.

43 Insecure tips

Any event (including any movement of material or any fire) which indicates that a tip to which Part 1 of the 1969 Act applies is or is likely to become insecure.

44 Locomotives

The bringing to rest of an underground locomotive by means other than its safety circuit protective devices or normal service brakes, when not used for testing purposes.

45 Falls of ground

Any fall of ground which—
- (a) results from a failure of an underground support system; and
- (b) prevents persons travelling through the area affected by the fall, or otherwise exposes them to danger, other than one which is part of the normal operations at a mine.

46 Accidents causing specified injuries

Any accident in which any person suffers a specified injury.

PART IV

DANGEROUS OCCURRENCES WHICH ARE REPORTABLE IN RELATION TO A QUARRY

47 Collapse of storage bunkers

The collapse of any storage bunker.

48 Sinking of craft

The sinking of any water-borne craft or hovercraft.

49 Projection of substances outside quarry

(1) Following a blasting operation, the projection of any material beyond the designated danger zone or the projection of any material which caused or might have caused injury.

(2) In this paragraph, "danger zone" means the area determined for each blast under the shotfiring rules required by regulation 25(2)(a)(i) and (b) of the 1999 Regulations.

50 Misfires

Any misfire, as defined by regulation 2(1) of the 1999 Regulations.

51 Insecure tips

Any event (including any movement of material or any fire) which indicates that a tip to which the 1999 Regulations apply is or is likely to become insecure.

52 Movement of slopes or faces

Any movement or failure of an excavated slope or face which—
- (a) could cause the death of any person; or
- (b) adversely affects any building, contiguous land, transport system, footpath, public utility or service, watercourse, reservoir or area of public access.

53 Explosion or fire in vehicles or mobile plant

Any explosion or fire in—
- (a) a dump truck with a load capacity of at least 50 tonnes; or

(b) an excavator with a bucket capacity of at least 5 cubic metres, which results in the stoppage of that vehicle or plant for more than 24 hours, and which affects—
(i) any place where persons normally work; or
(ii) the route of egress from such a place.

PART V
DANGEROUS OCCURRENCES WHICH ARE REPORTABLE IN RESPECT OF A RELEVANT TRANSPORT SYSTEM

54 Collision or derailment of passenger trains
Any collision between a passenger train and another train.

55
The derailment of the whole or part of a passenger train.

56 Collision or derailment not involving passenger trains
Any collision between non-passenger trains—
(a) on a running line, which causes damage to a train; or
(b) in a siding, which causes damage to a train and an obstruction to a running line.

57
The derailment of a non-passenger train—
(a) on a running line, except a derailment during shunting operations which does not obstruct any other running line; or
(b) in a siding, which causes an obstruction to a running line.

58 Accidents involving any train
Any collision between a train and a buffer stop which causes damage to the train, except a collision in a siding.

59
A train striking any cattle or horse, whether or not damage is caused to the train, or striking any other animal which causes damage necessitating immediate temporary or permanent repair (including damage to the windows of the driver's cab but excluding other damage consisting solely in the breakage of glass).

60
A train on a running line striking or being struck by any object which causes damage necessitating immediate temporary or permanent repair (including damage to the windows of the driver's cab but excluding other damage consisting solely in the breakage of glass) or which might have been liable to derail the train.

61
A train, other than one on a railway, striking or being struck by a road vehicle.

62
A passenger train, or a non-passenger train not fitted with continuous self-applying brakes, becoming unintentionally divided.

63 Failure of train parts
The failure of—
(a) an axle;
(b) a wheel or tyre, including a tyre loose on its wheel;
(c) a rope or the rope's fastenings;

(d) a winding plant or equipment involved in working an incline; or
(e) any part of a train which is likely to cause an accident to that or any other train, or to cause personal injury to any person, which occurs or is discovered whilst the train is on a running line.

64 Fire
Any fire—
(a) in or on any part of a passenger train or a train carrying dangerous goods within the meaning of the Carriage of Dangerous Goods and Use of Transportable Pressure Equipment Regulations 2009;
(b) in or on any part of a non-passenger train which was extinguished by a fire-fighting service;
(c) seriously affecting the functioning of signalling equipment;
(d) affecting the permanent way or works of a relevant transport system which necessitates the suspension of services over any line, or the closure of any part of a station or signal box or other premises, for a period—
 (i) of more than 30 minutes in the case of any part of a relevant transport system below ground; and
 (ii) in any other case, of more than 1 hour; or
(e) causing damage which could affect the running of a relevant transport system.

65 Severe electrical arcing or fusing
Severe electrical arcing or fusing—
(a) in or on any part of any train; or
(b) which seriously affects the functioning of signalling equipment.

66 Level crossings
Any train striking a road vehicle or gate at a level crossing.

67
Any train running onto a level crossing when not authorised to do so.

68
The failure of equipment at a level crossing which could cause a significant risk of personal injury to users of the road or path crossing the railway.

69 The permanent way and other works
The failure of a rail in a running line or of a rack rail, which results in—
(a) a complete fracture of the rail through its cross-section; or
(b) in a piece becoming detached from the rail which requires the immediate stoppage of traffic or the immediate imposition of a lower speed restriction.

70
The buckle of a running line which requires the immediate stoppage of traffic or the immediate imposition of a lower speed restriction.

71
An aircraft or vehicle of any kind either landing on, running onto or coming to rest across the line, or damaging the line, so as to cause damage—
(a) which obstructs the line; or
(b) to any railway equipment at a level crossing.

72
The runaway of an escalator, lift or passenger conveyor.

73

The following classes of accident where they are likely to cause an accident to a train or a significant risk of personal injury to any person—
- (a) the failure of a tunnel, bridge, viaduct, culvert, station or other structure or any part of it including the fixed electrical equipment of an electrified relevant transport system;
- (b) any failure in the signalling system which could cause a significant risk to the safe passage of trains other than a failure of a traffic light controlling the movement of vehicles on a road;
- (c) a slip of a cutting or of an embankment;
- (d) flooding of the permanent way;
- (e) the striking of a bridge by a vessel or by a road vehicle or its load; or
- (f) the failure of any other portion of the permanent way or works.

74 **Incidents of signals passed without authority**
Any train, travelling on a running line or entering a running line from a siding, passing a signal displaying a stop aspect without authority, unless the stop aspect was not displayed in sufficient time for the driver to stop safely at the signal.

PART VI

DANGEROUS OCCURRENCES WHICH ARE REPORTABLE IN RESPECT OF AN OFFSHORE WORKPLACE

75 **Release of petroleum hydrocarbon**
unintentional release of petroleum hydrocarbon on or from an offshore installation which—
- (a) results in—
   - (i) a fire or explosion; or
   - (ii) the taking of action to prevent or limit the consequences of a potential fire or explosion; or
- (b) could cause a specified injury to, or the death of, any person.

76 **Fire or explosion**
Any fire or explosion at an offshore installation, other than one caused by the release of petroleum hydrocarbon, which results in the stoppage of plant or the suspension of normal work.

77 **Release or escape of dangerous substances**
The unintentional or uncontrolled release or escape of any substance (other than petroleum hydrocarbon) on or from an offshore installation which could cause a significant risk of personal injury to any person.

78 **Collapses**
Any unintentional collapse or partial collapse of any offshore installation or of any plant on an offshore installation which jeopardises the overall structural integrity of the installation.

79 **Equipment**
The failure of equipment required to maintain a floating offshore installation on station which could cause a specified injury to, or the death of, any person.

80 **Dropping objects**
The dropping of any object on an offshore installation or on an attendant vessel or into the water adjacent to an installation or vessel which could cause a specified injury to, or the death of, any person.

81 Weather damage
Any damage to or on an offshore installation caused by adverse weather conditions and which could cause a specified injury to, or the death of, any person.

82 Collisions
Any collision between a vessel or aircraft and an offshore installation which causes damage to the installation, the vessel or the aircraft.

83
Any occurrence with the potential for a collision between a vessel and an offshore installation where, had a collision occurred, it might have jeopardised the overall structural integrity of the installation.

84 Subsidence or collapse of seabed
Any subsidence or collapse of the seabed likely to affect the foundations or the overall structural integrity of an offshore installation.

85 Loss of stability or buoyancy
Any incident which causes the loss of stability or buoyancy of a floating offshore installation.

86 Evacuation
The partial or complete evacuation of an offshore installation in the interests of safety.

87 Falls into water
Any fall of a person into water from more than 2 metres.

Asbestos

Andrea Oates

Introduction to asbestos

[A5001] Asbestos is the generic name for a group of naturally occurring fibrous minerals, metallic silicates, which have a wide range of industrial applications. They are divided into two sub-groups: serpentine (chrysotile – white asbestos), which is the most commonly used type of asbestos and amphiboles, which includes crocidolite (blue asbestos), amosite (brown asbestos), tremolite, actinolite and anthophyllite, of which crocidolite was the most commonly used in the past.

Mined in Canada, South Africa, Russia and elsewhere, the material has been manufactured into products which use its characteristics of heat and chemical resistance. Despite its contribution to fire protection, asbestos is now regarded as one of the most significant causes of occupational disease over the past 125 years. The Health and Safety Executive (HSE) says it is the single greatest cause of work-related deaths in the UK.

The use of asbestos has been extensive. As a result of UK prohibitions on new uses of asbestos-containing materials, products such as brake linings and other friction products containing asbestos are now unlikely to be found in most workplaces. However, its widespread application in materials used in construction has led to many buildings containing asbestos – it is present in around half a million non-domestic premises, including around 86% of schools according to the National Education Union (NEU), and around a million domestic premises. The current focus of asbestos controls is largely on the management of this legacy.

Asbestos diseases

[A5002] The HSE says asbestos can be found in any building built before the year 2000 (including houses, factories, offices, schools and hospitals) and asbestos-related diseases are responsible for around 5,000 deaths every year. There are four main diseases caused by asbestos: mesothelioma (which is always fatal), lung cancer (which is almost always fatal), asbestosis (which is not always fatal, but it can be very debilitating) and diffuse pleural thickening (which can be disabling and demonstrates exposure to asbestos, so people are at risk of developing other asbestos-related diseases including asbestosis and mesothelioma).

The diseases associated with asbestos arise when asbestos fibres are inhaled and penetrate deep into the lung. Extremely small fibres are capable of bypassing the body's defence mechanisms to achieve this. These are therefore called respirable fibres and have a diameter less than three micrometres. To illustrate how small these fibres are a human hair is approximately 50 micrometres in diameter.

Asbestosis is a thickening of the wall of the alveoli, the tiny lung sacs where oxygen passes into the blood. The HSE says it is generally recognised that heavy asbestos exposures are required in order to produce clinically-significant asbestosis within the lifetime of an individual. When millions of fibres are inhaled and reach the deep lung, the reduction in gas exchange capacity caused by thickening of the alveoli walls becomes significant. The disease of asbestosis is recognised when the patient has a measurable decrease in lung capacity and even notices a shortness of breath. Signs of the disease can also be detected by x-ray. If exposure is extensive and prolonged the damage to the lungs can be severe and eventually lead to death. It has occurred in industries where raw asbestos was handled in bulk – delivered in bales, manually cut open and fed into hoppers for the manufacturing process. Workers often went home white with asbestos fibres adhering to their body and clothes.

A condition associated with asbestosis is pleural plaques: areas of calcification or stiffening on the membrane on the outer surface of the lung. Pleural plaques do not generally cause symptoms. The current levels of asbestosis largely reflect the results of heavy exposures in the past. The HSE says that the best indication of the number of deaths where asbestosis contributed as a cause is to exclude death certificates mentioning mesothelioma. The HSE reports that deaths mentioning asbestosis (excluding those that also mention mesothelioma – see below) have increased substantially over a number of decades. There were 467 such deaths in 2015 compared with 109 in 1978. Typically, in recent years, the number of new cases of asbestosis assessed under the Industrial Injuries and Disablement Benefit (IIDB) scheme increased from 132 in 1978 to 1,050 in 2016. The number of cases of asbestosis has increased from 132 in 1978 to 1,050 in 2016.

Mesothelioma is a cancer occurring on the outer membrane of the lung. It is normally a very rare disease and so its occurrence among asbestos workers was quickly recognised. Typically, it is not detected in the early stages and, therefore, once diagnosed is advanced and usually progresses to death in months rather than years. However, the latency period from initial exposure to onset of the disease appears to be long, in the region of decades. The risk of contracting mesothelioma is based on exposure. The greater the concentration of respirable asbestos fibres inhaled, the greater is the risk of contracting the disease. In addition, the concentration of respirable asbestos fibres required to cause the disease is far lower than that associated with asbestosis. In the UK, mesothelioma was at first associated with crocidolite or blue asbestos and the control limit (the maximum level to which workers could legally be exposed) for blue asbestos was tightened. Historically it was recognised that all amphibole asbestos types were implicated, so amosite, brown asbestos, and crocidolite and the other amphibole asbestos types were assigned tighter control limits. Asbestos products are no longer manufactured in the UK and current exposure is to materials that contain mixtures of fibres rather than to a single asbestos type. Recent legislative updates have recognised this and all asbestos fibre types are now assigned a single lower control limit.

According to the latest HSE figures, there were 2,542 mesothelioma deaths in Great Britain in 2015, a similar number to the previous three years. The latest projections suggest that there will continue to be around 2,500 deaths per year for the rest of this decade before annual numbers begin to decline. The continuing increase in annual mesothelioma deaths in recent years has been driven mainly by deaths among those aged 70 and above. In 2015, there were 2,135 male deaths and 407 female deaths, similar to the annual numbers in among males and females in the previous three years. There were 2,170 new cases of mesothelioma assessed for IIDB in 2016, of which 240 were female, compared with 2,130 in 2015, of which 220 were female.

Lung cancer is now recognised as a disease associated with asbestos. The specific diagnosis of lung cancer being an asbestos-related disease is however occupational exposure. Since lung cancer occurring spontaneously, caused by smoking or caused by asbestos, is indistinguishable, it is the history of the patient which is used in the specific diagnosis. It is firmly established from epidemiology that there is an increased risk of lung cancer for smokers and that there is an increased risk of lung cancer for those exposed to asbestos. Indeed, for smokers who are also exposed to asbestos it is estimated that the risk of lung cancer is multiplied rather than the two risks being additive.

When asbestos fibres are inhaled, they have a very long half-life in the lungs. Their shape – long and thin – and robustness means that it is difficult for the usual lung-clearance mechanisms to remove them from the lungs (in the manner in which dust particles are cleared), and they do not readily dissolve in lung fluid. While the exact mechanisms which cause cancer remain obscure, the toughness and longevity of the fibres helps to explain why they are able to cause ill health many years after first exposure. Although the abrasive and chemical effect of the fibres in very high lung concentrations can cause asbestosis on a shorter timescale, this is largely an historical disease, while cancers continue to take their toll. If workers are exposed to airborne fibres, for example as a result of an incident in which asbestos has been inadvertently disturbed, there is no 'treatment' available to reduce the risk of future disease.

The overall scale of asbestos-related lung cancer deaths has to be estimated rather than counted – because it is difficult to tell these cancers apart from those due to other causes such as smoking. The HSE reports that research suggests there are probably about as many asbestos-related lung cancer deaths each year as there are mesothelioma deaths: around 2,500 deaths each year.

Historical exposures to asbestos – manufacturing

[A5003] In the 1950s and 1960s, the predominant concern arising from exposure to asbestos was for manufacturing workers contracting asbestosis following exposure to massive concentrations of respirable fibres. These workers were involved in manufacturing asbestos insulating boards and panels, asbestos cement products and friction products such as clutch plates, break shoes and gaskets. Anecdotally, they worked in 'snow storms' of the material, which because the fibres are soft and silky did not create any immediate discomfort, as would be the case with glass fibre which is immediately irritating to the eyes, throat and skin. In order to control this risk, and against the background of a well-funded lobbying campaign by the asbestos mining and manufacturing interests, a campaign which still functions, Asbestos Regulations were introduced imposing control limits on the maximum concentration of respirable asbestos fibres that workers could be exposed to. Where the concentrations could not be reduced below the control limits it was mandatory to provide suitable respirators. As the risk of lung cancer and mesothelioma was recognised and epidemiological data and analysis became available, the control limits for working with asbestos were gradually reduced. Originally, 'action levels' were adopted and defined in terms of concentrations of respirable airborne fibres measured over a twelve-week period. If the mean concentration of airborne fibres exceeded the action level then the area in the factory had to be designated an 'asbestos zone' and non-essential personnel excluded. If the airborne fibre concentration was likely to exceed the control limit then issuing and wearing respirators became mandatory. The original control regime included a lower control limit and action level for the amphibole asbestos fibre types than that of chrysotile asbestos.

The enactment of the *Control of Asbestos Regulations 2006 (SI 2006 No 2739)* enforced a single control limit that was set at 0.1 f/ml over a four-hour TWA (Time Weighted Average) covering all asbestos fibre types and the action levels were revoked.

On 6 April 2012, the *Control of Asbestos Regulations 2012 (SI 2012 No 632)* came into force revoking and replacing the *Control of Asbestos Regulations 2006 (SI 2006 No 2739)* (see below). The 2012 Regulations are still in force.

Asbestos prohibitions

[A5004] Historically, the *Asbestos (Prohibition) Regulations 1992 (SI 1992 No 3067)*, as amended, prohibited all imported materials to which asbestos had been added as part of a deliberate manufacturing process, but excluded naturally-occurring minerals which may contain traces of asbestos. Many producers and importers of mineral products carry out comprehensive testing to prevent materials containing significant quantities of asbestos getting into the supply chain. HSE advice on an amendment to the regulations advised suppliers that where asbestos was found, the material should not be sold unless the amount was trivially small. Even in those cases, suppliers had to inform their customers that trace quantities of asbestos may occasionally be found in their products. This allowed 'high-energy' users (for example traces of asbestos in blast-cleaning materials would release airborne fibres) to take appropriate precautions or consider alternative materials. The advice emphasised that even quite small amounts of asbestos in a material may give rise to a real risk from asbestos fibres. In addition to prohibitions on the sale of asbestos products, where they could be discovered in store rooms or warehouses, they could not be used and had to be disposed of safely. While there was no prohibition on asbestos already in place, new uses were effectively banned.

Asbestos removal

[A5005] As the manufacture and use of asbestos declined, accelerated by the *Asbestos (Prohibition) Regulations 1992 (SI 1992 No 3067)*, the risk to the health of manufacturing workers also decreased. However, it was recognised that there was a new exposure group and attention turned to those removing asbestos. The *Asbestos (Licensing) Regulations 1983 (SI 1983 No 1649)* were introduced to provide the enforcing authorities with information on which companies were involved in removing asbestos from buildings and where and when the work was being carried out. Approved Codes of Practice (ACOPs), aimed directly at the removal industry, provided information and instruction on meeting the requirements of these Regulations to those in charge of asbestos removal operations.

With the introduction of the *Control of Asbestos Regulations 2006 (SI 2002 No 2739)*, new editions of the ACOPs were published, including *Asbestos: The Licensed Contractors' Guide* (HSG 247) and *Asbestos: The analysts' guide for sampling, analysis and clearance procedures* (HSG 248). A full list of asbestos legislation, ACOPs and Guidance Notes are appended at the end of this chapter (see **A5043**).

Exposure of building maintenance workers

[A5006] As the *Asbestos (Licensing) Regulations 1983 (SI 1983 No 1649)* were enforced, and the asbestos removal industry became more thoroughly regulated, the risk to the health of the workers in that industry began to be more effectively controlled. However, in the late 1990s it was recognised that there was still a very large group of workers whose exposure to asbestos was not adequately controlled by the existing asbestos control regime. Maintenance workers and those engaged in building refurbishment were still exposed to asbestos because many building owners did not know that asbestos was present and many employers of maintenance workers did not consider that they would be working with asbestos and so took no precautions. For these reasons, Regulation 4 of the *Control of Asbestos at Work Regulations 2002 (SI 2002 No 2675)*, introduced a duty to manage asbestos within non-domestic premises. This became enforceable in 2004 and continues to be in force within the requirements of the *Control of Asbestos Regulations 2012 (SI 2012 No 632)*. The provisions require building owners and occupiers (duty holders) to ensure that a suitable and sufficient assessment is carried out as to whether asbestos is or is liable to be present in the premises and have a documented plan to control the risks to health presented by that asbestos. The aim is to reduce over time the number of asbestos-related deaths in the UK.

The Control of Asbestos Regulations 2012

[A5007] The Control of Asbestos Regulations 2006 repealed and replaced legislation previously governing asbestos within the UK:

— the *Asbestos (Licensing) Regulations 1983 (SI 1983 No 1649)*

[A5007] Asbestos

— the *Asbestos (Prohibitions) Regulations 1992 (SI 1992 No 3067)*
— the *Control of Asbestos at Work Regulations 2002 (SI 2002 No 2675)*

The *Control of Asbestos Regulations 2012 (SI 2012 No 632)* then revoked and replaced the *Control of Asbestos Regulations 2006 (SI 2006/2739)*. The requirements of the Control of Asbestos Regulations 2012 are summarised below.

Duties under the Control of Asbestos Regulations 2012

[A5007.1] Regulation 2(1) includes definitions of terms including 'asbestos cement', 'asbestos coating', 'asbestos insulation', 'asbestos insulating board', 'short duration work' and 'textured decorative coatings' 'relevant doctor' and 'licensable work with asbestos' as set out below:

- 'asbestos cement' means a material which is predominantly a mixture of cement and chrysotile and which when in a dry state absorbs less than 30% water by weight;
- 'asbestos coating' means a surface coating which contains asbestos for fire protection, heat insulation or sound insulation but does not include textured decorative coatings;
- 'asbestos insulating board' (AIB) means any flat sheet, tile or building board consisting of a mixture of asbestos and other material except:
  — asbestos cement; or
  — any article of bitumen, plastic, resin or rubber which contains asbestos, and the thermal or acoustic properties of the article are incidental to its main purpose;
- 'asbestos insulation' means any material containing asbestos which is used for thermal, acoustic or other insulation purposes (including fire protection) except:
  — asbestos cement, asbestos coating or asbestos insulating board; or
  — any article of bitumen, plastic, resin or rubber which contains asbestos and the thermal and acoustic properties of that article are incidental to its main purpose;
- 'the control limit' means a concentration of asbestos in the atmosphere when measured in accordance with the 1997 WHO recommended method, or by a method giving equivalent results to that method approved by the HSE, of 0.1 fibres per cubic centimetre of air averaged over a continuous period of four hours;
- 'licensable work with asbestos' is work:
  — where the exposure to asbestos of employees is not sporadic and of low intensity;
  — in relation to which the risk assessment cannot clearly demonstrate that the control limit (see above) will not be exceeded;
  — on asbestos coating; or
  — on asbestos insulating board or asbestos insulation for which the risk assessment:
    - demonstrates that the work is not sporadic and of low intensity;

- cannot clearly demonstrate that the control limit will not be exceeded; or
- demonstrates that the work is not short duration work.
- 'medical examination' includes any laboratory tests and x-rays that a relevant doctor may require;
- 'relevant doctor' means an appointed doctor or an employment medical adviser. In relation to work with asbestos which is not licensable work with asbestos and is not exempted by regulation 3(2), 'relevant doctor' also includes an appropriate fully registered medical practitioner who holds a licence to practice;
- 'textured decorative coatings' means decorative and textured finishes, such as paints and ceiling and wall plasters which are used to produce visual effects and which contain asbestos. These coatings are designed to be decorative and any thermal or acoustic properties are incidental to their purpose.

The HSE explains that certain types of work with asbestos-containing materials (ACMs) can only be done by those who have been issued with a licence by HSE. This is work which meets the definition of 'licensable work with asbestos' above. Short duration means the total time spent by all workers working with these materials does not exceed two hours in a seven-day period, including time spent in setting up, cleaning and clearing up, and no one person works for more than one hour in a seven-day period.

Guidance in L143 gives examples of licensable and non-licensable work:

- Work which requires a licence from HSE:
  — removing sprayed coatings (limpet asbestos);
  — removal or other work which may disturb pipe lagging;
  — any work involving loose fill insulation;
  — work on millboard;
  — cleaning up significant quantities of loose/fine debris containing ACM dust (where the work is not sporadic and of low intensity, the control limit will be exceeded or it is not short duration work);
  — work on AIB, where the risk assessment indicates that it will not be of short duration.
- Work which does not usually require a licence from HSE:
  — small, short duration maintenance tasks where the control limits will not be exceeded;
  — removing textured decorative coatings by any suitable dust-reducing method;
  — cleaning up small quantities of loose/fine debris containing ACM dust (where the work is sporadic and of low intensity, the control limit will not be exceeded and it is short duration work);
  — work on asbestos cement products or other materials containing asbestos (such as paints, bitumen, resins, rubber, etc) where the fibres are bound in a matrix which prevents most of them being released (this includes, typically, aged or weathered AC);
  — work associated with collecting and analysing samples to identify the presence of asbestos;

Regulation 2(2) sets out that work with asbestos includes: the removal, repair or disturbance of asbestos or materials containing asbestos; work that is ancillary to this; and the supervision of work with asbestos and ancillary work.

The HSE explains that 'ancillary work' includes maintenance work on equipment (eg class H vacuum cleaners (BS 8520-3:2009) and air extraction equipment (including 'negative pressure' units) which involves contaminated (or potentially contaminated) parts of that equipment. It also includes putting up and taking down scaffolding to provide access for licensable work, where it is foreseeable that the scaffolding activity is likely to disturb the asbestos.

Work with asbestos is not 'short duration work' if (in any seven-day period) that work, including any ancillary work liable to disturb asbestos in any seven-day period, takes more than two hours or any person carries out the work for more than an hour.

The ACOP to the regulations explains that for exposure to asbestos to be sporadic and of low intensity, the concentration of asbestos in the atmosphere should not exceed or be liable to exceed the concentration approved in relation to a specified reference period by the HSE. This is 0.6 fibres per cubic centimetre (f/cm3) in the air measured over a ten-minute period. Any exposure which exceeds or is liable to exceed this is not sporadic and of low intensity.

HSE guidance explains: 'The unit f/cm3 is the same unit as f/ml. This ten-minute limit is sometimes called the short-term exposure limit or STEL. It refers to the highest level of concentration for any ten-minute period of duration of the work. Note that this approved concentration for sporadic and low intensity exposure is not the same as the "control limit" defined in regulation 2.' The control limit is set out earlier in this section.

Regulation 3 sets out that the regulations apply to employers, employees and the self-employed. However, regulations 9 (notification of work with asbestos), 18(1)(a) (designated areas) and 22 (health records and medical surveillance) do not apply where:

- the exposure to asbestos of employees is sporadic and of low intensity (see above); and
- it is clear from the risk assessment that the exposure to asbestos of any employee will not exceed the control limit (see above);
- the work involves:
  — short, non-continuous maintenance activities in which only non-friable materials are handled;
  — removal without deterioration of non-degraded materials in which the asbestos fibres are firmly linked in a matrix;
  — encapsulation or sealing of asbestos-containing materials which are in good condition; or
  — air monitoring and control, and the collection and analysis of samples to ascertain whether a specific material contains asbestos.

The HSE points out that the requirements of regulations 9, 18(1)(a) and 22 apply to all licensable work. It also says: 'Although it does not require a licence issued by HSE, all non-licensable work with asbestos will still need to be carried out in accordance with the requirements contained in the Regulations. In particular, it needs to be carried out by trained and competent workers in accordance with a plan of work, using appropriate control measures to prevent exposure and the spread of asbestos.'

To decide if the exemption from the requirements in regulations 9, 18(1)(a) and 22 applies to non-licensable work, the employer needs to make an assessment of the work to be done and decide if it meets the conditions set out above.

Employer's duties under the regulations do not apply:

- under regulation 10 (information, instruction and training) with respect to people who are not employees of that employer unless they are on the premises where the work is being carried out; and
- under regulation 22 (health records and medical surveillance) with regard to people who are not employees of that employer.

In addition, regulation 17 (cleanliness of premises and plant) does not apply to fire and rescue services/authorities in respect of premises attended by its employees for the purpose of fighting a fire or in an emergency. The regulations do not apply to the master or crew of a ship.

Affected premises

[A5008] Regulation 4 sets out the duty to manage asbestos in non-domestic premises. The duty applies not only to non-domestic premises, but also to the common parts of commercial buildings, housing developments and leasehold flats. Given the widespread use of asbestos, particularly after the 1940s, many buildings still contain asbestos. A number of different parties with any measure of control need to take steps to comply with this duty.

Identifying the duty holder

[A5009] Those with a primary responsibility for a building, such as the owner, sole occupier or part occupier, have duties under the *CAR (SI 2012 No 632)*. Those with any responsibility for the maintenance of premises by virtue of a contract or tenancy also have a legal duty to manage the risks from asbestos and comply with the *CAR*. If no such formal agreement exists, the owner, sub-lessor or managing agent in charge of a workplace premise, or any employer or self-employed person who occupies commercial premises in which people work, will be subject to the same duty.

Risk assessment

[A5010] Under regulation 4, the duty holder is required to make a suitable and sufficient risk assessment as to whether asbestos is, or is liable to be, present in the premises. It should be noted that wherever a duty such as this applies, compliance requires that the duty holder ensures that it has been carried out. In this case, this means ensuring that a suitable and sufficient assessment has been conducted, documented and is available and used to inform decision-making, advise incoming occupiers, maintenance workers and others. Commissioning others to carry out such work is acceptable practice, so long as those employed are competent and allocated sufficient resources to carry out the work properly.

This includes taking into account the age of the premises, building plans and any other relevant information. In carrying out inspections it must be presumed that materials contain asbestos unless there is strong evidence to the contrary. At this stage it is necessary simply to identify whether asbestos may be present and, therefore, determine whether a more thorough investigation is required. The HSE advises that if a building was built in or after 2000, it is unlikely to have any asbestos. However, it warns that brownfield land, areas of land that was previously developed, often but not always for industrial and commercial purposes, may be contaminated and asbestos may lie buried. In addition, old equipment, including ovens, brakes, soundproofing, insulating mats, fire blankets, oven gloves, ironing surfaces should be presumed to contain asbestos.

The most likely asbestos materials to be found are:

- Sprayed asbestos and asbestos loose packing – typically used as fire breaks in ceiling voids and to protect steelwork, fire protection in ducts, firebreaks, etc.
- Moulded or preformed lagging – generally used in thermal insulation of pipes and boilers.
- Insulating board – fire protection, thermal insulation, panelling, partitioning, ducts.
- Ceiling tiles.
- Millboard and paper products – insulating electrical products and facing wood fibreboard.
- Woven products – boiler seals and old fire blankets.
- Cement products in flat or corrugated sheets.
- Textured coatings.
- Bitumen roofing materials.
- Vinyl and thermoplastic floor tiles.

Records

[A5011] If asbestos is present, an up-to-date record of the location and condition of asbestos-containing materials ('ACMs') or presumed ACMs in the premises must be made. This is often referred to as an 'asbestos register'. The asbestos register should record where asbestos is present, or presumed to be present, in the building, its form, eg asbestos insulating board, pipe lagging, ceiling tiles, asbestos cement sheets, and its condition. The health risk depends on the inhalation of respirable fibres, so the condition of the asbestos, as recorded in the asbestos register, should reflect the extent to which fibres may be released from the ACM during normal occupancy and during any maintenance work. Since maintenance work will be required at some stage, and it is necessary to determine the type of asbestos present before such work commences, it is a requirement to identify the type of asbestos present in the ACM and record it in the asbestos register. This requires samples of suspected ACM to be taken for laboratory analysis in accordance with HSG248 "The Analyst's Guide" by a suitably accredited laboratory in accordance with CAR regulation 21. The overall purpose of the asbestos register is to provide information regarding the locations of ACMs to assist in the planning of work to prevent incidental exposure to asbestos fibres. Selecting the correct format of an asbestos register is paramount in ensuring effective asbestos management – for example, electronic asbestos registers are easily updated and controlled and can provide easy access across multiple sites. However hard copy information may be more appropriate for single sites or where computer access is restricted.

Management plan

[A5012] The compilation of an asbestos register is only one part of the duty to manage asbestos. There must also be a documented plan to manage the health risks presented by the presence of ACMs. Where the health risk during normal occupancy is low, the plan may be to maintain the ACM in place and label it. There should then be written procedures for planned maintenance work and procedures for emergency work, and staff should be familiar with their duties in implementation. These procedures would normally include calling in licensed asbestos contractors to remove the ACM under controlled conditions before the maintenance work proceeded. If the presence of asbestos in the building represents a risk to health under normal occupancy conditions or it is envisaged that emergency situations would involve the removal of ACM, then the asbestos management plan may be to immediately encapsulate or remove ACMs for particular areas. This work requires the involvement of contractors licensed under the *Control of Asbestos Regulations (SI 2012 No 632) Regulation 8* working in accordance with the Approved Code of Practice (HSG 247) 'The Licensed Contractors Guide' (see below). Where asbestos removal operations are undertaken, it is a requirement under Regulation 20 for air testing to be undertaken by a United Kingdom Accreditation Service (UKAS) accredited laboratory under ISO 17025. It is considered best practice to retain the services of a UKAS accredited laboratory to assist in both the monitoring and management of asbestos removal work. The asbestos management plan should also contain the procedures for dealing with suspect ACM. Typically, such procedures would include the requirement to halt the work and call in a UKAS accredited laboratory to take samples and determine whether asbestos is present. The asbestos management plan should be reviewed regularly and the procedures and arrangements monitored to ensure that they are effective. It is important to test the procedures to ensure that they work. For instance, when routine maintenance is undertaken, checks that the tradespeople concerned were furnished with the appropriate information and that they followed the correct procedures should be made and recorded. In many commercial premises the effect of the *CAR's* duty to manage may create more than one duty holder. Here it is necessary to examine the nature and extent of the repair and maintenance obligation owed by each party, or the level of control in determining their relative contributions to comply.

[A5012] Asbestos

Asbestos

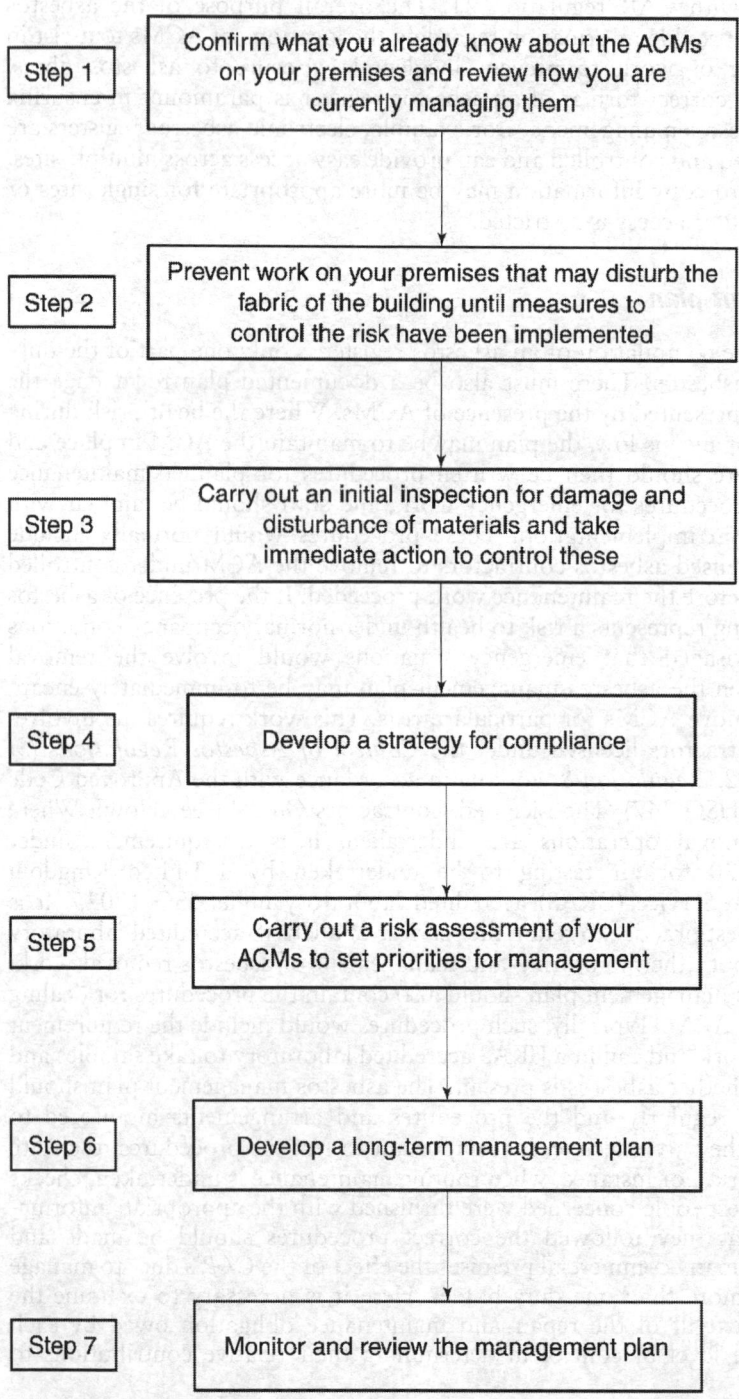

To ensure the effective compliance with the duty to manage asbestos in non-domestic premises, there is a simple seven step programme to follow.

Step 1: What ACMs do you know about and how are you managing them?

[A5013] Are there likely to be ACMs on your premises? If there are none, no further action is required. If ACMs are present, or there is uncertainty, a review of current management arrangements is required. Check that:

- Maintenance and building work is effectively controlled, with a system to prevent disturbance of asbestos and unknown material.
- Maintenance and building staff know the rules about ACMs, and how to work safely on or adjacent to them.
- The condition of known ACMs or materials presumed to contain asbestos is known, there is a system to check for damage and deterioration.
- Damaged and deteriorated ACMs are repaired, removed or isolated to prevent exposure to released fibres.
- There is a management plan for recording findings on ACMs, monitoring their condition and taking action as necessary.
- Everyone, including appointed contractors, knows their roles and responsibilities for the management of ACMs on the premises.
- A responsible person has been appointed, is competent and has the resources (premises or facilities manager, or for small businesses the owner/manager) to oversee the management of asbestos.
- There is access to competent advice from a qualified safety practitioner or hygienist with the necessary experience and qualifications for identifying and assessing asbestos. This is particularly important in buildings that are known or suspected to contain significant quantities of ACMs or have been constructed or refurbished prior to 2000.

Step 2: Control work on your premises

[A5014] In every building, there needs to be a system which prevents anyone doing building or maintenance work, even running new telephone cables or IT systems, without first checking whether ACMs may be disturbed (usually by reference to the Asbestos Register).

Step 3: Carry out an initial inspection

[A5015] After stopping new activities which could damage or disturb ACMs, it is essential to check on the condition of all existing known or suspected ACMs and take immediate action to prevent exposure to airborne fibres where such damage is identified. This check is looking for *any* damage to building materials. If any damaged materials are identified, they can be tested to see if they are ACMs, and kept isolated until the results are available and the material removed or repaired if asbestos is confirmed.

Step 4: Develop a strategy for compliance

[A5016] After taking immediate control (steps 1–3), the remaining steps are designed to manage occupancy and building works. The first such step is to develop a formal strategy to manage the long-term risks:

- Commission a survey to establish the extent of ACMs.
- Document the management controls to minimise risk of damage to ACMs, and to identify and act on damage and deterioration effectively.

Step 5: Risk assessment and priorities

[A5017] The HSE has developed a quantitative method for assessing the risks arising from asbestos materials. An asbestos surveyor will normally undertake an assessment of the materials condition in accordance with *HSG 264: Asbestos: The Survey Guide*. This will define the material's potential to release fibres. The second stage of the assessment deals with the likelihood of human exposure and is the responsibility of the duty holder in accordance with *CAR* Regulation 4. This risk assessment procedure is detailed within HSE Guidance document HSG 227 "A Comprehensive Guide to Managing Asbestos in Premises".

Step 6: Develop a long-term management plan

[A5018] The plan should contain:

- How the location and condition of known or presumed ACMs is recorded.
- Priority assessments and an action plan.
- Decisions about management options and their rationale.
- Action timetable.
- Monitoring arrangements.
- Employees and their responsibilities.
- Contractor procedures including selection, appointment, information and monitoring.
- Training arrangements.
- Procedures and plan for implementation, including controlling work on the fabric of the building.
- Method for passing on information about ACMs to those who need it.
- Responsibility for the management plan and its updating.
- Procedure and timetable for review of the management plan.

Step 7: Monitor and review the management plan

[A5019] Implementation of the management plan should be monitored and checks routinely carried out on the ACMs, and records should be kept up-to-date. Incidents and accidents require careful investigation and lessons learned should be used to improve the plan and its implementation. For most premises it is recommended that the plan is subject to at least a 12–monthly thorough review. Materials should be re-inspected according to the level of risk they present. For example, an ACM in a locked, disused area is less likely to become damaged than an ACM in a thoroughfare. Therefore, the material most likely to be disturbed should be monitored more routinely.

Duties of third parties

[A5020] The *CAR (SI 2012 No 632)* places any person under a duty to assist in the management of asbestos. Regulation 4(2) says: 'Every person must co-operate with the dutyholder so far as is necessary to enable the dutyholder to comply with the duties set out under this regulation.' Third parties are required to co-operate with the duty holder so far as is necessary to enable the duty holder to comply with the *CAR* by:

- making all information in relation to the premises available for inspection;
- providing them with information on the locations of asbestos, if known;
- allowing duty holders to identify asbestos or presumed asbestos for all parts of the premises repaired or maintained by the third party;
- assisting in the preparation and implementation of the asbestos management plan to control the health risk presented by ACMs;
- ensuring that everyone potentially at risk from ACMs receives information on the location and condition of the material so far as it is within the third party's control.

CAR contractual and tenancy agreements

[A5021] Organisations need to review the extent of their maintenance and repair obligations under any leases and be clear as to where the responsibility for asbestos control lies. Responsibility for dealing with any asbestos health risks, and therefore each party's contribution, should be considered in pre-contract lease negotiations and specific provisions should be made in the tenancy agreement. This will be especially important in determining who is responsible for common parts of the building where, for example, control over maintenance and repair activities in those parts could be shared.

Minimum requirements

[A5022] It will be a defence to show that an employer has done everything in their power to prevent exposure to asbestos. However, they will need to be able to produce an up-to-date asbestos register in a suitable format and a documented asbestos management plan containing risk assessments, procedures and arrangements to manage the risk from exposure to asbestos during normal occupancy, planned maintenance and emergency situations. Evidence will also be required to demonstrate that the asbestos management plan has been implemented and that it is effective.

Asbestos building surveys

[A5023] A second edition of HSG 264: *Asbestos: The Survey Guide* was published in 2012 and provides guidance for people carrying out asbestos surveys and people with specific responsibilities for managing asbestos in non-domestic premises under the Control of Asbestos Regulations 2012. It covers competence and quality assurance and surveys, including survey planning, carrying out surveys, the survey report and the dutyholder's use of the survey information. It can be downloaded free on the HSE website at www.hse.gov.uk/pubns/priced/hsg264.pdf.

Prevention and control of exposure to asbestos

Identification of the presence of asbestos

[A5024] Regulation 5 sets out that an employer must not carry out demolition, maintenance or other work which exposes, or is liable to expose, any employees to asbestos unless a suitable and sufficient assessment has been carried out.

This should identify whether asbestos is present and if so, what type of asbestos, contained in what material and in what condition is present, or liable to be present in the premises. If there is doubt as to whether asbestos is present, the employer must assume that it is present, and is not only chrysotile, and observes the requirement of the regulations.

The approved code of practice explains that before carrying out any work involving the potential disturbance of asbestos, employers should find out if the part of the building likely to be disturbed contains asbestos and, if so, its type and condition.

'This should include assessing relevant information, such as that contained in construction plans or provided by dutyholders responsible for the maintenance and repair of premises under regulation 4 of the Regulations (eg asbestos surveys or registers). If no records are available, or there are doubts about their accuracy/relevance, employers may need to arrange a survey and analysis of representative samples to determine the presence, type and condition of asbestos,' it says. 'Alternatively, employers should assume that the part of the building to be disturbed contains the most hazardous types of asbestos, crocidolite (blue) or amosite (brown), and apply the appropriate control measures required by the Regulations, using a licensed contractor if required.'

Assessment of work which exposes employees to asbestos

[A5025] Regulation 6 sets out that an employer must not carry out work which is liable to expose employees to asbestos unless he or she has:

- '(a) made a suitable and sufficient assessment of the risk created by that exposure to the health of those employees and of the steps that need to be taken to meet the requirements of these Regulations;
- (b) recorded the significant findings of that risk assessment as soon as is practicable after the risk assessment is made; and
- (c) implemented the steps referred to in sub-paragraph (a).'

The risk assessment must:

- identify the type of asbestos to which employees are liable to be exposed;
- determine the nature and degree of exposure which may occur in the course of the work;
  - consider the effects of control measures which have been or will be taken (see Regulation 11);
  - consider the results of monitoring of exposure (see Regulation 19);
  - set out the steps to be taken to prevent that exposure or reduce it to the lowest level reasonably practicable;

- consider the results of any necessary medical surveillance; and
- include any additional information as the employer may need in order to complete the risk assessment.

It must be reviewed regularly and immediately if there is reason to suspect that the existing risk assessment is no longer valid; there is a significant change in the work to which the risk assessment relates; or the results of any monitoring carried out under regulation 19 show it to be necessary.

Where changes to the risk assessment are required, these must be made and recorded.

Where the risk assessment shows that exposure to asbestos of employees may exceed the control limit (see above), the employer must keep a copy of the significant findings of the risk assessment at the premises while the work to which the risk assessment relates is being carried out.

The HSE advises employers to read regulation 6 with regulation 11(1), which places a duty on employers to entirely prevent the exposure of their employees to asbestos so far as is reasonably practicable. It says this should be the first consideration. The ACOP says the risk assessment must:

- be carried out in carry it out in time to comply with the Regulations and enable appropriate precautions to be taken before work begins;
- be job specific and consider the full scope of the work;
- establish the extent of potential risks and who could be affected; and
- identify the steps taken to remove the risk or, if that is not possible, to reduce the risk.

The significant findings must be recorded in writing (electronic or paper) and communicated to employees and others who could be affected in an understandable way in order to minimise risks or take appropriate precautions to reduce or remove the risk before work begins. The assessment must be regularly reviewed and updated as required.

The HSE also points out that employers have duties under the Safety Representatives and Safety Committees Regulations 1977 and the Health and Safety (Consultation with Employees) Regulations 1996 (as amended) to consult their employees.

The ACOP makes clear that employers must make sure that whoever carries out the risk assessment and provides advice on the prevention and control of exposure is competent to do this. To be suitable and sufficient, the risk assessment must include:

- for non-licensable work, a statement of why the work meets the criteria for non-licensable rather than licensable work, and whether it is notifiable non-licensed work (NNLW);
- a description of the work being carried out and the expected scale and duration;
- a description of the type(s) of asbestos and results of any survey or analysis or a statement that the assumption is that the asbestos is not chrysotile alone; and
- a description of the quantity, form, size, means of attachment, extent and condition of the ACMs present.

Plans of work

[A5026] Regulation 7 requires that an employer must not undertake any work with asbestos unless they have prepared a suitable written plan of work detailing how that work is to be carried out.

A copy of the plan of work must be kept at the premises while the work is being carried out.

In cases of final demolition or major refurbishment of premises, the plan of work must, so far as is reasonably practicable, specify that asbestos must be removed before any other major works begin, unless removal would cause a greater risk to employees than if the asbestos had been left in place.

The plan of work must include details of —

- the nature and probable duration of the work;
- the location of the place where the work is to be carried out;
- the methods to be applied where the work involves the handling of asbestos or materials containing asbestos;
- the equipment to be used for protection and decontamination of those carrying out the work, and the protection of others on or near the worksite;
- the measures the employer intends to take to comply with the requirements of regulation 11 (which deals with preventing and reducing exposure to asbestos); and
- the measures the employer intends to take in order to comply with the requirements of regulation 17 (which deals with cleanliness of premises and plant).

The work to which the plan relates must be carried out in accordance with the plan and any subsequent written changes to it.

The ACOP says that, as far as reasonably practicable, employers must make sure their employees follow the plan of work (also known as a method statement, plan or POW).

It also says:

> 'Where unacceptable risks to health and/or safety are discovered while work is in progress, eg disturbing hidden, missed or incorrectly identified ACMs, stop any work affecting the asbestos, except to put suitable controls in place and prevent further spread. Where there is extensive damage to ACMs which causes contamination of the premises, or part of the premises, the area should be immediately evacuated. Work should not restart until a new plan of work is drawn up or until the existing plan is amended. Some measures may need to be carried out by licensed contractors.'

Where necessary, the plan should include the site layout, a description of the location and nature of the asbestos present and which ACMs will be disturbed by the work. It should be a practical and useful document, describing a safe working method for site staff to follow and should be drawn up by a suitably competent person. Suitable and sufficient plans of work are a licence condition for any licensable work with asbestos and a legal requirement and a suitable and sufficient plan of work must have been prepared by the time of notification (see A5027 below).

Licensing and notification of work with asbestos

[A5027] Regulation 8 requires an employer to hold a licence before undertaking any licensable work with asbestos. Licenses are granted by the Health and Safety Executive (HSE) and applications must normally be made at least 28 days before the date the license is to run.

Regulation 9 sets out that notification of licensable work with asbestos to the enforcing authority is normally at least 14 days before the work is due to begin.

Information, instruction and training

[A5028] Regulation 10 requires employers to ensure that employees who are liable to be exposed to asbestos and their supervisors are given adequate information, instruction and training regarding:

- the properties of asbestos and its effects on health, including its interaction with smoking,
- the types of products or materials likely to contain asbestos,
- the operations which could result in asbestos exposure and the importance of preventive controls to minimise exposure,
- safe work practices, control measures, and protective equipment,
- the purpose, choice, limitations, proper use and maintenance of respiratory protective equipment,
- emergency procedures,
- hygiene requirements,
- decontamination procedures,
- waste handling procedures,
- medical examination requirements, and
- the control limit and the need for air monitoring,

in order to safeguard themselves and other employees.

This information, instruction and training must be given at regular intervals; adapted to take account of significant changes in the type of work carried out or methods of work used by the employer; and provided appropriately with regards to the nature and degree of exposure identified by the risk assessment and so that the employees are aware of the significant findings of the risk assessment, and the results of any air monitoring carried out with an explanation of the findings.

The ACOP sets out three main types of information, instruction and training employers should provide:

- asbestos awareness training to employees whose work could foreseeably disturb the fabric of a building and expose them to asbestos, or who supervise or influence the work. It should be given to those workers in the refurbishment, maintenance and allied trades where it is foreseeable that ACMs may become exposed during their work;
- additional task-specific information, instruction and training for those employees whose work will knowingly disturb ACMs, and which is defined as non-licensable work or NNLW (the ACOP contains detailed guidance); and

- additional task-specific information, instruction and training for those employees carrying out work defined as 'licensable work'. The ACOP contains detailed guidance and the licensed contractors' guide sets out the detailed content of the asbestos training modules for operatives, supervisors, managers and directors involved in licensable work.

Prevention or reduction of exposure to asbestos

[A5029] Regulation 11 deals with the prevention or reduction of exposure to asbestos and requires employers to prevent the exposure to asbestos of any employee so far as is reasonably practicable.

Where it is not reasonably practicable to prevent such exposure, they must work through a hierarchy of controls to:

- reduce exposure to asbestos of employees to the lowest level reasonably practicable by measures other than the use of respiratory protective equipment; and
- ensure that the number of any employees exposed to asbestos at any one time is as low as is reasonably practicable.

In order of priority these are:

- the design and use of appropriate work processes, systems and engineering controls and the provision and use of suitable work equipment and materials in order to avoid or minimise the release of asbestos; and
- the control of exposure at source, including adequate ventilation systems and appropriate organisational measures,

In addition, the employer must, so far as is reasonably practicable, provide any employee concerned with suitable respiratory protective equipment (RPE).

Where it is not reasonably practicable to reduce the exposure of an employee to asbestos to below the control limit (see **A5003** above) by these measures, they must also provide them with suitable RPE to reduce the concentration of asbestos in the air inhaled by that employee (after taking account of the effect of that respiratory protective equipment) to a concentration which is:

(a) below the control limit; and
(b) as low as is reasonably practicable.

Personal protective equipment (PPE) must be suitable for the purpose and comply with the requirements of the *Personal Protective Equipment Regulations 2002* or European Regulation 2016/425 which came into force in 2018. In the case of RPE, it must be of a type approved or conform to a standard approved by the HSE.

The employer must ensure that no employee is exposed to asbestos in a concentration in the air inhaled by that worker which exceeds the control limit; or if the control limit is exceeded they must:

- immediately inform any employees concerned and their representatives and ensure that work does not continue in the affected area until adequate measures have been taken to reduce employees' exposure to asbestos below the control limit,

- as soon as is reasonably practicable identify the reasons for the control limit being exceeded and take the appropriate measures to prevent it being exceeded again, and
- check the effectiveness of the measures taken by carrying out immediate air monitoring.

The ACOP sets out that 'where it is not reasonably practicable to prevent exposure to asbestos, employers must choose the most effective method or combination of methods to minimise fibre release and reduce exposure to the lowest levels reasonably practicable' and document this in the written risk assessment and/or plan of work. In addition, they should keep the number of both employees and others who might be exposed to asbestos at any one time as low as reasonably practicable.

Use of control measures

[A5030] Regulation 12 requires employers to ensure that control measures are properly used or applied and requires employees to make full and proper use of control measures, and where relevant, ensure it is returned after use to any accommodation provided for it; and to report any defect to the employer without delay.

The ACOP sets out that employers' procedures should not be made less effective by other work practices or other machinery, and should include regular checks, at least at the start of every shift, and prompt action when a problem is identified.

Employees should use any control measures properly and keep equipment in the places provided. This includes dust suppression and extraction equipment, RPE and protective clothing. They should carefully follow the procedures set out in the risk assessment and plan of work, including those for changing and decontamination, and comply with the use of control measures. They should also keep the workplace clean; eat, drink and smoke only in the designated places provided; and immediately report any defects concerning control measures to their supervisor/manager.

Maintenance of control measures

[A5031] Regulation 13 sets out that control measures, in the case of plant and equipment, must be kept clean and in good repair and must be maintained in an efficient state and in efficient working order. Systems of work and supervision and any other measure must be reviewed at suitable intervals and revised if necessary.

The employer must ensure that thorough examinations and tests of exhaust ventilation equipment or respiratory protective equipment (except disposable respiratory protective equipment) are carried out at suitable intervals by a competent person and records kept for at least five years.

The ACOP further explains that employers must draw up maintenance procedures for control measures and PPE which should also cover the equipment used for cleaning, washing and changing facilities and the controls used to prevent the spread of contamination. These should make clear which control measures require maintenance, when and how to carry this out and

[A5031] Asbestos

who is responsible for doing it. In particular, maintenance is required for: enclosures; hygiene facilities; vacuum cleaners; air extraction equipment; wet injection equipment; and RPE, including storage. The ACOP provides detailed guidance in relation to all these types of equipment.

Provision and cleaning of protective clothing

[A5032] Regulation 14 requires employers to provide adequate and suitable protective clothing for any employee who is exposed or is liable to be exposed to asbestos, 'unless no significant quantity of asbestos is liable to be deposited on the clothes of an employee while at work'. This must either be disposed of as asbestos waste or adequately cleaned at suitable intervals in a suitably equipped premises or laundry. Protective equipment which has been used must be suitably packed and labelled before being removed from the premises in accordance with Schedule 2 (which sets out how articles containing asbestos must be labelled) as if it were a product containing asbestos. Where failure or improper use of protective clothing results in a significant quantity of asbestos being deposited on personal clothing, this must be treated as protective clothing for the purposes of cleaning or disposal.

The ACOP says employers must decide whether protective clothing is required as part of their assessment and should start with the assumption that it will be needed, 'unless there is no potential for physical contamination and/or airborne exposures will be extremely slight and infrequent'. For licensable work employers will always need to provide a full set of PPE. Suitable protective clothing must: fit; be loose enough to avoid straining and ripping the seams; be comfortable; be suitable for cold environments; prevent penetration by asbestos fibres; be elasticated at the cuffs, ankles and on the hoods of overalls and designed to ensure a close fit at the wrists, ankles, face and neck; not have any pockets or other attachments which could attract and trap asbestos dust; and be easy to decontaminate or dispose of. The ACOP also contains detailed guidance on the removal of contaminated protective clothing and cleaning, maintenance and storage.

Arrangements to deal with accidents, incidents and emergencies

[A5033] Regulation 15 sets out that in the event of an accident, incident or emergency related to the unplanned release of asbestos at the workplace, the employer must ensure that immediate steps are taken to mitigate the effects of the event, restore the situation to normal, and inform any person who may be affected.

Only people responsible for repairs and other necessary work must be permitted in the affected area and they must be provided with appropriate RPE and protective clothing, and any necessary specialised safety equipment and plant, which must be used until the situation is restored to normal.

There are additional requirements set out with regard to licensable work with asbestos in the event of an accident, incident or emergency related to the use of asbestos. The employer must ensure that procedures, including safety drills, have been prepared and regularly tested and that information on emergency arrangements is available. This includes information about hazards and hazard identification arrangements and specific hazards likely to arise in an emergency.

They must provide suitable warning and communication systems to enable an appropriate response, including remedial action and rescue operations to be put into place. This requirement does not apply where: the risk assessment shows that there is only a slight risk to the health of employees because of the quantity of asbestos present at the workplace; and the measures the employer has taken to comply with the duty under regulation 11(1) – that is to prevent or reduce exposure to asbestos to the lowest level reasonably practicable by measures other than the use of RPE – are sufficient to control the risk.

The ACOP says employers must deal with all uncontrolled releases of asbestos into the workplace quickly and appropriately. The clean-up of any release that leads to potential exposures at or above the control limit, or that are not sporadic and of low intensity, such as releases of asbestos lagging, loose fill, asbestos coatings (not textured coatings) or largescale releases of AIB, must be carried out by a licensed contractor. If there is an uncontrolled release, employers should: warn people who may be affected; exclude people from the area who are not needed to deal with the release; identify the cause of the uncontrolled release; and regain adequate control as soon as possible.

In particular, anyone in the work area affected who is not wearing PPE, including RPE, must leave that area immediately; anyone who is contaminated with dust and debris must be decontaminated; any clothing or PPE must be decontaminated or disposed of as contaminated waste; and measures must be taken to contain and reduce fibre release.

A note that the exposure has occurred must be made on an employee's health record or personal record for any employee who was not wearing adequate RPE or has been potentially exposed to asbestos fibres in an incident.

Duty to prevent or reduce the spread of asbestos

[A5034] Regulation 16 requires employers to prevent or, where this is not reasonably practicable, reduce to the lowest level reasonably practicable the spread of asbestos from the workplace.

The ACOP says employers should select and use work methods that will reduce the disturbance and release of asbestos fragments and fibres to minimise the risk of spread, by removing items intact or whole and by using dust suppression techniques for example. Inside the work area, ACMs must never be left loose or in a state where they can be trampled on or spread. All asbestos waste should be bagged or wrapped promptly after removal and the waste should be removed from the work area regularly. For most licensable work with asbestos, a full enclosure is likely to be required, unless it is not reasonably practicable, and will be required for some types of non-licensable work. The ACOP provides detailed guidance in this area.

Cleanliness of premises and plant

[A5035] Regulation 17 requires employers who undertake work which exposes or is liable to expose any employees of that employer to asbestos to ensure that the premises and plant are kept in a clean state; and thoroughly cleaned on completion of the work.

Employers must choose work methods and equipment to prevent or reduce release of fibres and the build-up of asbestos waste on floors and surfaces in the

[A5035] Asbestos

working area and, wherever practicable, transfer waste directly into waste bags as workers remove the asbestos materials. Asbestos dust and debris must be cleaned up and removed regularly to prevent it accumulating (and drying out where wet removal techniques have been used), and at least at the end of each shift. Procedures for cleaning premises and plant should take account of the need for cleaning following an accidental and uncontrolled release of asbestos. Dustless methods of cleaning should always be used, including providing dedicated class H (BS 8520-3:2009) vacuum cleaning equipment fitted with suitable tools where practicable. The ACOP provides detailed guidance on cleaning and site clearance certification.

Designated areas

[A5036] Employers must ensure that where work liable to expose any employee to asbestos is being carried out, it is designated as an asbestos area. Where the risk assessment cannot clearly demonstrate that the control limit will not be exceeded, it must be designated as a respirator zone (regulation 18). Asbestos areas and respirator zones must be clearly and separately demarcated and identified by notices indicating that the area is an asbestos area or a respirator zone or both, and that in the case of a respirator zone, that the exposure of an employee who enters it is liable to exceed the control limit and RPE must be worn.

The employer must not permit any employee, other than an employee who is required for work purposes to be in an area designated as an asbestos area or a respirator zone, to enter or remain in any such area.

Only competent employees, who have received adequate information, instruction and training, must enter a respirator zone, and supervise any employees who enter a respirator zone.

Employers must ensure that employees do not eat, drink or smoke in an area designated as an asbestos area or a respirator zone; and they must make arrangements for such employees to eat or drink elsewhere.

All licensable work with asbestos should be carried out in an area designated as a respirator zone and an asbestos area.

Air monitoring

[A5037] Regulation 19 requires employers to monitor employees' exposure to asbestos by measuring asbestos fibres present in the air at regular intervals; and when a change occurs which may affect that exposure.

This is not required where the exposure of an employee is not liable to exceed the control limit; or the employer is able to demonstrate by another method of evaluation that the requirements of regulation 11(1) and (5) (see **A5029** above) have been complied with.

The employer must keep records of monitoring or the reason for the decision not to monitor. Records must be kept, in a case where exposure is such that a health record is required to be kept under regulation 22 (see below) for at least 40 years; or in any other case, for at least five years, from the date of the last entry made in it.

The employer must allow an employee access to the personal monitoring record (with reasonable notice), provide the appropriate authority (the HSE or Office for Nuclear Regulation (ONR)) with copies if they require them; and if the employer ceases to trade, notify the relevant enforcement authority without delay in writing and make available all monitoring records.

Personal sampling or air monitoring is required for a representative range of jobs and work methods and it should be done at regular intervals and when there is a change which may affect exposure. The ACOP provides detailed guidance on how to carry out and record the results of personal sampling and air monitoring.

Employers should consult employees, safety representatives or representatives of employee safety when making arrangements for monitoring.

Further information and guidance on the sampling strategy, the methods for sampling and analysis and the reporting of results of air monitoring can be found in *The analysts' guide* (see A5043B below).

Standards for air testing and site clearance certification

[A5038] Regulation 20 sets out standards for air testing and site clearance certification with reference to ISO 17025.

Standards for analysis

[A5039] Regulation 21 sets out the standards for the analysis of samples of materials to determine whether they contain asbestos with reference to ISO 17025. The United Kingdom Accreditation Service (UKAS) is currently the sole recognised accreditation body in Great Britain. The HSE recommends that employers who contract analysts should actively search the UKAS website to identify accredited organisations.

Health records and medical surveillance

[A5040] Regulation 22 deals with health records and medical surveillance.

For licensable work with asbestos employers must ensure that a health record is maintained and contains details approved by the HSE for all employees who are exposed to asbestos. This must be kept for at least 40 years and the employees must be kept under adequate medical surveillance by a relevant doctor.

This must include a medical examination not more than two years before the beginning of such exposure; and periodic medical examinations at intervals of at least once every two years or such shorter time as the relevant doctor may require while such exposure continues. It must include a specific examination of the chest.

For work with asbestos which is not licensable and not exempted by regulation 3(2), the requirements above apply and;

- a medical examination must take place not more than three years before the beginning of such exposure; and
- periodic medical examination must take place at intervals of at least once every three years, or such a shorter time as the relevant doctor may require while such exposure continues.

Regulation 3(20) concerns:

- asbestos exposure which is sporadic and of low intensity and, it is clear from the risk assessment, will not exceed the control limit and the work involves short, non-continuous maintenance activities in which only non-friable materials are handled;
- removal without deterioration of non-degraded materials in which the asbestos fibres are firmly linked in a matrix;
- encapsulation or sealing of asbestos-containing materials which are in good condition; or
- air monitoring and control, and the collection and analysis of samples to ascertain whether a specific material contains asbestos.

Where an employee has been examined under the regulations, the relevant doctor must issue a certificate to the employer and employee stating that the employee has been examined and the date, and the employer must keep that certificate, or a copy, for at least four years.

Examinations should be at the employer's cost, take place during working hours and employees are required to provide doctors with relevant information about their health. Employers must allow doctors to inspect records kept under these regulations, and ensure there are suitable facilities for medical surveillance where this is carried out on their premises. Again, they must allow employees access to their personal health records and provide the HSE or ONR with copies of personal health records on request. If the employer ceases to trade, they must notify the Executive, without delay and in writing, and make available all personal health records.

Where an employee is found to have an identifiable disease or adverse health effect which is considered by a relevant doctor to be the result of exposure to asbestos at work, the employer must —

- ensure that a suitable person informs the employee accordingly and provides the employee with information and advice regarding further medical surveillance;
- review the risk assessment;
- review any measure taken to comply with regulation 11 taking into account any advice given by a relevant doctor or by the HSE;
- consider assigning the employee to alternative work where there is no risk of further exposure to asbestos, taking into account any advice given by a relevant doctor; and
- provide for a review of the health of every other employee who has been similarly exposed, including a medical examination (which must include a specific examination of the chest) where such an examination is recommended by a relevant doctor or by the HSE.

The ACOP sets out that for licensable work the health record should contain the following:

- each employee's surname and first names, sex, date of birth, permanent address, post code and National Insurance number;

- a record of the types of work carried out involving asbestos, and where relevant, its location, with start and end dates, with the average duration of exposure in hours per week, exposure levels and details of any RPE used;
- a record of any work with asbestos before current employment, if the employer has been informed;
- dates of the medical examinations under the Regulations;
- a recording and planning system which brings forward the next required examination date for each individual.

For NNLW, the employer must enter the employees carrying out the work in a register or record, indicating the nature and duration of the activity and the exposure to which they have been subjected and have a recording and planning system which records the date of the last examination and brings forward the next required medical examination date for each individual.

Washing and changing facilities

[A5041] Regulation 23 requires employers to provide adequate washing and changing facilities, and where necessary, separate storage facilities for personal protective clothing, personal clothing not worn during working hours, and RPE.

The ACOP says the type and extent of washing and changing facilities provided should be determined by the type and amount of exposure indicated by the risk assessment. Suitable facilities should be provided, including: toilet facilities; facilities for washing and changing for non-licensable work; full hygiene facilities for licensable work (the ACOP provides detailed guidance on this); and an area to eat and drink (for licensable work these should be located as close as is reasonably practicable to the hygiene facilities).

Storage, distribution and labelling of raw asbestos and asbestos waste

[A5042] Regulation 24 requires employers who undertake work with asbestos to ensure that raw asbestos or waste which contains asbestos is not stored; received into or despatched from the workplace; or distributed within the workplace (unless it is in a totally enclosed distribution system) unless it is in a sealed receptacle or sealed wrapping, and clearly marked to show that it contains asbestos.

Raw asbestos must be labelled in accordance with Schedule 2 of the regulations.

Waste containing asbestos must be labelled where the Carriage of Dangerous Goods and Use of Transportable Pressure Equipment Regulations 2009 apply (see **P70** PRESSURE SYSTEMS) in accordance with those Regulations; and in any other case in accordance with the provisions of Schedule 2.

HSE guidance to the Regulations sets out that: 'Asbestos waste describes asbestos products or materials that are ready for disposal, including building materials, dust, rubble, disposable PPE, rags used for cleaning and used tools that cannot be properly decontaminated.'

It also advises that when packing asbestos waste:

[A5042] Asbestos

- it should be securely sealed in suitable, labelled bags, wrapping or packaging as it is produced;
- any bags, wrapping or packaging used must be designed, constructed and maintained to make sure that no asbestos fibres can be released during handling or transport;
- for most waste, double plastic sacks are suitable, provided they will not split during normal use;
- stronger packages must be used if the waste contains sharp metal fragments or other materials that could puncture plastic sacks; and
- any waste where the escape of hazardous quantities of respirable asbestos fibres can occur during carriage should be placed in UN-approved packaging. This is available in up to two tonnes capacity. (This does not apply to asbestos cement or textured decorative coatings.)

When filling bags:

- the inner bag should not be overfilled, especially when the debris is wet, and each bag should be securely tied or sealed;
- as far as possible, air from the bag should be excluded before sealing. Precautions will need to be taken as the exhaust air may be contaminated; and
- where practicable, the sealed packaging should be cleaned before it is removed from the work area or enclosure.

It further advises: 'If the asbestos waste is not to be disposed of immediately, the sealed bags and packages should be locked in a suitable and clearly marked storage area, ie a lockable skip.'

Detailed information on the disposal of asbestos waste is available on the HSE website at: www.hse.gov.uk/pubns/guidance/em9.pdf.

Prohibitions, exceptions and exemptions

[A5042.1] Part 3 and Part 4 of the Regulations set out various prohibitions and exceptions to and exemptions from the requirements of the regulations, as well as miscellaneous provisions.

Regulation 26 defines:

- 'asbestos spraying' – the application by spraying of any material containing asbestos to form a continuous surface coating;
- 'extraction of asbestos' – the extraction by mining or otherwise of asbestos as the primary product of such extraction, but does not include extraction which produces asbestos as a by-product of the primary activity of extraction; and
- 'supply' – supply by way of sale, lease, hire, hire-purchase, loan, gift or exchange for a consideration other than money, whether (in all cases) as principal or as agent for another.

Regulation sets out the prohibitions:

- a person must not undertake asbestos spraying or working procedures that involve using low-density (less than $1g/cm3$) insulating or soundproofing materials which contain asbestos;

- every employer must ensure that no employees are exposed to asbestos during the extraction of asbestos; and
- every employer must ensure that no employees are exposed to asbestos during the manufacture of asbestos products or of products containing intentionally-added asbestos.

Asbestos Compensation

[A5042.2] Because the period between exposure and development of asbestos-related diseases is very long – symptoms of asbestosis take around 20 to 30 years to develop while mesothelioma has a latency period of up to 50 years, for example – many workers were exposed to asbestos before the tighter controls contained in more recent legislation were introduced.

With regard to proving employer negligence in compensation cases, the Chief Inspector of Factories' 1938 annual report is often relied on to demonstrate that there was general awareness of the dangers of exposure to asbestos dust. According to Munkman on Employer's Liability this gave a very clear warning and 'means that it is no longer necessary to consider whether or not an employer is in breach of its duty to its workers if the level of dust may be described as "substantial"'.

The Compensation Act 2006 was enacted following a number of cases involving workers who had been exposed to asbestos by more than one employer. It states that where workers in this position have developed asbestos-related illnesses, they can claim compensation from all the employers concerned and do not have to prove which exposure was to blame – each employer is jointly and severally liable.

In addition, rulings over more recent years have confirmed that:

- an employer can be found to be liable in the case of 'low level' exposure to asbestos, as long as the asbestos made a 'material contribution' to the cause of death (*Sienkiewicz (Administratrix of the Estate of Enid Costello (deceased)) v Greif (UK) Ltd and Knowsley Metropolitan Borough Council v Willmore* [2011] UKSC 10);
- a worker 'sustains or contracts' mesothelioma at the point at which they are exposed to asbestos fibres, not when the symptoms appear (*Durham v BAI (Run Off) Ltd* [2012] UKSC 14);
- a parent company can be held liable for a personal injury case involving asbestos where the claimant was employed by the subsidiary rather than the parent company (*Chandler v Cape plc* [2012] EWCA Civ 525). This is particularly the case where the parent company has been directly involved in health and safety decision making at the subsidiary and had a high level of knowledge about the risks when the claimant was exposed to asbestos; and
- under the Factories Act 1937, the occupier of the premises is responsible for the welfare of people on the site, not just those it directly employs, and the *Asbestos Industry Regulations 1931 (SI 1931 No 1140)* applied to all factories using asbestos, not just those involved in the asbestos industry (*McDonald (Deceased) v National Grid Electricity Transmission plc* [2014] UKSC 53).

For more detailed analysis see Butterworths Personal Injury Litigation Service.

[A5042.2] Asbestos

In July 2018, the Court of Appeal ruled that Graham Dring, on behalf of the Asbestos Victims Support Group Forum UK, had a legitimate interest and should be given numerous copies of key documents held by global asbestos product manufacturer Cape Intermediate Holdings Limited. These which originated in separate litigation. Solicitors Leigh Day backed the case *Cape Intermediate Holdings Ltd v Graham Dring* [2018] EWCA Civ 1795 and said the ruling is highly significant as 'it upholds an essential part of an earlier decision which is believed to have been first time a non-party to a legal case has established the entitlement to copies of such a large number of documents used at trial'.

'The Court of Appeal judgment should provide access to a significant amount of documents, which Cape intermediate Holdings (Cape) have refused to disclose: documents which may assist mesothelioma sufferers and their families to claim compensation, and which may throw light on the way early knowledge of the dangers of asbestos was hidden from the public,' said Dring.

The government introduced the Diffuse Mesothelioma Payment Scheme to help those who have been diagnosed with mesothelioma and were exposed to asbestos either negligently or in breach of statutory duty by their employers, but are unable to claim compensation because their employer or employer's liability insurer is untraceable. Due to the length of time between exposure to asbFurther resources can be found on the HSE website at www.hse.gov.uk/asbestestos and diagnosis of cancer, many employers and their insurers no longer exist and so the liable successor organisations are often untraceable. Details of the government compensation scheme can be found at: www.mesoscheme.org.uk.

HSE guidance

[A5043] L143 *Managing and working with asbestos Control of Asbestos Regulations 2012 Approved Code of Practice and guidance* is available on the HSE website at: www.hse.gov.uk/pubns/books/l143.htm.

Other HSE guidance includes:

- HSG247 – Asbestos: The Licensed Contractors' Guide;
- HSG248 – Asbestos: The Analysts' Guide for Sampling, Analysis and Clearance Procedures;
- HSG 53 – Respiratory Protective Equipment at Work – A Practical Guide;
- HSG 210 – Asbestos Essentials – a task manual for building, maintenance and allied trades on non-licensed asbestos work;
- HSG 227 – A comprehensive guide to managing asbestos in premises;
- HSG 264 – Asbestos: The Survey Guide.
- INDG 223 – Managing asbestos in premises – a brief guide
- HSE 50 – Asbestos licence assessment, amendment and revocation guide (ALAARG) (2012)

Further resources can be found on the HSE website at www.hse.gov.uk/asbestos/index.htm.

Business Continuity

Lawrence Bamber

What is business continuity planning?

[B8001] Business continuity planning ('BCP') is not just about disaster recovery, crisis management, risk management or IT. It is a business risk management issue. It presents an opportunity to review the way an organisation performs its business processes, to improve procedures and practices, and increase resilience to interruption and loss.

To quote the Business Continuity Institute, the professional body for BCP:

> BCP is the act of anticipating incidents which will affect critical functions and processes for the organisation, and ensuring that it responds to any incident in a planned and rehearsed manner.

The Turnbull Committee 'Guidance for Directors on Internal Controls' sets out an overall framework of best practice for business, based upon an assessment and control of its significant risks. For many companies, BCP will address some of these key risks and help them to achieve compliance.

Hence BCP as an activity has two primary objectives:

- minimise the risk of a disaster befalling the organisation; and
- maximise the ability of the organisation to recover from a disaster.

It differs from disaster recovery by recognising that approximately 80 per cent of disasters which organisations suffer are generated internally. As a result, BCP places more emphasis on managing out potential causes of disasters, than on recovering from those incidents which may not be avoided.

The BCP is therefore a combination of disaster/risk avoidance and the ability to recover.

Business continuity planning in context

[B8002] Modern businesses cannot avoid all forms of corporate risk or potential damage. A realistic objective is to ensure the survival of an organisation by establishing a culture that will identify, assess and manage those risks that could cause it to suffer:

- inability to maintain customer services;
- damage to image, reputation or brand;
- failure to protect company assets;
- business control failure;

- failure to meet legal or regulatory requirements.

BCP therefore provides the strategic framework to achieve these objectives.

Why is business continuity planning of importance?

[B8003] Very simply the answer to the above question is that disasters kill businesses.

In the UK, 80 per cent of companies who suffered a major disaster and did not have some form of BCP capability went into liquidation within eighteen months. A further 10 per cent suffered the same fate within five years.

Without BCP companies have only a one in ten chance of survival. Also, simply because:

- Customers expect continuity of supply in all circumstances.
- Shareholders expect directors/managers to be fully in control, and to be seen to be in control of any crisis.
- Employees and suppliers expect the business to protect their livelihoods.
- The company's reputation and brand is at risk without BCP.
- It is implicit in good corporate governance and demonstrates best practice in business management.

All the above needs to be taken into account against a business background in which:

- The pressures on business generally are changing and increasing.
- Technology is transforming and underpinning the business environment.
- Consolidation, restructuring, re-engineering, take-overs, acquisitions and mergers etc are now a fact of life.
- New risks and exposures are continually being created.

Business continuity drivers

[B8004] The drivers or arguments that need to be understood by all involved in business continuity may be listed and grouped as follows:

- Commercial.
- Regulatory.
- Standards – Best Practice
- Financial.
- Imported disaster.
- Public relations.

Commercial drivers

[B8005] An organisation may operate in an intensively competitive market where any disruption to normal operation will be exploited by competitors.

For example, when Perrier were forced into a total product withdrawal following a benzene contamination at source, the gaps on supermarket shelves were promptly filled by 30–40 competing brands.

The organisation may depend on a high degree of public confidence to generate/maintain sales, e.g. air transport industry. The issue for many organisations is not merely one of lost revenue, but also of lost customers, who may prove very difficult to get back. On occasions, it has been observed that companies who have suffered a disaster have spent three times their normal PR budget in an attempt to reassure customers that they are still around and worth doing business with.

Conversely, a visible and demonstrable BCP capability can be used to enhance and differentiate products and services.

Quality and regulatory drivers

[B8006] The main regulatory driver within the UK is the *Civil Contingencies Act 2004* (see **B8044** below) which predominantly affects the role of local authorities (LAs) in the provision of business continuity support and advice.

The Act came into force in May 2006 and contains emergency powers that place duties and responsibilities on how organisations should respond to disasters.

Standards – Best practice

[B8007] The British Standards Institution (BSI) has published a two-part Standard on Business Continuity Management (BCM).

BS25999-1: 2006
- Business continuity management – Part 1: Code of practice

BS25999-2: 2007
- Business continuity management – Part 2: Specification

Part 1 essentially outlines the content of the Standard and Part 2 gives a steer on how the BCM process should be developed, implemented and maintained.

Part 1: Code of practice

This document presents an overview of the BCM process in section 1, 2 and 3 which are essentially narrative. The key sections are:

- Section 4: BCM policy
- Section 5: Programme management
- Section 6: Understanding the (ie your) organisation
- Section 7: Determining BCM strategy
- Section 8: Developing and implementing a BCM response
- Section 9: Exercising, maintaining and revising the BCM arrangements
- Section 10: Embedding BCM into the organisation's culture

Part 2: Specification for BCM

This document contains six sections. Sections 1 and 2 are descriptive but section 3, 4, 5 and 6 are auditable and follow the 'Plan, Do, Check, Act'

(PDCA) and 'Policy, Organisation, Plan, Implement, Monitor, Audit, Review' (POPIMAR) approaches to systems management. The key sections are:

Section 3: BCM planning
- General requirements;
- establishing, managing and maintaining the BCM system;
- embedding BCM into the organisational culture;
- BCM documentation and records.

Section 4: Implementation and operation
- Understanding the organisation;
- identification of critical activities/critical business processes;
- developing the BCM strategy;
- developing and implementing the BCM response;
- exercising, maintaining and reviewing the BCM arrangements.

Section 5: Monitoring and review
- internal audit: monitor and review of BCM system
- management review of BCM system

Section 6: Maintenance and improvement
- Preventive and corrective actions
- continual improvement.

A new International Standard for Business Continuity was published in 2012, namely BS ISO 22301: 2012 - Societal security: Business Continuity Management Systems. This gives guidance on how to move from BS 25999 to the new standard and follows the Plan-Do - Check-Act approach of other management systems in order to achieve continual improvement.

Sections include:

4. Context of the Organisation
5. Leadership
6. Planning
7. Support
8. Operation
9. Performance Evaluation
10. Improvement

Further requirements can be found on the British Standards Institute (BSI) website.

Financial drivers

[B8008] Traditional financial measures of mitigating or cushioning against risk, such as insurance, are an important protective element within the overall organisational business risk management framework. Insurance on its own however has limitations which do not protect the complete business cycle.

Despite the fact that business interruption insurance is a fairly common component in most commercial insurance packages, cover is normally limited to two years from the event, and certain consequential losses, which may contain significant unquantifiable costs, will be excluded and hence non-recoverable.

For example, BI (business interruption) insurance does not cover the risk of lost customers, lost productivity or the attrition costs of key staff following a disaster. In some cases, this could be as much as 40 per cent of the total loss.

In some organisations, the disaster itself may be caused by, or made worse because of, loss of control over finances and cash flow, e.g. Barings, AIB.

The imported disaster

[B8009] BCP is an approach which recognises that organisations do not operate in isolation of each other and that they rely on suppliers to continue to operate. Hence disasters elsewhere may have a knock-on effect within an organisation, and disasters within an organisation could adversely affect its customers.

With the concepts of single source supply and just-in-time delivery, there is an increasing need to undertake supplier audits to assess their BCP capability.

Also the risk of denial of access should be assessed, as in the case of the Manchester bomb explosion in 1996. Unaffected businesses were denied access by the emergency services for months after the explosion.

Natural disasters can also have the following knock-on effect:

Following the earthquake in Kobe, Japan, the world suffered a severe shortage of silicon chips (microprocessors). The price on the black market rose so dramatically, that it set off a spate of thefts worldwide. This led to any large office block thought to contain large numbers of PCs being hit.

The public relations disaster

[B8010] Increasingly, companies strive to raise their profile in the market place and increase their visibility by promoting themselves as aggressively and publicly as possible. The public awareness they seek, and need, in order to sell their products and services is a window through which to view their operations.

Just as members of the public will be more aware of the products and services the company offer, so they will be more aware of any mistakes, accidents, disasters, poor decisions, poor performance etc. for which the company is ultimately responsible. The ill feeling that can be generated against the company by the general public can be so damaging that confidence in the company is totally undermined for a long period of time.

For example, Gerald Ratner jokingly referred to his products as 'cr*p' and later compounded the felony by suggesting that there was more value in a prawn sandwich than in some of his products. The ultimate consequence of his action was loss of sales and the subsequent break-up of his jewellery business. He lost control of his company and a number of employees lost their jobs through the resulting re-organisation.

Hoover's handling of the massive over-subscription to its 'Free Flights' promotion became a national scandal which caused the company to be pilloried almost weekly on prime time TV consumer programmes. Not only did several directors lose their jobs, but the Hoover crisis severely damaged the Hoover brand name so much that the company even considered changing its name away from one of the most powerful and successful brand names/images in history, hence the 'free flights fiasco!'

More recently (January 2007) following the allegedly racist abuse of a Bollywood actress in the *Celebrity Big Brother* house by Jade Goody (and others), Jade herself effectively caused untold reputation damage to Jade Goody plc, which has an estimated value of £8 million. Indeed PR guru Max Clifford is quoted as saying: "in future when people think of PR blunders they will think of two people, Gerald Ratner and Jade Goody!"

Thus, the ability of a company to avoid, or recover from, a disaster effectively can have a tremendous and positive PR impact, thus vastly improving stakeholders' (investors, public, customers, suppliers, employees etc.) confidence.

This area is covered in more detail in the section on crisis communication (B8039 below).

Convincing the board – the business case

[B8011] A business case for BCP can be made by taking the following ten key points into account:

(1) Find out how much time you have for your presentation and make sure you rehearse to make full use of what may be a limited time allocation.
(2) Check the audience:
 - who will be there?
 - have non-executive directors previous experience of disaster and/or BCP?
(3) Determine, in advance, if there are any regulatory influences that the board may be particularly sensitive to.
(4) Achieve board buy-in/commitment to BCP by ensuring that they:
 - understand the negative impact of business interruption;
 - provide the necessary resources to reduce the risk of disaster;
 - appreciate the need to prepare and plan.
 - appoint a corporate champion, at board level, to take overall responsibility for the development and implementation of BCP.
(5) Stress the benefits of BCP:
 - peace of mind/sleep at night;
 - risk reduction.
(6) Be prepared for the 'It won't happen here' syndrome. Have some statistics (not too many) and some local or industry examples that the board may be aware of.
(7) Spell out that the board is ultimately responsible for business risk management. N.B. Turnbull/Corporate governance.
(8) Avoid being drawn into a detailed discussion about how the BCP capability is to be developed.

(9) Avoid being drawn into a discussion on actual or potential disasters, unless they are specifically relevant.

(10) Summarise and reinforce the need for BCP by highlighting the unique selling points from the above list.

The business continuity model

[B8012] The model is a phased and active process consisting of five main steps:

Stage 1: Understanding your business
- Business impact analysis ('BIA') and risk evaluation/assessment tools are used to: identify the critical business processes within the business; evaluate recovery priorities; and assess risks which could lead to business interruption and/or damage to the organisation's reputation.

Stage 2: Continuity strategies
- This involves determining the selection of alternative strategies available to mitigate loss, assessing the relative merits of these against the business environment; and deciding which are likely to be the most effective in protecting and maintaining the critical business processes.

Stage 3: Plan development
- The development of a proactive BCP plan is designed to improve the risk profile by upgrading optional procedures and practices; introducing alternative business strategies; and using risk financing measures; including insurance.

Stage 4: Plan implementation
- The introduction of BCP capability is achieved via education and awareness of all stakeholders, including employees, customers, suppliers and shareholders. The end result is to establish a BCP culture throughout the organisation. The BCPs should be widely communicated and published to all concerned so that everyone is aware of rules, responsibilities, accountabilities, timescales and KPIs (key performance indicator).

Stage 5: Plan testing and maintenance
- The plan is not the capability. The plan is the means by which the capability is realised.
  Failure to plan is planning to fail.
  BCP is therefore not a 'one-off' exercise. There is a need for ongoing plan testing, maintenance, audit and change in the management of the BCP and its processes. Regular communication and feedback on the results of testing and maintenance are imperative so as to ensure continual improvement of the BCP capability.

The process of developing a BCP capability is a cyclic and evolutionary one in order to take account of the ever-changing organisation.

Understanding your business (stage 1)

Risk evaluation

[B8013] Risk evaluation is the process by which an organisation determines the pattern of risk that is unique to its operation.

Risk is used to describe the relationship between the impact and likelihood for a particular event or scenario:

Risk = Impact × Likelihood

Within the BCP framework, risk evaluation is the process which identifies those aspects of the company's operations, which are essential to the survival of the business, and assesses how well they are protected against serious disruption and/or potential disaster.

As in the field of occupational safety and health ('OSH'), risk evaluation is a combination of identification and assessment. Within BCP, there are three commonly used approaches to risk evaluation:

- scenario driven;
- component;
- critical business processes.

All approaches assume that all concerned fully understand what the company is in business for and what is vital for its survival.

Scenario driven approach

[B8014] This approach involves the listing of all possible risk scenarios that can be thought of. Once the list has been compiled each scenario is considered in isolation with a view to planning specific counter-measures to tackle each risk scenario on an individual (and isolated) basis.

This approach has some serious limitations. It presupposes that risks exist in isolation and that they do not have a dynamic relationship with each other. This is clearly not the case. For example fire creates smoke, fumes and heat. The presence of smoke and fumes increases the risk of inhalation injury and may hinder evacuation or searching activities. If we then throw in the fact that the fumes are toxic or corrosive and also that they could adversely affect neighbouring sites, the single incidence of 'fire' has an exceptional number of associated risks.

Such an approach will have a high cost and also lead to complex planning implications.

Component approach

[B8015] This approach examines each department or unit within the organisation in order to develop their individual risk profiles. Whilst this approach does eventually cover the whole organisation, it fails to prioritise amongst departments and hence does not determine those departments which are important/critical.

Also it sometimes misses linkages and dependencies between departments, e.g. the internal customer. The approach as a whole suffers from a lack of organisation-wide perspective, and hence it is difficult to establish which departments should be reinstated first.

The business continuity model [B8017]

Critical business approach

[B8016] Business can be viewed as a set of linked processes. Some of these business processes, the critical ones, have a greater bearing on the core business or primary activity of the business. These are known as the critical business processes ('CBPs').

The fundamental question to be asked and understood is:

> What does my company actually do?

Developing a model may help to clarify thinking and understanding in this regard. Reference to the company's vision, and mission statement and goals will also demonstrate where the organisation is focused. It is on mission critical activities – the critical business processes – where BCP should be focused.

There are four basic questions to be asked:

(1) What is the business about?
(2) When are we to achieve our goals?
(3) Who is involved both internally and externally?
(4) How are the goals to be achieved?

An organisation has many dependencies, both internally and externally, that support the mission critical business processes. These may include suppliers, customers, shareholders, key employees, IT systems, and manufacturing processes. It is important to identify these at an early stage and to involve people from the CBPs in the development of the BCP.

External influences on the CBPs may include government departments, regulators, competitors, trade bodies, and pressure groups. These must also be taken into account with the BCP framework.

In summary, the critical business approach:

- Needs a common understanding of the company's primary business objective – the focus or mission of the company. This is ultimately what the capability is being designed to protect and recover.
- Some BCPs are doomed to failure because the participants cannot agree to what the company's primary business objective actually is.
- Determines the CBPs by asking how important are they in achieving the business objective. Again agreement amongst the participants regarding the CBPs is vital in developing a relevant BCP.
- Develops a BCP capability to protect all identified CBPs whilst keeping focused on the primary business objective will prevent participants getting sidetracked or bogged down in detail.

Business impact analysis

[B8017] Business impact analysis ('BIA') is the technique used to determine what the business stands to lose if its CBPs are disrupted.

BIA alerts the organisation to the likely cost of a business disruption and assists in the cost benefit analysis of risk reduction measures within the overall BCP capability. Indeed the approval and continuing support of your board for BCP may well hinge on the fact that the costs are justified in the light of potentially catastrophic business losses.

Quantification of business impact

[B8018] It is important that all concerned in the BCP process understand that disasters are characterised by time and not event. What causes the disruption is less important to the survival of the organisation than how long the disruption lasts. BIA therefore should consider the impact of a disruption in any of the CBPs spread over time. Traditionally, the impact of a disruptive event was derived merely by determining how much revenue would be lost over the period during which the company was not operational:

Loss of revenue – *Out for 1 day* –> **Impact** = Annual Revenue/ No of working days/year

Loss of revenue – *Out for a year* –> **Impact** = Annual Revenue

On its own, however, loss of revenue does not give the full picture of the overall impact on the business. Other factors, which need to be taken into account, include:

Erosion of customer base
- The organisation will lose customers if it cannot continue to service their requirements. Some will go simply because they perceive the service is unreliable and some will be forced to go for the sake of their own business continuity. Whatever the reason, it is generally estimated that it costs three times more to sell goods/services to a new customer than to an existing one. In the event of a prolonged disruption, the company could therefore have to treble its marketing, advertising, publicity and sales cost in order to restore its customer base.

Loss of key staff
- Following a disaster, some companies have reported that up to 40 per cent of their workforce have left as a result. The loss of intellectual capital and the cost of recruitment can be considerable.

Loss of reputation
- Most companies depend on their reputation as a reliable service provider to generate further sales. Also a great deal of resources, time and money, will have been spent over the years in cultivating a brand image linked to their reputation. Hence, any event which detracts from that image brings the company's reputation into question and not only jeopardises the previous investment but also seriously affects current and future sales. Such lack of confidence will also have a negative effect on the share price of publicly quoted companies.

Regulatory and contracted impact
- The company may be subject to penalty clauses/payments if it cannot meet contracted obligations. It may also be barred from operating by an inability to comply with regulatory (e.g. prohibition notice requirements), reporting or licensing obligations.

Loss of credit-worthiness
- The organisation may face increased costs of borrowing following a disruption to business, on the basis that it represents a greater investment risk to lenders as a result of the disaster.

Business impact Analysis: outputs

[B8019] BIA enables the organisation to focus on the CBPs and their negative impact on the business, rather than to conduct a global, risk-specific analysis. The process also takes time sensitivity into account, thus enabling the recovery objectives to be agreed.

It is essential to rate the impact of the loss of CBPs on the business. Risk rating may be qualitative: high, medium, low or quantitative, i.e. 1–9 rating scale. Where possible, financial values should be placed on the business impact.

It is also important to involve key personnel who work in the CBPs in the BIA process, especially as the CBPs are cross-functional/cross-departmental, and agreement must be reached on the relative ratings. Agreement from the board should then be obtained to ensure that the output from the BIA, i.e. the plan is put into place.

In summary therefore:

- Ensure involvement of appropriate functions/departments.
- Determine the impact on the business of the loss of CBPs.
- Apply risk ratings, including time dependencies.
- Obtain board approval for BIA outputs, i.e. the plan.

Continuity strategies (stage 2)

[B8020] Having identified those areas where the organisation is most at risk, a decision has to be made as to what approach is to be taken to protect the operation. With the introduction of the Turnbull Guidance on internal control (corporate governance) this decision must be taken at board level.

Many possibilities exist and it is likely that any strategy adopted will comprise a number of these approaches. Whichever are chosen, there are certain considerations to bear in mind, as indicated below:

- Do nothing (risk retention) – in some instances, the board may consider the risk to be commercially acceptable.
- Change or end the process (risk avoidance) – deciding to alter existing procedures have to be done, bearing in mind the organisation's key focus/mission.
- Insurance (risk transfer) – provides financial recompense/support following a loss but does not provide protection for brand, image and reputation.
- Loss mitigation (risk reduction) – tangible loss control programme to eliminate or reduce risk.
- Business continuity planning (risk reduction) – an approach that seeks to improve organisational resilience to interruption, allowing for the recovery of key business systems and processes within the recovery time frame objective whilst maintaining the organisation's critical business processes.

Any strategy must recognise the internal and external dependencies of the organisation and must be accepted by all members of management involved in the CBPs. Hence, in summary, strategy selection is as follows:

[B8020] Business Continuity

- Identify possible BCP strategies.
- Assess suitability of alternative strategies against the outputs of the BIA.
- Propose cost/benefit analyses of the various strategies.
- Present recommendations to the board for decision and approval.

Developing a business continuity plan (stage 3)

[B8021] As previously stated:

- Failing to plan is planning to fail.
- The plan is not the capability.

The BCP is the means by which the capability is realised and can be viewed as the route which takes the organisation from its present state (crisis/disaster) to its desired state (normal operation).

Too often organisations derive a false sense of security from the mere presence of a disaster recovery or BCP and lose sight of the fact that the plan is passive. It has no inherent capability to assist the organisation in any way until it is used. Its value to the organisation can only be assessed by how well it enabled the business to either avoid or recover from a disaster-threatening scenario, i.e. after the event.

The planning dilemma

[B8022] *Question*: How does one include enough detail to provide meaningful instruction and guidance without including so much data that individuals or departments find the plan difficult to use?

There is no simple answer to this question since requirements will vary from business to business. However, via the outputs from the BIA, from the experience of individuals involved in the BCP process, and from regular testing and maintenance of the plan, a balance that suits the organisation will be reached through plan refinement.

Plan components

[B8023] The objective of creating a BCP capability is firstly to design/engineer out as much of the operational risk from the identified CPBs as possible, i.e. risk avoidance.

Thereafter the BCP should provide a structural means of recovering those lost processes within a given timeframe.

Thus the plan will have two main components:

- *Proactive* – range of cost-justified risk reduction measures, procedural, physical, behavioural, processes, i.e. engineering.
- *Reactive* – disaster recovery plan.

Disaster recovery plan

[B8024] The vast majority of disaster recovery plans will comprise three stages:

Emergency response – Immediate
(1) This stage details the actions the company has predefined in immediate response to an emergency situation. Typically, it should include details of evacuation procedures, liaison with emergency services, how the BCP kicks in, notification procedures, damage assessment/control/mitigation, deployment of first aid and other emergency resources.

Fallback procedures – Short term
(2) This stage details the process by which the company starts to cope with the aftermath of the disaster, i.e. the fallback position. This may involve a physical move to alternative premises, alternative working methods, or the commencement of contractual agreements with third parties to undertake one or more of the CBPs on behalf of the company. A list of potential suppliers of such services, together with contract details is a 'must' in this regard.

Business resumption – Medium term
(3) This stage details the process which determines when and how the organisation gets on the road back to normal operation. Typically it will contain logistical, commercial, and political criteria which will be used by the company to decide where the business will be resumed, which CBPs will be reinstated first and where resources will be allocated to get the operations back to normality as quickly as possible.

Hierarchical planning structure

[B8025] In order to co-ordinate the recovery process, particularly when some of the decision-makers are not located at the site of the disaster, many companies make use of hierarchical management planning structure:

The command and control team
- This team is usually staffed by senior managers/decision-makers, supported by a business continuity professional. Their objective is to analyse the information being fed to them by the damage assessment team and to co-ordinate the overall response of the company to the disaster. It is their job to ensure that the recovery is on track; that they respond quickly and effectively to any developments in order to achieve restoration; to allocate emergency response resources; and to handle crisis communications both, internal and external.

The damage assessment team
- The damage assessment team is generally a multi-disciplinary group able to convey accurate information from the disaster site to the command and control team. They are also typically responsible for on-site OSH issues, site security, access control, liaison with emergency services, providing the command and control team with regular updates on restoration progress, and general tactical decision-making.

Departmental team(s)
- At some point within the BCP, those departments involved in CBPs will be instructed by the command and control team to incorporate their individual recovery plans. This may involve moving to an alternate location or developing an alternative way of working. Regular progress reports will be provided to the command and control team to enable them to keep track of the overall recovery. In this way, slippage against expected recovery timescales might be quickly identified, thus allowing additional resources to be allocated to that department, if available.

Plan development process

[B8026] The process for developing the BCP follows the criteria for project management methodology, increasingly employed by many organisations. It is essential that the BCP capability is developed in accordance with the following:

- Scope the development in terms of time and resources.
- Identify what skills are required to complete the recovery.
- Set milestones to demonstrate progress.
- Define reporting lines, roles, responsibilities, accountabilities.
- Establish a project tracking method.
- Set sign-off criteria.
- Move into maintenance phase.

Plan implementation (stage 4)

[B8027] The implementation process for the BCP requires the achievement of certain milestones in the following chronological order.

Securing buy-in

[B8028] Hopefully by this time, buy-in by the board will have been secured, using the techniques discussed above. It is vital, however, that this approval manifests itself in some way that allows it to be readily communicated throughout the organisation. A visible commitment to the BCP is required.

Policy

[B8029] Having a clearly stated BCP policy is one good way of visibly demonstrating board level commitment. The policy can be issued under the signature of the MD/CEO as an endorsement of the BCP process. This then provides a mandate to those involved to get the BCP implemented throughout the organisation.

Project authorisation

[B8030] The approval of the board should also include authorisation for the BCP project as a whole. This would generally be based on the information presented via the BIA, risk identification and risk evaluation exercises.

Individual responsibilities

[B8031] All key individuals should, as part of their job, have clearly defined and agreed responsibilities, accountabilities, timescales, KPIs and action plans, so as to ensure smooth plan implementation. In this way individual personal performance is geared to successful BCP implementation and an ongoing BCP capability.

Phased roll-out

[B8032] We have already determined that in order to provide the organisation with the most cost-beneficial BCP capability, there is a need to focus on the CBPs. In both the development and implementation stages, it is essential that the organisation stratifies its efforts to cover and protect those functions most directly involved in the core business (primary business objective) and the CBPs before moving on to less critical functions/areas.

This stratification will normally manifest itself as a phased roll-out of the BCP to the CBPs first, intermediate functions next, and non-business critical functions last, if at all.

Specialist services

[B8033] Quite apart from the emergency services, not all the resources required to implement the plan will be available in-house or on-site.

The following list is by no means exhaustive but serves to illustrate further possible components of the BCP. It is imperative that contact details for all such support services are kept readily available (in at least two known locations) and up to date:

- Data recovery/IT back-up.
- Emergency telecommunications.
- Salvage/decontamination.
- Cleaning/restoration.
- Buildings/facilities management.
- Security.
- Counselling (post-traumatic stress disorder (PTSD)).

Plan testing and maintenance (stage 5)

[B8034] Maintenance and testing are both vital components in ensuring that the BCP continues to support the disaster avoidance and recovery capability required by the organisation. Maintenance is essential to avoid the plan becoming unusable because it no longer reflects the organisation's structure, vision, mission or business priorities.

The objective of testing is to provide a high degree of assurance that the plan will work when it is needed, and also to highlight and rectify any shortcomings in the plan so that the BCP process is as streamlined as possible.

Maintenance should be an ongoing process throughout the lifetime of the BCP. Its primary focus is to change the plan to continuously reflect changes within the organisation. New business lines or processes, new reporting lines, reorganisations, changes of personnel, location, telephone numbers, user IDs etc all need to be reflected in the plan.

It is strongly advisable to allocate responsibilities for plan maintenance and testing to a group of individuals and make it part of their job descriptions, CYOs (current year objectives) and KPIs.

Testing strategies

The desk check/audit

[B8035] Many plans fall into disrepair due to a lack of accuracy and, consequently, integrity in the detail of the plan itself. This situation seriously undermines the effectiveness of the plan and its ability to protect the organisation in a disaster situation. The desk check/audit is a type of testing which seeks to verify that the factual detail contained in the plan is both accurate and current. As well as confirming the factual correctness of the plan, it also provides an excellent indication of how well the maintenance function is being performed.

The walkthrough test

[B8036] This test involves synthesising an incident or fictional set of circumstances, thereafter allowing participants to use the plan to resolve the described situation. Similar in concept to a war game, the rules of engagement are predefined and an external facilitator co-ordinates the introduction of extra information. The logic of the plan is tested conceptionally for robustness against a scenario designed to involve all sections of the plan.

Component testing

[B8037] Component testing normally involves single functions or departments testing in isolation their own parts of the plan by enacting their emergency, fallback and resumption arrangements. The objective is to test the practicality of the plan and to derive an estimate of how effective the plan is in terms of logistics and recovery times. It has the merit of eventually covering the whole organisation but it does not adequately test the linkages between functions and departments, which may more fully represent the flow of activity within the organisation.

Full simulation

[B8038] The organisation deliberately shuts down or denies access to CBPs, sometimes without warning, in order to get the most realistic feel for how groups and individuals react and perform in the event of a disaster.

Whilst this type of testing provides the highest possible degree of assurance that the plan works and the organisation can cope, it needs to be planned extremely carefully, otherwise it may well trigger off an alternative, real-life, disaster.

Normally a test of this type should not be undertaken until the organisation has successfully carried out the other types of test described with consistently favourable results.

Crisis communication and public relations

[B8039] In a crisis, many decisions have to be taken within a very short timescale. Some of those decisions may determine whether the organisation fails or survives. This fact represents one of the most powerful arguments for BCP and hence should ensure that the organisation has a proactive, up-to-date plan to communicate with its internal and external audiences throughout the crisis.

Employees, suppliers, customers, shareholders, the public and the media all expect and need to be kept informed in order for the organisation to retain its reputation and to maintain confidence in the business. An inability to communicate with such audiences before, during and after a crisis can easily cause a public relations disaster, thereby compounding the original incident.

Examples of companies who did not, or chose not to, communicate include:

- Hoffman-La Roche: Seveso (July 1976).
- Eli Lilly: Opren Withdrawal (August 1982).
- Union Carbide: Bhopal (December 1984).
- Delta Airlines: Plane Crash (August 1985).
- Sandoz: Rhine Pollution (November 1986).

In all cases, the share price fell within a short time of the disaster and did not get back up to the market average for at least a year after the event.

The converse of the situation faced by these companies is one where the organisation does communicate and manages to secure excellent PR from a difficult situation for example:

- British Midland: Kegworth Air Crash.
- Commercial Union: City IRA Bomb:

    We don't make a drama out of a crisis!

In order to minimise the risk of negative PR and the damage it can do, many companies include a communications plan within their BCP. This also enables the company to be portrayed in a positive and responsible light.

The communication plan

[B8040] The communications plan should ideally comprise three stages:

Stage 1: Pre-crisis – to minimise the risk of poor communications making the disaster worse

Stage 2: During crisis – to manage the volume and complexity of enquiries and information requirements

Stage 3: After crisis – to let shareholders know that we are back in business

Pre-crisis measures

[B8041] —

- Understanding which media are influential in your market and your location and developing a working relationship with them prior to any incident will tend to make them more sympathetic to your version of events in the reporting of a crisis.

- Press statements are an excellent way of ensuring that there is a single source of authorised information being disseminated to the media. They provide a consistent means of updating stakeholders. Having them prepared in advance, for most possible scenarios, will save the company valuable time in a crisis.
- In the event of a disaster, which generally excites the interest of the media, it is likely that directors/senior managers will be approached for comment in an interview format. The physical environment and the circumstances of the interview often conspire to generate a highly charged atmosphere, where the untrained individual may fail to portray the organisation in a good light, hence missing the opportunity to use this very powerful medium of communication to good effect. Training and rehearsal can dramatically improve individual performance on camera and enhance the organisation's ability to harness the power of the media.
- It is therefore advisable for the company to identify and train not just directors and senior managers, but any individuals who have some detailed knowledge or experience of the organisation's critical business processes. They then become the company spokespersons if problems occur within their area of expertise.
- In the event of a disaster, organisations experience a phenomenon known as call deluge. Organisations can expect to receive up to ten times as many calls and enquiries following a disaster than on a normal day.

Under such a weight of demand, the risk of providing inaccurate or misleading information is vastly increased.

One way of managing call deluge is to set up a dedicated communications office to which are channelled all disaster-related calls. In some cases a separate crisis hotline telephone number should be given out and all received calls be dealt with by trained operators.

From this centre the enquiry can be satisfactorily answered and passed on to one of the company spokespersons, if considered necessary. This approach enables the company to maintain integrity and consistency in the information being released concerning the disaster and its ongoing effects and mitigation.

During crisis

[B8042] —

- The communications office swings into action. It is advisable to include the setting up of a dedicated communications office within the testing and maintenance section of the BCP.
- Spokespersons needed for comment and/or interview should be readily available throughout the duration of the crisis. Ideally, they should be on site, or near to the site, working in tandem with the command and control team.
- Press/media statements should be released at regular and frequent intervals to update them on damage mitigation and recovery progress.

- Key stakeholders (e.g. customers) should be updated at regular and frequent intervals with regard to the recovery process and the estimated time for the reinstatement of their service.
- As the situation develops, all staff need to be kept informed, particularly if they are required to participate in the recovery process. In any event, all staff should be aware of the likely timescales for the business getting back into full operation.
- As the recovery from the disaster takes place, a crisis event log should be kept to enable significant events, actions or milestones to be recorded with the purpose of learning from them in the aftermath of the disaster, and possibly reviewing and amending the BCP in the light of unique experiences.

After the crisis

[B8043] —
- Use the crisis events log to review and analyse how well the organisation performed towards its agreed recovery targets. Does the BCP need amending in the light of unique experiences during actual recovery?
- Identify those aspects which worked well and those where there is room for improvement to the BCP, in order to make things work more smoothly if ever there is a next time.
- Publicise a summary of performance to all stakeholders in order to demonstrate that the BCP has actually worked in practice. This will breed confidence in the organisation and its ability to rise above and proactively manage any future potential disaster scenarios.
- And finally consider a dedicated marketing campaign to make capital of how the crisis was managed in a positive manner, and to share the lessons learnt with others operating in similar environments. This will again demonstrate the organisation's capability to cope before, during, and after any type of disaster or crisis.

Civil Contingencies Act, 2004

[B8044] This Act contains sweeping emergency powers that place duties and responsibilities on how organisations respond to disasters.

The act is accompanied by two documents:

Emergency Preparedness

This is statutory guidance under Part I of the Act, which sets out exactly what the legalisation requires and also gives best practise, advise to assist respondees in complying with their legal requirements.

Emergency Response and Recovery

This is non-statutory guidance that deals with the response and recovery phases of emergencies.

The Act in detail

[B8045] The CCA came into effect in May, 2006 and was primarily intended to rectify short comings in the UK civil defence and emergency powers legislation, and was therefore enacted to counter civil disasters such as foot and mouth diseases; widespread storm/flood damage/disruption; terrorist attacks/bombings etc.

Part 1 of the Civil Contingencies Act, 2004 covers local arrangements for civil protections by the emergency services — police, fire and rescue services, paramedics — and local authorities.

Part 2 covers emergency powers for wide scale, regional emergencies.

Part 3 contains technical details.

What is an emergency under the CCA?

[B8046] Part 1 of the CCA defines an emergency as:

- an event or situation which threatens serious damage to human welfare in a place in the UK;
- an event or situation which threatens serious damage to the environment of a pace in the UK
- war or terrorism which threatens serious damage to the security of the UK

Local Authorities: duty to provide business continuity advice

[B8047] The CCA places a duty on local authorities (LAs) to provide advice and assistance to businesses on their patch on aspects of business continuity, including emergency preparedness and disaster recovery. In other words, LAs have a duty to promote business continuity planning and management.

In order to comply with this duty, LAs have to demonstrate that they have taken reasonable steps to promote business continuity advice in their areas. This means that they have to identify what businesses need to know when they develop, implement or review their business continuity arrangements.

This implementation includes:

- the kinds of disruption that may occur and the impacts these may have on the business;
- what arrangements the LAs and emergency services have in place to respond to and recover from emergencies, such as evacuation or recovery plans;
- an awareness of sources or warnings, information and advice; and
- the steps that individual businesses can take to prepare for or mitigate the effects of an emergency, i.e. business continuity management procedures.

Government Business Continuity Tool Kit

[B8048] The Cabinet Office has now developed a national business continuity tool kit which is designed to help businesses prepare for potential disasters.

This BCM tool kit will be available on the '*Business Link*' website and also the '*Preparing for Emergencies*' website, www.preparingforemergencies.co.uk.

The tool kit includes a workbook which sets out the steps that businesses should go through to ensure they have sufficient BCM arrangements in place. It also features a BCM checklist, template and exercise scenarios.

The Civil Contingencies Secretariat, part of the Cabinet Office, has published a promotional leaflet for use by local authorities to encourage businesses to put BCM plans into practice.

Other Sources of Help

[B8049]

Business continuity bench-marking tool

This free bench-marking service is available on line from business advisory firm Deloitte.

ROBUST Business continuity toolkit

This free toolkit is available from RISC Authority, administered by the Fire Protection Association (FPA) at www.nfpa.org.

Business Continuity Institute (BCI) Good Practice Guidelines

These are available at www.thebci.org.

Cyber Security

[B8050] As part of the Health and Safety Executive's horizon scanning project, the threat to the safety of industrial processes from breaches in the security of safety critical electronic control systems has been identified.

Such breaches can result from targeted malicious attacks or from exposure to a wide range of threats/viruses ever present in an open IT environment, such as the Internet.

The increasing use of common, open operating systems - combined with wireless networking and greater interconnectivity of process control - safety-related systems, business management systems and external networks (such as the Internet) are all vulnerable, thereby leading to the likely occurrence of such problems increasing dramatically in the next few years.

Attacks by hackers, disgruntled employees, criminals etc are commonplace but are, to date, mainly being directed at activities such as spamming, denial of service, monetary gain (involving identity theft), fraud, extortion etc.

However, the same methods by which these known attacks are undertaken can readily be adapted to seriously disrupt computer-controlled hazardous processes and services in a way which could lead to major OSH risks for operators and the public at large.

The implications of the above are many and varied but include the accidental failure or malicious attack on process control systems which, in turn, could result in the loss of system - critical functions such as interlocking, emergency shutdown systems and disruption of overall process control. Such losses could thereafter result in serious risks to operators and the public.

While it is considered good practice to isolate safety-critical control or protection systems from any connectivity to the outside world, this approach is being challenged by the changing nature of plant electronic control and management systems.

This is leading to increased vulnerability of plant to electronic attack while, at the same time, the threat level is on the increase.

Cyber Risk Guidance

[B8051] The Institute of Risk Management (IRM) has published (February 2014) a detailed practical guide on Cyber risk entitled 'Cyber risk resources for practitioners'.

The aim of the publication is to help risk professionals understand and manage the Cyber risks facing their organisations. Tackling the risk of loss disruption and reputational damage arising from the use of IT systems has become a major issue for both private and public sector organisations.

The IRM guidance emphasises the importance of people, behaviour and processes rather than technological solutions.

Research by the IRM leading up to the publication of the guidance revealed that:

- although 82% of global organisations surveyed had an information security programme, less than 50% looked at the security practices of their supply chain;
- more than 90% of organisations were allowing staff to use mobile devices for business use but less than 40% required formal security configuration of such devices;
- nearly 40% of organisations reported using some sort of cloud-based facility but 33% of these have not developed a commensurate security policy;
- of those surveyed, 20% undertook no information security training;
- access to social media varied widely from the 20% of organisations that operated a complete lock-down with no social media access permitted from any business devices to the 9% who had no restrictions at all; and
- 10% reported that at least one breach of their online systems had taken place in the last three years with consequences ranging from regulatory fines to compensation costs, share price drops and reputational damage;

The IRM guidance outlines best practices in managing Cyber risks: these range from understanding the threat landscape and the iceberg impact of Cyber losses to balancing risks with opportunities, ensuring good corporate governance, effective internal Cyber risk audit, supply chain issues, and making training stick.

The guidance also looks in detail at the implications of cloud computing, social media, mobile devices and the essentials of secure systems.

The full report and an executive summary can be downloaded from the IRM's, hopefully, secure website: www.theirm.org.

Compensation for Work Injuries/Diseases

Andrea Oates

Introduction to compensation for work injuries/diseases

[C6001] Compensation for work injuries, diseases and death is payable under the welfare benefits system and in the form of damages for civil wrongs (torts). This chapter examines the two systems and the interaction between them.

The welfare benefits system is a form of public insurance, funded by employers/employees and taxpayers, and benefit is payable irrespective of liability on the part of an employer, i.e. 'no fault'. However, connection with employment must be established. The welfare benefits system (previously known as the social security system) was created by legislation, such as the various Social Security Acts, the *Social Security Administration Act 1992*, the *Social Security Contributions and Benefits Act 1992* ('*SSCBA 1992*'), the *Statutory Sick Pay Act 1994*, the *Social Security (Incapacity for Work) Act 1994*, the *Social Security (Recoupment of Benefits) Act 1997*, and other legislation. Over recent years, a series of Welfare Reform Acts have made changes to the benefits system. The *Welfare Reform Act 2012* particularly has introduced far-reaching changes, many of which take effect from 2013.

Liability in tort depends on proof of negligence or breach of statutory duty against an employer. Employers must be insured for such liability (see EMPLOYERS' LIABILITY INSURANCE). The current law relating to work injuries and diseases is to be found in a variety of Acts, including the *Law Reform (Personal Injuries) Act 1948*, the *Employers' Liability (Compulsory Insurance) Act 1969*, the *Employers' Liability (Defective Equipment) Act 1969*, the *Damages Act 1996*, various health and safety regulations and a wide body of case law.

A useful guides worth referring to is *Guidelines for the assessment of general damages in personal injury cases* (2012) 11th edition (ISBN 9780199664757) issued by the Judicial Studies Board. This aims to unify judicial approaches to awards of damages. These guidelines are not a legal document and a full examination of the law applicable to each case is required.

Industrial Injuries Scheme

[C6002] An employee who has an accident at work, contracts an industrial disease, or becomes deaf through their work may be able to claim through the Industrial Injuries Scheme. This provides non-contributory no-fault benefits for disablement because of an accident at work, or because of one of the prescribed diseases known to be a risk from certain jobs.

To qualify for industrial injuries disablement benefit (IIDB), the claimant must have been a paid employee (or "employed earner") when they suffered the injury or contracted the "industrial disease" – that is one of more than 70 "prescribed" diseases that are officially recognised as work-related by the Industrial Injuries Advisory Council (IIAC). These include:

- asthma;
- chronic bronchitis and emphysema;
- deafness;
- pneumoconiosis (lung disease from breathing in mineral dust); and
- prescribed disease A11 (formerly known as vibration white finger).

The Industrial Injuries Scheme Benefits are:

- Industrial Injuries Disablement Benefit (IIDB);
- Constant Attendance Allowance;
- Exceptionally Severe Disablement Allowance;
- Reduced Earnings Allowance (REA). (This is not covered in this chapter as there is no entitlement to REA for an accident which occurred on or after 1 October 1990); and
- Retirement Allowance (RA). RA replaces REA when you reach state pension age if REA is at least £2.00 a week and you are not in regular employment.

The amount paid depends on the extent of the disability resulting from the industrial injury or disease.

The claimant does not need to have paid any national insurance (NI) contributions in order to claim IIDB. Accidents or diseases which arise out of self-employment or service in HM forces are not included in the scheme. Normally benefits are only paid where the accident occurred, or the disease was contracted in Britain, although there are exceptions to this general rule.

The benefit is not paid for the first 15 weeks (90 days not including Sundays) after the accident.

Accident and personal injury provisions

[C6003] The employed earner must have suffered personal injury caused by an accident arising out of and in the course of their employment [SSCBA 1992, s 94]. The *Reporting of Injuries, Diseases and Dangerous Occurrences Regulations 1995 (SI 1995 No 3163)* contain detailed provisions which require reports to be made on prescribed forms to the Health and Safety Executive (HSE) after the occurrence of a reportable accident or on receipt of a report from a registered medical practitioner of their diagnosis of a prescribed disease. Records must be kept for three years containing prescribed details of accidents, or the date of diagnosis of the disease, the occupation of the person affected and the nature of the disease. For more information on these Regulations see ACCIDENT REPORTING AND INVESTIGATION.

Industrial accident and disease records may be kept for three years on:

- a B510 Accident Book;

- photocopies of completed form F2508; or
- computerised records.

Personal injury caused by accident

[C6004]–[C6005] An accident for IIDB purposes means any unintended happening or incident at work that has arisen out of and in the course of employment, and has resulted in a personal injury.

'Personal injury' includes physical and mental impairment, a hurt to body or mind, which includes nervous disorders or shocks (R(I) 22/52; R(I) 22/59). The injury must have been caused by an accident. Although this is usually an unintended and unexpected occurrence, such as a fall, if a victim is injured by someone else, may also be considered to be an accident (*Trim Joint District School Board of Management v Kelly* 83 LJPC 220, HL). A relevant accident may still have occurred where a series of accidents without separate definite times cause personal injury. An office worker was held to have suffered a series of accidents on each occasion she had been obliged to inhale her colleagues' tobacco smoke and this was held to have caused personal injury.

'Accident' must be distinguished from 'process', that is, bodily or mental derangement not ascribable to a particular event. Injuries to health caused by processes are not industrial injuries, unless they lead to prescribed industrial diseases (see **C6010** below). 'There must come a time when the indefinite number of so-called accidents and the length of time over which they occur, take away the name of accident and substitute that of process' (*Roberts v Dorothea Slate Quarries Co Ltd* [1948] 2 All ER 201, [1948] LJR 1409,HL). In *Chief Adjudication Officer v Faulds* [2000] 1 WLR 1035, HL, the House of Lords, in disallowing a claim for damages for psychological injury suffered by a fireman, held there must be at least one identifiable accident that caused the injury. The fact that an employee might develop stress from a stressful occupation would not satisfy the definition of 'accident'.

Accident arising out of and in the course of employment

[C6006] There is no need to show a cause for the relevant accident provided that the employee was working in the employer's premises at the time of the accident. An accident arising in the course of employment is presumed to have arisen out of that employment, in the absence of any evidence to the contrary.

What runs through all the case law is the common requirement giving rise to industrial injuries rights that the employee was doing something reasonably incidental to and within the scope of his employment, including extra-mural activities which the employee has agreed to do (R(I) 39/56). A male nurse who was injured in a football match watched by patients in the hospital grounds succeeded in a claim for benefit as this was reasonably incidental to and within the scope of his employment (R(I) 3/57), but a policeman who was injured whilst playing football for his force was unable to recover benefit despite the fact that his employers had encouraged him to play in the game (*R v National Insurance Commissioner, ex p Michael* [1977] 2 All ER 420). Also, in *Faulkner v Chief Adjudication Officer* [1994] PIQR P 244, a police officer who was injured whilst playing for a police football team was not entitled to industrial injuries benefit despite the benefit to the community resulting from his participation. He was not on duty at the time. Further guidance can be found in *Chief Adjudication Officer v Rhodes* [1999] ICR 178 where it was held that the two main questions to be asked are:

- what are the employee's duties; and
- was he discharging them at the time of the accident.

Supplementary rules

[C6007] Five statutory provisions establish rules under which the employee is deemed to be acting in the course of his employment duties. If the occurrence falls within these rules, the employee will be covered by industrial injuries benefit. These provisions are as follows:

- Illegal employment – if the employee was not lawfully employed, or his employment was actually void because of some contravention of employment legislation [*SSCBA 1992, s 97*].
- Acting in breach of regulations, or orders of the employer – if the employee is not acting outside his authority under his employment duties, and an accident occurs while the employee is doing something for the purposes of, and in connection with, the employer's business [*SSCBA 1992, s 98*], he will be covered. For instance, a kitchen porter hung up his apron to dry in a recess near to the ovens where he was forbidden to go. He was injured when he fell into a shallow pit. The hanging up of the apron was for the purposes of his employment, so the accident was deemed to have arisen out of, and in the course of, his employment duties (R(I) 6/55).
  Where a dock labourer was employed on loading a ship by the method of two slings, but he instead used a truck which he had not been authorised to use for this purpose, he was held to be acting in the course of his employment. (R(I) 1/70B). This contrasts with the earlier case of *R v D'Albuquerque, ex parte Bresnahan* [1966] 1 Lloyd's Rep 69, where a dock labourer was killed in an accident whilst driving a forklift truck, which he had no authority or permission to use to remove an obstruction. His widow was unable to recover industrial injuries benefit as her husband was held not to have been acting in the course of his employment.
- Travelling in an employer's transport – travel to and from work is not covered except where the employee is travelling in transport provided by the employer with his express or implied permission, whether or not the employee was bound to travel in this transport [*SSCBA 1992, s 99*]. Outside this express provision, the employee will, in most cases, both be required to be on the employer's premises doing what he was authorised to do, unless his work takes him off the premises. A postman was able to recover benefit when he was bitten by a dog on the street, as his job required him to walk along streets (R(I) 10/57). If the journey is preparatory to the start of timed itinerant duties, for example, as a

home help, there will be no entitlement to benefit in respect of injury sustained on the way to the first home, though if the employee has more discretion about his movements, he may be entitled to benefit on the way to his first call.
- An injury incurred while an employee is trying to prevent a danger to other people, or serious damage to property during an emergency [*SSCBA 1992, s 100*].
- Accidents caused by another's misconduct, boisterousness or negligence (provided that the claimant did not directly induce or contribute to the accident by his own conduct), the behaviour of animals (including birds, fish and insects). If these cause an accident, or if a person is struck by lightning or by any object, they respectively confer entitlement to industrial injuries benefit [*SSCBA 1992, s 101*].

Relevant employment

[C6008] 'Employed earner's employment' includes all persons who are gainfully employed in Great Britain under a contract of service, or as an office holder, and who are subject to income tax under Schedule E. Self-employed people and private contractors are thus not entitled to industrial injuries benefits.

Certain classes of person are expressly included for the purposes of industrial injuries benefits. They include unpaid apprentices, members of fire brigades (or other rescue brigades), first-aid, salvage or air raid precautions parties, inspectors of mines, special constables, certain off-shore oil and gas workers and certain mariners and air crew [*Social Security (Employed Earners' Employments for Industrial Injuries Purposes) Regulations 1975 Schedule 1*]. You must normally have had the accident or got the disease in Great Britain, but you may still get benefit if:

- you were a mariner, airman, worked on the continental shelf of the United Kingdom, or worked in an-other European Union (EU) country or Norway;
- your employer was paying Class 1 NI contributions for you while you were working out of the country;
- you were paying special Class 2 contributions as a volunteer development worker; or
- you worked in one of the countries with which Great Britain has an agreement covering industrial in-juries.

Certain types of employment are specifically excluded from cover [*SSCBA 1992, s 95; Social Security (Employed Earners' Employments for Industrial Injuries Purposes) Regulations 1975 (SI 1975 No 467), Regs 2–7* and *Sch 1, 2*].

Persons treated as employers

[C6009] The *Social Security (Employed Earners' Employments for Industrial Injuries Purposes) Regulations 1975 (SI 1975 No 467), Sch 3* also provide for cases where certain people who may not have a contract with the employee, or who may be an agency employer, to be the relevant employer for the person who has suffered the industrial injury, or who has developed a prescribed disease. An agency that supplies an office cleaner, or a typist, will be the relevant employer. For casual employees of clubs, the club will be the relevant employer.

Benefits for prescribed industrial diseases

[C6010] The rules for certain prescribed diseases (namely deafness, asthma and asbestos related diseases), differ in some respects from the provisions affecting prescribed diseases outlined below. (See **C6055** and **C6061** below.) To obtain the right to industrial injuries benefits for all other prescribed diseases, the claimant must show:

- that he is suffering from a prescribed disease;
- that the disease is prescribed for his particular occupation (where a disease is prescribed for a general activity, for example, contact with certain substances, he must clearly show that this was more than to a minimal extent); and
- that he contracted the disease through engaging in the particular occupation (there is a presumption that if the disease is prescribed for a particular occupation, the disease was caused by it, in the absence of evidence to the contrary).

(See also OCCUPATIONAL HEALTH AND DISEASES.)

Claims are made on Form BI 100PD and there is a right to, and it is advisable to, claim immediately after the disease starts. The 90-day waiting period for receipt of benefit applies as for accidents, as do the percentage disabilities and aggregated assessment rules, except in the case of loss of faculty resulting from diffuse mesothelioma when entitlement begins on the first day of the claim.

Assessments are made by one or possibly two experienced medical practitioners - doctors specially trained in industrial injuries disablement matters. They will decide on the percentage disability and how long the disability will last. Benefit will then be payable for the period stated in the assessment, but if the doctors are not sure of the period, benefit will be paid for a while with a further review. If the disease recurs during that period, there will be no need to make a further claim, but if the condition has worsened, the assessment may be reviewed. If there is a further attack after the period of the assessment, a fresh claim will have to be made, which will be subject to a further 90-day waiting period.

Industrial injuries disablement benefit

Entitlement and assessment

[C6011] A person is entitled to an industrial injuries disablement pension if:

- he suffers from a prescribed industrial disease (see **C6010** above);

- he suffers as a result of the relevant accident, a loss of physical or mental faculty such that the assessed extent of the resulting disablement amounts to not less than 14 per cent [SSCBA 1992, S 103, Sch 6]. See leaflet NI6 (July 1999) (update);
- 90 days (excluding Sundays) have elapsed since the date of the accident or onset of the prescribed disease or injury.

An assessment of the percentage disablement up to 100 per cent will be made by an adjudicating medical practitioner who, in the case of accidents, looks at the claimant's physical and mental condition, comparing him in those respects with those of a normal person of the same age and sex. No other factors are relevant. The assessment may cover a fixed period/or the life of the claimant. The degrees of disablement are laid down in a scale so that, for example, loss of one hand is normally 60 per cent and loss of both hands 100 per cent. Disfigurement is included even if this causes no bodily handicap.

Where the claimant suffered from a pre-existing disability before the happening of the industrial accident at the onset of the industrial disease, benefit will only be payable in respect of the industrial accident or disease itself, and the medical adjudicators will compare the original disability with the industrial disability for this purpose. Where one or more disabilities result from industrial accidents or diseases, the level of resulting disability may be aggregated, but not so as to exceed the 100 per cent disability and its corresponding rate of benefit. [*Social Security (General Benefit) Regulations 1982 (SI 1982 No 1408), Reg 11*].

Claims in accident cases

[C6012] Claims should be made on Form BI 100A obtainable from from the Regional Industrial Injuries Disablement Benefit Centre or the Disability benefits section of the GOV.UK website. The date of the claim is the date the fully completed claim form is received in an office of the Department for Work and Pensions (DWP).It is therefore important to carefully fill in all the details on the form and return it to the Regional Industrial Injuries Benefit Centre as soon as possible.

Rate of industrial injuries disablement benefit

[C6013] The amount of benefit depends on the extent of the disability caused by the accident. IIDB cannot be paid for the first 15 weeks (90 days not including Sundays) after the date of the accident and is not paid if the disablement is assessed at less than 14%. Benefit may be paid if the claimant has had more than one accident or disease and the total disablement, when the effects of all the accidents and diseases are added together, is 14% or more. This is known as aggregation. If disablement is at least 14% benefit will be paid as a weekly pension.

Constant attendance allowance and exceptionally severe disablement allowance

[C6014] Constant attendance allowance is available if:

- the claimant is receiving industrial injuries disablement benefit based on 100 per cent disablement, or aggregated disablements that total 100 per cent or more;
- the claimant needs constant care and attention as a result of the effects of an industrial accident or disease.

Furthermore, if the carer of the claimant spends at least 35 hours a week looking after them and is of working age and is not earning more than £100 per week, care allowance of £59.75 per week may be granted to that person. Claims for constant care allowance are made on Form BI 104. It is granted for a fixed period and may be renewed from time to time. There are four rates of payment:

- part-time (where full-time care is unnecessary);
- normal maximum rate (if the above conditions are fulfilled and full-time care is required);
- intermediate (if the claimant is exceptionally disabled and the degree of attendance required is greater, and the care necessary is greater than under the 'normal' classification, the benefit is limited to one and a half times the normal rate); and
- an exceptional rate (if the claimant is so exceptionally disabled as to be entirely dependent on full-time attendance for the necessities of life).

If the intermediate or exceptional rates are payable, an additional allowance (known as exceptionally severe disablement allowance) will be due if the condition is likely to be permanent. Constant attendance allowance may continue to be paid for up to four weeks if the claimant goes into hospital for free medical treatment.

Other sickness/disability benefits

[C6015] Most applicants for industrial injuries benefit will, as their incapacity results from accidents or diseases contracted during their employment, be able to claim statutory sick pay or Employment and Support Allowance (ESA)up to the date of commencement of industrial injuries benefit as 90 days is less than 28 weeks. The current rate of statutory sick pay (SSP) is £85.85 a week (as at May 2013). To be able to claim SSP you must be unable to work because you are sick or disabled and earn at least £109 a week (as at May 2013).

ESA replaced Incapacity Benefit and Income Support paid because of an illness or disability from January 2011. Self-employed workers, unemployed people and those whose earnings are too low to claim SSP can claim ESA if they have an illness or disability that affects their ability to work.

It is possible to claim extra for a spouse or person looking after children if the spouse is over 60, or, if younger, child benefit is being paid, and the claimant was maintaining the family to at least the extent of the dependency benefit being claimed. None of these restrictive rules apply to industrial injuries benefit claims.

Income support and other state benefits such as housing benefit and council tax benefit may be available if the claimant with or without dependants does not have sufficient to live on.

Disability Element of Working Tax Credit

[C6016] This benefit is for people who wish to work, but have a physical or mental disability which puts them at a disadvantage in securing a job under criteria set out in the regulations. The applicant must work for at least 16 hours a week to qualify. For a full analysis of the disability tax credit see www.hmrc.gov.uk/taxcredits/start/who-qualifies/workingtaxcredit/disability. htm.

Payment for work injuries/diseases

[C6017] For the financial year 2013/14 the weekly benefit rates payable for SSP, ESA, IIDB, Constant attendance allowance and Exceptionally severe disablement are set by the *Social Security Benefits Up-rating Order, SI 2013/574.*

Appeals

[C6018]–[C6019] Appeals relating to all industrial injuries are made to an appeal tribunal and should be made within one month from the date that the decision maker sends the decision to the applicant.

Change of circumstances and financial effects of receipt of benefit

- *Hospital*
  If a claimant enters hospital, industrial injuries disablement pension continues to be payable, as does exceptionally severe disablement allowance. Constant attendance allowance will stop after four weeks.
- *Taxation*
  Industrial injuries disablement benefit, constant attendance allowance, exceptionally severe disablement allowance and the disability element of working tax credit are not taxable. Contribution-based ESA is taxable. Statutory sick pay is taxable under the PAYE system.

Benefit overlaps

[C6020] IIDB does not affect any other National Insurance benefits such as:
- Incapacity Benefit

- Employment and Support Allowance (contribution-based)
- Contribution-based Jobseeker's Allowance
- Retirement Pension

However, it may affect income-related benefits (received by the disabled person or their partner) such as:

- Income Support
- Employment and Support Allowance (income-related)
- Income-based Jobseeker's Allowance
- Pension Credit
- Housing Benefit
- Working Tax Credit
- Child Tax Credit
- Universal Credit

(For recoupment of benefit after awards of damages see **C6044** below.)

Damages for occupational injuries and diseases

[C6021] When a person is injured or killed at work in circumstances indicating negligence or breach of duty on the part of an employer, he may be entitled to an award of damages. Damages are assessed by judges in accordance with previously decided cases; very exceptionally they may be assessed by a jury. Damages are also categorised as general and special damages, according to whether they reflect pre-trial or post-trial losses. Calculation of damages is often made by reference to Kemp and Kemp, *The quantum of damages*, and to the Judicial Studies Board's document, *Guidelines for the assessment of general damages in personal injury cases* (2012) (11th edition).

Damages normally take the form of a lump sum; however, 'structured settlements', whereby accident victims are paid a variable sum for the rest of their lives, are now a viable alternative [*Damages Act 1996, s 2*] (see **C6025** below). This may well involve greater reliance on actuarial evidence and a rate of return of interest provided by index-linked government securities.

The *Legal Aid, Sentencing and Punishment of Offenders Act 2012* has abolished legal aid for all personal injury (and clinical negligence) claims with the exception of "exceptional" cases where denying legal aid would amount to a breach of human rights. It has also abolished the recovery of success fees and After the Event (ATE) insurance premiums as part of costs (with limited exceptions related to mesothelioma cases) and has introduced damage-based agreements to replace conditional fees. The success fee will now be expressed as a percentage of the damages awarded. Referral fees are also barred. Further changes will be made through secondary legislation.

Accident Line, which is part of a scheme run by solicitors who are members of a specialist panel of personal injuries lawyers, provides a free half-hour consultation for claimants who have suffered personal injuries, including industrial accidents. The telephone number is 0500 192 939.

Basis of claim for damages for personal injury at work

[C6022] The basis of an award of damages is that an injured employee should be entitled to recoup the loss which he or she has suffered in consequence of the injury/disease at work. 'The broad general principle which should govern the assessment in cases such as this is that the court should award the injured party such a sum of money as will put him in the same position as he would have been in if he had not sustained the injuries' (per Earl Jowitt in *British Transport Commission v Gourley* [1956] AC 185). The *Damages Act 1996*, provides that periodical payments may be made in some cases.

It is not necessary that a particular injury be foreseeable, although it normally would be (see further *Smith v Leech Brain & Co Ltd* [1962] 2 QB 405). Moreover, if the original injury has made the claimant susceptible to further injury (which would not otherwise have happened), damages will be awarded in respect of such further injuries, unless the injuries were due to the negligence of the claimant himself (*Wieland v Cyril Lord Carpets Ltd* [1969] 3 All ER 1006). In this case, a woman, who had earlier injured her neck, was fitted with a surgical collar. She later fell on some stairs, injuring herself, because her bifocal glasses had been dislodged slightly by the surgical collar. It was held that damages were payable in respect of this later injury by the perpetrator of the original act of negligence. Conversely, where, in spite of having suffered an injury owing to an employer's negligence, an employee contracts a disease which has no causal connection with the earlier injury, and the subsequent illness prevents the worker from working, any damages awarded in respect of the injury will stop at that point, since the supervening illness would have prevented (and, indeed, has prevented) the worker from going on working.

[C6023]–[C6024] Listed below are the various losses for which the employee can expect to be compensated. Losses are classified as non-pecuniary and pecuniary.

- *Non-pecuniary losses*
  The principal non-pecuniary losses are:
  (i)   pain and suffering prior to the trial;
  (ii)   disability and loss of amenity (i.e. faculty) before the trial;
  (iii)   pain and suffering in the future, whether permanent or temporary;
  (iv)   disability and loss of amenity in the future, whether permanent or temporary;
  (v)   bereavement.
  (Damages for loss of expectation of life were abolished by the *Administration of Justice Act 1982, s 1(1)* (see further 'loss of amenity' at C6034 below).)
  There are four main compensatable types of injury, namely:
  (i)   maximum severity injuries e.g. irreversible brain damage, quadraplegia;
  (ii)   very serious injuries e.g. severe head injuries/loss of sight in both eyes/injury to respiratory and/or excretory systems;
  (iii)   serious injuries, e.g. loss of arm, hand, leg;
  (iv)   less serious injuries, e.g. loss of a finger, thumb, toe etc.

There is a scale of rates applicable to the range of disabilities accompanying injury to workers but it is nowhere as precise as the scale for industrial injuries disablement benefit. Damages for maximum severity cases can vary from several hundred thousand pounds to millions of pounds. In *Biesheuval v Birrell* [1999] PIQR Q40, the High Court awarded total damages of £9,200,000 (see **C6024** below) to a student who was almost completely paralysed in all four limbs after a car crash. In *Dashiell v Luttitt* [2000] 3 QR 4 a settlement of £5,000,000 was reached for brain damage sustained by a child aged 14 at the time of a school minibus crash. In *Cappoci v Bloomsbury Health Authority* (21 January 2000, unreported) the High Court awarded £2,275,000 damages to a 13-year-old boy who had been asphyxiated at birth and as a result suffered cerebral palsy and other severe physical and mental handicaps. Less serious injuries attract lower damages – for example, in *Williams v Gloucestershire County Council* (10 September 1999, unreported) an out of court settlement of £3,639 was reached in a case where a 6-year-old lost the top of her little finger in an accident.

- *Pecuniary losses*
  These consist chiefly of:
  (i) loss of earnings prior to trial (i.e. special damages);
  (ii) expenses prior to the trial, e.g. medical expenses;
  (iii) loss of future earnings (see below);
  (iv) loss of earning capacity, i.e. the handicap on the open labour market following disability.

In actions for pecuniary losses, employees can be required to disclose the general medical records of the whole of their medical history to the employer's medical advisers (*Dunn v British Coal Corporation* [1993] ICR 591).

General and special damages

- *General damages*
  General damages are awarded for loss of future earnings, earning capacity and loss of amenity. They are, therefore, awarded in respect of both pecuniary and non-pecuniary losses. An award of general damages normally consists of:
  (i) damages for loss of future earnings;
  (ii) pain and suffering (before and after the trial); and
  (iii) loss of amenity (including disfigurement).
- *Special damages*
  Special damages are awarded for itemised expenses and loss of earnings incurred prior to the trial. Unlike general damages, this amount is normally agreed between the parties' solicitors. When making an award, judges normally specify separately awards for general and special damages. A statement of special damage must be served with the statement of claim which should suffice to give the defendant a fair idea of the case he has to answer. More detailed information must be supplied after the exchange of medical and expert reports.

EXAMPLE

An example of the way in which damages awards are assessed and broken down is the case of *Biesheuval v Birrell* (referred to in C6023 above) where the High Court awarded £9,200,000 damages consisting of:

Pain and suffering and loss of amenity	£137,000
Interest on general damages	£6,617
Past loss of earnings	£80,700
Interest on past loss of earnings	£14,929
Other special damages	£360,113
Interest on special damages	£54,516
Tax on interest	£41,215
Loss of future earnings	£3,700,000
Loss of pension rights	£67,491
Initial capital expenditure	£551,803
Recurring costs and future care	£4,267,000

Structured settlements

[C6025] A structured settlement is an agreement for settling a claim or action for damages on terms that the award is made wholly or partly in the form of periodic payments. Such settlements are expected to become more usual in the case of larger awards of damages. Under the *Damages Act 1996*, these periodic payments must be payable in the form of an annuity for life, or for a specified period, and may be held on trust for the claimant if that should be necessary. Provision may be added to the settlements for increases, percentages or adjustments where the court or claimant's advisers secure these variations in his interest. Structured settlements in favour of claimants may be made for the duration of their life. Knowledge of the claimant's special needs is thus vital for the structure to be successful – it should be recognised that structured settlements will not be suitable for all cases. If the claimant dies while in receipt of periodic payments, they pass under his estate. Structured settlements may also be made in awards of damages in respect of fatal accidents. Tax-free annuities are payable directly to the claimant by the Life Office.

Structured settlement awards made by the Criminal Injuries Compensation Authority are also tax free.

When agreeing a settlement (whether structured or not), it is better to agree whether payments are net of repayable benefits. If a settlement offer is silent as to repayable benefits, then a deduction will have to be made in respect of them, possibly with unplanned results for the claimant.

Structured settlements are advisable where brain damage makes the injured person at risk and suggestible to pressure from relatives. They are also useful where the injured person has little experience or interest in investment, or dislikes the possibility of becoming dependent on the State or relatives should funds run out. Disadvantages are that annuities only last for the lifetime of the injured person. There is also a loss of flexibility to deal with changed circumstances and of the better return gained with the skilful investment of a lump sum.

Assessment of pecuniary losses

[C6026] Assessing pecuniary losses, i.e. loss of future earnings, can be a difficult process. As was authoritatively said, 'If (the claimant) had not been injured, he would have had the prospect of earning a continuing income, it may be, for many years, but there can be no certainty as to what would have happened. In many cases the amount of that income may be doubtful, even if he had remained in good health, and there is always the possibility that he might have died or suffered from some incapacity at any time. The loss which he has suffered between the date of the accident and the date of the trial [i.e. special damages (see above)] may be certain, but his prospective loss is not. Yet damages must be assessed as a lump sum once and for all [see 'Provisional awards' at C6040 below], not only in respect of loss accrued before the trial but also in respect of a prospective loss' (per Lord Reid in *British Transport Commission v Gourley* [1956] AC 185). Moreover, if, at the time of injury, a worker earns at a particular rate, it is presumed that this will remain the same. If, therefore, he wishes to claim more, he must show that his earnings were going to rise, for example, in line with a likely increase in productivity – a probable rise in national productivity is not enough.

When the court assesses loss of earnings, the claimant has to mitigate his loss by taking work if he can. In *Larby v Thurgood* [1993] ICR 66, the defendant applied to the court to dismiss an action brought by a fireman who was severely injured in a road traffic accident and who had taken employment as a driver earning £6,000 per annum, unless he agreed to be interviewed by an employment consultant who would then give expert evidence on whether the claimant could have obtained better paid employment. This application was refused. Evidence of whether the claimant could have obtained better paid employment depended partly on medical evidence of his capabilities and the present and future state of the job market where he lived, which could not be established by an employment consultant. His general suitability for employment, his willingness and motivation, were matters of fact for the judge; thus expert opinion was not required for that purpose.

A claimant may be earning practically as much as he was before his accident, but may be more at risk of losing his present job, of not achieving expected promotion, or of disadvantages in the labour market. Damages may be claimed for such prospective losses and are known as *Smith v Manchester* damages (after *Smith v Manchester City Council* (1974) 1118 Sol Jo 597).

Capitalisation of future losses

[C6027] Loss of future earnings, often spanning many years ahead, is awarded normally as a once-and-for-all capital sum for the maintenance of the injured victim. The House of Lords considered the way in which lump sums should be calculated in the leading case of *Wells v Wells; Thomas v Brighton*

Health Authority [1999] 1 AC 345. This also applies in Scotland (*McNulty v Marshall's Food Group Ltd* 1999 SC 195).

The award of damages is calculated on the basis of the present value of future losses – a sum less than the aggregate of prospective earnings because the final amount has to be discounted (or reduced) to give the present value of the future losses. Inflation is ignored when assessing future losses in the majority of cases (e.g. pension rights), since this was best left to prudent investment (*Lim Poh Choo v Camden and Islington Area Health Authority* [1980] AC 174). And where injury shortens the life of a worker, he can recover losses for the whole period for which he would have been working (net of income tax and social security contributions, which he would have had to pay), if his life had not been shortened by the accident. The present value of future losses can be gauged from actuarial or annuity tables. The net annual loss (based on rate of earnings at the time of trial) (the multiplicand) has to be multiplied by a suitable number of years (multiplier) which takes account of factors such as the claimant's life expectancy and the number of years that the disability or loss of earnings is expected to last. The multiplier normally ranges between 6 and 18, and is set out in actuarial tables. Both the multiplier and the multiplicand can vary for different periods; for example, where medical evidence shows that the need for care could increase or decrease over time – *Wells v Wells*; *Thomas v Brighton Health Authority* [1998] 3 All ER 481. In *McIlgrew v Devon County Council* [1995] PIQR Q66, the maximum multiplier of 18 was applied for permanent general losses, and the multiplier of 12 was applied to loss of earnings.

The multiplier is calculated on the assumption that the claimant will invest the lump sum prudently. Under *section 1(1)* of the *Damages Act 1996*, the Lord Chancellor may by order prescribe a rate of return which the courts must have regard to. Under this provision, the Lord Chancellor set the rate of return at 2.5 per cent.

EXAMPLE

In the case of a male worker, aged 30 at the date of trial, and earning £15,000 per year net, on a 2.5 per cent interest yield, the multiplier will be 22.80 (using Table 25 of the Ogden Tables). Hence, general damages will be about £342,000, assuming incapacity to work up to age 65.

In the case of a male worker, aged 50 at date of trial, earning £25,000 net, on a 2.5 per cent interest yield, the multiplier will be 12.06 (using Table 25 of the Ogden Tables). Hence, general damages will be about £301,500 assuming incapacity to work until age 65.

Deductions from awards under this head are made for the actual earnings of the injured claimant. Where the claimant takes a lighter, less well paid job, and thus suffers a loss of earnings, the courts have held that the fact that he gains more leisure through working shorter hours is not to be taken into account to reduce the amount of damages awarded for loss of earnings (*Potter v Arafa* [1995] IRLR 316).

Institutional care and home care

[C6028] Where they are provided on a commercial basis, expenses of medical treatment may be claimed. Alterations to a home, purchase of a bungalow accessible to a wheelchair, adaptations to a car and equipment (such as lifting equipment), which are necessary for care, may be claimed. Nursing and care requiring constant or less attendance may be claimed whether or not the carer is a professional or voluntary carer. It was confirmed by the House of Lords in *Hunt v Severs* [1994] 2 AC 350, that where an injured claimant is cared for by a voluntary carer, such as a member of his family, that damages could be recovered for this care, but the claimant should hold them in trust for the voluntary carer.

Where voluntary care is undertaken by relatives, compensation for the cost of this care is assessed as a percentage of the "Crossroad" rate agreed from time to time for community care by most local authorities; the actual percentage awarded being about two-thirds of that rate. A higher percentage of that rate will be awarded where the care being given is beyond the level of care normally provided by home helps (*Fairhurst v St Helens and Knowsley Health Authority* [1995] PIQR Q 1).

Damages for lost years

[C6029] Damages may be payable up to retirement age for lost earnings resulting from the shortening of the claimant's life expectancy by reason of the injury or disease. Estimated costs of living expenses are deductible from these damages. Damages under this head may also be awarded to dependants if the victim has died.

Non-pecuniary losses (loss of amenity)

[C6030] It is generally accepted by the courts that quantification of non-pecuniary losses is considerably more difficult than computing pecuniary losses. This becomes even more difficult where loss of sense of taste and smell are involved, or loss of reproductive or excretory organs. Unlike pecuniary losses, loss of amenity generally consists of two awards: an award for (i) actual loss of amenity and (ii) the impairment of the quality of life suffered in consequence (i.e the psychic loss).

Victims are generally conscious of their predicament, but in very serious cases they may not be, a distinction underlined in the leading case of *H West & Son Ltd v Shephard* [1964] AC 326. If a victim's injuries are of the maximum severity kind (e.g. tetraplegia) and he is conscious of his predicament, damages will be greater. However, where, as is often the case, the injuries shorten the life of an accident victim, damages for non-pecuniary losses will be reduced to take into account the fact of shortened life.

Types of non-pecuniary losses recoverable

[C6031] The following are the non-pecuniary losses which are recoverable by way of damages:

- pain and suffering;
- loss of amenity;
- bereavement.

Pain and suffering

[C6032] Pain and suffering refers principally to actual pain and suffering at the time of the injury and later. Since modern drugs can easily remove acute distress, actual pain and suffering is not likely to be great and so damages awarded will be relatively small.

Additionally, 'pain and suffering' includes 'mental distress' and related psychic conditions; more specifically (i) nervous shock, (ii) concomitant pain or illness following post-accident surgery and embarrassment or humiliation following disfigurement. Claustrophobia and fear are within the normal human emotional experience but are not compensatable, unless amounting to a recognised psychiatric condition, such as post-traumatic stress disorder (PTSD) (*Reilly v Merseyside Regional Health Authority* [1995] 6 Med LR 246).

In *Heil v Rankin and Another and joined appeals* [2001] QB 272, the Court of Appeal, reviewed the current awards for pain, suffering and loss of amenity. The court held that whilst payments under £10,000 should not increase, there would be a tapered increase for awards above that up to a maximum of 33 per cent for the highest level of damages.

Nervous shock

[C6033] Nervous shock refers to actual and quantifiable damage to the nervous system, affecting nerves, glands and blood; and, although normally consequent upon earlier negligent physical injury, an action is nevertheless maintainable, even if shock is caused by property damage (*Attia v British Gas plc* [1988] QB 304 where a house caught fire following a gas explosion. The claimant, who suffered nervous shock, was held entitled to damages).

Claimants fall into two categories, namely, (a) primary and (b) secondary victims, the former being directly involved in an accident and the latter are essentially spectators, bystanders or rescuers.

- *Primary victims*
  Primary victims can sue for damages for nervous shock/psychiatric injury, even if they have not suffered earlier physical injury (see *Page v Smith* [1996] AC 155 in which the appellant, who was physically uninjured in a collision between his car and that of the respondent, had developed myalgic encephalomyelitis and chronic fatigue syndrome, which became permanent. It was held by the House of Lords that the respondent was liable for this condition in consequence of his negligent driving). Foreseeability of physical injury is sufficient to enable a claimant directly involved in an accident to recover damages for nervous shock. Thus, in the case of *Bourhill v Young* [1943] AC 92 a pregnant woman, whilst getting off a tram, heard an accident some fifteen yards away between a motor cyclist and a car, in which the motor cyclist, driving negligently, was killed. In consequence, the claimant gave birth to a stillborn child. It was held on the facts of the case that the unknown motor cyclist could not have foreseen injury to the claimant who was unknown to him.
  In *Corr (administratrix of Corr deceased) v IBC Vehicles Ltd* [2006] EWCA Civ 331, [2007] QB 46, [2006] 2 All ER 929 a widow brought proceedings under the Fatal Accidents Act. Her husband was badly

injured at work while employed by the defendant and suffered PTSD, resulting in severe depression and him committing suicide six years after the accident. The Court of Appeal held that she did not have to establish that his suicide was reasonably foreseeable at the time of the accident. It flowed from the psychiatric illness for which the defendant had admitted responsibility.

In *Johnston v NEI International Combustion Ltd* [2007] UKHL 39, [2008] AC 281, [2007] 4 All ER 1047 the House of Lords rejected claims by workers who developed clinical depression after being negligently exposed to asbestos and developing pleural plaques. These are caused by exposure to asbestos; but having a pleural plaque does not increase the risk of developing asbestos diseases including asbestosis, mesothelioma or lung cancer. The workers argued that they should be considered primary victims, but the House of Lords said that their illness was caused by the fear of the possibility of something that had not actually happened, and was therefore not actionable.

- *Secondary victims*

As for secondary victims, defendants are taken to foresee the likelihood of nervous shock to rescuers attending to an injured person and to their close relatives, though the precise extent of nervous shock need not have been foreseen (*Brice v Brown* [1984] 1 All ER 997). People who witness distressing personal injuries, who are not related in either of these ways to the victim, are not only considered not to have been foreseen by the defendant, but are expected to be possessed of sufficient fortitude to be able to withstand the calamities of modern life. Only persons with a close tie with the victim or their rescuers who are within sight or sound of the accident or its immediate aftermath will be awarded damages for nervous shock as the law stands at the present. Husbands and wives will be presumed to have a sufficiently proximate tie whilst other relationships are considered on the evidence of the proximity of the relationship.

In *Chadwick v British Railways Board* [1967] 1 WLR 912, following a serious railway accident, for which the defendant was held to be liable, a volunteer rescue worker suffered nervous shock and became psychoneurotic. As administratrix of the rescuer's estate, the claimant sued for nervous shock. It was held that (i) damages were recoverable for nervous shock, even though shock was not caused by fear for one's own safety or for that of one's children, (ii) the shock was foreseeable, and (iii) the defendant should have foreseen that volunteers might well offer to rescue and so owed them a duty of care.

The class of persons who can sue for damages for nervous shock is limited, depending on proximity of the claimant's relationship with the deceased or injured person (*McLoughlin v O'Brian* [1983] 1 AC 410 where the claimant's husband and three children were involved in a serious road accident, owing to the defendant's negligence. One child was killed and the husband and other two children were badly injured. At the time of the accident, the claimant was two miles away at home, being told of the accident by a neighbour and taken to the hospital, where she saw the injured members of her family and heard that her daughter had been killed. In consequence of hearing and seeing the results of the road accident, the claimant suffered severe and recurrent shock. It was held that she was entitled to damages for nervous shock as she was present in the immediate aftermath). This approach was confirmed by the House of Lords in *Alcock v Chief Constable of South Yorkshire Police* [1992] 1 AC 310, where it was stated that the class of persons to whom this duty of care was owed as being sufficiently proximate, was not limited to particular relationships such as husband and wife or parent and child, but was based on ties of love and affection, the closeness of which would need to be proved in each case, except that of spouse or parent, when such closeness would be assumed. Similarly, in *Hinz v Berry* [1970] 2 QB 40 the appellant left her husband and children in a lay-by while she crossed over the road to pick bluebells. The respondent negligently drove his car into the rear of the car of the appellant. The appellant heard the crash and later saw her husband and children lying severely injured, the former fatally. She became ill from nervous shock and successfully sued the respondent for damages. On this basis, an employee who suffers nervous shock as a result of witnessing the death of a co-employee at work, will be unlikely to be able to claim damages against his employer for nervous shock, as being a 'bystander who happens to be an employee', as distinct from an active participant in rescue (*Robertson v Forth Bridge Joint Board* [1995] IRLR 251 where an employee was blown off the Forth Bridge in a high gale and fell to his death; a co-employee who watched this was unable to sue for damages for nervous shock).

(However, in *Young v Charles Church (Southern) Ltd* (1997) 39 BMLR 146 an employee who suffered psychiatric illness after seeing a workmate electrocuted close to him could recover damages as a primary victim (see above) because of the risk of physical injury to himself.)

In *Hunter v British Coal Corporation* [1999] QB 140, a claimant who was 30 metres away from the scene of a fatal accident and was told of the victim's death 15 minutes later was unable to recover damages for the psychiatric injury he suffered because he felt responsible for the accident. It was held that he was neither physically nor temporarily close enough to the accident to be a primary victim.

In any event, reasonable fortitude, on the part of the claimant, will be assumed, thereby disqualifying claims on the part of hypersensitive persons (see *McFarlane v EE Caledonia* [1994] 2 All ER 1 where the owner of a rig did not owe a duty of reasonable care to avoid causing psychiatric injury to a crew member of a rescue vessel who witnessed horrific scenes at the Piper Alpha disaster).

In *White v Chief Constable of South Yorkshire Police* [1999] 1 All ER 1, the House of Lords considered the question of psychiatric injury to police officers on duty during the Hillsborough football stadium disaster when 95 spectators were crushed to death. It was held that police officers were not entitled to recover damages against the Chief Constable for psychiatric injury suffered as a result of assisting with the aftermath of a disaster, either as employees or as rescuers. An employee who suffered psychiatric injury in the course of his employment had to prove liability under the general rules of negligence, including the rules restricting the recovery of damages for psychiatric injury.

- In *Monk v PC Harrington Ltd* [2008] EWHC 1879 (QB), [2009] PIQR P52, [2008] All ER (D) 20 (Aug) a self-employed foreman was working on a construction site when a temporary platform fell 60 feet onto two fellow workers, injuring one and killing the other. He tried to help both men and subsequently suffered symptoms of PTSD, and was unable to carry on working. He argued that he should be able to recover compensation as a rescuer. But although the court accepted that he could be regarded as a rescuer, he could not establish himself as a primary victim on the basis of his acts as a rescuer, because he could not show that he had reasonably believed that he was putting his own safety at risk. Neither could he show that he was a primary victim as an unwilling participant as he could not show that his injuries were induced by a genuine belief that he had causes another person's injury or death. It was not reasonably foreseeable that someone in his position would suffer psycatric injury as a result of such a belief.

Occupational Stress

Cases dealing with damages for mental ill health as a result of occupational stress are examined in **STRESS AT WORK**.

Loss of amenity

[C6034] Loss of amenity is a loss, permanent or temporary, of a bodily or mental function, coupled with gradual deterioration in health, e.g. loss of finger, eye, hand etc. Traditionally, there are three kinds of loss of amenity, ranging from maximum severity injury (quadraplegias and irreversible brain damage), multiple injuries (very severe injuries) to less severe injuries (i.e. loss of sight, hearing etc.). Damages reflect the actual amenity loss rather than the concomitant psychic loss, at least in maximum severity cases. Though, if a claimant is aware that his life has been shortened, he will be compensated for this loss. The *Administration of Justice Act 1982, s 1(1)* states:

(a) no damages shall be recoverable in respect of any loss of expectation of life caused to the injured person by the injuries; but

(b) if the injured person's expectation of life has been reduced by the injuries, the court, in assessing damages in respect of pain and suffering caused by the injuries, shall take account of any suffering caused or likely to be caused to him by awareness that his expectation of life has been so reduced."

Bereavement

[C6035] A statutory sum of £12,980 is awardable for bereavement by the *Fatal Accidents Act 1976, s 1A*. This sum is awardable at the suit of husband or wife, or of parents provided the deceased was under eighteen at the date of death if the deceased was legitimate; or of the deceased's mother, if the deceased was illegitimate. (In *Doleman v Deakin* (1990) Times, 30 January it was held that where an injury was sustained before the deceased's eighteenth birthday, but the deceased actually died after his eighteenth birthday, bereavement damages were not recoverable by his parents.)

Fatal injuries

[C6036] Death at work can give rise to two types of action for damages:

- damages in respect of death itself, payable under the *Fatal Accidents Act 1976* (as amended); and
- damages in respect of liability which an employer would have incurred had the employee lived; here the action is said to 'survive' for the benefit of the deceased worker's estate, payable under the *Law Reform (Miscellaneous Provisions) Act 1934*.

Actions of both kinds are, in practice, brought by the deceased's dependants, though actions of the second kind technically survive for the benefit of the deceased's estate. Previously paid state benefits are not deductible from damages for fatal accidents. Nor are insurance moneys payable on death deductible, e.g. life assurance moneys [*Fatal Accident Act 1976, s 4*].

Damages under the Fatal Accidents Act 1976

[C6037] 'If death is caused by any wrongful act, neglect or default which is such as would (if death had not ensued) have entitled the person injured to maintain an action and recover damages, the person who would have been liable if death had not ensued, shall be liable . . . for damages . . . '. [*Fatal Accidents Act 1976 s1*].

Only dependants, which normally means the deceased's widow (or widower) and children and grandchildren can claim – the claim generally being brought by the bereaved spouse on behalf of him/herself and children. [*Fatal Accidents Act 1976 s1(3)*]. The basis of a successful claim is dependency, i.e. the claimant must show that he was, prior to the fatality, being maintained out of the income of the deceased. If, therefore, a widow had lived on her own private moneys prior to her husband's death, the claim will fail, as there is no dependency. The fact that both the deceased and his partner were at the time of the accident on state benefits is irrelevant to the question of loss in assessing damages – *Cox v Hockenhull* [1999] 3 All ER 577. Contributory negligence on the part of the deceased will result in damages on the part of the dependants being reduced. [*Fatal Accidents Act 1976, s 5*].

Survival of actions

[C6038] Actions for injury at work which the deceased worker might have had, had he lived, survive for the benefit of his estate, normally for the benefit of his widow. This is provided for in the *Law Reform (Miscellaneous Provisions) Act 1934*. Any damages paid or payable under one Act are 'set off' when damages are awarded under the other Act, as in practice actions in respect of deceased workers are brought simultaneously under both Acts.

Assessment of damages in fatal injuries cases

[C6039] Damages in respect of a fatal injury are calculated by multiplying the net annual loss (i.e. earnings minus tax, social security contributions and deductions necessary for personal living (i.e. dependency)) by a suitable number of years' purchase. There is no deduction for things used jointly, such as a house or car. However, where a widow also works, this will reduce the dependency and she cannot claim a greater dependency in future on the ground that she and her deceased husband intended to have children.

It is possible to agree that fatal injury damages should be paid in the form of a structured settlement (see C6025 above).

Where both parents are dead as a result of negligence or a mother dies, dependency is assessed on the cost of supplying a nanny (*Watson v Willmott* [1991] 1 QB 140; *Cresswell v Eaton* [1991] 1 WLR 1113).

Provisional awards

[C6040] Because medical prognosis can only estimate the chance of a victim's recovery, whether partial or total, or alternatively, deterioration or death, it is accepted that there is too much chance and uncertainty in the system of lump sum damages paid on a once-and-for-all basis. Serious deterioration denotes clear risk of deterioration beyond the norm that could be expected, ruling out pure speculation (*Willson v Ministry of Defence* [1991] 1 All ER 638). Similarly in the case of dependency awards under the Fatal Accidents Act 1976 it can never be known what the deceased's future would have been, yet courts are expected and called upon to make forecasts as to future income. To meet this problem it is provided that provisional awards may be made,, where there is a chance that at some point in the future the person may develop a particular condition, and they will be allowed to return to court so that further damages may be awarded.

A claim in respect of provisional damages must be included in the statement of claim to entitle the claimant to such damages. The disease or type of deterioration in respect of which any future applications may be made must be stated (the *Civil Procedure Rules 1998 (SI 1998 No 3132), Part 41 rule 2(2)*). The defendant may make a written offer if the statement of claim includes a claim for provisional damages, offering a specified sum on the basis that the claimant's condition will not deteriorate and agreeing to make an award of provisional damages in that sum.

Interim awards

[C6041] In certain limited circumstances a claimant can apply to the court for an interim payment. This enables a claimant to recover part of the compensation to which he is entitled before the trial rather than waiting till the result of the trial is known – which may be some time away. This procedure is provided for in the *Civil Procedure Rules 1998 (SI 1998 No 3132), Part 25*, but it only applies where the defendant is either:

- insured,
- a public authority, or
- a person whose resources are such as to enable him to make the interim payment.

Prior to making an interim payment a judge is under an obligation to take into account any effect the payment may have on whether there is a 'level playing field' for the hearing – see *Campbell v Mylchreest* [1999] PIQR Q17.

Compensation recovery applies to interim payments as well as to payments into court. Care needs to be taken when applying for interim payments to avoid putting the claimant at a disadvantage. Capital of over £16,000, which could include an interim award, removes entitlement to means tested benefits, particularly income support, and there are also reductions on a sliding scale in such benefits for any capital above £6,000. In *Beattie v Department of Social Security* [2001] EWCA Civ 498, [2001] 1 WLR 1404 the Court of Appeal held that payments from structured annuity funds constituted 'income' for the purposes of the *Income Support (General) Regulations 1987 (SI 1987 No 1967)*.

Payments into court (Part 36 payments)

[C6042] A payment into court ('Part 36 payment') may be made in satisfaction of a claim even where liability is disputed. From the defendant's point of view, costs from the date of the Part 36 payment may be saved if the court does not order a higher payment of damages than that paid into court. The claimant may accept a Part 36 payment or Part 36 offer not less than 21 days before the start of the trial without needing the court's permission if he gives the defendant written notice of the acceptance not later than 21 days after the offer or payment was made. If the defendant's Part 36 offer or Part 36 payment is made less than 21 days before the start of the trial, or the claimant does not accept it within the specified period, then the court's permission is only required if liability for costs is not agreed between the parties. [*Civil Procedure Rules 1998 (SI 1998 No 3132), Part 36 rule 11*].

The defendant may, instead of making a Part 36 payment, make a Part 36 offer (formerly known as a Calderbank letter) in which he sets out his terms for settling the action

Interest on damages

[C6043] Damages constitute a judgment debt; such debt carries interest at 8 per cent (currently) up to date of payment. Courts have a discretion to award interest on any damages, total or partial, prior to date of payment (and this irrespective of whether part payment has already been made [*Administration of Justice Act 1982, s 15*]), though this does not apply in the case of damages for loss of earnings, since they are not yet due. Moreover, a claimant is entitled to interest at 2 per cent on damages relating to non-pecuniary losses (except bereavement), even though the actual damages themselves take into account inflation (*Wright v British Railways Board* [1983] 2 AC 773). Under *section 17* of the *Judgments Act 1838* (and *section 35A* of the *Supreme Court Act 1981* and *Schedule 1* to the *Administration of Justice Act 1982*), interest runs from the date of the damages judgment. Thus, where, as sometimes happens, there is a split trial, interest is payable from the date that the damages are quantified or recorded, rather than from the date (earlier) that liability is determined (*Thomas v Bunn, Wilson v Graham, Lea v British Aerospace plc*

[1991] 1 AC 362). Moreover, interest at the recommended rate (of 8 per cent) is recoverable only after damages have been assessed, and not (earlier) when liability has been established (*Lindop v Goodwin Steel Castings Ltd*, (1990) Times, 19 June).

Awards of damages and recovery of state benefits

[C6044] The *Social Security (Recovery of Benefits) Act 1997* and accompanying regulations made important changes to the rules for recoupment of benefit from compensation payments. One key change was that recoupment will not be taken from general damages for pain and suffering and for loss of amenity which it has been accepted should be paid in full. With respect to the other heads of compensation, namely loss of earnings, cost of care and compensation for loss of mobility, they are only to be subject to recoupment from specified benefits relevant to each of these heads of compensation. [*Social Security (Recovery of Benefits) Act 1997, s 8*].

Duties of the compensator

[C6045] Before the compensator makes a compensation payment, he must apply to the Secretary of State for a certificate of recoverable benefits. [*Social Security (Recovery of Benefits) Act 1997, s 4*]. The Secretary of State must send a written acknowledgement of receipt of the application and must supply the certificate within four weeks of receipt of the application. [*Social Security (Recovery of Benefits) Act 1997, s 4*].

He must supply the following information with his application:

- full name and address of the injured person;
- his date of birth or national insurance number, if known;
- date of accident or injury when liability arose (or is alleged to have arisen);
- nature of the accident or disease;
- his payroll number (where known), if the injured person is employed under a contract of service and the period of five years during which benefits can be recouped includes a period prior to 1994 [*Social Security (Recovery of Benefits) Regulations 1997 (SI 1997 No 2205), Reg 5*]; and
- the amount of statutory sick pay paid to the injured person for five years since the date when liability first arose, as well as any statutory sick pay before 1994, if the compensator is also the injured person's employer. The causes of his incapacity for work must also be stated.

The certificate of recoverable benefits will show the benefits which have been paid.

The compensator must pay the sum certified within 14 days of the date following the date of issue of the certificate of recoverable benefits. [*Social*

Security (Recovery of Benefits) Act 1997, s 6]. If the compensator makes a compensation payment without having applied for a certificate, or fails to pay within the prescribed fourteen days, the Secretary of State may issue a demand for payment immediately. A county court execution may be issued against the compensator to recover the sum as though under a court order. It is wise for the compensator to check the benefits required to be set off against the heads of compensation payment so that he is sure that the correct reduced compensation is paid to the injured person. Adjustments of recoupable benefits and the issue of fresh certificates are possible.

Where the compensator makes a reduced compensation payment to the injured person, he must inform him that the payment has been reduced. Statements that compensation has been reduced to nil must be made in a specific form. Once the compensator has paid the Secretary of State the correct compensation recovery amount and has made the statement as required, he is treated as having discharged his liability. [*Social Security (Recovery of Benefits) Act 1997, s 9*].

Complications concerning compensation for work injuries/diseases

Contributory negligence

[C6046] Where damages have been reduced as the result of the claimant's contributory negligence, the reduction of compensation is ignored and recovery of benefits is set-off against the full compensation sum.

Structured settlements

[C6047] The original sum agreed or awarded is subject to compensation recovery and for this purpose, the terms of the structured settlement are ignored and this original sum is treated as a single compensation payment. [*Social Security (Recovery of Benefits) Regulations 1997 (SI 1997 No 2205), Reg 10*].

Complex cases

[C6048] Where a lump sum payment has been made followed by a later lump sum, both payments are subject to recoupment of benefits where those benefits were recoupable. If the compensator has overpaid the Benefits Agency, he can seek a partial refund. [*Social Security (Recovery of Benefits) Regulations 1997 (SI 1997 No 2205), Reg 9*].

Information provisions

[C6049] Under *section 23* of the *Social Security (Recovery of Benefits) Act 1997*, anyone who is liable in respect of any accident, injury or disease must supply the Secretary of State with the following information within 14 days of the receipt of the claim against him:

- full name and address of the injured person;
- his date of birth or national insurance number, if known;
- date of accident or injury when liability arose (or is alleged to have arisen); and
- nature of the accident, or disease.

Where the injured person is employed under a contract of service, his employer should also supply the injured person's payroll number (if known) if the period of five years during which benefits may be recouped includes a period prior to 1994 and this is requested by the Secretary of State. [*Social Security (Recovery of Benefits) Regulations 1997 (SI 1997 No 2205), Regs 3, 5 and 6*].

If the Secretary of State requests prescribed information from the injured person, it must be supplied within 14 days of the date of the request. This information includes details of the name and address of the person accused of the default which led to the accident, injury or disease, the name and address of the maker of any compensation claim and a list of the benefits received from the date of the claim. If statutory sick pay was received by the injured person, the name and address of the employer who has paid statutory sick pay during the five-year period from the date of the claim or before 6 April 1994.

Appeals against certificates of recoverable benefits

[C6050] Appeals against the certificate of recoverable benefits must be in writing to the Compensation Recovery Unit. An appeal may only be made after final settlement of the compensation claim and payment of recoverable benefits and or lump sum payments have been made. It must be made within one month of the date on which the compensator makes the full payment of recoverable benefits and or lump sum payments to the Secretary of State. It is dealt with by an independent tribunal administered by the Appeals Service.

DWP leaflet Z1 – Recovery of benefits and or lump sum payments and NHS charges is available on the DWP website at www.dwp.gov.uk/docs/z1-print-version.pdf.

Treatment of deductible and non-deductible payments from awards of damages

[C6051] There are three well established exceptions to deductibility of financial gains:

- recovery under an insurance policy to which the claimant has contributed all or part of the premiums paid on the policy;
- retirement pensions;
- charitable or ex-gratia payments prompted by sympathy for the claimant's misfortune.

Appeals against certificates [C6053]

Deductible financial gains

[C6052] The courts, applying the principles outlined above, have ruled that the following financial gains are deductible:

- tax rebates where the employee has been absent from work as a result of his injuries (*Hartley v Sandholme Iron Co* [1975] QB 600);
- domestic cost of living expenses (estimated) must be set off against the cost of care (*Lim Poh Choo v Camden and Islington Area Health Authority* [1980] AC 174);
- estimated living expenses must be set off against loss of earnings (*Lim Poh Choo*, above);
- payment from a job release scheme must be set off against loss of earnings (*Crawley v Mercer*, (1984) Times, 9 March);
- statutory sick pay must be set off against loss of earnings (*Palfrey v GLC* [1985] ICR 437);
- sick pay provided under an insurance policy must be set off against loss of earnings not paid as a lump sum (*Hussain v New Taplow Paper Mills* [1988] AC 514);
- health insurance payment under an occupational pension plan paid before retirement must be set off against loss of earnings where no separate premium had been paid by the employee who had paid contributions to the pension scheme (*Page v Sheerness Steel plc* [1996] PIQR Q 26);
- reduced earnings allowance (not a disability benefit) must be set off against loss of earnings (*Flanagan v Watts Blake Bearne & Co plc* [1992] PIQR P 144);
- payments under *section 5* of the *Administration of Justice Act 1982* of any saving to the person who has sustained personal injuries through maintenance at the public expense must be set off against loss of earnings.

Non-deductible financial gains against loss of wages

[C6053]–[C6054] The following financial gains are non-deductible:

- accident insurance payments under a personal insurance policy taken out by the employee (*Bradburn v Great Western Railway Co* (1874) LR 10 Exch 1) (contributory);
- lump-sum wage-related accident insurance payment under a personal accident group policy payable regardless of the fault of the employee (*McCamley v Cammell Laird Shipbuilders* [1990] 1 WLR 963) (benevolent);
- incapacity pension (from contributory insurance scheme) both before and after retirement age (*Longden v British Coal Corporation* [1998] AC 653. It should be noted that even though the incapacity pension was triggered by the accident, benefit flows from the prior contributions paid by the injured party (contributory);
- private retirement pensions (*Parry v Cleaver* [1970] AC 1; *Hewson v Downs* [1970] 1 QB 73; *Smoker v London Fire and Civil Defence Authority* [1991] 2 AC 502) (contributory or benevolent);

- redundancy payment unconnected with the accident or disease (*Mills v Hassal* [1983] ICR 330). Where the claimant was made redundant because he was unfit to take up employment in the same trade, however, the redundancy payment was deductible from damages for lost earnings (*Wilson v National Coal Board* 1981 SLT 67);
- ex-gratia payment by an employer (*Cunningham v Harrison* [1973] QB 942; *Bews v Scottish Hydro-Electric plc* 1992 SLT 749) (benevolent);
- ill health award and higher pension benefits provided by the employer (*Smoker v London Fire and Civil Defence Authority* [1991] 2 AC 502) (benevolent);
- moneys from a benevolent fund, paid through trustees (not directly to the injured person or dependant) in respect of injuries;
- charitable donations (but not where the tortfeasor is the donor).

Non-deductible financial gains against cost of care

- Loss of board and lodging expenses awarded despite their being provided voluntarily by parents where the claimant had formerly paid such expenses herself (*Liffen v Watson* [1940] 1 KB 556).

Benefits and damages for prescribed diseases

[C6055] The claims for benefits are similar to those for personal injury. Claims for damages are also similar, with more latitude for late court applications because of recurrence of, or worsening of, industrial diseases. Lists of prescribed diseases are set out on the DWP website at: www.dwp.gov.uk/publications/specialist-guides/technical-guidance/db1-a-guide-to-industrial-injuries/appendix/appendix-1. The list of diseases is subject to amendment from time to time.

There are special provisions (including time limits) applicable to a number of prescribed diseases:

Osteoarthritis of the knee

To claim for osteoarthritis of the knee, you must have worked for at least 10 years underground in a coal mine. If this work was after 1986, it must have been in certain occupations known to cause osteoarthritis of the knee. From 30 March 2012, you can also claim IIDB for osteoarthritis of the knee if you have worked wholly or mainly fitting or laying carpets or floors (other than concrete floors), for a period of, or periods which amount in aggregate to, 20 years or more.

Primary carcinoma of the lung

From 1 August 2012, coke oven workers have been included in the list of those who can claim IIDB for this disease. You must have worked mainly as a coke oven worker for at least five years in top oven work, or at least 15 years in other oven work. If you worked fewer years on both types of oven work, then the time spent on both can be added together.

Chronic bronchitis and emphysema

In claims for chronic bronchitis and emphysema, it may be necessary to have a breathing test. If the doctor advises that you satisfy this test you will then have a medical examination. If not, your claim will be sent back to the decision maker who will consider whether to disallow it.

You must have worked underground in a coal mine for at least 20 years, or periods up to 40 years if you worked on the surface of a coal mine as a screen worker before 1983, or a mixture of the two such that two years on the surface equates to one year underground.

Pneumoconiosis

In claims for pneumoconiosis, you will normally have an X-ray of your chest. If the X-ray and other evidence shows that you may have the disease you will then have a medical examination. If the X-ray shows no trace of the disease your claim will be sent back to the decision maker who will consider whether to disallow it.

The *Pneumoconiosis, Byssinosis and Miscellaneous Diseases Benefit Scheme, and the Workmen's Compensation (Supplementation) Scheme* dealt with industrial accidents and disease exposure before 1948 and were both abolished on 5 December 2012. Those in receipt of payments were transferred to the main industrial injuries scheme and now receive IIDB.

Occupational deafness

[C6056] If the claimant's deafness was caused by an accident at work, then to qualify for disablement benefit his average hearing loss must have been at least 50 decibels in both ears due to damage of the inner ear. The claimant's disablement must be at least 20 per cent or more for him to qualify for disablement benefit (total deafness being 100 per cent). The claimant must have worked for at least the five years immediately before making his claim, or for at least ten years, in one of the listed jobs set out in the current edition of the leaflet. A claimant who has been refused benefit because the rules were not complied with will have to wait for three more years when he may qualify if he has worked in one of the listed occupations for five years. Benefit rates are published by the DWP each year at www.dwp.gov.uk/docs/dwp035.pdf

Work-related asthma

[C6057] As is well known, there has been an increase in the number of sufferers from asthma. To be able to claim disablement benefit because of asthma, a claimant must have worked for an employer for at least ten years and have been in contact with prescribed substances, and his disability must amount to at least 14 per cent. In the list of substances, item 23 includes 'any other sensitising agent encountered at work', so claims may be possible even if the offending sensitising substance is not yet on the list.

Repetitive strain injuries

[C6058] Repetitive strain injuries are included in the *Social Security (Industrial Injuries) (Prescribed Diseases) Regulations 1985 (SI 1985 No 967)*, Sch 1, as item A4 (see OCCUPATIONAL HEALTH AND DISEASES.) The difficulties in establishing causation can be seen from the case of *Pickford v Imperial Chemical Industries* [1998] 3 All ER 462, where the House of Lords reversed the decision of the Court of Appeal and upheld the trial judge's decision that a secretary was not entitled to damages for repetitive strain injury as she had failed to establish the cause of the injury and that the condition was not reasonably foreseeable. However, in many other cases damages for repetitive strain injury have been awarded – for example, *Ping v Esselte-Letraset Ltd* [1992] PIQR P74 where nine claimants who worked at a printing factory were awarded damages ranging from £3,000 to £8,000. In *Fish v British Tissues* (Sheffield County Court) (29 November 1994, unreported), damages of £57,482 were awarded to a factory packer for repetitive strain injuries to arms, hands and elbows.

A database expanding and updating existing information on important court judgments in repetitive strain injury (RSI) cases, funded by the HSE, has a long term aim of reducing the number of sufferers from this debilitating industrial injury. The work related upper limb disorder (WRULD) database has free access to users who register on www.humanetechnology.co.uk/wruldii/intro.php and provides details of judgments including the factors that courts considered important in reaching their decisions, the degree of care exercised by employers and the amount of damages awarded to claimants.

Claims for IIDB for prescribed diseases must be made on the form on the DWP website at www.dwp.gov.uk/advisers/claimforms/bi100pd_print.pdf.

For some conditions, there are extra sections to complete.

The 2008 Diffuse Mesothelioma Scheme

[C6059] The 2008 Diffuse Mesothelioma Scheme began on 1 October 2008 and provides compensation for sufferers of Diffuse Mesothelioma who have been exposed to asbestos in the UK but are unable to claim compensation from other sources – for example women who had washed their husband's clothes and self-employed people.

Diffuse Mesothelioma cancer is linked to asbestos exposure and most of those who develop the disease have worked in jobs or been in environments where they have been exposed to asbestos dust particles in the air. Where the illness is as a result of occupational exposure to asbestos and the person has been an employed earner he/she can claim under the Industrial Injuries Disablement Benefit (IIDB) Scheme (see above). However there are a number of cases of Diffuse Mesothelioma each year who have no occupational causation, who therefore cannot claim under the IIDB Scheme. The 2008 Diffuse Mesothelioma Scheme seeks to compensate this group of people.

The qualifying criteria for the scheme are:

- there must be evidence to show that the person suffers from Diffuse Mesothelioma;
- the claimant must have been exposed to asbestos in the UK; and
- the claim must have been made within one year of diagnosis or, in the case of a claim being made by a dependent, dependants can claim within a year of the date of death.

Claimants must provide evidence that they suffer from Diffuse Mesothelioma. If the claimant does not pro-vide any evidence at all then the claim will fail.

New support scheme for diffuse mesothelioma from 2014

In July 2012 the government announced a new support scheme for mesothelioma sufferers newly diagnosed with diffuse mesothelioma resulting from negligent exposure at work but unable to trace a liable employer or EL insurer. The scheme will be funded by a levy on current employers' liability insurers at an estimated cost of £25 to £35 million per year. Anyone diagnosed with mesothelioma from 25 July 2012 will be eligible to make a claim and the first payments are due to be made by July 2014.

Recent developments

[C6060] As set out in C6021 above, the *Legal Aid, Sentencing and Punishment of Offenders Act (LASPO) 2012* has introduced several changes to the funding regime for personal injury claims. The Act replaces a system of funding that was largely based on no-win, no-fee (NWNF) arrangements. Under these arrangements, legal representatives of successful claimants could fund the overall costs and risks of NWNF litigation by charging a "success fee" payable by losing employers. These arrangements were introduced after the government withdrew legal aid for personal injury.

Under LASPO, losing employers can no longer be required to pay a success fee, instead the employee will face deductions of up to 25% from their compensation payments.

In addition, the *Enterprise and Regulatory Reform Act 2013*, which gained Royal Assent on 25 April intro-duces further changes. It will repeal the law which makes employers liable if they breach health and safety regulations and mean that the burden of proof will now fall on the injured worker or the family of someone killed, rather than the employer. The TUC has set out that the change removes the right to claim compensation for injuries caused by a criminal breach of workplace health and safety regulations. Compensation will now be restricted to cases where employer negligence is established.

Community Health and Safety

Mike Bateman and Andrea Oates

Introduction to community health and safety

[C6501] Since the *Health and Safety at Work etc Act 1974 (HSWA 1974)* came into operation, every employer has had duties in respect of 'persons not in his employment', in addition to duties to their employees. These duties have come into closer focus through the risk assessment requirements contained in the *Management of Health and Safety at Work Regulations 1999* and other sets of regulations.

The term 'community' embraces a variety of different categories, such as:

- Neighbours and passers-by
- Trespassers on work premises
- Customers and users of facilities and services (including students and school children)
- Visiting individuals and groups
- Volunteers carrying out work

The Grenfell Tower fire has highlighted organisations' duties to tenants and other residents. By September 2017, the police investigation into the disaster, in which at least 72 people were killed, had identified 336 companies and organisations linked to the construction, refurbishment and management of the tower block.

Health and Safety Executive (HSE) statistics show 92 members of the public were killed due to work-related activities in 2016/17.

Meanwhile trade unions and safety campaigners calling for the removal of asbestos from school buildings say nearly 90% of schools still contain asbestos and this presents a serious risk to children as well as staff. They estimate between 200 and 300 adults are dying each year as a result of illnesses they developed after being exposed to asbestos in schools as children.

This chapter considers each of the categories above in some detail – who is at risk, what types of risk they might be exposed to and the types of precautions likely to be necessary to protect them. In addition, it deals with major events that can involve several community categories and may be subject to the scrutiny of Safety Advisory Groups (SAGs). These bodies are usually, but not exclusively, co-ordinated by a local authority (LA) and made up of representatives from the LA, emergency services and other relevant bodies. They meet at regular intervals, or when necessary, to review event applications and advise on public safety. The potential impact on the community from contractors used by businesses or hired in by event organisers is also considered in this chapter.

Relevant legal requirements relating to community health and safety

Health and Safety at Work etc Act 1974

[C6502] Section 3(1) of the *HSWA 1974* places duties on employers in respect of members of the public and other non-employees:

> It shall be the duty of every employer to conduct his undertaking in such a way as to ensure, so far as is reasonably practicable, that persons not in his employment who may be affected thereby are not exposed to risks to their health or safety.

As well as applying to various categories of members of the community (as summarised in C6501), these requirements also apply to others at work, such as contractors and their employees, agency staff, delivery personnel and members of the emergency services. These duties are qualified by the phrase 'so far as is reasonably practicable'. This means the level of risk must be considered together with the costs and practicalities of implementing related control measures.

Paragraph C6503 below contains examples of successful prosecutions under s 3 of the *HSWA 1974* in respect of risk or actual injury to members of the community.

£1 million plus fines following prosecutions under section 3 of HSWA 1974

[C6503] In 2005, a record fine of £15 million was handed down to Transco after it was convicted of a breach of s 3(1) of the *HSWA 1974*. This resulted from a gas explosion in 1999 which killed a family of four in their bungalow in Lanarkshire. The ductile iron gas main involved was badly corroded, arrangements for its inspection and maintenance were inadequate, and records of pipes in the area were inaccurate.

In March 2018, Southern Health NHS Foundation Trust was fined a total of £2 million after pleading guilty to two breaches of *s 3(1)* of the *HSWA 1974*. A series of management failings at the Trust had led to the deaths of two vulnerable patients. The HSE prosecution followed the deaths of 45-year-old Teresa Colvin at a Southampton Mental Health Hospital and 18-year-old Connor Sparrowhawk at a specialist unit in Oxford. Southern Health NHS Foundation Trust managed both centres. Oxford Crown Court heard both HSE investigations found failures to control risks and in planning.

Teresa was found slumped and unconscious at a telephone kiosk at Woodhaven Adult Mental Health Hospital in Southampton in April 2012. She died a short time later following treatment. During the HSE investigation it became clear that the Trust failed to act on the findings of assessments that it could better control the risks associated with the use of phones with cords. There had been a history of patients using phone cords as a ligature. Connor died in July 2013 after suffering an epileptic seizure in the bath at the Trust's specialist unit, Slade House in Oxford. The HSE investigation found despite his vulnerability and previous suspected seizures, he was allowed to use the bath alone with checks from staff taking place every 15 minutes.

In April 2017, Nottinghamshire County Council was fined £1 million after pleading guilty to breaches of ss 2(1) and 3(1) of the *HSWA 1974*. A disabled member of the public had been struck by a vehicle used for collecting branches in a country park. Nottingham Crown Court heard employees were working in Rufford County Park in June 2015, collecting branches and transporting them, using a tractor mounted grab attachment. At the same time a disabled man was on a guided walk in the park. The worker using the tractor could not see the member of public ahead and collided with him, causing severe bruising and injuries to his arms, legs and head.

An HSE investigation found the council had failed to implement a safe system of work by failing to segregate vehicle movements from the public. It also failed to train the workers to the required level to operate the mounted grab and act as banksman. The machine was not suitable for transporting materials long distances and the council also failed to supervise and adequately plan the work sufficiently in a public place. As a result, it put its own employees and members of the public at risk.

In September 2016, the owner of Alton Towers, Merlin Attractions Operations Ltd, was fined £5 million after pleading guilty to breaching s 3(1) of the *HSWA 1974* following a rollercoaster collision. Two young women on the Smiler ride who suffered leg amputations were among 16 people injured when their carriage collided with a stationary carriage on the same track in June 2015. Stafford Crown Court heard engineers overrode the ride's control system without the knowledge and understanding to ensure it was safe to do so. The HSE reported its investigation had found the root cause was a lack of detailed, robust arrangements for making safety critical decisions. It said the whole system, from training through to fixing faults, was not strong enough to stop a series of errors by staff when working with people on the ride.

In the same month, Network Rail was fined £4 million after pleading guilty to breaching s 3 of the *HSWA 1974* after pedestrian Olive McFarland was struck by a train and killed while using a pedestrian level crossing near Needham Market, Suffolk, in 2011. Ipswich Crown Court heard an investigation by the Office of Rail and Road (ORR) found Network Rail had failed to act on substantial evidence that pedestrians had poor visibility of trains when approaching the footpath crossing and were exposed to an increased risk of being struck by a train.

Corporate manslaughter convictions

[C6504] The *Corporate Manslaughter and Corporate Homicide Act 2007* created a new offence of corporate manslaughter. An organisation is guilty of the offence if the way in which it organises or manages its activities causes death and this amounts to a gross breach of a relevant duty of care it owed to the victim (see CORPORATE MANSLAUGHTER).

Several convictions under the 2007 Act have followed deaths of members of the public.

In November 2013, Princes Sporting Club was charged with corporate manslaughter and fined around £135,000 following the death of 11-year-old Mari-Simon Cronje. She died during a birthday celebration at the club in Bedfont, Middlesex on 11 September 2010. She died after she fell from an inflatable banana boat ride and was hit by the boat that had been towing it.

In December 2015, Cheshire Gates and Automation Ltd was handed down a £50,000 fine and given a Publicity Order after pleading guilty to the offence following the death of six-year-old Semelia Campbell. She died after becoming trapped in a faulty electric gate near her home in Manchester. The Crown Prosecution Service (CPS) said the company had left the gate without a proper limit on the force it exerted and it was unable to detect and pull back from any obstacles it encountered.

In February 2016, care home company Sherwood Rise Ltd was fined £300,000 following the death of 86-year-old resident Ivy Atkin, the first conviction against a care home under the Act. Mrs Atkin, who suffered from dementia, was found dehydrated and malnourished at Autumn Grange in Nottingham in 2012.

In June 2016, Monavon Construction Limited was fined £500,000 (£250,000 for offence) following the deaths of pedestrians Gavin Brewer and Stuart Meads. They died in October 2013 after falling through hoardings installed around renovation works to a flat. Edge protection to the lightwell of the basement flat was inadequate.

The police investigation into the Grenfell Tower fire is considering criminal offences in relation to organisations and individuals, including corporate manslaughter.

The Management Regulations

[C6505] Regulation 3(1) of the *Management of Health and Safety at Work Regulations 1999 (MHSWR) (SI 1999 No 3242)* states that:

> 'Every employer shall make a suitable and sufficient assessment of . . . the risks to the health and safety of persons not in his employment arising out of or in connection with the conduct by him of his undertaking.'

Risk assessments must therefore take account of relevant members of the community, the risks that they might be exposed to and the precautions necessary to control those risks to the extent required by the law – including the general requirements imposed the *HSWA 1974*.

SI 1999 No 3242, Reg 5 also requires employers to take effective steps to implement the necessary precautions identified through the risk assessment. It states:

> 'Every employer shall make and give effect to such arrangements as are appropriate, having regard to the nature of his activities and the size of his undertaking, for the effective planning, organising, control, monitoring and review of the preventive and protective measures.'

Appropriate and effective management techniques must be used to ensure the protection of members of the community, as well as employees and others who are at work. Detailed guidance is provided in the chapter entitled MANAGING HEALTH AND SAFETY.

Employers with five or more employees must record the significant findings of their risk assessments and their arrangements for implementing the related preventive and protective measures.

Other Specific Requirements of Regulations

[C6506] Several sets of regulations contain requirements relating to the protection of members of the wider community and other non-employees. These include:

- *Control of Substances Hazardous to Health Regulations 2002 (COSHH) (SI 2002 No 2677)*
  Regulation 3(1) states that:

  'Where a duty is placed by these Regulations on an employer in respect of his employees, he shall, so far as is reasonably practicable, be under a like duty in respect of any other person, whether at work or not, who may be affected by the work carried out by the employer . . .'

  While there are some exceptions to this requirement, in relation to health surveillance, monitoring, information and training and dealing with accidents, the main requirements of the *COSHH Regulations* therefore apply in respect of members of the community. COSHH assessments must take account of risks to the community, particularly the ways in which substances are used or stored.

- *Control of Major Accident Hazards Regulations 2015 (COMAH 2015) (SI 2015 No 483)*
  COMAH 2015 applies to workplaces where specified quantities of high-risk substances are stored, used or may be generated – through a chemical process for example. Employers must consider safety, health and environmental risks both inside and outside their own premises. The regulations contain requirements for both internal and external emergency plans and the provision of information to the public.

- *Construction (Design and Management) Regulations 2015 (CDM 2015) (SI 2015 No 51)*
  Under CDM 2015, duty holders are required to: manage the risks by applying the general principles of prevention; appoint the right people and organisations at the right time; make sure everyone has the information, instruction, training and supervision they need to carry out their jobs in a way that secures health and safety (of the wider community as well as workers); cooperate and communicate with each other and coordinate their work; and consult and engage with workers to promote and develop effective measures to secure health, safety and welfare.
  The Health and Safety Executive (HSE) guidance on these regulations makes clear that those with duties under the regulations, including clients, designers and contractors, must take into account the risks to

people other than workers, such as members of the public, who may be affected by the construction work. They must take measures to prevent or control the risks identified. Occupants of, and visitors to, premises where construction work is taking place must be considered, as must occupants of neighbouring property and passers-by (see the *Monavon Construction Ltd* case at **C6504** above for example).

- *Work at Height Regulations 2005 (SI 2005 No 735)*
  The regulations place duties on employers to prevent injuries caused by a fall from height and from falling objects. Members of the community must be protected as well as those at work. The duties include preventing materials and objects from falling, by using toe boards, brick guards, tool belts and securing stored items for example, and providing measures to prevent people being struck by falling items, using screens or nets for example, and ensuring materials and objects are not thrown or tipped from height in circumstances liable to cause injury.

The *Reporting of Injuries, Diseases and Dangerous Occurrences Regulations 2013 (RIDDOR) (SI 2013 No 1471)* also require those in control of premises, generally the employer, to report certain incidents, illnesses and dangerous events that occur there to the relevant safety enforcement authority. This is usually the HSE or local authority.

They must report all work-related fatalities resulting from work-related accidents (*Regulation 6*).

'Where any person not at work, as a result of a work-related accident, suffers—

(a) an injury, and that person is taken from the site of the accident to a hospital for treatment in respect of that injury; or

(b) a specified injury on hospital premises,

the responsible person must follow the reporting procedure . . . ' (*Regulation 5*).

Specified injuries include fractures, amputations, permanent loss of or reduction of sight, crush injuries leading to internal organ damage, serious burns, scalpings, and unconsciousness by head injury or asphyxia. Gas-related injuries are also reportable (for more information on RIDDOR see ACCIDENT REPORTING AND INVESTIGATION chapter).

The Regulatory Reform (Fire Safety) Order 2005 (SI 2005 No 1541)

[C6507] Since 1 October 2006, fire safety within workplaces and many other types of premises has been covered by the *Regulatory Reform (Fire Safety) Order 2005*. The Order places duties on the 'responsible person' as defined within the Order. For most workplaces this will be the employer while in other types of premises it will be the person who has control of the premises, or the owner.

Under Article 8 of the Order the responsible person must:

(a) take such general precautions as will ensure, so far as is reasonably practicable, the safety of any of his employees; and

(b) in relation to relevant persons who are not his employees, take such general fire precautions as may reasonably be required in the circumstances of the case to ensure that the premises are safe.

'Relevant persons' includes any person who is, or may be, lawfully on the premises and any person in the immediate vicinity who is at risk from a fire — this would include most of the members of the community referred to in C6501.

The Order is enforced by the fire and rescue authority for the area in most workplaces, although, for example, the Office for Nuclear Regulation (ONR) enforces fire safety in nuclear installations and the HSE enforces fire safety on ships and construction sites. The requirements under the Order are explained in more detail in FIRE PREVENTION AND CONTROL. A series of Fire Safety Guides provide more information concerning locations where the fire safety of members of the community is a major issue, such as premises providing sleeping accommodation, residential care, places of assembly, theatres and cinemas, educational premises, healthcare premises, transport premises and facilities, and open-air events (see C6528 for further information). A supplementary guide covers means of escape for disabled people.

This legislation is under scrutiny following the Grenfell Tower fire. The government appointed former HSE chair Dame Judith Hackitt to lead a review into building regulations and fire safety. The interim report was published in December 2017. The review found a lack of clarity around key roles and responsibilities. It found that 'responsible persons' under the Order are frequently not identified when the building is due to be handed over following construction. As a result, people are not aware of their responsibilities and often assume they are for someone else to do. There is also a confusing overlap during occupation between the Order and the *Housing Act 2004*.

The report also says fire and rescue service personnel may raise concerns about compliance with the Order which are not acted upon because of cost, because the building work is too far advanced to make changes or because their advice is ignored.

It also found fire and rescue services moved from a directive role in certain buildings to one of auditing following the introduction of the Order 2005. This abolished fire certificates with expectations placed firmly on 'responsible persons' to manage the risk in their buildings by completing a fire risk assessment.

The report can be found at: www.gov.uk/government/publications/independent-review-of-building-regulations-and-fire-safety-interim-report. The final report is due to be published in spring 2018.

Occupiers' Liability Acts 1957 and 1984

[C6508] Under the *Occupiers' Liability Act 1957*, occupiers must exercise a duty of care in respect of lawful visitors to their property. It requires them 'to take such care as in all the circumstances of the case is reasonable to see that the visitor will be reasonably safe in using the premises for the purposes for which he is invited or permitted by the occupier to be there'.

The occupier's duty of care was extended by the *Occupiers' Liability Act 1984* to also include people other than visitors (trespassers), in respect of any injury suffered on the premises, either because of any danger due to the state of the premises or things done, or omitted to be done, in the following circumstances:

(1) the occupier is aware of the danger or had reasonable grounds to believe that danger exists;
(2) they know or have reasonable grounds to believe that a trespasser (a person other than a visitor) is in the vicinity of the danger concerned, or may come into the vicinity of the danger; and
(3) the risk is one against which, in all the circumstances of the case, they may reasonably be expected to offer some protection.

The 1984 Act also states that 'no duty is owed . . . to any person in respect of risks willingly accepted by that person'.

Any case brought under this Act would be decided on its merits, but there are two important factors to consider:

- The Standard of Care Required
  A greater duty of care exists in relation to major risks (such as falls from significant height, presence of hazardous substances or deep water) than minor risks (such as uneven land or shallow water). The practicality of taking and maintaining precautions would also have to be taken into account. It may not be reasonable to expect to the fencing off of a large relatively low-risk workplace or site in its entirety. However, it may be reasonable to fence off individual high-risk areas within the site.
- The Type of Trespasser
  Case law has already established that a greater duty of care exists in respect of child trespassers, although parents are expected to assume the prime responsibility for controlling the whereabouts of very young children. Occupiers must also be aware of the presence of items on sites that may attract children, such as derelict machinery and opportunities to climb.
  The duty of care owed to adults will also vary. It will be much greater to those simply walking through a site, especially straying accidentally from a public footpath, than to those indulging in illicit activities such as theft, vandalism or fly-tipping. However, even these latter categories cannot be completely ignored, particularly if the occupier is already aware that significant risk exists.

In December 2015, gas distributor National Grid was fined £2 million after pleading guilty to breaching *s 3(1)* of the *HSWA 1974* in relation to the death of schoolboy Robbie Williamson. Eleven-year-old Robbie died while he and two friends were crossing the Leeds and Liverpool Canal, using a pipeline running on the outside of a canal bridge. He fell and died as a result of drowning and a head injury. Preston Crown Court heard the company had a procedure for inspecting this type of above-ground pipe-crossing and requirements for providing measures to prevent access on to these structures. (For more detailed information see Occupiers' Liability chapter.) However, its records incorrectly showed the pipe was buried within the bridge rather than exposed on the outside. As a result, the crossing had not been subject to inspection and had no access prevention measures fitted.

Adventure Activities Licensing Regulations 2004

Current arrangements require that providers of activities for under-18s in four areas – caving, climbing, trekking and water sports – have a licence issued by Adventure Activities Licensing Authority (AALA) before they can operate. The HSE is currently designated as the AALA. The licensing regime operates under the *Activity Centres (Young Persons' Safety) Act 1995* and the Adventure Activities Licensing Regulations 2004.

The AALA scheme is currently under review. Following an informal consultation, the HSE published a public discussion document in January 2018 setting out three options:

- retain the Adventure Activities Licensing Regulations (AALR) and current statutory scheme underpinned by the *HSWA 1974* and increase the fees;
- retain the AALR and current statutory scheme underpinned by the *HSWA 1974*, and increase fees and extend the activities covered; or
- remove the AALR and move to an industry-led, not-for-profit accreditation scheme, and underpinned by the HSWA 1974.

The discussion document, CD286 – Review of the Adventure Activities Licensing Authority (AALA), can be found at: www.hse.gov.uk/consult/condocs/cd286.htm. The consultation period ended in March 2018.

Licensing currently applies in England, Scotland and Wales but is a reserved matter in Scotland and Wales.

Neighbours and Passers-by

Who Must Be Taken Into Account?

[C6509] There are a number of categories of neighbours and passers-by who must be taken into account when carrying out risk assessments. Some of these will be at work, while others will be members of the wider community. They include:

- People within the workplace or site
- Those visiting or working within other parts of the premises, such as:
  — other businesses;
  — public buildings;
  — community facilities; and
  — voluntary groups.
- External Neighbours
  Occupants of property close to the workplace and those visiting them must be taken into account. In addition to the types of locations referred to above, this is also likely to include residential property.
- Passers-by

Pedestrians, drivers and users of recreational areas must be considered – whether or not the roads, footpaths or areas that they are using are designated for public access.

Risks to be considered

[C6510] Neighbours and passers-by may be affected by risks even if they remain outside the workplace but more so if they are able to stray into work areas. The types of risks most likely to be of relevance are:

- Vehicle traffic
  Vehicles are likely to be delivering materials to or collecting them from most workplaces. Heavy items of mobile plant may also be in circulation, particularly on construction sites.
- Lifting operations
  Lifting operations may involve objects being lifted over occupied premises or over areas where pedestrians or vehicles circulate.
- Work at height
  Work at height will inevitably involve a potential risk to neighbours and passers-by from falling objects. There is also a risk of unauthorised persons using access equipment to reach high levels from which they might fall.
- Hazardous and dangerous substances
  Substances may create risks to neighbours and passers-by through emissions or more substantial leakages. Fires and explosions within workplaces can also affect external areas. The explosion and subsequent fire at the Buncefield fuel storage depot in December 2005 illustrated the extent of the impact of such an event. Vapour from thousands of gallons of petrol ignited causing an explosion measuring 2.4 on the Richter scale. It was Britain's most costly industrial disaster and in July 2010, five companies were ordered to pay a total of £9.5 million in fines and costs.
  Waste substances or materials constitute a risk, particularly if left in unsecured areas.
  Outbreaks of Legionnaires' disease have resulted from inadequate cleaning and disinfectant procedures for water systems within workplaces. Two outbreaks in Edinburgh and Stoke-on-Trent in 2012 caused five deaths and left more than 120 members of the public ill. An investigation was unable to identify the source of the bacteria in the Edinburgh outbreak, but hot tub sales company JTF Wholesale pleaded guilty to an offence under the *HSWA 1974* and was fined £1 million after admitting responsibility for the deaths of two members of the public following the Stoke-on-Trent outbreak. The source was a hot spa pool on display at its premises.
- Work equipment
  Neighbours and passers-by may be at risk from contact with dangerous parts of equipment being used in areas to which they have access. They may also be hit by particles thrown from materials being worked on. Any tool, powered equipment or vehicle may be dangerous if it falls into the hands of unauthorised persons, particularly children. The

Provision and Use of Work Equipment (PUWER) Regulations 1998 contain a specific requirement that 'self-propelled work equipment' must have 'facilities for preventing it being started by an unauthorised person'.

- Dangerous activities
  There are a variety of activities, which might present risks to neighbours or passers-by, even those who are some distance away, such as pressure testing, the use of radioactive substances and the use of explosives. Potentially dangerous quarrying, agricultural and forestry activities are often carried out on land accessible to members of the public.
- Electrical equipment
  Uninsulated electrical cables and other unprotected electrical equipment are obvious risks to any unauthorised person who is able to gain access to them.

Precautions

[C6511] The precautions necessary will depend on the types of risks identified by the risk assessment, as described in **C6510** above.

For construction activities this will often be within the context of a CDM 'construction phase plan'. HSE advice on protecting the public from the risks of construction activities can be found on its website at: www.hse.gov.uk/construction/safetytopics/publicprotection.htm.

- Security of the workplace
  The level of risk may justify the whole workplace or parts of it being made secure against access by neighbours or passers-by. This may involve the use of fencing or other barriers, with locked access gates or doors. Periodic checks will ensure these measures are still in good condition and proving effective. In some cases, the presence of those carrying out the work (or a designated sentry) will be adequate to secure the work area against unauthorised access by neighbours or passers-by.
- Traffic management
  Controlling the routes by which traffic enters and leaves work premises may be important in reducing the risks to neighbours and passers-by. It may be possible to avoid high-risk locations, such as shopping areas and school entrances. The direction of approach may have to be specified in order to leave an adequate turning circle for larger vehicles. Controlling the timing of vehicle traffic can avoid times at which there will be large numbers of pedestrians or other vehicles in the vicinity. There are detailed requirements governing work being carried out on the public highway itself and it is important drivers are properly trained and sufficiently fit and healthy to drive safely and not put themselves and others at risk.
- Other traffic precautions
  Other precautions may be necessary to ensure the safe circulation of traffic where work activities are taking place, both on public and private roads. In dark conditions vehicles must have appropriate lights.

Adequate lighting must be provided for activities such as loading or unloading vehicles taking place on roads. Arrangements to clean roads of mud and other work-related debris may be necessary.
- Planning of lifting operations
Timing may also be important in minimising the possible presence of pedestrians and vehicles in areas where lifting operations must be carried out. Loads will often need to be lifted over buildings at times when they are unoccupied or when the removal of occupants causes minimum inconvenience. Measures may still be necessary to prevent any pedestrian or vehicle traffic which may be present from getting into risk areas.
- Storage locations and facilities
Materials and equipment should be stored in locations where the risks they present to neighbours and passers-by are minimised. Where non-workers have regular access to workplaces, the security of storage facilities will be important. Facilities storing hazardous or dangerous substances must meet relevant standards for the protection of both workers and members of the community.
- Planning of dangerous activities
As with traffic movements and lifting operations, the timing of other potentially dangerous work activities should be planned to minimise the risks to others. Precautions to protect those carrying out the work should also protect neighbours and passers-by, but further measures may be necessary to keep them away from such activities.
- Emergency procedures
Emergency procedures must take account of neighbours and passers-by and should consider:
  — how they will be notified of the emergency – through telephone messages, alarm sounds and loudspeaker announcements for example;
  — how they should respond – by closing all windows or evacuating to designated locations for example.
- Worker awareness
It is essential that workers are made fully aware of how their workplace or work activities may create risks for others. In particular they must be made aware of precautions involving them, such as designated traffic routes, permitted times for access or other activities, the importance of securing work areas, equipment or materials and their role in emergency procedures.
- Signs and notices
Signs and notices can play an important role in making workers aware of the need for precautions and in reinforcing their awareness. Neighbours and passers-by can also be made aware of risks and precautions, particularly the need to keep clear of areas where potentially dangerous activities are taking place.

The HSE points out that vulnerable groups such as the elderly, children and people with certain disabilities may need special attention.

Trespassers on Work Premises

Who must be taken into account?

[C6512] Employers have both statutory duties towards trespassers through the *HSWA 1974, s 3* and also a civil 'duty of care' (see C6508). These duties are much greater where the trespass is foreseeable, for example:

— pedestrians straying from a right of way;
— people recovering lost property (such as footballs or golf balls); and
— people recovering straying animals or children.

Potential trespass by children is a particular concern because of their lack of awareness of risks and their inability to read or understand warning notices. Children are also generally more inquisitive and adventurous than adults.

Deliberate trespass by adults, for theft, vandalism or other illicit purposes, presents particular difficulties. However, the actions of employers to protect their assets will often also protect this type of trespasser from any risks present.

Risks to be considered

[C6513] A number of the risks to neighbours and passers-by mentioned in C6510 may also affect trespassers. The following issues are of particular relevance in considering the need to protect trespassers:

- Unsafe access
  — derelict buildings or structures (unsafe floors, staircases etc);
  — fragile surfaces;
  — unstable ground or materials (eg waste tips); and
  — falls from significant height (shafts, cliffs, scaffolds, towers, gantries, etc).
- Water
  — deep water (including waste lagoons); and
  — fast-flowing water (both natural watercourses and artificial ones eg hydroelectric schemes).
- Railway lines
  — moving trains; and
  — power lines and live rails.
- Dangerous equipment or materials
  — abandoned equipment;
  — insecure items and those capable of toppling over; and
  — equipment capable of being used by unauthorised persons.
- Hazardous or dangerous substances
  — insecurely stored substances;
  — abandoned substances;
  — accessible waste products; and
  — toxic or asphyxiant atmospheres eg in sewers, pipes, conduits or other confined spaces.

Precautions

[C6514] Many of the precautions described in C6511 to protect neighbours and passers-by may also be appropriate in protecting trespassers.

- Fencing and other security measures
  The standard of measures such as fencing and locks, needed to prevent access by determined trespassers will generally need to be higher than those aimed at preventing more casual entry into work premises. Standards must relate to the level of risk inside the workplace. Fencing and other enclosures may need to be reinforced by periodic security patrols. They must also be inspected regularly to ensure they remain in a satisfactory condition.
- Signs and notices
  The use of signs and notices can be particularly important in making potential trespassers aware of risks within workplaces, particularly those which may not be readily apparent, such as fragile surfaces, deep water and high-voltage electrical equipment. The condition of signs and notices should be subject to regular inspection.
- Secure storage
  Potentially dangerous materials and equipment should be stored so they are not accessible to trespassers. The exact nature of the storage will depend on the items stored, the storage location and the anticipated level of approach by trespassers. Storage facilities on an unattended site accessible to the public will need to be much more robust and secure than in an occupied workplace where any trespass is likely to be less frequent and less prolonged.
- Emergency equipment
  Lifelines or lifebelts may be necessary to safeguard trespassers, as well as workers and any legitimate passers-by, against risks from water. These, too, should be subject to periodic inspection.

Customers and Users of Facilities and Services

Who must be taken into account?

[C6515] The need to protect people other than workers who may be present within workplaces was considered earlier in the chapter (see C6509). However, many workplaces involve direct contact by members of the community with work premises, work activities and people who are at work. Such members of the community can be described both as customers and as users of facilities and services.

- Customers
  Many businesses have customers who are members of the community:
  — shops, restaurants and other commercial premises;
  — indoor and outdoor markets;
  — transport undertakings (including railways, buses and airlines);

- utility suppliers (including companies supplying gas, electricity, water, sewage and telecommunications);
- the entertainment industry (including spectator sports and funfairs); and
- those providing domestic installation and repair services (gas equipment, electrical work, construction work, gardening and similar activities).
• Facility and service users
  This includes members of the community who use facilities or services such as:
  - hotels and similar accommodation;
  - hospitals and care homes;
  - social housing;
  - domiciliary care services;
  - gyms, sports centres and other sports facilities;
  - interactive attractions (including museums and galleries);
  - educational establishments (schools, colleges, universities); and
  - technical training services.
• Many other examples could be added to both of the above categories. (For brevity, all such persons will be referred to as 'customers' in the remainder of this chapter.)

Risks to be considered

[C6516] Many risks to 'customers' are very similar to those to neighbours and passers-by – sections C6510 and C6511 deal with such risks and the related precautions. However, some risks relate particularly to the 'customers' themselves – who they are, why they are present in the workplace and what they are doing there. Consideration must be given to the following:

• Access within the premises
  Ways in which 'customers' will access the premises must be taken into account:
  - how many people will be present?
  - what routes will they follow?
  - are those access routes safe?
  - what other risks might be present on access routes?
• Equipment
  What equipment will 'customers' be expected (or have the opportunity) to use and what other equipment (used by workpeople) might they have close contact with? Some situations may involve the use of temporary electrical installations and electrical generators which could represent a risk to 'customers'.
• Dangerous and hazardous substances
  Many portable electrical generators will run on petrol and will need additional supplies to be kept nearby. Temporary food outlets (eg in markets or at entertainment locations) will frequently use LPG cylinders for food preparation.
• Activities

What activities will 'customers' be involved in directly or come into close contact with? Some sporting and recreational activities (such as skiing, climbing or pole vaulting) involve an inevitable degree of risk. Activities may also involve risks from hazardous or dangerous substances.

- Accidents and emergencies
  In some workplaces, such as those in the healthcare sector, there are specific obligations placed on employers to make first-aid provision for their 'customers', and in others there is often a general expectation that they will do so. Fire and other emergencies can affect 'customers' as well as those at work.
- Risks from products
  Risks to 'customers' from food products is a specialist subject in its own right. There may be risks to 'customers' from other materials, substances or equipment supplied to them by a business or simply present on display. This will be of particular importance in plant hire businesses and other establishments offering potentially high-risk products. Gas and electricity present the greatest risks in the utilities sector. The gas supply industry is closely regulated.
- Capabilities of the 'customer'
  The capabilities of the 'customer' are important in considering the risks associated with all of the above. These may relate to their:
  — physical capabilities ('customers' may be elderly, or have disabilities or limited physical strength);
  — intellectual capabilities (these may be limited due to, for example, disability or the age of 'customers');
  — technical capabilities (some activities may require a particular level of skill, knowledge or experience, strength or intellectual ability); and
  — physical size (children in particular may be able to gain access to dangerous equipment or live electrical equipment through small apertures or may fall underneath guard rails).

Precautions

[C6517] The necessary precautions will relate to the types of 'customers' present and the foreseeable risks they will be exposed to in the workplace. Precautions are likely to include:

- Maintenance of safe access
  Access routes must be maintained in a safe condition for the benefit of staff as well as 'customers'. Particular consideration must be given to the numbers and types of 'customers' likely to be present. Special arrangements may be necessary for the elderly or those with disabilities (the partially sighted for example).
- Selection and maintenance of equipment
  Where equipment is expected to be used by 'customers', it must be suitable for the purpose. It may need to be simpler to use and more robust than equipment intended for use by workers. Arrangements must be in place for the maintenance of equipment, and some types will

require formal training of intended users (see below). Where equipment or utilities are supplied into customers' own premises these must be of a suitable construction and design, and arrangements are likely to be necessary for periodic inspection, testing or maintenance.
- Location and storage
Portable generators and liquid petroleum gas (LPG) powered equipment (as used at markets or entertainment venues) must be sited in locations where there is adequate ventilation and they are secure from interference by 'customers' and others. Additional petrol must be kept in suitable containers in safe locations, again secure from interference. Similarly, spare LPG cylinders must be stored safely and separately from petrol supplies.
- Assessment of 'customers'
In some situations, it will be necessary to assess 'customers' in order to establish their physical, intellectual or technical capabilities. This is particularly important where they are likely to be involved in potentially risky activities (sporting or other outdoor pursuits for example) or using dangerous equipment. Such assessment should establish the level of need for supervision and training and any special individual arrangements. In some cases, 'customers' may eventually be judged unsuitable to participate in the activity in question.
- Information and training
'Customers' may need to be made aware of risks and related precautions through the use of signs and notices or via customer information packs, in hotels and similar accommodation for example. Some situations may justify the use of formal briefings of 'customers', particularly where they are likely to be involved in higher risk activities or where there are significant risks present in the workplace. Certain activities or types of equipment will require 'customers' to receive formal training to ensure an adequate level of safety.
- Supervision and control
An appropriate number of properly trained and informed staff must be available to ensure 'customers' are adequately supervised. In some situations, in the retail sector for example, this supervision will be largely passive. Sporting and other outdoor activities will require more proactive forms of supervision, as will entertainment and other public events, through the presence of stewards. Those carrying out supervisory or stewarding duties must monitor the behaviour of their 'customers' and, in particular, restrict them from entering into higher risk areas.
- Monitoring
As well as monitoring 'customer' behaviour, the ongoing effectiveness of all arrangements for ensuring customer safety must be monitored. In particular, areas accessed by 'customers' and equipment used by them require periodic formal inspections to ensure they are still in satisfactory condition. Those in control of locations such as markets, where various small traders are present, will need to ensure those traders continue to maintain adequate health and safety standards.
- Emergency arrangements

In formulating emergency arrangements, account must be taken of 'customers'. Fire evacuation is an obvious case, but other foreseeable emergencies, such as chemical leaks or public disorder, may also need to be considered. 'Customers' must be informed about relevant emergency arrangements (see above regarding information and training). Staff must be fully briefed on their role in respect of 'customers'. This is particularly important where large numbers are present, at spectator events, in large retail outlets and in the educational and health care sectors for example, and where 'customers' with disabilities may be present. It may be necessary to identify 'customers' with disabilities so that appropriate action can be taken to protect them in an emergency — some disabilities, like deafness for example, may not be readily apparent.

Visiting Individuals and Groups

Who must be taken into account?

[C6518] Most of the immediately preceding sections (C6515–C6517) dealing with 'customers' and users of facilities and services is also of relevance to visitors. However, while dealing with 'customers' is an integral part of many work activities, other workplaces are less accessible to the community and may only occasionally have to deal with visitors. Among such visitors are likely to be:

- Schoolchildren and other students
  Individuals may visit as part of work experience and similar programmes. By virtue of the Health and Safety (Training for Employment) Regulations 1990, schoolchildren on formal work experience programmes have the status of employees in respect of health and safety legislation.
  HSE advice can be found online at: www.hse.gov.uk/youngpeople/workexperience/index.htm.
- Student and community groups
  Some workplaces are willing to accept groups of visitors from schools and other educational establishments. Similarly, special interest groups or members of the local community may make visits.
- Informal individual visits
  Individual friends and relations of workers may make visits. While employers will be aware of many such visits, others can pose more cause for concern, such as where a child is sent with a message for someone working outside normal working hours.

Risks to be considered

[C6519] Many of the risks referred to in C6516 in respect of 'customers' are also of relevance to visiting individuals and groups. Particular attention must be given to:

- Areas to be visited
  In order to gain benefit from their visit, it is likely that formal visitors will need to visit areas of the workplace where risks may be present. A balance must be struck between the intended benefits of the visit and not exposing visitors to an unacceptable degree of risk.
  If informal visitors are not allowed to get beyond a reception area, a canteen or a security lodge, there will be little risk. However, granting informal visitors free access to working areas can introduce considerable risks to safety and also give rise to security issues.
- Activities to be carried out
  Work experience students and similar longer-term visitors are likely to be involved in carrying out actual work or projects related to work activities. A risk assessment is likely to be necessary in each case to identify the risks involved in the work or project and the precautions which must be taken, taking into account the capabilities of the individual (see below).
- Capabilities
  The physical, intellectual and technical capabilities of potential visitors are an important consideration. A visiting group of engineers or safety specialists can be expected to be much more risk aware than a group of schoolchildren. The same applies to individual visitors, whether visiting formally or informally. This is where the potential for children to be making unauthorised visits to workplaces should be of particular concern. Any relevant physical disabilities, related to mobility, vision and hearing for example, should be identified in advance.

Precautions

[C6520] Inevitably the precautions necessary to protect visitors must relate to the areas they are visiting and the activities they will be involved in.

- Restricted areas and routes
  Consideration must be given to which areas individual visitors are allowed to visit, particularly those to which they are to be granted unaccompanied access. In planning routes for visiting groups, risks on or close to the proposed route must be taken into account. The composition of the group – their physical capabilities, awareness of risk, degree of responsibility, the size of the group and the abilities of those in charge to supervise it effectively must also be considered. In some situations, it may be necessary to provide additional signage or barriers to ensure the group keeps to its intended route.
- Restricted activities
  Where individuals are making extended visits, it must be made quite clear both to the visitor and to those providing immediate supervision, what the visitor is and is not permitted to do. These restrictions will relate to the capabilities of the individual and may be progressively removed as the individual acquires more knowledge and experience.
- Supervision

The level of supervision necessary for a group will need to relate to its size and type, as well as the areas it will visit. Support may be necessary from teachers or others connected with the visitors in order to ensure effective supervision. Someone must be made responsible for the overall supervision of long-term individual visitors, who may also need to be accompanied when visiting high-risk areas or carrying out high-risk activities. Short-term visitors are likely to need to be accompanied at all times, except when visiting low risk areas. This will be particularly the case for any children.

- Information and instruction

Visiting groups (and individuals) may need to be provided with information prior to their visit about such matters as:
— the level of supervision they are expected to provide;
— clothing and footwear to be worn; and
— any special health and safety issues for individuals with specific disabilities such as lack of mobility, deafness, visual impairment and asthma.

At the start of the visit further instruction and information will be necessary on topics such as:
— risks the visitors may be exposed to;
— behaviour during the visit;
— restricted areas, routes and activities;
— obeying instructions from visit guides and supervisors;
— emergency arrangements;
— use of Personal Protective Equipment (PPE) (see below); and
— what to do if they have questions or concerns.

During visits further emphasis may need to be given to the risks present or precautions necessary in individual areas. This might involve additional briefings from visit guides and/or the use of signs and notices.

- Personal Protective Equipment (PPE)

Some types of visitors are likely to be able to bring any PPE required for the visit, providing they are advised about this in advance. However, in other cases PPE will need to be made available by the organisation hosting the visit. This is relatively easy for individual visitors – long-term visitors will often, in effect, be treated as employees.

Provision of PPE for visiting groups can pose practical problems, such as hygiene, availability of suitable sizes, need for adjustment and training in use. Some types of PPE, such as safety helmets, high visibility clothing or disposable earplugs, are fairly simple to deal with. However, the issue of safety footwear or respiratory protection to visitors is much more problematic, and areas necessitating their use may need to be avoided. In some situations, it may be acceptable for visitors to be instructed to wear sturdy leather footwear for their visit, prohibiting the wearing of trainers, high heels or sandals for example.

Volunteers

Who must be taken into account?

[C6521] Volunteers carry out many types of work activities that might be carried out by paid staff in other locations. They are exposed to similar risks to workers but their capabilities may not be the same. Volunteers may have considerable knowledge and experience from previous work they have carried out but this may not always be relevant to their voluntary work. Older volunteers may also have diminishing physical (and even mental) capabilities. There is often a fine line between ensuring the health and safety of the volunteer and putting them off carrying out the voluntary work completely.

There are many activities in which volunteers are involved:

- Regular volunteers
  Volunteers might work regularly in sectors such as:
  — charity shops and other retail outlets;
  — care work, including home visits;
  — providing transport for clients or meals on wheels;
  — drop-in centres;
  — coaching or officiating at sporting activities;
  — acting as guides at museums and historical properties;
  — maintaining equipment such as vintage vehicles and steam locomotives; and
  — carrying out environmental work in country parks or on trails.
- 'One-off' volunteers
  Many people also volunteer to work on a one-off basis at major events, particularly those attracting large numbers, such as:
  — sporting or musical events;
  — carnivals; and
  — fundraising events.
  Their duties on such occasions often put them in the front line of dealing with other members of the community in roles such as:
  — traffic control;
  — car parking;
  — entry control;
  — staffing registration or enquiry desks;
  — crowd supervision; and
  — providing first-aid cover.

Risks to be considered

[C6522] Given the wide range of activities volunteers can become involved in, they are likely to be exposed to the same variety of risks as those who are at work – as described elsewhere in this publication. This section identifies some of the more common risks that must be taken into account during the risk assessment, including some which can be overlooked.

- Travel

Some volunteers need to travel to carry out their work. Apart from the normal risks associated with road traffic, there may be particular risks in the areas they are travelling through or visiting. Such risks will be greater during hours of darkness, particularly for staff who must travel late at night.
- Entry into clients' homes
Many people providing voluntary care services must enter clients' homes. There may be risks to their personal safety from the clients themselves, their relatives and friends and also from pets. Risks may be due to difficult access, due to household clutter, defective equipment, or dangerous electrical or gas installations.
- Other personal safety risks
Volunteers will also be prone to personal attack where they are expected to restrict entry to events, handle cash or deal with problems involving the public. Risks will be much higher at certain types of event, with alcohol consumption by members of the public often being a major factor.
- The work environment
In some places where volunteers work it is possible that no one will have been given specific responsibility for cleaning, tidying up or maintaining the premises. As a result, buildings themselves or electrical or gas installations may have deteriorated, possibly to a dangerous condition. Safe storage of equipment and the maintenance of safe access routes within premises will also merit particular attention.
- Equipment
Voluntary organisations often receive donations of equipment that is no longer required elsewhere. Arrangements must be put in place to screen out any potentially dangerous equipment and there must be suitable systems to ensure all equipment is properly maintained and regularly inspected. Some types of equipment, such as powered machinery, may be dangerous if not used correctly, while other types, such as storage or display racking and trestle tables, may pose risks if incorrectly assembled. Equipment may also be dangerous if it is left accessible for use by unauthorised persons.
- Manual handling
Some volunteers may be required to carry out significant amounts of manual handling of goods, materials and equipment. The normal criteria for carrying out manual handling assessments for workers should be applied. Account must be taken of the task, the load, the working environment and the capabilities of the individual, which is of particular relevance to many volunteers. For further information see the MANUAL HANDLING chapter.
- Fire and other emergencies
Volunteers must be capable of dealing with foreseeable emergencies both in respect of their own health and safety and that of clients and other members of the community they are working with. Volunteers may be responsible for assisting disabled clients in leaving buildings or shepherding members of the public away from fire and similar emergencies.

Precautions

[C6523] Some related precautions have already been identified in relation to the risks referred to in C6522 above, while others are detailed in other sections of this publication. Precautions of particular relevance to volunteers are as follows:

- Assessment
  Volunteers should be assessed in order to identify their existing knowledge and experience in relevant areas. Their physical and intellectual capabilities should also be assessed.
- Information and training
  It is likely that volunteers will need to be briefed on the health and safety aspects of their intended work and to be trained in the use of certain equipment or the carrying out of some activities. This needs to be done with sensitivity. While volunteers will accept they might require a certain level of information and training, they may react badly to spending time on subjects they regard as unnecessary or irrelevant. Sessions should be made as enjoyable as possible for the participants, with those who already possess relevant knowledge and experience encouraged to share this with others.
- Restrictions
  Some activities, or the use of certain types of equipment, may need to be restricted to volunteers who have undergone relevant training. In some cases, volunteers may need to be accompanied by a competent paid member of staff.
- Communication
  Effective means of communication need to be established for volunteers working within the community, particularly where they are remote from colleagues, or at major events. The use of radios, mobile phones and attack alarms may be necessary, depending on the situation. Peripatetic volunteers, such as those making home visits, should leave an intended itinerary back at base. There should be arrangements for them to log off on completing their duties and for following up any volunteers who fail to log off or do not keep in touch.
- Supervision and support
  As with paid workers, even very capable volunteers need to have someone they can communicate with, particularly when problems arise. Support should be readily available to volunteers who need it, especially in emergencies. This may be activated by a radio, telephone message or by sounding an alarm.
- Monitoring
  The working practices of volunteers should be monitored as those of paid workers are. However, a more diplomatic approach will need to be taken where corrective actions are necessary to modify the way volunteers behave. There should be regular inspections of equipment and premises used by volunteers and appropriate arrangements made for repair and maintenance work.
  The HSE provides advice and guidance on managing the health and safety of volunteers on its website at: www.hse.gov.uk/voluntary/index.htm.

Major Public Events

[C6524] Major events can create potential health and safety risks for many members of the communities already referred to in the chapter:

— participants in the event ('customers');
— volunteer workers (stewards and performers); and
— neighbours and passers-by.

Not only must event organisers take these members of the community into account, they must also consider contractors involved in the staging of the event and their potential impact on the community and on each other.

Many major events are now subject to Safety Advisory Groups (SAGs) (see C6501 above), at which the event is subject to scrutiny by relevant local authorities and the various emergency services. SAGs frequently ask organisers to prepare a properly documented risk assessment for their event so a full event-safety management plan can be developed.

HSE advice on event safety

[C6525] Public events can include music events, often requiring public entertainment licences, sporting events, carnivals, fun fairs, agricultural and county shows, firework displays, air shows, car rallies and religious gatherings.

The HSE event safety web pages (www.hse.gov.uk/event-safety/index.htm) have information on running events safely, with links to further information produced by other organisations and associations.

Risk Assessment

[C6526] The HSE website provides a list of health and safety topics to help event organisers with risk assessments. Each topic covers relevant safety planning, management and monitoring issues that may arise. These are:

- venue and site design;
- temporary demountable structures;
- managing crowd's safely;
- using barriers at events;
- transport;
- falls from height;
- electrical safety;
- fire safety;
- special effects;
- noise;
- amusements and attractions;
- employee welfare; and
- handling waste.

Of course, some of these topics will not be applicable to all events, while some events will require risk topics to be broken down in greater detail. For example, a running or cycling race taking place on public roads will need to take account of the direction of arrival and parking of competitors' vehicles as well as the interface between competitors and other road users during the event itself. This in turn will require consideration of the race route, barriers, signage, possible road closures, race marshals and the possible need for direct police assistance.

Use of Contractors

[C6527] Event organisers and others engaging contractors must consider the potential risks for members of the community posed by contractors' equipment and activities. Contractors themselves have many duties under the *HSWA 1974* and associated legislation, but it has been established for many years that those engaging contractors also have legal responsibilities for risks that the contractors may create, both for those at work and others who may be affected by their activities. Even where contractors already have good health and safety standards, they may not be fully aware of the risks associated with the environment they are working in and the members of the community who may be present there.

Advance checks on contractors' health and safety standards should include topics such as:

— insurance (particularly public liability);
— health and safety policy;
— risk assessments and method statements;
— competence to provide the services required;
— the competence and training of staff; and
— availability of relevant licences, certificates and inspection records.

Once contractors have been engaged, they must be informed about the precautions they will be expected to take in their work. Precautions of particular relevance in ensuring members of the community are adequately protected include:

— access routes and permitted vehicle access times;
— arrangements for parking;
— maintenance of safe pedestrian access;
— segregation of work areas;
— adequate control of work equipment;
— safe storage and use of hazardous and dangerous substances; and
— timing and control of high risk activities.

Their attention must be drawn to relevant characteristics of members of the community who may be present in or around the workplace. This will be of particular importance in relation to customers and other users of buildings and facilities. Special precautions will be necessary in many cases to ensure the safety of children, the elderly and those with disabilities.

In respect of major public events, contractors must be informed about relevant sections of the event safety management plan. In addition to the matters referred to above, they will need to be told about:

— the layout of the event site;

- availability of key services and facilities;
- communications, including with those managing the event and stewards; and
- emergency arrangements.

The performance of contractors should be monitored both at public events and in other work they carry out. Attention should be paid to their compliance with the health and safety standards set out in their own risk assessments and method statements, as well as to matters of importance in relation to the event or workplace concerned.

Further information on community health and safety

[C6528] The HSE website (www.hse.gov.uk) has sections dealing with school trips, societal risk and various other topics of relevance to community health and safety.

It also has an event safety micro-site at www.hse.gov.uk/event-safety.

The Fire Safety Guides referred to in C6507 can be found online at: www.gov.uk/government/collections/fire-safety-law-and-guidance-documents-for-business.

Confined Spaces

Angus Withington

Introduction to confined spaces

[C7001] Several people are killed or seriously injured each year as a result of work in confined spaces. These occur in a wide variety of situations and across a range of industries and commonly involve asphyxiation. Recent examples include:

- an employee who was fatally overcome by carbon monoxide and hydrogen sulphide whilst attempting the rescue of another employee similarly overcome in a hopper at a meat rendering plant (*R v John Pointon & Sons* (2008) 2 Cr App R (S) 82).
- two workers who were asphyxiated in a pit surrounding a hot isostatic press, which had filled with argon gas as a result of a leak (*R v HIP (Bodycote)* [2010] EWCA Crim 802, [2011] 1 Cr App Rep (S) 38).
- two workers who suffocated in the hold of a barge at a fish farm, where it was believed that rusting metal in a confined space had removed oxygen from the air.
- a fatal accident suffered when an employee was buried under 12 tonnes of limestone dust, as he was attempting to remove compacted material from the inside of a hopper (*R v Hanson Quarry Products Europe Ltd and another* (2011), unreported, Taunton Crown Court;
- a homeowner who died, having been overcome by carbon monoxide in a domestic wood pellet boiler storage tank; and
- three crew members who died on a general cargo vessel. Two of the crew members died in the attempted rescue of the first, who had collapsed as a likely result of oxygen depletion caused by the timber cargo. The vessel had no rescue plan or appropriate rescue equipment in place (2014).

The primary legislative control regulating the risks associated with such work is the *Confined Spaces Regulations 1997 (SI 1997 No 1713)* ("the Regulations"). The particular hazards which the Regulations seek to control arise as a result of the combination of substances or conditions with the confined nature of the place of work, which give rise to particular risks to health and safety. The most common hazards, which must be controlled are:

- flammable substances and oxygen enrichment
- toxic gases, fumes or vapours
- oxygen deficiency
- the ingress or presence of liquids
- free flowing solids or materials (eg. grain, flour, coal dust etc.)

- excessive heat (particularly if personal protective equipment is also required to be worn).

However, other risks (eg the difficulty of access, both generally and in the event of emergencies) also need to be taken into account in relation to confined space work – see **C7007** below.

A definition of the term 'confined space' is provided within the Regulations (see below). Confined spaces are enclosures with limited openings such as:

- storage tanks and silos
- reaction vessels
- sewers and enclosed drains
- pipes and ductwork
- rooms lacking in ventilation
- parts of structures under construction.

Relevant legal requirements relating to work in confined spaces

[C7002] The *Regulations* (see below) contain specific legal requirements relating to work in confined spaces. They provide a definition of 'confined space' and also define several 'specified risks', which must be addressed in the context of the regulations.

The *Management of Health and Safety at Work Regulations 1999 (SI 1999 No 3242)* require risk assessments to be carried out and effective risk management arrangements to be implemented in respect of all types of work activity. In the case of work in confined spaces, risk assessments must take account not just of 'specified risks' as defined in the regulations but also of other risks which may be present. Common examples of such risks are set out in **C7008** below, which also refers to several specific sets of regulations likely to be relevant to confined space work.

Given the broad definition of 'construction work' contained in the *Construction (Design and Management) Regulations 2015 (SI 2015 No 51)*, these regulations are also of considerable significance in relation to work in confined spaces. The CDM regulations place particular importance on ensuring that persons appointed or engaged in respect of construction work are competent for the purpose. Where work will involve entry into confined spaces, those who plan and manage such work must be competent as well as those actually carrying out the work. By reg 12(2) and Sch 3 of the CDM Regulations, the construction phase plan must include specific measures concerning works involving particular risks, some of which have obvious relevance for work in confined spaces, eg work which puts workers at risk from chemical or biological substances constituting a particular danger to the health or safety of workers.

Confined Spaces Regulations 1997

[C7003] *Regulation 1 of the Confined Spaces Regulations 1997 (SI 1997 No 1713) contains several definitions of which the most important are as follows:*

'confined space'	means any place, including any chamber, tank, vat, silo, pit, trench, pipe, sewer, flue, well or other similar space in which, by virtue of its enclosed nature, there arises a reasonably foreseeable specified risk;
'free flowing solid'	means any substance consisting of solid particles and which is of, or is capable of being in, a flowing or running consistency, and includes flour, grain, sugar, sand or other similar material;
'specified risk'	means a risk of (a) serious injury to any person at work arising from a fire or explosion; (b) without prejudice to paragraph (a)– 　(i) the loss of consciousness of any person at work arising from an increase in body temperature; 　(ii) the loss of consciousness or asphyxiation of any person at work arising from gas, fume, vapour or the lack of oxygen; (c) the drowning of any person at work arising from an increase in the level of liquid; or (d) the asphyxiation of any person at work arising from a free flowing solid or the inability to reach a respirable environment due to entrapment by a free flowing solid;
'system of work'	includes the provision of suitable equipment which is in good working order.

The definition of 'confined space' contains a list of examples of what might constitute a confined space.

The key words in the definition are in the latter part:

> . . . in which, by virtue of its enclosed nature, there arises a reasonably foreseeable specified risk.

It should be noted therefore that it is not necessary for the space to be entirely enclosed. The Regulations can therefore apply to open-topped tanks and inadequately ventilated silos, for example. The HSE Approved Code of Practice (ACOP) confirms that the identification of a confined space may not always be easy and that it should not be assumed that a confined space will necessarily be small or difficult to work in or to access. Grain silos, for example, can be very large. A confined space may also be a place where people regularly work, such as a car repair centre where the activities include spray painting.

Particular care should be taken where a location may become a confined space for the purposes of the Regulations if there is a change in the conditions inside the space or in the degree of confinement. For example, a roof space in a building would not come within the definition in the normal course of events. However, should significant quantities of a volatile and highly flammable adhesive or paint be used in the roof space, it would become a confined space under the Regulations. Similarly the Regulations would be likely to apply if work was to be carried out there in extremely hot conditions.

The list of 'specified risks' can be paraphrased as meaning a risk of:

- serious injury from fire or explosion;
- loss of consciousness from increase in body temperature;
- loss of consciousness or asphyxiation from gas, fume, vapour or lack of oxygen;
- drowning from an increase in liquid level; or
- asphyxiation or entrapment by a free flowing solid.

Paragraph 16 of the ACOP also highlights where a confined space is deliberately created, such as where oxygen levels are reduced so as to diminish the risk of fire (for example, in archives) or to delay oxidation for the purposes of assisting in the preservation of fresh food. The HSE has published specific guidance on Reduced Oxygen environments (Hypoxic Environments) on its website.

Therefore, as confirmed by paragraph 12 of the ACOP, a 'confined space' must therefore have both of the following defining features:

(a) It must be a space which is substantially (though not always entirely) enclosed; and
(b) One or more of the specified risks must be present or reasonably foreseeable.

A flow chart has been included at paragraph 19 of the ACOP to assist relevant decision makers to determine whether or not the Regulations apply to a space.

The Regulations do not apply to the crew of sea-going ships, mines or diving projects (regulation 2), as these are covered by separate legislation. The Regulations will, however, apply to work undertaken on a ship involving both its crew and shoreside workers (see paragraph 39 of ACOP).

Persons with duties under the Regulations

[C7004] *Regulation 3* of the Confined Spaces Regulations 1997 (SI 1997 No 1713) places duties on both employers and the self-employed.

Employers have duties to ensure compliance with the Regulations in respect of any work carried out by:

- their employees; and
- other persons (so far as is reasonably practicable), in relation to matters within their (the employers') control.

Self-Employed Persons have duties in respect of any work by:

- themselves; and
- other persons (so far as is reasonably practicable), in relation to matters within their (the self-employed persons') control.

This partially duplicates the position established by the *Health and Safety at Work etc. Act 1974 (HSWA 1974)*. In particular it places duties on employers who may bring in specialist contractors to work in confined spaces. Contractors and the self-employed also have duties towards each other.

Preventing the need for entry

[C7005] *Regulation 4 (1)* of the *Confined Spaces Regulations 1997 (SI 1997 No 1713)* imposes the primary obligation, namely a preference for avoiding the risk entirely. It states:

> No person at work shall enter a confined space to carry out work for any purpose unless it is not reasonably practicable to achieve that purpose without any such entry

There are many ways in which entry into confined spaces can be avoided. These should be identified as part of a risk assessment which should determine what is reasonably practicable. Examples include:

- testing atmospheres within confined spaces using long probes;
- cleaning or removing residues from outside by water jetting, steam or use of chemicals;
- integrating in-situ cleaning systems into the design of equipment (this also has hygiene and product quality advantages);
- clearing blockages using long tools or remotely-operated flails, vibrating devices or blasts of compressed air;
- providing sightglasses, openings etc. to see what is happening within vessels (internal glasses may need to be fitted with washers and/or wipers); and
- the use of CCTV cameras (additional internal lighting may be necessary).

It may also be reasonably practicable to remove sections of structures or vessels so that they no longer constitute a confined space for the duration of the work.

Designers, manufacturers, importers and suppliers of articles for use at work have duties under *section 6* of the *Health and Safety at Work etc. Act 1974 (HSWA 1974)* to ensure, so far as is reasonably practicable, that the article will be safe when being used, cleaned or maintained. This will include an obligation to ensure that plant and equipment containing confined spaces are properly designed, made or used to eliminate or minimise the need to enter such spaces. Even where it is not reasonably practicable to eliminate the need for entry, the design must still facilitate the establishment of safe systems of work and the provision of emergency arrangements.

Work in confined spaces

[C7006] Where it is not reasonably practicable to prevent entry, *Regulation 4(2)* of the *Confined Spaces Regulations 1997 (SI 1997 No 1713)* states that:

> Without prejudice to paragraph (1) above, so far as is reasonably practicable, no person at work shall enter or carry out work in or (other than as a result of an

emergency) leave a confined space otherwise than in accordance with a system of work which, in relation to any relevant specified risks, renders that work safe and without risks to health

Such safe systems of work are, of course, developed as a result of carrying out a risk assessment. Types of risks to be considered and the precautions likely to be necessary to control those risks are dealt with later in the chapter.

Emergency arrangements

[C7007] *Regulation 5* of the *Confined Spaces Regulations 1997 (SI 1997 No 1713)* sets down requirements for emergency arrangements:

(1) Without prejudice to regulation 4 of these Regulations, no person at work shall enter or carry out work in a confined space unless there have been prepared in respect of that confined space suitable and sufficient arrangements for the rescue of persons in the event of an emergency, whether or not arising out of a specified risk.

(2) Without prejudice to the generality of paragraph (1) above, the arrangements referred to in that paragraph shall not be suitable and sufficient unless–
    (a) they reduce, so far as is reasonably practicable, the risks to the health and safety of any person required to put the arrangements for rescue into operation; and
    (b) they require, where the need for resuscitation of any person is a likely consequence of a relevant specified risk, the provision and maintenance of such equipment as is necessary to enable resuscitation procedures to be carried out.

(3) Whenever there arises any circumstance to which the arrangements referred to in paragraph (1) above relate, these arrangements, or the relevant part or parts of those arrangements, shall immediately be put into operation.

The importance of this regulation cannot be under-stated as many fatalities unfortunately occur in the course of attempted rescue. It is estimated that approximately 60% of fatalities occurring in confined spaces are rescuers reacting to an emergency situation.

Several parts of this regulation require further emphasis:

- emergency arrangements must be in place for *any* type of emergency, not just those arising from specified risks – eg. a serious fall within the confined space;
- risks to rescuers must be reduced, so far as is reasonably practicable;
- resuscitation equipment may need to be provided and maintained (a worker who has been asphyxiated is likely to need to be resuscitated quickly (within a few minutes and therefore well before the emergency services can get there); and
- emergency arrangements must be immediately put into operation should an emergency arise.

Practical guidance on emergency arrangements is provided later in the chapter at **C7028** below.

Other risks in confined spaces

[C7008] Many other risks may be present when working in confined spaces. Some of these risks may be greater because of the enclosed nature of the space, whilst rescue of accident victims is likely to be much more difficult. Such risks include:

— restricted openings for entry into and exit from the space;
— restricted access within the space;
— potential for falls from height or into water;
— slips, trips and falls at the same level;
— falling objects;
— lack of visibility within the space;
— mechanical or electrical equipment within the space;
— chemical, biological or dust hazards;
— excessive noise (damaging hearing or interfering with communication);
— excessive cold; and
— the remote location of the confined space.

Even though these other risks are not covered by the *Confined Spaces Regulations 1997 (SI 1997 No 1713)*, other specific regulations will apply eg the *Workplace (Health, Safety and Welfare) Regulations 1992 (SI 1992 No 3004)*, the *Work at Height Regulations 2005 (SI 2005 No 735)*, the *Electricity at Work Regulations 1989 (SI 1989 No 635)*, the *Control of Noise at Work Regulations 2005 (SI 2005 No 1643)*, the *Provision and Use of Work Equipment Regulations 1998 (PUWER) (SI 1998 No 2306)* and the *Control of Substances Hazardous to Health Regulations 2002 (COSHH) (SI 2002 No 2677)*. The *Personal Protective Equipment at Work Regulations 1992 (SI 1992 No 2966*, as amended) will also be relevant in relation to PPE used in confined space work.

The general requirements of *Health and Safety at Work etc. Act 1974* will also apply (*HSWA 1974, s2 (2)(a)* contains a similar requirement for a safe system of work) and the *Management of Health and Safety at Work Regulations 1999 (SI 1999 No 3242)* still require a risk assessment to be carried out.

Figure 1 at **C7010** below provides a checklist to aid in the identification of both 'specified risks' and other risks.

Safe systems of work

[C7009] Both the specific requirements of the *Confined Spaces Regulations 1997 (SI 1997 No 1713)* and the general requirements of *Health and Safety at Work etc. Act 1974* mean that work in confined spaces must be carried out in accordance with a safe system of work. The ACOP makes a clear statement on how safe systems of work should be specified. It states (at paragraph 80):

> To be effective a safe system of work needs to be in writing. A safe system of work sets out the work to be done and the precautions to be taken. When written down it is a formal record that all foreseeable hazards and risks have been considered in

advance. The safe procedure consists of all appropriate precautions taken in the correct sequence. In practice a safe system of work will only ever be as good as its implementation.

Dependent upon circumstances it may be more appropriate to specify the safe system of work through a written model procedure (eg where a similar task is carried out on a regular basis) or through a permit to work (more appropriate for variable or infrequent tasks).

Permits to work

[C7010] The ACOP recommends the use of a 'permit to work' procedure as an extension to the general duty of providing a safe system of work where there is a reasonably foreseeable risk of serious injury for an employee entering or working in a confined space. Such a procedure supports the implementation of a safe system by providing the means of recording the results of the risk assessment process and the key elements of the safe system of work, as well as collating information which may be needed during an emergency. A correctly applied permit to work system should ensure that:

- people at work in the confined space are aware of the hazards involved;
- workers know the nature and extent of the work to be carried out;
- appropriate elements of a safe system of work are identified;
- there is a formal check that a safe system of work is in place before people enter or work in the confined space;
- suitable action is taken about other activities which may affect work in the confined space;
- any time limits on entry are specified;
- appropriate means of communication are established; and
- necessary respiratory protective equipment (RPE) and personal protective equipment (PPE) standards are identified.

Figure 2 (below) provides an example of a permit to work form for use in controlling entry into confined spaces. It identifies both the 'specified risks' defined in the Regulations and a range of other risks. However, those intending to use this form should ensure that it addresses fully the risks likely to be present in the confined spaces that they are responsible for – adaptation of the form may well be necessary to deal adequately with local circumstances.

Figure 3 (below) is intended as an aide-memoire to accompany the permit to work form provided in Figure 2. It provides the permit issuer with a list of the precautions which may be necessary to control the risks involved in the work.

Safe systems of work [C7010]

Figure 1: Confined Spaces – Identification of Risks

DEPARTMENT

TASK/LOCATION

SPECIFIED RISKS (Confined Spaces Regulations 1997) significant risk ? possible risk

	or ?	Notes
Serious injury from fire or explosion		
Loss of consciousness – increase in body temperature		
Loss of consciousness or asphyxiation due to: — gas, fume or vapour — lack of oxygen		
Drowning – increase in level of liquid		
Asphyxiation or entrapment – free flowing solid		
OTHER SIGNIFICANT RISKS		
Restricted entry/exit		
Restricted access within		
Falls from height/into water		
Slips, trips, falls – same level		
Falling objects		
Lack of visibility		
Mechanical equipment		
Electrical		
Chemical		
Biological		
Dust		
Noise		
Excessive cold		
Remote location		
Other (state)		

Should access be controlled by a permit to work?   **YES/NO**
Is a written task procedure required?   **YES/NO**

Signature _____   Name _____   Date _____

C70-9

[C7010] Confined Spaces

Figure 2: Permit to work – Entry into Confined Spaces

Confined Space _____   Purpose of Entry _____
Date   Can the work be carried out from outside? **YES/NO**

RISKS IDENTIFIED	√ or X	CONTROLS REQUIRED (inc. Tests) (if necessary, provide details on accompanying documents)
Fire or Explosion Increase in Body Temperature		
Gas, Fume, Vapour		
Lack of Oxygen		
Drowning		
Free-flowing Solids		
Restricted Entry/Exit/Access		
Falls, Slips, Trips		
Falling Objects		
Lack of Visibility		
Mechanical / Electrical		
Chemical / Biological / Dust		
Noise		
Excessive Cold		
Remote Location		
Other (provide details)		

DETAILS

Communication Methods

Emergency Equipment & Arrangements

Person(s) covered Name(s) by this PTW

Signature(s)

I confirm that the precautions detailed above have been/will be taken.
I authorise the persons named above to enter this confined space for the purpose stated.
They have been instructed on their roles and responsibilities.

PERMIT VALID   PERMIT CANCELLED AT
(Times) FROM   (Time)
TO   (Date)
Signature   Signature
(Permit Issuer)   Authorised Person

Safe systems of work [C7010]

FIGURE 3: CONFINED SPACES – CONTROL MEASURES AND OTHER PRECAUTIONS

RISKS	POSSIBLE CONTROLS
Fire or Explosion	Cleaning of residues. Atmospheric testing. On-going monitoring.
	Purging. Ventilation. Protected electrical equipment.
	Hot work restriction. No smoking. Fire fighting equipment.
Increase in Body Temperature	Delay before entry. Working time restrictions. Cooling fans.
Gas, Fume, Vapour	} Cleaning of residues. Atmospheric testing. Ongoing monitoring.
Lack of Oxygen (inc. presence of heavy gases nearby)	} Purging. Ventilation (natural or forced). Isolate incoming feeds.
	} Respiratory protection (inc. air-fed respirators or self-contained BA).
Drowning	} Isolate incoming feeds. Drain contents.
Free-flowing Solids	} Harness attached to secure lifeline.
Restricted Entry/Exit	Staff selection. Harness attached to secure lifeline. Instructions to staff.
Falls from height/into water	Harness attached to secure lifeline and inertia reel.
	Temporary scaffolding/platform.
Slips, trips, falls – same level	Clean up surfaces. Suitable footwear.
Falling Objects	Safety helmets. Protective screens.
Lack of visibility	Permanent or temporary lighting. Torches. BACK-UP ESSENTIAL.
Mechanical/Electrical	Secure isolation.
Chemical/Biological/Dust	Drain contents. Clean up. PPE eg gloves, body cover, eye protection.
	RPE – air-fed, self-contained BA, cartridge respirator, dust mask.
	Decontamination.
Noise	Hearing protection. Suitable communications (see below).
Excessive Cold	Suitable clothing and footwear.
Remote Location	Suitable communications (see below). Monitoring from base.
KEY PRECAUTIONS	**OPTIONS AVAILABLE TO CONSIDER**
Communications	Attendant outside. Word of mouth. Telephone. Portable phone.
	Radio. Air horn. Whistle.
	(Ensure that the equipment works in the work location.)

[C7010] Confined Spaces

RISKS	POSSIBLE CONTROLS
Emergency Arrangements (For all types of emergencies including falls, heart attacks etc.)	Attendant outside. Harness attached to secure lifeline. Recovery equipment eg winch (in position or available nearby). RPE/PPE worn by or readily available for rescuers. Resuscitation equipment. Stretcher. Other first aid equipment. Fire fighting equipment. Contact with emergency services/internal assistance. Medical staff on standby (high-risk situations only). Register of persons inside (large groups only). (Ensure relevant staff are trained in use of emergency equipment.)

An essential ingredient of any permit to work system is that those issuing and cancelling the permit are competent for the purpose. Competence and training are dealt with at **C7012** below.

A permit to work system is unlikely to be needed where:

- assessed risks are low and can be controlled easily;
- the system of work is very simple; and
- other work activities cannot affect safe working in the confined space.

However, that decision should also be made by a competent person, taking account of the findings of the risk assessment and the overriding obligation to ensure that a safe system of work is adopted.

Task procedures

[C7011] An alternative way of ensuring that a safe system of work is specified in writing may be to incorporate it into a formal 'task procedure' for carrying out the work. (The terms Job Safe Procedure' (JSP) or Standard Practice Instruction (SPI) are used by some to describe similar documents.) Such formal procedures are usually suitable where certain tasks are to be performed regularly within confined spaces. As with permits to work, they should result from a risk assessment of the task in question, and identify the risks which may be present and the various precautions which must be taken to control those risks to an adequate degree.

However, task procedures do not impose the same degree of formal checking (ie that the safe system of work is actually being implemented) as that provided by the permit to work system. They will usually be more suitable for providing a point of reference for a well-trained in-house work team who are already familiar with the environment, the work involved and the risks it entails, as well as the task procedure itself. A permit to work is more likely to be necessary for controlling outsiders (such as contractors), infrequently performed tasks and/or more hazardous work.

Competence and training

[C7012] Competence for work in confined spaces will be acquired through a combination of adequate training and relevant experience. Consideration should also be given to workers' physical and psychological suitability for work in confined spaces – they may need to be fit enough to wear breathing apparatus and conditions such as claustrophobia may be an issue. Training should include:

- an awareness of the Regulations and their requirements;
- an appreciation of the risks involved in work in confined spaces;
- the systems of work appropriate for different types of work;
- how to use relevant types of equipment eg for communications, access, testing, PPE; and
- emergency and rescue arrangements and equipment

An increasing number of organisations can provide general training on work in confined spaces. However, this will usually need to be augmented by additional training and/or experience to ensure that workers are:

- familiar with the confined spaces they will work in and the risks involved;
- aware of the precautions necessary in their types of work;
- capable of using the equipment required to ensure a safe system of work; and
- capable of using emergency equipment.

Where members of staff are required to issue and cancel permits to work for confined spaces they will need further training, including an appreciation of the purpose of permits to work. They must be familiar both with the range of risks which may be present and the wide variety of precautions which may be necessary to ensure a safe system of work eg isolations, ventilation, gas testing, RPE, PPE etc. There should be a formal system in place for their training, testing and authorisation (as should be the case for all permit to work systems).

Supervision

[C7013] All workers must be supervised effectively and this is particularly important for those working in confined spaces. Supervision will involve:

- identifying when work is to be carried out in confined spaces;
- arranging for necessary permits to work;
- ensuring that workers are aware of the risks involved and the precautions required;
- ensuring that they are capable of implementing the precautions and have all the equipment required;
- monitoring working practices (including the correct operation of the permit to work); and
- dealing with problems which may arise during the work.

For higher risk tasks, the supervisor is likely to remain present whilst the work is carried out. In other cases it may be acceptable to visit periodically, although the supervisor should still be readily available if required.

Precautions for work in confined spaces

[C7014] Figure 3 (see C7010 above) shows the range of precautions that may be necessary when working in confined spaces. Further detail is provided below on the types of precautions required to control 'specified risks'. Specific guidance is available in the ACOP (see in particular paragraphs 79-137). Other risks which may be present in confined spaces can be controlled using conventional techniques as set out in Figure 3.

Communications

[C7015] Effective communications must be provided between those inside a confined space, between those inside and those outside, and also in order to summon help in an emergency. A variety of means can be utilised, for example, speech, rope tugs, telephone, radio, so as to ensure that rapid and unambiguous communication can be achieved. Consideration should be given to risks of using communication equipment in flammable or explosive atmospheres, to the difficulties of communicating when wearing respiratory protection equipment (RPE) and to the practical difficulties of using telephones or radios in some situations. Air horns or whistles may provide a practical alternative for some environments. Communication equipment and methods should always be checked and tested at the work location before work starts.

Testing and monitoring the atmosphere

[C7016] Testing may be necessary in respect of hazardous gases, fumes or vapours (these may be hazardous to health, or flammable or explosive). This is particularly important where activities, such as welding, can displace the air inside enclosed spaces. The concentration of oxygen present in the atmosphere may also need to be checked. The HSE's Guidance is that the concentration of oxygen in the air should not be less than 19%. These test results will indicate possible needs for other precautions eg purging, ventilation, RPE. In some cases tests may need to be carried out periodically or prior to each further entry. Constant monitoring of levels may also be necessary - many monitoring devices can give alarms if concentrations of hazardous substances rise above a pre-determined level (or if oxygen levels fall). Those carrying out testing must be competent for the task and test results should be retained.

Gas purging

[C7017] Flammable or toxic gases or vapours may need to be removed by purging with an inert gas (such as nitrogen) for flammable substances, or by an inert gas and/or air for toxic contaminants. Where inert gases are used, further purging with air is likely to be necessary before entry can take place without

Precautions for work in confined spaces **[C7021]**

suitable RPE. Where purging has taken place, there is likely to be a need for further testing or monitoring of the atmosphere and the use of RPE may still be necessary. Consideration must also be given to the safe venting and disposal of the purged gases.

Ventilation

[C7018] Mechanical ventilation (or ventilation openings) may be necessary to remove contaminants already present within the confined space and/or to provide fresh breathing air for those inside the confined space and/or to remove gas, fume or vapour produced by the work. Consideration should be given to the layout of the confined space and the location of any contaminants in order to ensure effective fresh air circulation. Suitable positions should be chosen for air inlets and outlets to avoid possible dead spots and also to prevent contaminated extract air from re-entering. Oxygen must never be used to 'sweeten' any atmosphere, particularly in confined spaces, as oxygen enrichment (above the normal concentration in air) increases the combustibility of many materials and can also have a toxic effect, if inhaled.

Removal of contents and residues

[C7019] Work may only be able to be undertaken safely once the contents of the confined space, together with any remaining residues, have been effectively removed. Methods for removing residues include mechanical means (eg scraping), steam, high-pressure water, or use of chemicals. In some cases the removal work may require entry to be made into the confined space, necessitating its own safe system of work. Testing or inspection after the work may be needed, in order to ensure residues have been removed effectively.

Isolations

[C7020] Frequently the system of work will require the effective prevention of the ingress of substances into the confined space (gases, liquids and other free flowing materials), or the isolation of mechanical and electrical equipment within the space. Such isolations are likely to achieved by precautions such as the removal of sections of supply pipe, the insertion of blanks or the use of a locking valve. Consideration may have to be given to the need to maintain essential services eg lighting, power or water for the work to be carried out whilst preventing undesirable risks eg re-energisation of process equipment or possible flooding. Regular checks should obviously be made to ensure that the method of isolation adopted remains effective.

Electrical equipment

[C7021] Electrical equipment should be chosen in relation to risks within the confined space and the work to be carried out. It may need to be:

- suitable for use in a flammable or explosive atmosphere;
- suitable for use in damp conditions (or chemical exposure);

- low voltage because of electrocution risks (or at least protected by RCD's); or
- protected from mechanical damage or dropping.

Workers may need to be prevented from taking personal electrical equipment such as mobile phones or audio equipment into confined spaces where flammable or explosive atmospheres may be present.

Gas equipment and internal combustion engines

[C7022] Use of gas cylinders within confined spaces should be avoided if possible. If cylinders must be used then mechanical ventilation will usually be necessary. All gas equipment and pipelines should be checked for leaks before being taken into the confined space and removed to a well ventilated area at the end of every work period. This is to avoid possible contamination of the confined space by a slow leak from, for example, a gas cylinder. Where pipes and hoses cannot be removed they must be disconnected from the supply outside the confined space and their contents safely vented.

Petrol engines must not be used within confined spaces and use of diesel engines should be avoided whenever possible. Where the use of diesel engines or portable gas cylinders in confined spaces is unavoidable, adequate mechanical ventilation must be provided to ensure complete combustion takes place. Diesel engine exhausts must be vented well away from the confined space and downwind of any air intakes.

Personal protective equipment (PPE) and respiratory protective equipment (RPE)

[C7023] Although the law expresses a preference for other types of control measures the use of PPE and RPE in confined space work will often be unavoidable. Wearing of harnesses with safety lines is particularly likely to be necessary, as is the use of RPE providing higher levels of protection (self-contained breathing apparatus or air-fed respirators). Wearing PPE and RPE within a confined space may be an additional source of heat stress for workers.

Access and egress

[C7024] Safe routes into and out of the confined space must be provided and maintained. These must also be suitable for emergency access or escape. Sizes of openings must allow for PPE and/or RPE that may need to be worn. The minimum size of an opening to allow access (including in an emergency where self-contained breathing apparatus is used) is 575mm diameter. Some existing plant may have smaller openings (eg road tankers where the standard size is only 410mm). In such circumstances, further precautions should be taken, such as the availability of airline breathing apparatus. Larger workers may find it physically difficult to get through small openings or to move around within the confined space. Signs prohibiting unauthorised entry may also be required. It is recommended that practice drills should be undertaken in order to check that access and egress can be achieved, especially in an emergency situation.

Fire precautions

[C7025] Precautions such as removal of residues, gas purging, ventilation or the control of ignition sources may be necessary as described earlier. Where flammable substances must be present in a confined space (eg paints, cleaning materials) they must be kept to a minimum. Provision of suitable fire fighting equipment will be necessary in such cases. In larger structures consideration may need to be given to providing alternative means of escape in case of fire.

Lighting

[C7026] Adequate and suitable lighting must be provided as for all workplaces. Suitable back-up lighting (eg torches) must be available in the event of lighting failure. See **C7021** above for relevant criteria for the selection of electrical equipment.

Static electricity, hot work, smoking and sparks

[C7027] Risks from static electricity or other sources of ignition, such as hot work, may need to be considered. Some types of equipment, clothing or activities (eg. steam or water jetting) are prone to static build-up or discharge. As part of the health legislation enacted in England, Wales and Scotland, smoking is prohibited from most workplaces, including enclosed workplaces. However, paragraph 133 of the ACOP indicates it may also be necessary to set a smoking exclusion area a suitable distance beyond the confined space, if for example there is a risk of explosion. Non-sparking tools will be required where flammable or potentially explosive atmospheres may be present.

Limited working time and entry control

[C7028] Working within confined spaces may be physically demanding due to extremes of temperature or humidity, the difficulties of wearing RPE or PPE or cramped working conditions. Use of fans may be necessary to achieve a satisfactory thermal environment. Time limits for individuals to stay in the space and rest periods in between may need to be imposed. These may be enforced by a logging or tally system. Such systems could also be necessary in large confined spaces or for large work groups in order to provide a check as to who is inside at any point in time.

Emergency arrangements

[C7029] Suitable arrangements must be prepared for emergency rescue from a confined space. This may be due to a 'specified risk' or 'other risk' associated with the confined space or for other types of emergency, eg a worker inside suffering a heart attack or other medical condition. These arrangements must reduce risks to rescuers, so far as is reasonably practicable, and include the provision and maintenance of resuscitation equipment, where relevant due to 'specified risks'.

Emergency arrangements must be immediately put into operation should such a situation arise – a fall inside a confined space may allow time for medical and other assistance to be summoned, but a collapse due to gas, fume, vapour or lack of oxygen is likely to require immediate intervention by those on the spot eg. a rescue using a harness, lifeline entry into the space by rescuers using suitable respiratory protective equipment (RPE) and prompt use of resuscitation equipment. Consideration must also be given to lower levels of availability of outside assistance at certain times (eg nights or weekends) and to work taking place in more remote locations.

Examples of emergency equipment and arrangements which may be necessary are:

- an attendant immediately outside the confined space;
- a harness attached to a secure lifeline worn by those inside;
- recovery equipment eg a tripod and winch (either already in position or available nearby, if required);
- suitable RPE or PPE for rescuers (either already worn by them or readily available);
- appropriate resuscitation equipment readily available;
- presence of someone with first aid training;
- stretcher(s) and other appropriate first aid equipment near at hand;
- fire fighting equipment;
- suitable communication equipment (see **C7015**);
- means of contacting and liaising with the emergency services or other sources of assistance (some high risk work may justify informing the emergency services in advance);
- medical staff on standby nearby (only likely to be necessary for high risk situations);
- use of a register of persons inside the confined space (only necessary where large groups or large, complex confined spaces are involved); and
- arrangements to shut down nearby plant (which may be a source of noise or could increase the risks).

Relevant staff must be fully trained in the use of emergency equipment and physically capable of using it. This includes those working within confined spaces, acting as back-up attendants or intended to provide emergency assistance in other ways.

The emergency arrangements required should be identified during the risk assessment process and should normally be recorded – either on the permit to work form or within the task procedure.

Further Reading

[C7030] L101 Safe work in Confined Spaces, Confined Spaces Regulations 1997, 'Approved Code of Practice', (2014; 3rd edition) (HSE Books)

INDG 258 (rev1) Confined Spaces: A brief guide to working safely (HSE leaflet: 2013)

INDG273 (rev1) Working safely with solvents (HSE leaflet: 2014)

AIS 26 Managing confined spaces on farms (free HSE leaflet)

HSG250 Guidance on permit-to-work systems: A guide for the petroleum, chemical and allied industries (2005) (HSE Books)

There are also a number of short industry specific guidance documents concerning work in confined spaces including:

- Confined Spaces Off-shore (OCM1);
- Asphyxiation hazards in welding and allied processes (EIS45);
- Guidance on Confined Spaces in Ports (SIP015).

The HSE has also published on its website relevant updates on:

- Reduced Oxygen Environments (Hypoxic Environments);
- Wood Pellet Bio-fuels.

Construction and Building Operations

Andrea Oates

Introduction to construction and building operations

[C8001] The *Construction (Design and Management) Regulations 2015 (CDM 2015) (SI 2015 No 51)* came into force on 6 April 2015, replacing the *Construction (Design and Management) Regulations 2007 (CDM 2007) (SI 2007 No 320)*, subject to a number of transitional provisions which expired on 6 October 2015.

This chapter sets out the provisions of *Part 4 of CDM 2015* and is linked to the following CONSTRUCTION, DESIGN AND MANAGEMENT chapter which sets out the provisions of the other four parts: Parts 1, 2, 3 and 5.

Part 4 sets out general requirements for all construction sites and is largely unchanged from the 2007 Regulations.

Part 4, Construction (Design And Management) Regulations 2015

[C8002]–[C8017] The following is a summary of *Part 4 of CDM 2015*—

Person on whom duties are imposed

Regulation 16 sets out that *Part 4* of the Regulations "applies only to a construction site" and that:

- a contractor carrying out construction work must comply with the requirements set out in *Part 4*, so far as they affect the contractor or any worker under their control or relate to matters under their control (*Regulation 16(2)*); and
- a domestic client who controls the way in which construction work is carried out by workers must also comply with the requirements of *Part 4*, "so far as they relate to matters within the client's control".

CDM 2015 sets out a range of duty holders and their roles: clients, domestic clients, designers, principal designers, principal contractors, contractors and workers and these are examined in the linked CONSTRUCTION, DESIGN AND MANAGEMENT chapter.

Safe places of work

Regulation 17 sets out that, so far as is reasonably practicable—

— there must be suitable and sufficient safe access to and exit from every construction site to other places of work; and from every place construction work is being carried out to every other place to which workers have access within a construction site;
— every place of work must be made and kept safe for, and without risks to health of workers;
— action must be taken to ensure that no one uses access to or exit from, or gains access to any place, which does not comply with the requirements above; and
— a construction site must have sufficient working space and be arranged so it is suitable for people working there (or likely to be working there), taking into account any work equipment likely to be used. Ergonomic considerations should therefore be taken into account.

Good Order and site security

- *Regulation 18* links site security to good order. Every part of a construction site must, so far as is reasonably practicable, be kept in good order and those parts in which construction work is being carried out must be kept reasonably clean.
- Where necessary in the interests of health and safety, a construction site must, again so far as is reasonably practicable, and "in accordance with the level of risk posed" either have its perimeter identified by suitable signs so that the extent of the site is readily identifiable, and/or be fenced off.
(See also the requirements for barriered exclusion zones in the Work at Height Regulations (see Work at Height).)
- No timber or other material with dangerous projecting nails (or similar sharp objects) must be used in construction work or allowed to remain in place.

Stability of structures

- *Regulation 19* requires that, where necessary to prevent danger to any person, all practicable steps must be taken to ensure that any new or existing structure which could become unstable or is in a temporary state of weakness or instability due to construction work does not collapse.
- *Regulation 19(2)* sets out that any buttress, temporary support or temporary structure must be designed, installed and maintained so that it can withstand any foreseeable loads which may be imposed on it, and must only be used for the purpose for which it is designed, installed and maintained.

How this is to be achieved is not defined in the regulation. However, *Regulation 8* in *Part 3* requires anyone appointing a designer or contractor to work on a project to take reasonable steps to satisfy themselves that those who will carry out the work have the skills, knowledge, experience, and, where they are an organisation, the organisational capability, to carry out the work safely (see CONSTRUCTION, DESIGN AND MANAGEMENT).
- Scaffolding is frequently used to buttress or provide temporary support. European Standard BS EN 12811-1:2003, which is currently under review, specifies performance requirements and methods of structural and general design for access and working scaffolds. The requirements relate to scaffold structures that rely on adjacent structures for stability. In general, these requirements also apply to other types of working scaffolds.

This regulation also requires that structures must not be made unsafe through loading.

Demolition or dismantling

- *Regulation 20* requires that demolition or dismantling 'must be planned and carried out in such a manner as to prevent danger or, where it is not practicable to prevent it, to reduce danger to as low a level as is reasonably practicable'.
- The arrangements for carrying out demolition or dismantling must be recorded in writing before the work begins.

Explosives

- *Regulation 21* requires explosives to be stored, transported and used safely and securely, so far as is reasonably practicable.
- An explosive charge must only be used or fired if suitable and sufficient steps have been taken to ensure that there is no risk of injury as a result of the explosion, or from projected or flying material caused by the explosion.
- HSE information on establishing exclusion zones when using explosives in demolition can be found on the HSE website at: www.hse.gov.uk/pubns/cis45.pdf.

Excavations

- *Regulation 22(1)* states that; "All practicable steps must be taken to prevent danger to any person, including, where necessary, the provision of supports or battering to ensure that:
  (a) no excavation or part of an excavation collapses;
  (b) no material forming the walls or roof of, or adjacent to, any excavation is dislodged or falls; and

- (c) no person is buried or trapped in an excavation by material which is dislodged or falls."
- Suitable and sufficient steps must be taken to prevent any person, work equipment or accumulation of materials from falling into any excavation.
- The wording of *Regulation 22(4)* is as follows: "Construction work must not be carried out in an excavation where any supports or battering have been provided in accordance with *paragraph (1)* unless—
  - (a) the excavation and any work equipment and materials which may affect its safety, have been inspected by a competent person—
    - (i) at the start of the shift in which the work is to be carried out;
    - (ii) after any event likely to have affected the strength or stability of the excavation; and
    - (iii) after any material unintentionally falls or is dislodged; and
  - (b) the person who carried out the inspection is satisfied that the work can be safely carried out there".

  Regulation 22(5) prohibits construction work being carried out in an excavation until any matter found to be unsatisfactory during the inspection has been resolved.

  The risk assessed method statements covering the installation and removal of temporary supports will need careful checking by the person in control of the site. Systems of work that involve the use of trench boxes or drag boxes need to be checked to ensure that people working in the trench are adequately protected from danger in a manner equivalent or better to that specified by this regulation.
- Instability of an excavation caused by surcharge from work equipment (eg mobile plant) or materials (including material excavated or stockpiles for unrelated work) is covered by *Regulation 22(3)*: "Suitable and sufficient steps shall be taken, where necessary, to prevent any part of an excavation or ground adjacent to it from being overloaded by work equipment or material". The person responsible for making such judgement needs to be identified as part of the safe system of work.

Cofferdams and caissons

- This specialist area of work, which uses watertight structures to allow construction work to take place below the water level, is covered by *Regulation 23*.
- A cofferdam or caisson must be of suitable design and construction; provision must be made for workers to shelter or escape 'if water or materials' enter it and it must be properly maintained.

- A cofferdam or caisson must only be used to carry out construction work if it, and any work equipment and materials which may affect its safety, has been inspected by a competent person at the start of the shift and after any event likely to have affected its strength or stability and the person carrying out the inspection has the power to stop work if they are not satisfied as to its safety.

Reports of inspections

- A person carrying out inspections under *Regulation* 22 or 23 must, before the end of the shift within which the inspection is carried out, do the following:
    - Where he or she is not satisfied that the construction work can be carried out safely, inform the person for whom the inspection was carried out of any matters that could give rise to a risk to safety; and
    - Prepare a report. This must include:
        (i) the name and address of the person on whose behalf the inspection was carried out;
        (ii) the location of the place of construction work inspected;
        (iii) a description of the place of construction work or part of that place inspected (including any work equipment and materials);
        (iv) the date and time of the inspection;
        (v) details of any matter identified that could give rise to a risk to the safety of any person;
        (vi) details of any action taken as a result of any matter identified (such as alerting workers on site, fencing the area off or excluding access, for example);
        (vii) details of any further action considered necessary; and
        (viii) the name and position of the person making report.

The report (or a copy) must be provided to the person on whose behalf the inspection was carried out within 24 hours of completing the inspection.

Where the person who carries out an inspection works under the control of another person, whether as an employee or otherwise, the person in control must ensure that they comply with the requirements set out above (*Regulation 24(2)*).

Regulation 24(3) requires the person on whose behalf the inspection was carried out to keep a copy of the report available for inspection by an HSE inspector on-site until the construction work is completed and for three months following completion of the work. They must send the inspector copies or extracts of the report if required.

Where more than one inspection is carried out at the start of shifts within a seven-day period, more than one report does not have to be prepared (*Regulation 24(4)*).

Energy distribution systems

- The requirements in relation to services, buried or overhead, are covered by *Regulation 25*. It requires that energy distribution installations are suitably located, periodically checked and clearly indicated where this is necessary to prevent danger.
- The focus of the regulation is on electric power cables. Where there is a risk from these, they must be directed away from the area of risk or the power must be isolated, and where necessary earthed. If it is not reasonably practicable to do this, suitable warning notices must be provided, together with one or more of the following:
    - barriers suitable for excluding work equipment which is not needed;
    - suspended protections where vehicles need to pass beneath the cables; or
    - measures providing an equivalent level of safety.

No construction work which is liable to create a risk to health and safety from an underground service, or from damage to or disturbance of it, shall be carried out unless suitable and sufficient steps have been taken to prevent such risk, so far as is reasonably practicable.

Those in control of work where services may be damaged also need to be aware of the significant safety risks associated with water mains (volume of flow and pressure), gas mains and the cost of repairs and business interruption claims for damage to fibre optic and equivalent communication cables.

Prevention of drowning

- Where construction work involves the risk that people could fall into water or other liquid and drown, *Regulation 26* requires that: 'suitable and sufficient steps must be taken to (a) prevent, so far as is reasonably practicable, the person falling; (b) minimise the risk of drowning in the event of a fall; and, (c) ensure that suitable rescue equipment is provided, maintained and, when necessary, used so that such person may be promptly rescued in the event of a fall'.
- Suitable and sufficient steps must be taken to ensure the safe transport of any person conveyed by water to or from any place of work.
- Any vessel used for this purpose must not be overcrowded or overloaded.

The *Provision and Use of Work Equipment Regulations 1998 (PUWER 1998) (SI 1998 No 2306)* (see MACHINERY SAFETY) contains requirements concerning the construction and maintenance of vessels and their control by a competent person.

Traffic routes

- *Regulation 27* requires that construction sites are organised to allow pedestrians and vehicles to move safely and without risks to health, so far as is reasonably practicable.
- Segregation of people (*pedestrians*) from vehicles and mobile plant is a high priority area for enforcement due to the history of serious and fatal injuries. The planning of construction sites must consider the design of loading bays, separation, segregation, inter-visibility (*of pedestrians and vehicles*), choice of crossing points, siting of doors and gates, and the avoidance of obstructions that may cause people to be trapped or crushed.
- Traffic routes must be suitable for the people and vehicles using them, sufficient in number, in suitable positions and of sufficient size. Suitable and sufficient steps must therefore be taken to ensure that:
  - pedestrians and vehicles may use a traffic route without causing danger to the health and safety of people nearby;
  - any door or gate for pedestrians which leads on to a traffic route is sufficiently separated from the route so that pedestrians can see any approaching vehicle or plant from a place of safety;
  - there is sufficient separation between vehicles and pedestrians to ensure safety, or where this is not reasonably practicable, other means for the protection of pedestrians are provided and there are effective arrangements in place to warn anyone liable to be crushed or trapped by a vehicle of its approach;
  - any loading bay has at least one exit for the exclusive use of pedestrians; and
  - where it is unsafe for pedestrians to use a gate intended primarily for vehicles, at least one door for pedestrians is provided in the immediate vicinity, clearly marked and kept free from obstruction.
- Each traffic route must be indicated by suitable signs (where necessary for health and safety reasons) and traffic routes must be regularly checked and properly maintained.
- Vehicles must not be driven on a traffic route unless, so far as is reasonably practicable, the route is free from obstruction and permits sufficient clearance (*Regulation 27(5)*).
- The *Health and Safety (Safety Signs and Signals) Regulations 1996* (see WORKPLACES – HEALTH, SAFETY AND WELFARE chapter) require the use of road traffic signs, as prescribed in the Road Traffic Regulation Act 1984, to regulate road traffic within workplaces where necessary.

Vehicles

- The requirements of *Regulation 28* need to be read in conjunction with the *Provision and Use of Work Equipment Regulations 1998 (PUWER) (SI 1998 No 2306)*, regarding the selection, maintenance, inspection and use of mobile work equipment (see MACHINERY SAFETY).
- The unauthorised use and 'unintended movement' of unattended vehicles must be prevented or controlled.
- The driver or operator, or a person directing them, must be able to warn others that the vehicle is moving.
- Safe loading, operating, towing as well as driving must be considered.
- No one must ride, or be required to or permitted to ride, on any vehicle being used for construction work other than in a safe place provided for the purpose. Similarly, no one must remain, or be required or permitted to remain, on a vehicle during the loading or unloading of loose material unless a safe place of work is provided and maintained for them.
- Measures must be taken to prevent a vehicle from falling into an excavation or pit, or into water, or overrunning the edge of any embankment or earthwork. This is a particular issue when a number of different contractors are working on, or passing, an area of a site.

Prevention of risk from fire, flooding or asphyxiation

Regulation 29 requires that suitable and sufficient steps are taken to prevent, so far as is reasonably practicable, the risk of injury to a person during construction work as a result of fire or explosion, flooding or any substance liable to cause asphyxiation.

Emergency procedures

Regulation 30 requires that emergency procedures to deal with foreseeable emergencies are in place on construction sites. These should take account of:

- The type of work for which the construction site is being used;
- The characteristics and size of the site, and the number and location of workplaces;
- The work equipment being used;
- The number of people likely to be present on the site at any one time; and
- The physical and chemical properties of any substances or materials likely to be on the site.

The arrangements should be tested at suitable intervals and brought to the attention of everyone they are aimed at so that they are familiar with the arrangements.

Emergency routes and exits

Regulation 31 deals with emergency routes and exits and requires that there is a sufficient number of suitable emergency routes and exits in order to allow people to reach a place of safety quickly in the event of danger. Emergency routes and exits must lead as directly as possible to an identified safe area; be kept clear from obstruction; where necessary, be provided with emergency lighting; and be indicated by suitable signs. Traffic routes giving access to emergency routes and exits must also be kept clear and provided with emergency lighting where necessary.

Fire detection and fire-fighting

Regulation 32 requires that, where necessary in the interests of the health and safety of people working on construction sites, there is suitable and sufficient (and suitably located) fire-fighting equipment and fire detection and alarm systems.

The statutory duties in *Regulations* 29 and 32 aim to protect people from the risk of injury. Those in control of a site should also be aware of and take measures to protect against the commercial risk of damage to property and materials.

- For example, it is recommended that site rules need to cover the issuing of "hot work permits" for example, and site security to protect against the risk of arson must be a high priority.
- Where the works are being carried out in or adjacent to an existing workplace, fire risk assessment and management of emergency procedures need to take account of the requirements of the *Regulatory Reform (Fire Safety) Order 2005 in England and Wales*, or the *Fire (Scotland) Act 2005 and Fire Safety (Scotland) Regulations 2006*.
- The emergency procedures for any construction site are a key "site specific" component of the site induction for everyone entering that site.

HSE guidance on fire safety in construction is available at: www.hse.gov.uk/pubns/books/hsg168.htm.

Fresh air

- The provision of a safe place of work includes the environment of that workplace.
- *Regulation 33* requires that suitable and sufficient steps must be taken to ensure, so far as is reasonably practicable, that each construction site, or approach to it, has sufficient fresh or purified air to ensure that the site or approach is safe and without risks to health or safety. (Note: this supports the duties in the *Control of Substances Hazardous to Health (COSHH) Regulations 2002 (SI 2002 No 2677)* to suppress dust at source and avoid exposure to solvents for example.)
- Equipment providing fresh air must be fail safe or include an effective device to give visible and audible warnings of any failure.

Temperature and weather protection

Regulation 34 requires that suitable and sufficient steps must be taken to ensure, so far as is reasonably practicable that, during working hours, the temperature of a construction site that is indoors is reasonable, having regard to the purpose for which it is used. In addition, outdoor workplaces should provide protection from adverse weather, again so far as is reasonably practicable and where necessary in order to protect the health and safety of workers. Regard should be had to the purpose for which the workplace is used and to any protective clothing or work equipment provided.

Lighting

- *Regulation 35* requires that suitable and sufficient lighting is provided to construction sites, their approaches and traffic routes, so far as is reasonably practicable by natural light. The design of security shutters for site accommodation, cabins and huts, for example, needs to take this requirement for daylight into account.
- The colour of any artificial lighting provided shall not adversely affect or change the perception of any sign or signal provided for the purposes of health and safety.
- Lighting where IT and display screen equipment is to be used will need to be evaluated, as in any workplace.
- Suitable and sufficient secondary lighting must be provided in any place where there would be a risk in the event of a failure of the primary source of (artificial) lighting, ie for emergency evacuation.

Welfare facilities – Schedule 2

The requirements to provide welfare facilities on construction sites are set out in *Regulations 4(2)(b)*, requiring clients to make suitable arrangements to ensure that facilities are provided; *13(4)(c)*, requiring the principal contractor to ensure that facilities are provided throughout the construction phase; and *15(11)*, requiring contractors to ensure, so far as is reasonably practicable, that the requirements of *Schedule 2* are complied with so far as they affect the contractor or any worker under that contractor's control.

Schedule 2 of CDM 2015 sets out the welfare facilities that should be provided on construction sites.

They may be summarised as follows—

Sanitary conveniences—
- Suitable and sufficient sanitary conveniences must be provided or made available at readily accessible places. They should be adequately ventilated and lit and kept in a clean and orderly condition.
- Separate rooms containing sanitary conveniences must be provided for men and women, unless they are in a separate, lockable room.

Washing facilities—
- Suitable and sufficient washing facilities, including showers if required by the nature of the work or for health reasons, must be provided or made available at readily accessible places, so far as is reasonably practicable.
- Washing facilities must be provided in the immediate vicinity of every sanitary convenience, whether or not provided elsewhere, and of any changing rooms, again whether or not provided elsewhere.
- They must include—
  - a supply of clean hot and cold, or warm, water (which shall be running water so far as is reasonably practicable);
  - soap or other suitable means of cleaning; and
  - towels or other suitable means of drying.
- Rooms containing washing facilities must be sufficiently ventilated and lit.
- They must be kept in a clean and orderly condition.
- Separate washing facilities must be provided for men and women, except where they are provided in a lockable room intended to be used by only one person at a time.
- Privacy is not required for facilities which are provided for washing hands, forearms and faces only, eg a hand wash basin in food preparation or canteen area.

Drinking water—
- An adequate supply of wholesome drinking water must be provided or made available at readily accessible and suitable places.
- Drinking water supplies must be conspicuously marked by an appropriate sign where necessary for reasons of health and safety.
- Where a supply of drinking water is provided, there must also be a sufficient number of suitable cups or other drinking vessels unless it is in a jet from which persons can drink easily.

Changing rooms and lockers—
- Suitable and sufficient changing rooms must be provided or made available at readily accessible places if a worker has to wear special work clothing and cannot, for reasons of health or propriety, be expected to change elsewhere. This is particularly important where there is for example, significant contamination on site.
- Where necessary for reasons of propriety, separate rooms or 'secure' separate use of the same room must be available for men and women.
- Changing rooms shall be provided with seating and shall include, where necessary, facilities to enable a person to dry any special clothing and their own clothing and personal effects.

— People must be able to lock away any special clothing which is not taken home, their own clothing which is not worn during working hours, and their personal effects.

Facilities for rest
— Suitable and sufficient rest rooms or rest areas must be provided or made available at readily accessible places.
— They must be equipped with an adequate number of tables and adequate seating with backs for the number of persons at work likely to use them at any one time.
— Where necessary, they must include suitable facilities for pregnant women or nursing mothers to rest lying down.
— Rest rooms must include suitable arrangements to ensure that meals can be prepared and eaten. Food hygiene needs to be managed in terms of the maintenance and cleaning of these areas.
— Rest rooms must include the means for boiling water and be maintained at an appropriate temperature.

Typical issues occurring in building and construction projects

[C8018] Construction accounts for around 7% of the UK workforce, but more than a quarter (around 26%) of work-related deaths each year occur in the sector. In 2017/18, 38 construction workers died in incidents on construction sites in Britain. Almost half (48%) were caused by falls from height, 12% were due to being trapped by something collapsing and 11% were caused as a result of being struck by an object.

In February 2019, the All-Party Parliamentary Group on Working at Height published the results of its inquiry, which was launched in January 2018, *Staying Alive: Preventing Serious Injury and Fatalities while Working at Height*. The report says that the most effective way to improve the culture of working at height is to enhance the reporting of accidents and near misses and to investigate the introduction of civil enforcement. The APPG recommendations include the following:

- enhanced reporting through the Reporting of Injuries, Diseases and Dangerous Occurrences Regulations (RIDDOR) 2013 and as a minimum recording the scale of a fall, the method used and the circumstances of the fall;
- the appointment of an independent body to allow confidential, enhanced and digital reporting of all near misses and accidents that do not qualify for RIDDOR reporting. The data collected by this independent body should be shared with government and industry to inform health and safety policy;
- the extension of the *Working Well Together – Working Well at Height* safety campaigns to industries outside the construction sector;

- the extension of an equivalent system to Scotland's Fatal Accident Inquiry process to the rest of the UK to ensure employers are held to account for fatal injuries occurring as a result of workers falling from height and incidents are reported with sufficient information;
- the creation of a digital technology strategy, to include a new tax relief for small, micro and sole traders, to enable them to invest in new technology; and
- a major review of work at height culture. This should include an investigation into the suitability of legally binding financial penalties in health and safety, funds which could be used towards raising awareness and training, particularly in hard to reach sectors.

The report calls on the government to ensure that no individual working at height will be any less safe as a result of Brexit and examines the 'huge potential for ever-evolving digital technology', including drones, apps and virtual and augmented reality, to make work at height safer.

The report can be found at: www.workingatheight.info/wp-content/uploads/2019/02/Staying-Alive-APPG-REPORT.pdf

The *Construction (Head Protection) Regulations 1989*, were revoked by the *Health and Safety (Miscellaneous Repeals, Revocations and Amendments) Regulations 2013*. The *Personal Protective Equipment at Work Regulations 1992* were subsequently amended to cover the provision and use of head protection on construction sites in order to maintain the level of legal protection. It is the duty of the person in control of the site to establish and identify the zones or locations where head protection has to be worn and ensure that signing of the zones is clear and there is adequate supervision in place.

The individual's risk assessment-based method statement should identify the most suitable type of head protection to be worn. HSE guidance to the PPE Regulations, *Personal protective equipment at work* (L25), identifies several different types of widely used head protection, each designed to be most suitable for protecting against a different sort of injury.

A health and safety exemption that allowed Sikhs to wear a turban instead of a hard hat on construction sites was extended to all workplaces. Under the *Employment Act 1989*, turban-wearing Sikhs were exempt from the requirement to wear head protection when working on construction sites. The Deregulation Act 2015 extended this rule (with some limited exceptions including emergency response situations) to all workplaces from 1 October 2015.

The Lifting Operations and *Lifting Equipment Regulations 1998 (SI 1998 No 2307)*, requires that lifting operations be planned and supervised to ensure that they are not carried out above or over people.

Falls and falling objects are covered in detail by the *Work at Height Regulations 2005 (SI 2005 No 735)* (see WORK AT HEIGHT). The Personal Protective Equipment at Work Regulations 1992 are examined in more detail in PERSONAL PROTECTIVE EQUIPMENT. The *Lifting Operations and Lifting Equipment Regulations 1998* are covered in detail in LIFTING OPERATIONS.

Incidents in the workplace involving transport, moving vehicles and mobile plant, people being struck by excavators, lift trucks, dumpers and other work equipment, are also a major cause of death and serious injury in construction workplaces. In 2017/18, 9% of fatal injuries to workers in the sector occurred when they were struck by moving vehicles. *A guide to workplace transport safety* (HSG136) (www.hse.gov.uk/pubns/books/hsg136.htm) provides advice on all aspects of workplace transport operations.

Six per cent of fatal injuries in the sector were caused by contact with electricity in 2017/18. To support the *Electricity at Work Regulations 1989 (SI 1989 No 635)*, information is contained in *HSE guidance HSR 25 Memorandum of guidance on the Electricity at Work Regulations 1989: Guidance on Regulations*. Appropriate competence and experience proportionate to the task is a key factor in reducing or eliminating incidents when working with medium and high voltage electricity (see ELECTRICITY).

Construction workers are also at high risk of developing work-related diseases, including cancer. According to HSE statistics, construction accounts for over 40% of occupational cancer deaths, with past exposures causing more than 5,000 cases of cancer and around 3,700 deaths every year. The main causes are exposure to asbestos (70%), silica (17%), diesel exhaust fumes (6–7%) and working as a painter (6–7%). Exposure to hazardous substances including dusts, fumes, vapours and gases, causes significant breathing problems and lung diseases, and there are high rates of dermatitis in a number of construction occupations.

The *Control of Substances Hazardous to Health Regulations 2002 (SI 2002 No 2677) (as amended)* outline the principals of eliminating or reducing exposure to substances, including respirable dusts such as silica from cutting with abrasive wheels. Other hazardous substances such as asbestos and lead are covered by specific regulations: the *Control of Asbestos Regulations 2012 (SI 2012 No 632)* and the *Control of Lead at Work Regulations 2002 (SI 2002 No 2676)* (see ASBESTOS AND HAZARDOUS SUBSTANCES IN THE WORKPLACE).

The HSE report *Construction statistics in Great Britain, 2018* (www.hse.gov.uk/statistics/industry/construction.pdf) shows there were an estimated 82,000 new or longstanding work-related ill-health cases in the sector. Work-related illness (85%) and injury (15%) resulted in around 2.4 million full-time equivalent lost working days each year between 2015/16 and 2017/18. The total cost of workplace injury and new cases of work-related ill health in construction was estimated to be £1,062m in 2016/17 alone.

There were an estimated 58,000 work-related cases of injury. Thirty per cent involved over-three-days and 24% over-seven-days absence. The injury rate is about 50% above the 'All industries' rate and is statistically significantly higher.

There were almost 5,000 non-fatal injuries to employees reported and of these, almost 2,000 were specified (serious) injuries. Thirty per cent of these were caused by slips, trips and falls on the same level, 7% by handling, lifting or carrying, 33% by falls from a height and 13% resulted from being struck by moving, including flying or falling, objects.

The HSE campaign to reduce these injuries, *Shattered Lives*, can be found online at: www.hse.gov.uk/shatteredlives. Commonly referred to as housekeeping, the requirement for good order in *Regulation 18* of *CDM 2015* states that 'Each part of a construction site must, so far as is reasonably practicable, be kept in good order and those parts in which construction work is being carried out must be kept in a reasonable state of cleanliness'.

Skilled construction and building trades have some of the highest estimated prevalence of back injuries and upper limb disorders as a result of manual handling, and one of the highest rates of ill health caused by noise and vibration, according to the HSE.

Sixty two per cent of cases were musculoskeletal disorders, accounting for a higher proportion of ill-health cases in construction than in all industries (44%). The *Manual Handling Operations Regulations 1992 (SI 1992 No 2793)*, and associated guidance, outline the measures that need to be taken to minimise the costs to industry and individuals associated with this type of injury. It is not just about the weight of an object (see **MANUAL HANDLING**).

A quarter of ill-health cases involved stress, depression or anxiety. An Office for National Statistics research report shows the risk of suicide among low-skilled male labourers, particularly those working in construction roles, was three times higher than the male national average (see STRESS AT WORK).

Other conditions that can affect construction workers include occupational deafness and Hand Arm Vibration (largely made up of two conditions, Vibration White Finger and Carpal Tunnel Syndrome).

The *Control of Noise at Work Regulations 2005 (SI 2005 No 1643)*, outline the measures that need to be taken to reduce exposure to damaging levels of noise. Health surveillance is covered by SI 2005 No 1643, Regulation 9(1) which outlines when 'the employer shall ensure employees are placed under suitable health surveillance, which shall include testing of their hearing' (see NOISE AT WORK).

The *Control of Vibration at Work Regulations 2005, (SI 2005 No 1093)* specify the exposure limits for hand/arm vibration and whole-body vibration. However, as with manual handling, it is important to note the requirement of *SI 2005 No 1093, Regulation 6(1)* — 'The employer shall ensure that risk from the exposure of his employees to vibration is either eliminated at source or, where this is not reasonably practicable, reduced to *as low a level as is reasonably practicable*'. Regulation 7 identifies when health surveillance, intended to prevent or diagnose any health effect linked with exposure to vibration, is required (see VIBRATION).

Higher risk work is covered by separate Regulations. For example, entry into confined spaces is covered by the *Confined Spaces Regulations 1997 (SI 1997 No 1713)*, and work in compressed air is covered by the *Work in Compressed Air Regulations 1996 (SI 1996 No 1656)*. Formal safe systems of work, permits to work and effective arrangements for rescue and resuscitation are required.

The primary duty in the *Confined Spaces Regulations 1997 (SI 1997 No 1713)* is to avoid entry if the necessary work can be carried out without person entry.

CCTV survey of sewers and drains is a simple example. It is important to be able to correctly identify potential confined spaces, and the level of risk associated with them. Sewer access that requires horizontal traverse requires a different level of escape and rescue provision than a vertical entry into a manhole. A confined space can include, for example, an open trench or inspection chamber.

Similarly, the health risks of work in compressed air are well known and well documented – see below. The designer and principal designer under the *Construction (Design and Management) Regulations 2015 (SI 2015 No 51)* would need to justify why one of the many alternative safer methods of work are not being adopted (see CONSTRUCTION, DESIGN AND MANAGEMENT).

The Work in Compressed Air Regulations 1996 require principal contractors to appoint competent compressed air contractors, and the compressed air contractors to appoint contract medical advisers. They also contain requirements concerning fire prevention and protection measures (including, in particular, emergency means of escape and rescue).

Where workers (and their representatives) are consulted about and involved in health and safety arrangements, there is a measurable improvement in safety culture and performance. The *Safety Representatives and Safety Committees Regulations 1977 (SI 1977/500) (as amended)* set out in detail the functions of safety representatives and safety committees and employers' duties to consult with trade union appointed safety representatives. The *Health and Safety (Consultation with Employees) Regulations 1996 (SI 1996/1513)*, give non-union employees the same rights to consultation that union safety representatives have, but are more limited – for example there is not the right to carry out inspections of the workplace. They apply where there is no union or the union is not recognised in a workplace, or with regard to a particular group of workers.

Site inductions and site security

[C8019] The use of site-specific inductions has been shown to be effective in reducing the high rate of incidents that occur when people first start working on a new site. The induction can help to set the culture of the site, explaining what is expected in relation to health and safety. Key personnel can be identified. Practical details, where the welfare facilities are located, parking arrangements, access and egress routes, need to be covered. Emergency arrangements will include first aid, routes to the nearest accident and emergency department, as well as fire risk minimisation, sounding of the alarm and evacuation procedures. The induction gives an opportunity for the site manager to question those coming onto site to see if they understand their own risk assessments and method statements, and to assess whether their work will conflict with other parties working on site.

The HSE guidance, HSG151, "*Protecting the public – your next move*", stresses that hazardous construction work can be a fatal danger not only for those within the industry but also the wider public. It contains practical advice, especially for those designing, planning, maintaining or conducting on-site work, to prevent risks to those off-site and covers legalities, perimeter and boundary matters and premises requiring special attention.

Issues in building projects [**C8020**]

Under the *Occupiers' Liability Act 1957* and the *Occupiers' Liability Act 1984* a duty of care is imposed on occupiers of existing premises regarding visitors. The duty of care extends to children – it should be noted that a child is regarded as being at greater risk than an adult. The 1984 Act further extends the duty of an occupier to people other than lawful visitors, such as trespassers, to ensure that they are not injured whilst on the premises (see OCCUPIERS' LIABILITY).

Reference should also be made to the *Health and Safety (Safety Signs and Signals) Regulations 1996 (SI 1996 No 341)*. Signs are a useful tool to help reduce risk but are not a substitute for other methods of controlling or eliminating risk. Signboards should use the stipulated shape, colour and pictogram or symbol, with supplementary text added only if necessary. Pictogram signs are particularly important where speakers of other languages may be working. Their induction should ensure that they understand the meaning of any signs used.

Signs to regulate traffic off highway, but on site or similar workplaces are required to be as prescribed in the *Road Traffic Regulation Act 1984* and illustrated in the Highway Code.

In addition, reference should be made to the following related statutory provisions, which have aspects relating to public safety during construction or building operations—

(a) Part IX of the *Highways Act 1980*, which deals with lawful and unlawful interference with highways and streets – protection of public rights – contains several provisions relating to construction, including:—
 (i) Section 168 (building operations affecting public safety); and
 (ii) Section 169 (the control of scaffolding on highways).
(b) *New Roads and Street Works Act 1991*—
 (i) Section 50 (lays down particular safety requirements for work in the street and, specifically, the measures to be taken to minimise inconvenience to the disabled). This Act is supported by a Code of Practice, with additional specialist guidance for high speed and high-volume traffic routes.
(c) *Environmental Protection Act 1990*—
 (i) Section 79 as amended by the *Noise and Statutory Nuisance Act 1993* (noise or vibration emitted from buildings and from or caused by a vehicle, machinery or equipment in a street).

Waste management

[**C8020**] Environmental legislation governs the proper disposal of waste, ranging from low risk to hazardous waste and is enforced by environmental agencies and local authorities. The HSE provides information on materials storage and waste management on construction sites on its website at: www.hse.gov.uk/construction/safetytopics/storage.htm#waste.

Construction, Design and Management

Andrea Oates

Introduction to construction, design and management

[C8501]–[C8503] The *Construction (Design and Management) Regulations 2015 (CDM 2015) (SI 2015 No 51)* came into force on 6 April 2015, replacing the *Construction (Design and Management) Regulations 2007 (CDM 2007) (SI 2007 No 320)*. Transitional provisions expired in October 2015.

CDM 2015 cover the management of health, safety and welfare on construction projects. The Regulations set out a range of duty holders and their roles: clients, domestic clients, designers, principal designers, principle contractors, contractors and workers. Organisations or individuals can carry out the role of one or more duty holders for a project, provided they have the skills, knowledge, experience and (in the case of an organisation) the organisational capacity to carry out these roles safely.

The Regulations incorporate key elements the Health and Safety Executive (HSE) says are necessary to secure construction health and safety. These are:

- managing the risks by applying the general principles of prevention (see below);
- appointing the right people and organisations at the right time;
- making sure everyone has the information, instruction, training and supervision they need to carry out their jobs safely;
- duty-holders cooperating and communicating with each other and coordinating their work; and
- consulting workers and engaging with them to promote and develop effective health, safety and welfare.

CDM 2015 implement the provisions of European Directive 92/57/EEC on minimum safety and health requirements at temporary or mobile construction sites (with the exception of certain requirements which have been implemented by other health and safety regulations including the *Work at Height Regulations 2005 (SI 2005 No 735)* and the *Workplaces (Health, Safety and Welfare) Regulations 1992 (SI 1992 No 3004))*. They revoked and re-enacted, with modifications, CDM 2007.

HSE guidance on the 2015 Regulations, *Managing health and safety in construction. Construction (Design and Management) Regulations 2015. Guidance on Regulations (L153)*, is available on the HSE website (see www.hse.gov.uk/pubns/books/l153.htm) and is referred to and summarised throughout this chapter. The guidance provides a useful summary of the roles and main duties of the various duty holders under CDM 2015. Designers, principal designers, principal contractors and contractors are all required to take account of the general principles of prevention in carrying out their duties.

The HSE summarises these as follows:

(a) avoid risks where possible;
(b) evaluate those risks that cannot be avoided; and
(c) put in place proportionate measures that control them at source.

Structure of CDM 2015

[C8504] CDM 2015 is divided into 5 parts—

- Part 1 deals with the application of the Regulations and definitions;
- Part 2 covers client duties in relation to managing projects; appointing the principal designer and principal contractor; and notifying projects to the HSE (or other enforcing authorities). It also sets out how the Regulations apply to domestic clients;
- Part 3 contains health and safety duties and roles. It covers: general duties; designers' duties; designs prepared or modified outside Great Britain; the principal designer's duties; the construction phase plan and health and safety file; the principal contractor's duties; and contractors' duties;
- Part 4 contains general requirements that apply to all construction sites; and
- Part 5 deals with enforcement in respect of fire; transitional and saving provisions and revocations, amendments and review.

This chapter sets out Parts 1, 2, 3 and 5 of the Regulations. A linked chapter, CONSTRUCTION AND BUILDING OPERATIONS, sets out the requirements of Part 4.

Part 1 (application and definitions)

[C8505] The Regulations apply to all construction work in Great Britain and its territorial seas. They also apply to offshore construction work carried out in connection with, or in preparation for, the construction of any renewable energy structure in areas outside the territorial sea designated for the exploration or exploitation of energy from water or winds.

The term "construction work" is defined in regulation 2, and "means the carrying out of any building, civil engineering or engineering construction work and includes—

(a) the construction, alteration, conversion, fitting out, commissioning, renovation, repair, upkeep, redecoration or other maintenance (including cleaning which involves the use of water or an abrasive at high pressure or the use of corrosive or toxic substances), decommissioning, demolition or dismantling of a structure;

(b) the preparation for an intended structure, including site clearance, exploration, investigation (but not site survey) and excavation (but not pre-construction archaeological investigations), and the clearance or preparation of the site or structure for use or occupation at its conclusion;

(c) the assembly on site of prefabricated elements to form a structure or the disassembly on site of prefabricated elements which, immediately before such disassembly, formed a structure;

(d) the removal of a structure or of any product or waste resulting from demolition or dismantling of a structure or from disassembly of prefabricated elements which immediately before such disassembly formed such a structure; and
(e) the installation, commissioning, maintenance, repair or removal of mechanical, electrical, gas, compressed air, hydraulic, telecommunications, computer or similar services which are normally fixed within or to a structure".

It does not include mineral exploration and extraction.

A project includes planning, design, and management work.

Regulation 2(1) also defines the various duty holders: "client", "domestic client", "contractor", "principal contractor", "designer", "principal designer"; as well as documents including the "health and safety file" and "pre-construction information".

The client is defined as "any person for whom a project is carried out" and a domestic client is "a client for whom a project is being carried out which is not in the course or furtherance of a business of that client".

The HSE guidance explains that the definition includes both commercial and domestic clients and that "A domestic client is someone who has construction work done on their own home, or the home of a family member, which is not done in connection with a business". However, it says: "Local authorities, housing associations, charities, landlords and other businesses may own domestic properties, but they are not a domestic client for the purposes of CDM 2015."

A contractor is "any person (including a non-domestic client) who, in the course or furtherance of a business, carries out, manages or controls construction work."

A designer means "any person (including a client, contractor or other person referred to in these Regulations) who, in the course of furtherance of a business" prepares or modifies a design, or arranges for, or instructs a person under their control to do so relating to a structure or product or mechanical or electrical system intended for a particular structure.

Designers include architects, architectural technologists, consulting engineers, quantity surveyors, interior designers, temporary work engineers, chartered surveyors, technicians and anyone who specifies or alters a design.

However, while local authority or government officials may give advice and instruction on designs meeting statutory requirements, this does not make them designers.

A principal contractor is the contractor appointed under regulation 5(1)(b) to perform the duties set out in regulations 12–14 (see **C8513–C8515** below).

A principal designer is the designer appointed under regulation 5(1)(a) to perform the duties set out in regulations 11 and 12 (see **C8512–C8513** below).

Pre-construction information, the HSE guidance explains, is information already in the client's possession (such as an existing health and safety file, an asbestos survey, or structural drawings) or which is reasonable to obtain through sensible enquiry (regulation 2(1)). The information must be relevant to the project, have an appropriate level of detail and be proportionate to the nature of the risks.

Buried services, cables and pipes, previous modifications to buildings, the presence of hazardous materials, including asbestos and contamination due to industrial waste, are all examples of the type of information that needs to be assembled and passed on to those designing and managing construction projects.

A health and safety file is prepared under regulation 12 (see **C8513A** below) and must be revised from time to time in order to incorporate relevant new information and must be available for inspection. A health and safety file is required for projects involving more than one contractor.

A design includes drawings, design details, specifications and bills of quantities relating to a structure, and calculations prepared for the purpose of a design.

Any reference to a plan, rule, document, report or copy includes a copy or electronic version (regulation 2(2)).

Client duties (Part 2)

Client's duties to manage projects

[C8506] The client has an important role as they have a major influence over the way a project is procured and managed, according to the HSE. "Regardless of the size of the project, the client has contractual control, appoints designers and contractors, and determines the money, time and other resources available", it says.

CDM 2015 therefore makes the client accountable for the impact their decisions and approach have on health, safety and welfare on the project. While they are not usually experts in construction and are therefore not required to take an active role in managing the work, they are required to make "suitable arrangements for managing the project so that health, safety and welfare is secured."

As the HSE guidance to CDM 2015 explains: "Regulations 4 and 5 set out the client's duty to make suitable arrangements for managing a project and maintaining and reviewing these arrangements throughout, so the project is carried out in a way that manages the health and safety risks." Where the project involves more than one contractor, the client must appoint a principal designer and a principal contractor and make sure they carry out their duties.

While the Regulations apply in full to commercial clients, for domestic clients the effect of Regulation 7 (see **C8509** below) is to pass the client duties on to other duty-holders, including the principal designer and principal contractor.

Regulation 4 requires clients to make suitable arrangements for managing a project, including allocating sufficient time and other resources. This means ensuring that the construction work can be carried out, so far as is reasonably foreseeable, without risks to health and safety, and providing the welfare facilities set out in Schedule 2 (see **C8518** below) for construction workers. The client must ensure these arrangements are maintained and reviewed throughout the project.

The HSE guidance explains that the arrangements should include:

(a) assembling the project team – appointing designers (including a principal designer) and contractors (including a principal contractor);
(b) ensuring the roles, functions and responsibilities of the project team are clear;
(c) allocating sufficient resources and time for each stage of the project – from concept to completion;
(d) ensuring effective communication, cooperation and coordination between project team members;
(e) setting out how the client will take reasonable steps to ensure that the principal designer and principal contractor comply with their duties – at project progress meetings or via written updates for example;
(f) setting out how the health and safety performance of designers and contractors will be maintained throughout the project; and
(g) ensuring that workers are provided with suitable welfare facilities for the duration of construction work.

Clients must provide pre-construction information (see **C8505** above), to actual and potential designers and contractors, as soon as practicable. They must also ensure that before construction begins, the contractor (if there is only one contractor) or the principal contractor (if there is more than one contractor), draws up a construction phase plan and the principle designer prepares a health and safety file for the project (in compliance with regulation 12 – see **C8513A** below).

The HSE guidance explains: "The client must ensure that the principal designer prepares a health and safety file for their project. Its purpose is to ensure that, at the end of the project, the client has information that anyone carrying out subsequent construction work on the building will need to know about in order to be able to plan and carry out the work safely and without risks to health."

A client must also take reasonable steps to ensure that the principal designer and principle contractor comply with their duties under the regulations (as set out in regulations 11, 12, 13 and 14 – see **C8512, C8513, C8514** and **C8515** below).

Appointing the principal designer and principal contractor

[C8507] Where there is more than one contractor, or it is reasonably foreseeable that more than one contractor will be working on the project, the client must appoint (in writing) a designer with control over the construction phase as principle designer; and a contractor as principal contractor (Regulation 5). The appointments must be made as soon as practical and before the construction phase begins. If the client fails to appoint people to these roles, they must fulfil the duties themselves.

Notification

[C8508] A project is notifiable to the Health and Safety Executive, Office for Rail and Road or Office for Nuclear Regulation (Regulation 6) if the construction work is scheduled to last longer than 30 working days and have more than 20 workers working simultaneously at any point in the project; or exceed 500 person days.

The HSE guidance sets out that each day construction work is likely to take place (including weekends and bank holidays) counts towards the period of construction work.

Schedule 1 (see **C8518** below) specifies the particulars of the notice. It must be clearly displayed in the construction site office, be comprehensible to workers and periodically updated if necessary.

The HSE guidance makes very clear that the requirements of CDM 2015 apply whether or not the project is notifiable.

Application to domestic clients

[C8509] Where the client is a domestic client, Regulation 7 requires that the client duties (set out in Regulation 4) and notification requirements (set out in Regulation 6) must be carried out by the contractor (for projects where there is only one contractor), principle contractor (for projects where there is more than one contractor), or principle designer (where there is a written agreement that the principal designer will fulfil those duties). If a domestic client fails to appoint a principal designer, the designer in control of the pre-construction phase of the project becomes the principal designer. If a domestic client fails to appoint a principal contractor, the contractor in control of the construction phase of the project becomes the principal contractor.

Regulation 5(3) and (4), which set out that the client must fulfil the duties of principal designer and principal contractor if they do not appoint these roles, do not apply to domestic clients.

Health and Safety Duties and Roles (Part 3)

General Duties

[C8510] The HSE explains that Regulation 8 requires anyone appointing a designer or contractor to work on a project to take reasonable steps to satisfy themselves that those who will carry out the work have the skills, knowledge, experience, and, where they are an organisation, the organisational capability to carry out the work safely. They should make "sensible and proportionate enquiries" about their organisational capability to carry out the work and "due weight" should also be given to membership of an established professional institution or body.

The HSE guidance explains that: "Organisational capability means the policies and systems an organisation has in place to set acceptable health and safety standards which comply with the law, and the resources and people to ensure the standards are delivered."

A designer or contractor appointed to work on a project must have the skills, knowledge and experience (and in the case of an organisation the organisational capacity) necessary to safely fulfil the role that they are appointed to undertake. They must not accept an appointment to a project unless they fulfil these conditions. The HSE advises that there are companies that provide pre-qualification assessment services, including those who are members of the Safety Schemes in Procurement (SSIP) Forum (www.ssip.org.uk).

Regulation 8 also contains a duty of cooperation on everyone involved with the project or any project on an adjoining site, and a duty on those working under the control of others to report any health and safety concerns.

Any information or instruction provided under the Regulations must be comprehensible and provided as soon as is practicable.

The HSE guidance provides the following examples of types of information to be provided:

(a) pre-construction information the client is required to provide to designers and contractors;
(b) health and safety information about the design that designers are required to provide to other duty-holders;
(c) information the principal designer must provide to enable preparation of the construction phase plan;
(d) site rules that are part of the construction phase plan; and
(e) information that principal contractors must provide to workers (or workers' representatives).

Duties of designers

[C8511] Designers are important, says that HSE, because they have a strong influence during the concept and feasibility stage of a project and their earliest decisions can "fundamentally affect the health and safety of those who will construct, maintain, repair, clean, refurbish and eventually demolish a building". A designer should address health and safety issues from the very start, it says.

Regulation 9 sets out that a designer must not commence work on a project unless they are satisfied that the client is aware of their own duties under the Regulations. In addition, when preparing or modifying a design, they must take into account the general principles of prevention and any pre-construction information to eliminate, so far as is reasonably practical, foreseeable risks to health and safety of anyone carrying out or liable to be affected by construction work; maintaining or cleaning a structure; or using a structure designed as a workplace.

The general principles of prevention are set out in Schedule 1 of the Management of Health and Safety at Work Regulations 1999 and require employers to work through a hierarchy of prevention and control:

"(a) avoid risks;
(b) evaluate the risks which cannot be avoided;
(c) combat the risks at source;
(d) adapt the work to the individual, especially as regards the design of workplaces, the choice of work equipment and the choice of working and production methods, with a view, in particular, to alleviating monotonous work and work at a predetermined work-rate and to reducing their effect on health;

- (e) adapt to technical progress;
- (f) replace the dangerous by the non-dangerous or the less dangerous;
- (g) develop a coherent overall prevention policy which covers technology, organisation of work, working conditions, social relationships and the influence of factors relating to the working environment;
- (h) give collective protective measures priority over individual protective measures; and
- (i) give appropriate instructions to employees."

If the risks cannot be eliminated, they must, so far as is reasonably practicable, take steps to prevent or control the risks through the subsequent design process; provide information about those risks to the principal designer; and ensure appropriate information is included in the health and safety file.

The HSE guidance includes examples of questions that should be considered at the design stage:

- (a) Can I get rid of the problem (or hazard) altogether?
  For example, can air-conditioning plant on a roof be moved to ground level, so work at height is not required for either installation or maintenance?
- (b) If not, how can I reduce or control the risks, so that harm is unlikely or the potential consequences less serious?
  For example, can I place the plant within a building on the roof, or provide a barrier around the roof?

If risks cannot be totally eliminated, the HSE advises, a designer should apply the following principles in deciding how to reduce or control the remaining risks:

- (a) provide a less risky option, such as using lighter paving to reduce musculoskeletal disorders;
- (b) organise the work to reduce exposure to hazards, eg make provision for traffic routes so barriers can be provided between pedestrians and traffic; and
- (c) ensure that those responsible for planning and managing the work have the information they will need to manage any remaining risks, eg tell them about loads that will be particularly heavy, or about elements of the building that could become unstable.

Designers must also take reasonable steps to provide information about the design, construction and maintenance of the structure, with the design, to adequately assist other duty holders in complying with their duties under the regulations.

The HSE advises that designs for places of work must also comply with the Workplace (Health, Safety and Welfare) Regulations 1992, taking account of factors such as lighting and the layout of traffic routes (see WORKPLACES – HEALTH, SAFETY AND WELFARE).

Regulation 10 deals with designs prepared or modified outside Great Britain and requires those who commission a design, or the client, to ensure that Regulation 9 is complied with. This does not apply to domestic clients.

Duties of the principal designer

[C8512] A principal designer has control over the very earliest stage of a project – from concept design through to planning the delivery of the construction work. The HSE guidance explains that the role of principal designer is important as it involves coordinating the work of others in the project team to ensure that significant and foreseeable risks are managed throughout the design process.

The principal designer must plan, manage and monitor the pre-construction phase and coordinate health and safety matters in order to ensure, so far as is reasonably practicable, that the project is carried out without risks to health and safety (Regulation 11).

In doing so, they must take into account the general principles of prevention (see **C8511**), as well as pre-construction information and the content of any construction phase plan and health and safety file (see **C8513** and **C8513A** below).

The HSE guidance sets out that: "This information should be taken into account particularly when decisions are being taken about design, technical and organisational issues to plan which items or stages of work can take place at the same time or in what sequence; and when estimating the time needed to complete certain items or stages of work."

It also advises that "regular design meetings chaired by the principal designer are an effective way to:

(a) discuss the risks that should be addressed during the pre-construction phase;
(b) decide on the control measures to be adopted; and
(c) agree the information that will help prepare the construction phase plan."

The principal designer must identify and prevent or control foreseeable risks, so far as is reasonably practicable, to the health and safety of people carrying out or liable to be affected by construction work; maintaining or cleaning a structure; or using a structure as a workplace.

They must ensure that all designers comply with their duties under the regulations and everyone working in relation to the pre-construction phase cooperates with the client, the principle designer and each other.

They must help the client to provide the pre-construction information required under regulation 4 (see **C8506**) and provide it (promptly and in a convenient form) to designers and contractors appointed to the project (or being considered for appointment).

HSE guidance to the Regulations sets out that the client has responsibility for pre-construction information but that the principal designer must help the client bring together the information they hold (such as any existing health and safety file or asbestos survey). It adds:

"The principal designer should then:

(a) assess the adequacy of existing information to identify any gaps in the information which it is necessary to fill;

(b)   provide advice to the client on how the gaps can be filled and help them in gathering the necessary additional information; and
(c)   provide, as far as they are able to, the additional information promptly and in a convenient form to help designers and contractors who:
   (i)   are being considered for appointment; or
   (ii) have already been appointed, to carry out their duties."

The principal designer must also liaise with the principal contractor throughout the project and share information about the planning, management and monitoring of the construction phase and health and safety coordination with them. The HSE says that this can be done by holding regular progress meetings for example.

The construction phase plan

[C8513] Regulation 12 sets out that during the pre-construction phase, and before setting up the construction site, the principle contractor must draw up (or make arrangements for the drawing up of) a construction phase plan.

This must set out the health and safety arrangements and sites rules. In addition to taking account of any industrial activities taking place on the construction site, it must include specific measures concerning any work falling within one or more of the categories set out in Schedule 3 (see **C8518** below).

The principal designer must assist the principal contractor in preparing the construction phase plan by providing them with all the relevant information they hold. This includes pre-construction information obtained from the client and any information obtained from designers about the steps taken to reduce health and safety risks.

The principle contractor must ensure that the construction phase plan is reviewed, updated and revised as necessary.

Appendix 3 of L153 lists the following as topics that should be considered when drawing up the plan:

- a description of the project such as key dates and details of key members of the project team;
- the management of the work including: the health and safety aims for the project; the site rules; arrangements to ensure cooperation between project team members and coordination of their work, such as regular site meetings; arrangements for involving workers; site induction; welfare facilities; and fire and emergency procedures; and
- the control of any of the specific site risks listed in Schedule 3 (see below) where they are relevant to the work involved.

The health and safety file

[C8513A] During the pre-construction phase, the principle designer must prepare a health and safety file. This must contain information necessary to ensure the health and safety of people working on, or affected by, the construction work. It too must be appropriately reviewed, updated and revised.

The principal contractor must provide the principle designer with relevant information for the health and safety file throughout the project. At the end of the project, the health and safety file is passed to the client, and if the principal designer's appointment concludes before the end of the project, they must pass it to the principal contractor.

Appendix 4 of L153 provides information on preparing, providing and retaining a health and safety file. It sets out that the following information should be considered for inclusion:

- a brief description of the work being carried out;
- any hazards that have not been eliminated through the design and construction processes, and how they have been addressed (such surveys or other information concerning asbestos or contaminated land);
- key structural principles (such as bracing, sources of substantial stored energy – including pre- or post-tensioned members) and safe working loads for floors and roofs;
- hazardous materials used (such as lead paints and special coatings);
- information regarding the removal or dismantling of installed plant and equipment (such any special arrangements for lifting such equipment);
- health and safety information about equipment provided for cleaning or maintaining the structure;
- the nature, location and markings of significant services, including underground cables; gas supply equipment; and fire-fighting services; and
- information and as-built drawings of the building, its plant and equipment (such as the means of safe access to and from service voids and fire doors).

Health and safety duties of the principal contractor during construction

[C8514] Regulation 13 deals with the health and safety duties of the principal contactor during the construction phase. The HSE guidance emphasises the importance of having clarity over who is in control in any part of the site at any given time.

The principle contractor fulfils the role of safety and health coordinator for the project execution stage. Their role is important in managing risks and providing strong leadership. They must plan, manage and monitor the construction phase and coordinate health and safety matters, taking into consideration the general principles of prevention, particularly when:

"(a) decisions are being taken to plan which items or stages of work can take place at the same time or in sequence; and
(b) estimating the time certain items or stages of work will take to complete."

They must organise cooperation and coordination between contractors, ensure that employers apply the general principles of prevention in order to protect workers and self-employed people and follow the construction phase plan. They must also ensure that a suitable site induction is provided.

The HSE guidance lists the following issues that should be considered in site inductions:

(a) senior management commitment to health and safety;
(b) project outline;

(c) project management;
(d) first-aid arrangements;
(e) accident and incident reporting arrangements;
(f) arrangements for briefing workers on an ongoing basis, such as toolbox talks;
(g) arrangements for consulting the workforce on health and safety matters; and
(h) individual worker's responsibility for health and safety.

The principal contractor must ensure that the site is suitably secured against unauthorised access and that welfare facilities comply with the requirements of Schedule 2 (see C8518 below and Construction and Building Operations) throughout the construction phase.

They must also liaise with the principal designer and share information about the planning, management and monitoring of, and health and safety coordination during, the pre-construction phase. The HSE says that regular planning meetings between the principal contractor and contractors are an effective way of ensuring this.

The HSE guidance provides more detail regarding:

Planning:

This must take into account the risks to all those affected, including workers, members of the public and the client's employees. It must cover the risks likely to arise during construction and the measures needed to protect those affected. It includes providing and maintaining the right plant and equipment, providing the necessary information, instruction and training and the right level of supervision, and providing the resources, including time, necessary to organise and deliver the work.

Management:

Principal contractors must ensure that those engaged to carry out the work are capable of doing so; that effective, preventative and protective measures are put in place to control the risks; and that the right plant, equipment and tools are provided to carry out the work involved.

Monitoring:

Standards should be checked regularly because of the rapidly changing nature of a construction site. It says that effective monitoring involves time and effort; treating health and safety in the same way as other important aspects of the business; taking prompt action where necessary; and using a mix of active and reactive performance measures. Active measures include routine checks of site access and work areas, plant and equipment, or health risk management to prevent harm. Reactive measures include investigating near-miss incidents and injuries and monitoring cases of ill health.

Coordinating:

Principle contractors must ensure that contractors who start work at different stages of the construction phase cooperate with each other so any information and instruction relevant for a new contractor to carry out their work safely is provided. This can be done through regular planning meetings. In coordinating the work of employers and self-employed under their control, principal contractors must ensure they apply the general principles of prevention and follow the construction phase plan.

"This will involve the principal contractor liaising with those involved to establish a common understanding of the health and safety standards expected and gaining their cooperation in meeting these standards," says the HSE. The extent to which they should liaise will depend on the risks involved. The principal contractor should also work with the client to ensure cooperation outside the construction site. This includes coordinating the activities of contractors on the site with those on any neighbouring sites, particularly where hazards need to be jointly addressed.

Consulting and engaging with workers – principal contractor's duties

[C8515] Regulation 14 sets out the principal contractor's duties to consult and engage with workers and their representatives. They have a duty under CDM 2015 to involve the workforce in matters of health, safety and welfare. This is in addition to the duty on all employers to consult with their employees (or their representatives) on health and safety matters under separate legislation:

- Under the *Health and Safety at Work etc Act 1974* and the *Management of Health and Safety at Work Regulations 1999* to utilise the knowledge and information held by all workers on a construction project;
- Under the *Safety Representatives and Safety Committees Regulations 1977 (SI 1977/500)* (as amended) to consult with trade union appointed safety representatives. These regulations set out in detail the functions of safety representatives and safety committees and employers' duties; and
- Under the *Health and Safety (Consultation with Employees) Regulations 1996 (SI 1996/1513)*. These Regulations give non-union employees the same rights to consultation that union safety representatives have but are more limited – for example there is not the right to carry out inspections of the workplace. They apply where there is no union or the union is not recognised in a workplace, or with regard to a particular group of workers (see **JOINT CONSULTATION IN SAFETY** chapter).

Workplaces where workers are consulted and engaged in decisions about health and safety measures are safer and healthier. Research for the HSE has shown that where such arrangements are in place there is a measurable improvement in safety culture and performance. The TUC report *The Union Effect How unions make a difference on health and safety* brings together evidence showing that employers who had trade union health and safety committees had half the injury rate of those employers who managed safety without unions or joint arrangements. The arrangements that lead to the highest injury rates are those where management deals with occupational health and safety without consultation. The HSE guidance on CDM 2015 (L153) directs employers towards guidance contained in the *Leadership and Worker Involvement Toolkit* (www.hse.gov.uk/construction/lwit/).

The principal contractor must "make and maintain" arrangements to enable the principal contractor and construction workers to cooperate effectively in developing, promoting and checking the effectiveness of health, safety and welfare measures.

They must consult workers or their representatives, in good time, on health and safety and welfare measures (where they have not been consulted by their employer).

Regulation 14 also requires the principle contractor to ensure that workers or their representatives can inspect and take copies of health and safety information (with certain exceptions including information relating to national security, information relating specifically to individuals unless they have given consent for it to be disclosed, and information that has been obtained for the purpose of bringing, prosecuting or defending legal proceedings).

Contractors' duties

[C8516] Regulation 15 sets out the duties of contactors. The HSE says that contractors have an important role in planning, managing and monitoring the work, in liaison with the principal contractor where appropriate, to ensure risks are properly controlled.

"The key to this is the proper coordination of the work, underpinned by good communication and cooperation with others involved," says HSE guidance to CDM 2015 (L153).

Their main duty is to plan, manage and monitor the work under their control. They must also comply with directions given to them by either the principal designer or principal contractor and the relevant parts of the construction phase plan on sites where there is more than one contractor; and prepare a construction phase plan on sites where they are the only contractor. Contractors must not carry out construction work unless they are satisfied that the client is aware of their duties and they must plan, manage, and monitor construction work under their control in order to ensure that it is carried out safely.

Where there is more than one contractor working on a project, a contractor must comply with any directions given by the principal designer or principal contractor and relevant parts of the construction phase plan.

If there is one contractor, they must take account of the general principles of prevention (see **C8511** above) when deciding design, technical and organisational aspects in order to plan the various items or stages of work taking place simultaneously or in succession, and estimating the time needed to complete this work. They must also draw up a construction phase plan, or make arrangements for a plan to be drawn up, as soon as practicable after setting up a construction site. The plan must meet the requirements of Regulation 12(2) (see **C8513** above).

The people contractors employ must have (or be in the process of obtaining) the necessary skills, knowledge, training and experience to carry out the tasks they are allocated.

They must also provide appropriate supervision, instruction and information to workers under their control. This must include: a suitable site induction;

details of emergency procedures; information identified by the risk assessment carried out under regulation 3 of the Management of Health and Safety at Work Regulations 1999 (also see **RISK ASSESSMENT**); information about another contractor's activities that workers should be aware of; and any other necessary information.

> "Sole reliance should not be placed on industry certification cards or similar being presented to them as evidence that a worker has the right qualities," the HSE warns. "Nationally recognised qualifications (such as National Vocational qualifications (NVQs) and Scottish Vocational Qualifications (SVQs)) can provide contractors with assurance that the holder has the skills, knowledge, training and experience to carry out the task(s) for which they are appointed. Contractors should recognise that training on its own is not enough. Newly trained individuals need to be supervised and given the opportunity to gain positive experience of working in a range of conditions."

With regard to establishing whether a worker requires training, the HSE advises that contractors should assess the existing health and safety skills, knowledge, training and experience of their workers; compare these with the range of skills, knowledge, training and experience they will need for the job; and identify any shortfall between the two.

> "The difference between the two will be the 'necessary training'," it says.

> Assessing training needs should be an ongoing process and further training may be necessary if the risks to which workers are exposed alter due to a change in their working tasks; new technology or equipment is introduced; or the system of work changes, says the HSE.

> Skills that are not regularly used can decline, the HSE advises, so people who deputise for others on an occasional basis may need more frequent further training than those who do the work regularly.

> Contractors should also consider "softer skills", including "the ability to foresee risk, maintain sensitivity to risk, anticipate mistakes others might make and to communicate clearly, as well as the more technical skills workers require for their work."

The level of supervision contractors must provide depends on the risks involved, and the skills, knowledge, training and experience of the workers concerned.

The HSE guidance sets out that young, inexperienced and new workers will require closer supervision, and the level of individuals' safety awareness, education, physical agility, literacy and attitude will be relevant to the level of supervision required.

> "Even experienced workers may need an appropriate level of supervision if they do not have some or all of the skills, knowledge, training and experience required for the job and the risks involved," says the HSE. "Workers should always know how to get supervisory help, even when a supervisor is not present."

The HSE guidance sets out that the information provided by contractors to workers and employees must include:

(a) suitable site induction where this has not been provided by the principal contractor;

(b) the procedures to be followed in the event of serious and imminent danger to health and safety. These should make clear that any worker exposed to any such danger should stop work immediately, report it to the contractor and go to a place of safety. The procedures should:
  (i) include details of who to report to and who has the authority to take whatever prompt action is needed;
  (ii) take account of emergency procedures, emergency routes and exits and fire detection and fire-fighting;
(c) information on the hazards on site relevant to their work (eg site traffic), the risks associated with those hazards and the control measures put in place (eg the arrangements for managing site traffic).

Contractors must take steps to prevent unauthorised access before beginning work on a construction site. They must also ensure, so far as is reasonably practicable, that the welfare facilities set out in Schedule 2 (see **C8518** below) are provided for their workers and sub-contractors. The duties of contractors for all projects can be summarised as follows—

(i) To check that the client is aware of their duties under the regulations before they start work;
(ii) To plan, manage and monitor, not just turn up and "do";
(iii) To give sufficient time for any sub-contractor or other contractor they appoint to plan and prepare before they start work;
(iv) To ensure site induction appropriate to the work and site-specific risks is provided. The development of method statements based on site-specific risk assessment is a key part of this process;
(v) To ensure that emergency procedures, such as evacuation, first aid and fire safety are in place and all personnel know the arrangements and who is in control;
(vi) To protect the public and to protect the client and others by ensuring adequate security arrangements are in place. This may include fencing, access control, CCTV and other security arrangements. The likelihood of trespass by children (for example due to proximity of a school or play area) is information that will need to be taken into account;
(vii) To ensure that adequate welfare facilities for the numbers of people working on site and the type of work in progress, as laid out in Schedule 2 (see **C8518** below), are in place and are maintained. For example, preliminary site work when overall numbers on site are low may involve higher exposure to risk to health due to carrying out work in live sewers or the removal of asbestos for example.

Part 4 of the Regulations – General requirements for all construction sites

Part 4 is covered in the CONSTRUCTION AND BUILDING OPERATIONS chapter.

Part 5 General

[C8517] The final section of the Regulations deals with the enforcement of fire safety provisions, transitional and saving provisions, and revocations, amendments and review by the Secretary of State.

Information in the Schedules

[C8518] Schedule 1 sets out the particulars to be included in a notification (under Regulation 6). The information required is as follows:

- Date of forwarding.
- Address or precise location of the construction site.
- The name of the local authority where the site is located.
- A brief description of the project and the construction work that it entails.
- Contact details of the client, principal designer and principal contractor.
- Date planned for the start of the construction phase.
- The time allowed by the client for construction work.
- Planned duration of the construction phase.
- Estimated maximum number of people at work on the construction site.
- Planned number of contractors on the construction site.
- Names and addresses of any contractor and designer already appointed.
- A declaration signed by or on behalf of the client that he/she is aware of his/her duties under these Regulations.

Schedule 2 contains details of the requirements for welfare facilities (sanitary conveniences, washing facilities, drinking water, changing rooms and lockers and facilities for rest. See CONSTRUCTION AND BUILDING OPERATIONS and regulations 4, 13 and 15 above (see **C8506, C8514** and **C8516**). Schedule 3 sets out work involving particular risks (see regulation 12 in **C8513** above). These include:

- work with the risk of burial, engulfment or falling from height;
- work with the risk of exposure to chemical or biological substances;
- work with ionising radiation;
- work near high voltage power lines;
- work exposing workers to the risk of drowning;
- work on wells, underground earthworks and tunnels;
- work carried out by divers with an air supply system;
- work carried out in caissons with a compressed air atmosphere;
- work with explosives; and
- work involving the assembly or dismantling of heavy prefabricated components.

Schedule 4 sets out transitional and saving provisions.

CDM 2015 recognised that some construction projects would have started before they came into force on 6 April 2015 and continue beyond this date.

Transitional provisions regarding the appointment of a principal designer were therefore set out in Schedule 4 of the Regulations but expired on 6 October 2015.

Other transitional provisions set out that:

- pre-construction information, construction phase plans or health and safety files provided under CDM 2007 are recognised as meeting the equivalent requirements in CDM 2015;
- any project notified under CDM 2007 is recognised as a notification under CDM 2015; and
- a principal contractor appointed under CDM 2007 will be considered to be a principal contractor under CDM 2015.

In all other circumstances, the requirements of CDM 2015 apply in full from 6 April 2015.

Schedule 5 lists amendment to Acts (including the Factories Act 1961) and regulations, recording the extent of the amendments.

Information in the Appendices in the HSE guidance to the Regulations: Managing health and safety in construction

Construction (Design and Management) Regulations 2015. Guidance on Regulations (L153).

[C8519]

Appendix 1 sets out the General principles of prevention (see **C8511** above);
Appendix 2 gives guidance on the requirements for pre-construction information and the actions on each duty-holder;
Appendix 3 gives guidance on the requirements for the construction phase plan and the actions on each duty-holder;
Appendix 4 gives guidance on the preparation, provision and retention of a health and safety file and the actions on each duty-holder;
Appendix 5 provides a summary of how different types of information (pre-construction information, the construction phase plan and the health and safety file) relate to and influence each other in a construction project involving more than one contractor; and
Appendix 6 contains information about working for a domestic client.

The Corporate Manslaughter and Corporate Homicide Act 2007

[C8520] Under Section 8 of the *Corporate Manslaughter and Corporate Homicide Act 2007*, (see CORPORATE MANSLAUGHTER), in a prosecution under the Act, jurors may "consider the extent to which the evidence shows that there were attitudes, policies, systems or accepted practices within the organisation that were likely to have encouraged" any failure. These are matters specifically covered by the CDM 2015 Regulations. Jurors may also "have regard to any health and safety guidance that relates to the alleged breach". "Health and safety guidance" means any code, guidance, manual or similar publication that is concerned with health and safety matters and is made or issued (under a statutory provision or otherwise) by an authority responsible for the enforcement of any health and safety legislation. It therefore includes *Managing health and safety in construction. Construction (Design and Management) Regulations 2015. Guidance on Regulations* (L153).

Health and Safety Executive (HSE) prosecutions under the Construction (Design and Management) Regulation 2015

[C8521] A number of recent HSE prosecutions have arisen from failures to comply with the Construction (Design and Management) Regulations 2015, including the following:

In May 2019, R W Hill (Felixstowe) Limited was fined £15,000 and ordered to pay costs of nearly £14,000 after pleading guilty to breaching Regulation 15(2) and (8) of the Construction (Design and Management) Regulations 2015 following a fatal incident. The HSE reported that in April 2015, concrete worker Garry Louis was killed at a dockside apron-upgrade project on the Port of Felixstowe. He was employed by a concrete laying sub-contractor. The main contractor, R W Hill (Felixstowe) Limited, had sub-contracted specialist contractors to pump and lay concrete. A flexible delivery hose through which concrete was being pumped became momentarily blocked, then cleared under pressure, causing it to violently whip round, killing Mr Louis and injuring a second worker. An HSE investigation found R W Hill (Felixstowe) Limited had failed to effectively plan and manage the safe pumping of concrete by not enforcing an exclusion zone around the flexible delivery hose. The investigation also found the company did not adequately supervise, instruct, or provide suitable information to sub-contractors, and failed to monitor the pumping operations to ensure the ongoing safety of workers.

Also in May 2019, principal contractor Williams Homes was fined £60,000 and ordered to pay costs of more than £7,000 after pleading guilty to breaching Regulation 15(2) of the Construction (Design and Management) Regulations 2015. In this case, a self-employed joinery sub-contractor suffered serious facial injuries when he was struck by a falling beam during the construction of a timber frame home in November 2015. An HSE investigation found Williams Homes did not closely supervise the work to ensure it was properly planned, managed and monitored.

In April 2019, principal contractor S McMurray Ltd was fined £16,500 and ordered to pay costs of more than £1,200 after pleading guilty to breaching Regulation 13(1) of the Construction (Design and Management) Regulations 2015. An HSE investigation found S McMurray Ltd had failed to safely install joist hangers correctly. There was no other structural support arrangement in place, such as propping the first floor from underneath, and the floor was overloaded with blockwork.

In February 2019, Balfour Beatty Group Employment Limited pleaded guilty to breaching Regulation 13(1) of the Construction (Design and Management)

Regulations 2015 and was fined £600,000. In January 2016, an employee was killed when he was struck by a wheeled excavator at the construction site of the Third Don Crossing. Aberdeen Sheriff Court heard that as principal contractor, the civil engineering contractor had failed to ensure that the safe system of work for refuelling of all plant and equipment was fully implemented. An HSE investigation found that refuelling of plant and equipment was identified as a high-risk activity by the principal contractor. It had created a task briefing document detailing a safe system of work and had risk assessed the activity. However, although these procedures existed in documentary format, the safe system of work and its control measures were not fully implemented.

Corporate Manslaughter

Andrea Oates

Background to the Corporate Manslaughter and Corporate Homicide Act 2007

[C9001] After many years in the making, the *Corporate Manslaughter and Corporate Homicide Act 2007* (CMCHA 2007) finally came into force on 6 April 2008. Under the Act, companies, organisations and government bodies face an unlimited fine, as well as remedial and publicity orders, if they are found to have caused death due to their gross corporate health and safety failures.

At the time of its enactment, the Ministry of Justice (MoJ) said: 'The Corporate Manslaughter Act is a landmark in law and the culmination of ten years of campaigning by unions and other groups. Well-run businesses that already have effective systems in place for managing health and safety have nothing to fear from the new legislation. But employees of companies, consumers and other individuals will be offered greater protection against the worst cases of corporate negligence.'

Prior to the 2007 Act, a company could be prosecuted for the common law offence of gross negligence manslaughter – generally referred to as 'corporate manslaughter'. In order to be guilty of the offence, a company had to be in gross breach of a duty of care owed to the victim who was killed.

However, it proved virtually impossible to successfully prosecute a large company for corporate manslaughter under the common law, even where management failures lead to a death. Instead, prosecutions were nearly always taken under health and safety legislation. This led to calls, particularly from the trades unions and safety campaigners, for the law to be reformed to allow companies whose gross negligence lead to a death to be convicted of the more serious offence of corporate manslaughter, with corresponding harsher penalties.

The key problem with the law as it stood was known as the 'identification principle'. Before a company could be prosecuted for corporate manslaughter, a single individual at the top of the company – 'a directing mind' who could be said to embody the company in his or her actions and decisions – had to be shown to be personally guilty of manslaughter. If this could not be shown, the company escaped liability.

While 3,425 workers were killed in work-related incidents between 1992 and 2005, there were only six successful work-related corporate manslaughter prosecutions over the same period. These all involved small companies or sole traders.

[C9001] Corporate Manslaughter

In larger companies it proved extremely difficult to identify an individual who embodied the company and was culpable, due to often complex management structures and lines of control, and lack of clarity in overall responsibilities for safety matters.

Corporate manslaughter prosecutions were taken against a number of large companies following disasters in which many people died, but all were unsuccessful.

For example, on 6 March 1987 the roll-on roll-off P&O European ferry, the Herald of Free Enterprise, sank off the coast of Zeebrugge, Belgium, as a result of sailing with its bow doors open. One hundred and fifty passengers and 38 crew died. In the subsequent prosecution case against seven individuals and the company, the court found that there was insufficient evidence to convict any of the seven personal defendants with gross negligence manslaughter. The case against the company therefore also failed. The company was automatically acquitted at the same time as its directors.

The judge ruled there was insufficient evidence to show that the risk of the ferry leaving port with its bow doors open was 'obvious and serious'. He also ruled that there was insufficient evidence of wrongdoing by a director or senior manager. The case failed because the various acts of negligence could not be aggregated and attributed to any one individual who was a directing mind of the company.

No other criminal prosecution was taken. When the ferry sank, the *Health and Safety at Work Act 1974 (HSWA 1974)* did not apply to deaths at sea, and no appropriate charges under merchant shipping legislation were available. A new offence was created in the *Merchant Shipping Act 1988* in response to this situation.

The case was, however, a landmark decision in that it acknowledged that a company could properly be charged with corporate manslaughter, and the collapse of the case also triggered demands for reform of the law. These demands were strengthened by other unsuccessful corporate manslaughter prosecutions against large organisations.

They included the prosecution of Great Western Trains (GWT) after one of its high-speed passenger trains, travelling from Swansea to London on the 19 September 1997, went through a red signal and collided with an empty freight train at Southall, killing seven people and injuring 151. The Automatic Warning System that would normally have alerted the driver if he passed a signal at danger was malfunctioning in one of the train's two power units.

In a pre-trial hearing, the judge ruled that a company could not be prosecuted for manslaughter by gross negligence unless a named person deemed to be a 'controlling mind' of the company was also prosecuted, and no such person had been charged.

But he also said that even having a named director would have made no difference in this case, because no director was personally responsible for ordering the running of the train that crashed.

In July 1999, GWT pleaded guilty to breaches of *section 3(1)* of the *HSWA 1974*, for failing to ensure that members of the public were not exposed to risks to their health and safety. A fine of £1.5 million was imposed for 'a serious fault of senior management', but relatives of those killed called this 'derisory' in view of the company's £300 million turnover. The Crown Prosecution Service (CPS) dropped manslaughter charges against the driver of the train.

This case was also significant as the Attorney-General referred the legal issues to the Court of Appeal – the first time that an appeal court had considered the law relating to corporate manslaughter.

Manslaughter charges were also brought against both Network Rail (formerly Railtrack PLC) and Balfour Beatty, together with six managers from both companies, following another derailment. On 17 October 2000, a London-to-Leeds train travelling at 115 mph was derailed as a result of a defective rail near Hatfield. The rail had been identified as suffering from a form of metal fatigue, commonly found where track curves, 21 months earlier. Four people were killed and 70 injured.

Corporate manslaughter charges against Network Rail were subsequently dropped in September 2004, and in July 2005 the judge in the case against Balfour Beatty and five rail executives dismissed the manslaughter charges, five months into the trial.

Mr Justice Mackay commented 'This case continues to underline a long and pressing need for the long-delayed reform of the law in this area of unlawful killing'.

The Director of Public Prosecutions (DPP), head of the CPS, at the time explained: 'The ruling turned on the level of negligence that the jury was being invited to consider, and the judge took the view that it was not sufficient to invite the jury properly to conclude that it was grossly negligent. He therefore directed the jury to find the defendants not guilty.'

Network Rail and Balfour Beatty were instead found guilty of breaching health and safety law. Network Rail was fined a record £3.5 million, and Balfour Beatty was fined £10 million, although this was reduced on appeal to £7.5 million. They were also ordered to each pay £300,000 in costs. The five individuals facing health and safety charges were found not guilty.

The *CMCHA 2007* replaced the identification principle with a new test, focusing on the overall picture of how an organisation's activities are managed by its senior management, rather than focusing on the actions of one individual.

Instead of having to demonstrate that one or more individuals are guilty, liability for the new offence of corporate manslaughter, called corporate homicide in Scotland, depends on a finding of gross negligence in the way in which the activities of the organisation are run. As the explanatory notes to the Act set out—

> In summary, the offence is committed where, in particular circumstances, an organisation owes a duty to take reasonable care for a person's safety and the way in which activities of the organisation have been managed or organised amounts to a gross breach of that duty and causes the person's death. How the activities were managed or organised by senior management must be a substantial element of the gross breach.

The Ministry of Justice (MoJ) explained that the *Corporate Manslaughter and Corporate Homicide Act*—

— Makes it easier to prosecute companies and other large organisations when gross failures in the management of health and safety lead to death by delivering a new, more effective basis for corporate liability;
— Has reformed the law so that a key obstacle to successful prosecutions has now been removed. Before the Act came into force, a company could only be convicted of manslaughter if a 'directing mind' (such as a director) at the top of the company was also personally liable;
— Means that both small and large companies can be held liable for manslaughter where gross failures in the management of health and safety cause death, not just health and safety violations;
— Does not apply to individual directors, senior managers or other individuals: it is concerned with the corporate liability of the organisation itself. However, where there is sufficient evidence, individuals can already be prosecuted for gross negligence manslaughter and for health and safety offences. The Act does not change this position; and
— Lifts Crown immunity to prosecution (see below). The Act meant that, for the first time, crown bodies, such as government departments, were now liable to prosecution. The Act applies to companies and other corporate bodies in the public and private sector, government departments, police forces and certain unincorporated bodies, such as partnerships, where these are employers.

A number of government bodies, and quasi government bodies, such as government departments, had been able to claim immunity from prosecution because they were said to be acting as a servant or agent of the Crown – known as Crown immunity. In addition, many Crown bodies did not have a separate legal identity for the purposes of a prosecution and the 2008 Act deals with these issues.

The Act also applies to deaths in custody. This was a key demand of the House of Lords as the Bill progressed through parliament. After the Bill had gone through all its stages in the House of Commons and House of Lords, it then passed between the two houses (a process known as "ping-pong") until they could reach agreement on the issue as to whether deaths in custody should be included. Only once agreement was reached could the Bill receive Royal Assent and become an Act.

Eventually agreement was reached when the government tabled an amendment meaning that on the face of the bill, the new offence would apply to deaths in custody. At a later date, 1 September 2011, the *Corporate Manslaughter and Corporate Homicide Act 2007 Commencement (No 3) Order 2011* brought into force the custody provisions contained in section 2(1)(d) and 2(2) of the Act.

An organisation, including a government department, can now be convicted of the offence of corporate manslaughter if the way in which its activities were managed or organised caused a person's death and amounted to a gross breach of the duty of care owed to that person by virtue of them being held in custody.

The *Corporate Manslaughter and Corporate Homicide Act 2007 (Amendment) Order 2011* also came into effect (in England, Wales and

Scotland) in September 2011 and widened the scope of section 2(2) to include two categories of people not already covered by the Act: those detained in Service custody premises, which are the responsibility of the Ministry of Defence (MoD), and those detained for custody purposes in UK Border Agency offices.

Although it had been widely assumed since the reform of the law in this area was first discussed by the Law Commission back in 1994 that there would be separate legislation for Scotland, the Act applies to the whole of the UK. In October 2007, the MoJ published detailed guidance on the implementation of the Act, including *A guide to the Corporate Manslaughter and Corporate Homicide Act 2007*. This MoJ advice is referred to throughout this chapter.

The chapter sets out the main requirements of the *CMCHA 2007* and summarises the Sentencing Council guidelines for corporate manslaughter (and health and safety) offences which came into effect on 1 February 2016. It also summarises Sentencing Council guidelines on gross negligence manslaughter, which came into force in November 2018 (see **C9021**).

The main provisions of the *CMCHA 2007* are as follows:

A SUMMARY OF THE CORPORATE MANSLAUGHTER AND CORPORATE HOMICIDE ACT 2007

— The offence is called corporate manslaughter in England, Wales and Northern Ireland, and corporate homicide in Scotland;
— An organisation is guilty of the offence if the way in which it organises or manages its activities causes a death, and this amounts to a gross breach of a relevant duty of care it owed to the victim;
— The offence applies where the organisation owed a duty of care to the victim under the law of negligence;
— It applies to deaths in detention and custody;
— The offence applies to corporations, Government departments (and other bodies listed in a schedule to the Act), police forces, partnerships, trade unions and employer associations (where these are employers); but does not apply to corporations sole;
— Crown bodies are not immune from prosecution under the Act;
— Public policy decisions, exclusively public functions and statutory inspections are excluded from the scope of the offence;
— The Act includes a number of exemptions: for certain military activities and for police and other law enforcement bodies in particular situations (including terrorism and civil unrest); in relation to responding to emergencies; and for statutory functions relating to child protection and probation functions;
— The Act sets out factors for the jury to consider when deciding if there has been a gross breach of a relevant duty of care, including whether health and safety legislation was complied with; whether the culture of the organisation tolerated breaches; and any relevant health and safety guidance;
— Consent of the Director of Public Prosecutions (DPP) is needed for proceedings for the offence to be instituted;
— There is no secondary liability for the offence. Individuals will not be liable for aiding, abetting, counselling or procuring the commission of, or in Scotland, being art and part of, the offence. Individuals can still be prosecuted for gross negligence manslaughter under the common law;
— The Act abolishes the common law offence of gross negligence manslaughter in relation to corporations; and
— An organisation found guilty of corporate manslaughter or corporate homicide will face an unlimited fine, and the courts have the power to make remedial orders and publicity orders.

Corporate manslaughter: the offence

[C9002] The offence is called corporate manslaughter in England, Wales and Northern Ireland, and corporate homicide in Scotland (*CMCHA 2007 s 1(5)*).

An organisation is guilty of the offence of corporate manslaughter or corporate homicide if the way in which it manages or organises its activities causes a death and this amounts to a gross breach of a relevant duty of care it owed to the victim (*CMCHA 2007 s 1(1)*). Relevant duties of care are set out in *Section 2* of the Act (see below).

The MoJ explained in guidance published shortly before the Act came into force that: 'The offence is concerned with the way in which an organisation's activities were managed or organised. Under this test, courts will look at management systems and practices across the organisation, and whether an adequate standard of care was applied to the fatal activity.'

Although this test is not linked to a particular level of management – instead it considers how an activity was managed within the organisation as a whole – in order to be convicted of the offence, the way in which its activities are managed or organised by its senior management must be a substantial element in the breach (*CMCHA 2007 s 1(3)*).

Senior management is defined as the people who play a significant role in decision-making in how all, or a substantial part, of the activities are managed or organised; or actually managing or organising all, or a substantial part, of the activities (*CMCHA 2007 s 1(4)*). Those in the direct chain of management, as well as those in strategic or regulatory compliance roles, are therefore included in the definition.

The MoJ provided further guidance in this area:

> These are the people who make significant decisions about the organisation, or substantial parts of it. This includes both those carrying out headquarters functions (for example, central financial or strategic roles or with central responsibility for, for example, health and safety) as well as those in senior operational management roles.
>
> Exactly who is a member of an organisation's senior management will depend on the nature and scale of an organisation's activities. Apart from directors and similar senior management positions, roles likely to be under consideration include regional managers in national organisations and the managers of different operational divisions.

The offence does not require individual failings by senior managers to be identified. It is sufficient to show that senior management collectively were not taking adequate care and that this was a substantial part of the failure. In *R v Cornish* [2015] EWHC 2967 (QB), the court ruled that the Crown is not required to identify the individuals it believed failed to carry out their management functions properly. It must, however, identify the tier of management it considers to be the lowest level of the senior management team culpable of the offence.

The MoJ guidance also made clear that the offence cannot be avoided by senior management delegating responsibility for health and safety, and that inappropriately delegating health and safety matters will leave organisations vulnerable to a charge of corporate manslaughter or homicide. The courts will consider how responsibility was discharged at different levels of the organisation.

It advised: 'This does not mean that responsibility for managing health and safety cannot be made a matter across the management chain. However, senior management will need to ensure that they have adequate processes for health and safety and risk management in place and are implementing these.'

The MoJ referred to the guidance on directors' health and safety responsibilities produced by the Health and Safety Executive (HSE) and the Institute of Directors (IoD) (see C9009 below).

In order to amount to a 'gross' breach of duty of care, the conduct of the organisation must have fallen far below what can be reasonably expected in the circumstances (*CMCHA 2007 s 1(4)*), reflecting the threshold for the common law offence of gross negligence manslaughter.

The way in which the organisation's activities were managed or organised must also have caused the death. The usual principles of causation in criminal law apply – the management failure need not have been the sole cause of death; it need only be a cause. However, it must have made more than a minimal contribution to the death and an intervening act must not have broken the chain of events linking the management failure to the death. An intervening act will only break the chain of causation if it is extraordinary.

The MoJ provided further guidance on causation. It advised that although it will not be necessary for the management failure to have been the sole cause of death, '"but for" the management failure (including the substantial element attributable to senior management), the death would not have occurred'.

It also set out that the law does not recognise very remote causes, and in some circumstances, an intervening act may mean that the management failure is not considered to have caused the death.

Section 1(2) of the Act sets out that the offence applies to—

- Corporations — defined as a body corporate, whether incorporated in the United Kingdom or elsewhere;
- Government departments and other bodies listed in *Schedule 1*. These are Crown bodies which do not have a separate legal personality – *Section 11* (see below) sets out that there is no Crown Immunity under the Act;
- Police forces; and
- Partnerships, and trade union and employers' associations where these are employers, Partnerships within the *Partnership Act 1890* and limited partnerships registered under the *Limited Partnerships Act 1907* and similar firms and entities come within the scope of the Act (*CMCHA 2007 s 25*).

Companies incorporated under company law, and bodies incorporated under statute, as is the case with many non-Departmental Public Bodies and other bodies in the public sector, or by Royal Charter are all covered. But corporations sole, which cover a number of individual offices in England and Wales and Northern Ireland, are specifically excluded (*CMCHA 2007 s 25*).

[C9002] Corporate Manslaughter

The 2000 report, *Reforming the Law on Involuntary Manslaughter: The Government's Proposals*, provides the following definition of corporations sole: 'a corporation constituted in a single person in right of some office or function, which grants that person a special legal capacity to act in certain ways. Examples include many Ministers of the Crown and government officers, for example the Secretary of State for Defence and the Public Trustee and a bishop (but not a Roman Catholic bishop), a vicar, archdeacon, and canon.'

The list of organisations to which the offence applies can be further extended by the Secretary of State, for example to further types of unincorporated association. This is subject to approval in both Houses of Parliament before it would come into effect, known as the affirmative resolution procedure (*CMCHA 2007 s 21*).

The MoJ provided more guidance on how the offence applies, or does not apply, to organisations in a number of circumstances. It set out that:

- a parent company cannot be convicted because of failures within a subsidiary (although see CAV Aerospace Ltd and CAV Cambridge Ltd below): 'Companies within a group structure are all separate legal entities and therefore subject to the offence separately. In practice, the relevant duties of care that underpin the offence are more likely to be owed by a subsidiary than a parent';
- the offence applies to foreign companies: the offence 'applies to all companies and other corporate bodies operating in the UK, whether incorporated in the UK or abroad';
- where a company incorporated abroad is operating through a locally-registered subsidiary, it is the subsidiary that is likely to be investigated and prosecuted;
- where sub-contractors are involved, whether a particular contractor could be liable for the offence will firstly depend on whether they owed a relevant duty of care to the victim. The offence applies in respect of existing obligations on the main contractor and sub-contractors for the safety of worksites, employees and other workers they supervise;
- the offence applies to charities and voluntary organisations where these have been incorporated, as a company or as a charitable incorporated organisation under the Charities Act 2006, for example. It will also apply where a charity or voluntary organisation operates as any other form of organisation to which the offence applies, such as a partnership with employees.

The MoJ guidance does not mean that parent companies cannot be convicted. In July 2015 CAV Aerospace Ltd, the parent company of CAV Cambridge Ltd, was found guilty of corporate manslaughter following the death of Paul Bowers (see C9029). The CPS reported that: 'The senior management of CAV Aerospace Ltd [the parent company] received clear, unequivocal and repeated warnings over a sustained period of years prior to the fatal incident. Some of the most obvious solutions were rejected on cost grounds.' It failed to act on safety risks even though it had been warned ahead of the fatal accident that there were potentially disastrous consequences if nothing significant was done to address dangerously high levels of stock in the warehouse where Mr Bowers worked.

The meaning of 'relevant duty of care'

[C9003] The offence only applies where an organisation owed a duty of care, under the law of negligence, to the person who died, reflecting the common law offence of gross negligence manslaughter. Section 2(1) lists the following as a 'relevant duty of care':

- a duty owed to its employees or to other persons working for the organisation or performing services for it;
- a duty owed as occupier of premises;
- a duty owed in connection with—
  - the supply by the organisation of goods or services (whether for payment or not),
  - the carrying out by the organisation of any construction or maintenance operations,
  - the carrying out by the organisation of any other activity on a commercial basis, or
  - the use or keeping by the organisation of any plant, vehicle or other thing.

A relevant duty of care also arises where the organisation is responsible for the safety of person in custody or detention (*CMCHA 2007 s 2(1) and s 2(2)*).

There is provision for the categories of people to whom a duty of care is owed because they are in custody or detention to be extended by the Secretary of State.

As a result of the *Corporate Manslaughter and Corporate Homicide Act 2007 Commencement (No 3) Order 2011*, the *Corporate Manslaughter and Corporate Homicide Act 2007 (Amendment) Order 2011*, the *Immigration Act 2014 and the Criminal Justice and Courts Act 2015*, the Act now applies to deaths of people owed a duty of care by virtue of:

- being detained at a custodial institution, that is a prison, a young offender institution, a secure training centre, a secure college, a young offenders centre, a juvenile justice centre or a remand centre; in a custody area at a court or police station or UK Border Agency offices for customs purposes; and in service custody premises which are the responsibility of the Ministry of Defence;
- being detained at a removal centre, short-term holding facility, or pre-departure accommodation;
- transported in a vehicle or being held awaiting prison or immigration escort arrangements;
- living in secure accommodation in which they have been placed; or
- being a detained patient.

The following are 'relevant' duties under the Act—

- The duty to provide a safe system of work for employees. The breach of a duty owed to other people whose work the organisation controls or directs, but who are not formally employed, such as contractors, volunteers and secondees can also trigger the offence;
- Duties to ensure that buildings the organisation occupies are kept in a safe condition;
- Duties owed by organisations to their customers, such as those owed by transport providers to their passengers and by retailers for the safety of their products. It also covers the supply of services by the public sector, such as NHS bodies providing medical treatment;
- The duty of care owed by public sector bodies to ensure that adequate safety precautions are taken, when repairing a road for example, even where duties do not arise because they are not supplying a service or operating commercially; and
- The duty of care owed by organisations, such as farming and mining companies, for example, which are carrying out activities on a commercial basis, though not supplying goods and services.

The MoJ explained that statutory duties owed under health and safety law are not 'relevant' duties for the new offence – only a duty of care owed in the law of negligence. It explained that:

> In practice, there is a significant overlap between these types of duty. For example, employers have a responsibility for the safety of their employees under the law of negligence and under health and safety law (see for example section 2 of the Health and Safety at Work Act 1974 and article 4 of the Health and Safety at Work (Northern Ireland) Order 1978). Similarly, both statutory duties and common law duties will be owed to members of the public affected by the conduct of an organisation's activities.
>
> The common law offence of gross negligence manslaughter in England and Wales and Northern Ireland is based on the duty of care in the law of negligence, and this has been carried forward to the new offence. In Scotland, the concepts of negligence and duty of care are familiar from the civil law.

The offence applies where the duty of care owed under the common law of negligence has been superseded by statutory provision, including where this imposes strict liability. For example, the duty of care owed by an occupier, which is now owed under the *Occupiers' Liability Acts 1957* and *1984* and the *Defective Premises Act 1972 (CMCHA 2007 s 2(4))*, is included.

The MoJ guidance explained that in some cases where a person cannot be sued under the civil law of negligence, the offence may still apply. For example, this would be the case where a "no fault" scheme for damages has been introduced.

The offence is not affected by common law rules precluding liability in the law of negligence where people are jointly engaged in a criminal enterprise ("ex turpi causa non oritur actio") or because a person has accepted a risk of harm ("volenti non fit injuria") (*CMCHA 2007 s 2(6)*).

Whether a duty of care exists in a particular case is a matter of law for the judge to decide (*CMCHA 2007 s 2(5)*). Normally in criminal proceedings, questions of law are decided by the judge, while questions of fact, and the application of the law to the facts of the case, are for the jury, directed by the judge. However, because of the heavily legal nature of the tests relating to the existence of a duty of care in the law of negligence, the Act sets out that the judge will need to determine some facts, rather than the jury – for example whether the person killed was an employee of the organisation.

Public policy decisions, exclusively public functions and statutory inspections

[C9004] Public policy decisions, exclusively public functions and statutory functions are excluded from the scope of the offence – any duty of care owed by a public authority in respect of these is not a "relevant duty of care" (*CMCHA 2007 s 3(1)–(3)*).

Deaths alleged to have been caused by decisions of public authorities – the definition of which includes government departments, local councils and other public bodies, are therefore outside the scope of the offence. This would include, for example, decisions made by Health Service bodies about the funding of particular treatments.

Strategic funding decisions and other matters involving competing public interests are exempt, but decisions about how resources were managed are not.

An organisation will not be liable for a breach of any duty of care owed in respect of things done in the exercise of "exclusively public functions", unless the organisation owes the duty in its capacity as an employer or as an occupier of premises.

Any organisation, not just Crown or other public bodies, performing that particular type of function is excluded, although this does not affect individual liability. These functions continue to be subject to other forms of accountability such as independent investigations, public inquiries and the accountability of Ministers through Parliament.

Organisations with a duty of care owed in connection with the carrying out of statutory inspections, such as those carried out by health and safety enforcement authorities, will not be liable unless they owe duties as an employer or occupier of premises (*CMCHA 2007 s 3(3)*).

"Exclusively public functions" are those falling within the prerogative of the Crown, such as providing services in a civil emergency, and activities that require a statutory or prerogative basis and cannot be independently performed by private bodies, such as licensing drugs (*CMCHA 2007 s 3(4)*).

The MoJ explained that:

> This does not exempt an activity simply because statute provides an organisation with the power to carry it out (as is the case, for example, with legislation relating to NHS bodies and local authorities). Nor does it exempt an activity because it requires a licence (such as selling alcohol). Rather, the activity must be of a sort that cannot be independently performed by a private body. The type of activity involved must intrinsically require statutory or prerogative authority, such as licensing drugs or conducting international diplomacy.

It also explained that private companies carrying out public functions are broadly in the same position as public bodies: "Overall the Act is intended to ensure a broadly level playing field under the new offence for public and private sector bodies when they are in a comparable situation."

Military activities

[C9005] Certain military activities are exempt in respect of all categories of relevant duty of care (*CMCHA 2007 s 4*).

These include peacekeeping operations, operations for dealing with terrorism, civil unrest or serious public disorder, in the course of which members of the armed forces come under attack or face the threat of attack or violent resistance; as well as activities preparing for or supporting these operations. The exemption extends to training involving hazardous activities, and to the activities carried out by members of the special forces.

In July 2016, the government rejected proposals to remove Crown Immunity from the Ministry of Defence (MoD) and allow it to be charged with corporate manslaughter where military personnel are killed in training. The House of Commons Defence Committee recommended the move in its April 2016 report (*Beyond endurance? Military exercises and the duty of care* https://publications.parliament.uk/pa/cm201516/cmselect/cmdfence/598/598.pdf).

The previous month, the Health and Safety Executive (HSE) announced it would issue a Crown censure to the MoD over the deaths of three soldiers on a training exercise in the Brecon Beacons in early July 2013. Crown censure is an administrative procedure whereby the HSE may summon a Crown employer to be censured for a breach of the Act or a subordinate regulation which, but for Crown Immunity, would have led to prosecution with a realistic prospect of conviction.

Reservists Edward Maher, James Dunsby and Craig Roberts fell ill while on a training test march in temperatures of more than 31oC on the hottest day of the year. Lance Corporal Craig Roberts and Trooper Edward Maher died during the exercise. Corporal James Dunsby suffered multiple organ failure as a result of hyperthermia and died on 30 July 2013. The HSE investigation found a failure to plan, assess and manage risks associated with climatic illness during the training. These failings resulted in the deaths of the three men and heat illness suffered by ten other reservist casualties on the march.

Two officers who oversaw the SAS selection march were cleared of negligence in a military court in September 2018.

Policing and law enforcement

[C9006] Reflecting the law of negligence, the police and other law enforcement bodies are exempt in respect of all categories of relevant duty of care, including those as an employer or occupier, with respect to the following circumstances:

- operations dealing with terrorism, civil unrest or serious disorder in which an authority's officers or employees come under attack or the threat of attack; or
- where the authority in question is preparing for or supporting such operations; or
- where it is carrying on training with respect to such operations (*CMCHA 2007 s 5(1), (2)*).

A wider range of policing and law enforcement activities are excluded from the offence where the pursuit of law enforcement activities has resulted in a fatality to a member of the public, but not in respect of the duty of care owed as an employer or occupier (*CMCHA 2007 s 5(3)*).

Decisions about and responses to emergency calls, the manner in which particular police operations are conducted, the way in which law enforcement and other coercive powers are exercised, measures taken to protect witnesses, and the arrest and detention of suspects are not, for example, covered by the Act.

The exemption is not confined to police forces and extends to other bodies operating similar functions and to other law enforcement activity, such as action by the immigration authorities to arrest, detain, or deport an immigration offender. Note however, that the Act does not have any bearing on the question of individual liability.

For example, in March 2017, following a trial at Bristol Crown Court, a police sergeant and two custody detention officers were cleared of gross negligence manslaughter charges following the death of Thomas Orchard.

Mr Orchard, who had a history of serious mental illness, was arrested and restrained in Exeter on the morning of 3 October 2012. He was taken to a police station custody unit and removed from a police van into the holding area and then a cell. Emergency medical assistance was given and he was taken to hospital, but he later died.

In October 2018, the Office of the Chief Constable for Devon and Cornwall Police pleaded guilty to charges under the Health and Safety Act, in relation to the force's use of an Emergency Response Belt (ERB) to restrain him. It was the first ever guilty plea on health and safety charges from a police force in relation to a death in custody. In May 2019, the force was handed down a fine of £234,500 for health and safety breaches.

Emergencies and the emergency services

[C9007] The offence does not apply to the emergency services when responding to emergencies (*CMCHA 2007 s 6(1)*). However, the emergency services still have duties as employers and occupiers, so must still provide a safe system of work for their employees and secure the safety of their premises.

Organisations to which the exemption applies are—
- English and Welsh fire and rescue authorities;

- the Scottish Fire and Rescue Service;
- the Northern Ireland Fire and Rescue Service Board;
- any other organisation responding to an emergency either for one of the organisations listed above, or if not, otherwise than operating commercially;
- NHS bodies;
- an organisation providing ambulance services for a relevant NHS body or the Secretary of State or Welsh Ministers;
- an organisation providing services for the transport of organs, blood, equipment or personnel in pursuance of arrangements of the kind mentioned in the paragraph above;
- an organisation providing a rescue service, such as the Coastguard and the Royal National Lifeboat Institution (RNLI);
- the armed forces.

Emergency circumstances are defined as 'circumstances that are present or imminent' and are causing, or are likely to cause, serious harm or a worsening of such harm, or are likely to cause the death of a person (*CMCHA 2007 s 6(7)*). 'Serious harm' means serious injury to, or the serious illness (including mental illness) of, a person; serious harm to the environment (including the life and health of plants and animals); or serious harm to any building or other property.

Circumstances believed to be emergency circumstances are also covered (*CMCHA 2007 s 6(8)*).

Medical treatment, and decisions relating to this, other than those to establish the priority for treating patients, is not included in the exemption (*CMCHA 2007 s 6(3), (4)*). Nor does the exemption apply to duties that do not relate to the way in which a body responds to an emergency – so duties to maintain vehicles in a safe condition, for example, remain.

The MoJ advised that with regard to NHS trusts (including ambulance trusts) duties of care relating to medical treatment in an emergency, other than triage decisions (which determines the order in which injured people are treated), are not exempt.

In January 2016, a judge ruled that Maidstone and Tunbridge Wells NHS trust had no case to answer in the first corporate manslaughter prosecution brought against a health service trust. The case concerned the death of Frances Cappuccini, who died hours after giving birth to her second child at Tunbridge Wells hospital in Kent in October 2012. An anaesthetist accused of gross negligence manslaughter was also acquitted.

Child protection and probation functions

[C9008] The offence does not apply in relation to carrying out (or failing to carry out) statutory functions relating to child protection and probation (*CMCHA 2007 s 7(1)*). However, local authorities and probation services are covered by the offence with respect to their responsibilities to employees, and as occupiers of premises.

This section applies to any duty of care that a local authority or other public authority owes in respect of exercising its functions under:

- Parts 4 and 5 of the *Children Act 1989*;
- Part 2 of the *Children (Scotland) Act 1995*;
- the *Children's Hearings (Scotland) Act 2011*; or
- Parts 5 and 6 of the *Children (Northern Ireland) Order 1995*.

It also applies to any duty of care that a local probation board, a provider of probation services or other public authority owes in respect of exercising its functions under:

- Chapter 1 of Part 1 of the Criminal Justice and Court Services Act 2000;
- section 13 of the Offender Management Act 2007;
- section 27 of the Social Work (Scotland) Act 1968; or
- Article 4 of the Probation Board (Northern Ireland) Order 1982.

However, local authorities and probation services are covered by the offence with respect to their responsibilities to employees, and as occupiers of premises.

Factors for the jury to consider when deciding if there has been a 'gross' breach

[C9009] Where it has been established that an organisation owed a relevant duty of care to the victim and it falls to the jury to decide whether there was a gross breach of the duty, the jury must consider whether the evidence shows that the organisation failed to comply with health and safety legislation relating to the breach. If that is the case, it must look at how serious the failure was and how much of a risk of death it posed (*CMCHA 2007 s 8 (1), (2)*).

The jury may also consider the wider context, including cultural issues within the organisation, such as attitudes and accepted practices that tolerated breaches, and it may consider any relevant health and safety guidance.

Health and safety guidance does not provide an authoritative statement of required standards, and the jury is not therefore required to consider the extent to which this is not complied with. However, where breaches of relevant health and safety duties are established, health and safety guidance may assist a jury in considering how serious this was.

These factors are not exhaustive, and the jury can also have regard to any other matters they consider relevant (*CMCHA 2007 s 8 (4)*).

Relevant health and safety guidance includes statutory Approved Codes of Practice and other guidance published by regulatory authorities that enforce health and safety legislation.

The MoJ advised that: 'Employers do not have to follow guidance and are free to take other action. But guidance from regulatory authorities may be helpful to a jury when considering the extent of any failures to comply with health and safety legislation and whether the organisation's conduct has fallen far below what could reasonably have been expected.'

In addition to guidance from the HSE (and in Northern Ireland from the HSE Northern Ireland (HSENI)) and local authorities it advised: 'There are specific regulatory bodies, and in some cases separate legislation too, for certain sectors of industry (for example, in the various transport sectors: rail, marine, air and roads) and for dealing with particular safety issues (such as food and environmental safety). Further information about the standards that apply in these circumstances should be obtained from the relevant regulatory authority.'

For example, the Civil Aviation Authority (CAA) enforces health and safety provisions in respect of relevant civil aviation workers; the Driver and Vehicle Licensing Agency (DVLA) enforces provisions in respect of relevant road transport workers; and the Office of Rail and Road (ORR) and the Office for Nuclear Regulation (ONR) also have enforcement powers.

The MoJ added: 'Factors that might be considered will range from questions about the systems of work used by employees, their level of training and adequacy of equipment, to issues of immediate supervision and middle management, to questions about the organisation's strategic approach to health and safety and its arrangements for risk assessing, monitoring and auditing its processes. In doing so, the offence is concerned not just with formal systems for managing an activity within an organisation, but how in practice this was carried out.'

The HSE and IoD guidance on directors' responsibilities on health and safety *Leading health and safety at work – leadership actions for directors and board members* (INDG 417) (www.hse.gov.uk/leadership) is addressed to directors and their equivalents of corporate bodies and organisations in the public and voluntary sectors – including governors, trustees and officers.

Although the guidance is not obligatory, the HSE and IoD advise that it could be a relevant consideration for a jury depending on the circumstances of the particular case brought under the CMCHA 2007. It sets out that directors can be personally liable for breaches of health and safety law, and that members of the board have both collective and individual responsibility for health and safety.

Penalties and Sentencing

[C9010] Companies convicted of the offence of corporate manslaughter face very large fines, particularly as a result of the introduction of Sentencing Council guidelines for corporate manslaughter offences on 1 February 2016. These aimed to increase penalties for serious offending and ensure that fines are fair and proportionate to the seriousness of the offence and the means of offenders (see **C9011** below). They can also be given remedial and publicity orders (see **C9012** and **C9013** below).

Fines for corporate manslaughter

[C9011] An organisation found guilty of corporate manslaughter is liable on conviction on indictment to an unlimited fine (*CMCHA 2007 s 1(6)*) and the court may also make remedial orders and publicity orders (*CMCHA 2007 ss 9, 10*). The offence is triable in the Crown Court in England and Wales and the High Court of Justiciary in Scotland (*CMCHA 2007 s 1(7)*). Both involve trials by jury.

On 1 February 2016, Sentencing Council guidelines for corporate manslaughter (as well as health and safety and food safety and hygiene offences) came into force. The *Health and Safety Offences, Corporate Manslaughter and Food Safety and Hygiene Offences Definitive Guideline* sets out a range of sentences for both organisations and individuals (aged 18 and over), specifying the range of sentences appropriate for each type of offence; a number of categories reflecting varying degrees of seriousness within each offence range; and a starting point within each category for calculating the provisional sentence.

Once the provisional sentence has been calculated, they direct the court to take into account elements such as aggravating and mitigating factors in order to make a final adjustment to the sentence before considering a reduction for a guilty plea (of around a third).

Aggravating factors are:

- Previous convictions, having regard to the nature of the offence to which the conviction relates and its relevance to the current offence and the time that has elapsed since the conviction;
- Cost-cutting at the expense of safety;
- Deliberate concealment of any illegal nature of the activity;
- Breach of any court order;
- Obstruction of justice;
- A poor health and safety record;
- Falsification of documentation or licences;
- Deliberate failure to obtain or comply with relevant licences in order to avoid scrutiny by authorities; and
- The exploitation of vulnerable victims.

Mitigating factors are:

- No previous convictions or no relevant/recent convictions;
- Evidence of steps taken to remedy the problem;
- A high level of co-operation with the investigation, beyond that which will always be expected;
- A good health and safety record;
- Effective health and safety procedures in place;
- Self-reporting, co-operation and acceptance of responsibility; and
- Other events beyond the responsibility of the offender contributed to the death, although the actions of victims are unlikely to be considered contributory events.

For both health and safety and corporate manslaughter offences, the guidelines set out that the court must first determine the offence category, considering both 'culpability' and 'harm'. There is a very high level of culpability, for example, in cases involving deliberate breach or flagrant disregard for the law. In contrast, a low level of culpability is where the offender did not fall far short of the appropriate standard because, for example, they made significant efforts to address the risk but these were inadequate on this particular occasion, there was no warning or circumstances indicating a risk to health and safety, and the failings were minor and occurred as an isolated incident.

In order to categorise harm, courts must examine both the seriousness of the harm and the likelihood of it arising. They must also consider whether a number of people were at risk and also whether the offence was a significant cause of the actual harm.

Once the court has determined the offence category, the guidance refers them to tables setting out starting points and category ranges for organisations of different sizes. For organisations with a turnover of £50 million and over, which commit serious health and safety offences where there is a very high level of culpability and a high risk of harm, the starting point for the fine is £4 million and the category range is £2.6 million to £10 million. The guidelines also set out that: 'Where an organisation's turnover or equivalent very greatly exceeds the threshold for large organisations, it may be necessary to move outside the suggested range to achieve a proportionate sentence.'

In the case of corporate manslaughter offences, by definition the harm and culpability involved will be very serious and every case will involve death and corporate fault at a high level. In order to determine the seriousness of the offence the court will assess factors including:

- how foreseeable the serious injury was;
- how far short of the appropriate standard the offender fell;
- how common this kind of breach in this organisation is; and
- if there was more than one death, or a high risk of further deaths, or serious personal injury in addition to death.

The answers to these questions will determine whether the offence is category A (a high level of harm or culpability within the context of the offence) or B (a lower level of culpability).

For large organisations with a turnover of more than £50 million, where the offence category is A, the starting point for fines is £7.5 million and the range is £4.8 million to £20 million. Again, the guidelines state that where the turnover greatly exceeds the threshold, a higher fine may be necessary in order to achieve a proportionate sentence.

The guidelines also set out a range of penalties for individuals who have breached health and safety law, including prison sentences of up to two years for offences involving a very high level of culpability and high risk of harm. And they say that the court must consider whether to disqualify an offender from being a company director for a maximum of 15 years (if the case is being heard in the Crown Court) or five years (if it is being heard in the magistrates' court).

A Sentencing Council assessment of the guideline examined ten pre-guideline cases over the period from January to October 2015 and six post-guideline cases over the period from February to November 2016. It found the level of fines imposed on organisations sentenced for corporate manslaughter may have increased since the guideline came into force, as anticipated, but said the finding should be treated with caution due to low volumes.

Health and Safety Offences, Corporate Manslaughter and Food Safety and Hygiene Offences Definitive Guideline can be found on the Sentencing Council website at: www.sentencingcouncil.org.uk.

Assessing the impact and implementation of the Sentencing Council's Health and Safety Offences, Corporate Manslaughter and Food Safety and Hygiene Offences Definitive Guideline can be found at: www.sentencingcouncil.org.uk/wp-content/uploads/Health-and-safety-guideline-assessment.pdf.

Remedial Orders

[C9012] An organisation convicted of corporate manslaughter or corporate homicide may also be issued with a remedial order by the court, requiring it to take specific steps to remedy the breach; any matter the court believes to have resulted from the breach and caused the death; and any deficiencies in the organisation's health and safety policies, systems or practices of which the breach appear to be an indication (*CMCHA 2007 s 9(1)*).

Remedial orders can only be made on an application by the prosecution specifying the terms of the proposed order and after consultation with the health and safety enforcement authority, and there is the opportunity for the convicted organisation to make representations in court (*CMCHA 2007 s 9(2), (3)*).

The order must specify how long the organisation has to carry out the specified steps and may require that evidence showing the order has been complied is given to the enforcement authority (*CMCHA 2007 s 9(4)*). Failure to comply with the order is an offence and liable on conviction to an unlimited fine (*CMCHA 2007 s 9(5)*).

The MoJ explained that remedial orders would be used in relatively rare circumstances. The sanction was already available under *HSWA 1974 s 42* for health and safety offences but was not previously possible in relation to a manslaughter conviction. However, remedial orders for health and safety offences have only been used in a handful of cases.

The Sentencing Council guideline says that by the time of sentencing, an offender should have remedied any specific failings involved in the offence and if it has not, 'will be deprived of significant mitigation'. If, however, it has not, a remedial order should be considered if it can be made sufficiently specific to be enforceable.

No remedial orders have been handed down in relation to a corporate manslaughter conviction to date.

Publicity Orders

[C9013] The court may, after consulting the enforcing authorities and listening to representations from the prosecution, also require that an organisation convicted of corporate manslaughter or corporate homicide publicises the fact that it has been convicted of the offence; and specify the particulars of the offence, the amount of any fine, and the terms of any remedial order (*CMCHA 2007 s 10(1)*).

Again, the order must specify how long the organisation has to comply and may require that evidence showing the order has been complied is given to the enforcement authority (*CMCHA 2007 s 10(3)*). Failure to comply with the order is an offence and liable on conviction to an unlimited fine (*CMCHA 2007 s 10(4)*).

The Sentencing Council guidelines say that a publicity order should ordinarily be imposed in a case of corporate manslaughter and may require publication in a specified manner of:

- the fact of conviction;
- specified particulars of the offence;
- the amount of any fine;
- the terms of any remedial order (see **C9012**).

The order should normally specify the place where the public announcement is to be made, and the court should consider indicating the size of any notice or advertisement required. It should usually contain a provision designed to ensure that the conviction becomes known to shareholders, in the case of companies, and local people in the case of public bodies.

The court should also consider requiring a statement on the offender's website; although a newspaper announcement may not be necessary if the proceedings are certain to receive news coverage. If an order does require publication in a newspaper, it should specify the paper, the form of announcement to be made and the number of insertions required.

The guidelines also say that the prosecution should provide the court in advance of the sentencing hearing, and should serve on the offender, a draft of the form of order suggested. The judge should personally endorse its final form. The court should also consider stipulating that any comment placed by the offender alongside the required announcement should be separated from it and clearly identified as such.

Publicity orders have been handed out to several organisations convicted of corporate manslaughter offences, including Princes Sporting Club, Peter Mawson Ltd and Cheshire Gates and Automation Ltd for example (see **C9092**).

Application to Crown bodies

[C9014] Crown bodies are not immune from prosecution under the Act (*CMCHA 2007 s 11(1)*). Crown bodies that are either bodies corporate or are listed in Schedule 1 to the Act are subject to the offence. A Crown body is to be treated as owing the duties it would owe if it were a corporation that was not a servant or agent of the Crown (*CMCHA 2007 s 11(2)*). *Sections 11(3) and (4)* deal with the technicality which means that civil servants in government departments are employed by the Crown rather than the government department they work in, ensuring that the activities and functions of government departments and other Crown bodies can be attributed to the relevant body. Similarly, *section 11(5)* ensures the provisions apply to Northern Ireland departments in the same way as they apply to bodies listed in *Schedule 1*.

Application to the armed forces

[C9015] A duty of care is owed to personnel in the armed forces – including the Royal Navy, Army and Air Force – by the Ministry of Defence for the purposes of the offence (*CMCHA 2007 s 12*) (although see **C9005** above).

Application to police forces

[C9016] Police forces are not incorporated bodies. However, under the Act, a police force is treated as owing the duties of care it would owe if it were a body corporate (*CMCHA 2007 s 13*). Therefore, police officers, as well as police cadets and police trainees in Northern Ireland, and police officers seconded to the Serious Organised Crime Agency or the National Policing Improvement Agency are treated as the employees of the police force, or other organisation, for which they work and therefore owed the employer's duty of care.

Application to partnerships

[C9017] In the Act, partnerships are treated as though they owed the same duties of care as a body corporate for the purposes of the offence (*CMCHA 2007 s 14(1)*). Any proceedings will be brought in the name of the partnership rather than any of its members (*CMCHA 2007 s 14(2)*) and any fine imposed on conviction must be paid out of the funds of the partnership (*CMCHA 2007 s 14(2)*).

The MoJ explained that this approach reflects that taken under other legislation, such as the Companies Act 2006, and means that partnerships will be dealt with in a similar manner to companies and other incorporated defendants.

It took a cautious approach in extending the offence to unincorporated associations, since it represents a new extension of the criminal law to these organisations, explaining: "Extending the offence to partnerships will ensure that an important range of employing organisations, already subject to health and safety law, is within the offence and that large firms are not excluded because they have chosen not to incorporate.

The Act also makes provision for the range of organisations covered by the offence to be extended by secondary legislation (section 21).

Procedure, evidence and sentencing

[C9018] Any statutory provision that applies in relation to criminal proceedings against a corporation also applies, unless an order made by the Secretary of State has prescribed any adaptations or modifications, to the organisations listed in Schedule 1 of the Act (see below) and to police forces, partnerships, trade unions and employers' associations. This includes provisions in relation to procedure, evidence and sentencing (*CMCHA 2007 s 15(1)*).

An order made under this section is subject to the negative resolution procedure, which means that it would be laid before Parliament and become law, unless disapproved by Parliament (*CMCHA 2007 s 15(4)*).

Transfer of functions

[C9019] Where a death has occurred in connection with functions carried out by a government department, one of the bodies listed in Schedule 1, incorporated Crown bodies or police forces, and there has been a subsequent transfer of those functions within the public sector, prosecutions will be taken against the body with current responsibility for the relevant function. Where the function has been transferred out of the public sector – in the case of privatisation for example – proceedings will be taken against the public organisation which last carried out the function (*CMCHA 2007 s 16(1)–(3)*).

However, in some circumstances it may be appropriate for liability to lie with a different body, for example where a function transfers between government departments but there is no transfer of personnel. In this case, there is provision for the Secretary of State to make an order specifying that liability rests with a different body. Again, an order made under this section is subject to the negative resolution procedure (see above).

Director of Public Prosecutions (DPP) consent for proceedings

[C9020] Consent of the Director of Public Prosecutions (DPP) is needed for proceedings for the offence of corporate manslaughter to be instituted. This applies in England, Wales and Northern Ireland (*CMCHA 2007 s 17*). In Scotland, all proceedings on indictment must be instigated by the Lord Advocate.

No individual liability for corporate manslaughter or corporate homicide

[C9021] There is no secondary liability for the offence of corporate manslaughter or corporate homicide. An individual cannot be guilty of aiding, abetting, counselling or procuring the commission of, or in Scotland of being art and part of, the offence (*CMCHA 2007 s 18*). In addition, an individual cannot be guilty of an offence under Part 2 of the *Serious Crime Act 2007* – encouraging or assisting crime – by reference to an offence of corporate manslaughter.

However, an individual can still be directly liable for the offence of gross negligence manslaughter, culpable homicide, or health and safety offences. Under the common law offence of gross negligence manslaughter, individual officers of a company – directors or business owners – can be prosecuted for the offence if their own grossly negligent behaviour causes death.

The leading case on gross negligence manslaughter is *R v Adomako* [1995] 1 AC 171 which involved the death of a patient who was undergoing an eye operation. In this case, the House of Lords set down a four-stage test for the offence of gross negligence manslaughter:

- Did the defendant owe a duty of care towards the victim who has died?
- If so, has the defendant breached that duty of care?
- Has such breach caused the victim's death? and
- If so, was that breach of duty so bad as to amount, when viewed objectively, to gross negligence warranting a criminal conviction?

In addition, the MoJ made clear that although prosecutions under *CMCHA 2007* will be brought against organisations and not specific individuals, individual directors, managers and employees may be called as witnesses and stand in the dock.

According to the Home Affairs and Work and Pensions Committees joint report on the *Draft Corporate Manslaughter Bill*, 15 directors or business owners were personally convicted of manslaughter by gross negligence between April 1999 and September 2005.

Several people have been convicted of gross negligence manslaughter following work-related deaths since the Act came into force. The safety journal Health and Safety Bulletin (HSB) monitors work-related gross negligence cases and reported that three custodial sentences were handed down over the period from 13 September 2018 to 5 April 2019. Prison sentences of four years and six months, three years and two years were handed down. In total, HSB recorded 77 work-related gross negligence convictions to 5 April 2019.

A new Sentencing Council guideline on gross negligence manslaughter came into force at the beginning of November 2018. It recommends prison sentences of up to 18 years for employers found guilty of gross negligence manslaughter and whose 'long-standing and serious disregard for the safety of employees, motivated by cost-cutting, has led to someone being killed'. The new guideline came into effect from 1 November 2018 regardless of when the offence was committed.

Manslaughter definitive guideline can be found on the Sentencing Council website at: www.sentencingcouncil.org.uk/wp-content/uploads/Manslaughter-definitive-guideline-Web.pdf.

Convictions under the Act and under health and safety legislation

[C9022] A conviction for corporate manslaughter does not preclude an organisation being convicted for health and safety offences on the same facts (*CMCHA 2007 s 19*) where this is required in the interests of justice (see for example *CAV Aerospace Ltd* (C9029)).

An individual can be convicted on a secondary basis for an offence under *section 37* of the *HSWA 1974* (see for example *Pyranha Mouldings Ltd* (C9029)).

Abolition of liability of corporations for manslaughter at common law

[C9023] The application of the common law offence of gross negligence manslaughter is abolished in relation to corporations, and any application it has to other organisations to which section 1 applies ie unincorporated associations to which the offence applies (*CMCHA 2007 s 20*).

However, in Scotland, where the law on culpable homicide differs in certain respects from the law on gross negligence manslaughter, the common law continues to be in force. The Procurator Fiscal determines the appropriate charge in light of the circumstances of each individual case.

Section 27(4) deals with cases that occurred wholly or partly before the new offence came into force.

Extent and territorial application

[C9024] The Act applies to England and Wales, Scotland and Northern Ireland, and other locations where criminal jurisdiction currently applies (*CMCHA 2007 s 28*). For example, it applies where a death occurs as a result of an incident involving a British ship, or aircraft or hovercraft, but the victim was not on board because they were shipwrecked and drowned.

The MoJ set out the key areas where the offence applies:

'The Act applies across the UK.

— The new offence can be prosecuted if the harm resulting in death occurs:
- in the UK
- in the UK's territorial waters (for example, in an incident involving commercial shipping or leisure craft)
- on a British ship, aircraft or hovercraft

- on an oil rig or other offshore installation already covered by UK criminal law.'

It provided further guidance on jurisdiction: 'Harm resulting in death' will typically be physical injury that is fatal, and in most cases the injury and death will occur at the same time and in the same location. However, in some cases death may occur sometime after the injury or harm takes place. The courts have jurisdiction in cases where the relevant harm was sustained in the UK, even if the death occurs abroad.

In the case of fatalities related to ships, aircraft and hovercraft, the offence will apply where the death does not occur on board, as long as it relates to an on-board incident.

The Act does not apply to British companies responsible for deaths abroad. The harm leading to death must occur within the UK or one of the places described above.

Where a death occurs abroad, there are acute practical issues for investigators, since they will not have control over the crime scene or the gathering of evidence relating to the death, while the evidence will be a crucial part of the investigation.

Schedule 1: List of government departments etc

[C9025] —
- Attorney General's Office
- Cabinet Office
- Central Office of Information
- Crown Office and Procurator Fiscal Service
- Crown Prosecution Service
- Department for Business, Energy and Industrial Strategy
- Department for Culture, Media and Sport
- Department for Education
- Department for Environment, Food and Rural Affairs
- Department for International Development
- Department for Transport
- Department for Work and Pensions
- Department of Health and Social Care
- Export Credits Guarantee Department
- Foreign and Commonwealth Office
- Forestry Commission
- General Register Office for Scotland
- Government Actuary's Department
- Her Majesty's Land Registry
- Her Majesty's Revenue and Customs
- Her Majesty's Treasury
- Home Office
- Ministry of Defence
- Ministry of Housing, Communities and Local Government

[C9025] Corporate Manslaughter

- Ministry of Justice (including the Scotland Office and the Wales Office)
- National Archives
- National Archives of Scotland
- National Crime Agency
- National Savings and Investments
- National School of Government
- Northern Ireland Audit Office
- Northern Ireland Court Service
- Northern Ireland Office
- Office for National Statistics
- Office of Her Majesty's Chief Inspector of Education and Training in Wales
- Ordnance Survey
- Public Prosecution Service for Northern Ireland
- Registers of Scotland Executive Agency
- Royal Mint
- Scottish Executive
- Serious Fraud Office
- Treasury Solicitor's Department
- UK Trade and Investment
- Welsh Assembly Government

The offence also applies to Crown bodies that are incorporated, such as the Charity Commission, and fatalities caused by Executive Agencies come within the scope of the offence. Executive Agencies come under the responsibility of a parent department, which are all covered by the offence. The Schedule can be amended by an order made by the Secretary of State (*CMCHA 2007 s 22*).

Investigation and prosecution

[C9026]–[C9027] The Act did not change the responsibilities of the police to investigate, and the Crown Prosecution Service (CPS) in England and Wales, the Public Prosecution Service in Northern Ireland and the Procurator Fiscal in Scotland to prosecute, corporate manslaughter. The HSE and other health and safety authorities continue to use their expertise in investigations in order to look at whether there is liability under more specific legislation, and to provide advice and assistance to the police in investigating corporate manslaughter. The joint approach of the HSE and other enforcement authorities, the police and the CPS is set out in protocols, or in Northern Ireland in an agreement, on investigating work-related deaths in England and Wales, Scotland and Northern Ireland.

The Act does not change the role and powers of the independent accident investigation branches which investigate air, marine and rail accidents in order to establish the cause, independently of any criminal investigation.

HSE and Crown Prosecution Service (CPS) Guidance

[C9028] The HSE has guidance on the Act on its website at www.hse.gov.uk/corpmanslaughter.

The Crown Prosecution Service (CPS) guidance can be found on its website at: www.cps.gov.uk/legal-guidance/corporate-manslaughter.

Corporate manslaughter convictions

[C9029] By September 2019, 26 companies had been convicted of the offence of corporate manslaughter. They include the following:

Cotswold Geotechnical Holdings was the first company to be convicted of the offence in February 2011 and was fined £385,000. The case followed the death of 27-year-old geologist Alex Wright in September 2008. He was investigating soil conditions in a deep trench on a development plot in Stroud when it collapsed and killed him. The Crown Prosecution Service (CPS) said that Mr Wright was working in a dangerous trench because Cotswold Geotechnical Holdings' systems had failed to take all reasonably practicable steps to protect him from working in that way.

In convicting the company, the jury found its system of work in digging trial pits was wholly and unnecessarily dangerous. The company ignored well-recognised industry guidance that prohibited entry into excavations more than 1.2 metres deep, requiring junior employees to enter into and work in unsupported trial pits, typically from 2 to 3.5 metres deep. Mr Wright was working in just such a pit when he died.

Cotswold Geotechnical Holdings was a small company that employed eight people in 2008 and the company director, was in overall control of the way the company managed its affairs. He had been charged with gross negligence manslaughter and a health and safety offence, but a judge ruled that he was too unwell to stand trial. The company appealed against the fine but lost.

JMW Farms Limited (Co Armagh) was the first company in Northern Ireland to be convicted under the CMCHA 2007. On 8 May 2012, it was fined £187,500 plus £13,000 costs at Belfast's Laganside Crown Court for health and safety failings that led to the death of 45-year-old employee Robert Wilson. He died on 15 November 2010 while working in the meal-mixing plant on the company's pig farm at Tynan, Co Armagh after being struck by a metal bin which fell from a forklift being driven by one of the company's two directors. A joint police and HSE (NI) investigation found that the bin had not been attached or integrated with the forklift. In addition, it was not possible to insert the lifting forks into the sleeves of the bin as the forks were too large and incorrectly spaced.

In July 2012, Manchester-based company *Lion Steel Ltd* was convicted of corporate manslaughter and fined £480,000 (to be paid over four years) with £84,000 costs. The case was taken by the CPS following the death of employee Steven Berry. He died of his injuries after falling through a fragile roof panel at the company's Hyde site on 29 May 2008 while carrying out repairs. A joint investigation by Greater Manchester Police and HSE found there had been no risk assessment and Lion Steel had never offered health and safety training to Mr Berry who was not properly equipped and was working unsupervised. Three of the company's directors were also charged with gross negligence manslaughter; but these and a number of charges under *HSWA 1974* did not go before the jury.

In October 2013, County Down company *J Murray & Sons* pleaded guilty to the corporate manslaughter of casual worker Norman Porter who died in 2012 after either falling or being dragged into an animal feed mixing machine. The court heard that safety guards had been removed from the machine, that there were other health and safety breaches, and that any employee could have been entangled in the machine. The company was fined £100,000 plus £10,000 costs; but the court allowed the fine to be paid in £20,000 annual instalments in order to try to safeguard 16 jobs at the firm.

In November 2013 *Princes Sporting Club* was convicted under the Act after it pleaded guilty to the corporate manslaughter of 11-year-old Mari-Simon Cronje in 2010. She died at a friend's birthday party at the club in west London after she fell from a 20-foot inflatable boat and was then hit by the speedboat towing it. The court heard that the inflatable was not supervised by a competent adult in the speedboat to warn the driver if anyone fell into the water. The driver (from New Zealand) had no UK-recognised qualification despite having five years' experience as a ski-boat driver and staff reported a 'lax' attitude to health and safety. The company director was charged with breaching Section 37(1) of the *HSWA 1974* but the charge was subsequently dropped. The case was the first in which the court imposed a publicity order, in a magazine for speedboat enthusiasts, on the company. The fine, £134,579, was limited by the company's lack of assets.

In February 2014, road sweeping company *Mobile Sweepers (Reading)* was fined just £8,000 for corporate manslaughter; although the company director was also fined £183,000 and disqualified from acting as a director for five years for an offence under the *HSWA 1974*. Employee Malcolm Hinton died from crush injuries after working on a repair underneath a road-sweeping truck at Riddings Farm near Basingstoke in 2012. He had inadvertently removed a hydraulic hose which caused the back of the truck to fall on him. Hampshire Police and the HSE carried out a joint investigation into his death. The judge said that there was a lack of training for handling and servicing machinery, there was no servicing policy, and effectively no health and safety policy. While he said that he would have fined a larger company with a substantial turnover or large assets somewhere between £500,000 and £1 million, in this case he said that a fine would achieve nothing because the two defendant directors could not pay it.

Cavendish Masonry Limited was found guilty of the corporate manslaughter of stone mason's mate David Evans at Oxford Crown Court on 22 May 2014. The company had previously pleaded guilty to breaching the *HSWA 1974* and in November 2014 it was ordered to pay £237,117.69 (a £150,00 fine plus £87,117.69 costs). Mr Evans was killed as he was erecting a large wall at the Well Barn Estate in Moulsford, Wallingford in February 2010. A two-tonne limestone block fell off a concrete lintel and crushed him. He was taken to the John Radcliffe Hospital by air ambulance but was pronounced dead later the same day. Thames Valley Police and the HSE carried out a joint investigation into his death. Following the verdict, HSE Inspector Peter Snelgrove commented that drawings for the work were wholly insufficient, and the overall execution of the project fell significantly below the standard required and expected of a competent masonry company.

Recycling firm *Sterecycle* was fined £500,000 in November 2014 after being found guilty of corporate manslaughter following the death of waste processing operator Michael Whinfrey. He was killed at Sterecycle's Rotherham plant in January 2011 when an autoclave door failed and blew out under pressure. Another man suffered 'serious life-changing injuries'. A joint HSE and South Yorkshire Police investigation found 'systematic failings'. The company was aware of a longstanding issue with the autoclave doors but made no effort to repair the problem properly. A former maintenance manager was found not guilty of perverting the course of justice. A former operations manager and former operations director were also cleared of criminal health and safety breaches.

In January 2015, kayak manufacturing company Pyranha Mouldings was found guilty of corporate manslaughter after senior supervisor Alan Catterall was trapped in an industrial oven at the company's factory in Runcorn, Cheshire in 2010. The firm's technical director and designer of the oven was also found guilty of criminal safety breaches. In March 2015, the company received a £200,000 fine and the director was sentenced to nine months in prison, suspended for two years, and fined £25,000. The company and its director were also ordered to pay costs of £90,000 between them.

The same month, a Northern Ireland sawmill was fined £75,000 plus £15,832 costs after it pleaded guilty to corporate manslaughter. A Diamond and Son (Timber) Ltd was sentenced at Antrim Crown Court for criminal safety failings that led to the death of employee Peter Lennon. He was killed as he was carrying out a repair to a large automated machine in September 2012. The investigation found that power to the machine had not been disconnected and during the work the machine moved, crushing and fatally injuring Mr Lennon. It found that the repair could have been carried out safely and easily while the machine was isolated from all power sources. It showed that safety guards preventing access to dangerous parts of the machinery had been modified and were regularly bypassed for routine tasks, and that the electrical safety key for the safety gates had been disabled. In addition, operators did not know how to operate the machine in maintenance mode.

In February 2015, building firm Peter Mawson Ltd and its owner were sentenced at Preston Crown Court following the death of Jason Pennington. In October 2011, Mr Pennington had been working on a roof and had fallen through a skylight from a height of around 7.6 meters onto a concrete floor. The company had earlier pleaded guilty to corporate manslaughter and a breach of the HSWA 1974. It was fined £200,000 for the corporate manslaughter offence, and £20,000 for the Health and Safety breach. The director also pleaded guilty to a breach of the HSWA 1974 and was sentenced to the following: eight months in prison, suspended for two years; 200 hours unpaid work; a publicity order to advertise what happened on the company website, to take out a half-page spread in the local newspaper; and pay costs of £31,504.77.

In March 2015 the managing director of a caravan park, pleaded guilty on behalf of Nicole Enterprises to the corporate manslaughter of employee Thomas Houston, who died after being crushed by a static caravan on the Silvercover Caravan Park in County Down in February 2012. He also pleaded guilty, on behalf of the company, to breaching health and safety regulations. However, he pleaded not guilty to an individual charge of manslaughter and, as the managing director of Dieci Ltd, not guilty to corporate manslaughter and health and safety breaches. As a result, sentencing was delayed.

In July 2015, Greater Manchester-based Huntley Mount Engineering (HME) was convicted of corporate manslaughter and fined £150,000 following the death of 16-year-old apprentice Cameron Minshull. He was killed in January 2013 after being dragged into a lathe at the company. The sole director received an eight-month prison sentence and a supervisor received a four-month suspended sentence, 200 hours unpaid work and a £3,000 fine for breaches of the Health and Safety at Work etc Act 1974. Lime People Training Solutions, who placed Cameron at the company, was fined £75,000 for its health and safety failings.

Also in July 2015, aerospace company CAV Aerospace Ltd, the parent company of CAV Cambridge Ltd, was found guilty of corporate manslaughter following the death of Paul Bowers. The company had been warned ahead of the fatal accident that there were potentially disastrous consequences if it did not take action to deal with dangerously high levels of stock. Mr Bowers died in 2013 after a stack of metal sheets collapsed on top of him in a warehouse at Cambridge airport, trapping and crushing him as he made his way down a designated safe walkway. The Crown Prosecution Service (CPS) said: 'The senior management of CAV Aerospace Ltd. received clear, unequivocal and repeated warnings over a sustained period of years prior to the fatal incident. Some of the most obvious solutions were rejected on cost grounds.' The company was fined £600,000 for corporate manslaughter as well as £400,000 for breaches of the HSWA 1974, although the judge ordered the fines to be paid 'concurrently' meaning it was ordered to pay a total of £600,000 (plus £125,000 costs). This was an important case. Not only was it the first time a parent company was convicted of corporate manslaughter (see **C9002**), it also involved a full trial and the jury considered the role of collective failures by senior management. It is also the largest company, with 460 employees and an annual turnover of £73 million, to be convicted of the offence so far.

In September 2015, Linley Developments Ltd was fined £200,000 after pleading guilty to the corporate manslaughter of Gareth Jones at an earlier hearing. The bricklayer was killed in January 2013 when a wall collapsed on him on a site in Hertfordshire. St Albans Crown Court had heard that Gareth's manager had ignored a fellow worker's warning that a wall was unsafe as it was propped up only by the soil it leaned against. It collapsed when the soil was removed, crushing Gareth as he worked nearby. The company's managing director was given a six-month prison sentence, suspended for two years, fined £25,000 and ordered to pay court costs of £7,500. The project manager received a six-month prison sentence, suspended for two years, and ordered to pay court costs of £5,000. In addition, the company was ordered to publicise its conviction in the construction trade press.

In October 2015, Kings Scaffolding was fined £300,000 after earlier pleading guilty to corporate manslaughter at Preston Crown Court, following the death of Adrian Smith. He died as he was carrying out repairs on the roof of the company's head office in Liverpool in September 2012.

In December 2015, Cheshire Gates and Automation Ltd was handed down a £50,000 fine and given a publicity order after pleading guilty to the offence at an earlier hearing following the death of six-year-old Semelia Campbell in June 2010. She died after becoming trapped in a faulty electric gate near her home in Manchester. The Crown Prosecution Service (CPS) said the company had left the gate without a proper limit on the force it exerted and was unable to detect and pull back from any obstacles it encountered.

Also in December 2015, Baldwin's Crane Hire was found guilty of corporate manslaughter following the death of employee Lindsay Easton. He was killed in Lancashire in August 2011 when the brakes on his crane failed as he was driving down a steep road and it crashed into an earth bank. A joint investigation by Lancashire Police and the HSE found that the crash was caused by serious problems with the braking system, which had not been properly maintained.

The first corporate manslaughter conviction in the care sector followed the death of 86-year-old Ivy Atkin. She died in 2012, days after she was moved from the Autumn Grange care home in Nottingham. Sherwood Rise Ltd, the company that ran the home until it closed in 2012, pleaded guilty to the corporate manslaughter charge in December 2015. It was accused of failing to provide Mrs Atkin with adequate food and drinks and check she was taking fluids. The acting director pleaded guilty to a charge of gross negligence manslaughter and a health and safety offence while the deputy manager also admitted a health and safety offence. The company was fined £30,000. The director, who was in charge of the day-to-day operation of the Autumn Grange Residential Home, was sentenced to a three years and two months prison sentence and disqualified from being a company director for eight years. The manager was handed down a one-year prison sentence, suspended for two years, and was disqualified from being a company director for five years.

In June 2016, Monavon Construction Limited pleaded guilty to two counts of corporate manslaughter and a breach of *s 3(1)* of the *HSWA 1974* following the deaths of pedestrians Gavin Brewer and Stuart Meads. The company had been carrying out renovation works to a north London flat but did not provide adequate edge protection to a lightwell. The two men fell through hoardings to their deaths in the early hours of 19 October 2013. The company was fined £250,000 for each corporate manslaughter offence, £50,000 for breaching the Health and Safety at Work Act and costs of £23,000 were awarded.

In August 2016, Bilston Skips Limited was found guilty of the manslaughter of Jagpal Singh. The Crown Prosecution Service (CPS) reported that in June 2012, he fell eight feet from the top of a skip and died later in hospital having suffered severe injury to internal organs.

He had been working in and around two large skips, accessible only via a metal ladder on the side, in order to rearranging green garden waste for compression. The arm of a JCB excavator being used to compress the waste and said to have been in very poor condition was in very close proximity to where he was working. It is not known whether the arm made contact with his body before he fell. Its manager was operating the JCB excavator to compress the contents of the skip.

The company, which was in liquidation, was charged with one charge of corporate manslaughter and one of breaching s 2 of the *HSWA 1974*. It was convicted at Wolverhampton Crown Court and fined £600,000. In addition, the manager was convicted of gross negligence manslaughter and pleaded guilty to an offence under s 37 of the *HSWA 1974*. He was sentenced to two years imprisonment suspended for two years.

The CPS also said that there was no record of any health and safety qualified individual on site. The manager confirmed he had no health and safety training and there was no system in place to report or record any incidents.

In March 2017, construction company SR and RJ Brown admitted corporate manslaughter and was fined £300,000 following the death of Benjamin Edge, who fell from the roof of a shed being demolished at Fletcher Bank Quarry in Greater Manchester. A joint investigation by Greater Manchester Police and the HSE found that he was working in wet and windy conditions without safety equipment. In addition, two SR and RJ Brown directors were jailed for 20 months for health and safety offences and perverting the course of justice. Manchester Crown Court heard that they and an employee had conspired to cover up the events that led to his death. Groundworks firm MA Excavations had contracted the work to demolish the shed to SR and RJ Brown. Its director was also jailed for 12 months and the company was fined £150,000 for health and safety breaches.

In May 2017, construction firm Martinisation (London) Ltd was convicted of the corporate manslaughter of two workers who died after falling from a central London balcony. The trial at the Central Criminal Court heard how Tomasz Procko and Karol Symanski fell to their deaths after the railings at a flat in Cadogan Square gave way in November 2014. The men had been attempting to haul a heavy sofa up onto the balcony using ropes with only the Victorian railings for safety. The company director denied health and safety offences but was convicted of the charges against him. CPS spokesperson Nick Vamos said that the company and its director showed an appalling disregard for the safety of their employees, and the incident was not an isolated breach. He said 'evidence put forward by the prosecution clearly demonstrated to the jury how these tragic deaths were part of a pattern of serious neglect of basic health and safety'. In July 2017, the director was sentenced to 14-month imprisonment for each death (half in prison, half on licence), to run concurrently, and was disqualified from being a company director for four years. The company was fined £1.2 million for each death and £650,000 for health and safety breaches. The fines were concurrent, meaning a total fine of £1.2 million was payable.

Also in May 2017, two companies were convicted of corporate manslaughter and three company directors were jailed following the death of a man who fell while working at a warehouse in Essex.

Nikolai Valkov died in hospital after falling through the roof of a warehouse in Harlow on 13 April 2015.

At Chelmsford Crown Court, Koseoglu Metalworks Ltd, the company contracted to carry out the work, admitted a corporate manslaughter offence and its sole director, admitted an offence under the *HSWA 1974*. Ozdil

Investments Ltd, the owner of the warehouse, denied corporate manslaughter and a *HSWA 1974* offence but was convicted following a trial at Chelmsford Crown Court.

Ozdil Investments received a £500,000 fine for corporate manslaughter plus costs of more than £53,115 and was fined £160,000 for breaching the HSWA. One director was given a 12-month prison sentence and disqualified from being a director for ten years. A second director was given a ten-month prison sentence and disqualified from being a director for ten years. Koseoglu Metal Works Ltd was fined £300,000 for a corporate manslaughter offence plus costs of £21,236 and was fined £100,000 for breaching the *HSWA 1974*. Its director was given an eight-month prison sentence and disqualified from being a director for ten years.

In October 2017, a company and its director were sentenced for causing the death of a factory worker. Safi Qais Khan died at Master Construction Products (Skips) Ltd (MCPS) after he became entangled in a machine called a trommel, used to sort waste material. MCPS admitted the corporate manslaughter of Mr Khan, after an investigation found there was no safe system of work for the trommel and it was in a dangerous state. Essential guards to prevent entrapment were missing, there was no emergency stop button on the machine and it was surrounded by uneven and waste strewn ground. The company also admitted a health and safety breach of a duty owed to its employee in failing to ensure measures were in place to minimise risks of entrapment, crushing or falls whilst working at the trommel. The sole director pleaded guilty to a health and safety offence. He admitted that he was aware of the way in which the company operated the trommel. The company was given a fine of £255,000 for corporate manslaughter. The director was sentenced to 12 months' imprisonment, suspended for two years, and 300 hours of community service at Birmingham Crown Court. He was also disqualified as a company director for eight years and ordered to pay £11,500 in prosecution costs.

Other organisations have faced corporate manslaughter charges but have been acquitted.

PS & JE Ward: The first company to have been cleared of a charge of corporate manslaughter under the *CMCHA 2007* was Norfolk-based flower nursery operator PS & JE Ward. It was found not guilty of corporate manslaughter at Norwich Crown Court in April 2014. The trial followed the death of tractor driver Grzegorz Krystian Pieton, who was electrocuted in July 2010, when the trailer he was towing touched an overhead power line in Terrington St Clement while he was working at Belmont Nursery in the village near King's Lynn.

The company was found guilty of breaching the *HSWA 1974* and was fined £50,000 and ordered to pay £47,932 in costs in June 2014. The judge gave with two directors, three years to pay the fine.

MNS Mining Ltd: The mine manager and mine owner of the Gleision Mine, in which four miners drowned in September 2011, were found not guilty of manslaughter at Swansea Crown Court in June 2014. Miners Charles Breslin, Philip Hill, Garry Jenkins and David Powell were killed when the mine in which they were working was engulfed by an enormous inrush of water.

It charged the mine manager with four counts of gross negligence manslaughter, alleging that he caused the deaths of the four miners by mining into old, flooded mine workings in breach of health and safety regulations. In doing so, it alleged that he was grossly negligent. MNS Mining Ltd was also summonsed for four counts of corporate manslaughter. The prosecution alleged that because of the way in which its activities were managed or organised by its senior management, the company caused the deaths of the miners by failing to ensure a safe system of working was in place. It alleged that this failure amounted to a gross breach of duty of care owed by the company to each of the four mine workers.

Both the company and the manager denied manslaughter through gross negligence and were cleared of the charges by a jury at Swansea Crown Court in June 2014. Following the verdict, the solicitor representing the families of the mineworkers who died said: 'Getting a conviction on a charge of corporate manslaughter is very difficult, as the prosecution has to prove the manager's actions amounted to gross negligence, which is a hugely difficult legal burden. You can be careless, you can be negligent — and have men die as a result — but unless it is gross negligence, you walk free'.

Maidstone and Tunbridge Wells NHS Trust: In January 2016, the trust and an anaesthetist were acquitted of manslaughter after the judge said there was no case to answer. The case was brought following the death of Frances Cappuccini who died during an emergency caesarean in October 2012.

Clinton Devon Farms Partnership: In February 2019, the organisation was cleared of the offence of corporate manslaughter following a three and a half week-long trial at Exeter Crown Court. It was also found not guilty of failing to ensure the safety of Kevin Dorman, a 25-year-old farm worker who died in May 2014 while collecting silage. The tractor and trailer he was driving fell from a sloping field onto a sunken road and crushed him.

Grenfell Tower Fire: The Metropolitan Police is leading an ongoing criminal investigation into the fire which is considering health and safety, manslaughter and corporate manslaughter charges against individuals and organisations. In March 2019, the force explained that its investigation must take into account any findings or reports produced by the public inquiry, including its final report. It is therefore unlikely to submit any file to the CPS before the latter part of 2021 – more than four years after the fire. In June 2019, it said that a decision on whether charges will be brought could be at least two years away.

Culpable Homicide Bill Scotland: In Scotland, there has not been a single charge under the CMCH Act, and therefore no convictions. In November 2018, Labour MSP for Mid Scotland and Fife Claire Baker introduced a culpable homicide bill. She says this aims to 'plug the justice deficit' and amend the law of culpable homicide.

In the consultation paper on the bill – the consultation closed in April 2019 – she explained that if a medium-sized or larger company causes a death, it 'remains extremely difficult (and many experts would say practically impossible)' for the Crown Office and Procurator Fiscal Service (COPFS) to secure a conviction of corporate culpable homicide. This is 'even in circumstances where there was recklessness or gross negligence on the part of individuals operating in fairly senior positions within the company' She says the law of culpable homicide has always failed to have one clear set of rules that apply to all wrongdoers, individuals and organisations alike and that the CMHA Act 'has failed entirely in that objective'.

More information about the culpable homicide bill can be found at: www.parliament.scot/parliamentarybusiness/Bills/110169.aspx.

Dangerous Goods – Carriage

John Wintle

Introduction to dangerous goods – carriage

Scope

[D0401] This chapter sets out the legislation covering the carriage (transportation) of dangerous goods. The *Carriage of Dangerous Goods and Use of Transportable Pressure Equipment Regulations 2009 (SI 2009 No 1348)*, known as *CDG 2009*, cover all carriage of dangerous goods by road, rail and inland waterways, including radioactive materials. They revoked the previous *Carriage of Dangerous Goods and Use of Transportable Pressure Equipment Regulations 2007 (SI 2007 No 1573)*. As the carriage of dangerous goods can cross national boundaries, the UK legislation refers extensively to European Agreements and Directives, namely the *ADR* (Accord European relatif au transport international des marchandises dangereuses par route) for road transportation and *RID* (Reglement concernant le transport international ferroviare des marchandises dangereuses) for rail transportation.

The Health and Safety Executive (HSE) explains that the CDG Regulations were substantially restructured for 2009, with direct referencing to ADR for the main duties. The CDG Regulations now cross-refer almost totally to ADR, which contains the detailed requirements.

The European agreements and directives are derived from international standards recommended by the United Nations Committee of Experts through the UN Economic Commission for Europe. In general, these international standards apply to journeys wholly within the UK, but the UK legislation has certain national derogations (deviations) agreed with the European Commission for this type of journey. These are set out in the Department for Transport (DfT) Approved Document, *Carriage of Dangerous Goods: Approved Derogations and Transitional Provisions* which can be downloaded from the HSE website at: www.hse.gov.uk/cdg/regs.htm.

The scope of this chapter primarily refers to journeys by road; similar principles apply to journeys by rail and inland waterway. Journeys made by sea or air are covered by other legislation in line with international agreements (see **D0404** below). The chapter will be of use to those businesses involved with the carriage of dangerous goods. It will be relevant to businesses producing dangerous goods for carriage, haulage companies and specialist conveyors, and to businesses receiving dangerous goods. All aspects of the supply chain are covered: classification and identification, packaging and containment, consignment procedures (including marking, labelling, placarding, plating and documentation), and carriage (including loading, unloading and handling), the transport vehicle and equipment, and the training and duties of the crew.

The definition of dangerous goods for carriage is dealt with in some detail in the section on classification and identification. It covers goods whose release would cause a hazard to the general public, passengers, transport crews, equipment, property or the environment. This covers a wide range of goods, including those which are flammable, corrosive and toxic, as well as compressed gases and radioactive materials.

The aim of *CDG 2009* is to protect against the hazards and danger that release of these goods might cause within the transport system. Protection from substances within the workplace is covered by other regulations, including the *Control of Substances Hazardous to Health Regulations 1999 (COSHH)*, the *Dangerous Substances and Explosive Atmosphere Regulations 2006 (DSEAR)* and the *Pressure Systems Safety Regulations 2000 (PSSR)*. All these regulations are made under the general provisions of the *Health and Safety at Work etc Act 1974 (HSWA 1974)*.

The carriage of dangerous goods legislation is prescriptive and complex. It is not the aim of this chapter to provide a comprehensive guide to all the necessary detail; for this the reader is directed to *CDG 2009* and *ADR*, and to the various publications issued by the HSE and guidance on its website. This chapter will, however, enable businesses to gain an overview of the nature of their responsibilities and direct them to where more detail can be found.

Development of legislation in the UK

[D0402] Legislative control of dangerous substances began with the *Petroleum Act 1879* and the *Petroleum (Consolidation Act) 1928*. The latter remained (with subsidiary regulations) the main legislative control on the transport of all dangerous goods until the 1980's. In the 1980's, influenced by the 1978 tanker disaster in Spain, separate regulations were introduced regulating classification, labelling and packaging (1984), the transport by road of dangerous goods in tankers and tank containers (1981) and in packages (1986). The latter two sets of regulations were subsequently replaced by new regulations in 1992, together with regulations for the training of drivers of vehicles carrying dangerous goods.

In 1994, the classification, labelling and packaging regulations were replaced by a set covering these aspects for transport by road and rail. These regulations were themselves subsequently replaced in 1996, and aspects concerning the use of transportable pressure receptacles were added. In the same year the separate regulations for tankers and tank containers and for packages were combined into the *Carriage of Dangerous Goods by Road Regulations*, and the driver training regulations were updated. In 1999 various amendments were made, the most significant being the requirement by many duty holders to appoint safety advisers under the *Transport of Dangerous Goods (Safety Advisers) Regulations*.

In order to harmonise national and international regulations for the carriage of dangerous goods, the UK was a signatory to a European agreement. This

Introduction to dangerous goods – carriage [D0402]

necessitated aligning UK legislation with Directives concerning the international carriage of dangerous goods by road (*ADR*) and by rail (*RID*). As *ADR* and *RID* do not have provisions relating to national enforcement, this aspect had to be dealt with at the national level (see below).

In May 2004, a consolidating set of regulations came into force, the *Carriage of Dangerous Goods and Use of Transportable Pressure Equipment Regulations 2004 (SI 2004 No 568)*. These were substantially restructured in 2007 as the *Carriage of Dangerous Goods and Use of Transportable Pressure Equipment Regulations 2007 (SI 2007 No 1573)*, known as CDG 2007. Then, restructured *Carriage of Dangerous Goods and Use of Transportable Pressure Receptacles Regulations 2009 (CDG Regulations)* revoked the 2007 Regulations and create the majority of duties by direct reference to *ADR*. CDG 2009 reg 5 sets out that: 'No person is to carry dangerous goods, or cause or permit dangerous goods to be carried, where that carriage is prohibited by ADR or RID, including where that carriage does not comply with any applicable requirement of ADR or RID'. In addition, the *International Carriage of Dangerous Goods* by Inland Navigation (*ADN*) came into force in the European Union in February 2009. *ADN* only applies in the UK in relation to the training and examination system for safety advisers and the issuing and renewal of vocational training certificates and is not covered in detail in this chapter. The 2009 Regulations were amended in 2011, mainly to take account of changes to the Transportable Pressure Equipment Directive. This chapter does not currently cover the transportable pressure equipment aspects in *CDG 2009 Part 4* (see **P7001**).

CDG 2009 covers all road and rail carriage of dangerous goods, including radioactive materials. As set out above, the regulations directly reference *ADR* and *RID*, which contain the detailed prescriptive requirements. These documents are available in English and can be easily downloaded. *ADR* can be downloaded from the United Nations Economic Commission for Europe (UNECE) website at: www.unece.org/trans/danger/publi/adr/adr2017/17cont entse0.html. There are some differences in *CDG 2009* from *ADR* and *RID* for domestic transport. These are mainly for the carriage of explosives, but also to retain the UK system of marking road tankers – reg 6 sets out alternative placarding requirements that apply to certain national carriage. There are also some exemptions from the requirements of *ADR* and *RID* for purely domestic transport, but for international journeys the full requirements of *ADR* and *RID* apply. See the DfT Approved Document, *Carriage of Dangerous Goods: Approved Derogations and Transitional Provisions* (www.hse.gov.uk/cdg/regs.htm) and **D0439–D0444** below. *ADR* and *RID* do not contain provisions and arrangements for enforcement in any country. This comes under the national legislation *CDG 2009*.

The HSE is one of the enforcement authorities for many aspects of *CDG 2009*; the DfT is the 'competent authority' for most purposes; while suitably qualified and appointed police officers and Driver and Vehicle Standards Agency (DVSA) officers enforce the regulations 'on the road'. The Office for Nuclear Regulation (ONR) is also an enforcement authority.

Structure and content of ADR

[D0403] The HSE describes *ADR* as 'highly prescriptive but structured logically': each part is subdivided into chapters, with each chapter divided into sections, paragraphs and sub-paragraphs. For example, 2.1 introduces classification, 2.2 sets out the class specific provisions, 2.2.1 relates to class 1 (explosives), 2.2.2 to class 2 (gases) and so on. Following the introduction in Part 1, which sets out high level aims and duties and exemptions, including the need for a Dangerous Goods Safety Advisor (DGSA) (see **D0405** below), it is structured as follows:

- Part 2. Classification;
- Part 3. The dangerous goods list (including special provisions and exemptions related to limited quantities);
- Part 4. Packing and tank provisions;
- Part 5. Consignment procedures, including documentation and vehicle marking;
- Part 6. Construction and testing of packaging, intermediate bulk containers (IBC), large packaging and tanks;
- Part 7. Carriage, loading, unloading and handling;
- Part 8. Vehicle crews, equipment, operation and documentation (including driver training); and
- Part 9. Construction and approval of vehicles.

The HSE says 'the answer to most problems can be found' in *ADR* and 'for that reason there is little or no need for explanatory literature or guidance'.

Also note that a chapter dealing with pressure systems can be found further below (see 'Pressure Systems', para **P7001** onwards).

International and multi-modal context

[D0404] In order to maintain an international uniformity of approach to the transport of dangerous goods legislation, the United Nations has established guidelines on the regulation. These guidelines are published as the '*Recommendations on the Transport of Dangerous Goods: Model Regulations*' which is commonly referred to as the '*Orange Book*'. The *Orange Book* recommends example procedural systems for the classification of substances by types of danger, the identification of dangerous goods, containment system performance and use, documentation and training (see www.unece.org/trans/danger/publi/unrec/rev19/19files_e.html).

The UN Recommendations form the basis of a series of codes for the transport of dangerous goods by road, rail, sea and air. They have no legal force, but are adopted worldwide as the basis of national legislation, and are therefore of key importance to legislators in Europe and the UK. In Europe the transport of dangerous goods by road is covered by the *ADR* agreement.

In the UK, the HSE is the organisation most concerned with the regulation and enforcement of transport of dangerous goods by road and rail. Sea transport is subject to the international IMDG code which is enacted into UK law through the *Merchant Shipping (Dangerous Goods and Marine Pollution)*

Regulations 1990, which are enforced by the Department of Transport. Harbour areas are regulated by the *Dangerous Goods in Harbour Areas Regulations 2016*, enforced by HSE (or the ONR in some cases).

For air transport the international ICAO 'technical instructions' set the relevant standards, with UK enforcement by the Civil Aviation Authority. Airlines generally work to IATA rules which are based on ICAO technical instructions. Prohibited items are much more common for air transport than transport by road or rail due to the severe consequences of any loss of containment, and certain types of goods are only regarded as dangerous for specific modes of transport. For example, magnetised materials are dangerous goods for air transport but are not regulated for road or rail. In the UK, the DfT is the lead department and competent authority on the transport of dangerous goods in whatever mode.

Government guidance sets out that:

> 'If you consign goods that are classified as potentially dangerous when transported, you must arrange their packing and transportation by air, sea, road, rail or inland waterway according to international regulations.
>
> The UN Model Regulations harmonise the rules on the various methods of transportation into a classification system in which each dangerous substance or article is assigned to a class defining the type of danger which that substance presents. The packing group (PG) then further classifies the level of danger according to PG I, PG II or PG III [see **D0400** below]. Together class and PG dictate how you must package, label and carry dangerous goods, including inner and outer packaging, the suitability of packaging materials, and the marks and label they must bear.
>
> Other regulations define the training and qualifications that dangerous goods drivers and safety advisors must hold, and when you must use one.'

These requirements are set out in more detail throughout this chapter.

General requirements for carriage of dangerous goods

Dangerous goods safety adviser

[D0405] Businesses that handle, process or transport dangerous goods on a regular basis must appoint one or more Dangerous Goods Safety Advisers (DGSA) in line with *ADR section 1.8.3*. Government guidance explains that their role is to:

- monitor compliance with rules governing transport of dangerous goods;
- advise the business on the transport of dangerous goods;
- prepare an annual report to management on the business' activities in the transport of dangerous goods;
- monitor procedures and safety measures;
- investigate and compile records on any accidents or emergencies; and
- advise on the potential security aspects of transport.

Unless exempted (see below), the requirement applies to carriers, fillers, loaders and unloaders. This could include cargo consignors, freight forwarders, warehouse workers and manufacturers producing goods that will be collected from their factory. According to DfT guidance, 'final unloaders' (consignees) do not need to appoint a DGSA, but intermediate unloaders such as freight forwarders and consolidators and operators of in-transit storage facilities do need to appoint one.

ADR prescribes the training and certification regime required for the DGSA and their appointment and duties. They must obtain a vocational training certificate after receiving appropriate training and pass a written examination. The DfT approves the mandatory DGSA exams.

There are two exemptions in relation to the appointment of DGSAs:

- exemption (a) applies to carriage-related activities for small loads (see D0442 below), where *ADR subsection 1.7.1.4* applies for radioactive materials, and for limited and excepted quantities (see D0441 below);
- exemption (b) is where the main or secondary activity of the duty holder (or the business) is not the carriage of dangerous goods (or related activities) and where any engagement in such activity is only carried out occasionally and that carriage poses little or danger or risk of pollution. These terms pose practical difficulties of interpretation and if in doubt further advice should be sought.

These exemptions only apply to journeys in Great Britain and not to international carriage.

Classification and identification

[D0406] The HSE explains that 'ADR works in such a way that classification is the precursor for everything that follows'. Once a substance or article has been properly classified, table A *(ADR section 3.2.1)* – the Dangerous Goods List – allows every other requirement to be ascertained by working logically through the columns.

The consignor, that is the person or business shipping the goods, is responsible for classifying, marking and packaging the dangerous goods (see D0407). Government advice is that if you are involved in the processing, packing or transporting of dangerous goods, you will first need to classify them correctly so that all organisations in the supply chain, including the emergency authorities, know and understand exactly what the hazard is.

The United Nations classification/identification system

[D0407] The purpose of the classification is twofold; firstly, to determine which goods are dangerous during transport; secondly, to show what kind(s) of danger are to be found in a particular article or substance.

The UN Committee of Experts on the Transport of Dangerous Goods developed the system for classification. The UN guidelines, published as the *'Recommendations on the Transport of Dangerous Goods: Model Regulations'*, commonly referred to as the *'Orange Book'* (see D0404 above), has a classification system that splits dangerous goods into nine hazard classes according to the kind of danger inherent to the item. Some of these classes are further broken down into smaller groupings, which are known as divisions.

Various regional and national bodies have then modified this to produce the most appropriate controls for different modes of transport. All base their controls upon the UN system of nine classes, and follow the UN system of testing for the various classes. The nine classes and their divisions within *ADR* are as follows —

Class	Dangerous Goods	Division	Classification
1	Explosives	1.1–1.6	Explosive
2	Gases	2.1	Flammable gas
		2.2	Non-flammable, non-toxic gas
		2.3	Toxic gas
3	Flammable liquid		Flammable liquid
4	Flammable solids	4.1	Flammable solid
		4.2	Spontaneously combustible substance
		4.3	Substance which in contact with water emits flammable gas
5	Oxidising substances	5.1	Oxidising substance
		5.2	Organic peroxide
6	Toxic Substances	6.1	Toxic substance
		6.2	Infectious substance
7	Radioactive material		Radioactive material
8	Corrosive substances		Corrosive substance
9	Miscellaneous dangerous goods		Miscellaneous dangerous goods

In addition to indicating the kind of danger presented by a substance, the classification system also gives qualitative guidance as to the degree of danger of that substance in relation to other types of dangerous goods within the same class.

The packing group (PG) then further classifies the level of danger according to PG I, PG II or PG III.

UN Packing Group	Meaning
PG I	HIGH danger
PG II	MEDIUM danger
PG III	LOW danger

The packing group plays a wide role – from the determination of acceptable packaging and the definition of threshold limits, to stowage and segregation

[D0407] Dangerous Goods – Carriage

requirements. The packing group system does not apply to Class 1, 2, 5.2, 6.2 or 7, and other than self-reactive substances of class 4.1, as different specification criteria and special packaging requirements apply.

Together, class and PG dictate how dangerous goods must be packaged, labelled and carried, including inner and outer packaging, the suitability of packaging materials, and the marks and label they must bear.

[D0408] In an emergency the emergency services, vehicle drivers, other transport staff and the general public need to know what dangers they face from certain substances. This creates a need for a product reference that can be quickly interpreted for an effective response.

The UN has devised 2 such references—

- Proper Shipping Names ('PSNs') – These are recommended names for a wide range of commonly moved substances, listed in the UN Orange Book. However, they are often complex chemical names and the need for the identification to be achieved on a worldwide basis led to the development of a second system.
- UN Numbers – These are four-digit identification numbers directly linked to the PSN.

ADR classification procedures

[D0409] Dangerous substances (including articles) are very widely defined and classified according to the hazard they present. The rules for classification are in *ADR Part 2* with descriptions and criteria given in some detail. These largely conform to the UN system, including the nine hazard classes (see D0407 above) and the use of packing groups.

The consigner must assign a PSN and UN Number to the substances. Dangerous substances should be classified and named according to the properties of the predominant substance, and in general the presence of impurities is not relevant, although there are some specific exceptions to this.

All dangerous goods, not only those directly assigned UN 3077 (solids) or UN 3082 (liquids), meeting the relevant criteria will be regarded as environmentally hazardous substances and required to show the 'dead fish and tree' mark.

Apart from clinical wastes (see below), waste is classified in the same way as other substances.

The Dangerous Goods List

[D0410] The proper shipping name (the proper chemical name, not a trade-name) may also be qualified by the addition of descriptive terms such as Solution, Liquid, Molten or Solid (see *ADR subsections 3.1.2.3–3.1.2.7*).

[D0411] *ADR*, in line with the UN Recommendations, requires the use of the UN Proper Shipping Name (PSN) and UN Numbers. For consignment and transport, the PSN and corresponding UN Number can be found in the *ADR Dangerous Goods List*. This List is given in *Table A* of *ADR Part 3 chapter 3.2* ordered by UN Number and may be cross referenced using the PSN, or, alternatively, in the corresponding alphabetical PSN list in *Table B* of *chapter 3.2*.

The *ADR* Dangerous Goods List contains a large range of substances and generic groups (eg paints) that have been previously classified and assigned packaging groups. The list contains lists of all categories of dangerous goods authorised for road transport. If a substance is listed by name, or its hazardous properties are described, then it is subject to control. The classification and packing group of a substance can be determined by looking up the UN Number in list.

A solution or mixture that contains a dangerous substance and one or more non-dangerous substances is identified using the PSN for the listed one, with the addition of the word 'solution' or 'mixture' eg methanol solution.

This does not apply where—

- The solution or mixture is identified in the Dangerous Goods List.
- The entry in the Dangerous Goods List only applies to the pure substance.
- The class, physical state or packing group is different to that of the pure substance.
- There is a significant change in the measures to be taken in an emergency.

If this occurs, then the solution or mixture is identified under the appropriate generic or not otherwise specified (n.o.s.) entry in the list. The index contains a range of generic names that cover substances belonging to particular families (eg Alcohols n.o.s. and Ethers n.o.s.) and also involves more descriptive names that identify their danger (eg Flammable n.o.s. and Corrosive n.o.s.).

Each entry in the List provides information to be used by the consignor to fulfil classification, identification, package selection, labelling and documentation completion duties. For example—

[D0411] Dangerous Goods – Carriage

UNNo	Name and Description	Class	Classification Code	Packing Group	Labels	Special provisions	Limited and excepted quantities	Packaging		
								Packaging instructions	Special packing provisions	Mixed packing provisions
1106	Amylamine	3	FC	II	3 + 8		1L E2	P001 IBC02		MP19

UN Portable Tank Instructions	ADR Tank Special provisions	Tank code	Special provisions	Vehicle for tank carriage	Transport category	Special provisions for carriage			HIN
						Packages	Bulk	Loading, unloading and handling	
T7	TP1	L4BHTE1		FL	2			S20 / S20	38

UN No = United Nations Number
CLASS = Hazard Class
Classification Code = Hazard Classification Code
HIN = Hazard Identification Number

General requirements [D0413]

It is essential to check the implication of any classification codes in the table above as these can significantly affect consignor duties. These codes are described in *chapter 3.2* of *ADR*. The carrier of the dangerous goods can use the Dangerous Goods List to confirm the accuracy of the consignor's transport documentation, to determine the competence required by the driver carrying the load and to decide what vehicle markings are appropriate for the journey.

There is a hierarchy of classification (*ADR subsection 2.1.3.5.3*) and there are rules about choosing the most appropriate entry, and hence UN number, (*ADR section 3.1.2*).

Many preparations will not be found in *Table A* of *ADR*. In these cases, the *ADR* rules for classification from first principles need to be followed. Sometimes items may show danger characteristics of more than one class. In this case, the material is assigned to the hazard class with the greatest risk, and the lesser risks are assigned as subsidiary hazard classes.

Packing groups

[D0412] Some substances with the same name will have different degrees of danger (for example flash point). This is reflected in the packing group which is found in column 5 of *Table A*. The relevant part of *ADR* dealing with packing groups is *subsection 2.1.1.3*. Where a substance is not listed and its classification has been determined from first principles, its packing group is determined by its properties. *ADR subsection 2.2.3.1.3* shows, for example, how unlisted flammable liquids are assigned a packing group.

Some substances (for examples gases and explosives) are not assigned a packing group. These substances do have a transport category which is relevant with regard to limited load exemptions. Where it is not practicable to fully classify a substance before transport, *ADR Section 2.1.4* allows conservative over-classification on the basis of information already available, with a more stringent packing group than would otherwise be necessary.

Controls of radioactive materials follow a significantly different pattern and refer to further sections of *ADR*.

Low quantities of dangerous goods in small receptacles or small loads may be exempt from some or all of these regulations under limited quantities concessions (see **D0441** and **D0442**).

Packaging and containment

[D0413]–[D0414] The packaging requirements are in *Part 4* of *ADR*. *Chapter 4.1* covers the use of packaging, intermediate bulk containers and large packagings, while *chapters 4.2–4.5* cover various sorts of tanks. Class specific provisions are detailed in *Part 2*. The particular packaging requirements of each package type are identified in specific subchapters within *Part 4*.

The Dangerous Goods List allows all details of permitted packaging and containment to be accessed. Each packing instruction is identified in the list and describes a range of packing options for substances depending upon their Packing Groups. These must be carefully examined in order to determine any volume or mass limits that may apply to a particular package type.

Requirements for packages

[D0415] All packages and containment used for the transport of dangerous goods have to meet three criteria. In particular they must—

- Meet the general requirements established by the *ADR*.
- Be certified to UN specification standards.
- Meet the detailed directions for the packaging of the product.

All packages that meet these criteria are available for use to carry the product in the particular transport system. This means that, for most substances and modes of transport, a wide range of package types and materials are available. However, in the worst-case scenario only one system of containment will be authorised for transport use.

UN specification sets the standards for packages used in the transportation of dangerous goods and these are translated in *ADR Part 6* into detailed requirements for the construction and testing of packaging and tanks. The UN specification requires packaging design and materials to have proven their competence to a national competent authority by passing practical tests.

Approved packages are given markings that are clear, legible and readily visible. Such packages are marked in particular ways, prefixed by the UN logo and followed by codes, as detailed in *ADR Part 6*, which details the significance of the packaging code with examples (*section 6.1.3*). The HSE provides the following example for a paint product:

 1A1/Y1.6/270/**/GB/****

This is interpreted as follows (see ADR 6.1.3.1)

- **1A1** steel non-removable head drum
- **Y** for PG II, III
- **1.6** maximum relative density (formerly specific gravity) of contents. Not needed if 1.2 or less
- **270** Test pressure of drum in kPa
- ****** last two digits of year of manufacture
- **GB** country of certification
- ******** represents the number of the certificate (in GB this is all figures)

Requirements for Intermediate bulk container and tank systems

[D0416] Intermediate bulk containers ('IBCs') are designed for mechanical handling and have a general maximum capacity of $3m^3$ (3,000 litres). The selection procedures are similar to those for packages. IBCs that are acceptable packaging options in *ADR* are listed in *Volume II Ch 4.1*.

There are many types of tank used in the transport of dangerous goods, but they all are all built to tightly-controlled criteria. A range of descriptions are used to identify the different types, but unfortunately, terminology is not

General requirements [D0418]

consistent across the various transport modes. In each modal control, definitions will be found indicating what kind of equipment is being referred to by any particular name.

The various controls offer two avenues of control of tank traffic—

- Identification of how tanks must be built and equipped to meet the demands of the regulation
- Description of the type tank and ancillary equipment that must be used for the transport of a particular substance

ADR Ch 4.2–4.5 contain constructional and equipment details for the various types of *ADR* approved tanks, which detail requirements for portable tanks, fixed tanks, demountable tanks, tank containers, tank swap bodies, reinforced plastic tanks, welded tanks and vacuum waste tanks.

Consignment procedures

Basic principles

[D0417] To ensure that everyone involved in handling or moving dangerous goods are aware of the hazards, information warning of the dangers is required to be clearly displayed on the outside of packages, tanks, vehicles, etc. This is also vital for rapid and effective action from the emergency response teams faced with dangerous goods incidents.

Terminology is not yet fully harmonised and a distinction is drawn between packages, tanks and transport units (vehicles)—

- Package: drums, jerricans, boxes, gas cylinders, IBCs, etc.
- Transport unit: road freight or road tanker vehicles, rail wagons or rail tanks, multi modal freight containers and tank containers.

Requirements for all modes of transport are all based on the UN Recommendations, the Orange Book, which identify the need for—

- Marking and labelling on packages.
- Placarding and plating of tanks, containers, and transport units (vehicles).

CDG 2009

[D0418] *CDG 2009 reg 5* is the basis for implementing the requirements of *ADR* (including *Part 5* in respect of matters immediately prior to the transportation taking place). These include the marking and labelling of packages, and the placarding and marking (plating) of tanks, containers and vehicles. It also covers the documentation to be provided with the goods.

There are specific requirements for GB registered vehicles carrying tanks or tank containers on GB domestic journeys *CDG 2009 reg 6* and the associated Schedule. These are discussed under paragraph D0422 below. They are intended to retain elements of good custom and practice established by industry in GB.

Depending on circumstances, the duties under *reg 5* can fall on the packer, loader, consignor and/or the carrier. Marking and labelling of packages is typically the duty of packers and the consignor. Placarding and marking of containers, tanks and vehicles is typically the duty of the loader, consignor and carrier. Documentation is the duty of the consignor and the carrier. The consignor should exercise overall control over these aspects, while the carrier must ensure that the required documentation is carried and that the vehicle carries the necessary placards.

ADR requirements

[D0419] The detailed requirements under *ADR* are given in *Part 5*, with some linked requirements in *Part 8*. The *ADR* requirements for marking and labelling also apply to rail (*RID*).

Marking and labelling of packages

[D0420] Marking and labelling both apply to packages and are covered in *ADR Chapter 5.2*. Every package must be marked and labelled as follows. The marks and labels required for certain products can be established through the relevant substance entries in the first six columns of the Dangerous Goods List. Special packing provisions apply variations to the standard marking and labelling. Marking and labelling of packages mean something different as follows.

Marking on packages (*ADR section 5.2.1*) is the application of the UN Number and complete PSN in a readily visible and legible location on the outside of all packages. Other information is also required to be marked for packages containing explosives, gases and radioactive materials (hazard classes 1, 2, and 7). In addition, the UN package certification details are also to be marked.

Labelling, described in *ADR section 5.2.2*, is the application of diamond shaped warning labels representing both the class of primary danger and any subsidiary risks. Class labels are distinguished by the appropriate class number in the bottom corner, with the upper half containing a pictorial symbol representing the danger. Text indicating the nature of the hazard eg 'Corrosive' may also be shown in the lower half – this is mandatory for Class 7, radioactive materials. Division numbers must also be shown on Class 1 and Class 5 labels.

The standard size for such labels is 100mm x 100mm, with a border line of 5mm. Symbols, text and numbers must be in black, except those containing the Class 8, corrosive, number and any associated text, which must be in white. Labels with entirely green, red or blue backgrounds may also have white symbols, text and numbers. Each class is identified by means of a diamond label that indicates the nature of the hazard on packaging and vehicles carrying dangerous goods. These are reproduced in the figure below—

General requirements [D0420]

CLASS 1 HAZARD
Explosive substances or articles

(No.1)
Divisions 1.1, 1.2 and 1.3
Symbol (exploding bomb): black; Background: orange; Figure '1' in bottom corner

(No. 1.4)   (No. 1.5)   (No. 1.6)
Division 1.4   Division 1.5   Division 1.6

Background: orange; Figures: black; Numerals shall be about 30 mm in height
and be about 5 mm thick (for a label measuring 100 mm x 100 mm); Figure '1' in bottom corner

✶✶ Place for division - to be left blank if explosive is the subsidiary risk
✶ Place for compatibility group - to be left blank if explosive is the subsidiary risk

CLASS 2 HAZARD
Gases

(No.2.1)   (No.2.2)
Flammable gases   Non flammable, non-toxic gases
Symbol (flame): black or white;   Symbol (gas cylinder): black or white;
(except as provided for in 5.2.2.2.1.6 d))   Background: green; Figure '2' in bottom corner
Background: red; Figure '2' in bottom corner

CLASS 3 HAZARD
Flammable liquids

(No 2.3)   (No. 3)
Toxic gases   Symbol (flame): black or white;
Symbol (skull and crossbones): black;   Background: red; Figure '3' in bottom corner
Background: white; Figure '2' in bottom corner

[D0420] Dangerous Goods – Carriage

CLASS 4.1 HAZARD
Flammable solids, self-reactive substances, polymerizing substances and solid desensitized explosives

(No. 4.1)
Symbol (flame): black;
Background: white with seven vertical red stripes;
Figure '4' in bottom corner

CLASS 4.2 HAZARD
Substances liable to spontaneous combustion

(No. 4.2)
Symbol (flame): black;
Background: upper half white, lower half red;
Figure '4' in bottom corner

CLASS 4.3 HAZARD
Substances which, in contact with water, emit flammable gases

(No. 4.3)
Symbol (flame): black or white;
Background: blue;
Figure '4' in bottom corner

CLASS 5.1 HAZARD
Oxidizing substances

(No. 5.1)
Symbol (flame over circle): black;
Background: yellow;
Figure '5.1' in bottom corner

CLASS 5.2 HAZARD
Organic peroxides

(No. 5.2)
Symbol (flame): black or white;
Background: upper half red; lower half yellow;
Figure '5.2' in bottom corner

CLASS 6.1 HAZARD
Toxic substances

(No. 6.1)
Symbol (skull and crossbones): black;
Background: white; Figure '6' in bottom corner

CLASS 6.2 HAZARD
Infectious substances

(No. 6.2)
The lower half of the label may bear the inscriptions: 'INFECTIOUS SUBSTANCE' and 'In the case of damage or leakage immediately notify Public Health Authority';
Symbol (three crescents superimposed on a circle) and inscriptions: black;
Background: white; Figure '6' in bottom corner

General requirements [D0420]

CLASS 7 HAZARD
Radioactive material

(No. 7A)
Category I - White
Symbol (trefoil): black;
Background: white;
Text (mandatory): black in lower half of label:
'RADIOACTIVE'
'CONTENTS ......'
'ACTIVITY ......'
One red bar shall
follow the word 'RADIOACTIVE';
Figure '7' in bottom corner.

(No. 7B)
Category II - Yellow
Symbol (trefoil): black;
Background: upper half yellow with white border, lower half white;
Text (mandatory): black in lower half of label:
'RADIOACTIVE'
'CONTENTS ......'
'ACTIVITY ......'
In a black outlined box: 'TRANSPORT INDEX';
Two red vertical bars shall
follow the word 'RADIOACTIVE';
Figure '7' in bottom corner.

(No. 7C)
Category III - Yellow

Three red vertical bars shall
follow the word 'RADIOACTIVE';

(No. 7E)
Class 7 fissile material
Background: white;
Text (mandatory): black in upper half of label: 'FISSILE';
In a black outlined box in the lower half of the label:
'CRITICALITY SAFETY INDEX'
Figure '7' in bottom corner.

CLASS 8 HAZARD
Corrosive substances

(No. 8)
Symbol (liquids, spilling from two glass vessels
and attacking a hand and a metal): black;
Background: upper half white;
lower half black with white border;
Figure '8' in bottom corner

CLASS 9 HAZARD
Miscellaneous dangerous substances and articles

(No. 9)
Symbol (seven vertical stripes
in upper half): black;
Background: white;
Figure '9' underlined
in bottom corner

(No. 9A)
Symbol (seven vertical stripes
in upper half; battery group,
one broken and emitting flame
in lower half): black;
Background: white;
Figure '9' underlined
in bottom corner

Limited quantity concessions of dangerous goods are not subject to the standard marking and labelling duties (see *ADR sections 3.4.11–3.4.15*).

Placarding and marking (plating) of vehicles, tanks or containers etc

[D0421] Road vehicles, tanks and containers and other types of transport units must bear the appropriate placards, plates and markings to warn of the dangers of the goods being transported (*ADR Ch 5.3*).

Placards, described in *ADR section 5.3.1*, are larger versions of the diamond hazard warning labels used for packages, and are placed on the tank or container carrying the goods. Generally, placards must be at least 250 x 2,500mm with a border line of 12.5mm (*ADR 5.3.1.7*). Smaller placards can be used for small tanks and containers (*ADR subsection 5.3.1.7.3*).

Placards also have to be displayed on the vehicle according to the type of load (*ADR subsections 5.3.1.2–5.3.1.6*). For example, vehicles carrying packages containing explosive (Class 1) or radioactive (Class 7) substances must display placards on both sides and the rear of the vehicle. Where such packages are carried in freight containers, the container should display the relevant placards on all four sides. Marking (or plating) is described in *ADR section 5.3.2* refers to the plain orange plates carried at the front of vehicles (and on the back of vehicles carrying hazardous packages). Sometimes the plain orange plate can be divided by a horizontal black line and shows the UN and hazard identification number or emergency action code (see *ADR subsection 5.3.2.2.1*). Marking also applies to tanks and containers etc, and to other marks on the sides and backs of vehicles.

Examples of marking of UK vehicles carrying packages and a container are shown below—

General requirements [D0422]

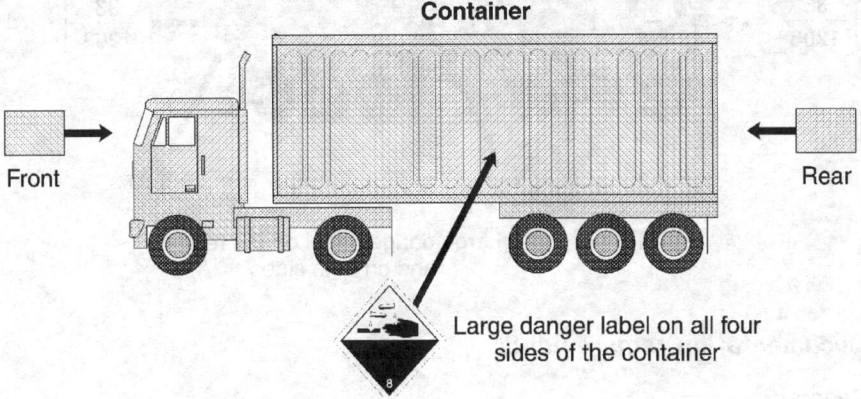

For certain journeys specific markings are required, the type, number and location of these vary depending on the transport mode and the nature of the journey – national or international. For example, the relevant UN number must be displayed on the outside of tank transport units. An elevated temperature mark is required for those units transporting liquids at temperatures of 100°C or more, or solids at 250°C or more. An emergency action code ('EAC'), which provides guidance for the emergency services in the event of an incident involving the load, along with other markings may also be needed for tank or bulk load transport, depending on the type of journey.

GB domestic and international journeys

[D0422] *Regulation 6* and the associated Schedule of *CDG 2009* require GB registered vehicles carrying tanks or tank containers on GB domestic journeys to be marked with 'Emergency Action Codes (EAC)' (sometimes called 'Hazchem codes'), and to include a telephone number for advice in the event of emergency. This is usually in the form of an integrated hazard warning panel, displaying the UN number, the EAC and telephone number, and the hazard placard, placed on both the sides and rear of the vehicle. This is in addition to the plain orange marking plate at the front of the vehicle.

[D0422] Dangerous Goods – Carriage

Vehicles carrying tanks or tank containers on international journeys (defined at *ADR subsection 1.1.2.4*) must carry a Hazard Identification Number (HIN - sometimes called the Kemler code) as shown at *ADR subsection 5.3.2.2.3*. The orange marking plate has a black border and is divided by a horizontal black line, above which is the HIN and below which is the UN number of the substance. These plates are placed at the rear and on both sides of vehicles, with the plain orange plate at the front, and are in addition to the diamond hazard placards. There is no requirement to display an emergency telephone number, although this would always be good practice.

An example of *ADR* road tanker vehicle marking and placarding is given below—

Road tanker

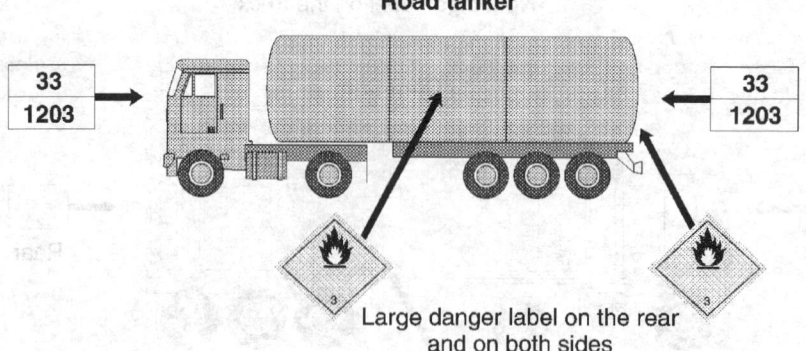

Large danger label on the rear and on both sides

Documentation requirements

General

[D0423] Documentation relating to a dangerous goods shipment describes important aspects of the cargo and its containment as well as indicating its point of origin and eventual destination. This information is required to ensure that all individuals involved in the transport process are fully aware of the dangers of the material that they are handling and employ suitable equipment and procedures to minimise any risks involved. The documentation also provides vital information in a concise form to emergency personnel. Description of packages etc should be as clear as possible to enable rapid identification, and substance descriptions must conform exactly to the requirements of *ADR* in order to minimise misunderstanding and errors during an emergency.

Key requirements

[D0424]–[D0425] The key requirements are that the documentation contains the following information, as set out in *ADR subsection 5.4.1.1*:

(1) the UN Number, preceded by the letters 'UN';
(2) the proper shipping name;
(3) depending on the article, substance or material, the classification code, class number or label model number;
(4) the packing group (where assigned);
(5) the number and description of packages;

(6) the total quantity (mass/volume) of each item of different UN number;
(7) the consignor name and address;
(8) the consignee names and addresses (where there are multiple consignees not known at the start of the journey the words 'Delivery Sale' may be used);
(9) declaration relating to any special agreement, where applicable (although this is uncommon);
(10) where assigned, the tunnel code, except where it is known that the journey will not involve passing through a relevant tunnel.

On the documentation, the first four items must appear in the order given (*ADR subsection 5.1.1.1.1*), but apart from this there is no requirement for all information to be on one document. There are special rules about documentation for wastes, salvage packaging, empty uncleaned packaging (*ADR subsections 5.4.1.1.3–5.4.1.1.6*) and for tanks and bulk (*ADR subsection 5.4.1.1.6*). There are other rules given under *ADR subsection 5.4.1.2* for class 1 (explosives), class 2 (gases), class 4.1 (flammable solids), class 5.2 (organic peroxides), class 6.2 infectious substances and class 7 (radioactive). Other special rules cover—

- Loads in a transport chain that includes air or sea (*ADR subsection 5.4.1.1.7*)
- Carriage in date expired IBCs (*ADR subsections 5.4.1.1.11, 4.1.2.2*)
- Multi compartment tanks or transport units with more than one tank (*ADR subsection 5.4.1.1.13*)
- Elevated temperature substances (*ADR subsection 5.4.1.1.14*)
- Substances stabilised by temperature control (*ADR subsection 5.4.1.1.15*)

In addition, a number of special provisions, including for waste and explosives are set out.

The language of the documentation should be that of the forwarding country and one of English, French or German if not already on the document.

The requirement to carry emergency information and instructions in writing is separate from the goods documentation and is covered under Crew and Vehicle below.

The ADR Agreement imposes specific duties relating to the provision and carriage of documentation when goods are moved by roads on international journeys. The consignor, vehicle operator and driver all have duties under the provisions of ADR.

The consignor has a duty to provide a transport document (information on the goods consigned), a declaration of compliance with ADR, written emergency instructions (actions to be implemented in the event of an accident) and a container packing certificate (only needed if the goods are loaded onto a freight container exceeding $3m^3$ capacity) to the vehicle operator.

It is the vehicle operator's responsibility to ensure that the appropriate documents are carried. These could include—

- The transport document and declaration of compliance;

- A container packing certificate, a vehicle approval certificate;
- The driver training certificate;
- The written emergency instructions; and
- A journey authorisation permit.

It is the driver's duty to ensure that the relevant documents are kept readily available and that any irrelevant documents are kept separate and clearly identified to avoid confusion.

Carriage, loading, unloading and handling

[D0426]–[D0427] The relevant sections are—

- *ADR 7.1* – general provisions
- *ADR 7.2* – special packaging provisions
- *ADR 7.3* – special bulk carriage provisions
- *ADR 7.4* – special carriage in tanks provisions
- *ADR 7.5* – covers mixed loading and includes special detail for explosives. It contains rules relating to foodstuffs, quantity limitations (mainly explosives and organic peroxides), handling and stowage, cleaning after packages have leaked, prohibition of smoking, precautions against electrostatic discharge (when carrying substances with a flash point of 61°C or lower), and special provisions 'CV' specific to particular substances or groups of substances.

The special provisions for packages, bulk carriage and tanks are contained in *ADR chapters* 7.2, 7.3 and 7.4 and require reference to the relevant column of *ADR Table A*.

Filling and loading

[D0428] All persons have a duty to ensure that the manner in which goods are loaded, filled or unloaded is not liable to create a significant risk or significantly increase any existing risk to the health and safety of any person. ADR makes provisions for particular hazard classes of dangerous goods. For example, where flammable goods are being transported by tank, the chassis must be properly earthed and flow rates controlled to prevent static discharges. It also provides more general safety requirements, such as prohibiting smoking in the vicinity of vehicles whilst loading and unloading (*ADR section 7.5.9*), and also cleaning tanks and freight containers after use (*ADR section 7.5.8*).

Stowage

[D0429] There are particular stowage requirements for certain dangerous goods. These requirements, detailed in *ADR section 7.5.7*, provide information on the various classes of dangerous goods. For example, flammable solids must be loaded in a fashion that ensures free air circulation and the maintenance of a uniform temperature. There is also general duty to ensure that all dangerous goods are suitably stowed and secured on vehicles to prevent load shifting during transit. This sets a high standard and supplements more general laws on the stowage of loads on goods vehicles.

General requirements [D0434]

Segregation

[D0430]–[D0431] From time to time a road traffic accident may occur, a tank valve may fail, or a fire may break out. These incidents could result in two substances reacting dangerously if they came into direct contact with each other. Segregation ensures that acceptable levels of safety are maintained in the transport system, in case one of these events does occur.

Dangerous goods must not be packed together in the same outer packaging with dangerous or other goods if there is a possibility that they will react dangerously together to cause: combustion and/or the evolution of considerable heat; the evolution of flammable, toxic or asphyxiant gases; or the formation of corrosive or unstable substances. Toxic or infectious substances must not be carried on the same vehicle as food, unless they can be 'effectively separated'.

It is the responsibility of the carrier to assess whether the loading arrangements are satisfactory. The circumstances under which the mixed packing of substances in a particular class are permitted, either with other items from the same class, or with products from other classes, are laid out in *ADR chapter 5.2* and *section 7.5.2*.

Crew and vehicle

Vehicle requirements

[D0432] The vehicle and its ancillary equipment is an integral part of the containment system, which must be effective in order to allow the goods to be carried in a safe manner.

[D0433] A range of requirements relating to the vehicle and its load carrying competence are laid down in the ADR Agreement. The information is presented at two levels: a set of duties that underpin all decision making on vehicle provision, equipment and use; and another set of directions related to each danger class and in some cases to specific item numbers within a class.

ADR Part 9 contains a number of general requirements regarding the physical competence of the vehicle and deals with any extra provisions or variations particular to a class, or a certain substance within a class. The degree of supplementary information varies between classes and covers a range of issues. For example, Class 1 has the greatest level of detail, ranging from vehicle construction requirements, to bodywork and exhaust systems, whereas Class 4 provisions cover aspects such as temperature control and the use of closed or sheeted vehicles. *Part 9* should therefore always be checked to confirm any additional requirements.

As a general requirement, a check is to be carried out on the vehicle, its equipment and the driver's documentation before loading the vehicle, to ensure that the journey can be undertaken in a compliant manner.

Operational procedures

[D0434] The vehicle crew (driver, assistant(s), and attendant(s)) is the primary channel of safety during road journeys involving the carriage of

[D0434] Dangerous Goods – Carriage

dangerous goods. The working procedures, duties, training and instruction of the crew are both a safeguard against the likelihood of an incident occurring involving their hazardous load and a primary means of minimising the consequences of any such incident.

Driver training

[D0435] When carrying dangerous goods, special training and certification requirements apply to many drivers. These requirements are covered in *ADR chapter 8.2*. *ADR subsection 8.2.2.8* sets out requirements concerning certificates of driver's training and a model certificate is shown at *ADR subsection 8.2.2.8.3*.

The ADR training requirements are summarised in the table below:

Vehicle/load	*Driver training*	*ADR reference*
All vehicles carrying dangerous goods except those carrying packages under the small load threshold	General training plus ADR training certificate. The certificate may be endorsed for different classes of goods or different modes of carriage	8.2.1
Any vehicle carrying packaged dangerous goods under the small load threshold	General training	8.2.3 *(refers to chapter 1.3)*, 1.1.36
Vehicle with a small tank (up to $1m^3$)	General training	8.2.1.3, 8.2.3

The general requirements for driver training are detailed in *ADR chapter 1.3*. Carriers must keep a record of training of their drivers as required by *ADR section 1.3.3*. Training should be carried out before a person assumes responsibilities in relation to dangerous goods. Otherwise, duties should be carried out only under the direct supervision of a trained person.

Under *CDG 2009* drivers are required to carry their training certificates (see *reg 24(6)*). These can be obtained by attending a government approved training course, (information may be found on the DfT website) and passing the examinations.

HSE inspectors and the police are entitled to inspect a driver's training certificate. They will ensure that the certificate is in the same name as the driving licence, the expiry date has not passed, the certificate is valid for the class of goods being carried and for the mode of carriage (eg tanks/other than tanks) and, where possible, that the certificate number matches that on the driving licence.

Duties of the crew

[D0436] There are various duties which must be carried out by the vehicle crew. In particular, the driver must—

- Ensure that any relevant transport documentation is kept readily available.

General requirements [D0438]

- Ensure that any relevant placards and markings required for the journey are displayed and removed when not relevant.
- If the load exceeds a specified limit then the special supervision and parking requirements must be satisfied.
- Ensure that any segregation, stowage or load safety instructions are followed.
- Ensure that the crew complies with any emergency information relating to the goods in the event of an accident and also ensure that the appropriate emergency services are contacted.
- Ensure that steps are taken to minimise the risk of fire or explosion.
- Ensure that the engine is shut off during loading and unloading.

ADR chapter 7.5 sets down the General Provisions Concerning Loading, Unloading and Handling.

Vehicle equipment requirements

[D0437] *ADR section 8.1.5* details the vehicle equipment that must be carried on journeys involving dangerous goods. Vehicle equipment requirements on dangerous goods journeys include—

- At least one suitable wheel chock for each vehicle.
- Two self-standing warning signs
- A suitable warning vest or warning clothing for each crew member
- One pocket lamp for each crew member (Note the pocket lamp has to be suitable for use in a flammable atmosphere in certain circumstances.)
- Other equipment that is needed according to the load eg eye rinsing liquid for loads with labels 3–9, emergency escape mask for labels 2.3 and 6.1; and a drain seal, shovel and plastic collecting container for labels, 4.1, 4.3, 8 and 9.

The table below sets out the current minimum requirements for fire extinguishers:

Vehicle	Minimum dry power fire extinguisher provision
Up to 3.5 tonne	2kg for cab plus another 2kg
Over 3.5 tonne and up to 7.5 tonne	2kg for cab – total 8kg (usually one 6kg but other provision is acceptable as long as there is one 6kg
Over 7.5 tonne	2kg for cab – total 12kg (including at least one 6kg)
Any vehicle carrying dangerous goods under the 'small load' limit or carrying only infectious substances	2kg only

Emergency information and other documents

[D0438] The Dangerous Goods Safety Advisor ('DGSA') must ensure the provision of proper emergency procedures and emergency information, and monitor their implementation in the event of any accident or incident that may affect the safety during the transport of dangerous goods.

ADR requires two types of documentation to be carried. The Transport Documents concerning the load are covered in paragraph **D0424** above. In addition, 'Instructions in Writing' containing emergency information must also be carried.

Emergency information, usually in the form of a Tremcard, comprises details of the measures to be taken by the driver in the event of an accident or an emergency and other safety information regarding the goods being carried. So long as the information that is set out in *ADR section 5.4.3* is carried and is valid for the substances being transported, then the requirement is satisfied. This information covers areas such as inherent dangers, measures to be taken if someone comes in to contact with the substance, measures to deal with a spillage and measures to be taken in the event of a fire. Additional requirements for specific types of material are given in *ADR 7.5.11*.

The driver of the vehicle must ensure that any documentation relevant to the load being carried is kept readily available on the vehicle, and must produce the documentation on request to any authorised person. The driver must also ensure that any documentation relating to dangerous goods not being carried is removed from the vehicle or is secured in a closed container and clearly marked to show that such information inside does not relate to the load currently on board. Once a journey is complete, the vehicle operator must maintain copies of transport documentation (but not emergency information) for three months (*CDG 2009 reg 31*).

Exemptions, revocations and radioactive materials

Types of exemptions

[D0439] Exemptions from some or all of the *CDG 2009* and *ADR* regulations can arise in three ways: either within *ADR* itself, or through provisions in *CDG 2009* or by Authorisation of the Secretary of State for Transport, which are applicable only to Great Britain.

ADR exemptions

[D0440] The exemptions in *ADR* are given in *section 1.1.3*. The main ones are summarised in the following list. For full detail, *ADR section 1.1.3* should be consulted.

(a) Private use of vehicles
(b) Carriage of machinery which happens to contain dangerous goods
(c) Carriage that is ancillary to the main activity - this is not easy to define. The HSE offers the following guidance:
- A driver taking dangerous goods with him for use with some machine or process that will be operated on arrival will be exempt.
- A journey taking dangerous goods to 're-supply' the above example will not be covered by this exemption (but other exemptions may apply).

Exemptions, revocations and radioactive materials [D0441.1]

(d) Carriage under the control of the emergency services
(e) Emergency transport intended to save life or protect the environment
(f) Uncleaned empty 'static' storage vessels that have contained certain gases, flammable liquids or class 9 (miscellaneous) substances. In this context static means not designed for the transport of dangerous goods
(g) Some carriage of gases (*ADR subsection 1.1.3.2*) and liquid fuels (*ADR 1.1.3.3*)
(h) Empty uncleaned packaging (*ADR subsection 1.1.3.5*), but not everything is exempt
(i) Limited quantity packages containing small receptacles
(j) Small load exemptions

The last two are probably the most frequently encountered and are explained as follows.

Limited quantity (LQ) packages exemptions

[D0441] Often dangerous goods are transported in such small receptacles within a package that there is no significant risk from them in the event of a serious incident. This exemption provides that, depending on the nature of the goods, small receptacles packed within a box or on a shrink-wrapped tray are subject to either simplified dangerous goods controls, or, in certain circumstances, are exempt from the scope of the regulations altogether – although there are requirements to mark certain transport units when carrying more than eight tonnes of LQ packages. This means that even if the substance is dangerous, the quantity in an individual receptacle is so small that the package of receptacles presents no particular danger for transport and is not considered to be a regulated package.

The definition of what is deemed to be a small receptacle varies greatly from class to class and packing group to packing group. Careful attention to detail is required to enable consignors to take advantage of limited quantity exemptions. Limited quantity concessions are available for shipments placed in combination packs, and there is usually a maximum weight specified for the package as well as a maximum size for the receptacles within the package.

The symbol for limited quantities of dangerous goods is a diamond (at least 100mm x 100mm unless package is too small in which case the minimum is 50mm x 50mm) with the UN Number within it where there is just one substance in a package, or, where there is more than one substance, all the UN numbers or the letters LQ within it. This must appear on the package and sometimes on the vehicle. Precise details are given at *ADR section 3.4.4(c)*.

Excepted quantities (EQ) exemption (see ADR 3.5)

[D0441.1] The HSE explains that 'excepted quantities' (EQ) is a relatively new concept for land transport of dangerous goods, although it has been commonly used in air transport. Like LQ it requires goods to be in combination packages (eg a bottle in a box). A code (E0 – E5) appears in column 7(b) of table A and links to paragraph 3.5.1.2 where what is allowed by the codes is set out. Unlike for LQ there are more prescriptive rules about packaging testing and for documentation. The packages must be marked with the 'EQ Symbol' and documents (where carried) must state 'dangerous goods in excepted quantities' and indicate the number of packages.

* (first) label number of the goods

** name of the consignor or consignee if not shown elsewhere on the package

Small load exemptions

[D0442] Small load exemptions relate to the total quantity of dangerous goods carried in packages by the vehicle (or trailer). The transport category of the dangerous goods determines the load limits for small load exemption. The transport categories are allocated on the basis of packing groups and class specific allocation in accordance with *ADR subsection 1.1.3.6.3*. The category for each specific material is listed in column 15 of Table A in *ADR*, (the Dangerous Goods List), and are summarised below: Note that small load exemption does not apply to tankers or bulk carriage.

Transport category		*Maximum mass/ volume threshold per transport unit*
0	Specific materials listed in *ADR Table 1.1.3.6.3* and in the Approved List	0
1	Packing Group I goods (unless otherwise specified) Specific materials listed in *ADR Table 1.1.3.6.3* and in the Approved List	20 litre/kilogram
2	Packing Group II goods (unless otherwise specified) Specific materials listed in *ADR Table 1.1.3.6.3* and in the Approved List	333 litre/kilogram

3	Packing Group III goods (unless otherwise specified) Specific materials listed in *ADR Table 1.1.3.6.3* and in the Approved List and any other dangerous goods not listed in other categories.	1000 litre/kilogram
4	Specific materials listed in *ADR Table 1.1.3.6.3* and in the Approved List. These materials are effectively exempt from *ADR* controls	Unlimited

If a vehicle is carrying goods under the small load threshold many of the requirements of *ADR* are not applicable. However, some are still applicable, depending on circumstances, and it is necessary to consult the relevant part of *ADR* for precise details. The relevant parts of *ADR* are *chapter 5.3* (placarding and marking), *section 5.4.3* (instructions in writing - emergency information), *chapter 7.2* (details of package requirements), *section 7.5.11* (prohibition of loading/unloading in a public place), *Part 8* (vehicle crews, equipment, documentation, operation and driver training), and *Part 9* (construction and approval of vehicles). In most cases the obligatory remaining requirements are—

- General driver training (*ADR section 1.3.2*) and training record (*ADR section 1.3.3*)
- Carrying one 2kg powder fire extinguisher or equivalent (*ADR subsection 8.1.4.2*)
- Proper stowage of the goods (*ADR section 7.5.7*)

It should be noted that any packages that meet the limited quantity criteria are not counted for the purposes of assessing small load exemption. Packages counting towards small load exception have to comply with the relevant standards. The use of small load exemption is optional and carriers may still wish to display the orange warning plates, provided the vehicle is actually carrying some dangerous goods. Orange plates must be removed when no dangerous goods are being carried.

CDG 2009 exemptions

Specific exemptions

[D0443] *CDG 2009* contains a number of further exemptions from *ADR* which are only applicable within Great Britain. The HSE sets out that:

- the main parts of the regulations do not apply where carriage is 'not undertaken by a vehicle'. The HSE says that in practical terms the regulations do not apply to: vehicles with a maximum design speed of 25 km/hour or less; vehicles that run on rails; mobile machinery (not defined but could include vehicles specially equipped for road construction purposes, such as white lining vehicles); agricultural or forestry tractors that do not travel at a speed exceeding 40 km/h when transporting dangerous goods; or any trailer being towed by such a vehicle;
- movement wholly within an enclosed area is exempt from the main parts of the regulations;

[D0443] Dangerous Goods – Carriage

- MoD vehicles will operate in line with ADR, but tankers will be placarded and plated to ADR practice rather than GB practice;
- a UK derogation makes changes to transport categories and load thresholds for many explosives and in effect these maintain older arrangements for domestic transport only; and
- crossing public roads: a UK derogation provides that, except for explosives and radioactive materials, the regulations do not apply to movements: between private premises and a vehicle in the immediate vicinity (for example loading a vehicle just outside the premises); and between private premises (in the immediate vicinity) occupied by the same person, including where separated by a road. The HSE says that 'Immediate vicinity' is not defined and there is scope for abuse. In addition, subject to any court decision, in the first case a journey of more than about 100 metres on the highway, and in the second case more than about 400 metres, should be regarded as not in the 'immediate vicinity'.

For further details and limitations, the regulations themselves should be consulted.

Exemption by 'authorisation'

[D0444] *CDG 2009 reg 12* empowers the HSE, the Secretary of State for Defence and the Secretary of State for Transport to authorise the carriage of dangerous goods contrary to the prohibitions or requirements of the regulations for the carriage of dangerous goods. (The power does not apply to transportable pressure equipment.) Authorisations are added or deleted as the need arises or recedes, and they are all time limited. A complete list of current authorisations is available as pdf files at: www.gov.uk/government/collections/transporting-dangerous-goods.

Radioactive materials

[D0445]–[D0447] *ADR* contains detailed requirements relating to radioactive materials. In particular, it refers to the need for a radiation protection programme (*ADR section 1.7.2*) and a management system (*ADR section 1.7.3*), and also provisions for special arrangements (*ADR 1.7.4*), subsidiary risk (*ADR section 1.7.5*), and non-compliance with applicable radiation or contamination levels (*ADR section 1.7.6*).

The *ADR* requirements derive from requirements set by the International Atomic Energy Authority (IAEA). General consignor duties are outlined in *ADR Part 7*, which provides the general requirements for transporting radioactive materials. These focus upon the need for the containment of radioactive contents, control of external radiation levels, prevention of criticality and heat damage. *ADR chapter 6.4* covers the requirements for the construction, testing and approval of packages and materials for containing radioactive materials.

In addition to radioactive and fissile properties, any subsidiary hazards of dangerous goods items such as flammability, corrosivity, etc, must be taken into account in the documentation, packaging, labelling, marking, placarding, stowage, segregation and carriage requirements.

Acknowledgement

[D0448] The information presented in this chapter has been derived from *CDG 2009, ADR* and supporting information issued by the government and HSE, in particular the *Carriage of Dangerous Goods Manual*, to which acknowledgement is made. It is a compilation and summary of complex regulations and is therefore not complete in all detail. For full detail and clarification, the reader is strongly advised to consult the regulations themselves (*CDG 2007* and *ADR*), the HSE or the DfT and their publications (see D0449 below).

Further Information

[D0449] *ADR* can be downloaded from the United Nations Economic Commission for Europe (UNECE) website at: www.unece.org/trans/danger/publi/adr/adr2017/17contentse0.html.

HSE information on the carriage of dangerous goods, including links to the Manual and other resources, can be found at: www.hse.gov.uk/cdg/index.htm.

DfT Approved Document, *Carriage of Dangerous Goods: Approved Derogations and Transitional Provisions* (www.hse.gov.uk/cdg/regs.htm).

Dangerous Substances and Explosive Atmospheres

John Wintle and William Delamare

Introduction to dangerous substances and explosive atmospheres

Background

[D0901] Until the *Dangerous Substances and Explosive Atmospheres Regulations 2002 (SI 2002 No 2776)* (*'DSEAR'*) came into force in December 2002, the regulation of substances that could give rise to the risks of fire and explosion had built up incrementally for different types of substances in different industries. There were separate regulations for highly flammable liquids and liquefied petroleum gases, celluloid in factories and workshops, dry cleaning, cinematograph film manufacture and petroleum. However, other substances and activities giving rise to these risks, such as flour dust and certain grinding operations, were not covered by health and safety regulations specific to the risks they created, and it was becoming increasingly impracticable to legislate on a substance or activity specific basis as technology advanced. There was a need for new legislation that would apply generically to all substances with these potential risks.

The new legislation that became DSEAR would remove the need to have to separate regulations for particular substances or groups of substances, and covered substances for which specific regulation did not previously exist. As a result, around 20 pieces of old health and safety legislation were repealed making a considerable simplification for duty holders and regulators alike. Significantly, the *DSEAR* replaced the *Highly Flammable Liquids and Liquefied Petroleum Gases Regulations 1972 (SI 1972 No 917)*.

European Directives

[D0902] The opportunity to modernise UK legislation, which led to the enactment of the *DSEAR,* arose from the need for the UK government to implement two European Union Directives into UK law: the Chemical Agents Directive and the ATEX 137 (99/92/EC) Directive. (The term ATEX is derived from the French 'Atmospheres Explosive').The Chemical Agents Directive ('CAD') requires employers to protect workers from fire, explosion and health risks arising from chemical agents present in the workplace. The health requirements of CAD have been implemented by changes to existing health legislation, mainly the *Control of Substances Hazardous to Health Regulations 2002 (SI 2002 No 2677)* ('*COSHH*'), with some changes to legislation on asbestos and lead.

The ATEX 137 (Explosive Atmospheres) Directive is concerned with workers health and safety in those workplaces where potentially explosive atmospheres may be present and requires employers to protect workers from the risks of fire and explosion caused by explosive atmospheres. It was decided that the safety (ie fire and explosion) aspects of the Chemical Agents Directive and the ATEX 137 Directive dealing with explosive atmospheres would be implemented together in one set of completely new safety regulations. Thus, *DSEAR* covers all substances that could give rise to fires, explosive atmospheres, explosions and similar energetic events and the management of places where explosive atmospheres may arise.

Another European Directive ATEX 95 (94/9/EC) is concerned with the supply of equipment, protective systems, components etc where these are for use in potentially explosive atmospheres. This is a trade Directive whose aim is to harmonise the manufacture and supply of equipment intended for use in potentially explosive atmospheres to common standards of safety. ATEX 95 was implemented in UK law under the Equipment and Protective Systems for Use in Potentially Explosive Atmospheres Regulations 1996 (EPS), by the Department for Trade and Industry, and now comes under the Department for Business Enterprise and Regulatory Reform (BERR), where further information can be obtained.

The DSEAR and EPS Regulations are complementary. DSEAR is concerned with the safe use of dangerous substances and requires employers to zone workplaces according to the hazard and to select equipment and protective systems appropriate for that zone that meets EPS requirements. The EPS puts a duty on the manufacturer and supplier to manufacture or supply equipment that is suitable and marked for use in different types of zone.

Guide to the chapter

[D0903] *Dangerous Substances and Explosive Atmospheres Regulations 2002 (SI 2002 No 2776) (DSEAR)* follow the trend of modern goal setting regulation based on the concept of risk assessment and management by the employer or the person responsible for the risk. This allows the Regulations to deal generically with the risks of fire and explosion, where prescriptive requirements relate directly to the risks, and not to the characteristics of any particular substance or activity. The amount of prescriptive requirement is thereby minimised.

The onus in law is placed squarely on those responsible for the risks to assess the risks within their situation and to manage them safely. It requires duty holders to think for themselves about safety and to take the necessary action. In support of the Regulations and to assist duty holders, the Health and Safety Commission has published a series of Approved Codes of Practice giving practical advice on how to comply with specific aspects of the Regulations and additional Guidance.

This chapter covers the requirements of the *DSEAR* in a form that should assist duty holders to understand easily what is needed to comply with the law.

The first part of the chapter is concerned with:

- aspects relating to the application of the Regulations;
- typical activities;
- who the duty holders are;
- what workplaces are covered;
- what dangerous substances and explosive atmospheres are; and

The rest of the chapter then deals with the requirements under the Regulations, and more specifically:

- the requirement for and considerations of a risk assessment;
- the principles of risk management: elimination, reduction, control and mitigation;
- the classification of places with explosive atmospheres into zones;
- equipment and protection systems;
- arrangements to deal with incidents and emergencies; and
- provision of information, instruction, and the marking and identification of containers.

Some examples and guidance are given on how to comply with the Regulations, but these should not be taken out of context.

The chapter will be useful to all employers and the self-employed where there is a possibility of dangerous substances or explosive atmospheres being present in their workplace. In practice, this will include a very wide range of businesses and public services, including manufacturing industries, service companies, some retailers, transport and shipping, garages, schools, universities and experimental research institutes, and local authorities. While all have a duty to comply, the manner of compliance can be commensurate with the risk. What is commensurate should be consistent with existing good practice and common sense, in addition current case law dating from 2009 is providing additional information. There are currently 21 cases highlighted in the HSE Prosecutions database.

Dangerous substances and explosive atmospheres: the regulations

Scope

[D0904] The *Dangerous Substances and Explosive Atmospheres Regulations 2002 (SI 2002 No 2776) (DSEAR)* set minimum requirements for the protection of persons from the risks of fire and explosion arising from dangerous substances and explosive atmospheres in the workplace. The Regulations assist duty holders to comply with the general requirement to manage risks from substances having these potential hazards under the *Management of Health and Safety at Work Regulations 1999 (SI 1999 No 3242) (MHSWR)*. While the *DSEAR* protect against substances with the potential to cause fire and explosion, they stand alongside the *Control of Substances Hazardous to Health Regulations (SI 2002 No 2677) (COSHH)* which protect against and control substances hazardous to health.

Many substances are classified as dangerous under the criteria given in the *DSEAR*. These include petrol, liquefied petroleum gas, welding gases, paints, varnishes and certain types of combustible and explosive dusts produced in, for example, machining and sanding operations. The Regulations are therefore relevant to the many businesses in Great Britain at which these substances may be present.

Typical activities

[D0905] The *Dangerous Substances and Explosive Atmospheres Regulations 2002 (SI 2002 No 2776) (DSEAR)* apply to many types of activities commonly found at work in businesses and public services, of which the following are typical:

- Storage and dispensing petrol as a fuel for cars, motor boats, horticultural machinery.
- Handling, storage and use of flammable gases, such as acetylene for welding.
- Handling and storage of waste dusts in a range of manufacturing industries.
- Handling and storage of flammable wastes such as fuel oils.
- Hot work on tanks or drums that have contained flammable material.
- Work activities that could release naturally occurring methane.
- Mining or processing of coal producing dusts.
- Use of flammable solvents such as ether in pathology and school laboratories.
- Storage/display of flammable goods, such as paints, in the retail sector.
- Filling, storage and handling of aerosols with flammable propellants such as LPG.
- Transport of flammable liquids in containers around the workplace.
- Deliveries from road tankers, such as petrol or bulk powders.
- Chemical manufacture, processing and warehousing.
- Extraction and processing of petrochemicals – onshore and offshore.

The examples given in the list above are only representative of the very wide range of relevant activities. All employers and the self-employed must consider if the *DSEAR* 2002 could apply to them.

General conditions for application

[D0906] The *Dangerous Substances and Explosive Atmospheres Regulations 2002 (SI 2002 No 2776) (DSEAR)* apply whenever:

(a) There is work being carried out by an employer or self-employed person.
(b) A dangerous substance is present or liable to be present at the workplace.

(c) The dangerous substance presents a risk to the safety of persons (as opposed to a risk to health).

Duty holders

[D0907] The responsibility for complying with the *Dangerous Substances and Explosive Atmospheres Regulations 2002 (SI 2002 No 2776) (DSEAR)* falls on the employer, or the self-employed person, whichever is appropriate, who is responsible for the workplace where dangerous substances or explosive atmospheres may be present [*SI 2002 No 2776, Reg 4*]. This is normally the owner, lessee or occupier of the workplace. (Note: references to the 'employer' throughout this chapter also apply to self-employed persons where appropriate.)

The duties extend not only to protecting employees but also to any other person, whether at work or not, who may be put at risk by dangerous substances and explosive atmospheres in the workplace. This includes:

- employees working for other employers;
- visitors to the site; and
- members of the public.

However, when making arrangements for dealing with accidents, incidents and emergencies and the provision of information, instruction and training, employers only have duties to persons who are normally at their workplace *SI 2002 No 2776, Reg 4(1)(b)*]. Further, their duties in law in respect of provision of suitable personal protective equipment and appropriate work clothing do not extend to persons who are not employees [*SI 2002 No 2776, Reg 4(1)(a)*]. Responsible employers will, however, ensure that the number of persons who need to be in hazardous places is minimised and that all those that are in hazardous workplaces are suitably equipped.

Sometimes two or more employers will share the same workplace (whether permanently or temporarily) where an explosive atmosphere may occur. Here the employer responsible for the workplace (normally the owner or lessee) has the duty to co-ordinate the implementation of the measures to protect employees from the risk of an explosive atmosphere. The duty for implementing the measures relating to dangerous substances would be expected to lie with the employer responsible for those substances [*SI 2002 No 2776, Reg 11*].

Workplaces

[D0908] The *Dangerous Substances and Explosive Atmospheres Regulations 2002 (SI 2002 No 2776) (DSEAR)* define 'workplace' widely as being any premises or part of premises used for or in connection with work [*SI 2002 No 2776, Reg 2*]. It includes any place to which an employee has access to while at work, and covers any room, lobby, corridor, staircase or road or other place used for access or where facilities are provided. Public roads are not workplaces within the definition of the Regulations. Premises include:

- all industrial and commercial sites;

- land based installations;
- vehicles and vessels;
- shared buildings, private roads and paths on industrial estates and business parks;
- houses and other domestic premises where there is a work activity.

A few types of workplaces and activities are specifically exempt from some or all of the requirements of the *DSEAR* because there is other legislation fulfilling these requirements, for example, at offshore installations [*SI 2002 No 2776, Reg 3*]. The Regulations do not in general apply to activities on board ships, except for construction, reconstruction, conversion and dry dock repairs carried out in Great Britain. The requirements in the *DSEAR* concerning zoning and co-ordination of safety in shared workplaces do not apply at workplaces in respect of the following substances, activities or types of equipment, where other controlling regulations exist:

(a) areas used directly for medical treatment;
(b) the use of gas appliances and fittings for cooking, heating, lighting etc;
(c) the manufacture, handling, use, storage and transport of explosives or chemically unstable substances;
(d) any activity at a mine quarry or borehole; and
(e) land transport regulated by international agreements.

The *DSEAR* apply within Great Britain, and apply outside Great Britain to the extent defined by virtue of the *Health and Safety at Work etc. Act 1974 (Application outside Great Britain) Order 2013* .

Dangerous substances

[D0909] The *Dangerous Substances and Explosive Atmospheres Regulations 2002 (SI 2002 No 2776) (DSEAR)* apply to any substance or preparation (or mixture of substances) with the potential to create a risk to persons from energetic events such as fire, explosions, or thermal runaway from exothermic reactions, etc. The definition of 'dangerous' includes all substances that meet the criteria for being explosive, oxidising, extremely flammable, highly flammable or flammable according to the *Chemicals (Hazard Information and Packaging for Supply) Regulations (SI 2009 No 716)* ('CHIP'). Substances classified as being explosive, oxidising, extremely flammable, highly flammable or flammable under the *CHIP Regulations* are therefore dangerous substances for the purposes of the *DSEAR* [*SI 2002 No 2776, Reg 2*].

The definition of 'dangerous' also includes other substances that are not classified under the *CHIP Regulations*, but meet the criteria by virtue of the way they are used or are present at the workplace. For example, the *DSEAR* apply to substances that decompose or react exothermically when mixed with other substances. Peroxides, wood, flour and many other dusts involved with grinding or machining can, when mixed with air, ignite and explode. While diesel (or other high flash point oils) at ambient temperature are not classified as flammable under the *CHIP Regulations*, if they are heated and used at sufficiently high temperatures in a process that creates a fire risk, they may become a dangerous substance for the purposes of the *DSEAR* [*SI 2002 No 2776, Reg 2*].

The Regulations [D0909]

Thus, in order to determine if a substance present is dangerous, two steps are necessary:

(a) a check to find out if the substance has been classified under the *CHIP Regulations*; and
(b) an assessment of whether the physical and chemical properties of the substance and the circumstances of its use in the workplace can create a safety risk to persons from a fire or explosion.

The *DSEAR* are only intended to protect persons from energetic events and the harmful physical effects from thermal radiation (burns), over-pressure (blast injuries) and oxygen depletion (asphyxiation) arising from fires and explosions. They implement the European Explosive Atmospheres Directive (99/92/EC) (ATEX 137) into UK law.

Many of the substances will, however, also create health risks and it should be noted that the *DSEAR* does not address these. Many solvents are toxic as well as being flammable. Health risks are dealt with by the *Control of Substances Hazardous to Health Regulations 2002 (SI 2002 No 2677) (COSHH)*, which have been amended to implement the health side of the European Chemical Agents Directive (98/24/EC).

The *Notification of Installations Handling Hazardous Substances Regulations 1982 (SI 1982 No 1357)* and the associated amendment *Regulations 2002 (SI 2002 No 2979)* were revoked on 6 April 2013. These changes were made under the *Health and Safety (Miscellaneous Repeals, Revocations and Amendment) Regulations 2013 (SI 2013 No 448)*.

A consequential amendment was also made to the *Dangerous Substances (Notification and Marking of Sites) (NAMOS) Regulations 1990 (SI 1990 No 304)* to bring into scope a requirement currently in the *NIHHS (Amendment) Regulations 2002 (SI 2002 No 2979)*.

The *NIHHS Regulations* were revoked because the provisions have been superseded by the requirements of the *Seveso II Directive* which is implemented in Great Britain through the *Control of Major Accident Hazards (COMAH) Regulations 1999 as amended*, and planning legislation. The revocation of NIHHS reduced the duplication of notification regimes for dangerous substances.

Changes to the NAMOS Regulations

The consequential amendment to the *NAMOS Regulations 1990* require operators to notify their local fire and rescue service (FRS) (not HSE) if they have on site 150 tonnes or more of ammonium nitrate (AN) and mixtures containing AN where the nitrogen content exceeds 15.75% of the mixture by weight. The threshold for all other dangerous substances in the *NAMOS Regulations* is 25 tonnes but for this particular type of AN the threshold has been lifted from the *NIHHS Amendment Regulations* (ie 150 tonnes) to avoid bringing into scope many small businesses, particularly farms, which would place an additional burden on those businesses.

There is not a requirement to mark the site for this particular mixture of AN because it is not included in the list of dangerous substances in the *Carriage of*

[D0909] Dangerous Substances and Explosive Atmospheres

Dangerous Goods and Use of Transportable Pressure Equipment Regulations 2009 on which the *NAMOS Regulations* are based.

Regulation 7 of the *NIHHS Regulations* made the HSE the enforcing authority for health and safety requirements at all notified sites. This no longer applies following revocation of the Regulations. Some sites will then be subject to Local Authority (LA) enforcement under the provisions of the *H&S (Enforcing Authority) Regulations* and will need to be identified and arrangements made to pass details of those sites to the relevant LA.

Local offices should be able to identify these sites from information held on sites with hazardous substances consent; those identified as sub and non-COMAH etc. Possible candidates are sites storing LPG above the *NIHHS* threshold of 25 tonnes but below the *COMAH* threshold of 50 tonnes.

A small number of former *NIHHS* sites where petrol is dispensed (eg for on-site vehicles rather than using a petrol filling station) which are not covered by the *COMAH Regulations*, are subject to the petroleum legislation and therefore the licensing regime. These may be HSE enforced if the main activity is not one assigned to LAs.

Schedule 1 of the *NAMOS Regulations* removed the requirement for sites to notify under *NAMOS* if they had notified under *NIHHS*, this will no longer applies following the revocation of *NIHHS*. This means that sites who would have notified under *NIHHS* are required to notify under *NAMOS* (subject to the type and amount of dangerous substances held on the site). It is suggested that this mainly applies to farms and a small number of other sites (eg docks).

There is not a requirement for operators who have already notified under *NIHHS* to re-notify under *NAMOS*. However, the number of sites affected is thought to be low. Therefore, as these sites are identified (ie former *NIHHS* sites which are not subject to *COMAH*) arrangements need to be put in place to transfer details to the relevant FRS (as details of *COMAH* sites should have already been shared as part. The commitment given to the Home Office to inspect all newly notified sites who notified under *NIHHS* will cease to apply from 6 April 2013. Field Operations Directorate (FOD) has advised that Operational Managers should use their judgement on the priority of any *NAMOS* notifications received and feed them, as appropriate, into local work programmes.

Explosive atmospheres and potentially explosive atmospheres

[D0910] An explosive atmosphere is a mixture, under atmospheric conditions, of air and one or more dangerous substances in the form of gases, vapours, mists or dusts in which, after ignition has occurred, combustion spreads to the entire unburned mixture. A potentially explosive atmosphere is one that could become explosive due to local and operational conditions. These would include maintenance activities and fault conditions such as leaks of volatile substances and gases. An atmosphere that is not explosive during the course of normal operations may be potentially explosive if dangerous substances are present.

Risk assessment [D0912]

Commencement

[D0911] The *Dangerous Substances and Explosive Atmospheres Regulations 2002 (SI 2002 No 2776) (DSEAR)* came fully into force on 30 June 2003, although certain transitional arrangements extended to 30 June 2006 [*SI 2002 No 2776, Regs 1* and *17*]. Since 30 June 2006 all workplaces must comply with the relevant requirements of DSEAR.

Risk assessment

Requirement and considerations

[D0912] When a dangerous substance is, or is liable to be present at the workplace, the employer is required to carry out a risk assessment before commencing any new work activity involving that substance [*Dangerous Substances and Explosive Atmospheres Regulations 2002 (SI 2002 No 2776) (DSEAR), Reg 5(1)*]. The risk assessment required by the *DSEAR* is identification and careful examination of the dangerous substance(s) that may be present, the work activities associated with the substance(s), and how the work activities might give rise to a risk of fire or explosion. The *DSEAR* require that the risk assessment considers certain aspects as follows [*SI 2002 No 2776, Reg 5(2)*]:

(a) the hazardous properties of the substance;
(b) information on safety or any safety data sheets that the supplier may provide;
(c) the circumstances of the work including:
  (i) the work processes and substances used and their possible interactions;
  (ii) the amount of substance involved;
  (iii) the risk when two or more dangerous substances are used in combination;
  (iv) the arrangements for the safe handling, storage and transport of dangerous substances and waste containing dangerous substances;
(d) activities such as maintenance where there is the potential for a high level of risk;
(e) the effect of measures taken or to be taken to comply with the Regulations;
(f) the likelihood that an explosive atmosphere will occur and persist;
(g) the likelihood that ignition sources, including electrostatic discharges, will be present and become active and effective;
(h) the scale of the anticipated effects of a fire or an explosion;
(i) any places which are or can be connected via openings to places in which explosive atmospheres may occur;
(j) such additional safety information as the employer may need to complete the risk assessment.

The requirements for a risk assessment and the management of risk are similar in principle to those required under the *Control of Substances Hazardous to*

Health Regulations 2002 (SI 2002 No 2677) (COSHH). A risk assessment may have already been carried out on dangerous substances under the *COSHH* Regulations if they were hazardous to health. However, because of the different nature of the hazards of fire or explosion, the risk assessment under the *DSEAR* has different considerations and must be treated separately. The management of dangerous substances will share aspects in common with the management of substances hazardous to health, but there will be individual elements of management required in each case.

The degree of detail in the risk assessment needs to be commensurate with the degree of risk. For a small workshop, a table showing the substances stored and the hazards that could be created (eg spillages) and the precautions in place (eg open windows when handling), with any recommended improvements (eg provision of an extra fire extinguisher) would probably suffice. For a large factory with many dangerous substances, a more sophisticated risk assessment would be expected that would analyse the threats, detail the protection systems and emergency procedures and the remedial actions in the event of an emergency.

A practical example of a fire risk assessment, and information about the risk assessment requirements contained in the *DSEAR* is given in *Tolley's Practical Risk Assessment Handbook*, 54th edition.

Recording findings

[D0913] The purpose of the risk assessment is to enable employers to decide what measures are needed to eliminate, or reduce as far as reasonably practicable, the safety risks from dangerous substances. An employer with five or more employees is required to record in writing the significant findings of the risk assessment as soon as possible after the assessment is made [*Dangerous Substances and Explosive Atmospheres Regulations 2002 (SI 2002 No 2776) (DSEAR), Reg 5(4)*], including:

- the measures (technical and organisational) that have been taken or will be taken to eliminate and/or reduce the risk;
- sufficient information to show that the workplace and work equipment will be safe during operation and maintenance, including details of hazardous zones, co-ordination of safety in shared workplaces, arrangements to deal with accidents, incidents and emergencies and measures to inform, instruct and train employees.

After completing the risk assessment, the *DSEAR* require that the measures identified by the risk assessment are implemented before any new work commences [*SI 2002 No 2776, Reg 5(5)*].

Review

[D0914] The risk assessment must be kept up to date and the employer is required to review it regularly [*Dangerous Substances and Explosive Atmospheres Regulations 2002 (SI 2002 No 2776) (DSEAR), Reg 5(3)*], particularly if:

(a) there is reason to suspect that it is no longer valid, or
(b) there has been a significant change in the workplace, work processes or organisation of work.

When changes to the risk assessment are required as a result of the review, the risk assessment must be updated and any resulting safety measures implemented before work commences again [SI 2002 No 2776, Reg 5(5)].

Management of risks

Safety principles

[D0915] Employers are required to ensure that the risks to the safety of employees and others from dangerous substances are either eliminated or reduced as far as reasonably practicable [*Dangerous Substances and Explosive Atmospheres Regulations 2002 (SI 2002 No 2776) (DSEAR), Reg 6(1)*]. Where it is not reasonably practicable to eliminate or reduce the risks, employers are required to take, so far as is reasonably practicable, measures to control the risks and measures to mitigate the detrimental consequences should a fire, explosion or similar event occur. The *DSEAR* therefore advocate the established safety management principles of elimination, reduction, control and mitigation of risk.

Elimination and reduction of risk

[D0916] The avoidance of risk by replacement of the dangerous substance with another substance or process that eliminates or reduces the risk is preferable if possible [*Dangerous Substances and Explosive Atmospheres Regulations 2002 (SI 2002 No 2776) (DSEAR)*]. When it is not reasonably practicable to eliminate the risk, it may be possible to reduce the risk by replacing the dangerous substance with another that is less dangerous (eg by replacing a low flashpoint solvent with a high flashpoint one). Alternatively, it may be possible to redesign the process so as to reduce the quantities of dangerous substances involved.

In support of these measures, there are guides published by the Health and Safety Executive on *7 steps to successful substitution of hazardous substances* HS(G) 110 (HSE) ISBN 0 7176 0695 3, which can be purchased from HSE Books) and by the Department of Trade and Industry on 'Process intensification' (which can be obtained from the DTI Publications Unit). When substituting other substances or redesigning the process, it is important to take care that the changes do not create new or increased safety or health risks from other sources.

Control measures

[D0917] When dangerous substances are present, the *Dangerous Substances and Explosive Atmospheres Regulations 2002 (SI 2002 No 2776) (DSEAR)* require employers to apply control measures, as far as reasonably practicable, consistent with the risk assessment and appropriate to the nature of the activity or operation [SI 2002 No 2776, Regs 6(3), (4)], including the following in priority order:

(a) reducing the quantity of dangerous substance to a minimum (eg limit volume stored);
(b) avoiding or minimising releases (eg keep in sealed containers);
(c) controlling the release at source (eg ensure that valves and stoppers are used);
(d) preventing the formation of an explosive atmosphere (eg apply appropriate ventilation);
(e) collecting, containing and removing any releases to a safe place (eg fume extraction);
(f) avoiding ignition sources (eg sparks, naked flames etc.);
(g) avoiding adverse conditions that could lead to danger (eg overheating, overpressure);
(h) keeping incompatible substances apart (eg by segregation).

Mitigation

[D0918] When dangerous substances are present, the *Dangerous Substances and Explosive Atmospheres Regulations 2002 (SI 2002 No 2776) (DSEAR)* require employers to apply mitigation measures, as far as reasonably practicable, consistent with the risk assessment and appropriate to the nature of the activity or operation [SI 2002 No 2776, Regs 6(3) and 6(5)], including the following:

(a) reducing the numbers of persons exposed to the risk (eg restricting access);
(b) providing plant that is explosion resistant (eg robust firmly mounted equipment);
(c) providing explosion suppression or relief equipment (eg pressure vents);
(d) taking measures to control or minimise the spread of fires or explosions (eg reducing combustible material, fire doors etc.);
(e) providing suitable equipment for the protection of personnel (eg helmets, visors etc.).

General safety measures

[D0919] In addition to the specific measures relating to dangerous substances and their immediate vicinity, the *Dangerous Substances and Explosive Atmospheres Regulations 2002 (SI 2002 No 2776) (DSEAR)* also specify a number of general safety measures that employers must take, as far as reasonably practicable [SI 2002 No 2776, Reg 6(8) and Sch 1]. These measures relate to the workplace and work processes as a whole and include:

(a) Ensuring the workplace is designed, constructed and maintained so as to minimise risk (eg use of fire resistant materials, installation of appropriate blast walls etc.).

(b) Providing work processes that are suitably designed, constructed, assembled, installed so as to reduce risk and ensuring that they are used properly and maintained in an efficient state, working order and good repair.
(c) Ensuring that equipment and protective systems meet the requirements for use in hazardous places (see below), and:
  (i) can be maintained in a safe state of operation independently of the rest of the plant in the event of a power failure;
  (ii) can be manually overridden when incorporated within automatic processes which deviate from the intended operating conditions;
  (iii) can dissipate accumulated energy quickly and safely on operation of emergency shutdown;
  (iv) contain measures to prevent confusion between connecting devices.
(d) Applying appropriate systems of work including written instructions, permits to work and other procedural systems of organising and controlling work.
(e) Identifying the hazardous contents of containers and pipes [*SI 2002 No 2776, Reg 10*]. Many will already be marked or labelled under existing legislation. For those that are not, 'identification' may require labelling, marking or warning signs. It could include training, information or verbal instruction.
(f) Arranging for the safe handling, storage and transport of dangerous substances and waste containing dangerous substances [*SI 2002 No 2776, Reg 6(6)*].

Classification of places with explosive atmospheres

Classification scheme

[D0920] Gases, vapours, mists and dusts can all form explosive atmospheres with air. The source of the gases, vapours, mists and dusts may be a consequence of the natural state of the substance (eg hydrogen gas), or it may be a result of the means of storage or handling (as in the case of a flammable liquid), or as a result of a work activity (eg grinding or powder processing). Where there is any prospect of an explosive atmosphere, a hazardous area classification should be carried out as an integral part of the risk assessment, and for those areas classified as hazardous, *special precautions* over the sources of ignition are needed to prevent fires and explosions.

In workplaces where an explosive atmosphere may occur, employers must first classify the workplace into hazardous and non-hazardous places [*Dangerous Substances and Explosive Atmospheres Regulations 2002 (SI 2002 No 2776) (DSEAR)*]. The criteria for deciding whether a place is hazardous is whether an explosive atmosphere could occur in such quantities as to require *special precautions* to protect the health and safety of workers concerned within the meaning of the *DSEAR*. Similarly, a non-hazardous place is one where an explosive atmosphere is not expected to occur in such quantities as to require special precautions to protect the health and safety of workers concerned [*SI 2002 No 2776, Sch 2 para 1*].

Thus, in the first instance, employers must assess the quantity of explosive atmosphere that could be present in a particular place. The term 'not expected to occur in such quantities' is deliberate and means that employers should also consider the likelihood or possibility of a release of an explosive atmosphere as well as the potential quantities of such a release. Places where it is credible that an explosive atmosphere could exist, even when this is not within normal operating limits should be classified as hazardous. If, on the other hand, a release is extremely unlikely to occur and/or if the quantities released are small, it may not be necessary to classify the area as hazardous.

As an example, a dangerous substance is being carried though a seamless pipe that has been properly installed, maintained and inspected. It is unlikely that the substance will be released. An explosive atmosphere would not be expected to occur from this source and the area surrounding the pipe would be non-hazardous. However, if the pipe contained flanged joints or fittings where there was the possibility of a leak or its condition was uncertain, a release would be a credible event and the area should be classified as hazardous.

With regard to quantity, a spillage from a small bottle of solvent used in a laboratory would release so little flammable vapour that no special precautions are needed other than general control of ignition sources (no smoking, for example) and cleaning and disposing of the spillage. The area would not be classified as hazardous. If, however, the vapour released also had a strong anaesthetic effect, and spillages were known to occur from time-to-time, then the area might be classified as hazardous. When deciding whether it is necessary to classify as 'hazardous' areas with small quantities of dangerous substances, the actual circumstances of use and any specific product guidance should be taken into account.

Where dangerous substances in small pre-packaged containers are stored or are on display for sale in retail premises, for example solvents or aerosols, the area would not normally need to be classified as hazardous. An exception to this might, however, be with storage in poorly ventilated basements. When such substances are held in large quantities, as in a warehouse, with the opportunity for a build up of vapours or a larger spillage, a hazardous classification would be expected. Procedures to control ignition sources and to clean up and dispose of any spillage/release would be needed.

In identifying hazardous and non-hazardous areas and assigning zones, the following matters should be considered:

- The hazardous properties of the dangerous substances involved.
- The amount of dangerous substances involved. The size of any potentially explosive atmospheres is, in part, related to the amount of dangerous substances present. Guidance is given in industry specific codes on the quantities of dangerous substances that may be safety stored. See, for example, the 'Code of Practice 7 on the storage of full and empty UK LPG cylinders and cartridges' produced by the LPG Association.

- The work processes, and their interaction, including any cleaning, repair or maintenance activities that will be carried out.
- The temperatures and pressure of the dangerous substances. This will affect the nature and extent of any release. Some substances do not form explosive atmospheres unless they are heated (eg diesel oil), while others if released under pressure will form a fine mist that can explode, even if there is insufficient vapour.
- The containment system and controls to prevent liquids, gases, vapours or dusts escaping into the general atmosphere (eg trays, seals).
- The possibility of an explosive atmosphere forming in an enclosed plant or storage vessel. A release of a dangerous substance into an enclosed space always increases the risk of an explosive atmosphere forming.
- The ventilation (natural or a fan extract system) or any other measures designed to dilute sources of release and ensure that any explosive atmosphere does not persist for an extended time. Well-designed ventilation may prevent the need for any zoned area or reduce it so it has a negligible extent.

When considering the potential for dangerous atmospheres, it is important to consider all substances that may be present in the workplace. Waste products, residues, cleaning agents, materials used for maintenance and fuel could all be potentially dangerous. The possibility that combinations of substances could react to create an ignition source or an explosive atmosphere should not be overlooked.

The refuelling of cars or loading and unloading of petrol from tankers intended for use on public roads involve the introduction of potential sources of ignition in an area where a spill is possible. This would normally meet the criteria for being a hazardous area. In these circumstances, safety can be achieved by isolating power sources (turning off engines) where a transfer of fuel is taking place and making suitable checks before and after transfer and before and after vehicles are moved into or out of the refuelling area. Garage owners have a particular responsibility to monitor the forecourt and to ensure that these practices are implemented.

Zoning

[D0921] For places classified as hazardous according to D0920, employers are required to further classify these places into zones on the basis of the frequency and duration of the explosive atmosphere occurring [*Dangerous Substances and Explosive Atmospheres Regulations 2002 (SI 2002 No 2776) (DSEAR), Reg 7(2)* and *Sch 2 para 2*]. The *DSEAR* give the following classification scheme for mixtures with air of dangerous substances (Zones 0, 1 and 2) and combustible dusts (Zones 20, 21, and 22). In decreasing order of risk:

- *Zone 0* – A place in which an explosive atmosphere consisting of a mixture with air of dangerous substances in the form of a gas, vapour or mist is present continuously or for long periods or frequently.

- *Zone 1* – A place in which an explosive atmosphere consisting of a mixture with air of dangerous substances in the form of a gas, vapour or mist is likely to occur in normal operation occasionally.
- *Zone 2* – A place in which an explosive atmosphere consisting of a mixture with air of dangerous substances in the form of a gas, vapour or mist is not likely to occur in normal operation, but if it does occur, will persist for a short period only.
- *Zone 20* – A place where an explosive atmosphere in the form of a combustible dust in air is present continuously or for long periods or frequently.
- *Zone 21* – A place where an explosive atmosphere in the form of a combustible dust in air is likely to occur in normal operation occasionally.
- *Zone 22* – A place where an explosive atmosphere in the form of a combustible dust in air is not likely to occur in normal operation, but if it does occur, will persist for a short period only.

Within this scheme, terms such as 'occasionally' or 'frequently' are not defined but are assumed to be that which a reasonable person would use. Normal operation is any situation when installations are used within their design parameters. Combustible dusts in the form of layers, deposits or heaps must be considered as a source of an explosive atmosphere and classified accordingly.

The basic principles of area classification are explained within the European standards, BS EN 60079/10 for gases and vapours and BS EN 61241/3 for dusts. These standards form a suitable basis for assessing the extent and type of the zone and can be used as a guide to complying with the requirements of the *DSEAR*. However, the guidance must be applied to the site-specific factors in order to determine the extent and type of zone in any particular case. It should be remembered that an explosive atmosphere may spread into areas away from the source of the hazard, for example through ducts, and these areas must also be included in the classification scheme.

Various organisations have published industry specific codes (eg Energy Institute). Providing they are applied appropriately, they are valuable in encouraging a consistent interpretation of the requirements.

Area classification

In areas where dangerous quantities and concentrations of flammable gas or vapour may arise, protective measures by zoning are required to reduce the risk of explosions. Essential criteria against which ignition hazards can be assessed and zone areas determined are given in Part 10 of International Standard ;IEC 60079-10-1:2008, which has been adopted as a European Standard and a British Standard, BS EN 60079-10-1:2009. This gives advice on design and control parameters which can be used to quantify the hazard in order to bound and limit such areas.

BS EN 60079-10-1:2009 gives guidance on the procedure for classifying areas where there may be an explosive gas atmosphere. The procedures are quite complex and should only be carried out by those who understand the relevance and significance of the properties of flammable materials and those who are familiar with the process and the equipment and the layout of the areas in

Places with explosive atmospheres [D0921]

which the equipment is located. Where the necessary electrical, mechanical and other qualified engineering personnel are not available, it is necessary to use consultants specialising in hazardous area classification.

This section is intended to provide an overview of the procedure but is in no way a substitute for reference to BS EN 60079-10-1:2009. It may assist users in determining whether they have the necessary capability and expertise to undertake the procedure themselves. Where consultants are used it will enable users to be better prepared to provide the necessary information.

For plant in design and construction, area classification should be carried out when the initial process and instrumentation line diagrams and layout plans are available so that the design and layout may be optimised to reduce zoned areas to a minimum. Area classification should be confirmed before start-up and reviewed during the life of the plant. For plants in service, area classification needs to take account of local conditions (eg ventilation and other equipment) and as these may change during service, classification needs to be reassessed as part of the management of change procedure.

The procedure comprises firstly of identifying sources, rate and grade of release and then considers factors that determine the type and extent of the zone including ventilation. The detail is provided in four Annexes. Annexes A and B provide information and examples on sources of release and release rate and the assessment of ventilation, while Annexes C and D give examples of areas classification and the treatment of flammable mists.

Sources of explosive atmospheres

Each item of process equipment containing a flammable gas or vapour (and flammable liquids and solids which may on release give rise to them) should be considered as a potential source of flammable material. Where process plant or containment (eg storage tanks) is not totally enclosed, or where air can enter the system, it is necessary to decide whether a flammable mixture with air can exist inside the equipment. For all items it is necessary to consider whether a release of flammable materials can create a flammable atmosphere outside the equipment. It is important to realise that explosive atmospheres can be generated by mists of leaking flammable liquid, even though the liquid may be below its flash point.

If the item cannot contain flammable materials under reasonably foreseeable circumstances it will not give rise to a hazardous area around it. The same applies if the item contains a flammable material but cannot release it to the atmosphere because the system is completely sealed. For example a continuous welded pipeline is not considered to be a source of release, however, valves, seals and flanges, which may potentially leak, are considered to be a potential source. If the total quantity of flammable material available for release is small (eg in some laboratory equipment), the area classification procedure may not be appropriate if other risk control measures are applied (eg high ventilation).

Having identified potential sources, these are then graded as continuous, primary or secondary according to the frequency and duration of the release. These normally give rise to zones 0, 1 and 2 as defined above. All times when a release may take place, including commissioning, normal operation, opening equipment, batch filling, cleaning, purging, start-up and shutdown cycles and when the equipment is mothballed and dormant, should be considered.

Extent of zone

The extent of zone is defined by the calculated or estimated distance over which a potentially explosive atmosphere from sources exists before it disperses to a concentration in air below its lower explosive limit with an appropriate safety factor. (The lower explosive limit is the concentration of flammable gas, vapour or mist in air below which an explosive gas atmosphere will not be formed.) The assessment of the spread of gas or vapour and its dispersion and dilution to below its lower explosive limit requires judgment and the advice of experts should be sought.

The key factors determine the extent of a zone are the release rate of gas or vapour, its lower explosive limit, ventilation, and whether the gas is heavier or lighter than air (relative density). Physical barriers (walls, sealed doors etc), and atmospheric barriers (maintaining an external over pressure and air purging) also need to be taken into account. Other parameters that should be considered include the climatic conditions (wind, air temperature) and topography (eg water courses where a flammable liquid may reside on the surface over a large area or trenches, pits and drains where an explosive atmosphere may not easily disperse and dilute).

Determining the rate of release of a gas or vapour can be a complex operation. In general, it depends on the geometry (area) and type of the source of release (eg open surface, leaking flange), the release velocity (related to the pressure differential, source geometry and the physical properties of the flammable material), and the concentration of flammable vapour or gas in the release (some liquids may comprise flammable and non flammable mixtures). After release of flammable liquid, the rate of gas or vapour formation depends on the volatility of the flammable liquid (determined by the vapour pressure as a function of liquid temperature and the enthalpy of vaporisation, to which the boiling point and flash point are roughly related), and the liquid temperature after release (which would need to take account of the ambient temperature and any hot/cold surfaces in the vicinity etc).

Application

Annex C of BS EN 60079-10-1:2009 gives ten worked examples that illustrate the principles of hazardous area classification. Examples range from mechanical pumps at ground level either indoors or outdoors pumping flammable liquid, a fixed process mixing vessel being opened regularly for operational reasons, a mixing room in a paint factory with multiple mixing vessels and pumps. Other examples include a flammable liquid storage tank situated outdoors with a fixed roof and no internal floating roof, a single tanker filling installation for gasoline, top filling with no vapour recovery and an oil water separator in a petroleum refinery.

The discussion on ventilation gives seven example calculations to ascertain the degree of ventilation under different circumstances. The calculations are based on determining a hypothetical volume over which the mean concentration of flammable gas or vapour will be typically between 0.25 to 0.5 times the lower explosive limit, depending on the factor of safety appropriate to the conditions. (The hypothetical volume is related to but not the same as the hazardous area dimensions.) The examples cover a range of flammable materials (eg toluene, propane gas, ammonia, methane) stored under different indoor and outdoor ventilation conditions.

Recent work by the Health and Safety Laboratory (HSE Research Report 630, 2008) has sought a more soundly based methodology as it applies to secondary releases from low pressure natural gas systems with the possibility of removing a significant amount of conservatism from the method given in BS EN 60079-10-1:2009. An extensive programme of theoretical and experimental work has lead to an alternative approach to measuring ventilation effectiveness in enclosures. This has the potential to assist the gas industry and businesses relying on supplies of natural gas for heating and other purposes.

Documentation

Area classification normally takes the form of drawings or plans identifying the hazardous areas and zones. Information about the dangerous substances that will be present, the work activities that have been considered, and other assumptions made will be given. These drawings or plans form part of the risk assessment record that the *DSEAR* require, and should be available for examination by those affected and regulators. They become particularly important when there is a change in the work activity and/or new equipment to be introduced into a zoned area.

Classification drawings or plans may need to be reviewed when maintenance is carried out. If the dangerous substances normally present have been removed it may be possible to treat the area as non-hazardous. Alternatively, if the maintenance creates a larger than normal risk of a release of a dangerous substance, for example if it is necessary to open a normally sealed pipe system or enter a tank where vapours could exist – a larger area may need to be treated as hazardous or the area may need to be to temporarily reclassified to a higher zone. It is not, however, normally necessary to create new classification drawings or plans for the duration of maintenance work, but the process needs to be carefully controlled.

Warning signs on entry

[D0922] Employers are required to ensure that places classified as hazardous are marked at their points of entry with warning signs [*Dangerous Substances and Explosive Atmospheres Regulations 2002 (SI 2002 No 2776) (DSEAR), Reg 7(3)* and *Sch 4*]. The warning sign at entry into places where explosive atmosphere may occur is specified in Annex III of the European Council Directive 99/92/EC and is:

(a) triangular in shape; and
(b) contains black letters 'EX' on a yellow background with black edging (the yellow part taking up at least 50 per cent of the area of the sign).

Verification of safety prior to first use

[D0923] Employers are required to ensure that before workplaces containing places classified as hazardous are used for the first time, they are confirmed as being safe (verified) by a person (independent or employed by an organisation) competent in the field of explosion protection [*Dangerous Substances and Explosive Atmospheres Regulations 2002 (SI 2002 No 2776) (DSEAR), Reg 7(4)*]. The person carrying out the verification must be competent to consider the particular risks at the workplace, the adequacy of the risk assessment, and the control and other measures put in place. As a result the person must have suitable experience, professional training or both.

Equipment and protection systems

Compliance with EPS Regulations

[D0924] Employers are required to take *special precautions* to ensure that equipment and protective systems in places classified as hazardous are selected so as to avoid sources of ignition from, for instance, sparks or flames [*Dangerous Substances and Explosive Atmospheres Regulations 2002 (SI 2002 No 2776) (DSEAR), Reg 7(2)* and *Sch 3*]. Specifically, equipment and protective systems must be selected on the basis of the requirements set out in the *Equipment and Protective Systems Intended for Use in Potentially Explosive Atmospheres Regulations 1996 (SI 1996 No 192)* (EPS) unless the risk assessment indicates otherwise.

EPS defines two Equipment Groups. Equipment Group 1 is equipment for use in underground parts of mines, and to those parts of surface installations of mines liable to be endangered by firedamp and/or combustible dust. Equipment Group II is intended for use in non-mining places liable to be endangered by explosive atmospheres. Within each Equipment Group different Categories of equipment are designated.

In a non-mining environment DSEAR specifies that the following categories of equipment (as defined within the EPS Regulations) must be used in the zones indicated, providing the category is suitable:

- in Zone 0 or Zone 20 – Group II Category 1 equipment;
- in Zone 1 or Zone 21 – Group II Category 1 or 2 equipment;
- in Zone 2 or Zone 22 – Group II Category 1, 2 or 3 equipment.

Equipment already in use before 1 July 2003 can continue to be used indefinitely providing the risk assessment shows that it is safe to do so.

The *DSEAR* provide definitions of the terms 'equipment', 'protective systems', 'devices', 'components'. A 'potentially explosive atmosphere' is an atmosphere that could become explosive due to local and operational conditions. These would include maintenance activities and fault conditions. This part of the *DSEAR* complements the *Equipment and Protective Systems Intended for Use in Potentially Explosive Atmospheres Regulations 1996*. For further information see the DTI website.

Manufacturers, suppliers and others supplying equipment for use in potentially explosive atmospheres have a duty under the EPS Regulations to mark their equipment to show which Group and Category it falls under, and therefore which zone it may be suitable for. These markings are in addition to the CE sand Ex markings required. Users have the duty to zone their workplaces and to use equipment suitable for that zone.

Equipment built to the requirements of the EPS Regulations can be identified as it will carry the explosion protection symbol 'Ex' in a hexagon, the equipment category (1, 2,or 3) and the letter 'g' or 'd' depending on whether it is intended for use in gas or dust atmospheres. In many cases a temperature rating, expressed as a 'T' marking, will also be included. These indicate the limitations to safe use.

For most electrical equipment designed for use in explosive atmospheres there will be little change required, except in the details of the marking on the equipment. However, the 1996 Regulations also apply to mechanical equipment as a potential ignition source. A harmonised European standard for category 3 mechanical equipment is now available as BS EN 13463 Part 1.

Equipment generating a potential explosive atmosphere

[D0925] Where the use of equipment may generate an explosive atmosphere, such as dust, then before installation and use the user has to make a risk assessment under Reg 5 of DSEAR. The assessment should consider the likelihood of explosive concentrations being generated under normal use and fault conditions. In addition, it should identify any additional precautions needed to prevent such concentrations and to mitigate the effects of any explosion to a safe level, such as local ventilation and filtration systems.

A milling machine is an example of equipment where an explosive dust may be generated. For the machine to be ATEX certified it has to be intended for use in a potentially explosive atmosphere and have its own ignition source. If the machine while containing an explosive dust is not operated or exposed to that dust (ie the dust is contained), it may not as a whole need to be ATEX certified.

However, the explosion risks from the processing of materials do have to be assessed and controlled by the machine's manufacturer to meet the essential health and safety requirements of the Machinery Directive (MD) (98/37/EC). Any equipment or protection systems installed as part of the machine to meet these requirements have, in turn, to meet the requirements of ATEX 95 (94/9/EC). For examples, if motors, switches etc could possibly be exposed to an explosive atmosphere of the dust generated, they would, individually, have to be ATEX certified. Explosion relief panels where these are fitted are protections systems and therefore also come within the scope of ATEX and must be certified.

Risks from existing mechanical equipment

[D0926] Equipment already in use before 1 July 2003 can continue to be used indefinitely providing the risk assessment shows that it is safe to do so. For most individual pieces of mechanical equipment there is frequently no documentation suggesting that anyone has previously considered the ignition risk at the time it was manufactured. The risk assessment therefore needs to demonstrate that the equipment is acceptably safe.

In general very little mechanical equipment is an ignition risk in normal operation, but may become a risk when a fault arises. The risk assessment would be a thorough assessment of the equipment under all credible conditions. The users own experience of operating the equipment is important, particularly if faults were known to occur and whether these faults created actual ignition or the obvious potential for ignition.

Contact with the manufacturer may still be possible. If so the manufacturer may be able to provide information about the risks, records of any ignitions and the types and frequency of faults that have been known to occur. The wider experience of the manufacturer may have identified faults that create an ignition risk that the user was not aware of.

The amount and adequacy of maintenance of the equipment are important considerations. Records of maintenance and the original maintenance instructions form an input to the risk assessment. There should be a reasonable basis for saying that the equipment remains as safe as when it was new.

Where new equipment of the same type has a relevant standard, a comparison of the existing equipment with this standard may identify risks that were not fully understood at the time when the old equipment was constructed. While there is no obligation to bring the existing equipment fully up to modern standards, simple changes that would improve safety where reasonably practicable should be adopted. This is the current HSE policy towards to continued use of old equipment.

Storage silos and bins

[D0927] Storage silo and bins that may be used for storing explosive products such a fine dusts are not ATEX 137/EPS equipment, unless they include, as an integral part, equipment or fittings that create an explosion risk. Where silos and bins have explosive vent panels for which there is no information about the standard to which they have been built and tested, nor any details about the operating pressure under normal and explosive conditions, the panels cannot be easily shown to be fit-for-purpose. A risk assessment under DSEAR would be likely to conclude that an assessment of the pressures was needed and new correctly rated panels carrying the CE and Ex mark should be fitted. Where the inside of such storage silos and bins is zoned under DSEAR, any equipment used in the zone would need to be ATEX/EPS rated.

Second hand equipment

[D0928] Both DSEAR and PUWER (Provision and Use of Work Equipment Regulations) are relevant to bringing second hand equipment into service. The EPS Regulations relate to the supply of new equipment. In general, the design, operation and condition of any new or second hand equipment brought into service needs to be reviewed and verified as fit for purpose. PUWER Regulation 10 makes clear that equipment brought into service at a new site where there is potential for an explosive atmosphere needs to comply with the APEX Directives or other relevant single market Directives that applied when it was first supplied or brought into service.

Work clothing

[D0929] Employers are required to ensure that appropriate work clothing that does not give rise to electrostatic discharges is provided to employees entering places classified as hazardous [*Dangerous Substances and Explosive Atmospheres Regulations 2002 (SI 2002 No 2776) (DSEAR), Reg 7(5)*]. Sparks from clothing could create a risk of igniting an explosive atmosphere. Advice should be sought from an industrial clothing supplier if the employer is in any doubt as to the suitability of any item.

Specific products and areas

Flammable substances in public and laboratory areas

[D0930] Where flammable substances are used in areas where the public is present, such as in nail treatments, it is difficult to maintain constantly an effective control over all sources of ignition, such as smoking or portable electrical equipment. Consequently hazardous area classification is not appropriate. It is better to limit the amount of the substance present to the amount used in half a day's work, to use containers that minimise the risk of spills and that will break if knocked over, and to ensure adequate ventilation.

A similar principle applies to areas that handle sacks of substances in the form of dusts, such as supermarkets with in-store bakeries that handle 25 kg sacks of flour sugar and custard powder that are listed as explosible substances in bulk storage areas. In shops there is usually no means to generate a large dust cloud unless a sack fails or tears during handling, and even then only a small proportion of the escaping product will be raised into a dust cloud. Unless there is some way that dust will be kept as a cloud for a longer period, the largest release foreseeable from a single failure is 25kg of dust, hazardous area classification is not normally appropriate.

Laboratories that handle flammable products but in small volumes. An HSE document 'Hazardous area classification and Laboratory operations' addresses this situation.

Liquefied Petroleum Gas (LPG) issues

[D0931] The DSEAR applies wherever a dangerous substance or explosive atmosphere is present or is liable to be present in a workplace, and it therefore applies wherever LPG is stored. Places for bulk LPG storage, LPG cylinder filling and cylinder storage areas therefore fall within the Regulations. The appropriate regulations apply, including the need for a risk assessment and a possible need for zoning.

The requirement to zone areas where LPG cylinders are stored outdoors in a ventilated area has been discussed with the LP Gas Association. Whilst an explosive atmosphere is not likely in normal operation, damage to cylinders has been known to occur during mechanical handling operations resulting in a leak and a short term build-up of gas. The LP Gas Association has published a Code of Practice 7 – The Storage of Full and Empty LPG Cylinders and Cartridges. Provided the outdoor storage area is designed, constructed and maintained fully in accordance with the Code of Practice, the HSE has accepted that it is not essential to designate the LPG cylinder outdoor storage area as Zone 2 as would otherwise be the case. Thus, ATEX compliant equipment and vehicles are not needed in such storage areas. Where the outdoor storage area does not comply with the conditions given in Code of Practice 7, a Zone 2 area may exist.

With regard to the storage of LPG cylinders indoors, Code of Practice 7 requires a full site specific risk assessment to be carried out. The zoning level would depend on a number of factors including ventilation provision. Where an area has been zoned, then suitable ATEX equipment (see below), including vehicles, would be required for work in that area.

Storage of aerosols

[D0932] Warehouses and other areas used for storing aerosols in bulk should be considered for zoning. Most aerosols use liquefied flammable gases as the propellant and may also contain flammable liquids as part of the product, but it is best to check with suppliers for details of the amounts and types in the products stored. There are examples where aerosols have been damaged during mechanical handling, they leak as a result and their gases have resulted in causing major fires. It is good practice to separate aerosols from other less hazardous products in a fire separated part of the warehouse.

The issue has been discussed with the British Aerosol Manufacturers Association, and they have published a guide to safe storage. Provided the conditions set out in the guide are followed closely, HSRE has accepted that it is not necessary to designate warehouse storing aerosols as zone 2, and ATEX compliant vehicles are not needed (see below). Where the conditions in the guide cannot be met, the warehouse should normally be classified as hazardous, zone 2.

Wood dust

[D0933] Wood dust can explode if dispersed in a cloud, and there may be a need to assign hazardous areas to a wood-working machine room if dense clouds can form in the case of some equipment fault or operator error. For example, a dense cloud can form if the local exhaust ventilation fails or machines continue to operate after the ventilation becomes ineffective. A dense cloud is also possible when dust deposits on horizontal surfaces are disturbed, for instance by a sudden blast of air.

IN many cases such as these there will be a need to assign zone 22 areas, but only a very limited need for zone 21. Where this is the case electrical equipment will need to be ATEX rated (see below). In zone 21 the machine or activity that generates the dust cloud must be stopped before any attempts are made to disperse the cloud or clean up the area.

Arrangements to deal with incidents and emergencies

[D0934] The *Dangerous Substances and Explosive Atmospheres Regulations 2002 (SI 2002 No 2776) (DSEAR)* require employers to make arrangements to protect employees (and other persons who are at the workplace) in the event of incidents and accidents involving dangerous substances and explosive atmospheres [*SI 2002 No 2776, Reg 8(1)*]. The provisions clarify what already needs to be done in relation to the safety management of dangerous substances and will not require any duties in addition to those already present in existing legislation. Specifically they build on existing requirements in *Regulation 8* of the *Management of Health and Safety at Work Regulations 1999 (SI 1999 No 3242)*.

The arrangements required are as follows:

(a) Giving suitable warnings (including audible and visual alarms) where necessary to alert employees (and other persons who are at the workplace) immediately of a release of a dangerous substance or the presence an explosive atmosphere before explosive conditions are reached, so they may be withdrawn to a safe place.

(b) Escape facilities (eg designated exit routes, stairways etc.) identified as necessary by the risk assessment to ensure that employees (and other persons who are at the workplace) can, in the event of danger, leave the endangered place promptly and safety.

(c) Warning and communication systems to enable an appropriate response to be made immediately when such an event occurs, including remedial actions and rescue operations to mitigate the effects and restore the situation to normal [*SI 2002 No 2776, Reg 8(3)(a)*].

(d) Emergency procedures to be followed in the event of an emergency, including defining responsibilities and actions to be taken.

(e) Providing equipment and clothing for essential personnel to deal with the incident [*SI 2002 No 2776, Reg 8(3)*].

(f) Organising relevant safety drills at regular intervals designed to test the emergency procedures.

(g) The provision of appropriate first-aid facilities in a safe place.

(h) Making information available to employees (eg through a display at the workplace) on the warnings, escape facilities, emergency procedures, relevant work hazards and hazard identification (eg zoning and marking), and on specific hazards likely to arise at the time of an incident [*SI 2002 No 2776, Reg 8(2)(b)*].

(i) Contacting the relevant accident and emergency services to let them know that information on emergency procedures is available (or providing them with any information that they consider necessary) [*SI 2002 No 2776, Reg 8(2)(a)*].

The scale and nature of these arrangements should be proportional to the risks, taking account of the frequency of such events and the possible consequences. When the results of the risk assessment show that because of the quantity of each dangerous substance at the workplace there is only a slight risk to employees and risk has been eliminated or reduced as far as reasonably practicable, specific emergency arrangements are not necessary [SI 2002 No 2776, Reg 8(4)]. In this case, the general provisions for safety at the workplace will apply.

Provision of information, instruction and training

[D0935] Employers are required to provide employees and other people at the workplace who might be at risk from dangerous substances with suitable information, instruction and training on precautions and actions they need to take to safeguard themselves and others so as to reduce the possibility of an incident taking place [Dangerous Substances and Explosive Atmospheres Regulations 2002 (SI 2002 No 2776) (DSEAR), Reg 9]. This should include:

- names of the substances in use and the risks they present;
- access to any relevant data sheets;
- any specific legislative provisions concerning the hazardous properties of the substance;
- the significant findings of the risk assessment.

Other aspects may include procedures for correct handling, protective clothing and equipment to be used, and the use of safe work processes. Information, instruction and training should be adapted if significant changes in the type or methods of work occur. Existing health and safety legislation already covers much of this.

Information, instruction and training need only be provided to non-employees where it is required to ensure their safety. Where it is provided, it should be in proportion to the level of risk. In general, non-employees should not be exposed to such risks.

Identification of hazardous contents of pipes and containers

[D0936] Employers are required to identify pipes, vessels, tanks and other containers of dangerous substances in accordance with the legislation listed below where this is applicable [Dangerous Substances and Explosive Atmospheres Regulations 2002 (SI 2002 No 2776) (DSEAR), Reg 10 and Sch 5].

- the Chemicals (Hazard Information and Packaging for Supply) Regulations 2009 (SI 2009 No 716);
- the Health and Safety (Safety Signs and Signals) Regulations 1996 (SI 1996 No 341);
- the Good Laboratory Practice Regulations 1999 (SI 1999 No 3106);
- the Classification and Labelling of Explosives Regulations 1983 (SI 1983 No 1140);

- the *Carriage of Dangerous Goods by Rail Regulations 1996 (SI 1996 No 2089)*;
- the *Packaging, Labelling and Carriage of Radioactive Material by Rail Regulations 1996 (SI 1996 No 2090)*;
- the *Carriage of Dangerous Goods (Classification, Packaging and Labelling) and Use of Transportable Pressure Receptacles Regulations 1996 (SI 1996 No 2092)*;
- the *Carriage of Dangerous Goods by Road Regulations 1996 (SI 1996 No 2095)*;
- the *Carriage of Explosives by Road Regulations 1996 (SI 1996 No 2093)*;
- the *Radioactive Material (Road Transport) (Great Britain) Regulations 1996 (SI 1996 No 1350)*.

Where containers of dangerous substances are not marked in accordance with this legislation, for reasons of applicability, employers must ensure that the contents of those containers are clearly identifiable together with the nature of the associated hazards.

Exemption regarding dangerous substances and explosive atmospheres

[D0937] The Health and Safety Executive (HSE) has the power to exempt any person or class of persons or any dangerous substance or class of substances from all or any of the requirements or prohibitions imposed by or under the *DSEAR* [*Dangerous Substances and Explosive Atmospheres Regulations 2002 (SI 2002 No 2776) (DSEAR), Reg 13*]. It is expected that any such exemption would only be used in exceptional circumstances. Such exemption is granted by the issue of a certificate in writing and may be subject to conditions and a limit of time and may be revoked at any time.

In exercising this power, the HSE must, having due regard to the circumstances of the case, be satisfied that the health and safety of persons likely to be affected will not be prejudiced. In addition, it must be satisfied that exemption will be compatible with the requirements of European Council Directives (98/24 and 99/92) relating to the protection of workers from risks from chemical agents and explosives, which take precedence.

The Secretary of State for Defence may, in the interests of national security, by a certificate in writing exempt from all or any of the requirements or prohibitions imposed by or under the *DSEAR* the following [*SI 2002 No 2776, Reg 14*]:

- any of Her Majesty's Armed Forces;
- any visiting force and members attached to a headquarters;
- any person engaged in work involving dangerous substances, if that person is under the direct supervision or control of a representative of the Secretary of State for Defence.

Where any such exemption is granted, suitable arrangements must be made for the assessment of risk to safety created by the work and for adequately controlling the risk to the persons to whom the exemption relates. This power of exemption could apply at defence establishments and centres of defence related research.

Amendments to modernise petroleum and other legislation

[D0938] Petroleum legislation is modernised as part of the *Dangerous Substances and Explosive Atmospheres Regulations 2002 (SI 2002 No 2776) (DSEAR)* by means of a number of amendments. Previously, the keeping of petrol in significant quantities was controlled by licences with conditions issued under the *Petroleum (Consolidation) Act 1928*. However, as petrol is a dangerous substance, the *DSEAR* apply to it and duplicate these controls. The *DSEAR* remove licensing requirements for holding petrol, except for petrol that is being kept for dispensing into vehicles (retail and non-retail) at, for example, garages. Accordingly the following Acts, Regulations and Orders are amended [*DSEAR (SI 2002 No 2776), Reg 15* and *Sch 6 Pts 1* and *2*]:

- the *Petroleum (Consolidation) Act 1928*;
- the *Petroleum Spirit (Motor Vehicles etc) Regulations 1929 (SI 1929 No 952)*;
- the *Petroleum (Liquid Methane) Order 1957 (SI 1957 No 859)*;
- the *Petroleum (Consolidation) Act 1928 (Enforcement) Regulations 1979 (SI 1979 No 427)*;
- the *Petroleum Spirit (Plastic Containers) Regulations 1982 (SI 1982 No 630)*.

In addition, the *DSEAR* amend a number of other pieces of legislation:

- the *Celluloid and Cinematograph Film Act 1922*;
- the *Dangerous Substances in Harbour Areas Regulations 1987 (SI 1987 No 37)*;
- the *Fire Precautions (Workplace) Regulations 1997 (SI 1997 No 1840)*;
- the *Fire Certificates (Special Premises) Regulations 1976 (SI 1976 No 2003)*;
- the *Carriage of Dangerous Good by Road Regulations 1996 (SI 1996 No 2095)*.

The amendment to the *Carriage of Dangerous Goods by Road Regulations 1996* prohibits the filling of fuel tanks for cars or other containers direct from a road tanker.

Repeals and revocations of existing legislation

[D0939] The *Dangerous Substances and Explosive Atmospheres Regulations 2002 (SI 2002 No 2776) (DSEAR)* repeal and revokes a large number of older Regulations and Orders relating to substances that now come under the general provisions of the 2002 Regulations. In this way the *DSEAR* are an important piece of modernising and reforming legislation. The following Acts

Repeals and revocations of existing legislation [D0939]

and statutory instruments are completely repealed and revoked [*SI 2002 No 2776), Reg 16* and *Sch 7 Pts 1* and *2*]:

- the *Celluloid etc. Factories and Workshops Regulations 1921 (SI 1921 No 1825)*;
- the *Manufacture of Cinematograph Film Regulations 1928 (SI 1928 No 82)*;
- the *Cinematograph Film Stripping Regulations 1939 (SI 1939 No 571)*;
- the *Petroleum (Carbide of Calcium) Order 1929 (SI 1929 No 992)*;
- the *Petroleum (Carbide of Calcium) Order 1947 (SI 1947 No 1442)*;
- the *Petroleum (Compressed Gases) Order 1930 (SI 1930 No 34)*;
- the *Magnesium (Grinding of Castings and other Articles) Special Regulations 1946 (SI 1946 No 2017)*;
- the *Dry Cleaning Special Regulations 1949 (SI 1949 No 2224)*;
- the *Dry Cleaning (Metrication) Regulations 1983 (SI 1983 No 977)*;
- the *Factories (Testing of Aircraft Engines and Accessories) Special Regulations 1952 (SI 1952 No 1689)*;
- the *Factories (Testing of Aircraft Engines and Accessories) (Metrication) Regulations 1983 (SI 1983 No 979)*;
- the *Highly Flammable Liquids and Liquefied Petroleum Gases Regulations 1972 (SI 1972 No 917)*;
- the *Abstract of Special Regulations (Highly Flammable Liquids and Liquefied Petroleum Gases) Order 1974 (SI 1974 No 1587)*.

All the Regulations (except for the final two relate) to particular industries or activities of a specialised kind. The repeal and revocation of Regulations relating to highly flammable liquids and liquefied petroleum gases has wider effect. The more specific requirements of these Regulations are covered by the general goal setting provisions of the *DSEAR* but remain good practice. Extensive guidance on the safe management of flammable liquids and liquefied petroleum gases is available from the LPG Association.

For completeness, the *DSEAR* also repeal and revoke particular parts of the following legislation:

- the *Petroleum (Consolidation) Act 1928*;
- the *Factories Act 1961*;
- the *Shipbuilding and Ship Repairing Regulations 1960 (SI 1960 No 1932)*;
- the *Dangerous Substances in Harbour Areas Regulations 1987 (SI 1987 No 37)*;
- the *Carriage of Dangerous Goods by Road Regulations 1996 (SI 1996 No 2095)*;
- the *Carriage of Dangerous Goods (Classification, Packaging and Labelling) and Use of Transportable Pressure Receptacles Regulations 1996 (SI 1996 No 2092)*;
- the *Workplace (Health, Safety and Welfare) Regulations 1992 (SI 1992 No 3004)*.

Dangerous substances and explosive atmospheres: enforcement

[D0940] The application of the *Dangerous Substances and Explosive Atmospheres Regulations 2002 (SI 2002 No 2776) (DSEAR)* is enforced through existing powers by:

- the HSE or local authorities depending on the allocation of premises under the *Health and Safety (Enforcing Authority) Regulations 1998 (SI 1998 No 494)*. In the main, the HSE will enforce at industrial premises and local authorities (environmental health officers) elsewhere, eg in retail premises.
- fire brigades at most premises subject to the *DSEAR* in relation to general fire precautions such as means of escape.
- petroleum licensing authorities at retail petrol filling stations in relation to the storage and dispensing of petrol, LPG and other fuel subject to the 2002 Regulations.

Regulatory impact assessment

[D0941] The Health and Safety Commission prepared a regulatory impact assessment (RIA) on the *Dangerous Substances and Explosive Atmospheres Regulations 2002 (SI 2002 No 2776) (DSEAR)*. RIA showed that businesses complying with current health and safety legislation would expect little or no additional cost as a result of the *DSEAR*. Businesses with work processes involving potentially explosive atmospheres that had not formally recorded zones where these could occur in their risk assessments would need to zone and mark zoned areas with a sign as a specific requirements of *DSEAR*.

Approved codes of practice and guidance etc.

General

[D0942] While *Unloading petrol from road tankers (L133)* continues as a separate ACOP, four DSEAR ACOP publications from 2003 have been merged into the original main ACOP L138:

- DSEAR – Design of plant equipment and workplaces (L134);
- DSEAR – Storage of dangerous substances (L135);
- DSEAR – Control and mitigation measures (L136);
- DSEAR – Safe maintenance, repair and cleaning procedures (L137);

ACoPs are published for the Health and Safety Commission by HSE Books. They give practical advice on how to comply with the Regulations and the law. Duty Holders are strongly recommended to have a copy and to have studied them and taken their advice.

Approved Codes of Practice have special legal status. By following the advice, the Duty Holder will be doing sufficient to comply with the Regulations on those matters to which the advice relates. If Duty Holders are prosecuted for an alleged breach of health and safety law, and it is shown that they did not follow the advice given in the ACoP, then the Duty Holder will need to demonstrate that they complied with the law in some other way. Otherwise, the court will find the Duty Holder at fault.

The Regulations and ACoPs are accompanied by Guidance that is more wide ranging than the ACoP. This does not form part of the ACoPs, and does not have the same legal status, but if Duty Holders follow the Guidance, then they will normally be doing enough to comply with the law. It is not compulsory to follow the Guidance, but Health and Safety Inspectors may refer to this Guidance as illustrating good practice as they ascertain compliance with the Regulations.

Aspects of the ACoPs and Guidance have been incorporated into this text, but this is by no means comprehensive or complete. Duty Holders are strongly advised to rely on the ACoPs and Guidance themselves. In some respects the scope of one ACoP overlaps with another, but the following gives a summary of the scope of each one.

Dangerous substances and explosive atmospheres (L138)

[D0943]–[D0947] This ACoP and Guidance provides an overview on how employers can meet their duties under the *Dangerous Substances and Explosive Atmospheres Regulations 2002 (SI 2002 No 2776) (DSEAR)*. It details all the DSEAR Regulations including the Schedules. A summary of the Regulations is provided and the relationship of DSEAR to other relevant and related health and safety legislation is set out. There is a comprehensive list of references to other Regulations and standards documents.

Guidance is given on the interpretation of terms used in the Regulations, such as dangerous substances, explosive atmospheres, physico-chemical or chemical property, combustion, fires and explosions, hazard, risk, and workplace and work processes. There is Guidance on the application of the Regulation in respect of activities including maritime activities, the medical treatment of patients, gas safety and fittings, the manufacture of explosives and mineral extracting industries, and use of means of transport. Advice is given on the duties of employers and self-employed under the Regulations.

There is substantial Guidance on *SI 2002 No 2776, Reg 5*, which is the requirement for employers to carry out a risk assessment and the nature of the risk assessment. It covers the large number of factors to consider when assessing risks from dangerous substances. The Guidance draws attention to a range of measures for assessing the overall risk presented by dangerous substances, both from the combination of factors as well as assessing each factor individually, and from more global risk management systems that may be in place.

The elimination or reduction of risks from dangerous substances is covered within *SI 2002 No 2776, Reg 6*. Within this document, elimination and reduction of risks are discussed within the context of the overall principle of 'so far as is reasonably practicable'. The elements of risk elimination or reduction by substitution and/or by control and mitigation measures are described, together with some specific issues.

The Guidance relating to SI 2002 No 2776, Reg 7 regarding places where explosive atmosphere may occur deals with the classification of areas containing explosive atmospheres, including zoning, the selection of equipment with an 'EX' mark and anti-static clothing for use in hazardous areas, and the marking and verification of areas containing explosive atmospheres.

SI 2002 No 2776, Reg 8 deals with accidents, incidents and emergencies, and the Guidance sets out the overall approach. Further Guidance covers the approach for assessing the detection of accidents, incidents and emergencies, assessing the requirements for emergency arrangements, including first aid and safety drills. There is Approved Practice and Guidance on warning and communication systems, escape facilities and mitigation measures and restoring the situation to normal. Additional Guidance is given on making information available to employees and emergency services, the need to review arrangements and the relationship with other legislation including fire safety.

There is Approved Practice on the information that employers must provide their employees to meet the requirements of SI 2002 No 2776, Reg 9. This covers the identity of dangerous substances that could be present and where they are used, the type and extent of the risks and the control and mitigation measures adopted and several other aspects. The Guidance draws attention to the need for information, training and instruction, particularly for new employees, and the level and means for delivering it so as to maximise effectiveness. Guidance is given on the need for training and information for non-employees who may be present on site, including members of the public. The Guidance reminds employers to update information, instruction and training whenever changes to the type of work or work methods are made.

The Guidance in support of SI 2002 No 2776, Reg 10 deals with the identification of the hazardous contents of containers and pipes. It considers means of identification such as labelling or colour coding etc. The duty of co-ordination where two or more employers share the same workplace is covered under SI 2002 No 2776, Reg 11, and the Guidance distinguishes between separate individual and shared workplaces.

Further Approved Practice and Guidance is provided covering the Transitional Provisions (SI 2002 No 2776, Reg 17), and General Safety Measures (Schedule 1). There is Guidance on the Amendments, including amendments to regulations covering the handling of the petroleum and other substances, and Repeals and Revocations (Schedule 7).

Unloading petrol from road tankers (L133)

[D0948] This ACoP and Guidance gives practical advice and details the necessary measures with regard to the safe unloading of petrol tankers at petrol filling stations, depots and other places. It is not intended to cover the unloading of liquefied petroleum gas (LPG), compressed natural gas (CNG) or liquid natural gas (LNG), but can be applied to the unloading of diesel. Nor is it intended to cover the unloading of petrol at major hazard sites or at sites other than filling stations where it is intended to store more than 100,000 litres.

The *Dangerous Substances and Explosive Atmospheres Regulations 2002 (SI 2002 No 2776) (DSEAR)* revokes, repeals or modifies the licensing controls under the *Petroleum Consolidation Act 1928*. The good practices of the old legislation are being maintained through the DSEAR and the ACoPs. DSEAR removes licensing requirements for petrol, except for petrol that is being kept for dispensing into vehicles. All parts of workplace premises, including non-workplace premises (such as boat or motor clubs etc) where dispensing of petrol takes place remain subject to licensing.

The ACoP and Guidance are concerned with *SI 2002 No 2776, Reg 6* on how the risk to safety from a dangerous substance, such as petrol, can be removed or eliminated. In this respect, they deal with preventing a fire through preventing the overfilling of a storage tank, controlling sources of ignition, during unloading and dealing with any spillages that may occur. However, in addition, the ACOP and Guidance also provide advice on preventing falls from petrol tankers unloading petrol at filling stations etc, in pursuance of the *Workplace (Health, Safety and Welfare) Regulations 1992, Regs 13(1)–(3)*.

The ACoP and Guidance highlights the general duty of everyone involved in the unloading of petrol from a road tanker under the *Health and Safety at Work etc. Act 1974*. It references duties on employers and the self-employed under the *Management of Health and Safety at Work Regulations 1999*, and duties for employers under the *Safety Representatives and Safety Committees Regulations 1977* and the *Health and Safety (Consultation with Employees) Regulations 1996*. All employers and the self-employed are required to carry out a risk assessment of activities that create a fire and explosion risk, such the unloading of petrol, under *SI 2002 No 2776, Reg 5*.

There is Approved Practice and Guidance detailing the responsibilities in respect of the unloading of petrol of the following parties:

(a) road tanker operators;
(b) site operators; and
(c) tanker drivers.

Additional requirements are given for pumped deliveries. There is a section on the prevention of falls from road tankers through the elimination of the need for routine access, or if this cannot be achieved, a safe means of access and a safe place to work.

Other information on dangerous substances and explosive atmospheres

General

[D0949] In addition to the ACoPs and Guidance, the Health and Safety Executive publish a free leaflet 'Controlling fire and explosion risks in the workplace', which provides a short guide to the *Dangerous Substances and Explosive Atmospheres Regulations 2002 (SI 2002 No 2776) (DSEAR)* and is aimed primarily and small and medium-sized businesses. Further information on DSEAR can be obtained on HSE's website: www.hse.gov.uk, which is updated regularly.

Safe handling of combustible dusts

[D0950] HSE has published fully revised and updated guidance aimed at industries dealing with combustible dusts, informing them of the safest way of handling them.

Safe handling of combustible dusts seeks to lower the risk of a dust explosion, present in industries such as food production (sugar, flour and custard powder), animal feed production and places handling sawdust, many organic chemicals, plastics, metal powders and coal.

The booklet is targeted at those who operate plants handling dusts which can explode. It describes in simple language the tests used on dusts to assess their explosive properties, the precautions used to control the risks and an outline of the health and safety law that applies.

Dust explosions occur when fine materials are disbursed to a certain concentration and an effective ignition source is present. If dust deposits around premises form a cloud an initial small explosion is often followed by a much larger one. Much can be done at the design stage to prevent such explosions, but some risks usually remain. There are many ongoing precautions that need to be followed by those who work in the plant. These are outlined in the booklet.

The booklet takes account of European Directive 99/92/EC on the Protection of Workers Potentially at Risk from Explosive Atmospheres (The ATEX Directive) which was implemented in the UK by the *Dangerous Substances and Explosive Atmospheres Regulations 2002 (SI 2002 No 2776)*. In particular, the guidance describes how the requirement for hazardous area classification, brought in by the Regulations, applies to dust handling plants, and explains which types of new equipment need to be 'ATEX compliant' ie be properly marked after undergoing specified tests and checks for their use in hazardous areas.

Copies of *Safe handling of combustible dusts: Precautions against explosions* (HSG103), ISBN 9 780717 627 264, price £10.95 are available from HSE Books, PO Box 1999, Sudbury, Suffolk, CO10 2WA (tel: 01787 881 165; fax: 01787 313 995).

Case histories

[D0951] The explosion and fire at the Buncefield Oil Depot on 11 December 2005 illustrated the power and devastating effect of dangerous substances and explosive atmospheres. The Final Report of the Major Incident Investigation Board was published by HSE on 11 December 2008.. The cause of the incident was attributed to overfilling a large petrol storage tank. Among its 78 recommendations are lessons for users of dangerous substances and owners and operators of workplaces with the potential for an explosive atmosphere. The effects of the incident in causing over a billion pounds of damage draw attention to the importance of the *Dangerous Substances and Explosive Atmospheres Regulations 2002 (SI 2002 No 2776)* (DSEAR) and the need for good management of such substances and places where explosive atmospheres may be present.

The Health and Safety Executive (HSE) and Sussex Police have warned motor vehicle repair garages about the importance of having a safe system in place for handling and storing petrol. The warning follows the conclusion of proceedings brought against the owner of a Sussex garage. Howard Hawkins, the owner, was found guilty of breaching *section 2(1) of the Health and Safety at Work etc. Act 1974 ('HSWA')*. He was sentenced at Lewes Crown Court receiving a fine of £10,000 with costs of £15,000.

The prosecution followed the death of an apprentice mechanic, Lewis Murphy (18), who died four days after becoming engulfed in flames in an explosion at the Anchor garage, Peacehaven, Sussex on 19 February 2004.

Passing sentence, Judge Richard Hayward, said of Howard Hawkins:

> To say that you were complacent about health and safety is an understatement. You regard health and safety as a tiresome intrusion into your business and a matter of common sense that you could leave to the experience of your mechanics. Being a dinosaur can sometimes be endearing but not on health and safety matters.

While the prosecution was made under *HSWA 1974*, DSEAR [*SI 2002 No 2776*] and the Approved Practice and Guidance would be have been relevant to the case.

Sources and acknowledgement

[D0952] In preparing this chapter, extensive use has been made of the *Dangerous Substances and Explosive Atmospheres Regulations 2002 (SI 2002 No 2776) (DSEAR)*, and of the commentary provided by the Safety Policy Directorate of the HSE, and the Approved Codes of Practice and Guidance. It should again be emphasised that this chapter is not a substitute for having and implementing these documents. It is primarily designed to increase awareness of the responsibilities under DSEAR and to draw attention to the detailed Approved Practice and Guidance that is available.

Disaster and Emergency Management Systems (DEMS)

Andrea Oates

The importance of Disaster and Emergency Management Systems (DEMS) has been highlighted by a series of events over the last year or so. They include the terrorist attacks on Westminster Bridge (March 2017), the Manchester Arena (May 2017), Borough Market and London Bridge (June 2017) and a tube train at Parsons Green (September 2017), the Grenfell Tower Fire (June 2017), and the nerve agent attack in Salisbury (March 2018).

Historical development of disaster and emergency management

[D6001] The terrorist attacks on the twin towers in New York on 11 September 2001 marked a paradigm shift in the thinking, planning and perception of 'man-made disasters'. Throughout the 1990s the US Federal Emergency Management Agency (FEMA), the authority dealing with natural and man-made disasters, was criticised by Congress and the media for being too resource consuming and inefficient. After '9/11' it was hailed for its expertise in emergency response management. Suddenly 'corporate America' and the democratic world began to take more seriously the importance of 'disaster and emergency management', 'civil defence management', 'civil protection management' and 'business continuity'.

In the UK, in June 2001 the 'civil contingencies' function transferred from the Home Office to the Cabinet Office, giving the Prime Minister's Office a greater strategic oversight of national 'major incidents'. In 2004, the *Civil Contingencies Act 2004* came into force, followed in 2005 by the *Civil Contingencies Act 2004 (Contingency Planning) Regulations 2005 (SI 2005 No 2042)*. The provisions put in place a clear framework for civil protection. The terrorist attacks on London's public transport system in July 2005 further highlighted the need for a co-ordinated, cross-organisational response to emergencies.

The UK experience shows disaster and emergency management (DEM) broadly evolved in three phases:

Phase 1: pre-Control of Industrial Major Accident Hazards Regulations 1984 (CIMAH) (SI 1984 No 1902)
- DEM was largely confined to local government, the emergency services, large organisations and civil protection agencies (environmental, military etc). This phase is marked by large-scale macro plans and detailed, sequential procedures to be followed in the event of a 'disaster' or 'emergency'. It is exemplified in government, United Nations, Red Cross and other guidance on 'incidents' involving radioactivity.

Phase 2: liberalisation phase
- With the advent of decentralisation of industry and the growth of privatisation, organisations (large or otherwise) began to focus on 'procedures' and 'business continuity planning'. British Telecom established a Disaster Recovery Unit and offered its expertise to customers. 'Major incidents' such as Chernobyl, *Piper Alpha*, Kings Cross, *The Marchioness* and *Challenger*, to name a few of those which hit headlines across the world in the 1980s, also focused planners' minds on effective DEM rather than isolated procedures alone.
Simultaneously the introduction of the *CIMAH* Regulations in 1984 required industrial sites with major accident potential to prepare on-site plans and co-operate with local authorities and other agencies in developing off-site plans for the area at risk around the installation.

Phase 3: holistic phase
- This has two dimensions including the role of Europe. The *Framework and Daughter Directives*, as transposed by the 1992 health and safety '*Six Pack*' of regulations', including the *Management of Health and Safety at Work Regulations 1992*, introduced for the first time explicit and strict duties on organisations in general to plan for 'serious and imminent danger'. The European Commission also began to take the risks of trans-national 'disasters' more seriously, as highlighted by the greater role given to the Civil Protection Unit in DG XI of the European Commission (European civil protection and humanitarian aid operations currently deal with civil protection).
This phase also saw domestic UK legislation becoming more aware of disaster and emergency issues:
    — the *Environment Act 1995* makes provisions for environmental emergencies;
    — the *Control of Major Accident Hazards Regulations 1999 (COMAH) (SI 1999 No 743)* replaces the *Control of Industrial Major Accident Hazards Regulations 1984 (CIMAH) (SI 1984 No 1902)*; and
    — the *Carriage of Dangerous Goods by Road Regulations 1996 (SI 1996 No 2095)* requires effective planning when carrying dangerous goods via road.

The most noticeable index of how DEM has changed is the availability of information and templates on 'disaster planning' and 'emergency planning' to organisations of all sizes.

Origin of disaster and emergency management as a modern discipline

[D6002] DEM is largely a fusion of four branches of knowledge:

Occupational safety and health (OSH)
- This relates to 'internal' or work-based causes of systems failures with major impact on life and property. It was exemplified by Heinrich in his publication *Unsafe Acts and Unsafe Conditions* in the early part of the

1900s, Bird and Loftus with their 'management failing' explanations of accidents and incidents, and Turner with his 'incubation' explanation of man-made disasters in the 1970s. OSH academics have been at the forefront of analysing the branch and root causes of major industrial disasters.

Security management
- In the 1970s, security threats in the UK such as terrorism and electronic surveillance failures have given insights to causation and the motivation behind man-made emergencies. Also, guidance from the Home Office and the Civil Contingencies Secretariat, as well as the emergency services, has enabled a practical understanding of how to cope with emergency situations.

Business management
- The late 1980s saw a shift in the academic paradigm in economics and business management from 'static' or closed business planning – where businesses were told to make the assumption of *cetaris paribus*, that is assume all things are constant with the business acting as if it was the only one in the market place – to 'dynamic' or open planning. The latter sees uncertainty and risk being factored into decision-making models. This influenced the development of 'business continuity management', developing strategies when the business faces major corporate uncertainty and crises as well as 'contingency planning'.

Insurance
- The fourth significant influence comes from insurance and loss control. The occurrence of disasters and accidents has involved loss adjusters and actuarial personnel. The former have developed methods of analysing the basic and underlying causes of an event, while the latter have developed statistical methods for calculating the chance of failure and the risk premiums needed to indemnify that failure.
  DEM uses both quantitative (including statistics, quantified risk assessments, hazard analysis techniques, questionnaires and computer simulation) and qualitative methods (including inspections, audits and case studies).

Definitions relating to disasters and emergencies

[D6003] Basic definitions can help to avoid confusion over the meaning given to core terms. Legislation and approved codes have not provided definitions of 'disaster' or 'emergency' for example, but dictionary definitions are as follows:

The *Oxford English Dictionary* provides the following primary meanings:
(a) Catastrophe: 'a sudden event causing great damage or suffering';
(b) Crisis: 'a time of severe difficulty or danger';
(c) Disaster: 'a sudden accident or natural catastrophe that causes great damage or loss of life';
A catastrophe and a crisis are types of disaster, namely more severe.

(d) Emergency: 'a serious and unexpected situation requiring immediate action';

While both disasters and emergencies can be sudden, the former has a macro, large scale impact while the latter requires an immediate response.

The definitions provided in Parts 1 and 2 of the *Civil Contingencies Act 2004* are key to understanding of the nature and consequences of emergencies and to ensuring the establishment of appropriate safeguards and responses.

(e) Accident: '1. an unpleasant and unexpected event, . . . 2. an event that happens by chance';

The *Control of Major Accident Hazards Regulations 1999 (COMAH) (SI 1999 No 743)* introduced the term 'major accident'.

The regulations currently in force, the *Control of Major Accident Hazards Regulations 2015 (SI 2015 No 483)*, define a 'major accident' as 'an occurrence such as a major emission, fire, or explosion resulting from uncontrolled developments in the course of the operation of any establishment to which these Regulations apply, and leading to serious danger to human health or the environment (whether immediate or delayed) inside or outside the establishment, and involving one or more dangerous substances'.

Given the potential for both human and property loss and the *COMAH* requirements for internal and external emergency plans, there seems to be little practical difference between a major accident and an emergency. Both are *response-based* concepts. However, if the legislators wished a major accident to be different from an emergency then they would have either said so or implied so. A major accident can be regarded as a type of emergency situation.

(f) Incident: '1. an event. 2. a violent event, such as an attack. 3. the occurrence of dangerous or exciting events';

The *Reporting of Injuries, Diseases and Dangerous Occurrences Regulations (RIDDOR) (SI 2013 No 1471)* do not define an accident or incident but classify the types, with requirements to report non-fatal injuries to workers, non-fatal injuries to non-workers, work-related fatalities, dangerous occurrences, occupational diseases and exposure to carcinogens, mutagens and biological agents.

(g) Major incident

The Lexicon of UK civil protection terminology (www.gov.uk/government/publications/emergency-responder-interoperability-lexicon) sets out that a major incident is an event or situation requiring a response under one or more of the emergency services' major incident plans.

So for example:

(1) two trains missing each other would be an 'incident' (near miss);
(2) an employee or member of the public being injured on a train would be an 'accident';
(3) an event at a *COMAH* site where dangerous substances ignite causing damage to the plant, injury to personnel and emissions into the local community, would be a 'major accident' under *COMAH*;

(4) the immediate event after a train collision and the response needed to the chaos – this would be an 'emergency'.
(5) if two trains collide causing multiple fatalities and immediate property and environmental damage, that would be referred to as a 'disaster'. For example, the Ladbroke Grove rail crash in 1999;
(6) if the collision, with the multiple fatalities and property damage is difficult to access, manage and control, this would be a 'crisis'. Ladbroke Grove fell short from being a 'crisis' in contrast to Clapham Junction in 1988 when a triple train crash caused major access and logistical problems;
(7) if the event generated major environmental, public and social harm that has 'longer term' implications, over and above the immediate human and property loss, this would be a 'catastrophe' For example, Kings Cross underground fire (1987), Chernobyl (1986), and *Piper Alpha* (1988) to name a few that had wider consequences over and above the immediate impact. More recent examples include the Fukushima Dai-ichi nuclear disaster in Japan and the Deepwater Horizon oil spill in the Gulf of Mexico.

Events (1)–(7) would be major incidents.

Figure 1 overleaf summarises the essential differences between the above events.

FIGURE 1: CLASSIFICATION OF INCIDENTS, ACCIDENTS AND MAJOR INCIDENT TYPES

Event:	Incident	Accident	Major Accident	Emergency	Disaster	Crisis	Catastrophe
Characteristic:			Types of emergency		Types of disaster		
1. MPL	Low						Very High
2. RISK:							
Severity	Near miss etc						Multiple Fatalities
Consequence	Minor						Major
3. NUMBERS	1–5						100 +
4. SOCIO-LEGAL IMPACT	No change likely						New Laws or Guidance
5. TIME	Short						Longer Impact
6. COST:							
Individual	Short Term						Irreparable
Social	None Usually						Irreparable
Environment	None Usually						Irreparable

'MAJOR INCIDENTS'

Key:
MPL=	Maximum Potential Loss (economic and property loss)
Risk=	Severity x Consequence (Severity refers to the quantum of harm generated by the event while Consequence measures the scale of impact)
Numbers=	The number of individuals affected
Socio-Legal Impact=	The impact on social attitudes and legislation/guidance as a result of the event
Time=	The length of the event
Cost=	The loss suffered by the individual, society or nature

Such classifications are important. From a philosophical perspective it is important to know how they differ and from a planning perspective, the resource allocation will differ accordingly. From a response perspective, the response to an incident differs from a disaster, with the organisation needing to define and clarify when an event is an incident and not a disaster.

The need for effective disaster and emergency management

[D6004] There are several reasons why DEM is needed: legal reasons (see D6005–D6029); insurance reasons; corporate reasons (see D6030); humanitarian and societal reasons (see D6032); and environmental reasons (see D6033).

Legal reasons

[D6005] In 2004, a fundamental change in the way the UK approached preparation for and management of emergencies took place. The *Civil Contingencies Act 2004* came into force and established a framework for civil protection capable of meeting the various challenges defined in the Act. It also repealed or revoked Acts or sections of Acts dealing emergency planning or response. In particular, the *Civil Defence Act 1948*; the *Civil Defence (Grant) Regulations 1953 (SI 1953 No 1777)*; the *Local Government Act 1972, s 138*; the *Civil Protection in Peacetime Act 1986*; and the *Emergency Powers Act 1920* all ceased to have effect.

The Civil Contingencies Act 2004

[D6006] The structure of the *Civil Contingencies Act 2004 (CCA 2004)* is in two substantive parts: *CCA 2004, Pt 1 (ss 1–18)*, focuses on local arrangements for civil protection and imposes a series of duties on local bodies in the UK which the Act classes as category 1 responders. Part 1 also establishes category 2 responders who are required to co-operate with and provide information to category 1 responders in pursuance of their civil protection duties. *CCA 2004, Pt 2 (ss 19–31)*, deals with emergency powers and allows for a Minister of State (there is provision for consultation with the devolved administrations), in certain defined situations, the power to make regulations if an 'emergency' has occurred or is about to occur. The definition of what constitutes an emergency is key to understanding the purpose and operation of *CCA 2004* and by extension the duties and responsibilities of the category 1 and category 2 responders.

The meaning of 'emergency' is defined in both substantive sections of *CCA 2004*. *CCA 2004, s 1(1)* defines 'emergency' for the purposes of *CCA 2004, Pt 1* as an event or situation which threatens serious damage to human welfare in a place in the UK, serious damage to the environment of a place in the UK or war or terrorism which threatens serious damage to the security of the UK. The meaning of these events or situations is contained in *CCA 2004, s 1(2)–(3)* and includes loss of human life, human illness or injury, homelessness, damage to property, disruption of a supply of money, food, water, energy or fuel, disruption to transport and communications, disruption to essential supplies, disruption to health services, contamination of land, water or air with biological, chemical or radio-active matter and disruption or destruction of plant and animal life.

The meaning of 'emergency' given in *CCA 2004, s 19(1)–(6)* for the purposes of *CCA 2004, Pt 2* are substantially the same as that given in *CCA 2004, Pt 1* with one exception, that of scale. In order to satisfy the definition given in *CCA 2004 Pt 2*, the threat must be serious and must apply to the UK, or a part or region of the UK. *CCA 2004, s 19(4)* provides the Secretary of State with the power to amend the list of events or situations to allow appropriate response should a system or service become so necessary that disruption of it fall within the given definition of 'emergency'.

Category 1 responders are listed in *CCA 2004, Sch 1, Pt 1* and are at the heart of emergency response: they include local authorities, the emergency services, health services, including the NHS Commissioning Board, NHS trusts and NHS foundation trusts, local health boards, and port health authorities constituted under the *Public Health (Control of Disease) Act 1984, s 2(4)*. Also included are the Environment Agency and the Secretary of State in respect of his/her obligations to respond to maritime and coastal emergencies as well as his or her functions with regard to the NHS and social care. The list of responders can be amended by a Minister of the Crown under *CCA 2004, s 13*.

Category 2 responders are listed in *CCA 2004, Sch 1, Pt 3* and include public utilities, transport operators, airport operators, harbour authorities, the Secretary of State in respect of matters for which he is responsible by virtue of the *Highways Act 1980, s 1* (c66) (highway authorities), clinical commissioning groups (CCGs), the Health and Safety Executive (HSE) and the Office for Nuclear Regulation. The role of category 2 responders is to be co-operating bodies. Although involved in the management of major hazard accidents within their own areas of influence they will be less directly involved in contingency planning for emergencies.

CCA 2004, s 2 establishes the primary duties of category 1 responders to assess, plan and advise in the context of contingency planning. In principle, category 1 responders are required to assess the risk of emergencies occurring and use this information to:

- inform contingency planning;
- put in place and maintain as necessary emergency plans, business contingency management plans, and arrangements to warn, inform and advise the public in the event of an emergency;
- share information with other local responders to enhance co-ordination; and
- co-operate with other local responders to enhance co-operation.

They are required to act to prevent the occurrence of an emergency or reduce, control or mitigate its effects and efficiency. The duties are given further form under the *Civil Contingencies Act 2004 (Contingency Planning) Regulations 2005 (SI 2005 No 2042) (as amended)* (see **D6007**). Local authorities are also required to provide advice about business continuity management to local businesses and voluntary organisations. *CCA 2004, s 2* is supported by *CCA 2004, s 4* which imposes a duty on responders to provide advice and assistance to the public.

Throughout *CCA 2004, Pt 1* the emphasis is on responsiveness with power given to Ministers to amend definitions and make regulations as required to give effect to the regulatory framework.

CCA 2004, Pt 2 continues this emphasis on responsiveness and provides, under *CCA 2004, s 20*, the power to make emergency regulations to meet the requirements of the situations defined in *CCA 2004, s 19* and to apply these regulations to specific areas or regions through the appointment of regional and emergency co-ordinators (*CCA 2004, s 24*).

The emergency regulations are made by Her Majesty by Order in Council or, in the event of urgency, by a 'senior Minister of the Crown'. A 'senior Minister of the Crown' is defined by *CCA 2004, s 20(3)* to mean the First Lord of the Treasury (the Prime Minister), any of Her Majesty's Principal Secretaries of State and the Commissioners of Her Majesty's Treasury.

Limitations are placed on this power by the definition of 'emergency' contained in *CCA 2004, s 19* and other provisions establishing limitations in respect of emergency regulations (*CCA 2004, s 23*), limitations on duration (*CCA 2004, s 26*) and Parliamentary scrutiny (*CCA 2004, ss 27 and 28*).

CCA 2004 seeks to be responsive to dynamic situations using specific, temporary legislation and to balance this power with safeguards. An emergency threatening serious damage to human welfare, the environment or national security must have occurred, or is about to occur. It is also demonstrably necessary to bring in legislation urgently because existing legislation is insufficient to deal with the situation and normal routes are too slow. The legislation provided must be proportional to the emergency faced.

The emergency regulations cannot be used to counter strike or other industrial action or to make substantive changes to existing legislation. They must be compatible with European legislation and the *Human Rights Act 1998 (HRA)* and open to challenge in the courts.

CCA 2004 provides for regulations to be made and has given rise to the *Civil Contingencies Act 2004 (Contingency Planning) Regulations 2005 (SI 2005 No 2042) (as amended)*.

The Act is accompanied by two documents:

- *Emergency Preparedness* – this is statutory guidance under Part 1 of the Act setting out what the legislation requires and providing best practice.
- *Emergency Response and Recovery* – this is non-statutory guidance dealing with the response and recovery phases of emergencies.

They are available on the government website at: www.gov.uk/government/publications/emergency-preparedness and www.gov.uk/government/publications/emergency-response-and-recovery.

Civil Contingencies Act 2004 (Contingency Planning) Regulations 2005 (SI 2005 No 2042)

[D6007]–[D6008] The Regulations are made under the *Civil Contingencies Act 2004 (CCA)*, *s 2(3)* and relate to the duties imposed on category 1 responders under CCA 2004, *ss 2* and *4* to assess and plan for emergencies and to provide advice and assistance to the public. The Regulations are split into 11 parts and comprise some 59 Regulations.

Part 2 of the Regulations, comprising *SI 2005 No 2042, Regs 4–12* inclusive, makes general provisions relating to the extent and the performance of the duties established in *CCA 2004*. *SI 2005 No 2042, Reg 4* relates to co-operation and local resilience forums and requires that category 1 responders co-operate with each other within a single forum, the local resilience forum, meeting at least once every six months and inform and invite category 2 responders as appropriate. *SI 2005 No 2042, Reg 5* makes the same provision for Scotland but referring to the forum for co-operation as the strategic co-ordinating group. *SI 2005 No 2042, Reg 6* provides the same requirement for co-operation in Northern Ireland. The co-operation required may be facilitated by the setting up of protocols (*SI 2005 No 2042, Reg 7*) or by establishing areas of joint responsibility (*SI 2005 No 2042, Reg 8*). *SI 2005 No 2042, Regs 9–11* relate to the establishment of a lead category 1 responder; *SI 2005 No 2042, Reg 9* allows for the identification of a lead category 1 responder in respect of a relevant civil protection duty under *CCA 2004, s 2* whilst *SI 2005 No 2042, Regs 10 and 11* relate to the duty to inform and co-operate falling upon the lead category 1 responder and the category 1 responders who do not a have a lead responsibility respectively. *SI 2005 No 2042, Reg 12* excludes responsibility for emergency planning duties covered under existing legislation: this includes major accident hazards under the *Control of Major Accident Hazard Regulations 2015 (SI 2015 No 483)*.

Part 3 of the Regulations, comprising *SI 2005 No 2042, Regs 13–18* inclusive, relates to the duties of category 1 respondents under *CCA 2004, s 2(1)(a)* and *(b)* to assess the risk of an emergency occurring. *SI 2005 No 2042, Reg 13* limits that duty to emergencies in areas where the category 1 responder functions. This duty extends under *SI 2005 No 2042, Reg 15* to collaborating with other category 1 responders to maintain a 'community risk register' of the risk assessments made by the local resilience forum. *SI 2005 No 2042, Regs 16–18* relate this duty to the sharing of information with other responders throughout England, Scotland and Wales. *SI 2005 No 2042, Reg 14* enables Ministers of State in England, Scotland and Wales to issue guidance on emergencies, the likelihood of emergencies occurring or their likely impact.

Part 4 of the Regulations, comprising *SI 2005 No 2042, Regs 19–26* inclusive, relates to the duty of category 1 responders under *CCA 2004, s 2(1)(c)* or *(d)* to maintain plans necessary to respond to an emergency made in respect of any risk assessment carried out under *CCA 2004, s 2(1)(a)* or *(b)* and, under *CCA 2004, s 2(1)(g)* with regard to the provision to maintain advice and information to the public *(SI 2005 No 2042, Reg 20)*. *SI 2005 No 2042, Reg 23* requires that in considering these plans that category 1 responders take into account the activities of voluntary organisations. In maintaining plans, category 1 responders should consider whether those plans can be generic or specific in respect of the emergency under consideration *(SI 2005 No 2042, Reg 21)* and *Reg 22* deals with multi-agency emergency plans. *SI 2005 No 2042, Reg 24* requires that a procedure be put in place to determine whether or not an emergency has occurred. *SI 2005 No 2042, Reg 25* requires that the effectiveness of plans be determined through appropriate arrangements being made for exercises and training. *SI 2005 No 2042, Reg 26* requires the consideration of plans and necessary revisions in light of any guidance issued by ministers of state in the UK.

Part 5 of the Regulations *(SI 2005 No 2042, Reg 27)* requires that in publishing any plans due care should be taken not to unnecessarily alarm the public.

Part 6 of the Regulations, comprising *SI 2005 No 2042, Regs 28–35* inclusive, relates to the duty of category 1 responders to maintain arrangements to warn, inform and advise members of the public if an emergency occurs or is likely to occur in respect of plans made under *CCA 2004, s 2(1)(g)*. As with *Part 4* of the Regulations the arrangement made may be either generic or specific as appropriate *(SI 2005 No 2042, Reg 29)*, must be supported by training and exercises *(SI 2005 No 2042, Reg 31)* and must not unduly alarm the public *(SI 2005 No 2042, Reg 30)*. *SI 2005 No 2042, Regs 32–34* require that arrangement be put in place to identify a lead category 1 responder to have lead responsibility to inform the public. *SI 2005 No 2042, Reg 35* requires that responders have due regard to information provided to the public by other bodies including the Meteorological Office and the Food Standards Agency.

Part 7 of the Regulations, comprising *SI 2005 No 2042, Regs 36–44* inclusive, relates to the duty owed by local authorities, as defined by *CCA 2004, Sch 1, para 1* referred to in *SI 2005 No 2042, Reg 36(a)* as a 'relevant responder', under *CCA 2004, s 4(1)* to provide advice and assistance to the public for the continuance of commercial activities *(SI 2005 No 2042, Reg 39)* or voluntary activities *(SI 2005 No 2042, Reg 40)* against which provision a charge may be levied *(SI 2005 No 2042, Reg 44)*. In common with other sections provision is made for co-operation between relevant responders and the establishment of a lead relevant responder *(SI 2005 No 2042, Regs 41–43)*, with *Regulation 42* relating to cross border communications with Scotland.

Part 8 of the Regulations, comprising *SI 2005 No 2042, Regs 44a–54* inclusive, relates to information and its control. *SI 2005 No 2042 Reg 44a* allows general responders to disclose information on request to another general responder (subject to regulations 45 to 53). *SI 2005 No 2042, Regs 45–46* define 'sensitive information' and provide for control of information where the release of such information may be prejudicial to national security or public safety, may be commercially sensitive or may be personal as defined by the *Data Protection Act 1998 (DPA), s 1(1)(a)–(d)*. *SI 2005 No 2042, Regs 47–50* allow category 1 and 2 responders and Scottish category 1 and 2 responders to seek information from other category 1 and 2 responders in connection with their duties under *CCA 2004*. It is for responders to determine whether the information requested is in fact sensitive. Where sensitive information has been received *SI 2005 No 2042, Reg 51* prohibits its disclosure except in exceptional circumstances. *SI 2005 No 2042, Reg 52* in the same context limits the use of such information to the specific function for which the user has requested it. *SI 2005 No 2042, Reg 53* requires that information once requested and received be kept secure. *SI 2005 No 2042, Reg 54* applies specifically to the HSE and amends the *Health and Safety at Work etc Act 1974 (HSWA), s 28(2)* and *28(7)* to allow disclosure of information under the requirements of *CCA 2004*. *Regulation 54A* deals with the disclosure of information by the Office for Nuclear Regulation (ONR) to another responder.

Part 9 of the Regulations, comprising *SI 2005 No 2042, Regs 55, 56*, applies the function of the regulations to London and provides for the London Fire Commissioner to take the lead role in all functions under the Regulations.

Part 10 of the Regulations, comprising *SI 2005 No 2042, Regs 57, 58*, relates to the performance of duties under *CCA 2004* in Northern Ireland.

Part 11 provides for a review of the regulations within five years from 1 April 2012.

The Health and Safety at Work etc Act 1974

[D6009] The intentions of the *Health and Safety at Work etc Act 1974* are captured by *HSWA 1974, s 1(1)(a)–(c)* which states that the Act is concerned with;

(a) securing the health, safety and welfare of persons at work;
(b) protecting persons other than persons at work against risks to health or safety arising out of or in connection with the activities of persons at work;
(c) controlling the keeping and use of explosive or highly flammable or otherwise dangerous substances, and generally preventing the unlawful acquisition, possession and use of such substances;"

Implicit at least is the intention that this Act will influence actions that contribute to disaster and emergency situations, whether industrial, chemical or environmental. The Act does give expressed powers to the HSE to investigate and to hold inquiries in relation to: ' . . . any accident, occurrence, situation . . . '. [*HSWA 1974, s 14(1)*].

The Management of Health and Safety at Work Regulations 1999 (MHSWR) (SI 1999 No 3242)

[D6010] These Regulations provide the main detail applicable to all organisations, for providing 'procedures for serious and imminent danger and for danger areas' [*Management of Health and Safety at Work Regulations 1999 (SI 1999 No 3242), Reg 8*] and 'contacts with external services'. [*SI 1999 No*

3242, Reg 9]. *Regulation 1* (Citation, commencement and interpretation) does not define 'procedure for serious and imminent danger'.

SI 1999 No 3242, Reg 8(1) says that a strict duty exists on all employers to:

(a) 'establish and where necessary give effect to appropriate procedures to be followed in the event of serious and imminent danger to persons at work in his undertaking'.
(b) 'nominate a sufficient number of competent persons to implement those procedures in so far as they relate to the evacuation from premises of persons at work in his undertaking'.
(c) 'ensure that none of his employees has access to any area occupied by him to which it is necessary to restrict access on grounds of health and safety unless the employee concerned has received adequate health and safety instruction'.

SI 1999 No 3242, Reg 8(2) says 'Without prejudice to the generality of *paragraph (1)(a)*, the procedures referred to in that sub-paragraph shall:

(a) 'so far as is practicable, require any persons at work who are exposed to serious and imminent danger to be informed of the nature of the hazard and of the steps taken or to be taken to protect them from it;
(b) enable the persons concerned (if necessary by taking appropriate steps in the absence of guidance or instruction and in the light of their knowledge and the technical means at their disposal) to stop work and immediately proceed to a place of safety in the event of their being exposed to serious, imminent and unavoidable danger; and
(c) save in exceptional cases for reasons duly substantiated (which cases and reasons shall be specified in those procedures), require the persons concerned to be prevented from resuming work in any situation where there is still a serious and imminent danger'.

SI 1999 No 3242, Reg 8(3) says: 'A person shall be regarded as competent for the purposes of paragraph *(1)(b)* where he has sufficient training and experience or knowledge and other qualities to enable him properly to implement the evacuation procedures referred to in that sub-paragraph'.

SI 1999 No 3242, Reg 8 can be summarised:

- Procedures need to be sequential, logical, documented and clear (clarity of procedures).
- Procedures need to be justified and authorised (legitimise procedures).
- A 'hierarchy of procedural control' seems to be advocated by the Regulation ie:
  — give information to those potentially affected by serious and imminent dangers;
  — take actions or steps to protect such people;
  — stop work activity if necessary to reduce danger; and
  — prevent the resuming of work if necessary until the danger has been reduced or eliminated.
- Appoint competent persons preferably from within the organisation who will assist in any evacuations. It is implied that the role and responsibility of such persons needs to be clearly demarcated.

- Generally prohibit access to dangerous areas (site management). If access to a 'danger area' is required ie a place which has an unacceptable level of risk but must be accessed by the employee, then appropriate measures must be taken as specified by other legislation (see below).

In addition, the Regulations imply:

- Risk assessments will identify foreseeable events that may need to be covered by procedures. Such assessments may also identify 'additional risks' that need additional procedures. The risk assessment, as discussed below, is therefore a vital tool to keep procedures in tune with current generic and specific risks.
- Procedures need to be dynamic – reflecting the fact that events can occur suddenly.
- There may be a need to co-ordinate procedures where workplaces are shared.
- Procedures should also reflect other legislative requirements (see below).

SI 1999 No 3242, Reg 9 states: 'Every employer shall ensure that any necessary contacts with external services are arranged, particularly as regards first-aid, emergency medical care and rescue work'.

It can be inferred that 'necessary contacts with external services' does not only relate to the emergency services but the organisation needs to identify both private sector and voluntary organisations that can be called on for assistance. The organisation needs to identify contact names, addresses and contact numbers of such bodies and develop relations with them.

The Control of Major Accident Hazards Regulations 2015 (SI 2015 No 483) (as amended)

[D6011]–[D6014] The *Control of Major Accident Hazards Regulations 2015 (SI 2015 No 483) (COMAH 2015)* came into force on 1 June 2015, repealing and replacing the *Control of Major Accident Hazards Regulations 1999 (SI 1999 No 743)* and implementing the so-called 'Seveso III' Directive (European Directive (EEC) 82/501 on the major-accident hazards of certain industrial activities).

COMAH 2015 aim to prevent major accidents occurring or mitigate the consequences should one occur.

The regulations apply to establishments having any dangerous substance specified in a schedule to the regulations present (actual or anticipated) at or above a particular quantity. These include materials classified as being dangerous to the environment. There are two thresholds, known as lower tier and upper tier.

The general duties under the *2015 Regulations* apply to all operators who control establishments where dangerous substances are present in specified quantities *(Regulations 5–7)*. Additional duties in relation to safety reports and emergency plans apply to the so-called 'upper tier establishments', which are those where the dangerous substances are present in higher quantities *(Regulations 8–16)*.

Operators must take all measures necessary to prevent major accidents, and to limit their consequences for human health and the environment (*Regulation 5*).

They must also prepare and keep a major accident prevention policy document which must be implemented by a safety management system (*Regulation 7*).

Upper-tier sites

The operator of an upper-tier site (that is of an establishment where dangerous substances are present in specified higher quantities) must send a safety report, to the competent authority (see below), and must not start construction of an establishment or commence operations until the competent authority has sent its conclusions (*Regulations 8 and 9*).

The safety report must be reviewed and, where necessary, revised at least every five years and:

- following a major accident at the establishment;
- where a review is justified because there are new facts or by technological knowledge about safety matters, including knowledge arising from analysis of accidents or near misses, or developments in knowledge concerning the assessment of hazards;
- before making any modifications to the establishment, process or the nature or physical form or quantity of dangerous substances which could have significant consequences for major accident hazards; and
- following any change to the safety management system which could have significant consequences (*Regulation 10*).

The operator of an upper site must also prepare an internal emergency plan adequate for:

(a) securing the containment and control of incidents, so as to minimise the effects and limit damage to persons, the environment and property;
(b) implementing the measures necessary to protect human health and the environment from the effects of major accidents;
(c) communicating the necessary information to the public and emergency services; and
(d) providing for the restoration and clean-up of the environment following a major accident (*Regulation 11*).

Internal emergency plans must be reviewed and tested at least every three years and revised when necessary (*Regulation 12*).

The operator of an upper-tier site must also ensure that people who are likely to be in an area liable to be affected by a major accident, and every school, hospital or other area of public use, are provided with information, in the most appropriate form and without their having to request it, on the safety measures at the establishment and on the behaviour required in the event of a major accident occurring (*Regulation 18*). The information must be made publicly available.

Competent authorities

The competent authority is (with the exception of nuclear establishments where it is the Office for Nuclear Regulation (ONR) and the appropriate agency acting jointly) the Health and Safety Executive acting jointly with the Environment Agency in England, the Scottish Environment Protection Agency in Scotland and the Natural Resources Body for Wales in Wales. The competent authority examines safety reports and can prohibit the operation of an establishment or installation if the measures for the prevention or mitigation of major accidents are seriously deficient.

The competent authority must identify groups of establishments ('domino groups') where the risk or consequences of a major accident may be increased because of their geographical position, their proximity to each other or the inventories of dangerous substances they hold (*Regulation 24*).

The competent authority must have an ongoing system of inspections and investigations (*Regulation 25*). If a major accident does occur, the operator is required to inform the authority and provide it with a range of information.

The authority must:

(a) ensure that urgent measures proved necessary are taken;
(b) collect any appropriate information;
(c) ensure that the operator takes remedial action;
(d) make recommendations for the future; and
(e) in the event of certain accidents, notify the European Commission

Local authorities

A local authority in whose area there is an upper-tier establishment is required to prepare an external emergency plan (subject to any exemption that may be granted by the competent authority (*Regulations 11* and *13*)).

The local authority must also review and test the external emergency plan and put it into effect in specified circumstances (*Regulations 14* and *16*).

Public information

The competent authority must make specified information, including by electronic means, available to the public. The information to be provided is:

(1) the name of the operator and the address of the establishment;
(2) confirmation that the *Control of Major Accident Hazards Regulations 2015* apply to the establishment and that the notification and the safety report have been sent to the competent authority;
(3) an explanation in simple terms of the activity or activities undertaken at the establishment;
(4) the hazard classification of the relevant dangerous substances involved at the establishment which could give rise to a major accident, with an indication of their principal dangerous characteristics in simple terms;
(5) general information about how the public will be warned, if necessary, and adequate information about the appropriate behaviour in the event of a major accident, or an indication of where that information can be accessed electronically;
(6) the date of the last site visit carried out, as part of a programme for routine inspections, and where more detailed information about the inspection and the related inspection plan can be obtained upon request; and

Effective disaster and emergency management **[D6015]**

(7) details of where further relevant information can be obtained (*Regulation 17*).

A guide to the *Control of Major Accident Hazards Regulations (COMAH) 2015* (L111) can be found on the HSE website at: www.hse.gov.uk/pubns/books/l111.htm.

The Control of Substances Hazardous to Health Regulations 2002 (SI 2002 No 2677) as amended

[D6015] Where emergency procedures drawn up to comply with the *Management of Health and Safety at Work Regulations 1999 (SI 1999 No 3242)* (see D6010) are insufficient to contain and control any risk to health that hazardous substances might pose during an emergency, then the *Control of Substances Hazardous to Health Regulations 2002, Reg 13* requires employers to extend their emergency procedures to ensure they are capable of:

(a) mitigating the effects of an incident;
(b) restoring the situation to normal as soon as possible; and
(c) limiting the risks to health of employees and anyone else likely to be affected.

The HSE publication *Control of Hazardous Substances* (L5), which contains the COSHH regulations together with the approved code of practice (ACOP) and guidance, provides considerable detail on the requirements. To deal with situations which could present significantly greater risks on account of substances hazardous to health, employers should extend their emergency procedures to include:

(a) the identity of the relevant substances present at the workplace, where they are stored, used, processed or produced, and an estimate of the amount in the workplace on an average day;
(b) the foreseeable types of accidents, incidents or emergencies which might occur involving those substances, and the hazards they could present. These could include failure of controls, spills, uncontrolled releases of vapours, dusts or fumes into the workplace, accidents with machinery transporting substances in the workplace, leaks and fire. Consider where such incidents might occur, what effect they might have, the other areas that might be affected by the incident spreading and any possible repercussions that might be caused;
(c) the special arrangements to deal with an emergency situation not covered by the general procedures and the steps to be taken to mitigate the effects;
(d) the safety equipment and personal protective equipment to be used in the event of an accident, incident or emergency, where it is stored, and who is authorised to use it. Judgements about the type of safety equipment and personal protective equipment, including respiratory protective equipment, to be used should be made with regard to the level and type of risk, and a worst-case estimate of the likely concentration of a hazardous substance in the air in the workplace;

(e) first-aid facilities sufficient to deal with an incident until the emergency services arrive, where the facilities are located and stored, and the likely effects on the workforce of the accident, incident or emergency such as burns, scalds, shock, the effects of smoke inhalation;
(f) the role, responsibilities and authority of the people nominated to manage the accident, incident or emergency and the individuals with specific duties in the event of an incident. These include the people responsible for checking specific areas have been evacuated, shutting down plant that might otherwise compound the danger, and contacting and liaison with the emergency services on their arrival and making sure they are aware of the hazardous substance(s) that are the cause of, or are affected by, the emergency;
(g) procedures for employees to follow and details of who should know these, how they should respond to an incident, what action they should take, and the roles of the people who have been assigned specific responsibilities;
(h) procedures for clearing up and safely disposing of any substance hazardous to health damaged or contaminated during the incident;
(i) regular safety drills. The frequency of practising emergency procedures will depend on the complexity of the layout of the workplace, the activities carried out, the level of risk, the size of the workforce, the amount of substances involved, and the success of each test;
(j) the special needs of any disabled employees, such as assigning other employees to help them leave the workplace in an emergency;
(k) If an incident results in the uncontrolled release of a carcinogen or mutagen into the workplace, the equipment the employer provides must always include suitable respiratory protective equipment which is capable of adequately controlling exposure.

The Control of Asbestos Regulations 2012 (SI 2012 No 632)

[D6016] Asbestos is one substance not included in the *Control of Substances Hazardous to Health Regulations 2002*. It has its own specific legislation, the *Control of Asbestos Regulations 2012*. Regulation 15 of these Regulations makes provisions for dealing with accidents, incidents and emergencies. The HSE publication *Managing and working with asbestos* (L143) sets out the regulations together with the ACOP and guidance and gives advice on dealing with uncontrolled releases of asbestos.

The ACOP sets out that:

> '"Employers must deal with all uncontrolled releases of asbestos into the workplace, quickly and appropriately. This applies to circumstances where asbestos is accidentally disturbed as a result of work or where asbestos is unintentionally released as a result of a failure of control measures, such as a leak from an enclosure.
>
> The steps required to clean up such releases must be appropriate for the scale of the release and the potential for further release and spread of fibres.
>
> The clean-up of any release that leads to potential exposures at or above the control limit or that are not sporadic and of low intensity, eg releases of asbestos lagging, loose fill, asbestos coatings (not textured coatings) or large-scale releases of [asbestos insulating board] AIB must be done by a licensed contractor." '

The Dangerous Substances and Explosive Atmospheres Regulations 2002 (SI 2002 No 2776)

[D6017] (See also D0934.)

Regulation 8 requires arrangements to deal with accidents, incidents and emergencies that closely mirror those required by the *Control of Substances Hazardous to Health Regulations 2002* (see **D6015** above).

Radiation (Emergency Preparedness and Public Information) Regulations 2001 (SI 2001 No 2975)

[D6018] The Radiation (Emergency Preparedness and Public Information) Regulations 2001 (SI 2001 No 2975) impose requirements on operators of premises where radioactive substances are present (in quantities exceeding specified thresholds). They also impose requirements on carriers transporting radioactive substances (in quantities exceeding specified thresholds) by rail or conveying them through public places, with the exception of carriers conveying radioactive substances by rail, road, inland waterway, sea or air or by means of a pipeline or similar means (see **D6019** below).

Essential guidance is contained in the HSE publication *Guide to the Radiation (Emergency Preparedness and Public Information) Regulations 2001 (L126)*.

The Regulations:

(a) impose a duty on the operator and the carrier to make an assessment as to hazard identification and risk evaluation and, where the assessment reveals a radiation risk, to take all reasonably practicable steps to prevent a radiation accident or limit the consequences should such an accident occur [*SI 2001 No 2975, Reg 4*];

(b) impose a duty on the operator and carrier to send the HSE a report of an assessment and empower the HSE to require a detailed assessment of such further particulars as may reasonably require [*SI 2001 No 2975, Reg 6 and Sch 5 and 6*];

(c) impose a duty on the operator and the carrier to make a further assessment following a major change to the work with the ionising radiation or within three years of the date of the last assessment, unless there has been no change of circumstances, and send the HSE a report of that further assessment [*SI 2001 No 2975, Regs 5 and 6*];

(d) where an assessment reveals a reasonably foreseeable radiation emergency arising, impose a duty on the operator or the carrier (as the case may be) and, in the case of an operator, the local authority in whose area the premises are situated, to prepare emergency plans [*SI 2001 No 2975, Regs 7, 8 and 9 and Sch 7 and 8*];

(e) require operators, carriers and local authorities to review, revise and test emergency plans at least every three years [*SI 2001 No 2975, Reg 10*];

(f) make provision as to consultation and co-operation by operators, carriers, employers and local authorities [*SI 2001 No 2975, Reg 11*];

(g) make provision as to charging by local authorities for performing their functions under the Regulations in relation to emergency plans [*SI 2001 No 2975, Reg 12*];

[D6018] Disaster and Emergency Management Systems (DEMS)

(h) in the event of a radiation emergency or an event which could reasonably be expected to lead to such an emergency, make provision for implementing emergency plans, and in the event of a radiation emergency, require both provisional and final assessments of the circumstances of the emergency [*SI 2001 No 2975, Reg 13*];

(i) where an emergency plan provides for the possibility of an employee receiving an emergency exposure, impose a duty on the employer to undertake specified arrangements for employees who may be subject to exposures, such as dose assessments, medical surveillance and the determination of appropriate dose levels. They impose further duties on employers in the event that an emergency plan is implemented [*SI 2001 No 2975, Reg 14*]; and

(j) impose requirements on operators and carriers, where an operator or carrier carries out work with ionising radiation which could give rise to a reasonably foreseeable radiation emergency, and on local authorities, where there has been a radiation emergency in their area, to supply specified information to the public [*SI 2001 No 2975, Regs 16 and 17* and *Sch 9 and 10*].

Public Information for Radiation Emergencies Regulations 1992 (SI 1992 No 2997)

[D6019] The *Public Information for Radiation Emergencies Regulations 1992 (SI 1992 No 2997)* were largely revoked by the *Radiation (Emergency Preparedness and Public Information) Regulations 2001 (SI 2001 No 2975)*. However, SI 1992 No 2997, Reg 3 continues in force to the extent that it applies in relation to the transport of radioactive substances by road, inland waterway, sea or air. Any other provisions of the 1992 Regulations continue in force so far as is necessary to give effect to *Reg 3*.

An employer (or self-employed person) must supply prior information where a radiation emergency is reasonably foreseeable from their undertaking in relation to the transport of radioactive substances by road, inland waterway, sea or air. They must:

(a) supply information to members of the public likely to be affected without their having to request it, concerning:
   (i) basic facts about radioactivity and its effects on persons/environment,
   (ii) the various types of radiation emergency covered and their consequences for people/environment,
   (iii) emergency measures to alert, protect and assist people in the event of a radiation emergency,
   (iv) action to be taken by people in the event of an emergency,
   (v) authority/authorities responsible for implementing emergency measures;
(b) make the information publicly available; and
(c) update the information at least every three years.

[Public Information for *Radiation Emergencies Regulations 1992 (SI 1992 No 2997), Reg 3 and Sch 2.*]

The Reporting of Injuries, Diseases and Dangerous Occurrences Regulations 2013 (SI 2013 No 1471)

[D6020] The *Reporting of Injuries, Diseases and Dangerous Occurrences Regulations (RIDDOR) 2013 (SI 2013 No 1471)* apply to 'Accidents' through to 'Major Incidents'. *RIDDOR* is a reporting requirement that must be followed by the employer or 'responsible person' in notifying the enforcing authority when there is an event resulting in a reportable injury, reportable dangerous occurrence or gas incident. In addition, work injuries lasting for seven days or more and reportable occupational diseases, are reported to the enforcing authority (see **A3001** for further details).

The types of incidents that must be reported are defined or listed in Schedules.

The Health and Safety (First-Aid) Regulations 1981 (SI 1981 No 917)

[D6021] These Regulations do not explicitly mention first-aid arrangements necessary in the event of an emergency or disaster. However, they will apply in all types of accidents and major incidents as defined above. The Regulations are a general statement of best practice for ensuring the existence of adequate medical equipment, competent and trained first-aiders or appointed persons and information on first-aid facilities and equipment to staff/others. The HSE has published guidance on the Regulations (L74).

The Construction (Design and Management) Regulations 2015 (SI 2015 No 51)

[D6022]–[D6023] *Regulation 30* requires that suitable and sufficient arrangements shall be prepared and implemented for dealing with any foreseeable emergency. HSE has published guidance on the regulations, *Managing health and safety in construction* (L153) which can be downloaded from the HSE website at: www.hse.gov.uk/pubns/books/l153.htm.

The Confined Spaces Regulations 1997 (SI 1997 No 1713)

[D6024] (See also C7029.)

Confined Spaces Regulations 1997 (SI 1997 No 1713), Reg 5 imposes duties on employers and others to make arrangements for emergencies in confined spaces. It should be noted that 'confined spaces' does not refer to merely a small area or congested place. The Regulations give a very specific definition based on a place where there arises a reasonably foreseeable specified risk listed as:

(a) serious injury to any person at work arising from a fire or explosion;
(b) without prejudice to paragraph (*a*):
 (i) the loss of consciousness of any person at work arising from an increase in body temperature;
 (ii) the loss of consciousness or asphyxiation of any person at work arising from gas, fume, vapour or the lack of oxygen;
(c) the drowning of any person at work arising from an increase in the level of a liquid; or
(d) the asphyxiation of any person at work arising from a free-flowing solid or the inability to reach a respirable environment due to entrapment by a free-flowing solid;

In summary:

- 'suitable and sufficient arrangements' need to be made for rescue of persons working in confined spaces, with no person accessing a confined space until such arrangements have been developed [SI 1997 No 1713, Reg 5(1)];
- so far as is reasonably practicable, the risks to persons required to put the arrangements into operation must also be considered as should resuscitation equipment in the event of it being required [SI 1997 No 1713, Reg 5(2)]; and
- there is also a duty to act, to implement the arrangements when an emergency results [SI 1997 No 1713, Reg 5(3)].

'Suitable and sufficient arrangements' for rescue and resuscitation should include appropriate equipment, rescue procedures, warning systems that an emergency exists, fire safety systems, first-aid, control of access and egress, liaison with the emergency services, training and competence of rescuers, etc.

The HSE publication *Safe work in confined spaces. Confined Spaces Regulations 1997 (L101)* (www.hse.gov.uk/pubns/priced/l101.pdf) contains an Approved Code of Practice and guidance on duties under the Confined Spaces Regulations applicable across all industry sectors (except offshore) and covers emergency procedures.

The Carriage of Dangerous Goods and Use of Transportable Pressure Equipment Regulations 2009 (CDG 2009) (SI 2009 No 1348) as amended

[D6025] Carrying goods by road or rail involves the risk of traffic accidents. If the goods carried are dangerous, there is also the risk of an incident, such as spillage of the goods, leading to hazards such as fire, explosion, chemical burn or environmental damage. Most goods are not considered sufficiently dangerous to require special precautions during carriage. Some goods, however, have properties which mean they are potentially dangerous if carried.

Dangerous goods are liquid or solid substances, and articles containing them, that have been tested and assessed against internationally agreed criteria – a process called classification – and found to be potentially dangerous (hazardous) when carried.

Carriage of dangerous goods by road or rail is regulated internationally by agreements and European Directives, with biennial updates of the Directives to take account of technological advances. New safety requirements are implemented by Member States via domestic regulations.

The *Carriage of Dangerous Goods and Use of Transportable Pressure Equipment Regulations 2009 (CDG 2009)* came into force on 1 July 2009. They succeed the 2007 Regulations and include the carriage of radioactive materials. *CDG 2009* implements the European Agreement concerning the International Carriage of Dangerous Goods by Road (ADR) (with a number of exceptions). They refer directly to ADR and there are some additional or alternative requirements to ADR.

The regulations place duties upon everyone involved in the carriage of dangerous goods to ensure that they know what they have to do to minimise the risk of incidents and guarantee an effective response to protect everyone either directly involved (such as consignors or carriers), or who might become involved (such as members of the emergency services and public).

[D6026]–[D6027] A duty is imposed on the operator of a 'container', 'vehicle' or 'tank' to provide information to other operators engaged/contracted, regarding the handling of emergencies or accident situations. Such information includes:

- the hazardous nature of the goods being carried and controls needed to make safe such hazards;
- actions required if a person is exposed/makes contact with the goods being carried;
- actions required to avert a fire and the equipment that should or should not be used to fight the fire;
- the handling of a breakage/spillage; and
- any additional information that should be given that can assist the operators or others.

(See also D0401.)

The Environment Act 1995

[D6028] While the *Environment Act 1995* (*EA 1995*) does not explicitly deal with DEM issues, its main objective is the mitigation or prevention of such events. For example, *EA 1995, ss 14–18* creates the flood defence committees to co-ordinate the prediction, consequence and response needed in the event of floods.

The Corporate Manslaughter and Corporate Homicide Act 2007

[D6029] Disasters such as Piper Alpha, Kings Cross, Clapham Junction, Herald of Free Enterprise and Ladbroke Grove involved multiple fatalities and allegations of managerial failure and negligence contributing to the disasters. However, health and safety law as it stood at the time did not permit the successful prosecution of large companies because of the difficulty in establishing that the individuals in charge were 'the embodiment of the company'. This situation was rectified through the *Corporate Manslaughter and Corporate Homicide Act 2007* which came into force in 2008 (see CORPORATE MANSLAUGHTER).

Insurance reasons

[D6030] There are also insurance-based reasons for effective DEM:

- given the positive correlation between risk and premium, the existence of DEM indicates hazard and risk control, consequently the premium ought to be less; and
- if insurers are not satisfied or are dissatisfied with the DEM system in place then they may not insure the operation, in turn increasing corporate risk as well as reducing corporate credibility. It may also prohibit the operation from tendering for contracts.

Corporate reasons

[D6031]

- The corporate experience of companies like P&O European Ferries (Dover) Ltd, indicates that proactive DEM systems would have saved the company considerable money, publicity and reputation. In March 1987, the *Herald of Free Enterprise*, the roll-on roll-off car ferry, left the Belgian port of Zeebrugge for Dover. It sank with the loss of 187 lives. The case against the company and five senior corporate officers collapsed for reasons cited above ('embodiment of the company' see **D6042**). With the subsequent introduction of the *Corporate Manslaughter and Corporate Homicide Act 2007*, there are now better chances for a successful prosecution of a company for corporate manslaughter. By April 2018, there had been 26 convictions under the Act.
- Failures to have sound DEM systems can also have dire financial consequences as disasters such as *Piper Alpha* indicate. Occidental Petroleum were generating 10 per cent of all the UK's North Sea output from *Piper Alpha*. Lord Cullen, who led the inquiry into the disaster said: 'The safety policy and procedures were in place: the practice was deficient'. *Piper Alpha* showed that the company had ineffective emergency response procedures resulting in workers being trapped, dying of smoke inhalation or jumping into the cold North Sea, something the procedures prohibited. A significant number of those that did jump into the sea survived. Such a disaster had consequences for Occidental's financial reputation, share value, ability to attract investment and growth potential.

Humanitarian and societal reasons

[D6032] The human cost for not planning for worst-case outcomes is the most significant. It is bad enough that incidents such as explosions claim victims, but when the death toll increases because the emergency response is inadequate, that may be seen as a worse crime. Society expects organisations to plan and prepare for the worst-case scenarios and when this does not happen, society seeks the closure, forfeiture or expulsion of the organisation from society, eg the legal and political costs to Union Carbide in India following the methyl isocyanate gas leak at its pesticide plant near Bhopal, which killed thousands of people.

Environmental reasons

[D6033] Failure to prepare and plan for disasters and emergencies will also have environmental costs, no better illustrated than the Sandoz warehouse fire in Basel, Switzerland in 1986, with its impact on the river Rhein not only in Switzerland, but also affecting Germany on the opposite bank, then France and Holland further downstream. Retention of the fire-fighting run-off water could have prevented much environmental damage.

Disaster and emergency management systems (DEMS)

[D6034] A DEMS is outlined below:

Figure 2: Disaster and Emergency Management Systems (DEMS)

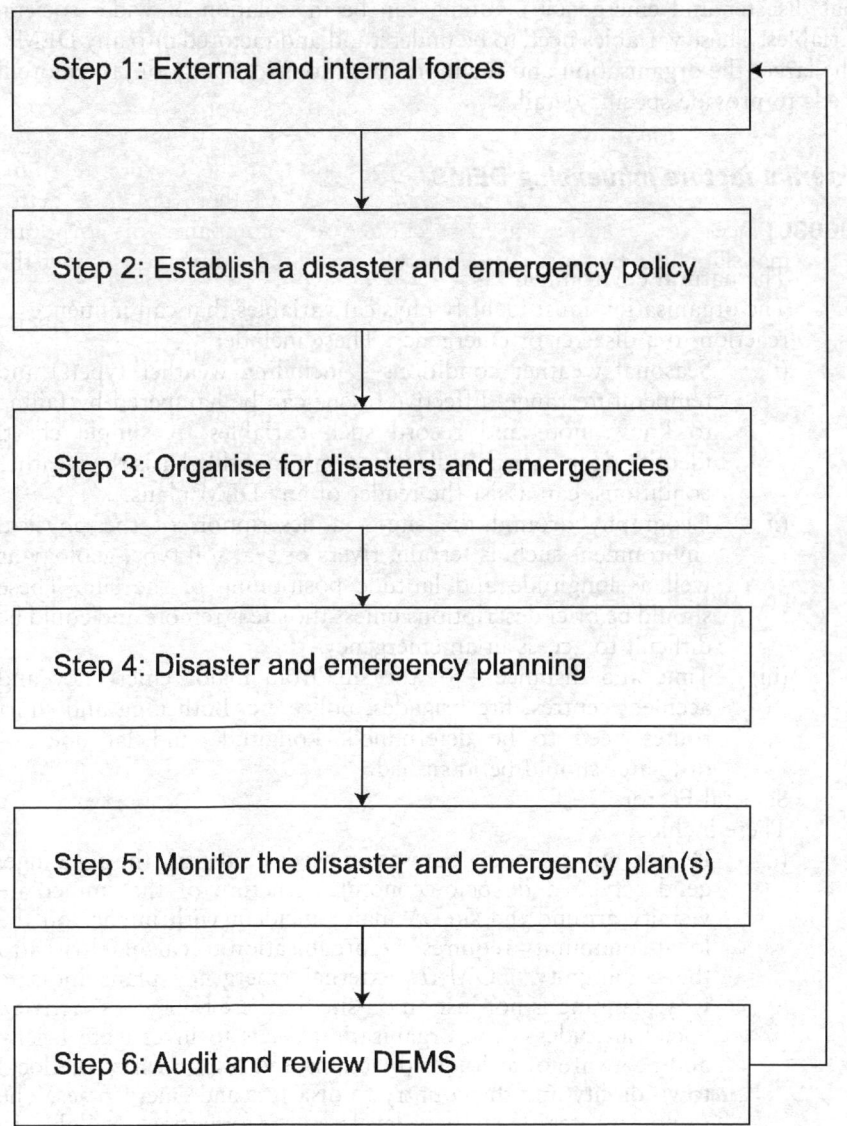

External and internal factors

[D6035] The starting point with DEMS is understanding the variables that can influence or affect them. These are both internal to the organisation and wider societal variables. It would be a mistake for an organisation to believe that disaster and emergency planning can be in isolation of wider societal variables. These variables need to be understood and factored into any DEMS. The larger the organisation and the more hazardous its operation, the more it needs to provide specific detail.

External factors influencing DEMS

[D6036]

- The natural environment
  The organisation must identify physical variables that can influence its reaction to a disaster or emergency. These include:
  (i) Seasonal weather conditions – including weather type(s) and temperature range. Effective rescue can be hampered by failing to know, note and record such variables. A simple chart, identifying in user-friendly terms on a monthly basis the weather conditions, can assist the reader of any DEM plans.
  (ii) Geography around the site – a description of the physical environment such as terrain, rivers or sea, soil type, geology as well as longitude and latitude positioning of the site. These should be brief descriptions unless the site is remote and could be difficult to access in an emergency.
  (iii) Time and distance – of the site from major emergency and accident centres, fire brigades, police etc. Both long and short routes need to be determined. Longitude and latitude co-ordinates should be identified.
- Societal Factors
  These include:
  (i) Demographics – the organisation needs to identify the age range, gender type and socio-economic structure of the immediate vicinity around the site. A major incident with impact on the local community requires the organisation to calculate risks to the community. *COMAH* external emergency plans indicate why planning is not just an on-site 'in these four walls' activity.
  (ii) Social attitudes – the organisation needs to investigate briefly and be aware of attitudes (reactions and responses) of the local town or city and the country to disasters and emergencies. The response rates, awareness levels and information available to different communities differs.
  (iii) Perception of risk – how does the local community view the site or operation? Is the risk 'tolerable' and 'acceptable' to them for having the operation in their community? Both Chernobyl and Bhopal show that economic necessity can alter the perception of risk.

(iv) History of community response – in brief the organisation needs to determine how many major incidents there have been in the past and how effectively the community has been assisted (emergency services and volunteers). Such support will be critical in a disaster.

(v) The built environment – a brief description of the urban environment, namely the street layout, urban concentration level, population density, access/egress to railways, motorways are useful. These issues could be covered by the inclusion of a map of the area.

- Government and political factors
Relevant factors are:
(i) The policy of national government to disasters and emergencies. Is it proactive in advising organisations? What information and guidance does it give? The organisation needs to identify the office of the local HSE, Environment Agency or similar body, make contact with officials and seek early input into any planning.

(ii) Local government – its plans, information and guidance. DEM's need to be aware of any local government restrictions on managing major incidents. In the case of *COMAH* sites, they will need to involve the local authority in preparing external emergency plans.

(iii) Committees and agencies – there are many legal and quasi-legal bodies which have guidance, information and templates on response. For example, the flooding committees under the *Environment Act 1995* (mentioned in **D6028** above).

(iv) Emergency and medical services – making contact, obtaining addresses/contact numbers and knowing the efficiency of the emergency services/accident and emergency are critical actions.

- Legal factors
The organisation needs to know the legal constraints it has to operate within, in order to comply with the law and follow the best advice contained in legislation as a means of preventing disaster and emergency.

- Sources of information issues
These include:
(i) The local and national media – will not only report any incidents but can be critical in relaying messages. Any planning requires the identification of local and national newspapers (including contact details), local/national radio details and any public-sector media (through local government).

(ii) Business associations – chambers of commerce, business links, training and enterprise councils should be identified. In the event of a major incident, they can convey and provide information on a local rescue or occupational health organisation.

(iii) Voluntary organisations – such as the British Red Cross and St John Ambulance should be identified and noted.

The organisation needs to build information networks locally and nationally. The experience of major disasters such as Chernobyl or Kings Cross indicate that letting others know of the incident is not a shameful or embarrassing matter – rather it can warn others and stop them from attending the site.

- Technological factors
  What technology and equipment exists in the local community and what does not exist? – for example, lifting equipment, rescue, stand-by computer facilities and monitoring and measuring equipment.
- Commercial factors
  These include:
  (i) Insurance – liaise with insurers from the beginning and ensure that all plans are drawn to their attention and if possible approved by them. This can avoid difficulties with any claim arising from a disaster or emergency and ensure advice from the insurer has been factored into the plans.
  (ii) Customer and supplier response – larger organisations should identify the responses from customers and suppliers, their willingness to work with the organisation in the event of a worst-case scenario and possible alternatives. Suppliers should be identified. Issues of customer convenience and loyalty should also be addressed.
  (iii) Attitudes of the bank – liquidity issues will arise when a major incident strikes. Most organisations do not and cannot afford to make financial provision for such eventualities. Therefore, if the operation is highly hazardous, establishing financial facilities with the bank beforehand is necessary to ensure availability of liquid cash.
  (iv) Strength of the economic sector and economy – larger organisations that are significant players in a sector or the economy need to be aware of the 'multiplier effect' that damage to their operation can do to the community and suppliers, employees and others.

Internal factors affecting DEMS

[D6037] DEMS must also account for various internal or organisational variables.

- Resource factors
  Cash flow and budgetary planning, whether annual or a longer period, must account for resource availability for major incidents. This includes physical, human and financial resources. This is wider than banking facilities and encompasses equipment, trained personnel and contingency funds.
- Design and architecture
  Is the workplace physically/structurally capable of withholding a major incident? What are the main design and architectural risks? Issues such as layout, access/egress, emergency routes and adequacy of space for vehicles should also be considered.

- Corporate culture and practice
  The organisation should identify its own collective behaviour and attitude, and its strengths and weaknesses in being able to cope with a major incident. Being a 'large organisation' does not mean it is a 'coping organisation' There is also a delay in response associated with hierarchical structures. In this case, the organisation needs to consider a small 'matrix' or project team cell in the hierarchy dedicated to incident response.
- Individual's perceived behaviour
  (a) The *Hale and Hale Model* is an attempt to explain how individuals internalise and digest perceived information of danger; make decisions/choices according to the cost/benefit associated with each decision/choice; and the actual actions they take as well as any reactions that result from their actions. In short, these five variables need to be understood in the organisation setting and a picture built up of the behavioural response of an individual.
  (b) The *Glendon and Hale Model* is a macro model of how the organisation (behaving like a system), being dynamic and fluid, with objectives and indeed limitations (systems boundary) can shape behaviour and in turn influence human error. Following Rasmussen, human error can be skill based (failing to perform an action correctly), rule based (have not learned the sequences to avoid harm) or knowledge based (breaching rules or best practice). If all three error types are committed then the danger level in the system is also greater. The model is a focus on how wider systems can contribute to human error and how that error can permeate into the organisation. The organisation must clearly define and communicate its intention and objectives and continuously monitor individual response.
  (c) *Reducing Error and Influencing Behaviour*, HSG 48, HSE, 1999 identifies the role of human error and human factors in major incident causation. The HSE says 'a human error is an action or decision which was *not intended*, which involved a deviation from an accepted standard, and which led to an undesirable outcome' (page 13). It classifies three types of error:
     (i) *slips* (unintended action);
     (ii) *lapses* (short term memory failure) with slips and lapses being skill based; and
     (iii) mistakes (incorrect decision) which are rule based.
  It also sets out that violations are deliberate breach of rules which are knowledge based.
  Different types of human errors contribute in different ways to major incidents. Organisations should identify from reported incidents the main types of human errors, why they are resulting and the negative harm generated.

The HSE says human factors is a distinct element which must be recognised, assessed and managed effectively in order to control risks. Human factors is a combination of understanding the person's behav-

iour, the job they do (ergonomics) and the wider organisational system. Major incidents can result if the organisation does not analyse and understand these three variables.
- Information systems at work
  The types of information systems, their effectiveness, and accuracy should be identified. Telephone, fax, email and mobile phone all need to be assessed for performance and efficacy during a worst-case scenario.

Establish a disaster and emergency policy

[D6038] The DEM policy is a concise document that highlights the corporate intent to cope with and manage a major incident. It will have three parts: statement of policy for managing major incidents (see **D6039**); arrangements for major incident management (see **D6040**); and command and control chart of arrangements (see **D6041**).

Statement of policy for managing major incidents

[D6039] This should be a short (maximum 1 page) mission and vision statement covering the following:

- Senior management commitment to be responsible for the co-ordination of major incident response.
- Compliance with the law, namely:
  — the protection of the health, safety and welfare of employees, visitors, the public and contractors;
  — compliance with the duties under the *Management of Health and Safety at Work Regulations 1999 (SI 1999 No 3242)*, Regs 8 and 9 (see **D6010**);
  — compliance with any other legislation that may be applicable to the organisation.
- Suitable arrangements to cope with a major incident and to be proactive and efficient in the implementation process.
- Application of the statement to all levels of the organisation and all relevant sites in the country of jurisdiction.
- Commitment of human, physical and financial resources to prevent and manage major incidents.
- Consultation with affected parties (employees, the local authority and others if needed).
- Review of the statement.
- Communication of the statement.
- It should be signed and dated by the most senior corporate officer.

Arrangements for major incident management

[D6040]

- This refers to what the organisation has done, is doing and will do in the event of a major incident and *how* it will react in those circumstances. The arrangements are a legal requirement under the *Management of Health and Safety at Work Regulations 1999 (SI 1999 No 3242), Regs 5, 8* and *9*.
- Arrangements should be realistic and achievable. They should focus on major actions to be taken and issues to be addressed rather than being a 'shopping list'. The arrangements will have to be verified (in particular for *COMAH* sites).
- Arrangements could be under the following headings with explanations under each. The larger the organisation and the more complex the hazard facing it, the more detailed the arrangements need to be. For example:

    (i) Medical assistance – including first-aid availability, first-aiders, links with accident and emergency at the medical centre, other specialists that could be called upon, rules on treating injured people, specialised medical equipment and its availability etc.

    (ii) Facilities management – the location, site plans and access to the main facilities (gas, electricity, water, substances etc), rendering safe such facilities, availability of water supply in-house and within the perimeter of the site etc.

    (iii) Equipment to cope – identification of safety equipment available and/or accessible, location of such equipment, types (personal protective equipment, lifting, moving, working at height equipment etc).

    (iv) Monitoring equipment – measuring, monitoring and recording devices needed, including basic items such as measuring tapes, paper, pens, tape recorders, intercom and loud-speakers.

    (v) Safe systems – procedures to access site, working safely by employees and contractors under major incident conditions (what can and cannot be done), hazard/risk assessments of dangers being confronted etc, risks to certain groups and procedures needed for rescue (disabled people, young people, children, pregnant women, elderly people).

    (vi) Public safety – preventing access to a major incident site by the public (in particular children, trespassers, the media and those with criminal intent), warning systems to the public etc.

    (vii) Contractor safety – guidance and information on working safely to contractors at a major incident.

    (viii) Information arrangements – the supply of information to staff, the media and others (insurers, enforcers) to inform them of the events. Where will the information be supplied from, when will it be provided and updated?

    (ix) The media – managing the media, confining them to an area, handling pressure from them, what to say and what not to say etc.

(x) Insurers/loss adjusters – notifying them and working with them at the earliest opportunity.

(xi) Enforcement agencies – notifying them of the major incident, working with eg HSE, Environment Agency (or Scottish or Welsh equivalent) as well as local authority (environmental health, planning, building control for instance).

(xii) Evidence and reporting arrangements – to cover strict rules on removal of evidence by employees or others, role and power of enforcers, incident reporting eg under the *Reporting of Injuries, Diseases and Dangerous Occurrences Regulations 2013 (SI 2013 No 1471)* etc.

(xiii) The emergency services – working with the police, fire, ambulance and other NHS services, and other specialists (British Red Cross, search and rescue), rules of engagement, issues of information supply and communication with these services.

(xiv) Specialist arrangements for specific major incidents such as bomb threats and explosions – issues of contacting the police, ordnance disposal, access and egress, rescue and search, economic and human impact for example.

(xv) Human aspects – removing, storing and naming dead bodies or seriously injured people during the incident. Informing next-of-kin, issues of religious and cultural respect. Issues of counselling support and person-to-person support during the incident.

This is not an exhaustive list. The arrangements should not repeat those in the safety policy, rather the latter can be abbreviated and attached as a schedule to the above, so the reader can have access to succinct and specific OSH arrangements such as fire safety, occupational health, safe systems at work and dangerous substances.

Command and control chart of arrangements

[D6041] This highlights who is responsible for the effective management of the major incident.

- It should be a graphical representation, preferably in a hierarchical format, clearly delineating the division of labour between personnel in the organisation and the emergency services/others.
- The chart should display three broad levels of command and control, namely strategic, operational and tactical. The first relates to the person or people in overall charge of the major incident. Will this be the person who signed the statement of policy for major incidents or will it be the disaster and emergency advisor or safety officer, for example? This person will make major decisions. The second relates to co-ordinators of teams. Operational level personnel need to have the above arrangements assigned to them in clear terms. The third refers to those in the front line of the major incident, such as first-aiders.

It is most important to note that internal command and control of arrangements does not mean *overall* command and control of the major incident. This will be vested with the appropriate emergency service,

normally the police or the fire authority. In the event of any conflict of decisions, the external body, such as the police, will have the final veto. Therefore, the chart and the arrangements must reflect this variable.
- The chart should list on a separate page contact details (names, addresses, emergency phone, email and mobile numbers) of those identified on the chart. It should also list contact details for the emergency services and others including the British Red Cross, specialist search and rescue, loss adjusters and enforcement authorities, for example.
- The chart should clearly ratify a principle of command and control: 'In-charge 1', 'In-charge 2', if the original person became unavailable.
- The chart and list of numbers should be accompanied by a set of 'rules of engagement' in short bullet points to remind personnel of the importance of command and control, safety, obedience, communication, accuracy and humanity for instance.

Summary
[D6042]
- The three parts of the DEM policy need to be in one document. Any detailed procedures can be separately documented as disaster and emergency procedures and could be an extensive source of information. However, unlike the safety policy and any accompanying safety manual, the same volume of information cannot apply to the DEM policy. For obvious reasons it must be concise, clearly written, very practical and without complex cross-referencing.
- The DEM policy must fit with the safety policy. There can be no conflict so the safety officer and the DEM officer need to cross check and liaise. The DEM policy must also fit with the broader corporate/business policy of the organisation.
- The DEM policy must be both proactive and reactive. The former is concerned with preventing/mitigating loss and the latter is concerned with managing the major incident, if and when it does arise, in a swift and least harmful manner.
- The DEM policy needs to be reviewed regularly. This could be when there is 'significant change' to the organisation, or as a part of an annual or biannual audit.
- It must be remembered that the DEM policy is a live document and must be accessible and up-to-date.
- Although accessibility is important, the policy should also have controlled circulation to core personnel only, those identified in the chart and the legal department for example. If the policy was to be accessed by individuals wishing to harm the organisation, it could enable them to pre-empt and reduce the efficacy of the policy.
- Finally, and most crucially, the core contents of the policy need to be communicated to all staff and others including contractors, temporary employees and possibly the local authority.

Organise for disasters and emergencies

[D6043] Once the establishment has accounted for external and internal factors and has produced a DEM policy taking account of such factors, it is then necessary to ensure personnel and others are aware of the issues raised. The '4' c's approach of the HSE publication, *Successful Health and Safety Management*, HSG 65, provides a logical framework to generate this: communication (see **D6044**); co-operation (see **D6045**); competence (see **D6046**); and control (see **D6047**).

Establish effective communication

[D6044] Communication is a process of transmission, reception and feedback of information, whether verbal, written, pictorial or intimated. Effective communication of the DEM policy is therefore not a matter of circulating copies, but requires the following:

(1) *Transmission*
- The whole policy should not be circulated as it will mean little to employees and others. An abridged version, possibly in booklet format or as an addition to any OSH documentation supplied, will make more sense. Such copies must be clear, user friendly and non-technical as possible. It should be aware of the end user's capability to digest the information, be logical/sequential in explanation and use pictorial representation as much as possible.
- Being aware of the audience is central to the effective communication of the DEM policy. It will consist of:
  — direct employees;
  — temporary employees;
  — contractors;
  — the media;
  — the enforcement agencies (*COMAH* sites);
  — the local authority (*COMAH* sites);
  — insurers;
  — emergency services (*COMAH* sites); and
  — the public (*COMAH* sites).
  This does not necessarily mean separate copies for each, but the abridged copy needs to satisfy the needs of all these groups.
- Transmission should start from the board, go through to departmental heads, and disseminate downwards and across.

(2) *Reception*
- What format will the end-user receive the abridged copy in – hard copy, electronic or via email for example?
- When will the copy be circulated – at induction, during training or ad hoc?

(3) *Feedback*
- Will the end-user have the opportunity to raise questions, make suggestions and be critical if they spot inconsistencies in the DEM policy?

- There should also be 'tool box talks' and other general awareness programmes to inform individuals of the policy. This could be combined with general OSH programmes or wider personnel programmes, so that a holistic approach is presented.
- Feedback allows the policy to be owned by all individuals – the single most important factor in successful pre-planning to prevent major incidents.

In general, it may be useful to retain copies of the abridged and full policy with other safety documentation in an in-house company library, in hard copy and electronic form. For smaller organisations, this could be one or two folders on a shelf. In larger organisations it could be a dedicated room. The abridged copy could also be available on an intranet site.

Co-operation

[D6045] Co-operation is concerned with collaborating, working together to achieve the following shared goal and objectives:

- Co-operation between strategic, operational and tactical level management is critical. This reflects the chart in the DEM policy as discussed above. This could be consolidated as part of a broader corporate meeting or preferably dedicated time to cover OSH and fatal incident issues. This could be a biannual event, with a dedicated day allotted for all grades of management to interact. This is not the same as a board level or management discussion.
- To give responsibility to either the safety committee or the safety group to co-ordinate, review, debate and carry out an assessment of the DEM policy or combine this function within a broader business/corporate review committee. The former has advantages as it is safety dedicated, while the latter would integrate DEM policy issues into the wider business debate.
- Involvement of safety representatives or representatives of employee safety (ROES), is both a legal requirement as well as inclusive safety management. They can be central in linking together management and workers. Trained trade union safety representatives are a knowledgeable resource. The TUC Diploma in Occupational Health and Safety at NOCN Level 3 is accredited to meet the academic requirement for safety technician (Tech IOSH) grade, for example.
- Co-operation should extend to contractors. The person responsible for liaising with contractors should up-date them and make them aware of the DEM policy and seek their support and suggestions. It may also be valuable to invite contractors to OSH/fatal incident awareness days or the general committee meetings as observers.
- Co-operation between the organisation and external agencies such as the local authority, insurers, enforcers and media is also important. This can be achieved by providing abridged copies of minutes or a short biannual newsletter informing them of the DEM policy and any changes, as well as other OSH issues. *COMAH* sites will have to demonstrate as a legal requirement that plans and policies are up-to-date and effective.

Competence

[D6046] Competence is a process of acquiring knowledge, skill and experience to enhance both individual and corporate response. It is about enhancing and achieving standards set by the organisation or others.

- Competence is important. Certain key people should be trained to understand the DEMS process. The above issues and their link to OSH in particular require personnel that can assimilate, digest and convey them. Training does not necessarily mean formal or academic training. It can be vocational or in-house. Neither does it mean the organisation spending vast sums. It can be a part of a wider in-house OSH awareness programme, one day each quarter for example.
- Larger organisations may be able to recruit competent persons to advise on OSH and fatal incident matters.
- In short, all employees need to be brought up to a minimum standard. *Piper Alpha* showed that while Occidental Petroleum had detailed procedures, the employees generally did not fully understand them and did not have an adequate understanding of major incident evacuation. The organisation had failed to impart sufficient knowledge, skill and experience to cope with fires and explosions on off-shore sites.

Control

[D6047] Control refers to establishing parameters, constraints and limits on behaviour and action. This ensures that on the one hand an effective DEM policy exists and on the other, personnel will act and react in a co-ordinated and responsible manner. Controls can be achieved through, for example:

- Contractual means – as a term of a contract of employment that instruction and direction on OSH and fatal incident matters must be followed by individuals.
- Corporate means – the organisation continuously makes individuals aware of following rules and best practice.
- Behavioural means – by establishing clear rules, training, supply of information, leading-through-example and showing top-level management commitment, for example.
- Supervisory means – ensuring supervisors monitor employee safety attitudes and risk perceptions.

The cumulative effect of the 4 C's should be a positive and proactive culture in which OSH issues and the DEM policy issues are understood. Factors that can mitigate against or prevent a positive and proactive culture developing include a lack of management commitment, lack of awareness of requirements, poor attitudes, misperception of the risk facing the organisation, lack of resources or the unwillingness to commit resources and fatalistic beliefs, for example.

Disaster and emergency planning

[D6048] Planning is a process of identifying a clear goal and objectives and pursuing the best means to achieve them. Although 'disaster' and 'emergency' are two separate but related terms, in the case of planning, the two need to be viewed together. This is because in practice the serious event (the disaster) cannot be divorced from the response to that serious event (emergency).

Disaster and emergency planning can be viewed in three broad stages:

- stage 1: before the event;
- stage 2: factors to consider during the event; and
- stage 3: after the event.

Stage 1: before the event

[D6049] Once the DEM policy has been established and a culture created where the policy has been understood and positively received, a 'state-of-preparedness' should be established, addressing issues, speculating on scenarios and developing support services if and when a major incident does strike.

The risk assessment of major incident potential and consequent contingency planning

[D6050] What is the probability of the major incident resulting? What would be the severity? What type of major incident would it be?

Figure 3 depicts a 'major incident matrix':

Figure 3: major incident matrix

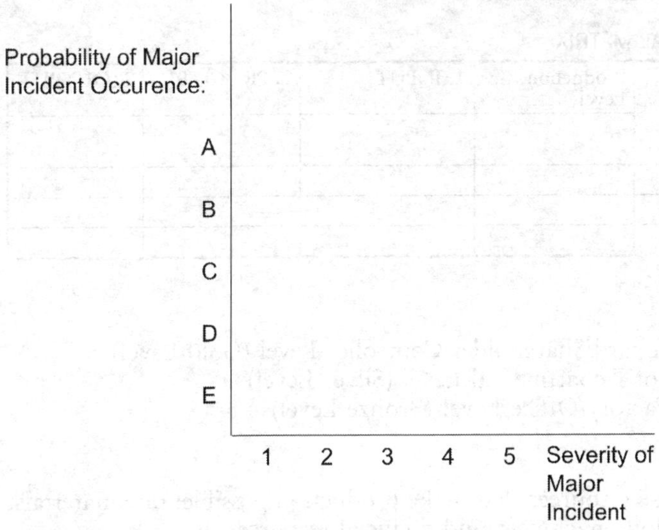

Probability of Major Incident Occurence:

A
B
C
D
E

1 2 3 4 5   Severity of Major Incident

Where:
A = Certainty of Occurrence (Probability = 1)
B = Highly Probable
C = 0.5 Probability of Occurrence
D = Low Probability

E= Most Unlikely

And:

1= Serious Injuries
2= Multiple Fatalities
3= Multiple Fatalities and Serious Economic/Property Damage
4= Fatal Environmental, Bio-Sphere and Social Harm
5= Socio-Physical Catastrophe

The planning responses will differ according to the particular cell. For example, cell 1E requires the least complex state of preparedness in terms of resources, detail of planning and urgency. 5A is a major societal event that requires macro and holistic responses, in which case the organisation's sole efforts at planning are futile. The organisation needs to assess which cell the major incidents it will confront will mainly fall in and develop plans accordingly. A conglomerate or an 'exposed operation' such as oil or gas refinery work may require planning at all 25 levels. This is 'contingency planning' – an analysis of the alternative outcomes and the provision of adequate responses to those outcomes.

Decisional planning

[D6051] Planning before the event also involves co-ordinating different levels of decision making.

Figure 4 depicts a decisional matrix:

FIGURE 4: DECISIONAL MATRIX

	Production Level	1. INPUT	2. PROCESS	3. OUTPUT
Management Level				
A. STRATEGIC				
B. OPERATIONAL				
C. TACTICAL				

Where:

A = Strategic or Board/Shareholder/Controller Level (Gold Level)
B = Operational or Departmental Level (Silver Level)
C = Tactical or Factory/Office Level (Bronze Level)

and

1 = Inputs or those resources that make production possible: raw materials, people, information, machinery and financial resources.
2 = Process or the manner in which the inputs are arranged or combined to enable production: method of production.
3 = Output or the product or service that is generated.

Decisional planning requires major incident risks to be identified for each cell. Essentially this is assessing where in the production process and at which management level problems can arise that can lead to major incidents. This matrix is not 'closed' but is 'open' and 'dynamic'. External factors, discussed above, as well as internal factors, need to be accounted for. For example, in a commercial operation to the board must consider:

- INPUTS: purchasing policy of any raw materials used (including safety), recruitment of stable staff, adequacy of resources to enable safe production, known and foreseeable risks of inputs used, reliability of supply, commitment to environmental safety and protection for example.
- PROCESS: external factors and their impact on production, production safety commitment, commitment to researching/investigating in the market for safer production processes for example.
- OUTPUT: boardroom commitment to quality assurance, environmental safety of the service and/or product produced, customer care and social responsibility for example.

The end result should be in each cell, major risks to production and their impact on a major incident are identified. This need not be a major, time consuming exercise. It can be determined through a brain-storming session or each management level completes the matrix during their regular meeting and it is then jointly co-ordinated in a short (2–5 page) document.

Testing

[D6052] This is a crucial aspect of pre-planning by all the emergency services. Testing is an objective rehearsal to examine the state of preparedness and to determine if the policy and the plan will perform as expected.

Training and exercising are both types of testing. Training is more personnel focused, aiming to assess how much the human resource knows about the policy and plan and means of enhancing the knowledge, skill and experience of that resource. Exercising is a broader approach, examining all aspects of the policy and plan, not just the human response, but also physical and organisational capability to deal with the major incident.

Training

[D6053] Carrying out a training needs analysis (TNA) to determine the quantity (in terms of time) and quality of training needed is essential. It may be that personnel and others adequately understand the policy and plan, therefore the training response should be proportionate. Excessive training is a motivational threat to interest and enthusiasm as is under-training. A TNA will consist of issues such as:

- Is the training 'necessary'? In other words, is there an alternative way of ensuring competence and awareness of the policy and plan, rather than just training? Could circulating the information be better? What about regular 'tool box talks'? It is important to weigh up the costs and benefits of each option.
- Is it 'needed'? Will the training meet personal, job or organisational needs thereby enhancing awareness and appreciation of the policy and plan?
- What will be the intended learning objectives of the training? Will it be 'this is a course on the content of the DEM policy and plan'?

- What type of training will be offered? Classroom, on-site or both? What are the costs and benefits associated with each?
- Management of the training – where will the training be conducted? Who will deliver it? When will it be delivered – day or evening? What training aids will be provided – if any?
- How will the effectiveness of the training be measured?

Exercising

[D6054] The Cabinet Office (www.gov.uk/guidance/emergency-planning-and-preparedness-exercises-and-training) cites three types of exercising:

- Discussion-based
  — a broad, brain-storming session assessing and analysing the efficacy of the policy and plan,
  — seminars need to be inclusive, bringing staff and others together in order to co-ordinate strategic, operational and tactical issues.
- Table-top exercises
  This is an attempt to identify visually, using a model of the production or office site and surrounding areas, what types of problems could arise, such as access/egress, crowd control and logistics for example.
- Live exercises
  This is a rehearsal of the major incident, actively and pro-actively testing the responses of the individual, organisation and possibly the community, including the emergency services and media for example. London Underground, the railway sector and the civil aviation sector tend to use such an approach.

There are costs and benefits associated with each of the approaches and a fourth category combines elements of all three. Training and exercising are both necessary to ensure human resources understand the policy and plan and ensure problems can be identified and competent responses developed.

HSE guidance to the *Control of Major Accident Hazards Regulations 2015 (SI 2015 No 483)* (L111) provides guidance on using table-top exercises, 'control post' exercises, which assess the physical and geographical posting of the personnel and emergency services during a major incident, and live exercises in relation to testing emergency plans.

Business continuity

[D6055] 'Business continuity' is a planning exercise to ensure critical facilities, processes and functions are operational and available during and immediately after a major incident, thereby enabling the organisation to function commercially and socially.

The Business Continuity Institute definition (2001) is as follows: 'A holistic management process that identifies potential impacts that threaten an organisation and provides a framework for building resilience with the capability for an effective response that safeguards the interests of its key stakeholders, reputation, brand and value-creating activities'.

There are many approaches to business continuity. Highlighted below is a 'generic approach'.

(1) *Generic approach*
Business continuity is a forecasting process of recovery, assessment and ensuring the adequacy of resources for the organisation to continue its operation.
  (a) Recovery
Recovery is a state of regaining or salvaging assets that otherwise would have been permanently lost. In the event of a major incident, the recovery phase needs to focus on:
    (i) human resource recovery:
- ensuring personnel are safely evacuated;
- others, such as lawful visitors and trespassers, are evacuated; and
- all persons in general are removed in the quickest and most practicable means;

    (ii) information resource recovery:
- essential documents and software; and
- private and confidential documents;

    (iii) physical resource recovery:
- primary electrical and mechanical facilities such as power supply, water and gas services; and
- essential work equipment, if possible and moveable;

    (iv) financial resource recovery; and
    (v) valuable assets recovery (if practicable).

The organisation must rate the recovery potential – how possible is recovery? This could be rated from 'certain' (1) through to 'not possible' (5). This calculation requires different recovery responses; if and when a major incident strikes, those on the scene need to decide if the rating is 1 or 5. Accordingly, the response will vary – if 5, there is little purpose risking life and resources to salvage assets.

  (b) Assessment
Upon the immediate recovery, the organisation must ask how much damage (deterioration, harm or erosion of value) has been inflicted to the operation by the major incident?
Damage assessment is a three-stage activity:
    (i) identify the type of damage:
- human resource damage: physical and/or behavioural;
- informational resource damage: primary, secondary and tertiary documentation;
- physical resource damage: work equipment, property, facilities, environment; and
- financial resource damage: cash, art, etc;

    (ii) identify severity of damage:
this can be a qualitative scale which says 'low' or 'high' or a quantitative scale, where the damage is rated on a scale from 1 to 5 for example. The severity needs to be forecasted for human, informational, physical and financial resources:

- human resources could be scaled from serious injury (1) to multiple fatality (5);
- informational: from minor harm (1) through to destroyed (5);
- physical: from reparable (1) to unsalvageable (5); and
- financial: from no impact on cash flow (1) to financial ruin (5).

(iii) consequence of damage:
how much does the damage affect the chance of the operation being resumed immediately or in the next few days? The longer it will take to resume, the worse the consequence. Consequence could also be rated in quantitative terms, such as a 1 for 'resume immediately' so the harm has been minimal through to 5 for 'resumption will take months' for example. The consequence needs to be examined in each case:
- human: how long will it take people to get back to work?
- informational: how long will it take for manual and electronic systems to be operational?
- physical: how long will it take for necessary facilities to be operational?
- financial: how long will it take for cash flow to become positive or to access financial facilities?

severity multiplied by consequence will give an index of forecasted potential loss, which can then be assessed as a spreadsheet over time. Loss adjustors use various detailed statistical models based upon these generic principles.

(c) Adequacy of resources

Once the organisation has forecast the potential for recovery and hypothesised the damage assessment, it needs to ensure it will have adequate resources to carry on operating in light of the resource losses identified by the assessment. Adequacy needs to consider:

(i) human resources:
- adequacy of competent and trained personnel at strategic, operational and tactical levels of the organisation;
- availability of key advisory support services such as lawyer and accountant for example; and
- if the major incident is classified as a 'crisis', then there is a chance some of the key personnel are not available. Pre-planning therefore requires liaison with recruitment and selection specialists;

(ii) physical resources:
- telephone, email, mobile connectivity and reliability;
- furniture and fittings;
- stationery;

- work equipment including computers and filing cabinets for example;
- working stock;
- working space;
- vehicles;
- safety equipment including personal protective equipment;

(iii) informational resources:
- legal documents (the organisation's certificates of incorporation and insurance liability certificates for example);
- personnel documents (such as PAYE, NIC and personnel records);
- financial documents (availability of bank books for example); and
- sales/marketing documents – this is at the heart of the operation and there will be a need to develop databases, contact potential customers and re-establish commercial functionality;

(iv) financial resources:
- adequacy of working capital.

See also **BUSINESS CONTINUITY**.

Stage 2: during the event

[D6056]

- Activate disaster and emergency policy and pre-planning
  Major incident response and management ought to be clear and effective if: the previous stages in the DEMS have been followed; external and internal factors and uncertainties have been accounted for; a chain of command as well as arrangements to cope have been developed; and the operation to cope with a major incident has been organised and pre-planned if the event actually arises. In the main, activating the arrangements made is the central action.
- Co-ordination
  This is the central action in ensuring efforts are in unison. This includes both liaising with external bodies such as the emergency services and media, and internally at strategic, operational and tactical levels as discussed above. The police will normally have overall command and control, so their guidance must be followed.
- Code of conduct
  A code of conduct is a set of 'golden rules' that staff and contractors need to follow if a major incident arises. This includes:
  (i) following instruction from those with command and control responsibilities;
  (ii) using initiative;
  (iii) facilities management (gas, electricity, water) – making safe, when and when not to switch on or off;

[D6056] Disaster and Emergency Management Systems (DEMS)

(iv) medical assistance – liaison with health service, first-aiders, when and when not to administer;
(v) rules of evacuation;
(vi) site access and egress;
(vii) site security;
(viii) work and rescue equipment – its safe use and logistics of use;
(ix) record keeping of the major incident (audio-visual, verbal and written);
(x) resources needed for effective major incident management;
(xi) media and public relations management; and
(xii) issues of care and compassion when dealing with injured people.

The code of conduct is a reflection of the disaster and emergency policy and the pre-planning. It should be reinforced verbally and in writing (one page of A4) and reiterated to the major incident team before and during the major incident. While this seems bureaucratic, if the core team (internal and external) fail to follow best practice and safeguard themselves, this increases the risk factor of the major incident and could even lead to a double tragedy. A five minute or so reiteration is a minor time cost.

Stage 3: after the event

[D6057]–[D6058] Major incident planning does not cease as soon as the major incident is physically over. There are continuous issues over time that need to be addressed (see below).

I. Immediate term (immediately after the major incident and within a few days)

Statutory investigation

[D6059] This will involve the HSE, local authority, one of the UK environment agencies and the police for example. These bodies have powers and duties to investigate a major incident.

- The organisation investigated must fully co-operate with these bodies and afford any assistance they require.
- There must be no removal or tampering with anything at the major incident site. They may require forensic or other data gathering.
- There must be provision for all documentation and information to be provided to these bodies.

Business continuity

[D6060]–[D6062] Implement plans (see above).

Insurers/loss adjusters

Assuming the insurance contract covers direct and consequential damage from a major incident (and not all will), the organisation needs to notify the insurer and ensure all paperwork is completed promptly. Most insurance contracts stipulate a time limit by which the paperwork has to be lodged with the insurer.

A major incident will divert attention from other issues, so ensuring that a person is appointed to carry out this insurance task is vital. This should be the organisation's lawyer.

The insurer in turn will notify their loss adjusters to investigate the basic and underlying causes of the event. Again, full disclosure and co-operation are implied insurance contractual requirements. The organisation must check all documents the loss adjusters complete to ensure they are accurate and cover all aspects of the event.

The police

If criminal neglect is suspected to have played a role in the event, the police will need to interview all core board members, senior management and others.

Building contractors

The organisation needs to plan for building contractors to visit the site and make it safe and secure. This may have to be done straight after the event, even if insurance issues have not been resolved. Adequate resources are therefore vital.

Visitors

Major incidents also result in public and media visits. The arrangements for handling these groups must extend to after the event.

Counselling support

Counselling support to affected employees and possibly contractors is not just a personnel management requirement showing 'caring management', but increasingly a legal duty of the organisation to provide such support. Issues of post-traumatic stress, nervous shock and bereavement mean the organisation has common law obligations to offer medical and psychological support. This should involve a medical practitioner and occupational nurse. The insurance policy can be extended before the event to cover the cost of such services.

II Short term (a week onwards after the event)

- Investigation and inquiry
  This can be both a statutory inquiry (although most are called within days) and/or the in-house investigation of the event and lessons to be learnt. Issues to consider include:
    — Basic causes: was it a fire, bomb, an explosion or a natural event?
    — Underlying causes: what led up to such an incident occurring? Examine the managerial, personnel, legal, technical, organisational and natural factors that could have caused the event.
    — Costs and losses involved.
    — Lessons for the future.
    — Did the disaster and emergency plan operate as expected? Were there any failings? What improvements are required? There needs to be complete debriefing and examination of the entire process involving internal personnel and external agencies.

[D6060] Disaster and Emergency Management Systems (DEMS)

> The organisation should weigh up the possibility of external people carrying out this exercise and consider whether in-house staff are objective and dispassionate enough to assess what went wrong.
> - Visit by enforcers
> The organisation also needs to be prepared for further visits from the enforcement authorities and the possibility of statutory enforcement notices being served to either regulate or prohibit the activity. Multiple notices are possible, from health and safety officers, fire authority, environment agencies, building control or planning officers. This will affect the operation, production process and have economic implications. This ought not to occur if the organisation has taken due care to pre-plan for the major incident and had continuous safety monitoring of the operation. Notices will be served if there is a failure to make the site safe.
> - Coping with speculation
> The public, media and employees will speculate about causation and there is the risk of adverse publicity. Public relations is a central activity. For legal and moral reasons, it is best practice to disclose all known facts unless the statutory investigation prohibits this.

III Medium term (a month plus)

[D6063]–[D6064] The 'normalisation process' will begin. The organisation needs to carry on the operation, be prepared for further visits from enforcement agencies and loss adjusters and reassess its corporate/financial health.

IV. Long term (six months onwards)

- Systems review
  — A review of the impact of the major incident and whether the organisation is recovering from it.
  — Impact on reputation.
  — Legal threats.
  — Any positive outcomes – learning from mistakes, improving technical know-how and wider industrial benefits from knowing the chain reaction of events for example.

V. Longer term (one year onwards)

[D6065] The organisation's memory and experience must be:

- included in in-house training programmes;
- factored into systems and procedures;
- included in review of the entire *espirit de corps* and corporate philosophy.

Disaster and emergency planning is an extensive exercise, being dynamic and accounting for a diverse range of phenomena as industrial, man-made, environmental, socio-technical, radiological and natural events.

Monitor the disaster and emergency plan(s)

[D6066] Monitoring is a process assessing and evaluating the value, efficiency and robustness of the disaster and emergency plan(s). This involves looking at all three stages as discussed above (before, during and after the event stages) and not just the core document, 'the plan'. It is comprehensive and holistic in its questioning of the entire planning process.

Monitoring can be classified as proactive or reactive. The former attempts to identify problems with the plan(s) before the advent of a major incident. It is a case of continuously comparing the plan(s) with even minor incidents and loop-holes identified in any training/exercising sessions. Reactive monitoring occurs after a major incident or occurrences that could have led up to a major incident, thereby reflecting back and assessing if the plan(s) need improvement. Both types are important.

FIGURE 5: MONITORING MATRIX

Stages in Planning	1. Before the Major Incident	2. During the Major Incident	3. After the Major Incident
Monitoring Types			
A. PROACTIVE	• Risk Assessments	• Inspections	• Inquiries
	• Testing via Training or Exercises	• Live Interviews	• Systems Review
	• Major Incident Assessment	• Feedback	• Counselling Reports
	• Facilities Inspections		
B. REACTIVE	• Incident Statistics	• Critical Assessments	• Brainstorming
	• Incident Reports	• Incident Levels	• Loss Assessments
	• Warnings	• Audio-Visual Assessment	• Enforcement
	• Notices		

The above cells are not strictly mutually exclusive. Many techniques are both proactive and reactive. A third dimension is added in *COMAH* cases, that of internal and external emergency plans (types of planning).

- Cell A1 (proactive before the major incident)
    — Risk assessments and hazard analysis techniques will identify significant hazards and their risk level. Monitoring such risk is an index of danger, which in turn is a variable in the type and potential of the major incident, outlined in Figure 1. The HSE's Five Steps approach (Identify the hazards; Decide who might be harmed and how; Evaluate the risks and decide on precautions; Record your significant findings; and Review your assessment and update if necessary) or Quantified Risk Assessment methodology provide outlines of assessing risk. Hazard techniques include hazard and operability studies, HAZard ANalysis (HAZANS) and fault and event tree analysis for example.

- Testing will identify any problems or concerns with the disaster and emergency plan. For example, a live exercise or a synthetic simulation could identify factors the plan has not considered, or which may not be practicable if a major incident was to arise. Enabling questioning, critical appraisal and comments should not be perceived as a threat but can instead provide vital information.
- Major incident assessment is a periodic overview of the plan(s), every quarter or biannually, by both internal and external persons. This could identify areas of concern. This assessment compares the plan(s) with the potential threat – can the former cope with the threat? Threats change as technology and know-how change, so such assessments become another vital source of information.
- Facilities inspections of gas, electricity, water, building structure and equipment available/not available, for example, can highlight issues of physical resourcing and adequacy of such resourcing.

- Cell B1 (reactive after the major incident)
  - Incident statistics will show the type of incidences, the type of injury, when and where they occurred. This enables the organisation to hypothesise/build a picture of the potential and severity of a bigger incident. Occupational health and safety data cannot be divorced from disaster and emergency management.
  - Analysing incident reports should enable issues of causation to be assessed. What type of occurrence could trigger a major incident? Identifying and developing a pattern of causes will enable an assessment of whether the plan(s) account for such causes.
  - Warnings from employees, contractors, enforcers, the public and others of possible and serious problems are to be treated seriously. All such warnings should be analysed and a common pattern and trend spotted.
  - Any notices served by enforcement authorities will identify failings in the operation and the remedial actions required. These can be factored into the plan(s).

- Cell A2 (proactive during the major incident)
  - Inspections will be made even as the event occurs. Inspections can range from the stability of the structure through to how personnel behaved and coped. As these inspections are made, the command and control team should evaluate whether any aspect of the plan(s), which is a live document, needs immediate changes.
  - Live interviews with internal personnel and emergency services personnel will enable a continuous appraisal of any difficulty with procedures, arrangements and instructions that emanate from the plan(s). Again, these can lead to immediate changes to the plan(s).

- Feedback is a proactive technique of requesting regular information on and off-site. This enables a picture to be constructed of what could happen next, trying to anticipate the next sequence and if the plan(s) can cope.
- Cell B2 (reactive during the major incident)
  - Critical assessments are carried out after some unexpected occurrence, which causes uncertainty and may even threaten the efficacy of the plan(s). The critical assessment is by the command and control team as a whole. Why did this happen? Why did we not account for it in the plan(s)?
  - Incident levels – in particular if serious injury or fatalities are increasing, then at a moral or philosophical level the organisation needs to ask whether the plan(s) have been overwhelmed by reality. All forms of planning, including the statutory *COMAH* planning must not be viewed with rigidity. If the plan(s) are failing, it is better to reappraise and replan. The *Control of Major Accident Hazards Regulations 2015 (SI 2015 No 483)* do not overtly allow for this, although they stress flexibility and continuous appraisal of the event. In such a case, there has to be quick and clear decision-making, with consequential command and control, as well as immediate communication of the alternative plan. Training and exercising sessions need to factor in this dimension and equip people with decisional techniques.
  - Audio-visual assessments can be a dramatic means of understanding the actual event. This can be video or photographic footage shot by the incident personnel or emergency services. This enables monitoring of the extent, potential and actual threat from the major incident.
- A3 (proactive after the major incident)
  - Inquiries are proactive. Even though the event has happened, the inquiry (whether internal or external) will identify strengths and weaknesses in the plan(s), which can lead to future improvements in planning.
  - Systems review is an overhaul of the entire reaction and holistic experience of the organisation to the trauma of a major incident. This involves developing future coping strategies for personnel and issues of how well the organisation responded. Were adequate resources in place to cope?
  - Counselling reports will identify the experiences, perceptual and cognitive issues that affected personnel and others. This can provide probably the most significant information on behavioural response of the command and control team and those who were injured. This in turn can be factored into training and exercise programmes, which will lead to personnel skill improvements.
- Cell B3 (reactive after the major incident)

— Brainstorming is an open-ended, participative and indeed critical analysis of what went wrong and what was right with the plan(s). Brainstorming should also be inclusive, involving emergency services and possibly enforcement authorities as well, so their guidance is factored in.
— The loss adjusters report will be a vital document as to the chain reaction that lead to the event and the consequences that followed. For legal reasons, their findings may have to be applied before insurance cover is available.
— Enforcement notices and enforcement agency reports will contain recommendations, which need to be viewed as lessons for the future.

Audit and review

[D6067] The final stage of the DEMS process is audit and review. An audit is a comprehensive and holistic examination of the entire DEMS process (Figure 2, see **D6034**). An audit will identify stages in this process that need improvement. A review is an act of 'zooming in' into that particular stage and carrying out those improvements.

Major incident auditing

[D6068]

- Major incident auditing can be qualitative or quantitative. The former adopts a 'yes' or 'no' response format to questions. The latter asks the auditor to rate the issue being examined from 0–5 for example.
- The audit must be comprehensive, assessing every aspect of the DEMS process. This means the audit will take time to be completed.
- Audits essentially benchmark (compare and contrast) performance. This can be against the DEMS process identified above or against legislation (eg *Control of Major Accident Hazards Regulations 2015 (SI 2015 No 483)*). The benchmarking could also be against another site or wider industry standards.
- Should audits be carried out in-house or rely on external consultants? There are costs and benefits associated with both, with no definitive answer. The *Management of Health and Safety at Work Regulations 1999 (SI 1999 No 3242)* in the UK, emphasises the need to develop and use in-house expertise in relation to OSH issues in general, with a reliance on external specialists as a last resort. This may be interpreted as best practice for DEM.
- Audits can be annual or biannual. The more complex the operation and risk it poses, the greater the need for biannual audits.
- Audits should be proactive, that is learning from weaknesses in the DEMS process and reducing or eliminating any weaknesses for the future.

- Finally, the results of the audit need to be fed back into the DEMS process and all those affected informed of any changes and risk management issues arising.
- Audits are holistic (assess anything associated with major incidents), systemic (assess the entire DEMS process, ie the 'system') and systematic (that is logical and sequential in analysis).

Review
[D6069]

(1) The review of any specific problems needs to be actioned by the organisation. The consultant will identify the areas of concern and make recommendations but the final discussion and implementation lies with the organisation. This needs to be led by senior officials in the organisation.

(2) Reviews are by definition 'diagnostic', meaning the organisation needs to look at causation and cure of the failure in any part of the DEMS process.

(3) Budgeting both in time and resource terms is critical in the review, as it will require management and external agency involvement.

(4) A review can be carried out at the same time as an audit. A review can also be a legal requirement, as with *Control of Major Accident Hazards Regulations 2015 (SI 2015 No 483)*, Regs 12 and 14, which require a review, and where necessary, a revision of the internal and external emergency plans for upper-tier establishments.

Conclusion to disaster and emergency management systems (DEMS)

[D6070] All organisations and societies need to prepare for worst-case scenarios. DEMS provide a logical framework to understand the main stages in effective preparation. The larger the organisation, the more detailed and analytical the preparation needs to be. DEMS is a live and open system requiring continuous monitoring.

Sources of information on disaster and emergency management systems (DEMS)

[D6071] These include:

(1) Legislation including:
- Civil Contingencies Act 2004
- Control of Major Accident Hazards Regulations 2015 (SI 2015 No 483)
- Control of Substances Hazardous to Health Regulations 2002 (SI 2002 No 2677)
- Control of Asbestos Regulations 2012 (SI 2012 No 632)

- Radiation (Emergency Preparedness and Public Information) Regulations 2001 (SI 2001 No 2975)
- Public Information for Radiation Emergencies Regulations 1992 (SI 1992 No 2997)
- Health and Safety (First-Aid) Regulations 1981 (SI 1981 No 917)
- Reporting of Injuries, Diseases and Dangerous Occurrences Regulations 2013 (SI 2013 No 1471)
- The Confined Spaces Regulations 1997 (SI 1997 No 1713)
- Carriage of Dangerous Goods and Use of Transportable Pressure Equipment Regulations 2009 (CDG 2009) (SI 2009 No 1348)

(2) Guidance
- Emergency Preparedness (statutory guidance)
- Emergency Response and Recovery (non-statutory guidance)
  These are available on the government website at: www.gov.uk/government/publications/emergency-preparedness and www.gov.uk/government/publications/emergency-response-and-recovery.
- A guide to the Control of Major Accident Hazards Regulations (COMAH) 2015 (HSE L111).
- Guide to the Radiation (Emergency Preparedness and Public Information) Regulations 2001 (HSE L126).
- The Health and Safety (First-Aid) Regulations 1981. Guidance on Regulation (L74), HSE Books, (2013) www.hse.gov.uk/pubns/books/l74.htm.

(3) The European Commission
DG Environment is the department responsible for environmental protection (ec.europa.eu/dgs/environment/index_en.htm), DG Energy deals with nuclear safety (https://ec.europa.eu/energy/en/home) and European civil protection and humanitarian aid operations deals with civil protection (ec.europa.eu/echo/index_en).

(4) Professional bodies such as:
- International Institute of Risk and Safety Management www.iirsm.org
- Institution of Occupational Safety and Health www.iosh.co.uk
- Business Continuity Institute www.thebci.org
- Fire Protection Association www.thefpa.co.uk

Display Screen Equipment

Andrea Oates

Introduction to display screen equipment

[D8201] The *Health and Safety (Display Screen Equipment) Regulations 1992 (SI 1992 No 2792)* ('the *DSE Regulations*') came into operation on 1 January 1993 as part of the so-called six pack of health and safety regulations which originated from the European Union (EU) 'framework' and 'daughter' directives. The aim of the DSE Regulations is to combat musculoskeletal disorders (MSDs) in the upper limbs (also known as repetitive strain injuries or RSI), eye and eyesight effects, together with general fatigue and stress associated with work at display screen equipment ('DSE').

In the 1980s, there was considerable public concern about whether electromagnetic radiation emissions from DSE were harmful, particularly in the early stages of pregnancy. This followed a number of reports of higher levels of birth defects and miscarriages among some groups of DSE workers. The Health and Safety Executive (HSE) states in its guidance to the Regulations, *Work with display screen equipment* (L26), that it does not consider there are any radiation risks from working with DSE or special problems for pregnant women.

It says that, taken as a whole, research has not shown any link between miscarriages or birth defects and work on DSE. It advises that if a woman is anxious about DSE work, or about work generally during pregnancy, they should be given the opportunity to talk to someone who is adequately informed of current scientific information and advice. The HSE also provides answers to frequently asked questions (FAQs) about new and expectant mothers on its website (www.hse.gov.uk/mothers/faqs.htm).

The Regulations require 'a suitable and sufficient analysis', to be carried out to identify any health and safety risks at workstations used by 'users' or 'operators', as defined in the Regulations (see D8202 below). In addition, 'users' have the right to an eye and eyesight test and any spectacles found to be necessary for their DSE work. These must be provided by their employer. The requirements of the Regulations are summarised and then set out in more detail in the following sections.

The Regulations summarised

Definitions

[D8202] *Regulation 1* of the *DSE Regulations (SI 1992 No 2792)* contains a number of important definitions which are summarised below:

- Display screen equipment
  - Any alphanumeric or graphic display screen, regardless of the display process involved. A European Court of Justice (ECJ) case involving a film cutter, *Dietrich v Westdeutscher Runddfunk*, C-11/99 [2000] ECR I-5589 ECJ, ruled that the term 'graphic display screen' had to be interpreted to include screens displaying film recordings. Screens used in work with television or film pictures are therefore included.
- Workstation
  An assembly comprising:
  - display screen equipment (whether provided with software determining the interface between the equipment and its operator or user, a keyboard or any other input device)
  - any optional accessories to the display screen equipment
  - any disk drive, telephone, modem, printer, document holder, work chair, work desk, work surface or other item peripheral to the display screen equipment, and
  - the immediate work environment around the display screen equipment.
- User
  - An employee who habitually uses DSE as a significant part of his normal work.
- Operator
  - A self-employed person who habitually uses DSE as a significant part of his normal work.

HSE guidance to the regulations set out in L26 says the main factors to consider in determining whether a person is a 'user' or 'operator' are:

- normal use of DSE for continuous or near-continuous spells of an hour or more at a time;
- use of DSE in this way more or less daily;
- the need to transfer information quickly to or from the DSE.

Other factors include:

- high levels of attention and concentration are required;
- high dependence on DSE with little choice about using it;
- special training or skills are needed to use the DSE.

Detailed examples are given in the HSE guidance booklet providing pen-portraits of definite users, possible users and those who are definitely not users. Examples of the type of employees likely to be classed as users include word-processing workers, secretaries and typists, data input operators and journalists. Those who are possibly users include scientists and technical advisers, client managers in large management accountancy consultancies, building society customer support officers, airline check-in clerks, community care workers and receptionists. Those who would not be users include, for example, senior managers who use DSE only for occasional monitoring purposes.

Homeworkers, teleworkers and agency workers all come within the scope of the Regulations, if they fulfil the definition of user or operator – see **D8213** below.

Exclusions

[D8203] The *DSE Regulations (SI 1992 No 2792)* do not apply to:

- drivers' cabs or control cabs for vehicles or machinery;
- DSE on board a means of transport;
- DSE mainly intended for public operation;
- portable systems not in prolonged use;
- calculators, cash registers or other equipment with small displays;
- window typewriters.

However, the *Health and Safety at Work etc Act 1974* and regulations made under it, such as the *Workplace (Health, Safety and Welfare) Regulations 1992 (SI 1992 No 3004)*, still apply to the use of such equipment. The HSE advises that particular attention should be paid to ergonomics in this context – that is the science of making sure that work tasks, equipment, information and the working environment are suitable for every worker, so that work can be done safely and productively.

Assessment of workstations and reduction of risk

[D8204] *Regulation 2* of the *DSE Regulations (SI 1992 No 2792)* requires employers to perform a 'suitable and sufficient' analysis of all workstations which:

- (regardless of who has provided them) are used for their purposes by 'users';
- have been provided by them and are used for their purpose by 'operators';

to assess the health and safety risks in consequence of that use. They must then reduce the risks identified to the lowest level reasonably practicable. As for other types of assessments, an assessment must be reviewed if there is reason to suspect it is no longer valid or there have been significant changes.

Further guidance on assessment, including an assessment checklist (at **D8217**), is provided later in the chapter.

Requirements for workstations

[D8205] *Regulation 3* of the *DSE Regulations (SI 1992 No 2792)* states that workstations must meet the requirements laid down in the *Schedule* to the

Regulations. Many of the requirements of the Schedule have been incorporated into the assessment checklist (at **D8217**). The full Schedule is contained in L26. The requirement must relate to a component present in the workstation concerned and be relevant in relation to the health, safety and welfare of workers. For example, there is no need to provide a document holder (referred to in the *Schedule*) if there is little or no inputting from documents. L26 provides further examples of where the requirements of the *Schedule* may not be appropriate.

Daily work routine of users

[D8206] Employers are required by *Regulation 4* of the *DSE Regulations (SI 1992 No 2792)* to plan the activities of 'users' so that their DSE work is periodically interrupted by breaks or changes of activity to reduce their workload on DSE. Breaks should be taken before the onset of fatigue and preferably away from the screen. Short frequent breaks are better than occasional longer breaks. It is best if users are given some discretion in planning their work and are able to arrange breaks informally rather than having formal breaks at regular intervals. Some software tools provide a means of ensuring that users take regular breaks but HSE guidance draws attention to the limitations of such software.

In terms of the frequency of breaks, the HSE advises that 5–10-minute breaks every hour are better than 20 minutes every 2 hours and it says that ideally, users should have some choice about when to take breaks.

Eyes and eyesight

[D8207] *Regulation 5* of the *DSE Regulations (SI 1992 No 2792)* requires that where a user (or a person who is to become a user) requests an appropriate eye and eyesight test, the employer must ensure that they have a test carried out by a competent person as soon as practicable after the request or, in the case of someone who is to become a user, before they become a user and at regular intervals thereafter.

The HSE guidance to the regulations says that employers should be guided by the clinical judgement of the optometrist or doctor on the regularity of testing and trade union advice is generally for repeat testing at specified intervals, such as two years, or on the advice of the optometrist.

'Users' are also entitled to tests on experiencing visual difficulties which may reasonably be considered to be caused by DSE work (on request).

Tests are normally carried out by opticians and involve a test of vision and an examination of the eye. Where companies have vision screening facilities, 'users' may opt for a screening test to see if a full eye test is needed. The HSE makes clear that vision screening tests are not an 'eye and eyesight test' and do not therefore satisfy the DSE Regulations, but it says that some employers may wish to offer them as an extra.

The Regulations set out that the employer should ensure that each user employed by them is provided with "special corrective appliances" appropri-

The Regulations summarised [D8209]

ate for the work being done where "normal corrective appliances" cannot be used where the result of the eye and eyesight test shows that this is necessary.

The HSE guidance to the Regulations sets out that: "'Special' corrective appliances (normally spectacles) provided to meet the requirements of the DSE Regulations will be those appliances prescribed to correct vision defects at the viewing distance or distances used specifically for the display screen work concerned. 'Normal' corrective appliances are spectacles prescribed for any other purpose."

The guidance also makes clear that the cost of the eye and eyesight tests and special corrective appliances is borne by the user's employer, even if the user works on other employers' workstations. However, they only have to pay for a basic appliance that is of a type and quality adequate for the user's work. Employees usually pay for extras if they want designer frames or tinted lenses for example.

Provision of training

[D8208] Under Regulation 6 of the *DSE Regulations (SI 1992 No 2792)* the employer must provide users with adequate health and safety training in the use of any workstation upon which they may be required to work. They must provide the training for those who are to become users before they become users. Training may also be required where workstations are substantially modified. The training should include:

- the causes of DSE-related problems, eg poor posture, screen reflections;
- the user's role in detecting and recognising risks;
- the importance of comfortable posture and postural change;
- equipment adjustment mechanisms, eg chairs, contrast, brightness;
- use and arrangement of workstation components;
- the need for regular screen cleaning;
- the need to take breaks and for changes of activity;
- arrangements for reporting problems with workstations, or ill health symptoms;
- information about the Regulations (especially eyesight tests and breaks); and
- the user's role in assessments.

The HSE has published a leaflet (*Working with display screen equipment (DSE) – A brief guide* (INDG36)) which provides a useful reference for training purposes.

Provision of information

[D8209] *Regulation 7 of the DSE Regulations (SI 1992 No 2792)* requires employers to ensure that operators and users at work within their undertaking are provided with adequate information. The table below shows the responsibility of the 'host' employer in this respect (and also gives a good guide to their responsibilities generally under the Regulations).

A Court of Appeal judgment found in favour of a worker whose repetitive strain injury (RSI) was made worse as a result of keyboard use. She was not required to make an excessive number of keystrokes and had opportunities during the working day to do other work away from the keyboard. However the judgment accepted that her work had aggravated the RSI. The Court found that the company involved had not complied with the DSE regulations, because having been informed of her condition, it provided no information or training or further reductions in her keyboard use (*Goodwin v Bennetts UK Ltd* [2008] EWCA Civ 1374, [2008] All ER (D) 220 (Dec)).

The 'host' employer must provide information about:	Regulation:	Own users:	Other users (ie agency staff):	Operators (self-employed):
DSE and workstation risks		YES	YES	YES
Risk assessment and reduction measures	2 and 3	YES	YES	YES
Breaks and activity changes	4	YES	YES	NO
Eye and eyesight tests	5	YES	NO	NO
Initial training	6(1)	YES	NO	NO
Training when workstation substantially modified	6(2)	YES	YES	NO

DSE workstation assessments

Decide who will carry out the assessments

[D8210] DSE workstation assessments within an organisation may be carried out by an individual or by members of an assessment team. Those responsible for carrying out assessments should have received appropriate training so that they are familiar with the requirements of the *DSE Regulations (SI 1992 No 2792)* and they should have the ability to:

- identify hazards (including less obvious ones) and assess risks from the workstation and the kind of DSE work being done;
- use additional sources of information or expertise as appropriate (recognising their own limitations);
- draw valid and reliable conclusions and identify steps to reduce risks;
- make a clear record and communicate the findings to those who need to take action; and
- recognise their own limitations so that further expertise can be utilised where necessary.

They may be health and safety specialists, or other in-house staff who have received appropriate training or who have the appropriate abilities to carry out assessments (see above). Safety representatives should also be involved in the risk assessment process. The HSE advises that safety reps should also be encouraged to report any problems in DSE work that come to their attention.

Identify the 'users'

[D8211] An important first step is to identify the users' (together with any 'operators') of DSE within the organisation. The definitions in *Regulation 1* of the *DSE Regulations (SI 1992 No 2792)* refer to habitual use of DSE. L26 provides considerable advice on the factors which must be taken into account, of which time spent using DSE is the most significant. Some organisations have adopted a rule of thumb that anyone spending more than 50 per cent of their time in DSE work is a DSE 'user'. However, the HSE guidance indicates that a less simplistic approach should be taken.

The assessment checklist provided at **D8217** includes reference to the other factors which the HSE says should be taken into account when deciding whether an individual is a 'user' or an 'operator'. It should be noted that employers have duties to assess workstations used for the purposes of their undertaking by *all* 'users' or 'operators'. This includes 'users' employed by others (eg agency-employed staff), 'operators' (eg self-employed draughtsmen or journalists), peripatetic staff (eg journalists, sales staff, careers advisors) and homeworkers or teleworkers (see **D8213** below).

In practice, most employers do not find it too difficult to decide who their 'users' and 'operators' are. Many have taken the approach of assessing the workstations of those individuals where doubt exists and, if necessary, making a final decision then. More time and expense can often be wasted debating a few borderline cases than would be involved in including them in the definition.

Decide on the assessment approach

[D8212] The HSE guidance to the regulations sets out the principal risks that may arise from DSE work relate to musculoskeletal problems, visual fatigue and mental stress. Ill health can result from poor equipment or furniture, work organisation, working environment, job design and posture, and from inappropriate working methods.

The known health problems associated with DSE work can be prevented in the majority of cases by good ergonomic design of the equipment, workplace and job, and by worker training and consultation.

Employers must assess the extent to which any of these risks arise for DSE workers using their workstations who are:

- users employed by them;
- users employed by others (for example agency employed 'temps'); or
- operators – self-employed contractors who would be classified as users if they were employees (for example self-employed agency 'temps', self-employed journalists).

It makes clear that individual workstations used by any of these workers must be analysed and the risks assessed. If employers require their employees to use workstations at home, these too will need to be assessed (see **D8213** below).

HSE guidance on assessing workstations sets out that users should be included in assessments. Their views are important and employees who are actively involved in the risk assessment process are also more likely to report any problems as they arise. Information from users can be through an ergonomic checklist, completed by users or with their input.

It also outlines other possible approaches. "For example more objective elements of the analysis (for example chair adjustability, keyboard characteristics, nature of work, etc) could be assessed generically in respect of particular types of equipment - or groups of workers performing the same tasks. Other aspects of workstations would still need to be assessed individually through information collected from users, but this could then be restricted to subjective factors (for example relating to comfort, adjustability of chairs, particularly where there is hot-desking, etc."

It advises that whatever type of checklist is used, employers should ensure that workers have received the necessary training before being asked to complete one.

A sample checklist, together with guidance on its completion, is provided at D8217. L26 provides a checklist covering similar topics and there is also a DSE workstation checklist on the HSE website at www.hse.gov.uk/pubns/ck1.htm.

Checklists should also be supported by other action including:

- an inspection of areas where DSE workstations are situated (evaluating general issues such as lighting, blinds, housekeeping, desk space and the standards of chairs and DSE hardware);
- providing employees with the option of an assessment of their individual workstation by a specialist;
- review of completed self-assessment checklists by a competent person;
- responding promptly to problems identified in completed assessment checklists.

Risk assessments should be reviewed if there is reason to suspect that it is no longer valid; or if there have been significant changes and the employer should implement any changes the review shows to be necessary (see **D8223**).

Homeworkers and teleworkers

[D8213] Homeworkers and teleworkers (working away from their employer's premises) are subject to the *DSE Regulations (SI 1992 No 2792)*, whether or not their workstation is provided by their employer. They face the same risks from DSE work, some of which may be increased because of their social isolation, the absence of supervision and the practical difficulties of carrying out risk assessments. Risks can be reduced by:

- training such staff to carry out workstation self-assessments (see above);
- requiring them to carry out a self-assessment of their main workstation;
- encouraging them to carry out ad hoc assessments of temporary workstations, eg hotel rooms;
- emphasising the importance of ensuring good posture and taking adequate breaks;

- providing clear communication routes for reporting equipment defects and possible health problems;
- responding promptly and effectively to reports of defects or problems.

Observations at the workstation

[D8214] Assessment checklist questions are intended to identify the principal factors that the assessor(s) need to look out for (as illustrated in the completed example).

Some of the more common problems identified are related to:

- Posture:
  — height of screen;
  — height of seat or position of backrest;
  — position of the keyboard or mouse;
  — keyboard and mouse technique;
  — need for footrest or document holder.
- Vision:
  — angle of screen;
  — position of lights or need for diffusers;
  — the availability, condition and effectiveness of blinds;
  — adjustment of brightness or contrast controls.

Discussions with 'users' and 'operators'

[D8215] An assessment provides an opportunity for a two-way dialogue between the assessor(s) and the 'user' and allows the assessor to evaluate whether any problems reported are related to deficiencies in the workstation.

The HSE also advises employers to encourage users to report any ill health that may be due to their DSE work. It says that this is a useful check that risk assessment and reduction measures are working properly and that reports of ill health may indicate that reassessment is required. DSE training should include the need to report and the organisational arrangements for making a report.

Common causes of problems are:
- Back, shoulders, neck:
  — the height or position of the screen;
  — positioning of the seat, including the backrest;
  — the need for a footrest or document holder.
- Hands, wrists, arms:
  — position of the keyboard or mouse;
  — keyboard or mouse technique.
- Tired eyes or headaches:
  — lack of regular breaks or activity changes;
  — reflected light (artificial or sunlight).

- Discussions with 'users' can also reveal other important pieces of information, eg:
  — there are problems with sunlight at certain times of day or periods of the year;
  — the 'user' does not know how to adjust their chair or brightness/contrast controls;
  — the 'user's' chair is broken and incapable of being adjusted;
  — the 'user' has never been offered an eye test.

  In addition, problems can arise from excessive workloads and high key stroke rates.

Reducing the risks

[D8216] The HSE advises that postural problems may be dealt with through simple adjustments to the workstation, such as repositioning equipment or adjusting the chair. They may also indicate that training may need to be reinforced with information about correct hand position, posture, or how to adjust equipment, for example. New equipment, such as a footrest or document holder, may be necessary.

Visual problems can often be tackled by straightforward methods such as repositioning the screen or using blinds to avoid glare, ensuring that the screen is at a comfortable viewing distance or by ensuring the screen is kept clean. In some cases, in appropriate lighting may be causing the problem.

Fatigue and stress may also be alleviated (in addition to the measures above) by ensuring that:

- software is appropriate to the task;
- the task is well designed;
- users have a degree of personal control over the pace and nature of their tasks; and
- proper provision is made for training and information, not only on health and safety risks but also on the use of software.

HSE guidance to the regulations says:

> 'It is important to take a systematic approach to risk reduction and recognise the limitations of the basic assessment. Observed problems may reflect the interaction of several factors or may have causes that are not obvious. For example backache may turn out to have been caused by the worker sitting in an abnormal position in order to minimise the effects of reflections on the screen. If the factors underlying a problem appear to be complex, or if simple remedial measures do not have the desired effect, it will generally be necessary to obtain expert advice on corrective action.'

Assessment records

[D8216A] There is no standard format for DSE workstation assessment records. The completed sample assessment checklists below, together with the guidance on completion are provided as examples of record formats that have been found successful in practice. There is a longer assessment checklist in L26 and a DSE workstation checklist can be found on the HSE website at www.hse.gov.uk/pubns/ck1.htm.

HSE guidance in L26 sets out that records may be stored in electronic as well as paper form. No guidance is provided on how long records should be kept. Prudent employers may prefer to retain them indefinitely, bearing in mind that civil claims for DSE-related conditions may be submitted many years after the condition first arose.

Display Screen Equipment Workstation Assessment		
USER'S NAME: *Christine Jones*		LOCATION: *Sales*
FACTOR		COMMENT
1	**WORK PATTERNS**	
1.1	Most time spent per day at DSE	*5 to 6 hours*
1.2	Average time per day at DSE	*4 hours*
1.3	Number of days per week at DSE	*5*
1.4	Longest spell without break	*2 hours*
1.5	Can breaks be taken?	*Yes – at Christine's discretion*
1.6	Concentration important?	*Accuracy is important*
1.7	Speed of operation important?	*Sometimes*
'USER' STATUS CONFIRMED		*Yes*
2	**PROBLEMS EXPERIENCED**	
	Has the user significant experience of problems with:	
2.1	Back, shoulders or neck	*Regular pains in shoulder and neck*
2.2	Hands wrists or arms	*Occasional aching wrists*
2.3	Tired eyes or headaches	*Sometimes*
2.4	Suitability of software	*Suitable for all tasks*
2.5	Other problems	*None*
3	**LIGHTING/ENVIRONMENT**	
3.1	Artificial lighting:	
	— adequate to see documents	*Yes*
	— any reflection or glare problems	*None – fittings recessed and diffused*
3.2	Sunlight:	
	— any reflection or glare problems	*On winter mornings from window behind*
	— suitable blinds available (if necessary)	*Effective vertical blinds provided*
3.3	Noise:	
	— hindering to communication	*No*
	— distracting or stressful	*Occasionally distracting*
3.4	Temperature and ventilation:	

	— satisfactory in summer and winter	*Office hot and stuffy in summer*
4	**SCREEN**	
4.1	Set at suitable height	*Screen height too low*
4.2	Stable image with clear characters	*Good. Able to vary colours*
4.3	Brightness and contrast adjustable	*Both have adjustable controls*
4.4	Swivels and tilts easily	*Yes*
4.5	Cleaning materials available	*In stationery cupboard*

5	**KEYBOARD**	
5.1	Separate and tiltable	*Yes*
5.2	Sufficient space in front	*Too near edge of desk (wrists bent)*
5.3	Keys clearly visible	*Yes*
6	**DESK AND CHAIR**	
6.1	Desk size adequate	*Satisfactory*
6.2	Sufficient leg room	*Materials being stored in desk well*
6.3	Desk surface low reflectance	*Yes – wooden surface*
6.4	Suitable document holder (if required)	*Not provided*
6.5	Chair comfortable and stable	*Yes*
6.6	Chair height adjustable	*Yes*
6.7	Back adjustable (height and tilt)	*Christine did not know how to adjust*
6.8	Footrest available (if required)	*Not required*

OTHER COMMENTS

Christine was advised to take short breaks more regularly.
Some lengthy jobs without breaks seem to be the cause of her tired eyes and headaches.
Repositioning of the screen and provision of a document holder should overcome the shoulder and neck problems.
Christine was shown how to adjust her chair back.

No.	Actions required	Responsibility
3.2	Close blinds when sunlight bright.	*C Jones*
3.4	Free up office windows (seized up by paint). Investigate whether additional ventilation is necessary.	*Facilities Manager*
4.1	Provide screen stand and keep top of screen at eye level.	*Office Manager/C Jones*
5.2	Keep space in front of keyboard – wrists horizontal.	*C Jones*

6.2	Remove items from desk well.	C Jones
6.4	Provide document holder.	Office Manager
6.7	Ensure induction includes chair adjustment mechanisms.	Training Department

Assessor: B WRIGHT   Signature: B Wright   Date: 11/12/17

PROGRESS WITH ACTIONS

All recommendations acted upon although Christine still needs to remember to take regular breaks on lengthy jobs. Pains in shoulder and neck have ceased and Christine only very occasionally experiences headaches and tired eyes.

B Wright 6/2/18

Planned date for assessment review: February 2019

DISPLAY SCREEN EQUIPMENT WORKSTATION CHECKLIST

USER'S NAME: LOCATION:

1	LIGHTING AND WORK ENVIRONMENT	COMMENTS
1.1	Is artificial lighting adequate?	
1.2	Does it cause any reflection or glare problems?	
1.3	Any reflection or glare problems from sunlight?	
1.4	Are suitable blinds available (if necessary)?	
1.5	Are temperature and ventilation satisfactory in summer and winter?	
2	SCREEN AND KEYBOARD	
2.1	Is the screen set at a suitable height?	
2.2	Stable image with clear characters?	
2.3	Brightness and contrast adjustable?	
2.4	Screen swivels and tilts easily?	
2.5	Cleaning materials available?	
2.6	Is the keyboard tiltable?	
2.7	Is there sufficient space in front of it?	
3	DESK AND CHAIR	
3.1	Is the desk size adequate?	
3.2	Is there sufficient leg room under it?	
3.3	Do you have a suitable document holder (if required)?	
3.4	Is the chair comfortable and stable?	

3.5	Can the seat height be adjusted?	
3.6	Can the height and tilt of the chair back be adjusted?	
3.7	Is there a footrest (if required)?	
4	**HAVE YOU HAD SIGNIFICANT EXPERIENCE OF PROBLEMS WITH:**	
4.1	Back, shoulders or neck?	
4.2	Hands, wrists or arms?	
4.3	Tired eyes or headaches?	
4.4	The suitability of the software?	
4.5	Other problems?	
ANY OTHER PROBLEMS OR COMMENTS?		
Is a further assessment of the workstation necessary? Yes/No		
Signature: Date:		
Name of person carrying out the assessment:		

Guidance on completing the DSE workstation self-assessment

Lighting and work environment

[D8217] In the work environment—

1.1 Artificial lighting should be adequate to see all documents.
1.2 Recessed lights with diffusers shouldn't cause problems. Lights suspended from ceilings might.
1.3 There may be problems in the early morning or afternoon, especially in winter when the sun is low.
1.4 Blinds provided should be effective in eliminating glare from the sun.
1.5 Strong sunlight may create significant thermal gain at times.

Screen and keyboard

[D8217A]

2.1 The top of the screen should normally be level with the user's eyes when sitting in a comfortable position.
2.2 There should be little or no flicker on the screen.
2.3 The user should know where the brightness and contrast controls are.
2.4 The screen should swivel and tilt to avoid reflections.
2.5 The user should know where to get cleaning items for the screen (and keyboard, if necessary).
2.6 Small legs at the back of your keyboard should allow its angle to be adjusted.

2.7 Space in front the keyboard allows wrists to be kept horizontal allows hands and wrists to be rested when not keying in.

Desk and chair

[D8217B]

3.1 The desk should have sufficient space to allow the screen and keyboard to be in a comfortable position and accommodate documents, document holder, phone etc.
3.2 There should be enough space under the desk to allow the user to move their legs freely.
3.3 If the user is inputting from documents, using a document holder helps avoid frequent neck movements.
3.4 Chairs with castors must have at least five (four is very unstable).
3.5 The user should be able to adjust the seat height to work in a comfortable position (arms approximately horizontal and eyes level with the top of the screen).
3.6 The angle and height of the back support should be adjustable so that it provides a comfortable working position. The angle and height of your back support should be adjustable so that it provides a comfortable working position.
3.7 DSE users who are shorter may need a footrest to help them keep comfortable when sitting at the right height for their keyboard and screen (see 3.5).

Possible problems

[D8217C]

4.1 Problems with back, shoulders or neck might indicate that the screen is at the wrong height, an incorrectly adjusted chair or need for a document holder.
4.2 Problems with hands, wrists or arms might indicate incorrect positioning of the keyboard or a poor keying technique. Hands should not be bent up at the wrist and a soft touch should be used on the keyboard, not overstretching the fingers.
4.3 Tired eyes or headaches could indicate problems with lighting, glare or reflections.
They may also indicate the need to take regular breaks away from the screen. Persistent problems might need an eye test - contact [eg Human Resources] to request one.
4.4 The software should be suitable for the work carried out.

Special situations concerning display screen equipment

[D8218] In L26, the HSE provides further guidance on several aspects of DSE work which have developed in recent years as information technology has changed.

Shared workstations

[D8219] 'Hot desking' arrangements mean that many workstations are shared, sometimes by workers of widely differing sizes. Assessments of such workstations should take into account aspects such as:

- whether chair adjustments can accommodate all the workers involved;
- the availability of footrests;
- adequacy of leg room for taller workers;
- arrangements for adjusting the heights of screens (eg adjustable mountings or stands).

In some situations the use of desks with adjustable heights may be appropriate.

Work with portable DSE

[D8220] Work with laptop computers as well as tablets, mobile phones and personal digital assistants (PDAs) which can be used to compose, edit or view text is becoming increasingly common. Where tablets, mobile phones or PDAs are used in this way for prolonged periods they are subject to the *DSE Regulations (SI 1992 No 2792)*, as is all work with laptops.

Mobile phones or PDAs will not generally be used for sufficiently long periods to require a workstation assessment but work involving laptops may be of much longer duration. The HSE publication L26 contains detailed guidance on the selection and use of laptop computers.

The design of laptops is such that postural problems are much more likely to result and there will be a far greater need to ensure that sufficient breaks are taken. Where the laptop is used in an office or at home for extended periods, the risks can be reduced considerably by the use of a docking station or the provision of separate monitors, keyboards etc which can be connected to the laptop. L26 provides further guidance on this. Other risks associated with the use of portable equipment (eg manual handling issues and possible theft involving assault) should also be considered.

Pointing devices

[D8221] Much work at DSE workstations involves the use of a mouse, trackball or similar device to move the cursor around the screen and carry out operations. The HSE booklet (L26) provides guidance on:

Choice of pointing devices
- Suitability of the device for:
  - the environment (space, position, dust, vibration);
  - the individual (right or left handed, physical limitations, existing upper limb disorder);
  - the task (some devices are better than others in respect of speed or accuracy).

Use of pointing devices
- Issues to be considered include:
  — positioning (close to the midline of the user's body, not out to one side);
  — technique (wrist straight, arm not stretched, forearm supported, do not grip too tightly);
  — work surface (particularly its height and degree of support for the arm);
  — mouse mats (smooth, large enough, without sharp edges);
  — software settings (suitable for the individual user);
  — task organisation (point device use mixed with other activities);
  — training (how to set up and use devices);
  — cleaning and maintenance.

The booklet also contains specific guidance on touch screens and speech interfaces.

Are smart phones covered by the DSE regulations?

[D8221A] Research carried out for Specsavers Corporate Eyecare found that almost a quarter (23%) of employers said all their employees use smart phones as part of their working role. The Yougov poll of more than 500 employers, carried out for the company in November 2017, also found that 60% said that at least half their employees used them as part of their working role and only 7% said their employees did not use a smartphone for work.

The results prompted Specsavers to suggest employers ensure their eyecare-at-work policies reflect the change in working patterns and the increasing use of smart phones and handheld devices for everyday work. The requirement to provide display screen users with an eye and eyesight test is set out earlier in the chapter.

'The reason the DSE regulations do not refer directly to smart phones is almost certainly down to the simple fact that they were not in wide use when the regulations were last amended in 2002,' said Specsavers Corporate Eyecare director of strategic alliances Jim Lythgow. 'The regulations do, however, only exclude portable DSE that is not in *prolonged* use (reg 1(4)(d)). So, it is the length of time employees are using screens that is more relevant than the type or size of screens they use.'

The company pointed to HSE guidance stating that 'portable DSE and handheld devices are subject to the Regulations if in prolonged use for work purposes'. It says that while there are no hard-and-fast rules on what constitutes 'prolonged' use, it is reasonable to conclude that portable equipment that is habitually in use by an employee for a significant part of his or her normal work, could well be regarded as covered by the DSE Regulations.

After the assessment

Review and implementation of recommendations

[D8222] The general guidance on the review and implementation of recommendations provided in the chapter on RISK ASSESSMENT is equally applicable to recommendations made as a result of DSE workstation assessments. Responsibility for implementing recommendations should be allocated clearly and all recommendations followed up. It should be noted that the sample assessment checklist includes a space which can be used as a follow-up of an assessment to describe 'Progress with actions'.

Assessment review/re-assessments

[D8223] Assessments must be kept up to date. Changes to the layout of office accommodation often take place with bewildering rapidity. Some of these changes have significant implications for DSE workstations. Both management and DSE 'users' should be aware that a review or reassessment should be requested from the assessor or assessment team whenever significant changes take place.

Because of the frequency of changes, a periodic review of DSE workstation assessments is beneficial. An HSE evaluation of the DSE regulations found that three quarters of businesses conducted risk assessments every 12 months. Significant changes in DSE equipment, software, furniture, lighting etc would obviously justify a review of relevant workstation assessments.

Further information

[D8224]

- HSE Publications
    - L26 Work with display screen equipment;
    - INDG36 Working with display screen equipment (DSE) – A brief guide.
- HSE Website (www.hse.gov.uk). Its content includes:
    - the DSE Regulations and guidance on their interpretation;
    - downloadable copies of the HSE publications listed above;
    - frequently asked questions;
    - a DSE case study (on reducing MSDs in keyboard users); and
    - links to further sources of information.

Electricity

Chris Buck and Andrea Oates

Introduction to electricity

[E3001] Electricity has become fundamental to our way of life – both at work and in the home. It provides an energy source to meet an ever-increasing range of needs and, used with care, is a good servant. However, treated with disrespect, it is a poor master and can result in serious injury or even death.

Every year, the Health and Safety Executive (HSE) receives many reports of incidents at work involving electric shock or burns. Most fatal injuries are caused by contact with overhead power lines.

The main causes of deaths and injuries are:

- the use of poorly maintained electrical equipment;
- work near overhead power lines;
- contact with underground power cables during excavation work;
- contact with live parts causing shock and burns – normal mains voltage, 230 volts AC, can kill;
- the use of unsuitable electrical equipment in potentially explosive or flammable areas such as car paint spraying booths; and
- fires started by poor electrical installations and faulty electrical appliances.

Understanding electricity

[E3002] A basic knowledge of electrical principles is important to an understanding of the dangers of electricity and, more importantly, the control measures that need to be put in place to avoid these dangers and prevent injury. The content of this section has been restricted to aiding understanding of the remainder of this chapter and a good textbook on electrical theory should be consulted for more extensive knowledge.

Because electricity is an unseen energy form, its nature is often difficult to appreciate. A simple analogy is to compare it with the flow of water in a pipe, where the water flow is determined by the water pressure and resistance to the flow arising from the size of pipe. Fundamentally, the flow of electricity (electric current) arises from the movement of electrons in an electrical circuit. The amount of current (I) flowing in a circuit is determined by the electrical pressure or voltage (V) and the resistance (R) in the circuit. These three basic parameters are related by an equation known as Ohm's Law:

$I = V/R$

Varying the electrical pressure will alter the current in direct proportion while varying the resistance will alter the current in inverse proportion.

Electric current is measured in amperes or 'amps' for short (A), voltage in 'volts' (V) and resistance in ohms (Ω). The range of values encountered for these basic parameters is great, often from millions to millionths of the basic unit. Standard prefixes are used to denoted multiples and sub-multiples of the basic unit. For example, 1 mA is one thousandth of an amp whilst 1 mΩ is one thousandth of an ohm. Similarly, 1 kA is one thousand amps, 1 kV is one thousand volts and 1 MΩ is one million ohms.

In any electrical circuit, resistance is encountered in two forms, as continuity or conductor resistance and as insulation resistance. Continuity resistance is the conductor resistance measured end-to-end. This should be as low as possible in order not to unduly restrict the flow of current. Typical values of continuity resistance will usually be less than an ohm, and are often quoted in milliohms (mΩ). Insulation resistance is the resistance between conductors and is therefore a measure of the quality of the electrical insulation surrounding the conductors. Because we do not want electricity to leak across or short-circuit from one conductor to another, or out of the circuit, the resistance value should be as high as possible. The insulation resistance of a circuit should therefore be a very large number of ohms, typically several million ohms (MΩ).

There are two forms of current, alternating (AC) and direct (DC) and both are likely to be encountered in the workplace. Alternating current is the most common because this is the form of 'mains' electricity. Direct current is the form provided by batteries and is therefore encountered in workplaces where battery-powered works vehicles are used. Certain specialist work processes also use DC obtained by conversion (rectification) of the AC mains supply. The process of rectification is not perfect and some oscillation generally appears on the DC output. This is referred to as ripple and has implications relating to electric shock. In the case of electrical circuits operating from the 'mains' the circuit resistance is more complicated because the oscillating waveform of the current brings into play additional characteristics, namely the magnetising and charging effects of the current. Suffice to say, the resistance of an AC circuit is referred to as its impedance (Z) which represents a fairly complex mathematical relationship involving the inductance and capacitance displayed by the circuit, in addition to its resistance. As with resistance, impedance is also measured in ohms.

The flow of an electric current represents power or energy. Electric power is the product of the current and voltage, measured in watts (W) or (kW), whilst electrical energy is the product of electric power and time, measured in watt seconds (Ws) or kilowatt hours (kWh).

Legal background and standards

[E3003] The fundamental piece of legislation regulating electrical safety at work is the *Electricity at Work Regulations 1989 (SI 1989 No 635)* ('*EWR*'). This chapter mainly explains the requirements of these Regulations, but **E3029** onwards includes information concerning other electrical legislation of relevance in specific circumstances.

Legal background and standards [E3003]

Reference is also made to electrical standards and approved codes of practice (ACOPs). A full list of commonly used electrical standards and ACOPs is set out on the HSE website at www.hse.gov.uk/electricity/standards.htm.

The *EWR* comprise 33 individual regulations grouped into three parts:

- *Part I*: Introductory matters (*Regulations 1–3*);
- *Part II*: General technical provisions (*Regulations 4–16*);
- *Part IV*: Miscellaneous and general provisions (*Regulations 29–33*).

Part III of the regulations contained provisions relating specifically to mines (*Regulations 17–28*) and was revoked by the *Mines Regulations 2014 (SI 2014/3248)*. These Regulations consolidated and modernised the law on health and safety in mines in Great Britain. They are not covered in detail in this chapter, but they contain requirements concerning the maintenance and testing of all electrical plant and equipment at a mine.

As the *Electricity at Work Regulations 1989* were made under the *Health and Safety at Work etc Act 1974* ('*HSWA*') they apply in the same circumstances as the parent Act, ie they are work-related rather than applying to specific categories of premises. They therefore apply to all types of work establishment.

Like the parent Act, they place responsibilities on employers, the self-employed and employees, in this case to avoid danger and prevent injury from electricity. The regulations also impose duties on operators of mines and quarries to ensure that all requirements or prohibitions imposed by or under the regulations are complied with. The responsibilities of duty holders are defined in *Regulation 3* and require action to be taken where the matter is under their control.

An example of matters under the control of an employer would be the preparation of an appropriate policy relating to electrical safety and the organisation and arrangements for putting it into effect, such as for the maintenance of electrical equipment. An example of corresponding matters under the control of an employee would be to comply with those electrical maintenance procedures and use any tools and equipment, provided in connection with the work, in the correct manner.

Some of the duties imposed by the regulations are qualified, ie 'so far as is reasonably practicable', as in the parent Act. The others are absolute. In the case of a prosecution for a breach of an absolute duty, it is available to the employer to utilise the defence provided by *Regulation 29*. This states that it shall be a defence for any person to prove that they took all reasonable steps and exercised all due diligence to avoid the commission of that offence.

Continuing with the example of electrical equipment maintenance, 'taking all reasonable steps' would involve establishing a maintenance policy and putting into place the organisation and arrangements for its implementation. 'Exercising all due diligence' would require on-going monitoring to ensure that the policy remained effective and that the maintenance was undertaken to the required standard.

The *EWR* apply to all electrical systems which, by definition (*Regulation 2*) means electrical equipment connected to a common source of energy. Electrical systems range from power stations and the national grid at one end of the spectrum to electrical installations within buildings and electrical equipment at the other. They require action to be taken only where there is danger to be avoided or injury prevented (see **E3004**).

When originally enacted, the *EWR* did not apply to offshore installations, ie outside territorial waters. However, one of the outcomes of the inquiry into the 1988 Piper Alpha oil rig fire was to extend the application of the EWR to such installations through the introduction of the *Offshore Electricity and Noise Regulations 1997 (SI 1997 No 1993)*.

Modern health and safety legislation tends to be goal-setting and therefore non-prescriptive. The *EWR* are no exception to this principle. They lay down the fundamental principles for achieving good standards of electrical safety but do not prescribe the specific means for compliance. There is a wide range of guidance published by the HSE, as well as British, European and International Standards, to assist employers in determining how to achieve compliance. A key HSE publication is HSR25 (www.hse.gov.uk/pUbns/priced/hsr25.pdf). This provides guidance on the general application of the Electricity at Work Regulations 1989 to workplaces.

Because of the higher level of risk relating to the use of electricity in mines, there is specific HSE guidance in this area: *Electrical safety in mines* (HSG278) (www.hse.gov.uk/pUbns/priced/hsg278.pdf) published in April 2015. There is also online guidance on electrical safety in higher risk areas in quarries at www.hse.gov.uk/quarries/hardtarget/electricity.htm.

In the introduction to HSR25 reference is made to the very important British Standard BS 7671 Requirements for Electrical Installations (also known as the IET Wiring Regulations). These are non-statutory regulations which 'relate principally to the design, selection, erection, inspection and testing of electrical installations, whether permanent or temporary, in and about buildings generally and to agricultural and horticultural premises, construction sites and caravans and their sites'. The HSE states that compliance with BS 7671 is likely to achieve compliance with the relevant aspects of the EWR. In addition, an electrical installation installed to an earlier edition of BS 7671, but current at that time, would not in itself mean that it did not comply with the EWR.

Although not a statutory document, the British Standards Institute (BSI) explains that BS 7671 sets the standard for how electrical installations should be done in the UK (and many other countries) and enables compliance with the law. The requirements in the 2018 version (18th edition) of 'the Regs' came into effect on 1 January 2019. Installations designed after 31 December 2018 must comply with BS 7671:2018.

The BSI provides a summary of the main changes introduced by BS 7671:2018. They include a new regulation recommending the installation of arc fault detection devices (AFDDs) to mitigate the risk of fire in AC final circuits of a fixed installation due to the effects of arc fault currents. There is also a new regulation on the requirements for the method of support of wiring systems in escape routes. More information can be found on the BSI website at: https://shop.bsigroup.com/.

Health and safety legislation of a more general application may also be of relevance to electrical safety. For example:

- the *Management of Health and Safety at Work Regulations 1999 (SI 1999 No 3242)*, which require the assessment of electrical risks and the provision of training relevant to electrical work activities;
- the *Provision and Use of Work Equipment Regulations 1998 (PUWER) (SI 1998 No 2306)*, which apply equally to electrical equipment, eg concerning the need for isolation from all energy sources prior to commencing work;
- the *Supply of Machinery (Safety) Regulations 2008 (SI 2008 No 1597)*; and
- the *Reporting of Injuries, Diseases and Dangerous Occurrences Regulations 2013 (SI 2013 No 1471)*.

Dangers of electricity

General

[E3004] *Regulation 2* of the *EWR (SI 1989 No 635)* gives the definitions of words used with special meaning in the Regulations. These include 'danger' and 'injury'. Danger is defined as risk of injury while 'injury' means 'death or personal injury from electric shock, electric burn, electrical explosion or arcing, or from fire or explosion initiated by electrical energy, where any such death or injury is associated with the generation, provision, transmission, transformation, rectification, conversion, conduction, distribution, control, storage, measurement or use of electrical energy'.

The 1989 Regulations require action to be taken to avoid danger or, where appropriate, injury. The reason for distinguishing between danger and injury is that in some situations, eg live working, the electrical danger is not eliminated because of the presence of live parts and, therefore, precautions must be put in place to prevent injury from that danger. Ideally, injury is prevented by completely removing or eliminating the danger, eg as in the case of equipment isolation prior to undertaking maintenance work.

It is important to appreciate that injury does not always require physical contact to be made with a live conductor. At high voltages, electricity can jump an air gap (flashover), the flashover distance increasing as the voltage increases. Environmental conditions are also relevant. Damp or wet conditions will reduce the effectiveness of air as an insulator. For this reason, the scope of the Regulations covers the prevention of injury from electrical danger arising from work near as well as on an electrical system or equipment.

The Regulations do not, per se, make a distinction between high and low voltage systems because system voltage is not necessarily the only factor in determining danger. A 1.5 V torch battery would not be expected to present a shock or short-circuit hazard in a normal work environment but might provide sufficient energy to cause ignition in a potentially explosive atmosphere.

The definition of injury effectively encompasses the dangers of electricity which may be summarised as electric shock, electric burn, fire, arcing and explosion.

Electric shock

[E3005] Electric shock may arise either through direct or indirect contact with electricity. The former involves direct contact with a live part, eg an exposed live terminal or conductor. The latter is a shock received from an exposed-conductive-part, eg the metal casing of electrical equipment, made live under a fault condition. An exposed-conductive-part is any conductive part of equipment which can be touched and which is not a live part, but which may become live under fault conditions. The *EWR (SI 1989 No 635)* specify control measures to deal with each of these circumstances (*Regulations* 7 and 8 respectively – see **E3014** and **E3015**).

The severity of an electric shock depends on a number of factors, in particular the magnitude and duration of the shock current passing through the body. The current magnitude will depend on the contact voltage and resistance or, to be more strictly correct in the case of a mains supply, which is an alternating current (AC), the impedance (Z) of the shock path. In the case of contact with mains electricity, therefore, the shock current is:

$I = V/Z$ (Ohms law)

The outcome from an electric shock will be very dependent on the circumstances in which it is received, in particular, the environment, as this will determine the shock path resistance or impedance. Indoors, standing on a floor covering acting as a good insulator, a hand-to-feet shock at mains voltage (230V nominal) may result in no more than a tingling sensation. Outdoors, however, with little insulation from earth, a shock at the same voltage would prove more severe, even fatal. A hand-to-hand shock, where one hand makes contact with a live part while the other is simultaneously in contact with earthed metal, such as the earthed case of electrical equipment, represents an even higher risk because the shock path impedance is only that of the body. At mains voltage such a situation is quite likely to result in a fatality.

Electric current is measured in amperes or 'amps' for short (A), voltage in 'volts' (V) and resistance in ohms (Ω).

Detailed technical information on electrical injury is outlined in the standard IEC 60479-1:2018 *Effects of current on human beings and livestock – Part 1: General aspects*.

In summary, the physiological effects of alternating current are:

Around 0.5 mA	Threshold of perception
Up to 10 mA	Unpleasant but usually no harmful physiological effects
Above 10 mA	Increasing likelihood of muscular contractions and, at higher currents, possibility of respiratory system failure

30 mA	Typical current trip rating for a residual current device (RCD)
Around 50 mA	Risk of ventricular fibrillation and cardiac arrest (increasing risk with increasing magnitude and duration of current flow)
250 mA	Current drawn by a 60 W light bulb

These values are only as a guide since the effect will vary from one person to another. With higher shock currents the effect on the body, in particular the probability of death, becomes increasingly dependent on how long the current flows for. Currents as low as 50mA (0.05A) can prove fatal, particularly if flowing through the body for several seconds. Higher currents will have a similar effect in a much shorter time.

While AC, per se, represents a high shock risk, contact with direct current (DC) is also hazardous and can disturb the normal heart rhythm. A direct current of approximately 3.75 times the AC value will have the same probability of inducing ventricular fibrillation. However, if ripple exists at more than 10 per cent of the nominal DC value due to imperfections in the process of rectification then it must be considered as hazardous as AC.

The HSE advises that a voltage as low as 50 volts applied between two parts of the human body causes a current to flow that can block the electrical signals between the brain and the muscles. It says that this could:

- stop the heart beating properly;
- prevent the person from breathing; and
- cause muscle spasms.

'The exact effect is dependent upon a large number of things including the size of the voltage, which parts of the body are involved, how damp the person is, and the length of time the current flows', it says. 'Electric shocks from static electricity such as those experienced when getting out of a car or walking across a man-made carpet can be at more than 10,000 volts, but the current flows for such a short time that there is no dangerous effect on a person. However, static electricity can cause a fire or explosion where there is an explosive atmosphere (such as in a paint spray booth)'.

Electric burn

[E3006] Body tissue burns may result from the passage of electric current. These burns tend to be deep seated and difficult to heal. Contact with high voltage is often characterised by body burn marks at the current entry and exit points, eg the palms of the hand and soles of the feet in the case of a hand-to-feet shock path. Electric burns may also arise through exposure to high-power sources of electro-magnetic radiation such as from radio-transmission antennae or induction-heating processes (see RADIATION chapter).

Fire

[E3007] Heat generation due to the overheating of cables or electrical equipment, which may occur for a number of reasons, can lead to fires. Examples are:

- overloading due to insufficient current carrying capacity for the connected load;
- cables bunched or covered with thermal insulation, significantly reducing current carrying capacity;
- lack of, or incorrectly rated, excess current protection (eg fuses or circuit breakers) for circuits and equipment;
- blocked ventilation apertures in equipment enclosures, eg motors;
- leakage currents due to deterioration of insulation surrounding live parts.

Arcing arising from short-circuit flashover or sparks, generated within electrical equipment such as motors or switching devices, may provide an ignition source for adjacent flammable materials (solids, gases, vapours or dusts).

Arcing

[E3008] Arcing or flashover will occur when short-circuiting or bridging between live parts at different voltages, eg between circuit conductors (phase to neutral or across phases in the case of a three-phase system) or from a live part to earth. As well as the risk of serious injury there is also the likelihood of damage to the equipment concerned. The severity of such a flashover will depend on the fault level at the point of short-circuit, ie the magnitude of the short-circuit current, which in turn will be determined by the upstream impedance in the electrical system back to the source of supply. Since this will be very low, the prospective short-circuit current will be very high (Ohm's law again). In a factory situation, prospective short-circuit currents as high as 10kA may well be encountered.

Short-circuit flashover may also result from equipment failure, eg failure of a circuit protective device to disconnect the supply safely following a fault. Where failure has arisen because of a malfunction during a switching operation, the operator is at risk of sustaining arc eye and severe burns.

Explosion

[E3009] In the case of a severe short-circuit flashover there is often little to distinguish between the effects of arcing and an explosion. Where the short-circuit occurs in a restricted space, such as within a busbar chamber or inside switchgear, the explosive forces are likely to result in considerable damage to the equipment involved, eg rupture of the external housing or covers blown off. If the switchgear is of the oil-filled type this may lead to the ejection of burning oil, a further hazard for anyone with the misfortune to be close by.

Electricity presents special problems in potentially explosive atmospheres (PEAs), where an electric spark may act as an ignition source. For this reason, particular attention must be given to the design and construction of electrical equipment intended for such environments – see E3027 below.

Assessing electrical risks

[E3009A] The assessment of electrical risks first requires an evaluation of the likelihood of electrical danger, following which the severity of injury can be predicted. The likelihood of danger will centre on the presence of hazards that may give rise to electric shock, burns, fire, etc. This will include situations where there may be exposed live parts, missing earth connections, incorrect fuse ratings, etc. The severity of the outcome will depend on factors such as the supply voltage, shock path and work environment, in the case of electric shock, and the fault level and bridging path in the case of short-circuit flashover. With electricity, there is a very thin dividing line between the accident being survivable and death. It is therefore best to work on the basis of endeavouring to reduce the likelihood of an electrical accident to as near zero as possible, which is the objective of the *EWR (SI 1989 No 635)*.

Fundamentals of controlling electrical risks

[E3010] The control measures aimed at preventing injury from electrical danger in general situations are prescribed in *Part II* of the *EWR (SI 1989 No 635)*, ie in *Regulations 4–16*. These cover:

- systems, work activities and protective equipment;
- strength and capability of electrical equipment;
- adverse or hazardous environments;
- insulation, protection and placing of conductors;
- earthing or other suitable precautions;
- integrity of referenced conductors;
- connections;
- means for protecting from excess of current;
- means for cutting off the supply and for isolation;
- precautions for work on equipment made dead;
- work on or near live conductors;
- working space, access and lighting;
- persons to be competent to prevent danger and injury.

Regulation 4 of the *EWR (SI 1989 No 635)* is effectively an 'umbrella' requirement, covering everything relating to the life of an electrical system, which includes electrical equipment as well as fixed electrical installations. The Regulation is in four parts, requiring:

(1) the construction of electrical systems such as to prevent danger;
(2) the maintenance of electrical systems so as to prevent danger;
(3) all work activities to be carried out in such a manner as not to give rise to danger;
(4) any equipment provided under the Regulations for the purpose of protecting persons at work on or near electrical equipment to be suitable for use, maintained in a suitable condition and properly used.

The first three parts of the Regulation are qualified duties, ie 'so far as is reasonably practicable'.

The subsequent regulations in *Part II* cover three important aspects relating to electrical safety:

- the need for competence, which underlies the effective implementation of the *EWR* duties (*Regulation 16*);
- the design, construction and specification of the hardware, eg cables, switchgear circuit protection, etc (*Regulations 5–12* and *15*);
- safe systems of work (*Regulations 13* and *14*).

These aspects will now be considered in turn in more detail.

Competence

[E3011] *Regulation 16* of the *EWR (SI 1989 No 635)* specifically requires that any person engaged in any work activity where technical knowledge or experience is necessary to prevent danger, or where appropriate injury, is competent to prevent danger and injury. Competence means the possession of such knowledge or experience as may be appropriate having regard to the nature of the work. If this is not the case, that person must be under an appropriate degree of supervision, again having regard to the nature of the work. This requirement for competence extends to non-electrical work activities in any situation where electrical danger may be present (see *EWR (SI 1989 No 635), Reg 4(3)*). The HSE publication HSR25, *The Electricity at Work Regulations 1989 Guidance on Regulations*, provides guidance concerning the scope of the phrase 'technical knowledge or experience' required under *Regulation 16*. This embraces:

- adequate knowledge of electricity;
- adequate experience of electrical work;
- adequate understanding of the system to be worked on and practical experience of that class of system;
- understanding of the hazards which may arise during the work and the precautions required to be taken;
- ability to recognise at all times whether it is safe for work to continue.

The list above effectively provides a framework training specification that can be used as a starting point for preparing a more detailed specification to suit a particular training need, eg to inspect and test portable electrical equipment. Competence therefore requires an understanding of the tasks to be performed, as well as of the equipment on which the work is to be undertaken, coupled with the necessary underpinning technical knowledge relating to electricity.

For example, an operative required to inspect and test portable electrical equipment would need a thorough knowledge of the different types and constructions of equipment likely to be encountered in order to know what to look for when carrying out an inspection and what tests would be appropriate. An understanding of the test instruments to be used would also be required, in particular their functions and methods of use. The underpinning technical knowledge would enable meaningful measurements to be taken, properly recorded and with the results correctly interpreted. The HSE guidance publication, *Maintaining portable electric equipment in low-risk environments* (INDG236) provides more guidance in this area and can be found at: www.hse.gov.uk/pubns/indg236.pdf.

A higher level of competence is required for live work since the dangers of electricity remain present. It is necessary for those involved to understand fully the shock and short-circuit hazards in relation to the equipment being worked on and the work being undertaken, in particular possible shock and bridging paths that might arise during the course of the work. The fact that such hazards are always present whilst the work is in progress also must be fully appreciated.

There are City and Guilds qualifications for certain electrical activities, eg installation work and electrical equipment inspection and testing, for which courses are offered by colleges and training organisations. Other routes to acquire the necessary competence, such as on-the-job apprenticeship training linked to NVQs based on engineering occupational standards, are also available. Many electricians belong to one or more trade organisations such as the Electrical Contractors' Association (ECA) for England and Wales or its Scottish equivalent (SELECT). The National Inspection Council for Electrical Installation Contracting (NICEIC) also exists, essentially as a consumer safety body, to maintain the standards of the electricians on its roll.

Online HSE advice on the training required to demonstrate competence is as follows:

> 'A person can demonstrate competence to perform electrical work if they have successfully completed an assessed training course, run by an accredited training organisation, that included the type of work being considered. As part of that course, this person should have demonstrated an ability to understand electrical theory and put this into practice. A successfully completed electrical apprenticeship, with some post-apprenticeship experience, is a good way of demonstrating competence for general electrical work. More specialised work, such as maintenance of high-voltage switchgear or control system modification, is almost certainly likely to require additional training and experience.'

Design and construction of electrical equipment and systems

Strength and capability

[E3012] *Regulation 5 of the EWR (SI 1989 No 635)* requires that: 'No electrical equipment shall be put into use where its strength and capability may be exceeded in such a way as may give rise to danger'.

HSE guidance on this regulation explains that 'the term "strength and capability" of electrical equipment refers to the ability of the equipment to withstand the thermal, electromagnetic, electrochemical or other effects of the electrical currents which might be expected to flow when the equipment is part of a system'.

Adverse or hazardous environments

[E3013] *Regulation 6 of the EWR (SI 1989 No 635)* requires that electrical equipment shall be constructed or as necessary protected to prevent danger, so far as is reasonably practicable, from reasonably foreseeable adverse or hazardous environments, namely:

- mechanical damage;
- the effects of the weather, natural hazardous, temperature or pressure;
- the effects of wet, dirty, dusty or corrosive conditions;
- any flammable or explosive substance, including dusts, vapours or gases.

Whilst exposure to earthquakes in the UK might not be reasonably foreseeable, the siting of electrical equipment outdoors means that it is highly likely to be exposed to wet and possibly dirty, dusty or even corrosive conditions, such as salt spray in coastal locations. An appropriate construction specification would be needed, therefore, to accommodate this type of environment.

Protection from electric shock by direct contact

[E3014] *Regulation 7 of the EWR (SI 1989 No 635)* deals with the prevention of electric shock from direct contact with live conductors. It requires that: 'All conductors in a system which may give rise to danger shall either:

(a) be suitably covered with insulating material and as necessary protected so as to prevent, so far as is reasonably practicable, danger; or
(b) have such precautions taken in respect of them (including, where appropriate, their being suitably placed) as will prevent, so far as is reasonably practicable, danger.'

The normal means for preventing direct contact with live conductors is:

- electrical insulation (eg cable insulation);
- enclosure (eg busbar chamber);
- use of barrier fencing or screening (eg high voltage test bay).

In some circumstances the use of an extra-low voltage system (eg 12 V) may provide an adequate safeguard because at such a low voltage any accidental contact could not give rise to a serious electric shock. Nevertheless, a short-circuit hazard may still exist, eg as with a 12 V high ampere-hour capacity battery.

The thickness of electrical insulation will depend on the conductor operating voltage. In some situations, additional protection may be required to prevent damage to the electrical insulation. In industrial premises, steel wire armoured (SWA) or mineral insulated copper sheathed (MICS) cables are often used. Alternatively, cables may be placed in steel conduit or trunking, or placed out of reach on cable trays, to provide additional protection. Where enclosures are used it is important to ensure adequate protection against the ingress of foreign objects or fingers, in accordance with British Standard BS EN 60529:1992+A2:2013 *Degrees of protection provided by enclosures* (also known as the IP code).

Electrical equipment and systems **[E3015]**

Barrier fencing is often used in test bays where there is a need to undertake live testing. Access gates must be interlocked with the test supply (eg key type) to ensure that entry cannot be gained while the supply is energised. In some circumstances, it might be impracticable to completely enclose all live parts. Examples are:

- for functional reasons, eg the supply to electric overhead travelling cranes;
- railway traction supplies and overhead power distribution systems; and
- in connection with operational requirements, eg fault finding and diagnostic live testing.

In such circumstances, it is then necessary for other appropriate precautions to be implemented to minimise the risk of contact. Placing out of reach, coupled with the posting of warning notices, is a commonly adopted safeguard. However, it is important to recognise that occasions may arise where it becomes necessary to encroach within the prescribed safety clearances. In these cases, it is then necessary to implement a safe system of work based on isolation of the supply to remove the electrical danger at source.

Protection from electric shock by indirect contact

[E3015] *Regulation 8* of the *EWR (SI 1989 No 635)* requires that precautions are taken, either by earthing or other suitable means, to prevent danger arising from any conductor (other than a circuit conductor) which may reasonably foreseeably become charged as a result of either the use of the system or a fault in the system. For example, an internal equipment fault may cause the metal case to be made live at the supply voltage, creating a shock risk. A range of techniques can be employed, either singly or in combination:

- earthing;
- equipotential bonding;
- double insulation;
- earth-free non-conducting environments;
- current/energy limitation;
- use of safe voltages;
- separated or isolated systems;
- connection to a common reference point on the system.

Earthing (for class I equipment) and double insulation (for class II equipment) are the techniques most commonly adopted. To achieve effective installation earthing it is most important that the earth fault loop impedance (the conductor continuity impedance/resistance from the supply source to the point of fault on the installation and back to source through the earth connection) is kept as low as possible. This is so sufficient fault current will flow to enable the circuit protective device to operate to disconnect the fault current quickly. In buildings, connections (referred to as main equipotential bonding) are also made from the main earthing terminal to other services, as well as to the metal structure of the building if applicable. Further connections (supplementary bonding) are made between conductive parts in high shock-risk situations (bathrooms, shower rooms, kitchens, etc). This bonding serves a different role to earthing. It is to minimise the risk of voltage differences arising between exposed-conductive-parts during the time that the fault exists and before the supply is automatically disconnected.

In high risk situations, a residual current device (RCD) may be used to supplement other protective measures, through very fast disconnection of the supply in the event of something going wrong (typically 20–40 ms). An RCD is a very sensitive device, capable of responding to the comparatively small currents typical of a shock situation. A common trip current rating is 30mA. An RCD will not prevent shock – a shock current must flow to create an out-of-balance situation, which is the basis of its operation – but the shock should be survivable.

Referenced conductors

[E3016] *Regulation 9 of the EWR (SI 1989 No 635)* requires that nothing shall be placed in the neutral conductor of an electrical system to cause it to become open-circuited while the phase conductor(s) remain live.

This is for two main reasons:

(i) on modern distribution supply networks, the neutral conductor is likely to be used also for earthing purposes, referred to as a protective multiple earthing (PME) system. An open-circuit in the neutral conductor would therefore result in loss of earthing;

(ii) the secondary (output) windings of three-phase LV distribution supply transformers are generally 'star' connected, with the star point earthed and serving as the system neutral. This provides a reference point for the phase voltages, to help maintain them within the prescribed statutory limits for supply. Loss of the neutral connection back to the supply transformer may result in undue voltage fluctuation depending on the loading on each phase.

Examples of prohibited devices are fuses and thyristors which can lead to danger by either becoming open-circuited or giving rise to a high conductor resistance. In the case of a three-phase four-pole switch, the switch must break the neutral conductor last and make it first when operated. This situation does not arise with a triple-pole and neutral (TP&N) switch, where only the phase conductors are switched and the neutral remains connected.

Connections

[E3017] *Regulation 10 of the EWR (SI 1989 No 635)* requires that all joints and connections should be mechanically and electrically suitable for use. This applies to both temporary and permanent connections and to connections in protective (earth) conductors as well as circuit conductors (phase and neutral). A broken phase connection could result in a short-circuit to a neutral conductor or to earth via an earthed metal equipment case. Failure of a joint or connection in a protective conductor could result in a loss of earthing and therefore loss of effective protection in the event of electric shock arising from indirect contact.

Electrical equipment and systems **[E3020]**

Excess current protection

[E3018] *Regulation 11* of the *EWR (SI 1989 No 635)* requires efficient means, suitably located, to be provided to protect every part of a system from excess current, as may be necessary to prevent danger. The abnormal conditions likely to give rise to excess current flow are:

- an overload, which can result in the overheating of cables or equipment and the possibility of fire;
- a short circuit, which can result in arcing or explosion, possibly accompanied by fire;
- an earth fault, giving rise to a shock risk through indirect contact.

In the case of electric shock protection, it is important that the protective device (eg a fuse or circuit breaker) disconnects the supply as quickly as possible. BS 7671 specifies maximum disconnection times for different system types at different voltages.

Isolation

[E3019] *Regulation 12* of the *EWR (SI 1989 No 635)* interfaces with *Regulation 19* of *PUWER (SI 1998 No 2306)*. It requires both a means for cutting off the supply as well as for isolation. In practice, switchgear often performs both functions. However, the fact that these are separately stated in this Regulation implies a difference. The distinction is made clear in the second part of the regulation, which explains the meaning of isolation in the electrical context. This amounts to a legal definition. The requirement for electrical isolation may be summed up as secure disconnection and separation from all supply sources. Key characteristics for compliance would be:

- adequate labelling, ie unambiguous and legible;
- adequate contact separation;
- means for securing the point of isolation, preferably by locking off.

Many accidents occur due to a failure to isolate plant correctly. Compliance with the requirements of *Regulation 12* is an important precursor to satisfying the requirements of the *Regulation 13* of the *EWR (SI 1989 No 635)*, which is concerned with work on de-energised systems (see **E3023**). The requirements of *Regulation 14* are set out in **E3024** below.

Working space, access and lighting

[E3020] *Regulation 15* of the *EWR (SI 1989 No 635)* requires adequate working space, access and lighting. This is relevant to safe systems of work, particularly in the case of live work. For example, live testing within control panels for fault-finding purposes requires accessibility, freedom of hand/arm movement, minimal risk of accidentally being pushed onto exposed live terminals and a lighting level sufficient to enable the live connections to be clearly observed, so that the testing can be undertaken safely. In terms of guidance for determining suitable accessibility, see Appendix 1 of HSR25, which reproduces a regulation from the former *Electricity (Factories Act) Special Regulations 1944 (SI 1944 No 739)*. The 1944 Regulations extended the provisions of the original 1908 Regulations, which were revoked by the EWR. As a guide, a clear distance of the order of one metre should be allowed in front of fixed equipment where access is required for live diagnostic testing and other such work.

Maintenance

[E3021] *Regulation 4(2) of the EWR (SI 1989 No 635)* states:

'(1) All systems shall at all times be of such construction as to prevent, so far as is reasonably practicable, danger.

(2) As may be necessary to prevent danger, all systems shall be maintained so as to prevent, so far as is reasonably practicable, such danger.'

This legal requirement sets an objective, to ensure that sufficient maintenance is carried out on all electrical systems to prevent danger. The duty is a qualified one and therefore the extent and frequency of the maintenance required should be proportionate to the level of risk. By virtue of the definitions contained in *Regulation 2 of the EWR (SI 1989 No 635)*, the term 'system' embraces electrical equipment of all forms. In the case of equipment, examples of risk factors would include:

- age;
- operating voltage;
- environment of use (eg indoor or outdoor);
- type of equipment (hand held, portable, fixed, etc);
- class of construction for protection from electric shock (class I, II or III);
- nature of use (or even abuse).

The high end of the risk spectrum would be a hand-held mains powered tool used on outdoor construction work at different sites, while the low end could be stationary IT equipment used in an office environment.

As a minimum, maintenance would include inspection to check for deterioration or damage and to verify safe condition for use by any user. However, some faults cannot be detected by inspection alone and therefore testing may also be required, eg to verify the effective earthing of exposed conductive parts such as a metallic equipment case.

HSE guidance *Maintaining portable electric equipment in low-risk environments* is available at www.hse.gov.uk/pubns/indg236.htm.

Further information on portable appliance testing can be found at www.hse.gov.uk/electricity/faq-portable-appliance-testing.htm.

Safe systems of work

General

[E3022] *Regulation 4(3) of the EWR (SI 1989 No 635)* imposes a general duty that all work activities, including work near an electrical system, shall be

Safe systems of work **[E3023]**

carried out in such a manner as not to give rise, so far as is reasonably practicable, to danger. The *EWR* provide for two basic approaches to achieving a safe system of work, covered respectively by *Regulations 13* and *14*:

(i) removal of any electrical danger by disconnection from all sources of supply and the implementation of a system of work aimed at verifying that the danger from electricity has been removed and remains removed for the duration of the work;
(ii) live working, recognising that the danger of electricity remains so that it is then necessary to implement a system of work aimed at preventing injury arising from the danger.

However, the second option is not an automatic choice since the intention to work live first must be properly justified, which forms the starting point for an assessment of electrical risks.

The HSE publishes guidance specific to safe working practices involving electricity (*Electricity at work – safe working practices* (HSG85) (www.hse.gov.uk/pubns/priced/hsg85.pdf).

Work on isolated equipment

[E3023] A model safe system of work would require consideration of the following:

- isolation from all points of supply;
- earthing to discharge any residual electrical energy or to prevent the build up of an induced charge;
- proving dead at the point of work;
- demarcation of the safe zone of work;
- safety documentation.

PUWER (SI 1998 No 2306), Reg 19 places duties on employers in respect of isolation from all forms of energy, which therefore includes electrical energy. These duties require that:

- where appropriate, the work equipment shall be provided with suitable means to isolate it from all sources of energy;
- the means of isolation shall be clearly identifiable and readily accessible;
- appropriate measures shall be taken to ensure that re-connection of any energy source to work equipment does not expose any person using the equipment to any risk to his health or safety.

Electrical isolation is defined in *EWR (SI 1989 No 635), Reg 12* (also discussed at **E3019** above) and means the secure disconnection and separation from all sources of electrical energy. The preferred means of securing points of isolation is by locking off, using special safety locks operated by unique keys. Where such facilities are not available, the HSE guidance (in HSG85) sets out, the removal of fuses or links and their being held in safe keeping can provide a secure arrangement if proper control procedures are used.

Where a number of different individuals or separate parties are working on the plant simultaneously a special locking device can be used to enable each party to apply its own safety lock, thereby maintaining personal control of the point of isolation for each person involved in the work.

Many accidents have arisen through incorrect plant isolation, coupled with a failure to prove dead. For example, this may be due to ambiguous, incorrect or missing circuit identification labelling or a failure to identify properly the isolation requirements appropriate to the intended work, particularly in the case of multiple supply sources. Confusion often arises concerning what has been made dead by the isolation of power and control circuits associated with factory production line equipment. For this reason, proving dead at the actual point of work is an essential element to achieving a safe system of work. Proprietary testers are available for this purpose and, because of the importance of using the right type of tester, the HSE publishes a guidance note on the subject *Electrical test equipment for use on low voltage electrical systems* (GS38) at www.hse.gov.uk/pUbns/priced/gs38.pdf.

It must also be recognised also that certain electrical equipment can retain energy, in the form of an electrical charge, for a period of time following its isolation. This applies to plant containing capacitors, eg power factor correction equipment, or systems exhibiting capacitive effects, such as a long run of high voltage cable. Similarly, in the case of high voltage systems, a charge may be induced into an isolated circuit from an adjoining live circuit. Earthing is used to safeguard from static or induced charge.

Demarcation of the safe zone of work may be required to minimise the risk of the worker inadvertently seeking to gain access to adjacent live plant. In complex situations it is often useful to confirm the isolation, the equipment to be worked on and the work to be undertaken through the issue of a permit-to-work (PTW). The PTW also can serve to demarcate the safe zone of work. A PTW does not constitute the safe system of work but serves to strengthen communication and act as a control document to hold plant out of service until signed off by the person in charge of the work. The HSE recommends that a PTW should be used for all high voltage work (ie above 3kV) as well as for low voltage work involving multiple points of isolation. Where PTWs are adopted it is important that a management procedure exists to specify and control their use.

The HSE makes clear (in HSG58) that: 'An electrical permit-to-work is primarily a statement that a circuit or item of equipment is safe to work on – it has been isolated and, where appropriate, earthed. You must never issue an electrical permit-to-work for work on equipment that is still live or to authorise live work.'

Live work

[E3024] *Regulation 14* of the *EWR (SI 1989 No 635)* states that no person shall be engaged in any work activity on or so near any live conductor (other than one suitably covered with insulating material so as to prevent danger) that danger may arise unless:

(a) it is unreasonable in all the circumstances for it to be dead;
(b) it is reasonable in all the circumstances for him to be at work on or near it while it is live; and

(c) suitable precautions (including where necessary the provision of suitable protective equipment) are taken to prevent injury.

Justification therefore centres around two issues. Firstly, there must be good reason for the equipment to remain live while the work is carried out, eg the nature of certain electrical testing may require this to be undertaken live. Secondly, the equipment must be safe to work on while it remains live. This relates to the equipment design and construction, eg working space, extent of exposed live parts, shock and bridging paths (particularly involving any earthed metal case). *Regulation 14* applies to all work activities, including testing, ie there are no exemptions. The criterion is whether electrical danger is present or could arise during the work activity. It is also relevant to non-electrical activities, eg work in the vicinity of underground cables or overhead lines (discussed at **E3028** below). The only exception to compliance with *Regulation 14* is where the live parts are suitably covered with insulating material and as necessary protected so as to prevent danger, therefore instead satisfying the requirements of *Regulation 7* of the *EWR (SI 1989 No 635)* for the prevention of electric shock by direct contact.

The HSE guidance (in HSR25) emphasises the importance of design of electrical equipment and, when ordering, purchasing and installing plant, considering the manner of operation, maintenance and repair of the electrical equipment which will be necessary during its life.

It says: 'The design of electrical equipment and of the installation should eliminate the need for live work which puts people at risk of injury. This can often be done by careful thought at the design stage of installations, for example by the provision of alternative power infeeds; properly laid out distribution systems to allow parts to be isolated for work to proceed; and by designing equipment housings etc which result in segregation of parts to be worked on and protect people from other parts which may be live.'

Provided that live working can be justified in accordance with the first two parts of *Regulation 14*, the remaining part of that Regulation requires that suitable precautions shall be taken to prevent injury. This is an absolute duty. Any accident occurring during the course of live work is therefore prima facie evidence that the precautions taken were not suitable, ie in breach of *Regulation 14*.

A safe system of work for live working must centre on the prevention of injury which, in turn, needs to recognise the types of injury and circumstances in which they might arise. The principal dangers are electric shock and short-circuit flashover. An understanding of shock and bridging paths, and how these might arise in the particular work activity, is therefore very important. Safeguards need to be devised and effectively implemented to prevent electric shock and short-circuit flashover.

A model safe system of work would require consideration of issues including the following:

- capability of those undertaking the work;
- provision of appropriate information concerning the proposed work;
- use of suitable tools and other protective equipment;

- use of suitable instruments and test probes;
- consideration of the need for accompaniment;
- control of the work area.

There are important considerations concerning the competence of personnel intending to undertake live work, mentioned previously (see **E3011** above). Adequate information concerning the work also needs to be provided so that those directly involved can be satisfied that it is within their capability and experience.

Protective equipment must satisfy the requirements of *Regulation 4(4)* of the *EWR (SI 1989 No 635)*, ie it must be:

- suitable for the use for which it is provided;
- maintained in a condition suitable for that use;
- properly used.

Suitability for use infers that management action has determined what protective equipment is needed and the circumstances in which it is to be used in contributing to the overall safe system of work. It is important that such equipment is of a suitable specification for the intended use, for which a number of British Standards exist. These include:

- BS EN 60900:2018 *Live working. Hand tools for use up to 1000 V AC and 1500 V DC*; and
- BS EN 60903: 2014 *Live working. Gloves of insulating material*.

These same standards include information to assist in the implementation of appropriate in-service care regimes for the protective equipment. Correct use infers the provision of relevant information and training to users, particularly concerning the limitations of such equipment.

While applied insulation may minimise the risk of electric shock, large areas of exposed metal, such as at tool working heads or test probe tips, may create a risk of short-circuit when used in a confined space with limited clearance between exposed live terminals or conductors.

The *EWR* requirements concerning protective equipment are supplemented by the *Personal Protective Equipment at Work Regulations 1992 (PPE) (SI 1992 No 2966)*. One important issue concerns compatibility between different items, eg insulated tools and gloves.

There is no automatic requirement for accompaniment in the case of live work. If it is considered necessary to have a second person present to render first aid in the event of an accident, this implies there could be doubt concerning the proposed system of work or its effective implementation. However, a second person may be necessary to enable safe working, eg by directly assisting with the job as 'another pair of hands' or to control the work area.

Special situations concerning electricity

Lightning protection systems

[E3025] Lightning is a natural phenomenon. However, *Regulation 6* of the *EWR (SI 1989 No 635)* requires electrical equipment which may reasonably foreseeably be exposed to the effects of the weather, such as lightning, to be of such construction or as necessary protected so as to prevent, so far as is reasonably practicable, danger arising from such exposure. In reality, it is not just electrical equipment that needs protection but buildings and their contents. A lightning strike onto a building may find its way to earth via the electrical installation. This risk can be minimised by careful attention to the design and installation of a lightning protection system for the building as a whole. Lightning protection is outside the scope of BS 7671 but is covered by another British Standard (BS EN 62305).

The standard comprises four parts:

- BS EN 62305-1:2011 *Protection against lightning. General principles*;
- BS EN 62305-2:2012 *Protection against lightning. Risk management*;
- BS EN 62305-3:2011 *Protection against lightning. Physical damage to structures and life hazard*; and
- BS EN 62305-4:2011 *Protection against lightning. Electrical and electronic systems within structures*.

Static electricity

[E3026] This is a common phenomenon. Most people have experienced the effects of static electricity as a short sharp shock when the body or another object appears to have become charged with electricity and discharge occurs as body contact, usually a hand, is made with the object. Frequently occurring examples are alighting from a vehicle and opening a metal filing cabinet in an office. The mechanism surrounding the build-up of static charge is not fully understood but involves interaction between different materials aided by particular atmospheric conditions. Certain work processes, involving the generation and movement of fine dusts or the flow of liquids through pipes and orifices, may give rise to the generation of static in work processes. The *EWR (SI 1989 No 635)*, in relating to electrical systems and associated equipment are not intended to apply to such situations. However, other legislation serves to ensure that appropriate safeguards are put in place. In particular, *PUWER (SI 1998 No 2306), Reg 12* requires that measures shall be taken to prevent or, where that is not reasonably practicable, to adequately control specified hazards associated with work equipment, such as the premature explosion of any article or substance produced, used or stored in it. Preventive measures might include redesigning the process or equipment or using different construction materials to prevent the build-up of static. A common practice is to electrically bond together all conductive parts, to reduce the possibility of charge build up between different surfaces.

Some work processes depend on electrostatic, eg electrostatic paint spraying and chimney flue gas filters. In these cases, *Regulation 6* of the *EWR (SI 1989 No 635)* is applicable.

Electricity in potentially explosive atmospheres

[E3027] The use of electricity in an area with a potentially explosive atmosphere (PEA) poses particular problems. An explosive atmosphere could be present in paint spray booths, near fuel tanks, in sumps, or places where aerosols, vapours, mists, gases, or dusts exist. As the use of electricity can generate hot surfaces or sparks, it can ignite an explosive atmosphere.

Ideally, the risk should be minimised either by siting any electrical equipment outside the area or the work location designed and constructed so that natural ventilation will ensure that a PEA will never arise within it. In practice, the latter option may be difficult to confirm and attention will then need to be given to the design and construction of the equipment, taking account of the likelihood of a PEA arising, to achieve the required level of safety. A British Standard provides a classification for potentially explosive atmospheres based upon the likelihood and duration of a PEA arising (BS EN 60079). BS EN 60079-10-1:2015 sets out the essential criteria against which the ignition hazards can be assessed, and gives guidance on the design and control parameters which can be used in order to reduce such a hazard.

Online HSE guidance on electricity in potentially explosive atmospheres advises that: 'Care should be taken to prevent static discharges in potentially explosive atmospheres. Measures such as earth bonding and the selection of antistatic work clothing and footwear can help to reduce the risk of static discharges.'

It says that electrical and non-electrical equipment and installations in PEAs must be specially designed and constructed to eliminate or reduce the risks of ignition. This can be done by sealing electrical equipment so that the explosive atmosphere cannot come into contact with electrical components, reducing the power of electrical equipment, and de-energising electrical equipment where a fault or an explosive atmosphere is detected.

Recently installed equipment should be marked with an 'Ex' to show that it is suitable for use in a PEA. All new equipment must comply with the *Equipment and Protective Systems Intended for Use in Potentially Explosive Atmospheres Regulations 1996* that implemented the European ATEX Directive.

This requires it to be assessed as suitable for a particular explosive atmosphere type and for this to be marked on the equipment along with CE and ATEX markings. Most new equipment being sold in the UK for use in potentially explosive atmospheres must have an ATEX certificate.

It also advises that equipment for use in explosive atmospheres should be regularly inspected and maintained to ensure it does not pose an increased risk of causing a fire or explosion, and maintenance of the equipment should only be carried out by people who are competent to do so. It points to BS EN 60079-17: *Explosive atmospheres. Electrical installations inspection and maintenance* as a source of further guidance as to the frequency and scope of maintenance required. BS EN 60079-10 *Explosive atmospheres. Classification of areas. Explosive dust atmospheres* provides guidance in respect of electrical equipment for use in the presence of flammable dusts.

In addition, in areas where it is possible that an explosive atmosphere may exist, the requirements of the *Dangerous Substances and Explosive Atmospheres Regulations 2002 (SI 2002 No 2776) (DSEAR)* apply. The regulations require that such areas be risk assessed before any new work is carried out in them and that measures be taken to control the risks. The HSE has produced guidance on *DSEAR* that explains how the Regulations can be complied with. This can be accessed via the HSE website at www.hse.gov.uk/fireandexplosion/dsear.htm. The *Dangerous Substances and Explosive Atmospheres Regulations 2002 (SI 2002 No 2776)* control fire and explosion risks in workplaces and are examined in more detail in DANGEROUS SUBSTANCES AND EXPLOSIVE ATMOSPHERES).

A common example of where a PEA may arise is in battery rooms. The charging of lead-acid batteries, even those that are described as maintenance free, generates hydrogen gas. Lack of ventilation may allow the build-up of a PEA since the lower explosive limit (LEL) is only four per cent. As hydrogen is lighter than air the PEA is likely to occur first at ceiling level where a light fitting could produce a spark to cause ignition. For this reason, care must be given to the ventilation requirements for battery rooms, with vents at both high and low level to achieve a good level of natural ventilation. Additionally, light fittings of an appropriate type, should be wall-mounted. Because of the high number of accidents involving electric storage batteries the HSE has produced a leaflet concerning safe charging and use *Using electric storage batteries safely* (INDG139) (www.hse.gov.uk/pubns/indg139.pdf).

Work near underground cables and overhead lines

[E3028] By means of *Regulation 4(3)* the *EWR (SI 1989 No 635)* regulate work carried out in close proximity to live conductors, whether overhead lines or buried cables, in situations where there is electrical danger to be avoided. Many accidents involving inadvertent contact with live overhead lines arise during construction or agricultural activities. In the case of the former, examples are the movement or operation of cranes, dumpers, excavators and access equipment too close to a line and for the latter, the use of crop spraying or other irrigation equipment often poses particular problems. With high voltage overhead lines, it is important to recognise that the electricity can flash across a gap and therefore actual contact does not have to be made for injury to occur. A safe system of work depends on establishing a safety clearance distance from the line and height clearances where plant is required to pass underneath. These measures are described in more detail in HSE guidance *Avoidance of danger from overhead electrical lines* (GS6) (www.hse.gov.uk/pubns/gs6.pdf) and, where there is any doubt, advice should be sought from the line owner.

Excavation work accounts for many cable damages, some of which result in serious injury or even death. Again, it is essential to implement a safe system of work founded on:

- establishing the presence of any buried cables by reference to plans (electricity distribution companies are required to make such information available);
- verifying the position of cables by the use of a cable locating device;
- employing safe excavation techniques involving the digging of trail holes.

The HSE has published guidance on avoiding danger from all types of underground service, which includes electricity cables: Avoiding danger from underground services (HSG47) (www.hse.gov.uk/pubns/gs6.htm).

Other relevant legislation on electricity

The Plugs and Sockets etc (Safety) Regulations 1994 (SI 1994 No 1768)

[E3029]–[E3030] These Regulations were made under the *Consumer Safety Act 1978* in response to concern that consumer safety was being compromised by the substantial quantity of counterfeit and unsafe electrical plugs and sockets being placed on the UK market and by the provision of electrical equipment without an appropriate means to connect it to the mains supply in the consumer's home. They contain requirements for the safety of plugs, sockets, adaptors and fuse links designed for use at a voltage of not less than 200 V. For example, under these Regulations, the circuit conductor pins (L and N) of 13 A plugs are required to be partially insulated.

Electrical Equipment (Safety) Regulations 2016 (SI 2016 No 1101)

[E3031] The *Electrical Equipment (Safety) Regulations 2016* implement European Directive 2014/35/EU, which harmonises laws relating to making available electrical equipment designed for use within certain voltage limits. They replace the previous *Electrical Equipment (Safety) Regulations 1994 (SI 1994 No 3260)*. 'Economic operators' – manufacturers, importers, distributors or authorised representatives – must ensure that electrical equipment made available on the market is safe.

Manufacturers must ensure that electrical equipment has been designed and manufactured in accordance with safety objectives set out in the regulations. They must have a conformity assessment procedure carried out before the equipment is placed on the market, affix the CE marking and label the equipment. Importers must check the manufacturer has carried out a conformity assessment procedure and labelled the electrical equipment correctly and indicate on the electrical equipment their name and address. Distributors must act with due care and check the equipment bears the CE marking and is labelled correctly.

The Electricity Safety, Quality and Continuity Regulations 2002 (SI 2002 No 2665)

[E3032] These Regulations cover electricity supply systems and the provision of electricity supplies to consumers. Their prime purpose is to ensure public safety. They are mainly directed at those with responsibilities for public supply networks, but are also of interest to organisations operating private distribution systems and highway power supply networks, particularly if these encroach on public land. Duties are placed mainly on generators, distributors, suppliers and meter operators. Some companies may be undertaking more than one of these roles. Some responsibilities are also placed on consumers. The Regulations prescribe the nominal supply voltage and frequency and the statutory limits of variation for these. For standard low voltage supplies this is 400/230 V +10 per cent/–6 per cent, 50 Hz ±1 per cent.

The Regulations specify important safety requirements concerning the design and construction of electricity networks, eg minimum ground clearances of overhead line conductors, the safeguarding of buried electricity cables and the protection of distribution equipment in substations, etc. Some of the requirements mirror those contained in the *EWR (SI 1989 No 635)*. The Regulations also specify fundamental requirements for safety concerning consumers' installations, prior to connection. These link in with the fundamental principles for the safety of electrical installations contained in chapter 13 of BS 7671. This is not only to ensure the safety of those using an installation but also to minimise the risk of the supply to other consumers being adversely affected, eg in terms of loss of supply or its quality, in consequence of how it is being used.

The regulations were amended by the Electricity Safety, Quality and Continuity (Amendment) Regulations 2006. The amendments:

- updated references to British Standard documents;
- more clearly exclude trams, trolleybuses and other forms of guided transport from the requirements of the Regulations on 'distributors';
- improved the resilience and reliability of overhead electricity distribution networks; and
- extended the scope of the Regulations in order to apply to offshore generating installations.

They were also amended by the *Electricity Safety, Quality and Continuity (Amendment) Regulations 2009*. Regulation 2 incorporated the latest revision to the British Standard Requirements for electrical installations (BS 7671) into the ESQCR. It enhanced the requirements for protection for people and livestock against injury, and property against damage caused by voltage disturbances and electromagnetic influences.

Building Regulations 2010 (SI 2010 No 2214)

[E3033] The Building Regulations require that building work must be carried out so that it complies with the requirements set out in Parts A to P of Schedule 1. Part P relates to electrical safety and sets out that: "Reasonable provision

[E3033] Electricity

shall be made in the design and installation of electrical installations in order to protect persons operating, maintaining or altering the installations from fire or injury."

The requirements concern electrical installations intended to operate at low or extra-low voltage and are:

- In or attached to a dwelling;
- In the common parts of a building servicing one or more dwellings but excluding power supplies to lifts;
- In a building that receives its electricity from a source located within or shared with a dwelling, or
- In the garden or in or on land associated with a building where the electricity is from a source located within or shared with a dwelling.

Emissions into the Atmosphere

Andrea Oates

Introduction to emissions into the atmosphere

[E5001] Pollution of the atmosphere is a subject that encompasses the whole range of distance scales (from personal to global) and can be important over minutes or many lifetimes. Regulation of emissions, therefore, tries to account for all these effects and, as a consequence, it has to take a variety of forms. The concept of the different distance scales is illustrated in Figure 1 below:

Figure 1: Distance Scales

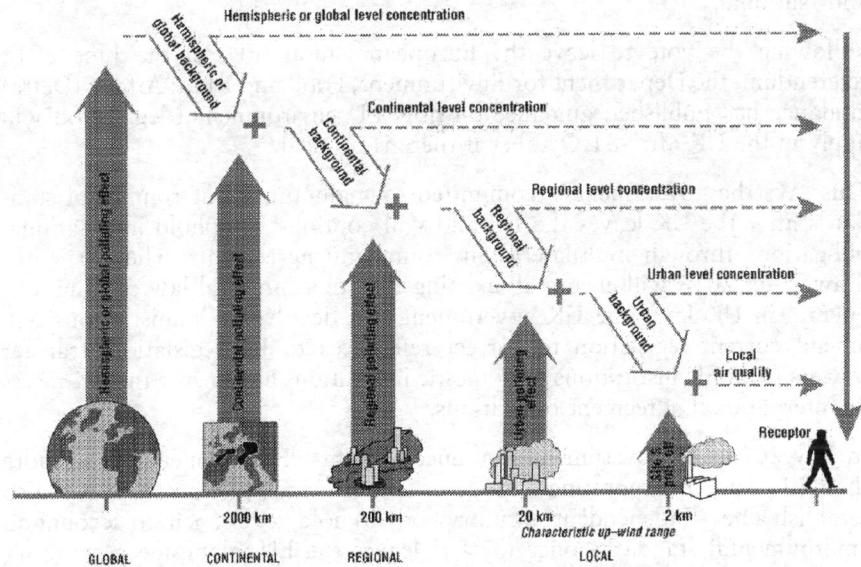

A 'receptor' (a human being, a population, an ecosystem) experiences air pollution as a sum of the contribution from various sources. The extent to which the sources contribute might depend on the pollutant of concern. A pollutant such as sulphur dioxide is relatively short lived in the atmosphere and has the capacity to affect the lungs of people within tens of kilometres of the point of emission. At distances of hundreds of kilometres, however, the sulphur dioxide will have been transformed into sulphate, which may be absorbed into cloud water and be deposited on the ecosystems of another country. This is an example of trans-boundary pollution and popularly known as 'acid rain'.

Pollutants that are chemically more stable than sulphur dioxide have the potential to be dispersed throughout the globe. Some organic pollutants are an excellent example of this. Poly-chlorinated biphenyls (PCBs) were used widely in the middle of the twentieth century as a component part of electrical transformers. They are highly toxic and accumulate in body fats, with animals (and humans) at the top of the food chain the most exposed. Hence, polar bears living in the Arctic Circle are far removed in distance and time from the original emission but still experience the impact. The manufacture, sale and use of products containing persistent organic pollutants (POPs), including PCBs, is now banned.

Regulation of emissions from industry must recognise all these effects – from the large coal-fired power station with sulphur emissions to the small manufacturer using solvents. Where the potential exists for an air pollution impact, regulators will be seeking to restrict emissions. This chapter examines the means by which this goal is currently being achieved.

There are three main sources of legislation and regulation to which industry has to respond and which this chapter seeks to cover: International, European and National.

Following the vote to leave the European Union (EU) in the June 2016 referendum, the Department for Environment, Food and Rural Affairs (Defra) guidance has published guidance on how EU environmental legislation will apply in the UK after 31 October if there is no deal.

This says the government is committed to maintaining environmental standards after the UK leaves the EU and will continue to uphold international obligations through multilateral environmental agreements. The EU Withdrawal Act 2018 will ensure all existing EU environmental law continues to operate in UK law. The UK government and devolved administrations will amend current legislation to correct references to EU legislation, transfer powers from EU institutions to domestic institutions and ensure the UK meets its international agreement obligations.

In July 2018, the government announced the first Environment Bill in more than 20 years, incorporating a range of issues including clean air. It will also establish a new, independent statutory body to hold government to account on environmental standards once the UK leaves the EU, alongside a statutory statement of environmental principles to guide future government policy making.

As at April 2019, the government was considering what interim measures may be necessary in a no deal scenario after 31 October and before the Environment Act is passed and comes into effect. The guidance also explains that the UK's legal framework for enforcing domestic environmental legislation by UK regulatory bodies or court systems is unaffected by leaving the EU and continues to apply. Environmental targets currently covered by EU legislation are already covered in domestic legislation. Permits and licences issued by UK regulatory bodies will continue to apply.

Regulatory drivers

Human health

[E5002]–[E5014] The effects of air pollution on human health have long been recognised in the UK. The often-quoted example is the prohibition of coal burning in London in 1273, on the grounds that it was 'prejudicial to health'.

Coal burning has been a major cause of air pollution since then and provoked most of the legislation aimed at improving air quality and human health. The infamous four day 'smog' of December 1952 was shown to have caused nearly 4,000 additional deaths through bronchitis and other diseases of the respiratory system. There was also an increase in heart disease.

Since then air quality has improved considerably as a result of cleaner-air technology and tighter environmental legislation, but air pollution is still an issue of major concern in the UK. Launching the government's clean air strategy in January 2019 (www.gov.uk/government/publications/clean-air-strategy-2019) health secretary Matt Hancock described it as a 'health emergency' and the 'biggest single environmental cause of death'. It is responsible for around 36,000 deaths each year and puts extra, preventable strain on the NHS through increased incidents of heart disease, stroke, lung cancer and child asthma.

Public Health England guidance published in 2018 (www.gov.uk/government/publications/health-matters-air-pollution/health-matters-air-pollution) sets out the health impacts of the key air pollutants. This is summarised below:

Particulate matter (PM)

PM is a generic term used to describe a complex mixture of solid and liquid particles of varying size, shape and composition. Primary particles are emitted directly while secondary PM is formed in the atmosphere through complex chemical reactions. The composition of PM varies greatly and depends on factors including geographical location, emission sources and weather.

The main sources of man-made PM are the combustion of fuels by vehicles, industry and in domestic properties, as well as other physical processes including tyre and brake wear. Natural sources include wind-blown soil and dust, sea spray particles, and fires involving burning vegetation.

PM_{10} is used to describe course particles that are less than 10 microns (μm) in diameter. $PM_{2.5}$ is used to describe fine particles that are less than 2.5 μm in diameter..$PM_{0.1}$ is used to describe ultrafine particles that are less than 0.1 μm in diameter.

Particles larger than 10 μm are mainly deposited in the nose or throat, but particles smaller than 10 μm pose the greatest risk because they can be drawn deeper into the lung. The strongest evidence for effects on health is associated with $PM_{2.5}$.

Long-term exposure to PM increases mortality and cardiovascular and respiratory diseases. Outdoor air pollution, particularly PM, has also been classified by the International Agency for Research on Cancer (IARC) as carcinogenic to humans (a Group 1 carcinogen) and causing lung cancer.

Nitrogen dioxide (NO$_2$)

NO$_2$ gas is produced with nitric oxide (NO) by combustion processes. Together they are often referred to as oxides of nitrogen (NOx). Defra estimates that 80% of NOx emissions in areas where the UK is exceeding NO$_2$ limits are due to transport, with emissions from diesel cars and vans the largest source. Other sources include power generation, industrial processes and domestic heating.

The independent Committee on the Medical Effects of Air Pollutants (COMEAP) has established that short-term exposure to NO$_2$, particularly at high concentrations, is a respiratory irritant that can cause inflammation of the airways, leading to shortness of breath for example. Studies have shown associations of NO$_2$ in outdoor air with reduced lung development and respiratory infections in early childhood and effects on lung function in adulthood.

Epidemiological studies have also shown associations of outdoor NO$_2$ with adverse health effects, including reduced life expectancy. These may be caused by NO$_2$ itself, or by other pollutants emitted at the same time by sources such as road traffic.

Other pollutants

Sulphur Dioxide (SO$_2$)

This is produced when sulphur-containing fuels, such as coal, are burned. Chemical reactions of SO$_2$ can also produce sulphates, which remain in the air as secondary particles, contributing to the PM mix (see above).

SO$_2$ has an irritant effect on the lining of the nose, throat and airways, and the effects are often felt very quickly. Most SO$_2$ in the UK now comes from industrial sources, such as coal and oil-burning power stations, as well as domestic sources such as boilers and stoves.

Ammonia (NH$_3$)

The main health impacts of NH$_3$ arise through its role in secondary PM$_{2.5}$ formation and health effects associated with exposure to PM, as described above. Agricultural emissions of NH$_3$ have been reported to be key contributor to some short-term episodes of high PM pollution in recent years.

Ozone (O$_3$)

Ozone gas occurs both in the earth's upper atmosphere and at ground level. Ground level, or tropospheric O$_3$, is not emitted directly into the air but is created by photochemical reactions involving the precursor pollutants NOx and volatile organic compounds (VOCs). Several epidemiological studies have reported adverse associations between short-term exposure to O$_3$ and human health.

The effects of exposure to O$_3$ are predominantly respiratory, but adverse effects on the cardiovascular system have also been reported.

Carbon Monoxide (CO)

CO is produced when fuels such as gas, oil, coal and wood burn without enough oxygen in household appliances, including boilers, central heating systems, gas fires, water heaters, cookers and open fires, for example. Burning charcoal, running cars and the smoke from cigarettes also produce CO gas. Exposure to high indoor levels can be fatal, while exposure to lower levels can result in symptoms that resemble flu, viral infections or food poisoning.

Non-Methane Volatile Organic Compounds (NMVOCs)

NMVOCs are emitted from a wide variety of products, industrial processes and agriculture. They also form a significant component of indoor air pollution emitted from household products. In the outside atmosphere, they react with NOx in the presence of sunlight to form tropospheric O_3, known to be harmful to health and the environment.

Indoors, VOCs emitted from consumer products are not thought to be a significant public health issue when homes are well-ventilated and products used according to the manufacturers' instructions. However, some sensitive people may suffer irritation of the eyes, nose and throat, headaches and dizziness if they are exposed.

Environmental consequences of air pollution

[E5015] Defra's February 2019 statistical release, *Emissions of Air Pollutants in the UK, 1970 to 2017 (htttps://assets.publishing.service.gov.uk/government/ uploads/system/uploads/attachment_data/file/778483/Emissions_of_air_p ollutants_1990_2017.pdf)*, explains that air pollution also damages ecosystems as a result of:

- acidification (SO_2, NOx and NH_3) – where chemical reactions involving air pollutants create acidic compounds. When deposited on land and aquatic systems, these can cause harm to soils, vegetation and buildings;
- eutrophication (NOx and NH3) – where nitrogen is deposited in soils or in rivers and lakes through rain, affecting the nutrient levels and diversity of species in sensitive environments, for example encouraging algae growth in lakes and water courses; and
- ground-level ozone (NOx and NMVOCs) – where chemical reactions involving NOx and NMVOCs produce the toxic gas ozone (O3). This can damage wild plants, crops, forests and some materials, and is a greenhouse gas contributing to global warming.

Air pollutants released in one country can be transported in the atmosphere, contributing to harmful impacts elsewhere.

Nuisance

[E5016]–[E5020] Air pollution can cause harm to the senses as well as harm to human health and the natural environment. Odour and dust nuisance can result in costly disputes and generally harm community relations.

[E5016] Emissions into the Atmosphere

Various emissions from small-scale processes (eg the outputs of local exhaust ventilation) are covered by statutory nuisance legislation (see E5062) and, depending on a range of factors, may be regarded as creating a local nuisance.

Regional and global issues

Trans-boundary pollution

[E5021] There are currently two main sources of controls on trans-boundary air pollution:

- the United Nations Economic Commission for Europe ('UNECE') Convention on Long-Range Transboundary Air Pollution ('CLRTAP'), which is an important framework for environmental assessment and policy in Europe (www.unece.org/) (see E5026 below); and
- the National Emissions Ceiling Directive (NECD).

The NECD sets national emission reduction commitments for Member States and the EU for five air pollutants: nitrogen oxides (NOx), non-methane volatile organic compounds (NMVOCs), sulphur dioxide (SO_2), ammonia (NH3) and fine particulate matter ($PM_{2.5}$). A new NECD came into force on 31 December 2016 and sets 2020 and 2030 emission reduction commitments for these five main air pollutants. The emission ceilings for 2010 set in an earlier directive remain applicable for Member States until the end of 2019.

The new directive transposes the reduction commitments for 2020 agreed by the EU and its Member States under the 2012 revised Gothenburg Protocol (see E5026) to the CLRTAP (see above). The more ambitious reduction commitments agreed for 2030 are designed to reduce the health impacts of air pollution by half compared with 2005. The Directive also requires Member States to draw up National Air Pollution Control Programmes to contribute to the successful implementation of air quality plans established under the EU's Air Quality Framework Directive (see E5033 below).

Climate change

[E5022] Climate change has come to be widely regarded as the most pressing global environmental problem. The natural presence of 'radiatively-active' gases in the atmosphere is essential for life. These gases trap heat in the lower atmosphere, creating a 'greenhouse effect'. The six main greenhouse gases are:

- carbon dioxide;
- methane;
- nitrous oxides;
- chlorofluorocarbons ('CFCs');
- PFCs; and
- ground level ozone.

Increasing concentrations of these greenhouse gases enable more infra-red radiation to be absorbed in the lower atmosphere, upsetting the earth's radiation balance. This increase allows the troposphere to be warmed more significantly, and the upper layers to be cooled. Without any greenhouse effect at all, the earth would be uninhabitable for humans. In geological history, the earth has variously been warmer and colder than at present. However, a fundamental environmental issue today is the rapid rate at which the climate appears to be changing.

The United Nations body for assessing the science related to climate change, the Intergovernmental Panel on Climate Change (IPCC), set out in a special report in October 2018 that limiting global warming to 1.5°C above pre-industrial levels will require urgent, far-reaching and unprecedented changes in all aspects of society. It also set out that countries have just 12 years (until 2030) to make massive changes to avoid the devastating consequences of allowing global warming to exceed this limit.

In November 2018, the government launched the UK Climate Projections 2018 (UKCP18) (www.metoffice.gov.uk/research/collaboration/ukcp) showing how the climate could change over the next century. Using the latest science from the Met Office and around the world, they illustrate a range of future climate scenarios until 2100 and show increasing summer temperatures, more extreme weather and rising sea levels, with urgent international action needed. The high emission scenario shows:

- summer temperatures could be up to 5.4°C hotter by 2070, while winters could be up to 4.2°C warmer;
- the chance of a summer as hot as 2018 is around 50% by 2050;
- sea levels in London could rise by up to 1.15 metres by 2100; and
- average summer rainfall could decrease by up to 47% by 2070, while there could be up to 35% more precipitation in winter.

Sea levels are projected to rise over the 21st century and beyond under all emission scenarios – meaning an expected increase in both the frequency and magnitude of extreme water levels around the UK coastline.

Even in the low emission scenario, the projections show the UK's average yearly temperature could be up to 2.3°C higher by the end of the century.

Consequences

[E5023] The consequences of rapid climate change are potentially massive and unpredictable. Regional sea currents may change dramatically due to polar melting. Vegetation will be in greater competition as optimum conditions become scarce. This may cause shifts of plant and animal life. Extra warmth can trigger insect plagues and plant diseases. Such consequences will affect agriculture, which is also directly affected by changes in rainfall. Changes to the land and coastal areas can lead to population and wildlife displacement, and land degradation.

The British Medical Association describes the potential adverse health consequences of climate change as including:

- an increase in frequency and severity of extreme weather events such as floods, droughts, extreme storms and heat waves;
- the spread of vector-borne diseases such as malaria and dengue to new locations;

- worsening nutrition resulting from decreased agricultural productivity and higher global food prices;
- rising sea levels and associated population displacement; and
- an exacerbation of poverty and inequalities.

UK Climate Change Risk Assessment

[E5023.1] In January 2017, a government report set out the six priority risk areas requiring further action in the UK over the next five years:

- from flooding and coastal change;
- to health and well-being from high temperatures;
- due to water shortages;
- to natural capital;
- to food production and trade; and
- from pests and diseases and invasive non-native species.

The report can be found on the government's website at: www.gov.uk/government/publications/uk-climate-change-risk-assessment-2017.

The depletion of stratospheric ozone

[E5024] The depletion of stratospheric ozone during the last few decades is a global problem. Stratospheric ozone protects the Earth's surface from harmful UV radiation. The loss is greatest nearer the poles and is sometimes referred to as the 'ozone hole'. The cause of this problem was the increase of industrially produced CFCs, halogens and other halogenated substances in the upper atmosphere, and although most emissions have stopped due to international bans on these substances, the ozone layer will take many years to recover and it may not recover to previous levels.

Consequences

[E5025] Any depletion of ozone means that certain wavelengths of ultraviolet solar radiation are able to penetrate through to the earth's surface, damaging human health and leading to an increase in skin cancer, as well as damaging ecosystems. This can also lead to a change in global circulation and climate, as the absorption of the radiation by ozone can lead to heat formation in the atmosphere.

International and European agreements

[E5026] Various agreements have been drawn up at international level. Key agreements include the Montreal Protocol on the Control of Ozone-depleting Substances (1988) (ozone-depleting substances), United Nations Framework Convention on Climate Change (1992) (carbon dioxide, unofficially known as the 'Climate Change Treaty'), Sulphur Protocol (1985) (sulphur), Nitrogen Oxides Protocol (1991), the VOC (Volatile Organic Compounds) Protocol (1994) and the Gothenburg Protocol (1999). Many of the protocols have then been implemented by European action.

There are two main European bodies that drive legislation and protocols, namely the European Union (EU) and the United Nations Economic Commission for Europe (UNECE). The latter body brings together 56 countries located in the European Union, non-EU Western and Eastern Europe, South-East Europe and Commonwealth of Independent States (CIS) and North America. Its chief role is to examine the consequences of trans-boundary pollution through the Convention on Long-Range Transboundary Air Pollution (CLRTAP) (see **E5021** above). The Convention states that all countries should;

> endeavour to limit and, as far as possible, gradually reduce and prevent air pollution, including long range trans-boundary pollution.

The Convention has been extended by eight protocols, including the following:

- *The 1985 Sulphur Protocol* required signatories to reduce national sulphur emissions by 30%, based on 1980 levels, on or before 1993. The Protocol was subsequently revised and signed in Oslo in June 1994 and came into force in August 1998. The UK ratified the Protocol (December 1996) and agreed to reduce SO2 emissions by 50% by 2000, 70% by 2005 and 80% by 2010, (again based on 1980 levels). This required 'fuel switching' and advances in abatement technology. In the main, these targets have most impact on the power generation and oil refining industries.
- The 1988 Nitrogen Oxides Protocol was adopted in Sophia in 1991. It froze emissions of nitrogen oxides by 1994 using 1987 as a baseline. The *Gothenburg Protocol to Abate Acidification, Eutrophication and Ground Level Ozone (1999)* also concerns nitrogen oxides as well as VOCs and ammonia.
- *The 1991 Volatile Organic Compounds Protocol* was signed in Geneva and came into force in September 1997. The UK ratified the Protocol in June 1994. VOCs are defined as *'all organic compounds of anthropogenic nature, other than methane, that are capable of producing photochemical oxidants by reactions with nitrogen oxides in the presence of sunlight.'* The Protocol required most parties to reduce overall VOC emissions by 30% on or before 1999, using a base year of 1988. Additionally, the protocol obliged signatories to control emissions from industries through new emission limits and to introduce abatement technologies as well as reduce solvent use and reduce VOC emissions through petrol distribution and refuelling.
- *The 1998 Heavy Metals Protocol* was signed in June 1998 at Aarhus. It requires emissions of cadmium, lead and mercury to be reduced below 1990 levels. Its aim is to reduce emissions through the use of stricter emissions limit values for industry and the use of best available technology (see **E5027** below).
- *The Persistent Organic Pollutants Protocol* was also signed in June 1998 at Aarhus. It aims to phase out the production of and use of a defined list of substances, as well as imposing requirements to eliminate discharges, emissions and losses and to ensure safe disposal methods. The list covers 16 substances in three categories, (pesticides, industrial chemicals and by products or contaminants, eg dioxins).

- *The Acidification, Eutrophication and Ground-level Ozone Protocol* – also referred to as the Multi-pollutant, Multi-effect Protocol, the 2nd NOx Protocol or the Gothenburg Protocol – was signed in December 1999. In Europe, it has a similar scope to EU directives in terms of pollutants, but applies to all European countries, not just EU Member States. When the Gothenburg Protocol is fully implemented, Europe's sulphur emissions should be cut by at least 63%, its NOx emissions by 41%, its VOC emissions by 40% and its ammonia emissions by 17% compared to 1990. This Protocol also sets limit values for specific emission sources (eg combustion plant, electricity production, transport) and requires 'best available techniques' to be used. VOC emissions from products as paints or aerosols will also have to be reduced.

The effects of these Protocols on industry are not usually seen immediately or directly. Normally they take effect through the subsequent actions of national governments and regulatory agencies who propose action and legislation designed to achieve the aims of the Protocols. In this sense, the European Commission is strongly interlinked with the CLRTAP Protocols. Not only is the Community a signatory in itself, but the Commission will frame directives so as to meet the aims of the Protocols.

The most active part of the Commission with regard to air quality legislation is Directorate General Environment ('DG Environment'). The units within this Directorate have direct responsibility for drafting new directives that set air quality standards and implement emissions reductions in key sectors.

DG Environment also has responsibility for some legislation relevant to industry and air pollution control, such as the implementation of Integrated Pollution Prevention and Control and the Waste Incineration Directive.

The Montreal Protocol

[E5026.1] The Montreal Protocol was drawn up in 1988 and signed by 50 countries. Since then, over 170 countries have signed up. The original protocol set targets for controlling CFCs, and halons. Carbon tetrachloride, 1,1,1 – trichloroethane, hydrochlorofluorocarbons (HCFCs) and methyl bromide and halons 1211, 1301 and 2402 were added to amendments to the protocol.

The Montreal Protocol affects many industries, eg refrigeration and air conditioning, aerosol manufacture and foam blowing. It is directly binding on the UK via European Regulations.

Hydrochlorofluorocarbons were prohibited from June 1995 except for use as solvents, refrigerants, insulating foams, carrier gas for various sterilising systems and feedstock for chemicals manufacture. Under the Beijing amendment (1999), production by developed countries was frozen in 2004 at 1989 levels. Exports to countries who have not signed up to the Montreal Protocol were banned from 2004.

From January 2000, the use of HCFCs was prohibited in new equipment in cold stores and warehouses, and for equipment of 150 kW and over.

The current UK Regulations are the *Ozone-Depleting Substances Regulations 2015 (SI 2015 No 168)* (as amended by the *Ozone-Depleting Substances and Fluorinated Greenhouse Gases (Amendment etc) (EU Exit) Regulations 2019 (SI 2019 No 583)*).

The UK regulations enforce EU Regulation (EC) No 1005/2009 which controls the production, placing on the market and use of substances that deplete the ozone layer and require that those who work on the recovery, recycling, reclamation or destruction of controlled substances and the prevention and minimising of leakages of controlled substances have minimum qualifications.

International climate change treaties

[E5027] The Climate Change Treaty was agreed in Rio de Janeiro in 1992.

The Kyoto Protocol (the *Kyoto Protocol to the 1992 United Nations Framework Convention on Climate Change (1997)*) was an initiative that came out of the 1992 Rio Earth Summit. At the 1997 Kyoto conference, cuts in the emission of six greenhouse gases: carbon dioxide, methane, nitrous oxide, hydrofluorocarbons, perfluorocarbons and sulphurhexafluoride were agreed. All developed countries agreed to reduce emissions by specified amounts to reduce the potential for global warming from sources such as road traffic. Ratification of the Protocol meant that the UK was legally required to reduce emissions of the six greenhouse gases by 12.5% below 1990 levels in 2008–2012.

Following the 2015 Paris climate conference, 195 countries adopted the first-ever universal, legally-binding global climate deal setting out an action plan to limit global warming to 1.5°C above pre-industrial levels. The agreement covers the period from 2020 onward, and in order to enter into force, 55 countries that make up at least 55% of global emissions had to have ratified it. The EU formally ratified the agreement on 5 October 2016, enabling it to come into force on 4 November 2016. At the international climate change conference in Katowice in Poland (COP 24) in December 2018, governments agreed to a 'rule book' setting out how they will deliver their promised cuts in carbon emissions to keep to the 1.5°C limit.

Climate change – emissions trading

[E5027.1] The European Commission describes the EU emissions trading system (EU ETS) as 'a cornerstone of the EU's policy to combat climate change and its key tool for reducing greenhouse gas emissions cost-effectively'.

The EU ETS works on a 'cap and trade' basis. A 'cap', or limit, is set on the total amount of certain greenhouse gases that can be emitted by each installation in the system. The cap is reduced over time so that total emissions fall. In 2020, emissions from sectors covered by the EU ETS will be 21% lower than in 2005. In 2030, under the revised system they will be 43% lower.

Within the cap, companies receive or buy emission allowances which they can trade with one another as necessary. They can also buy limited amounts of international credits from emission-saving projects around the world. The limit on the total number of allowances available ensures that they have a value. After each year, a company must surrender enough allowances to cover all its emissions or face heavy fines. If a company reduces its emissions, it can keep the spare allowances to cover its future needs or sell them to another company.

Directive 2003/87/EC came into force on 25 October 2003 and established a scheme for greenhouse gas (GHG) emission allowance trading within the European Union from 1 January 2005. The scheme (EU ETS) had two initial phases, 2005–07 and 2008–12 (the latter corresponds to the first Commitment Period under the Kyoto Protocol). The third phase, 2013–20, detailed in Directive 2009/29/EC is currently running:

The scheme covers:

- Carbon dioxide (CO_2) from power and heat generation, energy-intensive industry sectors including oil refineries, steel works and production of iron, aluminium, metals, cement, lime, glass, ceramics, pulp, paper, cardboard, acids and bulk organic chemicals, and commercial aviation;
- Nitrous oxide (NO) from the production of nitric, adipic, glyoxal and glyoxylic acids; and
- Perfluorocarbons (PFCs) from aluminium production.

Participation in the EU ETS is mandatory for companies in these sectors, but:

- in some sectors only plants above a certain size are included;
- some small installations can be excluded if governments put in place fiscal or other measures that will cut their emissions by an equivalent amount; and
- in the aviation sector, until 31 December 2023, the EU ETS will apply only to flights between airports located in the European Economic Area.

The EU ETS covers around 11,000 energy-intensive industrial installations throughout Europe, including power stations, refineries and large manufacturing plants and was expanded to the aviation industry on 1 January 2012.

From January 2005, all installations covered by the scheme were required to have a GHG emissions permit (non-tradable) enabling the installation to emit GHG from all or part of it. Each Member State was required to draw up national allocation plans (NAP), stating the total quantity of allowances it intended to allocate and how it proposed to allocate them to the installations covered.

A major revision approved in 2009 strengthened the system and means that the third phase is significantly different from and more harmonised than phases one and two. The main changes are:

- a single, EU-wide cap on emissions applies in place of the previous system of national caps;
- auctioning, not free allocation, is now the default method for allocating allowances;

- for those allowances still given away for free, harmonised allocation rules apply which are based on ambitious EU-wide benchmarks of emissions performance; and
- more sectors and gases are included.

The legislative framework of the EU ETS for phase 4 (2021–2030) was revised in early 2018 to enable it to achieve the EU's 2030 emission reduction targets in line with the 2030 climate and energy policy framework and as part of the EU's contribution to the 2015 Paris Agreement (see **E5027**).

Integrated pollution prevention and control ('IPPC') and the Environmental Permitting Regime

Background

Directive on Integrated Pollution Prevention Control 96/61/EC

[E5027.2] The IPPC directive (Directive on Integrated Pollution Prevention Control 96/61/EC) was designed to prevent, reduce, and eliminate pollution at source, through the careful use of natural resources. It applied to:

- energy industries (power, oil and gas);
- the production and processing of metals (ferrous and non-ferrous);
- mineral industries (cement and glassworks);
- chemical industries (organic, inorganic, and pharmaceuticals);
- waste management (landfill sites and incinerators); and
- other (slaughter houses, food/milk processing, paper, tanneries, animal carcass disposal, pig/poultry units and organic solvent users).

It applied to emissions to air, land and water, as well as heat, noise, vibration, energy efficiency, environmental accidents, site protection, and many more processes.

The IPPC directive required applications to show that installations were run in a way that prevented emissions, using the following principles:

- best available techniques ('BAT') to control emissions. Account must be taken of the relative costs and advantages of the available techniques;
- waste minimised and recycled where possible;
- energy conservation;
- accident prevention, and limitation of environmental consequences; and
- sites returned to a satisfactory state after operations cease.

The overall objective for the directive was a high level of environmental protection, and a system of permitting was set up to achieve this relating to:

- plant operating-conditions;
- emission limits for certain substances to air, land and water; and
- annual reporting of pollutant releases.

The *Pollution Prevention and Control Act 1999* paved the way for Regulations to be made which implemented the IPPC Directive in the UK (the *Pollution Prevention and Control (England and Wales) Regulations 2000 (SI*

2000 No 1973). These were superseded by the *Environmental Permitting (England and Wales) Regulations 2007* (as amended) as a result of the consolidating and more stringent IPPC Directive 2008/1/EC.

The current *UK regulations are the Environmental Permitting (England and Wales) Regulations 2016 (SI 2016/1154).*

IPPC Directive 2008/1/EC

IPPC Directive 2008/1/EC was implemented in the UK by the *Environmental Permitting (England and Wales) Regulations 2007 (SI 2007 No 3538)*, with separate regulations applying the IPPC Directive in Scotland, Northern Ireland and to the offshore oil and gas industries (see 5040 below). The 2007 Regulations replaced the previous pollution prevention control regime as well as the waste licensing system derived from the Environmental Protection Act 1990 Part II. They also replaced more than 40 sets of regulations.

The directive applied an integrated environmental approach to the regulation of particular industrial activities. Emissions to air, water and land, as well as a range of other environmental effects were to be considered together. IPPC aimed to prevent emissions (and waste production) and where that was not practicable, reduce them to acceptable levels. It also extended this integrated approach beyond permitting to the restoration of sites when industrial activities cease. Regulators set permit conditions to achieve a high level of environmental protection.

Industrial Emissions Directive (IED) (2010/75/EU)

In July 2010, the European Commission agreed the final text of the Industrial Emissions Directive (IED) (2010/75/EU). This superseded the Integrated Pollution Prevention and Control (IPPC) Directive and entered into force on 6 January 2011. Member States had two years to enact legislation to implement the IED, which also consolidated a further six directives into the new law:

- 1999/13/EC on solvent emissions;
- 2000/76/EC on waste incineration;
- 2001/80/EC on large combustion plants; and
- 78/176/EEC, 82/883/EEC and 92/112/EEC on the titanium dioxide industry.

Its requirements were implemented by amending regulations, the *Environmental Permitting (England and Wales) (Amendment) Regulations 2013 (SI 2013 No 390)* (and through the *Pollution, Prevention and Control Regulations 2012* in Scotland and the *Pollution Prevention and Control (Industrial Emissions) Regulations (Northern Ireland) 2013 (SI 2013 No 160)*).

The timetable was as follows:

- all new installations from 7 January 2013;
- existing installations from 7 January 2014;
- existing installations operating newly prescribed activities from 7 July 2015; and
- large combustion plants must meet specific requirements from January 2016.

New *Environmental Permitting (England and Wales) Regulations 2016 (SI 2016 No 1154)* came into force in January 2017, replacing the *Environmental Permitting (England and Wales) Regulations 2010 (SI 2010 No 675)* and transposing the provisions of 15 Directives which impose obligations required to be delivered through permits or capable of being delivered through permits – including the IED (see below).

Best Available Techniques (BAT)

[E5027.3] These permit conditions are based on the use of 'Best Available Techniques', which takes into account the costs to the operator and the benefits to the environment. It involves identifying options, assessing environmental effects and considering the economic factors.

Determining BAT involves comparing the techniques available to prevent or control emissions and then identifying those that will have the lowest overall impact on the environment. Once options have been identified, the environmental effects (particularly those which are significant) should be assessed. The environmental assessment should focus on identifying and quantifying the effects of releases. The option which minimises the environmental impact of the installation should be used unless economic factors rule this out.

BREF Notes

[E5028]–[E5032] Best available technique (BAT) reference documents, so-called BREFs have been drawn up (or are planned) as part of an exchange of information set out in Article 13(1) of the IED (and previously under the 2008 IPPC directive). These are published on the European IPPC Bureau (EIPPCB) website (eippcb.jrc.ec.europa.eu/reference/). The website provides information about the latest documents and where draft guidance is being prepared. The BREF Notes form the basis of guidance issued in the UK. The BREFs are not directly-applicable guidance notes and so they need to be supplemented by domestic guidance. This is produced by the Environment Agency, drawing on the information contained within BREFs. Where there is no domestic guidance available, operators and regulators should refer directly to the relevant BREFs.

European legislation on air quality

[E5033] Directive 96/62/EC on ambient air quality assessment, the Air Quality framework directive, provided a framework for the EU to use to set limit values for pollutants. This directive identified twelve pollutants for which limit or target values would be set in daughter directives:

- sulphur dioxide (SO_2);
- nitrogen dioxide (NO_2);
- particulate matter (PM_{10});
- lead;
- carbon monoxide;
- benzene;
- ozone;

- polycyclic aromatic hydrocarbons;
- cadmium;
- arsenic;
- nickel; and
- mercury.

This directive is particularly important, as it prepared the ground for a number of significant daughter directives on air quality standards and control measures, as set out below. These were prepared for sulphur dioxide, nitrogen dioxide, oxides of nitrogen, particulate matter, lead, carbon monoxide and benzene, ozone and arsenic, cadmium, mercury, nickel and polycyclic aromatic hydrocarbons.

Daughter directives

Daughter directive for limit values of SO2, oxides of nitrogen, particulate matter, and lead in ambient air

[E5034] The first daughter directive in 1998 established legally binding limit values for SO2, NO2 and PM10, to be achieved by 1 January 2005 and 2010. This directive was adopted in April 1999 and member states were required to implement it by July 2001.

Daughter directive to reduce the sulphur content of liquid fuels

[E5035] This daughter directive aimed to reduce further the emissions of SO2 resulting from the combustion of liquid fuels.

Daughter directive on limit values for benzene and carbon monoxide in ambient air

[E5036] Daughter Directive 2000/69/EC sets limit values for benzene and carbon monoxide.

National emissions ceilings directive and ozone daughter directive

[E5037]–[E5038] In June 1999 proposals were put forward for two further daughter directives. One was to set target values for ozone and the other emission ceilings for various pollutants. The latter is referred to as the *National Emissions Ceilings Directive* ('NECD'). This directive set ceilings for national emissions of SO2, NO2, NH3 and VOCs to be achieved by 2010 and was the main instrument for attaining air quality targets for ambient ozone.

As a result of these proposals, the following daughter directives were adopted:

- Directive 99/30/EC relating to limit values for sulphur dioxide, nitrogen dioxide and oxides of nitrogen, particulate matter and lead in ambient air;
- Directive 2000/69/EC relating to limit values for benzene and carbon monoxide in ambient air;
- Directive 2002/3/EC relating to limit values for ozone in ambient air; and
- Directive 2004/107/EC relating to arsenic, cadmium, mercury, nickel and polycyclic aromatic hydrocarbons in ambient air (known as the Fourth Daughter directive).

[E5039]–[E5040] The Ambient Air Quality and Cleaner Air for Europe Directive 2008/50/EC came into force in 2008 consolidating the Framework Directive and daughter directives into one piece of legislation and introducing new provisions for controlling $PM_{2.5}$, discounting natural sources. It also introduced options for time extensions to meet the PM_{10} and NO_2 deadlines.

The European Commission carried out a review of EU air policy in 2011–2013 and adopted a Clean Air Policy Package in December 2013. This consists of a new Clean Air Programme for Europe with new air quality objectives for the period up to 2030, a revised National Emission Ceilings Directive with stricter national emission ceilings for the six main pollutants, and a proposal for a new Directive to reduce pollution from medium-sized combustion installations.

The package also aligns with an international UNECE agreement to amend the Gothenburg Protocol (see **E5026** above). This sets national emission reduction targets, including for fine particulate matter, to be achieved by 2020. The EU as a whole is going to reduce its emissions of sulphur dioxide, nitrogen dioxide, ammonia, volatile organic compounds and PM2.5 by 59, 42, 6, 28 and 22%, respectively.

For the UK, the figures are: 59, 55, 8, 32 and 30% respectively.

(See also **E5002–E5007** and **E5016** above; **E5065–E5067** below.)

UK regulations relating to emissions into the atmosphere

Environmental Permitting (England and Wales) Regulations 2016 (SI 2016 No 1154)

[E5041]–[E5042] The Environmental Permitting Regime requires operators to obtain permits for particular facilities (and registration of exemptions for others) and aims to protect the environment while effectively and efficiently delivering permitting and compliance. It also encourages regulators to promote best practice in the operation of regulated facilities.

The current regime covers facilities previously regulated under the 2000 Pollution Prevention and Control Regulations, waste management licensing and exemptions and mining waste operations.

The regulations can be summarised as follows:

Part 1: General Provisions

- Regulation 8 defines the term 'regulated facility' and regulation 12 requires regulated facilities to be operated under the authority of an environmental permit.
- The combined classes of 'regulated facility' include (unless they are exempt or excluded) installations, mobile plant, waste operations, mining waste operations, radioactive substances activities, water discharge activities, groundwater and flood risk activities, whether or not carried on as part of the operation of another regulated facility. (See definitions in regulations 2(1) and Schedules 1, 9, 20–23 and 25).

- Regulation 5 defines the term 'exempt facility'. Schedule 2 contains details of the procedure in relation to exempt facilities, including registration requirements.

Part 2 (regulations 11–31) sets out the procedure in relation to environmental permits including the granting, variation, transfer, surrender and revocation of permits, the preparation of standard rules which can be incorporated into permits, and appeals.

Part 3 (regulations 32–35) sets out the functions of the regulator in relation to permits.

Part 4 sets out enforcement and penalties (regulations 36–44).

Part 5 (regulations 45–56) makes provision for public registers to be kept by the regulator.

Part 6 (regulations 57–69) confers powers on the regulator, the Secretary of State and Welsh Ministers and imposes duties on the regulator.

Part 7 contains miscellaneous provisions.

Schedules to the Regulations (including the requirements of various EU directives which must be delivered through the permitting regime) relate to:

- Part A installations: Industrial Emissions Directive (IED) (Schedule 7a);
- Part B installations and Part B mobile plant etc (Schedule 8);
- Waste Operations and Materials Facilities (Schedule 9);
- Landfill (Schedule 10);
- Waste motor vehicles (Schedule 11);
- Waste Electronic Electrical and Electronic Equipment (or WEEE) (Schedule 12);
- Waste Incineration: IED (Schedule 13);
- Solvents Emissions (Schedule 14);
- Large Combustion Plants: IED (Schedule 15);
- Asbestos (Schedule 16);
- Titanium Dioxide: IED (Schedule 17);
- Petrol Vapour Recovery (Schedule 18);
- Waste Batteries and Accumulators (Schedule 19);
- Mining Waste Operations (Schedule 20);
- Water Discharge Activities (Schedule 21);
- Groundwater Activities (Schedule 22);
- Radioactive Substances Activities (Schedule 23);
- Efficiency in heating and cooling energy: Energy Efficiency Directive (Schedule 24);
- Flood Risk Activities and Excluded Flood Risk Activities (Schedule 25);
- Medium Combustion Plants (Schedule 25A); and
- Specified Generators (Schedule 25B).

Regulated facilities

The regulations set out which "regulated facilities" require an environmental permit and provide for the exemption of some waste operations.

An environmental permit is required for a regulated facility that carries out:

- certain listed activities at installations or mobile plants, such as industrial, waste or intensive farming activities;
- waste operations, such as treating waste, metals recycling or operating a waste transfer station;
- water discharge activities, such as discharging pollutants to surface water (rivers, lakes, streams, etc);
- groundwater activities, such as directly or indirectly discharging pollutants to groundwater; and
- radioactive substances activities, such as keeping or using radioactive materials, or accumulating or disposing of radioactive waste.

An Environmental Permit can cover more than one regulated facility where both the regulator and operator are the same for each facility and (with exceptions) all the facilities are on the same site (Regulation 17).

Applications

Regulation 7 defines an operator as the person who has control over a regulated facility, and only an operator can obtain or hold an Environmental Permit. In most cases, a single operator will need to obtain a single environmental permit, but there are exceptions to this. Applications should normally be made at the design stage. The application procedure is set out in Schedule 5 to the Regulations. The application must:

The application procedure is set out in Schedule 5 to the Regulations. The application must:

- Be made by the operator;
- By the current operator and future operator in the case of a transfer application;
- Be made on the form provided by the regulator;
- Include information required by the application form (and any additional information required by the regulator); and
- Include the relevant fee.

The regulator must decide whether to grant or refuse the proposal in an application, and where applicable, what permit conditions to impose. Various periods for determination are set out in the regulations.

The application to the regulator will include an assessment of environmental risk (under normal and abnormal operating conditions) and the regulator must be satisfied that this assessment is robust.

When permitting facilities, the regulator must take into account the requirements of European directives which are set out in Schedules to the Regulations (see above).

Applications can be refused where:

- The operator is not competent to run the regulated facility;
- The environmental impact would be unacceptable;
- The information provided is not adequate for the regulator to determine the permit conditions; or
- The requirements of the European Directive cannot be met.

The regulations set out provisions to vary, transfer and surrender as well as apply for permits.

Standard rules

The regulations contain provisions for the Secretary of State, Welsh Ministers and the Environment Agency or NRBW to make Standard Rules consisting of requirements common to a particular class of facilities. These can be used instead of site-specific permit conditions. Standard rules do not require a site-specific assessment of risk for a standard facility, but they cannot be appealed against. It is the operator's decision as to whether they wish to operate under these rules.

Operator competence

In making the decision as to whether an operator is competent, the regulator will consider management systems, technical competence, including where relevant whether an approved scheme is being complied with, compliance with previous regulatory requirements and financial competence.

Consultation and public participation

The regulations require public consultation on some environmental permit applications A public consultee is anyone who could be affected by the application or has an interest in the application. Information on when and how the Environment Agency consults on permit applications and standard rules for environmental permits can be found at: www.gov.uk/government/publications/environmental-permits-when-and-how-we-consult.

Enforcement

Regulators can serve an enforcement notice if an operator has contravened, is contravening or is likely to contravene any permit conditions (regulation 36). These specify the steps necessary to remedy the problem and the timescale in which they must be taken.

Suspension notices (regulation 37) can be served where there is a serious risk of pollution, (or in the case of a flood risk activity, a risk of serious flooding; risk of serious detrimental impact on drainage; or risk of serious harm to the environment) whether or not an operator has breached a permit condition. This must describe the nature of the risk, the action necessary to remove the risk, and it must specify a deadline.

When a suspension notice is served, the permit ceases to authorise the operation of the facility or specified activities, although activities can continue unless their cessation is necessary to address the risk of pollution.

In the case of a prosecution, conviction (in the Crown court) can lead to an unlimited fine and up to five years imprisonment.

Operators have an emergency defence which provides a defence where operators can show that they have acted in an emergency in order to avoid danger to human health, they have taken all reasonable steps to minimise pollution and the regulator is informed promptly (regulation 40).

The details of any conviction or formal caution must be placed by the regulator on the public register.

The regulator can revoke a permit by serving a revocation order (regulation 22) where appropriate, for example where other enforcement tools have failed to protect the environment. Where a revocation notice is appealed, it does not come into effect until the appeal is determined (or withdrawn).

Where a regulated facility causes the risk of serious pollution, the risk can be removed by the regulator under regulation 57. This gives the regulator the power to arrange for steps to be taken to remedy the pollution at the expense of the operator if an offence that causes pollution is committed. Site protection must be addressed throughout the life of a permit rather that allowing an operator to contaminate a site during operation and decontaminating it when the facility closes.

The Crown is bound by the Regulations. While the Crown is not criminally liable if it contravenes the regulations, and the regulator cannot take legal proceedings if it does not comply with an enforcement or suspension notice, it may apply to the High Court to declare the contravention unlawful. Schedule 4 of the Regulations sets out the provisions relating to the Crown.

The conditions of the permit can be varied at any time by the regulator (regulation 20). Permits must be reviewed periodically (regulation 34) to check that the conditions remain adequate in the light of experience or new knowledge and that they continue to reflect current standards.

Appeals

An operator can appeal where an application for a permit (or for a variation, transfer or surrender of a permit) has been refused (or deemed to have been refused). The operator can also appeal where they disagree with the conditions set out in the permit, the regulator has deemed an application to be withdrawn. There are also provisions relating to appeals concerning closures, revocation, enforcement, and suspension and other notices relating to landfill, mining waste and flood risk activities. Decisions about the information that must be included on the public register can also be appealed. The time limits for making appeals varies according to the basis of the appeal. Appeals are made to the Secretary of State or Welsh Ministers.

Public registers and information

Regulators must maintain public registers for environmental permits containing information on all the regulated facilities for which they are responsible. Local authority registers must also contain information on facilities in their area regulated by the Environment Agency (other than mobile plant). Members of the public must be allowed to inspect the registers, free of charge and at all reasonable times and copies of any entry must be made available for a reasonable charge. Where information is not available due to confidentiality, the register must state that this information exists.

Registers can be internet or computer based. They must contain the information set out in paragraph 1 of Schedule 27 to the Regulations, including copies of permits, applications enforcement notices and monitoring information.

[E5041] Emissions into the Atmosphere

There is an exemption for information which the Secretary of State or Welsh Ministers deem would be contrary to the interests of National Security; and the regulator may judge information to be confidential and withhold it from the public register.

Information is also available under the *Freedom of Information Act* and the *Environmental Information Regulations 2004 (SI 2004 No 3391)* (see www.ico.gov.uk for further information).

Detailed guidance on Environmental Permitting is available on the government website www.gov.uk/topic/environmental-management/environmental-permits.

The regulators

The environment agencies

[E5043] In England, the Environment Agency is responsible for areas including:

- regulating major industry and waste;
- treatment of contaminated land; and
- water quality and resources.

Scottish Environment Protection Agency (SEPA)

[E5044] SEPA responsibilities include regulating:

- activities that may pollute water;
- activities that may pollute air;
- waste storage, transport, treatment and disposal;
- the keeping and disposal of radioactive materials; and
- activities that may contaminate land.

Natural Resources Body for Wales (NRBW)

[E5044.1] NRBW is the regulator for 'protecting people and the environment including marine, forest and waste industries, and prosecuting those who breach the regulations that we are responsible for'.

Northern Ireland

[E5045] The Department of Environment ('DoE') (Northern Ireland) undertakes activities to prevent, monitor and control pollution of the air, land and water.

Local authorities

[E5046] Local authorities have a wide range of responsibilities covering the whole spectrum of pollution control and environmental protection. The Local Government Association and the Environment Agency signed a Memorandum of Understanding covering those areas of environmental protection for which they have responsibility.

Local authorities regulate so-called 'medium-polluting' installations under the Local Authority Pollution Prevention and Control (LAPCC) regime – often referred to as Part B installations (and the processes they carry out are known as permitted Part B activities). The installations are classified into sectors, and a separate guidance note on air pollution standards is issued for each sector.

Useful websites

[E5047]–[E5060]

- Environment Agency: www.gov.uk/government/organisations/environment-agency
- Department for Environment, Food and Rural Affairs (DEFRA): www.gov.uk/government/organisations/department-for-environment-food-rural-affairs
- European Commission: eippcb.jrc.ec.europa.eu/reference/
- The UK-AIR (Air Information Resource) webpages (providing in-depth information on air quality and air pollution in the UK): https://uk-air.defra.gov.uk/
- Scottish Environment Protection Agency: www.sepa.org.uk/
- Natural Resources Wales: https://naturalresources.wales

The Clean Air Act 1993

[E5061] Historically, the *Clean Air Acts* of *1956* and *1968* were concerned very much with the control and reduction of smoke, largely through the burning of coal. These Acts, and other clean air legislation such as the *Control of Pollution Act 1974* and the *Control of Pollution (Amendment) Act 1989* were consolidated in the *Clean Air Act 1993*.

There are three parts to this Act that are relevant:

(1) *Part I: dark smoke*
 Prohibition of dark smoke from chimneys and from industrial or trade premises.
(2) *Part II: smoke, grit, dust and fumes*
 The Act controls the emissions of particles emitted from furnaces. Direction is also given on the height of chimneys.
(3) *Part V: Information about air pollution*
 The Act grants Local Authorities the power to provide information on emissions from specified premises, or to enter premises and make measurements.

DEFRA carried out a review of the *Clean Air Act (CAA) 1993* in order "to reduce burdens on business and Local Authorities whilst considering how the legislation can be modernised to make it more user friendly and relevant to current air quality challenges." A summary of the responses was published in July 2014 (see: www.gov.uk/government/consultations/clean-air-act-1993-review).

Statutory nuisance

[E5062]–[E5063] Air pollution control legislation has been rooted historically in the concept of nuisance to humans, rather than harm to the environment. The causes of such nuisance have traditionally been the burning of fuel or waste materials, giving rise to 'fumes' and 'smoke'. Odour is another form of nuisance.

The main legislation relating to nuisance is now in the *Environmental Protection Act 1990, Part III* as amended. It enables Local Authorities and individuals to secure the abatement of a statutory nuisance.

A nuisance is defined by the legislation, although only parts of the definition relate to air pollution. Others relate to noise and the state of premises that might be 'prejudicial to health'. In essence a nuisance can arise because of smoke or gases emitted from either private or business premises that are prejudicial to health or are a nuisance. A distinction is made between private dwellings and industrial premises in terms of the emissions. Private dwellings are exempt from causing a nuisance through emissions of 'dust, steam, smell or other effluvia'.

Unless the Secretary of State has granted consent, a local authority cannot bring a prosecution against a nuisance where a prosecution under the environmental permitting regime could be brought. However, activities not covered by environmental permitting, even if on the site of a regulated facility, may be regulated under the statutory nuisance provisions. In addition, private prosecutions can be brought and the 2012 Court of Appeal case, *Barr & others v Biffa Waste Services Ltd* [2012] EWCA Civ 312 showed that having an environmental permit does not prevent a company being sued for nuisance.

In *Anslow v Norton Aluminium* [2013] All ER (D) 03 the judge held that although local residents had failed to establish a legal nuisance by way of noise, smoke, fumes and dust, they had established an unreasonable interference with the use and enjoyment of their properties by reason of odour from an aluminium foundry. They succeeded with their private nuisance claim.

World Health Organisation air quality guidelines for Europe

[E5064] The World Health Organisation ('WHO') formally produced a set of air quality guidelines for Europe in 1987. This was a much-referenced document, despite having no legal status, and it was updated in 2005. The most recent WHO recommended air quality guidelines are used as a basis for setting standards in EU directives, and were also taken into account by the UK Expert Panel on Air Quality Standard (now merged into the Department of Health's Committee on the Medical Effects of Air Pollutants (COMEAP) when making recommendations for UK air quality standards. The table below at E5065 shows WHO air quality guidelines. The table below at E5067 shows the standards and guidelines that the UK has adopted to meet with the directive.

WHO air quality guidelines (including 2005 update)

[E5065]

Substances	Time-weighted average	Averaging time
nitrogen dioxide	200 µg m^{-3}	1 hour
	40 µg m^{-3}	annual
ozone	100 µg m^{-3}	8 hours
sulphur dioxide	500 µg m^{-3}	10 mins
	20 µg m^{-3}	24 hour
	50 µg m^{-3}	annual
carbon monoxide	100 mg m^{-3}	15 mins
	60 mg m^{-3}	30 mins
	30 mg m^{-3}	1 hour
	10 mg m^{-3}	8 hours
lead	0.5 µg m^{-3}	annual
cadmium	5×10^{-3} µg m^{-3}	annual
mercury	1.0 µg m^{-3}	annual
PM$_{10}$	20 µg m^{-3}	annual
	50 µg m^{-3}	24 hours

In addition to the guidelines expressed as a threshold concentration, guidelines are also given as risk factors for exposure to carcinogens such as benzene and benzo(a)pyrene.

Air quality standards in the UK

Air Quality Standards Regulations

[E5066] The *Air Quality Standards Regulations 2010 (SI 2010 No 1001)* apply to England and set air quality standards to protect human health and the environment. (Equivalent Regulations apply to Scotland, Wales and Northern Ireland). The *2010 Air Quality Standards Regulations* (which revoke earlier 2007 Regulations) implemented the requirements of the following Directives:

- Directive 2008/50/EC on ambient air quality and cleaner air for Europe (this Directive replaces Council Directive 96/62/EC on ambient air quality assessment and management
- Council Directive 1999/30/EC relating to limits for sulphur dioxide, nitrogen dioxide, oxides of nitrogen, particulate matter and lead in ambient air;
- Council Directive 2000/69/EC relating to limit values for benzene and carbon monoxide in ambient air;
- Council Directive 2002/3/EC relating to ozone in ambient air; and
- Directive 2004/107/EC relating to arsenic, cadmium, mercury, nickel and polycyclic aromatic hydrocarbons in ambient air.

The Regulations set down limit values for relevant pollutants. (It also sets down target values for particular pollutants, long term objectives for ozone,

[E5066] Emissions into the Atmosphere

information and alert thresholds, critical levels for the protection of vegetation and national exposure reduction targets for PM2.5.)

Schedule 2 containing the limit values is reproduced at **E5067**.

SCHEDULE 2 LIMIT VALUES

[E5067]–[E5068]

Sulphur Dioxide			
Averaging period	*Limit value*		
One hour	350 µg/m^3 not to be exceeded more than 24 times a calendar year		
One day	125 µg/m$_3$ not to be exceeded more than 3 times a calendar year		
Nitrogen dioxide			
Averaging period	*Limit value*		
One hour	200 µg/m^3 not to be exceeded more than 18 times a calendar year		
Calendar year	40 µg/m^3		
Benzene			
Averaging period	*Limit value*		
Calendar year	5 µg/m^3		
Lead			
Averaging period	*Limit value*		
Calendar year	0.5 µg/m^3		
PM$_{10}$			
Averaging period	*Limit value*		
One day	50 µg/m^3, not to be exceeded more than 35 times a calendar year		
Calendar year	40 µg/m^3		
PM$_{2.5}$			
Averaging period	*Limit value*	*Margin of tolerance*	*Date by which limit value is to be met*
Calendar year	25 µg/m^3	20% on 11 June 2008, decreasing on the next 1 January and every 12 months thereafter by equal annual percentages to reach 0% by 1 January 2015	1 January 2015
Carbon monoxide			
Averaging period	*Limit value*		
Maximum eight hour daily mean[1]	10 mg/m^3		

[1] The maximum daily eight hour mean concentration of carbon monoxide must be selected by

examining eight hour running averages, calculated from hourly data and updated each hour. Each eight hour average so calculated will be assigned to the day on which it ends, that is, the first calculation period for any one day will be from 17:00 on the previous day to 01:00 on that day, the last calculation period for any one day will be the period from 16:00 to 24:00 on that day.
(See also E5007–E5013, E5019 and E5033–E5039 above.)

Local air quality management

[E5069] Local Authorities have a major role to play in delivering cleaner air. They are responsible for land-use planning and traffic management, and for controlling industrial pollution sources. Local Authorities have a long history of local environmental control. They have responsibilities in controlling:

- industrial pollution;
- local pollution hotspots; and
- pollution from domestic sources.

In England and Wales local authorities control certain industrial processes, in Scotland SEPA, and in Northern Ireland the Industrial Pollution and Radiochemical Inspectorate control the equivalent. Air quality has been given increased consideration more recently when they fulfil their strategic planning and transport roles. This has been a natural progression for Local Authorities.

UK Climate Change Legislation

[E5070] In the UK, the *Greenhouse Gas Emissions Trading Scheme Regulations 2012 (SI 2012 No 3038)* came into force on 1 January 2013, implementing the European emissions trading scheme (see **E5027.1** above) in the UK.

The *Climate Change Act 2008* provides a broad framework for addressing climate change-related concerns. Among other provisions, it sets a target for the year 2050 for the reduction of targeted greenhouse gas emissions; provides for a system of carbon budgeting; establishes a Committee on Climate Change; confers powers to establish trading schemes for the purpose of limiting greenhouse gas emissions or encouraging activities that reduce such emissions or remove greenhouse gas from the atmosphere; paves the way for mandatory carbon reporting and makes other provision about climate change.

The *Companies Act 2006 (Strategic Report and Directors' Report) Regulations 2013* came into force on 1 October 2013 and require all UK quoted companies to report on their greenhouse gas emissions as part of their annual Directors' Report. Under the regulations, companies listed on the main market of the London Stock Exchange must report their greenhouse-gas (GHG) emissions alongside their financial reports. They must report total annual emissions of the six main greenhouse gases in tonnes of CO_2 equivalent, as well as a carbon intensity ratio. The directors' strategic report must also cover environmental risks and opportunities. DEFRA guidance in this area, *Environmental Reporting Guidelines: Including mandatory greenhouse gas emissions reporting guidance,* can be found on the government website at: www.gov.uk/government/publications/environmental-reporting-guidelines -including-mandatory-greenhouse-gas-emissions-reporting-guidance.

The guidance includes changes which take effect from 1 April 2019 and require all UK quoted companies to report on their global energy use as well as greenhouse gas emissions in their annual Directors' Report. Large unquoted companies and limited liability partnerships must also disclose their annual energy use and greenhouse gas emissions and related information. The government encourages all other companies to report similarly, although this is voluntary.

Future UK environmental policy, strategy and legislation

[E5071] In January 2018, the government launched its 25-year environment plan, *A Green Future: Our 25 Year Plan to Improve the Environment* (www.gov.uk/government/publications/25-year-environment-plan). This sets out how over the next quarter of a century the government will take action to improve air and water quality, create richer habitats for wildlife and 'curb the scourge of plastic in the world's oceans'. It includes commitments to consult on a new environmental watchdog to hold the government to account for environmental standards. The plan set out that the government will achieve clean air by:

- meeting legally binding targets to reduce emissions of five damaging air pollutants, which should halve the effects of air pollution on health by 2030;
- ending the sale of new conventional petrol and diesel cars and vans by 2040; and
- maintaining the continuous improvement in industrial emissions by building on existing good practice and the regulatory framework.

In December 2018, it published draft clauses on environmental principles and governance to be included in a new Environment Bill, set to be introduced in 2019 (www.gov.uk/government/publications/draft-environment-principles-and-governance-bill-2018). The government says the bill will introduce a new framework for environmental governance as the UK leaves the European Union (EU). The core elements published in the draft clauses are:

- environmental principles – including the 'polluter pays' principle and the principle that the public should be able to participate in environmental decision-making will act as guiding principles to help protect the environment from damage and will encourage decision-makers to further consider the environment in the development of government policy;
- the Office for Environmental Protection – a new 'green governance' body will uphold environmental law. It will be an independent, statutory environmental body to hold government and public bodies to account on environmental standards, including taking legal action to enforce the implementation of environmental law where necessary. Once the UK leaves the EU, this body will replace the current oversight of the European Commission;

- the 25-year environment plan – the draft bill proposes making it a legal requirement for the government to have a plan for improving the environment, to monitor and report annually to parliament on progress, and to update it at least every five years.

In January 2019, the government launched its new clean air strategy (www.gov.uk/government/publications/clean-air-strategy-2019). This includes:

- a long-term target to reduce people's exposure to particulate matter (PM), which the World Health Organization (WHO) has identified as the most damaging pollutant. It aims to halve the number of people living in areas breaching WHO guidelines on PM by 2025;
- a commitment to end the sale of conventional new diesel and petrol cars and vans from 2040; and
- a programme of work across government, industry and society to reduce emissions coming from a wide range of sources, including wood-burning stoves and agriculture – which is responsible for 88% of ammonia emissions.

Employers' Liability Insurance

Phil Grace

Introduction to employers' liability insurance

[E13001] Most employers carrying on business in Great Britain are under a statutory duty to take out insurance against claims for injuries/diseases brought against them by employees. When such a compulsory insurance policy is taken out the insurance company issues the employer with a certificate of insurance, and the employer must keep a copy of this displayed in a prominent position at his workplace, or supply copies in electronic form so that employees can see it. It is a criminal offence to fail to take out such insurance and/or to fail to display a certificate (see **E13013–E13015**); however, such a failure does not give rise to any civil liability on the part of a company director in England (see **E13008**). The above duties are contained in the *Employers' Liability (Compulsory Insurance) Act 1969* (referred to hereafter as the '1969 Act') and in the *Employers' Liability (Compulsory Insurance) Regulations 1998 (SI 1998 No 2573)* (referred to hereafter as the '1998 Regulations'). In addition, the requirements of the *1969 Act* extend to offshore installations but do not extend to injuries suffered by employees when carried on or in a vehicle, or entering or getting onto or alighting from a vehicle, where such injury is caused by, or arises out of use, by the employer, of a vehicle on the road. [*SI 1998 No 2573, Reg 9 and Sch 2, para 14*]. Such employees would normally be covered under the *Road Traffic Act 1988, s 145* as amended by the *Motor Vehicles (Compulsory Insurance) Regulations 1992 (SI 1992 No 3036)*. As from 1 July 1994, liability for injury to an employee whilst in a motor vehicle has been that of the employer's motor insurers.

Employees suffering from occupational diseases with long development periods such as those caused by exposure to asbestos face particular difficulties in making claims, especially with regard to tracking down the insurer of their old employer, who may, by now have ceased trading (see **E13026**).

In November 2005 the Government published its Compensation Bill. It forms part of a wider Government programme aimed at tackling the 'compensation culture' and also improving the system for obtaining compensation for those with a valid claim. The Lord Chancellor has said that the Bill will 'discourage false expectations that compensation is available for any untoward incident'. The first part of the Act came into effect in November 2006 (see **E13033**).

A key element of the Bill is the move to regulate the 'claims farmers' that is those firms who offer 'no win no fee' type arrangements. Such firms are thought to be fuelling the growing compensation culture in the UK.

But aside from this and one or two other specific changes the Bill has been described as adding little to existing legislation.

[E13001] Employers' Liability Insurance

After many years and numerous delays the Government introduced a Bill dealing with corporate killing in 2005. This was in response to public concerns that the majority of manslaughter cases against firms failed eg following various rail crashes. In contrast, prosecutions of small firms were often successful with fines and even custodial sentences being made against directors. The bill became law in 2008 with the passing of the Corporate Manslaughter and Corporate Homicide Act 2007. (see **E13032**).

Purpose of compulsory employers' liability insurance

[E13002] The purpose of compulsory employers' liability insurance is to ensure that employers are covered for any legal liability to pay damages to employees who suffer bodily injury and/or disease during the course of and as a result of employment. It is the liability of the employer towards his employees which has to be covered; there is no question of compulsory insurance extending to employees, since employers are under no statutory or common law duty to insure employees against risk of injury, or even to advise on the desirability of insurance; it is the employer's potential legal liability to employees which must be insured against (see **E13008**). Such liability is normally based on negligence or breach of statutory duty though not necessarily personal negligence on the part of the employer. Moreover, case law suggests that employers' liability is becoming stricter and there is a belief, in some quarters, that employers' liability claims are incapable of being defended, that the burden of proof has been reversed. While that is not strictly true, the task of defending claims is not made easier in those circumstances where the employer is unable to substantiate their contention that they have complied with legislation, supplied safe plant and equipment, trained employees etc. Good documentation is vital for a sound defence.

An employers' liability policy is a legal liability policy. Hence, if there is no legal liability on the part of an employer, the policy will not respond and no money will be paid out by the insurer. Moreover, if the employee's action against the employer cannot succeed, an action for damages cannot be brought against an employer's insurer (see **E13024**).

The policy indemnifies an employer against claims made by employees; an employee as such is not covered since he normally incurs no liability. Where an act of an employee causes injury to another employee and a claim results the employer is held to be liable for the acts of the employee through what is known as vicarious liability (see **E13027**). Employers' liability policies are unusual in that any compensation agreed or awarded by the court is paid direct to the employee rather than to the employer.

Although offering wide cover an employers' liability policy does not provide cover in respect of claims from third parties (eg independent contractors and members of the public). Such liability is covered by a public liability policy which, though advisable, is not compulsory.

This section examines:

— the general law relating to contracts of insurance (see E13003–E13006);

- the insurer's right of recovery (i.e. subrogation) (see E13007);
- the duty to take out employers' liability insurance (see E13008–E13012);
- issue and display of certificates of insurance (see E13013);
- penalties (see E13014, E13015);
- scope and cover of policy (see E13018–E13019);
- 'prohibition' of certain conditions (see E13020);
- trade endorsements for certain types of work (see E13021,);
- measures of risk and assessment of premium (see E13022)
- extensions to cover (see E13023)
- insolvent employers' liability insurers (see E13024);
- Limitation Act (see E13025);
- Long tail disease (see E13026);
- Vicarious liability (see E13027).
- Making the Market Work (see E3028)
- Woolf Reforms (see E13029)
- Benchmarking and Performance Indicators (see E13030)
- NHS Recoveries (see E13031)
- Corporate Manslaughter (see E13032)
- Compensation Act (see E13033)
- Employers' Liability Tracing Office (see E13034)
- Lord Young Report (see E13035)
- Lofstedt Review (see E13036)
- Fees for Intervention (see E13037)

General law relating to insurance contracts

[E13003] Insurance is a contract. When a person wishes to insure, for example, himself, his house, his liability towards his employees, valuable personal property or even loss of profits, he (the proposer) fills in a proposal form for insurance, at the same time making certain facts known to the insurer about what is to be insured. On the basis of the information disclosed in the proposal form, the insurer will decide whether to accept the risk and if so at what rate to fix the premium. If the insurer elects to accept the risk, a contract of insurance is then drawn up in the form of an insurance policy. The vast majority of commercial insurance is arranged through a broker. The broker advises the proposer on the nature and type of covers required in addition to advising on the levels of cover required, the Sum Insured for property insurance or Limit of Indemnity for liability insurance. Brokers are regulated by the Financial Services Authority and are required to act in a professional manner. Brokers are responsible for the advice they give and in the event that the advice proves wrong or incorrect they can be sued and thus brokers are required to take out Professional Indemnity insurance.

There is a growth of what is known as direct sales where companies, generally small firms, purchase insurance 'direct' using the telephone or internet. Although the vast majority of insurance for firms and businesses, what is known as commercial insurance, is arranged through a broker.

Incidentally, it seems to matter little whether the negotiations leading up to contract took place between the insured (proposer) and the insurance company or between the insured and a broker, since the broker is often regarded as the agent of one or the other, generally of the proposer (*Newsholme Brothers v Road Transport & General Insurance Co Ltd* [1929] 2 KB 356). However, a lot depends on the facts. If the broker is authorised to complete blank proposal forms, he may well be the agent of the insurer but this is rare. It is common practice for brokers to complete proposal forms and present them to the proposer for signing before return to the insurer.

Extent of duty of disclosure

[E13004] Most commercial contracts are governed by the doctrine of 'caveat emptor', which translates as 'let the buyer beware'. But the law relating to insurance contracts is based on the legal principle of 'uberrima fides'. This translates as 'utmost good faith' and places a higher duty on both parties but especially the proposer. Whilst the proposer can examine the insurance policy the insurer is unable to examine all aspects of the risk being presented. A proposer must disclose to the insurer all material facts within his actual knowledge. This does not extend to disclosure of facts which he could not reasonably be expected to know. 'The duty is a duty to disclose, and you cannot disclose what you do not know. The obligation to disclose, therefore, necessarily depends on the knowledge you possess. This, however, must not be misunderstood. The proposer's opinion of the materiality of that knowledge is of no moment. If a reasonable man would have recognised that the knowledge in question was material to disclose, it is no excuse that you did not recognise it. But the question always is – Was the knowledge you possessed such that you ought to have disclosed it?' (*Joel v Law Union and Crown Insurance Co* [1908] 2 KB 863 per Fletcher Moulton LJ).

The knowledge of those who represent the directing mind and will of a company and who control what it does, eg directors and officers, is likely to be identified as the company's knowledge whether or not those individuals are responsible for arranging the insurance cover in question (*PCW Syndicates v PCW Reinsurers* [1996] 1 All ER 774).

An element of consumer protection, in favour of insureds, was introduced into insurance contracts by the Statement of General Insurance Practice 1986, a form of self-regulation applicable to many but not to all insurers. This has consequences for the duty of disclosure, proposal forms (**E13005**), renewals and claims. In particular, with regard to the last element (claims), an insurer should not refuse to indemnify on the grounds of:

(a)   non-disclosure of a material fact which a policyholder could not reasonably be expected to have disclosed; or
(b)   misrepresentation (unless it is a deliberate non-disclosure of, or negligence regarding a material fact). Innocent misrepresentation is not a ground for avoidance of payment.

The trend towards greater consumer protection in (inter alia) insurance contracts is reflected in the *Unfair Terms in Consumer Contracts Regulations 1999 (SI 1999 No 2083)*. These Regulations apply in the case of 'standard form' (or non-individually negotiated) contracts [*SI 1999 No 2083, Reg 5*]. A contractual term in such a contract shall:

(i) be regarded as 'unfair' if contrary to the requirement of 'good faith' it 'causes a significant imbalance in the parties' rights and obligations arising under the contract, to the insured's detriment' [*SI 1999 No 2083, Reg 5(1), Sch 2*];

(ii) always be regarded as having been individually negotiated where it has been drafted in advance and the consumer has not been able to influence the substance of the term [*SI 1999 No 2083, Reg 5(2)*];

(iii) require a written contract term to be expressed in plain, intelligible language. If there is doubt about the meaning of terminology, a construction in favour of the insured will prevail [*SI 1999 No 2083, Reg 7*]; and

(iv) where the insurer claims that a term was individually negotiated, he must prove it [*SI 1999 No 2083, Reg 5(4)*].

Complaints (other than ones which are frivolous or vexatious) or which are to be handled by a qualified body [*SI 1999 No 2083, Sch 1*] relating to 'unfair terms' in standard form contracts, are considered by the Director General of Fair Trading, who may prevent their continued use [*SI 1999 No 2083, Reg 12*]. See also **P9056** PRODUCT SAFETY.

Filling in proposal form

[E13005] Generally only failure to make disclosure of relevant facts will allow an insurer subsequently to void the policy and refuse to compensate for the loss. The test of whether a fact was or was not relevant is whether its omission would have influenced a prudent insurer in deciding whether to accept the risk, or at what rate to fix the premium.

The effect or impact of 'uberrima fides' (i.e. utmost good faith) is significant. If, when filling in a proposal form, a statement made by the proposer is at that time true, but is false in relation to other facts which are not stated, or becomes false before issue of the insurance policy, this entitles the insurer to refuse to indemnify. In *Condogianis v Guardian Assurance Co Ltd* [1921] 2 AC 125 a proposal form for fire cover contained the following question: 'Has proponent ever been a claimant on a fire insurance company in respect of the property now proposed, or any other property? If so, state when and name of company'. The proposer answered 'Yes', '1917', 'Ocean'. This answer was literally true, since he had claimed against the Ocean Insurance Co in respect of a burning car. However, he had failed to say that in 1912 he had made another claim against another insurance company in respect of another burning car. It was held that the answer was not a true one and the policy was, therefore, invalidated.

Loss mitigation

[E13006] There is an implied term in most insurance contracts that the insured will take all reasonable steps to mitigate loss caused by one or more of the insured perils. Thus, in the case of burglary cover of commercial premises, this could extend to provision of security patrols, the fitting of burglar alarm devices and guard dogs. Again, in the case of fire cover, steps to mitigate the extent of the loss on the part of the insured might well extend to the installation of a sprinkler system, regular visits by the local fire and rescue service and/or securing competent advice on storage of products and materials. Such requirements may be made mandatory by means of policy conditions, especially for high risk trades where large losses are expected. Failure to observe these conditions may result in the insurer deciding not to pay the claim. However, such mandatory requirements cannot be placed on Employers' Liability policies (see **E13020**).

In the case of employers' liability, reasonable steps would be regarded as compliance with health and safety legislation, the management of risk and the securing of competent advice through the appointment of an accredited safety officer and/or occupational hygienist, either permanently or temporarily. This last aspect is a specific requirement of the *Management of Health and Safety at Work Regulations 1999 (SI 1999 No 3242)*.

Subrogation

[E13007] Subrogation is a legal principle that provides the right for one person to stand in place of another and avail themselves of all the rights and remedies of that other person. Subrogation enables an insurer to make certain that the insured recovers no more than exact replacement of loss (i.e. indemnity). 'It [the doctrine of subrogation] was introduced in favour of the underwriters, in order to prevent their having to pay more than a full indemnity, not on the ground that the underwriters were sureties, for they are not so always, although their rights are sometimes similar to those of sureties, but in order to prevent the assured recovering more than a full indemnity' (*Castellain v Preston* (1883) 11 QBD 380 per Brett LJ). Subrogation does not extend to personal accident insurance, policies whereby the insured (normally self-employed but who could in reality be any person) is promised a fixed sum in the event of injury or illness (*Bradburn v Great Western Railway* Co (1874) LR 10 Exch 1 where the appellant was injured whilst travelling on a train, owing to the negligence of the respondent. He had earlier bought personal accident insurance to cover him for the possibility of injury on the train. It was held that he was entitled to both damages for negligence *and* insurance moneys payable under the policy (see further COMPENSATION FOR WORK INJURIES/DISEASES C6001).

The right of subrogation does not arise until the insurer has paid the insured in respect of his loss, however, it should be noted that in respect of employer's liability cases payments are made direct to the injured party rather than the insured, the employer. Subrogation has been invoked infrequently in employers' liability cases. In some circumstances an employer may be entitled to an indemnity from employees for example where the act of an employee has resulted in an injury to another employee and the employer has been held to be vicariously liable. However, it is convention that such recoveries are not sought or pursued. (see **E13027**).

Duty of employer to take out and maintain insurance

[E13008] 'Every employer carrying on business in Great Britain shall insure, and maintain insurance against liability for bodily injury or disease sustained by his employees, and arising out of and in the course of their employment in Great Britain in that business' [*Employers' Liability (Compulsory Insurance) Act 1969, s 1(1)*].

Such insurance must be provided by an insurer authorised to transact Employers' Liability insurance under one or more 'approved policies'. An 'approved policy' is a policy of insurance not subject to any conditions or exceptions prohibited by regulations (see **E13020**) [*Employers' Liability (Compulsory Insurance) Act 1969, s 1(3)*]. This now includes insurance with an approved EU insurer [*Insurance Companies (Amendment) Regulations 1992 (SI 1992 No 2890)*].

There is no duty under the *1969 Act* to warn or insure the employee against risks of employment outside Great Britain (*Reid v Rush Tompkins Group plc* [1989] 3 All ER 228) although the *1998 Regulations* require the employer to insure employees employed on or from offshore installations or associated structures – see **E13009**.

The failure to take out Employers' Liability insurance does not give rise to any civil liability on the part of a company director in England. In *Richardson v Pitt-Stanley* [1995] 1 All ER 460 the plaintiff suffered a serious injury to his hand in an accident at work, and obtained judgment against his employer, a limited liability company, for breach of the *Factories Act 1961, s 14(1)* (failure to fence dangerous parts of machinery). Before damages were assessed, the company went into liquidation and there were no assets remaining to satisfy the plaintiff's judgment. The company had also failed to insure against liability for injury sustained by employees in the course of their employment, as required by the *Employers' Liability (Compulsory Insurance) Act 1969, s 1*. *Section 5* of that Act makes failure to insure a criminal offence. The plaintiff then sued the directors and company secretary who, he alleged, had committed an offence under *s 5*, claiming as damages a sum equal to the sum which he would have recovered against the company, had it been properly insured. His action failed. It was held that the *Employers' Liability (Compulsory Insurance) Act 1969* did not create a civil as well as criminal liability. In Scotland however the Sheriff Principal in *Quinn v McGinty* 1999 SLT 27, Sh Ct, on similar facts did not follow the Court of Appeal in *Richardson v Pitt-Stanley*.

Employees covered by the Act

[E13009] Cover is required in respect of liability to employees who:

(a) are ordinarily resident in Great Britain; or
(b) though not ordinarily resident in Great Britain, are present in Great Britain in the course of employment here for a continuous period of not less than 14 days; or
(c) though not ordinarily resident in the United Kingdom, have been employed on or from an offshore installation or associated structure for a continuous period of not less than seven days.

[*Employers' Liability (Compulsory Insurance) Regulations 1998 (SI 1998 No 2573), Reg 1(2)*].

Employees not covered by the Act

[E13010] An employer is not required to insure against liability to family members, that is employees who are:

(a) a spouse;
(b) father;
(c) mother;
(d) son;
(e) daughter;
(f) other close relative.

[*Employers' Liability (Compulsory Insurance) Act 1969, s 2 (2 a)*].

Those who are not ordinarily resident in the UK are not covered by the Act except as above. Nor are employees working abroad covered. Such employees can sue under English law in limited circumstances (*Johnson v Coventry Churchill International Ltd* [1992] 3 All ER 14 where an employee, working in Germany for an English manpower leasing company, was injured when he fell through a rotten plank. He was unable to sue his employer under German law; although he was working in Germany. A claim was thus made in England against the English firm on the basis that they were, in essence, the employer. Legal precedents stated that a claim of this type could not be made unless the case was of a type or nature that was covered by both overseas and English law. However, it was held this was an exception to the general principle, that England was the country with the most significant relationship with the claim because the injured person had made the contract in England; the firm had paid him, making deductions for National Insurance and held control over where and when he worked. The injured person was thus entitled to expect the firm to compensate him through their insurers for any personal injury sustained in Germany.

Degree of cover necessary

[E13011] The amount for which an employer is required to insure and maintain insurance is £5 million in respect of claims relating to any one or more of his employees, arising out of any one occurrence [*Employers' Liability (Compulsory Insurance) Regulations 1998 (SI 1998 No 2573), Reg 3(1)*].

Between 1 January 1972 (when the *Employer's Liability (Compulsory Insurance) Act 1969* came into force) and 1994, insurers provided unlimited cover under employers' liability policies. As from 1 January 1995, as a result of payments made in respect of claims exceeding the amount of premiums received during the period 1989–1993 across the entire insurance market, unlimited liability was withdrawn, but most insurers continued to offer a minimum of £10 million indemnity for onshore work. A consultative document issued by the Department of the Environment, Transport and the Regions entitled The Draft Employers' Liability (Compulsory Insurance) General

Regulations [C4857 September 1997] (hereafter referred to as 'the 1997 consultative document') which preceded the *1998 Regulations* assumed that this practice would continue.

Where an employing company has subsidiaries, there will be sufficient compliance if they insure/maintain insurance for themselves *and* on behalf of their subsidiaries for £5 million in respect of claims affecting any one or more of its own employees and any one or more employees of its subsidiaries arising out of any one occurrence [*Employers' Liability (Compulsory Insurance) Regulations 1998 (SI 1998 No 2573), Reg 3*].

Insurers and the courts have interpreted the legislation to mean that all injuries resulting from one incident (eg an explosion) are treated as one occurrence, and each individual case of gradually occurring injury or disease is treated as an individual occurrence – the only exception being a situation where a sudden and immediate outbreak of a disease amongst the workforce is clearly attributable to an identifiable incident (eg the escape of a biological agent). The introduction to the 1997 consultative document suggested that this interpretation might be challenged and set out a possible alternative regulation to be used instead of what is now *Reg 3* of the *1998 Regulations* if clarification was felt necessary. This was not adopted and no new regulations are currently proposed and therefore it would appear that the Government are now satisfied that the position is clear.

Exempted employers

[E13012] The following employers are exempt from the duty to take out and maintain insurance:

(a) nationalised industries;
(b) any body holding a Government department certificate that any claim which it cannot pay itself will be paid out of moneys provided by Parliament;
(c) any Passenger Transport Executive and its subsidiaries, London Regional Transport and its subsidiaries;
(d) statutory water undertakers and certain water boards;
(e) the Commission for the New Towns;
(f) health service bodies, National Health Service Trusts;
(g) probation and after-care committees, magistrates' court committees, and any voluntary management committee of an approved bail or approved probation hostel;
(h) governments of foreign states or commonwealth countries and some other specialised employers;
(i) Network Rail and its subsidiaries (the exemption ceasing when it is no longer owned by the Crown);
(j) the Qualifications & Curriculum Authority.

There are other types of employer specified in the Regulations, but these are the main exceptions [*Employers' Liability Compulsory Insurance Act 1969, s 3; Employers' Liability (Compulsory Insurance) Regulations 1998 (SI 1998 No 2573), Sch 2*].

The above situation was changed by the enactment of amending legislation in 2004 [*Employers' Liability (Compulsory Insurance) (Amendment) Regulations 2004 (SI 2004 No 2882)*]. With effect from 1 February 2005 certain small firms, known as sole-employee incorporated companies (SEICs), are exempt from the requirement to take out Employers' Liability insurance. An SEIC is defined as a company with one employee with that employee owning 51% or more of the issued share capital. This step has been taken to reduce the burden of unnecessary costs on small businesses. Such firms are not barred from taking out insurance if they wish to and if they were to take on extra employees, for example temporary staff to enable the company to undertake larger contracts, then insurance would be required.

Issue, display and retention of certificates of insurance

[E13013] The insurer must issue the employer with a certificate of insurance, which must be issued not later than 30 days after the date on which insurance was commenced or renewed [*Employers' Liability (Compulsory Insurance) Act 1969, s 4(1); Employers' Liability (Compulsory Insurance) Regulations 1998 (SI 1998 No 2573), Reg 4*]. Where there are one or more contracts of insurance which jointly provide insurance cover of not less than £5 million, the certificate issued by any individual insurer must specify both the amount in excess of which insurance cover is provided by the individual policy, and the maximum amount of that cover [*Employers' Liability (Compulsory Insurance) Regulations 1998 (SI 1998 No 2573), Reg 4(3)*].

A copy or copies of the certificate must be displayed at each place of business where there are any employees entitled to be covered by the insurance policy and the copy certificate(s) must be placed where employees can easily see and read it and be reasonably protected from being defaced or damaged [*Employers' Liability (Compulsory Insurance) Regulations 1998 (SI 1998 No 2573), Reg 5*]. The exception is where an employee is employed on or from an offshore installation or associated structure, when the employer must produce, at the request of that employee and within ten days from such request, a copy of the certificate [*Employers' Liability (Compulsory Insurance) Regulations 1998 (SI 1998 No 2573), Reg 5(4)*].

The situation changed in 2008 with the passing of *Employers' Liability (Compulsory Insurance) (Amendment) Regulations 2008 (SI 2008 No 1765)*. These Regulations made minor changes to the duties of employers. It is now possible for employers to make copies of their Certificate available to employees in an electronic format (amendment of Regulation 5). Whilst this may well prove of interest to those employers whose workforce is desk based or has ready access to computers, company intranets etc it is most likely that the majority of employees will find that posting a copy on notice boards remains the most economic and practicable approach.

An employer must, if a notice has been served on him by the Health and Safety Executive, produce a copy of the policy to the officers specified in the notice and he must permit inspection of the policy by an inspector authorised by the Secretary of State to inspect the policy [*Employers' Liability (Compulsory Insurance) Regulations 1998 (SI 1998 No 2573), Regs 7, 8*].

A change introduced by the *1998 Regulations* is that employers are now required by law to retain any certificate of employers' liability insurance (or a copy) for a period of 40 years beginning on the date on which the insurance to which it relates commences or is renewed [*Employers' Liability (Compulsory Insurance) Regulations 1998 (SI 1998 No 2573), Reg 4(4)*]. Companies may retain the copy in any eye-readable form in any one of the ways authorised by the *Companies Act 1985, ss 722 and 723* [*Employers' Liability (Compulsory Insurance) Regulations 1998 (SI 1998 No 2573), Reg 4(5)*].

This requirement was removed from the Regulations (deletion of paragraphs (4) and (5) from Regulation 4) in 2008 as part of a reduction in the administrative burden on business. However, it is strongly recommended that employers continue to retain their EL insurance certificates despite the removal of a specific legislative requirement to do so.

Penalties related to employers' liability insurance

Failure to insure or maintain insurance

[E13014] Failure by an employer to effect and maintain insurance for any day on which it is required is a criminal offence, carrying a maximum penalty on conviction of £2,500 [*Criminal Justice Act 1982, s 37(2)*].

It should be noted that prosecution for failure to insure is rare. The small number of prosecutions may be a result of the fact that the lack of insurance only becomes known during investigation into other, more serious matters, for example following an accident and where there are breaches of other regulations. In addition, breach of the legislation relating to compulsory insurance is a summary offence and a charge must be laid within a period of 6 months. This is difficult to achieve since an employer must be given time to produce a Certificate and this can delay verification that there is no insurance in place.

However, the enforcement agencies do prosecute from time to time. In 2009 a car wash firm was asked to produce evidence of having suitable EL insurance. They were unable to produce such evidence and the owner was fined £2,500 plus costs of £1,037. In 2010, following a number of warnings a public house was found guilty of not having EL insurance. The two landlords were fined £1,200 each with costs of £2,620. Other notable cases include:

- August 2013: A firm involved with the installation of solar panels was found not to have held EL ininsuranceor the period February to December 2012. The firm was fined £750 with £750 in costs.
- October 2013: A recycling firm was fined £300 plus £340 costs for a number of breaches including failing to hold EL insurance.
  Note: Whilst the fine seems low the firm was also found guilty of not having trained its forklift truck drivers. The Director received a 4 month prison sentence, suspended for 2 years and an application was made to have them suspended as a director for at least 5 years.

- April 2014: A café in Kent was found not to have taken out EL insurance and was fined £400 with £400 in costs. The firm's director declined to be interviewed but later admitted that they were unaware of the requirement to take out insurance.

Failure to display a certificate of insurance

[E13015] Failure on the part of an employer to display a certificate of insurance in a prominent position in the workplace is a criminal offence, carrying a maximum penalty on conviction of £1,000 [*Criminal Justice Act 1982, s 37(2)*].

In the 1997 consultative document, the Government suggests that penalties should be increased to become the same as those under the *HSWA 1974*, i.e. fines of £20,000 in a magistrates' court and unlimited in the Crown Court. To implement this change will require primary legislation and currently it would appear that the Government has no plans to introduce such legislation.

Cover provided by a typical policy

Persons

[E13016] Cover is limited to protection of employees. Independent contractors are not covered; liability to them should be covered by a public liability policy. Directors who are employed under a contract of employment are covered, but directors paid by fees who do not work full-time in the business are generally not regarded as 'employees'. Liability to them would normally be covered by a public liability policy. Similarly, since the judicial tendency is to construe 'labour-only' subcontractors in the construction industry as 'employees' (see CONSTRUCTION AND BUILDING OPERATIONS), employers' liability policies often contain an endorsement that will state: 'An employee shall also mean any labour master, and persons supplied by him, any person employed by labour-only subcontractors, any self-employed person, or any person hired from any public authority, company, firm or individual, while working for the insured in connection with the business'. The public liability policy should then be amended to exclude the insured's liability to 'employees' so designated.

It is commonplace, especially for smaller firms, for the same insurer to provide both the Employers' and Public Liability insurance and in such cases the wordings will be aligned. Current market practice is that few insurers provide Employers' Liability cover without also holding the corresponding Public Liability insurance. However, whenever different insurers are used to provide the two covers it will be necessary for the policyholder or their broker to check that the policy wordings are suitably aligned.

In the 1997 consultative document, the Government points out that the issue as to what constitutes 'an employee' cannot be completely resolved without primary legislation, but proposes to issue guidance on interpretation, although it was proposed that guidance on interpretation would be issued. Such guidance has not been issued and it must be assumed that the Government remains satisfied that the position is clear.

The definition of employee will usually be extended to include children of school age on Work Experience and young people employed under a Youth Training Scheme or similar. Persons who have been borrowed or hired by the company are also included. If the firm employs outworkers or home workers they are usually counted as employees.

The Health and Safety Executive have issued guidance to the effect that voluntary helpers or workers, eg persons who work in charity shops or on heritage railway lines, should be treated as employees. As described above most Employers' Liability policies include voluntary workers within the definition of employee.

The policy provides cover for the business, i.e. the declared trade of business of the company, but will in addition cover a number of other activities:

— repair and maintenance of buildings occupied by the insured;
— repair and maintenance of vehicles owned by the insured;
— provision of canteen and welfare services including sports and social clubs;
— private work for directors undertaken by employees with the consent of the insured.

There has been much debate in the recent past about the status of agency workers and in particular the question "Who is their employer?" since it may be either the agency or the employer on whose premises they are working that might be liable. There have been arguments presented for regarding both 'employers' as being responsible for the agency worker and this was considered in the case of *Brook Street Bureau [UK] Ltd v Dacas* [2004] EWCA Civ 217, [2004] ICR 1437. This case concerned a cleaner who had worked for a local authority for over a year, being placed with them by the employment agency Brook Street Bureau. Leaving aside the somewhat complex details of the case – it concerned employment rights rather than personal injury – the Court of Appeal decided that in view of the fact that she had been employed by the local authority for more than one year she was, in essence, their employee. This was contrary to previous case law and might change the view of some employers with respect to agency labour.

The more recent case of *James v London Borough of Greenwich Council* [2008] EWCA Civ 35, [2008] ICR 545 concerned an allegation of unfair dismissal. It was heard at an Employment Tribunal and an Employment Appeal Tribunal before going to the Court of Appeal. The finding was that there was no contract, explicit or implied. This finding contradicts previous case law which had suggested that the duration of a placement, exclusivity of services and control and supervision of agency workers were all factors suggesting that the end user (the firm or business using the agency worker) may be an implied employer, even in the absence of a contract. This case clarifies that an agency-supplied and contracted worker is **not** deemed to be the employee of the end-user, and a contract of employment could **not** be implied.

In brief, the claimant was employed by the Council as an asylum support worker. She began working again for them through an agency, from Sept 2001,

[E13016] Employers' Liability Insurance

and in 2003 moved to another agency which paid better. The claimant went off sick in 2004 and the agency provided another worker in her absence. When she returned, she was told she was no longer required as the agency had replaced her. The claimant tried to claim unfair dismissal from the council.

There was no express contract between the claimant and the council (i.e. end-user) but there was a contract between an agency and the worker and a separate contract between the end-user and the agency. However there was NO contract between the worker and the end-user and a contract was NOT implied to exist between the worker and the end-user.

Cases such as these will be judged on their own facts and concern the law of employment rather than personal injury and the law of negligence. The decision in this case is unlikely to be relevant in personal injury claims. There are well established tests that are used to establish a "master and servant" relationship which do not necessarily rely on the existence of a contract.

Scope of cover

[E13017] The policy provides for payment of:

(a) compensation for injury or disease, whether agreed or negotiated or awarded by a court
(b) costs and expenses of litigation, incurred with the insurer's consent, in defence of a claim against the insured (i.e. civil liability);
(c) claimants costs and expenses for which the insured is legally liable;
(d) solicitor's fees, and other legal costs incurred with the insurer's consent, for representation of the insured at proceedings in any court of summary jurisdiction (i.e. magistrates' court), coroner's inquest, or a fatal accident inquiry (i.e. criminal proceedings), arising out of an accident resulting in injury to an employee; it does *not* cover payment of a fine imposed by a criminal court;
(e) solicitor's fees, and other legal costs incurred with the insurer's consent, for representation of the insured at proceedings resulting from a prosecution for a breach of health and safety and related legislation, that is a breach of statutory duty; this is subject to certain provisos for example the breach must not have resulted from a deliberate act or omission by the insured and no indemnity will be paid if there is a legal expenses policy in force;
(f) compensation for any director, partner or employee attending court as a witness in a case for which the insurer is providing an indemnity.

The situation is changing with many insurers amending the wordings following the introduction of the Corporate Manslaughter legislation (see E13032.)

The policy may contain an excess negotiated between the insurer and employer, i.e. a provision that the employer pay the first £x of any claim. For the purposes of the *1969 Act*, any condition in a contract of insurance which requires a relevant employee to pay, or an insured employer to pay the relevant employee, the first amount of any claim or any aggregation of claims, is prohibited. Agreements which provide that the insurer will pay the claim in full and may then seek some reimbursement from the employer are permitted [*Employers' Liability (Compulsory Insurance) Regulations 1998 (SI 1998 No 2573), Reg 2*].

A contribution towards a claim is never sought from an employee; however, the amount of damages agreed or awarded may be reduced to reflect a degree of contributory negligence. If it is found that the employee was in some way to blame for the accident, for example they deliberately removed a guard, or failed to follow their training the damages will almost certainly be reduced.

The policy conditions require that as soon as a claim is received the insured, that is the employer, must notify their insurer as soon as possible. The insured must NOT make any admission of liability or offer or promise payment without the 'prior written consent' of the insurer. The insurer will take over the handling of the claim, communicating with the claimant's solicitors, investigating the circumstances and deciding on liability. The insured must supply the insurer with such particulars and information as may be required or requested. Any subsequent correspondence received by the insured must be forwarded to the insurer without delay.

The handling of claims has been dramatically influenced by the Woolf Reforms and the most recent changes introduced in 2014. These have considerably changed the way in which personal injury claims are handled by insurers and emphasise the need for employers to both pass onto insurers details of any circumstances that might give rise to a claim and immediately notify them of any letter of claim that is received (see **E13029**).

If the circumstances are clear cut an offer will be made and settlement negotiated. If a settlement cannot be agreed, if there is a dispute on facts or there is no legal precedent to follow then the claim may proceed to court. A judge will then be asked to decide on liability, the amount of damages that should be awarded or both. A very small percentage of claims go to court, in general less than 10%.

Geographical limits

[E13018]–[E13019] Cover is normally limited to Great Britain, Northern Ireland, the Channel Islands and the Isle of Man, in respect of employees normally resident in any of the above, who sustain injury whilst working in those areas. Cover is also provided for such employees who are injured whilst temporarily working abroad, but many policies limit this cover to periods not exceeding six months in any one year. This cover is usually granted on the basis that the action for damages is brought in a court of law of Great Britain, Northern Ireland, the Channel Islands or the Isle of Man – though even this proviso is omitted from some policies.

Employers must also have employers' liability insurance in respect of employees who, though not ordinarily resident in the United Kingdom, have been employed on or from an offshore installation or associated structure for a continuous period of not less than seven days; or who, though not ordinarily resident in Great Britain, are present in Great Britain in the course of employment for not less than fourteen days [*Employers' Liability (Compulsory Insurance) Regulations 1998 (SI 1998 No 2573), Reg 1(2)*].

Conditions which must be satisfied

(a) Cover only relates to bodily injury or disease; it does not extend to employee's property. This latter cover is provided by an employer's public liability policy. The only exception to this situation is where an employee's property is damaged in an event that gives rise to a claim under the policy. For example an employee could claim for clothing damaged in an incident that resulted in injury and if the employer's negligence was proven the employee would be entitled to compensation both for any injuries received and any damage to clothing or property. Such situations are rare and the property damage costs are generally far outweighed by any compensation for injury received.

(b) Injury must arise out of and during the course of employment (see below). If injury does not so arise, cover is normally provided by a public liability policy.

(c) Bodily injury must be caused during the period of insurance. Normally with injury-causing accidents there is no problem, since injury follows on from the accident almost immediately. Certain occupational diseases take many years to develop and may not manifest themselves until much later, eg asbestosis, mesothelioma, pneumoconiosis or deafness. Legal liability arises when the disease was caused i.e. when the exposure to the harmful substance or noise took place. Moreover, at least as far as occupational deafness is concerned, liability between employers can be apportioned, giving rise to contribution between insurers. This can be illustrated as follows:

— 20 years of employment spread across four employers A, B, C and D

— 15 years of negligent noise exposure involving employers A, B and C

— the compensation would be awarded against these three employers in proportion to the length of time the claimant was employed by each

— if there were five insurers who had variously insured the three employers A, B and C over the 15 years of negligent exposure they would contribute to the compensation in proportion to the length of time they had been on cover.

(see further NOISE AT WORK and VIBRATION) (see E13024)

(d) Claims must be notified by the insured to the insurer as soon as possible, or as stipulated by the policy (see E13017).

'Prohibition' of certain conditions

[E13020] Many policies of insurance contain restrictive conditions that may be used to reduce or limit the amount paid in settlement. For example

'Prohibition' of certain conditions [E13020]

household contents insurance may contain a condition that all windows must be fitted with a locking mechanism and that windows must be locked closed when the property in unoccupied. If the insurer were to establish that the windows were not locked in the closed position then they may choose to avoid a theft claim. All liability policies contain conditions with which the insured must comply if the insurer is to 'progress' his claim, eg the policyholder must notify the insurer of claims as soon as possible. Failure to comply with such condition(s) could jeopardise cover under the policy: the insured would be legally liable but without insurance protection. In the case of an employers' liability policy, an insurer is not entitled to avoid liability under the policy if the condition requiring the insured to take reasonable care to prevent injuries to employees was not complied with.

However, the insurer is not entitled to insist on conditions requiring compliance with the provisions of any relevant statutes/statutory instruments (eg *HSWA 1974; Ionising Radiations Regulations 1999 (SI 1999 No 3232)*), or to keep records, even though the keeping of records is vital to enable the insurer to mount a reasonable defence to a claim.

The object of the 1969 Act was to ensure that an employer who had a claim brought against him would be able to pay the employee any damages awarded. Regulations made under the Act, therefore, seek to prevent insurers from avoiding their liability by relying on breach of a policy condition, by way of 'prohibiting' certain conditions in policies taken out under the Act. More particularly, insurers cannot avoid liability in the following circumstances.

(a) Some specified thing being done or being omitted to be done after the happening of the event giving rise to a claim (eg omission to notify the insurer of a claim within a stipulated time) [*Employers' Liability (Compulsory Insurance) Regulations 1998 (SI 1998 No 2573), Reg 2(1)(a)*].

(b) Failure on the part of the policy-holder to take reasonable care to protect his employees against the risk of bodily injury or disease in the course of employment [*Employers' Liability (Compulsory Insurance) Regulations 1998 (SI 1998 No 2573, Reg 2(1)(b)*]. As to the meaning of 'reasonable care' or 'reasonable precaution' here, 'It is eminently reasonable for employers to entrust . . . tasks to a skilled and trusted foreman on whose competence they have every reason to rely' (*Woolfall and Rimmer Ltd v Moyle and Another* [1941] 3 All ER 304). The prohibition is therefore not broken by a negligent act on the part of a competent foreman selected by the employer. Where, however, an employer acted wilfully (in causing injury) and not merely negligently (though this would be rare), the insurer could presumably refuse to pay (*Hartley v Provincial Insurance Co Ltd* [1957] Lloyd's Rep 121, where the insured employer had not taken steps to ensure that a stockbar was securely fenced for the purposes of the *Factories Act 1937, s 14(3)* in spite of repeated warnings from the factory inspector, with the result that an employee was scalped whilst working at a lathe. It was held that the insurer was justified in refusing to indemnify the employer who was in breach of statutory duty and so liable for damages). This was confirmed in *Aluminium Wire and Cable Co Ltd v Allstate Insurance Co Ltd* [1985] 2 Lloyd's Rep 280.

(c) Failure on the part of the policy-holder to comply with statutory requirements for the protection of employees against the risk of injury [*Employers' Liability (Compulsory Insurance) Regulations 1998 (SI 1998 No 2573, Reg 2(1)(c)*] – the reasoning in *Hartley v Provincial Insurance Co Ltd* (see head (b) above), that wilful breach may not be covered, probably applies here too.

(d) Failure on the part of the policy-holder to keep specified records and make such information available to the insurer [*Employers' Liability (Compulsory Insurance) Regulations 1998 (SI 1998 No 2573, Reg 2(1)(d)*] (eg accident book or accounts relating to employees' wages and salaries (see ACCIDENT REPORTING)).

(e) By means of the use of an excess in policies [*Employers' Liability (Compulsory Insurance) Regulations 1998 (SI 1998 No 2573, Reg 2(2)*] – see **E13017**.

Trade endorsements for certain types of work

[E13021] The *Employers' Liability (Compulsory Insurance) Regulations 1998 (SI 1998 No 2573), Reg 2* does not prevent insurers, i.e. underwriters, from applying certain conditions in connection with intrinsically hazardous work; for instance, exclusion of liability for accidents arising out of demolition work, or in connection with the use of explosives. Insurers are also permitted to make decisions about which types of risks, that is trades, they wish to insure. There is a legal requirement for employers to take out insurance but no corresponding legal requirement for insurers to offer the cover. Thus it is possible for an insurer to underwrite, i.e. offer to provide insurance, for retail risks but not construction risks or for builders but not demolition firms. Trade endorsements are used frequently in underwriting employers' liability risks, and there is nothing in the 1969 Act to prevent insurers from applying their normal underwriting principles and applying trade endorsements where they consider it necessary, i.e. they will amend their standard policy form to exclude certain risks. Thus, there may be specific exclusions of liability arising out of types of work, such as demolition on a policy for a general builder, or the use of mechanically driven woodworking machinery, or work above certain heights for a window cleaner. Thus if a firm describes themselves as a builder the insurer is permitted to exclude demolition work from the policy. The declared trade, that of a builder, is used by the insurer to decide whether to underwrite the business i.e. offer insurance and also what rates to charge.

If that same firm were to undertake demolition work and an accident were to occur and a claim be made the insurer could choose to declare the policy void, that is cancel cover.

However, if a policy excludes certain activities, such as demolition this does not mean an employee will not obtain compensation from his employer in the event of an accident. Employers are required to obtain employer's liability insurance cover for such activities by means of another, additional policy taken out with another insurer. In practice such conditions are usually used to prevent or restrict an employer from developing their business into such areas without informing the insurer and paying the appropriate premium.

The business description is what is known as a material fact and must be declared to the insurer.

Measure of risk and assessment of premium

[E13022] Certain trades or businesses are known to be more dangerous than others. For most trades or businesses insurers have their own rate for the risk, expressed as a rate per cent on wages (other than for clerical, managerial or non-manual employees for whom a very low rate applies). This is known as the book rate and is derived from the insurer's experience i.e. the balance of premium against claims for all risks of that type. For small firms this rate is used as a guide and is altered upwards or downwards depending on:

(a) previous history of claims and cost of settlement;
(b) size of the company, generally taken to be indicated by the wage roll;
(c) whether certain risks are not to be covered, eg the premium will be lower if the insured elects to exclude from the policy certain risks, such as the use of power driven woodworking machinery;
(d) the insured's attitude towards safety.

For very small firms a minimum premium may be set.

For large firms that may have several claims each year it is possible to calculate the coming year's premium by basing it directly on the cost of claims in previous years, generally the preceding five years. The premium will be adjusted to take into account the firm's efforts to reduce risk, train staff, complete risk assessments etc. This is known as experience rating.

Many insurers survey premises with the object confirming the exact nature of the business. In addition the visit will provide an opportunity to improve the risk by offering guidance and advice or requiring that improvements are made, The completion of such improvements should minimise the incidence of accidents and diseases. This is an essential part of an insurer's service, and the risk surveyors will work in conjunction with the insured's own management and/or safety staff. However, it should be noted that this practice is restricted to the largest of risks and/or those with the greatest risk potential since insurers employ limited numbers of risk surveyors. Only a very small percentage of policyholders receive such visits.

Related covers

[E13023] In addition to employers' liability insurance, it is becoming increasingly common for companies to buy insurance in respect of directors' personal liability. Indeed, in the United States, some directors refuse to take up appointments in the absence of such insurance.

There is a growing trend to purchase legal expenses cover. This is primarily intended to cover the expenses incurred in defending employment tribunal claims such as arise from employment disputes eg relating to dismissal or to discrimination on the grounds of sex, gender or disability. (see E13032, C9001)

It should be noted that almost all liability insurance policies, both Employers' and Public, include an element of expenses cover. The costs and expenses of defending civil claims for compensation are always covered. If the claimant is successful, the policy will cover both the damages and the claimant's legal costs. However, there will usually be cover for the costs and expenses involved in the defence of criminal prosecutions arising from alleged breaches of health and safety and related legislation. Such prosecutions almost always take place before civil claims for compensation are settled. Thus insurers are interested in arranging for the defence of such claims since it provides early access to the facts of the accident, as determined by the enforcing authorities and enables an early view of potential liabilities to be formed.

Insolvent employers' liability insurers

[E13024] If the employee's action against the employer cannot succeed, the action for damages cannot be brought against an employer's insurer (*Bradley v Eagle Star Insurance Co Ltd* [1989] 1 All ER 961 where the employer company had been wound up and dissolved before the employer's liability to the injured employee had been established) (see below for the transfer of an employer's indemnity policy to an employee). The effect of this decision has been reversed by the *Companies Act 1989, s 141*, amending the *Companies Act 1985 (CA 1985), s 651* which allows the revival of a dissolved company within two years of its dissolution for the purpose of legal claims and, in personal injuries cases, the revival can take place at any time subject to the existing limitation of action rules contained in the *Limitation Act 1980*. For example, in the case of *Re Workvale Ltd (No 2)* [1992] 2 All ER 627, the court exercised its discretion under *section 33* of the *Limitation Act 1980* to allow a personal injuries claim to proceed after the three-year limitation period had expired. This meant that the company could also be revived under the provisions of *CA 1985, s 651(5)* and *(6)* (as amended). Thus, proceedings under the *Third Parties (Rights Against Insurers) Act 1930, s 1(1)(b)* may be brought in this manner.

If the employer becomes bankrupt or if a company becomes insolvent, the employer's right to an indemnity from his insurers is transferred to the employee who may then keep the sums recovered with priority to his employer's creditors. This is only so if the employer has made his claim to this indemnity by trial, arbitration or agreement before he is made bankrupt or insolvent (*Bradley v Eagle Star Insurance Co Ltd* [1989] 1 All ER 961). The employee must also claim within the statutory limitation period from the date of his injury (see E13007 for subrogation rights generally).

The problems that arise following the insolvency of an employer's liability insurer gained particular prominence following the much publicised liquidation of Chester Street Holdings Limited ('Chester Street') on 9 January 2001. It was initially feared that individuals whose claims pre-dated 1972 (the date when employers' liability insurance became compulsory) would be at risk of not receiving compensation. The situation has, however, largely been clarified following the establishment of the Financial Services Compensation Scheme ('the FSCS') which was created under the *Financial Services and Markets Act 2000* and acts as a 'safety net' for customers of finance sector companies who are unable to play claims against authorised companies. The FSCS came into effect on 1 December 2001. Under the FSCS policy holders (save for any period of cover when the policy holder was a nationalised industry which is specifically excluded from the provisions of the scheme) are eligible for protection if they are insured by an authorised insurance company under a contract of insurance issued in the UK, Channel Islands or Isle of Man. The FSCS scheme pays 100 per cent compensation for post-1972 compulsory employers' liability insurance claims. If the employer still exists and is solvent then the employer initially pays the compensation to the claimant although is able to recoup 100 per cent of the compensation outlay from the FSCS. The FSCS pays 90 per cent compensation for pre-1972 claims providing that the employer is not still in existence and solvent. If however the employer is still in existence – the employer must meet its liability to the claimant and is not compensated by the FSCS.

Special provisions apply to Chester Street for pre-1972 claims where:

- liability was established and quantified before insolvency on 9 January 2001, 90 per cent compensation is payable. Such claims are administered under the *Financial Services and Markets Act 2000* using the guidelines established by the Policyholders Protection Board;
- liability was agreed and quantified between 9 January and 30 November 2001. Such claims are subject to the requirements of a scheme set up by the Association of British Insurers who pay 90 per cent of the award.

Responsibility for claims handled by the Policyholder Protection Board (PPB) was transferred to the FSCS on 1 December 2001. PPB rules continue to apply to any claim arising from the insolvency of an insurance company where it occurred before 1 December 2001.

Further information can be obtained from FSCS, 10th Floor Beaufort House, 15 St Botolph Street, London EC3A 7QU, telephone 0800 678 1100 and 0207 741 4100.

Limitation Act

[E13025] Potential claimants have always been subject to a limitation on the time period within which they are able to bring a claim. Limitation is a practical approach to avoid the courts having to deal with a claim long after the event, when memories have faded, details have become uncertain and witnesses unreliable. Limitation is a defence against claims made after a lengthy delay and will be raised on behalf of the defendant, by the insurer acting on behalf of the employer. The law on limitation is contained in the *Limitation Act 1980*. There are many different periods dependent on the nature of the claim, a claim for personal injury through negligence, a claim under contract law etc but the main principle for Employers' Liability insurance is as follows.

An injured person has three years in which to make a claim for damages. This period runs from what is known as the 'date of accrual', usually the date on which the injury occurred. The date of accrual may also be the 'date of knowledge' which is taken to be the date on which the injured person realised they had a right to claim.

Where a specific injury has resulted, such as a broken bone, amputation of part of the body, a burn or serious flesh injury leading to scarring, there is no dispute. The three year period runs from the date of the accident. When a person is found to be suffering from a disease that might have an occupational cause the date of knowledge is usually regarded as being the date when a medical authority such as their doctor or specialist explains that their condition might have been caused by exposure to harmful substances encountered in the workplace.

There is a special case where the potential claimant has died. This may occur with mesothelioma which can result in death within a short period of time following diagnosis. The limitation period is three years from the date when the deceased might have been regarded as having knowledge. However, if the death occurs within the three year period the dependents have three years from the date of death or from the date of their knowledge.

Claims are usually initiated within a reasonable period after the accident but are sometime delayed and only reach insurers towards the very end of the three year period.

Long Tail Disease

[E13026] Employers' Liability policies provide cover in the event that an employee alleges that their disease was caused by exposure to harmful substances or situations in the working environment. Typically claims will arise from exposures to:

— asbestos giving rise to asbestosis, lung cancer or mesothelioma;
— harmful substances resulting in cancer, asthma etc;
— noise giving rise to noise induced deafness and/or tinnitus;
— vibration giving rise to hand arm vibration syndrome (vibration white finger).

Some diseases arise quickly after exposure and can be regarded as injuries, for example dermatitis, however some develop gradually over time eg deafness. There are others that have very long development periods or exhibit latency, not appearing for decades after exposure, mesothelioma which can take anything from 15 to 35 years to manifest itself. It should also be noted that for some diseases the extent of injury or disability is related to exposure. Thus the degree of deafness resulting from occupational noise exposure is related to the "dose" – a measure of both the level of exposure and the extent or time over which the exposure occurred. In contrast other diseases, especially cancers, are not related to dose. The condition can occur following small exposure, even of low level and for short periods of time.

Since the policy cover relates to the time when the harmful exposure took place the employee or ex-employee should make their claim against the employer they think is responsible and that employer will then have to identify the insurer for the relevant period. However, it is common for the claim to be made against the current employer (or the employer from which the employee retired) and that employer's insurer will undertake the task of determining which insurer is responsible. The task of determining which insurers are responsible is made especially difficult for diseases such as noise which develop over time as the employer may have changed insurer and the employee may have worked for several employers and been exposed to noise at some or all of them. Once liability is established and all responsible, i.e. liable, employers and insurers have been identified the cost of the claim will be apportioned between them according to the degree of liability agreed.

Employees suffering from occupational diseases with long development periods such as mesothelioma, may find it difficult to track down their previous employer. The firm may have ceased trading and in such circumstances tracing their employer's insurer may prove very difficult. In such cases claimants and their solicitor used to make use of a Code of Practice issued in November 1999. The Code was drawn up by the then Department of the Environment, Transport and the Regions, the Association of British Insurers, and the Non-Marine Association at Lloyds, and set out the procedures and standards of service required of insurers. The Code did not have statutory authority and was a voluntary code that committed member insurers to make a thorough search when requested of all records which exist. The Code obliges insurers to keep records of all policies issued for 60 years. No charge was levied for conducting a search.

There was widespread dissatisfaction with the performance of the Code and discussion took place between the Government and the insurance industry regarding the setting up of an Employers' Liability Insurance "database". There exists within the motor insurance market a database that holds details of all motor insurance policies, the details of the vehicle and policyholder. This performs several functions, including allowing the police to establish whether a vehicle is insured and also assists in the tracing of the insured party after accidents in which a vehicle has not stopped.

It was proposed that the creation of such a database for employers' liability would assist in the tracing of policyholders when a claimant seeks compensation for an occupational disease. Whilst this view has some merits there are considerable hurdles to be overcome before the full value of such a database could be realised. The database has been established and is known as the Employers' Liability Tracing Office (ELTO EL13034).

It should be noted that the real value or benefit from the creation of this database will be seen when those who are currently exposed to potentially harmful substance seek compensation at some time in the future, in perhaps 25 or 30 years. In order to secure an immediate benefit it would be necessary to "back load" the database with historic data in order for those currently seeking compensation who are finding it difficult to trace their employer's insurer to realise any value from the existence of such records. This could prove a mammoth task for insurers even if they had complete records dating back over many decades. It is already established that insurer's records are far from perfect eg there are incomplete records of subsidiary and associated companies, an essential aspect in proving which insurer held the cover and should pay the claim.

A further comparison was made with the Motor Insurers Bureau (MIB). The MIB is funded by contributions from all insurers who provide motor insurance to the public and its purpose is to provide compensation where there has been an accident resulting from the actions of an uninsured driver. A Private Members bill has been placed before Parliament suggesting that a corresponding body should be set up to compensate those who are unable to secure compensation for occupational disease. However, the bill faced opposition in Parliament and its second reading was delayed and it is not expected to make any further progress through Parliament.

Vicarious Liability

[E13027] Vicarious liability is where one person assumes liability for the acts and omission of another. It can arise in several ways but is most commonly encountered in Employers' Liability where the employer accepts liability for the acts of his employees. This is in accordance with the common law principle that sets out that an employee can expect their employer to provide competent fellow employees. There are some exceptions, eg if an employee was doing something expressly forbidden or carrying out a task in an unacceptable way, but generally the employer will be held liable. Also if one employee is responsible for injuring another then as a general rule the employer will be liable.

This principle can give rise to circumstances where an employer may be entitled to an indemnity from an employee. This right is set out in the Civil Liability (Contribution) Act 1978. This right was recognised in the case of *Lister v Romford Ice and Cold Storage* [1957] AC 555. In this case a father and son worked for the same firm and while driving a lorry the son knocked down and injured his father. The father's claim was successful and the employer's insurers then sought to recover from the son by use of the doctrine of subrogation. The insurance industry, through its trade body the British Insurance Association (now the Association of British Insurers), reached a market agreement that they would not sue an employee of an insured employer in respect of injury caused to a co-employee, unless there was either (a) collusion and/or (b) wilful misconduct on the part of the employee.

In *Morris v Ford Motor Co Ltd* [1973] 2 All ER 1084 Ford had subcontracted cleaning at one of its plants to X, for which Morris worked. While engaged on this work at the plant, Morris was injured owing to the negligence of an employee of Ford. Morris claimed damages and his claim was settled by Ford's insurers. However, the insurers sought to recover from Morris's employer on the basis of an indemnity clause in their contract with Ford. The clause bound X to indemnify Ford for all losses and/or claims arising out of the cleaning operations. However, X was not insured and sought to recover from the Ford employee who had injured Morris. Although accepting that it was bound by the terms of the indemnity, X argued that its liability should be subrogated against the employee who caused the accident, on the ground that the employee had carried out his work negligently. By a majority decision the Court of Appeal held that the claim for subrogation must fail. X had to borrow Ford's name in order to bring the claim and it was stated in the judgement that this was inequitable. And since subrogation was an equitable right the court had the discretion to refuse it.

A recent decision has changed the thinking about vicarious liability. It has been accepted for many years that only one person or organisation can be held to be vicariously liable. In other words the person who has acted negligently can only work for one employer. However, the outsourcing of activity by industrial firms and the growth of self employed contractors, agency workers and similar makes that view less clear.

Although the legal profession maintained the view that two employers could not both be found to be vicariously liable, other legal experts were coming to the opinion that this presumption should be tested in court. *Viasystems (Tyneside) Ltd v Thermal Transfer (Northern) Ltd and ors* [2005] EWCA Civ 1151, [2005] 4 All ER 1181 provided that opportunity. Viasystems (V) engaged Thermal Transfer (TT) to install air conditioning in its factory. TT subcontracted the work to SPD who in turn contracted with CMS to supply fitters and fitters' mates for the actual installation of the equipment. A self-employed fitter, H, contracted to SPD to act as supervisor, was supervising another fitter, M, and his mate, S, both employed by CMS. All three individuals were working in a roof space using crawling boards laid across the purlins. S was sent to get some fittings but in doing so he disturbed some of the already installed ducting which in turn caused damage to the sprinkler piping which leaked and flooded the factory. It was agreed that V could recover from TT in contract. In the first instance the Judge was asked to consider who was liable for the actions of S and decided that it was CMS but an appeal was lodged. The Court of Appeal had to consider whether both CMS and SPD could be jointly liable for the actions of the fitters mate. While it was obvious that the fitters mate had been negligent there was a question as to whether SPD or CMS were vicariously liable. The Court of Appeal considered who was entitled to give orders about how the work should be done. It was decided that H and M were both entitled and obliged to stop S's actions and thus both their employers were vicariously liable. After analysing the long standing assumption that two separate employers could not be vicariously liable for the actions of an employee the Court held that this could be possible in a modern context. Since both SPD and CMS were found to be vicariously liable there existed the question of contribution. The Court found that as both firms were liable they must be equally liable, and that each should thus contribute 50%.

Making the Market Work

[E13028] In an effort to facilitate the working of the Employers' Liability market and improve the flow of information between industry and insurers, the insurance industry set up the Making the Market Work Scheme. The Association of British Insurers set out which aspects of good health and safety management were of interest to insurers. Trade associations were invited to submit details of what actions they have taken and activities they offer to their members. The ABI has set up a standing committee to review the information received and provide constructive feedback to the trade associations.

This scheme is no longer in operation.

Woolf Reforms – the Pre-Action Protocol

[E13029] The Woolf Reforms, which were enacted as the *Civil Procedure Rules 1998 (SI 1998 No 3132)*, came into force on 26 April 1999 and apply to England and Wales. The aim of the new rules, also known as the CPR, was to speed up the processes by which people obtained compensation through the civil courts and make justice more accessible via management of costs.

Pre-action protocols are an important element of the new rules, which should ensure that more claims are settled without litigation. They are aimed at personal injury cases involving less than £15,000 in compensation. Should settlement not be reached through these protocols, the litigation process then goes through the courts. The reforms affect any class of insurance business where there is a third party injury, principally Motor, Public Liability and Employers Liability and occasionally Household.

The reforms were an attempt to resolve the dispute *before* court proceedings are even commenced; thus minimising both cost and court time. If the case is not settled via the pre-action protocols, it may be taken to court. The personal injury pre action protocol sets out the following stages:

— letter of claim;
— reply/acknowledgement;
— investigation;
— disclosure of documents;
— appointment of expert witnesses;
— negotiation and settlement;
— litigation in the courts.

Under the new rules all cases will be allocated to one of three routes:

— Small Claim: (previously the arbitration or Small Claims Court) property claims with a value of up to £5,000 or up to £1,000 for personal injury claims;
— Fast Track: claims with values between £5,000 and £15,000 (Property) or £1,000 and £15,000 (personal injury);
— Multi-track: all claims in excess of £15,000 and any complex claim of lower value.

Fast track cases will generally proceed quickly to a hearing, which is usually limited to one day, as most evidence has been provided in writing and agreed in advance.

The new rules affect policyholders in a number of ways. Any delay in notifying the claim may affect mounting a suitable defence and increase the cost of the claim. Policyholders must understand the reasoning behind decisions on liability; if there is no hard evidence insurers are unable to run a defence. And policyholders need to co-operate with claims handlers and solicitors by supplying information and documents. In particular policyholders must:

— promptly report all incidents likely to give rise to a claim to your insurer;
— carry out an investigation;

- retain all evidence and relevant documents (eg Accident Book or report form, accident investigation report, photos, plans and security videos, copies of risk assessments);
- forward any letter of claim to the insurer as soon as it is received but should not acknowledge its receipt.

Since 1998 the CPR have been regularly reviewed and amended/updated and the most recent update, No 68 came into effect on 1April 2014 Details of the amendments and updates can be found on the Ministry of Justice website www.justicegov.uk/courts/procedure-rules/civil. The most recent changes involve the extension of the "Portal" used for notification of Motor claims to include both Employers' Liability and Public Liability claims. The Portal is an electronic notification system that allows claimant's solicitors to identify the appropriate insurer and make an electronic notification of a claim. The insurer is identified using the Employers Liability Database (see **E13034**).

Insurers are given strict time limits within which they must admit liability and make an offer of settlement or deny liability. In addition to the new notification process a system of fixed costs has been introduced. The approach is intended to increase the speed of settlement of smaller, lower value claims. Factsheets and guidance on the new approach have been prepared by many insurers and solicitors, for example:

- www.qbeeurope.com/documents/casualty/risk/technical%20claims/QBE%20MOJ%20Claims%20Portal%20Extension%20June%202013.pdf
- www.hilldickinson.com/pdf/Fixed%20costs%20with%20some%20changes.pdf

Scotland

As stated above the Woolf Reforms only apply to England and Wales. However, following lengthy debate a voluntary Pre-Action Protocol has been agreed between insurers and the Law Society of Scotland and came into effect for all claims intimated after 1 January 2006. This voluntary arrangement remains in place as of April 2012.

This is a major step forward in Scottish pre-litigation procedure and is the culmination of much work between insurers and the Forum of Scottish Claims Managers. The main points to note about the Protocol are as follows:

- It is a voluntary protocol and does not have legal effect (although parties may seek to draw breaches in the protocol to the attention of the Court on the issue of costs).
- It applies only to claims intimated on or after 01/01/2006 and thereafter handled within the Protocol.
- It is primarily aimed at Motor claims with a value below £10,000 although can be agreed on **all** claims at all levels.
- Any existing claim notified prior to 01/01/2006 will be dealt with as per existing arrangements.
- The Protocol does not apply to property damage only claims — but will apply where a Property damage claim has a Personal Injury element.

— The Protocol does not apply to disease claims.

Admissions of Liability are binding where the value of the claim is below £10,000. Solicitors will send a Letter of Claim in similar format to the format of the letter of claim used in England / Wales and insurers will have 21 days to respond from date of receipt. Insurers then have 3 months to investigate liability and respond with a decision.

Medical evidence should be instructed by the Pursuer's (NB Scottish term for Claimant) solicitor within 5 weeks of a liability decision and medical experts will be instructed on an agreed basis (not joint instruction). Medical evidence should be disclosed by the Pursuer's solicitor within 5 weeks of receipt. Questions can be put to the expert by either side by mutual consent.

Pursuer's solicitors will serve a Statement of Valuation and insurers have 5 weeks in which to respond.

Where damages are agreed insurers have 5 weeks to issue cheques or else interest will be payable. The 2003 Fee Scale still applies but this now includes payment of the Investigation Fee.

Benchmarking and Performance Indicators

[E13030] The Health and Safety Executive (HSE), with the support of the DTI's Small Business Service, has launched a web-based tool in 2004. The tool was designed to assist SME's track and assess how well they are managing their own health and safety performance.

The Indicator was developed to help SME's regularly assess their health and safety performance eg from one year to the next. It is also intended to help companies tell their insurers how well they are managing health and safety so they can more accurately calculate insurance premiums based on individual performance.

Development of the Indicator involved HSE working closely with key stakeholders including the Department for Work and Pensions, the Small Business Service, the Association of British Insurers, the British Insurance Brokers Association and the Federation of Small Businesses.

The development of the Indicator arose in part from the Government's review of Employer's Liability Compulsory Insurance in 2003. In the report of the review Government called on HSE to develop a tool, for SME's in particular, which would enable insurers to determine levels of insurance premiums that better reflected how well those employers were managing risks to health and safety and show HSE's continued commitment in helping businesses to improve their health and safety performance.

The Indicator is an internet based tool. It is free to use and works by asking a series of questions on key hazards that the developers found most SME's encounter and incident frequency. A score out of ten is calculated for each – ten the best and zero the poorest. The results can be used in two ways. The questionnaire can be completed on a regular basis and the results used to monitor improvement over time. Alternatively, a user can benchmark their performance against their peers, for example firms in the same business or trade sector or firms of a similar size. As of May 2009 there were a total of nearly 10,000 records however this has fallen to just over 4,000 at April 2012.

SME's can also decide if they wish to share their results with anyone else, for example to let their insurers and brokers know how they are performing at health and safety, so this can be taken into account when insurance terms are set. For the main features of the indicator, go to www.businesslink.gov.uk.

There was a similar index designed for larger firms with greater than 250 employees. The Corporate Health and Safety Index (CHaSPI) focuses more on management systems than hazards.

At December 2008 there were 642 users of whom just 114 had completed the assessment. A review carried out in 2010 stated that opinion was divided as to the benefit and usefulness of both of these benchmarking tools. Neither HaSPI nor CHaSPI were available for use at the end of 2013.

NHS Recoveries

[E13031] The passing of the Road Traffic (NHS Charges) Act 1999 enabled hospitals to recover the costs of treatment of those injured in road traffic accidents. The collection of costs was centralised under the auspices of the Compensation Recovery Unit (CRU) that also recovers, from insurers of negligent employers, the benefits paid to personal injury claimants. It has been estimated that the recoveries channelled to the NHS totalled around £105m (2006 prices).

In 2002 the Department of Health investigated the possibility of extending the scheme to include all cases of personal injury where the injured person receives compensation. This would include any claimants seeking compensation from their employer or third parties such as visitors and customers seeking compensation from a shop owner, property owner or similar. This expansion of the scheme was included in the Health and Social Care (Community Health and Standards) Act 2003 and following further consultation the proposals were revised and amended and formed part of the Health Act 2006, coming into effect in January 2007.

The revised scheme is known as the NHS Injury Costs Recovery Scheme (ICR) and provides for the recovery of costs associated with the treatment of people who have suffered injuries as a result of:

— road traffic accidents (as already provided)
— workplace accidents
— other accidents eg those involving members of the public

and who subsequent make a successful claim for compensation. The recovery of costs is triggered when the compensation is paid to the injured person. The costs are recovered from the insurer who pays the compensation on behalf of the negligent party, the employer in the case of workplace accidents or the property owner, firm or other organisation where the injury involved a member of the public, customer or similar. The insurer will forward the appropriate amounts to the CRU who will then forward them to the appropriate NHS hospital, ambulance trust, primary care trust or similar.

A schedule of costs has been drawn up and as at April 2012 these are:
— Flat fee for treatment at hospital (without admission): £615
— Daily rate for treatment as an inpatient: £755
— Ambulance charges (per journey): £185

with a ceiling of £45,153 per incident. The scale of charges will be reviewed each April and increased in line with hospital cost inflation. The inflationary increase in April 2006 was 4.5% and this serves as an indication of likely future increases.

The ICR scheme will take account of any contributory negligence on the part of the claimant. Thus if an employee was held to be 30% to blame and their compensation reduced the costs recoverable would be reduced by the same amount.

The ICR scheme also recognises that some, mainly larger firms, may have excesses on their policy. Under such arrangements the employer, for example agrees to pay a proportion of each settled claim, perhaps the first £5,000 or £10,000. The usual arrangement is that the insurer will pay the claimant the full amount of damages agreed or awarded by the courts and the employer will then reimburse the insurer. The ICR requires that the insurer pay the full amount of any NHS costs owed irrespective of whether there is an excess in place. The insurer will then recover the insured's proportion of the NHS costs.

Corporate Manslaughter

[E13032] (Further information can be found at C9001)

Pressure groups and campaigning bodies have long held the view that the existing law on corporate manslaughter was inadequate and failed to deliver a suitable punishment following cases of death resulting from work activities, involving either employees or members of the public.

Whilst it was possible to bring a charge of corporate manslaughter against a company or corporate body there were difficulties in securing a conviction. It was an agreed legal principle that a corporate body does not possess a "mind" and is thus unable to commit a violent crime. Thus it was necessary to charge both the company and an individual employee with manslaughter. And unless the individual is successfully convicted the company cannot be found guilty. This situation resulted in few cases going to court and low conviction rates.

In the 2012/13 year there were 148 (173) deaths of workers/employees and 113 (90) work related deaths to members of the public (Figures for 2011/12 in brackets). Deaths of workers are fairly static but have dropped slowly over the last 10 years whilst deaths of members of the public are more volatile and do not display such a regular trend. In the period 1992 to 2006, a total of 160 cases of workplace death were referred to the Crown Prosecution Service, 45 of these proceeded to trial and a total of 10 convictions were achieved. The successful prosecutions also tend to involve smaller firms where it has proved easier to establish a link between the actions of senior management and the circumstances surrounding the accident that resulted in the death. Opponents of the existing arrangements thus maintain that the current approach is biased against smaller firms and allows larger firms that have caused deaths to "escape" justice.

In 1996 the Law Commission published a report on the subject of "involuntary manslaughter". This report was a response to growing public pressure for changes in the law. In the period following the Law Commission report the Government carried out various consultation exercises and published a draft bill in 2005. The final form of the Bill was published on late 2006 and after challenges in the House of Lords the Royal assent and into effect in April 2008.

The key elements of the Bill are as follows:

— A new offence of Corporate Manslaughter was created (Corporate Homicide in Scotland).
— For a charge of manslaughter to be made an organisation must have owed a duty of care to the victim and have breached that duty as the result of way in which its activities were managed or organised by senior manager(s).
— The duty of care can result from the employment of persons, ownership or occupation of premises, the supply of goods or services, construction or maintenance activities or the keeping/use of vehicles.
  Note: These duties match exactly the existing duties that are protected by liability policies.
— The breach of duty, termed a senior management failure, must have been gross and caused the death. In determining guilt consideration must be given to the risk of death presented by the breach, the seriousness of the breach and any breach of health and safety legislation.
— The gross breach must involve conduct falling far below what would be reasonably expected in the circumstances.
— Senior managers are described as being those who make decisions about the whole or a substantial part of the organisation's activities.
— All incorporated bodies are covered by the proposed legislation although certain classes of business are exempt eg partnerships, as are certain Government bodies.
— The offence will be tried in the Crown Court and the penalties available are unlimited fines or remedial measures.
— There is specific guidance that the jury should consider the firm's corporate safety culture, policies and procedures and any relevant safety information or guidance.

It should be noted that it is no longer necessary to secure a conviction of senior manager and thus it is thought that it will be easier to secure conviction of a firm.

The existing protocol for joint investigation of work related deaths by the Police and the HSE will remain in force.

The possibility of joint charges for breaches of health and safety legislation and corporate manslaughter remain – there is no change resulting from the proposed legislation. Thus there is the distinct possibility that a firm could be charged with breaches of health and safety legislation eg HSWA 1974 or specific legislation, and corporate manslaughter and senior managers or directors could be charged with a breach of HSWA 1974 s 37.

Guidance on sentencing and the level of fines was published for consultation in November 2007 - the consultation closed in February 2008 and guidelines were issued in February 2010 (http://sentencingcouncil.judiciary.gov.uk/)

The guidance set out the manner in which the evidence would be assessed and the factors that needed to be taken into account when judging the seriousness of the offence. For cases of Corporate Manslaughter it was stated that the fine would " . . . seldom be less than £500,000" and that it could run to "millions of pounds." For health and safety offences where a death has occurred it was stated that the fine would " . . . seldom be less than £100,000" and could run to hundreds of thousands of pounds. It was also stated that a fixed correlation between the level of fine and turnover or profit was not appropriate.

The guidelines also set out the details of Publicity Orders requiring the defendant to publicise the details of the incident and conviction, the extent of the fine and any Remedial Order that might have been made.

In addition to paying damages and compensation Employers' Liability insurance also covers a range of related costs and expenses (see E13017). Prior to 2008 policy wordings generally restricted the payment of defence costs to prosecutions made under health and safety and related legislation. However, it was common practice for insurers to agree to fund the costs of defending the few manslaughter prosecutions that did occur.

Following the introduction of the Corporate Manslaughter legislation the majority of insurers modified their wordings. In general the effect has been to explicitly state that the policy will cover the costs of defending any Corporate Manslaughter charge and pay any prosecution costs that may be awarded against the defendant employer. In addition the indemnity will extend, if the policyholder so requests, to any director, partner or employee. This recognises that such a person may be prosecuted for breaches of health and safety legislation at the same time as the company is prosecuted for Corporate Manslaughter.

Cotswold Geothechnical Holdings

In April 2009 the Crown Prosecution Service issued a press release stating that the first charge of Corporate Manslaughter had been made. An employee of Cotswold Geothechnical Holdings limited had died as a result of a trench collapse whilst working on a site in September 2008. The firm was charged with breaches of the HSWA 1974 in addition to the charge of Corporate Manslaughter. A director of the firm was charged with Gross Negligence Manslaughter in addition to breaches of the 1974 Act.

The case was complicated by the fact that the owner/director of this small firm was extremely ill. The charges against him were dropped in view of the fact that he was unable to attend court to give evidence in his defence. However, the case against the firm proceeded and it was found guilty of Corporate Manslaughter and fined £385,000. The fine represented 250% of the firm's annual turnover and was to be paid over a period of 10 years. An appeal was lodged but the Court of Appeal upheld both the original verdict and the fine.

JMW Farms

In May 2012 JMW Farms Ltd was fined £187,500 at Belfast's Langanside Crown Court for safety failings, which led to the death of its employee Robert Wilson.

In 2010 Mr Wilson, 45, was working at the firm's meal-mixing farm in Co Armagh, when he was struck by a metal bin, which fell from the raised forks of a forklift. The vehicle was being driven by one of the company's directors, Mark Wright. The bin had been raised off the ground to allow Mr Wilson to clean it. When the vehicle reversed the bin became unstable and fell from the forks of the fork lift truck. It struck Mr Wilson, causing fatal head injuries. The investigation into the circumstances of the accident revealed that a replacement vehicle being used while the usual truck was being serviced. It was also found that it was not possible to insert the forks of the replacement vehicle into the sleeves of the bin, as the forks were too large and incorrectly spaced.

JMW Farms pleaded guilty to breaching CMCHA 2007 and, in addition to the fine, it was ordered to pay £13,000 in costs. The company must pay both the fine and costs within six months.

Belfast Recorder Judge Tom Burgess said: "Yet again, the court is faced with an incident where common sense would have shown that a simple, reasonable and effective solution would have been available to prevent this tragedy. The very definition of the offence of corporate manslaughter is an acceptance of a gross breach of duty. That is a high and totally unacceptable breach in circumstances where the risks involved were high, with the more than foreseeable likelihood of serious injury, or death following if the proper steps were not taken."

The company may find it difficult to pay the fine since the previous year's turnover was just £1m.

A summary of the Judgement can be found at www.courtsni.gov.uk/en-GB/Judicial%20Decisions/SummaryJudgments/Documents/Summary%20of%20judgment%20-%20R%20v%20J%20M%20W%20Farm%20Limited/j_sj_R-v-JMW-Farm-Limited_080512.html.

Lion Steel

In July 2011 a firm called Lion Steel was charged under the CMCHA 2007 following the death of an employee after a fall through a fragile roof in 2008. Three directors of the firm were charged with Gross Negligence Manslaughter and offences under HSWA 1974.

The following notes are based on the detailed Sentencing Remarks prepared by the judge which can be found at www.judiciary.gov.uk.

[E13032] Employers' Liability Insurance

The Company—

- Lion Steel is a manufacturer of metal storage lockers. It was established in 1998 following a management buyout. It had around 140 employees at the time of the accident and operated two sites – in Hyde near Manchester and another near Chester.
- The company's turnover was of the order of £10m per annum. Following an examination of the Report and Accounts the judge commented that it seemed to be a company that was "holding its own" in difficult trading conditions but only by means of loans and overdrafts.
- The Sentencing Remarks contained a detailed description of the roof. There were various different areas of metal roofing consisting of sinusoidal and trapezoidal cross sections. Parts of the roof had been re-clad. There were the usual valley gutters although these were narrow and in some places made even narrower due to latter roof extensions or re-cladding projecting over or across the gutters. The roof sections included a number of skylights consisting of translucent fibreglass panels.
- There was evidence of holes or leaks which had been previously patched with strips of tape. From this the judge inferred that the roof needed repairing from time to time. Although he took pains to point out the case was not about standards of building maintenance.
- The issue of whether there had been access to the roof on other occasions was obviously not part of the prosecution and/or not discussed nor presented in evidence. However, from a risk management perspective it can be concluded that roof work must have been undertaken on previous occasions. It can only be a matter of conjecture who might have carried out such work.
- Access to the roof was via a door from part of the factory which allowed access to the valley gutter. The Sentencing Remarks made no comment about the interior access to the door but it can be assumed that it was reached by some form of internal stairs or similar. Such easy access does nothing to dissuade diligent employees who are seeking to carry out their work from accessing the roof. It also fails to discourage casual access by persons who have no need to be on the roof. It would have been prudent for such an access door to be locked in order to reinforce a clear ruling that employees should not carry out roof work (see below).

The accident—

- The deceased (B) was employed by Lion Steel as a general maintenance man – he was one of two such employees.
- These two employees were expected to carry out general maintenance work and small repairs around the factory premises.
- If they were in any doubt about their ability to carry out work they were instructed to ask for an independent outside contractor to be engaged.
- B was not a trained roofer, on the day of the accident B was dealing with water leaks. In order to deal with these leaks he went onto the roof.
- There was nothing to indicate exactly what B did when he gained access to the roof. It is not known where he walked, nor what route he took.

- However, what is known is that he stood on a fibreglass skylight which gave way under his weight and he fell 13m onto the factory floor.
- Evidence was presented that the Works Manager knew B was working on leaks that day.

The charges—

- There were 5 sets of charges—
  - Count 1: Corporate Manslaughter against Lion Steel
  - Count 2: Gross Negligence Manslaughter against three directors, P, W and C
  - Count 3: Breach of HSWA 1974 s 2
  - Count 4: Breach of HSWA 1974 s 37 by each of the three directors, namely that they were guilty of neglect
  - Count 5: Breach of Work at Height Regulations 2005
- In view of the closeness in time of the accident date (29 May 2008) to the commencement date of the CMCHA 2007 (6 April 2008) there were complex arguments about the presentation of evidence. In part because the CMCHA 2007 is not retrospective and thus it was not legally possible to introduce evidence of matters that related to events, actions or knowledge prior to 6 April 2008. As a result of these legal discussions Count 1 was set aside.
- The defence argued that Count 5 added nothing beyond that set out in Count 3 and should be dropped. Furthermore since a specific section of the WaH Regs had not been specified it was stated that the charge should be dropped. This argument as accepted by the judge and prosecution.
- As the trial progressed the judge stated that there was no case to answer in respect of the HSWA 1974 s 37 charges against the three directors and the charges were dropped.

The HSE involvement—

- There was evidence that the HSE had recommended the placing of signs warning of the fragile roof as far back as 2006.
- In the Sentencing Remarks the judge stated that there were various guidance documents and codes of practice relating to the prevention of falls from height and that such guidance was simply reinforced with the passing of the Work at Height Regulations in 2005.

The involvement of the Insurers—

- The role and participation of the insurer and brokers formed part of the prosecution case. The judge rejected what he saw as an unrealistic argument, namely that their involvement was causative of the accident.
- There was evidence that the insurer's surveyor visited in 2005 and 2007.
- In 2007 the survey report commented that large parts of the roof had been replaced and that the company was "acceptable subject to risk requirements being completed within the appropriate timescales."
- The judge specifically commented in his ruling:
  - no insurer had ever refused cover on the grounds that the precautions for working on the roof were inadequate;

- the insurer's requirements relating to roof work did not require a written risk assessment;
- none of the risk requirements dating from 2007 related to roof work; and
- at no point had the insurer, AXA, refused to insure the premises.
- The judge concluded by stating "In my judgement the insurers and insurance brokers were also less than thorough in their assessments about the roof".
- From the comments made and the line of questioning it would appear that the CPS/prosecution had little or no understanding about the role and involvement of an insurance surveyor. For a manufacturing risk involved in sheet metal work it would have been reasonable to expect the focus of the insurance company's surveyor's to have been machinery, vehicle movements, perhaps manual handling etc. Almost certainly, the accident history – if there were any – would have comprised accidents from such causes and not to have included any arising from falls from height, certainly not roof work.
- The judge's comments seem to indicate that he regarded the insurance surveyor's approach to have been more along the lines of an H&S consultant, a Competent Person undertaking a full hazard identification and analysis survey.

The prosecution case—

- The essence of the case was that–
  - the roof should have been assumed to be fragile unless tested to show otherwise;
  - the roof lights should have regarded as fragile without any need to test;
  - persons should not have been permitted to work on the roof without adequate precautions (to prevent the risk of fall);
  - suitable precautions could have included crawling boards or harness and arrest line/lanyard;
  - at the time of the accident there was no evidence of boards, lines, harness etc. all indicative of a lack of suitable precautions — nor were there any warning signs; and
  - any person expected to work on the roof should have been properly trained.
- In his Sentencing Remarks the judge stated that–
  - The risk of a fall from (or through) the roof was an obvious one.
  - He was of the opinion that Lion Steel had devised a way of working that was intended to keep B, the deceased off the fragile areas, but that what they had not done was to supply the deceased with training or protection.
  - From a risk management perspective it is reasonable to assume that the system of work was defective. It is not reasonable to expect a workman to make decisions about whether work is within his competence. There is the danger that they will tend to tackle work for fear of being regarded as "workshy", lacking in drive etc. With something as serious as work at height there

    should be a clear, plain ruling eg "No work at height to be undertaken". Or Lion Steel should have expressly stated that any work on the roof was only to be undertaken by specialist contractors.
  - What Lion Steel had done was to shift the responsibility for decisions about risk assessment onto the employee, and they had not provided the employee with either training to undertake such work nor the equipment to carry out the work safely.

The ultimate charge—

- Ultimately it was just the three charges of manslaughter against the director and the s 2 case against the company that remained.
- At this point the defence agreed to plead guilty to the Corporate Manslaughter charge against Lion Steel if the s 2 charge and the three manslaughter charges were dropped.

The fine—

- In his Sentencing Remarks the judge summarised his view of the seriousness of the case against Lion Steel. He stated–
    - the risk of accident was foreseeable;
    - the company fell short of the required standard and the measures required eg training, boards etc. "would not have been expensive to install";
    - there was a difference in the standards of health and safety between the two premises; and
    - the responsibility for the breach lies at the door of the director in charge of the Hyde premises.
- The judge placed the responsibility for the short comings squarely at the door of the director in charge of the premises – which is perfectly reasonable and it was encouraging that the judge commented that the costs of "simple" precautions such as training and the supply of crawling boards could hardly be regarded as expensive. It is also worthy of note that the judge commented on the difference in standards of H&S at the different locations.
- In outlining mitigating factors the judge stated that Lion Steel had as reasonable safety record. In addition the company had taken advice on health and safety, although mainly at its Chester premises. The judge said it could have done more at the Hyde factory.
- The judge drew attention to the Sentencing guidelines for the CM charge. Taking into account the guilty plea and the company's financial state the fine was set at £485,000 after a discount of 20%. The fine is to be paid in four instalments over the period September 2012 to September 2015.
- As regards costs the Defendant's QC asked for an order of no more than 25%. This was rejected by the judge although he did comment on the excessive length of time to bring the case to court and thus reduced the amount that the CPS was seeking. Lion Steel was ordered to pay 60% of the asked for costs – a figure of £84,000 to be paid within 2 years.

The Current Situation

The number of cases being referred to the Crown Prosecution Service has increased from 45 in 2011 to 63 in 2012. However, there is no evidence of a prosecution of a large firm, indeed the majority of prosecutions are of small firms and the fines have all been below the £500,000 threshold.

At January 2014 the following CM prosecutions have been launched:

- *PS & JE Ward (November 2012):* Market gardening firm – employee died following an overhead cable "strike" involving a tipping trailer
- *MSN Mining (January 2013):* Prosecution following the deaths of 4 miners at Gleison Colliery
- *Prince's Sporting Club (February 2013):* Death of an 11 year old girl in a "banana boat" incident
- *Mobile Sweepers (Rotherham) Ltd (March 2013):* Death as a result of employee working under-neath an unsupported vehicle body
- *Sterecycle Limited (October 2013):* Employee killed when door of pressurised autoclave flew off.

Compensation Act 2006

[E13033] The Compensation Act is part of the Government's programme to tackle the 'compensation culture' and in the case of those suffering from mesothelioma improve the system for obtaining compensation where there is a valid claim. The Act received the Royal Assent in November 2006 and further parts are expected in 2007. The Act contains a number of provisions and will enable the drafting of specific legislation to reform particular aspects and areas of the law relating to the claiming of compensation. The main elements of the Act are described as follows:

Part 1: Standard of Care
— Deterrent effect of potential liability
— Apologies, offers of treatment etc
— Mesothelioma
Part 2: Claims Management Services
Part 3: General

Potential Effect of Potential Liability (s 1): This section of the Act attempts to overcome the situation where, it is alleged, activities are not undertaken for fear of litigation in the event that something unforeseen occurs and legal action follows. The usual example is that of school trips being cancelled because of fear that teachers will be sued if a pupil is injured in an accident. The Act states that when a court is considering a claim in negligence or breach of statutory duty it should consider whether the taking of reasonable steps to meet a standard of care might prevent a desirable activity from being undertaken or discourage persons from undertaking functions in connection with a desirable activity. The effect of this text is dependent upon more detailed guidance to the courts about the exact meaning of words such as "desirable activity" and "standard of care".

Apologies (s 2): This part of the Act seeks to clarify that an apology or an offer of treatment or "other redress" shall not amount to an admission of negligence or breach of statutory. It is thought that in some cases injured parties are less concerned about claiming compensation and value an apology but that potential defendants are concerned that making an apology and offering treatment eg physiotherapy will be seen as an admission of guilt. This section of the Act seeks to overcome these perceptions and hopefully reduce claims and speed the delivery of rehabilitation.

Mesothelioma (s 3) this section of the Act will enable the making of Regulations to improve the delivery of compensation to those suffering from mesothelioma. It should be noted that this is the only part of the Act that applies in Scotland.

Claims Management Services (Part 2): This part of the Act provides for the introduction of regulations to control the activities of claims management services, often known as "claims farmers". It is thought that the activities of such firms, for example "door stepping" people in shopping malls and at railway stations only serves to increase claims numbers. The Government has already drafted regulations under this part of the Act.

Employers' Liability Tracing Office

[E13034] The Employers Liability Tracing Office (ELTO) was established in 2011 in response to dissatisfaction with the voluntary Code of Practice Act which had operated since 1999.

Insurers are providing information about companies and their EL insurance to ELTO and this information is being entered on a central database, the Employers Liability database (ELD). The information is being supplied as existing policies are renewed or new firms and businesses are underwritten. Details of cover from April 2011 are being recorded. In addition, when investigations related to specific long tail disease claims confirm details of historic cover eg dating back to the 1970s or 1980s this information is also being added.

The ELTO website is: www.elto.org.uk.

Preliminary results for 2013 are as follows:

- Policy Records: 12 million
- Enquiries: over 137,000 (more than double 2012)
- Success Rate: 77% (slight increase from previous year)

Lord Young Report

[E13035] In June 2010 the Prime Minister asked Lord Young to investigate and report on the compensation culture and the current low standing of health and safety. The report titled "Common Sense Common Safety" was published in October 2010. It was broad based and covered a number of areas and the report gave the impression of being driven by anecdotal concerns about a "compensation culture". The report gave support for the Jackson reforms of civil law. Despite providing no evidence of how widespread the practice was the report stated that insurers should stop requiring businesses to employ expensive health and safety consultants. The report also recommended consultation with the insurance industry" to ensure that worthwhile activities are not unnecessarily curtailed on health and safety grounds".

The report also recommended that insurers should prepare a "code of practice" for businesses and the voluntary sector. The intention was that such a code would provide an indication of what insurers required from their policyholders.

In response to this recommendation the Association of British Insurers coordinated the preparation of a Key Principles document titled "Health and Safety for Small Businesses and the Voluntary Sector" that sets out what insurers expect from their policyholders. The document was published in November 2011 and can be found here: www.abi.org.uk/.

Lofstedt Review

[E13036] In May 2011 the Government asked Professor Lofstedt of King's College London to chair a comprehensive review of health and safety regulations, enforcement and civil liability. Professor Lofstedt's report titled "Reclaiming Health and Safety for all" was published in November 2011. The full report and the Government's response, published simultaneously, can be found here: www.dwp.gov.uk.

In general terms the report concluded that there was no case for a radical overhaul of health and safety legislation. It concluded that much of legislation was beneficial in reducing risk and in turn workplace accidents. There were some recommendations for a review of HSE guidance and Approved Codes of Practice. In addition the legislation in sectors such as mining and quarrying was thought to be in need of revision and consolidation.

It was stated that there were some problems with interpretation and application of existing legislation and that businesses sometimes go beyond what is required by the law

Professor Lofstedt considered the usefulness and application of the Pre-Action Protocols, in particular the list of documents that can be requested or sought during disclosure. It was concluded that the manner in which these lists are used has perhaps deviated from the original intention. The report recommended that the original intention of the standard disclosure list be clarified and restated.

The report made numerous recommendations, however, there were none that related specifically to the provision of insurance or insurance companies. The HSE was recommended to continue to help businesses understand what is reasonably practicable and how to comply with the law in a proportionate manner. In early 2012 the Prime minister hosted a meeting of the Chief

Executives of several major insurance companies. Although much of the meeting focused on motor insurance, claims farmers and whiplash the subject of EL insurance was discussed. Following the meeting discussion took place between the Department of Work and Pensions and the ABI about how the insurance industry might be able to assist the HSE.

Fees For Intervention (FFI)

[E13037] In late 2011 the HSE launched a consultation on the approach to be used for the recovery of their costs. Existing legislation already provided the legal framework for the HSE to recover costs from firms for example for major hazard sites, offshore rigs etc. The legislation was amended to allow recovery of costs from firms that were, on inspection, found to have a "material breach" of legislation. The consultation exercise was intended to collect feedback about possible approaches to the recovery. The HSE undertook a pilot exercise involving key stakeholders and announced that the scheme would come into operation in April 2012.

In late March it was announced that cost recovery or FFI as it had become known, would not commence in April but was to be deferred until October. When this change was announced there was a slight change in that recovery of costs was to be triggered by a contravention rather than a material breach. This appears to be a lowering of the threshold which could result in more firms facing costs.

The proposal only extends to those firms that are subject to inspection by the HSE. Those firms that are subject to Local Authority inspection by Environmental Health Officers eg businesses in the retail, and hospitality sectors will not face FFI.

EL insurers have generally taken the view that any costs that the HSE is seeking to recover under FFI are not damages or compensation as defined by the policy but more akin to fines. Thus they would not be covered under an EL policy and the firm, the employer, would not be able to recover such costs from their insurer.

At the HSE's Board Meeting in June 2013 there was a report on the first 12 months of the operation of FFI. The report commented:

- Invoices issued: 5,766
- Total sums collected: £2.67m
- Average Invoice: £464

The majority of invoices were issued in the Manufacturing and Construction sectors, which together accounted for over 70% of all invoices.

Details of the Scheme can be found on the HSE website: www.hse.gov.uk.

Employment Protection

Andrea Oates

Introduction to employment protection

[E14001] The nature of employment protection is continually changing. At a domestic (national) level, legislation introduced towards the end of the 1990s and early 2000s extended certain employment protection rights to workers as well as employees. For example, *the Public Interest Disclosure Act 1998*, the *Working Time Regulations 1998 (SI 1998 No 1833)* and the *Part-Time Workers (Prevention of Less Favourable Treatment) Regulations 2000 (SI 2002 No 2035)* all apply to 'workers' – a wider category than 'employees' under traditional English employment law. New individual rights were also introduced. These include the right to be accompanied at a disciplinary or grievance hearing and the right to request flexible working. Several sets of regulations increased rights to parental leave and some collective rights were enhanced, with trade unions given the right to statutory recognition in certain circumstances.

UK discrimination legislation was simplified and consolidated into a single *Equality Act*. This came into force on 1 October 2010 and prohibits discrimination on the basis of a number of 'protected characteristics', including age, disability, religion or belief, race, pregnancy and maternity. It also banned (with some exceptions) pre-employment health checks on job applicants.

Over recent years, new legislation has, for example, doubled the qualifying period for unfair dismissal claims from one to two years for those starting work on or after 6 April 2012; introduced the requirement for ACAS early conciliation before a tribunal claim can go ahead; and reduced the consultation period in the case of large scale redundancies from 90 to 45 days. The 2016 Trade Union Act introduced new balloting, voting and notice restrictions in relation to taking industrial action as well as new restrictions on picketing.

In July 2013, the government introduced fees for lodging claims and hearings at employment tribunals. However, a judicial review challenge to their introduction in the Supreme Court by public services union UNISON saw the fees abolished with immediate effect – from July 2017 (*R on the application of UNISON v The Lord Chancellor* [2017] UKSC 51). Just as the introduction of the fees saw a huge decline in employment tribunal claims, so the abolition of the fees saw a year-on-year 165% increase in the number of single claims.

Membership of the Europe Union (EU) has also had a significant impact on the regulation of employment relations in the UK. For example, European directives have required the introduction of anti-discrimination legislation in areas including equal pay and equal treatment, part-time workers and fixed-term workers. Employers with more than 50 employees are required to set up national information and consultation procedures if requested by employees, and agency workers gained the right to equal treatment. At the time the chapter was being updated, negotiations on the terms of the UK's exit from the European Union (EU), scheduled for 31 October 2019, were on hold as the Conservative Party selected a new leader following Theresa May's resignation. The new leader will also become Prime Minister. The UK remains a member of the EU and will continue to comply with European legislation, including the rulings of the European Court of Justice (ECJ), until the withdrawal negotiations are complete and a new relationship defined, or the UK leaves without a deal.

The relationship between an employer and an employee is a contractual one and as such must have all the elements of a legally-binding contract to render it enforceable. In strict contractual terms an offer is made by the employer which is then accepted by the employee. As in the case of the offer, the acceptance may be oral, in writing, or by conduct, for example by the employee turning up for work. For a legal contract to exist, each party must bring something to it. The legal term given to this is "consideration". The consideration on the employer's part is the promise to pay wages and on the employee's part to provide his services for the employer. Once the employer's offer has been accepted, the contract comes into existence and both parties are bound by any terms contained within it (*Taylor v Furness, Withy & Co Ltd* (1969) 6 KIR 488).

The contractual analysis of the employment relationship is not entirely satisfactory in explaining the relationship between employee and employer. To fit the contract model, various elements comprising the reality of the employment relationship become part of the contract by implication.

A number of recent cases looking at employment status in the so-called gig economy have examined whether individuals described as being self-employed in their contracts are really 'workers' and therefore entitled to rights and protections including paid annual leave and rest breaks under the Working Time Regulations 1998. In most of these cases, including *Uber v Aslam* [2018] EWCA Civ 2748 and *Pimlico Plumbers v Smith* [2018] UKSC 29, the courts and tribunals have come down on the side of individuals arguing that they are workers, rather than self-employed or independent contractors.

An employment contract is unlike many other contracts, because many of the terms will not have been individually negotiated by the parties. The contract will contain the 'express' terms that the parties have agreed – most commonly hours, pay, job description – and there will be a variety of other terms which will be "implied" into the contract from other sources and which the parties have not agreed. Many of these are relevant to health and safety. If any of the express or implied terms in the contract are breached, the innocent party will have certain remedies. The fact that various employee rights, particularly in relation to health and safety, are implied into the contractual terms and conditions is important for this reason. In addition to terms implied into the contract by the common law, statute (such as the *Employment Rights Act 1996*) has created additional employment protection rights for employees, including some specific rights in relation to health and safety. These are in addition to detailed rights and duties arising from health and safety legislation and regulations which are discussed elsewhere in this publication.

An employer will often lay down health and safety rules and procedures. While the law allows an employer the ultimate sanction of dismissal as a method of ensuring that safety rules are observed, such dismissals should be lawful, that is generally with notice, and should be fair. Furthermore, statute has created specific protection from victimisation for employees who are protecting themselves or others against perceived health and safety risks. All of these provisions are the subject of this section.

As a specific health and safety protection measure, the *Health and Safety at Work etc Act 1974, s 2 (HSWA 1974)* lays down a general duty on all employers to ensure, so far as is reasonably practicable (for the meaning of this expression, see E15039 ENFORCEMENT), the health, safety and welfare of all their employees. An employer is also under a duty to consult about health and safety matters. The Health and Safety Executive (HSE) has also issued a number of codes of practice, under the *HSWA 1974*, relating to, for example, the functions of safety representatives and their rights to time off to train (see JOINT CONSULTATION IN SAFETY).

Sources of contractual terms

Express terms

[E14002] These are the terms agreed by the parties themselves and may be oral or in writing. Normally the courts will uphold the express terms in the contract because these are what the parties have agreed. However, if the term is ambiguous the court may be called upon to interpret the ambiguity, for example what the parties meant by 'reasonable overtime'.

Generally, the express terms cause no legal problems and the parties can insert such terms into the contract as they wish. There are, however, a number of restrictions which include the following:

(a) An employer cannot restrict his liability for the death or personal injury of his employees caused by his negligence. Further, they can only restrict liability for damage to their employee's property if such a restriction is reasonable (*Unfair Contract Terms Act 1977, s 2*).
(b) The terms in the contract cannot infringe the *Equality Act 2010*.
(c) The terms in the contract cannot infringe the *Part-time Workers (Prevention of Less Favourable Treatment) Regulations 2000 (SI 2000 No 1551)*. The Regulations provide that unless justified on objective grounds, a part-time worker has the right not to be treated less favourably than a comparable full-time worker on the ground that the worker is a part-timer in relation to the terms of the contract.

(d) The terms in the contract cannot infringe the *Fixed-term Employees (Prevention of Less Favourable Treatment) Regulations 2002 (SI 2002 No 2034)*. The Regulations provide that unless justified on objective grounds, a fixed-term employee has the right not to be treated less favourably than a comparable permanent employee on the ground that they are a fixed-term employee.

(e) The employer cannot have a notice provision which gives the employee less than the statutory minimum notice guaranteed by the *Employment Rights Act 1996, s 86*.

(f) Some judges have suggested that any express terms regarding hours are subject to the employer's duty to ensure their employee's safety and must be read subject to this. A term requiring an employee to work 100 hours a week will not be enforceable (see for example Stuart-Smith LJ in *Johnstone v Bloomsbury Health Authority* [1991] IRLR 118). More specifically, the provisions of the *Working Time Regulations 1998 (SI 1998 No 1833)* affect the contractual term in relation to working hours. The 1998 Regulations, save in the case of specified exemptions, set a maximum working week of 48 hours averaged over a 17-week reference period. They also provide for an obligatory daily rest period, weekly rest, rest breaks, limits on night work and minimum paid annual leave and otherwise regulate working time. In *Barber v RJB Mining UK Ltd* [1999] IRLR 308, the court decided the maximum imposed on weekly working time by the Regulations was part of the employees' contract. This decision gives some protection to employees who refuse to work beyond the statutorily stated maximum. It is possible under the 1998 Regulations for a worker to agree with his or her employer in writing to opt-out of the 48-hour working week, subject to complying with certain requirements.

(g) The terms in the contract cannot infringe the *Maternity and Parental Leave etc Regulations 1999 (SI 1999 No 3312)*. The Regulations entitle eligible employees who are still in work in the 15th week before their baby is due to a full year of maternity leave, regardless of their length of service.

(h) Confidentiality provisions should be subject to an employee's right to make a protected disclosure in accordance with the provisions set out in the *Employment Rights Act 1996*.

Common law implied duties – all contracts

[E14003] Both the employer and employee owe duties towards each other. These are duties implied into every contract of employment and should be distinguished from the implied terms discussed below which are implied into a particular individual contract. Although there are a number of different duties, three are of major importance in relation to health and safety:

(a) the duties on the part of the employee to obey lawful and reasonable orders;
(b) to perform their work with reasonable care and skill; and
(c) the duty on the part of the employer to ensure their employee's safety.

The duty to obey lawful and reasonable orders ensures the employer's safety rules can be enforced and, as it is a contractual duty, breach will allow the employer to invoke certain sanctions against the employee, the ultimate of which may be dismissal.

The same is true of the duty to perform work with reasonable care and skill. Should the employee be in breach of this duty and place their own or others' safety at risk, the employer may impose sanctions against them, including dismissal.

The imposition of the employer's duty is to complement the statutory provisions. Statutes such as the *HSWA 1974* provide sanctions against the employer should they fail to comply with the Act or any regulations made thereunder. The common law duty provides the employee with a remedy should the duty be broken, either in the form of compensation if they are injured, or, potentially, with a claim of unfair dismissal. The employer's duty to ensure their employees' safety is one of the most important aspects of the employment relationship. At least one judge has argued that it is so important that any express term must be read subject to it (see *Johnstone v Bloomsbury Health Authority* [1991] IRLR 118 at **E14002** above).

Breach of this duty can lead to the employee resigning and claiming constructive dismissal. In *Walton & Morse v Dorrington* [1997] IRLR 488, for example, an employee claimed that she had been constructively dismissed because her employer had breached the implied term of her contract of employment that it would provide, so far as reasonably practicable, a suitable working environment. The employee had been forced to work in a smoke-filled environment for a prolonged period of time, and her employer did not take appropriate steps to redress the problem when she raised the issue. The case arose before the Health Act 2006 came into force, banning smoking in most enclosed workplaces and public places. The Employment Appeal Tribunal (EAT) agreed she had been constructively dismissed because the employer had breached its duty to her.

The employer's duty to ensure their employee's safety has also been held to cover stressful environments resulting in mental injury to the employee. In *Walker v Northumberland County Council* [1995] IRLR 35, the High Court held that an employer was liable for damages on the basis that they owed their employee a duty not to cause him psychiatric damage by the volume and/or character of work that he was required to undertake.

In *Fraser v The State Hospitals Board for Scotland* 2001 SLT 1051, the Court of Session held that there was no reason to qualify an employer's duty to take reasonable care for the safety of employees so as to restrict the nature of the injury suffered to a physical one. The case involved a claim for damages for psychological injury as a result of disciplinary measures. It failed on the basis of lack of foreseeability.

In *Hatton v Sutherland* [2002] EWCA Civ 76, [2002] 2 All ER 1, [2002] IRLR 263 the Court of Appeal set out the principles for dealing with psychiatric injury arising from work-related stress. If the claimant has brought evidence of injury to the employer's attention, the claim is more likely to succeed. Failure on the part of the employer to provide support following a return-to-work following a period of sickness absence can also trigger liability (see *Barber v Somerset* CC [2004] for example). Employers have been found to have breached their duty of care if they knew, or should have known, about the risk to health, through bullying for example (see *Dickins v O2 plc* [2008] EWCA Civ 1144, [2009] IRLR 58 for example).

Common law implied terms – individual contracts

[E14004] The court may imply terms into the contract when a situation arises which was not anticipated by the parties at the time they negotiated the express terms. As such, the court is "filling in the gaps" left by the parties' own negotiations. The courts use two tests to see if a term should be implied, (i) the 'business efficacy' test (*The Moorcock* (1889) 14 PD 64) or (ii) the 'officious bystander' or 'oh of course' test (*Shirlaw v Southern Foundries Ltd* [1939] 2 KB 206). The second describes a term so obvious that it goes without saying that the parties must have intended it. Once the court has decided, by virtue of one of these tests, that a term should be implied, it will use the concept of reasonableness to decide the content of the term. Often this will involve looking at how the parties have worked the contract in the past. For example, if the contract does not contain a mobility clause, but the employee has always worked on different sites, the court will normally imply a mobility clause into the contract (*Courtaulds Northern Spinning Ltd v Sibson* [1988] IRLR 305). In *Aparau v Iceland Frozen Foods plc* [1996] IRLR 119, the EAT refused to imply a mobility clause on the basis that there were other ways of achieving the necessary flexibility.

In relation to dismissal, a term has been implied that, except in the case of summary dismissal, the employer will not terminate the employment contract while the employee is incapacitated where the effect would be to deprive the employee of permanent health insurance benefits (*Aspden v Webbs Poultry and Meat Group (Holdings) Ltd* [1996] IRLR 521). In a recent case, *Awan v ICTS UK Ltd* [2018] UKEAT/0087/18/RN, the EAT implied a contract term preventing the employer exercising its contractual right to dismiss the claimant for ill-health once he had become entitled to receive benefits under an ill-health retirement plan.

Terms can also be implied by the conduct of the parties or by custom and practice in a particular industry or area. The test for this is relatively difficult to fulfil – the term must be reasonable, notorious (well known) and certain and, in effect, everyone in the industry/enterprise must know that it is part of the contract. Arguments based on custom and practice come into play in relation to issues such as statutory holidays and redundancy policies.

Collective agreements

[E14005] Collective agreements are negotiated between an employer or employer's association and a trade union or unions. This means they are not contracts between an employer and individual employees because the employee was not one of the negotiating parties. Some terms of the collective agreement will be procedural and will govern the relationship between the employer and the union. Some, on the other hand, will impact on the relationship between the employer and each individual employee, for example a collectively-bargained pay increase.

As the employee is not a party to the collective agreement, the only way they can enforce a term which is relevant to them is if the particular term from the collective agreement has become a term of their individual employment contract. Procedural provisions, policy and more general aspirations are not suitable for incorporation into an individual contract of employment. Incorporation is important because the collective agreement is not a legally-binding contract between the employer and the union (*Trade Union and Labour Relations (Consolidation) Act 1992, s 179(1)*) and therefore needs to be a term of an employment contract to make it legally enforceable.

The two main ways a term from a collective agreement becomes a term of an employment contract is by express or implied incorporation. Until relatively recently, implied incorporation was the most common and was complex. It generally required the employee to be a member of the union which negotiated the agreement, to have knowledge of the agreement and of the existence of the term, and to have conducted themselves in such a way as to indicate that they accepted the term from the collective agreement as a term of his contract.

A decision in the EAT case *Healy v Corporation of London* (24 June 1999, unreported) illustrates that habitual acceptance of the benefits of a collective agreement does not, in itself, lead to the conclusion that the terms of that collective agreement have become contractually binding on an individual employee. There can be many reasons for an individual to accept the benefits of collective bargaining which do not amount to an acceptance that the underlying agreement forms part of his or her contract. Express incorporation means the employee has expressly agreed (normally in their contract) that any term collectively agreed would become part of their contract. This used to be unusual, but with the change made to the statutory statement which must be given to all employees (see below) employees must be told of collective agreements which apply to them, and this has been held as expressly incorporating those agreements into the contract.

Statutory statement of terms and conditions

[E14006] Under the *Employment Rights Act 1996, s 1* every employee no later than two months after starting employment, must receive a statement of their basic terms and conditions. The statement must contain:

(a) the names of the employer and employee;
(b) the date the employment began;
(c) the date the employee's continuous employment began;
(d) the scale or rate of remuneration and how it is calculated;
(e) the intervals when remuneration is paid;
(f) terms and conditions relating to hours;
(g) terms and conditions relating to holidays;
(h) terms relating to sickness or injury, including provision for sick pay (if any);

(i) terms and conditions relating to pensions (unless the employment is by a body or authority where pension terms are governed by separate legislation). If there is no pension scheme, the statement must say this;
(j) notice requirements;
(k) job description;
(l) title of the job;
(m) if the job is not permanent, the period of employment;
(n) place of work, or if various the address of the employer;
(o) any collective agreements which affect terms and conditions and, if the employer is not a party to the agreements, the persons with whom they were made;
(p) if the employee is required to work outside the UK for more than one month, the period he will be required to work, the currency in which he will be paid, any additional benefits paid to him and any terms and conditions relating to his return to the UK; and
(q) details of the employer's disciplinary and grievance procedures (or information about where to find them).

The terms in (a), (b), (c), (d), (e), (f), (g), (k), (l) and (n) must all be contained in a single document. In relation to pensions and sickness absence and sick pay, the employer may refer the employee to a reasonably accessible document. The employer can also refer the employee to a reasonably accessible collective agreement or to the *Employment Rights Act 1996, s 86* which contains provisions relating to minimum notice periods.

There is no duty on an employer to give details of any disciplinary or grievance procedures relating to health and safety. The Act (s 3(2)) specifically states that: Subsection (1) does not apply to rules, disciplinary decisions, decisions to dismiss, grievances or procedures relating to health or safety at work. However, given the employer's duties under the *HSWA 1974, s 2*, and the law relating to unfair dismissal, it is good industrial relations practice to ensure that all employees know of all the disciplinary procedures which could be invoked against them.

The employment service ACAS advises (see **E14008** below) that organisations may want to consider dealing with issues involving bullying, harassment or whistleblowing under a separate procedure to their main disciplinary and grievance procedure.

From 6 April 2020, the *Employment Rights (Employment Particulars and Paid Annual Leave) (Amendment) Regulations 2018 (SI 2018/1378)* and *Employment Rights (Miscellaneous Amendments) Regulations 2019 (SI 2019/731)* will make the right to a written statement of particulars of employment apply when an individual begins employment (a day 1 right) and extend the right to 'workers' as well as employees.

The statement will have to include additional information, including how long a job is expected to last, or the end date of a fixed-term contract, how much notice must be given to end the contract, and details of eligibility for sick leave and pay.

It will also require details of other types of paid leave, such as maternity and paternity leave, any terms and conditions on normal working hours, the days of the week the worker is required to work, whether or not those hours or days may vary and if so, how they vary and how that variation is decided.

Information about the duration and conditions of any probationary period, all remuneration, not just pay, and any training provided by the employer, any part of that training that is compulsory and any compulsory training that the employer will not be paying for, will also need to be included.

Works rules

[E14007] Works rules may or may not be part of the contract. If they are part of the contract and therefore contractual terms, they can be altered only by mutual agreement, that is, the employee must agree to any change. It is unusual, however, for such rules to be contractual. To be so, there would have to be some reference to them within the contract and an intention that they are terms of the contract. The more usual position with regard to the employer's rules was stated in *Secretary of State for Employment v ASLEF (No 2)* [1972] 2 QB 455 where Lord Denning said that they were merely instructions from an employer to an employee. This means that they are non-contractual and the employer can alter the rules without the consent of the employees. The fact that they are not contractual does not mean that they cannot be enforced against an employee. All employees have a duty to obey lawful and reasonable orders (see E14003 above) and thus failing to comply with the rules will be a breach of this duty and therefore a breach of contract. The only requirement that the law stipulates is that the order must be lawful and reasonable, and it is unlikely that an order to comply with any health and safety rules would infringe these requirements.

Disciplinary and grievance procedures

[E14008] It has already been noted that the employer must give details of grievance and disciplinary procedures to all employees. Failing to do so could lead to the employee resigning and claiming constructive dismissal (*W A Goold (Pearmak) Ltd v McConnell* [1995] IRLR 516).

The current ACAS *Code of practice on disciplinary and grievance procedures* (www.acas.org.uk/media/1047/Acas-Code-of-Practice-on-Discipline-a nd-Grievance/pdf/11287_CoP1_Disciplinary_Procedures_v1__Accessible.pdf) was issued under *section 199* of the *Trade Union and Labour Relations (Consolidation) Act 1992* and came into effect in March 2015, replacing the 2009 Code. More comprehensive advice and guidance on dealing with disciplinary and grievance situations is contained in the ACAS booklet, *Discipline and grievances at work: the Acas guide*, published in February 2019 (www.acas.org.uk/media/1043/Discipline-and-grievances-at-work-The-Acas -guide/pdf/DG_Guide_Feb_2019.pdf).

It is not legally binding and failure to follow the Code will not make a dismissal arising out of a disciplinary issue automatically unfair. But where an employer or employee has unreasonably failed to follow the Code, an employment tribunal can increase or reduce any compensation award (by up to 25%).

While many potential disciplinary or grievance issues can be resolved informally, it important that where they are pursued formally, they must be pursued fairly. The Code sets out the basic principles of fairness and recommends that:

- Clear and specific rules and procedures for handling disciplinaries and grievances are set down in writing. Disciplinary rules should give examples of what the employer regards as acts of gross misconduct, such as theft or fraud, physical violence, gross negligence or serious insubordination;
- Employees and, where appropriate, their representatives are involved in the development of the rules and procedures;
- Employees and managers are helped to understand what the rules and procedures are, where they can be found and how they are to be used;
- Employers and employees raise and deal with issues promptly and do not unreasonably delay meetings, decisions or confirmation of those decisions;
- Both employers and employees act consistently;
- Employers carry out any necessary investigations to establish the facts of the case;
- Employers inform employees of the problem and give them the opportunity to put forward their case in response before any decisions are made;
- Employers allow employees to be accompanied at any formal disciplinary or grievance meeting; and
- Employers allow an employee to appeal against any formal decision made.

Employees have the right to be accompanied at a disciplinary or grievance hearing under section 10 of the *Employment Relations Act 1999*. There is a statutory right to be accompanied by a companion where the disciplinary meeting could result in: a formal warning being issued; the taking of some other disciplinary action; or the confirmation of a warning or some other disciplinary action (appeal hearings). The companion can be a fellow employee, a trade union official, or a representative who has been certified by their union as being competent for this purpose. This right is enforceable in the employment tribunal and compensation is payable for any failure. It is worth noting that fellow employees are under no duty to perform the role of accompanying individual.

It is now a legal requirement, unless an exemption applies, for a claimant to have made an Early Conciliation notification to ACAS. Tribunal claims will not be accepted unless the complaint has been referred to ACAS and a conciliation certificate issued. This certificate confirms that the Early Conciliation requirements have been met.

The Code recommends that employers keep written records of any disciplinary or grievance cases they deal with, and consider dealing with issues involving bullying, harassment and whistleblowing under a separate procedure.

Subject to the above, employers can establish their own disciplinary procedures. Such procedures may become part of the contract. If, for example, the employer gives the employee a copy of the procedures with the contract, and the contract refers to the procedures and the employee signs for receipt of the contract and the procedures, it is likely that they will be contractual. Employers may prefer their disciplinary procedures not to be contractual. If the procedures are contractual, any employee will be able to claim that his or her contract has been breached if they are not followed. This possibility also applies to those employees who have been employed for less than the two-year qualifying period required to bring a claim for unfair dismissal. The period increased from one year to two years for those starting work on or after 6 April 2012. In response to a claim of breach of contract, a court may award damages against the employer. These damages are based on an assessment of the time for which, if the procedure had been followed, the employee's employment would have continued.

Employee employment protection rights

Right not to suffer a detriment in health and safety cases

[E14009] Under the *Employment Rights Act 1996, s 44* every employee has the right not to be subjected to a detriment, by any act or any failure to act, by his employer on the grounds that:

(a) having been designated by the employer to carry out activities in connection with preventing or reducing risks to health and safety at work, the employee carried out (or proposed to carry out) any such activities;

(b) being a representative of workers on matters of health and safety at work or a member of a safety committee –
 (i) in accordance with arrangements established under or by virtue of any enactment; or
 (ii) by reason of being acknowledged as such by the employer;
 the employee performed (or proposed to perform) any functions as such a representative or a member of such committee;

(ba) the employee took part (or proposed to take part) in consultation with the employer pursuant to the *Health and Safety (Consultation with Employees) Regulations 1996 (SI 1996 No 1513)* or in an election of representatives of employee safety within the meaning of those Regulations (whether as a candidate or otherwise);

(c) being an employee at a place where –
 (i) there was no such representative or safety committee; or
 (ii) there was such a representative or safety committee but it was not reasonably practicable for the employee to raise the matter by those means;
 he brought to his employer's attention, by reasonable means, circumstances connected with his work which he reasonably believed were harmful or potentially harmful to health or safety;

(d) in circumstances of danger which the employee reasonably believed to be serious and imminent and which he could not reasonably have been expected to avert, he left (or proposed to leave) or (while the danger persisted) refused to return to his place of work or any dangerous part of his place of work; or

(e) in circumstances of danger which the employee reasonably believed to be serious and imminent, he took (or proposed to take) appropriate steps to protect himself or other persons from the danger.

In considering whether the steps the employee took or proposed to take under (e) were reasonable, the court must have regard to all the circumstances including the employee's knowledge and the facilities and advice available to them (*Employment Rights Act 1996, s 44(2)*).

In *Kerr v Nathan's Wastesavers Ltd* (1995) IDS Brief 548, however, the EAT stressed that tribunals should not place too onerous a duty on the employee to make enquiries to determine if their belief is reasonable.

Danger under (d) does not necessarily have to arise from the circumstances of the workplace but can include the risk of attack by a fellow employee (*Harvest Press Ltd v McCaffrey* [1999] IRLR 778).

Danger under (e) can include danger to others as well as the employee themself (*Masiak v City Restaurants (UK) Ltd* [1999] IRLR 780).

Various actions by the employer can constitute a detriment to the employee, such as disciplining the employee. A failure to act on the part of the employer can also constitute a detriment, such as not sending the employee on a training course, for example. Furthermore, the section is not restricted to the health and safety of the employee or their colleagues. In *Barton v Wandsworth Council* (1995) IDS Brief 549 a tribunal ruled that an employee had been unlawfully disciplined when he voiced concerns over the safety of patients due to what he considered to be the lack of ability of newly-introduced escorts. This shows the legal protection is triggered in relation to any health and safety issue. It includes cases where the employee voices concerns, and is not limited only to those circumstances where the employee commits more positive action.

The *Employment Rights Act 1996, s 44(3)*, however, provides that an employee is not to be regarded as subjected to a detriment if the employer can show the steps the employee took or proposed to take were so negligent that any reasonable employer would have treated them in the same manner.

If the detriment suffered by the employee is dismissal, there is special protection under *s 100* (see **E14010** below).

If the employee should suffer a detriment within the terms of *s 44*, they may present a complaint to an employment tribunal (*Employment Rights Act 1996, s 48*). See **E14008** regarding the requirement for ACAS Early Conciliation. The complaint must be presented within three months of the act (or failure to act) complained of, or, if there is a series of acts, within three months of the date of the last act. The tribunal can waive this time limit if it was not reasonably practicable for the employee to present their complaint in time. If the tribunal finds the complaint well founded, it must make a declaration to that effect and may make an award of compensation to the employee, the amount of compensation being what the tribunal regards as just and equitable in all the circumstances (*Employment Rights Act 1996, s 49(2)*). The amount of compensation will take into account any expenses reasonably incurred by the employee in consequence of the employer's action and any loss of benefit caused by the employer's action. Compensation can be reduced because of the employee's contributory conduct.

The protection from being dismissed or subjected to a detriment on the health and safety grounds specified in *s 44* and *s 100* of the *Employment Rights Act 1996* has been reinforced by more general protection for whistle-blowers (see E14011.1 below).

Dismissal on health and safety grounds

[E14010] In addition to the normal protection against dismissal (see below), where an employee is dismissed (or selected for redundancy) and the reason or principal reason for the dismissal is one of the grounds listed in the *Employment Rights Act 1996, s 44*, the dismissal will be automatically unfair. The only defence available to an employer applies to a dismissal in relation to action taken by the employee to protect themselves or others from danger that the employee reasonably believed was serious and imminent (*Employment Rights Act 1996, ss 44(1)(e)* and *100(1)(e)*).

The employer can escape a finding of unfair dismissal if they can show the actions taken or proposed by the employee were so negligent that any reasonable employer would have dismissed the individual concerned. However, as long as the employee forms a genuine view of a risk they reasonably regard as serious and imminent, the fact that the employer disagrees with its seriousness or the appropriateness of the steps taken is irrelevant. Any dismissal for taking the steps will be automatically unfair (*Oudahar v Esporta Group Ltd* [2011] UKEAT/0566/10).

Where an employee genuinely and reasonably believes that a practice represents a health and safety risk or a breach of health and safety law, even where this turns out to have been mistaken, they are still protected. What matters is that they genuinely and reasonably held the view at the time (*Joao v Jurys Hotel Management UK Ltd* UKEAT0210/11).

In respect of a dismissal falling within *s 100*, the normal qualifying period of employment does not apply (*Employment Rights Act 1996, ss 108(3)(c)*). An employee who has only been employed for a few weeks can therefore claim unfair dismissal for a breach of *s 100*. A dismissal which is not automatically unfair under *s 100* may nevertheless be unfair under the general reasonableness test under *s 98*.

There is no limit on the amount of compensation that may be awarded for an unfair dismissal on health and safety (or whistleblowing) grounds.

Oudahar v Esporta Group Ltd [2011] UKEAT/0566/10 (see above) also provides guidance to tribunals on when a dismissal should be regarded as automatically unfair for a health and safety reason.

The EAT out a two-stage test:

(1) Did the employee reasonably believe that there was a serious and imminent danger and did they take steps to protect themselves and others?
(2) If so: Was the sole or main reason for dismissal because the employee took those steps?

Dismissal for assertion of a statutory right

[E14011] Under the *Employment Rights Act 1996, s 104*, an employee is deemed to be unfairly dismissed where the reason or principal reason for that dismissal was that the employee –

(a) brought proceedings against an employer to enforce a right of his which is a relevant statutory right, or

(b) alleged that the employer had infringed a right of his which is a relevant statutory right.

It is immaterial whether or not the employee has the right or whether or not the right has been infringed, as long as the employee made it clear to the employer what the right claimed to have been infringed was, and the claim is made in good faith.

A statutory right for the purposes of the section is any right under the *Employment Rights Act 1996* in respect of which remedy for infringement is by way of complaint to an employment tribunal; a right under *s 86* of the 1996 Act (minimum notice requirements); rights in relation to trade union activities under the *Trade Union and Labour Relations (Consolidation) Act 1992*; rights conferred by the *Working Time Regulations 1998 (SI 1998 No 1833)* (and other regulations relating to working time in particular industries); and rights conferred by the *Transfer of Undertakings (Protection of Employment) Regulations 2006 (SI 2006 No 246)*. This is an important right for employees. If, for example, after the employee has successfully claimed compensation from their employer for a breach of s 44 they are dismissed, the dismissal will be automatically unfair under *s 104*. Again if the employer unlawfully demotes or suspends without pay, as a disciplinary sanction for breach of health and safety rules, and after proceedings against them for an unlawful deduction from wages the employer dismisses the employee, this will be unfair under *s 104*. As with dismissal in health and safety cases under *s 100*, the normal qualifying period of employment does not apply.

Protection for whistleblowing

[E14011.1] The *Public Interest Disclosure Act 1998 (PIDA 1998)* provides protection to workers who blow the whistle on health and safety matters, as well as about criminal acts, failure to comply with legal obligations, miscarriages of justice, damage to the environment and deliberate concealment of any of these matters. Under the Act, which inserted new sections into the Employment Rights Act 1996 (see **E14009** and **E14010** above), a worker who makes a 'qualifying disclosure' which they reasonably believe shows one of these matters, may be protected against dismissal or being subjected to a detriment.

Disclosures are only protected if they are made to appropriate persons – which often means that the employer must be approached in the first instance. There are other possibilities available under the Act which include disclosure to a legal adviser and disclosure to a prescribed person (eg the Financial Services Authority or the Commissioners of the Inland Revenue).

In terms of health and safety risks, protection for qualifying disclosures is not limited to cases of imminent or serious danger – it can apply where the health and safety of any individual has been, is being or is likely to be, endangered. In all cases, the worker must have a reasonable belief and make the disclosure in good faith, except in the case of disclosure to a legal adviser. In *Parkins v Sodexho Ltd* [2002] IRLR 109, an employee successfully used the Act to complain about a lack of supervision that amounted to a breach of health and safety obligations.

If a disclosure is protected, and an employee is subjected to any detriment or dismissed as a result, it is unlawful. A dismissal in these circumstances is deemed to be automatically unfair and the tribunal is not required to consider whether or not the employer's actions were reasonable. There is no minimum qualifying period for entitlement to make an unfair dismissal claim for this reason and there is no limit on the compensation available to whistle-blowers who are unfairly dismissed because they have made a protected disclosure.

A PIDA claim can be brought where a disclosure relates to a previous employer *Elstone v BP plc* (2010) UKEAT/0141/09/DM, [2011] 1 All ER 718, [2010] IRLR 558. In *Hinds v Keppel Seghers UK Ltd* [2014] IRLR 754, a health and safety consultant who provided services via a personal service company was found to be a 'worker' entitled to protection under the whistleblowing provisions of the Employment Rights Act 1996.

The Court of Appeal has confirmed that a broad approach must be taken to the law to achieve parliament's aim of protecting whistle-blowers. In *Day v Health Education England* [2017] EWCA Civ 329, Dr Day alleged that he was victimised by the Trust where he worked and also by Health Education England, the body responsible for his training, after raising concerns about serious staffing problems affecting the safety of patients. The Court of Appeal ruled that he could bring his claim against both bodies. They were both his 'employers' for the purposes of whistleblowing law because both had a substantial role in deciding his terms of employment.

McTigue v University Hospital Bristol NHS Foundation Trust [2016] IRLR 742, [2016] All ER (D) 211 (Jul), established that some agency workers can claim they have been subject to a detriment as a result of making a protected disclosure against an 'end-user'. In this case, the claimant was employed by an agency and worked at a centre run by the NHS trust. The EAT found the tribunal should have considered whether both the agency and the Trust 'substantially determined' the terms of the claimant, so that she was in fact a 'worker' of both.

Protection under *PIDA 1998* was limited by the *Enterprise and Regulatory Reform Act 2013 (ERRA 2013)* which made the protection available to workers under the Act subject to a new public interest test. It also removed the requirement that certain disclosures be made in good faith, replacing this with a power to reduce compensation where disclosure is not made in good faith. It also introduced vicarious liability for employers if a worker is subjected to detriment by a co-worker for making a protected disclosure.

Employment service ACAS explains that vicarious liability refers to a situation where someone is held responsible for the actions or omissions of another person.

'In a workplace context, an employer can be liable for the acts or omissions of its employees, provided it can be shown that they took place in the course of their employment,' it explains.

A disclosure that is driven purely by self-interest will not be protected (*Parsons v Airplus International Ltd* [2018] UKEAT/0111/17). However, it is not for the tribunal, rather than the employer, to decide whether an employee is motivated by a public interest (*Beatt v Croydon Health Services v NHS Trust* [2017] EWCA Civ 401).

A dispute involving an individual worker's contract terms can be in the 'public interest' even if it only affects a small group of other people, such as co-workers, as long as the individual who makes the disclosure refers to the interests of those other people at the time of making it. If a disclosure is in the wider public interest, it does not matter that it is also in the whistle-blower's private interest, or in the private interests of everyone whose contract is affected (*Chesterton Global Ltd v Nurmohamed* [2017] EWCA Civ 314). This is an important case in which the Court of Appeal gave a broad interpretation of what 'public interest' means. It also provided guidance on the public interest test:

- it is not up to the tribunal to decide whether a disclosure really was in the public interest. All that matters is that the whistle-blower genuinely and reasonably believed this to be the case at the time of making it (whether or not they were correct);
- a disclosure will be protected even if public interest was not the worker's main motivation for making the disclosure;
- whether the worker held a reasonable belief that their intended disclosure was in the 'public interest' will depend on the facts of each case;
- a disclosure about someone's own employment contract or another issue affecting that person's individual interests, or the private interests of a group of co-workers, can be in the wider 'public interest' depending on the facts of the case.

The disclosure must convey information, not just make allegations (*Kilraine v London Borough of Wandsworth* [2018] EWCA Civ 1436). A person making generalised allegations without providing reasonably specific information to back them up is unlikely to be protected.

Individual managers can be held personally liable for losses as a result of whistleblowing. In *Royal Mail Group Ltd v Jhuti* [2017] EWCA Civ 1632, the Court of Appeal ruled that a manager could be held personally liable for whistleblowing detriment after he retaliated against a whistle-blower by engineering her dismissal by a different manager who knew nothing of the disclosures. In *Timis and Sage v Osipov* [2018] EWCA Civ 2321, individual senior managers in charge of deciding on a dismissal were held to be personally liable for all the resulting foreseeable losses, including uncapped compensation for future lost earnings and injury to feelings.

In April 2019, the European Parliament voted in favour of new rules setting EU-wide standards of protection for whistle-blowers. The law now needs to be approved by EU ministers and member states will have two years to comply with the rules (see E14001 above). The new rules will apply to areas including public procurement, financial services, money laundering, product and transport safety, nuclear safety, public health, consumer and data protection. They will allow whistle-blowers to disclose information either internally or directly to competent national authorities, as well as to relevant EU institutions, bodies, offices and agencies. In cases where no appropriate action is taken, or if they believe there is an imminent danger to the public interest or a risk of retaliation, they will still be protected if they choose to disclose information publicly. The UK is one of only ten EU countries that currently provides comprehensive legal protection for whistle-blowers.

Enforcement of safety rules by the employer

The rules

[E14012] Given the statutory duty on the employer, under the *HSWA 1974, s 2*, to have a written statement of health and safety policy, and their common law duty to ensure their employees' safety, the employer should lay down contractual health and safety rules, breach of which will lead to disciplinary action. These rules must be communicated to the employee and be clear and unambiguous so the employee knows exactly what they can and cannot do.

The employer's disciplinary rules will often classify misconduct, as minor misconduct, serious misconduct and gross misconduct for example. It is unlikely that a tribunal would uphold as fair a dismissal for minor misconduct. It will underline the importance of health and safety rules if the breach is deemed to be serious or gross misconduct. The tribunal will, however, look at all the circumstances of the case. It does not automatically follow that if an employer has stated that a breach of a particular rule will be gross misconduct, a tribunal will find a resultant dismissal fair.

In *Sarkar v West London Mental Health NHS Trust [2010] EWCA Civ 289, [2010] IRLR 508*, a consultant psychiatrist was dismissed following complaints that he bullied and harassed staff. In dismissing him, the employer used two different procedures. It first attempted to use a conflict resolution procedure to deal with the complaints. This could only result in a formal written warning. It then used a formal disciplinary procedure and found Mr Sarker guilty of gross misconduct and dismissed him. The Court of Appeal said that an informal procedure should not be used where the conduct could justify dismissal and found against the employer.

The procedures

Disciplinary issues

[E14013] Once an employer has laid down their rules, they must ensure that they have adequate procedures, which should be non-contractual, to deal with a breach. The procedures used by an employer are scrutinised by a tribunal in any unfair dismissal claim and past cases indicate that many employers have lost such claims due to inadequate procedures. As discussed below, an employer in an unfair dismissal claim must show the tribunal that they acted reasonably. This concentrates on the fairness of the employer's actions and not on the fairness to the individual employee (*Polkey v A E Dayton Services Ltd* [1987] IRLR 503). This means an employer cannot argue that a breach of procedures made no difference to the final outcome and that they would have dismissed the employee even if they had adhered to the procedure. Breach of procedures themselves by an employer is likely to render a dismissal unfair regardless of which rule was broken.

Essentially any disciplinary procedure should contain three elements: an investigation, a hearing (or meeting) and an appeal, and should observe the principles of natural justice.

(a) Investigation

The law requires that the employer has a genuine belief in the employee's 'guilt', and that the belief is based on reasonable grounds after a reasonable investigation (*British Home Stores v Burchell* [1978] IRLR 379). If the employer suspends the employee during the investigation, this suspension should be with pay and in accordance with the disciplinary procedure. The ACAS Code of practice on disciplinary and grievance procedures recommends that where a period of suspension with pay is considered necessary, it should be as brief as possible, be kept under review, and made clear that the suspension is not considered a disciplinary action.

An investigation is important because it may reveal defects in the training of the employee show, or that the employee was not told of the rules, or that another employee was responsible for the breach. In all these cases, disciplinary action against the suspended employee will be unfair. Any investigation should be as thorough as possible and documented. It should also take place as soon as possible since memories fade quickly and this is particularly important if other employees are to be questioned as witnesses. Likewise, taking too long to start an investigation may lead the employee to think that no action will be taken and to then discipline them may itself be unfair.

The ACAS Code says it may be necessary, in some cases, to hold an investigatory meeting with the employee before proceeding to any disciplinary hearing in order to establish the facts of the case without reasonably delay. If an investigatory meeting is held, this should not alone result in any disciplinary action.

In other cases, it says the investigatory stage will be the collation of evidence by the employer for use at any disciplinary hearing. It also recommends that where practicable, different people should carry out the investigation and disciplinary hearing in misconduct cases.

No disciplinary action should be taken until a careful investigation has been concluded.

(b) Hearing (or meeting)

Once the employer has investigated, they must conduct a hearing (the ACAS Code makes reference to a meeting) to make a decision as to the sanction, if any, they will impose. To act fairly, the employer must comply with the rules of a fair hearing. These are as follows.

(i) The employee must know the case against them so they can answer the complaint. This also means they should be given sufficient time before the hearing with copies of relevant documents to enable them to prepare their case.
(ii) The employee should have an opportunity to put their side of the case, ie the employer should listen to the employee's side of the story and allow them to put forward any mitigating circumstances.
(iii) The employee must be allowed to be accompanied at the hearing by a fellow employee or a trade union representative of their choice (*Employment Relations Act 1999*).
(iv) The hearing should be unbiased, ie the person chairing the hearing should come to it with an open mind and not have pre-judged the issue.
(v) The employee should be provided with an explanation as to why any sanctions are imposed.
(vi) The employee should be informed of their right to appeal (and the way in which they should go about it) to a manager who has not been involved in the first hearing. If the employee fails to exercise their right of appeal, however, they will not have failed to mitigate their loss, if ultimately a tribunal finds that they have been unfairly dismissed and therefore compensation will not be reduced (*William Muir (Bond 9) Ltd v Lamb* [1985] IRLR 95). Failing to allow an employee to exercise a right of appeal will almost certainly render any dismissal unfair (*West Midlands Co-operative Society Ltd v Tipton* [1986] IRLR 112).

The ACAS Code says that where it is decided there is a disciplinary case to answer, the employee should receive written notification of this. They should receive sufficient information about the alleged misconduct or poor performance and the possible consequences in order to prepare to answer the case at a disciplinary meeting. Although the meeting should be held without unreasonable delay, the employee should be given sufficient time to prepare their case.

At this point, copies of any written evidence, such as witness statements, should be provided to the employee. The employer should also inform the employee of the time and venue for the meeting and advise them that they have the right to be accompanied by a companion (see **E14008** above).

With regard to the conduct of the meeting, the Code sets out that:

"Employers and employees (and their companions) should make every effort to attend the meeting. At the meeting the employer should explain the complaint against the employee and go through the evidence that has been gathered. The employee should be allowed to set out their case and answer any allegations that have been made. The employee should also be given a reasonable opportunity to ask questions, present evidence and call relevant witnesses. They should also be given an opportunity to raise points about any information provided by witnesses. Where an employer or employee intends to call relevant witnesses they should give advance notice that they intend to do this."

Following the meeting, the employer should decide whether or not disciplinary or any other action is justified and inform the employee about any action to be taken in writing.

(c) Appeal

In an unfair dismissal case, a tribunal is required to consider the reasonableness of the employer's action, taking into account the resources of the employer and the size of the employer's undertaking. This means the tribunal will expect the employer to have provided an appeal for the employee. If an employer does not allow an appeal, they will not be complying with the ACAS Code of practice on disciplinary and grievance procedures, which sets out that employees should appeal against the decision if they feel the disciplinary action taken against them is wrong or unjust. All the rules of a fair hearing equally apply to an appeal. Only an appeal which is a complete rehearing of the case, rather than merely a review of the written notes of the disciplinary hearing, can rectify procedural flaws committed earlier on in the procedure (*Jones v Sainsbury's Supermarkets Ltd* (2000), unreported). An appeal, however, cannot endorse the sanction imposed by the earlier hearing for a different reason, unless the employee has had notice of the new reason and has been given an opportunity to put forward their argument in respect of it. Workers have a statutory right to be accompanied at appeal hearings.

Sanctions other than dismissal

[E14014]–[E14016] There are a variety of sanctions apart from dismissal that an employer may impose. It is important however that the 'punishment fits the crime'. The imposition of too harsh a sanction may entitle the employee to resign and claim constructive dismissal (see below).

(a) Warnings

The ACAS Code states that the usual first step in a formal procedure (where misconduct or unsatisfactory performance has been confirmed) is to give the employee a written warning. A further act of misconduct (or failure to improve performance) within a set period would then normally result in a final written warning. However, if the conduct or under performance is sufficiently serious, it may be appropriate to move directly to a final written warning.

In health and safety cases, a minor breach of a rule may justify a final written warning given the potential seriousness and consequences of breaches of such rules.

Dismissal is the final stage in the disciplinary process and should not be a penalty for a first offence, except in the case of gross misconduct. The Code sets out that some acts (termed gross misconduct) are so serious, or have such serious consequences, that they may call for dismissal without notice for a first offence. Breaches of health and safety rules have been held to be gross misconduct. Even so, a fair disciplinary process should still be followed before dismissing for gross misconduct and any decision to dismiss should only be taken by a manager who has the authority to do so.

(b) Fines or deductions

The employer must have contractual authority or the written permission of the employee before he can make a deduction from the employee's wages as a disciplinary sanction. Deducting without such authority is a breach of the *Employment Rights Act 1996, s 13* and gives the employee the right to sue for recovery in the employment tribunal. It will also lead to a potential constructive dismissal claim.

(c) Suspension without pay

Any suspension without pay will have the same consequences as a fine or deduction if there is no contractual authority or written authorisation from the employee to impose such a sanction.

(d) Demotion

Most demotions will involve a reduction in pay, and thus without written or contractual authority to demote the employer will be in breach of the *Employment Rights Act 1996, s 13* and liable to a constructive dismissal claim.

Employment protection: dismissal

[E14017] Dismissal is the ultimate sanction an employer can impose for breach of health and safety rules. All employees are protected against wrongful dismissal at common law, but, in addition, some employees have protection against unfair dismissal. Wrongful dismissal is based on a breach of contract by the employer and compensation will be in the form of damages for that breach – that is, the damage the employee has suffered because the employer did not comply with the contract.

Unfair dismissal, on the other hand, is statute based (under the *Employment Rights Act 1996*). It is not dependent on a breach of contract by the employer, but the employee must satisfy any qualifying criteria laid down by the statute before they can claim. Compensation for such dismissal is based on a formula within the statute (see **E14022** below). Given that the protection against wrongful and unfair dismissal rest alongside each other, an employee may claim for both, although they will not be compensated twice.

Wrongful dismissal

[E14018] A dismissal at common law is where the employer unilaterally terminates the employment relationship with or without notice. A wrongful dismissal is where the employer terminates the contract in breach, for example, by giving no notice or shorter notice than is required by the employee's contract and the employee's conduct does not justify this. An employer is entitled to dismiss without notice only if the employee has committed gross misconduct. In all other circumstances the employer must give contractual notice to end the relationship, or pay wages in lieu of notice.

However, this does require qualification. Firstly, the law decides what is gross misconduct and not the employer. Just because the employer has stated that certain actions are gross misconduct does not mean that the law will regard them as such. Only very serious misconduct is regarded by the law as gross, such as refusing to obey lawful and reasonable orders, gross neglect and theft. Secondly, contractual notice periods are subject to the statutory minimum notice provisions contained in the *Employment Rights Act 1996, s 86*. Any attempt by the contract to give less than the statutory minimum notice is void. These periods apply to all employees who have been employed for one month or more and are:

(a) not less than one week's notice if his period of continuous employment is less than two years;

(b) not less than one week's notice for each year of continuous employment if his period of continuous employment is two years or more but less than twelve years; and

(c) not less than twelve weeks' notice if his period of continuous employment is twelve years or more.

Where an employer terminates the contract and pays the employee in lieu of notice, in the absence of an express right to do so, this will be a technical breach of contract. The employee can waive their right to notice or accept wages in lieu of notice. If the contract gives notice periods which are longer than the statutory minimum, the contractual notice prevails. If an employer has an employee who has been employed for six years, and sacks them with four weeks' notice, the employee can sue for a further two weeks' wages in the employment tribunal.

Finally, if the employer fundamentally alters the terms of the employee's contract, without their consent, in reality the employer is terminating, or repudiating, the original contract and substituting a new one. The employee should therefore be given the correct notice before the change comes into effect. A unilateral change by the employer to a fundamental term of the contract, followed by resignation by the employee, will amount to constructive dismissal, that is repudiation, and compensation in the form of damages for breach of contract. The employee must take all reasonable steps to mitigate their loss by seeking other employment.

Unfair dismissal

Dismissal

[E14019] While all employees are protected against wrongful dismissal, generally employees must be employed for two years before they gain protection against unfair dismissal. In certain circumstances, however, an employee is protected immediately and does not need two years of employment. One of these is dismissal on certain health and safety grounds discussed in **E14009** and **E14010** above.

Once an employee is protected against unfair dismissal, the *Employment Rights Act 1996, s 95(1)* sets out the following situations which the law regards as dismissal:

- the contract under which he is employed is terminated by the employer (whether with or without notice);

- he is employed under a contract for a fixed term and that term expires without being renewed under the same contract, or he is employed under a limited-term contract and that contract terminates by virtue of the limiting event without being renewed under the same contract; or
- the employee terminates the contract under which he is employed (with or without notice) in circumstances in which he is entitled to terminate it without notice by reason of the employer's conduct.

In the first situation, the employer is unilaterally ending the relationship. Even if the employer gives the correct amount of notice so that the dismissal is lawful, it does not necessarily follow that the dismissal will be fair.

The second situation needs no explanation. If a fixed-term contract has come to an end and is not renewed, this is in effect the employer deciding to end the relationship. Employees cannot validly waive unfair dismissal rights in fixed-term contracts (*Employment Relations Act 1999*). The expiry of a 'limited-term contract' without renewal under the same contract is a dismissal. A 'limited-term contract' is where the employment under the contract is not intended to be permanent and there is provision for it to terminate by virtue of a 'limiting event', ie a specific fixed-term expires, a specific task is performed or an event occurs (or does not occur).

The last situation, constructive dismissal, is much more complex. On the face of it the employee has resigned. However, if the reason for their resignation is the employer's conduct, then the law treats the resignation as an employer termination. The action on the part of the employer which entitles the employee to resign and claim constructive dismissal is a repudiatory breach of contract. In other words, the employer has committed a breach which goes to the root of the contract and has, therefore, repudiated it. This means that not all breaches by the employer are constructive dismissals but that serious breaches may be.

It is also important to recognise that, as discussed in **E14005** above, the terms of the contract may include those which have not been expressly agreed by the parties and therefore rules, disciplinary procedures, terms collectively bargained, and all the implied duties discussed in **E14003** above, may all be contractual terms.

Breach of the health and safety duties owed to all employees is likely to give rise to a constructive dismissal claim (*Day v T Pickles Farms Ltd* [1999] IRLR 217).

In addition, the law requires as an implied term of the contract that both the employer and employee treat each other with mutual respect and do nothing to destroy the trust and confidence each has in the other. Breach of this duty may give rise to a constructive dismissal claim. In one case, a demotion imposed as a disciplinary sanction was held to be excessive by the EAT. Its very excessiveness was a breach of the duty of mutual respect which entitled the employee to resign and claim constructive dismissal.

It has also been held that an employer's disclosure about complaints against an employee in a reference, before first giving them an opportunity to explain, was a fundamental breach of the implied term of mutual trust and confidence which amounted to constructive and unfair dismissal (*TSB Bank v Harris* [2000] IRLR 157). In *(1) Reed (2) Bull Information Systems Ltd v Stedman* [1999] IRLR 299, the EAT held that where the employer was aware of an employee's deteriorating health, and the employee concerned had complained to colleagues at work about harassment, it was incumbent on the employer to investigate. Their failure to do so was enough to justify a finding of breach of trust and confidence and therefore constructive dismissal.

Where an employee with two years' continuous service (unless one of the specified reasons apply which render a dismissal automatically unfair) is constructively dismissed, they will also have the right to claim unfair dismissal. Obviously, the employee must resign before they can make a claim for unfair dismissal.

In the majority of cases, the repudiatory breach by the employer is a fundamental alteration of the contractual terms (for example hours). In this situation, the employer still wishes to continue the relationship, albeit on different terms. The employee has two choices: they can resign or continue to work under the new terms. If the employee continues to work and accepts the changed terms, the contract is mutually varied and no action will lie, provided the employee was given the correct notice before the change was implemented. If the employee resigns, however, they will have been dismissed. In *Walton & Morse v Dorrington* [1997] IRLR 488, the employee waited to find alternative employment before she resigned. The EAT decided that, in her circumstances, this was a reasonable thing to have done and agreed that she had not accepted her employer's breach of its duty to her and had been constructively dismissed (see **E14003** above).

Reasons for dismissal

[E14020] *Section 98* of the *Employment Rights Act 1996* gives the following potentially fair reasons for dismissal. These are:

- capability or qualifications;
- conduct;
- redundancy;
- contravention of statute;
- some other substantial reason.

Compulsory retirement was a potentially fair reason for dismissal until the law changed to abolish the default retirement age (DFA). Any retirement notified to the employer after 6 April 2011 no longer provides a fair reason for dismissal.

Dismissal on health and safety grounds could potentially fall within most of these reasons. It should, however, be remembered that where an employee is dismissed in circumstances where continued employment involves a risk to their health and safety, the employer may nevertheless face claims of unfair dismissal. Before terminating employment, an employer should consider all the circumstances of the case, assess the risk involved and take measures which are reasonably necessary to eliminate the risk.

Illness may make it unsafe to employ the employee; breach of health and safety rules will normally fall under misconduct; to continue to employ the employee may contravene health and safety legislation or it may be that the employer has had to reorganise the business on health and safety grounds and the employee is refusing to accept the change. This latter situation could be potentially fair under 'some other substantial reason'. However, an employer may be liable under the *Equality Act 2010* if found guilty of disability discrimination (see **EQUALITY ACT 2010**).

An employee who is suspended on medical grounds is entitled to normal remuneration for up to 26 weeks. If the employee unreasonably refuses an offer of suitable alternative work, no remuneration is payable for the period during which the offer applies. An employee may bring a complaint to an employment tribunal if an employer fails to pay the whole or any part of the remuneration to which they are entitled (*Employment Rights Act 1996, ss 64, 68, 70(1)*). In *British Airways Ltd v Moore* [2000] IRLR 296, the EAT upheld a purser's claim to a flying allowance on the basis that suitable work must be on terms and conditions not substantially less favourable.

Reasonableness

[E14021] Merely having a fair reason to dismiss does not mean that the dismissal is fair. *Section 98(4)* of the *Employment Rights Act 1996* requires the tribunal in any unfair dismissal case to consider whether the employer acted reasonably in all the circumstances, including the size and administrative resources of the employer's undertaking. This means that the tribunal will look at two things – (i) was the treatment of the employee procedurally fair, and (ii) was dismissal a reasonable sanction in relation to the employee's actions and all the circumstances of the case.

Procedures have already been discussed at **E14013** above. If the employer has complied with their procedures, they will not be found to have acted procedurally unfairly unless the procedures themselves are unfair. This is unlikely if the employer is following the ACAS Code (see **E14008** and **E14013–E14016** above). In respect of the fairness of the decision, the tribunal should consider whether the employer's decision to dismiss fell within the band of reasonable responses to the employee's conduct which a reasonable employer could adopt. The tribunal will look at three things – (i) has the employer acted consistently, (ii) have they taken the employee's past work record into account, and (iii) has the employer looked for alternative employment. The latter aspect is of major importance in relation to redundancy, incapability due to illness or dismissal because of a contravention of legislation but will not usually be relevant in dismissals for misconduct. It is important to note that special obligations also apply to an employer in the case of a disabled employee under the *Equality Act 2010* (see **E14020** above). When looking at consistency, the tribunal will look for evidence that the employer has treated the same misconduct the same way in the past. If the employer has treated past breaches of health and safety rules leniently, it will be unfair to suddenly dismiss for the same breach unless they have made clear to employees that their attitude has changed and breaches will be dealt with more severely in the future. Employees have to know the potential disciplinary consequences for breaches of the rules, and if the employer has never dismissed in the past, they are misleading employees unless they tell them things have changed.

The law only requires an employer to be consistent between cases which are the same. This is where a consideration of the employee's past work record is important. It is not inconsistent to give a long-standing employee with a clean record a final warning for a breach of health and safety rules and to dismiss another shorter-serving employee with a series of warnings behind them, as long as both employees know that the penalty for breach of the rules could be dismissal. The cases are not the same. It would, however, be unfair if both the employees had the same type of work record and length of service and only one was dismissed, and dismissal had never been imposed as a sanction for that type of breach in the past. In order for a misconduct dismissal to be fair, the employer must have had a reasonable belief in the employee's guilt of the misconduct in question on the basis of a reasonable investigation.

Remedies for unfair dismissal

[E14022] The remedies for unfair dismissal are:

(a) Reinstatement

The first remedy the tribunal is required to consider is reinstatement of the employee. When doing so the tribunal must take into account whether the employee wishes to be reinstated, whether it is practicable for the employer to reinstate them and, if the employee's conduct contributed to or caused their dismissal, whether it is just to reinstate them. In order to resist an order for reinstatement, an employer must provide evidence to show that it is not practicable because the implied term of mutual trust and confidence between employer and employee has broken down (*IPC Magazines Ltd v Clements (EAT/456/99)*; *Gentle v Perkins Engines Peterborough Ltd (formerly Perkins Group Ltd)* [2001] All ER (D) 360 (Jul), EAT). Reinstatement means the employee must return to their old job with no loss of benefits. If reinstatement is ordered and the employer refuses to comply with the order, or only partially complies, compensation will be increased. Reinstatement is rarely ordered by tribunals.

(b) Re-engagement

If the tribunal does not consider that reinstatement is practicable, it must consider whether to order the employer to re-engage the employee. In making its decision the tribunal looks at the same factors as when it considers reinstatement. Re-engagement is an order requiring the employer to re-employ the employee on terms which are as favourable as those they enjoyed before their dismissal, but it does not require the employer to give the employee the same job back. Failure on the part of the employer to comply with an order of re-engagement will lead to increased compensation, although again tribunals rarely make re-engagement orders. Less than one per cent of successful claimants are reinstated or re-engaged.

(c) Compensation

Compensation falls under a variety of different heads. In an unfair dismissal case, the employee will receive:

Basic award: This is based on his or her age, years of service and salary –

(i) one and a half weeks' pay for each year of service from the age of 41;

(ii) one week's pay for each year of service between 22 and 40;
(iii) half a week's pay for each year of service below the age of 22.

This is subject to a statutory maximum of £525 a week (from 6 April 2019), and a maximum of twenty years' service. The maximum basic award is therefore £15,750 (from 6 April 2019). Where the employee is unfairly dismissed for health and safety reasons under the *Employment Rights Act 1996, s 100*, the minimum basic award is £6,408 (from 6 April 2019).

Compensatory award: This is payable in addition to the basic award to compensate the employee for loss of future earnings, benefits etc, which are in excess of the basic award. As with the basic award the compensatory award can be reduced for contributory conduct. The present maximum compensatory award that can be made in most cases is £86,444 (from 6 April 2019).

Additional award: If the employer fails to comply with a reinstatement or re-engagement order, the tribunal may make an additional award. This will be between 26 and 52 weeks' pay (subject to a maximum of £525 per week (from 6 April 2019).

Note: There is no limit on the compensatory award for employees who are dismissed for health and safety reasons or in whistleblowing cases (see E14010). If the dismissal is discriminatory, compensation for lost earnings can be claimed under the Equality Act 2010 and again there is no limit on the amount that can be awarded.

In cases of discriminatory dismissal, compensation may include compensation for injury to feelings.

The recent case of *South Yorkshire Fire & Rescue v Mansell* [2018] UKEAT/0151/17/3001 established that compensation for injury to feelings can also be awarded in a claim for detriment for asserting working time rights (under sections 45A and 48 of the Employment Rights Act 1996).

Health and safety duties in relation to women at work

Sex discrimination

[E14023] Since health and safety issues may give rise to sex discrimination claims under the *Equality Act 2010*, it is appropriate to examine the particular considerations an employer needs to bear in mind in its relations with women workers.

The steps necessary to be taken by an employer, in order to comply with their duties under *HSWA 1974, s 2*, may differ for women.

While discrimination legislation removed some of the restrictions on women and their employment generally and health and safety specifically, a few restrictions and prohibitions on certain types of employment by women, in the interests of health and safety at work, remain.

Under the *Control of Lead at Work Regulations 2002 (SI 2002 No 2676)* an employer is prohibited from employing women of reproductive capacity (as well as young people) in particular activities relating to lead processes as follows:

(a) In the lead smelting and refining process:
    (i) handling, treating, sintering, smelting or refining any material containing 5 per cent or more of lead; or
    (ii) cleaning where any of the above activities have taken place.
(b) In the lead-acid battery manufacturing process:
    (i) manipulating lead oxides;
    (ii) mixing or pasting;
    (iii) melting or casting;
    (iv) trimming, abrading or cutting of pasted plates; or
    (v) cleaning where any of the above activities have taken place.

Pregnant workers, new and breastfeeding mothers

[E14024] Employers should have taken measures to guard against risks to new and expectant mothers in accordance with their general duties under *HSWA 1974, s 2*, and the *Management of Health and Safety at Work Regulations 1999 (SI 1999 No 3242)*. Employers are also required to protect new and expectant mothers in their employment from certain specified risks and to carry out a risk assessment of such hazards. If the employer cannot avoid the risk(s), they must alter the working conditions of the employee concerned or the hours of work, offer suitable alternative work and, if no suitable alternative work is available, suspend the employee on full pay (see **E14024** below).

Although the *Equality Act 2010* prohibits sex discrimination, as one of the protected characteristics set out in the Act, action taken to comply with health and safety legislation (eg *HSWA 1974*) will not amount to unlawful discrimination.

Case law under previous sex discrimination legislation includes *Page v Freight Hire (Tank Haulage) Ltd* [1981] IRLR 13. The complainant was an HGV driver. The employer, acting on the instructions of the manufacturer of the chemical dimethylformamide (DMF) refused to allow her to transport the chemical which was potentially harmful to women of child-bearing age. She brought a complaint of sex discrimination. It was held that the fact that the discriminatory action was taken in the interests of safety did not of itself provide a defence to a complaint of unlawful discrimination. However, the employer was protected by *s 51(1)* of the *Sex Discrimination Act (SDA) 1975* – the *Equality Act 2010* now applies - because the action taken was necessary to comply with the employer's duty under *HSWA 1974*. However, forcing a worker to accept a change of duties or suspension where the risk is low has been held to be sex discrimination (*New Southern Railway Ltd v Quinn* [2006] IRLR 266) (see **E14025** below).

The *Management of Health and Safety at Work Regulations 1999 (SI 1999 No 3242)* impose a duty on employers to protect new or expectant mothers from any process or working conditions or certain physical, chemical and biological risks at work (see E14025 below). The phrase 'new or expectant mother' is defined as a worker who is pregnant, who has given birth within the previous six months, or who is breastfeeding. 'Given birth' is defined as having delivered a living child or, after 24 weeks of pregnancy, a stillborn child.

Risk assessment

[E14025] The 1999 Regulations require employers to carry out an assessment of the specific risks posed to the health and safety of pregnant women and new mothers in the workplace and then to take steps to ensure that those risks are avoided. Risks include those to the unborn child or child of a woman who is still breastfeeding – not just risks to the mother.

A woman is not legally obliged to inform her employer she is pregnant or breastfeeding, but her employer has particular obligations once they have been notified in writing that she is a new or expectant mother. When an employee provides her employer with written notification, under regulation 18 of the Regulations, that she is pregnant, has given birth within the past six months, or is breastfeeding, the HSE advice is that the employer should immediately take into account any risks identified in their workplace risk assessment.

'Although it is not a legal requirement for employers to conduct another specific or further individual risk assessment for new and expectant mothers, employers may choose to do so as part of the process by which they reach a decision about what action should be taken,' it advises. 'An employer's risk assessment should have already considered any specific risks to new and expectant mothers when considering the rest of the workplace. This will enable employers to take immediate action, if and when necessary.' An interesting development in relation to this requirement is the case of *Day v T Pickles Farms Ltd* [1999] IRLR 217, where the employee suffered nausea when pregnant as a result of the smell of food at her workplace. As a result of the nausea, she was unable to work and was eventually dismissed after a prolonged absence.

The EAT found that she had not been constructively dismissed. However, it held that the obligation to carry out a risk assessment which considers possible risks to the health and safety of a pregnant female employee is relevant from the moment an employer employs a woman of childbearing age. The question of whether the applicant had been subjected to a detriment was remitted to the employment tribunal.

If an employer suspends a new or expectant mother to avoid health and safety risks, they must provide evidence of the risk and show that it could not otherwise have been avoided. In *New Southern Railway Ltd v Quinn* [2006] IRLR 266, a pregnant employee was removed from her duties as a duty station manager because of the risk of physical assault. Her salary was also reduced to reflect her change of duties and she brought a claim for sex discrimination. The EAT upheld her claim, finding that her employer had suspended her because of a 'paternalistic and patronising attitude' rather than for any real health and safety reasons.

In the European Court of Justice (ECJ) case, *Elda Otera Ramos v Servicio Galego de Saúde, Instituto Nacional de la Seguridad Social* Case C-531/15, the court confirmed that where a breastfeeding mother can show a risk assessment is inadequate, or not carried out at all, this can give rise to a potential discrimination claim.

The HSE website contains online guidance setting out the possible risks employers should consider as follows:

- Physical agents:
  - Movements and postures
  - Manual handling
  - Shocks and vibrations
  - Noise
  - Radiation (ionising and non-ionising)
  - Compressed air and diving
  - Underground mining work
- Biological agents:
  - Infectious diseases
- Chemical agents:
  - Toxic chemicals
  - Mercury
  - Antimitotic (cytotoxic) drugs
  - Pesticides
  - Carbon monoxide
  - Lead
- Working conditions:
  - Facilities (including rest rooms)
  - Mental and physical fatigue, working hours
  - Stress (including post-natal depression)
  - Passive smoking
  - Temperature
  - Working with visual display units (VDUs)
  - Working alone
  - Working at height
  - Travelling
  - Violence
  - Personal protective equipment
  - Nutrition

Where the assessment identifies a risk, affected employees or their representatives should be informed of the risk and the preventive measures to be adopted. The assessment should be kept under review.

In particular, employers must consider removing the hazard or seek to prevent exposure to it. If a risk remains after preventive action has been taken, the employer must take the following course of action:

(i) temporarily adjust her working conditions or hours of work (*Management of Health and Safety at Work Regulations 1999 (SI 1999 No 3242), Reg 16(2)*).

If it is not reasonable to do so or would not avoid the risk:

(ii) offer suitable alternative work for the same pay (*Employment Rights Act 1996, s 67*).

If neither of the above options is viable:

(iii) suspend her on full pay for as long as necessary to protect her health and safety or that of her child (*Management of Health and Safety at Work Regulations 1999 (SI 1999 No 3242), Regs 16(2) and 16(3); Employment Rights Act 1996, s 67*). This is known as a maternity suspension.

Night work by new or expectant mother

[E14026] Where a new or expectant mother works at night and has been issued with a certificate from a registered doctor or midwife stating that night work would affect her health and safety, the employer must first offer her suitable alternative daytime work, and suspend her as detailed at E14029 below if no suitable alternative employment can be found (*Management of Health and Safety at Work Regulations 1999 (SI 1999 No 3242), Reg 17*).

In *Gonzáles Castro v Mutua Umivale* [2018 IRLR 1142] the ECJ ruled that pregnant workers, workers who have recently given birth and those who are breastfeeding and who work shifts, some of which are at night, must be regarded as performing night work and therefore enjoy specific protection against the risks that night work is liable to pose.

Notification

[E14027] An employer is not required to alter a woman's working conditions or hours of work or suspend her from work under *Management of Health and Safety at Work Regulations 1999 (SI 1999 No 3242), Reg 16(2) or (3)* until she notifies them in writing that she is pregnant, has given birth within the previous six months or is breastfeeding. Any suspension or amended working conditions do not have to be maintained if the employee fails to produce a medical certificate confirming her pregnancy in writing within a reasonable time if the employer requests her to do so. The same applies once the employer knows that the employee is no longer a new or expectant mother or cannot establish whether she remains so (*Management of Health and Safety at Work Regulations 1999 (SI 1999 No 3242), Reg 18*).

However, an employer has a general duty under *HSWA 1974* and the *Management of Health and Safety at Work Regulations 1999* to take steps to protect the health and safety of a new or expectant mother, even if she has not given written notification of her condition.

Maternity leave

[E14028] All female employees have a right to 52 weeks' statutory maternity leave regardless of length of service and number of hours worked or pay as long as they follow statutory rules regarding notice and timing. At least 15 weeks before the due date, they must tell their employer when the baby is due and when they want to start their maternity leave.

If an employee is prohibited from working for a specified period after childbirth by virtue of a legislative requirement (eg under the *Public Health Act 1936, s 205* which sets out that women are not to be employed in factories or workshops within four weeks after birth of their child), her maternity leave period must continue until the expiry of that later period. The *Employment Rights Act 1996, s 72(1)* provides for two weeks of compulsory maternity leave more generally, which must be taken immediately following the birth.

Suspension from work on maternity grounds – suitable alternative work

[E14029] Where an employee is suspended from work on maternity grounds (ie she is pregnant, has recently given birth or is breastfeeding), the employer must offer available suitable alternative work. Alternative work will only be suitable if:

- the work is of a kind which is both suitable in relation to the employee and appropriate for the employee to do in the circumstances; and
- the terms and conditions applicable for performing the work are not substantially less favourable than corresponding terms and conditions applicable for performing the employee's usual work (*Employment Rights Act 1996, s 67*).

If an employer fails to offer suitable alternative work, the employee may bring a claim before an employment tribunal which can award 'just and equitable' compensation. Such complaint must normally be lodged within three months of the first day of the suspension (*Employment Rights Act 1996, s 70(4)*).

Remuneration on suspension from work on maternity grounds

[E14030] An employee who is suspended on maternity grounds, if no suitable alternative work is available, is entitled to normal remuneration for the duration of the suspension. However, if the employee unreasonably refuses an offer of suitable alternative work, no remuneration is payable for the period during which the offer applies. An employee may bring a complaint to an employment tribunal if an employer fails to pay the whole or any part of the remuneration to which the employee is entitled (*Employment Rights Act 1996, ss 64, 68, 70(1)*). See also *British Airways Ltd v Moore* [2000] IRLR 296 in E14020 above.

Pregnant and breastfeeding mothers working with radiation

[E14031] The *Ionising Radiations Regulations 2017 (SI 2017/1075)* apply in workplaces where there is a risk of exposure to ionising radiation from work with radioactive substances and other sources of ionising radiation. The regulations replaced earlier 1999 Regulations and removed the subsidiary dose limit for the abdomen of a woman of reproductive capacity.

When carrying out a risk assessment under Regulation 8 of the 2017 Regulations, the employer must take account of the risks arising from radiation exposure to those who are pregnant or breastfeeding and, in particular, the likely doses to the foetus or the breastfed infant.

Regulation 9 deals with restriction of exposure and Regulation 9(1) sets out that every employer must, in relation to any work with ionising radiation that it undertakes, take all necessary steps to restrict so far as is reasonably practicable the extent to which its employees and other persons are exposed to ionising radiation.

Under Regulation 9(6):

'Without prejudice to paragraph (1), an employer who undertakes work with ionising radiation must ensure that –

(a) in relation to an employee who is pregnant, the conditions of exposure are such that, after the employee's employer has been notified of the pregnancy, the equivalent dose to the foetus is as low as is reasonably practicable and is unlikely to exceed 1 mSv during the remainder of the pregnancy; and (b) in relation to an employee who is breastfeeding, that employee must not be engaged in any work involving a significant risk of intake of radionuclides or of bodily contamination.'

HSE guidance explains that the regulations do not prevent pregnant or breastfeeding employees from working with ionising radiation provided that exposures remain below the dose limit specified and are kept as low as is reasonably practicable. It adds that where exposure is to high-energy radiation, employers will need advice from a radiation protection adviser about an appropriate dose restriction.

Regulation 15 requires employers to ensure that employees working with ionising radiation are given appropriate training in radiation protection and receive suitable and sufficient information and instruction on the risks to health; the precautions to be taken; and the importance of complying with the medical, technical and administrative requirements of the regulations. They are also required to give adequate information to other people who are directly concerned with the work to ensure their health and safety.

They must inform female employees working with ionising radiation of the possible risk arising from ionising radiation to the foetus and to a nursing infant and of the importance of informing the employer in writing as soon as possible after becoming aware that they are pregnant or if they intend to breastfeed.

More detailed information can be found in the HSE publication *Work with ionising radiation* (L121) (see www.hse.gov.uk/pubns/priced/l121.pdf).

The Control of Electromagnetic Radiation at Work Regulations 2016 (SI 2016/588)

[E14032] The *Control of Electromagnetic Fields at Work Regulations* require employers to:

- make a suitable and sufficient assessment of the exposure limit values (ELVs) to which employees may be exposed to (with reference to action levels (ALs) and ELVs);
- ensure exposure is below a set of ELVs;
- when appropriate, assess the risks of workers' exposure and eliminate or minimise those risks. They must ensure they take into account workers at particular risk, including expectant mothers;
- when appropriate, devise and implement an action plan to ensure compliance with the exposure limits;
- provide information and training on the particular risks (if any) posed to employees by EMFs in the workplace and about any action to remove or control them;
- take action if employees are exposed to EMFs in excess of the ELVs; and

[E14032] Employment Protection

- provide health surveillance as appropriate.

More detailed information can be found in the HSE publication *A guide to the Control of Electromagnetic Fields at Work Regulations 2016* (HSG281) which can be found on the HSE website at: www.hse.gov.uk/pubns/books/hsg281.htm.

Enforcement

Malcolm Galloway and Alice Jarratt

Introduction to enforcement

[E15001]–[E15004] The *Health and Safety at Work etc Act 1974 ('HSWA 1974')* created new law and new bodies, the Health and Safety Executive (HSE) and the Health and Safety Commission, to deal with the management and enforcement of health and safety law in the UK. *HSWA 1974* provided a legal framework that now regulates nearly all work activity in the UK.

The HSE are responsible for regulating health and safety law across a wide range of work activities in Great Britain. Their aim is to protect the health, safety and welfare of people at work, and to safeguard others, including the public, who may be affected by work activities. The current approach to enforcement is shaped by the latest HSE Enforcement Policy Statement (dated October 2015), the Code for Crown Prosecutors in the context of prospective prosecutions, and the law in general. The Enforcement Policy Statement sets out the HSE's approach to enforcement, that is, where its inspectors take action to enforce the law when issues of non-compliance, hazard or serious risk have been identified. It is available to download from the HSE's website (www.hse.gov.uk/enforce/enforcepolicy.htm). The Code for Crown Prosecutors is available to download from the Crown Prosecution Service (CPS) website (www.cps.gov.uk/publications/code_for_crown_prosecutors). The HSE code follows the basic principles of prosecution that are set out therein. The current HSE Enforcement Policy Statement sets out that the purpose of enforcement is to prevent harm by requiring duty holders to manage and control risk effectively.

There have been important developments in the context of enforcement in recent years, notably with the introduction of the *Corporate Manslaughter and Corporate Homicide Act 2007* (which came into force on 6 April 2008) and the introduction of definitive guidelines by the Sentencing Council in respect of health and safety offences committed by organisations, individuals and a wholly-revised guideline in respect of corporate manslaughter. The guidelines came into force on 1 February 2016 and apply to all cases sentenced from that date onwards. They are therefore retrospective in nature. These guidelines are examined at **E15043** below.

The HSE cannot investigate or prosecute individual or corporate manslaughter. Whenever a work-related death occurs, and there is an indication that an offence of manslaughter (corporate or individual) may have been committed, the police will carry out an investigation and the CPS will bring the prosecution.

[E15001] Enforcement

Corporate manslaughter was first formally recognised as an offence in the prosecution that followed the Herald of Free Enterprise disaster in 1987. A number of prosecutions for corporate manslaughter followed, but convictions proved difficult to secure. The central difficulty was that it was incumbent on the prosecution to show that one of the company's 'directing' or 'controlling minds' had been guilty of gross negligence manslaughter. This meant that successful prosecution for corporate manslaughter at the hands of large companies was all but impossible. The 2007 Act created the statutory offence of corporate manslaughter. It now falls on the prosecution to show that the death resulted from serious management failures giving rise to a gross breach of a duty of care. Individual liability was precluded from the new statutory offence and individuals continue to be subject to the common law offence of gross negligence manslaughter.

The *Health and Safety (Offences) Act 2008*, which came into force on 16 January 2009, revised the mode of trial and maximum penalties applicable to certain health and safety offences. In particular, the 2008 Act increased the maximum fines that can be imposed by magistrates from £5,000 to £20,000 in respect of many health and safety offences, including breaches of sections 2 to 6. The range of offences punishable by a term of imprisonment was also broadened, and the magistrates' court is now entitled to send a larger number of offences to the Crown Court in order for more stringent penalties to be imposed. The *Legal Aid, Sentencing and Punishment of Offenders Act 2012 s 85 (LASPO 2012)* that came into force on 15 March 2015 introduced unlimited fines in the magistrates' courts for health and safety offences committed after this date with the intention that more cases will be disposed of in the magistrates' courts.

Currently, prosecutions and other enforcement procedures are the responsibility of the appropriate 'enforcing authority', either the HSE or a local authority (see **E15007–E15011** below).

This section deals with the following aspects of enforcement—

- The role of the Health and Safety Executive (see **E15005–E15006**).
- The 'enforcing authorities' (see **E15007–E15011** below).
- The 'relevant statutory provisions', which can be enforced under *HSWA 1974* (see **E15012** below).

Part A: Enforcement Powers of Inspectors'

- Improvement and prohibition notices (see **E15013–E15018** below).
- Appeals against improvement and prohibition notices (see **E15019**, **E15020** below).
- Grounds for appeal against a notice (see **E15021–E15025** below).
- Inspectors' investigation powers (see **E15026** below).
- Inspectors' powers of search and seizure (see **E15027** below).
- Indemnification by enforcing authority (see **E15028** below).
- Public register of notices (see **E15029** below).

Part B: Offences and Penalties

- Prosecution for contravention of the relevant statutory provisions (see **E15030, E15031** below).

- Main offences and penalties (see **E15032–E15038** below).
- Offences committed by particular types of persons, including the Crown (see **E15039–E15045** below).
- Sentencing guidelines (see **E15043** below).

The role of the Health and Safety Executive

[E15005] The *HSWA 1974, s 11* sets out the HSE's duties in respect of the "general purposes" (broadly, the securing of the health, safety and welfare of persons at work as defined in *HSWA 1974, s 1*) as follows—

In connection with the general purposes of this Part, the Executive shall—

(a) assist and encourage persons concerned with matters relevant to those purposes to further those purposes;

(b) make such arrangements as it considers appropriate for the carrying out of research and the publication of the results of research and the provision of training and information, and encourage research and the provision of training and information by others;

(c) make such arrangements as it considers appropriate to secure that the following persons are provided with an information and advisory service on matters relevant to those purposes and are kept informed of and are adequately advised on such matters—

    (i) government departments,

    (ii) local authorities,

    (iii) employers,

    (iv) employees,

    (v) organisations representing employers or employees, and

    (vi) other persons concerned with matters relevant to the general purposes of this Part.

The HSE also draws up proposals for regulations, replacing/updating existing statute and statutory instruments, and prepares approved codes of practice (ACoPs). Under *HSWA 1974, s 14* the HSE has the power to conduct and direct investigations and inquiries with a view to formulating new regulations. It also prepares draft regulations for approval by the Secretary of State.

The HSE is also tasked with the enforcement of health and safety legislation (unless responsibility lies with another enforcing body, such as a local authority). [*HSWA 1974, s 18(1)*].

The HSE's enforcement functions are principally performed by its inspectors and include the carrying out of routine inspections of premises, and the investigation of workplace accidents, dangerous occurrences or cases of ill health. Inspectors also provide advice to companies and individuals on the legal requirements under health and safety legislation and in respect of compliance with notices. The HSE also publishes guidance documents (see **E15006** below), carries out research, compiles statistics and licenses or approves certain hazardous operations.

The *Health and Safety and Nuclear (Fees) Regulations 2015 (SI 2015 No 363)* provide that the HSE may, in certain circumstances, charge for its services. In particular, fees are payable in relation to the following:

[E15005] Enforcement

- applications for approval under the *Agriculture (Tractor Cabs) Regulations 1974 (reg 3)*;
- applications for approval under the *Freight Containers (Safety Convention) Regulations 1984 (reg 4)*;
- various applications under the *Control of Asbestos Regulations 2012 (reg 5)*;
- examination or surveillance by an employment medical adviser *(reg 6)*;
- medical surveillance by an employment medical adviser under the *Control of Lead at Work Regulations 2002 (reg 7)*;
- in connection with the *Ionising Radiations Regulations 1999 and the Radiation (Emergency Preparedness and Public Information) Regulations 2001* (reg 8);
- in relation to the *Explosives Regulations 2014 and the Acetylene Safety (England and Wales and Scotland) Regulations 2014 (reg 9)*;
- in relation to the *Petroleum (Consolidation) Regulations 2014 (reg 10)*;
- application for or changes to an explosives licence under *Part 9 of the Dangerous Substances in Harbour Areas Regulations 1987 (reg 11)*;
- for notifications and applications under the *Genetically Modified Organisms (Contained Use) Regulations 2014 (reg 13)*;
- in respect of offshore installations *(reg 14)*;
- in respect of gas safety functions *(reg 15)*;
- in relation to nuclear installations *(reg 16)*;
- applications for approvals under the *Offshore Installations and Pipeline Works (First-Aid) Regulations 1989 (reg 18)*;
- notifications under the *Borehole Sites and Operations Regulations 1995 (reg 20)*; and
- for activities under the *Biocides Regulation (EU) No 528/2012* and the *Biocidal Products and Chemicals Regulations 2013 (reg 21)*.

Fee for Intervention Cost Recovery Scheme

[E15005.1] In addition, the HSE's 'fee for intervention' (FFI) cost recovery scheme came into effect on 1 October 2012, introduced by the *Health and Safety (Fees) Regulations 2012 (SI 2012/1652)* (which have since been repealed and replaced by the *Health and Safety and Nuclear (Fees) Regulations 2015 (SI 2015 No 363)*. The Regulations require the HSE to recover its costs for carrying out its regulatory functions from employers and other duty holders found to be 'in material breach' of health and safety law.

The HSE explains: 'A material breach is where you have broken the law and the inspector judges this is serious enough for them to notify you in writing. This will either be a notification of contravention, an improvement or prohibition notice, or a prosecution.'

In coming to a decision about what is a proportionate level of enforcement action, inspectors apply the principles of the HSE Enforcement Policy Statement and the HSE Enforcement Management Model (see below).

Examples of material breaches include not providing guards or other safety devices to prevent access to dangerous parts of machinery, and leaving material containing asbestos in a poor or damaged condition, resulting in the potential for asbestos fibres to be released.

FFI only applies where the HSE is the enforcing authority. Other health and safety regulators, including local authorities, cannot recover their costs under the scheme. It does not apply to the self-employed who do not put others at risk by their work. And it does not apply where businesses already pay fees to the HSE through other arrangements. For example it does not apply to licensable work with asbestos by those who hold a licence for work with asbestos under the Control of Asbestos Regulations 2012 as the licence fee contains an element to cover the costs of inspection.

The fee is payable for the costs that HSE reasonably incurs in relation to a material breach and includes:

- all work that is needed to identify a material breach. Where this is identified during a visit, costs for the whole visit are recoverable;
- all work to ensure that the breach is remedied; and
- any investigation or enforcement action, up to the point where HSE's intervention in relation to the material breach has been concluded, or a prosecution is started, or a report is submitted to the Procurator Fiscal in Scotland.

In its first year FFI brought in over £5.5 million in charges across all industries, with the manufacturing and construction sectors receiving the most fines. It generated £9.6 million in the financial year 2013/14.

Two reviews have examined the scheme since it was introduced in 2012. The first, a 2014 Triennial Review of the HSE, was critical of FFI and recommended that 'unless the link between "fines" and funding can be removed, or the benefits can be shown to outweigh the detrimental effects, and it is not possible to minimise those effects, FFI should be phased out'.

But the second, published in September 2014, found that cost recovery scheme has proven effective and should stay. The review was conducted by an independent panel chaired by Liverpool University professor of public policy, Alan Harding. It found that HSE inspectors have implemented the scheme 'consistently and fairly' since it began and found no evidence to suggest that its introduction had influenced enforcement policy decisions.

More information about the scheme is available on the HSE website at: www.hse.gov.uk/fee-for-intervention/index.htm.

Copies of the HSE Enforcement Policy Statement can be found at: www.hse.gov.uk/pubns/hse41.pdf.

Copies of the HSE Enforcement Management Model can be found at: www.hse.gov.uk/enforce/emm.pdf.

HSE publications

[E15006] The HSE also publishes information, advice and guidance for employers, local authorities, and trade unions etc on most aspects of health and safety.

Publications include *The health and safety toolbox – How to control risks at work*, which covers the most common workplace hazards and shows how most small to medium-sized businesses can put measures in place to control the risks, as well as those covering more detailed and specialised areas, such as gas safety in the context of the catering and hospitality industries.

HSE publications can be downloaded free from the HSE website.

Enforcing authorities

[E15007] While the HSE is the central body entrusted with the enforcement of health and safety legislation, in any given case enforcement powers rest with the body which is expressed by statute to be the 'enforcing authority'. Here the general rule is that the 'enforcing authority', in the case of industrial premises, is the HSE and, in the case of commercial premises within its area, the local authority (the enforcing authority in over a million premises), except that the HSE cannot enforce provisions in respect of its own premises, and similarly, local authorities' premises are inspected by the HSE. Each 'enforcing authority' is empowered to appoint suitably qualified persons as inspectors for the purpose of carrying into effect the 'relevant statutory provisions' within the authority's field of responsibility. [*HSWA 1974 s 19(1)*]. Inspectors so appointed can exercise any of the enforcement powers conferred by the *HSWA 1974* (see **E15013–E15018** below) and bring prosecutions (see **E15030**, **E15031** below).

The appropriate 'enforcing authority'

[E15008] The general rule is that the HSE is the enforcing authority, except to the extent that—

(a) regulations specify that the local authority is the enforcing authority instead; the regulations that so specify are the *Health and Safety (Enforcing Authority) Regulations 1998 (SI 1998 No 494)* or;
(b) one of the 'relevant statutory provisions' specifies that some other body is responsible for the enforcement of a particular requirement.

[*HSWA 1974 s 18(1), (7)(a)*].

Activities for which the HSE is the enforcing authority

[E15009] The HSE is specifically the enforcing authority in respect of the following activities (even though the main activity on the premises is one for which a local authority is usually the enforcing authority in accordance with the *Health and Safety (Enforcing Authority) Regulations 1998 (SI 1998 No 494) Sch 2* (see **E15010** below)—

(1) Any activity in a mine or quarry (other than a quarry in respect of which notice of abandonment has been given under *reg 45(1) of the Quarries Regulations 1999*);
(2) Any activity in a fairground;

(3) Any activity in premises occupied by a radio, television or film undertaking in which the activity of broadcasting, recording or filming is carried on;
(4) The following work carried out by independent contractors—
   (a) certain construction work including where the project includes work notifiable within the meaning of *reg 6(1)* of the *Construction (Design and Management) Regulations 2015 (SI 2015 No 51)*;
   (b) installation, maintenance or repair of gas systems or work in connection with a gas fitting;
   (c) installation, maintenance or repair of electricity systems;
   (d) most work with ionising radiations;
(5) Use of ionising radiations for medical exposure;
(6) Any activity in premises occupied by a radiography undertaking where work with ionising radiations is carried out;
(7) Agricultural activities, including agricultural shows where livestock or agricultural equipment is handled;
(8) Any activity on board a sea-going ship;
(9) Ski slope, ski lift, ski tow or cable car activities;
(10) Fish, maggot and game breeding (but not in a zoo);
(11) Any activity in relation to a pipeline within the meaning of *reg 3* of the *Pipelines Safety Regulations 1996*;
(12) The operation of a guided bus system or any other system of guided transport, other than a railway, that employs vehicles which for some or all of the time when they are in operation travel along roads; and
(13) The operation of a trolley vehicle system.

[*Health and Safety (Enforcing Authority) Regulations 1998 (SI 1998 No 494) reg 4(4)(b)* and *Sch 2*].

The HSE is the enforcing authority against the following, and for any premises they occupy, including parts of the premises occupied by others providing services for them. (This is so even though the main activity is listed in *Sch 1* of the *1998 Regulations*, see **E15010** below)—

(1) a county council;
(2) any other local authority (as defined in *reg 2*);
(3) a parish council in England or a community council in Wales or Scotland;
(4) a police authority, a local policing body, or the Receiver for the Metropolitan Police District;
(5) the Scottish Fire and Rescue Service;
(6) a fire and rescue authority under the *Fire and Rescue Services Act 2004*;
(7) a headquarters or an organisation designated for the purposes of the *International Headquarters and Defence Organisation Act 1964*; or a service authority of a visiting force within the meaning of *s 12* of the *Visiting Forces Act 1952*;
(8) the United Kingdom Atomic Energy Authority;
(9) the Crown, although not where the premises are occupied by the HSE itself

[*Health and Safety (Enforcing Authority) Regulations 1998 (SI 1998 No 494) reg 4(3)*].

The HSE is also the enforcing authority for the premises set out below, even if occupied by more than one occupier—

(1) the Channel Tunnel;
(2) an offshore installation
(3) a building or construction site;
(4) a campus or other premises occupied by an educational establishment;
(5) a hospital.

[*Health and Safety at Work (Enforcing Authority) Regulations 1998 (SI 1998 No 494) reg 3(5)*].

Finally, the HSE is the enforcing authority for the following—

(1) Common parts of domestic premises (*reg 3(1)*);
(2) Certain areas within an airport (*reg 3(4)(b)*);

Activities for which local authorities are the enforcing authorities

[E15010] Where the main activity carried on in non-domestic premises is one of the following, the local authority is the enforcing authority (ie the relevant county, district or borough council)—

(1) Sale or storage of goods for retail/wholesale distribution (including sale and fitting of motor car tyres, exhausts, windscreens or sunroofs), except—
   (a) at container depots where the main activity is the storage of goods which are of transit to or from dock premises, an airport or railway;
   (b) where the main activity is the sale or storage for wholesale distribution of dangerous substances or dangerous preparation;
   (c) where the main activity is the sale or storage for wholesale distribution of any hazardous substance or mixture; and
   (d) where the main activity is the sale or storage of water or sewage or their by-products or natural or town gas.
(2) Display or demonstration of goods at an exhibition, being offered or advertised for sale.
(3) Office activities.
(4) Catering services.
(5) Provision of permanent or temporary residential accommodation, including sites for caravans or campers.
(6) Consumer services provided in a shop, except—
   (a) dry cleaning;
   (b) radio/television repairs.
(7) Cleaning (wet or dry) in coin-operated units in laundrettes etc.
(8) Baths, saunas, solariums, massage parlours, premises for hair transplant, skin piercing, manicuring or other cosmetic services and therapeutic treatments, except where supervised by a doctor, dentist, physiotherapist, osteopath or chiropractor.

(9) Practice or presentation of arts, sports, games, entertainment or other cultural/recreational activities, save where the main activity is the exhibition of a cave to the public.
(10) Hiring out of pleasure craft for use on inland waters.
(11) Care, treatment, accommodation or exhibition of animals, birds or other creatures, except where the main activity is—
 (a) horse breeding/horse training at stables;
 (b) agricultural activity;
 (c) veterinary surgery.
(12) Undertaking, but not embalming or coffin making.
(13) Church worship/religious meetings.
(14) Provision of car parking facilities within an airport.
(15) Childcare, playgroup or nursery facilities.

[*Health and Safety (Enforcing Authority) Regulations 1998 (SI 1998 No 494) reg 3(1)* and *Sch 1*].

Transfer of responsibility between the HSE and local authorities

[E15011] Enforcement can be transferred (though not in the case of Crown premises), by prior agreement, from the HSE to the local authority and vice versa. Parties who are affected by such transfer must be notified. [*Health and Safety (Enforcing Authority) Regulations 1998 (SI 1998 No 494) reg 5*]. Transfer is effective even though the above procedure is not followed (ie the authority changes when the main activity changes) (*Hadley v Hancox* (1987) 85 LGR 402, decided under the previous Regulations).

Where there is uncertainty, these Regulations also allow responsibility to be assigned by the HSE and the local authority jointly to either body. [*Health and Safety (Enforcing Authority) Regulations 1998 (SI 1998 No 494) reg 6(1)*].

'Relevant statutory provisions'

[E15012] The enforcement powers conferred by *HSWA 1974* extend to any of the 'relevant statutory provisions'. These comprise—

(a) the provisions of *HSWA 1974 Part I* (ie *ss 1–53*);
(b) any health and safety regulations passed under *HSWA 1974*, eg the *Ionising Radiations Regulations 1999 (SI 1999 No 3232)*, the *Management of Health and Safety at Work Regulations 1999 (SI 1999 No 3242)* as amended and the *Workplace (Health, Safety and Welfare) Regulations 1992 (SI 1992 No 3004)* as amended; and
(c) the 'existing statutory provisions', ie all enactments specified in *HSWA 1974 Sch 1*, including any regulations etc made under them, so long as they continue to have effect.

National Enforcement Code

[E15012.1] The statutory National Enforcement Code came into force in 2013 and targets proactive local authority inspections on 'higher risk' activities in specified sectors, or when there is intelligence that workplaces are putting employees or the public at risk. The code explicitly outlaws proactive inspections outside 'high risk areas' by both the HSE and local authority regulators. Businesses are now only inspected if they are operating in high-risk areas, such as construction, or if they have a poor safety record.

In addition, if 'low-risk' businesses believe they are being unreasonably targeted they can complain to an independent panel, which will investigate and issue a public judgment. The HSE will also work with local authorities whose targeting of inspections fails to meet the standards set out. The National Enforcement Code can be found on the HSE website at: www.hse.gov.uk/la u/publications/la-enforcement-code.htm.

The March 2015 Department for Work and Pensions (DWP) report, *A final progress report on implementation of health and safety reforms*, sets out that the number of proactive inspections carried out by the HSE fell from 33,000 in 2010/11 to 22,000 (planned) in 2014/15. The trade union organisation TUC reported in April 2014 that local authority inspections had fallen by 93% since 2009–2010

Regulators' growth duty and the Regulators' Code

[E15012.2] In addition, a 'growth duty' for regulators (contained in *s 108* of the *Deregulation Act 2015*) requires the HSE (as a 'non-economic regulator') to take into account the impact of their activities on the economic prospects of firms they regulate, and the revised Regulators' Code, introduced in Spring 2014, requires 'methods of enforcement that are tailored to meet the needs of the business'.

Part A: Enforcement Powers of Inspectors

Improvement and prohibition notices

[E15013] It was recommended by the Robens Committee (the committee set up in 1970 and whose recommendations were substantially enacted in the *HSWA 1974*) 'that inspectors should have the power, without reference to the courts, to issue a formal improvement notice to an employer requiring him to remedy particular faults or to institute a specified programme of work within a stated time limit.' (*Cmnd 5034, para 269*). 'The improvement notice would be the inspector's main sanction. In addition, an alternative and stronger power should be available to the inspector for use where he considers the case for remedial action to be particularly serious. In such cases he should be able to issue a prohibition notice.' (*Cmnd 5034, para 276*). *HSWA 1974* put these recommendations into effect.

Crown bodies are exempt from the provisions relating to improvement and prohibition notices. Instead, Crown notices may be issued against a relevant body. The notice is not legally binding, and the Crown cannot be prosecuted for breach of the notice [see below].

Improvement notices

[E15014] An inspector may serve an improvement notice if he is of the opinion that a person—

(a) is contravening one or more of the 'relevant statutory provisions' (see E15012 above); or
(b) has contravened one or more of those provisions in circumstances that make it likely that the contravention will continue or be repeated.

[*HSWA 1974, s 21*].

In the improvement notice the inspector must—

(i) state that he is of the opinion in (a) and (b) above; and
(ii) specify the provision(s) in his opinion contravened; and
(iii) give particulars of the reasons for his opinion; and
(iv) specify a period of time within which the person is required to remedy the contravention (or the matters occasioning such contravention).

[*HSWA 1974 s 21*].

The period specified in the notice within which the requirement must be carried out (see (iv) above) must be at least 21 days – this being the period within which an appeal may be lodged with an Employment Tribunal (see E15019 below). [*HSWA 1974 s 21*]. In order to be validly served on a company, an improvement notice relating to the company's actions as an employer must be served at the registered office of the company, not elsewhere. Service at premises occupied by the company will only be valid if the notice relates to a contravention by the company in the capacity of occupier (*HSE v George Tancocks Garage (Exeter)* [1993] Crim LR 605 and *HSWA 1974 s 46*).

There is no requirement that inspectors give duty-holders written notice of an intention to serve an improvement notice. However, the HSE's leaflet entitled "What to expect when a health and safety inspector calls" states that inspectors will discuss with the duty-holder the contents of the notice so that any points of difference may be resolved.

Failure to comply with an improvement notice can have serious penal consequences (see **E15036** below).

Under *s 23(5)* an improvement notice can be withdrawn before the time for compliance has expired. This will be appropriate where the situation changed after the improvement notice was issued, or where further information has come to light which has made the notice inappropriate. If a new improvement notice is necessary the old one should be formally withdrawn.

Prohibition notices

[E15015] If an inspector is of the opinion that, with regard to any activities to which *s 22(1)* applies (see below), the activities involve or will involve a risk of serious personal injury, he may serve on the person carrying out the activities a notice (a prohibition notice). [*HSWA 1974 s 22(2)*]. Personal injury is defined in *HSWA 1974 s 53*.

A prohibition notice must—

(a) state that the inspector is of the opinion that the activities create a risk of serious personal injury;
(b) specify the matters which, in the inspector's opinion, create or will create the risk in question;
(c) where there is actual or anticipatory breach of provisions and regulations, state that the inspector is of the opinion that this is so and give reasons;
(d) direct that the activities referred to in the notice must not be carried out on, by or under the control of the person on whom the notice is served, unless the matters referred to in (b) above have been remedied.

[HSWA 1974, s 22(3)].

Failure to comply with a prohibition notice can have serious penal consequences (see E15028 below).

It is incumbent on an inspector to show, on a balance of probabilities, that there is a risk to health and safety (*Readmans Ltd and Another v Leeds CC* [1993] COD 419 where an environmental health officer served a prohibition notice on the appellant regarding shopping trolleys with child seats on them following an accident involving an eleven-month-old baby. The appellant alleged that the Employment Tribunal had wrongly placed the burden of proof on it by requiring it to show that the trolleys were not dangerous. It was held by the High Court (Roch J allowing the appeal), that it was for the inspector to prove that there was a health and safety risk; if the inspector succeeded in establishing that, the duty-holder would then have to prove that it had done all that was reasonably practicable).

The risk of personal injury does not need to be imminent (*Tesco v Kippax* (COIT No 7605, HSIB 180 p 8)

It should be noted however, that a company's temporary cessation of an activity does not preclude an inspector from issuing a prohibition notice. In the case of *Railtrack v Smallwood* [2001] EWHC Admin 78, [2001] ICR 714, Mr Justice Sullivan considered an appeal by Railtrack following the decision of an Employment Tribunal that a prohibition notice issued by an HSE inspector, Mr Smallwood, following the Ladbroke Grove train crash, was valid notwithstanding that activities on the part of the track to which the notice applied, had ceased. Mr Justice Sullivan stated that—

> 'In my judgment, it would be very surprising, and a significant lacuna in the Act if an Inspector, in the aftermath of such a serious accident, when operations have been suspended precisely because of the gravity of the accident, was unable to issue a prohibition notice upon the basis of a present risk of serious personal injury.'

He went on to consider the intention behind s 22 and concluded that—

> 'Looking at the words of s 22(1) in this context, and bearing in mind in particular that s 22 must have been intended to confer powers upon Inspectors, not merely prior to, but also in the aftermath of, the most serious accidents, I am satisfied that "activities" are (still) being carried on for the purposes of s 22 if they have been temporarily interrupted or suspended as a result of a major accident. I do not consider that such an interpretation does any violence to the ordinary meaning of the words in s 22, provided they are considered in a realistic context.'

In his judgment Mr Justice Sullivan stated that all the surrounding circumstances must be considered, including the reason for the temporary inactivity—

> 'There may well be a distinction to be drawn on the facts between a factory that is inactive because the employer has gone bankrupt and dismissed all of his employees, and a factory that is inactive because it has closed for the summer holidays. If, however, the factory has just closed, and the workers had been sent home, because of a tragic accident, I do not consider that activities will have ceased for the purposes of s 22. The position may be different if, not as a result of an accident, but pursuant to a pre-planned closure programme, for example, to replace outdated and unsafe machinery, the factory is at a standstill for some weeks or months.
>
> . . . I am satisfied that . . . even though there are temporary interruptions when nothing is actually taking place on the ground, so "activities" will continue for the purposes of s 22 even though they have been interrupted as a result of a serious accident.'

Railtrack's appeal was subsequently dismissed.

In *Chilcott v Thermal Transfer Ltd* [2009] EWHC 2086 (Admin), [2009] All ER (D) 94 (Aug) analogous circumstances produced a different outcome. Charles J considered *Railtrack v Smallwood* and affirmed the approach that is to be taken by a tribunal hearing an appeal from an improvement or prohibition notice (that being that a tribunal is at liberty to reach its own decision on the notice, rather than simply carrying out an assessment of its genuineness). In this case, the appellant company acted as a main contractor on a construction site, and had engaged sub-contractors to undertake platform steelwork to enable the installation of cooling towers. Risk assessments and method statements had been prepared which identified the potential risk to operatives of falls from height. Handrails were to be installed on the platforms to protect against this risk; the installation was to be done by operatives working on and from a mobile platform ('MEWP'). On 29 July 2008 one of the sub-contractor's employees, a Mr Campbell, was seen to be standing on the platform. He was not stopped because the appellant's representative was not aware that the method statement stated that all work was to be done from the MEWP without operatives standing on the platform. Later that day, Mr Campbell fell from the platform and broke both of his ankles. Mr Chilcott, the HSE's inspector, served a prohibition notice forbidding further 'work at height on the access platform for the new cooling towers' until contraventions of s 3 of the *HSWA 1974* and *Work at Height Regulations 2005* had been remedied. On 30 July 2008 Mr Chilcott indicated that he would be happy to 'lift' the notice, even though once made, a prohibition notice of immediate effect cannot be withdrawn.

The company appealed to an Employment Tribunal, which cancelled the notice on the basis that, with hindsight, there was no risk that the sub-contractor would go on to the platform that night, and that in those circumstances the inspector could simply have obtained an assurance that no work on the handrail would be done in the following 24 hours, and that all work thereafter would be carried out in accordance with the method statement. The tribunal deemed the service of the notice premature in the circumstances. The HSE appealed.

Charles J held that the Employment Tribunal had taken the wrong approach to the appeal that was before it and had effectively erred in law (see paragraphs 21 and 22 of the judgment). Charles J quoted with approval, at paragraph 5 of his judgment, the following passage from paragraph 44 of *Railtrack v Smallwood* [2001] EWHC Admin 78, [2001] ICR 714 in which Sullivan J stated:

> In light of those factors [as contained in *s 24 HSWA 1974*], and of the authorities cited in *De Smith Woolf & Jowell's Judicial Review of Administrative Law* (1999), pp 251 – 252, paragraph 6-010, I expressed the provisional view during the course of argument that a Tribunal hearing an appeal under *section 24* of the *1974 Act* was not limited to reviewing the genuineness and/or reasonableness of the Inspector's opinions. It was required to form its own view, paying due regard to the Inspector's expertise, see in particular *Sagnata Investments Ltd v Norwich Corporation* [1971] 2 QB 614.

Charles J went on to say that it was open to the Employment Tribunal to reach its own decision rather than simply to apply public law principles to the issue of whether the notice should have been served. This was recently reconfirmed in *MWH UK Ltd v Victoria Susan Wise (HM Inspector of Health & Safety)* [2014] EWHC 427 (Admin), a Court of Appeal case concerning the service, and modification by an Employment Tribunal, of an improvement notice served on a construction company.

In reaching its own decision, it was incumbent on the tribunal to focus its attention on the situation "on the ground" at the time that the notice was served, and specifically, to consider whether there was a risk flowing from an activity being carried out at that time. He said, at paragraph 19:

> The Employment Tribunal, on the basis of the evidence it has on the approach I have indicated, can and should itself, having due regard to the view of the Inspector and the Inspector's expertise, and indeed the expertise of the assessors of the Tribunal, assess the risk as at the relevant date. The assessment of risk is a multi-faceted exercise; some factors will be more important than others, and it seems to me inappropriate for me to seek to set guidelines in that context.

The notice in this case was cancelled by Charles J on the basis that the appellant had in place at the time that the notice was served a plan to protect against falls from height. Therefore, the risks from falls, as identified by the Inspector, had actually been considered and plans put in place by the appellant at the time of the notice. The notice consequently was not warranted.

Differences between improvement and prohibition notices

[E15016] Prohibition notices differ from improvement notices in two important ways—

(a) with prohibition notices, it is not necessary that an inspector believes that a provision of *HSWA 1974* or any other statutory provision is being or has been contravened. Instead, the inspector must consider that there are activities which are or may be capable of causing serious personal injury;

(b) prohibition notices are therefore served in anticipation of danger.

It is irrelevant that the hazard or danger perceived by the inspector is not mentioned in *HSWA 1974*; it can exist by virtue of other legislation, or even in the absence of any relevant statutory duty. In this way notices are used to enforce the later statutory requirements of *HSWA 1974* and the earlier requirements of the *Factories Act 1961* and other protective occupational legislation.

Unlike an improvement notice, where time is allowed in which to correct a defect or offending state of affairs, a prohibition notice can take effect immediately.

A direction contained in a prohibition notice shall take effect—

(a) at the end of the period specified in the notice; or
(b) if the notice so declares, immediately.

[*HSWA 1974, s 22(4)* as substituted by *Consumer Protection Act 1987 Sch 3*].

An improvement notice gives a person upon whom it is served time to correct the defect or offending situation. A prohibition notice, which is a direction to stop the work activity in question rather than put it right, can take effect immediately on issue; alternatively, it may allow time for certain modifications to take place (ie deferred prohibition notice). Both types of notice will generally contain a schedule of work which the inspector will require to be carried out. If the nature of the work to be carried out is vague, the validity of the notice is not affected. If there is an appeal, an Employment Tribunal may, within its powers to modify a notice, rephrase the schedule in more specific terms (*Chrysler (UK) Ltd v McCarthy* [1978] ICR 939).

Effect of non-compliance with notice

[E15017] If, after expiry of the period specified in the notice, or in the event of an appeal, expiry of any additional time allowed for compliance by the Employment Tribunal, an applicant does not comply with the notice or modified notice, he can be prosecuted. It is the HSE's policy, as stated in its Enforcement Guide (which provides legal guidance to its inspectors and members of staff), that prosecution should 'normally follow' a failure to comply with a notice. If convicted of contravening a prohibition notice, he may be imprisoned. [*HSWA 1974, s 33(1)(g), Sch 3A*]. In *R v Kerr; R v Barker* (1996, unreported) the directors of a company were each jailed for four months after allowing a machine which was subject to a prohibition notice – following an accident in which an employee lost an arm – to continue to be operated.

Prosecution for a failure to comply with a notice may fail if—

- the notice was not properly served
- it is incapable of being complied with because the wording is vague, confusing or unclear; or
- it was, in the opinion of the court, complied with.

If the court considers that there has been a failure to comply with an improvement or prohibition notice, it may order that the recipient of the notice to take steps to remedy the matters within his/her control within a specified time.

Service of notice coupled with prosecution

[E15018] Where an inspector serves a notice, he may at the same time decide to prosecute for the substantive offence specified in the notice. The fact that a notice has been served is not relevant to the prosecution. Nevertheless, an inspector will not normally commence proceedings until after the expiry of 21 days, ie until he is satisfied that there is to be no appeal against the notice or until the Employment Tribunal has heard the appeal and affirmed the notice, since it would be inconsistent if conviction by the magistrates were followed by cancellation of the notice by the Employment Tribunal. The fact that an Employment Tribunal has upheld a notice is not binding on a magistrates' court hearing a prosecution under the statutory provision of which the notice alleged a contravention; it is necessary for the prosecution to prove all the elements in the offence (see E15031 below).

Only a small proportion of Employment Tribunal cases heard relate specifically to health and safety. Moreover, they are not empowered to determine breaches of criminal legislation.

Appeals against improvement and prohibition notices

[E15019] A person on whom either type of notice is served may appeal to an Employment Tribunal within 21 days from the date of service of the notice. The Employment Tribunal may extend this time where it is satisfied, on application made in writing (either before or after expiry of the 21-day period), that it was not reasonably practicable for the appeal to be brought within the 21-day period. On appeal the Employment Tribunal may either affirm or cancel the notice and, if it affirms it, may do so with modifications in the form of additions, omissions or amendments. [*HSWA 1974, ss 24(2), 82(1)(c)*; *Employment Tribunals (Constitution and Rules of Procedure) Regulations 2013 (SI 2013 No 1237), Sch 1*].

The HSE's Enforcement Guide sets out the test the Employment Tribunal will apply, following the decision in *Chilcott*:

> Where the Inspector's decision is appealed the Tribunal is not limited to reviewing the genuineness and/or reasonableness of the Inspector's opinions. It was required to form its own view paying due weight to the Inspector's expertise. The test was not the judicial review test as to whether the decision was reasonable. The court should focus on the point at which the notice was served rather than look at the situation with the benefit of hindsight. Their task was to decide what they would have done at that point in time.

(See www.hse.gov.uk/enforce/ enforcementguide/notices/tribunals-tribunal.htm.)

Effect of appeal

[E15020] Where an appeal is brought against a notice, the lodging of an appeal automatically suspends operation of an improvement notice, but a prohibition notice will continue to apply unless there is a direction to the contrary from the Employment Tribunal. Thus—

(a) in the case of an improvement notice, the appeal has the effect of suspending the operation of the notice [*HSWA 1974, s 24(3)(a)*];

(b) in the case of a prohibition notice, the appeal only suspends the operation of the notice if the Employment Tribunal so directs, on the application of the appellant. The suspension is then effective from the time when the Employment Tribunal so directs.

[HSWA 1974, s 24(3)(b)].

Grounds for appeal

[E15021]–[E15023] The most common grounds for appeal involve assertions that—

(a) The inspector incorrectly interpreted the law or acted in excess of his powers;
(b) The notice is fundamentally flawed;
(c) The inspector did not genuinely hold the necessary opinion, or the opinion was based on unreasonable grounds;
(d) The inspector's opinion, although honestly held, is not one that should be endorsed by the tribunal;
(e) No contravention of the relevant statutory provision has occurred/will occur, or there is no risk of serious personal injury;
(f) Breach of law is admitted but the proposed solution is not "practicable" or not "reasonably practicable", or there are no means which would reduce the risk;
(g) Breach of the law is admitted but the breach is so insignificant that the notice should be cancelled.

Prior to the hearing of an appeal it will be open to the inspector to accept criticisms made of the notice by withdrawing it and serving a new notice in amended terms. Alternatively, the tribunal may affirm the notice and modify it to reflect the criticism.

Inspector's incorrect interpretation of the law

It is doubtful whether many cases have, or indeed would, succeed on this ground. Where regulations impose a strict duty (for example, the duty under *reg 5(1)* of the *Provision and Use of Work Equipment Regulations 1998 (SI 1998 No 2306)* to provide work equipment in good repair) there is no scope for argument by the employer. However, where the statute provides a defence, for example, it requires the duty to be carried out 'so far as reasonably practicable' or it provides for a due diligence defence, there is some scope to argue that the inspectors' interpretation of the law is incorrect.

For example, in *Canterbury City Council v Howletts and Port Lympne Estates Limited* (1996) 95 LGR 798, a prohibition notice was served on Howletts Zoo following the death of a keeper while he was cleaning the tigers' enclosure. It was Howletts' policy to allow their animals to roam freely; the local authority argued that the zoo's keepers could have carried out their tasks in the tigers' enclosure with the animals secured. The High Court affirmed the Employment Tribunal's decision to set aside the notice, holding that *s 2* of *HSWA 1974* was not intended to render illegal certain working practices simply because they were dangerous.

It can happen that an inspector exceeds his powers under statute by reason of misinterpretation of the statute or regulation (*Deeley v Effer* (COIT No 1/72, Case No 25354/77)). This case involved the requirement that 'all floors, steps, stairs, passages and gangways must, so far as is reasonably practicable, be kept free from obstruction and from any substance likely to cause persons to slip'. [*OSRPA 1963 s 16(1)*] (now replaced by equivalent duties under the *Workplace (Health, Safety and Welfare) Regulations 1992 (SI 1992 No 3004)*. The inspector considered that employees were endangered by baskets of wares in the shop entrance. The Employment tribunal ruled that the notice had to be cancelled, since the only persons endangered were members of the public, and *OSRPA 1963* was concerned with dangers to employees.

The HSE has recently appealed a High Court ruling. In *Rotary Yorkshire Ltd v Hague* [2014] EWHC 2126 (Admin), the judge decided that it was not reasonable for an inspector to issue a Prohibition Notice where 'lesser action was regarded as sufficient protection'. He said that the Employment Tribunal should have considered the inspector's other available means to deal with a situation. The inspector had served a prohibition notice after observing exposed conductors at the rear of a switchboard in a high voltage room. The company was not able to prove, during the unannounced inspection, that the equipment was dead, although they did confirm this the following day. The judge said the inspector could have instead used her powers (under *s 20(2)* of the *HSWA 1974*) and required the area to be undisturbed until the test could be carried out.

The HSE went on to appeal this decision to the Court of Appeal on three grounds:

- the court had been wrong to judge the enforcement decision with hindsight;
- the court should have found that any commercial disadvantage to a company caused by the registration of a prohibition notice was irrelevant; and
- the court had erred in holding that the use of the inspector's statutory power under *s 20(2)(e)* of the *HSWA 1974* could be an acceptable alternative to a prohibition notice where the inspector had concluded that there was a risk of serious personal injury.

The Court of Appeal (*Hague v Rotary Yorkshire Ltd* [2015] EWCA Civ 696) agreed with the inspector on all counts and reinstated the Employment Tribunal's original decision. Firstly, the question was whether there was a risk of serious personal injury on the facts known to the inspector at the time of the decision. Secondly, any commercial disadvantage to a company caused by the registration of a prohibition notice was irrelevant. Finally, where an inspector had already concluded that there was a risk of serious personal injury, there was no requirement under s 20(2)(e) to ask the contractor to remedy the error. No further appeals will follow.

Notice is fundamentally flawed

A clear instance of fundamental flaw will be where a notice fails to tell a recipient what is wrong, why it is wrong and what can be done to reduce or remove the risks posed by the activity (see Upjohn J in *Miller Mead v Minister of Housing and Local Government* [1963] 2 QB 196).

In *BT Fleet Ltd v McKenna* [2005] EWHC 387 (Admin), [2005] All ER (D) 284 (Mar) the need for the remedial instructions accompanying a notice to be clear was considered. The case concerned a perceived breach of the *Manual Handling Operations Regulations 1992 (SI 1992 No 2793)*. The inspector serving an improvement notice specified the need for lifting machinery to be installed, and the notice was expressed such that this was the only way that the breach could be remedied. BT Fleet appealed against the terms of an improvement notice which required it to provide mechanical lifting aids or 'any other equally effective measures' to ensure compliance with legislation. The Employment Tribunal hearing the appeal affirmed the notice, but added the words 'such as ensuring that adequate training and supervision is in place'. Evans-Lombe J, in cancelling the notice, held that where an inspector had elected to include the instructions in the notice (which were not of themselves compulsory) he was obliged to ensure that they were clearly worded. He said, at paragraph [18]:

> If the provisions of the relevant statute provide an option to proscribe how the recipient can comply with the notice, and that option is taken, then the specification of how compliance can be effected form part of the notice and, if confusing, may operate to make it an invalid notice.

Inspector did not hold the necessary opinion

While the decision in *Chilcott* has emphasised the need for the Employment Tribunal to make its own decision on the validity and need for a notice as at the date on which it was served, the wording of the judgment makes clear that there will still be circumstances where the Tribunal is required to make an assessment of the inspector's opinion at that time. In order to make this assessment, the Tribunal will need to address:

(a) Whether the inspector genuinely held the view stated in the notice; and
(b) If so, whether, at the date of the notice the activities complained of involved or would involve a contravention of the statutory provisions or a risk of serious personal injury.

The inspector will be required to satisfy the Tribunal that he did hold that view, and that it was based on reasonable grounds.

Inspector's opinion not to be endorsed

An example of the inspector holding an honest opinion which was ultimately deemed to be incapable of endorsement is to be found in *Chilcott v Thermal Transfer Limited* (see **E15015** above). In that case, the Inspector genuinely perceived there to be a risk of employees and subcontractors suffering serious personal injury such that the service of a prohibition notice was warranted. The Court of Appeal held that the notice was not, in fact, required because at the forefront of the Inspector's mind was the need for work at height to be planned to avoid or reduce the risks of falls from height. In actuality, a safe system of work and a risk assessment had been devised in that regard. The prohibition notice was therefore cancelled.

No contravention or risk

Employers should note that contesting a prohibition notice on the ground that there has been no legal contravention is pointless, since valid service of a prohibition notice does not depend on legal contravention (*Roberts v Day* (COIT No 1/133, Case No 3053/77)). All that the Inspector must believe is that the activities pose a risk of serious personal injury.

It is suggested that appealing a notice on the basis that the activities do not contravene statutory provisions or pose a risk of injury will require deployment of arguments similar to those outlined above in relation to asserting that an inspector has incorrectly interpreted the law. It is also suggested that it is likely to be extremely difficult to persuade a Tribunal that there has not been or will not be any contravention or any risk in circumstances where an Inspector has deemed it necessary to serve a notice (see *Railtrack v Smallwood*).

Proposed solution not practicable

[E15024] The position in a case where, although breach of the law is admitted, the proposed solution is not considered practicable, depends upon the nature of the obligation. The duty may be strict, or have to be carried out so far as practicable or, alternatively, so far as reasonably practicable. In the first two situations the cost of compliance is irrelevant; in the latter case, where a requirement has to be carried out 'so far as reasonably practicable', cost effectiveness is an important factor but has to be weighed against the risks to health and safety involved in failing to implement the remedial measures identified in the enforcement notice.

Where there is a real danger of serious injury the cost of complying with the notice is unlikely to be decisive. Thus in a leading case, the appellant was served with an improvement notice requiring secure fencing on transmission machinery. An appeal was lodged on the ground that the proposed modifications were too costly (£1,900). It was argued that because of the intelligence and integrity of the operators a safety screen costing £200 would be adequate. The Employment tribunal dismissed the appeal: the risk justified the cost (*Belhaven Brewery Co Ltd v McLean* [1975] IRLR 370).

The tribunal in *Associated Dairies v Hartley* [1979] IRLR 171 distilled the following principles from cases, including *Belhaven Brewery*, in respect of treatment of costs considerations—

(a) The questions for the Tribunal appear to be—
 (i) Is the requirement of the inspector practicable?
 (ii) Is the requirement reasonable taking into account all the circumstances?
(b) The question of whether, taking into consideration the whole circumstances, the requirement is reasonable, should not be determined having regard solely to the proportion which the risk to be apprehended bears to the sacrifice in money, time or trouble involved in meeting the risk.

Rather it is proper to consider whether the time, trouble and expense of the inspector's requirement are disproportionate to the risk involved to the employees if the Tribunal cancel the improvement notice or affirm it with modifications.

The cost of complying with a notice is likely to carry less weight in the case of a prohibition notice than an improvement notice, as there must be 'a risk of serious personal injury' for a prohibition notice to be served (*Nico Manufacturing Co Ltd v Hendry* [1975] IRLR 225, where the company argued that a prohibition notice in respect of the worn state of their power presses should be cancelled on the ground that it would result in a 'serious loss of production' and endanger the jobs of several employees'. The Employment tribunal dismissed this argument, having decided that using the machinery in its worn condition could cause serious danger to operators). Similarly, an undertaking by a company to take additional safety precautions against the risk of injury from unsafe plant until new equipment was installed was not sufficient (*Grovehurst Energy Ltd v Strawson (HM Inspector)* (COIT No 5035/90).

Where cost is a factor this is not to be confused with the current financial position of the company. A company's financial position is irrelevant to the question whether an Employment tribunal should affirm an enforcement notice. Thus in *Harrison (Newcastle-under-Lyme) Ltd v Ramsay (HM Inspector)* [1976] IRLR 135, a notice requiring cleaning, preparation and painting of walls had to be complied with even though the company was on an economy drive.

Employment tribunals have power under *HSWA 1974 s 24* to alter or extend time limits attaching to improvement and prohibition notices (*D J M and A J Campion v Hughes (HM Inspector of Factories)* [1975] IRLR 291, where even though there was an imminent risk of serious personal injury, a further four months were allowed for the erection of fire escapes as it was not practicable to carry out the remedial works within the time limits set). Extensions of time in which to comply with notices are most commonly granted where the costs of the improvements and modifications required by the notice are significant.

Breach of law is insignificant

[E15025] Employment tribunals will rarely cancel a notice which concerns breach of an absolute duty where the breach is admitted but the appellant argues that the breach is trivial: *South Surbiton Co-operative Society Ltd v Wilcox* [1975] IRLR 292, where a notice had been issued in respect of a cracked wash-hand basin, being a breach of an absolute duty under the *Offices, Shops and Railway Premises Act 1963*. It was argued by the appellant that, in view of their excellent record of cleanliness, there was no need for officials to visit the premises. The appeal was dismissed.

Inspectors' investigation powers

[E15026] Inspectors have wide ranging powers under *HSWA 1974* to investigate suspected health and safety offences. These include powers to—

(a) enter and search premises;

(b) direct that the premises or anything on them be left undisturbed for so long as is reasonably necessary for the purpose of the investigation;
(c) take measurements, photographs and recordings;
(d) take samples of articles or substances found in the premises and of the atmosphere in or in the vicinity of the premises;
(e) dismantle or test any article which appears to have caused or be likely to cause danger;
(f) detain items for testing or for use as evidence;
(g) interview any person;
(h) obtain information by way of a statement;
(i) require the production and inspection of any documents and to take copies; and
(j) require the provision of facilities and assistance for the purpose of carrying out the investigation.

[HSWA 1974, s 20].

During the course of an investigation, an inspector will generally seek to obtain documents that are required to be kept under any of the statutory provisions; and/or any other documents that are necessary for him to see to establish whether there has been any breach of the relevant statutory provisions. There is a restriction on obtaining material that is the subject of Legal Professional Privilege. Usually HSE inspectors will request information and documentation on a voluntary basis before seeking to invoke their powers under s 20. In *The Director of The Serious Fraud Office v Eurasian Natural Resources Corpn Ltd* [2017] EWHC 1017 (QB), the High Court reiterated the need for any pre-charge investigation by a defendant's solicitors to be in 'anticipation of litigation', either Criminal or Civil, to enable it to be protected by Litigation Privilege.

The ability of a duty holder to argue that documents he was compelled to produce should not be admissible against him at trial because of self-incrimination is limited: it only potentially applies in respect of documents the defendant was compelled to create and not to documents already in existence. In the wider context of criminal evidence, evidence that has been improperly obtained is not usually held to be inadmissible per se; rather the *Police and Criminal Evidence Act 1984 (PACE 1984), s 78* test is applied in the particular circumstances, the test being whether it would be in the interests of justice for the material to be admitted.

Interviews

Under these powers inspectors may require interviewees to answer such questions as they think fit and sign a declaration that those answers are true. [*HSWA 1974, s 20(2)(j)*]. However, evidence given in this way is inadmissible in any proceedings subsequently taken against the person giving the statement or his or her spouse. [*HSWA 1974, s 20(7)*]. Where prosecution of an individual is contemplated the inspector will, therefore, usually exercise his evidence-gathering powers under the *PACE 1984*. Evidence given in this way is admissible against that person in later proceedings. Interviews conducted under *PACE 1984* are subject to strict legal controls, the PACE Codes of Practice. For example, interviewees must be cautioned before the interview takes place, and after any breaks (Code C para 10.1).

'Interview' is widely defined by Code C (para 11.1A). An interview is the 'questioning of a person regarding their involvement or suspected involvement in a criminal offence or offences under which under para 10.1', a person whom there are grounds to suspect of an offence 'must be cautioned before any questions about an offence, or further questions if the answers provide the grounds for suspicion, are put to them if either the suspect's answers or their silence may be given in evidence to a court in a prosecution.'

Before any interview, there must be sufficient information provided to make it possible to understand the nature of the suspected offence and why any party is suspected of committing an offence in order to allow for the effective exercise of the rights of the defence (Code C para 11.1A). This disclosure obligation has been extended to reflect the requirements of the EU Directive 2012/13/EU, Article 6. The decision in relation to what should be disclosed rests with the investigating officer.

The HSE Enforcement Guide gives some guidance to inspectors, it says while inspectors may receive a request from a suspect's solicitor for disclosure of information prior to the suspect's attendance at an interview under caution, there is no obligation on an investigator to disclose the whole of the evidence against a suspect prior to interview. Inspectors are told they have a wide discretion in relation to disclosure of such information; but when a request is made, the suspect should be provided with some information to enable his solicitor to be able usefully to advise his client in relation to the interview under caution.

In any interview under *PACE 1984*, there is also a right to remain silent, although these are qualified by the *Criminal Justice and Public Order Act 1994*, in particular *s 34*, which deals with the circumstances in which an adverse inference can be drawn.

Inspectors' powers of search and seizure in case of imminent danger

[E15027] Where an inspector has reasonable cause to believe that there are on premises 'articles or substances ('substance' includes solids, liquids and gases – HSWA 1974 s 53(1)) which give rise to imminent risk of serious personal injury', he can—

(a) seize them; and
(b) cause them to be rendered harmless (by destruction or otherwise).

[*HSWA 1974, s 25(1)*].

Enforcing authorities are given similar powers regarding environmental pollution, under the *Environment Act 1995 ss 108, 109*.

Before an article forming 'part of a batch of similar articles', or a substance is rendered harmless, an inspector must, if practicable, take a sample and give to a responsible person, at the premises where the article or substance was found, a portion which has been marked in such a way as to be identifiable. [*HSWA 1974 s 25(2)*]. After the article or substance has been rendered harmless, the inspector must sign a prepared report and give a copy of the report to—

(i) a responsible person (eg safety officer); and
(ii) the owner of the premises, unless he happens to be the 'responsible person'. (See A3005 ACCIDENT REPORTING AND INVESTIGATION for the meaning of this term.)

[HSWA 1974 s 25(3)].

Indemnification by enforcing authorities

[E15028] Where an inspector has an action brought against him in respect of an act done in the execution or purported execution of any of the 'relevant statutory provisions' (see **E15012** above), and is ordered to pay damages and costs (or expenses) in circumstances where he is not legally entitled to require the enforcing authority which appointed him to indemnify him, he may be able to take advantage of the *HSWA 1974 s 26*. By virtue of that provision, the authority nevertheless has the power to indemnify the inspector against all or part of such damages where the authority is satisfied that the inspector honestly believed—

(a) that the act complained of was within his powers; and
(b) that his duty as an inspector required or entitled him to do it.

In practice there will be very few circumstances where an inspector is held liable for advice given or enforcement action taken as part of his statutory duties. The Court of Appeal has ruled (in *Harris v Evans* [1998] 3 All ER 522, recently applied in the context of the *Registered Homes Act 1984* (now no longer in force) in *Trent Strategic Health Authority v Jain* [2009] UKHL 4, [2009] 1 AC 853) that an enforcing authority giving advice which leads to the issue of enforcement notices does not owe a duty of care to the owner of the premises affected by the notice and a claim for economic loss arising from such allegedly negligent advice cannot therefore succeed. The court said that if enforcing authorities were to be exposed to liability in negligence at the suit of owners whose businesses are adversely affected by their decisions it would have a detrimental effect on the performance by inspectors of their statutory duties. *HSWA 1974* contains its own statutory remedies against errors by inspectors and the court was not prepared to add to those measures. The court did, however, suggest that a possible exception might arise if a requirement imposed by the inspector introduced a new risk or danger which resulted in physical damage or economic loss.

Public register of improvement and prohibition notices

[E15029] Improvement and prohibition notices relating to public safety matters have to be entered in a public register as follows—

(a) within 14 days following the date on which notice is served in cases where there is no right of appeal;
(b) within 14 days following the day on which the time limit expired, in cases where there is a right of appeal but no appeal has been lodged within the statutory 21 days.
(c) within 14 days following the day when the appeal is disposed of, in cases where an appeal is brought.

[*Environment and Safety Information Act 1988, s 3*].

Notices which impose requirements or prohibitions solely for the protection of persons at work are not included in the register. [*Environment and Safety Information Act 1988, s 2(3)*].

In addition, registers must be kept of notices served by—

(i) fire authorities, under the *Schedule* to the *Environment and Safety Information Act 1988* for the purpose of *Articles 29, 30 and 31* of the *Regulatory Reform (Fire Safety) Order 2005 (SI 2005 No 1541)*;
(ii) local authorities, under the Schedule to the *Environment and Safety Information Act 1988* for the purpose of *s 10* of the *Safety of Sports Grounds Act 1975*;
(iii) responsible authorities (as defined by the *Environment and Safety Information Act 1988 s 2(2)*);
(iv) enforcing authorities, under the *Radioactive Substances Act 1993* (see D1068 DANGEROUS SUBSTANCES I); and
(v) the Office for Nuclear Regulation, for the purposes of *Sch 8, paras 3 and 4* of the *Energy Act 2013*.

These registers are open to inspection by the public free of charge at reasonable hours and, on request and payment of a reasonable fee, copies can be obtained from the relevant authority. [*Environment and Safety Information Act 1988, s 1*]. Such records can also be kept on computer.

The HSE public register of enforcement notices gives details of enforcement notices issued by the Health & Safety Executive (HSE) and can be found at: www.hse.gov.uk/notices/.

Part B: Offences and Penalties

Prosecution for breach of the 'relevant statutory provisions'

[E15030] Prosecutions can follow non-compliance with an improvement or prohibition notice, but equally inspectors will sometimes prosecute without serving a notice. Service of notices remains the most usual method of enforcement. In 2013/14, a total of 13,790 enforcement notices were issued by all enforcing authorities. The HSE prosecuted 551 cases were prosecuted in England and Wales, while local authorities prosecuted 88 cases in England and Wales. In Scotland, the Procurator Fiscal prosecuted 35 cases.

Prosecutions are most commonly brought after a workplace accident or dangerous incident (such as a fire or explosion). Investigations by the enforcing authorities may take many months to complete and, in complex cases, it is not unusual for prosecutions to be commenced over a year after the original incident. Prosecutions take place in the magistrates' courts and the Crown Court.

In practice, the enforcing authorities usually only investigate the most serious workplace accidents; it continues to be the HSE's policy to investigate all

[E15030] Enforcement

work-related deaths which are reported to it. The HSE Enforcement Policy states: 'The enforcing authorities should carry out a site investigation of a reportable work-related death, unless there are specific reasons for not doing so, in which case those reasons should be recorded.'

Burden of proof

[E15031] Throughout criminal law, the burden of proof of guilt is on the prosecution to show that the accused committed the particular offence (*Woolmington v DPP* [1935] AC 462). The burden is a great deal heavier than in civil law, requiring a court to be sure of guilt. While not eliminating the need for the prosecution to establish general proof of guilt *HSWA 1974 s 40* makes the task of the prosecution easier by transferring the onus of proof to the accused for one element of certain offences. *Section 40* states that in any proceedings for an offence consisting of a failure to comply with a duty or requirement to do something so far as is practicable, or so far as reasonably practicable, or to use the best practicable means to do something, the onus is on the accused to prove (as the case may be) that it was not practicable, or not reasonably practicable to do more than was in fact done to satisfy the duty or requirement, or that there was no better practicable means than was in fact used to satisfy the duty or requirement. However, *s 40* does not apply to an offence created by *HSWA 1974 s 33(1)(g)* – failing to comply with an improvement notice.

Generally speaking, statute which requires a defendant to prove a fact is a prima facie affront to the presumption of innocence encapsulated in *Article 6(2)* of the *ECHR*. However, it has been held a reverse burden of proof may not necessarily be incompatible with a defendant's human rights. Such a view was articulated by Lord Bingham in *Sheldrake v DPP* [2004] UKHL 43, [2005] 1 AC 264, who stated that provided a reverse burden is reasonable and proportionate, it will not offend against the principle that it is for the prosecution to prove its case against the defendant.

The burden of proof and its application in prosecutions under *HSWA 1974* was considered by the House of Lords in *R v Chargot Ltd (t/a Contract Services)* [2007] EWCA Crim 3032, [2008] 2 All ER 1077; affd [2008] UKHL 73, [2009] 2 All ER 645. The case required the Court of Appeal and then the House of Lords to consider, amongst other things, the impact of *HSWA 1974 s 40*.

The first and second defendants, both companies, were members of the Ruttle Group, of which the third defendant was the managing director. Ruttle owned a farm at which a car park was being built. The deceased was driving a dumper truck at the farm during the course of the construction of a car park. He was not the usual dumper truck driver, but had been asked to drive it that day by the site foreman. As the deceased drove down a ramp, the dumper truck fell onto its side and he was buried under the soil and was killed. There were no witnesses to the accident, and its precise cause was never established.

The first defendant was charged with contravening *HSWA 1974 s 2(1)*. The second defendant was charged with contravening *HSWA 1974 s 3(1)*, and the third defendant, as an individual, was charged pursuant to *HSWA 1974 s 37* (it was alleged that, through his connivance, consent or neglect, he had caused the second defendant to breach *HSWA 1974 s 3(1)*). The defendants accepted that whilst there had been no risk assessments in relation to the use of the dumper truck, nor any training provided to the deceased, they had nonetheless done everything which was reasonably practicable to ensure his safety and that of the other workers.

The prosecution based its case against the first and second defendants on the proposition that it was sufficient for it to identify and prove a risk of injury arising from the state of affairs which had prevailed at the site, rather than from a series of specified acts and/or omissions. The case against the third defendant was that he had given the specific directions in respect of how works at the site were to be carried out. The prosecution submitted Routes to Verdict in questionnaire form to the jury in relation to the charges against the first and second defendants'. The defendants were duly convicted on this basis on 10 November 2006. They appealed against conviction and sentence.

The appeal centred on the effect of the burden placed on the defendants under *s 40* and on what the prosecution had to prove to establish the prime facie breach of duty which would trigger the need for the defence to prove that it was not reasonably practicable to do more than had in fact been done to satisfy the duty.

The Court of Appeal dismissed the defendants' appeal against conviction, holding that *HSWA 1974 ss 2, 3* imposed a duty to ensure a state of affairs (ie a risk-free work environment), so far as was reasonably practicable. *HSWA 1974 s 40* imposed an obligation on the defence to establish that they had done everything reasonably practicable to ensure the existence of that state of affairs. The policy behind *HSWA 1974* was clearly to impose a positive burden on employers rather than simply disciplining them for breaches of specific obligations. That being so, the prosecution was entitled to point to a state of affairs as amounting to a breach of the statutory duty. The prosecution had clearly established that a real risk of injury had been caused by driving the dumper truck (as evidenced by the fact that the accident had occurred). This was enough to justify the requirement that the first and second defendant should have had the burden of proving that they had done all that was reasonably practicable to protect against risk. It might have been different if the state of affairs which had been alleged to amount to a breach of duty by the employer or undertaker could not be shown to have any causal link to the employment or the undertaking in question, but that was not the case here.

Permission to appeal to the House of Lords was refused by the Court of Appeal but allowed by the House of Lords. Their Lordships delivered their decision on 10 December 2008.

The issues which fell to be determined were—

(a) What the prosecution had to prove in order to show that, subject to reasonable practicability, an offence had been committed;
(b) What the prosecution had to prove in order to demonstrate that the officer had committed an offence under *s 37*;
(c) What further particulars, if any, the prosecution had to produce in the interests of fairness when a prosecution was mounted.

The defendants again argued that the prosecution had to identify and then prove the acts and omissions which it asserted had led to the breach of duty, a generalised state of affairs being insufficient to establish culpability.

The House of Lords, in affirming the Court of Appeal's decision, held as follows (Lord Hope delivering the leading opinion)—

(a) That the duties contained in *HSWA 1974 ss 2, 3* were expressed in broad and general terms. The matters set out in *s 2(2)* were those to which employers had to have regard and could be taken as describing a result which employers had to achieve. If that result was not achieved an employer would have to show that it had not been reasonably practicable for it to achieve it. The prosecution had to show that the results set out in *ss 2(1), 3(1)* had not been achieved in order to establish a case. Once the breach had been established the onus was on the defendant to show that certain measures had not been reasonably practicable within the meaning of *s 40*.

(b) The allegations contained within the prosecution's case summary were not ingredients of the offence and so did not have to be proved. *HSWA 1974 ss 2(1), 3(1)* set out the various results that employers had to achieve, not the means by which they were to be achieved. It was the results with which the jury had to concern themselves.

(c) That reversing the burden of proof by placing it on the defendant pursuant to *HSWA 1974 s 40* was not disproportionate or incompatible with the presumption of innocence. The risks contemplated by *HSWA 1974 ss 2, 3* were risks which included serious injury or even death. Once a prima facie contravention of *ss 2, 3* has been established, it is entirely proportionate that the onus should pass to the defence to establish that certain preventative measures were not reasonably practicable.

(d) That fixed rules as to what the prosecution had to identify and then prove in relation to the charge under *s 37* (consent, connivance or neglect) were not capable of being laid down because of the extent to which each case differed on its facts. Where it was shown that a company had failed to achieve the results prescribed by *ss 2, 3* it would be a relatively short step for an inference to be drawn that there had been some connivance or neglect on the part of an officer who was controlling the workplace or the circumstances under which the relevant work activity was taking place.

(e) That in the present case, the particulars submitted by the prosecution had been sufficient, and it had not been necessary for it to specify the respects in which risk had been associated with the activity or to identify the cause of the accident. The fact that the accident had occurred at all was sufficient to show that the risk was real; Lord Hope commented that "where a person sustains injury at work, the facts will speak for themselves". The test of how much detail the particulars of the offence needed to go into was one of "fair notice". However, the particulars were not the ingredients of the offence and therefore did not need to be proved by the prosecution.

In relation to risk and what it was incumbent on the prosecution to prove, Lord Hope held as follows (at paragraph [27]):

The first point to be made is that when the legislation refers to risks it is not contemplating risks that are trivial or fanciful. It is not its purpose to impose burdens on employers that are wholly unreasonable. Its aim is to spell out the basic duty of the employer to create a safe working environment. This is intended to bring about practical benefits, bearing in mind that this is an all-embracing responsibility extending to all workpeople and all working circumstances: Robens report, para 130. The framework which the statute creates is intended to be a constructive one, not excessively burdensome . . . The law does not aim to create an environment that is entirely risk free. It concerns itself with risks that are material. That, in effect, is what the word "risk" which the statute uses means. It is directed at situations where there is a material risk to health and safety, which any reasonable person would appreciate and take steps to guard against.

The effect of the clarification given by the House of Lords in R v Chargot was considered and applied by the Court of Appeal in *R v EGS Ltd* [2009] EWCA Crim 1942, an appeal by the prosecution against the trial judge's finding that there was no case to answer in respect of certain charges on the indictment. After having set out at length passages from Lord Hope's decision, the Court of Appeal in EGS held that the trial judge had erred when he ruled that the prosecution was unable to establish a connection between the defendant's conduct and the accident, or that the risk was foreseeable. In allowing the prosecution's appeal, Dyson LJ held that the judge had confused himself on the role that foreseeability and causation played in the prosecution's formulation of its case and had fallen into error as a consequence. He stated that foreseeability was not relevant to the issue of whether the prosecution had proved the existence of a material risk (see paragraph [27]), and also stated that the judge had been wrong to require there to be a causal link between the defendants' conduct and the accident which eventuated.

HSWA 1974 ss 2 and 3 were considered by the Supreme Court within the context of a civil claim for damages for personal injury in *Baker v Quantum Clothing Group* [2011] UKSC 17, [2011] 4 All ER 223, [2011] 1 WLR 1003. The appeal concerned the liability of employers in the knitting industry for hearing loss shown by employees to have been suffered during the years prior to 1 January 1990 (which was then the *Noise at Work Regulations 1989* came into force). One of the issues which concerned the Supreme Court was whether liability arose under s 29 (1) of the *Factories Act 1961* (no longer in force) in relation to an employee who could establish that her hearing loss was caused by exposure to noise between 85 and 90 decibels. 90 decibels had been defined as a limit above which precautions were to be taken in a Code of Practice prepared by the Industrial Health Advisory Committee's Subcommittee on Noise in 1972. The Supreme Court therefore had to consider the whether employers could be said to have breached their statutory and common law duties in relation to exposure to noise below that limit.

Section 29 of the Factories Act 1961 provided:

(1) There shall, so far as is reasonably practicable, be provided and maintained safe means of access to every place at which any person has at any time to work, and every such place shall, so far as is reasonably practicable, be made and kept safe for any person working there.

(2) Where an person has to work at a place from which he will be liable to fall a distance more than six feet six inches, then, unless the place is one which affords secure foothold and, where necessary, secure hand-hold, means shall be provided, so far as is reasonably practicable, by fencing or otherwise, for ensuring his safety.

Lord Mance identified that consideration of s 29 required the Supreme Court to consider whether the safety of a workplace was an absolute and unchanging concept or a relative concept, the practical implications of which could change over time. The claimants' position was that safety was absolute, objective, unchanging and independent of any foresight of injury. This was accepted by Smith LJ in the Court of Appeal, who held that, under s 29, determination of whether a place was safe required it to apply "an objective test without reference to reasonable foresight", following Larner v British Steel plc [1993] 4 All ER 102, [1993] ICR 551, CA . Lord Mance rejected this approach: "Whether a place is safe involves a judgment, one which is objectively assessed of course, but by reference to the knowledge and standards of the time. There is no such thing as an unchanging concept of safety" (paragraph [64]). Lord Mance adverted to ss 2 (1), 3 (1) *HSWA* as the successors to s 29 of the *Factories Act*, and referred to the treatment given to the concept of safety by the House of Lords in Chargot. He quoted Lord Hope at paragraph [27] of Chargot, and noted that it would be strange if the "narrower formulation" in s 29 "had a more stringent effect". Lord Mance went on to hold that "[if] safety is a relative concept, then foreseeability must play a part in determining whether a place is or was safe" (paragraph [68]). The claimants had rejected this approach, arguing that foreseeability only became relevant when it came to consideration of reasonable practicability. Lord Mance held that the qualification "so far as is reasonably practicable" made it more rather than less likely that the concept of safety had to be judged by reference to what would "according to the knowledge and standards of the relevant time" have been considered safe (paragraph [78]). Lords Dyson and Saville (together the majority) agreed with this approach. The defendants' appeal was allowed.

Most recently, the Court of Appeal in *R v Tangerine Confectionary Ltd and Veolia ES (UK) Ltd* [2011] EWCA Crim 201 has identified the following questions as exercising Crown Courts following the decisions *Chargot*, *EGS* and *Baker v Quantum Clothing Group* [2011] UKSC 17, [2011] 4 All ER 223, [2011] 1 WLR 1003:

(1) What is the relationship between "safety" (*HSWA 1974 s 2*) and "risk" (to safety) (*HSWA 1974 s 3*)?

(2) Where there has been an injury, is the prosecution required to prove that the offence caused it?

(3) To what extent must the prosecution prove that the risk "derives" from the defendant's activities?

(4) What, if anything, is the relevance to these offences of foreseeability of injury or of an accident which has in fact happened?

In relation to the first question, the Court of Appeal held that the concepts of safety and "risk to safety" as contained in sections 2 and 3 respectively refer to the same thing: "Safety is not ensured, for the purposes of section 2, if there is a relevant risk to the safety of one or more employees" (paragraph 9). As to the second question, the Court referred to the judgments in Chargot and EGS as authority, if such were necessary, for the proposition that causation was not an ingredient of either offence. It adverted, however, to the potential for misunderstanding to arise from "too superficial a citation" from Lord Hope's speech at paragraph [22] of Chargot in which he held as follows:

> " . . . the statute prescribes the result which may be achieved. That is one thing. How the prosecution proposes to prove that this was so is another. The situation will vary from case to case. In cases such as the present, where a person sustains injury at work, the facts will speak for themselves. Prima facie, his employer or the person by whose undertaking he was liable to be affected, has failed to ensure his health and safety. Otherwise there would have been no accident."

The Court of Appeal cautioned against deploying this dicta "casually" or from relying on an injury to establish incontrovertibly a relevant risk: all that Lord Hope had been saying was that injury was evidence of risk. The potential for discussions as to whether the offence caused death or injury to hijack the proper course of the trial, or to become a "side-issue", was also articulated, with judges being advised to get proceedings back on track by making plain that the issue was whether or not the offence had been committed, not whether death or injury was the result (paragraph 17).

As to any need for the prosecution to establish that any risk had derived from the defendant's activities, the Court of Appeal branded this "potentially misleading" and identified circumstances where risk could exist concurrently with the defendant's undertaking and independently of it. The existence of the independent or external risk did not obviate or reduce the defendant's duty to avoid risk created by its undertaking.

On the fourth question, the Court of Appeal identified that the question that it was to determine was whether foreseeability was relevant to whether a material risk had been created – the Supreme Court had accepted in *Baker v Quantum Clothing Group* [2011] UKSC 17, [2011] 4 All ER 223, [2011] 1 WLR 1003 that foreseeability of danger was relevant to reasonable practicality per Chargot. The Court of Appeal also referred to its earlier decision in EGS and identified the error into which the trial judge in that case had fallen. Perhaps the most problematic reasoning behind his decision to accede to the half-time submission was his conclusion that the prosecution was unable to show that the accident was foreseeable. As it was not incumbent on the prosecution to establish that the accident had been caused by the offence so it could not be necessary for it to establish that the accident was foreseeable. It is material risk which must be prevented from eventuating; the justices in *Baker v Quantum Clothing Group* had confirmed that foreseeability was relevant to the assessment of risk and the objective determination of what constituted "safe". The Court of Appeal considered itself bound by that conclusion: "Whether a material risk exists or does not is, in these cases, a jury question and the foreseeability of risk (or lack of it) of some danger or injury is a part of the enquiry" (paragraph 36). The enquiry was not, however, concerned with whether the events in question were in fact foreseeable.

Main offences and penalties

[E15032] Health and safety offences are either (a) triable summarily (ie without jury before the magistrates), or (b) triable summarily and on indictment (ie triable either way), or (c) triable only on indictment. Most health and safety offences fall into categories (a) and (b). The main offences falling into these two categories are set out below. The *Health and Safety Offences Act 2008* has inserted a new *HSWA 1974 Sch 3A*, setting out the mode of trial for various offences under *s 33*.

Summary only offences

[E15033]

(a) Contravening any requirement imposed by or under regulations under *HSWA 1974 s 14* or intentionally obstructing any person in the exercise of his powers under that section;
(b) Intentionally obstructing an inspector in the exercise or performance of his powers or duties or obstructing a customs officer in the exercise of his powers under *HSWA 1974 s 25A*;
(c) Falsely pretending to be an inspector.

'Either way' offences

[E15034]

(a) Failure to carry out one or more of the general duties of *HSWA 1974 ss 2–7*;
(b) Contravening either—
 (i) *HSWA 1974 s 8* – intentionally or recklessly interfering with anything provided for safety;
 (ii) *HSWA 1974 s 9* – levying payment for anything that an employer must by law provide in the interests of health and safety (eg personal protective clothing);
(c) Contravening any health and safety regulations;
(d) Contravening a requirement imposed by an inspector under *HSWA 1974 s 25* (power to seize and destroy articles and substances);
(e) Contravening a requirement of a prohibition or improvement notice (including any notice which is modified on appeal);
(f) Intentionally or recklessly making false statements, where the statement is made—
 (i) to comply with a requirement to furnish information; or
 (ii) to obtain the issue of a document;
(g) Intentionally making a false entry in a register book, notice etc which is required to be kept;
(h) Using a document, with intent to deceive, which has been issued or authorised under any of the relevant statutory provisions or is required for any purpose thereunder, or making and possessing a document which resembles such a document so as to be calculated to deceive;
(i) failing to comply with a remedial court order made under *HSWA 1974 s 42*.

[*HSWA 1974 s 33(1)*].

In England and Wales there is no time limit for bringing prosecutions for indictable offences [see *Interpretation Act 1978 Sch 1* as to the definition of indictable offence]. Where an offence can only be tried summarily in the magistrates' courts the time limit for laying the information is 6 months from the date of the commission of the offence or from the time when the matters complained of became apparent [*Magistrates' Courts Act 1980 s 127(1)*]. In Scotland the 6-month time limit extends to 'either way' offences tried summarily. The period may be extended in the case of special reports, coroners' court hearings or in cases of death generally. [*HSWA 1974 s 34(1)*].

Summary trial or trial on indictment

[E15035] Many offences triable either way are tried summarily. However an increasing number of serious offences are being committed to the Crown Court, which has increased sentencing powers, for trial on indictment. The question for the court when deciding whether to commit an offence to the Crown Court is whether the case is of such seriousness that the Crown Court should have the power to deal with the offender in any way it could have dealt with him or her if convicted on indictment. The court should now have regard, principally, to the Sentencing Council Definitive Sentencing Guideline which detail the factors the court should consider when determining seriousness and the level of fine. The court should also consider the Sentencing Council Allocation Guideline which expressly states that either way offences should be tried summarily unless for reasons of unusual legal, procedural or factual complexity, the case should be tried by the Crown Court. It goes onto to state that for cases with no factual or legal complications the court should bear in mind its power to commit for sentence after trial and may retain jurisdiction notwithstanding that the likely sentence might exceed its powers. The obvious issue with this approach is that if a case is then committed for sentence the Crown Court may have to hear some of the evidence again to determine sentence, for example if the relevant sentencing category for harm or culpability is not agreed between the parties. In any event the defendant may elect trial on indictment.

For offences committed before 15 March 2015 the magistrates' court can impose a fine not exceeding £20,000 for an offence of failing to discharge one of the duties under *HSWA 1974 ss 2–6* and *s 7*. For offences committed after this date there is no statutory maximum to the fine they are able to impose. *Section 85* of *LASPO 2012* determines that there will now be unlimited fines available for the sentencing of health and safety offences. The effect of *s 85* is likely to be that more cases will be dealt with by the magistrates' court as it will consider its powers of sentence to be sufficient. This marks a huge increase in the sentencing powers of the magistrates' courts.

The Criminal Practice Direction ('CPD XIII') has been amended to incorporate guidance in respect of allocation of offences. CPD XIII sets out offences that must now be dealt with by either a District Judge in the magistrates' court or by a higher court. CPD XIII Annex 3 specifically deals with cases involving very large fines in the magistrates' court where *s 85* of *LASPO 2012* is invoked. An authorised District Judge must deal with any allocation decision, trial and sentencing hearing in the following types of cases that are triable either way:

(a) cases involving death or significant, life-changing injury or high risk of death or significant, life-changing injury;
(b) cases involving substantial environmental damage or polluting material of a dangerous nature;
(c) cases where major adverse effect on human health or quality of life, animal health, or flora resulted;
(d) cases where major costs through clean-up, site restoration, or animal rehabilitation have been incurred;
(e) cases where the defendant corporation has a turnover in excess of £10 million but does not exceed £250 million, and has acted in deliberate, reckless or negligent manner;
(f) cases where the defendant corporation has a turnover in excess of £250 million;
(g) cases where the court will be expected to analyse complex company accounts;
(h) high-profile cases or ones of an exceptionally sensitive nature.

The prosecution agency must notify the justices' clerk where practicable of any case listed in (a)–(h) above no less than seven days before the hearing. This is to ensure that the case is listed in front of an authorised District Judge as the matter must be adjourned if it is not listed in front of the designated tribunal.

In the Crown Court cases which fall into the below categories must be referred to the Resident Judge:

(a) any health and safety case resulting in a fatality or permanent serious disability;
(b) complex cases in which the defendant is a corporation (including cases for sentence as well as for trial);
(c) any case in which the defendant is a corporation with a turnover in excess of £1 billion (including cases for sentence as well as trial).

Once a case has been referred to the Resident Judge, the Resident Judge should refer the case to the Presiding Judge, who may retain the case for trial by a High Court Judge, or release the case back to the Resident Judge, either for trial by a named judge, or for trial by an identified category of judges.

Penalties for health and safety offences

[E15036] The maximum sentence for health and safety offences depends on the date that the offence was committed and the court that passes sentence. This is because the Health and Safety (Offences) Act 2008 increased penalties for some offences committed after 16 January 2009 (by increasing the maximum fine and introducing imprisonment for certain offences) and s 85 of LASPO 2012 (which came into force on 12 March 2015) had the effect of increasing the level of most fines available for magistrates' courts to an unlimited fine (previously £20,000 for most health and safety offences). However, the increase will only apply in respect of offences committed after 12 March 2015.

Part B: Offences and Penalties [E15036]

The *Health and Safety (Offences) Act 2008* set new penalties for certain offences and also increased the level of fines and terms of imprisonment that magistrates can impose.

The following penalties apply to offences committed after 16 January 2009 where a summary conviction is secured. Those committed before 12 March 2015 are limited to a £20,000 maximum fine in the magistrates' court. In the case of all offences listed below that are committed after 12 March 2015 the magistrates will have unlimited powers to fine.

(a) For failing to discharge a duty pursuant to *HSWA 1974 ss 2–6* – imprisonment for a term not exceeding 12 months or a fine or both;

(b) For failing to discharge a duty pursuant to *HSWA 1974 s 7* – imprisonment for a term not exceeding 12 months or a fine or both;

(c) For contravention of *HSWA 1974 s 8* – imprisonment for a term not exceeding 12 months, or a fine or both;

(d) For contravention of *HSWA 1974 s 9* – a fine;

(e) For an offence under *HSWA 1974 s 33 (1) (c)* - imprisonment for a term not exceeding 12 months, or a fine, or both;

(f) For an offence under *HSWA 1974 s 33(1)(d)* (includes contravention of regulations) – a fine not exceeding level 5 on the standard scale;

(g) For offences under *s 33(1)(e)–(g)* (includes contravention of a requirement contained in an improvement or prohibition notice - imprisonment for a term not exceeding 12 months, or a fine or both;

(h) For an offence under *s 33(1)(h)* (intentional obstruction of an inspector) – imprisonment for a term not exceeding 51 weeks in England and Wales or 12 months in Scotland, or a fine not exceeding level 5 on the standard scale or both;

(i) For an offence under *s 33(1)(i)* – a fine;

(j) For an offence under *s 33(1)(j)* – imprisonment for a term not exceeding 12 months or a fine not exceeding the statutory maximum, or both;

(k) For an offence under *s 33(1)(k), (l) or (m)* – imprisonment for a term not exceeding 12 months, or a fine, or both;

(l) For an offence under *s 33(1)(n)* (falsely pretending to be an inspector) – a fine;

(m) For an offence under *s 33(1)(o)* (failure to comply with a remedial order pursuant to *HSWA 1974 s 42*) – imprisonment for a term not exceeding 12 months or a fine, or both;

(n) For an offence under the existing statutory provisions for which no other penalty is specified – imprisonment for a term not exceeding 12 months, or a fine or both.

Except in respect of offences under *s 33(1)(b)* and *(i)*, and where offences can only be tried summarily, the Crown Court can impose a penalty of imprisonment for term not exceeding two years, or a fine, or both. Those penalties also apply where no penalty has been specified for an offence under existing statutory provisions.

The Definitive Sentencing Guidelines for Health and Safety Offences is the most significant recent event to have occurred in the field of health and safety. The guidelines came into force on 1 February 2016 and apply to all cases sentenced on or after that date (see **E15043** below).

The application of the recent guidelines has been tested in the Court of Appeal. The Court of Appeal has upheld a fine imposed on a local authority (*R (Health and Safety Executive) v Havering Borough Council* [2017] EWCA Crim 242, [2017] 2 Cr App Rep (S) 54, [2017] All ER (D) 86 (Mar). However, in *R (Health and Safety Executive) v MJ Allen Holdings Ltd* [2017] All ER (D) 58 (Jan), they found the sentence to be manifestly excessive.

Facts: The defendant was a holding company for a number of companies concerned with the iron industry. It was based in an industrial estate. At the relevant time, the defendant company employed a team of three maintenance workers; none of whom had received specific training regarding working from heights. In September 2014, the maintenance team were asked by the foundry manager to remove and replace a fan on the roof of one of the buildings on the site. The roof was made of sheeting containing asbestos and a number of thin perspex lights (the roof).

The roof was classified as "fragile" for the purposes of the Work at Height Regulations 2005, SI 2005/735. On September 19, 2014, the three maintenance employees began to do preparatory work with the aim of replacing the fan on the following day. During that preparatory work, one of the maintenance employees who was on the roof at the time slipped and his foot went through one of the asbestos sheets. There were employees underneath in the workshop, about eight to 10 metres away, who were potentially exposed to the debris and asbestos fibres released when the panel broke. No harm was done to the maintenance worker who had slipped.

The following day, the maintenance team (excluding the worker who had slipped) went onto the roof using the same method and replaced the fan. Two other roof fans had previously been replaced using the same methodology. In February 2016, the defendant company pleaded guilty to one count of failing to take suitable and sufficient measures to prevent any person falling a distance liable to cause personal injury, contrary to s.33(1)(c) of the Health and Safety at Work etc Act 1974. In sentencing, the Judge assessed the defendant company's culpability as "medium" under the Sentencing Council's Definitive Guideline: Health and Safety Offences, Corporate Manslaughter and Food Safety and Hygiene Offences (the guideline).

Further, the level of harm was assessed as level A and falling into category 2 on the basis that there was a risk of death or very serious injury, although that was balanced against the fact that no harm was actually caused. Furthermore, the Judge said that the defendant company fell into the "medium" category of the guideline given its annual turnover of £32m in the relevant year. On that basis, the Judge took a starting point of £240,000 under the guideline and gave a full deduction of one-third for the early guilty plea to arrive at a final sentence of a fine of £160,000. The defendant company appealed against sentence.

Among other things, the defendant company submitted that the Judge had taken too high a starting point of £240,000 and had failed to give any significant allowance for mitigating factors. In particular, consideration was given to the following sentencing remarks of the Judge:

> "The mitigating factors here are significant. The company has a very good record, indeed exemplary, record for health and safety, for which they are to be commended.

For over 50 years' trading since 1958, they have never been convicted of a health and safety offence. They have co-operated entirely with the Health and Safety Executive's investigation. They made their own investigation into the incident and changes were made and implemented. An updated risk assessment was made".

Holding: The appeal would be allowed.

In the present case, as the Judge herself had highlighted, there had been significant mitigating factors. Those should have driven the starting point down. Further, it was highly relevant to have had regard to the defendant company's small operating profit.

The net fine which the Judge had arrived at of £160,000, had represented some 23% of the defendant company's operating profit. Furthermore, since the United Kingdom European Union Membership Referendum in June 2016, the trading conditions for companies such as the defendant company had not been good. The defendant company employed some 240 people, most of whom worked in the foundry. Taking all those factors into account, the more appropriate starting point for the Judge to have taken in the present case would have been £120,000. Deducting a full one-third for the early plea of guilty, the net fine arrived at was £80,000.

Accordingly, the imposed fine of £160,000 would be quashed and substituted for a fine of £80,000. That fine met the objects of the legislation to mark and signify the breach that took place in the present case.

Time will tell how the further application of the guidelines will be viewed by the Court of Appeal, in particular if they will give further guidance in relation to 'very large companies'.

Defences

[E15037] Although no general defences are specified in the *HSWA 1974*, some regulations passed under the Act carry the defence of 'due diligence' (for example, *Regulation 21* of the *Control of Substances Hazardous to Health Regulations 2002 (SI 2002 No 2677)* as amended by *SI 2004 No 3386*. Currently, where a defence of 'due diligence' is available it is for the defendant to prove, on the balance of probabilities, that all appropriate steps to avoid the offence had been taken. Of course, in respect of *HSWA 1974 ss 2, 3*, it will be open to a defendant to assert that steps which would have enhanced health and safety were not reasonably practicable. However, this is not a defence as such, rather the words "so far as reasonably practicable" qualify the duties contained in *HSWA 1974 ss 2, 3*.

In *Polyflor Ltd v Health and Safety Executive* [2014] EWCA Crim 1522, [2014] ICR 1142 a company unsuccessfully appealed against its conviction on the basis that it could not guard against a careless employee doing something they knew was obviously unsafe. The case confirmed that an employer must assess and prevent or control the risks to the safety in these circumstances.

Manslaughter

[E15038] The *Corporate Manslaughter and Corporate Homicide Act 2007* came into force on 1 April 2008. It has re-formulated the test that must be satisfied by the prosecution before a conviction for corporate manslaughter can be secured. Cotswold Geotechnical Holdings was the first company to be convicted under the new legislation on 15 February 2011. Its appeal against a fine of £385,000 (which represented 250% of its annual turnover) was dismissed the following May.

Formerly, under the common law, in order to secure a conviction for corporate manslaughter it was necessary to establish gross negligence on the part of one or more of a company's directors or other 'controlling mind'. In large companies where there were several tiers of management below board level it was almost impossible to show gross negligence on the part senior executives who could be said to be 'controlling minds' and who were directly responsible for the commission of the offence. Successful prosecutions were therefore only really possible in respect of small companies.

The offence of corporate manslaughter (in England and Wales) or corporate homicide (in Scotland)

[E15038.1] The *Corporate Manslaughter and Corporate Homicide Act 2007* creates an offence whereby an organisation is guilty of corporate manslaughter (in England and Wales) or corporate homicide (in Scotland) if the way in which it is managed or organised by senior management causes a person's death and amounts to a gross breach of a relevant duty of care. As can be seen, this removes the need for a directing or controlling mind to be found guilty of gross negligence manslaughter. The offence applies to corporations and partnerships, government departments, police forces, and trade unions and employers' associations.

A duty of care is defined as any of the following duties owed under the law of negligence—

(a) a duty owed to employees or other persons working for a company or performing services for it (this is clearly designed to encapsulate relationships between a company and a deceased which fall beyond the usual employer-employee relationship);
(b) a duty owed as an occupier;
(c) a duty owed in connection with—
  (i) the supply by an organisation of goods and services;
  (ii) the carrying on by an organisation of any construction or maintenance operations;
  (iii) the carrying on by an organisation of any other activity on a commercial basis, or
  (iv) the use or keeping by an organisation of any plant, vehicle or other thing.
(d) a duty owed to a prisoner or a detained patient, for whose safety an organisation is responsible.

[*Corporate Manslaughter and Corporate Homicide Act 2007 s 2*].

The Act makes clear that reference to the "law of negligence" means a duty that would be owed in common law but for the existence of a statutory duty which is imposed. It is also clear in respect of the questions that fall to the jury: after the existence of a duty of care has been established, they will have to consider whether there has been a gross breach. The jury will also need to reach a decision on whether there has been a breach of health and safety legislation, how serious any breach was, and how great the risk of death the breach posed [s 8 of the *Corporate Manslaughter and Corporate Homicide Act 2007*].

Special provisions apply in respect of the Ministry of Defence [s 4 of the *Corporate Manslaughter and Corporate Homicide Act 2007*], police authorities [s 5 of the *Corporate Manslaughter and Corporate Homicide Act 2007*] and the emergency services [s 6 of the *Corporate Manslaughter and Corporate Homicide Act 2007*].

The Act created a new penalty imposable by the Court following a conviction – the publicity order [*s 10 of the Corporate Manslaughter and Corporate Homicide Act 2007*]. Application of this provision will compel an organisation to make public the fact of its conviction, specified particulars of the offence, the amount of any fine that has been imposed, and the terms of any remedial order.

Offences committed by particular types of persons

Corporate offences – delegation of duties to junior staff

[E15039] Companies cannot avoid liability for breach of general duties under *HSWA 1974* ss 2–6 by arguing that the senior management and/or the 'directing mind' of the company had taken all reasonable precautions, and that responsibility for the offence lay with a more junior employee or an agent who was at fault. *HSWA 1974* generally imposes strict criminal liabilities on employers and others (subject to the employer being able to establish that all reasonably practicable precautions had been taken) and it is not open to corporate employers to seek to avoid liability by arguing that their general duties have been delegated to someone lower down the corporate tree. In *R v British Steel plc* [1995] IRLR 310, British Steel was prosecuted under *HSWA 1974 s 3* after the death of a subcontractor who was carrying out construction work under the supervision of a British Steel engineer. British Steel argued that it was not responsible under s 3 for the actions of the supervising engineer as the engineer was not part of the 'directing mind' (following *Tesco Supermarkets Ltd v Nattrass* [1972] AC 153) of the company and all reasonable precautions to ensure the safety of the work had been taken by senior management. The Court of Appeal dismissed this argument; Steyn LJ commented that even passing negligence of an employee could give rise to liability under *HSWA 1974*. In *R v Gateway Foodmarkets Ltd* [1997] IRLR 189, the defendant was charged with breach of *s 2* when an employee fell through a trap door in the floor of a lift control room. The accident occurred while the store manager was manually attempting to rectify an electrical fault in the lift in accordance with a local practice which was not authorised by Gateway's head office. The Court of Appeal held that the failure at store manager level was attributable to the employer.

Regulation 21 of the *Management of Health and Safety at Work Regulations 1999 (SI 1999 No 3242)* makes clear that an employer cannot avoid liability in criminal proceedings by blaming an employee or "appointed person".

However, it does not follow that an employer will automatically be held criminally responsible for an isolated act of negligence by an employee performing work on its behalf. This is because it may still be possible for the employer to establish that it has done everything reasonably practicable in the conduct of its undertaking to ensure that employees and third parties are not exposed to risks to their health and safety by virtue of the way it has conducted its business (see *R v Nelson Group Services (Maintenance) Limited* [1999] IRLR 646, CA).

The definition of 'reasonably practicable' was authoritatively laid down in *Edwards v National Coal Board* [1949] 1 All ER 743 where it was said that: 'Reasonably practicable' is a narrower term than 'physically possible', and seems to imply that a computation must be made by the owner in which the quantum of risk is placed on one scale and the sacrifice involved in the measures necessary for averting the risk (whether in money, time or trouble) is placed in the other, and that, if it be known that there is a gross disproportion between them – the risk being insignificant in relation to the sacrifice – the defendants discharge the onus on them.' The role of foreseeability in determining what is reasonably practicable was considered in *Baker v Quantum Clothing Group* [2011] UKSC 17, [2011] 4 All ER 223, [2011] 1 WLR 1003 [see above].

The availability of a defence that a certain measure was not "reasonably practicable" was central *R v HTM Ltd* [2006] EWCA Crim 1156, [2007] 2 All ER 665, in which the Court of Appeal considered the previous decisions in *R v Gateway Foodmarkets* and *R v Nelson Group*. Two of the defendant company's employees had been fatally injured; it was consequently charged with breach of *HSWA 1974 s 2*. At trial, the company sought to adduce evidence that it had done all that was reasonably practicable to do, and that the accidents had resulted from the actions of the two employees who had perished. The company also asserted that it could not have been foreseen that the employees would behave as they did. The prosecution argued that foreseeabiliy was not relevant and that *reg 21* of the *Management of Health and Safety at Work Regulations 1999* [see above] precluded the company from relying on the act or default of its employees. The Court of Appeal held, in accordance with the judge at first instance, that the likelihood that the accident would happen was material in that it would assist in determining whether the risk would eventuate, and consequently what measures would be reasonable to put in place to prevent that. The Court of Appeal also held that the phrase "so far as is reasonably practicable" did not operate as a defence; instead it qualified the duty incumbent on employers. *Regulation 21* did not apply in the circumstances. It was therefore open to the defendant to attempt to demonstrate that the accident had been caused by the employees' actions.

Offences of directors or other officers of a company

[E15040] Where an offence is committed by a body corporate, senior persons in the hierarchy of the company may also be individually liable. Thus, where the offence was committed with the consent or connivance of, or was attributable to any neglect on the part of a director or officer of the company, that person is himself guilty of an offence and liable to be punished accordingly. Those who may be so liable are—

(a)　any functional director;
(b)　a manager (which does not include an employee in charge of a shop while the manager is away on a week's holiday (*R v Boal* [1992] QB 591, concerning s 23 of the *Fire Precautions Act 1971* – identical terminology to *HSWA 1974 s 37*));
(c)　a company secretary;
(d)　another similar officer of the company;
(e)　anyone purporting to act as any of the above.

[*HSWA 1974 s 37(1)*].

It is not sufficient that the company through its 'directing mind' (its board of directors) has committed an offence – there must be some degree of personal culpability in the form of proof of consent, connivance or neglect by the individual concerned. Evidence of this sort can be difficult to obtain and prosecutions under *s 37(1)* have, in the past, been rare compared with prosecutions of companies (although they are increasing in number). For example, 43 directors and/or senior managers and company secretaries were prosecuted under *s 37 in 2010/11*.

Directors, managers and company secretaries can be personally liable for ensuring that corporate safety duties are performed throughout the company (for example, a failure to maintain a safe system of work can give rise to personal liability). Liability may also arise as a result of a failure to perform an obligation placed on individuals by their employment contracts and job descriptions – for example, obligations imposed under a safety policy – not just in relation to duties imposed by law. In the case of *Armour v Skeen (Procurator Fiscal, Glasgow)* [1977] IRLR 310, an employee fell to his death whilst repairing a road bridge over the River Clyde. The appellant, who was the Director of Roads, was held to be under a duty to supervise the safety of council workmen. He had not prepared a written safety policy for roadwork, despite a written request that he do so, and was found to have breached *HSWA 1974 s 37(1)*.

Similar duties exist under the *Environmental Protection Act 1990 s 157* and the *Environment Act 1995, s 95(2)–(4)*. (See **E5023** emissions into the atmosphere.)

Directors convicted of a breach of *HSWA 1974 s 37* may also be disqualified, for up to 15 years, from being a director of a company, under the provisions of the *Company Directors Disqualification Act 1986 s 2(1)* as having committed an indictable offence connected with (inter alia) the management of a company. The question of whether an offence is in connection with the management of the company is to be decided by looking at whether it has some relevant factual connection with the management of the company, see *R v Goodman* [1993] 2 All ER 789. In *R v Chapman* (1992), unreported, a director of a quarrying company was disqualified and fined £5,000 for contravening a prohibition notice on an unsafe quarry where there had been several fatalities and major injuries.

More recently, a company director was disqualified from being a company director for the maximum of 15 years. He was also jailed for a total of 26 months: 16 months for a breach of the HSWA 1974; ten months for a breach of the Fraud Act 2006; and a total of eight months concurrent for four breaches of s 13 of the *Company Directors Disqualification Act 1986*.

The HSE and the Institute of Directors have produced a publication entitled "Leading Health and Safety at Work". This guidance sets out the duties and obligations incumbent upon directors, governors, trustees and equivalent officers in public, private and third sector organisations. Its primary aim is to ensure effective "top-down" leadership on health and safety. The HSE's inspectors will deploy the guidance in the context of routine inspections and in investigations.

Directors' insurance

[E15041] The *Companies Act 2006 s 232* prevents companies from 'immunising' their directors against liability by providing them with indemnities and similar arrangements. *Sub-section (1)* of the section states—'Any provision that purports to exempt a director of a company (to any extent) from any liability that would otherwise attach to him in connection with any negligence, default, breach of duty or breach of trust in relation to the company is void.'

The *Companies Act 2006 s 232* does allow companies to purchase and maintain for a director insurance against any of the liabilities mentioned in s 232. In practice, companies have since 1990 (under the terms of *Companies Act 1985 s 310*, now repealed) been able to buy insurance for their directors (usually known as Directors' and Officers', or D & O, insurance) which protects them in the event that they are named in civil or criminal proceedings.

Such insurance may (subject to the terms of the policy), protect directors against claims made during the policy period for breach of contract, negligence, misrepresentation and negligent misstatement caused by the person in his capacity as a director. Some policies also cover breach of statutory duty, employment claims, misfeasance, disqualification proceedings, proceedings by regulatory authorities (such as the HSE or Financial Services Authority) and corporate manslaughter claims. Policies typically allow directors to recover their legal costs and expenses incurred in defending or settling such claims, but fines and other penalties are usually excluded.

Offences due to the act of another person

[E15042] *Health and Safety at Work etc Act 1974, s 36(1)* makes clear that although provision is separately made for the prosecution of less senior corporate staff (eg safety officers and works managers) this does not prevent a further prosecution against the company itself. The section states that where an offence under *HSWA 1974* is due to the act or default of some other person, then—

(a) that other person is guilty of an offence; and
(b) a second person can be charged and convicted, whether or not proceedings are taken against the first-mentioned person.

Where the enforcing authorities rely on *HSWA 1974 s 36*, this must be made clear to the defendant. In *West Cumberland By Products Ltd v DPP* [1988] RTR 391, the conviction of a company operating a road haulage business for breach of regulations relating to the transport of dangerous substances was set aside as the offence charged related to the obligations of the driver of the vehicle and, in prosecuting the operating company, reliance was not placed on *HSWA 1974 s 36*.

Sentencing Guidelines

[E15043] The Definitive Sentencing Guidelines for Health and Safety Offences came into force on 1 February 2016. The guidelines apply to all organisations and offenders aged 18 and older, who are sentenced on or after 1 February 2016 regardless of the date of the offence. The guidelines replace the previous definitive guidelines for sentencing convictions for corporate manslaughter and of corporate bodies for health and safety offences which cause, or are one of the causes of, death. These guidelines did not apply where the death simply occurred, and was not causally related to the offence nor to any offences where death did not occur. The new guidelines apply to all health and safety offences. They provide a radical departure from the previous guideline case of *R v Howe & Son (Engineers) Ltd* [1999] 2 Cr App R (s) 37.

Section 125(1) of the *Coroners and Justice Act 2009* provides that when sentencing offences after 6 April 2010 every court must sentence an offender following any guidelines which are relevant, unless the court is satisfied that it would not be in the interests of justice to do so. The guidelines provide specific guidance, therefore, to both the magistrates and Crown Courts with nine steps that have to be followed to determine the level of any fine passed.

The level of fine imposed for health and safety offences has previously depended on the facts and circumstances of the case, including the gravity of the offence, whether the breach resulted in death or serious injury, and any mitigating evidence the defendant is able to put forward (including details of its means and ability to pay any fine imposed). The guidelines followed a consultation by the Sentencing Council on draft guidelines issued in November 2014, which closed in February 2015. The consultation found that the level of fines being imposed for offences in this area were, in some cases, too low to meet the objectives of sentencing. The Council also found that there were inconsistencies in the approach tribunals were taking to sentencing. The new guidelines are aimed at addressing these issues.

The guideline is divided into five main sections that deal with sentencing of health and safety offences in respect of organisations and individuals, corporate manslaughter and breach of food safety and food hygiene regulations in respect of both organisations and individuals. These latter two sections deal with breach of the *Food Safety and Hygiene (England) Regulations 2013 reg 19(1)*. The maximum penalties are the same as for breach of the *HSWA 1974* offences; maximum fine when tried on indictment and summarily is unlimited. In respect of the individual; trial on indictment two years custody and six months when tried summarily.

The guidelines for health and safety offences and corporate manslaughter are considered in more detail below.

Organisations

Step One: Culpability and harm

[E15043.1] The court must determine the offence category by making two separate assessments: culpability and harm. There are four categories for determining culpability:

Very high

Deliberate breach of or flagrant disregard for the law.

High

Offender fell far short of the appropriate standard; for example, by:

- failing to put in place measures that are recognised standards in the industry;
- ignoring concerns raised by employees or others;
- failing to make appropriate changes following prior incident(s) exposing risks to health and safety;
- allowing breaches to subsist over a long period of time.

Serious and/or systemic failure within the organisation to address risks to health and safety.

Medium

Offender fell short of the appropriate standard in a manner that falls between descriptions in 'high' and 'low' culpability categories.

Systems were in place but these were not sufficiently adhered to or implemented.

Low

Offender did not fall far short of the appropriate standard; for example, because:

- significant efforts were made to address the risk although they were inadequate on this occasion;
- there was no warning/circumstance indicating a risk to health and safety.

Failings were minor and occurred as an isolated incident.

Historically the HSE has brought prosecutions on the basis that the defendant fell 'far below' the required standard. Experience is that the HSE are asserting that culpability is high in its assessment of most cases under the new guidelines. The introduction to the culpability matrix states that where there are a number of factors present in a case that fall in different categories of culpability, the court should balance these factors to reach a fair assessment.

Harm

As set out at the outset of the guidelines in relation to harm; health and safety offences are offences of creating a risk of harm, there is no requirement for proof of any actual harm. The guidelines instruct that the below table is used to identify an initial category based on the risk of harm created by the offence.

	Level A	Level B	Level C
	• Death • Physical or mental impairment resulting in lifelong dependency on third party care for basic needs • Significantly reduced life expectancy	• Physical or mental impairment, not amounting to Level A, which has a substantial and long-term effect on the sufferer's ability to carry out normal day-to-day activities or on their ability to return to work • A progressive, permanent or irreversible condition	• All other cases not falling within Level A or Level B
High likelihood of harm	Harm category 1	Harm category 2	Harm category 3
Medium likelihood of harm	Harm category 2	Harm category 3	Harm category 4
Low likelihood of harm	Harm category 3	Harm category 4	Harm category 4

The guidelines state that the assessment of harm requires a consideration of both the seriousness of the harm risked by the offender's breach and the likelihood of that harm arising.

Seriousness of the risk is defined in Levels A, B and C. If death is a risk the correct category is A, if it is not, the correct category is B. In practice many health and safety offences that are prosecuted do present at least a risk of death, as many take place within the construction or manufacturing industries.

The likelihood of harm is assessed by determining how likely it was that the risk of harm present would be high, medium or low. In practice it is foreseeable that this may not always be a straightforward task as it may often require the court to speculate to make a determination. The court is likely to take into consideration, at this point, any risk assessment prepared by the defendant and the appropriateness of this risk assessment.

Having identified a harm category according to the table above the court must go onto consider two further factors before deciding whether to make an adjustment up or down the categories. The court must consider firstly whether the offence exposed a number of workers or members of the public to the risk of harm. The greater the number of people, the greater the risk of harm. Secondly the court must consider whether the offence was a significant cause of actual harm and the extent to which other factors contributed to the harm caused. In the footnotes of the guidelines a significant cause is one which more than minimally, negligibly or trivially contributed to the outcome. It does not have to be the sole or principle cause. Causation is an important consideration in health and safety offences in any sentencing exercise. If the breach caused the harm then the offence is likely to be viewed as more serious by the sentencing court. However the guidelines subsume this as an issue in that it is not dealt with as a distinct point but rather as part of the categorisation of the risk of harm.

It is of note that the guidelines go on to qualify its approach to significant cause by stating that the actions of victims are unlikely to be considered contributory events for sentencing purposes. It will therefore be difficult for a defendant to seek to establish that the actions of their employees should be taken into consideration. For example where an employee working at height on scaffolding falls due to the fact they chose not to wear personal protective equipment and an appropriate harness, despite this being company policy and the equipment being available. The guidelines go onto to say that offenders are required to protect workers or others who may be neglectful of their own safety in a way which is reasonably foreseeable. The case law pre-guidelines are littered with examples of how fines will not normally be reduced because an employee acted outside his instructions but the ability of the defendant to submit that the actions of an employee contributed to the harm caused cannot be entirely set aside. The court should not move up a category if actual harm was caused but to a lesser degree than the harm that was risked, as identified as per the table above.

Step two: Turnover

Once the court has determined the offence category it is required to determine the size of the organisation. This assessment is based on the annual turnover of the company. The guidelines define organisations by turnover or equivalent in the following way:

(a) Large: £50 million and over
(b) Medium: between £10 million and £50 million
(c) Small: between £2 million and £10 million
(d) Micro: not more than £2 million

The guidelines also refer to where a very large organisation's turnover very greatly exceeds the threshold for large organisations. In these cases it may be necessary to move outside the suggested range to achieve a more proportionate sentence. Very greatly exceeding £50 million is likely to be considered as a turnover or equivalent of £1 billion.

Once the size of the organisation is established there is starting point and category range of fine within each section for each level of culpability and harm category. Once these have been imputed the bracket of actual fine which the court will sentence from can be identified. The court should then consider further adjustment within the category range for aggravating and mitigating features. The guidelines contain a non-exhaustive list of factual elements providing the context of the offence and factors relevant to the offender. This list contains some matters that may have already been considered at the point that culpability was considered such as cost-cutting at the expense of safety. The guideline highlights that relevant recent convictions are likely to result in a substantial adjustment upwards.

Factors increasing seriousness

Statutory aggravating factor:

Previous convictions, having regard to a) the nature of the offence to which the conviction relates and its relevance to the current offence; and b) the time that has elapsed since the conviction.

Other aggravating factors include:

- cost-cutting at the expense of safety;
- deliberate concealment of illegal nature of activity;
- breach of any court order;
- obstruction of Justice;
- poor health and safety record;
- falsification of document or licences;
- deliberate failure to obtain or comply with relevant licenses in order to avoid scrutiny by authorities;
- exploitation of vulnerable victims.

Factors reducing seriousness or reflecting mitigation:

- no previous convictions or no relevant/recent convictions;
- evidence of steps taken voluntarily to remedy problem;
- high level of co-operation with the investigation, beyond that which will always be expected;
- good health and safety record;
- effective health and safety procedures in place;
- self-reporting, co-operation and acceptance of responsibility.

Steps three and four:

The court should 'step back', review and, if necessary, adjust the fine based on turnover to ensure that it fulfils its objectives of sentencing for the offences. The guidelines specifically state to check whether the proposed fine based on turnover is proportionate to the overall means of the offender. They go on to state that the fine must be sufficiently substantial to have a real economic impact which will bring home to both management and shareholders the need to comply with health and safety legislation. The court should examine the financial circumstances of the offender in the round to assess the economic realities of the organisation and the most efficacious way of giving effect to the purpose of sentencing. The guidelines go onto state that the court should have regard to the following specific factors:

- Profitability. If the organisation has a small profit margin relative to its turnover the Guidelines stipulate that downward adjustment may be needed and similarly upward adjustment may be necessary where organisation has a large profit margin.
- Quantifiable economic benefit derived from the offence, including through avoided costs or operating savings, should normally be added to the fine at step two.
- Where the fine will have the effect of putting the offender out of business, this will be a relevant consideration. The guidelines state that in some bad cases this may be an acceptable consequence.

When assessing the ability of an offender to pay any fine the court can order that the amount be paid in installments. There is no upper time limit for this and it can be deemed appropriate for an organisation to make payments over a number of years.

Step four requires the court to consider other factors that may warrant adjustment of the proposed fine. The court should consider any wider impacts of the fine within the organisation or on innocent third parties. The court should consider whether the fine impairs the offender's ability to make restitution to victims. It should consider the impact of the fine on the ability to improve conditions in the organisation to comply with the law. It should also consider the impact of the fine on employment of staff, service users and the local economy, but not shareholders or directors. The organisation may provide evidence to show how it will be affected, such as a letter from their accountants.

Where the fine will fall on public or charitable bodies, the fine should normally be substantially reduced if the offending organisation is able to demonstrate the proposed fine would have a significant impact on the provisions of its services.

Step five:

The guidelines at this point states that the court should consider any factors which indicate a reduction, such as assistance to the prosecution. This is assistance as an informer, pursuant to *ss* 73 and 74 of the *Serious Organised Crime and Police Act 2005*. It is not co-operating with the investigation itself, as is already dealt with in the list of mitigating factors at step two. It is difficult to foresee a situation when this is likely to occur in the everyday health and safety prosecution.

Step six: Reduction for guilty pleas

It is at this stage that the court is required to consider any potential reduction for a guilty plea in accordance with *s 144* of the *Criminal Justice Act 2003* and the guilty plea guideline. The reduction in sentence for a guilty plea has long been deemed appropriate as it avoids the need for a trial, allowing the case to be disposed of expeditiously, saving costs and witnesses and victims from having to give evidence. The Sentencing Guidelines Council Guideline, Reduction in Sentence for a Guilty Plea indicates that the level of reduction should be proportionate to the total sentence imposed, with the proportion calculated by reference to the circumstances in which the guilty plea was entered. The timing of the plea is deemed particularly important with the greatest reduction being given where the plea was indicated at the 'first reasonable opportunity'. In these circumstances this will be a recommended one-third.

Step seven: Compensation and ancillary orders

In all cases the court must go onto consider ancillary orders, including the following: compensation, remediation and forfeiture. In reality it is rare that these orders will be deemed necessary. In the case of compensation, it has long been recognised in the criminal courts that compensation is normally covered by insurance and should be dealt with through personal injury claims in the civil courts. Therefore orders are rarely made at this stage of proceedings.

Step eight: The principle of totality

Here the court should look at the total number of offences and consider whether the total sentence is just and proportionate to the offending behavior.

Step nine: Reasons

The court is under a duty to give reasons for, and explain the effect of, the sentence pursuant to *s 174* of the *Criminal Justice Act 2003*.

Individuals

[E15043.2] The guidelines in respect of health and safety offences committed by the individual mirrors the nine-step approach as set out above for sentencing organisations. For offences committed after 16 January 2009, individuals may be sentenced to imprisonment for health and safety offences, whether as employers or self-employed persons, for breaching one of the employer's general duties, either by virtue of secondary liability for a corporate health and safety offences under *s 26* or *s 27* of *HSWA 1974* or as an employee for a breach of *HSWA 1974 s 7*. It appears that *s 7* is not invoked as often as other offences, this may be due to the onus more readily falling on the organisation after investigation.

Step one: Culpability and harm

As with the guidelines for organisations the court must determine the offence category by making two separate assessments; culpability and harm. There are four categories for determining culpability:

Very High

Where the offender intentionally breached, or flagrantly disregarded the, the law.

High

Actual foresight of, or wilful blindness to, risk of offending but risk nevertheless taken.

Medium

Offence committed through act or omission which a person exercising reasonable care would not commit.

Low

Offence committed with little fault, for example, because:

- significant efforts were made to address the risk although they were inadequate on this occasion;
- there was no warning/circumstance indicating a risk to health and safety;
- failings were minor and occurred as an isolated incident.

Harm

The categorisation of harm is identical to that used for organisations as discussed above.

Steps two and three are also similar to those set out above for organisations. A table is provided to assist the court reach a sentence within a particular category range. These range from the most severe penalty of one to two years' custody down to a conditional discharge or a Band A fine. A Band A fine is 50% of relevant weekly income. The court should then consider further adjustment within the category range for aggravating and mitigating features. The aggravating features are identical to those in the organisation guideline, there are some additional mitigating factors that are relevant to the individual, such as mental disorder or learning disability, where linked to the commission of the offence.

Where the sentencing range includes a possible sentence of custody the guidelines state that the court should consider the custody threshold as follows:

- Has the custody threshold been passed?
- If so, is it unavoidable that a custodial sentence be imposed?
- If so, can that sentence be suspended?

Where the range includes a potential sentence of a community order, the court should consider the community threshold as follows:

- Has the community order threshold been passed?

The guidelines go onto state that even where the community order threshold has been passed, a fine will normally be the most appropriate disposal where the offence was committed for benefit.

In determining the level of fine, the court may conclude that the offender is able to pay any fine imposed unless the offender has supplied any financial information to the contrary. The guidelines state that if necessary, the court may compel the disclosure of an individual offender's financial circumstances pursuant to *s 162* of the *Criminal Justice Act 2003*. Where the court is not satisfied that it has been given sufficient reliable information, the court will be entitled to draw reasonable inferences as to the offender's means from evidence it has heard and from all the circumstances of the case which may include the inference that the offender can pay any fine.

At step three the court is required to step back and review the appropriate level of the fine, taking into consideration that the appropriate level of fine should be in accordance with *s 164* of the *Criminal Justice Act 2003*, which requires that the fine must reflect the seriousness of the offence and that the court must take into account the financial circumstances of the offender. In finalising the sentence the court should have regard to the factors relating to the wider impact of the fine on innocent third parties. Examples given are:

- impact of fine on offender's ability to comply with the law;
- impact of the fine on employment of staff, service users, customers and local economy.

Steps four and five are as per the organisation guidelines as discussed above.

Step six sets out that in all cases the court must consider whether to make ancillary orders. These are set out as per the organisation guidelines with the addition of the disqualification of a director. The guideline sets out that an offender may be disqualified from being a director of a company in accordance with s 2 of the *Company Directors Disqualification Act 1986* (see E15040 above).

Steps seven and eight are as per the organisation guidelines. Step nine is the consideration of time spent on bail. The court is required, pursuant to s *240A* of the *Criminal Justice Act 2003* to consider whether to give credit for time spent on bail.

Penalties for Corporate Manslaughter

The Definitive Guidelines for corporate manslaughter are incorporated into the new guidelines for health and safety offence that came into force on 1 February 2016 and are discussed above. They replace the guidelines that came into force on 15 February 2010 for corporate manslaughter and health and safety offences causing death. These guidelines addressed the question of seriousness following the guidance in the then leading case of *Howe*. They set out that where a conviction for corporate manslaughter is secured, the 'appropriate fine will seldom be less that £500,000 and may be measured in **millions of pounds**'. The guidelines went on to state that the 'range of seriousness in health and safety offences is greater than for corporate manslaughter. However, where the offence is shown to have caused death, the appropriate fine will seldom be **less than** £100,000 and may be measured in hundreds of **thousands of pounds or more**' (emphasis original).

The new guidelines are short in comparison to the health and safety offences guidelines. This is because in determining the seriousness of the offence each case will involve death and corporate fault at a high level. As is set out in the guidelines; by definition the harm and culpability involved in corporate manslaughter will be very serious.

Step one: Determining the seriousness of the offence

The court is required to assess factors affecting the seriousness of the offence by asking the following questions:

(a) How foreseeable was serious injury?
Usually, the more foreseeable a serious injury was, the graver the offence. Failure to heed warnings or advice from the authorities, employees or others to respond appropriately to 'near misses' arising in similar circumstances may be a factor indicating greater foreseeability of serious injury.

(b) How far short of the appropriate standard did the offender fall?
Where the offender falls far short of the appropriate standard, the level of culpability is likely to be high. Lack of adherence to recognised standards in the industry or the inadequacy of training, supervision and reporting arrangements may be relevant factors to consider.

(c) How common is this kind of breach in this organisation?
How widespread was the non-compliance? Was it isolated in extent or, for example, indicative of a systemic departure from good practice across the offender's operations or representative of systemic failings? Widespread non-compliance is likely to indicate a more serious offence.

(d) Was there more than one death, or a high risk of further deaths, or serious personal injury in addition to death?
The greater number of deaths, very serious personal injuries or people put at high risk of death, the more serious the offence.

The court must then decide whether this is a category A or a category B offence. This determination is simply to be made by assessing where the answer to questions (a)–(d) indicate a high level of harm or culpability within the context of the offence this is a category A offence and where there is a lower level of culpability it is a category B offence.

Step two: Turnover

Once the court has determined the offence category it is required to determine the size of the organisation. This assessment is based on the annual turnover. The new guideline provides starting points and category ranges as set out below:

Large organisations: turnover more than 50 million

Offence Category	Starting Point	Category Range
A	£7,500,000	£4,800,000–£20,000,000
B	£5,000,000	£3,000,000–£12,500,000

Medium organisations: turnover £10 million to £50 million

Offence Category	Starting Point	Category Range
A	£3,000,000	£1,800,000–£7,500,000
B	£2,000,000	£1,200,000–£5,000,000

Small organisations: turnover £2 million to £10 million

Offence Category	Starting Point	Category Range
A	£800,000	£540,000–£2,800,000
B	£540,000	£350,000–£2,000,000

Micro organisations: turnover up to £2 million

Offence Category	Starting Point	Category Range
A	£450,000	£270,000–£800,000
B	£300,000	£180,000–£540,000

As with the health and safety guideline in respect of organisations, it stipulates that where a defendant organisation's turnover or equivalent very greatly exceeds the threshold for large organisations, it may be necessary to move outside the suggested range to achieve a proportionate sentence. It is anticipated that this will be organisations with a turnover of 1 billion.

The offender will be expected to provide financial information, including comprehensive accounts for the last three years. If the court is not satisfied that it has been provided with sufficient reliable information, it will be entitled to draw reasonable inferences as to the offender's means from the evidence it has heard, which may include the inference that the offender can pay any fine.

It is of note that the guidelines state that normally only information relating to the organisation before the court will be relevant, unless it can be demonstrated to the court that the resources of a linked organisation are available and can properly be taken into account. The greater attention to the financial details of organisations that the new guidelines embrace may invite greater scrutiny in situations where the organisation does not stand alone and is part of a wider enterprise.

There is also specific guidance in the guidelines where the offender is a public body, local authority, health trust or charity. The guidelines stated that in respect of each of these, detailed analyses of specific expenditure or reserves is unlikely to called for unless there is a suggestion of unusual or unnecessary expenditure.

The court should then consider further adjustment within the category range for aggravating and mitigating features. The guidelines contain a non-exhaustive list of factual elements providing the context of the offence and factors relevant to the offender.

Factors increasing seriousness

Statutory aggravating factor:

Previous convictions, having regard to a) the nature of the offence to which the conviction relates and its relevance to the current offence; and b) the time that has elapsed since the conviction.

Other aggravating factors include:

- cost-cutting at the expense of safety;
- deliberate concealment of illegal nature of activity;
- breach of any court order;
- obstruction of Justice;
- poor health and safety record;
- falsification of document or licences;
- deliberate failure to obtain or comply with relevant licenses in order to avoid scrutiny by authorities;
- offender exploited vulnerable victims.

Factors reducing seriousness or reflecting mitigation:

- no previous convictions or no relevant/recent convictions;
- evidence of steps taken voluntarily to remedy problem;
- high level of co-operation with the investigation, beyond that which will always be expected;
- good health and safety record;
- effective health and safety procedures in place;
- self-reporting, co-operation and acceptance of responsibility;

- other events beyond the responsibility of the offender contributed to the death (however, actions of victims are unlikely to be considered contributory events. Offenders are required to protect workers or others who are neglectful of their own safety in a way which is reasonably foreseeable).

Steps four, five and six are identical to those set out above in respect of the guidelines for health and safety offences.

Step seven deals with ancillary orders where specific reference is made to publicity orders pursuant to *s 10* of the *Corporate Manslaughter and Corporate Homicide Act 2007*. A publicity order should ordinarily be imposed in a case of corporate manslaughter. It may require publication in a specified manner of the conviction. The object of a publicity order is deterrence and punishment. Any exceptional cost of compliance with such an order should be considered in fixing the fine.

Steps eight and nine are the same as those set out above in respect of the guidelines for health and safety offences.

Publication of convictions for health and safety offences and enforcement notices

[E15044] In an attempt to improve compliance with health and safety legislation, the HSE pursues an active policy of naming companies that breach the legislation. The HSE 'name and shame' website publicises all prosecution cases initiated by the HSE which resulted in a conviction as well as a register of improvement and prohibition notices issued by the HSE. This information can be accessed at: www.hse.gov.uk/enforce/prosecutions.htm. Generally, information relating to prosecutions and to the service of notices will remain on the public register for five years before being removed. Information relating to enforcement measures taken against companies will remain on the public register for longer.

Position of the Crown

[E15045] The general duties of *HSWA 1974* bind the Crown. [*HSWA 1974 s 48(1)*]. (For the position under the *Factories Act 1961*, see **W11031** WORKPLACES – HEALTH, SAFETY AND WELFARE.) However, improvement and prohibition notices cannot be served on the Crown, nor can the Crown be prosecuted [*HSWA 1974 s 48(1)*], although Crown employees can be prosecuted for breaches of *HSWA 1974* [*HSWA 1974 s 48(2)*]. Non-statutory procedures are in place for the issue of Crown improvement and prohibition notices, and for the censure of Crown bodies in circumstances in which a prosecution would otherwise have been brought. Crown censures are available to view at www.hse.gov.uk/enforce/prosecutions.htm.

Crown immunity is no longer enjoyed by health authorities, nor premises used by health authorities (defined as Crown premises) including hospitals (whether NHS hospitals or NHS trusts or private hospitals). [*National Health Service and Community Care Act 1990 s 60*]. Health authorities are also subject to the

Food Safety Act 1990. Most Crown premises can be inspected by authorised officers in the same way as privately run concerns, though prosecution against the Crown is not possible. [*Food Safety Act 1990 s 54(2)*].

Similarly, there is no Crown immunity under the *Corporate Manslaughter and Corporate Homicide Act 2007 (s 11(1))*.

Environmental Management

Mark Rutter

Introduction to environmental management

[E16001] The management of environmental performance is a critical issue for organisations in both the public and private sectors. Broadly speaking, environmental management refers to the controls implemented by an organisation to minimise the adverse environmental impacts of its operations. This includes the conservation of natural resources, minimising the production of waste, the protection of habitats and the control of hazardous substances. Historically, environmental management tended to be driven by a complex and interacting array of external pressures, to which organisations somewhat reluctantly responded. Now it is widely recognised that a positive and proactive approach towards environmental issues can benefit business performance.

Most large, as well as many medium and smaller-sized organisations, have responded by integrating policies that encourage good environmental management into their management systems and routines. External pressures are still important influences, but increasingly they tend to shape the nature and scope of environmental management practices, rather than triggering them in the first place. In addition, cooperative and constructive stakeholder dialogue is now commonly an element of effective and proactive environmental management. Furthermore, the ongoing development of new and increasingly innovative approaches to environmental management reflects the fact that there is commercial value to be gained from continuous improvement.

This positive approach to environmental considerations has become even more apparent in recent times, with the move towards a low carbon economy to combat fears of climate change. The most successful companies are likely to be those that show an early awareness of the risks and opportunities arising, and that face up to the changing economic values of businesses and assets under a carbon-constrained society. Even those organisations still refusing to face up to the threats of climate change, are now being subjected to numerous direct and indirect policies, laws and economic and fiscal instruments coming into force around the world. Parallels could be drawn with the recent banking and financial crisis, with the risk of a severe correction for those that have failed to respond to the impacts of conducting business in a low carbon world.

The sources of pressure on business to adopt more sustainable management practices can be divided into a number of areas:

- corporate social responsibility (CSR) expectations;
- management of information needs;
- employees;

[E16001] Environmental Management

- legislation;
- market mechanisms;
- the financial community;
- the supply chain;
- community and environmental groups;
- environmental crises;
- the business community;
- customers; and
- competitor initiatives.

Internal drivers

Corporate social responsibility (CSR)

[E16002] The concept of CSR, sometimes known as corporate responsibility (CR) has gradually crept into mainstream business practice, reflecting the changed conditions in which business operates. Environmental management is an integral part of effective CSR. The removal of trade barriers, the subsequent growth and political influence of multinational companies, and the opening of previously restricted markets have contributed to radical changes in the way business operates, including increasing the extent to which it controls its own performance. Consequently, the notion of CSR has also changed, with its increasing recognition by the public and its rise up the boardroom agenda. One reason for this has been a tendency for protest groups to target companies' wider impacts on global society, with environmental impact issues often to the fore. Society is looking less to government to control the social and environmental impacts of business activity, and instead is seeing business itself as being accountable for those impacts. Good corporate governance is no longer merely a reflection of responsible fiscal performance. As a result, the mandate of corporate directors and managers is expanding as they recognise the need to operate in a more transparent and inclusive manner. Proactive environmental management is a key aspect of that.

Governments have also taken an interest in CSR. In the UK, the Government first set out its interpretation of CSR in 2004, describing it as essentially the business contribution to sustainable development. It defined CSR as the action private and public sector organisations take voluntarily over and above the minimum legal requirements for social and environmental performance. This included environmental protection, as well as equal opportunities, employment terms and conditions, and health and safety. Business in the Community (BITC) took on the role of engaging organisations in CSR activities in 2007. BITC is a business-led charity with a membership of more than 800 institutions, which includes large multinational household companies, small local businesses and public sector bodies. It re-launched the new CR (Corporate Responsibility) Academy in 2008 as a 'one stop shop' offering training, support and advice on CSR for organisations of any size and sector to help incorporate CSR measures.

Another initiative is the BITCs' business-led CR Index. First launched in 2002, this has become one of the UK's leading benchmarks of responsible business. It is intended to help organisations to take a systematic approach to managing, measuring and reporting potential impacts arising from business operations, products and services. Such impacts may affect the local community, marketplace and workplace. The CR Index consists of an online survey that provides large companies with a framework for managing CR. This framework is designed to integrate and improve CR throughout the organisation by providing a systematic approach to managing, measuring and reporting on business impacts in society and on the environment. Originally participants were grouped into performance bands (Platinum, Gold, Silver and Bronze) according to the extent to which responsible practices are embedded within their corporate strategy and operations. The CR Index has now moved away from this approach and has replaced it with a five-star rating system, with half stars, to allow better differentiation of performance between companies and so improve comparability. The CR Index annual ranking is published on the BITC website. In 2014, the CR Index 2014 was made up of 98 companies, 30% of which were FTSE listed and nearly half reported on their global operations. Although sustainable production was identified as an area of significant progress, with 65% of participants taking steps to develop sustainable products and services, business processes or sourcing practices, the vast majority of investments focussed on responsible energy supply, rather than new products and services. Companies that are not ready to participate in the CR Public Index, can still use the management tool and benchmark themselves again others in the Private Index, but they will not be included in the annual CR Index public rankings.

Although some might predict a waning of interest in CSR during an economic downturn, the recent financial crisis has clearly shown that the business case is stronger than ever. As well as restoring public confidence in corporate activity, a recent study demonstrated a high correlation between responsible business practice and improved financial performance. The research, carried out by Ipsos MORI, showed that FTSE companies that actively managed their environmental and social impacts outperformed the FTSE 350 on total shareholder return by between 3.3% and 7.7% over the period 2002–07. Furthermore, returns from these companies recovered more quickly in 2009 after the dramatic downturn in 2008, compared with other FTSE listed companies.

The UK government is continuing to support CSR, and has stated that it is a key ingredient of a stronger, sustainable economy. It issued a consultation paper in 2013 to help inform the government's understanding of the importance of corporate responsibility to sustainable growth. From the 152 responses received from businesses and civil society it concluded that many UK-based companies are leading the way by putting CSR at the heart of their business plans. This it says not only leads to improved productivity by helping companies attract and retain the best staff, but also builds trust with investors and consumers.

At EU level, CSR first came under the scrutiny in the European Commission's Green Paper, published in July 2001. This document defined CSR in terms of both company internal management and its impact on society and argued for a European framework for CSR. The European Commission made a new commitment in March 2010 to renew the EU strategy for CSR. It issued a call for proposals aimed at encouraging the investment community to take more account of environmental, social and governance information, thereby increasing the incentives for businesses to incorporate more sustainable and responsible practices into their operations. This was followed by the publication of a new policy on CSR in October 2011, which stated that organisations should 'have in place a process to integrate social, environmental, ethical and human rights concerns into their business operations and core strategy in close collaboration with their stakeholders'. The new strategy was intended to encourage positive impacts, such as developing new products and services beneficial to enterprises and society, as well as preventing or minimising negative impacts. The Commission also put together an action agenda covering eight areas for 2011–2014. Included among these were measures such as enhancing the visibility of CSR, by for example creating an award, organising surveys on public trust in business, improving the self-regulation processes, and rewarding responsible business conduct through investment and public procurement measures.

An international standard providing guidelines for social responsibility was issued by the International Organisation for Standardisation (ISO) in November 2010. ISO 26000 is a voluntary standard suitable for public and private sectors in developed and developing countries. As it contains guidance, rather than strict requirements, it is not suitable for use as a management system standard or appropriate for certification purposes or regulatory use. The guidance is meant to assist organisations with contributing to sustainable development, and to encourage them to go beyond legal compliance. It is intended to complement other instruments and initiatives for social responsibility, rather than replace them.

Management information needs

[E16003] Effective business management relies on timely and reliable information on the multitude of factors that influence it. As managers' understanding of the relationship between environmental performance and business performance increases, so too does their requirement for information pertaining to environmental performance. Such information helps to increase their control over those factors.

This reflects the acceptance of the sustainability or sustainable development concept, which recognises that long-term business success requires environmental, social and economic factors to be balanced. While there is no clear guidance or agreement on how such a balance should be achieved, it is clear that it must be based on appropriate information on all three primary elements. Thus the recognition of the relevance of environmental performance to business performance, the value of controlling it, and the need for expanded management information, is an increasingly important driver of environmental management practices. Structured environmental management systems not only provide a means of controlling environmental performance *per se*, but also allow more informed strategic and operational decisions to be made by management.

Employees

[E16004] Employees have a potentially strong influence over the environmental management practices of an organisation. The desire to minimise staff turnover means that companies and other organisations have to be more responsive to employee enquiries and suggestions regarding environmental performance. Conversely, employers wishing to attract high calibre recruits must take on board the importance of maintaining a strong and positive corporate image, which is often dependent on environmental performance, amongst a number of other things. This highlights the need for a supportive organisation, where all employees are involved and where there is leadership from senior management, to ensure that the environmental management process becomes part of the work culture.

External drivers

Legislation

[E16005] UK companies are influenced by a range of international treaties, conventions and protocols; European regulations and directives; and domestic legislation. The latter may be a tool for implementing European directives, or they may have been enacted independently of any requirement of the European Union (EU).

Enforcement of the various legal instruments is primarily the responsibility of the Environment Agency within England, or Natural Resources Wales in Wales, although local authorities also play a role. In Scotland, responsibility lies with the Scottish Environment Protection Agency (SEPA).

Good corporate environmental management requires a thorough understanding of the legal requirements imposed on a company, in addition to evidence that the company has made reasonable attempts to ensure ongoing compliance with them, either through technological, procedural and/or administrative mechanisms. Many published corporate environmental policies now commit to going "beyond compliance", so that, legislative compliance is seen as the minimum standard.

This not only applies to large companies. It has often been assumed that due to their small size and lack of resources, few small and medium-sized enterprises (SMEs) would go beyond the bare minimum required to comply with environmental legislation. Indeed, a study published by the Institute of Environmental Management and Assessment (IEMA) in 2011 found low levels of knowledge, understanding and compliance with environmental legislation among SMEs. This was despite clear evidence that the impact on SMEs of complying with environmental legislation is over-estimated, and the high risk that they pose collectively to the environment as result of non-compliance. Other recent studies have shown that although most small businesses are engaged in some environmental initiatives, there is considerable variation between businesses. In particular, it was apparent that the smallest companies believed that there were significantly fewer benefits to be gained through engagement with environmental issues.

A Spanish study published in the *Journal of Environmental Management* found that a significant proportion of SMEs have proactive environmental policies. Their small size means that they have shorter lines of communication, close personal links, less bureaucracy and the ability to initiate change quickly. Moreover, the study went on to conclude that those with the most proactive environmental policies also had the best financial performance. Another study published in this Journal revealed that the engagement of Dutch SMEs in environmental management practices was influenced by the perceived financial benefits from energy conservation and innovative nature of the company, as well as its size.

Other initiatives have forced companies that have not voluntarily responded, to give more systematic consideration to environmental management and performance in making strategic and operational business decisions. Changes to company law have had an impact on business, and it is highly possible that further amendments will introduce new requirements in the future.

The latest Companies Act, which received Royal Assent in November 2006, replaced the Companies Act 1985 and introduced wide ranging changes to corporate responsibility and non-financial reporting. It requires that company directors act in a way most likely to promote the success of a company for the benefit of its shareholders. Implicit in this is that due regard should be taken of the impact of operations on the community and the environment, as these factors have the potential to significantly affect the long-term interests of a company.

Some companies have been obliged to prepare a Business Review from April 2005 as part of an 'enhanced directors' report' under the *EU Accounts Modernisation Directive (2003/51/EC)*. This Directive requires directors of quoted and large private companies to report on business practice to the 'extent necessary' for an understanding of a company's development, performance or position, in the Business Review. This requirement includes the disclosure of significant non-financial matters through the use of key performance indicators such as the impact of the company on the environment and the interests its employees. However, the Directive did not introduce a duty to discuss and assess future impacts. Medium private companies must also produce a Business Review, although they are not bound to produce performance indicators relating to non-financial information.

The Companies Act 2006 extended the scope of the Business Review for quoted companies. It must now be a forward-looking narrative to inform shareholders, and help them assess how the directors have performed their duty to promote the success of the company, and should include the company's environmental impacts and the risks. Despite the inclusion of these provisions, environmental groups were disappointed there was no statutory requirement for environmental reporting in the Act.

In 2011 BIS consulted on possible changes to the narrative reporting process to try to simplify company reports and make the information contained within them more accessible to shareholders. It suggested replacing the Business Review and Directors' Report with a Strategic Report and an Annual Directors' Statement. Vital environmental and social information would be included in the Strategic Reports of quoted companies, alongside financial information and a forward-looking analysis of future risks and opportunities. Companies would be encouraged to include additional voluntary disclosure on social and environmental matters in the new, more prescriptive Annual Directors' Statement. It also called for the environmental and social impacts of companies to be included in core business planning, and for the quality of reporting in these areas to be improved, in response to what it said was increasing interest from investors, consumers and wider society in these issues. Further consultation in 2012 led to the coming into force of the Companies Act 2006 (Strategic Report and Directors' Report) Regulations 2013 (SI 2013 No 1970) on 1 October 2013.

The original proposal to replace the Directors' Report was not implemented, and companies continue to prepare a Directors' Report as before. The regulations did though require the Business Review to be separated out from the rest of the Directors' Report as a Strategic Report, in which companies must divulge key environmental and social information, alongside information on the business model, and a forward-looking analysis of future risks and opportunities. In practice however, this resulted in little change in the information to be included, as these are already required by the UK Corporate Governance Code (see below), although not by the previous Companies Act. There was a new duty though for quoted companies to disclose some human rights and diversity information, and community issues. The 2013 regulations also incorporate provisions forcing all companies listed on the London Stock Exchange to report their greenhouse gas emissions in the Strategic Report, a measure first announced by the Government in June 2012. Directors have to measure, or calculate, and report on emissions of the six greenhouse gases covered by the Kyoto Protocol from those activities listed in the regulations. The measure is due to be reviewed shortly to decide whether to extend the approach to all large companies from 2016.

The *UK Corporate Governance Code* and the associated guidance replaced the amended version of the 2008 *Combined Code on Corporate Governance* in May 2010. The new Code contained broad principles and set out standards of good practice for FTSE 350 companies, as well as their effectiveness, remuneration, accountability and relations with shareholders. It included specific requirements for disclosure, as well as obligating companies to comment on how they have applied the Code in their annual report and accounts. Listed companies had to make a disclosure statement in two parts in relation to the Code. In the first part of the statement, the company was required to report on how it applies the main principles in the *Corporate Governance Code*. In the second part of the statement the company had either to confirm that it complies with the Code's provisions or, where it does not, to provide an explanation.

A review of the Code is conducted every two years to assess how companies are implementing it and to ensure that it adequately reflects the role of the board. The latest version of the Code was published by the Financial Reporting Council in September 2014 and applies to accounting periods beginning on or after 1 October 2014. It is applicable to all companies with a Premium listing of equity shares regardless of whether they are incorporated in the UK or elsewhere.

Principle C.2 of the 2014 Code states:

> 'The board is responsible for determining the nature and extent of the principal risks it is willing to take in achieving its strategic objectives. The board should maintain sound risk management and internal control systems.'

Such controls could include environmental issues. A new provision (C.2.2) introduced into the 2014 Code requires a broader statement about the board's reasonable expectation as to the company's viability based on a robust assessment of the company's principal risks and the company's current position. In addition, Supporting Principle E.1 on dialogue with shareholders now makes it the responsibility of the chairman to ensure that all directors are made aware of shareholders' concerns, and that all shareholders have equal access to information.

Guidance on assessing how the company has applied Principle C.2 is provided in an annexe to the Code. Originally published by the Institute of Chartered Accountants in England and Wales' Internal Control Working Party in 1999, this guidance has become known as the 'Turnbull Guidance', after the Working Party's chairman. Revised Turnbull Guidance on Internal Control was published in October 2005. It set out some questions that the board should consider and discuss with Management when reviewing reports on internal control. One of these questions asked whether the company communicates clearly to its employees what is expected of them and the scope of their freedom to act in a number of areas, including environmental protection. Alongside its September 2014 updates to the Code, the FRC issued new guidance to update the Turnbull Guidance and combined it with further guidance on going concern and liquidity risk. This new guidance was applicable for periods commencing on or after 1 October 2014.

At EU level there are a number of on-going developments that apply to corporate disclosure. A revised Accounting Directive (2013/34/EU), which entered into force on 20 July 2013 and which must be implemented by 20 July 2015, should lead to greater standardisation of reporting across member countries. This Directive merged the Fourth (78/660/EEC) and the Seventh (83/349/EEC) Accounting Directives on Annual and Consolidated Accounts, and consolidated several amendments to them. The requirements of the new Directive will apply to all companies listed on an EU regulated market. A further Directive (2014/95/EU) relating to disclosure of non-financial and diversity information by certain large undertakings and groups entered into force on 6 December 2014 and amends Directive 2013/34/EU. It requires companies to disclose in their management report, information on policies, risks and outcomes regarding environmental, social, human rights and employee matters, as well as anticorruption and bribery issues, and diversity in their board of directors. The new Directive applies to some large companies with more than 500 employees, including listed companies, public-interest entities such as banks, insurance companies, and other companies designated by an EU member country on the basis of their activities, size or number of employees. There is considerable flexibility however in the way that companies will be able to disclose the relevant information, with international, European or national guidelines all acceptable. EU Member States must transpose the directive into national legislation by November 2016. Although some countries, such as the UK, already have disclosure requirements on environmental issues that go beyond the new EU legislation, the European Commission estimates that fewer than 10% of the largest companies across the EU currently disclose the information necessary to comply with the new directives. Therefore, other EU countries could potentially be affected significantly, although some skeptics doubt the efficacy of the new rules. Citing opposition from some Member States during the passage of the legislation into EU law, they claim that the majority of large EU companies will be able to opt out of the new disclosure requirements.

Market mechanisms

[E16006] While the command and control approach embodied in early environmental legislation and regulations represents a significant pressure on business, policy makers have introduced new tools to encourage better management of environmental performance.

A range of measures has emerged designed to influence the economics of polluting activities. These so-called market-based or economic instruments impose costs on pollution-causing activities and provide incentives for companies to look for ways of minimising environmental damage. Further recent concerns over environmental damage, particularly in relation to greenhouse gas emissions, have been reflected in the Government's increasing focus on the use of fiscal instruments.

Such instruments include:

(a) Climate Change Levy
 The climate change levy ('CCL'), which first came into effect in April 2001, is a tax on the business use of fossil fuels and applies to energy supplied to the industrial, commercial, agricultural and public service sectors. It is designed to encourage energy conservation and a switch to cleaner fuels and renewable energy sources to help the UK reduce its greenhouse gas emissions. All revenues are recycled back to business through a 0.3 per cent cut in employers' National Insurance contributions and additional support for energy-efficiency measures and energy-saving technologies. The levy is applied as a specific rate per nominal unit of energy, depending upon which category the energy supplied falls into.

(b) Landfill Tax
 The Landfill Tax was introduced in October 1996 to encourage companies to reduce their volume of waste produced and since this time has increased greatly. The rate of tax in 2014–15 was £80 per tonne, and this increased to £82.60 from April 2015. The rate applying to inactive or inert waste increased from £2.50 per tonne in 2014–15 to £2.60 per tonne from 1 April 2015. The tax is intended to be revenue-neutral to business, with tax receipts going to fund the Land-fill Communities Fund (formerly the Landfill Tax Credit Scheme). This enables landfill site operators to claim tax credit for contributions they make to approved environmental bodies for spending on projects that benefit the environment.

(c) Aggregates Levy

An aggregates levy on extracting virgin aggregates, mainly sand, gravel and rock, was launched in 2002. It is was introduced to ensure that the environmental impact of aggregate extraction is reflected in the price, and to encourage more efficient use of aggregates and the development of alternatives such as waste glass, tyres and recycled construction and demolition waste. All the revenue raised is returned to business and the local communities affected by quarrying, through a 0.1 per cent cut in employers' National Insurance contributions and a sustainability fund. Between April 2002 and March 2008, the rate was £1.60 per tonne. This was increased to £1.95 per tonne from 1 April 2008, and to £2 per tonne from 1 April 2009, and has remained at this level since. Originally there were certain exemptions and reliefs from the levy in place eg for by-products resulting from extraction. These were suspended or limited from 1 April 2014 pending the results of an investigation by the European Commission into whether they gave rise to State aid.

(d) Vehicle and Fuel Duty

Lower levels of duty on cleaner fuels and changes to company car taxation and vehicle excise duty (VED) are designed to reduce pollution, in particular carbon dioxide, from vehicles. VED rates for cars registered on or after 1 March 2001 depend on CO_2 emissions and fuel type and are split into 13 bands. While there is no VED for vehicles with low CO_2 emissions (up to 100 g/km), the rate for the most polluting cars has increased significantly in recent years. For example the rate of VED for cars emitting over 255 g CO_2/km was £500 from 1 April 2014. In addition, differential First-Year Rates of VED were introduced to all new cars from 1 April 2010. With effect from April 2014, the nine highest bands were liable for additional VED, ranging from an extra £130 to an extra £1,090 for vehicles with emissions greater than 131 g CO_2/km and 255 g CO_2/km, respectively. The rates of VED for alternative fuel cars are £10 per year less than petrol and diesel cars in the equivalent band. There are also differential rates of VED for light goods vehicles, designed to encourage the use of more fuel-efficient transport.

(e) Emissions Trading

Companies that reduce their emissions below a quota can sell the unused part of their quota to other firms, thus providing an incentive to improve emissions performance. Such a scheme is included in the Kyoto Protocol, which was adopted in 1997 and came into force in February 2005. It sets targets for cuts in greenhouse gas emissions by developed countries. Article 17 of the Protocol allows developed countries that reduce their emissions by more than their assigned target to gain credits, which can be sold to other developed countries. EU wide emissions trading between companies started in January 2005 under a European Directive. The scheme has since been modified, with a second phase beginning January 2008, and lasting until December 2012. Phase III of the EU Emissions Trading System (previously known as the Emissions Trading Scheme) started on 1 January 2013 and will run to 31 December 2020. It introduced major changes including, additional greenhouse gases and emission sources, such as those from aviation, and has a more ambitious EU-wide cap on emissions.

(f) Air passenger duty (APD)
APD was originally introduced in 1994 as a flat charge of £5 or £10 per passenger depending on the class of seat purchased. In response to increasing concerns over the adverse impact of aviation on climate change, the rate of APD was increased significantly from 2007. On 1 November 2009, APD was structured to incorporate two rates according to the class of travel, and four distance bands (A to D), set at intervals of 2,000 miles from London, so that those flying further paid more. However successful lobbying from industry has seen a reduction in APD in certain cases. From 1 April 2015 there were only be two bands. Band A (0–2,000 miles from London) remained at £13 for the lowest class of travel £26 for other classes. All other destinations will be merged into band B with rates of £71 for the lowest class of travel £142 for other classes. A higher rate of APD applies to flights aboard aircraft of 20 tonnes and above with fewer than 19 seats. From 1 May 2015, children under the age of 12 years on the date of the flight in the lowest class of travel were exempt from APD. This exemption will be extended to children under the age of 16 from 1 March 2016. From 1 November 2011, direct long-haul rates for departures from Northern Ireland (bands B) were reduced to the short-haul rate (band A), irrespective of the destination.

The financial community

[E16007] The emerging realisation of the link between environmental performance and business performance has encouraged investors, shareholders and insurers to develop a direct interest in the environmental performance of companies. Investors are increasingly concerned that companies that fail to manage their social and environmental risk exposure will suffer. It is now standard practice for the financial community to seek information on how environmental issues will potentially impact on the long-term viability of the companies in which they have a commercial interest. In particular, they are concerned about the extent to which environmental risks are being controlled. Brand reputation accounts for an increasing proportion of stock market valuations. Therefore, they want reassurance that companies are not in breach of legal requirements with the consequent threats of fines, damage to reputation and the need for unanticipated expenditure. They need to be sure that assets, in the form of plant, equipment, property and brand value, against which they have lent money, are correctly valued. Raw materials, by-products and end products may need to be replaced, modified and/or discontinued, which may require provisions or contingent liabilities to be included in the corporate accounts. Such a situation may arise as a result of substances being phased out (for example, through legal controls on greenhouse gases), or it may reflect changing market attitudes such that the demand for environmentally damaging products begins to decline. Additional research and development costs are likely to be associated with such changes.

Consideration of environmental risks and performance is now an established component of acquisitions, mergers, flotations, buyouts or divestments. Management of companies involved in any of these deals must be able to demonstrate that environmental liabilities do not constitute an unacceptable risk for investors or insurers. Companies that emphasise their environmental performance are viewed by the City as forward-looking and actively managing their reputation. An increasing number of companies are also paying greater attention to their indirect impacts, such as the environmental implications of investment decisions.

Arguably the most striking example of progress in this area is the rapid growth in socially responsible investment (SRI). The Dow Jones sustainability indexes and FTSE4Good are two examples of indexes designed to track stocks and measure the performance of companies demonstrating strong environmental, social and governance practices for the purpose of SRI. They include only those companies that meet globally recognised corporate responsibility and environmental performance criteria.

Results from the BITC CR (see above) index have shown consistently that companies can gain competitive advantage and shareholder value by moving ahead of legislation in managing their environmental impacts. Furthermore, progressive organisations are now thinking strategically on environmental matters, such as supply chain, climate change and water consumption.

Hermes, the independent fund manager has published 'The Hermes Principles', setting out ten investment principles to address what owners should expect from UK public companies and what these companies expect from their owners. The Hermes Principles are designed to encourage companies to communicate clearly the plans they are pursuing and the likely financial and wider consequences of those plans. Principles 9 and 10 deal with social, ethical and environmental issues, and call for companies to manage effectively relationships with their employees, suppliers and customers, stating that they should behave ethically and have regard for the environment and society as a whole. They also require companies to support voluntary and statutory measures designed to minimise the externalisation of costs to the detriment of society.

The financial consequences of climate change have rapidly become a priority for many companies and investors. The risks, as well as the opportunities, posed by climate change are widespread and varied. They include physical risks such as asset damage and project delays resulting from changing weather patterns. Regulatory risks resulting from tighter legislation on greenhouse gas emissions have also come to the fore. In addition, there are possible competition risks due to a decline in consumer demand for energy-intensive products and a rise in costs for energy intensive processes. Risks to brand and reputation can arise through a perceived lack of action to implement measures to help counter climate change.

Several companies have been subjected to a boycott by environmental groups or have come under pressure from shareholders for their environmental policies. For example the oil company Exxon Mobil has suffered in this respect because of it opposition to the Kyoto Protocol. It has also experienced hostility from some investors and mainstream financial analysts concerned that the companies' environmental stance threatens shareholder value. As a result, a number of resolutions have been lodged at annual shareholder meetings urging the company to improve its environment disclosure and performance. Similarly, in 2012 the multinational engineering group Siemens faced a series of hostile shareholder resolutions, mainly from German activist groups, over its involvement in the in a large dam project in Brazil.

The Carbon Disclosure Project (CDP) works with the world's largest collaboration of institutional investors to annually survey 500 of the world's largest publicly listed companies (FTSE Global Equity Index Series) on the implications of climate change to their business. It collates and makes publicly available greenhouse gas emissions and climate performance data obtained from these companies, as well as advising investors on the risks and opportunities presented to them by climate change. This process is also aimed at driving improvements through more informed shareholder engagement. The CDP is now established as the gold standard for carbon disclosure methodology.

Since its earliest inception in 2003, the CDP survey has found a significant increase in the amount of climate change-related information communicated to investors and says that the majority of the Global 500 companies that reported cited regulation as a key risk factor. It believes corporations are making significant progress in understanding and disclosing the risks and opportunities associated with climate change, and that there is a growing determination to act on this. In April 2011, the CDP announced a new initiative of requesting detailed climate change action plans from the world's highest emitting companies. The Carbon Action scheme is aimed at getting organisations to deliver emissions cuts, and to identify and implement initiatives that give a positive return on investment. The CDP is trying to encourage companies to go beyond disclosure by providing information on how they are attempting to reduce their greenhouse gas emissions, as well as getting them to set a public emissions reduction target. The number of companies targeted in the fourth annual Carbon Action initiative undertaken in 2014 increased 7% from the previous year to 322, although the number reporting emissions reduction projects did not increase correspondingly. There was a positive trend though in the number of companies establishing emissions reduction targets, with an increase of 3% over three years to 79% of responding companies. Furthermore, for companies reporting all required data for projects, emissions reductions were increased by 8%, and investments in emissions-reduction projects were reported to have increased by 18%. However, 21% of responding companies still did not adopt any absolute targets. The Carbon Action survey was sent to over 1,300 companies covering 17 high emitting industries in 2015 to request various levels of information depending on the degree of previous engagement with the CDP.

A set of principles, known as the Equator Principles (EPs), has been developed for use by financial institutions when providing loans for large projects. These Principles are intended to ensure that projects being financed are developed in a socially responsible manner that reflects sound environmental management practices. They consist of a voluntary commitment based on the International Finance Corporation (IFC) performance standards on social and environmental sustainability and on the World Bank Group's environmental, health and safety general guidelines and provide the framework for identifying the issues that banks and borrowers need to be aware of. Most UK, and many other international banks have adopted the EPs, recognising their use in helping them to document and manage risk exposure. The Equator Principles Association launched a strategic review in October 2010 to look at the scope, reporting and transparency, and governance of the EPs. A revised framework (EP III) was finalised and published in 2013 and became mandatory for all new transactions from 1 January 2014. By the beginning of 2015, the EPs had been officially adopted by 79 financial institutions in 34 countries.

The United Nations launched its Principles for Responsible Investment (PRI), backed by world's largest institutional investors, in April 2006. This contains six overarching Principles relating to 35 possible actions that institutional investors can take to integrate environmental, social and corporate governance considerations into their investment decisions. Within these voluntary Principles, there is a recognition that while the global economy is driven by financial demands, better long-term investment returns and more sustainable markets are achieved through taking account of environmental and social considerations. By April 2015, around 1,325 investment institutions with around US$45 trillion of assets under management had signed up to the PRI. The PRI Initiative was created after the launch of the Principles to help investors to implement the Principles. The Initiative supports investors by sharing best practice, facilitating collaboration and managing a variety of activities.

Supply chain

[E16008] In recent times focus has gradually shifted so that the environmental credentials of organisations' supply chains are under greater scrutiny. Most companies are both purchasers and suppliers of a range of goods and services. Introducing environmental criteria into procurement decision-making processes emphasises the importance of issues other than price and quality in purchasing goods or services. Examples of environmental criteria are selecting materials, components or products that were manufactured using relatively less energy than alternatives, or that require relatively less energy in operations, or the substitution of chemical substances of concern with safer alternatives. Companies have worked to achieve more sustainable and ethical supply chains because it is in their long-term interests.

One example of the successful use of this scheme to make savings and improvement across the whole supply chain is the strategy employed by PepsiCo UK & Ireland. The company implemented a five year plan to reduce by 50% the carbon and water impacts of the oats, apples and potatoes from the UK that make their Quaker Oats, Copella Apple Juice and Walkers Crisps. It worked with farmers to identify their carbon 'hotspots' where the most efficient emissions reductions could be achieved. Another initiative was to use only 100% British potatoes to reduce food miles. In just two years, the carbon footprint of Walkers crisps was improved by 7%. Moreover, the measures resulted in savings of around £400,000.

The CDP Supply Chain Leadership Collaboration (SCLC) was launched in October 2007, in partnership with Wal-Mart. It is aimed at encouraging organisations to use a common methodology to measure and manage their supply chain emissions. Data is collected by incorporating an additional questionnaire on carbon emissions and climate strategy of the supply chain into its annual survey of major corporations. One of the main objectives of the SCLC was to better understand how supply chain companies were considering climate change and how they were working to reduce their greenhouse gas emissions.

The CDP estimate that over 50% of an average corporation's carbon emissions come from the supply chain. Furthermore, the results of its research, published in January 2013, identified that suppliers were considerably less prepared than their customers in responding to the challenges presented by climate change. Only 38% set emission reductions targets in comparison to 92% of purchasing companies. This gives rise to great vulnerability in the supply chain, which could lead to sudden and unexpected procurement problems at any time.

Pressure from government is also influencing the use of environmental considerations in procurement. Given that public authorities in EU countries spend trillions of euros annually, equivalent to around 16% of the whole region's GDP on procurement, the European Commission sees public purchasing as a useful policy tool for tackling environmentally damaging products and services. The process began in 2004 with the adoption of two directives aimed at simplifying and modernising existing European legislation on public procurement. In contrast with the earlier EU Directives governing procurement, Directive 2004/18/EC relating to public works, supply and service contracts, and Directive 2004/17/EC on procurement procedures in the water, energy, transport and postal services sectors, contain specific reference to the possibility of including environmental considerations in the contract award process. The European Commission's handbook for green public procurement (GPP) was launched at the same time in response to this revision of EU public purchasing rules. A second edition of the handbook, entitled 'Buying Green!', was published in October 2011. This handbook now forms the Commission's main guidance document to aid public authorities to buy goods and services with a lower environmental impact. It also contains sector specific GPP approaches for buildings, food and catering services, electricity and timber and acts as a reference for policy makers, and businesses responding to green tenders. The Commission also set up a help desk for GPP to promote and disseminate information about GPP and to respond to enquiries from stakeholders. There are currently 22 categories of products, services and works of EU GPP criteria published, each of which is comprised of recommended technical specifications, award criteria and contract clauses. These criteria form the basis for many sets of national GPP criteria across EU countries.

With an annual budget in the region of £150 billion, public sector goods and services in the UK has the potential to transform markets, encouraging much greater participation by householders and the private sector. In response to an EU initiative, the UK Government published a draft national action plan (NAP) on GPP 2006. Entitled 'Procuring the Future', it provided an analysis of the barriers to sustainable procurement and made six key recommendations, alongside details of the actions that should be taken, with clear target dates for the future. It concluded that the UK would gain significantly from being a leader in sustainable procurement. Among the benefits it said, would be better stewardship of taxpayers' money, environmental and social improvements and more support for environment-friendly technologies. The current Government policy follows EU recommendations that 50% of all tendering procedures should be green. While GPP is wholly voluntary, there are mandatory (and voluntary) Government Buying Standards (GBSs) in place for all central government departments and related organisations, which buy the equivalent of 9% of the UK's GDP. The GBS consist of a set of product specifications for public procurers. They are managed by DEFRA, with individual standards developed with input from across government and industry, and subject to review to ensure that they take account of the long-term cost effectiveness and state of the market. The government is currently working to try to ensure that the GBSs include and, where appropriate, match the relevant GPP criteria.

GPP has made some in-roads into the private sector in recent years. Research carried out by the Environment Agency in 2009 found that 95% of the UK's large construction firms give preference to subcontractors who can prove their environmental credentials. A large majority of these firms stated that they have more confidence in subcontractors with proven green policies and procedures in place, as they believed there was a reduced risk of prosecution. Many also believed that green policies would save subcontractors money. It is also quite common for purchasing companies to require their suppliers to demonstrate ongoing compliance with formal environmental management standards such as ISO 14001 or EMAS. Increasing green procurement opens up new opportunities for those with suitable green products and services. Conversely, it also poses a threat to traditional products and services that do not take into account environmental impacts.

The EU Eco-label voluntary award scheme has been in operation since 1993, when the first product groups were established, and was comprehensively revised in 2000. An Eco-label (the 'Flower symbol') is awarded to products possessing characteristics that contribute to improvements in environment protection. Ecolabel criteria are based on the impact of the product or service on the environment throughout its life-cycle, from raw material extraction through to production, distribution and disposal. The main objective of the scheme is to encourage business to market greener products by providing information to allow consumers to make informed environmental choices when purchasing. Although there has been a steady increase in both the number of applications from manufacturers and the number of eco-labelled products marketed since the inception of the scheme, the European Commission is continually striving to improve it.

In December 2006, it consulted on a range of issues including how the scheme could be improved to increase its uptake and what other product groups could be included. The consultation was also intended to collect views on how the scheme could be used to improve 'green' procurement and to support other environmental measures operated by the Commission and EU countries, such a the Eco-design Directive setting requirements for energy related products. The outcome of this was the replacement of *Regulation (EC)* No 1980/2000 by *Regulation (EC)* No 66/2010. This new Regulation was intended to increase the scope of product groups included in the scheme. It was also designed to speed up the development process by simplifying the assessment procedure and criteria documents and to provide better guidance for GPP. The Commission began reviewing the Eco-label Scheme again in 2014. In a two-stage process it is looking at the efficiency and effectiveness of the scheme across EU countries, and how it fits in with other policies. Secondly, it is assessing how the programme might perform in the future, and whether an alternative scheme will be required.

In the UK, the Carbon Trust has devised a label to provide information on a product's carbon footprint across its life cycle. As well as providing information to purchasers, it allows producers to demonstrate commitment to managing and reducing the carbon emissions of their products and services. The carbon reduction label is supported by a number of documents, including PAS 2050 - the standard method for the measurement of the lifecycle greenhouse gas emissions of goods and services. Originally published in 2008, this standard was developed by the BSI to be applicable to a wide range of sectors and product categories. It was updated in 2011, and published alongside 'The Guide to PAS 2050:2011' setting out how to carbon footprint products, identify hot spots and reduce emissions in the supply chain. The Guide is available to download free from the BSI website. A standard for sustainable procurement, *BS 8903: Principles and Framework for Procuring Sustainability*, was launched in 2010. This standard provides guidance to all sizes and types of organisation on good practice for adopting and integrating sustainable procurement principles and practices. It covers all stages of the procurement process and is applicable across industry, public, private and third sector organisations. Implementation of BS 8903 will enable a uniform approach to reporting and measuring sustainable procurement throughout the supply chains. It will also allow more accurate benchmarking of performance with others, therefore providing an opportunity for profile-raising and increasing competitive advantage.

Community and environment group pressure

[E16009] Public pressure on companies to improve their environmental performance is usually initiated by local communities that experience the direct effects of pollution. Most companies recognise the importance and value of working cooperatively with local communities. In fact many have established community liaison panels, which comprise representatives of the local community. Such panels interact with the company on a regular basis and provide input on environmental and other community issues.

Environmental pressure groups, often supported by a high level of legal and technical expertise, are also influential. Originally the relationship between environmental pressure groups and business was characterised by mutual mistrust and reactive criticism. The emergence of the concept of 'stakeholder engagement', whereby companies take a more inclusive approach to business management, means that environmental pressure groups can be expected to have more direct access to companies and management nowadays. Their opinions are actively sought by organisations that recognise the importance of constructive dialogue. While companies will not necessarily implement all suggestions made by external pressure groups, the trend towards more timely and constructive dialogue is likely to continue.

An example of the potential for corporate reputational damage that could occur when a company fails to engage with environmental pressure groups is Greenpeace's Stop Esso campaign. This campaign targeted Esso's parent company ExxonMobil worldwide, for its negative stance on climate change and lack of investment in renewable energy sources, by calling for a boycott of the company's products. Greenpeace claimed that its boycott campaign had resulted in around one million motorists, a quarter of the number of regular buyers of Esso petrol, boycotting its filling stations. This campaign also played a large part in Deutsche bank issuing a statement saying that Exxon Mobil was being labelled as 'environmental enemy number one' and that this posed a significant risk to its business.

Another tactic of pressure groups is to target customers of companies supplying goods or services that damage the environment. This is often seen as more effective when companies are proving difficult to engage with, particularly when the supplier is a private company, and therefore not accountable to shareholders. Greenpeace's lobbying of McDonald's and other key customers led to the signing a moratorium on Amazonian soya, which in turn influenced the suppliers such as Cargill to address the issue of illegal deforestation in the Amazon.

Legislation, for example EU Directive implementing the first objective of the Aarhus Convention concerning public access to information and the *Freedom of Information Act 2000*, has given the public stronger rights to access information on companies' impacts on the environment in recent times. Although this legislation only provides a right of access to recorded information held by public authorities, many industrial projects that could potentially affect the public would be included, thereby extending the rules to cover much more environmental and health and safety information. However, legally classified and commercially confidential information can still be excluded. The result is that pressure on companies to improve environmental performance has increased and this has spread into the international arena. Companies are finding it necessary to develop and adopt consistent environmental and social performance standards in all markets in which they operate, since their performance is increasingly subject to international scrutiny. The evolution of the internet, and in particular digital communication and social media, has significantly increased public awareness of and access to information about corporate environmental performance, as well as increasing the speed with which such information can be transferred and responded to.

The consultancy Accountability, whose members include leading companies, and public organisations and service providers, has developed the AA1000 series of standards designed to help organisations become more accountable, responsible and sustainable. These include a Stakeholder Engagement Standard (AA1000SES), which can be used by companies to improve and assure the quality of its communication with stakeholders. The AA1000SES is valid for a range of engagements, whatever their size, including customer care and human rights. It allows stakeholders to assess and comment on the quality of an engagement using set principles. Originally published in 2008, the engagement standard was updated in 2011. AA1000SES (2011) consists of four parts. These include a description of the standard's purpose and scope; how to integrate a commitment to stakeholder engagement in strategy and operations; how to define the purpose, scope, and stakeholders of the engagement; and what a quality stakeholder engagement process looks like. The new standard is part of a suite of stakeholder engagement tools that includes the AA1000SE Manual.

Environmental crises

[E16010] In many cases, high profile environmental crises are a catalyst for improved environmental management. There are two main types:

(a) crises generated as a result of the actions of an individual company, the effects of which are generally experienced at a local level; and
(b) crises generated by collective action or by natural forces, the effects of which are often experienced at the national or international level.

Individual companies that have been associated with environmentally damaging events such as oil spills generally find themselves exposed to intense pressure to improve their environmental management practices in the immediate future. The need to correct the damage to reputation caused by environmental crises is a further incentive to respond quickly. Obviously the costs of responding to pressures arising from catastrophic environmental incidents can be extremely high, and most companies seek to avoid those by incorporating environmental issues into their corporate risk management programmes. However, the impacts, in terms of clean-up costs, compensation, reputation damage and share price of such events can go far beyond those foreseen. This was the case with the BP Deepwater Horizon oil-rig explosion in 2010 and the devastating effects of the subsequent pollution. The situation was made worse for the company by its initial failings to manage its communication with outside agencies and the public, and was seen by many as a PR catastrophe.

Global or national environmental crises tend to emerge more gradually, and with considerably more debate about accountability and appropriate responses. Nevertheless, there are a number of examples of global environmental issues that have facilitated more systematic and intensive environmental management practices than may otherwise have occurred in the same time period. The most obvious examples are the depletion of the ozone layer and climate change. At a national level, issues such as water shortages, soil erosion and regional air pollution have triggered the adoption of improved environmental management practices on an extensive scale.

Business community

[E16011] There is a desire among employers and employees to improve the environmental impacts of the workplace. Trade associations and business groups such as the Confederation of British Industry (CBI) and the International Chamber of Commerce (ICC) have played a leading and effective role in encouraging businesses to adopt environmental management practices. Increasingly the importance of doing this within a sustainable development framework is being accepted. Sustainable development is generally recognised as the inter-relationship and inter-dependence of the three core elements of economic, environmental and social consideration, although it is a concept that is open to wide interpretation. However, it is generally understood to mean achieving a better quality of life with effective environmental protection. The most authoritative definition comes from the 1987 United Nations Brundtland report to the World Commission on Environment and Development. This definition states that it is 'development which meets the needs of the present without compromising the ability of future generations to meet their own needs'.

In the UK, the TUC has supported policies aimed at combating climate change. It has stated that the scientific evidence from the UN's Intergovernmental Panel on Climate Change (IPCC) requires urgent action. In 2011 the Congress formally recognised that decarbonising the economy is essential to promoting growth and employment, and endorsed government targets to reduce emissions of greenhouse gases. It also went further in stating that there is a need for more government intervention to provide incentives for investment in low-carbon energy generation. The TUC's Greenworkplaces projects cover a number of schemes, including energy saving at work, waste reduction, recycling, and green travel plans, and these are supported by an environmental education programme.

The Environment and Energy Commission represents ICC members in discussions on global initiatives and intergovernmental negotiations in these areas, including at the UN Framework Convention on Climate Change, the UN Commission on Sustainable Development, and the UN Environment Programme. It is made up of 350 members representing multinational corporations, industry associations, and the ICC national committees. Thousands of companies worldwide have been helped by the ICC voluntary business initiative 'Business Charter for Sustainable Development'. Launched in 1991 as a tool to help companies tackle the challenges and opportunities of the environmental issues that emerged in the 1980s and early 1990s, the Charter now contains a set of eight principles to guide company strategies and operations towards sustainable development. Its principle for environmental responsibility and management has provided for a global alignment of business to common objectives, and has helped thousands of companies worldwide establish the foundation on which to build their own integrated environmental management systems. The Charter also highlights such areas as transparency, communications and reporting, products and services, and responsibility towards people and societies. The World Business Council for Sustainable Development (WBCSD) was founded in 1992 at the Rio Earth Summit to provide a voice for the business community. Consisting of some 200 companies drawn from more than 35 countries and 20 major industrial sectors, it provides encouragement to the global business community to implement sustainable development policies and practices. The work of the WBCSD is carried out through a combination of advocacy, research, education, knowledge-sharing and policy development. It focuses on four key areas: energy and climate, development, the business role, and ecosystems. The organisation has a global network of about 60 national and regional business councils and regional partners. Members companies include Apple, Proctor & Gamble and Du Pont in the US; BP, BT plc and SABMiller in the UK; and Toyota and Toshiba in Japan.

The Responsible Care Programme is an example of an international initiative founded by a specific industry sector. It is monitored and coordinated by the International Council of Chemical Associations (ICCA), with member associations responsible for the detailed implementation of Responsible Care in their countries. The Programme is comprised of 54 national chemical manufacturing associations, including the Chemical Industries Association (CIA) in the UK. All of those associations have made acceptance of Responsible Care requirements compulsory for individual member companies. Responsible Care is designed to promote continuous improvement, not only in environmental management, but also health and safety. The companies must also adopt a policy of openness by releasing information about their activities. Adherence to the principles and objectives of Responsible Care is a condition of membership of CIA. Internationally, the Programme is at different stages of development in different countries, and its precise format will reflect the prevailing national conditions and requirements.

Individual companies have also contributed to the development of improved environmental performance standards. As the relevance of good environmental management to overall business performance has become more apparent, progressive companies have voluntarily adopted a number of innovative and unique approaches to corporate environmental management. This has had the effect of constantly moving the frontiers of "acceptable" environmental management practices and created substantial peer pressure which in turn has encouraged other companies to adopt similar or even more effective practices.

Customers

[E16012] In the late 1980s and early 1990s some of the major retailers started to market 'green' products, such as phosphate-free washing powders and biodegradable cleaning products, in response to consumer concern about high-risk chemicals.

In its 2008 document 'Framework for Pro-Environmental Behaviours', the Government reported the results of research carried out on public understanding of and behaviour in relation to environmental matters. It concluded that the most likely public actions are to reduce food waste and water usage, and the least likely to stop flying. The research also showed that there is potential for increasing the purchasing of low environment impact goods, particularly in relation to energy efficient products. Nearly half the people surveyed said they would be prepared to pay more in general for environmentally friendly products, with two thirds prepared to do so for appliances with high energy efficiency ratings. In addition, it was found that most people were aware of certification or assurance schemes such as Fair Trade and schemes for ensuring that timber originated from sustainable sources.

Retailers are coming under increasing scrutiny as a result of benchmarking by consumer groups. Consumer Focus (formerly part of BIS and the statutory consumer organisation for the UK, now known as Consumer Futures and part of Citizens Advice services), found in its 2009 study that although some UK supermarkets were making improvements in helping customers shop for green products, others were lagging well behind. Sainsbury's and Marks and Spencer were commended for making the biggest advances by achieving the first ever overall 'A' (excellent) score.

More recently a US survey conducted by Cone Communications in 2013 reported that a record high 71% of Americans consider the environment when they shop, up from 66% in 2008. Furthermore, it said that 71% of consumers wanted more information from companies to help them better understand environmental terms. The survey also warned that companies' honesty on green issues is an important factor for consumers, with some 78% stating they would boycott a product if they discover an environmental claim to be misleading. Therefore, accurate and clearly understandable information relating to claims about a benefit to the environment is essential to build consumer confidence. In the UK DEFRA has published a Green Claims Guidance in 2011 to provide advice to business on this matter. It also provides links to a range of tools and resources for both mandatory and voluntary schemes on its website.

A greater public awareness of environment issues will inevitably feed through into even more consumer demand, which will drive companies to produce 'greener' products and services. There is also evidence from academic studies that some consumers choose green products, eg hybrid cars because they believe it enhances their social status.

Environmental management guidelines

Standards for environmental management

[E16013] Various guidelines exist for responding to those many pressures to minimise damage to the environment and health. Effective environmental management, like quality management or financial management, requires:

(a) the setting of objectives and performance measures;
(b) the definition and allocation of responsibilities for implementing the various components of environmental management;
(c) the measurement, monitoring and reporting of information on performance; and
(d) a process for ensuring feedback on systems and procedures so that the necessary changes can be actioned.

There are two main approaches to environmental management. The first is the EU Eco-Management and Audit Scheme (EMAS) Regulation. The second is the International Standards Organisation (ISO) Series of Environmental Management Standards. The Acorn Scheme was set up by the Institute of Environmental Management and Assessment (IEMA) in the UK for organisations, in particular SMEs, where time and resource restraints prohibit implementation of EMAS or ISO 14001. A relatively recent development, it provides a more flexible approach as it can be implemented in small segments over a larger time-scale, while progressing to a fully functional environmental management system. All participants in the scheme must implement one or more phases of British Standard BS8555.

The Eco-Management and Audit Scheme (EMAS) Regulation

[E16014] This voluntary scheme came into force in July 1993 and has been open for participation by companies in all Member States since April 1995. It was initially established by *European Regulation* 1836/93, which has since been updated twice in order to attract more registrations and make the scheme more competitive. The most recent legislative instrument is *EU Regulation (EC)* No 1221/2009, which came into force in January 2010.

EMAS has three main aims:

(a) the establishment and implementation of environmental policies, programmes and management systems;
(b) the systematic, objective and periodic evaluation of these measures; and
(c) the provision of information to the public on environmental performance.

The Regulation sets out a number of elements of systematic environmental management. Sites that can demonstrate ongoing compliance with those requirements to an independent assessor have the right to be registered under EMAS.

Following the adoption of a company policy, an initial environmental review is made to identify the potential impacts of a site's operations, and an internal environmental protection system must be established. The system must include specific objectives for environmental performance and procedures for implementing them, and the results of the environmental review has to be described in an initial environmental statement. The statement must then be validated by an accredited external organisation before being submitted to nominated national authorities in individual Member States for registration of the site under the scheme.

There are a number of key points here:

(a) the first stage in developing environmental management is to carry out a thorough review of impacts on the environment;
(b) setting up an environmental management system is a prerequisite of registration under the scheme;
(c) external validation of the environmental statement is intended to ensure consistency in environmental management systems; and
(d) the description of the environmental management system within the statement will be on the public record.

Once a site has been registered under the scheme, it will require regular audits to review the effectiveness of the environmental management system as well as having available information on environmental impacts of the site. Here it is sufficient to note that the development of procedures for internal auditing is a crucial part of an environmental management system. In addition to the audit, the preparation and external validation of an environmental statement, submitted to the competent authority for continued registration and made public, are elements of an ongoing procedure.

The scheme is open to all sectors of the economy, including financial companies, transport and local and public bodies, although industry still accounts for most EMAS registrations. Registered organisations can use an official logo to publicise their participation in EMAS as well as gaining regulatory benefits. The scheme also encourages more involvement by employees in implementation and strengthens the role of the environmental statement, which helps to improve the transparency of organisations and their stakeholders.

The 2009 Regulation introduced revised audit cycles to make EMAS more applicable for small organisations, and offered the opportunity for a single corporate registration, rather than requiring registration for individual sites as under the old rules. It also sets out the 'environmental core indicators' that should be included in the environmental statement and used to describe the environmental performance. All the existing explanatory guidelines are now included in the EMAS Regulation. Some of the provisions are optional for EU countries. For example, the UK has chosen not to make use of the article that provides for organisations outside the EU to apply for EMAS registration in the UK. Neither will it appoint a separate licensing body in the UK, or allow the licensing of natural persons as environmental verifiers for the purposes of EMAS.

As of March 2015, there were a total of 8,150 EMAS registered sites owned by more than 4,500 organisations in a wide range of economic sectors worldwide. In the UK, registrations have fallen in recent years, and are at a relatively low level compared with the likes of Germany, Italy and Spain, the countries with the largest number of registrations. There were 48 organisations with an EMAS registration in the UK in early 2015.

International standards on environmental management (the ISO 14000 series)

[E16015] The International Standards Organisation (ISO) has developed a series of standards for various aspects of environmental management, which it keeps under review. All are designed to assist organisations with implementing more effective environmental management systems. Table 1 outlines the various standards and guidelines within the series.

The most high profile of these standards is ISO 14001, first published in June 1996, which sets out the characteristics for a certifiable environmental management system (EMS). ISO 14001 was based on the British Standard on Environmental Management Systems (BS 7750), although the latter has been superseded by the international standard.

The format of ISO 14001 reflects the procedures and manuals approach of the ISO 9000 quality management systems series. In practice this means that organisations that operate to the requirements of ISO 9000 can extend their management systems to incorporate the environmental management standard, although the existence of a certified quality management system is not a prerequisite for ISO 14001.

ISO 14001 requires an organisation to develop an environmental policy that provides the foundation for the rest of the system. The standard includes guidance on the development and implementation of other elements of an EMS, which is ultimately designed to allow an organisation to manage those environmental aspects over which it has control, and over which it can be expected to have an influence. ISO 14001 does not itself stipulate environmental performance criteria.

Revised versions of ISO 14001 and ISO 14004 were published in November 2004. ISO 14001:2004 is designed to be more closely aligned with ISO 9001, and to clarify a number of the requirements. ISO 14004:2004 is more consistent and compatible with ISO 14001:2004, to encourage their joint use, and to make it more accessible to SMEs. Although the new 14001 standard is very similar to the previous version in many respects, there are significant differences in the scope, impacts, and periodic evaluation of legal and other requirements.

ISO 14001 currently has over 250,000 users in at least 155 countries. The popularity of ISO 14001 over EMAS is due to the fact that, until recently, EMAS did not extend beyond Europe, and that it is much more exacting requirements than ISO 14001. This is despite the fact that the revised EMAS allows better integration of an existing ISO 14001 certification, which should allow a smoother transition and avoid duplication when upgrading from ISO 14001 to EMAS.

ISO 14001 is currently under revision with a final version planned for publication towards the end of 2015. The new standard should increase the focus on reducing environmental impacts across the whole life-cycle of products and services, as well as on the supply chain and outsourced processes. Additional changes, such as a greater consideration of the overall business strategy, should help organisations to more easily incorporate their environmental management system into core business processes, rather than just concentrating on operational issues. The new framework of terms, definitions, headings and text will be common to all management system standards so allow easier integration when implementing more than one management system. When implemented, ISO 14001:2015 should also result in improved knowledge of the factors that affect an organisation's environmental responsibilities and provide a stronger link to top management's responsibility for the management systems.

Table 1: The ISO 14000 environmental management series	
ISO 14001: 2004 ISO 14001:2004/Cor 1:2009	Environmental management systems — Requirements with guidance for use
ISO 14004:2004	Environmental management systems — General guidelines on principles, systems and supporting techniques
ISO 14005:2010	Environmental management systems — Guidelines for a staged implementation of an environmental management system, including the use of environmental performance evaluation
ISO 14006:2011	Environmental management systems — Guidelines for incorporating eco-design
ISO 14015: 2001	Environmental assessment — Sites and organisations
ISO 14020: 2000	Environmental labels and declarations — General principles
ISO 14021:1999 / Amd 1:2011	Environmental labels and declarations — Self-declared environmental claims (Type II environmental labelling)
ISO 14024:1999	Environmental labels and declarations — Type I environmental labelling — Principles and procedures
ISO 14025:2006	Environmental labels and declarations — Type III environmental declarations — Principles and procedures
ISO 14031:2013	Environmental management — Environmental performance evaluation — Guidelines
ISO/TR 14032:1999	Environmental management — Examples of environmental performance evaluation (EPE)
ISO 14033: 2012	Environmental management — Quantitative environmental information — Guidelines and examples
ISO/DIS 14034	Environmental management — Environmental technology verification (ETV) (under development)
ISO 14040: 2006	Environmental management — Life cycle assessment — Principles and framework
ISO 14044: 2006	Environmental management — Life cycle assessment — Requirements and guidelines
ISO 14045:2012	Environmental management — Eco-efficiency assessment of product systems — Principles, requirements and guidelines
ISO 14046:2014	Environmental management — Water footprint — Principles, requirements and guidelines
ISO/TR 14047:2012	Environmental management — Life cycle impact assessment — Examples of application of ISO 14044
ISO/TS 14048:2002	Environmental management — Life cycle assessment — Data documentation format

Table 1: The ISO 14000 environmental management series	
ISO/TR 14049:2012	Environmental management — Life cycle assessment — Examples of application of ISO 14044 to goal and scope definition and inventory analysis
ISO 14050:2009	Environmental management — Vocabulary
ISO 14051:2011	Environmental management — Material flow cost accounting — General principles and framework
ISO/CD 14052	Environmental management — Material flow cost accounting — Guidance for practical implementation in a supply chain (under development)
ISO/CD 14055-1 and ISO/CD 14055-2	Guidelines for establishing good practice for combating land degradation and desertification — Part 1: Guidelines and general framework; Part 2: Case studies
ISO/TR 14062:2002	Environmental management — Integrating environmental aspects into product design and development
ISO 14063:2006	Environmental management — Environmental communication — Guidelines and examples
ISO 14064:2006 (parts 1, 2 and 3)	Environmental management — Greenhouse gases — accounting and verification
ISO 14065:2013	Environmental management — Greenhouse gases — Requirements for greenhouse gas validation and verification bodies for use in accreditation or other forms of recognition
ISO 14066:2011	Environmental management — Greenhouse gases — Competence requirements for greenhouse gas validation teams and verification teams
ISO/TS 14067:2013	Carbon footprint of products — Requirements and guidelines for quantification and communication
ISO/TR 14069:2013	Greenhouse gases (GHG) — Quantification and reporting of GHG emissions for organisations — Guidance for the application of ISO 14064-1
ISO/TS 14071:2014	Life cycle assessment — Critical review processes and reviewer competencies — Additional requirements and guidelines to ISO 14044:2006
ISO/TS 14072:2014	Environmental management — Life cycle assessment — Requirements and guidelines for organisational life cycle
ISO 19011:2002	Guidelines for quality and/or environmental management systems auditing
ISO 26000:2010	Guidance on social responsibility
BS ISO 18634	Environmental management — Environmental technology verification (ETV) and performance evaluation (under development)

Differences between ISO 14001 and EMAS

[E16016] The key differences between ISO 14001 and EMAS are:

(a) ISO 14001 does not currently include a requirement for public reporting of environmental performance information, whereas sites registered under EMAS must produce an independently validated, publicly available environmental statement;

(b) Unlike EMAS, ISO 14001 does not include active involvement of employees;

(c) the level of control of contractors and suppliers required in EMAS is not matched in ISO 14001, which stipulates only that required procedures are communicated to them;

(d) EMAS specifies environmental core indicators to assess performance and to be included in the environmental statement. ISO 14001 does not include core indicators;

(e) a verified initial environmental review is required for EMAS but is only a recommendation in ISO 14001; and

(f) EMAS requires a demonstration that the implications of legal requirements relating to the environment have been identified, provide for legal compliance, and have procedures in place that enable the organisation to meet the requirements. ISO 14001 does not require organisations to demonstrate legal compliance, but to show commitment to comply.

Implementing environmental management

Practical requirements

[E16017] Both EMAS and ISO 14001 set a pattern for companies wishing to develop environmental management systems. The key steps in implementing such systems involve:

(a) conducting an initial review, designed to establish the current situation with respect to legislative requirements, potential environmental impacts and existing environmental management controls;

(b) developing an environmental policy which will provide the basis of environmental management practices as well as informing day to day operational decisions;

(c) establishing specific objectives and performance improvement targets;

(d) developing a programme to implement the objectives and establish operational control over environmental performance;

(e) ensuring information systems are adequate to provide management with complete, reliable and timely information; and

(f) auditing the system to compare intended performance with actual performance.

Review

[E16018] In order to be able to actively manage its interactions with the environment, an organisation needs to understand the relationship between its business processes and its environmental performance. An environmental review should therefore be conducted, which clarifies how various business activities could potentially affect, or be affected by, the quality of different components of the environment (air, land, water and the use of natural resources). This will allow an organisation to understand which of its activities have the greatest potential impact on the environment. Usually, but not always, these will relate to procurement and/or manufacturing processes. Therefore, the scope of the review should be broader than simply evaluating the direct impacts such as polluting emissions, discharges and wastes. ISO 14001 provides useful guidance on the range of issues that should be considered when reviewing the environmental aspects of an organisation's operations and activities. EMAS sets out more detailed advice on the content of an environmental review.

It is also important to consider which elements of environmental management and performance have the greatest potential impact on business performance, for example, high profile environmental prosecutions can have a significant impact on corporate reputation and brand value.

The initial and subsequent review should also consider the current and likely future legislative requirements that the company must comply with. It should consider the overall organisational strategy, any other relevant corporate policies, customer specifications and community expectations that could influence environmental management practices. The review also offers the opportunity to establish a baseline of actual management organisation, systems and procedures, its compliance record and the range of initiatives already in place to improve performance.

Ideally, an environmental review should involve input from a range of stakeholders, both internal and external to the organisation. This promotes a wider perspective on potential environmental impacts, and ensures that the resulting policies and programmes to be developed by the organisation, reflects a comprehensive range of issues and risks. Consequently, the chances of unidentified and therefore uncontrolled risks emerging will be minimised.

Policy

[E16019] The results of the initial review should inform the development of a written environmental policy. The policy directs and underpins the remainder of the EMS, and represents a statement of intent with regard to environmental performance standards and priorities.

The policy should be endorsed by the highest level of management in the company, and should be communicated to all stakeholders. There are many different types of environmental policy, ranging in length and detail included, and no established format. This is likely to be determined by nature and public position of the organisation.

Objectives

[E16020] It is important that the policy be supported by objectives, and both ISO 14001 and EMAS require this. These objectives should be both measurable and achievable. They should be cascaded throughout an organisation, so that at each level there are defined targets for each function to assist in the achievement of the objectives. A key part of ensuring continuous improvement in environmental performance is to review and update objectives in the light of progress and changing regulations and standards.

Objectives should be developed in conjunction with the groups and individuals who will have responsibility for achieving them. They should also be clearly linked to the overall business strategy and as far as possible with operational objectives. This ensures that environmental management is viewed as relevant and integral to business performance, rather than being seen as an isolated initiative.

The process of objective setting also needs to consider the most appropriate performance measures for tracking progress towards the ultimate objective. For example, if an objective is to reduce waste by 20 per cent over five years, a number of parameters could be used to reflect different aspects of the organisation's waste reduction efforts towards that, including volumes of waste recycled, efficiency with which raw materials are converted to product, and proportion of production staff that have received waste management training.

ISO 14031 provides useful guidance on the principles to be applied in the selection of appropriate environmental performance indicators. Performance, in whichever parameters are selected, should be measured regularly to enable corrective actions to be taken in a timely manner. As far as possible, performance measures and the achievement of quantified targets should be linked to existing appraisal systems for business units or individuals.

Responsibilities

[E16021] Allocation of responsibilities is vital for successful environmental management. Its implementation will typically involve changes in management systems and operations, training and awareness of personnel at all levels and in marketing and public relations.

A wide range of business functions will therefore need to be involved in developing and implementing environmental management systems. The commitment of senior personnel to introduce sound environmental management throughout the organisation, and to communicate it to all staff is vital.

Companies have adopted a range of organisational approaches as part of their environmental management systems. In some cases there is a single specialist function with responsibility for monitoring and auditing the system. An alternative is to have a central environmental function with only an advisory role, which can also undertake verification of the internal audits carried out by other divisions or departments. The approach needs to be adapted to the culture and structure of the organisation, but whatever system is adopted, there are a number of crucial elements:

(a) access to expertise in assessing environmental impacts and developing solutions;
(b) a degree of independence in the auditing function;
(c) a clear accountability for meeting environmental management objectives; and
(d) adequate information systems to help those responsible for evaluating performance against objectives, to identify problem areas and to ensure that action is taken to solve them.

Training and communications

[E16022] Both ISO 14001 and EMAS include training and communications as a key requirement of the EMS. In particular, they focus on ensuring that employees at all levels, in addition to contractors and other business partners, are aware of company policy and objectives; of how their own work activities impact on the environment and the benefits of improved performance; what they need to do in their jobs to help meet the company's environmental objectives; and the risks to the organisation of failing to carry out standard operating procedures.

This can involve a significant investment for companies, but if integrated with existing training modules and reinforced regularly, many hours of essential training can be achieved. An important benefit of training is that it can be a fertile ground for new ideas to minimise adverse environmental impacts.

Communication with, for example, regulators, investors, public bodies and local communities is an important part of good environment management and requires a preparedness to be open, honest and informative. It also requires clear procedures for liaison with external groups.

Operational controls

[E16023] The operational control elements of an EMS define its scope and essentially set out a basis for effective day-to-day management of environmental performance. Typically, such controls would include:

(a) a register of relevant legislation and corporate policies that must be complied with;
(b) a plan of action for ensuring that the policy is met, objectives are achieved and environmental management is continuously improved. This sets out the various initiatives to be proactively implemented and milestones to be achieved, and could be considered a 'road map' for guiding environmental performance;
(c) a compilation of operating procedures which define the limits of acceptable and unacceptable practices within a company and which incorporate consideration of the environmental interactions identified in the review and the objectives and targets that were defined subsequently;
(d) an emergency response plan to be implemented in the event of a sudden, unexpected and potentially catastrophic event which could potentially influence a company's environmental performance in an adverse manner; and

(e) a programme for monitoring, measuring, recording and reviewing environmental performance. This should incorporate a mechanism for regularly reporting back to senior management, since they are the key enablers of the EMS and because they retain ultimate responsibility of business performance.

These operational controls, and indeed all elements of an EMS, should be documented.

Information management and public reporting

[E16024] The critical factor in an EMS is the quality and timeliness of conveying information to internal and external users. EMAS goes beyond ISO 14001 in this respect, as it requires public access to company environmental reports, as well as independent verification of these reports. The importance of providing performance information to external users is also being influenced by increasing expectations of greater transparency and corporate responsibility (CR). Such expectations have been supported in the UK by strong encouragement from Government for companies to voluntarily and publicly report on their environmental performance. Regardless of whether they have adopted EMAS, ISO 14001, or neither, many companies have a statutory duty to report on some aspects of their environmental performance to demonstrate legislative compliance.

An increasing number of companies now provide information on their environmental performance, either in their annual report and accounts, or in a stand-alone document. In most cases, the information provided extends well beyond a demonstration of legislative compliance, and leans towards a general overview of all significant aspects of environmental management and performance.

Companies, recognising the importance of sustainability, now measure and report on their performance in terms of social impact. A requirement under the EU Accounts Modernisation Directive and the Companies Act 2006 for directors of quoted and large private companies to produce a Business Review, or Strategic Report form October 2013 (see [E16005]), has also raised the level of corporate environment and social disclosure for those companies affected. The CRC Energy Efficiency Scheme, which was introduced in April 2010, is a mandatory scheme designed to encourage large non-energy intensive public and private sector organisations to monitor their energy consumption and reduce their carbon dioxide emissions. Covering around 10% of the UK's greenhouse gas emissions, it requires organisations such as large supermarkets and water companies, as well as local authorities and Government Departments, to report on their carbon emissions.

Emissions reporting can benefit companies in terms of reputation and brand value. However, as stakeholders and communication media become more sophisticated, and our understanding of environmental interactions increases, so information requirements become more complex. This means that management information systems must incorporate database management, modelling, measuring, monitoring and flexible reporting. This trend also means that environmental reports that have been produced solely as a means of improving public relations and to defuse external pressures have become less acceptable to many stakeholders. The main reasons for this appear to be that they are seldom generated in response to internal management information needs, which in turn means that they tend to focus on statements of management intent, qualitative claims and descriptive anecdotes rather than actual performance. They are therefore less likely to include detailed and verifiable information.

Nevertheless, many organisations do recognise the value of maintaining a constructive and open dialogue with stakeholders, and external reporting is a major part of this. This has been encouraged by the setting up of award schemes that give recognition to the best reports. The likes of the Association of Chartered Certified Accountants (ACCA) and Corporate Register have been very active in this area in the past.

Institutional investors needing reassurance that companies are aware of their environmental risks are also driving companies into providing more environmental information. The Association of British Insurers (ABI) issued investment guidelines in October 2001 after consultation with a wide range of fund managers, corporate executives and NGOs. These guidelines, which were updated in 2007, aim to improve disclosure of companies' approach to external social, ethical and environmental risks. They set out the business case for CSR and recommend that it should be integrated into annual financial reports to allow for independent verification. The guidelines were intended to increase transparency and allow shareholders to engage with companies where they consider significant risks have not been assessed adequately. The introduction in July 2000 of a requirement for occupational pension funds to state the extent to which they consider environmental, social and ethical factors in investment decisions, also increased the pressure for environmental reporting among companies.

During 2012 many stock exchanges around the world announced new rules for the disclosure of non-financial information on environmental and social issues. Although the declaration at Rio+20 was widely expected to drive disclosures on environmental and social risks, as it would apply to 15,600 listed firms worldwide, weak enforcement of this requirement has meant that in practice it has had very little impact. Many listed companies are still able to easily evade financial, legal or reputational penalties for non-disclosure.

Reporting Guidelines

[E16024A] While there is no unified standard for environmental reporting, a number of sources of guidance are available. One of the first in the UK involved collaboration in the early 1990s between the United Nations Environment Program (UNEP) and SustainAbility an organisation that helps businesses improve their approach to sustainability issues. The Global Reporters Program was focused on advancing corporate reporting best practice worldwide. It was replaced in 1998 by the Engaging Stakeholders Program, a forum in which corporate members exchange ideas and best practice, as well as acquire expert guidance and advice on all aspects of corporate accountability. This Program is still in existence allowing members access to the latest developments in stakeholder engagement and transparency.

[E16024A] Environmental Management

The Global Reporting Initiative (GRI) provides the leading environmental and CSR reporting guidelines used currently. Established in 1997 to develop globally applicable guidelines for reporting on the economic, environmental and social performance of a range of organisations, it receives input from corporations, NGOs, accountancy organisations, business associations and other stakeholders from around the world. The GRI's Sustainability Reporting Guidelines, released in June 2000, provided an expanded model for voluntary non-financial reporting and reflected the move towards broader sustainability reporting. They included a core set of indicators, which, for the most part, were applicable to organisations in all sectors of commerce and included a list of performance indicators that took into account social and economic factors, rather than just concentrating on environmental issues. The guidelines also insisted that statements on a company's vision and strategy regarding sustainable development should be addressed before a report could be described as in accordance with GRI guidelines.

New guidelines (G3) were released in October 2006. Relevant to organisations of any size, sector, or location, they were made more user-friendly by incorporating a flexible outline of the core content of a report. Unlike previous revision cycles, when the entire set Guidelines were subject to revision, the G3 Guidelines are updated incrementally. This involves targeting certain portions of the Guidelines for revision on an annual basis. Stakeholders can make suggestions for specific amendments directly to the GRI. 'Sector-specific supplements' are also available to complement the core information in the general Guidelines. These cover a range of sectors including public authorities, financial services, mining and metals, tour operators, logistics and transportation, NGOs and the automotive and telecommunications industries. The improved G3 Guidelines have given rise to continued growth in reporting with over 1,500 companies, including sector leaders such as Barclays Bank, Rabobank, Royal Dutch Shell, Ford, Vodafone and Anglo American plc, having used them to produce voluntary standalone reports. Updated guidance, G3.1 Sustainability Reporting Guidelines, was launched by the GRI in March 2011, alongside guidance to help companies to determine what to measure and report. The 'Technical Protocol – Applying the Report Content Principles' should work for organisations looking to produce more relevant reports. Both versions of the guidance remain valid until the next generation of GRI Guidelines is in place, although the GRI recommends the use of G3.1. A fourth generation of Sustainability Reporting Guidelines (G4) was launched in May 2013. These incorporate recent developments in reporting, and are aimed at making the reporting process more focused and relevant. The latest guidance also provides generic format for disclosures on management approach and directions on how to select material topics. All reports published after 31 December 2015 have to be prepared in accordance with the G4 Guidelines, but the GRI will continue to recognise reports based on the G3 and G3.1 Guidelines until then.

Other international guidance on environmental reporting is available. The UN Global Compact (UNGC) provides a platform for the development, implementation and disclosure of responsible corporate policies and practices. Launched in 2000, it is the world's largest voluntary corporate sustainability initiative, with over 12,000 signatories from business and key stakeholder groups based in 145 countries. Business participants are required to issue an annual Communication on Progress (COP), a public disclosure to stakeholders on progress made in implementing its ten principles. The International <IR> Framework was founded in December 2013 by the International Integrated Reporting Council (IIRC). Its aim is to promote the inclusion in corporate reports of a long-term and integrated approach to resources and relationships used and affected by an organisation. It is used with ISO 26000, which provides guidance on social responsibility, to support companies in understanding and enhancing the value they create for society and for financial investors.

There is increasing cooperation between the different organisations involved in sustainability reporting. Nearly all of the various platforms and organisations promoting sustainability reporting have developed partnerships with the GRI. This includes the UN Global Compact, Accountability, OECD, UNEP, the Carbon Disclosure Project and many governments and sector organisations. ISO 26000 can also be used in conjunction with the GRI G4 Guidelines.

Online reporting is growing due to the availability of new communication tools and web-based applications, such as XBRL (extensible Business Reporting Language). This should improve the stakeholder feedback process, and the comparability of the information made available.

The first Guidelines intended to aid companies in the UK to report on their interactions and impacts on the environment, were issued by the Department for Environment, Food and Rural Affairs (DEFRA) in 2001. New voluntary guidelines were published in January 2006, partly in response to the need for preparing a Business Review under the Modernisation Directive and *Companies Act 2006*. They recommend using Key Performance Indicators (KPIs) as a tool for measuring environmental impacts. In theory this should allow costs to be reduced through the use of standard business data that may have been collected for other purposes. The DEFRA guidelines also contain information on how environmental impacts arising from the supply chain and from the use of products can be taken into account. They recommend that quantitative data should be provided where necessary, and provide a standard calculation method to help ensure consistent reporting. Businesses responded favourably to a consultation on the guidelines, commenting that they were relatively simple to follow. The government estimated that 80% of UK businesses would only have five or fewer of the 22 KPIs to report their performance against.

The guidance was updated by DEFRA, in partnership with the Department for Energy and Climate Change (DECC), in 2013 to include mandatory greenhouse gas emissions reporting required under the Companies Act 2006 (Strategic Report and Directors' Report) Regulations 2013 (SI 2013 No 1970). This guidance, which is aimed at all sizes of business as well as public and third sector organisations, contains methodology for measuring and reporting greenhouse gas emissions. It also gives guidance on calculating transport emissions from freight and work-related travel, as well as for both direct and indirect emissions from the supply chain.

Verification/assurance

[E16024B] In an attempt to increase the credibility of published reports and identify opportunities for improving management information systems, a number of companies are seeking independent third party assurance on the reliability, completeness and likely accuracy of information contained in their reports. While ISAE 3000 is used by public accounting firms and focuses on data quality and evidence gathering procedures, AA1000AS is a specialist tool for corporate public reporting on social, environmental and economic performance bodies. First published by the British organisation AccountAbility (AA) in March 2003, it emphasises the need for organisations to demonstrate effective stakeholder engagement, identify material sustainability issues and put in place a responsible business strategy to respond to those issues. The Standard placed certain demands on external auditors and verifiers of CSR and environmental reports. They are required to demonstrate their independence and impartiality by publicly disclosing commercial relationships with their clients, as well as proving their competency and commenting where a report has omitted information that could be important to stakeholders. The AA1000AS 2003 was replaced by 3 separate documents in 2008. AA1000AS 2008 requires the assurance provider to evaluate the extent of adherence to a set of principles rather than simply assessing the reliability of the data. In addition, the assurance provider must look at the underlying management approaches, systems and processes and how stakeholders have participated. It encourages integration of sustainability into organisations' routine operations, so that it is aligned with traditional financial accounting, reporting and auditing. The Assurance Standard is accompanied by Accountability Principles Standard 2008 (AA1000APS 2008), which provides a framework for an organisation to better identify, understand, prioritise and respond to its sustainability challenges. Guidance for the use of AA1000AS 2008 is also now available. The new standards and guidance documents are suitable for any organisation, from multinational businesses, to SMEs, governments and civil society organisations, although a license fee is now required for each commercial use of AA1000AS.

With recent advances in developing assurance standards for environmental and sustainability reports, external assurance is becoming the norm. KPMG reported in 2008 that formal third party assurance of Global 250 reports increased from 30% to 40% over three years. Developed in conjunction with the investment community, NGOs and business, and endorsed by the GRI and openly accessible on a non-commercial basis, AA1000AS is now used by hundreds of leading companies in the UK to assure their sustainability and corporate responsibility reports. All sustainability reports assured against the AA1000AS are contained in an online directory administered by Corporate Register.

However, there is evidence to suggest that third party assurance is not being applied as strictly across Europe. A recent study conducted by the Dutch Centre for Research on Multinational Corporations (SOMO) concluded that many companies claiming to report in accordance with the GRI guidelines often do not follow the indicators of environmental and social performance. It found that over half the claims in sustainability reports issued by 19 electric companies in Europe were false or misleading. Perhaps even more surprising was the fact that 10 of the 12 of these corporate reports purporting to have been externally assured, or checked by a third-party auditor or the GRI, contained some major discrepancies in their claims.

Auditing and review

[E16025] Auditing of environmental management systems, whether conducted by internal or external parties, is a means of identifying potential risk areas and can assist in identifying actions and system improvements required to facilitate ongoing system and performance improvement.

Benefits of environmental management

Effective environmental management

[E16026] Effective environmental management will involve changes across all business functions. It requires commitment from senior management and is likely to need additional human and financial resources initially. However, it can also offer significant benefits to businesses. These include:

(a) avoidance of liability and risk. Good environmental management allows businesses to choose when and how to invest in better environmental performance, rather than reacting at the last minute to new legislation or consumer pressures. Unforeseen problems will be minimised, prosecution and litigation avoided;

(b) gaining competitive advantage. A business with sound environmental management is more likely to make a good impact on its customers. The business will be better placed to identify and respond rapidly to opportunities for new products and services, to take advantage of 'green' markets and also respond to the increasing demand for information on supplier environmental performance;

(c) achievement of a better profile with investors, employees and the public. Increasingly, investors and their advisers are avoiding companies with a poor environmental record. The environmental performance of businesses and other organisations is an increasing concern for existing staff and potential recruits. Some are finding that a good environmental record helps to boost their public image;

(d) cost savings from better management of resources and reduction of wastes through attention to recovering, reusing and recycling; and reduced bills from more careful use of energy;

(e) an improved basis for corporate decision-making. Effective environmental management can provide valuable information to corporate decision-makers by expanding the basis of such decisions beyond financial considerations. Organisations that understand the interactions between their business activities and environmental performance are in a strong position for integrating environmental management into their business, thereby incorporating key elements of the principles of sustainable development.

Effective environmental management can turn environmental issues from an area of threat and cost, to one of profit and opportunity. As standards for environmental management systems are adopted and are applied widely, the question now is, as with quality management, can a company or large organisation afford not to adopt environmental management? The external pressures to improve environmental performance are unlikely to abate. Environmental management systems can help to respond to the pressures in a timely and cost-effective way.

It has often been suggested that companies that implement an environmental management system should be rewarded with lesser regulatory control. This policy has been implemented in the UK as a result of previous Government guidance to the Environment Agency that it should take account of robust environmental management systems, in particular EMAS and ISO 14001, in its regulatory approach to companies. The operational risk appraisal (OPRA) assessment provides a risk rating which the Agency uses to allocate its regulatory resources and to determine how much a business will be charged for regulation. Good performers are easier to regulate and are subject to less scrutiny from the Agency. Having a recognised EMS in place will help to improve performance and in turn enhance an organisation's OPRA score.

The future of environmental management

[E16027] Environmental management is now well recognised by many as an essential component of effective business management. Environmental management systems have been widely adopted and in most cases these have facilitated demonstrable and ongoing improvements in environmental performance. The involvement of a range of stakeholders in corporate environmental management is no longer the exception; the constructive contribution that they make is actively sought. There is continuing improvement in the quality and numbers of environmental reports published, with reporting becoming relatively common among leading companies. Most include a mechanism for obtaining feedback from external stakeholders, and an increasing number are independently assured in a similar way to annual financial reports and accounts.

Companies' annual reports have grown in length over the recent decades, largely due to the increased amount of narrative reporting, as well as greater legislative demands. This is also due in part to increased public scrutiny of business activities and the need for clearer explanations of company activity. Environmental pressures on businesses have grown dramatically in recent years. Stakeholders now routinely examine company websites and stand-alone social and environmental reports for information. Technological developments, globalisation, and the growth of the knowledge economy have radically increased expectations for comprehensive validated data on companies' environmental impacts to be made available. The relatively recent proliferation in social networking and blogging sites, has led to a steep rise in the speed at which information now travels around the world. The financial crisis of 2007–08 provoked a drop in public trust in the operation of the corporate world. An effective environment management system will allow business to respond rapidly to these pressures and demands. It is also more important than ever that the accuracy and quality of environmental reports is maintained, even in an economic downturn when finances are stretched. Developments in the quality and availability of clearer and more practical reporting guidelines, such as those produced by a WBCSD, UNEP and the GRI consortium, provide a very useful aid in this respect.

Business must also be prepared for new and emerging environmental concerns. Although these may pose a threat to some companies, they will provide opportunities for others, for example companies in the environmental technology sector. Water scarcity and other water-related issues have been identified as potential problem areas that could impact on business in the near future. Robust environmental management will help organisations to respond quickly to such emerging threats.

The trend towards sustainability management, whereby companies are attempting to systematically balance economic, environmental and social considerations in business strategies and operations continues to gather pace. The strong inter-relationships and inter-dependence between these three elements of sustainable development make it imperative for companies to move towards a more integrated approach to managing them.

There has been a real change in attitude relatively recently, with socially responsible investment (SRI) considerations becoming part of mainstream investment. Environmental and social factors are now more and more important in conventional investment, and fund managers are being more discerning about where their capital is invested. SRI factors are now commonly viewed not as a hindrance but as a critical tool in analysing corporate sustainability and long-term value. Continued development of this approach will be greatly influenced by shareholders, market analysts and financial institutions recognising the value of non-financial performance information as a basis for evaluating management competence and predicting business performance. Consequently they can increasingly be expected to insist on changes to the information presented to them by companies, thereby generating a radical shift in the traditional business paradigm.

Therefore, while environmental management will remain a critical issue for business to address, it is increasingly becoming integrated with other aspects of business management, reflecting a growing acceptance of the importance of sustainability. This is particularly the case in the UK. Each year the media company Corporate Knights produces a list of the world's 100 most sustainable companies. The latest list shows that 11 of these companies are based in the UK.

The Equality Act 2010

Alexander M S Green

Introduction to the Equality Act 2010

[E16501] Over the last forty years, a significant body of anti discrimination legislation has evolved covering individuals with a variety of 'protected characteristics' (e.g. race, age, disability, sex and sexual orientation etc). On 1 October 2010, The *Equality Act 2010* ('*EA 2010*') came into force thereby codifying the anti-discrimination legislation into one statute. Much of the case law on the earlier legislation and which pre-dates the EA will, however, remain relevant as the *EA 2010* draws heavily on the earlier legislation to which those cases refer.

Under the *EA 2010*, discrimination is unlawful if it is because of one or more of the 'protected characteristics' as defined by the statute. The 'protected characteristics' are:

- age;
- disability;
- gender reassignment;
- marriage and civil partnership;
- race;
- religion or belief;
- sex;
- sexual orientation;
- pregnancy and maternity.

The effect of the *EA 2010* is that employees and other workers are protected against direct discrimination, indirect discrimination, harassment, victimisation and discrimination arising from a disability and from a failure to make reasonable adjustments for disabled people. Employers and principals are generally liable for the discriminatory acts of their employees and agents.

The purpose of this chapter is to focus on the specific health & safety issues arising from individuals who have certain 'protected characteristics' under the *EA 2010*. Should readers require a general understanding of anti-discrimination law and employment rights they should consult the standard texts on the subject such as *Tolley's Employment Handbook* (27th Ed).

The HSE Stance on Equality and Diversity

[E16502] The HSE's stance on equality and diversity is expressed as follows:

For health and safety purposes, equality is concerned with breaking down the barriers that currently block opportunities for certain groups of people in the workplace, aiming to identify and minimise the barriers that exclude people and to take action to achieve equal access to all aspects of work for everyone. Eliminating discrimination is important in achieving equality, since it is not just the physical environment or poor policies that prevent equality from being achieved but also ways of working, attitudes and stereotypes about different groups of people.

Diversity is about recognising, valuing and taking account of people's different backgrounds, knowledge, skills, and experiences, and encouraging and using those differences to create a productive and effective workforce.

The workforce and working patterns are changing. The working population is getting older and there are more women and people from ethnic minorities at work.

Everyone has the right to be treated fairly at work and to be free of discrimination on grounds of age, race, gender, disability, sexual orientation, religion, pregnancy and maternity, gender reassignment, or belief.

Many employers have found that making adaptations to their working practices to accommodate a diverse workforce makes good business sense. It makes their business more attractive to both potential employees and customers and helps them recruit and retain the best people. This is not only good business sense but helps them meet the requirements of legislation.

Health and safety should never be used as a false excuse to justify discriminatory action.

(www.hse.gov.uk/diversity/discrimination.htm).

The HSE has stated that it is committed to the health and safety at work of all people in Britain and that it embraces the diversity of Britain's workforce and appreciates the roles it can play in eliminating discrimination and promoting equality of opportunity.

As part of its remit it believes that:

All people's health and safety in the workplace must be protected, whatever their race, gender, disability, age, religion or sexual orientation. It must be protected whatever their background and outlook on life.

The HSE is promoting equality and is tackling discrimination by:

- designing its interventions through an equality impact assessment tool;
- improving communications when reaching out to the diverse range of people and organisations it serves by:
  - providing a communication tool that helps HSE staff to understand the importance of the diverse audiences they are working with and the potential barriers to communication;
  - Responding positively to requests for alternative formats.
- building and making better use of research to provide information about diversity issues;
- where necessary, improving the diversity of our advisory bodies and its stakeholder networking.

(www.hse.gov.uk/diversity/discrimination.htm).

Prohibited Conduct

[E16503] The *EA 2010* prohibits certain types of conduct. The majority of the cases that are heard in the employment tribunal relate to direct discrimination but it is also important to note that prohibited conduct extends to cover: indirect discrimination, victimisation, harassment and, in the case of disability, discrimination arising from a disability.

Direct discrimination

[E16504] *EA 2010, s 13* defines direct discrimination as differential treatment because of a protected characteristic. A twofold test is applied:

- was a person treated less favourably than an actual or hypothetical comparator was or would have been treated in circumstances that were the same or not materially different?; and
- if so, was that less favourable treatment because of a protected characteristic?

Indirect discrimination

[E16505] Indirect discrimination arises where everyone is treated in the same way, but the consequences of that treatment are that people holding a protected characteristic are impacted more disparately than those who do not. *EA 2010, s 19* outlaws indirect discrimination. Indirect discrimination applies to all persons holding protected characteristics except pregnancy and maternity.

Discrimination arising from a disability

[E16506] *EA 2010, s 15* provides that disabled people are protected from unfavourable treatment arising in consequence of the disability where the same cannot be shown to be a proportionate means of achieving a legitimate aim. An example of this is given by the Equality and Human Rights Commission (EHRC):

> A woman is disciplined for losing her temper at work. However, this behaviour was out of character and is a result of severe pain caused by cancer, of which her employer is aware. The disciplinary action is unfavourable treatment. This treatment is because of something which arises in consequence of the worker's disability, namely her loss of temper. There is a connection between the 'something' (that is, the loss of temper) that led to the treatment and her disability. It will be discrimination arising from disability if the employer cannot objectively justify the decision to discipline the worker.

(www.equalityhumanrights.com.)

If a disabled person is treated less favourably than he or she would otherwise be treated by reason of something arising in consequence of the disability, an act of discrimination will have occurred (subject to the defence of justification). *EA 2010, s 15(1)(b)* sets out the test of justification. The treatment must be a proportionate means of achieving a legitimate aim.

An employer will also have a defence against this type of claim in circumstances where he did not know or could not reasonably have been expected to have known that the disabled person had a disability (*EA 2010, s 15(2)*).

Harassment

[E16507] *EA 2010, s 26* outlaws harassment. There are three separate species of harassment:

- harassment which is related to a protected characteristic;
- sexual harassment;
- less favourable treatment arising out of harassment.

Victimisation

[E16508] Victimisation is prohibited by *EA 2010, s 27*. This outlaws retaliation by employers where a person has done a protected act or where it is considered has done or may do a protected act.

Protected acts consist of:

- bringing proceedings under *EA 2010*;
- giving evidence or information in connection with proceedings under *EA 2010*;
- doing any other thing for the purposes of or in connection with proceedings under the EA 2010;
- making an allegation that a person has contravened *EA 2010*.

If the allegation, evidence or information is false and given or made in bad faith, it will not be a protected act (*EA 2010, s 27(3)*).

Health And Safety for Disabled People

Introduction

[E16509] The *Health and Safety at Work etc Act 1974, s 2* (*HSWA 1974*) imposes a general duty on all employers to ensure, so far as is reasonably practicable (for the meaning of this expression, see **E15039** Enforcement), the health, safety and welfare of all their employees. This includes disabled employees. The HSE estimates that 2% of the UK working age population becomes disabled every year. Furthermore, there are approximately 10 million disabled people in Great Britain who are covered by the *EA 2010*. This represents around 18% of the population (www.hse.gov.uk).

Employers should avoid discriminating against their disabled workers and carry out appropriate risk assessments to take account of an employee's disability which may entail making reasonable adjustments to accommodate the disability.

The meaning of 'disability'

[E16510] The starting point for understanding the health and safety implications of disability is to consider the meaning of 'disability'. Disability is defined by *EA 2010, s 6* as being where 'a person has a physical or mental impairment and that impairment has a substantial and long term adverse effect on the person's ability to carry out normal day to day activities'. Further information on the concept of disability is found in *Equality Act 2010 (Disability) Regulations 2010, (SI 2010/2128)* (www.legislation.gov.uk/uksi/2010/2128/contents/made) (the '2010 Regulations'). The Office for Disability Issues of HM Government has issued guidance on matters to be taken into consideration in determining questions relating to the definition of disability which can be found at:

www.equalityhumanrights.com.

These guidance notes provide a useful set of examples of the type of factual questions which are likely to be determined by employment tribunals in determining whether a person is or is not disabled. The Employment Appeal Tribunal ('EAT') has also provided guidance to employment tribunals as to the correct approach to establish whether a person is disabled or not. In *Goodwin v The Patent Office* [1999] IRLR 4, the EAT held that there were four questions which an employment tribunal should consider in determining whether a person was disabled or not:

- does the individual have a mental or physical impairment?
- does the impairment have an adverse effect of the ability of the individual to carry out normal day to day activities?
- is the adverse effect substantial?
- if the adverse effect is substantial, does it have a long term effect?

Whilst the concept of impairment is not defined by *EA 2010* the case law on the topic considers it to be a wide ranging term. A person may suffer from a physical impairment even where there is no underlying fault or defect (e.g. where the person suffers from the effects of an illness but it is not possible to identify the root cause of the illness) (*College of Ripon and York St John v Hobbs* EWCA Civ 1074). There is no longer a requirement for mental impairment to be linked to a clinically well recognised condition. Consequently, a person can be suffering from a mental impairment even though it does not amount to mental illness (*Dunham v Ashford Windows* [2005] ICR 1584).

Whether an impairment has an adverse effect on the ability of an individual to carry out normal day to day activities involves a consideration of what the individual is able to do and the way in which the individual is able to do it as against what would be the case if the individual did not have the impairment. In particular, the enquiry should focus on what the individual cannot do as opposed to what the individual can do (*Leonard v Southern Derbyshire Chamber of Commerce* [2001] IRLR 19). If a person can come to work and is able to do his job this does not mean that he is not necessarily disabled (*Law Hospitals NHS Trust v Rush* [2001] IRLR 611). It may also be necessary to consider what normal day to day activities an individual is able to carry out in the workplace if the working environment exacerbates the individual's impairment (*Cruickshank v VAW Motorcast Limited* [2002] IRLR 24). In *J v DLA Piper UK LLP* (UKEAT/0263/09/RN) [2010] ICR 1052, [2010] IRLR 936, EAT, the EAT confirmed that, particularly in cases where the mental impairment is disputed, the focus of the tribunal's enquiry should be on the effect the impairment has on an employee's day-to-day activities. If the tribunal finds a long-term substantial adverse effect, it will, in most cases, follow "as a matter of common sense inference" that the claimant is suffering from an impairment which has produced that effect. Having drawn that inference, the tribunal will not have to resolve the kinds of difficult medical issues which may otherwise arise. Any measures taken to treat or to correct an impairment are disregarded (e.g. medication) unless those measures mean that the individual is restored to being able to carry out normal day to day activities.

To be long term, the effects of the impairment must:

- have lasted for at least 12 months; or
- be likely to last for at least 12 months; or
- be likely to last for the rest of the life of the person.

If the impairment is likely to be recurrent it will be treated as though it is continuing.

EA 2010, Sch 1, para 3 provides that certain specified impairments are considered to be disabilities without more. These are:

- blindness, severe sight impairment, sight impairment and partial sightedness (provided this is certified by a consultant ophthalmologist);
- severe disfigurement;
- cancer;
- HIV;
- Multiple Sclerosis.

If a person suffers from a progressive condition which is likely to result in that person having impairment this will be deemed to be a disability.

The following conditions, set out in the 2010 Regulations are expressly stated not to be impairments (and hence are not disabilities) for EA 2010 purposes:

- Addiction to alcohol, nicotine or any other substance (although this does not apply where the addiction was originally the result of the administration of medically prescribed drugs or other medical treatment) (regulation 3). "Addiction" for this purpose "includes a dependency" (regulation 2);
- Tendency to set fires (regulation 4(1)(a));
- Tendency to steal (regulation 4(1)(b));
- Tendency to physical or sexual abuse of other persons (regulation 4(1)(c));
- Exhibitionism (regulation 4(1)(d));
- Voyeurism (regulation 4(1)(e));
- Tattoos and body piercings (regulation 5);
- Seasonal allergic rhinitis (usually known as hayfever). However, hayfever can be taken into account for EqA 2010 purposes where it aggravates the effect of any other condition (regulation 4(2) and (3)).

Health And Safety for Disabled People [E16511]

The duty to make reasonable adjustments

[E16511] *EA 2010, ss 20, 21, 22* and *Schedule 8* impose the duty to make reasonable adjustments in respect of disabled people. *Schedule 8* specifically deals with the duty to make reasonable adjustments in the workplace. If an employer fails to comply with his duty to make reasonable adjustments he is guilty of unlawful discrimination.

The duty to make reasonable adjustments consists of three requirements which are set out in *EA 2010, s 20(3), (4) & (5)*:

(3) The first requirement is a requirement, where a provision, criterion or practice of A's puts a disabled person at a substantial disadvantage in relation to a relevant matter in comparison with persons who are not disabled, to take such steps as it is reasonable to have to take to avoid the disadvantage.

(4) The second requirement is a requirement, where a physical feature puts a disabled person at a substantial disadvantage in relation to a relevant matter in comparison with persons who are not disabled, to take such steps as it is reasonable to have to take to avoid the disadvantage.

(5) The third requirement is a requirement, where a disabled person would, but for the provision of an auxiliary aid, be put at a substantial disadvantage in relation to a relevant matter in comparison with persons who are not disabled, to take such steps as it is reasonable to have to take to provide the auxiliary aid.

The first requirement addresses changing the way that things are done (e.g. changing a practice). The second requirement deals with making changes to the built environment (e.g. providing access to a building). The third requirement covers auxiliary and services (e.g. providing special computer software). In relation to the second requirement *EA 2010, s 20(9)* provides that a reference to avoiding a substantial disadvantage includes a reference to:

- removing the physical feature in question;
- altering it; or
- providing a reasonable means of avoiding it.

With regard to the third requirement, *EA 2010, s 20(11)* provides that reference to an auxiliary aid includes a reference to an auxiliary service.

EA 2010, Schedule 8 applies to the duty to make reasonable adjustments in the workplace. In this context, the duty is owed to an 'interested disabled person'. For the purposes of *Schedule 8*, an 'interested disabled person' is defined in relation to employers (referred to as 'employer A') as:

- an employee of employer A;
- an applicant for employment with employer A;
- a person who is or who has notified employer A that the person may be an applicant for employment.

The description of a disabled person is broad in that it also covers:

- principals in contract work;
- partnerships;
- limited liability partnerships;
- barristers/ advocates and their clerks;
- those making appointments to offices;

- qualification bodies;
- employment service providers;
- trade organisations;
- local authorities; and
- occupational pensions.

The EHRC have produced a Statutory Code of Practice which should be read in conjunction with the *EA 2010* (www.equalityhumanrights.com (the 'Code of Practice'). The Code of Practice sets out, in chapter 6, the principles and the application of the duty to make reasonable adjustments for disabled people who are in employment. The duty is referred to as a 'cornerstone' of the Act requiring employers to take positive steps to ensure that disabled people can access employment and progress in employment. The duty is not simply one of avoiding treating disabled workers, job applicants and potential job applicants unfavourably. The duty is broader and encompasses taking additional steps to which non-disabled workers and applicants are not entitled (*Archibald v Fife Council* [2004] IRLR 651).

The duty to make adjustments is qualified by the word 'reasonable'. The *Disability Discrimination Act 1995, s 18B* provides a useful list of what would be considered to constitute reasonable adjustments. These provisions have not been re-enacted by *EA 2010* but they do, nonetheless, provide a useful checklist for employers and the HSE have replicated these in their own guidance on the subject and have provided some examples of reasonable adjustments which include:

- adjustments to the workplace to improve access or layout;
- giving some of the disabled person's duties to another person, e.g. employing a temp;
- transferring the disabled person to fill a vacancy;
- changing the working hours, e.g. flexi-time, job-share, starting later or finishing earlier;
- time off, e.g. for treatment, assessment, rehabilitation;
- training for disabled workers and their colleagues;
- getting new or adapting existing equipment, e.g. chairs, desks, computers, vehicles;
- modifying instructions or procedures, e.g. by providing written material in bigger text or in Braille;
- improving communication, e.g. providing a reader or interpreter, having visual as well as audible alarms;
- providing alternative work (this should usually be a last resort).

(www.hse.gov.uk)

The duty to make adjustments is qualified by the word 'reasonable'. Examples of adjustments to working arrangements include:

- allowing a phased return to work;
- changing individual's working hours;
- providing help with transport to and from work;
- arranging home working, providing a safe environment can be maintained;

- allowing an employee to be absent from work for rehabilitation treatment.

Examples of adjustments to premises include:

- moving tasks to more accessible areas;
- making alterations to premises.

Examples of adjustments to a job include:

- providing new or modifying existing equipment and tools;
- modifying work furniture;
- providing additional training;
- modifying instructions or reference manuals;
- modifying work patterns and management systems;
- arranging telephone conferences to reduce travel;
- providing a buddy or mentor;
- providing supervision;
- reallocating work within the employee's team;
- providing alternative work.

(www.hse.gov.uk)

Further guidance on what would be constituted as reasonable adjustments can also be found in Chapter 6, paragraph 6.33 of the Code of Practice (www.equalityhumanrights.com/advice-and-guidance/guidance-for-employers/the-duty-to-make-reasonable-adjustments-for-disabled-people). An adjustment would be unlikely to be reasonable to an employer if would involve little benefit to the disabled person. The test of reasonableness is essentially an objective one. It may not be necessarily met by an employer who shows that he personally believed that the making of the adjustment would be too disruptive or costly (*Smith v Churchills Stairlifts plc* [2006] IRLR 41).

The employer's knowledge of a person's disability is relevant to the question of the duty to make reasonable adjustments. *EA 2010, Schedule 8, Part 3, para 20* states:

A is not subject to a duty to make reasonable adjustments if A does not know, and could not reasonably be expected to know—(a) in the case of an applicant or potential applicant, that an interested disabled person is or may be an applicant for the work in question; (b) in any other case referred to in *Part 2* of this *Schedule*, that an interested disabled person has a disability and is likely to be placed at a substantial disadvantage referred to in the first, second or third requirement.

In essence this means:

- did the employer know both that the employee was disabled and that his disability was liable to affect him in the manner set out in the definition of disability? If the answer to that question is 'no', then there is a second question, namely;
- ought the employer to have known both that the employee was disabled and this disability was liable to affect him in the manner set out in the definition of disability?

If the answer to the second question is 'no' then there is no duty to make reasonable adjustments.

Risk assessment and disability

[E16512] Risk assessment lies at the heart of health and safety management and consists of a careful examination of what could harm people and how likely this is to happen, so that employers can weigh up whether or not the steps they have taken are sufficient to comply with health and safety law. Clearly in relation to disabled people, the duty to make reasonable adjustments under the *EA 2010* interacts with an employer's duty to conduct risk assessments under general health and safety law and this fact should always be taken into consideration.

The quality of the risk assessment in relation to disabled people is critical. This point is emphasised by the EHRC who state that:

> Your employer may need to use specialist staff and the person doing the assessment must:
> - focus on you as an individual, not people with your condition in general;
> - consider the facts;
> - not make assumptions;
> - get individual specific medical advice; and
> - talk to you about how reasonable adjustments can be made.

The risk assessment should also consider the essential elements of the job; the length of time and frequency of any hazardous situations; and any reasonable adjustments that can be made to reduce the risk.

If there is still an unacceptable risk, even with adjustments, then the employer could lawfully dismiss or not employ the employee. The question is what is 'unacceptable'? This can only be tested in the courts. Increasing case law precedent is being set, which gives further guidance to employers.

The employer is the person who must take the decision about whether to employ or retain the employee in a job. If they seek expert advice from medical services or health and safety specialists, these are agents of the employer, and the employer has a duty to ensure that the specialists have considered all the facts.

(www.equalityhumanrights.com)

Risk assessments can be carried out in five steps:

- identify the hazards (what, in the work, could cause harm to people);
- decide who might be harmed and how;
- evaluate the risks and decide on precautions;
- record the findings and act on them;
- review the assessment and update if necessary.

(www.hse.gov.uk)

The HSE identify Step 2 as being about 'who might be harmed?'. To answer this question, employers must think about their whole workforce including disabled employees. There may be specific risks that arise in relation to a disabled employee, and in addressing these the employer should consult and involve the employee in the risk assessment. This helps avoid the following:

- people making assumptions about disabled people which can lead to poor practice or discrimination;

- people hiding an impairment that might have health and safety implications for fear they won't get or keep a job.

Examples of assumptions include:

- thinking a driver who loses an arm can't drive anymore. Steering wheels can be modified or replaced, e.g. with a joystick;
- believing a deaf person can't be warned of fire. Flashing lights can supplement sirens and bells.

Employers should:

- make sure they manage work risks for everyone;
- take account of disability, avoiding assumptions;
- involve disabled workers in doing risk assessments and making 'reasonable adjustments';
- consult others with appropriate expertise where necessary;
- review the situation if necessary, working with the disabled person and/or their representative.

Specific risks to disabled people arise in relation to fires in the workplace. The EHRC guidance on the topic is as follows:

> One thing that employers often worry about when thinking about employing a disabled person is what would happen in the event of a fire. This is often based on lack of information, with employers wrongly believing that wheelchair users should not be employed because they would not be able to escape from a building on fire where lifts were out of use. Deaf, hearing impaired blind people may also be discriminated against because people wrongly believe they won't know there is a fire alarm or may 'get in the way' of colleagues trying to escape.
>
> If you have a disability that may present a difficulty during a fire at your workplace, you need to have a discussion and draw up an agreed plan with your line manager and/or premises manager about this. There may be very simple changes that can be made to stop this from ever being a problem:
>
> - providing flashing lights as alarms, as well as things that make a noise;
> - making sure that colleagues working with deaf people have a basic awareness of sign language and deaf issues;
> - the establishment of a 'buddy' system to ensure wheelchair users are helped in an emergency;
> - providing a visually impaired person with named guides in case of fire;
> - providing temporary places of refuge for wheelchair users protected by fire resistant doors and from which there is a safe route to a final exit (the refuge may also have a means of communication to a central control point); or
> - training for staff in evacuation plans.
>
> These plans should be made known to all staff concerned and tested at regular intervals. A well thought through evacuation plan should take into account the needs of every individual in the building, including, for example, women in the later stages of pregnancy. Very often, this can help to improve procedures and safety overall. You may wish to discuss your needs in confidence, in which case only certain key individuals need know about your unique plan.

(www.equalityhumanrights.com/advice-and-guidance/your-rights/disability/disability-in-employment/health-and-safety-in-the-workplace)

Health and Safety in Relation to Age

Introduction

[E16513] Age is a protected characteristic (*EA 2010, s 5*). The concept of age includes a reference to a 'person of a particular age'. The protected characteristic encompasses the concept of an age group which is referred to in *EA 2010, s 5(2)* as a group of persons defined by reference to age whether by reference to a particular age or to a range of ages.

It is unlawful to discriminate against young workers as well as older workers. All workers and people who apply for work and vocational training are protected from: direct discrimination, indirect discrimination, harassment and victimisation. Discrimination because of age can be justified as a proportionate means of achieving a legitimate aim (*EA 2010, s 13(2)*).

Employers should not use 'health and safety' as an excuse for not continuing to employ older workers or for not receiving training.

Ageing workers

[E16514] It has been estimated that by 2020, approximately 1/3 of the UK's workforce will be over the age of 50. Some industries such as the UK offshore oil and gas industry have an ageing workforce. In 2008, the average age of people working offshore was 45.4 years (source: UK Oil and Gas www.oilandgasuk.co.uk)

In other industries such as manufacturing and construction older workers tend to be leaving at faster rates. Older workers are also likely to be self-employed. The HSE maintain that age is not an equivalent to personal capacity to work although it is accepted that certain cognitive functions such as memory abilities deteriorate with age.

Certain industries such as the agriculture and construction sectors have demonstrated persistent problems with workers over 65. In those sectors the fatality statistics show that the rate of fatal injuries is higher in workers aged 45 plus compared with younger workers (www.hse.gov.uk).

Since the abolition of the UK's statutory default retirement age of 65 in April 2011 combined with the general economic downturn and the 'pensions' crisis, the proportion of older people continuing to work beyond the age of 65 has increased. Furthermore, people are living longer and if they continue to work into their old age their employers will need to assess the health and safety risks which are associated with an ageing workforce.

Risk assessment

[E16515] As people grow older their ability to do their job may change as well as their perception of how they do their job. As part of their general duty to conduct risk assessments, employers should take account of their ageing employees to ensure that they can continue to perform their work safely. There are specific risks which need to be assessed in relation to older workers although each case should always to be assessed on its own particular circumstances. In its guidance on the matter the HSE states that employers should:

- carry out risk assessments routinely and not simply when an employee reaches a certain age;
- assess the activities involved in jobs and modify workplace design if necessary;
- make adjustments on the basis of individual and business needs, not age;
- consider modifying tasks to help people stay in work longer – with appropriate training;
- allow staff to change work hours and job content;
- not assume that certain jobs are too demanding for older workers. Decisions should be based on capability and objective risk and not age;
- encourage or provide regular health checks for staff, regardless of age;
- persuade staff to take an interest in their health and fitness;
- consider legislative duties such as the *EA 2010* or flexible working legislation. These could require businesses to make adjustments to help an employee with a health issue or consider a request to work flexibly.

(www.hse.gov.uk/vulnerable-workers/older-workers.htm).

Young workers

[E16516] Young workers (i.e. above school leaving age and under 18) are protected to at least the same level as adult workers. Furthermore, young people, especially those new to the workplace, will encounter unfamiliar risks from the jobs they will be doing and from the working environment. The Key risks for young people when starting work may arise because of their lack of experience or maturity and not having the confidence to ask for or knowing where they can get help. Employers should take account of this in their risk assessments.

The HSE provides guidance on the key risks which need to be assessed in relation to young workers (http://www.hse.gov.uk/youngpeople/risks/index.htm#ypkey). In summary these are as follows:

Prior to employing a young person, the health and safety risk assessment must take the following into account:

- the fitting-out and layout of the workplace and the particular site where they will work;
- the nature of any physical, biological and chemical agents they will be exposed to, for how long and to what extent;
- what types of work equipment will be used and how this will be handled;
- how the work and processes involved are organised;
- the need to assess and provide health and safety training; and
- risks from the particular agents, processes and work.

The Management of Health and Safety at Work Regulations require that young people are protected at work from risks to their health and safety which are a consequence of the following factors:

- physical or psychological capacity;
- pace of work;
- temperature extremes, noise or vibration;
- radiation;
- compressed air and diving;
- hazardous substances;
- lack of training and experience.

There are also risks to young people associated with specific industries or processes:

- agriculture;
- carriage of dangerous explosives and goods;
- shipbuilding and Ship-repairing Regulations;
- provision and use of work equipment;
- power presses;
- woodworking machines;
- mechanical lifting operations (including lift trucks).

The HSE state that there is no need for employers to carry out a new risk assessment each time they employ a young person, as long as their current risk assessment takes account of the characteristics of young people and activities which present significant risks to their health and safety.

The HSE recommend that employers may wish to consider developing generic risk assessments for young people as these could be useful when they are likely to be doing temporary or transient work, and when the risk assessments could be modified to deal with particular work situations and any unacceptable risks.

In all cases, employers will need to review the risk assessment if the nature of the work changes or when they have reason to believe that it is no longer valid.

In carrying out the risk assessment employers should identify the measures they need to take to control or eliminate health and safety risks.

Except in special circumstances, employers should not employ young people to do work which:

- is beyond their physical or psychological capacity;
- exposes them to substances chronically harmful to human health, e.g. toxic or carcinogenic substances, or effects likely to be passed on genetically or likely to harm the unborn child;
- exposes them to radiation;
- involves a risk of accidents which they are unlikely to recognise because of e.g. their lack of experience, training or attention to safety;
- involves a risk to their health from extreme heat, noise or vibration.

These restrictions will not apply in 'special circumstances' where young people over the minimum school leaving age are doing work necessary for their training, under proper supervision by a competent person, and providing the risks are reduced to the lowest level, so far as is reasonably practicable. Under no circumstances can children of compulsory school age do work involving these risks, whether they are employed or under training such as work experience.

Training includes Government-funded training schemes for school leavers, modern apprenticeships, in-house training arrangements and work qualifying for assessment for National/Scottish Vocational Qualifications, e.g. craft skills.

(http://www.hse.gov.uk/youngpeople/risks/index.htm#ypkey)

Further guidance on the management of health and safety of migrant workers can be found at: www.hse.gov.uk/migrantworkers/employer.htm#who).

Health and Safety in Relation to Race

Introduction

[E16517] Race is a protected characteristic. It is defined by *EA 2010, s 9(1)* to include:

- colour;
- nationality;
- ethnic or national origins.

EA 2010, s 9(2)(b) makes reference to persons who share a colour, nationality and/or ethnic origins as being persons of the same racial group. A racial group is a group of persons defined by reference to race (*EA 2010, s 9(3)*).

Examples of the concepts of colour, nationality, ethnic or national origins and a racial group are:

- colour includes being black or white;
- nationality includes being a British, Australian or Swiss;
- ethnic or national origin includes being from a Roma background or of Chinese heritage;
- a racial group could be 'Black Britons' which includes those people who are black and who are British citizens.

Jews and Sikhs have been held be members of a racial group. English and Scottish can be national origins even where the individual can describe his nationality as British.

Employers should not use 'health and safety' as an excuse not to employ people from different races or to continue to employ or to provide them with appropriate training.

Migrant workers

[E16518] The HSE have indentified that race can be a relevant factor in protecting the health and safety of workers. Race is a relevant factor to consider because of:

- differences in vulnerability;
- networks and communication channels for promoting health and safety; and
- the challenges of working with a different language.

These points are starkly illustrated in relation to the employment of migrant workers who are commonly from different racial backgrounds (e.g. Chinese).

In 2006, the HSE published its report 'Migrant Workers in England & Wales' (www.hse.gov.uk).

The Report was based on interviews with 200 migrant workers based in five regions of England & Wales and highlights the risks that migrant workers face and makes recommendations on how the health and safety of this group of workers can be protected. The Report found that in general, migrant workers are often over-qualified for the work that they do and they work long and anti social hours. The Report states that without migrant workers, in some areas of the country, hospitals and care homes would be without staff, construction would not boom, hotels and restaurants would be unable to service customers and farmers and food processing firms would not be able to distribute their produce to the shops.

The Report revealed that the vast majority of migrant workers were working with other migrant workers. In some cases a particular nationality might be dominant but in others the workforce could consist of workers from many different countries speaking different languages and with different skills and experience and knowledge of health and safety systems.

Few checks were made on migrant workers' skills and qualifications for undertaking the work that they were doing. There were cases where migrant workers were performing skilled and potentially dangerous work such as scaffolding who had no previous experience in the task. In food production and in catering most workers were not tested for their knowledge of food hygiene and only a minority were offered training in food hygiene.

Whilst many migrant workers were recruited directly by the employer, a significant number were supplied through recruitment agencies. Many of the migrant workers had bad experiences of dealing with recruitment agencies which included:

- being paid less;
- suffering unexplained deductions from wages;
- irregular work patterns;
- lack of clarity where health and safety responsibilities lay.

One third of the migrant workers interviewed had received no training in health and safety. The remaining two thirds had received cursory training, normally limited to a short session at induction. The problem was exacerbated where there was no common language amongst the work force which complicated effective communication of health and safety training.

Whilst personal protective equipment ('PPE') was generally provided, many migrant workers did not know how to use it properly.

25% of the migrant workers interviewed had either experienced an accident at work or had witnessed one involving a migrant co-worker. This suggested a higher level of accidents than would be expected by UK workers. Many of the accidents were associated with fatigue brought about by working excessively long or anti-social hours. Many migrant workers did not report accidents as they feared dismissal and deportation. Some of the migrant workers under estimated the risks that they faced in doing their job. The HSE attributed this to circumstances where the worker was undertaking work which they perceived to be below their qualifications and skills and they consequently tended to be less conscious of the risks associated with the jobs that they were doing. They took fewer measures to reduce health and safety risks. Risk assessment strategies should, therefore, take account of whether the migrant worker is engaged in work in which he or she has no previous experience.

Many of the workers suffered from discrimination at work because of their race or nationality. Many of the workers interviewed believed that they were allocated to the worst shifts and had less favourable terms and conditions. They frequently referred to name calling and harassment by supervisors and co-workers. The HSE hypothesised that that such discrimination might have an impact on worker health and safety in circumstances where discriminatory treatment combined not simply to contribute to stress at work but an inability to raise concerns about health and safety at work. Discrimination impacts on a worker's ability to challenge unfair and unsafe practices.

Lack of English was also a problem. For example, some of the migrant workers interviewed admitted that they were unable to follow the health and safety training that they were offered. Lack of English also impacted on issues such as supervision of work.

The Migrant workers also expressed a low level of knowledge of their health and safety rights and of how to enforce them. There was a generally held view that responsibility for health and safety lay with the individual worker and that accidents and incidents at work were the fault of individual workers. Consequently, they did not assume that their employers (or recruitment agencies or labour providers) had responsibility for their health and safety.

Risk assessment

[E16519] The HSE's Report on migrant workers highlights specific issues which should be risk assessed. The HSE have made the following recommendations:

- whilst there are no requirements in health and safety legislation for employers to ensure their staff are fluent in English the HSE recommends steps should be taken to ensure understanding of health and safety issues;
- the law requires that employers provide workers with comprehensible and relevant information about risks and about the procedures they need to follow to ensure they can work safely and without risk to health. This does not have to be in English;
- the employer may make special arrangements, which could include translation, using interpreters or replacing written notices with clearly understood symbols or diagrams;

- any health and safety training provided must take into account the worker's capabilities, including language skills;
- workers who do not speak English may need to understand key words and commands relating to danger, e.g. 'Fire' and 'Stop'. Employers will need to ensure that this is communicated clearly and simply, and check understanding afterwards;
- employers may wish to consider suitably tailored ESOL (English for Speakers of Other Languages) provision for longer-term workers.
- there will be few situations where health and safety considerations alone justify not employing workers with poor or no English.

(www.hse.gov.uk).

Health and Safety in Relation to Gender

Introduction

[E16520] Gender or sex is a protected characteristic. *EA 2010* defines references to sex as references to a man or to a woman. *EA 2010, s 13(6)* includes within the definition of sex (or more specifically less favourable treatment) the fact that a woman is breast feeding but excludes from the definition of sex the special treatment of women connected with pregnancy or childbirth.

Risk Assessment

[E16521] The HSE have estimated that women make up 42% of the employed population in the EU. Men and women are not the same and the jobs they do, their working conditions and how they are treated by society are not the same. These factors can affect the hazards they face at work and the approach that needs to be taken to assess and control them. The HSE states that factors to take into account include:

- women and men are concentrated in certain jobs, and therefore face hazards particular to those jobs;
- women and men face different risk to their reproductive health.

The impact of gender on both men's and women's occupational health and safety is generally under-researched and poorly understood. However, discrimination against new and expectant mothers is well known and the HSE has been working closely with other government departments to tackle this. There are specific provisions which address risk assessment in relation to pregnant workers which are dealt with elsewhere in this work [E14025].

The European Agency for Safety and Health at Work ('EASHW') have produced research into gender issues in relation to risk assessment which is summarised in its Fact Sheet 43 'Including gender in risk assessments'

(http://europa.eu/)

It is argued that taking a gender neutral approach to risk assessment and prevention can result in risks to female workers being underestimated or even being ignored altogether. It is further suggested that when we consider hazards at work there is a tendency to think about men working in high accident risk areas such as building site or a fishing vessel rather than women working in health and social care or call centres. It is argued that a careful evaluation of real work circumstances shows that men and women can face significant risks at work. Gender risks should be built into workplace risk assessments. The table below shows examples of hazards and risks which are found in female dominated work areas.

Work Area	Risk Factors and health problems include:			
Healthcare	Infectious diseases, e.g. blood borne, respiratory, etc.	Manual handling and strenuous postures; ionising radiation.	Cleaning, sterilising and disinfecting agents; drugs; anaesthetic gases.	'Emotionally demanding work'; shift and night work; violence from clients and the public.
Nursery workers	Infectious diseases, e.g. particularly respiratory.	Manual handling, strenuous postures.		'Emotional work'
Cleaning	Infectious diseases; dermatitis.	Manual handling, strenuous postures; slips and falls; wet hands.	Cleaning agents.	Unsocial hours; violence, e.g. if working in isolation or late.
Food production	Infectious diseases, e.g. animal borne and from mould, spores, organic dusts.	Repetitive movements, e.g. in packing jobs or slaughterhouses; knife wounds; cold temperatures; noise.	Pesticide residues; sterilising agents; sensitising spices and additives.	Stress associated with repetitive assembly line work.

Catering and restaurant work	Dermatitis.	Manual handling; repetitive chopping; cuts from knives and burns; slips and falls; heat; cleaning agents.	Passive smoking; cleaning agents.	Stress from hectic work, dealing with the public, violence and harassment.
Textiles and clothing	Organic dusts.	Noise; repetitive movements and awkward postures; needle injuries.	Dyes and other chemicals, including formaldehyde in permanent presses and stain removal solvents; dust.	Stress associated with repetitive assembly line work.
Laundries	Infected linen, e.g. in hospitals.	Manual handling and strenuous postures; heat.	Dry cleaning solvents.	Stress associated with repetitive and fast pace work.
Ceramics sector		Repetitive movements; manual handling.	Glazes, lead, silica dust.	Stress associated with repetitive assembly line work.
'Light' manufacturing		Repetitive movements, e.g. in assembly work; awkward postures; manual handling.	Chemicals in microelectronics.	Stress associated with repetitive assembly line work.
Call centres		Voice problems associated with talking; awkward postures; excessive sitting.	Poor indoor air quality.	Stress associated with dealing with clients, pace of work and repetitive work.

Education	Infectious diseases, e.g. respiratory, measles	Prolonged standing; voice problems.	Poor indoor air quality.	'Emotionally demanding work', violence.
Hairdressing		Strenuous postures, repetitive movements, prolonged standing; wet hands; cuts.	Chemical sprays, dyes, etc.	Stress associated with dealing with clients; fast paced work.
Clerical work		Repetitive movements, awkward postures, backpain from sitting.	Poor indoor air quality; photocopier fumes.	Stress, e.g. associated with lack of control over work, frequent interruptions, monotonous work.
Agriculture	Infectious diseases, e.g. animal borne and from mould, spores, organic dusts.	Manual handling, strenuous postures; unsuitable work equipment and protective clothing; hot, cold, wet conditions.	Pesticides.	

The EASHW recommend a model for making risk assessment more gender sensitive to take account of gender issues, differences and inequalities. They recommend a holistic approach to risk assessment to take account of the broad context of gender issues such as sexual harassment, discrimination, involvement in decision making in the workplace and conflicts between work and home. A further aim of a gender orientated risk assessment is to identify less obvious hazards and health problems that are more common with female workers.

A gender sensitive risk assessment should:

- have a positive commitment to gender issues;
- look at the real working situation;
- involve all workers, women and men, at all stages;
- avoid making assumptions about the hazards are and who is at risk.

It is suggested that hazard identification could include gender by:

- examining hazards prevalent in both male and female dominated jobs;
- looking for health hazards in addition to safety hazards;

- asking both female and male workers what problems they have in their work, in a structured way;
- avoiding making initial assumptions about what may be 'trivial';
- considering the entire workforce, e.g. cleaners, receptionists;
- not forgetting part-time, temporary or agency workers, and those on sick leave at the time of the assessment;
- encouraging women to report issues that they think may affect their safety and health at work, as well as health problems that may be related to work;
- looking at and asking about wider work and health issues.

Risk assessment could include gender:
- looking at the real jobs being done and the real work context;
- not making assumptions about exposure based purely on job description or title;
- being careful about gender bias in prioritising risks according to high, medium and low;
- involving female workers in risk assessment. Consider using health circles and risk mapping methods. Participative ergonomics and stress interventions can offer some methods;
- making sure those doing the assessments have sufficient information and training about gender issues in occupational safety and health (OSH);
- making sure instruments and tools used for assessment include issues relevant to both male and female workers. If they do not, adapt them;
- informing any external assessors that they should take a gender-sensitive approach, and checking that they are able to do this;
- paying attention to gender issues when the OSH implications of any changes planned in the workplace are looked at.

Once the risk assessment has been completed, it should be implemented to include gender by:
- aiming to eliminate risks at source, to provide a safe and healthy workplace for all workers. This includes risks to reproductive health;
- paying attention to diverse populations and adapting work and preventive measures to workers. For example, selection of protective equipment according to individual needs, suitable for women and 'non-average' men;
- involving female workers in the decision-making and implementation of solutions;
- making sure female workers as well as men are provided with OSH information and training relevant to the jobs they do and their working conditions and health effects. Ensure part-time, temporary and agency workers are included.

The risk assessment should be monitored and reviewed to include gender by involving female workers. Employers should be aware of new information about gender related occupational health issues.

Remedies and Enforcement in Relation to Unlawful Discrimination

[E16522] The employment tribunals have exclusive jurisdiction to hear claims of unlawful discrimination. Strict time limits apply for presentation of claims. *EA 2010, s 120* provides that for claims relating to discrimination in employment the application to the employment tribunal must be made:

- 3 months from the date when the discriminatory act was done (6 months for complaints within the armed forces);
- If a discriminatory act extends over a period as opposed to being a one off act, it is treated as done at the end of the period.

There is no qualifying period of service required for a claimant to make a complaint.

If an employment tribunal finds that a complaint of unlawful discrimination is well founded it must, if it considers it to be just and equitable, make one of the following orders:

- a declaration;
- an award of compensation; and/or
- a recommendation.

Compensation is assessed in the same manner as any other claim in tort or delict (Scotland) for breach of statutory duty. The award may include compensation for injury to feelings and, if appropriate injury to health and patrimonial loss flowing from the breach.

The award for injury to feelings ranges from a minimum band of between £750 and £7,000, to a middle band ranging from £7,000 to £18,000 to the maximum band of £18,000 to £30,000. The scale of compensation for injury to feelings depends on the gravity of the employer's unlawful act and the employee's reaction to it. These bands will be uprated to allow for inflation. If the discriminator acted in a high handed manner, malicious, insulting or oppressive manner the employment tribunal can award aggravated damages. The award must be compensatory and not punitive. Occasionally the employment tribunal will award 'exemplary damages' in circumstances where there has been oppressive, arbitrary or unconstitutional action by the state (or state employers). The employment tribunal may increase or decrease the award by up to 25% if it considers this just and equitable to do so if the claimant or the respondent have failed to comply with the ACAS Code of Practice.

In indirect discrimination cases, the employment tribunal may award compensation but only if the employment tribunal is satisfied that the 'provision, criterion or practice' was not applied with the intention of discriminating against the claimant.

Compensation may also be awarded in respect of financial loss flowing from the discriminatory act. Loss covers pecuniary loss, loss of benefits and expenses and are normally quantified according the same principles as in an unfair dismissal case.

The employment tribunal may also make a recommendation that the respondent takes action within a specified time for the purpose of obviating or reducing the adverse effect of any matter to which the proceedings relate both in relation to the claimant and to any other person. Failure to comply with a recommendation may lead to an increase in the award of compensation.

Unlike the position with unfair dismissal, there is no statutory limit to the amount of compensation which the employment tribunal may award.

Ergonomics

Margaret Hanson, Jill Cleaver and Andrea Oates

Introduction to ergonomics

[E17001] Ergonomics is concerned with the fit between people and the things they use. People vary considerably in a number of attributes and abilities, height, strength, visual ability, capacity to process information and so on. Ergonomics applies scientific information about human abilities, attributes and limitations to ensure the tools or equipment people use, the tasks they undertake, workstations, the working environment and work organisation are designed for people. Ergonomics adopts a people-centred approach, aiming to fit the work to the person rather than forcing the person to adapt to poorly-designed equipment, furniture, environments or work systems. Correct application of ergonomics will produce a work system that optimises human performance and minimises the risk to workers' health and safety. Ergonomics can be applied to any environment or system (work, leisure, travel, home, etc) with which people interact.

The word 'ergonomics' comes from the Greek words 'ergos', meaning 'work,' and 'nomos', meaning 'natural law'. Historically the term 'ergonomics' has been used in the UK and Europe, while 'human factors' has been used in North America, although this distinction is now blurring. Essentially, the terms 'ergonomics' and 'human factors' are synonymous, although in some quarters a distinction has been drawn between physical workplace design (which is referred to as 'ergonomics') and system design and human behaviour (which is referred to as 'human factors'). This chapter does not make that distinction and instead the term 'ergonomics' is used to cover all aspects of the design of equipment, environment or system in relation to people.

The benefits of applying ergonomics

[E17002] The benefits of ergonomically-designed equipment, furniture, workplaces and tasks are as follows:

- *Improved safety*, as people are less likely to make a mistake if equipment is designed to take account of their abilities (eg if characters on displays are of a suitable size and in a colour they can read accurately).
- *Reduced ill health and sickness absence*, as equipment and workstations are designed to fit people (eg seats are comfortable) and tasks are designed to take account of abilities (eg work rates and weights handled are within the person's capabilities).
- *Reduced fatigue and stress*, as tools, equipment and systems are easier to use.

- *Increased efficiency, performance, productivity and quality of product* (eg if tools are designed to fit the hands of the users, they will be easier to hold, users will be more comfortable and may work more efficiently and make fewer mistakes).
- *Increased job satisfaction*, as tasks are easier to perform.

Although people are highly adaptable and resourceful, and are often able to use poorly-designed equipment and systems, their use can lead to stress, errors, fatigue and injury. The consequences of not applying ergonomics can be enormous, potentially leading to human suffering through physical discomfort and disability, stress and accidents. It may also lead to reduced efficiency and productivity.

Ill health resulting from poorly-designed furniture, tasks, lifting and handling is also extremely costly.

Health and Safety Executive (HSE) statistics show that around 80 per cent of cases of work-related ill health are either musculoskeletal disorders (MSDs) or stress, depression or anxiety.

HSE annual statistics for 2017/18 show work-related MSDs accounted for 469,000 out of a total of 1,358,000 for all work-related illnesses, 35% of the total and a rate of 1,420 cases per 100,000 workers. They account for an estimated 6.6 million working days, an average of 14 days lost for each case. Not applying ergonomics can also be one of the factors that contribute to stress. Stress arises as a result of a mismatch between the demands of the job or situation and the abilities of the individual and results in a significant amount of ill health. The total number of cases of work-related stress, depression or anxiety in 2017/18 was 595,000, a prevalence rate of 1,800 per 100,000 workers. The total number of working days lost due to these conditions in 2017/18 was 15.4 million days, an average of 25.8 days lost per case. Stress, depression or anxiety accounted for 44% of all work-related ill health cases and 57% of all working days lost due to ill health.

Work-related stress may be caused by poor communication, lack of appropriate training, high workload, lack of control over work, inadequate feedback about work, or repetitive or boring work where tasks, jobs and systems are not designed with full consideration for the users. Individuals vary in their response to these factors but the employer has a responsibility to reduce the risks by designing work appropriately.

As well as increasing the potential for musculoskeletal injuries and increased stress, poor design that results in increased fatigue and error can contribute to accidents and system failures. Inadequate attention to human factor issues has been cited as a contributing element in many catastrophic failures including Chernobyl, Bhopal and Piper Alpha, see - Disaster and emergency management systems (**DEMS**)). These usually arise as a consequence of a series of errors resulting from inappropriate design and management. The prevention of accidents is an important aspect of ergonomics.

More information on human error is contained in the Health and Safety Executive (HSE) publication *Reducing error and influencing behaviour* (HSG48) (see [**E17059A**] below).

Core disciplines

[E17003] Ergonomics adopts a multi-disciplinary approach to an issue and draws on a number of other key disciplines, using this integrated knowledge to obtain a holistic view of the work system. The core disciplines include anatomy (the structure of the body), physiology (the function and capabilities of the body), biomechanics (the effect of movement and forces on the body), psychology (the performance of the mind and mental capability, including perception, memory, reasoning, concentration, etc) and anthropometry (the physical dimensions of the body).

The following text outlines:

- an overview of the ergonomic approach (E17004);
- design guidelines and principles relating to different aspects of the work system (E17005–E17043);
- musculoskeletal disorders (E17044–E17050);
- tools that can be used to assess work and assist with design (E17051–E17059);
- legislation that promotes ergonomic design (E17060–E17068).

The ergonomic approach

[E17004] The ergonomic approach considers all aspects of a person's interaction with their task, workstation, environment and work system, so that these can be designed to fit the abilities of the users and maximise ease of use. The model opposite can be used to illustrate the framework of this approach.

Figure 1: Ergonomic Approach

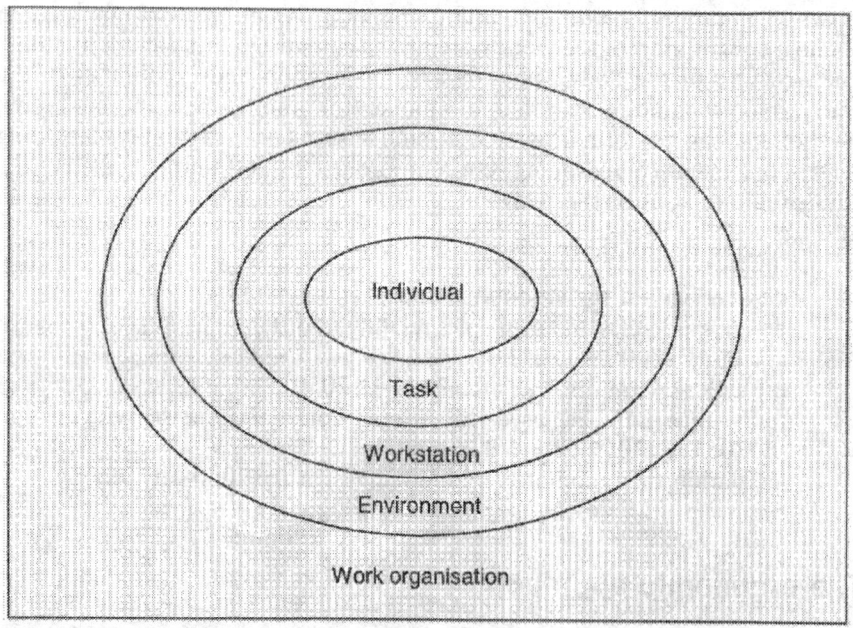

Ergonomics takes account of the physical and mental capabilities and needs of the potential users/workers/population (eg size, strength and ability to handle information), before considering the task and the context in which it will be undertaken. By establishing the physical and mental capabilities of the user group, certain criteria or constraints can be designed for the task. For example, most people can accurately remember up to seven numbers in their short-term memory. A picking task requiring order numbers to be remembered should therefore be designed so that the order number is seven digits or less.

The workstation should also be designed in relation to the physical requirements of the person and the operational requirements of the task. For example, the users' height and reach will be relevant to specifying the dimensions of the workstation; the sequence of use of tools will be relevant to their arrangement on the workstation.

The work environment should be designed in relation to the person, task and workstation requirements. For example, the lighting and noise level should be within the acceptable limits for human work but also appropriate to the requirements of the task (see also LIGHTING and NOISE AT WORK).

The work system is the organisation within which these elements sit and which ultimately affects the whole work. This includes the way in which the person, task and workstation are organised, as well as the requirements and constraints that affect the way work is conducted. It should complement and not constrain the effectiveness of the person, task, workstation and environment interactions.

Ergonomic criteria for these elements are discussed in more detail in paragraphs E17005–E17043 below.

Designing for people

[E17005] When adopting an ergonomic approach, the person is the central focus of the design of any system. Work should be designed to optimise the person's ability to complete the work, so it is within their mental and physical capabilities and they can work in a comfortable and effective manner – the equipment, furniture, environment and task activities should all support this. It is therefore important to understand the abilities and characteristics of the potential users.

It is clear there are wide differences between people in terms of their physical and psychological capabilities. It is important to understand the normal range of human abilities and potential variations within this in order to be able to design appropriately. In most work, transport and domestic situations it will not be possible or appropriate to select who uses equipment or systems, and therefore these should be designed to be appropriate for all potential users.

Physical characteristics of interest will include:

- the size and shape of people, so equipment and furniture can be designed to 'fit' the users;
- strength, so loads and tasks are designed to be within users' capabilities without leading to fatigue or injury;
- visual abilities, in terms of the items to be viewed, taking account of differences between people in terms of colour blindness, long and short sightedness and variations in this with age;
- hearing abilities, such that instructions and warnings can be clearly heard.

Psychological characteristics of interest will include:

- short- and long-term memory;
- uptake and processing of information;
- reaction time;
- motivation;
- concentration; and
- perception.

These physical and psychological characteristics will vary between people and will depend on individual factors such as age, gender, training and skills, health and fitness, and previous injury or disability. For any individual these may also vary over time (eg visual acuity tends to deteriorate with age).

Ergonomists are increasingly using digital human modelling (DHM) and simulation to help design work environments and tools. This incorporates data including anthropometric measurements (see **E17007** below) but allows the data to be applied to visual representations of humans within work environments (eg varying sizes of employees using a factory conveyor belt). It also incorporates other information gathered from academic research including biomechanical data (eg range of motion of a limb, posture analysis).

Simulation or DHM can help designers see if their designs might lead to potential harm without using real people to test that out.

Posture

[E17006] In order to work comfortably and reduce the risk of musculoskeletal injury or discomfort, people should be able to adopt 'neutral' postures. A neutral posture is one in which the joints of the body are at about the midpoint of their comfortable range of movement and with the muscles relaxed (eg arms hanging by the side of the body). Deviations from this place a strain on the muscles and soft tissue.

The trunk is in a neutral posture when the natural 'S' curve of the spine is maintained – as when standing. The lower part of the spine (lumbar region) flattens when in a sitting posture due to rotation of the pelvis, increasing pressure in the inter-vertebral discs. Suitable back support (with adequate lumbar support) should be provided when sitting. It should be recognised that the amount of curvature in the small of the back (lumbar area) varies from person to person and therefore one shape of backrest will not suit everyone.

The neck is in a neutral posture when the head is held upright and facing forward.

The shoulder is in a neutral posture when the upper arm is relaxed by the side of the body; the wrist is in a neutral posture when the muscles are relaxed and the hand is in line with the forearm.

Figure 2: The 'S' Shaped Curve of the Spine (S Pheasant *Bodyspace: Anthropometry, ergonomics and the design of work* (1996) p 69 fig. 4.1, reproduced with permission of Taylor and Francis)

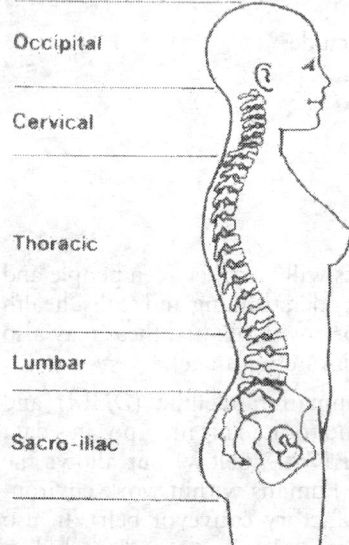

Depending on the degree of deviation and the duration of maintaining the posture, awkward postures (when the posture deviates significantly from the neutral) and static postures (postures in which there is no movement for a period of time) can both lead to fatigue, discomfort, strain and possibly injury. Awkward postures such as bending, twisting and stretching, reaching the arms behind the shoulders, and reaching above shoulder height or below knee height should be avoided as far as possible. Careful positioning and adjustment of the equipment and furniture used can achieve this. Introducing movement and changes in posture will help alleviate the discomfort that can arise from static postures.

Anthropometry

[E17007] Anthropometry (the measurement of body dimensions) can be used to ensure people will be able to reach, use and operate tools, equipment and workstations, by structuring and positioning items appropriately (eg so they are within easy reach, adjust sufficiently, and allow or prevent access).

The size of the body varies with age, gender and ethnicity. Tables of anthropometric data provide data for many different body dimensions, representing different population groups (see for example Stephen Pheasant and Christine M Haslegrave, *Bodyspace: Anthropometry, Ergonomics and the Design Of Work*, CRC Press (2005)). Variations in body size usually follow a normal Gaussian distribution, with most people falling within a central value for a given body dimension, and fewer people tailing off at either side. It is common when designing to take account of the middle 90 per cent of the population, for example to design for those who are taller than the smallest five per cent of the population, and smaller than the tallest five per cent of the population (namely to design within the fifth to ninety-fifth percentile data). However, in some situations it is necessary to design for the extremes of the population. For example, the height of a door should take account of the tallest potential user and machine guarding should take account of the finger width of the smallest potential user.

Population stereotypes

[E17008] Population stereotypes are cues and expectations which provide information regarding the meaning, state or operation of items. We have certain expectations concerning colours, directions of movement, shapes, sounds and abbreviations. For example, a red tap indicates that it supplies hot water. Turning a control dial clockwise turns the power up. Designs that take account of population stereotypes will result in increased accuracy and faster reaction times. Population stereotypes can vary by culture (eg pushing a light switch up turns it on in the US) and often by industry.

Examples of these stereotypes are shown below.

> Population stereotypes in the UK
> *Colour*
> - Red = hot, danger, warning, stop
> - Green = safe, go
> - Yellow = caution
> - Blue = cold, information
>
> *Shape of signs*
> - Round = enforceable
> - Triangle = caution/attention
> - Square = advisory
>
> *Direction*
> - Clockwise = on, tightens (screws), increases
> - Anti-clockwise = off, loosens, decreases
> - Upwards = switches off, increases
> - Downwards = switches on, decreases

Allocating tasks to people or machines

[E17009] In most systems people interact with machines. When designing a system, tasks can be allocated to either the person or the machine. Decisions on whether a task should be undertaken by a person or a machine will depend on an understanding of their relative capabilities.

In allocating functions between people and machines, economic or social factors also need to be considered alongside performance. In some situations, the 'best' performance may not be required, and contact with people may be preferable to dealing with machines. In order to optimise job design, retaining some 'interesting' tasks for people may be more important than mechanising.

Designing the task

[E17010] The task is the collection and sequence of activities and events that allow the work to be completed. The task should be designed to be within the capabilities of the person, but must take account of the constraints of the work process, environment or system. Mismatches between task requirements and individuals' capabilities increase the potential for human error.

Task design will influence the activities of the person, for example the postures adopted, the amount of force applied, repetition and duration of the activities. These may have an influence on performance outcome or on health (eg musculoskeletal disorders). Issues to consider include:

- *Duration of the task* – long periods of work without a break may lead to reduced performance and increased discomfort. Decline in concentration usually becomes evident after 30 minutes. Physical discomfort increases if static postures are maintained or movements are repeated without a break.
- *Breaks and changes in activity* – infrequent breaks can lead to reduced performance (eg on a vigilance task such as inspection) and increased discomfort. Regular changes in posture and movement help reduce physical discomfort and injury. Prolonged periods viewing a display screen without a break can also lead to visual discomfort. Varying the tasks that are undertaken will help to reduce boredom.
- *Amount of repetition of movements required* – highly repetitive work which involves the same muscle groups and movements can lead to physical discomfort and boredom.
- *Frequency of the task* – frequently performed tasks may lead to boredom and increased physical discomfort (as the same movements are made). Conversely, there may be training and information requirements for infrequently performed tasks.
- *Pacing* (if the person is required to work at a speed set by a machine or process) – this will dictate the speed of work and the amount of recovery the person has between tasks or operations. This can have an impact on discomfort and performance. A comfortable work rate should be established.
- *Work rate* – excessively busy periods or quiet periods can lead to stress and discomfort. Completing the task early should not result in benefits as this can lead to people rushing the task.
- *Complexity of the task* – a number of sub-tasks are usually required to complete the operation and more than one operator may be required to assist in the operation. Tasks can be divided so operators each undertake one sub-routine or a more complex combination of tasks. Jobs should be designed to facilitate the development of skills, interest, variety and commitment. Several sub-tasks may be undertaken by one operator, but to reduce boredom and prevent overload and the risk of musculoskeletal injury, these should vary in the demands they place on the physical and psychological capabilities of the workers.
- *Awkward or static postures required* – these may be dictated by task requirements or poor tool or workstation design. These can contribute to MSDs and should be avoided.
- *Force required to undertake the task* – application of frequent or excessive force, particularly in relation to the capabilities of the part of the body applying the force, can lead to discomfort. It may be appropriate to mechanise the task or use a tool or equipment to assist in generating force if required.
- *Sequence of use of tools/equipment* – appropriate layout of equipment and tools will facilitate ease of use so that, for example, those used most frequently are positioned closest to the user.
- *Levels of concentration and attention required* – the level of arousal (the state of consciousness) helps to maintain attention and affects performance. Low levels of arousal are found in undemanding jobs, where activities are either very mundane or very infrequent, eg produc-

tion line packing, night security work. High levels of arousal can be found in stressful and demanding jobs, eg air traffic control and ticket inspection. Both high and low levels of arousal can lead to poor performance. Tasks should be designed to maintain the vigilance and arousal necessary for the task. Methods of facilitating concentration and vigilance include reducing the task time and introducing frequent breaks, providing music or background noise for routine tasks, and exaggerating the size or colour of the stimulus.
- *Information and training required for the task* – some tasks will require personnel to be trained in how to undertake them. For other tasks, simple information or instructions may be adequate. The need for training and information will depend on the complexity of the task and the experience and abilities of potential users.

Task design will be influenced by the equipment and workstation design, the organisation of the work, and the abilities of the individuals. The application of anthropometric and biomechanical knowledge will help ensure the physical elements of the task are within the physical capabilities of the person. Understanding the psychological abilities of the person, such as their memory and concentration, will help to design a task within limits acceptable to the person (or group of people). The provision of any information or instruction should also account for the way in which people process information.

In general, it is those who are undertaking the task who have the most knowledge about the needs of the work system. Involving workers in evaluation and redesign is an important component in ensuring the design meets their needs and those of the system, and that any changes made are accepted by workers.

Design of equipment

[E17011] Equipment should be designed and selected for those who will use it, taking account of the task to be completed and the conditions under which it is used.

Hand tool design

[E17012] Tools enable the use of the hand to be extended, for example to extend its ability to apply force, manipulate items, make precise movements and so on. Appropriate design of hand tools can facilitate the task and reduce the risk of musculoskeletal injury.

Hand tools should be designed to allow neutral arm, wrist and finger postures in operation. The hand is in a neutral posture when the hand is in line with the wrist, the thumb facing upwards and the palm facing inwards. This can be seen when the shoulder is relaxed and the arm is allowed to hang relaxed by the side of the body. The maximum grip strength can be applied when the wrist is in a neutral posture; deviations from this (ulnar or radial, namely side to side, flexion or extension – up or down) will reduce the amount of force that can be applied, with the hand able to apply the least force in a flexed posture.

Working with the hand in a neutral posture also minimises the strain placed on joints and other soft tissue, which can be compressed or experience friction in awkward postures.

Angling either the work piece or the tool handle can help reduce the amount of wrist deviation required when using a hand tool. However, the potential for variation in orientation of the tool on the work piece may limit the benefit of this.

Key points in hand tool design are:

(1) Handles should be long enough to fit the whole hand – at least 100 mm, although a length of 120 mm may be preferable.

(2) In general, larger handles decrease the amount of muscular activity required when using the tool. A handle thickness of approximately 40 mm is generally recommended. Slightly thicker handles may be beneficial when applying torque (eg for screwdrivers).

(3) Tools with two handles which require a hand span (eg pliers, scissors) should have a span of approximately 60 mm. If the tool is used repetitively, an automatic spring opener will reduce the strain on the weaker finger extensor muscles (used to open the hand).

(4) Ideally the handle surface should be compressible, textured and non-conductive. It should be free from sharp edges and avoid ridges or finger contouring on the handle, as this can place pressure on the soft tissue of the hand. Avoid cold surfaces (eg metal), particularly if the tool is powered by compressed air.

(5) An excessively smooth handle surface requires the user to grip the handle more tightly, increasing the amount of force required, and this should be avoided. If the hands are sweaty or if the user is wearing gloves, they will also have to grip the handle more tightly. Tools used in these circumstances should therefore be made easier to grip.

(6) It should be possible to use the tool in either hand; if this is not possible, specific tools for left-handed workers should be provided.

(7) Power-assisted tools can greatly increase the speed of task completion, remove a large degree of force exertion from the operator and reduce some awkward postures, such as rotating the wrist. However, the user may experience vibration from the tool. Pneumatically powered tools can blow cold air exhaust over the hands, increasing the risk of discomfort and musculoskeletal problems such as Hand Arm Vibration Syndrome (see OCCUPATIONAL HEALTH AND DISEASES and VIBRATION). If power tools are used, ensure they are low-vibration and well maintained to reduce the vibration transmitted to the hand and arm. Mounting power tools in a jig can reduce vibration transmission.

(8) If force has to be applied through the tool, where possible use power tools rather than tools requiring manual application of force. However, power tools often weigh more than manually-operated tools and handling them can increase the risk of discomfort.

(9) In terms of weight, frequently-used tools should weigh as little as possible: a maximum of 0.5 kg is recommended. The distribution of weight in the tool should be even, with the centre of gravity as close to the hand as possible. Counterbalances may be required to support heavier tools or those with an off-set centre of gravity.

Design of controls

[E17013] Displays and controls are the mechanisms by which people and machines interact. The person gives instructions to the machine through the controls (eg knobs, switches, levers, buttons). Controls can be discrete, having a set number of conditions (eg on/off/standby), or continuous, where the condition varies along a scale (eg volume). Controls should be designed to facilitate the changes they allow and the amount of effort required to operate them. The following factors should be considered:

- The amount of force required to operate the control (and the amount of resistance to prevent accidental operation). Fingers and hands should be used for quick, precise movements, arms and feet for operations requiring force.
- The size of the control should be appropriate for the force required and part of the body used to activate it.
- The location of controls should facilitate their use. Hand-operated controls should be easily reached and grasped between elbow and shoulder height. Controls should be sufficiently far apart to allow space for the fingertips. Sequence and frequency of use, and importance, may also dictate the location of controls.
- Appropriate feedback should be provided to indicate activation of the control.
- Controls should be coded by colour, shape, texture or size (to allow identification by touch).

Design of displays

[E17014] The state of the machine is relayed to the person through a display. Displays should be designed to enable users to easily and accurately assess information from the machine. The following principles should be followed:

(1) Qualitative displays showing a small number of conditions should be used for discrete information (eg on/off). Quantitative displays should be used to present numerical or continuous information, eg temperature, speed etc. Qualitative displays may be lights or words etc. Quantitative displays may be scales, counters etc.

(2) Display scales should be clear, unobscured and concise. Scale intervals should increase left to right, bottom to top, preferably increasing in units of 10s, 100s, 1000s, etc as appropriate.

(3) Displays should conform to population stereotypes in terms of colours used (eg red = danger).

(4) Ensure labelling, if used, is clear.

(5) Limit the number of warning lights to avoid confusion and aid identification.

(6) On dials with pointers, avoid parallax (the difference in scale reading depending on the viewing angle) by keeping the dial and pointer close together.

(7) Text should be of a suitable size and font to allow easy reading. B and 8, O and 0 can be confused, and if resolution is not adequate, F and P can also be confused.

Grouping of controls and displays

[E17015] The result of activating a control is often indicated in a display. The relationship between displays and controls should be clear. This can be achieved through location, arrangement, text, shapes, colours and responsiveness. Grouping of controls and displays can improve their association with their function and facilitate their ease of use. Controls and displays can be grouped according to sequence of use (eg start, run, finish), by function (eg keeping all controls concerned with lights together) or by frequency of use, with those most frequently used within convenient reach and within the immediate viewing arc. Locating controls by importance of use can also aid their operation (eg emergency controls should be placed within convenient reach).

Text

[E17016] Text used as information, instruction or labelling should be clear and concise. Bold, italics, large fonts, underlining and colour can all be used to draw attention to or highlight information but should be used sparingly to preserve the meaning when they are used. Lower case should be used for phrases or sentences, as UPPER CASE DECREASES THE DIFFERENTIATION BETWEEN LETTERS AND IS SLOWER TO READ.

Positive instructions should be used, not negative or double negatives. For example: 'When the alarm sounds turn the machine off' – not 'When the alarm sounds do not leave the machine on'.

Software design

[E17017] The field of Human Computer Interaction (HCI) is a specialist area. In this context, it is suffice to say that software should be designed to make the use of a system intuitive and accessible (eg through the arrangement of icons, use of colours and text, structure of menus and commands).

Workstation design

[E17018] The workstation – the area where the task is conducted – should be designed so it is appropriate for the person using it. The workstation may be a desk or work surface, but may equally be at a conveyor belt or a driver's cab. Attention should be given to the physical dimensions of the workstation and the arrangement of any necessary equipment, tools or components to allow good posture and acceptable movements.

The issues discussed below cover the main points that should be considered in the design of workstations.

Work surface height

[E17019] Many tasks will involve workers using a work surface. The height of the work surface should permit a relaxed, upright posture and facilitate the use of equipment at the workstation. The surface height, whether designed for sitting or standing tasks, should be such as to avoid users stooping or reaching to its surface. In most situations, the height of the work surface is fixed. For seated tasks a height-adjustable chair should be provided so the working height can be set appropriately. A footrest may be required to support the feet when sitting at a comfortable height.

An appropriate work surface height is dependent on the work being done at that work surface. As a general rule, the top of the work surface should be level with the user's elbow height (the user may be sitting or standing, as required at the workstation). The height of the work surface may need to be reduced to take account of the thickness of the item being worked on, so this is at an appropriate height.

There may be a range of users utilising the same workstation, and unless easy height adjustment is provided (eg a step with appropriate grip covering), it will not be possible to obtain the optimum height for all potential users. For a fixed height workstation, some compromise will have to be made, based on the task requirements and the likely users.

The following working heights are recommended for different tasks in relation to the user's height (Stephen Pheasant and Christine M Haslegrave, *Bodyspace – Anthropology, Ergonomics and the Design of Work*, CRC Press Taylor and Francis (2005)).

Task	Height
Manipulative, requiring force and precision	50 mm–100 mm BELOW elbow height
Delicate (including writing)	50 mm–100 mm ABOVE elbow height
Heavy – requiring downward pressure	100 mm–250 mm BELOW elbow height
Lifting and handling	Between mid thigh and mid chest level, preferably close to waist level
Two-handed pushing/pulling	Hands just below elbow height
Hand-operated controls	Located between elbow and shoulder height

The variation in recommended working heights for different tasks arises from the aim of minimising the effort required to perform the tasks, by optimising the posture and muscle groups used.

Using this data, as an example tasks such as component assembly where pneumatic screwdrivers are used should be conducted at between 100 mm and 250 mm below elbow height, so downward pressure can be applied. If the unit being assembled stands 50 mm above the work surface, the work surface will need to be 150 mm–300 mm below elbow height. This assumes that all work at the workstation is of this nature. Workstations that are also used for writing tasks, for example, should be split-level, providing a writing surface at or just above elbow level. If a fixed-height workstation and height-adjustable chair are provided, users should adjust the height of the chair depending on the task.

For example, display screen equipment (DSE) users who also write at their workstation will find it more comfortable to sit higher when they are keying and lower when they are writing.

Height adjustable workstations can be beneficial, particularly for standing tasks and where a range of different users may work at the same workstation. Where height adjustable equipment is provided, users should be trained in how to adjust it and how to identify an appropriate working height.

Workstation characteristics

[E17020] Other characteristics recommended for the workstation include the following:

- The workstation should be a suitable size to allow all equipment to be located and arranged conveniently for the person to complete the task.
- The workstation surface properties should not present a risk to the user – they should be free from sharp edges and unprotected hot or cold surfaces. A front edge of 90° can cause discomfort if the arms are rested or pivoted on it.
- Surrounding workstations or other items in the environment, eg columns or posts, should not constrain the user's posture and ability to get into and out of the workstation easily.
- There should be sufficient legroom underneath the work surface to allow the user to sit comfortably, sufficiently close to the workstation, without their posture being constrained.

Workstation layout

[E17021] To facilitate a good working posture, the equipment and items used to complete the tasks should be within convenient reach of the user, so they do not have to stretch or lean. The zone of convenient reach and the normal working area can be used to establish a suitable layout of the workstation.

Zone of convenient reach

[E17022] The zone of convenient reach is defined by the area from the shoulder to the fingertips with the arm out-stretched (upward, downward and to the sides), which can be reached without any undue exertion. Items required for the task (eg control buttons, handles, work surfaces, tools) should be placed within this zone. The extent of this area obviously depends on the length of the arm and its arc. The fifth percentile arm length (95% of the population will be able to reach items if placed within this area) is 720 mm for British men and 655 mm for British women (Stephen Pheasant and Christine M Haslegrave, *Bodyspace – Anthropology, Ergonomics and the Design of Work*, CRC Press Taylor and Francis (2005)). If designing for both male and female users, the shorter dimension (655 mm) should be used.

Figure 4: Zone of Convenient Reach (E Grandjean Fitting the Task to the Man (1988) p 51 fig. 42 reproduced with the permission of Taylor and Francis)

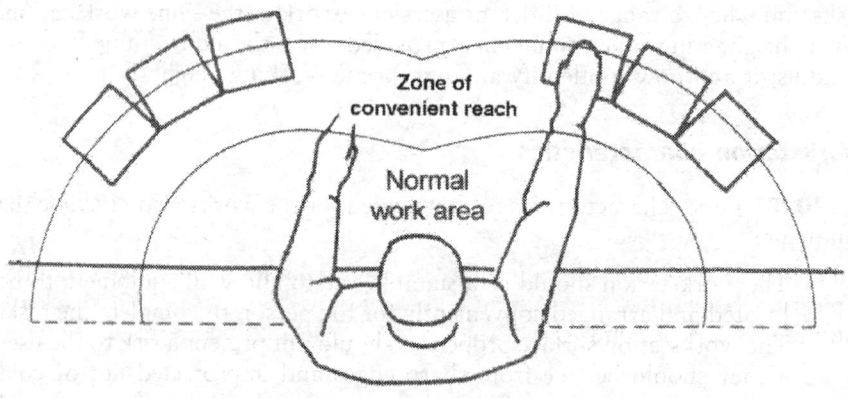

Normal working area

[E17023] The normal working area is defined by the comfortable sweep of the forearm with the elbow bent at 90° and the upper arm in line with the trunk. This provides an area that can be reached without extension of the arm at the shoulder and requiring no trunk movement. The extent of the area depends on the length of the arm from the elbow to the fingertips. The fifth percentile forearm length is 440 mm for men and 400 mm for women (Stephen Pheasant and Christine M Haslegrave, *Bodyspace – Anthropology, Ergonomics and the Design of Work*, CRC Press Taylor and Francis (2005)). Again, if designing for both male and female users, the shorter dimension (400 mm) should be used.

In a case highlighted by the retail union USDAW, a shop worker developed a repetitive strain injury (RSI) after the checkouts at the store she worked in were redesigned. Credit card readers and touch screens were installed, but cashiers had to stretch out of their chairs to reach these. The "chip and pin" machines were 535mm from the edge of the till and the touch screens 430mm away, making the overall workstation outside the comfortable reach of 95% of the women workers at the store.

One woman was particularly badly affected, because she was only four feet nine inches tall. The chip and pin machine was 23cm too far from her reach and within a month she developed tenosynovitis, a form of RSI.

Arrangement of items on the workstation

[E17024] Frequently used and critical items (eg emergency controls) should be placed in the normal working area. Occasionally-used items can be sited within the zone of convenient reach. Items can be grouped by function or sequence of use, or importance.

Sufficient space should be provided at the workstation to allow a flexible arrangement of all the equipment that is required for the task. Items should be arranged so that operators do not have to reach across their body, for example the phone should be placed on the left-hand side if it is answered with the left hand.

Visual considerations

[E17025] The following factors should be taken into consideration.

Up/Down Viewing

[E17026] Items can be viewed both by movement of the eyes and by movements of the head, neck and back.

Awkward neck and back postures may be required if the viewed item is low, high or to one side, and this may cause discomfort. Items to be viewed should be positioned such that no twisting or bending of the neck, head or back is required to see them, and so that the eye muscles are in a neutral position.

The neutral position of the eye is generally taken as about 15° below the horizontal line of sight. The eyes can comfortably move about 30–45° below the horizontal line of sight before the head must be inclined; an upward gaze of approximately 15° is achievable before the head must be tilted backwards to view further. An upward gaze is fatiguing when sustained for any length of time, as the muscles controlling upward movement of the eyes are not as strong as the muscles controlling downwards movement. The most comfortable eye position is an arc of 30° from the horizontal position of the eyes downward, and this is the acceptable viewing angle.

Figure 5: Viewing Angle (S Pheasant *Ergonomics, Work and Health* (1991) reproduced with permission of Palgrave Macmillan)

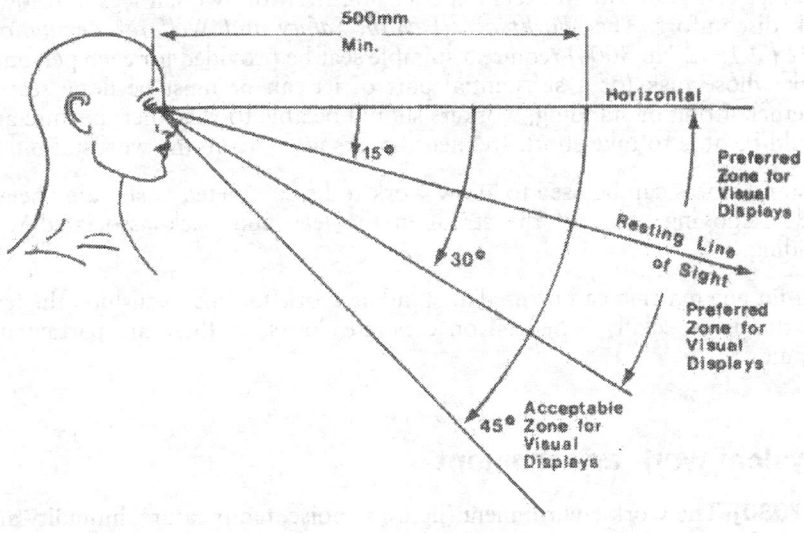

Side/Side Viewing

[E17027] Items outside about a 60° arc in front of the person will involve neck twist to view. Items to be viewed regularly or for long periods should not be positioned outside this area.

Viewing distance

[E17028] The required viewing distance generally depends on the person's visual acuity, the size of the item being viewed and the lighting levels. A viewing distance of between 500 mm and 750 mm is likely to be suitable for most items, but a shorter viewing distance may be required for fine work.

Sitting versus standing

[E17029] There are biomechanical benefits to both sitting and standing. Sitting can help reduce the development of fatigue in the lower limbs, as long as the chair provides adequate support and the feet are also supported. Tasks that require generally static postures (with limited or no back movement) will benefit from being undertaken while sitting. However, inappropriate seating and prolonged sitting can lead to discomfort and fatigue, and may prevent good posture (eg armrests may prevent the chair being brought sufficiently close to the workstation) (see WORKPLACES).

The HSE Workplace Health Expert Committee (WHEC) published a position paper on sedentary work and health in August 2018, based on a limited, but contemporary, review of the literature and the views of selected international experts. The paper can be found on the HSE website at: https://webcommunities.hse.gov.uk/connect.ti/WHEC/view?objectId=685957.

Standing is beneficial if the task requires a range of body movement or handling of heavy items. However, prolonged standing, particularly with little walking, can result in workers experiencing tired or swollen legs and lower back discomfort. The *Workplace (Health, Safety and Welfare) Regulations 1992 (SI 1992 No 3004)* require a suitable seat be provided for each person at work whose task (or a substantial part of it) can or must be done seated. Whether sitting or standing, workers should be able to vary their posture and should be able to take short, frequent breaks away from the workstation.

Sit/stand stools can be used to allow work to be completed at standing height while removing some of the strain in the legs and back associated with standing.

Anti-fatigue matting can be used at standing workstations to cushion the feet, and this is especially beneficial on concrete floors, as these are particularly fatiguing.

Physical work environment

[E17030] The work environment (lighting, noise, temperature, humidity and general layout of the place of work) should also be designed to suit the person

Physical work environment [E17031]

and the work. In general, the effects of the environment can impact on performance, subjective comfort, perception, attitudes, safety and health.

Lighting

[E17031] Appropriate levels of lighting should be provided in order to maintain acceptable levels of quality and performance, and to help avoid visual discomfort. Poor lighting can lead to visual fatigue (symptoms include red or sore eyes, blurred vision, headaches). Inappropriate lighting can also force users to adopt poor postures in order to view items more easily, and this can also lead to discomfort.

Issues to consider in designing visual environments are the brightness of the light sources, their position, the evenness of the light distribution, the contrast between the different work items and the potential for glare/reflections from other items in the work area.

The amount of illuminance (the amount of light energy reaching a given point) required should be determined by the task demands, for example visually-demanding tasks such as inspection or fine and precise work require a higher level of light than that required in walkways. Insufficient lighting can result in eye strain and discomfort, while too much light or a bright light source within the visual field will cause glare and result in visual discomfort and reduced visual performance. Where light levels are already high, it may be appropriate to improve the legibility of the source document or item being viewed (eg increasing size and contrast) rather than further increasing the lighting levels.

The location of lights will also affect visual performance and comfort – light sources within the visual field may cause glare (which may be visually disabling or cause discomfort). Excessive differences in lighting levels within the work areas should be avoided. Surfaces of a highly-reflective material should be covered or replaced to reduce glare. Directional lighting may be appropriate for some inspection tasks but diffused light is generally more satisfactory, as it reduces the amount of glare and shadows produced.

Light may come from a natural source (from windows) or an artificial source (electric lights). It is important to remember that lighting levels may vary according to the time of day or year and therefore appropriate adjustments may need to be made, for example provision of local lighting (lamps) or blinds.

The furniture and equipment in the room should be arranged so that the sources of light do not cause light to fall directly onto reflecting surfaces or into the eyes (eg sit at right angles to a window rather than facing or with the back to the window).

Colours should be chosen to provide contrast to aid detection but not to exaggerate glare or reflections. Colours with good contrast should be used for displays (eg black and white, blue and yellow, etc). Where lighting levels are high, dark characters on a light background are generally easier to read.

In general, individuals should have control over the lighting levels in their area, as individuals vary in what they find comfortable. In addition, having control over the work environment can help reduce stress. Lack of control over work and working methods is a psychosocial risk factor.

Increased lighting levels may be required for those with poorer eyesight (eg older people), although this should be done with care, as older people are generally more susceptible to glare. For further details see LIGHTING. Additional guidance is provided in the HSE Guidance Note *Lighting at Work* (HSG38).

Noise

[E17032] Noise can have an effect both on hearing ability (hearing loss) and on performance. Too much or too little noise can adversely affect concentration. Where work is monotonous, noise can be introduced to provide a stimulus; but where noise is excessive it should be reduced to facilitate concentration. Noise levels should not exceed approximately 55 dB(A) for tasks requiring concentration.

Noise can be used effectively as a warning or alarm, although excessive use of noise for warnings can be detrimental to their effectiveness. Hearing generally deteriorates with age. This, and the differences in hearing ability between people, should be taken into account when designing critical alarms.

There are several harmful effects of noise, which include:

(1) Physical damage to the eardrum and ossicles induced by excessively high noises eg explosives.
(2) Hearing damage due to exposure to high levels of noise. Damage can be divided into temporary threshold shift (reversible) or noise-induced hearing loss (irreversible).
(3) Annoyance and stress – which can lead to reduced concentration. Noises that are annoying are generally unexpected, infrequent, high frequency and those over which the listener has no control.
(4) Hindering communication – which can lead to an increased rate of accidents as well as stress.

The *Control of Noise at Work Regulations 2005 (SI 2005 No 1643)* require that where the noise level reaches 80 decibels, employers must assess the risk to workers' health and provide them with information and training. They must also ensure that the risk from exposure to noise is either eliminated at source or, where this is not reasonably practicable, reduced to as low a level as is reasonably practicable. If the level reaches 85 decibels (daily or weekly average exposure), they must provide hearing protection and establish hearing protection zones. There is also an exposure limit value of 87 decibels, taking account of any reduction in exposure provided by hearing protection, above which workers must not be exposed.

For further information see NOISE AT WORK.

Vibration

[E17033] Exposure to vibration can cause damage to the joints, muscles, circulation and sensory nerves and can also lead to reduced performance. There are two main categories of vibration transmission to the body: whole body and hand/arm (see N3027). Whole body vibration usually occurs via the feet or seat when the person is working on/in a vibrating environment (eg transport systems). The most widely reported health issue related to whole body vibration is back pain.

Hand-arm vibration affects only the hands and arms, and can be experienced when using power tools such as pneumatic screwdrivers and chainsaws. Exposure of the hands and arms to vibration is a known risk factor in the development of Hand Arm Vibration Syndrome (HAVS), a condition which affects the nerves and blood supply to the hand, resulting in tingling or numbness, impairment or loss of function and blanching of the fingers, notably on exposure to the cold. It is particularly higher frequencies of vibration (such as those associated with hand-held tools) that have been associated with muscle, joint and bone disorders affecting the hand and arm. Transmission of vibration has been found to increase as the gripping force increases and where tight gloves are worn.

Vibration, as with noise, should be tackled at its source. Larger machines can be isolated or the vibrations insulated, for example by introducing better mountings on the machine. Regular maintenance of machines can also reduce vibration. Relocation of the machine and insulation of the source should be considered. Floors, seats and handgrips can all be fitted with damping material. Hand tools should be designed to reduce the vibration created, the weight and the amount of force required, so that tools do not have to be gripped tightly. Where this cannot be achieved, exposure time should be limited and the hands kept warm.

The *Control of Vibration at Work Regulations 2005 (SI 2005 No 1093)* require employers to take action to prevent their employees from developing diseases caused by exposure to vibration at work from equipment, vehicles and machines.

For further information see **VIBRATION**.

Thermal environment

[E17034] The internal body temperature needs to be maintained within a relatively narrow range for health, and the ambient conditions that facilitate this are also relatively narrow. Deviations from comfortable temperatures can lead to reduced performance, fatigue, discomfort and ill health (hypo and hyperthermia as the body temperature rises or drops below those necessary for good health); in extreme circumstances this can be fatal.

Body temperature, and therefore comfort, is dependent on the ambient conditions (air temperature, humidity, any radiant heat and air velocity), the activity being undertaken (and therefore the heat produced within the body) and the clothing worn. Individuals' response to the thermal environment varies and some people are more affected by high or low temperature than others. Employees should be able to control the environment to suit themselves (eg opening windows etc), although this can be difficult in large, open plan offices.

As far as possible, environments should be designed and controlled to allow a comfortable working temperature (eg through heating, air conditioning, etc).

Where this is not possible (eg due to a work process or outdoor working), appropriate clothing can be used as a control measure or in extreme situations it may be necessary to limit exposure time. There is currently no legal upper temperature beyond which workers should not be exposed, although the Trades Union Congress (TUC) has long called for an action level of 24°C, which is the World Health Organisation (WHO) recommendation for maximum temperature for working in comfort, and an absolute maximum 30°C upper limit for indoor work areas (27°C for those doing strenuous work).

In its July 2018 report, *Heatwaves: adapting to climate change*, the parliamentary Environmental Audit Committee recommended the government consult on introducing maximum workplace temperatures, especially for work that involves significant physical effort. The government rejected the recommendation pointing to the existing legal obligation on employers, under the Workplace (Health, Safety and Welfare) Regulations 1992 (Workplace Regulations), to provide a 'reasonable' temperature in the workplace. It also highlighted a 2009 HSE review which found 'little evidence of significant numbers of cases of illnesses (long or short term, physical or psychological) caused or exacerbated by exposure to high temperatures at work'. In terms of a lower limit, the approved code of practice to the Workplace Regulations says the temperature in workrooms should normally be at least 16°C. If the work involves rigorous physical effort, the temperature should be at least 13°C. Further guidance is provided in the HSE publication L24, *Workplace health, safety and welfare Workplace (Health, Safety and Welfare) Regulations 1992 Approved Code of Practice and guidance*.

Low temperatures reduce manual dexterity, and where this is required the hands should be kept warm (eg through appropriate gloves). Manual handling injuries are more likely in extreme thermal environments and account should be taken of this when planning tasks.

Guidelines for control of employees' work in hot/humid/cold environments are shown below.

(1) Maintain temperatures as recommended by the Chartered Institute of Building Services Engineers for different working areas:
- Heavy work in factories: 13°C
- Light work in factories: 16°C
- Hospital wards and shops: 18°C
- Offices and dining rooms: 20°C

(2) Maintain humidity at between 40 and 70 per cent.

(3) Reduce air velocity (draughts) to less than 0.1 ms^{-1} in moderate and cold environments. Increasing air velocity can be an effective control measure in hot environments, provided the air temperature is not close to or above body temperature (37°C), when increased air velocity will increase the thermal load on the body.

(4) Decrease physical workload in hot environments.

(5) Schedule rest pauses or rotate personnel around more strenuous tasks in hot conditions.

(6) In hot environments allow rest periods to be taken in cooler environments.

(7) Schedule outdoor work to avoid periods of very high or low temperature.
(8) Permit gradual acclimatisation to hot or cold environments (7–10 days).
(9) Maintain hydration by consuming sufficient drinking water, particularly in hot environments.
(10) Avoid long exposure periods in cold environments.
(11) Avoid rest for long periods in cold environments.
(12) Provide protection from wind and adverse weather when working outside.

Physical hazards

[E17035] Other physical hazards may be present in the workplace (eg slipping, falling, tripping hazards). Ideally these should be eliminated at source through appropriate design (eg removing the need for an operator to enter a potentially dangerous area). Where hazards are still present, guarding, safety barriers, good housekeeping, appropriate lighting and good workplace layout will reduce the possibility of all types of accident. Raising awareness through education and training can be effective, although it should not be relied on as a first line of control. Personal protective equipment (PPE) should always be seen as the last resort control measure, due to issues related to fit, compatibility, impact on performance, loading on the body, etc, which may mean that it is not worn correctly or possibly at all.

Job design and work organisation

[E17036] The design and organisation of work can have a significant impact on health. The demands of the job should be matched to the skills and abilities of the individuals. Excessive demands or, conversely, under-utilisation of skills can both lead to a decrease in performance. The design of the job or work task is important for obtaining the maximum performance from the person for the minimum effort and facilitating a level of job satisfaction. Factors to consider include job rotation, job enlargement and job enrichment, as well as job scheduling, job demands and shift patterns.

Job rotation

[E17037] Job rotation, whereby people are moved from one task to another, allows changes in posture, movement and mental demands. This may lead to the acquisition of additional skills and may help the person identify more with the completed product or service. Job rotation can also help to overcome boredom, particularly in repetitive jobs, and help to prevent fatigue, loss of concentration and deterioration in performance. To be effective as a control measure for musculoskeletal discomfort, people should rotate to tasks that are significantly different in terms of postures, movements and forces required.

Job enlargement

[E17038] Job enlargement, whereby the number and range of tasks undertaken by the individual is increased, allows workers to become more flexible, develop other skills and have greater variety in their tasks.

Job enrichment

[E17039] Jobs can be enriched by incorporating motivating or growth factors, such as increased responsibility and involvement, opportunities for advancement and a greater sense of achievement. This should provide the workers with greater autonomy over the planning, execution and control of work.

Work scheduling

[E17040] Ideally workers should have control over how the work is scheduled. As far as is possible, work should be self-paced rather than machine-paced. Machine pacing may be too fast or too slow for an individual, both of which can lead to problems (lack of recovery time for muscles and soft tissue; frustration, etc). Machine pacing has been implicated as being a contributory risk factor in upper limb disorders, as well as stress and stress-related illnesses.

Rest breaks help alleviate both mental and physical fatigue. Allowing the worker to select when and how often they take breaks is preferable to fixed breaks, although there is some evidence that people work until fatigue occurs rather than stopping before this point. Therefore, under some circumstances (eg tasks that require high concentration or those requiring physical effort) it may be preferable to give fixed rest breaks before the person is likely to become fatigued.

Breaks should be taken away from the workstation, not only because this allows for physical movement and a change of scene, but also because it may facilitate reducing exposure to hazards such as noise. Short, frequent breaks are better than long, less frequent breaks.

Shift work

[E17041] It is well recognised that shift work affects health and safety at work. In particular, where shifts involve people changing their eating and sleeping patterns, symptoms such as irritability, depression, tiredness and gastrointestinal problems (eg indigestion, loss of appetite and constipation) can occur. These health problems are largely due to the disturbance of the body's physiological cycles. Although many shift workers adapt to the disruption, it is widely recognised that productivity levels are lower on night shifts than day shifts. In addition, studies have also linked shiftwork with an increased risk of conditions including diabetes and breast cancer.

The HSE says that poorly designed shift-working arrangements and long working hours that do not balance the demands of work with time for rest and recovery can result in fatigue, accidents, injuries and ill health. While all workers are potentially at risk from shift work, certain groups are more vulnerable than others, including:

- young workers;
- older workers;
- new and expectant mothers;
- workers with pre-existing health conditions, which may be made worse by shift work, such as those with gastro-intestinal problems, coronary heart disease and sleeping problems;
- workers taking time-dependent medication such as insulin;
- temporary and other workers, such as sub-contractors and maintenance workers, who may not be familiar with or be able to adhere to current shift work schedules, or who have been on a different schedule with a previous employer; and
- workers, who following a standard day's work, have remained on call through the subsequent night or weekend.

Job support

[E17042] Appropriate design of jobs, the structure of the organisation and the way it operates are important design criteria. The following points are recommended for providing adequate support in a job:

- Provide opportunity for learning and problem solving within the individual's competence.
- Provide opportunity for development and flexibility in ways that are relevant to the individual.
- Enable people to contribute to decisions affecting their jobs and their objectives.
- Ensure goals and other people's expectations are clear and provide a degree of challenge.
- Provide adequate resources including training, information, equipment and materials.
- Provide adequate supervision, support and assistance from contact with others.
- Provide feedback on performance and communication of objectives.

Organisational support

[E17043] The following factors are considered to be necessary to provide an acceptable background against which people can work:

- Employment relations policies and procedures should be agreed and understood, and issues handled in accordance with these arrangements.
- Payment systems should be seen as fair and reflect the full contribution of individuals and groups.
- Other personnel policies and practices should be fair and adequate.
- Physical surroundings and the health and safety provisions should be satisfactory.

Musculoskeletal disorders

[E17044] The term 'musculoskeletal disorders' (MSDs) refers to problems affecting the muscles, tendons, ligaments, nerves, joints or other soft tissues of the body. These disorders most commonly affect the back, neck and upper limbs.

Musculoskeletal disorders give rise to clinical effects in the individual, namely symptoms including pain, numbness, tingling, and signs including changes in the appearance of the limb which may be identified on medical examination. They usually also result in some functional changes including reduced ability to use the part of the body affected and restrictions to movement or strength. They can also have a general effect, such that there may be a resultant reduction in general health or quality of life, through restriction of activities. The February 2018 WHEC position paper, *Interrelationship between musculoskeletal problems and mental ill health*, sets out that mental ill health and musculoskeletal pain often coincide in affected individuals for example (see https://webcommunities.hse.gov.uk/connect.ti/WHEC/view?objectId=685925).

MSDs are frequently work related, but not necessarily caused by work. Some activities outside work may also present similar types of risk of MSDs to those experienced in work activities. However, non-work activities are usually not as repetitive, forceful or prolonged as work tasks, and the individual is likely to have more control over whether they undertake the activities and how long for. MSDs have been the subject of much civil litigation and employees have successfully brought a significant number of personal injury cases against their employers (see OCCUPATIONAL HEALTH AND DISEASES). The employer's duty of care to their employees with respect to upper limb disorders is now well established in the civil courts. This civil law duty runs parallel to the employer's statutory responsibility under health and safety legislation.

There is currently no specific legislation relating to MSDs. However, the *Manual Handling Operation Regulations 1992 (as amended) (SI 1992 No 2793)* and the *Health and Safety (Display Screen Equipment) Regulations 1992 (as amended) (SI 1992 No 2792)* were both introduced with the aim of reducing MSDs associated with these activities. There are also general duties under the *Health and Safety at Work etc Act 1974* and the *Management of Health and Safety at Work Regulations 1999 (SI 1999 No 3242)*. (For more details on the law see **E17060–E17068**.)

The Manual Handling Operations Regulations and Display Screen Equipment Regulations both stem from European directives introduced in 1990 by the European Commission (EC), both intended to address the problem of work-related MSDs. The impact of Brexit on UK health and safety law emanating from European directives will depend on the terms of the withdrawal agreement. The UK remains a member of the EU and is covered by all its provisions, including health and safety legislation, until the exit negotiations are complete, and a new relationship is defined.

Back disorders

[E17045] Back discomfort is extremely common in adults. Around 80 per cent of adults will experience an episode of back pain during their life, but the vast majority of these (around 90 per cent) will recover within six weeks, although recurrence of back pain is common. Back pain is rarely caused by a single injury but is generally cumulative in nature. However, a minor incident may be seen as the final straw that triggers discomfort.

Causes of back discomfort include manual handling activities; awkward postures, such as twisting, leaning, bending, stretching and prolonged static postures; poor seat design (including car seats); low physical fitness; and exposure to whole body vibration. The significance of psychosocial factors in experience of back pain is now well recognised.

The Back Book provides useful self-help guidance on how to cope with back pain (www.tsoshop.co.uk/bookstore.asp).

Upper limb disorders

[E17046] Upper limb disorders (ULDs) are a sub-category of MSDs. The term refers to injuries occurring in the upper limbs (fingers, hands, wrists, arms, shoulders and neck). It covers strains, sprains, injury and discomfort, including specific conditions such as tenosynovitis, tennis elbow and carpal tunnel syndrome, as well as non-specific disorders. These non-specific disorders have often been called 'RSI' (Repetitive Strain Injury) but could more accurately be referred to as Non-Specific Arm Pain. The cause of some of these disorders can be related specifically to the work undertaken and become known as work-related upper limb disorders (WRULDs). These disorders are usually cumulative in nature, arising from a series of micro-traumas, which lead to discomfort if the body does not have sufficient rest time to recover.

The main physical risk factors in the development of upper limb disorders are application of force, repetitive movements and awkward or static postures. It is usually the interaction of at least two of these risk factors which leads to disorders. The duration of exposure to these factors is obviously also significant, with long durations of exposure increasing the risk of injury.

(1) *Force* – application of excessive force in relation to the capabilities of the upper limb muscle group will place a strain on the muscles involved and may lead to injury. Different muscle groups are able to apply different levels of force (eg shoulder muscles can apply more force than finger muscles).

(2) *Repetition* – frequent, repetitive movements require the same muscles to contract and relax over and over again. If this activity is very rapid or continues for a long period of time the muscle can become fatigued and this can lead to discomfort.

(3) *Posture* – poor postures (both awkward and static postures) can lead to injury and discomfort by placing the muscles and joints under unnecessary strain.

(4) *Vibration* – exposure to hand/arm vibration is a recognised risk factor in the development of Hand Arm Vibration Syndrome (HAVS).

[E17046] Ergonomics

WRULDs are not confined to particular activities or industries but are widespread throughout the workforce. Some jobs are particularly associated with WRULDs, and these involve the recognised risk factors. Assembly line workers, construction workers, garment machinists, meat and poultry processors and display screen equipment workers (particularly those carrying out intensive data entry tasks) all suffer a risk of WRULDs. This list is not exhaustive and is provided for illustration only.

Common ULDs are listed in O1037. Legislation and guidance relating to ULDs is detailed in E17063 and E17067.

Lower limb discomfort

[E17047] Lower Limb Disorders (LLDs) affect the hips, knees and legs and usually happen because of overuse. Workers may report lower limb pain, aching and numbness without a specific disease being identified. Acute injury caused by a violent impact or extreme force is less common. Around 20% of all work-related musculoskeletal disorders affect the lower limbs.

Lower limb discomfort can be caused by prolonged standing, particularly on concrete floors; inappropriate (particularly hard-soled), heavy or ill-fitting footwear; operation of foot controls; and a lack of adequate foot support if sitting. Tasks that involve repetitive or prolonged kneeling (eg shelf stacking on low shelves) can lead to knee disorders.

Psychosocial risk factors

[E17048] As well as the physical factors that can lead to MSDs, there is strong evidence of the role played by work-related psychosocial factors (the worker's psychological response to work and workplace conditions) in the development of these disorders. Relevant factors include the design, organisation and management of work, the context of the work (overall social environment) and the content of the work (the specific impact of job factors).

Specifically, repetitive, monotonous tasks, excessive or undemanding workloads, lack of control over the task or organisation of the workplace, working in isolation, poor communication and lack of involvement in decision making have been implicated as psychosocial risk factors.

It is thought that many of the effects of these psychosocial factors occur via stress-related processes, which result in biochemical and physiological changes that can result in discomfort. Some can also have a direct impact on working practices and behaviours, for example work pressure may mean workers do not adjust the workstation to suit themselves at the start of the shift or may forego rest breaks.

Individual differences

[E17049] Some workers may be more likely to develop a musculoskeletal injury. They include new employees who may need time to develop the necessary skills or rate of work; those returning from holiday or sickness absence; older and younger workers; new and expectant mothers; and those with particular health conditions. Account should also be taken of differences in body size which may require awkward postures/reaches.

Prevention of WRMSDs

[E17050] The greatest risk reduction benefits will be achieved by tackling both physical and psychosocial risk factors in the workplace. These factors are best identified and tackled through consultation with the workforce. Appropriate design of tasks, workstations and tools, as outlined in E17005–E17043, can help to prevent injury to the musculoskeletal system by ensuring work is designed to be within the physical capabilities of people. Tasks or equipment may need to be modified for those returning to work following an absence related to a work-related musculoskeletal disorder.

Ergonomics tools

[E17051] The following techniques can be used to facilitate thorough and comprehensive investigation of all parts of the work so that these can be designed appropriately.

Risk assessments

[E17052] Ergonomic risk assessments can identify the risk of injury, accident or reduced performance due to ergonomic deficiencies in work or tasks. Any assessment should be systematic and comprehensive, accounting for all elements of the work, including irregular activities. Checklists can be used to ensure all elements within the task are covered, namely the person, the work task, the workstation, the work environment and work system.

All hazards should be identified and recorded and an assessment of the risk posed by the hazard made. All hazards must be addressed, eliminated where possible or reduced as far as practicable.

There are a number of risk assessment checklists and tools available on the HSE website including the following:

- Manual handling activities – the manual handling assessment charts (MAC tool) aims to help identify high-risk workplace manual handling activities and can be used to assess the risks posed by lifting, carrying and team manual handling activities. It can be used to categorise the level of risk of the various known risk factors associated with manual handling activities and incorporates a numerical and colour-coding score system to highlight high-risk manual handling tasks. There is also a V-MAC tool for assessing manual handling operations where load weights vary, which should be used in conjunction with the MAC tool. Information about the MAC tool can be found at: www.hse.gov.uk/msd/mac/. Information about the V-MAC tool can be found at: www.hse.gov.uk/msd/mac/vmac/index.htm.

- Display Screen Equipment workstations – The *Health and Safety (Display Screen Equipment) Regulations 1992 (as amended) (SI 1992 No 2792)* include a schedule which sets out the minimum requirements for workstations.
- Upper limb disorder risks – The HSE guidance *Upper limb disorders in the workplace (HSG60 rev)* contains a thorough risk assessment checklist for considering the risks associated with tasks involving upper limb movements (excluding display screen equipment activities).
- The Assessment of Repetitive Tasks (ART) Tool – is designed to help assess tasks that require repetitive movement of the arms and hands using a numerical score and a "traffic light" approach to indicate the level of risk for twelve factors. The ART Tool can be downloaded from the HSE website at: www.hse.gov.uk/msd/uld/art/index.htm.

The HSE has also developed a risk assessment tool for pushing and pulling operations (RAPP tool). This aims to help assess the key risks in manual pushing and pulling operations involving whole body effort. It is similar to the MAC tool (see above) and also uses colour-coding and numerical scoring to help identify high-risk pushing and pulling activities and help evaluate the effectiveness of risk-reduction measures. It can be used to assess the risks involved in moving loads using wheeled equipment, such as hand trolleys, pump trucks, carts or wheelbarrows; and moving items without wheels, involving dragging/sliding, churning (pivoting and rolling) and rolling. It can be found on the HSE website at: www.hse.gov.uk/msd/pushpull/index.htm.

The HSE warns that the tool is not sensitive to the level of risk in some tasks involving moving loads with hand pallet trucks equipment or similar with small wheels. Small irregularities including debris and small gradients in the floor surfaces, which would otherwise be assessed as low risk, can have a very significant effect on the manual forces required. A full, site-specific, pushing and pulling risk assessment may be more appropriate when assessing tasks (or parts of tasks) in locations with varying floor and environmental conditions, particularly outdoor tasks affected by the weather, such as deliveries or loading/unloading in yards. It also advises: 'Worker involvement in the assessment process is important as they have valuable knowledge of the specific risks of the task, particularly, for example, drivers who are experienced in delivery operations'.

See also Risk assessment.

Task analysis

[E17053] A task analysis provides information about the sequence and inter-relationship of activities undertaken in a task. It can be used to help in the evaluation of existing tasks, or the design of new ones, by identifying the demands on the individual, connections between activities and unplanned or unusual activities. From this, tasks can be allocated to different individuals or the decision may be taken to mechanise. Difficulties with equipment, tasks and workstations, and equipment and training needs can be identified.

Workflow analysis

[E17054] In a workflow analysis, the movements of an individual at the workstation, to the equipment that is used for example, or within a work area are assessed. This helps to identify the pattern of activities and the routes within the workplace to ensure they are optimised. Equipment can be arranged to ensure that items used together or in sequence are placed appropriately.

Frequency analysis

[E17055] Information on the frequency with which tools and equipment are used can be important in positioning items or controls. The frequency of use of tools or adoption of postures can be recorded regularly using a frequency count.

User trials

[E17056] Mock-ups of prototype equipment or workstations and trials of new furniture can be useful ways of assessing the suitability of new equipment. A representative sample of the user population should be used to evaluate the design. Rating scales can be used to allow people to score their opinion of a new item in terms of comfort, ease of use (clarity, adjustment etc) and so on. Rating scales should have a mid-point (neutral); either five- or seven-point rating scales are generally adequately discriminating.

Data collection

[E17057] Questionnaires and (formal or informal) interviews can also be useful means of collecting information and obtaining users' views of equipment, tasks or workplaces. Open (free response) or closed (series of options) questions can be used depending on the information that is required. Anonymity may be required in some situations (eg asking about health issues).

Discomfort surveys can be conducted to gain an understanding of the extent of any discomfort or disorders that are experienced by the workforce. In these, individuals report any discomfort they experience at the end of their shift according to the part of the body affected, and rate its severity (eg on a scale of one to five). The responses of individuals from a work area can be collated and this combined data can be used to identify where in the body operators are experiencing discomfort. It can also identify any trends in relation to work tasks or work areas. This information can similarly be used as a baseline against which the benefit of any intervention can be evaluated. It is a useful tool to use in relation to risk assessment, as it can assist in the interpretation of risks and prioritisation of risk-reduction measures. An example of a body map is shown in figure 6.

Figure 6: Body Map

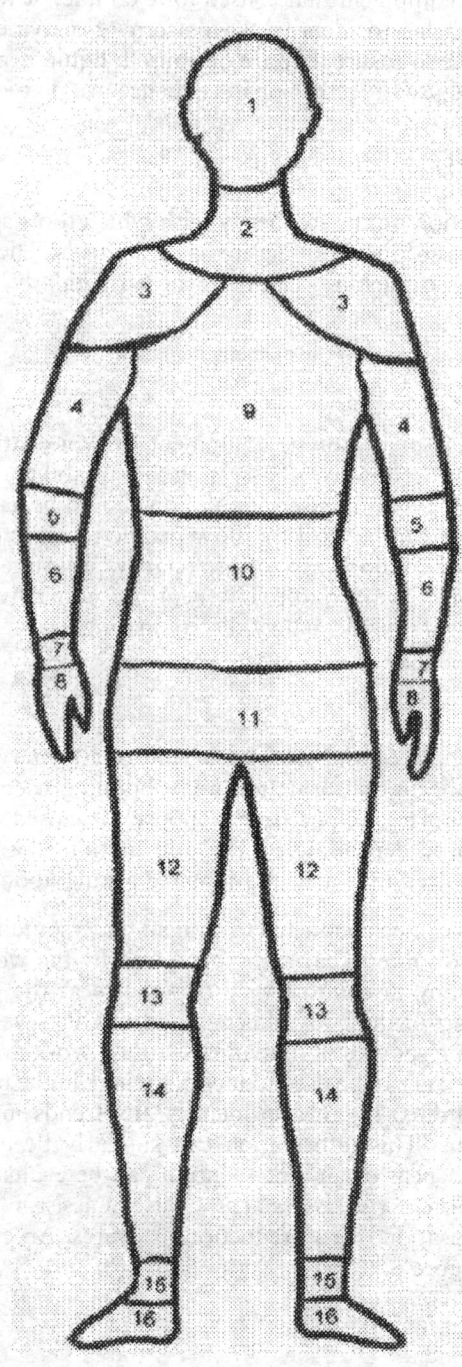

Alternative ways of using the body map are for groups of employees to apply stickers to a body map to indicate any discomfort experience, building up a pattern of where discomfort occurs.

Researchers at the University of Greenwich and Glasgow Caledonian University used body mapping to investigate MSDs among workers arising out of different recycling and waste collection systems. Their study included three surveys carried out with the same local authority workforce over four years. They reported that this was the first time researchers have used body mapping as a risk assessment tool for MSDs in waste collection.

The researchers created a unit of measurement, average pain count (APC), to examine the severity of the pain. Workers identified where they collectively felt pain or discomfort during their work activities and recorded the results through a chart or questionnaire. They reported the parts of the body with the highest APC were the lower back, shoulders, neck and upper spine. These decreased with a reduction in manual handling following the removal of boxes and baskets and an increase in the use of wheeled bins.

The workers also experienced less pain and reduced MSD risk when there was job rotation, variation in tasks and reduction in static loading for drivers. The research confirmed previously established links between awkward occupational postures and lower back pain, which can often be a result of bending, twisting, lifting boxes and sorting recycling into different components and bins.

The study, published in the Institution of Occupational Safety and Health (IOSH) Policy and Practice in Health and Safety journal, can be found online at: www.tandfonline.com/doi/abs/10.1080/14773996.2018.1491146?journalCode=tphs20.

Accurate assessment of tasks requires good observational skills and sufficient time to assess the task thoroughly. Observers should ensure a representative sample of the workforce or tasks undertaken is observed. Irregularly undertaken tasks (eg maintenance) should not be overlooked in assessments. In some tasks, movements may be rapid or only last for a short period. It can be useful, therefore, to collect video material of tasks to allow subsequent analysis.

Using existing data

[E17058] In some cases data will be available that may indicate ergonomic issues. Productivity, accident and ill health data may all be used to identify any particular issues and the extent of the problem. An accident book may also provide useful information on problems experienced. Data can be analysed by work area or type of injury in order to help prioritise areas for risk reduction measures.

Regular communication (eg through safety meetings) can also provide a useful route for identifying problem tasks or equipment.

Involving employees

[E17059] A number of people can contribute to the design of the workplace or task. These may include planners, procurers and human resources as well as those who undertake the task. Involvement of the relevant people in any proposed change is important to ensure it is acceptable. In particular, involvement of those who undertake the task in risk assessment, redesign and evaluation is a key element in ensuring that the task is thoroughly understood, any redesign is suitable and that any new equipment or changes are acceptable to the users. This can be achieved, for example, through questionnaires, discussions and trials of new equipment.

Reducing error and influencing behaviour

[E17059A] As set out earlier, the prevention of accidents is an important aspect of ergonomics. The HSE guidance *Reducing error and influencing behaviour* (HSG 48) highlights the importance of ergonomic principles in this area, making clear that tasks should be designed in accordance with ergonomic principles to take account of limitations and strengths in human performance, and that mismatches between job requirements and people's capabilities provide the potential for human error.

It sets out that failure to observe ergonomic principles can have serious consequences for individuals and organisations and challenges 'the commonly held belief that incidents and accidents are the result of a "human error" by a worker in the "front line"'. Instead, it says that: 'Organisations must recognise that they need to consider human factors as a distinct element which must be recognised, assessed and managed effectively in order to control risks.'

The guidance provides examples of what went wrong in a series of major incidents, including the Three Mile Island and Chernobyl nuclear incidents, the King's Cross fire and Clapham Junction train crash, the sinking of the Herald of Free Enterprise and the Bhopal chemical leak, to show how the failure of people at many levels within an organisation can contribute to a major disaster.

It lists the following circumstances in which errors are more likely to occur:

- work environment stressors, eg extremes of heat, humidity, noise, vibration, poor lighting, restricted workspace;
- extreme task demands, eg high workload, tasks demanding high levels of alertness, jobs which are very monotonous and repetitive, situations with many distractions and interruptions;
- social and organisational stressors, eg insufficient staffing levels, inflexible or overdemanding work schedules, conflicts with work colleagues, peer pressure and conflicting attitudes to health and safety;
- individual stressors, eg inadequate training and experience, high levels of fatigue, reduced alertness, family problems, ill health, misuse of alcohol and drugs; and
- equipment stressors, eg poorly designed displays and controls, inaccurate and confusing instructions and procedures.

It also sets out a number of steps to reduce human error:

- addressing the conditions and reducing the stressors which increase the frequency of errors;

- designing plant and equipment to prevent slips and lapses occurring or to increase the chance of detecting and correcting them (good ergonomic design can help with problems involving the layout of controls and displays which can influence the safety of a system);
- making certain that arrangements for training are effective;
- designing jobs to avoid the need for tasks which involve very complex decisions, diagnoses or calculations, eg by writing procedures for rare events requiring decisions and actions;
- ensuring proper supervision particularly for inexperienced staff, or for tasks where there is a need for independent checking;
- checking that job aids such as procedures and instructions are clear, concise, available, up-to-date and accepted by users;
- considering the possibility of human error when undertaking risk assessments;
- thinking about the different causes of human errors during incident investigations in order to introduce measures to reduce the risk of a repeat incident; and
- monitoring that measures taken to reduce error are effective.

One of the HSE's key messages is that accidents and injuries are the result of a combination of employee, employer and job factors and improved ergonomic design can influence safety behaviour.

The publication promotes the consideration of human elements as a key factor in effective health and safety management, provides practical advice on identifying, assessing and controlling risks arising from humans' interaction with the working environment, and provides guidance on further reading in this area.

HSG48 can be found on the HSE website at: www.hse.gov.uk/pubns/priced/hsg48.pdf.

Relevant legislation and guidance for ergonomics

[E17060] Ergonomics is mentioned in several pieces of legislation, although there is no single, specific piece of legislation concerning this subject. Requirements for ergonomic design, explicit or implicit, in legislation are summarised below. Specific pieces of legislation are discussed in more detail elsewhere in this publication. Selected relevant HSE guidance is also summarised.

Health and Safety at Work etc Act 1974

[E17061] Under this Act, employers have a general duty to ensure, so far as is reasonably practicable, the health, safety and welfare at work of their employees. This includes the provision of machinery and equipment that is without risks to health, the duty to keep workplaces in a safe condition and without risks to health, and to provide the information, instruction, training and supervision necessary to ensure employees' health and safety at work.

Management of Health and Safety at Work Regulations 1999 (SI 1999 No 3242)

[E17062] These Regulations require employers to carry out suitable and sufficient assessments of the risks to health and safety from work and to take appropriate preventative and protective measures to prevent or control those risks. Employees must be informed of the risks and preventative measures taken.

Further information on these Regulations is contained in RISK ASSESSMENT.

Manual Handling Operations Regulations 1992 (as amended 2002) (SI 1992 No 2793)

[E17063] These Regulations take an ergonomic approach to handling tasks. They apply to all manual handling tasks and place a duty on employers to avoid hazardous manual handling activities as far as is reasonably practicable. Where this is not practical, employers should assess the risk of injury and then take appropriate steps to reduce the risk of injury to the lowest level reasonably practicable through appropriate design of the task, load and environment, taking account of the capabilities of the individual.

Further information on these Regulations is contained in MANUAL HANDLING.

Health and Safety (Display Screen Equipment) Regulations 1992 (as amended) (SI 1992 No 2792)

[E17064] Particular ergonomic issues relate to work at display screen equipment (DSE). The *Health and Safety (Display Screen Equipment) Regulations 1992* require employers to assess the health and safety risks associated with DSE use and to reduce the risks identified. The health risks associated with DSE use include musculoskeletal disorders, eyestrain and stress. Specific requirements for the display screen equipment, workstation, software and task design are set out in the Schedule to the Regulations. These requirements ensure that it is possible to make certain adjustments to the equipment and furniture. It is advisable as part of the assessment to ensure that equipment is positioned for the user in such a way as to prevent awkward postures (eg twist to view screen) and that the adjustments are suitable for the user (eg seat height adjustment is adequate). The regulations and guidance are contained in the HSE publication *Work with display screen equipment* (L26), which contains a DSE risk assessment checklist.

Further information on the DSE Regulations is contained in DISPLAY SCREEN EQUIPMENT.

Provision and Use of Work Equipment Regulations 1998 (SI 1998 No 2932)

[E17065] The Regulations require employers to ensure that work equipment is suitable for the purpose and safe to use for the work being carried out so it does not pose any health and safety risk. General duties include the following:

When selecting work equipment, employers should take account of ergonomic risks . . . Operation of the equipment should not place undue strain on the user. Operators should not be expected to exert undue force or stretch or reach beyond their normal strength or physical reach limitation to carry out a task. This is particularly important for highly repetitive work.

Further information on these Regulations is contained in MACHINERY SAFETY M1014.

Personal Protective Equipment at Work Regulations 1992 (SI 1992 No 2966)

[E17066] These Regulations place a duty on employers to ensure that suitable personal protective equipment (PPE) is provided to employees who may be exposed to a risk to their health and safety while at work, in circumstances where such risks cannot be adequately controlled by other means. PPE should take into account the ergonomic requirements of the wearer and be capable of fitting them correctly.

PPE is designed to protect the wearer from a hazard. However, it may introduce other risks, such as reduced vision and hearing, which may restrict the ability to detect warnings. PPE may also restrict movement or the ability to perform a task (eg through reduced dexterity when wearing gloves). Some chemical protective clothing (particularly water vapour impermeable garments) can contribute to heat strain, as the wearer has limited potential to evaporate sweat from the body. In selecting PPE, the compatibility of different forms of PPE should be considered. It is often difficult to wear a hard hat with hearing defenders, or a chemical protective suit with a hard hat, although some integrated forms of PPE are available.

Reporting of Injuries, Diseases and Dangerous Occurrences Regulations 2013 (RIDDOR) (SI 2013 No 1471)

[E17067] Under the RIDDOR Regulations certain work-related accidents, diseases and dangerous occurrences have to be reported to the enforcing authorities. Regulation 8 states that:

"Where, in relation to a person at work, the responsible person receives a diagnosis of—

(a) Carpal Tunnel Syndrome, where the person's work involves regular use of percussive or vibrating tools;
(b) cramp in the hand or forearm, where the person's work involves prolonged periods of repetitive movement of the fingers, hand or arm; . . .

or

(f) tendonitis or tenosynovitis in the hand or forearm, where the person's work is physically demanding and involves frequent, repetitive movements,

the responsible person must follow the reporting procedure . . . "

Further information on these Regulations is contained in ACCIDENT REPORTING A3002.

Upper limb disorders in the workplace HSG60

[E17068]–[E17069] This guidance on upper limb disorders in the workplace provides a very useful and comprehensive approach to tackling these issues. It models a management approach, outlining seven stages in addressing these issues. These seven stages are:

(1) Understand the issues and commit to action on ULDs.
(2) Create the right organisational environment.
(3) Assess the risks of ULDs in the workplace.
(4) Reduce the risk of ULDs.
(5) Educate and inform the workforce of ULDs.
(6) Manage any episodes of ULDs.
(7) Carry out regular checks on programme effectiveness.

These stages are not necessarily sequential and different stages may well interact. For instance, educating the workforce concerning ULDs may be part of creating the right organisational environment where these issues are taken seriously and tackled supportively.

The guidance includes a two-stage risk assessment checklist. The first stage is a screening tool to help identify tasks where there may be a risk of injury and to assist in prioritising assessments. It contains five sections, asking a small number of questions concerning any signs and symptoms of ULDs, repetition, working postures, application of force and exposure to vibration. More detailed risk assessment worksheets are also included. These are divided into eight sections, concerning repetition, posture of the fingers, hands and wrist, posture of the arms and shoulders, posture of the head and neck, force, working environment, psychosocial factors, and individual differences. The risk assessment form encourages the identification of appropriate risk reduction measures for the risks identified.

It also contains useful suggestions for reducing the risk of injury, medical aspects of ULDs and case studies illustrating how organisations have successfully tackled these issues.

The HSE also publishes *Managing upper limb disorders in the workplace A brief guide* (INDG171) (www.hse.gov.uk/pubns/indg171.pdf).

Equality Act 2010

[E17070] The *Equality Act 2010* simplified discrimination law and enhanced workers' rights in this area. It consolidated over 100 pieces of anti-discrimination legislation including the Disability Discrimination Act 1995 and regulations outlawing discrimination on grounds of sexual orientation, religion or belief and age.

The Act protects job applicants and workers as well as employees. Former employees can take a claim for victimisation if a previous employer discriminates against them. The Act sets out harmonised definitions of discrimination which apply to the 'protected characteristics', one of which is disability.

The Act requires employers to make reasonable adjustments for disabled workers. This could include reallocating duties, altering their hours of work, providing technical aids such as adapted telephone or computer equipment or modifying the workplace. Taking account of the individual's abilities and needs when selecting equipment, furniture and designing the task and environment requires an ergonomic approach to ensure these suit the individual (see EQUALITY ACT 2010).

Europe – Health and Safety

Steve Granger

Introduction to Europe — health and safety

[E18001] *Have you ever thought about working abroad? Do you need to know how contracts are affected by different legal systems and business processes? Have you got the right qualifications and registration to work across the European Union (EU)?*

This chapter will outline the approaches to working in health and safety across the EU, including; qualifications, legal structures and how European law is both interpreted and applied. It will be useful to anyone who has business connections across the EU as well as contextualising the UK health and safety structures for those working in the UK. The chapter will include an overview of the harmonisation of professional qualifications across the EU and focus on the practical aspect of understanding how this is done in individual countries where regulation of the profession exists.

The chapter may be useful to consider from different perspectives;

- from the employer's perspective and the need to understand the regulatory framework that affects the business
- from a practitioners perspective in terms of any restriction or requirements imposed on the individual

Working in the European Economic Area

[E18002] The introduction of the Treaty of Rome in 1957 stated the intention to closer integration between people who live and work in the member nation states. The introduction of the Single European Act in 1987 and subsequent developments in harmonising with the Lisbon Treaty in 2009 further developed this to form what is now called the European Union. Laws relating to health and safety at work have been progressively harmonised to provide equal protection of workers and a balanced competitive market, in what is now referred to as the European Economic Area (EEA).

The Occupational Safety and Health Practitioner (OSH practitioner) who works in an international organisation will have to communicate with partner countries as part of their work. Other OSH practitioners who may wish to consider applying to work in another part of the EEA should also understand the relationship between their national legal framework and European law applied across the EEA, in order to integrate their skills and knowledge with what may exist elsewhere.

The actual process of making European law will not be discussed here; instead we will focus on a working application of these laws in relation to practising OSH and understanding where there are similarities and differences.

Law and Europe

[E18003] The basic structure of law remains the same, although there seem to be many ways of describing it. The reader is advised to just consider a few basic guidelines such as;

(1) who the parties are
(2) what type of dispute it is and where it occurred
(3) if sanctions are imposed and collected by the state – such as criminal penalties
(4) if sanction are awarded to an individual or organisation – such as compensation

How European law is both interpreted and applied

[E18004] The making of European law takes place in the official bodies of the EU. The process is not unlike the UK but obviously consultation, culture and language present challenges to the lawmakers. In many instances of OSH law the process is more about alignment than innovation.

Because of the different legal processes and maturity of OSH systems many EU countries will simply take the words and use them to apply new codes through their statutory processes. In other countries however, the choice might be to interpret new standards with existing requirements. The freedom to interpret into national law is determined by the level of societal risk in much the same way as UK law has always been made.

Apart from this, the UK generally is in a leading position for introducing law on OSH, so there are few surprises in what is required. For the OSH practitioner working in the EU it is more relevant to determine where to find such standards locally and who they are enforced by, than to know what the common details across the EU.

It is essential to understand the cultural perspective relating to OSH, the local workplace and industrial relations protocols and a general appreciation for the legal application of OSH law if the practitioner is to understand and communicate their competence effectively in other communities.

With the current debate on national identity by the UK and other nations, EU laws and its law making/enforcement process is under close scrutiny and health and safety may be a testing ground for wider politics. For example; the UK rationalisation programme of health and safety law by the government (known as the red tape challenge) might mean that laws introduced to comply with EU Harmonisation standards are now revoked – potentially leaving the UK in breach of its commitment. In reality the health and safety laws that are proposed or has been changed does not impact significantly on general day to day management of a very diverse subject area. Where gaps in the system might be formed other remedial measures can be adopted, for example; low risk self-employed workers might be exempt from the requirements to have health and safety documents but a client could invoke such a requirement through a tender or contract clause. In the UK the principle of 'less law' and consolidation into forward thinking and workable principles (goal setting standards) was actually established in the original *Health and Safety at Work Act* – so it seems that we are simply continuing to refine rather than remove what we have established.

EU legal structures

[E18005] The legal processes for each of the member states have been under development for centuries. Even if the laws relating to OSH are new, the way in which they are applied is not.

There are some fundamental differences in the approach to law that must be understood in order to appreciate how it is possible to have different means to the same end – reduced injury and ill health. This is clearly stated as one of the current strategies for the EU as a whole, as is the freedom to work across borders and mutual recognition of professional status.

Firstly, it is important to recognise that there is no unilateral agreement of terminology, despite attempts to provide definitions in legal articles. This often leads to confusion and discussion over the same words but may have different interpretation or meaning. For example the term Civil law is a much wider concept in global terms than a simplified relationship between individuals, as it is often taught to OSH practitioners and when compared with Criminal law.

Civil Law and Common Law systems

[E18006] In a broader sense EU *Civil law* describes the whole process of marshalling civil obedience and recording legal requirements precisely as standards in a structured way by the governing authority. The courts must follow and refer to these state written laws to the letter – even above any previous case examples. In this sense each case is measured against the standard – not on the basis of a previous court's example. There are many parts to the term Civil Law in this respect and a greater understanding of who made these laws, who enforces them and their jurisdiction is an important factor to establish in any country or region worked in. To Understand the relationship between Europe (through the European Commission), the nation state, internal regions of the nation and the type of industry it will be important to discern before even starting on the law, code or standard itself.

The UK has this concept of writing standards, or *codification* embedded in our statutes and regulations. But we also have the principles of *Common Law* which can be traced back over centuries. This courtroom-made law (i.e. not by a state authority) establishes *precedence*, and is subject to the influence of the court hierarchy. In a wider sense this may be considered (by some sceptics of the common law system) to put the influence of the Judge and Judiciary higher than that of democracy and the state, which is personified by Parliament.

Adversarial and Inquisitorial court proceedings

[E18007] The UK (Scotland has a hybrid system – see below), also uses the adversarial system in both our civil and criminal cases. This is where strength of argument over the presented facts to the court, is undertaken by qualified legal advocates. A trial to determine; innocence, guilt or liability depends upon the presentation and persuasion of one side of the argument over the other. A sceptical view might interpret that the judge has limited input and cannot go beyond the evidence presented to the court, toward the process of seeking the truth. In broad terms the Judges activity is more about the fair performance of the advocates within the rules of the court. However there are safeguarding mechanisms to facilitate (or deny) appeals on the grounds of fairness. The concept of presenting previous similar situations as comparators - or as a means of explaining key points in such a legal contest, is why previously established case law is so important for understanding common law and the UK application of UK and EU legislation for OSH.

The UK is almost exclusive in the EU in this respect because of the more prescriptive codification process across the rest of Europe, where previous case examples do not mean as much.

In contrast to the adversarial and common law process of the UK, in European systems the inquisitorial approach is used. This is based on the Roman principles of standards being explicit and codified and the focus is on determining if such standards had been breached. Also the inquisitorial system promotes greater enquiry and allows the judge more freedom to ascertain information. Generally the court is more open to discover the truth – and not necessarily restricted to the power to persuade through skilled advocates.

In the UK civil law refers to the relationship between individuals and organisations. This also introduces terms such as *tort* (wrong doing to another) including the tort of; *negligence, trespass and conversion*. Civil courts will also hear matters of contract dispute between two parties. In European systems this may also be known as the *law of obligations* and there are close parallels to be found.

Criminal law

[E18008] The concept of *Criminal law* — where punishment can be applied by the state as the state codes of conduct have been broken and the offense is against society at large is relatively similar across the EU – this is irrespective of an injured party. However, one major difference is that in the UK the civil and criminal courts systems are independent (although there are aspects of utilising the civil courts for efficiency such as appeals), whereas in some EU countries, such as Germany the courts structure may hear both Civil and Criminal cases and the access level is more about the size of the potential outcome in terms of compensation or penalty.

It has to be understood that wherever OSH matters arrive in a courtroom then; some criminal law becomes a civil issue and vice versa, and some law is open to interpretation and some isn't. We will avoid deepening the debate further and simply keep an open mind for now. Remember that law is said to be one of the founding 'arts' and not a precise science – but this may just be a UK perspective!

Harmonising legal systems

[E18009] Ultimately, and specifically for test cases in health and safety, the final outcome decision rests with the national court structure concerned. However, the European Court of Justice may be consulted for key decisions relating to interpretation of EU Regulation or Directives and equal application of EU law across the EEA. This is important for UK OSH laws, particularly when discussing terms in our legislation such as 'so far as reasonably practicable' – remembering that most EU countries codify the requirement more precisely.

Understanding the different structures and mechanics of the legal systems explains the confusion over the EU OSH requirements which dictate 'x&y will be done'.

The European Commission complained against the UK for not meeting the requirements of *Council Directive 89/391/EEC of 12 June 1989 on the introduction of measures to encourage improvements in the safety and health of workers at work* – also known as the Framework Directive on health and safety. The EC said that it was inappropriate to use qualifying instructions within the UK by using the term 'so far as is reasonably practicable' through regulations made under the HSWA1974. This was dismissed in 2007 by the Court of Justice of the European Communities. The court agreed that when using the UK legal system which is where the interpretation of the Directive would be applied (and not the EU based codified model yardstick of the complainant) then the application of the Directive within the UK jurisdiction meant the same thing.

We should therefore understand this cultural component of language is part of the problem of trying to achieve unilateralism across the EU and recognise the possible anomaly between words and legal processes when we work elsewhere. In Europe there will inevitably be a law or code to provide detail what should be done – go and look for it is the advice given here.

The reality is that most courts across Europe have an element of all of these principles, and not forgetting that the law and legal processes across the whole of the EU continues to develop. Where an individual considers they have a grievance against their own country then the European Court of Human Rights may be used. This might apply where for example an employee considers their country has not applied a European requirement fully.

We should not get out of our depth here; in all instances; the judge is to ensure that legal fair play is carried out; professional advocates represent the parties and the concept of court hierarchy for structured appeals. If the OSH practitioner ever ends up in a court it is important to know the different approach and who can ask questions. The need for legal counselling and advice cannot be over stated.

The global influence of European legal systems

[E18010] The influence of European culture and law has spread all over the world. The UK is not alone in the world in using common law as a legal basis. The USA, and Canada, India and Australia follow the same principles of common law, whereas Europe, Asia and much of Africa follow the civil codified approach. The Middle East and other parts of Africa use the Islamic principles of law and some regions or countries prefer to take the best of each of these systems and develop a hybrid approach. Scotland, for example incorporates part of the UK common law and part on the inquisitorial approach of the EU civil process.

Providing employers with OSH advice in the EEA

[E18011] The EU Framework Directive imposed the same standard of OSH management and competent advice to all member states, and required the member states to incorporate this within national legislative systems. It is clear therefore that all workplaces require someone to provide this advice and employers to ensure their business complies.

This advice can either be;

(1) Internal – an appointed employee, in the UK this is typically seen as a nominated OSH manager or other manager/director.
(2) External – such as an appointed consultant or contractor who are seeking commercial and enterprise opportunities.
(3) State provided - funded by the state or a subsidiary department. In terms of the UK state support for OSH information to business comes from the HSE and local authority Environmental Health professionals. This is provided as part of their remit under the HSWA1974. Other authorities such as the regional Fire Safety, Office of Rail Regulation also support this state provision of free OSH information, although the current trend for charging may change this relationship. In most cases the authority providing the advice may also be the prosecuting authority for non compliance – something which is certainly not universal across the EU.
(4) Not for profit - Across the EU there are different OSH bodies, some of who provide these services either free or on a cost recovery chargeable basis; some are associated with professional membership and some are simply fulfilling charitable objectives.

In all instances of the OSH practitioner providing advisory services on a personal or commercial level – either as an employee or a consultant, the individual has responsibilities under EU law to ensure they only do what they are capable of (although it is primarily the employers duty to ensure this happens). It should also be remembered that the Framework Directive protects 'workers' who identify potential risks or problems with safety processes.

In terms of protection it would be wise to ensure that European contracts include powers of the health and safety professional/consultant, sufficient explanation for the grounds of stopping work and reasons for breach of contract relating to advice given and how it is dealt with. This is especially important as many EU nations may impose personal liability on individual/consultant. In terms of personal liability for corporate health and safety performance it is worth looking at the country in question and asking yourself - 'can OSH responsibility be delegated to me for my any intended work role?'

It should be noted that the UK is also moving more towards personal liability (both civil and criminal) of the safety professional as an individual. The UK now has established case law and conviction of safety managers for failing in their duties.

The advice here is the same for both employed and contracted work; Remember that common law precedence is not there to explain what you thought might be the case in terms of your innocence, so protect yourself through your contract! Check that suitable employment policies, contracts and liability safeguards are put in place for your personal protection and importantly your potential legal costs.

This last consideration should now be included in incident management policies as existing policies and liability cover will usually be concerned with the highest levels of management and the corporate body as a priority – you cannot assume your legal costs will be covered automatically.

Corporate accountability – who is listening to the OSH practitioner?

[E18012] The relationship between organisational executive and the OSH practitioner will vary depending on the proximity between the organisational heads and their personal liability or exposure to prosecution. In the UK the debate on corporate accountability and manslaughter is still ongoing, even with relatively new legislation and case law on the subject. Understanding how executive responsibility and accountability differs may also make communication from the OSH practitioner easier.

One particular area OSH practitioners should be aware of is the growing trend towards Corporate Sustainability. This may also be referred to as Corporate Social Responsibility (CSR) or the 'triple bottom line which relates to social, economic and environmental impacts. Health and safety is already entrenched in these constructs – yet it has been significantly underplayed to date, usually to the benefit to the environmental element. Practitioners should make themselves aware of these high level corporate reporting processes and be ready to provide useful information on organisational cost/benefit of health and safety systems.

Worker participation

[E18013] The OSH practitioner will no doubt understand that the other end of the employment equation also needs balancing. The EU has information on consultation processes and employee rights regarding OSH. EU sponsored organisations such as; www.worker-participation.eu and the European Trade Union Institute www.etui.org are very useful websites which contain up to date comparisons and arrangements in each country for the requirements to consult and form safety forums or committees. This is particularly important when understanding the regional codification process described above in relation to consultation. ETUI published a very useful and free report in 2014 relating to the current situation regarding worker consultation requirements and achievements across the EU member states and the trend and effect of reducing social legislation such as this.

What is a regulated profession?

[E18014] The European Commission provides useful information on how the EU intends to enable professionals to work across borders;

> The term "regulated profession" as used in Directive 2005/36/EC of the European Parliament and the Council of 7 September 2005, is defined in section 3) 1.a) of the Directive: "an activity or group of professional activities, whose access, practice or conditions of practice are directly or indirectly dependent, in accordance with laws, regulations or administrative provisions, on possession of specific professional qualifications; the use of a professional title limited by laws, regulations or administrative provisions to holders of a given professional qualification constitutes a condition of practice."

Identifying countries where the OSH profession is regulated

[E18015] Europe is still in the process of harmonising work and qualifications in all vocational subjects. It is not yet possible to clearly identify a universal qualification or easily benchmark the role of the OSH practitioner. Each state (and part thereof) may have internal requirements concerning the provision of a 'business service'. Also, each individual OSH practitioner will have a portfolio of qualifications (safety and non safety) which might contribute to being accepted in one country or industry but not another. In short, it is the individual who needs to analyse their situation at the moment but it should be remembered that the codified systems in mainland Europe make it more likely that country specific recognised qualifications or a state issued license to practice are required.

There are some developments; The European Regulated Professions Database can point the direction you may need to follow in order to find out about your personal situation in the country you intend to work in. The database website states that;

A profession is said to be regulated when access and exercise is subject to the possession of a specific professional qualification.

It contains lists of regulated professions in the EU member states, EEA countries and Switzerland covered by the Directive 2005/36/EC.

Regulated professions are grouped together under headings called "generic professions". This is to help find regulated professions which are listed in the language of the country in which they are regulated, but several attempts using key words is advised as the list includes generic headings regulated in different Member States and these may cover different activities. The website also contains important details of the Contact Point in the future host country that are responsible for providing general information on recognition of professional qualifications and the competent authorities and national legislation governing the professions. This might be particularly useful as the question can be asked directly to the country representatives regarding necessary qualification, insurance and licensing.

At the time of publication the database is not comprehensive as a far as OSH is concerned, this may develop further following the EUSAFE project discussed below. For the UK, IOSH is listed as the professional body with Chartered Member being the standard professional level for the framework.

OSH Qualifications and competence

[E18016] The EUSAFE report was an EU funded project which provided a basis for co-operation by safety and health professional organisations and academics drawn from member states. The final report created a matrix of professional competencies and was accepted by the Education, Audio-visual and Culture Executive Agency in 2013. Further studies continue to determine what the essential skills knowledge and competencies are, and where state registration is required to deliver OSH advice.

Approximately half of the member states impose a statutory regulation of OSH practitioners whilst the remainder either have nothing in place or more likely impose a standard though alternative means such as professional membership and market forces on the profession, or as a result of economic requirements of the insurance processes.

Some countries are easily categorised – the UK for example does not have a mandatory registration scheme but it does have a voluntary register for the purpose of giving assurance to small businesses that those who are listed on the *Occupational Safety and Health Consultants Register* (OSHCR) have suitable qualifications and insurance. The scheme was set up following the release of the Governments report *Common Sense Common Safety*. It has a benchmark of qualification and professional accreditation to be registered, and as it is supported by the Health and Safety Executive and, IOSH the Chartered professional body for OSH practitioners, amongst others. OSHCR therefore, must carry some legitimacy, but this has yet to be tested in the UK courts.

State regulated OSH professionals

[E18017] Conversely, in many parts of the EEA either the state or regional authorities or the state funded schemes linked to state insurance processes will licence (or regulate by qualification) OSH practitioners work. These include;

Austria	Hungary	Slovakia
Belgium	Italy	Slovenia
Cyprus	Latvia	Spain
Germany	Luxembourg	Switzerland
Greece	Portugal	

Germany is an example where the state(s) have produced more comprehensive requirements to practice OSH, through the authority given to the state insurance coordination body; DGUV. From the DGUV website it explains their role and that of the OSH professional;

> The statutory accident insurance institutions in Germany and their umbrella association, the DGUV, have good reasons to be strongly involved in the area of qualification. Firstly, the German Social Code mandates them to conduct qualification measures in relation to all prevention issues; secondly, by virtue of the initial and further training of their own staff, they conduct continual quality assurance.
>
> Qualification measures addressing prevention issues are offered in particular for employers and management staff, OSH professionals, safety officers, teaching staff, trainers, company physicians, specialist medical personnel and other disseminators at plant level.

The DGUV statutory codes are a good example of the codified legal process above, and states;

> The statutory accident insurance institutions in Germany and their umbrella association, the DGUV, offer qualification measures in all areas of prevention, in accordance with their statutory mandate. Volume 7, Article 23 of the German Social Code (SGB VII) places the following requirements upon the statutory accident insurance institutions in this context:
>
> (1) The accident insurance institutions must assure the necessary initial and further training of persons within companies who are charged with performing measures for the prevention of occupational accidents, occupational diseases and work-related health hazards, and with first aid. The accident insurance institutions may also implement relevant measures for the company physicians and OSH professionals who are not company staff and who are to be appointed in accordance with the legislation governing company physicians, safety engineers and other OSH professionals. It is the task of the accident insurance institutions to encourage employers and insured individuals to attend courses of initial and further training

Therefore, in order to give advice on OSH the practitioner needs to be suitably qualified (approved by DGUV and registered accordingly as a business). If you are moving to work for an organisation based in Germany, then DGUV should be approached to determine what qualifications are required to supplement existing and home nation obtained accreditation.

In France a similar situation exists in that health and safety at work is closely linked with social security, health and insurance. Employers pay into the state insurance fund and this supports several departments including Caisse Nationale d'Assurance Maladie des Travailleurs Salariés (CNAMTS) -National Health Insurance Fund for Salaried Workers, and Caisses Régionales d'Assurance Maladie (CRAM)- regional health insurance funds. These organisations are responsible for enforcing the Labour code aspects on health and safety. The National Institute for Research and Safety (INRS) provide advice to individuals and employers and has a very useful website for resources.

There are other authorities such as the Occupational Injury and Disease Commission CaT/MP which has a national and regional structure of risk professionals and specialists. The French agency for Food, Environmental and Occupational Health Safety (ANSES) which is a public institution reporting to the Ministers for Health, Agriculture, the Environment, Labour and Consumer Affairs. The new health establishment became a legal and operational entity on 1 July 2010 and has incorporated the missions, resources and personnel of the French Food Safety Agency (AFSSA) and the French Agency for Environmental and Occupational Health Safety (AFSSET).

Other agencies such as the French Agency for the Safety of Health in the Environment and the Workplace improve knowledge of occupational risk prevention and are similar to the HSE. The National Agency for the Improvement of Working Conditions also provides information to organizations and promotes OSH.

Both France and Germany have dedicated services specifically working to reduce occupational ill health. It can be seen from these two examples that there are distinct similarities with the UK and the role of the HSE but also that the codified legislation has made provision for integrating OSH more with organisation of the state compensation allowances and benefits. This, in turn, has a vested interest to reduce accidents and ill health. In terms of the objectives for health and safety – to reduce accidents and ill health, this symbiotic relationship appears to work well. As discussed above – we are not judging the merits of either system here (the performance of UK plc is very favourable in terms of measuring safety performance), but we do have to understand how they work and how we work within them.

Until the (relatively recent) introduction for corporate and individual accountability in the UK both Germany and France stood out as examples where the law identified 'natural persons' rather than the ubiquitous 'employer' identified in UK law. If senior positions are gained it is strongly suggested that clarification on personal responsibility is sought and appropriate safeguards are taken such as individual Directors insurance for legal costs etc.

Development of a European Vocational Occupational Standard

[E18018] While vocational training was identified as an area of Community action in the Treaty of Rome in 1957, education was formally recognised as an area of European Union competency in the Maastricht Treaty 1992 (ratified by the Treaty of Lisbon in 2009) which established the European Community in 1992.

The European Parliament has expressed a desire to enable professionals to work across the European Economic Area more easily to improve economic development and sustainable economic recovery.

There are a number of European initiatives to harmonise qualifications and vocations. This is part of the drive towards an open marketplace and the development of transparent skills and qualifications to enable people to move around the EU more easily.

In particular, the EU is currently engaging on a number of schemes specifically focussed on professional harmonisation and professional recognition.

The EUSAFE project has been running since 2010 and was sponsored under the EU's Leonardo da Vinci *lifelong learning programme*. It aims to standardise a qualification for OSH practitioners and has produced a Vocational Occupational Standard for a generalist OSH practitioner at technician and manager levels.

There are currently a number of initiatives to develop an international competency framework. However it will be some years before this truly translates into a common profession due to the intricacy of national laws and requirements.

How the Vocational Occupational Standard VOS can be used by registered professional and those wishing to become registered

[E18019] The European Credit system for Vocational Education and Training (ECVET) is the European instrument to promote mutual trust and mobility in vocational education and training. Developed by Member States in cooperation with the European Commission, ECVET has been adopted by the European Parliament and the Council in 2009.

The European Credit Transfer and Accumulation System (ECTS) is similar to ECVET but concerns higher qualifications such as university based degree level. The approach here also provides flexible learning opportunities and crediting procedures for those studying at a higher level.

It should not be forgotten that the Framework Directive imposes a common standard (but through national interpretation) with regard to OSH advice; It identifies that; a) competent advice is necessary to achieve workplace safety and health, and that; b) the advice should be given by someone who has the necessary skills and aptitudes to ensure the advice will achieve the objective of worker safety and health.

> Article 7 – Protective and preventive services
>
> 8. *Without prejudice to the obligations referred to in Articles 5 and 6, the employer shall designate one or more workers to carry out activities related to the protection and prevention of occupational risks for the undertaking and/or establishment.*

And

In all cases:
- *the workers designated must have the necessary capabilities and the necessary means*
- *the external services or persons consulted must have the necessary aptitudes and the necessary personal and professional means*

and

8. *Member States shall define the necessary capabilities and aptitudes referred to in paragraph 5.*

If a competency framework such as EUSAFE is successful there will need to be a unilateral agreement on exactly what this means for the OSH practitioner. It is hoped that in the years following the publication of the EUSAFE report the state registration schemes will accept a wider range of qualifications to reduce the need to duplicate learning.

It should also be recognised that the OSH career market and vocational training providers influence 'market forces' and represents a significant element of 'who' or 'what' is referred to as 'competence'.

Monitoring OSH performance standards

[E18020] Like 'competence' there is no standard European answer to the value of organisational performance on OSH. EUROSTAT are responsible for all such information across the member states and have produced a valuable resource; Health and safety at work in Europe (1999–2007) a statistical survey.

It is useful not only for benchmarking but also highlights some of the differentials to be considered when trying to compare like for like frequency rates. Complications such as no standard definition for 'accident' or 'lost time' and the inclusion of highway statistics in some countries but not others mean that even this has to be viewed with caution.

The Health and Safety Executive produced a useful document; European Comparisons – summary of GB performance, as part of their statistical information service

Accident and ill health reporting

[E18021] Accidents and emergency actions are likely to be universal, but in terms of notification it is vital to understand that there may be significant differences on reporting and recording accidents and incidents. In much of France for example the information has to be with the insurance organisations within 24 hours. Knowing how to get hold of someone (and understanding them) in times of crisis will show the benefit of being prepared. It might be useful to establish a list of information you need to have ready to hand on arrival or task yourself with finding out without delay.

European organisations and resources

[E18022] As with OSH there are many resources to help you get started when working abroad.

Primary information sources and organisations

The UK foreign office and other government departments all offer resources associated with business and social integration within the EU.

Using the resources available from the European institutions is always worthwhile. The EU publishes most documents in at least 3 basic languages; English (predominantly the language of Europe), French and German. The European commission, Parliament and courts all have very easily accessed information which can often be a good starting point in looking for regional application.

Of course not everything is presented in English but internet translation proves very useful in getting to grips with most of the basic translation required to apply professional knowledge in a colloquial way.

UK organisations such as the HSE also contain web pages with information on different countries and working abroad.

Secondary information sources and organisations

There are other European OSH organisations where resources are freely available and produced with the OSH practitioner in mind. The European Agency for Safety and Health at Work or EU-OSHA and the European Network of Safety and Health Professional Organisations ENSHPO are two such bodies, each with links pages and news.

ENSHPO is useful in identifying which organisations future colleagues may be members of and a little research into membership qualifications might be a quick route to learning about statutory regulation processes.

Insurance organisations, manufactures of safety products and other professional organisations associated with occupational health, workers co-operatives and trades unions are also likely to yield some information to give a better perspective on OSH across the EEA.

Working essentials

[E18023] Individual circumstances will vary and the level of support will be determined by client/employer relationship. Irrespective of these it will pay to be prepared so that you know where you stand;

- Check the country you are going to and determine if there is requirement to complete a specific training scheme or obtain a licence to practice
- Check with your company insurance to see if your policy covers the whole of the EEA and the extent of liability you might carry – especially for legal assistance

- Check your contract and know exactly what you are responsible for in the small print
- Establish who and where are the local networks of OSH practitioners
- Determine the enforcing and advisory agents for the region and industry you are in
- Find the local interpretation of the Framework Directive and supporting requirements – it may be better to cite these that continually 'what we do'.

Facilities Management – An Overview

Robert Greenfield

Introduction to facilities management

[F3001] *Facilities management has gained so much momentum and recognition over the past few years and it is now a critical element of an organisation where FM's are being expected to deal with complex issues concerning the functionality of buildings that are capable of accommodating the needs of what is becoming a diverse and ever changing workforce within the workplace The facilities management structure of an organisation has become an essential element of the overall business strategy as this will play a major part in allowing the operational elements of the business deliver its product or services in order to meet its overall goal and objectives. In essence the modern day facilities manager has responsibility for the management of all aspects of an organisations non-core activities allowing the organisation to flourish within an effective workplace environment.*

The perception of facilities management and the services that may be provided has completely changed in a relatively short space of time and the facilities manager is now a widely respected and essential member of the senior management team within organisations. Facilities Management is now recognised in all areas of business, as providing a business edge within today's highly competitive world and is a highly developed strategic discipline which provides a well-managed and utilised built environment at best value for the company. Apart from the day-to-day operational running of activities Facilities Management is an essential element of the high level corporate end of a business in the delivery of an effective strategy. In the current economic climate Facilities Management can make or break an organisation and will no doubt move forwards in the future to gain even more recognition as the Facilities Management industry rises to the challenges posed in not only the current financially challenging world but also to the further move towards the use of modern game-changing technology. With organisations facing even more pressures to find ways of reducing the costs of the management of their buildings this is having a direct impact on the facilities management industry, which in turn is having to take a more innovative approach.

The current trend for facilities management providers is a multi-skilled approach where larger FM companies are offering clients an entire range of facilities management services under the one umbrella of Total or Integrated FM. More and more global household name types of organisations as well as much smaller ones have gone down this route and have in effect placed all their eggs in one basket. Although these types of contract models, especially the international delivery ones, are extremely complex the operating cost savings and performance benefits are planned to make a substantial difference to the profit margins of the client organisations. If these organisations do have to buy in specialist services then their buying power is so great that immediate cost savings in the region of 15% may be achieved.

The total FM model has now been taken a stage further by a number of the large global managing agent organisations, who up until recently have not had the ability to self-deliver some of the more technical elements of FM but that has changed through the acquisition of hard and soft FM organisations that has resulted in managing agent brand names being able to self-deliver a total FM package. This has its main advantages in that these types of organisations have a much better degree of management control over all of the technical elements of facilities management as well as eliminating the management structures and profit margins of what would have been supply partners or contractors thus again resulting in a streamlined structures and cost benefits.

There are now major facilities management organisations in many countries of the world including the British Institute of Facilities Management (BIFM) based in the United Kingdom, Facility Management Association of Australia (FMA), International Facility Management Association (IFMA) based in the USA, Hungarian Facility Management Society, Euro FM based in the Netherlands and Arseg based in France to name but a few. In addition, we now have Global FM which is based in Brussels and is an amalgamation of the facilities management organisations mentioned above and exists for the advancement of FM and in order to meet this objective organises educational workshops around the globe to provide information, prestigious awards for best practice and leadership to facilities managers. In addition, Global FM through its worldwide reach organises World FM Day where it encourages facilities managers and organisations to hold events on the same day to further raise the awareness of facilities management. In some instances more recently there have been financial challenges for some of these FM organisations and in order to survive they have had to call on all of their technical skills and knowledge. In some cases, this exercise has culminated in a steady surge of the membership and especially in the case of corporate members where organisations see membership as being essential in the cross pollination of best practice and has in some cases has helped them survive in a challenging market.

The definition of facilities management

[F3002] The British Institute of Facilities Management has adopted an official definition of Facilities Management developed by Central European Normalisation (CEN) the European Committee for Standardisation and which has also been ratified by the British Standards Institute (BSI), as follows:

> "Facilities management is the integration of processes within an organisation to maintain and develop the agreed services which support and improve the effectiveness of its primary activities".

An important element of this definition is the integrated management approach. Back in the mid 1990's there is no doubt that the sites and departments in an average organisation would most likely have been doing their own thing working in silos, setting their own strategy goal and objectives and managing themselves independently of each other with little cross pollination of ideas and best practice. Tasks and activities would have been duplicated due to the missing element of an overall integrated approach which is what makes facilities management such an important element of an organisations overall strategy.

> Therefore in summary the key function of a facilities management team are to ensure that all necessary resources and equipment are available and in place and that all elements of an organisations assets are maintained to preserve their value and functionality, so that an organisation's staff can carry on their core function in the most efficient manner possible.

Facilities management

[F3003] The modern organisation has to think strategically to survive and to support a diverse range of activities from its core business through to the provision of key activities such as hard FM eg technical maintenance and soft FM eg catering & cleaning, which enables a company to concentrate on its core business. However, the organisation may not have the expertise or the desire to manage such non-core functions and may not have the benefit of the most ideal workplace environment and it is under these circumstances that facilities management is able to effectively manage the non-core activities and use them to support the core business as well as adding efficiency and value. One way of achieving this is to outsource the non-core activities to an external contractor however such non-core activities would still be an essential part of the organisations activities so the conundrum is how to decide what element/s to outsource.

It is possible to divide the activities of an organisation into three streams, as follows:

- Core activities - are the key activities of an organisation which provide competitive advantage such as the provision of legal services, manufacture of pharmaceuticals. In other words if the organisation outsourced or contracted out these activities to an external party then it would in effect creating a competitor.
- Non-core, non-critical activities – these are activities such as the grounds maintenance, catering and cleaning which are still important but where even if they were of a low standard for a short period of time they are not likely to harm the organisation in the long term.
- Critical but non-core activities – these are activities such as supply of essential components to allow a production line to produce finished products, maintenance of critical assets that could lead to failure of a power supply to a main IT server, security leading to a serious breach. All of these may still be outsourced but if not performed to a required high standard would be a risk for the organisation and could place it at a disadvantage.

It is essential that organisations focus on their core business activities in order to fully utilise their skills and experience gained over the years in order to develop that competitive edge and from a facilities management point of view the modern strategy is that these non-core activities may be outsourced but with the proviso that particular attention be paid to the management of the non-core critical activities.

By outsourcing to specialist contractors outsourced activities may be managed by the facilities management team in such a way that they can place more certainty of costs which is essential for budgeting, save money or in most cases increase productivity through efficiencies gained and actually add value to the bottom line profits of the organisation.

The built environment and all associated assets are a major expense for a company and must suit the needs of the core business but also be flexible enough to accommodate change, especially in a dynamic work environment. Being such an expense it is important that all steps are taken to protect the assets of the organisation for the future and as a result facilities management may cover the management of a very broad spectrum of activities that could include the following:

- maintenance of technical plant and equipment;
- estate and property management;
- energy and environmental management;
- health, safety and environmental management;
- human resources management;
- waste management;
- sustainability/corporate responsibility;
- financial management;
- change management;
- domestic services eg cleaning, catering, security, post room, and reprographics;
- utility procurement; and
- lifecycle costing.

An illustration of good facilities management and the modern day challenges imposed may be seen in the following example.

The facilities team has seen the organisation expand in some areas of their activities but decrease in the production facility. The old head office building is of old construction with many fixed offices and as a result of the changing workforce there are now far too many staff attempting to work in the building with the additional problem of limited car parking spaces due to majority of plant equipment necessary to run the building being sited within the car parking area. The challenge was to develop a facilities management strategy to not only bring the building itself into the 21st century but also to reduce the space required within the building for each staff member. As a result a major budget was agreed for the refurbishment of the head office.

Key challenges for the facilities team were, as follows:

- Staff accommodation during refurbishment.
- Car Parking issues and resiting of plant equipment.
- Utilisation of the limited space available.

The staff accommodation issue was resolved by utilising the office space in the nearby production facility that was closing and reconfiguring some production areas into further open plan office and product demonstration and training space. As power, voice and IT network cabling was already installed and functioning plus existing furniture in excellent was transferred from head office to affect the move at a very low cost.

Plans had already been drawn up for the head office building in the past to increase the utilisation of the roof space and by strengthening some key areas of the roof, plant equipment housed within the car park was eventually moved to the roof thus creating a further thirty car parking spaces and improving access to the adjacent service yard for deliveries.

The internal space issue was a challenge for the facilities team and this was met in a number of ways, as follows:

- Open Plan environment with limited meeting rooms to provide maximum flexibility.
- Implementation of Webinar internet facility to complement conference calling and to negate the need for staff to attend office.
- Review of all staff at head office and identification of key staff essential to work full time with these members of staff being allocated full time desks.
- All other staff issued with laptop computers and secure system to access the organisations servers as part of an IT upgrade to correspond with the head office refurbishment.
- A majority of staff became home workers and head office became in effect a hot desk area with desks and car parking spaces only bookable in advance via an online internet based system. The result was that staff had to plan their days/weeks well in advance and either worked at home, on the road meeting clients or at other sites.
- Facilities Management team were set strict targets to police the new facilities model to achieve a minimum target rate of 95% occupancy for office space and car parks.

The net result for this organisation was a complete change in the culture of the organisation in that staff no longer expected there to be large amounts of available office space in head office and accepted that they had to plan their working days well in advance. The facilities management team used a variety of effective and proven change management techniques that were implemented to bring about a complete culture change within the organisation which has in turn has resulted in a flexible and highly efficient workplace resulting in cost savings for the business in what are proving to be challenging times.

Hard and soft facilities management

[F3004] The activities associated with facilities management and the support of the core business is normally split into the following two discipline categories.

Hard FM – which includes the provision of all activities involved in technical and statutory compliance maintenance services, energy management, refurbishment, life cycle management and costing, minor and/or additional works of the actual structure of the built environment. In addition, the provision of utilities and waste management could also fall into this category.

Soft FM – includes provision of all the remaining activities such as human resources, change management, health & safety, security, cleaning, catering, help desk, post room, reprographics and reception duties.

Managing

[F3005] Changes within our society and within the workplace have seen the previous ten years' experience a major upheaval. The technology available to all companies is now staggering and affects staff in a variety of ways. Modern buildings are designed and operated in a very high-tech manner which makes full use of a vast amount of computer software as well as the rapidly emerging application technology which allows numerous building management functions to be performed from handheld tablets. The same applies for manufacturing processes, which tend to be very automated and which rely on a massive amount of unseen infrastructure to support this process. The same is true for people management, due to the extensive range of well tested management techniques. During this period of change, which has seen technology and management techniques grow, the concept of facilities management has also grown. Gone are the days of the caretaker running the old school boilers, ordering paper towels and a few cleaning products whilst running his team of cleaners. Today's companies demand a holistic and more integrated approach to the provision of the built environment, workplaces and their management, and this responsibility would generally fall to a specialist facilities management team, depending on the size of the task. Failure to embrace an effective facilities management strategy for some organisations could lead to failure especially in the current challenging global financial climate and the challenges imposed on management to reduce costs.

In the modern organisation the facilities management team, responsible for say the ventilation and air-conditioning equipment is most likely to also be responsible for the design concept of the building, the selection and procurement of the ventilation and air-conditioning equipment, the maintenance and repair of the equipment and even the procurement of the energy actually powering the equipment. As can be seen facilities managers have a very broad, difficult and responsible role, as the impact upon the business of a company, if they get it wrong, can be considerable. As most companies are increasingly seen to outsource everything to a facilities management company, with the exception of their core business, means that the facility manager is being expected to manage an even more diverse range of activities.

If a company is to place facilities management at the forefront of its strategy, then the lead facilities manager will be required to be highly skilled and, ideally, have a seat on the board of directors. If a board placement is not possible then there should be an effective communications link directly up to a strategic level which is important as the company may have embarked upon a high level facilities management strategy and any decisions made at a board or operational FM level could have a detrimental effect upon the implementation and operation of the FM strategy.

Stoddart Review

[F3005A] The Stoddart Review has now become a very important periodical publication for the FM profession and raises the awareness of some key facilities management issues. Built on the back of the work by Christopher Stoddart, who passed away in 2014, a number of his colleagues and professionals from the facilities management industry have come together to continue his legacy and the vision of the importance that the workplace has to play in business and society in general.

The UK is facing a serious productivity gap when compared to 'The Group of Seven' (G7) and whilst there have been proposals to improve skills, infrastructure and investment, are we missing something here with our workplaces?

One of the issues currently being addressed is the complexity of buildings and their design, the time lag between the need for the workplace, the conception in design through to final construction and occupation. There is no doubt that by the time the building is occupied the actual needs have changed significantly leading to only 54% of employees in the UK agreeing that their workplace enables them to work efficiently. This is a serious problem as it was felt that the workplace actually held them back in their role rather than supporting it, which is not what facilities management is about.

Another worrying concern is that not enough organisations view the workplace as the key driver of performance which begs the question, do business leaders understand the value of the contribution of the workplace? Far too many organisations place emphasis on cost and cost reduction rather than broadening their outlook and considering value.

The Stoddart Review will bring together the thoughts of business leaders and aims to put the workplace back on the business agenda.

The project itself consists of a number of activities such as in-depth interviews with business leaders, an open view forum and finally data to provide statistical insights. As the project progresses the results will be included here in future issues.

Just maybe the incorporation of the workplace and facilities management could be the answer for UK plc!

Company approach to facilities management

[F3006] There are a number of ways in which a company could approach the introduction of facilities management and this could depend upon it size, location, types of buildings, core business and objectives, just as an example. It is important that the company develops a facilities management strategy tailored to the company requirements, which in turn must be capable of delivering its operational objectives.

It is important that the company is able to effectively distinguish between its core activities, eg production of engineering components or provision of a service and its non-core activities, eg security, catering, cleaning or maintenance. If the decision is made within the facilities management strategy to outsource, then it is the non-core activities that should be managed in this way, allowing the company to concentrate their efforts on their core business.

In order to manage the facilities of a company effectively the facilities strategy needs to be developed and agreed at senior management and board level. The company must decide where ideally it would like to position itself from both an organisational as well as facilities management perspective and then decide where is now in terms of achieving the strategy and then exactly how it is going to achieve its ultimate goal. Generally, it is good practice to set a clear mission statement together with objectives and targets to enable the development of the facilities management strategy. Scenario planning and business simulation may be useful in formulating a strategy where a series of 'what if?' questions are applied to produce a table of consequences enabling the senior management team of the organisation to form clearer decisions.

When determining the facilities management strategy, it is essential to build in flexibility and to have alternatives, should problems be encountered along the way as survival could depend upon the ability of a company to react to change quickly.

However the strategy is developed, the key is to take a systematic approach.

Facilities strategy

[F3007] *Analysis* of the company's present and future needs is important to look closely at objectives, policies, resources, processes, and procedures. It is important to know where the company stands in the market place and, if available, some idea as to its customer's perception. It is essential to review and take note of the *voice of the customer* to establish the image that the company is generating now and is this in alignment with its desired objectives? What needs to be changed? Where could improvements be made? In addition, a full understanding of how the company's resources and assets are being utilised needs to be made. It is important to understand all the business processes surrounding the company's core business to enable a facilities management strategist to establish where improvements can be made, so as to achieve cost savings, add value and increase efficiency by eliminating duplication. A really simple method for achieving this is to develop a series of process maps or flow diagrams to review the organisational processes and procedures to look for duplication and wastage. This can be taken to a higher degree of complexity if required by the use of Six Sigma learning techniques.

PESTLE Analysis is a really well used and proven management tool which allows a company to scan the external market place and is a very useful in the decision making process in the development of its facilities management strategy.

Let us examine each element of the PESTLE analysis, in brief as follows:

- Political – are there government decisions and emerging policies that may affect decisions made?
- Economic – what is the current trend in the economy within the company operating areas?

- Social – what are the social trends that may affect the demand for the company products?
- Technological – what advances and trends may affect the company product or services?
- Legal – is there emerging legislation that may affect your products or services?
- Environmental – are there any raw materials or sustainability issues that may affect the company?

SWOT Analysis is another useful management tool to utilise and generally is linked to and will take into account the findings of the PESTLE analysis as above. Again it is useful to examine each element in brief, as follows:

- Strengths – does the company have significant strengths and if so what are they? By carrying out this analysis in a group you will be surprised as to how you can build up a substantial list based on the combined opinions of the group.
- Weaknesses – in what areas is your company weak? Can these areas be improved?
- Opportunities – where are the key areas for improvement? Maybe by exploiting the company strengths and bolstering the weaknesses opportunities can be identified.
- Threats – are their peers or competitors who are a threat? Could they be about to follow a similar route to you based on market trends?

Development of Solutions to make up the facilities management strategy can then be made by careful interpretation of the data and resulting information gained within the method of analysis chosen. It is essential that solutions add flexibility, efficiency and are innovative eg by use of modern technology for communication of information. It can be useful to assemble a team who are capable of looking outside the 'box' and maybe have blue sky thinking, that is provided they are well managed. New ideas are crucial and whatever decisions are made they must be able to convince the senior management or board of the company that they will work, and ideally should be financially justified within any resulting report by use of a cost benefit analysis, in order to prove their case. Consider the use of a risk register to record the various scenarios and be sure to consider positives as well as negatives in this process. Any changes must ultimately be viewed as being cost effective and adding efficiency and value instead of being a cost burden.

Implementation of the solutions has to be carefully planned as without the use of effective 'change management' techniques will result in resistance to any changes being introduced and will probably result in staff problems taking up management time, thereby reducing efficiency. It is important to communicate with staff by way of group meetings and on a one-to-one basis to provide information and to allay any fears that they may have to proposed changes. Also listen to feedback from staff that may have some interesting ideas of their own and it is essential to engage with the actual staff undertaking the activities as they will be the ones who know the working practices. It is essential that the implementation plan is developed and includes time lines for each task with key milestones, performance measurement, proper ownership and accountability. It is important to ensure that the implementation plan includes all activities and elements of the company that will be affected by the change.

Also it is important to consider the integration of a change management process in order to ensure a seam-less transition in the implementation of the new strategy. Remember that we humans do not like change so consider all paths of potential resistance and plan to find ways of eliminating these.

A typical facilities management strategy should be capable of the effective delivery of the following:

(a) in-house/outsourced services – catering, cleaning, maintenance, security etc;
(b) specifications for required accommodation standards (RAS);
(c) procurement strategy;
(d) IT strategy;
(e) human resources strategy;
(f) provision of change management;
(g) service level agreements (SLA);
(h) performance measurement;
(i) financial planning;
(j) environmental health and safety management;
(k) accommodation/real estate strategy; and
(l) business systems and procedures.

In-house or outsourced facilities management?

[F3008] The decision taken is dependent upon the facilities management strategy of each individual company and may even be a combination of the following two options.

In-house – Also it is important to consider the integration of a change management process in order to ensure a seam-less transition in the implementation of the new strategy. Remember that we humans do not like change so as mentioned in the previous chapter, consider all paths of potential resistance and plan to find ways of eliminating these.

Outsourced – most companies consider that the elements of facilities management detract from their core business and is such a specialist skill that they would prefer to outsource, as all the problems and issues of running their building are left with a total facilities management provider or a contractor, and allows the company to concentrate on its core business. Although the facilities company, who will in effect be a contractor, will have to be managed, the added advantage is that facilities will be provided at a guaranteed price within an agreed service level for the contract period, which makes for easier budgeting. Consideration must also be given to the cost associated with the management of the facilities management contract to ensure that the agreed service is being delivered and that the service provider is complying with their health and safety duties.

The outsourced activities would be placed with a service provider, which could be one of the following:

- managing agent;
- contractor;
- cleaning/catering company;
- post room services company;
- maintenance company; or
- total FM provider.

There have been a number of changes in the outsourced facilities management marketplace over the last couple of years with a trend towards total facilities management (TFM) or Integrated Facilities Management (IFM) and more recently has seen the emergence of Managing Agents self-delivering hard FM technical services as well as soft FM through the acquisition of companies. The FM trend of large global blue chip organisations is definitely towards large bundled global TFM/IFM contracts.

Total Facilities Management (TFM)

[F3008A] Although this three-letter acronym has been around for a few years now the large global facilities management providers continue to innovate their delivery through the provision of a local more personalised approach but on a global basis. This approach is relatively new in its concept and very little data is available to establish its effect however as a marketing tool it certainly has some appeal.

The TFM solution will include all of the non-core service lines, which the TFM provider will self-deliver, typically over 70% in most cases and will achieve this through a One Team delivery model with the team having multi-functional responsibilities and cross skilling across all of the service lines delivered.

By adopting this approach to facilities management a number of service delivery improvements are being realised by customers, as follows:

- TFM providers are becoming well-versed in the change management techniques required when making the transition from in-house delivery to a TFM delivery model.
- TFM providers are becoming increasingly understanding of the TUPE arrangements required for the transfer of in-house staff to the TFM model.
- In general, the morale of staff improves as they see the benefits of working within a large TFM provider and the opportunities for promotion.
- By taking a more joint approach between client and customer there can be more effective risk management within a contract to deal with risks to the business such as disaster planning.
- Total transparency and team working may lead to the creation of trust and marked improvements in relationships long-term.
- In longer-term TFM contracts, teams are working in greater partnership and are becoming more creative resulting in far more innovations for contract delivery.

This goes to demonstrate that there is scope for not only cost savings to be realised but also improvements in service delivery.

Advantages to the company

[F3009] The advantages of using an effective integrated or TFM approach within an organisation are vast and are essential too, in some cases, for its very survival. Gone are the days where the various departments virtually managed themselves by doing their own thing in silos using practices which, had been handed down over the years. In such situations it was often found that tasks were carried out throughout the company with little or no communication, and certainly, no best practice with the result that tasks or operations overlapped and there was duplication thus costing the company time and money. Today the facilities management department is responsible for advising a company's board and will most probably have an essential role in setting the overall business strategy for the future. With facilities management encompassing all of the Hard and Soft elements of facilities management of a large organisation such a transition provides a totally joined-up management approach towards all elements of the business.

The facilities management holistic and integrated approach can add tremendous value to a company by identifying needs in all areas and looking at ways to satisfy such needs at a strategic level by innovation. For example, a department may need to bring in a temporary worker for short-term data entry. However, when viewed at strategic level it may be the case that other departments have administrative pressures at certain times and are also bringing in temporary staff. A view could be taken to either increase the head count of staff and use the new member of staff as a float to be used in fulfilling administrative requirements, or if this were not possible, the facilities management team could agree on more time for the temporary worker to act as the float in various departments and therefore negotiate a better price.

By giving consideration to the needs of the entire company for the supply of goods or a service and grouping the procurement of such goods and services together will result in added purchasing power to obtain the best deal from the suppliers and manufacturers. This concept could be taken further by establishing when, for example, a manufacturer is at their quietist. If a company's production run can coincide with a manufacturing quiet period, a more competitive rate could be negotiated with the added advantage to the manufacturer of not having to close the plant temporarily or be forced to lay off staff, thereby potentially losing skilled labour. If consideration is given to the service industry then a client could decide that for example the maintenance of an asset could be undertaken in say a three month period by the contracted specialist thus providing a high degree of flexibility. The contractor could maintain the asset when they happen to be in the area performing other work thus enabling an even more competitive price to be achieved.

Main advantages

[F3010] The main advantages are—

- Cost certainty – by inclusion within the facilities management contract of a detailed set of specifications and service level agreements, the desired services can be delivered at an agreed price aiding budgeting

within the company. By careful identification of the needs of the company, specifications may be tailored and delivered in a more efficient way adding value.
- **Reduced cost** – by the integrated management of all services, more efficient procurement by bundling of services, utilisation of company assets and staff management, huge savings can be realised which will lead to increased profits and ensure competitiveness and the long-term survival of the company. The claim that the facilities management department is an overhead, which perhaps costs a company more than it could ever save, can be totally refuted. It has been proven that good strategic facilities management can improve the efficiency of a company's systems, thereby reducing running costs and assisting in the delivery of the company core objectives. The use of Six Sigma and other management techniques can identify areas of duplication of activities within the organisation thus allowing even greater cost savings to be achieved.
- **Service delivery** – by careful facilities management contract negotiation and management of company expectations, services can be delivered to an agreed level which, if subject to the contract, may be measured periodically to ensure long-term delivery to the required standard. A system of continual improvement could form part of the initial contract and this could be implemented by both parties resulting in a proactive approach towards developing innovations in their working practices.
- **Improved communications** – will ensue following the integrated management of all services through one department. Issues can be quickly resolved or brought to senior management for a decision. Having all information collated in one department will ensure wide-ranging, accurate and meaningful information can be utilised in the decision making process.
- **Utilisation of support services** – by outsourcing to a total FM provider the client organisation can take advantage of a range of support services already *in situ* within the FM provider for example call centre, Computer Aided Facilities Management (CAFM) software, health and safety and administrative support to name but a few.

Responsibilities of facilities management

[F3011] As demonstrated earlier, the skills expected of the facilities manager are diverse and are defined by the British Institute of Facilities Management (BIFM) Professional Standards framework, which defines the competencies required at all levels of a facilities manager's career from trainee through to senior manager/director.

The Professional Standards framework may be downloaded from the BIFM website for use in your workplace at www.bifm.org.uk.

An overview of a range of competencies is shown as follows.

Development of a maintenance policy

[F3012] Maintenance of any building or process plant is crucial as without this equipment the organisations's environment will cease to function efficiently. One aspect of facilities management is to protect the assets of a company, such as the building fabric or ventilation system and to ensure they are utilised to maximum effect. The facilities management department would be responsible for the development and implementation of the maintenance policy for the company, which will depend on the culture of the company, the type of work carried out by the company, the age of the building, and type of plant equipment. For example, it would not be advisable to adopt a breakdown type of maintenance policy if the company was reliant on one piece of very complicated plant equipment which was business critical. Therefore, the development and implementation of the maintenance policy for the company is important in the provision of services to enable the company to function, but could also save money in the long term through the by selection of the most suitable policy.

Before any decision is made on the selection of a maintenance policy, each asset must be entered in an asset register and a condition survey completed. This will provide the facilities manager with information, which can be used when considering the policy to adopt. The type of company and business processes conducted will be taken into account during this decision making process, as will the available budget.

Companies always try to cut costs and maintenance is often seen as an area that is too expensive and that small reduction in the maintenance budget will hardly make any difference. It is for the facilities manager to demonstrate that cost cutting in this area could be counterproductive, leading to unreliability of plant equipment and general dilapidation of the assets.

It is important to make the correct choice of maintenance policy with current and future trends being considered, together with the long-term effects that a particular maintenance policy could have upon the company's assets and future capital expenditure. By adopting a reduced maintenance policy now, of say, reduced frequency, or perhaps even a breakdown policy, will no doubt have a detrimental effect upon the assets in the future and will perhaps reduce the life expectancy of the asset. This will either result in expensive maintenance in the future or complete replacement and therefore capital outlay before the expected life cycle of the asset.

A recent trend that has been born about by the need to trim even further costs from the budget of the company is that a vast number of facilities management teams have to reduce the maintenance of non-life safety and business critical assets to a breakdown maintenance strategy. It is too early to tell what effect this will have on the assets but theoretically this will no doubt reduce the performance of such assets and reduce their predicted life cycle. In some areas filters changes in air handling units has been reduced resulting in higher energy consumption caused by fan motors having to push air through partially blocked filters which really not only costs money in energy but generates more carbon and is therefore a false economy.

All components of a building and its plant equipment will require some form of maintenance to preserve its efficiency, reduce the risk of breakdown and to ensure that the equipment achieves it manufacturers designed and predicted life cycle for the equipment. There are various maintenance policies available and a combination of all could be implemented, as follows.

Planned preventative maintenance (PPM)

[F3013] Planned preventative maintenance (PPM) or Planned Maintenance (PM) refers to a scheduled visit to an item of plant equipment or asset to check the correct operation and to complete a maintenance task which is carried out by a competent person or engineer with the aim of preventing an unscheduled breakdown.

The advantages of this type of maintenance is that it is easier to plan and order spare or consumable parts, distribution of costs over a period of time and ensure optimum performance and efficiency.

In this policy, a maintenance schedule is developed for every item on the asset register and may include periodic inspection of the condition of assets, items of plant equipment being adjusted, and filters replaced or cleaned as appropriate. This is initially the most expensive form of maintenance, but will prevent unexpected breakdowns, will ensure that equipment runs efficiently and will prolong equipment life. This policy would be an option to maintain any 'business critical' assets within the company.

Planned corrective maintenance

[F3014] This is where the assets are inspected periodically but are not touched in anyway unless they are running very inefficiently or having an impact upon the running of other assets, or perhaps where there is a high risk of the asset failing in the near future. This maintenance policy is cheaper than planned preventative maintenance but will make it difficult to predict breakdown and an asset could run so far outside its design tolerance that long-term damage is inflicted. This is very much a hands off approach towards maintenance and although will achieve significant cost savings in the short term this will have a detrimental effect upon the life cycle of the asset.

Breakdown maintenance

[F3015] Assets are neglected and only touched when they fail. This is therefore purely a reactive system and by far the least expensive policy in the short term, but it's use makes it difficult to predict breakdowns. This policy does not protect assets for the future; dilapidation of equipment could prove costly.

Before deciding on the maintenance policy to be adopted the main aims and objectives of maintenance within the company need to be considered, some of which could include the following:

(1)   upkeep of the company's image;
(2)   to enable the built environment to fulfil its function;
(3)   compliance with legislation to provide a safe and healthy environment;

(4)   ensure plant runs efficiently to provide a comfortable environment;
(5)   minimum cost rather than reliability;
(6)   reliability of plant equipment; and
(7)   creation of a forward maintenance register to allow funds to be amortised for lifecycle replacement of equipment.

Reliability Centered maintenance (RCM)

[F3015A] Reliability-centered maintenance (RCM) is the process of preserving a system's function by selecting an appropriate planned maintenance regime which differs from other approaches to maintenance by focusing on the function of the equipment rather than merely the equipment itself.

RCM was developed as a result of the increase in size and complexity of systems within facilities such as buildings, chemical plants and power plants where the cost of adopting a more traditional planned preventative maintenance regime was ever becoming more difficult as new ways of trimming costs were required and innovation was sought. Typically, this is where assets are grouped into categories based upon the impact on the business if a particular asset were to fail. All assets that are operational or business critical or required for statutory compliance are maintained to manufacturer's specification or higher whilst other non-essential assets are run to failure. The advantage is that available budget is focused towards all of the critical assets to achieve 100% up-time and greater resilience whilst achieving overall cost savings.

Life cycle costing

[F3016] It is important to consider the building environment its condition and likely maintenance requirements for the future. The facilities management team can assist the company by using life cycle costing, also known as whole life costing. Life cycle costing provides the facilities manager with a cost profile, which can be broken down into three basic elements:

- acquisition cost;
- operational cost; and
- disposal cost.

This technique may be applied to services as well as actual assets.

All components, whether they are the outer protective layer of a building, a heating pump, or even a light bulb, will have been designed with a life expectancy which the manufacturers will quote provided that the component has been maintained to the manufacturers specification. Therefore, a whole building or section may be designed to provide a certain number of years' service before requiring total replacement, subject to the manufacturer's on-going maintenance being completed throughout its expected life. The cost element will also have to be considered as a cheaper product may have to be included in the design, which will require replacement earlier. However, the ongoing maintenance costs may be far higher than a more expensive product and this will also have to be taken into consideration. This highlights the fine balance a designer will have make to achieve the client's requirements and deliver within a prescribed budget. The compromise is that the client has to accept that although their initial capital expenditure is reduced at the construction phase, at some point in the future there will be an ongoing maintenance cost and further expenditure to replace the component.

In the case of some items of equipment such as a boiler its life expectancy could be 25 years however after say 15 years its efficiency may have reduced significantly and maintenance costs increased due to breakdowns. In this scenario there will come a time when a business case could be developed by the facilities management team to replace the boiler before the manufacturers stated life expectancy. Although there will be an initial capital expenditure over a period of time the cost savings gained through greater efficiency and reduced maintenance costs would far outweigh this expenditure.

Modern day LED lighting is becoming more efficient and the manufacturers are spending millions of pounds in research into the life expectancy of each lamp so that the new generation of lamp fittings have the ability to communicate with a buildings management system to inform the system when the lamps in that fitting will fail thus allowing the facilities team to plan the maintenance into their schedule and obtain maximum life from the product.

There are a number of computer software programs available to assist a facilities management team to create an asset register for the building. Such programs may include everything within the building from the very fabric of the construction to the fixed items of plant equipment, general office furniture and carpets. By inputting full details and the condition of each asset a lifecycle capital expenditure program can be produced to provide information indicating when assets should be replaced and will also provide an indicative guidance of costs to assist in budgeting.

Customer service

[F3017] One of the difficulties encountered when services are outsourced is the management of customer expectations, which comes from the desire to have everything work all the time, or for items to be delivered exactly on time. A key driver in the decision making process to outsource, may have been to save costs and it may be that the customer (eg the client company staff) could be expecting a gold service when in fact the company is only now only paying for a silver service. It is essential to manager customer expectations and to avoid confusion it is advisable to ensure that staff within the client company is provided with clear information as to the level of service that they should now expect.

The facilities manager will be able to measure customer service by use of metrics, which should be agreed within the initial contract. Such measures are normally referred to as Key Performance Indicators (KPI's) and will provide an ongoing indication of performance.

The KPI's that may be utilised for a typical facilities management help desk can be defined under the following three headings.

Timeliness – needs to be defined in a very precise way, but most probably that the help desk telephone is answered within five rings or perhaps within 15 seconds.

Appropriateness – when a help desk operator answers a call, was the customer's request understood or was the most appropriate action taken? Was the customer's expectation completely met or did the operator fail to establish a solution to the problem?

Accuracy – most probably the only acceptable standard in this category is 100%. In the example above, did the help desk operator enter the correct information into the system or were there inaccuracies that affected the resolution of the customer's problem?

The measuring and collating of this information is important to the facilities manager because it confirms whether or not the appropriate service is being delivered. With continuous performance measurement in this area it can be ensured that the service is delivered at the agreed level. This may be taken a stage further by publishing the service measures, or KPI's, to demonstrate how the help desk is performing and to encourage staff to provide suggestions on how service delivery may be improved. The performance may also be compared with those of help desks in other companies, provided the KPI measures are the same and if information is available from research or benchmarking clubs.

Service delivery is a key driver within the facilities management industry and is taken very seriously. In the example above, it would be expected that a department would be responsible for the completion of customer feedback surveys the results of which are then collated and measured.

Environmental and energy management

[F3018] Much emphasis is being placed on environmental protection and everybody has a duty to ensure that they use energy as efficiently as possible by carrying out basic tasks such as switching off lighting in areas that are not in use or lowering the temperature on a thermostat by a degree or so. Simple things that can be easily performed in everyday life, whether at home or in the workplace, and yet they can make such a difference. Expand this to all of the staff and processes contained within an average working environment and it can then be demonstrated that a company can make a huge difference. Energy consumption within a company burns fossil fuels, which constitute a significant repository of carbon and burning them to generate power results in the conversion of this carbon to carbon dioxide which is then released into the atmosphere. In global terms this results in billions of tons being released into the atmosphere every year resulting in an increase in the Earth's levels of atmospheric carbon dioxide, which enhances the greenhouse effect and contributes to global warming. Depending on the particular fossil fuel and the method of burning, other emissions can be produced such as ozone, sulfur dioxide, nitrous oxide and other gases are often released, as well as particulate matter. Sulfur and nitrogen oxides contribute to Smog and the resulting problems for asthma sufferers and people with chest problems and not to mention the effects of acid rain.

Facilities management can apply their skills not only to comply with legislation concerning environmental issues, but also to save the company costs on energy bills. A number of companies have also developed a corporate and social responsibility (CSR) strategy which will no doubt form part of the facilities management function to ensure compliance with this strategy.

The facilities management department would be able to assist energy management within a company at both a strategic and operational level. Typically, part of the environmental strategy would be for a company to agree upon its energy management strategy, which would normally consist of the following.

Strategic

[F3018A]

- policy statement;
- objectives;
- roles and responsibilities;
- process of review;
- arrangements for the following:
  - heating – space temperatures, heating periods, and
  - cooling – air conditioning policy – natural/mechanical ventilation.

It is important that all staff is made aware of and provided with training in the environmental and energy pol-icy, together with full details of the practical ways in which they can really make a difference. For the policy to be effective a clear message must be provided from the top down in the company, with directors and senior managers setting a good example and being supported by the facilities management team whose overall responsibility would be to manage energy issues on a day-to-day basis, as follows.

Operational

[F3018B]

- housekeeping issues;
  - switch off lighting,
  - doors and windows closed in winter,
  - switch off equipment not being used,
  - use water wisely,
  - wear sensible clothing for weather conditions,
- engineering issues;
  - install or replace existing lamps with energy efficient equipment,
  - ensure external lighting controlled by photo cell or time clock,
  - installation of waterless urinals in gents' toilets,
  - installation of internal lighting control by motion detectors (PIR),
  - use of improved ventilation rather than air-conditioning, and
  - installation of window film and or blinds to control solar gain.

The facilities management team would have a direct effect on the environmental impact of the company and would be able to reduce any adverse impact by careful management of the following service functions:

(a) products are procured from suppliers with environmental policies;
(b) all wood used is from sustainably managed sources;
(c) where possible, water is recycled in production or other processes;

(d) only use paints with a low volume of volatile organic compounds (VOC's);
(e) procure products made from recycled materials;
(f) recycle paper, cardboard, toner and printer cartridges, and mobile phones;
(g) manage ordering of marketing publications and leaflets to reduce waste;
(h) ensure that waste from refurbishment projects is recycled;
(i) procure energy from renewable energy sources; and
(j) procure energy efficient equipment and services.

These are some of the activities that could be undertaken by a skilled facilities management team, all of which will ensure compliance with current legislation and above all will realise considerable savings in the company's energy consumption and resulting expenditure on energy bills.

The organisation may also consider the installation of photovoltaic films on windows or installation of solar panels or wind turbines to generate power in order to reduce their consumption from the national grid thus saving carbon emissions.

In addition, the facilities manager can assist with the specification when designing a refurbishment or new building for the organisation in order to reduce energy consumption through the use of film or glass to reduce solar gain, use of ventilation as opposed to energy consuming air conditioning and even the orientation of the building, in the case of a new build, in order to reduce exposure of glass to the sun.

Sustainability

[F3018C] Sustainability within an organisation is a commitment to make ethical business decisions that benefit future generations, the environment and society in general and will encourage innovative working to ensure the conservation of natural resources. It can be demonstrated that every element of the operation of a company impacts on the overall product or service of the organisation and therefore will affect sustainability. The facilities manager will need to be aware of how to develop and implement sustainable strategy within the organisation as there are more and more reasons for organisations to go down this path. Case studies have proven time and again that true sustainable development can not only reduce wastage and increase profit but will also contribute towards compliance with legislation and improve links with the local communities in which the organisation works. Although there are many definitions of sustainability and sustainable development most probably the simplest was developed by the World commission on Environment and development which states the following:

> 'Sustainable development meets the needs of the present without compromising the ability of future generations to meet their own needs.'

Some typical keys drivers for sustainability are as follows:

- Legislation
- Customers

- Conservation of natural resources
- Economic value of the business is enhanced
- Staff motivation
- Shareholders demands and expectations
- Longevity of the organisation
- Financial stability

There are a number of business models that may assist the organisation towards a strategy for sustainable development such as the Five Capitals model, as follows:

(1) Natural – Purchasing of green/renewable energy or reduction in carbon emissions
(2) Human – Development of staff or recruitment of apprentices/trainees
(3) Social – Staff projects in the local community
(4) Manufactured – Reduction of waste to landfill
(5) Financial – Low risk financial decisions or sales and profit growth

An organisation can truly claim to be a sustainable organisation when it is living off the interest and not just the capital of the world.

Acquisition and disposal of buildings

[F3019] There are very few companies with buildings in their estate who could say that their buildings are entirely suitable for their operation. Compromises are often made as companies continually change and even if a building was designed specifically for a company, and even if there was a certain amount of flexibility built into the design to cope with future needs, there is no doubt that by the time the building was finally ready for occupation, the needs of the company would have changed.

In an effort to once again drive down costs the demand for smaller high quality office space is increasing where companies are striving towards zero waste of space in their organisations.

The facilities management team can use their skills in advising a company on the acquisition and disposal of buildings by taking an integrated approach, in which they would consider the company's objectives and current and future needs. The team may also advise on the suitability of a building the value of which may be severely affected by issues such as quality & flexibility of space, type of plant room assets, flexibility of use, poor energy efficiency or the presence of asbestos.

If the organisation for example required more space in which to conduct its operations, then areas for consideration could be made, as follows:

- Is this temporary or permanent?
- What has created the need?
- Can some staff be allowed to work from home with a laptop?
- Is it possible to move a department into a smaller satellite office?
- Is it financially viable to take short-term space nearby?
- Should the entire company relocate?
- What is the current lease situation on the present building?

- Is there a lease break clause or could the building be sub-let?
- What location would suit the company?
- Cost of any necessary refurbishment to new location?
- Maintenance and running costs?
- Energy efficiency?

There are so many areas to be considered that usually only a few of these options would be selected, each providing a range of advantages and disadvantages, although ultimately, senior management would most probably have to compromise.

Similar skills would have to be applied to the disposal of a building. It could be that only part of a building is no longer required, so consideration could be given to establish whether that particular area could be sub-let, or if there is a lack of a lease break, then it may be feasible to sub-let the entire building. If the building is no longer required but has to be retained due to the lease, then consideration has to be given to reducing building services to an absolute minimum whilst preserving the assets for future final disposal.

Space planning

[F3020] Further to the acquisition and disposal of buildings (at F3019 above) it can also be demonstrated that, due to the constantly changing company, space within a building will not be utilised to its full potential and facilities management can be called upon to find an appropriate solution.

A building occupied by an organisation is its largest overhead apart from the cost of the staff and thus if the facilities manager is able to reduce overheads by improving the utilisation of space, then profits will be increased, thus ensuring survival in the marketplace. If space planning is undertaken effectively then the company will further benefit from improved production created by better flow of work and processes through the building. Staff will be working more efficiently as they will have a better environment in which to work and with better management control, defining space requirements and reducing clutter by good housekeeping. A well-planned and managed workplace will enhance the company's image, which will provide a good impression to potential clients visiting the premises.

Project management

[F3021] A company will often have many projects running concurrently, such as the development and introduction of a new information technology system or the construction of an extension of a building. In most companies a facilities manager would normally manage small to medium sized projects, which would place further demands on the skills base required and of course workload. Often, it is unclear how to define what a project actually is. Each project will have its own unique set of co-ordinated activities being undertaken by an individual or a team brought together for this specific purpose. There will be defined start and finish points with objectives, time constraints and pre-set cost and performance parameters.

It is most likely that any construction type of project undertaken by the facilities management team will be governed by the *Construction (Design & Management) Regulations 2015*. From a health and safety aspect when undertaking a project then the facilities manager is advised to adhere to the specific duties of CDM 2015 placed upon clients, designers and contractors to rethink their approach to health and safety. It should be taken into account and then co-ordinated and managed effectively throughout all stages of the construction project and is applicable to all elements from conception, design and planning through to the execution of works on site and subsequent maintenance and repair, and even to final demolition and removal.

The key aim of CDM 2015 is to integrate health and safety into the management of the project and to encourage everyone involved to work together to:

- Identify hazards and risks as early as possible
- Focus efforts where it can do most good from the safety perspective
- Improve planning and management
- Discourage unnecessary bureaucracy and paperwork

A facilities manager would be aware, from the media, of the many larger projects that have failed elsewhere, so facilities managers must use their experience to prevent any similar failure from occurring by ensuring that the following criteria are managed at the outset of the project:

(1) agreeing a detailed project specification;
(2) timescales set are realistic;
(3) selection of appropriate staff in the team;
(4) management of customer expectations; and
(5) management of change when a project is implemented.

Successful projects are those that have the intended project sanctioned at a senior level within a company, containing a detailed project brief outlining the desired objective. It is important that time is spent defining a project brief, as this can save much effort and frustration at a later stage.

A typical project brief could include the following:

(a) goals and objectives – how will the project be deemed as being successful;
(b) scope of the project – remember to include areas out of scope;
(c) risks that will affect the project; and
(d) assumptions made.

Usually, a facilities manager will have to prepare a business case to put to senior management in order to obtain the desired budget. Part of the document would have to define cost, time limitations and benefits as well as operational risks of undertaking, or not undertaking, the project.

With project work it is advisable for the facilities manager to seek specialist health and safety advice to ensure all hazards are considered and thus ensure compliance with CDM 2015 and the accompanying ACOP.

During the project management of, say, a new IT system, the risks to a company in the changeover maybe considered too high until the new system has been proven. Both old and new systems could be run in parallel with the facilities management team analysing the performance of the new system until such a time as data is transferred completely and the old system discontinued.

Site Waste Management Plans (SWMP)

[F3021A] This is another challenge faced by the modern facilities manager who will need to be aware that the Site Waste Management Regulations 2008 legislation was revoked by the UK Government in December 2013 and the duty of care for waste and the requirements to record all waste movements by preparation of a SWMP is no longer a legal requirement.

Despite this change in the legal requirements the generation of SWMPs should still be considered as best practice on site as this will result in more effective waste management and cost reduction for waste disposal.

Contract management

[F3022] This is a very complex skill, but one in which it is essential that the facilities management team succeed, as a poorly defined contract will lead to poor performance and eventually a complete breakdown in communications between the two parties.

The contract management process itself is one that enables both parties to agree a contract that enables them to deliver the objectives required. It also involves building a good relationship between the parties in order to be able to work proactively. It is unhelpful having a relationship built around an adversarial approach, as the aim is not only to achieve delivery of the agreed services or products, but also to achieve value for money by looking at innovations in delivery and in some cases aiming for continuous improvement.

There are two types of contract that a facilities management team would normally be expected to deal with, which are: service delivery and construction contracts.

Service delivery contracts are generally divided into three areas of:

(1) service delivery management – to ensure that this is being delivered at the agreed level of performance;
(2) relationship management – to ensure that the relationship between the parties is maintained; and
(3) contract administration – to handle changes to the contract and to manage the more formal aspects.

The construction contract is very complex and the contract strategy will depend upon what suits the individual projects itself, and may be influenced by pricing and risk transfer. In this instance, the facilities manager will require much experience of the construction industry.

Change management

[F3023] Everything covered so far will involve an element of change. Humans tend to be creatures of habit and will often resist any changes, partly due to a fear or uncertainty of the unknown, created by a lack of knowledge of the intentions of the company and the reasoning behind the change.

It is important for the facilities management team to manage changes within a company, which is best carried out by effectively communicating with the staff affected. Such communication should be done at the earliest opportunity to reduce the risk of rumours escalating across the workforce which will only result in making the acceptance of change more difficult in the long run.

The first stage in the process is to analyse what effect the intended changes will have on staff, so that a chart can then be prepared to detail the likely resistance to change. Management can have in place a sequence of actions in situ to alleviate any areas of resistance, so ensuring the effective transition.

The second stage will be to hold regular staff meetings to keep the workforce informed of the changes and the drivers behind such changes, eg the introduction of a new management information system for improved efficiency in reporting. It is most likely that the staff most affected will need to be met on a one-to-one basis.

The final stage before the go live date will be for the company to agree and produce new procedures that can be communicated to staff through training.

Brand Strategy

[F3023A] This topic is becoming more of an important part of the facilities management role and a significant element involves brand protection. Imagine a factory unit or a retail outlet occupied by a well-known brand name in a prominent location and the damage to the brand if the building is looking dilapidated or the organisation does not treat the staff well. This can soon have a negative effect on the perception of the brand and cause damage. Just look at what has happened in the media in the past!

An organisation's brand can be exploited and damaged in a vast number of other ways such as counterfeiting of its products, resulting in the loss of revenue and erosion of margins, reputational loss, customer trust, exposure to legal threats and even paid for search traffic can be diverted to another organisation. A robust brand protection strategy can combat such losses as a result of what is often criminal activity.

Data Protection

[F3023B] In most cases the facilities management team will be responsible for the provision of Information and Communications Technology (ICT).

The facilities manager must be aware of impending changes in legislation that will no doubt have an impact on the protection and management of data within their organisation.

The new General Data Protection Regulation (GDPR) will apply from 25 May 2018 and applies to all organisations handling the personal data of European residents. Although 2018 seems to be a long way off it does contain a number of very onerous obligations which will take a long time to fulfill and therefore facilities managers should start their preparation now.

Once GDPR has kicked in the current Data Protection Directive 95/46/EC will be repealed and organisations that fail to abide by its requirements will face very tough penalties, which could have an impact on the very survival of the organisation itself. Typical penalties could be up to 4% of turnover so a considerable sum.

There are a number of obligations required to be implemented over the intervening period and in some cases organisations may find it more beneficial to work towards attainment of the ISO 27001 standard for Information Security Management, which will satisfy the requirements of GDPR.

The key changes of GDPR that will impact the facilities management department are:

- Organisations must prepare in advance for a data security breach and this is particularly necessary to combat cybercrime which it seems is constantly in the media at present. There should be clear policies and procedures to demonstrate preparation for dealing with any data breach.
- The GDPR requires that organisations adhere to strict data breach notification requirements to inform the data protection authority within 72 hours of becoming aware of the breach unless this is unlikely to be a risk to the individual data subjects concerned. If the data breach is high-risk then the data subjects must be informed however the GDPR does not provide timescales for this requirement.
- Organisations must develop a framework that ensures clear accountability with clear transparent policies and forms that must be simple to understand and demonstrate clear affirmative consent for processing data.
- There must also be an inbuilt culture of review and continual improvement to ensure policies and procedures are being met within the organisation and especially around the length of time that personal data is held.
- Privacy impact assessments (PIAs) will have to be prepared as a result of GDPR and should be at the heart of taking a privacy by design approach, which should be embraced. PIAs are mandatory for organisations with processes that are likely to results in a high risk to the rights of the data subjects. This in the long run is of benefit to the organisation and will likely save costs and reputational risk.
- A risk assessment should be undertaken if an organisation is seeking to transfer personal data to countries recognised as having poor or inadequate data protection legislation. There must a good legitimate business case in place before the transfer is completed.
- Parental consent will be necessary before the personal data of children under the age of 16 is processed and such consent must be clear, transparent and show details of how such consent was obtained.

This is clearly a complex area for facilities managers to deal with and we have no doubt that the impacts on some organisations will be considerable however with so much personal data being held within large organisations and the power of the media and potential reputational damage, should there be a breach, it is clear that this has to be a high priority for the facilities management team.

Four most common health and safety issues for facilities management

Asbestos

[F3024] The Duty to Manage Asbestos is a major activity for a facilities manager to deal with, especially in older buildings of pre-1981 construction, which potentially could contain large quantities of asbestos containing materials (ACM's). Buildings of pre-1999 construction can contain limited amounts of ACM's, which have also been found in recent new constructions where for example old gaskets have been fitted in pipe work.

The Control of Asbestos Regulations 2012 came into force on 6 April 2012, updating previous asbestos regulations to take account of the European Commission's view that the UK had not fully implemented the EU Directive on exposure to asbestos (Directive 2009/148/EC).

In practice though the changes are fairly limited they do require that some types of non-licensed work with asbestos now have additional requirements, ie notification of work, medical surveillance and record keeping.

All other requirements of the Control of Asbestos Regulations 2006 remain unchanged.

The *Control of Asbestos Regulations 2006* are often referred to as the 'duty to manage regulations' which impose a number of requirements upon the 'duty holder' who could be for example the facilities manager, landlord or managing agent but has been defined as the party who has 'control' of the building or part of a building.

The duties imposed on the duty holder which in effect will most likely be the facilities manager and may be summarised as follows.

- It is essential that, unless already completed, a competent asbestos surveyor must take reasonable steps to locate Asbestos Containing Materials (ACM's) within a building and to assess their condition. In effect this means the preparation of an asbestos survey.
- There is a further requirement to presume that materials contain asbestos unless there is strong evidence to prove that they do not. Instances where it could be presumed that there could be ACM's present is in areas where access was not gained, such as in lift shafts (fire compartmentalisation/suppression), locked areas, electrical switch gear (electrical flash protection) or in areas where samples are unable to be taken such as in listed buildings, where such damage would be prohibited.

- The findings of the asbestos survey must be maintained in a written record indicating the location and condition of known and presumed ACM's. This would normally be in the form of an asbestos register which should indicate the areas where known ACM's were located, together with non-asbestos samples and areas of no-access. The findings, ideally, would also be supported by the use of photographs and annotated drawings indicating locations where samples were taken.
- The final requirement is to produce an active asbestos management plan which must be kept up-to-date at all times. Most organisations believe that it is sufficient merely to carry out an asbestos survey but in effect this is the starting point for the preparation of an asbestos management plan which should indicate how known asbestos is to be managed and how you as the facilities manager is going to prevent exposure to the ACM eg removal, repair and encapsulation and periodic monitoring as well as persons responsible for management, emergency procedures and a record of all activities undertaken on the ACM's.

It is essential that ACM's are managed safely for the protection of all persons likely to come into contact with them and there is no doubt that this task would fall to the facilities management team.

Work at height

[F3025] Facilities managers will have to pay considerable attention to working at height. Working at height may appear to be relatively simple, from changing a lamp using a small pair of stepladders, to a major refurbishment project being undertaken from scaffolding. However, in each of these examples the inherent risk of falling from height may be high unless control measures are in place to mitigate such risk. It is important that the facilities manager is aware of the risks and is able to ensure that persons working at height are working in a safe manner and that they not only protect themselves, but also others who may be nearby.

The Health and Safety Executive (HSE) over recent years have implemented their Work at Height 'Shattered Lives' campaign and more recently launched a further toolkit for use when selecting access equipment for working at height called the *Work at Height Access equipment Information Toolkit - WAIT*. This online toolkit will provide information about which type of access equipment to use, It's important that the facilities management team assess the risks and select the right equipment for the job as every year on average 50 workers are killed in falls from height accidents and over 8,000 are injured every year.

The HSE has legislation to control working at height activities and the facilities manager will have to comply with various duties under the legislation. The *Work at Height Regulations 2005 (SI 2005 No 735)* came into force on 6 April 2005 and apply to all work at height where there is a risk of a fall liable to cause personal injury. As in the *Control of Asbestos Regulations 2006* there are duties placed on duty holders who are defined as employers, the self-employed and any person who controls the work of others eg a facilities manager.

Duty holders are required to ensure the following.

(a) Avoid work at height if at all possible.
(b) All work at height is planned and organised in a systematic way. Consider how long the work will take and if it can be successfully carried out with one hand from a ladder. Is there some other activity in the area that may affect the work?
(c) All persons working at height are competent. Ensure that if contractors are carrying out the work that their health and safety has been evaluated and that the operatives have been suitably trained.
(d) Risks from working at height have been assessed and that suitable control measures and work equipment are in place before work commences.
(e) Risks from working on fragile surfaces have been assessed and are properly controlled. Consider fitting moving bridges and gantries to span fragile surfaces requiring some form of maintenance.
(f) All equipment used for working at height is marked with a serial number or colour coding and is properly maintained and inspected. Operatives should also undertake their own inspection prior to using the equipment.

The facilities manager must ensure that there is an effective safe system of work in place such as a permit to work system for evaluating the competency of operatives, checking their risk assessments and method statements, checking that equipment to be used is suitable and has been maintained. Once work has commenced the facilities manager should supervise the work to ensure that the operatives are carrying out the work in the manner described in the risk assessment/method statement and that there are no other unforeseen hazards.

Water quality

[F3026] In the UK between 30 and 50 deaths are reported per annum, as being as a result of legionnaires' disease and these have been widely reported in the media over recent years, in spite of the introduction of the Code of Practice 'The Prevention or Control of Legionellosis' (including legionnaires' disease) L8 rev and its associated guidance document HS(G) 70. These deaths have resulted from poor maintenance of cooling towers and evaporative condensers and hot and cold water systems in buildings.

The legionella bacteria needs a number of components to proliferate such as Iron and L cysteine which is present in water, as well as the right temperature. Legionella grows at between 20 and 50 degrees centigrade and its optimum temperature seems to be 38 degrees centigrade. Therefore, if the facilities manager can maintain the temperature of the hot and cold water within the building outside of these temperatures then this will help prevent to prevent the growth of the bacteria. In addition, legionella will also build up within water outlets such as showers that are seldom used as well as dead legs in water pipework that are still connected to the system.

An essential component of the above L8 code of practice is to ensure that a water risk assessment (WRA) is undertaken for the company's building and to make recommendation to reduce the risk to the lowest possible. Within this document there will be an organisational chart indicating persons responsible for ensuring that water services are properly maintained. In a small to medium sized company the facilities manager will have responsibility for ensuring that all recommendation in the WRA are completed and that maintenance of the water system is carried out in accordance with the maintenance scheme shown in the WRA.

The following should be considered when dealing with the water services of a building:

(1) ensure that a WRA has been carried out by a competent person;
(2) ensure that all recommendations made in the WRA are completed and records maintained;
(3) ensure that the maintenance regime shown in the WRA is undertaken at the prescribed intervals by a competent person;
(4) ensure that all water samples taken are tested by a UKAS accredited laboratory;
(5) ensure that records are maintained for all maintenance and testing; and
(6) ensure that the WRA is reviewed on a regular basis or after any alterations have been made to the water system.

If the facilities manager adheres in essence to the above guidance and has maintained good, accurate and auditable records then the incidence of legionella should be avoided.

Management of contractors

[F3027] Any contractors working on site pose one of the most notable health and safety risks. Some senior managers share the mistaken belief that they are able to transfer any Health and Safety risk by contracting work out. In fact any such risk is shared because the company on whose premises work is undertaken also has a duty to others under Health and Safety at Work etc Act 1974 s 3(1) to ensure that persons not in their employ are working safely. This duty extends to contractors and their employees, visitors, customers, emergency services, neighbours, passers-by and the general public at large. It is highly probable that the facilities manager will be the person responsible for employing such contractors and will have to manage them whilst they are on site.

Before employing contractors, facilities management should ensure that the contractors have been thoroughly vetted by way of a health and safety evaluation questionnaire. Such a questionnaire can be designed to verify their areas of competency with points such as accident record, insurance, procedures and risk assessment systems being a minimum. In this era of outsourcing there are specialist procurement organisations out there that will not only source products and services for the facilities manager at best price but will also complete all of the health and safety evaluation checks that are deemed necessary. In addition, there are some organisations that also run the health and safety evaluation process within an online web-based system which improves communication and increases efficiency by allowing contractors to answer questions and upload documents directly into the online system. Some companies keep an 'approved contractors' list which shows all contractors who have been subjected to such an evaluation process. A word of warning, however, facilities management should check to ensure that the contractor has been approved for the type of skill required for the work activity to be undertaken. It could be that the contractor has been added to the approved list for some other activity.

Once facilities management is satisfied that the contractor is competent to undertake the allotted work activity, facilities management can allow them to conduct a site survey before holding a meeting to agree how the work should proceed. At this stage the contractor should be preparing their method statements and risk assessments while the facilities manager should inform the contractor of any known hazards connected with the site and should also provide rules or contractor's guidance notes. The next stage would be for the facilities manager to review the method statements and risk assessments to ensure that the work is to be undertaken safely and will not impact upon any other activity on site.

When the contractor arrives on site the facilities manager must be satisfied that the contractor has all the necessary control measures in place to protect not only the contractors but also company staff and anyone else who may be affected by the work activity, such as members of the public or visitors. It is important that work is undertaken as agreed and is recorded in the method statements and risk assessments with the facilities manager providing as much supervision as necessary. Provided there are good communications between both parties and with good preparation, any work should proceed smoothly. However, unforeseen circumstances can prevent work progress, in which case it is important that any changes in work procedure must be agreed and the facilities manager must ensure that any such changes will still result in safe working.

A good resource for use by facilities manager is a new publication that has been jointly produced and published by Westminster City Council and the British Institute of Facilities Management (BIFM) and is titled 'Management of the Health and Safety Performance of Contractors'.

How facilities management and building design can help a company

[F3028] The way in which facilities are managed within the work environment is important in creating the desired corporate image, whilst concurrently achieving its corporate objectives. It is pointless creating the desired corporate image if the efficiency of the company is undermined by bad building design or its location.

It is not uncommon to see many fine buildings, which are difficult to maintain with lights or air conditioning cassette units located high up in large atriums of buildings. In these instances the changing light bulbs or cleaning and replacing filters suddenly becomes a major part of the maintenance procedure as safe access becomes a more significant priority with, for example, the constructing scaffolding or mobile towers in order to undertake such a relatively simple task. It would be ideal if an experienced and skilled facilities manager were involved at the design stage, as they could consider the maintenance and operation of the building in a practical way. This situation has now improved as designers have duties, under the *Construction (Design and Management) Regulations 2015 (CDM 2015)*, where they have to consider the construction, maintenance and demolition of the building. However, it is still a problem in some areas.

A facilities manager is expected to be multi-skilled in a wide range of subjects, so that they are best able to advise a company how to create a safe and efficient workplace which provides a comfortable and positive atmosphere for all staff.

Most visitors will quickly form a first impression of a company by being affected by the synergy of the building and its design, location and its general condition. How a company's logo is displayed is all part of a carefully constructed corporate image. A skilled facilities manager with clever use of colours, furniture design and layout will display the company logo in such a way as to create the desired corporate image in order to help achieve the company's objectives.

Recognition of facilities management

[F3029] Facilities management of built environments is gaining greater recognition and credibility from a wide range of government departments, such as the Health and Safety Executive (HSE), the Office of the Deputy Prime Minister and the Sector Skills Council due, in part, to facility managers working in all types of companies who use a wide variety of skills. Consequently, facilities management is being recognised as an industry in its own right and facility managers are being used to provide feedback for government consultation documents, speak at conferences and attend innovation workshops. In fact, the facilities management industry has progressed to such an extent that it now employs more people than those working in the construction industry.

As mentioned in the introduction at the beginning of this chapter both the Central European Normalisation (CEN) and the British Standards Institute (BSI) recognise facilities management as an important industry and in fact this has established to such an extent that the British Institute of Facilities Management (BIFM) in conjunction with the Institute of Leadership and Management has developed a new qualification at award level 3 in Facilities Management which is also recognised by the Qualifications and Curriculum Authority (QCA).

BIFM has worked closely with the Office of the Deputy Prime Minister (ODPM) on a range of issues concerning energy and environmental management of the built environment and has produced a paper on Approved Document L2 of the *Building Regulations 2000 (SI 2000 No 2531)*. In addition, BIFM also has its own accredited qualification based around 20 core competencies which are endorsed by the Sector Skills Council, accredited by the Qualifications and Credit Framework and are also breaking across international borders, which further demonstrates the very diverse nature of facilities management.

Asset Skills, the Sector Skills Council for the places in which we live and work, has now created a facilities management sector board which is comprised of representatives from the facilities management industry and BIFM. The intention of the board is to:

- raise the standards of learning and professionalism;
- develop a National Vocational Qualification (NVQ) for facilities management;
- develop a research programme for facilities management (current research is being undertaken by Asset Skills Council, local councils and university estates management departments); and
- work closely with the facilities management industry and large companies to develop facilities management skills and best practice.

The Royal Institution of Chartered Surveyors (RICS) now has a very active facilities management faculty and a qualification of Chartered Facilities Management Surveyor. The institute is now actively undertaking research into best practice and is actively publishing facilities management articles on their website.

The Health & Safety Executive (HSE) recognise that the facilities management industry has considerable experience, retains broad skills and as a result holds regular update meetings with BIFM to discuss current trends and innovations. BIFM is also involved in the provision of feedback concerning government consultation documents by the collation of responses from specialists within its membership and is represented on the HSE lead Asbestos Liaison Group which meets on a quarterly basis.

Industry is also recognising facilities management as an essential element of their strategy to ensure that they remain competitive in the market place. Increasingly, more companies are joining BIFM as corporate members and are actively involved in providing input into the institute and also to putting forward representation on committees across the United Kingdom. Some of the larger blue-chip companies run their own facilities management training and, through their considerable experience, are laying down the foundations of best practice, which in some cases are being adopted throughout the facilities management industry.

Universities are offering degrees in facilities management with a number of the more traditional degrees eg architecture and surveying including a number of modules consisting of facilities management disciplines. The same applies to health and safety qualifications where a number of course providers are delivering the requisite course content but are focusing this towards facilities management using practical examples to demonstrate the operational implications within the facilities environment.

Innovations within the facilities management industry

[F3030] New products and methods of working are being developed constantly as manufacturing companies have seen the opportunities to be realised within facilities management, particularly within:

- information technology software manufacture;
- new construction products and materials;
- cleaning;
- security;
- sit stand;
- longer-term contracts and partnerships;
- big data;
- environmental; and
- energy management.

The development of telecommunications and software operating on the world-wide-web has provided an incredible choice of management systems to assist the facilities management industry. The large impact of such choice has led to the development of the 'virtual office'. A company has the opportunity to have staff based at home working from a laptop and phone line. Such working practices has advantages for both parties and can save the company costs in office space and staff equipment. Basically if an employee has a laptop, mobile telephone and a 3G card or Wi-Fi to link to the internet then within reason they are able to work from anywhere.

Travel can also be reduced these days through the use of Video and Teleconferencing as staff no longer has to travel vast distances to attend meetings. The same may be said of online interactive training courses that are widely available on the internet as a large volume of training is achievable in a short space of time at low cost and once again with zero travel requirements. A relatively new system called webinar is now in use to complement conference calling and is an abbreviated version of *Web-based seminar* which allows the chairperson of a meeting to give a presentation, training workshop or seminar over the World Wide Web. An additional feature is that the presenter has the ability to hand over control of the screen or presentation to others on the call. This innovation is a real plus for sustainability as apart from cost savings from reduction in travel there is the added bonus of fantastic carbon savings that can be made.

A key feature of a Webinar is its interactive elements — the ability to give, receive and discuss information. Contrast with Webcast, in which the data transmission is one way and does not allow interaction between the presenter and the audience.

LED lighting technology has come on in leaps and bounds and organisations are now investing in stripping out the old guard of heated element type lighting and are replacing with LEDs which are available in a vast range of products from general office strip lighting to high-intensity spotlights and high-bay lighting in warehouses and factories. This has a knock-on effect in the cost savings in energy costs, reduction in carbon produced in the power station, maintenance and replacement costs over the long-term. An additional feature of some lamp units is the addition of Wi-Fi and computer diagnostics to the unit which when linked to a management system can provide an update on condition, working hours, faults and can predict accurate information on replacement to enable accurate budgeting.

Within the cleaning industry, there have been considerable advances in cleaning products which are safe and easy to use and which have such a tolerance that they can be used by an unskilled operator and still provide a good finish. In other areas of cleaning, soap dispensers in toilets are being designed with large reservoirs to reduce the number of times they need to be refilled, helping reduce costs whilst dispensing soap over a longer period.

Floor cleaning equipment has progressed beyond large floor polishers with trailing leads and their inherent problems of trip, electrical and manual handling hazards. Modern machines tend to be smaller, lighter and are powered by maintenance-free batteries, thus eliminating the need for trailing cables. New flooring products manufactured from revolutionary materials are currently entering the marketplace and tend to be virtually maintenance-free. Flooring, especially in reception areas, can require much maintenance yet still look worn very quickly. Worn flooring may create a poor corporate image and as a result modern floor coverings are generally very durable and require little maintenance other than periodic vacuuming.

Technological advances have helped so-called 'intelligent buildings' realise much reduced running costs which have helped save energy and lower the environmental impact caused by the company. An intelligent building can detect when there is a lack of movement ie staff in an area, and can automatically switch off lighting and reduce the amount of heating and ventilation generated. Facilities and maintenance managers are being assisted by new hardware which can collate information from the central building management system (BMS) and transmit this information remotely to a smartphone or tablet, enabling facilities management real-time information. Recent developments in such building management software enable facilities management to pre-set parameters within the building and as soon as any readings extend outside those parameters a warning is sent to the Blackberry, warning the facilities manager, regardless of his location. This allows a facility manager to be proactive and take corrective action, hopefully before a problem arises.

Considered to be the most significant facilities management innovation is new technologies associated with the expansive use of smartphones and tablet technology by facilities managers and their teams the development of digital applications using really effective game-changing technology is now emerging rapidly to streamline the audit/inspection, health and safety, statutory compliance and reporting processes of the facilities management function.

The security industry has also seen much innovation with the introduction of new camera and software technology that can transmit pictures from remote, unmanned sites to a smartphone or tablet. The equipment runs in a hibernating mode until movement is detected, whereupon pictures and sound are transmitted back to a centrally manned site where they are then recorded. A mobile security patrol or the police may be called to deal with the situation. This equipment eliminates the need for a security guard to be exposed to lone working in a building and reduces security costs for the company.

Often you hear the term well-being in the workplace and one innovation that one can argue is improving the well-being of an organisation's staff is the new concept of Sit Stand Desks in the workplace. The theory is that by switching from sitting to standing can reduce discomfort and improves productivity.

Until recently these desks were viewed as a novelty in the workplace however with the release of recent research this type of desk is becoming more of the norm, where the use of computers is prevalent. Research has shown that standing at your computer screen reduces lower back pressure and relieves pressure on the buttocks and legs. By standing up the body tends to move and stretch more which in turn increases the blood flow. When staff are provided with the option to sit or stand they feel more comfortable working on their computer when standing, gain more benefits for their bodies and tend to take shorter breaks than if they were sitting, thus allowing them to feel healthier and more productive.

There is nothing worse than a short-term contract to stifle innovation especially if the contract is price-based and lacks any partnering or joint goal setting, which invariably leads to a lack of trust between the parties. New innovation in this field has seen the award of long-term contracts with total flexibility in service delivery and a joint partnering approach. In some cases, there is now evidence of the facilities management organisation providing investment to improve service delivery or improvements in the client's workplace to promote efficiency and eliminate hazards and improve safety.

With the onset of technology organisations are turning towards the analysis of data which can be captured in a number of ways. For example the installation of sensors on building services plant to measure run time, energy consumption and vibration in order to measure the condition of the equipment. That's not so bad if you are collating such data on say one asset but what if you have thousands within the portfolio? How do you analyse this data that may be changing or growing at an alarming rate? The ability to amass data on a large scale and process such data is called 'Big Data' and many organisations are turning to this in order to gain insight into their operations and to assist them in making informed decisions.

Facilities management help desks have benefited from software systems and the introduction of operative vehicle tracking modules, which not only show the location of the nearest operative, but their skills and competency. The helpdesk operative can locate and send the closest operative with the requisite skills for the work activity. Using smartphones hand-held tablet computers jobs can be sent to operatives electronically and once completed the operative is then able to close the job down in the system remotely which means that this data within the help desk system is live. A more recent innovation for the call centre is workforce scheduling and planning software which ensures that planned maintenance and reactive are allocated to engineers with the correct skills. The system is therefore automated and allows for the most efficient use of the available workforce thus achieving cost reductions and value add for clients.

The effects of safety, health and environmental legislation on facilities management

[F3031] A strong interrelationship exists between facilities management, safety, health and environmental (SHE) due to a large amount of compliance and legislation that is applicable to the built environment. Even just keeping abreast with ever changing legislation places a facilities manager under considerable pressure by not only having to acquaint themselves with new legislation, but by knowing how to apply it to the practical aspect of how and when it will impact upon a company and the budget required for any remedial works, training and implementation. The HSE completed a regulatory impact assessment (RIA) and has recognised that frequent changes in legislation caused problems in the past and now release legislative changes only twice a year in April and October.

Facilities management is becoming even more demanding in the field of SHE, as most small to medium sized companies will be unable to justify employing a SHE specialist or buying-in such advice from a consultant. A facilities manager is now expected to add health and safety to the ever-increasing range of facilities management skills. BIFM continually reviews its full range of facilities management competencies and the section concerning SHE legislation is forever expanding to include the working environment, environmental protection, energy efficiency, development of health and safety policy and risk management to name but a few subjects applicable.

As previously mentioned, building design is critical and if facilities management have some input at the design stage of a new build or refurbishment, then hopefully they can ensure that the building can be operated and maintained safely. A good, experienced facilities manager can make compromises when required and will take a proactive and flexible approach towards practical solutions to enable the safe operation of the building.

Every aspect of the facilities management role in the workplace, whether it is people management, or managing equipment, will at some stage be subject to an aspect of health and safety legislation. Staff has to be managed within an organisation to ensure that they have a safe and healthy place in which to work and that the actual way in which they work is also safe and will not endanger others. The facilities management role can influence the culture of a company in an effort to ensure that health and safety is viewed as a positive element of its business activities with benefits, rather than one that uses up valuable resources and impedes operations. As a result, it has been proven that a health and safety management system can identify operational issues before they become a problem, thus helping to create a more efficient workplace. Details of health and safety case studies from a whole range of industries are available on the hse.gov.uk website and you can keep up-to-date with these by signing up to the HSE eBulletin.

The EU also has an impact on health and safety legislation and although this is currently up for review with Brexit, there is no desire at this time to repeal any legislation so facilities managers should plan for any impacts on their operations made by legislative changes. A good example of this is the new Personal Protective Equipment (PPE) regulation, which replaces the 25-year-old PPE Directive. The new Regulation (EU) 2016/425 was approved by the EU on 31 March 2016 and became law on 21 April 2016 with a two-year transition period with enforcement commencing in April 2018. This legislation is more about the manufacture and import of PPE and will improve the adequacy and safety of the equipment. There is likely to be an impact on facilities managers as PPE such as hearing protection will be reclassified and could lead to the need for PPE to be upgraded.

Conversely, a building and all its associated assets, will be subject to legislation throughout its entire lifecycle, from design, use and then through to its final demolition or disposal, thus making it essential that a facilities manager is kept fully aware of such legislation especially during the building use element of the cycle. This maybe demonstrated by consideration of the legislation and best practice compliance concerning for example a passenger lift within a building at F3032–F3034 below. However, this three-stage lifecycle model could be applied to any asset within the built environment.

Design

[F3032] *Construction (Design and Management) Regulations 2015 (CDM 2015) (SI 2015 No 51)* – place duties on the principal designer of the lift to ensure there is a plan to manage and monitor the pre-construction phase to ensure that the equipment is safe to construct, use, maintain and finally de-commission.

Provision and Use of Work Equipment Regulations 1998 (PUWER) (SI 1998 No 2306) – place duties on the manufacturers and suppliers to ensure that the equipment is suitable for the purpose for which is has been designed and manufactured, and is safe for use and maintenance.

Construction

[F3033] *Construction (Design and Management) Regulations 2015 (CDM 2015)* – place duties on the client and principal contractor to ensure that the construction work is undertaken in a safe manner and is a safe system of work. There is also a requirement to keep a health and safety file, which is a record of all materials and equipment used and the methods of construction. This file must be kept in the building throughout its entire life as this contains valuable information.

Operation

[F3034] *Health and Safety at Work etc Act 1974* – places a duty on the owner/operator to ensure that the lift is safe.

Lifting Operations and Lifting Equipment Regulations 1998 (SI 1998 No 2307) – place duties upon owners and operators of lifts to ensure that they are subjected to preventative maintenance and also a thorough examination at regular intervals.

Construction (Design and Management) Regulations 2015 (CDM 2015) – ensure that the lift is demolished in a safe manner and information from the health and safety file would be used to assist in this process.

There are also environmental issues for a facilities manager to consider and increasingly more companies are developing and implementing environmental management systems (EMS) based upon the ISO 14001 model. Consequently, the suppliers to such companies are also expected to have an EMS.

Approved Document L2 of the *Building Regulations 2000 (SI 2000 No 2531)* deals with the conservation of fuel and power and therefore as a result has consequences for facilities managers as this requires a building logbook be maintained for all new or refurbished buildings. The logbook is designed to provide details of the plant equipment installed, it's controls and maintenance in order that energy consumption can be monitored and controlled. The object of such logbooks is to improve the efficiency of buildings by reducing energy costs and the emission of CO_2 gas into the atmosphere.

The Energy Performance of Buildings Directive is part of European legislation that is now being implemented by the use of Energy Performance Certificates and Display Energy Certificates. The construction of a building and the way in which it is insulated, heated and ventilated and the type of fuel used, all contributes to its energy consumption and carbon emissions and the facilities manager will no doubt be responsible in the day to day running of their portfolio, for the maintenance of this legislation.

The Energy Performance Certificate is one measure introduced to help improve the energy efficiency of our buildings. Other changes include requiring larger public buildings to display certificates showing the energy efficiency of the building and requiring inspections for air conditioning systems.

Please see below a summary of the implementation in England and Wales. Scotland and Northern Ireland have introduced their own regulations.

Energy Performance Certificates

[F3034A] When a commercial building is sold, built or rented an Energy Performance Certificate (EPC) is required.

The certificate provides energy efficiency A-G ratings and recommendations for improvement. The ratings - similar to those found on products such as fridges - are standard so the energy efficiency of one building can easily be compared with another building of a similar type.

Acting on an EPC is important to cut energy consumption, save money on bills and help to safeguard the environment.

You must display an EPC by fixing it to your commercial building if all these apply:

- the total useful floor area is over 500 square meters
- the building is frequently visited by the public
- an EPC has already been produced for the building's sale, rental or construction

It is essential for the facilities manager to ensure that EPCS are produced by accredited energy assessors and where appropriate are displayed.

If as a facilities manager you are responsible for the management of a public authority building which has a useful floor area in excess of $250m^2$, then after 9 July 2015 you will be required to display a Display Energy Certificate which is designed to raise the awareness of energy use and to inform visitors to public buildings about the energy usage of a building.

Summary of facilities management

[F3035] Facilities management continues to face challenges created by the impact of the ongoing need to reduce costs and the ever-emerging technological advances available to the facilities manager who as ever will rise to the challenge of generating even more cost reductions for organisations as a result of new FM delivery management models such as TFM and innovation in technology such as the use of tablet and digital application technology.

Facilities management continues to gain considerable recognition as an essential industry and many facilities managers are graduating from universities and colleges with FM qualifications, which is further proof that the profession is seen as a worthwhile and challenging career. It will be sometime before careers advisers are recommending facilities management to school leavers, but this will no doubt change in due course as more and more work is being carried out by the BIFM in its interaction with sixth form colleges and Universities.

There is strong collaboration between the British Institute of Facility Management in the UK and Global FM based in Brussels and the key facilities management organisations from around the world, including Mainland Europe, the United States, Hungary and Australia. These facilities management organisations have an agreement to undertake research, host exchange visits and share best practice, which is no doubt a powerful resource and is becoming invaluable to organisations on a global basis.

Once published, it will be interesting to review the results of the Stoddart Review and to analyse the findings, which will maybe just refocus the views and thinking of organisations, with regards to the workplace itself.

Future innovation and resultant advancements in facilities management will lead to improved efficiency and working conditions for staff with even greater value add and cost efficiencies for companies. In addition, if you include reduced energy consumption in buildings delivered through the use of new energy centric maintenance techniques and modern digital technology fitted to buildings and the resultant reduced impact to the environment then the facilities management industry is truly in for an exciting and rewarding future.

Fire Prevention and Control

Adair Lewis (updated by Andrea Oates)

Introduction

[F5001] In June 2017, a simple fault on a fridge freezer in a flat on the fourth floor of Grenfell Tower in west London sparked a fire that rapidly spread throughout the 24-storey residential block, killing 72 people. The disaster has resulted in intense scrutiny of both building control and fire safety legislation and its enforcement. There is an ongoing public inquiry and Metropolitan Police-led criminal investigation and former Health and Safety Executive (HSE) chair Dame Judith Hackitt has published the results of her independent review of building regulations and fire safety. She made 50 recommendations for the government on delivering a more robust regulatory system for high-rise homes. The government has responded to the Housing, Communities and Local Government Select Committee report on the independent review, setting out the steps it has taken, and plans to take (see **F5030.2** below).

Elements of fire

[F5001.1] Fire is a chemical reaction resulting in heat and light. For a fire to occur the following need to be combined in the correct proportions:

- oxygen (supplied by the air around us which contains about 21 per cent oxygen);
- fuel (combustible or flammable substances either solids, liquids or gases);
- source of heat energy (ignition sources such as open flame, hot surfaces, overheated electrical components).

For a fire to burn the fuel must be in a gaseous or vapour form. This means that solids and some liquids such as oils will need to be heated up until sufficient vapour is given off to burn. The form that the fuel takes will also influence the ease with which it ignites and the speed with which it burns. Generally, finely-divided materials are easier to ignite and burn more quickly than fuels in a solid form.

Once initiated, a fire will continue to burn as a result of the reinvestment of energy during the process provided sufficient fuel vapour, heat and oxygen are available and the process is not interrupted.

The 'triangle' of fire symbolises the fire process. Each side of the triangle represents fuel, heat or oxygen. Take away any one side and the fire will be extinguished or more importantly, stop all three sides from coming together and a fire can be prevented.

[F5001.1] Fire Prevention and Control

The first part of this chapter (F5001) considers generally the practical aspects of fire prevention and control. Special risks involving flammable or toxic liquids, metal fires or other hazards should be separately assessed for loss prevention and control techniques. For example, the HSE publications, *Storage of flammable liquids in containers* (HSG 51), *Safe use and handling of flammable liquids* (HSG 140) and *Storage of flammable liquids in tanks* (HSG 176) provide advice on precautions against fire hazards from flammable liquids. Control of these hazards is among the areas dealt with under the *Dangerous Substances and Explosives Atmospheres Regulations 2002 (SI 2002 No 2776)* (see **F5062–F5072** below).

Online HSE guidance makes clear that most fires are preventable and that those responsible for workplaces and other buildings to which the public have access can avoid them by taking responsibility for and adopting the right behaviours and procedures. Employers (and/or building owners or occupiers) must carry out a fire safety risk assessment, and based on the findings of the assessment, ensure that adequate and appropriate fire safety measures are in place in order to minimise the risk of death and injury in the event of a fire (www.hse.gov.uk/toolbox/fire.htm).

Considerable improvement can often be made immediately at little or no cost. Other recommendations which may require a financial appraisal must be related to loss effect values. In certain cases, however, due to high loss effect, special protection may be needed almost regardless of cost.

Modern developments in fire prevention and protection can now provide a solution to most risk management problems. However, it is a waste of time and money installing protective equipment unless it is designed to be functional, the purpose of such equipment is understood and accepted by all personnel, and the equipment is adequately inspected and maintained. The reasons for providing such equipment should be fully covered in any fire-safety training course. Fire routines should also be amended as necessary to ensure that full advantage is taken of any new measures implemented.

Fire statistics

[F5002] In the financial year 2017/18 (ending March 2018), 344 people died in fires in the UK, including 71 from the Grenfell Tower fire – one resident later died in the hospital bringing the total to 72 – compared with 263 in the previous year. This represents an increase of 27%. Fire and rescue services attended 167,150 fires during the year, three per cent more than the previous year.

Fire classification

[F5003] The classes of fire related to the fuel involved and the method of extinction are as follows.

- **Class A**
  This relates to fires generally involving solid materials, such as wood, paper and textiles, in which the combustion takes place with the formation of glowing embers.

- **Class B**
  This category relates to fires involving flammable liquids such as petrol, diesel or oils.

- **Class C**
  This relates to fires involving gases.

- **Class D**
  This relates to fires involving combustible metals, such as aluminium or magnesium.

- **Class F**
  This relates to fires involving cooking oils and fats.

There are also electrical fires, but these are not given a classification in the UK – see F5004 below. Also see F5011–F5015.1 below on the types of fire extinguisher that should be used for fighting different classes of fire.

Electrical fires

[F5004] Fires involving electrical apparatus must always be tackled by first isolating the electricity supply and then by the use of carbon dioxide or dry powder, both of which are non-conducting extinguishing mediums. Dry powder and carbon dioxide extinguish the fire by limiting or removing the oxygen by smothering.

EU Regulations, which apply in all EU member countries including the UK, prohibited the use of halon for extinguishing fires in most applications after 31 December 2003. Portable extinguishers and fixed systems using halon should no longer be available for use.

Fire extinction – active fire protection measures

[F5005]–[F5006] Extinction of a fire is achieved by one or more of the following:

- *Starvation* – this is achieved through a reduction in the concentration of the fuel. It can be effected by:
  (i)   removing the fuel from the fire;
  (ii) isolating the fire from the fuel source; and
  (iii) reducing the bulk or quantity of fuel present.

- *Smothering* – this brings about a reduction in the concentration of oxygen available to support combustion. It is achieved by preventing the inward flow of more oxygen to the fire, or by adding an inert gas to the burning mixture.
- *Cooling* – this is the most common means of fire-fighting, using water. The addition of water to a fire results in vaporisation of some of the water to steam, which means that a substantial proportion of the heat is not being returned to the fuel to maintain combustion. Eventually, insufficient heat is added to the fuel and continuous ignition ceases. Water in spray or mist form is more efficient for this purpose as the droplets absorb heat more rapidly than water in the form of a jet.

Passive and active fire protection

[F5007] Passive fire protection utilises inherently fire-resistant elements of the structure to divide a building into fire-resisting compartments. These serve two functions. One is to limit the spread of fire and can be a property protection measure as well as measure to protect life. The other is to protect escape routes by making the escape route a fire-resisting compartment. All internal fire escape stairways are also fire-resisting compartments.

In day-to-day work the integrity of these compartments should be maintained where there are doorways through the compartment walls. The doors in these openings should therefore be fire resisting door sets fitted with self-closing doors. It is essential that these doors are not obstructed and are allowed to self-close freely at all times. Any glazing in such doors must also be fire resistant and if damaged must be repaired to the appropriate standard. Guidance on the most common types of fire door is given in BS 8214:2016 *Timber-based fire door assemblies Code of Practice*.

The breaching of fire-resistant walls with services such as pipes, ducts and cables should be avoided. Where there is no alternative, it is important that all the openings are suitably protected to the same rating as the fire wall. Ducts should be fitted with fire dampers while openings around services should be suitably sealed or fire stopped. Guidance on the fire safety aspects of the design and construction of air handling duct work is given in BS 9999:2017 *Fire safety in the design, management and use of buildings Code of Practice*.

Active fire protection involves systems that are activated when a fire occurs, for example automatic fire detection or automatic sprinkler systems. As these systems are only required in an emergency it is essential that they are designed, installed, tested, inspected and maintained in accordance with acceptable standards and good practice.

Fire procedures and portable equipment

Fire procedures

[F5008] The need for effective and easily understood fire procedures cannot be over-emphasised and they should form a key element of the fire risk assessment for the premises. It may be necessary to provide a fire procedure manual, arranged so it can be used for overall fire defence arrangements, and sectioned for use in individual departments or for special risks.

It is essential that three separate procedures are considered:

- procedure during normal working hours;
- procedure during restricted staffing on shifts; and
- procedure when only security staff are on the premises.

All procedures should take into consideration absence of personnel due to sickness, leave, etc. The fire brigade should be called immediately if any fire occurs, irrespective of the size of the fire. Any delay in calling the fire brigade must be added to the delay before the fire brigade's actual arrival, which will be related to the traffic conditions or the local appliances already attending another fire.

In a large organisation, it may be necessary to identify the actions that should be taken by certain members of staff (such as security staff and till operators) when the alarm is raised. A person should be given the responsibility for ensuring pre-planned action is carried out and everyone in the premises can be accounted for when a fire occurs.

Large fires often result from a delayed call which may be due, not to delayed discovery, but to the wrong action being taken in the early stages following discovery of a fire. A pre-planned fire routine and suitable training to ensure all staff understand their responsibilities in the event of a fire is essential for fire safety. The fire brigade, when called, should be met on arrival by a designated person available to guide them directly to the area of the fire. It is essential that all fire routines, when finalised, be made known to the fire brigade.

Fire equipment

[F5009] The need to provide fire-fighting equipment is set out in Article 13 of the *Regulatory Reform (Fire Safety) Order 2005 (SI 2005 No 1541)* (see F5042 below). All areas of the workplace should be provided with a suitable number of appropriate fire extinguishers and staff should be trained in their use. (See *BS 5306–8:2012 Fire extinguishing installations and equipment on premises. Selection and positioning of portable fire extinguishers. Code of practice.*) Non-automatic fire-fighting equipment should be easily accessible, simple to use and indicated by signs.

There have been a number of cases where a person using an extinguisher has been seriously injured and investigations have shown that either the wrong type of extinguisher was being used or the operator had no training in the correct use of the appliance. The need for training staff cannot be over-

[F5009] Fire Prevention and Control

emphasised, particularly in areas of special risk, such as where deep fat fryers, furnaces, highly flammable liquids or gas cylinders are in use.

The following recommendations allow an evaluation of an existing problem and may need to be related to process risks:

- It is essential that people are trained in the use of extinguishers, especially in areas where special risks require a specific type of extinguisher to be provided.
- Workers should be clearly instructed that at no time should they jeopardise their own safety or the safety of others.
- People who may be wearing overalls contaminated with oil, grease, paint or solvents should not be instructed to attack a fire. Such contaminated materials may vaporise due to heat from the fire and ignite.

Types of fire extinguisher

[F5010] The type of extinguisher provided should be suitable for the hazard involved, adequately maintained and appropriate records kept of all inspections, tests etc. All fire extinguishers should be fitted on wall brackets or located in purpose-built floor stands. If this is not done, extinguishers may be removed and missing when required, or be knocked over and damaged. Extinguishers should be sited near exits or on the route to an exit. (See F5003 above for a description of different classes of fire.)

Water extinguishers

[F5011] Water extinguishers can only be used on Class A fires, although water extinguishers with additives can also be suitable for use on Class B fires – this will be indicated on the extinguisher where appropriate. Water extinguishers should not be used on liquid, electrical or metal fires.

Foam extinguishers

[F5012] Foam extinguishers can be used on Class A and B fires, particularly liquid fires such as petrol and diesel. However, only specially trained operators should use them on free-flowing liquid fires as these have the potential to rapidly spread to adjacent material. Foam extinguishers should not be used on electrical or metal fires.

Dry powder extinguishers

[F5013] This type of extinguisher can be used on most classes of fire and is capable of a quick 'knock down' of a fire. They can be used on electrical equipment (although it is likely the equipment will be rendered useless). Because power extinguishers do not significantly cool the fire, it can reignite. In addition, powder extinguishers can create loss of visibility and may affect people who have breathing problems. They are not generally suitable for enclosed spaces and should not be used on metal fires.

BCF extinguishers

[F5014] The extinguishing medium is a halon and manufacture of this class of chemicals is no longer permitted due to its adverse effect on the environment. New extinguishers of this type are no longer available, any that are still present in the workplace should be removed from service and returned to the supplier for safe disposal. Halon extinguishers may be replaced by dry powder or carbon dioxide extinguishers. EU Regulations prohibited the use of halon for extinguishing fires in most applications, except for some applications such as in aircraft and military applications, at the end of 2003.

Carbon dioxide extinguishers

[F5015] Carbon dioxide extinguishers are particularly suitable for fires involving electrical equipment. They can extinguish a fire without causing further damage to electrical equipment (except computer and other electronic equipment). The power should be disconnected if possible.

Carbon dioxide (CO_2) should not be used on metal fires. Training in the use of CO_2 extinguishers is essential; particular care should be taken not to hold the discharge horn on a CO_2 extinguisher as the surface gets very cold and can cause cold burns to unprotected skin. As carbon dioxide is an asphyxiant, it should not be used in a confined unventilated space.

Class 'F' extinguishers

[F5015.1] Class 'F' extinguishers can be used for fighting fires in catering establishments with deep-fat fryers. New fire extinguishers should comply with BS EN 3-7:2004+A1:2007 *Portable fire extinguishers. Characteristics, performance requirements and test methods.*

For detailed guidance on the number and type of fire extinguishers in different types of workplaces see: www.gov.uk/government/collections/fire-safety-law-and-guidance-documents-for-business.

Colour coding and distribution of portable fire extinguishers

[F5016] All new fire extinguishers for use throughout the EU are coloured red. Manufacturers are allowed to incorporate different coloured zones on or above the operation instructions label to help identify the extinguisher type – for example red for water, cream for foam, blue for dry powder and black for CO_2.

Extinguishers are distributed in accordance with BS 5306–8:2012 *Fire extinguishing installations and equipment on premises. Selection and positioning of portable fire extinguishers. Code of practice*, based on an extinguishers rating (capability) rather than by the type and size of the unit.

Fire alarms

[F5017] Every workplace should be provided with a fire alarm. This can range from a shouted warning to an electrical detection and warning system. As a result of the fire risk assessment, the alarm in a small workplace may take the form of a simple hand bell or air horn, provided that staff are aware of the sound and it can be clearly heard in all parts of the premises. In a larger workplace, however, a fire alarm system consisting of break glass call points and electric bells or sounders is normally installed (with the system often incorporating automatic fire detectors, see **F5039**).

The system should comply with the requirements of BS 5839-1:2017 *Fire detection and fire alarm systems for buildings. Code of practice for design, installation, commissioning and maintenance of systems in non-domestic premises.*

Fire alarm systems should be tested weekly, using the call points in rotation. Suitable records of the testing should be kept.

Good 'housekeeping'

[F5018] The need for good 'housekeeping' is the key to a safe working environment. The following are essential guidelines.

- Smoking in the workplace is prohibited. Legislation banning smoking in enclosed workplaces and public places has been in force across the UK for a number of years;
- Suitable 'smoking prohibited' notices should be displayed, in accordance with the *Smoke-free (Signs) Regulations 2012 (SI 2012 No 1536)*;
- Combustible waste and contaminated rags should be kept in separate metal bins with close fitting metal lids.
- Cleaners should, preferably, be employed in the evenings when work ceases. This will ensure that combustible rubbish is removed from the building to a place of safety before the premises are left unoccupied.
- Rubbish should not be kept in the building overnight, or stored in close proximity to the building.
- Where possible a 'clear desk' policy should be adopted, materials should not be stored on cupboard tops, and all filing cabinets should be properly closed, and locked if possible, at the end of the day.

Pre-planning of fire prevention

[F5019] A pre-planned approach to fire prevention and control is essential. Fire spreads extremely fast, and the temperature can rise to 1,000°C in only one minute. Smoke can be flammable and toxic. It is essential to review the fire risk assessment periodically, identify areas of high loss effect or high risk, and plan accordingly to meet requirements and legal responsibilities.

Means of escape

[F5020] Requirements for means of escape are set out in Article 14 of the *Regulatory Reform (Fire Safety) Order 2005 (SI 2005 No 1541)* (see **F5040** below). This indicates that in order to safeguard the safety of staff, visitors and others on the premises, the responsible person (see **F5033** below) must ensure that routes to emergency exits from premises and the exits themselves are kept clear at all times.

Pre-planning of fire prevention [F5020]

The following requirements (set out in Article 14) must be complied with in respect of premises where necessary (whether due to the features of the premises, the activity carried on there, any hazard present or any other relevant circumstances) in order to safeguard the safety of relevant persons:

(a) emergency routes and exits must lead as directly as possible to a place of safety;
(b) in the event of danger, it must be possible to evacuate the premises as quickly and as safely as possible;
(c) the number, distribution and dimensions of emergency routes and exits must be adequate having regard to the use, equipment and dimensions of the premises and the maximum number of people who may be present there at any one time;
(d) emergency doors must open in the direction of escape;
(e) sliding or revolving doors must not be used for exits specifically intended as emergency exits;
(f) emergency doors must not be locked or fastened so that they cannot be easily and immediately opened by anyone who may require to use them in an emergency;
(g) emergency routes and exits must be indicated by signs; and
(h) emergency routes and exits requiring illumination must be provided with emergency lighting of adequate intensity in the case of failure of their normal lighting.

The means of escape provided should be identified in the fire risk assessment for the workplace (required under Article 9 of the *Regulatory Reform (Fire Safety) Order (SI 2005 No 1541)* (see **F5037**).

Further guidance with regard to means of escape in different types of workplace is set out in a series of guides to the *Regulatory Reform (Fire Safety) Order (SI 2005 No 1541)*. These can be found online and downloaded free of charge at: www.gov.uk/government/collections/fire-safety-law-and-guidance-documents-for-business.

Any dimensions set out in guidelines or other forms of good practice should always be reviewed during the fire risk assessment for the workplace.

Care must always be taken when the arrangement of the work areas incorporates 'rooms within rooms' – an area where access is made to it from another room, not a corridor. The occupants of the inner rooms must be made aware of any threat to the integrity of their escape route at an earliest time as possible. In addition:

- the access room must not be classed as a high fire risk; and
- a clear glazed vision panel should be provided in the partition wall or the door between the inner room and the access room; or
- a suitable smoke detector should be provided in the access room; or
- the wall or partition between the two areas should not extend to within 0.5m from the ceiling.

The following are essential:

- all doors affording means of escape in case of fire should be maintained easily and readily available for use at all times that persons are on the premises;

- doors should be hung so as to open in the direction of travel;
- all doors not in continuous use, affording a means of escape in case of fire, should be clearly indicated;
- sliding doors should not be used on escape routes;
- doors on escape routes should not be locked or fastened in such a way that they cannot be easily and immediately opened without the use of a key; and
- all gangways and escape routes must be kept clear at all times.

Unsatisfactory means of escape

[F5021] Unsatisfactory means of escape that should not be used in the event of fire include the following:

- lifts (unless a specially designed evacuation lift for use in the event of fire);
- portable ladders;
- spiral staircases; and
- lowering lines.

Fire drills

[F5022] The *Regulatory Reform (Fire Safety) Order 2005 (SI 2005 No 1541)* (Article 15) (see also F5041 below) states that:

'(1) The responsible person must—
  (a) establish and, where necessary, give effect to appropriate procedures, including safety drills, to be followed in the event of serious and imminent danger to relevant persons;
  (b) nominate a sufficient number of competent persons to implement those procedures in so far as they relate to the evacuation of relevant persons from the premises; and
  (c) ensure that no relevant person has access to any area to which it is necessary to restrict access on grounds of safety, unless the person concerned has received adequate safety instruction.'

Employers should acquaint the workforce with the actions they should take in the event of fire. This consists of putting up a notice in a prominent place (normally next to each fire alarm call point) stating the action employees should take on:

- hearing the alarm, or
- discovering the fire.

In addition, employees should receive regular fire drills, as a minimum once a year, even though normal working is interrupted.

Employees should be designated and trained as fire wardens to ensure that everyone leaves the building safely and to assist members of the public or colleagues with a disability, as necessary. In addition, selected employees should be trained in the proper use of fire extinguishers.

Periodical visits by the local fire authority should be encouraged by employers, since this provides a valuable source of practical information on fire-fighting, fire protection and training and enables the fire brigade to become familiar with the layout of the premises and the hazards that may be present.

Typical fire action notice

[F5023]–[F5029] When the fire alarm sounds:

(1) Leave room, closing doors behind you.
(2) Walk quickly along the nearest available route to open air.
(3) Do not use lifts.
(4) Report to fire warden at assembly point.
(5) Do not re-enter building.

When you discover a fire:

(1) Raise alarm (normally by operating a break glass call point).
(2) Leave the room, closing doors behind you.
(3) Leave the building by the nearest available route.
(4) Report to fire warden at assembly point.
(5) Do not re-enter building.

Fire and fire precautions – legislation

[F5030] Fire safety legislation in the UK is concerned principally with protecting life, including the safety of firefighters and others who may need to enter a building on fire as part of their professional duties. There are two main sets of regulations that control this area: (*a*) the imposition of controls on the design and construction of buildings and (*b*) requirements leading to the safe management of buildings that are in use.

As set out in the introduction, both building regulations and fire safety are under scrutiny following the Grenfell Tower fire.

The provision of fire safety in new buildings relates to the physical provisions at the time of construction and is controlled through the Building Regulations in England and Wales (with equivalent legislation covering Scotland and Northern Ireland).

Buildings in use are currently controlled by a range of statutes relating to workplace fire risk assessment and licensing of premises for consumption of alcohol, entertainment and similar licensed activities. These statutes generally relate to physical provisions and management requirements.

The principle legislation relating to fire safety in the workplace is the *Regulatory Reform (Fire Safety) Order 2005 (SI 2005 No 1541)* (see **F5033** below). This legislation is concerned with the fire safety provisions and management of non-domestic premises.

Building regulations

[F5030.1] The *Building Regulations 2010 (SI 2010 No 2214)* lay down requirements for building work, including those related to fire safety. These

Regulations are minimum standards for design, construction and alterations to virtually every building and specify the following requirements:

Means of warning and escape

- The building shall be designed and constructed so that there are appropriate provisions for the early warning of fire, and appropriate means of escape in case of fire from the building to a place of safety outside the building capable of being safely and effectively used at all material times.

Internal fire spread

- To inhibit internal fire spread, internal linings must:
  (i)   adequately resist flame spread over surfaces; and
  (ii) if ignited, have a reasonable rate of heat release.
- The building must be designed and constructed so that, in the event of fire, its stability will be maintained for a reasonable period; and a common wall should be able to resist fire spread between the buildings.
- The building must be designed and constructed so that unseen fire/smoke spread within concealed spaces in its fabric and structure, is inhibited.

External fire spread

- External walls shall adequately resist fire spread over walls and from one building to another.
- A roof should be able to resist fire spread over the roof and from one building to another.

Access and facilities for the fire service

- The building shall be designed and constructed so as to provide reasonable facilities to assist firefighters in the protection of life and reasonable provision made within the site of the building to enable fire appliances to gain access.

The independent review of building regulations and fire safety

[F5030.2] Following the Grenfell Tower fire, the government commissioned former HSE chair Dame Judith Hackitt to carry out an independent review of building regulations and fire safety. She published her final report, *Building a safer future: independent review of building regulations and fire safety*, in May 2018 and called for a 'radical rethink of the whole system'. Her report makes over 50 recommendations for the government on delivering a more robust regulatory system for high-rise homes. She called for:

- a less prescriptive, outcomes-based approach to the regulatory framework overseen by a new regulator;

- clearer roles and responsibilities throughout the design and construction process and occupation;
- the consultation of residents and their involvement in decisions affecting the safety of their home;
- a more rigorous and transparent product-testing regime and a more responsible marketing regime; and
- an industry-led strengthening of competence of all those involved in building work and the establishment of an oversight body.

The report says the current system is far too complex and lacks clarity, while regulatory oversight and enforcement is inadequate. In her interim report, published in December 2017, she described the regulatory system covering high-rise and complex buildings as 'not fit-for-purpose'.

In her final report, she said regulations and guidance are not always read, and even when they are, they are misunderstood and misinterpreted. She identified the primary motivation as being to do things as cheaply and quickly as possible rather than to deliver quality, safe homes. She said that concerns are often ignored and some of those undertaking building work 'fail to prioritise safety, using the ambiguity of regulations and guidance to game the system'.

She identified a lack of clarity in the roles and responsibilities of those procuring, designing, constructing and maintaining buildings and inadequate regulatory oversight and enforcement, with penalties too low to act as an effective deterrent. She also found poor record keeping and 'change control', an opaque and insufficient regime for product testing, labelling and marketing, patchy competence across the system, and the voices of residents often going unheard even when they relate to safety concerns.

She called for a new and more robust system to strengthen regulatory oversight, clarify roles and responsibilities, raise and assure competence levels, and improve the quality and performance of construction products. The new system she recommended initially focuses on multi-occupancy higher-risk residential buildings (HRRBs) that are ten storeys or more in height, although the report says 'there would be merit in certain aspects of the new regulatory framework applying to a wider set of buildings'.

The recommendations include:
- a new Joint Competent Authority (JCA) comprising Local Authority Building Standards (the proposed new name for Local Authority Building Control), fire and rescue authorities and the HSE to 'oversee better management of safety risks in these buildings (through safety cases) across their entire life cycle';
- a mandatory incident reporting mechanism for dutyholders with concerns about the safety of HRRBs;
- a set of rigorous and demanding dutyholder roles and responsibilities, broadly aligned with those set out in the Construction (Design and Management) Regulations 2015;
- a series of 'robust gateway points' to strengthen regulatory oversight. Dutyholders would be required to show the JCA their plans are detailed and robust, their understanding and management of building safety is

appropriate, and they can properly account for the safety of the completed building in order to gain permission to move to the next phase of work and allow the building to be occupied;
- records of all changes to detailed plans previously signed off by the JCA. Regulatory oversight would be independent from clients, designers and contractors and there would be more rigorous enforcement powers and more serious penalties for those 'who choose to game the system and place residents at risk';
- a clear and identifiable dutyholder with responsibility for building safety of the whole building and a requirement on the dutyholder to present a safety case to the JCA at regular intervals to check building safety risks are being managed 'so far as is reasonably practicable'. The JCA would regulate fire and structural safety for the whole building;
- greater transparency of information on building safety;
- more resident involvement in decision-making, through the support of residents' associations and tenant panels and 'a no-risk route' for residents to escalate fire safety concerns through an independent statutory body;
- improved levels of competence in the construction and fire safety sector and moving towards a system 'where ownership of technical guidance rests with industry as the intelligent lead in delivering building safety';
- a more effective testing regime with clearer labelling and product traceability, including a periodic review process of test methods, underpinned by a more effective national market surveillance system; and
- a 'golden thread of information' about each HRRB to be created through a digital record, from initial design intent through to construction and including changes that occur throughout occupation. Dutyholders would use this to demonstrate to the regulator the safety of the building throughout its life cycle.

The report can be found on the government website at: www.gov.uk/government/publications/independent-review-of-building-regulations-and-fire-safety-final-report.

In September 2018, the government issued a response to the Housing, Communities and Local Government Select Committee report on the independent review of building regulations and fire safety and set out the action it will take. It has already taken a number of steps (see below) and is due to publish an implementation plan by the end of the 2018 providing more details on its response.

The Coroner's report into the fatal Lakanal House fire in 2009 recommended the government clarify statutory guidance on fire safety (known as Approved Document B (see below)), but this had not been completed by the time of the Grenfell Tower fire and was put on hold in order to consider the findings and recommendations of Dame Judith's review. She recommended consideration of 'presentational changes' to improve the clarity of the document.

In July 2018, the government published proposals to clarify. These aim to improve usability and reduce the risk of misinterpretation by those carrying out and inspecting building work (see https://assets.publishing.service.gov.uk/

government/uploads/system/uploads/attachment_data/file/727136/Clarificat
ion_of_ADB_con_doc_ECOMMS_MASTER.pdf).

The consultation document explains that technical requirements in Building Regulations are supported by statutory guidance set out in 'Approved Documents' which 'provide advice on approaches to compliance'. Guidance on fire safety is set out in Approved Document B in two volumes: volume 1 covers dwelling houses and volume 2 covers buildings other than dwelling houses. The consultation document contains revised guidance on restricting the use of assessments in lieu of tests (also known as desktop studies) and the use of combustible materials in the external walls of high-rise buildings, which have both been consulted on separately (see below).

At the beginning of October, the government announced that following a consultation exercise, it would ban the use of combustible cladding materials on the external walls of residential buildings more than 18 metres high. Secretary of State for Housing, Communities and Local Government James Brokenshire announced he would change the building regulations to ban the use of combustible materials for all new high-rise residential buildings, hospitals, registered care homes and student accommodation. The consultation began in July 2018 and closed in August 2018 (see https://www.gov.uk/government/consultations/banning-the-use-of-combustible-materials-in-t he-external-walls-of-high-rise-residential-buildings).

A separate consultation proposing amendments to Approved Document B would restrict the use of assessments in lieu of tests or desktop studies. This consultation ran in April and May 2018 (see www.gov.uk/government/cons ultations/approved-document-b-fire-safety-amendments-to-statutory-guida nce-on-assessments-in-lieu-of-tests). The outcome of this consultation was awaited in November 2018.

The *Government response to the Housing, Communities and Local Government Select Committee report on the independent review of building regulations and fire safety: next steps* can be found on the government website at www.gov.uk/government/publications/building-regulations-and-fire-safet y-government-response-to-select-committee-report.

For more information about the Building Regulations regime, see www.gov. uk/government/policies/building-regulation.

Fire Safety Legislation

[F5031] Until the introduction of the *Regulatory Reform (Fire Safety) Order 2005 (SI 2005 No 1541)* the statute law relating to fire and fire precautions was extensive and complex. The legislation included:

- *Fire Precautions Act 1971* ('FPA 1971') (and regulations and orders made thereunder);
- *Fire Safety and Safety of Places of Sport Act 1987*;
- *Fire Precautions (Workplace) Regulations 1997 (SI 1997 No 1840)* as amended by the *Fire Precautions (Workplace) (Amendment) Regulations 1999 (SI 1999 No 1877)*;

- *Health and Safety at Work etc Act 1974* ('HSWA 1974') (in the form of the *Fire Certificates (Special Premises) Regulations 1976 (SI 1976 No 2003)*);
- Petroleum Acts (and regulations made thereunder);
- *Public Health Acts 1936–1961*;
- *Building Act 1984* and the Building Regulations made thereunder; and
- *Fire Services Act 1947* and the *Fires Prevention (Metropolis) Act 1774*;

as well as regulations made under the *Factories Act 1961*.

The *Regulatory Reform (Fire Safety) Order 2005 (SI 2005 No 1541)*, introduced in October 2006, repealed a number of Acts and revoked a number of sets of Regulations. Other pieces of legislation were amended to streamline the legislative basis of fire safety in non-domestic premises. However, the *Dangerous Substances and Explosive Atmospheres Regulations 2002 (SI 2002 No 2776)* remain in force (see **F5062–F5072** below). Fire certificates are no longer issued by fire and rescue services but fire inspectors or duly authorised officers of the fire brigade have powers to visit and enter premises in order to carry out their duties (see **F5032** and **F5056**).

Fire certification

[F5032] Until the introduction of the *Regulatory Reform (Fire Safety) Order 2005 (SI 2005 No 1541)* in October 2006, one of the prime tools in respect of fire safety in the workplace was the *Fire Precautions Act 1971*. In the *Fire Precautions Act 1971* there was the requirement for 'designated' premises to be 'fire-certificated' – normally by the local fire authority, though in the case of exceptionally hazardous industrial premises (ie special premises) by the Health and Safety Executive (HSE). Two designation orders were made affecting factories, offices, shops and railway premises *and* hotels and boarding houses.

The fire certificate had to be kept on the premises to which it referred and this important document can still form a vital element when undertaking or reviewing the fire risk assessment and determining the fire safety policies and procedures for the building. Although fire certificates do not now have the legislative authority that they previously enjoyed they should be kept safely together with the fire safety risk assessment for the premises.

Fire certificates contained valuable information. For example, they specified:

- use/uses of premises;
- means of escape in case of fire;
- how means of escape can be safely and effectively used;
- alarms and fire warning systems; and
- fire-fighting apparatus to be provided in the building.

Moreover, at its discretion, the fire authority could, additionally, impose requirements relating to:

- maintenance of means of escape and fire-fighting equipment;
- staff training; and
- restrictions on the number of people within the building.

It should be remembered that statutory restrictions on the occupancy levels of certain areas of the premises may still be in force as a result of licenses issued by the local authority. Such restriction may apply, for example, to bars and other places of assembly.

The Regulatory Reform (Fire Safety) Order 2005

[F5033] The *Regulatory Reform (Fire Safety) Order 2005 (SI 2005 No 1541)* replaced fire certification under the *Fire Precautions Act 1971* with a general duty to ensure, so far as is reasonably practicable, the safety of employees, a general duty in relation to non-employees to take such fire precautions as may reasonably be required in the circumstances to ensure that premises are safe, and a duty to carry out a risk assessment. The Order imposes a number of specific duties in relation to the fire precautions to be taken and also provides for the enforcement of the Order. It amended or repealed other primary legislation concerning fire safety to take account of the new system and provided for minor and other consequential amendments, repeals and revocations. For example, safety certificates continue to be issued under the *Safety of Sports Grounds Act 1975* or the *Fire Safety and Safety of Places of Sport Act 1987* but are not allowed to require anyone to do anything that would cause them to contravene *SI 2005 No 1541*.

The Order applies to all non-domestic premises other than a restricted number of exceptions (see F5034).

The Order applies only in England and Wales. Similar legislation applies in Scotland and Northern Ireland.

The Regulatory Reform Order requires the 'responsible person' to undertake a fire risk assessment for the premises.

Article 3 sets out that the responsible person means the employer, if the workplace is under his control; otherwise the duties of the responsible person are extended to any person who has control of the premises.

There are a number of duties imposed on the responsible person, who must:

- take such general fire precautions as will ensure, so far as is practicable, the safety of his employees (Article 8);
- in relation to relevant persons who are not his employees, take such general fire precautions as may reasonably be required in the circumstances to ensure that the premises are safe (Article 8); and
- make a suitable and sufficient assessment of the risks to which relevant persons are exposed for the purpose of identifying the general fire precautions he needs to take (Article 9).

Article 4 sets out that general fire precautions are measures: to reduce the risk of fire and the spread of fire on the premises; in relation to the means of escape from the premises; for securing that the means of escape can be safely and effectively used; in relation to the means for fighting fires; in relation to the means for detecting fire on the premises and giving warning in case of fire; and in relation to the arrangements for action to be taken in the event of fire, including instruction and training of employees; and measures to mitigate the effects of the fire.

Application of the Regulatory Reform (Fire Safety) Order 2005

[F5034]–[F5036] The Order applies to all non-domestic premises other than a restricted number of exceptions. The Order does **not** apply in relation to:

- single private dwellings;
- an offshore installation within the meaning of the *Offshore Installation and Pipeline Works (Management and Administration) Regulations 1995 (SI 1995 No 738)*, Reg 3;
- a ship, in respect of the normal ship-board activities of a ship's crew which are carried out solely by the crew under the direction of the master;
- fields, woods or other land forming part of an agricultural or forestry undertaking but which is not inside a building and is situated away from the undertaking's main buildings;
- an aircraft, locomotive or rolling stock, trailer or semi-trailer used as a means of transport or a vehicle for which a licence is in force under the *Vehicle Excise and Registration Act 1994* or a vehicle exempted from duty under that Act;
- a mine within the meaning of the *Mines and Quarries Act 1954, s 180*, other than any building on the surface at a mine; and
- a borehole site to which the *Borehole Sites and Operations Regulations 1995 (SI 1995 No 2038)* apply.

Fire risk assessment

[F5037] The *Regulatory Reform (Fire Safety) Order 2005 (SI 2005 No 1541)* (Article 9) requires the responsible person for the premises to assess the risk a fire could pose to the employees and other persons who may be present in the workplace.

Although there is no prescriptive way in which the fire risk assessment should be carried out, it may involve five stages:

Stage 1 – identify the fire hazards;
Stage 2 – identify the people at risk;
Stage 3 – evaluate, remove or reduce, and protect from risk;
Stage 4 – record (a statutory requirement if more than five people are employed), plan, inform, instruct and train; and
Stage 5 – review.
The assessment should be reviewed periodically and when any significant changes occur to, for example, layout, management or staffing.

Carrying out the fire risk assessment

[F5038] The responsible person must make a suitable and sufficient assessment of the fire risk to which the staff and others on the premises are exposed. This should:

- identify any group of persons identified by the assessment as being particularly at risk;
- record the significant findings, including the measures which have, or will be taken by the responsible person;
- identify the general fire precautions needed to be taken; and
- include consideration of the presence, use and handling of any dangerous substances that are present.

The fire risk assessment should be recorded where there are five or more employees, there is a licence in force in relation to the premises and where an alterations notice requiring this is in force in relation to the premises (see F5057).

No new work involving a dangerous substance may commence unless a risk assessment has been made and the measures identified during that exercise have been implemented (Article 9).

In addition, the responsible person must not employ a young person unless he has made or reviewed a fire risk assessment, taking particular account of:

(a) the inexperience, lack of awareness of risks and immaturity of young persons;
(b) the fitting-out and layout of the premises;
(c) the nature, degree and duration of exposure to physical and chemical agents;
(d) the form, range, and use of work equipment and the way in which it is handled;
(e) the organisation of processes and activities;
(f) the extent of the safety training provided or to be provided to young persons; and
(g) risks from agents, processes and work listed in the Annex to Council Directive 94/33/EC on the protection of young people at work (Schedule 1, Part 2).

Where the responsible person implements any preventive and protective measures he must do so on the basis of the principles of prevention specified in Part 3 of Schedule 1 of the Order (Article 10). These are:

- avoiding risks;
- evaluating the risks which cannot be avoided;
- combating the risks at source;
- adapting to technical progress;
- replacing the dangerous by the non-dangerous or less dangerous;
- developing a coherent overall prevention policy which covers technology, organisation of work and the influence of factors relating to the working environment;
- giving collective protective measures priority over individual protective measures; and

- giving appropriate instructions to employees.

Controlling the residual risk

[F5039] Having eliminated and reduced fire hazards as far as practicable, there are a number of measures that should be taken to control the residual risks. These include the provision of:

- emergency routes and exits;
- fire safety policies and procedures;
- provisions for fire-fighting and fire detection; and
- additional measures in respect of dangerous substances.

Emergency routes and exits

[F5040] In order to safeguard the safety of employees and other persons on the premises in case of fire:

- routes to emergency exits must be kept clear at all times;
- emergency routes and exits must lead as directly as possible to a place of safety;
- in the event of danger it must be possible for persons to evacuate the premises as quickly and as safely as possible;
- the number, distribution and dimensions of emergency routes must be adequate regarding their use, equipment and dimensions of the premises and the maximum number of persons who may be present there at any one time;
- emergency doors must open in the direction of travel;
- sliding or revolving doors must not be used for exits specifically intended as emergency exits;
- emergency doors must not be so locked or fastened that they cannot be easily and immediately opened in an emergency;
- emergency routes and exits must be indicated by signs; and
- emergency routes requiring illumination must be provided with emergency lighting of adequate intensity in the case of failure of the normal lighting (Article 14).

Procedures for serious and imminent danger and danger areas

[F5041]

- The responsible person must establish appropriate procedures, including safety drills, to be followed in the event of serious and imminent danger to persons.
- They must nominate a sufficient number of competent persons to implement those procedures in so far as they relate to the evacuation of the premises.
- In addition, the responsible person must ensure that no person has access to any area to which it is necessary to restrict access on grounds of safety unless the person concerned has received adequate safety instruction (Article 15).

Provisions for fire-fighting and fire detection

[F5042] Where necessary to safeguard safety, the responsible person must:

- ensure that the premises is, to the extent that is appropriate, equipped with appropriate fire-fighting equipment and with fire detectors and alarms; and
- ensure that any non-automatic fire-fighting equipment so provided is easily accessible, simple to use and indicated by signs.

Article 13 sets out that what is appropriate will depend on the dimensions and use of the premises, the equipment present, the physical and chemical properties of the substances likely to be present, and the maximum number of people who may be present at any one time. Where necessary, the responsible person must:

- take measures for fire-fighting in the premises, adapted to the activities carried on there and the size of the undertaking and the premises concerned;
- nominate competent persons to implement these fire-fighting measures – the number of these persons, their training and the equipment available to them must be adequate, taking into account the size of the premises and the specific hazards there; and
- establish any necessary contacts with external emergency services, particularly as regards fire-fighting, rescue work, first aid and emergency medical care.

A competent person has sufficient training and experience or knowledge and other qualities to enable them properly to implement these measures (Article 13).

Additional emergency measures in respect of dangerous substances

[F5043] In order to safeguard staff and others in respect of an accident, incident or emergency arising from the presence of a dangerous substance, the responsible person must ensure (*Regulatory Reform (Fire Safety) Order 2005 (SI 2005 No 1541), art 16*) that:

- information on emergency arrangements is available. This should include details of relevant work hazards and hazard identification arrangements and specific hazards likely to arise at the time of an accident, incident or emergency. This information (along with information about procedures for serious and imminent danger and danger areas required under Article 15 (see **F5041** above) and information about warning systems and escape facilities (see below) should be displayed in the premises and also be made available to the emergency services to allow them to prepare their own response procedures.

This article also requires the responsible person to ensure that:

- suitable warning and communication systems are established to enable an appropriate response to be made when an incident occurs;

- where necessary visual or audible warnings are given before explosive conditions are reached and relevant persons are withdrawn from the area; and
- where the risk assessment indicates it to be necessary, escape facilities are provided and maintained to enable persons can leave the endangered areas promptly and safely.

In the event of a fire relating to the presence of a dangerous substance the responsible person must ensure that:

- immediate steps are taken to mitigate the effect of the fire, restore the situation to normal and inform the relevant person who may be affected; and
- only those persons essential for carrying out repairs and other necessary work are allowed into the affected area and that they are provided with appropriate personal protective clothing and equipment and any necessary specialised safety equipment and plant.

The above points do not apply, however, where the results of the risk assessment show that the quantity of the dangerous substances on the premises are such that there is only slight risk to relevant persons and the measures taken by the responsible person to comply with their duty under Article 12 are sufficient to control that risk.

Measures to be taken in respect of dangerous substances

[F5044]–[F5045] In applying measures to control risks the responsible person must, (*Regulatory Reform (Fire Safety) Order 2005 (SI 2005 No 1541), art 12*) in order or priority:

- eliminate or reduce the quantity of dangerous substances to a minimum;
- avoid or minimise the release of a dangerous substance;
- control the release of a dangerous substance at source;
- prevent the formation of an explosive atmosphere, including the application of appropriate ventilation;
- ensure that any release of dangerous substance which may give rise to risk is suitably collected, safely contained, removed to a safe place or otherwise rendered safe;
- avoid ignition sources, including electrostatic discharges and other adverse conditions that could result in harmful effects from a dangerous substance; and
- segregate incompatible dangerous substances.

The responsible person must ensure that mitigation measures are applied which include:

- reducing the number of persons exposed to a minimum;
- measures to avoid the propagation of fires and explosions;
- providing explosion relief arrangements;
- providing explosion suppression equipment;
- providing plant that is constructed to withstand the pressure likely to be produced by an explosion; and

- providing suitable personal protective equipment.

In addition, the responsible person must:

- ensure that premises are designed, constructed and maintained so as to reduce risk;
- ensure that mitigating measures are designed, constructed, assembled, installed, provided and used so as to reduce risk;
- ensure that special technical and organisational measures are maintained in an efficient working order and good repair;
- ensure that equipment and protective systems meet requirements concerning power failure, means for manual override for shutting down equipment and emergency shutdown; and
- ensure that appropriate systems of work are issued in writing and a suitable system of permits to work is instituted and maintained prior to work being carried out.

Fire fighters' switches for luminous signs

[F5046] In addition to the fire safety duties set out in *Regulatory Reform (Fire Safety) Order 2005 (SI 2005 No 1541)*, *Part 2*, there are additional miscellaneous requirements in *Part 5*; foremost among these are measures to protect fire fighters in the event of a fire.

SI 2005 No 1541, art 37 is concerned with the provision of switches for fire fighters' use for isolating luminous signs operating at a voltage in excess of the normal supply voltage. Where such apparatus is installed it must be provided with a cut-off switch so placed and marked as to be readily recognisable by and accessible to fire fighters.

Even though a switch may comply with the requirements of BS 7671 *Requirements for Electrical Installations IET Wiring Regulations*, the responsible person must give notice to the fire and rescue authority at least 42 days before work is to commence to install the apparatus. The proposal should be deemed to satisfy the requirements of the fire authority unless they serve a counter notice indicating that they are not satisfied with the proposals within 21 days from the date of the service of the notice.

This article does not apply to premises licensed under the *Licensing Act 2003* for the exhibition of a film or to premises where apparatus has already been installed in compliance with the *Local Government (Miscellaneous Provisions) Act 1982*.

In addition to the specific requirement for cut-off switches, *SI 2005 No 1541, art 38* requires the responsible person to safeguard the safety of fire fighters by ensuring that the premises and any facilities, equipment and devices provided for the use or protection of fire fighters under the Order be subject to a suitable maintenance regime and be maintained in efficient working order and good repair.

Maintenance

[F5047] Where necessary in order to safeguard the safety of relevant persons the responsible person must ensure that the premises and any facilities, equipment and devices provided in compliance with the *Regulatory Reform (Fire Safety) Order 2005 (SI 2005 No 1541)* are subject to a suitable system of maintenance and are maintained in an efficient state, in efficient working order and in good repair (*SI 2005 No 1541, art 17*).

Where the premises form part of a building then the responsible person may make arrangements with the occupiers of other parts of the building to ensure that the requirements outlined in the paragraph above are met. The occupiers of other parts of the building have a duty to co-operate in this respect.

Safety assistance

[F5048] *Regulatory Reform (Fire Safety) Order 2005 (SI 2005 No 1541), art 18* requires that the responsible person appoints one or more competent persons to assist them in undertaking the preventive and protective measures. The responsible person must make arrangements for ensuring adequate co-operation between themselves and safety assistant(s). Sufficient time must be made available to allow these safety assistant(s) to properly fulfil their functions.

Where there is a competent person in the responsible person's employment, that person must be appointed as safety assistant in preference to someone from outside the organisation. Where, however, the safety assistant appointed is not in the responsible person's employment, they must be provided with the information necessary for him to undertake his task. This will include access to relevant safety data relating to the dangerous substances present and the risk that they present.

Competent person

[F5049] A competent person is defined in *Regulatory Reform (Fire Safety) Order 2005 (SI 2005 No 1541), art 18(5)* as follows:

> 'A person is regarded as being competent where he has sufficient training and experience or knowledge and other qualities to enable him properly to assist in undertaking the preventive and protective measures.'

Provision of information to employees

[F5050] The responsible person must provide his employees with comprehensive and relevant information (*Regulatory Reform (Fire Safety) Order 2005 (SI 2005 No 1541), art 19*) on:

- the risks to them identified in the risk assessment;
- the preventive and protective measures;
- the procedures, including safety drills, to be followed in the event of serious and imminent danger;
- the identities of the staff nominated to assist with fire-fighting and the evacuation of relevant persons; and
- the risks in other parts of the building that have been notified to him.

With regard to dangerous substances on the premises, the responsible person must also provide his employees with:

- the details of such substances, including the name and the risk that it presents, access to relevant data sheets and information relating to any legislative provisions applying to those substances; and
- the significant findings of the risk assessment.

The information should be adapted to take into account the methods of work in use and be provided in a manner appropriate to the risk identified in the risk assessment.

Provision of information to other persons

[F5051] In addition to providing employees with information, the responsible person also has a duty (*Regulatory Reform (Fire Safety) Order 2005 (SI 2005 No 1541), art 20*) to inform non-employees in certain circumstances:

- the employer of employees working in or on the premises must be provided with comprehensible and relevant information on the risks to their employees and the preventive and protective measures that are in place. They must also be able to identify the competent people appointed to implement evacuation procedures; and
- non-employees working on the premises must be provided with appropriate instructions and comprehensible and relevant information regarding any risk to those persons. They too must be able to identify the competent people appointed to implement evacuation procedures.

Employment of a child

[F5052] Before employing a child the responsible person must (*Regulatory Reform (Fire Safety) Order 2005 (SI 2005 No 1541), art 19(2)*) provide a parent of the child with information on:

- the risks to that child identified by the fire risk assessment;
- the preventive and protective measures; and
- the risk in other parts of the building that have been notified to him.

Training

[F5053] The responsible person must (*Regulatory Reform (Fire Safety) Order 2005 (SI 2005 No 1541), art 21*) ensure that his employees are provided with adequate fire safety training:

- at the time when they are first employed; and
- on their being exposed to new or increased risks (such as being given a change of responsibilities or the introduction of new equipment, new work processes or new technology, or a new system of work or change in a system of work).

The training should:

- include suitable and sufficient instruction and training on the appropriate precautions and actions to be taken by the employee in order to safeguard themselves and other relevant persons on the premises;

[F5053] Fire Prevention and Control

- be repeated periodically where appropriate;
- be adapted to take account of new or changed risks to the safety of the employees;
- be provided in a manner appropriate to the risk identified by the risk assessment; and
- take place during working hours.

Co-operation and co-ordination

[F5054] Where two or more responsible persons share duties in respect of premises each must (*Regulatory Reform (Fire Safety) Order 2005 (SI 2005 No 1541), art 22*):

- co-operate with the other responsible persons(s) to enable them to comply with the requirements and prohibitions imposed on them by or under the Order;
- take reasonable steps to co-ordinate the measures taken; and
- inform the other responsible persons(s) of the risks to relevant persons in connection with their undertaking.

Where two or more responsible persons share premises where an explosive atmosphere may arise the responsible person with overall responsibility for the premises must co-ordinate the implementation of the measures to protect relevant persons from any risk from the explosive atmosphere.

Duties of employees at work

[F5055] In addition to the duties of the responsible person, every employee, while at work, must (*Regulatory Reform (Fire Safety) Order 2005 (SI 2005 No 1541), art 23*):

- take reasonable care for the safety of himself and other relevant persons who may be affected by his actions or omissions;
- co-operate with their employer to enable the employer to comply with his duties and requirements under the Order; and
- inform his employer or any other employee with specific responsibility for the safety of their fellow employees of any situation considered to be a serious and immediate threat to safety, or any matter which is considered to be a shortcoming in the employer's protection arrangements.

Enforcement

[F5056] *Regulatory Reform (Fire Safety) Order 2005 (SI 2005 No 1541), art 25* indicates that the enforcing authority for the Regulatory Reform Order is the fire and rescue authority for the area in which the premises are situated, apart from the following:

- The HSE enforces fire safety in:

(i) a ship in the course of construction, reconstruction, conversion or repair by persons other than the master and crew of the ship; and
(ii) construction sites (other than construction sites where the Office for Nuclear Regulation (ONR) is responsible for health and safety enforcement).
- The ONR enforces fire safety in nuclear installations (and construction sites where the ONR is the enforcing authority).
- The Defence Fire Service enforces fire safety in armed forces premises, premises occupied by any visiting force, international headquarters or defence organisations.
- The local authority enforces fire safety in sports grounds and stands.
- Fire inspectors enforce fire safety in Crown and United Kingdom Atomic Energy Authority premises, prisons and other types of premises used for detention or custody.

Every enforcing authority must (*SI 2005 No 1541, art 26*) enforce the provisions of the *Regulatory Reform (Fire Safety) Order 2005* and any regulations made under it. They may enforce the Order by the serving of one of three types of notices on the responsible person for the premises:

- alterations notices;
- enforcement notices; and
- prohibition notices.

Alterations notice

[F5057] An alterations notice (*Regulatory Reform (Fire Safety) Order 2005 (SI 2005 No 1541), art 29*) does not require the responsible person, or anyone else, to make alterations to the premises. It is served when the enforcing authority believes that the premises constitute a serious risk to relevant persons or there may be such a risk if a change is made to the premises or the use to which they are put.

The notice must state the matters which in the opinion of the enforcing authority constitute such a risk if the changes to the premises or their use is made.

Where such a notice has been served the responsible person must notify the enforcement authority of the proposed changes, and enclose a copy of the fire risk assessment and a summary of the changes he proposes to make to the existing general fire precautions if the alterations go ahead.

Enforcement notice

[F5058] If the enforcing authority is of the opinion that the responsible person has failed to comply with any provision of the Order or of any regulations made under it, the authority may serve on him an enforcement notice (*Regulatory Reform (Fire Safety) Order 2005 (SI 2005 No 1541), art 30*).

The enforcement notice must specify the provisions that have not been complied with and require the responsible person to take steps to remedy the failure within a stated period of time (but not less than 28 days).

An enforcement notice may include directions as to the measure that are considered necessary and give a choice between different ways of remedying the situation.

Prohibition notice

[F5059] If the enforcing authority is of the opinion that the use of the premises involves or will involve a risk to relevant persons so serious that use of the premises ought to be prohibited or restricted the authority may serve a prohibition notice (*Regulatory Reform (Fire Safety) Order 2005 (SI 2005 No 1541), art 31*) on the responsible person.

A prohibition notice must specify the matters which give rise to the risk and direct that the use to which the prohibition notice relates is prohibited or restricted to such extent as may be specified in the notice until the specified matters have been remedied.

A prohibition notice may include directions as to the measure which have to be taken to remedy the matters specified in the notice and any such measures should be framed so as to afford a choice between different ways of remedying the matters.

A prohibition notice takes effect immediately it is served if the enforcing authority is of the opinion that the risk of serious personal injury is sufficiently serious, otherwise it takes effect at the end of the time specified in the notice.

Offences and appeals

[F5060] It is an offence for a responsible person to fail to comply:

- with requirements or prohibitions, thereby placing one or more persons at risk of death or serious injury in case of fire;
- with any requirements of alteration notices, enforcement notices or prohibition notices; and
- in relation to the provision of luminous tube signs (*Regulatory Reform (Fire Safety) Order 2005 (SI 2005 No 1541), art 37*).

Any person guilty of an offence under points referred to above is liable:

- on summary conviction to a fine not exceeding the statutory maximum; or
- on conviction on indictment to a fine or to imprisonment for a term not exceeding two years, or to both.

It is an offence for any person to:

- fail to comply with the general duties of employees at work (*SI 2005 No 1541, art 23*) where that failure places one or more persons at risk of death or serious injury in case of fire;

- knowingly make a false entry into any register, book, notice or other document required to be kept under the Order;
- knowingly or recklessly give false information in response to any enquiry made under the Order;
- intentionally obstruct an inspector in the performance of his duties under the Order;
- fail without reasonable excuse to comply with any requirements imposed by an inspector under the Order;
- pretend, with intent to deceive, to be an inspector;
- fail to comply with the prohibition with regard to the charging of employees for things done or provided in connection with the Order (*SI 2005 No 1541, art 40*); and
- fail to comply with any prohibition or restriction imposed by a prohibition notice.

Any person guilty of other offences listed above is also liable on conviction to a fine, apart from failure to comply with a prohibition notice, which could result in a prison sentence.

Where an offence under this Order has been committed by a corporate body is proved to have been committed with the consent or connivance of any director, manager, secretary or similar officer of the body corporate he, as well as the corporate body is guilty of that offence and is liable to be proceeded against and punished accordingly.

Appeals against enforcement notices

[F5060.1] An appeal must be lodged against an alterations, enforcement or prohibition notice made under the *Regulatory Reform (Fire Safety) Order 2005 (SI 2005 No 1541)* within 21 days from the date of service. When an appeal is brought against an alterations or enforcement notice, the effect of an appeal is to suspend operation of the notice until the appeal is finally disposed of or withdrawn. When an appeal is made against a prohibition notice, the bringing of the appeal does not have the effect of suspending the operation of the notice unless the (magistrates) court so directs.

The court may either cancel or affirm the notice and may affirm it either in its original form or with such modifications as the court may see fit.

Application to the Crown

[F5061] As with other health and safety duties and regulations, generally speaking, statutory fire duties and fire regulations apply to the Crown and this continues with the introduction of the *Regulatory Reform (Fire Safety) Order 2005*, art 49. However, owing to Crown immunity in law, proceedings cannot be enforced against the Crown (see further E15045 ENFORCEMENT). This has the effect that the responsible person in Crown premises, that is, government buildings such as the Treasury and the Foreign Office as well as royal palaces, is required to comply with the requirements of the Order, including the preparation of a fire risk assessment. The articles relating to alterations and enforcement notices and consequent offences and appeals do not, however, apply.

Further guidance on specific types of premises

[F5061.1] A suite of guidance documents, 'Fire safety risk assessment guides', address the following occupancies:

- offices and shops;
- factories and warehouses;
- sleeping accommodation;
- residential care premises;
- educational premises;
- small and medium places of assembly;
- large places of assembly;
- theatres, cinemas and similar premises;
- open air events and venues;
- healthcare premises;
- transport premises and facilities; and
- animal premises and stables.

There are also guides on the hospitality sector and means of escape for disabled people and Fire safety risk assessment: five-step checklist. The guides can be found online and downloaded free of charge at www.gov.uk/government/collections/fire-safety-law-and-guidance-documents-for-business.

Regulatory Reform (Fire Safety) Order 2005 – a short guide to making your premises safe from fire is also available and can be downloaded at www.gov.uk/government/publications/making-your-premises-safe-from-fire. On 5 January 2016, responsibility for fire and rescue policies transferred from the Department for Communities and Local Government to the Home Office.

The dangerous substances and explosive atmospheres regulations (DSEAR) 2002

[F5062]–[F5072] The *Dangerous Substances and Explosive Atmospheres Regulations (DSEAR) 2002 (SI 2002 No 2776)* continue alongside the *Regulatory Reform (Fire Safety) Order 2005 (SI 2005 No 1541)* in that DSEAR requires employers and the self-employed to:

- carry out a risk assessment of any work activities involving dangerous substances;
- provide technical and organisational measures to eliminate or reduce, as far as is reasonably practical, the identified risks;
- provide equipment and procedures to deal with accidents and emergencies;
- provide information and training to employees; and
- classify places where explosive atmospheres may occur into zones and mark the zones where necessary.

Until the *Dangerous Substances and Explosive Atmospheres Regulations (DSEAR) 2002 (SI 2002 No 2776)* came into force, fire prevention measures for certain particularly dangerous processes were controlled by specific regulations. Most of these have been repealed or revoked and hazards are now controlled by the requirements of the *DSEAR*.

Regulations with specific fire safety requirements that are still on the statute book include:

- the *Offshore Installations (Prevention of Fire and Explosion, and Emergency Response) Regulations 1995 (SI 1995 No 743)* – to prevent and minimise the effects of fire and explosion on offshore installation;
- the *Construction (Design and Management) Regulations 2015 (SI 2015 No 51)*;
- the *Work in Compressed Air Regulations 1996 (SI 1996 No 1656)*, Reg 14; and
- the *Electricity at Work Regulations 1989 (SI 1989 No 635)*, Reg 6(d) – electrical equipment which may reasonably foreseeably be exposed to any flammable or explosive substance, must be constructed or protected so as to prevent danger from exposure.

Liability of occupier

[F5073] An occupier of premises where fire breaks out can be liable to:

- lawful visitors to the premises injured by the fire or falling debris (and is also liable to unlawful visitors, ie trespassers, as the principle of 'common humanity', enunciated in *Herrington v British Railways Board* [1972] 1 All ER 749 applies as does the *Occupiers' Liability Act 1984*, see OCCUPIERS' LIABILITY);
- fire fighters injured during fire-fighting operations; and
- adjoining occupiers.

In order to ensure therefore that occupiers and others involved may minimise their liability, insurance cover, though not compulsory, is highly desirable. The basic principles relating to fire cover are considered below (see **F5074**).

Fire insurance

[F5074] Insurance is intended to provide cover in the event of an accident that is not predictable. This applies to fire insurance and is reflected in the terms of insurance policies. If a fire occurs in a premises where it is thought that the insured might be involved in deliberately starting the fire, the insurance company may refuse to pay out under the policy. However, if the insured is innocent of any involvement, even if the fire was started deliberately, the insurer will pay out as arson is not predictable in the context of this issue. Where a high fire risk has been identified and adequate measures are not taken to manage the risk there may be a case for non-payment on the grounds of negligence. In practice insurance plays a very important role in the management of many risks in industry, including fire, and adequate appropriate fire cover is considered an essential part of modern business risk management.

First-Aid

Subash Ludhra

Introduction to first-aid

[F7001] The *Health and Safety (First-Aid) Regulations 1981 (SI 1981 No 917)*, which came into operation on 1 July 1982, require employers to have arrangements for the provision of first-aid in their place of work. The regulations were formally reviewed by the HSE in 2004/05, with agreement that no significant change was required at that time. However following further extensive consultation with employer duty holders, employees, first-aiders and first-aid training providers it was agreed that changes would be made to training requirements (shorter course duration for the first aid at work course and the introduction of the emergency first aid at work course) and a new approved code of practice and guidance was issued in 2009 and the new training requirement commenced on 1 October 2009.

The regulations and code of practice were further reviewed by the HSE and changes were made following a recommendation from 'Reclaiming Health and Safety for All: An independent review of health and safety legislation', by Professor Ragnar E Löfstedt, which was published in November 2011.

The main drivers for the change were to help businesses adopt proportionate first-aid arrangements suitable to their workplace. Significant changes from 1 October 2013 included the removal of the requirement for the HSE approval of first aid training and qualifications and the removal of the Approved code of practice status of the document to guidance only.

In 2018 the HSE made further minor amendments to clarify the significance of the 2013 amendment to regulation 3(2), which ended HSE's approval of first-aid training providers, updated the guidance on the use of automated external defibrillators AEDs, and blended learning in first-aid training and incorporated some additional amendments to take account of other previous legislative changes.

People can and do suffer injury or fall ill at work. This may or may not be as a result of work-related activity. However, it is important that they receive immediate attention.

Medical treatment should be provided at the scene promptly, efficiently and effectively before the arrival of any medical teams that may have been called. First-aid can save lives and can prevent minor injuries from becoming major ones. Employers are responsible for making arrangements for the immediate management of any illness or injury suffered by a person at work. First-aid at work covers the management of first-aid in the workplace – it does not include treating ill or injured people at work with medicines.

[F7001] First-Aid

However, the Regulations do not prevent specially trained staff taking action beyond the initial management of the injured or ill at work.

Interpretation

[F7001.1] First-aid means:

(1) In cases where a person will need help from a medical practitioner or nurse, treatment for the purpose of preserving life and minimising the consequences of injury and illness until such help is obtained and

(2) treatment of minor injuries which would otherwise receive no treatment or which do not need treatment by a medical practitioner or nurse.

The employer's duty to make provision for first-aid

[F7002] Employers must provide, or ensure that there is available equipment and facilities that are adequate and appropriate in the circumstances for enabling first-aid to be rendered to their employees if they are injured or become ill at work. Employers must also provide or ensure that there are an adequate and appropriate number of suitable persons (who can administer first aid to their employees if they become injured or ill) who:

(a) are trained and have such qualifications as may be appropriate in the circumstances of that case.

Where such a suitable person is absent in temporary and exceptional circumstances, an employer can appoint a person or ensure that a person is appointed:

(a) to take responsibility for first-aid in situations relating to an injured or ill employee who needs help from a medical practitioner or nurse;

(b) to ensure that equipment and facilities are adequate and appropriate in the circumstances;

(c) throughout the period of any such absence.

With regard to any period of absence of the first-aider, consideration must be given to:

(a) the nature of the undertaking;
(b) the number of employees at work; and
(c) the location of the establishment.

The assessment of first-aid needs

[F7003] Employers must assess their first-aid needs and requirements as appropriate to their particular circumstances (hazards and risks) relating to their workplace. Their principal aim must be to reduce the effects of injury or illness suffered at work, whether caused by the work itself or not. Adequate and appropriate first-aid personnel, equipment and facilities must be available at all times (taking into account alternative work patterns) for rendering assistance to persons with common injuries or illnesses and those likely to arise from specific hazards at work. Similarly, there must be adequate facilities for summoning an ambulance or other professional assistance.

Where first-aiders are provided in the workplace, the employer must ensure they have undertaken suitable training, have an appropriate first-aid qualification and remain competent to perform their role. First-aiders are likely to hold a valid certificate of competence in either first-aid at work (FAW) or emergency first-aid at work (EFAW). EFAW training enables a first-aider to give emergency first aid to someone who is injured or becomes ill while at work. FAW training includes EFAW and also equips the first-aider to apply first aid to a range of specific injuries and illnesses. Where the employer chooses to use qualifications other than FAW or EFAW to demonstrate workplace first-aid competence, it will be necessary for him to ensure that the common elements of the syllabus are taught. In addition, it may be appropriate for the employer to provide appointed persons or provide additional training above and beyond the FAW/EFAW where additional risks have been identified (see APPENDIX F). The four-layer framework for first aid provision is therefore:

- appointed person (AP);
- emergency first-aid at work (EFAW);
- first-aid at work (FAW);
- additional training.

The extent of first-aid provision in a particular workplace that an employer must make depends upon the circumstances of that workplace. There are no fixed levels of first-aid – the employer must assess what personnel and facilities are appropriate and adequate (see appendix B – assessment of first-aid needs). Employers with access to advice from occupational health services (internal or external to their organisation) may wish to use those resources for the purposes of conducting such an assessment and then take the advice given as to what first-aid provision would be deemed appropriate.

In workplaces that employ qualified medical doctors registered with the General Medical Council; nurses whose names are registered with the Nursing and Midwifery Council; or, paramedics registered with the Health Professions Council, those individuals can be utilised without additional first-aid training, provided that they can demonstrate current knowledge and skills in first aid.

Following the assessment, if the employer still decides a first-aider is not required in the workplace, a person should be appointed to take charge of the first-aid arrangements. The role of this appointed person includes looking after the first-aid equipment and facilities and calling the emergency services when required.

Although there is no legal requirement for the results of the first-aid risk assessment to be recorded in writing, it may be a useful exercise for the employer – for he may subsequently be asked to demonstrate how he came to the conclusion that the first-aid provision available is adequate or how he determined the number of first-aiders in place is appropriate for the workplace and its associated risk (see appendix C – record of first-aid provision).

When assessing first-aid needs, employers must consider the following:

- the nature of the work and workplace hazards and risks;
- the size of the organisation;
- the history of accidents in the organisation;
- the nature and distribution of the workforce;
- the distance from the workplace to emergency medical services;
- work patterns;
- travelling, distant and lone workers' needs and requirements;
- employees working on shared or multi-occupied sites;
- annual leave and other absences of first-aiders and appointed persons;
- first-aid provision for non employees.

Mental health first aid

[F7003.1] Many employers are training or considering training their staff in mental health first aid.

The MHFA programme is believed to have been developed in Australia in 2001 as a result of concerns that there was poor mental health literacy among the public, ie poor recognition of mental health disorders and lack of knowledge about appropriate responses and treatments. The purpose of the course was to equip members of the public to help others suffering with mental ill-health, or those experiencing a mental ill-health crisis.

MHFA came to England in 2007 and was launched under the Department of Health as part of a national approach to improving public mental health.

The MHFA training provided in England closely follows the process developed in Australia. The Department of Health subsequently encouraged employers in England to provide MHFA training as one of three steps in its 2012 'No Health Without Mental Health: Implementation Framework'.

MHFA is defined as:

> 'The help provided to a person developing a mental health problem or in a mental health crisis. The first aid is given until appropriate professional treatment is received or until the crisis resolves' (Kitchener and Jorm, 2002, cited in Kitchener and Jorm, 2008).

MHFA England state that their training courses teach people to spot in others symptoms of mental health, and to initiate help for these persons. 'We don't teach people to be therapists – but just like physical first aid, we teach people to listen, reassure and respond, even in a crisis'. The MHFA course in England includes the recognition of symptoms and risk factors in depressive, anxiety, psychotic and substance use disorders and associated mental ill-health and crisis situations; as well as suicidal thoughts and behaviours, panic attacks, experiencing a traumatic event, behaviour which is perceived as threatening, and issues surrounding drug overdosing. As in conventional first aid, an action plan is taught following ALGEE ('Assess risk of suicide or harm'; 'Listen non-judgementally'; 'Give reassurance and information'; 'Encourage the person to get appropriate professional help'; and 'Encourage self-help strategies'). Appropriate skills for these five actions are practised for each mental health disorder and crisis covered. In addition, employees undergoing the training are helped to recognise and to support colleagues experiencing mental ill-health or experiencing a crisis situation; including where to access further help, information, and additional professional support.

In 2018, the HSE issued their research report (RR1135), the summary of the evidence on the effectiveness of Mental Health First Aid (MHFA) training in the workplace, the report concluded that based on the published research, it is not possible to state whether MHFA training is effective in a workplace setting to improve the organisational management of mental-ill health. There is a lack of published occupationally-based studies, and the studies that have been conducted are limited in quality.

Based on the evidence reviewed, the following summary statements can be made:

- There is consistent evidence that MHFA training raises employees' awareness of mental ill-health conditions, including signs and symptoms.
- Those trained have a better understanding of where to find information and professional support, and are more confident in helping individuals experiencing mental ill-health or a crisis.
- There is no evidence from the published evaluation studies that the introduction of MHFA training in workplaces has resulted in sustained actions by those receiving the training or that it has improved the management of mental health in the workplace.
- There is limited evidence that the content of MHFA training has been considered for workplace settings.

The nature of the work and workplace hazards and risks

[F7004] *The Management of Health and Safety at Work Regulations 1999 (SI 1999 No 3242)* require employers to make a suitable and sufficient assessment of the risks to health and safety at work of their employees. The assessment must be designed to identify the measures required for controlling or preventing any risks to the workforce: highlighting what types of accidents or injuries are most likely to occur will help employers address such key questions as the appropriate nature, quantity and location of first-aid personnel and facilities.

Where the risk assessment conducted by an employer identifies a low risk to health and safety for example in a shop or an office, employers may only need to provide (i) a first-aid box clearly identified and suitably stocked, and (ii) an appointed person to look after first-aid arrangements and resources, and to take control in emergencies. However even in low risk environments it is still possible for an accident or sudden illness to occur and it is recommended that employers consider having a qualified first-aider available.

Where risks to health and safety are greater, employers may need to consider the following:

- the provision of an adequate number of trained first-aiders so that first-aid can be given immediately;
- the additional training of first-aiders in specialist skills to deal with specific risks or hazards;

- informing the local emergency services in writing of the risks and hazards on the site where hazardous substances or processes are in use;
- the requirement for additional first-aid equipment;
- the provision of one or more first-aid room(s).

Employers will need to consider the different risks within each part of their company or organisation. Where an organisation occupies large premises with different processes being performed in different parts of the premises or within different buildings, each area's or buildings risks must be assessed separately. It would not be appropriate to conduct a generic assessment of needs to cover a variety of activities – the parts of the building with higher risks will need greater first-aid provision than those with lower risk.

A list of common hazards found in workplaces can be found in appendix A.

The size of the organisation

[F7005] In general, the level of first-aid provision that is required will increase according to the number of employees present. Employers should be aware, however, that in some organisations there may be few employees but the risks to their health and safety might be high – and, as a result, their first-aid needs will be greater.

The history of incidents and accidents

[F7006] When assessing first-aid needs, employers might find it useful to collate data on accidents and near misses that have occurred in the past and then analyse them, for example, the numbers and types of accidents or near misses, their frequency and consequences. Organisations with large premises should refer to such information when determining the first-aid equipment, facilities and personnel that are required to cover specific areas.

The nature and distribution of the workforce

[F7007] The particular needs of young workers, trainees, temporary workers, students on work experience, pregnant workers, employees with disabilities or other special health problems should be addressed. Consideration must also be given to the gender of the employees and any ethnic or cultural needs that may need to be addressed including language barriers.

The employer should bear in mind that the size of the premises can affect the time it might take a first-aider to reach an incident. If there are a number of buildings on the site, or the building in question comprises several storeys, the most suitable arrangement might be for each building or floor to be provided with its own first-aiders.

The distance from the workplace to emergency medical services

[F7008] Where a workplace is far from emergency medical services, it may be necessary to make special arrangements for ensuring that appropriate transport can be provided for taking an injured person to the emergency medical services or providing additional facilities at the workplace to treat the injured person until the emergency medical services arrive.

In every case where the place of work is remote, the very least that an employer should do is to give written details to the local emergency services of the layout of the workplace, and any other relevant information, such as information on specific hazards.

Work patterns

[F7008.1] Where employees work shifts or work out of normal working hours it is important to ensure that sufficient provision is always available when they are at work. It may be necessary to have separate arrangements for each shift.

The needs and requirements of travelling, distant and lone workers

[F7009] Employers are responsible for meeting the first-aid needs of their employees whilst they are working away from their main company premises.

When assessing the needs of staff who travel long distances or who are constantly mobile, consideration should be given to what they are required to do, the hazards and risks they are typically exposed to and their medical condition to help decide whether they ought to be provided with a personal first-aid kit.

Organisations with staff working in remote areas or working alone must make special arrangements for those employees in respect of communications (this may include providing mobile phones), special training and arranging emergency transport.

Personnel on sites that are shared or multi-occupied

[F7010] Employers with personnel working on shared or multi-occupied sites can agree to have one employer on the site who is solely responsible for providing first-aid cover for all of the workers. It is strongly recommended that this agreement is written to avoid confusion and misunderstandings between the employers. It will highlight the risks and hazards of each company on the site and will ensure that the shared provision is suitable and sufficient. After the employers have agreed the arrangement, the personnel must be informed accordingly.

When employees are contracted out to other companies, their employer must ensure they have access to first-aid facilities and equipment. The host employer bears the responsibility for providing such facilities and equipment and making the employees aware of it.

First-aiders on annual leave or absent from the workplace

[F7011] Adequate provision of first-aid must be available at all times. Employers should therefore ensure that the arrangements they make for the provision of first-aid at the workplace are adequate to cover for any annual leave of their first-aiders or appointed persons. Such arrangements must also be able to cover for any unplanned or unusual absences from the workplace of first-aiders or appointed persons.

First-aid provisions for non employees

[F7012] Employers are not obliged by these regulations to provide first-aid cover or facilities for members of the public, as the regulations are aimed at employees. However, many organisations like health authorities, schools and colleges, places of entertainment, fairgrounds and shops do make first-aid provision for persons other than their employees, and this practice is strongly recommended by the HSE. In addition to its general guidance on first-aid at work the HSE has also produced more specific guidance, which incorporates first-aid, in relation to specific activities/sectors where there might be a large public presence, these include, Health and safety in swimming pools HSG179 2018 HSE Books ISBN 9780717666775, Fairgrounds and amusement parks – guidance on safe practice. Practical guidance on the management of health and safety for those involved in the fairgrounds industry, HSG175 2017 HSE Books ISBN 9780717666638 and Health and safety in care homes, HSG220 2014 HSE Books ISBN 9780717663682. The Department for children schools and families have produced guidance on first-aid provisions in schools and the *Road Traffic Act 1988* regulates first-aid provision on buses and coaches. The Events Industry Forum (EIF) produces the (purple) guide to staging events (which replaced the HSE's a guide to health, safety and welfare at music and similar events).

Where employers extend their first-aid provision to cover more than merely their employees, the provision for the employees must not be diminished and should not fall below the standard required by these regulations or the level of provision stipulated for non-employees under any other relevant legislation and guidance.

The compulsory element of employers' liability insurance does not automatically cover litigation resulting from first-aid given to non-employees. It is advised that employers check their public liability insurance policy on this issue.

Reassessing first-aid needs

[F7013] In order to ensure that first-aid provision continues to be adequate and appropriate, employers should periodically review the first-aid provision in the workplace, particularly when changes have been made to working practices. It is advisable to record details of any reviews made and their findings.

Duty of the employer to inform employees of first-aid arrangements

[F7014] Employers are under a duty to inform employees of the arrangements that have been made for first-aid in their workplace. This may be achieved by:

- Distributing guidance to all employees which highlights the key issues in the first-aid arrangements, such as listing the names of all first-aiders and describing where first-aid resources are located.
- Nominating key employees to ensure that the guidance is kept up to date and is distributed to all staff, and to act as information officers for first-aid in the workplace.
- Internal memos can be used as a method of keeping the personnel informed of any changes in the first-aid arrangements.
- Displaying announcements up on notice boards informing employees of the first-aid arrangements and of any changes to those arrangements.
- Providing new, or transferring, employees with the information as part of their induction training.

Any person(s) with reading or language difficulties must be provided the information in a way that they can understand.

First-aid and the self-employed

[F7015] The self-employed should provide, or ensure that there is provided, such equipment, if any, as is appropriate in the circumstances to enable them to render first-aid to themselves whilst at work. The self-employed who work in low-risk areas, for example at home, are required merely to make first-aid provision appropriate to a domestic environment.

Where self-employed persons are exempt from Health and Safety Regulations under the *Health and Safety at Work etc Act 1974 (General Duties of Self-Employed Persons) (Prescribed Undertakings) Regulations 2015*, there is no obligation to make any first-aid provision, however, they should consider making basic arrangements as stated in the previous paragraph.

When self-employed people work together on the same site, they are each responsible for their own first-aid arrangements. If they wish to collaborate on first-aid provision, they may agree a joint arrangement to cover all personnel on that particular site.

Number of first-aiders

[F7016] Sufficient numbers of first-aiders should be located strategically on the premises to allow for the administration of first-aid quickly when the occasion arises. The assessment of first-aid needs (see appendix B) may have helped to highlight the extent to which there is a need for first-aiders. Appendix E gives suggested numbers of first-aiders (FAW/ EFAW) or appointed persons who should be available at the workplace. The suggested numbers are not a legal requirement – they are merely for guidance.

Although there are no hard and fast rules on numbers, where 25 or more people are employed, even in low-hazard environments, at least one first aider should be provided. Employers will have to make a judgement, taking into account all of the relevant circumstances of their organisation. If the company is a long way from a medical facility or there are shift workers on site or the premises cover a large area, the numbers of first-aid personnel set out below may not be sufficient – the employer may have to make provision for a greater number of first-aiders to be on site.

Selection of first-aiders

[F7017] First-aiders must be reliable and of a good disposition. Not only should they have good communication skills, but they must also possess the ability and aptitude for acquiring new knowledge and skills, and must be able to handle stressful and physically demanding emergency incidents and procedures. Their position in the company should be such that they are able to leave their place of work immediately to respond to an emergency.

Employers should note that trained first-aiders are also more likely to promote health and safety awareness in the workplace and can have a positive impact in improving the overall safety culture within an organisation.

The training and qualifications of first-aid personnel

[F7018] Before taking up their first-aid duties, a first-aider should hold a valid certificate of competence in either;

- first-aid at work (FAW) issued by a training organisation; or
- emergency first-aid at work (EFAW) issued by a training organisation or a recognised awarding body of Ofqual/Scottish Qualifications Authority.

Appendix D details the contents of the FAW and EFAW courses.

Where possible organisations that are contracted to train first-aid personnel should be notified of any particular risks or hazards in a workplace so that the first-aid course provided can be tailored to include the risks specific to that workplace.

Additional special training may be undertaken to deal with unusual risks and hazards (the content of these additional training courses is not specified by the HSE. It may be undertaken as an extension to the FAW/EFAW training or as a standalone course and a certificate should be issued separately from the FAW/EFAW certificate). This will enable the first-aider to be competent in dealing with such risks.

It is important for employers to understand that FAW/EFAW certificates are valid for three years.

Employers need to arrange retraining before certificates expire. The FAW requalification course lasts two days (should cover the same content as the initial FAW course). If the first-aider does not retrain or re-qualify before the expiry date on their current certificate they are no longer considered competent to act as a first-aider in the workplace. They can re-qualify at any time after the expiry date by undertaking the two-day re-qualification course. However, it may be sensible to complete the three-day FAW course, especially where a considerable time period (in excess of one month) has elapsed since the FAW certificate expired. Employers must decide the most appropriate training course to re-qualify the first-aider. An EFAW requalification course should be of the same duration and content as the initial EFAW course.

In addition to re-qualification the HSE strongly recommended that employers provide their employees with annual refresher training for FAW/EFAW courses (see appendix D for typical content).

It is advisable for employers to keep a record of first-aiders in the company, together with their certification dates, in order to assist them in organising refresher/re-qualification training. Employers should develop a programme of knowledge and skills training for their first-aiders to enable them to be updated on new skills and to make them aware of suitable sources of first-aid information, such as occupational health services and training organisations qualified by the HSE to conduct first-aid at work training.

Bodies/Organisations concerned with the delivery of first-aid-at-work training

[F7018.1] Some first-aid training providers may choose to operate through voluntary accreditation schemes whose intention is to set and maintain standards in line with HSE requirements. These schemes are not mandatory and employers may decide to choose an independent training organisation. However, these bodies may help employers select training organisations who offer a standard of training with appropriate content, suitable trainers and assessors, and relevant and robust quality assurance systems.

The HSE does not verify the level of assurance that an employer can assume when a training organisation is a member of these voluntary accreditation schemes, with the exception of those offering 'regulated qualifications' (those that are nationally recognised and can be obtained from a training centre for an 'awarding organisation'). These awarding organisations (AOs) are recognised by qualification regulators (Ofqual, SQA or the Welsh Government), They have dedicated policies and quality assurance processes and must approve and monitor their training centres to ensure training meets a certain standard. Regulators stipulate that AOs and their training centres must work in compliance with the Assessment Principles for First Aid Qualifications and other key criteria, including the competence of trainers and assessors and the content of quality assurance systems.

Employers may obtain appropriate training from the Voluntary Aid Societies (St John Ambulance, British Red Cross and St Andrew's First Aid) who together are acknowledged by HSE as one of the standard-setters for currently accepted first-aid practice as far as they relate to the topics covered in FAW and EFAW training courses. The Voluntary Aid Societies work to similar principles of assessment and employ a similar hierarchy of policies and processes to AOs.

[F7018.1] First-Aid

Where an employer selects a training provider affiliated to other voluntary accreditation schemes, they will need to be confident that the provider will deliver training with appropriate content (ie content identified as being appropriate within the needs assessment, and/or content in line with Appendices 5 and 6), use suitable trainers and assessors, and has relevant and robust quality assurance systems in place. (Similar quality standards are also expected of all first-aid training providers including non-affiliated, independent first-aid training organisations.) All training providers should be able and prepared to demonstrate how they satisfy these criteria.

Selecting a training provider

[F7018.2] Employers must choose who they wish to engage when training their first aiders. The HSE provides guidance (GEIS3) for employers to assist them in selecting a competent first-aid training provider. Employers should satisfy themselves that any training providers engaged are using training material and teaching the first-aid management of injuries and illness as covered in FAW/EFAW training courses and in accordance with current guidelines published by the Resuscitation Council (UK); and the current edition of the first-aid manual of the Voluntary Aid Societies or other published guidelines provided they are in line with the two above or supported by a responsible body of medical opinion.

Where an employer's first-aid needs assessment identifies, or an employer chooses to use qualifications other than FAW or EFAW to demonstrate workplace first-aid competence, it will be necessary for an employer to ensure that common elements of the syllabus are taught in accordance with the same guidelines. Where an employer decides to provide this training in-house, they will need to establish that it is appropriate by ensuring that the content reflects the content of the FAW or EFAW qualifications listed in Appendix D and is delivered in accordance with currently accepted standards for first aid. In-house individuals acting as trainers/assessors should have the necessary skills, qualifications and competence expected of those working for an external training provider. A quality assurance system will be needed to ensure that the competence of trainers/assessors is regularly reviewed by competent 'verifiers'. These systems will need to be re-viewed on an annual basis by a competent person independent of those directly involved in the delivery/assessment of this training

Employers should consider maintaining records of training; they should be retained for a minimum of three years after the assessment process has been completed.

Although there is no requirement for the checks employers carry out when choosing a training provider to be formalised or written down, It may be useful for employers to retain a written record to confirm the competence of a training organisation, and retaining a record of those checks will help employers demonstrate to a HSE or local authority inspector how they selected a training provider.

Some training providers offer blended learning courses which are a combination of face to face and e-learning. Whilst this is an acceptable method of

Appointed persons [F7019]

training, employers must satisfy themselves that this method would be suitable for them, additional checks would include ensuring that:

- the individuals being trained are familiar with the technology being used and are able to use it;
- the training provider has an adequate means of supporting individuals during their training;
- the training provider has a robust system in place to prevent identity fraud;
- the training provider has an appropriate means of assessing the e-learning component of the training.

Irrespective of the way the training is delivered, employers must ensure that adequate time is set aside during the working day to undertake any first aid training employees receive.

Certificates

[F7018.3] In order for first aiders to be able to demonstrate that they have the necessary competence in first aid they will need to hold a certificate that contains all of the following (minimum) information:

- The name of training organisation;
- The name of qualification;
- Their name;
- A validity period for three years from date of course completion;
- An indication that the certificate has been issued for the purposes of complying with the requirements of the *Health and Safety (First-Aid) Regulations 1981*;
- A statement that the teaching was delivered in accordance with currently accepted first-aid practice;
- If the qualification is neither FAW nor EFAW, an outline of the topics covered should be stipulated (this may be on the reverse or as an appendix to the certificate).

Where an alternative qualification is identified in place of FAW/EFAW in the employer's needs assessment, the employer will need to seek assurance that the standard of training received and the competence of the organisation which delivered the training meet the necessary criteria.

Appointed persons

[F7019] An appointed person is an individual who takes charge of first-aid arrangements for the company, including looking after the facilities and equipment and calling the emergency services when required. The appointed person is allocated these duties when it is found through the first-aid needs assessment that a first-aider is not necessary. The appointed person is the minimum requirement an employer can have in the workplace. Clearly, even if the company is considered a low health and safety risk and, in the opinion of the employer, a first-aider is unnecessary, an accident or illness still may occur, therefore somebody should be nominated to call the emergency services, if required.

Appointed persons are **not** first-aiders - and therefore they should not be called upon to administer first-aid if they have not received the relevant training. However, employers may consider it prudent to send their appointed persons on an appropriate training course.

The only time an appointed person can replace a first-aider is when the first-aider is absent, due to circumstances that are temporary, unforeseen and exceptional. Appointed persons cannot replace first-aiders who are on annual leave. If the first-aid assessment has identified a requirement for first-aiders, they should be available whenever there is a need for them in the place of work.

Records and record keeping

[F7020] It is considered good practice to keep records of incidents that required the attendance of a first-aider and treatment of an injured person. It is advisable for smaller companies to have just one record book, but for larger organisations this may not be practicable and more than one may be needed, in these circumstances each book should be numbered to aid identification. Any such books should be kept in accordance with the Data Protection Act 1998.

The data entered in the record book should include the following:

- the date, time and place of the incident;
- the injured person's name and job title;
- a description of the injury or illness and of the first-aid treatment administered;
- details of where the injured person went after the incident, namely hospital, home or back to work;
- the name and signature of the first-aider or person who dealt with the incident.

This information may be collated to help the employer improve the environment with regard to health and safety in the workplace. It could be used to help determine future first-aid needs assessment and will be helpful for insurance and investigative purposes. The statutory accident book is not the same as the first-aid record book but they may be combined (provided the requirement of each are still met).

Remember employers, the self employed and controllers of premises have a duty to report some accidents and incidents at work under the Reporting of Injuries, Diseases and Dangerous Occurrences Regulations 2013 (RIDDOR) and any data or records held must be held in accordance with the *Data Protection Act 1998*.

First-aid resources

[F7021] Having completed the first-aid needs requirements assessment, the employer must provide the resources; that is the equipment, facilities, materials and time for first-aiders to carry out their duties, which will be needed to ensure that an appropriate level of cover is available to the employee's at all relevant times. First-aid equipment, suitably marked and obtainable, must be made available at specific places where working conditions require it.

First-aid containers

[F7022] First-aid equipment must be suitably stocked and contained in a properly identifiable container (white cross on a green background). At least one first-aid container with a sufficient quantity of first-aid materials must be made available for each worksite – this is the minimum level of first-aid equipment. Larger premises, for example, will require the provision of more than one container.

First-aid containers should be easily accessible and, where possible, near hand-washing facilities. The containers should be used only for first-aid equipment. Tablets and medications (potions, lotions, creams or sprays) should not be kept in them. The first-aid materials within the containers should be protected from damp and dust. It may be practical to provide first-aiders with their own individual containers and then make them responsible for their own specific containers.

Having completed the first-aid needs assessment, the employer will have a good idea as to what first-aid materials should be stocked in the first-aid containers. If there is no specific risk in the workplace, a minimum stock of first-aid materials would normally comprise the following (there is no mandatory list):

- a leaflet giving guidance on first-aid (eg HSE leaflet *Basic advice on first-aid at work*);
- 20 individually wrapped sterile plasters (assorted sizes), appropriate to the type of work being carried out and the needs of the employees);
- two sterile eye pads;
- two individually wrapped triangular bandages (preferably sterile);
- six safety pins;
- six medium-sized individually wrapped sterile unmedicated wound dressings – approximately 12 cm × 12 cm;
- two large sterile individually wrapped unmedicated wound dressings – approximately 18 cm × 18 cm;
- at least three pairs of disposable gloves (you may need to consider issues related to latex).

This list is a suggestion only – other equivalent materials will be deemed acceptable.

An examination of the first-aid kits should be conducted frequently (ie a recorded monthly inspection). Stocks should be replenished as soon as possible after use and ample back-up supplies should be kept on the company premises. Any first-aid materials found to be out of date should be carefully discarded.

Employers may also wish to refer to British Standard BS 8599 which provides further information on the contents of workplace first-aid kits. However irrespective of which kits are used their contents should reflect the outcome of the first-aid needs assessment.

Additional first-aid resources

[F7023] If the results of the assessment suggest a need for additional resources such as scissors, adhesive tape, disposable aprons, thermal blankets or individually wrapped moist wipes, they can be kept in the first-aid container if space allows. Otherwise they may be kept in a different container, as long as they are ready for use if required.

If the assessment highlights the need for such items as protective equipment, they must be securely stored next to first-aid containers or in first-aid rooms or in the hazard area itself. Only persons who have been trained to use these items may be allowed to use them.

If there is a need for eye irrigation and mains tap water is unavailable, at least a litre of sterile water or sterile normal saline solution (0.9%) in sealed, disposable containers should be provided. If the seal is broken, the containers should be disposed of and not reused. Such containers should also be disposed of when their expiry date has been passed.

There may be a need for items such as protective equipment in case first-aiders have to enter dangerous atmospheres, or calcium gluconate for the management of hydrofluoric acid burns. These items should be stored securely near the first-aid container, in the first-aid room or in the hazard area, as appropriate. Access to them should be restricted to people trained in their use.

As stated, first aid at work does not include giving tablets or medicines to treat illness. However, the only exception to this is where aspirin is used as first aid to a casualty with a suspected heart attack in accordance with currently accepted first-aid practice. It is recommended that tablets and medicines should not be kept in the first-aid container

Some employees may have their own medication that has been prescribed by their doctor. If an individual needs to take their own prescribed medication, the first-aiders role will be limited to helping them to do so and then contacting the emergency services as appropriate.

Automated External Defibrillators (AED)

[F7023.1] Whilst it is not a legal requirement to provide an AED, it is generally agreed that they can be invaluable in helping to save life. Modern AEDs are reliable and safe. If the employer decides to provide an AED, it is important to ensure that those people who are appropriately trained and authorised to use it do so. Employers will need to ensure that staff are appropriately trained to use the equipment provided. However the Resuscitation Council (UK) state that the lack of training (or recent refresher training) should not be a barrier to someone using a modern defibrillator, they state that 'If you are prepared to use the AED do not be inhibited from doing so'. The

HSE are also recommending that all first-aid training course providers include AED usage training as part of their course content and assessment process.

First-aid kits for travelling

[F7024] Based on the needs assessment, employers should consider issuing these types of kits to all mobile members of staff or, alternatively, placing them in vehicles used by mobile members of staff for business purposes. The assessment should also consider whether these employees should undergo a course of instruction in emergency first aid at work, particularly if they are involved in higher-hazard activities. First-aid kits for travelling may contain the following items:

- a leaflet giving general guidance on first-aid (eg HSE leaflet *Basic advice on first-aid at work*);
- six individually wrapped sterile plasters;
- one large sterile unmedicated dressing – approximately 18 cm x 18 cm;
- two triangular bandages;
- two safety pins;
- individually wrapped moist cleansing wipes;
- two pairs of disposable gloves.

This list is a suggestion only – it is not mandatory. However, if a kit is provided it must then be regularly inspected and topped up from a back-up store at the home site.

Rooms designated as first-aid areas

[F7025] A suitable room, or rooms, should be made available for first-aid purposes where the first-aid needs assessment deems such a room, or rooms, to be necessary. Such room(s) should have sufficient first-aid resources, be easily accessible to stretchers and be easily identifiable and where possible should be used only for administering first-aid.

First-aid rooms are normally necessary in organisations operating within high-risk industries. Therefore they would be deemed to be necessary on chemical, ship building and large construction sites or on large sites remote from medical services. A person should be made responsible for the first-aid room.

On the door of the first-aid room a list of the names and telephone extensions of all of the first-aiders should be displayed, together with details of how and where they may be contacted on site.

First-aid rooms should:

- have enough space to hold a couch with space in the room for people to work, a desk, a chair and any other resources found necessary;
- where possible, be near an access point in the event that a person needs to be taken to hospital;
- have heating, lighting and ventilation;

- have surfaces that can be easily washed;
- be kept clean and tidy; and
- be available and ready for use whenever employees are in the workplace.

The following is a list of resources that may be found in a first-aid room:

- a record book for logging incidents where first-aid has been administered;
- a telephone;
- a storage area for storing first-aid materials;
- a bed/couch with waterproof protection and clean pillows and blankets;
- a chair;
- a foot-operated refuse bin with disposable yellow clinical waste bags or some receptacle suitable for the safe disposal of clinical waste or sharps;
- a sink that has hot and cold running water – also drinking water and disposable cups;
- soap and some form of disposable paper towel.

Where first-aid rooms are provided, employers must also make provision for regular cleaning, emptying of bins (including sharp objects and bodily fluids) and for the laundering of any bed linen or blankets used. If the designated first-aid room has to be shared with the working processes of the company, the employer must consider the implications of the room being needed in an emergency and whether the working processes in that room could be stopped immediately. Can the equipment in the room be removed in an emergency so as not to interfere with any administration of first-aid? Can the first-aid resources and equipment be stored in such a place as to be available quickly when necessary? Lastly, the room must be appropriately identified and, where necessary, be signposted With signage complying with the Health and Safety (Safety Signs and Signals) Regulations 1996.

Appendix A

Common hazards found in the workplace

Hazard	Causes of accidents	Examples of injury requiring first-aid
Chemicals	Exposure during handling, spillages, splashing, leaks	Poisoning, loss of consciousness, burns, eye injuries
Electricity	Failure to securely isolate electrical systems and equipment during work on them, poorly maintained electrical equipment, contact with either overhead power lines, underground power cables or mains electricity supplies, using unsuitable electrical equipment in explosive atmospheres	Electric shock, burns
Machinery	Loose hair or clothing becoming tangled in machinery, being hit by moving parts or material thrown from machinery, contact with sharp edges.	Crush injuries, amputations, fractures, lacerations, eye injuries
Manual handling	Repetitive and/or heavy lifting, bending and twisting, exerting too much force, handling bulky or unstable loads, handling in uncomfortable working positions.	Fractures, lacerations, sprains and strains
Slip and trip hazards	Uneven floors, trailing cables, obstructions, slippery surfaces due to spillages, worn carpets and mats.	Fractures, sprains and strains, lacerations
Work at height	Over-reaching or over-balancing when using ladders, falling off or through a roof.	Head injury, loss of consciousness, spinal injury, fractures, sprains and strains
Workplace transport	Hit by, hit against or falling from a vehicle, being hit by part of a load falling from a vehicle, being injured as a result of a vehicle collapse or overturn.	Crush injuries, fractures, sprains and strains

Appendix B

Assessment of first-aid needs – checklist

The minimum first-aid provision for each worksite is:

- a suitably stocked first-aid container;
- an appointed person to take charge of first-aid arrangements;
- information for employees on first-aid arrangements.

This checklist below will help you assess what additional first-aid provision you need to make for your workplaces.

Factors to Consider	Impact on first-aid provision
Hazards – use the findings of your risk assessment and take account of any parts of your workplace that have different work activities/ hazards which may require different levels of first-aid provision	
Does your workplace have low hazards such as those that might be found in offices and shops?	The minimum provision is: (*a*) an appointed person to take charge of first-aid arrangements; (*b*) a suitably stocked first-aid box.
Does your workplace have higher hazards such as chemicals or dangerous machinery (see Table 1)? Do your work activities involve special hazards such as hydrofluoric acid or confined spaces?	You should consider: (*a*) providing first-aiders; (*b*) additional training for first-aiders to deal with injuries resulting from special hazards; (*c*) additional first-aid equipment (*d*) precise siting of first-aid equipment; (*e*) providing a first-aid room; (*f*) informing the emergency services in advance.
Employees How many people are employed on site?	Where there are small numbers of employees, the minimum provision is: (*a*) an appointed person to take charge of first-aid arrangements; (*b*) a suitably stocked first-aid box. Even in workplaces with a small number of employees, there is still the possibility of an accident or sudden illness so you should consider providing a qualified first-aider. Where there are large numbers of employees (ie more than 25) you should consider providing: (*a*) first-aiders; (*b*) additional first-aid equipment; (*c*) a first-aid room.

Assessment of first-aid needs – checklist

Are there inexperienced workers on site, or employees with disabilities or special health problems?	You should consider: (a) additional training for first-aiders; (b) additional first-aid equipment; (c) local siting of first-aid equipment. Your first-aid provision should cover any work experience trainees.

Record of accidents and ill health

What is your record of accidents and ill health? What injuries and illness have occurred and where did they happen?	Ensure your first-aid provision will cater for the type of injuries and illness that might occur in your workplace. Monitor accidents and ill health and review your first-aid provision as appropriate.

Working arrangements

Do you have employees who travel a lot, work remotely or work alone?	You should consider: (a) issuing personal first-aid kits; (b) issuing personal communicators to remote workers; (c) issuing mobile phones to lone workers.
Do any of your employees work shifts or work out of hours?	You should ensure there is adequate first-aid provision at all times people are at work.
Are the premises spread out, for example are there several buildings on the site or multi-floor buildings?	You should consider provision in each building or on each floor.
Is your workplace remote from emergency medical services?	You should: (a) consider special arrangements with the emergency services; (b) inform the emergency services of your location.
Do any of your employees work at sites occupied by other employers?	You should make arrangements with other site occupiers to ensure adequate provision of first-aid. A written agreement between employers is strongly recommended.
Do you have sufficient provision to cover absences of first-aiders or appointed persons?	You should consider: (a) consider special arrangements with the emergency services; (b) what cover is needed for unplanned and exceptional absences.

Non-employees

Do members of the public visit your premises?	Under the Regulations, you have no legal obligation to provide first-aid for non-employees but HSE strongly recommends that you include them in your first-aid provision. This is particularly relevant in workplaces that provide a service for others such as schools, places of entertainment, fairgrounds and shops.

Assessment of first-aid needs – checklist

Do not forget to allow for leave or absences of first-aiders and appointed persons. First-aid personnel must be available at all times when people are at work.

Appendix C

Record of First-aid Provision

First-aid personnel	Required yes / no	Number needed
First-aider with a first-aid at work certificate		
First-aider with an emergency first-aid at work certificate		
First-aider with additional training		
Appointed person		
First-aid equipment and facilities	**Required yes / no**	**Number needed**
First-aid container		
Additional equipment (specify)		
Travelling first-aid kit		
First-aid room		

Appendix D

Content of a first-aid at work course

Content of an Emergency first aid at work (EFAW) course

On completion of training, successful candidates should be able to:

- Understand the role of the first-aider, including reference to the importance of preventing cross infection, the need for recording incidents and actions and the use of available equipment;
- Assess the situation and circumstances in order to act safely, promptly and effectively in an emergency;
- Administer first aid to a casualty who is unconscious (including seizure);
- Administer cardiopulmonary resuscitation and use an automated external defibrillator;
- Administer first aid to a casualty who is choking;
- Administer first aid to a casualty who is wounded and bleeding;
- Administer first aid to a casualty who is suffering from shock;
- Provide appropriate first aid for minor injuries (including small cuts, grazes and bruises, minor burns and scalds, small splinters).

Content of a first aid at work (FAW) course

On completion of training, whether a full FAW course or a FAW requalification course, successful candidates should have satisfactorily demonstrated competence in all of the subject areas listed above and also to be able to:

- Administer first aid to a casualty with;
    - injuries to bones, muscles and joints, including suspected spinal injuries; chest injuries; burns and scalds; eye injuries; sudden poisoning; anaphylactic shock;
    - Recognise the presence of major illness and provide appropriate first aid (including heart attack, stroke, epilepsy, asthma, diabetes).

Typical content of a first aid refresher course

- Assess the situation and circumstances in order to act safely, promptly and effectively in an emergency.
- Administer first aid to a casualty who is unconscious (including seizure).
- Administer cardiopulmonary resuscitation.
- Administer first aid to a casualty who is wounded and bleeding.
- Administer first aid to a casualty who is suffering from shock.

Appendix E

Guide to the category and number of first-aid personnel to be available at all times people are at work

First-aid personnel Category of risk	Numbers employed at any location	Suggested number of first-aid personnel
Lower hazard – eg shops, offices, libraries	less than 25	at least one appointed person**
	25–50	at least one first-aider trained in EFAW
	more than 50	at least one first-aider (FAW) per 100 employees or part thereof
Higher Hazard – eg light engineering and assembly work, food processing, warehousing, extensive work with dangerous machinery or sharp instruments, construction, slaughterhouse, chemical manufacturer, work involving special hazards* such as hydrofluoric acid or confined spaces.	less than 5	at least one appointed person**
	5–50	at least one first-aider (EFAW or FAW)***
	more than 50	at least one first-aider (FAW) per 50 employees or part thereof

* additional training may be needed for first-aiders to deal with injuries resulting from special hazards.

** where first-aiders are shown to be unnecessary, there is still a possibility of an accident or sudden illness, so employers should consider providing qualified first-aiders.

*** the type of injuries that might arise in working with those hazards identified, will influence whether the first-aider should be trained in FAW or EFAW.

Appendix F

Table of examples of additional training needs

Additional training	When additional training may be relevant
Management of a casualty suffering from hypothermia or hyperthermia	Extensive exposure to the outdoor environment due to, for example, regular maintenance activity, eg trackside rail work, forestry
Management of a casualty suffering from hydrofluoric acid burns	Glass industry, chemical manufacture, or other industries using pickling pastes containing hydrofluoric acid
Management of a casualty suffering from cyanide poisoning	Chemical manufacture
Oxygen administration	Confined space work, for example tank cleaning operations and working in sewers. Also, where there is a risk of exposure to hydrogen cyanide
Management of a drowning casualty	Swimming pools, fish farms
Use of an automated external defibrillator	All sectors where you have decided that the presence of a defibrillator may be beneficial through a needs assessment
Recognise the presence of major illness and provide appropriate first aid (including heart attack, stroke, epilepsy, asthma, diabetes)	Wherever the environment is low hazard but you have identified a risk, either based on the known health profile, age and number of employees or a need to consider members of the public
Paediatric first aid, as required by the Department for Education or local authorities, which complies with the syllabus produced by OFSTED for first-aid provision for children in a school or other childcare setting	Schools and nurseries

Food Safety and Standards

Neville Craddock and Andrea Oates

Introduction

[F9001] Legislation has governed the sale of food for centuries. In Europe we can trace it back at least to the Middle Ages, and the ancient Hebrew food laws, found notably in the Book of Deuteronomy, show that it goes back even further. In the UK, the trade guilds were instrumental in the introduction of legislation aimed at stopping the adulteration of a wide variety of foods and the adulteration of tea, coffee and bread was prohibited by legislation as long ago as the early-mid 18th century.

Two themes have existed from the start:

- *Food Safety* – the protection of the health and wellbeing of anyone eating food; and
- *Food Standards* – the control of composition and adulteration of food for the prevention of fraud.

UK legislation is structured principally around the *Food Safety Act 1990* supported by a whole raft of more detailed regulations. Increasingly over the past 45 years Directives and Regulations from the European Union have determined UK legislation – with an almost total revision of food safety and consumer protection rules.

Devolution within the UK has seen the responsibility for food policy shifted to the Scottish parliament and regional assemblies. This chapter sets out food safety and hygiene law as it applies in England. For information on the law in Scotland, Wales and Northern Ireland, see www.foodstandards.gov.scot/; www.srs.wales/; and www.nidirect.gov.uk/contacts/contacts-az/food-standards-agency-northern-ireland.

Food safety

[F9002] The principal legislation in the UK is the *Food Safety Act 1990*, to which significant amendments have been introduced since January 2005 in order to align the technical details with the EU General Food Law Regulation 178/2002/EC. There have also been fundamental changes to UK food hygiene legislation. EU Regulation 178/2002/EC introduced new food safety requirements, new traceability requirements, and also measures to ensure effective product recall/withdrawals and notification to competent authorities. Although as an EU Regulation it was directly applicable in Member States, it was still necessary to amend the basic UK Act to introduce the new provisions and to ensure conformity with the EU law.

[F9002] Food Safety and Standards

The *Food Safety Act 1990 (Amendment) Regulations 2004 (SI 2004 No 2990)* aligned the definition of 'food' in the *Food Safety Act 1990* with the definition in the General Food Law Regulation (EC) 178/2002/EC, and also made some minor amendments to the *Food Safety Act 1990* in respect of public consultation requirements, to remove duplication with directly applicable provisions in Regulation 178/2002/EC.

The *General Food Regulations 2004 (SI 2004 No 3279)* provide the enforcement of certain provisions of Regulation 178/2002/EC and amended the *Food Safety Act 1990* to bring it into line with the Regulation.

Since the inception of the European Economic Community (EEC), now the European Union (EU), European 'hygiene' regulations have developed on the basis of specific measures for specific food sectors, the so-called 'vertical' regulations. A full complement of these was already in place by the 1990s covering the processing of most foods of animal origin such as meat, fish, eggs, milk and related products, even honey. These 'vertical' regulations did not generally cover businesses selling food direct to the ultimate consumer, for example catering or retail businesses. The 'horizontal' Food Hygiene Directive 93/43/EEC had previously established general principles for food hygiene and set hygiene standards for those businesses not covered by 'vertical' legislation, such as food factories processing non-animal products and also retail or catering outlets.

However, in 2004 the EU completed a radical overhaul of all food hygiene legislation, consolidating and recasting both the general requirements and those relating to animal-derived products into two principal Regulations (*Regulation 852/2004 on the Hygiene of Foodstuffs and Regulation 853/2004 laying down Specific Hygiene Rules for Products of Animal Origin*). These were transposed into UK legislation as the *Food Hygiene Regulations 2006 (SI 2006 No 14, amended by SI 2007 No 56)*.

The *Food Safety and Hygiene (England) Regulations 2013 (SI 2013 No 2996)* then revoked and re-enacted, with some minor changes, the *Food Hygiene (England) Regulations 2006 (SI 2006 No 14)* and certain provisions of the *General Food Regulations 2004 (SI 2004 No 3279)* as they apply in relation to England. Each of the devolved administrations has introduced their own set of hygiene legislation. For more information, see www.foodstandards.gov.scot www.srs.wales; and www.nidirect.gov.uk/contacts/contacts-az/food-standards-agency-northern-ireland.

The following account will be largely restricted to the provisions of the *Food Safety Act 1990 (as amended)* and the *Food Safety and Hygiene Regulations 2013 (SI 2013 No 2996), as amended*.

Food labelling

[F9003] The *European Food Information to Consumers (FIC) Regulation 1169/2011* brought together EU rules on general food and nutrition labelling into a single law. They are enforced by local authorities under the Food Information Regulations 2014 in England (and equivalent regulations in Northern Ireland and Wales. The Food Standards Authority (FSA) provides a

Introduction **[F9004]**

summary of the requirements of the Regulation, see www.food.gov.uk/business-guidance/packaging-and-labelling.

Pre-packed food

[F9004] All pre-packed food requires food labelling that displays certain mandatory information and is subject to general food labelling requirements. Any labelling provided must be accurate and not misleading. Certain foods are controlled by product specific regulations. These include: bread and flour; cocoa and chocolate; soluble coffee; evaporated and dried milk; honey; infant formula; jams; meat products – sausages, burgers and pies; natural mineral waters; spreadable fats; sugars; irradiated food; and foods containing genetic modification (GM).

Food labels must include:

- the name of the food;
- a list of ingredients;
- the ingredients or processing aids causing allergies or intolerances that are stated in the 14 allergens (see below);
- the quantity of certain ingredients or categories of ingredients;
- the net quantity of the food;
- the date of minimum durability or the 'use by' date;
- special storage conditions and/or conditions of use;
- the name or business name and address of the food business operator;
- the country of origin or place of provenance;
- instructions for use where it would be difficult to make appropriate use of the food in the absence of such instructions;
- the alcohol strength by volume for beverages containing more than 1.2% of alcohol, by volume; and
- nutritional declaration.

The 14 allergens are:

- cereals containing gluten, such as wheat (including spelt and khorasan wheat), rye, barley and oats;
- crustaceans, for example prawns, crabs, lobster, crayfish;
- eggs;
- fish;
- peanuts;
- soybeans;
- milk (including lactose);
- nuts (ie almonds, hazelnuts, pistachio nuts, pecan nuts, walnuts, Brazil nuts and macadamia or Queensland nuts);
- celery (including celeriac);
- mustard;
- sesame seeds;
- sulphur dioxide/sulphites, if they are more than 10 milligrams per kilogram or 10 milligrams per litre in the finished product;
- lupin, including lupin seeds and flour; and
- molluscs, for example mussels, oysters, snails and squid.

There are additional labelling requirements for certain food and drink products including for: foods containing certain gases, sweeteners, glycyrrhi-

[F9004] Food Safety and Standards

zinic acid or its ammonium salt, or with added phytosterols, phytosterol esters, phytostanols or phytostanol esters; beverages with high caffeine content or foods with added caffeine; or frozen meat, frozen meat preparations and frozen unprocessed fishery products.

Non pre-packed foods

[F9004.1] Food catering businesses do not have to label food in the same way that manufacturers and other food businesses do but are required to provide allergen and intolerance information to customers.

The FSA advises: 'Allergen information for non-prepacked food, or 'loose foods' can be communicated through a variety of means to suit how you display information in your business. The requirement is to provide information about the use of allergenic ingredients in a food. You are not required to provide a full ingredients list.'

Food labelling law came under scrutiny following the death of 15-year-old Natasha Ednam-Laperouse, who died in July 2016 after eating a baguette containing sesame from the sandwich chain Pret a Manger. FSA guidance, *Food allergen labelling and information requirements under the EU Food Information for Consumers Regulation No 1169/2011: Technical Guidance* (April 2015), explains that there are reduced labelling requirements for food produced on site:

> 'Prepacked foods for direct sale: This applies to foods that have been packed on the same premises from which they are being sold. Foods prepacked for direct sale are treated in the same way as non-prepacked foods in EU FIC's labelling provisions. For a product to be considered "prepacked for direct sale" one or more of the following can apply:
>
> It is expected that the customer is able to speak with the person who made or packed the product to ask about ingredients.
>
> Foods that could fall under this category could include meat pies made on site and sandwiches made and sold from the premises in which they are made.'

In September 2018, an inquest into her death heard that Pret a Manger was not required to include allergen advice on sandwich packaging. Instead, it was sufficient to post general allergen warnings around the shop and for specific advice to be given verbally by staff. In October 2018, the company announced that it would introduce full ingredient labelling, listing allergens, on all its freshly-made products, but there have been calls for the law to be changed.

Policy and enforcement

[F9005] Since 2000, UK policy on food safety and standards has derived principally from the Food Standards Agency (FSA) although much of the underlying policy and legislation is initiated from Europe. The FSA is a non-ministerial government department that was created and empowered by the Food Standards Act 1999 in the wake of BSE and other food 'crises' to combine the food policy functions of the then Department of Health (DH) and Ministry of Agriculture, Fisheries and Food (MAFF) into an organisation which operates at arm's length from the government with a primary role of protecting consumers' interests. Complementary FSA structures exist in Scotland, Wales and Northern Ireland.

Complex arrangements are in place for the enforcement of food legislation. Traditionally, local authorities have had the responsibility although central government officials have particular responsibilities in 'upstream' parts of the food chain. This is particularly true of 'meat hygiene' in abattoirs and cutting plants. This role was taken away from local authorities to a central Meat Hygiene Service (MHS) with the MHS itself becoming a branch of the FSA from April 2000. The FSA has also developed a role of local authority monitoring and plays a direct role in enforcement issues that cross local authority borders such as food fraud and contamination.

In April 2009, the FSA established a new Food Fraud Advisory Unit, specifically to support local authorities in their work to tackle food fraud in relation to any illegal activity relating to food or feed. This could cover, for example, the sale of food that is unfit and potentially harmful, or the deliberate misdescription of food in a way that, while not necessarily unsafe, deceives the consumer as to the nature of the product.

Most food enforcement at the production and retail level falls to two groups of enforcement officers employed by local authorities or Port Health Authorities ('authorised officers'):

- Environmental Health Practitioners (EHPs) have responsibility for food safety legislation. They are generally employed at district council (town hall) level.
- Trading Standards Officers (TSOs) take control of food standards issues, including weights and measures. Generally, they are employed at county council level.

In unitary authorities, TSO and EHP functions are usually combined in the same department.

Veterinary inspectors may also be involved in controlling and inspecting imports and exports of foods containing animal-based products. Specialist regimes also operate for wines and spirits, and for organic foods. Following numerous cases of excessive contamination of certain imported foods of non-animal origin, such as nuts, these are increasingly subject to enforcement inspection at officially-designated Border Inspection Posts.

Food Safety Act 1990

[F9006] The *Food Safety Act 1990,* as amended, defines the offences that may be committed if food legislation is contravened and specifies defences that may be available if charges are instigated. It sets up the mechanisms to enforce the laws and fixes penalties that may be levied. The Act also enables the government to make regulations that include more detailed food safety and consumer protection measures.

The Act is divided into four principle sections:

(i) Part I: Definitions and responsibilities for enforcement;
(ii) Part II: Main provisions;
(iii) Part III: Administrative and enforcement issues such as powers of entry; and
(iv) Part IV: Miscellaneous arrangements, notably the power to issue codes of practice.

EU Regulation 178/2002/EC (the '*General Food Law Regulation*') introduced new food safety requirements and new measures to ensure effective product traceability, recall/withdrawals and notification to competent authorities. Although, as an EU Regulation, 178/2002/EC is directly applicable in Member States, consequential changes to domestic legislation were necessary to introduce new enforcement provisions, and to ensure conformity with this EU law.

The *Food Safety Act 1990 (Amendment) Regulations 2004 (SI 2004 No 2990)* made minor changes to the definition of 'food' in the *Food Safety Act 1990* to align it with the new EU definition and also made some minor amendments to the Act itself in respect of public consultation requirements, and to remove duplication with the directly applicable provisions in Regulation 178/2002/EC. The *General Food Regulations 2004 (SI 2004 No 3279)* provide enforcement powers in respect of the food safety and traceability requirements introduced under Regulation 178/2002/EC, designate competent authorities, specify enforcement authorities, make provision for offences and penalties, and introduce some consequential amendments to the *Food Safety Act 1990*, in particular *sections* 7 and 8.

The following is a brief summary of the main provisions contained within the Act.

Sections 1–3: definitions

[F9007] Under the *Food Safety Act 1990*, as amended, 'food' has a wide meaning and includes drink, chewing gum and any substance, including water, intentionally incorporated into the food during its manufacture, preparation or treatment. It includes water after the point of compliance as defined in Article 6 of Directive 98/83/EC (Water Quality for Human Consumption) and without prejudice to the requirements of Directives 80/778/EEC (Natural Mineral Waters) and 98/83/EC.

'Food' does not include live animals, birds or fish (although shellfish that are eaten raw and alive such as oysters and similar are classed as 'food'). Neither does 'food' include animal feed, residues and contaminants, nor controlled drugs that might be taken orally, medicines or cosmetics that are covered under separate Community legislation.

The Medicines and Healthcare Products Regulatory Agency (MHRA) determines whether a given product is a 'medicinal product' or a food on a case by case basis, having regard to the overall presentation and function of the product. See https://www.gov.uk/guidance/decide-if-your-product-is-a-medicine-or-a-medical-device.

The scope of the *Food Safety Act 1990* is very wide and covers all commercial businesses, ranging from farmers, manufacturers, wholesalers, distributors, retailers to businesses such as canteens, clubs, schools, hospitals, care homes and so on, whether or not they are run for profit. Government establishments that once had 'crown immunity' are treated no differently from other food businesses. Charity events that sell food are also subject to the safety provisions of the Act but may be exempt from certain administrative requirements.

The Act therefore covers any premises from an abattoir to a retail superstore, from a street vendor to a five-star hotel. It includes any place (including premises used only occasionally for a food business such as a village hall), any vehicle, mobile stalls and temporary structures.

The *Food Safety Act 1990*, s 2 provides an extended, very wide meaning to the term 'sale' of food, and encompasses the European concept of 'placing on the market'. The Act does not apply only to sales where money changes hands and the business is run for profit. It also covers food given as a prize or as a reward by way of business promotion or entertainment. Entertainment includes social gatherings, exhibitions, games, sport and so on. Thus, if food, which turns out to be unfit, is given as a prize for a competition, it could be subject to an action under the 1990 Act. Similarly, a food business cannot avoid its obligations under the Act by giving away food with other non-food items for which payment is accepted.

The *Food Safety Act 1990*, s 3 establishes a presumption that food is intended for human consumption. If a food business has any food or substance capable of being used in the preparation of food in its possession, the Act presumes that the intention is to sell it, unless it can be proved otherwise. Where food raw materials, ingredients, additives or finished products are on trade premises but are either not ready for consumption or have been rejected, the presumption is that they are intended for manufacture or sale and it is prudent to keep them in separate rooms, areas or batches clearly identified as being 'not for human consumption'. For example, food past its 'use by' date should be segregated from food that is for sale and marked clearly, otherwise an enforcement officer could presume that it was for sale.

Sections 5 and 6: Food authorities, authorised officers and enforcement of the Act

[F9007.1] The *Food Safety Act 1990*, ss 5–6, defines the roles and responsibilities of local and central government and the officials authorised to act on their behalf.

The Food Law Code of Practice (England) (March 2017) and associated Food Law Practice Guidance (England) (November 2017) encompass enforcement of all aspects of food safety and hygiene legislation and the requirements of Regulation 6(1) of the *Official Feed and Food Controls (England) Regulations 2009*. Parallel documents are in place for the devolved administrations. The code of practice and guidance can be found online at: www.food.gov.uk/about-us/food-and-feed-codes-of-practice. Food Authorities must also have regard to the framework agreement on official feed and food controls by local authorities, which can be found online at: www.food.gov.uk/about-us/local-authorities.

Part II: main provisions of the Act

[F9008] The main provisions of the *Food Safety Act 1990* fall within ss 7–22. The offences are divided into two types: food safety and consumer protection, which each break down into two separate offences:

(i) Food Safety
- *Section 7*: Rendering food injurious to health.
- *Section 8*: Selling food not complying with food safety requirements.

(ii) Consumer Protection
- *Section 14*: Food not of the nature, substance or quality demanded.
- *Section 15*: Falsely presenting or describing food.

Section 7: rendering food injurious to health

[F9009] It is an offence to do anything intentional that would make food harmful to anyone that eats it. Even if one did not know that it would have that effect, an offence would still have been committed. For example, subjecting food to poor temperature control that allows bacteria to multiply could render food injurious to health. An example of a more deliberate offence would be the addition to, or removal from, food of substances or components so as to render the food injurious to health.

Section 7 was amended by the *General Food Regulations 2004 (SI 2004 No 3279)* to require regard to be had not only to the probable immediate and/or short-term and/or long-term effects of that food on the health of a person consuming it, but also on subsequent generations. It also requires the probable cumulative toxic effects and particular health sensitivities of a specific category of consumers, where the food is intended for that category of consumers, to be taken into account. Food would be injurious to health if it were likely to cause immediate harm to anyone that ate it, for example foods contaminants with pathogenic micro-organisms such as listeria, salmonella, etc. It would also be injurious to health if the harm were cumulative over a long period, for example fungal toxins in nuts or cereal products.

Section 8: selling food not complying with food safety requirements

[F9010] Whilst s 7 deals with the person who renders food injurious to health, the *Food Safety Act 1990*, s 8 made it an offence for them or anyone else in the food supply chain to sell food which 'fails to comply with food safety requirements' as defined in the EU Regulation. This offence is incorporated into the *General Food Regulations 2004 (SI 2004 No 3279)* which link the offence to contravention of Article 14(1) of EC Regulation 178/2002.

Food would be deemed to be unsafe if it was considered to be either 'injurious to health' or 'unfit for human consumption'. The concept of 'injurious to health' relates to safety, but the concept of 'unfit' relates to unacceptability. Unfit food is not necessarily unsafe in the ordinary sense of the word (eg sour milk and certain mouldy foods).

In determining whether any food is 'unsafe', consideration would be given to the normal use of the food and its handling at each stage of the chain, and to any information provided to the consumer, such as labelling, to help them avoid specific adverse health effects.

In respect of whether a food is 'injurious to health', regard would be had not only to the probable immediate and/or short-term and/or long-term effects of the food on the health of someone who had eaten it, but also on subsequent generations and to any probable cumulative toxic effects. The particular health sensitivities of a specific category of consumers would also be relevant if the food is specifically intended for them – for example sweets intended for consumption by young children where there is risk of choking. However, where a food is not specifically intended for such a group, the fact that it may harm someone in that group does not automatically mean that it is 'unsafe' for general consumption.

The Food Standards Agency advises that foods containing undeclared allergens are to be considered as 'unsafe', resulting in frequent and numerous recalls and withdrawals of such products from the market. Detailed advice on allergens and the law can be found on its website at: allergytraining.food.gov.uk/english/rules-and-legislation/.

Food would be 'unfit' (ie unacceptable) for consumption if it was putrid or contained serious foreign material, for example a dead mouse. This section gives wide scope to prosecute someone for selling food that fails to meet a customer's expectations. For example, if the food is mouldy or contains a rusty nail or there are excessive antibiotic residues in meat.

(The *Food Safety Act 1990*, ss 9–13 provide enforcement procedures for food safety offences. These sections are dealt with from **F9014–F9019**.)

Section 14: selling food not of the nature, substance or quality demanded

[F9011] It is an offence to supply to the prejudice of the purchaser food which is not of the nature, substance or quality demanded. The purchaser in this case does not have to be a customer in a retail store or catering outlet. One company, large or small may purchase from another and an offence is committed if the food is inferior in nature or substance or quality to that which they demanded. Once again this allows the law to be applied at any point of the food chain.

The purchaser does not have to buy the food for his or her own use to be prejudiced. They may intend to give it to someone else, for example members of their family.

The three offences created by this section are separate and a charge will be brought under one of the three according to the circumstances. These offences have provided a flexible and far reaching means of preventing the adulteration, contamination and mis-description of food and have been the foundation of food law in the United Kingdom for over 140 years. The wide application of this section has been the subject of many cases in law, and a detailed consideration of them is outside the scope of this brief overview.

If a purchaser asks for cod and gets coley, or beef mince contains a mixture of lamb or chicken, it would be not of the 'nature' demanded.

If a purchaser asked for diet cola and was served regular cola, or expected a sheep's milk cheese and it was made from cow's milk, it would be not of the 'quality' demanded.

Food that was not of the quality or substance demanded often formed the basis for complaints of mouldy food or foreign material. Such complaints are likely to be taken under the *Food Safety Act 1990, s 8* (ie that the food is so contaminated that it would be unreasonable to expect it to be eaten).

The above provision has also been used when a susceptible customer has requested information about peanuts or similar ingredients to which they may have a severe allergic reaction. If, notwithstanding such a request, the consumer is sold food that triggers a reaction, cases have been brought on the grounds that the food was not of the 'substance' demanded.

Section 15: falsely presenting or describing food

[F9012] The *Food Safety Act 1990, s 15* provides the principal protection for consumers in respect of food composition and description. It prohibits false and misleading descriptions of food by way of labelling and advertising. It is also an offence if material that is technically accurate is presented in such a way as to mislead the consumer. The scope of this section includes misleading pictorial representation. (See also para F9003 above on the *European Food Information to Consumers (FIC) Regulation 1169/2011*.)

Article 16 of Regulation 178/2002/EC (relating to 'Presentation') stipulates that the labelling, advertising and presentation of food shall not mislead consumers. This provision is not limited to 'food business operators'. It applies additionally to *Sections 14* and *15* of the *Food Safety Act 1990*. Unlike *sections 14* and *15*, Article 16 applies in the case of one-off supply free of charge, and also covers cases where a consumer is misled, whether or not it relates to the nature, substance or quality of the food.

The *Consumer Protection from Unfair Trading Regulations 2008 (SI 2008 No 1277)* impose a 'general prohibition' on the use of unfair commercial practices. A commercial practice will be unfair if it contravenes the requirements of professional diligence and it materially distorts, or is likely to materially distort, the economic behaviour of the typical consumer.

A commercial practice is a misleading action if it contains false information or in any way deceives or is likely to deceive the typical consumer in relation to a large number of matters specified in the legislation, and this causes or is likely to cause the consumer to take a transactional decision he would not otherwise have taken. If a commercial practice misleads in relation to a matter not directly specified in the regulations, then its unfairness will have to be assessed against the 'general prohibition'. The omission of relevant particulars may also be considered a misleading action.

Notwithstanding the requirements of food law and the procedures in place for their formal enforcement, food manufacturers, retailers and traders should also be aware of the role played by the Advertising Standards Authority (ASA) in matters related to misleading broadcast and non-broadcast advertising under the Advertising Standards Code for Broadcast Committee of Advertising Practice (BCAP Code) and the British Code of Advertising, Sales Promotion and Direct Marketing (CAP Code).

ASA is an independent, non-statutory body responsible for administering these Codes so that all broadcast and non-broadcast advertisements are perceived to be legal, decent, honest and truthful. The ASA receives and investigates complaints from the public and industry, and decisions on compliance are taken by the ASA Council; if a complaint is upheld, the advertisement must be withdrawn or amended. ASA adjudications are published and made available to the media.

BCAP is the regulatory body responsible for maintaining a number of standards Codes under a contracting-out agreement with the Office of Communications (Ofcom), the latter having statutory responsibility, under the Communications Act 2003, for maintaining standards in TV and radio advertisements.

CAP is the self-regulatory body that creates, revises and enforces the CAP Code, which covers UK-originated, non-broadcast marketing communications, such as advertisements placed in traditional and new media, sales promotions and direct marketing communications. Parties that do not comply with the CAP Code are liable to sanctions including the denial of media space and adverse publicity.

For more information, see www.asa.org.uk/.

Enforcement

[F9013] The *Food Safety Act 1990* gives enforcement officers strong powers, namely:

- To enter food premises to investigate possible offences.
- To inspect food.
- To detain or seize suspect food.
- To take action that requires a business to put things right.
- If all else fails:
    — to prosecute,
    — to prohibit the use of premises or equipment,
    — to bar an individual from working in a food business.

Under the *Food Safety Act 1990*, ss 32 and 33, enforcement officers have rights of access at any reasonable time and an offence will be committed if they are obstructed. They must be given reasonable information and assistance. They can inspect premises, processes and records. If records are kept on a computer, they are entitled to have access to them. They can take samples, photographs and even videos, and (under specific circumstances, eg where there is reason to believe an offence has been committed) copy records. If they are refused access to any of these, they can apply to a magistrate for a warrant. They must give 24-hours' notice to enter private houses used in connection with a food business.

Of course, none of these requirements to co-operate with officers and give them access to information cancel the basic right to avoid self-incrimination. In the extreme situation, one has the right to remain silent and to consult a solicitor.

Officers themselves commit an offence if they reveal any trade secrets learned in the course of official duties.

The *Food Safety Act 1990*, ss 29–31, allows officers to procure samples and control the sampling and testing procedures. Chapter 6 of the FSA Food Law Practice Guidelines covers 'Sampling and Analysis'. Samples that are taken carelessly may give unreliable results and may be not be acceptable as evidence in court.

The *Food Safety Act 1990*, ss 9–13, provides enforcement officers with a series of enforcement measures to control food or food businesses that present a health risk.

Section 9: detention and seizure of food

[F9014] Section 9 provides that, if an officer *suspects* that food is unfit or fails to meet food safety requirements, he can issue a detention notice. This requires the person in charge of the food to keep it in a specified place and not to use it for human consumption pending investigation. Chapter 7.2.10 of the Food Law Code of Practice describes the circumstances when the use of detention and seizure powers under the *Food Safety Act 1990, s 9*, as amended, is appropriate, including after food has been certified in accordance with Regulation 29 of the *Food Safety and Hygiene (England) Regulations 2013*. It also covers the procedures for serving and withdrawal of notices; voluntary surrender; and the destruction or disposal of food.

The detention notice must be written on a prescribed form, *Detention of Food (Prescribed Forms) Regulations 1990 (SI 1990 No 2614)*. It is an offence to ignore or contravene a detention notice even if the business may believe that it is unjustified. The officer has 21 days to complete his investigations but must clearly act quickly where the food is highly perishable. If to they conclude that the food was actually safe, they must withdraw the notice (using another prescribed form) and restore the food to the person in charge.

On the other hand, the authorised officer may conclude that the food is unsafe. they can seize it and take it before a justice of the peace (JP). The authorised officer may reach this conclusion as a result of tests performed during the 21 days after serving the detention notice. But in some circumstances, they may be confident that it is unsafe from the beginning and seize the food immediately without using a detention notice. Once again there is a form that must be used and the authorised officer must state exactly what is happening and when. After seizure the matter should be taken before the JP within two days, more quickly in the case of perishable food.

The hearing before the JP may not only decide the fate of the batch of food but may also be a forerunner to other legal proceedings. The accused is allowed to make representations at this hearing and to call witnesses. It is probably advisable to take specific legal advice from a solicitor when the first detention notice is issued. It is wise to have legal representation at a hearing following seizure.

If the JP decides that the food is unsafe, they can order it to be destroyed and he may order the offender to cover the costs of its disposal. If the seized food is only part of a larger batch, the court must legally presume that the whole batch fails unless evidence is provided to the contrary. Disposal must follow the requirements of the relevant waste disposal legislation, using licensed waste disposal contractors, and must also be in accordance with relevant animal by-products legislation.

If it turns out that the authorised officer was wrong, the offender may be entitled to compensation. If the food has deteriorated as a result, they may be entitled to compensation equal to the loss in value. If a reasonable sum cannot be agreed with the authority there is an arbitration procedure.

Section 10: Improvement Notices

[F9015] *Section 10* provides authorised officers with a range of powers to deal with unsatisfactory premises from informal advice through to prosecution and emergency prohibition. Chapter 7.2 of the Food Law Code of Practice describes the circumstances when it is appropriate to use Improvement Notices under the *Food Safety Act 1990, s 10* and the use of Hygiene Improvement Notices under Regulation 6 of the *Food Safety and Hygiene (England) Regulations 2013*. If an authorised officer has reasonable grounds for believing that a business does not comply with hygiene or processing regulations, he can issue a formal improvement notice demanding that it puts matters right. This provides a quicker and cheaper remedy than taking the business to court. An improvement notice should not be ignored. It is an offence to fail to comply with a notice and either matters must be put right as indicated on the notice or an appeal lodged against it.

An improvement notice must contain four elements:

(i) the reasons for believing that the requirements are not being complied with;
(ii) the ways in which regulations are being breached;
(iii) the measure which should be taken to put things right;
(iv) the time allowed for putting things right (this cannot be less than 14 days).

The business can choose to put things right in a different way to that suggested by the authorised officer provided that the authorised officer is satisfied that it will have the required effect.

If a party does not agree with an improvement notice, the *Food Safety Act 1990, s 37* gives the right to appeal to a magistrate within one month of its issue or within the time specified on the notice if shorter than one month.

The notice may contain a number of points but the whole notice does not have to be appealed. The proprietor may appeal just one point or simply ask for a time extension. The notice is effectively suspended during the appeal, but it is best to let the authority, as well as the court, know about the appeal. The *Food Safety Act 1990, s 39* allows the magistrate to modify or cancel any part of the notice.

If an authorised officer of an enforcement authority has reasonable grounds for believing that a food business operator is failing to comply with the *Food Safety and Hygiene Regulations*, he may serve a Hygiene Improvement Notice. The provisions for Hygiene Improvement Notices fall under the *Food Safety and Hygiene Regulations 2013 (SI 2013 No 2996), Reg 6*.

Food Authorities must continue to use the prescribed forms set out in the *Food Safety (Improvement and Prohibition - Prescribed Forms) Regulations 1991 (SI 1991 No 100)* when using powers under the *Food Safety Act 1990 s 10*.

Section 11: Prohibition Orders

[F9016] Chapter 7.2 of the Food Law Code of Practice describes the circumstances when it is appropriate to use Prohibition Orders under the *Food Safety Act 1990, s 11* and the use of Hygiene Prohibition Orders under the *Food Hygiene Regulations 2006, Reg 7*.

If a proprietor is convicted of food safety or processing offences, including failing to comply with an Improvement Notice, the Food Authority can also apply for a prohibition order. If the court is satisfied that any processing, or the construction or state of the premises, or use of equipment, poses a risk to public health, it must issue a prohibition order to the proprietor or any manager of the business. The order can deal with any of one three issues depending upon the actual risk:

- to prohibit a process or treatment,
- to deal with the *structure* of the premises or the use of equipment,
- to deal with a problem that relates to the *condition* of the premises or equipment.

In either of the last two cases, the order may prohibit the use of the equipment or the premises.

Any prohibition order must be served on the proprietor of the business. If it relates to equipment, it will also be fixed to the equipment and if it relates to the premises, it will be fixed prominently to the premises. It is automatically an offence to knowingly breach a prohibition order.

As soon as a party believes that they have put matters right and that the health risk no longer exists, they can apply to the Food Authority to have the prohibition lifted. The Food Authority must respond by reaching a decision within a fortnight and, if it believes that things are satisfactory, issue a certificate within another three days. The certificate will state that enough has been done to ensure that there is no longer an unacceptable risk to public health. If the Food Authority disagrees and refuses to issue a certificate, they must give their reasons and the business can appeal to a magistrate to have the order lifted. The Food Authority must give information on the right to appeal, and who to contact as well as the time scale (the appeal must be made within one month).

If a food business operator is convicted of an offence under the *Food Hygiene Regulations* and the court is satisfied that the health risk condition is fulfilled with respect to the food business concerned, the court must impose the appropriate prohibition. The provisions for Hygiene Prohibition Orders fall under the *Food Hygiene Regulations 2006 (SI 2006 No 14), Reg 7*.

Personal Prohibition under section 11

[F9017] Under the *Food Safety Act, s 11*, the court can also prohibit a person from running or managing a food business. An application to the court to lift a personal prohibition cannot be made within six months of its imposition and if an appeal is unsuccessful, it cannot be appealed again for a further three months.

In addition, if a business operator is convicted of an offence under the *Food Hygiene Regulations* and the court thinks it appropriate, taking into account all the circumstances of the case, it may impose a prohibition order on the food business operator, banning him from participating in the management of any food business, or any specified class of food business.

Section 12: Emergency Prohibition Notices and Orders

[F9018] Whereas *s 11* prohibitions are imposed only by a court following conviction, *s 12* permits, in an emergency, an authorised officer to act on his or her own authority, and with immediate effect. Where the Food Authority is satisfied that a business presents an imminent risk to health, it can serve an Emergency Prohibition Notice without prior reference to a court. Emergency Prohibition can apply to the whole premises or a specific part. The notice will be fixed in a prominent place and anyone removing it or deliberately ignoring it is guilty of an offence.

The authority must then take the matter before a magistrates' court within three days of serving the Emergency Prohibition Notice. At least one day before it goes to court, the authority must serve a notice on the proprietor of the business telling him of the court hearing. If an order is not issued by the court or if no application for one is made within the three-day period, the Emergency Prohibition Notice lapses. Once again, if a business finds itself faced with action of this kind, it is advisable to seek specific legal advice.

If the court agrees that the prohibition actions were justified, it will make an Emergency Prohibition Order which replaces the Emergency Prohibition Notice. The arrangements for lifting the Order are the same as for a *s 11* Order, above. If the court does not uphold an Emergency Prohibition Notice, the proprietor may seek compensation for loss or damages. An Emergency Prohibition Order cannot be made against an individual. Prohibition of a person can only be made under *s 11* following a conviction.

Similarly, and subject to the same procedures, if an authorised officer of an enforcement authority is satisfied that the health risk condition is fulfilled with respect to any food business, he may serve a 'Hygiene Emergency Prohibition Notice' on the relevant food business operator. The provisions for Hygiene Emergency Prohibition Notices and Orders fall under the *Food Safety and Hygiene Regulations 2013 (SI 2013 No 2996), Reg 8.*

Section 13: Emergency Control Orders

[F9019] *Section 13* empowers the Secretary of State or the Food Standards Agency to issue Emergency Control Orders where it appears that commercial food-related operations involve, or may involve, imminent risk of injury to health. In such circumstances either of the above may prohibit such operations. It is an offence to knowingly contravene an Emergency Control Order. If an Emergency Control Order is invoked, there is no right of appeal and no compensation arrangements.

It may be that an unsafe batch of food has already been distributed around the country. Emergency prohibition of the manufacturing site would stop further production but it would not control the risk from food already in the system. The *Food Safety Act 1990, s 13* allows an Emergency Control Order to require all steps to be taken that will remove the threat. This is a wide-ranging power that will only be used in exceptional circumstances. In most cases, any business told that it has received unsafe food from a supplier would stop selling it simply to protect their own reputation.

Voluntary arrangements are in place to co-ordinate food hazards and issue warnings and organise recalls when food is distributed more widely than the local area in which it is produced. Warnings are circulated by electronic mail to Food Authorities around the country. Where possible, trade organisations and the media are also informed and food hazard warnings and recalls are publicised, for example on the FSA website at: www.food.gov.uk/news-alerts/search/alerts or by direct emails to interested parties, including the public.

Remedial Action Notices

[F9019.1] In the case of an establishment subject to approval under Article 4(2) of Regulation 853/2004 (animal products hygiene legislation), an authorised officer may serve a Remedial Action Notice (Regulation 9) or Detention Notice (Regulation 10) where any of the requirements of the Food Safety and Hygiene Regulations is being breached or inspection under the Regulations is being hampered. This may:

- prohibit the use of any equipment or any part of the establishment (and must specify the reason and the action needed to remedy it);
- impose conditions upon or prohibit the carrying out of any process;
- require the rate of operation to be reduced to such extent as to ensure hygienic operation, or to be stopped completely;
- require the detention of any animal or food of animal origin for sampling and examination.

A Remedial Action Notice must be served as soon as practicable and state why it is being served. The notice must be withdrawn by a further written notice as soon as the authorised officer is satisfied that the necessary action has been taken.

Failure to comply with a Remedial Action Notice is an offence, although the right to appeal to a magistrate within one month is granted.

The use of Remedial Action or Detention Notices must be proportionate to the risk to public health and where immediate action is required to ensure food safety. A Remedial Action Notice may be used if a continuing offence requires urgent action owing to a risk to food safety or when corrective measures have been ignored by the food business operator and there is a risk to public health. As soon as the authorised officer is satisfied that the action specified in a Remedial Action Notice has been taken, the notice must be withdrawn by means a further notice in writing. Similarly, in respect of a Food Detention Notice, if the authorised officer is satisfied that the food need no longer be detained, the relevant notice must also be withdrawn by means a further notice in writing.

Sections 16 to 18: power to make regulations

[F9020] The *Food Safety Act 1990, ss 16–18* and 26 give the Secretary of State very wide-ranging powers to make regulations and orders for general food safety and consumer protection; for special provisions for particular foods; for the registration and licensing of food premises; for food hygiene training; and for facilities for sampling and the giving of information. For example:

- to prohibit specified substances in foods;
- to prohibit a certain food process;
- to require hygienic conditions in commercial food premises;
- to provide for a food hygiene information scheme;
- to make standards on food composition;
- to specify requirements for food labelling;
- to restrict and control claims made in the advertising of food.

They also provide the power to ban from sale food originating from potentially diseased sources; such powers were used to prohibit the sale of certain parts of beef cattle for human food on the basis that they may have been suffering from BSE.

Section 17 allows the government to bring EU provisions into UK law. *Section 18* regulations are intended to cover very particular circumstances, such as genetically modified and other novel foods, although in practice these are now harmonised at EU level and, while it remains a member of the EU, the UK has only very limited room in which to act independently.

The *Food Standards Act 1999, s 17*, provides limited powers to the FSA to make emergency orders. Under the *Food and Environment Protection Act 1985, ss 1* and *2* and the *Food Safety Act 1990, s 13* the Secretary of State may make emergency orders in response to circumstances or incidents which pose a threat to public health in relation to food. The Secretary of State retains these powers but may empower the FSA to make Emergency Orders itself on their behalf. This power does not give the FSA the ability to make legislation itself in other areas and, in practice, the FSA will only make orders in emergency situations where the Secretary of State is not available. The Secretary of State is ultimately answerable for emergency legislation made by the FSA on his behalf and anything done by the FSA is, in law, done by the Secretary of State.

Section 19: registration and licensing

[F9021] The *Food Safety Act 1990, s 19* permits the government to bring in regulations demanding licensing or registration of certain food businesses.

Registration and licensing confer significantly different powers on enforcement authorities. Registration is simply an administrative exercise, designed primarily to inform enforcement agencies what food businesses are operating in their area. It also provides some basic information on the size of the business and the type of food that they produce. No conditions are attached to registration, the Food Authority cannot refuse to register a business, there is no charge, and it does not have to be renewed periodically.

Licensing (or 'prior approval') is quite different. It is a control measure. Precise licensing criteria will be specified and there would normally be prior inspection and approval before a new business is allowed to open. There is often a charge for a licence, which must be renewed periodically. The enforcement authority will usually have the right to withdraw a licence if standards fall and the business would have to close.

Currently, licensing only applies to those businesses subject to the requirements of the EU 'vertical', animal product legislation, such as slaughter houses, cutting plants and dairy operations. Most other types of food businesses are subject only to registration. Article 6(2) of EC Regulation 852/2004 requires food business operators to register their establishments (ie each separate unit of their food businesses) with the appropriate Food Authority, unless they are subject to approval under EC Regulation 853/2004 (animal products hygiene) or fall outside the scope of the EC Regulation 852/2004 (in practice, very few food businesses fall outside the scope).

EC Regulation 882/2004, Article 31(1)(a) requires the Food Authority to establish procedures for food business operators to follow when applying for the registration of their establishments. Food business operators must provide the authority with full details of the activities to be undertaken at least 28 days before food operations commence. EC Regulation 882/2004, Article 31(1)(b) requires the authority to draw up and maintain a list of food establishments that have been registered. Any changes to details previously supplied, eg a change of business operator, a change to the activities carried out, the closure of an establishment etc. must be notified by the current operator. Notification of a change to the operator of a food establishment should be made by the new operator.

The principles of registration in accordance with the terms of the EC Regulation 852/2004 are maintained through the provisions of the *Food Safety and Hygiene Regulations 2013 (SI 2013 No 2996)* – see **F9026–F9032**.

Defences against charges under the Food Safety Act 1990

[F9022] The following paragraphs explain the defences available under the *Food Safety Act 1990*.

Section 21: defence of due diligence

[F9023] The so called 'due diligence' defence first appeared within food safety legislation when the *Food Safety Act 1990* was passed.

The *Food Safety Act 1990* creates offences of strict liability. The prosecution is not required to show that the offence was committed intentionally. However, the due diligence defence is intended to balance the protection of the consumer against the right of traders not to be convicted for an offence that they have taken all reasonable care to avoid committing. The intention of the due diligence defence is to encourage traders to take proper responsibility for their products and processes.

The *Food Safety Act 1990, s 21* states:

> 'In any proceedings for an offence under any of the preceding paragraphs of this Part . . . it shall . . . be a defence for the person charged to prove that he took all reasonable precautions and exercised all due diligence to avoid the commission of the offence by him or any person under his control.'

The onus is on the business to establish its defence on the balance of probabilities. Part of the due diligence defence may be to establish that someone else was at fault. If that is what is intended, the prosecutor must be informed at least seven days before the hearing or, if a business has been in court already, within one month of that appearance. Under the 1990 Act, reliance upon a supplier's warranty alone is not possible, although a warranty may be one element of a due diligence defence.

It is impossible to give precise advice on what any business would need to do to have a due diligence defence. Cases that have been decided by the courts serve only to demonstrate that each case is decided on its own facts and may also be determined by the size and resources otherwise available to a business. What is certain is that doing nothing will never provide a due diligence defence. Positive steps must be taken if a defence is to succeed.

The defence has two distinct parts:

- Taking *reasonable precautions* involves setting up a system of control. The business operator must consider the risks that threaten the operation and take *all reasonable precautions* that may be expected of a business of the size and type.
- Exercising *due diligence* means that those precautions are taken on a continuous basis.

All reasonable precautions and *all* due diligence are needed. The courts have decided consistently that if a precaution could reasonably have been taken, but was not taken, then the defence would not succeed.

The test is 'what is reasonable?' A business will be expected to take greater precautions with high risk, ready-to-eat foods than are needed for boiled sweets or biscuits. What is reasonable for a large-scale food business may not be reasonable for a smaller enterprise.

The business must ensure that the scope of 'due diligence' covers all aspects of food law requirements. It is not adequate to have excellent control of food safety hazards if there are no precautions in place to ensure that the composition and labelling of food complies with the regulations.

Documentation is important. Unless the precautions are written down, it will be difficult to persuade the authorised officer or the court that a proper system is in place. The same is true of records of any checks made to demonstrate that the system is followed diligently. The burden of proof lies with the business to establish that the defence is satisfied.

A good 'HACCP' plan (see **F9036**), as required by Article 5 of EC Regulation 852/2004, enforced in the UK by the *Food Safety and Hygiene Regulations 2013 (SI 2013 No 2996)* may be helpful in showing that a systematic approach to identifying the food safety precautions needed in the operation has been adopted. Cross-reference to industry guidelines or codes of practice may show that the system has a sound basis. Documentation from the hazard analysis will add to the evidence that precautions have been taken. Records such as specifications, cleaning schedules, training programmes and correspondence with suppliers or customers may all play a part in establishing the defence. An organisation chart and job descriptions that identify roles and responsibilities will also be useful evidence of the precautions taken.

Not all food businesses have the same defence. Manufacturers and importers must satisfy the full defence. Some traders within the retail and catering sectors may use a simpler defence of 'deemed due diligence' deriving from the *Food Safety Act 1990, s 21(3) and (4)*. Even then, 'deemed due diligence' is not available against the *s 7* offence of rendering food injurious to health. Larger businesses may be expected to take more precautions than smaller businesses. Those selling 'own label' packs may have greater responsibility for upstream activity than those selling manufacturers' 'branded' products. However, the basic responsibilities of retailers to ensure the accuracy of the labelling of all products sold by them was reinforced by the ECJ judgment in *Lidl Italia Srl v Comune di Arcole (VR)*: C-315/05 [2006] ECR I-11181, in which it was held that a retailer may be held liable for food labelling infringements by the producer, even if he simply markets a product as delivered by that producer.

Section 35: penalties under the Food Safety Act 1990

[F9024] The courts decide penalties on the merits of individual cases but they are limited by maximum figures included in the *Food Safety Act 1990, s 35*. For most offences, the crown court can send offenders to prison for up to two years and/or impose an unlimited fine. The magistrates' court, where most cases are heard, can set fines of up to £5,000 per offence and up to six months imprisonment, or both. For the two food safety offences in *ss 7* and *8* and the first of the consumer protection offences (*s 14*), the fine may be up to £20,000. Each set of regulations made under the Act will have its own level of penalties that will not exceed the levels mentioned here.

A new Sentencing Council guideline for use in courts in England and Wales was introduced from February 2016: *Health and safety offences, corporate manslaughter and food safety and hygiene offences: Definitive guideline*.

This recommends fines of up to £3 million for organisations guilty of breaching the *Food Safety and Hygiene (England) Regulations 2013* and custodial sentences for individuals.

See www.sentencingcouncil.org.uk/wp-content/uploads/HS-offences-definit ive-guideline-FINAL-web1.pdf.

Section 40: codes of practice

[F9025] The *Food Safety Act 1990, s 40* allows for the development of codes of practice. *Section 40* codes of practice are primarily intended to instruct and inform enforcement officers and related bodies or personnel, for example food analysts or examiners. In carrying out their duties in the execution and enforcement of the Act or Regulations laid under it, enforcement authorities and their authorised officers must 'have regard to' these *s 40* codes of practice. This means, in effect, that Food Authorities must follow and implement the provisions of the Code that apply to them. Any failure to follow a code requirement might result in their decisions or actions being successfully challenged, and evidence gathered during a criminal investigation being ruled inadmissible by a court, thus seriously jeopardising their ability to bring a successful prosecution. Twenty individual codes were originally made under s 40, some of which underwent several revisions. The subjects of these original codes were general matters, such as the demarcation of responsibility for enforcing different parts of the Act or general inspection procedures, and more specific information on the enforcement of particular hygiene-related regulations. The most significant was Code of Practice Number 9 on food hygiene inspections, which introduced detailed advice to enforcement officers on how to conduct such inspections and included a 'risk rating' system whereby inspectors could prioritise different food businesses and determine the necessary inspection frequency.

The 20 codes were subsequently consolidated by the FSA into a single Food Law Code of Practice but, to ensure a clear division between what Food Authorities must do in enforcing food law, which must be included in the statutory code, other advice is published as Practice Guidelines accompanying the Code.

The current code and guidance can be found on the FSA website at: https://signin.riams.org/files/display_inline/45497/Food-Law-CoP-Eng-01032017.pdf.

Section 40 codes should not be confused with 'industry guides'. These 'guides' provide more detailed practical explanation of the implications of the particular regulations in a food business sector. There are also numerous other guides and codes of practice published by industry, government, and other organisations.

Food premises regulations

[F9026] The principles of registration in accordance with the terms of the EU Regulation 852/2004, Article 6(2) have been maintained through the provisions of the *Food Safety and Hygiene Regulations 2013 (SI 2013 No 2996)*. Under EU food law, 'food business' is very widely defined as meaning any undertaking, whether for profit or not and whether public or private, carrying out any of the activities related to any stage of production, processing and distribution of food. Thus, all such businesses must be registered unless they are specifically exempted from the requirement. Food business operators must register their establishments (ie each separate unit of their food businesses) with the appropriate Food Authority, unless they are subject to approval under EC Regulation 853/2004 (animal products hygiene) or fall outside the scope of the EC Regulation 852/2004. In practice, very few food businesses fall outside the scope.

EC Regulation 882/2004, Article 31(1)(a) requires the authority to establish procedures that business operators must follow when applying for the registration of their establishments. Food business operators must provide the authority with full details of the activities to be undertaken at least 28 days before food operations commence. Any changes to details previously supplied, eg a change of business operator, a change to the activities carried out, the closure of an establishment, etc must be notified by the current operator. Notification of a change to the operator of a food establishment should be made by the new operator. The authority must, by law, draw up and maintain a list of food establishments that have been registered.

The following paragraphs cover the principal requirements.

[F9027] Food businesses must notify the local authority of all food premises under their control that carry out any production, processing and distribution of food (including retail and catering premises), unless they are already required by other legislation to be licensed (ie approved under 'vertical' legislation relating to handling products of animal origin). The legislation requires the notification of individual food *premises* and not food businesses. Thus, premises that may be used only occasionally by food businesses, even by different businesses, (eg a church hall, village hall or scout hut used from time to time for the sale of food) may have to be registered by the owner of the premises, depending on the frequency of use.

Similarly, wholesale markets or other premises used by more than one food business must also be registered. The market operator or the person in charge of the premises is responsible. If mobile food premises are used they may have to be registered. These can range from a handcart in a market to a forty-foot trailer used for hospitality. Privately-owned, moveable premises used in a market must be registered even though the market is registered by the operator. (If stalls provided by the market are used, registration is not necessary). If mobile food premises are operated, the premises at which they are normally kept must be registered with the food authority local to the base.

Staff restaurants must be registered. Contractors and clients should have a clear agreement as to who will action this.

New premises must notify the authorities at least 28 days before opening. This is designed to allow the Food Authority the opportunity to inspect a business before it opens. However, if the authority does not take that opportunity, it is not necessary for the business to wait for a visit or formal permission before opening, provided that the necessary registration form has been sent. To register, the operator must supply a range of information on a registration form obtained from the authority, which is normally contacted via the Town Hall. In case of doubt, it is best to keep a note of when the form was sent and, if possible check that it has been received.

Food business operators must also ensure that the authority always has up-to-date information on premises, including any significant change in food-related activities (eg change of product handled) and any closure of an existing establishment. The authority must be notified of the change within 28 days. Many of the other pieces of information on the application form may also change, for example the name of the manager, the phone number, the number of employees and so on but these changes do not need to be notified to the authority. Failing to register, giving false or incomplete information or failing to notify any of the changes mentioned above, can trigger prosecution proceedings.

Some food premises must by law be licensed, in which case additional notification/registration is not required. These premises are generally those (other than retail and catering) that handle raw/unprocessed products of animal origin, which fall into a number of categories under EU Regulation (EC) No 853/2004, and are approved ('licensed') by the competent authority, following at least one on-site visit.

Premises licensed or registered under other measures

Establishments handling raw products of animal origin

[F9028] The Meat Hygiene Service is responsible for the approval of meat-handling establishments where Regulation 853/2004 (animal products hygiene) requires control by an official veterinarian. These will include slaughterhouses, game handling establishments, cutting plants *and* any of these in which minced meat, meat preparations, mechanically separated meat, or meat products are *also* produced.

Local Authorities ('Food Authorities') are responsible for the approval of establishments where control does *not* fall to an official veterinarian. Such 'product-specific establishments' will be producing any, or any combination, of the following: minced meat; meat preparations; mechanically separated meat; meat products; live bivalve molluscs; fishery products; raw milk other than raw cows' milk; dairy products; eggs (not primary production) and egg products; frogs' legs and snails; rendered animal fats and greaves; treated stomachs; bladders and intestines; gelatine and collagen; and will include certain re-wrapping/re-packing establishments, cold stores and some wholesale markets.

Establishments producing 'composite products'

[F9028.1] Regulation 853/2004 does not apply to food comprising mixtures of plant origin and *processed* products of animal origin. However, establishments producing such 'composite products' which are also engaged in activities which would otherwise require *approval* will need to be approved in relation to those activities. Otherwise, Regulation 852/2004 will apply to the part of the establishment where the composite products are assembled. Only the notification process is required in respect of establishments engaged solely in assembling products originating from approved establishments to create 'composite products'.

Premises used only occasionally

[F9029] The EU hygiene legislation states that the requirements should apply only to undertakings, the concept of which implies a certain continuity of activities and a certain degree of organisation, but does not specify precisely how this concept is to be applied. In practice, this has historically been taken to include premises used for less than five days (not necessarily consecutive) in five consecutive weeks. However, the operators of all such premises are advised to seek advice on their specific circumstances from their local Food Authority.

Premises and areas exempt unless used for retail sales

[F9030] The following are examples of 'premises', areas and locations which are not generally subject to registration/licensing under the food hygiene legislation (although farms, smallholdings etc may be subject to measures under other legislation) but which may need to be so registered if the sale of products from them is other than the direct supply, by the producer, of small quantities of primary products to the final consumer or to local retail establishments directly supplying the final consumer. In general, however, the premises and the products obtained from these become subject to food hygiene controls as soon as they are harvested, slaughtered, caught etc if they are destined for sale to third parties.

- Where game is killed in sport (grouse moors).
- Where fish is taken for food (but not processed).
- Where crops are harvested, cleaned, stored or packed. However, if the crops are put into the final consumer pack the premises must be registered.
- Where honey is harvested.
- Where eggs are produced or packed.
- Livestock farms and markets.
- Shellfish harvesting areas (these are, however, subject to specific classification and related controls under Regulations 853/2004 and 854/2004).
- Places where no food is kept, for example an administrative office or headquarters of a food business.

Some domestic premises

[F9031] EU Regulation 852/2004 does not apply to primary production for private domestic use; or to the domestic preparation, handling or storage of food for private domestic consumption; or the direct supply, by the producer, of small quantities of primary products to the final consumer or to local retail establishments directly supplying the final consumer. In addition, the legislation states that the requirements should apply only to undertakings, the concept of which implies a certain continuity of activities and a certain degree of organisation but does not specify precisely how this concept is to be applied.

Thus, the use of domestic premises for a number of situations may, technically, bring them within the scope of "food businesses" and subject to registration under the legislation.

Other exemptions

[F9032]–[F9033] Other exemptions are:
- private cars;
- aircraft (airlines and in-flight caterers that are food businesses should register with the most appropriate food authority);
- ships unless permanently moored, eg as restaurants, or used for pleasure excursions in inland or coastal waters;

- food vehicles normally based outside the UK;
- vehicles or stalls kept at premises that are themselves registered or exempt;
- tents, marquees and awnings;
- places where the main activity has nothing to do with food but where light refreshments such as biscuits and drinks are served to customers without charge (for example hairdressers' salons);
- premises where food is sold only through vending machines. Vending machines themselves are subject to the relevant provisions of EU Regulation 852/2004 but, in practice, it is recognised that there is little practical value in the registration of individual vending machines or the premises on which they are sited if the only food-related activity on those premises relates solely to vending machines. However, all ancillary activities related to the vending machines, such as distribution centres where food for stocking them is stored and/or despatched must be registered. The delivery vehicles are subject to the general hygiene requirements for transport;
- places run by voluntary or charitable organisations, and used only by those organisations, provided that no food is stored on the premises except tea, coffee, dry biscuits etc. For example, some village or church halls, but not if they are used more than five days in five weeks by commercial caterers;
- crown premises exempted for security reasons;
- places supplying food and drink in religious ceremonies;
- stores of food kept for an emergency or national disaster.

Food hygiene regulations

[F9034] The *Food Safety and Hygiene (England) Regulations 2013 (SI 2013 No 2996)* revoked and re-enacted with minor changes the *Food Hygiene (England) Regulations 2006 (SI 2006 No 14)* and certain provisions of the *General Food Regulations 2004 (SI 2004 No 3279)* as they apply in relation to England. They provide for the implementation and enforcement of *European Regulation 178/2002/EC* and several EU Hygiene Regulations (set out in Regulation 2(1) and Schedule 1).

A detailed account of the hygiene requirements within the vertical Regulation 853/2004 is beyond the scope of this publication.

Food Hygiene Regulations 2013 (SI 2013 No 2996)

[F9035] For the purposes of the legislation, 'hygiene' is defined as the measures necessary to ensure fitness for human consumption of a foodstuff taking into account its intended use. The concepts of 'fitness' and food safety are now defined in the general food legislation, described above. There is no reference to 'wholesomeness', as was the case under previous legislation.

Since the UK legislation is implementing directly-applicable EU Regulations, it is necessary to refer to these in order to establish the precise requirements.

Regulation 852/2004 (Article 4, Part A of Annex I and the 12 chapters of Annex II) sets out the general principles to be followed.

Article 5 requires food business operators to put in place, implement and maintain a permanent procedure or procedures based on the Hazard analysis and critical control points (HACCP) principles and establishes the seven principles to be applied. The HACCP principles are established by Codex Alimentarius, the joint FAO/WHO Food Standards programme. The Regulation recognises the need to provide sufficient flexibility to avoid undue burdens for very small businesses by requiring that documentation and records need only be commensurate with the nature and size of the business to demonstrate the effective application of their HACCP measures.

Hazard analysis and primary production

[F9036] The requirement to establish procedures based on HACCP principles does not currently apply to primary production but the feasibility of doing so is under on-going review. Regulation 852/2004 specifies seven elements that must be included in the process of identifying steps in the activities of the business that are 'critical' to ensuring that food is safe by identifying and/or establishing:

- hazards that must be prevented, eliminated or reduced to acceptable levels;
- critical control points ('CCPs') at the step(s) at which control is essential to prevent, eliminate or reduce a hazard to acceptable levels;
- critical limits at CCPs to separate acceptability from unacceptability for the prevention, elimination or reduction of identified hazards;
- implementing effective monitoring procedures at CCPs;
- corrective actions that need to be taken when monitoring indicates that a CCP is not under control;
- procedures, to be carried out regularly, to verify that the measures are working effectively;
- documents and records (appropriate to the nature and size of the food business) to demonstrate the effective application of the measures.

HACCP is a powerful and sophisticated management tool that can be used by businesses to develop a food control plan. Full and formal HACCP requires detailed expert knowledge and training. A significant amount of literature has built up over the years aimed at larger businesses that employ staff with technical training.

Clear guidance for some specific food sectors has been developed by national governments and international agencies around the world. Critical control points (CCP) for microbiological food poisoning hazards will tend to focus on a small number of areas.

COOKING: In most processes, there is likely to be a risk of pathogens in raw ingredients, so cooking is frequently a CCP. Adequate cooking temperatures will be the control measure, and target temperatures should be established. These can be monitored either by checking the food temperature directly or by monitoring the process, for example by using a cooking time and temperature that you know will achieve a satisfactory product centre temperature.

CONTAMINATION: Process steps at which ready-to-eat food may become contaminated are also likely to be CCPs. Contamination could come from contact with raw food ('cross contamination'), or from contact with contaminated equipment or personnel. Separation of raw and ready-to-eat foods, and cleaning and disinfection routines will be appropriate control measures. In a good hazard analysis system, these procedures should be specified objectively to allow them to be monitored. Written cleaning schedules will help. Rapid test systems will assess effectiveness of cleaning.

TIME AND TEMPERATURE CONTROLS: Cooling of food after cooking is often a CCP, as are storage times and temperatures of ready-to-eat foods. In most cases, controls will be expressed as a combination of time and temperature. The critical limits may be just a few hours if the food is kept at warm room temperature, or many days if it is kept under good refrigeration. In this area, technology is providing increasingly sophisticated monitoring options including automatic alarms and real-time data monitoring. The temperature control requirements of the *Food Hygiene Regulations 2013 (SI 2013 No 2996)* also intervene by prescribing certain targets.

Documentation and records

[F9037] A full, Codex-defined HACCP system (see **F9035** above) would be fully documented. Any monitoring or verification procedures would be recorded and records kept. There is no explicit requirement for either of these in Regulation 852/2004, *Article 5* which states that Food Business Operators must put in place, implement and maintain a permanent procedure or procedures *based on* the HACCP principles. The introductory *Recital 15* explains that:

> 'The HACCP requirements should take account of the principles contained in the Codex Alimentarius. They should provide sufficient flexibility to be applicable in all situations, including in small businesses. In particular, it is necessary to recognise that, in certain food businesses, it is not possible to identify critical control points and that, in some cases, good hygienic practices can replace the monitoring of critical control points. Similarly, the requirement of establishing "critical limits" does not imply that it is necessary to fix a numerical limit in every case. In addition, the requirement of retaining documents needs to be flexible in order to avoid undue burdens for very small businesses.'

Most businesses would, nevertheless, be advised to keep some concise and succinct documentation. Without it, it may not be easy to demonstrate compliance with the legislation and it would be difficult to assemble a convincing due diligence defence if that should become necessary. However, the legislation recognises the need to avoid undue burdens for very small businesses by explicitly stating that documentation and records need only be commensurate with the nature and size of the business to demonstrate the effective application of their HACCP measures.

Hygiene 'pre-requisites'

[F9038] Codex Alimentarius, in its definitive *Guidelines for the Application of the HACCP System* (2003), says that any HACCP system must be supported by basic hygiene controls. These have become known as the pre-requisites to HACCP. EU Regulation 852/2004 takes a similar approach

[F9038] Food Safety and Standards

by establishing certain requirements for structures and services, covering the following subjects within twelve chapters:

Chapter I: General requirements for food premises.
Chapter II: Specific requirements for rooms where food is prepared.
Chapter III: Movable or temporary premises, etc.
Chapter IV: Transport.
Chapter V: Equipment.
Chapter VI: Food waste.
Chapter VII: Water Supply.
Chapter VIII: Personal hygiene.
Chapter IX: Protection of food from contamination.
Chapter X: Wrapping and packaging
Chapter XI: Heat treatment
Chapter XII: Training

Chapter I: General requirements for food premises, equipment and facilities

[F9039] Food premises must be kept clean, and in good repair and condition. The layout, design, construction, and size must permit good hygiene practice and be easy to clean and/or disinfect and should protect food against external sources of contamination such as pests, dirt, condensation etc. Cleaning agents and disinfectants are not to be stored in areas where food is handled.

Where necessary, suitable temperature-controlled handling and storage conditions should be available, with sufficient capacity for maintaining foodstuffs at appropriate temperatures and designed to allow those temperatures to be monitored and, where necessary, recorded.

Adequate sanitary and hand-washing facilities must be available and lavatories must not lead directly into food handling rooms. Sanitary conveniences are to have adequate natural or mechanical ventilation. Washbasins must have hot and cold (or better still mixed) running water and materials for cleaning and drying hands. Where necessary there must be separate facilities for washing food and hands. Drainage must be suitable and there must be adequate changing facilities.

Premises must have suitable natural or mechanical ventilation, and adequate natural and/or artificial lighting. Ventilation systems must be accessible for cleaning, for example to give easy access to filters.

Chapter II: Specific requirements in food rooms

[F9040] Food rooms should have surface finishes that are in good repair, easy to clean and, where necessary, disinfect. This would, for instance, apply to wall, floor and equipment finishes. The rooms should also have adequate facilities for the storage and removal of food waste. Of course, every food premises must be kept clean. How they are cleaned and how often, will vary. For example, it will be different for a manufacturer of ready-to-eat meals than for a bakery selling bread.

There must be facilities, including hot and cold water, for cleaning and where necessary disinfecting tools and equipment. There must also be facilities for washing food wherever this is needed, using potable water.

Chapter III: Temporary and occasional food businesses and vending machines

[F9041] Premises and vending machines must, as far as reasonably practicable, be sited, designed, constructed and kept clean and maintained in good repair so as to avoid the risk of contamination, in particular by animals and pests.

Most of the requirements outlined in F9040 and this paragraph apply equally to food businesses trading from temporary or occasional locations like marquees or stalls although for reasons of practicability, some requirements are slightly modified.

Chapter IV: Transport of food

[F9042] The design of containers and vehicles must allow cleaning and disinfection. Businesses must keep them clean and in good order to prevent contamination and place food in them so as to minimise risk of contamination. Precautions must be taken if containers or vehicles are used for different foods or, in particular, for both food and non-food products. Different products should be separated to protect against the risk of contamination, and vehicles or containers should be cleaned effectively between loads. In general, bulk foodstuffs in liquid, granular or powder form should be transported in receptacles and/or containers/tankers reserved for the transport of foodstuffs.

Chapter V: Equipment

[F9043] All articles, fittings and equipment with which food comes into contact must be designed, constructed and maintained in a way that permits them to be effectively cleaned and, where necessary, disinfected at a frequency sufficient to avoid any risk of contamination. One interpretation of this requirement is that wooden cutting boards are not deemed suitable for use with ready-to-eat foods. Equipment should be installed so to allow adequate cleaning of the equipment and the surrounding area.

Where necessary to ensure the safety of the food product, equipment must be fitted with appropriate control devices, eg temperature recorders. If chemical additives are used to prevent corrosion of equipment and containers, they should be used in accordance with good practice.

Chapter VI: Food waste

[F9044] Food and other waste must not accumulate in food rooms any more than is necessary for the proper functioning of the food business. Containers for waste must be kept in good condition and be easy to clean and disinfect. Waste storage containers should be lidded to keep out pests. The waste storage area must be designed so that it can be easily cleaned and prevent pests gaining access. There should be arrangements for the frequent removal of refuse and the area should be kept clean.

All waste must be disposed in accordance with relevant legislation, using licensed waste disposal contractors and, in the case of products containing animal-derived ingredients, relevant animal by-products legislation.

Chapter VII: Water supply

[F9045] There must be an adequate supply of potable (drinking) water. Ice must be made from potable water (there are special rules for ice and water used for cooling/washing fish). Recycled water used in processing or as an ingredient must be of potable standard, unless it can be shown not to affect the wholesomeness of the food.

Water used for cooling cans etc after thermal treatment must not be a source of contamination. Non-potable water may be used for fire control, steam production, refrigeration and other similar purposes but must be in a separate, duly-identified system and not connect with potable water systems.

Steam used directly in contact with food must not contain any substance that presents a hazard to health or is likely to contaminate the food.

Chapter VIII: Personal hygiene

[F9046] Anyone who works in a food handling area must maintain a high degree of personal cleanliness. The way in which they work must also be clean and hygienic. Food handlers must wear clean and, where appropriate, protective clothes. Anyone whose work involves handling food should—

- follow good personal hygiene practices;
- wash their hands routinely when handling food;
- never smoke or eat in food handling areas;
- be excluded from food handling if carrying an infection that may be transmitted through food if there is a risk of contamination of the food.

Nobody with infected wounds, skin infections, sores or diarrhoea, or suffering from, or being a carrier of a disease that is likely to be transmitted through food, is to be permitted to handle food or enter any food-handling area in any capacity if there is any likelihood of direct or indirect contamination. Personnel working in food handling areas are obliged to report any illness (like diarrhoea or vomiting, infected wounds, skin infections) and if possible their causes, immediately to the proprietor of the business. If a manager receives such notification, he or she may have to exclude them from food handling areas. Such action should be taken urgently. If there is any doubt about the need to exclude, it is best to seek urgent medical advice or consult the local authority.

These personal hygiene provisions are wide ranging. They apply not only to food handlers, but anyone working in food handling areas. There have been well-documented cases of cleaners contaminating working surfaces or equipment with pathogens that are later transmitted to food.

FSA guidance, *Food Handlers Fitness to Work* issued by the Food Standards Agency can be found on its website at: www.food.gov.uk/sites/default/files/media/document/fitnesstoworkguide.pdf.

Chapter IX: Preventing food contamination

[F9047] A business operator must not accept ingredients or other material for use in food products, if they are known or might be expected to be, contaminated with parasites, pathogenic microorganisms or toxic, decomposed or foreign substances to the extent that, even after sorting, preparation and processing, the final product would be unfit for consumption.

Food and ingredients must be stored in conditions that prevent harmful deterioration and protected against contamination that may make them unfit for human consumption or a health hazard. For example, raw poultry must not be allowed to contaminate ready-to-eat foods.

Products likely to support the growth of pathogenic micro-organisms or toxin formation must be kept at appropriate temperatures (more details about temperature controls and management are given below at F9050). Frozen materials must be thawed in a way that similarly minimises the risk of micro-organisms growth or toxin formation. Any run-off liquid from the thawing process that may present a health risk must be adequately drained.

Adequate procedures to control pests and to prevent access by domestic animals must be in place. Hazardous and/or inedible substances, including animal feed, must be adequately labelled and stored in separate, secure containers.

Chapter X: Wrapping and packaging

[F9047.1] Materials used for wrapping and packaging must be stored and used in a way which ensures that they are not to be a source of contamination. Where appropriate and, in particular, in the case of cans and glass jars, the integrity of the container's construction and its cleanliness must be checked prior to use.

Wrapping and packaging material re-used for foodstuffs must be easy to clean and, where necessary, disinfect.

Chapter XI: Heat processing

[F9047.2] Specific requirements apply to foods processed and sold in hermetically-sealed containers to ensure safe processing and to avoid defective containers or re-contamination.

Chapter XII: Training and supervising food handlers

[F9048] Whereas eleven of the chapters deal with structural and physical pre-requisites, the final chapter deals with the higher-level issue of competence of personnel to do their job. This provision is more subtle than it is often given credit for. The provision, carried forward from previous legislation, is still subject to occasional criticism that it is too lenient on training requirements but, in fact, it requires three complementary elements, instruction, training and supervision.

Food business operators must ensure that food handlers are supervised and instructed and/or trained in food hygiene matters commensurate with their work activity and that those responsible for the development and maintenance of the HACCP-based procedure or for the operation of relevant guides have received adequate training in HACCP principles. In practice, therefore, food handlers must be instructed on how to do their particular jobs properly and supervised to make sure that they follow instructions. The requirements will differ according to the nature of the business and the job of the individual staff member.

Businesses must also comply with any national legal requirements concerning training programmes for persons working in certain food sectors, eg the meat industry.

Industry guides to good hygiene practice

[F9049] EU Regulation 852/2004 retains the concept of voluntary industry guides to good hygiene practice. These provide more detailed guidance on complying with the Regulations as they relate to specific industry sectors. They are usually produced by trade associations and recognised by the Food Standards Agency. Importantly, enforcement officers are obliged to have regard for them when examining how businesses are operating and industry-demonstrated compliance is likely to contribute towards a potential defence of 'due diligence' should the need arise. Numerous guides have now been published in the UK.

Temperature Control Requirements *[Food Safety and Hygiene (England) Regulations 2013 (SI 2013 No 2996) Schedule 4]*

[F9050] Growth of pathogenic bacteria in food will significantly increase the risk of food poisoning. Indeed, growth of any micro-organisms in food will compromise its wholesomeness. Temperature controls at certain food holding or processing steps will be CCPs in most HACCP plans. The importance of good control of food temperatures is emphasised by the fact that some controls are prescribed in regulations.

The *Food Safety and Hygiene (England) Regulations 2013 (SI 2013 No 2996) Schedule 4* requires food business operators to observe certain temperatures during the holding of food if this is necessary to prevent a risk to health. The temperature requirements relate only to matters of food safety. For example, hard cheese that may go mouldy if kept at room temperature is not covered because it would *not* support the growth of *pathogens*.

The temperature control requirements in the *Food Safety and Hygiene (England) Regulations 2013 (SI 2013 No 2996) Schedule 4* do not apply to foods and establishments that fall under EU Regulation 853/2004 which sets more specific rules for most foods of animal origin during processing and handling prior to the retail or catering outlet.

Relationship between time and temperature

[F9051]–[F9052] Almost invariably the control of micro-organisms in food depends upon a combination of time and temperature. Temperature alone does not have an absolute effect. For example, destruction of micro-organisms can be effected in a very short time (seconds) at temperatures above 100°C or in a couple of minutes at around 70°C. A similar thermal destruction can be achieved even at temperatures as low as 60°C but exposure for around 45 minutes will be necessary. Similarly, in chilled storage, food can remain safe and wholesome for many days if the temperature can be kept close to freezing point at around −1°C. However, it will have a much shorter life at higher storage temperatures around 10°C.

The above remarks are generalisations and not all micro-organisms will react in the same way. Some organisms show greater resistance to heating; some are able to grow at storage temperatures as low as 3°C whereas others will show little growth at temperatures cooler than 15°C. But the general point that control is a function of time and temperature remains true.

Chilled storage and foods included in the scope of the regulations

[F9053] The principal rule for chilled storage is contained within the *Food Safety and Hygiene (England) Regulations 2013 (SI 2013 No 2996), Sch 4 para 2*:

'Subject to sub-paragraph (2) and paragraph 3, any person who keeps any food—
(a) which is likely to support the growth of pathogenic micro-organisms or the formation of toxins; and
(b) with respect to which any commercial operation is being carried out, at or in food premises at a temperature above 8°C commits an offence.'

Schedule 4 para 2 only applies to food that will support the growth of pathogenic micro-organisms. Such foods must be kept at 8°C or cooler. The types of food that will be subject to temperature control are indicated in the following table.

It is often good practice to keep foods at temperatures cooler than 8°C either to preserve quality or to allow longer-term storage and to allow a margin of error below the legal standard. Industry guides suggest a target food temperature of 5°C. This is especially important in cabinet fridges where there can be significant temperature rises during frequent door opening.

Food type	Comment
Cooked meats and fish, meat and fish products.	Includes prepared meals, meat pies, pates, potted meats, quiches and similar dishes based on fish.
Cooked meats in cans that have been pasteurised rather than fully sterilised.	Typically, large catering packs of ham or cured shoulder.
Cooked vegetable dishes.	Includes cereals, rice and pulses. Some cooked vegetables or dessert recipes may have sufficiently high sugar content[*] (possibly combined with other factors like acidity) to prevent the growth of pathogenic bacteria. These will not be subject to mandatory temperature control.
Any cooked dish containing egg or cheese.	Includes flans, pastries etc.
Prepared salads and dressings.	Includes mayonnaise and prepared salads with mayonnaise or any other style of dressing. Some salads or dressings may have a formulation (especially the level of acidity[**]) that is adequate to prevent growth of pathogens.

Food type	Comment
Soft cheeses/mould ripened cheeses (after ripening).	Cheeses will include Camembert, Brie, Stilton, Roquefort, Danish Blue and any similar style of cheese.
Smoked or cured fish, and raw scombroid fish.	For example, smoked salmon, smoked trout, smoked mackerel etc. Also raw tuna, mackerel and other scombroid fish.
Any sandwiches whose fillings include any of the foods listed in this Table.	
Low acid** desserts and cream products.	Includes dairy desserts, fromage frais and cream cakes. Some artificial cream may be 'ambient stable' due to low water activity and/or high sugar. Any product that does not support the growth of pathogenic micro-organisms does not have to be kept below 8°C. It may be necessary to get clarification from suppliers.
Fresh pasta and uncooked or partly cooked pasta and dough products.	Includes unbaked pies and sausage rolls, unbaked pizzas and fresh pasta.
Smoked or cured meats which are not ambient stable.	Salami, Parma hams and other fermented meats will not be subject to temperature controls if they are ambient shelf stable.

* Technically Aw (water activity) is the key criterion.
** Technically, pH 4.5 (or more acid) is the critical limit.

Mail order foods

[F9054] There is a controversial exemption from the specific 8°C requirement if the food is being conveyed by post or by a private or common carrier to the ultimate consumer. The sender still has a responsibility for the safety of the food but the specific 8°C temperature does not apply. This exemption for mail order foods cannot apply to supplies to caterers or retailers neither of whom meet the definition of 'ultimate consumers'.

Short 'shelf life' products

[F9055] Some perishable foods are allowed to be kept at ambient temperatures for the duration of their shelf life with no risk to health. This may be because they are intended to be kept for only very short periods (eg certain types of sandwiches). Bakery products like fresh pies, pasties, custard tarts can also be kept without refrigeration for limited periods. However, good practice is to keep all such products that contain animal-derived products such as meats and dairy ingredients under temperature-controlled conditions.

Canned foods and similar

[F9056] Sterilised cans or similar packs do not have to be stored under temperature control until the hermetically sealed pack is opened. After that, perishable food must be kept chilled (for example corned beef, beans, canned fish, and dairy products).

However, some canned meats are not fully sterilised (eg large catering packs of ham or pork shoulder) and must be kept chilled even before the can is opened.

High acid canned or preserved foods (some fruit, tomatoes, etc) do not have to be kept chilled after opening for safety reasons. However, it is advisable to remove these from the can for chilled storage after opening. It will inhibit mould growth and avoid any chemical reaction with the metal can body.

Raw food for further processing

[F9057] Raw food that is intended for further processing does not have to be kept at 8°C or cooler, providing the further processing will render the food fit for consumption. It is therefore not against the Regulations to keep raw meat, poultry and fish out of the fridge. Of course, for quality and safety reasons it is usually best kept in the fridge; exceptions would include game and similar meats being 'hung' to develop their characteristic maturity. Processed foods must be kept at 8°C or cooler, even if they are to be heated again.

Raw meat or fish that is intended to be eaten without further processing (for example beef for steak tartare, or fish for sushi) will not be exempted by this provision. It must be kept chilled. Raw scombroid fish (tuna, mackerel, etc) will also not be exempted by this provision. The 'scombrotoxin' is heat stable and processing will not render contaminated food fit for consumption. Scombroid fish (such as tuna and mackerel) must be kept at 8°C or cooler.

Variations from 8°C storage

[F9058] The *Food Hygiene Regulations 2013 (SI 2013 No 2996) Reg 32, Sch 4 para 4* allow the producer of a food to recommend that it can be kept safely at a temperature above 8°C. He must label the food clearly with the recommended storage temperature and the safe shelf life at that temperature, or convey the information by other written form. He must also have good scientific evidence that the food is safe at that temperature. In practice, little use has been made of this provision.

The four-hour rule

[F9059] The 'four-hour rule' allows cold food to be kept above 8°C when it is on display. This is crucial for many catering outlets, and some retailers, which depend upon this exemption for at least some of their operation. Without it, food could not be served on a buffet without refrigeration. The time the food is on display must be controlled. The maximum time allowed is four hours.

Only one such period of display is allowed, no matter how short. For example, if a dish of food is placed on display for one hour at the end of a service period, it cannot be displayed for a further three hours above 8°C at the next service. Food uneaten at the end of a display period does not have to be discarded provided that it is still fit for consumption. The food must be cooled to 8°C or cooler and kept at that temperature until it can be used safely.

The burden of proof is on the business, which must be able to demonstrate that the time limit is observed. Good management of food displays is therefore important. The amount of food on display must be kept to a minimum consistent with the pattern of trade.

There needs to be systems to help keep to the time limits and to demonstrate that they are being adhered to. These may include the labelling of dishes to indicate when they went onto display. Avoid topping-up of bulk displays of food. Food at the bottom of the dish will remain on display for much longer than four hours if topping-up is allowed.

Exemptions for other contingencies

[F9060] The Regulations allow for the fact that food may rise above 8°C for limited periods of time in unavoidable circumstances such as:

- transfers to or from vehicles;
- during handling or preparation;
- during defrost of equipment;
- during temporary breakdown of equipment.

The Regulations do not put specific figures on the length of time allowed or how warm the food might become. This tolerance is allowed as a defence and the burden of proof will be on the business to show that:

- the food was unavoidably above 8°C for one of the reasons allowed;
- that it was above 8°C for only a limited period;
- that the break in temperature control was consistent with food safety.

The acceptable limits will obviously depend upon the combination of time and temperature. Under normal circumstances, a single period of up to two hours is unlikely to be questioned.

All transfers of food must be organised so that exposure to warm ambient temperatures is reduced and rises in food temperature are kept to a minimum. For example, deliveries should be put away quickly and chilled food moved first, then frozen, and finally ambient grocery.

If food is being collected from a cash and carry warehouse, insulated bags or boxes should be used for any chilled foods to which the Regulations apply. The chilled food should be taken directly to the outlet and quickly put into chilled storage.

Risks of equipment breakdown should be minimised by ensuring planned and regular maintenance.

Hot food

[F9061] Hot food must be kept at 63°C or hotter if this is necessary for safety. This will apply to the same types of food described in the earlier table (see F9053 above). For example, the rule does not apply to hot bread or doughnuts.

Food must be kept at 63°C or hotter, whether:

- it is in the kitchen or bakery awaiting service or dispatch;
- or in transit to a serving point, no matter how near or far;
- or actually on display in the serving area.

Some tolerance or limited exemptions are allowed for practical handling reasons. As a parallel to the 'four-hour' rule for chilled food, hot food can be kept for service or display at less than 63°C for *one period* of up to *two hours*.

Again, the operator must be able to show that:

- the food was for service or on display for sale;
- it had not been kept for more than two hours;
- it had only had one such period.

There is a further general exemption if a well-founded scientific assessment can show that storage and safety at below 63°C does not present any risk to health.

Generally, there should be fewer problems with meeting the 63°C target in hot display equipment than encountered in achieving 8°C in 'chilled' display units. However, operators must be aware that hot display units are not designed to heat or cook foods from cold. They should be used for holding only, not heating.

Cooling

[F9062]–[F9069] UK industry guides recommend that cooling between 60°C and 10°C should be accomplished in a maximum of four hours, although research suggests that this may be conservative.

Note that rapid cooling is not only necessary for food that has been heated. Food that has become warm during processing should also be returned to chilled storage below 8°C as quickly as possible after the process is completed.

Food Safety Manual – food safety policy and procedures

[F9070] Both the *Food Safety Act 1990* and the Regulations made under it provide an obligation on proprietors of food businesses to operate within the law. In addition, the *Food Safety Act 1990, s 21* provides the defence of 'due diligence' (see **F9023**). One way for a business to demonstrate its commitment to operating safely and within the law is to document its management system, including the relevant policies and procedures. This documentation may form a significant part of a due diligence defence. The policy and procedures will illustrate the precautions that should be taken by the business and its staff in their day-to-day operations. The procedures will normally require the keeping of various records, and these will in turn help to demonstrate that the *'reasonable precautions'* are being followed *'with all due diligence'*.

It is often convenient to keep these policies and procedures together in a 'Food Safety Manual'. The following provides a brief outline of the contents of a Food Safety Manual for a typical food business.

Policy and procedures

[F9071] The 'policy' will be a statement of intent by the business. A Food Safety Manual may begin with a fairly broad statement of policy. Later, there may be more specific policy statements linked to particular outcomes, for example a policy to take all reasonable precautions to minimise the contamination of food with foreign material.

Procedures will detail the ways in which the business intends to put the policies into effect. A typical layout of any 'Food Safety Manual' is to begin each section or subject heading with a statement of policy, and to follow that with the procedures intended to deliver it. Procedures are often referred to by the acronym SOPs – 'Standard Operating Procedures'.

General policy and organisation

[F9072] A Food Safety Manual would normally begin with a top-level statement of the business's policy with regard to food safety. Recognising the wider scope of 'food safety' within the meaning of the 1990 Act and the EU Regulation 178/2002, it is advisable to add a statement about commitment to true and accurate labelling of food and adherence to compositional requirements.

The statement should be signed by the proprietor of the business, who is the person with ultimate legal responsibility. In a large group this may be the group chief executive.

The early part of the manual should also illustrate the organisational structures and management hierarchy that has responsibility for implementing the policy and procedures. There may be technical support services, (eg pest control, testing laboratories) either from inside or outside the business. Their roles and responsibilities can also be identified within the organisational structures in this part of the manual.

Hazard Analysis (HACCP)

[F9073] The law requires that food safety policy must be centred on a hazard analysis approach and the Food Safety Manual should include a policy statement to that effect. This will be followed by procedures to describe how the business establishes the necessary HACCP procedure and record its outcomes and implementation.

Pre-requisites to Hazard Analysis

[F9074] The modern approach to HACCP focuses on the small number of process steps that *must* be controlled (and monitored) to ensure that the food is safe. But this control must operate against a background of good hygiene practice. The so-called 'pre-requisites' to HACCP must all be in place and the Food Safety Manual should include policies and procedures for all of the following:

- Design – structures and layout

The Food Safety Manual should include a policy that the design, structure and layout will promote hygienic operation. The procedures should detail specific arrangements to achieve that. This section should also cover points such as the services, especially water and ventilation; provision for washing of hands, equipment and food; and removal of waste.

- Equipment specification

The policy should be that the business will only acquire and use equipment that is fit for its purpose. The focus should extend beyond just cleanability alone. In many cases, the ability of equipment to do the job can be even more important. For example, can the refrigeration equipment keep food at a specified temperature in operational conditions?

- Maintenance

A policy for maintenance should address both the premises in general and the equipment. Procedures should detail the arrangements for routine and emergency maintenance. There should be a record of any maintenance whether routine or emergency.

- Cleaning and disinfection

Detailed cleaning schedules for sections of the building or particular pieces of equipment should be available. Cleaning should be recorded.

- Pest Control

Employing a pest control contractor is only one small part of the pest control procedures. The prevention of pest access, the integrity of pest proofing, cleaning and the need for vigilance should also be included in the procedures. Records would normally include any sightings of or evidence of pests, positions of bait stations, visits by the contractor together with observations and action taken, and even details such as the changing of UV tubes on electronic flying insect killers.

- Personal hygiene

The business should have a very clear policy on personal hygiene. The arrangements will include a very clear statement of the dress code and personal hygiene requirements for all staff and contractors working in or passing through food areas; appropriate rules should also apply to visitors to such areas. There must also be very particular policies and procedures to deal with staff suffering illness that may potentially be food borne.

- Food purchasing

The policy should state that the business will purchase only from reputable suppliers. However, it may be more difficult to deliver such a policy, especially for smaller businesses that do not have the resources to inspect their suppliers. The Manual should, therefore, detail how this policy will be implemented. For assessments, it may be possible to rely on third party inspection systems; especially those operated by accredited inspection or certification bodies.

Supply specifications should also be drafted. Such specifications will usually cover food safety and 'quality' in its broadest sense. In addition to specific safety parameters such as chemical and microbiological

[F9074] Food Safety and Standards

criteria, they are the ideal route through which to state food safety conditions such as delivery temperature, shelf life remaining after delivery, packaging and so on.

- Storage

  Most businesses will have documented procedures for the storage of various categories of food as well as for stock checks to ensure proper storage temperature and stock rotation.

- Food handling and preparation

  The general policy should be to manage all steps in food handling and preparation in order to minimise contamination with any hazardous material or pathogenic micro-organisms. There should also be precautions to minimise the opportunities for growth of any micro-organisms that might contaminate the food. The procedures will be unique to the individual business, its particular range of foods and preparation methods.

- Foreign material

  The policy should require that all precautions are taken to minimise the risk of foreign material contamination, including chemical hazards such as cleaning materials. In general terms, procedures will cover three points:
  (a) Excluding or controlling potential sources of contamination;
  (b) Protecting food from contamination;
  (c) Taking any complaints seriously, tracking down what caused them and stopping them from happening again. Complaint trend analysis is important.

Allergy (Anaphylaxis)

[F9075] A responsible business will have a policy to protect susceptible customers from exposure to foods to which they may be allergic. In general terms procedures will rely upon:

(a) Awareness amongst staff of the types of food that may cause reaction.
(b) Labelling of the foods that include such ingredients.
(c) Careful handling, storage and use of these ingredients at all stages to avoid 'cross-contamination' to other products.
(d) An ability to supply reliable information about all ingredients on request from consumers.

Food labelling law requires comprehensive labelling of the known presence of 14 specified allergens to be declared when they or their derivatives are deliberately added to pre-packed food, including alcoholic drinks (see F9003). The detailed requirements of this legislation are outside the scope of this overview of food safety legislation.

Vegetarians and other dietary preferences

[F9076] Most food businesses today recognise the need to cater for customers with particular dietary preferences, such as vegetarian, vegan, kosher and Halal. There should be a policy to be able to supply vegetarian and/or other specified dishes properly formulated and accurately labelled. The FSA has issued guidance on the use of the terms 'vegetarian' and 'vegan' on food labels.

Consumer concerns

[F9077] From time to time particular issues will come to the forefront of consumer consciousness. For example, genetically modified ingredients, sustainability, food miles and carbon footprints or animal welfare issues, such as veal production, foie gras and intensive rearing of poultry. The Food Safety Manual could include a statement of the policy on these issues.

Procedures will be needed to deliver them. In the case of food and food ingredients derived from genetically modified organisms, comprehensive legislation requires the traceability and labelling of all such materials throughout the food supply chain. Numerous 'private' schemes cover areas including 'ethical trading', carbon footprint labelling etc but, as yet, no formal legislation specifically controls these (other than the general rules on misleading labelling and advertising).

Labelling and composition of foods and 'fair-trading'

[F9078] In view of the fact that the *Food Safety Act 1990* deals not only with health and hygiene, and in the light of growing interest in these areas, it may be considered desirable to include policies and procedures on topics such as food composition and nutrition labelling. Ethical and fair-trade issues might also be relevant to the company's product range and intended market; schemes and controls on these are as described under **F9077**.

Product recall

[F9079] Traceability and the requirement to initiate recall/withdrawal procedures if a company has reason to believe one of its products may be 'unsafe' are now legal requirements. Businesses involved in central production and distribution of food that will have several days of 'shelf life' (if not more) should have contingency plans for product recall by ensuring that products are labelled in such a way that individual batches can be identified and records kept to show where/when it might have been delivered.

A formal recall procedure should be an integral part of its food safety management policy.

Complaints

[F9080] Complaints are an important source of information. A business should have a policy and procedures to deal with them. These will normally include keeping a record of every complaint and making every attempt to track down its cause. Complaint trends should be analysed regularly as a means to identify developing problems in the manufacturing process or supply chain.

Training, instruction and supervision

[F9081] Cutting across all of the other policies and procedures will be a policy on staff training, instruction and supervision. Typically, a business will keep a personal training record for all members of staff.

Summary

[F9082] Modern businesses recognise the value of a properly documented management system. A Food Safety Manual can provide the focus of such a system. The design of a Food Safety Manual must be a fine balance between conflicting objectives. On the one hand, it must be reasonably comprehensive if it is to stand up as part of a 'due diligence' defence. On the other hand, it must be sufficiently concise to be usable as an effective working tool within the business. It must form an integral part of the business management's working tools – it must not be seen as a "nice-to-have" which, once developed languishes on an office shelf, gathering dust.

Gas Safety

Andrea Oates

Introduction to gas safety

[G1001] Explosions caused by escaping gas have been responsible for large-scale damage to property and considerable personal injury, including death. A number of people also die every year from carbon monoxide poisoning caused by gas appliances and flues that have not been properly installed, maintained or that are poorly ventilated.

The highest UK health and safety fine to date, totalling £15 million, was handed down to gas supply company Transco PLC in 2005 following a gas explosion which destroyed a house in Larkhall, Lanarkshire and killed the family of four who lived there.

A Health and Safety Executive (HSE) investigation showed there were holes in the iron pipe that ran through the front garden of the house. Gas leaking from the main found its way into the under-floor void and subsequently the kitchen of the property where it ignited.

The main pieces of legislation applying to the gas industry are as follows:

- the *Gas Act 1986* (which was considerably modified and extended by the *Gas Act 1995*);
- *Gas Safety (Installation and Use) Regulations 1998 (SI 1998 No 2451)*;
- *Gas Safety (Management) Regulations 1996 (SI 1996 No 551)*;
- *Pipelines Safety Regulations 1996 (SI 1996 No 825)*; and
- *Gas Safety (Rights of Entry) Regulations 1996 (SI 1996 No 2535)* (made under the authority of the *Gas Act 1986* (as amended)).

Some of this legislation concerns the safety responsibilities of gas processors, gas transporters and gas suppliers and is therefore outside the scope of this handbook. This chapter concentrates instead on the legislation which deals with the knowledge required by landlords, managing agents, health and safety managers, employers and other responsible persons who must ensure that gas appliances and fittings are installed safely and checked by a competent person on an annual basis.

The Health and Safety Executive (HSE) published a new (fifth) edition of its approved code of practice (ACOP) and guidance on the *Gas Safety (Installation and Use) Regulations 1998 (GSIUR) (as amended), Safety in the installation and use of gas systems and appliances*, (L56).

The ACOP and guidance provides advice on how to meet the requirements of GSIUR, which were amended on 6 April 2018. Those with duties under the Regulations include people who install, service, maintain and repair gas appliances and other gas fittings, as well as landlords.

The HSE explains that the 2018 amending regulations introduced some flexibility regarding the timing of landlords' annual gas safety checks. These can now be carried out in the two months before the due date, while retaining the existing expiry date. The aim is to prevent landlords waiting until the last minute and then experiencing access problems, or having to shorten the annual cycle to comply with the law.

The amending regulations also:

- allow alternative checks in situations where there is no meter to directly measure the heat input and it is not possible to measure the operating pressure;
- extend the scope slightly to include situations where the meter is not accessible or the meter display is not working; and
- disapply most of the requirements of GSIUR for installations fed by a dedicated gas supply, which are primarily used to supply compressed natural gas (CNG) to vehicles and which incorporate at least one gas compressor with an electric motor input power rating exceeding 5 kilowatts, bringing them in line with other industrial premises. The HSE says other more appropriate regulations already apply to these premises.

The definition of 'gas' under the GSIUR includes:

(a) methane, ethane, propane, butane, hydrogen and carbon monoxide;
(b) a mixture of two or more of these gases; and
(c) a combustible mixture of one or more of these gases and air.

The Regulations cover gas including natural gas, liquefied petroleum gas (LPG), methane from coal mines and landfill gas when these products are stored, supplied or used in situations within the scope of Regulations. Gas conveyed to premises through a distribution main (normally natural gas) and gas supplied from a storage vessel (mainly LPG) are both covered by the Regulations.

However, the Regulations do not cover most situations where LPG is likely to be used in liquid form, eg for automotive use or grain drying. The definition does not include gas consisting wholly or mainly of hydrogen when used in non-domestic premises, such as a laboratory in the industrial or education sector.

The use and storage of gas in any workplace is also subject to the *Dangerous Substances and Explosives Atmospheres Regulations 2002 (SI 2002, No 2776)*. In practical terms, the application of *SI 2002, No 2776* has the effect of bringing the standards of GSIUR into the workplace (see **DANGEROUS SUBSTANCES AND EXPLOSIVE ATMOSPHERES**).

Gas supply

The Gas Acts 1986 and 1995

[G1002] The Gas Act 1995 updated provisions in the Gas Act 1986, including new licensing arrangements for public gas transporters and permitting competition in the domestic gas market.

A number of safety issues are directly and indirectly addressed, including detailed provisions inserted into the *Gas Act 1986* under Schedule 2B – the 'gas code' – by the 1995 Act. These include duties to notify connection and disconnection of service pipes, and disconnection of meters in certain circumstances; maintenance of service pipes; duties on gas transporters in relation to certain gas escapes, and entry powers in specified circumstances.

With regard to safety, the *Gas Act 1986* (as amended) provides for:

- licensing and general duties [*Gas Act 1986, ss 4A, 5 and 8A*];
- security [*Gas Act 1986, s 11*];
- safety regulations regarding rights of entry for inspection of connected systems and equipment and the making safe and investigation of gas escapes [*Gas Act 1986, ss 18 and 18A*];
- pipeline capacity [*Gas Act 1986, ss 21, 22, 22A*];
- standards of performance [*Gas Act 1986, s 33A*].

Gas supply management – the Gas Safety (Management) Regulations 1996 (SI 1996 No 551)

[G1003] Principally, the *Gas Safety (Management) Regulations 1996 (SI 1996 No 551)* require an appointed person – the 'network emergency co-ordinator' – to prepare a safety case for submission to and acceptance by the HSE before a gas supplier can convey gas in a specified gas network [*SI 1996 No 551, Reg 3*].

The safety case

[G1004] The safety case document must contain:

— the name and address of the person preparing the safety case (the duty holder);
— a description of the operation intended to be undertaken by the duty holder;
— a general description of the plant, premises and interconnecting pipes;
— technical specifications;
— operation and maintenance procedures;
— a statement of the significant findings of the risk assessment carried out pursuant to the *Management of Health and Safety at Work Regulations 1999 (SI 1999 No 3242), Reg 3*;

— particulars to demonstrate the adequacy of the duty holder's management system to ensure the health and safety of his employees and of others (in respect of matters within his control);
— particulars to demonstrate adequacy in the dissemination of safety information;
— particulars to demonstrate the adequacy of audit and reporting arrangements;
— particulars to demonstrate the adequacy of arrangements for compliance with the duty of co-operation [*Gas Safety (Management) Regulations 1996 (SI 1996 No 551), Reg 6*];
— particulars to demonstrate the adequacy of arrangements for dealing with gas escapes and investigation [*Gas Safety (Management) Regulations 1996 (SI 1996 No 551), Reg 7*];
— particulars to demonstrate compliance with the content and characteristics of the gas to be conveyed in the network [*Gas Safety (Management) Regulations 1996 (SI 1996 No 551), Reg 8*];
— particulars to demonstrate the adequacy of the arrangements to minimise the risk of supply emergency;
— particulars of the emergency procedures [*Gas Safety (Management) Regulations 1996 (SI 1996 No 551), Reg 3(1) and Sch 1*].

Duties of compliance and co-operation

[G1005] The duty holder must ensure that the procedures and arrangements described in the safety case and any revision of it are followed. In addition, a duty of co-operation is placed upon specified persons including:

— a person conveying gas in the network;
— an emergency service provider;
— the network emergency co-ordinator in relation to a person conveying gas;
— a person conveying gas in pipes which are not part of a network;
— the holder of a licence issued under the *Gas Act 1986, s 7A*;
— the person in control of a gas production or processing facility [*Gas Safety (Management) Regulations 1996 (SI 1996 No 551), Regs 5 and 6*].

The *Gas Safety (Management) Regulations 1996 (SI 1996 No 551), Reg 7* deals with the duties and responsibilities of gas suppliers, gas transporters and 'responsible persons' regarding actions to be taken in the event of gas escape incidents. All incidents must be investigated including those resulting in an accumulation of carbon monoxide gas from incomplete combustion of a gas fitting.

Anyone discovering or suspecting a gas leak must notify British Gas plc immediately by telephone. British Gas is obliged to provide a continuously manned telephone service in Great Britain. The person appointed 'responsible person' for the premises must take all reasonable steps to shut off the gas supply.

The reporting of gas incidents to the HSE is laid down in the *Reporting of Injuries, Diseases and Dangerous Occurrences Regulations 2013 (SI 2013 No 1471) (RIDDOR)*. For further information regarding compliance with these

Regulations see the HSE publication L80: '*A guide to the Gas Safety (Management) Regulations 1996*' which can be downloaded from the HSE website at www.hse.gov.uk.

Rights of entry – the Gas Safety (Rights of Entry) Regulations 1996 (SI 1996 No 2535)

[G1006] Where an officer authorised by a gas transporter has reasonable cause to suspect an escape of gas into or from premises, he is empowered to enter the premises and to take any steps necessary to avert danger to life or property. [*Gas Safety (Rights of Entry) Regulations (SI 1996 No 2535), Reg 4*].

Inspection, testing and disconnection

[G1007] On production of an authenticated document, an authorised person must be allowed to enter premises in which there is a service pipe connected to a gas main for the purpose of inspecting any gas fitting, flue or means of ventilation. Fittings and any part of a gas system may be disconnected and sealed off when it is necessary to do so for the purpose of averting danger to life or property.

It is incumbent on the authorised officer carrying out any disconnection or sealing off activities to give a written statement to the consumer within five days. The notice must contain:

— the nature of the defect;
— the nature of the danger in question;
— the grounds and the manner regarding the appeal procedure available to the consumer.

Prominent and conspicuous notices must also be affixed at appropriate points of the gas system regarding the consequences of any unauthorised reconnection to the gas supply. [*Gas Safety (Rights of Entry) Regulations (SI 1996 No 2535), Regs 5–8*].

Prohibition of reconnection

[G1008] It is an offence for any person, except with the consent of the relevant authority, to reconnect any fitting or part of a gas system. This provision is qualified by the term 'knows or has reason to believe that it has been so disconnected'. [*Gas Safety (Rights of Entry) Regulations (SI 1996 No 2535, Reg 9*].

Pipelines – the Pipelines Safety Regulations 1996 (SI 1996 No 825)

[G1009] The *Pipelines Safety Regulations 1996 (SI 1996 No 825)* apply to all pipelines in Great Britain, both on and offshore, with the following exceptions:

- pipelines wholly within premises;
- pipelines contained wholly within caravan sites;
- pipelines used as part of a railway infrastructure;
- pipelines which convey water, air, water vapour or steam.

For the purposes of the Regulations, a pipeline for supplying gas to premises is deemed not to include anything downstream of an emergency control valve, that is, a valve for shutting off the supply of gas in an emergency, being a valve intended for use by a consumer of gas. [SI 1996 No 825, Reg 3(4)].

The Regulations complement the *Gas Safety (Management) Regulations 1996 (SI 1996 No 551)* and include:

- the definition of a pipeline [SI 1996 No 825, Reg 3];
- the general duties for all pipelines [SI 1996 No 825, Regs 5–14];
- the need for co-operation among pipeline operators [SI 1996 No 825, Reg 17];
- arrangements to prevent damage to pipelines [SI 1996 No 825, Reg 16];
- the description of a dangerous fluid [SI 1996 No 825, Reg 18];
- notification requirements [SI 1996 No 825, Regs 20–22];
- the major accident prevention document [SI 1996 No 825, Reg 23];
- the arrangements for emergency plans and procedures [SI 1996 No 825, Regs 24–26].

Generally, the Regulations place emphasis on 'major accidents' and the preparation of emergency plans by local authorities. As with the *Gas Safety (Management) Regulations 1996 (SI 1996 No 551)*, the detail relates to the specification and characteristics of gas and the design of the pipes to convey it. For further information on and guidance to the *Pipelines Safety Regulations 1996*, see the HSE Publication L82: 'A guide to the Pipelines Safety Regulations 1996'.

Gas systems and appliances

[G1010] The *Gas Safety (Installation and Use) Regulations 1998 (SI 1998 No 2451)* are supported by the HSE Approved Code of Practice and Guidance, L56: *Safety in the installation and use of gas systems and appliances Gas Safety (Installation and Use) Regulations 1998 as amended*. The Regulations deal with 'the safe installation, maintenance and use of gas systems, including gas fittings, appliances and flues, mainly in domestic and commercial premises, eg offices, shops, public buildings and similar places'.

They aim to protect gas consumers and cover natural gas, liquefied petroleum gas (LPG), landfill gas, coke, oven gas, and methane from coal mines when these products are 'used' by means of a gas appliance. Such appliances must be designed for use by a gas consumer for heating, lighting and cooking, or other purposes for which gas can be used.

Mobile or portable appliances, where gas is supplied from a cylinder, are not generally covered by the Regulations. However, the duty under *Reg 35*, which requires an employer or self-employed person to ensure that any gas appliance,

Gas systems and appliances [G1010]

flue or installation pipework installed at a place of work they control is maintained in a safe condition, does extend to certain mobile and portable appliances, such as LPG space heaters. In addition, they are subject to *Reg 3*, which requires anyone carrying out gas work to be competent, and *Reg 36*, which details landlords' duties to carry out annual safety checks on gas appliances and flues and carry out maintenance.

The Regulations do not cover gas appliances in domestic premises where a tenant is entitled to remove such appliances from the premises.

In addition to the general provisions governing the safe installation of gas appliances and associated equipment by HSE approved and competent persons, the 1998 Regulations lay down specific duties for landlords.

Except in the case of 'escape of gas' (*SI 1998 No 2451, Reg 37*), and certain types of valves to control pressure fluctuations, the Regulations do not apply to the supply of gas when used in connection with:

— bunsen burners in an educational establishment [*SI 1998 No 2451, Reg 2*];
— mines or quarries [see *Mines and Quarries Act 1954*] and the *Quarries Regulations 1999*];
— factories [see *Factories Act 1961 (FA61), s 175*] or electrical stations [*FA61, s 123*], institutions [*FA61, s 124*], docks [*FA61, s 125*] or ships [*FA61, s 126*];
— agricultural premises;
— temporary installations used in connection with any construction work within the meaning of the *Construction (Design and Management) Regulations 2015 (SI 2015 No 51)*;
— premises used for the testing of gas fittings; or
— premises used for the treatment of sewage.

Note: The Regulations would apply in relation to the above premises (or parts of those premises) if used for domestic or residential purposes or as sleeping accommodation, but not to hired touring caravans. [*SI 1998 No 2451, Reg 2*].

Generally, gas must be used in order to come within the scope of the Regulations. Therefore, the venting of waste gas from coal mines and landfill sites is excepted unless the gas is collected and intended for use. The Regulations do not cover gas used as automotive power or from grain drying.

'Work' in relation to a gas fitting is defined in the Regulations (Regulation 2) as including any of the following activities carried out by any person, whether an employee or not, that is to say:

(a) installing or reconnecting the fitting;
(b) maintaining, servicing, permanently adjusting, disconnecting, repairing, altering or renewing the fitting or purging it of air or gas;
(c) where the fitting is not readily movable, changing its position; and
(d) removing the fitting.

L56 provides further guidance on the meaning of work and states that 'work' for the purposes of these Regulations also includes do-it-yourself activities, work undertaken for friends and work for which there is no expectation of reward or gain, such as charitable work. [*L56, para 52*].

[G1011] Gas Safety

Duties and responsibilities

[G1011] Duties are placed on a wide range of people associated with domestic premises and on those concerned with the supply of gas to those premises and to industrial premises which have accommodation facilities.

Most of the legal requirements are 'absolute', ie they are not qualified by 'as far as is practicable' or 'so far as is reasonably practicable'; in all cases the duty must be satisfied to avoid committing an offence. Persons affected by the Regulations include:

— *individuals*. These include householders and other members of the general public;
— *responsible person*. This is the occupier of the premises or, where there is no occupier or the occupier is away, the owner of the premises or any person with authority for the time being to take appropriate action in relation to any gas fittings;
— *landlord*. In England and Wales, the landlord is defined as follows:
   (i) where the relevant premises are occupied under a lease, the person for the time being entitled to the reversion expectant on that lease or who, apart from any statutory tenancy, would be entitled to possession of the premises; and
   (ii) where the relevant premises are occupied under a licence, the licensor, save where the licensor is himself a tenant in respect of those premises.
   In Scotland, the landlord is the person for the time being entitled to the landlord's interest under a lease;
— *tenant*. In England and Wales, the tenant is defined as follows:
   (i) where the relevant premises are so occupied under a lease, the person for the time being entitled to the term of that lease; and
   (ii) where the relevant premises are so occupied under a licence, the licensee.
   In Scotland, the tenant is the person for the time being entitled to the tenant's interest under a lease. Duties are imposed on employers and the self-employed to ensure that all persons carrying out work in relation to gas fittings are competent to do so and are members of HSE approved organisations. [*Gas Safety (Installation and Use) Regulations 1998 (SI 1998 No 2451), Reg 3*].

Persons connected with work activities include:

— *supplier*. In relation to gas, a 'supplier' means:
   (i) a person who supplies gas to any premises through a primary meter; or
   (ii) a person who provides a supply of gas to a consumer by means of the filling or refilling of a storage container designed to be filled with gas at the place where it is connected for use, whether or not such container is or remains the property of the supplier; or
   (iii) a person who provides gas in refillable cylinders for use by a consumer whether or not such cylinders are filled, or refilled, directly by that person and whether or not such cylinders are or remain the property of that person.

Note: A retailer is not a supplier when he sells a brand of gas other than his own.
— *transporter.* This is defined as meaning a person, other than a supplier, who conveys gas through a distribution main.
— those carrying out 'work' in relation to a gas fitting. This is any person who installs, services, maintains, removes or repairs gas fittings whether he or she is an employee, self-employed or working on his or her own behalf (for example, a do-it-yourself activity).

Escape of gas

[G1012] It is the duty of the responsible person for the premises to take immediate action to shut off the gas supply to the affected appliance or fitting and notify the gas supplier (or the nominated gas emergency call-out office if different from the supplier) if they know or have reason to suspect that gas is escaping into those premises. All reasonable steps must be taken to prevent further escapes of gas, and the gas supplier must stop the leak within twelve hours from the time of notification. This may entail cutting off the gas supply to the premises. [*Gas Safety (Installation and Use) Regulations 1998 (SI 1998 No 2451), Reg 37(1)–(3)*].

An escape of gas also includes an emission of carbon monoxide resulting from incomplete combustion in a gas fitting. However, the legal duties regarding the action to be taken by the gas supplier are limited to advising of the need for immediate action by a competent person to examine, and if necessary, carry out repairs, to the faulty fitting or appliance. [*Gas Safety (Installation and Use) Regulations 1998 (SI 1998 No 2451), Reg 37(8)*].

Competent persons and quality control

[G1013] No work is permitted to be carried out on a gas fitting or a gas storage vessel except by a competent person. [*Gas Safety (Installation and Use) Regulations 1998 (SI 1998 No 2451), Reg 3(1)*].

Where work to a gas fitting is to any extent under their control – or is to be carried out at any place of work under their control – employers (and self-employed persons) must take reasonable steps to ensure that the person undertaking such work is registered with an HSE-approved body (that is the Gas Safe Register www.gassaferegister.co.uk) [*Gas Safety (Installation and Use) Regulations 1998 (SI 1998 No 2451), Reg 4*].

A gas fitting must not be installed unless every part of it is of good construction and sound material, of adequate strength and size to secure safety and of a type appropriate for the gas with which it is to be used. [*Gas Safety (Installation and Use) Regulations 1998 (SI 1998 No 2451), Reg 5(1)*].

Competent persons – qualifications and supervision

[G1014] To achieve competence in safe installation, a person's training must include knowledge of installing, purging, commissioning, testing, servicing, maintenance, repair, disconnection, modification and dismantling of gas systems, fittings and appliances.

[G1014] Gas Safety

This should include an adequate knowledge of:

(a) relevant associated services such as water and electricity;
(b) the potential for exposure to asbestos;
(c) the dangers these may give rise to; and
(d) the precautions to take.

In addition, a sound knowledge of combustion and its technology is essential, including:

— properties of fuel gases,
— combustion,
— flame characteristics,
— control and measurement of fuel gases,
— gas pressure and flow,
— construction and operation of burners, and
— operation of flues and ventilation.

To reach the approval standard required by *Gas Safety (Installation and Use) Regulations 1998 (SI 1998 No 2451), Reg 3*, gas installers and gas fitters should know:

(a) where/how gas pipes/fittings (including valves, meters, governors and gas appliances) should be safely installed;
(b) how to site/install a gas system safely, with reference to safe ventilation and flues;
(c) associated electrical work (e.g. appropriate electrical power supply circuits, that is, overcurrent and shock protection from electrical circuits, earthing and bonding);
(d) electrical controls appropriate to the system being installed/maintained/repaired;
(e) when/how to check the whole system adequately before it is commissioned;
(f) how to commission the system, leaving it safe for use.

In addition, they should know how to recognise and test for conditions that might cause danger and what remedial action to take, as well as being able to show consumers how to use any equipment they have installed or modified, including how to shut off the gas supply in an emergency. They should also alert customers to the significance of inadequate ventilation and gas leaks and the need for regular maintenance/servicing. L56 sets out that competence is a combination of practical skill, training, knowledge and experience to carry out the job in hand safely, and ensuring the installation is left in a safe condition for use. Knowledge must be kept up-to-date with changes in the law, technology and safe working practice. Industry guidance, *Standards of training in gas work* (available on the Institution of Gas Engineers and Managers website at www.igem.org.uk) provides criteria and guidance on the scope, standards and quality of training required to enable a gas engineer to achieve competence.

Individual gas fitting operatives must have their competence assessed at regular intervals by a certification body accredited by the United Kingdom Accreditation Service (UKAS).

Materials and workmanship

[G1015] It is incumbent on gas installers to acquaint themselves with the appropriate standards about gas fittings and to ensure that the fittings they use meet those standards. Most gas appliances are now subject to European Union (EU) Regulation 2016/426 and therefore should carry the CE mark.

The *Gas Appliances (Enforcement) and Miscellaneous Amendments Regulations 2018 (SI 2018/389)* provide for the enforcement of the EU regulation in the UK and revoked the *Gas Appliances (Safety) Regulations 1995 (SI 1995 No 1629)*. The 1995 Regulations continue to apply, as if they had not been revoked, to appliances and fittings placed on the market before 21 April 2018. It is an offence to carry out any work in relation to a gas fitting or gas storage vessel other than in accordance with the appropriate standards and in such a way as to prevent danger to any person. [*Gas Safety (Installation and Use) Regulations 1998 (SI 1998 No 2451), Reg 5(3)*]. Gas pipes and pipe fittings installed in a building must be metallic or of a type constructed in an encased metallic sheath and installed, so far as is reasonably practicable, to prevent the escape of gas into the building if the pipe should fail. Pipes or pipe fittings made from lead or lead alloy must not be used. [*Gas Safety (Installation and Use) Regulations 1998 (SI 1998 No 2451), Reg 5(2)*].

General safety precautions

[G1016] A general duty is imposed on all persons in connection with work associated with gas fittings to prevent a release of gas unless steps are taken which ensure the safety of any person [*SI 1998 No 2451, Reg 6(1)*].

The *Gas Safety (Installation and Use) Regulations 1998 (SI 1998 No 2451), Regs 6(2)–(6)* provide that the following precautions must be observed when carrying out work activities:

— a gas fitting must not be left unattended unless every complete gasway has been sealed with an appropriate fitting;
— a disconnected gas fitting must be sealed at every outlet;
— smoking or the use of any source of ignition is prohibited near exposed gasways;
— it is prohibited to use any source of ignition when searching for escapes of gas;
— any work in relation to a gas fitting which might affect the gas tightness of the installation must be tested immediately for gas tightness.

With regard to gas storage vessels it is an offence for any person intentionally or recklessly to interfere with a vessel, or otherwise do anything which might affect it so that the subsequent use of that vessel might cause a danger to any person. [*SI 1998 No 2451), Reg 6(9)*]. In addition:

— gas storage vessels must only be installed where they can be used, filled or refilled without causing danger to any person [*SI 1998 No 2451, Reg 6(7)*];
— gas storage vessels, or appliances fuelled by LPG which have an automatic ignition device or a pilot light, must not be installed in cellars or basements [*SI 1998 No 2451, Reg 6(8)*];

[G1016] Gas Safety

— methane gas must not be stored in domestic premises [*SI 1998 No 2451, Reg 6(10)*];
— the storage of dangerous substances such as LPG and natural gas should comply with the general safety measures set out in the *Dangerous Substances and Explosive Atmospheres Regulations 2002 (SI 2002 No 2776), Reg 6(8)*. The approved code of practice and guidance on the *Dangerous Substances and Explosive Atmospheres Regulations 2002* (L138) gives further details.

Gas appliances

[G1017] Precautions to be observed regarding gas appliances apply to everyone, not just gas installers. Generally, it is an offence for the occupier, owner or other responsible person to permit a gas appliance to be used if at any time he knows, or has reason to suspect, that the appliance is unsafe or that it cannot be used without constituting a danger to any person [*Gas Safety (Installation and Use) Regulations 1998 (SI 1998 No 2451), Reg 34*]. It is also an offence for anyone to carry out work in relation to a gas appliance which indicates that it no longer complies with approved safety standards [*SI 1998 No 2451, Reg 26(7), (8)*]. *Regulation 27* imposes similar safety requirements on persons who install or connect gas appliances to flues (see **G1019** below).

All gas appliances must be installed in a manner which permits ready access for operation, inspection and maintenance [*Reg 28*]. Manufacturer's instructions regarding the appliance must be left with the owner or occupier of the premises after its installation [*Reg 29*]. If the appliance is of the type where it is designed to operate in a suspended position, the installation pipework and other associated fittings must be constructed and installed as to be able to support the weight of the appliance safely [*SI 1998 No 2451, Reg 31*].

During installation, it is essential to ensure that all gas appliances are:

— connected (in the case of a flued domestic gas appliance) to a gas supply system by a permanently fixed rigid pipe [*SI 1998 No 2451, Reg 26(2)*];
— installed with a means of shutting off the supply of gas to the appliance unless it is not reasonably practicable to do so [*SI 1998 No 2451, Reg 26(6)*].

The *Gas Safety (Installation and Use) (Amendment) Regulations 2018* amended Regulation 26(9). It now sets out that:

'(9) Where a person performs work on a gas appliance he shall immediately thereafter examine—
(a) the effectiveness of any flue;
(b) the supply of combustion air;
(c) [subject to sub-paragraph (ca),] its operating pressure or heat input or, where necessary, both;
[(ca) if it is not reasonably practicable to examine its operating pressure or heat input (or, where necessary, both), its combustion performance;]
(d) its operation so as to ensure its safe functioning'.

Any defects must be rectified and reported to the appropriate responsible person as soon as is practicable; the defect must be reported to the appropriate gas supply organisation (or transporter) if a responsible person or the owner is not available [*SI 1998 No 2451, Reg 26(9)*].

Room-sealed appliances

[G1018] A room-sealed appliance has a combustion system sealed from the room in which it is located and obtains air for combustion from a ventilated uninhabited space within the premises or directly from the open air outside the premises. The products of combustion must be vented safely to open air outside the premises.

A room-sealed appliance must be used in the following rooms:

— a bathroom or a shower room [*SI 1998 No 2451, Reg 30(1)*];
— in respect of a gas fire, gas space heater or gas water heater (including instantaneous water heaters) of more than 14 kilowatt gross heat input, in a room used or intended to be used as sleeping accommodation [*SI 1998 No 2451, Reg 30(2)*].

Gas heating appliances which have a gross heat input rating of less than 14 kilowatt or an instantaneous water heater may incorporate an alternative arrangement, ie an approved safety control designed to shut down the appliance before a build-up of dangerous combustion products can occur [*Reg 30(3)*].

For the purposes of the *Gas Safety (Installation and Use) Regulations 1998 (SI 1998 No 2451), Reg 30(1)–(3)*, a room also includes:

— a cupboard or compartment within such a room; or
— a cupboard, compartment or space adjacent to such a room if there is an air vent from the cupboard, compartment or space into such a room [*SI 1998 No 2451, Reg 30(4)*].

Flues and dampers

[G1019] A flue is a passage for conveying the products of combustion from a gas appliance to the external atmosphere and includes the internal ducts of the appliance. Generally, it is prohibited to install a flue other than in a safe position, and, in the case of a power-operated flue, it must prevent the operation of the appliance should the draught fail [*Gas Safety (Installation and Use) Regulations 1998 (SI 1998 No 2451), Reg 27(4), (5)*].

A flue must be suitable and in a proper condition for the operation of the appliance to which it is fitted. It is prohibited to install a flue pipe so that it enters a brick or masonry chimney in such a manner that the seal cannot be inspected. Similarly, an appliance must not be connected to a flue surrounded by an enclosure, unless the enclosure is sealed to prevent spillage of products of combustion into any room or internal space other than where the appliance is installed. [*Gas Safety (Installation and Use) Regulations 1998 (SI 1998 No 2451), Reg 27(1)–(3)*].

Manually operated flue dampers must not be fitted to serve a domestic gas appliance, and, similarly, it is prohibited to install a domestic gas appliance to a flue which incorporates a manually operated damper unless the damper is permanently fixed in the open position [*Reg 32*].

In the case of automatic dampers, the damper must be interlocked with the gas supply so that the appliance cannot be operated unless the damper is open. A check must be carried out immediately after installation to verify that the appliance and the damper can be operated safely together and without danger to any person [*Reg 32*].

Where a gas appliance is connected to a flue, the installer must ensure that the flue and the means of connection are suitable and effective and complies with applicable building regulations. *Building Regulations* impose comprehensive requirements for the safe installation of heat-producing appliances in buildings (see www.planningportal.co.uk/info/200135/approved_documents/72/part _j_-_combustion_appliances_and_fuel_storage_systems).

Gas fittings

[G1020]–[G1021] Gas fittings are those parts of apparatus and appliances designed for domestic consumers of gas for heating, lighting, cooking or other approved purposes for which gas can be used (but not for the purpose of an industrial process occurring on industrial premises), namely:

— pipework;
— valves; and
— regulators, meters and associated fittings.

[*Gas Safety (Installation and Use) Regulations 1998 (SI 1998 No 2451), SI 1998 No 2451, Reg 2*].

An emergency control is a valve for use by a consumer of gas for shutting off the supply of gas in an emergency. Emergency controls must be appropriately positioned with adequate access, and there must be a prominent notice or other indicator showing whether the control is open or shut. A notice must also be posted either next or near the emergency control indicating the procedure to be followed in the event of an escape of gas [*Gas Safety (Installation and Use) Regulations 1998 (SI 1998 No 2451), SI 1998 No 2451, Reg 9*].

Meters and regulators

[G1022] The *Gas Safety (Installation and Use) Regulations 1998 (SI 1998 No 2451), Reg 12* provides that gas meters must be installed where they are readily accessible for inspection and maintenance. A meter must not be so placed as to adversely affect a means of escape or where there is a risk of damage to it from electrical apparatus. After installation, meters and other associated fittings should be tested for gas tightness and then purged so as to remove safely all air and gas other than the gas to be supplied [*SI 1998 No 2451, Reg 12(6)*].

Meters must be of sound construction so that, in the event of fire, gas cannot escape from them and, where a meter is housed in an outdoor meter box, the box must be so designed as to prevent gas entering the premises or any cavity wall, ie any escaping gas must disperse to the air outside.

Combustible materials must not be kept inside meter boxes. Where a meter is housed in a box or compound that includes a lock, a suitably labelled key must be provided to the consumer [*SI 1998 No 2451, Reg 13*].

Gas systems and appliances [G1023]

The *Gas Safety (Installation and Use) Regulations 1998 (SI 1998 No 2451), Regs 14–17* prescribe detailed precautions to be observed during the installation of the various types of meters. Most of these are for compliance by gas installers and should be included in work procedures. In essence, a meter must not be installed in service pipework unless:

(a) there is a regulator to control the gas pressure [*SI 1998 No 2451, Reg 14(1)(a),(b)*];

(b) a relief valve or seal is fitted which is capable of venting safely [*SI 1998 No 2451, Reg 14(1)(c)*]; and

(c) the meter contains a prominent notice (in permanent form) specifying the procedure in the event of an escape of gas [*SI 1998 No 2451, Reg 15*].

Similar requirements are imposed regarding gas supplies from storage vessels and re-fillable cylinders [*SI 1998 No 2451, Reg 14(2)–(4)*].

Where gas is supplied from a primary meter to a secondary meter, a line diagram must be provided and prominently displayed showing the configuration of all meters, installation pipework and emergency controls. It is incumbent on any person who changes the configuration to amend the diagram accordingly [*SI 1998 No 2451, Reg 17*].

Installation pipework

[G1023] For the purposes of this chapter, 'installation pipework' covers any pipework for conveying gas to the premises from a distribution main, including pipework which connects meters or emergency control valves to a gas appliance, and any shut-off devices at the inlet to the appliance. The term 'service pipe' is used to describe any pipe connecting the distribution main with the outlet of the first emergency control downstream from the distribution main. Similarly, 'service pipework' are those pipes which supply gas from a gas storage vessel. All service pipes normally remain the property of the gas supplier or transporter and their permission must be obtained before any work associated with such pipes can be performed.

Installation pipework must not be installed:

— where it cannot safely be used, having regard to other pipes, pipe supports, drains, sewers, cables, conduits and electrical apparatus;

— in or through any floor or wall, or under any building, unless adequate protection is provided against failure caused by the movement of these structures;

— in any shaft, duct or void, unless adequately ventilated;

— in a way which would impair the structure of a building or impair the fire resistance of any part of its structure;

— where deposit of liquid or solid matter is likely to occur, unless a suitable vessel for the reception and removal of the deposit is provided. (It should be noted that such clogging precautions do not normally need to be taken in respect of natural gas or LPG as these are 'dry' gases.)

[*Gas Safety (Installation and Use) Regulations 1998 (SI 1998 No 2451), Regs 18–21*].

Following work on installation pipework, the pipes must immediately be tested for gas tightness and, if necessary, a protective coating applied; followed, where gas is being supplied, by satisfactory purging to remove all unnecessary air and gas. Except in the case of domestic premises, the parts of installation pipework which are accessible to inspection must be permanently marked as being gas pipes. [*Gas Safety (Installation and Use) Regulations 1998 (SI 1998 No 2451), Regs 22, 23*].

Testing and maintenance requirements

[G1024] According to the *Gas Safety (Installation and Use) Regulations 1998 (SI 1998 No 2451), Reg 33(1)*, a gas appliance must be tested when gas is supplied to premises to verify that it is gastight and to ensure that:

(a)   the appliance has been installed in accordance with the requirements of the 1998 Regulations;
(b)   the operating pressure is as recommended by the manufacturer;
(c)   the appliance has been installed with due regard to any manufacturer's instructions accompanying the appliance; and
(d)   all gas safety controls are in proper working order.

If adjustments are necessary in order to comply with (a) to (d) above, but cannot be carried out, the appliance must be disconnected or sealed off with an appropriate fitting [*SI 1998 No 2451, Reg 33(2)*].

A general duty is imposed on employers and self-employed persons to ensure that any gas appliance, installation pipework or flue in places of work under their control is maintained in a safe condition so as to prevent risk of injury to any person [*SI 1998 No 2451, Reg 35*]

A similar duty is imposed on landlords of premises occupied under a lease or a licence to ensure that relevant gas fittings and associated flues are maintained in a safe condition [*SI 1998 No 2451, Reg 36(2)*]. A relevant gas fitting includes any gas appliance (other than an appliance which the tenant is entitled to remove from the premises) or installation (not service) pipework installed in such premises.

It is the responsibility of landlords to ensure that each gas appliance and flue is checked by a competent person (HSE approved person) on an annual basis [*SI 1998 No 2451, Reg 36(3)*].

Such safety checks shall include, but not be limited to, the following:
(a)   the effectiveness of any flue;
(b)   the supply of combustion air;
(c)   [subject to sub-paragraph (ca),] its operating pressure or heat input or, where necessary, both;
[(ca)   if it is not reasonably practicable to examine its operating pressure or heat input (or, where necessary, both), its combustion performance;]
(d)   the operation of the appliance to ensure its safe functioning.

A new Regulation 36A on determining the date when the next safety check is due, introduced by the Gas Safety (Installation and Use) (Amendment) Regulations 2018 (SI 2018/139), sets out that:

'[(1) Where a safety check of an appliance or a flue made in accordance with regulation 36(3)(a) or (b) is or was completed within the period of 2 months ending with the deadline date, that check is to be treated for the purposes of regulation 36(3)(a) and (b) as having been made on the deadline date.
(2) Subject to paragraph (3), the landlord may ensure that an appliance or flue is checked for safety within the 2 month period beginning with the deadline date, instead of checking it within the 12 month period ending with that date.
(3) The discretion conferred by paragraph (2) may be exercised—
   (a) only once in relation to each appliance or flue in the relevant premises; and
   (b) only in order to align the deadline date in relation to the next safety check of that appliance or flue with the deadline date in relation to the next safety check of any other appliance or flue in the same relevant premises.
(4) In this regulation "the deadline date", in relation to a safety check for an appliance or flue, means the last day of the 12 month period within which the check is or was required to be made under regulation 36(3)(a) or (b).]'

Nothing done or agreed to be done by a tenant can be considered as discharging the duties of a landlord in respect of maintenance except in so far as it relates to access to the appliance or flue for the purposes of carrying out such maintenance or checking activities [*SI 1998 No 2451, Reg 36(10)*].

In addition, it is the duty of landlords to ensure that appliances and relevant fittings are not installed in any room occupied as sleeping accommodation such as to cause a contravention of *Reg 30(2)* or *(3)* (see **G1018** above).

Inspection and maintenance records

[G1025] The *Gas Safety (Installation and Use) Regulations 1998 (SI 1998 No 2451)* marked a significant increase in the obligations placed upon landlords.

A landlord is under a duty to retain a record of each inspection of any gas appliance or flue until there have been two further checks of the appliance or flue or, in respect of an appliance or flue that is removed from the premises, for a period of two years from the date of its last check.

— the date on which the appliance or flue is checked;
— the address of the premises at which the appliance or flue is installed;
— the name and address of the landlord of the premises at which the appliance or flue is installed;
— a description of, and the location of, each appliance and flue that has been checked together with details of safety defects and any remedial action taken.

The person who carries out the safety check must enter on the record their name and the particulars of registration. Then they should sign the record confirming that they have examined the following:

— the effectiveness of any flue;
— the supply of combustion air;
— the operating pressure or heat input or, where necessary, both; or if it is not reasonably practicable to examine the operating pressure or heat input (or, where necessary, both), the combustion performance;

— the operation of the appliance to ensure its safe functioning. [SI 1998 No 2451, Reg 36(3)].

[SI 1998 No 2451, Reg 36(3)].

Within 28 days of the date of the check, the landlord must ensure that a copy of the record is given to each tenant of the premises to which the record relates [SI 1998 No 2451, Reg 36(6)(a)]. In addition, before any new tenant moves in, the landlord must provide any such tenant with a copy of the record – however, where the tenant's right to occupation is for a period not exceeding 28 days, a copy of the record may instead be prominently displayed within the premises [SI 1998 No 2451, Reg 36(6)(b)]. A copy of the inspection record given to a tenant under Reg 36(6)(b) need not include a copy of the signature of the person who carried out the inspection, provided that it includes a statement that the tenant is entitled to have another copy, containing a copy of such signature, on request to the landlord at an address specified in the statement. Where the tenant makes such a request, the landlord must provide the tenant with such a copy of the record as soon as is practicable [SI 1998 No 2451, Reg 36(8)].

The *Gas Safety (Installation and Use) Regulations 1998 (SI 1998 No 2451), Reg 36(7)* provides that where any room occupied, or about to be occupied, by a tenant does not contain any gas appliance, the landlord may – instead of giving the tenant a copy of the inspection record in accordance with *Reg 36(6)* (see above) – prominently display a copy of the record within the premises, together with a statement endorsed upon it that the tenant is entitled to have his own copy of the record on request to the landlord at an address specified in the statement. Where the tenant makes such a request, the landlord must provide the tenant with a copy of the record as soon as is practicable.

Pressure fluctuations

[G1025A] The *Gas Safety (Installation and Use) Regulations 1998 (SI 1998 No 2451), Reg 38*, provides that where gas is used in plant which is liable to produce pressure fluctuations in the gas supply, such as may cause danger to other consumers, the person responsible for the plant must ensure that any directions given to him by the gas transporter to prevent such danger are complied with.

If it is intended to use compressed air or any gaseous substance in connection with the consumption of gas, at least 14 days' written notice must be given to the gas transporter.

Any device fitted to prevent pressure fluctuations or to prevent the admission of a gaseous substance into the gas supply must be adequately maintained.

Liquid Petroleum Gas (LPG)

[G1025B] Lord Gill's July 2009 report of the public inquiry (https://assets.publishing.service.gov.uk/government/uploads/system/uploads/attachment_data/file/229279/0838.pdf) into the fatal explosion at ICL Plastics Ltd in Maryhill, Glasgow in May 2004, which was caused by an LPG leak, found many weaknesses in the safety regime in existence at the time. It set out a new safety regime for the use of LPG in industrial and commercial premises. At the ICL factory, LPG had leaked from an on-site underground metal pipe into the basement of the factory and ignited, causing an explosion which led to the collapse of the four storey Victorian factory.

As a result of the explosion and recommendations made by Lord Gill, the HSE agreed a comprehensive programme with UK LPG suppliers for buried metal pipe work to be replaced with newer and more robust plastic pipes. Businesses with buried metallic service pipework, which can corrode over time, are required to replace it with more durable materials, such as polyethylene. The oldest buried metallic service pipework in the least well-maintained condition and located in the most corrosive soils were targeted first. The programme continues until 2025.

The HSE publishes guidance on inspecting and maintaining or replacing buried metallic pipework carrying LPG vapour and other resources on health and safety in the LPG industry on its website at www.hse.gov.uk/gas/lpg/resources.htm.

HSE Guidance on gas safety

[G1025C] The Health and Safety Executive (HSE) has produced a series of leaflets providing guidance for landlords and important safety advice for gas consumers.

The leaflet *Gas appliances: Get them checked, keep them safe* (INDG238), provides information on the need to obtain urgent medical advice if a person suspects that they or their family have been exposed to carbon monoxide (CO) poisoning. The doctor would need to take a blood or breathe sample but if delayed for more than four hours after exposure has ceased the test results may be inaccurate as CO rapidly leaves the blood.

Symptoms of CO poisoning can mimic many common ailments and may be confused with flu or simple tiredness. If in doubt consult your doctor.

The leaflet also advises on how to ensure that a gas installer is registered with Gas Safe and provides new easy-to-understand diagrams showing the difference between safe and dangerous gas appliances. It also mentions the potential contribution of CO alarms as a useful back-up precaution, but emphasises that they must not be regarded as a substitute for proper installation and maintenance by a Gas Safe registered installer.

A guide to landlords' duties: Gas Safety (Installation and Use) Regulations 1998 as amended Approved Code of Practice and guidance is a 2018 publication and explains landlord's duties to ensure the safety of gas appliances, gas fittings and flues, and carry out annual safety checks. This leaflet also tells landlords how to ensure their gas installer is registered with Gas Safe.

HSE's Domestic Gas Safety web pages at: www.hse.gov.uk/gas/domestic/index.htm, offers printable versions of these information leaflets, as well as useful advice and information for members of the public.

The HSE Catering Information Sheet No 23 *Gas safety in catering and hospitality* advises the installation of a suitable device or system to interlock the gas supply with the mechanical ventilation system in commercial kitchens so that in the event of failure of the flue gas extract system the gas supply is automatically shut down. See: *BS 6173: 2009 Specification for the installation of gas-fired catering appliances for use in catering establishments (2nd and 3rd family gases)*. The standard was current but under review in February 2019.

More HSE information, advice and guidance is available at www.hse.gov.uk/gas.

Interface with other legislation

[G1026] Although wide-ranging in their scope the *Gas Safety (Installation and Use) Regulations 1998 (SI 1998 No 2451)* are not concerned directly with product safety. Certain premises, gas fittings and uses are also subject to exceptions. However, in many of these instances where exceptions are made, similar requirements for gas safety are found in the *Health and Safety at Work etc Act 1974* and supporting legislation. The most important of these are:

- Regulation (EU) 2016/426
  The Regulation lays down essential safety requirements and 'type-approval' rules for appliances and fittings burning gaseous fuels. Appliances can only be placed on the market or brought into service on condition that they do not represent a risk to the safety of people, domestic animals or property (see **G1015** above).
- *Management of Health and Safety at Work Regulations 1999 (SI 1999 No 3242)* (see also Risk Assessment)
  These Regulations require employers and relevant self-employed people to undertake a suitable and sufficient assessment of the risks to the health and safety of their employees and others who may be affected by the work that they do. Employers, including those who install or supply gas or gas appliances or who undertake gas maintenance work, should ensure controls are in place and kept under review.
- *Provision and Use of Work Equipment Regulations 1998 (SI 1998 No 2306) as amended (SI 2002, No 2174)* (see also Machinery Safety)
  These Regulations impose health and safety requirements on work equipment including any gas appliance, apparatus, fitting or tool used or provided for use at work. Equipment must be suitable for its intended use, be maintained and inspected and users provided with appropriate health and safety information and where necessary written instructions.
- *Pressure Systems Safety Regulations 2000 (SI 2000 No 128)*
  The objective of these regulations is the prevention of serious injury from the failure of a pressure system or its components, including those that contain gas in excess of 0.5 bar above atmospheric pressure and pipelines used to convey gas at 2 bar above atmospheric pressure. Under these Regulations duties are placed on those who design, manufacture, use and maintain gas systems overlapping with similar duties under the *Gas Safety (Installation and Use) Regulations 1998 (SI 1998 No 2451)*.

- *Dangerous Substances and Explosive Atmospheres Regulations 2002 (SI 2002 No 2776)*
  These Regulations apply to all workplaces where dangerous substances are used or generated irrespective of the quantity involved. Such dangerous substances include gases and liquefied gases such as methane, propane and butane (LPG) which are extremely flammable and may form explosive atmospheres if mixed in air within the explosive limits. Employers are required to eliminate or reduce the risk to safety from fire and explosion arising from the use or occurrence of such substances and to have in place a combination of control and mitigation measures to ensure the safety of employees and others. *Regulation 7* (classification into hazardous and non-hazardous areas) and *Reg 11* (co-ordination between employers sharing a workplace) do not however apply to workplaces where gaseous fuels are used for cooking, heating, hot water production, refrigeration, lighting or washing; and have, where applicable, a normal water temperature not exceeding 105°C.
- *Construction (Design and Management) Regulations 2015 (CDM 2015) (SI 2015 No 51)*
  These Regulations place wide-ranging duties on clients, designers, and contractors to take health and safety matters into account and manage them effectively from the planning stages of a construction project through commission and any future construction work, including dismantling or demolition.
- *Building Regulations 2010 (SI 2010 No 2214)* (see also WORKPLACES – HEALTH, SAFETY AND WELFARE)
  These Regulations were made under the authority of the *Building Act 1984*. They are supported by non-mandatory approved documents that provide practical guidance on for example the safe installation of gas appliances used for heating – Approved Document J is concerned with combustion appliances and fuel storage systems.

Recent Significant Gas Safety Prosecutions

[G1026A] In January 2019, an Essex-based builder was handed down a suspended prison sentence after putting three people at risk of carbon monoxide poisoning. He was contracted to build a single storey extension to the rear of a house. The homeowner told him the boiler flue exited the rear of the property where the extension was to be built. He advised that this would not be a problem, and that he would arrange for a plumber to move the flue so it exited the side of the property.

An HSE investigation found that he failed to ensure the gas boiler flue was moved to a safe place (out of the side of the property) before this extension was built. The gas flue was therefore releasing the products of combustion into the finished extension, which the homeowner was alerted to by her carbon monoxide alarm.

The landlord pleaded guilty to breaching Regulation 8(1) of the Gas Safety (Installation and Use) Regulations 1998 and was sentenced to four months in prison, suspended for 24 months, 30 days rehabilitation and 150 hours of community unpaid work. He was also ordered to pay £3,000 in costs over six months.

Also in January 2019, two self-employed fireplace installers were given suspended prison sentences after installing new gas fires in three domestic properties, despite not being Gas Safe registered. An HSE investigation found that substandard, and in some cases dangerous, gas work was undertaken at the properties. Gas Safe inspectors attended all three properties, and found that two had installation defects that were immediately dangerous, one of which was spilling excessive carbon monoxide into the room.

The pair pleaded guilty to breaching Regulations 3(1), 3(3) and 5(3) of the Gas Safety (Installation and Use) Regulations 1998 and were each handed down a ten-month suspended sentence, 150 hours unpaid community work and ordered to pay £1,000 and £2,200 respectively toward the costs.

In November 2018, a landlord was given a suspended prison sentence after failing to ensure proper gas safety checks were carried out at his tenanted property. The former Gas Safe Register gas engineer contracted a fitter who was not a member of Gas Safe Register to undertake a landlord's gas safety check at one of his tenanted properties.

An HSE investigation found that the landlord had failed to make any checks on this individual, including checking if he was registered with the Gas Safe Register. The investigation also found that the gas safety certificate used false Gas Safe Register engineer details. The landlord later admitted that he had produced the fraudulent certificate. It was also found he tried to bribe a prosecution witness before the trial by offering them £300 to change their evidence.

He was found guilty of breaching Regulation 36(4) of the Gas Safety (Installation and Use) Regulations and was sentenced to 26 weeks prison, suspended for two years, and ordered to undertake 240 hours unpaid work. He was also ordered to pay costs of £5,330.76.

Also in November 2018, a former gas engineer was given a suspended prison sentence after conducting gas work he was no longer registered or competent to do and leaving it in a dangerous condition. Hours after he had completed the installation, a boiler developed faults and the homeowners reported these faults to him. He attended the address on numerous occasions but was unable to resolve the issues. A Gas Safe Registered gas engineer later inspected the work and found it to be of poor standard, classing it as 'At Risk'.

An HSE investigation found that the former gas engineer's membership of the Gas Safe Register had expired and he was no longer registered to undertake gas work. He used his old Gas Safe Register number on the commissioning document supplied to the home owners and did so knowing this was no longer valid. The investigation also found he had left the gas boiler flue he fitted in a dangerous state that allowed fumes to leak into the property and could have caused carbon monoxide poisoning.

He pleaded guilty to breaching Regulations 3(3), 3(7) and 26(1) of the Gas Safety (Installation and Use) Regulations 1998 and was sentenced to eight months prison, suspended for 18 months for each offence, to run concurrently. He was also fined £500, and ordered to pay the homeowners £500 compensation and £1,000 in prosecution costs.

The same month, a major UK gas distribution company and construction company were fined after a gas main ignited as it was being repaired, injuring two workers. Southern Gas Network Plc (SGN) employees were called to a gas escape caused by employees of Cliffe Contractors Ltd damaging a gas main during construction work. During the repair by SGN, the gas ignited causing the injuries to two of its employees. One worker suffered severe burns while the other sustained cuts and bruises.

An HSE investigation found Cliffe Contractors Ltd had not followed safe digging techniques when excavating around the pipeline resulting in the gas main being damaged by a mechanical excavator. This led to a significant amount of gas being released. Subsequently, SGN did not follow their own procedures or recognised safe systems of work when repairing the main.

SGN pleaded guilty to breaching Section 2(1) of the Health and Safety at Work Act 1974 and was fined £1.2 million and ordered to pay costs of £18,975.43.

Cliffe Contractors Ltd pleaded guilty to breaching Sections 2(1) and 3(1) of the Health and Safety at Work Act 1974 and was fined £60,000 and ordered to pay costs of £12,689.13.

In September 2018, Willmott Partnership Homes Ltd (WPHL) was fined £1.25 million after exposing members of the public to carbon monoxide fumes. A number of gas installations were found to be either immediately dangerous or at risk after a householder reported smelling gas.

An HSE investigation found that WPHL built the flats several years before the incident and in 2014 some remedial was work needed to be carried out on an external wall. During the demolition and reconstruction of the wall, many live flues of gas boilers were removed damaged and blocked, exposing the residents to a risk from carbon monoxide poisoning.

WPHL as the principal contractor had not ensured that an adequate system of work was in place to manage the risks from working around the live flues. The company pleaded guilty to breaching Section 3(1) of the Health and Safety at Work etc Act 1974 and was fined £1.25 million and ordered to pay cost of £23,972.33.

In August 2018, a landlord was sentenced for failing to maintain gas appliances at a rental property and repeatedly failing to provide tenants with a Landlords Gas Safety Certificate. HSE and Gas Safe Register inspectors found a gas oven to be at risk and the gas central heating boiler to be unsafe to use.

The subsequent HSE investigation found the landlord had failed in his duty to have the gas appliances regularly inspected or maintained, and failed to provide a Gas Safety Certificate for a number of years. He also failed to comply with an Improvement Notice which required he take action to deal with these issues.

He pleaded guilty to breaching Section 21 of the Health and Safety Work etc Act 1974 and breaching Regulation 36(2) and Regulation 36(3) of the Gas

Safety (Installation and Use) Regulations 1998. He received a 20-week custodial sentence, suspended for two years, and was ordered to carry out 100 hours of unpaid community work and to pay full costs of £4,146.34.

In July 2018, a self-employed gas fitter was sentenced to 16 months in prison after he carried out unsafe gas work while falsely pretending to be Gas Safe Registered. He was prosecuted following HSE investigations at seven different locations where the Gas Safe Register had been alerted to unsafe work.

He had produced Gas Safety Certificates and falsely claimed to be Gas Safe Registered by using the registration number of another business who had never heard of him. He left customers with faulty installations that presented risks of gas leaks and dangerous accumulations of the products of combustion. He pleaded guilty to 21 offences under the Gas Safety (Installation and Use) Regulations 1998 and was sentenced to 16 months in prison and a concurrent 28-day sentence for not attending the court hearing.

Harassment in the Workplace

Nick Humphreys (Partner, Penningtons Manches Cooper LLP)

Introduction to harassment in the workplace

[H1701] The law relating to harassment in the workplace was substantially amended in the period following the passing of the Equal Treatment Directive (2002/73/EC). It was responsible for, inter alia, extending the categories of unlawful discrimination which are actionable under EU law, and providing a free standing claim of harassment under various aspects of national law (these being the *Sex Discrimination Act 1975 (SDA 1975)*, the *Race Relations Act 1976 (RRA 1976)*, the *Disability Discrimination Act 1995 (DDA 1995)*, the *Employment Equality (Religion and Belief) Regulations 2003 (SI 2003 No 1660) (Religion and Belief Regulations 2003)*, the *Employment Equality (Sexual Orientation) Regulations 2003 (SI 2003 No 1661) (Sexual Orientation Regulations 2003)* and the *Employment Equality (Age) Regulations 2006 (SI 2006 No 1031) (Age Regulations 2006)*(collectively "the Old Discrimination Legislation")).

Prior to the changes introduced into the Old Discrimination Legislation by the Equal Treatment Directive (2002/73/EC), the law relating to harassment was but a specific example of the general law relating to discrimination in the workplace under the various strands of the Old Discrimination Legislation, albeit an area of such significant importance that in the context of sex discrimination, the European Commission was moved to publish a Recommendation (OJ 1992, L49/1) and annexed Code of Practice on the Protection of the dignity of men and women at work (92/131/EEC).

The law has now moved on with the implementation of the *Equality Act 2010 (EA 2010)* in October 2010.

The core architecture of the *EA 2010* in the field of employment law is that *EA 2010, s 4* sets out the protected characteristics to which the *EA 2010* applies (which will include all those covered by the Old Discrimination Legislation), and as regards harassment, this is now dealt with under *EA 2010, s 26* as a freestanding code.

Under the *EA 2010, s 26* there are three types of harassment, as follows.

The first type of harassment will apply to all the protected characteristics (apart from pregnancy and maternity, and marriage and civil partnership), and involves unwanted conduct which is related to a relevant circumstance (which includes employment) and has the purpose or effect of creating an intimidating, hostile, degrading humiliating or offensive environment for the claimant or violating the claimant's dignity.

The second type is sexual harassment which is unwanted conduct of a sexual nature where this has the same purpose or effect as the first type of harassment.

The third type is treating someone less favourably because they have either submitted to or rejected sexual harassment, or harassment related to sex or gender reassignment.

History

[H1702] The law has also moved on in the way that harassment is considered as a concept.

Historically, harassment was considered as a form of standard discrimination law. The first significant case on the issue was under the *Sex Discrimination Act 1975*, *Porcelli v Strathclyde Regional Council* [1986] IRLR 134. Here, the claimant was a female laboratory assistant working at a school alongside two male colleagues. The male assistants did not like the employee and they subjected her to a campaign of harassment in the form of brushing against her in the workplace and making sexually suggestive remarks to her and about her in order to make her leave her job. The campaign was a 'success' in that it did indeed force the claimant to apply for a transfer to another school. Having done so, the claimant then brought a claim against her employer claiming that the actions of her former colleagues amounted to direct sex discrimination for the purposes of the *SDA 1975, s 1(1)(a)*, in that she had been subjected to a detriment. The employer tried to defend the claim by calling in evidence the male employees responsible for the course of harassment. During their evidence, they ventured to suggest that the conduct that had been meted out to the claimant would also have been displayed towards a hypothetical male comparator employee that they did not like and, consequently, the employer could not have committed sex discrimination against the claimant since the treatment was gender neutral.

The Scottish Court of Session refused to accept the defence. Lord Emslie stated of sexual harassment that it was:

> . . . a particularly degrading and unacceptable form of treatment which it must be taken to have been the intention of Parliament to restrain.

The court graphically likened the use of sexual harassment as a 'sexual sword' which had been 'unsheathed and used because the victim was a woman'. It added that, although an equally disliked male employee may well have been subjected to a campaign of harassment, the harassment would not have had as its cutting edge the gender of the victim. The court then went on to give general guidance as to what amounted to harassment. It stated that sexual harassment amounted to:

> . . . [u]nwelcome acts which involve physical contact which, if proved, would also amount to offences at common law, such as assault or indecent assault; and also conduct falling short of such physical acts which can be fairly described as sexual harassment.

The concept of what amounted to harassment was then refined following the *Porcelli* decision, most notably in the re-statement by the Employment Appeal Tribunal (EAT) in the case of *(1) Reed (2) Bull Information Systems Ltd v Stedman* [1999] IRLR 299 and *Macdonald v Advocate General for Scotland; Pearce v Governing Body of Mayfield Secondary School* [2003] UKHL 34,

[2003] IRLR 512. The *MacDonald* and *Pearce* cases showed that whilst the judicial construct of harassment was effective, it was also limited in its scope since comparison under the relevant constituent of the Old Discrimination Legislation was a necessary component of a claim for harassment.

Following the passing of the Race Directive (2000/43/EC), the General Framework Directive (2000/78/EC), and the Equal Treatment Directive (2002/73/EC), the law of harassment has been amended to where it now stands under the *EA 2010, s 26* covering the following:

- harassment on grounds of sex, the law having been amended so as to comply effectively with the definition of harassment under the Equal Treatment Directive (76/207/EC) (this occurring initially from 6 April 2008 following changes made by the *Sex Discrimination Act 1975 (Amendment) Regulations 2008 (SI 2008 No 656)* in the light of *Equal Opportunities Commission v DTI* [2007] EWHC 483 (Admin), [2007] IRLR 327, which held that the definition of 'harassment' under the *Sex Discrimination Act 1975* should be recast so as to provide that it arises from 'association' with sex, not causation by it);
- harassment on grounds of sex and gender reassignment (with effect from 1 October 2005 following changes made by the *Employment Equality (Sex Discrimination) Regulations 2005 (SI 2005 No 2467)* to the *Sex Discrimination Act 1975*, by introducing to it *s 4A*;
- harassment on grounds of race, this being from 19 July 2003 following changes made by the *Race Relations Act 1976 (Amendment) Regulations 2003 (SI 2003 No 1626)* to the *Race Relations Act 1976*;
- harassment on grounds of religion or religious belief, under the *Employment Equality (Religion or Belief) Regulations 2003 (SI 2003 No 1660), Reg 5*;
- harassment on grounds of sexual orientation, under the *Employment Equality (Sexual Orientation) Regulations 2003 (SI 2003 No 1661), Reg 5*; and
- harassment on grounds of age, under the *Employment Equality (Age) Regulations 2006 (SI 2006 No 1031), Reg 6*.

All provisions of the Old Discrimination Legislation insofar as they relate to harassment were repealed by the *EA 2010* with effect from 1 October 2010. However, the provisions of the Old Discrimination Legislation remain good law for acts of harassment which occurred before 1 October 2010 and which were not continuing act cases (albeit if the cases have not been commenced in the Employment Tribunal there was likely to be a limitation issue); cases of harassment which are continuing act cases will now be governed by the principles set out in the *EA 2010* under the transitional provisions set out in the *Equality Act 2010 (Commencement No 4, Savings, Consequential, Transitional, Transitory and Incidental Provisions and Revocation) Order 2010, Art 7*.

The effect of these changes (under both the Old Discrimination Legislation and also under the *EA 2010, s 26*) was considered by the EAT in *Richmond Pharmacology v Dhaliwal* [2009] IRLR 336. In the *Richmond Pharmacology* case, Underhill P confirmed the codifying nature of the changes to the law of harassment and held that harassment claims should be treated in a broadly consistent fashion, regardless of the particular protected characteristic of discrimination in a given case. Accordingly, 'harassment' will now, for cases both before and after the commencement of the EA 2010, focus on three elements, these being:

(1) unwanted conduct;
(2) where the same has the purpose *or* effect of either:
    (a) violating the claimant's dignity; or
    (b) creating an adverse environment for the claimant; and
(3) in case before 1 October 2010 on, or in cases after 1 October 2010, related to, the protected statutory grounds.

The EAT recommended that Employment Tribunals in harassment cases should consider each of these points (stating that to do so would be a 'healthy discipline'), and should ensure that factual findings are made on each of the points.

The EAT then went on to add four general heads of guidance for Employment Tribunals in harassment cases, as follows:

(1) Case law decided before the amendments made by various European Directives on discrimination is 'unlikely to be helpful', and it is more likely to hinder. Further, still less assistance can be sourced from the 'entirely separate provisions' of the Protection from Harassment Act 1997.
(2) Element (2) above of Underhill P's analysis shows that there are two separate heads of liability, purpose or effect (albeit the EAT acknowledged that Employment Tribunals will commonly focus on cases where harassment is the effect rather than the purpose).
(3) It is not enough that harassment within the statutory test is alleged to have been suffered by a claimant; it must also be reasonable that the conduct complained of should have that effect. This therefore addresses the situation of the overly sensitive claimant who, although it is stated by the claimant that the claimant has suffered harm is acting unreasonably. The EAT held that whether a claimant is found to be unreasonably prone to take offence is 'quintessentially a matter for the factual assessment of the tribunal'.
(4) Finally, element (3) of Underhill P's test above, ('on the grounds that' or 'by reason that', ie "the reason why" the alleged harasser acted) is a separate element and has to be determined by the Employment Tribunal.

Accordingly, it is necessary now to ensure when pleading a claim for harassment that it is expressly pleaded as a separate statutory head of claim setting out the matters in the preceding paragraphs and which type of harassment is being alleged.

Elements of a harassment claim

[H1703] The EA 2010, s 26 contains a definition of harassment in the following terms:

Harassment

(1) A person (A) harasses another (B) if—
    (a) A engages in unwanted conduct related to a relevant protected characteristic, and
    (b) the conduct has the purpose or effect of—
        (i) the conduct has the purpose or effect of—
        (ii) creating an intimidating, hostile, degrading, humiliating or offensive environment for B.

(2) A also harasses B if—
    (a) A engages in unwanted conduct of a sexual nature, and
    (b) the conduct has the purpose or effect referred to in subsection (1)(b).

(3) A also harasses B if—
    (a) A or another person engages in unwanted conduct of a sexual nature or that is related to gender reassignment or sex,
    (b) the conduct has the purpose or effect referred to in subsection (1)(b), and
    (c) because of B's rejection of or submission to the conduct, A treats B less favourably than A would treat B if B had not rejected or submitted to the conduct.

(4) In deciding whether conduct has the effect referred to in subsection (1)(b), each of the following must be taken into account—
    (a) the perception of B;
    (b) the other circumstances of the case;
    (c) whether it is reasonable for the conduct to have that effect.

(5) The relevant protected characteristics are—
    age;
    disability;
    gender reassignment;
    race;
    religion or belief;
    sex.

The consolidated definition of harassment under the *EA 2010 s 26* now provides a broad view of harassment, with the conduct in relation to which a complaint is made being conduct "related to" a relevant protected characteristic (for the purposes of the *EA 2010, s 26(5)*) rather than the characteristic being a personal one of the claimant.

The definition is divided into three types of harassment being:

- a general head which applies to all "relevant protected characteristics" (defined in *EA 2010 s 26(5)* so as to exclude marriage and civil partnership, and pregnancy and maternity). This head relates to "conduct" which is unwanted by the claimant, the same being related to a relevant characteristic which has the purpose or effect of creating an intimidating, hostile, degrading, humiliating or offensive atmosphere for the claimant, or which violates the claimant's dignity (an "unlawful purpose or effect") (*EA 2010, s 26(1)*);
- unwanted conduct of a sexual nature which has an unlawful purpose or effect (*EA 2010, s 26(2)*); or

- treatment of a person less favourably because that person has either submitted to or rejected sexual harassment or harassment related to sex or gender reassignment (*EA 2010, s 26(3)*).

Further, although marriage (including civil partnership) and pregnancy and maternity are not covered as relevant protected characteristics for the purposes of a claim of harassment under the *EA 2010, s 26*, where conduct is meted out to a claimant in respect of any of marriage, civil partnership, pregnancy or maternity, there is a possibility that a general discrimination claim might be raised, For example, in *Nixon v Ross Coates Solicitors and another* UKEAT/0108/10ZT, the EAT held that workplace rumour concerning an employee's pregnancy was capable of being both pregnancy discrimination and sexual harassment.

Yet further, the *EA 2010* has made some changes to the Old Discrimination Legislation in that there is now protection against colour and nationality discrimination (EA 2010, s 9), and protection for all of the relevant protected characteristics under the *EA 2010, s 26(5)* in respect of perceived and associative discrimination thereby extending and harmonising the spread of case law (see, for example: *Saini v All Saints Haque Centre* [2009] IRLR 74 in relation to religion/belief discrimination; *EBR Attridge Law LLP v Coleman (No 2)* [2010] IRLR 10 in relation to disability discrimination; and, *Showboat Entertainment Centre Limited v Owens* [1984] IRLR 7, *Weathersfield Limited v Sargent* [1999] IRLR 94 and *Redfearn v Serco Limited* [2006] IRLR 623 in relation to race discrimination).

Commission of harassment

[H1704] It follows from the foregoing that a claim for harassment can arise in a variety of different formats. However, whichever format is argued as the claim, all of the elements of the claim must be present, as follows.

(i) The unwanted nature of the conduct giving rise to a claim

[H1705] The conduct giving rise to a claim must be unwanted by the victim of the harassment. Accordingly, if the victim participates willingly in the conduct, the conduct cannot be unwanted.

To this end, in the case of *Thomas Sanderson Blinds Limited v English* UKEAT/0316/10/JOJ, a case involving allegations of perceived sexual orientation discrimination, the EAT had to determine whether an Employment Tribunal had directed itself correctly in looking at the claimant's own perceptions and feelings in order to decide whether the alleged unwanted conduct had the effect of violating his dignity or creating an intimidating, hostile, degrading, humiliating or offensive environment for him in a situation where the Claimant willingly participated in banter about his alleged sexual orientation, and, indeed, responded in kind, writing articles for a work newsletter which the Tribunal found to "riddled with sexist and ageist innuendo" (the Tribunal also recorded an occasion when the Claimant had been obliged to apologise to a woman for an offensive remark about her breasts). The Tribunal concluded that the claimant had not been discriminated against.

Each case turns on its own facts and it is up to the Tribunal to determine whether the conduct alleged is lawful or unlawful: one act of conduct may be found not to be discrimination, even where the conduct is unwanted, but repeated conduct may be found to overstep the mark; conversely, conduct on one occasion may be extreme and unwelcome, and therefore harassment. However, the intention of the party engaged in the conduct is irrelevant in determining whether the conduct amounts to harassment.

Distinguishing between whether conduct is lawful or not can cause real problems especially where the parties might view the conduct in different ways. Further, drawing the line can be difficult; a one off act, such as asking a colleague out for a drink, may not be harassment, but could become so if the rejected party then repeats the invitation. This type of situation was commented upon by HH Judge McMullen QC in *Fearon v The Chief Constable of Derbyshire* UKEAT/0445/02/RN, 16 January 2004.

It is no defence to a claim for harassment that the protected characteristic is shared by both perpetrator and victim, or that neither of them has the protected characteristic: it is entirely possible for an employee of the same gender to treat another employee of that gender in a way that is sexually harassing fashion notwithstanding that the harasser would not have treated a different gender employee similarly to the harasser's actual victim (see, for example, *Walker v BHS Ltd* [2005] All ER (D) 146 (May), EAT under the old law, and note the *EA 2010, s 24* which provides that the characteristics of the alleged discriminator will be irrelevant to the question of liability under the new law).

(ii) The nature of harassing conduct

[H1706] There is no definition of what amounts to conduct under the *EA 2010*. Accordingly, it falls to an Employment Tribunal to determine what conduct is on a case by case basis; it can be anything, whether written, oral or physical.

The older case law provides various examples of conduct related cases in three broad categories, these being course of conduct cases, single incident cases and failure to investigate cases.

Course of conduct cases

As the heading suggests, the hallmark of this genus of cases is that an employee is subjected to repeated harassment in the course of her employment. The following are reported examples of such cases, albeit decided under the Old Discrimination Legislation and therefore needing to be treated with some caution.

(a) *Porcelli v Strathclyde Regional Council* [1986] IRLR 134. The claimant was a female laboratory assistant and was the victim of a campaign of harassment by fellow male employees to make her leave her job. The male employees would frequently brush past the claimant in a sexually suggestive manner and direct comments of a sexual nature at her. The Court of Session held that the harassment the claimant had suffered amounted to sex discrimination by way of a detriment.

(b) *Tower Boot Company Limited v Jones* [1995] IRLR 529. The claimant was a black employee who was subjected to such acts as being whipped with a leather belt, 'branded' with a red-hot screwdriver and subjected to racist name calling. The Court of Appeal held that, although employees were not employed to harass other employees (the employer's defence being that it could not be vicariously liable on that basis), the acts had been committed during the course of the employees' employment and therefore the employer was liable for the discrimination by way of detriment through the harassment.

(c) *Reed and Bull Information Systems v Stedman* [1999] IRLR 299. The claimant was a junior secretary responsible to the marketing manager of the employer. The claimant catalogued fifteen separate incidents ranging from the telling of dirty jokes by the manager to other colleagues in her presence to the pretence of looking up her skirt in the office smoking room. The EAT upheld the findings of the Employment Tribunal that, although the incidents by themselves would have been insufficient to amount to a detriment, collectively they amounted to a course of conduct of innuendo and general sexist behaviour amounting to harassment.

(d) *Ministry of Defence v Fletcher* [2010] IRLR 25. This case concerned the bullying of a lesbian soldier by a male superior with the bullying comprising conduct and comments made to the victim to the effect that: the victim's partner was "ugly"; and, that the victim should have sex with the superior. Further, the superior sent text messages of a sexual nature to the victim. The victim presented a grievance and was then subsequently discharged from the army for being temperamentally unsuitable, notwithstanding that her grievances were upheld. An Employment Tribunal found that the complaint of sex discrimination under the *SDA 1975* was proven (albeit the EAT held on appeal that the Tribunal had erred in relation to the amount of compensation ordered by the Tribunal).

(e) *Wilton v Timothy James Consulting Ltd* (2015) UKEAT/82/14, [2015] ICR 765, [2015] IRLR 368, EAT. The claimant began a personal relationship with one of the co-owners of the respondent, Mr O'Connell, and subsequently became a director of the respondent. In 2012, the claimant's relationship with Mr O'Connell ended. Subsequently, a Ms Docker joined the respondent and began working in the claimant's team under the claimant's management. Mr O'Connell and Ms Docker began a personal relationship. The claimant became aware of that relationship and raised concerns with Mr O'Connell about its effect on the claimant's team and the claimant's ability to manage Ms Docker. Several months later, in a meeting between Mr O'Connell and the claimant, Mr O'Connell subjected the claimant to a 30-minute highly personal tirade of criticism. Mr Connell referred to the claimant's management style, and called her 'a green-eyed monster'. Mr O'Connell stated that he believed the claimant to be jealous of Ms Docker. In a further meeting a month later, Mr O'Connell again became aggressive towards the claimant. During the meeting, Mr O'Connell insisted that the claimant should apologise to Ms Docker in respect of alleged bullying of Ms Docker. The claimant denied the allegation of bullying and refused to apologise. Several days later, Mr O'Connell called the claimant into a yet further meeting and told the claimant that he wanted to move on from the matter. The claimant asked Mr O'Connell whether he believed the allegation of the bullying of Ms Docker. Mr O'Connell did not answer. The claimant was left with the impression that Mr O'Connell believed in the truth of Ms Docker's allegations even though no proper investigation had taken place. The claimant as a result lost all trust and confidence in Mr O'Connell and the respondent. The following day she left work and did not return. She gave notice of termination of employment. The EAT held that Mr O'Connell's conduct towards the claimant, his colleague and former partner, had been sexual harassment. It was important to give effect to the words that Parliament had used in the *Equality Act 2010, s 26* and not to substitute alternative words for them. The words used by Parliament were that the conduct had to be 'related to' the relevant protected characteristic, and on the facts, the conduct of Mr O'Connell towards the claimant was harassing in nature and related to the claimant's sex; the green-eyed monster comment by Mr O'Connell was made with reference to alleged jealousy of Miss Docker and the Tribunal hearing the case was entitled to be satisfied that a man would not have been subjected to such treatment by Mr O'Connell. The claimant received that treatment because she had previously had a relationship with Mr O'Connell, and because she was a woman to whom he ascribed jealousy. The conduct was therefore related to the protected characteristic of sex. It should be noted that the EAT also held that a resignation which amounts to a constructive dismissal does not fall within the meaning of harassment; subsequently, the EAT in *Urso v Department for Work & Pensions* [2017] IRLR 304 held that whilst it is correct that a constructive dismissal does not amount to an act of harassment, an actual dismissal may amount to affront to an employee's dignity such as to constitute harassment.

Single incident cases

There is no requirement that conduct should be a course of conduct and it is perfectly possible that conduct could arise from a single incident. Examples of such single incidents cases are as follows.

(a) *Bracebridge Engineering Ltd v Darby* [1990] IRLR 3. The claimant was physically manhandled into the office of the works manager by both a chargehand and the works manager, and then indecently assaulted. She was warned not to complain because no one would believe her. The claimant subsequently complained but the complaint was then not properly investigated by the employer (in respect of which, see below). It was held by the EAT that the assault was sufficiently serious to amount to sexual harassment.

(b) *Insitu Cleaning Co Ltd v Heads* [1995] IRLR 4. The claimant was a cleaning supervisor who attended a company meeting. Also present was the son of one of the directors of the company who greeted her by saying 'Hiya, big tits'. The claimant found the remarks particularly distressing as the director's son was approximately half her age. The

EAT held that the remark was capable of subjecting the claimant to discrimination and rejected the employer's contention that the remark was of the same effect as a statement made to a bald male employee about his head.

(c) *Chief Constable of the Lincolnshire Police v Stubbs* [1999] IRLR 81. The claimant was a Detective Constable in the Lincolnshire Constabulary. She attended an off-duty drink with some colleagues in a public house one evening after work. During the course of the evening a male colleague pulled up a stool next to her, flicked her hair and re-arranged her collar so as to give the impression that there was a relationship between himself and the claimant. The claimant found this attention to be distressing and moved away from him. On another occasion at a colleague's leaving party that she had attended with her boyfriend, she was accosted on her way to the toilet by the same officer who stated to her 'Fucking hell, you look worth one. Maybe I shouldn't say that it would be worth some money'. The claimant found the remark to be humiliating and brought proceedings for sex discrimination. The EAT upheld the decision of the Employment Tribunal that the statement amounted to harassment and a detriment.

(d) *Dhaliwal v Richmond Pharmacology* [2009] IRLR 336. The comment made "We will probably bump into each other in the future unless you are married off in India", was found not to be harassment due to the unreasonable perception of the victim of the allegedly harassing nature of the comment.

(e) *Quality Solicitors CMHT v Tunstall* (UKEAT/0105/14), [2014] EqLR 679. A statement made by a manager to the effect that that the Claimant was "Polish, but very nice" (the implication allegedly being that Poles were usually not very nice), made in respect of a Polish employee who it was found in evidence was a "sensitive woman who took things to heart", was found on appeal, not to be an act of harassment on grounds of race because it could not truly be said to have violated the Claimant's dignity or created a proscribed environment for her.

Failures to investigate or protect

Where an employee raises an allegation of harassment by another employee, their employer has a duty to carry out a proper investigation into the complaint and to take such further steps as are necessary in the light of the findings of the investigation. A failure to carry out such a proper investigation can amount to a further acts of harassment in its own right. The following cases are examples of alleged failures to investigate or protect employees.

(a) *Bracebridge Engineering Ltd v Darby* [1990] IRLR 3. The claimant was sexually assaulted by male staff at the factory where she worked. The employer failed properly to investigate the claimant's complaint (the personnel manager simply accepted that it was the word of two employees against one and that there was insufficient evidence to warrant further investigation), whereupon she resigned and claimed in respect both of constructive dismissal and unlawful sex discrimination. The EAT held that where serious allegations of sexual harassment are made, employers have a duty to investigate them properly.

(b) *Burton v De Vere Hotels* [1996] IRLR 596. The case took place under the provisions of the Old Discrimination Legislation, specifically the *SDA 1975* and the *RRA 1976*. The case was brought by a waitress at a hotel and concerned the detriment of suffering sexual and racial harassment by a third party not connected to the employer but whom over the employer was able to exercise control (at a function at which the claimant was working, she was subjected to sexist and racial abuse, masquerading as comedy, from the third party, a well known comedian). Although the employer apologised to the claimant for the actions of the comedian, the claimant presented proceedings for sexual and racial harassment and succeeded in the claim. The EAT held that where an employer was in a position to exercise control over a third party so that discrimination would either not occur, or the extent of the harassment would be reduced, the employer would be subjecting the claimant to a detriment by failing to take the steps necessary to control the third party.

The reasoning in *Burton* was overturned by the House of Lords in the joined appeal in *MacDonald v Advocate General for Scotland and Pearce v Governing Body of Mayfield Secondary School* [2003] UKHL 34, [2003] IRLR 512, where it was held that Burton was wrongly decided, and had gone too far. Their Lordships held that the failure to take reasonable steps to prevent an employee from racial or sexual abuse was discrimination only where the reason for that failure to act amounted to race or sex discrimination. Their Lordships held that the Burton decision was vulnerable because it treated an employer's inadvertent failure to take such steps to protect their employees from racial or sexual abuse as discrimination, even though the failure had nothing to do with the sex or race of the employees.

Following *MacDonald*, the Old Discrimination Legislation was amended in consequence of the Equal Treatment Directive (2002/73/EC), (the amendment carrying over into the *EA 2010*). One of the first recorded decisions under this amended legislation was *May & Baker Limited v Okerago* [2010] IRLR 394, a case where there was a statement by an agency worker that a full time employee should "go back to [her] own fucking country". It was held that an employer, without more, would not be responsible for the racially harassing actions of such third parties.

In the subsequent case of *Sheffield City Council v Norouzi* [2011] UKEAT/0497/10/RN, the EAT held that the employer (a local authority) was liable for acts of racial harassment carried out by a child in a care home against one of the employer's employees. This decision was arrived at following a consideration of the decision of Burton J in *R (Equal Opportunities Commission) v Secretary of State for Trade and Industry* [2007] ICR 1234, on the basis that an employer can be liable for the conduct of a third party in circumstances where there is a continuing course of offensive conduct about which the employer is aware but does nothing to safeguard against.

Further, following the introduction of the *EA 2010, s 40*, employers were made liable for harassment arising due to sex, race, disability, age, gender reassignment, religion or belief, or sexual orientation arising as a result of a third party's conduct. Such liability arose where the third party subjects an employee to harassment in the course of the employee's employment and the employee's employer failed to take such steps as would have been reasonably practicable to prevent the third party from doing so. The scope of this head of liability was limited by the *EA 2010, s 40(3)*, which provided that the employer would only be liable where the employer knew that the employee had been subjected to harassment in the course of the employee's employment on at least two other occasions by a third party. The *EA 2010, s 40* extended substantially the scope of such third party liability, since, prior to its implementation, liability existed only for sex discrimination under this head (under the (now repealed) *SDA 1975, s 6(2)(b)* (as amended by the *Sex Discrimination Act 1975 (Amendment) Regulations 2008 (SI 2008 No 656)*).

However, on 7 September 2013, the *Enterprise and Regulatory Reform Act 2013 (Commencement No 3, Transitional Provisions and Savings) Order 2013 (SI 2227/2013)* took effect and commenced the *Enterprise and Regulatory Reform Act 2013, s 65*. This section provides that with effect from 1 October 2013 the *EA 2010, s 40* was repealed subject to contraventions of *s 40* occurring before 1 October 2013 still being action-able.

Accordingly, with effect from 1 October 2013, new cases will be dealt with under the general position set out in *MacDonald*, whilst cases prior to that date will be dealt with under *EA 2010, s 40*.

Finally, conduct does not even have to be conduct which is specific to the claimant; it can comprise conduct of which a claimant becomes aware (for example, widespread downloading of pornography within the workplace on screens to which the Claimant has access *Moonsar v Fiveways Express Transport Ltd* [2005] IRLR 9, EAT).

(iii) Purpose or effect

[H1707] Under the *EA 2010, s 26(1)(b)*, it is necessary for the conduct of which the claimant has raised a complaint to have either the purpose or the effect of violating the claimant's dignity or creating an intimidating, hostile, degrading, humiliating or offensive environment.

Given the distinction purpose based claims and effect based claims, it is necessary to consider what Employment Tribunals will look at to prove such claims.

In purpose based claims, Employment Tribunals will examine the motive or intention of the person who is alleged to have carried out the act(s) of harassment.

Such an examination can involve an Employment Tribunal having to draw inferences as to what that true motive or intent actually was, and the Tribunal will also be required to take into account the transferring burden of proof under the *EA 2010, s 136*.

Where the harassment is purpose based harassment, it does not matter whether the conduct was undertaken against the victim due to the victim's protected status under the *EA 2010*: all that is necessary is that, objectively in all the circumstances, the wrongdoer has committed the act.

Conversely, in order to establish a claim of harassment based on the effect of conduct, the Employment Tribunal must consider the perception by the victim of what was done. Although decided before the changes to the Old Discrimination Legislation by the Equal Treatment Directive (2002/73/EC), the guidance provided by Morison P in *Reed and Bull Information Systems v Stedman* [1999] IRLR 299, is still useful for Employment Tribunals when determining whether harassment may have been committed. In that case, his Lordship stated of harassment (on grounds of sex) that—

(i) A characteristic of harassment is that it undermines the victim's dignity at work and constitutes a detriment . . . lack of intent is not a defence.
(ii) The words or conduct must be unwelcome to the victim and it is for her to decide what is acceptable or offensive. The question is not what (objectively) the Tribunal would or would not find offensive.
(iii) The Tribunal should not carve up a course of conduct into individual incidents and measure the detriment from each; once unwelcome interest has been displayed, the victim may be bothered by further incidents which, in a different context, would appear unobjectionable.
(iv) In deciding whether something is unwelcome, there can be difficult factual questions for a Tribunal; some conduct . . . may be so clearly unwanted that the [victim] does not have to object to it expressly in advance. At the other end of the scale is conduct which normally a person would be unduly sensitive to object to, but because it is for the individual to set the parameters, the question becomes whether that individual has made it clear that she finds that conduct unacceptable. Provided that that objection would be clear to a reasonable person, any repetition will generally constitute harassment.

Ultimately though, since it is a question of objective fact for an Employment Tribunal to determine whether there may have been an act of harassment committed, it is still possible that a victim may be perceived as what has been termed 'hypersensitive' (*Driskel v Peninsula Business Services Limited* [2000] IRLR 151 at 155); it is not enough under the statutory definition of harassment under the *EA 2010, s 26* that a claimant personally suffers a sense of grievance arising from what he or she felt to be unlawful conduct if that is not objectively borne out by the facts.

Further, harassment may exist without proof of the elements needed for direct discrimination. In particular, no requirement exists now to see how a person in what would otherwise be a comparator group is or would be treated. Under the old case law (ie, *Porcelli v Strathclyde Regional Council* [1986] IRLR 134), it was necessary to show that less favourable treatment had been meted out to a victim on the grounds of his or her protected status.

Now, harassment must arise within the scope of the statutory definition under the relevant head of the *EA 2010, 26*: a claim may not arise where a person carries out generalised indiscriminate workplace bullying (sometimes referred to as the 'bastard defence', ie, 'I am a bastard to everyone'), although see in this regard the possible claims which may arise under the *Protection from Harassment Act 1997*; there is also the fact that such generalised indiscriminate bullying will be likely to amount to a breach of the implied term not without reasonable cause to undermine the relationship of trust and confidence: *Horkulak v Cantor Fitzgerald* [2003] EWHC 1918 (QB), [2003] IRLR 716.

To succeed, the claim must be founded on one of the protected forms of harassment under the EA 2010, s 26.

In this regard, the decision in *Brumfitt v Ministry of Defence* [2005] IRLR 4 (which pre-dates the amendments under the Equal Treatment Directive (2002/73/EC) to the *SDA 1975*), which was brought by a woman in connection with abusive language made to a mixed-sex group of which she was a member, would in all likelihood still stand today.

Further, in the case of *English v Thomas Sanderson Blinds Limited* [2008] IRLR 342, a case brought under the *Employment Equality (Sexual Orientation) Regulations 2003*, the importance of a claim being within the ambit of the relevant aspect of the discrimination legislation was brought home. The claimant, a heterosexual male, had been made the subject of workplace bullying based upon him having certain personality traits which were apparently stereotypical of homosexual males in the minds of those bullying him. His claim for sexual orientation-based discrimination failed, with the EAT holding that the treatment was not meted out on grounds of the claimant's actual sexual orientation.

However, on appeal, the Court of Appeal held that the perceived discrimination of the claimant's possession of homosexual characteristics could give rise to a claim for harassment ([2008] EWCA Civ 1421, [2009] IRLR 206, CA).

Furthermore, the *EA 2010, s 26(1)* makes it clear that the definition of harassment applies to both of associative and perceived discrimination in that it covers conduct "related to a relevant protected characteristic".

(iv) Violation of claimant's dignity or the creation of an intimidating, hostile, degrading, humiliating or offensive work environment

[H1708] The final hurdle to be crossed in order to bring a successful claim is to show that a claimant has suffered, what under the previous case law on the point was, a detriment. Now, following the changes made by the Equal Treatment Directive (2002/73/EC), although a detriment does not need to be shown, there still must be some form of adverse consequence which the claimant suffers, either in the form of a violation of the claimant's dignity, or by the creation of an intimidating, hostile, degrading, humiliating or offensive work environment.

Unsurprisingly, an unjustified sense of grievance will not give rise to a claim (*Barclays Bank plc v Kapur (No 2)* [1995] IRLR 87). The point was considered further in *Pemberton v Inwood* [2018] EWCA Civ 564, a case involving alleged harassment by the Church of England in relation to the same-sex marriage of a priest. The claimant was unable to take up the post of Chaplaincy and Bereavement Manager at the Kingsmill Hospital after the respondent, the Acting Bishop of Southwell and Nottingham, revoked his Permission to Officiate within the Diocese and declined to grant him an Extra Parochial Ministry Licence, consequent to that marriage. This was because there was an express provision prohibiting priests from entering into same-sex marriages (which identified the consequences if that provision was broken) under the doctrines of the Church of England. The claimant's appeal to the Court of Appeal was rejected with the court holding that 'If you belong to an institution with known, and lawful, rules, it implies no violation of dignity, and is not cause for reasonable offence, that those rules should be applied to you, however wrong you may believe them to be. Not all opposition of interests is hostile or offensive.' The point was subsequently affirmed by the EAT in *Ahmed v The Cardinal Hume Acadamies* UKEAT/0196/18/RN, with the EAT holding in relation to alleged harassing conduct that if it is not reasonable for the conduct to be regarded as violating a claimant's dignity or creating an adverse environment for him, then it should not be found to have done so.

However, beyond this, there are many situations in which a claim for harassment can be triggered. In *Reed and Bull Information Systems v Stedman* [1999] IRLR 299, the claimant had resigned claiming that the behaviour of her manager amounted to sexual harassment. The employee cited a list of fifteen alleged incidents that she considered amounted to a course of harassment. The Employment Tribunal concentrated on four of the allegations, these being such matters as telling the employee that sex was a beneficial form of exercise, telling dirty jokes to colleagues in her presence, making a pretence of looking up the employee's skirt and then laughing when she angrily left the room, and stating to her when she was listening to a trial presentation that 'You're going to love me so much for my presentation that when I finish you will be screaming out for more and you will want to rip my clothes off'. The Employment Tribunal stated that, of themselves, each of the allegations was insufficient to amount to sexual harassment. However, if the claimant was able to prove the incidents, they would amount to a course of conduct that would itself be sufficient to amount to harassment. The Employment Tribunal went on to find that the acts had occurred and that the employer and the manager had therefore discriminated against the claimant. The employer and manager appealed to the EAT. The EAT dismissed the appeal agreeing that the actions by the manager, although not meant to be sexually harassing, were indeed discriminatory.

That said, not every utterance or act in the workplace which prima facie appears to be harassing on prohibited grounds will give rise to a claim of harassment. For example, the EAT had to consider in *Dhaliwal v Richmond Pharmacology* [2009] IRLR 336 whether a comment made ("We will probably bump into each other in the future unless you are married off in India"), by an employee's line manager to an Indian employee who had tendered her resignation, to emphasise that they both worked in a small business world, was harassing to the extent necessary to give rise to a claim in law. The EAT held that an employer will only be liable for harassment where a claimant feels or perceives that his or her dignity has been violated and that it is reasonable for that perception to arise. To that end, it is necessary to consider all the relevant circumstances including the purpose for which an act was undertaken or comment made. On the facts of the case, it was not reasonable for the employee to have felt that she was being harassed given the reason for the making of the statement.

In *Kelly v Covance Laboratories Ltd* [2016] IRLR 338, the EAT again held that it was necessary to look at all the circumstances of the case. Kelly involved a Russian national claimant who was placed on a performance review by her employer and told not to speak in Russian on her personal mobile telephone whilst in the workplace. Part of the workplace involved the use of animals for testing products, in consequence of which the employer had been the subject of attention by those involved in the animal rights movement, which had included violent assaults on some of the employer's employees. From early on in her employment, concerns arose relating to the claimant's conduct and performance; during the early weeks of her employment, the claimant's conduct had been sufficiently unusual for a new employee in her position that her line manager began to wonder whether she was in fact an animal rights activist who had infiltrated the employer.

The Employment Tribunal rejected the claimant's complaint relating to the instruction that she should not speak Russian in the workplace; there was no reason to believe that another employee, of a different national origin to the claimant but seeking to speak a language other than English in the workplace, would have been treated any differently. The claimant relied on the fact that two Russian speaking Ukrainian colleagues had not been subjected to the same instruction. However, on the facts, the Tribunal found that the claimant's Line Manager had told the line managers of the Ukrainians to impose a similar prohibition on them, albeit that instruction had not been carried out. The Tribunal also held that the correct comparator had been another employee speaking some language other than English in circumstances that had given the claimant's Line Manager reasonable cause for concern. It was that that caused him to give the instruction to the claimant and, as a matter of fact, it was found that the Line Manager would have given the same instruction in respect of any other employee where those concerns had arisen. Specifically with regard to the harassment claim, the Tribunal allowed that the instruction had been unwanted conduct, but it was not satisfied that it related to the claimant's nationality. Whilst (as a Russian national) Russian was the claimant's mother tongue, the test was not one of causation but was a subjective test as to why a person had acted as they had; a matter to be determined in the context within which the comment was made. On appeal, the EAT agreed with the Tribunal, holding that a reference to someone's own or native language might mean that it was related to that person's race. However, the Tribunal had then found that the employer had provided an alternative explanation for its conduct, which the Tribunal had accepted. Ultimately, it had not been the claimant's race or national origin that had caused the Line Manager to give the instruction not to speak in Russian, but the claimant's behaviour, given the context in which the employer operated and the risks it faced.

Limitations to the definition of harassment

[H1709] The *EA 2010* does have its limitations in relation to what amounts to harassment.

First, as stated above, it is possible that an employer may argue the 'bastard defence' and attempt to show that the treatment meted out to a victim is indiscriminate and that the employer would always behave in that fashion, regardless of any protected status of a claimant (albeit that such a defence has its limitations).

Second, it is entirely possible that although a victim may have been subjected to what he or she believes to be unpleasant conduct, the conduct is found not to be unlawful under the *EA 2010*.

For example, in *Balgobin v London Borough of Tower Hamlets* [1987] IRLR 401, complaints were made by women employees that they had been subjected to the detriment of being required to work with their alleged harasser after an investigation into their allegations of harassment were found to be inconclusive. The EAT refused to accept that the claimants had been treated less favourably on grounds of sex since it was possible that if a male employee had been harassed, the employer would have treated the male in a similar way. This decision would probably be decided in the same way under the *EA 2010*.

Third, under the EA 2010, there are three heads of claim which are not allowed to present a free-standing claim for harassment, these being claims arising from a victim's:

(i)   marital status;
(ii) pregnancy; or
(iii) maternity leave.

In respect of these types of claim, it would be necessary to attempt pursue claims in the form of a claim of direct discrimination under the *EA 2010 Part II, Chapter II*.

Employer's liability for harassment

Vicarious liability

[H1710] A problem that occurs in connection with harassment claims is that, often, it is not the employer who directly subjects a victim to harassment; rather it is a fellow employee of the victim/claimant who is responsible. The employer may then be held responsible for the employee's actions on the grounds of vicarious liability.

The problem here is that employers do not employ staff to harass other members of the workforce. Therefore, it is necessary to determine what amounts to 'the course of employment' in order to see when an employer will be liable for the harassment of its employees by other members of staff.

A further issue arises, which is whether an employer can be liable for the acts of putative agents of the employer.

The course of employment

[H1711] The first case to take as an employer's line of defence the fact that a harasser was not acting in the course of his employment when committing extreme acts of harassment was *Tower Boot Co Ltd v Jones* [1997] IRLR 168.

The case concerned proceedings for racial harassment brought under the provisions of the now repealed *RRA 1976* prior to its amendment by the Equal Treatment Directive (2002/73/EC). The applicant had been subjected to a particularly vicious campaign of name-calling and physical assaults. An Employment Tribunal found that this constituted racial harassment and that the employer was vicariously liable for the acts of the employees engaged in the course of conduct. The employer appealed to the EAT which found for the employer on the ground that the common law test for vicarious liability had not been satisfied. The EAT took the view that for this test to be met the aggressor employees had to be engaged in doing the business of the employer, albeit in an improper manner, ie the act had to be a different mode of doing what the aggressor employees were employed to do. The EAT then held that on the facts, the actions of the aggressor employees could not 'by any stretch of the imagination' be found to be an improper method of doing what the aggressors were properly employed to do.

The EAT decision was widely condemned on the grounds that the more serious that harassment was, the less likely that an employer would be found to be vicariously liable for it. Indeed, the logical conclusion to the EAT's decision was that, since people were not employed to harass other members of staff, employers would never be liable, a surprising result indeed. Unsurprisingly, the decision was appealed to the Court of Appeal which readily found for the employee. In doing so, the court held that the words, 'in the course of his employment' (the test under the now repealed *RRA 1976, s 32(1)* and its replacement the *EA 2010, s 109(1)*), were not subject to the old common law rules on vicarious liability which would artificially restrict the natural everyday sense that every layman would understand of the wording in the *EA 2010, s 109(1)*.

Consequently, harassment by an aggressor employee did not have to be connected with acts authorised to be done as part of the aggressor's work. In so deciding, the Court of Appeal attacked the reasoning of the EAT, stating that its interpretation of the section creating liability:

> . . . cut across the whole legislative scheme and underlying policy which was to deter racial and sexual harassment through a widening of the net of responsibility beyond the guilty employees themselves by making all employers additionally liable for such harassment.

Requirement of causal link with employment

[H1712] It is not every act of harassment that is committed by an employee that will create liability on the part of the employer: there must still be a causal link with employment. A good example of this point is the case of *Waters v Commissioner of Police of the Metropolis* [1997] IRLR 589. In this case, the complaint was one of sexual harassment in the form of an alleged sexual assault by a male police officer on a female police officer in her section house room after they had returned from a late night walk together. Given that the attack took place outside the normal working hours of the claimant, the Employment Tribunal refused to follow the decision in the *Tower Boot* case and held that the claimant was in the same position as if the alleged assailant had been a total stranger to the employer. The alleged assailant did not live in the section house, both officers were off duty and there was nothing linking the alleged assailant to the employer other than the fact that he was a police officer. The Court of Appeal ultimately upheld the decision of the Tribunal on appeal.

This said, the decision in *Waters* does not provide an absolute defence to harassment undertaken after employees leave work for the day. Whether there is a causal link between the harassment and the employment, thereby attaching liability to the employer, will be a question of fact in each particular situation. In the subsequent proceedings brought by Ms Waters in respect of bullying and failure to investigate her complaint, the House of Lords held that an employer owes a duty of care to investigate allegations of rape, bullying and harassment (*Waters v Commissioner of Police for the Metropolis (No 2)* [2000] 4 All ER 934). An example of this point is *Chief Constable of the Lincolnshire Police v Stubbs* [1999] ICR 547, [1999] IRLR 81. The claimant had been the subject of distressing personal comments of a sexual nature from a fellow police officer made at a colleague's leaving party. The Employment Tribunal at first instance held that since the party was an organised leaving party, the incidents were 'connected' to work and the workplace. It stated that the incident:

> . . . would not have happened but for [WPC Stubbs'] work. Work-related social functions are an extension of employment and we can see no reason to restrict the course of employment to purely what goes on in the workplace.

The EAT emphasised, however, that it would have been different 'had the discriminatory acts occurred during a chance meeting between the employee and the aggressor outside the workplace.

The decisions give a wide degree of discretion to Employment Tribunals on the facts of a particular case to decide whether harassment has been committed in the course of employment. If a Tribunal finds that it is not committed in such circumstances, the effect is that, subject to *Waters (No 2)*, whilst an employee may be able to sue the harasser in tort, the remedy that is provided may well be empty if the harasser does not have the means to pay for the wrongdoing, let alone meet the costs of the claimant's action.

Employer's defence

[H1713] A defence is provided to employers in relation to allegations that they are vicariously liable under the *EA 2010, s 109(4)*. The defence provides that:

> In proceedings against A's employer (B) in respect of anything alleged to have been done by A in the course of A's employment it is a defence for B to show that B took all reasonable steps to prevent A—
>
> (a) from doing that thing, or
> (b) from doing anything of that description.

The problem for employers with the defence is that, whilst it provides a defence where an employer has a full procedure dealing with harassment claims (including such matters as monitoring, complaints handling and disciplinary proceedings), the defence requires that the employer must pay more than mere lip service to the existence of such procedures and must also be able to show that its employees were aware of the procedures and, where relevant, that the employer had fully implemented the procedures in practice.

A good example of a case where the defence succeeded was *Balgobin v London Borough of Tower Hamlets* [1987] IRLR 401. Here, the alleged harasser was suspended following the initial allegations, his conduct was fully investigated and it was found that there was insufficient evidence against the cook to justify the claims. The EAT was moved to comment that it was difficult to see what further steps the employer could have taken to avoid the alleged discrimination.

Ultimately, if an employer has gone to the expense of having policies and procedures drafted then they should not be kept in the office safe. Employees should be left under no illusions as to what will happen in cases of harassment.

Liability of employees

[H1714] Although an employer may be vicariously liable for acts of harassment, primary responsibility for such action will remain with the employee who has committed the discrimination (*EA 2010, s 110*).

Further, the *EA 2010, s 112* provides that a person who knowingly aids another person to do an act made unlawful by the relevant element of the Discrimination Legislation is to be treated for the purposes as himself doing an unlawful act of the like description.

The scope of these provisions was given a liberal interpretation in *AM v WC & SPV* [1999] IRLR 410. Here, the EAT held that an Employment Tribunal had erred in not allowing a police officer to bring a complaint of harassment against another individual officer. The case arose under the now repealed *SDA 1975* and the EAT considered that the construction of the *SDA 1975, s 41(1), (3)* (*EA 2010, ss 109(1)* and *(4)*), and *SDA, s 42* (*EA 2010, s 112*), was to make an employer vicariously liable for acts of discrimination committed by employees *EA 2010, s 109(1)*), to provide a conscientious employer with a defence for the actions of its employees where the actions were effectively a 'bolt from the blue' (now *EA 2010, 109(4)*) and to allow an employee who has committed acts of discrimination against another to be joined as a 'partner' to proceedings commenced against an employer (now *EA 2010, s 112*). The EAT added that even where a defence did exist for an employer under what is now *EA 2010, s 109(4)*, there is nothing in the *EA 2010* that prevented the victim of discrimination from launching proceedings against a discriminator employee alone.

That said, in order for an employer to be responsible, there must be a genuine employer or principal relationship; if there is no such relationship, there can be no liability (see *May & Baker Limited v Okerago* [2010] IRLR 394, in this regard).

The employer's defence has also been considered by the House of Lords in *Anyanwu v South Bank Student Union* [2001] IRLR 305, where it was held that the words 'knowingly aids' means that the person providing aid must have given some kind of assistance to the person committing the discriminatory act which helps the discriminator to do the act. The amount or value of the help is of no importance. All that is needed is an act of some kind, done knowingly, which helps the discriminator to do the unlawful act.

Narrowing the extent of employers' liability

[H1715] Where an employee is able to establish that he or she has been the victim of unlawful harassment, there can be circumstances where the damages that would otherwise be awarded to a victim of such discrimination will be dramatically reduced. In the context of sexual harassment, this will be where the victim's conduct shows that she would not have suffered a serious detriment as a consequence of the harassment. For example, in *Snowball v Gardner Merchant Ltd* [1987] IRLR 397, the claimant complained that she had been sexually harassed by her manager. The allegations were denied and the claimant was subjected to questioning at the Tribunal hearing as to her attitude to matters of sexual behaviour so as to show that even if she had been harassed, her feelings had not been injured. In particular, evidence was called as to the fact that she referred to her bed as her 'play pen' and that she had stated that she slept between black satin sheets. On appeal, the EAT held that the evidence had been correctly admitted so as to show that the claimant was not likely to be offended by a degree of familiarity having a sexual connotation.

Likewise, in *Wileman v Minilec Engineering Ltd* [1988] IRLR 144, the EAT refused to overturn an award of compensation in the sum of £50 that had been made for injury to feelings. In this case, the claimant had made a complaint of sexual harassment against one of the directors of her employer. The award had been set at this figure on account of the fact that the victim had worn revealing clothes and it made it inevitable that comment would be passed. The extent of the compensation awarded in the *Wileman* case would now have to be considered in the light of the guidelines given in *Vento v Chief Constable of West Yorkshire Police (No 2)* [2002] EWCA Civ 1871, [2003] IRLR 102 (which are updated from time to time, most recently, following the case of *De Souza v Vinci Construction (UK) Ltd* [2017] EWCA Civ 879, and subsequently, on 25 March 2019, when a Second Addendum to the Joint Presidential Guidance was issued by the Presidents of the Employment Tribunals for Scotland and for England and Wales as to the bands of compensation to be awarded for injury to feelings).

With respect to the EAT, these cases, whilst reflecting that harassment can still occur in cases where a claimant has a high degree of confidence in sexual banter, unfortunately appear to send out the wrong message in that they can be construed as being a 'harasser's charter' and do not seem to recognise that a person may be particularly upset by the unwelcome attentions of a specific person.

Liability of employer as a principal for a putative agent

[H1715A] A further possible area of liability for employers is that they may be alleged to be liable for the acts of putative agents pursuant to *EA 2010, s109(2)*.

Whether such liability exists was considered in the case of *Yearwood v Comr of the Police of the Metropolis* (UKEAT 0310/03/2805) [2004] ICR 1660, EAT.

Yearwood was one of a series of conjoined cases which had to determine whether the Commissioner was responsible, on an agency basis, for the discriminatory acts of respectively, police officers and civilian officers (the latter being employed by the Commissioner). The case held that when considering whether a person was an agent of an employer, common law agency principles had to be applied.

Following *Yearwood*, the point was subsequently considered by the Court of Appeal in *Kemeh v Ministry of Defence* [2014] EWCA Civ 91 (2014) Times, 11 March. In *Kemeh*, the Court had to determine whether the Claimant, a cook employed by the British Army and working in the Falkland Islands, could establish principal/agent liability of the employee of a third party (Sodexo) who was supervised by an NCO. The facts were that the Claimant asked the Sodexo employee for some chicken pieces to make fresh soup for a large number of soldiers. The Sodexo employee gave him just two pieces and when he asked for more she replied, "Why should I trust you? First you are a Private in the British Army and then you are black", a matter which upset the Claimant. Shortly afterwards, the Claimant was also told by his immediate line manager (a senior NCO), to "shut up you dumb black bastard" during the course of a conversation about a football match. The Claimant presented proceedings against the Ministry of Defence ("MoD") in respect of both incidents, arguing that the MoD was responsible on the principal/agent basis for the Sodexo employee.

It was argued on behalf of the Claimant before the Court that the general approach to liability for the principal/agent basis for the purposes of the *EA 2010* must be that set out in Tower Boot v Jones; that is, there should be a broad, purposive approach.

However, the Court held that: " . . . *it cannot be appropriate to describe as an agent someone who is employed by a contractor simply on the grounds that he or she performs work for the benefit of a third party employer. She is no more acting on behalf of the employer than his own employees are, and they would not typically be treated as agents.*"

The Court went on to add that agency might be established in appropriate cases, but that it would require "very cogent evidence" to do so.

Accordingly, whilst each turns on its own facts, the rebuttable presumption is against there being an agency relationship.

In *Unite the Union v Nailard* [2016] IRLR 906, [2017] ICR 121, [2016] All ER (D) 43 (Oct), the EAT held that two elected officials of the union, who had bullied and harassed the Claimant, could be considered to be agents of the union. The EAT stated that in determining whether a person is an agent of a principal for the purposes of the *EA 2010, s 109(2)*, the legal concept of agency cannot be disregarded. An investigating tribunal must therefore have regard to what, if anything, the putative agent was authorised to do. It is not necessary that the putative agent has the authority to bind the principal contractually; it is also not enough that the putative agent merely performs some work for the principal. In the instant case, the two officials were elected to perform certain duties for union, and they were performing those duties as regards the claimant, who was a member of the union, which was what gave rise to the agency. As regards the elected officials, the Court of Appeal [2018] IRLR 730 confirmed that a principal will be liable wherever an agent discriminates in the course of carrying out the functions he or she is authorised to do. That formulation effectively equates the circumstances in which a principal may be liable for the acts of an agent with the 'course of employment' test governing the liability of employers for the acts of their employees. The court added that it may well extend the scope of the liability beyond what would apply at common law; but there is no reason why Parliament should not have chosen to effect such an extension in discrimination cases.

Legal action and remedies for harassment

[H1716] Where a person unlawfully harasses another, the consequences of such action can be extremely far-reaching and potentially spread across the full range of actions in both civil and criminal law.

Unlawful discrimination

Elements of the tort

[H1717] The most obvious form of action that a victim of harassment may consider is a claim for harassment under the relevant provision of the old Discrimination Legislation.

Remedies

[H1718] The remedies that are open to a victim of unlawful harassment are:

(a) a declaration;
(b) a recommendation; or
(c) compensation.

A declaration is a formal acknowledgement from an Employment Tribunal that discrimination in the form of harassment has occurred. It will be granted in every successful case brought before a Tribunal.

Recommendations are used where an claimant to an Employment Tribunal remains in the employment of her employer. The purpose of a recommendation is for a Tribunal to give guidance to an employer relating to its business for the purposes of eliminating discrimination.

However, the main remedy that victims of harassment will seek is compensation for the loss that has been suffered as a result of the harassment including, particularly, injury to feelings (NB, compensation awarded for injury to feelings is awarded as general damages for personal injury in tort and is therefore not subject to deductions for income tax under the *Income Tax (Earnings and Pensions) Act 2003, s 406* (per *Wilton v Timothy James Consulting Ltd* (2015) UKEAT/82/14, [2015] ICR 765, [2015] IRLR 368, EAT)). Compensation for injury to feelings is awarded in accordance with the bands set in the case of *Vento v Chief Constable of West Yorkshire Police* [2002] EWCA Civ 1871 and depends on the extent of the injury to feelings. The Vento bands are updated from time to time, most recently in the Second

[H1718] Harassment in the Workplace

Addendum to the Joint Presidential Guidance, the same being issued on 25 March 2019, following the case of *De Souza v Vinci Construction (UK) Ltd* [2017] EWCA Civ 879. This guidance provides that for cases arising on or after 6 April 2019, the lower band will be £900 to £8,800 (less serious cases); there is a middle band of £8,800 to £26,300 (cases that do not merit an award in the upper band); and an upper band of £26,300 to £44,000 (the most serious cases), with the most exceptional cases capable of exceeding £44,000.

Unfair dismissal

[H1719] Unfair dismissal can arise as a claim where an employer fails to control unlawful harassment in the workplace or fails to investigate properly complaints of unlawful harassment. Such failures may (and probably will) also act to undermine the employee's trust and confidence in the employer. The result of such failures is that an employee may be able to claim that the employer has constructively dismissed the employee, thereby amounting to a dismissal for the purposes of the *Employment Rights Act 1996, s 95(1)(c)*. Indeed, where a unlawful harassment claim has been brought on the back of an employee's departure from his or her employment, most cases will also include a claim for unfair dismissal as well.

Proving constructive dismissal

[H1720] The first hurdle that an employee has to overcome is to prove that he or she has been dismissed for the purposes of the *Employment Rights Act 1996 (ERA 1996), s 95(1)(c)*. It is trite law that for an employee to show that the employee has been constructively dismissed the acts of his or her employer must show that the employer, by its conduct, evinced an intention not to be bound by the terms and conditions of the contract of employment (see, to this effect, *Western Excavating (ECC) Limited v Sharp* [1978] IRLR 27). The problem that an employee may face in this regard is that if an employer has in place a full harassment procedure that is properly implemented, the employee may fail to establish that the employer has acted in breach. For an example of this occurring in practice see *Balgobin v London Borough of Tower Hamlets* [1987] IRLR 401.

Likewise, an employee may also fail to show that an employer has breached the implied term of trust and confidence where the act of harassment is a serious one-off act of harassment that the employer was unable to predict would happen and where the employer properly deals with the matter immediately upon a complaint being made. To the extent that the employer does not investigate properly a complaint in such circumstances it is almost inevitable that the employer will breach the contract of employment (see *Bracebridge Engineering Ltd v Darby* [1990] IRLR 3). Further, the High Court confirmed in the case of *Horkulak v Cantor Fitzgerald* [2003] EWHC 1918 (QB), [2003] IRLR 756 that workplace bullying (bullying being a form of harassment) of an employee may amount to grounds for claiming that the employee has been constructively dismissed.

Remedies

[H1721] The remedies that can be provided for unfair dismissal are those set out in *ERA 1996, ss 113–118*, namely:

(a) an order for reinstatement;
(b) an order for re-engagement; or
(c) an order for compensation.

The practical consequence of an employee claiming that he or she has been constructively dismissed as a result of an employer's breach of the implied duty of trust and confidence is that an Employment Tribunal is likely to be unwilling to make an order of reinstatement or re-engagement and therefore the reality is that most cases will provide for compensation to be awarded. The downside to claiming unfair dismissal as opposed to bringing a claim for unlawful discrimination is that, whereas the compensation for discrimination is both unlimited as regards the amount that can be awarded and can contain an award for injury to feelings, the compensatory award that is available for unfair dismissal (in addition to the basic award) is the lower of the statutory cap of £80,541 or 52 weeks' pay and there is no right to an award of compensation for injury to feelings.

Actions in tort

[H1722] Broadly speaking, torts are the category of actionable civil wrongs not including claims for breach of contract or those arising in equity. Acts of unlawful harassment, depending upon the particular characteristics of the harassment concerned, are capable of amounting to a variety of different torts.

Assault and battery

[H1723] Assault and battery are forms of trespass to the person. They are separate torts and occur where a person fears that either he or she will receive or has been the subject of unlawful physical contact. As was stated of the torts in *Collins v Wilcock* [1984] 3 All ER 374:

> An assault is an act which causes another person to apprehend the infliction of immediate, unlawful, force on his person; a battery is the actual infliction of unlawful force on another person.

For the tort of assault to be committed it is irrelevant that the victim is not in fear of harm. Consequently, it is possible for an assault to occur where a victim of harassment is threatened with the possibility that he or she will receive unwanted physical contact. All that must exist for the tort to be committed is that the victim must have a reasonable apprehension that she will imminently receive unwanted contact.

Further, for either tort, the use of the word 'force' in the formulation of the tort does not mean physical violence. It is not necessary that the victim is subjected to a full-scale attack (as occurred, for example, in *Bracebridge Engineering Ltd v Darby* [1990] IRLR 3 and *Tower Boot Company Limited v Jones* [1995] IRLR 529). It is possible for the torts to be committed where a person brushes past another suggestively (as in *Porcelli v Strathclyde Regional Council* [1986] IRLR 134) or even kisses another without consent. The torts therefore cover the full range of possible forms of contact related harassment from physical injury to unwanted molestation and are capable of providing a cause of action even though no actual physical harm has been occasioned to the victim of the particular tort.

Negligence

[H1724] While negligence may imply that an act committed against a person has been committed carelessly, it is perfectly possible for the tort to be committed intentionally by a person. The classic phrasing of the tort is derived from the case of *Donoghue v Stevenson* [1932] AC 562, where Lord Atkin stated of the elements of the tort:

> You must take reasonable care to avoid acts or omissions which you can reasonably foresee would be likely to injure your neighbour . . . [who are] persons who are so closely and directly affected by my act that I ought reasonably to have them in contemplation as being so affected when I am directing my mind to the acts or omissions which are called in question.

The test thus framed ensures that liability will be created where there is reasonable foresight of harm being inflicted upon persons foreseeably affected by an actor's conduct. Consequently, the conduct can be intentional or reckless as well as careless.

Where it is the case that an employee harasses a fellow employee, the harasser and victim are likely to be within the relationship of proximity required by the law (it being trite law that fellow employees owe a duty of care to each other).

Once it has been established that a duty is owed, two other factors become relevant these being whether harm has been sustained by a victim and whether the magnitude of the harm suffered by the victim is far greater than the actor intended.

(a)　*Requirement for loss*. For the tort of negligence to be committed, the victim must suffer loss or damage flowing from the negligent act, an obvious form of which is stress related illnesses that can occur from prolonged harassment (for example, nervous shock). It is well recognised in law that where a person suffers from medically treatable stress related illnesses that are caused by the negligence of an employer, it is possible to recover damages for the injuries so inflicted (see to this end the non-harassment case of *Walker v Northumberland County Council* [1995] ICR 702).

(b)　*'Eggshell-skulls'*. The extent of the injury suffered, provided that a victim could foreseeably have suffered a particular form of harm (eg distress) from the commission of the unlawful act, is irrelevant. This principle is known as the 'eggshell-skull' rule. The law recognises that it is possible for 'eggshell-personalities' to exist (see, for example, *Malcolm v Broadhurst* [1970] 3 All ER 508). It is irrelevant that another person possessed of the same characteristics of the claimant would not have suffered injury as a consequence of harassment being meted out provided that the hypothetical person would have been distressed by it.

Given that it is possible to recover damages for injury to feelings occasioned as a consequence of unlawful discrimination, it may well be the case that an employee will not want to pursue a claim for negligence. This said, it is not possible to obtain an injunction in respect of unlawful harassment under *EA 2010*.

Remedies in tort

[H1725] Where an employee commits an actionable tort two remedies will commonly be available to the victim. These will be:

(a) damages (which is available as of right for a tort); and
(b) an injunction (where damages are not adequate either in whole or in part).

In tort, the correct measure of damages is that the victim should be given such a sum as would put the victim in the position as if the tort had not been committed (*Livingstone v Rawyards Coal Co* (1880) 5 App Cas 25).

As to how much this will be is a question of fact in each case and, potentially, in the case of a course of harassment that induces, say, nervous shock preventing an employee from working again, loss of career damages could run into hundreds of thousands of pounds.

However, the problem incurred with the common law solution of awarding damages to the victim of a tort in the context of unlawful harassment is that the victim may not be properly compensated for on-going acts. In the circumstances, damages may not be adequate in such a situation and the better remedy may be to sue for an injunction as well.

Further, the use of an injunction to stop on-going harassment may be of more practical benefit to a currently distressed employee than would a future award of damages. This is because injunctions can be applied for and obtained relatively speedily as an interim remedy pending the full trial of an action.

The problem with injunctions is that, assuming that an interim injunction is applied for whilst both the harasser and the victim are still in the employment of their employer, the victim's trust and confidence in his or her employer may well have become exhausted by the employer's failure to take preventative action in relation to the aggressor employee. Consequently, in many cases, the better course of action may well be to leave employment on the grounds of constructive dismissal and sue for compensation under *EA 2010*.

Protection from Harassment Act 1997

[H1726] An offence and tort of harassment was created by the *Protection From Harassment Act 1997* ('*PHA 1997*'). The *PHA 1997* was passed to protect victims of 'stalking' but this has not proved a barrier to it being used in the workplace as a means of seeking to prevent workplace bullying (see for example the case of *Majrowski v Guy's and St Thomas's NHS Trust* [2005] EWCA Civ 251, [2005] IRLR 340). Not all cases of alleged harassment will be successful, and although a standard of criminal conduct had been held necessary to establish liability for harassment claims in order to establish liability under the *PHA 1997* (according to the Court of Appeal in *Conn v Sunderland City Council* [2007] EWCA Civ 1492, [2008] IRLR 324), this was reviewed in *Veakins v Kier Islington Ltd* [2009] EWCA Civ 1288, [2010] IRLR 132, where the Court of Appeal developed this proposition in holding that the court must focus on whether the conduct complained of is oppressive and unacceptable, as opposed to merely unattractive, unreasonable or regrettable, albeit the court must keep in mind that the conduct must be of an order that would sustain criminal liability. Further, in *Marinello v City of Edinburgh Council* [2010] CSOH 17, the Scottish Court of Session, following a full review of the authorities held that what is required for conduct to become criminal is that the conduct amounts to harassment.

Harassment is not specifically defined by the *PHA 1997*, although *PHA 1997, s 7* provides that harassment must amount to a course of conduct causing alarm or distress to a person.

The phrase 'a course of conduct' is defined to be conduct on two or more occasions and 'conduct' includes words (*PHA 1997, s 7(4)*). It has been held in the case of *Lau v DPP* [2000] All ER (D) 224 that where there are few incidents separated by a long period of time, it will be difficult to establish a course of conduct.

It is possible to obtain an injunction against a harasser under the *PHA 1997 s 3* and also to claim damages.

Breach of contract

[H1727] Where an employee harasses a fellow employee, the harasser will almost certainly breach one or more of the terms of his or her employment contract. In the first instance, most disciplinary procedures provide that harassment of fellow employees on grounds of sex, race or disability amounts to gross misconduct. Given that many contracts of employment provide that a disciplinary procedure is contractual, harassment *per se* may provide grounds for summary dismissal. In any event, gross misconduct would act to undermine the implied term not without lawful and reasonable cause to undermine the relationship of trust and confidence between employer and employee.

Even if an employer fails to provide a contractual disciplinary procedure, an employee may well be in breach of several of the implied terms under his contract of employment if he or she is engaged in unlawfully harassing a fellow employee. This is because the employee will owe a duty of care to the employer that requires the employee to take reasonable care for fellow employees. Deliberate unlawful harassment will in all likelihood breach this term and fix the employer with potential vicarious liability for the harassing employee's wrongdoing.

Remedies

[H1728] As with remedies provided for torts, the remedies for breach of contract will be either damages or an injunction.

The law places a different emphasis on the method of computing damages for breach of contract to that provided in relation to torts. Practically though, in the context of a contract of employment, the final outcome will usually be the same. In the law of contract, the victim is to be placed so far as possible in the position as if the contract had been properly performed by the employer (*Robinson v Harman* (1848) 1 Exch 850). The amount of damages payable will vary according to the loss that has been suffered by a victim of a breach of contract. Ordinarily, if the employee leaves his/her employment claiming that he or she has been constructively dismissed, the employee will be able to claim damages for loss of contractual benefits during her notice period (subject to the employee's duty to mitigate). However, it seems that in spite of the decision of the House of Lords in *Malik v BCCI* [1997] IRLR 462, following the subsequent decision of the House of Lords in *Johnson v Unisys Ltd* [2001] UKHL 13, [2001] IRLR 279, it is not possible to recover damages for injury to feelings occasioned as a result of the humiliating manner of a dismissal.

This said, if the claim is framed by an employee as one going to the employer's duty of care to the employee during employment, it is well established that an employee will be able to claim damages for the illness so caused (see, to this effect, *Walker v Northumberland County Council* [1995] ICR 702).

Criminal action

[H1729] The conduct of a person who harasses another may also amount to one or more of the following crimes:

(a) battery;
(b) harassment under the *Protection from Harassment Act 1997*;
(c) intentional harassment under the *Public Order Act 1986*; or
(d) racially aggravated harassment under the *Crime and Disorder Act 1998*.

Battery

[H1730] Battery is a common law offence. The elements of the offences are that the harasser intentionally and unlawfully inflicts personal violence upon the victim. In *Collins v Wilcock* [1984] 3 All ER 374 the Court of Appeal held that there was a general exception to the crime which embraced all forms of physical contact generally acceptable. However, it is clear from this that if the victim makes known to his/her harasser the fact that the victim does not welcome physical contact from him, the harasser will commit the offence simply by touching the victim.

The offence can only be tried summarily.

Harassment under the Protection from Harassment Act 1997

[H1731] The *PHA 1997* can act to criminalise a course of conduct pursued by a person who 'knows or ought to know' that the conduct amounts to harassment. The *PHA 1997, s 1(2)* provides that a person will be deemed to have knowledge where a reasonable person in possession of the same information as the harasser would think that the course of conduct amounted to harassment.

The offence is one that is triable either way. If the offence is tried on indictment, the maximum penalty is five years imprisonment or a fine or both. If tried summarily, the maximum penalty is a term of imprisonment not exceeding six months, a fine subject to the statutory maximum or both.

Harassment under the Public Order Act 1986

[H1732] The offence exists under the *Public Order Act 1986 (POA 1986), s 4A* and criminalises threatening, abusive or insulting words or behaviour thereby causing harassment, alarm or distress. The offence can be committed in a public or private place (other than a dwelling). Consequently, sexual harassment in the workplace is capable of being covered by the *POA 1986*.

The offence can only be tried summarily and a fine on level 3 of the standard scale can be levied upon conviction.

Racially aggravated crimes under the Crime and Disorder Act 1998

[H1733] The offence of racially aggravated harassment arises under the *Crime and Disorder Act 1998 (CDA 1998), s 32*. The offence is committed where a person commits an offence against another and the offence is racially aggravated (as defined by *CDA 1998, s 28*).

The offence in either case is triable either way. In the event that a person is convicted on indictment of racially aggravated harassment under *CDA 1998, s 32(1)(a)*, the person can be sentenced to imprisonment for a term of up to two years, a fine or to both, or for putting people in fear of violence under *CDA 1998, s 32(1)(b)*, the person can be sentenced to imprisonment for a term of up to seven years, a fine or to both.

Where a person is summarily convicted of racially aggravated harassment under *CDA 1998, s 32(1)(a)*, the person can be sentenced to imprisonment for a term of up to six months, or a fine not exceeding the statutory maximum or to both, or for putting people in fear of violence under *CDA 1998, s 32(1)(b)*, the person can be sentenced to imprisonment for a term of up to six months or a fine not exceeding the statutory maximum or to both.

Preventing and handling claims of harassment

Harassment policy

[H1734] Codes of practice covering unlawful harassment under the headings of gender, race and disability in respect of the Discrimination Legislation have been issued by the Equality and Human Rights Commission.

In order for an employer's policy to be effective and properly used, it must have the full support of senior managers within the business and must be used fairly and properly on each occasion that its provisions become relevant. So as to achieve this, it is important that the policy is clear and made known to all employees. A harassment policy is an essential feature of a proper equal opportunities policy and should exhibit the following characteristics:

(a) a complaints procedure;
(b) confidentiality;
(c) a method for informal action to be taken; and
(d) a method for formal action to be taken.

Complaints procedure

[H1735] The hallmark of an effective policy is that provision is made for a complaints procedure that both encourages justifiable complaints of harassment to be made and protects the maker of the complaint. By way of guidance, in the context of sexual harassment, the European Commission ('EC') Code of Practice states in this regard (at section B paragraph (iii)) that a formal complaints procedure is needed so as to:

> . . . give employees confidence that the organisation will take allegations of sexual harassment seriously.

The EC Code also states that such a procedure should be used where employees feel that informal resolution of a dispute is inappropriate or has not worked.

In order for the complaints procedure to be effective, the EC Code recommends that employees should know clearly to whom to make a complaint and should also be provided with an alternative source of person to whom to complain in the event that the primary source is inappropriate (eg because that person is the alleged harasser).

The EC Code adds that it is good practice to allow complaints to be made to a person of the same sex as the victim.

Confidentiality

[H1736] In order for complaints to be made effectively, it is necessary that some measure of confidentiality be maintained in the complaint making procedure. The problem that arises here is that there is an inevitable trade-off between the right of the victim of harassment to be protected against recrimination and the right of an alleged harasser to know the case that he/she is meeting for disciplinary purposes. Indeed, to the extent that complaints are made against a person who is not told where, when and whom he/she is alleged to have harassed and the alleged harasser is then disciplined as a result of the complaint, there is a chance that the employer will be acting in breach of the duty to maintain the alleged harasser's trust and confidence.

To this end, the requirement for confidentiality exists to the extent that although a harasser may have the right to know the full nature of the complaint that he/she has to meet, the victim should be protected from recrimination in the workplace by the employer ensuring that all details surrounding the complaint are suppressed from other employees in the workplace save to the extent that it is necessary to involve them to either prove or disprove the allegations of harassment.

Further, the requirement of confidentiality also extends to ensuring that the allegations, to the extent that they are made in good faith, are not subsequently used against the claimant for any other purposes connected with the claimant's employment (thereby tying in with the duty of the employer to ensure that there is no victimisation occasioned to the claimant — see *Nagarajan v London Regional Transport* [1999] IRLR 572 dealing with the issue of victimisation).

Informal action

[H1737] Most victims of harassment simply want the unwanted attention or conduct towards them to stop. Whilst the employer's policy should leave employees in no doubt as to the fact that deliberate harassment will amount to a case of gross misconduct being brought against an employee, it may be the case that the form of the harassment is innocent in nature and can be remedied by informal action.

In the context of sexual harassment, it may be the case that the victim of harassment can simply explain to the harasser that the conduct that he/she is receiving is unwelcome and makes the victim feel uncomfortable in the workplace. Where the victim finds it embarrassing or uncomfortable to address the harasser, it may be possible for a fellow employee (in the case of a larger organisation possibly a confidential counsellor) to make the requests to the harasser. Equal opportunities policies should reflect this.

This said, in some cases it simply will not be possible or appropriate for the matter to be dealt with informally and where this is the case, the only possible resolution will be through formal channels.

Formal action

[H1738] In the event that formal action is required to end harassment, both harassers and their victims should be made fully aware as to the procedures that an employer will adopt in relation to a claim of harassment.

Equal opportunities policies should include details as to:

(a) how investigations are to be carried out;
(b) the time frame for investigations (which should be as short as possible for the benefit of both the victim of actual harassment and the alleged harasser);
(c) whether there is to be a period of suspension whilst a claim is investigated; and
(d) the penalty for an employee who is found to have engaged in harassment.

It may be necessary to require employees to be transferred to different duties or sites (in some cases even where a case is not proven against an alleged harasser so as to prevent the employees being forced to work with each other against their will). Indeed, where they are required to do so in the face of an unproved allegation, unless the matter has been thoroughly investigated by the employer, it may be a recipe for litigation as the case of *Balgobin v London Borough of Tower Hamlets* [1987] IRLR 401 illustrates.

Training

[H1739] In the context of unlawful discrimination based harassment, there should training of the workforce is essential in order to ensure that harassment does not occur in the first place. Specifically, there should be training for supervisory staff and managers. The recommendations for training include:

(a) identifying the factors contributing to a working environment free of sexual harassment;
(b) familiarisation of employees with the employer's harassment policy;
(c) specialist training of those employees required to implement the harassment policy; and

(d) training at induction days for new employees as to the employer's policy on harassment.

Monitoring and review

[H1740] The acid test of an employer's equal opportunity policy is not how it looks on paper but whether and how it is implemented and monitored in practice. As to monitoring, much will depend upon the facts relating to particular employers.

In the case of smaller employers, it may simply require the employer to be aware of any recommendations that are proposed by bodies such as Equality and Human Rights Commission and/or the Chartered Institute of Personnel and Development. However, where employers are larger, the monitoring function may require a full comparative analysis of complaints made in order to determine what matters need to be implemented with regard to training and whether any changes need to be made to the employer's policy for the purposes of such matters as investigating complaints and tightening up the policy generally.

In short, once the policy is adopted, it must be kept up to date with regard to what is acceptable within the workplace.

Hazardous Substances in the Workplace

Andrea Oates

Introduction to hazardous substances in the workplace

[H2101] For most employers, the main requirements relating to hazardous substances in the workplace are those contained in the *Control of Substances Hazardous to Health Regulations SI 2002 (SI 2002 No 2677)* ('*COSHH Regulations*'). This chapter is therefore mainly concerned with these regulations.

The *COSHH Regulations* were first introduced in 1988 but have been subject to a number of changes and amendments since. Currently the *Control of Substances Hazardous to Health Regulations 2002*, after as amended, are in force.

There are many other legal requirements affecting what would generally be regarded as hazardous substances, even though some of these do not fall within the definition of 'substance hazardous to health' contained in the *COSHH Regulations*. Relevant regulations include:

- The European Regulation on the Classification, Labelling and Packaging of Substances and Mixtures (the CLP Regulation) which adopts the Globally Harmonised System for the Classification and Labelling of Chemicals (GHS) and the Registration, Evaluation, Authorisation and restriction of Chemicals (REACH) regime (see **H2137**). The directly-acting European Regulation replaced the *Chemicals (Hazard Information and Packaging for Supply) Regulations 2009 (SI 2009 No 716)*, commonly known as 'CHIP';
- *Control of Asbestos at Work Regulations 2012 (SI 2012 No 632)* (see **H2138**);
- *Control of Lead at Work Regulations 2002 (SI 2002 No 2676)* (see **H2140**);
- *Control of Major Accident Hazards Regulations 2015 (SI 2015 No 483)* (known as 'COMAH') (see **H2139**);
- *Dangerous Substances and Explosive Atmospheres Regulations 2002 (SI 2002 No 2776)* (known as 'DSEAR') (see **H2141**);
- *Dangerous Substances (Notification and Marking of Sites) Regulations 1990 (SI 1990 No 304)*;
- *Ionising Radiations Regulations 2017 (SI 2017 No 1075)* (see **H2145**).

The requirements of these regulations are summarised at the end of this chapter and are dealt with more fully in other chapters.

How hazardous substances harm the body

Routes of entry

[H2102] Hazardous substances can be present in the workplace in a variety of forms – solids, dust, fumes, smoke, liquids, vapour, mists, aerosols and gases. In deciding what risks are posed by these substances, three potential entry routes into the body must be considered:

- *Inhalation* – the hazardous substance may cause damage directly to the respiratory tract or the lungs. Inhalation also allows substances to enter the bloodstream via the lungs and thus affect other parts of the body.
- *Ingestion* – accidental ingestion of hazardous substances is always possible, particularly if containers are not correctly labelled. Eating and drinking in the workplace introduce the risk of inadvertently ingesting small quantities of hazardous substances. Legislation banning smoking in enclosed workplaces and public places was introduced across the UK in 2006 and 2007.
- *The skin* – some substances can damage the skin (or eyes) through their corrosive or irritant effects. Some solvents can enter the body by absorption through the skin. Damage is more likely when there are cracks or cuts in the skin.

Occupational ill health

[H2103] The adverse effects of hazardous substances are many and varied and are a field of study in their own right (see OCCUPATIONAL HEALTH AND DISEASES – AN OVERVIEW). Some types of occupational diseases must be reported to the enforcing authorities under the requirements of the *Reporting of Injuries, Diseases and Dangerous Occurrences Regulations 2013 (SI 2013 No 1471)* ('RIDDOR'):

- occupational dermatitis where the person's work involves significant or regular exposure to a known skin sensitiser or irritant;
- occupational asthma, where the person's work involves significant or regular exposure to a known respiratory sensitiser;
- any cancer attributed to an occupational exposure to a known human carcinogen or mutagen (including ionising radiation); and
- any disease attributed to an occupational exposure to a biological agent.
A carcinogen is a cancer-causing substance. A mutagen is a physical or chemical agent that alters genetic material. A biological agent is a micro-organism, cell culture, or human endoparasite which may cause infection, allergy, toxicity or otherwise create a hazard to human health.

See ACCIDENT REPORTING AND INVESTIGATION.

In addition, others are eligible for social security payments under government schemes. Employees suffering from occupational ill health can also bring civil actions for damages against their employers, former employers or others they consider responsible for their condition. See COMPENSATION FOR WORK INJURIES/DISEASES for more information.

How hazardous substances harm the body [H2103]

Some examples of occupational ill-health caused by hazardous substances are:
- Respiratory problems
  — *Pneumoconiosis* – eg asbestosis, silicosis, byssinosis, siderosis.
  — *Respiratory irritation* – caused by inhaling acid or alkali gases or mists.
  — *Asthma* – sensitisation of the respiratory system which may be caused by many substances, eg isocyanates, flour, grain, hay, animal fur, wood dusts and diesel engine exhaust emissions (DEEEs).
  — *Respiratory cancers* – eg those caused by certain chemicals, asbestos, pitch, tar or DEEEs for example.
  — *Metal fume fever* – a flu-like condition caused by inhaling zinc fumes.
- Poisoning
  Acute (short-term) or chronic (long-term) poisoning may be caused by:
  — *Metals and their compounds* – particularly lead, manganese, mercury, beryllium, cadmium.
  — *Organic chemicals* – which may affect the nervous system, the liver, the kidneys or the gastro-intestinal system.
  — *Inorganic chemicals* – as above plus possible asphyxiant effects such as those caused by carbon monoxide.
- Skin conditions
  — *Dermatitis*:
    (i) caused by primary irritants where skin tissue is damaged by acids or alkalis or natural oils are removed by solvents, detergents etc.
    (ii) due to the effects of sensitising agents, eg isocyanates, solvents, some foodstuffs.
  — *Skin cancer* – caused by contact with pitch, tar, soot, mineral oils etc.
- Biological problems
  From contact with animals, birds, fish (including their carcasses and products) or with micro-organisms from other sources, eg:
  — *Livestock diseases* – eg anthrax or brucellosis.
  — *Allergic Alveolitis* – caused by inhaling mould or fungal spores present in grain etc.
  — *Viral Hepatitis* – usually due to contact with blood or blood products.
  — *Legionnaires' Disease* – caused by inhaling airborne water droplets containing the legionella bacteria (which can occur extensively in water at certain temperatures).
  — *Leptospirosis* – Weil's disease, caused by contact with urine from small mammals, particularly rats.

Information on the harmful effects associated with individual products may be available on packaging but certainly should be contained in a material safety data sheet provided by the manufacturer or supplier.

The COSHH Regulations

What are substances hazardous to health?

[H2104] Under the *COSHH Regulations 2002 (SI 2002 No 2677), Reg 2(1)*, a substance hazardous to health is defined as a substance, including a mixture:

- which is listed in Part I of the approved supply list as dangerous for supply in the CLP Regulation (see above) and for which an indication of danger specified for the substance is very toxic, toxic, harmful, corrosive or irritant;
- which meets the criteria for classification as hazardous within any health hazard class laid down in the CLP Regulation whether or not the substance is classified under that Regulation;
- for which the Health and Safety Executive has approved a workplace exposure limit (see **H2112** below);
- which is a biological agent;
- other dust of any kind when present at a concentration in air equal to or greater than 10 mg/m3, as a time-weighted average over an eight-hour period, of inhalable dust, or 4 mg/m3, as a time-weighted average over an eight-hour period, of respirable dust; or
- any other substance creating a risk because of its chemical or toxicological properties and the way it is used or is present at the workplace.

HSE advice is as follows: 'COSHH covers substances that are hazardous to health. Substances can take many forms and include:

- chemicals
- products containing chemicals
- fumes
- dusts
- vapours
- mists
- nanotechnology
- gases and asphyxiating gases
- biological agents (germs). If the packaging has any of the hazard symbols then it is classed as a hazardous substance and
- germs that cause diseases such as leptospirosis or legionnaires disease and germs used in laboratories.'

Given this widely drawn definition, in any cases of doubt it is always advisable to treat the Regulations as applying, although they do not cover lead, asbestos or radioactive substances as these are all covered by their own specific regulations as set out at the end of this chapter.

Who has duties under the COSHH Regulations?

[H2105] The *COSHH Regulations* place duties on employers in relation to their employees. In addition, they place duties on employees themselves. *Regulation 8* requires employees to make full and proper use of any control measure, other thing or facility provided in accordance with these Regulations. 'Relevant' self-employed people have duties as if they were both employer and employee.

Only 'relevant' self-employed people are now covered by health and safety law. The HSE explains that if a self-employed person's work activity is specifically mentioned in the *Health and Safety at Work etc Act 1974 (General Duties of Self-Employed Persons) (Prescribed Undertakings) Regulations 2015*, or if their work activity poses a risk to the health and safety of others, then health and safety law applies. The activities mentioned are agriculture (including forestry), work with asbestos, work on a construction site, an activity to which the *Gas Safety (Installation and Use) Regulations 1998* applies, work with genetically modified organisms and work on the railways. A 'risk to the health and safety of others' is the likelihood of someone else – members of the public, clients or contractors for example – being harmed or injured as a consequence of the work activity. For health and safety law purposes, 'self-employed' means that a person does not work under a contract of employment and works only for themselves. If a person is self-employed and employs others, then the law applies.

The HSE guidance on how health and safety legislation applies to self-employed people can be found on its website at: www.hse.gov.uk/self-employed/index.htm.

Regulation 3(1) extends most employer duties under the *COSHH Regulations*, 'so far as is reasonably practicable', to 'any other person, whether at work or not, who may be affected by the work carried on'. (There are exceptions with regard to health surveillance, monitoring, information, training and dealing with accidents.)

Consideration must be therefore be given to:

- others at work in the workplace, eg contractors or co-tenants;
- visitors;
- members of the public, particularly in public places and buildings;
- customers and service users; and
- members of the emergency services.

Some of these (such as children) cannot be expected to behave in the same way as employees or contractors. An early prosecution under the *COSHH Regulations* was of a doctor's surgery when a young child gained access to an insecure cleaner's cupboard and drank the contents of a bottle of carbolic acid, suffering serious ill effects.

Full details of the *COSHH Regulations* together with associated Approved Codes of Practice ('ACoPs') and guidance are contained in a single HSE booklet – *Control of substances hazardous to health* (L5).

The COSHH Regulations summarised

[H2105.1] In summary, the COSHH Regulations set out:

- *Prohibitions relating to certain substances* [SI 2002 No 2677, Reg 4] – the manufacture and use of certain specified substances is prohibited (full details are contained in *Schedule 2* to the Regulations).

- *Application of Regulations 6 to 13 [SI 2002 No 2677, Reg 5]* – exceptions are made from the application of these Regulations where other more specific regulations are in place. These relate to lead and asbestos. There are also exceptions where substances are hazardous solely due to their radioactive, explosive or flammable properties; because they are at high or low temperatures or high pressure; or where they are administered in the course of medical treatment.
- *Assessment [SI 2002 No 2677, Reg 6]* – an assessment of health risks must be made to identify steps necessary to comply with the Regulations and these steps must be implemented before any work which is liable to expose employees to any substance hazardous to health is carried out.
- *Prevention or control of exposure [SI 2002 No 2677, Reg 7]* – this must be achieved preferably through elimination or substitution or alternatively the provision of adequate controls, eg enclosure, local exhaust ventilation (LEV), general ventilation, systems of work, and as a last resort where adequate control cannot be achieved by other means, through the use of personal protective equipment ('PPE'). This regulation requires a 'hierarchical' approach to be taken – this is explained in **H2107** below.
- *Use of control measures etc [SI 2002 No 2677, Reg 8]* – employees are required to make 'full and proper use of any control measure', such as PPE, and employers must 'take all reasonable steps' to ensure they do.
- *Maintenance, examination and testing of control measures [SI 2002 No 2677, Reg 9]* – control measures, including PPE, must be maintained in an efficient state. Controls must also be examined and tested periodically.
- *Monitoring exposure at the workplace [SI 2002 No 2677, Reg 10]* – monitoring (eg through occupational hygiene surveys) may be necessary in order to ensure adequate control or to protect employees' health.
- *Health surveillance [SI 2002 No 2677, Reg 11]* – surveillance of employees' health may be required in some circumstances.
- *Information, instruction and training [SI 2002 No 2677, Reg 12]* – must be provided for persons who may be exposed to hazardous substances (and also for those carrying out COSHH assessments).
- *Arrangements to deal with accidents, incidents and emergencies [SI 2002 No 2677, Reg 13]* – emergency procedures must be established where significant risks exist, and information on emergency arrangements provided.

Carrying out COSHH risk assessments

Assessing the risk to health

[H2106] Sub-paragraph (2) of *Regulation 6* of the *COSHH Regulations 2002 (SI 2002 No 2677)* details many specific factors which must be taken into account during the assessment. The Regulation states:

(1) An employer shall not carry on any work which is liable to expose any employees to any substance hazardous to health unless he has—
  (a) made a suitable and sufficient assessment of the risk created by that work to the health of those employees and of the steps that need to be taken to meet the requirements of these Regulations; and
  (b) implemented the steps referred to in sub-paragraph (*a*).
(2) The risk assessment shall include consideration of—
  (a) the hazardous properties of the substance;
  (b) information on health effects provided by the supplier, including information contained in any relevant safety data sheet;
  (c) the level, type and duration of exposure;
  (d) the circumstances of the work, including the amount of the substance involved;
  (e) activities, such as maintenance, where there is the potential for a high level of exposure;
  (f) any relevant workplace exposure limit or similar occupational exposure limit;
  (g) the effect of preventive and control measures which have been or will be taken in accordance with regulation 7 [which sets out requirements concerning the prevention or control of exposure to substances hazardous to health];
  (h) the results of relevant health surveillance;
  (i) the results of monitoring of exposure in accordance with regulation 10 [which sets out requirements concerning monitoring exposure at the workplace];
  (j) in circumstances where the work will involve exposure to more than one substance hazardous to health, the risk presented by exposure to such substances in combination;
  (k) the approved classification of any biological agent; and
  (l) such additional information as the employer may need in order to complete the risk assessment.
(3) The risk assessment shall be reviewed regularly and forthwith if—
  (a) there is reason to suspect that the risk assessment is no longer valid;
  (b) there has been a significant change in the work to which the risk assessment relates; or
  (c) the results of any monitoring carried out in accordance with regulation 10 show it to be necessary,
  and where, as a result of the review, changes to the risk assessment are required, those changes shall be made.
(4) Where the employer employs 5 or more employees, he shall record—
  (a) the significant findings of the risk assessment as soon as practicable after the risk assessment is made; and
  (b) the steps which he has taken to meet the requirements of regulation 7.'

In carrying out assessments, it should be noted that *Regulation 3(1)* also requires employees to take account, so far as is reasonably practicable, of others who may be affected by the work, eg visitors, contractors, customers, passers-by and members of the emergency services.

Prevention or control of exposure

[H2107] The *COSHH Regulations 2002 (SI 2002 No 2677), Reg 6(1)(a)* state that the assessment must take into account the steps needed to meet the requirements of the Regulations. *Regulation 7* of the *COSHH Regulations 2002* sets out a clear hierarchy of measures which must be taken to prevent or control exposure to hazardous substances. It states:

(1) Every employer shall ensure that the exposure of his employees to substances hazardous to health is either prevented or, where this is not reasonably practicable, adequately controlled.

(2) In complying with his duty of prevention under paragraph (1), substitution shall by preference be undertaken, whereby the employer shall avoid, so far as is reasonably practicable, the use of a substance hazardous to health at the workplace by replacing it with a substance or process which, under the conditions of its use, either eliminates or reduces the risk to the health of the employees.

(3) Where it is not reasonably practicable to prevent exposure to a substance hazardous to health, the employer shall comply with his duty of control under paragraph (1) by applying protection measures appropriate to the activity and consistent with the risk assessment, including in order of priority—

    (a) the design and use of appropriate work processes, systems and engineering controls and the provision and use of suitable work equipment and materials;

    (b) the control of exposure at source, including adequate ventilation systems and appropriate organisational measures; and

    (c) where adequate control of exposure cannot be achieved by other means, the provision of suitable personal protective equipment in addition to the measures required by sub-paragraphs (*a*) and (*b*).

This establishes a clear order of preference:

(1) Prevention of exposure (by elimination or substitution).
(2) Appropriate work processes, systems and engineering controls (eg enclosure) and suitable work equipment and materials.
(3) Adequate control at source (by ventilation or organisational measures).
(4) Adequate control through PPE (as a last resort where adequate control measures cannot be achieved by any other means and in addition to the measures above).

(*Sub-paragraph (4)* of *Regulation 7* contains further details of what the measures referred to in *paragraph (3)* must include – these are referred to in H2109.)

[H2108] The first preference should always be to prevent exposure to hazardous substances. This could be achieved by:

- changing work methods, such as adopting ultrasonic techniques or high-pressure water jets for cleaning purposes rather than using solvents;
- using non-hazardous (or less hazardous) alternatives, such as replacing solvent-based paints or inks with water-based ones (or ones utilising less hazardous solvents);
- using the same substances but in less hazardous forms, such as substituting powdered materials with granules or pellets, using more dilute concentrations of hazardous substances; or

- modifying processes, such as eliminating the production of hazardous by-products, emissions or waste by changing process parameters such as temperature or pressure.

Exposure must be prevented 'so far as is reasonably practicable' – a definition of this qualifying phrase is provided in the chapter on RISK ASSESSMENT. It has been interpreted in the courts as meaning that the costs of carrying out the health and safety measures must not be grossly disproportionate to the benefits. Technical means and financial considerations can be taken into account, but while an employer may be able to argue that the cost of a particular safety measure is not justified by the reduction in risk, they cannot avoid their responsibilities simply by claiming that they cannot afford improvements.

The level of risk will be determined by the type and severity of the hazards associated with a substance and its circumstances of use.

Considerations weighing against the prevention of exposure being reasonably practicable might be the costs associated with using alternative methods or materials, the detrimental effect of alternatives on the product or additional risks which may be introduced by the alternatives.

For example:

- ultrasonic cleaning may not achieve the required standards of cleanliness;
- use of high pressure water can involve additional risks;
- water-based paints may result in increased problems from corrosion; and
- alternative solvents may be more flammable.

Control measures, other than PPE

[H2109] The *COSHH Regulations 2002 (SI 2002 No 2677), Reg 7(4)* states that the control measures required by *Reg 7(3)* shall include:

(a) arrangements for the safe handling, storage and transport of substances hazardous to health, and of waste containing such substances, at the workplace;
(b) the adoption of suitable maintenance procedures;
(c) reducing, to the minimum required for the work concerned—
 (i) the number of employees subject to exposure,
 (ii) the level and duration of exposure, and
 (iii) the quantity of substances hazardous to health present at the workplace;
(d) the control of the working environment, including appropriate general ventilation; and
(e) appropriate hygiene measures including adequate washing facilities.

Examples of such control measures include:

- enclosed or covered process vessels;
- enclosed transfer systems, such as the use of pumps or conveyors, rather than open transfer;

- use of closed and clearly labelled containers;
- plant, processes and systems of work which minimise the generation of, or suppress and contain, spills, leaks, dust, fumes and vapours. This often involves a combination of partial enclosure and the use of LEV;
- adequate levels of general ventilation;
- appropriate maintenance procedures;
- minimising the quantities of hazardous substances stored and used in workplaces;
- restricting the numbers of people in areas where hazardous substances are used;
- prohibiting eating and drinking in areas of potential contamination. Legislation banning smoking in enclosed workplaces and public places came into effect across the UK from 2006;
- providing showers in addition to normal washing facilities;
- regular cleaning of walls and other surfaces;
- using signs to indicate areas of potential contamination; and
- safe storage, handling and disposal of hazardous substances (including hazardous waste).

Paragraphs (5) and *(6)* of *Regulation 7* set out measures required to control exposure to carcinogens, mutagens and biological agents – these are dealt with later in **H2113**.

Control using PPE

[H2110] Control of exposure using PPE should be the last resort. The ACoP to the *COSHH Regulations* (see L5) sets out that: 'PPE must be used where it is not reasonably practicable to achieve adequate control of exposure by other control measures alone, and then only in addition to these control measures . . .'

PPE necessary to control risks from hazardous substances might include:

- respiratory protective equipment (RPE);
- protective clothing;
- hand or arm protection (usually gloves or gauntlets);
- eye protection; and
- protective footwear.

The *COSHH Regulations 2002 (SI 2002 No 2677), Reg 7(9)* state that:

Personal protective equipment provided by an employer in accordance with this regulation shall be suitable for the purpose and shall—

(a) comply with any provision in the Personal Protective Equipment Regulations 2002 which is applicable to that item of personal protective equipment; or

(b) in the case of respiratory protective equipment, where no provision referred to in sub-paragraph (a) applies, be of a type approved or shall conform to a standard approved, in either case, by the Executive.'

The HSE provides detailed guidance on PPE generally and on RPE in particular on its website at www.hse.gov.uk/coshh/basics/ppe.htm and www.hse.gov.uk/respiratory-protective-equipment/.

Adequate control

[H2111] The *COSHH Regulations 2002 (SI 2002 No 2677), Reg 7(7))* (as amended) and *(11)* define what is meant by the term 'adequate control' in respect of risks from hazardous substances:

'(7) Without prejudice to the generality of paragraph (1), where there is exposure to a substance hazardous to health, control of that exposure shall only be treated as adequate if—

(a) the principles of good practice for the control of exposure to substances hazardous to health set out in Schedule 2A are applied;

(b) any workplace exposure limit approved for that substance is not exceeded; and

(c) for a substance—
  (i) which carries the hazard statement H340, H350 or H350i, or for a substance or process which is listed in Schedule 1; or
  (ii) which carries the hazard statement H334, or which is listed in section C of HSE publication "Asthmagen? Critical assessments of the evidence for agents implicated in occupational asthma" as updated from time to time, or any other substance which the risk assessment has shown to be a potential cause of occupational asthma, exposure is reduced to as low a level as is reasonably practicable.'

Schedule 2A sets out the principles of good practice for the control of substances hazardous to health.

Hazard statements describe the nature of the hazard in the substance or mixture and replaced the 'risk or R-phase' used in CHIP:

- Hazard statement H340: May cause genetic defects;
- H350: May cause cancer;
- H350i: May cause cancer by inhalation;
- H334: May cause allergy or asthma symptoms or breathing difficulties if inhaled.

Regulation 7(11) goes on to say:

'In this regulation, "adequate" means adequate having regard only to the nature of the substance and the nature and degree of exposure to substances hazardous to health and "adequately" shall be construed accordingly.'

Workplace exposure limits

[H2112] Workplace exposure limits (WELs) for substances hazardous to health are approved by the HSE and are set in relation to a specified reference period. The HSE explains that many thousands of substances are used at work but only around 500 substances have WELs, which are listed in EH40 *Workplace exposure limits*. This is legally-binding guidance. The current (2011) version of EH40 was updated to include new and revised WELs introduced by the 2nd and 3rd Indicative Occupational Exposure Limit Values (IOELV) Directives.

In November 2017, the HSE launched a consultation on the implementation of new and revised indicative occupational exposure limit values (IOELVs) for 31 chemical substances. The consultation document, *CD 283 Consultation on implementing new and revised Workplace Exposure Limits*, sets out proposals

[H2112] Hazardous Substances in the Workplace

for establishing WELs for the substances listed in the 4th IOELVs directive (2017/163/EU) by amending EH40. The consultation document can be found on the HSE website at www.hse.gov.uk/consult/condocs/cd283.htm. The consultation ends on 2 February 2018 (see **H2088**). EH40 can be found on the HSE website at www.hse.gov.uk/pubns/books/eh40.htm.

The HSE advises that ensuring that exposures are below the WEL relies on monitoring – measuring the substance in the air that workers breathe in during the work activity. For more detailed guidance, it points to the guidance sheet Exposure measurement: air sampling (G409) which sets out what employers can expect from a competent consultant providing monitoring services. This publication can also be found on the HSE website at www.hse.gov.uk/pubns/guidance/g409.pdf.

Carcinogens, mutagens and biological agents

[H2113] The *COSHH Regulations 2002 (SI 2002 No 2677), Reg 7(5)* and *(6)* set out a number of specific measures which must be applied where it is not reasonably practicable to prevent exposure to carcinogens, mutagens and biological agents. These are in addition to the measures required by *Reg 7(3)*.

Measures specified for controlling carcinogens and mutagens are:

- totally enclosing process and handling systems (unless not reasonably practicable);
- prohibiting eating, drinking and smoking in areas that might be contaminated by these substances. Legislation banning smoking in enclosed workplaces and public places across the UK came into effect from 2006;
- cleaning floors, walls and other surfaces regularly and whenever necessary;
- designating areas of potential contamination by warning signs;
- storing, handling and disposing of carcinogens or mutagens safely, using closed and clearly labelled containers.

Measures for controlling biological agents are:

- use of warning signs (a biohazard sign is shown in *Schedule 3* of the Regulations);
- specifying appropriate decontamination and disinfection procedures;
- safe collection, storage and disposal of contaminated waste;
- testing for biological agents outside primary physical confinement areas;
- specifying work and transportation procedures;
- where appropriate, making available effective vaccines;
- using appropriate hygiene measures including providing adequate washing and toilet facilities and prohibiting eating, drinking, smoking and applying cosmetics in working areas where there is a risk of contamination by biological agents; and
- control and containment measures where human patients or animals are known (or suspected to be) infected with a Group 3 or 4 biological agent.

A Group 3 biological agent can cause severe human disease and may be a serious hazard to employees. It may spread to the community, but there is usually effective treatment available. A Group 4 biological causes severe human disease and is a serious hazard to employees. It is likely to spread to the community and there is usually no effective treatment available.

The *COSHH Regulations 2002 (SI 2002 No 2677), Reg 7(10)* require work with biological agents to be in accordance with *Schedule 3* to the Regulations. This sets out several detailed requirements, including standards for containment measures.

Further guidance on the control of carcinogens, mutagens and biological agents is provided in the HSE publication L5.

In May 2016 and January 2017, the European Commission proposed two sets of amendments to the carcinogens and mutagens directive to add more carcinogens to the list of controlled substances. The main effect of the first proposal would be to set new exposure limits for respirable crystalline silica (RCS) dust, to which workers in the construction and quarrying sectors are frequently exposed, and for hardwood dust, which is common in woodworking and construction.

The second proposal would bring exposure to used engine oils within the scope of the directive and assign a 'skin notation' to these oils. It would add five carcinogenic substances to the list of chemicals covered by the Directive. It also extends the scope of the Directive to a new category of Polycyclic Aromatic Hydrocarbons (PAHs).

In July 2017, the European Parliament and Council of Ministers reached agreement on first of the two proposals, which also included lower exposure limit values for Chromium (VI) compounds and hardwood dust and an extension of the existing requirement for employers to conduct health surveillance of their workers.

In terms of how the amendments will affect UK exposure levels, there is a two-year implementation period for the directive plus a five-year transition period. It is possible the UK will implement the amendments even though the likely dates of its application fall after March 2019 when the UK is due to leave the European Union. The text of the so-called Brexit deal negotiated in December 2017, which allowed negotiations to progress to the next stage, suggests that the whole UK will remain in 'regulatory alignment' with the EU in order to avoid the need for customs checks on the Irish border. In addition, existing and new EU laws will continue to apply during a two-year transition period after the UK leaves the EU on 29 March 2019.

Practical aspects of COSHH assessments

The requirements summarised

[H2114] COSHH assessments must involve the following:
- identification of risks to the health of employees;

- consideration of whether it is reasonably practicable to prevent exposure to those hazardous substances which create risks (by elimination or substitution);
- if prevention is not reasonably practicable, identification of the measures necessary to achieve adequate control of exposure, as required by the *COSHH Regulations 2002 (SI 2002 No 2677), Reg 7* (these control measures are described at **H2109** and **H2110**);
- identification of other measures necessary to comply with the *COSHH Regulations 2002 (SI 2002 No 2677), Regs 8–13*, eg:
    — use of control measures;
    — maintenance, examination and test of control measures etc;
    — monitoring of exposure at the workplace;
    — health surveillance;
    — provision of information, instruction and training;
    — arrangements for accidents, incidents and emergencies.

(These are described more fully at **H2127–H2135** below).

The assessment must be 'suitable and sufficient', and its complexity will vary considerably depending on the circumstances. The planning, preparation and implementation of a COSHH assessment will be very similar to other types of risk assessment and much of the guidance given in the chapter on RISK ASSESSMENT will be valid here.

Who should carry out COSHH assessments?

[H2115] *Regulation 7* of the *Management of Health and Safety at Work Regulations 1999 (SI 1999 No 3242)* requires employers to appoint a competent person or persons to assist them in complying with their duties, which will include carrying out COSHH assessments. *Regulation 12(4)* of the *COSHH Regulations 2002 (SI 2002 No 2677)* also states:

> Every employer shall ensure that any person (whether or not his employee) who carries out any work in connection with the employer's duties under these Regulations has suitable and sufficient information, instruction and training.

The degree of knowledge and experience required will depend upon the circumstances. An assessment in a workplace containing a few simple uses of hazardous substances may be carried out by someone without any specialist qualifications but with the capability to understand and apply the principles contained in this chapter. However, such a person may need access to someone with greater health and safety knowledge and a more detailed understanding of the *COSHH Regulations*. Some qualified and experienced individuals may be capable of carrying out detailed assessments of relatively high-risk situations themselves, although they are likely to need to consult others during the assessment process.

HSE guidance provided in L5 is as follows: "The person carrying out the assessment may have access to people who can help to deliver the assessment and the implementation of the risk management measures. Employers should ensure everyone appointed to assist in meeting this requirement is competent to do the job and there is formal, written specification for the work that is being planned (see www.hse.gov.uk/business/competent-advice.htm)."

Practical aspects of COSHH assessments [H2117]

Higher risk situations, with much more complex and significant use of hazardous substances are more likely to require multi-disciplinary teams, some of whom have specialist knowledge and experience. Possible candidates for such a team might be:

- health and safety officers or managers;
- occupational hygienists;
- occupational health nurses;
- physicians with occupational health experience;
- chemical or process engineers; or
- maintenance or ventilation engineers.

The ACOP sets out that: "Where more than one person is involved in carrying out the assessment, the employer should nominate someone to co-ordinate, consult, compile, assure quality, record, communicate and implement the risk management measures and monitor their effectiveness, as well as consider the need for reviewing the assessment.

The competent person carrying out the assessment should:

- know how the work activity uses, produces or creates substances hazardous to health;
- have the knowledge, skills, training and experience to make sound decisions about the level of risk and the measures needed for prevention or adequate control of exposure;
- have the ability and the authority of the employer to collate all the necessary, relevant information."

How should the assessments be organised?

[H2116] In larger workplaces it is often best to divide them into manageable assessment units which could be based on:

- departments or sections;
- buildings or rooms;
- process lines; or
- activities or services.

Where an assessment team is used, those individuals most suited for a particular situation can be selected to assess that unit.

Pooling of knowledge can allow, for example, a supervisor's experience of a process to be combined with the technical and legal knowledge of a health and safety manager.

Gathering information together

[H2117] Prior to the assessment, information should be gathered together, particularly about which hazardous substances are present and what hazards are associated with them. Substances to be taken into account include:

- Raw materials used in processes.
- The contents of stores and cupboards.

- Substances produced by processes:
  - intermediate compounds;
  - products;
  - waste and by-products;
  - emissions from the process.
- Substances used in maintenance and cleaning work.
- Buildings and the work environment:
  - surface treatments;
  - pollution or contaminants;
  - bird droppings, animal faeces etc;
  - possible legionella sources.
- Substances involved in the activities of others, such as contractors.

Potential sources of information about the hazards associated with these substances include:

- Manufacturers' or suppliers' safety data sheets.
- Information provided on containers or packaging (including the symbols required by the CLP Regulation (see **H2101**).
- HSE publications and the HSE website.
- Other reference books and technical literature.
- Direct enquiries to specialists or suppliers.

This information will primarily concern the hazards associated with the substance but is also likely to contain recommendations on control methods. However, the substance's actual circumstances of use may constitute a greater or lesser degree of risk than the manufacturer or supplier anticipated – it is for the assessor(s) to determine what control measures are necessary in each situation.

Other useful information may be available from previous assessments or surveys. Even if previous COSHH assessments were inadequate or are now out of date, they may still contain information of value in the assessment process. The results of previous occupational hygiene surveys (eg dust, gas or vapour concentrations) are also likely to be useful as are the collective results of any health surveillance work which has been carried out.

Details of existing control measures may be contained in operating procedures, health and safety rule books or listings of PPE requirements or may be referred to within training programmes. Specification information may be available for some control measures, such as design flow rates for ventilation equipment or performance standards for PPE such as respiratory protection or gloves. Some of this information may also be sought during the assessment itself or prior to the preparation of the assessment record.

Observations in the workplace

[**H2118**] Time spent in making COSHH assessments can be reduced and made more productive by good planning and preparation. Nevertheless, observations of workplaces and work activities must still be an integral part of the COSHH assessment process.

There must be sufficient observations in order to arrive at a conclusion on the level of risk and the adequacy of control measures for all the hazardous substances in the workplace – whether raw materials, products, by-products, waste or emissions. This should include the manner and extent to which employees (and others) are exposed and an evaluation of the equipment and working practices intended to achieve their control.

Aspects to be considered include:

- Are specified procedures being followed?
- Does local exhaust ventilation or general ventilation appear effective?
- Is PPE being used correctly?
- Is there evidence of leakage, spillage or dust accumulation?
- What equipment is available for cleaning?
- Are dusts, fumes or strong smells evident in the atmosphere?
- Are there any restrictions relating to eating or drinking? Legislation banning smoking in enclosed workplaces and public places across the UK came into effect from 2006.
- What arrangements are there for storage, cleaning or maintenance of PPE?
- Are there suitable arrangements for washing, showering or changing clothing?
- Are special arrangements for laundering clothing required?
- What arrangements are in place for storage and disposal of waste?
- Is there potential for major spillages or leaks?
- What emergency containment equipment or PPE is available?

Discussions with staff

[H2119] It is also important to talk to those working with hazardous substances, their safety representatives and those managing or supervising their work. Questions might relate to working practices, awareness of risks or precautions, the effectiveness of precautions or experience of problems. These might include:

- Why are tasks done in particular ways?
- Are the present work practices typical?
- How effective is the LEV/general ventilation?
- What happens if it ever breaks down?
- What types of PPE are required for the work?
- Are there any problems with the PPE?
- Have workers experienced any health problems?
- How often is the workplace cleaned?
- What methods are used for cleaning?
- How is waste disposed of?
- Is there any possibility of leaks occurring or emergencies arising?
- What would happen in such situations?
- What are the arrangements for washing/cleaning/maintaining PPE?

Further assistance

[H2120] Guidance on COSHH assessments is available in the HSE booklet *A step by step guide to COSHH assessment* (HSG97). The HSE has also developed a range of *COSHH essentials* publications and there are COSHH essentials web pages at www.hse.gov.uk/coshh/essentials.

The HSE has also compiled a set of example COSHH risk assessments for particular industries and workplaces and has examples of general risk assessments on its website at: www.hse.gov.uk/coshh/riskassess/index.htm.

Further investigations

[H2121] Assessment in the workplace may reveal the need for further tests on the effectiveness of control measures before the assessment can be concluded. The tests may be in the form of occupational hygiene surveys – usually to measure the airborne concentrations of dust, fume, vapour or gas and compare them with the relevant WEL. Alternatively, there may be a need to measure the effectiveness of local exhaust ventilation or general ventilation.

Further investigations may also be necessary into matters such as:

- hazards associated with substances which were not identified during the preparatory phase of the assessment;
- whether the specifications for PPE found in use are adequate for the exposure involved;
- the feasibility of using alternative work methods to prevent exposure or alternative methods of control to those currently in place.

COSHH assessment records

Preparation of assessment records

[H2122] Much of the guidance on note taking and the preparation of assessment records contained in the chapter on RISK ASSESSMENT is relevant to COSHH assessments. The essential components which must be included in COSHH assessment records are:

- the work activities involving risks from hazardous substances;
- information as to the hazardous substances involved (and their form, eg liquid, powder, dust etc);
- who might be harmed and how;
- control measures which are (or should be) in place;
- improvements identified as being necessary.

Although employers with less than five employees are exempt from the requirement to record assessments, it is strongly advised that keeping records is good practice. In all other cases the significant findings must be recorded and kept readily accessible to those who may need to see them. Employees or their representatives should be informed of the results of COSHH assessments.

The amount of information should be proportionate to the risks posed by the work. In some cases it will be appropriate to state that, in the circumstances of use, there is little or no risk and no more detailed assessments are necessary.

Review and implementation of recommendations

[H2123] Once the assessment has been completed and the record has been prepared it is important that recommendations are reviewed, implemented and followed up. The review process is likely to involve people who were not involved in the original assessments – senior managers, or staff with relevant technical expertise. While the assessment team may be prepared to make changes to their assessment findings or recommendations, they should not allow themselves to be pressurised into making changes they find unacceptable. If senior management choose not to implement COSHH assessment recommendations then they must accept the responsibility for their actions.

Responsibilities for implementing each recommendation within a designated timescale must be clearly allocated to individuals, possibly as part of an overall action plan. Each of these recommendations must then be followed up to ensure that they have actually been carried out and also to check that additional risks have not been inadvertently created. The assessment record should then be annotated to indicate that the recommendation has been completed (alternatively a new record could be prepared reflecting the improved situation).

Example assessment records

[H2124] The HSE provides examples of COSHH risk assessments for different types of workplace on its website at: www.hse.gov.uk/coshh/riskassess/index.htm. These are provided for a paving company, feed mill, die caster, electronics, fruit farm, warehouse, office, DIY shop, workshop, engineer, sandwich maker and a garden centre. Examples of general risk assessments are also provided.

Ongoing reviews of assessments

[H2125] The *COSHH Regulations 2002 (SI 2002 No 2677), Reg 6(3)* requires COSHH assessments to:

> be reviewed regularly and forthwith if—
>
> (a) there is reason to suspect the risk assessment is no longer valid;
> (b) there has been a significant change in the work to which the risk assessment relates; or
> (c) the results of any monitoring carried out in accordance with regulation 10 show it to be necessary.

- *Assessments no longer valid*
  Assessments might be shown to be no longer valid because of:
  — new information received about health risks, eg information from suppliers, revised HSE guidance, changes in the WEL;
  — results from inspections or through examinations or tests (*Regulation 9*), eg indicating fundamental flaws in engineering controls;
  — regular reports or complaints about defects in control arrangements;

- the results from workplace exposure monitoring (*Regulation 10*), eg showing the WEL is regularly being exceeded or approached;
- results from health surveillance (*Regulation 11*), eg demonstrating an unsatisfactory or deteriorating position;
- a confirmed case of an occupational disease.

- *Significant changes*
  Changes necessitating a review of an assessment might involve:
  - the types of substances used, their form or their source;
  - equipment used in the process or activity (including control measures);
  - altered methods of work or operational procedures;
  - variations in volume, rate or type of production;
  - staffing levels and related practical difficulties or pressures.

- *Regular review*
  The periods elapsing between reviews should relate to the degree of risk involved and the nature of the work itself. HSE guidance on how often COSHH assessments should be reviewed is as follows:
  - an assessment should be regularly revisited to ensure that it is kept up to date;
  - the date of the first review and the length of time between successive reviews will depend on type of risk, the work, and the employer's judgement on the likelihood of changes occurring; and
  - it should be reviewed immediately if there is any reason to suppose that the original assessment is no longer valid, eg evidence from the results of examining and testing engineering controls, reports from supervisors about defects in control systems; or any of the circumstances of the work should change significantly and especially one which may have affected employees' exposure to a hazardous substance.

A review would not necessarily require a revision of the assessment – it may conclude that existing controls are still adequate despite changed circumstances. However, where changes are shown to be required, *Reg 6(3)* requires that these be implemented.

Further requirements of the COSHH Regulations

[H2126] Some recommendations from COSHH assessments could relate to the ongoing maintenance of effective control measures and many of these aspects are covered by further specific requirements of the *COSHH Regulations*.

Use of control measures

[H2127] The *COSHH Regulations 2002 (SI 2002 No 2677)*, *Reg 8* place duties on both employers and employees in respect of the proper use of control measures, including PPE.

(1) Every employer who provides any control measure, other thing or facility in accordance with these Regulations shall take all reasonable steps to ensure that it is properly used or applied as the case may be.
(2) Every employee shall make full and proper use of any control measure, other thing or facility provided in accordance with these Regulations and where relevant shall—
   (a) take all reasonable steps to ensure it is returned after use to any accommodation provided for it; and,
   (b) if he discovers a defect therein, report it forthwith to his employer.'

Employers have duties under the *Management of Health and Safety at Work Regulations 1999 (SI 1999 No 3242)* to monitor all of their health and safety arrangements. Areas where monitoring of COSHH control measures are likely to be necessary include the following:

- correct use of LEV equipment;
- compliance with specified systems of work;
- compliance with PPE requirements;
- storage and maintenance of PPE;
- compliance with requirements relating to eating and drinking. Legislation banning smoking in enclosed workplaces and public places across the UK came into effect from 2006;
- condition of washing and showering facilities and personal hygiene standards; and
- whether defects are being reported by employees.

The ACOP sets out that:

"Employers should establish procedures to ensure that control measures, including PPE and any other item or facility, are properly used or applied and are not made less effective by other work practices or by improper use. The procedures should include:

- visual checks and observations at appropriate intervals;
- ensuring that where more than one item of PPE is being worn, the different items are compatible;
- supervising employees to ensure that the defined methods of work are being followed;
- monitoring systems for the effectiveness of controls and prompt remedial action where necessary.

In addition, it sets out that:

"Employees should use the control measures in the way they are intended to be used and as they have been instructed. In particular they should:

- use the control measures provided for materials, plant and processes;
- follow the defined methods of work;
- wear PPE provided, including any RPE, correctly and in accordance with the manufacturer's instructions;
- store the PPE, when not in use, in the accommodation provided;
- remove any PPE, which could cause contamination, before eating, drinking or smoking;
- maintain a high standard of personal hygiene, and make proper use of the facilities provided for washing, showering or bathing and for eating and drinking;

- report promptly to the appointed person, eg 'foreman', supervisor or safety representative any defects discovered in any control measure, including defined methods of work, device or facility, or any PPE, including RPE."

Maintenance, examination and testing of control measures

[H2128] The *COSHH Regulations 2002 (SI 2002 No 2677), Reg 9* contain both general and specific requirements for the maintenance of control measures –

'(1) Every employer who provides any control measure to meet the requirements of regulation 7 shall ensure that -
   (a) in the case of plant and equipment, including engineering controls and personal protective equipment, it is maintained in an efficient state, in efficient working order, in good repair and in a clean condition; and
   (b) in the case of the provision of systems of work and supervision and of any other measure, it is reviewed at suitable intervals and revised if necessary.
(2) Where engineering controls are provided to meet the requirements of regulation 7, the employer shall ensure that thorough examination and testing of those controls is carried out—
   (a) in the case of local exhaust ventilation plant, at least once every 14 months, or for local exhaust ventilation plant used in conjunction with a process specified in Column 1 of Schedule 4, at not more that the interval specified in the corresponding entry in Column 2 of that Schedule; or
   (b) in any other case, at suitable intervals.
(3) Where respiratory protective equipment (other than disposable respiratory protective equipment) is provided to meet the requirements of regulation 7, the employer shall ensure that thorough examination and, where appropriate, testing of that equipment is carried out at suitable intervals.
(4) Every employer shall keep a suitable record of the examinations and tests carried out in pursuance of paragraphs (2) and (3) and of repairs carried out as a result of those examinations and tests, and that record or a suitable summary thereof shall be kept available for at least 5 years from the date on which it was made.

Regulation 9(5), (6) and *(7)* contains requirements specifically relating to PPE – these are examined later (see **H2131** below). Also see **H2146** on future changes.

General maintenance of controls

[H2129] HSE guidance is that "All control measures in use should be visually checked, where possible, at appropriate intervals and without undue risk to maintenance staff. In the case of LEV and work enclosures, such checks should be carried out at least once a week."

Such checks may simply confirm that there are no apparent leaks from vessels or pipes and that LEV or cleaning equipment appear to be in working order. It is good practice (and prudent) to keep a record of such checks.

Reg 9(2) requires that thorough examinations and tests of engineering controls are carried out. Requirements relating to LEV are reviewed below at **H2130**

Further requirements of the COSHH Regulations [H2131]

but for other engineering controls such examinations and tests must be 'at suitable intervals', and suitable records must be kept for at least five years. The nature of examinations and tests will depend upon the engineering control involved and the potential consequences of its deterioration or failure. Examples might involve:

- detailed visual inspections of tanks and pipelines;
- inspections and non-destructive testing of critical process vessels;
- testing of detectors and alarm systems;
- planned maintenance of general ventilation equipment; and
- checks on filters in vacuum cleaning equipment.

Those carrying out maintenance, examinations and testing must be competent for the purpose in accordance with the *COSHH Regulations 2002 (SI 2002 No 2677), Reg 12(4)*: 'Every employer shall ensure that any person (whether or not his employee) who carries out work in connection with the employer's duties under these Regulations has suitable and sufficient information, instruction and training.'

Local exhaust ventilation (LEV) plant

[H2130] Most LEV systems must be thoroughly examined and tested at least once every 14 months, although *Schedule 4* of the *COSHH Regulations 2002 (SI 2002 No 2677)*, requires increased frequencies for LEV used in conjunction with a handful of specified processes. The examination and test might involve visual inspection, air flow or static pressure measurements or visual checks of efficiency using smoke generators or dust lamps, depending on the design and purpose of the LEV concerned. In some cases, air sampling to confirm efficiency levels, filter integrity tests or checks on air flow sensors, may be appropriate.

Personal Protective Equipment (PPE)

[H2131] All types of PPE are subject to the general maintenance requirements contained in the *COSHH Regulations 2002 (SI 2002 No 2677), Reg 9(1)* whilst *Regulation 9(3)* contains specific requirements relating to non-disposable respiratory protective equipment ('RPE').

Further requirements in respect of PPE are contained in *Regulation 9(5), (6)* and *(7)* of the 2002 Regulations that state:

(5) Every employer shall ensure that personal protective equipment, including protective clothing, is:

    (a) properly stored in a well-defined place;
    (b) checked at suitable intervals; and
    (c) when discovered to be defective, repaired or replaced before further use.

(6) Personal protective equipment which may be contaminated by a substance hazardous to health shall be removed on leaving the working area and kept apart from uncontaminated clothing and equipment.

(7) The employer shall ensure that the equipment referred to in paragraph (6) is subsequently decontaminated and cleaned or, if necessary, destroyed.

Some types of PPE can easily be seen to be defective by the user. In other cases, such as gloves or clothing providing protection against strongly corrosive chemicals, it may be appropriate to introduce more formalised inspection systems.

For non-disposable RPE the COSHH ACoP states that:

> "Thorough maintenance examinations and, where appropriate, tests of items of RPE, other than disposable respirators, should be made at suitable intervals. The frequency should increase where the health risks and conditions of exposure are particularly severe.
>
> In situations where RPE is used only occasionally, an examination and test should be made before their next use and maintenance carried out as appropriate. The person who is responsible for managing the maintenance of RPE should determine suitable intervals between examinations. Emergency escape-type RPE should be examined and tested in accordance with the manufacturer's instructions.
>
> Suitable arrangements should be made to ensure that no employee uses RPE previously used by another person, unless it has been thoroughly washed and cleaned in accordance with the manufacturer's instructions."

An HSE booklet *Respiratory protective equipment at work: a practical guide* provides detailed guidance in this area (www.hse.gov.uk/pubns/priced/HSG53.pdf).

Monitoring exposure at the workplace

[H2132] Reference was made earlier in the chapter at **H2111** and **H2121** to the possible need to carry out air testing in the workplace as part of the COSHH assessment process in order to determine the adequacy of control measures. Such testing may also be necessary in order to ensure that adequate control continues to be maintained. *COSHH Regulations 2002 (SI 2002 No 2677), Reg 10* sets out that:

(1) Where the risk assessment indicates that—
   (a) it is requisite for ensuring the maintenance of adequate control of the exposure of employees to substances hazardous to health; or
   (b) it is otherwise requisite for protecting the health of employees,
   the employer shall ensure that the exposure of employees to substances hazardous to health is monitored in accordance with a suitable procedure.
(2) Paragraph (1) shall not apply where the employer is able to demonstrate by another method of evaluation that the requirements of regulation 7(1) have been complied with.
(3) The monitoring referred to in paragraph (1) shall take place—
   (a) at regular intervals; and
   (b) when any change occurs which may affect that exposure.
(4) Where a substance or process is specified in Column 1 of Schedule 5 [see below], monitoring shall be carried out at least at the frequency specified in the corresponding entry in Column 2 of that Schedule [see below].
(5) The employer shall ensure that a suitable record of any monitoring carried out for the purpose of this regulation is made and maintained and that record or a suitable summary thereof is kept available—
   (a) where the record is representative of the personal exposures of identifiable employees, for at least 40 years; or

(b) in any other case, for at least 5 years, from the date of the last entry made in it.
(6) Where an employee is required by regulation 11 to be under health surveillance, an individual record of any monitoring carried out in accordance with this regulation shall be made, maintained and kept in respect of that employee.
(7) The employer shall—
   (a) on reasonable notice being given, allow an employee access to his personal monitoring record;
   (b) provide the appropriate authority with copies of such monitoring records as the appropriate authority may require; and
   (c) if he ceases to trade, notify the Executive forthwith in writing and make available to the Executive all monitoring records kept by him.

Schedule 5 refers to specific substances and processes for which monitoring is required. These are vinyl chloride monomer, which must be continuously monitored or monitored in accordance with the HSE, and spray given off from vessels at which an electrolytic chromium process is carried on (except trivalent chromium) and must be monitored every 14 days while the process is being carried on.

A number of techniques are available for monitoring air quality in the workplace. Most of these involve the use of chemical indicator tubes, direct reading instruments, or sampling pumps and filter heads. An HSE booklet *Monitoring strategies for toxic substances* provides further guidance on the subject (see www.hse.gov.uk/pubns/priced/HSG173.pdf).

Health surveillance

[H2133] The *COSHH Regulations 2002 (SI 2002 No 2677), Reg 11(1)* and *(2)* contain the main requirements relating to the need for health surveillance:

(1) Where it is appropriate for the protection of the health of his employees who are, or are liable to be, exposed to a substance hazardous to health, the employer shall ensure that such employees are under suitable health surveillance.
(2) Health surveillance shall be treated as being appropriate where—
   (a) the employee is exposed to one of the substances specified in Column 1 of Schedule 6 [see below] and is engaged in a process specified in Column 2 of that Schedule, and there is a reasonable likelihood that an identifiable disease or adverse health effect will result from that exposure; or
   (b) the exposure of the employee to a substance hazardous to health is such that—
      (i) an identifiable disease or adverse health effect may be related to the exposure,
      (ii) there is a reasonable likelihood that the disease or effect may occur under the particular conditions of his work, and
      (iii) there are valid techniques for detecting indications of the disease or the effect,
   and the technique of investigation is of low risk to the employee.

Schedule 6 sets out substances for which medical surveillance is appropriate when they are used in particular processes.

The remaining parts of *Regulation 11 (paragraphs (3)–(11))* relate to the manner in which health surveillance is conducted and used, together with the maintenance of and access to surveillance records.

The decision as to whether health surveillance is appropriate to protect the health of employees is one that would normally be taken at the time of a COSHH assessment or during its subsequent review. However, for those processes and substances specified in *Schedule 6* to the Regulations, surveillance must be carried out.

Normally health surveillance programmes would be initiated and carried out under the overall supervision of a registered medical practitioner, and preferably one with relevant occupational health experience. However, the surveillance itself may be carried out by an occupational health nurse, a technician or a responsible member of staff, providing that individual is competent for the purpose.

There are many different procedures available for health surveillance including:

- biological monitoring, eg tests of blood, urine or exhaled air;
- biological effect monitoring, eg through lung function testing;
- medical surveillance, eg through physical examinations; and
- interviews about possible symptoms, eg skin abnormalities or shortness of breath.

For health surveillance to be 'appropriate' there must firstly be a significant enough risk to justify it and there must also be a valid technique for detecting indications of related occupational diseases or ill-health effects. HSE provides guidance on health surveillance on its website (www.hse.gov.uk/health-surveillance/index.htm) - and has also published guidance on the practical application of biological monitoring to chemical exposure: *Biological monitoring in the workplace* (see www.hse.gov.uk/pubns/priced/hsg167.pdf). Records of health surveillance must contain information specified in the COSHH ACoP and must be retained for at least 40 years.

The ACoP sets out that employers must keep an up-to-date health record for each individual employee placed under health surveillance. It should contain at least the following:

- identifying details:
  - surname;
  - forename(s);
  - gender;
  - date of birth;
  - permanent address and postcode;
  - national insurance number;
  - date when present employment started;
  - an historical record of jobs in this employment involving exposure to identified substances requiring health surveillance;
- results of all other health surveillance procedures and the date on which, and by whom, they were carried out. The conclusions should relate only to the employee's fitness for work and will include, where appropriate:

- a record of the decisions on an employee's fitness for continued exposure or restrictions made by the appointed doctor, or by the registered medical practitioner, occupational health nurse or other suitably qualified or responsible person;
- whether the results require increased health surveillance.

The health record should not include confidential clinical data.

Information, instruction and training

[H2134] The *COSHH Regulations 2002 (SI 2002 No 2677), Reg 12* contain requirements relating to information, instruction and training for persons who may be exposed to substances hazardous to health:

(1) Every employer who undertakes work which is liable to expose an employee to a substances hazardous to health shall provide that employee with suitable and sufficient information, instruction and training.

(2) Without prejudice to the generality of paragraph (1), the information, instruction and training provided under that paragraph shall include—
   (a) details of the substances hazardous to health to which the employee is liable to be exposed including—
      (i) the names of those substances and the risk which they present to health,
      (ii) any relevant workplace exposure limit or similar occupational exposure limit,
      (iii) access to any relevant safety data sheet, and
      (iv) other legislative provisions which concern the hazardous properties of those substances;
   (b) the significant findings of the risk assessment;
   (c) the appropriate precautions and actions to be taken by the employee in order to safeguard himself and other employees at the workplace;
   (d) the results of any monitoring of exposure in accordance with regulation 10 and, in particular, in the case of any substance hazardous to health for which a workplace exposure limit has been approved, the employee or his representatives shall be informed forthwith, if the results of such monitoring show that the workplace exposure limit is exceeded;
   (e) the collective results of any health surveillance undertaken in accordance with regulation 11 in a form calculated to prevent those results from being identified as relating to a particular person; and
   (f) where employees are working with a Group 4 biological agent [see H2113 above] or material that may contain such an agent, the provision of written instructions and, if appropriate, the display of notices which outline the procedures for handling such an agent or material.

(3) The information, instruction and training required by paragraph (1) shall be—
   (a) adapted to take account of significant changes in the type of work carried out or methods of work used by the employer; and
   (b) provided in a manner appropriate to the level, type and duration of exposure identified by the risk assessment.

(4) Every employer shall ensure that any person (whether or not his employee) who carries out work in connection with the employer's duties under these Regulations has suitable and sufficient information, instruction and training.

(5) Where containers and pipes for substances hazardous to health used at work are not marked in accordance with any relevant legislation listed in Schedule 7 [see below], the employer shall, without prejudice to any derogations provided for in that legislation, ensure that the contents of those containers and pipes, together with the nature of those contents and any associated hazards, are clearly identifiable.'

Schedule 7 lists the legislation concerned with the labelling of containers and pipes.

The general requirements of *Reg 12(1)* match those found in various other regulations but the contents of *Reg 12(2)* are much more prescriptive than the requirements of previous versions of the COSHH Regulations. It is important that workers are aware of the substances they are exposed to and the risks that they present. A good awareness of the risks will mean employees are more likely to take the appropriate precautions. Instruction and training will be particularly relevant in relation to:

- awareness of safe systems of work;
- correct use of LEV equipment;
- use, adjustment and maintenance of PPE (especially RPE);
- the importance of good personal hygiene standards;
- requirements relating to eating and drinking. Legislation banning smoking in enclosed workplaces and public places across the UK came into effect in 2006 and 2007;
- emergency procedures;
- arrangements for cleaning and the disposal of waste;
- contents of containers and pipes (see *Regulation 12(5)* above).

Employers must inform employees or their representatives of the results of workplace exposure monitoring straightaway if the WEL is shown to have been exceeded. They must also inform employees of the collective results of health surveillance – the average results within a department or on a particular shift from biological monitoring, or the numbers of employees referred for further investigation following a skin inspection for example.

The type of information, instruction and training provided must be appropriate for the level, type and exposure involved and adapted to take account of significant changes.

Any person carrying out work on the employer's behalf (whether or not an employee) is required by *Reg 12(4)* to have the necessary information, instruction and training. A consultant carrying out a COSHH assessment or an occupational hygienist conducting exposure monitoring must be verified by the employer as being competent for the purpose and be provided with the information necessary to carry out their work effectively.

The Occupational Safety and Health Consultants Register (OSHCR) provides details of consultants who have met the highest qualification standard of recognised professional bodies and who are bound by a code of conduct that requires them to only give advice which is sensible and proportionate. The register is freely accessible and searchable for employers at www.oshcr.org.

Further requirements of the COSHH Regulations **[H2135]**

Accidents, incidents and emergencies

[H2135] *Regulation 13* requires employers to ensure that arrangements are in place to deal with accidents, incidents and emergencies related to hazardous substances. This is in addition to the general duty to have procedures for serious and imminent danger and for danger areas contained in *Regulation 8* of the *Management of Health and Safety at Work Regulations 1999 (SI 1999 No 3242.)* Based on HSE guidance in the COSHH ACoP booklet (L5) such emergencies might include:

- a serious process fire posing a serious risk to health;
- a serious spillage or flood of a corrosive agent;
- a failure to contain biological, carcinogenic or mutagenic agents;
- a failure that could lead (or has led) to a sudden release of chemicals;
- a threatened significant exposure over a WEL, eg due to a failure of LEV or other controls.

The requirements of the regulation are summarised below.

Arrangements should include:

- emergency procedures, such as:
    — appropriate first aid facilities;
    — relevant safety drills (tested at regular intervals);
- providing information on emergency arrangements:
    — including details of work hazards and emergencies likely to arise;
    — made available to relevant accident and emergency services;
    — displayed at the workplace, if appropriate;
- suitable warning and other communications:
    — to enable an appropriate response, including remedial actions and rescue operations.

In the event of an accident, incident or emergency, the employer must:

- take immediate steps to:
    — mitigate its effects;
    — restore the situation to normal;
    — inform employees who may be affected;
- ensure only essential persons are permitted in the affected area and that they are provided with:
    — appropriate PPE;
    — any necessary safety equipment and plant;
- in the case of a serious biological incident, inform employees or their representatives, as soon as practicable of:
    — the causes of the incident or accident;
    — the measures taken or being taken to rectify the situation.

Regulation 13(4) states that such emergency arrangements are not required where the risk assessment shows that because of the quantities of hazardous substances the risks are slight and control measures are sufficient to control that risk. This 'exception' does not apply in the case of carcinogens, mutagens or biological agents.

Regulation 13(5) requires employees to report possible releases of a biological agent which could cause severe human disease, forthwith to their employer (or another employee with specific responsibility for health and safety).

The ACoP sets out that:

> 'Employers need not extend the scope of their general emergency procedures drawn up under regulation 8 of the MHSW Regulations if they are satisfied that:
>
> - the quantity and type of hazardous substance(s) at the workplace would either individually or cumulatively create no more than a slight risk because they have a low toxic effect;
> - existing control measures and emergency arrangements are sufficient to contain and control any risk to health that the substances might pose during an emergency, and that they are capable of quickly restoring the situation to normal.
>
> If the conditions described above do not apply, employers must extend their emergency procedures as required by regulation 13 and ensure that they are capable of limiting the extent of any risks to health of employees and, [so far as is reasonably practicable] SFARP, the health of anyone else likely to be affected by the incident, eg people living in the neighbourhood.'

The ACoP provides further details of what emergency procedures might need to include:

- the identity, location and quantities of hazardous substances present;
- foreseeable types of accidents, incidents or emergencies;
- special arrangements for emergencies not covered by general procedures;
- emergency equipment and PPE, and who is authorised to use it;
- first-aid facilities;
- emergency management responsibilities eg emergency controllers;
- how employees should respond to incidents;
- clear up and disposal arrangements;
- arrangements for regular drills or practice;
- dealing with special needs of disabled employees.

Full details of the Regulations together with associated Approved Codes of Practice ('ACoP') and guidance are contained in a single HSE booklet – *Control of substances hazardous to health* (L5).

Other important regulations on hazardous substances in the workplace

[H2136] The introduction to this chapter referred to several other legal requirements relating to hazardous substances in the workplace and these are summarised below.

Other important regulations **[H2137]**

European Regulation No 1907/2006 concerning the Registration, Evaluation, Authorisation and Restriction of Chemicals (REACH) and European Regulation 1272/2008 on the classification, labelling and packaging of substances and mixtures (the CLP Regulation)

[H2137] The requirement to provide safety data sheets is contained in the European REACH Regulation – REACH stands for **R**egistration, **E**valuation, **A**uthorisation and restriction of **CH**emicals.

The HSE leaflet *REACH and Safety Data Sheets* explains to manufacturers, importers, downstream users and distributors supplying substances or mixtures when a safety data sheet has to be provided (www.hse.gov.uk/reach/resources/reachsds.pdf).

This is where substances or mixtures have been classified as hazardous, persistent, bioaccumulative or toxic (or very persistent and very bioaccumulative) or are included in the European Chemicals Agency's Candidate List of substances of very high concern for reasons other than these.

The HSE advises that safety data sheets also have to be provided by suppliers where the customer requests one for certain other mixtures that have not been classified as dangerous under the CLP Regulation. Suppliers of a product listed as a 'special case' in paragraph 1.3 of Annex 1 of the CLP Regulation for which there are labelling derogations, for example gas containers intended for propane, butane or liquefied petroleum gas, must also provide a safety data sheet.

The HSE sets out that the safety data sheet must be dated and contain information under the following headings:

(1) Identification of the substance/mixture and of the company/undertaking;
(2) Identification of the hazards;
(3) Composition or information about the ingredients;
(4) First-aid measures;
(5) Fire-fighting measures;
(6) Accidental release measures;
(7) Handling and storage;
(8) Exposure controls or personal protection;
(9) Physical and chemical properties;
(10) Stability and reactivity;
(11) Toxicological information;
(12) Ecological information;
(13) Disposal considerations;
(14) Transport information;
(15) Regulatory information; and
(16) Other information.

The European Regulation on the Classification, Labelling and Packaging of Substances and Mixtures (the CLP Regulation) introduces into Europe the Globally Harmonised System for the Classification and Labelling of Chemicals

(GHS) (see H2137). The same system is now used throughout the world. It also replaced the *Chemicals (Hazard Information and Packaging for Supply) Regulations 2009 (SI 2009 No 716)*, commonly known as 'CHIP'.

The aim of the CLP Regulation is to protect human health and the environment and allow the free movement of substances, mixtures and articles by harmonising the criteria for the classification of substances and mixtures and the rules on labelling and packaging for hazardous substances and mixtures.

Manufacturers, importers and downstream users are required to classify substances and mixtures before placing them on the market. They must carry out an identification and examination of the available information on substances and mixtures, carry out an evaluation of the hazard information and decide on the classification.

Packaged hazardous substances and mixtures must be labelled with specified information. Suppliers are also required to label and package substances and mixtures placed on the market. Manufacturers and importers must notify the European Chemicals Agency where they are required to register substances under the European Regulation concerning the registration, evaluation, authorisation and restriction of chemicals ('REACH') (see above).

The CLP Regulation also lays down requirements for competent authorities and contains enforcement provisions.

Suppliers of hazardous substances must classify and label them using CLP symbols (black symbols on white with a red diamond-shaped border) rather than CHIP symbols (black on orange squares). More information can be found on the HSE website at: www.hse.gov.uk/chemical-classification/legal/clp-regulation.htm.

The COSHH essentials e-tool includes the incorporation of Hazard (H) statements to comply with the CLP Regulation. This can be found on the HSE website at: www.hse.gov.uk/coshh/essentials/coshh-tool.htm.

Further information can be found in the chapter REACH: REGISTRATION, EVALUATION AND AUTHORISATION OF CHEMICALS IN EUROPE.

Control of Asbestos Regulations 2012 (SI 2012 No 632)

[H2138] The *Control of Asbestos Regulations 2012* require an assessment of work which exposes employees to asbestos and the preparation of a suitable written plan of work which must prevent or reduce exposure to asbestos. Many other requirements also mirror COSHH requirements, eg information, instruction and training; use of control measures; maintenance of control measures; requirements for air monitoring (plus standards for air testing and analysis); and health records and medical surveillance.

There are also a number of other specific requirements including those relating to the notification of work with asbestos (to HSE), to the storage, distribution and labelling of raw asbestos and asbestos waste and to the labelling of products containing asbestos.

More information concerning asbestos, including the important 'duty to manage asbestos in non-domestic premises', are contained in the chapter dealing with ASBESTOS.

Other important regulations **[H2140]**

Control of Major Accident Hazard (COMAH) Regulations 2015

[H2139] The *Control of Major Accident Hazards (COMAH) Regulations 2015 (SI 2015 No 483)* aim to prevent major accidents involving dangerous substances and limit the consequences to people and the environment of any accidents which do occur. The *COMAH Regulations 2015* implement the majority of the provisions contained in the so-called 'Seveso III' European Directive, although the land-use planning requirements are implemented through planning legislation.

The regulations apply to establishments where dangerous substances are present (or likely to be present) in quantities equal to or exceeding the quantities set out in a schedule to the Regulations. Establishments may be either 'lower tier' or 'upper tier' depending on the quantities of dangerous substances present, and both must make information available to the public.

An operator must take all necessary measures to prevent major accidents and to limit their consequences for human health and the environment and demonstrate to the competent authority (such as the HSE or Environment Agency) that it has taken these measures. Operators must also prepare and keep up to date a major accident prevention policy and operators of upper tier establishments must send a safety report to the competent authority demonstrating that all the necessary measures have been taken to prevent major accidents and to limit their consequences to people and the environment. Safety reports must be kept up to date and an internal emergency plan must be reviewed and tested.

A local authority in whose area there is an upper tier establishment must prepare an external emergency plan containing specified information. The authority must review and test the external emergency plan and put it into effect in specified circumstances.

An operator of an upper tier establishment must regularly send information to people liable to be affected by a major accident occurring there (without them having to request it) and the competent authority to adopt a procedure to deal with requests for information. Competent authorities have a range of powers, including imposing duties on groups of establishments identified as 'domino' groups to co-operate.

Further details are provided in the chapter MAJOR ACCIDENT HAZARDS.

Control of Lead at Work Regulations 2002

[H2140] Like asbestos, previous regulations on lead preceded the original *COSHH Regulations* and the *Control of Lead at Work Regulations 2002 (SI 2002 No 2676)* continue to follow similar principles. The Regulations together with an ACoP and related guidance are contained in an HSE booklet *Control of Lead at Work Regulations 2002. Approved code of practice and guidance* (see www.hse.gov.uk/pubns/books/l132.htm) which is supported by other HSE guidance material.

Dangerous Substances and Explosive Atmospheres Regulations 2002 (DSEAR)

[H2141]–[H2143] The *DSEAR Regulations (SI 2002 No 2776)* are concerned with protecting against risks from fire and explosion. The *DSEAR Regulations (SI 2002 No 2776)* contain a detailed definition of a 'dangerous substances':

'(a) a substance or mixture which meets the criteria for classification as hazardous within any physical hazard class laid down in the CLP Regulation, whether or not the substance is classified under that Regulation;
(b) a substance or mixture which because of its physico-chemical or chemical properties and the way it is used or is present in the workplace creates a risk, not being a substance or mixture falling within subparagraph (a) above; or
(c) any dust, whether in the form of solid particles of fibrous materials or otherwise, which can form an explosive mixture with air or an explosive atmosphere, not being a substance or mixture falling within subparagraphs (a) or (b) above;'

An explosive atmosphere is defined as 'a mixture, under atmospheric conditions of air and one or more dangerous substances in the form of gases, vapours, mists or dusts, in which, after ignition has occurred, combustion spreads to the entire unburned mixture'.

The *DSEAR Regulations (SI 2002 No 2776)* require employers to:

- Carry out a risk assessment of work activities involving dangerous substances [*SI 2002 No 2776, Reg 5*].
- Eliminate or reduce risks as far as is reasonably practicable [*SI 2002 No 2776, Reg 6*].
- Classify places where explosive atmospheres may occur in zones, and ensure that equipment and protective systems in these places meet appropriate standards, marking zones with signs, where necessary [*SI 2002 No 2776, Reg 7*]
- Provide equipment and procedures to deal with accidents and emergencies [*SI 2002 No 2776, Reg 8*].
- Provide employees with suitable and sufficient information, instruction and training [*SI 2002 No 2776, Reg 9*].

Regulation 6 requires control measures to be implemented according to a specified order of priority which can be summarised as:

- Avoid the presence or use of a dangerous substance at the workplace by replacing it with a substance or process which either eliminates or reduces the risk.
- Reduce the quantity of dangerous substances to a minimum.
- Avoid or minimise releases of dangerous substances.
- Control releases at source.
- Prevent the formation of an explosive atmosphere (including providing appropriate ventilation).
- Collect, contain and remove any releases to a safe place or otherwise render them safe.
- Avoid the presence of ignition sources (including electrostatic discharges).

- Avoid adverse conditions that could lead to danger (eg by temperature or other controls).
- Segregate incompatible dangerous substances.

The regulation also requires measures to mitigate the detrimental effects of a fire or explosion (or other harmful physical effects). These include minimising the number of employees exposed and the provision of explosion pressure relief arrangements or explosion suppression equipment.

Further information on DSEAR is provided in the chapter **DANGEROUS SUBSTANCES AND EXPLOSIVE ATMOSPHERES** and detailed guidance on the subject is available on the HSE website at: www.hse.gov.uk/fireandexplosion/index.htm.

Dangerous Substances (Notification and Marking of Sites) Regulations 1990

[H2144] The *Dangerous Substances (Notification and Marking of Sites) Regulations 1990 (SI 1990 No 304)* are primarily for the benefit of the fire service and other emergency services. Both the fire authority and HSE must be provided with notification of dates when it is anticipated that a total quantity of 25 tonnes of 'dangerous substances' will be present on the site. (The definition of 'dangerous substances' is contained in the Regulations.) Further notifications must be made where significant changes occur in substances present, including cessation or reductions. Sites must also be marked with standard signs as specified in *Schedule 3* to the Regulations. A booklet providing detailed guidance on the regulations is available from HSE (see www.hse.gov.uk/pubns/priced/hsr29.pdf).

Ionising Radiation Regulations 2017

[H2145] *Regulation 8* of the *Ionising Radiation Regulations 2017 (SI 2017 No 1075)* require employers to carry out a risk assessment before commencing any new activity involving work with ionising radiation. The assessment must be:

. . . sufficient to demonstrate that—

(a) all hazards with the potential to cause a radiation accident have been identified; and
(b) the nature and magnitude of the risks to employers and other persons arising from those hazards have been evaluated.

Where such radiation risks are identified, all reasonably practicable steps must be taken to:

- prevent any such accident;
- limit its consequences should such an accident occur;
- provide employees with necessary information, instruction, training and equipment necessary to restrict their exposure.

The Regulations contain many detailed requirements on the prevention of exposure to radiation and related control measures. An HSE booklet contains the Regulations and an ACoP and guidance on the regulations (see www.hse.gov.uk/pubns/books/l121.htm) and further guidance is available on the HSE website in relation to specific aspects of radiation safety. Further details are provided in the chapter dealing with RADIATION.

Future changes: HSE review of the Control of Substances Hazardous to Health Regulations 2002, the Control of Lead Regulations 2002 and the Dangerous Substances and Explosive Atmospheres Regulations 2002

[H2146] The HSE initiated a fundamental review of the regulatory framework for chemicals in the workplace at the beginning of 2016. It considered three sets of regulations: the *Control of Substances Hazardous to Health Regulations 2002*; the *Control of Lead Regulations 2002*; and the *Dangerous Substances and Explosive Atmospheres Regulations 2002*.

The review aimed to simplify and modernise the regulatory framework and included an online survey and a number of focus groups. A discussion document was planned for June 2017 but this did not go ahead due to the general election.

The HSE has analysed the findings of the review and is taking forward work to address the key issues identified. It will revise guidance in this area and develop case studies to aid understanding of what is expected to achieve compliance when using and controlling hazardous substances in various sectors and activities.

In addition, a key finding of the review was that the current requirement in COSHH for thorough examination and testing of LEV every 14 months does not ensure that LEV plant is contributing to control of dust or fume (see H2128 above). It is possible for an LEV plant to pass an examination or test as it is operating as originally intended, but not effectively contributing to control – because it is not properly designed for the purpose it is used for or it has not been correctly commissioned, for example. The HSE is to undertake further research in order to 'strengthen the evidence base to inform any potential future legislative change in this area'.

It is also exploring potential amendments to COSHH as part of any future amendment to the Regulations, such as transposition of the amended carcinogens and mutagens directive (see **H2113** above). Changes may include simplification of the definition of dust and removal of *Schedule 6*, which relates to medical surveillance requirements for certain substances and processes (see **H2133** above).

International Health and Safety

Steve Granger

Introduction to international health and safety

[I2001] In this chapter we will discuss the framework for Occupational Safety and Health in North America and how the practitioner can quickly integrate with their colleagues by being aware of the differences and similarities between; continents, countries, states and the workplace industries.

Legal framework

[I2002] In the chapter *European Health and Safety* the different legal systems across Europe were discussed. As the legal history of the USA and Canada stem from a mixture of European countries the foundations of American and Canadian law will obviously be heavily influenced by both EU Civil (codified) and Common law processes. The passage of time and self governance inherently meant that the traditional processes would become refined and personalised by those people that were using them. As with Europe, the distances, regionalisation and cultures represented within the USA and Canada mean that a process of rationalisation is necessary to bring the subject of OSH to a common understanding.

Societal differences across the continent also exist – both socially and within business. In this respect the European challenge of 'harmonisation' is equally relevant to OSH practitioners.

Health and safety in the USA

[I2003] The United States of America is a *Federation* of member *States*. Federal Law applies across the whole of the USA and is made by a process involving the National Government called Congress. Congress is made up of both the Senate (upper house – 2 elected senators from each state) and the House of Representatives (state elected allocation number is determined by population size) in much the same way as the UK.

In discussing OSH it is important to understand the relationship and history between the member States, the centralised Congress (National Government) and the bodies who apply the legislation across the individual States.

Since the Declaration of Independence, the US Constitution supported the hard-fought doctrine of 'self governance' for each of the (now) 50 States. Although the US Congressional Government (Congress) can make *Federal*

Law, there are subjects where *State law* is authoritative. To confuse further, Federal Law can be supplemented by additional rules made either centrally or regionally –providing the enactment facilitates this. This means that the Roman based *codified* process of law making and rule setting is an inherent part of American law.

However, unlike mainland Europe, America has also preserved the *common law* principles of *case law* and court hierarchy. It is clear that Federal law is supreme over either State laws or Common Law, and in this respect OSH is covered in very much the same way as the UK. The *Occupational Safety & Health Act of 1970* provides the fundamental structure for all US OSH.

The Federal Regulator for OSH is the Occupational Safety and Health Authority, OSHA. This a federal government body with a similar role to the HSE. In some states OSHA will delegate to the state (approximately half the states have been delegated full authority for the enforcement of OSHA) certain functions such as accident monitoring. The OSHA website is an excellent resource and provides a good example of codified legislation under the 'Regulation' tab of the website. The full text of the updated Occupational Safety and Health Act 1970 plus all of the current regulations can be found. Part 1910 describes general OSH requirements. The page also provides information on proposals and the process of rule making.

OSHA can also make criminally enforceable instruments (OSHA cannot make laws – it develops regulations to implement laws that Congress creates) through the powers given to them to define OS standards — unlike the HSE who require ministerial power to give authority to regulations. These standards are applicable across the whole of the USA.

The courts and legal cases operate in much the same way as the UK – criminal and civil law objectives for defendants and plaintiff, inclusion of common law and the tort of negligence, using evidence of what has been decided before. One main difference between UK civil disputes and the USA is the ability of the US court to award *punitive damages* to a litigant. This is often the 'talking point' between the two systems as it is not permissible in Europe to profit from compensation. This is currently under debate in the USA but considering the powerful lobby of the legal profession and the way in which litigation is funded in the USA it is unlikely to change significantly.

States

[I2004] The OSHA website provides details of its regions and where, within those regions, individual states have been approved to run their own OSH programmes and systems to meet the overall objectives. For anyone working in the USA it is necessary to know which authority coves the workplace.

Like the HSE, OSHA provides free guidance and tools for employers. One incentive programme provided by OSHA offers a free workplace inspection, which is not linked to the enforcement activity. Small business who achieve good results an be exempt from OSH inspections. This innovative approach provides both assurance to small businesses and cost saving to the regulators. Another Federal body that provides information and research is the National

Institute for Occupational safety and Health (NIOSH). Whereas OSHA is within the Department of Labour NIOSH falls within the Department of Health and Human Services providing a valuable link between occupational health and the nations wellbeing.

Professional recognition

[I2005] There are no requirements for registering as an OSH practitioner in the USA. The market is generally controlled by insurance requirements which will probably require a *certified* professional. There are federal and state recognised and approved certification schemes, it is worth enquiring which one is suited to individual and industry needs. The professional Societies, such as those below are probably the best source of advice if you are choosing a qualification or study pathway. Regional branches may also be able to help with choosing a provider and a valuable support network.

In terms of professional organisations and competency registration the USA has a number of organisations where professionals associated with OSH can be found; The American Society of Safety Engineers is the largest covering all subjects and geographic regions. It utilises a regulated professional qualification pro-gramme - the Certified Safety Practitioner (CSP) (which is actually run by separate organisation not affiliated with ASSE – The Board of Certified Safety Professionals BCSP). ASSE has a reciprocal agreement to re-duce the burden of initial qualification for IOSH Chartered Safety and Health professionals to join at member level.

Like ASSE, The American Society of Industrial Hygiene, AIHA also has close relationships with other international professional bodies such as IOSH, the British Occupational Health Society.

Both have excellent free advice available and should be a priority resource for preparing to work in the USA.

OSH Management and Performance

[I2006] Although the codification of safety is comprehensive it does not include everything. For example there is no statutory requirement to manage OSH in the same way as required by the EU Framework Directive or the Management of Health and Safety at Work Regulations in the UK.

However more supportive documentation does appear in the form of standards which (because of their inherent sense) become adopted as the 'norm'. At present the *American National Standards Institute* has endorsed the ANSI/AIHA Z10 standard on Occupational health and safety management systems. This is comparable to OHSAS 18001 (soon to be replaced with ISO 45001) so the fundamental principles are common to both nations –one being statutory and the other driven by business expectation and market forces.

In terms of culture and perception of OSH American society are generally more positive towards the safety and wellbeing in the workplace than the UK (although this public perception is not limited to the USA nor is it necessarily a validated difference). Regulation by the state is taken more seriously and with less scepticism, but this does not necessarily mean improved performance. One reason for this might be explained by the ability to issue immediate fines by the inspectorate, which they do frequently.

In 2010; 4690 fatalities occurred in the workplace and the reality of OSH is that application of the law varies considerably wherever it is examined. Measuring performance is not simply a matter of comparing statistics. The declared incidence rate is 3.6 per 100,000 full time equivalents for the USA, compared to the UK 0.6 deaths per 100 000 workers. However, information presented like this can be misleading. Variables such as proportion of employees in high risk industry, definition of reporting categories, inclusion of different categories of fatalities (homicide and road deaths are included) etc.

On the positive side, the cautious approach to litigation and liability in the USA means that workplace rules, conditions of contract and contract performance monitoring are valued, lessening the need to convince organisational management for the OSH practitioners input into planning and management.

Working with the workforce

[I2007] Like any country the relationship between employers and workers depends upon variables such as; the economy, industry and corporate values. In terms of worker representation and Trade Unions, the USA does have a history of extremes of exceptionally good to destructive influence and turbulence (although this is much less so now). But this has left a mark on the business society and how some employers feel about consultation and co-operation. It is advisable to learn a little history of this relationship and how it may have matured in the relatively recent past.

Validity of contracts

[I2008] Before you go... Determining the legality of working and living in the USA, obtaining visas and paying tax will not be covered here. But there are some things the practitioner should be aware of;

If the worst happens and you are involved in litigation or prosecution you need to know that your contract of engagement is valid, or determine exactly who is responsible for what within a contract. Do not be hesitant about commenting or requesting alteration to agreements or contract proposals before signing if you think you might face liability alone. For those that are working under an employment contract you need to ensure it is valid to protect you in the same way, or make contingency arrangements.

Professional Insurance

[I2009] It is most unlikely that any Professional Indemnity or Public Liability insurance held in the UK or Europe will cover the USA without special arrangement. Although this may sound obvious it is in your personal interest to ensure that advice or work done by you, on behalf of your employer or agent is suitably covered. It is advisable to take a copy of policy document along with other schemes such as healthcare which should also be checked for validity and the duration of visit. If the trip is of short duration include repatriation cost and it is also worth researching individual State policy with respect to cover that may be available for treatment. This essential element may well be worth negotiating with prospective employers or including in specifications rather than face an unexpected, and very considerable cost.

Health and safety in Canada

Government and law

[I2010] Canada is a *Federal Democracy* with the Queen still as the recognised Monarch. Like the USA each of the ten *Provinces* is self governing but subject to federal law. The three additional *Territories* rely on federal jurisdiction for the simple reason that the population is low and would not sustain a fully functioning self government. Acts – referred to as *Codes*, are made in a similar way to the UK and parliamentary make up is similar to the USA with Province representation. Canada, like the USA incorporates both the codified system of law and the principles of *Common Law*. One significant difference between the USA and Canada is that because of the constitutional link between the UK and Canada the *common law precedence* established in Canada can influence the UK courts and vice versa. Health and Safety is a provincial domain, and therefore each province has its own Occupational Health and Safety Act for their particular province. Federally regulated companies fall under the Canada Labour Code. A simple way to differentiate between a provincially regulated and federally regulated firm, is that if they cross provincial boundaries, then they are regulated. The list includes the following:

- banks
- marine shipping, ferry and port services
- air transportation, including airports, aerodromes and airlines
- railway and road transportation that involves crossing provincial or international borders
- canals, pipelines, tunnels and bridges (crossing provincial borders)
- telephone, telegraph and cable systems
- radio and television broadcasting
- grain elevators, feed and seed mills
- uranium mining and processing
- businesses dealing with the protection of fisheries as a natural resource
- many First Nation activities
- most federal Crown corporations
- private businesses necessary to the operation of a federal act

Source: www.hrsdc.gc.ca/eng/labour/employment_standards/regulated.shtml.

A useful resource for legislation across the whole of Canada is CanOSH –a website which conveniently brings together the different laws for easy referencing, particularly useful for multinational that span several agencies. From the website;

Government agencies in all Canadian jurisdictions have a role in workplace health and safety. The regulatory framework outlines the general rights and responsibilities of the employer, the supervisor and the worker. Each of the ten provinces, three territories and the federal government has its own OSH legislation.

The federal government has responsibility for the health and safety of its own employees and federal corporations, plus workers in certain industries such as inter-provincial and international transportation (e.g., railways and air transport), shipping, telephone and cable systems, etc. Approximately 10% of the Canadian workforce falls into the federal jurisdiction. The remaining 90% of Canadian workers fall under the legislation of the province or territory where they work.

Canada's federal *Labour Code 1985* provides an excellent example of the codification process and integration of laws relating to employment. Part II specifically covers Health and Safety and it is supported by Regulations in much the same way as the HSAWA1974. Enforcement is also provisioned by the appointment of *Health and Safety Officers* who have similar powers to UK HSE and Local Authority Inspectors.

The Code also establishes the acceptance of any 'standard' deemed suitable to support the objectives of the Act – a significant improvement on other legislative systems which do (including the UK) which do not properly address the need to use best practice effectively.

Regional application of OSH law will need to be looked at but generally follows the same principles as both Federal and UK expectations.

Like the USA one omission is the general duty to manage health and safety required in the EU. As with the similar codified system of America this duty is expected through compliance of all parts collectively.

Workers Compensation systems

[I2011] Canada operates a 'no fault' compensation system, which began around 1915 and was adopted across the country. William Meredith provided the framework for workers' compensation in Canada in the early 1900's. The worker gave up the right to sue the employer, with the employer also giving up the right to sue and is responsible to fund the insurance system. Employment contributions fund a provincially run scheme that people can claim from following injury or illness - without requiring recourse to litigation. A great summary of the different workers' compensation plans is available on the Association of Workers Compensation Boards of Canada at www.awcbc.org/en. The list may be found at www.awcbc.org. This has several benefits; the monies collected can be reinvested in health and safety promotion as there is a vested interest in the scheme looking after its 'investment' in peoples safety and health, the scheme can be used to either incentivise or penalise for OSH performance and as the scheme is not restricted to workplace accidents and ill health then OSH is not marginalised as a drain on business. Each jurisdiction has an experience rating program that reimburses employers a portion of their premiums if their costs are lower than the average for their industry or rate group. The net result seems to work very favourably in the promotion of OSH as part of business and societal development – making the work of the OSH practitioner less contentious! Such schemes are not without their problems, but overall Canada seems to be successful in properly integrating the various components of OSH to make it easy to understand, follow and apply as a result of some early joined up thinking.

Many provinces have additional incentive programs, that support prevention activities and provide financial incentives for better than average health and safety performance. A useful summary of the programmes may be found at www.awcbc.org.

Professional development and recognition

[I2012] The Canadian Society of Safety Engineers CSSE is a health and safety association for the health, safety and environment professional.

The CSSE has the Certified Health and Safety Consultant designation that provides recognitions of competence as a safety consultant. More information is available at www.csse.org/chsc_designation/about_chsc.htm. There is an application process for membership.

There is an application and it requires on-going professional development to retain the designation.

The Board of Canadian Registered Safety Professionals or BCRSP issues the Canadian Registered Safety Professional (CRSP) designation. CRSP is an individual who has met the requirements for registration established by the Governing Board. A CRSP applies broad based safety knowledge to develop systems that will achieve optimum control over hazards and exposures detrimental to people, equipment, material and the environment. A CRSP is dedicated to the principles of loss control, accident prevention and environmental protection as demonstrated by their daily activities. IOSH has a memorandum of understanding with BCRSP to recognise these qualifications and fast track to attain the CRSP or CMIOSH designations.

Like many professional bodies, CSSE is an excellent resource to provide a support network for anyone considering a move to Canada.

Contract, work permits, and insurances

As with the USA above, the issue of contract validity (look for a jurisdiction clause or limitation) and insurance cover need to be checked, carefully, as does the application for appropriate work permits or extended stay visa.

Resources

[I2013] As well as the federal and provincial government websites the *Canadian Centre for Occupational Health and Safety* offers a variety of resources and information. The Canadian Department of Justice and the Provincial equivalents offer very good online documentation of OSH law.

The professional bodies above and the *Canadian Occupational Health Nurses Association* and the Institut de recherche Robert-Sauvé en santé et en sécurité du travail *IRSST* is a scientific research organization, also have useful resources.

Aspects of work permit, insurance and contract is through the Citizenship and Immigration Canada, at their website www.cic.gc.ca/english/work/apply-how.asp.

Other useful websites

These include —

- www.healthandsafetyontario.ca/HSO/Home.aspx
- www.worksafebc.com/
- www.csse.org/links
- awcbc.org/
- www.ccohs.ca/oshlinks/

Periodicals and magazines

Useful periodicals and magazines include —

- www.ohscanada.com/
- www.cos-mag.com/

Joint Consultation in Safety – Safety Representatives, Safety Committees, Collective Agreements and Works Councils

Nick Humphreys, Partner, Thomas Cooper LLP

Introduction to joint consultation in safety

[J3001] The provision of 'information, consultation and participation' and of 'health protection and safety at the workplace' are outlined as two of the 'fundamental social rights of workers' in accordance with the Community Charter of Fundamental Rights of 1989. Combined with Article 138 of the Consolidated Version of the Treaty Establishing the European Community (ie Treaty of Rome 1957), the Health and Safety Directive 1989 (resulting in 'the six pack' collection of regulations) and the United Kingdom's signing of the Social Chapter in 1998, the EU has provided great stimuli for the introduction of extensive obligations to consult on health and safety issues.

At national level, there has already been change to the obligations owed by employers to their workers. The *Working Time Regulations 1998 (SI 1998 No 1833)* ('*Working Time Regulations*') were introduced in October 1998 and have subsequently been amended.

Furthermore, the obligations owed under the European Works Councils Directive 1994 have been transposed into national law under the *Transnational Information and Consultation of Employees Regulations 1999 (SI 1999 No 3323)* ('*Works Councils Regulations*').

Yet further, the Information and Consultation Directive (Directive 2002/19/EC) (transposed into national law under the *Information and Consultation of Employees Regulations 2004 (SI 2004 No 3426)*), was introduced, on a phased basis, into all undertakings with 50 or more employees.

The Health and Safety Commission ('HSC') is responsible for seeking to promote change in health and safety at work in relation to the UK's regulatory framework. A list of HSC consultations can be found at www.hse.gov.uk/consult/live.htm.

Regulatory framework for joint consultation in safety

[J3002] The legal requirements for consultation are more convoluted than might be expected, principally because there are two groups of workers who are treated as distinct for consultation purposes. Until 1996, only those employers who recognised a trade union for collective bargaining purpose were obliged to consult the workforce through safety representatives. The *Safety Representatives and Safety Committees Regulations 1977 (SI 1977 No 500) (as amended)* ('*Safety Representatives Regulations*'), which were made under the provisions of the Health and Safety at Work etc Act 1974 ('HSWA 1974'), s 2(6), came into effect in 1978 and introduced the right for recognised trade unions to appoint safety representatives. The *Safety Representatives Regulations* were amended in 1993 – by the *Management of Health and Safety at Work Regulations 1992 (SI 1992 No 2051)* (now superceded by the *Management of Health and Safety at Work Regulations 1999 (SI 1999 No 3242)*) – to extend employers' duties to consult and provide facilities for safety representatives.

The duty to consult was extended on 1 October 1996 by virtue of the *Health and Safety (Consultation with Employees) Regulations 1996 (SI 1996 No 1513)* ('*Consultation with Employees Regulations*'), which have subsequently been amended. The *Consultation with Employees Regulations* were introduced as a 'top up' to the *Safety Representatives Regulations*, extending the obligation upon employers to consult all of their employees about health and safety measures, including those who are not represented by a trade union. The *Consultation with Employees Regulations* expanded the obligation beyond just trade union appointed representatives. When introduced, the *Consultation with Employees Regulations* addressed the general reduction of union recognition over the preceding five years. However, the more recent trend in workforces, driven by the twin forces of Europe and the current Government, is for greater worker participation in employers' undertakings and the current framework facilitates this whether or not a union is the conduit for worker consultation.

The obligation of consultation, its enforcement and the details of the role and functions of safety representatives and representatives of employee safety, whether under the *Safety Representatives Regulations* or the *Consultation with Employees Regulations*, are almost identical but, for the sake of clarity, are dealt with separately below. The primary distinction is that different obligations apply depending on whether the affected workers are unionised or not. The respective Regulations also cover persons working in host employers' undertakings.

There are specific supplemental provisions which apply to offshore installations, which are governed by the *Offshore Installations (Safety Representatives and Safety Committees) Regulations 1989 (SI 1989 No 971)*. Likewise, in the case of the education sector, the HSC has published the 'Safety Representatives' Charter' ('the Charter'). The Charter has been developed to:

- Promote and emphasise the rights, roles and functions of safety representatives and to actively promote the involvement of safety representatives in the education sector's efforts to improve health and safety.
- Motivate employers, safety representatives and employees to work in partnership to develop a positive safety culture throughout the education sector.

- Raise awareness amongst employers of the important contribution of safety representatives towards the development of such a culture.
- Encourage employers to demonstrate their full commitment towards consulting and involving safety representatives in matters of health, safety and welfare.
- Increase the participation of education sector employees and their safety representatives in health and safety activities.
- Contribute towards an improved health and safety performance which aims to reduce accidents and ill health in the education sector.

In particular, the Charter sets out the legal obligations that employers in the education sector owe under the *Safety Representatives Regulations* and the *Consultation with Employees Regulations*.

While it is not legislation itself, the Charter is evidence of best practice within the education sector.

Although the *Consultation with Employees Regulations* and the *Safety Representatives Regulations* are based upon good industrial relations practice, there is still the possibility of a prosecution by HSE inspectors of employers who fail to consult their workforce on health and safety issues. There is substantial overlap, however, with employment protection legislation and the obligations in respect of consultation – as can be seen from the protective rights which are conferred upon safety representatives, such as the right not to be victimised or subjected to detriment for health and safety activities, together with the consequential right to present a complaint to an employment tribunal if they are dismissed or suffer a detriment as a result of carrying out their duties (arising under the *Employment Rights Act 1996, ss 100* and *44*, respectively).

The *Consultation with Employees Regulations* also extend these rights to the armed forces. However, armed forces representatives are to be appointed rather than elected, and no paid time off is available. The *Consultation with Employees Regulations* do not apply to sea-going ships.

The Working Time Directive 1994 has resulted in the *Working Time Regulations 1998 (SI 1998 No 1833)* (subsequently amended) under national law. The *Working Time Regulations* allow collective modification of the night working requirements in *Reg 6*, the daily rest provisions in *Reg 10*, the weekly rest provisions in *Reg 11*, the rest break provisions in *Reg 12* and also allow modification of the averaging period for calculating weekly working time under *Reg 4* where the same is either for objective, technical or organisational reasons. The method of modification is either by collective agreement in the case of a unionised workforce or by workforce agreement (as defined in the *Working Time Regulations 1998 (SI 1998 No 1833), Sch 1*) in the case of non-unionised employees.

In relation to pan-European employers, obligations are owed to workers under the European Works Councils Directive 1994. The United Kingdom adopted the Directive on 15 December 1997 and provisions were enacted into national law by the *Works Councils Regulations (SI 1999 No 3323)* and came into force on 15 January 2000. While the *Works Councils Regulations* do not specifically encompass health and safety obligations, these issues are within the remit of a European Works Council.

National level consultation requirements have also been introduced under the *Information and Consultation of Employees Regulations 2004 (SI 2004 No 3426)*, derived from the Information and Consultation Directive (Directive 2002/19/EC). As with European Works Councils, health and safety is not specifically included in the scope of the information and consultation requirements under those Regulations, but may be included under the general remit of an employee information and consultation procedure.

Consultation obligations for unionised employers

[J3003] Under the *Safety Representatives Regulations (SI 1977 No 500)*, *Regs 2* and *3*, an independent trade union recognised by an employer has the right to appoint an individual to represent the workforce in consultations with the employer on all matters concerning health and safety at work, and to carry out periodic inspections of the workplace for hazards. Every employer has a duty to consult such union-appointed safety representatives on health and safety arrangements (and, if they so request him, to establish a safety committee to review the arrangements – see **J3017** below) [*HSWA 1974, s 21*].

General duty

[J3004] The general duty of the employer under the *Safety Representatives Regulations (SI 1977 No 500)* is to 'consult with safety representatives with regard to both the making and maintaining of arrangements that will enable the employer and its workforce to co-operate in promoting and developing health and safety at work, and monitoring its effectiveness'.

Employers are also required under *Safety Representatives Regulations (SI 1977 No 500), Reg 4A(2)* to provide such facilities and assistance as safety representatives may require for the purposes of carrying out their functions under the *Heath and Safety at Work etc Act 1974, s 2(4)*.

General guidance has been issued by HSC in the form of the Codes of Practice *'Safety Representatives and Safety Committees'* and *'Time Off for the Training of Safety Representatives'* to which regard should be had generally.

Appointment of safety representatives

[J3005] The right of appointment of safety representatives was, until 1 October 1996, restricted to independent trade unions who are recognised by employers for collective bargaining purposes. The *Consultation with Employees Regulations (SI 1996 No 1513)* extended this to duly elected representatives of employee safety, as detailed above. In a case where there are no Safety Representatives under the *Safety Representatives Regulations (SI 1977 No 500)* and where the employees do not elect representatives under the *Consultation with Employees Regulations (SI 1996 No 1513)*, the obligation will fall on employers to consult with the entire workforce as to the matters within the scope of the *Consultation with Employees Regulations (SI 1996 No 1513)* (per *Reg 3*).

The terms 'independent' and 'recognised' are defined in the *Safety Representatives Regulations (SI 1977 No 500)* and follow the definitions laid down in the *Trade Union and Labour Relations (Consolidation) Act 1992, ss 5* and *178(3)* respectively. The *Safety Representatives Regulations* make no provision for dealing with disputes which may arise over questions of independence or recognition (this is dealt with in the *Trade Union and Labour Relations (Consolidation) Act 1992, ss 6* and *8* – the amendments that have been made to the *Trade Union and Labour Relations (Consolidation) Act 1992* by the *Employment Relations Act 1999* in relation to recognition of unions do not help as *Sch A1* is confined to recognition disputes concerning pay, hours and holiday).

Safety representatives must be representatives of recognised independent trade unions, and it is up to each union to decide on its arrangements for the appointment or election of its safety representatives [*Safety Representatives Regulations (SI 1977 No 500), Reg 3*]. Employers are not involved in this matter, except that they must be informed in writing of the names of the safety representatives appointed and of the group(s) of employees they represent [*Safety Representatives Regulations (SI 1977 No 500), Reg 3(2)*].

The *Safety Representatives Regulations (SI 1977 No 500), Reg 8*, state that safety representatives must be employees except in the cases of members of the Musicians' Union and the Actors' Union Equity. In addition, where reasonably practicable, safety representatives should have at least two years' employment with their present employer or two years' experience in similar employment. The HSC guidance notes advise that it is not reasonably practicable for safety representatives to have two years' experience, or employment elsewhere, where:

- the employer is newly established;
- the workplace is newly established;
- the work is of short duration; or
- there is high labour turnover.

The same general guidance is followed for employee safety representatives under the *Consultation with Employees Regulations (SI 1996 No 1513)*.

[J3005] Joint Consultation in Safety – Safety Representatives etc

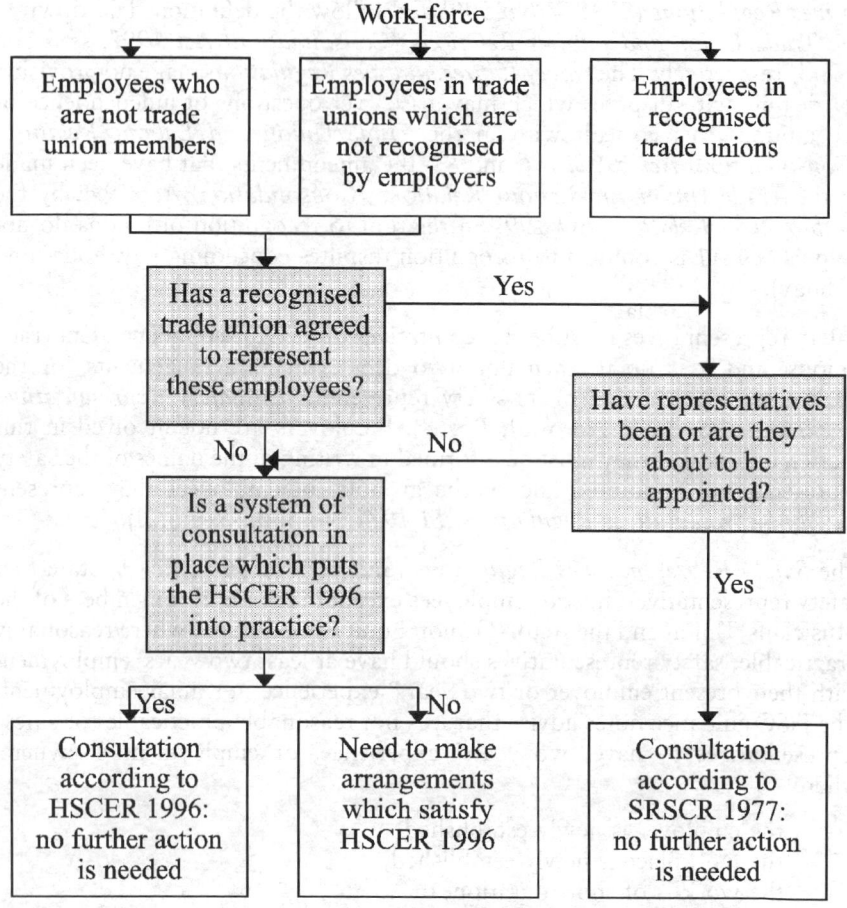

Number of representatives for workforce

[J3006] The *Safety Representatives Regulations 1977 (SI 1977 No 500)* do not lay down the number of safety representatives that unions are permitted to appoint for each workplace. This is a matter for unions themselves to decide, having regard to the number of workers involved and the hazards to which they are exposed. The HSE's view is that each safety representative should be regarded as responsible for the interests of a defined group of workers. This approach has not been found to conflict with existing workplace trade union organisation based on defined groups of workers. The size of these groups varies from union to union and from workplace to workplace. While normally each workplace area or constituency would need only one safety representative, additional safety representatives are sometimes required where workers are exposed to numerous or particularly severe hazards; where workers are distributed over a wide geographical area or over a variety of workplace locations; and where workers are employed on shiftwork.

Role of safety representatives

[J3007] The *Safety Representatives Regulations 1977 (SI 1977 No 500), Reg 4(1)* (as amended by the *Management of Health and Safety at Work Regulations 1999 (SI 1999 No 3242)*) lists a number of detailed functions for safety representatives:

(a) to investigate potential hazards and causes of accidents at the workplace;
(b) to investigate employee complaints concerning health etc at work;
(c) to make representations to the employer on matters arising out of (*a*) and (*b*) and on general matters affecting the health etc of the employees at the workplace;
(d) to carry out the following inspections (and see **J3008** below):
- of the workplace (after giving reasonable written notice to the employer – see *Safety Representatives Regulations, Reg 5*),
- of the relevant area after a reportable accident or dangerous occurrence (see accident reporting) or if a reportable disease is contracted, if it is safe to do so and in the interests of the employees represented (see *Safety Representatives Regulations 1977 (SI 1977 No 500), Reg 6*),
- of documents relevant to the workplace or the employees represented which the employer is required to keep (see *Safety Representatives Regulations 1977 (SI 1977 No 500), Reg 7*) – reasonable notice must be given to the employer; and
(e) to represent the employees they were appointed to represent in consultations with HSE inspectors, and to receive information from them (see **J3016** below); and
(f) to attend meetings of safety committees.

These functions are interrelated and are to be implemented proactively rather than reactively. Safety representatives should not just represent their members' interests when accidents or near-misses occur or at the time of periodic inspections, but should carry out their obligations on a continuing day-to-day basis. The *Safety Representatives Regulations* make this obligation clear by stating that safety representatives have the functions of investigating potential hazards and members' complaints *before* accidents, as well as investigating dangerous occurrences and the causes of accidents *after* they have occurred. These functions are only assumed when the employer has been notified in writing by the trade union or workforce of the identity of the representative.

Thus safety representatives may possibly be closely involved not only in the technical aspects of health, safety and welfare matters at work, but also in those areas which could be described as quasi-legal. In other words, they may become involved in the interpretation and clarification of terminology in the *Safety Representatives Regulations*, as well as in discussion and negotiation with employers as to how and when the regulations may be applied. This would often happen in committee meetings.

Under the *Safety Representatives Regulations 1977 (SI 1977 No 500), Reg 4A(1)* (introduced by the *Management of Health and Safety at Work Regulations 1992*), the subjects on which consultation 'in good time' between employers and safety representatives should take place are:

- the introduction of any new measure at a workplace which may substantially affect health and safety;
- arrangements for appointing competent persons to assist the employer with health and safety and implementing procedures for serious and imminent risk;
- any health and safety information the employer is required to provide; and
- the planning and organisation of health and safety training and health and safety implications of the introduction (or planning) of new technology.

The safety representative's terms of reference are, therefore, broad, and exceed the traditional 'accident prevention' area. For example, the *Safety Representatives Regulations 1977 (SI 1977 No 500), Reg 4(1)* empowers safety representatives to investigate 'potential hazards' and to take up issues which affect standards of health, safety and welfare at work. In practice it is becoming clear that four broad areas are now engaging the attention of safety representatives and safety committees:

- health;
- safety;
- environment; and
- welfare.

These four broad areas effectively mean that safety representatives can, and indeed often do, examine standards relating, for example, to noise, dust, heating, lighting, cleanliness, lifting and carrying, machine guarding, toxic substances, radiation, cloakrooms, toilets and canteens. The protective standards that are operating in the workplace, or the lack of them, are now coming under much closer scrutiny than hitherto.

Workplace inspections

[J3008] Under the *Safety Representatives Regulations 1977 (SI 1977 No 500), Regs 5* and *6*, safety representatives are given a power to undertake health and safety inspections. Arrangements for three-monthly and other more frequent inspections and reinspections should be by joint arrangement.

The TUC advises that the issues to be discussed with the employer can include:

- more frequent inspections of high risk or rapidly changing areas of work activity;
- the precise timing and notice to be given for formal inspections by safety representatives;
- the number of safety representatives taking part in any one formal inspection;
- the breaking-up of plant-wide formal inspections into smaller, more manageable inspections;
- provision for different groups of safety representatives to carry out inspections of different parts of the workplace;
- the kind of inspections to be carried out, eg safety tours, safety sampling or safety surveys; or

- the calling in of independent technical advisers by the safety representatives.

Although formal inspections are not intended to be a substitute for day-to-day observation, they have on a number of occasions provided an opportunity to carry out a full-scale examination of all or part of the workplace and for discussion with employers' representatives about remedial action. They can also provide an opportunity to inspect documents required under health and safety legislation, eg certificates concerning the testing of equipment. It should be emphasised that during inspections following reportable accidents or dangerous occurrences, employers are not required to be present when the safety representative talks with its members. In workplaces where more than one union is recognised, agreements with employers about inspections should involve all the unions concerned. It is generally agreed that safety representatives are also allowed under the *Safety Representatives Regulations 1977 (SI 1977 No 500)* to investigate the following:

- potential hazards;
- dangerous occurrences;
- the causes of accidents; and
- complaints from their members.

This means that imminent risks, or hazards which may affect their members, can be investigated right away by safety representatives without waiting for formal joint inspections. Following an investigation of a serious mishap, safety representatives are advised to complete a hazard report form, one copy being sent to the employer and one copy retained by the safety representative.

Rights and duties of safety representatives

Legal immunity

[J3009] Ever since the *Trade Disputes Act 1906*, trade unions have enjoyed immunity from liability in tort for industrial action, taken or threatened, in contemplation or furtherance of a trade dispute (although such freedom of action was subsequently curtailed by the *Trade Union and Labour Relations (Consolidation) Act 1992, s 20*). Not surprisingly, perhaps, this immunity extends to their representatives acting in a lawful capacity. Thus, the *Safety Representatives Regulations 1977 (SI 1977 No 500)* state that none of the functions of a safety representative confers legal duties or responsibilities [*Safety Representatives Regulations 1977 (SI 1977 No 500), Reg 4(1)*]. As safety representatives are not legally responsible for health, safety or welfare at work, they cannot be liable under either the criminal or civil law for anything they may do, or fail to do, as a safety representative under the *Safety Representatives Regulations*. This protection against criminal or civil liability does not, however, remove a safety representative's legal responsibility as an employee. Safety representatives must carry out their responsibilities under the *HSWA 1974, s 7* if they are not to be liable for criminal prosecution by an HSE inspector. These duties as an employee are to take reasonable care for the health and safety of one's self and others, and to co-operate with one's employer as far as is necessary to enable him to carry out his statutory duties on health and safety.

Time off with pay

General right

[J3010] Under the *Safety Representatives Regulations 1977 (SI 1977 No 500)*, safety representatives are entitled to take such paid time off during working hours as is necessary to perform their statutory functions, and reasonable time to undergo training in accordance with a Code of Practice approved by HSC.

Definition of 'time off'

[J3011] The *Safety Representatives Regulations 1977 (SI 1977 No 500), Reg 4(2)* provides that the employer must provide the safety representative with such time off with pay during the employee's working hours as shall be necessary for the purposes of:

- performing his statutory functions; and
- undergoing such training in aspects of those functions as may be reasonable in all the circumstances.

Further details of these requirements are outlined in the Code of Practice attached to the *Safety Representatives Regulations* and the HSC Approved Code of Practice on time off for training. The Code of Practice is for guidance purposes only – its contents are recommendations rather than requirements. However, it is guidance that an employment tribunal can and will take into account if a complaint is lodged in relation to an employer's unreasonable failure to allow time off.

The combined effect of the ACAS Code No 3: *'Time Off for Trade Union Duties and Activities'* (2010) and the HSC Approved Code of Practice on time off is that shop stewards who have also been appointed as safety representatives are to be given time off by their employer to carry out both their industrial relations duties and their safety functions, and also paid leave to attend separate training courses on industrial relations and on health and safety at work – this includes a TUC course on COSHH (*Gallagher v The Drum Engineering Co Ltd, COIT 1330/89*).

An employee is not entitled to be paid for time taken off in lieu of the time he had spent on a course. This was held in *Hairsine v Hull City Council* [1992] IRLR 211 when a shift worker, whose shift ran from 3 pm to 11 pm, attended a trade union course from 9 am to 4 pm and then carried out his duties until 7 pm. He was paid from 3 pm to 7 pm and he could claim no more.

A similar decision was arrived at in *Calder v Secretary of State for Work and Pensions*(UKEAT/0512/08/LA) [2009] All ER (D) 106 (Aug) in relation to an employee who worked part-time Tuesdays, Wednesdays and Thursdays. The employee was a health and safety secretary for a major trade union in the public sector. In December 2006, the employee applied to the employer to attend the stage 3 health and safety representatives training course provided by the Trade Union Congress. The course was held on Fridays for a period of 36 weeks. The employer refused permission for the employee to attend on a paid basis since because of the nature of the employee's duties and the nature of the course curriculum, it was not reasonable for her to go on the course nor necessary for her duties. In the employee's application to the Employment Tribunal for her health and safety representatives paid time off claim, and the employee's appeal to the EAT, the employee lost; it was held that unless there was a refusal to permit time off from working hours (which Fridays were not for the employee), the *Safety Representatives Regulations 1977 (SI 1977 No 500)* did not come into effect.

However, where more safety representatives have been appointed than there are sections of the workforce for which safety representatives could be responsible, it is not unreasonable for an employer to deny some safety representatives time off for fulfilling safety functions (*Howard and Peet v Volex plc (HSIB 181)*).

A decision of the EAT seems to favour jointly sponsored in-house courses, except as regards the representational aspects of the functions of safety representatives, where the training is to be provided exclusively by the union (*White v Pressed Steel Fisher* [1980] IRLR 176). Moreover, one course per union per year is too rigid an approach (*Waugh v London Borough of Sutton* (1983) HSIB 86).

Definition of 'pay'

[J3012] The amount of pay to which the safety representative is entitled is contained in the *Safety Representatives Regulations (SI 1977 No 500), Sch 2*.

Recourse for the safety representative

[J3013] Where the employer's refusal to allow paid time off is unreasonable, he must reimburse the employee for the time taken to attend [*Safety Representatives Regulations 1977 (SI 1977 No 500), Reg 4(2), Sch 2*]. In the case of *Scarth v East Herts DC (HSIB 181)*: the test of reasonableness is to be judged at the time of the decision to refuse training.

Safety representatives who are refused time off to perform their functions or who are not paid for such time off are able to make a complaint to an employment tribunal [*Safety Representatives Regulations 1977 (SI 1977 No 500), Reg 11*].

Facilities to be provided by employer

[J3014] The type and number of facilities that employers are obliged to provide for safety representatives are not spelled out in the Regulations, Code of Practice or guidance notes, other than a general requirement in the *Safety Representatives Regulations 1977 (SI 1977 No 500), Reg 5(3)* which states, *inter alia*, that 'the employer shall provide such facilities and assistance as the safety representatives shall require for the purposes of carrying out their functions'. Formerly, the requirement to provide facilities and assistance related only to inspections.

Trade unions consider that the phrase 'facilities and assistance' includes the right to request the presence of an independent technical adviser or trade union official during an inspection, and for safety representatives to take samples of substances used at work for analysis outside the workplace. The TUC has recommended that the following facilities be made available to safety representatives:

- a room and desk at the workplace;
- facilities for storing correspondence;
- inspection reports and other papers;
- ready access to internal and external telephones;
- access to typing and duplicating facilities;
- provision of notice boards;
- use of a suitable room for reporting back to and consulting with members; and
- other facilities should include copies of all relevant statutes, regulations, Approved Codes of Practice and HSC guidance notes; and copies of all legal or international standards which are relevant to the workplace.

Disclosure of information

[J3015] Employers are required by the *Safety Representatives Regulations* to disclose information to safety representatives which is necessary for them to carry out their functions [*Safety Representatives Regulations 1977 (SI 1977 No 500), Reg 7(1)*]. A parallel provision exists under the *Management of Health and Safety at Work Regulations 1999, Reg 10(2)* in relation to the information that is to be provided to the parents of a child to be employed by an employer.

Regulation 7 is consolidated by paragraph 6 of the Code of Practice which details the health and safety information 'within the employer's knowledge' that should be made available to safety representatives. This should include:

- plans and performance and any changes proposed which may affect health and safety;
- technical information about hazards and precautions necessary, including information provided by manufacturers, suppliers and so on;
- information and statistical records on accidents, dangerous occurrences and notifiable industrial diseases; and
- other information such as measures to check the effectiveness of health and safety arrangements and information on articles and substances issued to homeworkers.

The exceptions to this requirement are where disclosure of such information would be 'against the interests of national security'; where it would contravene a prohibition imposed by law; any information relating to an individual (unless consent has been given); information that would damage the employer's undertaking; and information obtained for the sole purpose of bringing, prosecuting or defending legal proceedings [*Safety Representatives Regulations 1977 (SI 1977 No 500), Reg 7(2)*].

However, the decision in *Waugh v British Railways Board* [1979] 2 All ER 1169 established that where an employer seeks, on grounds of privilege, to withhold a report made following an accident, he can only do so if its dominant purpose is related to actual or potential hostile legal proceedings. In this particular case, a report was commissioned for two purposes following the death of an employee:

- to recommend improvements in safety measures; and

- to gather material for the employer's defence.

It was held that the report was not privileged.

This was followed in *Lask v Gloucester Health Authority* (1995) Times, 13 December where a circular *'Reporting Accidents in Hospitals'* had to be discovered by order after an injury to an employee while he was walking along a path.

Where differences of opinion arise as to the evaluation or interpretation of technical aspects of safety information or health data, unions are advised to contact the local offices of HSE, because of HSE expertise and access to research.

Technical information

[J3016] HSE inspectors are also obliged under the *HSWA 1974, s 28(8)*, to supply safety representatives with technical information – factual information obtained during their visits (ie any measurements, testing and results of sampling and monitoring), notices of prosecution, copies of correspondence and copies of any improvement or prohibition notices issued to their employer. The latter places an absolute duty on an inspector to disclose specific kinds of information to workers or their representatives concerning health, safety and welfare at work. This can also involve personal discussions between the HSE inspector and the safety representative. The inspector must also tell the representative what action he proposes to take as a result of his visit. Where local authority health inspectors are acting under powers granted by the *HSWA 1974* (see ENFORCEMENT), they are also required to provide appropriate information to safety representatives.

Safety committees

[J3017] There is a duty on every employer, in cases where it is prescribed (see below), to establish a safety committee if requested to do so by safety representatives. The committee's purpose is to monitor health and safety measures at work [*HSWA 1974, s 2(7)*]. Such cases are prescribed by the *Safety Representatives Regulations 1977 (SI 1977 No 500)* and limit the duty to appoint a committee to requests made by trade union safety representatives.

Establishment of a safety committee

[J3018] If requested by at least two safety representatives in writing, the employer must establish a safety committee [*Safety Representatives Regulations 1977 (SI 1977 No 500), Reg 9(1)*].

When setting up a safety committee, the employer must:

- consult with both:
  — the safety representatives who make the request,
  — the representatives of recognised trade unions whose members work in any workplace where it is proposed that the committee will function; and

- post a notice, stating the composition of the committee and the workplace(s) to be covered by it, in a place where it can easily be read by employees; and
- establish the committee within three months after the request for it was made.

[*Safety Representatives Regulations 1977 (SI 1977 No 500), Reg 9(2)*]

Function of safety committees

[**J3019**] In practical terms, trade union appointed safety representatives are now using the medium of safety committees to examine the implications of hazard report forms arising from inspections, and the results of investigations into accidents and dangerous occurrences, together with the remedial action required. A similar procedure exists with respect to representatives for tests and measurements of noise, toxic substances or other harmful effects on the working environment.

Trade unions regard the function of safety committees as a forum for the discussion and resolution of problems that have failed to be solved initially through the intervention of the safety representative in discussion with line management. There is, therefore, from the trade unions' viewpoint, a large measure of negotiation with its consequent effect on collective bargaining agreements.

If safety representatives are unable to resolve a problem with management through the safety committee, or with HSE, they can approach their own union for assistance – a number of unions have their own health and safety officers who can, and do, provide an extensive range of information on occupational health and safety matters. The unions, in turn, can refer to the TUC for further advice.

The 'Brown Book', which contains the *Safety Representatives Regulations*, Code of Practice and guidance, was revised in 1996 to include the amendments made in 1993 by the *Management of Health and Safety at Work Regulations 1992* (now superceded by the *Management of Health and Safety at Work Regulations 1999 (SI 1999 No 3242)*) (see above) and the *Consultation with Employees Regulations* (see **J3021**).

Non-unionised workforce – consultation obligations

[**J3020**] A representative of employee safety is an elected representative of a non-unionised workforce who is assigned with broadly the same rights and obligations as a safety representative in a unionised workforce.

General duty

[**J3021**] The *Consultation with Employees Regulations 1996 (SI 1996 No 1513)* introduced a duty to consult any employees who are not members of a group covered by safety representatives appointed under the *Safety Represen-*

tatives Regulations. Under the *Consultation with Employees Regulations*, employers must consult either with elected employee representatives or in the absence of such representatives, directly with the whole workforce (the *Consultation with Employees Regulations, Reg 3*). The obligation imposed on such employers is to consult those employees in good time on matters relating to their health and safety at work.

Number of representatives for workforce

[J3022] Guidance notes on the *Consultation with Employees Regulations 1996 (SI 1996 No 1513)* state that the number of safety representatives who can be appointed depends on the size of the workforce and workplace, whether there are different sites, the variety of different occupations, the operation of shift systems and the type and risks of work activity. A DTI Workplace Survey concluded that in non-unionised workplaces which have appointed worker representatives, it is usual for there to be several representatives, with the median number of such representatives being three. The survey estimated that there are approximately 218,000 representatives across all British workplaces with 25 or more employees.

Role of safety representatives

[J3023] The functions of representatives of employee safety are:

- to make representations to the employer of potential hazards and dangerous occurrences at the workplace which affect or could affect the group of employees the representative represents;
- to make representations to the employer on general matters affecting the health and safety at work of the group of employees the representative represents, and in particular on such matters as the representative has been consulted about by the employer under the *Consultation with Employees Regulations*; and
- to represent that group of employees in consultations at the workplace with inspectors appointed under the *HSWA 1974*.

Rights and duties of representatives of employee safety

Time off with pay

[J3024] Under the *Consultation with Employees Regulations 1996 (SI 1996 No 1513), Reg 7(1)(b)*, the right that a representative of employee safety has to take time off with pay is generally the same as that for safety representatives.

Definition of 'time off'

[J3025] An employer is under an obligation to permit a representative of employee safety to take such time off with pay during working hours as shall be necessary for:

- performing his functions; and
- undergoing such training as is reasonable in all the circumstances.

A candidate standing for election as a representative of employee safety is also allowed reasonable time off with pay during working hours in order to perform his functions as a candidate [*SI 1996 No 1513, Reg 7(2)*].

Definition of 'pay'

[J3026] The *Consultation with Employees Regulations (SI 1996 No 1513), Sch 1* deal with the definition of pay, and generally the definition is the same as that for union safety representatives.

Provision of information

[J3027] The employer must provide such information as is necessary to enable the employees or representatives of employee safety to participate fully and effectively in the consultation. In the case of representatives of employee safety, the information must also be sufficient to enable them to carry out their functions under the *Consultation with Employees Regulations 1996 (SI 1996 No 1513)*.

Information provided to representatives must also include information which is contained in any record which the employer is required to keep under *Reporting of Injuries, Diseases and Dangerous Occurrences Regulations 2013 (SI 2013 No 1471)* and which relates to the workplace or the group of employees represented by the representatives. Note that there are exceptions to the requirement to disclose information under the *Consultation with Employees Regulations, Reg 5(3)*, these being similar to those under the *Safety Representatives Regulations 1977 (SI 1977 No 500), Reg 7* (see **J3015** above).

Relevant training

[J3028] Under the *Consultation with Employees Regulations 1996 (SI 1996 No 1513), Reg 7(1)*, representatives of employee safety must be provided with reasonable training in respect of their functions under the *Consultation with Employees Regulations*, for which the employer must pay.

Remedies for failure to provide time off or pay for time off

[J3029] A representative of employee safety, or candidate standing for election as such, who is denied time off or who fails to receive payment for time off, may make an application to an employment tribunal for a declaration and/or compensation. As in the case of safety representatives, the remedies obtainable (set out in the *Consultation with Employees Regulations 1996 (SI 1996 No 1513), Sch 2* are similar to those granted to complainants under the *Trade Union and Labour Relations (Consolidation) Act 1992, s 168* (time off for union duties).

Recourse for safety representatives

[J3030] Safety representatives (whether they are appointed under the *Safety Representatives Regulations 1977 (SI 1977 No 500)* or the *Consultation with Employees Regulations 1996 (SI 1996 No 1513)* are provided with statutory protection for the proper execution of their duties.

Recourse for safety representatives [J3030]

An employee who is:

- designated by his employer to carry out a health and safety related function;
- a representative of employee safety; or
- a candidate standing for election as such,

has the right not to be subjected to any detriment or unfairly dismissed on the grounds that:

— having been designated by the employer to carry out a health and safety related function, he carried out, or proposed to carry out, the function;
— he undertook, or proposed to undertake, any function(s) consistent with being a safety representative or member of a safety committee;
— he took part in, or proposed to take part in, consultation with the employer; or
— he took part in an election of representatives of employee safety.

In relation to a detriment claim, where the employer infringes any of these rights, the employee has the right to make a complaint to an employment tribunal under the *Employment Rights Act 1996, s 44(1)* (as amended by the *Consultation with Employees Regulations 1996 (SI 1996 No 1513), Reg 8*. An employment tribunal can make a declaration and also award compensation [*Employment Rights Act 1996, ss 48–49*]. There is neither a minimum qualifying period of service nor an upper age limit for bringing such a claim.

If the employer unfairly dismisses such an employee for one of the above reasons, or where it is the principal reason for the dismissal, that dismissal shall be deemed automatically unfair [*Employment Rights Act 1996, s 100*, as amended by the *Consultation with Employees Regulations 1996 (SI 1996 No 1513), Reg 8*]. Conversely, if the dismissal is not connected with the health and safety issues which have arisen, the dismissal will not be automatically unfair [*Dunn v Ovalcode Limited* [2003] ALL ER (D) 241 (Apr), EAT].

It will also be an automatically unfair dismissal to select a representative or candidate for redundancy for such a reason [*Employment Rights Act 1996, s 105*]. However, there is no presumption of a right to positive discrimination for such representatives where the employer is undertaking a redundancy programme [*Shipham & Co Limited v Skinner* [2001] All ER (D) 201 (Dec), EAT].

The normal minimum qualifying period of service, the normal upper age limit and the cap on the compensatory award for unfair dismissal claims do not apply [*Employment Rights Act 1996, ss 108, 109 and 124(1A)* as amended by the *Employment Relations Act 1999, s 37(1))*].

Under the *Employment Rights Act 1996, s 103A* a right to automatically claim unfair dismissal would exist for a worker who is dismissed for making a protected disclosure. A right also exists under the *Employment Rights Act 1996, s 47B* for workers to claim that they have been subjected to a detriment for making a protected disclosure. 'Protected disclosures' in this regard are capable of covering the situation where the health and safety of an individual has been or is likely to be endangered [*Employment Rights Act 1996, s 43B(1)(d)*].

European developments in health and safety

Introduction

[J3031] Health and safety law will continue to be subject to change in the future with the implementation of further European directives and HSC programme of modifying and simplifying health and safety law. At a European level the most important legislation is the directives made under Art 138 of the Treaty of Rome (as amended). The Framework Directive, from which the *Consultation with Employees Regulations 1996 (SI 1996 No 1513)* were derived, will continue to drive forward developments in UK health and safety law.

Further developments have occurred under European law which are having an impact at national level these being in the form of the Working Time Directive (93/104) and the European Works Councils Directive (94/45). These Directives have been implemented into national law under the *Working Time Regulations 1998 (SI 1998 No 1833)* (as amended) and the *Works Councils Regulations 1999 (SI 1999 No 3323)* respectively.

The Working Time Regulations 1998 (as amended)

[J3032] A full discussion of the impact of the *Working Time Regulations 1998 (SI 1998 No 1833)* is beyond the scope of this chapter (see WORKING TIME).

That said, the *Working Time Regulations* have introduced a joint consultation function into the operation of these Regulations by use of collective, workforce and relevant agreements. The *Working Time Regulations* provide that it is possible to vary the extent to which the *Working Time Regulations* must be strictly complied with through the use of these devices. The various types of agreements can be described as follows:

- collective agreements – these are defined by *s 178 of the Trade Union and Labour Relations (Consolidation) Act 1992* as being agreements between independent trade unions and employers;
- workforce agreements – these were created by the *Working Time Regulations* and are defined in *the Working Time Regulations 1998 (SI 1998 No 1833), Reg 2 and Sch 1*. They amount to agreements between an employer and either duly elected worker representatives of the employer or, in the case of an employer employing less than 20 workers, a majority of the individual workers themselves, where the agreement concluded:
  — is in writing;
  — has effect for a specified period not exceeding five years;
  — applies to either:
     (i)   all of the relevant members of the workforce; or
     (ii) all of the relevant members of the workforce who belong to a particular sub-group;
  — is signed:
     (i)   by the worker representatives or by the particular group of workers; or

(ii) in the case of an employer having less than 20 employees on the date on which the agreement is first concluded, either by appropriate representatives or by a majority of the workers working for the employer; and
— before being made available for signature, was provided in copy form to all of the workers to whom the agreement was intended to apply, together with such guidance as the workers might reasonably require in order to understand the draft agreement;
- relevant agreements – these are workforce agreements that cover a worker, any provision of a collective agreement that is individually incorporated into the contract of employment of a worker, or any other agreement in writing between a worker and his employer that is legally enforceable (eg a staff handbook).

By the *Working Time Regulations 1998 (SI 1998 No 1833), Reg 23(a)* it is possible to modify the provisions relating to:

- the length of night work (see *Working Time Regulations 1998 (SI 1998 No 1833), Reg 6*); and
- the minimum daily and weekly rest periods and rest breaks (see *Working Time Regulations 1998 (SI 1998 No 1833), Regs 10–12* respectively).

Modification must be by way of a collective or workforce agreement and such an agreement must make provision for a compensatory rest period of equivalent length [*Working Time Regulations 1998 (SI 1998 No 1833), Reg 24*].

Further, by the *Working Time Regulations 1998 (SI 1998 No 1833), Reg 23(b)* it is possible for an employer and its workers to agree by collective or workforce agreement to vary the reference period for calculating the maximum working week from the usual 17 weeks to a 52-week period if there are objective or technical reasons relating to the organisation which justify such a change.

The variation provisions allow a degree of flexibility where it is necessary for the interests of an employer's business to effect such change for operational reasons whilst still ensuring the protection of the health and safety of the workforce. The provisions also ensure that any change that is to be made must survive collective scrutiny of the employer's workforce.

European Works Councils and information and consultation procedures

[J3033] The provisions relating to European Works Councils are at present confined to large pan-European entities.

As stated at J3001, the Information and Consultation Directive 2002/14/EC will impose information and consultation obligations on any employer employing more than 50 workers. Transitional provisions will apply initially limiting the impact of Directive 2002/14/EC to employers employing more than 150 employees. All employers in the United Kingdom employing 50 are required to comply. National legislation to implement Directive 2002/14/EC was introduced in the form of the *Information and Consultation of Employees Regulations 2004 (SI 2004 No 3426)*. From 6 April 2008, employers employing 50 or more employees are subject to its provisions. The *Information and Consultation of Employees Regulations 2004* do not apply directly to health and safety consultation, although such consultation could be voluntarily covered under an information and consultation procedure under those Regulations.

The European Work Council Directive (94/45) was incorporated into national law by the *Works Councils Regulations 1999 (SI 1999 No 3323)*. A full discussion of the operation of the *Works Councils Regulations* is beyond the scope of this chapter, which instead focuses upon the health and safety aspect of the Regulations.

The *Works Councils Regulations* govern employers employing a total of 1,000 or more workers where at least 150 workers are so employed in each of two or more member states.

Their main purpose is procedural. *Part IV* of the *Works Council Regulations 1999 (SI 1999 No 3323)* creates machinery between workers and their employer for the purpose of establishing either a European Works Council ('EWC') or an Information and Consultation Procedure ('ICP'). *Regulation 17(1)* of the *Works Councils Regulations 1999 (SI 1999 No 3323)* provides that the central management of the employer and a special negotiating body (defined in *Part III* of the Regulations) are bound to:

> . . . negotiate in a spirit of co-operation with a view to reaching a written agreement on the detailed arrangements for the information and consultation of employees in a Community-scale undertaking or Community-scale group of undertakings.

Regulation 17(3) of the *Works Councils Regulations 1999 (SI 1999 No 3323)* leaves the choice of whether to proceed with an EWC or an ICP to the parties.

The parties are free to include in the agreement reference to whatever matters are likely to affect the workers of the employer at a trans-national level. Health and safety is clearly such an issue.

If:

- the parties fail to agree the content of the agreement; or
- within six months of a valid request being made to the central management of an employer the employer fails to negotiate so as to create either an EWC or ICP; or
- after three years of negotiation to produce an agreement for an EWC or ICP the parties cannot agree as to the constitution,

default machinery is provided by the *Schedule* to the *Works Councils Regulations* [*Works Councils Regulations 1999 (SI 1999 No 3323), Reg 18(1)*]. *Paragraph 6* of the *Schedule* provides in relation to EWCs that:

> The competence of the European Works Council shall be limited to information and consultation on the matters which concern the Community-scale undertaking or Community-scale group of undertakings as a whole or at least two of its establishments or group undertakings situated in different Member States.

In relation to ICPs, *para 7(3)* of the *Schedule* to the *Works Councils Regulations 1999 (SI 1999 No 3323)* provides that meetings of ICPs:

> . . . shall relate in particular to the structure, economic and financial situation, the probable development of the business and of production and sales, the situation and probable trend of employment, investments, and substantial changes concerning organisation, [and] introduction of new working methods or production processes . . .

Although not expressly providing for discussion of health and safety issues, the provisions relating to both EWCs and ICPs will, by implication, include debate of health and safety matters.

Lifting Operations

Kevin Chicken

Introduction to lifting operations

[L3001] General requirements for the provision and maintenance of safe plant and equipment are set out in the Health and Safety at Work etc Act 1974 sections 2(2)(a) and (b).

In addition, the *Provision and Use of Work Equipment Regulations 1998 (PUWER) (SI 1992 No 2932)* place duties on people and companies who own, operate or have control over work equipment and on organisations whose employees use work equipment, whether owned by them or not. Under PUWER, the Health and Safety Executive (HSE) sets out that: equipment provided for use at work must be suitable for its intended use; safe for use, maintained in a safe condition and inspected to ensure it is correctly installed and does not subsequently deteriorate; used only by people who have received adequate information, instruction and training; accompanied by suitable health and safety measures, such as protective devices and controls, including emergency stop devices, adequate means of isolation from sources of energy, clearly visible markings and warning devices.

More specific requirements came into force on 5 December 1998 in the *Lifting Operations and Lifting Equipment Regulations 1998 (LOLER) (SI 1998 No 2307)* which encompass all lifting equipment for use at work.

LOLER demands that four main principals are complied with:

- the initial integrity of the equipment is suitable for its intended use;
- the lifting operation is correctly planned and assessed;
- the operation is undertaken in a safe manner; and
- thereafter, the equipment's safe integrity is maintained.

There is a duty on the employer to select suitable lifting equipment based on a fitness for purpose criteria in both *Regulation 4* of PUWER (see **M1004**) and its sister legislation, LOLER.

Statutory requirements relating to the manufacture of lifting equipment

[L3002] The law relating to the manufacture of lifting equipment includes:

- the *Lifts Regulations 2016 (SI 2016 No 831)*; and
- the *Supply of Machinery (Safety) Regulations 2008 (SI 2008 No 1597)* and the *Supply of Machinery (Safety) (Amendment) Regulations 2011 (SI 2011 No 2157)*

(which implemented the requirements of the European Union Machinery Directive). The *Supply of Machinery (Safety) (Amendment) Regulations 2011 (SI 2011 No 2157)* cover the manufacturer's duties for all lifting equipment and accessories other than those detailed within the *Lifts Regulations 2016 (SI 2016 No 831)*. The relationship between the manufacturers and the owner/users duties are shown in Figure 1.

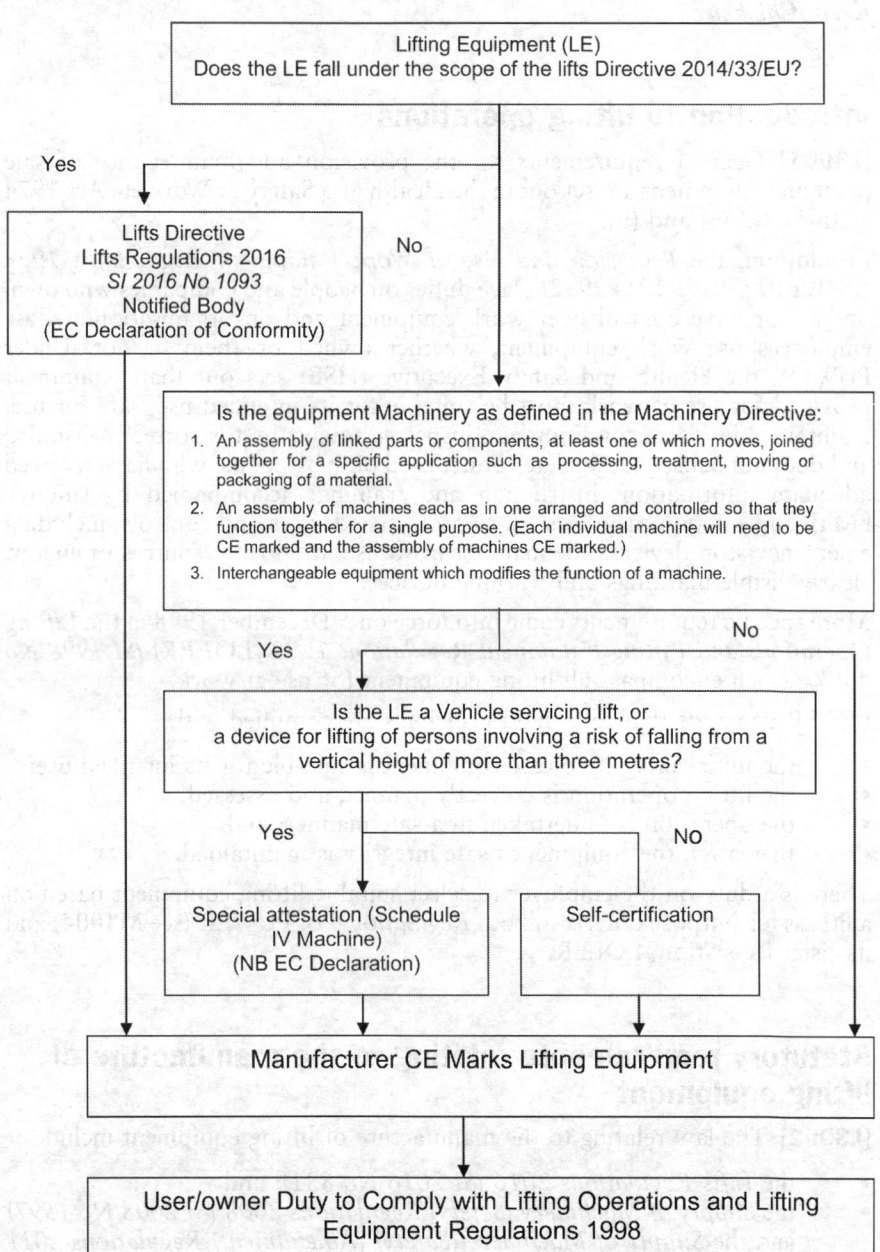

In addition to the specific requirements above, lifting equipment incorporating electrical equipment must also comply with:

- *Electrical Equipment (Safety) Regulations 2016 (SI 2016 No 1101)* implementing the Low Voltage Directive; and
- *Electromagnetic Compatibility Regulations 2016 (SI 2016 No 1091)*.

Lifts Regulations 1997

[L3003] 'Lift' – this means an appliance serving specific levels, having a carrier moving along guides which are rigid – or along a fixed course even where it does not move along guides which are rigid (for example, a scissor lift) – and inclined at an angle of more than 15 degrees to the horizontal and intended for the transport of (there is difference in terminology when LOLER refers to Lifts as Carriers):

- persons;
- persons and goods; or
- goods alone if the car is accessible, that is to say, a person may enter it without difficulty, and fitted with controls situated inside the car or within reach of a person inside.

The Regulations

[L3004] These Regulations apply to lifts permanently serving buildings or constructions; and safety components for use in such lifts. The *Lifts Regulations 2016* implement Directive 2014/33/EU (and replace the previous *Lifts Regulations 1997* which implemented Directive 96/16/EC, as amended by 2006/42/EC).

These Regulations do not apply to:

(a) the following lifts – and safety components for such lifts (see *Sch 2* of the regulations which remains unchanged from the *Lifts Regulations 1997*):
— Lifting appliances whose speed is not greater than 0.15m/s.
— Construction site hoists.
— Cableways, including funicular railways.
— Lifts specially designed and constructed for military or police purposes.
— Lifting appliances from which work can be carried out.
— Mine winding gear.
— Lifting appliances intended for lifting performers during artistic performances.
— Lifting appliances fitted in means of transport.
— Lifting appliances connected to machinery and intended exclusively for access to work-stations including maintenance and inspection points on the machinery.
— Rack and pinion trains.
— Escalators and mechanical walkways.
[*SI 1997 No 831, Reg 4, Sch 14*].

(b) any lift or safety component which is placed on the market (ie when the installer first makes the lift available to the user, but see below) and put into service before 1 July 1997 [*SI 1997 No 831, Reg 5*];
(c) any lift or safety component placed on the market and put into service on or before 30 June 1999 which complies with any health and safety provisions with which it would have been required to comply if it was to have been placed on the market and put into service in the United Kingdom on 29 June 1995.
This exclusion does not apply in the case of a lift or a safety component which;
— unless required to bear the CE marking pursuant to any other European Union obligation, bears the CE marking or an inscription liable to be confused with it; or
— bears or is accompanied by any other indication, howsoever expressed, that it complies with the Lifts Directive.
[*SI 1997 No 831, Reg 6*]; and
(d) any lift insofar as and to the extent that the relevant essential health and safety requirements relate to risks wholly or partly covered by other EU directives applicable to that lift [*SI 1997 No 831, Reg 7*].

General requirements

General duty relating to the placing on the market and putting into service of lifts

[L3005]

(i) Subject to the *Lifts Regulations 2016* implement Directive 2014/33/EU (*SI 2016 SI 1093*), *Reg 12* (see **L3010** below), no person who is a responsible person shall place on the market and put into service any lift unless the requirements of paragraph (ii) below have been complied, see *Reg 8(1)*.
(ii) *Regulation 8(2)* provides that the requirements in respect of any lift are that:
— it satisfies the relevant essential health and safety requirements and, for the purpose of satisfying those requirements:
– where a transposed harmonised standard covers one or more of the relevant essential health and safety requirements, any lift constructed in accordance with that transposed harmonised standard shall be presumed to comply with that (or those) essential health and safety requirement(s);
– by calculation, or on the basis of design plans, it is permitted to demonstrate the similarity of a range of equipment to satisfy the essential safety requirements;
— the appropriate conformity assessment procedure in respect of the lift has been carried out in accordance with *SI 2016 No 1093, Reg 7(1)* (see **L3011** below);
— the CE marking has been affixed to it by the installer of the lift in accordance with *SI 2016 No 1093, Reg 8(1)*;
— a declaration of conformity has been drawn up in respect of it by the installer of the lift *Reg 8(1)*; and

(iii) — it is in fact safe.
Any technical documentation or other information in relation to a lift required to be retained under the conformity assessment procedure used must be retained by the person specified in that respect in that conformity assessment procedure for any period specified in that procedure [*SI 2016 No 1093, Reg 7*].

In these Regulations, 'responsible person' means:

— in the case of a lift, the installer of the lift;
— in the case of a safety component, the manufacturer of the component or his authorised representative established in the European Union; or
— where neither the installer of the lift nor the manufacturer of the safety component nor the latter's authorised representative established in the European Union, as the case may be, have fulfilled the requirements of *SI 2016 No 1093, Reg 8(2)* or *SI 2016 No 1093, Reg 9(2)* (see **L3006** below), the person who places the lift or safety component on the market.

General duty relating to the placing on the market and putting into service of safety components

[L3006] Safety components include devices for locking landing doors; devices to prevent falls; overspeed limitation devices; and electric safety switches.

(i) Subject to the *Lifts Regulations 2016* implement Directive 2014/33/EU (*SI 2016 No 1093, Reg 48* (see **L3010** below), no person who is a responsible person shall place on the market and put into service any safety component unless the requirements of paragraph (a)–(c) below have been complied with in relation to it [*SI 2016 No 1093, Reg 48*].

(a) the model of the safety component for lifts must be submitted for EU type examination set out in Part A of Annex IV to the Directive (as amended from time to time) and the conformity to type must be ensured with random checking of the safety component for lifts set out in Annex IX to the Directive (as amended from time to time);

(b) the model of the safety component for lifts must be submitted for EU type examination set out in Part A of Annex IV to the Directive (as amended from time to time) and be subject to conformity to type based on product quality assurance in accordance with Annex VI to the Directive (as amended from time to time);

(c) conformity based on full quality assurance set out in Annex VII to the Directive (as amended from time to time).

Note: *SI 2016 No 1093, Reg 49* provides that a safety component – or its label – which bears the CE marking, and is accompanied by an EC declaration of conformity, is taken to conform with all the requirements of the *Lifts Regulations 2016, Reg 50* (see **L3011** below), unless reasonable grounds exist for suspecting that it does not so conform.

(ii) Any technical documentation or other information in relation to a safety component required to be retained under the conformity assessment procedure used must be retained by the person specified in that respect in that conformity assessment procedure for any period specified in that procedure [*SI 2016 No 1093, Reg 33*].

General duty relating to the supply of a lift or safety component

[L3007] Subject to the *Lifts Regulations 2016 (SI 2016 No 1093), Reg 12* (see below) any person who supplies any lift or safety component but who is not a person to whom *Reg 8* or *9* applies (see **L3005** and **L3006** above) must ensure that that lift or safety component is safe [*SI 2016 No 1093, Reg 10*].

Penalties for breach of Regs 8, 9 or 10

[L3008] A person who is convicted of and found guilty of a breach of Reg 70 of the *Lifts Regulations 2016 (SI 2016 No 1093)* for contravening or failing to comply with any requirement of any of Regs 6–12, 13(2), 14–22, 23(2), 25–33, 34(2), 36–40, 41(2), 44–45 or who contravened or failed to comply with any requirement of a withdrawal or recall notice served on that person by an enforcing authority under these regulations is liable on summary conviction:

(a) in England and Wales, to a fine or imprisonment for a term not exceeding three months or both;
(b) in Scotland and Northern Ireland, to a fine not exceeding level 5 on the standard scale or imprisonment for a term not exceeding three months or both.

A person guilty of an offence under *Reg 9, 13(2), 18, 23(2), 34(2)* or *41(2)* is liable on summary conviction:

(a) in England and Wales, to a fine;
(b) in Scotland or Northern Ireland, to a fine not exceeding the level 5 on the standard scale.

Specific duties relating to the supply of information, freedom from obstruction of lift shafts and retention of documents

[L3009]

(i) The person responsible for work on the building or construction where a lift is to be installed and the installer of the lift must keep each other informed of the facts necessary for, and take the appropriate steps to ensure, the proper operation and safe use of the lift. Shafts intended for lifts must not contain any piping or wiring or fittings other than that which is necessary for the operation and safety of that lift (*SI 2016 No 1093, Reg 14*).
(ii) Where, in the case of a lift, for the purposes of *Lifts Regulations 2016 (SI 2016 No 1093)*, the person responsible for the design of the lift must supply to the person responsible for the construction, installation and testing all necessary documents and information for the latter person to be able to operate in absolute security (*Reg 23*).
(iii) A copy of the declaration of conformity must:

— in the case of a lift, be supplied to the Enforcing Authority, on request, by the installer of the lift together with a copy of the reports of the tests involved in the final inspection to be carried out as part of the appropriate conformity assessment procedure referred to in *SI 2016 No 1093, Reg 23*; and

— be retained, by the person who draws up that declaration, for a period of ten years – in the case of a lift, from the date on which the lift was placed on the market; and in the case of a safety component, from the date on which safety components of that type were last manufactured by that person.

A person who fails to supply or keep a copy of the declaration of conformity, as required above, is liable on summary conviction (see L3005) they may however rely on a defence of due diligence: ie the defendant must show that he took all reasonable steps and exercised all due diligence to avoid committing the offence.

Exceptions to placing on the market or supply in respect of certain lifts and safety components

[L3010] For the purposes of *Lifts Regulations 2016 (SI 2016 No 1093), Reg 8, 9 or 10*, a lift or a safety component is not regarded as being placed on the market or supplied:

(i) where that lift or safety component will be put into service in a country outside the EU; or is imported into the EU for re-export to a country outside the EU – but this paragraph does not apply if the CE marking, or any inscription liable to be confused with such a marking, is affixed to the lift or safety component or, in the case of a safety component, to its label; or

(ii) by the exhibition at trade fairs and exhibitions of that lift or safety component, in respect of which the provisions of these Regulations are not satisfied, if:

— a notice is displayed in relation to the lift or safety component in question to the effect that it does not satisfy those provisions; and that it may not be placed on the market or supplied until those provisions are satisfied; and

— adequate safety measures are taken to ensure the safety of persons.

[*SI 2016 No 1093, Reg 12*].

Conformity assessment procedures

[L3011] For the purposes of *Lifts Regulations 2016 (SI 2016 No 1093), Reg 8 or 9* (see above), the appropriate conformity assessment procedure is as follows:

For lifts – one of the following procedures:

(1) For the assessment of conformity of a lift, the installer must carry out one of the following procedures—

    (a)   if the lift is designed and manufactured in accordance with a model lift that has undergone an EU-type examination set out in Part B of Annex IV to the Directive (as amended from time to time)—
- (i) final inspection for lifts set out in Annex V to the Directive (as amended from time to time);
- (ii) conformity to type based on product quality assurance for lifts set out in Annex X to the Directive (as amended from time to time);
- (iii) conformity to type based on production quality assurance for lifts set out in Annex XII to the Directive (as amended from time to time);

    (b)   if the lift is designed and manufactured under a quality assurance system approved in accordance with Annex XI to the Directive (as amended from time to time)—
- (i) final inspection for lifts set out in Annex V to the Directive (as amended from time to time);
- (ii) conformity to type based on product quality assurance for lifts set out in Annex X to the Directive (as amended from time to time);
- (iii) conformity to type based on production quality assurance for lifts set out in Annex XII to the Directive (as amended from time to time);

    (c)   conformity based on unit verification for lifts set out in Annex VIII to the Directive (as amended from time to time);

    (d)   conformity based on full quality assurance plus design examination for lifts set out in Annex XI to the Directive (as amended from time to time).

(2) Where one of the procedures in paragraph (1)(a) or (1)(b) is carried out and the person responsible for the design and manufacture of the lift and the person responsible for the installation and testing of the lift are not the same person, the former must supply to the latter all the necessary documents and information to enable the latter to ensure the correct and safe installation and testing of the lift.

(3) Where one of the procedures in paragraph (1)(a) is carried out, the installer must ensure that all permitted variations between the model lift and the lifts derived from the model lift are clearly specified (with maximum and minimum values) in the technical documentation referred to in regulation 7(b).

(4) When using the procedure in paragraph 9(1)(a), in order to demonstrate the conformity of a lift with the essential health and safety requirements, the installer may demonstrate the similarity of a range of equipment—
- (a) by calculation;
- (b) on the basis of design plans; or
- (c) using both of the methods specified in sub-paragraphs (a) and (b).

For safety components – one of the following procedures:

(a) the model of the safety component for lifts must be submitted for EU type examination set out in Part A of Annex IV to the Directive (as amended from time to time) and the conformity to type must be ensured with random checking of the safety component for lifts set out in Annex IX to the Directive (as amended from time to time);

(b) the model of the safety component for lifts must be submitted for EU type examination set out in Part A of Annex IV to the Directive (as amended from time to time) and be subject to conformity to type based on product quality assurance in accordance with Annex VI to the Directive (as amended from time to time);

(c) conformity based on full quality assurance set out in Annex VII to the Directive (as amended from time to time).

Requirements fulfilled by the person who places a lift or safety component on the market

[L3012] Where in the case of a lift or a safety component, any of the requirements of *Lifts Regulations 1997 (SI 1997 No 831), Regs 8, 9, 11* and *13* to be fulfilled by the installer of the lift or the manufacturer of the safety component or, in the case of the latter, his authorised representative established in the EU, have not been so fulfilled such requirements may be fulfilled by the person who places that lift or safety component on the market [*SI 1997 No 831, Reg 14*].

This provision, however, does not affect the power of an enforcement authority to take action in respect of the installer of the lift, the manufacturer of the safety component or, in the case of the latter, his authorised representative established in the EU in respect of a contravention of or a failure to comply with any of those requirements.

Notified bodies

[L3013] For the purposes of these Regulations, a notified body is a body which has been appointed to carry out one or more of the conformity assessment procedures referred to in *Lifts Regulations 2016 (SI 2016 No 1093), Reg 13* which has been appointed as a notified body by the Secretary of State in the United Kingdom or by a member State.

[*SI 2016 No 1093, Reg 51*].

Lifting Operations and Lifting Equipment Regulations 1998 (LOLER) (SI 1998 No 2307)

[L3014] The *Lifting Operations and Lifting Equipment Regulations 1998 (LOLER) (SI 1998 No 2307)* uses the term employer rather than duty-holder and duties specifically assigned to the employer can be assumed to apply to the duty holder, if they have any control over lifting operations.

Where a company provides personnel to undertake work, which will involve the use of lifting equipment then that company is regarded as an employer and they have a duty under LOLER to provide persons competent to undertake the work.

[L3014] Lifting Operations

Duties under LOLER apply to anyone with control, to any extent, of:
— lifting equipment;
— a person at work who uses or supervises or manages the use of lifting equipment; or
— the way in which lifting equipment is used,

and to the extent of his control.

The following general definitions apply to LOLER:

- 'lifting equipment' – work equipment for lifting or lowering loads and includes its attachments for anchoring, fixing or supporting it;
- 'accessory for lifting' – lifting equipment for attaching loads to machinery for lifting (pendant, sling, shackle, etc);
- 'load' – includes material or people lifted by the lifting equipment;
- 'examination scheme' – suitable scheme drawn up by a competent person for such thorough examination of lifting equipment at such intervals as may be appropriate for the purpose described in *SI 1998 No 2307, Reg 9*; and
- 'thorough examination' – means a thorough examination by a competent person including such testing as is appropriate for the purpose.

In its publication *Safe Use of Lifting Operations and Equipment Regulations 2008 Approved code of practice and guidance* (L113), the Health and Safety Executive (HSE) provides the following examples of the type of equipment which should be assessed for the application of LOLER. This is a non-exhaustive list:

(a) cranes;
(b) lift trucks and telescopic handlers;
(c) hand pallet trucks, specifically those that have the ability to raise the forks;
(d) goods lifts or passenger lifts, for example in an office block, hospital etc which are provided for those at work;
(e) simple systems such as a rope and pulley used to raise a bucket of cement on a building site, a construction site hoist, a gin wheel, or a dumb waiter in a restaurant or hotel;
(f) pull-lifts;
(g) vacuum lifting equipment;
(h) a vehicle inspection hoist;
(i) a scissor lift or a mobile elevating work platform (MEWP);
(j) ropes used for climbing or work positioning during arboriculture, climbing telecommunication towers and structural examination of a rock face or external structure of a building;
(k) a paper roll hoist on a printing machine;
(l) an automated storage and retrieval system;
(m) a front-end loader on a tractor used for raising and lowering loads such as a bale of hay;
(n) an excavator (or other earth-moving machinery) adapted to be used for lifting using lifting attachments (eg forks, grabs, lifting magnets), but not when used for normal earth-moving operations;

(o) a hoist or sling used for lifting people from, for example, a bed or a bath;
(p) a loader crane fitted to a lorry, eg used to raise bins for delivery duties;
(q) a refuse vehicle loading arm, eg used to raise bins for tipping;
(r) an air cargo elevating transfer vehicle;
(s) a car transporter or vehicle recovery equipment;
(t) a skip collection vehicle; and
(u) vehicle tail lifts.

It advises that lifting accessories would include such items as slings, removable eyebolts, chains, ropes, shackles, grabs, magnets, vacuum lifters, crane forks, lifting beams and spreaders.

Some of these specific items of lifting equipment are looked at in more detail in L3024 below onwards.

General requirements

Strength and stability

[L3015] Every employer must ensure that lifting equipment is of adequate strength and stability for each load, having regard in particular to the stress induced at its mounting or fixing point – and that every part of a load and anything attached to it and used in lifting it is of adequate strength [*Lifting Operations and Lifting Equipment Regulations 1998 (LOLER) (SI 1998 No 2307), Reg 4*].

HSE guidance here is that employers should:

- assess whether the lifting equipment has adequate lifting capacity for the proposed use;
- take account of the combination of forces to which the lifting equipment will be subjected as well as the weight of any associated accessories used in the lifting operation and how they have been configured together; and
- assess foreseeable events such as loads snagging during use, eg on other structures.

The lifting equipment selected should not be unduly susceptible to any of the foreseeable failure modes likely to arise in service, for example fracture, wear or fatigue it advises, and it should provide an appropriate margin of safety against failure under foreseeable failure modes.

Although the load does not fall within LOLER it is incumbent upon the employer to ensure that any lifting points on the load are of adequate strength.

Lifting equipment for lifting persons

[L3016] *Lifting Operations and Lifting Equipment Regulations 1998 (LOLER) (SI 1998 No 2307), Reg 5(1)(b)* does contain the exception in the applied duty under LOLER, in this case the standard is qualified **by so far as is reasonably practicable (SFAIRP)**, in that the employer shall SFAIRP prevent a person being injured whilst carrying out activities from a carrier eg mobile elevated work platform, suspended cradle etc. All other parts of this regulation demand an **absolute duty**.

Every employer must ensure that lifting equipment for lifting persons:

(i) is such as to prevent a person using it being crushed, trapped or struck or falling from the carrier;
(ii) is such as to prevent so far as is reasonably practicable a person using it, while carrying out activities from the carrier, being crushed, trapped or struck or falling from the carrier;
(iii) has suitable devices to prevent the risk of a carrier falling, and, if the risk cannot be prevented for reasons inherent in the site and height differences, the employer must ensure that the carrier has an enhanced safety coefficient suspension rope or chain which is inspected by a competent person every working day; and
(iv) is such that a person trapped in any carrier is not thereby exposed to danger and can be freed.

[SI 1998 No 2307, Reg 5].

Common examples of equipment NOT designed for lifting persons are forklift trucks and telescopic handlers.

Positioning and installation

[L3017] Employers must ensure that lifting equipment is positioned or installed in such a way as to reduce to as low as is reasonably practicable the risk of the lifting equipment or a load striking a person, or the risk from a load:

— drifting;
— falling freely; or
— being released unintentionally,

and that otherwise it is safe. Further consideration should be given to:

- the need to lift loads over people, and were this unavoidable minimise the risk;
- crushing is prevented at extreme operating positions;
- loads moving along a fixed path are suitably protected to minimise the risk of the load or equipment striking a person; and
- trapping points are prevented or access limited on travelling or slewing equipment.

Employers must also ensure that there are suitable devices for preventing anyone from falling down a shaft or hoistway and that lifting equipment cannot be unintentionally released during a loss of power to the lifting equipment or through the collision of equipment or their loads.

The use of hooks with safety catches, motion limiting devices and safe systems of work are possible means of minimising these risks. [*Lifting Operations and Lifting Equipment Regulations 1998 (LOLER) (SI 1998 No 2307), Reg 6*].

Marking of lifting equipment

[L3018] Employers must ensure that:

(i) machinery and accessories for lifting loads are clearly marked to indicate their safe working loads;

(ii) where the safe working load of machinery for lifting loads depends on its configuration, either the machinery is clearly marked to indicate its safe working load for each configuration, or information which clearly indicates its safe working load for each configuration is kept with the machinery;
(iii) accessories for lifting are also marked in such a way that it is possible to identify the characteristics necessary for their safe use;
(iv) lifting equipment which is designed for lifting persons is appropriately and clearly marked to this effect; and
(v) lifting equipment which is not designed for lifting persons but which might be mistakenly so used is appropriately and clearly marked to the effect that it is not designed for lifting persons.

[Lifting Operations and Lifting Equipment Regulations 1998 (LOLER) (SI 1998 No 2307), Reg 7].

Organisation of lifting operations

[L3019] Every employer must ensure that any lifting or lowering of a load which involves lifting equipment is properly planned by a competent person, appropriately supervised and carried out in a safe manner.

The *Lifting Operations and Lifting Equipment Regulations 1998 (LOLER) (SI 1998 No 2307)* demands as a minimum standard that all lifting operations are carried out safely under adequate supervision and following a lifting plan. The competent person planning the operation should have adequate practical and theoretical knowledge and experience of planning lifting operations.

The plan will need to address the risks identified during a risk assessment and should identify all resources, procedures and responsibilities necessary for safe operation. The degree of planning will vary considerably depending on the type of lifting equipment and complexity of the lifting operation and degree of risk involved. This regulation does allow for routinely repeated lifting operations to be assessed in the first instance and thereafter as appropriate.

There are two elements to the plan: the suitability of the lifting equipment as per *Regulation 4* of the *Provision and Use of Work Equipment Regulations 1998 (PUWER) (SI 1998 No 2306)* and the individual lifting operation to be performed.

SI 1998 No 2307, Reg 8 requires that the employer or controller of lifting operations ensures that suitable persons are appointed for planning and supervising of such operations.

For any lifting operation it is necessary to:

(a) carry out a risk assessment under the *Management of Health and Safety at Work Regulations 1999 (SI 1999 No 3242)*;
(b) select suitable equipment for the range of tasks; and
(c) plan the individual lifting operation.

The term 'Competent Person' is not prescriptively described in LOLER and is used to identify a number of different roles under the regulations. In practical terms the competency of a person may be confirmed by formal, vocational qualification or through first-hand knowledge of planning or supervising the lifting operations.

The approved code of practice (ACOP) and guidance to the regulations, *Safe use of lifting equipment – Lifting Operations and Lifting Equipment Regulations 1998 (L113)* (available on the HSE website at www.hse.gov.uk/pubns/priced/l113.pdf) states that the competent person carrying out the planning task is unlikely to be the same person as the competent person undertaking the thorough examination and testing:

> 'The term "competent person" required to carry out the planning means the person must have the skills, knowledge and experience to make the relevant assessment of the requirements of the lifting equipment being used and the type of task being carried out. It does not have the same meaning as, and is unlikely to be, the same competent person referred to in regulation 9 (thorough examination and inspection).'

This planning must take into account the location, the load to be lifted, the duration and the specific operation to be carried out.

Regulation 4 of *(PUWER) (SI 1998 No 2306)* requires suitable work equipment to be provided for the task. There is therefore a close link between this regulation and the requirement for planning. Factors to be considered when selecting lifting equipment so that it is suitable for the proposed task should include:

- the load to be lifted;
- its weight, shape and centre of gravity and availability of lifting points;
- positioning of the load before and after lifting;
- how often the lifting equipment will be used to carry out the task;
- where the lifting equipment will be used; and
- the personnel available and their knowledge, training and experience (ie competence).

Planning of individual operations

[L3020] Planning for routine operations will usually be a matter for the persons undertaking the task, such as a trained and certified forklift operator. The competent person carrying out this exercise will have to have appropriate knowledge and experience.

The HSE provides the following example of a simple plan for routine use of an overhead travelling crane:

(a) assess the weight and size of the load;
(b) choose the right accessory for lifting, eg depending upon the nature and weight of the load and the environment in which it is to be used;
(c) check the anticipated path of the load to make sure that it is not obstructed;
(d) prepare a suitable place to set down the load;
(e) fit the sling to the load (using an appropriate method of slinging);
(f) make the lift (a trial lift may be necessary to confirm the centre of gravity of the load; tag lines may be necessary to stop the load swinging);
(g) release the slings (boards or similar may be necessary to prevent trapping of the sling); and
(h) clear up.

It says that the same principles could be applied to other routine lifting operations involving other types of lifting equipment, eg forklift truck, use of an electric hoist etc.

For routine operations the initial plan will only be required for the first lifting operation. This will however need to be reviewed to ensure that nothing has changed and that the plan has remained valid.

A standard plan can be used for routine similar lifting operations as long as it is periodically reviewed to make sure that nothing has changed and the 'plan' remains valid.

Examples where standard lifting plans could be provided for routine lifting operations may include:

- fork-lift trucks in a warehouse;
- a construction site hoist;
- a mobile elevated work platform (MEWP) used for general maintenance; and
- a suspended cradle used for window cleaning;
- a vehicle tail lift; and
- a patient hoist.

[*Lifting Operations and Lifting Equipment Regulations 1998 (LOLER) (SI 1998 No 2307, Reg 8*].

Hemel Hempstead-based Parker Hannifin Manufacturing Ltd was fined £1 million after a worker was crushed to death by falling machinery. Colin Reddish was involved in moving a large CNC milling machine within the company's Grantham factory in April 2015 when it overturned and killed him. The machine had been lifted using jacks and placed onto skates in order to give Mr Reddish access to use an angle grinder to cut and remove the bolts that had secured it to the floor. He was working alone at the time of the incident. Lincoln Magistrates Court heard how the company had not ensured that workers who were tasked with lifting and moving the machine were sufficiently trained and had the right experience and training for carrying out such a potentially dangerous activity.

An HSE investigation found that the work had not been properly planned, the centre of gravity of the machine had not been properly assessed and taken into account before the move took place, and that this resulted in an unsafe system of work being used for the job, with fatal consequences.

The company pleaded guilty to breaching *Reg 3(1) of Management of Health and Safety at Work Regulations 1999* and s 2(1) of the Health and Safety of Work etc Act 1974 and was also ordered to pay full costs of £6,311 and a victim surcharge of £120.

HSE Inspector Martin Giles said the death was entirely preventable.

'Parker Hannifin Manufacturing Ltd had already tried unsuccessfully to lift the machine using a fork lift truck but instead of learning from this failure they carried on. Their ad hoc approach to managing dangerous tasks resulted in one of their workers losing his life,' he said. 'All companies can learn from this incident and make sure they have properly risk assessed the situation before they start and that they have trained staff with the right type of experience to carry out the task in hand safely. Taking an extra few minutes to properly think through a problem could save a worker's life.'

Thorough examination and inspection

[L3021] An employer is under a duty to ensure:

(i) before lifting equipment is put into service for the first time by him, that it is thoroughly examined for any defect unless:
— the lifting equipment has not been used before; and
— in the case of lifting equipment for which an EC declaration of conformity could or (in the case of a declaration under the *Lifts Regulations 1997 (SI 1997 No 831)* should have been drawn up, the employer has received such declaration made not more than twelve months before the lifting equipment is put into service; or
— if obtained from the undertaking of another person, it is accompanied by physical evidence referred to in (iv) below;
[*Lifting Operations and Lifting Equipment Regulations 1998 (LOLER) (SI 1998 No 2307), Reg 9(1)*].

(ii) where the safety of lifting equipment depends on the installation conditions, that it is thoroughly examined – after installation and before being put into service for the first time and after assembly and before being put into service at a new site or in a new location, to ensure that it has been correctly installed and is safe to operate [*SI 1998 No 2307, Reg 9(2)*];

(iii) that lifting equipment which is exposed to conditions causing deterioration which is liable to result in dangerous situations is: thoroughly examined;
— at least every six months, in the case of lifting equipment for lifting persons or an accessory for lifting;
— at least every twelve months, in the case of other lifting equipment; or
— in either case, in accordance with an examination scheme; and
— whenever exceptional circumstances which are liable to jeopardise the safety of the lifting equipment have occurred, and
if appropriate for the purpose, is inspected by a competent person at suitable intervals between thorough examinations,
to ensure that health and safety conditions are maintained and that any deterioration can be detected and remedied in good time;
[*SI 1998 No 2307, Reg 9(3)*].

(iv) that no lifting equipment leaves his undertaking; or, if obtained from the undertaking of another person, is used in his undertaking, unless it is accompanied by physical evidence that the last thorough examination required to be carried out under this regulation has been carried out. [*SI 1998 No 2307, Reg 9(4)*].

Competent person

[L3022] The competent person responsible for undertaking the thorough examination must remain objective in their approach to this activity and be totally independent of any commercial or other conflicts of interest in the item of lifting equipment that they are to examine.

External competent persons eg Engineering Insurance Companies are expected to be accredited to the relevant BS EN ISO/IEC 17020 *Conformity Assessment. Requirements for the operation of various types of bodies performing inspection standard.*

Thorough examination

[L3023] The employer (owner of the equipment) must determine the level of thorough examination based on manufacturer's information and other factors. The employer may need to seek specialist advice to comply with this requirement. Factors that must be taken into account by the employer should include the work being carried out, any specific risks at the location that may affect the condition of the equipment, and the intensity of use of the equipment.

A thorough examination may include visual examination, a strip down of the equipment and functional tests. Advice should be sought from manufacturer's instructions, and a competent person for guidance on what an inspection should include for each piece of equipment.

A thorough examination must be carried out under the following circumstances.

- When the equipment is put into service for the first time. If it is new equipment that has not been used before there should be a declaration of conformity, which confirms that the equipment has undergone a thorough examination.
- Where safety depends upon the conditions after installation and before being used for the first time.

If the equipment is obtained from another undertaking, as in the case of hired equipment, then a copy of the previous certificate of thorough examination must accompany the equipment.

Installed equipment applies to lifting equipment erected or built on site, such as tower cranes, hoists or gantry cranes ie lifting equipment which is intended to be there for a period of time. It would not apply to equipment such as a mobile crane as it is not 'installed'. For equipment such as a mobile crane, there must be a copy of the previous thorough examination certificate.

Thorough examination periods

[L3024]

(i) Lifting equipment for lifting persons must be thoroughly examined at least every six months.
(ii) Lifting accessories must be thoroughly examined at least every six months.
(iii) All other lifting equipment must be thoroughly examined at least every twelve months
(iv) Each time that exceptional circumstances that are liable to jeopardise the safety of the lifting equipment have occurred, a thorough examination is required. These may include:

- any accident involving lifting equipment, eg collapse of a crane boom;
- after extended periods without proper servicing or maintenance; and
- after long periods out of use after substantial modifications and repairs.

In general **below the hook** items are examined at six monthly intervals with **above the hook** items examined at twelve monthly intervals unless the equipment is used for lifting persons in which case the lifting equipment defaults to a six monthly interval.

The employer may decide to examine specific items of lifting equipment at different intervals in accordance with an examination scheme approved by a competent person eg an occasionally used overhead crane could be examined depending on the number of lifting operations, stipulated in the scheme of examination which may be greater than the previous frequency.

In June 2015, Wrexham-based coach company GHA Coaches was fined £90,000 after repeatedly failing to comply with legal notices to get its lifting equipment examined. Mold Magistrates' Court heard that, between April 2014 and August 2015, the company failed to have its lifting equipment thoroughly examined within the required timescales to ensure that health and safety conditions were maintained and that any deterioration could be detected and remedied in good time.

An inspection carried out in 2015 found overdue LOLER examinations on at least 14 items. An improvement notice was served, and extended twice, and still resulted in a failure to comply. An HSE investigation found that a previous improvement notice was served in 2011. The company pleaded guilty to breaching *Reg 9(3)(a)(ii)* of the *Lifting Operations and Lifting Equipment Regulations 1998 (LOLER)* and failing to comply with an Improvement Notice.

Reports and defects

[L3025]

(i) A person making a thorough examination for an employer under *Lifting Operations and Lifting Equipment Regulations 1998 (LOLER) (SI 1998 No 2307), Reg 9* must:
- immediately notify the employer of any defect in the lifting equipment which in his opinion is or could become a danger to anyone;
- write a report of the thorough examination (see **L3022** below) to the employer and any person from whom the lifting equipment has been hired or leased;
- where there is in his opinion a defect in the lifting equipment involving an existing or imminent risk of serious personal injury send a copy of the report to the relevant enforcing authority ('relevant enforcing authority' means, where the defective lifting

equipment has been hired or leased by the employer, the Health and Safety Executive – and otherwise, the enforcing authority for the premises in which the defective lifting equipment was thoroughly examined).

(ii) A person making an inspection for an employer under *SI 1998 No 2307, Reg 9* must immediately notify the employer of any defect in the lifting equipment which in his opinion is or could become a danger to anyone, and make a written record of the inspection.

(iii) Every employer who has been notified under (i) above must ensure that the lifting equipment is not used before the defect is rectified; or, in the case of a defect which is not yet but could become a danger to persons, after it could become such a danger.

[SI 1998 No 2307, Reg 10].

Prescribed information

[L3026] *Lifting Operations and Lifting Equipment Regulations 1998 (LOLER) (SI 1998 No 2307), Sch 1* specifies the information that must be contained in a report of a thorough examination, made under *SI 1998 No 2307, Reg 10* (see **L3021** above). This information can be stored electronically in a secure manner eg read only with write access protected, and available only to the competent person.

The information within the report of thorough examination must include the following.

(i) The name and address of the employer for whom the thorough examination was made.
(ii) The address of the premises at which the thorough examination was made.
(iii) Particulars sufficient to identify the lifting equipment including its date of manufacture, if known.
(iv) The date of the last thorough examination.
(v) The safe working load of the lifting equipment or, where its safe working load depends on the configuration of the lifting equipment, its safe working load for the last configuration in which it was thoroughly examined.
(vi) In respect of the first thorough examination of lifting equipment after installation or after assembly at a new site or in a new location:
— that it is such thorough examination; and
— if in fact this is so, that it has been installed correctly and would be safe to operate.
(vii) In respect of all thorough examinations of lifting equipment which do not fall within (vi) above:
(a) whether it is a thorough examination under *SI 1998 No 2307, Reg 9(3)*:
— within an interval of six months;
— within an interval of twelve months;
— in accordance with an examination scheme; or
— after the occurrence of exceptional circumstances;
(b) if in fact this is so, that the lifting equipment would be safe to operate.

(viii) In respect of every thorough examination of lifting equipment:
 (a) identification of any part found to have a defect which is or could become a danger to anyone, and a description of the defect;
 (b) particulars of any repair, renewal or alteration required to remedy a defect found to be a danger to anyone;
 (c) in the case of a defect which is not yet but could become a danger to anyone:
  — the time by which it could become such a danger;
  — particulars of any repair, renewal or alteration required to remedy the defect;
 (d) the latest date by which the next thorough examination must be carried out;
 (e) particulars of any test, if applicable; and
 (f) the date of the thorough examination.
(ix) The name, address and qualifications of the person making the report; that he is self-employed or, if employed, the name and address of his employer.
(x) The name and address of a person signing or authenticating the report on behalf of its author.
(xi) The date of the report.

Keeping of information

[L3027] An employer who obtains lifting equipment to which the 1998 Regulations apply, and who receives an EC declaration of conformity relating to it, must keep the declaration for so long as he operates the lifting equipment [*Lifting Operations and Lifting Equipment Regulations 1998 (LOLER) (SI 1998 No 2307), Reg 11(1)*].

SI 1998 No 2307, Reg 11(2)(a) provides that the employer must ensure that the information contained in every report made to him under *SI 1998 No 2307, Reg 10(1)* (see **L3025** above) is kept available for inspection:

— in the case of a thorough examination under *SI 1998 No 2307, Reg 9(1)* (see **L3021** above) of lifting equipment other than an accessory for lifting, until he stops using the lifting equipment;
— in the case of a thorough examination under *SI 1998 No 2307, Reg 9(1)* (see **L3021** above) of an accessory for lifting, for two years after the report is made;
— in the case of a thorough examination under *SI 1998 No 2307, Reg 9(2)* (see **L3021** above), until he stops using the lifting equipment at the place it was installed or assembled;
— in the case of a thorough examination under *SI 1998 No 2307, Reg 9(3)* (see **L3021** above), until the next report is made under that paragraph or the expiration of two years, whichever is later.

The employer must ensure that every record made under *SI 1998 No 2307, Reg 10(2)* (see **L3025** above) is kept available until the next such record is made [*SI 1998 No 2307, Reg 11(2)(b)*].

Lifting operations and equipment failure

General

[L3028] Lifting equipment and accessories are not subjected to frequent failure but should failure occur it is often catastrophic. Lifting equipment is subject to statutory inspection to ensure that a fitness for purpose examination has been carried out. This examination must be current throughout the equipment's working life. The *Lifting Operations and Lifting Equipment Regulations 1998 (LOLER) (SI 1998 No 2307)* have placed further demands than those that previously existed under repealed lifting legislation in that *SI 1998 No 2307, Reg 9(3)(b)* requires that if appropriate, a competent person undertakes an inspection between thorough examinations (See **L3021** above). The specific requirements of this intermediate inspection are based on risk exposure and subsequent determination of the equipment.

Some failures do however still occur and they are often attributed to a combination of procedural and hardware faults.

Examples of specific causes of failure are examined in the following chapters. These examples look at possible causations, but they are by no means exhaustive and human error will almost certainly be a factor in most failures.

Cranes

[L3029] Cranes are widely used in lifting/lowering operations in construction, dock and shipbuilding works. The main hazard, generally associated with overloading or incorrect slewing, is collapse or overturning. Contact with overhead power lines is also a danger. In such cases, the operator should normally remain inside the cab and not allow anyone to touch the crane or load; the superintending engineer should immediately be informed. (For reportable dangerous occurrences in connection with cranes etc. see ACCIDENT REPORTING at **A3027**.)

There are a number of different types of cranes available to perform varying tasks and while there are variants they generally fall into the following categories.

(i) Fixed crane.
(ii) Tower crane.
(iii) Mobile crane.
(iv) Overhead travelling crane.

There are in addition, a hybrid mobile tower crane (also referred to as self-erecting tower cranes) that has seen increasing popularity in recent years, on construction sites, rough terrain cranes as well as crawler and wheeled cranes. Other variants include ship to shore cranes (portainers) in use at container terminals.

A statutory tower crane registration scheme was introduced in April 2010 but was subsequently revoked by the *Health and Safety (Miscellaneous Repeals, Revocations and Amendments) Regulations 2013*.

Safe lifting operations by mobile cranes

[L3030] A mobile crane is a crane which is capable of travelling under its own power.

British Standard BS 7121-:2016 *Code of practice for the safe use of cranes - Inspection, maintenance and through examination - Mobile cranes* includes recommendations for pre-use checks, in-service inspection, maintenance, thorough examination (in service and following exceptional circumstances) and supplementary testing of mobile cranes, including truck cranes, wheeled rough terrain cranes, wheeled yard cranes, crawler mounted cranes, mini-cranes and vehicle mounted self-erecting tower cranes. This BS is broken into several parts, with Parts 1, 3, 4, 5, 7 and 9 being current.

I. BS 7121-2-1:2016 Code of practice for safe use of cranes. General
II. BS 7121-2-3:2017 Code of practice for safe use of cranes. Mobile cranes
III. BS 7121-2-4:2013 Code of practice for the safe use of cranes. Inspection, maintenance and thorough examination. Loader cranes
IV. BS 7121-2-5:2006 Code of practice for safe use of cranes. Tower cranes
V. BS 7121-2-7:2012+A1:2015. Code of practice for the safe use of cranes. Inspection, maintenance and thorough examination. Overhead travelling cranes, including portal and semi-portal cranes, hoists, and their supporting structures
VI. BS 7121-2-9:2013 Code of practice for the safe use of cranes. Inspection, maintenance and thorough examination. Cargo handling and container cranes

The important areas addressed within BS 7121 include safe lifting operations depend on co-operation between supervisor (or appointed person), slinger (and/or signaller) and crane driver:

Appointed person
(a) Overall control of lifting operations rests with an 'appointed person', who can, where necessary, stop the operation. Failing this, control of operations will be in the hands of the supervisor (who, in some cases, may be the slinger). The appointed (and competent) person must ensure that:
  (i) lifting operations are carefully planned and executed;
  (ii) weights and heights are accurate;
  (iii) suitable cranes are provided;
  (iv) the ground is suitable;
  (v) suitable precautions are taken, if necessary, regarding gas, water and electricity either above or below ground;
  (vi) personnel involved in lifting/lowering are trained and competent; and
  (vii) access within the vicinity at where the lifting operation is undertaking is minimised.

Supervisor
(b) Supervisors must:
  (i) direct the crane driver where to position the crane;
  (ii) provide sufficient personnel to carry out the operation;
  (iii) check the site conditions;
  (iv) report back to the appointed person in the event of problems;

(v) supervise and direct the slinger, signaller and crane driver; and
(vi) stop the operation if there is a safety risk.

Crane driver
(c) Crane drivers must:
(i) erect/dismantle and operate the crane as per the manufacturer's instructions;
(ii) set the crane level before lifting and ensure that it remains level;
(iii) decide which signalling system is to apply;
(iv) inform the supervisor in the event of problems; and
(v) carry out inspections/weekly maintenance relating to:
— defects in crane structure, fittings, jibs, ropes, hooks, shackles; and
— correct functioning of automatic safe load indicator, over hoist and derrick limit switches.

Drivers should always carry out operations as per speeds, weights, heights and wind speeds specified by the manufacturer, mindful that the weight of slings/lifting gear is part of the load. Such information should be clearly displayed in the cab and not obscured or removed. Windows and windscreens should be kept clear and free from stickers containing operational data.

Any handrails, stops, machinery guards fitted to the crane for safe access should always be replaced following removal for maintenance; and tools, jib sections and lifting tackle properly secured when not in use.

After a load has been attached to the crane hook by the lifting hook, tension should be taken up slowly, as per the slinger's instructions, the latter being in continuous communication with the driver. In the case of unbalanced loads, drivers/slingers should be familiar with a load's centre of gravity - particularly if the load is irregularly shaped. Once in operation, crane hooks should be positioned directly over the load, the latter not remaining suspended for longer than necessary. Moreover, suspended loads should not be directed over people or occupied buildings.

Slingers
(d) Slingers must:
(i) attach/detach a load to/from the crane;
(ii) use correct lifting appliances;
(iii) direct movement of a load by correct signals. Any part of a load likely to shift during lifting/lowering must be adequately secured by the slinger beforehand. Spillage or discharge of loose loads (eg scaffolding) can be a problem, and such loads must be properly secured/fastened. Nets are useful for covering palletised loads (eg bricks).

Signallers
(e) Quite frequently, signallers are responsible for signalling in lieu of slingers. Failing this, their remit is to transmit instructions from slinger to crane driver, when the former cannot see the load.

Crane failure

[L3031] Crane failure may be attributed to:

(a) failure to lift vertically, eg dragging a load sideways along the ground before lifting;
(b) 'snatching' loads, ie not lifting slowly and smoothly;
(c) exceeding the maximum permitted moment, ie the product of the load and the radius of operation;
(d) excessive wind loading, resulting in crane instability;
(e) defects in the fabrication of the crane, eg badly-welded joints;
(f) incorrect crane assembly in the case of tower cranes;
(g) brake failure (rail-mounted cranes); or
(h) in the case of mobile cranes:
  (i) failure to use outriggers,
  (ii) lifting on soft or uneven ground, and
  (iii) incorrect tyre pressures.

In March 2016, national crane hire company Falcon Crane Hire Ltd was fined £750,000 and ordered to pay costs of £100,000 for failings that led to the death of two men when a crane collapsed in London in 2006. The company was sentenced for breaches of ss 2 and 3 of the Health and Safety at Work etc Act 1974.

Southwark Crown Court heard that crane operator Jonathan Cloke died after falling from the crane as it collapsed, also killing a member of the public Michael Alexa. The court heard that sections of the tower crane, which was on a housing development in Battersea, separated when 24 bolts failed due to metal fatigue. The 24 bolts were a significant safety feature on the crane's slew ring, which connected the mast (tower) to the slew turret, allowing the arms of the crane (jib) to rotate through 360 degrees. When the bolts failed the slew turret and jib separated from the mast and fell to the ground.

The HSE investigation into the incident found that Falcon Crane Hire Ltd did not investigate a similar incident which happened nine weeks before, when the bolts failed on the same crane and had to be replaced. It found that the company had an inadequate system to manage the inspection and maintenance of its fleet of cranes and that its process to investigate the underlying cause of components' failings was also inadequate. The court heard how the particular bolts were a safety critical part of the crane and the bolts failing previously was an exceptional and significant occurrence, which should have been recognised by the company. HSE Head of Operations Mike Wilcock highlighted Falcon Crane Hire's failure to take decisive action when the bolts failed the first time.

Hoists and lifts

[L3032] The safe use and maintenance of hoists and lifts is governed primarily by the *Lifting Operations and Lifting Equipment Regulations 1998 (SI 1998 No 2307)*.

Examples include: goods lifts; man hoists, scissors lifts and passenger lifts.

Powered working platforms are commonly used for fast and safe access to overhead machinery/plant, stored products, lighting equipment and electrical installations as well as for enabling maintenance operations to be carried out on high-rise buildings. Their height, reach and mobility give them distinct advantages over scaffolding, boatswain's chairs and platforms attached to fork lift trucks. Typical operations are characterised by self-propelled hydraulic booms, semi-mechanised articulated booms and self-propelled scissors lifts.

Platforms should always be sited on firm level working surfaces and their presence indicated by traffic cones and barriers. Location should be away from overhead power lines – but, if this is not practicable the safe system of work in operation needs to control this additional risk. A key danger arises from overturning as a consequence of overloading the platform. Maximum lifting capacity should, therefore, be clearly indicated on the platform as well as in the manufacturer's instructions, and it is inadvisable to use working platforms in high winds (ie above Force 4 or 16 mph). Powered working platforms should be regularly maintained and only operated by trained personnel.

Hoist and lift failure

[L3033] Hoist and lift failure can be attributed to:

(a) excessive wear in wire ropes;
(b) excessive broken wires in ropes;
(c) mechanical failure (eg worm and gearing);
(d) failure of the overload protection device;
(e) failure of the overrun device;
(f) failure of the speed governor;
(g) safety gear failure; or
(h) failure of the landing door interlocks.

Fork lift trucks

[L3034] Fork lift trucks are the most widely used item of mobile mechanical handling equipment. There are several varieties which are as follows:

(1) *Pedestrian-operated stackers – manually-operated and power-operated*
Manually-operated stackers are usually limited in operation, for example, for moving post pallets, and cannot pick up directly from the floor. Power-operated stackers are pedestrian-operated or rider-controlled, operate vertically and horizontally and can lift pallets directly from the floor.

(2) *Reach trucks*
Reach trucks enable loads to be retracted within their wheel base. There are two kinds, namely, (a) moving mast reach trucks, and (b) pantograph reach trucks. Moving mast reach trucks are rider-operated, with forward-mounted load wheels enabling the carriage to move within the

wheel base – mast, forks and load moving together. Pantograph reach trucks are also rider-operated, reach movement being by pantograph mechanism, with forks and load moving away from static mast.

(3) *Counterbalance trucks*
Counterbalance trucks carry loads in front counterbalanced to the weight of the vehicle over the rear wheels. Such trucks are lightweight pedestrian-controlled, lightweight rider-controlled or heavyweight rider-controlled.

(4) *Narrow aisle trucks*
With narrow aisle trucks the base of the truck does not turn within the aisle in order to deposit/retrieve load. There are two types, namely, side loaders for use on long runs down narrow aisles, and counterbalance rotating load turret trucks, having a rigid mast with telescopic sections, which can move sideways in order to collect/deposit loads.

(5) *Order pickers*
Order pickers have a protected working platform attached to the lift forks, enabling the driver to deposit/retrieve objects in or from a racking system. Conventional or purpose-designed, they are commonly used in racked storage areas and operate well in narrow aisles.

Forklift trucks are involved in far too many incidents although manufacturers have increased the number of safety devices in an attempt to limit human error. These include:

- speed limiters with the forks elevated;
- pressure sensitive seats, that cut off the fuel supply when the driver disembarks;
- TV monitors when the truck has obscured vision;
- intelligent start up systems, swipe cards or similar;
- lap straps; and
- falling object protection.

Forklift trucks are mobile work equipment and with the exception of rough terrain vehicles all are protected from rolling over beyond 90 degrees should an overturn occur by their masts. Lap straps should be used when fitted although the requirement for their use is risk based.

Forklift truck failure

[L3035] Forklift truck failure can be attributed to:

(a) uneven floors, steeply inclined ramps or gradients, ie in excess of 1:10 gradient;
(b) inadequate room to manoeuvre;
(c) inadequate or poor maintenance of lifting gear;
(d) the practice of driving forwards down a gradient with the load preceding the truck;
(e) load movement in transit;
(f) sudden or fast braking;
(g) poor stacking of goods being moved;
(h) speeding;
(i) turning corners too sharply;

(j) not securing load sufficiently eg pallets stacked poorly;
(k) hidden obstructions in the path of the truck;
(l) use of the forward tilt mechanism with a raised load; or
(m) generally bad driving, including driving too fast, taking corners too fast, striking overhead obstructions, particularly when reversing and excessive use of the brakes.

Patient/bath hoist

[L3036] Patient/bath hoists are used for the lifting of persons and require a six-month thorough examination.

Patient hoist failure:

(a) mechanical damage to pulleys sheaves and drums;
(b) anchor point failure;
(c) breaking mechanism failure;
(d) boom arm distortion or corrosion;
(e) electrical faults; and
(f) hook catch failure.

Vehicle lifting table

[L3037] The primary function is as a vehicle lift but personnel are at risk when the vehicle is descending. By definition a twelve-monthly thorough examination would normally satisfy the inspection frequency laid down in the *Lifting Operations and Lifting Equipment Regulations 1998 (SI 1998 No 2307)*, but most of industry has adopted a six-monthly examination frequency. Lifting tables are normally either wire suspension type or hydraulic actuated.

Vehicle lifting table failure:

(a) mechanical damage to pulleys sheaves and wires;
(b) anchor point failure;
(c) breaking mechanism failure;
(d) electrical faults;
(e) hydraulic failure;
(f) foundation bolt failure;
(g) catching device failure causing vehicles to fall from table; and
(h) carrying arm failure.

Lifting accessories

[L3038] 'Accessory for lifting' – is defined in LOLER as lifting equipment for attaching loads to machinery for lifting and includes pendants, slings and shackles.

[L3038] Lifting Operations

It has become common practice within industry to use a simple colour-coded cable tie to identify that the accessory is within its inspection cycle. This colour tie will normally be displayed at the lifting equipment store for ease of recognition.

Lifting accessories refer to:

(a) ropes;
(b) chains;
(c) slings;
(d) shackles etc.

Ropes

[L3039] Ropes are used quite widely throughout industry in lifting/lowering operations, the main hazard being breakage through overloading and/or natural wear and tear. Ropes are either of natural (eg cotton, hemp) or man-made fibre (eg nylon, terylene). Natural fibre ropes, if they become wet or damp, should be allowed to dry naturally and kept in a well-ventilated room. They do not require certification prior to service, but six-monthly examination thereafter is compulsory (all lifting equipment accessories require a six-monthly examination to meet the requirements under the *Lifting Operations and Lifting Equipment Regulations 1998*). If the specified safe working load (SWL) cannot be maintained by a rope, it should be withdrawn from service immediately. Of the two, man-made fibre ropes have greater tensile strength, are not subject so much to risk from wear and tear, are more acid/corrosion-resistant and can absorb shock loading better.

Fibre rope failure

[L3040] Fibre rope failure can be attributed to:

(a) bad storage in wet or damp conditions resulting in rot and mildew;
(b) inadequate protection of the rope when lifting loads with sharp edges;
(c) exposure to direct heat to dry, as opposed to gradual drying in air;
(d) chemical reaction; or
(e) Ultra Violet degradation.

Wire (steel) ropes

[L3041] Wire ropes are much in use in cranes, lifts, hoists, elevators etc and as such become part of the lifting equipment. They are also commonly utilised as accessories and they need to be lubricant-impregnated to minimise corrosion and reduce wear and tear unless they are manufactured from corrosive resistive materials for specialist applications. BS EN 12385-1:2002+A1:2008 sets out general safety requirements for steel wire ropes.

Wire rope failure(s)

[L3042] There have been instances where ropes have had to be changed due to visible damage to the wires. Ropes showing no external visible sign of damage have failed during normal operations with catastrophic consequences as a result of fatigue failure, often brought about by reverse cycle leading which significantly reduces the ropes operational life.

Wire rope failure can be attributed to:

(a) excessive broken wires;
(b) failure to lubricate regularly;
(c) fatigue failure of the rope;
(d) frequent knotting or kinking of the rope; or
(e) bad storage in wet or damp conditions which promotes rust.

Chain

[L3043] A chain is a classic example of a lifting accessory, in spite of an increase in the use of wire ropes. There are several varieties, including mild, high-tensile, and alloy steel chains. The principal risk is breakage, usually occurring in consequence of a production defect in a link, or through application of excessive loads. A crucial part of the thorough examination is to measure the chain wear. A simple gauge which allows measurement of elongation and link/pin wear is used for this.

Chain failure

[L3044] Chain failure can be attributed to:

(a) mechanical defects in individual links;
(b) application of a static in excess of the breaking load;
(c) snatch loading; and
(d) hydrogen embrittlement (manufacturing welding defects refer HSE publication PM 39, *Hydrogen cracking of grade T and grade 8 chain and components*).

Slings

[L3045] Slings are manufactured in natural or man-made fibre or chain. The safe working load of any sling varies according to the angle formed between the legs of the sling.

The relationship between sling angle and the distance between the legs of the sling is also important. (See Table 1 below)

Table 1 Safe working load for slings	
Sling Angle	*Distance between legs*
30°	$\frac{1}{2}$ leg length
60°	1 leg length
90°	$1\frac{1}{3}$ leg length
120°	$1\frac{2}{3}$ leg length

For a one tonne load, the tension in the leg increases as shown in Table 2 below.

Table 2
Safe working load for slings – increased tension

Sling leg angle	Tension in leg (tonnes)
90°	0.7
120°	1.0
151°	2.0
171°	6.0

Causes of failure in slings include:

(a) cuts, excessive wear, kinking and general distortion of the sling legs;
(b) failure to lubricate wire slings;
(c) failure to pack sharp corners of a load, resulting in sharp bends in the sling and the possibility of cuts or damage to it; and
(d) unequal distribution of the load between the legs of a multi-leg sling.

Eyebolts

[L3046] Eyebolts are mainly for lifting heavy concentrated loads; there being several types, namely, dynamo, collar and eyebolt incorporating link.

Eyebolt failure

[L3047] Eyebolt failure can be attributed to:

(a) mechanical defects;
(b) incorrect thread (metric and imperial threads, cross threaded);
(c) dynamo eyebolt used where collar eyebolt was required;
(d) eyebolt not hardened down to the shoulder; or
(e) oil remaining in the tapped hole receiving the eyebolt, thus destroying the receiving housing under hydraulic pressure.

Shackles

[L3048] Shackles come in two main types 'Dee' shackle and 'Bow' so named because of their shapes. The bow shackle has the capability to accept a larger number of ropes, slings etc. The shackle and pin are matched pairs and should not be separated and mixed up.

Shackle failure

[L3049] Shackle failure can be attributed to:

(a) mechanical defects;
(b) interchange of pin and shackle;
(c) nut and bolt used instead of the shackle pin; or
(d) wear on the shackle.

Once only use accessories

[L3050] Also included are 'one off' accessories which are disposed of after their use and examples of these are:

- industrial bulk carriers (IBC's) bags that are commonly used in the building and chemical industry where it would not be possible certify integrity following their first use;
- health care, under-body sling used with patient hoists, the problems associated with sterilising such equipment through autoclaving when it would deteriorate and the cost of examination by a competent person have made it both a practical solution and an economic proposition to discard such equipment after use; and
- dynamo eyebolt, are designed for a single lift only.

Mobile lifting equipment

[L3051] Mobile work equipment is any equipment which carries out work whilst travelling. It may be self-propelled, towed or remotely controlled and it may incorporate attachments. An example of mobile lifting equipment is an excavator involved in digging tasks.

The risk of mobile lifting equipment overturning is another cause for concern, and the problem was addressed by the additional requirements introduced in the *Provision and Use of Work Equipment Regulations 1998 (SI 1998 No 2306)*. These implement into UK legislation the amending directive to the *Use of Work Equipment Directive (AUWED)*. Under this legislation, workers must be protected from falling out of the equipment and from unexpected movement eg overturning.

To prevent such equipment overturning, the following action should be taken:

— increase stability when reasonably practicable (this may be done by reducing the equipment's speed)
— fit stabilisers (eg outriggers or counterbalance weights);
— ensure that roll-over protection structure is installed which ensures that it does no more than fall on its side;
— ensure the structure gives sufficient clearance to anyone being carried if it overturns more than on its side;
— provision of harnesses or a suitable restraining system to protect against being crushed;
— use only on firm ground; and
— avoid excessive gradients.

Further information regarding lifting operations

[L3052] More information on lifting equipment and lifting operations can be found on the HSE website at: www.hse.gov.uk/work-equipment-machinery/loler.htm.

Lighting

Nicola Coote

Introduction to lighting

[L5001] Increasingly, over the last decade or so, employers have come to appreciate that indifferent lighting is both bad economics and bad ergonomics, not to mention potentially bad industrial relations. Conversely, good lighting uses energy efficiently and contributes to general workforce morale and profitability – that is, operating costs fall whilst productivity and quality improve. Alternatively, poor lighting reduces efficiency, thereby increasing the risk of stress, denting workforce morale, promoting absenteeism and leading to accidents, injuries and even deaths at work.

Ideally, good lighting should 'guarantee' (*a*) employee safety, (*b*) acceptable job performance and (*c*) good workplace atmosphere, comfort and appearance. This is not just a matter of maintenance of correct lighting levels. Ergonomically relevant are:

(a) horizontal illuminance;
(b) uniformity of illuminance over the job area;
(c) colour appearance;
(d) colour rendering;
(e) glare and discomfort;
(f) ceiling, wall, floor reflectances which magnify lighting levels;
(g) job/environment illuminance ratios;
(h) job and environment reflectances; and
(i) vertical illuminance.

Increasingly there are growing financial pressures to review lighting more from an energy conservation point of view. This can have many financial and environmental benefits, provided the energy efficient system being used does not detriment the lighting levels necessary to complete the work safely and effectively. The Chartered Institute of Building Service Engineers (CIBSE www.cibse.org) *Guide F: Energy efficiency in buildings* (chapter 9) contains further information about energy efficiency lighting.

Statutory lighting requirements

General

[L5002] Adequate standards of lighting in all workplaces can be enforced under the general duties of the *Health and Safety at Work etc Act 1974*, which require provision by an employer of a safe and healthy working environment (see EMPLOYERS' DUTIES TO THEIR EMPLOYEES). *Regulation 8* of the *Workplace (Health, Safety and Welfare) Regulations 1992 (SI 1992 No 3004)* requires that all workplaces have suitable and sufficient lighting and that, so far as is reasonably practicable, the lighting should be natural light.

More particularly, however, people should be able to work and move about without suffering eye strain (visual fatigue) and having to avoid shadows. Local lighting may be necessary at individual workstations and places of particular risk. Outdoor traffic routes used by pedestrians should be adequately lit after dark. Lights and light fittings should avoid dazzle and glare, and be so positioned that they do not cause hazards, whether fire, radiation or electrical. Switches should be easily accessible and lights should be replaced, repaired or cleaned before lighting becomes insufficient. Moreover, where persons are particularly exposed to danger in the event of failure of artificial lighting, emergency lighting must be provided (*Workplace (Health, Safety and Welfare) Regulations 1992 (SI 1992 No 3004), Reg 8(3)* (and ACOP)).

To accommodate these 'requirements', refresh rates (flickering), a combination of general and localised lighting (not to be confused with 'local' lighting, eg desk light) is often necessary. Moreover, state of the art visual display units have thrown up some occupational health problems addressed by the *Health and Safety (Display Screen Equipment) Regulations 1992 (SI 1992 No 2792) as amended by the Health and Safety (Miscellaneous) Regulations 2002* (see O5005 *et seq* OFFICES AND SHOPS).

The Chartered Institute of Building Services Engineers (CIBSE) produces a wide range of technical and industry-specific guidance on design of lighting in various environments, commissioning and installation, etc. As an example, *Lighting Guide 7: office lighting* covers all aspects of the lighting of offices, from board room through general offices to the post room. This organisation provides guidance on considerations to be taken when designing energy reduction systems for lighting in various work areas as well as recommended luminescence for different types of work environment. There are several lighting guides for a wide range of working environments and further industry specific guidance can be obtained via the CIBSE website. They provide nine main lighting guides tailored for specific industry sectors, and taking into account the nature of products, equipment and environment. They also provide guidance notes on design features, surface reflectance and colours etc. A list of the guides is below:

- SLL Lighting Guide 1: The Industrial Environment (2012);
- SLL Lighting Guide 2: Hospitals and Health Care Buildings (2008);
- SLL Lighting Guide 4: Sports (2006);
- SLL Lighting Guide 5: Lighting for Education (2011);
- SLL Lighting Guide 6: The Exterior Environment (2016);
- SLL Lighting Guide 7: Office Lighting (2005) – (including Addendum);
- SLL Lighting Guide 8: Lighting for Museums and Art Galleries (2015);
- SLL Lighting Guide 9: Lighting for Communal Residential Buildings (2013);
- SLL Lighting Guide 10: Daylighting – A Guide to Designers (2014);
- SLL Lighting Guide 11: Surface Reflectance and Colour (2001);

Statutory lighting requirements **[L5005]**

- SLL Lighting Guide 11 Chart: Reflectance Sample Chart;
- SLL Lighting Guide 12: Emergency Lighting NEW (2015);
- SLL Lighting Guide 13: Lighting for Places of Worship (2014);
- Guide to Limiting Obtrusive Light (2012); and
- Guide to the Lighting of Licensed Premises (2011).

The Society of Light and Lighting is a part of the Chartered Institute of Building Service Engineers and acts as the professional body for lighting in the UK. It operates worldwide and offers a range of publications and advice on specific areas of lighting.

HSE Guidance document HSG38 Lighting at Work provides the good practice standards that should be maintained in a workplace in relation to lighting levels.

Specific processes

Lighting and VDUs

[L5003] Annex A to BS EN 9241-6-1999 provides guidance on illuminance levels, specifically in relation to use with display screen equipment (BS EN 9241 – Ergonomic requirements for office work with visual display terminals (VDTs); Part 6 – Guidance on work environment). BS EN 12464-1–2002 (Light and lighting. Lighting of work places, Indoor work places) also provides advice and guidance in this respect. The other key professional bodies who advise on lighting design are the Professional Lighting Designers' Association (PLDA) and the International Association of Lighting Designers (IALD).

Machinery and work equipment – Provision and Use of Work Equipment Regulations 1998 (SI 1998 No 2306)

[L5004] General lighting requirements for work with machinery comes under the *Provision and Use of Work Equipment Regulations 1998 (SI 1998 No 2306), Reg 21*. This Regulation is very broad and merely requires that suitable and sufficient lighting is provided, taking into account the operations to be carried out. To ascertain what is required to meet this duty, the HSE guidance has been produced. This refers to occasions when additional or localised lighting may be needed such as:

- when the task requires a high perception of detail;
- when there is a dangerous process;
- to reduce visual fatigue;
- during maintenance operations.

Additional guidance is provided in the HSE Guidance Note HS(G) 38, 'Lighting at Work' and in the CIBSE Lighting Guide 1 – The Industrial Environment 2012 (SLL LG1).

Lighting requirements on construction sites – Construction (Design & Management) Regulations 2015 (SI 2015 No 51) (CDM Regulations)

[L5005] Lighting requirements for construction sites can be found in Regulation 35. These regulations emulate those requirements given in the Workplace (Health, Safety & Welfare) Regulations 1992 (SI 1992 No 3004) and require that every place of work, including traffic routes around the construction site and the approaches to the workplace must be provided with suitable and sufficient lighting, which shall be, so far as is reasonably practicable, by natural light. The colour of any artificial lighting provided must not adversely affect or change the perception of any sign or signal provided for the purposes of health and safety.

Suitable and sufficient secondary lighting shall be provided in any place where there would be a risk to the health or safety of any person in the event of failure of primary artificial lighting. The welfare requirements under the CDM Regulations (Schedule 2) further stipulate that washing and sanitary facilities must be adequately lit.

Lighting requirements for electrical equipment

[L5006] So as to prevent injury, adequate lighting must be provided at all electrical equipment on which or near which work is being done in circumstances that may give rise to danger (*Electricity at Work Regulations 1989 (SI 1989 No 635), Reg 15*). The most recent HSE guidance on these regulations (HSR25 (Third edition) published 2015) stipulates that the requirement to provide adequate lighting is not restricted to those circumstances where live conductors are exposed, but applies where any work is being done in circumstances which may give rise to danger. It further states that lighting is not required to be provided at times other than when work is being done, so this would allow for portable lighting to be used and then removed when a dangerous electrical activity is being completed.

Sources of light

Natural lighting

[L5007] Daylight is the natural, and cheapest, form of lighting, but it has only limited application to places of work where production is required beyond the hours of daylight, at all seasons, and where daytime visibility is restricted by climatic conditions. But however good the outside daylight, windows can rarely provide adequate lighting alone for the interior of large floor areas. Single storey buildings can, of course, make use of insulated opaque roofing materials, but the most common provision of daylight is by side windows. There is also the fact that the larger the glazing area of the building, the more other factors such as noise, heat loss in winter and unsatisfactory thermal conditions in summer must be considered.

Modern conditions, where the creation of a pleasant building interior environment requires the balanced integration of lighting, heating, air conditioning, acoustic treatment, etc, are such that lighting cannot be considered in isolation. At the very least, natural lighting will have to be supplemented for most of the time with artificial lighting, the most common source for which is electric lighting.

Sources of light [L5009]

Solar lighting

[L5008] This technology is still fairly new but is beginning to expand rapidly, and government grants are now frequently available to assist with installation. However, they still provide quite limited results in a number of workplace environments. There are some environments when use of solar lighting is useful and effective. Examples include isolated satellite work premises that are occupied for short periods, some outdoor areas for additional evening lighting and areas where low levels of luminance are sufficient. CIBSE issued guidance on this (guide J: *Weather, Solar and Illuminance Data*) which was withdrawn in 2015 and is currently being updated.

Electric lighting

[L5009] Capital costs, running costs and replacement costs of various types of electric lighting have a direct bearing on the selection of the sources of electric lighting for particular application. Such costs are as important a consideration as the size, heat and colour effects required of the lighting. The efficiency of any type of lamp used for lighting is measured as light output, in lumens, per watt of electricity. Typical values for various types of lamp are as follows (the term 'lumen' is explained in the discussion of standards of illuminance in L5010 below).

Type of lamp	Lumens per watt
Incandescent lamps	10 to 18
Tungsten halogen	22
High pressure mercury	25 to 55
Tubular fluorescent	30 to 80 (depending on colour)
Mercury halide	60 to 80
High pressure sodium	100
LED	45

In general, the common incandescent lamps (coiled filament lamps, the temperature of which is raised to white heat by the passage of current, thus giving out light) are relatively cheap to install but have relatively expensive running costs. The manufacture of incandescent bulbs was phased out in the UK from 2011 but some remain in use. A discharge or fluorescent lighting scheme (which works on the principle of electric current passing through certain gases and thereby producing an emission of light) has higher capital costs but higher running efficiency, lower running costs and longer lamp life.

In larger places of work the choice is often between discharge and fluorescent lamps. The normal mercury discharge lamp and the low pressure sodium discharge lamp have restricted colour performance, although newly developed high pressure sodium discharge lamps and colour corrected mercury lamps do not suffer from this disadvantage.

LED (light emitting diodes) are bulbs without a filament. They are low in power consumption and have a long life span. LEDs are just starting to rival conventional lighting, but unfortunately they just do not provide the output (lumen) needed to completely replace incandescent, and other type, bulbs at the present time. They are particularly efficient and safe where localised and direct lighting is required.

Standards of lighting or illuminance

The technical measurement of illuminance

[L5010] The standard of illuminance (the amount of light) required for a given location or activity depends on a number of variables, including general comfort considerations and the visual efficiency required. The unit of illuminance is the 'lux', which equals one lumen per square metre; this unit has now replaced the 'foot candle' which was the number of lumens per square foot. The term 'lumen' is the unit of luminous flux, describing the quantity of light received by a surface or emitted by a source of light.

Light measuring instruments

[L5011] For accurate measurement of the degree of illuminance at a particular working point, a reliable instrument is required. Such an instrument, suitable for most measurements, is a pocket light meter, often referred to as a lux meter; this incorporates the principle of the photo-electric cell, which generates a tiny electric current in proportion to the light at the point of measurement. This current deflects a pointer on a graduated scale measured in lux. Manufacturers' instructions should, of course, be followed in the care and use of such instruments. Lux meters with 10% margin of error are now available on the market and can provide a quick and relatively cheap method for identifying the range of light levels within a workspace.

Average illuminance and minimum measured illuminance

[L5012] HSE Guidance Note HS(G) 38 'Lighting at work' (1998) relates illuminance levels to the degree or extent of detail which needs to be seen in a particular task or situation. Recommended illuminances are shown in Table 1 at L5013 below.

This guidance note makes recommendations both for average illuminance for the work area as a whole and for minimum measured illuminance at any position within it. As the illuminance produced by any lighting installation is rarely uniform, the use of the average illuminance figure alone could result in the presence of a few positions with much lower illuminance, which pose a threat to health and safety. The minimum measured illuminance is therefore the lowest illuminance permitted in the work area taking health and safety requirements into account.

The planes on which the illuminances should be provided depend on the layout of the task. If predominantly on one plane, for example horizontal, as with an office desk, or vertical, as in a warehouse, the recommended illuminances are recommended for that plane. Where there is either no well-defined plane or more than one, the recommended illuminances should be provided on the horizontal plane and care taken to ensure that the reflectances of surfaces in working areas are high.

Illuminance ratios

[L5013] The relationship between the lighting of the work area and adjacent areas is significant. Large differences in illuminance between these areas may cause visual discomfort or even affect safety levels where there is frequent movement, for example, forklift trucks. This problem arises most often where local or localised lighting in an interior exposes a person to a range of illuminance for a long period, or where there is movement between interior and exterior working areas exposing a person to a sudden change of illuminance. To reduce hazards and possible discomfort, specific recommendations shown in Table 1 below should be followed.

Where there is conflict between the recommended average illuminances shown in Table 1 and maximum illuminance ratios shown in Table 2, the higher value should be taken.

Table 1
Average illuminances and minimum measured illuminances for different types of work

General activity	Typical locations/types of work	Average illuminance (Lx)	Minimum measured illuminance (Lx)
Movement of people, machines and vehicles*	Lorry parks, corridors, circulation routes	20	5
Movement of people, machines and vehicles in hazardous areas; rough work not requiring any perception of detail	Construction site clearance, excavation and soil work, docks, loading bays, bottling and canning plants	50	20
Work requiring limited perception of detail**	Kitchens, factories, assembling large components, potteries	100	50
Work requiring perception of detail	Offices, sheet metal work, bookbinding	200	100
Work requiring perception of fine detail	Drawing offices, factories assembling electronic components, textile production	500	200
Notes			

*	Only safety has been considered, because no perception of detail is needed and visual fatigue is unlikely. However, where it is necessary to see detail to recognise a hazard or where error in performing the task could put someone else at risk, for safety purposes as well as to avoid visual fatigue the figure should be increased to that for work requiring the perception of detail.
**	The purpose is to avoid visual fatigue: the illuminances will be adequate for safety purposes.

Table 2
Maximum ratios of illuminance for adjacent areas

Situations to which recommendation applies	Typical location	Maximum ratio of illuminances		
		Working area		Adjacent area
Where each task is individually lit and the area around the task is lit to a lower illuminance	Local lighting in an office	5	:	1
Where two working areas are adjacent but one is lit to a lower illuminance than the other	Localised lighting in a works store	5	:	1
Where two working areas are lit to different illuminances and are separated by a barrier but there is frequent movement between them	A storage area inside a factory and a loading bay outside	10	:	1

The CIBSE (Chartered Institute of Building Services Engineers) produces a Code for Interior Lighting which gives lighting requirements for different areas. This is also replicated in BS EN 12464-1:2002 Light and lighting – Lighting of work places – Part 1: Indoor work places.

A sample is shown below.

Illuminance (lux)	**Activity**	**Area**
100	Casual seeing	Corridors, changing rooms, stores.
150	Some perception of detail	Loading bays, switch rooms, plant rooms.
200	Continuously occupied	Foyers, entrance halls, dining rooms.
300	Visual tasks moderately easy	Libraries, sports halls, lecture theatres.
500	Visual tasks moderately difficult	General offices, kitchens, laboratories, retail shops.

Illuminance (lux)	Activity	Area
750	Visual tasks difficult	Drawing offices, meat inspection, chain stores.
1000	Visual tasks very difficult	General inspection, electronic assembly, paintwork, supermarkets.
1500	Visual tasks extremely difficult	Fine work and inspection, precision assembly.
2000	Visual tasks exceptionally difficult	Assembly of minute items, finished fabric inspection.

Maintenance of light fitments

[L5014] The lighting output of a given lamp will reduce gradually in the course of its life but an improvement can be obtained by regular cleaning and maintenance, not only of the lamp itself, but also of the reflectors, diffusers and other parts of the luminaire. A sensible and economic lamp replacement policy is called for (eg it may be more economical, in labour cost terms, to change a batch of lamps than deal with them singly as they wear out).

Qualitative aspects of lighting and lighting design

[L5015] While the quantity of lighting afforded to a particular location or task in terms of standard service illuminance is an important feature of lighting design, it is also necessary to consider the qualitative aspects of lighting, which have both direct and indirect effects on the way people perceive their work activities and dangers that may be present. The quality of lighting is affected by the presence or absence of glare, the distribution of the light, brightness, diffusion and colour rendition.

Glare

[L5016] This is the effect of light which causes impaired vision or discomfort experienced when parts of the visual field are excessively bright compared with the general surroundings. It may be experienced in three different forms:

(a) disability glare – the visually disabling effect caused by bright bare lamps directly in the line of vision;
(b) discomfort glare – caused by too much contrast of brightness between an object and its background, and frequently associated with poor lighting design. It can cause discomfort without necessarily impairing the ability to see detail. Over a period it can cause visual fatigue, headaches and general fatigue;
(c) reflected glare – the reflection of bright light sources on shiny or wet work surfaces, such as plated metal or glass, which can almost entirely conceal the detail in or behind the object that is glinting.

NB The Illuminating Engineering Society of North America (IES) publishes a Limiting Glare Index for each of the effects in (*a*) and (*b*) above. This is an index representing the degree of discomfort glare which will be just tolerable in the process or location under consideration. If exceeded, occupants may suffer eye strain or headaches or both. This could be considered in line with general guidance for the particular type of work environment produced by CIBSE.

Distribution

[L5017] Distribution is concerned with the way light is spread. The British Zonal Method classifies luminaires (light-fittings) according to the way they distribute light from BZ1 (all light downwards in a narrow column) to BZ10 (light in all directions). However, a fitting with a low BZ number does not necessarily imply less glare. Its positioning, the shape of the room and the reflective surfaces present are also significant. Light distribution, for simplicity, is often categorised into five main types, with type I giving a lateral diffusion of 15 degrees to type V giving a circular distribution that has the same intensity at all angles. This type of lighting is often used in large car parks, the centre of roadways or areas where evenly distributed light is necessary.

The actual spacing of luminaires is also important when considering good lighting distribution. To ensure evenness of illuminance at operating positions, the ratio between the height of the luminaire and the spacing of it must be considered. The IES spacing: height ratio provides a basic guide to such arrangements. Under normal circumstances, for example offices, workshops and stores, this ratio should be between $1\frac{1}{2}:1$ and $1:1$ according to the type of luminaire.

Brightness

[L5018] Brightness or 'luminosity' is very much a subjective sensation and, therefore, cannot be measured. However, it is possible to consider a brightness ratio, which is the ratio of apparent luminosity between a task object and its surroundings. To ensure the correct brightness ratio, the reflectance (the ability of a surface to reflect light) of all surfaces in the working area should be well maintained and consideration given to reflectance values in the design of interiors. Given a task illuminance factor (the recommended illuminance level for a particular task) of 1, the effective reflectance values should be 0.6 for ceilings, 0.3 to 0.8 for walls and 0.2 to 0.3 for floors.

Diffusion

[L5019] This is the projection of light in all directions with no predominant direction. The directional flow of light can often determine the density of shadows, which may prejudice safety standards or reduce lighting efficiency. Diffused lighting will reduce the amount of glare experienced from bare luminaires.

Colour rendition

[L5020] Colour rendition refers to the appearance of an object under a specific light source compared to its colour under a reference illuminant, for example natural light. Good standards of colour rendition allow the colour appearance of an object to be properly perceived. Generally, the colour rendering properties of luminaires should not clash with those of natural light, and should be just as effective at night when there is no daylight contribution to the total illumination of the working area.

It is worth noting that colour of walls, ceiling and work surfaces can play a major part in exacerbating or reducing effects of reflection and glare. Matt surfaces are always preferential in a workplace to minimise glare, and to select soft colour hues rather than harsh primary colours.

Stroboscopic effect

[L5021] One aspect of lighting quality that formerly gave trouble was the stroboscopic effect of fluorescent tubes, which gave the illusion of motion or even the illusion that a rotating part of machinery was stationary. With modern designs of fluorescent tubes, this effect has largely been eliminated.

Machinery Safety

Andrea Oates

Introduction to machinery safety

[M1001] The introduction of harmonised European Union ('EU') regulations and standards in health and safety changed the approach to the prevention of accidents and ill health at work. In the place of a prescriptive set of standards and regulations, the harmonised regulations and standards represented a remarkable breakthrough in the identification, assessment and control of machinery and work equipment risks. This approach is found in the EU Machinery and Use of Work Equipment Directives as well as the Transposed Harmonised EU Machinery Safety Standards which resulted in UK Regulations controlling the manufacture and use of machinery and work equipment.

Currently, statutory duties are dually (but separately) laid on both manufacturers of machinery for use at work and employers and users of such machinery and equipment.

General and specific duties are imposed on manufacturers by the *Supply of Machinery (Safety) Regulations 2008 (SI 2008 No 1597) (as amended)*.

Duties on employers are incorporated into the *Provision and Use of Work Equipment Regulations 1998 (SI 1998 No 2306)*; as well as the *Lifting Operations and Lifting Equipment Regulations 1998 (SI 1998 No 2307)* (which are examined in LIFTING OPERATIONS).

Legal requirements relating to machinery

[M1002] The introduction of the *Health and Safety at Work etc Act ('HSWA 1974')* in 1974, in addition to placing general duties on employers under *section 2* of the Act, placed responsibility for machinery safety on designers, manufacturers and suppliers. This is still applicable in the UK, in addition to the EU requirements.

Section 6(1)(a) of the *1974 Act* requires that:

> It shall be the duty of any person who designs/supplies . . . any article for use at work to ensure, so far as is reasonably practicable, that the article is so designed . . . as to be safe and without risks to health . . . when properly used.

Machinery safety – the risk based – approach

[M1003] In May 1985 European Ministers agreed a New Approach to Technical Harmonisation and Standards to overcome trade barriers. The current Machinery Directive (2006/42/EC) (as amended) is one of several Product Safety Directives and sets out the essential health and safety requirements ('EH&SR's') for machinery which must be met before machinery is placed on the market anywhere in the EU.

EH&SRs are expressed in general terms and it was intended that the European Harmonised Standards should fill in the detail so that machinery designers and suppliers have clear guidance on how to achieve conformity with the Directive.

The Use of Work Equipment Directive (89/655/EEC) ('UWED'), repealed by the Use of Work Equipment Directive 2009/104/EC, was also introduced to outline the responsibilities of employers for the protection of workers from the risks arising from the use of [machinery and work equipment]. In the UK this became the *Provision and Use of Work Equipment Regulations 1992 (SI 1992 No 2932)* ('PUWER'92'). These regulations were replaced, as a result of amendments to EU Directive 89/655/EEC, by *the Provision and Use of Work Equipment Regulations 1998 (SI 1998 No 2306)* ('PUWER'98').

The 1998 Regulations remain in force and have been amended a number of times by legislation including:

(a) the *Police (Health and Safety) Regulations 1999 (SI 1999 No 860)*;
(b) the *Mines Regulations 2014 (SI 2014 No 3248)*;
(c) the *Work at Height Regulations 2005 (SI 2005 No 735)*;
(d) the *Construction (Design and Management) Regulations 2015 (SI 2015 No 51)*;
(e) the *Deregulation Act 2015 (Health and Safety at Work) (General Duties of Self-Employed Persons) (Consequential Amendments) Order 2015 (SI 2015 No 1637)*.

The Provision and Use of Work Equipment Regulations 1998 (PUWER'98)

[M1004] The Health and Safety Executive (HSE) explains that the primary objective of PUWER'98 is to ensure that work equipment should not result in health and safety risks, regardless of its age, condition or origin.

Structure of PUWER'98

[M1005] The Regulations are structured in five main parts. These are set out as follows:

Part I: Regulations 1–3 (Introduction)
This part includes interpretation, definitions and application of *PUWER'98*.
Part II: Regulations 4–10 (General – management duties)
This part sets out the 'management' duties of *PUWER'98* covering: selection of suitable equipment, maintenance, inspection, specific risks, information, instruction and training. *Regulation 10* covers the conformity of work equipment with the requirements of EU Directives on product safety.
Part II: Regulations 11–24 (General – 'hardware' aspects of dangerous machinery)

This part deals with the hardware aspects of *PUWER'98*. It covers the guarding of dangerous parts of work equipment, protection against specified hazards, high or very low temperatures, controls for starting or making a significant change in operating conditions, the provision of appropriate stop and emergency stop controls, controls and control systems, isolation from sources of energy, stability, lighting maintenance operations, markings and suitable warnings or warning devices.

Part III: Regulations 25–30 (Mobile work equipment)
This part deals with specific risks associated with the use of self-propelled, towed and remote controlled mobile work equipment and set outs requirements in relation to employees carried on mobile work equipment, rolling over of mobile work equipment, overturning of forklift trucks, self-propelled work equipment, remote-controlled self-propelled equipment and drive shafts.

Part IV: Regulations 31–35 (Power presses)
This deals with the management requirements for the safe use and thorough examination of power presses including thorough examination and inspection, reports and the keeping of information.

Part V: Regulations 36–39
This part covers miscellaneous issues including an exemption for the armed forces, transitional provisions, repeal of Acts and revocation of instruments.

What is work equipment?

[M1005.1] The definition of 'work equipment' covers any machinery, appliance, apparatus, tool or installation for use at work (Regulation 2). Although pre-dating these Regulations, in *Knowles v Liverpool City Council [1993] IRLR 588*, it was held that a flagstone used for repairing a pavement was 'equipment'.

In *Smith v Northamptonshire County Council* [2008] EWCA Civ 181, [2008] 3 All ER 1054, a claim was brought by a carer/driver who was injured while pushing a wheelchair down a wooden ramp outside the home of the wheelchair user. The court found that the ramp, which had given way, was not work equipment. But in the House of Lords case, *Spencer-Franks v Kellogg Brown and Root Ltd* [2008] UKHL 46, [2009] 1 All ER 269, a door closer on the door of the central control room of an offshore oil platform was held to be work equipment.

In *Coia v Portavadie Estates Ltd* [2015] CSIH 3, 2015 SC 419, 2015 SCLR 479, the Scottish Court of Session held that a metal pole in a wardrobe in accommodation provided to a chef working at a hotel by his employer (and usually used by guests) was not 'work equipment'. The pole was dislodged and fell as he was moving his belongings out of a lodge, injuring his foot.

In *Johnson v University of Bristol (2017) (Civ) 17 October 2017 (unreported) Court of Appeal*, a faulty kitchen cupboard in student accommodation (being repaired by a carpenter) was held not to be work equipment under *Regulation 2(1) of PUWER'98 (SI 1998 No 2306)*.

The term 'use' covers starting, stopping, modifying, programming, setting, transporting, maintaining, servicing and cleaning.

Risk assessment and PUWER'98

[M1006] While there is no requirement in *PUWER'98 (SI 1998 No 2306)* to carry out a risk assessment, the general risk assessment requirements under *Regulation 3(1) of the Management of Health and Safety at Work Regulations 1999 (SI 1999 No 3242)* apply.

The HSE publication *Safe use of work equipment: Provision and Use of Work Equipment Regulations* 1998 (L22) contains specific guidance on risk assessment in relation to work equipment.

The guidance indicates that the factors to be considered in a risk assessment to meet the requirements of *PUWER'98* should include: type of work equipment, substances and electrical or mechanical hazards to which people may be exposed. Action to eliminate/control any risk might include, for example, during maintenance: disconnection of power supply, supporting parts of the work equipment which could fall, securing mobile equipment so that it cannot move, removing or isolating flammable or hazardous substances, and depressurising pressurised systems.

It advises employers to consider environmental conditions such as lighting, problems caused by weather conditions, other work being carried out which may affect the operation and the activities of people who are not at work. They should also identify groups of workers that might be particularly at risk, such as young or disabled people not English

Summary of main requirements of PUWER'98

Part II: General (Regulations 4–24)

[M1007] *Part II* of *PUWER'98 (SI 1998 No 2306)* applies to all work equipment, including machinery, mobile plant and work equipment.

Regulation 4 –   Suitability

Equipment must be suitable, by design, construction or adaptation, for the actual work it is provided to do. In selecting the work equipment, the employer must take account of the environment in which the equipment is to be used and must ensure the equipment is used only for the operations, and under the conditions, for which it is suitable.

The *Police (Health and Safety) Regulations 1999 (SI 1999 No 860)* amended *Regulation 4* with regard to offensive weapons provided for use as self-defence, or as deterrent equipment, and work equipment provided for use for arrest or restraint by police officers and cadets.

HSE guidance provided in L22 sets out that the risk assessment carried out under *Regulation 3(1)* of the Management Regulations will help in selecting work equipment and assessing its suitability for particular tasks. It says most duty holders will be capable of making the risk assessment themselves using expertise within the organisation. Where there are complex hazards or equipment, they may need the help of external health and safety advisors appointed under *Regulation 7 of the Management Regulations*.

Regulation 5 – Maintenance

Equipment must be maintained in an efficient state, in efficient working order and in good repair. The term 'efficient' relates to how the condition of the equipment might affect health and safety, not productivity. There is no requirement to keep a maintenance log but where one exists, and this is good management practice, it must be kept up to date.

Regulation 6 – Inspection

Regulation 6 of *PUWER'98 (SI 1998 No 2306)* includes a requirement for the inspection of work equipment:

> '(1) Every employer shall ensure that, where the safety of work equipment depends on the installation conditions, it is inspected—
> (a) after installation and before being put into service for the first time; or
> (b) after assembly at a new site or in a new location,
>
> to ensure that it has been installed correctly and is safe to operate.'

Regulation 6(2) requires employers to ensure work equipment exposed to conditions causing deterioration liable to result in dangerous situations is inspected at suitable intervals and each time exceptional circumstances liable to jeopardise the safety of the work equipment have occurred, to ensure health and safety conditions are maintained and any deterioration can be detected and remedied in good time.

The purpose of an inspection is to identify whether the work equipment can be operated, adjusted and maintained safely and that any deterioration, eg defects, damage or wear can be detected and repaired before it results in unacceptable risks.

The ACoP to the regulations sets out that where the risk assessment under *Regulation 3* of the *Management Regulations* has identified a significant risk to the operator or other workers from the installation or use of the work equipment, a suitable inspection should be carried out. Guidance on the regulations then explains that a significant risk is one which could result in an imminent failure, which could lead to a major injury, as a result of incorrect installation or re-installation, deterioration, or exceptional circumstances which could affect the safe operation of the work equipment.

The extent of the inspection required will depend on the level of risk associated with the work equipment, and should include, where appropriate visual and functional checks and testing to the more comprehensive inspection which may require some dismantling and/or testing.

Guidance to the regulations sets out that:

> 'An inspection will vary from a simple visual external inspection to a detailed comprehensive inspection, which may include some dismantling and/or testing. An inspection should always include those safety-related parts necessary for safe operation of equipment, for example overload warning devices and limit switches. The extent of the inspection required will depend on:
> (a) the type of equipment;
> (b) where it is used;
> (c) how it is used.'

The nature and frequency of inspections made under *Regulation 6* is to be decided by a 'competent person', the ACoP makes clear.

A fundamental requirement under Regulation 6 is that every employer must ensure that the result of an inspection is recorded and kept until the next inspection is recorded. Although records do not have to be kept in a particular form, the example shown in FIGURE 1 below identifies the main items to be recorded on an inspection record sheet.

Components which must receive particular attention as part of an inspection are the safety-related and safety-critical parts, for example interlocking devices, safeguards, protection devices, controls and control systems.

Regulation 6 also states that no work equipment should leave an undertaking, or be obtained from an undertaking of another person (for example by hiring), unless it is accompanied by physical evidence that the last inspection has been carried out.

Regulation 6(5) was amended by the *Work at Height Regulations 2005 (SI 2005 No 735)* to exclude from the regulations the inspection of work equipment where regulation 12 of those regulations, which also deals with inspection, applies; and by the *Construction (Design and Management) Regulations 2015 (SI 2015 No 51)* to exclude work equipment required to be inspected under those regulations.

In 2016, the HSE launched a review of the requirements for the inspection or thorough examination of work plant and equipment contained in the *Lifting Operations and Lifting Equipment Regulations 1998; Provision and Use of Work Equipment Regulations 1998; Pressure Systems Safety Regulations 2000 (PSSR); and Work at Height Regulations 2005.*

This included the requirement for inspections set out in *Regulation 6* of *PUWER'98* and the requirement for the thorough examination of power presses set out in *Regulation 32* (see Part IV: Power Presses at **M1008.1** below).

The review's conclusion was that the current framework for equipment inspection remains proportionate and fit for purpose. The HSE Board agreed in March 2017 that no action should be taken at this time and the review should be brought to a close.

Figure 1: Example of an inspection record sheet

INSPECTION RECORD

Type and model of equipment: Press Brake-Komato . . .

Serial number/identification mark: KP/012 . . .

Normal location of equipment: Press Shop . . .

Date inspection carried out: 08/09/2018 . . .

> **Type of inspection:**
>
> > Visual check:
> > Functional check:
> > Dismantling/testing:
>
> **Who carried out inspection:** John Smith . . .
>
> **Item and safety-related part inspected** . . .
> 1. Guard-interlocking Device . . .
> 2. Photo-electric Trip Device . . .
> 3. Fixed and Interlocking Safeguard . . .
> 4. Pneumatic Clutch and Brake System . . .
> 5. Machine Control and Electrical System . . .
>
> **Fault and defect found:**
> 1. Small air leak near clutch . . .
> 2. Interlocking device wiring loose . . .
>
> **Action/corrective measure taken:**
> 1. Seals replaced for pneumatic system . . .
> 2. Re-wiring of interlocking system . . .
>
> **To whom fault have been reported:** Kevin Taylor . . .
>
> **Date repair/action were carried out:** 12/09/2018 . . .

Regulation 7 -   Specific risks

Regulation 7 of *PUWER'98 (SI 1998 No 2306)* requires employers to ensure that, where the use of work equipment is likely to involve a specific risk to health and safety, the use, repair, modification, maintenance or servicing of such equipment is restricted to those persons who have been specifically designated to perform operations of that description. It is further required that employers must provide adequate training to perform such tasks safely.

The Approved Code of Practice ('ACoP') and Guidance to *PUWER'98* does not define what is meant by 'specific risks', or what makes such risks 'specific'. The guidance does however explain that: 'Specific risks can be common to a particular class of work equipment. There can also be a specific risk associated with the way a particular item of work equipment is repaired, set or adjusted as well as with the way it is used.'

The guidance also makes clear that rather than focusing only on safety risks, employers must also consider risks to the health of workers from manual handling, dust, fumes, noise and hand/arm vibration in their risk assessment.

The ACoP relating to *Regulation 7* requires employers to ensure that 'risks are always controlled by (in the order given):

(a) eliminating the risks, or if that is not possible;
(b) taking engineering (physical) measures to control risks such as the provision of guards; but if risks cannot be adequately controlled;
(c) taking appropriate management measures to deal with the residual risks, such as following safe systems of work and the provision of information, instruction and training.'

Regulation 8 –   Information and instructions

Managers, supervisors and users of work equipment must be provided with adequate health and safety information and, where appropriate, written instructions relating to work equipment. This must include the conditions in which, and the methods by which, the equipment may be used, the limitations on its use together with any foreseeable difficulties that might arise and the required actions.

Regulation 9 –   Training

Managers, supervisors and all users of work equipment must receive adequate health and safety training in the methods which may be adopted when using equipment, any risks which may arise from such use, and precautions to be taken. The level of training will depend on the circumstances of use and the competence of the employee. Particular attention must be paid to the training and supervision of young persons. Self-propelled work equipment, including any attachments or towed equipment must only be driven by workers who have received the appropriate training in the safe driving of the equipment, and the ACOP provides specific guidance regarding the training and certificates of competence in relation to the use of chainsaws.

The case of *Allison v London Underground Ltd* [2008] EWCA Civ 71, [2008] IRLR 440 provided important guidance on the duty to ensure that all persons who use work equipment receive adequate health and safety training. It established that the duty to provide training is mandatory, and that the test for adequacy is what training is needed in the light of what the employer ought to have known about the risks arising from the activities of the business.

Regulation 10 –   Conformity with Community requirements

Regulation 10 of *PUWER'98 (SI 1998 No 2306)* (amended by the *Health and Safety (Miscellaneous Amendments) Regulations 2002 (SI 2002 No 2174)*), applies to items of work equipment provided for use in the premises or undertaking of the employer for the first time after 31 December 1992. *Regulation 10(1)* requires employers to ensure that work equipment complies at all times with any essential requirements that applied to that kind of work equipment at the time of its *first supply* or when it is *first put into service*. It places a duty on the employer that complements those on manufacturers and suppliers in other jurisdictions regarding the initial integrity of equipment.

Regulation 10(2) defines the term 'essential requirements' – as meaning 'requirements relating to the design and construction of work equipment of its type in any of the instruments listed in Schedule 1 (being instruments which give effect to Community directives concerning the safety of products)'. (See below.)

The employer can demonstrate compliance with *Regulation 10* by checking the equipment bears a CE marking and asking for a copy of the EU Declaration of Conformity; checking that adequate operating instructions have been provided and that there is information about residual hazards such as noise and vibration; and, more importantly, checking the equipment for obvious faults.

The HSE booklet '*Buying new machinery*' (INDG271) contains two checklists:

Checklist A - *What should I talk to a supplier (or manufacturer) about?*

Checklist B - *What do I have to do when I have bought new machinery?*

The guidance is aimed at small/medium-sized enterprises and signposts employers to further information. It can be found on the HSE website at www.hse.gov.uk/pubns/indg271.pdf.

Regulation 11 - Dangerous parts of machinery

Regulation 11(1) of *PUWER'98 (SI 1998 No 2306)* requires employers to ensure that access to dangerous parts of the machine is prevented, or if access is needed to ensure that the machine is stopped before any part of the body reaches a danger zone.

Regulation 11(2) sets out the following hierarchy of preventive measures (see also **M1027** onwards):

(a) fixed enclosing guards;
(b) other guards or protective devices;
(c) protection appliances, eg jigs, holders, push sticks, etc.

Employers are also required to provide information, instruction, training and supervision.

Regulation 11(3) describes the need for safeguards or safety devices which are robust, difficult to defeat, well maintained and meet the requirements of relevant European (EN) standards.

Illustrated examples of the options given in Regulation 11 are shown in an appendix to the guidance to PUWER'98.

Regulation 12 - Protection against specified hazards

Employers must take appropriate measures to prevent or adequately control exposure to 'specified hazards' arising from the use of work equipment. These measures should be other than the provision of personal protective equipment or information, instruction, training and supervision, so far as is reasonably practicable.

The 'specified hazards' are: falling or ejected articles or substances; rupture or disintegration of parts of equipment; work equipment catching fire or overheating; and unintended or premature discharges or explosions.

Regulation 13 – High or very low temperature

Employers must ensure work equipment, parts of work equipment and any article or substance produced, used or stored in work equipment which is at high or low temperature is protected so as to prevent burns, scalds or sears.

Regulations 14 to 18 – Controls and control systems

Employers must ensure work equipment is provided with:

- One or more controls for starting the equipment or for controlling the operating conditions (speed, pressure etc), where the risk after the change is greater than or of a different nature from the risks beforehand.
- One or more readily accessible stop controls that will bring the equipment to a safe condition in a safe manner and which operate in priority to any control that starts or changes the operating conditions of the equipment.
- One or more readily accessible emergency stop controls – unless by the nature of the hazard, or by the adequacy of the normal stop controls (above), this is unnecessary. Emergency stop controls must operate in priority to other stop controls.

Controls should be clearly visible and identifiable (including appropriate marking). The position of controls should be such that the operator can establish that no person is at risk as a result of the operation of the controls. Where this is not possible, safe systems of work must be established to ensure that no person is in danger as a result of equipment starting or where this is not possible, there should be an audible, visible or other suitable warning whenever work equipment is about to start. Sufficient time and means shall be given to a person to avoid any risks due to the starting or stopping of equipment.

Control systems should not create any increased risk, which ensure that any faults or damage in the control systems or losses of energy supply do not result in additional or increased risk, and which do not impede the operation of any stop or emergency stop control.

Regulation 18 of PUWER'98 (SI 1998 No 2306) requires that:

(a) Every employer shall—
    (a) ensure, so far as is reasonably practicable, that all control systems of work equipment are safe; and
    (b) are chosen making due allowance for failures, faults and constraints to be expected in the planned circumstances of use.

A control system can be defined as: a system which responds to input signals and generates an output signal which causes the equipment to perform controlled functions.

The input signals may be made by the operator (manual control), or automatically controlled by the equipment through sensors or protection devices, eg guard interlocking devices, photoelectric device, emergency stop or speed limiters.

The objective of *Regulation 18* is to prevent or reduce the likelihood that a single failure in the interlocking device, wiring, contactors, hardware, software could cause injury or ill health.

The subject of control systems generally and the safety related functions of control systems in particular are complex, even for engineers. HSE guidance in L22 refers to national, European and international standards providing guidance on the design of control systems so as to achieve high levels of performance related to safety. Aimed at new machinery, the HSE advises that they may also be used as guidance for existing work equipment.

Regulation 19 - Isolation from sources of energy

Employers must ensure equipment is provided, where appropriate, with identifiable and readily accessible means of isolating it from all its sources of energy. Reconnection of any source of energy must not expose anyone to any health and safety risk.

Regulation 20 - Stability

Employers must ensure all work equipment is stabilised where necessary – by clamping, tying or fastening.

Regulation 21 - Lighting

Employers must ensure suitable and sufficient lighting, taking account of the operations carried out, is provided. This may entail additional lighting where the ambient lighting is insufficient for the required tasks.

Regulation 22 - Maintenance operations

Employers must ensure work equipment is constructed or adapted in such a way that maintenance operations, which involve a risk to health or safety, can be carried out while the equipment is shut down. If this is not possible, the work should be carried out in such a way as to prevent exposure to risk and appropriate measures taken to protect those carrying out the maintenance work.

Such measures might include, for instance, functions designed into the equipment that limit the power, speed or range of movement of dangerous parts during maintenance. The HSE advises that these could include providing temporary guards; limited movement controls; crawl speed operated by hold-to-run controls; or using a second low-powered visible laser beam to align a powerful invisible one.

Regulation 23 - Markings

Employers must ensure equipment must be marked for the purposes of ensuring health and safety with clearly visible markings.

Regulation 24 - Warnings

Employers must ensure equipment incorporates appropriate warnings or warning devices, which must be unambiguous, easily perceived and easily understood.

A warning is normally in the form of a notice and can be positive instructions, eg "Hard hats must be worn", prohibitions, eg "Not to be operated by people under 18 years", or restrictions eg "Do not heat above 60oC". A warning device is an active unit giving a signal, typically audible or visible.

Second-hand machinery

[M1007.1] The Health and Safety Executive (HSE) advises:

> 'Where the second-hand equipment is in scope of one of the European product supply Directives and has not previously been put into service in the European Economic Area (EEA), or placed on the market of the EEA, the person importing it into the EEA must meet the conformity assessment requirements of the relevant European Directives, including CE marking. The term "sold as seen" or similar cannot be used to avoid these legal responsibilities.'

It should also be subject to a suitable and sufficient risk assessment and should comply with *Regulations 11 to 24 of PUWER'98 (SI 1998 No 2306)*.

Second-hand machinery supplied from outside the EU should be treated as new machinery and will be subject to CE marking requirements.

It is a management responsibility to ensure that second-hand machinery and any modification work comply with the requirements of the *PUWER'98*.

For more detailed information see the HSE website at www.hse.gov.uk/work-equipment-machinery/second-hand-products.htm

Part III: Mobile work equipment

[M1008] HSE guidance to the regulations explains that 'mobile equipment is any work equipment which carries out work while it is travelling or which travels between different locations where it is used to carry out work'. Mobile work equipment may be self-propelled, towed or remote controlled and may incorporate attachments. Part III of *PUWER'98, Regulations 25 to 30,* sets out additional requirements for mobile work equipment.

Regulation 25 – Employees carried on mobile work equipment

Regulation 25 of PUWER'98 (SI 1998 No 2306) contains a general requirement relating to the risks to people (drivers, operators and passengers) carried by mobile equipment, when it is travelling. This includes risks of people falling from the equipment or from unexpected movement while it is in motion or stopping.

It requires every employer to ensure that no employee is carried by mobile work equipment unless:

(a) it is suitable for carrying persons; and
(b) it incorporates features for reducing to as low as is reasonably practicable risks to their safety, including risks from wheels or tracks.

Mobile work equipment can be made suitable for carrying persons by means of operator stations with seats or work platforms to provide a secure place.

Features for reducing risks may include the following:

- Falling object protective structures ('FOPS'). This may be achieved by a strong safety cab or protective cage.

- Restraining systems – this can be full-body seat belts, lap belts or purpose-designed restraining systems. This should provide adequate protection against injury through contact with or being flung from the mobile equipment if it comes to a sudden stop moves unexpectedly or rolls over.
- Speed adjustment – when carrying people, mobile equipment should be driven within safe speed limits to ensure stability when cornering and on all surfaces and gradients.
- Guards and barriers fitted to mobile work equipment to prevent contact with wheels and tracks.

Regulation 26 –   Rolling over of mobile work equipment

In addition to the main general requirements of *Regulation 25, Regulation 26* of *PUWER'98 (SI 1998 No 2306)* covers the measures necessary to protect employees where there are risks from roll-over while travelling. The following measures should be considered:

- Stabilisation – this might include fitting appropriate counterbalance weights, wider wheels and locking devices.
- Structures to prevent rolling over by more than 90 degrees, eg boom of a hydraulic excavator, when positioned in its recommended travel position.
- Roll-over protective structures ('ROPS') – normally fitted on mobile equipment to withstand the force for roll over through 180 degrees.

Regulation 27 –   Overturning of forklift trucks

The HSE guidance explains that *Regulation 27 of PUWER'98 (SI 1998 No 2306)* applies to FLT's fitted with vertical masts, which effectively protect seated operators from being crushed between the FLT and the ground in the event of roll-over. FLT's capable of rolling over 180° should be fitted with ROPS.

FLT's with a seated ride-on operator can roll-over in use. There is also a history of accidents on counterbalanced, centre control, high lift trucks that have a sit-down operator. Restraining systems will normally be required on these trucks.

Regulation 28 –   Self-propelled work equipment

Regulation 28 of *PUWER* sets out that every employer shall ensure that, where self-propelled work equipment may, while in motion, involve risk to the safety of persons there are/is:

- Measures to preventing unauthorised start-up – this could be achieved by a starter key or device, which is issued only to authorised personnel.
- Measures in place to minimise consequences of a collision of rail-mounted work equipment – this can be achieved by introducing safe systems of work, buffers or automatic means to prevent contact.
- Devices for stopping and braking – all self-propelled work equipment should have brakes to enable it to slow down and stop in a safe distance and park safely.
- Emergency braking and stopping facilities – a secondary braking system may be required if risk is significant.

[M1008] Machinery Safety

- Adequate devices for improving vision, so far as is reasonably practicable, where the driver's direct field of vision is inadequate to ensure safety – mirrors or more sophisticated visual or sensing facilities may be necessary.
- Lighting for use in the dark – to be fitted to equipment if outside lighting is inadequate.
- Carriage of appropriate fire-fighting equipment – where escape from self propelled work equipment in the case of fire is not easy.

Regulation 29 – Remote-controlled, self-propelled work equipment

This is self-propelled work equipment that is operated by controls which have no physical link with it, for example radio control. *Regulation 29* of *PUWER* requires employers to ensure that where remote-controlled self-propelled work equipment involves a risk to safety while in motion, it stops automatically once it leaves its control range. In addition, where there is a risk of crushing or impact, it incorporates features to guard against such risk unless other appropriate devices are able to do so.

The HSE advises employers, in its guidance to the regulations, that they may need to consider:

- Alarms or flashing lights – to be considered for the prevention of collision with others.
- Sensing or contact devices – to prevent or reduce risk of injury following contact.
- Hold-to-run control devices – so that any hazardous movement can stop when the controls are released.
- Control range – once equipment leaves its control range should stop and remain in a safe state.

Regulation 30 – Drive shafts

A 'drive-shaft' is a device which conveys the power from the mobile work equipment to any work equipment connected to it. In agriculture these devices are known as power take-off shafts.

Where the seizure of the drive shaft between mobile work equipment and its accessories or anything towed is likely to involve a risk to safety, Regulation 30 of PUWER sets out that employers shall:

- ensure that the work equipment has a means of preventing such seizure; or where such seizure cannot be avoided,
- take every possible measure to avoid an adverse effect on the safety of an employee.
- Where mobile work equipment has a shaft for the transmission of energy between it and other mobile work equipment; and it could become soiled or damaged by contact with the ground while uncoupled, employers must ensure that the work equipment has a system for safeguarding the shaft.

Part IV   Power presses

[M1008.1] Part IV of the regulations deals with power presses; although it does not apply to types set out in Schedule 2. These are:

- a power press for the working of hot metal;
- a power press not capable of a stroke greater than 6 millimetres;
- a guillotine;
- a combination punching and shearing machine, turret punch press or similar machine for punching, shearing or cropping;
- a machine, other than a press brake, for bending steel sections;
- a straightening machine;
- an upsetting machine;
- a heading machine;
- a riveting machine;
- an eyeletting machine;
- a press-stud attaching machine;
- a zip fastener bottom stop attaching machine;
- a stapling machine;
- a wire stitching machine; and
- power press for the compacting of metal powders.

Regulation 32 requires the thorough examination of power presses, guards and protection devices.

Power presses must not be put into service for the first time after installation, or after assembly at a new site or in a new location, unless it has been thoroughly examined to ensure that it has been correctly installed, would be safe to operate and any defect has been rectified.

A power press with fixed guards only must also be thoroughly examined at least every 12 months and in other cases they must be examined at least every 6 months. In addition, examination is required where exceptional circumstances have occurred which could jeopardise the safety of the power press or its guards or protection devices. Any defects must be remedied before the power press is used again.

Regulation 33 concerns the inspection of guards and protection devices and requires employers to ensure that power presses are not used after the setting, re-setting or adjustment of its tools (apart from trying out its tools or in die proving) unless every guard and protective device has been inspected and tested while in position by a competent person – or someone undergoing training and acting under the immediate supervision of a competent person.

It also requires that:

> 'Every employer shall ensure that a power press is not used after the expiration of the fourth hour of a working period [ie working day, night or shift] unless its every guard and protection device has been inspected and tested while in position on the power press by a person appointed in writing by the employer who is—
>
> (a) competent; or
> (b) undergoing training for that purpose and acting under the immediate supervision of a competent person'

In both cases, the person carrying out the inspection must sign a certificate confirming that inspection and testing has been carried out.

Regulation 34 requires that the person carrying out the examination must notify the employer of any defects, provide a written report (Schedule 3 sets out the information this must include) and where there are potentially dangerous defects, send a copy of the report to the relevant enforcing authority.

Employers must keep reports for two years and ensure that certificates of inspection are kept at or near the power press (Regulation 35).

The HSE publication *Safe use of power presses* (L112) gives advice on precautions that can be taken to ensure the safe use of power presses. It contains information on thorough examination and inspection and on reports and the keeping of information (see www.hse.gov.uk/pubns/books/l112.htm).

The HSE publication *Safe use of woodworking machinery* (L114) gives practical advice on the safe use of woodworking machinery and covers the provision of information and training, as well as aspects of guarding (see www.hse.gov.uk/pubns/books/l114.htm).

The Supply of Machinery (Safety) Regulations 2008 (as amended)

[M1009] The Health and Safety Executive (HSE) explains:

> 'Since 1995 all new machinery in scope of the Machinery Directive has to be designed and constructed to meet common minimum European requirements for safety
>
> The outward signs of compliance are CE marking on the equipment and a document (Declaration of Conformity) issued by the Responsible Person (normally the manufacturer) declaring the product's conformity.
>
> To achieve compliance the Responsible Person must undertake a conformity assessment process to meet the Directive's obligations. This includes meeting all relevant essential health and safety requirements (EHSRs) for the product, producing comprehensive user instructions, and showing how compliance has been achieved in the technical file.
>
> For certain higher risk products the conformity assessment process will normally require the use of an independent Notified Body.'

In the UK, the *Supply of Machinery (Safety) Regulations 2008 (SI 2008 No 1597)* as amended by the *Supply of Machinery (Safety) (Amendment) Regulations 2011 (SI 2011 No 2157)* implemented these requirements.

The *2008 Regulations* set out in *Part 1* and 2 the types of product which constitute 'machinery' and 'partly completed machinery' and are subject to their requirements.

Part 3 sets out the key obligations on those who place machinery or partly completed machinery on the market or put it into service. These are referred to in the Directive as 'manufacturers or their authorised representatives' and in the *Regulations* they are referred to as 'responsible persons'.

These include:

- ensuring the safety of products (by reference to the essential health and safety requirements set out in *Part 1* of *Schedule 2*);

- following a 'conformity assessment procedure' (set out in *regulations 10 to 12*); and
- documenting their compliance with the requirements of the Directive in various ways (particularly by drawing up an 'EC declaration of conformity' and affixing the CE marking to the product).
- *Part 4* contains further provisions regarding CE marking.
- *Part 5* is concerned with the activities of 'notified bodies'. These bodies assess conformity of products with the Regulations.
- *Part 6* is concerned with enforcement.

The requirements of the *Supply of Machinery (Safety) Regulations 2008* are summarised below.

The Regulations extend to machinery and products including safety components, lifting tackle and partly-completed machinery, but exclude, for example, domestic electrical machines and fairground equipment. Regulation 4 sets out which products the regulations are concerned with and define machinery.

Regulation 5 sets out that the regulations do not apply to a product if (or to the extent that) other European Union directives make more specific provisions with regard to its particular hazards.

The Regulations require that all machinery:

- is designed and constructed to be safe, meeting the essential health and safety requirements (EHSRs) listed in the *Regulations* (see below);
- is CE marked;
- is supplied with instructions in English; and
- has a Declaration of Conformity (or, in the case of partly completed machinery, a Declaration of Incorporation) (see below).

Essential Health and Safety Requirements

[M1010] All New Approach product safety Directives impose essential requirements that products covered by the Directive must meet. In most cases these are health and safety requirements. Before the product may be placed on the market, or brought into use for the first time, designers and manufacturers must meet all the essential requirements relevant to the particular product.

The *Supply of Machinery (Safety) Regulations 2008, SI 2008/1597* sets out the EHSRs of the machinery directive in full in part 1 of Schedule 2.

They are set out in six sections:

(1) General – including safety principles, design to facilitate handling, ergonomics, control systems, protection against mechanical hazards, other hazards such as from electricity and other forms of energy, temperature, fire and explosion, emissions from noise, vibration, radiation and hazardous substances, maintenance, cleaning and information (including sales literature) and markings/warnings;

(2) Those relating to certain classes of machines – including machinery for foodstuffs, cosmetics and pharmaceuticals, for portable hand-held and hand-guided machinery, and machinery for working wood and materials with similar characteristics;

(3) Requirements relating to the mobility of work equipment - including operating/seating positions, controls, roll-over, turning over, falling object protection, means of access, power transmission between machinery, and towing;
(4) Requirements relating to lifting operations – including stability, lifting accessories, information and marking;
(5) Requirements for machinery intended to work underground, and
(6) Requirements for machinery lifting people, including for greater strength/safety margins, and risks to persons being carried.

While not compulsory the harmonised standard BS EN ISO 12100:2010 *Safety of machinery – General principles for design – Risk assessment and risk reduction* (available from the British Standards Institute (BSI)) provides fundamental guidance and an overall framework to enable designers to design machines that are safe for their intended use. The international standard for machinery safety outlines the general principles of machinery safety and risk assessment and management in order to help identify and eliminate hazards at different stages of the machinery's lifecycle and avoid accidents.

CE Marking

[M1011] The HSE explains that a CE mark is required for all new products subject to one or more European product safety Directives. It shows that the manufacturer of the product is declaring conformity with all relevant product supply law.

It is the responsibility of the person who places the product on the market, or puts it into service, for the first time – the Responsible person. In most cases this is the manufacturer (or the manufacturer's representative who has been authorised in writing). It could also be someone who imports non CE marked products into Europe; any user in Europe who makes a product for their own use; or those who modify existing products already in use to such an extent they are considered 'new' products.

CE marking is the final stage of the conformity assessment process.

Conformity Assessment

[M1012] This is the process by which persons can legally place safe and compliant products onto the European market (or bring them into use) for the first time.

Some products can be self-certified, but others must undergo one of a number of specific conformity assessment procedures involving third parties known as Notified or Conformity Assessment Bodies, before the manufacture can declare conformity.

Technical File

[M1013] Manufacturers of new machinery must collect and be able to assemble comprehensive information covering the design, construction, conformity assessment and use of the product to demonstrate how it complies with all the relevant directives. This is known as a technical file and should be kept available for at least 10 years since last production of the product range.

Declaration of Conformity

[M1014] Most new products must be supplied to end users with a Declaration of Conformity. This is a certificate containing key information including:

- The name and address of the organisation taking responsibility for the product;
- A description of the product; and
- A list of which product safety directives it complies with.

It may include details of the relevant standards used and it should be dated and signed by a representative of the organisation placing on the European market.

The HSE explains that a declaration of conformity is 'a formal declaration by a manufacturer, or the manufacturer's representative, that the product to which it applies meets all relevant requirements of all product safety directives applicable to that product. It is a sign that a product has been designed and constructed for compliance with relevant essential requirements, and has been through the appropriate conformity assessment processes.'

It advises purchasers of machinery that they should retain Declarations of Conformity as evidence that a product complied with the safety requirements when it was first placed the market, or brought into use, and have met their duty under *Regulation 10* of the *Provision and Use of Work Equipment Regulations 1998 (PUWER)* (see **M1007** above).

The *Supply of Machinery (Safety) Regulations 2008* apply to manufacturers or their 'authorised representatives'. An authorised representative is defined in the regulations as someone who has received a written mandate from the manufacturer to perform, on the manufacturers behalf, all or part of the obligations and formalities imposed on them by the regulations.

In some cases, they apply to others – such as importers of non-CE-marked equipment from outside the EU, and those who design and construct machinery for their own use. However, they do not usually apply to intermediate suppliers of CE-marked machinery, which is covered by section 6 of the HSW Act. The Health and Safety Executive (HSE) advise that the non-application of the Regulations to intermediate suppliers may change in the next few years, when the Machinery Directive is brought into line with other EU legislation.

The regulations are enforced by both the HSE and local trading standards officers, depending on the field of use of the equipment. Where machinery is for use at work, the HSE leads on enforcement.

More information is available on the HSE website at www.hse.gov.uk.

Requirements of other EU directives

[M1015]–[M1017] Machinery suppliers need to identify, apart from the Machinery Directive, all other EU Directives relevant to their products. It is then necessary to identify specific legislation, which implement them. Usually more than one EU directive may apply to particular machinery. For example, in addition to the Machinery Directive, an item of electrically driven machine would be subject to the following Directives:

- Electromagnetic Compatibility Directive (2014/30/EU) implemented in the UK by the *Electromagnetic Compatibility Regulations 2016 (SI 2016 No 1091)*;
- Low Voltage Directive (2014/35/EU) implemented in the UK by the *Electrical Equipment (Safety) Regulations 2016 (SI 2016 No 1101)*;
- Other EU directives, which may be relevant to some machinery include the Simple Pressure Vessel Directive (2014/29/EU) and the Potential Explosive Atmosphere Directive (2014/34/EU).

More information is available on the HSE website at www.hse.gov.uk/work-equipment-machinery/european-community-law-supply-new-products.htm.

The machine is in fact 'safe'

[M1018] The key objective of the EU Machinery Directive is that the onus is placed on suppliers to demonstrate that a machine complies with relevant EH&SRs before it is placed for sale within the market place of the EU. The requirement for the machine to be in fact safe is a UK rather than EU requirement.

'Safe' is defined by the *Supply of Machinery (Safety) Regulations 1998* as meaning:

> 'in relation to machinery, that when it is properly installed and maintained, and used for the purposes for which it is intended, or under conditions which can reasonably be foreseen, it does not –
> (a) endanger the health of, or result in death or injury to any person; or
> (b) where appropriate –
>   (i) endanger the health of, or result in death or injury to domestic animals; or
>   (ii) endanger property [or the environment].'

Failure to comply with relevant EU requirements will mean that the machinery cannot legally be supplied in the UK and could result in prosecution and penalties on conviction of a fine or in some cases, of imprisonment for up to three months, or of both.

The same rules apply everywhere in the EU, so machinery complying with the EU regime may be supplied in any Member State.

The Transposed Harmonised European Machinery Safety Standards

[M1019] Application of the European Transposed Harmonised Machinery Safety Standards is fundamental to the risk-based approach. These standards have now replaced corresponding British Standards relating to machinery safety, as well as other EU member countries national standards.

The HSE sets out that the use of standards in complying with European product safety directives is not compulsory but can be useful when designing products.

It also explains that some European standards have a special legal status and define minimum acceptable levels for health and safety by supporting the essential requirements of these directives, and if followed fully 'may give a presumption of conformity to the relevant Directive's essential requirements'.

Status and scope of the Harmonised Standards

[M1020] It goes on to explain: 'In the field of European product safety transposed harmonised standards are a group of standards with special status. Their status is confirmed by their listing in the Official Journal of the European Union, although usually an indication of the status of any particular standard is given in the standard, and any limits on its coverage in its own Annex Z. Although their use remains voluntary, if a transposed harmonised standard is followed in full by a product designer it can confer presumption of conformity with one or more essential (health and safety) requirements, provided the product is within scope of the standard and the standard supports the relevant product safety Directive without qualification.'

Structure of the Harmonised Standards

[M1021] The structure of safety standards for machinery is as follows:

- 'A' type Standards provide general requirements applicable to all machines. The key standard relating to the EU Machinery Directive is BS EN ISO 12100:2010 *Safety of machinery. General principles for design. Risk assessment and risk reduction*. This sets the international standard for machinery safety.
- 'B' type Standards are those on techniques or components, which could be applicable to a large number of machines such as BS EN 574 *Two-hand control devices*. The B type or horizontal group safety standards deal with only one safety aspect or one safety-related device at the time.
- 'C' type Standards are drawn up to cover a particular type or class of machines. An example of a C Standard is BS EN 1493 Vehicle Lifts.

Machinery risk assessment

[M1022] The risk-based approach involves a structured and systematic risk assessment. Risk assessment is essentially a proactive means to foresee how accidents can happen and to decide what corrective and preventive actions are needed in advance to prevent such accidents.

Risk assessment can be defined as ' . . . A structured and systematic procedure for *identifying hazards* and *evaluating risks* in order to prioritise decisions to reduce risks to a *tolerable* level'.

This section will provide guidance on the following key elements of a structured and systematic risk assessment:

- techniques for hazard identification and analysis;

- criteria for the estimation and evaluation of risks;
- the concept of 'ALARP' and risk tolerability, and
- the preferred hierarchy for risk control options.

In addition to demonstrating compliance with relevant legislation and standards, risk assessment is crucial in aiding the consistency of decision making and the cost effectiveness in allocation of resources for health, safety and environmental issues.

Risk assessment is vital for two reasons. Firstly, the level of risk determines the priority which should be accorded to, and the selection of appropriate health, safety and environmental control measures to deal with, different hazards. Secondly, the standard of integrity of corrective/preventive measures will depend on the level of risk.

Risk assessment techniques are complementary to the more pragmatic ways of problem identification and assessment. They highlight systematically how hazards can occur and provide a clearer understanding of their nature and possible consequences, thereby improving the decision-making process for the most effective way to prevent injury and damage to health or the environment.

The techniques range from relatively simple qualitative methods of hazard identification and analysis to the advanced quantitative methods for risk assessment, in which numerical values of risk frequency or probability are derived.

Hazard and risk – definitions

[M1023] A *hazard* can be defined as 'a potential to cause harm'.

Harm can be defined as:

- injury or damage to health;
- damage to the environment;
- economic losses (interruption to production/asset damage).

Risk can be defined as:

- the chance (probability) of the harm being realised, combined with its consequences; or
- Risk = Chance of exposure to the hazard x Consequences (severity).

FIGURE 3 below illustrates the concept of risk. If for ,example, the chance of an accident involving a collision between a forklift truck and an operator is estimated as being 'remote', and the resulting injury severity as fatal 'catastrophic', then the Risk Matrix shows this as risk level 'A' or high, which should warrant urgent attention. If on the other hand the chcance of slipping on a wet surface in part of the workplace was estimated as being 'occasional', which would result in 'minor' injury, the Risk Matrix shows this to be risk level 'C' or low.

Figure 3: The Risk Matrix

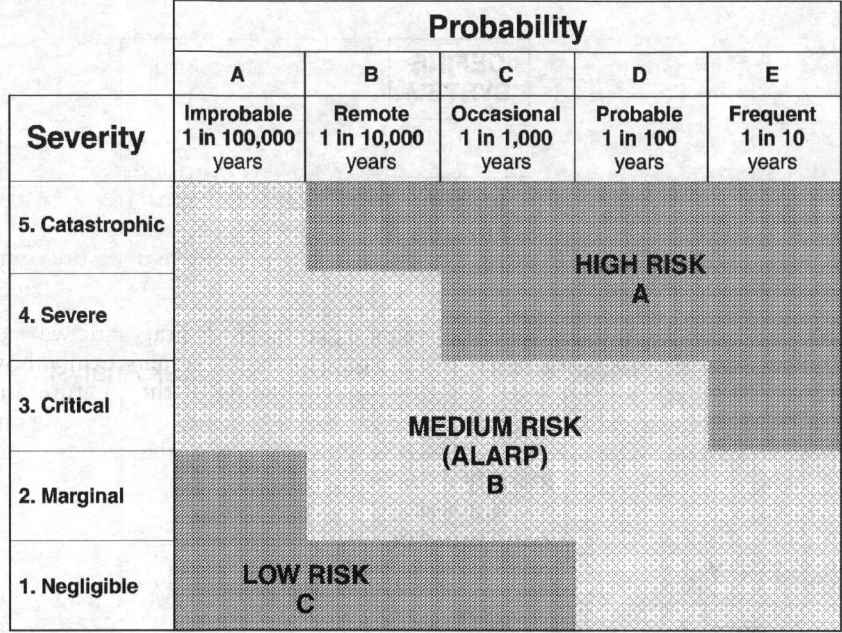

Risk assessment framework

[M1024] Hazard analysis is an essential part of the overall risk assessment framework. The flow-chart shown in FIGURE 4 outlines elements of the overall framework. The effectiveness of hazard analysis and risk evaluation processes involves a number of practical considerations.

Figure 4: Risk assessment framework

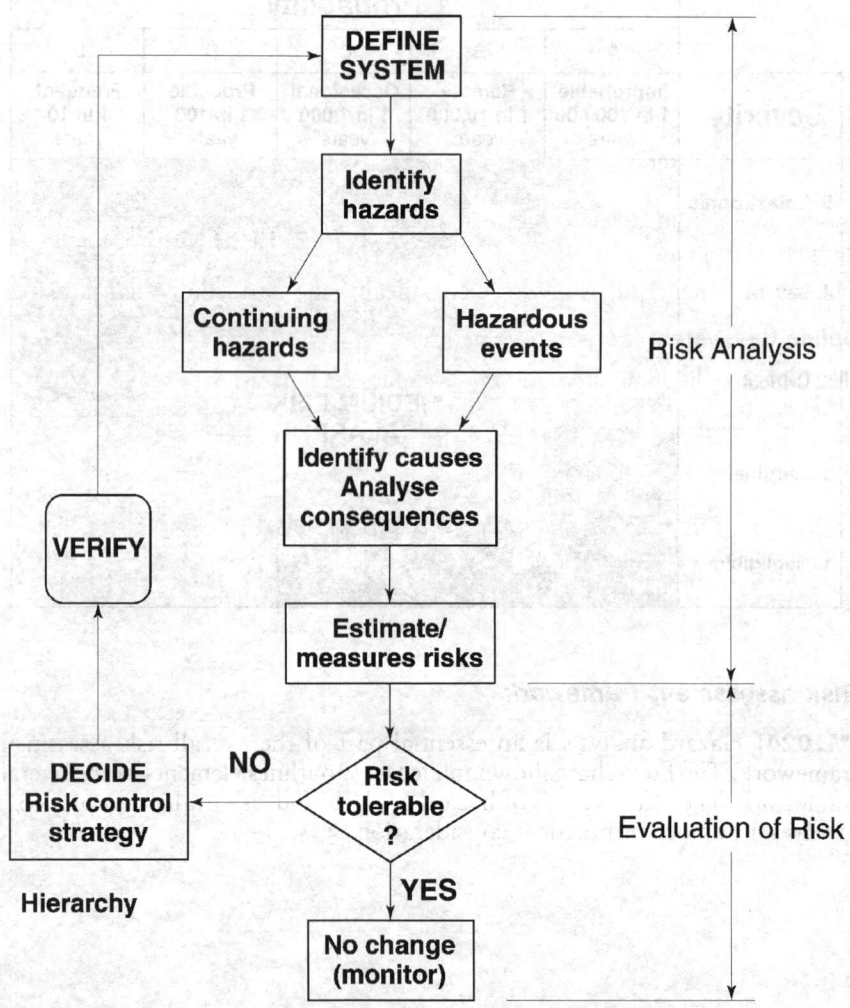

Before commencing the study, its objectives and scope must be defined explicitly. The purpose of the study and the fundamental assumptions made must be clearly stated. In order to limit the analysis, the human/system boundaries must also be defined, as well as its intended design, use, operation and layout.

The risk assessment procedure contains two essential elements:

(1) *Risk analysis* – where hazards and hazardous situations are systematically identified and their consequences are analysed. The level of risk associated with each hazardous situation is estimated or measured.

(2) *Risk evaluation* – in which a judgement is made as to whether the level of risk is acceptable/tolerable or whether corrective/preventive action is needed.

Following any modifications in the plant/process design or operation/maintenance procedures, there is a need to verify that the original hazards/hazardous situations are controlled, and that new hazards are not introduced as a result of these modifications. It is crucial that the risk assessment is updated as a result of design/procedural evolution. The safety management system developed by the employer as a result of risk assessment is also a living system, capable of learning from experience of the people who run it, and should reflect what is actually taking place, and not what should happen.

The main elements of risk assessment, ideally, should include the following:

Define the system/activity

[M1024.1] The first step in any risk assessment is to provide terms of reference to the detailed description of the process/activity being studied. This clear description is needed at this stage in order to avoid confusion at a later stage and to define the boundary and interfaces with other processes/activities. This step also involves the identification of the design objectives, material being handled/processed and exposure patterns for individuals involved in operation and maintenance.

Define the study objective

[M1024.2] Because risk assessment is used as a means of demonstrating compliance with a number of codes/regulations, it is important to limit the study to the range of hazards and potential consequences in relation to specific requirements, eg COSHH (control of substances hazardous to health), noise at work, the environment, machinery safety or economic risks.

Identify the hazards

[M1024.3] Having defined the scope and objectives, the identification of hazards is the third, but the most important, step in any risk assessment study because any hazard omitted at this stage will result in the associated risk not being assessed.

It is important in this respect to distinguish between continuing hazards (those inherent in the work activity/equipment/machinery/substances under normal conditions), eg noise, toxic substances, mechanical hazards, and hazards which can result from failures/error, eg hardware/software failures as well as foreseeable human error.

The checklist shown in TABLE 1 at [[M1025]] may be used to identify continuing hazards.

The range of continuing hazards outlined in the checklist is based on the EU Machinery Directive and is intended to be used as an aid-memoire. This approach to hazard identification will focus attention on parts of the work activity/equipment/machinery to identify the sources of each relevant hazard, who may be exposed to them, eg the operator, maintenance, and the task involved, eg loading, removing, tool setting, cleaning, adjusting.

[M1024.3] Machinery Safety

Hazards resulting from equipment failure and human error require an open-ended type of analysis, based on brain storming. This approach normally involves the following:

(1) A detailed hazard identification to be carried out for all stages of the process life-cycle. Ideally, this should include the identification of contributory causes for each hazard.
(2) Analysis of systems of work and established procedures to identify who might be exposed to a hazard and when, eg operator under normal conditions, maintenance/tool fitter/cleaner, etc.

Structured and systematic techniques which could assist in the identification of hazardous events include the following:

- *Hazard and Operability Study* ('HAZOP') – a qualitative technique to identify hazards resulting from hardware failures and human error. This involves potential causes and consequences
- *Failure Modes and Effects Analysis* ('FM&EA') – an inductive technique to identify and analyse hardware failures. This technique can be used to quantify risks.
- *Task Analysis* – an inductive technique to identify the opportunity for human error.

Machinery hazards identification

[M1025] A person may be injured at machinery as a result of one or more of the following:

(a) contact or entanglement with the machinery;
(b) crushing between a moving part of the machine and fixed structures;
(c) being struck by ejected parts of the machinery;
(d) being struck by material ejected from the machinery.

These are regarded as mechanical hazards.

Figure 5: Examples of mechanical hazards

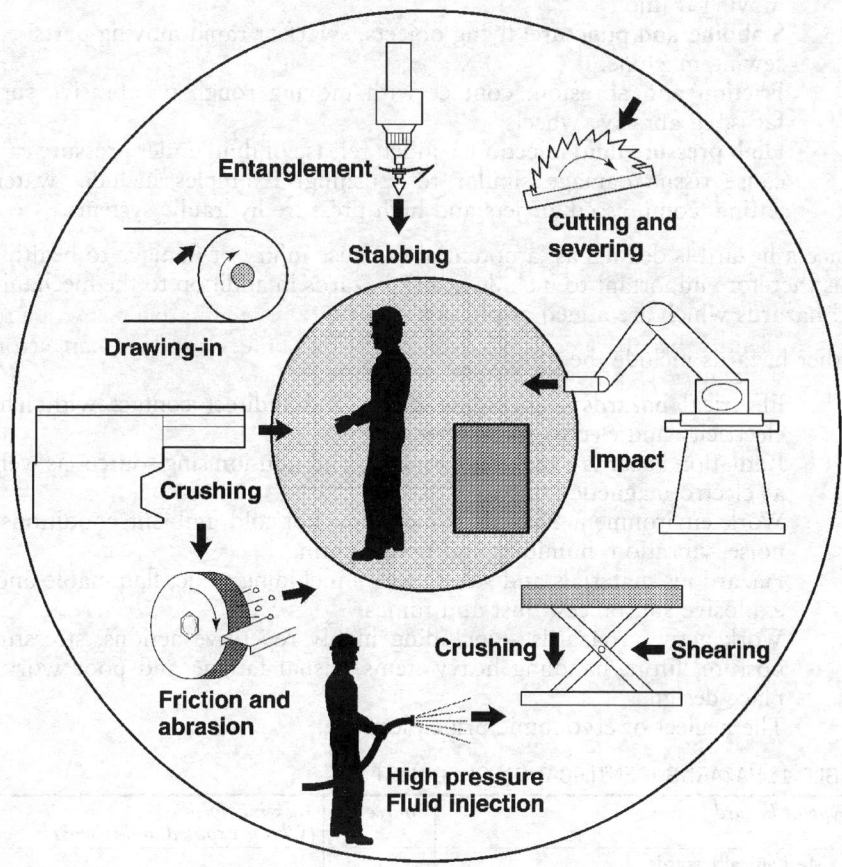

The main types of mechanical hazards shown in FIGURE 5 are described as:

- **Crushing**: Occurs when part of the body is caught between a moving part of a machine against a fixed object, eg underneath scissor lift or between the tools of a press.
- **Shearing**: Parts of the body may be sheared by scissor action caused by parts of the machine, eg mechanism of scissor lift or oscillating pendulum.
- **Cutting and severing**: Cutting hazards include contact with circular saws, guillotine knife, rotary knives or moving sheet metal.
- **Entanglement**: Occurs as a result of clothing or hair contact with rotating objects or catching on projections or in gaps, eg drills, rotating workpiece or belt fasteners.
- **Drawing-in or trapping**: Occurs when part of the body is caught between two counter-rotating parts, eg gears, mixing mills or between belt and pulley or chain and chain wheel.

- **Impact:** Impact hazards are caused by a moving object striking the body without penetrating it. Examples include: being hit by a robot arm or by moving traffic.
- **Stabbing and puncture:** flying objects, swarf or rapid moving parts, eg sewing machine.
- **Friction and abrasion:** contact with moving rough or abrasive surfaces, eg abrasive wheels.
- **High pressure fluid injection:** sudden release of fluid under pressure can cause tissue damage similar to crushing. Examples include: water jetting, compressed air jets and high pressure hydraulic systems.

Since a hazard is defined as 'a potential to cause injury or damage to health', it is therefore important to include health hazards in addition to the mechanical hazards which are aimed at physical injury.

Other hazards include the following:

- Electrical hazards – including direct and indirect contact with live electricity and electro-static phenomena.
- Radiation hazards – including ionising and non-ionising sources as well as electro-magnetic effect.
- Work environment hazards – including hot/cold ambient conditions, noise, vibration, humidity and poor lighting.
- Hazardous materials and substances – including toxic, flammable and explosive substances, dust and fumes.
- Work activity hazards – including highly repetitive actions, stressful posture, lifting/handling heavy items, visual fatigue and poor workplace design.
- The neglect of ergonomic principles.

TABLE 1: HAZARDS IDENTIFICATION CHECKLIST

Type of hazard	Source	Task involved (Who is exposed and when?)
1. Mechanical hazards		
1.1 Crushing 1.2 Shearing 1.3 Cutting/severing 1.4 Entanglement 1.5 Drawing-in/trapping 1.6 Impact 1.7 Stabbing/puncture 1.8 Friction/abrasion 1.9 High pressure fluid injection 1.10 Slips/trips/falls 1.11 Falling objects 1.12 Other mechanical hazards		
2. Electrical hazards		
2.1 Direct contact 2.2 Indirect contact 2.3 Electrostatic phenomena 2.4 Short circuit/overload 2.5 Source of ignition 2.6 Other electrical hazards		

Type of hazard	Source	Task involved (Who is exposed and when?)
3. Radiation hazards		
3.1 Lasers 3.2 Electro-magnetic effects 3.3 Ionising/non-ionising radiation 3.4 Other radiation hazards		
4. Hazardous substances		
4.1 Toxic fluids 4.2 Toxic gas/mist/fumes/dust 4.3 Flammable fluids 4.4 Flammable gas/mist/fumes/dust 4.5 Explosive substances 4.6 Biological substances 4.7 Other hazardous substances		
5. Work activities hazards		
5.1 Highly repetitive actions 5.2 Stressful posture 5.3 Lifting/handling heavy items 5.4 Mental overload/stress 5.5 Visual fatigue 5.6 Poor workplace design 5.7 Other workplace hazards		
6. Work environment hazards		
6.1 Localised hot surfaces 6.2 Localised cold surfaces 6.3 Significant noise 6.4 Significant vibration 6.5 Poor lighting 6.6 Hot/cold ambient temperature 6.7 Other work environment hazards		

Options for machinery risk reduction

[M1026] The most effective risk control measures are those implemented at the machine/work equipment design stage. It is a legal requirement that all machinery supplied in the EU should carry a 'CE' mark. This demonstrates the machine complies with all relevant EU machinery directives, including meeting the essential health and safety requirements (ES&HRs). Fundamental to this approach is manufacturers and suppliers carrying out a risk assessment and demonstrating that all risks are adequately controlled by design, rather than by procedures. The following risk control options reflect the preferred order of priority:

Technical measures to be taken at the design stage

[M1026.1]

(1)   To make the machinery inherently safer (design out hazards) as a priority:
— by eliminating hazards at the design stage;
— by substituting hazardous substances used by the machine or process by less hazardous ones.

[M1026.1] Machinery Safety

(2) To provide protection from the hazards by the machine design, eg to avoid the need for access to hazardous parts of the machine.
(3) To improve machinery reliability by reducing the chance for fail-to-danger of critical components/systems.
(4) To provide protection from the hazards by adequate safeguarding.
(5) To reduce ease of access to danger zones (eg safeguards or safety devices which allow safe access but prevent access at other times).
(6) To reduce the chance of potential human error:
  - by improving the ergonomics of machine control layout to reduce the chance of unintended errors;
  - by reducing the chance of violations by making maintenance, cleaning, adjustments, etc tasks convenient, easy and safe.

Procedural measures (make work tasks safer)

[M1026.2] These may be achieved by the following measures:

- Planned preventive maintenance and regular inspection of machines and safeguards/safety devices.
- Safe systems of work (which would minimise the need for access into the danger zone).
- Permit-to-work procedures (to formalise precautions in the face of a hazard).
- Adequate personal protection equipment ('PPE') and planning for emergencies.

Behavioural measures (develop safer people)

[M1026.3] This involves the following:

- Selection and certification of personnel against specific work tasks.
- Training: basic skills, systems and procedures, as well as knowledge of hazards.
- Adequate instructions, warnings and supervision.
- Improve safety culture within the business.
- Ask: why should safeguards/safety devices be violated/defeated?

Types of safeguards and safety devices

[M1027] The main types of safeguards and safety devices can be classified as follow:

(1) Fixed guards.
(2) Fixed guards with adjustable element.
(3) Automatic guards.
(4) Interlocked guards.
(5) Safety devices.
   These are described as:
   5.1 Trip devices, eg photoelectric light curtains, pressure sensitive devices.

5.2 Two-hand control devices.

Machinery safeguards – general considerations

Figure 6: Design/selection of safeguards [M1028]

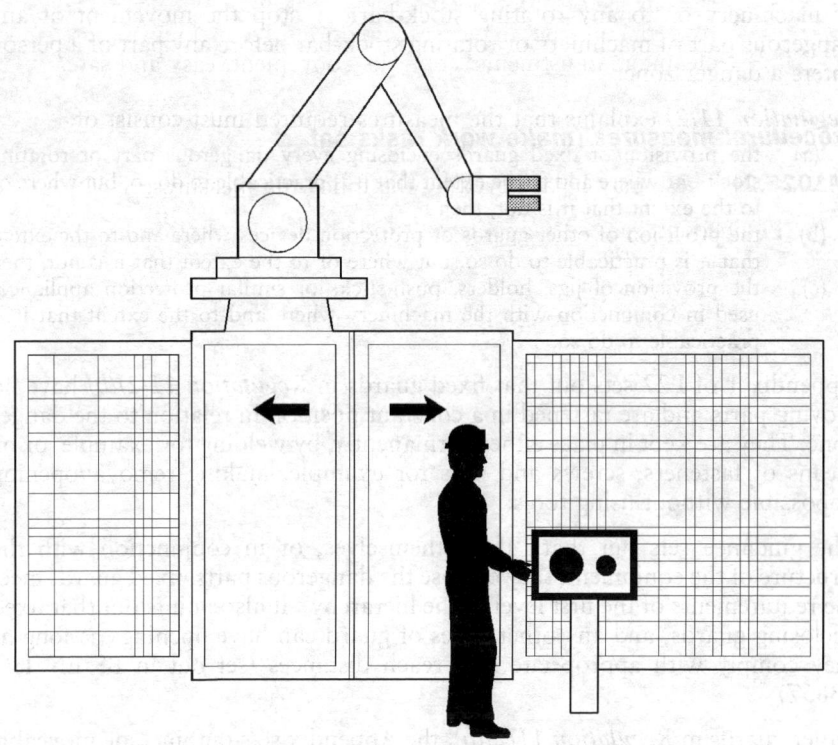

It is a general requirement that all machinery safeguards and safety devices must have the following characteristics:

- Be of robust construction: strength, stiffness and durability to prevent ejected parts of the machine/components or material penetrating the guard.
- Not give rise to additional hazards.
- Not be easy to bypass or render non-operational.
- Be located at an adequate distance from the danger zone.
- Cause minimum obstruction of view for machine operators.
- Enable essential work (eg maintenance, cleaning) to be done without safeguard removal.

The two most applicable EU Standards relevant to safeguards are *BS EN ISO 14120:2015 Safety of machinery. Guards. General requirements for the design* and construction of fixed and movable guards and BS EN ISO 13857 *Safety of machinery. Safety distances to prevent hazard zones being reached by upper and lower limbs.*

Fixed guards

[M1029] *Regulation 11* of *PUWER'98*, which deals with dangerous parts of machinery, sets out that employers must prevent access to any dangerous part of machinery or to any rotating stock-bar; or stop the movement of any dangerous part of machinery or rotating stock-bar before any part of a person enters a danger zone.

Regulation 11(2) explains that the measures required must consist of:

'(a) the provision of fixed guards enclosing every dangerous part or rotating stock-bar where and to the extent that it is practicable to do so, but where or to the extent that it is not, then
(b) the provision of other guards or protection devices where and to the extent that it is practicable to do so, but where or to the extent that it is not, then
(c) the provision of jigs, holders, push-sticks or similar protection appliances used in conjunction with the machinery where and to the extent that it is practicable to do so, . . . "

Appendix 1 of L22 sets out that fixed guards in *Regulation 11(2)(b)* have no moving parts and are fastened in a constant position in relation to the danger zone. They are kept in place either permanently, by welding for example, or by means of fasteners, screws and nuts for example, making removal/opening impossible without using tools.

The guidance sets out that: 'If by themselves, or in conjunction with the structure of the equipment, they enclose the dangerous parts, fixed guards meet the requirements of the first level of the hierarchy'. It also points out that fixed enclosing guards, and any other types of guard can have openings as long as they comply with appropriate safe reach distances (set out in BS EN ISO 13857).

Other guards in *Regulation 11(2)(b)*, the Appendix sets out, include moveable guards which can be opened without the use of tools, and fixed guards that are not fully enclosing. These allow limited access through openings or gates, for example, to allow feeding materials, making adjustments and cleaning.

A fixed guard should, when fitted, be incapable of being displaced casually, so the method of fixing is important. The ideal method of fixing the guard would be of a captive type, which would make unofficial access difficult.

It is clear from experience that if there is a need for access into hazardous parts of a machine fitted with a fixed guard, there will be a tendency not to replace the guard back onto the machine. Therefore, an interlocked guard is more suited in this situation.

Figure 7: Example of a fixed guard

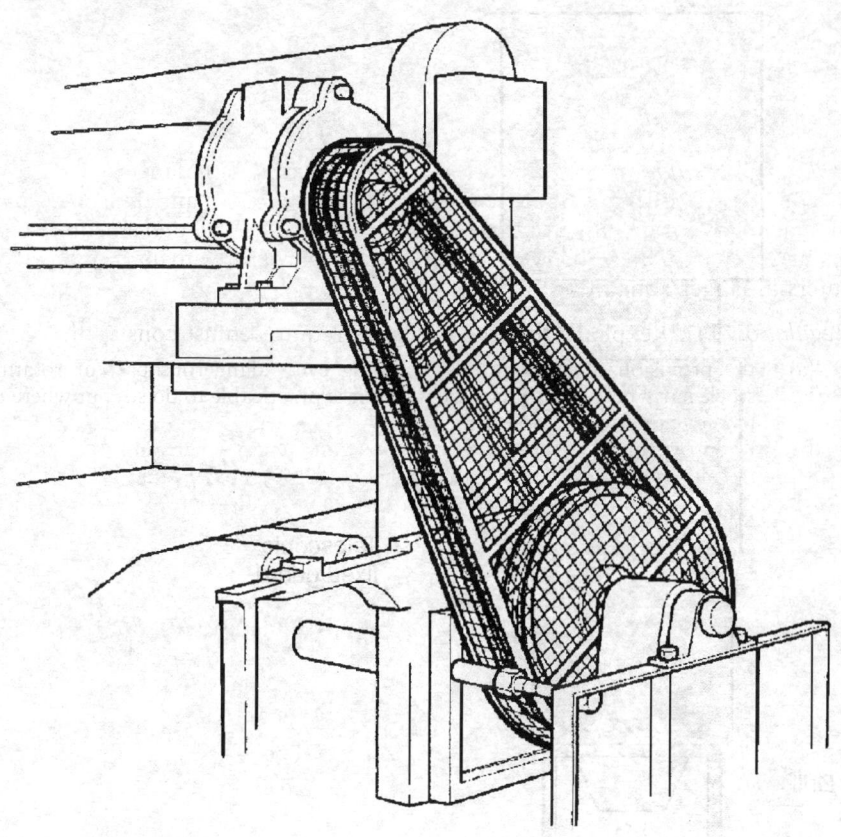

Adjustable guards

[M1030] Adjustable guards comprise a fixed guard with adjustable elements that the setter or operator has to position to suit the job being worked on. They are widely used for woodworking and tool-room machines. Where adjustable guards are used, the operators should be familiar with, and trained in, how to adjust them so that full protection can be obtained.

FIGURE 8 shows the application of a telescopic guard to a heavy duty-drilling machine. The idea here is that as the drill descends into the workpiece, the fixed guard will always enclose the drill. The approach to selecting the most appropriate safeguard should start by hazard identification. Hazards associated with the use of the drilling machine include the following:

- entanglement with the drill;
- stabbing/puncture by the swarf;
- ejection of broken drill bit;
- stabbing by the drill bit; and
- possible ejection of the work-piece.

Figure 8: Fixed guard with adjustable element

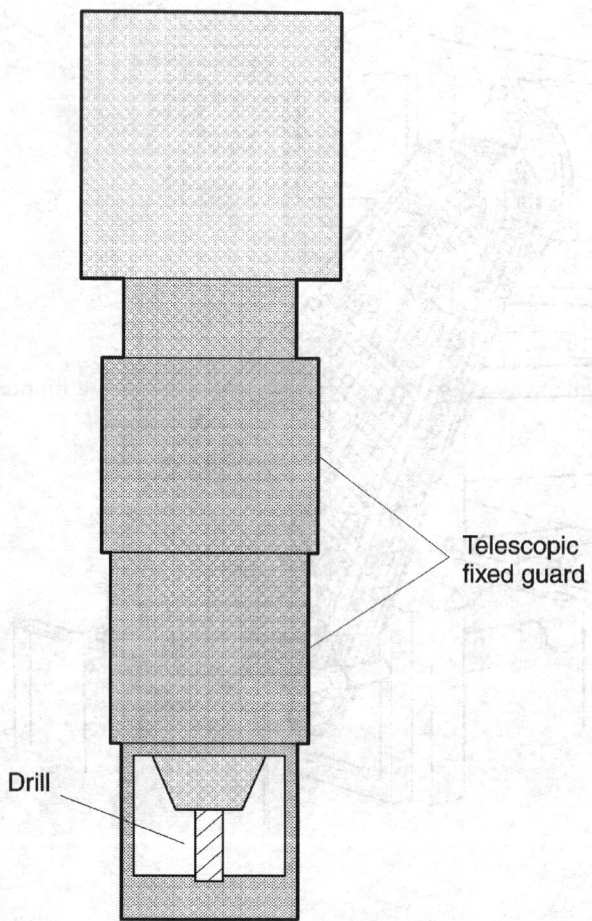

It is noted that the telescopic guard should provide protection against most of the above hazards apart from stabbing/puncture by the drill and ejection of the work-piece.

Self-adjusting guards are also classified as a fixed guard with adjustable element. This guard prevents access to the hazard except when the guard is forced open by the passage of the work. They usually incorporate a spring-loaded pivoted element. These types of guards are mainly used on woodworking machinery.

Automatic guards

[M1031] An automatic guard is moved into position automatically by the machine, thereby removing any part of a person from the hazardous area of

Fixed guards [M1032]

the machine. This is sometimes known as a 'sweep away' guard and is used on some paper guillotines and large mechanical or hydraulic presses.

There are however many factors that could limit the effectiveness of these types of guards.

Some possible hazards associated with automatic guards include the following:

- shearing hazard between the moving and fixed parts of the guard;
- impact with the moving part of the guard;
- crushing hazard against fixed structures nearby as well as the ergonomics of the guard, eg height of guard in relation to operators and the size of guard gaps.

Interlocking guards

[M1032] Interlocking guards are usually movable, eg they could be hinged, sliding or removable. These guards are used, where frequent access to hazardous parts of a machine may be needed, and are connected to the machine controls by means of 'position sensors'. These interlocking elements could be electrical, mechanical, magnetic, hydraulic or pneumatic. The position sensors interlock the guard with the power source of the hazard. When the guard is open the power is isolated, therefore allowing safe access into the relevant part of the machine.

Figure 10: Main types of electrical interlocking switches

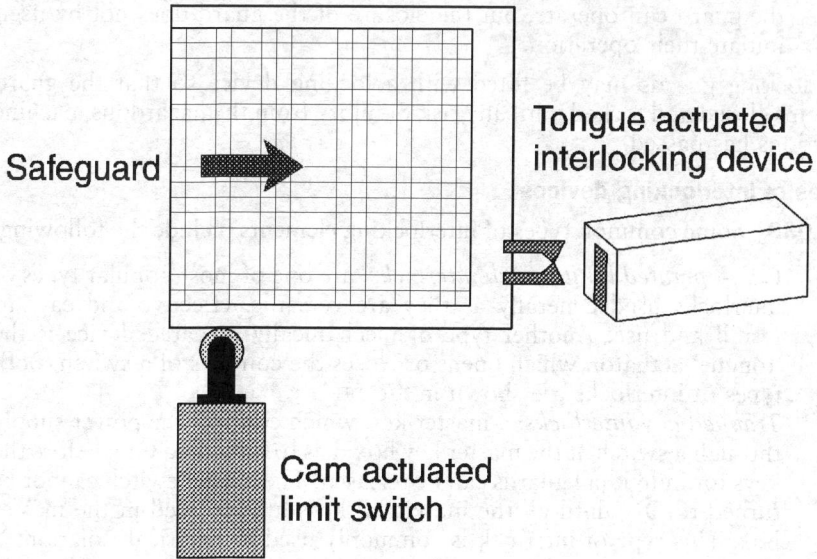

BS EN ISO 14119 *Safety of machinery. Interlocking devices associated with guards. Principles for design and selection* provides detailed guidance on the design and selection of the appropriate type of devices. Risk assessment plays a vital role in the selection of type and level of integrity on interlocking devices.

The integrity of the interlocking mechanism must ensure that the device is reliable, capable of resisting interference, difficult to defeat and the system should not, as far as possible, fail-to-danger. Fail-to-danger of an interlocking device means that the machine can be operated when the guard is open.

Interlocking guards are convenient as they give ready access into the danger zone while ensuring the safety of the operator. However, some of these guards may be defeated to carry out necessary tasks, eg maintenance, setting, fault finding etc. Therefore, additional measures such as locking off or isolation may need to be introduced under a safe system of work. It is, however, a requirement that the machine designer should consider how these tasks can be safely performed at the design stage.

Interlocking guards can be categorised into two broad categories:

- Guard is interlocked with the source of power.
- Guard is interlocked with respect to the motion itself.

There are clear advantages to interlocking the guard with the moving parts rather than the source of power, as the power might fail-to-danger due to wiring/earth faults or failures of the interlocking switch. The main requirements for the movable interlocking guards are:

(a) the hazardous machine functions covered by the guard cannot operate until the guard is closed;
(b) if the guard is opened while hazardous machine functions are operating, a stop instruction is given;
(c) when the guard is closed, the hazardous machine functions covered by the guard can operate, but the closure of the guard does not by itself initiate their operation.

Interlocking guards may be fitted with a locking device so that the guard remains closed and locked until any risk of injury from the hazardous machine functions has passed.

Types of interlocking devices

[M1033] Some common types of interlocking elements include the following:

- *Cam-operated limit switch interlocks*: are one of most popular types of interlocks used generally as they are versatile, effective and easy to install and use. Another type of mechanically actuated device is the 'tongue' actuator, which opens or closes the contacts of a switch. Both types of interlocks are shown in FIGURE 13.
- *Trapped key interlocks*: a master key, which controls the power supply through a switch at the master key box, has to be turned OFF before the keys for individual guards can be released. The master switch cannot be turned to ON until all the individual keys are replaced in the master box. This type of interlock is commonly used in electrical isolation.
- *Captive-key interlocking*: involves a combination of an electrical switch and a mechanical lock in a single assembly.
- *Direct manual switch or valve interlocks*: where a switch or valve controlling power source cannot be operated until a guard is closed, and the guard cannot be opened at any time the switch is in run position (position sensors).

- *Mechanical interlocks*: provide a direct linkage from guard to power or transmission control. In other words, the switch cannot be reached when the guard is open.
- *Magnetic switches* with the actuating 'coded' magnet attached to the guard: have the disadvantage that they can be defeated easily. However, types using a shaped magnet have been used with removal guards on screw conveyors. Normal reed switches are not acceptable unless they incorporate special current limiting features. Reed switches are often encapsulated and have application in flammable atmospheres. A similar type of switch relies on inductive circuits.
- *Electro-mechanical device*: can provide a time delay where the first movement of the bolt trips the machine, but the bolt has to be unscrewed a considerable distance before the guard is released. Time delay arrangements are necessary when the machine being guarded has a long run-down time. A solenoid-operated bolt can also be used in conjunction with a time delay circuit.
- *Mechanical restraints (scotches)*: are used as a back-up to other forms of interlocks on certain types of presses to protect the operator against crushing injury between platens.

Trapped key interlocking (Power isolation)

[M1034] The electrical interlocking switches interrupt the power source of the hazard by switching of a circuit which controls the power-switching device. This is known as control interlocking.

Power interlocking, on the other hand, involves the direct switching of the power supply to the hazard, eg isolation of power. The most practical means of power interlocking is a trapped key system.

An example of such a system is illustrated in FIGURE 11. The power supply isolation switch is operated by key 'A' which is trapped in position while the isolation switch is in the ON position. When the key is turned and released, the isolation switch contacts are locked open thus isolating the power supply. The machine safeguard door in the meantime is locked closed. To open the guard, the isolation key 'A' is inserted into the guard-locking unit and turned to release the door. The key is then trapped in position and cannot be removed until the guard is closed and locked again.

One of the drawbacks of this system is that unknown to others, a person may still be inside the machine while another person decides to start it up by closing and locking the guard. This foreseeable problem can be overcome by using a personal key 'B' which is also trapped in the guard locking unit and is released when the guard is open. Key 'A' cannot be released from the guard unless key 'B' is inserted and turned into the guard locking unit. Key 'B' can also be used for other tasks, eg inch mode for presses or programming mode for robotics.

Figure 11: Trapped key system for power interlocking

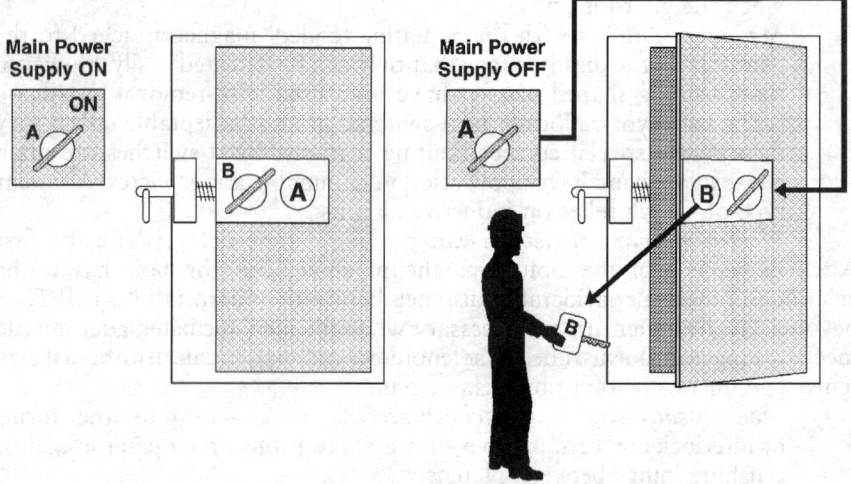

This type of system is reliable and does not require electrical wiring to the guard. The main disadvantage is that because it requires the exchange of the keys every time, it may not be suitable if access is frequently required. Strict control of the keys is essential.

Mechanical interlocking

Figure 12: Example of mechanical interlocking system

[M1035]

The system shown in FIGURE 12 is known as guard-inhibited interlocking. This is achieved in the above example by a sliding guard. When open the guard retains the lever of a switch or a control valve in the power OFF position. When the guard has been closed, the lever or switch can be moved to the ON position, thus inhibiting the guard from being open.

This system however is easy to defeat by removing one of the door-stops, or even by removing the guard.

Other forms of mechanical interlocks include mechanical restraints (scotches) to prevent gravity fall of vertical machinery, eg scissors lifts and presses.

Safety devices

[M1036] A safety device is a protective appliance, other than a guard, which eliminates or reduces risk, alone or associated with a guard. There are many forms of safety device available.

(a) Trip device

A trip device is one which causes a machine or machine elements to stop (or ensures an otherwise safe condition) when a person or a part of his body goes beyond a safe limit. Trip devices take a number of forms – for example, mechanical, electro-sensitive safety systems and pressure-sensitive mat systems.

Trip devices may be:

(i) mechanically actuated – eg trip wires, telescopic probes, pressure-sensitive devices; or
(ii) non-mechanically actuated – eg photo-electric devices, or devices using capacitive or ultrasonic means to achieve detection.

(b) Enabling (control) device

This is an additional manually-operated control device used in conjunction with a start control and which, when continuously actuated, allows a machine to function.

(c) Hold-to-run control device

This device initiates and maintains the operation of machine elements for only as long as the manual control (actuator) is actuated. The manual control returns automatically to the stop position when released.

(d) Two-hand control device

A hold-to-run control device which requires at least simultaneous actuation by the use of both hands in order initiate and to maintain, while hazardous condition exists, any operation of a machine. It therefore affords a measure of protection only for the person who actuates it. BS EN 574 *Safety of machinery. Two-hand control devices. Functional aspects. Principles for design* specifies the safety requirements relating to two-hand control devices.

(e) Limiting device

Such a device prevents a machine or machine elements from exceeding a designed limit (eg space limit, pressure limit).

(f) Limited movement control device

The actuation of this sort of device permits only a limited amount of travel of a machine element, thus minimising risk as much as possible; further movement is precluded until there is a subsequent and separate actuation of the control.

(g) Deterring/impeding device

This comprises any physical obstacle which, without totally preventing access to a danger zone, reduces the probability of access to this zone by preventing free access.

Major Accident Hazards

Andrea Oates

Background to major accident hazards

[M1101]–[M1102] The flammable, oxidising, explosive or toxic properties of some of the substances used and produced in the chemical, petrochemical and other industries are such that accidents involving them may have the potential to cause harm to people, property or the environment. Where the substances are present in the scale associated with the major chemical industries, the consequences of accidents may cause loss of life, injury and damage to property over a wide area. Communities where such accidents have occurred have paid a high price for the hazards on their doorstep, for example in Seveso in Italy, Bhopal in India, Mexico City and in the UK, Flixborough. Twenty-eight workers were killed and a further 36 injured in an explosion at the Nypro (UK) site at Flixborough in Lincolnshire on 1 June 1974 after a large quantity of cyclohexane escaped and subsequently ignited.

It was the incident on 10 July 1976 in the Italian town of Seveso that prompted European legislation to prevent and control major industrial accidents. A bursting disc on a chemical reactor at the Icmesa chemical company ruptured, releasing a cloud of vapour which included a small amount of the highly toxic material dioxin or TCDD. Although no one was killed, many people became ill and 26 pregnant women who had been exposed to the release had abortions. In addition, many thousands of animals in the contaminated area died or were slaughtered in order to prevent dioxin from entering the food chain.

The original 'Seveso' Directive (Directive 82/501/EEC) was amended following incidents including the Bhopal disaster at the Union Carbide India Ltd plant. In the early morning of 3 December 1984, a cloud of highly toxic methyl isocyanate (MIC) was released and drifted over nearby housing in the Indian town of Bhopal. It killed around 2,000 people within a short period, injuring many more, and thousands more people have continued to die over the years since the incident.

Seveso II (Directive 96/82/EC) was itself superseded in 2012 by Seveso III (Directive 2012/18/EU). The latest directive takes account of the changes in European legislation on the classification of chemicals and increased rights regarding access to information. Seveso III applies to more than 10,000 industrial establishments across the European Union (EU) where dangerous substances are used or stored in large quantities, mainly in the chemical, petrochemical, logistics and metal refining sectors.

In the UK, the *Control of Major Accident Hazards Regulations 2015 (SI 2015 No 483) (COMAH 2015)* came into force on 1 June 2015, implementing the requirements of Seveso III.

[M1101] Major Accident Hazards

Following the so-called 'Brexit' vote to leave the EU in the June 2016 referendum, the UK will remain a member of the EU and is covered by all its provisions, including health and safety legislation, until the exit negotiations are complete, and a new relationship is defined. However, by the beginning of May 2019, the terms under which the UK will leave the EU were still not clear, with negotiations ongoing.

In July 2018, the government laid the *Health and Safety (Amendment) (EU Exit) Regulations 2018* to ensure that EU-derived health and safety protections will continue to be available in domestic law after the UK has left the EU. The Regulations do not make any policy changes beyond the intent of ensuring continued operability of the relevant legislation and maintain the status quo.

The HSE is contributing to cross-government work being coordinated by the Department for Exiting the European Union (DExEU), including 'work to support the Government's commitment to protect workers' rights as the UK leaves the EU by ensuring that health and safety regulation continues to provide a high level of protection in the workplace'. It is also contributing to joint work on chemicals regulation and workplace product safety.

These deal with areas including regulating chemicals (see [H20102.1]) and can be found on the DExEU website (www.gov.uk/government/collections/how-to-prepare-if-the-uk-leaves-the-eu-with-no-deal).

The Regulations replaced the earlier *Control of Major Accident Hazards Regulations 1999 (SI 1999/743)* and the HSE explained that although the main duties stayed the same, the 2015 Regulations introduced a number of important changes. These relate to how dangerous substances are classified and to the information that has to be made available to the public. Lower tier operators must now provide public information about their site and its hazards, and operators of both upper tier (formerly referred to as top tier) and lower tier establishments must now provide public information electronically and keep it up to date.

The HSE together with the Environment Agency (EA) in England, Natural Resources Body for Wales (NRBW) and the Scottish Environment Protection Agency (SEPA) are jointly responsible, as the Competent Authority (CA), for enforcing COMAH 2015 (regulations 2(1) and 4(b)), except in relation to nuclear establishments. Here, the Office for Nuclear Regulation (ONR), rather than the HSE, acts jointly with the appropriate environmental protection enforcement agency (the 'appropriate agency') (regulation 4(a)). Northern Ireland makes separate legislative and administrative arrangements.

[M1103] COMAH 2015 impose requirements where dangerous substances are present in quantities equal to or exceeding those specified in *Schedule 1*, irrespective of whether they are raw materials, products, by-products, residues or intermediates (regulation 2(1)).

The Regulations apply to dangerous substances including those which it is reasonable to foresee may be generated during the loss of control of a process, including storage activities, in these quantities. The emphasis is on controlling risks to both people and the environment through demonstrable safety management systems, which are integrated into the routine of business.

An occurrence is regarded as a major accident if:

— it results from uncontrolled developments in the course of the operation of an establishment to which the Regulations apply;
— it leads to serious danger to people or to the environment, on- or off-site (whether immediate or delayed); and
— it involves one or more dangerous substances defined in the Regulations.

Major emissions, fires and explosions are the most typical major accidents (regulation 2(1)).

The duties placed on operators by *COMAH* fall into two categories, defined by the amounts of dangerous substances: these are known as 'lower tier establishments' and 'upper tier establishments'.

General duties fall upon all operators with inventories exceeding the lower threshold – these are set out in regulations 5, 6 and 7. Where inventories exceed the higher threshold, operators are subject to additional upper tier duties, set out in regulations 8–14.

There are also requirements relating to the provision of information (regulations 17–21). Part 6 of the Regulations sets out the functions of the competent authority (regulations 22–25) and regulation 26 sets out what action must be taken following a major accident.

An 'operator' is defined (in regulation 2(1)) as a person who is in control of the operation of an establishment or installation. Where the establishment or installation is yet to be constructed or operated, the 'operator' is the person who proposes to control its operation and where that person is not known, the operator is the person who has commissioned its design and construction.

Once it has been established that the Regulations apply to an establishment, the duties for the appropriate tier then apply to all dangerous substances present, not just those that are in excess of the threshold.

The Regulations (regulation 28) provide for fees payable by the operator to the competent authority for the performance of specified functions by the HSE, ONR or environmental protection enforcement agency. Details of the current charging regime may be found on the HSE website at: www.hse.gov.uk/charging/comahcharg/comahch1.htm.

HSE guidance on the Regulations is contained in the Control of Major Accident Hazards Regulations Guidance on Regulations (L111, Third edition, 2015) which can be downloaded from the HSE website at: www.hse.gov.uk/pubns/books/l111.htm.

COMAH 2015 in brief

In summary, *COMAH 2015* impose a duty on operators to:

(a) take all measures necessary to prevent major accidents and to limit their consequences for human health and the environment (regulation 5(1));
(b) demonstrate to the competent authority that they have taken all the necessary measures (regulation 5(2));

(c) provide the competent authority with all assistance necessary to enable it to perform its functions under the Regulations, including inspections, investigations and gathering information (regulation 5(3) and (4));
(d) send the competent authority a notification containing specified information (regulation 6); and
(e) prepare and retain in writing a major accident prevention policy (MAPP) and revise it in specified circumstances (regulation 7 and Schedule 2).

With regard to operators of an upper tier establishment, COMAH 2015 requires them to:

(a) send (at specified times) a safety report to the competent authority (for the purposes set out in regulation 8) containing specified information, and not to start construction or operation of the establishment or permit modifications leading to a change in the inventory of dangerous substances, until the authority's conclusions of its examination of the safety report have been received (regulation 9 and Schedule 3);
(b) review and revise the safety report in specified circumstances (regulation 10); and
(c) prepare an internal emergency plan (regulation 11), containing specified information and review and test the plan (regulation 12 and Schedule 4).

In addition, COMAH 2015 require:

(a) a local authority in whose area there is an upper tier establishment to prepare an external emergency plan (regulation 11), containing specified information (regulation 13 and Schedule 4), subject to any exemption that may be granted under regulation 15;
(b) a local authority to review and test the external emergency plan (regulation 14);
(c) an operator or local authority who has prepared an emergency plan to put it into effect in specified circumstances (regulation 16);
(d) the competent authority to ensure that specified information is made available to the public, including by electronic means (regulation 17);
(e) the operator of an upper tier establishment to send specified information regularly to people liable to be affected by a major accident occurring at the establishment in the most appropriate form without them having to request it (regulation 18);
(f) the competent authority to adopt a specified procedure in dealing with a request for information (regulation 19); and
(g) the competent authority to provide a potentially affected Member State with sufficient information where an upper tier establishment presents a major accident hazard with possible trans-boundary consequences (regulation 20).

COMAH 2015 also:

(a) give the competent authority power to accept information in another document (regulation 21);
(b) impose functions on the competent authority with respect to—

(i) its examination of the safety report sent by the operator (regulation 22);
(ii) prohibiting the operation of an establishment (regulation 23);
(iii) its identification of domino groups of establishments (see [M1113A]), and impose cooperation duties on the operators of such establishments (regulation 24);
(iv) inspections and investigations (regulation 25);
(c) impose requirements as regards action to be taken following a major accident on the operator of the establishment concerned, the competent authority and the local authority in whose administrative area the accident has occurred (regulation 26);
(d) impose functions on the competent authority with respect to enforcement and penalties (regulation 27);
(e) provide for fees payable by the operator to the competent authority for the performance of specified functions by the HSE, ONR or environmental protection enforcement agency (regulation 28); and
(f) provide for fees payable by the operator to the local authority for the preparation, review and testing of the external emergency plan (regulation 29).

COMAH 2015 also amended several pieces of older legislation. These are set out in *Schedule 6* (regulation 30).

Application

[M1104]–[M1108] The *Control of Major Accident Hazards Regulations 2015* (COMAH 2015) apply to lower tier and upper tier establishments (regulation 3(1)) which are defined (in regulation 2(1)) with reference to specified quantities of dangerous substances (set out in Schedule 1) either present, where their presence is anticipated, or where it is reasonable to believe they may be generated during the loss of control of an industrial chemical process.

An 'establishment' refers to the whole area under the control of the same person where dangerous substances are present in one or more installations, including common or related infrastructures or activities. 'Dangerous substances' are substances or mixtures, including raw materials, products, by-products, residues and intermediates, which are:

(a) listed in column 1 of Part 2 of Schedule 1; or
(b) in a category listed in column 1 of Part 1 of Schedule 1.

These include substances with a health hazard category of 'acute toxic', explosives, flammable gases and liquids and substances that are hazardous to the aquatic environment. They include bromine, chlorine, formaldehyde, sulphur dichloride and propylamine when present in particular quantities.

A 'lower tier establishment' means 'an establishment where a dangerous substance is present in a quantity equal to or in excess of the quantity listed in the entry for that substance in column 2 of Part 1 or in column 2 of Part 2 of Schedule 1, but less than that listed in the entry for that substance in column 3 of Part 1 or in column 3 of Part 2 of Schedule 1, where applicable using the rule laid down in note 4 of Part 3 of that Schedule'.

An 'upper tier establishment' means 'an establishment where a dangerous substance is present in a quantity equal to or in excess of the quantity listed in the entry for that substance in column 3 of Part 1 or in column 3 of Part 2 of Schedule 1, where applicable using the rule laid down in note 4 of Part 3 of that Schedule'.

Part 1 of Schedule 1 specifies the qualifying quantities of dangerous substances for the application of lower tier (in Column 2) and upper tier requirements (in Column 3) in relation to the following hazard categories (in accordance with the European classification, labelling and packaging (CLP) Regulation): health hazards; physical hazards (explosives, flammable gases, flammable aerosols, oxidising gases, flammable liquids, self-reactive substances and mixtures, pyrophoric liquids and solids and organic peroxides, and oxidising liquids and solids); environmental hazards (acute and chronic hazards to aquatic environments); and other hazards.

Part 2 specifies the qualifying quantities of named dangerous substances for the application of lower tier (Column 2) and upper tier (Column 3) requirements.

Note 4 of Part 3 sets down rules governing the addition of dangerous substances or categories of dangerous substances.

The lists of categories and named dangerous substances are not reproduced here, but the Schedule can be found, together with detailed guidance, (from page 84 onwards) in the HSE publication the *Control of Major Accident Hazards Regulations Guidance on Regulations (COMAH) 2015* (L111, Third edition, 2015). This can be downloaded from the HSE website at: www.hse.gov.uk/pubns/books/l111.htm.

Exclusions

[M1109]–[M1110] The Control of Major Accident Hazards Regulations (SI 1999 No 743) does not apply to:

— Ministry of Defence establishments;
— extractive industries exploring for, or exploiting, materials in mines and quarries (with certain exceptions);
— offshore exploration and exploitation of minerals;
— the storage of gas at underground offshore sites;
— waste landfill sites (with certain exceptions);
— hazards created by ionising radiation originating from substances and substances nuclear establishments which create a hazard from ionising radiation (see below); and
— sites used for the storage of metallic mercury.

HSE guidance explains that substances which emit ionising radiation are outside the scope of the COMAH Regulations if they are within a nuclear establishment as the hazards are already addressed by stringent nuclear legislation which ensures at least an equivalent level of safety. However, dangerous substances which do not emit ionising radiation would bring a nuclear establishment into the scope of COMAH if the quantity met or exceeded the appropriate threshold set out in the Regulations.

Provision of Information to the Public

[M1111] Under the general information about public warnings and appropriate behaviour in the event of a major accident, or an indication of where that information can be accessed electronically; *Control of Major Accident Hazards Regulations 2015 (SI 2015 No 483), Reg 17*, the CA must make the following information available to the public, including by electronic means, in relation to every establishment:

- the name of the operator and the address of the establishment;
- confirmation that the Regulations apply to the establishment and that the required notification and safety report have been sent to the CA;
- a simple explanation of the activity or activities undertaken at the establishment;
- the hazard classification of the relevant dangerous substances which could give rise to a major accident, with a simple indication of their principal dangerous characteristics;
- general information about how the public warnings and appropriate behaviour in the event of a major accident, or an indication of where that information can be accessed electronically;
- the date of the last site visit carried out, as part of a programme for routine inspections, and where more detailed information about the inspection and the related inspection plan can be obtained upon request; and
- details of where further relevant information can be obtained (regulation 17(1)).

The CA must also make additional information about upper tier establishments publicly available, including electronically. This includes information about:

(a) the nature of the major accident hazards, including their potential consequences on human health and the environment, and the control measures to address them;
(b) liaison with the emergency services;
(c) co-operating with any instructions or requests from the emergency services;
(d) whether the establishment is close to the territory of another Member State with the possibility of a major accident with trans-boundary consequences (regulation 17(2)).

The CA must keep this information up to date and operators must provide the CA with up-to-date information and comply with reasonable requests for information from the CA (regulation 17(5)).

Regulation 18 requires operators of upper tier establishments to provide information to people (without them having to request it) who are likely to be affected by a major accident, as well as to every school, hospital or other 'area of public use' within an area where people are liable to be affected by a major accident occurring at the establishment.

This must include publicly-available information, be in 'the most appropriate form', and contain clear and intelligible information about safety measures and what to do in the event of a major accident. The operator must review, and where necessary revise, the information at least every three years. They must resend the information if it is revised following a review and otherwise at least every five years.

The COMAH Regulations are linked to the *Environmental Information Regulations 2004 (SI 2004/3391) (EIR 2004)* in England and Wales and the *Environmental Information (Scotland) Regulations 2004 (SSI 2004/520)* in Scotland and set out the exceptions to the duty for disclosure (regulation 19).

Regulation 20 contains requirements for upper tier establishments concerning trans-boundary consequences; but the HSE advises that as an island nation, this requirement is unlikely to be applicable in Great Britain and is therefore dealt with on a case-by-case basis.

Regulation 21 allows CAs to accept information that operators are required to provide in another document sent to the appropriate agency.

Action to be taken following a major accident

[M1112] Following a major accident, the operator of the establishment where the accident has occurred must, as soon as practicable, inform the CA of the accident and provide the following information:

- the circumstances of the accident;
- the dangerous substances involved;
- the data available for assessing the consequences on human health, the environment and property; and
- the emergency measures taken.

They must also inform the CA of the steps required to mitigate the medium and long-term damage and prevent the recurrence of such an accident and update the information they have provided if necessary (regulation 26(1)).

The CA must take any necessary urgent, medium and long-term measures, carry out inspections and investigations necessary to analyse the accident, take appropriate action to ensure that the operator takes any necessary remedial measures and make recommendations on future preventative measures (regulation 26(2)).

In certain circumstances (set out in Schedule 5) they must provide information about the accident to the European Commission (regulation 26(3)). The *Health and Safety (Amendment) (EU Exit) Regulations 2018* which come into force on exit day (see **[M1101]–[M1102]** above) amend COMAH to make reference to international organisations rather than European institutions.

The local authority in whose administrative area the accident has occurred must inform people likely to be affected about the accident and, where relevant, the measures taken to mitigate the effects (regulation 26(5)).

Prohibiting operation

[M1113] The CA must, within a reasonable period of time after it has received a safety report, tell the operator about its conclusions having examined the report; or if necessary, prohibit the operation of the establishment or any part of it (regulation 22).

The CA must prohibit the operation of any establishment, installation or storage facility (or part of it) where the measures taken by the operator for the prevention and mitigation of major accidents are seriously deficient (regulation 23(1)). It may serve a notice on the operator to prohibit operation if the operator has failed to submit a required notification, safety report or other information within the specified time.

A notice must give reasons for the prohibition and specify the date when it is to take effect, may specify measures to be taken and can be withdrawn in writing. [*Control of Major Accident Hazards Regulations 2015 (SI 2015 No 483), Reg 23(2), (3).*]

Domino effects and domino groups

[M1113A] Under some circumstances, a major accident at one establishment might be triggered by an incident at another (not necessarily adjacent) establishment – the so-called 'domino effect'.

Under regulation 24, the CA must identify 'domino groups' of establishments presenting an increased risk of a major accident (or increased consequences) because of their geographical position; their proximity to one another; or the inventories of dangerous substances they hold. In order to identify domino groups, the CA may use notifications, safety reports, planning information, and information from inspections and investigations. They may also request additional information from any operator in a domino group.

Where the CA has information, in addition to that provided by any operator of an establishment which is part of a domino group, about the immediate environment of the establishment, or factors which are likely to cause a major accident or aggravate its consequences, including details of neighbouring establishments, sites of operation that fall outside the scope of the COMAH Regulations, or areas and developments that could be the source of or increase the risk or consequences of a major accident and of domino effects, it must provide that information to each operator in the group.

Where the CA identifies a domino group, it must notify the operators of the name and address of each establishment within the group. Operators who have been notified that they are in a domino group are under a duty of co-operation and must exchange information with each other. This includes sharing information in their major accident prevention policy (MAPP) and safety management systems. Upper tier operators must also share information in their safety reports, internal emergency plans and the information provided to people likely to be affected by a major accident. Where there are neighbouring sites outside the scope of COMAH 2015, operators must provide these with suitable information.

Inspections and investigations

[M1113B]–[M1114] The Seveso III Directive strengthens requirements for inspection and COMAH 2015 (regulation 25) requires the CA to organise a system of inspections to ensure that operators have taken appropriate measures to prevent major accidents and provided appropriate means to limit the consequences of major accidents. The CA must regularly review and revise the inspection plan and prepare programmes for routine inspections of all establishments. It must also investigate serious complaints, serious accidents or near misses and any significant non-compliance with the Regulations.

Following an inspection or investigation, the CA must tell the operator about its conclusions and the necessary action they must take, within four months of the date of the inspection or investigation, and ensure that the operator takes this action. If an 'important case of non-compliance' is identified it must re-inspect within six months.

General duties of operators of all establishments

[M1115] Under the *Control of Major Accident Hazards Regulations 2015 (SI 2015 No 483), Reg 5*, every operator, of both lower tier and upper tier establishments, is under a general duty to take all *measures necessary* to prevent major accidents and limit their consequences to human health and the environment. They must also demonstrate to the CA that they have taken these measures (reg 5(2)).

HSE guidance to the Regulations sets out that this is the general duty on all operators and underpins the whole of the Regulations. It is a high standard which applies to all establishments within the scope of the Regulations, by requiring measures both for prevention and mitigation. The wording of the duty recognises that risk cannot be completely eliminated. There must be some proportionality between the risk and the measures taken to control the risk. The phrase 'all measures necessary' will be interpreted to include this principle. Where hazards are high, high standards will be expected by the enforcement agencies to ensure that risks are acceptably low. The ideal should always be, wherever possible, to avoid a hazard altogether. This is known as inherent safety. Where reliance is placed on people as part of the necessary measures, human factor issues (including human reliability) should be addressed with the same rigour as technical and engineering measures.

Operators should look at how activities can be made safer by reducing hazards, for example by reducing the inventory. The required standard is that risks have been reduced to a level as low as is reasonably practicable (ALARP) and this is generally done by adopting good practice. (Note: the CA view of ALARP is explained briefly in Appendix 4 of the HSE guidance *Preparing Safety Reports: Control of Major Accident Hazards Regulations 1999*, HSG 190 (www.hse.gov.uk/pubns/books/hsg190.htm).

Good practice represents a consensus on what constitutes proportionate action to control a given hazard. Among other things it takes account of what is technically feasible and the balance between the costs and benefits of the measures taken. Sources of good practice include Approved Codes of Practice and standards produced by organisations such as the British Standards Institution (BSI) and Comité Européen de Normalisation (CEN).

In most cases, good practice will mean adopting sound engineering design principles and good operating and maintenance practices.

Operators may employ a risk management approach to prevention and mitigation based on first principles as an alternative to compliance with established good practice, but the competent authority will require this to be thoroughly justified. In cases where no suitable standard for good practice exists, this may be the only possible course of action.

Risk management systems typically include the following elements:

(a) identifying the hazards and risks;
(b) examining the control options available and their merits, including the human factor aspects;
(c) adopting decisions for action informed by the findings of (a) and (b) above;
(d) implementing the decisions; and
(e) evaluating the effectiveness of the actions taken and revising where necessary.

Based on this system, operators must be able to demonstrate that the control measures adopted are adequate for the risks identified.

'All measures necessary' includes measures for mitigating the effects of major accidents. This includes land-use planning which helps to mitigate the effects of major accidents by ensuring adequate separation of people and the environment from their consequences. The hazardous substances authority and planning authorities deal with land-use planning but information from the operator in the hazardous substances consent application is essential for them to perform this function (see **DANGEROUS SUBSTANCES AND EXPLOSIVE ATMOSPHERES**).

Notifications

[M1116] Notification requirements are set out in *Control of Major Accident Hazards Regulations 2015 (SI 2015 No 483), Reg 6*.

Within a reasonable period of time before the start of construction of an establishment, and before the start of operation the operator must send the competent authority (CA) a notification containing the following information:

(a) the name of the operator and the full address of the establishment;
(b) the registered place of business of the operator;
(c) the name or position of the person in charge of the establishment;
(d) information about the dangerous substances or category of dangerous substances present;
(e) the quantity and physical form of these dangerous substances;
(f) a description of the activity or proposed activity of the installations or storage facilities; and
(g) a description of the immediate environment of the establishment, and factors likely to cause a major accident or to aggravate its consequences including, where available, details of neighbouring establishments; sites of operation that fall outside the scope of the COMAH Regulations; and areas and developments that could be the source of or increase the risk or consequences of a major accident and of domino effects (see **[M1113A]** above) (regulation 6(1), (2)).

Operators of lower tier or upper tier establishments in operation before the Regulations came into force were required to send a notification with this information to the CA by 1 June 2016 (regulation 6(4)).

Operators of other establishments are required to send a notification within a year of becoming an 'other establishment'.

The HSE website has forms that may be used for the notification process (see www.hse.gov.uk/Comah/notification-forms-free-leaflets.htm).

Guidance on the amount of information required under each heading is given in the HSE's publication *The Control of Major Accident Hazards Regulations (COMAH) 2015 Guidance on Regulations* (L111).

Notification is a continuing duty and *SI 2015 No 483, Reg 6(6)* requires operators to further notify the CA in advance of:

— any significant increase or decrease in the quantity of dangerous substances notified;
— any significant change in the nature or physical form of the dangerous substances notified, or the processes employing them.
— any modification of the establishment or an installation which could have significant consequences in terms of major accidents hazards;
— the permanent closure of the establishment or its decommissioning; or
— any changes in names and addresses provided under regulation 6(1)(a)–(c).

If an operator submits a safety report or a revised safety report before submitting a notification, they must still send a separate notification to the CA. The HSE advises that this is a change from the 1999 Regulations, which did not require a notification if a safety report had been submitted.

There are particular notification requirements for 'liquefied flammable gases' and 'petroleum products'.

Major accident prevention policy (MAPP)

[M1117] The *Control of Major Accident Hazards Regulations, Reg 7(1)* requires that every operator prepares and retains a written major accident prevention policy (MAPP).

The MAPP must:

(a) be designed to ensure a high level of protection of human health and the environment;
(b) be proportionate to the major accident hazards;
(c) set out the operator's overall aims and principles of action; and
(d) set out the role and responsibility of management, and its commitment towards continuously improving the control of major accident hazards (regulation 7(2)).

Operators of new establishments must prepare a MAPP within a reasonable period of time prior to construction or operation or to modifications leading to a change in the inventory of dangerous substances at the establishment. Operators of establishments in operation when the Regulations came into force were required to do so by 1 June 2016. Operators of any other establishments must prepare a MAPP within a year from the date it first becomes an 'other establishment' (see above) (regulation 7(3)).

Where a MAPP was prepared under the 1999 Regulations and immediately before the 2015 Regulations and is still valid, the operator is not required to prepare a further MAPP (regulation 7(4)).

A MAPP must be reviewed in the event of a significant increase or decrease in the quantity of dangerous substances notified under regulation 6; significant changes in the nature or physical form of the substances or the processes employing them. In any event, a MAPP must be reviewed every five years (regulation 7(6)).

An operator must implement its MAPP by a safety management system (regulation 7(7)).

SI 2015 No 483, Sch 2 sets out that a safety management system must:

A safety management system must:

- be proportionate to the hazards, industrial activities and complexity of the organisation in the establishment;
- be based on assessment of the risks; and
- include the general management system, including the organisational structure, responsibilities, practices, procedures, processes and resources for determining and implementing the major accident prevention policy.

It must address the following:

(a) *organisation and personnel* – the roles and responsibilities of personnel involved in the management of major hazards at all levels in the organisation, together with the measures taken to raise awareness of the need for continuous improvement; training; and the involvement of employees and, where appropriate, sub-contractors;

(b) *identification and evaluation of major hazards* – procedures for systematically identifying major hazards arising from normal and abnormal operation, including any subcontracted activities, and the assessment of their likelihood and severity;

(c) *operational control* – including procedures and instructions for safe operation, including maintenance of plant, processes, and equipment and for alarm management and temporary stoppages;

(d) *management of change* – procedures for planning modifications to, or the design of, new installations, processes or storage facilities;

(e) *planning for emergencies* – procedures to identify foreseeable emergencies by systematic analysis and to prepare, test and review emergency plans to respond to such emergencies; specific training for everyone working on the establishment, including sub-contractors, on the procedures to be followed in the event of emergencies;

(f) *monitoring performance* – procedures for the on-going assessment of compliance with the major accident prevention policy and safety management system, and the mechanisms for investigation and taking corrective action in the case of non-compliance. The procedures must cover the operator's system for reporting major accidents or near misses, particularly those involving failure of protective measures, and their investigation and follow-up on the basis of lessons learnt; and

(g) *audit and review* – procedures for periodic systematic assessment of the major accident prevention policy and the effectiveness and suitability of the safety management system; the documented review of performance of the policy and safety management system and its updating by senior management.

The MAPP is a key document for operators which sets out the framework within which adequate identification, prevention/control and mitigation of major accident hazards is achieved. Its purpose is to compel operators to provide a statement of commitment to achieving high standards of major hazard control, together with an indication that there is a management system covering all the issues set out in (a)–(g) above.

Not only must the MAPP be kept up-to-date (*SI 1999 No 743, Reg 7(6)*), but also the safety management system described in it must be put into operation (*SI 1999 No 743, Reg 7(7)*).

Upper tier duties

Purpose of safety reports

[M1118]–[M1119] Operators of upper tier sites are required to produce a safety report. The key requirement is that operators must show that they have taken all necessary measures for the prevention of major accidents and for limiting the consequences to people and the environment of any that do occur.

Safety reports are required before construction as well as before start-up. The HSE publication '*Preparing safety reports: Control of Major Accident Hazards Regulations 1999:*' HSG 190 (www.hse.gov.uk/pubns/books/hsg190.htm) gives practical and comprehensive guidance to site operators in the preparation of a *COMAH* safety report.

In detail, the operator must send a safety report to the CA:

- where the establishment is new, within a reasonable time prior to construction or the start of operation or before any modifications leading to a change in the inventory of dangerous substances;
- where it was an existing establishment, either on or before 1 June 2016, or where a review of the safety report would, had the 1999 Regulations not been revoked, have been required to be carried out before the 1 June 2016, no later than five years after the relevant date; and
- where it is 'an other' establishment, within two years from when it first became an other establishment (regulation 9(2)).

Control of Major Accident Hazards Regulations 2015 (SI 2015 No 483), Reg 8(1) requires every operator of an upper tier establishment to prepare a safety report that must:

(a) demonstrate that a major accident prevention policy and a safety management system have been put into effect. Schedule 3 (see [M1120] below) sets out the minimum data and information a safety report must include;

(b) demonstrate that the major accident hazards and possible major accident scenarios have been identified and the necessary measures have been taken to prevent such accidents and to limit their consequences for human health and the environment;
(c) demonstrate that adequate safety and reliability have been taken into account in the design, construction, operation and maintenance of any installation, storage facility, equipment and infrastructure connected with the establishment's operation which are linked to major accident hazards inside the establishment;
(d) demonstrate that an internal emergency plan has been prepared in accordance with regulation 12, which includes sufficient information to enable an external emergency plan to be prepared; and
(e) provide sufficient information to the CA to enable decisions to be made regarding the siting of new activities or developments around establishments.

Requirements relating to the preparation of safety reports

[M1120] The minimum information to be included in safety reports is set out in Schedule 3 of the 2015 Regulations:

(1) Information on the management system and on the organisation of the establishment with a view to major accident prevention.
This information must contain the elements set out in *Control of Major Accident Hazards Regulations 2015 (SI 2015 No 483), Sch 2* in relation to the safety management system (see **[M1117]** above).
(2) The environment of the establishment:
    (a) a description of the establishment and its environment including the geographical location, meteorological, geological, hydrographic conditions and, if necessary, its history;
    (b) identification of installations and other activities of the establishment which could present a major accident hazard;
    (c) on the basis of available information, identification of neighbouring establishments, and sites outside the scope of the COMAH Regulations, areas and developments that could be the source of, or increase the risk or consequences of a major accident and of domino effects; and
    (d) a description of areas where a major accident may occur.
(3) The establishment:
    (a) a description of the main activities and products of the parts of the establishment which are important from the point of view of safety, sources of major accident risks and conditions under which such a major accident could happen, together with a description of proposed preventive measures;
    (b) a description of processes, in particular the operating methods, where applicable, taking into account available information on best practices;
    (c) a description of dangerous substances:
        (i) an inventory of dangerous substances including–

[M1120] Major Accident Hazards

- the identification of dangerous substances: chemical name, the number allocated to the substance by the Chemicals Abstract Service (CAS), name according to International Union of Pure and Applied Chemistry (IUPAC) nomenclature;
- the maximum quantity of dangerous substances present or likely to be present;

(ii) the physical, chemical, toxicological characteristics and indication of the hazards, both immediate and delayed, for human health and the environment;

(iii) the physical and chemical behaviour under normal conditions of use or under foreseeable accidental conditions.

(4) Identification and accidental risks analysis and prevention methods:
   (a) a detailed description of the possible major accident scenarios and their probability or the conditions under which they occur including a summary of the events which may play a role in triggering each of these scenarios, the causes being internal or external to the installation. In particular, operational causes, external causes such as those related to domino effects, and natural causes;
   (b) an assessment of the extent and severity of the consequences of identified major accidents including maps, images or, as appropriate, equivalent descriptions, showing areas which are liable to be affected by such accidents arising from the establishment;
   (c) a review of past accidents and incidents with the same substances and processes used, consideration of lessons learned and explicit references to specific measures taken to prevent such accidents; and
   (d) a description of technical parameters and equipment used for the safety of the installations.

(5) Measures of protection and intervention to limit the consequences of an accident:
   (a) a description of the equipment installed in the plant to limit the consequences of major accidents;
   (b) the organisation of alert and intervention;
   (c) a description of mobilisable resources, internal or external; and
   (d) a description of any technical and non-technical measures relevant for reducing the impact of a major accident.

[*Control of Major Accident Hazards Regulations 2015 (SI 2015 No 483), Sch 3.*]

Safety reports must contain, as a minimum, the data and information set out in Schedule 3 and identify the organisations involved in preparing it (regulation 9(1)).

Review of safety reports

[M1121] The operator must review, and where necessary revise, a safety report (where regulation 9(6) does not apply), no later than five years after the date it was last sent to the CA or, where it was not required to be sent to the CA, it was last reviewed by the operator. Regulation 9(6) applies where an operator had, immediately before 1 June 2015, sent the competent authority a safety report in relation to an establishment under regulation 7 or 8 of the 1999 COMAH Regulations. The information in the report must be materially unchanged and be in compliance with the current Regulations. In these circumstances, the operator is not required to send a further safety report.

In addition, the operator must review, and where necessary revise, a safety report:

- following a major accident at the establishment;
- where new facts or technical knowledge, or developments in knowledge concerning the assessment of hazards, justify a review;
- before making modifications that could have significant consequences for major accident hazards; and
- following any change to the safety management system that could have significant consequences for major accident hazards.

The revised safety report must be sent to the CA without delay (or in the case of the modifications above, in advance) (regulation 10).

The HSE provides guidance in L111 on issues to consider at the five-year stage and about the 'trigger events':

At least every five years – regulation 10(1)

'It is important that the review not only details the changes that have occurred, but also assesses the significance of the changes in terms of the identification, prevention, control and mitigation of major accidents. Examples include: changes in the land use of areas surrounding the establishment, including changes in population; or changes in the conservation designations of the surrounding area. Particular attention should be given to the cumulative effects of any minor changes that have taken place over the period,' the guidance says. 'One of the purposes of the review is to see whether the standards, both technical and procedural, remain appropriate in the light of new knowledge and technological developments. For example, it may not always be sufficient for the operator to maintain the plant and systems in an "as built" condition. If the review reveals that further measures are necessary, these additional measures should be taken.'

Trigger events – regulation 10(2)(a)–(c)

The HSE gives the following as examples to illustrate the kinds of things that would trigger a review:

- a substance which is present on-site, but not previously classified as a dangerous substance, is reclassified as dangerous, or the reverse;
- incidents which reveal potentially hazardous reactions or loss of control scenarios not previously considered;
- recommendations made following a major accident or public inquiry;
- changes in surrounding land use or the environment, eg a change in environmental designation; and
- lessons learned from worldwide incidents.

Trigger event – a modification – regulation 10(2)(d)

The HSE gives the following as examples of the sorts of changes which may have significant consequences:

- a change in the quantity of a dangerous substance;
- changes of phase of a dangerous substance, eg a change from liquid to gaseous chlorine;
- the introduction of new, or removal of existing, dangerous substances;
- new processes;
- changes to storage facilities;
- changes to a safety instrumented system;
- changes to the mode of delivery or transport of dangerous substances, eg a change from daily road tanker deliveries to weekly ship deliveries;
- changes to the design or location of control rooms and/or the number of people present within them;
- changes to the location of occupied buildings and/or the number of people present within them; and
- changes to the original design parameters such as process operating conditions or practices, changed throughput, design life extensions or removal of safety-critical plant.

Trigger event – changes to the SMS – regulation 10(2)(e)

The HSE provides the following examples of the types of changes which may be considered significant by the competent authority include:

- changes in use of contractors, management structure and workforce numbers and competences in relation to the operation or maintenance of the establishment; and
- changes in health, safety and environment policy, procedures, standards, aims, objectives or priorities, including changes to the MAPP or SMS.

Key to the effective review of a safety report are well-developed change-management procedures. If implemented properly and in a timely manner, these procedures will enable identification of changes that could have significant repercussions with respect to the prevention of major accidents or the limitation of their consequences.

An operator must notify the CA when they have reviewed the safety report even if it has not been necessary to revise it (regulation 10(6)).

Emergency plans for upper tier establishments

[M1122] Objectives of emergency plans

Regulation 11 sets out that internal and external emergency plans must have the following objectives:

(a) containing and controlling incidents to minimise the consequences and limit damage to human health, the environment and property;

(b) implementing the necessary measures to protect human health and the environment from the consequences of major accidents;
(c) communicating the necessary information to the public and relevant services or authorities; and
(d) providing for the restoration and clean-up of the environment following a major accident.

Preparation, review and testing of internal emergency plans

[M1123] *Control of Major Accident Hazards Regulations 2015 (SI 2015 No 483), Reg 12(1)*, requires operators of upper tier establishments to prepare an internal emergency plan.

The timescale is as follows:

- for new establishments, within a reasonable period of time before operations begin or modifications that lead to a change in the inventory of dangerous substances are made;
- existing establishments had to do this by 1 June 2016; and
- other establishments were given two years from when they first became an other establishment (regulation 12(2)).

Operators of establishments in existence when the Regulations came into force in June 2015 are not required to prepare an internal emergency plan if the on-site emergency plan prepared under the 1999 Regulations, immediately before 1 June 2015, remains materially unchanged; and it complies with the requirements of regulation 11 (see **[M1122]** above). In these circumstances, it will be treated as an internal emergency plan prepared under regulation 12 (regulation 12(3)).

The internal emergency plan must contain the information specified in *SI 2015 No 483, Sch 4, Pt 1*, namely:

— the name or position of any person authorised to set emergency procedures in motion and the person in charge of and co-ordinating the mitigatory action within the establishment;
— the name or position of the person with responsibility for liaison with the local authority responsible for preparing the external emergency plan (also see **DANGEROUS SUBSTANCES AND EXPLOSIVE ATMOSPHERES**);
— for foreseeable conditions or events which could be significant in bringing about a major accident, a description of the action which should be taken to control the conditions or events and to limit their consequences, including a description of the safety equipment and the resources available;
— the arrangements for limiting the risks to persons within the establishment including how warnings are to be given and the actions people are expected to take;
— arrangements for providing early warning of the incident to the local authority responsible for setting the external emergency plan in motion, the type of information which should be contained in an initial warning and the arrangements for the provision of more detailed information as it becomes available;

- where necessary, the arrangements for training staff in the duties they will be expected to perform, and where appropriate, co-ordinating this with the emergency services; and
- the arrangements for providing assistance with mitigatory action outside the establishment.

[SI 2015 No 483, Sch 4, Pt 1.]

When preparing an internal emergency plan, the operator must consult:

- persons working in that the establishment;
- the appropriate agency (ie EA, NRBW or SEPA);
- the emergency services;
- the health authority for the area where the establishment is situated;
- the NHS Commissioning Board and Public Health England (if it is in England); and
- the local authority, unless it has been exempted from the requirement to prepare an external emergency plan in respect of the establishment (under regulation 15 – see below).

The operator must, at least every three years, review, and where necessary revise, the internal emergency plan, and test it (regulation 12(6)).

Preparation of external emergency plans

[M1124]–[M1125] Subject to regulation 15 (see below) the local authority in whose area an upper tier establishment is located must prepare a written external emergency plan for dealing with off-site consequences of possible major accidents (regulation 13). The operator must provide the information the local authority needs to prepare the plan before the date on which it is required to prepare the internal emergency plan under regulation 12(2) (see above). The local authority must prepare it no later than six months (or such longer period not exceeding nine months agreed by the CA in writing) after it has received the necessary information from the operator.

There is an exception (regulation 13(5)) where the CA has exempted the LA under regulation 15 from the requirement to prepare an external emergency plan. This is where, taking into account the information contained in the safety report, the CA is of the opinion that the establishment is not capable of creating a major accident hazard beyond the site.

The plan must contain the following information, specified in Part 2 of Schedule 4:

- names or positions of any person authorised to set emergency procedures in motion and of any person authorised to take charge of and co-ordinate action outside the establishment;
- arrangements for receiving early warning of incidents;
- arrangements for co-ordinating resources necessary to implement the external emergency plan;
- arrangements for providing assistance with mitigatory action within the establishment;
- arrangements for mitigating actions outside the establishment, including responses to major accident scenarios as set out in the safety report and considering possible domino effects;

— arrangements for providing the public and any neighbouring establishments or sites that fall out the scope of the Regulations (in accordance with regulation 24 regarding domino effects and domino groups) with specific information relating to the accident and the behaviour they should adopt; and
— arrangements for the provision of information to the emergency services of other member states in the event of a major accident with possible transboundary consequences. This will be amended to countries after the UK leaves the EU (see [M1011]–[M1102] above).

In preparing the off-site emergency plan, the local authority must consult:

— the operator;
— the appropriate agency;
— the designated authorities who are liable to be required to respond to an emergency at the establishment;
— if the establishment is situated in England, the local authority shall also consult the National Health Service Commissioning Board and Public Health England, an executive agency of the Department of Health and Social Care; and
— such members of the public and other people as it deems appropriate.

[SI 2015 No 483, Reg 13(7).]

Reviewing and testing external emergency plans

[M1126] *Control of Major Accident Hazards Regulations 2015 (SI 2015 No 483), Reg 14*, requires local authorities to review and, where necessary, revise and test external emergency plans, at least every three years.

Reviewing is a key process for addressing the adequacy and effectiveness of the components of the emergency plan – it should take into account:

— changes occurring in the establishment to which the plan relates;
— any changes in the emergency services relevant to the operation of the plan;
— advances in technical knowledge;
— knowledge gained as a result of major accidents either on-site or elsewhere; and
— lessons learned during the testing of emergency plans.

These tests will help assess the accuracy, completeness and practicability of the plan: if the test reveals any deficiencies, the relevant plan must be revised. The operator, emergency services and local authority should agree on the scale and nature of the emergency plan testing to be carried out. Regulation 14(4) requires local authorities to consult with the people/organisations set out in regulation 13(7) (see above) where it is of the opinion that an external emergency plan requires substantial revision.

Where there have been any modifications or significant changes to the establishment, operators should not wait for the three-year review before reviewing the adequacy and accuracy of the emergency planning arrangements.

[M1126] Major Accident Hazards

Operators and local authorities must put emergency plans into effect without delay in the event of a major accident, or an uncontrolled event which could lead to a major accident (*SI 2015 No 483, Reg 16*).

Local authority charges

[M1127]–[M1129] A local authority may charge the operator a fee for preparing, reviewing and testing off-site emergency plans (regulation 29).

The charges can only cover costs that have been reasonably incurred. In presenting a fee to an operator, the local authority must provide a statement of work done and costs incurred.

Statement of policy for managing major incidents

[M1130] This should be a short (maximum 1 page) mission and vision statement covering the following:

- Senior management commitment to take responsibility for the co-ordination of major incident response.
- To comply with the law, namely:
  - the protection of the health, safety and welfare of employees, visitors, the public and contractors;
  - to comply with the duties under the Management of Health and Safety at Work Regulations 1999 (SI 1999 No 3242), Regs 8 and 9 (see **D6010**);
  - to comply with any other legislation that may be applicable to the organisation.
- To make suitable arrangements to cope with a major incident and to be proactive and efficient in the implementation process.
- That this statement applies to all levels of the organisation and all relevant sites in the country of jurisdiction.
- The commitment of human, physical and financial resources to prevent and manage major incidents.
- To consult with affected parties (employees, the local authority and others if needed).
- To review the statement.
- To communicate the statement.
- Signed and dated by the most senior corporate officer.

Arrangements for major incident management

[M1131] —

- This refers to what the organisation has done, is doing and will do in the event of a major incident and *how* it will react in those circumstances. The arrangements are a legal requirement under the *Management of Health and Safety at Work Regulations 1999 (SI 1999 No 3242), Regs 5, 8* and *9*.
- Arrangements should be realistic and achievable. They should focus on major actions to be taken and issues to be addressed. The arrangements will have to be verified (in particular for *COMAH* sites).

- Arrangements could be under the following headings with explanations under each. The larger the organisation and the more complex the hazard facing it, the more detailed the arrangements need to be. For example:

  (i) Medical assistance – including first-aid availability, first-aiders, links with accident and emergency services, other specialists that could be called upon, rules on treating injured people, specialised medical equipment and its availability etc.

  (ii) Facilities management – the location, site plans and accessibility to the main facilities (gas, electricity, water, substances etc), rendering safe such facilities, availability of water supply in-house and within the perimeter of the site etc.

  (iii) Equipment to cope – identification of safety equipment available and/or accessible, location of such equipment, types (personal protective equipment, lifting, moving, working at height equipment etc).

  (iv) Monitoring equipment – measuring, monitoring and recording devices needed, including basic items such as measuring tapes, paper, pens, recorders, intercom and loud-speakers.

  (v) Safe systems – procedures to access site, working safely by employees and contractors under major incident conditions (what can and cannot be done), hazard/risk assessments of dangers being confronted etc, risks to certain groups and procedures needed for rescue (disabled people, young persons, children, pregnant women, elderly people).

  (vi) Public safety – ensuring non-access to major incident site by the public (in particular children, trespassers, the media and those with criminal intent), warning systems to the public etc.

  (vii) Contractor safety – guidance and information to contractors at the major incident on working safely.

  (viii) Information arrangements – the supply of information to staff, the media and others (insurers, enforcers) to inform them of the events. Where will the information be supplied from, when will it be done and updates?

  (ix) The media – managing the media, confining them to an area, handling pressure from them, what to say and what not to say etc.

  (x) Insurers/loss adjusters – notifying them and working with them at the earliest opportunity.

  (xi) Enforcers – notifying them of the major incident, working with them eg HSE, Environment Agency (or NRBW or SEPA) as well as local authority (environmental health, planning, building control for instance).

  (xii) Evidence and reporting arrangement – to cover strict rules on removal or evidence by employees or others, role and power of enforcers, incident reporting eg under the *Reporting of Injuries, Diseases and Dangerous Occurrences Regulations 2013 (SI 2013 No 1471)* etc.

(xiii) The emergency services – working with the police, fire, ambulance/NHS, and other specialists (British Red Cross, search and rescue), rules of engagement, issues of information supply and communication with these services.

(xiv) Specialist arrangements for specific major incidents such as bomb explosions – including for example contacting the police, ordnance disposal, access and egress, rescue and search, economic and human impact.

(xv) Human aspects – removing, storing and naming dead bodies or seriously injured persons during the incident. Informing the next-of-kin, issues of religious and cultural respect. Issues of counselling support and person-to-person support during the incident.

This is not an exhaustive list.

Command and control chart of arrangements

[M1132] This highlights who is responsible for the effective management of the major incident.

- It should be a graphical representation preferably in a hierarchical format, clearly delineating the division of labour between personnel in the organisation and the emergency services/others.
- The chart should display three broad levels of command and control, namely strategic, operational and tactical. The first relates to the person(s) in overall charge of the major incident. Will this be the person who signed the statement of policy for major incidents or will it be another (disaster and emergency advisor, safety officer, others)? This person will make major decisions. The second relates to co-ordinators of teams. Operational level personnel need to have the above arrangements assigned to them in clear terms. The third refers to those at the front end of the major incident, for instance first-aiders.

  It is most important to note that internal command and control of arrangements does not mean *overall* command and control of the major incident. This can (will be) vested with the appropriate emergency service, normally the police or the fire authority. In the event of any conflict of decisions, the external body such as the police will have the final veto. Therefore, the chart and the arrangements must reflect this variable.
- The chart should list on a separate page names/addresses/emergency phone, email and mobile numbers of those identified on the chart. It should also list the numbers for the emergency services as well as others (British Red Cross, specialist search and rescue, loss adjusters, enforcing body).
- The chart should also clearly ratify a principle of command and control, as to who would be 'In-charge 1', 'In-charge 2', if the original person became unavailable.

- The chart and the list of numbers should also be accompanied by a set of 'rules of engagement' in short 'bullet points' to remind personnel of the importance of command and control eg safety, obedience, communication, accuracy, humanity for instance.

Testing and validation

[M1133] The emergency services consider this to be a crucial aspect of pre-planning. Testing is an objective rehearsal to examine the state of preparedness and to determine if the policy and the plan will perform as expected.

Training and exercising are both types of testing. Training is more personnel focused, aiming to assess how much the human resource knows about the policy and plan and means of enhancing the knowledge, skill and experience of that resource. Exercising is a broader approach, examining all aspects of the policy and plan, not just the human response but also physical and organisational capability to deal with the major incident.

Training and exercising are both necessary to ensure that human resources understand the policy and plan, as well as to ensuring that problems can be identified and competent responses developed.

[M1134] Four types of exercising can be identified:
- Seminar exercises
  — a broad, brain-storming session, assessing and analysing the efficacy of the policy and plan;
  — seminars need to be 'inclusive', that is bring staff and others together in order to co-ordinate strategic, operational and tactical issues.
- Table-top exercises
  — This is an attempt to identify visually using a model of the production site and surrounding areas, what types of problems could arise (access/egress, control, logistics, etc). It involves a limited number of people, an advantage in terms of time commitment, but a disadvantage in that others do not have the opportunity to learn from it. It is perhaps most valuable in validation of new plans.
- Live exercises
  — This is a rehearsal of the major incident, actively and proactively testing the responses of the individual, organisation and possibly the community (emergency services, the media, etc) to a simulated event.
- Synthetic simulation
  — This applies the techniques used in flight simulation and military simulation, to major incidents. Thus, using computers one can model the site and operation and in 3-d format move around the screen. This then enables various eye-points around the site, achieved by moving the mouse. Some advanced systems enable 'computer generated entities' to be included into the data-set.

For example, a collision can be simulated outside the production plant or the rate of noxious substance release modelled and calculated across the local community.

There are costs and benefits associated with each of the four approaches. The need is for developing exercise budgets which enable all four to be used in proportion and to complement each other, though synthetic simulation may well be out of reach for smaller organisations.

Land-use planning

[M1135] Much has already been achieved through controls under *COMAH* to reduce the likelihood of major accidents and to mitigate the consequences of those which do occur, by emergency planning and information to people in the surrounding area. But there is a further and very important way in which the consequences of major accidents can be mitigated: land-use planning. Land-use planning allows decisions to be made about the siting of major hazard installations and about the development of land around existing installations.

In an ideal world, industries using large quantities of hazardous substances would be located far away from centres of housing and other developments that could be affected in case of accident. In reality, the situation is very different. Factories, housing, schools and shops have developed close to each other; indeed, in many cases these industries provide the economic heart of the local community.

Greater control is possible when planning the location of new hazardous activities, but even here the options may be limited. There are few locations where new hazardous installations can be 'shoehorned' into place without creating some risk to an existing community. The UK is a small, densely populated island and such undeveloped areas as do exist are often so remote or of such environmental value as to be unsuitable for industrial use. Since the early 1970s, arrangements have existed for local planning authorities (PAs) to obtain advice from the HSE about risks from major hazard sites and the potential effect on populations nearby. The Advisory Committee on Major Hazards (ACMH), set up in the aftermath of the Flixborough disaster in 1974 (see **[M1101]**–**[M1102]**), laid down a framework of controls which included a strategy of mitigating the consequences of major accidents by controlling land-use developments around major hazard installations.

Land-use planning legislation

[M1136]–**[M1140]** Land-use planning is covered by planning legislation in Great Britain, in particular the *Planning (Hazardous Substances) Act 1990* and the *Planning (Hazardous Substances) Regulations 2015 (SI 2015 No 627)*. The *Planning (Hazardous Substances) Regulations 2015 (SI 2015 No 627)* implement the land-use planning obligations (Articles 13 and 15) in Directive 2012/18/EU the 'Seveso III Directive' and consolidate, with amendments, existing regulations made under the *Planning (Hazardous Substances) Act*

1990. They revoked the *Planning (Hazardous Substances) Regulations 1992* and the *Planning (Control of Major-Accident Hazards) Regulations 1999*.

Enforcement of the *Planning (Hazardous Substances) Act 1990* and the Regulations are the responsibility of the appropriate hazardous substances authority (usually the local planning authority).

Land-use planning legislation is not examined in detail in this chapter.

Sources of information on major accident hazards

[M1141] These include:

(1) Legislation
- *Control of Major Accident Hazards Regulations 2015 (SI 2015 No 483)*;
- *The Planning (Hazardous Substances) Regulations 2015 (SI 2015 No 627)*.

(2) Guidance
- A guide to the Control of Major Accident Hazards Regulations (COMAH) 2015 (L111);
- 'Preparing safety reports: Control of Major Accident Hazards Regulations 1999' (HSG 190);
- 'Emergency Planning for Major Accidents: Control of Major Accident Hazards Regulations 1999' (HSG 191);
- 'Major accident prevention policies for lower-tier COMAH establishments' (HSE Information Sheet, Chemical Sheet No 3);
- HSE website: www.hse.gov.uk/comah/index.htm.

(3) Professional bodies such as:
- International Institute of Risk and Safety Management: www.iirsm.org;
- Institution of Occupational Safety and Health (www.iosh.co.uk), particularly the Hazardous Industries group.

Managing Absence

Lynda Macdonald

Introduction to managing absence

[M1501] It is advisable for all employers, irrespective of their size or industry sector, to take active steps to manage employee absence, both long and short-term. The causes of absences should be recorded and analysed so that the employer can review whether any patterns exist or whether factors in the workplace are causing or contributing to employee absence. Records should also be kept of all employees' sickness absences when they occur. In this way, employers may be able to minimise the incidence of employee absence, especially short-term absence.

Where, despite good management, employees are nevertheless absent from work, the employer should adopt good practice techniques to manage those absences. This chapter aims to help managers to understand the various employment laws that impact on employee sickness absence and to apply sound management techniques to the management of both long-term and short-term absences.

Research published by the Chartered Institute of Personnel and Development (CIPD) in 2016 (CIPD Annual Survey Report on Absence Management) shows that employee absence during 2016 was running at an average of 6.3 days per employee per year (down from 6.9 days per employee per year in 2015), at a median cost to employers of £522 per employee per year. According to CIPD, average absence levels are highest in the public services and non-profit sectors and lowest in the private services sector. Absence levels in the public sector were highest at an average of 8.5 days per employee (down slightly from 8.7 days the previous year). In the private sector, the average level was 5.2 days whilst in manufacturing and production, it was 5.4 days. Non-profit sector organisations experienced an average of 6.9 days absence per employee per year. Larger organisations tend on the whole to have higher average levels of absence than smaller organisations.

CIPD reported that the most common cause of long-term absence amongst manual workers was musculoskeletal injuries (for example neck strains and repetitive strain injury). The next most common causes were acute medical conditions (for example stroke, heart attack and cancer), stress, back pain and mental ill-health (for example clinical depression and anxiety). For non-manual workers, the most common cause of long-term absence was acute medical conditions, followed by stress, mental ill-health and musculoskeletal injuries. By far the most common cause of short-term absence for both manual and non-manual workers was minor illness such as colds, flu and stomach upsets. For manual employees, the next most common cause of short-term absenteeism was musculoskeletal injuries, followed by back pain, stress, home/family/carer responsibilities and mental ill-health. Stress remains the second most common cause of short-term absence among non-manual workers, followed by mental ill-health, musculoskeletal injuries and home/family responsibilities.

CIPD reported in 2016 that 24% of organisations found that non-genuine absence was one of the top causes of short-term absenteeism. For manual workers, the figure was 30% whilst for non-manual workers, it was 20%.

Managing sickness absence effectively is likely to result in:

- a reduction in the potential duration of some employees' sickness absence;
- an increase in the likelihood of long-term sick employees returning to work;
- a reduction in the likelihood of repeated spells of absence after employees have returned to work; and
- a decrease in the costs to the business of employee absence.

Managing long-term sickness absence

[M1502] Long-term absence is usually defined as absence of more than four weeks (although there is no legal definition).

There are two stages to managing an employee's long-term sickness absence. The first is to manage the employee's absence from work and the second to manage their return to work.

The management of long-term sickness absence requires a pro-active approach from management in order that:

- the employee's absence can be regularly reviewed;
- appropriate steps can be taken to support the absent employee;
- actions can be taken to facilitate the employee's return to work;
- the employee's return to work can be effectively managed; and
- where the employee does not recover sufficiently to resume working, the employer can fairly dismiss the employee.

Reviewing the employee's absence

[M1503] Employers should adopt a 'case management' approach when dealing with employees on long-term sickness absence. This means regularly reviewing the employee's absence and state of health/fitness to see whether there is any improvement, and whether the employer can do anything to facilitate the employee's recovery and return to work.

Case reviews would normally be held quarterly or even monthly, and these would involve the employee's line manager, health and safety personnel, HR department, medical specialists and of course the employee who is absent.

Managing long-term sickness absence [M1504A]

Keeping in touch with an employee who is absent

[M1504] Where an employee is absent from work due to a long-term condition, it is good practice for the employer to nominate an appropriate person to keep in touch with them. This has a number of advantages:

- the employee who is absent will feel reassured of their employer's ongoing support and is less likely to feel 'abandoned' or gain the impression that their employer does not care about their welfare;
- the employer will be able to keep up-to-date on the employee's state of health and progress and the likelihood of a return to work;
- the employer will be able to organise temporary cover more effectively;
- the employer can keep the employee informed of any developments in the business and the work of their department; and
- the employee is more likely to return to work sooner, if they feel confident that their employer is adopting a supportive attitude towards them.

Naturally, managers should take care to ensure that contact is not viewed as unwelcome or intrusive. Equally, it is important to make sure that no unfair pressure is put on the employee through such contact and that the employee understands that this is not the case. It is, however, a common assumption that contacting an employee who is off sick will be inappropriate, whilst in fact many employees who are absent long-term would welcome regular contact. The absent employee may feel very isolated and excluded whilst off sick and regular contact may help to reduce these feelings and make the employee feel that they are still viewed as part of the team even though they are not able to work.

The type of contact and its frequency should be agreed directly with the employee. This can be instigated by writing to the employee explaining that the manager would like to maintain contact and asking the employee whether they would prefer telephone calls, occasional visits at home (perhaps by a workmate), e-mail communication or a combination of these. The letter should make clear the manager's interest and concern about the employee's welfare and progress and offer any support that is reasonable and practicable. The employee's views on how contact should be maintained should always be sought and respected.

The employee's line manager should take responsibility for arranging and maintaining such contact, but he or she will not always be the best person to maintain the contact. It may be that a close colleague of the absent employee can be nominated instead, in particular where visits to the employee's home are agreed.

Medical statements

[M1504A] From 6 April 2010, the traditional GP's medical certificate, or 'sick note', was replaced by a new 'statement of fitness for work', also known as a 'fit note'. The fit note allows doctors to state either that the employee is 'not fit for work' or that he/she 'may be fit for work' taking account of advice specified by the doctor. There is no 'fit for work' option as it was considered that GP's do not have sufficient knowledge about individual employees' roles or the risks inherent in them to assess this properly.

The purpose of the fit note scheme is to facilitate return to work in circumstances where adjustments by the employer would help the employee resume working sooner than might otherwise be the case.

The fit note system gives doctors the opportunity to highlight one (or more) of four options to help facilitate the employee's return to work. These are:

- a phased return;
- amended job duties;
- altered hours of work;
- workplace adaptations.

The GP may also write in any other option that they consider may be appropriate in the circumstances.

There is no legal obligation on an employer to comply with any recommendation made on a GP's fit note and what is written on the fit note may therefore be regarded as advisory rather than obligatory. Equally, any changes to employees' hours or job duties, whether temporary or permanent, should be made only with the agreement of the employee. Nevertheless, managers should take what the employee's doctor has written seriously and give fair consideration – in consultation with the employee – as to whether any of the changes recommended by the doctor can be accommodated.

If the employer is unable to facilitate the change(s) recommended by the doctor, this will be evidence that the employee remains unfit to carry out their normal job, ie there is no need for the employee to obtain a further fit note to confirm this.

Fit for Work service

[M1504B] In 2015, a new, independent health and work advisory and assessment service (known as 'Fit for Work') was introduced. This service is state-funded and is therefore free to employers and employees. The aim of the service is to facilitate the return to work of employees who have been absent from work for four weeks or more and to give employers support to help them manage long-term sickness absence and reduce their costs. The service is provided by Health Matters Ltd (a private organisation) in England and Wales, and by NHS Scotland in Scotland.

The scheme consists of two main elements: an advice service (by telephone and online) for employers, employees and GP's and an employee assessment service provided by occupational health professionals.

An employee can be referred for assessment either by their GP or by their employer once they have been off sick for four weeks or more (earlier in some cases). An employer can refer as many employees as it wishes, but an employee can only be referred once a year. The scheme is voluntary and so no employee can be compelled to agree to an assessment.

Assessments are usually conducted by telephone, although a face-to-face assessment can be requested. Following the assessment, an occupational health specialist will (where appropriate) produce a return-to-work plan which will make recommendations as to how the employer can help the employee return to work and how any obstacles to a return (whether health-related or not) can be overcome.

It is not the case that the Fit for Work service should replace any occupational health provision that an employer has in place. 'Fit for Work' may, however, be particularly useful for those smaller employers who do not have access to an occupational health provider.

Employment laws that impact on long-term sickness absence

[M1505] There are a number of laws that impact on long-term sickness absence:

- *Equality Act 2010*;
- *Employment Equality (Repeal of Retirement Age Provisions) Regulations 2011*;
- *Access to Medical Reports Act 1988*;
- *Employment Rights Act 1996, s 1* (the duty on employers to provide employees with a written statement of employment particulars);
- *Social Security Contributions and Benefits Act 1992, Pt XI* (the duty to pay statutory sick pay);
- *Working Time Regulations 1998*;
- *Social Security (Medical Evidence) and Statutory Sick Pay (Medical Evidence) (Amendment) Regulations 2010*.

When an employee's condition may amount to a disability

[M1506] The definition of disability in the *Equality Act* is very wide and includes both physical and mental impairments. An individual who has a serious disfigurement may qualify as disabled under the Act, depending on the severity and positioning of the disfigurement.

Definition of disability

[M1507] Specifically, a person is disabled if he or she has a physical or mental impairment that has a substantial and adverse long-term effect on his or her ability to carry out normal day-to-day activities. 'Long-term' in this context means twelve months or more. A condition will be long-term if it has already lasted 12 months, is likely to last 12 months or more or is likely to last for the rest of the person's life (ie in cases of terminal illness). 'Substantial' is defined as 'not minor or trivial' (which is a low hurdle). There is no need for an employee to be 'registered' as disabled to qualify for protection under the *Equality Act 2010*.

'Normal day-to-day activities' stands to be interpreted according to the lay-person's understanding of the phrase, in other words any activity that people do on a reasonably regular basis in their every-day lives will be a normal day-to-day activity for the purpose of the Act.

Although 'normal day-to-day activities' was not originally intended to mean an individual's specific job duties, the EAT ruled in *Banaszczyk v Booker*

UKEAT/0132/15, [2016] IRLR 273, [2016] All ER (D) 173 (Feb) that an employee whose job duties involved manual lifting and moving cases weighing up to 25kg was disabled on account of a long-term back condition that prevented him from carrying out these duties.

In *Gallop v Newport City Council* [2013] EWCA Civ 1583, the Court of Appeal made it clear that it is an employer's responsibility to form its own view of whether or not a particular employee is disabled - based on all the relevant facts. The Court ruled that it is not appropriate for an employer to rely unquestionably or automatically on a medical adviser's opinion as to whether or not an employee is disabled, although the employer should take any advice received from professional medical advisers into account. The Court also suggested that an employer should ask its medical advisers specific questions directed to the particular circumstances of the employee's supposed disability and then judge for itself, based on the answers to these questions, whether or not the criteria for disability are satisfied.

Scope of the Act

[M1508] All workers are protected by the disability discrimination provisions of the *Equality Act 2010* and all employers, with the exception of the armed forces, are under a duty to comply with the Act's provisions.

The *Equality Act 2010* provides that workers are protected against 'discrimination by association'. This means that individuals who, for example, care for a disabled child, parent or partner are protected against discriminatory treatment (including harassment) on any grounds related to the fact that they have an association with or responsibility towards the disabled person. For this type of claim to succeed, however, the claimant would have to show that the alleged discrimination against him/her occurred specifically because the associated person had a disability.

There is no qualifying period of employment for an employee to bring a claim of disability discrimination to an employment tribunal, and no ceiling on the amount of compensation that can be awarded if a claim is successful.

The main duties in respect of disability under the Act are:

- the duty not to discriminate directly, ie the duty not to treat a disabled worker (or job applicant) unfavourably directly because of a disability;
- the duty not to discriminate indirectly, ie not to apply any provision, criterion or practice that puts, or would put, people who share a particular type of disability at a particular disadvantage, unless the provision, criterion or practice is objectively justified;
- the duty not to discriminate for a reason 'arising in consequence of' a disability (unless the particular treatment of the person is objectively justified), for example discrimination against someone because of sickness absence that is occasioned by their disability;
- the duty to make reasonable adjustments to any provision, criterion or practice, or to physical features of premises, that may place a disabled worker at a substantial disadvantage.

Each of these is considered separately below.

Conditions that may amount to a disability

[M1509] A wide range of physical and mental conditions and illnesses may amount to disabilities, depending always on whether the effect of the condition on the person is substantial and long-term. Conditions that are intermittent, or that fluctuate in their effects, will entitle the person to protection under the disability discrimination provisions of the *Equality Act 2010* at all times (provided the condition is likely to recur – 'likely' has been defined by the House of Lords as meaning 'could well happen'), even if at a particular point in time the condition is in remission. A further important point is that a condition will amount to a disability even if the person, as a result of medication or other form of support, experiences no adverse effects on a day-to-day basis. The question that determines whether an employee is disabled is how the condition would affect them if they did not take their medication or use the support. The one exception to this is in respect of sight impairments. If a sight impairment is capable of correction by spectacles or contact lenses, then the question of whether the individual is disabled will stand to be judged against the assessment of the person's sight whilst wearing the spectacles or lenses.

Progressive conditions such as Alzheimer's, Parkinson's disease and muscular dystrophy (for example) are covered under the definition of disability as soon as the condition is diagnosed and the person's normal day-to-day activities are affected in any way, whether substantial or not provided the effects of the condition are likely to become substantial in the future. Since October 2005 (when changes to the legislation were implemented), any employee diagnosed with cancer, multiple sclerosis or HIV is automatically deemed disabled whether or not they are experiencing any symptoms at the time in question and whether or not it is likely that the effects of the condition will, in time, become substantial.

In the Danish case of Fag og Arbejde (FOA), acting on behalf of *Karsten Kaltoft v Kommunernes Landsforening (KL)*, acting on behalf of Billund Kommune: C-354/13 (2014) ECLI:EU:C:2014:2463, [2015] All ER (EC) 265, [2015] 2 CMLR 500, the ECJ ruled that obesity may in certain circumstances amount to a disability if it 'entails a limitation resulting in particular from long-term physical, mental or psychological impairments which in interaction with various barriers may hinder the full and effective participation of the person concerned in professional life on an equal basis with other workers'. The cause of an illness or other condition is irrelevant to the assessment of whether it amounts to a disability. In the Kaltoft case (above), the ECJ emphasised that the key issue is whether the person's disability hinders full and effective participation in working life (for example where mobility is severely impaired).

Equally, there is no need for an employee to prove a specific medical diagnosis in order to claim protection under the Act, nor to prove that some kind of illness underlies their impairment. Instead it will be sufficient for them to show that the adverse effects of the condition are substantial and long-term. There are some exceptions to the Act, ie conditions that are specifically deemed not to be disabilities. These include alcohol addiction, addiction to illegal drugs and hay fever. A small number of mental conditions are also excluded.

The duty not to discriminate

[M1510] Employers are under a duty not to treat a disabled employee unfavourably:

- directly because they have a particular disability;
- indirectly, eg by applying a policy or provision which has a disproportionate adverse impact on people with a particular type of disability and which is not justified;
- for a reason arising from disability, unless justified.

Direct disability discrimination, which can never be justified, is where an employer treats a disabled employee or job applicant unfavourably just because they have a disability, or a particular type of disability. An example could be the automatic rejection of a job applicant just because they had disclosed that they had a history of mental illness.

In contrast, it is possible for an employer to justify indirect disability discrimination if the provision in question can be shown to be appropriate and necessary with a view to achieving a legitimate business aim. This defence must, however, relate not only to the provision in question, but also to its application to the individual. This was confirmed in *Buchanan v Commissioner of Police of the Metropolis* (2016) UKEAT/0112/16, [2016] IRLR 918, [2017] ICR 184 in which the EAT held that, although an employer may well have a legitimate aim underpinning the operation of an absence management policy, its legitimacy will not, on its own, necessarily mean that its rigid application to a disabled employee is justified.

It is also possible to justify discriminating against an employee for a reason arising in consequence of their disability, for example if an existing employee had, following an accident, become completely unable to perform the job for which he or she was employed. In these circumstances it might be justifiable to terminate the employee's employment, provided the employer had first explored whether any reasonable adjustments could be made (see below), for example whether the employee could reasonably be transferred to different job duties. To bring a claim of this type, the individual is not required to demonstrate that their treatment was less favourable than that afforded to anyone else. Justification will be made out if the employer can show that the treatment of the employee represented a proportionate means of achieving a legitimate business aim.

Unfavourable treatment can take many forms and may include:

- rejection for employment or promotion;
- exclusion from training opportunities;
- being afforded less favourable terms and conditions;
- removal of job duties or transfer to less interesting or less responsible work;
- exclusion from contractual benefits or non-contractual perks;
- harassment, which can be physical, verbal or non-verbal; or
- dismissal or selection for redundancy.

The list is not exhaustive.

The duty to make reasonable adjustments

[M1511] The *Equality Act 2010* places a duty on employers to make reasonable adjustments to any provision, criterion or practice applied by the employer and to physical features of premises in order to accommodate the needs of disabled workers and job applicants. There is also a duty to provide an 'auxiliary aid' (eg specialised computer software) for a disabled employee where it is reasonable for the employer to do so. The duty to make reasonable adjustments arises whenever any aspect of the employer's working practices or premises, puts a disabled person at a substantial disadvantage or where the disabled person would, but for the provision of an auxiliary aid, be put at a substantial disadvantage.

In *Hainsworth v Ministry of Defence* [2014] EWCA Civ 763, the Court of Appeal confirmed that the employer is not under a duty to make reasonable adjustments where the disabled person is someone associated with the employee, rather than the employee him-or-herself.

An employer who fails to comply with this duty will be acting in contravention of the *Equality Act 2010* unless they can show that it was not reasonable for them to make adjustments generally, or to make a particular adjustment.

The employer will only be subject to the duty to make reasonable adjustments if the manager in question knows, or ought reasonably to have known, that the individual is disabled. This was confirmed by the EAT in *Gallop v Newport City Council* (2016) UKEAT/0118/15, [2016] IRLR 395, [2016] All ER (D) 218 (Mar) in which it was ruled that because the manager who took the decision to dismiss the employee (on account of extensive sickness absence) did not personally know that he was disabled, his dismissal could not amount to disability discrimination. This was despite the fact that the employer's medical adviser did know that the employee's condition amounted to a disability. The EAT held that this knowledge could not be imputed to the manager who took the decision to dismiss him (unless the decision to dismiss had been taken jointly).

Furthermore, in *Secretary of State for the Department for Work and Pensions v Alam* [2010] ICR 665, [2010] IRLR 283, the EAT held that the duty to make reasonable adjustments did not apply because management did not know, and could not reasonably have known, the extent of the effects of the employee's condition (even though they were aware that he had a disability). An important point to bear in mind is that the duty to make reasonable adjustments places the employer under a positive duty to take the initiative and consider what adjustments would be possible and practicable for a particular disabled employee (or job applicant). It is advisable for employers to make a full assessment of the employee's condition, its effects on their ability to perform their job duties and the steps that might reasonably be taken to support them.

Although the duty to make reasonable adjustments is on the employer, it is nevertheless advisable for managers to consult the employee concerned about possible adjustments as the employee will, in any event, know more about their condition and its effects than the manager, and will often be able to suggest adjustments that would be helpful for them. Managers should be open-minded and willing to take on board any reasonable suggestions from the employee for adjustments.

Adjustments can include:

- transferring the employee to another job, for example lighter work, provided the employee consents to such a move;
- adjusting the duties of the job, for example exempting an employee with a back condition from doing any heavy physical work;
- changing the method of doing the job, for example allowing an employee who cannot drive on account of a medical condition to use taxis to travel on business;
- adjusting working hours, for example allowing someone with a mental illness a later or flexible start time, or more frequent rest breaks, or exempting an employee with a debilitating condition from a requirement to work overtime;
- changing the place of work, for example moving someone with limited mobility to a ground floor location or allowing home-working for part of the working week;
- adjusting procedural requirements, for example allowing an employee with a recurring illness to take more time off work than would normally be acceptable;
- providing additional or tailored training, coaching or mentoring, for example for someone with learning difficulties;
- providing special equipment, for example voice activated software for someone with a visual impairment;
- modifying instructions or reference manuals, for example providing them in Braille;
- providing a reader or interpreter, for example for someone who is profoundly deaf;
- providing an auxiliary aid, for example specialised computer software;
- providing additional or special supervision; and
- modifying premises, for example widening a doorway for a wheelchair user, or relocating door handles or shelves for someone who has difficulty reaching.

The Employment Appeal Tribunal said, however, (*in O'Hanlon v HM Revenue and Customs* [2006] IRLR 840) that it is unlikely to be a reasonable adjustment for an employer to adjust their sick pay scheme so as to allow a disabled employee to receive full pay indefinitely when absent from work for a disability-related reason, unless there are exceptional circumstances. The Court of Appeal subsequently upheld the EAT's decision in *O'Hanlon v Commissioners of HM Revenue and Customs* [2007] EWCA Civ 283, [2007] IRLR 404.

Unless an employer has done all they reasonably can to make adjustments for a disabled employee, any subsequent unfavourable treatment (for example dismissal) will always be discriminatory and unlawful.

In *Muzi-Mabaso v Revenue & Customs Comrs* (2014) UKEAT/0353/14/DA, unreported, the EAT ruled that even if a particular policy or practice is objectively justified in a general sense (and so a claim for indirect disability discrimination can be successfully defended), that does not mean that the employer is no longer under a duty to make reasonable adjustments for the disabled individual. The employer's duty will be to consider the discriminatory impact of the particular policy or practice on the individual concerned and review whether there are any reasonable adjustments they can make to accommodate him or her in light of the particular disability.

Insurance benefits

[M1511A] Under the *Employment Equality (Repeal of Retirement Age Provisions) Regulations 2011*, an exemption was introduced (on 6 April 2011) for employers who offer or provide their employees with certain types of insurance benefits on a group risk basis (ie through third party insurers). Examples include life assurance, income protection schemes, sickness/accident insurance and private medical insurance.

The exemption permits employers who offer or provide these benefits to their employees to stop providing them in respect of employees who have reached the age of 65 (or state pension age, if higher, ie when it is raised above age 65 in the future). Such a policy would otherwise be age discriminatory.

Access to medical reports

[M1512] The *Access to Medical Reports Act 1988* places certain restrictions on organisations that wish to obtain medical information about someone from his or her doctor. The Act also gives individuals a variety of rights in relation to any such medical reports.

The Act applies whenever a medical report is 'prepared by a medical practitioner who is or has been responsible for the clinical care of the individual'. The effect of this is that the Act applies to medical reports prepared by an employee's own doctor, specialist or consultant, but would not normally apply to reports prepared by an occupational doctor.

Employees have a number of rights under the Act:

- to be informed in writing if the employer wishes to seek their consent to contact their doctor for a medical report;
- (at the same time) to be informed by the employer of their rights under the Act;
- to decline to give their consent for the employer to apply to their doctor for a medical report;
- (if they have given consent) to ask their doctor for a copy of the report once it has been prepared;
- to ask the doctor to amend the report, if, in their opinion, it contains anything inaccurate or misleading; and
- to refuse to allow the report to be released to the employer.

The employee's right to refuse to give consent for their employer to apply to their doctor for a medical report and their right to refuse to allow a report, once prepared, to be released to the employer, are statutory rights. This means that the employer cannot override these rights by inserting a clause in employment contracts to the effect that employees are required to give such consent. Any contractual term to this effect will be unenforceable.

[M1512] Managing Absence

Since the Act does not normally apply to reports prepared by occupational doctors, it is open to an employer to include clauses in employees' contracts to the effect that they must agree, on request by the employer, to be examined by an occupational doctor, and allow the doctor to provide a report to the employer.

General Medical Council Guidance

[M1512A] Although the *Access to Medical Reports Act 1988* does not normally apply to reports prepared by occupational doctors, Guidance from the General Medical Council, published in October 2009, requires all doctors (including occupational doctors) preparing reports for employment purposes to obtain the employee's written consent before passing any report to the employer. The doctor is also required to be satisfied that the employee concerned is fully informed of the purposes and likely results of disclosing the report to the employer, and offer to show the employee the report before passing it to the employer. This means, amongst other things, that employees may in effect be able to refuse to consent to an occupational doctor's report being disclosed to the employer as it is unlikely a doctor would be prepared to disclose a report in breach of the General Medical Council's Guidance.

Employees' contractual rights

[M1513] Under the *Employment Rights Act 1996, s 1*, employers are under a duty to provide a written statement of key terms and conditions of employment to all employees whose employment is to continue for more than one month. The 'written statement' as it is usually known must include information on a number of defined issues, including any terms and conditions relating to 'incapacity for work due to sickness or injury, including any provision for sick pay'.

There is no duty on an employer to pay an employee 'occupational sick pay', ie their normal wages or salary, if they are absent from work due to sickness or injury. That is a matter for each employer to decide for themselves. Employers must, however, inform each employee in writing (as part of the written statement) what their entitlement to sick pay will be in the event of sickness absence, and of any conditions attached to the payment of sick pay.

Many employers pay occupational sick pay for defined periods of time, whilst others make the payment of sick pay discretionary. Employers should bear in mind, however, that a discretionary policy carries with it the danger that employees who are not paid sick pay will perceive the policy (and management) to be unfair and that this may lead to grievances. Where the employer wishes to retain discretion over the payment of sick pay, it is advisable therefore to devise clear and transparent management guidelines on how discretion should be exercised to ensure consistency and fairness across different departments. In *Clark v Nomura International plc* [2000] IRLR 766, the High Court held that management discretion in relation to pay must not be exercised perversely or irrationally. It would also be normal practice to make the payment of occupational sick pay dependent on certain conditions being fulfilled, for example to require employees to:

- telephone a designated person as soon the employee knows that they will be unable to attend work;
- provide a self-certificate for the first week of absence and doctors' statements for all periods of absence thereafter;
- maintain contact with the employer at regular intervals (for example weekly) with a view to advising the employer how long they expect to be absent;
- agree to be examined by an occupational doctor on request and to the doctor providing a report to the employer; and
- attend a 'return to work' interview with a designated person (usually the line manager) on return to work.

These conditions allow the employer to manage absence and sick pay effectively.

Statutory sick pay (SSP) is an entirely different matter and is dealt with at M1514 below.

Statutory sick pay

[M1514] Although the employer may decline to make payment of wages or salary during sickness absence, they cannot contract out of their obligation to pay SSP.

There is no length of service required for an employee to qualify for SSP, and so both temporary and permanent employees can, potentially, qualify. However, complex provisions are in place with regard to when payment is due and there is a requirement for certain record keeping.

Both full-time and part-time employees will qualify, but an employee who earns less than the equivalent of the lower limit for the purposes of national insurance contributions is excluded. The lower limit is £113 per week as from April 2017. Where an employee is paid occupational sick pay, SSP must be offset against this, ie SSP cannot be paid over and above normal wages or salary.

To qualify for SSP, an employee must comply with their employer's conditions for notification of absence and provide suitable evidence of incapacity.

SSP is a fixed weekly rate, payable to an employee who is absent from work due to personal sickness for up to 28 weeks during any 'qualifying period'. The first three days of absence during any qualifying period are 'waiting days' and no SSP is payable for these days. The rate in force (as from 6 April 2017) is £89.35 per week (or the appropriate pro-rata daily rate).

When the rules on SSP were first introduced, employers were able to recoup amounts paid out by offsetting payments against their national insurance contributions. These provisions were, however, subsequently abolished, which means that employers now foot the bill for SSP themselves.

Additionally, the SSP percentage threshold scheme, which allowed employers whose sickness absence costs exceeded 13% of their gross NICs for any particular month, was abolished on 6 April 2014. The record-keeping requirements for the percentage threshold scheme were also abolished in respect of SSP paid out after 6 April 2014. Employers are nevertheless still required to maintain PAYE records which demonstrate that they are meeting their SSP obligations.

Accrual of annual leave during periods of sickness absence

[M1514A] In January 2009, the European Court of Justice (ECJ) published an important judgement (*Stringer v HM Revenue and Customs* [2009] IRLR 214) concerning the question of whether statutory annual holidays continue to accrue whilst an employee is absent from work due to sickness. It has since been established (in *Neidel v Stadt Frankfurt am Main* C-337/10 [2012] 3 CMLR 92, [2012] IRLR 607, ECJ) that this principle applies only in respect of the first four weeks of statutory annual holiday leave, ie the portion of leave governed by the EC Working Time Directive.

The ECJ ruled as follows:

- employees who are absent from work due to sickness continue to accrue statutory annual holiday entitlement in the usual way;
- this is the case irrespective of whether the employee is absent for a whole holiday year or only part of it (and irrespective of whether or not the employee is being paid whilst on sick leave);
- the employer may agree, if the employee requests it, to convert a period of sickness absence into paid annual holiday leave;
- if an employee has been unable to take holidays as a result of sickness absence, the employer must permit them to take their full statutory entitlement after they have returned to work;
- if the employee's absence has spanned more than one holiday year (ie if they have accrued annual leave whilst off sick during one holiday year and return to work only during the next holiday year), they must be permitted to carry forward any untaken holidays to the next holiday year and take them during that year (irrespective of whether the employee has made any request to carry the holidays forward);
- if an employee is dismissed whilst absent from work, the employer must make a payment in lieu of all accrued statutory annual holiday which has not been taken, even in circumstances where the employee was on sick leave for one or more full holiday years.

In a subsequent case, *Pereda v Madrid Movilidad SA* [2009] IRLR 959, the ECJ ruled that where an employee goes off sick just before a period of planned statutory annual holiday, the employer must permit them, on request, to reschedule the period of annual holiday to another time once they have recovered and returned to work. The same principle applies to employees who become incapacitated whilst they are on holiday. Employees have the right to choose to take their annual holiday leave at a time other than during a period that overlaps with a period of sickness absence.

In *Plumb v Duncan Print Group Ltd* (2015) UKEAT/0071/15, [2015] IRLR 711, EAT, the EAT ruled that in light of EU law and ECJ case law, any holiday leave that an employee has carried over as a result of sickness absence must be taken within (at most) 18 months from the end of the holiday year in which the leave accrued. Employers are not legally required to allow carry-over for unlimited periods.

Rehabilitation of an employee who has been absent from work

[M1515] Where an employee has been absent from work for a lengthy period of time, it will be essential to manage their return to work carefully. The employee's line manager should take responsibility for discussing the employee's return to work directly with them and for agreeing a detailed plan to support the employee on their return.

Many employees will feel anxious about the prospect of returning after a lengthy period away from the workplace and they may be worried about how they will cope with being back at work, or concerned about their colleagues' reactions.

The first action point will be for the manager to speak to the employee (preferably prior to their return if that is possible, but if not on the employee's first day back) to discuss the details of their return. One way of dealing with this would be to suggest to the employee that they might make a social visit to the workplace shortly before their agreed return date to allow a meeting with their line manager and/or HR department and also allow them to chat informally with their colleagues about any ongoing relevant matters.

The discussion with the line manager should include:

- the employee's opinion about their capabilities, for example whether the employee is confident that they are capable of full job performance or only partial performance;
- whether the employee's return should be to full-time duties or whether a phased return would be beneficial;
- whether there are any recommendations from the employee's GP or from the Fit for Work service regarding adaptations that might be made to facilitate their return to work (ie recommendations contained in a doctor's 'fit note');
- whether the employee will still be taking any medication after their return to work that might have side effects, for example tiredness;
- any special arrangements, additional support or adjustments to the employee's duties or environment that would help the employee to reintegrate into the workplace; and
- whether a tailored induction programme is desirable or necessary, for example if a number of organisational or procedural changes have taken place during the employee's absence, a 'mini-induction' may be very helpful.

The employer should also arrange for the employee to be medically examined by an occupational doctor to confirm that they are genuinely capable – both physically and mentally – of returning to work. The doctor should also be asked to comment on whether the employee should be exempted initially from any particular job duties and if any special arrangements are necessary or advisable.

After the employee's return, the line manager should:

- monitor the situation over the first few weeks to ensure the employee is coping with their work and the day-to-day pressures of working life;
- make sure the employee is not 'thrown in the deep end', for example by requiring them to deal with a huge backlog of work caused by their absence; and
- take all reasonable steps to ensure the employee's reintegration into the workplace.

When dismissal on the grounds of long-term absence can be fair

[M1516] Long-term sickness absence can be a fair reason for dismissal for the purposes of the *Employment Rights Act 1996, s 98* under the heading of 'capability'. 'Capability' is defined as having reference to 'skill, aptitude, health or any other physical or mental quality'. For a dismissal to be fair, however, the employer would have to show that the employee's long-term absence was sufficient to justify dismissal and that they had acted reasonably in dismissing the employee for this reason, taking into account the size and administrative resources of the organisation. Employment tribunals, in practice, will assess, based on all the available evidence, whether the dismissal for the stated reason was within what the Courts have termed the 'band of reasonable responses'.

In the event of a claim to tribunal, the burden of proving the reason for the dismissal is on the employer.

All employees are entitled to bring a claim for unfair dismissal to an employment tribunal provided that they:

- are employed under a contract of employment;
- have a minimum of two years' continuous service with the employer.

There are no upper or lower age limits on the right to claim unfair dismissal.

Before contemplating dismissal, the employer should review the circumstances to establish whether grounds for dismissal exist. In doing so, the employer should consider:

- the effect of the employee's past and likely future absences on the organisation, for example to what extent colleagues are being burdened by the employee's absence;
- the degree to which it is possible (if at all) to cover the absent employee's work, for example by using temporary agency workers;
- how similar situations have been handled in the past;
- medical advice;
- the likelihood of the employee being able to return to work in the reasonably near future; and
- the availability of suitable alternative work which the employee would be capable of doing, ie whether there is any alternative to dismissal.

In the case of *Perth & Kinross Council v Gauld* [2014] EATS 0046/13, the EAT held that if an employee presents his or her own medical report to the employer and the employer has some concerns about the quality of that medical evidence, it should obtain its own medical report. A failure to do so could lead an employment tribunal to rule that any subsequent dismissal of the employee for sickness absence is unfair.

Procedure for ensuring a dismissal is fair

[M1517] Over and above having a potentially fair reason for dismissal (ie ill-health absence), an employer who is challenged in an employment tribunal must also be able to demonstrate that they:

- acted reasonably towards the employee in an overall sense; and
- followed their own in-house procedures fully.

Courts and tribunals have, over the course of many years, developed tests of fairness in cases of dismissal. In relation to dismissals on the grounds of long-term ill-health absence, the employer should, as a minimum, ensure that they:

- review the employee's absence record to assess whether it is sufficient to justify dismissal;
- set time limits for review and inform the employee of these;
- consult the employee regularly;
- obtain up-to-date medical advice;
- advise the employee as soon as it is established that termination of employment has become a possibility;
- advise the employee that they have the right, if they attend any formal meetings with the employer to discuss their ongoing employment, to bring a colleague or trade union representative of their choice along with them to the meeting if they wish;
- review whether there are any other jobs that the employee could do prior to taking any decision on whether to dismiss; and
- act reasonably throughout.

In *BS v Dundee City Council* [2013] CSIH 91, the Court of Session listed some of the questions that employers should address before dismissing an employee on the grounds of ill-health absence. These were:

- can the employer be expected to wait any longer for the employee to recover and return to work, and if so, how much longer?
- has the employee been properly consulted and have his or her views been taken into account?
- has the employer taken reasonable steps to establish the employee's condition and likely prognosis?
- have the employee's representations (if any) been properly balanced against the opinions of medical practitioners?

There is no time limit in law after which it is fair to dismiss an employee who is absent from work. The key question is whether in all the circumstances the employer can reasonably be expected, in light of the requirements of their business, to wait any longer for the employee to recover and return to work. This will depend, amongst other things, on the size and resources of the business and the degree of disruption or difficulty that the employee's long-term absence is causing.

A court or tribunal will judge the fairness of a dismissal in light of the circumstances at the time the employee's employment came to an end, and not at the time when management took the decision to dismiss. It follows that in a situation where an employee is given notice that their employment will terminate at a defined future date, and where circumstances change or new information comes to light during the notice period, the employer should reconsider their decision. In *Fox v British Airways plc* [2013] EWCA Civ 972, [2013] ICR 1257, [2013] IRLR 812, a case in which the employee was dismissed with notice on account of long-term sickness absence, the Court of Appeal ruled that the dismissal was unfair because the employee's condition had improved during the notice period and his consultant had provided a report referring to the possibility of his being able to return to work. The fairness of a dismissal stands to be judged in line with the House of Lords' ruling in the case of *Polkey v A E Dayton Services Ltd* [1987] IRLR 503. In this case, the House of Lords said that any procedural shortfall will make a dismissal unfair, even in circumstances where the employer had a legitimate reason to dismiss the employee. This means that any failure of the employer to adhere to its own in-house procedures when dismissing a member of staff (for any reason) is likely, in almost all cases, to render the employee's dismissal unfair.

In April 2009, ACAS published a revised Code of Practice on Disciplinary and Grievance Procedures. The Code provides statutory guidance to employers (and employees) on matters of discipline, performance and grievances. The Code does not specifically address termination of employment on the grounds of sickness absence, nevertheless many of its provisions may be broadly relevant to such dismissals as they relate to fair treatment in a general sense. The Code is accompanied by a comprehensive non-statutory Guide, 'Discipline and Grievances at Work', which contains a section titled 'Dealing with Absence' (in appendix 4).

Managing short-term absences

[M1518] Frequent, persistent short-term absences can be extremely disruptive for an employer. Where an employee is absent long-term, the employer can at least make plans, as they will have some idea about how long the employee is likely to be absent. The nature of short-term absenteeism, however, is that it tends to be unpredictable and uncertain.

The CIPD, in their 2016 annual survey on absence management, reported that by far the most common cause of short-term absence was minor illness such as colds, flu and stomach upsets. Other common causes of short-term absence included stress, musculoskeletal injuries (particularly for manual workers), back pain, home/family/carer responsibilities, mental ill-health (for example clinical depression and anxiety) and acute medical conditions (such as stroke, heart attack and cancer). Employees' rights in relation to short-term absence will be determined according to the terms of their contracts of employment. Contracts or supplementary documentation should state clearly what

Managing short-term absences **[M1520]**

employees' obligations and entitlements are when they are unable to attend work, for example what they will be paid when they are absent and how they should notify the employer.

To manage short-term absenteeism effectively, employers should:

- require employees who are unable to come to work personally to notify a designated person of their absence as soon as possible;
- require employees to 'self-certify' for all periods of absence of up to one week and to provide doctors' statements for all periods of absence beyond the first week;
- conduct regular 'return-to-work interviews' after every period of absence, however short;
- investigate the possible causes of frequent absenteeism;
- look for patterns in absenteeism to see whether there is, for example, a problem in a particular department;
- implement a policy with defined trigger points which, when reached, activate a formal review of the employee's attendance;
- set targets and time limits for improvement; and
- keep detailed records of all employee absences.

These issues are explored further below.

Sound management techniques for dealing with short-term absenteeism

[M1519] There is much an employer can do to manage short-term absenteeism effectively with a view to reducing the frequency of its occurrence.

Requirement to notify

[M1520] Employers should have in place a set procedure under which employees are required personally to notify a designated person (usually their line manager) by telephone as soon as they know that they will be unable to come to work. The line manager should record the date and time of the call, the reason given by the employee for inability to attend work and the employee's view of how long they expect to be absent. This creates the start of an absence record. A pre-printed questionnaire could be devised for this purpose, thus ensuring consistency in approach. The questionnaire could also be used as a means of instigating payment of statutory sick pay.

If the manager is unavailable when the employee telephones, they should telephone the employee at home soon afterwards to obtain the necessary information. Accepting a voicemail or e-mail message is unlikely to be sufficient and will not act as a deterrent to casual absence.

The information should subsequently be passed to the HR department (if the organisation has one) or a named senior manager to whom responsibility has been allocated for collecting and analysing sickness absence data across the whole organisation.

Certification

[M1521] Employers should routinely require employees who are absent from work due to sickness to complete a self-certification form when they return to work (at the return-to-work interview – see below). The employee should be asked to complete and sign the form in front of their manager who should counter-sign the form, and the obligation to do so should apply to every absence of one day or more. The form should require the employee to write down the reason for their absence, the dates on which their sickness began and ended and whether they consulted a doctor, rather than just tick a series of boxes. This makes the process of self-certification a more active process.

Line managers should check that self-certification forms have been properly completed, for example by ensuring that the reason given for the absence is clearly stated. Any vague, woolly reasons for absence such as 'sickness' or 'debility' should be questioned (although the manager should not ask intrusive questions about the employee's state of health).

It is a sound idea to make the payment of occupational sick pay dependent on the completion of a self-certification form (and on the provision of doctors' statements in the case of longer periods of absence).

Return to work interviews

[M1522] As well as requiring the employee to complete a self-certification form, employers should adopt a policy whereby their managers routinely conduct 'return-to-work interviews' every time an employee has been absent from work due to sickness.

The interview should be:

- informal (but should be more than just a 'corridor chat')
- private and confidential;
- structured and factual;
- carried out in a positive and supportive way; and
- recorded in writing (with a copy of the record being provided to the employee).

The dual purpose of such an interview will be to establish the reason for the employee's absence (without requiring disclosure of any sensitive medical details) and to assess whether the employer can do anything to help or support the employee. It is important to note that this type of interview should not be confused with a disciplinary interview and that fact should be made clear to the employee. It may assist the process of communication if the manager explains to the employee that the purpose of return-to work interviewing is to manage and monitor employees' absences so that any problem areas can be identified and support offered where appropriate.

Because the interview is informal, the right for the employee to be accompanied will not apply.

Adopting this procedure will let employees know that management is serious about monitoring absences and may therefore deter casual absences.

CIPD reported in their 2016 survey on absence management that the most effective approaches to the management of short-term absenteeism were (in order of effectiveness): conducting return-to-work interviews, trigger mechanisms for reviewing attendance, the provision of sickness absence information to line managers, the use of disciplinary procedures to deal with unacceptable absence, giving line managers primary responsibility for managing absence, providing leave for family circumstances (such as carer/emergency/dependant leave) and flexible working.

Considering the possible causes of frequent absenteeism

[M1523] Most employers have one or more employees who are absent from work on a regular basis. If the employee has an underlying medical condition that is causing these absences, then the employer will know what they are dealing with and medical advice can be sought to support any employment decisions that the employer might contemplate making. The employer should also consider whether the employee might qualify as disabled under the *Equality Act 2010* (see **M1507** above) and if so review what reasonable adjustments might be made to the employee's job or working environment.

There are, however, many other possible causes of frequent absenteeism. Apart from ill-health, the employee may:

- have personal or family difficulties, for example childcare problems;
- have another job which is causing a conflict of interest;
- be de-motivated and disinclined to attend work; or
- have a specific problem in the workplace (see **M1524** below) that is causing or contributing to their frequent absences.

It will be essential for the manager to remain open-minded and to refrain from jumping to the conclusion that the employee is taking time off work without good reason. It may be that the cause of the employee's absences is something outside their control. The manager should speak to the employee to establish whether there is a specific underlying cause and whether anything can be done to support the employee. Understanding the root cause of the absence problem is important because, until the cause is correctly pinpointed, it will not be possible to identify an appropriate course of action to remedy it.

Reviewing whether factors in the workplace are causing absence

[M1524] Whilst a certain level of absenteeism is inevitable due to minor illnesses, there may also be occasions when employees take time off work for reasons associated with factors in the workplace.

Managers should be constantly alert to the possibility that an employee's absences might be caused by, or exacerbated by, factors in the workplace. For example, the problem may be related to:

- the volume of work being too much for the employee to cope with;
- deadline dates or other pressures that the employee feels unable to face;
- unsatisfactory working relationships or outright conflict with colleagues;

- bullying or harassment;
- de-motivation caused by ineffective management or an authoritarian management style;
- inability to cope with change;
- a feeling of inadequacy or loss of control over the job; or
- other factors at work causing dissatisfaction, for example ineffective procedures or equipment, having no clear goals or targets, etc.

Issues such as those identified above may well lead to absences and managers should be alert to signals that the employee may be suffering from stress to an extent that he or she is not coping. Medical certificates that state 'stress', 'depression' or 'anxiety' should put the manager on notice that there may be a workplace problem that needs to be addressed.

Line managers should always be alert to such issues and should make enquiries of any employee who has frequent short-term absences as to whether there is anything in the workplace causing or contributing to their absences. Such enquiries should be made sympathetically, with the manager reassuring the employee that if he or she is experiencing problems at work, it would be their genuine wish to provide support with a view to resolving those problems.

If a workplace problem is identified, the manager should of course take steps to remove or reduce the factor that is causing the problem, if that is at all possible. Once the cause of the employee's absences is removed, the employee's attendance may well improve. A failure to take steps to support an employee who is experiencing health problems as a result of factors in the workplace may have serious consequences as the employer could be held liable in law if the employee subsequently had a mental breakdown as a result of factors in the workplace and the breakdown was reasonably foreseeable.

A number of court decisions demonstrate the dangers of doing nothing when an employee complains of serious workplace stress. In *Intel Corporation (UK) Ltd v Daw* [2007] IRLR 355, the employer was held liable in negligence when Ms Daw had a complete breakdown, which had been caused by overwork. The employee had, at her manager's request, provided a written account of the problems she was experiencing, but despite this, nothing was done to reduce her workload. The Court held that it would have been clear to the employer, following the written account, that Ms Daw was at risk of personal injury. It followed that the injury to her health as a result of an excessive workload had been reasonably foreseeable. Management should have taken immediate action to reduce Ms Daw's workload upon receiving the written account of her problems, as this was the only course of action that would have properly addressed the problem.

Similarly, in *Dickins v O2 plc* [2008] EWCA Civ 1144, [2009] IRLR 58, the employer was held liable for having caused the depressive illness of an employee who had repeatedly complained of overwork and of being 'at the end of her tether'. The Court of Appeal held that management's failure to reduce Ms Dickins' workload amounted to a breach of the duty of care, and that once they became aware of the extent of her difficulties, they should have sent her home pending urgent investigation by occupational health, even though, at that time, she had not been signed off by her GP.

Looking for patterns

[M1525] Employers who are experiencing high levels of short-term absenteeism should use the records created by line managers about individuals' absences to review the absenteeism, and whether any patterns exist, for example:

- whether absenteeism is higher than average in any particular department or section;
- whether there are any particular times of the year, month or week when absenteeism is noticeably higher than at other times;
- whether the rates of absence of employees performing any particular type of work are disproportionate when compared to other employees; or
- whether there is any particularly common cause of absence (for example back pain).

In each case, the employer should investigate in order to try to establish the root cause of the particular pattern. Following identification of the likely cause, steps can be taken to remedy the situation. For example if absenteeism is unusually high in a particular manager's section, it could possibly be that the manager is the cause. Well-designed training in supervisory skills or interpersonal skills, or individual coaching may improve the manager's people-management skills and subsequently lead to a reduction in levels of absenteeism in his or her department.

On a more local level, line managers should, when reviewing individual employees' absence records, look to see whether there is a pattern amongst the absences. Examples could include the frequent Monday morning absentee or the employee whose absences tend to occur at particular times, for example just before a detailed month-end report is due.

If such a pattern is identified, the manager should speak to the employee about it and ask the employee if they can explain the pattern. Managers should, of course, not make assumptions, nor blame or accuse the employee, but instead should remain open-minded. The main aim in speaking to the employee about a pattern of absences will, once again, be to try to establish the underlying reasons(s) for the frequent absences. It is only when the underlying cause is identified that it will be possible to decide what to do. For example, if absence is due to tiredness and the reason the employee is tired is because they have a second job, then the manager might proceed by explaining to the employee that the situation is unsatisfactory and seek to reach a compromise on how many hours the employee works in the other job.

Even if the employee is unable to put forward any explanation, such a meeting will have the advantage of alerting the employee to the fact that the line manager has noticed the pattern. This, in turn, may deter further casual absences.

Defined trigger points

[M1526] It can be very effective for employers to devise and implement an attendance procedure with defined trigger points that act to instigate a formal review of the employee's level of absence and its causes. Trigger points are usually defined by reference to:

- the total number of days' absence the employee has had in a defined period (for example in any twelve-month period); or
- the number of occasions when the employee has been absent from work during the same period.

A typical example of trigger points could be three separate absences (whatever their length) or absences totalling 15 days in any twelve-month period.

Setting targets and time limits for improvement

[M1527] Where an employee's level of short-term absences has reached a stage where their level of attendance is unsatisfactory, the line manager should seek to agree with the employee reasonable targets and time limits for improvement in attendance. The manager should try to ensure that the employee is committed to achieving the agreed level of improvement. The manager should of course follow up at the agreed time and review the employee's level of attendance again. If it has not improved to a satisfactory level, formal action may have to be instigated according to the employer's procedures.

Keeping records

[M1528] Full records should always be kept of employees' absences and of all discussions held with the employee about absence and attendance. Self-certificates and medical statements should also be retained. Such records should, of course be held confidentially, preferably by the organisation's HR department (if they have one). The terms of the *Data Protection Act 1998* will have to be adhered to in holding such records.

Giving formal warnings for unsatisfactory attendance

[M1529] Although it is appropriate for managers to be supportive in the first instance towards employees who, for genuine reasons, have frequent absences from work, managers also need to ensure the work of their department is done efficiently and effectively. Even where the reason for an employee's absences is undisputedly genuine, that does not change the fact that the employee's work still needs to be done on a regular and reliable basis. If, therefore, informal measures have not led to an improvement in the employee's attendance, it may be that formal procedures need to be instigated. This will be appropriate when the employee's absences have become excessive or where they are beginning to cause serious disruption or dissatisfaction.

Some employers use their standard disciplinary procedure to deal with this type of situation, whilst others have separate capability procedures which follow a similar format. The key action steps of any such procedure will be that the employer would:

- write to the employee, explaining that their level of absence has reached a level that is causing concern, enclosing a note detailing the dates and duration of all absences over the last six or twelve-month period (as appropriate);
- invite the employee to attend a meeting to discuss the matter, informing them that they have the right to bring a colleague or trade union official along, either to represent them or simply to provide moral support;
- hold the meeting, and explain to the employee that their absence has reached a level that is considered unsatisfactory and the reasons why this is the case;
- give the employee a full and fair opportunity once more to explain their absences and put forward any mitigating factors or other representations;
- decide after the meeting whether it is appropriate to issue a formal warning, following the terms of the employer's own procedure;
- make it clear in any warning that if the employee's level of attendance does not improve to a defined standard within a defined time period, further formal action will be taken and that continuing absences could, eventually, lead to dismissal; and
- set down a date for a further review, typically in three or six months' time.

This type of procedure may be instigated irrespective of whether or not the employee's absences are for genuine reasons. If some of the absences are not genuine, the action taken would of course fall under the banner of discipline. Even where the employee's absences are for genuine reasons (for example genuine ill-health, family problems, etc), formal action can be justified under the banner of 'capability' or 'attendance'.

Fair dismissal on the grounds of short-term absence

[M1530] If, following formal action, the employee's level of attendance does not improve to a satisfactory degree within the defined time-scale, the employer would move to the next stage of the disciplinary or capability procedure. It is usual for two written warnings to be given following which the employee may ultimately be dismissed. Dismissal should not, of course, be undertaken lightly and should normally only ever be a last resort after all other possible courses of action have been fully explored.

The reason for dismissal will be either 'lack of capability' (ie ill-health that has led to the employee being unable to perform their job to a satisfactory standard) or 'some other substantial reason' (ie unsatisfactory attendance, whatever the causes). Both of these are stated in the *Employment Rights Act 1996* as potentially fair reasons for dismissal.

For a dismissal to be fair, however, the employer also has to show that the employee's level of absence was sufficient to justify dismissal, that they had acted reasonably in dismissing the employee for this reason and that they followed their own in-house procedures fully.

Conclusion

[M1531] Many employers are wary of tackling employee absence and uncertain as to what action they can lawfully take. The basic underlying principle is that, if an employee is incapable of performing their job, whether due to genuine ill-health or not, the employer may take action up to and including dismissal. Long before the dismissal stage is reached, however, active management intervention can often help to:

- identify the cause(s) of an employee's absences;
- provide support to the employee, if appropriate;
- deter casual absences;
- identify whether there are any problems inherent in the workplace that are contributing to rates of absenteeism, for example bullying, and ensure these are addressed;
- establish whether or not an employee's level of attendance is likely to improve within a reasonable time-frame;
- improve morale and motivation; and
- lead to a reduction in rates of absenteeism within the organisation and an associated reduction in costs.

Managing Health and Safety

Abigail Cohen

Introduction to managing health and safety

[M2001] There are many reasons why employers need to manage health and safety well. These are usually grouped under three main headings:

Humanitarian

Statistics produced by the Health and Safety Executive report that in 2013/2014 around 203 people in the UK (both workers and members of the public) were fatally injured in work-related accidents. Approximately 646,000 people suffered an accident at work in 2012/2013. Each year around 30,000 people suffer major injuries (fractured limbs, amputations etc), well over 2,000 people die as a result of occupational diseases and at least 10,000 other deaths are partly attributable to occupational ill health. Employers would obviously not wish to be involved in these accidents and cases of ill health. The psychological effects of individual incidents on employers, employees and others are sometimes quite traumatic.

In recent years greater attention has been placed on corporate responsibility with the European Agency for Safety and Health at Work publishing a report in 2004 entitled *'Corporate social responsibility for health and safety at work'*, which contains many examples of good practice together with related conclusions and recommendations. Further, the *Corporate Manslaughter and Corporate Homicide Act 2007* came into force on 6 April 2008 creating the offence of corporate manslaughter in England, Wales and Northern Ireland and corporate Homicide in Scotland.

As at the beginning of 2014 there had been 6 convictions under the *Corporate Manslaughter and Corporate Homicide Act 2007* (1 trial and 5 guilty pleas) with at least 5 further cases pending this year. Figures from 2013 indicate that the number of cases opened and under investigation is far higher, with 141 corporate manslaughter cases having been opened since records began in 2009 and 56 cases being investigated for prosecution.

Legal

Society has been increasingly unwilling to accept such a toll of accidents and ill health. In many respects the UK has led the world in health and safety law and now European directives result in a steady flow of health and safety legislation. Many regulations (particularly the *Management of Health and Safety at Work Regulations 1999 (SI 1999 No 3242)* – see **M2005** below), specifically require the application of appropriate management techniques. The *Corporate Manslaughter and Corporate Homicide Act 2007* requires corporate entities to ensure that their activities are safely managed so as to prevent fatalities.

Financial

Accidents and ill-health can have a direct financial impact on a business through fines and compensation claims. Fines for breaches of health and safety legislation have increased considerably in recent years, with the present record fine of £15 million for a single offence being set in 2005 following the prosecution of Transco after an explosion at its Larkhall plant in Lanarkshire. The Sentencing Guidelines Council issued Guidelines which took effect from 15 February 2010 which provide that for corporate manslaughter The appropriate fine will seldom be less than £500,000 and may be measured in millions of pounds and for health and safety offences causing death the appropriate penalty will seldom be less than £100,000 and may be measured in hundreds of thousands of pounds or more. Even in health and safety cases where there has not been a fatality, the Court of Appeal has recently held that there is no ceiling on the fine to be imposed and that where a corporate defendant has a significant turnover, the court must scrutinise the structure, turnover and profitability of the company when determining the appropriate fine.

In *R v Sellafield Ltd & R v Network Rail Infrastructure* [2014] EWCA Crim 49 (a decision of the Court of Appeal, with Judgment by the Lord Chief Justice), Network Rail Infrastructure Ltd and Sellafield Ltd had pleaded guilty in the Magistrates' Court for breaches of health and safety, following a collision at an unmanned level crossing, and environmental protection legislation, following the disposal of radioactive waste, respectively. The cases had been committed to the Crown Court for sentence. Both companies had an annual turnover in excess of GBP 1 billion. Sellafield was fined £700,000 and Network Rail was fined £500,000 and both sought to appeal those fines on the basis that they were manifestly excessive. Neither case involved a fatality. The Court of Appeal laid down principles for sentencing Courts and stressed that, in determining the appropriate level of fine, the Court must consider not only the seriousness of the offence but also the financial means of the offender, which will serve to increase or reduce the level of any fine. This applies whether the case involves a small local business or a multi-million pound corporation. The Court had called for the provision of company accounts and stated that accounts or financial information must be made available in good time. Where a corporate defendant has a turnover in excess of one billion pounds, the Court must carefully scrutinise the structure, turnover and profitability of the company and the remuneration of the directors. The Court rejected submissions that, on early guilty pleas in environmental and health and safety cases, fines in excess of GBP 500,000 are only appropriate where there has been a disaster or fatality. The Court stated there was no ceiling on the amount of a fine which could be imposed, although each case will turn upon its own facts and the financial means of the defendant.

There may also be significant indirect costs associated with the unavailability of injured employees or damaged equipment, or due to the impact of legal sanctions such as prohibition notices. In 2012/13, over a third of all injuries suffered by workers led to over 3 days absence from work and almost a quarter led to absences over 7 days. Loss of reputation following serious accidents has had a significant adverse impact on a number of companies, particularly in the rail sector, and supply-chain pressure can result in those whose health and safety standards are found lacking, finding it increasingly difficult to gain business.

The cost to employers of workplace injuries and work-related ill health was estimated at £2.8 billion in 2010/11. These costs comprise expenditure relating to sick pay payments, production distribution costs, administrative and legal costs and insurance premiums, of which the latter make up the largest proportion of approximately 50%.

The total cost associated with workplace injuries and ill health (excluding occupational cancers) in Great Britain in 2010/11 was estimated to be some £13.8 billion which falls on individuals, employers and Government.

Figure 1: The costs of accidents

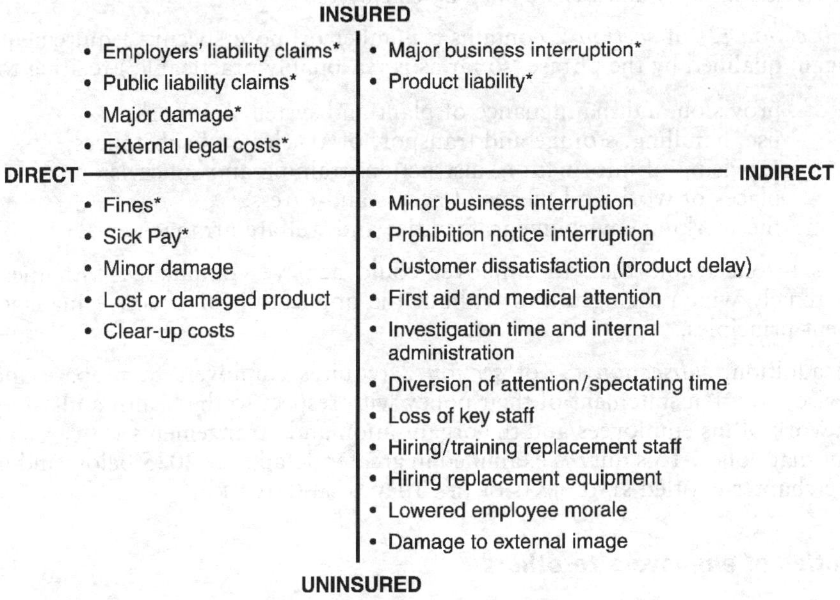

INSURED

DIRECT		INDIRECT
• Employers' liability claims*	• Major business interruption*	
• Public liability claims*	• Product liability*	
• Major damage*		
• External legal costs*		
• Fines*	• Minor business interruption	
• Sick Pay*	• Prohibition notice interruption	
• Minor damage	• Customer dissatisfaction (product delay)	
• Lost or damaged product	• First aid and medical attention	
• Clear-up costs	• Investigation time and internal administration	
	• Diversion of attention/spectating time	
	• Loss of key staff	
	• Hiring/training replacement staff	
	• Hiring replacement equipment	
	• Lowered employee morale	
	• Damage to external image	

UNINSURED

* Only these costs can be quantified without considerable effort

(Expanded version of diagram previously published in HSE booklet HSG 96 (no longer available).)

Legal requirements relating to managing health and safety

[M2002] Almost all recent health and safety legislation contains provisions relating to effective management, often requiring the application of risk assessment and control techniques. Set out below are the requirements of *sections 2* and *3* of the *Health and Safety at Work etc. Act 1974* ('*HSWA 1974*'), together with those contained in the *Management of Health and Safety at Work Regulations 1999 (SI 1999 No 3242)* (the '*Management Regulations*').

The *Construction (Design and Management) Regulations 2007 (SI 2007 No 320)* increased the requirements for effective health and safety management of all construction work – see **M2022** below

Duties of employers to their employees

[M2003] *Section 2* of *HSWA 1974* sets out a number of general duties that employers hold towards their employees. *Subsection (1)* of *section 2* contains an all-embracing requirement:

> It shall be the duty of every employer to ensure, so far as is reasonably practicable, the health, safety and welfare of all his employees.

Subsection (2) of *section 2* contains a number of more specific requirements (again qualified by the phrase 'so far as is reasonably practicable') relating to:

- provision and maintenance of plant and systems of work;
- use, handling, storage and transport of articles and substances;
- provision of information, instruction, training and supervision;
- places of work and means of access and egress;
- the working environment, facilities and welfare arrangements.

It is inconceivable that an employer could achieve compliance with these extremely wide-ranging duties without the application of effective management principles.

In addition, *subsection (3)* of *section 2* requires employers to prepare and revise a written statement of their policy with respect to the health and safety at work of his employees and the organisation and arrangements for carrying out that policy. This duty is examined in greater detail at **M2025** below and in the chapter entitled STATEMENTS OF HEALTH AND SAFETY POLICY.

Duties of employers to others

[M2004] *Section 3(1)* of *HSWA 1974* places a general duty on employers in respect of persons other than their employees:

> It shall be the duty of every employer to conduct his undertaking in such a way as to ensure, so far as is reasonably practicable, that persons not in his employment who may be affected thereby are not exposed to risks to their health or safety.

Persons protected by this requirement include contractors (and their employees), visitors, customers, members of the emergency services, neighbours, passers-by and other members of the general public – including trespassers, although the 'so far as is reasonably practicable' qualification is of particular relevance to them. (*Subsection (2)* of *section 3* places a similar duty on the self-employed.)

Once again, effective health and safety management provides the only way of complying with this duty.

The Management Regulations

[M2005] The *Management Regulations 1999 (SI 1999 No 3242)* came into operation on 29 December 1999. They modified the 1992 Management Regulations, which in turn had developed many of the principles already established by the *HSWA* in 1974. The Regulations make it quite clear that health and safety must be managed systematically, like any other aspect of an organisation's affairs. The Approved Code of Practice (L21) for the 1999 Regulations has been withdrawn but guidance remains available on the HSE Website and in particular in '*Managing for Health and Safety*' (also known as HSG65) which is available online. Some of the main requirements of the Regulations relate to:

- *Risk assessments* – to identify risks and related precautions (*Regulation 3*).
- *Management systems* – to ensure that precautions are implemented (*Regulation 5*).
- A *competent source of health and safety advice* being appointed (*Regulation 7*).
- *Emergency procedures* being developed (*Regulation 8*).

The 1999 Regulations incorporated previous amendments to the 1992 Regulations, relating to young persons, new or expectant mothers and fire risks.

The requirements of a number of the regulations (eg risk assessment) are dealt with more fully elsewhere in this publication (see RISK ASSESSMENT) and therefore receive only limited attention here.

Breach of the Regulations had given rise to civil liability in the majority of cases following the amendment to the Regulations made by the *Management of Health and Safety at Work Fire Precautions (Workplace) (Amendment) Regulations 2003* which came into effect on 27 October 2003.

Note however that as of 1 October 2013, Section 47 of the *Health and Safety at Work Act 1974*, as amended by section 69 of the *Enterprise and Regulatory Reform Act 2013*, removed the standard of strict liability from certain health and safety regulations.

Section 69 amends s 47 to provide that Breach of duty imposed by a statutory instrument containing (whether alone or with other provision) health and safety regulations shall not be actionable except to the extent that regulations under this section so provide Breach of duty by an existing statutory provision shall not be actionable except to the extent that regulations under this section so provide (including by modifying any of the existing statutory provisions).

Going forward, no civil claim may be brought for breach of statutory duty unless a regulation expressly provides for it; this effectively reverses the previous position. In almost all cases where the accident occurs after this provision came into force it will be for the injured employee to rely on common law negligence and prove that his injuries were caused by the employer's negligence (though failure to comply with statutory duties may often be re-lied upon as evidence of negligence). The regulatory position is not affected and employers are still liable for prosecution if statutory duties are breached.

Risk assessment (Regulation 3)

[M2006] Employers (and the self-employed) must make 'suitable and sufficient' assessments of the risks to their employees and others who may be affected by their work activities, in order to identify the measures necessary to comply with the law. The 'significant findings' of the assessment must be recorded by those with five or more employees.

Given the widely drawn nature of the requirements of *HSWA 1974* and other legislation, all risks must be included in this exercise. Note however that in order for breach of the Regulation 3 duty to be the basis of a claim by an injured employee, the risks need to have been such as ought to have been identified by a suitable and sufficient assessment and thus to have given rise to a duty to combat those risks: *Spencer v Boots The Chemist Ltd* [2002] EWCA Civ 1691, [2002] All ER (D) 465 (Oct).

There is no need to repeat other 'assessments' required under more specific regulations, provided these are still valid. The *Management Regulations 1999 (SI 1999 No 3242)* require special consideration to be given to risks to young persons (under 18s) (*Reg 3(4)* and *(5)*) and to new and expectant mothers and their babies by virtue of *Regulation 16*.

Principles of prevention to be applied (Regulation 4)

[M2007] Preventive and protective measures must be implemented on the basis of principles specified in *Schedule 1* to the *Management Regulations 1999 (SI 1999 No 3242)*. These include:

- avoiding risks;
- evaluating risks which cannot be avoided;
- combating risks at source;
- replacing the dangerous by the non-dangerous or less dangerous;
- giving collective protective measures priority over individual measures, i.e. following good principles of health and safety management.

Health and safety arrangements (Regulation 5)

[M2008] Employers must make and give effect to appropriate 'arrangements' for the effective planning, organisation, control, monitoring and review of preventive and protective measures. Records should be kept of these arrangements by those with 5 or more employees.

While *Regulation 3* of the *Management Regulations 1999 (SI 1999 No 3242)* requires precautions to be identified, *Regulation 5* requires them to be put into practice effectively, utilising a systematic management cycle (see FIGURE 2).

Figure 2

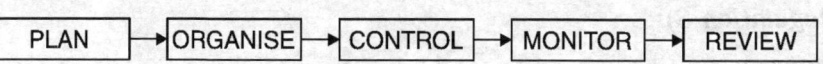

The risk assessment cannot remain a meaningless piece of paper – *Regulation 5* of the *Management Regulations (SI 1999 No 3242)* requires it to be put into effect. Much of the remainder of this chapter is concerned with the practical application of these management principles.

Health surveillance (Regulation 6)

[M2009] Employers must ensure that employees are provided with appropriate health surveillance where this is identified by the risk assessment as being necessary. In most cases surveillance should already have been identified as being necessary to comply with other specific regulations (eg COSHH, asbestos). However, there may be other types of health risk not adequately covered by other legislation, eg colour blindness for train drivers or electricians; risks of vibration white finger.

'Health surveillance' is not defined. In *Paterson v Surrey Police Authority* [2008] EWHC 2693 (QB), [2008] All ER (D) 85 (Nov), it was held that the words cover a spectrum from keeping an eye on an employee to insisting that he has a medical check up: in each case the requirements are conditioned by what has been identified in the risk assessment conducted in accordance with regulation 3.

Health and safety assistance (Regulation 7)

[M2010] Employers must appoint one or more 'competent' persons to assist them in complying with the law. The numbers of such persons and the time and resources available to them must be adequate for the size of and the risks present in the undertaking.

Competence in relation to this requirement might involve:

- knowledge and understanding of the work activities involved;
- knowledge of the principles of risk assessment and prevention;
- understanding of relevant legislation and current best practice;
- the capability to apply this in a practical environment;
- an awareness of one's own limitations (in experience or knowledge);
- the willingness and ability to learn.

Although formal health and safety qualifications give a good guide to competence, they are not an automatic requirement. Small employers are allowed to appoint themselves under this Regulation, providing they are competent. The *Management Regulations 1999 (SI 1999 No 3242)* state (in

Regulation 7(8)) that there is a preference for the competent person to be an employee rather than someone from outside the organisation such as a consultant.

Procedures for serious and imminent danger and for danger areas (Regulation 8)

[M2011] Procedures must be established 'to be followed in the event of serious and imminent danger to persons at work'. These procedures must be implemented where appropriate. Access must also be restricted to areas which are known to be dangerous.

Most employers are familiar with the process of preparing a fire evacuation procedure. Similar procedures must be established for dealing with other types of emergency which may have been identified through the process of risk assessment as being foreseeable. Such emergencies might include bomb threats, leaks or discharges of hazardous substances, runaway processes, power failure, violence, animal escape, severe weather etc. Competent persons must be appointed to implement evacuation where necessary.

Access to danger areas must be restricted. A danger area is an area of the work environment identified as being dangerous because the level of risk is unacceptable without taking special precaution. Examples include, the presence of electrical risks, chemical risks, potentially dangerous animals (or humans). Access to such areas must be restricted to those who have received adequate instruction on the risks present and the precautions to be taken.

The ACOP had provided that employers should establish procedures for any worker to follow if situations presenting serious and imminent danger were to arise or for the police and emergency services in case of an outbreak of public disorder. The procedure for any worker to follow in serious and imminent danger has to be clearly explained by the employer so that employees and others at work know when they should stop work and how they should move to a place of safety.

Contacts with external services (Regulation 9)

[M2012] Necessary contacts must be arranged with emergency services (and others if appropriate) regarding first aid, emergency medical care and rescue work, particularly in respect of the types of emergencies referred to in M2011 above.

The Guidance in HSG 56 provides that where a number of employers share a workplace and their employees face the same risks, it is possible for one employer to arrange contacts on behalf of themselves and the other employers. In these circumstances it would be for the other employers to ensure that the contacts had been made. In hazardous or complex workplaces, employers should designate appropriate staff to routinely contact the emergency services to give them sufficient knowledge of the risks they need to take into account in emergencies, including those likely to happen outside normal working hours. This will help these services in planning for providing first aid, emergency medical care and rescue work, and to take account of risks to everyone involved, including rescuers. Contacts and arrangements with external services should be recorded, and should be reviewed and revised as necessary, in the light of changes to staff, processes and plant, and revisions to health and safety procedures.

Information for employees (Regulation 10)

[M2013] Employers must provide employees with comprehensible and relevant information on:

- risks to their health and safety identified by risk assessment;
- the related preventive and protective measures;
- emergency procedures (and evacuation co-ordinators);
- risks notified to them by other employers in shared workplaces.

The Guidance provides that the information provided should be pitched appropriately, given the level of training, knowledge and experience of the employee. It should be provided in a form which takes account of any language difficulties or disabilities. Information can be provided in whatever form is most suitable in the circumstances, as long as it can be understood by everyone. For employees with little or no understanding of English, or who cannot read English, employers may need to make special arrangements. These could include providing translation, using interpreters, or replacing written notices with clearly understood symbols or diagrams.

This regulation applies to all employees, including trainees and those on fixed-duration contracts. Additional information for employees on fixed-duration contracts is contained in regulation 15.

Regulation 10(2) of the *Management Regulations 1999 (SI 1999 No 3242)* requires parents or guardians of children (under the minimum school leaving age) to be provided with information on risks the child may be exposed to and the precautions identified as necessary by the risk assessment. This is of relevance in relation to school work experience programmes and part-time work by schoolchildren.

Co-operation and co-ordination (Regulation 11)

[M2014] Where two or more employers share a workplace (whether temporarily or permanently) each employer must:

- co-operate with the other employers where necessary for them to comply with the law;
- take all reasonable steps to co-ordinate their precautions with others;
- take all reasonable steps to inform the others of risks arising out of their work.

(Self-employed persons are also covered by this requirement.)

This type of co-operation and co-ordination was in effect already required under *sections* 2 and 3 of *HSWA 1974*. In many cases a main employer may already control the worksite with the other employers assisting. However, in some circumstances it may be necessary to identify an individual or organisation to carry out a co-ordinating role. Even those without employees in the workplace (eg landlords) may still be required to co-operate in order to meet their obligations under *section 4* of *HSWA 1974*.

Persons working in host employers' or self-employed persons' undertakings (Regulation 12)

[M2015] Host employers must ensure that employers whose employees are working in their undertaking are provided with comprehensible information on:

- risks arising out of or in connection with the host's activities; and
- measures being taken by the host to comply with the law.

This Regulation may apply for example to a contractors' employees carrying out cleaning, repair, or maintenance work under a service contract or to employees in temporary employment businesses, hired to work under a host employer's control.

Hosts have an additional duty to ensure that visiting employees are given instructions and information on the risks present and any emergency or evacuation procedures.

(Under *Regulation 12* of the *Management Regulations 1999 (SI 1999 No 3242)* the self-employed are treated as both employers and employees).

Over and above the requirements of *Regulation 11* of the *Management Regulations 1999 (SI 1999 No 3242)* and *HSWA 1974*, hosts are given twin duties of informing visiting employers and of *ensuring* that visiting employees are instructed and informed. Whether they do this themselves or require others to do it on their behalf will depend on local circumstances. *Regulation 12* and *Regulation 11* inter-relate closely with requirements of the *Construction, Design and Management Regulations. 2007 (SI 2007 No 320)* – see **M2023**.

The ACOP stated that "information may be provided through a written permit-to-work system. Where the visiting employees are specialists, brought in to do specialist tasks, the host employer's instructions need to be concerned with those risks which are peculiar to the activity and premises. The visiting employee may also introduce risks to the permanent workforce (eg from equipment or substances they may bring with them). Their employers have a general duty under section 3 of the HSW Act to inform the host employer of such risks and to co-operate and co-ordinate with the host employer to the extent needed to control those risks."

Capabilities and training (Regulation 13)

[M2016] Employers must take into account employees' capabilities as regards health and safety in entrusting tasks to them. The intellectual and physical capabilities of individual employees must be considered as well as the demands of the task and the knowledge and experience required to perform it. Adequate health and safety training must be provided to employees:

- on recruitment into the undertaking;
- on exposure to new or increased risks because of:
  - transfer or a change in responsibilities;
  - new or changed work equipment;
  - new technology;
  - new or changed systems of work.

Training must:

- be repeated periodically where appropriate;
- be adapted to take account of new or changed risks;
- take place during working hours.

The requirements on training stem from the cornerstone duty in *section 2* of *HSWA 1974*. Many other sets of regulations also have specific training requirements. Even where employees have been trained in the same way their capabilities may differ and *Regulation 13* of the *Management Regulations 1999 (SI 1999 No 3242)* requires employers to take ability into account. It also identifies the possible need for refresher training.

The Guidance provides that training needs are likely to be greatest for new employees on recruitment. Employees should receive basic induction training on health and safety, including arrangements for first-aid, fire and evacuation. Particular attention should be given to the needs of young workers. The risk assessment should identify further specific training needs. In some cases, training may be required even though an employee already holds formal qualifications. Training and competence needs should be reviewed if the work activity a person is involved in or the working environment changes, such as a change of department or the introduction of new equipment, processes or tasks.

Employees' duties (Regulation 14)

[M2017] Employees must use machinery, equipment, dangerous substances etc in accordance with the training and instructions they have been provided with. They must also inform their employer (or his representative) of situations representing serious and immediate danger and of any shortcomings in the employer's protection arrangements. To a large extent, this requirement duplicates those already contained in *section 7* of *HSWA 1974* and other legislation. A failure to report a dangerous situation is likely to be considered a serious 'omission' under *section 7*.

Temporary workers (Regulation 15)

[M2018] *Regulation 15* of the *Management Regulations (SI 1999 No 3242)* creates several detailed requirements in relation to the use of fixed-term contract or agency staff and the provision of information to such people. It should always be remembered that temporary workers are still affected by most of the other requirements of these Regulations.

[M2019] Managing Health and Safety

Risk assessment in respect of new or expectant mothers (Regulations 16, 16A, 17 and 18)

[M2019] Where there are:

- women workers of childbearing age, and
- the work could involve risk to the mother or baby, eg processes, working conditions, physical, biological or chemical agents (see Annexes I and II of EC Directive 92/85/EEC),

the risk assessments made under *Regulation 3(1)* of the *Management Regulations 1999 (SI 1999 No 3242)* must take account of such risks.

For individual employees the employer must, *if necessary*:

- alter working conditions or hours of work, or *if necessary*,
- suspend the employee from work,

in order to avoid such risks.

Regulation 17 of the *Management Regulations 1999 (SI 1999 No 3242)* allows doctors (or midwives) to require employers to suspend women from night work where necessary for health and safety reasons. *Regulation 18* of the 1999 Regulations states that employers need not take action under *Regulations 16* and *17* in respect of individual employees unless the employee has notified them in writing of her pregnancy or that she has given birth within the previous 6 months or is breast feeding. (The employer may request confirmation of pregnancy through a certificate from a doctor or midwife.)

NB *Regulation 16* is exempt from the effect of the amendment to *section 47 HSWA 1974* as to civil liability.

A failure to carry out a risk assessment in respect of a pregnant woman, as required by these Regulations is capable of amounting to unlawful sex discrimination; *Hardman v Mallon (t/a Orchard Lodge Nursing Home)* [2002] IRLR 516, EAT.

The *Agency Regulations 2010 (SI 2010/93)* amended the Regulations to add Regulations 16A, 17A, 18A and 18AB with effect from 1 October 2011. Regulation 16A imposes duties in respect of the alteration of working conditions in respect of new or expectant mothers who are agency workers. Regulations 17A and 18A mirror Regs 17 and 18 but apply to agency workers.

Regulation 18AB provides that:

(1) Regulations 16A, 17A and 18A shall not apply where the agency worker—
    (a) has not completed the qualifying period, or
    (b) is no longer entitled to the rights conferred by regulation 5 of the Agency Workers Regulations 2010 pursuant to regulation 8(a) or (b) of those Regulations.
(2) Nothing in regulations 16A or 17A imposes a duty on the hirer or temporary work agency beyond the original intended duration, or likely duration of the assignment, whichever is the longer.

Protection of young persons (Regulation 19)

[M2020] Employers must ensure young persons are protected from risks which are a consequence of their:

- lack of experience;
- absence of awareness of risks;
- lack of maturity,

eg by appropriate training, supervision, restrictions (see also *Regulation 3* of the *Management Regulations 1999 (SI 1999 No 3242)*).

Young persons may not be employed for work:

- beyond their physical or psychological capacity;
- involving harmful exposure to toxic, carcinogenic, genetically damaging agents etc;
- involving harmful exposure to radiation;
- involving risks they cannot recognise or avoid (due to insufficient attention, lack of experience or training);
- in which there are health risks from extreme cold or heat, noise or vibration.

Such work may be carried out if it is necessary for the training of young persons who are not children, providing they will be supervised by a competent person and risks are reduced to the lowest level reasonably practicable.

This topic is dealt with more fully in the chapter VULNERABLE PERSONS.

Other requirements

[M2021] The remaining requirements under the *Management Regulations 1999 (SI 1999 No 3242)* cover the following:

Regulation 20–Exemption certificates
- (For the armed forces in the interests of national security)

Regulation 21–Provisions as to liability
- (Liability of employers for breaches of employees in relation to criminal proceedings. Note see *R v HTM Ltd* [2006] EWCA Crim 1156, [2007] 2 All ER 665, [2006] ICR 1383 as to the interpretation and effect of this Regulation on the general health and safety duties under the *HSWA 1974*.)

Regulation 22–Restriction of civil liability for breach of statutory duty
- (A breach of a duty imposed on an employer by these Regulations does not confer a right of action in civil proceedings insofar as that duty applies for the protection of persons not in his employment. The previous exclusion of right of action under the Regulations by employees was removed in 2003.)

Regulation 23–Extension outside Great Britain
- (The Regulations apply to various offshore activities)

Regulations 23–27
- (Contain minor amendments in respect of first aid, offshore installations and pipeline works, mines and construction)

Regulations 28–30
- (Deal with administrative matters, revocations and transitional arrangements)

Construction (Design and Management) Regulations 2007

[M2022] The future of the *CDM 2007 Regulations* is currently unclear. In December 2011 the HSE presented the findings of an evaluation of *CDM 2007* to the HSE Board. It was clear from the evaluation that the model of risk management embedded within *CDM 2007* had become standard practice in the more organised part of the construction industry, A clear case had emerged, however, to make changes which would embed this approach at the smaller end of the industry.

The HSE Board agreed, therefore, that a significant revision to *CDM 2007* was needed rather than minor amendments. It directed HSE to develop revised regulations based on copy-out of the Temporary or Mobile Construction Sites Directive (TMCSD) as a starting point, but supported an approach which retained duties going beyond Directive requirements where these were objectively justifiable.

The Board agreed in 2013 to publication of a Consultation Document (CD) which included draft Regulations. The key proposals described in the CD were:

- removal of the ACOP and its replacement with a suite of sector-specific guidance aimed principally at SMEs;
- removal of the detailed requirements on competence and their replacement with a more generic framework;
- replacement of the CDM co-ordinator role with the Principal Designer;
- addressing shortcomings in the transposition of TMCSD, with removal of the domestic client exemption and the alignment of the threshold for the appointment of co-ordinators and for formalised health and safety plans with that in the Directive.

The revised draft Regulations were much more closely aligned with TMCSD than are the *CDM 2007 Regulations*, simpler, with a more linear and considerably shorter structure, yet maintaining the same set of core 'hardware' requirements as *CDM 2007*.

Following further detailed Government scrutiny the consultation opened on 31 March 2014 for a period of ten weeks. The results of the consultation were presented in an HSE Report dated August 2014. In terms of next steps the Report provides that:

> "The construction industry is one of the largest sectors in the UK economy and is exceptionally diverse in its makeup. What has emerged from the consultation is a

strong degree of agreement on the broad proposals tempered by a diverse range of often opposing views on the detail of the proposals.

HSE considers that there is a strong case to proceed with the revision to CDM broadly as proposed in the CD. HSE proposes, however, subject to Board agreement, to develop an ACOP in a reduced form . . .

The Board is therefore invited to agree that the proposed amendments to the Regulations proceed to the relevant Government clearances to allow the draft CDM 2015 Regulations to be included in the Statement of New Regulation 9, with a view to CDM 2015 coming into force in April 2015."

A new set of Regulations therefore seems likely to be forthcoming. The requirements of CDM 2007 as currently on the statute books are summarised below but full details of the regulations, together with an ACOP and guidance are contained in HSE book L144 whilst it remains in force.

Current requirements

The *Construction (Design and Management) Regulations 2007 (SI 2007 No 320)* (CDM 2007) came into operation on 6 April 2007, replacing the 1994 CDM Regulations, the Construction (Health, Safety and Welfare) Regulations 1996 and the Construction (General Provisions) Regulations 1961. The Regulations are in five parts:

(1) Introduction (dealing with interpretation of terms and application of the regulations)
(2) General Management Duties applying to construction projects
(3) Additional Duties where project is notifiable
(4) Duties relating to health and safety on construction sites (dealing with a variety of practical health and safety topics relating to construction work)
(5) General (covering civil liability, enforcement etc.)

The definition of 'construction work' contained in the regulations is extremely broad and would include such minor work as replacing a broken window or installing a new electrical socket outlet in a building. The management duties in Part 2 of the regulations apply to all such work, as also do the practical requirements set out in Part 4.

Where a project is notifiable to the enforcing authorities (i.e. the construction phase is likely to involve more than 30 days or 500 person days of construction work) the additional management duties of Part 3 apply.

Management Duties applying to all construction projects

[M2023] Part 2 of CDM 2007 contains requirements applying to all construction projects, whether notifiable or not.

Competence

Sub-section (1) of Regulation 4 states that:

No person on whom these Regulations place a duty shall—

(a) appoint or engage a CDM co-ordinator, designer, principal contractor or contractor unless he has taken reasonable steps to ensure that the person appointed or engaged is competent;
(b) accept such an appointment or engagement unless he is competent;
(c) arrange for or instruct a worker to carry out or manage design or construction work unless the worker is—
    (i) competent, or
    (ii) under the supervision of a competent person.

(This requirement relates to competence in respect of compliance with health and safety legislation.)

Consequently all parties managing construction activities must ensure that all those they appoint, engage or instruct in respect of construction work are competent — whether they employ them directly or obtain their services from elsewhere.

The ACOP (appendices 4,5,6) provide guidance on assessing competence.

Cooperation and Coordination

Regulations 5 and 6 require persons concerned in construction work to co-operate with each other and co-ordinate their activities, largely duplicating requirements in the Management Regulations (see **M2014**). Similarly Regulation 7 requires that the General Principles of Prevention contained in Schedule 1 of the Management Regulations are applied in relation to construction projects (See **M2007**).

Where there is more than one client involved in a project, Regulation 8 allows a written agreement between clients so that only one (or more) client carries out the majority of the duties of the client in respect of the regulations.

Specific duties are placed on clients, designers and contractors in relation to all construction projects, of whatever size.

Clients

[M2024] Clients are required by Regulation 9 to ensure that suitable management arrangements are in place for the project and that these arrangements are maintained and reviewed.

The ACOP and Guidance provides more details as to what the arrangements should include and how clients should comply with the Regulation 9 duty in both notifiable and non notifiable projects.

Regulation 10 requires clients to provide designers and contractors with relevant information, particularly about the site and about the proposed use of the structure as a workplace.

In addition to these specific duties, clients are also subject to the general duties contained in regs 4–7 as to competence, cooperation and coordination.

Designers

[M2025] Regulation 11 imposes duties on designers. Designers must not commence work on a project unless the client is aware of his duties under the regulations. In preparing or modifying designs for construction work designers must, so far as is reasonably practicable, avoid foreseeable risks to the health and safety of any person:

- carrying out construction work;
- liable to be affected by such work;
- cleaning windows or transparent or translucent surfaces;
- maintaining permanent fixtures and fittings;
- using a structure as a workplace.

Designers must provide relevant information about their designs to clients, other designers and contractors to assist them in complying with their duties. Where designs are prepared or modified abroad, Regulation 12 places the responsibilities set out in Regulation 11 with the person commissioning the design or the client.

The ACOP and Guidance provides a list of who may be a designer and provides practical examples of what a designer is and is not required to do in order to discharge the duties imposed by this Regulation.

Contractors

[M2026] Regulation 13 provides contractors must not carry out any construction work unless the client is aware of his duties. They must plan, manage and monitor construction work in a way which ensures that, so far as is reasonably practicable, it is carried out without risks to health and safety.

Contractors must also provide workers under their control with information and training necessary for the work to be carried out safely and without risks to health. They must not begin work on a site unless reasonable steps have been taken to prevent access by unauthorised persons.

Additional duties where the project is notifiable

[M2027] The duties in Part 3 of CDM 2007 come into operation where a project is notifiable (see M2022 above). Full details of these requirements are contained in the HSE book L144, but because of their importance to the management of larger construction projects, these regulations are summarised below.

Clients

- Regulation 14 requires that for notifiable projects the client appoints in writing a CDM co-ordinator and a principal contractor (see below). If no such appointment is made the client is deemed to hold either or both of these roles.

  Clients are required by Regulation 15 to provide the CDM co-ordinator with relevant information. They must ensure that the construction phase does not start unless the principal contractor has prepared a satisfactory 'construction phase plan' (Regulation 16). Clients must also provide the CDM co-ordinator with information in their possession necessary for the 'health and safety file' for the structure and subsequently keep the 'health and safety file' available for those who may need it and revise it as appropriate (Regulation 17).

Designers
- Under Regulation 18 designers must not commence work on a project (other than initial design work) unless a CDM co-ordinator has been appointed. They must also provide the CDM co-ordinator with sufficient information to assist him in complying with his duties.

Contractors
- Contractors have additional duties under Regulation 19, particularly in respect of providing the principal contractor with information about relevant parts of their risk assessments and about any sub-contractors they may engage. Additionally they must comply with directions from the principal contractor and any site rules, and take reasonable steps to work in accordance with the construction phase plan.

CDM Co-ordinator
- The CDM co-ordinator has slightly enhanced duties compared to the Planning Supervisor under the 1994 CDM Regulations. These are set out in Regulation 20 and include:
  — giving advice and assistance to the client (particularly in respect of whether the 'construction phase plan' is at a satisfactory stage for construction work to start);
  — ensuring suitable arrangements for co-ordinating health and safety during planning and preparation activities;
  — liaising with the principal contractor;
  — identifying and collecting pre-construction information;
  — providing this in a convenient form to designers and contractors engaged by the client;
  — taking reasonable steps to ensure designers comply with their duties and co-operation between designers and the principal contractor;
  — preparing (or reviewing and updating) the 'Health and Safety File';
  — passing the 'Health and Safety File' to the client at the end of the construction phase.

Regulation 21 requires the CDM Co-ordinator to ensure that the project is notified to the HSE (or the Office of Rail Regulation).

Principal Contractor
- The principal contractor has many specific duties set out in Regulation 22, particularly in planning, managing and monitoring the construction phase to ensure it is carried out without risk to health and safety, so far as is reasonably practicable. He is required by Regulation 23 to prepare a suitable 'construction phase plan' to ensure that work is planned, managed and monitored so that the construction phase can be started. This plan must be updated, reviewed, revised and refined as the construction work progresses and the principal contractor must ensure that it is implemented.

Regulation 24 places duties on the principal contractor in respect of co-operation and consultation with workers.

(Guidance on the content of the 'Health and Safety File' and 'construction phase plan' is provided in HSE book L144.)

Accident causation

[M2028] The domino model of accident or loss causation has been used by many to demonstrate how accidents result from failures in health and safety management. The model is based on five dominos standing on their edges in a line:

- The *fifth and final domino* in the line represents all the *potential losses* which can result from an accident or incident – personal injury, damage to equipment, materials, premises, disruption of business etc;
- The *fourth domino* in the line represents the specific *accident or incident* which occurred – for example a failure of an item of lifting equipment;
- This failure could have a number of *immediate causes* which are represented by the *third domino* – overloading or misuse of the lifting equipment, an undetected defect etc;
- Behind these immediate causes are a variety of *basic or underlying causes*, represented by the *second domino* – an absence of training, a lack of supervision, failure to conduct regular inspections and thorough examinations of the lifting equipment;
- These in turn are indicative of a *lack of management control* which is represented by the *first domino* – in a well-managed workplace appropriate training is provided, there is effective supervision, and inspection and thorough examination procedures are in place.

If there is effective management control then all the dominoes stay in position. However, if management control breaks down then the other dominoes will topple, eventually causing an accident or incident and the resultant losses.

Health and safety management systems

[M2029] The vast majority of occupational safety and health ('OSH') management systems (and environmental management systems) have followed the guidance given in HSE's publication 'Successful health and safety management', (HSG65) and embellished in BS 8800:2004, Occupational Health and Safety Management Systems Guide, which outlined a system based on:

- establishing a policy with targets and goals;
- organising to implement it;
- setting forth practical plans to achieve the targets/goals;
- measuring performance against targets/goals; and
- reviewing performance.

The whole process is overlaid by auditing. This outline may be remembered via the use of the mnemonic: POPIMAR:

Policy
Organising

Planning
Implementing
Measuring/monitoring
Auditing
Reviewing

These key elements are linked in the form of a continual improvement loop (see FIGURE 3).

Figure 3: Continual improvement loop

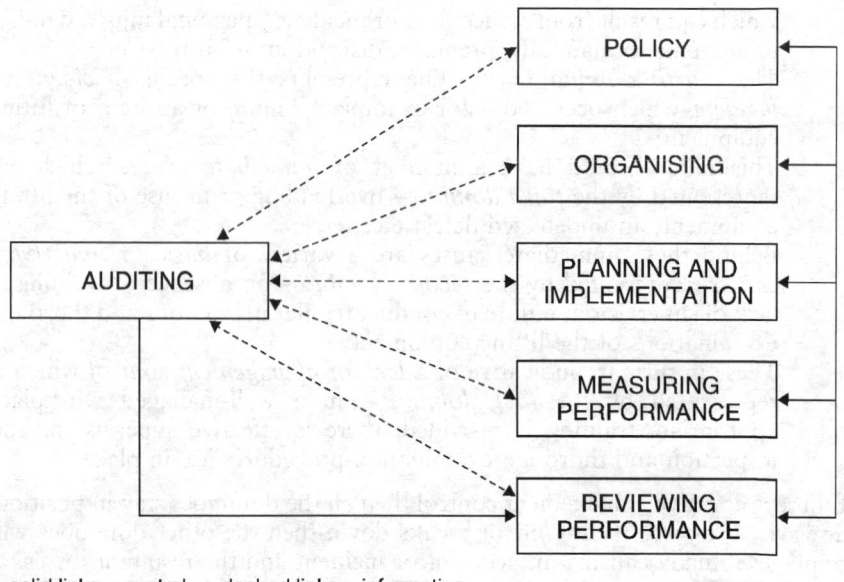

solid links = control   dashed links = information

The more times organisations go round the loop, the better will be their OSH management performance.

Within BS 8800, two alternative approaches are suggested:

- the HSG65 approach;
- the ISO 14001 approach (BSI BS EN ISO 14001: 2004, Environmental Management Systems – Requirements with guidance for use).

BS 8800 itself is merely a guide; it is *not* a certifiable standard; whereas ISO 14001 *is* a certifiable standard. To date, there is no equivalent in health and safety to the ISO 9000 standard for quality systems or the ISO 14001 standard for environmental management systems. OHSAS 18001 is not a recognised certifiable ISO standard, but an assessment specification for occupational health and safety management systems.

BS 8800 builds on the HSG65 approach (outlined above) but starts off the loop with an initial status review in order to establish 'Where are we now?' BS 8800 also includes an alternative approach aimed at those organisations wishing to base their OSH management system not on HSG65 but on BS EN ISO 14001, the environmental management systems standards which again incorporates an initial status review.

The HSE have however overhauled their guidance and moved away from the POPIMAR approach. HSG65 is no longer a detailed manual based on POPIMAR but is now a micro-site (www.hse.gov.uk/managing/index.htm) that can be referred to for guidance and reference links.

The management system now advocated by the HSE is "Plan, Do, Check, Act" or PDCA referred to further below. Further information on this system can be accessed on the HSE website at www.hse.gov.uk/managing/delivering/index.htm and there is a 2013 publication entitled *"Plan, Do, Check, Act; An introduction to managing for health and safety"* which can be downloaded. The content of the system is addressed below. References to the approach under the POPIMAR system below remain useful as guides for effective management systems but it should be borne is mind that PDCA is now the favoured system and approach of the HSE.

A review checklist

[M2030] In order to undertake an initial status review, use may be made of the following checklist:

- Does a current OSH policy exist for the organisation/location?
- Is the policy up to date (i.e. not more the three years old)?
- Has the policy been signed (and dated) by a director/senior manager who has (site) responsibility for OSH?
- Does the policy recognise that OSH is an integral and critical part of the business performance?
- Does the policy commit the organisation/location to achieve a high level of OSH performance with legal compliance seen as a minimum standard?
- Does this commitment include the concept of continual improvement?
- Does the policy state that adequate and appropriate resources, i.e. time, money, people, will be provided to ensure effective policy implementation?
- Does the policy allow for the setting and publishing of OSH objectives for the organisation/location and/or for individual directors, managers and supervisors?
- Does the policy clearly place the prime responsibility for the management of OSH on to line management, from the most senior executive to first line supervision?
- Has the policy been effectively brought to the attention of *all* employees/agency staff/temporary employees?
- Are copies of the policy on display throughout the organisation/location?
- Is the policy understood, implemented and maintained at all levels within the organisation/location via the use of suitable arrangements?
- Does the policy ensure that employee involvement and consultation takes place in order to gain their commitment to it and its implementation?

- Does the policy require that it gets periodically reviewed (at least every three years) in order to ensure that management and compliance audit systems are in place?
- Does the policy require that all employees at all levels, including agency staff and temporary employees, receive appropriate training to ensure that they are competent to carry out their duties?
- Does the policy contain a section dealing with the organisational framework – people and their duties – so as to facilitate effective implementation?
- Does the policy contain a section dealing with the need for risk assessments to be undertaken – both general and specific – their significant findings acted upon, and an assessment record-keeping system maintained?
- Does the policy contain a section dealing with the arrangements – systems and procedures – by which the policy will be implemented on a day-to-day basis?
- Does the policy contain a section dealing with the monitoring/measurement of OSH performance?
- Does the policy contain a section dealing with planning for and reviewing the organisation's policy implementation and overall OSH performance? (This would normally be in the arrangements section.)
- The HSE has produced a free leaflet (INDG 343) 'Directors' responsibilities for health and safety' which sets out the critical role directors play in the effective management of health and safety. Further publications are in preparation in anticipation of legislation on 'Corporate Manslaughter' see **M2001**.

Policy

Requirements under HSWA 1974

[M2031] *Section 2(3)* of *HSWA 1974* states:

> Except in such cases as may be prescribed, it shall be the duty of every employer to prepare and as often as may be appropriate revise a written statement of his general policy with respect to the health and safety at work of his employees and the organisation and arrangements for the time being in force for carrying out that policy, and to bring the statement and any revision of it to the notice of his employees.

Regulation 2 of the *Employers' Health and Safety Policy Statements (Exception) Regulations 1975 (SI 1975 No 1584)* excepts from these provisions any employer who carried on an undertaking in which for the time being he employs less than five employees. In this respect regard is to be given only to employees present on the premises at the same time (*Osborne v Bill Taylor of Huyton Ltd* [1982] IRLR 17).

Contents of policy statements

[M2032] The policy is the first stage in the POPIMAR management system and the statement must reflect that. Reference to its contents was made in the checklist provided in M2030 above. Some contents will be common to all organisations, whatever their size or work activity. Typical of these are the following:

- A commitment to achieving high standards of health and safety (not only in respect of employees but also others who may be affected by the organisation's activities).
- Reference to the importance of people to the organisation and the organisation's moral obligations.
- A statement that compliance with legal obligations must be achieved (although these should be regarded as a minimum standard).
- Effective management of health and safety must be a key management objective (it must continue right through the line of management responsibility).
- Reference to the provision of resources necessary to implement the policy.
- A commitment to consulting employees on health and safety matters.

Where the organisation is fully committed to the type of management system represented by POPIMAR this should be reflected in the policy statement by:

- A reference to achieving continuous improvement in health and safety standards.
- Reference to the setting of health and safety objectives and the allocation of responsibility for their implementation (whether on an annual basis or otherwise).

The statement must be signed by an appropriate director or senior manager to reflect a high level commitment to the policy and must be dated. Review of the policy as a whole, where appropriate, is a legal requirement, as well as a key component of the POPIMAR system. An out of date policy statement signed by someone who has long since left the organisation is not indicative of a real commitment to effective health and safety management.

The HSE provide examples and templates for a health and safety policy on the '*Health and Safety Made Simple, Basics for your Business*' area of its website.

Organisation

[M2033] The HSE booklet 'Successful health and safety management' (HSG65) (now a more slim line guide and reference website as referred to above) referred to the need to establish a 'health and safety culture' which it defines as 'the product of individual and group values, attitudes, perceptions, competencies and patterns of behaviour that determine the commitment to, and the style and proficiency of, an organisation's health and safety management.' Put in other words it is 'the way we do things here'. In order to achieve such a culture, activities are necessary on four fronts (often called the four Cs):

- establishing and maintaining *control* in respect of health and safety matters;
- securing *co-operation* between individuals and groups;
- achieving effective *communication*;
- ensuring the *competence* of individuals to make their contribution;

In effect the four Cs encompass both the organisation and arrangements which are required to implement the health and safety policy statement.

Control

[M2034] Management must take responsibility and provide clear direction in respect of all matters relating to health and safety. Key elements of control include:

Clear definitions of responsibilities
- Responsibilities must be allocated to all levels of line management as well as persons with specialist health and safety roles. These responsibilities should be set out in the health and safety policy (see the chapter STATEMENTS OF HEALTH AND SAFETY POLICY) and also included in individuals' job descriptions.

Nomination of a senior person to control and monitor policy implementation
- This should ensure that health and safety is seen to be important to the organisation. The senior figure must be involved in the development of plans and particularly in monitoring their implementation. They may also be involved in formal means of consultation with the workforce such as health and safety committees.

Establishment of performance standards
- The old adage of 'what gets measured gets done' applies just as much to health and safety as to other aspects of management. Performance standards must be incorporated into health and safety plans and also into health and safety related procedures and arrangements. Standards must specify:
  — what needs to be done;
  — who is responsible for carrying it out;
  — when (or how often) the task must be carried out;
  — what is the expected result,
  — For example, an investigation must be made into all accidents requiring first aid treatment. This must be carried out by the injured employee's supervisor and a report submitted on the company report form within 24 hours.

Adequate levels of supervision
- The precise level of supervision will depend on the risks involved in the work and the competence of those carrying it out. Supervisors are key players in informing, instructing and, to an extent, training the members of their teams. They also act as an important point of reference on health and safety matters, particularly on whether it is safe to continue with a task.

Co-operation

[M2035] Health and safety must become everyone's business – there must be real commitment to health and safety throughout the organisation. Commitment comes where there is a genuine consultation and involvement. This can partially be achieved through formal means of consultation, such as safety representatives and health and safety committees. Hazard report books and suggestion schemes can also play their part. However, much can also be achieved informally by line managers and health and safety specialists by simply 'walking the job' and talking to those at work about the health and safety aspects of their job.

The HSE have taken a number of initiatives to encourage the involvement of workers in health and safety management. Their website contains a 'worker involvement' site which provides tools to aid businesses in increasing the ways in which employers and workers can co-operate in improving health and safety standards.

Communication

[M2036] Good communication is a key element of all aspects of running any organisation. Employees need to be made aware of what is being done to improve health and safety standards (and why), and also of their own role in achieving this improvement. Information needs to move in various directions.

Incoming information
- The organisation must keep abreast of developments in the outside world, eg:
  — potential and actual changes in legal requirements;
  — technical developments in respect of control measures;
  — awareness of new or changed risks;
  — good practices in health and safety management.

Information circulation within the organisation
- Information will need to be passed down from management levels but there must also be means of passing information up the line and across the organisation. Important information to circulate includes:
  — the health and safety policy document itself;
  — senior management's commitment to implementing it;
  — current plans for achieving improvement;
  — progress reports on previous plans;
  — other performance reports, eg accident frequency rates;
  — proposed changes to health and safety procedures and arrangements;
  — employees' views on such proposals;
  — comments and ideas for improvement;
  — details of accidents and incidents, together with lessons learned as a result.

Information can be communicated in written form (eg policy documents, booklets, newsletters, notices, e-mails etc.) but face to face communication is also important. This can include health and safety committee meetings, health and safety discussions or briefings at other meetings, tool box talks. Once again, 'walking the job' by management can be invaluable in achieving two way communication in an informal setting.

Increasing use of company 'Intranets' provides a means of ensuring that health and safety related procedures and documents are readily available to those affected by them.

Outgoing information
- Health and safety related information may also need to be communicated outside the organisation, for example:
    — formal reports to enforcing authorities (eg as required by RIDDOR);
    — technical information about products supplied or manufactured (eg material safety data sheets, noise and vibration levels);
    — information to other organisations involved in the same business (about potential risks or successful control measures);
    — emergency planning information (to the emergency services, local authorities and neighbours who may be involved).

Competence

[M2037] All employees need to be capable of carrying out their jobs safely and effectively. What is required to achieve this will vary depending upon the individual's role in the organisation. Important elements include:

Recruitment and appointment arrangements
- Employees must have appropriate levels of intellectual and physical abilities to carry out their jobs. Whilst knowledge and experience can be acquired subsequently, certain levels may be a pre-requisite for some posts.

Training
- Systems should be in place to ensure that staff receive adequate training for their work, particularly following their recruitment, transfer or promotion. Arrangements will also be necessary to train those given particular responsibilities, eg first aiders, fire wardens. Other training needs may also be identified as a result of audits, job appraisals, changes in technology, legislative changes etc. Refresher training may also be necessary for some staff.

Succession and contingency planning
- Account must be taken of the need to have sufficient competent staff available at all times in some safety-critical roles. Staff may be absent for short periods, eg holidays or sickness, for longer periods due to serious injury or illness, or leave permanently, whether for other jobs or on retirement.

CDM 2007 (see **M2023**) place specific responsibilities on duty holders under the regulations to ensure that those they appoint or engage to carry out or manage construction work (and related design work) are competent for the purpose.

The chapter TRAINING AND COMPETENCE IN OCCUPATIONAL SAFETY AND HEALTH provides more information on competence generally.

Planning

[M2038] -

> 'Failure to plan is planning to fail'.

Effective planning results in an occupational safety and health management system ('OSHMS') which controls risks, reacts to changing demands and sustains a positive OSH culture. Such planning involves designing, developing and installing risk control systems and workplace precautions commensurate to the risk profile of the organisation. It should also involve the operation, maintenance and improvements to the OSHMS to suit changing needs, hazards and risks.

Planning is the first stage of the PDCA cycle: Plan, Do, Check, Act, which is the cornerstone of the vast majority of management systems.

Plan:	Policies, organisation, hazard identification, risk assessments, change management
Do (Implement):	The implementation of the management plans and processes
	Worker consultation is a crucial aspect of planning and doing which helps to create a positive culture
Check (Monitor):	Inspections, audits, exposure monitoring, health surveillance, attendance/absence monitoring, accident reporting and investigation, statistical/causal analyses, testing of emergency arrangements/procedures
Act (Audit/Review):	Take control action based on monitoring activities, keep track of developments/corrective measures, continually improve processes and plans

The planning process

[M2039] The planning process involves:

- Setting objectives.
- Risk control systems.
- Legal and other requirements.
- OSH management arrangements.
- Emergency plans and procedures.

Setting objectives

This relies on having an annual plan for the organisation with clearly outlined individual and collective 'SMART' OSH targets. SMART refers to the need for the targets to be: specific, measurable, achievable, realistic, trackable.

The annual OSH plan should:

- outline individual and collective action plans for the current year;
- target continual improvement in OSH performance;
- fix accountabilities via SMART targets;
- be linked to job descriptions/performance appraisals;
- be adequately resourced – time, money, people.

Risk control systems

These should ideally evolve from an ongoing review of all general and specific risk assessments in place within the organisation. Indeed, as part of the annual plan, all current risk assessments should be reviewed, updated, and recommunicated to all concerned.

Risk should be reduced to tolerable or acceptable levels via an ongoing process of hazard identification and risk assessment within the working environment. Suitable risk control systems should be developed and implemented with the goal of hazard elimination and risk minimisation. As part of the process all significant findings/risk assessments should be recorded in writing.

Legal and other requirements

These need to be incorporated into the annual plan in order to establish and maintain arrangements which ensure that all current and emerging OSH legal and other requirements are complied with and understood throughout the organisation.

Bench-marking within the organisation's sector can ensure that best practice and performance measures are adopted and communicated.

OSH management arrangements

These should reflect both the needs of the business and the overall risk profile and should cover all elements of the POPIMAR model. Specific arrangements should include:

- plans, objectives, personnel, resources;
- operational plans to control risks;
- contingency plans for emergencies;
- planning for organisational arrangements;
- plans covering change management (positive OSH culture);
- plans for interaction with third parties (eg contractors);
- planning for performance measurement, audits and reviews;
- implementing corrective actions.

Emergency plans and procedures

These should be in place and reviewed as part of the annual planning processes. Specific serious and imminent dangers should have been identified through risk assessments and procedures for dealing with them must be developed in accordance with the *Management Regulations (SI 1999 No 3242), Reg 8* (see **M2011**).

Documented emergency procedures must have been communicated effectively to all those affected by them, and necessary emergency equipment must be readily available. Situations requiring such precautions might include fire, flood, structural collapse/subsidence, chemical leak or spillage, transport accident, bomb scare, explosion, etc.

In addition some businesses may be under specific duties to prepare emergency plans given the nature of their operation, for example the duty imposed on operators under Regulation 9 of the *Control of Major Accident Hazards Regulations 1999*.

Implementation

[M2040] In order to ensure smooth implementation of the OSH policy and management system, it is imperative to have sufficient health and safety related records and documentation as this is a vital element in organisational communication and continual improvement. However, in order to ensure efficiency and effectiveness of the OSHMS, documentation should be kept to realistic proportions and should be tailored to suit organisational needs. The detail should be in proportion to the level and complexity of the associated hazards and risks.

Specifically, some of the key risk areas requiring implementation of commensurate risk control systems are:

- employee selection and training;
- general/specific risk assessments;
- safe systems of work/permit to work systems;
- management of third parties on site;
- workplace and safe access;
- work equipment;
- fire precautions;
- hazardous and dangerous substances;
- noise and vibration;
- first aid;
- issue and use of personal protective equipment.

These are generally referred to as 'arrangements' in most organisational OSH policies.

As outlined in the section on Planning above at **M2032**, this 'doing' section requires the use of action plans and SMART targets in order to ensure that the written down standards and procedures are actually achieved in practice on a continual basis. The phrase 'Mind the gap(s)' springs to mind!

Most organisations split their arrangements into general and specific. General arrangements require implementation throughout the organisation whereas specific arrangements only operate at certain locations or for certain tasks.

Within the implementation section of the OSHMS, the risk control systems relevant to the organisation should be briefly described, together with who is responsible for their implementation and who should be involved in the implementation process.

In larger organisations the detail may well be contained in a separate OSH manual which is referred to but is not part of the OSH policy. Intranet systems allow documents to be linked directly.

Monitoring

What gets measured gets done!

[M2041] Monitoring of the OSHMS is vital to ensure continual improvement. Monitoring procedures fall into two distinct categories:

- proactive;
- reactive.

Proactive monitoring takes place before an unwanted event takes place (accident, disease, damage etc.) and includes audits, inspections and specific surveys or checks.

Reactive monitoring is always after the event and includes accident and incident reporting and investigation, numerical/causal analysis, compilation of accident and incident statistics, and adherence to the RIDDOR '95 requirements on accident notification.

OSH inspections – proactive monitoring
- This should involve competent individuals going around the workplace on a regular (ideally monthly) basis in order to identify hazards – unsafe acts as well as unsafe conditions – with a view to their rectification.
  Such inspections should preferably be carried out jointly by management and workforce representatives and should always involve talking to the people at risk in the workplace being inspected. This may help to identify unsafe practices, near misses and shortcomings in the established systems and procedures. In most cases use will be made of a specific inspection checklist which ideally should be compiled by the people undertaking the inspection.
  In all cases there needs to be a post-inspection action plan prepared which clearly indicates who is going to take remedial action and by when. There should also be full communication of the findings and recommendations from all such inspections.

Specific surveys and checks – proactive monitoring
- Such surveys and checks may involve: housekeeping inspections; checks on adherence to agreed safe systems of work, permit to work systems, site safety rules; utilisation of personal protective equipment; documentation checks – risk assessments, statutory inspection records, maintenance and test records; training records, etc..

Accident and incident reporting and investigation systems – reactive monitoring
- This should encompass all accidents, damage and near-miss incidents and occupational ill-health in order to ensure RIDDOR '95 compliance and an effective accident database with which to plan further accident reduction strategies. However, even more importantly, all accidents and incidents should be promptly investigated in order to identify their causes and to implement commensurate control measures so as to prevent a recurrence. Most organisations will have set up an internal reporting and investigation system to capture accident and incident data. Any investigation culture should be positive, i.e. not blame-apportioning or fault-finding.

  The main purposes of accident and incident investigation are therefore to:
  — discover the immediate and underlying causes (note: plural!);
  — prevent recurrences;
  — minimise future legal liability;
  — collect sufficient data on which to base future plans;
  — identify accident/illness/absence trends;
  — ensure RIDDOR '95 compliance;
  — maintain records for organisational purposes;
  — continually improve the OSHMS performance.

Audit

[M2042] Formal OSH audits should be undertaken at regular intervals, (possibly annually), by trained and competent auditors, sometimes external to the organisation. The main objectives of any audit are to reduce risks, with avoidance being the best strategy, and to ensure continual improvement of the OSHMS.

All audit findings (successes as well as shortcomings) should be highlighted and communicated to all concerned, including worker representatives. Any recommendations for improvement should be prioritised and allocated to named individuals for action within agreed, finite time-scales. The progress on action completion should also be followed up and communicated.

The audit needs to compare actual OSH performance in the workplace against established standards – mind the gaps! – such as:

- legislative compliance;
- HSE guidance;
- best practice (bench-marking);
- national/international standards.

Essentially the OSH audit, as with financial audits, should be a deep and critical appraisal of all elements (POPIMAR) of the OSHMS and should be seen as being additional to routine monitoring and reviews. Preferably, the auditors need to be independent of the area/activity/location being audited and the audit checklist should be tailored to suit the activities, risks and needs of the organisation.

The audit scope should address key areas such as:

- is the overall OSHMS capable of achieving and maintaining the required standards?
- is the organisation fulfilling all OSH obligations?
- what are the strengths and weaknesses of the OSHMS?
- is the organisation actually doing and achieving what it claims or believes?
- are the audit results acted upon, communicated and followed up?
- are the audit reports used in management reviews?

In 2006 the HSE introduced their Corporate Health and Safety Performance Index (CHaSPI) which through a series of structured questions allows organisations to evaluate their OSHMS. CHaSPI could be accessed via the HSE website. However, with effect from 1 June 2012, CHaSPI has been withdrawn as an interactive benchmark tool. The HSE have stated that following a review it has decided to withdraw the tool as a cross sector benchmark aid.

The decision is said to have been informed by a number of factors, including:

- A reluctance of some users to publish their performance ratings despite the option to publish anonymously, undermining its usefulness as a bench mark tool.
- A recent trend towards industry taking a greater lead role in producing sector focused health and safety performance assessment tools, with a move towards tools that provide feedback on company specific performance, and clarity in identifying strengths and weaknesses to inform improvement actions.
- Following the withdrawal on 1 June 2012 the HSE is to consider providing a simple checklist of performance related questions, extracted from CHaSPI, in a PDF format that may be useful to employers as a stand alone self assessment aid.

The HSE state that these developments suggest that CHaSPI has been superseded by a range of industry specific performance assessment tools that offer an alternative approach that appears to have greater appeal and success in attracting interest and take up.

Following the withdrawal on 1 June 2012 the HSE is to consider providing a simple checklist of performance related questions, extracted from CHaSPI, in a PDF format that may be useful to employers as a stand alone self assessment aid.

Review

[M2043] Reviews of the performance of the OSHMS generally fall into two categories:

- periodic status review ('PSR');
- management review.

The main purposes of the PSR are to make judgements on the adequacy of performance and to ensure that the right decisions are made and taken in connection with the nature and timings of the remedial actions.

Specifically, the PSR should consider:

- overall OSHMS performance;
- performance of individual elements of the management system;
- audit findings;
- internal and external factors such as changes in organisational structure, changing business activities and demands, pending legislation, introduction of new technology;
- anticipated future changes;
- information gathered from proactive and reactive monitoring.

The PSR should be a continual process which includes management and supervisory responses to system shortcomings such as non-implementation of agreed risk control systems, substandard performances, or failure to assess the impact of forward plans and objectives on their part of the organisation. Successes should also be highlighted.

The effectiveness of the PSR can be enhanced by clearly establishing responsibilities for implementing the remedial actions identified and by setting deadlines for action completion, using the SMART approach — see **M2039**. The PSR needs to be auditable, trackable, fully communicated and followed up.

The management review should be undertaken annually by the board/senior management. It should ensure that the OSHMS is suitable, adequate, effective and efficient, and that it satisfies and achieves the aims and objectives as stated in the OSH policy.

The review should be based on the audit results and on the periodic status review reports. Specifically it should:

- establish new OSH objectives for continual improvement;
- develop OSH annual plans;
- consider changes to OSHMS arrangements;
- review findings

which should then be documented, communicated to stakeholders via annual reports and followed up to ensure implementation and continual improvement.

Continual improvement

[M2044] Continual improvement is at the heart of the OSHMS. It essentially is a fundamental commitment to manage OSH risks proactively so that accidents and ill-health are continuously reduced (system effectiveness) and the system achieves its desired aims using less resources (system efficiency).

The International Labour Office Guidelines on Occupational Safety and Health Management Systems (ILO, 2001) includes the following model which clearly illustrates the importance of the goal of continual improvement.

Figure 4

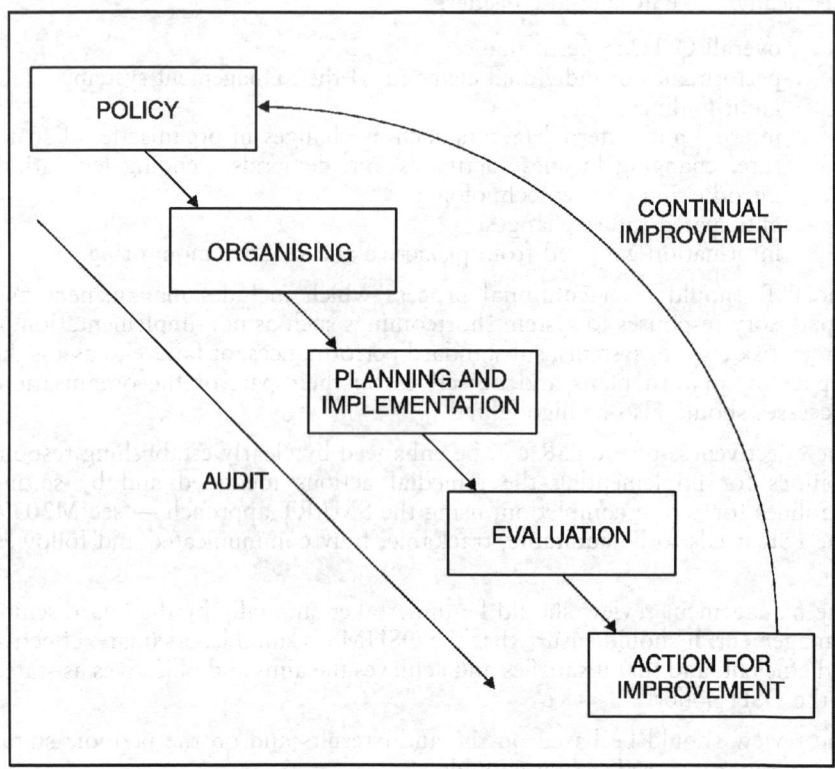

The route to continual improvement has been highlighted throughout the course of this chapter. It involves the use of SMART targets, the setting of OSH Key Performance Indicators ('KPIs'), the use of performance appraisals which include individual and collective OSH performance reviews, and clearly establishing OSH responsibilities at all levels within the organisation.

The goal of continual improvement should manifest itself in a number of ways:

- better year on year results – less injuries, diseases, damage, near misses;
- steadily improving results using less resources;
- better targeted and more efficient and effective OSHMS;
- indications of breakthrough performances;
- a more positive, proactive OSH culture;
- OSHMS elements which are more comprehensive and easier to understand;

In order to achieve continual improvement, the following factors should be taken into account:

- regular OSH audits;

- statistical information;
- bench-marking;
- industry/sector guidelines;
- ownership of health and safety matters:
  — managers, supervisors;
  — workers' representatives;
- results of proactive monitoring in relation to targets;
- effective and efficient operation via workforce involvement;
- generation and evaluation of ideas for improvement(with regular feedback to originators);
- fully resourced, implemented and monitored action plans (with SMART objectives);
- widespread and effective OSHMS training;
- appropriate OSHMS documentation and records.

POPIMAR in action

[M2045] The example below shows how the POPIMAR and PDCA principles can be followed in practice. It relates to the introduction into an organisation of a permit to work ('PTW') system to control certain high risk activities.

Policy

The PTW system will become part of the 'arrangements' for implementing the health and safety policy statement. The PTW procedure (see below) is likely to contain a general statement about the importance of ensuring that appropriate controls are in place to minimise the risks associated with potentially high risk activities.

Organisation

This will primarily centre around the development of a formal PTW procedure which will involve:

Control
- The procedure must identify:
  — what activities will be controlled by the PTW;
  — who will issue (and cancel) PTWs;
  — the timing sequence for issuing and canceling individual PTWs;
  — the format of the PTW form.

Co-operation
- The draft procedure must be reviewed with those likely to be involved in issuing and working under PTWs. This could involve an ad hoc consultative group in addition to safety representatives and the health and safety committee.

Communication
- A wider section of the workforce must be informed about the proposed procedure and the reasons for its introduction.

Competence
- Those who are to issue the PTWs must receive formal training in the procedure and use of the PTW form. There should also be a formal appointment process to ensure that they have successfully assimilated the training and can relate it to their practical work situations. Arrangements for this should be referred to in the procedure.

Planning and implementation

Once the procedure has been finalised the timescales and responsibilities must be identified for achieving the various component parts of its implementation. This will include:

- delivery of training to PTW issuers;
- formal appointment of PTW issuers (after an appropriate test process);
- printing and distributing the PTW forms;
- obtaining any necessary equipment (eg isolation locks, atmospheric testing equipment);
- announcement of the formal introduction of the PTW procedure.

Measuring performance

Monitoring of the PTW can be carried out in various ways:
- a survey soon after the procedure's introduction;
- ongoing monitoring during formal health and safety inspections;
- periodic mini-audits of the procedure (eg by health and safety specialists or managers).

Auditing

The operation of the PTW procedure would be included in any general OSH auditing activities.

Reviewing performance

Both monitoring and audit findings must be reviewed by appropriate individuals and groups such as:

- health and safety specialists;
- the health and safety committee;
- senior management;
- ad hoc review teams.

Where satisfactory standards are not being achieved, the reviewers must consider how the POPIMAR sequence can best be applied to taking remedial action.

Managing health and safety: references

[M2046] (All HSE publications)

L 144 Managing health and safety in construction

HSG 65 Managing for Health and Safety

Plan, Do, Check, Act, An introduction to managing for health and safety

INDG 343 Directors' responsibilities for health and safety

www.hse.gov.uk/statistics

Managing Work-related Road Safety

Roger Bibbings and Andrea Oates

Overview of Managing Work-Related Road Safety

[M2101] According to the Trades Union Congress ('TUC'), work-related road accidents are the biggest cause of work-related accidental death in the UK. Department for Transport ('DfT') statistics released in November 2018 (www.gov.uk/government/collections/road-accidents-and-safety-statistics) show there were 1,793 reported road deaths in 2017. There were also 24,831 serious injuries in road traffic accidents (RTAs) reported to the police that year and 170,993 casualties of all severities in reported RTAs.

The Health and Safety Executive ('HSE') estimates that 'more than a quarter of all road traffic incidents may involve somebody who is driving as part of their work at the time'.

The Royal Society for the Prevention of Accidents ('RoSPA') says between one-quarter and one-third of reported road casualties are estimated to occur in road accidents involving someone who was driving, riding or otherwise using the road for work purposes. It says police road accident data shows that every year over 500 people are killed (almost a third of all road deaths), 5,000 are seriously injured and more than 40,000 are slightly injured in collisions involving drivers or riders who are driving for work. This includes other road users as well as the at-work drivers and riders themselves.

The Parliamentary Advisory Council for Transport Safety (PACTS) says that after a period of rapid decline, the number of road deaths has remained unchanged in statistical terms since 2011. In August 2018, the UCL Centre for Transport Studies published the findings of research, funded by the Road Safety Trust charity, highlighting concerns that modern forms of work could increase the risk of RTAs.

A recent survey of gig economy drivers, riders and their managers, based on 48 in-depth interviews and 200 responses to an online survey, found the majority of those surveyed (63%) were not provided with safety training on managing risks on the road. Forty-two per cent of drivers and riders reported that their vehicle had been damaged as a result of a collision while working, and a further one in ten said someone had been injured.

Researcher Heather Ward said the findings highlighted 'that the emergence and rise in the popularity of gig work for couriers could lead to an increase in risk factors affecting the health and safety of people who work in the gig economy and other road users'.

The human and financial costs of work-related RTAs to families, businesses and the wider community are massive.

Employers have clear duties under health and safety legislation to manage occupational road-risk in the same way that they manage other health and safety risks. Essentially this means organisations need to:

- communicate clear messages to their staff about their approach to road safety;
- set up systems and allocate duties to key members of staff (particularly managers);
- carry out 'suitable and sufficient' risk assessments;
- use these to check that they are doing all that is 'reasonably practicable' to avoid risk on the road or to ensure safe driving;
- provide driver training where necessary; and
- monitor and review performance.

The joint HSE/DfT publication 'Driving at Work' (INDG 382) (www.hse.gov.uk/pubns/indg382.pdf) emphasises that employers' duties under health and safety legislation extend to at work driving – or 'on-the-road activities'.

Organisations should integrate Work-Related Road Safety (WRRS) within their general policy, organisation and arrangements for managing health and safety at work. They should undertake suitable risk assessments (covering journey, vehicle and driver risk factors) and ensure safe journey planning and other safety measures are in place. Top of the list of possible risk reduction measures is 'meeting without moving' (for example, by being video enabled). Next best is going by a safer means of transport, such as train.

Vehicles should be fit-for-purpose and properly maintained with additional safety features where necessary.

Managers should avoid systems of work which cause people to speed such as 'just in time' delivery, payment by the number of calls made, 'job and finish', and unrealistic guaranteed call-out or delivery times. They should avoid asking staff to drive while tired and at times of day when falling asleep at the wheel is more likely. They also need to consider employees' sleep deprivation at home (caused by looking after sick children and dependents for example) and avoid introducing driver distractions like making and receiving mobile phone calls while driving, even with 'hands free' – No mobile while mobile!.

As well as being illegal to drive a vehicle or ride a motorbike while using a handheld mobile phone or similar device, hands-free phones can be a distraction, and drivers risk prosecution for not having proper control of a vehicle if the police see them driving poorly while using one.

New penalties for driving while using a mobile phone came into force in March 2017. The police can give drivers a fixed penalty notice for less serious traffic offences including using a mobile phone while driving. They can be fined up to £200 and get penalty points on their licence as a result. They could also be taken to court and banned from driving and/or be fined up to £1,000. The maximum fine is £2,500 for bus and lorry drivers. Drivers who passed their driving test in the last two years will lose their licence if they are caught illegally using a mobile phone will driving. Government guidance on this can be found at: www.gov.uk/using-mobile-phones-when-driving-the-law. Managers should also be aware of potential health impairments, driver fitness issues, and issues like alcohol and drugs which can affect people's ability to drive safely.

Drugs and driving – recent developments

Scotland's road safety laws will be strengthened by the introduction of drug-driving limits and roadside testing on 21 October 2019. The Scottish government says there will be a zero-tolerance approach to eight drugs most associated with illegal use, with limits set at a level where any claims of accidental exposure can be ruled out. A list of other drugs associated with medical use will have limits based on impairment and risk to road safety.

The drugs which will have a zero-tolerance limit are benzoylecgonine, cocaine, delta–9–tetrahydrocannabinol (cannabis and cannabinol), ketamine, lysergic acid diethylamide (LSD), methylamphetamine, methylenedioxymethaphetamine (MDMA – ecstasy), and 6-monoacetylmorphine (6-MAM – heroin and diamorphine). The drugs with medical uses which will have limits based on impairment are clonazepam, diazepam, flunitrazepam, lorazepam, methadone, morphine, oxazepam, and temazepam. A separate approach will be taken to amphetamine, balancing its legitimate use for medical purposes against its abuse.

The new limits in Scotland will follow those introduced in England and Wales from March 2015.

People who take medicines and are not sure if they are safe to drive should check with their pharmacist or doctor.

The limits set for the prescription drugs exceed normal prescribed doses. The government advises that the vast majority of people can drive as they normally would, so long as:

- they are taking their medicine in accordance with the advice of a healthcare professional and/or as printed in the accompanying leaflet; and
- their driving is not impaired.

There is a medical defence if a driver has been taking medication as directed and is found to be over the limit, but not impaired. Government advice for drivers who are taking prescribed medication is that 'it may be helpful' to keep evidence of with them in case they are stopped by the police.

Planning for safety

Drivers and managers should always plan the safest routes, avoiding congestion, crash sites and adverse weather. In November 2018, the Scottish government published a new charter encouraging employers to treat workers fairly and be mindful of their safety if they are unable to get to work during extreme weather. Following the 'beast from the east' snow storm with in 2018, the Fair Work Charter for Severe Weather provides guiding principles to help employers prepare for similar circumstances in the future. It was developed jointly by the Scottish Government and the Scottish Trades Union Congress (STUC) and sets out fair work practices – including the recommendation that all employers have a severe weather policy. For more information see: https://news.gov.scot/news/severe-weather-employer-guidelines.

Staff may need to set off the night before and stay over in the case of long journeys. And managers should be aware of the effects of stress and fatigue on road safety arising from poor work/life balance.

Organisations should assess their drivers' attitudes and driving competence, follow their crash and penalty points histories, and analyse and learn from their crashes and 'near-misses'. Driver assessment should be used to target training at those with the greatest need.

Above all, organisations should train their line managers, consult their safety representatives, require their senior managers to lead by example and recognise, celebrate and reward safe driving achievement. A good approach is to set up a multi-disciplinary team, with driver and safety representative involvement, to ask 'where are we now?', and to develop an action plan with clear targets.

Managing Work-Related Road Safety – employers' duties

[M2102] It is estimated that up to a third of all road crashes involve someone who is working at the time. Driving as part of their job is one of the riskiest activities many people will undertake while at work. Road crashes wreck families and no amount of money can compensate for the loss of a loved one or for a severe disabling injury.

Although most drivers tend to believe that they are safe, crashes still happen. Clearly as road users, employee drivers have a direct responsibility to obey road traffic law. Employers' duties under health and safety law are also applicable to at-work driving. They should therefore manage occupational road risk within the framework they should already have in place for managing other workplace health and safety risks. For example, organisations should ensure their approach to safety at work on the road dovetails with arrangements for safe use of vehicles on their work sites.

The following section describes legal responsibilities.

Legal responsibilities

[M2103] As well as duties employers may have under specific road traffic law, their duties under the *Health and Safety at Work etc Act 1974* and the *Management of Health and Safety at Work (MHSW) Regulations 1999 (SI 1999 No 3242)* also apply when their employees are engaged in on-the-road work activities. This does not extend to commuting journeys, except where an employee is travelling from their home to a location that is not their usual place of work, according to HSE advice (see *Driving at work Managing work-related road safety* (INDG 382) www.hse.gov.uk/pubns/indg382.pdf).

Legal responsibilities [M2103]

Under the Act, employers have a general duty of care to ensure, so far as is reasonably practicable, the health, safety and welfare of their employees at all times. They also have a similar responsibility to ensure that the safety of others, such as contractors' staff and members of the public, is not put at risk by their employees' activities.

The driver is responsible for the way a vehicle is driven on the road and should not put themselves or others at risk. They must also co-operate with their employer. But employers can have a significant influence over what the driver does. For example, they could increase the risk of crashes by:

- imposing unrealistic schedules which increase pressure to break speed limits;
- requiring drivers to work excessive hours – particularly when driving is combined with other tasks – which may lead them to become dangerously fatigued;
- failing to provide drivers with adequate or suitable training; and
- failing to provide drivers with the right vehicle for the job and to maintain it in a safe condition.

Under the MHSW Regulations, employers have a duty to establish a suitable policy, organisation and arrangements to manage health and safety at work, and to have access to competent advice.

The Regulations require employers to:

- carry out an assessment of the risks to health and safety of their employees while at work and others who may be affected by their activities (and record the results where the organisation employs more than five people);
- monitor health and safety performance, review risk assessments and keep them up-to-date; and
- ensure that employees are competent and receive any necessary training, information and supervision.

They also have a duty to consult with employees over health and safety matters, and where they are appointed, their trade union health and safety representatives.

Failure to comply with these duties could lead to enforcement action being taken. In extreme circumstances it could result in a prosecution leading to a fine and/or imprisonment.

In addition to the *Health and Safety at Work etc Act and the Management Regulations*, the following pieces of legislation apply:

- The *Workplace (Health, Safety and Welfare) Regulations 1992* (see **W110 WORKPLACES – HEALTH, SAFETY AND WELFARE**) contain requirements concerning traffic routes for vehicles within the workplace.
- Road Traffic laws – the RoSPA publication, *An introduction to managing occupational road risk*, explains that 'road traffic law focuses mainly on individual driver behaviour and the vehicle owner. However, the various Road Traffic Acts and regulations also require employers to ensure that vehicles used for work purposes are safe and legal to be on

the road, and that drivers are properly licensed and insured. Employers can be held liable for various "cause or permit" road traffic offences'. It is an offence to cause or permit someone to: use a handheld mobile phone while driving; use a vehicle in a dangerous condition; or drive on the road without a valid driving licence or motor insurance.

- *The Road Transport (Working Times) Regulations 2005* and European rules on driving hours, domestic drivers' hours rules and tachograph rules – under these rules and regulations, both drivers and employers are responsible for ensuring compliance with drivers' hours and tachograph regulations (see W100 WORKING TIME). The impact on UK health and safety law emanating from European directives as a result of the so-called 'Brexit' vote to leave the EU in the June 2016 referendum will depend on the terms of the withdrawal agreement. The UK remains a member of the EU and is covered by all its provisions, including health and safety legislation, until the exit negotiations are complete, and a new relationship is defined.
- The *Road Vehicles (Construction and Use) Regulations 1986* include a wide range of requirements concerning the construction and maintenance of vehicles including the dimensions and weights of vehicles used on British roads and the safety of loads.
- The Corporate Manslaughter and Corporate Homicide Act 2007 introduced a new offence of corporate manslaughter (or corporate homicide in Scotland) for very serious senior management failures which result in fatality (see *C90 Corporate Manslaughter*).

In December 2015, Baldwins Crane Hire Limited was convicted of corporate manslaughter and ordered to pay £700,000, after employee Lindsay Eaton was killed when the brakes on a 130-tonne crane failed as he was driving down a steep road and it crashed into an earth bank. A joint investigation by Lancashire Police and the HSE found that several wheel brakes were inoperable, worn and contaminated. Of its four auxiliary braking systems, three had been disconnected altogether, causing the vehicle to crash while attempting to negotiate a steep bend on an access road.

Although Labour Force Survey – 'LFS' figures show that there are between 52,000 and 83,000 non-fatal work-related RTA injuries each year, work-related RTAs are not reportable under the *Reporting of Injuries, Diseases and Dangerous Occurrences Regulations 2013* – RIDDOR.

In December 2018, the general Unite union called for urgent reforms in the way work-related RTAs are recorded and investigated after a Freedom of Information (FoI) request revealed a 50% increase in lorry driver deaths, from 14 in 2016 to 21 in 2017. The union says that because employers do not have to report RTAs to the HSE under RIDDOR, these deaths are 'hidden from official statistics on workplace fatalities' and 'not properly investigated'.

The HSE annual work-related fatality statistics for 2017/18, published in July 2018, show that 144 workers were killed at work in Great Britain. However, as the *Workplace fatal injuries in Great Britain 2018* report explains, this figure only includes those the relevant enforcing authority (the HSE, local authorities or the Office of Rail and Road) have judged as reportable under RIDDOR.

Certain types of work-related injury are not reportable under RIDDOR and are therefore excluded from the figures. This includes fatal RTAs involving workers travelling on a public highway. These are instead enforced by the police and reported to the DfT. Those killed while travelling to and from home to work are also excluded from the statistics.

The HSE/DfT guidance explains that in most cases, the police will take the lead on investigating road traffic incidents on public roads. HSE will usually only take enforcement action where the police identify that serious management failures have been a significant contributory factor to the incident.

The business case for action

[M2104] In addition to meeting duties under health and safety law, managing work-related road safety (WRRS) should also deliver a wide range of business benefits, including financial benefits. As with other kinds of work-related accidents, the true costs of crashes to organisations are nearly always considerably higher than the resulting repairs and insurance claims. For small businesses and the self-employed, the consequences of a major road crash are likely to be particularly severe. Even relatively minor road crashes can lead to time off work due to whiplash or psychological trauma, and some of the effects can be long lasting.

By improving WWRS and reducing crashes, the benefits to a business can include:

- fewer lost days due to injury;
- reduced risk of work-related ill health;
- reduced stress and improved morale;
- less need for investigation and paperwork;
- less lost time due to work rescheduling;
- fewer vehicles off the road for repair;
- reduced running costs through better driving standards – less wear and tear and lower fuel costs;
- fewer missed orders and business opportunities;
- less risk of damage to the organisation's reputation and image, to both the general public and potential customers;
- reduced risk of losing the goodwill of customers; and
- preventing key employees from being banned from driving.

In addition to these business benefits, organisations will also have better control over costs, including:

- wear and tear and fuel payments;
- insurance premiums and legal fees; and
- claims from employees and third parties.

Although they should be primarily interested in the safety of their employees while driving for work purposes, from a business perspective, organisations also have an interest in promoting the safety of their staff when they are commuting by road or when they are driving for domestic reasons. Whether staff are injured when driving for work purposes or at other times, the resulting costs to an organisation are likely to be very similar. By promoting a safety culture in their business, and encouraging staff to take improved road safety home, businesses can offer a wider societal benefit by influencing their employees' private driving behaviour and in turn by employees influencing family members and friends.

The following section shows how businesses can extend their existing health and safety management system to ensure that it addresses WRRS issues.

Extending health and safety management systems to cover Work-Related Road Safety (WRRS) hse management systems to cover WRRS

[M2105]–[M2110] To manage work-related road safety (WRRS) the HSE and DfT advise employers to follow a 'Plan, Do, Check, Act' approach:

Plan – *Describe how you manage health and safety in your organisation and plan to make it happen in practice*

- Assess the risks from work-related road safety.
- Produce a health and safety policy covering, for example, organising journeys, driver training and vehicle maintenance.
- Make sure there is top-level commitment to work-related road safety.
- Clearly set out everyone's roles and responsibilities for work-related road safety. Those responsible should have enough authority to exert influence and be able to communicate effectively to drivers and others.

Do – *Prioritise and control your risks, consult your employees and provide training and information*

- In larger organisations, make sure departments with different responsibilities for work-related road safety co-operate with each other.
- Make sure there are adequate systems to allow work-related road safety to be managed effectively. For example, ensure vehicles are regularly inspected and serviced according to manufacturers' recommendations.
- Involve workers or their representatives in decisions.
- Provide training and instruction where necessary

Check – *Measure how you are doing*

- Monitor performance to ensure the work-related road safety policy is effective and has been implemented.
- Encourage employees to report all work-related road incidents or near misses.

Act – *Review your performance and learn from your experience*

- Collect enough information to allow informed decisions about the effectiveness of existing policy and the need for changes to be made, for example targeting those more exposed to risk.
- Regularly revisit the health and safety policy to see if it needs updating.

Source: Driving at work Managing work-related road safety (INDG382).

Figure 1 The Plan, Do, Check, Act cycle

Plan, Do, Check, Act should not be seen as a once-and-for-all action:

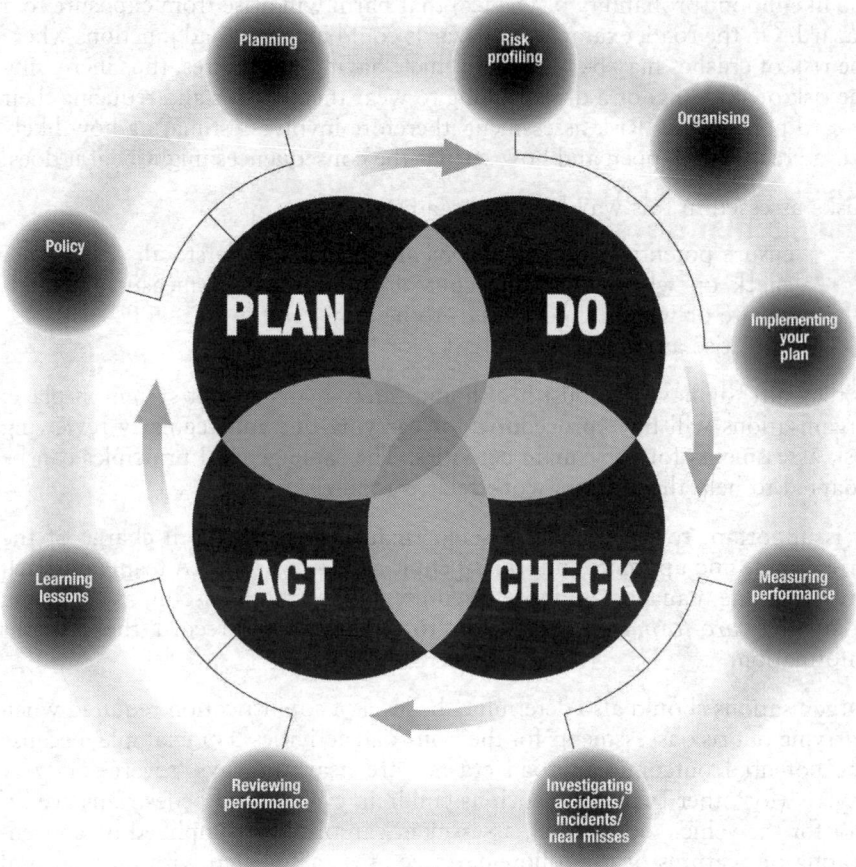

Source: *HSE guide: HSG65 (Third edition, published 2013)*. Contains public sector information licensed under the Open Government Licence v2.0.

Assessing risks on the road

[M2111] Risk assessment has many meanings in a wide range of different contexts. In the context of WRRS, risk assessment is an aid to understanding and making judgements about safety. It needs to:

- be appropriate to the circumstances of the organisation;
- not be over-complicated and bureaucratic;
- be carried out by competent people with a practical knowledge of the work activities being assessed;

- generate 'an inventory of actions'; and
- encourage managers, supervisors, work allocators and drivers to think of themselves as road risk assessors.

A 'hazard' is something with potential to cause harm. Risk is an estimation of the likelihood (probability or chance) that harm will arise from exposure to a hazard. On the road, examples of hazards could include: road junctions where the risk of crashes may be higher; a vehicle having worn tyres, thus increasing the risk of skidding; or a driver failing to wear their glasses and reducing their hazard perception. 'Risk assessment' therefore involves estimating how likely it is a crash will happen and how serious the consequences might be if it does.

Risks assessed in this way help an organisation to:

- ensure potential safety problems are properly understood;
- check on whether existing control and emergency measures are adequate or whether more needs to be done; and
- prioritise any unacceptable risks for further action.

If they already have a robust health and safety management system in place, organisations will have procedures for carrying out and regularly reviewing risk assessments for their main activities. The same general principles can be adapted to help them assess work-related road risks.

It is important to ensure that a senior manager is in overall charge of the programme and appoints and trains people as risk assessors. A team approach involving line managers and safety representatives can be useful, as can using a suitable *pro-forma* (or software) to organise and record the relevant information.

Organisations should also determine the level of sophistication required when carrying out risk assessments for their on-road activities. For example, because the potential outcomes of road crashes are nearly always 'severe' or 'very severe' (for other road users such as children, cyclists and pedestrians even if not for the vehicle occupants), assessment can often be simplified by concentrating on features of the following three areas associated with an increased probability of crashes occurring:

(1) the journey;
(2) the vehicle; and
(3) the driver.

By estimating the combined effect of such factors, organisations can assign a risk rating (for example, 'high', 'medium', or 'low') to particular driving tasks, prioritise them for attention and then consider what risk reduction measures might be taken.

Figure B: Suggested risk assessment framework

Factors to consider If any of the factors listed below are likely to increase the likelihood of an incident then tick the box and consider the level of risk	Risk Level			Existing control/ Action required
	Low	Medium	High	
The journey – enhancement of risk associated with: • road types • distances to be covered • reasonable time allocation • allowance for suficient breaks • traffic density (urban or rural) • areas with a high pedestrian density • driving at night/darkness • poor weather conditions	☐ ☐ ☐ ☐ ☐ ☐ ☐ ☐			
The vehicle – enhancement of risk associated with: • maintenance to a suitable standard • performance (power, all-terrain) • crash resistance • other safety features (e.g. ABS, traction control, air bags) • distractions (e.g. mobile phones) • driver familiarity with vehicle • loads to be carried	☐ ☐ ☐ ☐ ☐ ☐ ☐			
The driver – enhancement of risk associated with: • age • experience • driving competence • associated skills (loading, checks) • health and fitness • stress and fatigue • attitude • crash/enforcement history	☐ ☐ ☐ ☐ ☐ ☐ ☐ ☐			
Remedial Action. What remedial action should be taken in order of priority? 1 _____ 2 _____ 3 _____ 4 _____ 5 _____				

This sort of specific risk assessment is likely to be most relevant where journeys follow a regular and predictable pattern, for example planned deliveries or regular sales calls. Line managers, drivers and safety representatives can review risk factors such as routes, traffic conditions, time of day, seasonal conditions, vehicle specifications, and fitness and driver profiles to see if there are areas of potential concern and if there is scope for improving safety.

Much can be achieved by applying thorough management practice, but organisations can also use sources of national crash data and analyse their own claim/incident records and investigation findings to focus particularly on specific factors such as start times or time of day, speed, fatigue or specific tasks.

Where driving is a clear job requirement, they can include questions on a candidate's driving record at interview. Line managers can regularly inspect drivers' licences for entitlements – to drive a particular type of vehicle, drivers need an 'entitlement' for that category on their licence – and penalty points, track their annual mileages and review their crash involvement. They can also arrange for in-vehicle competence assessments or use computer or questionnaire-based techniques to assess knowledge, attitudes and hazard-perception skills.

Risk assessment is not just a matter of assessing drivers. The aim should be to make an overall assessment of all the various factors that can increase the chances of a crash happening.

Where journeys have many factors which are identical or broadly similar (such as same vehicles, same routes and same times) it may be possible to develop 'generic' risk assessments to help identify safety requirements.

Where drivers have considerable autonomy over key features of their journey tasks, they can be trained to apply the risk assessment system themselves ('dynamic assessment'). A simple driver checklist highlighting risk factors and indicating relevant standards or limits to be observed can be a highly effective tool to assist drivers make safe choices about the safest routes, timing, breaks and any other risks.

A very simple approach is to identify those drivers exposed to the highest risks (such as high mileages, night-time driving and congested urban areas), with the most penalty points on their licences and the worst crash histories. This enables the organisation to focus on options for reducing risk, for example, by journey task re-design and improving driver competence.

Whatever approach is adopted, the overall aim is to make informed choices about where, when and how to take action to reduce the chances of crashes occurring and to reduce the consequences if they do occur.

The HSE guidance, *Driving at Work – Managing work-related road safety*, provides a step-by-step guide to assessing risks on the road:

- Identifying the hazards – look for hazards and ask employees (both those who drive extensively and occasionally) or their representatives as they will have first-hand experience of what happens in practice. Consider the driver, vehicle and journey;
- Who might be harmed? – Decide who might be harmed and how. Not only the driver, but also passengers and other road users and pedestrians. Those new to the job, working long hours or driving long distances could be groups particularly at risk;

- Evaluate the risks – Having identified the hazard, decide how likely it is that harm will occur. While employers do not have to eliminate every risk, they must do everything practicable to protect people from harm;
- Record your findings – (this is required by law where there are five or more employees and is good practice in any case); and
- Regularly review your assessment – ensure you know about road incidents, drivers and vehicle history in order to be able to do this effectively. Changes such as new routes, new equipment or a change in vehicle specification may all prompt a review.

The following section discusses some of the options for reducing risks.

Road risk control measures

[M2112] Suitable risk assessment will help to identify where further action needs to be taken and the kinds of risk control measures the organisation needs to consider. Employers may need to consider what they can do to improve the safety of:

- the journey task;
- the vehicle; and
- the driver.

The following list is not exhaustive and is intended as an introductory overview of risk management options. The control measures which are implemented should be appropriate to the organisation's particular needs and operations.

The journey task

[M2113] Eliminating unnecessary journeys by road

- Can the organisation justify the need to travel by road?
- Rather than face-to-face contact, could the organisation use other means of communication, such as teleconferencing?

Changing mode of travel

- Could alternative, safer methods of transport such as train or plane be used?

Avoiding driving in adverse conditions

- Where possible, are drivers discouraged from driving at night and in bad weather conditions?
- Where possible, are drivers encouraged to stay overnight rather than continue a journey back in bad conditions?

Controlling driver hours

- Does the organisation set in-house limits for driving hours and distances?
- Does this include daily, weekly and monthly driving hours and distances?

Setting safe schedules

- Does the organisation take into account traffic and weather conditions and speed limits when planning journey times and schedules?
- Does the organisation minimise the need for driving during high risk hours for fatigue (between 2am and 6am and 2pm and 4pm) when planning schedules?
- Are the organisation's schedules contributing to driver fatigue and stress?

Encouraging the use of the 'safest' routes

- Does the organisation ensure that drivers take the safest routes (ie: using motorways and dual carriageways and avoiding crash 'black spots')
- If using a SatNav, are drivers instructed to input their destination before setting off, and if they need to change it, stop in a safe place to do so?

Avoiding incentives to speed

- Has the organisation eliminated incentives to speed (such as 'job and finish' contracts, payment by number of visits made, or unrealistic customer service promises)?

Discipline

- Are significant and persistent breaches of speed limits dealt with in normal disciplinary procedures?

The vehicle

[M2114] Specifying appropriate vehicles

- Is it the most appropriate vehicle for the job (eg where load carrying is involved)?
- Does it have undesirable features such as 'bull bars' that increase injury severity to vulnerable road users?
- Does it incorporate desirable additional safety features? (for example, primary features such as Anti-lock Braking System (ABS), high level brake lights, traction control, headway information systems and alcohol ignition interlocks to help prevent crashes happening; and secondary safety features such as crash resistance, side impact bars, air bags and seatbelt interlocks to reduce consequences if crashes do occur).
- Is it ergonomically suited to the driver, particularly to help minimise the risk of musculo-skeletal disorders?

Ensuring effective vehicle maintenance

- Are vehicles properly and regularly maintained to ensure they are in a safe condition?
- Are vehicles maintained by skilled, knowledgeable and qualified mechanics?
- Where staff provide their own vehicles, does the organisation require them to have them serviced in accordance with the manufacturer's instructions and to provide a current MoT certificate?

Road risk control measures [M2115]

- Does the organisation require daily and weekly driver/line manager checks of safety critical features such as tyres, mirrors and lights? If any defects are found, are they reported and repaired immediately?

The driver

[M2115] Ensuring health and fitness

- Does the organisation monitor employee well-being through periodic medical and eyesight checks? (Note that the *Equality Act 2010* introduced prohibitions regarding pre-employment health checks — see EQUALITY ACT 2010 E16501.)
- Does the organisation pay particular attention to musculo-skeletal problems such as back pain?

Tackling driver impairment

- Does the organisation have clear policies and procedures on alcohol, drugs and medicines (prescription and non-prescription)?
- Does the organisation address impairment factors such as stress (both work-related and domestic) and poor sleep?
- Does the organisation advise drivers on strategies to manage fatigue?
- Does the organisation prohibit the use of hand-held or hands-free mobile phones while driving? (Both are distracting).

Enhancing driver competence

- Has the organisation put in place a needs/risk-based driver assessment and training programme?
- Does driver assessment and training address specific needs such as safe reversing?
- Does it focus on factors such as attitude, domestic and occupational stress, fatigue and poor preparation for work?

Providing effective information and supervision

- Does the organisation have a comprehensive and frequently updated handbook for drivers?
- Does the handbook explain the organisation's work-related road safety policy and make clear that the organisation requires all employees to comply with key aspects of road traffic law such as compliance with speed limits?
- Does the handbook explain procedures for incident reporting, dealing with emergencies such as crashes or breakdowns and ensuring the personal safety of staff?
- Are the organisation's staff coached and assessed on their understanding and application of the content?
- Does the organisation use internal communications (such as bulletins, the intranet and award schemes) to update drivers on WRRS?
- Is there adequate supervision of drivers by line managers trained in WRRS?

The HSE guidance *Driving at work – Managing work-related road safety* also sets out a work-related road safety checklist under Safe driver, Safe Vehicles and Safe Journey headings.

Implementing an organisation's road risk control measures

[M2116] The "Plan, Do, Check, Act" approach described at **M2105** above, is the framework that needs to be in place for tackling risks in a planned and methodical way, informed by the results of appropriate risk assessments (at M2111 above) and control measures such as those suggested above. Management systems by themselves, however, will not deliver continuous improvements in performance unless they are underpinned by proactive management and a positive 'health and safety culture'. Organisations need to build on this to develop a 'corporate road safety culture' in which WRRS becomes a common talking point and is taken seriously by every employee. The following steps can aid this process.

- Avoid focussing solely on drivers and driving, but aim to eliminate or reduce risks at source wherever possible. There is no point in training drivers and then requiring them to use unfit vehicles or to carry out unsafe, unsatisfactory or badly designed driving tasks.
- Consult key players such as managers, safety representatives and drivers (who are likely to have differing points of view) in order to get consensus.
- Use limited resources most appropriately to arrive at the most cost effective combination of measures by trading off the costs and benefits of different options.
- Check to see whether the expected safety gain resulting from a given 'spend' in one direction – say, upgrading to 'safer' vehicles – is justified when compared with the expected safety gain from the same 'spend' on other measures such as awareness raising or setting safer journey standards.

The following section highlights the importance of monitoring and evaluation in the process of improving WRRS. This should be built into programmes at the beginning of the process rather than as an afterthought.

Monitoring and evaluation

[M2117] To manage WRRS successfully organisations need appropriate systems to be able to monitor performance and learn lessons that can help an organisation improve all aspects of its management system.

Traditionally organisations have monitored road safety performance 'reactively' by focusing on 'lagging' indicators such as insurance claims data and costs but it is also important to undertake 'active' monitoring by looking at 'leading' indicators which help managers and safety reps to understand how well the management system is working.

Monitoring and evaluation [M2121]

Active monitoring

[M2118] Some of the options for 'active' monitoring include:

- tracking risk assessment records and journey plans to monitor driving patterns and adherence to standards;
- fitting 'black box' technologies to vehicles (after informed consultation with drivers and their representatives) to monitor excessive acceleration/deceleration forces, speeding and compliance with other journey standards such as time and distance limits;
- checking vehicle servicing schedules and records;
- spot checking vehicle conditions and fuel usage;
- monitoring key performance indicators (KPIs) such as near hits, complaints and driving at blackspots;
- quarterly monitoring of driving licences (with informed consent) to identify any penalty points and details of the offences for which they were incurred;
- regular end of shift or journey debriefs for posts with substantial driving;
- monitoring drivers' attitudes, for example through surveys; and
- checking during staff appraisals to see how well managers and drivers are meeting their WRRS responsibilities.

Reactive monitoring

[M2119] Organisations also need be able to assess the number, type, causes and cost of at-work crashes in order to learn lessons to help it improve prevention and track changes in performance.

Gathering data

[M2120]

- Are procedures for collecting crash data at the scene explained in all driver training programmes, internal communications and driver handbooks?
- Are there clear claim and incident reporting procedures in place, for example, suitable 'bump cards', report and investigation forms?
- Are cameras or miniature tape recorders available to help drivers capture crash/incident information at the scene?
- Is there a similar procedure for reporting and sharing information on 'near hits' or WRRS problems?
- Is there a 'no blame' reporting policy to make it clear that the organisation is more interested in reducing crashes/incidents than in penalising those involved?

Investigation

[M2121]

- Are there criteria and procedures in place to investigate crashes, incidents and near hits in depth to determine immediate and underlying causes?

- Does the organisation adopt a team approach led by a senior manager (and involving managers, driver trainers, senior drivers and safety representatives) to share insights and expertise?
- Is it policy to concentrate most investigation effort on those events that are likely to yield the most significant lessons?
- Are the results of crash/near hit investigations considered as part of periodic performance review?

Evaluation

[M2122] Has the organisation set KPIs to help it evaluate the effectiveness of its WRRS programme?

Is it policy, for example, to track:

- the number of crashes and claims?
- the claims rate per vehicle or per 100,000 kilometres per year?
- the number and severity of injuries?
- the type and location of crashes?
- severities (including 'write-offs' and air bag 'go-offs')?

Does the organisation seek to evaluate crash costs (including vehicle, driver, third party and other costs)?

Are there regular discussions with the organisation's insurer, brokers or accident management company about how they can analyse and provide effective feedback from claims?

Have baselines been established from which to measure progress and/or change over time, and to provide a basis for performance benchmarking, for example with other organisations?

Key questions:

- Does the organisation monitor drivers' day-to-day behaviour (for example through incident rates, fuel/tyre/brake pad use, free phone schemes, black box technology, end-of-shift debriefs or other methods)?
- Are the organisation's drivers actively encouraged to report all incidents, and near hits, no matter how small?
- Does the organisation investigate all, but particularly significant, crashes to establish immediate and underlying causes?
- Does the organisation set appropriate KPIs?
- Does the organisation monitor insured and uninsured crash costs?
- Does the organisation feed back results to key managers and other staff (for example, through the staff intranet, email, notice-boards and in-house magazines)?

Monitoring and evaluation is a key element in improving WRRS and helps the organisation to understand 'where it is now', which is discussed next.

Where is the organisation now?

[M2123] If the importance of WRRS is still not fully recognised or understood in an organisation, its full extent needs to be quantified and costed to raise awareness and convince everyone, from board level down, of the need for a comprehensive approach. The most important first step is to undertake an initial status review (ISR) to answer the following questions:

- 'where is the organisation now (and why is WRRS important to it)?'
- 'where does the organisation want to be?' and
- 'how is the organisation going to get there?'.

A useful starting point is the collation of information on:

(1) key fleet parameters such as vehicles, miles and drivers;
(2) safety incidents such as injuries, fatalities, airbag 'go-offs', insurance claims, near hits, wear and tear, maintenance repairs, violations, fuel use and costs; and
(3) policies, people and procedures which may already be in place to manage WRRS.

This should also help the organisation to focus on the nature and scale of its actual and potential WRRS costs and opportunities and its capacity to manage them systematically. In turn, this should help to focus attention on the importance of WRRS and provide a baseline from which progress can be measured.

The results of an ISR should be considered at board level, with the proposal that a WRRS 'action plan' be developed and implemented throughout the organisation. Everyone needs to be convinced that WRRS is an important issue affecting costs, business effectiveness and employee well-being. Simply counting crashes, injuries and costs for the first time and talking about the impact of crashes on individuals, their families and on the business can be extremely motivating.

To develop an 'action plan' a small team can be set up, for example, under the leadership of the board level health and safety 'champion'. It can involve managers and safety representatives and professionals from areas such as health and safety, fleet management, insurance risk management, training and human resources. Others such as insurers, brokers, WRRS service providers, the police and local authority road safety officers can be invited to participate. Key elements of the plan can include: establishing a clear corporate policy; developing a framework for risk assessment; setting some indicative WRRS standards; improving crash/incident reporting and investigation procedures; and setting a target date for reviewing performance. It is very important that, before investing in driver assessment and training or other safety control measures, appropriate training is provided for all line managers of staff who drive or work on the road.

If an organisation has already developed a modern, proactive risk management approach and has a strong health and safety culture, it will have few difficulties in embracing such an approach to WRRS. Colleagues will also be keen to learn from good practice developed by other businesses by benchmarking through their trade associations and other networks. For those that have yet to find this path, tackling WRRS can often provide the organisational learning experience which will help them establish a cycle of continuous improvement in all areas of health and safety risk management.

The next section describes where to get support.

Where to get help

[M2124] There are many bodies that can provide organisations with help and advice (either free or on a fee-for service basis). Some of these include:

- employer/trade associations;
- professional bodies;
- trades unions;
- the police;
- national road safety organisations;
- national occupational safety organisations;
- local authority road safety departments;
- local health and safety or advanced drivers group;
- driver training providers;
- specialist fleet risk management consultants;
- motoring organisations;
- insurers, brokers or accident management companies; and
- vehicle suppliers/hirers.

There is also much useful information available, for example on the HSE's WRRS home page on its website (www.hse.gov.uk/roadsafety/index.htm), the Department for Transport's road safety website (www.dft.gov.uk/topics/road-safety) and on the resources section of the Occupational Road Safety Alliance website (www.orsa.org.uk). Time spent visiting these sites can pay dividends in helping to identify other sources of guidance and information relevant to an organisation's specific needs.

Useful publications

[M2125] The HSE recommends the following publications for more information on work-related road safety:

The health and safety toolbox: How to control risks at work HSG268 HSE Books 2013 www.hse.gov.uk/pubns/books/hsg268.htm

IOSH Management of occupational road risk information www.iosh.co.uk/Books-and-resources/IOSH-management-of-occupational-road-risk-policy.aspx

Management of work-related road safety RR018 HSE Books 2002 www.hse.gov.uk/research/rrpdf/rr018.pdf

Managing occupational road risk associated with driver fatigue: a good practice guide RSSB 2013 www.rssb.co.uk/Library/improving-industry-performance/2013-good-practice-guide-t997-managing-occupational-road-risk.pdf

RoSPA Road safety advice, information and resources www.rospa.com/roadsafety

RoSPA Road Safety Resources for Employers www.rospa.com/roadsafety/resources/employers (including *An introduction to managing occupational road risk* www.rospa.com/rospaweb/docs/advice-services/road-safety/employers/introduction-to-morr.pdf)

Work-related road safety case studies www.drivingforbetterbusiness.com/casestudies/default.aspx

Useful websites:

[M2126] The HSE recommends the following websites as sources of further useful information:

Brake: The road safety charity: www.brake.org.uk

Fleet Safety Benchmarking Website: www.fleetsafetybenchmarking.net/main

Driver and Vehicle Licensing Agency: www.gov.uk/government/organisations/driver-and-vehicle-licensing-agency

Freight Transport Association: www.fta.co.uk

Occupational Road Safety Alliance: www.orsa.org.uk

Road Haulage Association: www.rha.uk.net

Roadsafe: www.roadsafe.com

'Think!' Department for Transport's road safety site for road users: 'Think!' Department for Transport's road safety site for road users: www.think.gov.uk

Manual Handling

Andrea Oates

Introduction to manual handling

[M3001] The *Manual Handling Operations Regulations 1992 (SI 1992 No 2793), as amended,* came into operation on 1 January 1993. The Health and Safety Executive (HSE) had long been concerned at the large number of injuries resulting from manual handling activities at the time of their introduction, particularly as many manual handling injuries have cumulative effects and eventually result in physical impairment or even permanent disability.

Work-Related Musculoskeletal Disorders (WRMSDs) continue to be one of the two most common causes of occupational ill health in Great Britain, affecting around a million people a year.

Estimates from the Labour Force Survey show that in Great Britain the total number of WRMSDs cases (prevalence) in 2016/17 was 507,000 out of a total of 1,299,000 for all work-related illnesses. This represents 39% of the total number of work-related illnesses and a rate of 1,550 cases per 100,000 workers. The number of new cases of WRMSDs (incidence) in 2016/17 was 159,000, an incidence rate of 480 cases per 100,000 workers.

Also, an estimated 8.9 million working days were lost due to WRMSDs in 2016/17, an average of 17.6 days lost for each case, and work-related musculoskeletal disorders account for 35% of all working days lost due to work-related ill health. Although working days lost per worker due to self-reported work-related musculoskeletal disorders showed a generally downward trend until around 2010/11, since then the rate has remained broadly flat. WRMSDs therefore remain an important priority area for the HSE.

The fourth edition of the HSE publication *Manual Handling, Manual Handling Operations Regulations 1992, Guidance on Regulations* (L23) was published in September 2016. It has been simplified and restructured, with the regulations and guidance in Part 1 and more detailed guidance to assist in carrying out risk assessments and control risks in Parts 2–4. The risk filter which helps to identify those tasks that do not require a detailed assessment (formerly in Appendix 3 of the third edition) has been adapted so that HSE's assessment tools, including the 'MAC' and 'RAPP' tools (see **M3003.1**) can be used as part of the risk assessment process. The full risk assessment checklists are now online at www.hse.gov.uk/pubns/ck5.pdf.

L23 also includes changes to Regulation 2 to reflect an amendment to section 3(2) of the Health and Safety at Work etc Act 1974 relating to the duties of self-employed people.

What the Regulations require

[M3002] Various terms used in the Manual Handling Operations Regulations 1992 (SI 1992 No 2793) are defined in Regulation 2:

> '"manual handling operations" means any transporting or supporting of a load (including the lifting, putting down, pushing, pulling, carrying or moving thereof) by hand or by bodily force.
>
> "Injury" does not include injury caused by any toxic or corrosive substance which –
>
> (a)   has leaked or spilled from a load;
> (b)   is present on the surface of a load but has not leaked or spilled from it; or
> (c)   is a constituent part of a load;
>
> and "injured" shall be construed accordingly;'

This in effect establishes a demarcation with the *Control of Substances Hazards to Health (COSHH) Regulations 2002 (SI 2002 No 2677)*. For example, where there is oil on the surface of a load making it difficult to handle that is a matter for the *Manual Handling Operations Regulations*. Any risk of dermatitis is covered by the *COSHH Regulations*.

> '"Load includes any person and any animal"'.

This is very important in a number of work sectors including health and social care, agriculture and veterinary work.

HSE guidance accompanying the Regulations, L23, states that an implement, tool or machine, such as a chainsaw or breathing apparatus, being used for its intended purpose is not considered to constitute a load – in effect establishing a demarcation with the requirements of the *Provision and Use of Work Equipment Regulations 1998 (PUWER'98) (SI 1998 No 2306)*. However, when such equipment is being moved before or after use it is a load for the purpose of these Regulations.

Regulation 2(2) imposes the duties of an employer to their employees under the Regulations on a 'relevant' self-employed person 'in respect of himself'. The wording reflects changes introduced by the *Health and Safety at Work etc Act 1974 (General Duties of Self-Employed Persons) (Prescribed Undertakings) Regulations 2015 (SI 2015/1583)* and the *Deregulation Act 2015 (Health and Safety at Work) (General Duties of Self-Employed Persons) (Consequential Amendments) Order 2015 (SI 2015/1637)*. Both came into force on 1 October 2015.

Health and safety law applies to 'relevant' self-employed persons – whose work activity is specifically mentioned in the *Health and Safety at Work etc Act 1974 (General Duties of Self-Employed Persons) (Prescribed Undertakings) Regulations 2015 (SI 2015 No 1583)*, or poses a risk to the health and safety of others. The activities mentioned are agriculture (including forestry), work with asbestos, work on a construction site, an activity to which the Gas Safety (Installation and Use) Regulations 1998 applies, work with genetically modified organisms and work on the railways.

For health and safety law purposes, 'self-employed' means a person does not work under a contract of employment and works only for themselves. If a person is self-employed and employs others, then the law applies. A risk to the health and safety of others is the likelihood of someone else being harmed or injured (eg members of the public, clients, contractors etc) as a consequence of their work activity. HSE advice on the law can be found on its website at: www.hse.gov.uk/self-employed/index.htm.

Regulation 3 excludes the normal ship-board activities of a ship's crew under the direction of the master from the application of the Regulations. These are subject to separate merchant shipping legislation.

The duties of employers under the Regulations are all contained in *Regulation 4* which establishes a hierarchy of measures that employers must take:

'(1) Each employer shall—
- (a) so far as is reasonably practicable, avoid the need for his employees to undertake any manual handling operations at work which involve a risk of their being injured; or
- (b) where it is not reasonably practicable to avoid the need for his employees to undertake any manual handling operations at work which involve a risk of their being injured–
  - (i) make a suitable and sufficient assessment of all such manual handling operations to be undertaken by them, having regard to the factors which are specified in column 1 of Schedule 1 to these Regulations and considering the questions which are specified in the corresponding entry in column 2 of that Schedule,
  - (ii) take appropriate steps to reduce the risk of injury to those employees arising out of their undertaking any such manual handling operations to the lowest level reasonably practicable, and
  - (iii) take appropriate steps to provide any of those employees who are undertaking any such manual handling operations with general indications and, where it is reasonably practicable to do so, precise information on–
    - (a) the weight of each load, and
    - (b) the heaviest side of any load whose centre of gravity is not positioned centrally.

(2) Any assessment such as is referred to in paragraph (1)(b)(i) of this Regulation shall be reviewed by the employer who made it if –
- (a) there is reason to suspect that it is no longer valid; or
- (b) there has been a significant change in the manual handling operations to which it relates,

and where as a result of any such review changes to an assessment are required, the relevant employer shall make them.

(3) In determining for the purpose of this regulation whether manual handling operations at work involve a risk of injury and in determining the appropriate steps to reduce that risk regard shall be had in particular to–
- (a) the physical suitability of the employee to carry out the operations;
- (b) the clothing, footwear or other personal effects he is wearing;
- (c) his knowledge and training;
- (d) the results of any relevant risk assessment carried out pursuant to regulation 3 of the Management of Health and Safety at Work Regulations 1999;

(e) whether the employee is within a group of employees identified by that assessment as being especially at risk; and
(f) the results of any health surveillance provided pursuant to regulation 6 of the Management of Health and Safety Regulations 1999.'

The hierarchy of measures which must be taken by employers where there is a risk of injury can be summarised as follows:

- Avoid the operation (if reasonably practicable);
- Assess the remaining operations;
- Reduce the risk (to the lowest level reasonably practicable);
- Provide information and training to employees.

Reference will be made later in the chapter to a variety of measures which can be taken to avoid or reduce risks. The factors to which regard must be given during the assessment (as specified in *Schedule 1* to the Regulations) are the *tasks, loads, working environment* and *individual capability* together with other factors referred to in *Regulation 4(3)*. These factors and the questions set out in *Schedule 1* are also covered later in the chapter.

Regulation 5 places a duty on employees:

'Each employee while at work shall make full and proper use of any system of work provided for his use by his employer in compliance with Regulation 4(1)*(b)*(ii) of the Regulations.'

Regulation 6 provides for the Secretary of State for Defence to make exemptions (in the interests of national security) from the requirements of the Regulations in respect of home forces and visiting forces and their headquarters.

Regulation 7 extends the Regulations to apply to offshore activities such as oil and gas installations, including diving and other support vessels. Previous provisions relating to manual handling were replaced or revoked by *Regulation 8*.

The Regulations themselves are extremely brief but they are accompanied by considerable HSE guidance, much of which will be referred to during this chapter. A number of legal cases have provided further interpretation on the requirements set out in the regulations. These include the following:

In *Swain v Denso Martin Ltd (2000)*, The Times, 24 April, the Court of Appeal ruled that the fact than an employer had failed to carry out a risk assessment did not mean that he was not under a duty to take steps to give information about the weight of a load where there was a risk of injury.

In *King v Sussex Ambulance NHS Trust* [2002] EWCA Civ 953, the court took into account the difficulties an ambulance service has in reconciling its duties towards patients with its duties towards its employees. The case involved an ambulance technician who had injured his back while carrying a patient downstairs from his bedroom. However, in *Knott v Newham Healthcare NHS Trust* [2002] EWHC 2091 (QB), a High Court judge ruled that the employer's arrangements for lifting were inadequate and breached the Regulations by failing to provide sufficient lifting equipment and effectively expecting nurses to lift patients manually.

In a 2002 Court of Appeal case, *O'Neill v DSG Retail Ltd* [2002] EWCA Civ 1139, it was held that an employer required to take appropriate steps to reduce a risk of injury to the lowest level reasonably practicable must, in assessing the risk, consider the particular task, the context of where it is performed and the employee required to perform it.

In *Ghaith v Indesit Co UK Ltd* [2012] EWCA Civ 642, the Court of Appeal held that the employer had failed to discharge its duty to undertake a risk assessment and was therefore liable for the claimant's injury sustained while carrying out manual handling during a stock take. His claim was initially dismissed but on appeal it was allowed. The judge said the risk assessments the employer was using were not suitable or sufficient as they did not deal with stock taking. In addition, the employer had not taken appropriate steps to reduce the risk to the lowest reasonable practicable level. It had not taken the obvious precaution of providing regular breaks of reasonable length in the stock taking operation. Mr Gaith had been working all day with just four short breaks.

In *Sloan v Governors of Rastrick High School* [2014] EWCA Civ 1063, a learning support assistant's personal injury claim against the school failed. Shortly after beginning work she experienced a pain in her shoulder and back after pushing a student in her wheelchair. She returned to work after taking time off, but handed in her notice shortly after and claimed damages for personal injury. She claimed that the school had failed to carry out its duties under *Regulation 4* of the *Manual Handling Operations Regulations 1992*. Her claim initially failed and her appeal was dismissed for reasons including the following:

- it was not reasonably practicable to avoid the use of manual wheelchairs. The NHS and parents rather than the school provided the students' wheelchairs;
- the school had carried out a suitable and sufficient assessment of manual handling operations. It prepared an annual risk assessment for students using wheelchairs and staff supporting them. Although the person carrying out the assessment was not a health and safety officer, she had attended a five-day training course and attended annual refresher training. Manual wheelchairs were not considered to be specialised plant and equipment; and
- in terms of reducing the risk of injury to the lowest level reasonably practicable, the school regularly rotated staff and trained them. The wheelchairs were pushed for short periods and the gradient of slopes in the school were acceptable.

Assessing the risk of injury from manual handling

Risk assessment filters

[M3003] In L23, the HSE makes clear that while weight is an important factor, there are many other risk factors employers must take into account.

'Medical and scientific knowledge stress the importance of an ergonomic approach to manual handling, taking into account the nature of the task, the load, the working environment and individual capability, and this approach requires worker participation,' it says. An ergonomic approach tries to fit the operations to the individual rather than the other way round.

Psychosocial risk factors, including high workloads, tight deadlines and lack of control of the work and working methods are also important as they may affect workers' psychological response to their work and workplace conditions.

As set out earlier, the current (fourth) edition of the HSE guidance on the Regulations (L23) has been simplified and restructured (see **M3001**). The Appendix in L23 sets out ways to help employers decide how detailed manual handling assessments should be. These are:

- simple weight filters which assume that handling is infrequent, symmetrical and takes place in favourable working conditions;
- a more detailed assessment of the most common manual handling risk factors, carried out using the HSE's 'MAC tool' for example (see **M3003.1**). Employers can choose to use other tools not published by the HSE;
- a full risk assessment taking account of all the factors is set out in Schedule 1 of the Regulations. The HSE website includes blank and example checklists to help carry out full assessments at www.hse.gov.uk/pubns/ck5.pdf.

The filters referred to above help to identify low-risk manual handling operations.

HSE advice is as follows: 'If the task is within the filter values you do not normally need to do any other form of risk assessment unless you have individual employees who may be at significant risk, eg pregnant workers, young workers, those new to the job, or those with a significant health problem or a recent injury. If you are not sure that a task is low risk, you should do a more detailed risk assessment.'

It must be stressed that the filters do not contain weight limits. They must not be regarded as safe weights nor must they be treated as thresholds which must not be exceeded.

The HSE guidance states that application of the guidelines set out in the filters will provide a reasonable level of protection to around 95 per cent of working men and women. However, it adds: 'Even for a minority of fit, well-trained individuals working under favourable conditions, operations which exceed the filter values by more than a factor of about two may represent a serious risk of injury. Always make these operations high priority for carrying out full risk assessments and implementing appropriate risk reduction measures.'

There are filters for four types of manual handling operations:

Lifting and lowering

The guidelines for lifting and lowering operations as set out in L23 are shown in the accompanying diagram (see Figure 1 on page 8). They assume the load can be easily grasped with both hands and the operation takes place in reasonable working conditions, using a stable body position. Where the hands enter more than one box, the lowest weights apply. A more detailed assessment, using the MAC tool or a full risk assessment, should be made if the weight guidelines are exceeded or if the hands move beyond the box zones.

Carrying

The guidance sets out that the filter weights for lifting and lowering above apply to carrying operations where the load:

- is held against the body;
- is carried no further than about 10 m without resting;
- does not prevent the person from walking normally;
- does not obstruct the view of the person carrying it; and
- does not require the hands to be held below knuckle height or much above elbow height (owing to static loading on the arm muscles).

Where the load can be carried securely on the shoulder without first having to be lifted, when unloading sacks from a lorry for example, the filter values can be applied to carrying distances up to 20 m.

Again, if the weight lifted exceeds the filter weight or the assumptions above are not met, a more detailed assessment, using the MAC tool or a full risk assessment, should be carried out.

Pushing and pulling

In pushing and pulling operations, the load may be slid, rolled or moved on wheels. The HSE guidance sets out that the task is likely to be low risk if:

- the force is applied with the hands;
- the torso is largely upright and not twisted;
- the hands are between hip and shoulder level; and
- the distance involved is no more than about 20 m.

The task is also likely to be low risk if the load can be moved and controlled easily with only one hand.

If it requires significant forces for pushing and pulling, as indicated by the posture while the operation is being carried out, a more detailed assessment, using the RAPP tool (see **M3003.1**), or a full risk assessment should be carried out.

Even where the task is within the filter, the HSE advises that a more detailed risk assessment will be necessary if there are risk factors such as slopes, uneven floors, confined spaces or trapping hazards.

Handling when seated

The accompanying diagram (see Figure 2 on the next page) illustrates the guideline figures for handling while seated.

If handling beyond this box zone is unavoidable, a full assessment should be carried out.

Fig. 1: Lifting and lowering

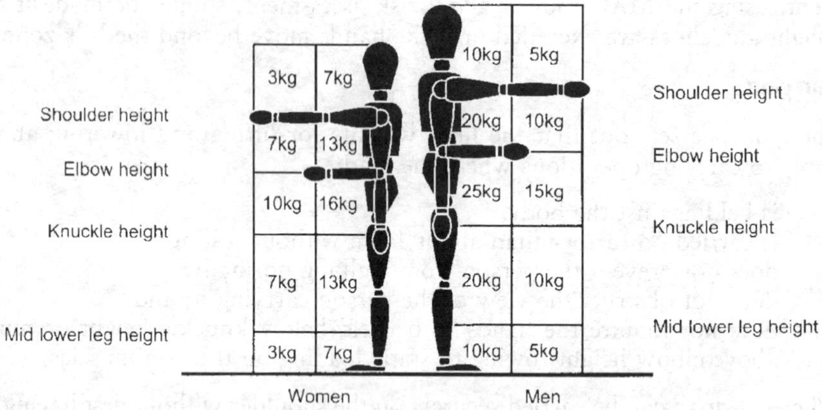

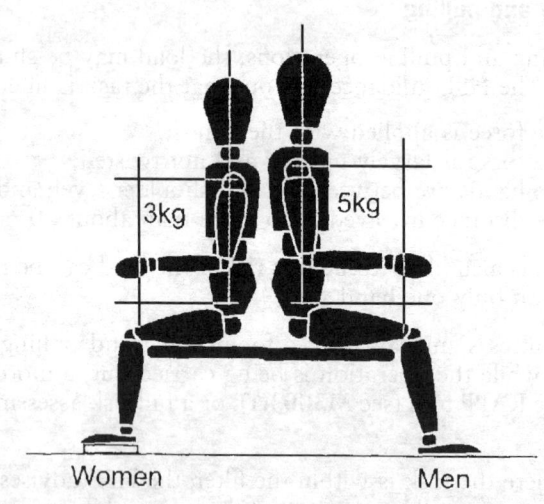

Fig. 2: Handling while seated

Carrying out more detailed risk assessments using the HSE manual handling assessment tools

[M3003.1] The HSE has developed an MSD toolkit which includes the *Manual handling assessment charts* (MAC) and *Risk assessment of pushing and pulling* (RAPP) tool as well as the Assessment of repetitive tasks of the upper limbs (ART tool) and the Variable MAC tool (V-MAC). These aim to provide logical processes to identify high-risk manual handling operations for which urgent further action is needed to reduce risk.

The MAC tool

The *Manual Handling Assessment Charts (the MAC tool)* (INDG 383) helps to assess the most common risk factors in lifting and lowering, carrying and team handling operations. The V-MAC (see below) can be used with the MAC tool for lifting operations where load weights or handling frequencies vary.

It uses a 'traffic light' approach, from green, through to amber, red and purple to categorise risks as low, medium, high and very high respectively. Numerical values are provided for each level of risk for each risk factor, with risk levels based on published data in ergonomics literature.

The MAC tool is available online at: www.hse.gov.uk/pubns/indg383.pdf.

An interactive score sheet can be downloaded at: www.hse.gov.uk/msd/mac/scoresheet.htm.

The RAPP tool

The risk assessment tool for pushing and pulling operations (the RAPP tool) helps assess the key risks in manual pushing and pulling operations involving whole body effort. Like the MAC tool it uses colour-coding and numerical scoring. It helps to identify high-risk pushing and pulling activities and to evaluate the effectiveness of any risk-reduction measures.

Two types of pulling and pushing operations can be assessed using the RAPP:

- moving loads using wheeled equipment, such as hand trolleys, pump trucks, carts or wheelbarrows; and
- moving items without wheels, involving dragging/sliding, churning (pivoting and rolling) and rolling.

For each type of assessment there is a flow chart, an assessment guide and a score sheet. For more information see: www.hse.gov.uk/msd/pushpull/index.htm.

The Variable MAC tool (V-MAC)

The V-MAC is a tool for assessing manual handling operations where load weights vary and should be used in conjunction with the MAC tool (see above). It was developed to help assess jobs where load weights are variable, such as in order picking, parcel sorting, trailer loading/unloading, and parts delivery in manufacturing. It is therefore best suited for order picking and distribution systems which can automatically generate the data for importing into V-MAC. Significant background knowledge is needed to use the V-MAC successfully.

This can be gained through the HSE website support pages or through training.

For more information see: www.hse.gov.uk/msd/mac/vmac/index.htm.

The ART tool and other HSE guidance

The Assessment of Repetitive Tasks (ART) tool is designed to help risk assess tasks that require repetitive movement of the upper limbs (arms and hands). It

[M3003.1] Manual Handling

helps to assess some of the common risk factors in repetitive work that contribute to the development of Upper Limb Disorders (ULDs).

It is aimed at people with responsibility for the design, assessment, management and inspection of repetitive work, such as assembly, production, processing, packaging, packing and sorting work, as well as work involving regular use of hand tools.

It also uses a numerical score and a traffic light approach to indicate the level of risk for 12 factors, grouped into four stages:

- A: Frequency and repetition of movements;
- B: Force;
- C: Awkward postures of the neck, back, arm, wrist and hand;
- D: Additional factors, including breaks and duration.

The factors are presented on a flow chart which leads, step-by-step, to evaluate and grade the degree of risk. The tool is supported by an assessment guide, providing instruction to help score the repetitive task observed. There is also a worksheet to record the assessment. Training is recommended to help use the ART tool reliably and appropriately.

For more information see: www.hse.gov.uk/msd/uld/art/whatis.htm.

The HSE says that where the risk assessment identifies a task may be harmful to a worker's upper limbs (hands, arms, shoulders) or neck, the HSE publications *Managing upper limb disorders in the workplace*, *Upper limb disorders in the workplace* as well as the ART tool may be useful. Information on manging back pain can be found on the HSE's Musculoskeletal disorders web pages and The *back book* (see **M3028**).

HSE guidance on carrying out a full risk assessment

[M3003.2] Part 3 of L23 provides additional information on what to look for when making risk assessments of manual handling activities under regulation 4(1)(b)(i). It looks at task factors aspects of the load and the working environment and includes action to reduce the risks in relation to the duties set out in regulation 4(1)(b)(ii).

The structure of Part 3 follows paragraphs 1 to 3 in *Schedule 1* to the *Manual Handling Operations Regulations 1992 (SI 1992 No 2793)*: 'The tasks', 'The loads' and 'The working environment'.

Paragraphs 4 and 5 of *Schedule 1*: 'Individual capability' and 'Other factors', such as clothing and personal protective equipment (PPE) are covered in the guidance to regulation 4(3) in Part 1 of L23.

Schedule 1 contains a list of questions which must be considered during assessments of manual handling operations.

The Tasks

Holding or manipulating loads at a distance from the trunk?

[M3004] Stress on the lower back increases as the load is moved away from the trunk. Holding a load at arms' length imposes around five times the stress as the same load when held very close to the trunk. Any change that allows the load to be lifted or held closer to the body is likely to reduce the risk of injury.

Unsatisfactory body movement or posture?

The risk of injury increases with poor feet or hand positions, eg feet too close together, body weight forward on the toes, heels off the ground.

Twisting the trunk while supporting a load increases stress on the lower back – the principle should be to move the feet, not twist the body.

Stooping also increases the stress on the lower back while *reaching upwards* places more stress on the arms and back and lessens control of the load.

Excessive movement of loads?

Risks increase the further loads must be *lifted or lowered*, especially when lifting from floor level or from above head height. Movements involving a *change of grip* are particularly risky. Carrying a load for an excessive distance prolongs physical stresses, leading to fatigue and increased risk of injury. The HSE guidance sets out that as a rough guide, if a load is carried further than around 10m, demands on the heart and lungs, and muscle fatigue from carrying the load, will become the limiting factors rather than the strength needed for lifting and lowering.

Excessive pushing or pulling?

Here the risks relate to the forces which must be exerted, particularly when starting to move the load, and the quality of the grip of the handler's feet on the floor. Risks also increase when pushing or pulling below waist height or above shoulder height.

Risk of sudden movement

Freeing a box jammed on a shelf or a jammed machine component can impose unpredictable stresses on the body, especially if the handler's posture is unstable. Unexpected movement from a client in the health care sector or an animal in agricultural work can produce similar effects.

Frequent or prolonged physical effort

Risk of injury increases when the body is allowed to become tired, particularly where the work involves constant repetition or a relatively fixed posture. The periodic inter-changing of tasks within a work group can do a lot to reduce these risks.

Insufficient rest and recovery

The opportunity for rest and recovery will also help to reduce risks, especially in relation to tasks which are physically demanding.

Rate of work imposed by a process

In such situations workers are often not able to take even short breaks, or to stretch or exercise different muscle groups. More detailed assessments should be made of assembly line or production line activities. Again, periodic role changes within a work team may provide a solution.

Load related factors

Is the load too heavy?

[M3005] The guideline figures provided earlier are relevant here but only form part of the picture. Task-related factors (eg twisting or repetition) may also come into play or the load may be handled by two or more people. Size, shape or rigidity of the load may also be important.

Is the load bulky or unwieldy?

Large loads will hinder close approach for lifting, be more difficult to get a good grip of or restrict vision whilst moving. The HSE guidance (in L23) suggests that any dimension exceeding 75cm will increase risks and the risks will be even greater if it is exceeded in more than one dimension. The positioning of suitable handholds on the load may somewhat reduce the risks.

Other factors to consider are the effects of wind on a large load, the possibility of the load hitting obstructions or loads with offset centres of gravity (not immediately apparent if they are in sealed packages).

Is the load difficult to grasp?

Where loads are large, rounded, smooth, wet or greasy, inefficient grip positions are likely to be necessary, requiring additional strength of grip. Apart from the possibility of the grip slipping there are likely to be inadvertent changes of grip posture, both of which could result in loss of control of the load.

Is the load unstable?

The handling of people or animals was referred to earlier. Not only do such loads lack rigidity, but they may also be unpredictable and protection of the load as well as the handler is an important consideration. Other potentially unstable loads are those where the load itself or its packaging may disintegrate under its own weight or where the contents may shift suddenly, eg a stack of books inside a partially full box.

Is the load sharp, hot etc?

Loads may have sharp edges or rough surfaces or be extremely hot or cold. Direct injury may be prevented by the use of protective gloves or clothing but the possibility of indirect risks (eg where sharp edges encourage an unsuitable grip position or cold objects are held away from the body) must not be overlooked.

Factors relating to the working environment

Space constraints

[M3006] Areas of restricted headroom (often the case in maintenance work) or low work surfaces will force stooping during manual handling and obstructions (eg in front of shelves) will result in other unsatisfactory postures. Restricted working areas or the lack of clear gangways will increase risks in moving loads, especially heavier or bulkier items.

Assessing the risk of injury from manual handling [M3007]

Uneven, slippery or unstable floors

All of the above can increase the risks of slips, trips and falls. The availability of a firm footing is a major factor in good handling technique – such a footing will be a rarity on a muddy construction site. Moving workplaces (eg trains, boats, elevating work platforms) will also introduce unpredictability in footing. The guideline weights should be reduced in such situations.

Variation in levels of floors or work surfaces

Risks will increase where loads have to be handled on slopes, steps or ladders, particularly if these are steep. Any slipperiness of their surface, eg due to ice, rain or mud will create further risks. The need to maintain a firm handhold on ladders or steep stairways is another factor to consider. Movement of loads between surfaces or shelving at different levels may also need to be considered, especially if there is considerable height change, eg floor level to above shoulder height.

Extremes of temperature or humidity

High temperatures or humidity will increase the risk of fatigue, with perspiration possibly affecting the grip. Work at low temperatures (eg in cold stores or cold weather) is likely to result in reduced flexibility and dexterity. The need for gloves and bad weather clothing is another factor to consider.

Ventilation problems or gusts of wind

Inadequate ventilation may increase fatigue while strong gusts of wind may cause considerable danger when handling larger loads, eg panels being moved on building sites.

Poor lighting

Lack of adequate lighting may cause poor posture eg by encouraging stooping or an increased risk of tripping. It can also prevent workers identifying risks associated with individual loads, eg sharp edges or corners, offset centres of gravity.

Part 3 of the HSE guidance (L23) provides detailed guidance on all the areas above.

Individual capability

[M3007] As set out in **M3002** above, Regulation 4(3) of the *Manual Handling Operations Regulations 1992 (SI 1992 No 2793)* (as amended) requires that:

> 'In determining for the purposes of this regulation whether manual handling operations at work involve a risk of injury and in determining the appropriate steps to reduce that risk regard shall be had in particular to:
>
> (a)   the physical suitability of the employee to carry out the operations;'

Under *Schedule 1*, employers must consider the following questions:

Does the job:

- require unusual strength, height, etc?
- create a hazard to those who might reasonably be considered to be pregnant or to have a health problem?
- require special information or training for its safe performance?

Manual handling capabilities will vary significantly between individual workers. Reference was made earlier in the chapter to the assessment guideline weights providing a reasonable level of protection to around 95 per cent of workers and those guideline figures differ for men and women. *Regulation 4(3)* specifically refers to factors which must be taken into account in respect of individual workers (see **M3002** above). However, HSE guidance to the Regulations (L23) sets out that while the ability to carry out manual handling safely varies between individuals, these variations 'are less important than the nature of the handling operations in causing manual handling injuries. Assessments which concentrate on individual capability at the expense of task or workplace design are likely to be misleading.'

Employers also need to be aware of their duties under the *Equality Act 2010* – for more information see EQUALITY ACT Chapter).

Disabled workers

[M3007.1] In *Banaszczyk v Booker Ltd* [2016] UKEAT/0132/15/RN, [2016] IRLR 273, the Employment Appeal Tribunal (EAT) ruled in favour of a worker who was dismissed after a spine injury he developed following a car accident affected his ability to do his job. This involved lifting and moving heavy cases.

Mr Banaszczyk was employed as a picker at a Booker distribution centre. After his accident, he became unable to meet the hourly 'pick rate' of 210 25kg cases per hour and was dismissed. He issued a claim for unfair dismissal and disability discrimination. An employment tribunal found that because his spinal injuries were not severe enough to have an adverse effect on his day-to-day activities, such as shopping, eating and driving, he could not be considered legally disabled. However, the EAT upheld his appeal. It said that moving and lifting cases of up to 25kg is a normal day-to-day activity for many workers and ruled that under the *Equality Act 2010*, Mr Banaszczyk could be classified as disabled. His dismissal was therefore unfair and discriminatory.

The HSE says employers should not assume someone with a disability is not able to do a job involving manual handling. The *Equality Act 2010* requires employers to make 'reasonable adjustments' for disabled workers and this could include limiting the number, size or weight of loads handled.

Further HSE advice on individual capability and employees at particular risk is as follows:

'Take account of the particular requirements of employees who:
- are or have recently been pregnant;
- have a disability which may affect their manual handling capability;
- have recently had a manual handling injury or have a history of back, knee or hip trouble, hernia or other health problems which could affect their manual handling capability;
- are young workers or new to the job;

Assessing the risk of injury from manual handling [M3007.2]

- are older workers.'

Gender

In general, the lifting strength of healthy women is less than that of healthy men. Their body sizes and amounts of muscle are, on average, smaller than those of men. However, for both men and women, the range of individual strength and ability is large, and there is considerable overlap. Some women can safely handle greater loads than some men.

Age

Physical capability also varies with age and tends to decline as part of the ageing process, but again there is considerable variation between individuals. Employers should not assume that certain jobs are physically too demanding for older workers and should not overlook the benefits of their judgement, experience and knowledge.

Young workers may be more at risk of manual handling injuries as their muscle strength may not be fully developed. They may be less skilled in handling techniques or in pacing their work. Those new to the job may lack experience or be unfamiliar with the workplace.

Health status

Care should be taken when considering placing an individual with a history of back pain, particularly recent back pain, in a job which involves heavy manual handling. But individuals should not be excluded from work unless there is a good medical reason for restricting their activity. The HSE also points out that making changes to reduce the physical demands of a heavy job can allow a wider range of individuals to be employed.

New and expectant mothers

[M3007.2] Special consideration should be given to new and expectant mothers whose capabilities may be affected by hormonal and physical changes.

The HSE says:

- there is some evidence linking physically demanding work to premature birth, particularly where the work includes long periods of standing and/or walking;
- during the last three months of pregnancy there is an increased risk of musculoskeletal symptoms when heavy or repeated lifting is undertaken due to hormonal changes affecting the ligaments that support the joints;
- as pregnancy progresses it may become more difficult to achieve and maintain good postures; and
- a shift of the centre of gravity can increase the risk of back pain for pregnant women.

Under the Management Regulations, when an employee tells her employer in writing she is pregnant, has given birth within the past six months, or is breastfeeding, the employer should immediately take into account any risks identified in the workplace risk assessment and take appropriate action. The HSE recommends that employers have a well-defined plan on how to respond when pregnancy is confirmed. This could include the following steps:

[M3007.2] Manual Handling

- reassess the handling operations (positioning of the load and feet, frequency of lifting etc) and consider improvements;
- provide training to recognise ways in which the work may be altered to help with changes in posture and physical capability, including the timing and frequency of rest periods;
- consider job-sharing, alternative work or suspension on full pay where the risk cannot be reduced by a change to the working conditions;
- take into account any medical advice provided by the GP.

Are manual handling operations unusual for a group of workers?

[M3007.3] The HSE advises employers to consider whether the physical demands imposed by manual handling operations are unusual for a particular group of workers. For example, demands that would be considered unusual for a group of employees engaged in office work might not be out of the ordinary for those normally involved in heavy physical labour.

Other factors

Clothing, footwear etc

Personal protective equipment (PPE) should be used only as a last resort, when engineering or other controls do not provide adequate protection against hazards. If wearing PPE cannot be avoided, its implications for the risk of manual handling injury should be considered. It may significantly affect the user's mobility or dexterity or increase the possibilities of fatigue, dehydration etc. Similarly uniforms or costumes, in the entertainment industry for example, may also be a factor to be taken into account.

Avoiding or reducing risks

[M3008] The *Manual Handling Operations Regulations 1992 (SI 1992 No 2793)* require employers to avoid the need for employees to undertake manual handling operations involving a risk of injury, so far as is reasonably practicable. Measures for avoiding such risks can be categorised as:

- Elimination of handling
- Automation or mechanisation
- Reduction of the weight or size of the load

Elimination of handling

[M3009] The first step is to consider whether manual handling can be avoided. It may be possible for employers to eliminate manual handling altogether by means such as:

Redesigning processes or activities

For example, can activities such as machining or wrapping be carried out in situ, rather than a product being manually handled to a position where the activity should be takes place? Can a treatment be taken to a patient rather than vice versa?

Using transport better

For example, allowing maintenance staff to drive vehicles carrying tools or equipment up to where they are working, rather than manually handling them into place.

Requiring direct deliveries

For example, requiring suppliers to make deliveries directly into a store rather than leaving items in a reception area. Suppliers may be better equipped in terms of handling aids and their staff may be better trained in respect of manual handling.

Automation or mechanisation

[M3010] The automation or mechanisation of a work activity is best considered at the design stage although there is no reason in principle why such changes cannot be made later, as the result of a risk assessment for example. In making changes it is important to avoid creating additional risks. The use of fork lift trucks for a task in an already cramped workplace, for example, may not be the best solution. However, as employers have become more aware of manual handling risks and the requirements of the Regulations, a much greater range of mechanical handling solutions to manual handling problems have become available. These include:

Mechanical lifting devices

Fork lift trucks, mobile cranes, lorry mounted cranes, tail lifts, mobile elevating work platforms, vacuum devices and other powered handling equipment are commonly used to eliminate or greatly reduce the manual handling of loads.

Manually-operated lifting devices

Pallet and stacker trucks, manually-operated chain blocks or lever hoists can all be used to move heavy loads using very limited physical force.

Powered conveyors

As well as fixed conveyors (both belt and driven roller types), mobile conveyors are used particularly for loading and unloading vehicles, such as stacking bundles of newspapers in delivery vans for example.

Non-powered conveyors, chutes etc

Free-running roller conveyors, chutes or floor-mounted trolleys allow loads to move under the effects of gravity or to be moved manually with little effort. Such devices may operate between different levels or be set into the workplace floor.

Raw material transfer systems

Raw materials such as powders or liquids can be transferred using gravity feeds or pneumatic transfer equipment, avoiding handling of bags or other containers.

Trolleys and trucks

Trolleys and trucks can be used to greatly reduce the manual handling effort required in transporting loads. Some types also incorporate manual or mechanically-powered lifting mechanisms. Specialist trolleys are designed for carrying drums or other containers while a triple-wheel system can be lifted to some trucks to aid them in climbing or descending stairs.

Lifting tools

Special lifting tools are available to reduce the manual handling effort required for lifting or lowering certain loads, eg paving slabs, manhole covers, drums or logs.

These and many other types of devices are illustrated in HSE booklet INDG398 *Making the best use of lifting and handling aids?* which can be downloaded from the HSE website at www.hse.gov.uk/pubns/books/indg398.htm.

Part 4 of L23 also provides detailed guidance on mechanical assistance and good handling technique.

Reduction of the weight or size of the load

[M3011] There is considerable scope to avoid the risk of injury by reducing the sizes of the loads which have to be handled. In some cases, the initiative for this has come from suppliers of a particular product whereas in others the impetus has come from customers or from users of equipment, often as a result of their manual handling assessments. Examples include the following:

- Limiting individual bag weights on airlines.
- Reducing the weight of packaged building materials (eg cement).
- Supplying photocopying paper in five ream boxes instead of 10 ream boxes as was previously commonplace.
- Reducing the weight of newspaper bundle sizes. Printers reduce the number of copies in a bundle as the number of pages in the paper increases.
- Customers specifying maximum container weights they will accept from suppliers, eg weights of brochures from commercial printers.
- Equipment being separated into component parts and assembled where it is to be used, eg emergency equipment used by fire services.

The HSE advises that materials like liquids and powders may be packaged in smaller containers, while it may be possible to specify lower package weights where loads are brought in.

However, it advises, breaking down loads will not always be the safest course of action as it will increase the handling frequency.

'When the frequency of handling is low and the weights are relatively high, it is preferable to reduce the weight and increase the frequency of handling,' it says. 'However, increasing the frequency too much will cause fatigue. Another option is to make the load so big that it cannot be handled manually.'

Where it is not reasonably practicable to avoid a risk of injury from a manual handling operation, the employer must carry out an assessment of the operation with the aim of reducing the risks to the lowest level reasonably practicable. The four main factors to be taken into account during the assessment are *the task, the load, the working environment* and *individual capability*. Possible risk reduction measures related to each of these factors are described below.

Task-related measures

Mechanical assistance

[M3012] The types of mechanical assistance described earlier can be used to reduce the risk of injury, even if such risks cannot be avoided entirely. The use of roller conveyors, trolleys, trucks or levers can reduce considerably the amount of force required in manual handling tasks.

Task layout and design

The layout of tasks, storage areas etc should be designed so that the body can be used in its more efficient modes, eg:

- heavier items stored and moved between shoulder and mid-lower leg height (preferably around waist height);
- allowing loads to be held close to the body;
- avoiding reaching movements, eg over obstacles or into deep bins;
- avoiding twisting movements or handling in stooped positions, eg by repositioning work surfaces, storage areas or machinery;
- providing resting places to aid grip changes while handling;
- reducing lifting and carrying of loads, eg by pushing, pulling, sliding or rolling techniques;
- utilising the powerful leg muscles rather than arms or shoulders;
- avoiding the need for sustaining fixed postures, eg in holding or supporting a load;
- avoiding the need for handling in seated positions.

Work routines

Work routines may need to be altered to reduce the risk of injury using such measures as:

- limiting the frequency of handling loads (especially those that are heavier or bulkier);
- ensuring workers are able to take rest breaks (either formal breaks or informal rest periods as and when needed);
- introducing job rotation within work teams (allowing muscle groups to recover while other muscles are used or lighter tasks undertaken).

Team handling

Tasks which might be unsafe for one person might be successfully carried out by two or more. HSE guidance (L23) explains that the load a team can handle safely can be less than the sum of the loads that the individual team members could cope with when working alone. As an approximate guide, it advises using 85% of the total of individual capabilities to allow a safety margin. This can be applied to teams of different sizes.

[M3012] Manual Handling

Other factors to take into account in team handling include the following:

- the availability of suitable handholds for all the team;
- whether steps or slopes have to be negotiated;
- possibilities of the team impeding each other's vision or movement;
- availability of sufficient space for all to operate effectively;
- the relative sizes and capabilities of team members; and
- good communication between team members.

Load-related measures

Weight or size reduction

[M3013] The types of measures described earlier in the chapter might be adopted to reduce the risk of injury even if it is not possible to avoid the risk entirely. However, reducing the individual load size will increase the number of movements necessary to handle a large total load. This could result in a different type of fatigue and also the possibility of corners being cut to save time. Sizes of loads may also be reduced to make them easier to hold or bring them closer to the handler's body.

Making the load easier to grasp

Handles, grips or indents can all make loads easier to handle (as demonstrated by the handholds provided in office record storage boxes). It will often be easier handling a load placed in a container with good handholds than handling it alone without any secure grip points. The positioning of handles or handholds on a load can also influence the handling technique used, eg help avoid stooping. Handholds should be wide enough to accommodate the palm and deep enough to accommodate the knuckles (including gloves where these are worn).

The use of slings, carrying harnesses or bags can all assist in gaining a secure grip on loads and carrying them in a more comfortable and efficient position.

Making the load more stable

Loads which lack rigidity themselves or are in insecure packaging materials may need to be stabilised for handling purposes. Use of carrying slings, supporting boards or trays may be appropriate. Partially full containers may need to have their contents wedged into position to prevent them moving during handling.

Reducing other risks

Loads may need to be cleaned of dirt, oil, water or corrosive materials for safe handling to take place. Sharp corners or edges or rough surfaces may have to be removed or covered over. There may also be risks from very hot or very cold loads. In all cases, the use of containers (possibly insulated), other handling aids, or suitable gloves or other PPE may need to be considered.

Improving the work environment

Providing clear handling space

[M3014] The provision of adequate gangways and sufficient space for handling activities to take place will be linked with general workplace safety issues. Low headroom, narrow doorways and congestion caused by equipment and stored materials are all to be avoided. Good housekeeping standards are essential to safe manual handling.

Floor condition and design

Even in temporary workplaces such as construction sites, every effort should be made to ensure that firm, even floors are provided where manual handling is to take place. Special measures may be necessary to allow water to drain away or to promptly clear away any potentially slippery materials (eg food scraps in kitchens). Equipment is available to measure the slipperiness of floors. Where risks are greater, special slip-resistant surfaces may need to be considered. In outdoor workplaces the routine application of salt or sand to surfaces made slippery by ice or snow may need to be introduced.

Differing work levels

Measures may be necessary to reduce the risks caused by different work levels. Slopes may need to be made more gradual, steps may need to be provided or existing steps or stairways made wider to accommodate handling activities. Work benches may need to be modified to provide a uniform and convenient height.

Thermal environment and ventilation

Unsatisfactory temperatures, high humidity or poor ventilation may need to be overcome by improved environmental control measures or transferring work to a more suitable area. For work close to hot processes or equipment, in very cold conditions (eg refrigerated storage areas) or out of doors the use of suitable PPE may be necessary.

Gusts of wind

Where gusts of wind (or powerful ventilation systems) could affect larger loads, extra precautions in addition to those normally taken may be necessary eg:

- using handling aids (eg trolleys);
- utilising team handling techniques;
- adopting a different work position;
- following an alternative transportation route.

Lighting

Suitable lighting must be provided to permit handlers to see the load and the layout of the workplace and to make accurate judgements about distance, position and the condition of the load.

Individual capability

Health surveillance and health monitoring

[M3015] In ensuring that individuals are capable of carrying out manual handling activities in the course of their work, account must be taken both of factors relating to the individual and those relating to the activities they are to perform.

The HSE guidance (L23) explains there is no duty in the Regulations to carry out health surveillance. However, 'valuable information can be obtained from less formal systems of reporting, monitoring and investigating symptoms, known as "health monitoring"'. It is good practice to put in place systems that allow individuals to make early reports of manual handling injuries or back pain. Where appropriate these can be supplemented, for example by monitoring sickness absence records and annual health checks.

Training and information

[M3015.1] The HSE sets out that section 2 of the HSW Act and regulations 10 and 13 of the *Management Regulations* require employers to provide their employees with health and safety information and training. Employers should supplement this, as necessary, with more specific information and training on manual handling injury risks and prevention as part of the steps to reduce risk required by regulation 4(1)(b)(ii) of the *Manual Handling Operations Regulations*.

The HSE does not publish prescriptive guidance on what a 'good' manual handling training course should include or how long it should last. Courses should be suitable for the individual, tasks and environment involved, use relevant examples, relate to what workers actually do and last long enough to cover all the relevant information. This information is likely to include advice on:

- manual handling risk factors and how injuries can occur;
- how to carry out safe manual handling, including good handling technique;
- appropriate systems of work for the individual's task and environment;
- safe use of lifting and handling aids; and
- practical work to allow the trainer to identify and put right anything the trainee is not doing safely.

According to new HSE advice published in March 2018, research has found that general training in lifting techniques is an ineffective way of controlling the risks of manual handling in businesses. It says off-the-shelf manual handling training should become a thing of the past. Employers should instead get help to change the way they work, reduce manual handling risks and avoid paying for ineffective or unnecessary training.

HSE research shows simplistic training involving 'bending your knees to lift a cardboard box' is a waste of time and money and does not make any difference. The overall aim is to avoid and reduce manual handling and that is where employers should start if their workforce faces manual handling risks. Rather than starting with training, they should start with re-organising and redesigning working practices.

If they do need staff training for residual risks, this should be customised and professionally delivered. It should be based on observations of current working practices and should be informed by the views and experience of the workforce.

The new advice can be found on the HSE website at: www.hse.gov.uk/msd/external-help.htm.

Task-specific training and instruction

Training and instruction should relate to the safety of specific manual handling tasks, eg:

- how to use specific handling aids;
- specific PPE requirements;
- specific handling techniques.

General principles of handling technique are summarised in the HSE leaflet *Manual handling at work A brief guide* (INDG 143) and also in L23. Many other HSE publications deal with specific occupational areas and often contain guidance on training (see 'HSE Guidance' at **M3019**).

HSE guidance to Regulation 4(1)(b)(iii) makes clear that the requirement to provide 'general indications' of the weight and nature of the loads to be handled should form part of any job induction or basic training so employees have sufficient information to carry out the operations they are likely to be asked to do.

Regulation 4(3)(c) of the *Manual Handling Operations Regulations 1992 (SI 1992 No 2793)* (as amended) requires employees' level of knowledge and training to be considered during the risk assessment.

Planning and preparation

[M3016] The principles of planning and preparation for manual handling assessments are no different from those for any other types of risk assessments.

Who should carry out the assessments?

[M3017] Responsibility for carrying out an assessment lies with the employer (or relevant self-employed person). HSE advice is that whoever carries it out must be competent, whether they are in-house personnel or external assessors. They must involve the workforce (see **M3020**) and understand when specialist help is required.

They should be familiar with the requirements of the Regulations and be able to:

- identify manual handling hazards and assess risks;
- use additional sources of information on risks, as appropriate;
- draw valid and reliable conclusions from assessments and identify steps to reduce risks;

- clearly record the assessment;
- communicate the findings to workers and those who need to take action; and
- recognise their own limitations, calling on further expertise as necessary.

How should the assessments be carried out?

[M3018] HSE advice is that assessments should be carried out in consultation with employees and should use practical experience of how the work is done. They should identify foreseeable problems likely to arise and the necessary measures to deal with them, including training.

L23 provides advice on carrying out generic assessments – for assessing risks common to a number of broadly similar operations; to individuals rotating between similar tasks or to groups of workers carrying out similar jobs. In multi-stage manual handling assessments; generic assessments can be supplemented with assessments of individual tasks.

The findings should be available to employees and safety representatives (see M3020 below).

It also contains specific advice on generic and individual person assessments for moving and handling people and more specialist advice is available (see below).

Gathering information

[M3019] It is important to gather information prior to the assessment in order to identify potentially high-risk manual handling activities and precautions which should be in place to reduce the risks. Relevant sources may include:

Accident investigation reports

Dependent upon the requirements of the organisation's accident investigation procedure, these may only include the more serious accidents resulting from manual handling operations.

Ill health records

These may reveal short absences or other incidents due to manual handling which may have escaped the accident reporting system. Enquiries may need to be made in respect of the health surveillance records of individuals (see **M3015** above).

Accident books

Accident books or other treatment records may identify regularly recurring minor accidents which do not necessarily result in any absences from work, eg cuts or abrasions from sharp or rough loads.

Operating procedures etc

Reference to manual handling activities and precautions which should be adopted may be contained in operating procedures, safety handbooks etc.

Work sector information

Trade associations and similar bodies publish information on manual handling risks and how to overcome them.

HSE sector guidance

As well as HSE guidance already referred to above, HSE publications on manual handling in specific work sectors include the following:

HSG 196	Moving food and drink: Manual handling solutions for the food and drink industries (2014)
INDG 269	Managing musculoskeletal disorders in checkout work A brief guide (2013)
INDG 318	Manual handling solutions in woodworking (2013)
INDG 390	Choosing a welding set. Make sure you can handle it (2012)
INDG 439	Reducing manual handling risks in carpet retail (2015)
INDG 470	Handling newspaper and magazine bundles (2015)

HSE advice and information on moving and handling in health and social care can be found on the HSE website at: www.hse.gov.uk/healthservices/moving-handling.htm.

HSE advice and guidance on manual handling in the paper industries can be found on the HSE website at: www.hse.gov.uk/paper/manual-handling.htm.

HSE advice and guidance on manual handling in the textiles industry can be found on the HSE website at: www.hse.gov.uk/textiles/textiles-and-manual-handling.htm.

Picking up the pieces: Prevention of musculoskeletal disorders in the ceramics industry (1996)

Further general sources of guidance on handling are listed in **M3028**.

Consulting and involving workers

[M3020] As with other types of assessments, visits to the workplace are an essential part of manual handling assessments. Manual handling operations need to be observed (sometimes in some detail) and discussions with those carrying out manual handling work, their safety representatives and their supervisors or managers need to take place.

HSE advice is as follows: 'You must consult all your employees, in good time, on health and safety matters. In workplaces where a trade union is recognised, this will be through union health and safety representatives. In non-unionised workplaces, you can consult either directly or through other elected representatives. Consultation involves employers not only giving information to employees but also listening to them and taking account of what they say before making health and safety decisions.'

Employees will know about manual handling risks in the workplace and can probably offer practical solutions to controlling those risks. They can raise specific problems caused by, for example, the size, shape or weight of loads, how often loads are handled, the order in which tasks are carried out, and the environment in which handling operations are carried out, including space constraints which make it difficult to manoeuvre the load, unsuitable storage and uneven flooring, for example.

The assessment checklist provided on the following pages provides guidance on the factors which may need to be taken into account during the observations and discussions and also on possible measures to reduce the risk. The checklist utilises the four main factors specified in the Regulations – task, load, working environment and individual capability. In practice each of these will have greater or lesser importance, depending upon the manual handling activity.

Consulting and involving workers [M3020]

MANUAL HANDLING ASSESSMENT CHECKLIST
THE TASK

ASSESSMENT FACTORS	REDUCING THE RISK
DISTANCE OF THE LOAD FROM THE TRUNK	TASK LAYOUT
BODY MOVEMENT / POSTURE	USE THE BODY MORE EFFECTIVELY
• Twisting • Reaching • Stooping • Sitting	eg SLIDING OR ROLLING THE LOAD
DISTANCE OF MOVEMENT	SPECIAL SEATS RESTING PLACE / TECHNIQUE
• Height (? grip change) • Carrying (? over 10m) • Pushing or pulling	USE OF TROLLEYS etc TECHNIQUE / FLOOR SURFACE
RISK OF SUDDEN MOVEMENT	
FREQUENT / PROLONGED EFFORT	AWARENESS / TRAINING IMPROVED WORK ROUTINE
	eg Job rotation
CLOTHING / FOOTWEAR ISSUES RATE OF WORK IMPOSED BY A PROCESS	ADEQUATE REST OR RECOVERY PERIODS

THE LOAD

ASSESSMENT FACTORS	REDUCING THE RISK
HEAVY	MAKE THE LOAD LIGHTER TEAM LIFTING MECHANICAL AIDS
BULKY (Any dimensions above 75cm)	MAKE THE LOAD SMALLER
UNWIELDY (Offset centre of gravity)	INFORMATION / MARKING
DIFFICULT TO GRASP (Large, rounded, smooth, wet, greasy)	MAKE IT EASIER TO GRASP
	• Handles • Handgrips • Indents • Slings • Carrying devices • Clean the load
UNSTABLE • Contents liable to shift • Lacking rigidity	WELL-FILLED CONTAINERS USE OF PACKING MATERIALS SLINGS / CARRYING AIDS
SHARP EDGES / ROUGH SURFACES	AVOID OR REDUCE THEM GLOVES OR OTHER PPE
HOT OR COLD	CONTAINERS (? insulated)

THE WORKING ENVIRONMENT

ASSESSMENT FACTORS	REDUCING THE RISK
SPACE CONSTRAINTS AFFECTING POSTURE	ADEQUATE GANGWAYS, FLOORSPACE, HEADROOM
UNEVEN, SLIPPERY OR UNSTABLE FLOORS	WELL MAINTAINED SURFACES PROVISION OF DRAINAGE SLIP-RESISTANT SURFACING GOOD HOUSEKEEPING PROMPT SPILLAGE CLEARANCE
VARIATION IN WORK SURFACE LEVEL *eg steps, slopes, benches*	STEPS NOT TOO STEEP GENTLE SLOPES UNIFORM BENCH HEIGHT
HIGH OR LOW TEMPERATURES HIGH HUMIDITY POOR VENTILATION	BETTER ENVIRONMENTAL CONTROL RELOCATING THE WORK SUITABLE CLOTHING
STRONG WINDS OR OTHER AIR MOVEMENT	RELOCATING THE WORK ALTERNATIVE ROUTE HANDLING AIDS TEAM HANDLING
LIGHTING	SUFFICIENT WELL-DIRECTED LIGHT
MOVING WORKPLACE • Boat • Train • Vehicle	

INDIVIDUAL CAPABILITY

ASSESSMENT FACTORS *INDIVIDUAL FACTORS*	REDUCING THE RISK
	BALANCED WORK TEAMS 'SELF SELECTION' MAKE ASSISTANCE AVAILABLE ENCOURAGE FITNESS
PROBLEMS WITH CLOTHING ETC. PREGNANCY INJURY / HEALTH PROBLEM • Back trouble • Hernia • Temporary injury	SPECIAL ASSESSMENTS FORMAL RESTRICTIONS CLEAR INSTRUCTIONS

TASK FACTORS OPERATIONS OR SITUATIONS REQUIRING PARTICULAR • Awareness of risks • Knowledge • Method of approach • Technique	FORMALISED WORK PROCEDURES CLEAR INSTRUCTIONS ADEQUATE TRAINING • Recognising potentially hazardous operations • Dealing with unfamiliar operations • Correct use of handling aids • Proper use of PPE • Environmental factors • Importance of housekeeping • Knowing one's own limitations • Good technique

Observations

[M3021] Sufficient time should be spent in the workplace to observe a sufficient range of the manual handling activities taking place and to take account of possible variations in the loads handled (raw materials, finished product, equipment etc). Particular aspects that should be considered include:

- comparison of actual methods used with those specified in operating procedures, industry standards etc;
- the level and manner of use of handling aids and their effectiveness;
- handling techniques adopted;
- physical conditions, eg access, housekeeping, floor surfaces, lighting;
- variations between employees, eg size, strength, technique adopted;
- damage to products, containers or equipment.

Discussions

[M3022] In discussing manual handling issues with workers an enquiring approach should be taken, utilising open-ended questions wherever possible. Aspects which might be discussed include:

- whether normal conditions are being observed;
- workers' awareness of procedures, rules etc;
- workers' training in the use of handling aids or handling techniques;
- possible variations in work activities, eg product changes, seasonal variations, differences between day and night shift;
- what happens if handling equipment or handling aids break down or are unavailable;
- whether assistance is available if required;
- which manual handling operations are difficult to perform;
- what manual handling problems have been experienced;
- whether there have been any manual handling injuries;
- reasons precautions are not being taken.

Notes

[M3023] Notes should be taken during the assessment of anything which might be of relevance such as:

- further detail about manual handling risks found;
- handling risks which are discounted as insignificant;
- handling aids available;
- other control measures in place and appearing effective;
- problems identified or concerns expressed;
- alternative precautions worthy of consideration;
- related procedures, records or other documents.

These rough notes will eventually form the basis of the assessment.

After the assessment

Assessment records

[M3024] The significant findings of the assessment should be recorded. The HSE says the record should include the hazards, how people may be harmed by them, what is in place to control the risks and details of any groups identified as being particularly at risk. Employers with five or more employees are required by law to record the results of the assessment.

The HSE website includes sample record formats together with worked examples for lifting and carrying and pushing and pulling (www.hse.gov.uk/pubns/ck5.pdf). It includes checklists of assessment factors under the four main headings (task, load, working environment, individual, capability) in a similar listing to the assessment checklist provided in this chapter. Alongside each checklist item the assessor has space to:

- identify whether the risk from this factor is low, medium or high;
- provide a more detailed description of the problems;
- identify possible remedial action.

Some will no doubt find this checklist format meets their needs. However, many assessment factors often have little or no relevance to a specific manual handling operation. As a result, the eventual record contains a lot of blank paper, with only occasional comments. Consequently, it is often preferable to keep a separate checklist (as with the checklist earlier in the chapter — see **M3020**), and only include within the record a reference to the risks, precautions and recommended improvements which are relevant to the operation being assessed. Two worked examples of this type of record are included below.

After the assessment [M3024]

MANUAL HANDLING ASSESSMENT		BROADACRES DEVELOPMENT	DATE: 4 April 2018	ASSESSMENT BY: A Backworth
No.	ACTIVITY	RISKS	EXISTING PRECAUTIONS	RECOMMENDATIONS
1	HANDLING OF STATIONERY AND PRINTED MATERIALS	Heaviest stationery items are 5 ream boxes of paper (approximately 12 kg). Some packs of printed promotional material weigh over 20 kg.	Suppliers deliver items directly to main store.	Require printing suppliers to supply materials in packs weighing no more than 12 kg.
		Items are moved from main store on the ground floor to smaller stores on other floors.	Trolley and lift used for transporting between main store and smaller stores.	Ensure paper and printed materials are stored on shelves between knee and shoulder height.
2	HANDLING OF ARCHIVED RECORDS	Records are normally stored in file boxes weighing no more than 12 kg. They are moved into and out of the record store on the third floor.	File boxes are kept below shoulder height on substantial racking. They are moved around using the trolley and lift.	Access to the accounts section of the archive store racking was obstructed by various items which must be removed.
		Older records are removed for storage off-site by a specialist company.	Movement to off-site storage is carried out by facilities staff who have had manual handling training.	Some records were stored in oversize boxes — this practice must cease.
3	MOVEMENT OF EQUIPMENT PC monitors, TV sets, projection equipment etc.	These items (of varying weights) are moved between the equipment store on the first floor and locations where they are required.	The trolley and lift are used where appropriate. IT staff have been trained in handling techniques.	Training section assistants should also receive handling technique training. Provide dust coats (so that staff can carry dirty items close to their bodies).
			Some racking is provided in the equipment store but this is inadequate and many items are stored on the floor.	Install additional racking in the equipment store and keep heavier items on more accessible shelves.
4	MOVEMENT OF FURNITURE Desks, seating etc.	Desks and seating are moved as required for individual workstations.	Facilities section staff have had manual handling training.	Training section assistants require training (see above).
		Major moves are carried out by contractors.	Team lifting is used for larger items, eg desk tops. The lift is used as appropriate.	Suitable racking must be provided in the equipment store for desk and table tops.

[M3024] Manual Handling

MANUAL HANDLING ASSESSMENT		BROADACRES MENT	DEVELOP-	DATE: 4 April 2018		ASSESSMENT BY: A Backworth
No.	ACTIVITY	RISKS		EXISTING PRECAUTIONS		RECOMMENDATIONS
		Seating and tables are moved by facilities section staff when required for major meetings and training events.		Desk and table tops are kept in the equipment store.		

GOLD STAR ENGINEERING MAINTENANCE STORES Assessment by: T Wrightson		Date: 26 February 2018
MANUAL HANDLING ASSESSMENT		
ACTIVITIES/ RISKS	EXISTING PRECAUTIONS	RECOMMENDATIONS
Handling activities involve the movement of equipment and materials within the stores building and the yard outside, together with two storage containers and a flammable liquid store situated within the yard. Both equipment and materials must also be moved to locations within the factory.	A fork lift truck is used to handle heavier items within the yard. A hand-operated pallet truck and a lightweight lifting truck are available. Several trucks and trolleys are provided including a stair-climbing truck, gas cylinder and fire extinguisher trolleys and a drum transporter. Skates, bogie units and dollies are also available for transporting heavy items.	
	Most internal and external floor surfaces are in good condition.	A section of the floor just inside the main stores entrance requires repair.
There is considerable variation in the sizes and weights of items handled. Some items are palletised, some are supplied in boxes and some are kept in bins and cages.	There is a gradual ramp into the main stores building but not into the storage containers or the flammable liquid store (the latter having a sill at its entrance).	Suitable permanent or removable ramps should be provided to allow handling aids to be used in the storage containers and flammable liquid store.
Drums, gas cylinders and large fire extinguishers must also be moved.	Lighting standards are good in the main stores building, the flammable liquid store and the yard, but there are no lights in the storage containers.	Lighting must be provided within the storage containers.
Racking and cupboards are provided within the building and inside the outside storage containers.	Heavier items are generally stored either on the floor or on accessible levels on racking or in cupboards.	Ensure walkways between racking are kept clear.

GOLD STAR ENGINEERING MAINTENANCE STORES	MANUAL HANDLING ASSESSMENT Assessment by: T Wrightson	Date: 26 February 2018
ACTIVITIES/ RISKS	EXISTING PRECAUTIONS	RECOMMENDATIONS
Some items to be handled have sharp or rough edges.	Latex coated knitted gloves (for grip) and gloves with chrome leather palms are available (as are gloves for chemical risks). Most stores' staff have attended a one-day course in manual handling techniques tailored to their needs.	Develop a formal training programme to ensure that all stores staff receive training in manual handling technique and the correct use of the various handling aids available.

Review and implementation of recommendations

[M3025] As for other types of risk assessments, recommendations should be reviewed and action plans for their implementation prepared. Recommendations will need to be followed up in order to ensure that they have actually been implemented and also to assess any new risks which may have been introduced. This latter point is particularly relevant to manual handling as changes in working practice may have resulted in different techniques being adopted (with their own attendant risks) or the introduction of handling aids may have created additional training needs.

Assessment review

[M3026] *Regulation 4(2)* of the *Manual Handling Operations Regulations 1992* requires a manual handling assessment to be reviewed if:

(a) there is reason to suspect that it is no longer valid; or
(b) there has been a significant change in the manual handling operations to which it relates.

As is the case for other assessments, a review would not necessarily result in the assessment being revised, although a revision may be necessary. A significant manual handling related injury should automatically result in a review of the relevant assessment. Assessments should be reviewed periodically in any case in order to take account of small changes in work practices which will eventually have a cumulative effect.

Training

[M3027] The need for specific manual handling technique training may be one of the recommendations made as a result of a manual handling assessment (see **M3015** above).

Further sources of guidance

[M3028] In addition to the work sector specific guidance in **M3019**, many other HSE publications provide guidance of relevance to manual handling. This includes:

L 23	Manual Handling. Manual Handling Operations Regulations 1992 (as amended). Guidance on Regulations. (2004)
INDG 383	Manual handling assessment charts (2014) Variable manual handling assessment chart (V-MAC) tool
INDG 478	Risk assessment of pushing and pulling (RAPP) tool (2016)
INDG 438	Assessment of repetitive tasks in the upper limbs (the ART tool) (2010)
HSG 60	Upper limb disorders in the workplace (2002)

INDG 171	Managing upper limb disorders in the workplace (2013) (www.tsoshop.co.uk/Medicine/?DI=346223&CLICKID=002289)
INDG 143	Manual handling at work A brief guide (2012)
INDG 398	Making the best use of lifting and handling aids? (2013)

Sections of particular relevance on the HSE website (www.hse.gov.uk) include:

- Moving goods safely;
- Musculoskeletal disorders (MSDs).

The HSE also recommends *The Back Book* which can be found on The Stationery Office (TSO) website www.tsoshop.co.uk.

Noise at Work

Andrea Oates

Introduction to noise at work

[N3001] This chapter is aimed at company directors, legal advisors, health and safety officers, safety representatives and others who need a comprehensive, concise reference source on the issue of noise at work.

It will be helpful to anyone who needs to commission a workplace noise assessment or to undertake a noise control project. It will also assist those who wish to devise long-term noise reduction and equipment-buying strategies.

Despite decades of warnings and legislation, noise-induced hearing loss is still a huge but underrated problem, with a slow, cumulative and irreversible effect. A particular problem with noise-induced hearing loss is the long time-lag between exposure and the damage becoming apparent.

This chapter discusses the scale of problems arising from workplace noise exposure, the legal obligations on employers to reduce the risk of noise-induced hearing damage (**N3012**), and the risks in terms of compensation claims (**N3022**). It also provides an overview of industrial injuries disablement benefit (IIDB) as it applies to occupational deafness (**N3018**).

It includes a short introduction to the perception of sound and the various indexes used to measure noise (**N3004**), provides general guidance on methods of assessing and reducing noise (see **N3029** and **N3035**), and suggests sources of more detailed guidance (see **N3023**).

Scale of the problem

[N3002] The charity Action on Hearing Loss (previously the RNID) sets out that:

- there are more than 11 million people in the UK with some form of hearing loss, around one in six of the population;
- by 2035, it is estimated that there will be 15.6 million people with hearing loss in the UK;
- more than 40% of over 50-year olds have some form of hearing loss; and
- around one in ten adults in the UK have tinnitus.

(www.actiononhearingloss.org.uk/about-us/our-research-and-evidence/facts-and-figures/)

The Health and Safety Executive (HSE) estimates that around 20,000 workers have work-related hearing problems, including those with new and longstanding hearing problems, based on estimates using three years of the Labour Force Survey from 2014/15 to 2016/17.

In order to qualify for industrial injuries disablement benefit (IIDB) for occupational deafness, the disabled person must suffer an average hearing loss of at least 50 decibels in both ears due to damage to the inner ear. This level of hearing loss is roughly equivalent to listening to the television or a conversation through a substantial brick wall. In at least one ear this must be due to noise at work and claimants must have been employed for at least ten years in a specified noisy occupation.

There were just 90 new claims for work-related deafness under the Industrial Injuries Disablement Benefit Scheme in 2016.

Overview of the Control of Noise at Work Regulations 2005

[N3002A] The *Control of Noise at Work Regulations 2005 (SI 2005 No 1643)* came into force in April 2006 and gave effect to the European Physical Agents (Noise) Directive (2003/10/EC). Since 6 April 2008 they have also applied to the music and entertainment industry, where there has been a recent important case involving a personal injury claim for acoustic shock (see N3027.1 below).

With less than a year to go before the UK is scheduled to leave the European Union (EU) on 29 March 2019, the impact on UK health and safety law emanating from European directives as a result of Brexit is not yet clear. Negotiations are ongoing with the hope that there will be an agreement on the outline of the future relationship between the EU and UK in October 2018. At present, the UK remains a member of the EU and is covered by all its provisions, including health and safety legislation, until the exit negotiations are complete and a new relationship is defined.

The government's European Union (Withdrawal) Bill will end the primacy of EU law in the UK when it leaves the EU. This incorporates all EU legislation into one UK law which the government will be able to retain, amend or revoke post-Brexit.

The *Control of Noise at Work Regulations 2005* apply to places outside Great Britain where the *Health and Safety at Work, etc Act 1974* applies, such as offshore installations. However, the regulations do not normally apply to the master and crew of a seagoing ship.

The Regulations, which are fully described later in this chapter, emphasise a risk-based approach. This means that in some circumstances noise measurements may not be needed, while in other circumstances more stringent precautions may be necessary, including ongoing health surveillance (which includes hearing tests). Where hearing damage is identified, the employer must ensure a doctor examines the employee concerned to consider whether the damage is likely to have been caused by noise exposure, and if so, they must put in place a range of measures to reduce the risk to them and other employees.

Assessments must be carried out by a person who has 'sufficient information, instruction and training'. The Regulations require exposure to be assessed when it exceeds 80 dB(A) averaged over eight hours (in some cases, averaged over five eight-hour days). In addition, hearing protection must be made available at 80 dB(A) personal daily exposure, and information and training must also be provided. Health surveillance must be made available where exposure reaches 85 dB(A) and hearing protectors must be worn. There is an 87dB (A) limit on exposure and the exposure must be calculated including the effect of any hearing protection that is provided.

Hearing damage

[N3003] We all lose hearing acuity with age, and this loss is accelerated and worsened by exposure to excessive noise. Fortunately, in early life, we have much greater hearing acuity than modern life demands, and a small loss is readily compensated by turning up the volume of the radio or television.

Loud noises cause permanent damage to the nerve cells of the inner ear in such a way that a hearing aid is ineffective. At first, the damage occurs at frequencies above normal speech range, so the sufferer may have no inkling of the problem, although it could be identified easily by an audiogram which measures the sensitivity of the ear at a number of frequencies across its normal range. As exposure continues, the region of damage progresses to higher and lower frequencies. Damage starts to extend into the speech range, making it difficult to distinguish consonants, so that words start to sound the same. Eventually, speech becomes a muffled jumble of sounds.

Some people are much more susceptible to hearing damage than others. A temporary dullness of hearing or tinnitus (a ringing or whistling sound in the ears) when emerging from a noisy place are both indicative of damage to the nerve cells of the inner ear. Because these symptoms tend to disappear with continued exposure, sufferers may think their ears have become 'hardened' to the noise, whereas in fact they have lost their ability to respond to the noise. Even when hearing has apparently returned to normal, a little of the sensitivity is likely to have been lost. Therefore, anyone who experiences dullness of hearing or tinnitus after noise exposure must take special care to avoid further exposure. They may also be advised to seek medical advice and an audiogram.

Very loud impulsive sounds, such as those that may arise from cartridge-operated tools, firearms and explosives, can be particularly damaging to hearing, as they occur too suddenly for the ear's defensive mechanisms to operate. These can cause instantaneous damage to hearing.

Apart from such catastrophic exposures, the risk of hearing loss is closely dependent on the 'noise dose' received, and especially on the cumulative effect over a period of time. The noise exposure level (often referred to as the 'noise dose') takes account of both the sound pressure level (see below) and how long it lasts. British Standard *BS ISO 1999:2013 Acoustics. Estimation of noise-induced hearing loss* (which is currently under review) provides guidance in this area.

The perception of sound

[N3004] Sound is caused by a rapid fluctuation in air pressure. The human ear can hear a sensation of sound when the fluctuations occur between 20 times a second and 20,000 times a second. The rate at which the air pressure fluctuates is called the frequency of the sound and is measured in Hertz (Hz). The loudness of the sound depends on the amount of fluctuation in the air pressure. Typically, the quietest sound that can be heard (the threshold of hearing) is zero decibels (0 dB) and the sound becomes painful at 120 dB. Surprisingly perhaps, zero decibels is not zero sound. The sensitivity and frequency range of the ear vary somewhat from person to person and deteriorate with age and exposure to loud sounds.

The human ear is not equally sensitive to sounds of different frequencies (ie pitch): it tends to be more sensitive in the frequency range of the human voice than at higher or lower frequencies (peaking at about 4 kHz). When measuring sound, compensation for these effects can be made by applying a frequency weighting, usually the so-called 'A' weighting, although other weightings are sometimes used for specific purposes.

The ear has an approximately logarithmic response to sound: for example, every doubling or halving in sound pressure gives an apparently equal step increase or decrease in loudness. In measuring environmental noise, sound pressure levels are therefore usually quoted in terms of a logarithmic unit known as a decibel (dB). To signify that the 'A' weighting has been applied, the symbol of dB(A) is often used. However, current practice tends to prefer the weighting letter to be included in the name of the measurement index. For example, dB(A) L_{eq} and dB L_{Aeq} both refer to the 'A' weighted equivalent sound level in decibels.

Depending upon the method of presentation of two sounds, the human ear may detect differences as small as 0.5 dB(A). However, for general environmental noise the detectable difference is usually taken to be between 1 and 3 dB(A), depending on how quickly the change occurs. A 10 dB(A) change in sound pressure level corresponds, subjectively, to an approximate doubling or halving in loudness. Similarly, a subjective quadrupling of loudness corresponds to a 20 dB(A) increase in sound pressure level (SPL). When two sounds of the same SPL are added together, the resultant SPL is approximately 3 dB(A) higher than each of the individual sounds. It would require approximately *nine* equal sources to be added to an original source before the subjective loudness is doubled.

Noise indices

[N3005] The sound pressure level of industrial and environmental sound fluctuates continuously. A number of measurement indices have been proposed to describe the human response to these varying sounds. It is possible to measure the physical characteristics of sound with considerable accuracy and to predict the physical human response to characteristics such as loudness, pitch and audibility. However, it is not possible to predict *subjective* characteristics such as annoyance with certainty. This should not be surprising: one would not expect the light meter on a camera to be able to indicate whether one was taking a good or a bad photograph, although one would expect it to get the physical exposure correct. Strictly speaking, therefore, a meter can only measure sound and not noise (which is often defined as sound unwanted by the recipient): nevertheless, in practice, the terms are usually interchangeable.

Equivalent continuous 'A' weighted sound pressure level, L_{Aeq}

[N3006] This unit takes into account fluctuations in sound pressure levels. It can be applied to all types of noise, whether continuous, intermittent or impulsive. L_{Aeq} is defined as the steady, continuous sound pressure level which contains the same energy as the actual, fluctuating sound pressure level. In effect, it is the energy-average of the sound pressure level over a period which must be stated, eg $L_{Aeq, (8-hour)}$.

Levels measured in L_{Aeq} can be added using the rules mentioned earlier. There is also a time trade-off: if a sound is made for half the measurement period, followed by silence, the L_{Aeq} over the whole measurement period will be 3 dB less than during the noisy half of the period. If the sound is present for one-tenth of the measurement period, the L_{Aeq} over the whole measurement period will be 10 dB less than during the noisy tenth of the measurement period. This is a cause of some criticism of L_{Aeq}: for discontinuous noise, such as may arise in industry, it does not limit the maximum noise level, so it may be necessary to specify this as well.

Daily personal noise exposure level, $L_{EP,d}$

[N3007] This is used in the *Control of Noise at Work Regulations 2005 (SI 2005 No 1643))* as a measure of the total sound exposure a person receives during the day. It is formally defined in a *Schedule* to the Regulations.

$L_{EP,d}$ is the energy-average sound level (L_{Aeq}) to which a person is exposed over a working day, disregarding the effect of any ear protection which may be worn, adjusted to an eight-hour period. If a person is exposed to 90 dB L_{Aeq} for a four-hour working day, their $L_{EP,d}$ is 87 dB, but if they work a 12–hour shift, the same sound level would give them an $L_{EP,d}$ of 91.8 dB.

Weekly personal noise exposure level, $L_{EP,w}$

[N3007.1] This is the daily personal noise exposure level averaged over a five-day working period. For example, if a person has a daily personal noise exposure of 90 dB $L_{EP,d}$ on two and a half days, but had no noise exposure for the rest of the week, their $L_{EP,w}$ would be 87 dB. However, if they worked for six days a week, their $L_{EP,w}$ would be 90.8 dB.

Maximum 'A' weighted sound pressure level, $L_{Amax,T}$

[N3008] The maximum 'A'-weighted root-mean-square (rms) sound pressure level during the measurement period is designated L_{Amax}. Sound level meters indicate the rms sound pressure level averaged over a finite period of time. Two averaging periods (T) are defined, S (slow) and F (fast), having averaging times of 1 second and 1/8th second. It is necessary to state the averaging period. Maximum SPL should not be confused with peak SPL which is a measure of the instantaneous peak sound pressure.

Peak sound pressure

[N3009] Instantaneous peak sound pressure is used in the assessment of explosive sounds, such as from gunshots, cartridge-operated tools and blasting, and in hearing damage assessments of this type of sound. For the *Control of Noise at Work Regulations 2005 (SI 2005 No 1643)*, it is measured in C-weighted decibels. C-weighting is used because it responds equally to sounds over the wide range of frequencies that can occur in an explosive event. It can only be measured with specialist instruments that have special circuitry designed to respond to, and capture, very short, intense events.

Background noise level, L_{A90}

[N3010] L_{A90} is the level of sound exceeded for 90 per cent of the measurement period. It is therefore a measure of the background noise level, in other words, the sound drops below this level only infrequently. The term 'background' should not be confused with 'ambient', which refers to *all* the sound present in a given situation at a given time.

Sound power level, L_{WA}

[N3011] The sound output of an item of plant or equipment is frequently specified in terms of its sound power level. This is measured in decibels relative to a reference power of 1 pico-Watt (dB re 10^{-12} W), but it must not be confused with sound pressure level. The sound pressure level at a particular position can be calculated from a knowledge of the sound power level of the source, provided the acoustical characteristics of the surrounding and intervening space are known. As a crude analogy, the power of a lamp bulb gives an indication of its light output, but the illumination of a surface depends on its distance and orientation from the source, and the presence of reflecting and obstructing objects in the surroundings.

To give a rough idea of the relationship between sound power level and sound pressure level, for a noise source which is emitting sound uniformly in all directions close above a hard surface in an open space, the sound pressure level 10m from the source would be 28 dB below its sound power level.

General legal requirements relating to noise at work

[N3012] Statutory requirements relating to noise at work are contained in the *Control of Noise at Work Regulations 2005 (SI 2005 No 1643)*. The Regulations require employers to assess the risk of exposure to noise at work affecting the health and safety of their employees and to identify the measures needed to meet the requirements of the Regulations.

There are specific legal provisions relating to vibration exposure which are covered in the VIBRATION chapter. There are also provisions contained in the *Social Security (Industrial Injuries) (Prescribed Diseases) Regulations 1985 (SI 1985 No 967)* and the *Reporting of Injuries, Diseases and Dangerous Occurrences Regulations 2013 (SI 2013 No 1471)*. Occupational deafness is a prescribed industrial diseases for which disablement benefit is payable (although the 14 per cent disablement rule limits the number of successful claimants (see also **C6011** COMPENSATION FOR WORK INJURIES/DISEASES). Damages may also be awarded against the employer (see further **N3022** below).

Control of Noise at Work Regulations 2005 (SI 2005 No 1643)

[N3013] The following duties are laid on employers:

(1) To make and keep under review a suitable and sufficient assessment of the risk to the health and safety of employees who are liable to noise exposure at or above any **Lower Exposure Action Values** which are:
 (a) 80 dB(A) daily or weekly personal noise exposure; and
 (b) a peak sound pressure of 135 dB(C).
(2) To eliminate or reduce noise exposure to as low a level as is reasonably practicable. Where any employee is likely to be exposed above any **Upper Exposure Action Value (UEAV)**, they must reduce noise exposure to as low a level as is reasonably practicable by means of organisational and technical measures other than personal hearing protectors. The UEAVs are:
 (a) 85 dB(A) daily or weekly personal noise exposure; and
 (b) a peak sound pressure of 137 dB(C).
(3) To ensure that employees are not exposed to noise above any **Exposure Limit Values**, which are:
 (a) 87 dB(A) daily or weekly personal noise exposure; and
 (b) a peak sound pressure of 140 dB(C).
(4) If an Exposure Limit Value is exceeded, to immediately reduce exposure to below the Exposure Limit Value. The reason for exceeding the value must be identified and organisational and technical changes made to prevent it happening again.

In calculating exposure to an Exposure Limit Value, the effect of any personal hearing protectors provided by the employer may be taken into account. However, the effect of personal hearing protectors cannot be taken into account when assessing exposure to the Lower or Upper Exposure Action Values.

The risk assessment shall be conducted by assessing the noise exposure of employees and relating this to the Action and Limit values. The assessment shall be made by:

(a) observing specific working practices;
(b) reference to probable levels of noise from the equipment and particular working conditions;
(c) noise measurements, if necessary.

The risk assessment shall include consideration of the level, type and duration of exposure, including the peak sound pressure and the effect of exposure on people whose health is at particular risk from noise exposure.

Consideration must also be given to interactions such as between noise and vibration, noise and the use of substances at work that may be ototoxic (poisonous to the hearing mechanism). Indirect effects on health and safety such as the ability to hear audible warning systems or other sounds must also be considered.

The risk assessment must consider any information provided by the equipment manufacturers; the availability of quieter equipment or noise-reducing equipment; exposure to noise at the workplace beyond normal working hours, such as in rest-rooms; appropriate information from health surveillance, including published information; and the availability of adequate personal hearing protectors.

Employees or their representatives should be consulted on the risk assessment.

A record of the risk assessment must be kept. The record must include the significant findings of the risk assessment and the measures to be taken in order to eliminate or control the risks. This must include details of the provision of hearing protection and the provision of information and training to employees on the detection and prevention of hearing damage.

The risk assessment should be kept under regular review and also reviewed immediately where:

(a) there is reason to suspect that the risk assessment is no longer valid; or
(b) there has been a significant change in the work to which the assessment relates.

Changes should be made where the review shows that these are required.

The regulations require health surveillance (ie hearing checks) for employees who may have a health risk arising from their noise exposure.

The Health and Safety Executive (HSE) advises that health surveillance for hearing damage usually means carrying out regular hearing checks in controlled conditions; telling employees about the results of their hearing checks; keeping health records; and ensuring employees are examined by a doctor where hearing damage is identified.

It says: "Ideally, you would start the health surveillance before people are exposed to noise (ie for new starters or those changing jobs), to give a baseline. It can, however, be introduced at any time for employees already exposed to noise. This would be followed by a regular series of checks, usually annually for the first two years of employment and then at three-yearly intervals (although this may need to be more frequent if any problem with hearing is detected or where the risk of hearing damage is high). The hearing checks need to be carried out by someone who has the appropriate training. The whole health surveillance programme needs to be under the control of an occupational health professional (for example a doctor or a nurse with appropriate training and experience)."

The employer has responsibility for making sure the health surveillance is carried out properly.

Hearing protection is to be made available on request to any employee exposed above the lower exposure action value, and must be provided to any employee exposed at or above the upper exposure action level. Any area where an employee is likely to be exposed at or above the upper action level must be designated as a hearing protection zone, and no employee should enter that area unless they are wearing ear protectors.

Personal hearing protectors must be selected to eliminate the risk to hearing or reduce it to as low a level as is reasonably practicable and after consultation with employees or their representatives. In addition, it must comply with the relevant requirements of the *Personal Protective Equipment Regulations 2002* or the directly acting European Regulation (EU) 2016/425 on personal protective equipment, which came into force in April 2018. A Hearing Protection Zone should be demarcated and identified by means of the sign specified for the purpose of indicating that ear protection must be worn in the *Health and Safety (Safety Signs and Signals) Regulations 1996* (see image below):

(From the *Health and Safety (Safety Signs and Signals) Regulations 1996 (SI 1996 No 341)*)

Employers must ensure that personal ear protection is fully and properly used and that noise control equipment and ear protectors are maintained in efficient, working order and in good repair.

Where employees are exposed to noise which is likely to be at or above a lower exposure action value, employer must provide those employees and their representatives with 'suitable and sufficient information, instruction and training'. This should include:

- the nature of risks from exposure to noise;

- the organisational and technical measures taken to comply with the requirements of the regulations;
- the exposure limit values and upper and lower exposure action values;
- the significant findings of the risk assessment, including any measurements taken, with an explanation of those findings;
- the availability and provision of personal hearing protectors and their correct use;
- why and how to detect and report signs of hearing damage;
- the entitlement to health surveillance and its purposes;
- safe working practices to minimise exposure to noise; and
- the collective results of any health surveillance undertaken.

Legal obligations of designers, manufacturers, importers and suppliers of plant and machinery

[N3014] The *Supply of Machinery (Safety) Regulations 2008 (SI 2008 No 1597* as amended) require manufacturers and suppliers of noisy machinery to design and construct such machinery so that the risks from noise emissions are reduced to the lowest level taking account of technical progress. Information on noise emissions must be provided when specified levels are reached.

The regulations require that instruction manuals include at least the following information on airborne noise emissions:

— "the A-weighted emission sound pressure level at workstations, where this exceeds 70 dB(A); where this level does not exceed 70 dB(A), this fact must be indicated,
— the peak C-weighted instantaneous sound pressure value at workstations, where this exceeds 63 Pa (130 dB in relation to 20µPa),
— the A-weighted sound power level emitted by the machinery, where the A-weighted emission sound pressure level at workstations exceeds 80 dB(A).

These values must be either those actually measured for the machinery in question or those established on the basis of measurements taken for technically comparable machinery which is representative of the machinery to be produced.

In the case of very large machinery, instead of the A-weighted sound power level, the A-weighted emission sound pressure levels at specified positions around the machinery may be indicated.

Where the harmonised standards are not applied, sound levels must be measured using the most appropriate method for the machinery. Whenever sound emission values are indicated the uncertainties surrounding these values must be specified. The operating conditions of the machinery during measurement and the measuring methods used must be described.

Where the workstation(s) are undefined or cannot be defined, A-weighted sound pressure levels must be measured at a distance of 1 metre from the surface of the machinery and at a height of 1.6 metres from the floor or access platform. The position and value of the maximum sound pressure must be indicated."

A 'harmonised standard' means a non-binding technical specification adopted by one of a number of European standardisation bodies. The *Supply of Machinery (Safety) Regulations 2008 (SI 2008 No 1597* as amended) also set out that where specific directives lay down other requirements for the measurement of sound pressure levels or sound power levels, the provisions of those directives must be applied and the corresponding provisions of this section shall not apply.

Specific legal requirements regarding noise at work

Agriculture

[N3015] The *Agriculture (Tractor Cabs) Regulations 1974 (SI 1974 No 2034)* (as amended) provide that noise levels in tractor cabs must not exceed 90 dB(A) or 86 dB depending on which annex is relevant in the certificate under Directive 77/311/EEC. [*SI 1974 No 2034, Reg 3(3)*].

Equipment for construction sites and other outdoor uses

[N3016] The *Noise Emission in the Environment by Equipment for Use Outdoors Regulations 2001 (SI 2001 No 1701)* apply to a very wide range of powered equipment used outdoors. They do not apply to most non-powered equipment, equipment for the transport of goods or persons by road, rail, air or on waterways, or equipment for use by military, police or emergency services.

Each type of equipment is subject to a maximum permissible sound power level, dependent on the power of the equipment's engine or other drive system.

Examples of the type of equipment covered include: cranes, hoists, excavators, dozers, dump trucks, compressors, generators, pumps, power saws, concrete breakers, compactors, glass recycling containers, lawnmowers and other gardening equipment.

The Regulations define the noise testing methods in great detail and also require a standardised marking of conformity indicating the guaranteed sound power level.

Code of practice for noise and vibration control from construction and open sites

[N3017] British Standard BS 5228-1:2009+A1:2014 (which was amended in 2014) *Code of practice for noise and vibration control on construction and open sites. Noise* provides information on protection against noise and vibration for anyone living near or working on a building site.

Compensation for occupational deafness

Industrial injuries disablement benefit

[N3018] The most common condition associated with exposure to noise is occupational deafness. Occupational deafness is prescribed occupational disease A10 (see **OCCUPATIONAL HEALTH AND DISEASES**). It is defined as: "Sensorineural hearing loss amounting to at least 50 dB in each ear, being the average of hearing losses at 1, 2 and 3 kHz frequencies, and being due in the case of at least one ear to occupational noise (occupational deafness)" (*Social Security (Industrial Injuries) (Prescribed Diseases) Regulations 1985 (SI 1985 No 967) (as amended)*).

Conditions for which deafness is prescribed

[N3019] Occupational deafness is prescribed for a wide range of occupations involving the use of, or work wholly or mainly in the immediate vicinity of the use of, a:

- a band saw, circular saw or cutting disc to cut metal in the metal founding or forging industries, circular saw to cut products in the manufacture of steel, powered (other than hand-powered) grinding tool on metal (other than sheet metal or plate metal), pneumatic percussive tool on metal, pressurised air arc tool to gouge metal, burner or torch to cut or dress steel-based products, skid transfer bank, knock out and shake out grid in a foundry, machine (other than a power press machine) to forge metal including a machine used to drop stamp metal by means of closed or open dies or drop hammers, machine to cut or shape or clean metal nails, or plasma spray gun to spray molten metal;
- a pneumatic percussive tool to drill rock in a quarry, on stone in a quarry works, underground, for mining coal, for sinking a shaft, or for tunnelling in civil engineering works;
- a vibrating metal moulding box in the concrete products industry, or circular saw to cut concrete masonry blocks;
- a machine in the manufacture of textiles for weaving man-made or natural fibres (including mineral fibres), high speed false twisting of fibres, or the mechanical cleaning of bobbins;
- a multi-cutter moulding machine on wood, planing machine on wood, automatic or semi-automatic lathe on wood, multiple cross-cut machine on wood, automatic shaping machine on wood, double-end tenoning machine on wood, vertical spindle moulding machine (including a high speed routing machine) on wood, edge banding machine on wood, bandsawing machine (with a blade width of not less than 75 millimetres) on wood, circular sawing machine on wood including one operated by moving the blade towards the material being cut, or chain saw on wood;
- a jet of water (or a mixture of water and abrasive material) at a pressure above 680 bar, or jet channelling process to burn stone in a quarry;

Compensation for occupational deafness [N3021]

- a machine in a ship's engine room, or gas turbine for performance testing on a test bed, installation testing of a replacement engine in an aircraft, or acceptance testing of an Armed Service fixed wing combat aircraft;
- a machine in the manufacture of glass containers or hollow ware for automatic moulding, automatic blow moulding, or automatic glass pressing and forming;
- a spinning machine using compressed air to produce glass wool or mineral wool;
- a continuous glass toughening furnace;
- a firearm by a police firearms training officer; or
- a shot-blaster to carry abrasives in air for cleaning.

The Regulations are detailed and should be consulted for precise definitions of these occupations. [*Social Security (Industrial Injuries) (Prescribed Diseases) Regulations 1985 (SI 1985 No 967)* (as amended)].

Conditions under which benefit is payable

[N3020] For a claimant to be entitled to disablement benefit for occupational deafness, the following conditions under the *Social Security (Industrial Injuries) (Prescribed Diseases) Regulations 1985 (SI 1985 No 967)* (as amended) currently apply:

- they must have been employed:
  (i) at any time on or after 5 July 1948, and
  (ii) for a period or periods amounting (in the aggregate) to at least ten years.
- there must be "sensorineural hearing loss amounting to at least 50 dB in each ear, being the average of hearing losses at 1, 2 and 3 kHz frequencies, and being due in the case of at least one ear to occupational noise (occupational deafness). [*SI 1985 No 967, Sch 1, Pt 1*]; and
  (There is a presumption that occupational deafness is due to the nature of employment where the person has worked in one of the occupations/types of work listed [*SI 1985 No 967, Reg 4(2)*]); and
- the claim must be made within five years of the last date when the claimant worked in an occupation prescribed for deafness [*SI 1985 No 967, Reg 25(2)*].

Assessment of disablement benefit for social security purposes

[N3021] The extent of disablement is the percentage calculated by:

(a) determining the average total hearing loss due to all causes for each ear at 1, 2 and 3 kHz frequencies; and
(b) determining the percentage degree of disablement for each ear; and then
(c) determining the average percentage degree of binaural disablement.

[Social Security (Industrial Injuries) (Prescribed Diseases) Amendment Regulations 1989, Reg 34(2)].

The following chart shows the scale for all claims made on/after 3 September 1979.

Table 1
Percentage degree of disablement in relation to hearing loss

Hearing loss	Percentage degree of disablement
50–53 dB	20
54–60 dB	30
61–66 dB	40
67–72 dB	50
73–79 dB	60
80–86 dB	70
87–95 dB	80
96–105 dB	90
106 dB or more	100

[*Social Security (Industrial Injuries) (Prescribed Diseases) Regulations 1985, Reg 34, Sch 3, Pt 2,* as amended]

Any degree of disablement, due to deafness at work, assessed at less than 20 per cent, must be disregarded for benefit purposes. [*Social Security (Industrial Injuries) (Prescribed Diseases) Amendment Regulations 1990 (SI 1990 No 2269)*].

Action against employer at common law

[N3022] There is no separate action for noise at common law; liability comes under the general heading of negligence. It was not until as late as 1972 that employers were made liable for deafness negligently caused to employees (*Berry v Stone Manganese Marine Ltd* [1972] 1 Lloyd's Rep 182).

In *Carragher v Singer Manufacturing Co Ltd* 1974 SLT (Notes) 28 reference was made to the *Factories Act 1961, s 29*: 'every place of work must, so far as is reasonably practicable, be made and kept safe for any person working there', to the effect that this was wide enough to provide protection against noise).

Noise should be regarded as an aspect of the working environment (*McCafferty v Metropolitan Police District Receiver* [1977] 2 All ER 756. Here the plaintiff, who was a ballistics expert, suffered ringing in the ears as a result of the sounds of ammunition being fired from different guns in the course of his work. When he complained about ringing in the ears – the ballistics room had no sound-absorbent material on the walls and he had not been supplied with ear protectors – he was advised to use cotton wool. It was held that his employer was liable since it was highly foreseeable that the employee would suffer hearing injury if no steps were taken to protect his ears, cotton wool being useless.

Although there are specific statutory requirements to minimise exposure to noise (in agriculture and offshore operations and construction operations, see **N3015, N3016** above), these have generated little or no case law.

The main points established at common law are as follows.

Action against employer at common law [N3022]

(a) As from 1963, the publication date by the (then) Factory Inspectorate of 'Noise and the Worker', employers have been 'on notice' of the dangers to hearing of their employees arising from over-exposure to noise (*McGuinness v Kirkstall Forge Engineering Ltd* (1979), unreported). Hence, consistent with their common law duty to take reasonable care for the health and safety of their employees, employers should 'provide and maintain' a sufficient stock of ear protectors.

This was confirmed in *Thompson v Smiths*, etc (see (*d*) below). However, more recently, an employer was held liable for an employee's noise-induced deafness, even though the latter's exposure to noise, working in shipbuilding, had occurred *entirely before* 1963. The grounds were that the employer had done virtually nothing to combat the *known* noise hazard from 1954–1963 (apart from making earplugs available) (*Baxter v Harland & Wolff plc* [1990] IRLR 516, NI CA). This means that, as far as Northern Ireland is concerned, employers are liable at common law for noise-induced deafness as from 1 January 1954 – the earliest actionable date. Limitation statutes preclude employees suing prior to that date (*Arnold v Central Electricity Generating Board* [1988] AC 228).

(b) Because the true nature of deafness as a disability has not always been appreciated, damages have traditionally not been high (*Berry v Stone Manganese Marine Ltd* [1972] 1 Lloyd's Rep 182 – £2,500 (halved because of time limitation obstacles); *Heslop v Metalock (Great Britain) Ltd* (1981) Observer, 29 November – £7,750; *O'Shea v Kimberley-Clark Ltd* (1982) Guardian, 8 October – £7,490 (tinnitus); *Tripp v Ministry of Defence* [1982] CLY 1017 – £7,500).

(c) Damages will be awarded for exposure to noise, even though the resultant deafness is not great, as in tinnitus (*O'Shea v Kimberley-Clark Ltd*, (1982) Guardian, 8 October).

(d) Originally the last employer of a succession of employers (for whom an employee had worked in noisy occupations) was exclusively liable for damages for deafness, even though damage (ie actual hearing loss) occurs in the early years of exposure (for which earlier employers would have been responsible) (*Heslop v Metalock (Great Britain) Ltd* (1981) Observer, 29 November). More recently, however, the tendency is to *apportion* liability between offending employers (*Thompson, Gray, Nicholson v Smiths Ship Repairers (North Shields) Ltd*; *Blacklock, Waggott v Swan Hunter Shipbuilders Ltd*; *Mitchell v Vickers Armstrong Ltd* [1984] IRLR 93). This is patently fairer because some blame is then shared by the original employer(s), whose negligence would have been responsible for the actual hearing loss.

(e) Because of the current tendency to apportion liability, even in the case of pre-1963 employers (see *McGuinness v Kirkstall Forge Engineering Ltd* above), contribution will take place between earlier and later insurers.

(f) Although judges are generally reluctant to be swayed by scientific/statistical evidence, the trio of shipbuilding cases (see (*d*) above) demonstrates, at least in the case of occupational deafness, that this trend is being reversed (see Table 2 below, the 'Coles-Worgan classification'); in particular, it is relevant to consider the 'dose response' relationship published by the National Physical Laboratory (NPL), which relates long-term continuous noise exposure to expected resultant hearing loss. This graph always shows a rapid increase in the early years of noise exposure, followed by a trailing off.

(g) Current judicial wisdom identifies three separate evolutionary aspects of deafness, ie (i) hearing loss (measured in decibels at various frequencies); (ii) disability (ie difficulty/inability to receive everyday sounds); (iii) "social handicap" (attending musical concerts/meetings etc). That "social handicap" is a genuine basis on which damages can be (*inter alia*) awarded, was reaffirmed in the case of *Bixby, Case, Fry and Elliott v Ford Motor Group* (1990), unreported.

The case of *Baker v Quantum Clothing Group* [2011] UKSC 17, [2011] 4 All ER 223, [2011] 1 WLR 1003 was an appeal to the Supreme Court and concerned the liability of employers in the knitting industry of Derbyshire and Nottinghamshire for hearing loss suffered by employees prior to 1 January 1990. The central issue was whether liability existed at common law in negligence and/or under s 29(1) of the *Factories Act 1961* towards an employee who suffered noise-induced hearing loss due to exposure to noise levels between 85 and 90dB(A) LEP,d.

The court ruled that liability did not exist at common law and/or under s 29(1) of the Act, towards an employee who could establish that an employer had not implemented measures to protect their employees in respect of noise exposure at levels below 90dB(A) Lep,d prior to 1 January 1990 and the decision of the Court of Appeal [2009] All ER (D) 205 (May) was reversed.

In *Goldscheider v Royal Opera House* [2018] EWHC 687 (QB), the High Court ruled that the Royal Opera House had failed to protect the hearing of its musicians and had caused acoustic shock to a former professional viola player. The case is important because the court examined the music industry's legal obligations to protect the hearing of musicians under the Control of Noise at Work Regulations 2005 and because it recognised acoustic shock as a compensatable condition for the first time (see **N3027.1** below).

General guidance on noise at work

Guidance on the Control of Noise at Work Regulations 2005

[N3023] The HSE provides guidance on the Control of Noise at Work Regulations in the publication *Controlling noise at work (L108)*. It covers the following areas:

- the management and control of workplace noise;
- legal duties of employers to prevent damage to hearing;
- guidance on the assessment and management of noise risks;
- practical advice on noise control;
- selection of quieter tools and machinery;
- selection use, care and maintenance of hearing protection; and
- health surveillance.

Further resources, including free leaflets and priced publications that can also be downloaded free can be found on the HSE noise microsite at www.hse.gov.uk/noise/index.htm.

Reducing noise in specific working environments

[N3024]–[N3026] Although many noisy industries are in decline, noise is now becoming an issue in some newer industries and service sectors. In particular, the application of the Regulations to the music and entertainment industries posed new challenges, not least because operators have two conflicting needs: their clients may want loud music, but a safe working environment must be provided for staff.

Music and Entertainment

Pubs and clubs

[N3027] The HSE's Sound Advice website (www.soundadvice.info) provides guidance on the steps employers can take to reduce exposure to noise in pubs and clubs where live amplified music is performed or loud amplified recorded music is played:

These include:

- Acoustic controls: to help to absorb reverberant noise;
- Physical separation: to keep employees away from noisy areas;
- Direction controls: to point the sound where it is wanted - the dance floor or performance area, and away from areas such as bars;
- Volume control: to keep all equipment in good working condition and make it clear who can use the volume controls;
- Managing exposure: to reduce the length of time to which individuals are exposed;
- Information, instruction and training: to make sure everyone understands what you are doing about noise, and why it is important;
- Hearing protection: but only after all other ways of controlling noise have been considered or while you are implementing a more permanent solution; and
- Hearing health checks: to make sure people at risk are regularly monitored.

More information can be found on the Sound Advice website at: www.soundadvice.info/pubsandclubs/pubsandclubs-step2.htm.

Concert halls and theatres

[N3027.1] The HSE advises that musicians playing in an orchestra or a band are exposed to high levels of noise, whether they are on or off stage or in a pit, and that exposure to live music can cause hearing damage.

Employers and employees working in concert halls and theatres have responsibilities to protect the hearing of all employees, including musicians, performers and crew, including guest performers. Actors and other workers on stage can be exposed to high levels of noise caused by on-stage bands and choruses.

The length of exposure is as important as noise level and, says the HSE, noise exposures in orchestra pits can quickly exceed safe levels.

It sets out a number of control measures that employers can put into place to reduce exposure to harmful levels of noise. These include:

- Concert, show and tour planning: including making sure the repertoire or show is appropriate for the venue and varying the repertoire of concerts;
- Acoustic controls: including screens, drapes and floor coverings;
- Layout and position: allowing enough space between performers and using risers so that the sound projects over other players' heads;
- Information, instruction and training: regular discussions with musicians, crew and managers; and
- Hearing protection: but only after all the other ways of controlling noise have been considered, or while a more permanent solution is being implemented;
- Hearing health checks: make sure people at risk are regularly monitored;
- Designing, altering and building venues: noise can be controlled by careful design of the premises; and
- Venue operators: can take steps to control noise levels.

In *Goldscheider v Royal Opera House* [2018] EWHC 687 (QB) the High Court ruled the Royal Opera House had failed to protect the hearing of its musicians and had caused acoustic shock to former professional viola player Chris Goldscheider. The case is important not only because the court examined the music industry's legal obligations to protect the hearing of musicians under the Control of Noise at Work Regulations 2005, but also because it recognised acoustic shock as a compensatable condition for the first time.

Acoustic shock is caused by the sudden onset of unexpected noise, often delivered at a very intense frequency. Expert evidence presented to the court set out that 'it is important to distinguish acoustic shock from acute acoustic trauma that is experienced with exposure to extremely loud sounds, over 140dB' and that 'acoustic shock is unrelated to noise-induced hearing loss, in which repeated exposure to sounds of an intensity greater than 85dB causes cochlear damage'. The court heard that a range of 82dB to 120dB is sufficient to cause acoustic shock.

Goldscheider claimed damages after the noise levels he was exposed to while seated in front of the brass section of the orchestra during a rehearsal of Wagner's Ring Cycle in September 2012 were so high that his hearing was immediately and irreversibly damaged. He was unable to continue working because of his injuries.

The court found the Royal Opera House had breached the Control of Noise at Work Regulations 2005 in a number of ways, including: failing to carry out a suitable and sufficient risk assessment; failing to take reasonably practicable measures to reduce the risk of noise at the rehearsal; and failing to properly inform and instruct musicians about the 'imperative nature' of the need to wear hearing protection in what should have been a designated area. Damages were to be decided at a later date.

Outdoor music venues

[N3027.2] Outdoor music venues require considerable thought and attention, as the correct balance between audience sound levels and environmental noise must be obtained, at the same time as ensuring that noise exposure of staff is controlled. This will require the various staff duties to be identified and the noise exposure of each duty to be separately assessed. External staff must be included in the assessment. It will be important to ensure that everyone is made aware of the risks and the control measures necessary to manage the risks.

Detailed HSE guidance on controlling and reducing exposure to noise at events is available on its website at: www.hse.gov.uk/event-safety/noise.htm.

Call centres

[N3028] Call centres can be very noisy simply because of the large number of people talking on the telephone. If poorly designed, each operator will inevitably raise their voice so as to be heard above the noise produced by their neighbours. This becomes self-defeating as others then raise their voices further. They may also need to raise the volume of their earpiece, and this could lead to excessive sound exposure. There have been successful cases of compensation for deafness (sometimes only in one ear) suffered by people using telephone and radio headsets, and employers should ensure that staff do not adjust them to be excessively loud.

A number of techniques must be used to achieve a satisfactory acoustic environment. Firstly, the basic acoustics of the call centre must be satisfactory. Working areas should be as sound-absorbent as feasible, with acoustic ceiling tiles and a thick, sound absorbent carpet. Acoustic screens should be used between workstations unless they are widely-spaced. The screens need to have thick absorbent surfaces, and should be high enough to screen operators when they are sitting at their workstations.

Staff training must include discussion of appropriate voice and earpiece levels, and the risks of excessive levels. They need to be encouraged to keep their voices down: it should be explained that it is not necessary for them to raise their voices in order for the customer to hear them clearly.

The HSE has reported concerns regarding acoustic shock – a term used in connection with incidents involving exposure to short duration, high frequency, high-intensity sounds through a telephone headset (see **N3027.1**). It says that it has not been established whether the reported symptoms are caused directly by exposure to these unexpected sounds. There is no clear single cause of these incidents, but one cause may be interference on the telephone line.

It advises that although call handlers may be shocked or startled by the sounds, exposure to them should not cause hearing damage as assessed by conventional methods. It reports that since 1991, major manufacturers have incorporated an acoustic limiter in the electronics of their headsets to ensure that any type of noise (eg conversation, short duration impulses) above 118 dB is not transmitted through the headset.

The HSE considers that, in general, call handlers' daily personal noise exposure is unlikely to exceed the 80 dB lower exposure action value defined in the Control of Noise at Work Regulations 2005, provided good practice in the management of noise risks is followed. It advises that call handlers should be encouraged to report to management exposure to acoustic shock incidents and management should keep a record of these reported events.

Although *Goldscheider v Royal Opera House* [2018] EWHC 687 (QB) (see N3027.1 above) is the first case in which the High Court has allowed a claim for compensation for acoustic shock, there have been out-of-court settlements in the telecoms sector. A communications CWU union official estimated the total compensation figure had reached around £2 million by the end of September 2012. Goldscheider was reported to be seeking £750,000 in damages.

Workplace noise assessments

Competent person

[N3029] The *Control of Noise at Work Regulations 2005 (SI 2005 No 1643)* require that any person who carries out work in connection with the employer's duties under the Regulations has 'suitable and sufficient information, instruction and training' (Regulation 10(4)). The Regulations do not stipulate a particular form of certification for such people, although the Institute of Acoustics (www.ioa.org.uk/education-and-training) and other organisations provide formal training and certification intended to meet these requirements.

Sound level meters

[N3030] It may be more economical to employ an acoustics consultant for workplace noise assessments than to send a member of staff on a training course and to hire a sound level meter to do the measurements in-house. Suitable sound level meters are expensive. A precision meter is required and these cost thousands of pounds, depending on their capabilities, and they must be sent away for laboratory calibration at regular intervals. Those meters that can be bought cheaply from electronics suppliers are not suitable: they could be misleading even as a 'screening' tool. This is because cheap meters do not have an accurate response to rapid or impulsive sounds, and usually do not contain an L_{Aeq} function as this requires an on-board computer chip for the complex averaging process.

HSE guidance (in L108) is as follows: "Anyone who helps you comply with your duties under the Noise Regulations (eg by making noise measurements, determining exposures or planning for control of risk through changes to industrial processes or working practices) must be competent to undertake the task. Whether you employ a consultant or use members of your staff for these purposes you must satisfy yourself of their competence and provide them with any information on the work necessary for them to undertake the tasks."

It also advises that the sound level meter should be an integrating sound level meter and should meet at least Class 2 of BS EN 61672-1 (the current instrumentation standard for sound level meters), or at least Type 2 of BS EN 60804:2001 (the former standard).

(The current standard is BS EN 61672-1:2013 *Electroacoustics. Sound level meters. Specifications.*)

Assessing the daily personal noise exposure

[N3031] It is important to appreciate that noise assessments must determine the daily personal noise exposure of each individual member of staff – it is not sufficient simply to measure the noise level in various parts of the workplace and to mark the noisy areas as 'hearing protection zones'.

In some workplaces, this is not particularly difficult to do: if the employee has a fixed workstation and fixed working routine, it will only be necessary to measure at a location representative of their head position, for a sufficient period of time to cover a few work cycles.

However, it is quite common for a worker to use a variety of machines or to move between different parts of the works during a normal day. Some workers, such as crane or lorry drivers, may have a work environment where noise levels vary a great deal and it is not practicable for the noise surveyor to accompany them throughout a representative period.

These considerations mean that a workplace noise assessment needs planning before it is made. First, establish the different types of work that are being done on the site. This will need a meeting with the site manager to determine how many staff there are, the jobs they do, shift-working patterns, and their hours of work, including when and where meal-breaks are taken. It is also important to involve the workforce. It is helpful for the surveyor to explain what they are doing and to confirm what the workforce does. They will also want to know the results, but the surveyor should explain this cannot be provided immediately as the daily personal noise exposure needs to be calculated. Where they are present in the workplace, trade union safety representatives should be consulted.

A scale plan of the workplace will be needed so that measurement locations, noise sources and noisy areas can be recorded for the report. The surveyor should ensure all machinery and plant is tested, and note anything that is out of use during the survey.

Dosemeters

[N3032] Where there are people such as lorry drivers, who the surveyor cannot accompany, it will probably be best to ask them to wear a dosemeter.

This is a small electronic device about the size of a mobile phone that clips into the top pocket, with a microphone that clips to the lapel or some convenient part of the clothing near to the ear. The dosemeter does not record the actual sound, only the sound level. It is not a 'spy in the cab'. Dosemeters usually contain some protection to deter tampering or falsification and to prevent the worker obtaining the readout.

Where workers move between various workstations during their working day, it is usually acceptable to measure the noise level at each workstation and calculate the daily personal noise exposure from an estimate of the proportion of the day each worker typically spends at each one.

Although it may seem that this is less accurate than using a dosemeter, it is important to recognise that each of the work-cycles will usually produce a slightly different noise exposure, as there are so many things that affect the noise levels. Moreover, it is necessary to obtain actual workstation noise levels in order to identify any ear protection zones that need to be marked out. In some borderline cases, dosemetry may settle the matter, but it would be better to look at ways of reducing noise levels, since the effect on any one individual's hearing is difficult to determine, even though the Regulations define action levels precisely.

Hazardous environments

[N3033] Special care is required in hazardous working environments, especially those with an explosion risk (such as gas and petroleum processing plants). Many sound level meters use a high 'polarisation voltage' for the microphone, which could produce a spark. 'Intrinsically safe' meters can be obtained for such instances, although they can be less convenient for ordinary use.

Requirements for the report

[N3034] A written report of the survey must be produced and kept as a record at least until the next survey is made. The report should state all the information used in the survey, and should contain a plan showing the areas surveyed and the hearing protection zones identified. It must identify individual staff or jobs where noise exposure exceeds any of the action levels. The report should also contain a statement of the employer's duties resulting from the survey findings. However, it is not usual for the report to include detailed specifications for any noise remediation that might be needed. This is usually a separate study, as the issues cannot be known prior to the survey and there will usually be a number of options which will require detailed discussion with the employer.

Noise reduction

[N3035] The HSE advises that noise-reduction programmes are only likely to be effective if they include a positive purchasing policy which makes sure noise is taken into account when selecting machinery.

HSE guidance (in L108) advises employers that when buying, hiring or replacing equipment they should ask potential suppliers for information on the noise emission of machines under the conditions of intended use, and use that information to compare machines.

Where it is found to be necessary to purchase machinery which causes workers to be exposed over the action levels set out in the regulations, keeping a record of the reasons for the decision will help guide future action, eg by providing those responsible for future machine specifications with information on improvements that are needed.

Selecting low-cost tools and machinery through a positive purchasing and hire policy can avoid the need to apply retrofit noise control and 'could be the single most cost effective, long-term measure' an employer can take to reduce noise, according to the HSE.

It suggests such a policy could include:

- preparing a machine specification;
- drawing suppliers' attention to the requirements of the *Supply of Machinery (Safety) Regulations 2008* (see above);
- introducing a company noise limit – a realistic low-noise emission level that the company is prepared to accept from incoming plant and equipment given the circumstances and planned machine use;
- comparing the noise information declared by the manufacturer to identify low-noise machines;
- requiring a statement from all companies who are tendering or supplying, saying if their machinery will meet the company noise limit specification;
- discussing noise issues with the supplier of the machine, which may influence the design of future low-noise machines;
- where it is necessary to purchase noisy machinery, keeping a record of the reasons for decisions made to help with the preparation of future machine specifications with information on where improvements are necessary;
- using an agreed format for the presentation of results by suppliers; and
- discussing machinery needs and noise emission levels with safety or employee representatives.

Hearing protection

[N3036] There are two main categories of hearing protection:

(i) circumaural protectors (known as ear muffs) which fit over and surround the ears and seal to the head by cushions filled with soft plastic foam or a viscous liquid; and
(ii) ear plugs, which fit into the ear canal.

Hearing protection will only be effective if it is in good condition, suits the individual and is worn properly. It can be uncomfortable, especially if it presses too firmly on the head or ear canal, or cause too much sweating. Some users may be tempted to bend the head-band so the muffs do not press so tightly, which reduces the effectiveness of the seal. Ear muffs are also less effective for people with thick spectacle frames, long hair or beards that prevent the muff from sealing fully against the head.

The amount of protection differs between different designs, and usually gives more protection at middle frequencies and less protection at low frequencies. Where the noise has strong tones (often described by words such as rumble, drone, hum, whine, screech) then particular care is needed to ensure the protection gives adequate attenuation at those frequencies. However, it can be counterproductive to choose protection that gives far more attenuation than necessary, since it will be heavier, with greater head-band or insert pressure and so less comfortable to wear. Hearing protection that is not worn does not give any protection at all.

Ear plugs should not be used by people with ear infections and certain other ear conditions, and they may need to obtain medical advice.

Because individuals differ greatly in their preferences, the employer should select more than one type of suitable protector and offer the user a personal choice where possible. British Standard BS EN 458: 2016 gives guidance and recommendations for the selection, use, care and maintenance of hearing protectors.

Ear protectors need to be maintained in a clean and hygienic condition. Insert types and muffs should not be shared between people. With ear muffs, it is necessary to check the sealing cushion is not damaged, as this will seriously reduce its effectiveness. On some makes, it is possible to change a damaged cushion. The ear cups have a foam lining which must not be discarded. If it gets dirty, it can usually be taken out and washed in detergent, then dried and replaced.

If the headband tension becomes weak, the muffs should be discarded.

Some earplugs are designed for one-off use, but others are reusable. These need to be kept in a protective container when not in use, and regularly washed according to the manufacturer's instructions. If the seals become damaged or hardened, the plugs must be replaced.

The employer should ensure workers know how to check the condition of their ear protection and how to obtain replacements as soon as necessary.

Acoustic booths

[N3037] In some workplaces, it may be possible for operators to be located in a control booth from which they can monitor the plant, only occasionally entering the noisy area. They may still need to wear ear defenders when leaving the acoustic booth, since it may require only a few seconds or minutes to exceed the allowable daily noise dose.

Even where it is not practical to have a separate control booth, it may be possible to use an acoustic refuge – a small cabin with an open side – that will give some respite from the noise. However, this should not be expected to give more than a nominal sound reduction in most cases.

Work rotation

[N3038] Work rotation can sometimes be a method of reducing the daily personal noise exposure. The staff work part of the day in a noisy area and the rest of the day in a quiet area. If a person works for half their day in a place with a noise level of 88 dB L_{Aeq} (four hours) and the other half of their day in a place with a noise level of 70 dB L_{Aeq} (4 hours) they would just meet a noise exposure of 85 dB. However, this is not a very satisfactory method as it can easily go wrong. They may spend longer than expected in the noisy workplace, or there may be a higher than expected noise level in the quiet area, for example. Again, while this approach could satisfy the legal requirements, some people may suffer hearing loss from shorter exposure to loud noise.

Noise barriers

[N3039] Noise barriers often give very limited noise reduction within buildings. This is because sound can be reflected over the barrier from the ceiling, pipe-work, and other overhead fittings. It is usually necessary to use noise barriers in conjunction with sound baffles suspended from the ceiling. Suspended sound baffles on their own have very limited application. They can sometimes be helpful in very reverberant work-spaces where the staff are not exposed to the direct sound of noisy machines.

PA and music systems

[N3040] Some workplaces use public address and music systems. One study found that they were a major source of excessive noise in lorry cabs. Care must be taken to ensure that in noisy workplaces these systems are not so loud that they create excessive noise in themselves. This also applies to people who wear radio or telephone headsets.

Sound systems for emergency purposes

[N3041] Voice messages can be superior to bells or sirens to convey warnings and instructions in emergencies. However, these must be properly audible in noisy places. British Standard BS EN 50849:2017 gives specifications for sound systems for emergency purposes.

Occupational Health and Diseases – An Overview

Leslie Hawkins and Andrea Oates

What is occupational health?

[O1001] Occupational health is concerned with the prevention, identification and management of health issues relating to employment. It is very broad and covers not only ill-health caused by work and the conditions of work, but also the management of employees who have a health problem which may impact on their work or their employment. One definition of occupational health is that it concerns the *effects of work on health and the effects of health on work*. The primary aim of good occupational health practice should always be to prevent ill health. Occupational health practitioners (see **O1010** below) are health professionals who are also trained to understand the complex relationships between work and the impact this may have on the health of the individual. Occupational ill-health is caused by exposure to a wide variety of chemical, physical and organisational aspects of work and the working environment. To be effective, an occupational health practitioner must have a good understanding of how the environment at work (in its broadest context) might affect a person's health.

Why is it important?

[O1001.1] The trade union organisation TUC says that even using the most conservative estimates, at least 20,000 people die prematurely every year because of occupational disease. But very few workers have access to a fully comprehensive occupational health service. While two million people suffer from a work-related condition, it says that only one in eight UK workers having access to an occupational physician.

This situation is set to get even worse. In November 2017, the Department for Work and Pensions (DWP) announced that the national Fit for Work occupational health service, which aimed to help employers to access services when a person was ill for over a month, would close from 31 March 2018 after less than three years of operation (see **O1015.1** below).

Meanwhile, the Health and Safety Executive (HSE) signalled in its health and work strategy that it will have an increased focus on occupational health, particularly those conditions with widespread prevalence, the largest lost-time and economic cost consequences, and life-limiting or life-altering impacts: work-related stress, musculoskeletal disorders (MSDs) and occupational lung disease.

Sickness absence

[O1002] Occupational health is an important, but often neglected, aspect of an organisation's management. It is important to the organisation for a number of reasons. Employees are the most important, and often the most expensive, asset any organisation has. Sickness absence can therefore be very costly. In March 2017, the Office for National Statistics (ONS) reported the lowest recorded rate of sickness absence since the series began in 1993. The figure dropped from 7.2 days per worker to 4.3 days per worker, but the ONS estimates that this still amounted to 137 million working days lost due to sickness or injury in the UK in 2016. The HSE estimates that in 2015/16, workplace injury and illness cost £14.9 billion. A good occupational health service should be able to considerably reduce the level of avoidable sickness absence and hasten the successful return to work of those who are absent long term, with significant financial savings for the organisation (see **R2001 REHABILITATION**).

Personnel support

[O1003] An occupational health service can be an invaluable support to the Human Resources/Personnel function in managing the many aspects of employment that relate to an individual's medical condition and their consequent fitness to work. As well as managing sickness absence (see **O1002** above), the occupational health service can advise on fitness to work at the point of recruitment or placement. This will reduce the risks to the organisation of recruiting someone to a position for which an existing medical condition renders them vulnerable to further illness or injury or for which they may be simply unfit to undertake the proposed work.

Employers should note, however, guidance produced by the Equality and Human Rights Commission (EHRC) on the *Equality Act 2010*. The Act came into effect on 1 October 2010 and replaced several anti-discrimination laws, including the *Disability Discrimination Act 1995*, with a single Act.

The guidance sets out that: 'Except in very restricted circumstances or for very restricted purposes, employers are not allowed to ask any job applicant about their health or any disability until the person has been:

- offered a job either outright or on a conditional basis, or
- included in a pool of successful candidates to be offered a job when a position becomes available (for example, if an employer is opening a new workplace or expects to have multiple vacancies for the same role but doesn't want to recruit separately for each one).'

The EHRC guidance makes clear that this includes:

- asking such a question as part of the application process or during an interview; or
- sending them a questionnaire about their health for them to fill in before the employer has offered them a job.

The guidance also clarifies that questions relating to previous sickness absence count as questions that relate to health or disability. No-one else can ask these questions on behalf of the employer. An applicant cannot be referred to, or asked to fill in a questionnaire provided by, an occupational health practitioner before the offer of a job is made or before they are included in a pool of successful applicants, except in very limited circumstances. The purpose of this is to ensure that job applicants are not ruled out just because of issues related to or arising from their health or disability, which may well say nothing about whether they can do the job. It is only once the job offer has been made, or an applicant is included in a group of successful candidates, that the employer can make sure that someone's health or disability would not prevent them from doing the job and they must also consider whether reasonable adjustments to enable them to do the job can be made.

Employers can only ask questions about health or disability when:

- They are asking the questions to find out if any applicant needs reasonable adjustments for the recruitment process, such as for an assessment or an interview;
- They are asking the questions to find out if a person (whether they are a disabled person or not) can take part in an assessment as part of the recruitment process, including questions about reasonable adjustments for this purpose;
- They are asking the questions for monitoring purposes to check the diversity of applicants;
- They want to make sure that an applicant who is a disabled person can benefit from any measures aimed at improving disabled people's employment rates;
- They are asking the question because having a specific impairment is an occupational requirement for a particular job;
- The questions relate to a requirement to vet applicants for the purposes of national security; or
- The question relates to a person's ability to carry out a function that is absolutely fundamental to that job, where a health or disability-related question would mean that they would know if a person could carry out that function with reasonable adjustments in place.

The EHRC advises that there will be very few situations where a question about a person's health or disability needs to be asked.

Other medical aspects of employment on which the occupational health service can advise include retirement on grounds of ill-health, temporary restrictions on work or working times to aid recovery or rehabilitation, keeping at-risk employees under health surveillance, and advice on managing medical conditions which are impacting on the capacity of the employee to undertake expected duties.

Legal duties

[01004] Another important reason for having access to occupational health advice is to ensure that statutory and civil legal obligations are being met. Several regulations place a legal requirement on the employer to place at-risk employees under health surveillance or to have regular medical examinations (usually by a doctor 'appointed' or 'approved' for the particular requirements

of the regulation). The *Ionising Radiations Regulations 2017 (SI 2017 No 1075)*, the *Control of Lead at Work Regulations 2002 (SI 2002 No 2676)* and the *Control of Asbestos at Work Regulations 2012 (SI 2012 No 632)* are examples of regulations which contain requirements for medical surveillance by a relevant doctor. In addition, other areas of work, such as working with respirators and working offshore, may require medical examinations either under statute or industry standards.

Risk assessment

[01005] The occupational health department may also need to be involved in risk assessments where these involve health considerations. Under the *Management of Health and Safety at Work Regulations 1999 (SI 1999 No 3242)* there is a general requirement to undertake an assessment of the risks to health and safety of employees. In some circumstances the risk may be to the employee's health (a classic example being the risk assessment for work-related stress (see O1020.8 below) and the occupational health department can play an important role in helping to determine those aspects of the job that may be stressful. In many instances the risk assessment will have to take account of an individual's vulnerability (perhaps because of age, pregnancy, or previous history of injury or illness) and the occupational health department is in a unique position to translate a previous medical history (or current condition) into the additional risks that this may pose to the employee. As importantly, occupational health can also advise on what needs to be done to avert the risk either by additional protection or modifications to the job or work routine. Examples of where there may be a need to take account of individual vulnerability, in addition to pregnancy, include a history of upper limb disorder, a history or earlier episode of back pain, earlier periods of stress-related disorder, a history of a skin allergy and existing respiratory disorder (such as asthma). There are of course many other examples, but it is extremely important to take such vulnerabilities fully into account in the risk assessment; failure to do so may increase the chances of successful litigation for damages for failure to exercise an appropriate level of duty of care.

Other regulations such as the *Control of Substances Hazardous to Health Regulations 2002 (SI 2002 No 2677)*, the *Health and Safety (Display Screen Equipment) Regulations 1992 (SI 1992 No 2792)*, and the *Manual Handling Operations Regulations 1992 (SI 1992 No 2793)*, may require occupational health input into the risk assessment and may require employees to be placed under health surveillance in order to reduce the risk of injury or ill health to the lowest level reasonably practicable.

Relationships between occupational health and safety

[01006] In general, only larger organisations have an occupational health department, although many smaller employers have access to occasional occupational health or medical advice through buying in a service from a local provider.

In addition, most organisations will have some sort of safety provision, either a dedicated safety department (again mostly in larger organisations) down to a safety person who fits in safety with their main job. Occupational health and safety are two different, but complementary, means of achieving the same goal – ensuring the well-being of both the organisation and its employees.

Prevention is better than cure

Being reactive

[O1007] Inevitably there are cases of employees becoming ill, either as a result of work or for reasons completely unrelated to work. Either way, the role of the occupational health service is to advise management on the employment issues associated with the individual. If the illness is work related or thought to be so, then the advice relates to what can be done to ensure that there is no exacerbation of the condition when the employee returns to work. If it is unrelated to work then the advice may be concerned with how a person's newly acquired health condition may impact on work, either by affecting the way he or she can perform the expected duties or the way in which the medical condition may now render that person more susceptible to the conditions of work. The employer may require advice on employees who are taking frequent periods of short-term sick leave or a period of long-term sick leave and how to manage the sickness absence or rehabilitation and return to work. They may need advice on medical retirement on grounds of permanent ill health. All of these are examples of occupational health, albeit playing an important role, but being *reactive* – reacting to the needs of the organisation in managing the consequences of staff being unfit for work.

A good occupational health service should also be *proactive* – prevention being better than cure. Being proactive involves anticipating the possibility of ill health and putting into place measures which are aimed at preventing or minimising the risk of this occurring. Prevention starts with ensuring that employees are not recruited to jobs for which they are medically unfit or making sure that where preventative measures need to be taken (eg vaccinations) then these are undertaken before commencement of work (see O1003 above regarding the requirements of the *Equality Act 2010*, concerning the use of pre-employment medicals and questionnaires). During employment, prevention is then exercised through a number of initiatives that a good occupational health service will be engaged in. Health education and health and well-being programmes are a good way of improving and maintaining the health of the workforce. This may be strictly confined to those health issues related to the work, (for example how to look after your skin, how to prevent hearing damage, or how to prevent musculoskeletal disorders) or it may be concerned with general health and well-being such as smoking cessation, women's or men's health, look after your heart campaigns and exercise, diet and obesity. It is good practice for an organisation to be actively involved in promoting health generally, and maintaining a healthy workforce will bring business benefits.

Prevention and risk assessment

[O1008] Prevention is also achieved by involving occupational health in the risk assessment process in order to ensure that relevant health issues are fully included. As discussed at O1005 above, the occupational health practitioner can help ensure the medical aspects of the risk assessment are fully included and individuals' vulnerabilities, where relevant, are taken into account. Occupational health input can also ensure where required, appropriate health surveillance is set up and carried out as a means of early detection and hence prevention of work-related disorders.

Health surveillance

[O1009] Health surveillance is another important aspect of prevention in which the occupational health department has an important role to play. *Regulation 6* of the *Management of Health and Safety at Work Regulations (SI 1999 No 3242)* requires that:

> 'Every employer shall ensure that his employees are provided with such health surveillance as is appropriate having regard to the risks to their health and safety which are identified by the assessment.'

The HSE advises that health surveillance is required if all the following criteria are met:

- there is an identifiable disease/adverse health effect and evidence of a link with workplace exposure;
- it is likely the disease/health effect may occur;
- there are valid techniques for detecting early signs of the disease/health effect; and
- these techniques do not pose a risk to employees.

HSE guidance also sets out how and who should carry out health surveillance: 'In its simplest form, health surveillance could involve employees checking themselves for signs or symptoms of ill health following a training session on what to look for and who to report symptoms to. For example, employees noticing soreness, redness and itching on their hands and arms, where they work with substances that can irritate or damage the skin.'

It goes on to advise that: 'A responsible person can be trained to make routine basic checks, such as skin inspections or signs of rashes and could, eg, be a supervisor, employee representative or first aider. For more complicated assessments, an occupational health nurse or an occupational health doctor can ask about symptoms or carry out periodic examinations.'

HSE industry-specific guidance includes advice on the type of jobs that may require health surveillance. There are several high-hazard substances and agents where the law requires health surveillance programmes include statutory medical surveillance. This involves a medical examination and possibly tests by a doctor with appropriate training and experience who has been appointed by the HSE.

For example, where surveillance includes physiological or biochemical measurements, such as hearing function, lung function, or urine or blood tests, then an occupational health doctor or nurse would usually be necessary. What is most important, however the health surveillance is carried out, is that there is a recognised means of referral for expert help if the health surveillance reveals someone with early signs or symptoms of a work-related condition.

In summary, while occupational health has an important role to play in investigating and managing illness and the consequences of ill health that occur during the course of employment, the core of occupational health practice should be to maintain a healthy workforce and prevent as many work-related conditions as possible from arising in the first place.

The occupational health team

[01010] The occupational health 'team' varies enormously from one organisation to another depending on the complexity of the working environment and the needs of the organisation. It can vary from no provision at all up to a multi-professional team. The majority of organisations that make provision for occupational health do so with the core service being provided by an occupational health nurse who may be full or part time depending on the size of the organisation. Occupational health nurses, working on their own, will usually have a network of external support, including a physician. The external support should include other occupational health nurses who can participate in clinical audit.

Occupational health nurses

[01011] Experienced occupational health nurses can undertake a wide range of duties including contributing health considerations to the risk assessment process, undertaking health surveillance, and advising the organisation on health issues relating to individuals, such as sickness absence management and return to work. Occupational health nurses also usually play a key role in preventing work-related illness and promoting health and well-being among the workforce. Many occupational health nurses are also competent to undertake environmental assessments such as simple measurements of noise and lighting, and carrying out workstation assessments.

Competent occupational health nurses also recognise their limitations and will be prepared to refer an individual for a medical opinion from a doctor or refer a problem that is outside of his or her competence for expert help.

Where other advice or practical help is required this may be from any one of a number of sources:

- occupational health physicians;
- occupational hygienists;
- ergonomists;
- physiotherapists; and
- counsellors.

Occupational physicians

[01012] *Occupational physicians* are doctors who have specialised in occupational medicine. Their role is to undertake medical examinations and to diagnose work-related ill health, to determine the work-related causes of ill-health and to advise on the management of health problems which impact on the person's ability to work. Pension funds may require a doctor to make the decision regarding permanent ill-health retirement or admission to pension schemes, although they do not necessarily have to be specialist occupational physicians.

The Health and Safety Executive (HSE) advises employers that when appointing an occupational health doctor or nurse, it is important to find someone with experience in their industry, or a related industry. It also advises them to ask to see their registration/personal identification number (PIN) and their certificates/diplomas relevant to the duties they will undertake. Relevant qualifications can be confirmed with the appropriate governing body: the General Medical Council and the Faculty of Occupational Medicine (for doctors); or the Nursing and Midwifery Council. Nurses holding specialist qualifications in occupational health will be registered as specialist community health nurses (OH).

Occupational health technicians

[01012.1] Occupational health technicians are trained and qualified in specific areas, such as spirometry or audiology.

Occupational hygienists

[01013] Occupational hygienists are employed to measure chemical and physical hazards in the workplace. Information on levels of exposure to chemicals, dusts, fumes and noise for example, is often required in order to determine the risks to health. There is also sometimes a statutory need to determine exposure levels and to keep records of individual exposures. Occupational hygienists are also expert in advising on protective equipment such as the suitability of hearing defenders and dust masks.

Ergonomists

[01014] *Ergonomists* have specialist skills, which enable them to design work systems that optimise the match between the person and their work and work environment. The important skill of the ergonomist is to understand the mental and physical capabilities of people and the wide range of human variability, and to design work systems which impact least on the person's health (see **ERGONOMICS**).

Treatment services

[01015] *Physiotherapists, chiropractors and counsellors* Occupational health is primarily concerned with the prevention of work-related disorders and the management of health problems that affect employees at work and the organisation employing them. Although it is not primarily the function of occupational health to provide treatment, many organisations consider it a worthwhile investment to provide some types of treatment in order to assist the employee in recovering and returning to work more quickly. Such treatment services typically involve those aimed at recovery from musculoskeletal disorders and psychological illness which together form the two most frequent reasons why employees are off sick. Physiotherapy and counselling are therefore sometimes provided by the occupational health service, or organised with external providers through the occupational health service.

Closure of the fit for work service

[01015.1] The national Fit for Work occupational health service allowed employers and GPs to refer employees who were, or were likely to be, off work sick from work for four weeks or more for a voluntary occupational health assessment to help them return to work.

The service was set up following a recommendation in the 2011 government-commissioned review, *Health at Work – an independent review of sickness absence*. This found that while much sickness absence ended in a swift return to work, a significant number of absences lasted longer than they needed to, and each year over 300,000 people were falling out of work and onto health-related state benefits.

The review recommended that the government fund a new independent service to provide an in-depth assessment of an individual's 'physical and/or mental function' and provide advice about how people on sickness absence could be supported to return to work.

The November 2017 paper *Improving Lives The Future of Work, Health and Disability*, which sets out the government's strategy on the future of work, health and disability, reported that the DWP-commissioned service for offering free occupational health assessments had very low take-up and, as a result, the assessment service would close at the end of March 2018. Employers, employees and GPs will continue to be able to use the Fit for Work advice helpline, website and web chat, which offer general health and work advice as well as support on sickness absence.

The strategy paper sets out that the government is now looking at how to shape, fund and deliver effective occupational health services. It says that: 'Good OH [occupational health] advice can improve the health of the UK's working population by preventing work-related illness, and unnecessary sickness absence.'

However, the *Improving Lives* consultation found that:

- the current model of occupational health provision does not meet the needs of employers or individuals;
- those at greatest risk of job loss generally lack access to occupational health;

- the funding model for occupational health provision is unclear with a lack of consensus about responsibility for providing and funding occupational health services – NHS, government, or employers;
- larger employers and firms with skilled employees are more likely to pay for occupational health provision while SMEs tend to offer little or no access;
- Fit for Work had very low take-up;
- there is a lack of availability and capability of occupational health services, characterised by too few qualified OH professionals;
- there is a lack of conversations and collaboration between GPs, employers, Fit for Work service, other healthcare professionals, and Jobcentre Plus work coaches; and
- there are limitations of the evidence on the effectiveness of services, and inconsistent use of evidence-based interventions where they are available.

By 2019/20 the government aims to be in a position to set out a clear direction and strategy for future reform and will set up an expert working group to look at:

- building the evidence base;
- potential funding models and where responsibility for occupational health support should fall;
- methods for improving quality of existing provision, for example accreditation of services, staff and training;
- emerging new models of provision (in primary, secondary health and across sectors) and local models to integrate work and health support;
- exploiting the potential of technology as a way to help people access the advice they need easily and quickly; and
- workforce development.

The *Improving Lives* strategy document can be found at: www.gov.uk/government/uploads/system/uploads/attachment_data/file/663399/improving-lives-the-future-of-work-health-and-disability.PDF.

GP fit notes

[01015.2] In April 2010, 'fit notes' replaced 'sick notes'. GPs can use fit notes either to indicate that a person is 'not fit for work' or that they 'may be fit for work taking account of the following advice'. The GP can recommend: a phased return to work; altered hours; amended duties; or workplace adaptations. Department for Work and Pensions (DWP) advice on GP fit notes aimed at employers and line managers, employees and patients, and GPs can be found at: www.dwp.gov.uk/fitnote, together with guidance for occupational health professionals.

The *Improving Lives* strategy document (see O1015.1 above) reports that the government has conducted an internal review of the fit note and aims to reform the system over the next two to three years. It wants the fit note to 'become an enabler for conversations about health and work, focussing on what people can do, rather than what they cannot do. It should facilitate returns to work and help people stay in work where appropriate, by providing information to the employer about what support might enable that to happen.'

It reports that while employers and GPs support the idea of helping people to return to work when they can, recently published fit note statistics show that only 6.6% of fit notes used the 'may be fit for work' option. In response, the government is:

- starting development work to legislate for the extension of fit note certification powers to other healthcare professionals, along with the design and development of a set of competencies for those completing fit notes;
- conducting a feasibility test to investigate whether, for the purposes of Statutory Sick Pay, employers could use the Advisory Fitness for Work report (which can be completed by some Allied Health Professionals) as an alternative to the fit note;
- integrating fit note training into GP undergraduate and postgraduate education;
- commissioning the feasibility of clinical guidelines for workplace adjustments for the top five clinical reasons people are off work sick or are on health-related benefits; and
- exploring whether changes to the way GPs complete fit notes could support better return to work conversations.

Health records

Health records and confidentiality

[01016] It is sometimes necessary to obtain the medical records of an employee in order to make a judgement about the current state of the person's health and their fitness to work. If this is done then it is important that these are obtained by a heath professional employed by the company (or engaged by the company under contract), who can take responsibility for the confidentiality of the records. Medical records should always be kept confidentially under the guardianship of a registered medical practitioner or registered nurse, and should not form part of the person's personnel file.

The advice, conciliation and advisory service ACAS says: 'Employers can seek to collect information regarding an employee's health if the employee freely gives consent. Employers should consider why they need the information and exactly what information is needed. This information once collected should be held securely, this could be allowing only one or two people access to the information or by password protecting it. Employers should check that the information collected can be justified.'

Obtaining records and reports

[01017] Requesting medical records or a medical report requires the informed consent of the individual and is subject to the legal requirements of the

Access to Medical Reports Act 1988 (in the case of medical reports) and the Data Protection Act 1998, (in the case of health and medical *records*).

Under the *Access to Medical Reports Act 1988, s 1* an individual has the right of access to any medical report relating to him or her which is to be, or has been, supplied by a medical practitioner for employment purposes or insurance purposes.

Subject access to health records under the *Data Protection Act 1998* is much wider than under the *Access to Medical Reports Act 1988, s 68(2)*;

""health record" means any record which—

(a) consists of information relating to the physical or mental health or condition of an individual, and
(b) has been made by or on behalf of a health professional in connection with the care of that individual."

Health professionals include doctors, nurses, midwives, dentists, opticians and optometrists. An applicant for access to a health record can be the patient or another person that the patient has authorised in writing to make application on his behalf. In the case of an employer seeking access to an employee's health records, the employer is acting on behalf of the patient (their employee) and therefore must seek written informed consent to make an application. Under this Act the individual does not have the automatic right to see the record before it is released to the applicant, but does have the right to subsequently request that any misleading or inaccurate entry in the health records is corrected.

The normal way in which an employer seeks to obtain a medical report or access to health records is to explain to the employee what the purpose is in requesting a medical report or copies of health records and asking the employee to sign a consent form. The consent form must clearly spell out the legal rights of the individual under the *Access to Medical Reports Act 1988* and *Data Protection Act 1998*.

The General Data Protection Regulation comes into force on 25 May 2018 and will bring in stricter obligations that all employers must follow. The Information Commissioner's Office (ICO) has published an overview of the regulation and a checklist of 12 steps employers can take in preparation for the new law. This can be found at: https://ico.org.uk/media/1624219/preparing-for-the-gdpr-12-steps.pdf.

Keeping and making use of the occupational health records.

[01018] Some occupational health records have to be kept for a statutory period of time. Under the *Control of Asbestos Regulations 2012 (SI 2012 No 632)*, for example, the health records of employees who are exposed to asbestos have to be kept for a minimum of 40 years. Under the *Control of Substances Hazardous to Health Regulations 2002 (SI 2002 No 2677)*, health records pertaining to individuals must also be kept for a minimum of 40 years.

Health records are kept for management purposes and these obviously relate to individuals and must be kept confidentially. These records are used firstly to provide a record of any health surveillance, referrals for medical opinion and the opinions received, the results of tests and vaccination history etc. They form an important record of the duty of care afforded by the employer and might be necessary to defend possible claims for damages. Health records are also kept for health management purposes such as recalling employees for routine health checks and tests or making sure that vaccinations are kept up-to-date.

Health records on individuals are confidential and that confidentiality must be preserved. However, if the records are kept in such a way as to allow the analysis of collective and anonymised data then they can be a useful means of detecting trends in illness or detecting the emergence of a new illness or condition that has not previously been recognised in the organisation. For example, by analysing the health surveillance records of all the employees working in a particular department (perhaps by producing monthly statistics), it might become apparent that since a new solvent cleaner was introduced there has been an increase in the incidence of dermatitis. Health records should therefore not simply be a passive record of what was done and what was found, but be used positively to promote the health of the individual and employees collectively and to protect the liability of the organisation.

Buying services and priorities

[O1019] The priority for any manager contemplating buying in an occupational health service is to determine exactly what the organisation needs and what it wants from its occupational health provider. The needs are determined firstly as a 'must have' list. 'Must have' would include any statutory requirement for medical examination or health surveillance and whatever else the organisation determines it must have to manage the business effectively and support the personnel function. This would include advice on the management of sickness absence and rehabilitation, and any requirement of the pension fund regarding ill-health retirement (see O1003 above). In addition, depending on the organisation's resources and intention to promote health and be proactive, non-statutory health surveillance, occupational health involvement in risk assessment, and health promotion and health education activities may also be required.

NHS *Health at Work* provides information for businesses seeking occupational health advice and support at www.nhshealthatwork.co.uk.

It is important to ensure that the provider is able to offer competent and properly qualified personnel for the work expected from them. Ask to see CV's of the personnel being put forward and seek confirmation of the professional registration of the medical and nursing staff.

Occupational diseases and disorders

[O1020] The following section describes some of the main diseases and disorders that may be caused or worsened by working conditions.

Musculoskeletal disorders

What are they?

[01020.1] The term musculoskeletal disorders (MSDs) refers to a range of conditions which affect the skeletal system and its associated muscles, tendons and ligaments. These conditions are usually separated into those affecting the upper limb (from the shoulder to the tips of the fingers), the lower limb (from the hip to the toes) and the back. The most common conditions are those affecting the upper limb (work-related upper limb disorders or WRULDs) and the back (backache, acute and chronic back pain). Conditions which affect the lower limb are less common but in some sectors (such as carpet laying) the knee can be damaged by repeated kneeling, causing a condition called beat knee (bursitis). The other part of the body often affected is the neck, especially in sedentary jobs, (at computer workstations for example) where postural problems result in neck pain.

The common conditions affect the tendons of the arm and hand. These tendons run from the muscles of the forearm through the wrists and hand to insert on the individual bones of the wrist and fingers. In places, as they pass across the wrist and the back and front of the hand, they pass through lubricated sheaths (tendon sheaths). Repetitive and forceful hand-arm movements can damage almost any part of this system.

Musculoskeletal disorders have consistently been one of the most commonly reported type of work-related illness. According to Health and Safety Executive (HSE) statistics, 507,000 workers were suffering from work-related musculoskeletal disorders (new or long-standing) in 2016/17, out of a total of 1,299,000 for all work-related illnesses. They account for 39% of all work-related ill-health and led to 8.9 million lost working days in 2016/17, an average of 17.6 days lost for each case. Working days lost per worker due to self-reported work-related musculoskeletal disorders showed a generally downward trend up to around 2010/11, but since then the rate has remained broadly flat.

There are a range of WRMSDs:

Upper limb disorders. These are also called repetitive strain injury (RSI) and 'work-related upper disorders' (WRULDs). There are a number of recognised conditions that make up this collection of conditions, but their precise nature is uncertain and drawing a diagnostic distinction between them is difficult.

- *Tenosynovitis.* An inflammatory and painful condition of the tendons as they pass through the tendon sheaths of the wrist and in the hand. The condition may also involve inflammation and dryness of the tendon sheath itself.
- *Tendinitis.* Inflammation of the tendons but not involving the tendon sheaths. Tendinitis can occur in the wrists and hand and also in the upper arm and shoulder.
- *Peritendinitis.* Inflammation at the point where the tendon joins the muscle in the arm.
- *De Quervain's disease.* Tenosynovitis with pain and swelling at the back of the wrist, particularly involving the tendon to the thumb.

- *Epicondylitis*. Pain at the point where the muscles of the arm join the bone at the elbow (the epicondyles). On the inner side of the elbow this is called medial epicondylitis (golfers elbow). On the outer side this is called lateral epicondylitis (tennis elbow).
- *Carpal tunnel syndrome*. Entrapment or compression of the median nerve as it passes through the front of the wrist (the carpal tunnel). This causes numbness and tingling in the part of the hand supplied by the median nerve – the palm side of the thumb, the second and third fingers and half the fourth (ring) finger.
- *Chronic (non-specific) upper limb pain*. A chronic widespread pain in the upper limb with tenderness to which no certain diagnosis can be made. Linked with repetitive use of the hand and arm at work. Some doctors will diagnose this as RSI, despite it not being a clinical entity, simply because no clearer diagnosis can be made.
- *Beat hand and beat elbow*. Repeated use of picks, shovels and hand tools in labourers, quarrymen and other manual jobs, can cause chronic pain, tenderness, redness and swelling of the hand and the elbow.
- *Cramp*. A condition that appears to affect people in occupations that involve repetitive movements of the hand and arm. There is some dispute about its true nature but it is a prescribed occupational disease. It causes spasm, tremor and pain in the affected muscles when the familiar action is attempted. It is typified by writer's cramp.

Back pain. Most back pain is caused by mechanical damage to the spine and associated muscles, ligaments and tendons. Acute (sudden onset) pain is usually associated with a torn ligament or muscle around the spine caused by sudden forceful lifting or movement associated with awkward posture. Around 90% of people with acute back pain will recover within six weeks. However, these conditions are cumulative and repeated episodes may take longer to recover and recovery may be progressively less complete. The condition therefore becomes recurrent. Back pain can also be caused by rupture (prolapse or herniation) of the intervertebral disc. These soft tissue discs act as 'shock absorbers' between each of the vertebral bones. Severe force on the spine can cause them to rupture and compress the nerve roots as they leave the spine. The most common part of the spine for this to occur is in the lumbar region where the nerves leave to supply the leg – compression of these nerves causes pain to be referred down the leg (sciatica). Chronic back pain occurs where there is incomplete recovery from repeated trauma. Disc rupture is often associated with subsequent chronic back pain.

Osteoarthritis. Osteoarthritis is a very common condition in the population as whole, but there is evidence that its prevalence is higher in some occupations. Osteoarthritis is a degeneration and loss of the articular cartilage in some joints particularly of the hand, knee and hip. Articular cartilage forms a smooth lubricated surface between the joint surfaces – its loss means that bone rubs against bone causing severe damage to the joint. Osteoarthritis of the knee is more common in people involved with heavy manual labour and especially if the work involves kneeling. Osteoarthritis of the hip is less clearly associated with occupation although prolonged heavy lifting may be a risk factor.

What causes them?

[O1020.2] The following are risk factors associated with most musculoskeletal disorders.

- **Repetition.** Frequent repeated movements over a relatively long period of time.
- **Force.** The amount of force put into the movement such as pulling, pushing or lifting. The greater the force required, the greater the risk.
- **Duration.** The longer someone is engaged in the same work, the greater is the risk of cumulative damage.
- **Lack of rest.** The risk of musculoskeletal damage is increased if the work is carried out for long periods of time without rest or time for recovery.
- **Posture.** Work carried out in awkward postures greatly increases the risk. This is true of sedentary jobs such as typing or checkout work, or in manual jobs such as lifting.
- **Working in adverse conditions.** Manual work where the hands are cold, the floor uneven or slippery, or where loads are difficult to handle, increase the risk. Working in uncomfortable conditions at computer workstations increases the risk of upper and lower limb disorders.
- **Stress.** Jobs carried out while under stress from time pressures etc increase the risk of musculoskeletal disorder.
- **Individual factors.** Factors such as age and sex, height, weight and strength, a history of such conditions (or a pre-existing condition), and pregnancy will have an influence on the risk of musculoskeletal injury.

How are they detected?

[O1020.3] Musculoskeletal disorders can normally only be detected by self-reporting or by causes of sickness absence. Employees in any at-risk job should be encouraged to report symptoms. Intervention at an early stage can often result in preventing the problem becoming worse. Acute conditions will result in sudden disabling pain and disability, which require medical intervention. The onset of the chronic condition may be insidious and occur over a period of time. However, during this time there will be early symptoms, which should not be ignored. With upper limb conditions or slowly developing back pain there may be episodes of acute pain or numbness, which quickly resolve. Recurrent episodes or early signs of disorder such as pain, numbness or tingling, should be referred to an occupational physician (or GP) for diagnosis.

How are they managed?

[O1020.4] Heavy manual work may carry a more obvious risk but sedentary jobs are no less likely to be associated with problems. In an office the unplanned task of carrying or moving something heavy may be all that is required to cause an acute back injury. All jobs should be considered as inherently capable of giving rise to a musculoskeletal condition, although of course some jobs, which carry the known risk factors (see O1020.2 above), will be much more at risk. All jobs should, therefore, have a risk assessment for

MSD that takes into account all of the known and likely factors associated with the job and any individual vulnerabilities. This must be done by a competent person who recognises all the risk factors and has a good understanding of what control measures should be put in place. Having recognised the risks and implemented the necessary controls it is important that these are adhered to – this has to be done partly by effective supervision (eg if rest pauses are built into the job, make sure they are taken) and partly by providing good information, education and training of the people involved.

Employees must be made aware of the symptoms to report and to whom they should be reported. A system of referral to an occupational health nurse or doctor (or if that is not available to the person's GP) must be put in place. Referral to the health professional should ask for advice on what needs to done in the workplace or to the job to ensure that the condition does not worsen. Referral to an occupational health professional is much more likely to result in practical advice, which many GPs may not have the experience to offer.

Conditions that have become chronic or disabling to the point where the individual is not able to carry out their expected duties must be referred to an occupational physician with the view that permanent ill-health retirement may be necessary. However, the *Equality Act 2010* (see **O1003** above) will almost certainly apply to chronic disabling musculoskeletal disorder. Under the Act, a person has a disability if they have a physical or mental impairment which has a substantial and long-term adverse effect on their ability to perform normal day-to-day activities. 'Substantial' means more than minor or trivial; 'long-term' means that the effect of the impairment has lasted or is likely to last for at least twelve months (there are special rules covering recurring or fluctuating conditions); and 'normal day-to-day activities' include eating, washing, walking and going shopping.

In *Banaszczyk v Booker Ltd* [2015] UKEAT/0132/15/RN, the Employment Appeal Tribunal (EAT) ruled that, for a warehouse operative, lifting and moving goods weighing up to 25kg, in part manually and in part using a pallet truck, was a day-to-day activity.

'No-one with any knowledge of modern UK working life could doubt that large numbers of people are employed to work lifting and moving cases of up to 25kg across a range of occupations, including in particular occupations concerning warehousing and distribution' noted the EAT judge. Mr Banaszczyk, who suffered from chronic back pain, was a disabled person and therefore entitled to the protection of the *Equality Act 2010*, according to the ruling.

What are the legal requirements?

[O1020.5] General duties of risk assessment, risk management, information and training, competency of those undertaking risk assessment and health surveillance come under the *Management of Health and Safety at Work Regulations 1999 (SI 1999 No 3242)*. These regulations apply where the risk arises from adverse conditions at work such as slipping and tripping hazards, poor environmental conditions (heat, wet and cold), where the primary hazard is poor or awkward posture and where stress may be a contributing factor.

Where the risk arises specifically from manual handling (lifting, pushing or pulling loads) the *Manual Handling Operations Regulations 1992 (SI 1992 No 2793)* apply. Working with display screens (including computer workstations, and video display screens) falls within the *Health and Safety (Display Screen Equipment) Regulations 1992 (SI 1992 No 2792)*.

(See also MANUAL HANDLING and the EQUALITY ACT 2010 chapters.)

Psychological Disorders (Stress, Anxiety and Depression)

What are they?

[01020.6] The HSE defines stress as 'the adverse reaction people have to excessive pressure or other types of demand placed on them'. Pressure is a normal part of work, indeed a normal part of life, but stress occurs when the pressures are perceived by the individual as being greater than their ability to cope. Stress is therefore a very individual phenomenon; people vary greatly both in their perception of the amount of pressure on them and their ability to cope. Stress often leads to anxiety and depression, but anxiety and depression can occur for many reasons without the underlying cause being stress. Stress also occurs in our everyday lives outside of work and it is important for the employer to understand what contribution, if any, the work has in causing an individual's stress or other psychological problem. An employer can only be responsible for that part of an individual's health complaint that is related to work, but of course most employers will be concerned with an employee who is suffering a psychological illness, whatever the cause, because it is likely to have a major impact on the person's performance at work. Stress is not by itself an illness, but a set of symptoms which result from the mental and physiological reaction to an environment in which the perceived demands exceed the perceived ability to cope. The resulting psychological and physiological illness may be classified as diseases. And stress may not only cause psychological illness but may be related to mild to serious physical illness including high blood pressure, gastrointestinal complaints, disturbed sleep and fatigue.

Someone suffering from stress may exhibit any of the following signs, some of which will be obvious to work colleagues and management and should trigger action to discover, and hopefully remedy, the cause.

- Irritability with colleagues.
- Expressed feeling of inability to cope.
- Frequent periods of sickness absence.
- Difficulty concentrating.
- Tearfulness.
- Mood changes.
- Undue tiredness/lethargy.
- Headaches.
- Insomnia.

- Palpitations.
- Appetite change either loss of appetite or over-eating.
- Abdominal and chest pains.
- Diarrhoea or constipation.

It is possible too that individuals under stress will drink more alcohol, smoke more cigarettes and indulge in 'comfort' eating. These are attempts at reducing the symptoms of stress but will, particularly in the long term, have serious adverse health effects.

Stress has consistently been one of the most commonly reported types of work-related illness. HSE statistics show that the total number of cases of work related stress, depression or anxiety in 2016/17 was 526,000, of which 236,000 were new cases. In 2016/17 stress, depression or anxiety accounted for 40% of all work-related ill health cases and 49% of all working days lost due to ill-health. The total number of working days lost was 12.5 million days, an average of 23.8 days lost per case. Working days lost per worker due to self-reported work-related stress, depression or anxiety has remained broadly flat but has shown some fluctuations over recent years.

What causes them?

[01020.7] There are very many factors causing stress at work. Stress may be caused, or the response to it influenced, by any of the following.

- **The work itself and the work environment.** This includes a huge number of possible factors including the workload, time pressures, and complexity (and adequate training for the work expected). It also includes managerial and organisational aspects, how well-supported the individual feels, how valued and rewarded they feel, how managers and work colleagues treat them. The physical environment can also be stressful – noise which distracts and interferes with communication and concentration, high or low temperatures which cause physical discomfort, poor lighting etc. Unreliable equipment, frequent system failures, and the pressures brought about by modern technology (including emails) are 'technological' causes of stress in the modern workplace. Bullying, harassment and violence at work are also causes of stress for some. In some jobs stress arises because of the nature of the job – working in the emergency services, in health care and social work brings emotional stress including traumatic stress. Individual factors that are capable of causing stress are called 'stressors'; stress is the adverse reaction people may have to exposure to these stressors. Stress is often caused by an additive effect when several of these stressors are present together.
- Modifying factors. The way individuals respond to stressors will depend on a number of modifying factors. These include the support they receive (both at work and outside of work) and their ability to learn successful coping strategies. The effect stressors have will also depend on the person's previous history of stress and their 'stress proneness'. It should usually be taken as true that someone with a history of stress-related disorder is likely to be more vulnerable to

continuing or additional levels of stress. Stress frequently arises outside of work and one important modifying factor is the reduced ability to cope with the stressors of work if the person is already suffering stress from non-work-related causes.

The HSE has categorised a number of causes of stress which form part of its stress management standards.

- **Demands.** Issues like workload, work pattern and the work environment.
- **Control.** How much say the person has in the way they do their work.
- **Support.** The encouragement and support provided to employees by management and colleagues.
- **Relationships.** Conflict and unacceptable behaviour.
- **Role.** People should understand their role within the organisation and the organisation should avoid creating conflicting roles.
- **Change.** How organisational change is managed and communicated.

How are they managed?

[O1020.8] As with any occupational health issue the primary aim is to prevent stress happening. Where individuals do become stressed, the organisation, usually with professional support, will need to carefully manage the individual's recovery and return to productive working and address any work-related factors which may have caused or contributed to the illness.

The key to prevention is to undertake a stress risk assessment. Identify any areas of the work or workplace which might place undue pressures on people. Identify aspects of the organisation which might create stress. The list of factors identified in the HSE management standards (at O1020.7 above) serves as a good checklist for undertaking the risk assessment although there may be other factors particular to the organisation. The risk assessment will have to be kept under constant review as things within the organisation change. In particular, it is important to reconsider the risk assessment for an individual who has been diagnosed with 'stress' or a stress-related disorder. Failure to take account of an individual's known vulnerability to stress has led to several successful civil claims for damages.

Consider having a stress management policy. This would detail the way in which stress will be identified in the organisation. It will also detail individual responsibilities and the support services available to employees. An important aspect of a stress management policy is that it shows a concern by the management and a commitment to address the issues, but only if it is put into practice and is seen to be working.

Even with the best stress management policy there may be individuals who will become stressed and it is important to have a support system in place to deal quickly and effectively with these cases. If the cause is clearly rooted within the organisation (excessive workload, poor working relationships, bullying, harassment, etc) then it is a management problem that will need to be addressed and resolved. Individuals may need support during this time and the organisation should have available to it a professional support service (occupational health or a counsellor experienced in stress counselling). If the causes are partly or wholly outside the organisation, but it is nevertheless impacting on the work (causing excessive sickness absence for example), then it is also important for the company to provide the necessary support.

What are the legal requirements?

[01020.9] The legal requirements fall under the *Management of Health and Safety at Work Regulations 1999 (SI 1999 No 3242)*. Under these regulations there is a requirement to assess the risks of stress occurring and the source of the stress (the stressors). The risk assessment should identify who is at risk (taking into account any existing vulnerability), what the possible adverse outcome might be and what controls need to be put in place to prevent or minimise the risk. The HSE Stress Management Guidelines and the associated Indicator Tool are a useful guide to undertaking a risk assessment in this area (www.hse.gov.uk/stress/index.htm).

(See also STRESS AT WORK chapter.)

Noise and occupational hearing loss

What is it?

[01021] Noise is defined as unwanted sound. Sound is created when an object that is vibrating or impacting causes a wave of pressure to move through the air. The magnitude of the air pressure wave is a measure of the intensity of the sound which our ears and brain detect as volume or sound level. The number of times the wave repeats itself in a given time is the frequency of the sound. The ear and the brain interpret this as the pitch of the sound. By utilising combinations of intensity and pitch we can interpret the meaning of sounds such as speech or music for example.

Sounds from the environment are an important way for us to communicate and be aware of dangers. However, both the ear and the brain can be affected by sounds that are harmful. Sound can be harmful in two ways:

- when the sound level is sufficiently high to cause damage to the mechanism of the ear; and
- when the level and pitch of the sound is such that it interferes with concentration, communication or the peaceful enjoyment of our surroundings.

Noise is therefore a level of sound sufficient to cause hearing damage or sound of sufficient intensity or frequency to interfere with expected levels of communication or mental well-being. The latter effect of noise can have very important consequences at work when, for example, noise in a busy office causes annoyance, distraction, loss of concentration and stress. These psychological effects are important to note but will not be dealt with in further detail here. This section looks in particular at the effects of noise on hearing loss.

How is it caused?

[01022] The ear receives the changes in air pressure, by a vibration of the eardrum (tympanic membrane). The vibrations of the eardrum are passed through a chain of small bones (the ossicles), and transmitted into the inner ear (the cochlea). In the cochlea the vibrations distort microscopic hairs attached to hair cells, and it is this distortion of the hairs that triggers nerve impulses which pass to the brain where frequency and sound level are interpreted as recognisable sounds. The cochlea is able to separate the sounds into individual frequency and sound level is interpreted by the magnitude of the distortion of the hair cells. Hearing loss is caused when the sound levels are so great that the hair cells are distorted so much that they fracture. Once damaged in this way the hair cells do not recover. Also, because the cochlea separates frequencies, the part of the cochlea that is damaged will depend on the frequency of the sound that has caused the damage. It is for this reason that hearing loss is sometimes confined to a small range of frequencies (see figure 1 at **O1023** below). The young human ear can normally detect frequencies in the range of about 20 to 20,000Hz (1Hz (Hertz)) is one cycle per second), although this ability declines significantly with age and the upper limit may fall to 16,000Hz or less as we get older. The ear can detect sound pressure from as little as 2×10^{-5} Newtons per square metre ($N.m^{-2}$), (the threshold of hearing), and is most sensitive at around 2 to 4kHz (kilo Hertz) which roughly corresponds with human speech frequencies.

Under most circumstances, hearing loss is gradual with the damage being cumulative over many years. Hearing loss, even in the early stages, may also be accompanied by tinnitus.

How is it detected?

[01023] Sound level is measured using a sound level meter which measures the sound pressure level according to an agreed international standard. Conventionally the sound intensity is measured using a logarithmic scale, the decibel (dB) scale. International and British Standards use reference levels that define 0dB as the threshold of hearing. On this scale 0dB is the average threshold of hearing and 130dB is the pain threshold; this roughly defines the range of human hearing ability (although not safe levels of exposure). One other characteristic of sound level meters is the convention to relate the measured sound to the sensitivity of the ear at different frequencies. The A weighted scale is most often used (although others exist) as this is weighted to correlate most closely between measured and perceived sound. It also correlates best with the ability of sound to cause hearing damage. When the A weighted scale is used, the letter A is placed in brackets eg 60dB(A).

Hearing loss is measured using an audiometer; the resulting graphical expression of the ability to perceive sounds is the audiogram. The audiometer measures the sound level at which the individual can just perceive the sound, at a range of individual frequencies, usually between 125Hz to 8,000 or 12,000Hz. As this uses the dB(A) scale, a normal audiogram would show a fairly straight line across this frequency range at around 0dB(A) (defined as the threshold of hearing). If there is hearing loss, greater levels of sound will be required to elicit hearing. Hearing loss may occur at any frequency or across the whole range of measured frequencies, but most often starts with a loss at around 3–4,000Hz, the frequency at which the ear is most sensitive.

Figure 1 (a) shows a normal audiogram with the responses close to 0dB across the frequency range. Figure 1(b) shows the typical drop in hearing ability, in this case of between 50 and 60dB, in the frequencies around 4,000Hz. This is a case of occupational hearing loss.

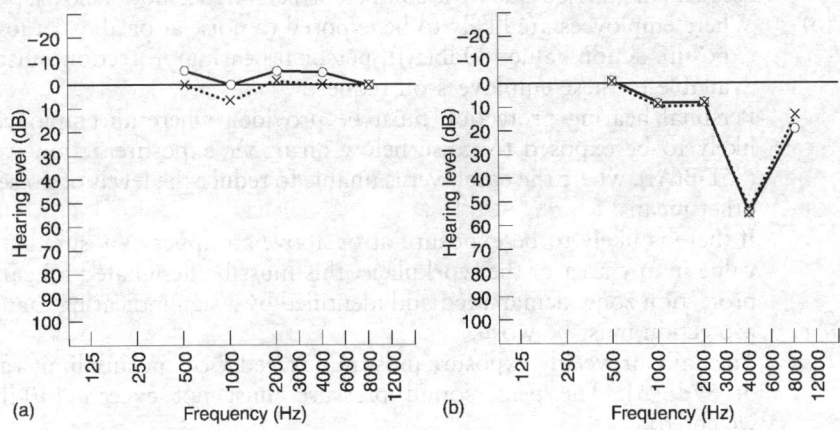

How is it managed?

[01024] Hearing loss (occupational deafness) is a serious risk in many industries which involve the use of machinery, or physical percussion or impact. To ensure that the risk of hearing loss is minimised (and legal requirements are being met) the following steps should be taken.

- Identify any areas or work tasks which may exceed the legal threshold for noise at work (see **O1025** below).
- Identify individuals who may be exposed to these noisy areas or types of work.
- Obtain a baseline audiogram on any exposed individuals. Audiometry should be undertaken in an approved soundproof booth to avoid erroneous results due to interference with the test from background sounds.
- Have a noise assessment undertaken on the noisy areas by a competent person and act on the report and advice given (see note under **O1025** below).
- Ensure that every reasonable effort is taken to reduce the noise at source. Protect the hearing of exposed employees by issuing recommended hearing protection.
- Ensure relevant employees are properly trained in and informed about the measures to protect themselves from the damaging effects of noise and how to properly wear and maintain the hearing protection provided.
- Have periodic audiometry undertaken and keep the records.

What are the legal requirements?

[01025] The primary legal requirements are contained in the *Control of Noise at Work Regulations 2005 (SI 2005 No 1643)*. The following are the key duties under these regulations:

(a) Employers must survey the workplace for areas likely to come under the Regulations. Any area likely to be affected must have a formal 'noise assessment' carried out by a competent person (see note below).

(b) Where employees are likely to be exposed to noise at or above a lower exposure action value (80db(A)), personal hearing protection must be available to these employees on request.
Personal hearing protection must be provided where an employee is likely to be exposed to noise below an upper exposure action value (85DB(A)), where the employer is unable to reduce the levels of noise by other means.
If there is likely to be exposure at or above an upper exposure action value in any area of the workplace, this must be designated a hearing protection zone, demarcated and identified by a sign indicating that ear protection must be worn.

(c) The daily or weekly exposure must not exceed the exposure limit value of 87db(A). The peak sound pressure must not exceed 140dB(C weighted).

(d) Any area of the workplace that is likely to create exposure at or above the upper exposure value (85bB(A)) must be demarcated and designated as a 'Hearing Protection Zone'.

(e) Risk of hearing damage must be reduced to the lowest reasonably practicable level.

(f) Noise control measures must be fully maintained and properly used.

(g) Information and training – employees must know the risks from noise and how to properly use the means to protect themselves from these risks.

[01026] Note. Under these regulations it is necessary to assess the cumulative dose of noise during an eight-hour working day. This is done by measurement of the sound levels in the work area (or personal noise dose), and by extrapolation to derive a daily personal exposure over an eight-hour working day (the LEP,d). The values of 85 and 90dB(A) quoted above are LEP,d values. Under the 2005 regulations it is permissible to use weekly exposure, in place of daily exposure, if the noise exposure varies markedly from day to day. It is a legal requirement to have the noise assessment performed by a competent person, which usually means an occupational hygienist with experience of noise assessment work (see **O1013** above).

Regulation 9 requires the employer to place an employee who is likely to be exposed to noise under health surveillance. The choice, issue and maintenance of hearing protection is also covered by the *Personal Protective Equipment at Work Regulations 1992 (SI 1992 No 2966)*.

(See also **NOISE AT WORK** chapter)

Hand arm vibration syndrome (HAVS) and whole body vibration

What is it?

[01027] Hand arm vibration syndrome (HAVS) is a set of symptoms suffered by people whose work exposes them to vibrating hand tools. These include drills, de-scalers, grinders, jackhammers, riveting guns, chain saws, brush cutters etc. Whole body vibration causes a different set of symptoms and is potentially caused by driving or being a passenger on a variety of vehicles or work platforms, especially those driven off-road and over rough terrain.

Vibration describes the oscillatory movements of an object. Vibration can be characterised by its magnitude (the amount of displacement measured as the distance the object moves during the course of its motion) or by the speed it moves or the rate of change of speed (the acceleration). Vibration, either of the hand-arm or whole body, is also characterised by the direction of movement. This can be in the fore-aft direction, or the horizontal or vertical plane or a complex combination of these axes of movement. Although all of these characteristics can be measured, the value that most closely correlates with the health risk is the acceleration. The acceleration, measured in the unit of metres per second per second ($m.s^{-2}$) is the most usual way of describing the magnitude of the vibration in terms of the likely health risk. It is also usual to measure the acceleration in each of the three axes of movement (fore-aft, vertical and horizontal).

The common symptoms of HAVS are:

- painful finger blanching attacks triggered by cold or wet conditions (vibration white finger);
- loss of sense of touch and temperature;
- numbness and tingling;
- loss of grip strength; and
- loss of manual dexterity.

HAVS is a progressive condition, which starts with minor sensory changes to the fingers and progresses either slowly or rapidly to become a serious and debilitating condition. It may cause the sufferer to be unable to continue working.

Whole body vibration causes a number of symptoms, which commonly include:

- back pain, displacement of intervertebral discs, degeneration of the vertebrae, osteoarthritis;
- abdominal pain, digestive disorders;
- increased urinary frequency, prostatitis;
- visual disturbance;
- sleeplessness; and
- headache.

How is it caused?

[01028] The causes of HAVS and symptoms of whole body vibration are not fully understood. HAVS can be due to changes in the blood vessels of the arm and hand, the nerves supplying the blood vessels or in the muscles and bones. The most likely cause of the condition in the majority of cases is damage to the nerves supplying the blood vessels caused by the repeated trauma of the shaking of the upper limb. Vibration white finger is caused by a severe reduction in the blood supply to the fingers but may actually be due to damage of the nerves controlling the blood vessels. Once damaged, the nerves do not recover. HAVS is often triggered, and the condition worsened, by exposure to the cold at the same time as using the vibrating tool or equipment.

The symptoms of whole body vibration are also caused by the trauma of repeatedly being shaken and may involve more generalised symptoms such as fatigue, headache and sleeplessness.

How is it detected?

[01029] HAVS and symptoms of whole body vibration will need to be diagnosed by an occupational physician. Anyone exposed to vibrating equipment should be kept under health surveillance. Health surveillance does not need to be done by an occupational health professional, but the person responsible for undertaking the surveillance must know what to look for and there must be in place a means of referral for expert occupational health advice for anyone suspected of showing early signs or symptoms of the condition.

Vibration white finger is usually quantified and recorded using the Stockholm Scale.

THE STOCKHOLM SCALE

Stage	Grade	Description
0		No attacks
1	Mild	Occasional attacks only affecting the tips of one or more fingers
2	Moderate	Occasional attacks affecting the distal and middle phalanges of one or more fingers
3	Severe	Frequent attacks affecting all the phalanges of most fingers
4	Very severe	As in stage 3 but with trophic skin changes at the finger tips

How is it managed?

[01030] The health risks from vibration exposure, especially hand arm vibration, are serious and debilitating. The risk of these conditions must be effectively managed and competent advice taken if there is thought to be a need for exposure assessment, diagnosis or management of the health effects.

Jobs which expose people to the risk of vibration illness and the individuals who are potentially affected must be identified. A competent risk assessment must be made and the control measures identified by the risk assessment must be put in place. This may include measures such as job rotation to reduce the risk, improved maintenance of equipment to reduce vibration, replacement of old equipment by new, wearing vibration absorbing gloves, making sure that the hands are protected from the cold, and placing at-risk individuals under health surveillance.

Individuals identified by the health surveillance as having early signs or symptoms of the condition must be referred for expert occupational health advice.

What are the legal requirements?

[01031] The EU Physical Hazards (Vibration) Directive (2002/44/EC) gave rise to the *Control of Vibration at Work Regulations 2005 (SI 2005 No 1093)*. Under these regulations there is a requirement to carry out a health risk assessment and to eliminate or control the risks to health arising from vibration. For hand arm vibration there is a daily exposure *limit value* of $5m/s^2 A(8)$ and an *action level* of $2.5m/s^2 A(8)$. For whole body vibration the daily exposure *limit* is set at $1.15m/s^2 A(8)$ and the daily exposure *action value* is $0.5m/s^2 A(8)$. These values all refer to the exposure during a working day normalised to an eight-hour day. Whereas older guidance set limits for vibration in the dominant axis of movement only, the current regulations require the level of vibration to be determined as the sum of each of the three axes of movement (see **O1027** above). The measurement and calculation of daily dose is therefore technically quite complex and will need to be done by a competent person.

The provision of personal protection, which may include gloves to reduce exposure to the cold or to reduce the transmission of vibration, fall under the *Personal Protective Equipment at Work Regulations 1992 (SI 1992 No 2966)*.

In addition, the *Provision and Use of Work Equipment Regulations 1998 (SI 1998 No 2306)*, apply insofar as equipment must be suitable for the work, and must be properly maintained in good repair. *Reg 7* of these regulations requires 'specific risks' in the use of the equipment to be eliminated or adequately controlled. The regulations also require users of equipment to receive adequate health and safety training. Under the *Supply of Machinery (Safety) Regulations 2008 (SI 2008 No 1597)*, manufacturers and suppliers of machinery are obliged to reduce risks to a minimum, to provide data on vibration, and information on risks to health and their control.

(See also **VIBRATION** chapter)

Diseases and disorders of the eye

What are they?

[01032] Diseases and disorders of the eye relate to:

- disorders, including injuries, which are *caused* by the conditions of work; and
- conditions of the eye, such as poor visual acuity or defects in colour vision which affect safety at work or affect the health of the worker.

How are they caused?

[01033] Eye injuries. Injuries are caused by a number of hazards such as:

- physical penetration of the eye with sharp objects (such as metal chips or swarf and wood splinters);
- splashes with acid, alkali or other caustic substances; and
- exposure to high flux of UV light, typically in arc welding operations (arc eye).

Irritation. The surface of the eye (the conjunctiva) can be irritated by exposure to irritant fumes or chemicals (such as formaldehyde, some pesticides and dusts etc, causing conjunctivitis. Conjunctivitis can also be caused by *allergy* to particles in the atmosphere including animal dander (in veterinary work or animal husbandry), plant material including pollens (in horticultural and agricultural workers). Frequently, allergic conjunctivitis is accompanied by irritation of the nasal passages and the respiratory system generally in a condition commonly called hay fever. Employers have a particular responsibility if the employee's work brings them in contact with known allergens, which cause eye or respiratory allergy. However, these are common conditions in the population and many employees will, at certain times of the year, have to work while suffering severe irritation of the eye or have to take time off sick if the condition prevents them doing their work or driving to work. In some cases, the employer may have to determine if the employee is safe to continue working if, for example, they work at heights, drive on the roads or drive forklift trucks. The medication taken to alleviate these conditions may itself cause a safety issue.

Dryness and low humidity. The surface of the eye is also made sore by low humidity. This causes the surface of the eye to dry and become irritant. This is often seen in air-conditioned buildings where the air has low humidity or in cases of 'sick building syndrome'.

Photokeratitis and photoconjunctivitis. Exposure to UV light in particular can cause damage to the tissues of the cornea (photokeratitis) and to the conjunctiva (photoconjunctivitis). Photokeratitis (arc eye) is now rare among arc welders but is, nevertheless, a risk if proper precautions are not observed. Other causes are where there is high reflectivity (snow blindness), in the use of sun beds and in laboratories using UV light. The possibility that with increased outdoor UV radiation in recent years, normal outdoor exposure in the summer may now be sufficient to cause photokeratitis and photoconjunctivitis remains uncertain, but there is sufficient evidence that it may be a risk to suggest that outdoor workers should protect their eyes with suitable sunglasses.

Cataract. Cataracts are caused by the lens of the eye becoming opaque. This may be a natural age-related change or may be caused or hastened by exposures at work. Cataracts may be caused by excessive exposures, or

Diseases and disorders of the eye [O1034]

exposures over a period of time, to ultraviolet light, infra-red light or microwave radiation. Sun lamps, sun beds, welding arcs, lasers and furnaces (including glass blowing), are sources of these radiations which may give rise to the risk of cataract. With increasing levels of natural UV in sunlight, there is growing concern that workers who spend long periods of time outdoors may also be at risk. There is sufficient evidence that this may be the case to suggest wearing suitable sunglasses as a precaution.

Visual acuity. The other eye disorders, which have great importance in occupational health, relate to visual acuity. Visual acuity is the ability of the eye to sharply focus on objects from near to far distance. The young eye can accommodate (ie change its focal length) from about 10cms up to far distance. However, as we get older the ability to focus on near objects diminishes, so that by the age of 60, the near point of vision has receded to about 83cms. Visual acuity is very important in some jobs where the ability to see accurately may have important implications for safety. Driving, for example, has standards determined by the Driver and Vehicle Licensing Authority (www.gov.uk/government/organisations/driver-and-vehicle-licensing-agency).

Visual acuity is important in display screen workers because inability to focus sharply at the average distance of the computer screen is sometimes not corrected by glasses or contact lenses, which correct for near or far sight visual defects. Correction for the intermediate distance of the display screen equipment (DSE) screen may require specific lenses. It is for this reason that there is provision in the *Health and Safety (Display Screen Equipment) Regulations 1992 (SI 1992 No 2792)* for eyesight and vision tests and for the employer to supply suitable corrective lenses (see **O1036** below).

Colour vision. In certain jobs accurate colour perception may be a safety issue. About 8% of males and 0.4% of females are colour blind. The most common form is red-green colour blindness, which makes the distinction between these colours difficult. People working in jobs that require accurate colour perception, or where safety is dependent on colour perception, should have their colour vision tested.

How are they detected?

[O1034] Defects in visual acuity are measured by a number of standard tests. The basic test is the Snellen Chart which requires a person to read letters of diminishing size from a distance of 6 meters (20 feet). The normal visual acuity is quoted as 6/6 (or 20/20 in imperial) although many people have better acuity than this. The UK driving standard is equivalent to a Snellen's score of 6/12.

Other tests may measure binocular vision, the ability of the two eyes to work together to give proper depth and distance perception. Vision tests should normally be carried out by a registered optician or ophthalmic practitioner. However, screening tests, using a vision-screening instrument, are often carried out by occupational health advisors to detect individuals who may have visual difficulties. Anyone with a suspected visual impairment would be referred to an optician for full examination.

Tests for colour blindness can be carried out by occupational health advisors using a vision screening instrument or standard colour charts (Ishihara charts).

Other eye conditions or suspected eye conditions, such as injuries to the eye, allergic inflammation, cataracts, photokeratits or photoconjunctivitis must be referred to a doctor or medically qualified ophthalmic practitioner for diagnosis and treatment.

How are they managed?

[01035] The risk of eye injury must always be anticipated in those work environments where there are hazards such as metal swarf, wood dusts or splinters, mineral and cement, brick and concrete dusts or chips, or when handling chemicals. A risk assessment (see O1036 below) will reveal where such risks may exist and appropriate controls must be put in place. These normally involve the establishment of systems of work (ways of working and adequate supervision), which should prevent such injuries occurring. However good the system of work is, there is always likely to be a residual risk and further safeguards will be necessary. These include the provision and use of eye protection (safety glasses and goggles), and suitable first aid facilities (eye wash bottles) and first aid provision.

Jobs in which visual acuity or colour perception is critical should have eye tests on recruitment and periodically thereafter. The frequency of these will depend on the condition, its likely rate of change and the age of the individual. The occupational health advisor or optician (if referred to one) will normally advise on the frequency of tests. DSE users should have their visual acuity measured at the normal viewing distance of the display screen and, if necessary, be provided with suitable corrective lenses (see O1036 below).

What are the legal requirements?

[01036] There is a duty under the *Management of Health and Safety at Work Regulations 1999 (SI 1999 No 3242)*, for every employer to undertake a suitable and sufficient assessment of the risks to the health and safety of their employees to which they are exposed while they are at work. Where there is a risk of eye injury or eye disease, the risk assessment should identify who is at risk (taking into account any existing vulnerability), what the possible eye injury or condition might be and what controls need to be put in place to prevent or minimise the risk.

Where personal eye protection is considered necessary then the *Personal Protective Equipment at Work Regulations 1992 (SI 1992 No 2966)* apply.

The *Health and Safety (Display Screen Equipment) Regulations 1992 (SI 1992 No 2792)* gives employees who are designated as DSE 'users' the right to have an eye and eyesight test carried out by a competent person. Employers have a legal duty to supply corrective lenses, if needed, for the work at the display screen.

The *Control of Artificial Optical Radiation at Work Regulations 2010 (SI 2010 No 1140)* implemented the requirements of a European Union (EU) Directive on the protection of workers from hazardous sources of artificial light such as UV radiation and powerful lasers, which have the potential to cause harm to the skin or eyes, when they came into force in April 2010.

They apply to organisations using hazardous light sources as part of their work activities. HSE guidance provides examples of hazardous sources of light that present a "reasonably foreseeable" risk of harming the eyes (and skin) of workers and where control measures are needed:

- Metal working – welding (both arc and oxy-fuel) and plasma cutting;
- Pharmaceutical and research – UV fluorescence and sterilisation systems.
- Hot industries – furnaces;
- Printing – UV curing of inks;
- Motor vehicle repairs – UV curing of paints and welding;
- Medical and cosmetic treatments – laser surgery, blue light and UV therapies;
- Intense Pulsed Light sources (IPLs);
- Industry, research and education, for example, all use of Class 3B and Class 4 lasers, as defined in British Standard BS EN 60825-1: 2007; and
- Any Risk Group 3 lamp or lamp system (including LEDs), as defined in British Standard BS EN 62471: 2008, for example search lights, professional projections systems.

It advises that less common hazardous sources are associated with specialist activities – for example lasers used during the manufacture or repair of equipment, which would otherwise not be accessible.

The HSE advice can be downloaded at www.hse.gov.uk/radiation/nonionising/employers-aor.pdf.

Diseases of the skin

What are they?

[01037] The HSE reported in 2017 that there were an estimated 6,000 new cases of self-reported 'skin problems' each year caused or made worse by work, according the Labour Force Survey (LFS), over the last three years. There were an estimated 16,000 people currently or recently in work (within the last year) with skin problems they regard as caused or made worse by work.

Estimated numbers of annual case reports of skin disease by dermatologists within the EPIDERM scheme, part of The Health and Occupation Reporting (THOR) network, are much lower that estimates based on the LFS, and include only those cases serious enough to be seen by a skin disease specialist.

In 2016, there were an estimated 1,207 individuals with new cases of occupational skin disease within EPIDERM. There were 1,261 new diagnoses among these individuals and of these, 74% were contact dermatitis, 9% were other non-cancerous dermatoses (mainly contact urticarial and nail conditions), and the remaining 16% were skin cancers.

Of the occupational dermatitis diagnoses in 2016, 33% were among men and 67% among women. Contact dermatitis often occurs at a young age, particularly among female workers, with 57% of reports to EPIDERM among women were aged less than 35 years compared with 42% among men.

Contact dermatitis is so called because the conditions are caused by contact with substances in the workplace. Contact dermatitis is further classified as:

- irritant contact dermatitis; and
- allergic contact dermatitis.

In addition to the above, other, but much less common diseases of the skin include oil folliculitis, chloracne, leucoderma, and ulcerations.

How are they caused?

[01038] *Irritant contact dermatitis.* Contact with a wide variety of agents that are irritant to the skin can cause this disease. The hands are usually affected because they are most likely to be in contact with the irritant agent. These are either chemical substances or physical agents that cause damage to the cells of the outer layer of the skin. Well known highly irritant substances (strong acids and alkalis for example) are normally well protected against and are not the usual causes of contact irritant dermatitis, although accidental splashes may cause skin burns. It is usually the milder substances, which may not cause an immediate reaction, that are the cause. Irritant contact dermatitis often occurs after a lengthy period of contact with agents that do not immediately cause any problem. The onset is therefore insidious and not usually predictable in any particular individual. There is a large individual variation in susceptibility and the only probable risk factor is a history of eczema as a child affecting the hands.

The agents most frequently encountered as causing irritant contact dermatitis are soaps, detergents, mild acids and alkalis, cutting oils, organic solvents, animal and plant products, and oxidising and reducing agents. Physical agents include friction and low humidity. It is often the case that the dermatitis occurs after exposure to more than one irritant source.

Allergic contact dermatitis. This disease is caused by contact with materials that can penetrate the skin and provoke an immunological response in the skin cells. The type of response, which characterises dermatitis, is called a type IV allergy. The allergic reaction occurs after a period of sensitisation, which may occur after repeated exposure over a prolonged period (many years in some cases) or as little as a single previous contact. Allergic contact dermatitis is much less easy to control than irritant contact dermatitis because once sensitised the person will react to minute quantities of the substance. Substances known to be capable of causing contact allergy are called contact allergens. Hundreds of contact allergens are now known including chromates, methacrylates, formaldehyde, dyes, epoxy resins, and some woods.

Urticaria. An alternative allergic reaction is called urticaria (nettle rash). This is a typical wheal and flare reaction, which occurs within 20–30 minutes of contact with the allergen. This is not classified as an allergic contact dermatitis because it involves a different immune response (a type I reaction) and may involve a whole-body response, including asthma and anaphylaxis (whole-body shock). Reactions to substances causing urticaria may therefore be more serious and even life-threatening. The most common substance at work causing urticaria is latex rubber. Other allergens include proteins from animal skin and hair, and foodstuffs. People working in the food industry, with animals and in the health care industry (using latex gloves) may be at risk.

Other conditions. Oil folliculitis (oil acne), chloracne, leucoderma, ulcerations and skin cancers are caused by contact with (or ingestion of) a variety of substances found in the workplace. **Oil folliculitis** is an irritation of the hair follicles caused usually by contact with petroleum oils. **Chloracne** is a serious skin condition caused by ingestion of certain polychlorinated aromatic hydrocarbons. **Leucoderma** is a chemical depigmentation of the skin caused by exposure to a number of chemicals including phenols, catechols and hydroquinones. Such chemicals are commonly found in products such as adhesives. The skin depigmentation may occur at sites remote from the site of skin contact. **Skin ulcers** are usually caused by contact with wet cement or chromium compounds (chrome ulcers).

Occupational **skin cancers** were once a common industrial disease as a result of exposure to coal tar and machine oils. Outdoor exposure to sunlight may now be the most likely cause of occupational skin cancers. According to the findings of a 2015 study carried out by Imperial College London, skin cancers caused by sun exposure at work now affect nearly 50 workers every year. The research looked at those working outdoors in industries including construction, agriculture, leisure and entertainment. It found that 240 new cases are being registered each year.

A separate study carried out by the University of Nottingham the same year, found a lack of awareness of the risks of solar or ultra violet (UV) radiation in the construction industry. The study into work attitudes to sun safety found that two thirds of construction workers working outside for an average of nearly seven hours a day thought that they were not at risk or were unsure about the risks. Nearly six in ten reported having sunburn, a major contributor to skin cancer, at least once a year. And it found a 'macho culture' in parts of the industry and misconceptions about the risk in the UK climate.

How are they detected?

[O1039] Skin conditions, of whatever type, are easily detected by visual examination. Workers identified by the risk assessment as being at risk of skin disease (ie because they are working with known irritants or allergens), should be placed under health surveillance (see **O1009** above). It is important that the person undertaking the periodic skin inspection (and the people themselves who are at risk) should know exactly what to look for and be able to recognise early signs of a skin complaint. Equally important is to have a system in place for rapid referral to a skin specialist or occupational physician. Although abnormal reactions in the skin are easily detected, the diagnosis of the precise nature of the disease, what is causing it and how to manage it will require expert advice from an occupational physician or dermatologist experienced in occupational skin diseases.

How are they managed?

[01040] Occupational skin diseases can be very debilitating and may render the individual unable to continue their work. Allergic contact dermatitis in particular is difficult to manage once someone has become sensitised because it may be difficult to avoid the very small levels of exposure needed to bring about a reaction. Prevention is therefore always better than cure. The risk assessment should be able to identify those employees who are at risk from exposure to substances and physical agents, which could possibly cause skin conditions. Avoidance of contact with such agents will prevent irritant contact dermatitis and will greatly reduce the possibility of sensitisation and prevent allergic contact dermatitis. Protection against exposure should ideally start with elimination of the substances (through substitution with safer alternatives). If this is not reasonably practicable, then engineering solutions should be considered. Enclosure of processes, local exhaust ventilation, and remote working which avoids handling irritant substances are possible solutions. Personal hygiene is also very important. Staff must be encouraged to wash their hands after handling irritant material and be provided with easily accessible means to do so. Remember though that soaps and detergents may themselves be a cause of dermatitis. Protection may require the use of gloves (but again remembering that latex gloves themselves may be a risk), or barrier creams. Barrier creams may have a place if gloves are difficult to wear but do not afford the same level of protection.

What are the legal requirements?

[01041] Since the major causes of dermatitis and other skin complaints are chemical substances, the primary legal obligations fall under the *Control of Substances Hazardous to Health Regulations 2002 (COSHH) (SI 2002 No 2677)*. Under these regulations there is a requirement to undertake a risk assessment of the health effects and to devise and implement effective control measures including, if necessary, personal protection. The COSHH regulations also require health surveillance under certain circumstances (see the criteria for health surveillance set out in **O1009** above). In the case of skin disease these criteria are met, so health surveillance is mandatory for people at risk. Health records pertaining to individuals have to be kept for 30 years (see **O1016** above). Under the COSHH regulations there are requirements relating to information, instruction and training.

The use of personal protection falls under the *Personal Protective Equipment at Work Regulations 1992 (SI 1992 No 2966)*.

Physical risks, such as friction or low humidity are subject to the general requirements for risk assessment, control, health surveillance, information, instruction and training of the *Management of Health and Safety at Work Regulations 1999 (SI 1999 No 3242)*.

The Health and Safety Executive (HSE) published revised 'simplified and streamlined' guidance on preventing work-related dermatitis and urticaria in 2015. The leaflet explains what the law requires and which jobs present most risk. While those working in health care, hairdressing, the beauty industry, printing, cleaning, catering, construction and metalworking are at greater risk, the guidance makes clear that contact dermatitis and urticaria can occur in just about any workplace. It describes the signs and symptoms of contact dermatitis and contact urticaria and sets out what employers can do to prevent them. The guidance can be found on the HSE website at: www.hse.gov.uk/pubns/indg233.htm.

Occupational respiratory diseases

What are they?

[01042] Respiratory diseases are those that affect the respiratory tract from the nose down to, and including, the lung. The majority of occupational diseases of the respiratory system involve the lower respiratory tract within the lung. Historically, respiratory diseases have been a major cause of industrial illness and death. Diseases associated with dust inhalation in the coalmines (coal worker's pneumoconiosis), in the metal and pottery industries (siderosis and silicosis) and in the cotton industry (byssinosis) were scourges of the industrial revolution. Collectively, diseases of the lung associated with dust inhalation are called the pneumoconioses. Some of these diseases are also associated with development of cancer of the lung. Chronic bronchitis and emphysema (chronic obstructive pulmonary disease – COPD) was also very prevalent. Later, diseases associated with asbestos, (asbestosis, mesothelioma, other asbestos-related lung cancers and pleural thickening), became the respiratory lung disease of major concern. Today, most of the older diseases associated with traditional industries have not completely gone way. Asbestos-related lung cancers can have a very long lag time between exposure and the onset of disease. Asbestos still kills around 5,000 workers each year and the number of deaths due to mesothelioma is still rising. There were 2,542 mesothelioma deaths in 2015, a similar figure to the 2,538 deaths in 2013, but substantially higher than the 2,312 deaths in 2011. There were also 467 asbestosis deaths in 2015.

Other common work-related respiratory diseases are allergies of the upper respiratory tract (allergic rhinitis), and the lung (extrinsic allergic alveolitis and asthma). *Extrinsic allergic alveolitis* is an inflammatory immune response in the lung to the inhalation of a wide variety of both organic and inorganic materials. Examples include bird fancier's lung (from inhaling proteins in avian excreta), farmer's lung (from mould spores in mouldy hay or grain), and mushroom worker's lung (from mushroom spores). Some inorganic substances of industrial importance such as those used in spray painting and polyurethane foam, (diisocyanates and diphenyl methane) can cause the same reaction. Asthma is caused by narrowing of the airways, which is reversible usually over a short period of time. *Occupational asthma* is defined as attacks, which are initiated by inhalation of substances at work. Initiators can include irritants (irritant-induced asthma) or agents which bring about an immune response (hypersensitivity-induced asthma).

Tuberculosis (TB) is also worth noting as an occupational respiratory disease. In 2016, there were 5,664 TB cases notified in England, down from 5,727 in 2015. Following a sustained annual decline of at least 10% in the number of TB cases since 2012, the decline slowed to just 1% in 2016, according to Public Health England. Although it is rarely caused by work, TB has great importance for those working in the health care industry and others, who may come in close contact with vulnerable groups (community social workers, teachers etc). There is a risk in these cases for infected staff to pass on the infection to those in their care. Other infectious diseases of the lung have occupational importance (see O1043 below).

How are they caused?

[O1043] Respiratory diseases are normally caused for one of three main reasons:

- inhalation and retention in the lung of dust and particles;
- inhalation of substances that bring about an allergic reaction; and
- inhalation of infectious organisms.

The resulting pathology is, however, very complex. Inhaled and deposited dusts and particles will have different effects depending on the chemical properties of the substance. Some will be fairly inert and be retained in the lung as patches of dust accumulation. Simple coal worker's pneumoconiosis (the early stages of the disease) is an example of this. Some will trigger the formation of protective fibrous capsules, for example iron dust (siderosis), and some will stimulate wider fibrous changes in the lung leading to serious impairment of lung function and eventually death. Silicosis is an example. Some substances are carcinogenic and their retention in the lung significantly increases the risk of lung cancer. Asbestos has been mentioned as one such substance, others include arsenic, beryllium and cadmium compounds, silica, tars, soots, diesel-engine exhaust particulates, welding fumes, formaldehyde, vinyl chloride monomer and radioactive gases and particles (including radon). This list is of course not exhaustive. Exposure to inhaled carcinogens not only cause cancer in the lung but may also affect the nose and upper airway.

Allergic responses in the lung occur after a period of sensitisation. The allergic reaction releases substances into the lung which either cause inflammation in the lowest parts of the lung (the bronchioles and alveoli) or may cause constriction of the slightly higher parts of the airway. The inflammatory reaction, which also causes fluid build-up in the lower parts of the lung, is the cause of extrinsic allergic alveolitis. The bronchial constriction results in an asthmatic attack. Asthma may also be triggered by irritant substances which do not involve an allergic reaction.

Work-related *infectious disease*s of the lung are not common but can have serious occupational implications. Tuberculosis may be caught at work, but the occupational health risk is far more likely to be that an infected employee transmits the infection to someone else. Animal workers are also at risk from catching a variety of animal-borne lung infections (zoonoses). These include brucellosis, psittacosis and Q fever, which are pneumonia-like illnesses. Legionnaire's disease is a serious type of pneumonia caused by the legionella bacterium infecting water systems. To cause infection the water has to be released as fine spray capable of being inhaled.

How are they detected?

[01044] This depends on the particular disease. It is important for employers to identify individuals who may be at risk from any of the occupational lung diseases. Employees must be made aware of the risks and what symptoms might arise if they become affected. They should be encouraged to report any symptoms such as chest tightness, breathlessness, cough or wheeze and the organisation must have a system in place for immediate referral to an occupational health doctor or nurse (or advise the employee to go to their own doctor with a note of what they are exposed to at work). It may be necessary to place some employees under health surveillance and this typically involves routine lung function testing. Lung function tests are performed using a spirometer, which measures a number of functional capacities. The most useful predictors of early changes that might indicate the onset of lung disease are the forced vital capacity (FVC) and the forced expiratory volume in one second (FEV1). Monitoring the health of someone with asthma is best done with another measure, the peak flow (PF). Modern electronic spirometers calculate the lung function values and indicate abnormal values by comparison with 'predicted' values for the person's age, sex and height. However, this cannot by itself be relied on to show early significant changes in lung function. It is better to compare individuals with themselves over time. It is best practise to have lung function tests done by an occupational health practitioner who can interpret the findings and ask other relevant questions about the individual's respiratory health at the same time.

Other tests for respiratory diseases include chest x-rays.

How are they managed?

[01045] Lung disease is usually serious, debilitating and sometimes life-threatening. As with any occupational disease every effort should be put into prevention. This starts with the risk assessment which, by law, must be thorough, systematic and take into account all the relevant factors. Anyone working with any substance or potentially exposed to infectious organisms, capable of causing respiratory illness, must have a risk assessment undertaken on their work. If it is reasonably practicable, then the first action must be to eliminate the substance (through substitution with a safer alternative for example), then prevent exposure at source. This may mean suitable engineering solutions (such as enclosure or local exhaust extraction), to prevent dust, particulates or fumes from escaping into the work area. It may be necessary, because of the nature of the work, to use respiratory protection, but this should only be used if engineering solutions are not possible or do not provide complete protection. A wide variety of respiratory protection is available. The correct type and specification must be provided for the type of substance it is intended to protect against. It may be necessary to seek the advice of an occupational hygienist.

Individuals at risk of respiratory illness should be placed under health surveillance. The criteria for health surveillance under the *Management of Health and Safety at Work Regulations 1999 (SI 1999 No 3242)* and the *Control of Substances Hazardous to Health Regulations 2002 (SI 2002 No 2677)* are probably met in most cases which means that health surveillance is a legal requirement (see **O1009** above). Health surveillance might include self-surveillance for symptoms, respiratory health questionnaires and lung function tests (see **O1009** and **O1044** above), and should usually be devised and carried out by an occupational health advisor.

What are the legal implications?

[01046]–[01055] Occupational lung disease usually results from inhalation of 'substances' at work and so the primary legal duty falls under the *Control of Substances Hazardous to Health Regulations (COSHH) 2002 (SI 2002 No 2677)*. Infectious organisms are also defined as 'substances' under these regulations. Under these regulations there may be a requirement to monitor the levels of exposure to ensure that they do not fall over any prescribed limits. (Current Workplace Exposure Limit (WEL) values are listed in the guidance document 'HSE Workplace Exposure Limits', (www.hse.gov.uk/pubns/books/eh40.htm).

If the risk assessment under the COSHH regulations determines a need for personal respiratory protection then the choice of this, its suitability, issue, maintenance and instruction and training all fall within the *Personal Protective Equipment at Work Regulations 1992 (SI 1992 No 2966)*.

If the respiratory risk specifically arises from working with asbestos or potentially coming into contact with asbestos, then the *Control of Asbestos at Work Regulations 2012 (SI 2012 No 632)* apply.

Other occupational health concerns

[01056] This short review has focussed on some of the main occupational health issues that confront today's workforce and employers. Inevitably it is not complete and many additional occupational health issues will from time to time arise.

Some of these will involve potentially serious health problems and must be addressed proactively through the usual practice of risk assessment and risk management. The legal requirement for any of these is much the same – assessment of the risk, devising and implementing suitable controls, education and training, monitoring and review, and keeping at-risk individuals under health surveillance.

Other conditions range from the serious and life threatening to those that affect the well-being and comfort of the employee. The more serious (as well as cancers of the lung and skin already mentioned) included cancers of the mouth, the nose, the liver, the bladder and blood cancers. A very large number of chemicals and other agents (such as radiation) used in industry and in manufacturing are known or suspected carcinogens and anyone working with these should be well protected.

Other conditions of note are disorders of the nervous system including the brain. Many commonly used substances such as solvents, metals, insecticides, styrene and carbon monoxide are neurotoxic and capable of causing nerve damage. Radiation poses a special risk. Ionising radiations are well known hazards potentially capable of causing cancers and must be well controlled. Protection against the risk of exposure to ionising radiation comes under the *Ionising Radiations Regulations 2017 (SI 2017 No 1075)* (see RADIATION). However, the risk from non-ionising radiations, such as extra low frequency (from mains electricity), radio frequency and microwaves has been more controversial.

In July 2016, the *Control of Electromagnetic Fields at Work Regulations 2016 (CEMFW Regulations)* came into force, implementing the requirements of a European Union (EU) directive on electromagnetic fields. The directive takes account of recommendations published by the International Commission on Non-Ionizing Radiation Protection (ICNIRP) and introduces a system of exposure limit values (ELVs) 'on the basis of biophysical and biological considerations, in particular on the basis of scientifically well-established short-term and acute direct effects, i.e. thermal effects and electrical stimulation of tissues'.

The directive does not address suggested long-term effects of exposure to electromagnetic fields, including possible carcinogenic effects, although it does require the European Commission to take into account new scientific knowledge in this area.

The CEMFW Regulations are the first specific UK EMF regulations and transpose the requirements of the directive which go beyond, or are more specific than, those in existing UK legislation, that is the *Management of Health and Safety at Work Regulations 1999* and corresponding legislation in Northern Ireland, supported by a Public Health England recommendation that the international ICNIRP guidelines be followed.

Under the CEMFW Regulations, employers must:

- assess the levels of EMFs to which employees may be exposed;
- ensure that exposure is below a set of ELVs;
- when appropriate, assess the risks of workers' exposure and eliminate or minimise those risks. The employer must ensure they take workers at particular risk, including expectant mothers and workers with active or passive implanted or body-worn medical devices, into account;
- when appropriate, devise and implement an action plan to ensure compliance with the exposure limits;
- provide information and training on the particular risks (if any) posed to employees by EMFs in the workplace and details of any action they are taking to remove or control them. This information should also be made available to safety representatives;
- take action if employees are exposed to EMFs in excess of the ELVs; and
- provide health surveillance as appropriate.

See RADIATION for more information.

There is also mounting concern about the health impact of working at night, particularly the link between night work and cancer. For example, researchers at Sichuan University in China reviewed 61 studies on the effects of night work on the health of women published between 1996 and 2017. The data cover 115,000 cases of cancer and four million participants living in the US, Australia, Asia and Europe. They concluded that the long-term risk of breast cancer increases by 32%, that of skin cancer by 41%, and that of gastrointestinal cancers by 18% for those working nights. The risk of breast cancer increases by 3.3% for every five years of night work. In addition, a study published in the *Scandinavian Journal of Work Environment & Health* found that male workers carrying out shift or night work had a higher risk of contracting prostate cancer. They found the risk is double that for workers who have never worked shifts, while men who have worked shifts or nights for more than 20 years are three or four times more likely to get prostate cancer.

Occupational cancer kills 100,000 people every year across the European Union and is the most common work-related cause of death. In May 2016 and January 2017, the European Commission proposed two sets of amendments to the carcinogens and mutagens directive to add more carcinogens (cancer causing substances) to the list of controlled substances. The main effect of the first proposal would be to set new exposure limits for respirable crystalline silica (RCS) dust, to which workers in the construction and quarrying sectors are frequently exposed, and for hardwood dust, which is common in woodworking and construction.

The second proposal would bring exposure to used engine oils within the scope of the directive and assign a 'skin notation' to these oils. It would add five carcinogenic substances to the list of chemicals covered by the Directive. It also extends the scope of the Directive to a new category of Polycyclic Aromatic Hydrocarbons (PAHs).

In July 2017, the European Parliament and Council of Ministers reached agreement on first of the two proposals, which also included lower exposure limit values for Chromium (VI) compounds and an extension of the existing requirement for employers to conduct health surveillance of their workers.

This review is not a complete description of the many and varied diseases and disorders that may be caused or worsened by the conditions of work, nor has it dealt in detail with the various legal requirements. Readers who believe they may have a health issue in their workplace should seek advice on what they may need to do to manage such problems and avoid the inevitable subsequent cost and disruption to work. Occupational health problems rarely go away on their own; many are cumulative and will progress to more serious conditions the longer they are ignored. The key to good occupational health management is to anticipate what might cause ill health and to avert the risk of serious harm to the employee and to the organisation.

Many occupational health conditions will require reporting to the HSE under the *Reporting of Injuries, Diseases and Dangerous Occurrences Regulations 2013 (RIDDOR) (SI 2013 No 1471)*.

Where a worker is diagnosed with any of the following occupational diseases:

(a) Carpal Tunnel Syndrome, where the person's work involves regular use of percussive or vibrating tools;

(b) cramp in the hand or forearm, where the person's work involves prolonged periods of repetitive movement of the fingers, hand or arm;
(c) occupational dermatitis, where the person's work involves significant or regular exposure to a known skin sensitizer or irritant;
(d) Hand Arm Vibration Syndrome, where the person's work involves regular use of percussive or vibrating tools, or the holding of materials which are subject to percussive processes, or processes causing vibration;
(e) asthma, where the person's work involves significant or regular exposure to a known respiratory sensitizer; or
(f) tendonitis or tenosynovitis in the hand or forearm, where the person's work is physically demanding and involves frequent, repetitive movements,

their employer must report this to the relevant enforcement authority as set out in the regulations.

In addition, employers must report workers diagnosed with:

(a) any cancer attributed to an occupational exposure to a known human carcinogen or mutagen (including ionising radiation); or
(b) any disease attributed to an occupational exposure to a biological agent.

The legal duties cited in this chapter apply mainly to regulations which come under the *Health and Safety at Work etc Act 1974*. These should be interpreted together with the general provisions of the Act itself.

Occupiers' Liability

Alison Newstead

Introduction to occupiers' liability

[O3001] Inevitably, by far the greater part of this work details the duties of employers (and employees) both at common law, and by virtue of the *Health and Safety at Work etc Act 1974* (*HSWA 1974*) and other related legislation. Duties are also laid on persons including companies and local authorities who merely occupy premises which others visit or carry out work activities on, eg repair work or servicing. More particularly, persons in control of premises (see **O3004** for meaning of 'control'), that is occupiers and employers, where others work (although not their employees), have a duty under *HSWA 1974* s 4 to take reasonable care towards them. Failure to comply with this duty can lead to prosecution and a fine on conviction (see further **ENFORCEMENT**). This duty applies to premises not exclusively used for private residence, eg lifts and electrical installations in the common parts of a block of flats, and exists for the benefit of workmen repairing or servicing them (*Westminster City Council v Select Managements Ltd* [1985] 1 ALL ER 897). Moreover, a person who is injured while working on or visiting premises may be able to sue the occupier for damages, even though the injured person is not an employee. Statute law relating to this branch of civil liability (ie occupiers' liability) is to be found in the *Occupiers' Liability Act 1957* (*OLA 1957*) and, as far as persons other than visitors (ie trespassers) are concerned, in the *Occupiers' Liability Act 1984* (*OLA 1984*). (In Scotland, the law is to be found in the *Occupiers' Liability (Scotland) Act 1960*.)

Duties owed under the Occupiers' Liability Act 1957

[O3002] An occupier of premises owes the same duty, the 'common duty of care', to all his lawful visitors. [*OLA 1957, s 2(1)*]. 'The common duty of care is a duty to take such care as in all the circumstances of the case is reasonable to see that the visitor will be reasonably safe in using the premises for the purposes for which he is invited or permitted by the occupier to be there' [*OLA 1957, s 2(2)*]. Thus, a local authority which failed, in severe winter weather, to see that a path in school grounds was swept free of snow and treated with salt and was not in a slippery condition, was in breach of s 2, when a schoolteacher fell at 8.30 am and was injured (*Murphy v Bradford Metropolitan Council* [1992] 2 All ER 908). Similarly, the Court of Appeal held in *Brioland Limited v Searson* [2005] All ER (D) 197 (Jan) that the owner of a hotel had breached s 2 by not displaying a sign warning guests of the unusual presence of a 2.8cm high upstand in the doorway of the entrance to the hotel over which the claimant tripped whilst looking ahead and sustained severe injuries.

The duty extends only to requiring the occupier to do what is 'reasonable' in all the circumstances to ensure that the visitor will be safe. Some injuries occur where no one is obviously to blame. For example, in *Graney v Liverpool County Council* (1995), (unreported) an employee who slipped on icy paving slabs could not recover damages against his employer. The Court of Appeal held that many people slip on icy surfaces in cold weather – it did not follow that in this particular case the employer was to blame. In the case of *Clare v Perry* [2005] EWCA Civ 39, [2005] ALL ER (D) 67 (Jan), the claimant deliberately jumped over the edge of a wall running alongside the defendant's hotel car park, to access the road below (even though a safe exit was provided). The claimant sustained serious injuries. The Court of Appeal dismissed the claimant's argument that the defendant had breached s 2 by failing to fence off the highest part of the wall and held that the standard of care imposed on the defendant did not extend to protecting people against their 'foolhardy and unexpected conduct' which was, in this case, unforeseeable. Similarly, in *Tacagni v Cornwall County Council* [2013] EWCA Civ 702, [2013] All ER (D) 182 (Apr), the claimant fell from a two-metre raised footpath into the road and sustained head injuries. The accident happened at night, on a badly lit road and the claimant had been drinking. The claimant used a fence (put up as a result of a landslip) as a guiderail. She fell where the fence ended. The Court of Appeal held that although the path had presented a real possibility of danger, the erection of the fence had been a proportionate reaction to the landslip and there had been no suggestion that further fencing was required. Furthermore, it had not been necessary to consider the scenario where someone would use the fence as a guiderail. As a result, there was no breach of s 2; the judge considering that a highly material factor was what could be expected of an ordinary visitor.

In *Esdale v Dover District Council* [2010] EWCA Civ 409, [2010] All ER (D) 73 (Oct) a tenant was injured when she tripped and fell on a pathway to her block of flats. The council was the owner and occupier of the flats and the pathway and was aware of the defect. The tenant tripped as a result of a small change of level in the path (between ¾ and one inch. The Court held that the relevant test did not depend on what standards of safety the council had set for itself. The test was objective. Consideration had to be given as to whether the authority took such steps as were reasonable in all the circumstances to keep the visitor to the premises reasonably safe. The judge found that the path was reasonably safe and the council had complied with its duty of care by regularly inspecting it.

Nature of the duty

[03003] The *OLA 1957* is concerned only with civil liability. 'The rules so enacted in relation to an occupier of premises and his visitors shall also apply, in like manner and to like extent as the principles applicable at common law to an occupier of premises and his invitees or licensees would apply . . . ' [*OLA 1957, s 1(3)*]. This contrasts with the *HSWA 1974*, in which the obligations are predominantly penal measures enforced by the HSE (or some other 'enforcing authority'). The OLA 1957 cannot be enforced by a state agency – nor does it give rise to criminal liability. Action under the OLA 1957 must be brought in a private suit between parties.

The liability of an occupier towards lawful visitors at common law was generally based on negligence; so, too, is the liability under the OLA 1957. It is never strict. In *McGivney v Golderslea Ltd* (1997), (unreported) the defendant owned a block of flats which had been built in 1955. The glass in the door at the foot of the communal stairs had been installed at the same time. When the claimant, who was visiting a friend on the third floor, came down the stairs, he slipped and his hand went through the glass panel on the front door resulting in personal injury. The claimant claimed that the defendant was in breach of the OLA 1957 s 2(2) in that the glass was unsafe and should have been replaced with thicker safety glass. However, at the time the flats were built, the glass satisfied the building regulations then in force. There was therefore no breach of any statutory duty. The Court of Appeal stressed the words in s 2(2) "to take such care as in all the circumstances of the case is reasonable", and held that it was impossible to find the defendant in breach of s 2(2) of the OLA 1957 in failing to remove a pane of glass which complied with the regulations when it was installed and to replace it with more up-to-date safety glass.

This can be contrasted with the case of *Moloney v Lambeth Borough Council* (1966) 64 LGR 440, in which the child of a tenant of a block of flats tripped on the stairs, fell through the balustrade and was seriously injured. There was no evidence that the child was misbehaving. In this case, the local authority contended that the balustrade conformed with all reasonable standards of safety under the Act since no other accidents had occurred. The Court found that whilst the evidence of absence of other accidents supported the local authority's contention that the balustrade conformed with all reasonable standards of safety, the local authority should have envisaged that a child would use the stairs unaccompanied and, taking into account the position and gaps in the balustrade, it was a matter of common sense that the staircase did not comply with the occupier's common law duty of care under OLA 1957.

In *Butcher v Southend-on-Sea* [2014] EWCA Civ 1556, [2014] All ER (D) 87 (Dec), the claimant exited the back door of a local authority premises and rather than following a tarmac path diagonally, stepped off the path to take a straight line course. She lost her balance due to a drop of 2½ inches at the edge of the path and was injured. The Court of Appeal upheld the finding of the judge at first instance in that it had been reasonably foreseeable that someone would take such a route and the local authority was in breach of its duty under the OLA 1957.

In *Stevens v Blaenau Gwent County BC* [2004] EWCA Civ 715, [2004] All ER (D) 116 (Jun), a case which proceeded on the basis of common law negligence and not under the OLA 1957, the Court of Appeal held that the defendant local authority was not liable in negligence for the injuries suffered by a child who had fallen out of the window of a local authority house. The defendant had refused to comply with the claimant's mother's request to install safety locks on the windows of her house to protect her children, on the basis of fire regulations. The Court of Appeal held that the claimant had not demonstrated that the local authority owed him a duty of care. The authority were entitled to apply their policy on locks arrived at having assessed the risks – in the absence of any evidence that it presented exceptional risk to the claimant's family.

The relationship of occupier and visitor is less immediate than that of employer and employee, and is not, as far as occupiers' liability is concerned, underwritten by compulsory insurance as would be the case with employers' liability insurance. Such a relationship would be covered by public liability insurance, which is not generally obligatory but advisable.

Who is an 'occupier'?

[03004] 'Occupation' is not defined in OLA 1957, which merely states that the 'rules regulate the nature of the duty imposed by law in consequence of a person's occupation or control of premises . . . but they shall not alter the rules of the common law as to the persons on whom a duty is so imposed or to whom it is owed . . . ' [OLA 1957, s 1(2)]. The meaning of 'occupation' must, therefore, be gleaned from common law which focuses primarily on the issue of control. Determining who is an occupier is a question of fact.

Where premises, including factory premises, are leased or subleased, control may be shared by lessor and lessee or by sublessor and sub-lessee.

> Wherever a person has a sufficient degree of control over premises that he ought to realise that any failure on his part to use care may result in injury to a person coming lawfully there, then he is an "occupier" and the person coming lawfully there is his "visitor" and the "occupier" is under a duty to his "visitor" to use reasonable care. In order to be an occupier it is not necessary for a person to have entire control over the premises. He need not have exclusive occupation. Suffice it that he has some degree of control with others.

(Per Lord Denning in *Wheat v E Lacon & Co Ltd* [1965] 2 All ER 700.)

In *Jordan v Achara* (1988) 20 HLR 607, the claimant, a meter reader, was injured when he fell down stairs in the basement of a house. The defendant landlord, the owner of the house, had divided it into flats. Because of arrears of payment, the electricity supply had been disconnected. The local authority, having arranged for it to be reconnected for the tenants, took over responsibility for the payment of electricity charges and the right to have rents payable to them. However, it was held that the landlord was liable for the injury, under OLA 1957, since he was the occupier of the staircase and the passageway where the injury occurred and maintained his control of the common parts (and therefore his duties under OLA 1957) in spite of the local authority being in receipt of rents.

Liability associated with dangers arising from maintenance and repair of premises will fall upon the person responsible for maintenance and/or repair under the lease or sublease:

> The duty of the defendants here arose not out of contract, but because they, as the requisitioning authority, were in law in possession of the house and were in practice responsible for repairs . . . and this control imposed upon them a duty to every person lawfully on the premises to take reasonable care to prevent damage through want of repair.

(Per Denning LJ in *Greene v Chelsea BC* [1954] 2 All ER 318, concerning a defective ceiling which collapsed, injuring the appellant, a licensee.)

Significantly managerial control constitutes 'occupation' (*Wheat v Lacon*, see above, concerning an injury to a lodger at a public house owned by a brewery and managed by a manager. Both were held to be 'in control'). The case of *Fischer v CHT Ltd (No 2)* [1966] 1 All ER 88, further demonstrates that two or more people may be occupiers of the same land, each under an obligation to use such care as is reasonable in relation to his degree of control. In this case, the owner of a club and the defendant who ran a restaurant in the club were both found to be occupiers. The more recent case of *Rhind v Astbury Water Park Ltd and another* [2003] EWHC 1029 (QB), [2003] All ER (D) 217 (May) confirms this principle. In this case, the first defendant, *Astbury Water Park Ltd*, obtained a ten-year licence from freeholders of a mere and surrounding land. They in turn entered into an agreement with the second defendant, *Maxout Ltd (Astbury Sailing School)*. Although the case was dismissed on the basis that there was no breach of duty on the part of the defendants, it was held that—

> Two or more people may be in occupation of the same premises and are under a duty of care to visitors. Whether a particular occupier is in breach of the common duty of care to a particular visitor will depend on the circumstances [. . .] Thus the circumstances that may give rise to liability will include the extent and purpose of a particular occupier's occupation and the nature of his ability on the premises and his relationship with the visitor.

(Per Morland J)

More recently, the case of *Furmedge v Chester-Le-Street District Council* [2011] EWHC 1226 (QB), [2011] All ER (D) 162 (May) considered the question of whether a number of people could be constituted as an occupier. The case related to apportionment of liability between two parties following a fatal accident involving a public art installation. Both defendants were involved in the organisation of the event. The Court considered that:

> 'the issue of who is occupier (whether sole or one of a number) is in fact a sensitive issue. It normally requires some degree of physical control over the premises, even if it is not entire or exclusive. An appreciation that a failure to take care could result in injury to someone coming onto the premises is another factor'.

If a landlord leases part of a building but retains other parts, such as the roof, common staircase, or lifts, he remains liable for that part of the premises (*Moloney v Lambeth BC* see above). The Court of Appeal has since held (*Ribee v Norrie* [2001] PIQR P 8), that the proper question was not what actual control the landlord did or did not exercise but what power of control he had, ie whether or not he had the authority to act and was in a position where he could have taken steps to prevent certain risks.

Where the owner of land has no actual knowledge of ownership, he may still be deemed an occupier. In *Harvey v Plymouth City Council* [2010] EWCA Civ 860, 154 Sol Jo (no 31) 30, the council granted a licence to a supermarket to use some land. An accident occurred sometime later, after the licence had expired. While the council's Park Department did not realise it owned the land or that it had a duty to maintain the land, the Court considered that actual position would have been known to the council's legal department and thus the council conceded it was the occupier.

However, it is not the occupier's duty to make sure that the visitor is *completely safe*, only reasonably so. For example, in *Berryman v Hounslow London Borough Council* [1997] PIQR P 83, the claimant injured her back when she was forced to carry heavy shopping up several flights of stairs because the lift in the block of flats, maintained by the landlord, was broken. The landlord's obligation to take care to ensure that the lift he provided was reasonably safe was discharged by employing competent contractors. There was no causal connection between the claimant's injury on the stairs and the absence of a lift service. In other words, the injury to the claimant was not a 'foreseeable consequence' of the broken lift.

Premises

[03005] *OLA 1957* regulates the nature of the duty imposed by law in consequence of a person's occupation of premises [*OLA 1957 s 1(2)*]. This means that the duties are not personal duties but depend on occupation of premises and extend to a 'person occupying, or having control over, any fixed or movable structure, including any vessel, vehicle or aircraft' [*OLA 1957 s 1(3)(a)*] (eg a car, in *Houweling v Wesseler* [1963] 40 DLR(2d) 956-Canada or a sea wall in *Staples v West Dorset District Council* (1995) 93 LGR 536 (see below)) or even a 'dreamspace' public art installation in *Furmedge v Chester-Le-Street District Council* [2011] EWHC 1226 (QB), [2011] All ER (D) 162 (May) (see **O3004**). In *Bunker v Charles Brand & Son Ltd* [1969] 2 ALL ER 59 the defendant was a contractor, digging a tunnel for the construction of the London Underground Victoria Line. The contractor used a large digging machine owned by London Transport which moved forward on rollers. The claimant was injured when he slipped on the rollers. The defendant was held to be the occupier of the tunnel, even though it was owned by London Transport, as it retained control of the tunnel and the machine.

To who is the duty owed?

Visitors

[03006] Visitors to premises entitled to protection under the Act are both (*a*) invitees and (*b*) licensees. This means that protection is afforded to all lawful visitors, whether the visitors enter for the occupier's benefit (clients or customers) or for their own benefit (factory inspectors, policemen), though not to persons exercising a public or private way over premises [*OLA 1957 s 2(6)*] (see **O3007** Countryside and Rights of Way Act 2000). A person entering premises in exercise of rights under this Act is not considered a visitor of the occupier of those premises for the purposes of the *OLA 1957*.

Nevertheless, occupiers are under a duty to erect a notice warning visitors of the immediacy of a danger (*Rae v Mars (UK) Ltd* [1990] 03 EG 80). (In this case, a deep pit was situated very close to the entrance of a dark shed. There was no artificial lighting and a visiting surveyor fell, sustaining injury. It was held that the occupier should have erected a warning). In some instances it may be necessary not only to provide proper warning notices but also to arrange for the supervision of visitors at the occupier's premises (*Farrant v Thanet District Council* (1996), unreported). However, there is no duty on an employer to light premises at night, which are infrequently used by day and not occupied at night. It is sufficient to provide a torch (*Capitano v Leeds Eastern Health Authority* (Current Law, October 1989). In this case a security officer was injured when he fell down a flight of stairs at night while checking the premises following the sounding of a burglar alarm. The steps were formerly part of a fire escape route but were now infrequently used). However, landowners need not necessarily warn against obvious dangers (see O3017).

Harvey v Plymouth City Council [2010] EWCA Civ 860 154 Sol Jo (no 31) 30 confirmed that a landowner owes a duty under the *OLA 1957* only to those who use its land within the terms of the license granted to them. In this case, a young man had injured himself on the council's land after running from a taxi without paying. He had run across parkland late at night, tripped on a section of chain link fence and fallen on to a car park below. The court held that the council had licensed the public to use the parkland for normal recreational purposes carrying normal risks, not for any activity, however reckless. He was therefore not the council's "visitor" under the *OLA 1957*.

It is not the occupier's duty to make sure that the visitor is *completely safe*, only reasonably so. (See *Berryman v Hounslow London Borough Council* (1996) 30 HLR 567, [1997] PIQR P 83, CA, at O3005.)

However, in considering whether a duty of care is owed to a visitor under *s 2(2)* of the *OLA 1957*, the occupier is also required to take account of any known vulnerability of the visitor. In *Pollock v Cahill* [2015] EWHC 2260 (QB), [2015] All ER (D) 330 (Jul), the claimant – a blind man – fell from an open second floor window at the defendant's home. The Court held that reference to 'such a visitor' in *s 2(3)* of the *OLA 1957* required the occupier to have regard to any known vulnerability:

'If [the claimant] had been a sighted person, the open window would not have rendered the premises unsafe. It was the fact that he was blind that made them so'

(per Davis J).

Trespassers

[03007]

' . . . a trespasser is not necessarily a bad man. A burglar is a trespasser; but so too is a law-abiding citizen who unhindered strolls across an open field. The statement that a trespasser comes upon land at his own risk has been treated as applying to all who trespass, to those who come for nefarious purposes and those who merely bruise the grass, to those who know their presence is resented and those who have no reason to think so.'

(*Railways Comr (NSW) v Cardy* (1961) 104 CLR 274, Aus HC.)

Common law defines a trespasser as a person who:

(a)   goes onto premises without invitation or permission; or

(b) although invited or permitted to be on premises, goes to a part of the premises to which the invitation or permission does not extend; or
(c) remains on premises after the invitation or permission to be there has expired; or
(d) deposits goods on premises when not authorised to do so.

The *Countryside and Rights of Way Act 2000* (*CRWA 2000*) gives the public the right to roam across the open countryside in England and Wales. *CRWA 2000 s 13* amends *OLA 1957 s 1(4)* so that: 'A person entering any premises in exercise of [the right to roam] is not, for the purposes of this Act, a visitor of the occupier of the premises.' Therefore, ramblers under the *CRWA 2000* are also classed as 'trespassers'. Indeed, *CRWA 2000* amends *OLA 1984 s 1* so that no duty of care is owed to a person exercising the right to roam under the Act in respect of: 'a risk resulting from the existence of any natural feature of the landscape, or any river, stream, ditch or pond whether or not a natural feature . . . or a risk of that person suffering injury when passing over, under or through any wall, fence or gate, except by proper use of the gate of or a style.' The 2000 Act therefore explicitly states what the position is in law – whilst 'visitors' falling under the 1957 Act must refer to case law.

Duty owed to trespassers, at common law and under the Occupiers' Liability Act 1984

Common law

[03008] The position under common law used to be that an occupier was not liable for injury caused to a trespasser, unless the injury was either intentional or done with reckless disregard for the trespasser's presence (*R Addie & Sons (Collieries) Ltd v Dumbreck* [1929] AC 358). This is no longer true and in more recent times, the common law has adopted an attitude of humane conscientiousness towards simple (as distinct from aggravated) trespassers. ' . . . the question whether an occupier is liable in respect of an accident to a trespasser on his land would depend on whether a conscientious, humane man with his knowledge, skill and resources could reasonably have been expected to have done, or refrained from doing, before the accident, something which would have avoided it. If he knew before the accident that there was a substantial probability that trespassers would come, I think that most people would regard as culpable failure to give any thought to their safety' (*Herrington v British Railways Board* [1972] 1 All ER 749). The effect of this House of Lords case was that it became possible for a trespasser to bring a successful action in negligence under the common law. However, the precise nature of the occupier's duty to the trespasser remained unclear, leading to Parliamentary intervention culminating in the *OLA 1984*.

Occupiers' Liability Act 1984

[03009] *OLA 1984* introduced a duty on an occupier in respect of trespassers, that is to persons other than visitors, in respect of any risk of them

suffering injury on the premises (either because of any danger due to the state of the premises, or things done or omitted to be done on them) [OLA 1984 s 1(1)(a)]. Such a duty arises in respect of the occupier if—

(a) he was aware of the danger or had reasonable grounds to believe that it exists;
(b) he knows or has reasonable grounds to believe that a trespasser is in the vicinity of the danger concerned, or that he may come into the vicinity of the danger (whether the trespasser has lawful grounds for being in that vicinity or not); and
(c) the risk is one against which, in all the circumstances of the case, he may reasonably be expected to offer some protection.

[OLA s 1(3)]

Where under *OLA 1984*, the occupier is under a duty of care to the trespasser in respect of the risk, the standard of care is 'to take such care as is reasonable in all the circumstances of the case to see that he does not suffer injury on the premises by reason of the danger concerned' [*OLA 1984, s 1(4)*]. The duty can be discharged by issuing a warning, giving warning of the danger or to discourage persons from incurring the risk. One way to do this would be to post notices warning of hazards. However, these must be explicit and not merely vague. Thus, 'Danger' might not be sufficient whereas 'Highly Flammable Liquid Vapours – No Smoking' would be [*OLA 1984 s 1(5)*] (see **O3015**). Moreover, under *OLA 1984* there is no duty to persons who willingly accept risks (see **O3016**). However, the fact that an occupier has taken precautions to prevent persons going on to his land where there is a danger, does not mean that the occupier has reason to believe that someone would be likely to come into the vicinity of the danger, thereby owing a duty to the trespasser under the *OLA 1984 s 1(4)* (*White v St Albans City and District Council*, (1990) Times, 12 March).

In *Revill v Newbery* [1996] 1 All ER 291, the court considered the duty of an occupier to an intruder. It found that the intruder could recover damages despite being engaged in unlawful activities at the time the injury occurred. The defendant was held to have injured the intruder by using greater force than was justified (by shooting the intruder through a locked door) in protecting himself and his property. The doctrine of *ex turpi causa non oritur actio* (ie no cause of action may be founded on an immoral or illegal act) cannot, without more, be relied upon by the defendant to escape liability. A person is entitled to use reasonable and necessary force to protect himself, others and his property (*R v Owino* [1995] Crim LR 743). Violence may be returned with necessary violence, but the force used must not exceed the limits of what is reasonable in all the circumstances. Courts have recognised that in considering what is 'reasonable' a person being attacked may overreact with the measure of his response (*Palmer v R* [1971] 1 All ER 1077). For the extent of the intruder's contributory negligence, see **O3011**.

It should be noted that dangers arising from the trespasser's own activities rather than the premises themselves will not render the occupier liable (See *OLA 1984 s 1(1), (9)* and *Keown v Coventry Healthcare NHS Trust* [2006] EWCA Civ 39, [2006] 1 WLR 953 in which a twelve year old fell from a fire escape, having climbed up the underside. The judge held that the fire escape itself was not dangerous, and any danger was due to the claimant's own actions on the premises and not due to the state of the premises themselves. See also *Siddorn v Patel* [2007] EWHC 1248 (QB), [2007] All ER (D) 453 (Mar) in which the claimant rented a flat from the defendants. The claimant sustained injuries having fallen through a skylight on a flat roof (which did not form part of the demised premises) and which had been accessed via a lounge window during the course of a party. The Court held that the claimant had failed to establish that the flat roof and skylights were dangerous within the meaning of *OLA 1984 s 1(1)(a)*; the danger had arisen out of the activities of those on the roof rather than the state of the premises. In *Kolasa v Ealing Hospital NHS Trust* [2015] EWHC 289 (QB), [2015] All ER (D) 262 (Feb), a man fell from a wall outside a hospital A&E Department, which he left after being told to wait for treatment. The man was intoxicated at the time. The Court held that at the time of the accident, the individual was a trespasser, or phrasing it alternatively, *'no longer a visitor acting within the scope of his permission'*.

Children

[03010] 'An occupier must be prepared for children to be less careful than adults' [*OLA 1957 s 2(3)(a)*]. Where an adult would be regarded as a trespasser, a child is likely to qualify as an implied licensee, in spite of the stricture that:

> it is hard to see how infantile temptations can give rights however much they excuse peccadilloes.

(Per Hamilton LJ in *Latham v Johnson & Nephew Ltd* [1913] 1 KB 398).

If there is something or some state of affairs on the premises (eg machinery, a boat, a pond, bright berries, a motor car, forklift truck, scaffolding), this may constitute a 'trap' to a child. If the child is then injured by the 'trap', the occupier will often be liable. (See *Dornan v Department of the Environment for Northern Ireland* [1993] NI 1, in which the concept of what action may constitute a trap was considered). In the case of *Phipps v Rochester Corporation* [1955] 1 All ER 129 the Court noted that the existence of a trench dug for the purposes of laying a sewer did not constitute an allurement for a child, or a danger concealed from adults, but that it was a danger imperceptible to a child of five years of age, because he did not realise that care should be taken to avoid the trench and, in addition, children did not frequently venture into the area unaccompanied. The Court therefore held that the 'licence' allowing the claimant to enter onto the land was a condition that he be accompanied by an adult. As he was not, the claim failed. See also *Southern Portland Cement v Cooper* [1974] AC 623 in which the Court held that the defendants (limestone quarriers) could have expected the presence of children at the quarry (a sandhill being an allurement to the children) and it would have been easy for them to take steps to prevent the dangerous situation and the 13–year old claimant's injuries.

Although under *OLA 1984* a duty of care will normally only arise when an occupier knows that a trespasser may come upon a danger on his premises that is latent (see **O3017**, *Titchener v British Railways Board*), the judge held in *Young v Kent County Council* [2005] EWHC 1342 (QB), [2005] All ER (D)

217 (Mar) that an occupier may also owe a duty under the Act to protect children against an obvious risk. In this case, the claimant, who was 12 years old at the time of the accident, had climbed on the roof of a school building and had fallen through a skylight. The judge held that as the claimant was a child, and that it was probable on the evidence available that the defendant knew that children were climbing on the roof of the building, it was foreseeable that children might injure themselves. The defendant should therefore have taken the cheap, precautionary measure of fencing off the access to the roof. The notices warning the children against the dangers of going on the roof were not sufficient to discharge this duty. The defendant therefore breached its duty of care under the Act. However, the judge also held that the claimant was 50 per cent contributory negligent as he would have been aware of the warning notices and of the fact that he was misbehaving when climbing on the roof.

The case of *Jolley v Sutton London BC* [2000] 3 All ER 409, concerned an old wooden boat that had been dumped on a grassed area belonging to a block of local authority flats. It became rotten and decayed, but was not covered or fenced off. The council placed a danger sticker on the boat, warning people not to touch it. This did not deter two teenage boys, who decided to repair it so they could take it sailing. Using a car jack and some wood from home, they jacked up the front of the boat so that they could repair the hull. One of the boys was lying on the ground underneath the boat when the prop gave way. The boat fell on him and caused very serious spinal injuries.

The Court of Appeal ruled that although it was foreseeable that children who were attracted by the boat might climb on it and be injured by the rotten planking giving way, it was not reasonably foreseeable that an accident could occur as a result of the boys deciding to work under a propped-up boat. However, the House of Lords decided this was wrong and that the council was liable; the trial judge was right in saying that the wider risk was that children would meddle with the boat at the risk of some physical injury, and what the boys did was not so different from normal play and was reasonably foreseeable.

The case of *Pierce (a child by his litigation friend Annette Pierce) v West Sussex County Council* [2013] EWCA Civ 1230, [2013] All ER (D) 166 (Oct), considered the case of a child who had injured his thumb on the underside of a newly installed water fountain at his school. He had been larking about when the accident happened. The judge considered that harm must be reasonably foreseeable and that there should be consideration of what a reasonable defendant would and should do in light of the harm that can be foreseen. Schools had an obligation to keep children safe, but not safe in all circumstances. See also *Hufton v Somerset County Council* [2011] EWCA Civ 789, [2011] All ER (D) 72 (Jul), which concerns a pupil who slipped on a wet school floor on a rainy day. The Court of Appeal held that the school was not expected to take absolute measures which would have prevented any accident from ever occurring. All that was required was the exercise of reasonable care.

Contributory negligence

[03011] In cases brought under *OLA 1957*, the question of contributory negligence is often considered and impacts considerably on the amount of damages eventually awarded. This may be because the duty on the occupier is only to take such care as is reasonable to see that the visitor will be reasonably safe in using his premises. Therefore, it is possible to imagine that, even where an occupier is found to have breached his duty of care, the injured party may also, by his own conduct, have contributed to the circumstances giving rise to the injury and may have exacerbated the severity of the injuries sustained. The *Law Reform (Contributory Negligence) Act 1945 s 1(1)* provides that 'where any person suffers damage as the result partly of his own fault and partly of the fault of any other person or persons, a claim in respect of that damage shall not be defeated by reason of the fault of the person suffering the damage, but the damages recoverable in respect thereof shall be reduced to such extent as the court thinks just and equitable having regard to the claimant's share in the responsibility of the damage'. There is nothing to suggest that the above provision does not apply to children or trespassers. However, Lord Denning in *Gough v Thorne* [1966] 3 All ER 398 (a case in which it was held that a 13½ year old girl could not reasonably be expected to stop and check for herself whether it was safe to cross a road, when a lorry driver had beckoned her across) held that—

> a very young child cannot be guilty of contributory negligence. An older child may be. But it depends on the circumstances.

(See *Young v Kent County Council* (at **O3010**).)

In *Revill v Newbery* (see **O3009**), although the defendant's conduct was not reasonable, the plaintiff intruder was found to be two-thirds to blame for the injuries he suffered and his damages were reduced accordingly. In *Betts v Tokley* [2002] All ER (D) 99 (Jan), the Court of Appeal upheld a ruling that the claimant was 60 per cent contributorily negligent after she fell on some unlit steps whilst leaving her employer's premises by using the back exit (as the front exit was locked). The rationale behind the decision was that the claimant had pressed on to find the back exit despite the fact that she was fully aware that she had to take great care if she continued in the dark, there was nothing to stop her returning through the lit building to the front door and the steps to the front of the building should have alerted her to the risk of further steps at the back. The claimant's suggestion that the defendant had created a trap was rejected.

In 2002 the Court held that a claimant cannot be guilty of 100 per cent contributory negligence (*Anderson v Newham College of Further Education* [2002] EWCA Civ 505, [2003] ICR 212) if evidence showed that the entire fault was to lie with the claimant then there could be no liability on the defendant. This should be compared to the dicta of the Court of Session in the Scottish case of *McEwan v Lothian Buses Plc* 2006 SCLR 592 in which the court, in stark contrast to the approach in *Anderson*, considered the case of *Jayes v IMI (Knoch) Ltd* [1985] ICR 155 to be much more persuasive in that there was no logical reason why 100 per cent contributory negligence should not apply when the degree of fault on the part of the defendant could be too small to warrant an apportionment.

Assessing a claimant's contributory negligence will often involve a careful examination of the facts. In *Wattleworth v Goodwood Road Racing Co Ltd* [2004] EWHC 140 (QB), [2004] All ER (D) 51 (Feb) the claimant's husband, an experienced driver, died when his car crashed into a tyre wall during an amateur track day. Although the occupier of the premises, as well as the official bodies which had inspected the track prior to the incident, were not held liable for the driver's death, the judge went on to consider whether the driver would have been held contributory negligent had the defendants been liable for his death. Evidence showed that the driver had committed two mistakes prior to the incident. Firstly, he had taken the wrong driving line when approaching the bend where he crashed. Although the mistake was probably due to a lack of concentration, the margin of error was very small and therefore the judge decided that he could not have been held contributory negligent by virtue of this mistake. However, following his first error, he maintained his foot on the throttle and did not attempt to brake despite having 'ample time' (per Davis J at para 178) to do so (he had travelled a distance of over 100 metres before crashing against the tyre wall). The judge decided that had it been necessary, he would have held the driver 20 per cent contributory negligent on the basis of his second mistake.

Dangers to guard against

[03012] The duty owed by an occupier to his lawful visitors is a 'common duty of care', so called since the duty is owed to both invitees and licensees, ie those having an interest in common with the occupier (eg business associates, customers, clients, salesmen) and those permitted by regulation or statute to be on the premises, eg factory inspectors or policemen. That duty requires that the dangers against which the occupier must guard are twofold: (*a*) structural defects in the premises; and (*b*) dangers associated with works or operations carried out for the occupier on the premises.

Structural defects

[03013] As regards structural defects in premises, the occupier will only be liable if either he actually knew of a defect or foreseeably had reason to believe that there was a defect in the premises. Simply put, the occupier would not incur liability for the existence of latent defects causing injury or damage, unless he had special knowledge in that regard (eg a faulty electrical circuit, unless he were an electrician); whereas, he would be liable for patent (ie obvious) structural defects, eg an unlit hole in the road. This duty now extends to 'uninvited entrants', eg trespassers, under the *OLA 1984 s 1*.

Workmen on occupier's premises

[03014] 'An occupier may expect that a person, in the exercise of his calling, will appreciate and guard against any special risks ordinarily incident to it, so far as the occupier leaves him free to do so' [*OLA 1957 s 2(3)(b)*]. This has generally been taken to imply that risks associated with systems or methods of work on customer premises are the exclusive responsibility of the employer (not the customer or occupier) (*General Cleaning Contractors Ltd v Christmas* [1952] 2 All ER 1110, concerning a window cleaner who failed to take proper precautions in respect of a defective sash window: it was held that there was no liability on the part of the occupier, but liability on the part of the employer). In *Hood v Mitie Property Services (Midlands) Limited* [2005] All ER (D) 11 (Jul), the claimant, the employee of an independent contractor hired by the occupier, fell through a skylight whilst working on a roof. The Court of Appeal said that there was no reason why the occupier should not expect a visitor in the position of the claimant to guard against the risks normally incident to this job, especially as the independent contractor had held itself out as able to do roofing work and as having a proper approach to safety. However, where a window was so designed to be cleaned from the inside, and not from the sill, the employer was held to be liable (*King v Smith* [1995] ICR 339).

Further, there is the requirement to ensure that the independent contractor is competent. This is particularly true where an occupier wishes to organise a hazardous event on his land. In *Bottomley v Todmorden Cricket Club* [2003] EWCA Civ 1575, [2003] All ER (D) 102 (Nov) the claimant, who had volunteered to help two independent contractors with a pyrotechnic display held on the defendants' premises, had been filling a mortar tube with gunpowder when the contents of the tube exploded prematurely, causing him severe burns. Although the defendants were occupiers and not employers of the claimant, the Court of Appeal held that the defendants were liable for the claimant's injuries as they had not taken all the necessary precautionary steps to ensure that the independent contractors were competent (in particular, the defendants failed to ensure that the contractors had public liability insurance and had prepared a written safety plan). The rationale behind the Bottomley decision was that it was reasonable to impose liability on the Club because it did not do what it ought to have done before it allowed the dangerous event to take place on its land.

The Court of Appeal decisions in the cases of *Gwilliam v West Hertfordshire Hospital NHS Trust* [2002] EWCA Civ 1041, [2003] 3 WLR 1425 (see judgment of Lord Woolf CJ), *Bottomley v Todmorken Cricket Club* [2003] EWCA Civ 1575, [2003] 48 LS Gaz R 18 and *Naylor (trading as Mainstreet) v Payling* [2004] EWCA Civ 560, [2004] All ER (D) 83 (May) suggest that when an occupier wishes to organise a hazardous or dangerous activity on its premises, part of their duty to ensure the competence of an independent contractor will involve ensuring that they hold public liability insurance. When delivering the first instance judgment in *Bottomley*, the judge held that enquiring about a contractor's insurance cover was important, as insurers would wish to reduce the risk of having to meet liability claims and would therefore want to ask questions to clarify the extent of the risk which was to be covered. Further, the judge added that if a contractor was unable to provide evidence of insurance cover, there could be a good reason for this and this could highlight a contractor's incompetence. It must be pointed out that this rule is limited to hazardous activities such as a 'Splat Wall' funfair (*Gwilliam*) or a pyrotechnic display (*Bottomley*) and does not apply to more regular activities such as, for example, those of a bouncer providing security outside a nightclub (*Naylor*). In *Naylor*, the Court held that there was not, save in special circumstances, a duty on an employer to check that his independent contractor had insurance cover or was capable of meeting any liabilities which might arise.

'Unusual' Dangers

The occupier will ordinarily only incur liability for a structural defect in premises which the oncoming workman would not normally guard against as part of a safe system of doing his job, ie against 'unusual' dangers.

> And with respect to such a visitor at least, we consider it settled law that he, using reasonable care on his part for his own safety, is entitled to expect that the occupier shall on his part use reasonable care to prevent damage from unusual danger, which he knows or ought to know.

(Per Willes J in *Indermaur v Dames* (1866) LR 1 CP 274, concerning a gasfitter testing gas burners in a sugar refinery who fell into an unfenced shaft and was injured.)

Firemen

The law is not entirely settled in the case of firemen. In *Sibbald v Sher Bros*, (1981) Times, 1 February, the Court examined the issue of liability when a number of firemen were killed following a fire at a warehouse which had an untreated hardboard ceiling which had suddenly burst into flames. It was held that the occupier's duty in relation to firemen was not the same as a duty to other visitors, such as employees. The firemen were, in fact, considered to be the occupier's 'neighbours'.

In *Hartley v British Railways Board* (1981) 125 Sol Jo 169 (a case in which a railway station caught fire and the defendant wrongly indicated the station was staffed, leading to the claimant unnecessarily entering the station and being injured), Waller LJ held—

> It (is) . . . very unlikely that the duty of care owed by the occupier to workers was the same as that owed to "firemen" (per Lord Fraser of Tullybelton), (but) it is arguable that a "fireman" (is) a "neighbour" of the occupier in the sense of Lord Atkin's famous dictum in *Donoghue v Stevenson* [1932] AC 562, so that the occupier owes him some duty of care, as for instance, to warn firemen of an unexpected danger or trap of which he knew or ought to know.

A neighbour has been defined as —

> persons who are so closely and directly affected by my act that I ought reasonably to have them in contemplation as being so affected when I am directing my mind to the acts or omissions which are called in question.

(Per Lord Atkin in *Donoghue v Stevenson* [1932] AC 562.)

The case of *Simpson v A I Dairies Farms Limited* [2001] EWCA Civ 13, [2001] All ER (D) 31 (Jan) reinforces this principle, as a farmer who failed to warn a fireman of the presence of an uneven drain in a yard which was concealed by water was deemed liable when the fireman was injured. The water covered drain constituted an unexpected danger or trap which gave rise to a duty to issue a warning. (See also *Bermingham v Sher Bros* 1980 SC 67, 1980 SLT 122, HL.)

The case of *Ogwo v Taylor* [1987] 3 All ER 961 concerned a fire caused by the occupier's negligence when using a blowtorch to remove paint from his house. The fireman was injured by steam created when fighting the fire. The Court considered the question of whether the duty of care to fireman was limited to special or exceptional risks. The Court held that the defendant was responsible for the injuries, regardless of whether the injuries were sustained as a result of exceptional or ordinary risks.

Dangers associated with works being done on premises

[03015] Where work is being done on premises by a contractor, the occupier is not liable if he—

(a) took care in selecting a competent contractor; and
(b) satisfied himself that the work was being properly done by the contractor [*OLA 1957 s 2(4)(b)*].

An examination of the apportionment of liability for the negligence of contractors was provided by the Court of Appeal in *Clark v Hosier & Dickinson Ltd* [2003] EWCA Civ 1467, [2003] All ER (D) 225 (Oct). In this case, a theatre company commissioned the claimant to install some gates on its premises. The theatre company also engaged building contractors to carry out ground works where the gates were to be placed. During the course of the ground works the building contractor came across a mains electric cable. He drew this to the attention of the occupier. The occupier left the building company to deal with this matter. The building company provided an inaccurate drawing as to the location and depth of the mains cable to the claimant, who subsequently struck the cable whilst drilling and sustained serious injuries.

The Court of Appeal held that the building company had been in breach of its contractual obligations to carry out its work safely and competently. The Court considered that the building company should have acted on its knowledge, by at least contacting an architect or the Electricity Board. The theatre company was also criticised for failure to seek advice from these parties, but only one third of the liability was attributed to the theatre company occupier, the building contractor being held two thirds responsible on the basis of the inaccurate drawing being the cause of the claimant's injuries.

As regards the checking of works, it may be highly desirable (indeed necessary) for an occupier to delegate the 'duty of satisfaction', especially where complicated building or engineering operations are being carried out, to a specialist, eg an architect or a geotechnical engineer. Not to do so, in the interests of safety of visitors, is probably negligent.

> In the case of the construction of a substantial building, or of a ship, I should have thought that the building owner, if he is to escape subsequent tortious liability for faulty construction, should not only take care to contract with a competent

contractor . . . but also cause that work to be supervised by a properly qualified professional . . . such as an architect or surveyor . . . I cannot think that different principles can apply to precautions during the course of construction, if the building owner is going to invite a third party to bring valuable property on to the site during construction.

(Per Mocatta J in *AMF International Ltd v Magnet Bowling Ltd* [1968] 2 All ER 789.)

Waiver of duty and the Unfair Contract Terms Act 1977 (UCTA 1977)

[03016] Where damage was caused to a visitor by a danger of which he had been warned by the occupier (eg by notice), an explicit notice, eg 'Highly Flammable Liquid Vapours – No Smoking' (as distinct from a vague notice such as 'Fire Hazards') used to absolve an occupier from liability. Now, however, such notices are ineffective (except in the case of trespassers, see O3007–O3009) and do not exonerate occupiers. Thus: 'a person cannot by reference to any contract term or to a notice given to persons generally or to particular persons exclude or restrict his liability for death or personal injury resulting from negligence' [*UCTA 1977 s 2*]. Additionally, in terms of consumer contracts and notices, *s 65(1)* of the *Consumer Rights Act 2015* states 'a trader cannot by a term of a consumer contract or by a consumer notice exclude or restrict liability for death or personal injury resulting from negligence'. Such explicit notices are, however, a defence to an occupier when sued for negligent injury by a simple trespasser under *OLA 1984*, and the *Consumer Rights Act 2015*, provided they are appropriately worded.

As regards negligent damage to property, a person can restrict or exclude his liability by a notice or contract term, but such notice or contract term must be 'reasonable', and it is incumbent on the occupier to prove that it is in fact reasonable [*UCTA 1977 ss 2(2), 11*]. In respect of consumer contracts, an unfair consumer notice is not binding on the consumer [*s 62(2)* of the *Consumer Rights Act 2015*]. Under this legislation, 'a notice is unfair if, contrary to the requirement of good faith, it causes a significant imbalance in the parties' rights and obligations to the detriment of the consumer' [*s 62(6)*] and 'whether a notice is fair to be determined by taking into account the nature of the subject matter of the notice' [*s 62(7)(a)*].

The question of whether the defendant did what was reasonable to bring to the notice of the claimant the existence of a particularly onerous condition regarding his statutory rights, was considered in *Interfoto Picture Library Ltd v Stiletto Visual Programmes Ltd* [1989] QB 433. This decision reaffirmed the pre-*UCTA 1977* position as set out in *Parker v South Eastern Rly Co* (1877) 2 CPD 416, CA and *Thornton v Shoe Lane Parking Ltd* [1971] 2 QB 163, [1971] 1 All ER 686, CA. In *Interfoto* it was held that for a condition seeking to restrict statutory rights (eg under *OLA 1957*) to be effective it must have been fairly brought to the attention of a party to the contract by way of some clear indication which would lead an ordinary sensible person to realise, at or before the time of making the contract, that such a term relating to personal injury was to be included in the contract. In most instances this would be covered by clear signs indicating hazards.

The test is objective and whether the particular individual was aware of the sign or not is not relevant. It should be noted, however, that the more onerous the term the greater the notice which must be given and the more effort an occupier must make to bring it to someone's attention (see Lord Denning in *Spurling (J) Ltd v Bradshaw* [1956] 2 All ER 121, [1956] 1 WLR 461, CA). (See also the Court of Appeal case of *O'Brien v MGN Ltd* [2001] EWCA Civ 1279, [2001] All ER (D) 01 (Aug) which considered this issue, albeit in a contractual context and not in respect of personal injury.)

Risks willingly accepted

[03017] 'The common duty of care does not impose on an occupier any obligation to a visitor in respect of risks willingly accepted as his by the visitor' [*OLA 1957 s 2(5)*]. However, this 'defence' has generally not succeeded in industrial injury claims. This is similarly the case with occupiers' liability claims. See, for example, *Burnett v British Waterways Board* [1973] 2 All ER 631 where a lighterman working on the River Thames was held not to be bound by the terms of a notice erected by the respondent, (stating that the Board accepted no liability for loss, damage or injury arising from the lighterman's use of a lock) even though the lighterman had seen the notice many times and understood it. This decision has since been reinforced by the *Unfair Contract Terms Act 1977*: 'Where a contract term or notice purports to exclude or restrict liability for negligence a person's agreement to or awareness of it is not of itself to be taken as indicating his voluntary acceptance of any risk' [*UCTA 1977 s 2(3)*]. Further, in the case of *Wattleworth v Goodwood Road Racing Co Ltd* (see O3011), it was held that although a driver who died as a result of a crash during an amateur track day, would have accepted the risks inherent in racing, he would only have done so on the basis that those responsible for the circuit's safety had taken all due steps to see that the circuit was reasonably safe. Therefore, had it been necessary to consider the defendants' defence of *volenti non fit injuria* (the defendants were held not to be liable for the driver's injuries), the judge would have rejected the defence.

The *OLA 1957* makes clear that there may be circumstances in which even an explicit warning will not absolve the occupier from liability. Nevertheless, the occupier is entitled to expect a reasonable person to appreciate certain obvious dangers and take appropriate action to ensure his safety. In those circumstances no warning would appear to be required. It was not necessary, for example, to warn an adult of sound mind that it was dangerous to go near the edge of a cliff in *Cotton v Derbyshire Dales District Council* (1994) Times, 20 June. Likewise in *Staples v West Dorset District Council* [1995] PIQR P439 no duty of care was imposed on a local authority to warn a visitor of the obvious danger of slipping on visible algae that had formed on a wall; the visitor being able to evaluate the danger. In *Darby v The National Trust* [2001] EWCA Civ 189, [2001] All ER (D) 216 (Jan) it was held there was no duty on the defendant to warn against the risk of swimming in a pond as the dangers of drowning were no more than an adult should have foreseen. The risk to a competent swimmer was an obvious one. There was therefore no relative causative risk; a sign would have told the deceased no more than he already knew. In *Tomlinson v Congleton Borough Council* [2003] UKHL 47, [2004] 1 AC 46, the claimant sustained severe injuries after diving into the shallow water at the edge of a lake (in a country park owned by the defendant) and striking his neck on the bottom. The House of Lords held that the defendant, who had prohibited swimming and erected notices and distributed leaflets warning against the dangers of swimming in the lake, was under no duty under *OLA 1984* to offer protection to the claimant against a risk which was perfectly obvious (the claimant was held to be a trespasser by virtue of the prohibition on swimming). Further, two Law Lords, Lord Hoffman and Lord Hobhouse, said that there was an important question of freewill at stake and that—

> it is not and should never be the policy of the law to require the protection of the foolhardy or reckless few to deprive, or interfere with, the enjoyment by the remainder of society of the liberties and amenities to which they are rightly entitled.

(Per Lord Hobhouse at para 81.)

Therefore, it would have been unjust for the defendant to be under a legal obligation to block access to the lake. Although this case was concerned with whether a duty arose under *OLA 1984*, the House of Lords made clear that had the defendant owed a duty under *OLA 1957 s 2(2)*, this duty would not have required him to take any steps to warn the claimant against dangers which were 'perfectly obvious' (per Lord Hoffman at para 50).

See also *Ratcliff v McConnell* [1999] 1 WLR 670, [1999] 03 LS Gaz R 32, CA in which the Court of Appeal held that there was no duty on the occupier of a swimming pool to warn an adult trespasser of the dangers of diving into shallow water. If the trespasser willingly accepted the risk as his, there was no duty owed by the occupier.

In the case of *Evans v Kosmar Villa Holidays Plc* [2007] EWCA Civ 1003, [2008] 1 All ER 530, the Court of Appeal considered the issue of whether a tour operator had a duty to protect against obvious risk. In this case, the claimant booked a holiday with a tour operator. Whilst on holiday, he dived into the shallow end of a swimming pool in the early hours of the morning, sustaining serious injuries. The claimant's case was that the positioning and size of the "no diving" signs were inadequate and that the signage relating to the closure of the pool at night should have been explicit and enforced. The claimant accepted that he knew that diving into the shallow end of the pool posed a dangerous risk. The defendant tour operator said that it had no duty to warn of the obvious risk of diving into shallow water or water of an unknown depth. The Court held that the principle that there is no duty to protect against obvious risks, which had been applied in the cases of trespassers and lawful visitors to whom a duty of care was owed under the *OLA 1957*, also applied to those to whom a duty of care was owed in contract. The defendant's duty of care did not extend to warning about the obvious risk, the claimant being able to make a genuine and informed decision.

In a more recent case, *Geary v JD Weatherspoon plc* [2011] EWHC 1506 (QB), [2011] All ER (D) 97 (Jun), a claimant failed to establish liability against an occupier after she attempted to slide down the banister of a pub and fell around four metres to a marble floor below, suffering severe injuries. The judge found that she had deliberately and freely chosen to do something that she knew was dangerous.

OLA 1984 s 1(6) regulates the position with respect to trespassers. *OLA 1984 s 1(6)* states, 'No duty is owed . . . to any person in respect of risks willingly accepted as his by that person . . . '. This applies even if the trespasser is a child (see **O3010**) (*Titchener v British Railways Board* [1983] 3 All ER 770 where a 15-year-old girl and her boyfriend aged 16 had been struck by a train. The boy was killed and the girl suffered serious injuries. They had squeezed through a gap in a fence to cross the line as a short-cut to a disused brickworks which was regularly used. It was held by the House of Lords that the respondent did not owe a duty to the girl to do more than they had done to maintain the fence. It would have been 'quite unreasonable' for the respondent to maintain an impenetrable and unclimbable fence. However, the Court held that—

> 'the duty (to maintain fences) will tend to be higher with a very young or a very old person than with a normally active and intelligent adult or adolescent.

(Per Lord Fraser)

Scott and Swainger v Associated British Ports and British Railways Board [2000] All ER (D) 1937, concerned schoolboys who on separate occasions were both severely injured whilst attempting to 'train surf' ie hitch rides on slow passing freight trains. The case was decided on the narrow issue of causation. The judge found as a question of fact that the provision of a fence would not have deterred them. In this situation, the court seemed to conclude that defendants may not do more than fence in dangers in order to protect children and that by scaling a fence the claimants would have willingly consented to the risk of injury. This position might well be otherwise if anti-trespasser devices were not satisfactorily maintained (*Adams v Southern Electricity Board* (1993) Times, 21 October where the electricity board was held to be liable to a teenage trespasser for injuries sustained when he climbed on to apparatus by means of a defective anti-climbing device). In the curious case of *Arthur v Anker* [1996] 3 All ER 783 a car was parked without permission in a private car park. It was clamped but the driver removed his vehicle with the clamp still attached. The owner of the car park sued the driver for the return of the wheel clamp and payment of the fine for illegal parking. The question of *volenti* was considered in detail. The judge held that the owner parked his car in full knowledge that he was not entitled to do so and that he was therefore consenting to the consequences of his action (payment of the fine); he could not complain after the event. The effect of this consent was to render conduct lawful which would otherwise have been tortious. By voluntarily accepting the risk that his car might be clamped, the driver accepted the risk that the car would remain clamped until he paid the reasonable cost of clamping and de-clamping.

Actions against factory occupiers

[03018] The *Workplace (Health, Safety and Welfare) Regulations 1992 (SI 1992 No 3004)*, as amended apply with respect to the health, safety and welfare of persons in a defined workplace. The application of the Regulations is not confined to factories and offices – they also affect public buildings such as hospitals and schools, but they do not apply to, for example, construction sites and quarries. Employers, persons who have control of the workplace and occupiers are all subject to the Regulations, the provisions of which address the environmental management and condition of buildings, such as the day-to-day considerations of ventilation, temperature control and lighting and potentially dangerous activities such as window cleaning. In many ways the Regulations reflect the obligations imposed by the old *Factories Act 1961*. For example, *Regulation 12(3)* requires floors and the surfaces of traffic routes, 'so far as is reasonably practicable', to be kept free from obstacles or substances which may cause a person to 'slip, trip or fall'. These Regulations also closely mirror the general obligation placed on employers by *HSWA 1974 s 2* to ensure so far as is reasonably practicable, the health, safety and welfare of their employees whilst at work. Breach of these Regulations gives rise to civil liability by virtue of *HSWA 1974 s 47(2)* (for criminal liability, see O3019 and ENFORCEMENT). An Approved Code of Practice and guidance to the Regulations is available from the Health and Safety Executive.

The case law on these Regulations is mostly concerned with the interpretation of *Regulations 5* and *12*. *Regulation 5* applies in particular to maintenance of the workplace and workplace equipment failure which is likely to result in a breach of the Regulations. Under *Regulation 5(1)*, the workplace, equipment, devices and systems to which the Regulation applies, must be maintained in 'an efficient state, in efficient working order and in good repair'. Under *Regulation 12(1)*, every floor in a workplace and the surface of every traffic route in a workplace must be of a construction 'so that the floor or surface of the traffic route is suitable for the purpose for which it is used'. *Regulation 12(3)* requires that, so far as is reasonably practicable, floors and traffic routes be kept free from obstructions which may cause a person to 'slip, trip or fall'. It must be pointed out that although *Regulation 12* imposes specific obligations, this does not prevent other obligations from arising under *Regulation 5*, as the duties imposed by both provisions are not mutually exclusive (see the first instance decision in *Irvine v Metropolitan Police Commissioner* [2004] EWHC 1536 (QB), [2004] All ER (D) 79 (May)). Further, the Court of Appeal held in *Ricketts v Torbay* [2003] EWCA Civ 613 that the duties imposed by *Regulation 5* were not absolute as this would be difficult to reconcile with the fact that the duties owed under *Regulation 12(3)* are subject to the qualification of reasonable practicability.

In *Lewis v Avidan* [2005] EWCA Civ 670, [2005] All ER (D) 136 (Apr), an employee slipped on a patch of water in a corridor during a night shift. The water had leaked from a concealed pipe that had burst shortly before the incident. The Court of Appeal held that although the pipe could be considered as 'equipment' under *Regulation 5(3)*, the fault in the pipe did not amount to a breach of *Regulation 12(3)*. Such a breach would only have occurred if the employer had failed to take reasonable steps to prevent the floor from being slippery, which the Court considered was not the case here. Further, the Court was of the opinion that the definition of a workplace was limited to areas of work to which employees had access, which would therefore include the floor but not an enclosed pipe. In addition, the Court held that the word 'maintaining' implied the doing of an action to the floor, such as cleaning or repairing it. The court therefore held that an 'entirely unexpected and unpredictable flood' did not mean that the floor was not maintained in an efficient state and the employee's claim under *Regulation 5(1)* was therefore dismissed.

In *Jaguar v Coates* [2004] EWCA Civ 337, [2004] All ER (D) 87 (Mar), the Court of Appeal held that the defendant was not liable under *Regulation 5(1)* for the injuries sustained by his employee who fell whilst going up four regular steps which were not equipped with a handrail. The Court held that *Regulation 5(1)* did not apply as this Regulation was concerned with 'maintenance' and the fitting of a handrail on the side of the steps was not a maintenance issue.

However, in the Court of Appeal's decision in *Gitsham v CH Pearce & Sons Plc* [1992] PIQR P57, Stocker LJ submitted that a duty to maintain should not necessarily be limited to the state of a structure but could also extend to taking precautions against snow and ice. This case was concerned with s 29 of the *Factories Act 1961* which imposes on employers a duty, 'so far as reasonably practicable', to provide and maintain 'safe means of access to every place at which any person has at any time to work'. As this duty is similar to those imposed by the Regulations, literature suggests that this decision also applies to the meaning of 'maintaining' under Regulation 5. It must be pointed out that following the decisions of the Court of Appeal in *Ricketts v Torbay* and of the Scottish Inner House (the appeal court of the Scottish Court of Session) in *Donaldson v Hays Distribution Services Ltd* 2005 SLT 733, the duty owed by employers under the Regulations apply only to persons at work in the workplace and not to visitors coming into the premises.

Occupier's duties under HSWA 1974

[03019] In addition to civil liabilities under *OLA 1957* and at common law, occupiers of buildings also have duties, under the *HSWA 1974*. Failure to carry out such duties can lead to criminal liability. In accordance with the provisions of *HSWA 1974*, 'each person who has, to any extent, control of premises (ie 'non-domestic' premises) or the means of access thereto or egress therefrom or of any plant or substance in such premises' must do what is reasonably practicable to ensure that the premises, means of access and egress and plant/substances in the premises or provided for use there, are safe and without health risks [*HSWA 1974 s 4(2)*]. This section applies in the case of (*a*) non-employees; and (*b*) non-domestic premises [*HSWA 1974 s 4(1)*]. In other words, to a large extent, it places health and safety duties on persons and companies letting or sub-letting premises for work purposes, even though the people working in those premises are not employees of the lessor/sublessor. However, it should be noted that the duty also extends to people who are not at work eg children at a play centre (*Moualem v Carlisle City Council* (1994)

158 JP 1110). As with most other forms of leasehold tenure, the person who is responsible for maintenance and repairs of the leased premises is the person who has 'control' (see, by way of analogy, O3004) [*HSWA 1974 s 4(3)*]. Included as 'premises' are common parts of a block of flats. These are 'non-domestic' premises (See *Westminster City Council v Select Managements Ltd* [1985] 1 All ER 897 which held that being a 'place' or 'installation on land', such areas are 'premises'; and they are not 'domestic', since they are in common use by the occupants of more than one private dwelling).

The reasonableness of the measures which a person is required to take to ensure the safety of those premises is to be determined in the light of his knowledge of the expected use of the premises and of the extent of his control and knowledge, if any, of the use.

More particularly, if premises were not a reasonably foreseeable cause of danger to anyone acting in a way a person might reasonably be expected to act, in circumstances that might reasonably be expected to occur during the work, further measures against unknown and unexpected events would not be required (*Mailer v Austin Rover Group* [1989] 2 All ER 1087 where an employee of a firm of cleaning contractors, while cleaning a paint spray booth and the sump underneath it, was killed by escaping fumes. The contractors had been instructed by the appellants not to use paint thinners from a pipe in the booth (which the appellants had turned off but not capped) and only to enter the sump (where the ventilator would have been turned off) with an approved safety lamp and when no one was working above. Contrary to those instructions, an employee used thinners from the pipe, which entered a sump where the deceased was working with a non-approved lamp, and an explosion occurred. It was held that the appellant was not liable for breach of the *HSWA 1974 s 4(2)*). (See also dicta in *R v HTM Ltd* [2006] EWCA Crim 1156, [2007] 2 All ER 665 where the Court of Appeal considered the issue of foreseeability in determining the issue of reasonable practicability.)

When considering criminal liability of an occupier under *HSWA 1974* in relation to risks to outside contractors, *s 4* is likely to be of less significance as a result of the decision of the House of Lords in *R v Associated Octel* [1996] 4 All ER 846, a decision which in effect enables the *HSWA 1974 s 3* to be used against occupiers. *HSWA 1974 s 3(1)* provides that 'It shall be the duty of every employer to conduct his undertaking in such a way as to ensure, so far as is reasonably practicable, that persons not in his employment who may be affected thereby are not thereby exposed to risks to their health or safety'.

The decision establishes that an occupier can still be deemed to be conducting his undertaking by engaging contractors to do work, even where the activities of the contractor are separate and not under the occupier's control. *HSWA 1974 s 3* therefore requires the occupier to take steps to ensure the protection of contractors' employees from risks not merely arising from the physical state of the premises but also from the process of carrying out the work itself.

However, the judgment of *Makepeace v Evans Brothers (Reading) (A firm)* [2000] BLR 737 should be noted in that it was held that an occupier would not usually be liable to an employee of a contractor employed to carry out work at the occupier's premises if the employee was injured as a result of any unsafe system of work used by the employee and the contractor. It was not generally reasonable to expect an occupier of premises, having engaged a contractor whom he had reasonable grounds to regard as competent, to supervise the contractor's activities in order to ensure that he was discharging his duties to ensure a safe system of work for his employees. Even if the occupier may have reason to suspect that the contractors may have been using an unsafe system of work (see also *McGarvey v EVE NCI Ltd* [2002] EWCA Civ 374, [2002] All ER (D) 345 (Feb), this would not, in itself, be enough to impose upon him liability under *HSWA 1974*, or in negligence.

In *Hampstead Heath Winter Swimming Club v Corp of London* [2005] 1 WLR 2930, the defendant refused to grant the members of a swimming club permission to swim unsupervised in a pond which it owned on the basis that it would then be liable to prosecution under *HSWA 1974 s 3*. The judge held that although the defendant had been correct to consider that its power to regulate the admission of swimmers to the pond constituted an undertaking under *HSWA 1974 s 3*, the defendant had made a legal error by refusing admission to the pond to unsupervised swimmers. The judge referred to the concept of freewill which played an important role in the House of Lords' decision in *Tomlinson v Congleton Borough Council* (see **O3017**) and held that the risks incurred by adult swimmers with knowledge of the dangers of swimming unsupervised, could only be attributed to their decision to swim unsupervised and not to the permission to do so given to them by the defendant as part of its undertaking. The defendant could therefore not be liable to conviction under *HSWA 1974 s 3*.

Establishing breaches of *ss 3* and *2 of the HSWA 1974* should be considered in light of the case of *R v Chargot Ltd (t/a Contract Services)* [2008] UKHL 73, [2009] 2 All ER 645, [2009] 1 WLR 1. In this case, the House of Lords considered the issue of whether it was necessary to prove merely that a risk of injury existed or that particular acts or omissions gave rise to a breach of duty. The House of Lords concluded that the nature of the duties in *ss 2* and *3 of the HSWA 1974* are general and 'results based'. Thus if a result is not achieved, then the duty has been breached ie once there has been an injury, then *ipso facto* there is a relevant risk).

Offshore Operations

Fred Osliff, Ian Wallace and Andrea Oates

Introduction to offshore operations

[O7001] 'Offshore operations' take place, broadly, in three specific stages:

(a) exploration for recoverable reserves of oil and gas;
(b) production of those reservoirs which prove to be economic; and
(c) decommissioning and/or removal of the facilities once the reservoir has been depleted to the economic level.

In addition to offshore oil and gas operations, the *Health and Safety at Work etc Act 1974 (Application outside Great Britain) Order 2013* consolidated two earlier orders and introduced new articles associated with so-called emerging energy technologies (EETs). The order maintains the Health and Safety Executive's (HSE's) ability to regulate work activities associated with 'energy structures' (wind farms), which are being built beyond the territorial sea within the Renewable Energy Zone. It also provides legal clarity about the HSE's power to regulate work activities with new EETs (such as combustible gas storage and underground coal gasification) and climate change technologies (such as carbon dioxide storage) (see **O7009** below).

The HSE has also made clear that the existing health and safety regulatory regime applies to shale gas operations (or 'fracking'). For example, the *Offshore Installations and Wells (Design and Construction, etc) Regulations 1996 (DCR) (SI 1996 No 913)* (see **O7024** below) apply to all wells drilled with a view to the extraction of petroleum, regardless of whether they are onshore or offshore.

Activities connected with petroleum exploration and production have a history of almost 100 years when, by an enactment of 1918, the searching for petroleum onshore was made the subject of Government licence. This was intended to be a temporary measure, but the *Petroleum (Production) Act 1934* (now repealed) vested property rights in the Crown, and gave new powers to the Board of Trade to make regulations and impose conditions based on 'model clauses' which were incorporated into exploration/exploitation licences.

Model clauses were drafted in conjunction with the Institute of Petroleum, and matters of safety were covered by the phrase 'licensees must execute all operations in accordance with good oilfield practice' and comply with any instructions which might be given by the Board. Good oilfield practice was not, however, defined. The relevant provisions of earlier legislation governing the issue of licences were amended by the *Petroleum Act 1998*, which repealed the *1989 Petroleum Act*.

[O7001] Offshore Operations

In the 1950s petroleum exploration activities were being undertaken at sea. The rights of coastal countries to exploit the natural resources of their continental shelves were first formally recognised in international law by the Geneva Convention on the Continental Shelf 1958. The Convention was ratified by the United Kingdom in 1964 by the *Continental Shelf Act 1964* (*CSA 1964*). This vested in the Crown all rights exercisable in designated areas outside territorial waters in relation to the sea bed and subsoil in those areas. The *CSA 1964* also provided that the model clauses for offshore petroleum licences should make provision for the health, safety and welfare of employees. Enforcement of the Act and subordinate legislation continued to be the responsibility of the Department of Energy under the Safety Directorate of the Petroleum Engineering Division. However, the only sanction available for the enforcement of the safety procedures was withdrawal of a licence.

Following an accident in 1965 involving the exploration rig Sea Gem, which capsized and sank with the loss of 13 lives, the government accepted recommendations for strengthening the statutory provisions and the *Mineral Workings (Offshore Installations) Act 1971* (*MWA 1971*) was enacted specifically to regulate offshore safety matters. *MWA 1971* and the regulations made under it remained in force until 1992 (see **O7003** below).

The *Offshore Safety Act 1992* applied *Pt 1* of the *Health and Safety at Work etc Act 1974* (*HSWA 1974*) for offshore purposes.

Today, most of the provisions of the *MWA 1971* have been repealed and replaced by health and safety regulations made under the *HSWA 1974*. The provisions of the 1934 Act have been re-enacted with modifications by the *Petroleum Act 1998* (*PA 1998*) to cover:

- petroleum (rights vested in Her Majesty, licensing, development fees, mines and workings):
  — 'petroleum' is defined to include any mineral oil or relative hydrocarbon and natural gas existing in its natural condition in strata. It does not include coal or bituminous shales or other stratified deposits from which oil can be extracted by destructive distillation;
- offshore activities:
  — application of criminal and civil law etc;
  — prosecutions for offences with respect to carriage of passengers and goods by helicopter (*Civil Aviation Act 1982* – *in part*); infringement of safety zones (*Petroleum Act 1987, s 23*); any offence from remaining provisions of the *Mineral Workings (Offshore Installations) Act 1971*;
- submarine pipelines (see also **O7034** below):
  — construction and use of pipelines; authorisations; compulsory modifications; rights to use pipelines; and
  — pipeline inspectors and procedures;
- abandonment of offshore installations (see also **O7014** below):
  — abandonment programmes – duties and default procedures.

In addition, the *Petroleum Act 1998* makes amendments to the following statutes associated with offshore (safety) operations:

Introduction to offshore operations [07003]

- *Continental Shelf Act 1964;*
- *Prevention of Oil Pollution Act 1971;*
- *Offshore Petroleum Development (Scotland) Act 1975;*
- *Sex Discrimination Act 1975* and *Race Relations Act 1976* (both repealed by the *Equality Act 2010*);
- *Energy Act 1976;*
- *Food and Environment Protection Act 1985;*
- *Oil and Pipelines Act 1985;*
- *Gas Act 1986;*
- *Petroleum Act 1987;*
- *Territorial Sea Act 1987;*
- *Food Safety Act 1990;*
- *Aviation and Maritime Security Act 1990;*
- *Social Security Contributions and Benefits Act 1992;*
- *Offshore Safety Act 1992;*
- *Trade Union and Labour Relations (Consolidation) Act 1992;*
- *Merchant Shipping Act 1995;*
- *Employment Rights Act 1996.*

Application of criminal and civil law to offshore installations

[07002] The *Petroleum Act 1998* provides that any act or omission constituting an offence under United Kingdom law which takes place on, under or above an offshore installation in United Kingdom waters or within 500 metres of such an installation (the safety zone) shall be treated as taking place in the United Kingdom. Police powers are granted accordingly. The application of criminal law to offshore installations also extends to corporate bodies, including the provisions relating to ' . . . the consent or connivance of, or to be attributable to any neglect on the part of, any director, manager . . . ' (*HSWA 1974, s 37*).

Matters which may be heard under civil law regarding offshore activities are determined in accordance with the law in force in such part of the United Kingdom as may be specified in an Order in Council. Any jurisdiction conferred on a court under this provision cannot prejudice any jurisdiction exercisable by any other court. The application of civil law also extends to mobile installations in transit.

The Cullen Report

[07003] *MWA 1971* and the regulations made under it remained in force until 1992 when, following the Piper Alpha tragedy in 1988 (in which 167 lives were lost), the fatal accident inquiry (the Cullen Report) made 76 recommendations regarding offshore health and safety. During the inquiry Lord Cullen made reference to the Robens Report (1972) and the Burgoyne Report (1980) and in particular referred to the significance of the following:

— the Robens Report (which led to the enactment of *HSWA 1974*) had implications for safety in onshore petroleum exploitation but it did not explicitly consider offshore safety;

- the decision of the Robens Committee to exclude *MWA 1971* because of the similarities in fact and in relevant (marine) law between offshore installations and ships which are not covered by *HSWA 1974*;
- the subsequent decision of the government to extend *HSWA 1974* to cover oil and gas workers offshore [*Health and Safety at Work etc Act 1974 (Application outside Great Britain) Order 1977 (SI 1977 No 1232)* – now revoked] but that its enforcement should remain the responsibility of inspectors appointed by the Department of Energy under the powers provided under *MWA 1971*;
- the ineffectiveness of inspections as compared with those of HSE inspectors regulating similar major hazards onshore, under the *Control of Industrial Major Accident Hazards Regulations 1984 (CIMAH 1984) (SI 1984 No 1902 as amended) (since revoked and replaced by Control of Major Accident Hazards (COMAH) Regulations)*.

Regarding *CIMAH 1984* Lord Cullen, although influenced by the Health and Safety Commission/Health and Safety Executive (HSC/HSE) enforcement policy, recognised that due to the close proximity of working and living areas and the impracticability of rapid evacuation from offshore installations in an emergency, the following additional provisions were required which had no direct counterpart in *CIMAH 1984*:

(a) that the 'duty holder's' standards for management of health and safety and the control of major hazards must be demonstrated by the submission of a Safety Case and subject to formal acceptance by the regulatory body. Any installation without an accepted Safety Case will not be allowed to operate;

(b) that measures to protect the workforce must include arrangements for temporary refuge from fire, explosion and associated hazards until evacuation can be effected from an installation;

(c) that suitable use should be made of quantitative risk assessment (QRA) techniques as an aid to demonstrating the adequacy of the preventive and protective measures;

(d) that arrangements should be put in place relating to safety management and audit.

Offshore safety statistics

[07003.1] Offshore safety statistics are published on an annual basis in the HSE publication *Offshore Statistics & Regulatory Activity Report 2016*.

The report sets out the headline health and safety statistics for the offshore oil and gas industries and provides an important indicator of health and safety performance in the sector. The headline statistics for 2016 include the following:

- there was one fatal injury in 2016; there have been six fatalities in the last ten years;
- there were 20 specified injuries, with a rate of 66 per 100,000 full-time equivalent (FTE) workers;
- there were 78 over-seven-day injuries, with a rate of 257 per 100,000 FTE workers;

- there were ten occupational diseases reported;
- there were 263 dangerous occurrences reported;
- there were 104 hydrocarbon releases; the hydrocarbon release rate has fluctuated over the last ten years; and
- 816 non-compliance issues were raised with operators; 37 enforcement notices were issued (35 improvement notices and two prohibition notices); and there was one prosecution case instituted, which resulted in a conviction.

In 2016, the Department for Energy and Climate Change (DECC) reported that there were 302 installations in the United Kingdom Continental Shift (UKCS), of which 261 were operational and 143 were manned.

Copies of *Offshore Statistics & Regulatory Activity Report 2016* can be viewed on the HSE website at: www.hse.gov.uk/offshore/statistics/hsr2016.pdf.

In 1997, the oil and gas industry trade associations founded the *Step Change in Safety* initiative which aimed to reduce the UK offshore injury rate by 50%. By 2002, a new vision was created: to make the UK the safest place to work in the global oil and gas industry by 2010. It is now an independent tripartite organisation, separate from Oil and Gas UK and representing the workforce, regulators and employers. More information can be found on the Step Change website: www.stepchangeinsafety.net.

Repeal and revocation of outdated legislation

[07004] Offshore health, safety and welfare is now regulated by a specialist division of the HSE – the Energy Division (ED). The regulatory regime promotes risk-based and goal-setting legislation made under *HSWA 1974*.

The *Offshore Safety Act 1992 (OSA 1992)* and the *Offshore Safety (Repeals and Modifications) Regulations 1993 (SI 1993 No 1823)* expressly extended the application of *Part I (General Duties)* of *HSWA 1974* to include:

(a) securing the safety, health and welfare of persons on offshore installations or engaged on pipe-line works;
(b) securing the safety of such installations and preventing accidents on or near them;
(c) securing the proper construction and safe operation of pipe-lines and preventing damage to them; and
(d) securing the safe dismantling, removal and disposal of offshore installations and pipe-lines. [OSA 1992, s 1(1).]

At the same time the following statutes were expressed to come within the ambit of the enforcement provisions of *HSWA 1974*:

— the *Pipe-lines Act 1962*;
— the Mineral Workings (Offshore Installations) Act 1971;
— the Petroleum and Submarine Pipe-lines Act 1975 (now repealed); and
— the Petroleum Act 1987.

Subsequent regulations made under *HSWA 1974* have repealed most of the other sections of *MWA 1971* and its subordinate legislation. The *Offshore*

[07004] Offshore Operations

Installations and Pipeline Works (Management and Administration) Regulations 1995 (SI 1995 No 738) replaced:

— the *Offshore Installations (Logbooks and Registration of Death) Regulations 1972 (SI 1972 No 1542)*;
— the *Offshore Installations (Managers) Regulations 1992 (SI 1992 No 703)*;
— the *Offshore Installations (Registration) Regulations 1992 (SI 1992 No 702)*;
— the *Offshore Installation (Included Apparatus and Work) Order 1989 (SI 1989 No 978)*;
— the *Offshore Installations (Application of Employers Liability (Compulsory Insurance) Act 1969) Regulations 1979 (SI 1979 No 1289)*;
— the *Offshore Installations (Inspectors and Casualties) Regulations 1973 (SI 1973 No 1842)*; and
— the *Submarine Pipe-lines (Inspectors etc) Regulations 1977 (SI 1977 No 835)*.

It also amended the meaning of Manager and logbook in a number of regulations.

The *Offshore Installations (Safety Case) Regulations 1992*, enacted to implement Lord Cullen's recommendations, have been repealed and replaced by the *Offshore Installations (Safety Case) Regulations 2005 (SI 2005 No 3117)*. These regulations also amend the following:

— the *Offshore installations (Safety Representatives and Safety Committees) Regulations 1989*;
— the *Offshore Installations and Pipeline Works (Management and Administration) Regulations 1995*;
— the *Offshore Installations (Prevention of Fire and Explosion and Emergency Response) Regulations 1995*;
— the *Offshore Installations and Wells (Design and Construction etc) Regulations 1996*; and
— the *Diving at Work Regulations 1997*.

The regulations also implement Article 3(2) of Council Directive 92/91/EEC (OJ No L348 28.11.92, p9) concerning minimum requirements in the mineral-extracting industries through drilling.

The Deepwater Horizon disaster

[07004.1] The *Offshore Installations (Offshore Safety Directive) (Safety Case etc) Regulations 2015 (SI 2015 No 398)* (SCR 2015) implement the requirements of a European Offshore Directive adopted following the Deepwater Horizon explosion in the Gulf of Mexico in 2010. Eleven people died as a result of the explosion, which also caused the worst oil spill in US history.

SCR 2015 came into force in July 2015 and applies to oil and gas operations in external waters, that is, the territorial sea adjacent to Great Britain and any designated area within the UK continental shelf. They replaced the *Offshore*

Introduction to offshore operations **[O7005]**

Installations (Safety Case) Regulations 2005 (SCR 2005) in these waters, but activities in internal waters (eg estuaries) continue to be covered by SCR 2005 (see **O7009.1** below).

The primary aim of the 2015 Regulations is to reduce the risks from major accident hazards to the health and safety of the workforce employed on offshore installations or in connected activities. They also aim to increase the protection of the marine environment and coastal economies against pollution and ensure improved response mechanisms in the event of such an incident. The regulations build on the central recommendation of Lord Cullen's report on the public inquiry into the Piper Alpha disaster (see **O7003** above): that the operator or owner of every offshore installation should be required to prepare a safety case and submit it to the regulator for acceptance.

This safety case now incorporates the additional requirements of Directive 2013/30/EU on the safety of offshore oil and gas operations and amending Directive 2004/35/EC. The Regulations also implement aspects of Directive 92/91/EEC3 concerning the minimum requirements for improving the safety and health protection of workers in the mineral-extracting industries through drilling.

Safety cases under *SCR 2015* are submitted to the Competent Authority (CA) for assessment. This is the Offshore Safety Directive Regulator (OSDR) and comprises the HSE and DECC working in partnership.

SCR 2015 is set out in more detail in **O7010** onwards.

What is an offshore installation?

[O7005] Although the HSE Energy Division is responsible for applying the health and safety legislation to offshore installations, many offshore oil and gas facilities are not classified as an offshore installation at all times. Examples of these are mobile drilling rigs, flotels, floating and production units etc when they are not on station. At these times the units are deemed to be vessels and are controlled by the Maritime and Coastguard Agency under maritime law including the *Marine Safety Act 2003*.

The *Offshore Installations and Pipeline Works (Management and Administration) Regulations 1995 (SI 1995 No 738 as amended)* specify that for a structure to be classed as an 'installation' it must be a structure which is, or is to be, or has been used, while standing or stationed in relevant waters, or on the foreshore or other land intermittently covered with water:

(a) for the exploitation, or exploration with a view to exploitation, of mineral resources by means of a well;
(b) for the storage of gas in or under the shore or bed of relevant waters or the recovery of gas so stored;
(c) for the conveyance of things by means of a pipe;
(d) for undertaking activities that involve mechanically entering the pressure containment boundary of a well; or
(e) mainly for the provision of accommodation for persons who work on or from a structure falling within any of the provisions of this paragraph,

[07005] Offshore Operations

together with any supplementary unit which is ordinarily connected to it or any part of it.

Excepted structures are:

(a) a structure which is connected with dry land by a permanent structure providing access at all times and for all purposes;
(b) a well;
(c) a structure or device which does not project above the sea at any state of the tide;
(d) a structure which has ceased to be used for any of the purposes specified above, and has since been used for a purpose not so specified;
(e) a fixed structure which has ceased to be used for any of the purposes specified above, for so long as it is used for a purpose not so specified;
(f) a mobile structure which has been taken out of use and is not for the time being intended to be used for any of the purposes specified above eg a mobile drilling rig when it is off station; and
(g) any part of a pipeline.

Where two or more structures are, or are to be, connected permanently above the sea at high tide they shall for the purposes of these Regulations be deemed to comprise a single offshore installation. Offshore installations are divided into three types.

Drilling installations which—

(a) carry out exploration for oil and gas reserves;
(b) appraise the potential productivity of and reserves contained in an identified accumulation of oil or gas;
(c) complete wells for production or abandon wells;
(d) workover wells to improve productivity or correct a downhole problem,

Drilling installations can be either jackup or floating rigs and floating rigs can semi submersibles or drillships.

Production installations which—

(a) extract petroleum from beneath the seabed by means of a well;
(b) store gas in or under the shore or bed of relevant waters and recover gas so stored;
(c) are used for the conveyance of petroleum by means of a pipe; and
    (i) includes a:
        — non-production installation converted for use as a production installation for so long as it is so converted,
        — production installation which has ceased production for so long as it is not converted to a non-production installation; and production installation which has not come into use, and
    (ii) does not include an installation which, for a period of no more than 90 days, extracts petroleum from beneath the seabed for the purpose of well testing.

Non-production installations which—

(a) are an installation other than a production installation or drilling rig.

Because of the complexity of defining offshore installations, appropriate advice should be sought in any case of doubt.

Parts of offshore installations

[07006]–[07008] The *Offshore Installations and Pipeline Works (Management and Administration) Regulations 1995, Reg 3(3)* states that the following offshore equipment is deemed to be part of an offshore installation:

'(a) any well for the time being connected to it by pipe or cable;
(b) such part of any pipeline connected to it as is within 500 metres of any part of its main structure;
(c) any apparatus or works which are situated—

  (i) on or affixed to its main structure; or
  (ii) wholly or partly within 500 metres of any part of its main structure and associated with a pipe or system of pipes connected to any part of that installation.'

Which health and safety law applies to offshore installations?

[07009] HSE guidance explains that the purpose of the *Health and Safety at Work etc Act 1974 (Application outside Great Britain) Order 2013* is to apply ss 1–59 and 80–82 (the 'prescribed provisions') of the *HSWA 1974* beyond the mainland of Great Britain to specified offshore areas and work activities. These offshore areas include the 'territorial sea' and 'designated areas'.

The 2013 Order applies these sections to:

(a) offshore installations and most activities on or in connection with them (including certain diving projects) (*Article 4*);
(b) wells and activities in connection with wells (*Article 5*);
(c) pipelines, pipeline works and activities in connection with pipeline works (*Article 6*);
(d) mines and activities in connection with them (*Article 7*);
(e) activities connected to gas unloading, storage and recovery, the conversion of any natural feature for the permanent or temporary storage of gas and related exploration activities (*Article 8*);
(f) production of energy from water or wind and activities in connection with their exploration and exploitation (*Article 9*);
(g) underground coal gasification and any activities in connection with it (*Article 10*); and
(h) various other activities within the territorial sea (*Article 11*).

Determining whether or not health and safety regulations apply to offshore operations can be complex. Advice should be sought from company legal representatives or the HSE Energy Division in any cases of doubt. The following comments, however, may be useful:

Firstly, regulations made under *HSWA 1974* apply offshore only if they contain a clause in the current Application Order which explicitly states that they do apply. Each set of new regulations issued should detail the extent of their application by referring to the relevant Articles in the Application Order.

Secondly, regulations which contain an 'Application outside Great Britain' clause but which do not refer to any specific Article in the Application Order should be taken as applying to all the categories of offshore descriptions and activities specified in the Application Order.

To understand how a particular set of regulations is applied offshore, it is necessary to take account of several legally defined terms and expressions that may be included in the 'Application outside Great Britain' clause including the following:

— *territorial waters* – ie United Kingdom territorial waters adjacent to Great Britain.
[*The Health and Safety at Work etc Act 1974 (Application outside Great Britain) Order 1995 (SI 1995 No 263), Reg 2(1)*].

— *within territorial waters* includes on, over or under United Kingdom territorial waters adjacent to Great Britain.
[*The Health and Safety at Work etc Act 1974 (Application outside Great Britain) Order 1995 (SI 1995 No 263), Reg 2(1)*].

— *territorial sea* – ie the breadth of the territorial sea adjacent to the United Kingdom (the 12-mile limit).
[*Territorial Sea Act 1987, s 1*].

— *relevant waters* – internal waters (eg estuaries) and external waters (that is the territorial sea adjacent to Great Britain and any designated area within the UKCS).
[*The Offshore Installations and Pipeline Works (Management and Administration) Regulations 1995 (SI 1995 No 738 as amended), Reg 2(1)*].

— *associated structure* – ie in relation to an offshore installation a 'vessel', aircraft or hovercraft attendant on the installation or any floating structure used in connection with the installation.
[*The Offshore Installations and Pipeline Works (Management and Administration) Regulations 1995 (SI 1995 No 738 as amended), Reg 2(1)*].

— *vessel* – ie a hovercraft and any floating structure which is capable of being staffed.

— *jurisdiction* – ie (a) the law in force in England and Wales, (b) the law in force in Scotland, (c) the law in force in Northern Ireland,
and the authority given to the High Court for the determination of any questions of law arising in a designated area made under the *Continental Shelf Act 1964, s 1(7)*, and in particular the application of law as it would apply in those countries by virtue of the *Civil Jurisdiction (Offshore Activities) Order 1987 (SI 1987 No 2197)* and the *Criminal Jurisdiction (Offshore Activities) Order 1987 (SI 1987 No 2198)*.

— *fixed installation* – means an installation which cannot be moved from place to place without major dismantling or modification, whether or not it has its own motive power.

— *mobile installation* – No longer defined.

— *activities within territorial waters* – Article 11 specifies the following activities which may be carried out by non-offshore workers in accordance with the provisions of *HSWA 1974* and its subordinate legislation:

(a) the construction, reconstruction, alteration, repair, maintenance, cleaning, demolition and dismantling (or any preparatory activities) of any building or other structure (other than a vessel);
(b) loading, unloading, fuelling or provisioning of a vessel (which includes a 'stacked' installation, ie one which is not in use or intended for use at that time);
(c) the construction, reconstruction, finishing, refitting, repair, maintenance, cleaning or breaking up of a vessel (including a 'stacked' installation) except when carried out by the master or any officer or member of the crew of that vessel;
(d) diving operations;
(e) maintaining a 'stacked' installation on station.

The current Application Order is the *Health and Safety at Work etc Act 1974 (Application outside Great Britain) Order 2013 (SI 2013 No 745).*

Generally, if the most onerous of the regulations were complied with then the requirements of other health and safety legislation may be deemed to be satisfied.

In addition to health and safety regulations that are extended to apply offshore, the following regulations apply specifically to the Offshore Oil and Gas Industry and are discussed in the following sections:

— *Offshore Installations (Offshore Safety Directive) (Safety Case etc) Regulations 2015 (SI 2015 No 398)*;
— *Offshore Installations (Safety Case) Regulations 2005 (SI 2005 No 3117)*;
— *Offshore Installations and Wells (Design and Construction etc) Regulations 1992 (SI 1992 No 913)*;
— *Offshore Installations Prevention of Fire and Explosion and Emergency Response Regulations 1995 (SI 1995 No 743)*;
— *Offshore Installations and Pipeline Works (Management and Administration) Regulations 1995 (SI 1995 No 738)*;
— *Pipeline Safety Regulations 1996 (SI 1996 No 825)*;
— *Offshore Installations (Safety Representatives and Safety Committees) Regulations 1989 (SI 1989 No 971)*; and
— *Offshore Installations and Pipeline Works (First Aid) Regulations 1989 (SI 1989 No 1671).*

The *Borehole Sites and Operations Regulations 1995 (SI 1995 No 2038)* do not apply to offshore installations or activities carried out from an offshore installation.

Offshore Installations (Offshore Safety Directive (Safety Case Etc) Regulations 2015 (SI 2005 No 398)

[07009.1] The *Offshore Installations (Offshore Safety Directive) (Safety Case etc) Regulations 2015 (SCR 2015)* came into force in July 2015 and apply to oil and gas operations in external waters, that is the territorial sea adjacent to Great Britain and any designated area within the UKCS. They replaced the

SCR 2005 in these waters (see **O7009.13** below), subject to certain transitional arrangements. The *SCR 2005* continue to apply to oil and gas activities in internal waters (eg estuaries) and their requirements are set out in **O7010** onwards.

The HSE has published guidance on *SCR 2015* in the *Offshore Installations (Offshore Safety Directive) (Safety Case etc) Regulations 2015, Guidance on Regulations* (L154). This is aimed at those with duties under *SCR 2015*, including licensees, production installation operators, non-production installation owners and well operators.

This explains that: 'The primary aim of SCR 2015 is to reduce the risks from major accident hazards to the health and safety of the workforce employed on offshore installations or in connected activities. The Regulations also aim to increase the protection of the marine environment and coastal economies against pollution and ensure improved response mechanisms in the event of such an incident'.

The regulations build on Lord Cullen's central recommendation following the Piper Alpha disaster (see above) that the operator or owner of every offshore installation should be required to prepare a safety case and submit it to the regulator for acceptance.

A 'safety case' is defined in the Regulations as a document containing specified information relating to the management of health and safety and the control of major accident hazards (and containing the relevant particulars specified in *Sch 6* or *7* to the Regulations – relating to production and non-production installations).

This safety case now incorporates the additional requirements of Directive 2013/30/EU on the safety of offshore oil and gas operations (and amending Directive 2004/35/EC). (The *Offshore Petroleum Licensing (Offshore Safety Directive) Regulations 2015 (SI 2015 No 385)* and the *Merchant Shipping (Oil Pollution Preparedness, Response and Co-operation Convention) (Amendment) Regulations 2015 (SI 2015 No 386)* also implement some of the requirements of the Offshore Safety Directive.) *SCR 2015* also implements aspects of Directive 92/91/EEC3 concerning the minimum requirements for improving the safety and health protection of workers in the mineral-extracting industries through drilling.

Safety cases under *SCR 2015* are submitted to the CA for assessment. This is the joint HSE and DECC body the OSDR.

In summary, *SCR 2015*:

- require a licensee to ensure that any operator or well operator they appoint is capable of satisfactorily carrying out the functions and discharging the duties of an operator (*Reg 5*);
- require the CA to inform the licensing authority (which is established in the *Offshore Petroleum Licensing (Offshore Safety Directive) Regulations 2015*) where it determines that an operator or well operator no longer has the capacity to meet the requirements of the relevant statutory provisions (as defined in *Reg 2(1)*) (*Reg 6*);
- require operators and owners to prepare and implement a corporate major accident prevention policy (CMAPP) (*Reg 7* and *Schs 1* and *2*);

- require an operator and owner to prepare a document setting out its safety and environmental management system (SEMS) and to integrate that system with its overall management system (*Reg 8* and *Schs 2* and *3*);
- impose requirements with respect to the creation, revision and continuing effect of a verification scheme in respect of an installation (*Regs 9, 10* and *14* and *Pt 1* of *Sch 4*);
- impose requirements with respect to the creation, revision and continuing effect of a well examination scheme (*Regs 11, 12* and *14* and *Pt 2* of *Sch 4*);
- require an operator to prepare and send to the CA a design notification for a production installation which is to be established (*Reg 15(1)* and *Sch 5*) and a relocation notification for a production installation that is to be moved to a new location (*Reg 15(3)* and *Sch 5*);
- prohibit the operation of a production installation unless a safety case has been sent to and accepted by the CA (*Regs 16* and *17* and *Sch 6*);
- prohibit the movement of a non-production installation in external waters with a view to its being operated there unless a safety case has been sent to and accepted by the CA (*Regs 16* and *18* and *Sch 7*);
- require a design notification to be sent to the CA in respect of the conversion of a non-production installation to a production installation (*Reg 19(1)*) and prohibit operation unless a safety case has been sent to and accepted by the CA (*Reg 19(7)*);
- prohibit the dismantling of a fixed installation unless a revised safety case has been sent to and accepted by the CA (*Reg 20* and *Sch 8*);
- prohibit the commencement of a well operation unless a notification has been sent to the competent authority (*Reg 21(1)* and *Sch 9*) or where the CA objects to the notification (*Reg 21(7)*);
- prohibit the engagement of an installation in a combined operation with another installation or installations unless a notification has been sent to the CA (*Reg 22* and *Sch 10*);
- require an owner or operator to review their safety case at intervals of five years and at such other times as the CA may direct (*Reg 23*);
- require a safety case to be revised when appropriate and when directed by the CA (*Reg 24*);
- grant to the CA powers in respect of safety cases and related documents (*Reg 25*);
- grant to the CA a power to prohibit operations where measures for preventing or limiting the consequences of a major accident proposed in a safety case or in a notification of well operations or combined operations are insufficient (*Reg 26*);
- impose requirements with respect to the making and keeping of documents (*Reg 27*);
- require any procedures or arrangements in safety cases and plans stated in a notification of well operations or a notification of combined operations to be followed (*Reg 28*);

- require an operator, owner and well operator to take suitable measures to reduce risk, including where necessary suspending operations, where an activity carried out significantly increases the risk of a major accident (*Reg 29(1)* and *(2)*) and to report to the CA when such measures have been taken (*Reg 29(3)*);
- require the operator or owner to perform certain duties under the *Offshore Installations (Prevention of Fire and Explosion, and Emergency Response) Regulations 1995* (the duties are set out in *Reg 30(14)*) consistent with the external emergency response plan (as defined in *Reg 30(13)*) and taking into account the risk assessment undertaken during the preparation of the safety case (*Reg 30(1)*);
- require the operator, owner and well operator to communicate to their employees, contractors and contractors' employees the arrangements for confidential reporting of safety concerns (*Reg 31*);
- require duty holders to cooperate with the CA in developing, preparing and revising standards and guidance on major accident prevention (*Reg 32*);
- require the operator, well operator or owner to notify the CA of any major accident or situations where there is an immediate risk of such an accident (*Reg 33*);
- require UK-registered companies to provide the CA with information about accidents outside the European Union in which they or their subsidiaries are involved as licensees, operators or well operators (*Reg 34*);
- provide for the granting of exemptions from the Regulations by the CA (*Reg 35*);
- make specific provision for enforcement of the Regulations and penalties for offences (*Regs 36 and 40*); and
- provide for an appeal to the Secretary of State against certain decisions of the CA (*Reg 37* and *Sch 12*).

The HSE sets out the main changes between the 2005 and 2015 regulations. These are as follows:

Competent Authority

[07009.2] The competent authority for *SCR 2015* is the joint HSE/DECC OSDR as set out above.

Application

[07009.3] *SCR 2015* applies to oil and gas operations in external waters: the territorial sea adjacent to Great Britain and any area designated within the continental shelf. They replace *SCR 2005* in these waters, subject to certain transitional arrangements (see **O7009.13** below). Activities in internal waters (eg estuaries) will continue to be covered by *SCR 2005*.

Definitions

[07009.4] Several new definitions include 'oil and gas operations':

'Offshore oil and gas operations' are all activities associated with an installation relating to exploration and production of petroleum, including the design, planning, construction, operation and decommissioning of the installation, but excluding the conveyance of petroleum from one coast to another.

Existing definitions have been updated, including the definitions of 'major accident' and 'production installation'.

A 'major accident' is:

'(a) an event involving a fire, explosion, loss of well control or the release of a dangerous substance causing, or with a significant potential to cause, death or serious personal injury to persons on the installation or engaged in an activity on or in connection with it;

(b) an event involving major damage to the structure of the installation or plant affixed to it or any loss in the stability of the installation causing, or with a significant potential to cause, death or serious personal injury to persons on the installation or engaged in an activity on or in connection with it;

(c) the failure of life support systems for diving operations in connection with the installation, the detachment of a diving bell used for such operations or the trapping of a diver in a diving bell or other subsea chamber used for such operations;

(d) any other event arising from a work activity involving death or serious personal injury to five or more persons on the installation or engaged in an activity on or in connection with it; or

(e) any major environmental incident resulting from any event referred to in paragraph (a), (b) or (d), and for the purposes of determining whether an event constitutes a major accident under paragraph (a), (b) or (e), an installation that is normally unattended is to be treated as if it were attended.'

A 'production installation' is an installation which:

'(a) extracts petroleum from beneath the seabed by means of a well; or
(b) is used for the conveyance of petroleum by means of a pipe, and—

   (a) includes a—
      (i) non-production installation converted for use as a production installation for so long as it is so converted;
      (ii) production installation which has ceased production for so long as it is not converted to a non-production installation; and
      (iii) production installation which has not come into use; and
   (b) does not include an installation which, for a period of no more than 90 days, extracts petroleum from beneath the seabed for the purposes of well testing.'

In addition, the definitions of licensee, operator (in relation to a production installation) and well operator now refer to the definitions outlined in the *Offshore Petroleum Licensing (Offshore Safety Directive) Regulations 2015 (OPLR)*.

Scope

[07009.5] The scope of *SCR 2015* is greater than *SCR 2005*, as *SCR 2015* covers the management and control of environmental major accident hazards. This means that environmental protection issues need to be addressed in

CMAPPs, SEMS, verification schemes, emergency response arrangements and demonstrations related to the management and control of major accident hazards (see below). There are also additional requirements to provide information on environmental issues within the safety case and notifications sent to the CA.

Corporate major accident prevention policies (CMAPPs)

[07009.6] Every duty holder and well operator must now have a CMAPP. This should provide a high-level overview of how the management and control of major accident hazards will be implemented throughout an organisation.

Regulation 7 applies only to a duty holder which is a body corporate or unincorporate. It sets out that the duty holder must prepare in writing a CMAPP which establishes the overall aims and arrangements for controlling the risk of a major accident and how those aims are to be achieved and those arrangements put into effect by the officers of the duty holder. It covers the duty holder's installations in external waters; and outside the European Union.

The policy must address at least the particulars set out in *Sch 1* and must be prepared in accordance with the matters set out *Sch 2*. It may also outline the commitment of the duty holder to mechanisms for effective tripartite consultation.

A duty holder must implement the policy throughout its offshore oil and gas operations and set up appropriate monitoring arrangements to assure effectiveness of the policy.

Safety and Environmental Management Systems

[07009.7] Every duty holder and well operator must have a documented safety and environmental management system (SEMS) in operation within its organisation. This must be integrated with its overall management system.

Regulation 8 sets out that in the case of a body corporate or unincorporate, the SEMS must include the organisational structure, responsibilities, practices, procedures, processes and resources for determining and implementing the CMAPP. It must be integrated with the overall management system of the duty holder and must address the particulars in *Sch 3* and be prepared in accordance with the matters set out in *Sch 2*.

The document setting out the SEMS must include a description of:

(a) the organisational arrangements for the control of major hazards;
(b) the arrangements for preparing and submitting documents under the relevant statutory provisions; and
(c) the verification scheme (which description must comply with *Reg 13(1)* – see below).

The regulation applies to well operators and well examination schemes.

The arrangements within the SEMS must include arrangements for determining and implementing the CMAPP and address various matters relevant to the management and control of major hazards set out in *Schs* 2 and 3. An adequate description of the SEMS must be included in the duty holder's safety case (see *Schs 6(5)* and *7(5)*), and for well operators, in a well notification, if appropriate (see *Sch 9(16)*).

Verification

[07009.8] The *SCR 2015* verification scheme requirements have been extended to cover environmental, as well as safety-critical elements. The verification requirements in *SCR 2015* have also been updated to accommodate new duties (eg relating to the competence of verifiers and the sharing of information between verifiers and duty holders).

Regulation 9 sets out that the duty holder must establish a verification scheme for ensuring that the safety and environmental-critical elements and the specified plant are suitable and in good repair and condition. They must do this through:

- examination, including testing where appropriate, of the safety and environmental-critical elements and the specified plant by a verifier;
- examination of any design, specification, certificate, CE marking or other document, marking or standard relating to those elements or that plant by a verifier;
- examination by a verifier of work in progress;
- the creation of reports by a verifier on the examination and testing carried out, the findings, and any remedial action recommended;
- the taking of appropriate action by the duty holder following a report;
- the making of a note of action taken by the duty holder following a report; and
- the reporting by a verifier to the duty holder of any instances of non-compliance of the duty holder with the standards of the scheme.

The duty holder must ensure that the verification scheme is drawn up by or in consultation with the verifier and recorded in writing, and that a note is made of any reservation expressed by the verifier as to the content of the scheme in the course of drawing it up.

The duty holder must produce a written record of the safety and environmental-critical elements and the specified plant, invite comment on the record by a verifier, and make a note of any reservation expressed by a verifier as to the contents of the record.

These duties must be completed:

(a) in the case of a production installation, before completion of its design; and

(b) in the case of a non-production installation, before it is moved into external waters with a view to its being operated there.

Regulation 10 sets out that a verification scheme must provide for the matters contained in *Pt 1* of *Sch 4*. The duty holder must:

(a) ensure that where tasks under a verification scheme are allocated by the verifier to personnel of the verifier they are appropriately allocated to personnel qualified to undertake them;

(b) make suitable arrangements for the communication of information between the duty holder and the verifier; and
(c) give the verifier suitable authority to carry out the functions under the verification scheme effectively.

The duty holder must also ensure that the verification scheme is reviewed as often as may be appropriate and, where necessary, revised or replaced by or in consultation with the verifier, and that a note is made of any reservation expressed by the verifier in the course of drawing up the verification scheme.

Where there is a material change to a design notification, a relocation notification, the safety case or a notification of combined operations (that is combined between two units that have safety cases) the duty holder must refer the material change to the verifier for further comment in accordance with the verification scheme.

If the competent authority requests this, the duty holder must communicate the outcome of the referral of the material change to the CA.

The duty holder must ensure that the verification scheme is put into effect from the time it is established and that effect continues to be given to the scheme, or any revision or replacement of the scheme, while the installation remains in existence.

Regulation 13 sets out that a description of the verification scheme complies if it includes:

(a) a description of the criteria for selection of the verifier to carry out functions under the scheme;
(b) a description of the means of verifying that the safety and environmental-critical elements and any specified plant remain in good repair and condition; and
(c) details of the arrangements to carry out the functions under the scheme including—
 (i) the examination and testing of the safety and environmental-critical elements by the verifier;
 (ii) the verification of the design, standard, certification or other system of conformity of the safety and environmental-critical elements;
 (iii) the examination of work in progress;
 (iv) the taking of remedial action by the duty holder;
 (v) the reporting of any instances of non-compliance of the duty holder with the standards of the scheme; and
 (vi) the review of the scheme throughout the lifecycle of the installation.

A description of the well-examination scheme complies if it includes—

(a) a description of the criteria for selection of the well examiner to carry out functions under the scheme;
(b) a description of the means of verifying that the well is designed and constructed, and is maintained in such repair and condition, that:
 (i) so far as is reasonably practicable, there can be no unplanned escape of fluids from the well; and

Offshore Installations (Offshore Safety Directive etc) Regulations 2005 **[07009.12]**

(ii) risks to the health and safety of persons from it or anything in it, or in strata to which it is connected, are as low as is reasonably practicable; and
(c) details of the arrangements to carry out the functions under the scheme including:
(i) the examination of the well, or a similar well, by the well examiner;
(ii) the examination of information required under *Reg 11(2)(a)(ii)*;
(iii) the examination of work in progress;
(iv) the taking of remedial action by the well operator;
(v) the reporting of any instances of non-compliance of the well operator with the standards of the scheme; and
(vi) the review of the scheme.

Well examination

[07009.9] Well examination requirements for wells in external waters have been revoked from the *Offshore Installations and Wells (Design and Construction etc) Regulations 1996*, updated to accommodate new principles and duties (for example in relation to the competence of well examiners and the transference of information from well examiners to well operators), and placed in *SCR 2015*.

Internal emergency response arrangements

[07009.10] Certain duties under the *Offshore Installations (Prevention of Fire and Explosion and Emergency Response) Regulations 1995* are now designated 'internal emergency response duties'. A description of these duties together with the Oil Pollution Emergency Plan produced under the *Merchant Shipping (Oil Pollution Preparedness, Response Co-operation Convention) Regulations 1998* (OPRC 1998) deliver the description of the internal emergency response arrangements (see *Regs 2(10)* and *30* of the *SCR 2015*). This description must be included in the safety case.

Well notification

[07009.11] A statement must now be included in a well notification, made after considering reports by the well examiner, that the risk management relating to well design and its barriers to loss of control are suitable for all anticipated conditions and circumstances. Well notifications, or a material change to a well notification, must also be accompanied by the relevant report of the well examiner (see *Reg 21*).

Other changes

[07009.12] These include the following:
- a requirement for the CA to inform the licensing authority (which is defined in *Offshore Petroleum Licensing (Offshore Safety Directive) Regulations 2015* (OPLR)) where it determines that an operator of a

[07009.12] Offshore Operations

production installation or well operator no longer has the capacity to meet the requirements of the relevant statutory provisions (as defined in *Reg 2(1)*) (see *Reg 6* of the *SCR 2015*);
- a power for the CA to prohibit operations where measures for preventing or limiting the consequences of a major accident proposed in a safety case, in a notification of well operations or a notification of combined operations are insufficient (*Reg 26*);
- a specific requirement for plans stated in a notification of well operations or a notification of combined operations to be followed (*Reg 28*);
- a requirement for a duty holder and well operator to take suitable measures to reduce risk, including where necessary suspending operations where an activity carried out significantly increases the risk of a major accident (*Reg 29(1)* and *(2)*), and to report to the competent authority when such measures have been taken (*Reg 29(3)*);
- requirements for duty holders to perform certain duties under PFEER (the duties are set out in *Reg 30(14)* of the *SCR 2015*) consistent with the external emergency response plan (as defined in *Reg 30(13)*) and taking into account the risk assessment undertaken during the preparation of the safety case (*Reg 30(1)*);
- requirements for duty holders and well operators to communicate to their employees, contractors and contractors' employees the arrangements for confidential reporting of safety and environmental concerns (*Reg 31*);
- requirements for duty holders and well operators to co-operate with the competent authority in developing, preparing and revising standards and guidance on major accident prevention (*Reg 32*);
- a requirement for the duty holder or well operator to notify the CA of any major accident or situations where there is an immediate risk of such an accident (*Reg 33*); and
- a requirement for UK-registered companies acting as licensees or operators to provide the competent authority, when requested to do so, with information about accidents outside the European Union in which they or their subsidiaries are involved (*Reg 34*).

Transitional provisions

[07009.13] Oil and Gas UK (the trade body for the UK offshore oil and gas industry) provides the following transitional timeline for the new law:

- existing installations are those that existed on 18 July 2013;
- operators of production installations that came into existence, or owners of non-production installations without a UK safety case and working in UK waters for the first time, after the 19 July 2016 the new laws apply immediately;
- owners of other non-production installations, and production installations which came into existence after 18 July 2013, must comply by the earlier date of 19 July 2016 or the date of their next thorough review;

- operators of existing production installations (other than executing well operations) which existed prior to 18 July 2013, must comply by the earlier date of the 19 July 2018 or the date of their next thorough review;
- operators executing well operations, the new laws apply on 19 July 2016, unless this involves an existing installation with an earlier implementation date.

Schedule 14 of the *SCR 2015* sets out the transitional provisions in detail.

The *Offshore Installations (Offshore Safety Directive) (Safety Case etc) Regulations 2015, Guidance on Regulations* (L154) can be downloaded from the HSE website.

Offshore Installations (Safety Case) Regulations 2005 (SI 2005 No 3117)

[07010] As Lord Cullen recommended, new offshore safety regulations, the *Offshore Installations (Safety Case) Regulations 1992*, were introduced within the *HSWA 1974* framework.

These regulations have been repealed and replaced by the *Offshore Installations (Safety Case) Regulations 2005 (SI 2005 No 3117)*, which have in turn been replaced (subject to transitional arrangements (see O7009 above)) by the *Offshore Installations (Offshore Safety Directive) (Safety Case etc) Regulations 2015* in relation to oil and gas operations in external waters, that is the territorial sea adjacent to Great Britain and any designated area within the UK continental shelf. Activities in internal waters (eg estuaries) continue to be covered by *SCR 2005*.

The 2005 Regulations require:

- licensees to ensure anyone they appoint as an operator is capable of fulfilling their legal responsibilities for safety and carries out those responsibilities;
- well operators to notify the HSE before commencing well operations;
- duty holders to send an early design notification, instead of a design safety case, to the HSE when establishing a new production installation;
- duty holders to send a relocation notification before moving a production installation to a new location (whether from outside relevant waters or not).
- safety cases for production installations and non-production installations;
- safety cases be revised and resubmitted before dismantling an offshore installation;
- duty holders to carry out a thorough and fundamental review of their safety cases at least every five years, or as directed by HSE;
- the revision and resubmission of a safety case before making a material change to an installation; and
- duty holders to ensure compliance with the safety case.

The regulations also introduced an appeals procedure and amended the *Offshore Installations (Safety Representatives and Safety Committees) Regulations 1989 (SI 1989 No 971)* (see **O7000** below) in order to extend consultation with safety representatives to reviewing and revising a safety case, as well as preparing one.

The primary aim of the regulations is to reduce risks to the health and safety of the workforce employed on offshore installations or in connected activities. The principal matters to be demonstrated in safety cases and submitted to the HSE in compliance with *SI 2005 No 3117*, while taking account of the requirements of the above regulations, are that:

(a) the management system is adequate to ensure compliance with all statutory health and safety requirements;
(b) adequate arrangements have been made for audit, and the preparation of audit reports, and for the management of arrangements with contractors and sub-contractors;
(c) all hazards with the potential to cause a 'major accident' have been identified, their risks evaluated, and measures taken to control those risks; and
(d) adequate arrangements are in place for ensuring the suitability of safety critical elements and emergency escape and evacuation.

The definition of 'major accident' embraces fire, explosion or the release of a dangerous substance involving death or serious personal injury, an event involving major damage to the structure of the installation or plant or loss in the stability of the installation, collision of a helicopter with the installation, the failure of life support systems for diving operations in connection with the installation, the detachment of a diving bell used for such operations or the trapping of a diver in a diving bell or other subsea chamber used for such operations; or any other event arising from a work activity involving death or serious personal injury to five or more persons on the installation or engaged in an activity in connection with it.

The Offshore Installations and Wells (Design and Construction, etc) Regulations 1996 (SI 1996 No 9130) introduced the concept of Safety Critical Elements. These are defined as;

> Such parts of an installation and such of its plant (including computer programmes), or any part thereof—
>
> (a) the failure of which could cause or contribute substantially to; or
> (b) a purpose of which is to prevent or limit the effect of,
>
> a major accident.

For convenience the expression 'duty holder' is used as a shorthand way of referring to the person on whom duties are placed by the Regulations, for a production installation that is the operator and for a non-production installation the owner.

The case for safety is required to relate to all activities carried out on, or in connection with, an installation including, for example, relevant activities involving divers, pipelaying or the movement of vessels or helicopters. The various people concerned with these activities are under a duty to co-operate with 'duty holders' in the preparation of installation safety cases.

The 2005 Regulations allow the use of electronic communication and storage of information. *The Guide to the Offshore Installations (Safety Case) Regulations 2005* (L30) is available on the HSE website along with a significant number of documents giving guidance on how to comply with the legal requirements and HSE's procedures for handling and accepting safety cases and other submissions. The HSE is empowered to recover costs associated with the assessment of safety cases and the enforcement of relevant statutory provisions (see *Cost recovery for offshore activities – A guide* on the HSE website at: www.hse.gov.uk/charging/offshore/chgoffsh.htm and O7022 below).

Regulations to underpin the Safety Case Regulations

[07011] In addition to the Safety Case Regulations, three further goal-setting regulations were enacted to implement the recommendations made by Lord Cullen; completing offshore implementation of the Extractive Industries (Boreholes) Directive and supporting and complementing the objectives of the *Safety Case Regulations*:

— *Offshore Installations and Pipeline Works (Management and Administration) Regulations 1995 (SI 1995 No 738)*;
— *Offshore Installations (Prevention of Fire and Explosion, and Emergency Response) Regulations 1995 (SI 1995 No 743)*; and
— *Offshore Installations and Wells (Design and Construction etc) Regulations 1996 (SI 1996 No 913)*.

Previously, under *section 5* of the now defunct *Mineral Workings (Offshore Installations) Act 1971*, a general duty was placed on the manager of an offshore installation for all matters affecting safety, health or welfare on his offshore installation. This duty also included the maintenance of order and discipline over all people on or about the installation. This authority and duty was continued by the *Offshore Installation and Pipelines (Management and Administration) Regulations 1995 (SI 1995 No 738)*.

Verification of safety-critical elements of installations

[07012] The *Offshore Installations (Safety Case) Regulations 2005 (SI 2005 No 3117)* impose a duty on the duty holder to set up arrangements for the scrutiny of safety-critical elements of the offshore installation to be verified as suitable by an independent and competent person. This requirement (continued from earlier legislation) replaced the previous certification regime established by *the Offshore Installations (Construction and Survey) Regulations 1974 (SI 1974 No 289)* which were revoked by the *Offshore Installations and Wells (Design and Construction etc) Regulations 1996*.

The purpose of such scrutiny is to obtain assurance of the achievement of the satisfactory condition of the safety critical elements of the installation. The scheme is to be carried out throughout the installation's life cycle (from design

[07012] Offshore Operations

through fabrication, construction, hook-up and commissioning, and the whole operating life to decommissioning and dismantling). For the duty holder, the scheme will contribute to the duty holder's ability to demonstrate that the installation is 'fit for purpose'. A summary of the scheme must be included in the installation's Safety Case.

Before commencing operations

[07013] (Note: The HSE has the discretion to allow a shorter notification period in appropriate cases.)

(a) Production installations to be established or relocated

[07013.1] *Reg 6(1)* requires the operator of a production installation to submit a Design Notification (see *Sch 1*) by a date which is early enough to allow them to take account in the design of any matters raised by the HSE within three months of the submission of a field development programme to the Oil and Gas Authority (OGA). The OGA's role is to regulate, influence and promote the UK oil and gas industry in order to maximise the economic recovery of the UK's oil and gas resources.

Regulation 6(2) requires the operator of a production installation which is to be moved to a new location within internal waters (whether from outside internal waters or not) to submit a relocation notification (see *Sch 1*) at such time before the submission of a field development programme to the OGA to allow him to take account in the design of any matters raised by the HSE within three months of the submission.

The HSE will indicate in writing any changes to the design or any considerations which it would wish to be taken into account at detailed design, construction or commissioning stages, and which would be likely in due course to assist in acceptance of the full operational Safety Case required under *Reg 4(2)*. No operations may be commenced until six months have elapsed since the operator sent the Safety Case to the HSE and it has accepted it. Commencement of an operation includes the first well drilling operation which may require the release of hydrocarbons beneath the sea bed or when hydrocarbons are brought on to the site for the first time.

(b) Production installations

[07013.2] The Operator of a production installation must ensure that it is not operated within internal waters unless a Safety Case has been sent to the HSE at least six months before commencing the operation of the installation, and it has accepted the Safety Case [*Reg 7* and *Sch 2*].

(c) Non-production installation

[07013.3] The Operator of a non-production installation must ensure that it is not moved in internal waters with a view to it being operated there unless a safety case has been sent to the HSE at least three months before the movement of the installation and it has accepted the Safety Case [*Reg 8* and *Sch 2*].

The Operator of a non-production installation must ensure that it is not converted to enable it to operate as a production installation unless a Design

Notification of any particulars not contained in the current Safety Case has been sent to the HSE by a date which is early enough to allow him to take account in the design of any matters raised by the HSE within three months of the submission [*Reg 9* and *Sch 1*].

Decommissioning fixed installations

[07014] The Operator of a fixed installation must not decommission the installation unless he has:

(a) revised the current Safety Case in line with Sch 5;
(b) sent it to the HSE at least three months before the decommissioning starts; and
(c) the HSE has accepted the revision to the Safety Case [*Reg 11*].

The submission required by *Reg 11* is intended to ensure that the method and arrangements for abandonment of a fixed installation provide for adequate control of risks to persons on the installation or engaged in connected activities. It is an offence to begin decommissioning before the Safety Case has been accepted.

The HSE's acceptance of a Safety Case for decommissioning of an installation is required, and will be given, independently of any other approval such as the obligation under the *Petroleum Act 1998* to submit an abandonment programme.

The abandonment programme

[07015] The process of abandonment of a fixed offshore installation or submarine pipeline is subject to the acceptance of a detailed formal document known as the abandonment programme by the Secretary of State which must:

(a) contain an estimate of the cost of the measures proposed in the programme;
(b) specify the periods/times when the programme is to be put into effect, or provide information relevant to determining how the specified times are to be determined;
(c) include provisions regarding the continuing maintenance requirements of installations or pipelines (including those not fully demolished) which are to be left in position.

Part IV of the *Petroleum Act 1998* provides that when an abandonment programme has been approved by the Secretary of State, it becomes the duty of the persons who submitted the programme to ensure that it is carried out and that any conditions are complied with.

Oil and Gas UK Code OP071 *Guidelines for the Abandonment of Wells*, Issue 5, July 2015, *Guidelines on qualification of materials for the abandonment of wells*, Issue 2, October 2015 and *Guidelines on Well Abandonment Cost Estimation*, Issue 2, July 2015 can be purchased as a package or separately from the Oil and Gas UK website at: https://oilandgasuk.co.uk/product/op071/ HSE guidance, Offshore Information Sheet 2/2008 *Verification during Decommissioning and Dismantling* can be found on the HSE website at: www.hse.gov.uk/Offshore/verification.htm.

Safety Case specifications

[07016] The particulars to be included in all Safety Cases are required to demonstrate that:

(a) there is a management system that will make sure that the statutory provisions relating to the installation are complied with and that there is satisfactory management of arrangements with contractors and sub-contractors;
(b) adequate arrangements have been made for audit and reporting of the management system;
(c) all hazards that could cause a major accident have been identified;
(d) all major accident risks have been evaluated and measures have been or will be taken to keep these risks to the lowest level that is reasonably practicable [*Reg 12*].

Matters to be considered include:

(i) the plant and arrangements for the control of well operations;
(ii) a description of any pipeline with the potential to cause a major accident and how it is intended to secure safety and how he will ensure compliance with the *Pipelines Safety Regulations 1996 (SI 1996 No 825), Reg 11*;
(iii) how compliance with the PFEER regulations will be achieved;
(iv) how personnel will be protected from toxic gas;
(v) how the duty holder will ensure compliance with the *Offshore Installations and Wells (Design and Construction etc) Regulations 1996 (SI 1996 No 913)* and the suitability of the safety critical elements; and
(vi) where appropriate, the arrangement, methods and procedures for dismantling the installation and connected pipelines.

Verification of safety-critical elements

[07017] The duty holder (fixed and mobile installations) must ensure that the installation is not operated unless a suitable verification scheme covering the 'safety-critical elements' of the installation has been put in place. These elements include such parts of an installation and of its plant (including computer programs) the failure of which could cause or contribute substantially to a major accident (*Reg 19* and *Sch 7*).

To be suitable, the written scheme, prepared by a competent person, must demonstrate that the safety-critical elements are in good repair and condition and are verified as such by competent persons carrying out the following actions:

— examination, including testing where appropriate;
— examination of any design, specification, certificate, CE marking or other document, marking or standard relating to those elements;
— examination of work in progress;
— the taking of appropriate action following reports by such persons.

Competent persons must be independent, that is, their functions will not involve the consideration of elements for which they bear or have borne such responsibility as might compromise objectivity. In addition, it must be shown that they are sufficiently independent of a management system to ensure that they will be objective in discharging their functions.

Guidance on the verification of Safety Critical Elements is available on the HSE website, *Guide to the Integrity Environmental and Misc Aspects of the Offshore Installations and Wells (Design and Construction etc) Regs L85* and the Oil and Gas UK publication HS007 *Guidelines on the Management of Safety Critical Elements Issue 2 (2015)* (see https://oilandgasuk.co.uk/product/guidelines-for-the-management-of-safety-critical-elements/).

Review, revision and recording of verification schemes

[07018] The duty holder must ensure that as often as may be appropriate:

— the verification scheme is reviewed and, where necessary, revised or replaced by or in consultation with an independent and competent person;
— a note is made of any reservation expressed by such person in the course of drawing it up;
— verification schemes must be kept (at an address notified to the HSE) for a minimum of six months following cessation of the scheme along with the following documents;
— all examinations and testing carried out;
— the findings;
— remedial actions recommended; and
— remedial actions performed [*Reg* 20 and *Sch* 7].

Notification of hazardous activities

[07019] Duty holders must ensure that the following operations are not commenced until the HSE has been notified. However, unlike a Safety Case, notification does not require acceptance by HSE before the activities can be commenced. Should the information provided give cause for concern, HSE may intervene by, for example, requesting further information, inspecting, or by issuing an enforcement notice.

When considering whether intervention is appropriate HSE will take account of the following factors (among others):

— the major accident potential of the operations being undertaken;
— the assessed adequacy of the management and other risk control systems; and
— whether the operations will take place in conditions approaching the operating limits described in the installation Safety Case (for example, extreme or unusual environmental conditions or well operations).

Combined operations

[07019.1] Operators of installations must not engage in a combined operation unless they agree a combined operations notification and one of them sends it to the HSE 21 days before the combined operation is due to start [*Reg* 10 and *Sch* 4]. See www.hse.gov.uk/osdr/guidance/combined-operations.htm.

Well operations

[07020]–[07021] Well operations require notification of the HSE at least 21 days before commencing the operation. The notification must contain the information listed in *Sch 6*. However, for well operations on a production installation involving the insertion of a pipe into a well or altering the construction of a well the notice period is reduced to ten days [*Reg 17* and *Sch 6*].

Summary of contents of a Production Safety Case

[07022] The Production Safety Case which is required to be submitted under *Reg 7* must include a comprehensive account of the following matters:

— a description of the structure, plant, pipelines, wells and any connections to other installations and any limits to safe operation;
— how compliance with the *Offshore Installations and Wells (Design and Construction) Regulations 1996 (SI 1996 No 913)* and the suitability of the safety critical elements will be achieved;
— management of health and safety, control of major hazards and the auditing of these systems;
— compliance with the PFEER regulations and the protection of personnel from toxic gas;
— all foreseeable activities which are intended to be undertaken, or which may need to be undertaken, during the operating lifetime of the installation;
— any occasional activities such as major maintenance projects or diving work and any planned construction or alteration projects;
— activities which may involve other vessels (for example, nearby diving support vessels, supply vessels, floating storage units), aircraft or other installations should be addressed as far as they are foreseeable (*note*: where not foreseeable, further details may need to be provided in a revised Safety Case (under *Reg 9(2)*) or a combined operations Safety Case (as required by *Reg 6*);
— well operations so far as they are foreseeable and, in particular, a description of the intended number of wells and the purpose of each (that is, whether for production, gas injection or water injection);
— the drilling facilities and the general drilling programme;
— the expected performance characteristics of the reservoir and individual wells (including well drive mechanisms);
— well control procedures;
— any simultaneous drilling and production operations from the installation and the arrangements for controlling such activities;
— procedures for the suspension or abandonment of any well;
— the safety management system for the operation, control and emergency procedures of the pipelines connected to the installation;
— a summary of the design philosophy for ensuring the safe operation of the installation; and
— management of arrangements with contractors and sub-contractors.

The *Health and Safety and Nuclear (Fees) Regulations 2016 (SI 2016 No 253)* include provisions allowing the HSE to charge for the assessment and acceptance of safety cases. The regulations set the fees payable by an operator or owner for the performance by the HSE of the following functions under the *Offshore Installations (Safety Case) Regulations 2005 (SI 2005 No 3117)* and the *Offshore Installations (Safety Directive) (Safety Case) Regulations 2015* as follows:

Function	*Person by whom fee is payable*
Assessing a design notification (sent to the Executive pursuant to *regulation 6(1)* or *9(1)* of the 2005 Regulations) for the purpose of deciding whether to raise matters relating to health and safety and raising such matters	The operator or owner who sent the design notification to the Executive pursuant to that provision
Assessing a relocation notification (sent to the Executive pursuant to *regulation 6(2)* of the 2005 Regulations) for the purpose of deciding whether to raise matters relating to health and safety and raising such matters	The operator who sent the relocation notification to the Executive pursuant to that provision
Assessing a safety case or a revision to a current safety case (sent to the Executive pursuant to any provision of the 2005 Regulations) for the purpose of deciding whether to accept that safety case or revision and accepting any such safety case or revision	The operator or owner who sent the safety case or revision to the Executive pursuant to that provision
Providing advice with respect to the preparation of a safety case or a revision to a current safety case which is proposed to be sent to the Executive pursuant to any provision of the 2005 Regulations	The operator or owner who has requested that advice
Assessing whether to grant an exemption pursuant to *regulation* 23 of the 2005 Regulations and granting any such exemption	The operator or owner who has requested the exemption
Assessing a design notification (sent to the competent authority pursuant to *regulation 15(1)* or *19(1)* of the 2015 Regulations) for the purpose of deciding whether to raise matters relating to health and safety and raising such matters	The operator or owner who sent the design notification to the competent authority pursuant to that provision
Assessing a relocation notification (sent to the competent authority pursuant to *regulation 15(3)* of the 2015 Regulations) for the purpose of deciding whether to raise matters relating to health and safety and raising such matters	The operator who sent the relocation notification to the competent authority pursuant to that provision

Function	Person by whom fee is payable
Assessing a safety case or a revision to a current safety case (sent to the competent authority pursuant to any provision of the 2015 Regulations) for the purpose of deciding whether to accept that safety case or revision and accepting any such safety case or revision	The operator or owner who sent the safety case or revision to the competent authority pursuant to that provision
Providing advice with respect to the preparation of a safety case or a revision to a current safety case which is proposed to be sent to the competent authority pursuant to any provision of the 2015 Regulations	The operator or owner who has requested that advice
Assessing whether to grant an exemption pursuant to regulation 35 of the 2015 Regulations and granting any such exemption	The operator or owner who has requested the exemption

Record keeping

[07023] Copies of the current Safety Case and any summary of any review along with every audit report must be kept at an address in Great Britain and on the installation for at least three years. Documents relating to the verification scheme (see O7018 above) must be kept for six months.

Reports must be made, and the actions recorded, in respect of audits carried out under *Reg 8(1)(b)*. These requirements are intended to ensure, among other matters, that the reports and action records will be conveniently available for examination by the HSE during the auditing of the duty holder's safety management and audit system.

Documents required to be kept on the installation are normally kept by the installation manager. Arrangements for keeping the documents and revising them as required should be considered as part of the safety management system.

Offshore Installations and Wells (Design and Construction, etc) Regulations 1996 (SI 1996 No 913)

[07024] The *Offshore Installations and Wells (Design and Construction, etc) Regulations 1996* (DCR) aim to ensure (*a*) the structural integrity of offshore installations is maintained throughout their normal life cycle; (*b*) that risks to installation workers from structural failure or loss of stability are minimised to reasonably practicable levels; (*c*) the safe condition of wells at all stages of their life cycle; and (*d*) to implement the relevant aspects of the Extractive Industries (Boreholes) Directive (EID) (92/91/EEC).

DCR complements the requirements of the *Pipeline Safety Regulations 1996 (SI 1996 No 825)*, *Provision and Use of Work Equipment Regulations 1998 (SI 1998 No 2306)* *(PUWER)*, marine and aviation regulations. The regulations have been amended by the 2005 and 2015 Safety Case Regulations.

They are supported by the following guides;
- A Guide to the Well Aspects of the Offshore Installations and Wells (Design and Construction etc) Regulations 1996 [L84].
- A Guide to the Integrity, Workplace Environment and Misc Aspects of the Offshore Installations and Wells (Design and Construction etc) Regulations 1996 [L85].

Integrity of installations – duties of duty holders

[07025] So far as reasonably practicable, duty holders must ensure that:

(a) installations are able to withstand such adverse forces as are reasonably foreseeable [*Reg 5(1)(a)*];
(b) the structural integrity of an installation is not prejudiced by (i) its layout and configuration, and (ii) its fabrication, transportation, construction, operation, modification, maintenance, repair [*Reg 5(1)(b), (c)*];
(c) in the event of reasonably foreseeable damage, the installation will retain sufficient integrity to enable action to be taken to safeguard the health and safety of personnel on or near the installation [*Reg 5(1)(e)*];
(d) installations are capable of being decommissioned and dismantled safely [*Regs 5(1)(d), 10*];
(e) installations are composed of materials sufficiently proof against and/or protected from anything likely to prejudice their integrity [*Reg 5(2)*];
(f) work to an installation does not impair its integrity [*Reg 6*];
(g) installations are (i) operated within appropriate limits, and (ii) that environmental conditions in which they may safely operate have been recorded. Such records must be kept on the installation and readily available for inspection by personnel [*Reg 7*];
(h) installations are subject to periodic maintenance, and arrangements made for (i) periodic assessment, and (ii) execution of any remedial work [*Reg 8*];
(i) within ten days after the appearance of evidence of a significant threat to the integrity of an installation, a report is sent to HSE, identifying the threat and specifying any remedial action [*Reg 9*];
(j) every helicopter landing area is large enough for landing and departure purposes and is of adequate design and construction [*Reg 11*]; and
(k) the additional requirements set out in *Sch 1* are complied with.

Duty holders have a duty to ensure that temporary equipment operated on an offshore installation also meet the requirements of the Safety Case and supporting regulations. The HSE has issued guidance on temporary equipment offshore which can be found on its website at: www.hse.gov.uk/foi/internalops/hid_circs/technical_osd/spc_tech_osd_25.htm.

Integrity of wells – duties of well-operators

[07026] Well-operators must ensure that:

(a) wells are designed, constructed and operated so that (i) so far as reasonably practicable, there is no unplanned escape of fluids, and (ii) risks to the health and safety of persons are as low as reasonably practicable [Reg 13];
(b) prior to design of a well, assessments have been made of geological strata and formations and fluids therein, including any hazards [Reg 14];
(c) wells are designed and constructed so that, so far as reasonably practicable,
 (i) they can be suspended or abandoned safely, and
 (ii) following suspension or abandonment, there is no unplanned escape of fluids [Reg 15];
(d) wells are constructed of suitable materials [Reg 16];
(e) suitable well control equipment is provided [Reg 17];
(f) prior to commencement or adoption of design of a well, written arrangements are made for examination, by independent and competent persons, of a well or parts thereof [Reg 18];
(g) in the case of:
 (i) escape of fluids,
 (ii) completion of a well,
 (iii) abandonment operations,
 (iv) drilling operations, and
 (v) workover operations;
 a report is sent to the HSE at such intervals as are agreed (or otherwise at weekly intervals following commencement), containing:
 — identifying number of well,
 — name of installation,
 — summary of activity,
 — diameter and true vertical and measured depths of (a) any hole drilled, and (b) any casing installed,
 — drilling fluid density,
 — in the case of an existing well, its current operational state [Reg 19]; and
(h) personnel carrying out well operations have received appropriate information, instruction and training and are properly supervised [Reg 21].

Duties of other personnel

[07027] Every person concerned (or to be concerned) in whatever capacity, in an operation in relation to a well, must co-operate with the well-operator, so far as is necessary, to enable him to discharge his duties [Reg 20].

Offshore Installations (Prevention of Fire and Explosion, and Emergency Response) Regulations 1995 (SI 1995 No 743)

[07028] The regulations have been amended by the *Offshore Installations (Safety Case) Regulations 2005 (SI 2005 No 3117)* (see **O7010** onwards) and the *Offshore Installations (Offshore Safety Directive) (Safety Case etc) Regulations 2015 (SI 2015 No 398)* (see **O7009.1** onwards) to bring the definitions into line with those of the 2005 and 2015 Safety Case Regulations and to delete the requirements for maintaining the equipment since these are now covered by the verification requirements of the safety case regulations.

These regulations, applicable to both production and non-production installations, require operators, owners and duty holders to:

(a) protect persons on the installation from fire and explosion, and secure effective emergency response [*Reg 4*];
(b) carry out, update, where necessary, and keep a record of an assessment of measures (including performance standards) for effective:
  (i) evacuation,
  (ii) escape,
  (iii) recovery, and
  (iv) rescue [*Reg 5*];
(c) establish the organisation/arrangements to be implemented in an emergency, and ensure provision of:
  (i) training/instruction on necessary action, and
  (ii) written information on use of emergency plant [*Reg 6*];
(d) ensure availability of suitable equipment in the event of an accident involving a helicopter [*Reg 7*];
(e) prepare and update the emergency response plan:
  (i) ensure its availability and that its contents are known, and
  (ii) ensure that it is tested from time to time [*Reg 8*];
(f) take effective measures to prevent fire/explosion, eg
  (i) identifying/designating areas where there is a risk of occurrence of flammable/explosive atmospheres,
  (ii) controlling the carrying on of hazardous activities there,
  (iii) prohibiting/limiting use of plant (in such areas) unless suitable for use, and
  (iv) controlling placement (in such areas) of electrical fixtures/sources of ignition [*Reg 9*];
(g) take steps to detect fire and communicate information to places where control action can be taken [*Reg 10*];
(h) provide suitable emergency response arrangements:
  (i) ensure a warning is given of an emergency (see below), and
  (ii) communicate the purpose of the emergency response.
  These arrangements should, so far as is reasonably practicable, remain effective during an emergency [*Reg 11(1)*].
  Warnings take the form of (A) illuminated signs and (B) acoustic signals.
  (A) Illuminated signs
    These are:

- in the case of warning of toxic gas, a red flashing sign, and
- in other cases, a yellow flashing sign.

(B) Acoustic signals

These are:
- in the case of warning of evacuation, a continuous signal of variable frequency,
- in the case of warning of toxic gas, a continuous signal of constant frequency, and
- in other cases, an intermittent signal of constant frequency [*Reg 11*];

(i) control/limit extent of emergencies (including fire and explosions), eg by provision for remote operation of plant and ensuring that it is capable of remaining effective in an emergency [*Reg 12*];

(j) protect persons from the effects of fire and explosion [*Reg 13*];

(k) provide muster areas and:
 (i) provide safe egress from accommodation and work areas, and
 (ii) provide safe access to muster areas/evacuation and escape points which must be kept unobstructed, adequately lit with emergency lighting, marked with suitable signs and, so far as is reasonably practicable, remain passable in an emergency [*Reg 14(1), (2)*];

Notes:
- doors for use in an emergency must open in the appropriate direction, or, if this is not possible, be sliding doors, and must not be fastened,
- accommodation areas must be provided with at least two means of egress situated suitably apart at each level [*Reg 14(3)*], and
- each person on an installation must be assigned to a muster area; and muster lists must be kept up-to-date and clearly displayed [*Reg 14(4)*];

(l) arrange for safe evacuation of persons and their conveyance to a place of safety [*Reg 15*];

(m) ensure provision of means of escape, where arrangements for systematic evacuation fail [*Reg 16*];

(n) arrange for:
 (i) recovery of persons following evacuation/escape,
 (ii) rescue of persons near the installation, and
 (iii) conveyance of rescued/recovered persons to a place of safety [*Reg 17*];

(o) provide emergency protective equipment and prepare and operate a written scheme for its examination and testing by a competent person [*Reg 18*] (see further Personal Protective Equipment);

(p) provide verifications of the 2005 and 2015 Safety Case Regulations see **O7012** and **O7009.8** (not of (SI 2005 No 3117);

(q) provide life-saving appliances, eg survival craft, life-rafts, life-buoys, life-jackets, which are:
 (i) of conspicuous colour,
 (ii) suitably equipped (where necessary), and
 (iii) kept available in sufficient numbers [*Reg 20*];

(r) provide information to all persons on the installation, in relation to:

(i) areas in which there is a risk of occurrence of flammable/explosive atmospheres,
(ii) non-automatic plant for fire-fighting purposes,
(iii) plant connected with personal protective equipment and life-saving appliances [Reg 21].

The HSE published a third edition of *Prevention of fire and explosion, and emergency response on offshore installations Offshore Installations (Prevention of Fire and Explosion, and Emergency Response) Regulations 1995, Approved Code of Practice and guidance* (L65) in 2016. This is aimed at all those who own, operate or work on offshore installations and looks at how to prevent fires and explosions as well as how to protect people working on offshore installations should they occur. It also looks at how to respond to emergencies, considering issues such as escape, evacuation, rescue and recovery. It can be downloaded from the HSE website.

Offshore Installations and Pipeline Works (Management and Administration) Regulations 1995 (SI 1995 No 738)

[07029] These regulations impose managerial and administrative duties on the duty holder who is defined as:

(a) operators of production installations, and
(b) owners of non-production installations;

not on licensees or managers (except the duty to co-operate). The installation manager is the delegate of such duties and as such has a limited liability, whilst ultimate residual liability rests with the duty holder (see *McDermid v Nash Dredging and Reclamation Co Ltd* [1987] 2 All ER 878).

Chain of command

Licensee

Duty holder

Installation manager

Duties on operators of fixed installations and owners of mobile installations – duty holders

[07030] Duty holders must:

(1) before the date on which an offshore installation is due to enter or leave relevant waters, notify the HSE in writing of the date of entry/departure of an installation into/out of relevant waters. Where there is a change of duty holder, an installation must not be operated until the HSE has been notified of:
 (i) the date of change, and
 (ii) the name and address of the new duty holder [Reg 5];
(2) appoint a competent installation manager, who must be known or readily ascertainable by everyone on the installation [Reg 6];

(3) keep both an offshore and onshore record of details of all persons on and working on the installation [*Reg 9*];
(4) where necessary, introduce a permit to work system [*Reg 10*];
(5) issue written instructions on procedures to be observed, bringing them to the attention of all personnel [*Reg 11*];
(6) establish a sufficient system for communicating health and safety arrangements between installation and shore, vessels and aircraft [*Reg 12*];
(7) appoint a competent helicopter landing officer to ensure safe helicopter landings and take-offs [*Reg 13*];
(8) collect and keep meteorological/oceanographic data and information concerned with the movement of the installation [*Reg 14*];
(9) ensure that the address and telephone number of the local HSE is known or readily ascertainable by personnel on the installation [*Reg 15*];
(10) provide employees with health surveillance [*Reg 16*];
(11) ensure an adequate supply of clean wholesome drinking water at suitable locations [*Reg 17*];
(12) ensure that all food is fit for human consumption [*Reg 18*];
(13) ensure that the installation is readily identifiable by sea or from the air [*Reg 19*];
(14) ensure that the *Employer's Liability (Compulsory Insurance) Act 1969* is fully complied with by all employers of employees who work on any offshore installation under his control [*Reg 21*];
(15) notify and register deaths and other losses from offshore installations [*Regs 21A–21D*].

Certificates of exemption may be issued by the HSE in respect of these duties [*Reg 20*].

Duty of installation manager

[07031] Where necessary, in the interests of health and safety, an installation manager can restrain a person and put him ashore [*Reg 7*].

Duties of all persons on the installation

[07032] Everyone on the installation must co-operate with the installation manager to enable him to discharge his statutory safety duties and functions; similarly, with the helicopter landing officer [*Reg 8*].

Powers of inspectors

[07033] The powers of inspectors of offshore installations, in relation to pipelines and the duty to provide accommodation and subsistence for inspectors are set out in *Regs 21F* and *21G*.

Safety Zones

[07033.1] *Regulation 21H* deals with safety zones around offshore installations and sets out that:

'The prohibition under section 23(1) of the Petroleum Act 1987 on a vessel entering or remaining in a safety zone established around an installation by virtue of that Act does not apply to a vessel entering or remaining in the safety zone—

(a) in connection with the laying, inspection, testing, repair, maintenance, alteration, renewal or removal of any submarine cable or pipe-line in or near that safety zone;
(b) to provide services for, to transport persons or goods to or from, or under the authority of a government department to inspect, any installation in that safety zone;
(c) if it is a vessel belonging to a general lighthouse authority (within the meaning given in section 193 of the Merchant Shipping Act 1995) performing duties relating to the safety of navigation;
(d) in connection with the saving or attempted saving of life or property;
(e) owing to stress of weather;
(f) when in distress; or
(g) if there is consent from the duty holder.'

Temporary exclusion zones

[07033.2] In addition to the set safety zones around each offshore installation or sub-sea installation, the Secretary of State for Transport is empowered under marine law to establish temporary exclusion zones around a wrecked ship or structure. In determining whether such a casualty, including those in distress, requires the establishment of an exclusion zone, the Secretary of State will take into account the 'significant harm' factor, that is, in terms of pollution or damage to persons or property. [*Merchant Shipping Act 1995, s 100A* – as introduced by the *Merchant Shipping and Maritime Security Act 1997*].

Construction and use of submarine pipelines

[07034] The *Petroleum Act 1998* prohibits the construction of a pipeline (or use of a pipeline constructed after 1975) without the specific written authorisation of the Secretary of State. Once authorised, the pipeline becomes a '*controlled pipeline*' and any work affecting the design, route and boundaries is subject to any terms and conditions specified by the Secretary of State. In particular, it is necessary to ensure that sufficient funds are available regarding liability cover in the event of damage attributable to the release or escape from the pipeline of 'permitted substances', that is, anything or any substance authorised to be transported in that pipeline.

The Secretary of State is granted power to order the compulsory modifications of a pipeline for the purposes of improving flow through that pipeline. A person other than the owner of the pipeline is permitted to make an application to the Secretary of State for a modification order [*PA 1998, s 16*].

Pipelines Safety Regulations 1996 (SI 1996 No 825)

[07035] The *Pipelines Safety Regulations 1996* aim to secure the initial and continuing integrity of pipelines throughout their life cycle. The Operator, in relation to a pipeline means;

(a) the person who has control over the conveyance of fluid in the pipeline;
(b) until that person is known, the person who commissions, or is to commission, the design and construction of the pipeline; or
(c) where the pipeline is no longer in use the person last having control over the conveyance of fluid in it.

To this end, the operator of a pipeline ensure that:

(a) it is designed so as to be safe within range of all operating conditions [*Reg 5*], provided with appropriate safety systems [*Reg 6*], can be examined and maintained safely [*Reg 7*] and is composed of suitable materials [*Reg 8*];
(b) it is constructed and installed so that it is fit for the purpose which it was designed [*Reg 9*];
(c) it is modified and maintained so that its fitness for purpose is not prejudiced [*Reg 10*] and operated within safe operating limits [*Reg 11*];
(d) adequate arrangements have been made to deal with any accidental loss of fluid, discovery of a defect or other emergency [*Reg 12*];
(e) the pipeline is maintained in an efficient state, efficient working order and good repair [*Reg 13*];
(f) abandoned at the end of their life cycle so as not to become a source of danger [*Reg 14*];
(g) steps are taken to inform persons of its existence and where to ensure that no damage is caused to the pipeline [*Reg 16*]; and
(h) other persons have a duty not to cause damage to a pipeline as may give rise to danger to persons [*Reg 15*].

Moreover, in the case of pipelines conveying dangerous fluids (eg flammable in air, very toxic, oxidising, reacting violently with water):

(i) the pipeline must be fitted with emergency shut-down valves [*Reg 19*];
(ii) details relating to construction/fluid to be conveyed must be notified to the HSE at least six months prior to commencement of construction [*Reg 20*];
(iii) the HSE must be notified at least 14 days before fluid is conveyed in a major accident pipeline [*Reg 21*] or within 14 days of a change in operator [*Reg 21*];
(iv) the HSE must be notified of specified particulars 3 months before any event listed in schedule 5 takes place [*Reg 22(2)*];
(v) a major accident prevention policy must be prepared (and revised) prior to completion of design [*Reg 23*];
(vi) appropriate emergency organisation and arrangements must in place and details of emergency procedures must be finalised [*Reg 24*]; and
(vii) local authorities, in whose area the pipeline is to pass, must be notified that construction is to take place, and then prepare and update an emergency plan [*Reg 25*], for which it may charge a fee [*Reg 26*].

The HSE has published guidance on the regulations, *A guide to the Pipelines Safety Regulations 1996 Guidance on Regulations* (L82). The guidance is designed to give pipeline operators and others involved with pipeline activities or who may be affected by the regulations, an understanding of what is required.

The Diving at Work Regulations 1997

[07036] The *Diving at Work Regulations 1997 (SI 1997 No 2776)*, which revoked the *Diving Operations at Work Regulations 1981*, impose requirements and prohibitions relating to health and safety upon persons who work as divers. Under the 1997 Regulations, a person 'dives' if he enters water or any other liquid; or a chamber in which he is subject to pressure greater than 100 millibars above atmospheric pressure and in order to survive in such an environment he breathes in air or other gas at a pressure greater than atmospheric pressure. The HSE has issued advice on many aspects of diving safety and how to comply with the regulations; see the HSE website.

Environments such as scientific clean rooms or submersible craft subjected to a pressure less than 100 millibars above atmospheric pressure are not covered by the Regulations. The regulations do not cover work carried out in air that is compressed to prevent the ingress of ground water or the treatment of patients not under the control of a diving contractor [*Reg 3(1)*].

The regulations do apply offshore [*Reg 3(2)*].

Every person who is engaged in a diving project or is responsible for, or has control over, such a project must take such measures as it is reasonable for a person in his position to take to ensure that the provisions of the 1997 Regulations are complied with [*Reg 4*].

There could be a number of diving projects taking place on one site. Each of these projects could be separate from each other and may have different 'diving contractors' in charge of them.

The HSE are authorised to approve diving qualifications [*Reg 14(1)*] and medical examiners of divers [*Reg 15(6)*].

Diving personnel responsibilities

The diving contractor

[07037] One person and only one person shall be appointed to be diving contractor for a diving project [*Reg 5(1)*].

The diving contractor must ensure, so far as is reasonably practicable, that the diving project is planned, managed and conducted in a manner which protects the health and safety of all persons taking part in that project. This requires the preparation of a diving project plan which must be updated as necessary during the continuance of the project.

Other duties placed on the diving contractor include:

- appointing a person to supervise a diving operation and ensuring that the person is competent and suitably qualified to act as supervisor for the operation;
- making a written record of that appointment;
- ensuring that the person appointed is supplied with a copy of any part of the diving project plan which relates to that operation;
- ensuring that there are sufficient people with suitable competence to carry out safely and without risk to health both the diving project and any action (including the giving of first-aid) which may be necessary in the event of a reasonably foreseeable emergency connected with the diving project;
- ensuring that suitable and sufficient plant is available whenever needed to carry out safely and without risk to health both the diving project and any action (including the giving of first-aid) which may be necessary in the event of a reasonably foreseeable emergency connected with the diving project;
- ensuring that all diving equipment is maintained in a safe working condition;
- ensuring, so far as reasonably practicable, that any person taking part in the diving project complies with the requirements and prohibitions imposed on him by or under the relevant statutory provisions and observes the provisions of the diving project plan; and
- ensuring that a record containing the required particulars, given in *Sch 1*, is kept for each diving operation and retained for at least two years after the date of the last entry in it [*Reg 5–9*].

The diving supervisor

[07037.1] Although the main duty for the whole project is placed on the diving contractor, supervisors also have a duty to manage the diving operation safely and give such reasonable directions to any person to enable him to do so. Supervisors should not participate in a diving operation which they consider to be unsafe because insufficient supervisors have been appointed [*Reg 10–11*].

The diver

[07037.2] A diver has the responsibility not to dive unless he has an approved valid qualification [*Reg 12(1)*] and is competent to carry out the activities he may reasonably expect to carry out whilst taking part in the diving project [*Reg 13*]. A diver must maintain a daily record of his diving [*Reg 12(3)*] and have a valid medical certificate to dive [*Reg 12(1)*] and must not dive if he knows of anything which makes him unfit to dive [*Reg 13(1)*].

Dive planning and risk assessment

[07038] The diving contractor must ensure that a diving project plan is prepared before commencement of diving and the plan must be updated as necessary [*Reg 6(2)*]. The diving plan must be based on an assessment of the risks to any person taking part in the diving project [*Reg 8*].

Similar duties regarding the safe operation of diving projects are also placed on 'clients' who raise diving contracts. They should be involved in the risk assessment elements of the 'diving project plan' required by *Reg 8*.

Further information on regulations covering diving at work safely and guidance to help employers meet their duties under the law can be found on the HSE website at: www.hse.gov.uk/diving.

All five diving Approved Codes of Practice (ACOPs) were revised following industry consultation and came into force on 8 December 2014, including *Commercial diving projects offshore, The Diving at Work Regulations 1997 Approved Code of Practice and Guidance* (L103).

Other offshore specific legislation

[07039]–[07045] The following Regulations were enacted under the *Mineral Workings (Offshore Installations) Act 1971* and apply only to offshore operations:

— *Offshore Installations (Safety Representatives and Safety Committees) Regulations 1989 (SI 1989 No 971)*;
— *Offshore Installations and Pipeline Works (First-Aid) Regulations 1989 (SI 1989 No 1671)*.

Offshore Installations (Safety Representatives and Safety Committees) Regulations 1989

[07046] Following the Piper Alpha disaster, the *Offshore Installations (Safety Representatives and Safety Committees) Regulations 1989 (SI 1989 No 971)* were issued by the then enforcing authority, the Department of Energy. The purpose of these Regulations, which were made under the authority of the *Mineral Workings (Offshore Installations) Act 1971 and not the Health and Safety at Work etc Act 1974*, is to ensure that the whole workforce is formally involved in promoting health and safety through freely-elected safety representatives and the safety committee.

Since the publication of the Cullen Report in 1990, much has been written regarding the identification and control of major accident hazards and protection of the workforce from such risks. The Cullen inquiry made it quite clear that high standards of safety and health performance, necessary on an offshore installation, cannot be achieved without the positive and informed commitment of the workforce. Everyone involved with offshore operations must have a safety management system which describes the arrangements for securing this commitment – the provision of information and advanced consultation – whenever it is appropriate to do so.

Information must be given in a form and manner which ensures that it is both comprehensible and relevant to those who are to receive it.

The HSE has published guidance to the Regulations: '*A Guide to the Offshore Installations (Safety Representatives and Safety Committees) Regulations 1989*' (L110). The third edition was published in 2012.

The definitions in the safety representatives regulations have been updated by the *Offshore Installations (Safety Case) Regulations 2005 (SI 2005 No 3117)* and the *Offshore Installations (Offshore Safety Directive) (Safety Case etc) Regulations 2015 (SI 2015 No 398)*.

Establishment of constituencies

[07047] The first stage in the appointment of safety representatives is the establishment of a system of constituencies by the installation manager in consultation with the safety committee. Thus, unlike onshore safety representatives appointed by recognised trade unions, offshore safety representatives are elected by all the workers in a constituency; in other words, by voluntary decision of the workforce. Their principal role is to act as a conduit on safety from workforce to management. Constituencies will take account of:

(a) areas of the offshore installation;
(b) activities undertaken on or from the installation;
(c) employees of the workforce; and
(d) other objective criteria which appear to the installation manager to be appropriate to the circumstances of the installation.

There must be, at least, two constituencies and every worker can be assigned to one, but subject to a maximum of forty and a minimum of three. It is the duty of the installation manager to post particulars of a constituency in suitable places on the installation and, if necessary, in appropriate languages. This done, the workers comprising the constituency can elect a safety representative. Any worker can stand for election to represent his constituency as a safety representative, provided that he:

(i) is a member of that constituency;
(ii) is willing to stand as a candidate;
(iii) has been nominated by a second member; and
(iv) his nomination is seconded by a third member.

A list of duly nominated candidates must be posted within a week after nominations have expired [*Regs 5–10*].

The purpose of dividing an installation into constituencies is to provide for appropriate groupings of the workforce from which safety representatives can be elected. Each installation, no matter how small the permanent workforce, must have at least two constituencies. If an installation manager wishes to create new constituencies, the proposals must be agreed by the safety committee, or both safety committees if two installations are temporarily bridge-linked (for example, production and accommodation platforms). Every new arrival to an installation must be given written notification and details of their constituency. Exceptions to this rule may be permitted where a person is not expected to remain on the installation for longer than 48 hours.

Each installation manager should actively encourage the filling of a constituency vacancy. HSE inspectors may wish to discuss such matters with other safety representatives during visits to the installation [*L110, paras 12–33*].

Functions of offshore safety representatives

[07048] Safety representatives on offshore installations have similar functions to safety representatives generally. More particularly, they can:

(a) investigate potential hazards/dangerous occurrences and examine causes of accidents;
(b) investigate health and safety complaints from their members;
(c) draw matters arising from (*a*) and (*b*) to the attention of the installation manager and any employer;
(d) approach the installation manager/any employer about general health and safety matters;
(e) attend meetings of the safety committee;
(f) represent members in consultation with the Inspectorate;
(g) consult members on any part of the above matters;
(h) inspect accident sites/equipment;
(i) alert the HSE if an imminent risk of serious injury is considered to exist;
(j) carry out site inspections;
(k) consult with HSE inspectors;
(l) receive copies of statutory health and safety documentation.

Exercise of these functions cannot give rise to criminal and/or civil liability on the part of a safety representative [*Reg 16*].

It is important to distinguish between functions and duties – safety representatives cannot be held to account for not carrying out their functions. However, since the functions are the principal basis of the safety representative system, safety representatives should endeavour to carry them out as fully as possible. Although not normally a formal part of the installation's management team, safety representatives should be able to take matters of safety to the installation manager without delay and without the need for a formal written approach. [*L110, paras 52 55*].

Inspection of equipment

[07049] A safety representative can inspect any part of the installation or its equipment, provided that part of the installation or equipment has not been inspected in the previous three months, and that he has given reasonable notice in writing to the manager and, if his employer is not the duty holder, also his employer.

More frequent inspections can take place, by agreement with the manager [*Reg 17*].

Inspection of safety documents and Safety Case

[07050] A safety representative is entitled to see occupational health and safety at work documents which are required by law to be kept on the installation. This entitlement does not extend to a document consisting of or relating to any health record of an identifiable individual.

Regarding the Safety Case for the installation, safety representatives must be supplied with a written summary of the main features of the Safety Case including any particulars concerning remedial work and the time by which it will be done. Appropriate facilities must be made available on the installation for reading the information if it is kept on film or in electronic form [*Regs 18, 18A*].

The *Health and Safety at Work etc Act 1974, s 28(8)* requires inspectors appointed under that Act to give certain factual and other information concerning their actions to employees or their representatives. Information is normally provided in the form of letters from inspectors, and the Regulations allow for the safety representatives to receive such information. It is recommended that the safety representatives agree that one of their number will act as a contact point with inspectors for this purpose. This representative can then inform his fellow safety representatives, but such arrangements should not preclude any of the representatives establishing working relationships with inspectors and meeting them when they visit.

One of the functions of safety representatives is the carrying out of formal inspections, both on a regular basis and following an incident. When determining the cause of a 'notifiable incident', safety representatives are empowered to inspect relevant parts of the installation and equipment, provided that the installation manager is notified and confirms:

(a) there has been a notifiable incident or over three-day injury;
(b) it is safe for an inspection; and
(c) the interests of his members might be involved.

A 'notifiable incident' is any death, injury, disease or dangerous occurrence which is required to be reported under the *Reporting of Injuries, Diseases or Dangerous Occurrences Regulations 2013*.

An 'over three day injury' means an injury required to be recorded in accordance with *Reg 12(1)(b)* of the *Reporting of Injuries, Diseases and Dangerous Occurrences Regulations 2013*.

In addition, a safety representative can inspect any part of the installation or its equipment, provided that the relevant part of the installation or equipment has not been inspected in the previous three months, and that he has given reasonable notice in writing to the manager and, if his employer is not the duty holder, also his employer.

More frequent inspections can take place by agreement with the manager [*Reg 17*].

The main purpose of the examination should be to determine the cause so that the possibility of action to prevent a recurrence can be considered. For this reason, it is important that the approach to the problem should be a joint one by the installation manager or his representative, relevant employers and the safety representatives. Whilst it may be necessary, following a notifiable incident, for the installation manager to take immediate steps to safeguard against further hazards he should involve the safety representatives at the earliest opportunity. If this is not possible he should notify the safety representatives of the action he has taken and confirm this in writing.

OF (SR and SC) Regs 1989 **[07052]**

Examinations may include visual inspection, photography (subject to any work permit requirements) and discussions with persons who are likely to be in the possession of relevant information and knowledge regarding the circumstances of the notifiable incident. The examination must not, however, include interference with any evidence or the testing of any machinery, plant, equipment or substance which could disturb or destroy the factual evidence before a HSE inspector has had the opportunity to investigate in accordance with his powers.

The number of safety representatives taking part in a formal inspection should be a matter for agreement with the installation manager. It may be appropriate for an installation safety officer or a specialist to provide advice during an inspection. Where any remedial action has been taken, safety representatives should be given the opportunity to make a re-inspection to satisfy themselves in order to report to the safety committee. [*L110, paras 63–70*].

Risk of serious personal injury

[07051] If two or more safety representatives think that there is an 'imminent risk of serious personal injury', arising from an installation activity, they can:

(a)   make representations to the manager, who must send a written report to an inspector, and
(b)   themselves send a written report to an inspector [*Reg 17(4)*].

Safety representatives are not directly empowered to stop activities which they consider involve imminent risk of serious personal injury. However, where two or more of them consider that there is such imminent risk, they are required by *Reg 17(4)(a)* to make representations on the matter to the installation manager. The manager is then required to report the matter in writing to an inspector as soon as is reasonably practicable.

In addition, the safety representatives may make their own written report on the matter to an inspector, exercising good judgment in invoking this function which should be reserved for genuinely imminent danger. In exercising their powers, a safety representative is entitled to seek advice and guidance from persons on the installation or elsewhere. This may usually be informally sought, for example from management, the safety officer, inspectors or from his trade union health and safety adviser. However, if a safety representative considers that there is a need for formal external independent advice, for example a noise survey or a hazard analysis, it is recommended that the matter be raised in the safety committee which may direct the request to the owners [*L110, paras 71–73*].

Safety committees

[07052] Owners of offshore installations, with one or more safety representatives, must establish a safety committee [*Reg 19*].

This must consist of:

(a)   the installation manager (as chairman),

(b) one other person (to be appointed by the owner or manager),
(c) all safety representatives, and
(d) persons who may be co-opted by unanimous vote, eg safety officer(s) [*Reg 20*].

The person who may be appointed to the committee should preferably be a representative of onshore management not directly involved in the day-to-day operation of the installation. Such an appointment would ensure a direct and continuous link with onshore management for the purpose of pursuing any improvements which cannot be made with the resources directly at the disposal of the installation manager. It would also enable solutions to common problems encountered on other installations owned by the same company to be relayed directly to the committee, ensuring consistency of approach.

The Regulations also allow for the safety committee, by unanimous vote, to co-opt on to the committee such additional persons as they wish, but these co-opted individuals are not entitled to vote on the co-option of any further individuals. It is expected that one of them should normally be the installation safety adviser, whilst others might include specialists in particular activities on the installation. Co-option of a minutes secretary may assist the committee [*L110, paras 78–83*].

Safety committee meetings

[07053] A safety committee must be convened by the chairman (normally the installation manager) within six weeks of the date of its establishment, and thereafter at least once every three months. If the committee wishes to meet more frequently, it may establish a policy on the circumstances under which such meetings may be called, for example to consider an urgent unforeseen matter. A safety representative who is unable, for whatever reason, to attend a meeting can nominate another member of his constituency to attend in his stead. The nominee should act on his representative's behalf at the meeting but has no other functions or powers under the Regulations.

Generally, safety committee representatives should cover certain basic items, for example:

— all incidents involving the installation since the committee last met;
— any feedback from onshore management including any incidents on other installations which may have implications for their installation;

and any other matter which could possibly affect the health and safety of people working on the installation. Safety committees on bridge-linked installations may decide to co-ordinate their business but it is a legal requirement for each installation to have a safety committee which must meet at least every three months. [*L110, paras 84–89*].

Functions of safety committees

[07054] Safety committees have the following functions:

(i) to monitor health and safety at the workplace;
(ii) to keep under review the constituency system;
(iii) to monitor arrangements for the training of safety representatives;

(iv) to monitor the frequency of safety committee meetings;
(v) to consider representations from any member of a safety committee;
(vi) to consider the causes of accidents, dangerous occurrences and cases of ill- health;
(vii) to consider any statutory documents relating to occupational health and safety, eg registers (but not medical records);
(viii) to prepare and maintain a record of its business, with a copy being kept for one year from the date of the meeting; and
(ix) to review safety representative training arrangements [Reg 22].

Installation owners/managers must co-operate with safety committees and safety representatives to enable them to perform their functions and they must disclose relevant information to safety representatives [Regs 23, 24]. This includes the revision, review or preparation of a safety case (see O7010).

The general philosophy behind the Regulations is that the safety committee will promote co-operation on all matters affecting occupational health and safety between all parties on the installation and will seek to promote and develop measures to ensure the occupational health and safety of the workforce. However, the committee should not attempt to replace the normal day-to-day channels of communication on specific health and safety issues between the individual and his immediate supervisor and ultimately the installation manager, but should serve as a back-up in the event of continuing concern.

The safety committee decides upon the format of its meetings. For example, the installation manager may be required to:

— report on any feedback from shore management on previous meetings;
— make a report on all notifiable incidents involving the installation since the committee last met;
— report any incidents or hazards to health which happened on other installations operated by the same operator which may have repercussions for his own installation (for this he would need briefing from shore);
— report any major changes in the equipment or method of operation planned for the installations; and
— report on anything else which could possibly affect the health and safety of persons working on the installation.

The importance of consultation to satisfy the requirements of *Reg 23* cannot be overemphasised. The installation manager should keep the safety committee informed of progress on agreed actions to be taken. [*L110, paras 90–96*].

Training

[07055] Employers must permit safety representatives time off from work, without loss of pay, to enable them:

(a) to perform their functions, and
(b) to undertake training,

and employers must meet the cost of training, travel and subsistence expenses. [*Regs 26, 27*].

Safety representatives must be permitted time off without loss of pay (including any appropriate offshore allowances) during normal working hours for basic training. This should be as soon as possible after their election. The length of training required is not prescribed but basic training should take into account:

— the functions of safety representatives;
— the training necessary for providing an understanding of the role of the safety representatives and safety committees;
— the legal requirements relating to the health and safety of persons on the installation, particularly the constituency members they directly represent;
— the nature and extent of workplace hazards, and the measures necessary to eliminate or minimise them; and
— the health and safety policy of the employer and the organisations and arrangements for fulfilling that policy.

A major element of safety representative training should be devoted to his representational role and the desirability that the representative should be looked upon as a part-time safety officer. Training should also include inspection and investigation techniques, an appreciation of the appropriate level of response to problems and how to pursue any action with management.

The Trades Union Congress (TUC) has a long history of providing training in the onshore sector, particularly in the representative skills that are central to a safety representative's task and has developed an education and training course for offshore oil and gas industry safety representatives, which includes distance learning.

The Offshore Petroleum Industry Training Organisation (OPITO) is the offshore industry lead body with the responsibility for approving training standards.

Further training will be needed where the safety representative has special responsibilities, or in order to meet changes in circumstances or relevant legislation.

Installation owners and managers may make allowances in contractual arrangements in order that contractors' employees may be elected as safety representatives. The responsibility for training then rests with the contractor who must ensure not only that training is provided to any employee who is a safety representative, but also that any such employee is given the necessary time off both for training and for the performance of his role and duties as safety representative.

One of the functions of the safety committee is to keep the training requirements of safety representatives under review. All training providers and courses should be considered in order to ensure that safety representatives have been provided with suitable opportunities to become competent for their role, including the requisite interpersonal skills of communication, presentation and representation. Follow-up training may need to be arranged according to developments affecting the installation, new legislation, changing technologies or differing working practices. [*L110, paras 103–119*].

Offences/defences

[07056] Definitions of 'offence' and 'defence' are as follows:

(a) *Offence* – If a duty holder, manager or employer fails to comply with any of his statutory duties, he commits an offence (punishable as with mainline *HSWA 1974* offences – see ENFORCEMENT) [*Reg 28(1)*].
(b) *Defence* – It is a defence for the accused to prove:
 (i) that he exercised all due diligence to prevent the commission of the offence, and
 (ii) that the relevant failure to comply was committed without his consent, connivance or wilful default [*Reg 28(2)*].

Under the Employment Rights Act 1996, all employees (both onshore and offshore) irrespective of length of service or hours of work, are entitled to complain to an employment tribunal if they are dismissed, selected for redundancy or subjected to any other detriment by their employer. So far as offshore workers are concerned, this means that they can expect not to be disciplined, dismissed, 'blacklisted' or 'not required back' (NRB), transferred to other duties or to another installation, or disadvantaged in any way, because of any actions they take on health and safety grounds [*L110, Appendix 1*].

First-aid – the Offshore Installations and Pipeline Works (First-Aid) Regulations 1989

[07056.1] The Regulations requires duty holders to ensure adequate first-aid and basic healthcare provision for all personnel, including visitors, who are injured or become ill while on offshore installations or pipeline works. They have been amended several times since they were enacted, including in 2015. The third edition of *Health care and first aid on offshore installations and pipeline works, Offshore Installations and Pipeline Works (First-Aid) Regulations 1989, Approved Code of Practice* (L123) was published in 2016 and can be found on the HSE website.

Duty to provide first-aid facilities and personnel

[07057] Installation managers/owners and those in control of pipeline works must make adequate arrangements for first-aid for persons at work. This includes provision of suitably trained personnel under the direction of a registered medical practitioner, and employees must be informed of these arrangements. [*Offshore Installations and Pipeline Works (First-Aid) Regulations 1989 (SI 1989 No 1671), Reg 5(1)*.] The Regulations do not apply to diving operations offshore, which are subject to the requirements of the *Diving at Work Regulations 1997*. However, when a diving operation ceases, any members of a diving team who remain present on an installation or barge will be covered by the duty of care which the 1989 Regulations impose upon the person in control.

'Suitable persons'

[07058] The 'suitable persons' required by *Reg 5(1)(a) and (b)* may be offshore medics or offshore first-aiders. These are defined in the Code of Practice as follows:

- an 'offshore first-aider' means a person who holds a current Offshore First Aid Certificate issued by an organisation approved by HSE to train, examine and certify offshore first-aiders;
- an 'offshore medic' means a person who holds a current Offshore Medic Certificate issued by an organisation approved by HSE to train, examine and certify offshore medics.

The ACOP also sets out that a 'registered medical practitioner' means a fully registered person within the meaning of the *Medical Act 1983* who holds a licence to practise under that Act (ie someone who holds a current General Medical Council (GMC) registration and has a licence to practise in the United Kingdom).

The assessment of needs prepared under *Reg 5(1)(a)* and *(b)* must include an assessment of how many offshore medics and offshore first-aiders are required. The ACOP sets out that this will normally indicate that an offshore medic needs to be available at all times. If 25 or fewer people are regularly present, or if the installation or vessel has access to onshore medical services at all times, then continuous cover by an offshore medic may not be required. However, there must always be an adequate number of offshore first-aiders, both where an offshore medic is available and where there is no need for one. Arrangements should ensure cover for absence, especially of the offshore medic.

Although not required by the regulations, HSE guidance is that all offshore workers who are not qualified offshore medics or offshore first-aiders should receive training in the basic principles of first aid, including resuscitation, the control of bleeding and management of unconsciousness.

Sick bays and first aid equipment

[07059]–[07062] In particular, manned offshore installations and all pipe-laying barges and barges used in offshore construction, repair, maintenance, cleaning, demolition or dismantling activities must have a sick bay.

The sick bay should be clearly identifiable and in the charge of a suitable person (eg offshore medic or first-aider).

HSE guidance sets out that the location of the sick bay and of first-aid and medical equipment should be clearly identified with notices posted in conspicuous positions, including the sick bay, giving the locations of first-aid and medical equipment and facilities, the names and, as far as possible, locations of the offshore medic and offshore first-aiders.

The size, layout, equipment, medications and facilities of the sick bay should be sufficient for the number of people regularly present at one time on the installation or vessel, and appropriate for the type of activity carried out. In addition to equipment kept in the sick bay, the HSE advises that offshore first-aiders should be provided with appropriate first-aid and medical equipment, which should be a type they are familiar with. This equipment also needs to be provided at convenient locations on the installation or vessel where working conditions require.

The Offshore Electricity and Noise Regulations 1997

[07063] These Regulations *(SI 1997 No 1993)* amend the *Electricity at Work Regulations 1989 (SI 1989 No 635)* (see E3001 ELECTRICITY), by applying them in their entirety to offshore installations.

Legislation for other hazardous offshore activities

[07064] Generally, an installation's Safety Case will not be considered for acceptance unless it contains all the particulars specified in the Safety Case Regulations (see O7009.1 onwards and O7010 onwards). In some cases, with respect to hazardous activities carried out on offshore installations, there will be a degree of overlap with the particulars required in accordance with the provisions of associated legislation.

Set out below are brief references to various regulations which relate to hazards that potentially could cause a major accident.

Working in confined spaces

[07065] The *Confined Spaces Regulations 1997 (SI 1997 No 1713)* impose requirements and prohibitions with respect to the health and safety of persons carrying out work in confined spaces.

Although the regulations do not cover offshore installations, the regulations would apply to certain activities aboard installations which are 'stacked' out of use in territorial waters (see O7009 above) and which are not currently classed as offshore installations, such as the activities of shore-based workers undertaking repair, maintenance or cleaning.

A 'confined space' is defined in *Reg 1(2)* as 'any place, including any chamber, tank, vat, silo, pit, trench, pipe, sewer, flue, well or other similar space in which, by virtue of its enclosed nature, there arises a reasonably foreseeable specified risk'.

Under these regulations a confined space has two defining features. Firstly, it is a place which is substantially (though not always entirely) enclosed and, secondly, there will be a reasonably foreseeable risk of serious injury from hazardous substances or conditions within the space or nearby.

In addition, the regulations and associated ACOP and guidance provide valuable advice for activities carried out in confined spaces (see *Safe work in confined spaces Confined Spaces Regulations 1997, Approved Code of Practice and guidance* (L101) (third edn, published 2014).

Radiation hazards

[07066] There are a number of activities carried out on offshore installations that involve radioactive substances and radiation sources. They include radioactive scale deposited on well tubulars and oil separation equipment, radioactive sources used in instrumentation and X-ray and radioactive sources used for weld inspection and wall thickness measurement. All these activities must be carried out with total compliance with the *Ionising Radiations Regulations 1999 (SI 1985 No 1333)* the *Radioactive Substances Act 1993* in terms of the designation of competent persons, authorised persons and classified workers, medical supervision, personal monitoring procedures following excessive exposure, control of sealed sources of radiation, equipment tests and the maintenance of records. The HSE have issued advice specific to offshore operations in *Operations Notice 4 Radioactive Substances Regulation*, and *Operations Notice 34 Ionising Radiations Regulations 1999: notification of offshore site radiography work*. Other useful information is given in *Ionising Radiation Protection Series No 8 Control of Radioactive Substances*. New Ionising Radiation Regulations 2017 will come into force on 1 January 2018 and will replace the 1999 Regulations.

Oil pollution

[07067] In addition to the environmental aspects, oil and volatile hydrocarbons spilled into the sea in the vicinity of an offshore installation pose a real threat of fire to that installation. It is beyond the scope of this chapter to provide a detailed overview of legislation in this area. A summary of environmental regulations and guidance on offshore oil and gas exploration and production, offshore gas unloading and storage and offshore carbon dioxide storage activities can be found on the government website at: www.gov.uk/guidance/oil-and-gas-offshore-environmental-legislation.

Carriage of dangerous substances

[07068] In addition to the Regulations relating to the labelling, packaging and transportation of dangerous goods by road or rail (see DANGEROUS GOODS – CARRIAGE), carriage to the installation must comply with the following legislation:

— *Merchant Shipping (Dangerous Goods and Marine Pollutants) Regulations 1997 (SI 1997 No 2367)*; and
— *Air Navigation (Dangerous Goods) Regulations 2002 (SI 2002 No 2786)*.

Reference to these regulations is given for guidance only and must not be regarded as comprising a complete list.

More detailed advice is contained in the *Oil & Gas UK Best Practice for the Safe Packing & Handling of Cargo to & from Offshore Locations Issue 6 2015*.

Written procedures and training programmes for dealing with hazards

[07069] Detailed plans, procedures, safe systems of work and written instructions, including those covering the operation of permit to work systems, are a standard feature of offshore safety operations. Such procedures and safe systems of work may include the use of access equipment, eg scaffolding, boatswain's chairs, ladders, etc, work over the sea, the safe use of lifting equipment, helicopter operations and general fire and emergency procedures.

High standards of operator training, prior to arrival on a working platform and at frequent intervals during their employment, are crucial to ensure safe operation. See *Offshore Petroleum Industry Training Organisation (OPITO) Minimum Industry Safety Training Standard (MIST)*.

Induction training should incorporate the following aspects:

(a) *General safety* – dealing with the hazards of offshore work and the personal precautions necessary, in particular the correct use of all forms of personal protective equipment, ie foul weather protection, safety helmets, safety harnesses.

(b) *Safe working procedures* – the use of permit to work systems and safe systems of work covering electricity, welding and ionising radiation in particular; the use of scaffolding, working platforms, the use of plant and equipment, maintenance and inspection procedures and record keeping (see further 43 INTRODUCTION).

(c) *Health precautions* – including precautions against toxic dusts, fumes, gases and vapours, asbestos, the risk of asphyxiation and anoxia, and physical hazards associated with noise, electricity and radiation.

(d) *Survival* – instruction in the use of life jackets, survival capsules, life boats and rafts; emergency platform evacuation procedures, particularly in the event of fire or explosion.

(e) *Helicopter operations* – procedures for safe movement of personnel by helicopter to or from the platform.

(f) *Fire protection* – the correct use of fire appliances, platform evacuation procedure in the event of fire, platform fire alarm system.

(g) *First-aid* – elementary first-aid procedures.

The RIDDOR regulations apply offshore (see **A3002**) using Form OIR/9B. In addition, there is a requirement to report hydrocarbon releases using Form OIR/12. Guidance on this requirement has been provided by the HSE in *OTO 96 956* and *Oil & Gas UK Supplementary Guidance For Reporting Hydrocarbon Releases*. This data is input by the HSE into the hydrocarbon releases (HCR) system database, analysed and published as non-attributable failure nature, cause and rate data.

[07070] Offshore Operations

Hazards in offshore activities

Cranes and lifting equipment

[07070] The potential for crane and lifting equipment failures, including the use of helicopters in lifting operations during the initial work stage, represents a serious hazard in offshore work. All lifting operations and equipment must comply with the *Lifting Operations and Lifting Equipment Regulations 1998* (see **L3023** LIFTING MACHINERY AND EQUIPMENT).

The HSE has published technical guidance on the safe use of lifting equipment offshore as part of a key initiative to reduce the incidence of lifting-related injuries and deaths offshore.

The guidance, *Technical guidance on the safe use of lifting equipment offshore* (HSG221), provides technical information for those who are involved in the operation and control of lifting equipment offshore. It has been written for duty holders, OIMs, managers and people who are directly involved in using lifting equipment offshore, although others, such as safety representatives, manufacturers, suppliers and verification bodies will also find it useful. The guidance covers all frequently used lifting equipment and accessories. It includes a section on personnel carriers. It can be downloaded from the HSE website.

IMCA and Step Change in Safety have published useful guidance on many aspects of lifting operations and cargo handling.

Burns and eye injuries

[07071] These are the two most common forms of injury to operators, in many cases associated with personal protective equipment failures (see PERSONAL PROTECTIVE EQUIPMENT REGULATIONS at **P3002**).

Access hazards and working at height

[07072] Falls from scaffolds and ladders and whilst working 'over the side' are common, emphasising the need for good systems of scaffold erection, inspection and maintenance and adherence to the *Work at Height Regulations 2005 (SI 2005 No 735)* which apply offshore (see **W9001** WORKING AT HEIGHT).

Gassing accidents and incidents

[07073] Precautions against asphyxiation, oxygen enrichment, hydrogen sulphide and various fumes, dusts, gases and vapours are now a standard feature of the industry.

Permit to work

[07074] As with many other industries non-standard operations are controlled by permit to work systems. Guidance on these systems has been issued by the International Association of Oil & Gas Producers and the HSE.

Hazards in offshore activities [07075]

Sources of Further Advice

[07075] The following are sources of further information;

- HSE – general health and safety issues
- The Offshore Oil and Gas section of the HSE website (www.hse.gov.uk/offshore/index.htm) – offshore specific health and safety; offshore operations notices and offshore information sheets
- Institution of Occupational Safety and Health (IOSH) – general health and safety issues
- IOSH Offshore Group – offshore specific health and safety issues
- Oil and Gas UK and Step Change in Safety – offshore specific health and safety issues including design guidance
- International Association of Drilling Contractors – drilling specific design and operational health and safety
- Offshore Contractors Association
- International Maritime Organization
- Energy Institute
- American Petroleum Institute
- British Standards Institute
- Offshore Petroleum Industry Training Organisation
- Society of Petroleum Engineers
- Offshore Coordinating Group of offshore unions (www.offshoreworkers.org.uk/)

Personal Protective Equipment

Nicola Coote

Introduction to personal protective equipment

[P3001] Conventional wisdom suggests that 'safe place strategies' are more effective in combating health and safety risks than 'safe person strategies'. Safe systems of work, and control/prevention measures serve to protect everyone at work, whilst the advantages of personal protective equipment are limited to the individual(s) concerned and their ability or cooperation in using it correctly. Given, however, the fallibility of any state of the art technology in endeavouring to achieve total protection, some level of personal protective equipment is inevitable in view of the obvious (and not so obvious) risks to head, face, neck, eyes, ears, lungs, skin, arms, hands and feet.

Current statutory requirements for employers to provide and maintain suitable personal protective equipment are contained in the *Personal Protective Equipment at Work Regulations 1992 (SI 1992 No 2966)* as amended by *the Health and Safety (Miscellaneous Amendments) Regulations 2002 (SI 2002 No 2174)*. This legislation requires the employer to carry out an assessment to determine what personal protective equipment is needed and to consider employee needs in the selection process. Adjunct to this, common law insists, not only that employers have requisite safety equipment at hand, or available in an accessible place, but also that management ensure that operators use it (see *Bux v Slough Metals Ltd* [1974] 1 All ER 262). That the hallowed duty to 'provide and maintain' has been getting progressively stricter is evidenced by *Crouch v British Rail Engineering Ltd* [1988] IRLR 404 to the extent that employers could be in breach of either statutory or common law duty (or both) in the case of injury/disease to a member of the employee's immediate family involved, say, in cleaning protective clothing; however, the injury/disease must have been foreseeable (*Hewett v Alf Brown's Transport Ltd* [1992] ICR 530).

This section deals with the requirements of the *Personal Protective Equipment at Work Regulations 1992*, the *Personal Protective Equipment (EC Directive) Regulations 1992 (SI 1992 No 3139)* (now revoked and consolidated by *SI 2002 No 1144*), the duties imposed on manufacturers and suppliers of personal protective equipment, the common law duty on employers to provide suitable personal protective equipment, and the main types of personal protective equipment and clothing and some relevant British Standards.

The Personal Protective Equipment at Work Regulations 1992

[P3002] The *Personal Protective Equipment at Work Regulations 1992*, require all employers to make a formal assessment of the personal protective equipment needs of employees and provide ergonomically suitable equipment in relation to foreseeable risks at work (see further **P3006** below). However, employers cannot be expected to comply with these statutory duties, unless manufacturers of personal protective equipment have complied with theirs under the requirements of the *Personal Protective Equipment (EC Directive) Regulations 1992 (SI 1992 No 3139)* and *SI 1994 No 2326* (now revoked and consolidated by *SI 2002 No 1144*) – that is, had their products independently certified for EU accreditation purposes. Although imposing a considerable remit on manufacturers, EU accreditation is an indispensable condition precedent to sale and commercial circulation (see **P3012** below).

It is important that personal protective clothing and equipment should be seen as 'last resort' protection – its use should only be prescribed when engineering and management solutions and other safe systems of work do not effectively protect the worker from the danger, and where risk of injury is still foreseeable after other controls have been implemented (*Health and Safety (Miscellaneous Amendments) Regulations 2002 (SI 2002 No 2174)*) further require employers to ensure it is appropriate for the risk or risks involved, the conditions at the place where exposure to the risk may occur, and the period for which it is worn, and to ensure that it takes account of ergonomic requirements and the state of health of the person or persons who may wear it, and of the characteristics of the workstation of each such person (*Gerrard v Staffordshire Potteries Ltd* [1994] EWCA Civ 31, [1995] ICR 502, [1995] PIQR P 169).

Employees must be made aware of the purpose of personal protective equipment, its limitations and the need for on-going maintenance. Thus, when assessing the need for, say, eye protection, employers should first identify the existence of workplace hazards (e.g. airborne dust, projectiles, liquid splashes, slippery floors, inclement weather in the case of outside work) and then the extent of danger (e.g. frequency/velocity of projectiles, frequency/severity of splashes). Selection can (and, indeed, should) then be made from the variety of CE-marked equipment available, in respect of which manufacturers must ensure that such equipment provides protection, and suppliers ascertain that it meets such requirements/standards [*HSWA s 6*] (see further PRODUCT SAFETY). Typically, most of the risks will have already been logged, located and quantified in a routine risk/safety audit (see RISK ASSESSMENT), and classified according to whether they are physical/chemical/biological in relation to the part(s) of the body affected (e.g. eyes, ears, skin). Good practice dictates that any such assessment should include the input from the users, as simply providing equipment that will meet a technical specification for protection will not meet requirements for a "suitable and sufficient" assessment. To achieve this, a deeper analysis of environment in which it is being used, comfort, fit etc will be needed.

Selection of personal protective equipment is a first stage in an on-going routine, followed by proper use and maintenance of equipment (on the part of both employers and employees) as well as training and supervision in personal protection techniques. Maintenance presupposes a stock of renewable spare parts coupled with regular inspection, testing, examination, repair, cleaning and disinfection schedules as well as keeping appropriate records. Depending on the particular equipment, some will require regular testing and examination (e.g. respiratory equipment), whilst others merely inspection (e.g. gloves, goggles). Generally, manufacturers' maintenance schedules should be followed.

Equipment must be appropriate for the risk or risks involved, the conditions at the place where exposure to the risk may occur, and the period for which it is worn.

Suitable accommodation must be provided for protective equipment in order to minimise loss or damage and prevent exposure to cold, damp or bright sunlight, e.g. pegs for helmets, pegs and lockers for clothing, spectacle cases for safety glasses.

On-going safety training, often carried out by manufacturers for the benefit of users, should combine both theory and practice. Where it is appropriate, and at suitable intervals, the employer should organise demonstrations in the wearing of personal protective equipment. (*Health and Safety (Miscellaneous Amendments) Regulations 2002 (SI 2002 No 2174)*.)

Work activities/processes requiring personal protective equipment

[P3003] Examples abound of processes/activities of which personal protective equipment is a prerequisite, from construction work and mining, through work with ionising radiations, to work with lifting plant, cranes, as well as handling chemicals, tree felling and working from heights. Similarly, blasting operations, work in furnaces and drop forging all require a degree of personal protection, as does work in healthcare, research laboratories and many agricultural activities.

Statutory requirements in connection with personal protective equipment

[P3004] General statutory requirements relating to personal protective equipment are contained in the *Health and Safety at Work etc Act 1974*, ss 2, 9, the *Personal Protective Equipment at Work Regulations 1992*, and the *Health and Safety (Miscellaneous Amendments) Regulations 2002 (SI 2002 No 2174)* 'Every employer shall ensure that suitable personal protective equipment is provided to his employees who may be exposed to a risk to their health or safety at work except where, and to the extent that such risk has been adequately controlled by other means which are equally or more effective'. [*Reg 4(1)*]. More specific statutory requirements concerning personal protective equipment also exist in the in sundry other regulations applicable to particular industries/processes, eg asbestos, noise, construction (see **P3008** below). Where personal protective equipment is a necessary control measure to meet a specific statutory requirement or following a risk assessment, this must be provided free of charge [HWSA 1974 s 9]. However, where it is optional or non essential, there is nothing to stop the employer from seeking a financial contribution.

A financial contribution by the employee may also be appropriate if the individual requests a more expensive type, purely for their own personal reasons and aesthetic benefits. However, where a more expensive type of equipment is needed to address a specific need of an individual due to health or other relevant conditions then it is reasonable for the employer to meet the additional costs following medical evidence being submitted by the individual.

General statutory duties – HSWA 1974

[P3005] Employers are under a general duty to ensure, so far as is reasonably practicable, the health, safety and welfare at work of their employees – a duty which clearly implies provision/maintenance of personal protective equipment. [*HSWA s 2(1)*].

Specific statutory duties – Personal Protective Equipment at Work Regulations 1992 (SI 1992 No 2966 as amended by SI 2002 No 2174)

Duties of employers

[P3006] Employers – and self-employed persons in cases (*a*), (*b*), (*c*), (*d*) and (*e*) below, must undertake the following:

(a) Formally assess (and review periodically) provision and suitability of personal protective equipment. [*Reg 6(1), (3)*].
 The assessment should include [*Reg 6(2)*]:
 (i) risks to health and safety not avoided by other means;
 (ii) reference to characteristics which personal protective equipment must have in relation to risks identified in (i);
 (iii) a comparison of the characteristics of personal protective equipment having the characteristics identified in (ii);
 (iv) an assessment as to its compatibility with other personal protective equipment in use at the same time.
 The aim of the assessment is to ensure that an employer knows which personal protective equipment to choose to effectively protect against the identified risks; it constitutes the first stage in a continuing programme, concerned also with proper use and maintenance of personal protective equipment and training and supervision of employees.

(b) Provide suitable personal protective equipment to his employees, who may be exposed to health and safety risks while at work, except where the risk has been adequately controlled by other equally or more effective means. [*Reg 4(1)*].
 Personal protective equipment is not suitable unless:

(i) it is appropriate for risks involved and conditions at the place of exposure;
(ii) it takes account of ergonomic requirements, the state of health of the person who wears it and the duration of its use;
(iii) it is capable of fitting the wearer correctly; and
(iv) so far as is practicable (for meaning, see ENFORCEMENT), it is effective to prevent or adequately control risks involved without increasing the overall risk.

[Reg 4(3)].

Where it is necessary to ensure that personal protective equipment is hygienic and otherwise free of risk to health, every employer and every self-employed person shall ensure that personal protective equipment provided is used only by the person to whom it has been issued. [Reg 4(4)]

(c) Provide compatible personal protective equipment – that is, that the use of more than one item of personal protective equipment is compatible with other personal protective equipment. [Reg 5].
(d) Maintain (as well as replace and clean) any personal protective equipment in an efficient state, efficient working order and in good repair. [Reg 7].
(e) Provide suitable accommodation for personal protective equipment when not being used. [Reg 8].
(f) Provide employees with information, instruction and training to enable them to know:
(i) the risks which personal protective equipment will avoid or minimise;
(ii) the purpose for which and manner in which personal protective equipment is to be used;
(iii) any action which the employee might take to ensure that personal protective equipment remains efficient.

[Reg 9].

(g) Ensure, taking all reasonable steps, that personal protective equipment is properly used. [Reg 10].

Summary of employer's duties

- duty of assessment;
- duty to provide suitable PPE;
- duty to provide compatible PPE;
- duty to maintain and replace PPE;
- duty to provide suitable accommodation for PPE;
- duty to provide information and training;
- duty to see that PPE is correctly used.

Duties of employees

[P3007] Every employee must:

- use personal protective equipment in accordance with training and instructions [Reg 10(2)];
- return all personal protective equipment to the appropriate accommodation provided after use [Reg 10(4)]; and

- report forthwith any defect or loss in the equipment to the employer [*Reg 11*].

Specific requirements for particular industries and processes

[P3008] In addition to the general remit of the *Personal Protective Equipment at Work Regulations 1992*, the following regulations impose specific requirements on employers to provide personal protective equipment up to the EU standard.

- *Control of Lead at Work Regulations 2002 (SI 2002 No 2676), Regs 6, 8* – supply of suitable respiratory equipment and where exposure is significant, suitable protective clothing and equipment (see APPENDIX 4);
- *Control of Asbestos Regulations 2012 (SI 2012/632), Reg 14* – provision and cleaning of protective clothing (see APPENDIX 4);
- *Control of Substances Hazardous to Health Regulations 2002 (SI 2002 No 2677 as amended by SI 2003 No 978) (COSHH), Regs 7, 8, 9* – supply of suitable protective equipment where employees are foreseeably exposed to substances hazardous to health;
- *Noise at Work Regulations 2005 (SI 2005 No 1643), Reg 7* – personal ear protectors;
- *Ionising Radiations Regulations 1999 (SI 1999 No 3232), Reg 9* – supply of suitable personal protective equipment and respiratory protective equipment;
- *Shipbuilding and Ship-repairing Regulations 1960 (SI 1960 No 1932), Regs 50, 51* – supply of suitable breathing apparatus, belts, eye protectors, gloves and gauntlets (amended in 1994).
- (These regulations have not been revoked, despite them being included in a government review as part of its deregulation strategy).

Increased importance of uniform European standards

[P3009] To date, manufacturers of products, including personal protective equipment (PPE), have sought endorsement or approval for products, prior to commercial circulation, through reference to British Standards (BS), HSE or European standards, an example of the former being Kitemark and an example of the latter being CEN or CENELEC. Both are examples of *voluntary* national and international schemes that can be entered into by manufacturers and customers, under which both sides are 'advantaged' by conformity with such standards. Conformity is normally achieved following a level of testing appropriate to the level of protection offered by the product. With the introduction of the *Personal Protective Equipment (EC Directive) Regulations 1992* as amended by *SI 1994 No 2326* (now revoked and consolidated by *SI 2002 No 1144*) (see **P3012** below), this voluntary system of approval was replaced by a statutory certification procedure.

Towards certification

[P3010] Certification as a condition of sale is not exactly a legal requirement, but there are requirements within the Blue Guide on the Implementation of EU Product Rules (DOI: 10.2769/9091) http://ec.europa.eu/CEmarking. Which states inter alia: " Union harmonisation legislation applies to all forms of selling. A product offered in a catalogue or by means of electronic commerce has to comply with Union harmonisation legislation when the catalogue or website directs its offer to the Union market and includes an ordering and shipping system".

The EN 45000 series of standards helps to meet this requirement as it establishes how test houses and certification bodies are to be established and independently accredited. In addition, a range of European standards on PPE have been implemented in BS EN format, for example, on eye protection, fall arrest systems and safety footwear.

Interim measures

[P3011] Almost every reputable manufacturer has now elected to comply with the current CEN standard when seeking product certification. Alternatively, should a manufacturer so wish, the inspection body can verify by way of another route to certification - this latter might well be the case with an innovative product where standards did not exist. Generally, however, compliance with current CEN (European Standardisation Committee) or BS ISO 9001:208 *Quality Assurance* has been the well-trodden route to certification. To date, several UK organisations, including manufacturers and independent bodies, have established PPE test houses, which are independently accredited by the National Measurement Accreditation Service (NAMAS). Such test houses will be open to all comers for verification of performance levels.

Personal Protective Equipment (EC Directive) Regulations 1992 (SI 1992 No 3139 as amended by SI 1994 No 2326)

[P3012] The *Personal Protective Equipment (EC Directive) Regulations 1992 (*as amended by *SI 1994 No 2326)* required most types of PPE (for exceptions, see P3019 below) to satisfy specified certification procedures, pass EU type-examination (i.e. official inspection by an approved inspection body) and carry a 'CE' mark both on the product itself and its packaging before being put into commercial circulation. Indeed, failure on the part of a manufacturer to obtain affixation of a 'CE' mark on his product before putting it into circulation, is a criminal offence under the *Health and Safety at Work etc Act 1974*, s 6 carrying a maximum fine, on summary conviction, of £20,000 (see ENFORCEMENT).

The *Personal Protective Equipment Regulations 2002 (SI 2002 No 1144)* implement the Product Safety Directive 89/696/EEC, on the approximation of the laws of the Member States relating to personal protective equipment. These Regulations came into force on 15 May 2002.

[P3012] Personal Protective Equipment

The 2002 Regulations place a duty on responsible persons who put personal protective equipment (PPE) on the market to comply with the following requirements:

- the PPE must satisfy the basic health and safety requirements that are applicable to that type or class of PPE;
- the appropriate conformity assessment procedure must be carried out;
- a CE mark must be affixed on the PPE.

The Regulations will be enforced by the Weights and Measures Authorities (District Councils in Northern Ireland).

CE mark of conformity

[P3013] The *Personal Protective Equipment Regulations SI 2002 No 1144*) specify procedures and criteria with which manufacturers must comply in order to be able to obtain certification. Essentially this involves incorporation into design and production basic health and safety requirements. Hence, by compliance with health and safety criteria, affixation of the 'CE' mark becomes a condition of sale and approval (see *fig. 1* below). Moreover, compliance on the part of manufacturers with these regulations, will enable employers to comply with their duties under the *Personal Protective Equipment at Work Regulations 1992*. Reg 4(3)(e) of the *Personal Protective Equipment at Work Regulations 1992* requires the PPE provided by the employer to comply with this and any later EU directive requirements.

CE accreditation is not a synonym for compliance with the CEN standard, since a manufacturer could opt for certification via an alternative route. Neither is it an approvals mark. It is rather a quality mark, since all products covered by the regulations are required to meet formal quality control or quality of production criteria in their certification procedures. [*Arts 8.4, 11, 89/686/EEC*].

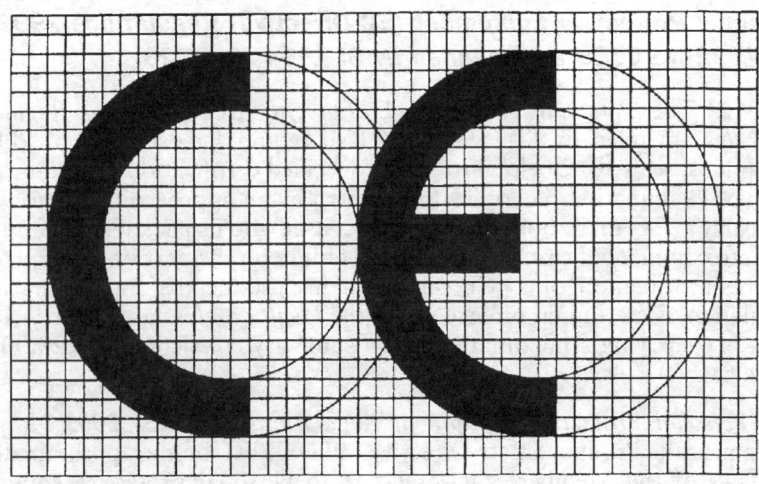

Certification procedures

[P3014] The *Personal Protective Equipment Regulations SI 2002 No 1144*, identify three categories of PPE:

- PPE of simple design;
- PPE of non-simple design;
- PPE of complex design.

Each of these have different requirements for manufacturers to follow.

In order to obtain certification, manufacturers must submit to an approved body the following documentation (except in the case of PPE of simple design, where risks are minimal).

(1) **PPE – simple design**
This includes use of PPE where risks are minimal e.g. gardening gloves, helmets, aprons, thimbles. It is used where the designer assumes that the user can himself assess the level of protection provided against the minimal risks concerned the effects of which, when they are gradual, can be safely identified and managed. This type of equipment must still be subject to a manufacturers declaration of conformity.

(2) **PPE – non-simple design**
In order to obtain certification, manufacturers must submit to an approved body the following documentation:
 (a) Technical file, i.e.
  (i) overall and detailed plans, accompanied by calculation notes/results of prototype tests; and
  (ii) an exhaustive list of basic health and safety requirements and harmonised standards taken into account in the model's design.
 (b) Description of control and test facilities used to check compliance with harmonised standards.
 (c) Copy of information relating to:
  (i) storage, use, cleaning, maintenance, servicing and disinfection;
  (ii) performance recorded during technical tests to monitor levels of protection;
  (iii) suitable PPE accessories;
  (iv) classes of protection appropriate to different levels of risk and limits of use;
  (v) obsolescence deadline;
  (vi) type of packaging suitable for transport;
  (vii) significance of markings.

This must be provided in the official language of the member state of destination.

(3) PPE – complex design

This includes use of PPE for protection against mortal/serious dangers where the user cannot identify the risk in sufficient time. It requires compliance with a quality control system. Examples of PPE coming within the complex design category include: firemen's helmets, breathing air apparatus for spray painters and fall arrest systems for roof / scaffold workers.

This category of PPE, including:

(a) filtering, respiratory devices for protection against solid and liquid aerosols/irritant, dangerous, toxic or radiotoxic gases;
(b) respiratory protection devices providing full insulation from the atmosphere, and for use in diving;
(c) protection against chemical attack or ionising radiation;
(d) emergency equipment for use in high temperatures, whether with or without infrared radiation, flames or large amounts of molten metal (100°C or more);
(e) emergency equipment for use in low temperatures (–50°C or less);
(f) protection against falls from heights;
(g) protection against electrical risks; and
(h) motor cycle helmets and visors;

must satisfy either

(i)
— an EU quality control system for the final product, or
— a system for ensuring EU quality of production by means of monitoring;
and in either case,
(ii) an EU declaration of conformity.

Basic health and safety requirements applicable to all PPE

[P3015]–[P3016] All PPE must:

- be ergonomically suitable, that is, be able to perform a risk-related activity whilst providing the user with the highest possible level of protection;
- preclude risks and inherent nuisance factors, such as roughness, sharp edges and projections and must not cause movements endangering the user;
- be kept in clean and hygienic condition and be of a design that enables good hygienic standards, unless it is disposable;
- provide comfort and efficiency by facilitating correct positioning on the user and remaining in place for the foreseeable period of use; and by being as light as possible without undermining design strength and efficiency; and
- be accompanied by necessary information – that is, the name and address of the manufacturer and technical file (see **P3014** above).

[Annex II].

General purpose PPE – specific to several types of PPE

- PPE incorporating adjustment systems must not become incorrectly adjusted without the user's knowledge;
- PPE enclosing parts of the body must be sufficiently ventilated to limit perspiration or to absorb perspiration;
- PPE for the face, eyes and respiratory tracts must minimise risks to same and, if necessary, contain facilities to prevent moisture formation and be compatible with wearing spectacles or contact lenses;
- PPE subject to ageing must contain date of manufacture, the date of obsolescence and be indelibly inscribed if possible. If the useful life of a product is not known, accompanying information must enable the user to establish a reasonable obsolescence date. In addition, the number of times it can be cleaned before being inspected or discarded must (if possible) be affixed to the product; or, failing this, be indicated in accompanying literature;
- PPE which may be caught up during use by a moving object must have a suitable resistance threshold above which a constituent part will break and eliminate danger;
- PPE for use in explosive atmospheres must not be likely to cause an explosive mixture to ignite;
- PPE for emergency use or rapid installation/removal must minimise the time required for attachment and/or removal;
- PPE for use in very dangerous situations (see **P3015** above) must be accompanied with data for exclusive use of competent trained individuals, and describe procedure to be followed to ensure that it is correctly adjusted and functional when worn;
- PPE incorporating components which are adjustable or removable by user, must facilitate adjustment, attachment and removal without tools;
- PPE for connection to an external complementary device must be mountable only on appropriate equipment;
- PPE incorporating a fluid circulation system must permit adequate fluid renewal in the vicinity of the entire part of the body to be protected;
- PPE bearing one or more identification or recognition marks relating to health and safety must preferably carry harmonised pictograms/ideograms which remain perfectly legible throughout the foreseeable useful life of the product;
- PPE in the form of clothing capable of signalling the user's presence visually must have one or more means of emitting direct or reflected visible radiation;
- multi-risk PPE must satisfy basic requirements specific to each risk.

Additional requirements for specific PPE for particular risks

[P3017]–[P3018] There are additional requirements for PPE designed for certain particular risks as follows.

PPE protection against

- *Mechanical impact risks*
  - (i) impact caused by falling/projecting objects must be sufficiently shock-absorbent to prevent injury from crushing or penetration of the protected part of the body;
  - (ii) falls. In the case of falls due to slipping, outsoles for footwear must ensure satisfactory adhesion by grip and friction, given the state and nature of the surface. In the case of falls from a height, PPE must incorporate a body harness and attachment system connectable to a reliable anchorage point. The vertical drop of the user must be minimised to prevent collision with obstacles and the braking force injuring, tearing or causing the operator to fall;
  - (iii) mechanical vibration. PPE must be capable of ensuring adequate attenuation of harmful vibration components for the part of the body at risk.
- *(Static) compression of part of the body* – must be able to attenuate its effects so as to prevent serious injury of chronic complaints.
- *Physical injury* – must be able to protect all or part of the body against superficial injury by machinery, e.g. abrasion, perforation, cuts or bites.
- *Prevention of drowning* – (lifejackets, armbands etc.) must be capable of returning to the surface a user who is exhausted or unconscious, without danger to his health. Such PPE can be wholly or partially inherently buoyant or inflatable either by gas or orally. It should be able to withstand impact with liquid, and, if inflatable, able to inflate rapidly and fully.
- *Harmful effects of noise* – must protect against exposure levels of the Noise at Work Regulations 2005 (i.e. 80 and 85 dB(A)) and must indicate noise attenuation level.
- *Heat and/or fire* – must possess sufficient thermal insulation capacity to retain most of the stored heat until after the user has left the danger area and removed PPE. Moreover, constituent materials which could be splashed by large amounts of hot product must possess sufficient mechanical-impact absorbency. In addition, materials which might accidentally come into contact with flame as well as being used in the manufacture of fire-fighting equipment, must possess a degree of non-flammability proportionate to the risk foreseeably arising during use and must not melt when exposed to flame or contribute to flame spread. When ready for use, PPE must be such that the quantity of heat transmitted by PPE to the user must not cause pain or health impairment. Second, ready-to-use PPE must prevent liquid or steam penetration and not cause burns. If PPE incorporates a breathing device, it must adequately protect the user. Accompanying manufacturers' notes must provide all relevant data for determination of maximum permissible user exposure to heat transmitted by equipment.
- *Cold* – must possess sufficient thermal insulating capacity and retain the necessary flexibility for gestures and postures. In particular, PPE must protect tips of fingers and toes from pain or health impairment and

prevent penetration by rain water. Manufacturers' accompanying notes must provide all relevant data concerning maximum permissible user exposure to cold transmitted by equipment.

- *Electric shock* – must be sufficiently insulated against voltages to which the user is likely to be exposed under the most foreseeably adverse conditions. In particular, PPE for use during work with electrical installations (together with packaging), which may be under tension, must carry markings indicating either protection class and/or corresponding operating voltage, serial number and date of manufacture; in addition, date of entry into service must be inscribed as well as of periodic tests or inspections.
- *Radiation*
    (i) Non-ionising radiation – must prevent acute or chronic eye damage from non-ionising radiation and be able to absorb/reflect the majority of energy radiated in harmful wavelengths without unduly affecting transmission of the innocuous part of the visible spectrum, perception of contrasts and colour differentiation. Thus, protective glasses must possess a spectral transmission factor so as to minimise radiant-energy illumination density capable of reaching the user's eye through the filter and ensure that it does not exceed permissible exposure value. Accompanying notes must indicate transmission curves, making selection of the most suitable PPE possible. The relevant protection-factor number must be marked on all specimens of filtering glasses.
    (ii) Ionising radiation – must protect against external radioactive contamination. Thus, PPE should prevent penetration of radioactive dust, gases, liquids or mixtures under foreseeable use conditions. Moreover, PPE designed to provide complete user protection against external irradiation must be able to counter only weak electron or weak photon radiation.
- *Dangerous substances and infective agents*
    (a) respiratory protection – must be able to supply the user with breathable air when exposed to polluted atmosphere or an atmosphere with inadequate oxygen concentration. Leak-tightness of the facepiece and pressure drop on inspiration (breath intake), as well as in the case of filtering devices' purification capacity, must keep the contaminant penetration from the polluted atmosphere sufficiently low to avoid endangering health of the user. Instructions for use must enable a trained user to use the equipment correctly;
    (b) cutaneous and ocular contact (skin and eyes) – must be able to prevent penetration or diffusion of dangerous substances and infective agents. Therefore, PPE must be completeely leak-tight but allow prolonged daily use or, failing that, of limited leak-tightness restricting the period of wear. In the case of certain dangerous substances/infective agents possessing high penetrative power and which limit duration of protection, such PPE must be tested for classification on the basis of efficiency. PPE conforming with test specifications must carry a mark indicating

names or codes of substances used in tests and standard period of protection. Manufacturers' notes must contain an explanation of codes, detailed description of standard tests and refer to the maximum permissible period of wear under different foreseeable conditions of use.
- *Safety devices for diving equipment* – must be able to supply the user with a breathable gaseous mixture, taking into account the maximum depth of immersion. If necessary, equipment must consist of:
  (a) a suit to protect the user against pressure;
  (b) an alarm to give the user prompt warning of approaching failure in supply of breathable gaseous mixture;
  (c) life-saving suit to enable the user to return to the surface.

Excluded PPE

- PPE designed and manufactured for use specifically by the armed forces or in the maintenance of law and order (helmets, shields);
- PPE for self-defence (aerosol canisters, personal deterrent weapons);
- PPE designed and manufactured for private use against:
  (i) adverse weather (headgear, seasonal clothing, footwear, umbrellas),
  (ii) damp and water (dish-washing gloves),
  (iii) heat (e.g. gloves);
- PPE intended for the protection or rescue of persons on vessels or aircraft, not worn all the time.

Main types of personal protection

The main types of personal protection covered by the *Personal Protective Equipment at Work Regulations 1992* are (*a*) head protection, (*b*) eye protection, (*c*) hand/arm protection, (*d*) foot protection and (*e*) whole body protection.

In particular, these regulations do not cover ear protectors and most respiratory protective equipment – these areas are covered by other existing regulations and guidance, for example the *Noise at Work Regulations 2005*, and HSE's guidance booklet HS(G) 53, 'Respiratory protective equipment: a practical guide for users'.

The Regulations do not cover the use of protective equipment by professional sports people during competitive sports eg wearing shin guards, racing helmets etc. However, those engaged in sports-related occupations such as stable staff or canoeing instructors will be covered by the Regulations, and their protective equipment must comply with the Regulations.

The Regulations do not cover the use of PPE used for protection while travelling on a road as far as the Road Traffic Acts are concerned, although where protective equipment is required for occupational pur-poses, such as by farm workers riding all-terrain vehicles, the Regulations will apply.

Main types of personal protection [P3020]

Head protection

[P3019] This is likely to come under the non-simple or complex categories of PPE and takes the form of:

- crash, cycling, riding and climbing helmets;
- industrial safety helmets – to protect against falling objects;
- scalp protectors (bump caps) – to protect against striking fixed obstacles;
- caps/hairnets – to protect against scalping,

and is particularly suitable for the following activities:

(i) building work – particularly on scaffolds;
(ii) civil engineering projects;
(iii) blasting operations;
(iv) work in pits and trenches;
(v) work near hoists/lifting plant;
(vi) work in blast furnaces;
(vii) work in industrial furnaces;
(viii) ship-repairing;
(ix) slaughterhouses;
(x) tree-felling;
(xi) suspended access work, e.g. window cleaning.

Eye protection

[P3020] The majority of these are likely to come under non-simple category but some will come under the complex category. This takes the following forms:

- safety spectacles – these are the same as prescription spectacles but incorporating optional sideshields; lenses are made from tough plastic, such as polycarbonate – provide lateral protection;
- eyeshields – these are heavier than safety spectacles and designed with a frameless one-piece moulded lens; can be worn over prescription spectacles;
- safety goggles – these are heavier than spectacles or eye shields; they are made with flexible plastic frames and one-piece lens and have an elastic headband – they afford total eye protection; and
- faceshields – these are heavier than other eye protectors but comfortable if fitted with an adjustable head harness – faceshields protect the face but not fully the eyes and so are no protection against dusts, mist or gases.

Eye protectors are suitable for working with:

(i) chemicals;
(ii) power driven tools;
(iii) molten metal;
(iv) welding;
(v) radiation;
(vi) gases/vapours under pressure;

(vii)   woodworking.

Hand/arm protection

[P3021] Gloves that protect against low risk cleaning fluids etc will come under the simple category but the majority will come under the non-simple or complex categories. They are used to provide protection against:

- cuts and abrasions;
- extremes of temperature;
- skin irritation/dermatitis;
- contact with toxic/corrosive liquids,

and are particularly useful in connection with the following activities/processes:

(i)   manual handling;
(ii)   vibration;
(iii) construction;
(iv)   hot and cold materials;
(v)   electricity;
(vi)   chemicals;
(vii) radioactivity.

Foot protection

[P3022] This type of protection will range from simple to complex and takes the form of:

- safety shoes/boots;
- foundry boots;
- clogs;
- wellington boots;
- anti-static footwear;
- conductive footwear,

and is particularly useful for the following activities:

(i)   construction;
(ii)   mechanical/manual handling;
(iii) electrical processes;
(iv)   working in cold conditions (thermal footwear);
(v)   chemical processes;
(vi)   forestry;
(vii) molten substances.

Respiratory protective equipment

[P3023] The majority of these come under the non-simple or complex categories. This takes the form of:

- half-face respirators,

- full-face respirators,
- air-supply respirators,
- self-contained breathing apparatus,

and is particularly useful for protecting against harmful:

(i) gases;
(ii) vapours;
(iii) dusts;
(iv) fumes;
(v) smoke;
(vi) aerosols.

The general advice given in the PPE Regulations relating to the selection and maintenance of suitable protective equipment will usefully apply to the above types of respiratory equipment, although in most cases where respiratory apparatus is required specific Regulations apply. In particular, employers should be aware of the *Control of Asbestos at Work Regulations 2012* and the *Control of Substances Hazardous to Health Regulations 2002*.

Whole body protection

[P3024] This takes the form of:

- coveralls, overalls, aprons;
- outfits to protect against cold and heat;
- protection against machinery and chainsaws;
- high visibility clothing;
- life-jackets,

and is particularly useful in connection with the following activities:

(i) laboratory work;
(ii) construction;
(iii) forestry;
(iv) work in cold-stores;
(v) highway and road works;
(vi) food processing;
(vii) welding;
(viii) fire-fighting;
(ix) foundry work;
(x) spraying pesticides.

Some relevant British Standards for protective clothing and equipment

[P3025] Although not, strictly speaking, a condition of sale or approval, for the purposes of the *Personal Protective Equipment Regulations SI 2002 No 1144*), compliance with British Standards (e.g. ISO 9001:2008 'Quality Assurance') may well become one of the well-tried verification routes to obtaining a 'CE' mark. For that reason, if for no other, the following non-exhaustive list of British Standards on protective clothing and equipment should be of practical value to both manufacturers and users.

Table 1
Protective clothing

EN 340	Protective clothing - general requirements
EN 342	Protective clothing against cold
EN 343	Protective clothing against rain
EN 381	Protective clothing for users of hand-held chain saws
EN 469	Protective clothing for firefighters. Performance requirements for protective clothing for firefighting
EN 471	High-visibility warning clothing for professional use. Test methods and requirements
EN 510	Specification for protective clothing for use where there is a risk of entanglement with moving parts
EN 943-1	Protective clothing against liquid and gaseous chemicals, aerosols and solid particles. Performance requirements for ventilated and non-ventilated "gas-tight" (Type 1) and "non-gas-tight" (Type 2) chemical protective suits
EN 943-2	Protective clothing against liquid and gaseous chemicals, including liquid aerosols and solid particles - Part 2: Performance requirements for "gas-tight" (Type 1) chemical protective suits for emergency teams (ET)
EN 1073-1	Protective clothing against radioactive contamination. Requirements and test methods for ventilated protective clothing against particulate radioactive contamination
EN 1073-2	Protective clothing against radioactive contamination. Requirements and test methods for non-ventilated protective clothing against particulate radioactive contamination
EN 1149-5	Protective clothing. Electrostatic properties. Material performance and design requirements
EN 1150	Protective clothing. Visibility clothing for non-professional use. Test methods and requirements
EN 1486	Protective clothing for fire-fighters. Test methods and requirements for reflective clothing for specialized firefighting
EN ISO 11611	Protective clothing for use in welding and allied processes
EN ISO 11612	Protective clothing. Clothing to protect against heat and flame
EN 13034	Protective clothing against liquid chemicals. Performance requirements for chemical protective clothing offering limited protective performance against liquid chemicals (Type 6 and Type PB [6] equipment)
EN 13982-1	Protective clothing for use against solid particulates. Performance requirements for chemical protective clothing providing protection to the full body against airborne solid particulates (type 5 clothing)
EN ISO 13998	Protective clothing. Aprons, trousers and vests protecting against cuts and stabs by hand knives

Table 1
Protective clothing

EN ISO 14116	Protective clothing. Protection against heat and flame. Limited flame spread materials, material assemblies and clothing
EN 14605/A1	Protective clothing against liquid chemicals. Performance requirements for clothing with liquid-tight (Type 3) or spray-tight (Type 4) connections, including items providing protection to parts of the body only (Types PB [3] and PB [4])
EN ISO 14877	Protective clothing for abrasive blasting operations using granular abrasives
EN 14126+Ac	Protective clothing. Performance requirements and tests methods for protective clothing against infective agents
EN 1082-2	Protective clothing. Gloves and arm guards protecting against cuts and stabs by hand knives. Gloves and arm guards made of material other than chain mail
EN 13356	Visibility accessories for non-professional use
EN 13758-2	Textiles. Solar UV protective properties. Classification and marking of apparel
EN 13911	Protective clothing for firefighters. Requirements and test methods for fire hoods for fire-fighters
EN 14058	Protective clothing. Garments for protection against cool environments
EN 14404	Personal protective equipment. Knee protectors for work in the kneeling position
EN 14786	Protective clothing. Determination of resistance to penetration by sprayed liquid chemicals, emulsions and dispersions. Atomizer test
EN 15614	Protective clothing for firefighters. Laboratory test methods and performance requirements for wildland clothing
ISO 13688	Protective clothing - General requirements
ISO 15384	Protective clothing for firefighters -- Laboratory test methods and performance requirements for wildland firefighting clothing
IEC 61482-2	Live working - Protective clothing against the thermal hazards of an electric arc - Part 2: Requirements

Protective footwear

BS 2723: 1956 (1995)	Specification for fireman's leather boots
BS 4676: 2005	Protective clothing. Footwear and gaiters for use in molten metal foundries.
BS 5145: 1989	Specification for lined industrial vulcanized rubber boots
BS 6159: Part 1: 1987	Polyvinyl chloride boots Specification for general and industrial lined or unlined boots

Table 1
Protective clothing

BS EN	Safety, protective and occupational footwear for professional use
BS EN	Safety footwear for professional use
ISO 20345:2004	Personal Protective Equipment: Footwear

Head protection

BS 6658: 1985 (1995)	Specification for protective helmets for vehicle users
BS 3864: 1989	Specification for protective helmets for firefighters
BS 4033: 1978	Specification for industrial scalp protectors (light duty)
BS 4423: 1969	Specification for climbers helmets
BS EN 1384: 1997	Specification for helmets for equestrian activities.

Face and eye protection

BS 679: 1989	Specification for filters, cover lenses and backing lenses for use during welding and similar industrial operations
BS 1542: 1982 (1995)	Specification for eye, face and neck protection against non-ionising radiation arising during welding and similar operations
BS EN 166:2002	Specification for personal eye protectors for industrial and non-industrial users
BS EN 167: 2002	Personal eye protection. Optical test methods
BS EN 168: 2002	Personal eye protection. Non-optical test methods
BS EN 169: 2002	Personal eye-protection. Filters for welding and related techniques.
BS EN 170: 2002	Personal eye-protection. Ultraviolet filters.
BS EN 171: 2002	Personal eye-protection. Infrared filters
BS EN 172: (2000)	Specification for sunglare filters used in personal eye-protectors equipment for industrial use

Respiratory protection

BS EN 132: 1999	Respiratory protective devices. Definitions of terms and pictograms
BS EN 133: 2001	Respiratory protective devices. Classification
BS EN 134: 1998	Respiratory protective devices. Nomenclature of compounds

Main types of personal protection [P3025]

Table 1
Protective clothing

BS EN 135: 1999	Respiratory protective devices. List of equivalent terms
BS EN 136: 1998	Respiratory protective devices. Full face mask requirements for testing and marking
BS EN 137: 1993	Specification for respiratory protective devices self-contained open-circuit compressed air breathing apparatus
BS EN 138: 2000	Respiratory protective devices. Specification for fresh air hose breathing apparatus for use with full face mask, half mask or mouthpiece assembly
BS EN 141: 1991	Specification for gas filters and combined filters used in respiratory protective equipment
BS EN 143: 1991	Specification for particle filters used in respiratory protective equipment
BS EN 144	Respiratory protective devices. Gas cylinder valves
BS EN 145: 1998	Respiratory protective devices. Self-contained closed-circuit compressed oxygen breathing apparatus
BS EN 148	Specification for thread connection
BS EN 149: 2001	Respiratory protective devices. Filtering half masks to protect against particles
BS EN 371: 1992	Specification for AX gas filters and combined filters against low boiling organic compounds used in respiratory protective equipment
BS EN 372: 1992	Specification for SX gas filters and combined filters against specific named compounds used in respiratory protective equipment
BS EN 269: 1995(2000)	Respiratory protective devices. Specification for powered fresh air hose breathing apparatus incorporating a hood.
BS EN 270: 1995	Respiratory protective devices. Compressed air line breathing apparatus incorporating a hood. Requirements, testing, marking
Hearing protection	
BS EN 352:	Hearing protectors. Safety requirements and testing
Part 1: 2002	Ear muffs
Part 2: 2002	Ear plugs
Part 3: 2002	Ear muffs attached to an industrial safety helmet
Hand protection	
BS EN 374:	Protective gloves against chemicals and microorganisms
Part 1: 2003	Terminology and performance requirements
Part 2: 2003	Determination of resistance to penetration
Part 3: 2003	Determination of resistance to permeation by chemicals
BS EN 511: 1994	Specification for protective gloves against cold

Table 1	
Protective clothing	
BS EN 421: 1994	Protective gloves against ionising radiation and radioactive contamination
BS EN 407: 2004	Protective gloves against thermal risks (heat and/or fire)
BS EN 388: 2003	Protective gloves against mechanical risks

Note on revisions to current British Standards

Some standards in the above list are current but obsolescent eg: BS 3664 : 1963 which means that despite the techniques being old there is still a small level of use. Some standards have been withdrawn altogether eg BS 4001: Part 1 1989, while others have been replaced eg: BS 3864 : 1989 has been replaced with BS EN 443: 1997. In some cases a standard has simply been revised eg: BS EN 511 : 1994 is now known as BS EN 511: 2006.

To ensure up to date information contact BSI Knowledge Centre at www.bsigroup.com.

Common law requirements relating to personal protective equipment

[P3026] In addition to specific statutory requirements, the duty on employers at common law to provide and maintain a safe system of work, including appropriate supervision of safety duties, still applies to protection against foreseeable risk of eye injury (*Bux v Slough Metals Ltd* [1974] 1 All ER 262) and dermatitis/facial eczema (*Pape v Cumbria County Council* [1991] IRLR 463). *Pape* related to an office cleaner who developed dermatitis and facial eczema after using Vim, Flash and polish. The employer was held liable for damages at common law, since he had not instructed her in the dangers of using chemical materials with unprotected hands and had not made her wear rubber gloves.

Consequences of breach

[P3027] Employers who fail to provide suitable personal protective equipment, commit a criminal offence, under the *Health and Safety at Work etc Act 1974* and the *Personal Protective Equipment at Work Regulations 1992* and sundry other regulations (see **P3006–P3008** above). In addition, if, as a result of failure to provide suitable equipment, an employee suffers foreseeable injury and/or disease, the employer will be liable to the employee for damages for negligence. Conversely, if, after instruction (where necessary) training, an employee fails or refuses to wear and maintain suitable personal protective equipment, he too can be prosecuted and/or dismissed. If he is injured or suffers a disease in consequence, he will probably lose all or, certainly, part of his damages.

"Not excused boots"

Despite some clear case law decisions, employers still have to wrestle with difficult situations in the work-place when employees do not or will not wear PPE. It may be on religious grounds, hygiene grounds or simply due to management uncertainty.

An EAT decision regarding an employee's dismissal (upheld) over the non-wearing of safety footwear is a case in point. A warehouse worker was dismissed because his employer could not find an alternative to the safety boots that aggravated his skin condition (psoriasis), and no other work deployment was available. [*Lane Group plc v Farmiloe, Court of Appeal,* (UKEAT/0352/03/2201) [2004] All ER (D) 08 (Mar)].

Another ruling has implications for deciding on who must pay for PPE. A recruitment agency won an appeal that meant temporary workers may be asked to contribute to the wholesale price of work equipment, where a key issue is that the PPE is not exclusive to one workplace [*Miles Recruitment Ltd v Gangmasters Licensing Authority* (2007), unreported].

As the PPE Regulations themselves contain no explicit exemptions on religious, medical or other grounds (with the exception mentioned above regarding Sikhs), the upshot is likely to be that if employers demonstrate sufficient resolve to overcome employees' objections to wearing PPE they may in the end be justified in dismissing them. In essence, one cannot opt out of health and safety requirements and these take precedency over the Equality Act and employment law.

Pressure Systems

Kevin Chicken

Introduction to pressure systems

[P7001] Legislation concerning the safety of pressure systems applied in its infancy to steam generating boilers and is traceable back to the original *Boiler Explosions Act 1882*. At this time statutory requirements were placed on the users and owners of the equipment for the safety of steam generating boilers and little or no consideration was given to the safety of the system into which it was incorporated. Periodic examination of boilers became mandatory following the *Factories and Workshops Act 1901* when requirements were also extended to include other items of pressure equipment such as steam and air receivers (storage vessels).

The legislation governing the construction and use of pressure equipment is divided into two areas where the initial duty falls onto those manufacturing, installing or supplying pressure equipment and thereafter to users or owners of the equipment.

There are three principal statutes that cover the manufacture, installation and operation of pressure equipment:

- The *Simple Pressure Vessels (Safety) Regulations 2016 (SI 2016 No 1092)*, this replaced the *Simple Pressure Vessel Regulations 1991 (SI 1991 No 2749)* as amended.
- The *Pressure Equipment (Safety) Regulations 2016 (SI 2016 No 1105)*, this replaced the Pressure *Equipment Regulations 1999 (SI 1999 No 2001)* as amended.
- *Pressure Systems Safety Regulations 2000 (SI 2000 No 128)*.

These regulations provide a synergistic approach to the management of pressure systems based on a cradle to grave concept.

The main changes to the *Simple Pressure Vessels (Safety) Regulations 2016* and the *Pressure Equipment (Safety) Regulations 2016* relate to alignment of the New Legislative Framework (NLF) principles. The NLF is a set of legislative acts (including the Regulation (EC) No 765/2008 and the Decision No 768/2008/EC) that create a more coherent and consistent legal framework for the marketing of products in the European Union across all sectors. The new content of the 2016 Regulations, amongst others, relate to definitions and detailed obligations of economic operators (manufacturers, authorised representatives, importers or distributors); definitions of 'placing on the market' and 'making available on the market'; market surveillance procedures including the Union safeguard procedures and enforcement penalties applicable in the UK against offences committed.

These Regulations set minimum standards of safety and quality for the manufacture of new pressure systems and equipment in the European Union so that non-tariff trade barriers are removed and they present a 'level playing field' for all member states. The Health and Safety Executive 're responsible for the enforcement of the Legislation and the Department for Innovation, and Skills (formerly known as The Department of Trade and Industry) have the responsibility for the assessment of and the appointment of conformity assessment bodies.

The *Pressure Systems Safety Regulations 2000 (SI 2000 No 128)* set minimum safety standards upon organisations that install, repair, examine use and manage pressure systems within the UK.

Pressurised transportable gas containers are legislated for under the *Carriage of Dangerous Goods and Use of Transportable Pressure Equipment Regulations 2009* (CDG Regs) and the European agreement ('Accord europÈen relatif au transport international des marchandises dangereuses par route', known as ADR).

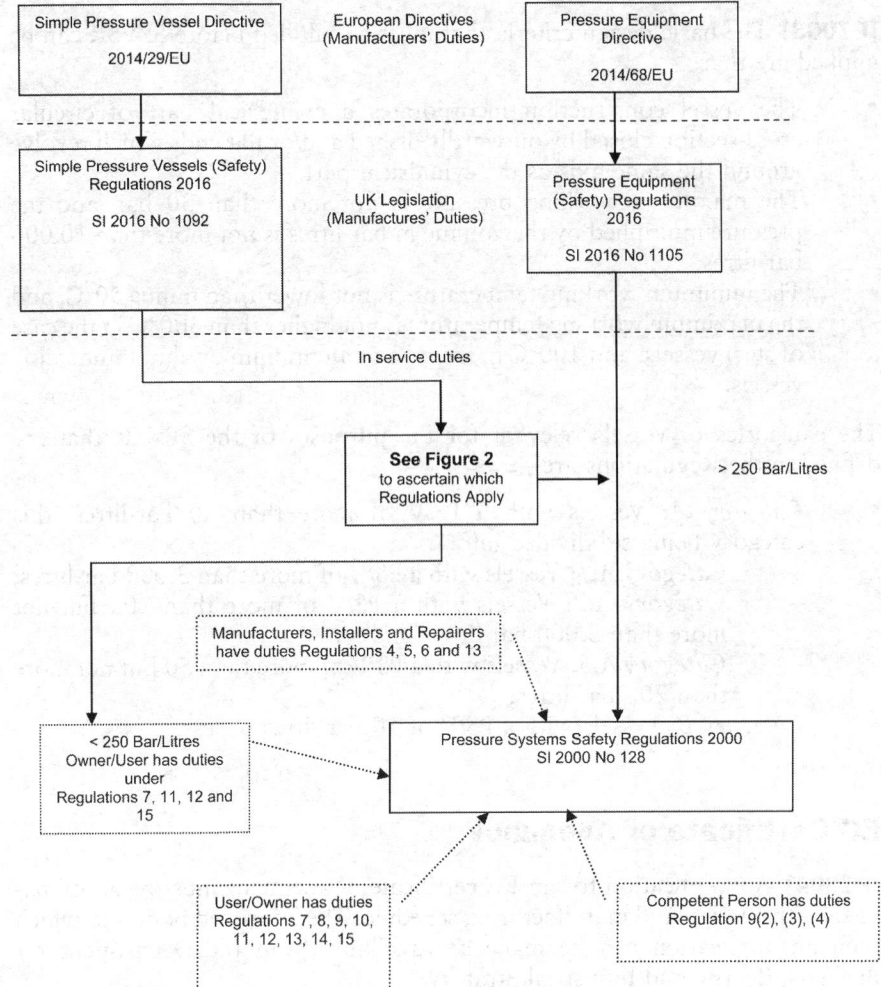

Figure 1. – Relationship between Manufactures' Duties

Simple pressure vessels

[P7002] Simple pressure vessels are those that are constructed under the criteria of the *Simple Pressure Vessels (Safety) Regulations 2016 (SI 2016 No 1092)*.

A simple pressure vessel is a welded vessel intended to contain air or nitrogen at a gauge pressure greater than 0.5 bars but one that is not intended for exposure to a flame.

The Simple Pressure Vessels (Safety) Regulations 2016 (SI 2016 No 1092)

[P7003] The basic design criteria that must be fulfilled before SPVSR can be applied are that:

- The vessel construction incorporates a cylindrical part of circular cross-section closed by outwardly dished and/or flat ends which revolve around the same axis as the cylindrical part.
- The maximum working pressure is not more than 30 bar, and the pressure multiplied by the volume in bar litres is not more than 10,000 bar litres.
- The minimum working temperature is not lower than minus 50°C, and the maximum working temperature is not higher than 300°C in the case of steel vessels, and 100°C in the case of aluminium or aluminium alloy vessels.

The categories of vessels relevant for the purposes of the SPVSR that are defined in the Regulations are—

- *Category A*: Vessels with a PS/V of more than 50 bar-litres, this category being subdivided into:
  - *Category A.1*: Vessels with a PS/V of more than 3,000 bar-litres;
  - *Category A.2*: Vessels with a PS/V of more than 200 but not more than 3,000 bar-litres; and
  - *Category A.3*: Vessels with a PS/V of more than 50 but not more than 200 bar-litres;
- *Category B*: Vessels with a PS/V of 50 bar-litres or less.

EC Certificate of Adequacy

[P7004] An application for an EC certificate of adequacy must be accompanied by the design and manufacturing schedule. The approved bodies to which such an application can be made are available from the Department for Business, Energy and Industrial Strategy.

The schedule must contain all the required information; and additionally provide evidence that the vessel has been manufactured in accordance with the schedule and that it conforms to a relevant national standard eg BSI EN 13445.

A certificate of adequacy can then be issued by the approved body which allows the simple pressure vessel to be placed on the market as being 'safe'.

If the approved body is not so satisfied and refuses to issue a certificate of adequacy it must inform the applicant in writing the reasons for the refusal.

EC Type-Examination Certificate

[P7005] EC type-examination is the process whereby an approved body ascertains, and certifies by means of an EC type-examination certificate, that a prototype representative of the production envisaged satisfies the requirements of the Directive.

The prototype may be representative of a family of vessels in which case the Approved Body is required to examine the prototype, and perform such tests as it considers are appropriate, and once it is satisfied that the prototype is manufactured in conformity with the schedule, and meets the essential safety requirements specified in Schedule 1, declare that it is safe.

The Pressure Equipment (Safety) Regulations 2016 (SI 2016 No 1105)

[P7006] The Directive was taken into UK legislation as the *Pressure Equipment (Safety) Regulations 2016 (SI 2016 No 1105)*, and covers pressure equipment and assemblies with a maximum allowable pressure greater than 0.5 bar. Pressure equipment includes vessels, piping, safety accessories and pressure accessories.

Assemblies are several pieces of pressure equipment assembled to form an integrated, functional whole.

The intention of legislation is to remove technical barriers to trade by harmonising national laws of legislation regarding the design, manufacture, marking and conformity assessment of pressure equipment. It does not deal with in-use requirements, which may be necessary to ensure the continued safe use of pressure equipment.

The Directive covers a wide range of equipment such as, reaction vessels, pressurised storage containers, heat exchangers, shell and water tube boilers, industrial pipework, safety devices and pressure accessories.

There are some specific exclusions such as: equipment within the scope of other directives (eg simple pressure vessels, aerosols, parts for the functioning of vehicle braking system); and equipment covered by the international conventions for the transport of dangerous goods and equipment for which the pressure hazards are low and are adequately covered by other directives (such as the Machinery Directive).

Essential safety requirements

[P7007] The principle adopted in the *Pressure Equipment (Safety) Regulations 2016* is similar to other European Directives, and requires the manufacturer to consider hazards associated with pressure equipment and accordingly take these hazards into consideration when considering the design and construction of the equipment.

Compliance with the Essential Safety Requirements (ESRs) provides the necessary safety elements for protecting public interest. Essential safety requirements for design, manufacture, testing, marking, labelling, instructions and materials, are provided in general terms, compliance is mandatory and must be met before products may be placed on the market in the European Community. The main areas for assessment are:

- General requirements.
- Design requirements:
  - Adequate strength;
  - Loading considerations;
  - Simultaneous loading considerations;
  - Calculation methods;
  - Experimental design methods;
  - Provisions to ensure safe handling and operation;
  - Means of examination;
  - Means of draining and ventilating;
  - Corrosion or other chemical attack;
  - Wear;
  - Assemblies;
  - Provisions for filling and discharge;
  - Protection against exceeding the allowable limits of pressure equipment;
  - Safety accessories;
  - Pressure limiting devices;
  - Temperature monitoring devices;
  - External fire.
- Manufacturing:
  - Manufacturing procedures;
  - Preparation of the component parts;
  - Permanent jointing;
  - Non-destructive tests;
  - Heat treatment;
  - Traceability;
  - Final assessment;
  - Final inspection;
  - Proof test;
  - Inspection of Safety devices;
  - Marking and labelling;
  - Operating instructions.
- Materials.
- Fired or otherwise heated pressure equipment with a risk of overheating.
- Piping.
- Specific quantitative requirements for certain pressure equipment:
  - Allowable stresses;
  - Joint coefficients;
  - Pressure limiting devices, particularly for pressure vessels;
  - Hydrostatic test pressure;
  - Material characteristics.

The requirements set out above may be met by applying one of the following methods, as appropriate, if necessary as a supplement to or in combination with another method:

— design by formula;
— design by analysis; or
— design by fracture mechanics.

Conformity assessment procedure

[P7008] Schedule IV of the *Pressure Equipment (Safety) Regulations 2016*, details the procedures and responsibilities for the Notified Body Requirements. This is a detailed document and it consists of a product classification table and nine classification charts.

A Conformity Assessment must be undertaken by the manufacturer or notified body, depending on the category of the equipment, in order to demonstrate that the essential safety requirements are met.

In order to declare pressure equipment to be safe the manufacturer or their authorised representative must, before supplying any equipment, satisfy the Certification. The revised Regulations continue the modular approach to conformity assessment, thereby subdividing it into a number of independent activities. Modules differ according to the type of assessment (eg documentary checks, type approval, design approval, quality assurance) and the organisation carrying out the assessment (ie the manufacturer or a third party).

- Draw up a technical document which will allow an assessment to be made of the conformity of the pressure equipment, containing:
    — General description;
    — Conceptual design and manufacturing drawings;
    — Descriptions and explanations necessary to understand the drawings;
    — A list of harmonised standards applied and the solutions adopted to meet the essential safety requirements were harmonised standards have not been applied;
    — Results of design calculations made;
    — Test reports.
- Ensure manufacturing process complies with technical documentation.
- Affix CE Mark.
- Draw up written declaration of conformity.
- Retain declaration of conformity and technical documentation for ten years.

Published Harmonised (European) Standards, list are a specific subset of European Standards (EN, produced by CEN and available from the national Standards Institutes) with particular consideration of the Essential Safety Requirements the reference number of which is published in the Official Journal of the European Commission. The use of a Published Harmonised Standard in the design and manufacture of a product will give the presumption of conformity to those ESRs listed in the particular Harmonized Standard.

There are a number of different conformity assessment modules available for the different categories of equipment; eg the boiler manufacturer's codes:

- BS EN 12592 Water Tube Boilers and Auxiliary Installations;
- BS EN 12593 Shell Boilers.

The unfired pressure vessel manufacturer's new code:

- BS EN 13445 Unfired pressure vessels.

To satisfy the requirements of the *Pressure Equipment (Safety) Regulations 2016*, manufacturers will need to:

- meet essential safety requirements covering design, manufacture and testing;
- satisfy appropriate conformity assessment procedures; and
- undertake the CE marking.

By affixing a CE mark the manufacturer declares the completion of conformity assessment and that the equipment or assembly complies with the provisions of the Directive and meets the essential safety requirements.

Pressure equipment and assemblies below the specified pressure / volume thresholds must that needs to conform to "Sound Engineering Practice (SEP)" (SEP applies to equipment that is not subject to conformity assessment but must be designed and manufactured in accordance with the sound engineering practice of a Member State in order to ensure safe use. That the design and manufacture takes into account all relevant factors influencing safety during the intended lifetime. The equipment must be accompanied with adequate instructions for use and must bear the identification of the manufacturer. The responsibility for compliance with the PED lies solely with the manufacturer):

To be safe and meet compliance with SEP the equipment must;

- be designed and manufactured according to sound engineering practice; and
- bear specified markings (but not the CE marking).

Notified bodies

[P7009]–[P7010] The Secretary of State for the Department for Innovation, and Skills is responsible for the appointment of recognised third-party notified bodies, who must satisfy all the criteria laid down in the directive.

The Notified Body is a semi-official or private technical organisation appointed by Member States, either for approval and monitoring of the manufacturers' quality assurance system or for direct product inspection. A Notified Body may be specialised for certain products/product categories or for certain modules.

Pressure Systems Safety Regulations 2000

Duty holders

[P7011] The *Pressure Systems Safety Regulations 2000* impose statutory duties on individuals and organisations who own, use or manage pressure systems.

The *Pressure Systems Safety Regulations, SI 2000/128* ('the Regulations') came into force on 14th February 2000. They are the United Kingdom's statutory controls over the operation and safety of in-service pressure systems. The aim of the Regulations is to prevent serious injury from the hazard of stored energy as a result of failure of the system or one of its component parts.

- **Users and owners** of pressure systems have primary and wide responsibilities for ensuring the safety of their systems.
- **'Competent persons'** are responsible for drawing up or certifying suitable written schemes of examination, for the examination in accordance with the written scheme, and for action in the case of imminent danger.
- **Designers, manufacturers, importers, suppliers and installers** have responsibilities relating to design, materials and construction, and information and marking.
- Employers of **persons, who install,** modify or repair pressure systems have duties to ensure that their work does not give rise to danger.
- **The enforcing authority** (Health and Safety Executive or Environmental Health Department of the local authority), has powers under the Regulations to enforce and exempt the application of the Regulations.

In cases where pressure systems are supplied by way of lease, hire or other arrangements, the supplier is allowed to discharge the duties of the user with regard to examinations, operation, maintenance and record keeping. [*Pressure Systems Safety Regulations 2000 (SI 2000 No 128), Reg 3(5) and Sch 2.*]

A duty holder charged with an offence under the Regulations is allowed a defence if the offence was due to an act or default of another person (not being an employee) and that all reasonable precautions and due diligence had been taken to avoid the offence taking place. [*Pressure Systems Safety Regulations 2000 (SI 2000 No 128), Reg 16.*]

Systems and equipment covered by the regulations

[P7012] The Regulations are concerned with systems containing relevant fluids under pressure (See Figure 2 below). A relevant fluid is:

- Steam (above atmospheric pressure);
- A gas or a mixture of gases;
- A liquid which would evaporate as a gas if system failure occurred.

The Regulations therefore cover systems containing compressed air, other compressed gases (eg nitrogen, oxygen or acetylene), hot water contained above its normal boiling point, and a gas dissolved under pressure in a solvent contained in a porous substance.

Except in the case of steam (which is always a relevant fluid when stored above atmospheric pressure), a fluid is relevant if it exerts a pressure or vapour pressure greater than 0.5 bar above atmospheric pressure.

The Regulations cover both installed (fixed) pressure systems and mobile systems that can be readily removed between and used in different locations. Transportable pressure receptacles such as a tanker used for carrying dangerous goods by rail or road, are covered under the *Carriage of Dangerous Goods and Use of Transportable Pressure Equipment Regulations 2004* (if these Regulations cease to apply and a relevant fluid is carried, then the *Pressure Systems Safety Regulations 2000* will apply). Steam locomotives, including model steam engines, are classed as installed systems for the purposes of the Regulations. Within the Regulations, an installed system is the responsibility of the user, while the owner has responsibility for a mobile system.

A pressure system comprises of one or more pressure vessels of rigid construction, any associated pipework, and protective devices. Pipework used in connection with transportable pressure receptacles and pipelines with their associated protective devices are also pressure systems. The terms 'pipework', 'pipeline' and 'protective devices' are defined by the Regulations and cover all items of pressure equipment including valves, pumps, hoses, compressors, bursting discs and pressure relief devices. [*Pressure Systems Safety Regulations 2000 (SI 2000 No 128), Reg 2.*]

Users and owners should decide, seeking suitable advice if required, whether equipment for which they are responsible is covered by these Regulations. Examples of equipment for which the Regulations are likely to apply in full include:

- Steam boilers and associated pipework and protective devices.
- Compressed air systems with a receiver of a stored capacity exceeding 250 Bars/litres
- Portable hot water steam cleaning units fitted with a pressure vessel.
- Steam sterilising autoclaves and associated pipework and protective devices.
- Steam pressure cookers.
- Gas loaded hydraulic accumulators if the free gas capacity exceeds 250 Bars/litres
- Steam locomotives including model and narrow gauge steam engines.
- Vapour compression refrigeration systems where the power exceeds 25KW.
- Fixed liquid petroleum gas (LPG) storage systems eg supplying fuel for heating in a workplace.

Systems and equipment covered by the regulations [P7012]

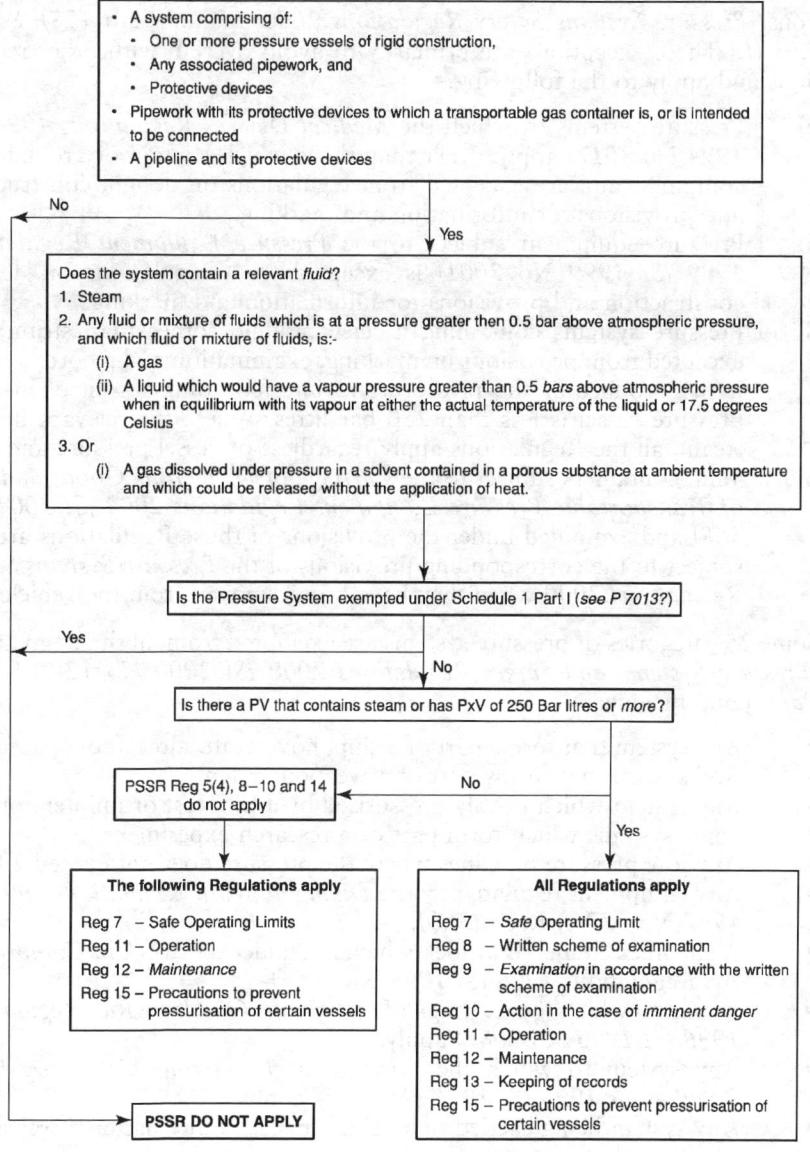

Figure 2. Pressure Systems Safety Regulations 2000 – Duties Decision Flow Chart

Systems and equipment excepted from some or all the regulations

[P7013] Some types of pressure systems are excepted from certain Regulations [*Pressure Systems Safety Regulations 2000 (SI 2000 No 128), Sch 1, Part II.*] These exceptions are primarily to avoid overlap with other regulations and apply to the following.

(i) Pressure systems to which the *Medical Devices Regulations 1994 (SI 1994 No 3017)* apply, other than those which contain or are liable to contain steam, are excepted from regulations on design, construction and provisions for information and marking.

(ii) Pressure equipment subject to the *Pressure Equipment Regulations 1999 (SI 1999 No 2001)* is excepted from regulations on design, construction and provisions for information and marking.

(iii) Pressure systems containing a relevant fluid (other than steam) are excepted from provisions on marking, examination and record keeping if the product of the pressure and internal volume control in each pressure vessel is less than 250 bar litres. Where the relevant fluid is steam, all the Regulations apply regardless of vessel pressure and size.

(iv) Tank containers subject to the *Carriage of Dangerous Goods and Use of Transportable Pressure Equipment Regulations 2004 (SI 2004 No 568)* and examined under the provisions of those Regulations are not subject to the corresponding provisions of the *Pressure Systems Safety Regulations 2000* when they have been removed from the vehicle.

Some 25 categories of pressure systems are exempted from all the Regulations [*Pressure Systems and Safety Regulations 2000 (SI 2000 No 128), Sch 1, Part I*] and include:

- Any system that forms part of a ship, hovercraft, aircraft or spacecraft.
- Any system that forms part of a weapon system.
- Any system which is only pressurised by a leak test or unintentionally.
- Some systems which form part of a research experiment.
- Any low pressure pipelines where the pressure does not exceed 2 bars.
- Any equipment regulated by the *Diving at Work Regulations 1997 (SI 1997 No 2776), Reg 6(3)(b)*.
- A chamber, tunnel or airlock which fall under the *Work in Compressed Air Regulations 1996 (SI 1996 No 1656)*.
- A tank to which the *Carriage of Dangerous Goods by Rail Regulations 1996 (SI 1996 No 2089)* apply.
- Any system to which the *Carriage of Dangerous Goods by Road Regulations 1996 (SI 1996 No 2095)* apply.
- Any system being carried in a vehicle engaged international transport.
- A system that forms part of the braking, control or suspension system of a vehicle.
- Any water cooling system on an international combustion engine or compressor.
- Any tyre used or intended to be used on a vehicle.
- Any system fitted to a vehicle within the meaning of the *Road Traffic Act 1988, s 185*.

- Any portable fire extinguisher with a working pressure of less than 25 bars at 60°C.
- Any part of a tool or appliance designed to be held in the hand.

Some of these categories may, however, be subject to other regulations.

Approved code of practice and guidance

[P7014] The Regulations are largely non-prescriptive goal-setting regulations. For example, the Regulations state that pressure systems 'shall be properly designed and properly constructed from suitable material so as to prevent danger'. The goal is to prevent danger; what is 'proper' and 'sufficient' to achieve this goal is left to duty holders and ultimately the courts to decide.

The Health and Safety Commission's (HSC) Approved Code of Practice ('the ACoP') to the *Pressure Systems Safety Regulations 2000 (SI 2000 No 128)* also includes guidance on compliance with the Regulations (*'Safety of Pressure Systems, Approved Code of Practice'*, L122, ISBN 07176 1767 X, published by HSE Books). This document is essential reading for duty holders. It has special legal status in relation to compliance with the Regulations.

The ACoP gives practical advice on how to comply with and exceed minimal compliance with the regulations by following the approved Code of Practice, which is addressed at all duty holders under the Regulations, eg users, owners, competent persons, designers, manufacturers, importers, suppliers and installers of pressure systems. It reflects pertinent issues such as: design and construction; provision of information and marking; installation; safe operating limits; matters concerning the written scheme of examination; imminent danger action; maintenance; modification/repair; record keeping; and specific precautions preventing pressurisation of certain vessels.

Following the advice in the ACoP should be enough to ensure compliance with the law on specific matters, although alternative methods may be used. If the relevant provisions of the code are not followed, then it is necessary, in any prosecution for breach of the law, to show compliance with the law in other ways.

The additional guidance material is not compulsory but illustrates good practice. The guide sets out the law and explains what it means in simple terms.

Interpretation

[P7015] The following words and phrases have a specific defined meaning within the context of the Regulations:
- Competent person.
- Danger.
- Data.
- Examination.

[P7015] Pressure Systems

- The Executive.
- Installed system.
- Mobile system.
- Owner.
- Pipeline.
- Pipework.
- Pressure system.
- Protective devices.
- Relevant fluid.
- Safe operating limits.
- Scheme of examination.
- Ship.
- System failure.
- Transportable pressure receptacle.
- User.

[*Pressure Systems Safety Regulations 2000 (SI 2000 No 128), Reg 2.*]

Design and construction

[P7016] The following persons have duties for ensuring that the design and construction of pressure systems and component parts meet the requirements of the Regulations:

(i) Designers.
(ii) Manufacturers.
(iii) Importers.
(iv) Suppliers.

[*Pressure Systems Safety Regulations 2000 (SI 2000 No 128), Reg 4.*]

The Regulations require that pressure systems shall be properly designed and properly constructed from suitable material so as to prevent danger. The design and construction must enable necessary examinations to be carried out for preventing danger from defects etc., and that where access to the inside of a pressure system is provided, it can be gained without danger to the inspector. Pressure systems must be provided with suitable protective devices (safety relief valves, bursting discs etc.) to prevent danger, and such devices must be capable of releasing the contents safely.

Pressure systems supplied with a CE marking under the *Pressure Equipment Regulations 1999 (SI 1999 No 2001)* are presumed to have complied with this Regulation (see **P7008**.)

Employers of persons installing a pressure system have a duty to ensure that there is nothing about the way it is installed that could give rise to danger from the system or otherwise impair the operation of protective devices or the facility to examine the equipment. The ACoP draws attention to a number of items that the employer should consider relating to the siting and access to and around the pressure system and the means and manner of installation. Special consideration needs to be given by the manufacturer to any in-service examinations that may require access to constricted areas and the specific requirements of *the Confined Space Regulations 1997 (SI 1997 No 1713)*.

The provision of information and marking

[P7017] The following information and documentation is required for each pressure system.

(a) *Information concerning the design, construction, examination, operation and maintenance*
 The Regulations require that pressure systems (or any component parts) are provided with written information concerning the design, construction, examination, operation and maintenance. Responsibility for providing this information falls on the person who designs or supplies (whether as manufacturer, importer or in some other capacity) the system. When a pressure system is modified or repaired, the responsibility to provide information about the modification or repair falls on the employer of the person who carried out the work or any self-employed individual. [*Pressure Systems Safety Regulations 2000 (SI 2000 No 128), Reg 5.*]
 The Regulations do not specify the information to be provided, but set the objective that it must be sufficient enough to comply with their provisions. The information must be provided with the design or with the supplied pressure system. When systems are modified or repaired, the information must be provided immediately afterwards. Although it is not possible to give a complete list of all the information which might be needed, the ACoP highlights items that should be considered.

(b) *A plate on each pressure vessel with the information given in Schedule 3 of the Regulations*
 Manufacturers of pressure vessels must ensure that before the vessel is supplied, it is marked or has a plate attached containing the information specified in the *Pressure Systems Safety Regulations 2000 (SI 2000 No 128), Sch 3*. This information includes the manufacturer's name, standard of construction, and the design pressure(s) and temperature(s). Similarly, no one may import a vessel unless it is marked or has a plate containing this information. The Regulations prohibit the removal or falsification of information contained on the plate or on any other mark on the vessel.

(c) *A statement of the safe operating limits*
 The user of an installed pressure system or the owner of a mobile system will have established *safe operating limits* of the system before the system is operated. The nature of the safe operating limits is not specified by the Regulations and will depend on the system. They should, however, be such as to ensure that there is a suitable margin of safety before failure is liable to occur. Typically, operating limits may be set on pressure, temperature, number of cycles, or time in operation, but it is for the duty holder to determine what is appropriate.

[P7017] Pressure Systems

In the case of mobile systems, the owner, if not the user, must supply the user with a written statement specifying the safe operating limits or ensure that the system is marked with this information in a legible, durable and visible way.

(d) *A written scheme of examination* (see **P7019** below);
(e) *A report on each examination carried out under the written scheme* (see **P7024** below);
(f) *Instructions for safe operation and action in emergencies* (see **P7026** below).

Under certain circumstances, the following additional documentation is required:

(g) *An agreement to postpone the date of examination;*
(h) *A report in case of imminent danger* (see **P7025** below);
(i) *Information required by the Pressure Equipment Regulations 1999 (SI 1999 No 2001);*
(j) *A certificate exempting application of the requirements/prohibitions of the Regulations.*

Competent persons

[P7018] Whilst users and owners have responsibility for the management of their pressure systems, the Regulations require and define the role and responsibilities of a 'competent person'. The role of the competent person is primarily in connection with the examination and action in the case of imminent danger. The term 'competent person' is used in connection with two distinct functions:

- drawing up or certifying schemes of examination;
- carrying out examinations and making the examination report.

A single individual may carry out these functions, but equally, separate individuals can also carry them out. The user or owner may, however, also use the competent person in an advisory capacity in conjunction with other matters relating to the Regulations. For example, the competent person may assist users/owners in deciding the scope of the examination.

The level of expertise needed by the competent person depends on the size and complexity of the pressure system in question. The ACoP distinguishes between minor, intermediate and major systems and the attributes required of the competent person and back up specialist services and organisation required in each case. In all but minor systems, the competent person must be a chartered engineer.

For larger and more complex systems, it is unlikely that one individual will have sufficient knowledge and expertise. In this case the competent person should be part of a corporate body with a team of employees which have the necessary expertise and experience. The requirements for 'competence' then applies to the company and not to individual employees.

The following bodies may provide competent person services:

- A user company with its own in-house inspection department.
- An inspection organisation providing such services to clients.
- A self employed person.

The Competent Person should be either an individual or company with sufficient knowledge and experience to undertake the work. As a minimum standard the Competent Person should be able to demonstrate sufficient knowledge and experience to undertake the work required. A national certification scheme has been developed for bodies that provide competent person services known as the RG2, UKAS – Accreditation for In-Service Inspection of Pressure Systems/Equipment who will accredit competent persons to EN 45004:1995.

This standard provides guidance to those requirements in BS EN ISO/IEC 17020:2004 *General criteria for the operation of various types of bodies performing inspection* which need interpretation when applied by Inspection Bodies carrying out in-service inspection of pressure systems / pressure equipment. BS EN ISO/IEC 17020:2004, as applied by UKAS in accordance with IAF/ILAC-A4: 2004.

The following criteria is provided by UKAS as suitable qualifications for competent persons:

Category 1 Chartered Engineer as defined by the Engineering Council or equivalent (eg appropriate degree with relevant experience, NVQ Level V Engineering) including at least 3 years experience within an engineering discipline associated with in-service inspection of pressure systems.

Category 2 Incorporated Engineer as defined by Engineering Council or equivalent (eg appropriate HNC with relevant experience, NVQ Level IV Engineering) including at least 5 years experience within a relevant engineering discipline of which at least one year* shall have been spent working within an engineering discipline associated with in-service inspection of pressure systems.

Category 3 Engineering Technician as defined by Engineering Council or equivalent (eg appropriate ONC with relevant experience, NVQ Level III) having a minimum of 5 years experience within a relevant discipline of which at least one year shall have been spent working within an engineering discipline associated with the in service inspection of pressure systems.

Category 4 Person trained in a relevant engineering discipline with a recognised and documented engineering apprenticeship with a minimum of 5 years experience within a relevant discipline of which at least one year shall have been spent working within an engineering discipline associated with the in-service inspection of pressure systems.

Category 5 Person with less than tradesman's apprenticeship but with a minimum of 5 years spent working with or within the industry associated with pressure systems and has general knowledge of pressure systems and its operating environment. Personnel shall be placed on recognised training courses with appropriate documented tests in in-service inspection of pressure systems. The minimum age for this Category is 21 years.

Persons in Categories 4 and 5 shall pass a qualifying test, established by the Inspection Body, associated with the particular inspection activities relating to pressure systems / equipment and this should cover relevant knowledge of the law, codes of practice and inspection techniques.

* Where a person meets the minimum requirement for a specific discipline and is to be trained in a second discipline, it may not be necessary to have experience of at least one year in the second discipline provided that the required competence can be demonstrated.

Safe operating limits

[P7018A] No system shall be operated unless the safe operating limits have been established by the user/owner.

They may comprise:

- For steam and air systems — maximum safe working pressure
- Refrigeration plant — pressure and temperature
- Complex plant — pressure, temperature, external loads, operating time, nature of content, etc

For mobile plant the owners must provide the user with a written statement detailing the safe operating limit or ensure that the information is clearly marked on the equipment. For long term hire both a written statement and durable markings are preferred.

Written scheme of examination

Scope

[P7019] At the heart of the Regulations is the requirement for users of installed pressure systems or owners of mobile systems to have a 'suitable' written scheme of examination for every pressure system that they use or own. The user or the owner must ensure that the scheme is drawn up or certified as suitable by a competent person. The written scheme must specify:

- the parts to be examined (the scope);
- the maximum intervals between examinations (the frequency);
- the types of examinations necessary (the nature);
- any special measures necessary to prepare the system for safe examination.

Written schemes of examination may be held as electronic or other forms of data providing that the data is capable of being printed as a written copy and is kept secure from loss or unauthorised interference.

Responsibility for defining the scope of the written scheme of examination rests with the user or owner. The user or owner is likely to be most conversant with the equipment and relevant boundaries between systems. The competent person may be useful in assisting the user or owner to define the scope. The written scheme of examination must cover the relevant parts of the system which are specified by the Regulations as:

(a) All protective devices.
(b) Every pressure vessel and every pipeline in which a defect may give rise to danger.
(c) Those parts of pipework where a defect may give rise to danger.

The user or owner should first identify and establish which parts of the pressure system are pressure vessels, protective devices or pipework as defined by the Regulations. The user or owner will then decide which parts of the system should be included for examination. Where the decision is made to exclude parts from regular examination, users and owners should take advice from a person with appropriate and relevant technical expertise and experience (who need not be the competent person) and be able to justify the decision.

A written scheme of examination must be prepared before the pressure system is operated for the first time. It will specify the examination necessary before the pressure system enters service. The user or owner with the competent person will take account of fabrication and installation quality and inspection reports.

The user or owner must ensure that the scheme is reviewed at appropriate intervals by a competent person to determine whether it is still suitable for the current conditions of use of the system. This normally occurs after every examination. If, as a result of the review, the competent person recommends modifications to the content of the scheme, then the user or owner must ensure that the scheme is modified.

Schemes of examination must be suitable for the purposes of preventing danger arising from the system. Whilst it is for the duty holders to decide how to meet the goal of suitability and whether the scheme achieves it, the ACoP gives detailed advice about the factors that should be considered.

Further explanation regarding the preparation of written schemes of examination can be found in the Safety Assessment Federation (SAFed) Guidelines for the Production of Written Schemes of Examination and the Examination of Pressure Vessels Incorporating Openings to Facilitate Ready Internal Access Ref: PSG4

Frequency of examination

[P7020] The written scheme will specify a suitable frequency of examination ie the maximum intervals between successive examinations. The ACoP and Guidance give advice about the factors that should be taken into account when deciding the frequency between examinations. Separate considerations apply to the initial examination before entry to service, the first examination after entering service, and the regular examinations thereafter.

The Regulations do not prescribe intervals between examinations and therefore allow flexibility and freedom to exercise good engineering judgement in each case. Different parts of the system may be examined at different intervals depending on the degree of risk associated with each part. Much will depend on the extent and degree of confidence in the knowledge that is available about the predicted condition of the equipment, degradation mechanisms and rates, the operating conditions and operating history.

Advice on the frequency of examinations is available. The *Factories Act 1961* sets limits on the intervals between examinations of boilers, steam receivers and air receivers. These limits are still relevant good practice. The Institute of Petroleum Model Code of Practice (Parts 1 and 2) provides advice on examination periods for chemical pressure vessels, heat exchangers and other equipment depending on the results of earlier examinations. The Safety Assessment Federation (SAFed), which represents the major engineering insurers, has published recommendations on the periodicity of examination of pressure equipment (*'Pressure Systems Guidelines on the Periodicity of Examinations'* (PSG1), available from the Safety Assessment Federation, ISBN 1901212106). (safed.co.uk/technical-guides/pressure-equipment/)

Protective device

Means a device that protects control or measuring equipment which is essential to prevent a dangerous situation from arising. Protective devices are often categorised as:

- Category 1: Devices designed to protect the system against system failure eg safety valve;
- Category 2: Devices that warn that a system failure could occur eg pressure gauge.

Examples of protective devices commonly found in use include:

For steam systems

- Safety (or relief) valves;
- Pressure gauges;
- Water level gauges;
- Low water level devices;
- Pressure limiting devices;
- High pressure cut out devices;
- Flame failure protection systems;
- Pressure reducing vales.

For compressed air systems

- Safety (or relief) valves;
- Fusible plugs;
- Bursting discs;
- Pressure gauges;
- Pressure limiting or unloading devices.

For bulk LPG installations

- Safety (or relief) valves;
- Pressure gauges;
- Bursting discs.

Written scheme of examination [P7022]

The HSE Guidance to the Regulations (L122) deals with the scheduling of examinations of several identical vessels on the same site. The overriding principle is that each vessel must be examined within the interval specified by the written scheme for that vessel. Within this limit, a form of staged examination may be used.

Nature of examination

[P7021] The written scheme must specify the nature of the examination. This may vary, depending on the system, from a simple visual examination to a requirement for the use of non-destructive testing methods. Pressure vessels within a system will need to be examined from the inside when their accessibility permits this. The Safety Assessment Federation (SAFed), have a number of publications covering specialist examinations that provide the explanation and guidance that the competent person needs to follow for:

- Guidelines for the examination of shell-to-endplate and furnace-to-endplate welded joints Ref: SBG1 (ISBN 1 901212 05);
- Guidelines for the examination of longitudinal seams of shell boilers Ref: SBG2 (ISBN 1 901212 30 0).

Some pressure systems will require an examination both with the system stripped down and out of service and with the system running under normal operating conditions so that the systems integrity and protective devices can be verified as operational to protect the equipment under fault conditions, eg Pressure relief valve lifts under an overpressure condition.

Other guidance documents produced by the Safety Assessment Federation (SAFed) that cover these areas of the thorough examination include:

- Pressure Systems: Guidelines on Periodicity of Examinations Ref: PSG1 (ISBN 1 901212 10 6);
- Guidelines for the Operation of Steam Boilers Ref: PSG2.

Examination

[P7022] The Regulations require users and owners to ensure that their pressure systems are examined by a competent person as specified by the written scheme of examination. [*Pressure Systems Safety Regulations 2000 (SI 2000 No 128), Reg 9.*] The examination must take place within the maximum time interval specified by the scheme. There is a clear duty on users/owners to ensure that the equipment is not operated/supplied beyond the date specified in the last examination report until it has been examined again.

Users and owners must ensure that systems are adequately prepared in time for examination and that appropriate precautions for safety are taken. Adequate advance notice should be given to the competent person.

Conoco Phillips Limited was fined a total of £895,000 and ordered to pay £218,854 costs at Grimsby Crown Court, after pleading guilty to breaching health and safety legislation at an earlier hearing. The case follows an investigation by the HSE into two incidents: a fire and explosion at the

Humber Refinery, South Killingholme, North Lincolnshire on 16 April 2001 and a release of liquefied petroleum gas (LPG) at the Immingham Pipeline Centre, Immingham Dock, on 27 September 2001.

At an earlier hearing at Grimsby Crown Court, ConocoPhillips pleaded guilty to seven breaches of *HSWA 1974* and the *Pressure Systems and Transportable Gas Containers Regulations 1989*.

The first incident happened on 16 April 2001 when 170 tonnes of highly flammable LPG was released from ConocoPhillips' (then Conoco Ltd) Saturate Gas Plant at its Humberside oil refinery. The gas cloud ignited causing a massive explosion and fire. As the fire burned it caused failures of other pipework resulting in another explosion and fireball.

The fire burned for approximately two-and-a-half hours. There were no serious injuries but considerable damage was caused to other processing plants and buildings on the refinery and to properties off-site. HSE's investigation found that the initiating event was the failure of a 15 cm diameter pipe at an elbow, due to corrosion and erosion. The most likely source of the ignition was a gas-fired heater in an adjacent processing unit.

During HSE's investigation a second incident occurred at the company's nearby Immingham Pipeline Centre. Over the night of 27 September 2001, approximately 16 tonnes of LPG leaked from a road tanker and the liquid pool and gas cloud dispersed without ignition. As a result, HSE launched a detailed investigation into the cause of the release.

The company pleaded guilty to seven breaches, the first six applying to its Humberside plant:

- Section 2(1) of HSWA 1974 in not ensuring the safety of its employees, for which it was fined £400,000;
- Section 3(1) of the same Act applying to non- employees. It was fined a further £400,000 on this charge;
- Four breaches of reg 9(1)(a) of the Pressure Systems and Transportable Gas Containers Regulations 1989 for not ensuring parts of the pressure system were examined by a competent person and for which it was fined £5000 on each charge; and
- Section 2(1) of HSWA 1974 in not ensuring the safety of employees at the Immingham Pipeline Centre, for which it was fined £75,000.

After the hearing, Kevin Allars, head of HSE's Chemical Industries Division, said:

> The incident at the Humber refinery was possibly the most serious chemical incident in Britain since the Flixborough disaster in 1974 and it is fortunate that there were no deaths or very serious injuries. This was mainly because the incident occurred on a Bank Holiday and during a shift change when the limited staff on site were away from the plant. The potential for loss of life was great. However, the extent of the damage to the site and to properties in the nearby village of South Killingholme indicates the violent nature of the explosion. The severity of the events at the Humber Refinery have been reflected in the penalties imposed by the court today.
>
> Allars added: 'Our investigation revealed a systematic failure by the company to inspect the pipework in certain parts of the refinery. It is vital that companies who

operate high-hazard sites – such as oil refineries and chemical plants – put rigid and robust systems in place for inspecting pipework to detect corrosion or other defects.'

There is provision to postpone the due date of examination once, by agreement in writing between the user or owner and the competent person. The date of examination may be postponed to a later date provided that the following conditions are satisfied:

(a) The postponement does not give rise to danger.
(b) Only one postponement is made for any one examination.
(c) Such a postponement is notified in writing by the user or owner to the enforcing authority before the date set for the next examination in the previous report.

Where the user or owner is the competent person, such an agreement is unnecessary, but the notification to the enforcing authority must contain a declaration that postponement will not give rise to danger. Any postponement must be agreed by the Competent Person who undertook the previous examination. No periods of deferment are quoted within the Regulations, ACoP and Guidance (L122) but when justified the Competent Persons are unlikely to agree more than 25% of the frequency period.

The competent person must take responsibility for all examinations. Often the competent person will carry out the examination, but a user or owner and the competent person may agree that some examinations are carried out by specialist non-destructive testing technicians who are not classed as competent persons within the meaning of the Regulations. The competent person will accept responsibility for ensuring that the examinations are carried out properly according to the scheme and for the results, interpretation and reporting.

Fitness for service

[P7023] After obtaining the results of the examination, the competent person will make an assessment of the continued fitness for service of the system. If the competent person is not satisfied about continued fitness for service under the current condition or safe operating limits of the system, then s/he will specify in the report any repairs, modifications, or changes to the safe operating limits deemed necessary.

The competent person will then, having regard to the assessment of continued fitness for service, specify the next date, within the limits set by the written scheme, beyond which the system may not be operated without a further examination. The competent person will also consider whether the written scheme remains suitable or whether it should be modified.

Examination report

[P7024] The Regulations require the competent person to make a written report of the examination and to sign or otherwise authenticate it. [*Pressure Systems Safety Regulations 2000 (SI 2000 No 128), Reg 9.*] The competent person must send the report to the user or owner as soon as is practicable after completing the examination (or for integrated systems, the last examination in the series). In any event, the report must arrive within 28 days of the examination (or the last in the series). The report must also arrive before the date specified by the previous examination report. Where the user or owner is also the competent person, the report must also be made by these times.

The examination report must contain the following information:

(a) A statement of the parts examined, the condition of those parts and the results of the examination.

(b) Specification of any repairs, modifications or changes to the operating limits necessary to prevent danger or ensure the continued effective working of protective devices. The date by which any repairs, modifications or changes are to completed must also be specified. The user or owner must ensure that the system is not operated after the date specified for completion of the repairs, modifications or changes until these have been made.

(c) The next date (within the limits of the scheme of examination) after which the pressure system may not be operated without a further examination under the scheme. The owner of a mobile system must ensure that the date of the next examination is legibly and durably marked on the system and is clearly visible.

(d) A statement of whether the current written scheme of examination is suitable for preventing danger or whether it should be modified (with reasons).

Imminent danger

[P7025] After the examination of a pressure system and assessment of fitness for service, the competent person must consider whether the condition of the system, or any part thereof, is such that continued operation in its current condition would give rise to imminent danger. If this is the case, then the competent person will immediately make a written report. [*Pressure Systems Safety Regulations 2000 (SI 2000 No 128), Reg 10.*] The report will specify the system and the parts in dangerous condition, and the repairs, modifications or changes to the operating conditions needed in order to make the system safe from imminent danger. A pressure system in a state of imminent danger usually points to wider failings in general management of the user's/owner's organisation.

The competent person will give the report to the user of an installed system or the owner and user of a mobile system. In addition, within 14 days of the examination, the competent person will send a report to the enforcing authority for the place where the pressure system is situated. This will normally be the Health and Safety Executive or the Environmental Health Department of the local authority.

On receipt of this report, the user of an installed system must ensure that the system (or the part effected) is not operated until the repairs, modifications or changes have been carried out. Owners of mobile systems must take all practicable steps to ensure that the system is not operated. Where the competent person is the user or owner, s/he must take responsibility to prepare

Operations, maintenance, modification and repair

the report, inform the enforcing authority and ensure that the system is not operated until the repairs, modifications or changes have been carried out.

Safe operation

[P7026] Users and owners have a duty to provide adequate and suitable operating instructions for the safe operation of the system and the action to be taken in an emergency. Users must ensure that their pressure systems are not operated except in accordance with the instructions.

Users of pressure systems must provide adequate training to persons who operate the systems as required by the *Provision and Use of Work Equipment Regulations 1998 (SI 1998 No 2306), Reg 9.*

Precautions to prevent overpressure

[P7027] Users have a duty to take precautions to prevent overpressure to vessels. [*Pressure Systems Safety Regulations 2000 (SI 2000 No 128, Reg 15.*] The Regulation applies to vessels constructed with a permanent outlet to the atmosphere (or to a space where the pressure does not exceed atmospheric pressure) which could become pressurised if the outlet were obstructed. In this case, the user must ensure that the outlet is kept open at all times and is free from obstruction when the vessel is in use.

Maintenance

[P7028] Users and owners must ensure that their systems are properly maintained so as to prevent danger. [*Pressure Systems Safety Regulations 2000 (SI 2000 No 128), Reg 12.*] This should not be confused with the requirement for examinations under the written scheme. Maintenance might include the periodic replacement of parts, flange tightening, and checks on vulnerable areas of pipework systems not included for examination under the written scheme. Advice on maintenance programmes is given in the ACoP.

Systems and equipment covered by the regulations

The HSE has warned of the need for effective operation, inspection and maintenance regimes to ensure the safe use of high-pressure equipment.

Example 1

The warning comes after Giancarlo Coletti, 48, a worker at Clariant Life Science Molecules (UK) Limited's plant in Sandycroft, Flintshire, sustained serious injuries to his right arm in October 2003. The accident happened when the clamping system on a pressure vessel lid he was operating failed, causing the lid to fly off and hit him.

Clariant Life Science Molecules (UK) Limited was fined £100,000, split equally between *Regulations 11* and *12* of the *Pressure Systems Safety Regulations*

2000 (SI 2000 No 128), with £24,474 costs, at Mold Crown Court on 17 December 2004 after pleading guilty at an earlier hearing. The company admitted failing to provide workers operating the pressure vessel with adequate and suitable instructions for its safe operation and also failing to ensure the vessel was adequately maintained.

HSE Inspector, Dr Stuart Robinson, who investigated the incident, said:

> Our investigation revealed serious deficiencies in Clariant's safe systems of work. Opening and closing of the lid was regarded as a simple process, carried out three times a day, but the hazards had been overlooked. Although the clamping system was designed with a substantial margin of safety, it had been allowed to deteriorate to such an extent that the risk of injury became unacceptably high.

In particular, the HSE's investigation found that Clariant had failed to put in place adequate operating procedures to ensure the system was used correctly. For example, clamps were regularly over-tightened, occasionally causing them to break and the system was allowed to operate with less than its full complement of eight clamps.

In addition, at the time of the accident, one of the clamps was missing and others showed excessive wear and tear, or inadequate repair. Furthermore, it had become common practice for leaks to be nipped up with the system under pressure because the operating procedures failed to state that the system should be depressurised first.

Robinson added:

> The investigation also revealed that although Clariant had arranged for an independent competent person to examine the pressure system periodically, this was insufficient due to the frequent operation of the clamps, and the high level of wear and tear they showed. Instead, Instead, the firm should have introduced a more frequent system of inspection and maintenance.

Example 2

On 12 June 2007, a 44-year-old Loughborough man employed at Authentic World Cuisine Limited, which produced a range of sachet Asian meals for supermarkets, sustained multiple injuries when the door of an autoclave (known as a retort, or pressure cooker) exploded under pressure midway through a cooking cycle.

The detached circular steel door struck him and sent him eight metres across the factory causing multiple fractures. He also sustained burns from the hot contents of the vessel. As a result of the incident, his leg was amputated, he spent more than five months in hospital and is now looked after at home whilst he rehabilitates.

The company which manufactured at Castle Business Park, Pavilion Way, was fined £4,000, and ordered to pay this within 28 days*, at Loughborough Magistrates Court today, after pleading guilty to breaching Regulation 12 of the Pressure Systems Safety Regulations 2000 for failing to ensure that the company's pressure vessel (autoclave) was maintained in a state of good repair.

Roger Amery of the HSE who deals with Leicestershire said:

Related health and safety legislation **[P7031]**

"Major injuries in the food industry arising from the sudden or explosive release of pressure are, I am pleased to say, infrequent events. But this incident shows just how serious the consequences can be when things do go wrong and just how much energy is contained within industrial pressure systems. There might easily have been one or several fatalities in this explosion. The injuries that did result were very severe with long term consequences for the man who had the misfortune of being in the path of this door when the system exploded."

"I hope this will serve as a reminder to employers and to managers who have specific responsibility for plant and equipment, that as well as ensuring periodic statutory examinations are made, it is equally important to ensure that all maintenance on pressure vessels is undertaken competently and that the protection systems are kept in good working order at all times."

Modification and repair

[P7029] Employers of persons who modify or repair pressure systems have to duty to ensure that the manner in which modifications or repairs are carried out does not give rise to danger, or otherwise impair the operation of any protective devices or inspection facility. [*Pressure Systems Safety Regulations 2000 (SI 2000 No 128), Reg 13).*]

Record keeping

[P7030] The user of an installed system or the owner of a mobile system must keep the originals or copies of the following documents:

(a) The report of the last examination made by the competent person.
(b) Reports of previous examinations if they contain information that can assist in assessing whether the system is safe to operate or whether any repairs can be carried out safely.
(c) Any documents where reports of the last or previous examinations are recorded as data.
(d) Any agreement, declaration or notification made in relation to the postponement of the date of the next examination.

[*Pressure Systems Safety Regulations 2000 (SI 2000 No 128), Reg 14.*]

For an installed system, these documents must be kept at the premises where the system is installed or at other premises approved by the enforcing authority (normally the Health and Safety Executive). For a mobile system, the documents must be kept at the premises from which the deployment of the system is controlled.

Where the user or owner of a pressure system changes, the previous user or owner must give to the new user or owner all originals or copies of documents relating to the system as soon as practicable.

Related health and safety legislation

[P7031] The Regulations are concerned with the hazard of stored energy. The control of hazards associated with the release of hazardous substances, for example toxic or flammable materials, from containers and systems falls under the *Control of Major Accident Hazards Regulations 2015 (COMAH) (SI 2015 No 483)*. Controls to prevent the release of hazardous materials cover the management of storage systems, containers and tanks.

Other legislation relating to pressure equipment includes:

- *The Carriage of Dangerous Goods and Use of Transportable Pressure Equipment Regulations 2009 (SI 2009 No 1348)*;
- *Pipelines Safety Regulations 1996 (SI 1996 No 825)*.

The main legislation covering health and safety at work in general is:

- *Health and Safety at Work etc Act 1974*;
- *Workplace (Health and Safety) Regulations 1992 (SI 1992 No 3004)*;
- *Management of Health and Safety at Work Regulations 1999 (SI 1999 No 3242)*;
- *Provision and Use of Work Equipment Regulations 1998 (SI 1998 No 2306)*.

Product Safety

Alison Newstead

Introduction to product safety

[P9001] With the expansion of EU directives and other legislation aimed at manufacturers, designers, importers and suppliers, product safety has emerged as a fast growth area over recent years. Indeed, there has been an identifiable trend towards placing responsibilities on producers and those involved in commercial circulation as well as on employers and occupiers (see MACHINERY SAFETY and PERSONAL PROTECTIVE EQUIPMENT).

Although there is no question of removal of duties from employers and occupiers and users of industrial plant, machinery and products, there is a realisation that the *sine qua non* of compliance on the part of employers and users with their safety duties is compliance with essential health and safety requirements on the part of designers and manufacturers. More products have been brought within the scope of the EU's 'new approach' regime which promotes supply of products throughout the single market provided they meet essential requirements which can be demonstrated by compliance with harmonised European standards.

More recently, the European Commission's Product Safety and Market Surveillance Package (adopted in February 2013) sets out increased obligations for manufacturers, importers and national authorities to improve the safety of products on the EU market and strengthens market surveillance activities. The proposed Regulation on Consumer Product Safety will replace Directive 2001/95/EC (the General Product Safety Directive) and Directive 87/357/EEC on dangerous imitations. The proposed Regulation on Market Surveillance of Products will unify current fragmented EU legislation, in particular bringing together enforcement rules for all non-food consumer products into one Regulation. The proposals are currently being considered by the European Parliament and are expected to come into force in 2015.

More information about the Package of legislative and non-legislative measures can be found on the European Commission website at: ec.europa.eu/index_en.htm.

Background

[P9001A] It was thalidomide (prescribed for easing morning sickness in pregnancy in the 1950s and 1960s with devastating consequences) that first focused serious attention on the legal control of product safety, leading to the *Medicinal Products Directive (65/65/EEC)* and the *Medicines Act 1968*, the

[P9001A] Product Safety

first comprehensive regulatory system of product testing, licensing and vigilance. Since then a host of regulations has appeared, covering areas as diverse as pencils, aerosols, cosmetics and electrical products. The General Product Safety Directive (92/59/EEC), introduced the general concept that producers of consumer goods must place only safe products on the market and undertake appropriate post-marketing surveillance. The significance of the introduction of these Regulations lies not in their substantive requirements, but rather in the completion of an all-embracing consumer product safety regime across the European Union (EU) covering any regulatory gaps in the consumer protection network.

The General Product Safety Directive was amended and re-issued as Directive 2001/95/EC and implemented in the UK by way of the *General Product Safety Regulations 2005 (SI 2005 No 1803)*. The 2005 Regulations came into force in the UK on 1 October 2005 and incorporated additional post-marketing obligations and enforcement powers in respect of consumer products, even those already covered by sector-specific regulatory requirements.

In contrast, the regulation of products for use in the workplace has developed along separate lines (although there is an increasing overlap by virtue of the so-called 'new approach' directives, such as those for machinery or electro-magnetic compatibility, which apply uniform requirements for consumer and workplace products albeit with separate enforcement regimes).

Under the *Health and Safety at Work etc Act 1974 (HSWA 1974)*, s 6 all articles and substances for use at work are required to be as safe as is reasonably practicable, and manufacturers and other suppliers are under obligation to provide information and warnings for their safe use. Various product-specific regulations have also been made under *HSWA 1974*, for example the *Chemicals (Hazard Information and Packaging for Supply) Regulations 2009 (SI 2009 No 716)*.

This section deals with criminal and civil liabilities for unsafe products for consumer and industrial use as well as the duties imposed on sellers and suppliers of consumer products and after-sales personnel as far as contract law is concerned.

Consumer products

General Product Safety Regulations 2005

Duties of producers and distributors

[P9002] The *General Product Safety Regulations 2005 (SI 2005 No 1803)* lay down general requirements concerning the safety of products intended for or likely to be used by consumers. The General Product Safety Regulations 2005 implement Directive 2001/95/EC concerning consumer products which are not covered by any other sector specific Directive.

The Regulations significantly increased the obligations on producers and distributors which were formerly in place under the General Product Safety Regulations 1994.

The 2005 Regulations define a consumer product as:

> . . . a product which is intended for consumers or likely, under reasonably foreseeable conditions, to be used by consumers even if not intended for them, and which is supplied or made available, whether for consideration or not, in the course of a commercial activity, and whether new, used or reconditioned and includes a product which is supplied or made available to consumers for their own use in the context of providing a service [. . .].

The scope of the 2005 Regulations extended the previous definition of 'product' found under the 1994 Regulations in that:

(1) It encompasses products which are made available to consumers, as well as supplied to them.
(2) It includes products supplied or made available in the context of providing a service.

Products used solely in the workplace are not covered by the 2005 Regulations, but the fact that a product has a wider commercial application will not prevent it being a consumer product for the purposes of these Regulations, if it is supplied to consumers or it is likely, under reasonably foreseeable conditions, to be used by consumers, even if not intended for them.

The Directive makes it clear that, although business products are not included *per se* it does include products which are designed exclusively for professional use but have subsequently migrated to the consumer market, such as laser pens. The 2005 Regulations also cover products which were originally intended for professional use, but which subsequently 'migrate' onto the consumer market (eg power tools).

'Migration' does not necessarily mean that a product is unsafe for a consumer, but if it is foreseen that such migration may occur, instructions for consumer use and warnings of non-obvious risks must be provided. In addition, in circumstances in which a professional product could never be safely used by a consumer, supply should be strictly controlled and labelling a product 'strictly for professional use only' is unlikely to be sufficient on its own to render the product safe.

The Regulations apply to all products (new and second-hand) used by consumers whether intended for use by consumers or not.

The Regulations do not apply to:

— second-hand products which are antiques;
— second-hand products supplied for repair or reconditioning before use, provided that the purchaser is clearly informed accordingly;
— products exported direct by the UK manufacturer to a country outside the European Union; or
— equipment used by service providers themselves to supply a service to consumers, in particular equipment on which consumers ride or travel which is operated by the service provider.

The Regulations do not apply in relation to any product where there are specific provisions in European law governing *all aspects* of the safety of a product. Where there are product-specific provisions in European law which do not cover all aspects, the 2005 Regulations apply to the extent that specific provision is not made by European law.

[*General Product Safety Regulations 2005 (SI 2005 No 1803), Reg 3(1).*]

The main requirements of the Regulations are that:

(a) no producer may place a product on the market unless it is safe [*General Product Safety Regulations 2005 (SI 2005 No 1803), Reg 5(1)*];

(b) producers provide consumers with relevant information, so as to enable them (ie consumers) to assess risks inherent in products throughout the normal or foreseeable period of use, where such risks are not immediately obvious without adequate warnings, and to take precautions against those risks; [*General Product Safety Regulations 2005 (SI 2005 No 1803), Reg 7(1)(a) and (b)*];

(c) producers update themselves regarding risks presented by their products and take appropriate action (if necessary recall and withdrawal), for example by:
 (i) identifying products/batches of products by marking;
 (ii) sample testing;
 (iii) investigating and, if necessary, keeping a register of complaints concerning the safety of a product;
 (iv) keeping distributors informed of the results of such monitoring where a product presents a risk or may present a risk;
 [*General Product Safety Regulations 2005 (SI 2005 No 1803), Reg 7(3), (4)*;]

(d) in order to enable producers to comply with their duties, distributors must act with due care and in particular:
 (i) must not supply products to any person which are known or presumed (on the basis of information in their possession and as professionals) to be dangerous products, and
 (ii) must participate in monitoring safety of products placed on the market – in particular, by passing on information on product risks, keeping (and if necessary, producing) the documentation necessary for tracing the origin of the product and co-operating in action taken by a producer or an enforcement authority to avoid risks.

[*General Product Safety Regulations 2005 (SI 2005 No 1803), Reg 8(1)(a), (b).*]

The new proposed European Regulation on Consumer Product Safety – if implemented as currently drafted – will introduce some important changes for manufacturers and distributors. There is a focus on measures to ensure traceability of products, new obligations to label country of origin on products (or on packaging or documentation accompanying the product), a new obligation on manufacturers to prepare and retain technical documentation, including a documented risk assessment, expanded obligations to retain documentation identifying parties throughout the supply chain, explicit obligations to label products with type, batch number and serial number, enhanced obligations to label products with contact details of the manufacturer and importer and new explicit obligations on importers to ensure that the manufacturer has complied with its documentation and labelling requirements.

The 2001 Directive introduced a significant additional obligation on producers and distributors to notify the relevant competent authorities immediately if they know or ought to know, on the basis of information in their possession and as professionals, that a product they have placed on the market poses risks to the consumer that are incompatible with the general safety requirement, ie that it is not safe, and of any action taken to prevent such risks [General Product Safety Directive 2001/95/EC, Article 5.3]. It is not necessary for there to have been an incident involving personal injury or property damage. Equally, 'isolated circumstances or products' do not need to be notified [*General Product Safety Regulations 2005 (SI 2005 No 1803) Reg 9(2)(b)*]. The new proposed Regulation on Consumer Product Safety sets out a further exemption if '*the manufacturer, importer or distributor can demonstrate that the risk can be fully controlled and cannot any more endanger the health and safety of persons*'.

The UK government has published guidance on when notification is appropriate (Notification Guidance for Producers and Distributors (DTI, September 2005)). This refers to the European Commission's methodological framework for assessing risk contained in its published Guidelines for the Notification of Dangerous Consumer Products (2004) for the purposes of the GPSD. However, these risk-assessment guidelines have been superseded by Decision 2010/15/EU, which sets out revised risk-assessment guidelines. The aim of the new guidelines is to provide a practical and transparent risk-assessment method for use by Member States' competent authorities when they assess risk in non-food products. The risk-assessment methodology looks at the product itself, the product hazard, the abilities and behaviour of the consumer (in particular vulnerable consumers), injury scenarios, the severity and probability of injury and the determination of risk. The number of products supplied or users potentially affected is not a relevant consideration for notification, although it may be taken into account in determining what action to take to address the risk.

The Regulations require producers and distributors to put in place sophisticated procedures to capture and assess safety information. Deciding whether new information means that a product is no longer legally safe (and therefore notification is necessary) can be a major issue in practice and is best undertaken against a background of previous safety data.

A risk assessment will need to be carried out to determine whether notification is necessary. Independent expert advice from engineers and lawyers may be necessary prior to and during the risk-assessment process. For certain products, at least, it is necessary to undertake a documented risk assessment before a product is placed on the market, and subsequently to keep it up to date (eg with monitoring and sample testing), so as to define the level of anticipated risks with a product's use, to determine that these are acceptable and that the general safety requirement is met, and to provide a background and baseline statement of acceptable risks against which any increase in risk may be assessed.

The new proposed Consumer Product Safety Regulation, if implemented in its current draft form, sets out new obligations on manufacturers to prepare and retain technical documentation, including a risk assessment. Technical documentation which is used to put together the risk assessment must also be retained for ten years and presented to the market surveillance authorities on request.

The obligation to notify applies to both the producer and the distributor, although discussion between the producer and the distributor is encouraged when a potential safety issue arises to avoid duplication of notifications.

The 2005 Regulations stipulate that written notification to the relevant authority should be made 'forthwith' [*General Product Safety Regulations 2005 (SI 2005 No 1803) Reg 9(1)*]. In contrast, the General Product Safety Directive stipulates that notification should be *'immediately' (Article 5(3))*. Guidance from the European Commission states notification must be *'as soon as the information on the dangerous product has become available and in the case of a serious risk, within 3 days and in other cases within 10 days in any event'*.

Notification in the UK should be made to the Trading Standards Department of the local authority for the area where the decision-making function of the business is located. Notification should not be delayed on the basis that the form is incomplete. Incomplete notifications can be supplemented with information as and when it is received. The reporting form for general consumer products is available at www.bis.gov.uk.

Notification is not required in respect of isolated circumstances or products [*General Product Safety Regulations 2005 (SI 2005 No 1803) Reg 9(2)(b)*].

Producers and distributors are required to co-operate with the competent authorities, on request, on action taken to avoid risks posed by products that they supply or have supplied [*General Product Safety Regulations 2005 (SI 2005 No 1803) Reg 9(4)*].

In addition, producers (as part of their duty to update themselves regarding risks and to take appropriate action) are obliged to take appropriate action when necessary including withdrawal, adequately and effectively warning customers as to the risks, or as a last resort recall. (The recall obligation includes products already supplied or made available to the consumer/user.)

[*General Product Safety Regulations 2005 (SI 2005 No 1803) Reg 7(1)(b).*]

Recalls are only required as a last resort where other measures would not suffice to prevent the risks involved, where the producer considers it necessary or where they are obliged to do so by a Member State. Practical issues which may arise include: whether adequate systems are in place to trace the products when a safety issue arise, whether a system is in place to record the number and percentage of products returned in a recall and what level of returns would be acceptable. In practice, a recall should never 'close' (unless all products are identified and recalled, which rarely happens).

Complementary to the obligation to recall products from consumers is a power for enforcement authorities to serve suspension notices on producers (in order to prevent the supply of products whilst safety evaluations are being carried out), to require producers to mark products which could pose risks in certain conditions, to warn consumers of risks that products may post and to order or organise the immediate withdrawal of any dangerous product already on the market and alert consumers to the risks it presents and to order or co-ordinate or, if appropriate, to organise with producers and distributors its recall from consumers and its destruction in suitable conditions [*General Product Safety Regulations 2005 (SI 2005 No 1803) Regs 11–15*].

The Directive also includes a provision to prohibit the export from the European Community any products which have been the subject of a European Commission-initiative decision (after consulting the Member States and, whenever it proves necessary also a Community scientific committee) that requires Member States to take measures in relation to that particular product on the grounds that it presents a serious risk (unless the decision specifically provides that the products may still be exported outside the Community).

Presumption of conformity with safety requirements

[P9003] A 'safe' product is a product which, under normal or reasonably foreseeable conditions of use (as well as duration), does not present any risk (or only minimal risks) compatible with the product's use, considered as acceptable and consistent with a high level of protection for the safety and health of consumers, with reference to:

(a) product characteristics (eg composition, packaging, instructions for assembly and, where applicable, installation and maintenance instructions);

(b) the effect on other products, where use with other products is reasonably foreseeable;

(c) product presentation (ie labelling, any warnings and instructions for use and disposal and any other indication or information concerning the product); and

(d) categories of consumers at serious risk – in particular, children and the elderly.

[*General Product Safety Regulations 2005 (SI 2005 No 1803), Reg 2.*]

Where a product conforms with specific UK safety requirements (eg the *Plugs and Sockets etc (Safety) Regulations 1994 (SI 1994 No 1768)*) there is a presumption that the product is safe, until the contrary is proved. In addition, where a product conforms to a voluntary national standard in the UK, giving effect to a European Standard (which has been published in the Official Journal of the European Union in accordance with the 2001 Directive) the product is presumed to be safe so far as concerns the risks and categories of risk covered by that national standard [*General Product Safety Regulations 2005 (SI 2005 No 1803) Reg 6(1), (2)*].

In the absence of these two particular circumstances where a presumption of safety exists, the assessment of the safety of a product is to be made by taking into account:

(i) voluntary national standards implementing European standards other than those listed in the Official Journal with reference to the General Product Safety Directive;

(ii) standards drawn up in the UK in which the product is marketed;
(iii) European Commission recommendations setting guidelines on product safety assessment;
(iv) product safety codes of good practice in the relevant sector;
(v) state of the art and technology; and
(vi) the safety that consumers may reasonably expect.

[General Product Safety Regulations 2005 (SI 2005 No 1803), Reg 6.]

Market surveillance and enforcement

[P9004] The 2001 Directive increased the strength of market surveillance and enforcement provisions introducing additional obligations on Member States. The European Commission also has an obligation to promote and take part in the operation of a European network of authorities of the Member States.

The obligations and enforcement powers of Member States were expanded to include the following elements:

(a) a detailed definition of the tasks, powers, organisation and cooperation arrangements for market surveillance of competent authorities;
(b) establishment of sectoral surveillance programmes by categories of products or risks and the monitoring of surveillance activities, findings and results;
(c) the follow-up and updating of scientific and technical knowledge concerning the safety of products;
(d) periodic review and assessment of the functioning of the control activities and, if necessary, revision of the approach and organisation of surveillance;
(e) procedures to receive and follow-up complaints from consumers and others on product safety, surveillance or control activities;
(f) exchange of information between Member States on risk assessment, dangerous products, test methods and test results, recent scientific developments and other aspects relevant for control activities;
(g) joint surveillance and testing projects between surveillance authorities;
(h) exchange of expertise and best practices, and cooperation in training activities; and
(i) improved collaboration at Community level on tracing, withdrawal and recall of dangerous products.

[General Product Safety Directive 2001/95/EC, Articles 6–10]

An extended list of enforcement powers is included in the 2001 Directive and includes the power for a Member State to order or organise the issuance of warnings about, or recall of, dangerous products.

[General Product Safety Directive 2001/95/EC, Article 8.1.]

The new proposed EU Regulation on Market Surveillance of Products, will transfer the current Market Surveillance rules from Directive 2001/95/EC into a new Regulation. The new Regulation sets out the process of market surveillance in chronological order with the aim of making it more user-friendly and effective. Enforcement is also more rigorous, with the objective of protecting consumers from non-compliant products.

The new proposed Market Surveillance provisions are far-reaching. The activities and obligations of the market surveillance authorities will apply irrespective of the end-user of the product ie they are applicable to consumer and industrial products. This is in contrast to the new proposed Regulation on Consumer Product Safety which is concerned solely with consumer products. There are four key themes set out in the proposed Market Surveillance Regulation:

- Increased and new powers given to market surveillance authorities.
- New obligations on market surveillance authorities, such as the need to report their activities.
- Market surveillance authorities are required to carry out more active surveillance across the EU via coordinated and targeted activities, with positive obligations to carry out 'appropriate checks' with 'adequate frequency' and on an 'adequate scale'.
- Border controls will be strengthened.

Other significant changes introduced by the 2001 Directive

[P9005] The 2001 Directive also introduced the following significant changes:

(1) Member States are required to take due account of the precautionary principle when they are taking enforcement measures. The precautionary principle is that Member States should use a precautionary approach to risks where there is scientific uncertainty as to the level of risk

[General Product Safety Directive 2001/95/EC, Article 8.2].

(2) Information which is available to the authorities relating to consumer health and safety, in particular on product identification, the nature of any risk and on measures taken, shall in general be available to the public, in accordance with the requirements of transparency. However, information which is obtained by the authorities shall not be disclosed if the information, by its nature, is covered by professional secrecy in duly justified cases.

[General Product Safety Directive 2001/95/EC, Article 16.1].

Offences, penalties and defences

[P9006] Contravention of Regulations regarding provision of information to consumers, risk assessment, monitoring and notification is an offence, carrying, on summary conviction, a maximum penalty of:

(a) imprisonment for up to three months, or
(b) a fine; or
(c) both.

Offering or agreeing to supply or supplying an unsafe product is an offence and liable to conviction, on indictment to:

(a) imprisonment for a term not exceeding 12 months, or
(b) a fine; or
(c) both; or

[P9006] Product Safety

[General Product Safety Regulations 2005 (SI 2005 No 1803), Reg 20.]

There is a proposal in the new Regulation on Consumer Product Safety that penalties should have regard to the size of business and whether the economic operator has committed a previous similar infringement.

Offence by another person

[P9007] As with breaches of *HSWA* and similarly-oriented legislation, where an offence is committed by 'another person' in the course of his commercial activity, that person can be charged, whether or not the principal offender is prosecuted. Similarly, where commission of an offence is consented to or connived at, or attributable to neglect, on the part of a director, manager or secretary, such persons can be charged in addition to or in lieu of the body corporate. *[General Product Safety Regulations 2005 (SI 2005 No 1803), Reg 31.]*

Defence of 'due diligence' – and exceptions

[P9008] It is a defence for a person charged under the Regulations to show that he took all reasonable steps and exercised all due diligence to avoid committing the offence. *[General Product Safety Regulations 2005 (SI 2005 No 1803), Reg 29(1).]* (See further ENFORCEMENT.)

The exceptions to the above are:

(a) Where, allegedly, commission of the offence was due to:
 (i) the act or default of another, or
 (ii) reliance on information given by another,
 a person so charged cannot, without leave of the court, rely on the defence of 'due diligence', unless he has served a notice, within, at least, seven days before the hearing, identifying the person responsible for the commission or default or who gave the information. *[General Product Safety Regulations 2005 (SI 2005 No 1803), Reg 29(2), (3).]*
(b) A person so charged cannot rely on the defence of 'information supplied by another', unless he shows that it was reasonable in all the circumstances for him to have relied on the information. In particular, whether he took reasonable steps to verify the information or whether he had any reason to disbelieve the information. *[General Product Safety Regulations 2005 (SI 2005 No 1803), Reg 29(4).]*

A mere recommendation on the part of an importer that labels should be attached to boxes by retailers does not constitute 'due diligence' for the purposes of *Reg 29(1)*. (In *Coventry County Council v Ackerman Group plc* [1995] Crim LR 140, an egg boiler imported by the defendant failed to contain instructions that eggs should be broken into the container and yolks pricked before being microwaved. The defendant had learned of the problem and had printed instructions which were sent to all retailers, recommending that they be fixed to the boxes.)

Compliance with recognised standards (such as British Standards) will not amount to the defence of due diligence if a product is nevertheless unsafe for the user. In *Whirlpool (UK) Ltd and Magnet Ltd v Gloucestershire County Council* (1993) 159 JP 123 cooker hoods which were intrinsically safe and which met applicable standards for the purposes of the *Low Voltage Electrical Equipment (Safety) Regulations 1989* failed to meet the general safety requirement contained in the consumer protection legislation that applied at the time because they were liable to result in fires when used in conjunction with certain gas hobs.

Industrial products

[P9009] The *HSWA 1974* was the first Act to place a general duty on designers and manufacturers of industrial products to design and produce articles and substances that are safe and without health risks when used at work. Prior to this date legislation, such as the *Factories Act 1961*, had tended to avoid this approach on the premise that machinery could not be made design safe (see **MACHINERY SAFETY**). Statutory requirements had tended to concentrate on the duty to guard and fence machinery, and with the placement of a duty upon the user or employer to inspect and test inward products for safety. There is a general residual duty on the employers/users of industrial products to inspect and test them for safety under *HSWA 1974, s 2*.

These duties notwithstanding, the trend of more recent legislation has been towards safer design and manufacture of products for industrial and domestic use. Thus, *HSWA 1974, s 6* as updated by the *Consumer Protection Act 1987, Sch 3* imposes general duties on designers, manufacturers, importers and suppliers of products ('articles and substances') for use at work, to their immediate users. Contravention of s 6 carries with it (on summary conviction) imprisonment for a term not exceeding 12 months, or a fine not exceeding £20,000, or both (see Enforcement).

In addition, a separate duty is laid on installers of industrial plant and machinery. Because of their involvement in the key areas of design and manufacture, more onerous duties are placed upon designers and manufacturers of articles and substances for use at work than upon importers and suppliers, who are essentially concerned with distribution and retail of industrial products. However, under *HSWA 1974, s 6(8A)* importers are made liable for the first time for the faults of foreign designers/manufacturers.

A 2003 case concerned the interaction between *HSWA 1974*, European directives concerning the placing on the market and putting into service of certain product types, and the UK legislation which implements the European directives. The House of Lords case of *R (on the application of Junttan Oy) v Bristol Magistrates' Court* [2003] UKHL 55, [2004] 2 All ER 555, concerned a judicial review of a decision of the Magistrates' Court involving a prosecution brought by the HSE based on offences under *HSWA 1974, ss 3 and 6* regarding machinery. It was questioned whether the HSE should have brought a prosecution based on *HSWA 1974, s 6* where the equipment's safety requirements were covered by the *Supply of Machines (Safety) Regulations 1992, Reg 29(a)* (implementing the Machinery Directive 98/37/EC). The House of Lords ruled that in this particular instance there was nothing in the Regulations to prevent the HSE bringing a prosecution under *HSWA 1974, s 6* instead, and nothing in the Directive that affected the pre-existing rights of a

Member State to take action against machinery that it believed to be unsafe. It should be noted that fines are higher under the *HSWA 1974* than that relevant Regulations, and prosecutions under the Regulations are not subject to time limits.

Regulations made under the Health and Safety at Work etc Act 1974

[P9010] Various safety regulations are made under the umbrella of the *HSWA 1974* and these often implement European health and safety directives. The *Provision and Use of Work Equipment Regulations 1998 (SI 1998 No 2306)* are such Regulations and implemented European Directive 89/655/EEC.

The 1998 Regulations place obligations on employers and on certain persons having control of work equipment, or who use or supervise or manage the use of work equipment or the way in which the equipment is used. A key requirement is to ensure that work equipment is maintained in an efficient state, in efficient working order and in good repair. [*Provision and Use of Work Equipment Regulations 1998 (SI 1998 No 2306), Reg 5(1)*.]

It was confirmed in *Stark v The Post Office* [2000] All ER (D) 276, CA that this imposes strict liability on the employer and that, even where a workplace product develops a hidden fault (in this case metal fatigue in a bicycle), the employer would be liable even in situations where through wear and tear the product has become dangerous. At this time, the employer's duty to maintain translated into a duty to ensure that there are no latent defects.

The ability to assert a claim for breach of statutory provision without any degree of fault or negligence (as in *Stark v The Post office*) is called into doubt for causes of action after 1 October 2013 by the *Enterprise and Regulatory Reform Act 2013*. *Section 69* of the *Enterprise and Regulatory Reform Act 2013* amended *Section 47* of the *Health and Safety at Work act 1974*. The effect of *Section 69* is to remove strict liability in health and safety regulations, effectively removing the right of injured people to rely upon breaches of health and safety regulations when pursuing a civil claim. A claimant must now show that the employer was negligent.

Criminal liability for breach of statutory duties

[P9011] This refers to duties laid down in *HSWA 1974* as revised by the *Consumer Protection Act 1987, Sch 3*. The duties exist in relation to articles and substances for use at work and fairground equipment. (For the specific requirements now applicable to machinery for use at work, see MACHINERY SAFETY.)

Definition of articles and substances

[P9012] An article for use at work means:

(a) any plant designed for use or operation (whether exclusively or not) by persons at work; and

(b) any article designed for use as a component in any such plant.

[*HSWA 1974, s 53(1)*.]

A substance for use at work means 'any natural or artificial substance (including micro-organisms), whether in solid or liquid form or in the form of a gas or vapour'. [*HSWA 1974, s 53(1)*.]

An article upon which first trials/demonstrations are carried out is not an article for use at work, but rather an article which *might* be used at work. The purpose of trial/demonstration was to determine whether the article could safely be later used at work (*McKay v Unwin Pyrotechnics Ltd* [1991] Crim LR 547, DC where a dummy mine exploded, causing the operator injury, when being tested to see if it would explode when hit by a flail attached to a vehicle. It was held that there was no breach of *HSWA 1974, s 6(1)(a)*).

Duties in respect of articles and substances for use at work

[P9013] *HSWA 1974, s 6* (as amended by the *Consumer Protection Act 1987, Sch 3*) places duties upon manufacturers and designers, as well as importers and suppliers, of (*a*) articles and (*b*) substances for use at work, whether used exclusively at work or not (eg lawnmower, hair dryer).

Articles for use at work

[P9014] Any person who designs, manufactures, imports or supplies any article for use at work (or any article of fairground equipment) must:

(a) ensure, so far as is reasonably practicable (for the meaning of this expression see ENFORCEMENT), that the article is so designed and constructed that it will be safe and without risks to health at all times when it is being (i) set, (ii) used, (iii) cleaned or (iv) maintained by a person at work;

(b) carry out or arrange for the carrying out of such testing and examination as may be necessary for the performance of the above duty;

(c) take such steps as are necessary to secure that persons supplied by that person with the article are provided with adequate information about the use for which the article is designed or has been tested and about any conditions necessary to ensure that it will be safe and without risks to health at all such times of (i) setting, (ii) using, (iii) cleaning, (iv) maintaining *and* when being (v) dismantled, or (vi) disposed of; and

(d) take such steps as are necessary to secure, so far as is reasonably practicable, that persons so supplied are provided with all such revisions of information as are necessary by reason of it becoming known that anything gives rise to a serious risk to health or safety.

[*HSWA 1974, s 6(1)(a)–(d) as amended by the Consumer Protection Act 1987, Sch 3*.]

(See **P9016** and **P9017** below for further duties relevant to articles for use at work.)

In the case of an article for use at work which is likely to cause an employee to be exposed to 80 dB(A) or above, or to peak sound pressure of 135dB(c) or above, adequate information must be provided about noise likely to be generated by that article. [*Control of Noise at Work Regulations 2005 (SI 2005 No 1643), Reg 10.*] (See further NOISE AT WORK).

Substances for use at work

[P9015] Every person who manufactures, imports or supplies any substance must:

(a) ensure, so far as is reasonably practicable, that the substance will be safe and without risks to health at all times when it is being (i) used, (ii) handled, (iii) processed, (iv) stored, or (v) transported by any person at work or in premises where substances are being installed;

(b) carry out or arrange for the carrying out of such testing and examination as may be necessary for the performance of the duty in (*a*);

(c) take such steps as are necessary to secure that persons supplied by that person with the substance are provided with adequate information about:
 (i) any risks to health or safety to which the inherent properties of the substance may give rise;
 (ii) the results of any relevant tests which have been carried out on or in connection with the substance; and
 (iii) any conditions necessary to ensure that the substance will be safe and without risks to health at all times when it is being (*a*) used, (*b*) handled, (*c*) processed, (*d*) stored, (*e*) transported and (*f*) disposed of; and

(d) take such steps as are necessary to secure, so far as is reasonably practicable, that persons so supplied are provided with all such revisions of information as are necessary by reason of it becoming known that anything gives rise to a serious risk to health or safety.

[*HSWA 1974, s 6(4)* as amended by the *Consumer Protection Act 1987, Sch 3.*]

Additional duty on designers and manufacturers to carry out research

[P9016] Any person who undertakes the design or manufacture of any article for use at work (or any article of fairground equipment) must carry out, or arrange for the carrying out, of any necessary research with a view to the discovery and, so far as is reasonably practicable, the elimination or minimisation of any health or safety risks to which the design or article may give rise. [*HSWA 1974, s 6(2).*]

Duties on installers of articles for use at work

[P9017] Any person who erects or installs any article for use at work in any premises where the article is to be used by persons at work (or who erects or installs any article of fairground equipment), must ensure, so far as is reasonably practicable, that nothing about the way in which the article is

Criminal liability for breach of statutory duties **[P9022]**

erected or installed makes it unsafe or a risk to health when it is being (a) set, (b) used, (c) cleaned, or (d) maintained by someone at work. [*HSWA 1974, s 6(3)* as amended by the *Consumer Protection Act 1987, Sch 3.*]

Additional duty on manufacturers of substances to carry out research

[P9018] Any person who manufactures any substance must carry out, or arrange for the carrying out, of any necessary research with a view to the discovery and, so far as is reasonably practicable, the elimination or minimisation of any health/safety risks at all times when the substance is being (a) used, (b) handled, (c) processed, (d) stored, or (e) transported by someone at work. [*HSWA 1974, s 6(5)* as amended by the *Consumer Protection Act 1987, Sch 3.*]

No duty on suppliers of industrial articles and substances to research

[P9019] It is not necessary to repeat any testing, examination or research which has been carried out by designers and manufacturers of industrial products, on the part of importers and suppliers, in so far as it is reasonable to rely on the results [*HSWA 1974, s 6(6).*]

Custom built articles

[P9020] Where a person designs, manufactures, imports or supplies an article (for use at work or an article of fairground equipment and does so for or to another) on the basis of a written undertaking by that other to ensure that the article will be safe and without health risks when being (a) set, (b) used, (c) cleaned, or (d) maintained by a person at work, the undertaking will relieve the designer/manufacturer etc from the duty specified in *HSWA 1974, s 6(1)(a)* (see **P9013** above), to such extent as is reasonable, having regard to the terms of the undertaking. [*HSWA 1974, s 6(8)* as amended by the *Consumer Protection Act 1987, Sch 3.*]

Importers liable for offences of foreign manufacturers/designers

[P9021] In order to give added protection to industrial users from unsafe imported products, the *Consumer Protection Act 1987, Sch 3* introduced a new subsection (*HSWA 1974, s 6(8A)*) which, in effect makes importers of unsafe products liable for the acts/omissions of foreign designers and manufacturers. *Section 6(8A)* states that nothing in (*inter alia*) *s 6(8)* is to relieve an importer of an article/substance from any of his duties, as regards anything done (or not done) or within the control of:

(a) a foreign designer; or
(b) a foreign manufacturer of an article/substance.

[*HSWA 1974, s 6(8A).*]

Proper use

[P9022] The original wording of *HSWA 1974, s 6(1)(a), 6(4)(a)* and *6(10)* concerning 'proper use' excluded 'foreseeable user error' as a defence. This had the consequence of favouring the supplier. If a supplier could demonstrate a degree of operator misuse or error, however reasonably foreseeable, the question of initial product safety was side stepped. Moreover, 'when properly used' implied, as construed, that there could only be a breach of *s* 6 once a product had actually been *used*. This was contrary to the principle that safety should be built into design/production, rather than relying on warnings and disclaimers. The wording in *s* 6 was amended so that only *unforeseeable* user/operator error will relieve the supplier from liability; he can no longer rely on the strict letter of his operating instructions. [*Consumer Protection Act 1987, Sch 3.*]

Powers to deal with unsafe imported goods

[P9023] *HSWA 1974* does not empower enforcing authorities to stop the supply of unsafe products at source or prevent the sale of products by foreign producers after they have been found to be unsafe, but enforcement officers have the power to act at the point of entry or anywhere else along the distribution chain to stop unsafe articles/substances being imported by serving prohibition notices (see ENFORCEMENT).

Customs officers can seize any imported article/substance, which is considered to be unsafe, and detain it for up to two (working) days. [*HSWA s 25A (incorporated by Schedule 3 to the Consumer Protection Act 1987).*]

In addition, customs officers can transmit information relating to unsafe imported products to HSE inspectors. [*HSWA 1974, s 27A (incorporated by the Consumer Protection Act 1987, Sch 3).*]

The new proposed EU Regulation on Market Surveillance of Products is due to come into force in 2015 and its provisions are far-reaching. The activities and responsibilities of the market surveillance authorities will apply irrespective of the intended user of the products, so will apply to products commonly described as 'industrial' and for use at work. The accompanying Regulation on Consumer Product Safety only applies to consumer products.

The proposed Regulation on Market Surveillance of Products gives increased and new powers to market surveillance authorities, encourages the authorities to carry out more active surveillance across the EU with 'appropriate checks', of 'adequate frequency' and on an 'adequate scale' and strengthens border controls.

In accordance with Article 9 of the draft Regulation, where a market surveillance authority believes a product may present a risk, the authority will have an obligation to carry out a risk assessment in respect of that product. In addition, when a product is considered a 'serious risk' the market surveillance authority will be required to take 'all necessary measures' and may do so without requiring the economic operator to take corrective action first or providing the opportunity to be heard beforehand. This includes, ultimately, recall (Article 10). Articles 14–18 of the draft Regulation set out measures to control products entering the EU, including checks at borders, suspension of release of products and refusal to release products.

Civil liability for unsafe products – historical background

[P9024] Originally at common law where defective products caused injury, damage and/or death, redress depended on whether the injured user had a contract with the seller or hirer of the product. This was often not the case, and in consequence many persons, including employees repairing and/or servicing products, were without remedy. This rule, emanating from the decision of *Winterbottom v Wright* (1842) 10 M & W 109, remained unchanged until 1932, when *Donoghue v Stevenson* [1932] AC 562 was decided by the House of Lords. This case was important because it established that manufacturers were liable in negligence if they failed to take reasonable care in the manufacturing and marketing of their products, and in consequence a user suffered injury when using the product in a reasonably foreseeable way. More particularly, a 'manufacturer of products, which he sells in such a form as to show that he intends them to reach the ultimate consumer in the form in which they left him with no reasonable possibility of intermediate inspection, and with the knowledge that the absence of reasonable care in the preparation or putting up of the products will result in an injury to the consumer's life or property, owes a duty to the consumer to take reasonable care' (per Lord Atkin). In this way, manufacturers of products which were defective were liable in negligence to users and consumers of their products, including those who as intermediaries repair, maintain and service industrial products, it being irrelevant whether there was a contract between manufacturer and user (which normally there was not).

Donoghue v Stevenson is a case of enormous historical importance in the context of liability of manufacturers, but although the principle has become well established and is still widely applied it did not give a reliable remedy to injured persons, who still had to satisfy the legal burden of proving that the manufacturer had not exercised reasonable care, a serious obstacle to overcome in cases involving technically complex products. In some instances it became possible to avoid this obstacle, by establishing liability on other bases.

Defective equipment supplied to employees

[P9025] At common law an employer obtaining equipment from a reputable supplier was unlikely to be found liable to an employee if the equipment turned out to be defective and injured him (*Davie v New Merton Board Mills Ltd* [1959] AC 604). Although it represented no bar to employees suing manufacturers direct for negligence, the effect of this decision was reversed by the *Employers' Liability (Defective Equipment) Act 1969* rendering employers strictly liable (irrespective of negligence) for any defects in equipment causing injury. In such circumstances the employer would be able to claim indemnity for breach of contract by the supplier of the equipment.

Breach of statutory duty

[P9026] Whilst legislation and regulations dealing with domestic and industrial safety are principally penal and enforceable by state agencies (Trading Standards officers and Health and Safety Executive inspectors) if injury, death or damage occurs in consequence of a breach of such statutory duty, it may be possible to use this breach as the basis of a civil liability claim. Here there may be strict liability if there are absolute requirements, or the duty may be defined in terms of what it is practicable or reasonably practicable to do (see **E15031** ENFORCEMENT).

Section 69 of the *Enterprise and Regulatory Reform Act (ERRA) 2013* amended *Section 47* of the *Health and Safety at Work act 1974*. The effect of Section 69 is to remove strict liability in health and safety regulations, effectively removing the right of injured people to rely upon breaches of health and safety regulations when pursuing a civil claim. A claimant must now show that the employer was negligent.

The *Consumer Protection Act 1987* and the safety regulations made under the main Act are unaffected by ERRA.

No provision is made in the *General Product Safety Regulations 2005 (SI 2005 No 1803)* for any breach thereof to give rise to civil liability for the benefit of an injured person. Given that the Regulations stem from European law it is unlikely that they would be construed in such a way as to give more extensive rights than those contained in the Product Liability Directive (85/374/EEC) (see **P9031**).

Consumer Protection Act 1987

[P9027] At European level it was deemed necessary to introduce a degree of harmonisation of product liability principles between Member States, and at the same time to reduce the importance of fault and negligence concepts in favour of liability being determined by reference to 'defects' in a product. Thus the nature of the product itself would become the key issue, not the conduct of the manufacturer. After protracted debate the Product Liability Directive (85/374/EEC) was adopted.

Introduction of strict product liability

[P9028] The introduction of strict product liability is enshrined in the *Consumer Protection Act 1987, s 2(1)*. Thus, where any damage is *caused* wholly or partly by a defect (see **P9031** below) in a product (eg goods, electricity, a component product or raw materials), the following may be liable for damages (irrespective of negligence):

(a) the producer;
(b) any person who, by putting his name on the product or using a trade mark (or other distinguishing mark) has held himself out as the producer; and
(c) any person who has imported the product into a Member State from outside the EU, in the course of trade/business.

[Consumer Protection Act 1987, s 2(1), (2).]

Producers

[P9029] Producers are variously defined as:

(a) the person who manufactured a product;
(b) in the case of a substance which has not been manufactured, but rather won or abstracted, the person who won or abstracted it; or
(c) in the case of a product not manufactured, won or abstracted, but whose essential characteristics are attributable to an industrial process or agricultural process, the person who carried out the process.

[Consumer Protection Act 1987, s 1(2)(a)–(c).]

Liability of suppliers

[P9030] Although producers are principally liable, intermediate suppliers can also be liable in certain circumstances. Thus, any person who supplied the product is liable for damages if:

(a) the injured person requests the supplier to identify one (or more) of the following:
   (i) the producer,
   (ii) the person who put his trade mark on the product,
   (iii) the importer of the product into the EU; and
(b) the request is made within a reasonable time after damage/injury has occurred *and* it is not reasonably practicable for the requestor to identify the above three persons; and
(c) within a reasonable time after receiving the request, the supplier fails either:
   (i) to comply with the request, or
   (ii) identify his supplier.

[Consumer Protection Act 1987, s 2(3).]

Importers of products into the EU and persons applying their name, brand or trade mark will also be directly liable as if they were original manufacturers *[Consumer Protection Act 1987, s 2(2)(b), (c)]*.

Defect – key to liability

[P9031] Liability presupposes that there is a defect in the product, and indeed, existence of a defect is the key to liability. Defect is defined in terms of the absence of safety in the product. More particularly, there is a 'defect in a product . . . if the safety of the product is not such as persons generally are entitled to expect' (including products comprised in that product) *[Consumer Protection Act 1987, s 3(1)]*.

Defect can arise in one of three ways and is related to:

(a) construction, manufacture, sub-manufacture, assembly;
(b) absence or inadequacy of suitable warnings, or existence of misleading warnings or precautions; and
(c) design.

The definition of 'defect' implies an entitlement to an expectation of safety on the part of the consumer, judged by reference to *general* consumer expectations not individual subjective ones. (The American case of *Webster v Blue Ship Tea Room Inc* 347 Mass 421, 198 NE 2d 309 (1964) is particularly instructive here. The claimant sued in a product liability action for a bone which had stuck in her throat, as a result of eating fish chowder in the defendant's restaurant. It was held that there was no liability. Whatever her own expectations may have been, fish chowder would not be fish chowder without some bones and this is a general expectation.)

The claimant must prove that the product is defective and that the defect caused damage. Following the case of *Hufford v Samsung Electronics (UK) Ltd* [2014] EWHC 2956 (TCC), [2014] All ER (D) 60 (Sep), a claimant need not prove the exact cause of the defect or identify it with precision. He only needs to prove in general terms that the defect exists and that it caused damage:

> 'in relation to a claim under the 1987 Act, a claimant does not have to specify or identify with accuracy or precision the defect in the product he seeks to establish, and thus prove. It is enough for a claimant to prove the existence of a defect in broad and general terms, such as "a defect in the electrics of the Lexus (motor car)"'.

(per Grant HHJ).

Consumer expectation of safety – criteria

[P9032] The general consumer expectation of safety must be judged in relation to:

(a) the marketing of the product, ie:
  (i) the manner in which; and
  (ii) the purposes for which the product has been marketed;
  (iii) any instructions/warnings against doing anything with the product; and
(b) by what might reasonably be expected to be done with or in relation to the product (eg the expectation that a sharp knife will be handled with care); and
(c) the time when the product was supplied (eg a product's shelf-life).

A defect cannot arise retrospectively by virtue of the fact that, subsequently, a safer product is made and put into circulation.

[Consumer Protection Act 1987, s 3(2).]

The definition of 'defect' was discussed extensively by the Court in *A v National Blood Authority* [2001] 3 All ER 289. The Claimants in this case had contracted Hepatitis C from infected blood transfusions. The Court found that the blood was defective; the judge concluding that the Claimants were entitled to expect that the blood being supplied to them was free from infection. In *Pollard v Tesco Stores Ltd* [2006] EWCA Civ 393, [2006] All ER (D) 186 (Apr), the Court gave further consideration as to the definition of defect and the link between the legitimate expectation of the public and national safety standards.

In *Ide v ATB Sales Ltd* [2008] EWCA Civ 424; [2008] PIQR P13, the Court of Appeal up-held a decision in favour of a cyclist who had fallen when the handlebar of their bicycle had broken. It held that the judge was entitled to find that the handlebar had caused the fall even though there was no evidence of a specific defect. The judge had rejected expert evidence from the defendant that the handlebar had broken as a result of the fall, rather than being the cause of the fall. (See also *Hufford v Samsung Electronics (UK) Ltd* [2014] EWHC 2956 (TCC), [2014] All ER (D) 60 (Sep) which extended this principle.)

The European Court of Justice's ruling in the joined cases of *Boston Scientific Medizintechnik GmbH v AOK Sachsen-Anhalt - Die Gesundheitskasse: Joined Cases C-503/13 and C-504/13* (2015) ECLI:EU:C:2015:148, 144 BMLR 225, [2015] All ER (D) 88 (Mar), ECJ looked at whether a risk of failure could constitute a defect. The cases involved a pacemaker and a cardioverter-defibrillator which were explanted due to them belonging to a series of products which could potentially fail. The Court held that for products such as pacemakers and cardioverter-defibrillators which carry a high risk, the potential lack of safety constituted a defect. Furthermore, the Court held that where a product belongs to the same group or production series of products which had a potential defect, such a product may be classified as defective. There was no need to show that the product in question had such a defect.

These cases concerned medical devices (a defibrillator and a pacemaker) which the Court considered to create a particularly high consumer expectation test, due to their function and the particular vulnerable situation of patients. It is therefore not certain that the decision will necessarily transcend all other categories of product liability cases.

Time of supply

[P9033] Liability attaches to the *supply* of a product (see **P9041** below). More particularly, the producer will be liable for any defects in the product existing at the time of supply (see **P9049**(d) below and *Piper (Terence) v JRI (Manufacturing) Ltd* [2006] EWCA Civ 1344, 92 BMLR 141, 150 Sol Jo LB 1391; and where two or more persons collaborate in the manufacture of a product, say by sub-manufacture, either and both may be liable, that is, severally and jointly (see **P9043** below).

Contributory negligence

[P9034] A person who is careless for his own safety may risk a reduction in his damages due to his contributory negligence. [*Consumer Protection Act 1987, s 6(4)*.] However, in the product liability context, carelessness of the user may mean that there is no liability at all on the part of the producer. If, for example, clear instructions and warnings provided with the product had been disregarded, when compliance would have avoided the accident, it is highly unlikely that the product would be found to be defective for the purposes of the Act. No off-setting of damages for contributory fault of the claimant would arise.

Absence or inadequacy of suitable/misleading warnings

[P9035] The common law required that the vendor of a product should point out any latent dangers in a product which he either knew about or ought to have known about. Misleading terminology/labelling on a product or product container could result in liability for negligence.

The duty, on manufacturers, to research the safety of their products, before putting them into circulation is even more necessary and compulsory now, given the introduction of strict product liability. This includes safety in connection with instructions for use on a product. A warning refers to something that can go wrong with the product; instructions for use relate to the best results that can be obtained from products, if the instructions are followed. In order to avoid potential product liability actions, manufacturers should provide both warnings, indicating the worst results and dangers, and directions for use, indicating the best results; the warning, in effect, identifying the worst consequences that could follow if instructions for use were not complied with.

In the case of *Worsley v Tambrands Ltd (No 2)* [2000] MCR 0280, it was held that a tampon manufacturer had done what was reasonable in all the circumstances to warn a woman about the risk of toxic shock syndrome from tampon use. They had placed a clear legible warning on the outside of the box directing the user to the leaflet. The leaflet was legible, literate and unambiguous and contained all the material necessary to convey both the warning signs and the action required if any risk were present. The manufacturer could not cater for lost leaflets or for those who chose not to replace them. This gives valuable guidance as to the extent that manufacturers are expected to warn users of the risks associated with their products.

Defect must exist when the product left the producer's possession

[P9036] This situation tends to be spotlighted by alteration or modification to, or interference with, a product on the part of an intermediary, for instance, a dealer or agent. If a product leaves an assembly line in accordance with its intended design, but is subsequently altered, modified or generally interfered with by an intermediary, in a manner outside the product's specification, the manufacturer is probably not liable for any injury so caused. In *Sabloff v Yamaha Motor Co* 113 NJ Super 279, 273 A 2d 606 (1971), the claimant was injured when the wheel of his motor-cycle locked, causing it to skid, then crash. The manufacturer's specification stipulated that the dealer attach the wheel of the motor-cycle to the front fork with a nut and bolt, and this had not been done properly. It was held that the dealer was liable for the motor-cyclist's injury (as well as the assembler, since the latter had delegated the function of tightening the nut to the dealer and it had not been properly carried out). Similarly, in *Piper (Terence) v JRI (Manufacturing) Ltd* [2006] EWCA Civ 1344, 92 BMLR 141, 150 Sol Jo LB 1391, which concerned an allegedly defective hip prosthesis, the Court found that the prosthesis had not been defective at the time it was supplied to the hospital by the defendant. As a result, the defence had been established.

Role of intermediaries

[P9037] If a defect in a product is foreseeably detectable by a legitimate intermediary (eg a retailer in the case of a domestic product or an employer in the case of an industrial product), liability used to rest with the intermediary rather than the manufacturer, when liability was referable to negligence

(*Donoghue v Stevenson* [1932] AC 562). This position does not duplicate under the *Consumer Protection Act 1987*, since the main object of the legislation is to fix producers with strict liability for injury-causing product defects to users. Nevertheless, there are common law and statutory duties on employers to inspect/test inward plant and machinery for use at the workplace, and failure to comply with these duties may make an employer liable. In addition, employers, in such circumstances, can incur liability under the *Employers' Liability (Defective Equipment) Act 1969* and so may seek to exercise contractual indemnity against manufacturers.

Comparison with negligence

[P9038] The similarity between product liability and negligence lies in causation. Defect must be the material *cause* of injury. The main arguments against this are likely to be along the lines of misuse of a product by a user (eg knowingly driving a car with defective brakes), or ignoring warnings (a two-pronged defence, since it also denies there was a 'defect'), or – as is common in chemicals and pesticides cases – a defence based on alternative theories of causation of the claimant's injuries.

Product liability differs from negligence in that it is no longer necessary for injured users to prove absence of reasonable care on the part of manufacturers. All that is now necessary is proof of (*a*) defect (see **P9035** above) and (*b*) that the defect caused the injury. It will, therefore, be no good for manufacturers to point to an unblemished safety record and/or excellent quality assurance programmes, or the lack of foreseeability of the accident, since the user is not trying to establish negligence. How or why a defect arose is immaterial; what is important is the fact that it exists.

Liability in negligence still has a role to play in cases where liability under the *Consumer Protection Act 1987* cannot be established because, for example, the defendant is not a 'producer' as defined, or because the statutory defence or time limit would bar a strict liability claim (see **P9044** and **P9048**). There is greater scope under the law of negligence for liability to be established against distributors and retailers who may be held responsible for certain defects, especially where inadequate warnings and instructions have been provided (eg *Goodchild v Vaclight*, [1965] CLY 2669).

Joint and several liability

[P9039] If two or more persons/companies are liable for the same damage, the liability is joint and several. This can, for example, refer to the situation where a product (eg an aircraft) is made partly in one country (eg England) and partly in another (eg France). Here both partners are liable (joint liability) but in the event of one party not being able to pay, the other can be made to pay all the compensation (several liability) [*Consumer Protection Act 1987, s 2(5)*].

For injuries arising out of occupational exposures, the traditional approaches to causation and joint and several liability have been subject to revision. The asbestos cases of *Fairchild v Glenhaven Funeral Services Ltd* [2002] UKHL 22, [2003] 1 AC 32, *Barker v Corus (UK) plc* [2006] UKHL 20, [2006] 2 AC 572 and *Sienkiewicz v Greif (UK) Ltd* [2011] UKSC 10, [2011] 2 AC 229, [2011]

2 All ER 857 were concerned with mesothelioma injuries. The Claimants had been exposed to asbestos by various (negligent) defendants but it was impossible to show which of these exposures had caused the disease. It was held that in such circumstances, the Defendants were nevertheless liable in spite of the lack of proof of causation. Causation was established where the claimant demonstrated that the defendant's wrongdoing materially increased the risk of injury. In balancing fairness between the parties, each Defendant was only liable (severally, not joint) for a share of the liability apportioned according to its contribution to the risk (this is likely to be according to the relative times and intensity of exposure). It is of note that in *Sienkiewicz* the Court expressed the view that it was unlikely that this exceptional approach to causation was unlikely to extend to other types of claim. This view was confirmed in *AB v Ministry of Defence* [2012] UKSC 9, [2013] 1 AC 78, [2012] 3 All ER 673.

Parameters of liability

[P9040] For certain types of damage including (*a*) death, (*b*) personal injury and (*c*) loss or damage to property liability is included and relevant to private use, occupation and consumption by consumers [*Consumer Protection Act 1987, s 5(1)*]. However, producers and others will not be liable for:

(a) damage/loss to the defective product itself, or any product supplied with the defective product [*Consumer Protection Act 1987, s 5(2)*];
(b) damage to property not 'ordinarily intended for private use, occupation or consumption' eg car/van used for business purposes [*Consumer Protection Act 1987, s 5(3)*]; and
(c) damage amounting to less than £275 (to be determined as early as possible after loss) [*Consumer Protection Act 1987, s 5(4)*].

Defences

[P9041] The following statutory defences are open to producers:

(a) the defect was attributable to compliance with any requirement imposed by law/regulation or a European Union rule/regulation [*Consumer Protection Act 1987, s 4(1)(a)*];
(b) the defendant did not supply the product to another (ie did not sell/hire/lend/exchange for money/give goods as a prize etc (see below 'supply')) [*Consumer Protection Act 1987, s 4(1)(b)*];
(c) the supply to another person was not in the course of that supplier's business [*Consumer Protection Act 1987, s 4(1)(c)*];
(d) the defect did not exist in the product at the relevant time (ie it came into existence after the product had left the possession of the defendant). This principally refers to the situation where for example a retailer fails to follow the instructions of the manufacturer for storage or assembly [*Consumer Protection Act 1987, s 4(1)(d)*];
(e) that the state of scientific and technical knowledge at the relevant time was not such that a producer 'might be expected to have discovered the defect if it had existed in his products while they were under his control' (ie development risk) [*Consumer Protection Act 1987, s 4(1)(e)*] (see **P9049** below);
(f) that the defect:

(i) constitutes a defect in a subsequent product (in which the product in question was comprised); and
(ii) was wholly attributable to:
(A) the design of the subsequent product; or
(B) compliance by the producer with the instructions of the producer of the subsequent product.

[*Consumer Protection Act 1987, s 4(1)(f)*.] This is known as the component manufacturer's defence.

The meaning of 'supply'

[P9042] Before strict liability can be established under the *Consumer Protection Act 1987*, a product must have been '*supplied*'. This is defined as follows:

(a) selling, hiring out or lending goods;
(b) entering into a hire-purchase agreement to furnish goods;
(c) performance of any contract for work and materials to furnish goods (eg making/repairing teeth);
(d) providing goods in exchange for any consideration other than money;
(e) providing goods in or in connection with the performance of any statutory function/duty (eg supply of gas/electricity by public utilities); and
(f) giving the goods as a prize or otherwise making a gift of the goods.

[*Consumer Protection Act 1987, s 46(1)*.]

Moreover, in the case of hire-purchase agreements/credit sales the effective supplier (ie the dealer), and not the ostensible supplier (ie the finance company), is the 'supplier' for the purposes of strict liability [*Consumer Protection Act 1987, s 46(2)*].

Building work is only to be treated as a supply of goods in so far as it involves provision of any goods to any person by means of their incorporation into the building/structure, eg glass for windows [*Consumer Protection Act 1987, s 46(3)*].

No contracting out of strict liability

[P9043] The liability to person who has suffered injury/damage under the *Consumer Protection Act 1987*, cannot be (*a*) limited or (*b*) excluded:

(i) by any contract term; or
(ii) by any notice or other provision.

[*Consumer Protection Act 1987, s 7*.]

Time limits for bringing product liability actions

[P9044] No action can be brought under the *Consumer Protection Act 1987, Part I* (ie product liability actions) after the expiry of ten years from the time when the product was first put into circulation (ie the particular item in question was supplied in the course of business/trade etc) [*Limitation Act 1980, s 11A(3); Consumer Protection Act 1987, Sch 1*]. In other words, ten years is the cut-off point for liability. An action can still be brought for common law negligence after this time.

However in *SmithKline Beecham plc and Another v Horne-Roberts* [2001] EWCA CIV 2006, [2002] 1 WLR 1662 the court allowed a claimant to substitute a new defendant for an existing one in a strict liability claim despite the fact that the substitution was outside the ten year cut-off period.

The European Court of Justice case, *O'Byrne v Sanofi Pasteur MSD Ltd*: C-127/04 [2006] ECR I-1313 cast further doubt over the finality of the ten-year long-stop. In this case, the Court had to consider whether the passing of a product between two different companies (and therefore legal entities) constituted 'putting the product in circulation'. The European Court concluded that:

> A product must be considered as having been put into circulation [. . .] when it leaves the production process and enters a marketing process in the form in which it is offered to the public in order to be used or consumed.

Switching from the production to the marketing process therefore appears to be key. In addition, the European Court also considered whether, if an individual had sued the wrong defendant, the Product Liability Directive allowed proceedings to continue against the correct defendant outside the ten-year long-stop limit. The Court indicated that this was an issue for national courts to decide.

All actions for personal injury caused by product defects must be initiated within three years of whichever event occurs later, namely:

(a) the date when the cause of action accrued (ie injury occurred); or
(b) the date when the injured person had the requisite knowledge of his injury/damage to property.

[*Limitation Act 1980, s 11A(4); Consumer Protection Act 1987, Sch 1.*]

However, if during that period the injured person died, his personal representative has a further three years from his death to bring the action. (This coincides with actions for personal injuries against employers, except of course in that case there is no overall cut-off period of ten years.) [*Limitation Act 1980, s 11A(5).*]. Of course, the limitation period for a child to bring an action does not commence until that child reaches 18 years of age.

Development risk

[P9045] The development risks defence was initially considered as potentially being the most important defence to product liability actions. However, its use has not been extensive to date. Manufacturers have argued that it would be wrong to hold them responsible for the consequences of defects which they could not reasonably have known about or discovered. The absence of this defence would have the effect of increasing the cost of product liability insurance and stifle the development of new products. On the other hand, consumers maintain that the existence of this defence threatens the whole basis of strict liability and allows manufacturers to escape liability by, in effect, pleading a defence associated with the lack of negligence. For this reason, not all EU states have allowed this defence; Finland and Luxembourg excluded the defence entirely from national law, for example. The burden of proving development risk lies on the producer and government guidance set out that: 'It will not necessarily be enough to show that he (the producer) has done as many tests as his competitor, nor that he did all the tests required of him by a government regulation setting a minimum standard.'

Additionally, the fact that judgments in product liability cases are 'transportable' could have serious implications for the retention of development risk in the United Kingdom (see **P9050** below).

Transportability of judgments

[P9046] The so-called 'Brussels Regulation' (Regulation 44/2001 of 22 December 2000 on *Jurisdiction and the Recognition and Enforcement of Judgments in Civil and Commercial Matters*) requires judgments given in one of the Member States (except Denmark which will continue to apply the 1968 Brussels Convention) to be enforced in another. The United Kingdom became bound by this Regulation as of 1 March 2002, which, by virtue of the *Civil Jurisdiction and Judgments Order 2001 (SI 2001 No 3929)* (which amends the *Civil Jurisdiction and Judgments Act 1982*), is part of UK law. Where product liability actions are concerned, litigation can be initiated in the state where the defendant is based or where injury occurred and the judgment of that court 'transported' to another Member State. This could pose a threat to retention of development risk in the United Kingdom and other states in favour of it from a state against it.

Regard also needs to be given to the European Court of Justice's decision in *Andreas Kainz v Pantherwerke AG: C-45/13* (2014) ECLI:EU:C:2014:7, [2015] QB 34, [2014] 3 WLR 1292 where an Austrian national bought a German manufactured bike from an Austrian supplier and had an accident whilst riding the bike in Germany. The Court held that in product liability claims, the place of the event giving rise to the damage (and therefore the place where a person may be sued) was the place where the event which damaged the product itself occurred. In this case, the harmful event (both the damage and the event giving rise to the damage) occurred in Germany. If the place where the damage occurred is different from the place where the product was manufactured, the claimant can bring proceedings in that country.

Contractual liability for sub-standard products

[P9047] Contractual liability is concerned with defective products which are substandard (though not necessarily dangerous) regarding quality, reliability and/or durability. In respect of goods supplied between companies and businesses liability is predominantly determined by contractual terms implied by the *Sale of Goods Act 1979* and the *Sale and Supply of Goods Act 1994*, in the case of goods sold; the *Supply of Goods (Implied Terms) Act 1973*, where goods are the subject of hire purchase and conditional and/or credit sale;

[P9047] Product Safety

and the *Supply of Goods and Services Act 1982*, where goods are supplied but not sold as such, primarily as a supply of goods with services; hire and leasing contracts are subject to the 1982 Act as well. The new *Consumer Rights Act 2015* deals with sale of goods from a 'trader' to a consumer.

Exemption or exclusion clauses in commercial contracts may be invalid by virtue of the *Unfair Contract Terms Act 1977*, which also applies to such transactions. Part 2 of the *Consumer Rights Act 2015* governs unfair terms in consumer contracts.

Contractual liability is strict. It is not necessary that negligence be established. For instance, in *Frost v Aylesbury Dairy Co Ltd* [1905] 1 KB 608 the defendant supplied typhoid-infected milk to the claimant, who, after its consumption, became ill and required medical treatment. It was held that the defendant was liable, irrespective of the absence of negligence on his part.

Sale of Goods Acts and Consumer Rights Act

[P9048] Conditions and warranties as to fitness for purpose, quality and merchantability were originally implied into contracts for the sale of goods at common law. Those terms were then codified in the *Sale of Goods Act 1893*. However, this legislation did not provide a blanket consumer protection measure, since sellers were still allowed to exclude liability by suitably worded exemption clauses in the contract. This practice was finally outlawed, at least as far as consumer contracts were concerned, by the *Supply of Goods (Implied Terms) Act 1973* and later still by the *Unfair Contract Terms Act 1977*, the current statute prohibiting contracting out of contractual liability and negligence. The protection for consumers was further bolstered by the *Sale and Supply of Goods to Consumers Regulations 2002 (SI 2002 No 3045)* (hereafter the 2002 Regulations) (implementing Directive 1999/44/EC on certain aspects of the sale of consumer goods and associated guarantees). These Regulations extended the definition of satisfactory quality, giving a statutory right to repair or replacement, amending the law relating to the passing of risk in consumer contracts, and further amending UCTA to make the position more favourable to consumers.

The *Consumer Rights Act 2015* (which received Royal assent on 26 March 2015) has now revoked and superseded the 2002 Regulations in governing the law on the protection of consumers. The 2015 Act makes a number of amendments to the *Sale and Supply of Goods Act 1979*. Part 1 of the Act deals with consumer contracts for goods, digital content and services.

The initial law relating to sale and supply of goods was updated by the *Sale of Goods Act 1979* and the *Sale and Supply of Goods Act 1994*, the latter replacing the condition of 'merchantable quality' with 'satisfactory quality'. The difference between the 'merchantability' requirement, under the *1979 Act*, and its replacement 'satisfactory quality', under the *1994 Act*, is that, under the former Act, products were 'usable' (or, in the case of food, 'edible'), even if they had defects which ruined their appearance; now they must be free from minor defects as well as being safe and durable. Further, under the previous law, a right of refund disappeared after goods had been kept for a reasonable time; under the 1994 Act, consumers benefited from a right of examination for a reasonable time after buying.

Regulation 5 of the 2002 Regulations further amended the *Sale of Goods Act 1979* by allowing consumers who ordered goods which do not conform to contract at the time of delivery, the right to require the seller to repair or replace the product. If a defect is discovered in the product during the first six months after delivery and that defect amounts to non-conformity with the contract of sale, then it will be regarded as having been in existence at the time of delivery, subject to certain minor exceptions.

The *Consumer Rights Act 2015* also gives a number of remedies for breach of contract, such as the right to reject (s 20), the partial rejection of goods (s 21), the right to a repair and replacement (s 23) and the right to price reduction or final right to reject (s 24).

As stated above, in 1979 a consolidated *Sale of Goods Act* was passed and the law on quality and fitness of products for non-consumer contracts is contained in that Act. Another equally important development has been the extension of implied terms, relating to quality and fitness of products, to contracts other than those for the sale of goods, that is, to hire purchase contracts by the *Supply of Goods (Implied Terms) Act 1973*, and to straight hire contracts by the *Supply of Goods and Services Act 1982*. In addition, where services are performed under a contract, that is, a contract for work and materials, there is a statutory duty on the contractor to perform them with reasonable care and skill. In other words, in the case of services liability is not strict, but it is strict for the supply of products. This is laid down in the *Supply of Goods and Services Act 1982, s 4*. This applies whether products are simultaneously but separately supplied under any contract, eg after-sales service, say, on a car or a contract to repair a window by a carpenter, in which latter case service is rendered irrespective of product supplied.

Products to be of satisfactory quality – sellers/suppliers

[P9049] The *Sale of Goods Act 1979* (as amended) writes two quality conditions into all contracts for the sale of products, the first with regard to satisfactory quality, the second with regard to fitness for purpose.

Where a seller sells goods in the course of business, there is an implied term that the goods supplied under the contract are of satisfactory quality, according to the standards of the reasonable person, by reference to description, price etc. [*Sale of Goods Act 1979, s 14(2)* as substituted by the *Sale and Supply of Goods Act 1994, s 1(2)*.] The 'satisfactory' (or otherwise) quality of goods can be determined from:

(a) their state and condition;
(b) their fitness for purpose (see **P9056** below);
(c) their appearance and finish;
(d) their freedom from minor defects;
(e) their safety; and
(f) their durability.

However, the implied term of 'satisfactory quality' does not apply to situations where:

(i) the unsatisfactory nature of goods is specifically drawn to the buyer's attention prior to contract; or

(ii) the buyer examined the goods prior to contract and the matter in question ought to have been revealed by that examination. *[Sale of Goods Act 1979, s 14(2) as substituted by the Sale and Supply of Goods Act 1994, s 1(2A), (2B) and (2C).]*

The *Consumer Rights Act 2015* confirms that goods supplied under a contract where sale is made to a consumer must be of satisfactory quality (s 9):

> '9(2) *The quality of goods is satisfactory if they meet the standard that a reasonable person would consider satisfactory, taking account of –*
>
> (a) *any description of the goods,*
> (b) *the price or other consideration for the goods (if relevant), and*
> (c) *all the other relevant circumstances (see subsection (5)).*
>
> (5) *The relevant circumstances mentioned in subsection (2)(c) include any public statement about the specific characteristics of the goods made by the trader, the producer or any representative of the trader or the producer.*
>
> (6) *That includes, in particular, any public statement made in advertising or labelling'.*

As with the *Sale of Supply of Goods Act 1979*, the *Consumer Rights Act 2015* also carves out a defence for the trader [s 9(4)] in circumstances where:

> 'anything which makes the quality of the goods unsatisfactory –
>
> (a) which is specifically drawn to the consumer's attention before the contract is made,
> (b) where the consumer examines the goods before the contract is made, which that examination ought to reveal, or
> (c) in the case of a contract to supply goods by sample, which would have been apparent on a reasonable examination of the sample'.

Sale by sample

[P9050] In the case of a contract for sale by sample, there is an implied condition that the goods will be free from any defect making their quality unsatisfactory, which would not be apparent on reasonable examination of the sample. [*Sale of Goods Act 1979, s 15(2) as substituted by the Sale and Supply of Goods Act 1994, s 1(2).*]

The *Consumer Rights Act 2015* further protects the consumer when they have had the chance to inspect a sample. *Section 13 states: (a) the goods will match the sample except to the extent that any differences between the sample and the goods are brought to the consumer's attention before the contract is made; and (b) the goods will be free from any defect that makes their quality unsatisfactory and that would not be apparent on a reasonable examination of the sample.*

Conditions implied into sale – sales by a dealer

[P9051] The condition of satisfactory quality only arises in the case of sales by a dealer to a consumer, not in the case of private sales. The *Sale of Goods Act 1979, s 14(2)* also applies to second-hand as well as new products. It is not necessary, as it is with the 'fitness for purpose' condition (see **P9052** below), for the buyer in any way to rely on the skill and judgment of the seller in selecting his stock, in order to invoke *s 14(2)*. However, if the buyer has examined the products, then the seller will not be liable for any defects which the examination should have disclosed. Originally this applied if the buyer had been given opportunity to examine but had not, or only partially, exercised it. In *Thornett & Fehr v Beers & Son* [1919] 1 KB 486, a buyer of glue examined only the outside of some barrels of glue. The glue was defective. It was held that he had examined the glue and so was without redress.

Products to be reasonably fit for purpose

[P9052]

> Where the seller sells goods in the course of a business and the buyer, expressly or by implication, makes known:
>
> (a) to the seller, or
> (b) where the purchase price or part of it is payable by instalments and the goods were previously sold by a credit-broker to the seller, to that credit-broker,
>
> any particular purpose for which the goods are being bought, there is an implied condition that the goods supplied under the contract are reasonably fit for that purpose, whether or not that is a purpose for which such goods are commonly supplied, except where the circumstances show that the buyer does not rely or that it is unreasonable for him to rely, on the skill or judgment of the seller or credit-broker.
>
> *[Sale of Goods Act 1979, s 14(3).]*

Reliance on the skill/judgment of the seller will generally be inferred from the buyer's conduct. The reliance will seldom be express: it will usually arise by implication from the circumstances; thus to take a case of a purchase from a retailer, the reliance will be in general inferred from the fact that a buyer goes to the shop in the confidence that the tradesman has selected his stock with skill and judgment (*Grant v Australian Knitting Mills Ltd* [1936] AC 85). Moreover, it is enough if the buyer relies partially on the seller's skill and judgment. However, there may be no reliance where the seller can only sell goods of a particular brand. A claimant bought beer in a public house which he knew was a tied house. He later became ill as a result of drinking it. It was held that there was no reliance on the seller's skill and so no liability on the part of the seller (*Wren v Holt* [1903] 1 KB 610).

Even though products can only be used normally for one purpose, they will have to be reasonably fit for that particular purpose. A claimant bought a hot water bottle and was later scalded when using it because of its defective condition. It was held that the seller was liable because the hot water bottle was not suitable for its normal purpose (*Priest v Last* [1903] 2 KB 148). But, on the other hand, the buyer must not be hypersensitive to the effects of the product. A claimant bought a Harris Tweed coat from the defendants. She later contracted dermatitis from wearing it. Evidence showed that she had an exceptionally sensitive skin. It was held that the coat was reasonably fit for the purpose when worn by a person with an average skin (*Griffiths v Peter Conway Ltd* [1939] 1 All ER 685, CA).

Like *s 14(2)*, *s 14(3)* extends beyond the actual products themselves to their containers and labelling. A claimant was injured by a defective bottle containing mineral water, which she had purchased from the defendant, a retailer. The bottle remained the property of the seller because the claimant had paid the seller a deposit on the bottle, which would be returned to her, on return of the empty bottle. It was held that, although the bottle was the property of the seller, the seller was liable for the injury caused to the claimant by the defective container (*Geddling v Marsh* [1920] 1 KB 668).

Like s 14(2) (above), s 14(3) does not apply to private sales.

Section 10 of the *Consumer Rights Act 2015* reaffirms the principle laid down in s 14(3) of the *Sale and Supply of Goods Act 1987* and provides that if a consumer acquires goods for a specific purpose, and has made this purpose known to the trader beforehand, the goods must be fit for that purpose unless the consumer does not rely, or it would be unreasonable for the consumer to rely, on the skill or judgment of the trader.

Strict liability under the Sale of Goods Act 1979, s 14

[P9053] Liability arising under the *Sale of Goods Act, s 14* is strict and does not depend on proof of negligence by the purchaser against the seller (*Frost v Aylesbury Dairy Co Ltd* (see **P9051** above)). This fact was stressed as follows in the case of *Kendall (Henry) & Sons (a firm) v William Lillico & Sons Ltd* [1969] 2 AC 31, [1968] 2 All ER 444:

> If the law were always logical one would suppose that a buyer who has obtained a right to rely on the seller's skill and judgment, would only obtain thereby an assurance that proper skill and judgment had been exercised, and would only be entitled to a remedy if a defect in the goods was due to failure to exercise such skill and judgment. But the law has always gone further than that. By getting the seller to undertake his skill and judgment the buyer gets . . . an assurance that the goods will be reasonably fit for his purpose and that covers not only defects which the seller ought to have detected but also defects which are latent in the sense that even the utmost skill and judgment on the part of the seller would not have detected them.

(Per Lord Reid).

Dangerous products

[P9054] As distinct from applying to merely substandard products, both s 14(2) and (3) of the *Sale of Goods Act 1979* can be invoked where a product is so defective as to be unsafe, but the injury/damage must be a reasonably foreseeable consequence of breach of the implied condition. If, for instance, therefore, the chain of causation is broken by negligence on the part of the user, in using a product knowing it to be defective, there will be no liability. In *Lambert v Lewis* [1982] AC 225, manufacturers had made a defective towing coupling which was sold by retailers to a farmer. The farmer continued to use the coupling knowing that it was unsafe. As a result, an employee was injured and the farmer had to pay damages. He sought to recover these against the retailer for breach of *s 14(3)*. It was held that he could not do so.

Credit sale and supply of products

[P9055] Broadly similar terms to those under the *Sale of Goods Act 1979, s 14* exist, in the case of hire purchase, credit sale, conditional sale and hire or lease contracts, by virtue of the Acts described above at **P9048** having been modified by the *Sale and Supply of Goods Act 1994*.

Sale by description

[P9056] Both the *Sale and Supply of Goods Act 1979* (s 13) and the *Consumer Rights Act 2015* (s 11) confirm that the goods must correspond with their description.

Exclusion of liability of implied terms by agreement or practice

[P9057] Under s 55(i) of the *Sale and Supply of Goods Act 1979* it is possible for parties to vary or opt out of some of the implied terms discussed above. This can be done by express agreement, or by the course of dealing between the parties, or by such usage as binds both parties to the contract.

However, s 31 of the *Consumer Rights Act 2015* states that it is not possible for a seller to contract out of a number of implied terms with a consumer including, that goods will match description, that goods will match the sample, that goods will be of satisfactory quality and that they will be fit for purpose.

Product safety: unfair terms

Unfair Contract Terms Act 1977 (UCTA 1977)

[P9058] In spite of its name, this Act is not directly concerned with 'unfairness' in contracts, nor is it confined to the regulation of contractual relations. It should be noted that following the enactment of *Consumer Rights Act 2015*, UCTA no longer applies to consumer contracts. The main provisions are as follows:

(a) Liability for death or personal injury resulting from negligence cannot be excluded by warning notices or contractual terms. *[UCTA 1977, s 2.]* It should be noted, however, that this does not necessarily prevent an indemnity of any liability to an injured person being agreed between two other contracting parties. For instance, if A hires plant to B on terms that B indemnifies A in respect of any claims by any person for injury, the clause may be enforceable (see *Thompson v T Lohan (Plant Hire) Ltd (JW Hurdiss Ltd, third party)* [1987] 2 All ER 631, [1987] 1 WLR 649.

(b) In the case of other loss or damage, liability for negligence cannot be excluded by contract terms or warning notices unless these satisfy the requirement of reasonableness and the party seeking to rely on the notice takes sufficient steps to bring it to the injured party's attention.

(c) In all cases where the reasonableness test applies, the burden of establishing that a clause or other provision is reasonable will lie with the person who is trying to rely on the term in question. The consequence of contravention of the Act is, however, limited to the offending term being treated as being ineffective; the courts will give effect to the remainder of the contract so far as it is possible to do so.

Unfair Terms under the Consumer Rights Act 2015

[P9059] Part 2 of this Act replaces existing consumer legislation on unfair terms. The Act made a number of amendments to UCTA (in that it no longer applies to consumers) and it repealed the Unfair Terms in Consumer Contracts Regulations 1999.

In accordance with s 62 of *Consumer Rights Act 2015*, terms used in contracts and notices will only be binding upon the consumer if they are fair.

The Act states:

'62(4) A term is unfair if, contrary to the requirement of good faith, it causes a significant imbalance in the parties' rights and obligations under the contract to the detriment of the consumer.

(5) Whether a term is fair is to be determined –

(a) taking into account the nature of the subject matter of the contract, and

(b) by reference to all the circumstances existing when the term was agreed and to all of the other terms of the contract or of any other contract on which it depends.

(6) A notice is unfair if, contrary to the requirement of good faith, it causes a significant imbalance in the parties' rights and obligations to the detriment of the consumer.

(7) Whether a notice is fair is to be determined –

(a) taking into account the nature of the subject matter of the notice, and

(b) by reference to all the circumstances existing when the rights or obligations to which it relates arose and to the terms of any contract on which it depends.'

Section 65 of the Act confirms the previous UCTA position that a trader cannot by a term of a consumer contract or by a consumer notice exclude or restrict liability for death or personal injury resulting from negligence. In addition, written terms and written notices must be transparent ie legible and in plain and intelligible language (s 68).

As had been the position under UCTA, if a term in a consumer contract, or a consumer notice, could have different meanings, the meaning that is most favorable to the consumer is to prevail (s 69).

Examples of 'unfair terms'

[P9060] Schedule 2 of the *Consumer Rights Act 2015* sets out a non-exhaustive list of unfair terms. These are:

1. A term which has the object or effect of excluding or limiting the trader's liability in the event of the death of or personal injury to the consumer resulting from an act or omission of the trader.

2. A term which has the object or effect of inappropriately excluding or limiting the legal rights of the consumer in relation to the trader or another party in the event of total or partial non-performance or inadequate performance by the trader of any of the contractual obligations, including the option of offsetting a debt owed to the trader against any claim which the consumer may have against the trader.

3. A term which has the object or effect of making an agreement binding on the

consumer in a case where the provision of services by the trader is subject to a condition whose realisation depends on the trader's will alone.

4. A term which has the object or effect of permitting the trader to retain sums paid by the consumer where the consumer decides not to conclude or perform the contract, without providing for the consumer to receive compensation of an equivalent amount from the trader where the trader is the party cancelling the contract.

5. A term which has the object or effect of requiring that, where the consumer decides not to conclude or perform the contract, the consumer must pay the trader a disproportionately high sum in compensation or for services which have not been supplied.

6. A term which has the object or effect of requiring a consumer who fails to fulfil his obligations under the contract to pay a disproportionately high sum in compensation.

7. A term which has the object or effect of authorising the trader to dissolve the contract on a discretionary basis where the same facility is not granted to the consumer, or permitting the trader to retain the sums paid for services not yet supplied by the trader where it is the trader who dissolves the contract.

8. A term which has the object or effect of enabling the trader to terminate a contract of indeterminate duration without reasonable notice except where there are serious grounds for doing so.

9. A term which has the object or effect of automatically extending a contract of fixed duration where the consumer does not indicate otherwise, when the deadline fixed for the consumer to express a desire not to extend the contract is unreasonably early.

10. A term which has the object or effect of irrevocably binding the consumer to terms with which the consumer has had no real opportunity of becoming acquainted before the conclusion of the contract.

11. A term which has the object or effect of enabling the trader to alter the terms of the contract unilaterally without a valid reason which is specified in the contract.

12. A term which has the object or effect of permitting the trader to determine the characteristics of the subject matter of the contract after the consumer has become bound by it.

13. A term which has the object or effect of enabling the trader to alter unilaterally without a valid reason any characteristics of the goods, digital content or services to be provided.

14. A term which has the object or effect of giving the trader the discretion to decide the price payable under the contract after the consumer has become bound by it, where no price or method of determining the price is agreed when the consumer becomes bound.

15. A term which has the object or effect of permitting a trader to increase the price of goods, digital content or services without giving the consumer the right to cancel the contract if the final price is too high in relation to the price agreed when the contract was concluded.

16. A term which has the object or effect of giving the trader the right to determine whether the goods, digital content or services supplied are in conformity with the contract, or giving the trader the exclusive right to interpret any term of the contract.

17. A term which has the object or effect of limiting the trader's obligation to respect commitments undertaken by the trader's agents or making the trader's commitments subject to compliance with a particular formality.

18. A term which has the object or effect of obliging the consumer to fulfil all of the consumer's obligations where the trader does not perform the trader's obligations.

19. A term which has the object or effect of allowing the trader to transfer the trader's rights and obligations under the contract, where this may reduce the guarantees for the consumer, without the consumer's agreement.
20. A term which has the object or effect of excluding or hindering the consumer's right to take legal action or exercise any other legal remedy, in particular by —

    (a) requiring the consumer to take disputes exclusively to arbitration not covered by legal provisions,
    (b) unduly restricting the evidence available to the consumer, or
    (c) imposing on the consumer a burden of proof which, according to the applicable law, should lie with another party to the contract.

Public & Products Liability Insurance

Phil Grace

Introduction to public & products liability insurance

[P9501] Insurance is a way by which firms and businesses protect themselves against claims that could adversely affect their continued profitability or even viability and existence. Claims for compensation or damages may not be capable of being covered by income or profits. Most employers carrying on business in Great Britain are under a statutory duty to take out insurance against claims for injuries/diseases brought against them by employees.

Firms and businesses should also take out insurance to protect themselves from claims made against them by people who are not its employees or by other firms or trading entities. Such insurance is called Public Liability Insurance. Where a firm manufacturers, imports or sells products then it is advisable for it to take out Products Liability insurance. Whilst such insurances are not compulsory (with one or two specific exceptions) failure to protect the firm could threaten its continued existence in the event that it is found legally liable and required to pay compensation or damages to a third party.

This section examines:

- General law and practice relating to Liability insurance policies
- Public Liability Insurance
- Products Liability Insurance
- Miscellaneous Matters Relating to Public and Products Liability Insurance

General law and practice relating to liability insurance policies

[P9502] Insurance is a contract. When a person or firm (the proposer) wishes to insure, for example, themselves, their house, their liability towards employees, valuable personal property or even loss of profits, they complete a proposal form for insurance, at the same time making certain facts known to the insurer about what is to be insured. On the basis of the information disclosed in the proposal form, the insurer will decide whether to accept the risk and if so what premium to charge. If the insurer decides to accept the risk, a contract of insurance is then drawn up in the form of an insurance policy.

The vast majority of commercial insurance is arranged through a broker. The broker advises the proposer on the nature and type of covers required in addition to advising on the levels of cover required, the Sum Insured for property insurance or Limit of Indemnity for liability insurance. Brokers are regulated by the Financial Conduct Authority and are required to act in a professional manner. Brokers are responsible for the advice they give and in the event that the advice proves wrong or incorrect they can be sued and thus brokers are required to take out Professional Indemnity insurance. (**P9513**)

Incidentally, it seems to matter little whether the negotiations leading up to contract took place between the insured (proposer) and the insurance company or between the insured and a broker, since the broker is often regarded as the agent of one or the other, generally of the proposer (*Newsholme Brothers v Road Transport & General Insurance Co Ltd* [1929] 2 KB 356). However, a lot depends on the facts. If the broker is authorised to complete blank proposal forms, he may well be the agent of the insurer but this is rare. It is common practice for brokers to complete proposal forms and present them to the proposer for signing before return to the insurer.

There is a growth of what is known as direct sales where companies, generally small firms, purchase insurance 'direct' using the telephone or internet. However the vast majority of insurance for firms and businesses, what is known as commercial insurance, is still arranged through a broker.

Proposal forms

[**P9503**] Traditionally a firm or business seeking insurance would complete a proposal form, a standard document prepared by insurers on which the proposer, would describe their business and provide relevant details for the insurers. Failure to declare or reveal certain information, known as material facts, placed the proposer at risk of having any future insurance cancelled or declared void. The test of whether a fact was or was not material is whether its omission would have influenced a prudent underwriter in deciding whether to accept the risk, or what premium to charge. Generally only failure to make disclosure of relevant facts will allow an insurer subsequently to void the policy and refuse to compensate for the loss.

However, the use of proposal forms has almost died out. It is generally the case that brokers, who act on behalf of, as an agent of the firm seeking insurance make a "presentation" to the insurance company. This presentation will summarise the activities of the firm and contain relevant information such as the nature of the business activity, its size, the products manufactured or sold, turnover etc. The broker knows the key information required by the underwriter and will ensure it is included in the presentation. On the basis of the presentation the underwriter will then makes a decision whether to provide cover, that is offer terms, and what premium to charge,

Duty to take out insurance

[**P9504**] Unlike Employers' Liability where there is an express legal requirement to take out insurance there is no corresponding duty to take out Public and Products Liability insurance. There are two specific exceptions to this where it is compulsory to take out Public Liability Insurance. (**P9516**)

Two cases illustrate the complexity of this matter.

An individual attending a fair held in the grounds of a hospital was injured when they made use of a "splat wall" device. The equipment had been negligently set up by an independent contractor who had been given permission to attend the fair by the hospital trust, the organisers of the event. The injured person sued both the contractor and the hospital board. It was found that the operator's Public Liability insurance had expired. The Trust stated that when the booking had been made, several months prior to the event, the operator had confirmed that suitable insurance was in place.

A claim was made under the Occupiers' Liability Act 1957, against both the Trust and the operator, the independent contractor. The Trust was not held liable but the case was appealed. On appeal the Trust was held not to have been negligent. They had taken steps to ensure that the contractor was competent and enquired about insurance although they had not requested a copy of the policy. It was stated that the Trust had no reason to believe the insurance had lapsed or was no longer in force and that it would be unreasonable to expect the Trust to check the policy.

Thus the Trust escaped liability and the operator was required to pay compensation even though they had no insurance.

Gwilliam v West Hertfordshire Hospital NHS Trust [2002] EWCA Civ 1041, [2003] QB 443, [2003] 3 WLR 1425

Note: Despite this ruling any organisation that employs contractors is strongly recommended to enquire about insurance AND request a copy of the policy document.

In another case concerning a Local Authority a person attending a horse fair held in the streets of a town was kicked by an untethered horse owned by a third party. Although there was no single organising body the fair was organised by a "committee" that included representatives of the local authority. It was held that it was reasonable to expect the local authority to have arranged suitable insurance. Since the fair was held on their land, they took a leading role in the organisation and were aware of the risks.

The injured person sued the local authority stating that they should have taken out Public Liability insurance to protect themselves against the risk of being sued following accidents such as had occurred. The court agreed.

However, the local authority appealed stating that there was no special relationship between themselves and the injured person and as such they did not owe them a duty of care either to take out insurance themselves or to ensure any other persons took out suitable insurance. The Court of Appeal agreed commenting that even in the absence of insurance there would have still have been negligence in that there was a failure to properly segregate visitors and horses.

Glaister v Appleby-in-Westmorland Town Council [2009] EWCA Civ 1325, [2010] 1 LS Gaz R 14, [2009] All ER (D) 79 (Dec)

A similar, more recent case resulted from a 2010 fatal accident that occurred when a horse and carriage ran through spectators at a "Fair". In 2012 the inquest returned a verdict of accidental death and the coroner commented on inadequate risk assessments. Both the operator of the carriage rides and the Local Authority were prosecuted by the HSE for breaches of s 3(1) of the *Health and Safety at Work etc Act 1974*. The operator was found guilty of providing less than adequate training for his staff and using an unsafe system of work and were sentenced to community service. The Local Authority was found not guilty.

Duty of disclosure

[P9505] Most commercial contracts are governed by the doctrine of 'caveat emptor', which translates as 'let the buyer beware'. But the law relating to insurance contracts is based on the legal principle of 'uberrima fides'. This translates as 'utmost good faith' and places a higher duty on both parties but especially the proposer. Whilst the proposer can examine the insurance policy the insurer is unable to examine all aspects of the risk being presented. A proposer must disclose to the insurer all material facts within his actual knowledge. This does not extend to disclosure of facts which the proposer could not reasonably be expected to know. 'The duty is a duty to disclose, and you cannot disclose what you do not know. The obligation to disclose, therefore, necessarily depends on the knowledge you possess. This, however, must not be misunderstood. The proposer's opinion of the materiality of that knowledge is of no moment. If a reasonable man would have recognised that the knowledge in question was material to disclose, it is no excuse that you did not recognise it. But the question always is — Was the knowledge you possessed such that you ought to have disclosed it?' (*Joel v Law Union and Crown Insurance Co* [1908] 2 KB 863 per Fletcher Moulton LJ).

The impact of 'uberrima fides' (i.e. utmost good faith) is significant. If, when filling in a proposal form, a statement made by the proposer is at that time true, but is false in relation to other facts which are not stated, or becomes false before issue of the insurance policy, this entitles the insurer to refuse to indemnify. In *Condogianis v Guardian Assurance Co Ltd* [1921] 2 AC 125 a proposal form for fire cover contained the following question: 'Has proponent ever been a claimant on a fire insurance company in respect of the property now proposed, or any other property? If so, state when and name of company'. The proposer answered 'Yes', '1917', 'Ocean'. This answer was literally true, since he had claimed against the Ocean Insurance Co in respect of a burning car. However, he had failed to say that in 1912 he had made another claim against another insurance company in respect of another burning car. It was held that the answer was not a true one and the policy was, therefore, invalidated.

The knowledge of those who represent the directing mind and will of a company and who control what it does, e.g. directors and officers, is likely to be identified as the company's knowledge whether or not those individuals are responsible for arranging the insurance cover in question (*PCW Syndicates v PCW Reinsurers* [1996] 1 All ER 774, [1996] 1 Lloyd's Rep, CA).

An element of consumer protection, in favour of insureds, was introduced into insurance contracts by the Statement of General Insurance Practice 1986, a form of self-regulation applicable to many but not to all insurers. This has consequences for the duty of disclosure, proposal forms (E13005), renewals and claims. In particular, with regard to the last element (claims), an insurer should not refuse to indemnify on the grounds of:

(a) non-disclosure of a material fact which a policyholder could not reasonably be expected to have disclosed; or
(b) Misrepresentation (unless it is a deliberate non-disclosure of, or negligence regarding a material fact). Innocent misrepresentation is not a ground for avoidance of payment.

The trend towards greater consumer protection in (inter alia) insurance contracts is reflected in the *Unfair Terms in Consumer Contracts Regulations 1999 (SI 1999 No 2083)*. These Regulations apply in the case of 'standard form' (or non-individually negotiated) contracts [SI 1999 No 2083, Reg 5]. A contractual term in such a contract shall:

(i) be regarded as 'unfair' if contrary to the requirement of 'good faith' it 'causes a significant imbalance in the parties' rights and obligations arising under the contract, to the insured's detriment' [SI 1999 No 2083, Reg 5(1), Sch 2];
(ii) always be regarded as having been individually negotiated where it has been drafted in advance and the consumer has not been able to influence the substance of the term [SI 1999 No 2083, Reg 5(2)];
(iii) Require a written contract term to be expressed in plain, intelligible language. If there is doubt about the meaning of terminology, a construction in favour of the insured will prevail [SI 1999 No 2083, Reg 7]; and
(iv) Where the insurer claims that a term was individually negotiated, he must prove it [SI 1999 No 2083, Reg 5(4)].

Complaints (other than ones which are frivolous or vexatious) or which are to be handled by a qualified body [SI 1999 No 2083, Sch 1] relating to 'unfair terms' in standard form contracts, are considered by the Director General of Fair Trading, who may prevent their continued use [SI 1999 No 2083, Reg 12]. See also **P9056 PRODUCT SAFETY**.

The importance attached to the duty to disclose is confirmed by the standard policy condition relating to alteration of risk. If there is change in the nature of the business after the commencement of the policy that increases the risk exposure the insurer retains the right to cancel the policy if they are not informed of the changes. This has the effect of requiring the policyholder to inform or notify the insurer of any changes, whenever they may occur during the period of the policy.

Claims Procedures

[P9506] Insurers prefer to handle any claims that may arise under policies. Thus there is a policy condition that requires policyholders to notify the insurer of any loss or event that might give rise to a claim. (**P9523**) The policy conditions require that as soon as a claim is received the insured, that is the firm, must notify their insurer as soon as possible. The insured must NOT make any admission of liability or offer or promise payment without the 'prior written consent' of the insurer. The insurer will take over the handling of the claim, communicating with the claimant's solicitors, investigating the circumstances and deciding on liability. The insured must supply the insurer with such particulars and information as may be required or requested. Any subsequent correspondence received by the insured must be forwarded to the insurer without delay. Finally, the condition requires that the policyholder allow the insurer to take over the conduct of the claim in the name of the policyholder.

When policyholder receives correspondence that indicates either loss has occurred or a claim might be made they are required to forward such correspondence to the insurer immediately. This might be a letter from a commercial customer complaining of a problem with a component, a letter from solicitors acting on behalf of a customer who slipped over in a shop or representing a firm whose property has been damaged by the construction activities on an adjacent building site.

Thus construction firms should notify their insurer if items fall from the scaffold and injure a passer-by or damage a car parked near the site. Retail firms should notify their insurer of any slips involving customers, either in the shop or the car park especially if the customer is injured or if they are taken to hospital. Firms that sell products should notify their insurer when they become aware of any incident where it seems that the product has resulted in injury to the user or damage to property.

All of the above serve to protect the interests of the both the insurer and policyholder and ensure that any claim is handled in the best and most appropriate manner.

Policyholders must not make any admission of liability nor repudiate any claims, nor make any payment or offers of payment without the written consent of the insurer. Policyholders often regard the requirement not to admit liability in a strict or extreme manner. There is nothing to prevent, for example a retailer from offering sympathy, arrange for first aid treatment, a cup of tea, transport to hospital (provided an ambulance is not required) or home. Such action, perhaps accompanied by a bunch of flowers, replacement of spoilt shopping e.g. frozen goods may well help to deflect a claim. However, statements such as "You are the xth person to have slipped over . . . " or "Lots of people have fallen over here." are best avoided. Such comments can create the wrong impression and encourage the injured person to make a claim.

If the circumstances are clear cut an offer will be made and settlement negotiated. If a settlement cannot be agreed, if there is a dispute on facts or there is no legal precedent to follow then the claim may proceed to court. A judge will then be asked to decide on liability and the amount of damages that should be awarded or both. A very small percentage of claims go to court, in general less than 10%.

Discharge of Liability

[P9507] Whilst insurers will endeavour to represent the policyholder and settle claims in a timely fashion and with the minimum cost there may be circumstances where this proves difficult if not impossible.

The most likely scenario is where the costs of a claim are estimated to be higher than the policy Limit of Indemnity. In such circumstances the policyholder is liable for any amount over and above that provided by their insurer. They may have an excess of loss policy that would provide additional funds but if not they must pay any compensation over the Limit of Indemnity from their own funds. In such circumstance the insurer has no interest in the final settlement and will generally pay the Limit of indemnity to the insured

Whilst not common it is possible that a breakdown in the relationship between the insurer and the policyholder could arise. In such a situation the insurer may decide to discontinue the handling of the claim. This is permitted through the Discharge of Liability condition. This allows the insurer to pay the Limit of Indemnity or any smaller amount that the claim can be settled for and then allow the policyholder to settle the claim themselves. This is a rare event but the condition allows for this extreme action if it were necessary. In practice this condition is rarely used.

Reasonable Precautions

[P9508] Policyholders are expected to maintain their business premises in good order, to keep plant and machinery well maintained and take reasonable precautions to prevent injury to third parties and/or loss or damage to their property. They are also expected to comply with all legal requirements, such as health and safety legislation and keep records of their business, for example a record of purchases and sales.

Policies contain a condition that allows the insurer to avoid indemnifying, that is settling a claim if there is evidence that the policyholder has not complied with any of the above. The condition does not place an absolute duty on the policyholder but simply requires that reasonable care be taken. Thus, policyholders are expected to have the fixed wiring in their premises inspected since faulty wiring is known to cause serous fires. Similarly property owners are expected to maintain their buildings and property in good condition. If a piece of masonry were to fall from a building the expectation is that the property owner would be found to have been negligent through not arranging for suitable and sufficient inspections of the premises and carrying out maintenance as necessary. It is likely that most property owners would have some systems in place but they had simply not been sufficient. It would only be if there was evidence of wilful neglect of the property that an insurer might seek to invoke this condition and avoid settling a claim. Such circumstances are rare and infrequent.

Subrogation

[P9509] Subrogation is a legal principle that provides the right for one person to stand in place of another and avail themselves of all the rights and remedies of that other person. Subrogation enables an insurer to make certain that the insured recovers no more than exact replacement of loss (i.e. indemnity). 'It [the doctrine of subrogation] was introduced in favour of the underwriters, in order to prevent their having to pay more than a full indemnity, not on the ground that the underwriters were sureties, for they are not so always, although their rights are sometimes similar to those of sureties, but in order to prevent the assured recovering more than a full indemnity' (*Castellain v Preston* (1883) 11 QBD 380 per Brett LJ). Subrogation does not extend to personal accident insurance, policies whereby the insured (normally self-employed but who could in reality be any person) is promised a fixed sum in the event of injury or illness (*Bradburn v Great Western Railway* Co (1874) LR 10 Exch 1 where the appellant was injured whilst travelling on a train, owing to the negligence of the respondent. He had earlier bought personal accident insurance to cover him for the possibility of injury on the train. It was held that he was entitled to both damages for negligence *and* insurance moneys payable under the policy (see further **COMPENSATION FOR WORK INJURIES/DISEASES C6001**)

The right of subrogation does not arise until the insurer has paid the insured in respect of his loss, however, it should be noted that in respect of employers' liability cases payments are made direct to the injured party rather than the insured, the employer. Subrogation has been invoked infrequently in Employers' Liability cases. In some circumstances an employer may be entitled to an indemnity from employees for example where the act of an employee has resulted in an injury to another employee and the employer has been held to be vicariously liable. However, it is convention that such recoveries are not sought or pursued. (see **E13027**).

Policies contain a condition that allows the insurer to take action in the name of the policyholder to enforce any rights or remedies and recover any monies that might be owed by third parties.

Policy Exceptions

[P9510] The term exception means something that is not covered under the policy, for which the insurer will not indemnify the policyholder. There are a number of policy exceptions that apply across all and any covers that may exist. Typical exceptions that are applied to Public and Products Liability policies include:

- War Risks: The policy will not cover any loss that arises as a result of war, civil disturbance, nationalisation or any seizure by the Government or any action relating to such actions
- Radioactive Contamination: All losses arising from any form of release or exposure to nuclear waste or nuclear fuel. There is a "pool" arrangement that covers such losses
- Any claim relating to a loss of computer data, virus attacks, "denial of service" events and unauthorised access to computers i.e. hacking: Insurers do not wish to include loss of data and similar events within the definition of injury loss or damage.

In addition to the above there will be exceptions that operate at a lower level, limiting the cover for Public Liability or Products Liability.

Limitation Act

[P9511] Potential claimants have always been subject to a limitation on the time period within which they are able to bring a claim. Limitation is a practical approach to avoid the courts having to deal with claims long after the event, when memories have faded, details have become uncertain and witnesses unreliable. Limitation is a defence against claims being made after a lengthy delay and will be raised on behalf of the defendant, by the insurer. The law on limitation is contained in the *Limitation Act 1980*. There are many different periods dependent on the nature of the claim, a claim for personal injury through negligence, a claim under contract law etc but the main principle for liability insurance is as follows.

An injured person has three years in which to make a claim for damages. This period runs from what is known as the 'date of accrual', usually the date on which the injury occurred. The date of accrual may also be the 'date of knowledge' which is taken to be the date on which the injured person realised they had a right to claim.

Where a specific injury has resulted, such as a broken bone, amputation of part of the body, a burn or serious flesh injury leading to scarring, there is no dispute. The three year period runs from the date of the accident. When a person is found to be suffering from a disease that might have an occupational cause the date of knowledge is usually regarded as being the date when a medical authority such as their doctor or specialist explains that their condition might have been caused by exposure to harmful substances encountered in the workplace.

There is a special case where the potential claimant has died. This may occur with mesothelioma which can result in death within a short period of time following diagnosis. The limitation period is three years from the date when the deceased might have been regarded as having knowledge. However, if the death occurs within the three year period the dependents have three years from the date of death or from the date of their knowledge. For minors the situation is somewhat complex. A claim for an injury to a minor can be commenced by a parent or guardian. However, the legal 3 year period does not commence until the minor reaches the age of 18.

Claims are usually initiated within a reasonable period after the accident but are sometime delayed and only reach insurers towards the very end of the three year period.

Where the claim is for damage to property arising from negligence the period in which a claim must be made is six years. The period is also six years for claims made from an alleged breach of contract.

Vicarious Liability

[P9512] Vicarious liability is where one person assumes liability for the acts and omission of another. It can arise in several ways but is most commonly encountered in Employers' Liability where the employer accepts liability for the acts of his employees. This is in accordance with the common law principle that sets out that an employee can expect their employer to provide competent fellow employees. There are some exceptions, e.g. if an employee was doing something expressly forbidden or carrying out a task in an unacceptable way, but generally the employer will be held liable. Also if one employee is responsible for injuring another then as a general rule the employer will be liable.

However, vicarious liability can also arise under Public Liability. The most common example would be where an employee of the firm causes injury to a third party, for example where a worker on a building site drops something from the scaffold resulting in an injury to a passer-by. In such circumstances the employer would be vicariously responsible for the actions of their employee and if the passer-by made a claim against the firm then it would be regarded as a Public Liability claim and the employer would most likely be regarded as having been negligent for not establishing working practices that prevented such an event. A typical example would be scaffolders who, when dismantling a scaffold throw the clamps and fittings from the scaffold into the vehicle that is parked adjacent to the footpath. The risk of injury to passers-by is high, as is the risk of damage to a passing car.

Where there has been physical or sexual abuse then the perpetrator's employer is generally regarded as being vicariously liable for the actions of their employee. Victims of such abuse are entitled to sue the employer for damages and compensation.

Professional Indemnity

[P9513] A professional or expert simply offers advice, unlike a manufacturing firm that makes and sells a physical product or a retailer who in addition to selling products invites the customer into their premises. It is possible that advice given by a professional could result in some form of loss and be regarded as having been negligently given. Such a liability is called Professional Indemnity (PI).

This form of liability is very different from other forms of liability, and is usually excluded from Public Liability policies. Experts who offer advice such as solicitors, surveyors and estate agents, engineers, architects and consultant must purchase a PI policy to protect them from the effect of their advice being deemed negligent.

It is possible that manufacturers may offer advice to their customers. They may provide guidance and advice regarding which of their products is most suitable and will meet the customer's requirements. Alternatively, a manufacturer may obtain details of a customer's specific needs and develop a product that is designed to meet those needs. Similarly, a supplier of steel frame buildings may design buildings for customers.

In all these examples the advice and professional guidance is being given in connection with the manufacture, sale, supply or installation of a product. In such cases the Public Liability policy is extended to include any professional errors or omissions. However, the policy will clearly state that the cover does not extend to advice given "for a fee". Thus if the policyholder performs a design service in return for a fee and does not supply the product then there is no cover. A separate PI policy must be purchased to protect against the risk of claims from such work.

Public liability insurance

Purpose

[P9514] The purpose of Public Liability insurance is to ensure that firms are covered for any legal liability to pay damages to third parties that is persons who are not employees who suffer bodily injury and/or disease as a result of the activities of the firm. Such liability is normally based on negligence or breach of statutory duty on the part of the firm itself. Liability may also result from a breach of contract, for example to supply goods or services to an agreed standard or that they have a defect that might cause injury or damage.

A Public Liability policy is a legal liability policy. Hence, if there is no legal liability on the part of a firm, the policy will not respond and no compensation or damages will be paid out by the insurer. The policy indemnifies a firm against claims made by any person except employees and any claims made by any firm or other body.

Case law suggests that Employers' Liability is becoming stricter and there is a belief, in some quarters, that Employers' Liability claims are incapable of being defended, that the burden of proof has been reversed. There have been concerns about a similar trend in Public Liability however; recent court decisions indicate that the courts are attempting to ensure a balance of responsibility.

Duty to take out insurance

[P9515] Employers' Liability insurance is a compulsory class of insurance. 'Every employer carrying on business in Great Britain shall insure, and maintain insurance against liability for bodily injury or disease sustained by his employees, and arising out of and in the course of their employment in Great Britain in that business' [*Employers' Liability (Compulsory Insurance) Act 1969, s 1(1)*].

There is no corresponding duty to take out public liability insurance with just a few very specific exceptions (see **P9516**). However, failure to take out such insurance exposes the firm to an unacceptable risk of failure through being required to pay damages or compensation of such value that it would bankrupt the firm.

In some cases it is necessary to take out Public Liability insurance in order to comply with contractual requirements for example in the Construction industry or when tendering for Local Authority contracts.

Compulsory Public Liability Insurance

[P9516] There are very few situations where a business is required to take out Public Liability insurance

Riding Establishments Act 1970

Riding schools must be licensed by the Local Authority. The Act states that all licensed riding schools must be insured against any liability for injury to any person sustained whilst riding a horse hired from the School or riding a horse in connection with tuition being given by the school.

There is no mention in the Act of the cover required, the limit of indemnity that must be purchased, nor is there any reference to a certificate of insurance.

Dangerous Wild Animals Act 1976

This Act requires that any person who wishes to keep certain prescribed wild animals must hold a licence that is issued by the Local Authority. The granting of the licence is conditional upon the prospective licence holder having insurance against the risk of any damage caused by the animals. As with riding establishments there is no mention of certificates or minimum levels of indemnity.

In addition to the above there are certain very specific requirements relating to operators of nuclear installations and aircraft operators. However, these are outside of the scope of normal business activity.

Cover provided

[P9517] The insurer will pay both compensation and any costs and expenses provided they arise from an accidental

- Personal Injury
- Damage to property
- Any obstruction, trespass, nuisance or interference with any right of way, air, light, or water

The event must happen within the period of insurance and occur within the Territorial Limits as set out in the policy. The maximum that will be paid is the Limit of Indemnity together with any costs and expenses. However, for any claim raised or made in certain overseas locations, primarily North America, the costs will be included within the Limit of Indemnity.

The policy provides for payment of:

(a) compensation for personal injury or disease, or loss of or damage to third party property whether agreed or negotiated or awarded by a court

(b) costs and expenses of litigation, incurred with the insurer's consent, in defence of a claim against the insured (i.e. civil liability);

(c) claimants costs and expenses for which the insured is legally liable;

(d) solicitor's fees, and other legal costs incurred with the insurer's consent, for representation of the insured at proceedings in any court of summary jurisdiction (i.e. magistrates' court), coroner's inquest, or a

(e)   fatal accident inquiry (i.e. criminal proceedings), arising out of an accident resulting in injury to an employee; it does *not* cover payment of a fine imposed by a criminal court;
(e)   solicitor's fees, and other legal costs incurred with the insurer's consent, for representation of the insured at proceedings resulting from a prosecution for a breach of health and safety and related legislation, that is a breach of statutory duty; this is subject to certain provisos for example the breach must not have resulted from a deliberate act or omission by the insured and no indemnity will be paid if there is a legal expenses policy in force;
(f)   compensation for any director, partner or employee attending court as a witness in a case for which the insurer is providing an indemnity.

Persons

[P9518] With Employers' Liability insurance the classes of persons who are able to make a claim against their employer has to be strictly defined. The indemnity or benefit from an Employers' Liability policy is provided to the employers in respect of claims from their employees and a specified number of other persons such as volunteers and persons who are on loan to the employer who are regarded as employees.

However, this approach cannot be used on Public Liability policies. It is not possible to make any restrictions other than the exclusion of employees or any other persons who are treated or regarded as employees. If employees have been injured as a result of the negligence or breach of statutory duty by their employer then they must make a claim against their employer's Employers' Liability policy.

Any other person who has been injured or suffered a loss of or damage to property as a result of the activities of the firm can make a claim against the firm's Public Liability policy. Typical classes of persons who make such claims include:

- Customers to retail premises; including both the shop itself and surrounding paths, walkways and car parks
- Visitors to offices or other premises
- Tenants in rented accommodation
- Members of the public who are injured as a result of a slip or trip on the pavement
- Passers-by who may injured by materials falling from a vehicle delivering to a construction site

It is commonplace, especially for smaller firms, for the same insurer to provide both the Employers' and Public Liability insurance and in such cases the wordings will be aligned. Current market practice is that few insurers provide Employers' Liability cover without also holding the corresponding Public Liability insurance. However, whenever different insurers are used to provide the two covers it will be necessary for the policyholder or their broker to check that the policy wordings are suitably aligned.

Rating

[P9519] Certain trades or businesses present a greater risk than others. For most trades or businesses insurers have their own rate for the trade or business occupation, expressed as a rate per mille on turnover. This is known as the book rate and is derived from the insurer's experience i.e. the balance of premium against claims for all firms of that type. For small firms this rate is used as a guide and is altered upwards or downwards depending on:

(a)   previous history of claims and cost of settlement;
(b)   size of the company, generally taken to be indicated by the turnover;
(c)   whether certain risks are not to be covered, e.g. the premium will be lower if the insured elects to exclude certain types of risk exposure from the policy, or has accepted an excess
(d)   the insured's attitude towards safety.

For very small firms a minimum premium may be set.

For large firms that may have several claims each year it is possible to calculate the coming year's premium by basing it directly on the cost of claims in previous years, generally the preceding five years. The premium will be adjusted to take into account the firm's efforts to reduce risk, train staff, complete risk assessments etc. This is known as experience rating.

There are a number of different methods of rating of Public Liability insurance but all are related to a measure of the businesses size or activity.

Turnover:

Work away: applied to the wage roll of employees who work away from the premises

Premises: a flat charge applied to the policyholder's premises

Contract Price: where the activities of the policyholder are based on specific contracts

Payments to sub-contractors:

Footfall or Customer Numbers: where the activities of the business involve large numbers of visitors or customers e.g. theme parks, stadia and cinemas

The most appropriate measure or group of measures is then used as the basis of the premium calculation. For most businesses the turnover is the most appropriate measure of the business activity. Some firm have more than a nominal amount of work away from the premises. The premium will be modified according to the actual business activity carried out. For example, a retail business with a number of shops and many thousands of customers visiting each working day presents a greater risk of accidents and claims from third parties than a factory, which may have very few if any visitors' apart from in the offices. A leisure risk that has swimming pools, saunas, climbing walls and gym facilities presents a significant risk if injury to third parties.

For firms with multiple premises, for example a chain of retail shops, the premium may be based upon the number of premises since each location represents a "unit" of risk, a place where customers may slip or trip and be injured.

Some firms in the retail sector may have substantial numbers of employees engaged in delivery work. Any work away from the premises presents an additional risk of Public Liability claims. Thus the amount of work away from the premises, denoted by the wageroll may be a Public Liability rating factor.

In the leisure sector firms that operate cinemas or attractions such as theme parks may be rated on the basis of customer numbers, footfall, since this may better present the risk exposure than financial turnover.

Some firms will have teams of engineers installing or servicing the products that have been sold or supplied or simply providing a maintenance service. For such firms insurers will be interested in the nature of the contracts and the type of work undertaken. It will also be necessary for the firm to confirm the type of premises visited. Insurers seek confirmation of any work undertaken at "high risk premises" such as petrochemical plants, power stations etc. Work at such premises presents a risk of more serious claims.

If firms make use of sub-contractors then they will need to declare the type of contractor employed, whether they are bona fide firms that are assumed to hold their own insurances or labour only sub-contractors where a measure of responsibility may fall upon the employing firm. Such information will need to be disclosed to the insurer together with the amounts paid to such contractors.

In addition to the above other trades may well be rated on specific measure that are particular to them e.g. acreage for farms. Thus it is possible for the premium of a firm to comprise a number of elements: thus for a retailer there might be premium derived from:

- The premises – based on the number of locations
- Turnover – to represent the sales of the firm
- Work away – to reflect work undertaken on installation, for example of electrical appliances
- Drivers – to reflect the risks associated with delivery work

Many insurers survey premises with the object of confirming the exact nature of the business and the insured's attitude towards and management of risk. In addition the visit will provide an opportunity to improve the risk by offering guidance and advice or requiring that improvements are made, the completion of such improvements should minimise the incidence of accidents and diseases. This is an essential part of an insurer's service, and the risk surveyors will work in conjunction with the insured's own management and/or safety staff. However, it should be noted that this practice is restricted to the largest of risks and/or those with the greatest risk potential since insurers employ limited numbers of risk surveyors. Only a very small percentage of policyholders receive such visits.

Limits of Indemnity

[P9520] Whilst the legislation stipulates the amount for which an employer is required to insure and maintain insurance relating to Employers' Liability (£5 million in respect of claims relating to any one or more of his employees) there is no corresponding legal requirement for Public Liability.

[P9520] Public & Products Liability Insurance

The usual level of indemnity is either £2m or £5m. This is generally described as being the limit of indemnity for "any one event or series of events arising from one cause. The effect of this is to limit the amount payable irrespective of the number of claimants. Thus, for example if a firm carrying out the shot blasting of a bridge caused damage to many cars the total amount payable would be the limit of indemnity no matter how many claimants there were and the amount that was claimed.

The Limit of Indemnity does not include any cost and expenses – these are paid in addition to the compensation and damages.

The indemnity may be extended at the request of the policyholder. If the holding insurer will not provide an increased indemnity the policyholder will have to purchase what is known as Excess of Loss insurance (XoL). This is a separate insurance policy that agrees to pay an additional indemnity over and above the underlying policy. This if the first policy had a limit of indemnity of £5m but the claim resulted in damages of £10m the XoL policy would pay the additional £5m.

The policy may contain an excess negotiated between the insurer and firm, and this is much more common in Public Liability policies than Employers' Liability i.e. a provision that the employer pay the first £x of any claim.

Territorial Limits

[P9521] Cover under Public Liability policies is normally limited to

Great Britain, Northern Ireland, the Channel Islands and the Isle of Man and offshore installations

Other member countries of the European Community

The rest of the world in respect of any injury, loss of or damage to properly arising from non-manual work

Insurers are generally not interested in providing cover for companies that have overseas manufacturing operations and may further limit the cover to business activities carried out in premises in Great Britain, Northern Ireland, the Channel Islands or the isle of Man. The Public Liability exposure of overseas premises or locations should be insured locally in that country.

The cover is usually granted on the basis that any action for damages is brought in a court of law of Great Britain, Northern Ireland, the Channel Islands or the Isle of Man — though even this proviso is omitted from some policies. This is known as the jurisdiction clause.

Strict Liability

[P9522] In some circumstances, Public Liability insurance is subject to the doctrine of strict liability. Strict Liability does not mean that the policyholder is automatically liable and must pay damages – that is known as absolute liability. Strict Liability involves a reversal of the burden of proof. The claimant is not required to prove that the policyholder has been negligent or that there has been a breach of contract. It is for the person who is alleged to have been negligent to prove that they were not, that they acted reasonably. There are a limited number of defences available, including:

- Act of God
- Act of a stranger or third party (that in turn resulted in the loss)
- Statutory Authority (whereby the policyholder had statutory authority to act in the manner hey did.
- Negligence of the claimant

However, it is difficult for the policyholder to avoid being held at fault. Strict Liability was established by the legal case of *Rylands v Fletcher* (1868) LR 3 HL 330, 33 JP 70, HL. In this case the defendant constructed a reservoir on their land to hold water for use in their mill. Through some fault or negligence of the engineers and contractors who constructed the reservoir water escaped, enter a neighbours mine and caused damage. When the claim came to be heard certain difficulties were realised:

- The case did not involve trespass since the damage was not immediate
- It was not a case of nuisance since the damage was not the result of a recurring problem and
- Since the defendant had not themselves been negligent it was not a case of negligence

However, the defendants were held to be strictly liable for the damage. The judgement centred on the judge's view that any person who brings something onto their land, which although harmless whilst remaining there becomes dangerous if it escapes is strictly liable for the damage caused. In a later hearing in the House of Lords it was further clarified that there must be a "non-natural use of land" for the strict liability to exist.

Over the years there have been many cases that have relied upon the "rule of Rylands V Fletcher" and judges have given much thought to what substances fall within the original judgement's view of substances that become dangerous if they "escape" and what constitutes non-natural use of land.

Certain substances and things are held to be always dangerous, for example, chemicals and electricity. However, water trees and motor cars may or may not be regarded as dangerous. Thus whilst the water in the original case of Rylands V Fletcher was held to be a dangerous thing in a later case water stored in a cistern for use by the occupants was not held to be dangerous and nor was the use of the land regarded as non-natural

Strict Liability also existed in much health and safety legislation as a result of s 40 of the *Health and Safety at Work etc Act 1974*. This section required the defendant in any criminal proceeding taken under the Act or related regulations to prove that it was not practicable or reasonably practicable for them to have done what was necessary to satisfy the particular requirement or that there was no better means available.

In 2013 an amendment to the Health and Safety at Work etc Act 1974 deleted the relevant section. The effect was to remove the ability to bring a civil claim for damages alleging breach of Statutory Duty unless the Act or Regulation specifically allowed for civil action.

Policy Conditions

[P9523] All Public Liability policies will have a number of standard terms and conditions that serve to modify the cover provided in some way. The following are some of the most common.

Reasonable Precautions Condition

The Employers' Liability (Compulsory Insurance) Act prohibits the placing of conditions on the policy that would mean that the insurer was able to avoid the policy if the policyholder had been negligent.

There is no such limitation for Public Liability insurance and thus most policies contain a "reasonable precautions" condition. The effect of this is to require that the policyholder takes precautions against loss or to mitigate or minimise the effects of a loss. There is no requirement for the policyholder to act upon risks or situations that are not foreseeable – that would be unreasonable.

Suspension of Cover

The insurer may decide that they no longer wish to continue providing cover. An example might be if the policyholder's premises were subject to a visit and inspection by the insurer's risk engineer or surveyor and the conditions were found to be unacceptable. For example risks were not being managed, housekeeping was poor, the risk of injury to customer or visitors was high etc. In such circumstance the insurer can issue formal notice of cancellation of the policy. However, in practice the insurer would set out what needed to done to improve the situation and allowed reasonable time for the necessary changes to be made.

Discharge of Liability

This rarely used condition gives the insurers the right to pay the limit of indemnity and cease to be involved. For example if the policy limit of indemnity was £1M but the claim involved far greater sums, say £2.5M the insurer is entitled to pay the limit of indemnity and leave the policyholder to continue with the handling of the claims.

The rationale for this is that in this example the policyholder's interest far exceeds that of the insurer and so it is hardly appropriate for the insurer to continue handling the claim.

Contractual Liability

The majority of Public Liability policies contain a condition relating to contractual liabilities. A typical wording might read as follows:

> "liability assumed by the Insured under contract or agreement and which would not have attached in the absence of such contract or agreement shall be the subject of indemnity under this section only if the conduct and control of any claims so relating is vested in the Insurers and subject to the exceptions and extensions of this section".

This means that the policyholder will be indemnified if they are liable for injury or damages to property solely as a result of some contractual liability but only if the conduct and control of the claim is given to the insurer.

A recent case has clearly demonstrated the operation of this clause. Tesco were building a store which was to be located above a railway line. This was to be achieved by building a culvert for the train tracks and then covering them to create land for the building of the store. During the work there was a collapse and the train operating company suffered losses through not being able to run their trains. The train operating company and Tesco had entered into a deed of covenant whereby Tesco agreed to pay the company any losses arising out of or in connection with the works. Tesco had failed to disclose the existence of the deed to their insurer. Tesco settled with the train operating company and then attempted to recover their outlay from their insurer. Tesco's insurance policy contained a contractual liability clause as described above and the insurer declined to indemnify.

The case went to court and the judge found in favour of the insurer. Tesco were given leave to appeal but the Court of Appeal did not overturn the original decision, agreeing that the decision not to indemnify was correct. This judgment confirms that a typical public liability wording will not cover a "standalone" contractual liability unconnected with any liability owed to the third party in tort.

A policy could be amended to cover contractual liability for pure financial or economic loss — provided it says so clearly. *Tesco Stores V Constable and Others* [2008] EWCA Civ 362

Claims Notification

This condition requires that the policyholder notify the insurer as soon as they are aware of any circumstances that might give rise to a claim and to allow the insurer to take over the handling of the claims. This condition is expressed as condition precedent meaning that the insurer can avoid making a payment if there has been a breach of the condition or example the policyholder filed to notify the insurer of a possible claim. Since it is a condition precedent to liability the insurer doe not even have to have been prejudiced by the lack of or late notification.

Pollution

All Public Liability polices contain a condition limiting the insurer's liability for damage caused by or resulting from pollution. (see **P9526**)

Other conditions relating to the firm's trade include the following:

Use of Heat:

The term "use of heat away from the premises" is used by insurers to denote any work involving the use of heat. Such work has been the cause of many costly claims under Public Liability policies and insurers thus identify those activities and trades that might use heat and require them to conform to certain, minimum good standards in order that the risk is acceptable. This is done by applying a condition to the policy that gives the insurer the right to not pay any claim if it is found that the policyholder breached the condition.

Work involving the use of heat is taken to include welding and flame cutting of metal, the use of soldering equipment, the use of angle grinders to cut metal or smooth welds, the use of hot air guns to join plastic materials and the use of tar boilers e.g. for flat roofing work. Thus very many trades fall within the scope of this condition, from jobbing engineers to steel erector, from plumbers to roofing contractors.

Work at Height:

The fact that a particular firm undertakes work at height is generally self evident to the insurer from the business description. Insurers may make further enquiries via the broker as to the exact nature of the work, the height at which it is undertaken and what proportion of the firm's activities involve work at height etc. From a Public Liability perspective work at height carries a particular risk. The chance that something may happen that could cause injury to a third party is increased if the work is undertaken at height. Tools and equipment may be dropped causing injury to persons working below. It is possible that the machinery or equipment that is being worked on may have to be lifted into position with the risk that it may be dropped or damaged. Or the work may involve the erection of buildings e.g. steel erection, in which case there is the risk that the structure may collapse before it is fully erected.

Underground Services:

Certain trades whose work involves excavation face the risk of striking underground cables. Firms that operate as ground workers, or are involved with the laying of utilities pipe work or installing cables for TV or broadband services may work in locations where there are existing cables or services. There is well documented guidance e.g. from the Health and Safety Executive describing the risk management approach that should be adopted. The checking of plans, securing the advice of utility companies and the use of cable detection devices (CAT scanning) are all extremely useful in reducing the risk of damage to underground services. Coupled with careful planning and the use of safe digging practices i.e. hand digging, it is possible to excavate with a very low risk of striking an underground pipe or cable. In an effort to ensure that sound risk management practices are adopted insurers place a condition on the policies of those firms that undertake digging and excavation work.

The condition makes it possible for the insurer to avoid paying the claim if the policyholder did not follow best practice.

Policy Endorsements and Extensions to Cover

[P9524] There are a number of endorsements and extensions that are commonly used to limit or restrict cover, or to add in covers but unlike Conditions they do not set out exactly how the insurer expects work to be carried out.

Defective Premises

The Defective Premises Act 1972 imposes a liability on persons who dispose of or sell property making them liable for injury or damage arising out of work done on the premises. It is common for Public Liability policies to be extended to provide cover in the event that the policyholder should sell property or move from one premises to another.

Motor Contingent Cover

This is a standard extension that protects against the eventuality that a motor vehicle that is not owned by the policyholder but is being used in connection with the business causes injury, loss of or damage to property. The extensions will usually exclude any damage to the vehicle itself and will not be effective if there is insurance under another policy.

Excess Clause

It is sometimes the case that policyholders retain some of the risk for themselves in much the same way that motorists have an excess placed on their private car insurance. For example, the policyholder may choose to pay the first £10,000 of each personal injury claim and perhaps a greater amount of property damage claims.

The effect of this is that the insurer pays less in claim and this has the effect of reducing the premium. In addition, where the policyholder has several premises e.g. a chain of shops or subsidiary companies it is possible to allocate the excess to the different premises or companies where the claims have occurred. This will have the effect of focussing their activity towards risk management. If, for example each shop is regarded as a profit centre then having to pay the excess of any claim will reduce profits. generally motivates the store manager to identify, manage and reduce risks.

Health and Safety at Work Act

Public Liability policies are usually extended to provide cover for defence costs. In the event that the policyholder is prosecuted under the Health and Safety at Work etc Act or any related health and safety regulations the insurer will cover the costs of defending such an action. The cover is conditional upon the policyholder requesting such an indemnity; however, insurers are very keen that they be allowed to fund a defence.

By arranging for the policyholder to be represented it is possible for insurers to gain valuable information about the circumstances of the accident or event that could prove useful in any subsequent civil compensation claim.

The cover will extend to pay any prosecution costs that the court may order the policyholder to pay. Generally, the cover will extend to the funding of any appeal against conviction, if such an appeal is of merit. However, there is no cover for the payment of any fines, since this is regarded as being against public policy. Fines and penalties cannot be insured.

The cover is often extended to include prosecutions under the Food Safety Act.

Corporate Manslaughter Prosecutions

With the enacting of the Corporate Manslaughter Act it became necessary for insurers to amend their cover in respect of defence costs. This Act is not part of health and safety legislation. Thus a further extension to cover was added to policies to ensure that policyholders were entitled to have their defence funded.

Policy Exclusions

[P9525] There are a number of standard exclusions that appear within all Public Liability policies. Generally they serve to clarify the cover that is being provided or to ensure that there is not dual insurance. The following are some typical examples

Employee accidents: Claims arising from accident involving employees should be made under the Employers' Liability policy.

Aircraft, Watercraft and Motor Vehicles: These are normally insured under more specific policies, for example a marine policy.

Property owned by the insured: This would usually be insured under a Material Damage policy.

Pollution: Some policies exclude all cover for pollution, thereby allowing the insurer to make a full assessment of the potential risk and then add cover back into the policy in a controlled fashion.

Penalty payments, Liquidated Damages, Fines etc: in the construction sector failure to complete contract on the agreed date can lead to a firm having to pay pre-determined penalties. These are not accidental and are regarded as penalties, not the result of an accidental event. Fines are a form of punishment and it would be against public policy to be able to insure against them.

Punitive and Exemplary Damages: These are more commonly found in North America where the court may require firms to not only pay compensation but also the pay additional monies to represent a punishment. Such payments are not insured.

Pollution

[P9526] If harmful substances were to escape from the premises and cause damage to neighbouring property it is possible that the firm or business could be sued for compensation. Insurers regard this risk as unattractive and generally the cover provided is restricted. Insurers generally recognise two types of pollution: gradual and "sudden and accidental".

The following wording is taken from the Public Liability policy of a major general insurer.

1 Pollution clause

This Policy excludes all liability in respect of pollution or contamination other than caused by a sudden identifiable unintended and unexpected incident which takes place in its entirety at a specific time and place during the Period of Insurance.

All pollution or contamination which arises out of one incident shall be deemed to have occurred at the time such incident takes place.

For the purpose of this clause 'pollution or contamination' shall be deemed to mean:

a) all pollution or contamination of buildings or other structures or of water or land or the atmosphere and

b) all loss or damage or personal injury directly or indirectly caused by such pollution or contamination.

The effect of this wording is to exclude from the cover provided by the policy all pollution other than that which occurs "suddenly". Thus, if a forklift truck were to knock over a diesel storage tank and in doing so cause the tank to rupture then any pollution caused by the leak of diesel would be covered. And if a third party were to make a claim for compensation following damage to their property then the policy would respond. However, if the tap or valve was leaking and diesel dripped onto the yard surface over an extended period then this would be regarded as gradual pollution and would not be covered. Insurers describe pollution resulting from poor control of the process, of drips and minor leaks that continue for long periods of time as gradual and exclude it from the policy.

From a risk management perspective the Environment Agency guidance on the bunding of tanks not only ensures that any leaks or drips are continued within the bund wall but also provides physical protection against the majority of sudden and accidental events that could lead to pollution.

Damage arising from pollution that is caused by a sudden and accidental event is covered. If there was a failure of equipment designed to clean and filter gases before they were discharged and this resulted in the escape of harmful gases then cover would be provided.

A Public Liability policy only provides an indemnity where there has been damage to third party property and there is no cover for damage to the environment. Thus in the above examples cover would be provided if:

The escaping diesel killed fish that had been placed in the river by a fishing club

or

The escaping gases caused damage to cars in the car park

However, if the Environment Agency carried out a clean–up and sought to recover their costs there would be no cover. Such statutory clean up costs are not damages or compensation as defined by the policy. This was confirmed by the case of Bartoline

In this case there was a fire at a factory that manufactured paints and wood stains. The flammable components generated a serious fire and the Fire and Rescue Services had to use considerable quantities of water to contain and extinguish the fire. Some of this fire water run-off caused damage to surrounding water courses and the Environment Agency had to carry out emergency clean up operations to mitigate the harmful effects and placed enforcement notices requiring the firm to carry out other preventative work. The Environment Agency sought to recover its costs.

The firm sought to recover from their Public Liability insurer both the costs the Environment Agency sought to recover and the costs of complying with the notices. The insurers refused to indemnify on the grounds that the statutory clean-up costs were not damages within the meaning of the policy.

Bartoline v Heath Lambert and RSA [2006] EWHC 3598 (QB).

Following this case some insurers are prepared to extend their policies to provide an indemnity for such costs. This is usually restricted to certain trades, will provide a lower level of cover than the main policies Limit of Indemnity and may require payment of an additional premium.

Legislation relating to Public Liability Insurance

[P9527] There are a number of pieces of legislation that apply to the insurance of Public Liability. In addition to the duties under sec 3 of the Health and Safety at Work etc. Act 1974 there is legislation relating to fire safety and the duties placed on occupiers of premises. Some of the most commonly encountered pieces of legislation are as follows:

- Health and Safety at Work etc. Act 1974
- Control of Asbestos Regulations 2012
- Defective Premises Act 1972
- Fire Safety (Regulatory Reform) Order 2005
- Gas Safety and Use Regulations 1998
- Hotel Proprietors Act 1956
- Occupiers' Liability Acts 1957/1984
- Riding Establishments Act 1964/1970

Common Public Liability Risk Exposures

[P9528] Some of the following liabilities have been established by criminal law. However, it is more likely that there was no prosecution and the precedent has been set by a civil case. Since relatively few civil cases proceed to court and are reported the legal position can often only by estimated.

Risks arising from the Ownership and Control of Premises:

There are many risks that arise from the ownership and control of premises. The simplest of risks relate to the condition of the premises and policyholders should ensure that the premises are free of slip and trip hazards. Flooring and floor coverings should be in good order and replaced or repaired if damaged. Floors and particularly stairs should be clear of trip hazards and in good order e.g. stair treads should ideally be of a non-slip nature and stair nosings should be in good condition.

In retail premise the displays should not present any risk to customers e.g. lighting displays with hot bulbs should be out of reach of small children and stacked materials in a DiY store should not be at risk of falling against customers.

The failure to maintain buildings can give rise to the risk of injury through, for example, falling debris. In Edinburgh, in 2000 a waitress was killed when she was struck by a part of the building that fell from roof level. Just two years later there was a similar incident a matter of yards from that described above but which did not result in injury since the loose part of the structure was retained by a parapet. However, the Council instituted a check of some 16,000 buildings in the city centre.

In some cases bad weather can result in parts of buildings become loose or detached and cause injury to persons and passers-by;

- in January 2013 a metal shop sign fell onto a man as he exited a betting shop in North London the man had a cardiac arrest and died in hospital;
- in February 2013 woman who suffered a mild brain injury and concussion when she was hit by a falling shop front sign received £61,000 in an out of court settlement; and
- in February 2014, a taxi driver was killed almost instantly when a piece of masonry feel from a building onto the taxi she was driving.

The award of compensation described in the last example illustrates the importance of having Public Liability insurance.

All those with responsibility for the maintenance and upkeep of buildings should have in place measures for inspection and checking so as to ensure the structural integrity of the building and roof.

Such dramatic incidents attract press comment, however, minor accidents involving the public and giving rise to a liability can occur anywhere, inside or outside a public place and at any time of day. Most commonly they happen inside business premises, schools, grocery stores and shopping centres. But accidents are just as likely to take place outside in public parks, playgrounds, and on the high street and side roads, where broken or uneven pavements and walking surfaces are a continuing major cause of personal injury claims.

Spread of Fire to Third Party Property

All businesses are at risk of causing damage to neighbours property if there is a fire on the premises that spreads and causes damage. It is possible that a firm could be held to have been negligent and held liable for the damage to the neighbouring property.

Fire has sometimes been regarded as a dangerous "substance" and claims for damage caused by fire originating on a neighbours land been the subject of claims under Rylands v Fletcher. However, there are separate precedents regarding liability for the escape of fire although the liability is also strict.

The earliest case dates back to 1401 where it was established that if a householder places a candle in their window and ignites their thatch they are responsible for the spread of fire to neighbouring properties. However, if the thatch had been ignited by a stranger or third party then they could not be held responsible. This approach has been maintained up to the present day.

A case of fire spread that was decided under *Rylands V Fletcher* concerned a factory that manufactured packaging from polystyrene. During the process of cutting a block of polystyrene using a hot wire (a standard method for carrying out such cutting) a fire started that in turn spread to neighbouring property. It was held that use of the premises for the manufacture of polystyrene was non-natural use of land and that the cutting process carried a risk of fire. The firm was held strictly liable for the spread of fire. In addition there was negligence in that the employee carrying out the work was not trained, there were no safety devices on the machine nor was there a fire detection system.

LMS International Ltd v Styrene Packaging and Insulation Ltd [2005] EWHC 2065 (TCC), [2006] BLR 50, [2005] All ER (D) 171 (Sep)

In a more recent case a claim made under the rule of *Rylands v Fletcher* failed. A tyre fitting firm had stored many surplus tyres in an alley or passageway between their premises and those of a neighbour. Due to an electrical fault a fire developed and engulfed the tyres whereupon it destroyed the defendant's property and that of their neighbour. At the first hearing an allegation of negligence was dismissed but the defendant was held to have a strict liability under *Rylands v Fletcher*. The case went to appeal where the claimant lost their case. It was decided that:

- The carrying out of a tyre fitting business did not amount to non-natural use of land given that the premises were located on an industrial estate.
- Tyres had been brought onto the premises but they were not, in themselves, dangerous items and
- It was the fire that escaped not the tyres and the defendant had not brought the fire onto their land.

Stannard (t/a Wyvern Tyres) v Gore [2012] EWCA Civ 1248, [2013] 1 All ER 694, [2012] NLJR 1289

Use of Contractors

Most firms make use of contractors from time to time. Some firms make use of builders, electricians and plumbers to carry out routine maintenance. Work such as extensions to the premises or the installation of new plant and machinery might require the use of specialist contractors.

There have been well documented examples of firms being involved following an accident resulting from or caused by the contractor's activities or following an accident involving a contractor's employee. Criminal prosecutions have resulted in the employing firm being fined and civil claims for compensation have resulted in the damages and compensation being shared between the employing firm and the contractors.

Firms should take steps to ensure that any contractors they use are competent and experienced, adopt good health and safety practices and have suitable and sufficient insurance cover. There are many examples of situations where the sub-contractor did not have insurance and the employing firm became involved and had to settle the claim under their own insurance.

In September 2008 a contractor's employee working on the roof of Tullis Russell's paper mill in Fife died after falling through the roof. The Health and Safety Executive investigated and found that Tullis Russell had instructed the contractor to clean the roof and gutters. Although the contractor had said that crawling boards would be used they were not supplied to the workers. Tullis Russell failed to ensure that the work was properly planned and organised and they did not monitor, control or review the manner in which the work was carried out. They failed to check that crawling boards were made available and used and they breached their own procedures that required daily monitoring of contractor's activities. Tullis Russell were prosecuted and fined £260,000. The contractor who operated as a self-employed person was fined £20,000.

In October 2009 a contractor's employee fell through a skylight whilst cleaning the gutters on a factory roof. The Health and Safety Executive investigation determined that the contractor's employees were not trained to carry out work at height in safety. The employees were not supplied with any safety equipment nor were there any measures in place to prevent falls from height. The owners of the factory had failed to assess the competency of the contractor and had not requested a risk assessment. They also failed to supervise the work or carry out any check to ensure that appropriate safety equipment was being used.

The factory owners were fined £50,000 with costs of £19,300 whilst the contractor was fined £65,000.

These two very similar accidents clearly demonstrate the extent to which firms employing contractors need to ensure that they employ competent contractors and that they carry out suitable and sufficient checks to ensure that the contractors are working safely.

Work away from the Policy holder's Premises

Some firms carry out work away from their premises. A typical example is firms that install their product e.g. manufacturers of machinery or kitchen installers. Such work represents a significant increase in risk of loss or damage. Insurers will want to know about any such work and how much of the businesses activity derives from work away. Other examples of work away from the premises include delivery and servicing or maintenance of products supplied. Perhaps the most serious work carried out away from the premises is that which involves the use of heat.

Use of Heat:

Insurers will want to know whether there is any work involving the use of heat away from the premises. This could be a specific part of the trade activity e.g. painters and decorators who might use heat to remove paint, plumbers who might make joints using heat to roofers who may repair flat roofs using hot bitumen. Similarly jobbing welders are expected to use heat on third party premises. For such trades insurers will require full details of the nature of the activity and the percentage of time or amount of such work undertaken.

For other trades there may be less obvious use of heat. For example maintenance engineers who undertake servicing work may use heat to free seized bolts or may even cut off rusted nuts in order to allow equipment to be removed or replaced. Such activities have resulted in serious fires and given rise to expensive claims.

There should be robust procedures in place to control any use of heat. The procedure is essentially a risk assessment that requires the insured to identify the risks e.g. the presence of flammable materials and then take appropriate precautions. Such procedures should assess the risk with the aim of avoiding the use of heat if possible. If the use of heat cannot be eliminated then suitable precautions should be taken. Public Liability policies are generally endorsed along the following lines:

2 Use of heat clause

It is a condition precedent to the liability of the Insurer under this policy that the following precautions are complied with on each occasion of the use or application of heat (as defined below) by or on behalf of the Insured taking place elsewhere than on the Insured's own premises.

a) Application of heat by means of electric oxyacetylene or other welding or cutting equipment or angle grinders, blow lamps, blow torches, hot air guns or hot air strippers.
    i) The area in the immediate vicinity of the work (including in the case of work carried out on one side of a wall or partition, the opposite side of the wall or partition) must be cleared of
all loose combustible material; other combustible material must be covered by sand or over-lapping sheets or screens of non-combustible material.
    ii) At least two adequate and appropriate portable fire extinguisher, in proper working order, must be kept in the immediate area of the work being undertaken and used immediately smoke or smouldering or flames are detected.
    iii) A fire safety check of the working area must be made approximately 60 minutes after the completion of each period of work and immediate steps taken to extinguish any smouldering or flames discovered.
    iv) Blow lamps and blow torches must be filled in the open and must not be lit until immediately before use and must be extinguished immediately after use.
    v) A person must be appointed by the Insured to act as an observer to watch for signs of smoke or smouldering or flames.
Sub-paragraph v) does not apply to the application of heat by means of blow lamps, blow torches, hot air guns or hot air strippers.
b) Use of asphalt, bitumen, tar, pitch or lead heaters
    i) The heating must be carried out in the open in a vessel designed for the purpose and, if carried out on a roof, the vessel must be placed on a non-combustible heat insulating base.

Gas Safety:

Every year a number of people are exposed to carbon monoxide generated by faulty gas appliances. Such exposures can cause death and each year there are around 20 deaths. Landlords in control of domestic premises or letting agents that act on behalf of landlords are required to ensure the safety of tenants. They must ensure installation work is only carried out by suitably competent persons. Gas equipment such as fires and water heaters must be maintained in a safe condition. An annual safety check must be carried out and a record of such inspections and checks kept and passed to any tenants.

The competence of gas engineers is confirmed by their registration with the regulatory authority GasSafe.

There are many cases of landlords being prosecuted for a failure to comply with the relevant legislation, including failure to use competent contractors to carry our installation, servicing or annual safety checks or for failing to supply tenants with the require gas safety certificate. In the worst cases such prosecutions follow incidents in which there has been an escape of carbon monoxide. In such cases the prosecution will generally be followed by a civil claim for compensation.

Asbestos:

Landlords and those firms that own or have control of premises have specific duties to identify, assess and manage the risks of exposure to asbestos and asbestos continuing materials. These duties are set out in Regulation 4 of the Control of Asbestos Regulations 2012. Firms should have an Asbestos Management Plan that sets out they will manage the risk of exposure to asbestos.

Following work at its Reading store in 2006 Marks & Spencer were successfully prosecuted and fined £1m. It was found that the retailer had placed too much emphasis on keeping their stores operating and placed very strict demands on contractors, requiring them to work at night and return the stores for trading after each night shift. This resulted in corners being cut and asbestos control failing. The retailer was found to have failed to properly address the risk and to have failed to follow its own written procedures.

The contractors employed by the retailer had failed to follow their own procedures but the retailer had failed to properly supervise them and ensure that the safety rules were complied with. Several contractors were also prosecuted and fined for breaches of asbestos legislation.

Pollution

Any business that stores, handles, manufacturers or processes chemicals and similar potentially harmful substances faces the risk of pollution from an escape of such materials. Similarly any firm that has oil fired heating will have the potential for an escape of oil that could lead to pollution. (see **P9526**)

Fire Safety and Fire Risk Assessments

Fire safety is a wide subject that is concerned with both the protection of property and life safety. There were many pieces of legislation that set out what firms and businesses. The Regulatory Reform (Fire Safety) Order (RR(FS)O) established a coherent framework that introduced the concept of risk assessment to fire safety.

The Fire and Rescue Service are the enforcing authority for the RR(FS)O and have been actively pursuing those in control of premises where a breach of the Order has been established. There have been a number of high profile prosecutions for a range of offences including well known High Street retailers:

New Look: Fined £400,000 plus costs after a fire at their Oxford Street store in April 2007. New Look were found guilty of a number of offences including the re-setting of a fire alarm after the initial outbreak.

Cooperative Stores: Fined £250,000 after a number of breaches discovered at stores in East Sussex in 2006. Problems included a lack of fire alarms, locked and blocked fire exits and fire doors that were wedged open.

B&M Retail: Mansfield retailer fined £33,000 after pleading guilty to six fire safety offences under the Regulatory Reform (Fire Safety) Order 2005. The offences found included the following:

- Lack of a suitable and sufficient Fire Risk Assessment

- Fire exits and exit routes obstructed
- An exit route that led to locked enclosed area
- Lack of adequate safety training for employees
- Fire doors that were locked and unable to be used in an emergency

In addition there have been successful prosecutions of hotel owners and landlords, primarily those operating houses in multiple occupations (HMOs).

- Following a prosecution taken by Oxfordshire County Council on behalf of the Fire and Rescue Service a landlord was prosecuted for failing to comply with a fire safety prohibition notice. The individual was fined £2,250 and ordered to pay costs totalling £1,300.
  The fire safety arrangements were so poor that people living there were at risk of death or serious injury in the event of fire. The charges included failures to comply with the prohibition notice, as well as on-going failures to provide an adequate fire detection and alarm system, together with on-going failures to provide a safe means of escape, such that people were at risk of death or serious injury if there were a fire.
- The former owner of a Rochdale takeaway received a suspended prison sentence for 11 fire safety failings and putting lives at risk in 2011. Greater Manchester Fire and Rescue Service officers inspected the premises in February 2011 and as a result of the inspection a prohibition notice was issued.
  The officers found no suitable fire alarm in the building and the takeaway and the living accommodation were not separated by fire doors. In the basement there were overloaded extension leads and sockets powering a fridge, kettle and microwaves – all increasing the risk of fire. Combustible materials were stacked in front of a gas meter and this area was directly opposite the bottom of the stairs.
- In a case taken by Cheshire Fire and Rescue Service a Widnes landlord pleaded guilty to failing to carry out an adequate fire risk assessments and a further eight counts of breaching the Fire Safety Order 2005 in his three-storey property where the top two floors had been transformed into a number of flats. The landlord who was ordered to pay £8,500.

Whilst there has been no injury in any of the above cases if people had been injured by a fire there would have been a strong case for compensation. From an insurer's perspective if the policy has been extended to cover the costs of prosecution under the FS(RR)O then these cases will have resulted in payment of costs:

Electrical Installations

Electricity is one of the most common causes of fire. Insurers expect firms to keep their electrical equipment in good order. The fixed wiring should be inspected regularly, generally every 5 years but more frequently where the environment is particularly harmful e.g. is corrosive or dusty. Insurers may ask for evidence that such an inspection has been carried us and if not may make it a condition of continued cover that such an inspection is carried out. This will be a matter of interest to both the Property insurer, who will have to pay out if the building, contents, machinery and stock is damaged and the provider of the Public Liability policy how might have to pay any compensation if neighbouring property is damaged.

There is a risk that faulty electrical equipment or installations could give rise to injury or death and a subsequent claim for compensation, for example:

- A tenant could be injured by the electrical installation in a flat
- A customer could be injured by a faulty electrical installation
- Following work by a contractor such as an electrician the customer could be injured by faulty electrical equipment or a fire could break out causing damage to property

All of the above have resulted in Public Liability claims.

Risks arising from the activities of the firm

The most obvious hazards arise from the ownership and control of premises. However, other aspects of a firm's activities could give rise to damage to third party property or injury of persons. As mentioned under Rating (see **P9519**) the following factors all have a bearing on the likely frequency that is number of claims and their severity, the amount that they might cost.

- The nature of the business i.e. Manufacturing, Retail, Construction, Leisure etc
- Ownership of property
- The extent to which the firm or business interfaces or interacts with the public
- The nature and extent of work away from the premises
- The location of the premises and the nature of surrounding property
- The location of work away from the premises
- The use of sub-contractors

Examples of Public Liability Claims

[P9529] Claims made under Public Liability policies can be based on a number of different "headings".

Breach of Statutory Duty:

This is most commonly encountered when a claim is made under legislation such as the Defective Premises Act or the Occupiers' Liability Acts. A liability could arise, for example, if a visitor to the premises were to slip or trip due to some defect in the property.

Negligence:

The is the basis of the vast majority of claims made under legal liability policies and is the most frequently encountered basis for claims under Public Liability policies.
Thus a visitor to premises who has been injured through tripping on some defect in the premises, for example an uneven paving slab could allege negligence or breach of the Occupiers Liability Acts. The negligence claim would be based upon the presumption that:

- The owner or occupier of the premises owed a duty of care to the visitor
- That the duty was breached or broken and
- That the breach of the duty had resulted in the injury

It is necessary for the outcome or result of the breach to be foreseeable and not remote. Thus it is obvious that an uneven paving slab could result in a tripping accident and that this could lead to a broken ankle. If whilst staying in hospital the claimant contracted an infectious disease such as MRSA it could be alleged that this was a direct result of the initial accident and that but for the accident they would not have been in hospital and not have been at risk. It would be possible to further argue that the broken ankle and the resultant MRAS infection resulted in the claimant's business failing. However, it would be for the courts to decide whether the loss of the claimant's business was too remote a consequence.

Nuisance:

Nuisance is the tort or civil wrong that relates to the use or enjoyment of land. Thus if the activities of a business unreasonably interferes with person's enjoyment of their land this is regarded as nuisance. The purpose of the law of nuisance is to protect a person's right to use and enjoy their land without interference. An example would be if the operation of a factory gave rise to noise or emissions of dust such that the owners of adjacent or neighbouring houses were unable to sit in their gardens or hang out their washing. Unlike negligence which is based on a duty owed by one party or person to another, nuisance is based upon reasonableness. Not all activities are unreasonable even though they may result in damage.

There are two forms of nuisance; A Public Nuisance is one that affects an entire community, group of class of person. It is a criminal offence and if for example the public highway is obstructed no one individual person can bring a civil action for nuisance.

In contrast Private Nuisance is specific and relates to an interference with one person's right to enjoy their land. An example would be that of a factory that emits dust and smoke that affects a neighbouring property. However, it is necessary to prove that damage has occurred in order that an action can be commenced. It is worthy of note that the emission of smoke or dust that causes damage is nuisance whereas to walk onto someone's land is trespass. To hang a sign over a neighbours land cannot be nuisance since no damage has been caused; however, it does constitute trespass.

The remedy for an action based in nuisance is generally that of an injunction requiring the acitivty that is giving rise to nuisance to cease. Actions for nuisance and trespass are rare.

Products liability insurance

Introduction

[P9530] Product liability insurance covers firms and businesses against the risk that damages may be awarded against them if a product that has been sold or supplied causes damage to property or personal injury.

Much of the legal basis of liability for products including both the criminal and the civil law, starting with the case of *M'Alister (or Donoghue) v Stevenson* [1932] AC 562, 101 LJPC 119 can be found in the chapter on Product Safety [P9001]

Purpose

[P9531] The purpose of Products Liability insurance is to ensure that firms are covered for their legal liability to pay damages to third parties, that is persons who are not employees, who suffer bodily injury and/or disease as a result of the use of products that have been sold or supplied to them. The policy indemnifies the policyholder against claims made by a firm or any person except employees.

Such liability is normally based on negligence or breach of statutory duty. Liability may also result from a breach of contract, for example to supply goods or services to an agreed standard.

A Products Liability policy is a legal liability policy. Hence, if there is no legal liability on the part of a firm, the policy will not respond and no compensation or damages will be paid by the insurer.

Compulsory Products Liability Insurance

[P9532] Products Liability insurance is not compulsory in the same way that Employers' Liability is compulsory. However, legislation such as the Consumer Protection Act which effectively reverses the burden of proof and requires the defendant manufacturer to prove that their product was not defective means that those firms and businesses that manufacture, sell or supply goods take out Products Liability insurance as a matter of course.

Cover provided

[P9533] A Products Liability policy protects the policyholder against the risk that they might be required to pay compensation or damages to someone who has been injured or suffered some loss or damage to their property caused by a product sold or supplied.

The policy provides for payment of:

(a) compensation for personal injury or disease, or loss of or damage to third party property whether agreed or negotiated or awarded by a court

(b) costs and expenses of litigation, incurred with the insurer's consent, in defence of a claim against the insured (i.e. civil liability); and also any appeal against convection
(c) claimants costs and expenses for which the insured is legally liable;
(d) solicitor's fees, and other legal costs incurred with the insurer's consent, for representation of the insured at proceedings resulting from a prosecution for a breach of health and safety and related legislation, that is a breach of statutory duty; this is subject to certain provisos for example the breach must not have resulted from a deliberate act or omission by the insured and no indemnity will be paid if there is a legal expenses policy in force; and this cover is generally extended to include prosecutions under Part II of the Consumer Protection Act 1987 and part II of the Food Safety Act 1990
(e) compensation for any director, partner or employee attending court as a witness in a case for which the insurer is providing an indemnity.

Persons

[P9534] The approach adopted with regard to the specifying the classes of person who are entitled to make a claim against a Public Liability policy has been described (see **P9518**). The situation is similar with Products Liability claims. Any person who has suffered injury or loss of or damage to property caused by a product is entitled to make a claim.

Thus, unlike Employers' Liability policies it is not necessary or even possible to define the classes of person who can make a claim.

Rating

[P9535] The rating of Products Liability insurance is universally based on the turnover of the business. This provides a good measure of the volume of activity, of the number of items sold and in turn the exposure that is faced.

Limits of Indemnity

[P9536] Whilst the legislation stipulates the amount for which an employer is required to insure and maintain Employers' Liability insurance (£5 million in respect of claims relating to any one or more of his employees) there is no corresponding legal requirement for the amount of Products Liability cover that must be purchased.

Generally the Limit of Indemnity for Product Liability will be the same as for the Public Liability with the difference being that the indemnity for the Products Liability will be "in the aggregate" for the period of insurance. This means that if the Limit of Indemnity is, say, £5m then no matter how many claims there might be the maximum amount that the insurer will pay out in damages or compensation will be £5m. However many claimants and no matter how much is claimed in total the maximum amount the insurer will pay is the Limit of Indemnity together with any related costs.

This limitation provides protection for the insurer against the risk that a defective product could be sold in large numbers or distributed very widely and could, if defective, result in many claims being made. There can be protracted arguments between insurer and policyholder as to the cause of a product failure. It may well be that a defect in a batch of products is the result of a single failure on the production line and this would be regarded as a single cause. Thus all claimants would be grouped into single claim.

A Products Liability policy may contain an excess negotiated between the insurer and firm i.e. a provision that the firm will pay the first £x of any claim. This is much more common in Public and Products Liability policies than Employers' Liability.

The purchase of excess protection by means of excess of loss insurance is much more common with Products Liability than Public Liability. Claims under Products Liability policies are rare, described as being of low frequency, but can be extremely expensive, described as being of high severity. Thus it is prudent for firms in certain high risk sectors such the manufacture of pharmaceuticals or products for use in the aviation or aerospace industries to secure additional protection, up to £25m or even £50m.

As with Public Liability the costs and expenses are not included and are in addition to the Limit of Indemnity. However, in respect of any claim brought in the United States of America or Canada the total amount payable is inclusive of any costs and expenses. The effect of this is to reduce the amount of compensation that is available within the Limit of Indemnity. The reason for this limitation is the extremely high legal fees associated with claims that are heard in North American courts.

Territorial Limits

[P9537] Cover for Products Public Liability policies cannot be limited or restricted in the same manner as Public Liability. A business might sell its products in many countries and they may be used in countries where they are not even sold. Thus cover under the Products Liability policy is unlimited or worldwide provided that the activities of the firm or business, the policyholder, are conducted from an address or location within Great Britain, Northern Ireland, the Channel Island or the Isle of Man. These are generally specified in the policy and are often known as the Defined Territories.

It should be noted that if a company has overseas subsidiaries then the products supplied by such subsidiaries are not covered. Appropriate insurance should be obtained locally.

Product Liability cover is usually granted on the basis that any action for damages is brought in a court of law of Great Britain, Northern Ireland, the Channel Islands or the Isle of Man – though even this proviso is omitted from some policies.

Policy Endorsements and Extensions to Cover

[P9538] Insurers have a range of standard endorsements that may be applied to a Products Liability policy in order to exclude certain aspects of cover. Some typical endorsements include:

- Products Supplied Restriction:
  This has the effect of completely removing any cover for products supplied. It is used if the policyholder does not manufacture or sell/supply products. The restriction does not apply to a number of low risk activities including:
    - The sale or supply of food and drink
    - The disposal of office furniture and equipment previously used in the business i.e. sale of second-hand office equipment
- Products supplied to North America:
  The purpose of this endorsement is to remove all cover for any products supplied to North America. Claims in North America are extremely costly and the purpose of this restriction is to protect the insurer against claims for injury, loss or damage resulting from the use of the policyholder's products in North America. This endorsement is applied automatically unless the policyholder declares that they make sales to North America and thus need cover. Then the insurer will assess the risk apply any terms and condition that may be necessary, amend the policy and collect a suitable and sufficient premium.
  Failure by the policyholder to declare that they have sales to North America would be regarded as a failure to disclose a material fact and could result in the insurer voiding the policy.
- Damage to Products Supplied:
  This endorsement is designed to limit the amount paid where the product causes damage and must itself be replaced. Thus if the policyholder manufactured and supplied a concrete beam for a block of flats it is possible that a failure of the beam could cause damage to the building. The Products Liability policy would cover the costs of repairing the building but not the costs associated with removing and replacing the defective beam. Similarly if the policyholder supplied a patient hoist to a hospital or care home the policy would respond if there was a failure that caused injury to a patient. However, the policyholder would have to bear the costs of repairing or replacing the defective hoist.

Efficacy

[P9539] There are certain products where the correct operation of the device is an essential matter. Example include fire alarms where a failure to operate as required when a fire is small could lead to a considerable loss. Similarly, a burglar alarm that failed to sound when the premises were broken into would be considered as not having functioned as expected. In neither case has there been any damage caused by the failure of the device to operate. Thus there is no claim under a Products Liability policy.

If there had been an electrical fault in an alarm that had caused a fire then the product would be regarded as defective and a there is a valid basis for a claim.

However, the owner of a warehouse that was burnt down because of a non-functioning fire alarm would rightly consider that they should be able to make a claim against the manufacturers or supplier of the alarm.

In order to protect their position and clarify exactly what cover is required insurers generally add what is known to efficacy exclusion to policies for manufacturers that manufacture or sell/supply products that are required to operate in a specific manner. A typical wording for an efficacy exclusion is as follows

" . . . . the failure of any Product to correctly fulfil its intended use or function and/or meet the level or performance, quality, fitness or durability warranted or represented by the Insured."

Thus policies for manufacturers or suppliers of fire and security related products will almost certainly have such an exclusion placed on their policies.

Financial Loss

[P9540] A Products Liability policy protects the policyholder against claims arising from any injury, loss or damage to property that might occur as a result of a defect in their product. Thus, if negligence or breach of statutory duty were proven, the policy would respond and provide compensation if:

- A person were injured
- or
- A person's property was damaged.

If the property damage also resulted in other financial losses, often called consequential losses, these would be paid if they were directly related to and flowed from the damage to property. However, a Products Liability policy will not pay for financial losses resulting from a defect in the product if there was no damage to property. Such losses are often described as pure financial losses. The following examples illustrate how this situation could arise.

Example 1: A machinery manufacturers supplies a piece of equipment, for example a conveyor, which is incorporated into the production line. As a result of a fault in the machine a fire starts and the other parts of the production line are damaged. The purchaser of the product can claim for

- The damage to the production line and the costs of repairing it.
- Lost production
- Costs of arranging for production to be continued e.g. at a competitors' plant
- Loss of profit due to the loss of a contract – since they were not able to fulfil the orders

It is possible that the last of these might be ruled as being too remote and the clamant would not receive compensation for that part of their claim.

Example 2: The machinery manufacturer supplies the same piece of equipment, the conveyor, which is incorporated into the production line. As a result of a fault in the machine the production line has to be stopped whilst repairs are carried out. The only loss is the value of the product that could not be produced whilst the production line was stopped.

This is a pure financial loss and will not be covered by a normal Products Liability policy.

[P9540] Public & Products Liability Insurance

It is possible to request an extension of the Products Liability policy to include cover for Financial Loss. Insurers adopt a strict approach to requests for such cover.

There will be lists of manufacturers/products for which cover will not be provided under any circumstances. The insurer will require additional information about the product, where and by whom it will be used, and the likelihood of pure financial loss occurring, details of any past losses or claims etc. Once in possession of this information the insurer will determine a premium and may well require the policyholder to accept an excess or co-insurance. Usually a lower Limit of Indemnity will be applied for this extension of cover.

Products Guarantee

[P9541] A Products Liability policy provides insurance against the risk that the product that has been manufactured or sold/supplied is defective and causes injury or loss of/damage to property. But such a policy does not cover the costs of replacement of any product supplied or any financial loss resulting from its failure to perform. A Products Liability policy provides an indemnity against a legal liability and is not intended to provide any form of guarantee. Thus if the product proves ineffective and does not perform as expected there are no grounds for a claim under the policy. For example, if a paint manufacturer sells paint and claims that it possesses anti-corrosion properties and will protect metal structures against rust and corrosion for 10 years there is no claim if rust is found after 7 years. The paint manufacturer may be asked to arrange for the bridge to be repainted but the costs of this cannot be recovered from the insurer.

If a manufacturer wished to obtain protection against such a risk they would need to purchase Product Guarantee insurance. This can be purchased but is not commonly available and will thus be expensive. It can generally only be obtained from specialist insurers.

Products Guarantee insurance will cover the cost of replacing or reworking the product that has failed to perform its intended function for whatever reason. The cover may extend to include any financial loss that has been incurred by the policyholder's customer as a result of the products failure to perform. Products Guarantee is more concerned with covering the consequences of an unfulfilled contractual responsibility than a legal liability.

Product Replacement insurance is a similar form of cover and would cover for example, the costs of digging up and replacing a car park if there had been swelling and cracking due to defective hardcore laid under the tarmac surface.

Products Recall

[P9542] If a product fails to perform as intended it may need to be recalled. Similarly if a product is found to be defective and liable to cause injury or damage it is essential that customers are alerted and told not to use the product and perhaps return it to the manufacturer. Such an exercise is known as a product recall and is a key component of mitigating any potential loss. Policyholders are expected to take all reasonable steps to reduce or mitigate the extent of losses and this could include undertaking a product recall.

Under Product Safety legislation regulatory and enforcing bodies may initiate a product recall programme in advance of any such action by the manufacturers. Manufacturers are advised to prepare product recall plans in exactly the same way that emergency plans are drawn up. Such plans should allocate key responsibilities and contain a clear description of what action should be taken and when. Plans should be rigorously tested, ideally by means of mock exercises.

Products Liability policies do not cover any costs associated with replacement of defective products or any costs associated with notifying customers and recalling the product. It may be possible to extend a Products Liability policy to include such cover or it may be purchased as part of a combined Products Guarantee/Products Recall policy.

Although Products Recall cover does not generally form part of a Products Liability policy the insurer may seek confirmation that a recall plan is in place as a demonstration of the firm's ability to cope with a potentially costly situation and mitigate costs by recalling defective products.

Legislation relating to Products Liability Insurance

[P9543] Consumer Protection Act 1987

General Product Safety Regulations 2005

Sale & Supply of Goods Act 1994

Supply of Goods and Services Act 1982

Unfair Contract Terms Act 1977

Food Safety Act

Common Products Liability Risk Exposures

[P9544] For firms that manufacture products the risk of an injury or loss will generally fall under one of the following headings

Design: A fault in the design results in a product that is unsafe e.g. a claim that the design of the product is inherently unsafe. The most memorable example is the large number of accidents involving Ford Pinto car in the 1970's. A serious design error resulted in the risk that the vehicle could burst into flames if hit from behind. More recently there have been many injuries to small children due to a design fault in McLaren buggies.

Manufacture or Production Faults – It is possible that a fault in some part of the production process created an unreasonably unsafe defect in the resulting product. In recent years there have been a number of withdrawal and recalls of products manufactured in China relating to the possibility that small parts might become detached and swallowed by small children or the presence of dangerous chemicals in their products. Such situations could result in claims.

Quality Control/Assurance: A failure in the systems and procedure designed and intended to ensure that products are made to a satisfactory standard and that any manufacturing faults or failures are detected before the faulty product is dispatched from the factory.

Defective Warnings or Instructions: Claims can be based on the allegation that the product was not properly labelled or had insufficient warnings for the consumer to understand the risk. The claims for burns and scalds from "hot coffee" are an example where the defence rests on the adequacy of warnings.

Examples of Products Liability Claims

[P9545] There are no significant differences between the approach to handling Products Liability claims and that adopted for Public Liability claims. Products Liability claims are rare but may result in substantial damages, either as awards to single claimants or multiple, smaller awards made to many persons. It is also possible that high profile press coverage of problems with products only result in the product being recalled and either repaired or replaced with no actual loss or injury. This is often the case with consumer products. In such cases Products Liability policies will not respond although a manufacturer may be able to claim under a Products Recall policy if they have one in place

In the absence of specific regulations or standards the safety of products will be assessed taking the following into account in turn:

- voluntary European standards;
- Community technical specifications;
- national standards (i.e. British standards which are not UK versions of European standards);
- industry codes of good practice; and
- state of the art and technology, and the standard of safety which consumers may reasonable expect.

The following examples illustrate some common features of Products Liability claims.

Balding v Lew-Ways Ltd (1995) 159 JP 541, [1995] Crim LR 878, DC

Lew Ways manufactured a child's tricycle that was fitted with a tipping bucket somewhat like a dumper truck. When the bucket was tipped a sharp metal "catch" was exposed and this caused injury.

The manufacturer stated in their defence that the toy complied with the relevant British/European Standards for safety of toys. However, the existence of the sharp protrusion and the resultant injury meant that the products failed to satisfy the Toy Safety Regulations 1995. The firm was successfully prosecuted.

This case illustrates that if a product is determined not to conform to relevant standards then if an accident or injury occurs the manufacturers may still face prosecution in addition to a civil claim for compensation. If a prosecution is successful then the chances civil clam for compensation being successful are increased.

Maclaren Buggies

In 2009 Maclaren, the manufacturer of children's buggies and pushchairs became aware of injuries to children's fingers in the United States. The accidents happened when the buggies were being opened or closed not when they were being wheeled along with a child. It was possible for children's fingers to be in such a position that as the catch mechanism "sprang closed" the tips of their fingers could be amputated. The firm took steps to reduce the risk of further injuries by issuing a product recall and making available a shroud or cover that could be retrofitted to the buggies. Around 1 million buggies were affected. (Note the term "product recall" has a somewhat different meaning in the United States)

The firm initially took no action in the UK despite the fact that the buggies and pushchairs sold in the UK were of an identical design to those marketed in the United States. Social media meant that the problem became known to owners of buggies in the UK. News of similar accidents involving UK children began to circulate. Maclaren came under pressure from Trading Standards and the consumer organisation Which and eventually was forced to take action. Notices were placed in the press and information, including a video, was placed on the firm's website. Covers were made available on request and also at major stores that sold the buggies. These steps eliminated the risk of further injuries. However, a number of children had already been injured and claims for compensation have been made against the firm and are being dealt with by insurers.

This situation illustrates the risk faced by manufacturers, especially of consumer products that can be placed on the market in great volume. Prompt action in the UK once the problem was identified in the United States may have reduced the numbers of UK injuries. The firm's delay in taking action in the UK may well have adversely affected their image and public profile.

Leather Sofas Litigation

More than 4,500 claimants issued claims against a number of retailers alleging they suffered dermatological injuries from exposure to Chinese-manufactured leather furniture containing a mould inhibitor DMF. Sachets of the substance were placed in the furniture at the point of manufacture. The sofas were sold but without any instruction regarding the substance, whether it should be removed etc. Many users began to suffer from a range of symptoms.

In April 2012 Mr Justice MacDuff approved plans for a compensation fund for claimants pursuing claims against Argos, Land of Leather and Walmsleys the retailers who had sold the majority of the leather sofas.

Although the court found that the insurer, Zurich could not be held liable for claims against Land of Leather, which had gone into administration the insurer agreed to pay those claims against Land of Leather where sofas had been supplied by Chinese company Eurosofa. For all other claims the court ruled,

Zurich would not be held liable for Land of Leather's losses. That strand of the case will be taken to the Court of Appeal and will be heard in January 2013.

Defective Garden Furniture

A consumer was injured whilst using a plastic sun lounger purchased from their local DIY store.

The consumer followed the instructions for assembly of the sun lounger and placed it in the sitting position. After sitting on the sun lounger for a short time it suddenly collapsed from beneath them and the individual fell heavily onto the floor. As a result they sustained injuries to their lower back and knee, requiring treatment from their GP.

The DIY store and/or manufacturer of the sun lounger were alleged to have acted negligently in several ways:-

(1) They had failed to take any or any adequate care in the design, development and manufacture of the goods;
(2) They had sold the defective goods;
(3) They had caused or permitted the goods to be or to become or to remain defective and dangerous;
(4) They had failed to institute or enforce any or any adequate system of quality control or to inspect, check or test the goods before dispatch;

Liability for the Defendant's negligence was admitted and compensation of £6,250 was awarded to the claimant.

General matters relating to public and products insurance

Woolf Reforms

[P9546] The Woolf Reforms, which were enacted as the *Civil Procedure Rules 1998 (SI 1998 No 3132)*, came into force on 26 April 1999 and apply to England and Wales. The aim of the new rules, also known as the CPR, was to speed up the processes by which people obtained compensation through the civil courts and make justice more accessible via management of costs.

Pre-action protocols are an important element of the new rules, which should ensure that more claims are settled without litigation. They are aimed at personal injury cases involving less than £15,000 in compensation. Should settlement not be reached through these protocols, the litigation process then goes through the courts? The reforms affect any class of insurance business where there is a third party injury, principally Motor, Products Liability and Employers Liability and occasionally Household.

The reforms were an attempt to resolve the dispute *before* court proceedings are even commenced; thus minimising both cost and court time. If the case is not settled via the pre-action protocols, it may be taken to court. The personal injury pre action protocol sets out the following stages:

- letter of claim;
- reply/acknowledgement;
- investigation;
- disclosure of documents;
- appointment of expert witnesses;
- negotiation and settlement;
- Litigation in the courts.

Under the new rules all cases will be allocated to one of three routes:

- Small Claim: (previously the arbitration or Small Claims Court) property claims with a value of up to £5,000 or up to £1,000 for personal injury claims;
- Fast Track: claims with values between £5,000 and £15,000 (Property) or £1,000 and £15,000 (personal injury);
- Multi-track: all claims in excess of £15,000 and any complex claim of lower value.

Fast track cases will generally proceed quickly to a hearing, which is usually limited to one day, as most evidence has been provided in writing and agreed in advance.

The new rules affect policyholders in a number of ways. Any delay in notifying the claim may affect mounting a suitable defence and increase the cost of the claim. Policyholders must understand the reasoning behind decisions on liability; if there is no hard evidence insurers are unable to run a defence. And policyholders need to co-operate with claims handlers and solicitors by supplying information and documents. In particular policyholders must:

- promptly report all incidents likely to give rise to a claim to your insurer;
- carry out an investigation;
- retain all evidence and relevant documents (e.g. Accident Book or report form, accident investigation report, photos, plans and security videos, copies of risk assessments);
- forward any letter of claim to the insurer as soon as it is received but should not acknowledge its receipt.

Since 1998 the CPR have been regularly reviewed and amended/updated and the most recent update, the 78th will come into effect in early 2015. Details of the amendments and updates can be found on the Ministry of Justice website.

www.justice.gov.uk/courts/procedure-rules/civil

Scotland

[P9547] As stated above the Woolf Reforms only apply to England and Wales. However, following lengthy debate a voluntary Pre-Action Protocol has been agreed between insurers and the Law Society of Scotland and came into effect for all claims intimated after 1 January 2006. This voluntary arrangement remains in place as of April 2012.

This is a major step forward in Scottish pre-litigation procedure and is the culmination of much work between insurers and the Forum of Scottish Claims Managers. The main points to note about the Protocol are as follows:

- It is a voluntary protocol and does not have legal effect (although parties may seek to draw breaches in the protocol to the attention of the Court on the issue of costs).
- It applies only to claims intimated on or after 01/01/2006 and thereafter handled within the Protocol.
- It is primarily aimed at Motor claims with a value below £10,000 although can be agreed on all claims at all levels.
- Any existing claim notified prior to 01/01/2006 will be dealt with as per existing arrangements.
- The Protocol does not apply to property damage only claims — but will apply where a Property damage claim has a Personal Injury element.
- The Protocol does not apply to disease claims.

Admissions of Liability are binding where the value of the claim is below £10,000. Solicitors will send a Letter of Claim in similar format to the format of the letter of claim used in England / Wales and insurers will have 21 days to respond from date of receipt. Insurers then have 3 months to investigate liability and respond with a decision.

Medical evidence should be instructed by the Pursuer's (NB Scottish term for Claimant) solicitor within 5 weeks of a liability decision and medical experts will be instructed on an agreed basis (not joint instruction). Medical evidence should be disclosed by the Pursuer's solicitor within 5 weeks of receipt. Questions can be put to the expert by either side by mutual consent.

Pursuer's solicitors will serve a Statement of Valuation and insurers have 5 weeks in which to respond.

Where damages are agreed insurers have 5 weeks to issue cheques or else interest will be payable. The 2003 Fee Scale still applies but this now includes payment of the Investigation Fee

Corporate Manslaughter

[P9548] (Further information can be found at C9001)

Pressure groups and campaigning bodies have long held the view that the existing law on corporate manslaughter was inadequate and failed to deliver a suitable punishment following cases of death resulting from work activities, involving either employees or members of the public.

Whilst it is possible to bring a charge of corporate manslaughter against a company or corporate body there are difficulties in securing a conviction. It is currently an agreed legal principle that a corporate body does not possess a "mind" and is thus unable to commit a violent crime. Thus it is necessary to charge both the company and an individual employee with manslaughter. And unless the individual is successfully convicted the company cannot be found guilty. This situation has resulted in few cases going to court and low conviction rates.

In the 2007/08 year there were 229 deaths of workers/employees and 358 work related deaths to members of the public, including 263 due to acts of suicide or trespass on the railways. Deaths of workers are fairly static and have dropped slowly over the last 10 years whilst deaths of members of the public are more volatile and do not display such a regular trend. In the period 1992 to 2006, a total of 160 cases of workplace death were referred to the Crown Prosecution Service, 45 of these proceeded to trial and a total of 10 convictions were achieved. The successful prosecutions also tend to involve smaller firms where it has proved easier to establish a link between the actions of senior management and the circumstances surrounding the accident that resulted in the death. Opponents of the existing arrangements maintained that the approach is biased against smaller firms and allows larger firms that have caused deaths to "escape" justice.

In 1996 the Law Commission published a report on the subject of "involuntary manslaughter". This report was a response to growing public pressure for changes in the law. In the period following the Law Commission report the Government carried out various consultation exercises and published a draft bill in 2005. The final form of the Bill was published on late 2006 and after challenges in the House of Lords the Royal assent and into effect in April 2008.

The key elements of the Bill are as follows:

- A new offence of Corporate Manslaughter has been created (Corporate Homicide in Scotland).
- For a charge of manslaughter to be made an organisation must have owed a duty of care to the victim and have breached that duty as the result of way in which its activities were managed or organised by senior manager(s).
- The duty of care can result from the employment of persons, ownership or occupation of premises, the supply of goods or services, construction or maintenance activities or the keeping/use of vehicles.
  Note: These duties match exactly the existing duties that are protected by liability policies.
- The breach of duty, termed a senior management failure, must have been gross and caused the death. In determining guilt consideration must be given to the risk of death presented by the breach, the seriousness of the breach and any breach of health and safety legislation.
- The gross breach must involve conduct falling far below what would be reasonably expected in the circumstances.
- Senior managers are described as being those who make decisions about the whole or a substantial part of the organisation's activities.
- All incorporated bodies are covered by the proposed legislation although certain classes of business are exempt e.g. partnerships, as are certain Government bodies.
- The offence will be tried in the Crown Court and the penalties available are unlimited fines or remedial measures.
- There is specific guidance that the jury should consider the firm's corporate safety culture, policies and procedures and any relevant safety information or guidance.

It should be noted that it is no longer necessary to secure a conviction of senior manager and thus it is thought that it will be easier to secure conviction of a firm.

There have been a number of prosecutions (12 as at October 2014) since the Act came into force and all have involved the death of employees. The first two prosecutions involved small firms. The third involved a larger firm with two factories. The most recent prosecution is of a small market garden/nursery. A driver tipping some waste oil with a tipping trailer drove off before lowering the raised body and came into contact with overhead electrical cables.

However, the majority of the prosecutions have involved smaller firms and have not provided any better indications of how the Act might work against larger firms. There has been just one prosecution involving a person who was not an employee, this involved a water sports firm (Princes Sporting Club) that was prosecuted after a young girl died in a power boat accident.

Compensation Act

[P9549] The Compensation Act is part of the Government's programme to tackle the 'compensation culture' and in the case of those suffering from mesothelioma improve the system for obtaining compensation where there is a valid claim. The Act received the Royal Assent in November 2006 and further parts are expected in 2007. The Act contains a number of provisions and will enable the drafting of specific legislation to reform particular aspects and areas of the law relating to the claiming of compensation. The main elements of the Act are described as follows:

Part 1: Standard of Care

- Deterrent effect of potential liability
- Apologies, offers of treatment etc
- Mesothelioma

Part 2: Claims Management Services

Part 3: General

Potential Effect of Potential Liability (sec1): This section of the Act attempts to overcome the situation where, it is alleged, activities are not undertaken for fear of litigation in the event that something unforeseen occurs and legal action follows. The usual example is that of school trips being cancelled because of fear that teachers will be sued if a pupil is injured in an accident. The Act states that when a court is considering a claim in negligence or breach of statutory duty it should consider whether the taking of reasonable steps to meet a standard of care might prevent a desirable activity from being undertaken or discourage persons from undertaking functions in connection with a desirable activity. The effect of this text is dependant upon more detailed guidance to the courts about the exact meaning of words such as "desirable activity" and "standard of care".

Lord Young Report

[P9550] In June 2010 the Prime Minister asked Lord Young to investigate and report on the compensation culture and the current low standing of health and safety. The report titled "Common Sense Common Safety" was published in October 2010. It was broad based and covered a number of areas and the report gave the impression of being driven by anecdotal concerns about a "compensation culture". The report gave support for the Jackson reforms of civil law. Despite providing no evidence of how widespread the practice was the report stated that insurers should stop requiring businesses to employ expensive health and safety consultants. The report also recommended consultation with the insurance industry" to ensure that worthwhile activities are not unnecessarily curtailed on health and safety grounds".

The report also recommended that insurers should prepare a "code of practice" for businesses and the voluntary sector. The intention was that such a code would provide an indication of what insurers required from their policyholders.

In response to this recommendation the Association of British Insurers coordinated the preparation of a Key Principles document titled "Health and Safety for Small Businesses and the Voluntary Sector" that sets out what insurers expect from their policyholders. The document was published in November 2011 and can be found on the ABI website.

Lofstedt Review

[P9551] In May 2011 the Government asked Professor Lofstedt of King's College London to chair a comprehensive review of health and safety regulations, enforcement and civil liability. Professor Lofstedt's report titled "Reclaiming Health and Safety for all" was published in November 2011.

The full report and the Government response can be found here: www.gov.uk/government/organisations/department-for-work-pensions.

In general terms the report concluded that there was no case for a radical overhaul of health and safety legislation. It concluded that much of legislation was beneficial in reducing risk and in turn workplace accidents. There were some recommendations for a review of HSE guidance and Approved Codes of Practice. In addition the legislation in sectors such as mining and quarrying was thought to be in need of revision and consolidation.

It was stated that there were some problems with interpretation and application of existing legislation and that businesses sometimes go beyond what is required by the law

Professor Lofstedt considered the usefulness and application of the Pre-Action Protocols, in particular the list of documents that can be requested or sought during disclosure. It was concluded that the manner in which these lists are used has perhaps deviated from the original intention. The report recommended that the original intention of the standard disclosure list be clarified and restated.

The report made numerous recommendations; however, there were none that related specifically to the provision of insurance or insurance companies. The HSE was recommended to continue to help businesses understand what is reasonably practicable and how to comply with the law in a proportionate manner. In early 2012 the Prime minister hosted a meeting of the Chief

Executives of several major insurance companies. Although much of the meeting focussed on motor insurance, claims farmers and whiplash the subject of EL insurance was discussed. Following the meeting discussion took place between the Department of Work and Pensions and the ABI about how the insurance industry might be able to assist the HSE.

Fees for Intervention

[P9552] In late 2011 the HSE launched a consultation on the approach to be used for the recovery of their costs. Existing legislation already provided the legal framework for the HSE to recover costs from firms for example for major hazard sites, offshore rigs etc. The legislation was amended to allow recovery of costs from firms that were, on inspection, found to have a "material breach" of legislation. The consultation exercise was intended to collect feedback about possible approaches to the recovery. The HSE undertook a pilot exercise involving key stakeholders and announced that the scheme would come into operation in April 2012.

In late March 2012 it was announced that cost recovery or FFI as it had become known, would not commence in April but was to be deferred until October. When this change was announced there was a slight change in that recovery of costs was to be triggered by a contravention rather than a material breach. This appears to be a lowering of the threshold which could result in more firms facing costs.

The proposal only extends to those firms that are subject to inspection by the HSE. Those firms that are subject to Local Authority inspection by Environmental Health Officers e.g. businesses in the retail, and hospitality sectors will not face FFI.

EL insurers have generally taken the view that any costs that the HSE is seeking to recover under FFI are not damages or compensation as defined by the policy but more akin to fines. Thus they would not be covered under an EL policy and the firm, the employer, would not be able to recover such costs from their insurer. The same situation would apply with regard to Public and Products Liability policies. Whilst it is expected that the most common cause of a charge will be a material breach of legislation relating to employee safety it is possible that the breaches of legislation that impact on the safety of others could lead to an Improvement or Prohibition Notice and subsequent recovery of fees and costs. An example would be poor working practice on a construction site that could create a risk for passers-by and members of the public.

Details of the Scheme can be found on the HSE website.

Long Tail Disease

[P9553] Employers' Liability policies provide cover in the event that an employee alleges that their disease was caused by exposure to harmful substances or situations in the working environment. Typically claims will arise from exposures to:
- asbestos giving rise to asbestosis, lung cancer or mesothelioma;

- harmful substances resulting in cancer, asthma etc;
- noise giving rise to noise induced deafness and/or tinnitus;
- Vibration giving rise to hand arm vibration syndrome (vibration white finger).

Some diseases arise quickly after exposure and can be regarded as injuries, for example dermatitis, however some develop gradually over time e.g. deafness. There are others that have very long development periods or exhibit latency, not appearing for decades after exposure, mesothelioma which can take anything from 15 to 35 years to manifest it. It should also be noted that for some diseases the extent of injury or disability is related to exposure. Thus the degree of deafness resulting from occupational noise exposure is related to the "dose" – a measure of both the level of exposure and the extent or time over which the exposure occurred. In contrast other diseases, especially cancers, are not related to dose. The condition can occur following small exposure, even of low level and for short periods of time.

Many of the workplace risks faced by employees are not an issue for the classes of person that might make a claim under a Public and Products Liability policy, such as a visitor to the premises, a customer or user of a product. Thus whilst claims for long tail disease are rare under Public Liability policies they are not unknown. There are perhaps two main classes of claim that might be made.

It is possible that a user of a product may suffer from some form of illness or disease. But usually these are short term in nature, appear quickly and would not fall under the definition of a long tail disease.

By far the most common long tail disease that generates claims under Public Liability policies is one relating to or caused by exposure to asbestos. There are well documented cases of people suffering from mesothelioma who have no employment history of exposure to asbestos. Wives and children have developed mesothelioma from exposure to asbestos containing dust from their husband/father's overalls. In the past, when the risks were less well know or understood and standards of workplace hygiene less strict it was often the case that employees went home in their overalls, which were shaken and washed. This has given rise to a popular description for claims from wives as "shake out" claims.

There have also been claims from people who lived next to factories that processed asbestos who have contracted mesothelioma following exposure to asbestos containing dust that escaped from the factory premises.

Whether there was a Public Liability policy in force at the time would need to be established and whether it would respond are both complex matters that can only be decided on a case by case basis.

It is also possible that the failure of the directors of a firm to take out suitable and sufficient insurance could be regarded as a breach of their duties to the business. This could even generate a claim against the Directors and Officers insurance, if such a policy existed.

Recovery of NHS Costs and State Benefits

[P9554] A unit under the control of the Department of Work and Pensions called The Compensation Recovery Unit (CRU) works with insurance companies, solicitors and DWP customers, in order to recover:

- Social security benefits paid as a result of an accident, injury or disease, where a compensation payment has been made. This is known as the Compensation Recovery Scheme.
- Costs incurred by NHS hospitals and Ambulance Trusts for treatment from injuries from road traffic accidents and personal injury claims. This is known as Recovery of NHS Charges.

Insurers notify the CRU of all personal injury claims in order to initiate the process of recovery and once the claim is settled in favour of the claimant the insurer refunds any benefits that have been paid.

Radiation

Andrea Oates

Non-ionising and ionising radiation

[R1001] The new *Ionising Radiations Regulations 2017 (IRR 2017) (SI 2017 No 1075)* came into force in January 2018, repealing and replacing the *Ionising Radiations Regulations 1999 (SI 1999 No 3232)* and implementing the requirements of the European Basic Safety Standards Directive (2013/59/Euratom) (BSSD). The main changes and requirements of these new regulations are summarised in **R1010** below.

Non-Ionising Radiation – What is it?

[R1001.1] Ionising radiation has sufficient energy to displace tightly bound electrons from atoms, causing the atom to become electrically charged or *ionised*. These ions, although short-lived, can be very reactive, causing tissue damage and changes in DNA, which lead to cancer (see **R2001** onwards). *Non-ionising radiation* is that part of the electromagnetic spectrum which has insufficient energy to cause ionisation. Its interaction with body tissues, and hence the potential harm it can cause, is quite different from ionising radiation.

All radiation can be thought of as waves of energy radiating away from the source that produces it. In the case of non-ionising radiation, this energy can be characterised by both an *electrical* field (the E-field) and a *magnetic* field (the B or H field – see **R1002**). Both of these characteristics can be measured and can be used to quantify the strength of the field.

In addition, the energy wave has a *frequency*: the number of times it repeats itself in a given time (a frequency of 1 cycle per second is 1 Hertz, or Hz). The *wavelength* (which is directly related to the frequency) is the distance between the waves measured in the unit of metres (or down to nanometres (nm) for the very short wavelengths in the visible and ultraviolet region). The frequency is an important way to characterise non-ionising radiation, since the effects it can have on the body are determined as much by frequency as by the strength of the field. The frequency of non-ionising radiation extends from just above zero Hz (if it was zero and therefore had no frequency, it would be a *static* electrical or magnetic field) to between 10^{15} and 10^{16} Hz in the ultraviolet region. At this frequency the non-ionising region gives way to the ionising region.

The complete range of energies is divided into bands, which are described by the electromagnetic spectrum (see TABLE **1** below).

[R1001.1] Radiation

TABLE 1. THE ELECTROMAGNETIC SPECTRUM

Region			Wavelength (metres)	Approximate frequency range (Hz)*
		Static	0	0
Non – Ionising Radiation	Radiofrequencies	ELF	10^8–10^5	0.3 Hz to 3 kHz
		VLF	10^5–10^4	3 kHz to 30 kHz
		LF	10^4–10^3	30 kHz to 300 kHz
		MF	10^3–10^2	300 kHz to 3 MHz
		HF	10^2–10	3 MHz to 30 MHz)
		VHF	10–1	30 MHz to 300 MHz)
	Micro-waves	UHF	1–10^{-1}	300 MHz to 3GHz)
		SHF	10^{-1}–10^{-2}	3 GHz to 30 GHz)
		EHF	10^{-2}–10^{-3}	30 GHz to 300 GHz)
	Infrared	IR	10^{-3}–10^{-6}	10^{11}–10^{14}
	Visible **	Red	740–625 nm	10^{14}–10^{15}
		Orange	625–590 nm	
		Yellow	590–565 nm	
		Green	565–520 nm	
		Cyan	520–500 nm	
		Blue	500–435 nm	
		Violet	435–400 nm	
	Ultraviolet	UVA	400–315 nm	10^{15}–10^{17}
		UVB	315–280 nm	
Ionising Radiation		UVC	280–100 nm	
	X ray	X-ray	10^{-8}–10^{-11}	10^{17}–10^{20}
	Gamma ray	γ- ray	10^{-11}–10^{-14}	10^{20}–10^{22}

*	1 Hz is 1 cycle per second: kHz = one thousand Hertz (kilo Hertz); MHz = one million Hertz (Mega Hertz); GHz = one thousand million Hertz (Giga Hertz)
**	The visible band is a small range of frequencies that the retina of the eye is able to detect and which gives us the sense of vision. The wavelength of the visible part of the spectrum is usually expressed in terms of nanometres (nm; 1 nm is 10^9 of a metre). Visible light extends from 740 nm (perceived as red) to 400 nm (perceived as violet). Above the violet part of the visible spectrum is the ultraviolet region, which is broken down into three bands (A, B and C). The transition from non-ionising radiation to ionising radiation occurs around the wavelength of UVC.

[R1002] Electromagnetic fields (EMFs) can be characterised by their frequency and/or wavelength (see TABLE 1). The frequency is measured in the unit of Hertz (Hz). One Hz equals one cycle per second. The wavelength is measured in the unit of a metre (or for very high frequency radiations, down to nanometres (nm)). One nm is equal to one thousand-millionth of a metre (10^{-9} m). The strength of the field is measured by either the strength of the electrical component of the field or the strength of the magnetic component. The electrical field (the E-field) is measured as volts per metre ($v.m^{-1}$). The magnetic field can be measured as the magnetic field strength in units of Amps per metre ($A.m^{-1}$; the H-field) or the magnetic flux density in units of Tesla (or more usually as micro Tesla – µT; the B field). 1 µT is one-millionth of a Tesla.

EMFs above 100kHz in the radio frequency part of the spectrum have the ability to heat body tissues, but the risk depends on the amount of radiation received. This is measured as the amount of power falling on a given area of the body surface in the units of Watts per square metre ($W.m^{-2}$). This measure is the power density. The rate at which energy is absorbed in the body tissues is measured in the units of Watts per kilogram ($W.kg^{-1}$). This measure, the Specific Absorption Rate (SAR), is a fundamental measure of absorbed dose, which relates to the health and safety risk and is used as the basis for guidelines to ensure safe limits of exposure (see R1003).

How Does Non-ionising Radiation Interact with the Body?

[R1003] Non-ionising radiation poses acute risks and potentially chronic risks to health. Acute risks relate to the way in which the radiation is absorbed by the body and causes the heating of body tissues, possible electrical shock and skin burn. The chronic health effects relate to a possible risk of cancers and other diseases that may be linked to exposure to non-ionising radiation.

Acute Health Risks

[R1004] The acute health effects are separated into direct effects and indirect effects. Direct effects are those that result from a direct interaction between the body and the EMF to which the body is exposed. Indirect effects occur as a result of an interaction between the EMF, an external object (usually a metallic structure) and the human body.

Direct Effects

[R1005] The direct effects are usually considered differently for frequencies above and below 100kHz. Below 100kHz the hazard is one of induced electrical current in the body. Induced electrical charge on the surface of the body can be detected by a small proportion of people as a tingling sensation or as a perception of the raising or the vibration of the skin hairs. About 10 per cent of people can detect surface charges in this way, but only when the electrical field to which they are exposed is in the order of $10-15kV.m^{-1}$ (at ELF frequencies – see TABLE 1; 50Hz is the frequency of the electrical mains distribution and would be the usual source of exposure to fields causing this effect). This is not a health hazard but is annoying. A lot of research has focused on the physiological effects induced currents may have on the body. These have included effects on the nervous control of breathing and the heart, the effects of induced electrical signals in the brain and spinal cord, effects on performance and behaviour, effects on vision and hearing, and effects on the blood and on circadian rhythms (the body's daily cycle)). For frequencies above 100 kHz, the major effects are body heating. Healthy people can tolerate short-term rises in body temperature of about 1°C without significant adverse effects. However, a rise of 1°C will cause increased sweating, increased heart rate and a fall in blood pressure. A body temperature rise of 1°C may also be associated with changes in circulating hormone levels (such as thyroxin and adrenaline), and rises of more than this are considered potentially hazardous. As well as the hazard of whole body heating, frequencies above 100 kHz can cause localised heating, which may present a risk of damage to particular tissues. The lens of the eye, the testes and the embryo and foetus are the tissues most often considered to be at particular risk. These localised effects are most likely to be a problem with microwave radiations, but in the case of damage to the lens of the eye, infrared and ultraviolet radiations are also a significant risk (see TABLE 1). In the case of the eye, localised heating may cause cataracts, an opacity in the lens causing visual impairment and blindness. The testes may be vulnerable because they are usually maintained at a few degrees below body temperature, so that a rise in testicular temperature can cause reduced sperm counts and transient infertility. The embryo and foetus may be particularly sensitive to heating, which has been linked to birth defects.

Indirect Effects

[R1006] Indirect effects result from an accumulation of charge in the body, which is then discharged to a conductive (usually metallic) object. The discharge current may be sufficient to cause significant electric shock and can have serious consequences, such as paralysis of the respiratory muscles and ventricular fibrillation (an erratic beating of the heart, causing collapse and death). At lower levels of current the discharge may cause pain at the point of exit of the discharge (the point where the body is in contact with the metallic object) and sufficient localised heating to result in burns to the skin. Electrical discharges may be sudden or continuous. Sudden discharges occur where the body is effectively insulated from the ground, allowing the accumulation of charge, which is then suddenly discharged when contact is made with a metallic object. This is most likely to happen if the body is in an EMF of less than 100 kHz but is also extremely common with static electrical fields. This is a common occurrence in indoor environments and in cars where static electricity causes an accumulation of charge sufficient to cause a discharge on subsequent contact with a metallic object. The discharge current in such circumstances will not be sufficient to cause serious effects but will result in pain and discomfort, and is particularly distressing if repeated frequently. Alternatively, discharge may occur if the person is in continuous contact with the metallic object (or other grounded conductor) during the time of exposure to the EMF. The person effectively becomes a conduit for the electrical current to pass to earth. This is most likely to occur when working with fields of below 100 kHz and at the same time in contact with a grounded conductor. The resulting current may be sufficient to cause electric shock if the field strength is sufficient.

Implanted Medical Devices

[R1007] Both direct and indirect exposure to EMF sources may interfere with implanted devices such as cardiac pacemakers, automatic implantable cardioverter defibrillators (AICDs) and cochlea implants. It is impossible to be precise about what levels of exposure are safe for people fitted with such devices, because each device varies, but there is a risk that induced currents will interfere with their function. In the case of pacemakers and AICDs, the consequence of malfunction could be serious. It is also possible that individuals fitted with metallic implants, such as hip or knee replacements, will experience localised heating in certain circumstances.

Chronic Health Effects

[R1008]–[R1009] These effects result from chronic (namely, long-term) exposure to EMFs, and may be caused by relatively low levels of exposure. The risks cited mainly include cancers of various types, including cancers of the breast, brain and nervous system, leukaemia, degenerative brain and spinal cord conditions such as Amyotrophic Lateral Sclerosis (ALS), and behavioural conditions.

The cancer risk is the most frequently cited and the most researched. The possibility of a cancer risk from chronic exposure to EMFs first appeared in 1979 (Wertheimer and Leeper, 1979). This study, from America, demonstrated a two- to three-fold risk in childhood leukaemia, lymphomas and nervous system tumours in houses that had a high level of magnetic fields from the use of mains electricity (in the USA these are 60 Hz magnetic fields).

Since then, scores of epidemiological studies have been done to attempt to confirm these findings. Epidemiological studies attempt to look for statistical differences in disease prevalence in people exposed to a particular factor, compared with people who are not (or are less) exposed. In these studies, the cancer prevalence rates are compared for people living near, for example, power transmission lines, compared to people who have never lived near a power line. The difficulty encountered by such studies is that the numbers of people exposed to power lines is fairly small and the number of those with relevant cancers is also small. The statistical power of many of the studies is therefore poor.

One of biggest studies ever carried out, the UK Childhood Cancer Study, failed to find evidence that exposure to magnetic fields associated with electricity supply and transmission increases the risk of childhood cancers (UKCC, 1999). However, other studies with similar statistical validity have concluded that there is a risk between two to three times higher when the mean exposure exceeds about 0.2–0.3 µT (see, for example, Savitz et al, 1988; Feychting and Ahlbom, 1993.)

The Public Health England (PHE) website currently refers to an extract from Health Protection Agency (HPA) advice (based on National Radiation Protection Board (NRPB) 2004 advice) which says: ' . . . the overall evidence for adverse effects of EMFs on health at levels of exposure normally experienced by the general public is weak. The least weak evidence is for the exposure of children to power frequency magnetic fields and childhood leukaemia'.

The NRPB became part of the HPA in 2005 which in turn became part of PHE in 2013.

Much of the research over the last 30 years, especially on leukaemia and nervous system tumours, has been concerned with exposure to electrical power fields at 50/60 Hz. However, considerable work has also been done on exposures to radiofrequency fields. To some extent this has been because of concerns about living near to radio or television transmitter masts, but probably in greater part because of the concern that the transmitted power from mobile phones may be a health risk. There is also concern that living near to mobile phone base stations poses a health risk.

In the United Kingdom AM radio uses frequencies between about 180 kHz and 1.6 MHz; FM radio between 88 and 108 MHz; and television from 470 to 854 MHz. Mobile phone networks operate between 872 and 960 MHz or 1710 and 1875 MHz (see **TABLE 1**). In addition, other frequencies, particularly in the microwave region, are used by the emergency services, for civilian and military radar, for satellite communications and for digital radio and television transmission.

There have been a number of epidemiological studies to investigate the health risk of living near to television and radio transmitter sites with almost the same equivocal results as exist for living near to power lines. One UK study found that there was no relationship between living within two kilometres of TV and radio transmitter sites and the prevalence of leukaemia and lymphoma, although the prevalence did decline with increasing distance from the sites (Dolk et al, 1997a). In addition, the same authors have shown that the incidence of leukaemia and lymphoma around a single transmitter site in the Midlands was increased by a factor of 1.8 (Dolk et al, 1997b). Other studies in Sydney and Hawaii have shown an increased risk of leukaemia to those living close to television and radio masts, (Maskarinec et al, 1994 and Hocking et al, 1996), whilst a study in San Francisco around a single microwave transmitter failed to find any excess incidence of disease (Selvin et al, 1992).

A 2012 report by the HPA's independent Advisory Group on Non-ionising Radiation (AGNIR) concluded that there was still no convincing evidence that mobile phone technologies cause adverse effects on human health. The report updated a 2003 review and considered the scientific evidence on exposure to radiofrequency (RF) EMFs, which are produced by mobile phone technologies and other wireless devices, such as Wi-Fi, as well as television and radio transmitters. The report can be found at: www.gov.uk/government/publicat ions/radiofrequency-electromagnetic-fields-health-effects.

One area that has received much attention is the suggestion that exposure to electromagnetic radiation may increase the risk of breast cancer. A theoretical link between the two has been suggested for over 20 years. This theory suggests that electromagnetic radiation can influence the production of a hormone, melatonin, and that this in turn increases the risk of breast cancer.

The AGNIR concluded that EMFs do not influence the production or action of melatonin and that there is no evidence that EMF exposure is associated with an increased risk of breast cancer (HPA 2006) (webarchive.nationalarchives. gov.uk/20140714110131/http://www.hpa.org.uk/webc/HPAwebFile/ HPAweb_C/1204286180274).

Mobile phone cancer risks have also been the focus of numerous research studies. Animal and cellular studies suggest that exposure to the power levels received by the brain during a mobile phone conversation can cause changes in the DNA of cells, which, theoretically, could be associated with cancer (Lai and Singh, 1996).

A study of the mortality rate and leukaemia and brain tumour prevalence among 250,000 mobile phone users did not show any increased risk (Rothman et al, 1996). An important study published in 2006 concluded that the use of mobile phones was not associated with an increased risk of glioma (the most common type of brain cancer), (Hepworth et al. 2006). The Independent Expert Group on Mobile Phones (IEGMP), in a wide ranging and detailed review of the evidence linking mobile phones (and other RF sources) to health risks, concluded that RF exposure below the NRPB and International Commission on Non-ionising Radiation Protection (ICNIRP) guidelines do not cause adverse health effects (IEGMP 2000).

However, this report recommended that, as a precautionary measure, the ICNIRP guidelines should be adopted in preference to those of the NRPB. The IEGMP report was subsequently updated by two NRPB Reports (NRPB 2003, NRPB 2004). In 2006, the World Health Organisation (WHO) similarly concluded that in relation to exposure to mobile phone base station emissions, there was to date no scientific evidence that these cause adverse health effects (WHO 2006).

Since then there have been a number of further studies and in May 2010 the results of the important multi-national INTERPHONE study were published. The authors concluded: "Overall, no increase in risk of glioma or meningioma was observed with use of mobile phones. There were suggestions of an increased risk of glioma at the highest exposure levels, but biases and error prevent a causal interpretation. The possible effects of long-term heavy use of mobile phones require further investigation."

The AGNIR commented on the findings of the study saying: "Because mobile phones have only been in widespread use for less than 20 years, the Interphone study could not address the possibility of longer term risks to health. Given the enormous scale on which the technology has been adopted, there is therefore a need for continuing epidemiological surveillance to ensure that any adverse effects are detected at the earliest possible stage. The recently launched COSMOS study, which is being funded in the UK as part of the Mobile Telecommunications and Health Research (MTHR) programme, will contribute importantly to meeting this need."

In May 2011, the World Health Organisation (WHO) International Agency for Research on Cancer (IARC) classified radiofrequency EMFs as possibly carcinogenic to humans (Group 2B), based on an increased risk for glioma, a malignant type of brain cancer, associated with mobile phone use.

The main conclusion from the 2012 AGNIR Report on the Health Effects from Radiofrequency Electromagnetic Fields is that, although a substantial amount of research has been conducted in this area, there is no convincing evidence that RF field exposures below guideline levels cause health effects in adults or children. These "guideline levels" are those of the ICNIRP, which already form the basis of public health protection in many countries.

Several studies have also looked specifically at the risks of working with significant sources of electromagnetic energy. Such jobs are found in electronics and electrical engineering, engineering (welding), in the electrical supply industry, on the railways, in radio and television transmission and transmitter maintenance, and in the military. Some of these studies on occupational exposures have concluded that there is a significant risk of brain cancers, leukaemia and breast cancer (in both men and women) in those occupations in which there is significant exposure to electromagnetic radiation. In most cases the excess risk is between two and four times higher. However, there are also studies in which no association has been found. A number of studies have looked at paternal occupational exposure to electromagnetic fields and the risk of brain tumour in their offspring. While several showed an increased risk of brain tumour in the children of exposed fathers others failed to show any relationship.

With regard to current and future research, the WHO website reports that much effort is currently being directed towards the study of EMFs in relation to cancer, with studies looking at possible carcinogenic (cancer-producing) effects of power frequency fields continuing, although at a reduced level compared to that of the late 1990's. It also, it says that the long-term health effects of mobile telephone use is another topic of much current research. It reports that while no obvious adverse effect of exposure to low level radiofrequency fields has been discovered, given public concern regarding the safety of mobile phones, further research aims to determine whether any less obvious effects might occur at very low exposure levels.

In conclusion, the health risks from chronic exposure either to community or occupational sources of low level electromagnetic radiation remain uncertain. The vast amount of research in this area has failed to uncover any consistent or reliable evidence that there is a health risk. Many of the studies are too weak to reach definite conclusions on a health risk, but the underlying body of evidence that there might be a health risk cannot be ignored. Brain tumours, leukaemia and lymphomas are the most consistent health effects noted. In the absence of conclusive evidence for a health risk, the 'precautionary principle' should be applied.

On 1 July 2016, the *Control of Electromagnetic Fields at Work (CEMFAW) Regulations 2016 (SI 2016 No 588)* came into force, implementing the requirements of a European Union (EU) directive on EMFs. The directive, *Directive 2013/35/EU on the minimum health and safety requirements regarding the exposure of workers to the risks arising from physical agents (electromagnetic fields)*, takes account of the ICNIRP recommendations. It does not address the possible long-term effects of exposure to EMFs including cancer, although it does require the European Commission to take into account new scientific knowledge in this area. The European Trade Union Confederation (ETUC) criticised the directive at draft stage, saying that it 'short-changed' workers and put them at 'deadly risk' because it disregarded the long-term effects.

The CEMFAW Regulations are the first specific EMF regulations to be introduced into UK health and safety law. The Health and Safety Executive (HSE) explains that they transpose the requirements of the Directive which go beyond, or are more specific than those covered by, UK legislation already in place: the *Management of Health and Safety at Work Regulations 1999* supported by a PHE recommendation that the ICNIRP guidelines be followed. The requirements of the new regulations are set out at **R1010** below.

Legal Requirements: The Control of Electromagnetic Fields at Work Regulations 2016

[R1010] The *Control of Electromagnetic Fields at Work (CEMFAW) Regulations 2016* require employers to:

- make a suitable and sufficient assessment of the levels of EMFs to which employees may be exposed with reference to action levels (ALs) and exposure limit values (ELVs);
- ensure that exposure is below a set of ELVs;
- when appropriate, assess the risks of employees' exposure and eliminate or minimise those risks, ensuring that workers at particular risk, such as expectant mothers and workers with active or passive implanted or body-worn medical devices, are taken into account;
- when appropriate, devise and implement an action plan to ensure compliance with the exposure limits;
- provide information and training on the particular risks (if any) posed to employees by EMFs in the workplace and details of any action being taken to remove or control them. This information should also be made available to safety representatives;
- take action if employees are exposed to EMFs in excess of the ELVs; and
- provide health surveillance or medical examination as appropriate.
  As regulation 3 sets out, the Regulations do not apply to the shipping industry.

Definitions

Action Levels (ALs) and Exposure Limit Values (ELV)

[R1010.1] ALs and ELVs are set out in a Schedule to the Regulations, which can be found at: www.legislation.gov.uk/uksi/2016/588/pdfs/uksi_20160588_en.pdf.

HSE guidance to the Regulations explains that the requirements in the Regulations are based on these two sets of values related to EMFs which in turn are based on the ICNIRP recommendations (see above). It sets out that:

- ELVs are the legal limitations on the exposure of employees to EMFs and primarily relate to the levels of exposure to EMFs within the body.
- Health effect ELVs prevent possible harm from the heating of tissue and electrical stimulation of nerve and tissue from exposure to EMFs.
- Sensory effect ELVs prevent effects such as magnetophosphenes (a flickering sensation) or a feeling of nausea, vertigo or a metallic taste caused by static magnetic fields.

However, it points out that: 'These are often impossible or difficult and expensive to measure directly. For this reason, a separate set of values, known as ALs, have been produced, relating to quantities which can be measured more easily.'

The guidance explains that certain ALs may be used to demonstrate that EMF levels are below particular ELVs. If the AL is not exceeded, exposure cannot exceed the corresponding ELV. If the AL is exceeded, further consideration and assessment is required to determine whether the corresponding ELV may be exceeded. It is still possible, and it is often the case, that the corresponding ELV will not be exceeded.

Simple measures to reduce exposure, such as moving the worker further away from the EMF source or installing screening, the HSE advises, may be the easiest way to ensure that exposure is below the relevant ELV.

Indirect-effect ALs are not tied to a particular ELV and instead detail the EMF levels above which particular indirect effects may take place, such as interference with pacemakers, or the risk of ferromagnetic objects becoming projectiles in the vicinity of strong magnets. However, indirect-effect ALs do not cover every possible indirect effect. The HSE advises employers that they are already expected to know what these risks are and factor them in to their risk assessment.

The Regulations set out (Regulation 2) that an 'electromagnetic field' means a static electric, static magnetic and time-varying electric, magnetic and electromagnetic field with a frequency of up to 300 GHz.

An 'employee at particular risk' is an employee who has declared to their employer a condition which may lead to higher susceptibility to the potential effects of exposure to EMFs. This includes expectant mothers and workers who use active implanted medical devices (AIMDs), passive implanted medical devices (PIMDs) or body-worn medical devices (BWMDs), and employees who work in close proximity to electro-explosive devices, explosive materials or flammable atmospheres.

A 'health effect' is a direct biophysical effect (which in turn is defined as an effect on human body tissue caused by its presence in an EMF) which is potentially harmful to health; an 'indirect effect' is an effect, caused by the presence of an object or substance in an EMF, which may present a safety or health hazard; and a 'sensory effect' is a direct biophysical effect involving a transient disturbance in sensory perception or a minor and temporary change in brain function.

Exemptions

[R1010.2] Regulation 4 requires employers to ensure that employees are not exposed to EMFs in excess of the ELVs.

However, HSE guidance to the Regulations, *Electromagnetic Fields at Work A guide to the Control of Electromagnetic Fields at Work Regulations 2016 (HSG281)*, explains that:

> the Regulations allow the 'sensory effects' ELVs to be exceeded when certain safety conditions are met. These are set out in the Schedule to the Regulations. In these circumstances, an employer will not need to produce an exposure action plan (see below) or carry out further risk assessment unless there are employees at particular risk in the workplace or exposure exceeds any of the direct-effect ALs.
>
> In addition, the HSE may exempt work activities from the exposure limits stated in the CEMFAW Regulations, with exemption conditions laid down. These conditions are:

- exposure must be as low as reasonably practicable; even though exposure is permitted to be above the ELVs, employers are still required to bring exposure down to the lowest levels, ie they must ensure the ELVs are exceeded to the lowest extent reasonably practicable; and
- employees must be protected against the health effects and safety risks posed by that exposure.

Employers must still carry out risk assessments and provide suitable information and training (see below).

The exposure limit requirements do not apply during the development, testing, installation and use and maintenance of, or research related to, MRI equipment for patients in the health sector subject to the conditions set out above.

Nor do they apply to any activity where a suitable and sufficient alternative exposure limitation system is in place and where the activity is carried out by a member of Her Majesty's armed forces or a visiting force; by any civilian working with such a person; or on any Ministry of Defence premises or those of the services authorities of a visiting force. The other requirements of the CEMFAW regulations, with the exception of the requirement to produce an action plan, apply in full.

Low exposure equipment and work activities or equipment which may exceed the ELVS

[R1010.3] When the new regulations came into force in 2016, the HSE advised that most employers would not need to take any additional action to reduce the risks from EMFs, because in most workplaces EMFs are already at safe levels or because where employees may be exposed to higher levels of EMFs, the levels and risks should already have been assessed and managed.

The guidance provides a non-exhaustive list of 'low exposure equipment' (in Table 2). Where workplaces contain only equipment on this list, the ELVs and indirect-effect ALs will not be exceeded and employers do not need to take further action under the CEMFAWR unless there are employees at particular risk and/or they have five or more employees, and must therefore make a record of their findings.

It advises employers that determining that their work activity/equipment is on this list will be their exposure assessment. Employers with five or more employees must record that they checked all equipment against this list.

The list includes:

- wireless communications: including landlines and mobile phones, faxes and wireless communications devices such as Wi-Fi or Bluetooth;
- office equipment: including TVs, DVDs, computer and IT equipment, electric fans and room heaters, photocopiers, printers and shredders;
- buildings and grounds: including alarm systems, electrical room heating equipment, electric garden appliances, electric handheld and transportable tools, and lighting including desk lamps;
- electrical supply: including some overhead lines and electrical circuits and installations;
- light industry: coating and painting equipment; control equipment not containing radio transmitter; and measuring equipment and instrumentation not containing radio transmitters; and

- miscellaneous: including equipment placed on the European market as compliant with Council Recommendation 1999/519/EC or harmonised EMF standards; battery chargers, non-inductive coupling designed for household use; battery powered portable equipment that does not contain radio frequency transmitters; hydraulic ramps; and workplaces containing electrical handheld, portable tools.

The HSE guidance also provides a non-exhaustive list of work activities or equipment which may exceed the ELVs (see Table 3) and therefore potentially pose a risk to workers to which employers need to give further consideration. This list includes:

- Infrastructure (buildings and grounds): broadcast and telecoms base stations inside operator's designated exclusion zone; radio frequency or microwave energised lighting equipment; and radio and TV broadcasting systems and devices;
- Electrical supply: including particular electrical circuits or installations;
- Light industry: dielectric heating and welding; resistance welding: manual spot and seam welding; induction heating; induction soldering; magnetic particle inspection (crack detection); industrial magnetiser and demagnetisers, eg tape erasers; microwave heating and drying; RF plasma devices including vacuum deposition and spluttering;
- Heavy industry: industrial electrolysis, furnaces, arc and induction melting;
- Construction: microwave drying in the construction industry;
- Medical: MRI equipment; medical diagnostic and treatment equipment using EMFs eg diathermy and transcranial magnetic stimulation;
- Transport: electrically powered trains and trams; radar, air traffic control, weather and long range; and
- Military activities: maintenance of radar or high-powered communications systems.

Employers' Duties

Exposure assessment

[R1010.4] The first step employers must take is to assess the potential level of EMFs to which workers may be exposed (Regulation 5).

Assessments must demonstrate that where employers are required to ensure that employees are not exposed to EMFs in excess of the ELVs (Regulation 4(1)), the assessment must demonstrate whether that regulation is complied with, if necessary through the use of calculations and measurements and may, in accordance with the Schedule, assess exposure against the ALs in order to determine that specific ELVs are not exceeded.

The HSE guidance explains that employers should carry out exposure assessments using information in the HSE guidance, evidence such as records of reports of ill-health; emission information and other safety data provided by manufacturers or distributors of equipment; any sector or industry standards and guidelines; the EU non-binding guide to good practice for implementing Directive 2013/35/EU (see below); any exposure databases; information provided by trade associations and other industry bodies; and Medicines and Healthcare products Regulatory Agency (MHRA) guidance.

It further advises: 'Employers may permit employees to be exposed in excess of the sensory effect ELVs where the safety conditions stated in the schedule to the CEMFAW Regulations are met. Measurement or calculations should only be necessary for those employers where no exemption applies and information already available is insufficient to determine that the health-effect ELVs are not exceeded'.

Employers who employ five or more employees must keep a suitable record of the significant findings of the most recent exposure assessment and where required, the most recent action plan and the significant findings of the most recent risk assessment. They must also keep the exposure assessment under review and make any changes to ensure that it remains suitable and sufficient.

Action plans

[R1010.5] Regulation 7 requires employers to make and implement a suitable and sufficient action plan to ensure compliance with regulation 4(1). HSE guidance makes clear that employers must 'devise and implement an action plan to ensure compliance with the exposure limits' unless:

- the exposure assessment shows that the ELVs are not exceeded; or
- the exposure limits are only exceeded during: work activities where the applicable safety conditions stated in the Schedule to the CEMFAW Regulations 2016 are met (this only allows the sensory effect ELVs to be exceeded); and/or work activities covered by the MRI or military exemption; or work activities exempted from the exposure limits by the HSE.

The HSE advises that action plans must include the consideration of:

- other working methods that entail less exposure to EMFs;
- the choice of equipment emitting less intense EMFs, taking account of the work to be done;
- technical and/or organisational measures that limit the duration and/or intensity of emission of EMFs, such as the use of interlocks or screening. The HSE says that in many situations ELVs are only exceeded where the worker is close to the EMF source and that this can be easily remedied by moving the person further away from the EMF source or installing screening. However, screening may not be effective for low-frequency work activities;
- using signs, access controls and floor markings;
- exposure to electric fields; measures and procedures to manage spark discharges and contact currents through technical means and through the training of workers;
- maintenance and selection of equipment and design of workplaces and, when replacing or hiring equipment, consideration of selecting equipment which emits less intense EMFs; and
- providing personal protective equipment (PPE) such as insulating shoes, gloves and other protective clothing, where appropriate.

Trade union safety representatives, or other worker representatives, should be consulted when employers are deciding risk control measures.

Risk Assessments

[R1010.6] Regulation 8 requires the employer to make a suitable and sufficient assessment of the risks to employees arising from their exposure to EMFs. HSE guidance explains that risk assessments are required where the exposure assessment demonstrates that the ELV may be exceeded; there is a risk of indirect effects; and or there are workers at particular risk.

The HSE advises that risk assessments must consider the following where these are relevant:

- the ALs and ELVs;
- the frequency of the EMFs, level, duration and type of exposure, including the distribution over the employee's body and the variations between areas in the workplace;
- direct effects and indirect effects;
- employees at particular risk;
- simultaneous exposure to multiple frequency fields;
- multiple sources of exposure;
- information available from the manufacturer of relevant equipment;
- information obtained from any appropriate health surveillance undertaken;
- the existence of replacement equipment designed to reduce the level of exposure to EMFs; and
- other health and safety-related information.

Where a risk assessment is required and there are five or more employees, the significant findings must be recorded. The most recent action plan must also be recorded.

Regulation 9 sets out the employer's obligation to ensure that, so far as is reasonably practicable, the risks identified in the most recent risk assessment under regulation 8 are eliminated or reduced to a minimum.

Employees at particular risk

[R1010.7] The HSE guidance sets out that employers must give special consideration to the safety of employees at particular risk. These are expectant mothers who have advised employers of their condition and workers who have declared their use of AIMDs (such as cardiac pacemakers, cochlea implants and neural stimulators), PIMDs (such as orthopaedic implants or joints, stents and metallic contraceptive implants) and BWMDs (such as insulin pumps or hearing aids). In addition, an employee who works in close proximity to electro-explosive devices, explosive materials or flammable atmospheres is also at particular risk.

Employers should encourage workers to advise them in writing if they become pregnant as working with certain levels of EMFs could result in a greater risk to an expectant mother. The HSE advises employers that they may wish to take a practical approach and limit the exposure of expectant mothers to the public exposure limits set out in Council Recommendation 1999/519/EC (ec.europa.eu/health/electromagnetic_fields/docs/emf_rec519_en.pdf).

Sources of EMF which may pose specific risks to expectant mothers (see Table 5 in the HSE guidance) include work where workers need to be in close proximity to cables carrying high currents; automated induction heating systems: fault-finding and repair involving close proximity to the EMF source, automated welding systems, fault-finding: repair and teaching involving close proximity to the EMF source and MRI equipment.

Some levels of EMFs could cause medical devices to malfunction or cause injuries to workers using them as a result of EMFs interacting with them. The HSE gives the example of very strong static magnetic fields creating turning forces that move ferromagnetic PIMDs, while intermediate frequencies may cause them to heat up, which could cause injury to surrounding tissues.

Again, the HSE advises employers to encourage workers to advise them if they think they may be affected and obtain information and instructions from the manufacturer or advice from the medical professional who implanted the device.

There are two British Standards relating to the specific assessment of the exposure to EMFs of workers with AIMDs:

BS EN 50527-1:2010: Procedure for the assessment of the exposure to electromagnetic fields of workers bearing active implantable medical devices – General.

BS EN 50527-2-1:2011: Procedure for the assessment of the exposure to electromagnetic fields of workers bearing active implantable medical devices – Part 2-1: Specific assessment for workers with cardiac pacemakers.

The HSE guidance provides a non-exhaustive list of sources of EMF which may pose a risk to workers with PIMDs (see Table 6). These are similar to those for expectant mothers (see above) but also include hand-held induction heating coils.

The non-exhaustive list of sources of EMF which could interfere with ACMDs and ABWDs (see Table 7 in the HSE guidance) includes wireless communications; audio-visual equipment containing radiofrequency transmitters; the use of electric garden appliances; article surveillance equipment and radio frequency identification, tape or hard drive erasers and metal detectors; work on generators or emergency generators and where workers need to be in close proximity to cables carrying high currents, inverters, including photovoltaic systems; arc welding including MIG, MAG and TIG, industrial and large professional battery chargers, use of heat and glue guns, automated welding systems, automated induction heating systems and machine tools; MRI equipment; construction equipment; motor vehicles and plant, maintenance of inverters used on mainline trains; professional inductive cooking equipment and maintenance of radar or high powered communications systems.

Information and training

[R1010.8] Regulation 10 requires the employer to provide relevant information and training to any employees who are likely to be subjected to risks identified in the most recent risk assessment under Regulation 8.

The HSE guidance sets out that this should also be provided to their representatives and should include:

- an explanation of ALs and ELVs;
- details of possible health, sensory or indirect effects and what to do if these are experienced;
- the safe working practices the employer will adopt to eliminate or reduce the risks;
- an explanation of any safety signs used;
- details of appropriate personal protective equipment;
- information for workers at particular risk such as expectant mothers, workers using AIMDs, PIMD or BWMD and employees who work in close proximity to electro-explosive devices, explosive materials or flammable atmospheres; and
- the circumstances in which they may be entitled to an appropriate medical examination and/or health surveillance.

Health surveillance

[R1010.9] Regulation 11 requires the employer to ensure that health surveillance and medical examinations are provided as appropriate to any employee who is exposed to EMF levels in excess of the health effect ELVs and reports experiencing a health effect to that employer.

The HSE guidance explains that because the regulations only relate to short-term effects resulting from exposure to EMFs, and while it is possible to incur health effects, there is no well-established scientific evidence of long-term effects, health surveillance is only likely to be necessary in very limited circumstances.

Any health surveillance or medical examination must be provided during hours chosen by the employee and a suitable record kept of any health surveillance and medical examinations provided.

It is beyond the scope of this chapter to reproduce the Schedule contained in the CEMFAWR. Readers who need to undertake a risk assessment should refer to the Regulations and the Schedule it contains, and be prepared to understand the complexities and nuances of how they should be interpreted. It may be necessary to seek expert advice on how to determine levels of exposure, how to interpret the guidelines and how to devise suitable means of risk control.

Further information

[R1010.10] The *Control of Electromagnetic Fields at Work Regulations 2016 (SI 2016 No 588)* can be found at: www.legislation.gov.uk/uksi/2016/588/pdfs/uksi_20160588_en.pdf.

The HSE guidance, *Electromagnetic Fields at Work A guide to the Control of Electromagnetic Fields at Work Regulations 2016* (HSG281) is available on the HSE website at: www.hse.gov.uk. This includes a list of references and further reading on EMFs.

European Commission guidance on the Directive can be found at: ec.europa.eu/social/main.jsp?catId=738&langId=en&pubId=7845&type=2&furtherPubs=yes.

European Trade Union Institute (ETUI) guidance, Electromagnetic fields in working life. A guide to risk assessment, can be found at: www.etui.org/Publications2/Guides/Electromagnetic-fields-in-working-life.-A-guide-to-risk-assessment.

Control of Artificial Optical Radiation at Work Regulations 2010

[R1010.11] These Regulations implemented the requirements of an EU Directive on the protection of workers from hazardous sources of artificial light such as UV radiation and powerful lasers, which have the potential to cause harm to the skin or eyes, came into force in April 2010.

Common sources of light in workplaces like office lights, photocopiers and computers are not affected by the regulations. They do apply to organisations using hazardous light sources as part of their work activities.

HSE guidance provides examples of safe light sources and sources of light that have the potential to cause harm if inappropriately used, but which are safe under normal conditions of use.

It also sets out examples of hazardous sources of light that present a 'reasonably foreseeable' risk of harming the eyes and skin of workers and where control measures are needed:

- Metal working – welding (both arc and oxy-fuel) and plasma cutting;
- Pharmaceutical and research – UV fluorescence and sterilisation systems.
- Hot industries – furnaces;
- Printing – UV curing of inks;
- Motor vehicle repairs – UV curing of paints and welding;
- Medical and cosmetic treatments – laser surgery, blue light and UV therapies;
- Intense Pulsed Light sources (IPLs);
- Industry, research and education, for example, all use of Class 3B and Class 4 lasers, as defined in British Standard BS EN 60825-1:2014; and
- Any Risk Group 3 lamp or lamp system (including LEDs), as defined in British Standard BS EN 62471: 2008, for example search lights, professional projections systems.

It advises that less common hazardous sources are associated with specialist activities – for example lasers used during the manufacture or repair of equipment, which would otherwise not be accessible.

More information: The Control of Artificial Optical Radiation at Work Regulations 2010 (SI 2010 No 1140) can be found at www.opsi.gov.uk/si/si2010/uksi_20101140_en_1. HSE advice can be downloaded at www.hse.gov.uk/radiation/nonionising/employers-aor.pdf.

Ionising Radiation – What is it?

[R1011] Ionising radiation is radiation with sufficient energy to electrically charge or *ionise* material that the radiation strikes. Ionising radiation arises from a wide range of natural and man-made radioactive sources, and its properties have been harnessed extensively for industrial and medical use. Ionising radiation can, however, affect living tissue and its use must always be assessed and monitored carefully to prevent harmful effects. Types of radiation that cause ionisation include:

- α (alpha) particles;
- β (beta) radiation;
- γ (gamma) radiation;
- x rays;
- neutrons.

Units of Measurement: Radioactivity and Radiation Dose

[R1012] Radiation is emitted from radioactive sources and this radioactivity is measured in units known as *bequerels*. Radiation dose to humans is defined as the amount of energy absorbed by the body and the *gray (Gy)* is the unit of dose. Different types of radiation have, however, different effects on tissue, and in order to take account of these differences, a further unit of dose, the *sievert*, has been developed.

The *sievert (Sv)* is the most commonly used measure of radiation in occupational and environmental exposures. Doses, however, are usually measured in smaller fractions of the sievert, either *millisieverts* or *mSv* (one-thousandth of a sievert), or *microsieverts* or μSv (one-thousandth of a millisievert).

Natural Radiation

[R1013] On earth there are two main sources of natural or background radiation: cosmic radiation from space and natural radiation from the ground. Most cosmic radiation is filtered by the upper layers of the atmosphere, but ionising radiation from space can be significant for those in the space and aviation industry. A typical radiation dose for a flight from the UK to Spain would be about 10 µSv (NRPB, 1994).

Certain rock types, such as granite, emit weak ionising radiation but the most significant source of background radiation from the ground in the UK occurs from radon, a naturally occurring gas, which largely accounts for the difference in background radiation across the UK. Typical annual doses from background radiation in the UK range from 2.1 mSv in London, where radon levels are low, to 7.8 mSv in Cornwall, where they are much higher (NRPB, 1994).

Man-Made Radiation

[R1014] The development of ionising radiation for industrial and medical purposes has produced significant benefits. Major uses include:

- Medicine – Ionising radiation has a number of medical applications including the diagnosis of disease, using techniques such as computed tomography (CT) scanning or X-ray radiography, and in the treatment of cancer.
- Nuclear Power – around 20 per cent of UK electricity is generated by nuclear energy.

- Radiography in industry – Radiographic techniques are used in monitoring the integrity of metal structures such as pipelines and bridges and X-ray machines are used widely at ports and airports for baggage security.
- Consumer products – Radioactive materials are used in a number of consumer products including smoke detectors and in luminous dials.

Harmful Effects

[R1015] Ionising radiation can damage essential proteins in cells, and this damage can give rise to both short and long-term health effects:

- Short-term effects are generally related to the capacity of ionising radiation to kill cells, and this is proportional to the dose of radiation. Doses above 3 Gy to the skin will produce redness brought about by the death of skin tissue. In humans whole body doses of greater than about 4–5 Gy can be fatal.
- Lower radiation doses can damage but not kill living cells. The long-term effects of ionising radiation arise as a result of imperfect repair of damage to cells leading to abnormal cell growth. In turn, this abnormal cell growth can lead to cancer. These effects are random, but in a working population the possibility of developing cancer is *increased* by about 4 per cent for every sievert of radiation dose received (NRPB, 1995).

Irradiation and Contamination

[R1016] In considering the possible harmful effects of radiation, it is useful to differentiate between *irradiation* and *contamination*. Exposure to penetrating radiation from an external source, *irradiation*, may cause an individual to sustain damage to their body depending upon the dose, but the individual would not remain radioactive and would not become a hazard to others. Once the source of radiation is removed, the individual would not receive a further dose. An example of this would be a chest X-ray.

By contrast, if an individual has radioactive material on them or in their body, they are *contaminated*. As the *contamination* continues to emit radiation, they will continue to receive a radiation dose until the contamination is removed or declines naturally, even if they leave the area where the contamination occurred. They may also transfer the contaminated material to others, who may become at risk.

Protection against Radiation

[R1017] The key to radiation protection lies in understanding the nature of the hazard. Different types of radiation can be blocked or *shielded* by different types of material. For instance, α (alpha) particles can be stopped by a sheet of paper, whereas β (beta) radiation requires a sheet of metal and γ (gamma) radiation a thickness of lead.

Other properties of radiation can be used when considering protection. The effect of radiation, for example, is reduced by distance from the source, and if the distance is doubled, the dose is reduced to a quarter. This is the *inverse square* phenomenon. The duration of exposure is also important. The less time an individual is exposed to a source, the less the received radiation dose will be.

PHE's Centre for Radiation, Chemical and Environmental Hazards provides expert advice on radiological protection (www.gov.uk/guidance/radiation-products-and-services).

Legal Requirements

[R1018] In the UK work with ionising radiation is now covered by the *IRR 2017 (SI 2017 No 1075)* which came into force in January 2018.

Other legislation covering ionising radiation can be accessed via the UK Government Cabinet Office website www.opsi.gov.uk, and includes:

- The *Radioactive Substances Act 1993*, the legislation by which the environment agencies regulate the use of radioactive materials and the disposal of radioactive waste.
- The *High-activity Sealed Radioactive Sources and Orphan Sources Regulations 2005 (SI 2005 No 2686)* which implement the EU High-Activity Sealed Sources and Orphan Sources (HASS) Directive.
- The *Radiation (Emergency Preparedness and Public Information) Regulations 2001 (SI 2001 No 2975)*, which lay down basic safety standards for the protection of the health of workers and the general public against the dangers arising from ionising radiation and impose requirements on the operators of premises where radioactive substances are present in quantities exceeding specified thresholds. They also impose requirements on carriers transporting radioactive substances (again in quantities exceeding specified thresholds) by rail or conveying them through public places.
- The *Carriage of Dangerous Goods and Use of Transportable Pressure Equipment Regulations 2009 (SI 2009 No 1348)*.
- The *Justification of Practice Involving Ionising Radiation Regulations 2004 (SI 2004 No 1769)*, by which the *Department of Environment, Food and Rural Affairs* assesses new processes involving ionising radiation.

Medical exposures to ionising radiation are governed by:

- The *Ionising Radiation (Medical Exposure) Regulations 2017 (SI 2017 No 1322)*.
- The *Medicines (Administration of Radioactive Substances) Regulations 1978 (SI 1978 No 1006)*.

In relation to work with ionising radiation, health and safety managers may also have to consider the *Management of Health and Safety at Work Regulations 1999 (SI 1999 No 3242)* (MHSWR) and the *Personal Protective Equipment Regulations 2002 (SI 2002 No 1144)*.

Key Features of the Ionising Radiations Regulations 2017

[R1019] Employers with staff working with ionising radiation, or those who are considering work with ionising radiation, should carefully study the *IRR 2017*, together with the Approved Code of Practice and Guidance (ACOP).

HSE guidance to the Regulations explains that work with ionising radiation means:

- a practice, which involves the production, processing, handling, use, holding, storage, transport or disposal of artificial radioactive substances and some naturally-occurring sources;
- the operation of electrical equipment emitting ionising radiation and containing components operating at energies above 5 kV; and
- work in places where the radon gas concentration exceeds the values set out in the Regulations.

The new regulations came into effect on 1 January 2018, implementing the European Basic Safety Standards Directive 2013/59/Euratom (BSSD) in the UK and repealing and replacing the previous *Ionising Radiations Regulations 1999 (IRR 1999)*.

The main changes are that the dose limit for exposure to the lens of the eye is reduced from 150 mSv to 20 mSv in a year. Dose limits are the radiation exposure levels that must not be exceeded and are set out in *Schedule 3* of *IRR 2017*. In addition, there is now flexibility for five-year averaging for the dose limit to the lens of the eye, subject to HSE-specified conditions.

The other key change is that, depending on the level of risk, employers are required to notify, register or obtain consent from the HSE, even if they have already told the HSE that they work with ionising radiation. There is now a three-tier system of notification, registration and consent, depending on the size and likelihood of exposure. This replaces the previous requirement under *IRR 1999* for notification and prior authorisation.

The HSE summarises the changes as including the following:
- lowering the dose limit to the lens of the eye;
- flexibility for five-year averaging for dose limit to lens of the eye, subject to conditions specified by HSE;
- changing the radon reference level. The *IRR 1999* radon reference level was over a 24-hour period, while the BSSD expresses the reference level on an annual basis. The *IRR 1999* reference level is broadly equivalent to the annual average reference level in *IRR 2017*;
- introducing a three-tier system of notification, registration and consent that replaces the previous requirement for notification and prior authorisation;
- changing the requirement for notification, which for some radionuclides is at a lower threshold than in *IRR 1999*;
- broadening the scope of the definition of an outside worker to include both classified and non-classified workers (see Designation of classified persons and Outside workers and passbooks below);
- introducing new weighting factors for dosimetry;
- changing the dose record retention period from 50 years to not less than 30 years after the last day of work;
- introducing a requirement to put procedures in place to estimate doses to members of the public;
- removing the requirement for a registered medical practitioner to be appointed 'in writing' for the purposes of the Regulations;

- introducing a requirement for authorisation of the annual whole-body dose limit in special cases: the HSE or the Office for Nuclear Regulation (ONR) may authorise the application of an effective dose limit of 100 mSv over five years (with no more than 50 mSv in a single year) rather than duty holders only giving prior notification;
- recording and analysis of significant events, ie radiation accidents;
- removing the subsidiary dose limit for the abdomen of a woman of reproductive capacity; and
- removing references to 'radiation employers', a term that has previously caused confusion, and replacing it with employers.

Regulations 5, 6 and 7 set out the three-tier system of notification, registration and consent which has replaced the previous requirement under *IRR 1999* for notification and prior authorisation.

The regulations set out the following dose limits in *Schedule 3*:

- For employees and trainees of 18 years of age or above the limit on the dose of radiation to the whole body (the effective dose) is 20 mSv in any calendar year. The limit on equivalent dose for the lens of the eye is 20 mSv in a calendar year; or 100 mSv in any period of five consecutive calendar years subject to a maximum equivalent dose of 50 mSv in any single calendar year. The limit on equivalent dose for the skin is 500 mSv in a calendar year as applied to the dose averaged over any area of 1 cm^2 regardless of the area exposed. The limit on equivalent dose for the extremities is 500 mSv in a calendar year.
- For trainees under 18 years, the limit on effective dose is 6 mSv in any calendar year. The limit on equivalent dose for the lens of the eye is 15 mSv in a calendar year; the limit on equivalent dose for the skin is 150 mSv in a calendar year as applied to the dose averaged over any area of 1 cm^2 regardless of the area exposed; and the limit on equivalent dose for the extremities is 150 mSv in a calendar year.
- The limit on effective dose for any person other than an employee or trainee referred to above, including any person below the age of 16, is 1 mSv in any calendar year. This does not apply in relation to any person (not being a carer and comforter) who may be exposed to ionising radiation resulting from the medical exposure of another. In these cases, the limit on effective dose is 5 mSv in any period of five consecutive calendar years. The limit on equivalent dose for the lens of the eye is 15 mSv in any calendar year; the limit on equivalent dose for the skin is 50 mSv in any calendar year averaged over any 1 cm^2 area regardless of the area exposed; and the limit on equivalent dose for the extremities is 50 mSv in a calendar year.

Key sections of the regulations for the non-specialist include the following:

- Radiation risk assessment – *Regulation 8* requires employers, before commencing any new activity with ionising radiation, to undertake a 'suitable and sufficient' risk assessment. The HSE guidance explains that this was previously known as a prior risk assessment, and complements the requirements of *regulation 3* of the *Management of Health and Safety at Work Regulations 1999*.

- Restricting exposure – *Regulation 9* requires employers to restrict exposure to ionising radiations so far as is reasonably practicable. HSE guidance explains that: 'Firstly, in any work with ionising radiation, employers should take action to control the doses received by their employees and other people by means of engineering controls. Only after these have been applied should they consider using supporting systems of work. Lastly, employers should provide PPE to further restrict exposure where this is necessary and reasonably practicable'.
- Under *regulation 10* employers must ensure that any PPE provided is suitable and complies with the *Personal Protective Equipment Regulations 2002*, or in the case of some respiratory protective equipment, be of a type approved or conform to a standard approved in either case by the Executive. Employers must also ensure that adequate facilities are provided for the storage of that equipment.
- *Regulation 11* requires employers to maintain engineering controls and personal protective equipment and examine and test them at appropriate intervals where appropriate.
- *Regulation 12* sets out requirements in relation to dose limits (see above) and the HSE explains that while employers must make sure that exposures arising from the work are kept as low as reasonably practicable, complying with dose limits is an absolute requirement.
- *Regulation 13* sets out requirements in relation to contingency plans (see below).
- Management of radiation protection – *Regulation 14* requires employers engaged in work with ionising radiation to consult a suitable Radiation Protection Advisor (RPA) to advise on the Regulations. Employers must ensure that the RPA fulfils the HSE's criteria of competence but will find that an RPA will be particularly useful in providing practical guidance in applying the Regulations. The HSE website, www.hse.gov.uk/radiation/rpnews/rpa.htm, is helpful for employers new to work with ionising radiation, and the site has a list of RPA bodies recognised by HSE under *IRR 2017*.
- Information, instruction and training – *Regulation 15* requires employers to ensure that employees working with ionising radiation are given appropriate training in the field of radiation protection and receive suitable and sufficient information and instruction on the risks to health; the precautions to be taken; and the importance of complying with the medical, technical and administrative requirements of the regulations. In addition, adequate information should be given to other people who are directly concerned with the work with ionising radiation carried on by the employer to ensure their health and safety. Female employees working with ionising radiation should be informed of the possible risk arising from ionising radiation to the foetus and to a nursing infant and of the importance of informing the employer in writing as soon as possible after becoming aware that they are pregnant or if they intend to breastfeed. Employees engaged in work in a controlled area (as designated under *regulation 17*) must be given specific training in connection with the characteristics of the workplace and the activities within it; and this must be repeated at appropriate intervals and documented by the employer.

- Co-operation between employers – *Regulation 16* deals with work with ionising radiation undertaken by one employer likely to give rise to the exposure to ionising radiation of an employee of another employer. The employers concerned must co-operate, by exchanging information for example, so they all have access to information on the possible exposure of their employees to ionising radiation and can comply with the requirements of the Regulations. This reflects the general obligation of the *Management of Health and Safety at Work Regulations 1999 (SI 1999 No 3242, Reg 11* of employers sharing the same workplace to co-operate.
- Designated areas – *Regulations 17–20* cover the requirements of employers to designate and monitor areas in which employees may receive a significant radiation dose. These areas may be designated as *controlled* or *supervised*, and employers are obliged to have in place written *local rules* to identify key working instructions to restrict radiation exposure in these areas and the steps to be taken in the event of a radiation accident. Employers are also required to appoint *Radiation Protection Supervisors* to ensure compliance with the Regulations and local rules.
- Designation of classified persons – *Regulation 21* requires employers to designate as *classified* persons employees who are likely to receive a radiation dose greater than 6 mSv per year or an equivalent dose greater than 15 mSv per year for the lens of the eye or greater than 150 mSv per year for the skin or the extremities. They must immediately inform those employees that they have been designated. Classified persons must be over 18 years of age and a relevant doctor must have certified in the health record that that employee is fit for the work with ionising radiation which that employee is to carry out.
- Dose assessment and recording – *Regulation 22* sets out the requirement for employers to appoint an *approved dosimetry service* (ADS) to monitor the radiation dose and to keep the dose records. Dosimetry services have to be approved by the HSE for certain functions, and employers should confirm that the chosen dosimetry service has an appropriate certificate of approval. The HSE website, www.hse.gov.uk/radiation/ionising/dosimetry/ads.htm, lists approved dosimetry services.
- Medical surveillance – *Regulation 25* sets out the requirements concerning medical surveillance. HSE guidance to the Regulations explains that:
    - employers of anyone exposed to ionising radiations as a result of work activities must decide when such an employee needs to be designated as a classified person;
    - before that person is classified, employers must make sure that the employee has been certified as fit for the intended type of work within the previous 12 months. This may require a medical examination;
    - employers will then need to make arrangements with the relevant doctor (that is an appointed doctor or an employment medical adviser) for continuing medical surveillance; and

- employers must arrange for adequate medical surveillance for any employee who has received an overexposure, whether or not that employee has been designated as a classified person.

The medical surveillance of classified persons must be undertaken by a registered medical practitioner who has been appointed in accordance with the Regulations. A list of appointed doctors can be found on the HSE website at: https://webcommunities.hse.gov.uk/connect.ti/appoint eddoctors/viewdatastore?DSID=7172&ADV=S&search_0=LIKE&sea rch_0=&search_7_4=Y&search_allcolumns=.

A *health record* has to be maintained to record the outcome of annual medical surveillance for each classified person, and the requirements of this record are set out in Schedule 6 of the Regulations. Form F2067 is the standard form available from the HSE. Note that this health record is separate from the employee's medical file and should be kept available for scrutiny by a Health and Safety Executive (HSE) inspector. The HSE advises that after the initial medical examination conducted before designation as a classified person, periodic reviews of health should take place at least once every year. The relevant doctor may specify a shorter period between reviews.

HSE guidance sets out that employees can ask for a review of any decision by the doctor if they do not agree with any condition recorded in the health record or a finding that they are not fit for the work with ionising radiation.

- Pregnant and breast-feeding employees – *Regulation 9(6)* sets out that an employer who undertakes work with ionising radiation must ensure that after an employee has notified them of the pregnancy, the equivalent dose to the foetus is as low as is reasonably practicable and is unlikely to exceed 1 mSv during the remainder of the pregnancy. They must also ensure that an employee who is breastfeeding must not carry out any work involving a significant risk of intake of radionuclides or of bodily contamination.
- Outside workers and passbooks – HSE guidance explains that an outside worker is any person who is carrying out services in a controlled area or supervised area (see above) but who does not have an individual contract of employment with the employer responsible for that area (as distinct from any contract for service between their own employer and the employer responsible for the area). The intention of the definition is that all outside workers, including non-classified outside workers, have the same level of protection as other employees (those formally employed by the organisation) in relation to training, instruction, protective equipment, dose monitoring and entry to controlled and supervised areas. A key feature of outside workers is their requirement to hold a *radiation passbook (Regulation 22(5))*, which is provided by the ADS. This is a document personal to them and cannot be transferred to other employees. *Schedule 5* sets out the particulars to be entered in the radiation passbook.
- Overexposures – *Regulations 26* and *27* detail the action to be taken in the event of an overexposure. This includes the employer notifying the HSE and the appointed doctor as soon as practicable and taking reasonable steps to notify the suspected overexposure to the person

affected. Care should also be taken to notify other employers in the event that their employees are involved. A special medical examination should be undertaken by the appointed doctor in the event that the effective dose of radiation to an employee is greater than 100 mSv in a year. In practice it is recommended that employees are seen by the RPA and the appointed doctor at an early opportunity and provided with as much information and support as possible.

The new regulations are set out together with the approved code of practice and guidance in a revised version of the HSE publication L121: *Work with ionising radiation Ionising Radiations Regulations 2017 Approved Code of Practice and guidance* (www.hse.gov.uk/pubns/priced/l121.pdf). The revised document includes a summary of other changes introduced as a result of the new regulations.

The ACOP was still in draft form in March 2018. The *Ionising Radiation (Medical Exposure) Regulations 2017 (IR(ME)R 2017)* amended *IRR 2017* when they came into force on 6 February 2017. *IR(ME)R 2017* impose duties on employers and those with responsibilities for administering ionising radiation to protect people undergoing medical exposures whether as part of their own medical diagnosis or treatment, as part of research, as asymptomatic individuals, as those undergoing non-medical imaging using medical radiological equipment or as carers and comforters of persons undergoing medical exposures. The HSE was due to review the draft ACOP and determine a final publication date after *IR(ME)R 2017* came into force.

Key Features of Other Legislation relating to work with Ionising Radiation

[R1020] Much other legislation relating to work with ionising radiation is specialised and beyond the scope of this overview. In the event, however, that an employer believes that any of this legislation might apply, the advice of an appropriate RPA should be sought at an early opportunity.

Emergencies and Unforeseen Incidents

[R1021] The *IRR 2017* place an obligation on the employer to prepare a contingency plan if the risk assessment shows that an accident is reasonably foreseeable (*Regulation 13*).

The *Radiation (Emergency Preparedness and Public Information) Regulations 2001* set out the requirements for emergency planning and the obligation for information to be made available to the public in the event that the public could be affected.

The UK has a scheme, the National Arrangements for Incidents involving Radioactivity (NAIR). The scheme provides assistance to the police and other emergency services where there is no radiation expert available. Assistance is provided in two stages and is drawn from hospitals, the nuclear industry and government departments.

In the first stage, a radiation expert with simple monitoring equipment will provide assistance. They will be able to tell whether a hazard exists and advice the police on appropriate action. If necessary they will advise the police to obtain stage 2 assistance, providing more sophisticated resources – a small team of experts with readily available transport, monitoring and decontamination equipment and special clothing.

Those requiring assistance from the NAIR scheme should in the first instance contact the Communications Centre of the Civil Nuclear Constabulary on 0800 834153.

Nuclear industry

[R1021.1] It is beyond the scope of this chapter to provide detailed guidance on health and safety legislation applying to nuclear installations.

Legislation regulating safety in the nuclear industry covers health, safety and radioactive waste management at nuclear sites, aimed at protecting both the public and workers. In addition, there is also legislation to protect nuclear and radioactive materials on licensed civil nuclear sites, sensitive nuclear information and those employed in the civil nuclear industry, against 'criminal or malevolent acts that threaten national security, the environment or public safety'. The Nuclear Installations Act 1965 (NIA 1965), as amended, requires that a site cannot have nuclear plant on it unless a site licence has been granted. Safety in the nuclear industry is regulated by the Office for Nuclear Regulation (ONR), an HSE agency.

Following the 2011 Fukushima Dai-ichi nuclear accident in Japan, an ONR analysis revealed that there are no fundamental safety weaknesses in the UK's nuclear industry but concluded that by learning lessons it can be made even safer.

Final report on the Japanese earthquake and tsunami: Implications for the UK nuclear industry, by Her Majesty's Chief Inspector of Nuclear Installations and Executive Head of the Office for Nuclear Regulation, Mike Weightman, is available at www.hse.gov.uk/nuclear/fukushima/final-report.pdf.

Rehabilitation

Andrea Oates

What is rehabilitation?

[R2001] The term vocational rehabilitation is widely used to broadly describe the process of getting people back to work. However different stakeholders have different views of what is meant by vocational rehabilitation. One view is that it refers to getting people into employment when, for some reason, they have been out of work for a long period of time or perhaps have never worked. This is indeed one important aspect of rehabilitation. The reasons why people may not have worked for a long time, or indeed not have worked at all, are complex and not always due to a disability or long-term health impairment.

The other view is that vocational rehabilitation relates to the successful return to work of an employee who has been ill, or injured, or for some other reason is taking significant periods of absence. The employee may eventually return to their original job, a modified job, or even a new job within the organisation. This is good for the employer as they will reduce the costs of absence and retain a valued employee.

The two views therefore refer in the one instance to getting people into employment who are currently unemployed, and in the other instance to getting people who do have a job back into the workplace, usually following a period of sickness or injury. It is this latter we will explore in more detail in this chapter since it concerns the relationships between employers and their employees and concerns the successful management of the organisation.

Why should we be concerned?

The business argument

[R2002] All organisations should seek to maximise the use of resources and minimise their operational costs. People are usually an organisation's most valuable and most costly asset and it makes good business sense, therefore, to ensure employees are absent as little as possible.

Sickness absence figures published by the Office for National Statistics (ONS) in July 2018 show that sickness absence has fallen to the lowest rate on record. The average number of sickness absence days that UK workers take has almost halved since 1993. Employees took an average of 4.1 sickness absence days in 2017, compared with 7.2 days in 1993. However, the ONS says there may be an increase in presenteeism, where people go to work even though they are ill (see **R2007**).

More than a quarter (26.2%) of days lost through sickness absence in 2017 were attributed to minor illness such as coughs and colds. This accounts for around 34.3 million days per year. Musculoskeletal problems, such as back and joint pain, was a reason for sickness absence for 21% of 50- to 64-year-olds and 19% of 35- to 49-year-olds. There has been an increase in the proportion of younger workers aged 25 to 34, who attribute their sickness absence to mental health conditions, rising from around 7% in 2009 to around 9.5% in 2017. The government's November 2017 policy paper, *Improving Lives The Future of Work, Health and Disability* (see **R2008**), puts the costs to the economy of ill-health that prevents people from working at an estimated £100bn a year.

The advice and conciliation service (ACAS) says that most absence is genuine, and clearly a proportion is unavoidable, with much short, self-limiting periods of absence resulting from common causes such as colds, headaches and stomach upsets.

In addition, the trade union organisation (TUC) has consistently warned about the effects on productivity of workers feeling obliged or pressurised to come into work when they are ill and has reported a growing trend of "presenteeism" over recent years (see **R2007** below).

Around 80% of sickness absence is estimated to be short term. The remaining 20% is long term – employees who are absent for weeks or months with a condition that renders them unable to travel to work and/or to carry out their expected duties. This long-term sickness absence has been estimated to account for around 70% of the total costs of sickness absence. From a business point of view therefore:

- reducing the impact of long-term sickness absence;
- controlling any non-genuine reasons for absence; and
- controlling frequent short-term absences

are essential in reducing costs. The processes of rehabilitation can play an important role in helping an organisation to reduce the impact of these types of absence.

There is no agreed definition of short term and long term. Often short term refers to absences of seven days or less which coincides with the point at which an employee who is absent from work due to ill health or injury must obtain a 'fit note' (Form Med3) from their doctor (see **R2008** below). However, long term is not normally taken to be more than seven days and more usually is defined as three weeks or more. Many organisations have their own definition of short and long-term absence set out within their absence policy.

In addition to single short-term absences, there will be people who take very frequent one or two-day absences which over the course of a year add up to a significant amount of lost time. Frequent short-term absence has different causes from continuous long-term absence but nevertheless the problem may need to be addressed.

Absence costs a great deal of money both in terms of direct salary loss, loss of business or business inefficiency and replacement costs and needs to be managed like any other aspects of an organisation. It is important to focus attention to where efforts to manage absence are most likely to pay dividends. It is pointless devoting energy to unavoidable short-term absence. However, having control over frequent short-term absence, avoidable short-term absence and long-term absence can be very cost effective. Rehabilitation can be seen to be the process by which people can be enabled back to work or to a more reliable work pattern (see **R2007** below).

The legal argument

[**R2003**] Although it is important to get people back to work as soon as possible and to limit the amount of sickness absence time taken, there is no legal duty to do so. The Health and Safety Executive (HSE) produces information and advice for employers on rehabilitation, but this is guidance and contains no legal obligation. Rehabilitation often involves the intervention of Occupational Health Services (OHS), but many organisations do not have not have access to Occupational Health provision (see **R2010** below). Rehabilitation therefore is something employers will want to provide because of the benefits to the business rather than any direct legal necessity for it, although see **R2004** below.

Common law

[**R2004**] Although there is no legal obligation to provide for rehabilitation, there are a number of important legal ramifications. In the case of *Young v The Post Office* [2002] EWCA Civ 661, [2002] IRLR 660, for example, the Court of Appeal ruled that the employers were in breach of their duty of care in failing to ensure arrangements made for the claimant's return to work were adhered to. In this case the claimant had returned after an absence of four months due to a nervous breakdown brought on by stress at work. The arrangements for his return to work were not adhered to with the result that, within seven weeks of returning to work, the claimant suffered a recurrence of his psychiatric illness. An important aspect of this judgement is that the court rejected the suggestion that it was up to the claimant to speak out if he felt he was under stress. While this case involved stress, the principle would equally apply to return to work following other illnesses or injury. The employer is liable for the subsequent injury to the employee if adequate arrangements are not made for the return to work and/or if the arrangements made are not adhered to. It is up to the employer to monitor the employee and not to rely on his or her reporting of any worsening of their condition.

This case and others like it (see *Walker v Northumberland County Council* [1995] 1 All ER 737 for example), have established an important principle. If an employer knows an employee has been off sick and will be returning with a vulnerability, they have a legal duty to ensure the employee's residual condition is not going to be made worse on return to work. However, the employer's liability will be subject to the usual principles of law. These are as follows.

Foreseeability	The employee must show that their employer should have been able to reasonably foresee the conditions of work on their return would have made their condition worse or prevented their recovery. Foreseeability will depend on what an employer knows or ought reasonably to know about the employee's condition. For an employee who has been off sick for reasons that are well documented by Medical Statements (known as 'fit notes') (see **R2008** below), it would be difficult for an employer to argue they did not know of the employee's condition. If the employer has any doubts about the vulnerability the employee may suffer from on their return to work and how this vulnerability might be affected by the conditions of work, it would be considered reasonable for them to have taken steps to find out. This would normally mean referral to an Occupational Health professional and a request for an opinion on what adjustments might need to be made to ensure any risks inherent in the employee's return to work were minimised (see **R2010** below).
Breach of duty	An employer is in breach of their duty of care to their employee if they have failed to take reasonable steps to prevent the employee being harmed. What is reasonable will depend on factors including the size of the undertaking, its resources and what is feasible. It may eventually be necessary for an employer to argue in court that what they did, or failed to do, was reasonable. It is important to note here the amendment to section 47 of the Health and Safety at Work etc Act 1974 (via section 69 of the Enterprise and Regulatory Reform Act 2013), which came into effect on 1 October 2013. The effect of the amendment is that employers no longer have a strict liability for the health and safety of their workers. 'Strict liability' means an employer is automatically responsible for a breach of health and safety law regardless of whether the breach was their fault. It referred to a few workplace regulations where, if the employer breached those regulations, they would be liable in a civil claim even if they could argue they had taken reasonably practical steps to try and prevent the incident. As a result of the amendment, workers can no longer rely on an employer's breach of health and safety law to win a personal injury claim but must provide proof of negligence.
Causation	The employee will need to show the breach of duty has caused, or has materially contributed to, the harm they have suffered. It would not be sufficient to show that the work by itself caused an exacerbation of the residual illness or injury someone returned to work with. It must be shown that the employer did not do all that was reasonably practicable, and this breach of duty was responsible for the harm caused.
Apportionment and quantification	Where a worsening of a condition on return to work may have more than one cause, the courts may have to decide on apportioning the damages. The courts may also have to decide on any contribution the claimant may have made to the worsening of their condition – by not complying with the risk controls put in place by the employer for example. They may also have to decide whether the deterioration in a person's condition may have been going to happen anyway – because, for example, it is a progressive condition. The principle is that the employer should only pay for the proportion of the damage caused by their breach of duty.

The Equality Act 2010

[R2005] Another aspect of the legal argument for ensuring rehabilitation is firmly on an organisation's agenda, is the requirements of the *Equality Act 2010 (EA 2010)*.

The Act replaced a raft of previous anti-discrimination legislation, including the *Disability Discrimination Act 2005 (DDA 2005)*. Disability is a 'protected characteristic' under the *EA 2010*, which covers direct discrimination, indirect discrimination, discrimination by association, perceptive discrimination and protection against victimisation and harassment. The Act also introduced a new offence of 'discrimination arising from disability' (see the **EQUALITY ACT 2010** chapter).

The Act requires employers not to treat people with disabilities less favourably than those without that disability, and to make reasonable adjustments to the work or the workplace to enable someone with a disability to work. The Act applies to all aspects of employment including at the point of recruitment, but in the context of rehabilitation its importance is that it may need to be applied when someone returns to work following an illness or injury. Much depends on the definition of disability and what adjustments might be considered reasonable under *EA 2010*. If an employee claims breach of the Act, redress is generally through an employment tribunal rather than the courts.

Under *EA 2010* a disability is regarded as any mental or physical impairment that has a substantial and long-term adverse effect on a person's ability to carry out their normal day-to-day activities. The Act abolished a list of specified "day-to-day" activities which included mobility, manual dexterity and physical coordination. This was widely regarded as unnecessarily restrictive. Employment tribunals are now required to take a broad view when interpreting "day-to-day activities".

In *Banaszczyk v Booker Ltd* [2015] UKEAT/0132/15/RN, the Employment Appeal Tribunal (EAT) ruled that lifting and moving goods weighing up to 25kg was a day-to-day activity – in this case for a warehouse operative. The judge commented: "No-one with any knowledge of modern UK working life could doubt that large numbers of people are employed to work lifting and moving cases of up to 25kg across a range of occupations, including in particular occupations concerning warehousing and distribution."

If an employee returns to work with a residual condition following an illness, injury or operation, and the condition is likely only to be temporary before they return to normal full capacity, this would not fall within the requirements of *EA 2010*. However, if the condition is expected to be long term – the Act defines this as lasting at least twelve months – then adjustments have to made. If the effects are likely to recur beyond 12 months after the first occurrence, they are to be treated as long-term.

Examples of adjustments that may be necessary and would be considered reasonable include making adjustments to premises, redeploying someone to a different job or different job location, altering the person's work hours, allocating some of the person's job to another person or allowing time off for rehabilitation or treatment. This is not an exhaustive list and many other types of adjustment, such as providing new work equipment or modifying existing equipment, would need to be considered.

Even where an employee is not disabled, if their illness or injury is work-related, the employer should "go the extra mile" in terms of investigating all the alternatives before moving to a dismissal (*McAdie v Royal Bank of Scotland* [2007] EWCA Civ 806, [2007] IRLR 895, [2008] ICR 1087.)

Health and safety and the Equality Act 2010

[R2006] There is a potential conflict between an employer's duties under health and safety law and disability discrimination law. This arises when an employer may decide it is unsafe to employ a person with a disability or, when considering rehabilitation, to retain someone in the workplace or allow them back to work. The employer may come to the decision that a person has become a danger to themselves or to other people in the workplace as a result of a disability and use health and safety law to justify not allowing someone back to work.

However, an HSE research report, *The extent of use of health and safety requirements as a false excuse for not employing sick or disabled persons* (www.hse.gov.uk/research/rrpdf/rr167.pdf), suggested that in 40% of cases heard at employment tribunals this justification was not upheld and the tribunal ruled that adequate adjustments were possible but had not been made.

What happens if an employer decides it is unsafe for an individual to continue in employment and no reasonable adjustments are possible, but the employee wants to continue working? Case law on this is contradictory. In the case of *Withers v Perry Chain Co Ltd* [1961] 3 All ER 676 it was concluded that there was no common law duty to dismiss in these cases. A similar judgement was made in the Appeal Court in the case of *Hatton v Sutherland and other appeals* [2002] EWCA Civ 76, [2002] 2 All ER 1.

However, in *Coxall v Goodyear Great Britain Ltd* [2002] EWCA Civ 1010 [2002] IRLR 742, Mr Coxall brought a claim for damages against his employer for allowing him to continue to work in an environment which had caused his asthma, even though he had wanted to continue working. The Court of Appeal held that an employer may be under a duty to stop an employee from working if he is put at risk by continuing to do so, even though it was the employee's wish to continue.

When an employee has been off sick with an illness, whether physical or mental, injury or develops a chronic illness, and appears ready to return to work, the implications of disability discrimination law and health and safety law have to be weighed up very carefully.

If there is a residual disability which appears to be both long term and has substantial adverse effects on the person's ability to carry out day-to-day activities (see **R2005** above), the *EA 2010* applies and the question of reasonable adjustment to avoid discrimination has to be considered.

If there appears to be a health and safety reason why the person should not continue in that employment, it may be necessary to cease their employment.

However, this step should only be made after a thorough examination of possible adjustments has been made, including additional health and safety precautions. Examples of where a particular job or task puts a worker's health and safety at such risk that it is not reasonable for their employer to allow them to perform it are rare and health and safety should never be used as a false excuse for not employing, or continuing to employ an individual.

HSE advice in *Health and Safety for Disabled People and their Employers* says: 'There are very few cases where health and safety law requires the exclusion of specific groups of people from certain types of activity. While work in hazardous situations cannot always be eliminated, it can often be substantially reduced with comparatively little cost. With reasonable adjustments and plans to review if circumstances change, risks can be managed. This might be achieved by reallocating responsibilities or rescheduling duties to more suitable times.'

The publication also provides examples of good practice showing practical steps, including involving disabled workers, to promote disability equality when managing health and safety. The guide can be found online at: www.hse.gov.uk/disability/largeprint.pdf.

Managing sickness absence

[R2007] A number of terms are used to describe systems to reduce absence through sickness. These include:

- absence management;
- sickness absence management;
- attendance management.

Absence management addresses the whole problem of non-attendance at work, some of which include:

- genuine absence for sickness and injury;
- absence for other reasons such as routine dental and doctors' appointments;
- staying at home with sick children;
- compassionate leave; and
- unauthorised absence.

Many organisations have a policy on such issues and rules by which employees will be expected to conform as part of their contract of employment. When referred to specifically, sickness absence means absence from work because of mental or physical ill-health, injury or recovery from treatments such as surgery, chemotherapy or radiotherapy. Within the boundaries of sickness absence lies the problematic question of absence for non-genuine sickness (see **R2002** above). Attendance management is not dissimilar to absence management but places the emphasis on the more positive aspects of encouraging people to stay at work or return to work rather than the sometimes negative and heavy-handed methods employed to discourage sickness absence. There has been a drive over recent years to reduce the level of short-term sickness absence in the public services. This has translated into policies to discourage employees from taking sick leave even though the reasons may be genuine. Public service union UNISON says that all too often workers are being forced back to work through punitive policies that punish them for being sick. It is calling for fair sickness absence policies that help staff get well, and not punish them for being sick. It is also calling for more action to tackle the underlying causes of sickness absence such as stress, mental health and MSDs.

A 2017 employment tribunal judgement (www.gov.uk/employment-tribunal-decisions/mrs-k-allen-powlett-v-cardiff-vale-university-health-board-1600619-2016) said a health service trust had unfairly dismissed a nurse for frequent short-term sickness absences at a time when she was experiencing domestic violence. The tribunal ruled against Cardiff and Vale University Health Board and said that Karen Allen-Powlett must be allowed to return to work and be paid compensation for her loss of earnings. She had taken 15 short-term absences from work in her first two years' employment with the trust, a number of which were migraines caused by the mental stress of her husband's violence. Following the end of the relationship, she returned to better health, but after she took one day off with flu during the following seven months, the board dismissed her, arguing that her attendance had not shown "significant improvement". In this case, she had told her supervisor six months earlier that the underlying reason for her previous ill health was her violent husband.

The reverse of absenteeism is presenteeism. Presenteeism is the trend, encouraged by stigmatising absenteeism, of people being at work who should be off for reasons of ill health. Kivimaki et al (2005) found twice the risk of coronary heart disease episodes (non-fatal myocardial infarction and fatal coronary heart disease) in men who took no sickness absence despite being ill, compared with those who took moderate levels of sickness absence. This seems to suggest that encouraging people to remain at work, even though they are ill, has damaging health effects.

The conditions that people often endure and continue working include:

- asthma;
- migraine;
- allergies;
- back pain and other musculoskeletal disorders;
- depression; and
- irritable bowel syndrome.

Consider too the huge cost of having people at work who are not well. There could be safety implications, increased risk of accident and reductions in efficiency and productivity. Two studies by Stewart and his colleagues in America showed that presenteeism with depression cost US companies $31 billion (£15.8 billion) per year and with common pain conditions $61 billion (£31.2 billion) per year. Pain conditions include headache, back pain, arthritis and other musculoskeletal complaints. The extremely common complaint of headache loses companies on average 3.5 hours per affected employee per week (Stewart et al 2003).

In 2015, drawing on research from the Chartered Institute of Personnel and Development (CIPD) annual Absence Management report, Professor Sir Cary Cooper said that the annual cost of presenteeism (in the UK) is "twice that of absenteeism".

The October 2017 report *Thriving at Work*, an independent review of mental health and employers by Lord Dennis Stevenson and Paul Farmer, included a quote from Chartered Institute of Personnel and Development (CIPD) senior policy adviser Rachel Suff stating: 'Presenteeism is particularly common in organisations where a culture of long working hours is the norm and where operational demands take precedence over employee wellbeing. Also, in periods of job insecurity, people may be more likely to go into work when they are ill, rather than take a day off sick, for fear their commitment to their job will be doubted. It is this culture and these fears that need to be addressed in order to reduce presenteeism at work.' *Thriving at work* also reported that various studies suggest presenteeism is increasing year on year. It estimated the cost of presenteeism related to poor mental health at between £17bn and £26bn a year, compared to £8bn as a result of absenteeism.

Sickness absence management must therefore tread a very careful line between exerting too much pressure on genuinely sick people to remain at work and allowing those who are not genuinely ill to take time off without question.

The key to managing sickness absence is to have a well-developed policy which is written, transparent and made available to everyone. The *Employment Rights Act 1996* requires employers to provide staff with information on "any terms and conditions relating to incapacity for work due to sickness or injury, including any provision for sick pay".

In *Managing sickness absence and return to work An employers' and managers' guide (HSG 249)* (see www.hse.gov.uk/pUbns/priced/hsg249.pdf) the HSE identifies six key elements involved in effectively managing sickness absence and return to work. These are:

- Recording sickness absence: Knowing which employees are off sick and why will help to identify patterns and high-level causes of short and long-term sickness absence, identify work-related and other causes, plan cover for absent employees and benchmark the organisation's performance.
- Keeping in contact: This can be a sensitive topic as some employees may feel they will be pressed to come back to work too early, but without contact, those who are absent may feel increasingly out of touch and undervalued.
- Planning and undertaking workplace adjustments: The aim is to return the employee to their job with any modifications needed, or to an alternative job if no adjustments are possible; retain valuable skills; and remove any obstacles to returning to work.
- Using professional or other advice and treatment (see **R2010** below).
- Agreeing and reviewing a return-to-work plan: HSE advice is that the best time to prepare a plan is generally three to four weeks into an absence as discussing it too soon may put pressure on the employee, but leaving it too late may mean the employee loses confidence in being able

to return. The plan should be tailored to the individual and include, for example, its goal; the time scale; information about alternative working arrangements and changes to terms and conditions; the checks that will be made to make sure the plan is put into practice; and review dates.
- Coordinating the return-to-work process: It may be useful to appoint a coordinator where a number of advisors are involved to ensure information is available on time, arrangements are smooth and everybody knows what is expected from them. The coordinator should be familiar with the employee's job and work environment, be able to communicate and negotiate with staff, at all levels, and be sensitive to the needs of the employee concerned. A more formal, case-management approach may be needed in complicated cases or when input is needed from a wide variety of sources. A case manager is typically someone who is professionally qualified in a relevant medical area and may be involved in treating the employee. They can also mediate in cases where communications have broken down or help is needed to move things on. To control sickness absence, it is essential to have an effective system of recording absence (of all types) and to have clear policies of when interventions are required. In larger organisations, this will require the full co-operation of line managers and supervisors who will have to ensure that absence is recorded in a central location (usually in HR).

Employers must be careful not to breach the General Data Protection Regulation (GDPR) which came into force on 25 May 2018 and replaced the Data Protection Act 1998. Employment service ACAS online guidance (www.acas.org.uk/index.aspx?articleid=3717#whatisgdpr) explains the GDPR regulation contains six principles:

- personal data should be processed fairly, lawfully and transparently;
- data should be obtained for specified and lawful purposes and not further processed in a manner that is incompatible with those purposes;
- the data should be adequate, relevant and not excessive;
- the data should be accurate and where necessary kept up to date;
- data should not be kept for longer than necessary; and
- data should be kept secure.

The Information Commissioner's Office (ICO) has produced a *Guide to the General Data Protection Regulation (GDPR)*. This can be found on its website at: https://ico.org.uk/for-organisations/guide-to-data-protection/guide-to-the-general-data-protection-regulation-gdpr/.

The ICO Employment Practices Data Protection Code (part 4) deals with information about workers' health. The code is not legally-binding but clarifies the law and establishes standards which employers are expected to follow and covers sickness records. However, the code has not been updated since the GDPR became law.

Local managers should be able to conduct return-to-work interviews. They are likely to need training to do this sensitively and confidentially and to be able to recognise when a case that gives concern, for example when a work-related illness or injury appears to be developing or when there appears to non-genuine causes or persistent short-term absences. There should be clear guidance to managers on who to report problem cases to – normally a more senior manager or human resources. Someone should also have responsibility for monitoring the central absence records to pick up particular patterns of absence (eg persistent Monday absence), to identify particular departments or sections of the workforce with high absence rates, and to detect when someone has reached the threshold for long-term absence – usually three weeks. Managers should be given regular absence information for their department. Appropriate interventions for long-term absence should then be put in place. This might include considering rehabilitation when the individual returns to work.

The CIPD reports that while employee absence is a significant cost for many organisations, research suggests that only a minority of employers monitor its cost.

The most common measure of absence is the lost-time rate. This expresses the percentage of the total time available which has been lost to absence in a given period, usually an annual lost time rate.

It is calculated from:

Total Absence (hours or days)/Possible total hours or days worked × 100.
If someone has 20 days absence in a year and could possibly have worked 220 days then the annual lost time rate is:
$$20/220 \times 100 = 9.1\%$$

It is very useful as part of an absence policy to have a target for lost time rate and some organisations have set this as low as 2%.

GP Fit Notes

[R2008] GP fit notes replaced sick notes in April 2010 and these medical statements can either indicate that a person is "not fit for work" or that they "may be fit for work taking account of the following advice". The GP can recommend: a phased return to work; altered hours; amended duties; or workplace adaptations.

Department for Work and Pensions (DWP) guidance on GP fit notes to employers and line managers advises:

> 'The advice in the fit note is about your employee's fitness for work in general, and not specifically about their current job", it says. "This gives you maximum flexibility to discuss possible changes to help them return to work (which may include changing their duties for a while).'

The advice can be found online at: www.gov.uk/government/uploads/system/uploads/attachment_data/file/578032/fit-note-guidance-for-employers-and-line-managers.pdf.

The government's November 2017 policy paper, *Improving Lives The Future of Work, Health and Disability* set out plans to 'transform employment prospects for disabled people and those with long term health conditions over the next 10 years'.

This includes reforming fit notes so they 'become an enabler for conversations about health and work, focussing on what people can do, not what they cannot do'. It says they should facilitate returns to work and help people stay in work where appropriate, by providing information to the employer about what support might enable that to happen.

It reported that recently published fit note statistics show that only 6.6% of fit notes used the 'may be fit for work' option, rather than 'not fit for work'.

Some of those responding to a government consultation said the fit note process could be improved by including more detailed information from employers on their workplace and the patient's job role, and from specialists involved in their care, such as physiotherapists, occupational therapists or psychiatrists.

There was general agreement that additional OH training was required for GPs and that other healthcare professionals should only take on this role with both comprehensive training and ongoing support.

The government has completed an internal review of the operation of the fit note. This concluded that the fit note remains an important tool but is not always used effectively across the system to support people staying in or returning to work. It says too many fit notes say 'not fit for work' when people 'may be fit for work' as long as appropriate workplace adjustments are made.

It reported it would:

- start development work to legislate for the extension of fit note certification powers to other healthcare professionals, along with the design and development of a set of competencies for those completing fit notes;
- conduct a feasibility test to investigate whether, for the purposes of Statutory Sick Pay, employers could use the Advisory Fitness for Work report (which can be completed by some Allied Health Professionals) as an alternative to the fit note;
- integrate fit note training into GP undergraduate and postgraduate education;
- commission the feasibility of clinical guidelines for workplace adjustments for the top five clinical reasons people are off work sick or are on health-related benefits; and
- explore whether changes to the way GPs complete fit notes could support better return to work conversations.

Improving Lives The Future of Work, Health and Disability can be found at: https://assets.publishing.service.gov.uk/government/uploads/system/uploads/attachment_data/file/663399/improving-lives-the-future-of-work-health-and-disability.PDF.

Long-term absence and rehabilitation

[R2009] Long-term absence may need a carefully thought through plan for return to work. The potential legal pitfalls of not having a comprehensive return-to-work plan, or of not adhering to it, were discussed earlier (see **R2004** above). The longer an employee is off sick, the more likely a plan for rehabilitation will be needed. However, the need for a return-to-work plan will also depend on the nature of the illness or injury for which the person has been off sick and may be essential even if they have only been absent for a relatively short time. This will be important if the employee has a condition such as stress or a musculoskeletal injury and particularly if the condition may have been caused by the conditions of work. Indeed, any condition which may be work-related, such as a developed asthma or dermatitis, neither of which are likely to entail excessively long-term absence, will need to be carefully managed on the person's return to work.

It is important to keep in touch with the long-term absent employee on a regular basis during their absence. This can be done by, say, a weekly telephone call from HR or the immediate line manager to discuss how they are progressing and, at an appropriate time, to start to discuss when they might return to work, what the return-to-work plan might consist of, and what adjustments might need to be made.

Return-to-work options include the following:

- A phased return – a gradual re-introduction to the workplace starting with part-time and building up over a planned number of weeks to full-time. A phased return is an extremely important way of getting someone back to work but may be resisted by some managers as being too disruptive and difficult to manage. However, a relatively short period of difficulty will likely ensure a successful return to full productivity. Without it they may never return.
- Altering the employee's hours, either temporarily or permanently. Moving the employee from shift work to permanent days. Allowing flexible-working.
- Moving the employee to another job or another department in the organisation. This might be necessary to avoid lifting and carrying following a back injury, to avoid exposure to a chemical that has caused an allergy, for example. This might also be necessary if the job entails driving and if the employee has developed a condition which would now preclude them from holding a licence. If someone has been bullied or harassed, employers should address the problem and not transfer the person making the complaint unless they ask for such a move.
- Considering what adjustments might accommodate a new condition the employee now suffers. This might be, for example, epilepsy following brain surgery, partial paralysis or a speech defect following a stroke, or reduced physical capacity following a heart attack. In these cases, there may be no need to redeploy the person. They may be able to carry out their normal work, but with some adjustments made to their workplace, work equipment or working practices. Remember that if the condition falls within the meaning of the *Equality Act 2010*, there is a legal requirement to make reasonable adjustments so as not to discriminate against the person (see *EA 2010* at **R2005** above).

Whatever the agreed return-to-work plan it is essential that someone, probably the line manager, is made responsible for ensuring that it is adhered to, reviewing it if necessary and keeping the employee fully consulted. The legal consequences of not seeing through a plan were highlighted in the case of *Young v Post Office* [2002] EWCA Civ 661, [2002] IRLR 660, (see **R2004** above).

An analysis by the Centre for Economics and Business Research, reported in *Improving Lives The Work, Health and Disability Green Paper*, indicates that employees who have access to early intervention and rehabilitation services and use them tend to have shorter long-term absences than those that do not. On average, the duration is shorter by 16.6%.

Using professional advice

[R2010] A good occupational health practitioner will be invaluable in helping to manage absence generally and ensuring that rehabilitation is successful. Alternatively, and it sometimes becomes the case, an occupational health physician may need to give advice on the necessity for retirement on the grounds of ill-health.

Many different practitioners contribute to the overall practice of occupational health but the two most relevant are occupational health nurses and occupational health physicians. It is possible that in individual circumstances the advice or input of an ergonomist, an occupational hygienist, physiotherapist or safety practitioner may be required. However, the occupational health nurse or occupational health physician will be able to advise on the need for additional expert help and one of these two should always be the first point of reference.

Occupational health nurses (also called occupational health advisors), are registered nurses who have specialised in occupational health. The occupational health nurse is likely to be the key professional advising on the rehabilitation of employees and advising on suitable return-to-work options and any necessary work adjustments. The OH nurse will also play an important role in monitoring the progress on return to work.

Occupational health nurses are qualified and registered nurses, many of whom have chosen to gain additional training and qualifications as specialist community public health nurses. They must be registered on part 1 of the Nursing and Midwifery Council (NMC) register.

The Faculty of Occupational Medicine (FOM) publishes *An employer's guide to engaging an occupational health physician*. This advises that specialists in occupational medicine will have had in depth training and experience in occupational medicine and be Members or Fellows of the Faculty of Occupational Medicine and will have the letters MFOM or FFOM after their name. The Associate of the Faculty of Medicine (AFOM) qualification is being phased out. Those who have the qualification have core knowledge in occupational health, but are not 'specialists', the FOM advises.

'If a medical practitioner does not have any of these qualifications (DOccMed, AFOM, MFOM or FFOM), it is likely that they only have the minimal training in occupational medicine which most, but not all, doctors get at medical school (a few hours' worth). They do not have the qualifications, experience or validation that would enable an employer or a worker to be confident of their competency in this field,' according to the FOM guide.

In addition, it advises that the term "specialist" has a specific legal meaning in medicine and a doctor can only call themselves a specialist if they are on the General Medical Council (GMC) Specialist Register.

Occupational Health Service Standards for Accreditation are available on the FOM website at: www.fom.ac.uk/wp-content/uploads/standardsjan2010.pdf.

In addition, the SEQOHS (Safe, Effective, Quality Occupational Health Service) Accreditation Scheme for Occupational Health Services is a set of standards and a voluntary accreditation scheme for occupational health services in the UK (and beyond). SEQOHS accreditation is the formal recognition that an occupational health service provider has demonstrated that it has the competence to deliver against the measures in the SEQOHS standards. The scheme is managed by the Royal College of Physicians of London on behalf of the Faculty of Occupational Medicine. For more information see: www.seqohs.org/.

Occupational Health Services

[R2011] Evidence shows effective occupational health provision can help protect and promote employee health and wellbeing and prevent unnecessary sickness absence. However, there is no legal requirement for employers to have an occupational health service (OHS), and many organisations in the UK do have not have access to occupational health (OH) provision.

According to a recent government consultation, the current model of OH provision does not meet the needs of employers or individuals. It also found:

- those at greatest risk of job loss generally lack access to OH;
- the funding model for OH provision is unclear, with a lack of consensus about responsibility for providing and funding OH services – NHS, government, or employers;
- larger employers and firms with skilled employees are more likely to pay for OH provision, while small and medium-sized companies ('SMEs') tend to offer little or no access;
- Fit for Work, the Department for Work and Pensions (DWP)-commissioned service for offering free OH assessments, has had very low take-up (see below);
- unavailability and capability of OH services, characterised by too few qualified OH professionals;
- a lack of conversations and collaboration between GPs, employers, Fit for Work service, other healthcare professionals, and Jobcentre Plus work coaches; and
- limitations of the evidence on the effectiveness of services, and inconsistent use of evidence-based interventions where they are available.

The government closed the national occupational health service, Fit for Work, at the end of March 2018 after less than three years of operation. There is still a Fit for Work advice helpline, website and web chat, offering general health and work advice and support on sickness absence.

It is now looking at how to shape, fund and deliver effective OHSs and 'aims to be in a position to set out a clear direction and strategy for future reform by 2019-20'.

An Expert Working Group on occupational health will look at:

- building the evidence base;
- potential funding models and where responsibility for OH support should fall;
- methods for improving quality of existing provision, for example accreditation of services, staff and training;
- emerging new models of provision (in primary, secondary health and across sectors) and local place-based models to integrate work and health support across the health pathway;
- exploiting the potential of technology as a way to help people access the advice they need easily and quickly; and
- workforce development so there is expert capacity in future.

In April 2019, the Work and Health Unit, which leads the government's strategy to support disabled people and people with long-term health conditions to enter and stay in work, published research on how and why employers use OHSs and why they do not buy OHSs. A summary of the report, *Employers' motivations and practices: A study of the use of occupational health services*, can be found on the government website: www.gov.uk.

The government is also working partners including the Greater Manchester Combined Authority and the Scottish government to test new elements of the OH offer.

In July 2019, the government launched a consultation, *Health is everyone's business: proposals to reduce ill health-related job loss* (www.gov.uk/government/consultations/health-is-everyones-business-proposals-to-reduce-ill-health-related-job-loss).

This includes proposals to improve access to high quality, cost-effective OH services for employers and self-employed people and seeks views on ways to reduce the cost for SMEs through potential co-funding of OH, for example through a direct subsidy or voucher scheme.

Web-based advice and details of the occupational health services available to businesses via the NHS can be found at the NHS Health at Work website at www.nhshealthatwork.co.uk/.

Further reading and information

[R2012] Kivimaki M, Head J, Ferrie JE, Hemingway H, Shipley MJ, Vahtera J, and Marmot MG (2005) Working While Ill as a risk factor for Serious Coronary Events: The Whitehall II Study. American Journal of Public Health, 95 (1) 98–102.

Stewart WF, Ricci JA, Chee E, Hahn SR and Morganstein D (2003) Cost of Lost Productive Work Time Among US Workers with Depression. JAMA 289: 3135–3144

Stewart WF, Ricci JA, Chee E, Morganstein D and Lipton R (2003). Lost Productive Time and Cost Due to Common Pain Conditions in the US Workforce. JAMA 290: 2443–2454

Wardell G, Kim Burton A, Kendall NAS, Vocational rehabilitation What works for whom, and when? www.gov.uk/government/publications/vocational-rehabilitation-scientific-evidence-review.

Advisory, Conciliation and Arbitration Service (ACAS) advice on managing sickness absence is available at: www.acas.org.uk.

Chartered Institute of Personnel and Development (CIPD) information on absenteeism, the causes of absence, absence monitoring and management, attendance, sickness and long-term illness, return to work, fit notes and sick pay can be found at: www.cipd.co.uk.

Health and Safety Executive (HSE) advice on managing sickness absence and return to work is available at: www.hse.gov.uk.

Improving Lives The Work, Health and Disability Green Paper can be found online at: www.gov.uk/government/uploads/system/uploads/attachment_data/file/564038/work-and-health-green-paper-improving-lives.pdf.

REACH – Registration, Evaluation and Authorisation of Chemicals in Europe

John Wintle and Andrea Oates

Introduction to REACH

[R2501] A European Regulation on chemicals came into force across Europe on 1 June 2007 and was hailed as one of the most significant and far-reaching legislative reforms of the last 20 years. Known as REACH, the Regulation provides for the Registration, Evaluation and Authorisation of CHemicals manufactured or imported in quantities of one tonne per year or more.

Registration has been phased in over ten-year period (from 1 June 2008), with the final registration deadline, for substances that are manufactured in, or imported into, the European Union or European Economic Area (EU/EEA) in quantities of between 1 tonne and 100 tonnes per year, on 31 May 2018 (REACH Registration 2018). The deadlines for substances manufactured or imported in higher volumes have already passed.

A guidance note produced by the UK Chemicals Stakeholder Forum, which advises the government on managing risks to the environment and human health that may result from the production, distribution and use of chemicals (www.gov.uk/government/groups/uk-chemicals-stakeholder-forum), warns that 'the clock is ticking', while the Health and Safety Executive's (HSE's) message is 'No data, no market'.

'If you do not register your substances, then the data on them will not be available and as a result, you will no longer be able to manufacture or supply them legally, i.e. no data, no market!' the HSE warns.

The HSE also says that, based on the timescale for negotiating the UK's exit from the EU (or Brexit), those UK-based companies affected by REACH still have the legal obligation to register their substances by the 2018 deadline.

REACH affects manufacturers, importers, suppliers, distributors, and 'downstream users' of chemicals, substances, and articles containing substances. They all have responsibilities under the Regulation, as do the regulatory authorities. Downstream users are businesses that use chemicals in the course of their activities. The HSE says that this probably includes most businesses in some way.

Under REACH, all businesses in Europe that manufacture or import more than one tonne of any given substance each year are responsible for registering information about that substance with the European Chemicals Agency (ECHA) based in Helsinki, Finland (https://echa.europa.eu/).

Registrants have a duty to inform their supply chain and downstream users of the hazards and appropriate risk management measures for particular uses of the substance. Downstream users have a duty to apply the risk management measures and a responsibility to report back any new hazards or uses.

REACH replaced what the trade union organisation TUC described as 'a confusing range' of European Directives and Regulations with a single system, helping to harmonise the information on substances and their control across the EU. It provides an opportunity to consider the latest data available on the properties of substances so as to classify them accordingly, and to exchange and improve information for users. For substances that pose a particular threat and are of high concern, the Regulation provides for authorising or restricting their availability and use.

Purpose of REACH

[R2502] The purpose of REACH is to make those businesses and organisations that place substances on the European market responsible for understanding and managing the risks associated with their use. They are responsible for collecting, collating and registering information about their chemicals with the ECHA. The ultimate aim of REACH is to ensure that employees, users, the public and the environment in the EU have a high level of protection from any properties of substances that may be hazardous to human and biological health.

The Regulation is intended to promote the free movement of substances on the EU market by removing any specific national restrictions and introducing common regulation across the EU, ensuring that substances imported into the EU are registered to the same standards as substances manufactured within the EU. The registration process gives wide-ranging consideration to the properties of substances and aims to encourage the development of new or alternative methods by which the hazardous properties of substances may be assessed. It also aims to promote the development of less hazardous substitutes. In these ways, REACH aims to stimulate the European chemicals industry into more innovation and increased competitiveness in world markets.

The need for REACH

[R2503] There are thousands of substances supplied by businesses in quantities of over one tonne that are on the market in the EU. In general, there has been very little publicly-available information about the hazards and risks these substances pose to human health and the environment, and the previous system of regulation was slow to produce results, particularly for substances existing prior to the requirement for notification. The *Notification of New Substances Regulations 1993 (NONS Regulations)* were replaced by REACH from 1 June 2008 and responsibilities under the *NONS Regulations* were repealed. The number of these substances that were properly assessed over the 30 years prior to 2008 was estimated to be less than 200, while the number of notifications of new substances was around 1,000. At a time of increasing public concern over the health risks from chemicals, there was therefore a limited evidence base of information to address this concern and needed to be improved.

At a regulatory level, REACH was intended to simplify the system for the control of substances that was considered confusing for industry to understand and for authorities to administer.

Chemicals covered and exempted

[R2504] Under REACH, 'chemicals', usually referred to as 'substances', are defined as chemical elements and their compounds in the natural state or obtained by any manufacturing process. This includes any additive necessary to preserve stability and any impurity derived from the process used. The term excludes any solvent that may be separated without affecting the stability of the substance or changing its composition.

REACH applies to individual substances on their own, in preparations, or in articles. A preparation is a mixture or solution of two or more substances. An article is an object which, during production, is given a special shape, surface or design, which determines its function to a greater degree than its chemical composition. The HSE gives the examples of a car, a battery and a telephone.

It explains: 'It is only chemicals on their own that are registered, not deliberate mixtures of chemicals (formulations/preparations). Where a chemical is supplied or used as part of a deliberate mixture, for example, it is produced and used in a solvent or as part of a product, (e.g., paint or glues, etc.), it is the individual ingredients that are registered. In some cases, substances in articles need to be registered.'

Generally, REACH applies to substances manufactured or imported into the EU in quantities of one tonne per year or more. The amount of a given substance supplied in excess of one tonne affects the timetable for its registration (see below). A distinction is made for substances supplied at 100 tonnes or more per year and 1,000 tonnes or more per year.

Some substances that may not be properly on the market or that are transitory in nature are specifically excluded:

- substances under customs supervision;
- substances undergoing transport;
- radioactive substances;
- non-isolated intermediates – the HSE explains that a non-isolated substance is manufactured solely for the purpose of being transformed into another substance (or synthesis) and is used up within this reaction. This type of intermediate is not intentionally removed from the synthesising equipment (except for sampling);
- waste;
- some naturally occurring low hazard substances.

Some substances, which are covered by more specific legislation, have tailored provisions within the REACH legislation:

- substances intended as human and veterinary medicines;
- food and foodstuff additives;
- plant protection products and biocides.

Other substances have tailored provisions within the REACH legislation, as long they are used in specified conditions:

- isolated intermediates – the HSE explains that an on-site isolated intermediate is a substance manufactured for or used for chemical processing in order to be transformed into another substance, the synthesis of which takes places on the same site which is operated by one or more legal entities;
- substances used for research and development.

Duty Holders within REACH

[R2505] Many businesses have responsibilities as duty holders under REACH. The UK Chemicals Stakeholder Forum (see **R2501** above) says that REACH 'has the potential to impact all UK business sectors from chemicals manufacturers, metal industry, furniture makers and retailers to builders, food companies and printers. Most businesses use chemicals, in some form, in their day-to-day operations.'

There are three main types of REACH duty holder, each with different roles and responsibilities. These types of duty holder are:

- manufacturers or importers (registrants);
- distributors or suppliers;
- downstream users.

Businesses that manufacture or import more than one tonne of any given substance in Europe each year are duty holders as Registrants under REACH. Because substances in articles also count, it is possible that some producers and importers of articles will also be Registrants, if the substances contained in them are intended to be released during use and exceed the threshold. Where downstream users are also importers of substances, then they will also be Registrants.

Distributors are businesses or persons established within Europe that buy, store, place on the market and sell a substance, on its own, or in a preparation or article, but are not the importer, manufacturer or final user. A retailer who only stores and places on the market substances and articles for sale to third parties is a distributor.

'You are a distributor under REACH and CLP [see **R2510** below] if you source a chemical substance or a mixture within the EEA, store it and then place it on the market for someone else (also under your own brand without changing its chemical composition in any way,' the ECHA explains. 'For example, retailers and wholesalers are distributors under REACH and CLP.'

Downstream users include any business or persons established within Europe, other than the manufacturer or importer who uses a substance, either on its own or in a preparation, in the course of their industrial or professional activities. A consumer buying substances for use at home is not a downstream user.

The ECHA set out:

'You are a downstream user, if:
- You are established in the EEA, and
- The supplier of a substance or mixture you use is also established in the EEA, and
- You use chemicals in your industrial or professional activities, as described above

You are not a downstream user, whenever:
- The supplier of a substance you use is established outside the EEA and has not designated an only representative to register the substance and ensure its access on the EU market. If this is the case, you are an importer
- You only store chemicals and place them on the market (without changing their composition or packaging). Then you are a distributor
- You buy a product and you use it yourself outside an industrial/professional setting. If so, you are a consumer.'

Information on why and how to appoint an only representative can be found on the ECHA website at: https://echa.europa.eu/support/getting-started/enquiry-on-reach-and-clp.

Manufacturers and importers have two key responsibilities. Firstly, they are responsible for registering with the ECHA information about the properties and hazards of each substance that they manufacture or import and the uses to which that substance may be put. Secondly, registrants are responsible for providing information directing downstream users to the appropriate risk management measures, and responding to information about the use and effects passed back by distributors and downstream users.

Suppliers have specific duties to pass information provided by manufacturers and importers down the supply chain to the downstream users. They also have duties to pass information back up the supply chain to their own suppliers when their own clients ask them to do so.

Downstream users have a duty to use substances on their own or in preparations in a safe way, and according to the information on risk management measures passed down the supply chain from manufacturers and importers for that use. Where they gain new information about the hazardous properties of a substance, they have a responsibility to pass the information back up the supply chain to the manufacturer or importer and, where appropriate, to the ECHA.

Where a Safety Data Sheet (SDS) (see **R2510** below) has an attached exposure scenario (see **R2509** below) that details how chemicals may be used, the HSE advises that users should implement the required risk management measures (or use equivalent measures).

A downstream user chemical safety report is prepared by a downstream user to document the assessment of the conditions of safe use of a substance where a use (including conditions of use) is not covered in the exposure scenarios received from the supplier. ECHA guidance, *How to prepare a downstream user chemical safety report (Practical guide 17)*, is available on the Agency's website at: https://echa.europa.eu/documents/10162/13655/pg17_du_csr_final_en.pdf.

If the downstream user wishes to keep its use of the substance confidential, they must provide information on the use with the ECHA independently of the original manufacturer or importer.

Preparation for REACH registration 2018

[R2506] Businesses that think they are potential duty holders under REACH should prepare for their responsibilities depending on whether they are a potential registrant (manufacturer or importer), supplier, or downstream user, or more than one of these.

The HSE advises companies that are new to REACH to compile an inventory of the substances that enter, become part of, or leave the business. This should include feedstocks, intermediates (isolated or otherwise), and substances in preparations, articles and products that are manufactured.

All substances should be identified. For preparations and mixtures, it is necessary to know all the ingredient substances and the proportions by weight of each. A list of the substances can then be drawn up with the estimated tonnage of each substance that enters, is created by, and leaves the business per year.

If the business produces or imports articles, they must determine if any substance which is intended to be released under normal or reasonably foreseeable conditions of use is present in these articles in quantities which might total more than one tonne per year. The HSE gives the example of a fragranced bin liner as an article that intentionally releases substances – it intentionally releases the substances that create the fragrance. These substances should to be added to the list.

When this inventory has been established, businesses should consider their supply chain and determine which type(s) of REACH duty holder they are, taking account of alternatives or variations in substances, processes and supply. Potential registrants (manufacturers and importers) should establish the typical uses of their product, while downstream users can be more specific about the uses to which substances are put, and whether their suppliers are likely to be aware of these uses. Users should consider whether they would wish to share information about their use with their suppliers, or where this might be sensitive whether they would prefer to compile a risk assessment themselves.

In some areas, REACH can have commercial impact. Certain substances may be subject to authorisation and/or restrictions of use or supply. Businesses should consider the possible impact of any such measures on their business, particularly for substances that are especially hazardous, which should be specifically identified.

Having determined what type(s) of duty holder it is, a business may wish to understand the role and responsibilities for that duty holder, and the roles of responsibilities of other duty holders in its supply chain. It may be helpful for the supply chain to discuss the implications and expectation of REACH for each actor. A focal point for REACH within each business would be useful so as to provide an interface with suppliers and customers on REACH responsibilities.

The April 2017 UK Chemicals Stakeholder Forum (see **R2501** above) guidance note, *What SMEs need to know about the 2018 REACH Regulation deadline*, says that it is concerned that many companies, particularly smaller ones, are not aware of, or are not preparing for, registration. In the worst case, it says, if nobody registers a substance, it will no longer be available on the market.

The Registration process

[R2507] Registration is the requirement on manufacturers and importers to collect and collate specific sets of information on the properties of the substances that they manufacture or supply at or above one tonne a year. An assessment of the hazards and risks a substance may pose and how those risks can be managed is also required. All registrants must register their substances and the necessary information and assessment with the ECHA.

For the thousands of substances being manufactured in or imported into Europe, registration has been taking place over three phases spread over ten years. For any given substance, a single collation of the information and assessment is produced. Where there is more than one manufacturer or importer for a given substance, as will often occur, there are arrangements to put all like registrants into contact with each other for generating the information and making the assessment jointly.

The ECHA provides information for businesses registering the same chemical to come together and form a Substance Information Exchange Forum (SIEF). Through this forum they can collectively share their available data and knowledge, and generate a single collation of the information. All registrants of that substance then share the information and hazard assessment for the purposes of registration.

It is for the members of the SIEF to negotiate the sharing of their data and the costs of generating any new data required. Where there is information that is business specific or sensitive, companies can choose to submit information separately to the ECHA. This process is intended to help industry reduce the costs of registration.

More information on SIEFs can be found on the ECHA website at: https://echa.europa.eu/support/registration/finding-your-co-registrants.

Timescale for Registration

[R2508] Phased registration applies to the thousands of existing substances, known as 'phase-in substances', that meet at least one of the following criteria.

(a) A substance that is listed on the European Inventory of Existing Commercial Chemical Substances (EINECS) (see https://echa.europa.eu/information-on-chemicals/ec-inventory), or
(b) A substance that was manufactured in the Community (or in the countries acceding to the European Union on 1 January 1995 or 1 May 2004) but not placed on the market at least once in the 15 years prior to REACH entering to force, or
(c) A substance that was placed on the European market by the manufacturer or importer before REACH came into force and was considered to be notified in accordance with the first indent of Article 8(1) of Directive 67/548/EEC, but does not meet the definition of a polymer as set out in REACH.

Manufacturers or importers should have documentary evidence to support criteria (b) and (c). In order to benefit from phased registration, manufacturers and importers should have pre-registered their substances between 1 June and 30 November 2008. The registration of phase-in substances has been taking place over three phases, with the date for registration according to their tonnage and/or hazardous properties.

(a) Substances that had to be registered by 1 December 2010 are:
 — substances supplied at 1,000 tonnes or more per year; or
 — substances classified as very toxic to aquatic organisms that are supplied at 100 tonnes or more per year; or
 — substances classified as Category 1 or 2 carcinogens, mutagens or reproductive toxicants hat are supplied at 1 tonne or more per year.
(b) Substances that had to be registered by 1 June 2013 are substances supplied at 100 tonnes or more per year.
(c) Substances that must be registered by 1 June 2018 are substances supplied at 1 tonne or more per year.

The ECHA advises that companies planning to register a non-phase-in (new substances not covered by the definition of a phase-in substance) substance or phase-in (existing) substance that has not been pre-registered, have a duty to inquire with ECHA whether a registration has already been submitted for that substance. To submit an inquiry, potential registrants must prepare an electronic inquiry dossier with substance identify information clearly describing the manufactured or imported substance (see https://echa.europa.eu/support/substance-identification). They must wait for the result of the inquiry before submitting the registration or beginning any animal testing.

Information required for Registration

[R2509] The information required for registration depends on the amount of the chemical produced. For substances produced in quantities of more than one tonne but less than ten tonnes per year, a Technical Dossier is required. For substances produced in amounts greater than ten tonnes per year, a more extensive Chemical Safety Report is required.

Information required for Registration [**R2509**]

The Technical Dossier contains information on the properties, uses and classification of the chemical, as well as guidance on its safe use. The ECHA sets out that manufacturers and importers must collect 'all existing available information on the intrinsic properties of a substance as well as on its manufacture, uses and exposure'. It explains that the technical dossier contains information about:

- the identity of the substance;
- information on its manufacture and use;
- the classification and labelling of the substance;
- guidance on its safe use;
- 'robust' study summaries of the information on the intrinsic properties;
- proposals for further testing, if relevant; and
- for substances registered in quantities between one and ten tonnes, the technical dossier also contains exposure-related information for the substance (main use categories, type of uses, significant routes of exposure).

Under REACH, the classification of substances according to their characteristics follows an established system. For example, established classification characteristics include those that are corrosive, or toxic to fish (see **R2510** below).

In contrast to the Technical Dossier, a Chemical Safety Report (CSR) details the classification and hazardous effects of the chemical. It includes an assessment as to whether the chemical is persistent (P), bio-accumulative (B) and toxic (T) or very persistent and very bio-accumulative (vPvB).

In addition, the CSR also describes 'Exposure Scenarios' within a Chemical Safety Assessment (CSA). The ECHA explains that a CSR contains a detailed summary of information on a substance's environmental and human health hazard properties and an assessment of exposure and risk where required. It documents the CSA carried out as part of the REACH registration process and is the key source from which the registrant provides information to users of chemicals through the exposure scenarios. It also forms a basis for other REACH processes including substance evaluation, authorisation and restriction.

Exposure Scenarios describe the conditions under which the substance is manufactured or is to be used during its life. They specify the controls that the manufacturer or importer has or recommends in order to limit exposure to humans and the environment. These should include the appropriate risk management measures and operational restrictions, limits and controls to ensure that, when these are properly applied, the risk from exposure to the substance is sufficiently low.

The CSA should address all the identified and known uses of the substance on its own (but including any major impurities and additives), in a preparation and in an article. All stages of the life-cycle of the substance must be considered, from manufacture to final disposal. A connection is made between the potential adverse effects of the substance with the known or reasonably foreseeable Exposure Scenarios for humans and/or the environment.

Safety Data Sheets and downstream use

[R2510] A key feature of REACH is to ensure good transmission of information up and down the supply chain. Downstream users need to know about the hazards involved in using substances and how to control the risks from the knowledge of manufacturers and importers. In addition, suppliers need to know from users about new uses of the substances that they supply, and the experience of users so that this information can be passed to other users.

Safety Data Sheets (SDSs) provide the supply chain and downstream users with appropriate safety information on a substance. It is the responsibility of manufacturers and importers to prepare a SDS and to include it when supplying the substance. Where a CSR is required for registration (see **R2509** above), the SDS should include the information given in the CSR and be consistent with information provided for registration.

REACH replaced the previous system under the *Chemicals (Hazard Information and Packaging for Supply) Regulations 2002 (CHIP Regulations)* and takes account of the requirements of the *Classification, Labelling and Packaging Regulation (CLP Regulation)* which fully replaced *CHIP Regulations* when the transitional provisions ended in June 2017. The *CLP Regulation* introduces into the EU the globally harmonised system (GHS) for classifying and labelling chemicals, so the same system is now used throughout the world.

The ECHA advises that a SDS should be provided when:

- a substance or a mixture is classified as hazardous;
- a substance is persistent, bioaccumulative and toxic (PBT) or very persistent and very bioaccumulative (vPvB); or
- a substance is included in the Candidate List for Authorisation (see **R2513** below) according to REACH for other reasons than the above.

In addition, mixtures which are not classified as hazardous, but which contain specified concentrations of certain hazardous substances, require a safety data sheet to be provided on request.

The HSE advises that SDSs must be dated and contain the following headings:
1. Identification of the substance/mixture and of the company/undertaking;
2. Hazards identification;
3. Composition/information on ingredients;
4. First-aid measures;
5. Fire-fighting measures;
6. Accidental release measures;
7. Handling and storage;
8. Exposure controls/personal protection;
9. Physical and chemical properties;
10. Stability and reactivity;
11. Toxicological information;
12. Ecological information;
13. Disposal considerations;
14. Transport information;
15. Regulatory information;
16. Other information.

It also advises that, where appropriate, exposure scenarios should be included in an Annex; or for mixtures, included in the main body (under Section 8 above – exposure controls/personal protection for example). It says that the information in the SDS should be consistent with the information in the CSR for that substance, or mixture if a CSA for the mixture is available. Further guidance on how to compile a SDS is provided in Annex II of REACH.

Downstream users have a duty to follow the information and guidance provided on the SDS. Where a downstream user intends a use with an exposure scenario outside those listed on the SDS, a dialogue with the supplier on the safety of the intended use might be the first step. Where the downstream user does not wish to do this, or is advised against the use by the supplier, the downstream user should normally prepare a CSR unless specifically exempted. This should include a risk assessment and risk management measures for the use, and must be available to the ECHA if required.

ECHA guidance in this area, *Guide on Safety data sheets and Exposure scenarios*, can be found on its website at https://echa.europa.eu/documents/10162/22786913/sds_es_guide_en.pdf/b5e90791-68a0-4ad3-8769-6b3a17e61c36.

HSE guidance, *REACH and Safety Data Sheets*, can be found online at www.hse.gov.uk/reach/resources/reachsds.pdf.

European Chemicals Agency

[R2511] The ECHA was founded in 2007 and is based in Helsinki, Finland. It describes itself as 'the driving force in implementing the EU's groundbreaking chemicals legislation for the protection of human health and the environment'.

Companies register the information on the hazards, risks and safe use of chemical substances they are required to provide under the REACH Regulation with the ECHA. This information is freely available on the website. As a result of REACH, thousands of the most hazardous and most commonly used substances have been registered. In March 2017, the Agency announced that key information on around 15,000 chemicals can now be downloaded and used by researchers, regulatory authorities and businesses in order to improve the safe use of chemicals, enable innovation and help avoid the unnecessary testing of chemicals on animals. The information can be downloaded from the ECHA website at https://iuclid6.echa.europa.eu/reach-study-results.

Companies must also notify ECHA of the classification and labelling that they use for their chemicals. The Agency has received millions of these notifications and again this information is freely available on its website.

More information about the ECHA can be found on its website at https://echa.europa.eu/about-us.

Evaluation

[R2512] Registration dossiers submitted under REACH by industry to the ECHA can be evaluated in different ways:

Compliance checking

A proportion of registration dossiers are subject to a full compliance check of the quality and accuracy of the information and assessment submitted.

Dossier evaluation

For substances registered at 100 tonnes per year or more, the ECHA carefully considers any proposals for further animal toxicity tests to obtain missing information to complete registration dossiers before deciding whether to approve the proposed tests. This is to ensure that no unnecessary animal testing is carried out.

Substance evaluation

National competent authorities (see **R2515** below) evaluate substances that have been prioritised for potential regulatory action because of concerns about their hazardous properties. This could lead to restrictions on the manufacture, supply or use of a substance, lead to a substance being added to the priority list for authorisation (see **R2513** below), or result in a proposal to change the classification and labelling of a substance.

Authorisation

[R2513] The ECHA explains that under the REACH Regulation, the route to authorisation starts when a Member State or the Agency, at the request of the Commission, proposes a substance to be identified as a substance of very high concern (SVHC). A substance identified as an SVHC is included in the 'Candidate List' for eventual inclusion in the 'Authorisation List'. There is a 45-day public consultation period for this SVHC identification process.

SVHCs include those identified as: carcinogenic, mutagenic or toxic to reproduction (CMRs); persistent, bio-accumulative and toxic (PBTs); very persistent and very bio-accumulative (vPvBs); and substances demonstrated to be of equivalent concern, such as endocrine disruptors.

Companies manufacturing or importing articles containing SVHCs in a concentration above 0.1% weight of the article are required to inform recipients of the articles about the presence of the substance and how to use it safely. They are also required to provide this information to consumers if they request it.

Substances placed on the Candidate List can move to the Authorisation List. Where substances are admitted to the Authorisation List, after a given date, they cannot be supplied or used unless an authorisation for that supply or use has been applied for and granted. Any company wishing to supply or use a substance on the list must apply to the ECHA for an authorisation, which may be granted or refused.

The HSE explains that:

- the use is considered safe as long as the risks are adequately controlled, and the conditions of the authorisation are met; or
- the use is considered safe as above and the substance can be demonstrated to be so important on socio-economic grounds that its continued use outweighs the risks to human health and the environment.

In both cases, those applying for authorisations must provide information on the availability of alternatives, and if there is a suitable alternative, a plan for phasing out the SVHC in question – known as the 'substitution plan'.

More information on authorisation can be found in *UK REACH CA Information Leaflet Number 19 – Authorisation* available on the HSE website at www.hse.gov.uk/reach/resources/19authorisation.pdf.

The ECHA explains that one of the main aims of authorisation is to phase out SVHCs where possible. The Candidate List is updated every six months and is based on proposals made by national authorities or the ECHA. A *Roadmap on Substances of Very High Concern* aims to have all relevant currently known SVHCs included in the Candidate List by 2020.

In June 2017, the European Commission added 12 SVHCs to the Authorisation List bringing the total to 43 substances. The Commission adopted a moratorium on additions in August 2014. Organisations including trade unions and NGOs have criticised the authorisation process on the basis that too few substances have been added to the list and that safer alternatives are being ignored. They have established their own lists of substances. For example, the second version of the European Trade Union Confederation (ETUC) Trade Union Priority List for REACH authorisation includes 334 substances or group of substances in order of priority. The ETUC says that most of these substances are identified as causative agents for recognised occupational diseases in the EU and that it 'is convinced that including the union-listed chemicals in the REACH Authorisation List would cut the incidence of chemical-related occupational diseases and the attendant costs for the community, workers and industry itself' and that it 'will also be a strong incentive for companies to innovate and replace them by safer alternatives'.

The Candidate List can be found online at https://echa.europa.eu/candidate-list-table.

The Authorisation List can be found online at https://echa.europa.eu/addressing-chemicals-of-concern/authorisation/recommendation-for-inclusion-in-the-authorisation-list/authorisation-list.

Information about the Trade Union Priority List for REACH authorisation can be found on the ETUC website at: www.etuc.org/trade-union-priority-list.

Restrictions

[R2514] REACH enables restrictions on the supply and/or use of substances to be placed across the EU, where this is shown to be necessary. Restrictions may limit or ban the manufacture, placing on the market or use of a substance. The ECHA says that a restriction applies to any substance on its own, in a mixture or in an article, including those that do not require registration and that it can also apply to imports.

Either Member States or the ECHA, at the request of the European Commission, can propose restrictions on any particular chemical. The ECHA can also propose a restriction on articles containing substances that are in the Authorisation List (see **R2513** above). When approved, the form of restrictions can vary. This aspect of REACH takes over the provisions of the Marketing and Use Directive.

UK Competent Authority and Helpdesk

[R2515] In the UK, the Competent Authority is hosted by the HSE, working with the Environment Agency and other government departments. It lists its responsibilities under REACH as including the following:

- to provide advice to interested parties on their responsibilities under REACH through a Helpdesk;
- to enforce compliance with registration;
- to conduct in-depth evaluation of selected prioritised substances and prepare draft decisions;
- to consider the identification of substances of very high concern for authorisation;
- to consider whether substances need to be managed through restrictions;
- to liaise with relevant enforcing organisations in relation to downstream responsibilities under REACH, such as the implementation by users of the risk management measures designated by suppliers and adherence to restrictions or to the refusal of authorisation.

The UK CA helpdesk can be contacted via email: UKREACHCA@hse.gov.uk or by post:
UK REACH CA Helpdesk
5S.1 Redgrave Court
Bootle
Merseyside
L20 7HS

Enforcement of REACH

[R2516] The *REACH Enforcement Regulations 2008 (SI 2008/2852)* came into force on 1 December 2008. They allocate responsibility for REACH enforcement to a number of enforcing authorities, including the HSE, environment agencies, local authorities and the Office for Road and Rail (ORR), for example, provide the authorities with the powers they need to carry out their role, require them to co-operate and share information with other enforcement agencies, and set down the offences and penalties for contravening REACH requirements.

In addition, a Forum for Exchange of Information on Enforcement provides for co-operation and co-ordination across the European Union.

Under the *REACH Enforcement Regulations 2008*, it is an offence for a person to contravene a 'listed REACH provision' or to cause or permit another person to do so. The maximum penalties are a fine of up to £5,000 and/or up to three months' imprisonment following summary conviction (in a Magistrates' Court); and an unlimited fine and/or up to two years' imprisonment following conviction on indictment (in the Crown Courts). Criminal offences include obstruction of inspectors, providing false statements, failing to comply with enforcement notices, and so on. These additional offences are also the subject of penalties, which are the same as those above.

The REACH Compliance Team can be contacted by email: CRDCompliance@hse.gov.uk or by post:
REACH Compliance
2.3 Redgrave Court
Merton Road
Bootle
Merseyside
L20 7HS.

In May 2013, the news service Chemical Watch (https://chemicalwatch.com) used a freedom of information (FOI) request as part of an investigation into how REACH requirements were being enforced by national enforcement authorities. It found that since 2010, the HSE had serviced 21 notices. These included three enforcement notices, requiring companies to immediately remove articles containing restricted substances; four improvement notices, all to the same company, each requiring it to supply a safety data sheet for a particular substance or mixture; and 14 improvement notices requiring companies to register substances. In June 2013, Chemical Watch reported that since it had made the FOI request, the number of REACH-related notices on the HSE's public register of enforcement had increased to 54 (over the period from 2009 to 2013).

The restriction offences related to the presence of the phthalate plasticisers DINP, DEHP and DBP in toys supplied by one company, and the presence of chrysotile asbestos fibres in sky lanterns supplied by another.

There were no HSE prosecutions pending in relation to these notices, but Chemical Watch reported that there had been at least three prosecutions by local authority trading standards departments for the supply of goods containing a substance restricted under REACH. These included two cases where a company director was prosecuted for supplying 'popper' recreational drugs containing isobutyl nitrite. In one case, the director was sentenced to two years conditional discharge and ordered to pay £1,500 costs; in the other, they received an eight-week prison sentence suspended for 12 months, plus 200 hours community service. In another case, a company director was fined £1,000 plus £3,000 costs for supplying toys containing a banned substance.

Further information related to REACH

[R2517] HSE information on REACH can be accessed from its website at www.hse.gov.uk/reach/index.htm.

The ECHA website is at https://echa.europa.eu.

UK Chemicals Stakeholder Forum publications can be downloaded from the government website at www.gov.uk/government/groups/uk-chemicals-stakeholder-forum.

The full text of the consolidated version of the REACH Regulation including amendments can be found online at eur-lex.europa.eu/legal-content/EN/TXT/?uri=CELEX:02006R1907-20140410.

Acknowledgement

[R2518] In preparing this chapter, use has been made of official publications and information provided by the HSE, the Department for Food and Rural Affairs (DEFRA) and the ECHA.

Risk Assessment

Andrea Oates

Introduction to risk assessment

[R3001] Although the term 'risk assessment' probably first came into common use as a result of the *Control of Substances Hazardous to Health Regulations 1988* (since replaced by the *Control of Substances Hazardous to Health Regulations 2002 (SI 2002 No 2677)* or 'COSHH Regulations'), similar requirements were contained previously in both the *Control of Lead at Work Regulations 1980* and the *Control of Asbestos at Work Regulations 1987*. The regulations currently in force are the *Control of Asbestos Regulations 2012 (SI 2002 No 632)*.

In practice a type of risk assessment had already been necessary for some time, particularly as a result of the use of the qualifying clause 'so far as is reasonably practicable' in a number of the important sections of the *Health and Safety at Work etc Act 1974* ('HSWA 1974').

Risk assessment is now at the heart of effective health and safety management, with many regulations containing specific risk assessment requirements — see **R3006** to **R3016** below.

HSWA 1974 requirements

[R3002] *HSWA 1974, s 2* contains the general duties of employers to their employees with the most general contained within *s 2(1)*:

> It shall be the duty of every employer to ensure, so far as is reasonably practicable, the health, safety and welfare at work of all employees.

Other more specific requirements are contained in *s 2(2)* and these are also qualified by the term 'reasonably practicable'.

HSWA 1974, s 3 places general duties on both employers and the self-employed in respect of persons other than their employees. *Section 3(1)* states:

> It shall be the duty of every employer to conduct his undertaking in such a way as to ensure, so far as is reasonably practicable, that persons not in his employment who may be affected thereby are not exposed to risks to their health or safety.

Employers therefore have duties to contractors and their employees, visitors, customers, members of the emergency services, neighbours, passers-by and the public at large. This may to a certain point extend to include trespassers. Self-employed people are put under a similar duty by virtue of *HSWA 1974, s 3(2)*. As a result of a change in the law on 1 October 2015, brought in by the

Deregulation Act 2015, health and safety law now applies to self-employed people whose work activity is specifically mentioned in the *Health and Safety at Work etc Act 1974 (General Duties of Self-Employed Persons) (Prescribed Undertakings) Regulations 2015*, or their work activity poses a risk to the health and safety of others. The activities specifically mentioned in these regulations are agriculture (including forestry), work with asbestos, work on a construction site, an activity to which the *Gas Safety (Installation and Use) Regulations 1998* applies, work with genetically modified organisms and work on the railways. The new wording of *s 3(2)* is as follows:

'It shall be the duty of every self-employed person [who conducts an undertaking of a prescribed description] to conduct [the undertaking] in such a way as to ensure, so far as is reasonably practicable, that he and other persons (not being his employees) who may be affected thereby are not thereby exposed to risks to their health or safety.'

HSWA 1974, s 4 places duties on each person who has, to any extent, control of non-domestic premises used for work purposes in respect of those who are not their employees. *HSWA 1974, s 6* places a number of duties on those who design, manufacture, import or supply articles for use at work, or articles of fairground equipment and those who manufacture, import or supply substances. Many of these obligations also contain the 'reasonably practicable' qualification.

What is reasonably practicable?

[R3003] The Health and Safety Executive (HSE) says that the concept of 'reasonably practicable' lies at the heart of the British health and safety system and is a key part of the general duties of *HSWA 1974* and many sets of health and safety regulations. Lord Justice Asquith provided a definition of the term in his judgment in *Edwards v National Coal Board* [1949] 1 All ER 743 in which he stated:

> '"Reasonably practicable" is a narrower term than "physically possible" and seems to me to imply that a computation must be made by the owner in which the quantum of risk placed on one scale and the sacrifice involved in the measures necessary for averting risk (whether in money, time or trouble) is placed in the other, and that, if it be shown that there is a gross disproportion between them – the risk being insignificant in relation to the sacrifice – the defendants discharge the onus on them. Moreover, this computation falls to be made by the owner at a point in time anterior to the accident.'

HSWA 1974, s 40 places the burden of proof in respect of what was or was not 'reasonably practicable' (or 'practicable', see **R3004** below) on the person charged with failure to comply with a duty or requirement. Employers and other duty holders must establish the level of risk involved in their activities and consider the various precautions available in order to determine what is reasonably practicable. In other words they must carry out a form of risk assessment.

The HSE website section dealing with Risk Management (www.hse.gov.uk/risk/theory/alarp1.htm) provides guidance on the principles and guidelines which are used by the HSE itself in determining what is 'reasonably practicable'. Both 'risk' and 'sacrifice' are considered in some detail. In respect of the latter the HSE states: 'Individual duty-holders' ability to afford a control measure or the financial viability of a particular project is not a legitimate factor in the assessment of its costs. HSE must present duty-holders with a level playing field. Thus HSE cannot take into account the size and financial position of the duty-holder when making judgements on whether risks have been reduced ALARP [as low as reasonably practicable].'

Further guidance is provided on the influence of 'societal concerns', 'transfer of risks' (eg to other employees or members of the public) and 'good practice'. Additional information on material available on the HSE website is contained in **R3053** below.

Practicable and absolute requirements

[R3004] Not all health and safety law is qualified by the phrase 'reasonably practicable'. Some requirements must be carried out 'so far as is practicable'. 'Practicable' is a tougher standard to meet than 'reasonably practicable' – the precautions must be possible in the light of current knowledge and invention (*Adsett v K and L Steelfounders & Engineers Ltd* [1953] 2 All ER 320). Once a precaution is practicable it must be taken even if it is inconvenient or expensive. However, it is not practicable to take precautions against a danger which is not yet known to exist (*Edwards v National Coal Board* [1949] 1 All ER 743), although it may be practicable once the danger is recognised.

Many health and safety duties are subject to neither 'practicable' nor 'reasonably practicable' qualifications. These absolute requirements usually state that something 'shall' or 'shall not' be done. However, such duties often contain other words which are subject to a certain amount of interpretation, such as 'suitable', sufficient', 'adequate', 'efficient' and 'appropriate'.

In order to determine whether requirements have been met 'so far as is practicable' or whether the precise wording of an absolute requirement has been complied with, the employer must carry out a proper evaluation of the risks and the effectiveness of the precautions.

The Management of Health and Safety at Work Regulations 1999

[R3005] The *Management of Health and Safety at Work Regulations* were introduced in 1992 and replaced by the current *Management of Health and Safety at Work Regulations 1999 (the 'Management Regulations') (SI 1999 No 3242)*. They implement the requirements of the European Framework Directive (89/391) on the introduction of measures to encourage improvements in the safety and health of workers at work. *Regulation 3* requires employers and 'relevant' self-employed people (see **R3002**) to make a suitable and sufficient assessment of the risks to employees (or themselves in the case of relevant self-employed people) and persons not in their employment. The purpose of the assessment is to identify the measures needed 'to comply with the requirements and prohibitions imposed ... by or under the relevant statutory provisions ...' ie identifying what is needed to comply with the law.

Given the extremely broad obligations contained in *HSWA 1974, ss 2, 3, 4* and 6, all risks arising from work activities should be considered as part of the risk assessment process (although some risks may be dismissed as being insignificant). Compliance with more specific requirements of regulations must also be assessed, whether these are absolute obligations or subject to 'practicable' or 'reasonably practicable' qualifications. The requirement for risk assessments introduced in the Management Regulations simply formalised what employers (and others) should have been doing all along ie identifying what precautions they need to take to comply with the law. The more detailed requirements of the Management Regulations are covered later in the chapter.

Any such additional precautions must then be implemented (the *Management of Health and Safety at Work Regulations 1999 (SI 1999 No 3242), Reg 5* contains requirements relating to the effective implementation of precautions). Practical guidance is provided in the chapter on MANAGING HEALTH AND SAFETY.

Common regulations requiring risk assessment

[R3006] An increasing number of regulations contain requirements for risk assessments. Several of these regulations are of significance to a wide range of work activities and are dealt with in more detail elsewhere in this publication. The most significant regulations requiring risk assessment include the following:

Control of Substances Hazardous to Health Regulations 2002 (COSHH)

[R3007] The *Control of Substances Hazardous to Health Regulations 2002 (COSHH) (SI 2002 No 2677 as amended), Reg 6* require employers to make a suitable and sufficient assessment of the risks created by work liable to expose any employees to any substance hazardous to health and of the steps that need to be taken to meet the requirements of the regulations. See HAZARDOUS SUBSTANCES IN THE WORKPLACE, which provides more details of the COSHH requirements.

The Health and Safety Executive (HSE) booklet L5, contains both the regulations and the associated Approved Code of Practice (ACOP) and guidance. There are many other relevant HSE publications including HSG97 '*A step by step guide to COSHH assessment*'.

Control of Noise at Work Regulations 2005

[R3008] Regulation 5 of the *Control of Noise at Work Regulations 2005 (SI 2005 No 1643 as amended)* requires employers carrying out work which is liable to expose any employees to noise at or above a 'lower exposure action value' to make an assessment of the risk from that noise. The lower exposure action values are:

- a daily or weekly personal noise exposure of 80 dB (A-weighted); and
- a peak sound pressure of 135 dB (C-weighted).

The measures necessary to meet the requirements of the regulations must be identified during the assessment.

See the chapter dealing with NOISE AT WORK and also the HSE booklet L108 'Controlling Noise at Work' which contains guidance on the interpretation of the regulations and on the assessment process.

Manual Handling Operations Regulations 1992

[R3009] Employers are required by *Reg 4* of the *Manual Handling Operations Regulations 1992 (SI 1992 No 2793*as amended), to first of all, so far as is reasonably practicable, avoid the need for employees to undertake any manual handling operations at work which involve a risk of their being injured. Where it is not reasonably practicable to avoid the need for employees to undertake such work, the employer must make a suitable and sufficient assessment of all manual handling operations at work which involve a risk of employees being injured, and take appropriate steps to reduce the risk to the lowest level reasonably practicable. See MANUAL HANDLING which provides further details of the Regulations and the carrying out of assessments. HSE booklet L23, '*Manual Handling: Manual Handling Operations Regulations 1992* (as amended)' provides detailed guidance on the Regulations and on carrying out manual handling assessments.

Health and Safety (Display Screen Equipment) Regulations 1992

[R3010] The *Health and Safety (Display Screen Equipment) Regulations 1992 (SI 1992 No 2792* as amended), *Reg 2* require employers to perform a suitable and sufficient analysis of display screen equipment (DSE) workstations for the purpose of assessing risks to 'users' or 'operators' as defined in the Regulations. Risks identified in the assessment must be reduced to the lowest extent reasonably practicable.

The HSE booklet L26 '*Work with display screen equipment*' contains guidance on the Regulations and on carrying out workstation assessments.

Personal Protective Equipment at Work Regulations 1992

[R3011] Under the *Personal Protective Equipment at Work Regulations 1992 (SI 1992 No 2966* as amended), *Reg 6*, employers must ensure that (before choosing any personal protective equipment or PPE) an assessment is made to determine risks which have not been avoided by other means and identify PPE which will be effective against these risks. The Regulations also contain other requirements relating to the provision of PPE; its maintenance and replacement; information, instruction and training; and the steps which must be taken to ensure its proper use.

See PERSONAL PROTECTIVE EQUIPMENT which provides further details of the requirements under the Regulations. HSE booklet L25, '*Personal protective*

[R3011] Risk Assessment

equipment at work' provides detailed guidance on the Regulations and on carrying out assessments of PPE needs.

Regulatory Reform (Fire Safety) Order 2005

[R3012] The *Regulatory Reform (Fire Safety) Order 2005 (SI 2005 No 1541)* requires the 'responsible person' (a definition is provided in the Order and means the employer, if the workplace is under their control; otherwise the duties of the responsible person are extended to any person who has control of the premises) to ' . . . make a suitable and sufficient assessment of the risks to which relevant persons are exposed for the purposes of identifying the general fire precautions he needs to take' (Article 9).

Fire safety legislation has come under particular scrutiny following the Grenfell Tower fire in June 2017, which killed 72 people. In June 2019, the government issued a call for evidence as a first step in ensuring the Order is fit for purpose for all regulated premises. Details of the consultation, which closed at the end of July 2019, can be found at: www.gov.uk/government/consultations/the-regulatory-reform-fire-safety-order-2005-call-for-evidence.

More information is provided in the chapter on FIRE PREVENTION AND CONTROL.

Control of Asbestos Regulations 2012

[R3013] The *Control of Asbestos Regulations 2012 (SI 2012 No 632 as amended)* require employers to make a risk assessment before carrying out work which is liable to expose employees to asbestos. *Regulation 6* contains several specific requirements about what the risk assessment must involve. Other regulations detail the many precautions which must be taken in asbestos work. Some types of asbestos work can only be done by companies licensed by the HSE.

The 2012 Regulations contain a 'Duty to manage asbestos in non-domestic premises'— in *Reg 4*. Duty holders are normally the owner or leaseholder, the employer occupying the premises or a combination of these, depending upon the terms of any lease or contract or who controls the premises.

Duty holders must carry out an assessment of their premises in order to identify any asbestos-containing material ('ACM') and assess its condition. They must then prepare a written plan for managing the risks from ACM, including such measures as regular inspections of confirmed and presumed ACM, restricting access to or maintenance in ACM areas and, where necessary, protection, sealing or removal of ACM.

Further details on these requirements are contained in the chapter on ASBESTOS.

The HSE publication L143, *Managing and working with asbestos*, contains the *Control of Asbestos Regulations 2012*, the Approved Code of Practice (ACOP) and guidance for employers about work which disturbs, or is likely to disturb, asbestos, asbestos sampling and laboratory analysis. Other important HSE references worth referring to are:

- HSG210 asbestos essentials. A task manual for building, maintenance and allied trades on non-licensed asbestos work;

- HSG264 Asbestos: The survey guide;
- HSG227 A comprehensive guide to managing asbestos in premises; and
- INDG223 Managing asbestos in buildings: A brief guide.

Dangerous Substances and Explosive Atmospheres Regulations 2002

[R3014] These Regulations (known sometimes as 'DSEAR') are concerned with protecting against risks from fire and explosion. The definition of 'dangerous substances' contained within the regulations includes petrol, many solvents, some types of paint, and liquid petroleum gas (LPG), as well as potentially explosive dusts.

Employers are required by the *DSEAR (SI 2002 No 2776), Reg 5* to carry out a risk assessment of work involving dangerous substances. *Regulation 6* requires a range of specified control measures to be implemented in order to eliminate or reduce the risks as far as reasonably practicable. Places where explosive atmospheres may occur must be classified into zones and equipment and protective systems in these zones must meet appropriate standards (*DSEAR (SI 2002 No 2776), Reg 7*). *Regulation 8* requires equipment and procedures to deal with accidents and emergencies to be put into place, while *Reg 9* states that employees must be provided with information, instruction and training. Further details are contained in HAZARDOUS SUBSTANCES IN THE WORKPLACE.

L138 *Dangerous substances and explosive atmospheres, Dangerous Substances and Explosive Atmospheres Regulations 2002* Approved Code of Practice and guidance provides more detailed guidance in this area.

Work at Height Regulations 2005

[R3015] Although the *Work at Height Regulations 2005 (SI 2005 No 735)* do not introduce a separate requirement for risk assessment, *Reg 6* sets out that:

- in identifying the measures required by this regulation, every employer shall take account of a risk assessment under *Reg 3* of the Management Regulations;
- every employer shall ensure that work is not carried out at height where it is reasonably practicable to carry out the work safely otherwise than at height; and
- where work is carried out at height, every employer shall take suitable and sufficient measures to prevent, so far as is reasonably practicable, any person falling a distance liable to cause personal injury.

In February 2019, the all-party parliamentary group on working at height published its inaugural report on how to improve the safety environment for work at height, *Staying Alive: Preventing Serious Injuries and Fatalities While Working at Height*. The report can be found at: https://workingatheight.info/wp-content/uploads/2019/02/Staying-Alive-APPG-REPORT.pdf.

See further information in the chapter WORK AT HEIGHT. Various HSE publications provide further guidance, including *Working at height – A brief guide* (INDG401) at www.hse.gov.uk/pubns/indg401.pdf.

Specialist regulations requiring risk assessment

[R3016] Some regulations requiring risk assessments are of more specialist application and are covered in more detail elsewhere in this publication. References to relevant chapters of this publication and to key HSE publications of relevance are included below. These regulations include:

- *Control of Lead at Work Regulations 2002 (SI 2002 No 2676)*
  See HAZARDOUS SUBSTANCES IN THE WORKPLACE.
  L132 Control of lead at work.
- *Supply of Machinery (Safety) Regulations 2008 (SI 2008 No 1597)*
  The Regulations include a variety of procedures which must be followed in assessing conformity of machinery with essential health and safety requirements set out in the Machinery Directive.
  See MACHINERY SAFETY.
  INDG270 *Supplying new machinery: A short guide to the law and your responsibilities when supplying machinery for use at work.*
  INDG271 *Buying new machinery: A short guide to the law and your responsibilities when buying new machinery for use at work.*
- *Control of Major Accident Hazard Regulations 2015 (SI 2015 No 483) (COMAH 2015)*
  The Regulations apply to sites containing specified quantities of dangerous substances.
  See DISASTER AND EMERGENCY MANAGEMENT SYSTEMS (DEMS).
  L111 *A guide to the Control of Major Accident Hazards Regulations 2015 (COMAH 2015).*
- *Ionising Radiations Regulations 2017 (SI 2017 No 1075)*
  L121 *Work With Ionising Radiation: Ionising Radiations Regulations 2017 – approved code of practice and guidance.*
  Control of Electromagnetic Fields at Work Regulations 2016 (SI 2016 No 588).
  HSG 281 *A guide to the Control of Electromagnetic Fields at Work Regulations 2016*
  See RADIATION.
- *Control of Artificial Optical Radiation at Work Regulations 2010 (SI 2010 No 1140)*
  Guidance for Employers on the Control of Artificial Optical Radiation at Work Regulations (AOR) 2010
- *Control of Vibration at Work Regulations 2005 (SI 2005 No 1093)*
  See VIBRATION.
  L140 *Hand-arm vibration. Control of Vibration at Work Regulations 2005.*
  INDG175 *Hand-arm vibration – A brief guide.*
  L141 *Whole-body vibration, Control of Vibration at Work Regulations 2005.*

INDG242 *Control back pain risks from whole-body vibration. Advice for employers.*

Related health and safety concepts

[R3017] Risk assessment techniques are an essential part of other health and safety management concepts.

- *Safe systems of work*
  Employers are required under *HSWA 1974, s 2(2)(a)* to provide and maintain 'systems of work that are, so far as is reasonably practicable, safe and without risks to health'.
  Both the *Confined Spaces Regulations 1997 (SI 1997 No 1713)* and the *Lifting Operations and Lifting Equipment Regulations 1998 (LOLER) (SI 1998 No 2307)* contain similar requirements for safe systems of work. A safe system of work can only be established through a process of risk assessment.
  See the chapter on SAFE SYSTEMS OF WORK for more information on this concept.
- *'Dynamic risk assessment'*
  The term 'dynamic risk assessment' is often used to describe the day-to-day judgements that employees are expected to make in respect of health and safety. However, employers must ensure that employees have the necessary knowledge and experience to make such judgements. The employer's 'generic risk assessments' must have identified the types of risks which might be present in the work activities, established a framework of precautions, such as procedures and equipment, which are likely to be necessary and provided guidance on which precautions are appropriate for which situations.
- *Permits to work*
  A permit-to-work system is a formalised method for identifying a safe system of work, usually for a high-risk activity, and ensuring that this system is followed. The permit issuer is expected to carry out a dynamic risk assessment of the work activity and should be more competent in identifying the risks and the relevant precautions than those carrying out the work.
  Permits to work are dealt with in more detail in the chapter on SAFE SYSTEMS OF WORK.
- *CDM Construction Phase Plans*
  The construction phase plan required for 'notifiable projects' under the *Construction (Design and Management) Regulations 2015 (CDM) (SI 2015 No 51), Reg 12*, is in effect a risk assessment in relation to the construction project. The various risks associated with the project must be identified, together with the means for eliminating or controlling those risks. Further detail on the regulations is provided in CONSTRUCTION, DESIGN AND MANAGEMENT and also in the HSE booklet *L144 'Managing health and safety in construction'* in which Appendix 3 sets out topics for inclusion in a construction phase plan.
- *Method statements*

[R3017] Risk Assessment

Method statements usually involve a description of how a particular task or operation is to be carried out and should identify all the components of a safe system of work arrived at through a process of risk assessment.

Management Regulations requirements

[R3018] The general requirement concerning risk assessments is contained in the *Management of Health and Safety at Work Regulations 1999* (the '*Management Regulations*') *(SI 1999 No 3242)*, *Reg 3*. The risk assessment must take specific account of risks to both young persons (under 18's) (see **R3025** below) and new and expectant mothers (see **R3026** below). Other regulations require more specific types of risk assessment, including hazardous substances (COSHH), noise, manual handling operations, display screen equipment and personal protective equipment (see **R3006–R3016** above).

Regulation 3(1) of the *Management Regulations* states:

every employer shall make a suitable and sufficient assessment of:

(a) the risks to the health and safety of his employees to which they are exposed whilst they are at work; and
(b) the risks to the health and safety of persons not in his employment arising out of or in connection with the conduct by him of his undertaking,

for the purpose of identifying the measures he needs to take to comply with the requirements or prohibitions imposed upon him by or under the relevant statutory provisions.

Regulation 3(2) of the *Management Regulations* imposes similar requirements on 'relevant' self-employed persons (see **R3002**). *Regulation 3(3)* requires a risk assessment to be reviewed if:

- there is reason to suspect that it is no longer valid; or
- there has been a significant change in the matters to which it relates.

Regulation 3(4) requires a risk assessment to be made or reviewed before an employer employs a young person, and *Regulation 3(5)* identifies particular issues which must be taken into account in respect of young persons, including their inexperience, lack of awareness of risks and immaturity. Further requirements in respect of young persons are contained in *Regulation 19*.

Regulation 16 contains specific requirements concerning the factors which must be taken into account in risk assessments in relation to new and expectant mothers. These relate to processes, working conditions and physical, biological or chemical agents.

Assessments in respect of young persons and new or expectant mothers are dealt with in more detail later in the chapter (see **R3025–R3027** below).

Regulation 3(6) requires employers who employ five or more employees to record:

— the significant findings of their risk assessments; and

— any group of employees identified as being especially at risk.

Methods of recording assessments are described later in this chapter (see **R3042** below).

HSE online guidance on managing health and safety at work includes the following:

- *Health and safety made simple: The basics for your business* provides basic information for those getting started in managing for health and safety;
- *The health and safety toolbox: How to control risks at work* provides guidance on controlling risks from specific topics; and
- *Managing for Health and Safety (HSG65)* is divided into four sections covering: Core elements of managing for health and safety; Deciding if you are doing what you need to do; Delivering effective arrangements; and Resources. It is aimed at those who need to put in place or oversee their organisations arrangements for health and safety, as well as workers and their representatives.

Hazards and risks

[R3019] The HSE explains that:

- a **hazard** is anything that may cause harm, such as chemicals, electricity, working from ladders, an open drawer etc; and
- the **risk** is the chance, high or low, that somebody could be harmed by these and other hazards, together with an indication of how serious the harm could be.

The *extent of the risk* will depend on:

- the likelihood of that harm occurring;
- the potential severity of that harm (resultant injury or adverse health effect);
- the population which might be affected by the hazard ie the number of people who might be exposed.

As an illustration, work at height includes the hazard of those below being struck by falling objects. The extent of this risk might depend on factors such as the nature of the work being carried out, the weight of objects which might fall, the distance to the ground below and the numbers of people in the area.

Evaluation of precautions

[R3020] Risk assessment involves identifying the hazards present in any working environment or arising out of work activities, and evaluating the extent of the risks involved, taking into account existing precautions and their effectiveness.

The evaluation of the effectiveness of precautions is an integral part of the risk-assessment process. This is overlooked by some organisations who concentrate on the identification, and often the quantification, of risks without checking whether the intended precautions are actually being taken in the workplace and whether these precautions are proving effective.

Using the illustration in the previous paragraph, the risk assessment must take account of what precautions are taken, such as:

- use of barriers and/or warning signs at ground level;
- use of tool belts by those working at heights;
- provision of edge protection on working platforms; and
- use of head protection by those at ground level.

'Suitable and sufficient'

[R3021] Risk assessments under the *Management Regulations 1999 (SI 1999 No 3242)* (and several other regulations) must be 'suitable and sufficient', but the phrase is not defined in the Regulations themselves.

HSE guidance contained in *Managing for health and safety* (HSG65) sets out that for a risk assessment to be 'suitable and sufficient', it should show that: 'a proper check was made; you asked who might be affected; you dealt with all the obvious significant risks, taking into account the number of people who could be involved; the precautions are reasonable, and the remaining risk is low; you involved your workers or their representatives in the process'.

It further advises: 'The level of detail in a risk assessment should be proportionate to the risk and appropriate to the nature of the work. Insignificant risks can usually be ignored, as can risks arising from routine activities associated with life in general, unless the work activity compounds or significantly alters those risks.'

However, in practice, a risk can only be concluded to be insignificant if some attention is paid to it during the risk assessment process and, if there is any scope for doubt, it is prudent to state in the risk assessment record which risks are considered insignificant.

Winter weather, including rain, ice, snow and wind, may be considered to pose a routine risk to life. However, driving a fork-lift truck in an icy yard or carrying out agricultural or construction work in a remote location may involve a far greater level of risk than normal and require additional precautions to be taken.

The case of *Kennedy v Cordia (Services) LLP* [2016] UKSC 6, [2016] 1 WLR 597, [2016] ICR 325 involved a claim from a home care worker who was injured when she slipped on an icy, ungritted path on her way to visit a client. It included a consideration of whether the company had breached its statutory duty to carry out a suitable and sufficient risk assessment with regard to this 'routine' risk. In this case, the worker was exposed to a risk of slipping and falling on snow and ice. The risk was obvious and within the knowledge of the company. It had previous knowledge of home carers suffering such accidents. The risk had been identified in assessments in 2005 and 2010 – the incident occurred in December 2010 – but no consideration had been given to the possibility of personal protective equipment. Advice to wear appropriate footwear was given, but this did not specify what might be appropriate. The

Supreme Court set out that the Lord Ordinary (the judge who first heard the case in the Outer Court of Session) was entitled to conclude there had been a breach of *Reg 3(1)* of the Management Regulations.

The risk assessment must take into account workers and other people who might be affected by the undertaking.

For example, a construction company would need to consider risks to and from their employees, sub-contractors, visitors to their sites, delivery drivers, passers-by and even possible trespassers on their sites.

Similarly, a residential care home should take into account risks to and from their staff, visiting medical specialists, visiting contractors, residents and visitors.

Reviewing risk assessments

[R3022] The *Management of Health and Safety at Work Regulations 1999 (SI 1999 No 3242), Reg 3(3)* require a risk assessment to be reviewed if:

(a) there is reason to suspect it is no longer valid; or
(b) there has been a significant change in the matters to which it relates;

and where as a result of any such review changes to an assessment are required, the employer or relevant self-employed person concerned shall make them.

The HSE guidance *Managing for health and safety* (HSG65) advises: 'Your risk assessment should only include what you could reasonably be expected to know – you are not expected to anticipate unforeseeable risks.'

However, what is foreseeable can be changed by subsequent events. An accident, a non-injury incident or a case of ill-health may highlight the need for a risk assessment to be reviewed because:

- a previously unforeseen possibility has now occurred;
- the risk of something happening, or the extent of its consequences, is greater than previously thought; or
- precautions prove to be less effective than anticipated.

A review of the risk assessment may also be required because of significant changes in the work activity, such as changes to the equipment or materials used, the environment where the activity takes place, the system of work used or to the numbers or types of people carrying out the activity. The need for a risk assessment review may be identified from information received from elsewhere, such as from others involved in the same work activity, through trade or specialist health and safety journals, from the suppliers of equipment or materials, or from the HSE or other specialist bodies. Routine monitoring activities, such as inspections and audits, or consultation with employees may also identify the need for an assessment to be reviewed. A review of the risk assessment does not necessarily require a repeat of the whole risk assessment process, but it is quite likely to identify the need for increased or changed precautions.

In practice, workplaces and the activities within them are constantly subject to gradual changes and HSE advice in *Risk assessment – A brief guide to controlling risks in the workplace* (INDG163) is as follows:

'Few workplaces stay the same. Sooner or later, you will bring in new equipment, substances and procedures that could lead to new hazards. So it makes sense to review what you are doing on an ongoing basis, look at your risk assessment again and ask yourself:

- Have there been any significant changes?
- Are there improvements you still need to make?
- Have your workers spotted a problem?
- Have you learnt anything from accidents or near misses?

Make sure your risk assessment stays up to date.'

There are many activities where the nature of the work or the workplace itself changes constantly. Examples are construction work or peripatetic maintenance or repair work. Here it is possible to carry out 'generic' assessments of the types of risks involved and the types of precautions which should be taken. However, some reliance must be placed upon workers themselves to identify what precautions are appropriate for a given set of circumstances or to deal with unexpected situations. Such workers must be well informed and well trained so they are competent to make what are often called 'dynamic' risk assessments (see **R3017** above).

Related requirements of the Management Regulations

[R3023] A number of other requirements within the *Management Regulations (SI 1999 No 3242)* are closely related to the risk assessment process.

Regulation 4: Principles of prevention to be applied

> The *Management of Health and Safety at Work Regulations 1999 (SI 1999 No 3242), Reg 4* and *Sch 1* require preventive and protective measures to be implemented on the basis of a number of general principles of prevention:

- avoiding risks;
- evaluating risks that cannot be avoided;
- combating the risks at source;
- adapting the work to the individual, particularly with regard to workplace design, the choice of work equipment and working and production methods, with a view, in particular, to alleviating monotonous work and work at a predetermined work rate and to reducing their effect on health;
- adapting to technical progress;
- replacing the dangerous with safe or safer alternatives;
- developing a coherent overall prevention policy covering technology, work organisation, working conditions, social relationships, and the working environment;
- giving priority to collective over individual protective measures; and
- giving appropriate instructions to employees.

Regulation 5: Health and safety arrangements

> Under the *Management of Health and Safety at Work Regulations 1999 (SI 1999 No 3242), Reg 5*, employers are required to have appropriate arrangements for the effective planning, organisation, control, monitoring and review of preventive and

protective measures. Those employers with five or more employees must record these arrangements. This application of the 'management cycle' to health and safety matters is dealt with in greater detail in **R3024** below.

Regulation 6: Health surveillance

Under the *Management of Health and Safety at Work Regulations 1999 (SI 1999 No 3242), Reg 6*, where risks to employees are identified through a risk assessment, they must be 'provided with such health surveillance as is appropriate having regard to the risks to their health and safety which are identified by the assessment'. Surveillance is also likely to be necessary to comply with the requirements of more specific regulations such as *COSHH (SI 2002 No 2677)*, the *Control of Asbestos Regulations 2012 (SI 2012 No 632)*, and the *Ionising Radiations Regulations 2017 (SI 2017 No 1075)*. HSE guidance on health surveillance can be found at: www.hse.gov.uk/health-surveillance/index.htm.

Regulation 8: Procedures for serious and imminent danger and for danger areas

The *Management of Health and Safety at Work Regulations 1999 (SI 1999 No 3242), Reg 8* require every employer to 'establish and where necessary give effect to appropriate procedures to be followed in the event of serious and imminent danger to persons at work in his undertaking.' It also refers to the possible need to restrict access to areas 'on grounds of health and safety unless the employee concerned has received adequate health and safety instruction.'

The need for emergency procedures or restricted areas should of course be identified through the process of risk assessment. Situations 'of serious and imminent danger' might be due to fires, bomb threats, escape or release of hazardous substances, out-of-control processes, personal attack, or the escape of animals for example. Areas may justify access to them being restricted because of, for example, the presence of hazardous substances, unprotected electrical conductors, particularly high voltage, or potentially dangerous animals or people.

Application of the management cycle

[R3024] The time spent in carrying out risk assessments will be wasted unless those precautions identified as being necessary are implemented. The *Management of Health and Safety at Work Regulations 1999 (SI 1999 No 3242), Reg 5* require the application of a five stage 'management cycle' to this process.

- Planning
  The employer must have a planned approach to health and safety involving such measures as:
  — a health and safety policy statement;
  — annual health and safety plans; and
  — development of performance standards.
- Organisation
  Organisation must be put in place to deliver these good intentions, such as:
  — individuals allocated responsibility for implementing the health and safety policy or achieving elements of the health and safety plan;

- arrangements for communicating with and consulting employees, using health and safety committees, newsletters and notice boards for example; and
- the provision of competent advice and information.

- Control

  Detailed health and safety control arrangements must be established. The extent of such arrangements will reflect the size of the organisation and the degree of risk involved in its activities, but they are likely to involve:
  - formal procedures and systems for accident and incident investigation, fire and other emergencies, the selection and management of contractors for example;
  - the provision of health and safety related training at induction, for the operation of equipment or certain activities, and for supervisors and managers for example; and
  - provision of appropriate levels of supervision.

- Monitoring

  The effectiveness of health and safety arrangements must be monitored using methods such as:
  - health and safety inspections, both formal and informal;
  - health and safety audits; and
  - accident and incident investigations.

- Review

  The findings of monitoring activity must be reviewed and, where necessary, the management cycle applied once again to rectify shortcomings in health and safety arrangements.

  The review process might involve:
  - joint health and safety committees;
  - management health and safety meetings; and
  - activities of health and safety specialists.

The management cycle can also be applied to ensure the effective implementation of individual health and safety management procedures, for the investigation and reporting of accidents and non-injury incidents, for example, or to specific types of health and safety precautions such as guarding of machinery or precautions for working at heights.

HSE guidance is moving away from using the 'POPMAR' (Policy, Organising, Planning, Measuring performance, Auditing and Review) model for managing health and safety to a Plan, Do, Check, Act approach. The HSE says this achieves a better balance between the systems and behavioural aspects of management and treats health and safety management as an integral part of good management generally, rather than as a stand-alone system.

A diagram explaining how the two models work, with POPMAR in the right-hand column and the new approach on the left, is provided on the HSE website at: www.hse.gov.uk/managing/delivering.htm.

The chapter MANAGING HEALTH AND SAFETY also provides more information on management systems.

Children and young people

[R3025] Previous Acts and regulations identified many types of equipment that children and young people were not allowed to use, or processes or activities that they must not be involved in. Many of these 'prohibitions' were revoked by the *Health and Safety (Young Persons) Regulations 1997 (SI 1997 No 135)*, since incorporated into the *Management of Health and Safety at Work Regulations 1999 (SI 1999 No 3242)*. A few of these 'prohibitions' remain, but generally the emphasis has changed to restrictions on the work which children and young persons are allowed to do, based upon the employer's risk assessment.

The *Health and Safety (Training for Employment) Regulations 1990 (SI 1990 No 1380)* have the effect of giving students on work experience training programmes and trainees on training for employment programmes the status of 'employees'. The immediate provider of their training is treated as the 'employer'. There are exceptions for courses at educational establishments, such as universities, colleges and schools. Employers therefore have duties in respect of all children and young people at work in their undertaking: full-time employees, part-time and temporary employees and students or trainees on work placement with them.

The term 'child' and 'young person' are defined in the *Management of Health and Safety at Work Regulations 1999 (SI 1999 No 3242)*, Reg 1(2).

- A *child* is defined as a person not over compulsory school age in accordance with:
  — the *Education Act 1996, s 8* (for England and Wales); and
  — the *Education (Scotland) Act 1980, s 31* (for Scotland).
  In practice this is just under or just over the age of sixteen.
- A *young person* is defined as 'any person who has not attained the age of eighteen'.

Some of the prohibitions remaining from older health and safety regulations use different cut off ages.

See the chapter on VULNERABLE PERSONS which provides more detail on the requirements of the Management Regulations and the factors to be taken into account when carrying out risk assessments in respect of children and young persons. These factors centre around their:

- lack of experience;
- lack of awareness of existing or potential risks; and
- immaturity in both the physical and psychological sense.

That chapter also contains a checklist of 'Work presenting increased risks for children and young persons' (see **V12016**).

New or expectant mothers

[R3026] The *Management of Health and Safety at Work Regulations 1999 (SI 1999 No 3242)* require employers to consider risks to new or expectant mothers in their risk assessments.

Regulation 1 contains two relevant definitions:

- *'New or expectant mother'* means an employee who is pregnant; who has given birth within the previous six months; or who is breastfeeding.
- *'Given birth'* means 'delivered a living child or, after twenty-four weeks of pregnancy, a stillborn child.'

The requirements for 'risk assessment in respect of new and expectant mothers' are contained in the *Management of Health and Safety at Work Regulations 1999 (SI 1999 No 3242), Reg 16* which states in *paragraph (1)*:

(1) Where —
 (a) the persons working in an undertaking include women of child-bearing age; and
 (b) the work is of a kind which could involve risk, by reason of her condition, to the health and safety of a new or expectant mother, or to that of her baby, from any processes or working conditions, or physical, biological or chemical agents, including those specified in Annexes I and II of Council Directive 92/85/EEC on the introduction of measures to encourage improvements in the safety and health at work of pregnant workers and workers who have recently given birth or are breastfeeding [as amended by Directive 2014/27/EU],
 the assessment required by regulation 3(1) shall also include an assessment of such risk.

Regulation 16(4) states that in relation to risks from infectious or contagious diseases, an assessment must only be made if the level of risk is in addition to the level of exposure outside the workplace. (The types of risk which are more likely to affect new or expectant mothers are described in **R3027** below.) *Regulation 16(2)* and *(3)* set out the actions employers are required to take if these risks cannot be avoided. *Paragraph (2)* states:

Where, in the case of an individual employee, the taking of any other action the employer is required to take under the relevant statutory provisions would not avoid the risk referred to in paragraph (1) the employer shall, if it is reasonable to do so, and would avoid such risks, alter her working conditions or hours of work.

Consequently, where the risk assessment required under *Reg 16(1)* shows that control measures would not sufficiently avoid the risks to new or expectant mothers or their babies, the employer must make reasonable alterations to their working conditions or hours of work.

In some cases, restrictions may still allow the employee to substantially continue with her normal work but in others it may be more appropriate to offer her suitable alternative work.

Any alternative work must be:

— suitable and appropriate for the employee to do in the circumstances;
— on terms and conditions which are no less favourable.

Regulation 16(3) states:

'If it is not reasonable to alter the working conditions or hours of work, or if it would not avoid such risk, the employer shall, subject to section 67 of the 1996 Act, suspend the employee from work for so long as is necessary to avoid such risk.'

The *1996 Act* referred to is the *Employment Rights Act 1996* which provides that any such suspension from work on the above grounds is on full pay.

However, payment might not be made if the employee has unreasonably refused an offer of suitable alternative work.

Regulation 16A deals with the alteration of working conditions in respect of new or expectant mothers who are agency workers and the duties of hirers:

'(1) Where, in the case of an individual agency worker, the taking of any other action the hirer is required to take under the relevant statutory provisions would not avoid the risk referred to in regulation 16(1) the hirer shall, if it is reasonable to do so, and would avoid such risks, alter her working conditions or hours of work.

(2) If it is not reasonable to alter the working conditions or hours of work, or if it would not avoid such risk, the hirer shall without delay inform the temporary work agency, who shall then end the supply of that agency worker to the hirer.'

The *Management of Health and Safety at Work Regulations 1999 (SI 1999 No 3242), Reg 17* deals specifically with night work by new or expectant mothers and states:

'Where —

(a) a new or expectant mother works at night; and
(b) a certificate from a registered medical practitioner or a registered midwife shows that it is necessary for her health or safety that she should not be at work for any period of such work identified in the certificate,
the employer shall, subject to section 67 of the 1996 Act, suspend her from work for so long as is necessary for her health or safety.

Such suspension, on the same basis as described above, is only necessary if there are risks arising from work. HSE advice is that a new or expectant mother may work nights, provided this presents no risk to her health and safety. However, if a specific work risk has been identified – or her GP/midwife has provided a medical certificate stating she must not work nights – her employer must offer suitable alternative day work, on the same terms and conditions. If that is not possible, the employer must suspend her from work on paid leave for as long as is necessary to protect her health and safety and/or that of her child.

Regulation 17A contains similar provisions in respect of new and expectant who are agency workers:

'Where—

(a) a new or expectant mother works at night; and
(b) a certificate from a registered medical practitioner or a registered midwife shows that it is necessary for her health or safety that she should not be at work for any period of such work identified in the certificate,
the hirer shall without delay inform the temporary work agency, who shall then end the supply of that agency worker to the hirer.'

The requirements placed on employers in respect of altered working conditions or hours of work and suspensions from work only take effect when the employee has formally notified the employer of her condition. The *Management of Health and Safety at Work Regulations 1999 (SI 1999 No 3242), Reg 18* states:

'Nothing in paragraph (2) or (3) of regulation 16 shall require the employer to take any action in relation to an employee until she has notified the employer in writing that she is pregnant, has given birth within the previous six months, or is breastfeeding.'

Regulation 18(2) states that the employer is not required to maintain action taken in relation to an employee once the employer knows that she is no longer a new or expectant mother; or cannot establish whether this is the case.

Similar provisions apply in relation to hirers and new and expectant mothers who are agency workers (*Reg 18A*).

Case law has established that, for example:

Whether or not a separate risk assessment is required will depend on the particular circumstances of the individual worker and the kind of work she is doing (*O'Neill v Buckinghamshire County Council* (2010) UKEAT/0020/09, [2010] All ER (D) 15 (Feb), EAT).

The HSE advises: 'Although it is not a legal requirement for employers to conduct another specific or further individual risk assessment for new and expectant mothers, employers may choose to do so as part of the process by which they reach a decision about what action should be taken. An employer's risk assessment should have already considered any specific risks to new and expectant mothers when considering the rest of the workplace. This will enable employers to take immediate action, if and when necessary.'

Not taking steps to avoid any health risks will amount to a 'detriment' and 'sex discrimination'.

In *Bunning v GT Bunning and Sons Ltd* [2005] EWCA Civ 983 a risk assessment concluded that Suzanne Bunning's welding job was not high risk and that she should return to that role. She took a claim following a miscarriage.

Where a breastfeeding mother can show a risk assessment is inadequate, or not carried out, this can give rise to a potential discrimination claim (*Elda Otera Ramos v Servicio Galego de Saúde, Instituto Nacional de la Seguridad Social* Case C-531/15).

It would be sex discrimination to force a worker to accept a change of duties or suspension where the risk is low and does not require such a drastic response (*New Southern Railway Ltd v Quinn* [2006] IRLR 266).

There is no requirement for a pregnancy risk assessment to be in writing (*Stevenson v J M Skinner & Co* UKEAT/0584/07). In this case the employer had carried out a risk assessment in the form of meetings addressing particular concerns, used a generic pregnancy risk assessment, and kept a record of the risk assessment.

Pregnant workers, workers who have recently given birth or are breastfeeding who work shifts, some of which are at night, must be regarded as performing night work and therefore enjoy specific protection against the risks that night work is liable to pose (*Gonzáles Castro v Mutua Umivale* [2018] IRLR 1142). The chapter VULNERABLE PERSONS also provides information on these requirements.

Risks to new or expectant mothers

[R3027] The HSE website www.hse.gov.uk/mothers provides guidance on those risks which may be of particular relevance to new or expectant mothers. It advises that possible risks include:

Physical agents
- Movements and postures
- Manual handling
- Shocks and vibrations
- Noise
- Radiation (ionising and non-ionising)
- Compressed air and diving
- Underground mining work

Biological agents
- Infectious diseases

Chemical agents
- Toxic chemicals
- Mercury
- Antimitotic (cytotoxic) drugs
- Pesticides
- Carbon Monoxide
- Lead

Working conditions
- Facilities (including rest rooms)
- Mental and physical fatigue, working hours
- Stress (including post-natal depression)
- Passive smoking
- Temperature
- Working with visual display units (VDUs)
- Working alone
- Working at height
- Travelling
- Violence
- Personal protective equipment
- Nutrition

It also makes reference to the non-exhaustive list that can be found in the annexes of the Pregnant Workers Directive 92/85/EEC and European Commission guidelines.

HSE advice on working with display screen equipment (DSE) is that pregnant women can safely work with DSE. Scientific studies have not shown any link between miscarriages or birth defects and working with DSE.

Other factors associated with pregnancy may also need to be taken into account, such as morning sickness, backache, difficulty in standing for extended periods or increasing size. These may necessitate changes to the working environment or work patterns and the HSE advises that pregnant workers are entitled to more frequent rest breaks.

Despite the legal requirements, a 2016 Women and Equalities parliamentary committee inquiry into pregnancy and maternity discrimination heard that as many as 21,000 women leave their jobs each year because employers are failing to tackle the pregnancy and maternity risks they face at work. It also found that pregnant women and mothers were reporting more discrimination and poor treatment at work than they did a decade ago.

An Equality and Human Rights Commission survey of 3,000 mothers found two in five (41%) respondents felt there was a risk to, or impact on, their health or welfare. However, 38% said that their employer did not initiate a conversation about risks when they informed them of their pregnancy; almost one in five said they had identified risks that their employer had not; and 10% said that their employer had identified risks and had not tackled them.

Following its findings, the HSE revised its guidance for new and expectant mothers. A new risk assessment action flowchart outlines the action an employer must take (www.hse.gov.uk/mothers/flowchart.htm):

Stage 1: General Risk Assessment—Assess the risk to the health and safety of your employees, including females of child-bearing age and new and expectant mothers;
Stage 2: After notification—revisit your general risk assessment.

Even if no risk is identified, it advises employers they must still 'monitor and review this assessment regularly as circumstances may change'.

In May 2019, the TUC and campaign group Maternity Action warned that employers are still not doing enough to protect pregnant women at work and published new guidance, *Pregnancy, breastfeeding and health and safety—A guide for workplace representatives*.

This outlines the potential risks that can affect new and expectant mothers and their babies, the law and what employers should do, and provides examples practical measures employers can take.

The new guide can be found on the TUC website at: www.tuc.org.uk/resource/pregnancy-breastfeeding-and-health-and-safety.

The Women and Equalities Committee inquiry report, *Pregnancy and maternity discrimination*, can be found on the parliament website at: https://publications.parliament.uk/pa/cm201617/cmselect/cmwomeq/90/90.pdf.

Further information on the risks all of the above pose to new or expectant mothers is contained in the chapter dealing with **VULNERABLE PERSONS**.

Other vulnerable persons

[R3028] As well as children and young persons and new or expectant mothers, there are other vulnerable members of the workforce who merit special consideration during the risk assessment process. These include:

Other vulnerable persons [R3028]

- **Lone workers**
  Some staff members may work alone continuously or only part of the time. They may be working on the employer's own premises or outside, within the community. The level of risk may relate to the worker's location, the activities they are involved in or the type of people they might encounter.
  Among lone workers will be:
  — security staff;
  — cleaners;
  — isolated receptionists or enquiry staff;
  — some retail staff;
  — some maintenance workers;
  — staff working late, at nights or at weekends;
  — keyholders, particularly at risk if called to suspected break-ins;
  — delivery staff;
  — sales and technical representatives; and
  — property surveyors.

- **Disabled people**
  Regulation 25A of the *Workplace (Health, Safety and Welfare) Regulations 1992 (SI 1992 No 3004)* sets out that:

  > Where necessary, those parts of a workplace (including in particular doors, passageways, stairs, showers, washbasins, lavatories and workstations) used or occupied directly by disabled persons at work shall be organised to take account of such persons.

  In considering access in general, and also specific risks which may affect disabled people, account must be taken of all possible types of disability, including:
  — people lacking mobility, including wheelchair users;
  — those lacking physical strength, particularly in relation to manual handling;
  — people with relevant health conditions, especially on COSHH-related matters;
  — blind or visually impaired people;
  — those who have hearing difficulties; and
  — people with learning disabilities.

  While employers have particular duties towards their own employees, the risk assessment must also consider risks to disabled persons who are:
  — on work placements;
  — visitors;
  — service users; and
  — neighbours or passers-by.

  All types of risks must be considered but particular note must be taken of:
  — access posing particular dangers to disabled people;
  — the presence of hazardous substances;
  — potentially dangerous work equipment;
  — hot items or surfaces; and

- the need for special emergency arrangements for disabled people.

The HSE website has a section dealing with risk assessment and disability at www.hse.gov.uk/disability/. This makes clear that: 'Health and safety legislation should not prevent disabled people finding or staying in employment and should not be used as a false excuse to justify discriminating against disabled people.'

See also the EQUALITY ACT 2010.

- **Inexperienced workers**
  Some adult workers may, like young people, be lacking in experience of some types of work or lack awareness of existing or potential risks of particular work activities. Some may also lack a suitable degree of maturity. This must be taken into account during the risk assessment process, especially in situations such as:
  — training or retraining programmes;
  — work activities with rapid staff turnover; and
  — activities involving people with special needs.
  More information on the above is available in the chapter VULNERABLE PERSONS.

Who should carry out the risk assessment?

[R3029] The *Management of Health and Safety at Work Regulations 1999 (SI 1999 No 3242)*, Reg 7 require employers to appoint competent health and safety assistance. Those appointed must have sufficient training and experience or knowledge together with other qualities, in order to be able to identify risks and evaluate the effectiveness of precautions to control those risks. (*Regulation 7(8)* expresses a preference for such people to be employees as opposed to others eg consultants.) HSE guidance in *Managing for health and safety* (HSG65) is that: 'A risk assessment should be completed by someone with a knowledge of the activity, process or material that is being assessed. Workers and their safety representatives are a valuable source of information. If an adviser or consultant assists with the risk assessment, managers and workers should still be involved.'

And in the online guidance, *Health and Safety Made Simple*, it advises: 'As an employer, you must appoint someone competent to help you meet your health and safety duties. A competent person is someone with the necessary skills, knowledge and experience to manage health and safety. If you run a low-risk business, health and safety is something you can manage without needing to buy in expert help. Here you could appoint yourself as a competent person or one or more of your workers. However, if you are not confident of your ability to manage all health and safety in-house, or if you are a higher-risk business, you may need some external help or advice.'

HSE guidance on the training and qualifications required to carry out a risk assessment is as follows:

'You do not necessarily need specific training or qualifications to carry out a risk assessment . . . . You could appoint one or a combination of: yourself, one or more of your workers, someone from outside your business. You may need extra help or advice if you do not have sufficient experience or knowledge in-house. You may also need extra help if the risks are complex.'

In selecting people to carry out risk assessments, larger employers may set up risk assessment teams who might be drawn from managers, engineers and other specialists, supervisors or team leaders, health and safety specialists, safety representatives and other employees. Some team members may require some additional training or information about the principles of risk assessment and possibly technical aspects of the work activities to be assessed.

Many employers continue to utilise consultants to co-ordinate or carry out their risk assessments. A national register of occupational safety consultants, the Occupational Safety and Health Consultants Register (OSCHR), has been developed by the HSE and a network of professional bodies representing safety consultants.

This provides details of consultants who have:

- a status recognised by the professional bodies participating in the scheme;
- assessed experience and qualifications;
- demonstrated adequate continuing professional development;
- a commitment to abiding by their code of conduct;
- a commitment to providing sensible and proportionate advice; and
- appropriate insurance.

To be eligible to join the register, individual consultants must meet one of the following criteria:

- Chartered status with IOSH (Institution of Occupational Safety and Health); Chartered Member of the BPS (British Psychological Society); CIEH (Chartered Institute of Environmental Health); or REHIS (Royal Environmental Health Institute of Scotland) with health and safety qualifications;
- Fellow status with IIRSM (International Institute of Risk and Safety Management) with degree level qualifications;
- Chartered Member or Chartered Fellow of BOHS (British Occupational Hygiene Society) Faculty of Occupational Hygiene;
- Registered Member or Fellow status with IEHF (Institute of Ergonomics and Human Factors.
  The register is freely accessible and searchable for employers at www.oshcr.org.

The OSHCR Board announced in April 2019 that a strategic review of the register is being carried out. This will include 'an assessment of the skills, knowledge and behaviours required of consultants to ensure the integrity of the register and the quality of advice businesses can access, and options to both increase awareness of the register amongst businesses and enhance the customer service offered to consultants who apply to be included on the register'.

Whether the assessments are to be carried out by an individual or by a team, others will need to be involved during the process. Managers, supervisors, employees, specialists etc will all need to be consulted about the risks involved in their work and the precautions that are (or should be) taken.

Assessment units

[R3030] In a small workplace it may be possible to carry out a risk assessment as a single exercise but in larger organisations it will usually be necessary to split the assessment up into manageable units. If done correctly this should mean that assessment of each unit should not take an inordinate amount of time and also allows the selection of the people best able to assess an individual unit. Division of work activities into assessment units might be by departments or sections, buildings or rooms, processes or product lines, or services provided.

As an illustration a garage might be divided into:
- Servicing and repair workshop.
- Body repair shop.
- Parts department.
- Car sales and administration.
- Petrol and retail sales.

Relevant sources of information

[R3031] There is a wide range of documentation which might be of value during the risk assessment process including:
- *Previous risk assessments* – including risk assessments carried out to comply with specific regulations eg COSHH or the Control of Noise Regulations.
- *Operating procedures* – where these include health and safety information.
- *Safety policies and handbooks*
- *Training programmes and records*
- *Accident and Incident records*
- *Health and safety inspection or audit reports*
- *Relevant Regulations and Approved Codes of Practice* – Risk assessment involves an evaluation of compliance with legal requirements and therefore an awareness of the legislation applying to the workplace in question is essential.
- *Relevant HSE publications*
  The HSE publishes online advice and information providing guidance on a wide range of health and safety topics. Some of these relate to the specific requirements of regulations, others deal with specific types of risks or sectors of work activity.
  All these can be of considerable value in identifying which risks the HSE regards as significant and in providing benchmarks against which precautions can be measured. The guidance on specific types of

workplaces, which includes engineering workshops, motor vehicle repair, warehousing, kitchens and food preparation, and many others, should form an essential basis for those carrying out assessments in those sectors.

Details of HSE publications are available via the HSE Books website (https://books.hse.gov.uk/bookstore.asp).

- *HSE website*
  The HSE website at www.hse.gov.uk contains an ever-growing resource of guidance material. The Risk Management section provides further links to example risk assessments for a wide range of workplaces and an online risk assessment web tool for office and other environments which prompts employers to answer a series of questions that generates their risk assessment and action plan (www.hse.gov.uk/risk/assessment.htm).

  The tool was developed by the HSE in response to a recommendation in the 2010 Young Review of health and safety legislation, *Common Sense: Common Safety* (www.gov.uk/government/publications/common-sense-common-safety-a-report-by-lord-young-of-graffham) to simplify the risk assessment procedure for 'low-hazard' workplaces such as offices, classrooms and shops.

  Safety organisations including the Chartered Institute of Environmental Health (CIEH) and the Royal Society for the Prevention of Accidents (RoSPA) pointed out that while specific sectors may appear to present low hazard workplaces, even seemingly low hazard settings, such as offices, can have significant hazards associated with them.
- *Trade Association codes of practice and guidance*
- *Information from manufacturers and suppliers* – Equipment handbooks or substance data sheets.
- *General health and safety reference books*

Consider who might be at risk

[R3032] The risk assessment process must take account of all those who may be at risk from the work activities ie both employees and others. It is important that all of these are identified.

- *Employees*
  Different categories of employees to take account of might include:
  — production workers;
  — maintenance workers;
  — administrative staff;
  — security officers;
  — cleaners;
  — delivery drivers;
  — sales representatives;
  — others working away from the premises;
  — temporary employees.

  Some of these employees may merit special considerations:
  — children and young people (see **R3025**);

- women of childbearing age – potential new or expectant mothers (see **R3026** and **R3027**); and
- other vulnerable persons (see **R3028**);
- *Contractors and their staff*
  Contracted services might involve:
  - construction or engineering projects;
  - routine maintenance or repair;
  - hire of specialist plant and its operators;
  - support services including catering, cleaning, security and transport workers;
  - professional services including architects, engineers, trainers; and
  - supply of temporary staff.

  Contractors also have duties to carry out risk assessments in respect of their own staff.
- *Others at risk*
  The types of people who might be put at risk by the organisation's activities will depend upon the nature and location of those activities. Groups of people to be considered include:
  - volunteer workers;
  - co-occupants of premises;
  - occupants of neighbouring premises;
  - drivers making deliveries;
  - visitors (both individuals and groups);
  - residents – in the care or hospitality sectors for example;
  - passers-by;
  - users of neighbouring roads;
  - trespassers; and
  - customers or service users.

Identify the issues to be addressed

[R3033] The issues to be addressed during the risk assessment process should be identified. This may be done for the assessment overall or separately for each of the assessment units. It will be based upon the knowledge and experience of those carrying out the assessment and the information gathered together from the sources described earlier. It will create an initial list of headings and sub-headings for the eventual record of the risk assessment, although in practice this list is likely to be amended along the way.

These headings are likely to consist of the more common types of risk, such as fire, vehicles and work at height, together with some specialised types of risk associated with the work activities such as violence, lasers or working in remote locations. A checklist of possible risks to be considered during risk assessments is provided in **R3035** below. This includes some of the risks which have specific regulations associated with them.

In some situations, particular notes may also be made to check on the effectiveness of the precautions which should be in place to control the risks, such as standards of machine guarding, compliance with personal protective equipment (PPE) requirements or the quality and extent of training.

Variations in work practices

[R3034] Consideration should also be given at this stage to possible variations in work activities which may create new risks or increase existing risks. Such variations might involve:

- Fluctuations in production or workload demands.
- Reallocation of staff to meet changing workloads.
- Problems introduced by staff absences.
- Seasonal variations in work activities.
- Abnormal weather conditions.
- Alternative work practices forced by equipment breakdown or unavailability.
- Urgent or 'one-off' repair work.
- Work carried out in unusual locations.
- Difference in work between days, nights or weekends.

Further variations may emerge later in the assessment process.

Checklist of possible risks

[R3035] This checklist is intended to assist in identifying which issues need to be addressed during risk assessments. In some cases, the headings and/or sub-headings might be used in the form shown, in other cases it may be more appropriate to combine them or modify the titles.

Work Equipment
- Process machinery
- Other machines
- Powered tools

- Hand tools

- Knives/blades
- Fork-lift trucks
- Cranes

- Lifts
- Hoists
- Lifting equipment
- Vehicles

Other factors
- Vibration
- Lasers
- Ultraviolet/infrared radiation
- Work related upper limb disorders
- Stress
- Bullying and harassment
- Working Time (eg long hours/shift work/fatigue)
- Workloads

Services/power sources
- Electrical installation
- Compressed air
- Steam
- Hydraulics

Access
- Vehicle routes
- Rail traffic
- Pedestrian access
- Glazing

Work at height
- Scaffolding
- Mobile elevating work platforms
- Ladders and stepladders
- Falling objects
- Roof work

Work activity
- Burning or welding
- Entry into confined spaces
- Electrical work
-
- Handling cash/valuables
- Use of compressed gases
- Molten metal

External factors
- Violence or aggression
- Robbery
-
- Large crowds
- Animals
- Clients' activities

- Other pressure systems
- Buried services
- Overhead services

Storage
- Shelving and racking
- Stacking
- Silos and tanks
- Waste

Fire and explosion prevention
- Flammable liquids *
- Flammable gases *
- Storage of flammables *
- Hot work

* DSEAR requirements (see **R3014**)

Work locations
- Heat
- Cold
- Severe weather
- Deep water
- Tides
- Remote locations
- Work alone
- Homeworking
- Poor hygiene
- Infestations
- Work abroad
- Work in domestic property
- Clients' premises
- Site security
- Work-related driving

Risks/issues subject to separate/additional assessment requirements	
Hazardous substances (COSHH)	Lead
Noise	Asbestos work
Manual handling	Asbestos in premises
Display screen equipment workstations	Conformity of machinery
PPE needs	Major accident hazards (COMAH)
Fire risks and precautions	Ionising and electromagnetic radiation
Young workers New and expectant mothers	Artificial optical radiation
Dangerous Substances and Explosive Atmospheres (DSEAR)	Vibration

Carrying out the risk assessment

[R3036] Good planning and preparation can reduce the time spent in carrying out the risk assessment as well as enabling that time to be used much more productively. However, it is essential that time is spent in work locations, seeing how work is actually carried out, as opposed to how it should be carried out.

Observation

[R3037] Observation of the work location, work equipment and work practices is an essential part of the risk assessment process. Where there are known to be variations in work activities, a sufficient range of these should be observed in order to form a judgement on the extent of the risks and the adequacy of precautions. Evaluations can be made of the effectiveness of fixed guards, the suitability and condition of access equipment, compliance with PPE requirements, the observance of specified operating procedures or working practices and many other aspects. It should also be borne in mind that work practices may change once workers realise they are under observation. Initial or undetected observation of working practices may be the most revealing.

Discussions

[R3038] Discussions with people carrying out work activities, their safety representatives and those responsible for supervising or managing them are also an essential part of risk assessment. Among aspects of the work that might be discussed are possible variations in the work activities, problems that workers encounter, workers' views of the effectiveness of the precautions available, the reasons some precautions are not utilised and their suggestions for improving health and safety standards.

Tests

[R3039] In some situations, it may be appropriate to test the effectiveness of safety precautions such as the efficiency of interlocked guards or trip devices, the audibility of alarms or warning devices, or the suitability of access to remote workplaces including crane cabs and roofs. When carrying out such tests, care must be taken by those carrying out the assessment not to endanger themselves or others or disrupt normal activities.

Further investigations

[R3040] Frequently further investigations will need to be made before the assessment can be concluded. Such investigations may involve detailed checks on standards or records, or enquiries into how non-routine situations are dealt with. Examples of further checks or enquiries which might be appropriate are:

- Detailed requirements of published standards such as design of guards and thickness of glass.
- Contents of operating procedures.
- Maintenance or test records.
- Contents of training programmes or training records.

Once again, the potential list is endless although lines of further enquiry should be indicated by the observations and discussions during the initial phase of the assessment.

Notes

[R3041] Rough notes should be made throughout the assessment process. It will be on these notes that the eventual assessment record will be based. It will seldom be possible to complete an assessment record 'on the run' during the assessment itself. The notes should relate to anything likely to be of relevance, such as:

- Risks discounted as insignificant.
- Further detail on risks, or additional risks identified during the assessment.
- Risks which are being controlled effectively.
- Descriptions of precautions which are in place and effective.
- Precautions which do not appear to be effective.
- Alternative precautions which might be considered.
- Problems identified or concerns expressed by others.
- Related procedures, records or other documents.

Assessment records

[R3042] This phase of the risk assessment process concludes with the preparation of the assessment records. It may be preferable to record the findings in draft form initially, with a revised version produced once the assessment has been reviewed more widely and/or recommended actions have been completed.

Assessment records [R3043]

- *Decide on the record format*
  Many alternatives are available. Different types of records may be appropriate for different departments, sections or activities. Use of a 'model' assessment might be relevant for similar workplaces or activities. Some or all of the assessment record might be integrated into documented operating procedures.
- *Identify the section headings to be used*
  During the preparatory phase of the assessment a list of risks and other sources to be addressed in each assessment unit was prepared as an aide memoire. This list should now be converted into section headings for the assessment records. As a result of the assessment, some of the headings may have been sub-divided into different headings while others may have been merged.
- *Prepare the assessment records*
  The rough notes made during assessment must then be converted into formal assessment records. Preparing the records will normally require at least half the time spent in the workplace carrying out the assessment and sometimes might even take longer. It is wise not to allow too long to elapse between assessing in the workplace and preparation of the assessment record. Notes will seem much more intelligible and memory will often be able to 'colour in' between the notes.
- *Identify the recommendations*
  The assessment will almost inevitably result in recommendations for improvements and these must be identified. Some assessment record formats incorporate sections in which recommendations can be included but in other cases separate lists will need to be prepared. The use of risk rating matrices (see **R3043**) may assist in the prioritisation of recommendations.

Risk rating matrices

[**R3043**] Some organisations use risk-rating systems to help identify priorities. Most involve matrices, utilising a simple combination of the likelihood of a hazard having an adverse effect and the severity of the consequences if it did. Some use numbers to produce a risk rating, as in the example below:

Risk Assessment Matrix			Likelihood of adverse effect		
			Unlikely	Possible	Frequent
			1	2	3
Severity of consequences	Minor	1	1	2	3
	Moderate	2	2	4	6
	Severe	3	3	6	9

The numbers can be replaced by descriptions of the level of risk as shown in the next example:

Risk Assessment Matrix		Likelihood of adverse effect		
		Unlikely	Possible	Frequent
Severity of consequences	Minor	Low	Low	Medium
	Moderate	Low	Medium	High
	Severe	Medium	High	Very High

The danger of using such matrices is that time is often spent considering risk values at the expense of evaluating the effectiveness of the controls which ought to be in place. The HSE has urged businesses to spend less time in dotting 'i's and crossing 't's and more time on putting practical actions into effect.

After the assessment

[R3044] While the recording of the risk assessment is an important legal requirement, it is even more important that the recommendations for improvement identified during the assessment are implemented. This is likely to involve several stages.

Review and implementation of the recommendations

[R3045] There may be a need to involve others outside, and probably senior to, the risk assessment team. The reasoning behind the recommendations can be explained and various alternative ways of controlling risks can be evaluated. Changes to the assessment findings or recommendations may be made at this stage but the risk assessment team should not allow themselves to be browbeaten into making alterations they do not feel can be justified. Similarly, they should not hold back from making recommendations they consider are necessary just because they believe senior management will not implement them. The assessment team should carry out their duties to the best of their abilities in identifying the precautions necessary to comply with the law. The responsibility for achieving compliance rests with their employer.

Some recommendations may need to be costed in respect of the capital expenditure or staff time required to implement them. It is unlikely that all recommendations will be able to be implemented immediately. There may be a significant lead time for the delivery of materials or the provision of specialist services from external sources. Once costings and prioritisation have been agreed, the recommendations should be converted into an action plan with individuals clearly allocated responsibility for each element of the plan, within a defined timescale. Some members of the risk assessment team, particularly health and safety specialists, may have responsibility for implementing parts of the plan or providing guidance to others.

Recommendation follow-up

[R3046] Even in well-intentioned organisations, recommendations for improvement that have been fully justified and accepted are often still not implemented. It is essential that the risk assessment process includes a follow-up of the recommendations made. As well as establishing the improvements have been carried out, consideration should also be given as to whether any unexpected risks have inadvertently been created.

Once the follow-up has been carried out, the assessment record should be annotated or revised to take account of the changes made. If recommendations have not been implemented, there is a clear need for the situation to be referred back to senior management for them to take action to overcome the obstacles to progress.

Assessment review

[R3047] The *Management of Health and Safety at Work Regulations 1999 (SI 1999 No 3242)* state that assessments must be reviewed in certain circumstances (see **R3022** above). A frequency for reviews should be established which relates to the extent and nature of the risks involved and the likelihood of creeping changes (as opposed to a major change which would automatically justify a review).

The review may, however, conclude that the risks are unchanged, the precautions are still effective and that no revision of the assessments is necessary. The process of conducting regular reviews of assessments and, where appropriate, making revisions may be aided by the application of document control systems of the type used to achieve compliance with the quality management system ISO 9001 and similar standards.

Content of assessment records

[R3048]–[R3051] The *Management of Health and Safety at Work Regulations 1999 (SI 1999 No 3242), Reg 3(6)* states:

> Where the employer employs five or more employees, he shall record —
>
> (a) the significant findings of the assessment; and
> (b) any group of his employees identified by it as being especially at risk.

HSE advice on risk assessment records is as follows: Any record produced should be simple and focused on controls and any paperwork produced should help to communicate and manage the risks in the business.

'Record the significant findings', it advises. 'These should include a record of the preventive and protective measures in place to control the risks, and what further action, if any, needs to be taken to reduce risk sufficiently, for example health surveillance.'

It recommends using the HSE risk assessment template (www.hse.gov.uk/risk).

The essential content of any risk assessment record should be:

- Hazards or risks associated with the work activity.
- Any employees identified as especially at risk.
- Precautions which are or should be in place to control the risks with comments on their effectiveness.

- Improvements identified as being necessary to comply with the law.

Other important details to include are:

- Name of the employer.
- Address of the work location or base.
- Names and signatures of those carrying out the assessment.
- Date of the assessment.
- Date for next review of the assessment.

These can be provided as an introductory sheet.

Sensible risk management

[R3052] 'Health and safety' still gets blamed for nonsensical decisions which have nothing to do with the correct application of risk assessment and risk management principles. Often such decisions are made by people with no real knowledge of health and safety, sometimes just as an excuse for a particular course of action or inaction.

The HSE sets out that sensible risk assessment is about:

- ensuring that workers and the public are properly protected
- enabling innovation and learning not stifling them.
- ensuring that those who create risks manage them responsibly and understand that failure to manage significant risks responsibly is likely to lead to robust action
- providing overall benefit to society by balancing benefits and risks, with a focus on reducing significant risks - both those which arise more often and those with serious consequences
- enabling individuals to understand that as well as the right to protection, they also have to exercise responsibility

It is not about:

- reducing protection of people from risks that cause real harm
- scaring people by exaggerating or publicising trivial risks
- stopping important recreational and learning activities for individuals where the risks are managed
- creating a totally risk-free society
- generating useless paperwork mountains.'

See www.hse.gov.uk/risk/principles.htm.

References

HSE publications

[R3053] Many specific references are contained earlier in this chapter. The following general references are relevant in risk assessment and/or risk management.

INDG 163	Risk assessment – A brief guide to controlling risks in the workplace (2014)
INDG 449	Health and safety made simple – The basics for your business (2014)
INDG 417	Leading health and safety at work (2013)
—	The online resource *The health and safety toolbox: How to control risks at work* builds on the basics laid out in *Health and safety made simple* and provides the next level of advice to help businesses identify, assess and control common risks in their workplace.
HSG 65	Managing for health and safety (2013)

HSE website

[R3054] The HSE website contains a section dealing with risk management (www.hse.gov.uk/risk/index.htm).

HSE publications can be downloaded from the web pages.

Safe Systems of Work

Lawrence Bamber

Introduction to safe systems of work

[S3001] This chapter examines the need for safe systems of work ('SSWs') within the framework of an occupational safety and health management system ('OSHMS').

Indeed, the latest OSHMS standard to feature SSWs is the proposed ISO 45001: *Occupational health and safety management systems — requirements with guidance for use*, which is progressing through the ISO committee stages which includes a consultative process. It is likely that the final version of this standard will appear in late 2016 / early 2017.

The standard follows the "Plan, Do, Check, Act", approach which demonstrates an iterative process to be used by organisations to achieve continued improvement in OSH management.

Written safe systems of work are a vital part of the system's documented information. It is important to keep the level of complexity of such a documented information at the minimum level possible to assure effectiveness, efficiency and simplicity of the system so as to enhance understanding.

The extent of documented information — including written SSWs — contained in an OSHMS will differ from organisation to organisation depending on:

- the size of the organisation;
- the type of activities, processes, products and services;
- the complexity of the processes and their hazardous interactions; and
- the competence of persons involved.

In essence SSWs outline the control measures needed to avoid or reduce risk in the workplace and are usually one of the main results of a suitable and sufficient risk assessment.

This chapter considers:

- the components of a safe system of work – both legal and others (S3002–S3006);
- appropriate safe system of work required for level of risk involved (S3007);
- development of SSWs, including job safety analysis; preparation of job safety instruction and safe operating procedures (S3008–S3010);
- permit to work systems (S3011–S3022);
- isolation procedures (S3023).

This particular chapter will be of use to guide those managers and supervisors who have the responsibility to develop and utilise SSWs, including permit to work systems, within their areas of control. It will also be useful to occupational safety and health ('OSH') practitioners who have to advise line management on practical development and implementation of such systems.

Essentially, a safe system of work is written down, and chronologically lists the methods of doing a particular job in such a way so as to avoid or minimise risk within the workplace. The safe system of work is therefore a very important workplace precaution or control measure which should be known by all concerned, agreed with employees via joint consultation, understood via regular communications/toolbox talks, adhered to by all involved in the job concerned, and rigorously enforced by supervision within the workplace.

All employers should therefore utilise suitable and sufficient safe systems of work commensurate with the degree of risk highlighted in a risk assessment, especially when the legal requirement for SSWs are taken into account (see S3003 and S3004).

The pitfalls of not utilising SSWs within the risk control system include:

- an increase in the number of accidents/injuries/diseases;
- the increased risk of prosecution;
- increased employers' liability insurance premiums;
- a more dangerous workplace.

Remember – SAFE SYSTEMS OF WORK MEAN FEWER ACCIDENTS! To illustrate the down side of not utilising SSWs as part of an OSHMs, consider the case of a fatality in a Nottingham paint shop, outlined below:

Nottingham paint shop Leadmaster Ltd was fined £20,000 at Newark Magistrates' Court on 18 April 2005.

The HSE prosecution followed an investigation into the circumstances surrounding the death of Robert Fountain, who had been spray-painting a 300kg fabricated steel grid hanging from a forklift truck in the company's Newark paint shop on 17 June 2004. He was found by his son, Jason Fountain, trapped against a steel table by the grid, which had fallen on him.

Leadmaster Ltd is based on the Colwick Industrial Estate, Nottingham, and has a factory at Beaconside Works, Beacon Hill, Newark, which employs eight people and trades as Parsons (Newark).

The company admitted one charge under *HSWA 1974 s 2(1)*. It was fined £20,000, the maximum at magistrates' court, and ordered to pay costs of £2,922.

HSE Inspector David Appleton said:

> 'This was a tragic incident to a newly recruited employee. It would have been simple and cheap to prevent this death by using a rope or strap to secure the fabrication, or by placing it on trestles for painting. The incident underlines the need for a safe system of work to be devised for all tasks and for everybody involved to understand how to do the job safely.'

More recently, further prosecutions by HSE have highlighted how accidents could and should have been prevented via the use of written SSWS.

9th December 2008: the HSE warned construction companies to ensure they provide written safe systems of work after part of a man's leg was amputated when a tipper truck fell onto him, trapping him against a pile of brick rubble.

M J Curle Ltd of Shifnal were fined £5,000 and ordered to pay costs of £7,500 at Shrewsbury Crown Court after pleading guilty to breaching *section 3(1)* of the *HSW Act 1974*.

Principal Contractor, Anthony Wilson Homes Ltd of Shifnal pleaded guilty to breaching *sections 2(1)* and *3(1)* of *HSW Act 1974* and *Regulation 3(1)(a)* of the *Management of Health and Safety at Work Regulations 1999* (MHSWR).

The court imposed a fine of £25,000 with costs of £10,000 on Anthony Wilson Homes Ltd for its greater responsibility for the lack of site safety, including a lack of SSWS.

14th January 2009: The HSE have called on employers in the removal and haulage business to ensure that proper training and written safe systems of work are in place, even for routine tasks.

This message follows an accident in Louth, Lincolnshire in which a worker was crushed between a seventeen-tonne removals van and a brick wall.

Fox Group (Moving and Storage) Ltd were fined £3,515 and ordered to pay £2,000 costs by Skegness Magistrates Court after pleading guilty to breaching *MHSW 1999* and contravening *Regulation 9(1)* of the *Provision and Use of Work Equipment Regulations 1998* (PUWER) for failing to undertake sufficient risk assessment and training in the resulting SSWS for employees.

29th March, 2011: James Paterson Haulage Ltd appeared at Inverness Sheriff Court and pleaded guilty to breaching *section 2(1)* of the *HSWA 1974* and were fined £13,000. Mackay Steelwork and Cladding Ltd appeared at the same hearing and pleaded guilty to breaching *section 3(1)* of *HSWA 1974* and were fined £40,000.

These prosecutions followed a fatal accident to one of James Paterson's lorry drivers who was killed when two steel gates fell off his vehicle and landed on him during an inadequately planned lifting operation which took place during the delivery of twenty steel gates from Mackay's yard at Belney, Invergordon to a garden centre in Inverness.

When the delivery driver arrived at the garden centre, he removed the securing straps from his load and began to assist a forklift truck (flt) driver in unloading the gates. He directed the flt driver to ensure that the forks were positioned underneath one of the two stacks of gates. It is believed he then walked around to the far side of the lorry to place the straps inside a storage box located next to the vehicle's fuel tank.

As he was bending over to open the box, the flt prongs extended beyond the first stack of gates and struck the second stack, causing four of the gates to fall off the lorry. Two of the gates landed on the lorry driver who died at the scene as a result of fatal neck injuries.

HSE Inspector, Graeme McMinn, explained that both companies involved had failed to adequately liaise with each other or obtain enough information so as to ensure that a written safe system of work was in place, particularly in relation to the role that the lorry driver would play in the unloading of the gates. He went on to say that one method of doing this would have been to create a segregated, clear zone, designed to prevent workers from standing around the lorry as it was being unloaded.

Inspector McMinn said: 'Those involved in arranging and carrying out deliveries should exchange and agree information to ensure lorries can be loaded and unloaded in a safe manner. They must make sure a safe way of working is in place and that workers have clear responsibilities so everyone involved in the lifting operation knows what everyone else is meant to be doing and where they are meant to be.'

Components of a safe system of work

[S3002] The components of a safe system of work comprise of legal elements – common law (S3003) and statute law (S3004) – and other components which explain where SSWs fit into the overall framework of an occupational safety and health management system ('OSHMS') (S3006). Indeed SSWs are considered to be the cornerstone of all successful OSHMSs.

Legal components – common law

[S3003] The need for SSWs can be traced back to 1905 when Lord McLaren in *Bett v Dalmey Oil Co* (1905) 7F (Ct of Sess) 787 judged that:

> the obligation is threefold: the provision of a competent staff of men, adequate material, and a proper system and effective supervision.

In 1938, the common law duty of care was highlighted in the case of *Wilsons and Clyde Coal Co Ltd v English* [1938] AC 57 where Lord Wright said that:

> the whole course of authority consistently recognises a duty which rests on the employer, and which is personal to the employer, to take reasonable care for the safety of his workmen, whether the employer be an individual, a firm or a company, and whether or not the employer takes any share in the conduct of the operations.

This duty of care implies the need for SSWs and also that delegation to an employee does not remove the employer's duty to provide a safe system of work.

The question of what constitutes a safe system of work was further considered by the Court of Appeal in 1943 by Lord Greene, Master of the Rolls, in the case of *Speed v Thomas Swift Co Ltd* [1943] 1 All ER 539, who said:

> I do not venture to suggest a definition of what is meant by a system. But it includes, or may include according to circumstances, such matters as the physical layout of the job, the setting of the stage, the sequence in which the work is to be carried out, the provision of warnings and notices, and the issue of special instructions.

He also said that: 'a system may be adequate for the whole course of the job, or it may have to be modified or improved to meet circumstances which arise: such modifications or improvements appear to me equally to fall under the heading of system.'

He added that: 'the safety of a system must be considered in relation to the particular circumstances of each particular job.'

The conclusion here is that the safe system of work must be tailored to the job to which it applies by co-ordinating the work, planning the arrangements and layout, stating the agreed method of work, providing the appropriate equipment, and giving employees the necessary information, instruction, training and supervision. All the above are designed to prevent harm to persons at work.

Indeed, safe systems of work have aided the improvement in health and safety performance of a number of industries in recent years.

The Health and Safety Commission's, Paper and Board Industry Advisory Committee (PABIAC), welcome the evident improvement in health and safety performance in the paper industry. Statistics provided by the Confederation of Paper Industries (CPI) display a downward trend in the workplace accident rate and significant reductions in major injuries.

On 20 February 2007 PABIAC Chair and HSE Head of Manufacturing Sector, James Barrett, said: 'It is very encouraging to see the positive affect the PABIAC strategy has had on health and safety standards in the paper industry. By continuing our partnership approach between employer's, trade unions and HSE, we are confident that there will be further reductions in the rate of accidents across the industry.'

Director General of the CPI and PABIAC member Martin Oldman said: 'The papermaking industry has been working diligently over a number of years to significantly improve its health and safety performance and, whilst more can be done, we are heartened by these results and anticipate further progress throughout the industry'.

Since March 2004, members of the CPI Paper and tissue making sector have seen a 33% reduction in accidents in their businesses. The amount of major injuries has been reduced by 32% and the number of slips and trips is down by 30%. There has also been a 17% improvement in the number of manual handling accidents.

CPI Corrugated sector members have successfully reduced their accident rate by 48%. In the key areas of manufacturing and manual handling accidents have been reduced by 53% and 48% respectively.

CPI Recovered Paper Sector, who only recently signed up to the PABIAC strategy have already seen a fall of 25% in major industries between December 2004 and August 2006.

Legal components – statute law

[S3004] The above common law precedents (at S3003) were taken into account during the drafting of the *Health and Safety at Work etc. Act 1974* ('*HSWA 1974*').

Section 2(2)(a) of *HSWA 1974* requires 'the provision and maintenance of plant *and systems of work* that are, so far as is reasonably practicable, safe and without risks to health'.

Furthermore, *section 2(3)* of *HSWA 1974* requires employers having five or more employees to prepare a *written* health and safety policy statement, together with the organisation and *arrangements* for carrying it out.

These 'written arrangements' should include all relevant written safe systems of work.

This general duty has been tightened up and more clearly defined since the enactment of the *Management of Health and Safety at Work Regulations 1999 (SI 1999 No 3242)* ('MHSWR').

Specifically, *Regulation 3* of MHSWR (SI 1999 No 3242) requires 'suitable and sufficient' risk assessments. A safe system of work should be the end result of the vast majority of risk assessments.

Regulation 4 of MHSWR (SI 1999 No 3242) requires the 'Principles of Protection' to be applied in connection with the introduction of 'any preventive and protective measures'. These measures are control measures and are 'arrangements' within the written health and safety policy. These principles of protection are listed in *Schedule 1* to the Regulations and should be referred to when developing SSWs.

Further clarification of 'arrangements' is given in *Regulation 5* of MHSWR (SI 1999 No 3242): 'arrangements for the effective planning, organisation, control, monitoring and review of the preventive and protective measures'. This infers that as SSWs are part of the preventive and protective (i.e. control) measures, then they should be planned, monitored and reviewed.

An example of a company falling found of the duty to have a known, understood and adhered to safe system of work is outlined below:

A managing director was sentenced on 7 January 2005 to a 16-month custodial sentence following a prosecution brought by the Crown Prosecution Service (CPS). The case, heard at Manchester Crown Court, followed a police led, joint investigation with the HSE into the death of Mr Daryl Arnold on 11 June 2003.

Mr Arnold, aged 27, and several others had been employed by Mr Lee Harper of Cannock, Staffordshire, to remove and replace the roof of a warehouse on the Lynton industrial estate in Salford. No safe system of work had been prepared before the work began and no safety precautions were in place at the time of the incident. Mr Arnold had never worked on a roof before.

Whilst working on the roof, Mr Arnold stepped backwards onto a fragile roof light on an adjoining warehouse, which gave way. Mr Arnold fell approximately 6.75 metres landing on the ground floor directly below. He died as a result of his injuries.

Mr Harper, managing director of Harper Building Contractors Ltd of Cannock Staffordshire pleaded guilty to charges of manslaughter and a breach of *HSWA 1974 s 2*.

Pam Waldron, HSE's Head of Construction for Scotland and the North West, said:

> 'No penalty can make up for the loss of a loved one. However, Lee Harper's sentence properly reflects the seriousness of his failure to ensure that Daryl Arnold was safe

and HSE is pleased that the matter has been concluded. There was a fundamental failure to recognise that the roof included fragile roof lights that will not bear a man's weight. Moreover, the equipment to prevent people falling through fragile materials is readily available and relatively cheap. A sensible, straightforward approach to health and safety in managing risks on this job should have prevented this tragic death.'

HSWA 1974 s 2 states: 'It shall be the duty of every employer to ensure, so far as is reasonably practicable, the health, safety and welfare at work of all his employees.'

The breach of *HSWA 1974 s 2* by Mr Harper was brought against him by virtue of *HSWA 1974 s 37*. Mr Harper was the sole Director of Harper Building Contractors Ltd. *HSWA 1974 s 37* states: 'Where a "body corporate" commits a health and safety offence, and the offence was committed with the consent or connivance of, or was attributable to any neglect on the part of, any director, manager, secretary or other similar officer of the body corporate, then that person (as well as the body corporate) is liable to be proceeded against and punished.'

The arrangements for liaison between the police, CPS and HSE are set out in a '*Work-related deaths: A protocol for liaison*', which is available from HSE website at www.hse.gov.uk/pubns/misc491.pdf.

HSE's free leaflet '*Working on Roofs*' includes this advice and gives more information about preventing falls from a variety of types of roof work. '*Health and Safety in Roof Work*', HSG33 gives more detailed information and includes advice about other issues encountered whilst working on roofs.

The construction industry is one where there is a need for specific written safe systems of work to be in place. In the majority of cases, they may well be labelled as Method Statements. However, an absence of such systems increases the risks to construction workers.

"Nearly 1 in 3 construction refurbishment sites inspected put the lives of workers at risk", Stephen Williams, HSE Head of Construction said on 11 September 2007.

This startling figure comes after The HSE carried out over 1500 inspections as part of its rolling inspection programme, resulting in enforcement action on 426 occasions in just two months.

Stephen Williams said: 'We stopped work on site immediately during 244 inspections because we felt there was a real possibility that life would be lost or ruined through serious injury. It is completely unacceptable that so many lives have been put at risk. Our inspectors were appalled at the apparent willingness to ignore basic safety precautions The simple fact is that despite knowing what they should be doing, too many people are prepared to allow bad practices to continue, even though last year 39 people died on refurbishment, repair and maintenance sites. We are determined to tackle this issue head on and will continue to take enforcement action against those rogues who flout safety precautions. Let me be clear to all those who put lives at risk – we will continue to carry out further inspections and will take all action necessary to protect workers, including closing sites and prosecution'.

Work at height remains the biggest concern. Over half of the enforcement action taken during this inspection initiative was against dangerous work at height, which last year led to the death of 23 workers.

Stephen Williams continued: "My advice to those who work in the refurbishment sector is to plan work, use competent workers and if working at height use the right equipment and use it safely".

Welcoming the Secretary of State's decision to hold a construction forum to discuss safety standards in the construction industry, HSE confirmed that inspectors will continue to target falls and trips in the refurbishment sector as part of their ongoing work.

More guidance and advice is available at hse.gov.uk/construction/index.htm.

In order to ensure SSWs are fully implemented within the workplace, it is imperative that they are communicated to all concerned and enforced via adequate supervision.

Companies are reminded of the need to ensure proper training and supervision is given to employees following the death of an untrained demolition worker.

On 22 December 2006 HSE successfully brought criminal charges against five different parties after the death of Mr David Moran. Between them they were fined a total of £87,000 and ordered to pay £57,228 costs at Manchester's Minshull Street Crown Court.

David fell eight metres to his death when he stepped on a fragile roof light at 17 Chesford Grange in Warrington on 20 September 2002. David and another untrained demolition worker, Anthony Harris from Collyhurst, Manchester, were using the roof to access another roof on the site.

HSE construction inspector for Cheshire Nic Rigby who brought the cases to court said: 'This prosecution follows the tragic death of a young man on a site in Warrington. Unfortunately, his death is not unique: on average, one person is killed on a construction site in Great Britain every five or six days, and many more are seriously injured.'

David's employer Elmsgold Haulage Ltd of Clayton House, Piccadilly, Manchester and John McSweeney, of Willow Road, Prestwich, the Managing Director of Elmsgold Haulage Ltd – pleaded guilty to two charges under *HSWA 1974 s 2(1)* in that they failed to provide a safe system of work and failed to ensure that people working on site were properly trained and supervised, and a third charge under *Regulation 9(3) of the Lifting Operations and Lifting Equipment Regulations 1998* in that they failed to ensure that lifting equipment was properly examined and inspected.

Elmsgold Haulage was fined £10,000 for each charge and ordered to pay total costs of £9,756. Mr McSweeney was fined £5,000 for each charge and ordered to pay total costs of £5,000.

Demolition contractor Excavation & Contracting (UK) Ltd of Sandringham Avenue, Denton in Manchester, the principal contractor for the Chesford Grange project, and the company's former Managing Director Bernard O'sullivan, now living in Australia, pleaded guilty to a charge under *HSWA 1974 s 3(1)* in that they each failed to ensure that risks to non employees were adequately controlled.

Excavation & Contracting (UK) Ltd was fined £35,000 and ordered to pay £9,972 costs. Bernard O'sullivan was fined £20,000 and ordered to pay £30,000 costs.

Dennis O'Connor, of St James' Road, Orrell, Wigan, Elmsgold Haulage's site foreman pleaded guilty to a charge under *HSWA 1974 s 7* in that he failed to ensure the safety of other employees. He was fined £2,500 and ordered to pay £2,500 costs.

At an earlier hearing at Warrington Magistrate's Court on 31 January 2006, John Edge of Knight Frank, a property management company acting for the owner of Chesford Grange and planning supervisor for the project, pleaded guilty to two charges under Regulation 15 of the Construction (Design and Management) Regulations 1994 for which Knight Frank was fined a total of £7,000 plus full prosecution costs of £4,500.

Knight Frank operated as a partnership, which does not constitute a legal entity for the purposes of prosecution, and therefore the case was taken against one of the partners, Mr John Edge.

Another recent example of a heavy fine being levied because of a lack of a safe system of work was highlighted in the June 2008 edition of Health and Safety at Work journal.

City of York Council was fined £20,000 plus £20,423 costs after pleading guilty to a breach of section 2(1) of the Health and Safety at Work Act 1974 for failing to ensure employee safety.

This successful prosecution followed the death of a council worker which occurred when his ride-on lawnmower overturned on sloping ground. It was held that the Council had not properly assessed the risk and was also not following the mower manufacturer's safe operating instructions, which should have been referred to in any written SSW.

Frank Smith, aged 54, died on 19 May 2005, whilst using a Hayter ride-on mower to cut long, wet grass on an embankment at Water End in York City Centre. His job was to cut the top of the embankment and then to turn the mower and tidy up an area near a bridge.

There were no witnesses to the accident but the HSE believes that in turning the mower, Mr Smith lost control and he and the mower slid down the sloping bank, hitting a low retaining wall at the bottom. This caused the mower to flip over, throwing him out of the driving seat.

HSE Principal Inspector, Keith King, stated that the Council had no site specific assessment of the risks and has failed to implement a written safe system of work. It has a generic assessment – dated 1997 – which was considered to be 'quite old'. However, although this generic assessment identified the possibility of a mower turning over with fatal consequences, the Council had failed to act. The mowers were not equipped with either seat belts or roll-over protection.

Mr King stated that the main issue was the Council's failure to send anyone competent to assess the degree of the slope of the embankment. At 25 degrees, this was significantly greater than the 19 degree upper limit specified in the manufacturer's instructions for safe use of the mower.

Mr King continued by saying that the Council should have identified all areas of sloping ground so as to ensure that the right equipment was in use and it should also have provided proper supervision and guidance for employees. The only warning provided by the Council was the instruction to take extra care when working on slopes.

In May 2010 another *HSWA 1974 s 2(1)* breach involved the family partnership operating Brunton Farms of Brunton, Adberdeenshire, where a guilty plea resulted in a fine of £8,000 after a farm worker's toe was severed when his foot became trapped in a fertiliser spreading machine.

Two workers were spreading lime on a field when they noticed fertiliser was jamming the spreader. One stood inside the hopper of the spreader and used a broom handle to keep the fertiliser flowing smoothly. As the vehicle was driven over uneven ground, Mr Walis slipped and his foot got trapped in the mechanism of the spreader. The HSE investigation found the farm did not provide or maintain a safe system of work in relation to the spreading of the lime.

HSE Inspector, John Radcliffe, said that the farm worker's life will always be affected by this totally avoidable incident. He should never have been asked to put his safety at risk by standing inside the machine so close to the moving parts. If his employers had used their common sense, it would have been clear that they were asking their employee to do something very unsafe and he was likely to be injured.

Specific Legal Requirements for SSWs [S30044/2]

The *Confined Spaces Regulations, SI 1997/1713* require that work in confined spaces must be carried out in accordance with a safe system of work. The ACoP makes a clear statement on how safe systems of work should be specified (see CONFINED SPACES).

The ACoP recommends the use of a 'permit to work' procedure where there is a reasonably foreseeable risk — ie high, or very high risk — of a serious injury for an employee entering or working in a confined space (see CONFINED SPACES for further information).

The *Electricity at Work Regulations, SI 635/1989* provides for two basic approaches to achieving a safe system of work — regs 13 and 14 (see ELECTRICITY for further information).

The HSE has published guidance on safe systems of work in connection with electricity: HS(G)85 *Electricity at Work — Safe Working Practises*, 3rd Edition (2013).

Summary of legal components of SSWs

[S3005] From the common and statute law references above (S3004), it may be seen that the key elements of SSWs are:

- adequate plant and equipment:

- the right tools for the job;
- plant designed for safe access and isolation.
- competent staff:
  - properly trained and experienced in the work they do;
  - instructed clearly in the work to be done;
  - provided with the necessary information on substances, safe use of equipment etc.
- proper supervision:
  - to monitor the work as it progresses;
  - to ensure the safe system of work is adhered to.

Other components

[S3006] The bulk of the SSWs component parts are derived from the common and statute law considerations above.

The safe system of work may be therefore seen as the centre piece of a jigsaw which links together all the other component parts including:

- competent staff;
- premises;
- safe design;
- plant safeguards;
- safe installation;
- safe access/egress;
- tools and equipment;
- instruction;
- supervision;
- training;
- information;
- safety rules;
- planned maintenance;
- monitoring;
- PPE;
- environment.

Hence SSWs need to take account of all of the above components, and they need to be in writing.

The relative importance of each of the components will be task-dependent. Some tasks will be heavily dependent on the intrinsic safety built into the plant at the design stage. Other jobs will rely on the skills and expertise of the competent employee(s).

It may be of use therefore to consider the SSWs components under the sub-headings: hardware and software.

The 'hardware' includes: design, installation, premises/plant, tools and equipment, and working environment – heating, lighting, ventilation, noise control.

The 'software' comprises: planned maintenance, competent employees, adequate supervision, safety rules, training and information, correct use of PPE, and correct use of tools and equipment.

Which type of safe system of work is appropriate for the level of risk?

[S3007] Different jobs will require different systems, depending on the levels of risk highlighted via the risk assessment process.

A very low risk job may require adherence to safety rules or a previously agreed guide, whereas a very high risk job may well require a formal written permit to work system. Both types qualify as safe systems of work.

The following matrix may prove useful:

Risk level	Type of SSW
Very high	Permit to Work
High	Permit (or written SSW)
Medium	Written SSW
Low	Written SSW
Very low	Verbal (with written back-up, such as safety rules)

Once the SSWs are in place, there is a need to ensure that they are reviewed at least annually, so as to ensure:

- continued legislative compliance (N.B. any new legislation);
- continued compliance with most recent risk assessment;
- SSWs still work in practice;
- any plant modifications are incorporated;
- substituted (safer) substances are allowed for;
- new work methods are incorporated;
- advances in technology are exploited;
- control measures are improved in the light of accident experience; and
- continued involvement in, and awareness of the importance of, all relevant SSWs.

It is vital that regular feedback is needed to all concerned following any updates of existing SSWs.

In essence, SSWs are one of the most important foundations of any OSHMS. They need to be:

- known, i.e. brought to the attention of all relevant employees and third parties;
- well understood and agreed with by all concerned;
- adhered to by all concerned; and
- enforced by management and supervision.

Development of safe systems of work

[S3008] The steps in the development of SSWs of all types are as follows:

- Hazard identification.

- Risk assessment.
- Define safety operating procedures/methods ('SOP's).
- Implement agreed system.
- Monitor system performance.
- Review system (at least annually).
- Provide feedback on system updates.

Job safety analysis ('JSA')

[S3009] JSA is also known as job hazard analysis and/or task analysis. It is a technique that has evolved from the work study techniques of method study and work measurement which, in turn, formed part of the scientific management approach in the early twentieth century.

The work study engineer analysed jobs in order to improve methods of production by eliminating unnecessary job steps or changing those steps that were inefficient.

From an occupational safety and health ('OSH') viewpoint, JSA assists in the elimination of hazardous steps, thus eliminating or avoiding risky operations.

In essence, JSA is a useful risk assessment tool as it identifies hazards, describes risks and assists in the development of commensurate control measures.

The original work study approach made use of the 'SREDIM' principle:

Select	=	work to be studied
Record	=	how work is done
Examine	=	the total job
Develop	=	the best method for doing the work
Install	=	the preferred method into the company's way of working
Maintain	=	the agreed method via training, supervision etc.

Work study is used to break the total job down into its component parts and, by measuring (work measurement/time and motion) the quantity of work done in each of the component parts, the total job time could be established. If any of the component parts could be modified to make them more efficient or productive usually by saving time these improvements would be made as part of the overall study.

From measuring a number of similar tasks undertaken by different employees, standard times for each job component – and hence the total job-were derived. These were subsequently used in the development of payment and bonus systems.

JSA uses the SREDIM principle, but measures/assesses the risk content, as opposed to the work content, in each of the component parts of the job. From this detailed breakdown a safe system of work or safe operating procedure can then be compiled.

The basic JSA procedure is as follows:

(1) Select the job to be analysed (Select).
(2) Break the job down into its component parts in an orderly and chronological sequence of job steps (Record).
(3) Critically observe and examine each job step/component part to determine whether there is any risk involved (Examine).
(4) Develop control measures to eliminate or reduce the risk (Develop).
(5) Formulate written SSWs and job safety instructions ('JSIs') for the job (Install).
(6) Review SSWs at regular intervals to ensure continued utilisation (Maintain).

Use may be made of the following three-column layout in order to record the JSA:

Sequence of job steps	Risk factors	Controls advised

It is important to work vertically down the columns in their entirety, rather than to list one or two job steps and then work horizontally across the rows. Experience has shown that the former approach is much less time-consuming!

Once the job to be analysed has been selected, the next stage is to break the job down into its component parts or job steps. On average there should be approximately ten to twenty job steps. If there are more than twenty, too large a job for analysis has probably been chosen and one should therefore split it into two separate jobs. If there are less than ten job steps, then combine two or more jobs together.

The job steps should follow an ordered and chronological approach. Consider the following job – *changing a car wheel*:

The scenario is that you have come out of work and gone to your car where you left it on the works car park. It has a flat tyre. You are not allowed to call for assistance but you have to replace the wheel yourself! The sequence of job steps will be as follows:

(1) Check handbrake on and check wheels.
(2) Remove spare tyre from boot, check tyre pressure.
(3) Remove hub cap/wheel trim.
(4) Ensure jack is suitable and located on solid ground.
(5) Ensure jacking point is sound.
(6) Jack car up part way, but not so wheels leave ground.
(7) Loosen wheel nuts.
(8) Jack car up fully.
(9) Remove nuts, then wheel.
(10) Fit spare.
(11) Replace nuts and tighten up.
(12) Lower car.
(13) Remove jack and store in boot with replaced wheel.

Development of safe systems of work [S3009]

(14) Re-tighten wheel nuts.
(15) Replace trim/hub cap.
(16) Ensure wheel is secure before driving off.

Once column 1 has been agreed and completed, then commence listing the associated risk factors by completing column 2. Thereafter, assign commensurate control measures in column 3.

The complete JSA should look like this:

Sequence of job steps	*Risk factors*	*Controls advised*
1. Check handbrake, check wheels	Strain to wrist/arm	Avoid snatching, rapid movements
2. Remove spare tyre, check tyre pressure	Strain to back	Use kinetic handling techniques
3. Remove hub cap/trim	Strain, abrasion to hand	Ensure correct lever used
4. Ensure jack on solid ground	Vehicle slipping, jack sinking into ground	Check jack and ground
5. Ensure jacking point is sound	Vehicle collapse	Place spare wheel under car floor as secondary support
6. Jack car up part way, wheels still on ground	Strain, bumping hands on car/jack	Avoid snatching, rapid movements. Use gloves
7. Loosen wheel nuts	Hands slipping-bruised knuckles, strain	Ensure spanner/brace in good order; avoid snatching, rapid movements. Use gloves
8. Jack car up fully	Strain, bumping hands on car/jack	Avoid snatching, rapid movements
9. Remove wheel	Strain to back, may drop onto foot	Use kinetic handling techniques. Use gloves to improve grip
10. Fit spare	Strain to back	Use kinetic handling techniques
11. Tighten nuts	Hands slipping-bruised knuckles, strain	Use gloves, avoid snatching, rapid movements
12. Lower car	Strain, bumping hands on jack/car	Avoid snatching, rapid movements
13. Remove wheel and store with jack in boot	Strain to back	Use kinetic handling techniques
14. Re-tighten nuts	Hands slipping-bruised knuckles	Use gloves, avoid snatching rapid movements
15. Replace hub cap	Abrasion to hand	Use gloves
16. Ensure wheel secure	Vehicle collapse, loose wheel	Check wheel and area around car before driving off

The above example serves to illustrate what a typical JSA should look like. Any job/task within the workplace should be able to be recorded in this way and, ideally, displayed in close proximity to where the job is being undertaken.

The ultimate aim must be to undertake JSA on all jobs within the workplace and to communicate the findings and resultant control measures to all concerned. However, to assist in prioritising which jobs to analyse initially, the following criteria should be taken into account:

- past accident/loss experience;
- worst possible outcome/maximum potential loss ('MPL');
- probability of recurrence;
- specific legal requirements;
- newness of the job;
- number of employees at risk.

JSA may be activity-based or job-based.

Activity-based JSAs may cover:

- all work carried out at height;
- all driving activities:
  — internal (works transport);
  — external (driving on public highways);
- loading and unloading of vehicles.

Job-based JSAs may cover:

- the activities of a maintenance engineer repairing a roof;
- the activities of a chemical process worker obtaining samples for analysis.

The third column 'Controls advised' is in essence the job safety instructions ('JSI') and forms the basis of the written safe system of work.

Preparation of job safety instruction and safe operating procedures

[S3010] The purpose of job safety instructions ('JSIs') and/or safe operating procedures ('SOPs') is to communicate the safe system of work to all concerned – supervisors, employees and contractors.

For each job step (column 1 in table in **S3009**) there should be a corresponding control action (column 3) designed to reduce or eliminate the risk factor (column 2) associated with the job step. The chronological ordering of the control measures listed in column 3 becomes the JSI for that particular job.

Such JSIs should be utilised in OSH training and communication, both formal (classroom style) and informal (on the job contact sessions).

All managers and supervisors should be fully knowledgeable and aware of all JSIs and SSWs that are operational within their areas of control and they should ensure that they are adhered to at all times.

JSIs may be:

- verbal (for very low risk tasks) – a word in the ear from a supervisor reminding the employee of the do's and don'ts;

- written (for low, medium and some high risk tasks) – as part of an agreed safe system of work;
- written and incorporated into safe operating procedures (usually for medium and some high risk tasks); or
- as part of an all singing, all dancing formal permit to work system (usually for 'high' or 'very high' risk tasks).

From a practical viewpoint, to aid communication and use of JSIs and SSWs, they should be listed on encapsulated cards and posted in the areas where the jobs are to be carried out. They should also be individually issued to all relevant employees, who should receive training in their application and use. All such training and communication – whether formal or informal – should be recorded on individual training records.

The use of the three-column format to communicate JSIs and SSWs is well proven and is an excellent method of revitalising what in most organisations has become a rather moribund set of documents. As stated above, all jobs/tasks should be able to be fitted onto one or two sheets of A4 and use of the technique is a good way to regenerate interest in SSWs via a review.

Most organisations only review SSWs when things have gone wrong, e.g. as part of an accident investigation. This is negative and a form of reactive monitoring.

It is much better to be proactive and to review SSWs at least annually. This can be achieved by setting targets for managers and supervisors to review one safe system of work per month, together with relevant safety representatives, employees and/or contractors. Any findings or changes to the existing SSWs must rapidly be fed back to all concerned, so as to ensure continual improvement in OSH performance.

Permit to work systems

[S3011]

'A permit to work system (PTW) is a formal written system of work used to control certain types of work that are potentially hazardous. A permit to work is a document which specifies the work to be done and the precautions to be taken. Permits to work form an essential part of a safe system of work for many maintenance activities. They allow work to start only after safe procedures have been defined and they provide a clear record that all foreseeable hazards have been considered. A permit is needed when maintenance work can only be carried out if normal safeguards are dropped or when new hazards are introduced by the work. Examples are: entry into vessels, hot work, and pipeline breaking.'

(HSE, 1997)

When may PTW systems be required?

[S3012] A PTW system may be required in the following instances:

- where the risk assessment indicates 'very high' or 'high' risk;
- entry into confined spaces;

[S3012] Safe Systems of Work

- working at heights;
- working over water;
- asbestos removal;
- high voltage electrical work;
- complex maintenance work – usually involving mechanical/electrical/chemical isolation;
- demolition work;
- work in environments which present considerable health hazards, such as:
  — radiation work;
  — thermal stress (work in hot or cold environments);
  — toxic dusts, gases, vapours;
  — oxygen enrichment or deficiency;
  — flammable atmospheres;
- lone working;
- work on or near overhead travelling cranes (7 metre rule);
- work involving contractors/third parties on site.

Essential features of PTW systems

[S3013] HSE Guidance (1997) states that employers should incorporate the following essential features into bespoke PTW systems:

- provision of information;
- selection and training;
- competency-use of competent persons;
- description of work to be undertaken;
- hazards and precautions;
- procedures.

Provision of information in PTW systems

[S3014] The PTW system should clearly state how the system works in practice, the job(s) it is to be used for, the responsibilities and training of those involved, and how to check that it is working effectively.

There must be clear identification in the system of those persons who may authorise particular jobs and also a note as to the limits of their competency.

There should also be a clear identification of who is responsible and competent for specifying the necessary precautions, e.g. isolation, use of RPE/PPE, emergency arrangements etc.

The PTW system paperwork should be clear, unambiguous and not misleading.

The system should be so designed that it can be used in unusual circumstances, especially to cover contractors.

Selection and training for PTW systems

[S3015] Those persons authorised to issue permits must be sufficiently knowledgeable concerning the hazards and precautions associated with the plant/equipment/substances and the proposed work. They should have sufficient imagination and experience to ask enough 'what if' questions to enable them to identify all potential hazards. Lateral thinking is required here, especially if more than one trade is involved; or when own employees are working alongside contractors.

Ideally formal training should be required for all:

- permit issuers/authorisers;
- permit receivers/users;
- contractors' supervision;
- line managers/supervisors.

All staff and contractors should fully understand the importance and assurance of safety associated with PTW systems and should be 'competent persons' within the framework of the system.

What is a 'competent person'?

[S3016] Recent legislation, codes of practice and guidance refer to a competent person in connection with permit to work systems in particular, and the OSHMS in general.

Regulation 3 of the *MHSWR (SI 1999 No 3242)* requires employers to undertake suitable and sufficient risk assessments and *Regulation 7* of the same Regulations requires employers to appoint one or more competent persons to assist them in undertaking such assessments.

MHSWR '99 defines 'competent' as follows: 'a person shall be regarded as competent for the purposes of these Regulations where he has sufficient training and experience or knowledge and other qualities to enable him properly to assist in undertaking the measures referred to in these Regulations.'

This is especially important in connection with the authorisation and receipt of permit to work documentation. Specifically, permit authorisers and receivers need:

- a knowledge and understanding of:
  — the work involved;
  — the principles of risk assessment;
  — the practical operation of commensurate control measures;
  — current OSH best practice.
- the ability to:
  — identify OSH problems;
  — implement prompt solutions;
  — evaluate the PTW system effectiveness;
  — promote and communicate the workings of the PTWS to all involved/concerned.

PTW system: work description

[S3017] The permit documentation should clearly identify the work to be done, together with the associated hazards. Plans and line diagrams may be used to assist in the description, location and limitations of the work to be done.

All plant and equipment should be clearly tagged/identified/numbered, so as to assist permit issuers and users in selecting and/or isolating the correct piece of kit. All isolation switchgear should be similarly numbered or tagged. An asset register listing all discrete numbering should be maintained.

The permit should also include reference to a detailed method statement for the more complicated tasks. This should form part of the overall PTW system.

Hazards and precautions

[S3018] The PTW system should require the removal of hazards – risk avoidance being the best strategy. Where this cannot be achieved then effective control measures designed to minimise the risk should be incorporated.

Any control measures should ensure legislative compliance is achieved as an absolute minimum standard. Specific regulations that may apply include: *COSHH 2002*, *Confined Spaces Regulations 1997*, *Control of Asbestos Regulation, 2006*. Permit authorisers and users should be aware of all relevant legislation.

The permit should state those precautions that have already been taken before work commences, as well as those that are needed whilst work is in progress. This may include what physical, chemical, mechanical and electrical isolations are required and how these may be achieved and tested. Also, the need for different types of PPE and/or RPE should be specified.

Reference should also be made to any residual hazards that might remain, together with any hazards that might be introduced during the work – e.g. welding fume, vapour from cleaning solvents.

PTW system procedures

[S3019] The permit should contain clear rules about how the work should be controlled or abandoned in the case of an emergency.

The permit should have a hand-back procedure which incorporates statements that the maintenance work has finished and that the plant has been made safe prior to being returned to production staff.

Time limitations and shift changeovers should be built into the PTW system. Ideally the permit should only remain valid for the time that the authorisation/issuer is present on site.

There should be clear procedures to be followed if the work has to be suspended for any reason.

It is vitally important that there is a method for cross-referencing when two or more jobs (at least one of which is subject to a PTW system) may adversely affect each other.

A copy of the permit must be displayed as close as possible to where the work is being undertaken.

All jobs subject to a PTW system should be checked at regular intervals during the duration of the job, so as to ensure that the system is being followed, is still relevant and is working properly.

General application of PTW systems

[S3020] In order for the PTW system to be applied to risk reduction in the workplace there needs to be a well understood communication system in place that everyone concerned is aware of and becomes involved in.

Most organisations make use of PTW system forms which have been designed to take account of individual site conditions and requirements. In some cases there will be general permits; in other cases special permits will be required for specific jobs, e.g. hot work, entry into confined spaces. This enables sufficient emphasis to be given to the particular hazards present and the precautions advised.

Outline of a PTW system

[S3021] There are many and varied PTW system documents available in the OSH literature. The key headings/subject areas on PTW system documentation should include the following as a minimum requirement:

- permit title;
- permit number;
- date and time of issue;
- reference to other permits;
- job location;
- plant/equipment identification;
- description of work to be done;
- hazard identification;
- precautions necessary – before, during and after job/work;
- protective equipment;
- authorisation;
- acceptance;
- shift handover procedures;
- hand back;
- permit cancellation – work satisfactorily and safety completed.

Operating principles

[S3022] In operating a PTW system the following principles should be observed:

- Title of permit.
- Permit number (including reference to other relevant permits or isolation certificates).
- Precise, detailed and accurate information concerning:
  — job location;
  — plant identification;
- Clear description of work to be done and its limitations.
- Hazard identification – including residual hazards and hazards introduced by the work.
- Specification of plant already made safe, together with an outline of what precautions have been taken, e.g. isolation.

- Specification of what further precautions are necessary prior to the commencing of work-e.g. rpe/ppe
- Specification of what control measures should be in place for the duration of the work.
- Clear statement of when the PTW system comes into effect, and for how long it remains in effect. A re-issue or extension should take place if the work is not completed within the allocated time. Generally, the PTW system should only be valid for the time the authoriser is on site.
- The permit document should be recognised as the master instruction which, unless it is cancelled, overrides all other instructions.
- The authorised/competent person issuing the permit must assure himself/herself – and the persons undertaking the work – that all the precautions/controls specified as necessary to make the job, plant and working environment safe and healthy have in fact been taken. This inevitably should involve an inspection of the job location by the authoriser. The authoriser's signature on the permit document gives this assurance of safety to the people doing the work.
- The person who accepts the permit becomes responsible for ensuring that all specified safety and health precautions continue in force and that only permitted work is undertaken within the location specified on the permit. At this time, confirmation that all permit information has been communicated to all concerned should be obtained. A signature of the accepting person should also be on the permit document.
- A copy of the permit should be clearly displayed in the work area.
- If the permit needs to be extended beyond the initially agreed duration, e.g. because of shift changeover, then signatures of oncoming authorised/competent persons – together with that of the original authoriser – should be added to the appropriate section of the permit to confirm that checks have been made – again involving a visual walk-round inspection to ensure that the plant/equipment remains safe to be worked on. The new acceptor and all involved are then made fully aware of all hazards and associated precautions. A new expiry time is agreed. This should be built into relevant shift hand-over procedures.
- The hand-back procedure should involve the acceptor signing that the work has been satisfactorily completed. The authoriser of the permit also signs off certifying that the work is complete and the plant/equipment is ready for testing and recommissioning.
- The permit is then cancelled, certifying that the work has been tested and the plant/equipment has been satisfactorily completed. The copy permit should also be removed from the work location once the work has been completed.
- It is advisable to keep copies of each PTW system for a period of five years as these constitute maintenance records.

Isolation procedures

[S3023] SSWs and PTW systems frequently require the effective isolation of plant and associated services, including:

- electrical power;
- pneumatic/hydraulic pressure;
- mechanical power;
- steam;
- chemical lines.

The principles of effective isolation require four steps to be taken:

1. Switch off — at the appropriate operating control panels. It is a good idea to post warning signs on the panel to alert other workers that it is switched off.
2. Isolate — by creating a physical break or gap between the source of energy and the part to be worked on. This can be achieved by operating an electrical isolator, removing a section of pipe or blanking off with a plate. N.B. Removing pipe sections is preferable to blanking off.

   Reliance on control valves alone has been a frequent cause of accidents and is not considered to be effective isolation. From a mechanical isolation viewpoint, drive belts and linkages should be physically removed.
3. Secure — the means of isolation by locking off. Multiple hasp padlock devices (e.g. Isolok) can be used when several people are working on the same piece of kit. Each puts his own padlock on the isolator so that the power cannot be reinstated until all personal padlocks are removed. It is therefore imperative to have a suite of different keys/padlocks.
4. Check — the isolation is effective by trying all available operating controls.

Many maintenance procedures will justify either a written safe systems of work or a formal PTW system; both can have the isolation procedure incorporated into the documentation.

For further information on electrical safe systems of work and permit to work systems, see ELECTRICITY.

Further reading

[S3024] The following publications are available either via the HSE website www.hse.gov.uk or from HSE Books, Sudbury, telephone 01787 881165.

- HS G 250 Guidance on permit to work systems '*A guide for the petroleum, chemical and allied industries*' (2005).
- HS G 253 '*The safe isolation of plant and equipment*' (2006).
- '*Permit to work systems*', INDG98 (rev) (1997).
- '*Safe work in confined spaces: Confined Spaces Regulations 1997: Approved Code of Practice, Regulations and Guidance (3rd Edition)*, L101 (2014).
- '*A brief guide to working safely in confined spaces*', IND G 258 (rev. 1) (2013).

- 'Management of health and safety at work: Management of Health and Safety at Work Regulations 1999: Approved Code of Practice and guidance', L21 (rev) (2000).
- 'Emergency isolation of process plant in the chemical industry', CHIS 2 (1999).
- 'Work at Height Regulations 2005 (as amended), IND G 401 (rev. 2) (2014).
- 'Electricity at work. Safe working practices (3rd Edition) HS G 85 (2013).
- 'Safe working with flammable substances', IND G 227 (1996).
- 'Dangerous Substances and Explosive Atmospheres Regulations 2002'. Approved Code of Practice and guidance, L138 (2nd Edition) (2013).
- 'A guide to the Control of Major Accident Hazards, 1999 (as amended)'. Guidance on Regulations L111 (2006).
- 'Developing process safety indicators: A step by step guide for chemical and major hazard industries', HS G 254 (2006).
- 'Safe use of work equipment, Provision and Use of Work Equipment Regulations 1998'. Approved Code of Practice and guidance (rev), L22 (2008).
- 'Working alone: Health and safety guidance on the risks of working alone', IND G 73 (rev 3) (2013)
- 'Managing and working with asbestos, Control of Asbestos Regulations 2012' Approved Code of Practice and guidance, L143 (2nd Edition) (2013)

Statements of Health and Safety Policy

Andrea Oates

Introduction to statements of health and safety policy

[S7001] Employers with five or more employees have a statutory duty under s 2(3) of the *Health and Safety at Work etc Act 1974 (HSWA 1974)* to prepare a written statement of their health and safety policy, including the organisation and arrangements for carrying it out, keep it up to date, and bring the statement to the attention of their employees. Employers with less than five employees are excepted from this requirement, but it is important to note that all employers have a duty of care to protect their employees and others from harm arising from work activities.

This section sets out the legal requirements relating to health and safety policies, explains the essential ingredients that a policy should contain and provides checklists to assist in the preparation of policies. It also refers to further sources of guidance on the subject.

The legal requirements of statements of health and safety policy

[S7002] *Section 2(3)* of the *HSWA 1974* states:

> 'Except in such cases as may be prescribed, it shall be the duty of every employer to prepare and as often as may be appropriate revise a written statement of his general policy with respect to the health and safety at work of his employees and the organisation and arrangements for the time being in force for carrying out that policy, and to bring the statement and any revision of it to the notice of his employees.'

The *Employers' Health and Safety Policy Statements (Exception) Regulations 1975 (SI 1975 No 1584)* provide an exception from these provisions for 'Any employer who carries on an undertaking in which for the time being he employs less than five employees'.

Only employees present on the premises at the same time can be included (*Osborne v Bill Taylor of Huyton Ltd* [1982] IRLR 17).

This test case asked both what is an 'undertaking' and what does 'for the time being' mean? It involved a prosecution against employers running a chain of 31 betting shops in respect of one of their premises. The Divisional Court ruled that the issue was whether there was a single undertaking being carried on in 31 separate places, or 31 separate undertakings. It said that this was a matter of fact and degree. With regard to the interpretation of who is employed 'for the time being', the court ruled that this means 'at any one time' and therefore occasional relief workers in this case should not be included.

Directors, managers and company secretaries can be personally liable for a failure to prepare or to implement a health and safety policy (*Armour v Skeen* [1977] IRLR 310).

In this first case in which a manager was held personally liable for offences under HSWA 1974, the High Court found that a local authority director's failure to develop a written general safety policy for his department amounted to 'neglect' and he was therefore personally liable to prosecution.

Developing a health and safety policy

[S7002.1] A health and safety policy statement should be written by people within the organisation who know best how it operates. Although employers may seek external assistance and advice, they should also involve staff in developing the policy as they have day-to-day experience of the job. They are more likely to be committed to carrying out its aims if they have been involved in its development.

The policy should set out clearly to everyone what is expected of them in order to comply with its requirements. The aims of the policy should be linked to the level of risk, with risk assessment determining how explicit the policy needs to be in the 'arrangements' section of the policy (see S7003 below).

In large organisations with multiple sites or activities, having an overarching or corporate policy that covers general issues, with more detailed policies relating to individual sites or activities, can allow employers to tailor the 'organisation' section of the policy to the individual management of each site (see S7003 below).

The policy should be brought to the attention of all employees. It should also be monitored, through spot checks or safety inspections using checklists, audits and reviewing management reports and accident investigations for example, to ensure it remains effective.

Guidance produced for small businesses by the Health and Safety Executive for Northern Ireland (HSE NI) includes an eight-step flow chart showing where the health and safety policy statement fits into the overall management of health and safety. This is summarised below:

1. Assess the risks from the organisation's work activities, record the significant findings, develop an action plan and let staff know the outcome.
2. Develop a written health and safety policy which outlines the plan for managing health and safety. This should give details of who will be responsible for putting the policy into practice and set out the health and safety arrangements that are in place.
3. Decide on user-friendly rules and procedures for areas such as fire safety and evacuation in an emergency, manual handling, using work equipment, using hazardous substances and electrical safety.
4. Decide who will co-ordinate and manage the health and safety policy and procedures day-to-day – the business owner or director or a senior member of staff for example. Whoever has the responsibility for health and safety

5. Communicate with staff to make sure they know about the health and safety policy and the arrangements that have been put into place and share the findings of risk assessments and controls with all staff who may be affected. Not all staff may come into contact with the same hazards. Staff meetings are an ideal place for passing on this information and feedback on the improvements that could be made should be encouraged.
6. Monitor the arrangements against those you have set in your policy. Monthly workplace or equipment inspections and weekly fire equipment inspections can identify and deal with possible hazards before they become a problem and cause injury or ill-health.
7. Carry out thorough accident or 'near-miss' investigations to learn from the experience and make changes where necessary.
8. Regularly review health and safety arrangements (policy, procedures and risk assessments) to make sure they are still effective and all staff understand them. This should be done every six to 12 months or if significant changes are introduced.

Source: *Protect Your Profit* (www.hseni.gov.uk/sites/hseni.gov.uk/files/publications/%5Bcurrent-domain%3Amachine-name%5D/protect-your-profit.pdf).

Content of the policy statement

[S7003] The policy statement should have three main parts:

— *The Statement of Intent*
This should involve a statement of the organisation's overall commitment to good standards of health and safety and compliance with relevant legislation. While s *2(3)* of the *HSWA 1974* only requires the statement to relate to employees, most organisations also refer to others who may be affected by their activities eg contractors, clients and members of the public. This may be of particular relevance if the policy is being reviewed by other parties such as clients. In order to demonstrate commitment at a high level, the statement should be signed by the chair, chief executive or someone in a similar position of seniority.

— *Organisation*
It is vitally important that the individuals responsible for putting the policy into practice are clearly identified. This may be relatively simple in a small organisation but larger employers are likely to need to identify the responsibilities held by those at different levels in the management structure as well as those for staff in more specialised roles eg health and safety officers. In all cases the competent person (or people) who are to assist in complying with health and safety requirements should be identified [*Management of Health and Safety at Work Regulations 1999 (SI 1999 No 3242), Reg 7*]. The requirement to appoint competent people under the Management Regulations is set out in more detail in MANAGING HEALTH AND SAFETY.

— *Arrangements*

The practical arrangements for implementing the policy, such as emergency procedures, consultation mechanisms, and health and safety inspections should be identified in this section. It may not be practical to detail all the arrangements in the policy document itself but the policy should identify where they can be found eg in a separate health and safety manual or within risk assessment records, organisational procedures or similar documents.

More detailed guidance on what should be included in each of these parts is contained in the following sections, together with guidance on two other important issues:

- Communication of the policy to employees (see S7008).
- Revision of the policy (see S7009).

The statement of intent

[S7004] The statement should emphasise the organisation's commitment to health and safety. The detailed content must reflect the philosophical approach within the organisation and also the context of its work activities. Typically, it might include:

— a commitment to achieving high standards of health and safety in respect of its employees;
— a similar commitment to others involved or affected by the organisation's activities (possibly specifying them eg contractors, visitors, clients, tenants, members of the public);
— a recognition of the organisation's legal obligations under the *HSWA 1974* and related legislation and a commitment to complying with these obligations;
— a reference to the importance of people to the organisation and its moral obligations towards them;
— references to specific obligations eg to carry out risk assessments and provide safe and healthy working conditions, equipment and systems of work;
— a commitment to providing the resources necessary to implement the policy;
— a reference to the importance of health and safety as a management objective;
— a commitment to consultation with the organisation's employees – employees or their representatives should be involved in decisions that affect their health and safety;
— a reference to achieving continuous improvement in health and safety standards.

Most employers keep the statement of intent relatively short – less than a single sheet of paper. This allows it to be prominently displayed (eg in reception areas or on noticeboards), or to be inserted easily into other documents such as employee handbooks.

EXAMPLE:

The Midshires Housing Association (MHA) is committed to achieving high standards of health and safety not only in respect of its own employees but also in relation to tenants of the Association's property, contractors working on the Association's property or projects, visitors and members of the community who may be affected by the Association's activities. MHA recognises the importance to the Association of its employees and its moral obligations towards ensuring their health and safety.

The Association is also well aware of its obligations under the *Health and Safety at Work etc Act 1974* and related legislation and is fully committed to meeting those obligations. MHA will provide the resources necessary to implement this Policy and regards the successful management of health and safety as a key management objective. An annual health and safety plan will be prepared and implemented, with the aim of continuously improving health and safety standards. MHA supports the concept of consultation with its staff on health and safety matters and has established a Health and Safety Committee to provide a forum for such consultation.

The organisation and arrangements for the implementation of this Policy are set out in the full Health and Safety Policy document, which is available on the Association's Intranet. The Policy will be reviewed annually and staff will be informed of any revisions that are made.

A Champion

Chief Executive

October 2018

Organisation

[S7005] The statement of intent is of little value unless the organisation for implementing these good intentions is clearly established. This usually involves identifying the responsibilities of people at different levels in the management organisation and also those with specific functions eg, the health and safety officer. How these responsibilities are allocated will depend upon the structure and size of the organisation. The responsibilities of the managing director of a major company will be quite different from those of a managing director in a small business. The latter will quite often have responsibilities held by a supervisor in a larger concern eg ensuring staff are provided with and make proper use of suitable personal protective equipment (PPE).

Further advice on leadership for directors, governors, trustees, officers and their equivalents in the private, public and third sectors is available in the joint Health and Safety Executive (HSE) and Institute of Directors (IOD) publication, *Leading Health and Safety at Work* (INDG417), which is available to download from the HSE website at www.hse.gov.uk/pubns/indg417.pdf.

The document sets out the following essential principles:

- Strong and active leadership from the top, including visible, active commitment from the board; effective 'downward' communication systems and management structures; and the integration of good health and safety management with business decisions;
- Worker involvement: including engaging the workforce in the promotion and achievement of safe and healthy conditions; effective 'upward' communication; and providing high-quality training; and

- Assessment and review: including identifying and managing health and safety risks; accessing (and following) competent advice; and monitoring, reporting and reviewing performance.

Most organisations prefer to identify the responsibilities through job titles rather than by name as this avoids the need to revise the document when individuals change jobs or leave. Responsibilities might be allocated to levels in the organisation eg:

— Managing Director/Chief Executive
— Senior Managers
— Line Managers
— Supervisory Staff
— Employees

and also to those with specialist roles eg:

— Health and Safety Manager or Officer
— Occupational Hygienist
— Occupational Health Staff
— Personnel/Human Resources Specialists
— Maintenance and Project Engineers
— Training Manager or Officer

Where health and safety advice is provided from outside (eg a consultant or a specialist at a different location), this should be stated and the means of contacting this person for advice should be identified. The following examples are provided as an illustration of how responsibilities might be allocated in a medium-sized organisation. The allocation will be different for businesses of greater or lesser size.

EXAMPLES:
- Managing Director
  The Managing Director has overall responsibility for health and safety and in particular for:
    — ensuring that adequate resources are available to implement the health and safety policy;
    — ensuring health and safety performance is regularly reviewed at board level;
    — monitoring the effectiveness of the health and safety policy;
    — reviewing the results of health and safety audits and arranging the implementation of appropriate action;
    — reviewing the policy annually and arranging for any necessary revisions.
- Works Manager
  The Works Manager is primarily responsible for the effective management of health and safety within the workplace. In particular this includes:
    — delegating specific health and safety responsibilities to others;
    — monitoring their effectiveness in carrying out those responsibilities;
    — ensuring the company has access to adequate competent health and safety advice;
    — ensuring that safe systems of work are established for activities within the workplace;
    — ensuring that premises and equipment are adequately maintained;
    — ensuring that risk assessments are carried out;
    — ensuring that employees receive suitable and sufficient information and instruction;
    — ensuring that adequate health and safety training is provided;

Content of the policy statement [S7006]

- — ensuring that appropriate remedial action is taken following accident and incident investigations;
- — ensuring that contractors working within the workplace have satisfactory health and safety standards;
- — chairing the health and safety committee;
- — preparing an annual health and safety plan and co-ordinating its implementation.

• Supervisors
Each Supervisor is responsible for the effective management of health and safety within his or her own area or function. In particular this includes:
- — ensuring that safe systems of work are implemented;
- — enforcing personal protective equipment (PPE) requirements and appropriate standards of behaviour;
- — ensuring that staff are adequately trained for the tasks they perform;
- — monitoring premises and work equipment, reporting faults where necessary;
- — monitoring any contractors or visitors who may be present;
- — identifying and reporting health and safety related problems and issues;
- — identifying training needs;
- — investigating and reporting on accidents and incidents;
- — participating in the risk assessment programme;
- — setting a good example on health and safety matters.

• All employees
All employees have a legal obligation to take reasonable care for their own health and safety and for that of others who may be affected by their actions eg colleagues, contractors, visitors and delivery staff. Employees are responsible for:
- — complying with company procedures and health and safety rules;
- — complying with PPE requirements;
- — behaving in a responsible manner;
- — identifying and reporting defects and other health and safety concerns;
- — reporting accidents and near miss incidents to their supervisor;
- — suggesting improvements to procedures or systems of work;
- — co-operating with the company on health and safety matters.

• Health and Safety Officer
The Health and Safety Officer is responsible for co-ordinating health and safety activities and for acting as the primary source of health and safety advice within the organisation. These responsibilities specifically include:
- — co-ordinating the company's risk assessment programme;
- — acting as secretary to the Health and Safety Committee;
- — administering the accident investigation and reporting procedure;
- — liaising with the HSE, the company's insurers and other external bodies;
- — submitting reports as required by *Reporting of Injuries, Diseases and Dangerous Occurrences Regulations 2013* (RIDDOR);
- — co-ordinating the health and safety inspection programme;
- — arranging annual health and safety audits;
- — identifying health and safety training needs;
- — providing or sourcing health and safety training;
- — providing health and safety induction training to new staff;
- — identifying the implications of changes in legislation or Health and Safety Executive (HSE) guidance;
- — preparing and submitting progress reports on the annual health and safety action plan;
- — sourcing additional specialist health and safety assistance when necessary;
- — assisting the Works Manager in the preparation of the annual health and safety plan;
- — preparing regular (eg quarterly) progress reports on the health and safety plan.

Arrangements

[S7006] The duty to state the arrangements for carrying out the health and safety policy overlaps with the duty imposed by *Regs* 3 and 5 of the

[S7006] Statements of Health and Safety Policy

Management of Health and Safety at Work Regulations (SI 1999 No 3242) to carry out and record the results of risk assessments and to make, give effect to and record the arrangements 'for the effective planning, organisation, control, monitoring and review of the preventive and protective measures'. Rather than providing all the detail on arrangements within the health and safety policy, some employers prefer to provide information on where these details can be found, for example in:

— risk assessment records;
— health and safety manuals, handbooks or other documents.

This has the benefit of keeping the policy document itself relatively short – which helps to effectively communicate the policy to employees. Some key aspects of health and safety arrangements which should be included or referred to in this section of the policy (where appropriate) are those for:

— conducting and recording general risk assessments;
— specific risk assessments eg fire, control of substances hazardous to health (COSHH), noise, manual handling, display screen equipment (DSE), workstations, asbestos-containing materials;
— assessing and controlling risks to particular groups of workers, such as young workers and new or expectant mothers;
— establishing PPE standards and providing PPE;
— operational procedures (where these relate to health and safety);
— permit-to-work systems;
— routine training eg induction, operator training;
— specialised training eg first aiders, fork lift truck drivers;
— health surveillance;
— statutory examinations and inspections;
— consultation with employees eg health and safety committees, staff meetings, shift briefings;
— health and safety inspections;
— auditing health and safety arrangements;
— investigation and reporting of accidents and incidents;
— selection and management of contractors;
— control of visitors;
— fire prevention, control and evacuation;
— dealing with other emergencies eg terrorist threats and incidents, chemical leaks, flooding;
— first-aid arrangements;
— stress, bullying, harassment and violence;
— maintaining and repairing premises and work equipment;
— welfare facilities;
— the working environment;
— staff working elsewhere eg sales representatives, delivery staff, installation engineers, homeworkers, secondees, peripatetic staff; and
— work abroad.

If there is a safety committee, its constitution and terms of reference should be included in this section.

Ensuring arrangements work effectively

[S7007] *Regulation 5* of the *Management of Health and Safety at Work Regulations 1999 (SI 1999 No 3242)* requires employers to have appropriate arrangements for the effective planning, organisation, control, monitoring and review of preventive and protective measures ie to have a suitable 'management cycle' in place.

If a health and safety policy is to be meaningful and to have credibility with employees and others, these arrangements must actually work in practice.

In order to ensure this, the management cycle must be applied to the arrangements themselves. New procedures, systems etc must be properly thought through and introduced effectively, with particular attention being paid to communicating them to those affected. The monitoring and review of health and safety arrangements can be carried out through health and safety inspections and audits, and the activities of health and safety committees.

The chapter MANAGING HEALTH AND SAFETY provides more detail on how to ensure that health and safety arrangements work effectively while the chapter on RISK ASSESSMENT provides further guidance on the application of the management cycle.

Communicating the policy

[S7008] Employers must bring the policy (and any revision of it) to the notice of their employees. The most important time for doing this is when new employees join the organisation and communication of the policy should be an integral part of any induction programme. Young people, for example, may need to be reminded of their personal responsibilities at this stage, as well as any restrictions on what they can and can't do. Even experienced workers may need to be made aware that they are joining an organisation that takes health and safety seriously, as well as knowing what the PPE rules are and where they can obtain the PPE from.

Many employers provide their employees with a personal copy of the policy while others include it in a general employee handbook or display it prominently eg on noticeboards as well bringing it to employees' attention electronically. In all cases communication will be aided by keeping the policy itself relatively brief and providing the detail (particularly in respect of 'arrangements') in other related documents.

Intranet systems provide an excellent means of communicating the health and safety policy itself and giving links to related procedures, forms, risk assessments and other important health and safety documents. Such systems make it relatively easy to communicate changes to the policy and related documents and can also be used to record the fact that individual employees have been informed about the changes.

If there are workers who do not read English, employers will need to ensure that measures are in place to communicate the policy and any relevant supporting documents to them. The HSE advises that employers have a duty to provide comprehensible information to workers, although this does not have to be in writing, or necessarily in English. Other options it sets out include providing written information in relevant languages, ensuring that a competent translator familiar with any technical terms is used, or using non-verbal communications, such as DVDs or videos for example (see www.hse.gov.uk/migrantworkers/employer.htm#language).

Review

[S7009] The policy must be revised 'as often as may be appropriate' but the need for revision can only be identified through a review. The HSE does not stipulate a frequency for review but has in the past suggested carrying one out annually. This should not be too onerous even for small employers, for whom it should be a fairly simple affair. The HSE NI advises small businesses to regularly review health and safety arrangements to make sure they are still effective and all staff understand them every six to 12 months or if significant changes are introduced (see S7002 above). In larger organisations, the review could be an annual task of the health and safety committee. Alternatively, the review could be made the responsibility of an individual (eg the Managing Director or the Health and Safety Officer), either formally through specifying this within the organisational responsibilities or informally through a simple diary entry.

The review may well conclude that there is no need for change but it may identify the need for revision. For example:

— A new Managing Director should make their personal commitment to the statement of intent by signing and re-issuing it.
— Changes in the management structure necessitate a reallocation of responsibilities for health and safety.
— New 'arrangements' for health and safety have been established (or existing ones have been altered) and the policy needs to be amended to match this.
— Changes in health and safety legislation or HSE guidance.

A policy which is clearly well out of date does not reflect well on the organisation's commitment to health and safety, nor on the effectiveness of its health and safety management.

Contractors' Approval Systems

[S7010] Being able to demonstrate effective standards of health and safety management to others is of increasing importance to employers supplying services to others (see Introduction [IN15]: Supply-chain pressure). Client organisations such as major construction companies and local authorities increasingly require those wishing to work for them to submit details of their health and safety management arrangements as part of an approval system.

The High Court case *General Building and Maintenance plc v Greenwich Borough Council* (1993) 92 LGR 21, [1993] IRLR 535 looked at the extent to which an authority is entitled to take into account health and safety records in determining who should be permitted to tender. Dismissing an application by a company that did not win a tender because its health and safety policy was said not to have met the required standard, the case confirmed that health and safety matters could be considered in the selection of contractors putting in tenders for public works.

These, sometimes complex, questionnaires require information to be submitted on the contractor's health and safety policy — the three key elements (policy statement, organisation and arrangements), usually accompanied by details of how the policy is communicated to employees and how it is reviewed and revised.

An example of guidance to assist small businesses complete paperwork and comply with the health and safety requirements that are necessary for inclusion on approved contractors lists, *Guidance on completing a Contractor Appraisal Questionnaire*, can be found on the Health Scotland website at www.healthscotland.com/uploads/documents/15798-contractorAppraisal.pdf.

Checklist for writing and reviewing a health and safety policy statement

[S7011] Does the statement clearly express the organisation's commitment to health and safety?

Does it make clear its obligations to employees (and others who may be affected by its work activities)?

Does it set out who is responsible for implementing the policy and keeping it under review and how they will do this?

Has someone at the top of the organisation signed and dated the policy?

Have managers, supervisors, safety representatives and staff been involved in developing the policy and have their views been taken into account?

Are the people with duties and responsibilities set out in the statement clear about their duties and responsibilities?

Do they have sufficient resources (in terms of both time and money) to be able to carry out their roles?

Are the mechanisms for worker involvement clearly set out?

Does the policy clearly set out who has responsibilities in relation to specific areas of health and safety (such as first aid, reporting, investigating and recording injuries, ill-health and dangerous occurrences, carrying out risk assessments and health and safety inspections, providing health and safety information and training, fire safety, work equipment, PPE, electrical and gas safety, hazardous substances, noise, manual handling, stress, bullying, harassment and violence, manual handling, work at height, premises and the work environment, and welfare facilities)?

Further Guidance

[S7012] The free HSE leaflet *Health and safety made simple The basics for your business* (INDG 449) provides guidance on many aspects of health and safety. One section of the leaflet contains a template of a health and safety policy into which a small business can insert its own details as appropriate. As well as a statement of general policy, it provides sections into which details of both responsibilities and action and arrangements for the following can be inserted:

- Preventing accidents and cases of work-related ill health by managing the health and safety risks in the workplace.
- Providing clear instructions and information, and adequate training, to ensure employees are competent to do their work.
- Engaging and consulting with employees on day-to-day health and safety conditions.
- Implementing emergency procedures – evacuation in case of fire or other significant incident. You can find help with your fire risk assessment at: www.gov.uk/workplace-fire-safety-your-responsibilities.
- Maintaining safe and healthy working conditions, providing and maintaining plant, equipment and machinery, and ensuring safe storage/use of substances.

Where a small business needs to refer to important topics not contained in the above list, it will often be easier to deal with these through accompanying sections, rather than prepare a policy from scratch.

Important references in ensuring that adequate arrangements are in place for carrying out the policy are the HSE publication '*Managing for health and safety*' (HSG 65) (which can be downloaded at www.hse.gov.uk/pubns/books/hsg65.htm), the principles of which are summarised in the leaflet *Plan, Do, Check, Act – An introduction to managing for health and safety* (ING 275) (which can be downloaded at www.hse.gov.uk/pubns/indg275.pdf). These principles are also explained in more detail in the chapter MANAGING HEALTH AND SAFETY.

The Healthy Working Lives website, which provides employers in Scotland with workplace health, safety and wellbeing information, includes advice on writing a health and safety policy statement at www.healthyworkinglives.com/advice/Health-and-Safety-Quick-Start-Guide/Health-and-safety-policy-statement.

Several local authorities have also published advice for employers on writing health and safety policies.

Stress at Work

Nicola Coote

Introduction to stress at work

[S11001] Stress continues to be a concern for employers and employees alike. More seems to be demanded by organisations in less time. At work, life is becoming more competitive, with more people chasing the same goals.

People are very complex and the way in which they react to events within their lives varies considerably. Sometimes, a single event may cause a crisis in someone's life. How they respond will depend upon a number of factors, including other circumstances that may exist within their personal or professional life, their own reaction to the event and the length of time it takes to deal with the event. There are people who, when stressed, will make sure that everyone around them knows it. They will become withdrawn, and internalise their stress. Physically retreating from stressful situations, such as by taking frequent smoking breaks, is one way that some people cope. Others will become angry, they care little about anyone except themselves, and their tension is obvious. This may result in behaviour that appears to be aggressive or threatening. Indeed, in some cases the stress may build to an extent that someone resorts to expressing their feelings by using violent behaviour. For others, stress will be a result of them being exposed to violent or threatening situations. Examples where this may occur include: care workers, retailers, police and other emergency support workers, and call centre staff.

Definition of stress

[S11002] Taken in its simplest form, the Oxford English dictionary defines stress as 'demand upon physical or mental energy'. The HSE's definition of stress is given as 'the adverse reaction people have to excessive pressures or other types of demand placed upon them'. This has been seen by some as a rather negative view, as it implies that stress produces only adverse effects.

Stress often occurs when the pressures placed upon a person exceed their capacity or perceived capacity to cope. This can be negative or positive. For some people, the pressure of a tight deadline will 'positively stress' them, with the end result being that it enhances their performance. This may be due to their fear of failure or the consequences of not meeting the deadline. This is particularly true of people working in competitive sport and those people who set themselves challenges eg to increase sales figures by 50 per cent in the next 12 months. In these cases, the pressures placed upon them can have a positive effect in that it produces drive, challenge and excitement. Research has shown, however, the body suffers under prolonged periods of stress due to physical and mental overload. This is discussed below.

Understanding the stress response

[S11003] People's ability to respond to difficult or stressful situations has evolved over many million years. The mind recognises a particular need to respond to an external source (both physical and mental). The alert message is sent to the brain which then initiates dramatic changes in the hypothalamus. This is the control centre which integrates our reflex actions and coordinates the different activities within our bodies. When the hypothalamus is activated, the muscles, the brain lungs and the heart are given priority over any other bodily activity. The chemical adrenaline is produced and the message of a threat is passed on to other parts of the body which prepares itself for vigorous physical action – known as the fight or flight instinct. When the threat has abated, the parasympathetic nervous system takes over again and the body returns to a state of equilibrium, (known as homeostasis) where all our functions are in balance. This is demonstrated in the stress response curve.

Figure 1: Stress Response Curve

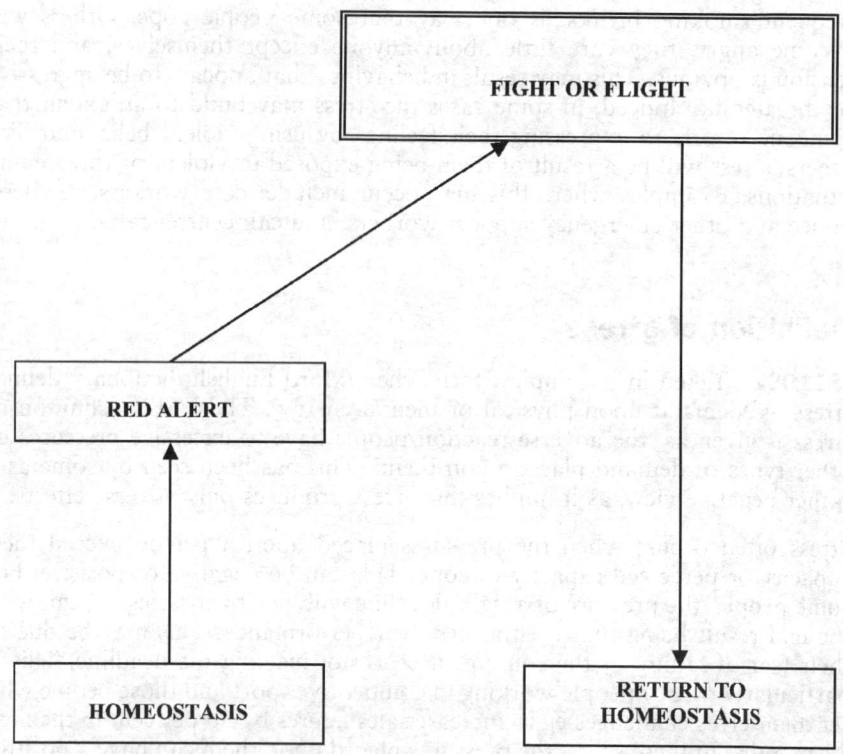

To summarise, given below are the physical responses to the 'fight or flight' mechanism:

- brain goes on red alert and stimulates hormonal changes including the production of adrenalin and noradrenaline (stress hormones);
- muscles tense, ready for action;
- eye pupils dilate to enable the danger to be seen more clearly;
- the heart beat increases to provide more blood to the muscles, therefore blood pressure increases;
- breathing increases to provide extra oxygen for the tensed muscles;
- liver releases glucose to provide extra energy for the muscles;
- digestive system shut down, making the mouth go dry and the sphincters close;
- sweating occurs in anticipation of expending extra energy; and
- immune system slows down.

These reactions are highly complex, but demonstrate the increased demands that are placed upon the body when feelings of stress occur. It is easy to see, therefore, how this response can have a detrimental effect upon well being and health if the stress is prolonged.

Why manage stress?

[S11004] The following paragraphs explain the importance of managing stress.

Legal issues

[S11005] There is no legislation that relates specifically to stress but the following Regulations touch upon the issue.

Health and Safety at Work etc Act 1974

[S11006] Under the *Health and Safety at Work etc Act 1974, s 2(1)* employers have a general duty to ensure, so far as is reasonably practicable, the health, safety and welfare of employees at work. This implies that measures should be in place to ensure that employees do not suffer stress-related illness as a result of their work.

Management of Health and Safety at Work Regulations 1999

[S11007] Under the *Management of Health and Safety at Work Regulations 1999 (SI 1999 No 3242), Reg 3* employers are required to undertake risk assessments, and this implies that they must also take into consideration the risk of stress-related ill health or unsafe working practices that may be caused or exacerbated by stress. Stress therefore needs to be considered when completing general risk assessments (eg a risk assessment of the work environment, or of a particular task) as people's responses may differ when they are feeling the symptoms of stress, or overload. Some organisations may decide to undertake a specific assessment of occupational stress within their workplace, ie factors associated with the work that are known to affect the way people feel and respond.

Working Time Regulations 1998

[S11008] The *Working Time Regulations 1998 (SI 1998 No 1833)* seek to reduce the effects on employees who are subjected to working long hours, or who have insufficient rest periods during and between work shifts (these issues are known to be occupational stressors). The Regulations are detailed, and cover minimum rest times during a shift (20 minutes every six hours), minimum rest time between shifts (11 hours) and maximum number of hours that can be worked within a weekly period (48 hours). The maximum number of hours may be calculated over a 17-week reference period.

There are a number of exclusions that apply to certain categories of personnel under the Regulations, and these continue to be under review in the European Courts. Detailed requirements apply to employers to maintain a record of hours worked for each employee who is covered by this legislation.

For further information see WORKING TIME.

Criminal and civil liability

[S11009] Whilst there is potential for prosecution in criminal law for failure to meet the above duties, the likelihood is that compliance will be driven by civil litigation. To date, there have been no prosecutions under criminal law for failure to manage stress at work. Enforcement of the general duty of care under health and safety legislation has been due to successful claims being made for compensation following allegations by an employee that they have suffered illness or ill-health as a result of being exposed to unacceptable levels of stress at work.

Indeed, there have been some large insurance settlements for employees who have successfully claimed for stress at work. The most notable was *Walker v Northumberland County Council* [1995] IRLR 35. In this case, Mr Walker, a senior social worker, had been employed by the council for 17 years. During the 1980s his workload had increased, and as a result, he suffered a nervous breakdown. Upon his return to work he told his employers that his workload must be reduced. This was not done, and Mr Walker suffered another nervous breakdown which resulted in him being dismissed for permanent ill health. He claimed compensation from his employer, and was successful on the grounds that the employer had been aware that he was under extreme pressure at work. A key question in the case was whether Mr Walker's risk of mental illness was materially higher than that which would normally affect a senior social worker with a heavy workload.

As regards the second breakdown, the employer should have foreseen a risk that Mr Walker's career would end and they should have appreciated that Mr Walker was more vulnerable to damage than he had been before his first breakdown. Because of this, it was deemed reasonable for Mr Walker to have received assistance, but in not providing such assistance, the council was in breach of their common law duty of care.

In February 2002, a landmark decision by the Court of Appeal in *Hatton v Sutherland and Others* [2002] EWCA Civ 76, [2002] 2 All ER 1 overturned three out of four stress claims and quashed the awards for damages initially given to the workers in the High Court. The Appeal Court ruling stated that for a compensation claim to succeed it must first be 'reasonably foreseeable' to the employer that the employee concerned would suffer a psychiatric injury as a result of stress in the workplace. When deciding whether a risk is 'foreseeable' the following needs to be taken into account:

- nature and extent of the work undertaken by the employee (eg is the workload more than is normal for that particular job?); and
- signs from the employee themselves of impending harm to their health as a result of work-related stress (eg has an employee already suffered harm to their health from work-related stress?).

The Appeal Court ruling further stated that employers are entitled to assume that an employee can withstand the normal pressures of a job, except where they are aware of a particular problem or vulnerability (eg the employee has warned the employer that they are suffering from work-related stress). In addition, employers are entitled to take at face value what they have been told by their employees about the effects of work-related stress on their mental health. For example, if an employee has had sickness absence due to stress but does not inform their employer of the cause of the absence, then it would be reasonable for the employer to expect that the worker would be fit to return to their normal work.

The Court of Appeal set out a number of factors that must be taken into account by the courts once it has been established that it was reasonably foreseeable that an employee would suffer harm as a result of work-related stress. These include:

- an employer is only in breach of their duty of care if they fail to take reasonable steps to prevent the employee from suffering as a result of stress;
- indications of impending harm to health must be plain enough for any reasonable employer to realise;
- when considering what actions are reasonable for an employer to take to reduce risk of harm from stress, the size and resources of the employer's operation may be taken into account;
- it is not enough for a claimant to show they are suffering harm from a result of occupational stress – they must demonstrate that their 'injury' was caused by the employer's breach of duty of care by failing to take reasonable steps to protect the worker;
- an employer will not be deemed to be in breach of their duty of care if they allow a willing employee to continue with their job, despite them stating they have been suffering from stress, if the only reasonable alternative is to dismiss the employee; and
- if an employer offers a confidential staff counselling service for those who experience stress at work with referral to treatment services, they will be unlikely to be found in breach of their duty of care.

The court ruling places more responsibility on the employee to let an employer know if they are suffering from symptoms of occupational stress, rather than leaving it totally to the employer to identify a 'stressed' employee. In practice, therefore, an employee is unlikely to have a valid claim unless they can demonstrate that they have told their employer that they are suffering from work-related stress. Note that this does not take away the employer's duty to be proactive and seek to identify stress risk factors in the workplace, and to implement suitable precautions to minimise these risks.

This ruling was used in the case of *Pratley v Surrey County Council* [2003] EWCA Civ 1067, [2004] ICR 159 when a stress compensation claim made by Ms Pratley failed because she had not told her County Council employer that she was feeling under pressure.

Another significant ruling emerged from the case of *Barber v Somerset* [2004] UKHL 13, [2004] 2 All ER 385. Mr Barber was a teacher in East Bridgewater Community School and made a claim for the damages caused to him by suffering a mental breakdown which he attributed to the pressures and stresses of his workload. He was successful in his claim and was awarded £91,000 for loss of earnings plus £10,000 for pain, suffering and loss of amenity. Somerset County Council took the case to the Court of Appeal on the advice on their insurers and were successful in overturning the decision. However, it was referred to the House of Lords who overturned the decision on 1 April 2004 but reduced the damages by almost £30,000. It is important to note that the House of Lords broadly supported the guidance given by the Court of Appeal in dealing with cases concerning stress at work and that the principles established by the Court of Appeal were still valid. The main rulings from this lengthy case were:

- unless the employer knows of some particular problem or vulnerability he is usually entitled to assume that the employee is up to the normal pressures of the job;
- employers are generally entitled to take what they are told by or on behalf of the employee at face value; and
- an employee who returns to work after a period of sickness without making further disclosure or explanation to his employer is usually implying that he believes himself fit to return to the work which he was doing before.

The House of Lords decided on the basis of facts in this particular case that Mr Barber's employer was in breach of its duty of care by failing to take steps to lessen job-related stress that could lead to psychiatric illness, and thus made a compensatory award of £72,500.

The Court of Appeal's judgment gave useful guidance to employers, with practical steps that could be taken in case of complaints relating to psychiatric illness brought about by stress at work.

For an employee to succeed with such a claim, the employer must be able to reasonably foresee whether the employee would suffer from psychiatric harm which could be attributable to stress at work. This of course depends on what the employer knows (or ought to reasonably know) about the individual employee and their illness.

In general, the employer is entitled to take what they are told by the employee at face value, unless they have good reason to think to the contrary. They do not generally have to make searching inquiries of the employee or seek permission to make further enquiries of the employee's medical advisers. However, the employer should always seek further medical advice to clarify the nature of the illness and any recommended actions.

Another case tested the defence of reasonable foreseeability. In *Hartman v South Essex Mental Health and Community Care NHS Trust* [2005] EWCA Civ 6, [2005] ICR 782 the Court of Appeal reviewed the general principals that were considered in Barber v Somerset. In the *Hartman v South Essex* case, the court did not accept that the employer knew the claimant was vulnerable because she had disclosed the fact that she had suffered a nervous breakdown and was taking medication when completing a confidential questionnaire produced by the occupational health service. The court decided that it was not appropriate to attribute to the employer knowledge of the contents of information submitted by the employee in a confidential medical document to the occupational health department. In addition, the Court of Appeal commented that the fact that an employer offers an occupational health service does not mean that the employer had foreseen the risk of psychiatric injury owing to stress at work to any individual or class of employee.

The case of *Hone v Six Continents Retail Ltd* [2005] EWCA Civ 922, [2006] IRLR 49 is one of many decided cases in this area. The Court of Appeal found unanimously in favour of the Claimant and dismissed his employer's appeal. In this case the claimant was a pub manager who had been working for excessive numbers of hours with little support.

The Court of Appeal held in the case that the test laid down in the *Hatton* case (see above) which further stated that a duty to take steps would be triggered when 'the indications of impending harm to health arising from stress at work must be plain enough for any reasonable employer to realise that he should have done something about it'. The case therefore supports the notion that a successful claim for occupational stress can be mounted from a breach of the *Working Time Regulations*.

In recent years there have been a number of cases where claimants have alleged that occupational stress induced by their employer has caused or triggered psychiatric health conditions requiring inpatient treatment in a mental health hospital. Other cases have related to pre-existing health conditions which claimants have alleged have been triggered by their employment. The case of *Daniel v Secretary of State for the Department of Health* [2014] EWHC 2578 (QB), [2014] All ER (D) 290 (Jul) demonstrates these points. In this case the claimant had a pre-existing history of bipolar disorder (which carries a very high rate of recurrence) and she claimed that she suffered occupational stress caused by victimisation, bullying and work overload which resulted in her becoming an inpatient at a mental health hospital. It was contended by the defendant that there was no bullying and that the defendant had willingly undertaken additional work responsibilities. They further defended that there was no foreseeable risk of injury, that if a duty by the employer had arisen there was no breach, and that causation had not been established. In this case the principal issues were whether there was a foreseeable risk that the claimant would injure her mental health arising from her work and whether impending harm to her health from the stress was plain enough for any reasonable employer to realise that something had to be done about it. The claim was dismissed as no breach of duty of care arose. Established principles from earlier cases resulted in the decision that the conduct being complained about, when taken properly in context, had not amounted to conduct which was either genuinely offensive or oppressive and unacceptable or otherwise to amount to bullying. Rather, it was the pressures of her job and her own perception of the conflict which was one of the causes of the stress she had suffered at work.

An interesting case arose in 2012 *(MacLennan v Hartford Europe Ltd* [2012] EWHC 346 (QB), [2012] 11 LS Gaz R 23, [2012] NLJR 363) where the claimant argued that working long stressful hours lowered her immune system to such an extent that she developed Chronic Fatigue Syndrome. The Queen's Bench Division threw out the claim as the claimant was unable to establish a causal link between the stress and the medical condition. The case was a useful demonstration of the relevant legal points that must be established and acts as a reminder that unless a job is identified as putting a person at greater risk of injury, or if that person is particularly vulnerable to occupational stress, the employer can assume that the employee can cope with the normal stresses of the work.

There have also been a number of Employment Tribunal cases relating to stress at work. An example occurred in 2012 *(Santos v Disotto Foods Ltd* (2014) UKEAT/0623/12, [2014] All ER (D) 139 (Apr), EAT). In this case the worker (Mr Santos) had worked for the company since its infancy in 1994. He was a factory manager by 2009, working 12 hours per day. His contractual weekly working hours were 57. He had not signed an opt-out of the maximum 48-hour working week. He fell ill with stress in April 2009 and had two weeks off work.

On his return to work, Mr Santos agreed to change his job to warehouse manager, as part of a reorganisation of the expanding business. This involved a reduction in pay from £43,000 to £35,000 and his contractual hours were reduced from 57 to what was described as "40 plus".

From June 2009, the employer embarked on disciplinary action over a number of issues. Mr Santos was dismissed after an incident in November 2010. He was accused of not following a management instruction to load a vehicle with what appears to have been four banoffee pies. He had been at work from 6am that day, going home at 5pm for an hour and a half to see his children. He returned to work at 6.30pm. At 10pm, he failed to load a vehicle as instructed, claiming later that he was unsure of which vehicle he was supposed to put the pies on. He completed the loading the next morning, which the tribunal assumed he did when he started work at 6am.

The employment tribunal upheld Mr Santos's unfair dismissal claim. It decided that the employer's decision to dismiss was one that no reasonable employer would have taken. Mr Santos was in a position of responsibility and had chosen the best way to carry out the instruction he was given. He was tired, having worked 13 or 14 hours that day. Despite his reduction in pay, he was still often required to work 11 hours per day. The incident for which Mr Santos was dismissed was "so slight a matter that no reasonable employer could reasonably contemplate dismissing an employee because of that matter even when the previous disciplinary issues, which, in themselves, had not justified dismissal, were taken into account".

The employment tribunal awarded Mr Santos £59,315. This included a basic award of £7,200 and loss of earnings of £32,270. The compensation was "grossed up" to take account of the tax that would have to be paid on the award.

See also OCCUPATIONAL HEALTH AND DISEASES O1052.

Recent research on occupational stress

[S11009.1] There have been numerous studies in recent years to identify if workplace stress exists and to what extent it is affecting the health and wellbeing of the workforce. There are too many to include in this chapter but details of some of the reports can be found on the HSE website (www.hse.gov.uk/stress). This website includes a range of useful theoretical and practical information, including statistical information (the most recent research on this website is from 2004 to 2007), HSE Management Standards introduced in 2009 and an occupational stress assessment tool.

One of the more notable pieces of research is the *Beacons of Excellence* report which describes the authors' work to identify good practice in stress prevention and then to identify organisations within the UK that could be called 'beacons of excellence' in comparison with this model. The length of programmes reviewed varied from a few hours to a ten-year longitudinal study. The report summarises and draws conclusions from all of the substantive academic studies on stress prevention over the last decade and uses this information, as well as advice from a panel of international experts, to develop a comprehensive stress prevention model. This model is then used to describe examples of stress prevention practices within a wide range of UK organisations. The full report can be viewed via www.hse.gov.uk/research/rrhtm/rr133/htm.

The *Beacons of Excellence* report states that organisational strategies to date have concentrated on employers providing access to specific services, with the intention of assisting employees during stressful periods. These strategies have been referred to as primary, secondary and tertiary levels of stress intervention (Murphy 1988).

Primary interventions seek to eliminate sources of stress by changing the physical and socio-political environment so it meets more with employee needs and provides more control over individuals' work. They also focus on improving communication, involving employees in decision-making processes and redesigning jobs to suit the needs of individuals.

Secondary interventions seek to help employees manage their stress but do not try to eliminate or modify workplace stressors. Such stress management programmes will help workers to identify stress within themselves and to cope it with more effectively.

Tertiary prevention strategies seek to help people who are experiencing on-going problems either from their working life or work environment. The idea behind these strategies is to modify individual lifestyle and behaviour but to make little or no change to organisational practices.

According to Kompier and Cooper (1999) stress intervention practice is currently focusing more on secondary and tertiary strategies rather than working on the primary prevention techniques, and the *Beacons for Excellence* study reiterated this view. Reasons for this are given in detail in the report. It states that the benefits of implementing stress prevention and management strategies are likely to vary according to the management commitment in implementing programmes that are suitably developed, and with the involvement of its employees.

Another study of note is the research project that the British Occupational Health Research Foundation (BOHRF) completed to provide evidence-based answers to some of the key questions related to mental ill health in the workplace. The questions were:

- What is the evidence for preventative programmes at work, and what are the conditions under which they are most effective?
- For employees who are identified as at risk, what interventions most effectively enable them to remain in work?
- For those employees who have experienced bouts of mental health related sickness, what systems most effectively support their rehabilitation and return to work?

The study looked at over 15,000 references in published research worldwide and identified 111 papers that were critically appraised. Based on these it was found there was moderate evidence to suggest that stress prevention strategies can have a beneficial and practical impact. There was stronger evidence to suggest that retention strategies are effective, ie properly targeted interventions to individuals can help to prevent long-term illness. The most effective programmes focused on personal support, individual social skills and coping skills training, and using a combination of these suggested that the most long lasting effects were likely. The study also found that rehabilitation of an affected worker can be very effective, and the most effective treatment was Cognitive Behaviour Therapy (CBT). This is a generic term that is used to describe a range of approaches designed to modify the individual's behaviour. The evidence suggested that this treatment was most suitable for those who had a reasonable level of control over their work.

In addition to the research a series of discussions were held with over 200 senior managers, HR professionals, health professionals and policy makers from both the private and public sectors. Again, one of the key findings was that good management was seen as being crucial to the effectiveness of a stress management programme. There was strong support for the training of managers and supervisors in how to be pre-emptive in dealing with stress.

Full details of the BOHRF research study can be found at the BOHRF website: www.bohrf.org.uk.

The most recent research study (*Kerr et al 2009*) can be found on the HSE website. The conclusion from this study provided empirical evidence to support the use of the Management Standards approach in tackling workplace stress that was adopted by the HSE and which is referred to further in this chapter.

Management standards on occupational stress set by the Health and Safety Executive (HSE)

[S11009.2] The most recent published statistics by the HSE from the Labour Force Survey (LFS) have identified that the total number of cases of stress in 2014/2015 was 440,000 cases, a prevalence rate of 1,380 per 100,000 workers. The report identified that the number of new cases was 234,000 which results in an incidence rate of 740 per 100,000 and these figures have remained broadly flat for more than a decade. This survey also identified:

- The total number of working days lost due to occupational stress in 2014/15 was 9.9 million days, equating to an average of 23 days lost per case.
- During 2014/15 stress accounted for 35% of all workplace ill-health cases and 43% of all working days were lost due to ill health attributed to stress.
- Stress continues to be more prevalent in public service industries, eg education, health and social care and public administration and defence.
- By occupation, the higher levels of stress occur in jobs such as healthcare, teaching and public service professionals.
- The main occupational factors given by respondents as causing stress, anxiety or depression (LFS, 2009/10–2011/12) were tight deadlines, too much responsibility and a lack of managerial support.

The full report can be found at: www.hse.gov.uk/statistics/causdis/stress/stress.pdf.

The HSE's Management Standards on Occupational Stress seek to tackle the absence and ill health caused by occupational stress. Compliance with these is not a legal requirement but will assist the employer in reducing stress at source, ie by using primary interventions, and by having systems in place that will support a worker who is experiencing stress problems related to their work. Following the guidance in these standards and having evidence of doing this will also support an employer's defence that they have taken all reasonably practical measures to eliminate or mitigate against occupational stress in their workplace.

The Management Standards are based around six key areas; these are:

- demands, which includes issues such as workload, work patterns and the work environment (standard 1);
- control, which concentrates on how much involvement or autonomy a person has in the way they work (standard 2);
- support, which considers the level of encouragement and assistance provided by the organisation and the resources made available to provide support where additional assistance is needed (standard 3);

- relationships, which promotes positive behaviours and cooperation at work, and identifies if there are systems in place to respond to unacceptable behaviour (standard 4);
- role, which considers how clearly job descriptions and organisational structures are defined, whether there are conflicts in workers' roles and whether there is a clear and unambiguous reporting line (standard 5); and
- change, which assesses how organisation change is managed and communicated to the workforce (standard 6).

The standards seek to enable employers to identify the gaps between their current performance and conditions that are set in the standards, and to develop their own solutions where problems are identified. Note that the standards are not numbered by the HSE, but have been numbered for easy reference when reading the next sections in this chapter.

Further information on these standards can be found on the HSE's website: www.hse.gov.uk

Occupational stress risk factors

[S11010] The following paragraphs discuss in detail the most common occupational stress risk factors.

The physical environment

[S11011] The physical environment will come under HSE Management Standard No 1 (Demands).

The physical environment can be described as the general conditions in which people have to work. The work environment is often seen as a contributory cause for stress at work, ie something that compounds a problem that is being experienced at work. For others the work environment may be the main cause of stress amongst the workforce, especially if the conditions prevent work targets from being achieved, or detract workers from undertaking their work effectively.

Our general work conditions have improved drastically over the past century, but so have our expectations. The environments in which people work will vary considerably, and will be dependant upon the type of work that is being undertaken. For example, a die casting factory is traditionally a hot, dirty, and malodorous environment. Any investments are usually spent upon the equipment to produce the product rather than to enhance the work environment, and managers may be reluctant to spend a heavy outlay on a workplace that is likely to become dusty and dirty within a short time. These types of environment often have less time and effort spent on them in maintaining high standards of cleanliness, good housekeeping or ensuring that lighting remains in good order throughout the premises.

This can, however, have a knock-on effect for staff, both in their mental attitude towards the workplace, in their general conduct and in their general well-being. It has been found, following several studies of the workplace, that an environment that is poorly maintained produces a negative attitude amongst the staff – 'if management don't care then why should we' type approach. If the workplace is allowed to become dirty, untidy or cramped, then workers will respond accordingly with the 'why should we tidy up – management don't seem to mind' type attitude. A workplace that is dull, untidy, badly designed or scruffy is likely to make employees feel negative about coming to work, and they are more likely to respond negatively to issues that arise at work.

The condition of the workplace

[S11012] The condition of the workplace can be described as its general standard of maintenance and decoration, both internal and external. Housekeeping standards within the workplace could be included under this heading. To understand how a workplace condition can affect how a person feels and responds, consider the following examples.

- *Scenario 1*
  An old hospital building was first constructed during the early 1900s. The hospital management has struggled to keep up with the increase in usage over the years and the structural changes that have been necessary to accommodate more patients and more services. The hospital is crowded and cluttered. Government financial cutbacks have been a common setback and funding has focussed on provision of medical care, rather than on maintenance.
  The walls are grey and flooring is mainly grey concrete. It is cracked in many areas, causing ruts and trip hazards. Lighting is poor and several overhead tubes are not working. Paint on the wall is dirty, marked in places and chipped in others. Some of the windows are cracked.
  Doors are damaged due to the constant thoroughfare of hospital trolleys being pushed against them.
  Sacks of dirty laundry, and black sacks containing rubbish, seem to be constantly accumulating in corners awaiting collection.

- *Scenario 2*
  This scene depicts another hospital that was constructed in the early 1900s. Similar challenges have been faced within the hospital to provide increased services and a higher level of care. Budgets restraints also present a challenge.
  The walls are still grey but have been painted within the last few years. However, different parts of the hospital have received a splash of colour by way of painting the support pillars, door surrounds etc with bright contrasting colours. The doors are less damaged as they have metal plates fitted to reduce damage from the trolleys. Windows are cracked in some places, but have been cleaned, which has resulted in more light coming into the hospital.
  Rubbish bags and dirty laundry bags are out of sight. The floor looks cleaner and there are less pot holes and cracks in the flooring.

- *Scenario 3*
  This scene depicts a hospital that was also built in the early 1900s. However, despite the hospital facing the same challenges as the hospitals in the first two scenarios, the management team have adopted a much more proactive approach to general decoration. Staff have been encouraged to participate in choosing colour schemes when new services have been introduced and the patients' views have been taken into account when planning any refurbishment. This has resulted in local colleges being invited to practise their artistic skills in main areas at little or no cost to the hospital. Waiting rooms have brightly coloured murals along the walls. Corridors are coloured differently for each separate unit (eg cardiology, rehabilitation) and collages and other pictures have been displayed along the walls. These have resulted in interesting schemes for people to enjoy as they walk through the corridors.

Some extra money has been made available each year for ongoing repairs which has resulted in a higher standard of maintenance. A proactive maintenance schedule is in place whereby all lighting tubes are replaced annually, thereby significantly minimising lighting failure. Staff have been encouraged to share their requirements or preferences for lighting in different areas, and for reporting to management when problems are occurring. Windows are kept clean.

Wards are brightly coloured, and curtains have been fitted at windows which are pleasant to look at. Plants are situated around the hospital and a water feature has been installed in the main reception waiting area.

Waste collection points have been clearly defined. Laundry collection/delivery areas are clearly marked and waste is not allowed to accumulate.

An employee working in any one of these environments, doing work which is known to be emotionally and physically demanding, would obviously prefer to be working in an environment depicted in scenario 3.

People feel better and therefore respond more positively when their environments are pleasant. This positive factor is increased when staff feel that they have had some level of control over their own environment, even if it is something as basic as deciding on the best areas for storage or waste collection points.

The workplace layout

[S11013] The workplace layout will be considered in HSE Management Standard 1 (Demands) and 2 (Control).

The workplace layout can make a significant impact upon a worker's efficiency, safety and general well being. This is where the study of ergonomics can have a major input into reducing stress levels at work. Ergonomics is a specialist area, which looks at the interaction of the worker, the work equipment and the work environment. The use of an expert may be needed at some stage in workplace layout, but the first step should always be to discuss employee requirements and to see how these can be accommodated within current restraints. See ERGONOMICS E17021.

Simple changes to a work layout can make a drastic improvement in a person's efficiency, safety and general well being, especially when the idea has emanated from the workforce itself. The worker is the expert in the particular task that is being undertaken, and whilst an employer does not want to provide a 'wish list' to the workforce, valuable information can be ascertained merely by finding out if and what causes frustration within a work layout.

An employer might consider asking its staff the following questions.

Question	Yes/No	Comments
Are worksurfaces large enough for the work and equipment that are needed?		
Are worksurfaces of a suitable height (too high or too low) for the worker(s)?		
Is there sufficient space for the worker to get to, from and around their own work area?		
Are storage facilities in close proximity and easy to access?		

Question	Yes/No	Comments
Is there sufficient storage?		
Is storage being used to full effect?		
Do workers have somewhere to store their own personal items?		
Are seats or suitable rest areas provided, at or near to the workstations?		
Are walkways wide enough to enable people and equipment to pass through (eg trolleys)?		
Are work areas in large areas clearly identifiable, signed and labelled?		
Is the layout designed to minimise environmental hazards (eg glare from windows, noise from nearby machines)?		

Poorly designed layouts can induce other health conditions such as musculoskeletal disorders eg back problems from having to stoop, stretch or crouch to undertake a task. Stress creates physical tension which can contribute to these health conditions. Some proposed new EU legislation will require that stress and work pressure be considered as part of risk assessments for back pain and other musculoskeletal disorders.

Staff facilities

[S11014] Staff facilities will need to be considered under HSE's Management Standard 1 (Demands) and perhaps under Support (Standard 3).

Everyone occasionally feels a need to take a break from their work, and this may be more of a need than a desire in some work situations that are known to incur high levels of stress eg dealing with the public, working in the care profession. Having nowhere to go to take a break or let off steam may result in stress levels increasing to a point where a worker becomes distressed. This will be compounded when other aspects of the work are problematic, for example, the shift patterns are badly designed or working hours are excessive.

Staff facilities should include basic facilities such as sanitary conveniences and washing facilities, and somewhere from where drinks can be prepared or obtained. Employers should ensure that such facilities:

- accommodate the numbers of people using them;
- take into consideration the need for privacy (especially if workers need to change clothes or shower before or after work);
- are maintained to a good state of hygiene and cleanliness;
- provide basic equipment such as soap, drying equipment, toilet paper and sanitary bins;
- enable workers to obtain both hot and cold drinks, including provision to clean spillages or clean utensils if drinks are being prepared by the workers; and
- provide a ready supply of hot/warm and cold water, and potable water for drinking purposes.

The *Workplace (Health, Safety and Welfare) Regulations 1992 (SI 1992 No 3004)* and supporting guidance contain detailed minimum standards for sanitary conveniences, washing facilities, drinking water facilities for changing clothes.

These Regulations also contain provisions for rest facilities, and require that somewhere that is suitable and sufficient is made available for people to eat meals where meals are regularly eaten in the workplace. Resting and eating facilities must ensure that people who do not smoke are protected from the effects of tobacco smoke.

This legislation also requires the provision of rest facilities for workers who are pregnant, or who are nursing mothers. Some employers have expressed concern at the practicalities of meeting this requirement. However, the female workforce is increasing as more mothers are returning to work earlier and employers have a legal as well as moral duty to ensure that these provisions are available. These types of issue do have an impact upon the stress levels of employees.

Since July 2007 it has been against the law to smoke in virtually all enclosed public places, workplaces and public/work vehicles.

The *Working Time Regulations 1998 (SI 1998 No 1833)* contain requirements for rest periods. During working hours the maximum that someone is allowed to work is six hours, following which they are entitled to a 20 minute break. Rest facilities should be in line with the standards stipulated above.

For further information see WORKPLACES – HEALTH, SAFETY AND WELFARE W11001.

Environmental conditions

[S11015] Environmental conditions will be considered in HSE Management Standard 1 (Demands), but will also need to be considered under Standard 2 (Control), ie how much control do they have over their local environment (eg temperature, space etc), and Standard 3 (how much support is available to address concerns that are raised).

When considering stress induced by environmental conditions, an employer should consider issues such as lighting, heating, ventilation, humidity and space.

In some cases, the environmental conditions may be the sole or primary cause of stress, in others it may be a factor that exacerbates other stressful conditions within the workplace. Poor environmental conditions are the cause of other workplace issues such as sick building syndrome.

Several regulations contain specific duties for provision of environmental standards. These include the *Workplace (Health, Safety and Welfare) Regulations 1992 (SI 1992 No 3004)*, the *Schedule* to the *Health and Safety (Display Screen Equipment) Regulations 1992 (SI 1992 No 2792)* and the *Provision and Use of Workplace Regulations 1998 (SI 1998 No 2306)*.

An employer should take the following into account when deciding on the standards that should be achieved in the workplace:

- legal requirements contained in various statutes;
- official guidance issued by the HSE, trade associations or other recognised bodies such as the Chartered Institute of Building Service Engineers;
- the nature of the work (work requiring intricate detail will need a higher level of lighting than one undertaking general tasks);
- the amount of physical effort in a task (a cooler background temperature will be needed where workers exert physical effort);
- the need to combine workers' needs with other stakeholders (eg customers, patients);
- the level of humidity in the environment (work in a humid environment will reduce a worker's ability to control their own internal temperature by sweating, thus increasing the risk of heat exhaustion); and
- the amount of space for the work components, the work equipment, and the worker's need for movement.

There are several sources of guidance which will help an employer to interpret conditions that will be deemed as 'reasonable' to meet statutory requirements for a work environment. A good starting point, other than a specific trade association, is the Chartered Institute of Building Service Engineers. This institute can provide detailed guidance on recommended environmental standards, and should be referred to when deciding optimum work conditions for different types of work activity.

Employers should take into account, however, that any such guidance should not be used in isolation. The most effective way of reducing physical and emotional stress from work environments is to involve the staff. It is not possible to please everyone, but by obtaining a general consensus an employer has an opportunity of ascertaining if any problem exists, where the problem might be, and the extent that the environment may be cause or adding to a problem.

Another important key to reducing stress induced by environmental conditions is to provide as much control as possible over the local environment. For example, it is usual in large open workspaces for the lights to be controlled by only a few switches. The lighting circuits could be changed so that a smaller area is controlled by each switch, enabling more control over an employee's immediate working vicinity. The strength and colour of the tubes could also be altered to further meet individual preferences.

The work

[S11016] There are many aspects of work that may induce pressure. Pressure at work is not necessarily bad as it helps to generate enthusiasm and can produce energy. However, prolonged pressure is likely to create a state of nervous tension as the person has no respite from the constant rush of adrenaline that occurs when stress is induced. Unless this is relieved, the end result can lead to health problems.

The trick is to try to achieve a balance between providing a challenge to staff, and overloading them.

The following areas are known to be occupational stressors, and should be reviewed when any stress management strategy is being considered.

Targets, deadlines and standards

[S11017] Targets and deadlines clearly come under HSE's Management Standard 1 (Demands). They may also need to be considered under Standard 2 (Control) and 3 (Role). For example, if a worker is given conflicting roles and job functions this may well have a negative impact on their ability to meet their targets.

Targets are necessary at work. This is because an employee who is not given a target will set their own target which may not be in line with organisational expectations. However, problems arise when the standard that has been set by the employer exceeds the individual's ability to meet it.

When a deadline or target is unachievable, a sense of helplessness may be experienced by the person. Similar emotions may be experienced when each time the target is met the employer increases the next target. Whilst for some employees this will represent a sense of excitement, there will inevitably be a time when an employee begins to feel defeated, undermined or useless. This problem becomes more serious when there is no mechanism for the employee to let the employer know how they are feeling, or when there is no system for monitoring staff behaviour and performance once the target has been set. If the employee feels that their job depends upon meeting the target then the feeling of pressure is increased, as the consequences of failure are more significant to that individual.

There are various reasons why a deadline or target may be unachievable. Managers should consider the following questions, and perhaps agree with the employee what is reasonable when all the restraints have been taken into consideration. This will help the employee to perceive a sense of involvement and control over the amount of pressure being placed upon them.

Assessment of targets			
Question	Yes/No	Comments	Agreed action
Is the employee sufficiently trained/experienced to do their work effectively?			
Are there adequate tools/equipment to meet the deadline/target?			
What other demands are being made which may affect the standard/deadline being met?			
Is there any mechanism for the employee to report early if he/she feels they are not likely to meet the standard?			
To what extent can machine failure etc affect the target being met?			

Assessment of targets			
Question	Yes/No	Comments	Agreed action
Is the time frame for the amount of work being set reasonable (ie is this being achieved reasonably by others)?			
What are the consequences of meeting or failing the target?			
Are there any other factors that the employee feels may detrimentally affect the target/deadline being achieved?			

People who have experienced continual pressure from unrealistic deadlines or targets often explain their feelings with comments such as not being able to see the light at the end of the tunnel and being on a treadmill that they cannot get off. When these feelings become excessive, a common response is for the employee to take time off for sickness absence, or even to leave the company. Early warning signs that an employee may be feeling excessive pressure may include erratic behaviour or mood swings, reduced or erratic work performance, looking worn out or tired or even compulsive behaviour (increased smoking, drinking or eating). Some people, of course, will respond with the opposite effects, ie by losing their appetite, becoming very quiet or seeming to become particularly conscientious.

Time demands

[S11018] Time demands clearly fall under HSE Standard No 1 (Demands) and No 5 (Role).

Stress from conflicting demands on time may be a sign of 'role conflict' or 'role ambiguity', ie which task is more important. It may be that the time available is insufficient to deal with the extent of work being required. If time demands being made upon the employer are such that everything is expected to be undertaken at the same time, then the worker will suffer.

Different people have different ways of handling their work. Some people prefer to complete one task before giving any attention to another, whilst others appear to thrive on dealing with numerous matters all at the same time. Problems arise when the manager operates to one of these principles whilst the subordinates preferred style is totally different. The end results can be disastrous, with neither party feeling that a job is being completed satisfactorily and both parties feeling that they are being misunderstood. Two-way dialogue between management and employees is needed to ensure that each party understands the other's needs, and to agree parameters for working.

A commonly reported cause of stress is knowing that several other tasks are building up whilst only one task is being addressed. The employee will either rush through all the work, to try to reduce the mountain that is piling, or select one task that they feel is the most important at the expense of others which are perceived to be less important. Neither response is likely to result in staff feeling satisfied that they are coping and doing a good job. A key to managing this type of problem is to:

- review work loads;
- take into account deadlines;
- agree priorities for each task that is building up;
- ensure the most suitable type of person is allocated (some people deal better with a high throughput that requires low levels of accuracy whilst others prefer work that require less output but a higher level of attention); and
- encourage and develop self-organisation for the individual to manage their work load.

Working hours and shift patterns

[S11019] Working hours and shift patterns will be reviewed under a number of the HSE Management Standards, including Demands, Control, Support and maybe even Relationships and Role. This will be particularly relevant for people who are covering more than one job role eg during periods of short staffing.

Management of working hours is a key feature in reducing stress levels at work. When considering working hours, the employer should take into consideration:

- the length of the shift (how many hours worked each day);
- the shift pattern, especially if rotating shifts are used;
- rest times between shifts; and
- rest and break times during the shift.

As mentioned previously (see **S11008**) the *Working Time Regulations 1998 (SI 1998 No 1833)* contain requirements on the number of hours that should be worked each week, the minimum rest time between shifts and the minimum rest breaks during shifts. Regardless of this whether an employee is excluded from the Regulations or chooses to 'opt out', the employer has a duty of care to ensure that health, safety and welfare are not detrimentally affected by the work.

Problems with working hours seem to crystalize when the time spent at work impinges on employees' home and social activities. Someone who habitually works long hours often feels too tired to enjoy their private time. An employee will be prepared to 'trade off' a well-designed shift pattern in order meet social demands being made at home. It is clear, therefore, that a balance must be struck between shift patterns that have been deemed to be the most effective, against the social needs of the workers.

The most frequently reported source of stress among shift workers who work night shifts is the interruption in their sleep, or difficulty in being able to sleep during the day. Possible sources of interruptions are numerous and include doorbells, telephones, increased traffic noise, family, pets and neighbours. Workers who operate to rotating shifts are more likely to feel the effects of sleep disturbance.

The need for variety during night shifts in terms of the range of tasks, and the mental and physical output necessary, is more important for night workers. The attention span will reduce, typically between 3.00 am – 4.00 am and this will be exacerbated if the work is monotonous, requiring high levels of concentration or is physically demanding. Fatigue is unavoidable during night work, but will be reduced if the interest of the shift worker is maximised.

Travel during work

[S11020] Time spent on travelling and the modes of transport permitted within the organisation's policy or resources can have a significant impact upon a worker's wellbeing. This is likely to be considered under HSE Management Standards 1 (Demands), 2 (Control), and 3 (Support).

Many employees enjoy travelling as part of their work as it provides a sense of variety, change in environment and a level of autonomy about their work pattern (eg by not feeling that they are being watched). Problems arise when the amount of time travelling exceeds the amount of time that can be spent in the workplace, or impinges upon the time available for the remainder of work that needs to be completed. Travelling is also exhausting, both physically and mentally, and can reduce a person's work performance. Other problems associated with travelling include feelings of isolation from colleagues, family and friends, especially when extended stays away from home occur.

Apart from drivers of heavy goods or public services vehicles, there are no maximum times set in legislation that a worker may travel on the road without taking a break. Therefore an employer should undertake a risk assessment, and agree with the workers what travel arrangements are considered reasonable. Factors that need to be taken into account include:

- preferred method of travel;
- manual handling issues of the traveller (cases, files etc);
- the length of the journey (including times for delays);
- how much time the employee has been given to reach their destination (ie work time versus private travelling);
- travel routes (especially if using the road);
- the need for regular two-way communication with the worker whilst away from the workplace; and
- suitability of overnight accommodation, including location of accommodation.

Organisations such as ROSPA (Royal Society for the Prevention of Accidents) have produced detailed guidance and literature detailing results of their research into occupational road risk, and this may be a useful source of reference when considering the physical and psychological risks associated with travelling.

Technology and work equipment

[S11021] This section will be considered in HSE's Management Standards on Control, Support and maybe Change (if the technology changes).

There is nothing more frustrating that having a mismatch between the tools necessary to complete a task and the task itself. If the problem is simply being given the wrong tool to complete the task, then this can quickly be rectified (provided the employer is informed). However, stress levels increase when equipment constantly fails to meet its design criteria, breaks down due to overload or poor maintenance, or when there is a skill's shortage amongst the workers.

To overcome the problem the employer should undertake a task analysis to identify what tools and equipment are needed, and to ensure that they are provided. Ongoing maintenance may be needed, including an emergency call-out service where the job is dependant upon the work being completed within a certain time. This needs to be followed by a training needs' analysis, to identify if and what type of training is required to ensure workers can use the equipment effectively and safely.

Most people are instinctively sociable, and respond to both verbal and non-verbal behaviour and communication whenever they meet. Someone making a light-hearted comment can lighten a dull or pressurised meeting, a smile will indicate whether an ambiguous comment is meant as a joke. This type of behaviour is missing from computerised communication ie e-mail or even telephone. Messages received without the additional benefit of non-verbal communication may come across as blunt, aggressive or hostile and may make the recipient feel threatened when the opposite was the intention. Stress will result from the perception of aggression or threat, especially if that person feels helpless to respond or defend themselves. This needs to be taken into account when transmitting difficult messages, to avoid inducing unnecessary anxiety. In extreme cases, e-mail communication has been known to cause severe disruption to workplace harmony and cause individuals to become insecure or even sick.

Role conflict or ambiguity

[S11022] Very clearly, this will be predominantly reviewed under HSE's Management Standard 5 (Role), but also under Standard 3 (Support).

People can work much more effectively if they know what they are supposed to do, how it is supposed to be undertaken, why it needs to be undertaken in a particular way and how their role fits into the corporate objective.

When job roles conflict or are ambiguous the end result is confusion and frustration for everyone involved. If this becomes a frequent event then those feelings of frustration may increase into anger and resentment. This may be compounded when the person's level of authority is limited and they have no access or authority to change what is causing the problem. The following scenarios may be indicative of conflicting or ambiguous roles.

- *Scenario 1*
  A secretary is told by her manager to manage the diary and to keep all the clients satisfied when they ring. Several potential clients have telephoned to book meetings which the secretary has entered into the book, but her boss is very displeased when he finds out and tells her to cancel them. Nearly all other bookings made into the diary by the secretary get changed or altered, causing the secretary additional work and

upset. When challenged by the manager about why all these bookings have been made, the secretary defends herself by stating that she is doing what she was told to do. The boss responds by saying that nothing should be booked without checking with him first.

- *Scenario 2*
  A departmental manager has responsibility for reducing expenditure in his area of authority. However, he has also been given responsibility for managing health and safety. The health and safety standards in the department are poor. The manager identifies a need for health and safety training which will improve standards but this will take him over the agreed spending threshold which will directly affect profitability and his requirement to reduce expenditure.

Clerical/administration support

[S11023] Many employees now undertake their own basic administration tasks, using basic computer software packages. This is acceptable provided they have the basic skills to use the equipment, and the time allocated within their work patterns to complete the administration. Common indicators that completion of paperwork is becoming problematic may include:

- workers regularly taking work home in the evenings/weekends;
- employees not using their holiday entitlement;
- increase in errors within written correspondence (typographical, grammar or content);
- tasks being half completed;
- paperwork piling up on workstations;
- messages not being answered; and
- comments from workers about bureaucracy.

Employers should ensure that time is made available to enable workers to complete any necessary paperwork, and that the equipment is accessible. This is particularly relevant where the worker is based in a non-office environment, and would not automatically have access to a desk, paper, filing system etc readily to hand.

Most people do not like excessive paper-work, and some do not like any paper-work. Employers should therefore seek to streamline their systems so that administration is minimised, and simplified.

Cover during staff absence

[S11024] This is an area which will cross through a number of the Management Standards, but is likely to be predominantly considered under Standard 4 (Relationships) and Standard 5 (Role).

Administration often becomes a stress issue when staff are constantly required to cover for colleagues. The additional pressures this causes are unlikely to be problematic if it is for only short duration, eg sickness or holiday. However, staff who feel that they are frequently undertaking additional work to cover for shortages will begin to feel the effects of the burden. Early symptoms to look for may include reduced efficiency, increased error rates, lower patience or attention thresh-hold, or increased sickness absence.

Managers should seek to avoid relying on employee goodwill as the sole method for dealing with the workload when a colleague is absent. Use of temporary staff should only be used as a stopgap. Often their effectiveness is limited as they are unfamiliar with the work and work systems. It also takes time to train a temporary worker, and if this becomes a regular exercise then the use of temporary workers will be seen as a hindrance rather than a help.

Employers should therefore adopt a more proactive stance, and consider what alternatives would be suitable to deal with staff shortages. Examples may include use of multi-skilled workers who could be moved around, or 'floating' staff who are employed specifically to cover during other staff absences.

Company structure and job organisation

[S11025] The following paragraphs discuss the relevance of company structure and job organisation on stress levels.

Hierarchy of control

[S11026] The following areas will focus on HSE's Management Standard 2 (Control) and 5 (Role) but might also impact on Standard 4 (Relationships). Where changes in company organisation occur then Standard 6 (Change) will be predominant.

Ambiguity over job roles and the level of control over a task is a major stressor at work.

Problems occur when employees are given responsibility for a task but have no authority to implement necessary systems, training etc to effectively deal with the task. Colleagues and fellow team members will judge both the corporate reaction and the reaction of the particular manager to any decisions that are made within an organisation, and will build their own attitudes as a result when showing commitment back to the organisation. It is therefore very important that everyone knows what their level of responsibility is, and the authority that they have to deal with a given situation. When a challenge is presented at work the employee's physiological state will alter to deal with the event, (increased blood pressure, etc). Any nervousness or heightened state of alertness to deal with the situation will be better controlled by the individual if they know exactly how they are supposed to respond, and the parameters in which they can deal with the given situation.

The interface between two overlapping parts of an organisation can also be a cause of stress – each will have their own priorities and their own way of working. Good communication and clearly defined lines of authority are vital to identify the essential patterns of overlap and to enhance the organisation's efficiency.

To overcome such problems the employer should review where any overlap in processes or responsibilities may occur, and where any gaps or areas that have not been clearly defined may be present. These should be cross-checked against the recipients (ie employee's) perceptions and interpretations of the responsibilities to ensure that misunderstandings do not occur.

Organisational changes and arrangements

[S11027] Any change within the organisation will need to be considered under HSE Standard 6 (Change). Change is stressful to many people but symptoms can be alleviated by good communication and a feeling that they are involved.

People generally do not like change. It upsets the person's comfort factor and invariably leads to a period of uncertainty whilst they are adapting to the changes. Some people adapt better to changes in their lives than others. Having an awareness that any change is likely to cause stress is the first step to managing the problem.

It is necessary to communicate as clearly as possible the nature of any changes, and the consequences of these to the workers who are affected. Even bad news can be dealt with positively if it is communicated honestly, openly and with sensitively.

Responsibility

[S11028] Responsibility links with the level of control an individual has (HSE Standard 2). How clearly this responsibility is defined will need to be reviewed under HSE Standard 4 (Role).

Once a responsibility has been allocated to an individual, there should be sufficient scope allocated to enable the person to carry out the task. This does not mean that the employee is then left to deal with a given responsibility without any mechanism for support or assistance if the responsibility becomes more extensive or complex than first anticipated. Being given a responsibility without any access to support when needed will inevitably lead to stress.

The feeling of a lack of appreciation, through non-realisation of an individual's potential, often arises from being given insufficient responsibility. Many people, especially those in jobs which do not enjoy high status in their organisation, feel frustration through lack of challenge, or feel frustrated when they can see things going on around them when they do not have an opportunity to be involved in making a contribution. Giving people autonomy, ie increasing the degree of self-control over their work, will help to counteract this.

Developing a system for employee recognition and staff development will help to identify where an employee needs more or less responsibility within their work. This will also provide a mechanism for a manager to explain why more responsibility has not been given to an individual who feels they want or need it, and for managers and employees to work together to reach a level which is mutually agreeable.

Individual perception of company

[S11029] How people feel about their organisation will reflect in their work attitude and performance, so this area may be considered under Standard 4 (Relationships), 5 (Role) and perhaps also 2 (Control).

If staff have a negative view of the company, they will respond accordingly. Employees who feel that the company is going nowhere, or does not care about its image, will not care about the company's future or its image. This will give them no sense of achievement or pride in their employment. The end result will be a higher staff turnover and the associated problems that go with this. Output will not be as effective as a motivated team of workers, and standards of work are likely to be lower.

Involving staff in the company's vision, or business plan will help employees to see what the company is hoping to achieve and the goals that have been set. Even if the employees do not fully understand or agree with the vision they will at least have an understanding of what is happening when changes begin to occur.

Perception of one's self-esteem and role within the company

[S11030] Everyone needs to feel that they have a valid role to play, in their life and in their work; everyone needs to feel valued and needed. When this does not occur in the workplace, the result is an employee who feels worthless, undervalued and misunderstood. The problem will become compounded if the individual concerned is someone who is ambitious or is trying to achieve promotion or a more fulfilling role.

Previous experience

[S11031] Someone who has experience of the environment will be more aware of the penalties and rewards associated with the work. They are likely to have been exposed to any associated work stressors before and may be more adept at managing them. However, this can sometimes have the opposite effect, in that if someone has been exposed to high levels of stress in a similar type job previously then this may well have reduced their tolerance to pressures.

Personal factors

[S11032] The way in which someone responds to their work can be reviewed under most of the Management Standards, but is perhaps more prevalent under Standard 1 (Demands). The compatibility of the demands being made upon the person with their own personality traits needs close review and monitoring to ensure the job and the person have the right fit.

Everyone responds differently to events that occur in their lives. This very much depends upon their basic personality traits, their upbringing and their experiences in life.

Various studies have been undertaken in behaviour patterns. The research undertaken by Friedman and Rosenman *Type 'A' behaviour and your heart* (1975) (ISBN 0704501589) gave some insight into why some people are more prone to stress-related disease. In these studies, two main risk behaviours were identified, which were called Type A for high risk, and Type B for low risk. In general, Type B behaviours are the opposite of Type A behaviours. These categories should only be considered as a general yardstick as there are no absolutes, and most people will fall somewhere between the two extremes identified.

Type A people were said to be:

- autonomous, dominant and self-confident;
- very self-driven, busy and self-disciplined;
- aggressive, hostile and impatient;
- ambitious and striving for upward social mobility;
- preoccupied with competitive activities, enjoying the challenge of responsibility; and
- believing that their behaviour pattern is responsible for their own success.

Type A people were said to work longer hours and spend less time in relaxation and recreational activities. They received less sleep, communicated less with their partners and tended to derive less pleasure from socialising.

In contrast, type B people tend to:

- find it easier and are more effective at delegating;
- work well in a team;
- allow subordinates some flexibility in how a task is undertaken;
- avoid, or do not set, unrealistic targets;
- keep a sense of balance in the events that occur in their lives; and
- be more accepting of a given situation.

Type B people are often more secure within themselves and more accepting of their strengths and limitations. A very simplistic summary of this type of character could be that they are simply more 'laid back'.

The level of responsibility and control over work

[S11033] Clearly the following sections are most relevant to Standards 2 and 5 (Control and Role).

These are major factors as to whether someone is likely to suffer the effects of stress. Karasek and Theorell *Healthy Work* (1996), a comprehensive study of job demand and autonomy, showed that stress is a function of the combination of psychological demand and 'decision latitude', or autonomy.

The term 'executive stress' has been widely used, insinuating that perhaps people with high levels of responsibility are more likely to suffer from occupational stress. Research has shown, in fact, that this is not true. Whilst the level of responsibility increases with 'white collar' roles, so does the level of autonomy about how the work is to be carried out. Material rewards are usually higher, which helps to enhance self-esteem within the worker.

Karasek and Theorell's studies showed that the greatest risks are among those occupations with least autonomy and of a lower status. Examples include bus drivers, production line workers and care staff. These are classic examples of the conclusion reached by Karasek and Theorell that it 'is the bossed, and not the bosses' who are most likely to suffer from stress. They often have little or no control over how the work is to be undertaken but have to deal with the problems and difficulties that arise when a job is badly designed or managed. They also have less material reward for the work undertaken.

Qualifications and competencies

[S11034] There is a rather cynical saying that people are promoted to their level of incompetence. Whilst this may be an exaggerated view there is some truth in the concept that people who are newly promoted may not immediately be competent. The same could apply to almost any role at work, if there is a mismatch between the task that is required and the skills that are possessed by the people undertaking the work. People need to grow and develop into their roles at work, to a point that they feel comfortable and competent in what they do without becoming bored or complacent. An employee who works beyond their competence is likely to cause problems for the organisation in that work standards may reduce and production may be affected. From the workers' point of view they are not likely to feel confident or able that they can achieve what is required, and this will have a knock-on effect upon their self-esteem. The end result is likely to be that such an employee would feel frustrated and dissatisfied, and maybe very unhappy. The manager is likely to feel equally frustrated and dissatisfied because the work is not being completed satisfactorily.

Personal issues

[S11035] Separating personal problems from work issues is difficult for some people, and impossible for others. Both work and personal issues have an indirect effect upon each other and if an employee is experiencing a stressful situation in their home life then this may impact upon work performance.

An employer cannot be responsible for personal issues in an employee's life, but an awareness that a problem exists for the individual is useful knowledge for the employer. Systems should be developed that enable a manager to offer support and sympathy whilst the difficulty exists. For example, if a family member is in hospital perhaps working hours could be adapted slightly to fit in with the hospital's visiting times. The reaction to an individual who is going through a crisis, in terms of the level of support and flexibility shown by the manager, can stay with that person for a long time. Genuine help being offered by the manager can be an investment which promotes commitment from that person at a later stage.

A pre-agreed workload level may be acceptable in normal circumstances, but intolerable when a long term problem is occurring in someone's personal life. Therefore, the events going on in the employees' lives outside work will, at some stage, have an effect on their ability to cope with work. If these events are positive then the ability to cope will improve, with the opposite effect happening when the personal events are negative.

Managers need to take a balanced approach when responding to an individual's personal issues. An employee who is considered to have a good 'credit' balance in terms of their commitment and enthusiasm to the organisation may be treated more sympathetically by the organisation if a crisis develops in their lives. However, this could in itself have a significant knock-on effect to other employees who feel that they have not been treated so considerately, and this can induce feelings of being undervalued, misinterpreted or simply overlooked by those colleagues.

Stress management techniques

[S11036] The following paragraphs provide detailed guidance and advice on managing stress.

Stress auditing/surveys

[S11037] Stress audits or surveys are a good starting point, as this will give the employer an indication of whether a stress issue exists in the workplace, and the areas where this may be most problematic.

There are several approaches to audits and surveys, ranging from a simple checklist that employees are asked to complete, to detailed analyses with scoring systems that are undertaken by specialists. The type of audit that is used will depend upon the organisational culture and the extent to which an employer feels that a stress problem exists.

A survey in its simplest form may merely be a short questionnaire that asks employees to list their three most positive and three most negative aspects about their work. Guided headings may be included, perhaps with a tick box and a space for the employee to write comments explaining why it is seen as a positive/negative aspect of work. Headings that are used could include areas that the employer suspects are causing pressures upon the employees. This type of approach is often used as a pre-audit review, to enable employers to gain a feel as to any patterns which might be obvious to the workers.

Stress assessments are more detailed, and require planning, thought and sensitivity. They usually take the form of a guided list of questions or areas that the assessor wishes to cover about the work. They can include all members of staff, or selected members of staff from certain job titles to act as a representation. The interview needs to be undertaken by someone who is sufficiently competent to complete such a task. Competencies necessary to achieve this may include:

- counselling skills.
- polished interview techniques.
- knowledge of the organisation and its management structure.
- knowledge of procedures and practices.
- understanding of health and safety issues, including stressors at work.
- understanding knowledge of human behaviour (ie human factors at work).
- the ability to gain the confidence of the interviewee.
- the ability to listen.

- understanding of the HSE's Management Standards and what to cover in each topic.

Points to remember:

- Reassure staff that all comments will be confidential.
- Respect the confidentiality of your staff.
- Tell your staff what you are planning to do, what you hope to achieve, why you are doing it, and the benefits that may be in it for them.
- Provide a mechanism by which anyone can discuss any queries or concerns about the audit process beforehand.
- Involve *all* levels of employment, from senior board members to junior staff.
- Consider issuing a pre-interview questionnaire which you can review before your interview (eg three best and three worst aspects of their job).
- Design a simple framework around which you can develop your audit/survey format.
- Ask simple, open questions.
- Give people time to consider questions asked, or points that have been raised.
- Allow plenty of time for the interview, and sufficient time afterwards to write up the key findings.
- Tell your staff what you plan to do with any information collected.
- Involve them as much as possible in any further decisions or developments arising from the survey.

After the audit:

- Inform staff and/or representatives about the key findings and what the company plans to do in response.
- Provide (where possible) an indication of time frames as to when and how these changes can expect to occur.
- Involve staff and/or representatives.
- Explain why any suggestions may not be taken forward by the company, don't just refuse.
- Follow up any changes you make to ensure they are having the effect you intend.
- Keep staff and representatives involved at all times during the design and implementation of any systems or procedures resulting from the audit.
- Lead by example – managers can communicate important signals about the importance of managing stress.

Stress management policy

[S11038] A proactive management strategy for occupational stress should include development of a policy. This should highlight the company's philosophy and recognition of the issue, and clearly state the procedures in place to eliminate or manage stress associated with work.

The policy should include:

- the company's definition of occupational stress;
- recognition that stress is a work issue that will be taken seriously and sympathetically by management;
- acknowledgement that stress is not an illness or a weakness;
- explanation of the procedures the company will implement to identify occupational stressors and implement risk reduction control systems;

- details of personnel who have been allocated with responsibilities for implementing and monitoring risk reduction measures for stress;
- explanation of employees' responsibilities and procedures for reporting to management when they feel they are suffering from stress; and
- training and support packages, including any counselling services that may be available.

On-going monitoring

[S11039] Like any other health and safety system it is necessary to monitor systems and procedures to ensure they are effective. Monitoring will have the added benefit of helping to identify a stressed employee who is unwilling to report they have a problem. Monitoring can be undertaken in many ways including:

- monitoring during team meetings, by encouraging a general discussion about positive and negative aspects of work that have occurred;
- monitoring during appraisals and other one-to-one meetings with staff;
- implementing an anonymous stress report box in which people who are finding a work situation to be continuously problematic can report this without any fear of recrimination;
- circulating staff attitude or opinion surveys which will encourage feedback from workers;
- observation of workers' behaviour patterns and relationships with colleagues (looking for any obvious differences from the norm, or spotting relationships/atmospheres that appear to be tense); and
- monitoring accident and incident statistics, and sickness absence records.

Information and training

[S11040] Provision of information and training is a general duty under the *Health and Safety at Work etc Act 1974, s 2(2)(c)*. This will include information and training in occupational stress. Information should be given to employees to enable them to understand the organisation's views on occupational stress, its policies and its procedures for dealing with stress and the part that employees can play in managing occupational stress for themselves.

Training of managers and supervisors in how to recognise symptoms of stress, and how to be pre-emptive when addressing any problems, was identified as a key benefit in the research project undertaken by the British Occupational Hygiene Research Foundation.

Training can be used to underpin any findings from surveys or assessments that may have been undertaken, and to help the policy on stress to be understood and effectively implemented. Training should be provided to managers and to staff in general. Managers will need to know:

- what stress is and it contrasts with or effects occupational stress;
- symptoms of stress and how to recognise someone who may be suffering from stress;

- health risks associated with stress;
- occupational factors which cause or contribute towards stress;
- identifying, separating and managing personal and management stressors;
- outline of company policy and procedures on stress;
- identifying measures to identify stress; and
- understanding of measures to manage/reduce occupational stress.

Employees will need to understand for themselves exactly what stress is, and how to recognise for themselves when they or a colleague may be experiencing symptoms. The training will enable the organisation to demonstrate that it is a caring employer.

The Chartered Institute of Environmental Health (CIEH) has developed a one day accredited course called 'Stress Awareness'. This provides a detailed programme that covers the key issues associated with occupational stress. The fact that is accredited by a recognised body will help to increase its profile to employees who may be more cynical about the subject itself, or the organisation's commitment to employee welfare.

Other stress management techniques

[S11040.1] As part of the above measures, employers should seek to implement stress management techniques that focus on primary interventions, but also have secondary and tertiary interventions to underpin the management system and deal with workers at all stages of a stressful situation. To recap, components of these stages are as follows:

Primary interventions

These seek to eliminate occupational stress symptoms at source, rather than support someone who is experiencing a stressful situation at work. Development of a cohesive stress prevention and management policy is a key to such a pro-active approach, supported by training and other mechanisms to implement the policy objectives.

Specific pro-active interventions that employers could take include:

- developing, or further improving, communication channels with employees so that they feel involved and have a better understanding of workplace issues and changes;
- including aspects of stress assessment and review during job appraisals and personal feedback sessions;
- redesigning jobs to improve efficiency, comfort and productivity;
- involving staff in any changes to their working pattern or equipment, and in the work process;
- keeping up to date with technology and equipment available, and ensuring that tools and equipment do not become too outdated or cumbersome;
- monitor and review working hours and environmental conditions to ensure the workplace is as pleasant and conducive to work as it can reasonably be;

- encourage early reporting of stress by workers, with a cohesive policy of supporting actions following the report being made; and
- more attention to the person/job fit during the recruitment process.

Secondary interventions tend to help people to manage stress without seeking to eliminate or modify workplace stressors. Examples may include:

- training of employees in how to recognise stress in themselves;
- promoting self-help or coping mechanisms when someone is reporting stress;
- introduction of mentoring or 'buddy' systems amongst the workforce when someone is feeling under stress;
- provision of 'quiet rooms' or other such places where someone can go if they are feeling under excess pressure; and
- grievance procedures for staff who feel that they are not being treated fairly.

Tertiary interventions seek to assist workers who are experiencing on-going problems from their working life, without making any changes to the source of the problem:

- counselling service for someone who needs support and assistance in helping themselves to cope;
- access to various specialist treatments such as Cognitive Behavioural Therapy (CBT), Psychoanalysis; or
- yoga, meditation (including sessions on mindfulness) or other relaxation classes offered during lunch breaks.

In reality an effective stress management system should include elements of all the interventions mentioned above. The most effective system will focus predominantly on primary interventions, and will have the full commitment of senior management which is visible to all the workforce.

Training and Competence in Occupational Safety and Health

Steve Granger

Introduction to training and competence in occupational safety and health

[T7001] This chapter considers the subject of training and competence. The revised edition reflects the current and ongoing developments in organisation competency frameworks – including that at the Institution of Occupational Safety and Health (IOSH), the International Network of Safety Professional Organisations (INSHPO) and several major organisations who have influence over their supply chain.

It will provide details of the training and competence of employees and the safety profession. We will discuss the legal matters relating to health and safety training and competence and look closely at the structure of the current UK and European marketplace for training products, as well as the role of government and industry to ensure that peoples training and qualifications are suitable and sufficient for their needs.

We will illustrate the people and processes with respect to the training and competence requirements of the workforce and identify how to select the right process from the wide range of products on the market. We will also look at comparisons across borders and look at the latest developments in harmonising standards of training and competence. The descriptions below apply equally to manual and managerial skills and tasks.

What is competence?

Defining competence

[T7002] Looking at the concept of competence from a plain English perspective, The New Oxford Dictionary of English defines 'competence' as 'the ability to do something successfully or efficiently and 'competent' as having the necessary ability, knowledge, or skill to do something successfully.'

The (UK) National Vocational Standards define competence as 'the ability to perform to the standards required in employment across a range of circumstances and to meet changing demands'.

In UK case law a competent person is viewed as 'one who is a practical and reasonable man who knows what to look for and how to recognise it when he sees it' (Gibson v Skibs A/S Marina and Orkla Grobe A/B and Smith and Coggins Ltd [1966] 2 All ER 476). It could also be added that they should also know what to do with the acquired knowledge once they have found it.

It is difficult to find a universal definition of competence, but training, skill and appointment have featured in the UK definition for many years. The European definition of competence varies slightly to the UK in that it is described as 'a combination of knowledge, skills and attitudes', instead of the UK term 'experience'. The European version incorporates the individual's autonomy in decision making, thus introducing an indication of responsibility.

Is it just about the law and liability?

[T7003] All of these definitions basically amount to the same thing, perhaps with a slightly different slant, that competence is a combination of knowledge, experience and skills to be able to do something (note; this may be a manual or managerial activity). This concept has been enshrined in the legislature through the European Framework Directive and subsequently into UK law through the *Health and Safety at Work Act* via the *Management of Health and Safety at work Regulations*. The supporting guidance to these regulations also added the important aspect of 'recognising one's own limitations' of competence which has transcended into professional codes of conduct.

Training is specifically mentioned within *section 2* of the *Health and Safety at Work etc Act 1974* ('*HSWA 1974*') which prescribes the general duties of employers. *Section 2(2)* of the 1974 Act states that in addition to other requirements, it is an employer's duty to ensure 'the provision of such information, instruction, training and supervision as is necessary to ensure, so far as is reasonably practicable, the health and safety at work of his employees'. This in fact does not place an absolute duty on an employer to provide training, but recognises that having carried out generic and specific risk assessments of the workplace, an employer may well believe that training would be an effective measure of control. Many employers take this approach in addition to other risk control methods (Note the disparity between some subordinate regulations which adopt a more strict application in higher risk situations).

The term 'competence' is not specifically identified within the *HSWA 1974* other than an allusion in *section 19(1)* which refers to the appointment of inspectors. Here it relates to 'such persons having suitable qualifications as it thinks necessary for carrying into effect the relevant statutory provisions within its field of responsibility'.

Perhaps the most important part of the *HSWA 1974* in relation to training and competence is in *section 15* which relates to the power to make health and safety regulations which may provide more detail of what is required. The *Safety Representatives and Safety Committees Regulations, SI 1977 No 500*, require an employer to allow a trade union appointed safety representative to take time off 'to undergo such training as may be reasonable in the circumstances'. Other regulations describe training content and even assessment processes necessary before appointing competent people, or refer to industry schemes. For example; UK gas safety law and the combined effort between the

Institution of Gas Engineers and Managers (IGEM) and the industry skills advisory body; Energy and Utility Skills, as being the benchmark of acceptability. The release of their report; Standards of Training in Gas Work 2015, provides a good example of an industry which recognises future training and competence development needs to be properly co-ordinated and managed by its organisational stakeholders and individual members. The process used also links with education and assessment frameworks such as the Qualifications and Curriculum Framework (QCF) described in more detail below.

Regulating for health and safety training and competence

[T7004] The *Management of Health and Safety at Work Regulations, SI 1999 No 3242* (MHSWR) also make several statements particularly in relation to the requirement for training.

These Regulations are the main focus when considering competence in health and safety. Specifically in the context of safety management they state that:

> 'Every employer shall, appoint one or more competent persons to assist him in undertaking the measures he needs to take to comply with the requirements and prohibitions imposed upon him by or under the relevant statutory provisions.'

Regulation 13 specifically relates to capabilities and training for all employees. *Regulation 13(2)* particularly states that every employer shall ensure that his employees are provided with adequate health and safety training, and very specifically relates to the various times that this should be given, ie—

- on recruitment;
- on being exposed to new or increased risks because of being transferred or having a change of responsibility;
- on the introduction of new work equipment or a significant change to it;
- on the introduction of new work equipment;
- on the introduction of a new system of work or a change to an existing system.

Regulation 13 also says that this training should be repeated periodically and be adapted to take account of any changes or new risks and should also take place during working hours.

Employers should make themselves aware of any specifically cited requirements in their sphere of business, in addition to the more general management requirements described in the MHSWR. They should give the necessary training required that will lead their workforce to perform competently with regards to health and safety. Simple examples of the statutory duty to train can also be found in the *Provision and Use of Work Equipment Regulations, SI 1998 No 2306* (PUWER) which covers almost any item of equipment used in the workplace. These regulations cover information provision on how to use equipment safely, abnormal situations and, importantly, that the information should be given in a way that the recipient understands it. They also require employers to ensure that only those authorised (competent in the employer's view), that have received appropriate training which is not only for the worker but also for those who supervise them—

9.—(1) Every employer shall ensure that all persons who use work equipment have received adequate training for purposes of health and safety, including training in the methods which may be adopted when using the work equipment, any risks which such use may entail and precautions to be taken.

(2) Every employer shall ensure that any of his employees who supervises or manages the use of work equipment has received adequate training for purposes of health and safety, including training in the methods which may be adopted when using the work equipment, any risks which such use may entail and precautions to be taken.

The principles of what is outlined in both of the regulations above needs to be understood by workers in relation to their own safety and health and that of those who are affected by their work, and is found in other references to safety training. Most health and safety regulations and their supporting literature now contain considerable sections on the subject. It is usually the case that when regulations describe a level of responsibility, the issue of competence and 'suitable and sufficient training' is either directly referenced, or implied by general requirements of health and safety law.

Societal risk and personal safety

[T7005] In most work situations the law relates to the employer providing training and the employee following that training. There are a few situations where personal responsibility is placed on the employee to apply this principle, especially where there is a significant risk to people other than themselves. Statutory trade skills training and safety registration schemes can be found in key hazard legislation, which may also provide named training programmes. Activities worth particular noting in this area are radiation protection advisors, gas fitting, mining, the carriage of dangerous goods and diving.

Legislation relating to major industrial hazards has also introduced the statutory duty to develop both the processes and the people who can either prevent, or at least control, the impact of incidents. To obtain an operator's licence, it is necessary to provide information on how those who work in the high hazard areas will have their competence developed and maintained.

In all other instances a training needs analysis is required to be undertaken. This commences with a risk assessment and investigation of the necessary provision and training standard that are expected in either statutory or contractual responsibilities. In many cases, either the HSE or the industry Sector Skills Council/Body (SSC/B) will have developed a Sector Qualification Strategy (SQS) for that industry's needs, and employers should look to this for guidance rather than trying to remember all of the regulations that apply.

From training to competence

[T7006] On the basis that competence includes knowledge and skills gained through appropriate training and sufficient experience, it is important to recognise that in an employment situation it is the responsibility of the employer to ensure these are in place before they appoint someone to a job, hence the legislature usually making it an employer's *'absolute'* duty without any *'reasonable'* qualification or softening.

Additionally it may be argued within the expectation of contractual terms when supplying services as a contractor or consultant, that such a person; a) knows what they are doing, b) is capable of doing it, and c) will not try to do anything they are not properly able to do.

Equally where those skills are not present – or not kept up to date, the issue may be raised as a *'capability'* issue if those skills were once held, or it was reasonable (or lawful) to expect them to be held.

Additionally, someone may be qualified or able to undertake a task to a certain level or in a specified area, so they could be considered competent either generally at a lower level or in part. This particular dilemma is one that has faced most professions and occupations during their development stages. It has led to the formation of professional bodies and trade associations who, as one of their major reasons for existence, have set standards of practice for their own particular disciplines.

This is perhaps where competent performance related to health and safety is very different to other professions and trades, in so much that everyone, whether at work or not, needs to have knowledge of how to remain safe and healthy and from many different angles and perspectives.

It becomes particularly relevant in workplaces where both employers and employees have moral and legal responsibilities both for themselves and others in terms of maintaining a safe working environment.

Additionally, specialist advisers in health and safety will need to have much higher and broader levels of competence in occupational health and safety issues, so that the advice they give to both the management and the workforce in organisations is clearly understood and of a standard expected from a professional practitioner.

But we should not just consider the safety practitioner when discussing health and safety competence. Everyone needs to have a level of competence in terms of looking after themselves or those around them. This applies to those who actually carry out work activities and those who manage them. It also applies to the supply chain and contractors, as their work (or products) may be the cause of incidents in the workplace.

In order to develop or select training products with a view to establishing competency schemes, it is useful to have an understanding of:

- The learning process, assessment and certification as a means to establish competence
- The organisations who work in the field of education, to ascertain what they do and how to compare them
- The terminology used in education to determine the training product quality and value to the consumer and the end user
- The current status of Occupational Standards in occupational safety and health
- How to compare qualifications and understand the terminology and framework for qualification development in the UK and Europe and see how it compares in the global market

The education framework

How do we acquire knowledge, skills and experience?

[T7007] It is not possible to cover all of the learning theories or types of training that are available to develop the knowledge and skills required for today's workforce and management. However, it is possible to discuss some of the concepts that have been shown to work when faced with the task of developing people.

Studies have shown that 'doing' is better than 'not doing', particularly for adult learning. They have also proved that engaging trainees with innovative problem solving is more likely to develop useable skills that will stay with someone, than the ability to recite something they have studied deeply.

We have also learned that not everyone is suited to one type of learning or assessment process – so variety in terms of training delivery methods, taking advantage of technological advances and on-the- job training has been introduced as 'blended learning' for Vocational Qualifications (VQ). The European project for lifelong learning and the development of transferable skills means that training today needs to be portable and pertinent to the employee, as well as relevant to the employer.

The last few years have seen the emergence of new or revised education frameworks with the aim of achieving these objectives. We have also seen a significant shift in the equivalence and standardisation process in the European Union. Although not perfect, such transparency within Europe also helps when benchmarking across the globe as by either design or default, there are many similarities.

This is particularly relevant when considering the employability of health and safety professionals in international organisations whose activities span several countries.

Who are UKCES, Ofqual, SSC's and AO's and what is a SQS, NOS, QCF, SCQF ?

[T7008] When seeking, designing, or delivering health and safety training it is useful to understand something about the validation processes and schemes of recognition. Whether it is simple 'in house' training session or an internationally recognised brand name of accreditation, there are a few concepts to understand so that the right choice can be made when selecting what is needed. The training world has just as much jargon as the health and safety community. This section outlines the different bodies that accredit qualifications and what that actually means in terms of outputs.

We will only concern ourselves with Vocationally Related Qualifications (VRQ), although health and safety may appear on the National Curriculum for schools in the future. One body which is particularly involved in the development of workplace skills is the UK Commission for Employment and Skills (UKCES), a social partnership, led by Commissioners from large and small employers, trade unions and the voluntary sector.

The term VRQ's can relate to any type of employment and any level of employment. There have been attempts for many years to create a system whereby qualifications could be benchmarked against each other to deliver 'transferable skills' between employers and even industry sectors. This might be useful for example when determining pay bands and management levels across an organisation. When looking at a particular trade or industry it is useful to refer to the Sector Qualification Strategy (SQS), which is a plan developed for the industry sector, by those in the industry sector. We will discuss the SQS for health and safety in the workplace later.

The Qualifications and Credit Framework (QCF) is the national credit transfer system for education qualification in England, Education in Northern Ireland and Wales. Scotland has its own system – the Scottish Credit and Qualifications Framework (SCQF).

In 2010 the QCF replaced the National Qualifications Framework (NQF), in England as the latest move towards this goal. Similar systems were established in Wales (Credit and Qualifications Framework for Wales, CQFW), Scotland (Scottish Credit and Qualifications Framework, SCQF) and Northern Ireland (National Framework of Qualifications for Ireland, NFQ). These frameworks are similarly constructed but unfortunately have a different number of levels attached to them.

To see a comparison of these frameworks and see how cross border qualifications are compared follow the link or visit the Ofqual website or any of the bodies identified above.

To add to the confusion they must all link back to the European Credit Framework (EQF) which will be discussed further below.

The UK established these frameworks as part of the European harmonisation programme regarding the development of lifelong learning and skills. The intention being to create a common standard of skill sets across the European Union, to enable free movement of people and work. In the UK the Office of Qualifications and Examinations Regulation (Ofqual) is the independent regulator of qualifications, examinations and assessments in England and of vocational qualifications in Northern Ireland. The Ofqual website and of those organisations above provide the following description of Vocational Qualifications;

> 'VQs are vocational qualifications which are designed to allow learners to learn in a way that suits them, and give learners the skills that employers are looking for. Occupational VQs are designed to meet the national occupational standards (NOS) for a particular sector/work place and employers rely on these qualifications for evidence that an employee is competent to carry out the job.'

> 'NOS are set by Sector Skills Councils (SSCs) and industry and they define the competency requirements employers in a particular sector, such childcare or retail, expect employees to demonstrate when carrying our particular jobs at work. VQs are often designed to prepare learners to be able to carry out a job role or to confirm competence of doing that role in the workplace.'

> 'Vocationally-related qualifications (VRQs) may not be based on the national occupational standards and can be designed to allow learners access to further/higher education and/or the workplace. Some VRQs are technical certificates which assess the knowledge requirements of apprenticeships.'

How does the Qualifications Credit Framework work?

[T7009] Although employers are not compelled to use it, the QCF does provide a suitable structure to initiate discussion and understanding of good practice in the provision of training and development of competence. Therefore it is important to understand the basis of QCF and the technical jargon that goes with it so that we can understand training processes and products that fall both within and outside of the framework.

The QCF is built using 'Units', which are described as 'the building blocks of all qualifications'

These units describe what someone should be able to do if they have successfully undergone the process of receiving and understanding the related knowledge, and where appropriate, they have developed the skills in order to be able to utilise this knowledge properly.

> 'Every unit and qualification has a credit value and a level. One credit represents ten notional hours of learning, showing how much time the average learner would take to complete the unit or qualification. Levels indicate difficulty and vary from a basic entry 1 to level 8. In addition to level, the depth of knowledge on any given level is also considered (consider a cube of knowledge with both height and depth). There are three types of qualification at every level:
>
> - Awards (1 to 12 credits)
> - Certificates (13 to 36 credits)
> - Diplomas (37 credits or more)'

Once units have been determined and put onto the framework they can be used individually as a standalone subject, or using 'rules of combination' they can be assembled in any way the user wishes. To safeguard the quality of units and qualifications that are derived from them, only Ofqual approved Awarding Organisations (AO) can access the system. The process of building Units and developing Qualifications is undertaken by the Awarding Organisation using the online system known as the Regulatory Information Technology System (RITS).

Providing the rules are followed, commercial and employers training providers can certificate against them by registering with the Awarding Organisations. This further cascades responsibility for maintaining the standards.

Using this approach it is possible to create either standard or a unique qualification for a particular group/industry. In theory the QCF was intended to reduce the mass of training titles that was confusing industry, but in reality the new framework has only just started on this issue and it will be an ongoing process of refinement that is needed to redress the problem.

The QCF is now available online and anyone can search The Register of Regulated Qualifications for either units or qualifications. The register provides details of who owns the title and details of the curriculum.

What does a 'recognised qualification' mean?

[T7010] Although this sounds simple in reality there is still confusion and disparity in the training market. For example; use of the terms 'certificate' and 'diploma' may mean something when applied within the official framework described above, but of course these terms are also found in everyday language and as such, in unaccredited training programmes. That is not to say that unaccredited training is not suitable or fit for purpose, but it does mean that the buyer or user should know what they want when searching the myriad of suppliers and courses available. The term 'approved course' is similarly misleading; for example; approved by whom, and suitable for whom? Generally this means that it is Government approved and possibly funding may be available in certain circumstances.

This may give confidence in some cases but may be less relevant in others where, for example a trade body may take the lead in determining standards. And remember the system described above is UK centric.

The term caveat emptor, or let the buyer beware – might usefully be supplemented in the situation of training procurement with variation 'let the buyer know what they want and understand the best place to get it'.

As well as the QCF which relate predominantly to vocational training, there exists a similar framework known as the Framework for Higher Education Qualifications (FHEQ) which is managed by the Qualifications Assurance Agency (QAA). This works in a similar way to the QCF but has its own levelling system and generally holds University programme details. When discussing education systems it is important to be able relate to both of these frameworks, as either can be a route for the qualification of safety professionals.

Together these two frameworks cover most of the government accredited training and education products for employees at different levels of skill, knowledge and understanding. There may be other forms of 'accreditation', for example by professional organisations. Once again it is for the buyer of these products to be wise about their choice and its intended use.

The main benefit of accredited courses is that they have been benchmarked to a standard and regulated by Ofqual. They may also be used in conjunction with other parts or units on the framework to devise specific development programmes which lead to one of the achievement levels; Award, Certificate or Diploma described above. Traditionally, the other main difference related to part funding by the government so that certain types of skills training were at a lower cost. In reality this is disappearing with the exception perhaps of qualifications from the framework which applies to apprenticeships.

Provided that the employer understands what they are required to provide in terms of training, they are at liberty to choose what best suits their risk control strategy. The only caveat is when certain types of health and safety training are regulated (for example; gas safety described at the top of this chapter), either in terms of content or by whom the provider or certification is underwritten. Details on such limitations would be found within Regulations, Approved Codes produced by the HSE or industry-approved processes.

In reality, the training products market has established generally good practice in the formation of either accredited or unaccredited training and qualification products and, irrespective of the QCF, many of these common templates of skills already existed.

Foreseeable difficulties of achieving the complete harmonisation across the UK or Europe for that matter are varied and include; differences in national legislation, differing sector standards, commercial confidentiality and copyright to name a few. This has meant that the reality of creating this standardised training is, and will remain, more of an ongoing and extremely complex project rather than a completed one.

For the purpose of illustration we will focus on two types of skills – firstly those of employees relating to their health and safety training and secondly, those of the health and safety profession itself.

A case study; Using the QCF for health and safety qualifications

[T7011] The QCF has an Entry Level plus 8 levels. Health and safety qualifications will start at the entry level, or more likely at level 1 for all employees. This would include hazard awareness and the necessary risk control features of 'looking after oneself and of those around you'. A level 3 type qualification introduces leadership and supervisory skills. For example, being responsible for how others work safely and ensuring the workforce and workplace follow appropriate safety procedures – but without necessarily requiring a skill to develop these procedures in the first instance. Technician entry into the Institution of Occupational Safety and Health (IOSH), the Professional body for health and safety practitioners, can be made with a level 3 qualification, providing the qualification has been approved by the Institution (not the QCF described above) who set the standard for the industry through their membership structure.

At a level 6 qualification in safety management, competencies involving professional judgement and management skills are necessary. This is the current level at which IOSH sets its membership qualifications for Graduate Members (GradIOSH) and provides the route for further Initial Professional Development (IPD) and the route towards Chartered member status (CMIOSH). The completion of a qualification can be by various means and should be suited to the learner. Some people favour examination whilst other favour on-the-job training and work experience, known as the National Vocational Qualification NVQ route.

As with a level 6 'QCF accredited qualification' which has been recognised by IOSH, an approved University Bachelor's or Master's Degree falls into the same membership catchment criteria as a qualification on the QCF.

IOSH utilises the high standards of the QCF and FEHQ to establish the quality of training, assessment and accreditation in this respect. By doing so it has assurance that the Awarding Organisation (AO) who actually issue the certificates, will have already been scrutinised and had to follow strict criteria in order to get their qualification accredited. However, IOSH has also produced its own criteria for such qualifications so that it may accredit those courses which are not on the QCF. This is particularly relevant when considering international education programmes – especially those outside Europe.

The SCQF is similar to the above but the levels differ in that this framework has 12 levels instead of 8. For practitioner entry level the SCQF is set at level 10. Similarly the National Qualifications Framework for Ireland has 10 levels.

The website of Ofqual provides comprehensive plan and details of cross border comparison.

Using the FHEQ for developing other professions

[T7012] It has been a long-standing request from the health and safety community that health and safety should be included in MBA programmes, so that potential future leaders of industry should have health and safety included within their academic programmes. In recent years the approach to developing competency frameworks has been adopted by many professions. One of the advantages of using a framework approach is that until recently, graduate programmes in the non-safety professions had very little input of workplace safety management. Consequently managers and executives have always seen safety as an 'add-on', not integral to their chosen professional skills.

With the framework approach it makes the updating of such programmes much easier, as bespoke units can be added to enhance existing programmes and the work required to fulfil them can be integrated into the existing learning process. For example, a design-based degree might require the candidate to complete a 'design, build, maintain and de-construct risk assessment' and also to furnish supplementary research on the necessary risk control measures might be used.

The competency framework for any given profession needs to integrate a number of elements; the level of responsibility (worker to executive), the technical knowledge (protection of self, to design and implementation of complex safety procedures) and practical skill in both 'information processing' and 'physically doing' (using Personal Protective Equipment to completing a high risk safety case) can be represented by a three-dimensional shape. A cube is the conventional model to use – but in reality life is seldom so symmetrical.

What makes a safety and health qualification?

[T7013] The National Occupational Standards for health and safety (NOS) for occupational safety and health relate to all employees, not just safety practitioners. They are grouped into general and practitioner standards. There is also a section for the regulatory profession such as HSE or local authority inspectors.

They were established in the mid 1980's, and have been through several revisions. They now reside under the guardianship of the Sector Skills Body; Proskills, and were last revised in 2011/12. In principle, these standards are the foundation on which safety education of the workforce is built. They relate to the actual task that the person is required to do in the workplace. The standards are grouped by subject area; within each area performance criteria are described and these are defined by measurable learning objectives that are outputs that the candidate should be able to do. The next review of this process is likely in 2016/17. Development of competency frameworks should refer to the NOS and the Sector Qualification Strategy so that objective and valid development processes and products (ie training courses) can be suitably accredited in both a national and international context. The European situation of qualification parity is actually regulated for and explained further below – but the same principles can (and should) be applied when benchmarking and considering equality of status and competency of practitioners who belong to different professional bodies.

The verbs that describe these competency outputs are important and part of the Grade/Credit Descriptors. For lower level qualifications simple outputs are expected, conversely for higher levels and thus more complex outputs, the verbs change accordingly.

For example; a level 3 qualifications would ask the candidate to; list items of personal protective equipment required to control the risk of exposure to a particular hazard'. Whereas a level 6 qualification might expect a candidate to; determine the suitability of the same equipment and recommend a more suitable strategy for risk control.

Those seeking more on the use of such key words are advised to look under Bloom and Krathwohl's 'A taxonomy for learning teaching and assessing' 1956 and particularly useful is the revision and model of Blooms Taxonomy presented by Rex Heer at Iowa State University in 2012 or the Southern England Consortium for Credit Accumulation (SECC) Credit Level Descriptors for Higher Education released in 2010 which is particularly useful in the development of the professional level competency processes.

Having determined what the output expectation is, it is necessary to derive a curriculum that involves the necessary knowledge to enable the learner to comprehend the subject sufficiently for them to be able to perform the desired skills. Again we should consider the context and variance from one extreme; following a set of simple instructions for carrying out relatively simple physical task such as 'close the machinery guard', to the other which requires cognitive output such as 'design the safety systems for a chemical process plant by undertaking a HAZard and OPerability Study'.

Assessment

[T7014] It is also necessary to determine how the evidence will be collected and presented by a candidate in order to determine if the skill and knowledge has been learned and applied. This is achieved by designing the assessment process. Although this sounds simple, it is probably the most complex part of the qualification development. It has to be fair, equitable to the described outcome, applied consistently across learners and be open to appeal and scrutiny. The assessment process should consider the needs of both the standard and the learner, and specifically the individual needs of members of society who might be taking it.

Literacy skills, physical and visual impairment, language, location, transparency and application of examination rules all play their part here. Of note in recent years is the problems of plagiarism – especially in the use of the internet by candidates delivering assignments and projects. Larger assessment bodies and universities use fraud detection programmes, such as 'Turnitin', to identify where this may have occurred in electronic submissions – indeed this is one of the reasons it may be asked for in this format.

When considering skill based qualifications (ie doing something) the necessity to undertake practical assessments is crucial. Consider the competency (knowledge and skills) required to wear respiratory protective equipment. There are practical elements relating to a) selecting it and checking it, b) putting it on, and c) knowing what to do if it malfunctions.

Although it is possible to obtain written answers relating to the above, these do not give any indication of the person's ability to actually do it – especially in an emergency. An assessment process is required to combine both the knowledge relating to this task and of course a demonstration that it can be done. The only rules relating to how these are done are generally concerned with collecting evidence of satisfactory performance – and of course, the safety of the assessment process itself.

Assessment opportunities

[T7015] With the digital age it is no longer necessary to provide large volumes of text to either prove or record that learning has taken place and that the individual can deliver the particular skill. One of the early means of determining hazard perceptiveness can be seen in the DVLC theory driving test and the use of a simulated driving environment. With today's technology the ease and capability to take film clips and upload them instantly, the process of recording a skill demonstration becomes simple and easy to store on personal data files. If we add sound to this any additional test relating to the candidate's underpinning knowledge can be added by the assessor, providing evidence of both the candidate's performance and the fairness of the assessment process.

In only a few years we have seen computer games progress from simple table tennis to actually participating in 3D movies. The training world is still limited in the use of this technology, but the use of hand-held game devices and the simulated application of hazard control mechanisms would seem very adaptable to the health and safety market.

These have served the aircraft industry very well in their pilot simulator training. Until recently such thinking has largely been limited by cost. We now see e-learning tools that do provide realistic experience and extreme flexibility in the learning and assessment process. The safety industry needs to take advantage of such technology. Examples of possible applications could be: multi-lingual delivery (on-site inductions), practical and physical assessment (banksman's signalling in response to video streaming) and management problem solving (virtual incident management), all employing the gaming technology that is commonly available today.

One of the current shortfalls of employee training is the reluctance of trainers to denote that 'competence' has been shown. Perhaps using the techniques described here (ie digital recording and video capture of the assessment process) might give trainers less concern about this, since it can be shown (and replayed) in their assessment process, and hence reduce their largely perceptual liability. Indeed such compelling evidence of employee performance may be a valuable investment and simplify the requirement of employers to prove how and when training of their employees took place, and importantly whether such training was successful.

Achieving and maintaining competence in safety practice

[T7016] As described above; the National Occupational Standards for OSH Practice are recognised by the profession as representing the core competence requirements for those in health and safety practice. There are, however, many ways of achieving these standards through the various routes of qualification, coupled with the development of skills and experience. The standards, which are based on safety management models reflect the current management-based approach to safety which has gone a long way to improving health and safety within companies in the UK.

Candidates for an NVQ are assessed against these occupational standards. This type of qualification is an assessment of the ability of a person to perform to the specified standards. It involves an assessor who will monitor and lead the candidate through the process of developing a practice portfolio. For a VQ at this higher level, it is also necessary for the assessor, backed up by the verification processes required by the awarding bodies, to determine that a candidate has the requisite domain knowledge to carry out the competence-based tasks required.

It should be clarified here that academic qualifications (such as those from the HE sector, or commercial market) are not assessments of competence in the same way as an NVQ qualification. They are a demonstration that a programme of academic work, leading to the development of knowledge and some skills, has been undertaken and formally assessed by examination and/or other suitable form of academic assessment. This by itself would not be sufficient to demonstrate competence to do the job, but with suitable experience and further development of skills in a practical setting, the knowledge gained is a vital part of the competence equation.

However, it should also be recognised that although the achievement of an NVQ actually demonstrates competence, this is only in the workplace setting in which the person undertaking the qualification is employed. This means that the broader knowledge base required of a competent practitioner has not necessarily been covered or assessed to the same extent to a person who has taken the academic qualifications.

Employee representatives training

[T7017] Many people who perform the function of safety representatives in the workplace, as defined by the Safety Representatives and Safety Committees Regulations 1977 (SI 1977 No 500), actually train in health and safety. There is a specific programme run by the Trade Union Congress (TUC) under the auspices of the National Open College Network (NOCN), which has been specifically designed to meet the requirements of this group. This programme, the TUC Certificate in Occupational Health and Safety, is a level 3 programme that builds on the TUC basic stage 1 and stage 2 health and safety courses or other union's equivalent courses. The assessment of the course is by a competence-based type of assessment similar to the NVQ qualifications with the addition of workplace-biased projects of practical value to the candidate's workplace and union.

The course is available for union representatives on either a part-time day-release basis or by distance learning. The course additionally gives access to higher education for those who wish to progress further.

The TUC Education Service normally pays for the course, provided that the candidate is nominated by a trade union and it is a valuable contribution towards competent safety advice in the workplace. This is recognised by IOSH who currently offer Tech IOSH membership access for those with this qualification and who are able to demonstrate at least two years pro-rata full-time equivalent experience.

Where there are representatives of workplace safety who are not members of a trade union, the employer is responsible for determining how they should be trained.

The training and competence of non-employees

[T7018] Leading on from the requirement of the development of a workforce that is competent to perform its role in a healthy and safe manner is the requirement to look at all those people who also may be present in a workplace.

Although organisations have the ability to control the training and competence of individuals that they have directly employed by developing programmes which meet their specific requirements, there is one area that they need to control but do not necessarily have complete control over - outside contractors coming into their workplaces. This is particularly relevant in some industrial sectors, such as construction and related industries, where the use of contracted labour is extensive. However, at some point all employing organisations may have people other than their own employees within their workplaces. Many organisations control this situation by having 'permit to work' systems that ensure that anyone entering a workplace has sufficient knowledge of the health and safety arrangements that are in place.

Passport schemes

[T7019] Within the last few years there has been a move to a more transferable skills type of scheme. This can allow those who do move between workplaces to show that they have sufficient underpinning knowledge of health and safety matters to allow briefer induction programmes to be given, when moving between different workplaces and sites. These are the safety passport schemes.

These passports are available from several industry-related schemes and allow employers to determine who may be given access to their workplaces. The employing organisation makes their workplace a passport-controlled environment. The passports themselves generally take the form of a credit-card sized plastic card, containing the name, photograph and other relevant details of an individual; they also verify what training a person has undertaken towards receiving this passport. The passport usually also has an expiry date.

The advantages of passport schemes are—

- they can save time and money through reduced induction training as they are the starting point for workers training
- they demonstrate an organisation's commitment to a safe and health working environment;
- they can create good relationships with the supply chain between organisations and their suppliers;
- they can be used by employees of companies as well as sub-contractors;
- contractors can move between companies and demonstrate that they have the necessary health and safety awareness;
- they can be used to control access to worksites;
- probably most importantly, they can increase a positive safety culture within organisations, which can lead to a reduction in accidents or ill health;

Helpfully the HSE have also produced a core syllabus containing all the items that the schemes should include within their training and assessment programmes.

From incompetence to competence and back

[T7020] As we have seen competence is the development of sufficient knowledge, skills and experience to be able to perform a task to the laid-down standards in a workplace. These standards may be prescribed by a legislative requirement, national standards or by best practice in the workplace. However competence is prescribed or developed - there is a cycle that needs to be recognised in the attainment and then in the subsequent maintenance of the desired performance level.

Competence is actually a developed attribute and passes through several stages, these are recognised as distinct states:

- unconscious incompetence--a person is unaware of the requirements to perform in a competent manner;
- conscious incompetence--a person begins to undertake training and becomes aware of what they do not know;
- conscious competence--a person has undergone sufficient training to be able to complete a task in a competent manner and is aware of this;
- unconscious competence--a person now performs tasks in a fully competent manner and has now developed to become unaware of this as it is part of their behavioural patterns.

At this point in the cycle there can be two pathways, a person can continue to perform in a competent manner or they can unconsciously develop bad habits (possibly through overconfidence and familiarity), which can lead to a diminishment in their performance. Any level of competence is time-bounded.

Skills and knowledge become out-dated very quickly in the light of new technologies and practices, and is a particular aspect that needs to be considered in maintaining the health and safety of the workforce. Continuing Professional Development, re-certification, re-licensing are just some of the issues relating to sustaining a competent workforce.

There is a need for those responsible for the workforce to revisit both their own current knowledge and that of their employees on a regular basis, to ensure that performance relating to health and safety is in fact to the level they believe it to be.

There can be a number of triggers that can identify that there may be a need for refresher or further training. This is not necessarily simply because of a time lapse since the initial training took place - other issues such as the introduction of new equipment, changes to methods of use, the re-site of equipment, changes of shift patterns etc. should be considered. Also safety audits may identify that there have been behavioural changes that need addressing.

Many of the external agencies, which play a role in the quality control of training programmes, will specify an expiry date for their accreditation. The safety passport schemes have particular expiry dates usually 3-5 years built into the actual passport, whilst organisations such as IOSH have a three-year expiry term on their Managing and Working Safely series of programmes.

Continuing Professional development (CPD)

[T7021] For professionals who practice in health and safety there is a particular need for the maintenance and development of their skills. The arena of health and safety is constantly changing, both in terms of legislative requirements and in the knowledge of best practice in health and safety issues. Most members of IOSH are now required to maintain CPD records via the on-line system. Affiliate Members are also recommended to undertake CPD as many of them are in a developmental stage of their careers

The benchmark for the amount of CPD is estimated at five days per year, but as CPD is very subjective and based on reflective practice, this will vary depending on the stage a person is at in their career.

Employers looking to recruit either an employee or consultant are advised to ask to seek evidence of CPD as an indication of continuing competence. This is an extension of the concept of life-long learning which is now being incorporated into the initial training programmes for those in occupational health and safety practice, and in fact features within the national vocational standards.

As an example of good practice, the IOSH CPD online system allows members of the Institution to maintain their CPD as efficiently as possible without creating unnecessary burdens or paperwork. Also as part of this revision activity, the practitioner takes overall ownership of their own CPD requirements and this allows them to be more flexible in the type of activities that they can use to demonstrate CPD. Although there are a number of well-designed

CPD taught courses available, these form only part of the requirement for CPD and practitioners are being encouraged to think beyond attendance at events.

The CPD categories in the system above include: maintaining core skills of the professional, developing new professional skills and developing transferable skills such as management activities.

To sum up, it is critical for good performance in health and safety, that the knowledge and skills acquired initially are developed and constantly maintained. This can only be achieved by regular checking of the competence of the entire workforce. Some industry sectors have formalised this approach within their safety management systems and here competence frameworks have been developed and monitored on a formal basis. This is good practice and should be considered by those who have not yet taken this approach. A revised health and safety competency framework for health and safety is being developed by IOSH with a release due in 2015.

Working in the global market

[T7022] The reality of applying the principles of safety management in more than one country and of having a truly global workforce presents new opportunities for the development of competence, both for the workforce and safety managers.

Existing recognition schemes, memoranda of understanding and the use of international standards rather than regional ones, all indicate that there is a desire to develop portable skills around the globe. An example can be seen in the mutual recognition procedures written between IOSH and the Board of Canadian Registered Safety Professionals (BCRSP).

The EU Directive 2005/36/EC of the European Parliament and of the Council of 7 September 2005 on the recognition of professional qualifications required members to work together to harmonise professional qualifications. The European Union sponsored 'EUSAFE' project was created in response to this and delivered its final report to the European Commission in January 2013. This project provided the framework for developing a revised and harmonised competency scheme for OSH practitioners which was based on the vocational activities for technician and manager levels.

One of the complications presented is that of compulsory regulation and registration of those who practice safety management. In some EU states there is a requirement to enrol on the state register, whereas in others; such as the UK, it has only recently been introduced as a voluntary option with the Occupational Safety and Health Consultants Register (OSHCR). This was set up following Government publication of 'Common Sense, Common Safety', and under the scrutiny of the HSE. Its declared vision is;

> The aim of this register is to help businesses find advice on managing their general health and safety risks. The register is only open to those health and safety consultants who have met certain standards within their professional bodies.

Along with language and cultural differences there is a need to understand the processes and management styles of the regions of the world of work. Personal development plans take on another dimension and professional development may be more pertinent in terms of learning a language than those directly concerned with the workplace safety. This is an opportunity to enrich and develop the professional which is just as valid as acquiring a new technical discipline, and is certainly classed as a 'transferable skill' in the IOSH Continuing Professional Development process.

Conclusions

[T7023] Health and safety training is required by law at all levels of employee and is determined by the hazards encountered and the degree of involvement with the management of safety within an organisation.

Standards for qualifications are currently under review, both in the UK and across Europe. The objective is to develop more transferable and transparent competencies.

Commercial and cultural barriers cause resistance to change within the training industry. This has resulted in confusion over the use of terminology. Buyers need to understand exactly what they want from training products and how it links to their employees work related development activities.

Organisations that develop and regulate training standards can provide information to assist in the use of accredited qualifications, but the commercial training market is already established in many industries so introducing any change will be difficult. Essentially employers need to understand the current situation so that they can make an informed choice.

To be successful in developing people appropriate methods should be used in the training and assessment process. There are opportunities to utilise many different forms of communication and technology in the transfer of knowledge and assessment of individuals.

Competence is a mixture of several components. Developing, assessing and maintaining competence can be achieved in many ways, including work based and academic routes. If it is to remain current Continuing Professional Development is required to be undertaken.

Ventilation

Andrea Oates

Introduction to ventilation

[V3001] Ventilation involves the movement and supply and/or removal of air. This can be by natural ventilation, from doors and windows, or mechanical means, by a powered fan for example, to or from an enclosed space or workplace. In protecting the health and safety of people at work, ventilation provides fresh air to workplaces and controls exposure to hazardous airborne contaminants from work processes.

Ventilation provides fresh air for:

— breathing, providing oxygen and diluting and removing carbon dioxide;
— diluting unpleasant odours;
— removing excess heat from the work environment to maintain thermal comfort; and
— controlling harmful airborne contamination from work activities.

Industrial processes and activities generate a very wide range of airborne contaminants, including dusts, fumes and vapours, many of which represent an inhalation health hazard to the workers exposed to them. Health and safety legislation requires employers to manage these risks. In general, this requires identifying hazards and assessing risks, followed by action to prevent, so far as is reasonably practicable, or adequately control those risks.

In workplace contaminant control, ventilation can be divided into two categories: local exhaust ventilation (LEV) and dilution or general ventilation. These ventilation approaches represent options in the wider portfolio of control measures that may be chosen in a particular risk situation. A hierarchy of control options is available where prevention is not reasonably practicable. These measures include:

— substituting a substance for a less toxic one;
— substituting a process for a safer method;
— changing a process (for example, reducing operating temperatures to decrease fume emission; or handling substances remotely);
— segregation – total enclosure of the process;
— LEV to control at source;
— dilution ventilation to reduce the concentration of a contaminant;
— as a last resort, personal protective equipment (PPE) including respiratory protection.

By controlling airborne contaminant exposure at source, LEV represents a powerful tool to control the work environment. Its application represents the major focus of this chapter. Legal requirements relating to the provision and use of ventilation systems are also considered where appropriate.

In an LEV system, the collection point (enclosure, hood or slot) is positioned as close to the source of the contaminant as possible. Air is drawn into the system ductwork by means of a suitable fan, passes through an air cleaning device, and is finally expelled at a point outside the workplace.

With dilution or general ventilation, fresh air is induced to enter the workplace air and this causes dilution of the airborne contaminants present. This air flow may be induced either by natural means (by opening windows or roof vents for example) or by mechanical means (by electrical fans in walls or roofs for example). With dilution ventilation there is no special control at the source of emission.

Under the *Health and Safety at Work etc Act 1974* employers have a duty to protect employees and others against risks to their health and safety, so far as is reasonably practicable. This includes the risks arising from the use, handling, storage and transportation of substances together with the provision and maintenance of safe plant and systems of work, which includes ventilation systems.

Under the *Management of Health and Safety at Work Regulations 1999 (SI 1999 No 3242)*, the employer has a duty to carry out a suitable and sufficient assessment of the risks to the health and safety of their employees and others for the purpose of identifying the measures needed to control the risks. They must then take steps to eliminate the hazards identified or adequately control exposure.

In workplaces where people are exposed to airborne contaminant hazards, statutory regulations require hazard identification, risk assessment and prevention or, where this is not reasonably practicable, adequate measures to control exposure. This requirement is set out in the *Control of Substances Hazardous to Health Regulations 2002 (SI 2002 No 2677)* (COSHH Regulations), the *Control of Asbestos Regulations 2012 (SI 2012 No 632)* and the *Control of Lead at Work Regulations 2002 (SI 2002 No 2676)*.

LEV may also be applied as a control measure where there is exposure to ionising radiation and dangerous substances as set out in the *Ionising Radiations Regulations 2017 (SI 2017 No 1075)* and the *Dangerous Substances and Explosive Atmospheres Regulations 2002 (SI 2002 No 2776)*. This is not discussed further in this chapter.

Where ventilation is used as an air contaminant control measure, the various regulations require competent maintenance and, for LEV, statutory examination and testing at specified intervals. This is to ensure ventilation systems continue to work effectively. In addition, the *Workplace (Health, Safety and Welfare) Regulations 1992 (SI 1992 No 3004)* require workplaces to be supplied with sufficient quantities of fresh air. These legal aspects are further discussed below.

This chapter will focus on the principles of operation, design features and the application of LEV systems as well as the requirements for maintenance, examination and testing. The statutory requirements for the provision and use of these systems is considered, as is the Health and Safety Executive (HSE) guidance on LEV: *Controlling Airborne Contaminants at Work – A Guide to Local Exhaust Ventilation* (LEV) (2017, HSG 258, HSE Books). Other aspects of ventilation, including dilution ventilation and the provision of fresh air to workplaces, are also discussed.

Sources of contamination

[V3002] Many industrial operations and tasks, including machining metals, welding, paint spraying, charging reactor vessels and heating processes, create sources of airborne contamination. Airborne contaminants can be broadly divided by their physical state into aerosols and gases, which in turn can be classified into various groups:

An aerosol can be defined as any disperse system of liquid or solid particles suspended in a gas (usually air). These include the following:

— Dusts – solid particles made airborne by the mechanical disintegration of bulk solid material;
— Spray – large liquid droplets (of a few microns upwards) produced during condensation or atomisation;
— Mist – fine liquid droplets (up to a few microns) produced during condensation or atomisation – a micron (1 μm) is one millionth of a meter;
— Fume – small solid particles (usually the aggregates of much smaller primary particles) produced by condensation of vapours or gaseous combustion products;
— Smoke – solid or liquid particles resulting from incomplete combustion, aggregates of very small particles;
— Bioaerosol – solid or liquid particles containing biologically viable organisms (viruses, bacteria, fungal spores), ranging from submicron (ie less than a micron) to over 100μm.

Aerosol particles above 100μm do not remain airborne and therefore are unlikely to be an inhalation hazard. Particles in the range 3–20μm are known as the 'thoracic fraction' and if inhaled they may be deposited beyond the larynx in the respiratory tract. Particles in the range 0.5–7μm are very fine particles. They are invisible under normal lighting conditions but are hazardous as they can avoid respiratory defences and penetrate to the deepest parts of the lungs. Particles in this size range are known as the 'respirable fraction'. When particles reach the deep lung, they may be absorbed into the systemic blood circulation and cause toxic harm by accumulating in target organs. For example, exposure to cadmium oxide fume can cause irreversible kidney damage. Alternatively, particles may be deposited as insoluble particles in the deep lung and provoke serious tissue reactions, for example, crystalline silica particles and silicosis.

These fine particles move with the air in which they are suspended and only sediment out slowly. It is important to emphasise it is the aerodynamic

[V3002] Ventilation

diameter of aerosols that determines how they move in the atmosphere and where they will deposit in the respiratory system.

Gases include substances which are normally in the gaseous state at room temperature and vapours, the gaseous form of a substance which is normally a liquid at room temperature, such as toluene and benzene vapours for example. Toxic vapour/air mixtures have a density virtually identical to that of air and disperse with movement of the air in which they are located.

Harmful aerosols and gases disperse with the movement of the air in which they are located, so they can be controlled at source by LEV. The system collection point is positioned close to the source of emission allowing the contaminated air to be drawn into the system and removed.

Local exhaust ventilation (LEV)

[V3003] The objective of an LEV system is to remove contaminant efficiently using the minimum volume of exhaust air possible. Unfortunately, there are many examples of poor design which do not achieve this objective. There is a wide variation in the size and complexity of LEV systems and the appropriate system will be determined by the particular risk situation and other factors such as building constraints. For example, a small hood can be used to extract fume from a manual soldering iron, while a large walk-in booth may be used for paint-spraying large components. With an LEV system, there are important energy cost implications since air expelled from the workplace has normally to be replaced with warmed clean air. It is therefore important that the system is well designed to achieve satisfactory control with a minimum volume of exhaust air.

The basic components of an LEV system are:

— a collection point, an enclosure, hood or slot to collect and remove contaminant close to its source;
— ducting to transport the contaminant and exhaust air to the air cleaner and ultimately to the discharge point;
— an air-cleaning device to remove the contaminant from the airstream. This is normally placed between the hood and the fan to protect the integrity of the fan from corrosive or hot airborne contaminants;
— a fan which provides the power to move a sufficient volume of air through the system; and
— a discharge point to ensure efficient removal of contaminated air from the system.

LEV contaminant collection points can be divided into partial enclosures, hoods and slots. The collection point is a key part of the system and requires careful design. It is crucial it is positioned as close as possible to the source of emission. The type of system collection point chosen in a specific situation will depend on the particular industrial process and access requirements for materials and people to carry out tasks. Where the work process allows this, it is desirable to enclose the contaminant-emitting process as far as possible in a box-like structure. However, it may not be possible to enclose a particular process at all, and in this case hoods and slots may be used as the inlet point for the local removal of contaminant.

A simple LEV system is shown below and the next section will consider the various components.

Inlet ⟩ Hood ⟩ Ducting ⟩ Air cleaner ⟩ Air mover ⟩ Discharge ⟩

Partial enclosures

[V3004] A partial enclosure is a box-like structure which surrounds a process emitting airborne contaminant. This type of collection point is used where operator access or the entry/egress requirements for process materials only permit a partial enclosure rather than a total enclosure of the process. The HSE explains that partial enclosure is a compromise between containment and accessibility.

The amount of exhaust air needed for a partial enclosure depends on the size of the openings required in the enclosure and the velocity of the air needed to overcome the tendency of the contaminated air inside to escape. Turbulent air conditions both inside and outside the enclosure can influence the likelihood of contaminant escape.

Face velocities ranging from 0.5–2.0 ms^1 are used and the total flow volume can be calculated from the equation $Q = VA$. (A is open face area in metres (m); V is air velocity in metres per second (ms^1); Q is total volume flow rate in cubic metres per second ($m^3 s^1$). The volume flow rate equals the area of the opening multiplied by the face velocity necessary to prevent contaminant escaping. The face velocity required depends on the toxicity of a contaminant and its momentum on release. For example, a higher face velocity would be needed in paint spraying than in a paint dipping operation. With toxic substances, a face velocity in excess of 1.5 ms^1 will be required. Both the size of the enclosure and the size of the openings should be kept to a practicable minimum. A three-sided booth, such as a laboratory fume cupboard or paint spray booth, is a common type of partial enclosure.

In a fume cupboard, there is a sliding front panel which the operator can adjust. Modern fume cupboards are fitted with a by-pass mechanism or variable fan arrangement which allows the same face velocity to be achieved for all positions of the panel. The contaminant source is located within the fume cupboard and the sliding panel not only provides access for the operator but also provides an opportunity for the contaminant to escape. Therefore, in order to prevent escape, sufficient air velocity is required across the face of the cupboard to induce the contaminant to enter the system ducting. A fume cupboard should be aerodynamically designed to allow a smooth pattern of inward air flow and it should be carefully sited in the workplace to avoid external local turbulent conditions (for example, busy passageways). The minimum face velocity will depend on the nature of the contaminant, the operator's required movement and local air conditions near the partial enclosure.

In large ventilated enclosures, as used in paint spraying for example, the operator may carry out work entirely inside the booth. Paint spray aerosol particles rebound from sprayed surfaces at high speed and a sufficient air flow rate is required to overcome this in order to prevent escape and to induce these particles to enter the system ducting. In this case, a face velocity of 0.7 ms^1 is recommended with an even and stable face velocity across the whole booth. Often a perforated plate is placed at the back of the booth to ensure an even laminar air flow inside the booth. It is important that the operator is not placed between the contaminant source and the booth exhaust point. Where possible the worker should stand at the side of the work. In some situations, a rotating turntable for the work piece is used to prevent this occurring.

An average air velocity of 0.5 ms^1 is normally required for a partial enclosure. Some processes are more toxic and emit particles at high velocity, in which case a higher face velocity of at least 1.5 ms^1 may be required.

Hoods

[V3005] Often it is not possible to enclose a process because of operator and/or process material access needs. In such cases, a hood or slot collection point may be required, but these are not as effective as enclosures.

Hoods vary from small apertures to extensive canopies above, below or at the side of the emission source. They should not be placed above if operators have to lean over and put themselves between the contaminant source and the ducting exhaust point. Hoods should be located as close as practicable to the contaminant source. In a large hood, the face velocity distribution may vary across the hood (ie higher close to the duct entrance and lower at the extremities). In some work processes, contaminant is generated with considerable energy and a receptor hood may be placed in the path of the contaminant to collect and remove it. For example, a receptor hood may be placed above a source that is emitting a hot contaminant, which then rises and enters the hood of its own volition.

On the other hand, a captor hood pulls contaminant into the system collection point and here the air flow needs to be sufficient to overcome any forces for the contaminant to disperse. Capture velocity is the air velocity required at the source of contaminant emission, which is sufficient to move contaminant to the mouth of the extract to be successfully captured by the system. Small aerosol particles and gases and vapours will move with the air flow while larger particles will be more difficult to divert from their natural pathway. However, captor systems can be designed to intercept such particles. A minimum capture velocity of 0.25–1.0msec^1 is required for low velocity particle release with up to 2.5–10msec^1 needed to capture high speed particles. For example, a capture velocity of 0.25msec^1 would be needed to capture the gentle evaporation of solvent from a degreasing tank while a capture velocity of around 10msec^1 would be required to capture high velocity particles from a grinding process.

Other factors relevant to hood design are the toxicity of the contaminants, the quantities emitted and the required size of the hood. The objective, as stated earlier, is to achieve sufficient capture velocity utilising the minimum volume of exhaust air. Hood size is based on the size of the source or what is practicable and convenient. Applying flanges (ie adding raised edges) to the entrance of a hood reduces the amount of air drawn in from behind. This improves air entry conditions, giving an even velocity distribution across the hood and a lower entry pressure loss (see **V3007** below). Captor systems should not be sited in areas of local air turbulence – for example, next to busy passageways. Captor hoods require a greater total volume flow than enclosures for equivalent effectiveness and every effort should be made to convert captor hoods into partial enclosures with the use of flanges and general boxing in.

Low-volume high-velocity (LVHV) systems employ very small hoods, with capture velocities from $50-100 msec^1$, placed very close to the source of emission. They are used for high speed particles, those generated by grinding or sanding for example, or for gases and vapours generated during welding and soldering. For successful application, such hoods need to be integrated into the design of the tool or they may be cumbersome and difficult to use.

Slots

[V3006] Where the aspect ratio (length:breadth) of a capture hood is less than 0.2, it is called a captor slot. Hoods may be replaced by slots where there are limitations on space and access or process reasons. For example, slots are used on degreasing and electroplating tanks to remove vapours produced at the surface of these tanks. Where there is a wide surface area of potential contaminant release, two slots are often employed to minimise capture distance. According to FS Gill (*Ventilation* in *'Occupational Hygiene'* eds Gardiner, K and Harrington, JM, Blackwell, 2005), it is difficult to pull air into a slot from more than 750mm.

Ducting

[V3007] Enclosures, hoods and slots are connected to air cleaners, fans and the discharge point by ductwork. This consists of straight pieces, bends and changes in cross-section, dampers and other fittings which connect the inlet with the discharge point. The volume flow rate of the system is determined by the air velocity required at the system inlet and its cross-sectional area. The size of the ducting will be determined by the air velocity required or the building space available.

Energy losses due to friction are expressed as a pressure loss and this is proportional to velocity squared (v^2). The larger the cross-section of the duct, the lower the air velocity and the pressure loss will be. Conversely, the smaller the cross-section of the duct the higher the air velocity, pressure loss and noise levels will be.

The pressure loss for each section of ductwork, including any dust collection device, is calculated and the total pressure loss of the system is determined. The duty of the system fan is calculated based on the volume flow rate at the total pressure loss for the system. If particles of powder or dusts are to be transported by an LEV system, a sufficient transport velocity needs to be maintained in the system to keep these particles suspended in the airstream and prevent them from settling out in the ducting.

If only gases and vapours are transported by an LEV system, transport velocity is less important. The values for suitable transport velocities vary from 5–10 ms^1 for gases, vapours, smoke and fumes to up to 25 ms^1 for heavy dusts. If a particular transport velocity is required, once the desired volume flow rate has been determined for the system, the duct diameter can be fixed at a size to allow that transport velocity to be achieved. Ducting is normally made out of galvanised sheet steel unless corrosive gases or vapours are to be transported, in which case stainless steel or plastics are used. The design of the system should avoid abrupt changes in direction and changes in section should be smooth to minimise energy losses. There must be adequate access for internal cleaning of ductwork and inspection ports for monitoring equipment.

Fans

[V3008] These provide the motive power for moving air through the system.

A fan must be matched with the volume flow rate required and the total pressure drop of the system. With fans of similar design, relative performance depends on their diameter and speed of rotation. A fan has two fundamental components: the impeller which rotates on a shaft and the casing which guides air to and from the impeller.

Two basic types of fan are used, axial and centrifugal. Axial fans are installed in-line in the contaminant airstream. They have a cylindrical casing, are compact and fit readily into ducting. Axial fans can overcome only low resistance to flow and they are generally used in wall and roof-mounted units. The fan motor is positioned in the airstream and therefore should not be used with high temperature or corrosive contaminant exhausts.

With centrifugal fans, air is drawn into the centre of the impeller and thrown off at high velocity into the fan casing and discharge. Centrifugal fans can deliver the required airflow against high system resistance and are the most common type of fan found in LEV systems. Centrifugal fans can be used with abrasive dusts and corrosive chemicals and, although the blades wear out quickly, they can be readily replaced. Fans should be sited outside the building or as near to the discharge as possible as when a leak occurs in front of the fan, the ductwork will be under negative pressure and contaminant is less likely to escape.

Air cleaning devices

[V3009] Before exhausted air is discharged to the atmosphere or re-circulated it needs to pass through an appropriate air cleaning device. Usually such devices are located in front of the fan to give the fan some protection from damage by contaminants. A range of devices are available depending on the contaminant in question. Dry dusts may be removed from the airstream by air filters, contaminants with larger particles may be removed by cyclones, and those with smaller particles by bag filters. Dust particles that will easily take an electrical charge can be removed by electrostatic precipitators as long as there is no fire and explosion hazard. Wetted particles can be removed by a venturi scrubber method. Organic vapours can be removed by activated charcoal or a filter with a suitable adsorbent.

Discharge

[V3010] The behaviour of an airstream close to the entry and exit of the LEV system is very different. At the suction end the influence of the system will be very localised, while at the discharge end the system will have a much greater influence and cause considerably more disturbance in the air. It is therefore important the discharge point is located away from any air inlets so as to avoid re-circulation of exhaust air. Discharge stacks should be positioned so as to ensure efficient mixing with the atmosphere and this may mean extending them to at least 3m above roof level.

Further information on the fundamentals of LEV systems can be found in the HSE publication *Controlling airborne contaminants at work: A guide to local exhaust ventilation* (LEV) HSG 258.

The LEV design process

[V3011] The basic components of LEV systems have been outlined above. Unfortunately, there are many examples of inadequate systems. For example:

- poorly designed hoods placed in the wrong position based on misunderstandings about how aerosols and vapour/air mixtures behave in air;
- captor hoods placed too far from the emission source;
- inadequate total volume flow; and
- lack of provision for ease of maintenance.

Designing LEV systems that will efficiently and reliably collect airborne contaminants on a continuing basis is a difficult and complex task. A suggested sequence is as follows:

(1) From knowledge of the health effects of the contaminant(s), decide on tolerable exposure level. Consider Workplace Exposure Limits (WELs) and application of the principles of good hygiene practice. For carcinogens and asthmagens, exposure must be reduced to as low a level as is reasonably practicable.
(2) Identify all sources of emission sources and rank in order of magnitude.
(3) Examine process and operator work methods.
(4) Perform occupational hygiene measurements (static and personal monitoring) and re-rank emission sources in terms of contribution to operator personal exposure.
(5) Consider control by methods which do not involve ventilation systems; consider how the number of sources and/or emission rates can be reduced.

Control by local exhaust ventilation

(6) Decide on the shape and size of the hood, slot or enclosure. Consider possible changes to the work process itself.
(7) Decide on appropriate capture velocities and distances for captor hoods and face velocities for receptor hoods.
(8) Calculate the required volume flow-rate for the system.

(9) Design ductwork to transport exhaust air in the most energy efficient manner (to minimise losses due to friction); consider required duct transport velocity where this is relevant.
(10) Determine what air cleaning device will be required.
(11) Consider how replacement air will be provided to the work area.
(12) Choose an appropriate fan taking into account to the total volume flow rate required and the total pressure drop for the system.
(13) Plan for commissioning and maintenance, examination and testing of the LEV system to comply with the *Control of Substances Hazardous to Health Regulations (as amended) 2002 (SI 2002 No 2677)*.

Maintenance, examination and testing of LEV systems

[V3012] In addition to well-designed LEV there is a requirement for a high and continuing level of performance. LEV systems, as with any workplace plant and equipment, must therefore be appropriately maintained with specified tasks carried out at defined frequencies by designated personnel. However, maintenance is more than the tasks carried out by maintenance workers and should include visual checks, inspection, servicing, reviewing systems of work and any remedial work to ensure system effectiveness. A thorough examination normally includes a visual check, quantitative measurement of performance and an assessment of control.

To evaluate performance, among other things, the following tasks need to be carried out: the measurement of air velocity, volume flow rate and pressure at various points in the system, together with observation of the characteristics of local air movement. The following questions should be answered:

(1) Are airborne contaminants successfully captured near to their point of release?
(2) How do the current measured parameters differ from the design specification values?
(3) What are air velocities at different points in the system?
(4) What pressure or suction is the fan developing?
(5) How much pressure (energy) is lost in the different parts of the system? This may be particularly important at filters or dust collectors.
(6) Is the exposure of operators under adequate control?

Ventilation system pressure

[V3013] Monitoring ventilation system pressure and its change is an important measure of a system's performance. A pressure difference is required for air to flow in a ventilation system and it will flow from a region of high pressure to one of low pressure. Such pressure differences can be induced naturally or artificially, by a chimney and a fan for example. Pressure appears in two forms: static pressure and velocity pressure. Static pressure is the pressure exerted in all directions by a stationary fluid (gas or liquid). In a moving fluid, static pressure is measured at right angles to the direction of flow to eliminate the effects of velocity. Static pressure can be negative or positive with respect to normal atmospheric pressure. Velocity pressure is a measure of the kinetic energy of a fluid in motion. It provides the driving force of 'wind'. Total pressure is the sum of velocity pressure and static pressure.

Comparing static pressure measurements at different points with those normally found in an LEV system can be used for checking performance and fault finding. The duct pressures on both sides of a fan are illustrated in Figure 1. A U-tube is used to measure static pressure (p_s), velocity pressure (p_v) and total pressure (p_t).

Figure 2: Duct pressures

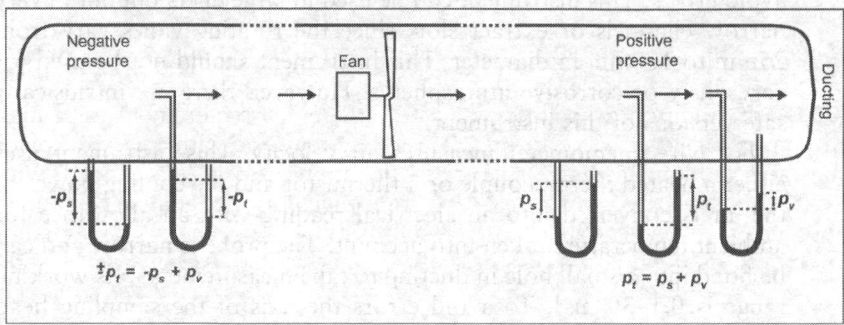

At the suction side of a fan static pressure is negative, and on the discharge side it is positive. Velocity pressure gives the push to move air through the system and this is always positive.

Making some assumptions, for air, this can be represented by the simple equation:

$$Pv = 0.6v^2$$

(v is expressed in metres per second (ms^{-1}), velocity pressure (Pv) is in newtons per square metre (Nm^2) or pascals (Pa)).

Total volume flow

When a quantity of air is moving within a system or duct, the volume flow can be calculated by the following formula:

$$Q = vA$$

(v is the average air velocity across the cross-section of the duct (or face velocity) in ms1; A is the cross-sectional area of the ducting or face in square metres (m2); and Q is in cubic metres per second (m3s1)).

Monitoring equipment

Monitoring equipment used in evaluating the performance of a ventilation system includes the following:

(1) A pitot-static tube measures system pressures (and air velocity indirectly). It measures velocity pressure inside a ventilation system irrespective of the static pressure at the point of measurement. Velocity

pressure can be converted into air velocity using the equation above. Measurements should be carried out in a straight section of ducting with streamlined air flow and minimal turbulence. This instrument is sensitive to duct air velocities above $3 ms^1$. The instrument should only be used in ducting and it is not suitable for use at the face of hoods or booths. It can also be used to measure static pressure.

(2) A rotating vane anemometer measures air velocity. The rotating vane is connected electrically or mechanically to a display dial. The axis of the rotating vane should be positioned parallel to the airstream contours to avoid errors. This instrument can be used in large ducts but not in very narrow channels or extract slots since the rotating vanes vary from 25mm to 100mm in diameter. This instrument should not be used for very dusty or corrosive atmospheres. However, there are intrinsically safe versions of this instrument.

(3) A hot wire anemometer measures air velocity. This instrument uses either a heated thermocouple or a thermistor and the cooling power of the air is converted into an electrical reading with an allowance for ambient temperature taken into account. The probe is narrow and can be fitted into a small hole in ducting to take measurements. Its working range is $0.1-30$ ms^1. To avoid errors the axis of the sampling head should be parallel to the airstream contours. Regular calibration of anemometers is required to avoid measurement errors.

(4) A dust lamp enables visualisation of air flow patterns. This allows study of the movement and distribution of dust aerosols. Many harmful airborne dust particles are too small to be seen by the naked eye. When a parallel beam of light is shone through a dust cloud the particles reflect light by forward light scattering to an observer whose eyes are shielded from the main light source. This is similar to visualising dust particles floating in a ray of sunlight. This is not a quantitative method but does allow direct observation of the performance of a ventilation system. Other direct reading dust monitors can be used to make observations.

(5) Smoke tracers allow visualisation of air flow patterns. These devices generate a plume of smoke and can be used to visualise air movement patterns generated by ventilation systems.

(6) Air monitoring equipment enables evaluation of environmental levels and personal exposure of operators indicating whether adequate control of exposure is being achieved. Static and personal monitoring using air collection pumps and sampling devices (as appropriate to the specific contaminants) can be carried out. In addition, direct monitoring equipment may also be used to assess control. Infrared spectrophotometers may be used for gases and vapour which absorb infra-red radiation for example.

In checking system performance, static pressure is measured behind each enclosure or hood and at various other points in the system. Air velocity is measured at the face of the enclosure or point of emission and at various points in the ducting. The key question is whether the system is performing as was intended in the design specification.

Dilution ventilation

[V3014] Insufficient fresh air can lead to tiredness, lethargy, headaches, dry or itchy skin or eye irritation. Dilution or general ventilation refers to the movement of air into and out of a work area. A minimum standard of ventilation is required in all enclosed workplaces. Fresh air must be supplied, and stale, hot or humid air replaced at a reasonable rate. Dilution ventilation may also be used in some situations where it can remove small quantities of low toxicity airborne contaminants generated by work activities.

Careful positioning of inlets is required to ensure that fresh air is contaminant free and sometimes incoming air may need to be filtered to remove particulates. In some cases, natural ventilation with doors and windows may be sufficient, but often mechanical systems will be required to provide fresh air and dilution and removal of contaminants. Natural ventilation is generated by temperature and pressure changes in buildings where differences can produce an upward movement of air known as the stack effect. This is where cooler air enters at a low level and, as it is warmed and mixes with contaminant, it rises and finally escapes through a roof vent or high-level opening.

Dilution ventilation can be used to control exposure to airborne contaminants where:

— LEV is not practicable;
— the contaminant is of low toxicity;
— only small quantities of contaminant are released at a uniform rate;
— contaminants can be sufficiently diluted to control operator exposure adequately.

In recirculation systems there needs to be adequate filtration to remove impurities, and there should also be a significant fresh air component. With air-conditioning, both the temperature and humidity of recirculated air must be adjusted, since the aim is to provide comfortable conditions for workers. With natural ventilation there is limited scope to control air pressure and temperature differences. Where mechanical systems are used, however, these parameters are more controllable. As with LEV, mechanical systems including air conditioning should be cleaned, maintained, examined and tested. Where a mechanical system provides dilution ventilation to dilute and remove airborne contaminants, a warning device is required in case of breakdown.

HSE guidance recommends the fresh air supply rate should be at least 5–8 litres per second per building occupant. Relevant factors to be considered are the amount of floor space for each building occupant, the processes and equipment involved and whether the work is strenuous.

Chartered Institute of Building Services Engineering (CIBSE) Guide A: "Environmental Design" recommends an outdoor air supply rate of 10 litres/second per person for offices.

If gas, coal or oil-fired equipment is used, fresh air requirements will depend on the type of flue arrangement in use. Where dilution ventilation is used to control airborne contaminants from work processes, the correct air supply rate must be determined. This requires calculation of the rate of release of the contaminant into the space to be ventilated.

Legal requirements relating to ventilation

[V3015] There are statutory requirements regarding the application and maintenance, examination and testing of LEV and dilution or general ventilation systems. An outline of the principal duties is set out below.

The Health and Safety at Work etc Act 1974 (ch 37)

Under *section* 2 of the Act, an employer must provide and maintain a working environment that is, so far as is reasonably practicable, safe and without risk to health.

In addition, the HSE publication *Controlling airborne contaminants at work A guide to local exhaust ventilation* (LEV) HSG 58 sets out: "It is important to remember that companies who sell LEV or provide related services are also subject to duties under health and safety law (eg sections 3, 6 and 36 HSWA Act). This means that anyone who, for example, supplies, installs, commissions or tests LEV, has health and safety duties with respect to the people who use it (or are meant to be protected by it). Consequently, it is not just the owner of an LEV system who has responsibilities".

The Control of Substances Hazardous to Health Regulations (as amended) 2002 (SI 2002 No 2677)

HSG 58 sets out that:

- employers must assess the degree of exposure and the risks to their employees, devise and implement adequate control measures, and check and maintain them;
- employees must use these control measures in the way they are intended to be used and as they have been instructed;
- employers must ensure that the equipment necessary for control is maintained 'in an efficient state, in efficient working order, in good repair and in a clean condition';
- employers must ensure that thorough examination and testing of their "protective" LEV (LEV may have been required for reasons other than COSHH, eg nuisance) is carried out at least every 14 months (unless otherwise stipulated), other engineering controls at 'suitable intervals' and must 'review and revise' ways of working so that controls are being used effectively;
- the frequency of examination and tests should be linked to the type of engineering control in use, the size of the risk if it failed or deteriorated and how likely it is to fail or deteriorate;
- employers and employees should give the person carrying out the thorough examination and test all the co-operation needed for the work to be carried out correctly and fully;
- any defects should be put right as soon as possible or within a time laid down by the person who carries out the examination;
- the person carrying out the thorough examination and test should provide a record, which needs to be kept by the employer for at least five years.

'Assessment of risk' (reg 6) and 'Adequate control' (reg 7)

The COSHH Regulations *(reg 6)* require an assessment of the risks to health created by work involving 'substances hazardous to health' (as defined in *reg 2*). Such risks must be prevented (ie the hazard eliminated) or, where this is not reasonably practicable, adequately controlled.

Where appropriate to the activity and the risk assessment, control of exposure at source is required as part of a hierarchy of control and this includes the provision of adequate ventilation systems and their associated maintenance *(regs 7(3)(b), 7(4)(b))*. Measures to control the working environment including appropriate general ventilation are also required under *Control of Substances Hazardous to Health Regulations (as amended) 2002 (SI 2002 No 2677), reg 7(4)(d)*.

'Adequate control' is achieved only if:

— the principles of good practice for control as specified in *Sch 2A* are applied *(reg 7(7)(a))*;
— WELs are not exceeded *(reg 7(7)(b))*; and
— for specified groups of carcinogens and asthmagens, exposure levels are reduced as low as is reasonably practicable *(reg 7(7)(c))*.

WELs are intended to prevent exposure exceeding a set level and represent the personal maximum exposure concentration averaged over a reference time period – the sampling device is placed in the 'breathing zone' of the person – of either 8 hours or 15 minutes. WELs can be used to evaluate the application and effectiveness of ventilation control. They are not to be regarded as the boundary between 'safe' and 'unsafe' conditions.

Current WEL values are listed in guidance document '*HSE Workplace Exposure Limits*', EH40/2005 (2011) www.hse.gov.uk/pubns/books/eh40.htm. This statutory guidance will be amended to include WELs for 31 chemical substances listed in the 4th Indicative Occupational Exposure Limit Values (IOELVs) Directive 2017/164/EU, which must be implemented in the UK by August 2018.

In view of the uncertainties relating to the health effects of many substances hazardous to health the general philosophy should always be to reduce exposures as low as is reasonably practicable utilising the principles of good practice specified in *Sch 2A* to the Regulations.

Use of control measures (reg 8)

The employer has a duty to ensure that LEV and other ventilation systems are properly used and applied and employees have a duty to ensure that any defects are immediately reported to the employer.

Maintenance, examination and testing of control measures (reg 9)

Any engineering control including local exhaust ventilation and other ventilation systems must be: "maintained in an efficient state, in efficient working order, in good repair and in a clean condition" *(reg 9(1)(a))*.

Engineering controls must also be subjected to thorough examination and testing and in the case of local exhaust ventilation this must be at least once every 14 months *(reg 9(2))*.

Similarly, the provision of associated systems of work and supervision must be reviewed at suitable intervals (*reg 9(1)(b)*).

Maintenance

The aim of maintenance is to ensure that an LEV system continues to perform as originally intended. Maintenance includes visual checks, inspection, servicing, observations of systems of work and any remedial work to ensure system effectiveness. Deterioration in function must be detected and remedied and the frequency of checks will depend on the nature of the risks involved and the importance of the particular system.

For LEV and work enclosures, visual checks should be carried out at least once per week (see para 172 *L5 Control of Substances Hazardous to Health Regulations 2002 (as amended): Approved Code of Practice and Guidance*, 2013). The employer should ensure that the person carrying out the maintenance, examination and testing is competent to do so in accordance with *reg 12(4)*. People carrying out examinations and tests on control measures such as LEV and PPE must have adequate knowledge, training and expertise in examination methods and techniques (para 175 ACOP).

Schedule 3 contains additional provisions relating to work with biological agents with containment measures specified in *Parts II* and *III*.

Examination and testing

Thorough examination and testing is specifically required for LEV systems every 14 months, unless the process is specified in *Sch 4* and requires more frequent examination and testing (*reg 9(2)(a)*). A suitable record of the examination and test of the local exhaust ventilation system must be kept available for at least five years (*reg 9(4)*). For all LEV systems, whether portable or fixed and including biological safety cabinets, the examination and test is to ensure that the system is still functioning at the level as originally intended. Guidance is given on the maintenance, examination and testing of LEV systems in the 2017 HSE publication HSG 258 *Controlling Airborne Contaminants at Work – A Guide to Local Exhaust Ventilation (LEV)* referenced at **V3001**.

A thorough examination will normally include a visual check, a measurement of LEV performance and an assessment of control. Where air is recirculated there must be an assessment of the air cleaning device. The minimal content of a suitable examination and test record is given in para 186 of the COSHH ACOP and the type of information required is outlined in para 189. This includes information on:

— enclosures/hoods: maximum number in use; position; static pressure behind each hood or extraction point; face velocity;
— ducting: dimensions; transport velocity; volume flow;
— filter/collection point: specification; volume flow; static pressure at inlet; outlet and across filter;
— fan/air mover: specification; volume flow; static pressure at inlet; direction of rotation;
— systems which recirculate air to the workplace: filter efficiency; concentration of contaminant in returned air.

In 2016, the HSE initiated a fundamental review of the regulatory framework for chemicals in the workplace and considered the COSHH Regulations, as well as the *Control of Lead Regulations 2002* and the *Dangerous Substances and Explosive Atmospheres Regulations 2002*.

A key finding was that the current COSHH requirement for thorough examination and testing of LEV every 14 months does not ensure that LEV plant is contributing to control of dust or fume. The HSE says it is possible for an LEV plant to pass an examination or test as it is operating as originally intended, but not effectively contributing to control. This is because, for example, it is not properly designed for the purpose it is used for or it has not been correctly commissioned. The HSE will undertake further research to "strengthen the evidence base to inform any potential future legislative change in this area" and is exploring potential amendments to COSHH.

Control of Lead at Work Regulations (as amended) 2002 (SI 2002 No 2676)

Similar duties to those found in the COSHH Regulations are found in these regulations and in the *Control of Asbestos Regulations 2012 (SI 2012 No 632)* as discussed below.

'Assessment of risk' (reg 5) and 'Adequate control' (reg 6)

The *Control of Lead at Work Regulations (as amended) 2002 (SI 2002 No 2676)* require an assessment of risks to health arising from work involving exposure to lead *(reg 5)* with a requirement for the prevention and control of those risks *(reg 6)*. Protective measures must be applied as appropriate to the situation and the risk assessment and, for control at source amongst other things, LEV must be considered *(reg 6(3)(b))*. Suitable maintenance procedures must be adopted for LEV systems *(reg 6(4)(b))* and control of the environment by 'appropriate general ventilation' is also required *(reg 6(4)(d))*.

The occupational exposure limit for lead should not be exceeded or, where it is, the employer should identify the reasons and take immediate action to remedy the situation *(reg 6(6))*. The occupational exposure limit for inorganic lead is an airborne concentration of 0.15 mg/m^3 (averaged over an eight-hour work period) *(reg 2(1))*. The limit for organic lead is 0.10 mg/m^3.

Maintenance, examination and test of control measures (reg 8)

LEV systems must be: 'maintained in an efficient state, in efficient working order, in good repair and in a clean condition' *(reg 8(1))*.

In addition, thorough examination and testing of control measures is required *(reg 8(2))* and for local exhaust ventilation systems this must be carried out at least every 14 months *(reg 8(2)(a))* with a suitable record kept for at least five years *(reg (8)(4))*. Thorough examination and testing should include the following:

— A thorough examination both internally and externally, where appropriate, of the condition of all parts of the system, ie exhaust openings, collection hoods or suction points, ductwork, dust collection and filtration units, fans or air movers;

- Measurements of static pressure at a point immediately behind each exhaust opening, collection hood or suction when the equipment is simultaneously extracting from all points;
- Measurements of air velocity at the plane of openings to enclosures, collection hoods or suction points for which the standard velocities have been specified;
- An assessment of whether lead dust, fume or vapour is being effectively controlled at each opening, collection hood or suction point. This would be complemented by lead in air monitoring.

Paragraph 185 of *Control of Lead at Work Regulations 2002: Approved Code of Practice* (L132), HSE, 2002.

The Control of Asbestos Regulations 2012 (SI 2012 No 632)

Where work is liable to result in exposure to asbestos the employer must carry out a suitable and sufficient assessment of risks to health that considers the effects of control measures which have been or will be taken *(reg 6(2)(a)* and *(2)(c))*. Where prevention of exposure is not reasonably practicable, adequate control is required. Measures to avoid or minimise the release of asbestos fibres must be applied together with control at source, which includes 'adequate ventilation systems' *(reg 11(2)(b))*.

Where maintenance of air extraction plant contaminated with asbestos is carried out, appropriate precautions should be taken and it may be necessary to use LEV (paras 298 and 299, *Managing and working with asbestos* (L143, HSE 2013).

There are similar requirements to the COSHH Regulations regarding the use of controls. Where controls are provided, the employer has a duty to ensure they are used *(reg 12(1))* and employees have a duty to make use of controls provided and report defects *(reg 12(2))*.

In addition, controls must be 'maintained in an efficient state, in efficient working order, in good repair and in a clean condition' *(reg 13(1)(a))*.

Systems of work and supervision must be reviewed at suitable intervals and revised where necessary *(reg 13(1)(b))*. For local exhaust ventilation systems, thorough examination and testing must be carried out at 'suitable intervals' by a competent person *(reg 13(2))*. Portable equipment and enclosures may require more examination and testing than static systems (for example, during asbestos removal operations). Suitable records of examination, test and repair of LEV systems must be kept available for five years *(reg 13(3))*.

HSE guidance *Managing and working with asbestos* (L143, HSE 2013) sets out that for most licensable work with asbestos, it is likely that a full enclosure will be required, unless it is not reasonably practicable. While full enclosures will not normally be required for non-licensable work with asbestos cement or other bonded materials, or for some work outside or in remote areas, it says a partial enclosure should be used for removing external asbestos cement soffits (ACOP paras 384 and 385). A thorough visual inspection and check on the integrity of the enclosure, airlocks and ducting from air extraction equipment must be carried out at least at the beginning of each shift (ACOP para 316).

Air extraction equipment must be operated while work is being carried out, during breaks and for at least one hour after each shift (ACOP para 316).

Appropriate air monitoring outside the enclosure must be carried out, for example where the air exhausted from the enclosure is discharged into an occupied building because it is not reasonably practicable to discharge externally (ACOP para 316).

All necessary air extraction equipment (including air movers and negative pressure units) should be visually inspected daily when in use and thoroughly examined and tested every six months by a competent person to make sure it is working properly to its design (ACOP para 326).

Reference is made to BS 8520 *Equipment used in the controlled removal of asbestos-containing materials: negative pressure units — specification.*

Workplace (Health, Safety and Welfare) Regulations (as amended) 1992 (SI 1992 No 3004)

Every enclosed workplace must be ventilated by a sufficient quantity of fresh or purified air *(reg 6(1))*. The aim is that stale, hot or humid air should be replaced at a reasonable rate. Careful positioning of air inlet points is required and sometimes filtration may be needed to remove particulates. In some cases natural ventilation may be sufficient but often mechanical ventilation systems will be required to achieve sufficient quantities of fresh air. Where air is re-circulated adequate filtration to remove impurities will be required.

Mechanical ventilation systems (including air-conditioning systems) should be regularly and adequately cleaned, and properly tested and maintained, to ensure that they are kept clean and free from anything which may contaminate the air (para 52, *Workplace (Health, Safety and Welfare)* (L24), HSE, 2013).

A warning device in case of system breakdown is required where this is necessary for health and safety reasons, for example, where a mechanical system is employed to dilute airborne contaminant concentrations *(reg 6(2))*. It is recommended that the fresh air supply rate is not below 5–8 litres per minute per occupant (para 57, *Workplace (Health, Safety and Welfare)* (L24), HSE, 2013).

Provision and Use of Work Equipment (PUWER) Regulations 1998 (SI 1998 No 2306)

The PUWER Regulations apply to LEV systems and their components when used at work. LEV as work equipment should be suitable for its intended purpose, maintained for safety and conform at all times with any essential requirements that applied when it was first put into service. Many LEV systems have dangerous parts (motors, fans, rotary valves etc) for which adequate safety measures must be taken.

Other relevant legislation

The *Supply of Machinery (Safety) Regulations 2008* require that machinery placed on the market, or put into service, is safe. The *Dangerous Substances and Explosive Atmosphere Regulations 2002* (DSEAR) set out user obligations relating to fire and explosion risks associated with LEV systems. Under REACH (the European Union regulation on the Registration, Evaluation, Authorisation and restriction of Chemicals), companies that use chemicals have a duty to use them safely and information on risk management measures, including LEV, should be passed down the supply chain via safety data sheets.

Ventilation: summary

[V3016] This chapter has addressed ventilation in the workplace in general and has focused in particular on LEV as a very important workplace environmental control. Where prevention of exposure is not reasonably practicable, control at source may be possible using LEV to limit inhalation exposure to hazardous substances.

The constituent components of LEV systems have been discussed illustrating how careful design is crucial. This applies particularly to the design of the system collection point, the enclosure, hood or slot. It is very important that such systems continue to perform as intended in their original design specification. This requires regular checks and maintenance by competent personnel with thorough examination and testing at appropriate intervals. In addition, the role of dilution or general ventilation in providing sufficient fresh air to workplaces and in contaminant control is important in maintaining healthy and comfortable enclosed working environments.

Statutory obligations regarding the application of LEV systems and the provision of maintenance, examination and testing have also been reviewed. There are broadly similar statutory requirements in the *Control of Substances Hazardous to Health Regulations (as amended) 2002 (SI 2002 No 2677)* the *Control of Lead at Work Regulations (as amended) 2002 (SI 2002 No 2676)* and the *Control of Asbestos Regulations 2012*. The statutory requirement to provide a healthy general environment in the workplace has also been discussed with reference to the ventilation requirements of the *Workplace (Health, Safety and Welfare) Regulations (as amended) 1992 (SI 1992 No 3004)*.

Vibration

Andrea Oates

Vibration: the scope of this chapter

[V5001] This chapter deals with the effects of vibration on people and the obligations of employers to assess, monitor and reduce the risks of the adverse health effects of hand-arm and whole-body vibration on their workforce. This area emerged from relative regulatory obscurity following EC Directive (2002/44/EC), the Physical Agents (Vibration) Directive, which was implemented in the UK by the *Control of Vibration at Work Regulations 2005 (SI 2005 No 1093)*. At the time of updating this chapter, December 2018, the impact of the Brexit vote to leave the European Union (EU) on UK health and safety law emanating from European directives was not clear.

The regulations came into operation in July 2005 but with transitional provisions for work equipment first provided before 6 July 2007 and for the agriculture and forestry sectors. These expired in 2010 and 2014 respectively and the regulations are now fully in force.

The *Control of Vibration at Work Regulations 2005* are covered in detail in a later section of this chapter, which also provides some basic information on vibration and general advice on reducing vibration exposure at work.

An introduction to the effects of vibration on people

[V5002] The effects of vibration on people can be divided into three broad classes:

- building vibration perceptible to occupiers;
- the exposure of people in vehicles and industrial situations to whole-body vibration (WBV); and
- the exposure of people operating certain tools or machines to hand-arm vibration (HAV).

This chapter considers these separately, as the cause, effect and method of control is different in each case.

Hand-arm vibration

[V5003] The effects of industrial vibration on workers received relatively little attention until an award of damages to seven ex-British Coal miners in July 1998 in the landmark Court of Appeal decision *Armstrong v British Coal Corporation* (1998), unreported. The former miners received compensation ranging from £5,000 to £50,000 for the effects of vibration white finger ('VWF'), a form of hand-arm vibration (HAV) characterised by the fingers becoming numb and turning white. In its early stages, the disease is reversible, but continued exposure leads to permanent damage and even gangrene, resulting in the loss of fingers or even a complete hand. The decision was important not only because of the size of the awards, but also in setting out the terms under which British Coal was negligent and the standards of exposure that would be reasonable (see **V5012** below). The Labour government in office at the time set up a compensation scheme for miners to avoid the need for individual litigation. The scheme ran until 1 May 2009. By June 2011, 169,609 of a total of 169,611 claims had been settled, with £1.7 billion being paid out in compensation. The number of claims and the amount of compensation were far higher than had originally been expected.

More recently, in *Bowe v Mersey Rewinds Engineering Ltd* [2018] EWCA Civ 72, 168 NLJ 7768, [2018] All ER (D) 11 (Feb), the Court of Appeal allowed an appeal by three companies who had employed the claimant over his working life against a finding that they were in breach of their duty of care in relation to exposure to vibration (see **V5012** below).

In addition, the Health and Safety Executive (HSE) has prosecuted several organisations for breaches of the *Control of Vibration at Work Regulations 2005*.

For example, in June 2018, construction contractor Balfour Beatty received a £500,000 fine after workers were left at risk of developing Hand-Arm Vibration Syndrome (HAVS) from 2002 to 2011. The workers were regularly exposed to hand-arm vibration (HAV) while operating handheld power tools including hydraulic breakers.

Health and Safety Executive (HSE) research in the 1990s estimated that around five million British workers were exposed to hand-arm vibration in the workplace and approximately 1.7 million were believed to be exposed at levels above the exposure action value set out in the *Control of Vibration at Work Regulations 2005* (see below) and around 900,000 exposed above the exposure limit value (see below). Around 288,000 people were estimated to have a particularly severe form of HAVS, vibration white finger.

There were 270 new claims for Industrial Injuries Disablement Benefit for hand-arm vibration syndrome in 2017 and an annual average of 455 over the last three years. In 2014, there were 610 new claims.

Carpal Tunnel Syndrome

[V5004] Use, either frequent or intermittent, of hand-held vibratory tools, can result in injury to the wrist. Carpal tunnel syndrome ('CTS') is one example of this. It is thought to arise in part from the trapping or compression of nerves in the wrist. It can arise from repetitive twisting or gripping movements of the hand as well as from the use of vibrating tools. CTS arising from the use of vibrating tools was prescribed as occupational disease A12 in April 1993 [*Social Security (Industrial Injuries) (Prescribed Diseases) Amendment Regulations 1993 (SI 1993 No 862), Reg 6(2)*].

Effects of vibration on people [V5006]

The number of new cases of CTS recognised by the Department for Work and Pensions (DWP) rose steadily from 1993 until 2002/03 when the number peaked at 1030. While as set out above, Carpal Tunnel Syndrome may have other occupational causes, such as repetitive twisting or gripping movements of the hand, such cases do not qualify for compensation by the DWP. There were 145 new claims for industrial injuries disablement benefit for CTS in 2017 and an annual average of 215 over the last three years. In 2014, there were 220 new claims.

Key messages

[V5004.1] The HSE's key messages are as follows:

- HAVS is preventable, but once the damage is done it is permanent;
- HAVS is serious and disabling, and nearly two million people are at risk;
- damage from HAVS can include the inability to do fine work and cold can trigger painful finger blanching attacks;
- the costs to employees and to employers of inaction could be high;
- there are simple and cost-effective ways to eliminate risk of HAVS;
- the *Control of Vibration at Work Regulations 2005* focus on the elimination or control of vibration exposure;
- the long-term aim is to prevent new cases of HAVS occurring and enable workers to remain at work without disability;
- the most efficient and effective way of controlling exposure to hand-arm vibration is to look for new or alternative work methods which eliminate or reduce exposure to vibration; and
- health surveillance is vital to detect and respond to early signs of damage.

Whole-body vibration

[V5005] The four principal effects of whole-body vibration ('WBV') are:

- health effects;
- impaired ability to perform activities;
- impaired comfort; and
- motion sickness;

Exposure to WBV causes a complex distribution of oscillatory motions and forces within the body. These may cause unpleasant sensations giving rise to discomfort or annoyance, resulting in impaired performance (eg loss of balance, degraded vision) or present a health risk. The most widely reported WBV injury is back pain.

Buildings and structures

[V5006] Vibration can affect buildings and structures, but a detailed description of these effects are beyond the scope of this chapter. People are more sensitive to vibration than are buildings or structures, so vibration within a building is likely to become unacceptable to the occupants at values well below

[V5006] Vibration

those which pose a threat to a structurally-sound building. For detailed technical guidance on the measurement and evaluation of the effects of vibration on buildings, see British Standard BS ISO 4866:2010 *Mechanical vibration and shock. Vibration of fixed structures. Guidelines for the measurement of vibrations and evaluation of their effects on structures* and BS 7385 Part 2: 1993 *Evaluation and measurement for vibration in buildings. Guide to damage levels from groundborne vibration.*

Vibration measurement

Measurement units

[V5007] Vibration is the oscillatory motion of an object about a given position. The rate at which the object vibrates (ie the number of complete oscillations per second) is called the frequency of the vibration and is measured in Hertz (Hz). The frequency range of principal interest in vibration is from about 0.5 Hz to 100 Hz, ie below the range of principal interest in noise control.

The magnitude of the vibration is generally measured in terms of the acceleration of the object, in metres per second squared (ie metres per second, per second) denoted $m.s^{-2}$ or m/s^2. In mathematical notation, the minus sign in front of the index, such as in s^{-2}, means that s^2 is on the bottom line. Therefore, in $m.s^{-2}$, m is 'divided by' s^2. In vibration work, other values of index may be encountered, such as 1 and 1.75.

Vibration magnitude can also be measured in terms of peak particle velocity, in metres per second (denoted $m.s^{-1}$ or m/s), or in terms of maximum displacement (in metres or millimetres).

For simple vibratory motion, it is possible to convert a measurement made in any one of these terms to either of the other terms, provided the frequency of the vibration is known, so the choice of measurement term is to some extent arbitrary. However, *acceleration* is now the preferred measurement term because modern electronic instruments generally employ an *accelerometer* to detect the vibration. As the name suggests, this responds to the acceleration of the vibrating object and hence this characteristic can be measured directly.

Direction and frequency

[V5008] The human body has different sensitivity to vibration in the head to foot direction, the side to side direction, and in the front to back direction. This sensitivity varies according to whether the person is standing, sitting or lying down. In order to assess the effects of vibration, it is necessary to measure its characteristics in each of the three directions and to take account of the recipient's posture. The sensitivity to vibration is also highly frequency-dependent. BS 6841: 1987 *Guide to measurement and evaluation of human exposure to whole-body mechanical vibration and repeated shock* provides a set of six frequency-weighting curves for use in a variety of situations. The curves may be considered analogous to the 'A'-weighting used for noise measurement (see NOISE AT WORK).

Vibration dose value

[V5009] In most situations, vibration magnitudes do not remain constant, and the concept of vibration dose value ('VDV') has been developed to deal with such cases. This is analogous to the concept of noise dose (see NOISE AT WORK).

The Control of Vibration at Work Regulations 2005 set a daily exposure limit value of 1.15 m/s^2 A(8) and the daily exposure action value is 0.5 m/s^2 A(8) for whole body vibration (see **V5015** below).

Structural vibration of buildings can be felt by the occupants at low vibration magnitudes. It can affect their comfort, quality of life and working efficiency. Low levels of vibration may provoke adverse comments, but certain types of highly-sensitive equipment (eg electron microscopes) or delicate tasks may require even more stringent criteria.

Adverse comment regarding vibration is likely when the vibration magnitude is only slightly in excess of the threshold of perception and, in general, criteria for the acceptability of vibration in buildings are dependent on the degree of adverse comment rather than other considerations such as short-term health hazard or working efficiency.

Vibration levels greater than the usual threshold may be tolerable for temporary or infrequent events of short duration, especially when the risk of a startle effect is reduced by a warning signal and a proper programme of public information.

Detailed guidance on this subject may be found in BS 6472-1:2008 *Guide to evaluation of human exposure to vibration in buildings. Vibration sources other than blasting* and BS 6472-2:2008 *Guide to evaluation of human exposure to vibration in buildings. Blast-induced vibration.*

Causes and effects of hand-arm vibration exposure

[V5010] Intense vibration can be transmitted to the hands and arms of operators from vibrating tools, machinery or work materials. Examples include the use of pneumatic, electric, hydraulic or engine-driven chain-saws, percussive tools, grinders or sanders. The vibration may affect one or both arms, and may be transmitted through the hand and arm to the shoulder.

The vibration may be a source of discomfort, and possibly reduced proficiency. Habitual exposure to hand-arm vibration (HAV) has been linked to various diseases affecting the blood vessels, nerves, bones, joints, muscles and connective tissues of the hand and forearm, most commonly VWF and Carpal Tunnel Syndrome (CTS).

VWF arises from progressive loss of blood circulation in the hand and fingers, sometimes resulting in necrosis (death of tissue) and gangrene, for which the only solution may be amputation of the affected areas or complete hand. Initial signs are mild tingling and numbness of the fingers. Further exposure results in blanching of the fingers, particularly in cold weather and early in the morning. The condition is progressive to the base of the fingers, sensitivity to attacks is reduced and the fingers take on a blue-black appearance. The development of the condition may take up to five years, according to the degree of exposure to vibration and the duration of such exposure.

CTS is a relatively common condition that causes pain, numbness and a burning or tingling sensation in the hand and fingers. The carpal tunnel is a small tunnel that runs from the bottom of the wrist to the lower palm. Several tendons that help to move the fingers pass through the carpal tunnel, as does the median nerve which controls sensation and movement of the hand. It can be caused by, among other things, exposure to vibration.

CTS and HAV as reportable occupational diseases

[V5011] Regulation 8 of the *Reporting of Injuries, Diseases and Dangerous Occurrences Regulations 2013 (SI 2013 No 1471)*, includes: (a) Carpal Tunnel Syndrome, where the person's work involves regular use of percussive or vibrating tools; and (d) Hand Arm Vibration Syndrome, where the person's work involves regular use of percussive or vibrating tools, or the holding of materials which are subject to percussive processes, or processes causing vibration; in the list of reportable occupational diseases.

VWF is also prescribed occupational disease (A11). It is described as:

'(a) Intense blanching of the skin, with a sharp demarcation line between affected and non-affected skin, where the blanching is cold-induced, episodic, occurs throughout the year and affects the skin of the distal with the middle and proximal phalanges, or distal with the middle phalanx (or in the case of a thumb the distal with the proximal phalanx), of —

    (i) in the case of a person with 5 fingers (including thumb) on one hand, any 3 of those fingers, or

    (ii) in the case of a person with only 4 such fingers, any 2 of those fingers, or

    (iii) in the case of a person with less than 4 such fingers, any one of them or, as the case may be, the one remaining finger, where none of the person's fingers was subject to any degree of cold-induced, episodic blanching of the skin prior to the person's employment in an occupation described in the second column in relation to this paragraph, or

(b) significant, demonstrable reduction in both sensory perception and manipulative dexterity with continuous numbness or continuous tingling all present at the same time in the distal phalanx of any finger (including thumb) where none of the person's fingers was subject to any degree of reduction in sensory perception, manipulative dexterity, numbness or tingling prior to the person's employment in an occupation described in the second column in relation to this paragraph, where the symptoms in paragraph (a) or paragraph (b) were caused by vibration."

It is prescribed in relation to the following occupations:

(a) the use of hand-held chain saws on wood; or

(b) the use of hand-held rotary tools in grinding or in the sanding or polishing of metal, or the holding of material being ground, or metal being sanded or polished, by rotary tools; or
(c) the use of hand-held percussive metal-working tools, or the holding of metal being worked upon by percussive tools, in riveting, caulking, chipping, hammering, fettling or swaging; or
(d) the use of hand-held powered percussive drills or hand-held powered percussive hammers in mining, quarrying, demolition, or on roads or footpaths, including road construction; or
(e) the holding of material being worked upon by pounding machines in shoe manufacture.

[Social Security (Industrial Injuries) (Prescribed Diseases) Regulations 1985 (SI 1985 No 967) Schedule 1]

Carpal tunnel syndrome is occupational disease A12. It is prescribed in relation to:

'(a) The use, at the time the symptoms first develop, of hand-held powered tools whose internal parts vibrate so as to transmit that vibration to the hand, but excluding those tools which are solely powered by hand; or
(b) repeated palmar flexion and dorsiflexion of the wrist for at least 20 hours per week for a period or periods amounting in aggregate to at least 12 months in the 24 months prior to the onset of symptoms, where "repeated" means once or more often in every 30 seconds."

[Social Security (Industrial Injuries) (Prescribed Diseases) Regulations 1985 (SI 1985 No 967) Schedule 1]

Action at common law

[V5012] In July 1998, the Court of Appeal upheld an award of damages to seven employees of British Coal claiming damages for VWF. The court determined that after January 1976, British Coal should have implemented a range of precautions, including training, warnings, surveillance and job rotation where exposure to vibration was significant.

The court went on to consider what degree of exposure would be reasonable. They supported the standards set out in the HSE Guidance booklet *Hand-arm vibration* published in 1994, which established limits for vibration dose measured as an eight-hour average or A8. The court considered that an exposure of 2.8 m/sec^2 A8 would be an appropriate level for prudent employers to use. This level of exposure would lead to a 10 per cent risk of developing finger blanching (the first reversible stage of VWF) within eight years.

The court also suggested a form of wording to warn employees: 'If you are working with vibrating tools and you notice that you are getting some whitening or discolouration of any of your fingers then, in your own interests, you should report this as quickly as possible. If you do nothing, you could end up with some very nasty problems in both hands.'

In *Docherty v Rugby Machinery (UK) Ltd* [2004] EWCA Civ 147 the Court of Appeal upheld a decision that by 1991 at the latest a woodworking industry owner should have been aware of the health effects of using hand-powered tools. In *Brookes v South Yorkshire Passenger Services Executive* [2005] EWCA Civ 452, the Court of Appeal held that the owners of garages should have known about the risks of workers suffering vibration injuries caused by powered hand tools from 1987 onwards. The Control of Vibration Regulations 2005 have since been enacted and set out, for hand-arm vibration, a daily exposure action value of 2.5 m/s^2 A(8), and a daily exposure limit value of 5 m/s^2 A(8) which must not be exceeded. (See **V5015** below.)

In *Bowe v Mersey Rewinds Engineering Ltd* [2018] EWCA Civ 72, 168 NLJ 7768, [2018] All ER (D) 11 (Feb), the Court of Appeal allowed an appeal by three companies who had employed the claimant over his working life against a finding that they were in breach of their duty of care in relation to exposure to vibration. The court at first instance had found he had been occasionally exposed to vibrating tools and that, from time to time, his exposure would exceed threshold levels. However, the Court of Appeal found that transitory and occasional exposure levels above the threshold did not automatically lead to a finding of breach of duty. It also found that failing to provide warnings and monitoring were only a breach of duty when there was frequent exposure to vibrating tools above the threshold limit.

Stages of vibration white finger

[V5013] Damages are awarded according to the stage of the disease at the time of the action from the date the employer should have known of the risk. The Taylor-Pelmear Scale is usually used to describe these stages as follows.

Taylor-Pelmear Scale System		
Stage	Grade	Description
0	—	No attacks
1	Mild	Occasional attacks affecting the tips of one or more fingers
2	Moderate	Occasional attacks affecting the tips and the middle of the fingers (rarely the base of the fingers) on one or more fingers
3	Severe	Frequent attacks affecting the entire length of most fingers
4	Very severe	As in stage 3, with damaged skin and possible gangrene in finger tips

Prescription of hand-arm vibration syndrome

[V5014] Foreseeably, prescription may extend to hand-arm vibration syndrome ('HAVS') instead of VWF. The former would cover recognised neurological effects as well as the currently recognised vascular effects of vibration. Neurological effects will include numbness, tingling in the fingers and reduced sensibility. Moreover, the current list of occupations, for which VWF is prescribed, may be replaced by a comprehensive list of tools/rigid materials against which such tools are held, including:

- percussive metal-working tools (eg fettling tools, riveting tools, drilling tools, pneumatic hammers, impact screwdrivers);
- grinders/rotary tools;
- stone working, mining, road construction and road repair tools;
- forest, garden and wood-working machinery (eg chain saws, electrical screwdrivers, mowers/shears, hedge trimmers, circular saws); and
- miscellaneous process tools (eg drain suction machines, jigsaws, pounding-up machines, vibratory rollers, concrete levelling vibratables).

The Control of Vibration at Work Regulations, 2005 (SI 2005 No 1093)

[V5015] The Control of Vibration at Work Regulations 2005 implemented EC directive 2002/44/EC, the Physical Agents (Vibration) Directive on the protection of workers from exposure to vibration, and came into force on 6 July 2005. The Regulations included transitional provisions but are now fully in force.

The Regulations apply to work activities on mainland Britain (including work on moored vessels), on offshore workplaces and in aircraft in flight over Britain, although exemption certificates may be granted to emergency services, air transport and the Ministry of Defence. Separate regulations were introduced in Northern Ireland, The *Control of Vibration at Work Regulations (Northern Ireland) 2005*, and for sea transport, the *Merchant Shipping and Fishing Vessels (Control of Vibration at Work) Regulations 2007*.

The *Control of Vibration at Work Regulations 2005* require that an employer shall make a suitable and sufficient assessment of the risk to health and safety of employees arising from exposure to vibration at work, and if an exposure action value is likely to be exceeded, the employer shall reduce the exposure to the lowest level that is reasonably practicable by appropriate technical and organisational means.

If exposure exceeds an exposure limit value, the exposure must be reduced below the limit value 'forthwith', the reason for exceeding the limit must be identified and appropriate technical and organisational means used to ensure that the limit is not exceeded again.

For *hand-arm vibration*, the daily exposure action value is 2.5 m/s^{2}A(8), and the daily exposure limit value is 5 m/s^{2}A(8) which must not be exceeded.

For *whole-body vibration*, the daily exposure action value is 0.5 m/s^2 A(8), and the daily exposure limit value is 1.15 m/s^2 A(8), which must not be exceeded.

The daily personal exposure values are to be adjusted (or normalised) to an eight-hour reference period. This can be averaged over a seven-day period where vibration exposure varies considerably from day to day.

The risk assessment must assess whether any employees are likely to be exposed to vibration at or above an exposure action value or limit value by means of:

(a) observation of specific working practices;
(b) reference to information on the probable magnitude of vibration from the equipment used in the particular working conditions;
(c) if necessary, measurement of the magnitude of vibration exposure.

The risk assessment should consider the magnitude, type and duration of the exposure, including the effect of intermittent vibration or repeated shocks. Any health or working conditions that might worsen the risk must also be considered, such as working in low temperatures. In the case of whole body vibration, consideration should be given to exposure beyond normal working hours, including vibration in rest facilities.

The effect of vibration on the ability to perform the task is also relevant, including the effect on handling of controls, reading of indicators and the stability of structures and joints.

Risk assessments must be recorded and updated as necessary. Details should include significant findings and measures taken to control or eliminate vibration exposure and the provision of information, instruction and training to employees, including entitlement to health surveillance.

Health surveillance is required for all employees where there is a risk to health, which includes all employees likely to be exposed to vibration at or above an exposure action value. The health surveillance shall prevent or diagnose any health effect linked with vibration exposure and if such an effect is detected, then it will be necessary to prevent further exposure of the person concerned, to review the vibration protection measures in force and to check whether any other employee is affected.

Other relevant legislation

[V5016] In addition to the requirements imposed by *Control of Vibration at Work Regulations 2005* there are specific obligations on employers contained in other health and safety regulations, in addition to the general duty of care under the *Health and Safety at Work etc. Act 1974*.

Such obligations include, among others, those under:

- the *Reporting of Injuries, Diseases and Dangerous Occurrences Regulations 2013 (SI 2013 No 1471)*;
- the *Provision and Use of Work Equipment Regulations 1998 (SI 1998 No 2306)*; and
- the Supply of Machinery (Safety) Regulations 2008 (SI 2008 No 1597).

In addition, the *Management of Health and Safety at Work Regulations 1999 (SI 1999 No 3242)* set out in *regulation 16* that where there are women of child-bearing age in the workforce, risk assessments must cover the risks that are specific to new and expectant mothers.

Annexes 1 and 2 of the EC Directive on Pregnant Workers (92/85/EEC) include physical risks where they are considered as agents causing foetal lesions and/or likely to disrupt placental attachment, and include particular reference to shocks and vibration. More information on the health and safety of new and expectant mothers can be found on the Health and Safety Executive (HSE) website at: www.hse.gov.uk/mothers.

Regulation 19 requires that the employer ensures that young workers are protected from any risks to their health or safety which arise from their lack of experience, or awareness of existing or potential risks, or their lack of maturity. Work that young people must not be employed to carry out includes that where there is a risk to health from extreme cold or heat, noise or vibration.

Advice and obligations in respect of hand-arm vibration

Vibration magnitudes

[V5017] The energy level of the hand tool is significant. Percussive action tools, such as compressed air pneumatic hammers, operate within a frequency range of 33-50 Hz. These cause considerable damage whereas rotary hand tools, which operate within the frequency range 40-125 Hz, are less dangerous.

The Health and Safety Executive (HSE) publishes guidance on the vibration magnitudes of a variety of tools in a selection of typical working conditions. Vibration can vary not only with the type and design of the tool, but also with other things such as the task, the operator's technique and the material being worked (see www.hse.gov.uk/vibration/hav/advicetoemployers/assessrisks.htm).

For example, angle grinders create vibration of between 4 and 8 m/s^2, giving between three hours' and 45 minutes' use before the exposure action value is reached, and between 12 hours' and three hours' use before the exposure limit value is reached. Road breakers create between 5 and 20 m/s^2, giving between two hours' and ten minutes' use, with 12 m/s^2 being typical, giving 20 minutes' use before the exposure action value is reached, and between eight hours and 30 minutes before the exposure limit value is reached.

It is clear from this data that most of these tools cannot be used for a whole working day without exceeding the vibration limit value, and few can be used without exceeding the exposure action value. Most currently-available vibrating hand tools will exceed the vibration exposure action value, and are very likely to exceed the exposure limit value if used for a whole working day.

Managing vibration exposure

[V5018] The HSE has produced general guidance aimed at finding alternatives to the use of hand-held vibratory tools. It has been found that methods of working that produce less vibration also produce more accurate work and less user fatigue, so multiple benefits can flow from a change of working methods.

In the construction industry, work can be planned better, for example, casting-in ducts, and detailing box-outs, so that there is less need to break through new masonry. Machine-mounted hydraulic breakers, floor saws, diamond core drilling, hydraulic crushers and hydraulic bursters can replace hand-held hammers and breakers in demolishing concrete or masonry.

The HSE has issued definitive advice that the removal of pile caps using hand-held breakers is not acceptable. It advises on various alternatives of hydraulic bursting and hydraulic cropping. It also advises that concrete scabbling for purely aesthetic effect is not acceptable and alternatives should be used.

Where the use of a vibratory tool cannot be avoided, then it is necessary to assess and reduce the risk. Where a tool is used for a well-defined process, it might be helpful to control the vibration exposure by limiting the number of at-risk tasks in a day, rather than the number of hours use. For example, it may be found that an operator can drill 50 to 60 holes 100 mm deep in concrete before reaching the exposure action value, and 200 to 230 such before reaching the exposure limit. This is a theoretical example.

Many work processes involving hand-held power tools will require employees to undertake a variety of tasks, possibly using different tools, each with different vibration exposures. In this case, it could be more difficult to set a limit on the number of operations that are allowable. An alternative is to use a 'points' system.

For example, an employer might allocate points to vibration exposure such that 100 points per day is equal to the exposure action value. The employer assesses the vibration 'dose' of an angle grinder as 30 points per hour, and of an impact drill as 120 points per hour. This means that the grinder gives an exposure of five points every 10 minutes and the impact drill gives 20 points for every 10 minutes' use. Thus 20 minutes' use of the drill plus 120 minutes' use of the grinder would cause the exposure action value to be reached.

The allocation of points to each tool will have to be calculated using a formula which depends on the vibration magnitude of the tool and process. Note that the system described here is only intended to be indicative, as vibration dose does not strictly obey the rules of simple arithmetic.

The HSE guidance L140, *Hand-arm vibration*, provides advice on using an exposure points system and ready-reckoner to calculate daily vibration exposures.

Personal protection against hand-arm vibration

[V5019] The risk of injury to workers from hand-arm vibration can be reduced by ensuring that they can keep the blood flowing while working, for example by keeping warm, and exercising the hands and fingers. Gloves and other warm clothing should be provided when working in cold conditions. Tools should be designed for the job, both to lessen vibration and to reduce the strength of grip and amount of force needed. The equipment should be used in short bursts rather than long sessions. Workers should be encouraged to report any symptoms to the employer.

There are a number of 'anti-vibration' gloves on the market but, these may seriously limit dexterity and can in some circumstances actually increase the risk of vibration injury.

The HSE advises: "Gloves marketed as 'anti-vibration', which aim to isolate the wearer's hands from the effects of vibration, are available commercially. There are several different types, but many are only suitable for certain tasks, they are not particularly effective at reducing the frequency-weighted vibration associated with risk of HAVS and they can increase the vibration at some frequencies. It is not usually possible to assess the vibration reduction provided in use by anti-vibration gloves, so you should not generally rely on them to provide protection from vibration. However, gloves and other warm clothing can be useful to protect vibration-exposed workers from cold, helping to maintain circulation."

Manufacturers' warning of risk

[V5020]–[V5021] The *Supply of Machinery (Safety) Regulations 2008 (SI 2008 No 1597)* established essential health and safety requirements for machinery supplied in the European Economic Area and set out that: "Machinery must be designed and constructed in such a way that risks resulting from vibrations produced by the machinery are reduced to the lowest level, taking account of technical progress and the availability of means of reducing vibration, in particular at source."

Machinery manufacturers and suppliers have a duty to warn whenever a machine carries a vibration risk, and to provide information for hand-held or hand-guided machines where vibration emissions exceed 2.5 m/s^2. This is the vibration emission of the machine, not the daily exposure of the user, which will depend on the amount of use of this and other machines. It should be noted that 2.5 m/s^2 is not to be regarded as an acceptable target for the amount of vibration a machine may emit – there is an overriding duty to reduce vibration emission as far as possible within the current state of the art.

Manufacturers are required to test vibration emissions in accordance with test procedures aimed at obtaining a reproducible value, and they must declare both the typical measured value and the amount of uncertainty (variation) of measurements from the typical value. Thus, the typical value plus the uncertainty value should give the 'worst case', but it may still not represent the total amount of vibration in real use. This can complicate the choice of tool, to the extent that sometimes a tool that gives slightly more vibration than another tool may still be better if it is more comfortable, appropriate or quicker for the task.

Moreover, tools that are incorrectly assembled, used wrongly, with worn or loose parts can give many times the amount of vibration than normal.

Formal assessments

[V5022] It is quite difficult to measure the vibration emission of equipment in actual use, as the measuring device must be attached to the equipment in such a place and in such a manner as to represent the vibration being transmitted to the operator. The measuring system must be capable of determining the amount of vibration over a range of frequencies and in different directions. It is therefore usually necessary to rely on general advice.

The first priority should be to try to find ways to eliminate the use of vibrating tools, to use the most appropriate equipment to get the job done quickly and safely, and to make sure that the tool is correctly assembled and used, with the correct cutting bit kept sharp.

Advice and obligations in respect of whole-body vibration

Incidence of whole-body vibration

[V5023] Prior to the introduction of the *Control of Vibration Regulations 2005*, the HSE estimated that around nine million people in Britain were exposed to whole-body vibration (WBV) at work. Although most of these were users of road vehicles and unlikely to be at risk of WBV injury, more than 1.3 million workers were estimated to be exposed above the WBV exposure action value set in the Regulations; and more than 20,000 men were exposed in excess of the WBV exposure limit value.

WBV can cause or worsen back pain, and the effect is aggravated when the vibration contains severe shocks and jolts. These conditions occur particularly in off-road driving, such as in farming, construction and quarrying, but can also occur in small, fast boats at sea and in some helicopters. Old railway vehicles running on track work in poor condition can also give rise to high levels of WBV.

Back pain can arise from many causes other than WBV, and can be aggravated by poor driving posture, sitting for long periods without being able to change position, awkwardly placed controls that require the operator to stretch or twist to reach them, manual lifting, and repeatedly climbing into or jumping out of a high cab.

Jobs which combine two or more of these factors can increase the risk further, for example, driving over rough ground while twisting around to check on the operation of equipment being operated behind the vehicle.

Assessing the risk of WBV problems

[V5024] Road transport drivers are not usually at risk unless there are other aggravating factors such as long hours of driving, or use of unmade or poor road surfaces. However, if there is a history of back pain in the job, this could be indicative of a problem, and it may be wise to consider regulations and guidance on manual handling of materials as well as the *Control of Vibration Regulations 2005*. Operators of off-road vehicles are at greater risk. Indicators of risk include the operator being thrown around in the cab, or receiving shocks and jolts through the seat, or if the manufacturer warns of a risk. A history of back pain is again an indicative factor.

Measurements of whole-body vibration in construction machines have shown that smaller loaders and excavators can produce WBV levels that exceed the exposure action value over a typical working day, although it is unlikely that the exposure limit value would be exceeded. However, it is possible to exceed the limit in smaller machines that are intensively used for long periods of the day.

Keeping records

[V5025] Risk assessments should be recorded. These should show those jobs or processes identified as being at risk from WBV; an estimate of the amount of exposure; comparison of exposure with the action and limit values; the available risk controls; the plans to control and monitor risks; and the effectiveness of the control measures.

Reducing the risk of WBV problems

[V5026] When a risk is identified the obvious initial actions are to ensure that the machine is suitable for the job, is used correctly and driven at safe speeds for the ground conditions, and that seats are correctly adjusted for the driver's size and weight. Access routes on work sites should be maintained in good condition, for example by using a scalping blade and roller.

Exposure assessments

[V5027] The HSE advises: "In most cases it is simpler to make a broad assessment of the risk rather than try to assess exposure in detail, concentrating your main effort on introducing controls."

It advises employers of drivers and operators to observe work tasks and talk to managers, employees and others in order to collect basic information to allow them to make a broad assessment of the risk and introduce simple control measures to reduce the risk.

Exposures may be high where:

- machine or vehicle manufacturers' handbook warns of risks from whole-body vibration;
- machines or vehicles are unsuitable for the tasks for which they are being used;
- operators and drivers are using poor techniques – driving too fast or operating the machine too aggressively for example;
- employees are operating or driving, for several hours a day, machines or vehicles likely to cause high vibration exposures;
- employees are being jolted, continuously shaken or, when going over bumps, rising visibly in the seat;
- vehicle roadways or work areas are potholed, cracked or covered in rubble;
- road-going vehicles are regularly driven off-road or over poorly-paved surfaces for which they are not suitable; or
- operators or drivers report back problems.

It advises: "This kind of broad risk assessment can be done without needing to estimate or measure vibration exposure. Most employers of drivers or operators will not need to do any measurements or employ vibration specialists to help with the risk assessment."

Health surveillance

[V5028] In *Whole-body vibration the Control of Vibration at Work Regulations 2005 Guidance on Regulations* (L141) the HSE advises that health surveillance is not appropriate for WBV 'because it is considered that no methods currently exist for detecting changes in peoples' backs which can reliably indicate the early onset of changes (which may cause low back pain) that are specifically related to workplace factors'. Instead, it advises employers to use other methods to monitor the health of employees exposed to the risks, such as encouraging symptom reporting and checking sickness records:

"Valuable information can, however, be obtained from less precise measures than those provided for by a formal health surveillance approach, such as reporting and monitoring of symptoms. This is generally known as 'health monitoring'. It is good practice to put in place this type of system to allow individuals to make early reports of low back pain in order to assess the need for action on WBV, manual handling or posture."

HSE guidance on hand-arm and whole-body vibration

[V5029] HSE resources include:

Hand-arm vibration – the Control of Vibration at Work Regulations 2005 Guidance on Regulations (L140) at: www.hse.gov.uk/pubns/books/l140.htm;

Whole-body vibration – Control of Vibration at Work Regulations 2005 (L141) at: www.hse.gov.uk/pubns/books/l141.htm;

Vibration solutions: Practical ways to reduce the risk of hand-arm vibration injury, (HSG 170) which includes 45 case studies dealing with vibration at: www.hse.gov.uk/pubns/books/hsg170.htm;

An online exposure points system and ready-reckoner at: www.hse.gov.uk/vibration/hav/readyreckoner.htm;

An online vibration calculator to assist in calculating exposures for hand-arm vibration can be found at: www.hse.gov.uk/vibration/hav/vibrationcalc.htm;

Advice on vibration exposure monitoring which is available at: www.hse.gov.uk/vibration/hav/advicetoemployers/vibration-exposure-monitoring-qa.pdf#?eban=rss-vibration.

Guidance leaflets include:

- *Hand-arm vibration at work: A brief guide* (INDG175) www.hse.gov.uk/pubns/indg175.pdf
- *Control back-pain risks from whole-body vibration* – Advice for employers on the Control of Vibration at Work Regulations (INDG242) www.hse.gov.uk/pubns/indg242.pdf

- *Hand-arm vibration* – advice for employees (INDG296) (pocket card) www.hse.gov.uk/pubns/indg296.pdf
- *Drive away bad backs* (INDG404) www.hse.gov.uk/pubns/indg404.pdf

HSE research reports include the 2015 publication: *A critical review of evidence related to hand-arm vibration syndrome and the extent of exposure to vibration (RR1060)* at: www.hse.gov.uk/research/rrhtm/rr1060.htm.

Violence in the Workplace

Andrea Oates

Introduction to violence in the workplace

[V8001] Work-related violence, which the Health and Safety Executive (HSE) defines as "Any incident in which a person is abused, threatened or assaulted in circumstances related to their work", continues to be a significant and problem with wide-ranging consequences. Whereas violence was once seen exclusively as a problem for occupations such as police officers and prison staff, it is now widely acknowledged that people working across many occupations and economic sectors can be exposed. Those at particular risk of exposure include workers who come into contact with the public, those who handle money, those who, on occasions must deny rather than provide a service, and those who deal with people in pain and distress, or under the influence of alcohol or drugs. There are many jobs to which some, or all, of these dimensions apply, including teachers, nurses and other healthcare workers, taxi drivers and other transport staff, emergency service workers, shop workers, public house staff, social workers and 'helpline' staff.

The HSE emphasises that the process of managing the risk of violence is the same as for any other health and safety risk. It acknowledges, however, that violence is a problem within society and the risk of violence in the workplace depends on the interaction of a range of factors, some of which lie outside of the employer's ability to control. There are no easy solutions or short cuts and there are no universal blueprints for dealing with the risk of violence. What works for one organisation may not work for another. When developing a workplace violence prevention programme, it is vital that organisations make discernible judgements based on the circumstances in their particular working environments.

HSE guidelines suggest the adoption of a risk management framework with a recursive cycle of activities to ensure continuous improvement in the assessment and management of risk, beginning with an appraisal of the nature and extent of the problem, and followed by the design, implementation and evaluation of appropriate preventive measures.

The nature and extent of the problem of workplace violence

[V8002]–[V8003] The HSE uses figures from the Crime Survey for England and Wales (CSEW) as well as its own sources to estimate the extent of workplace violence.

[V8002] Violence in the Workplace

The annual CSEW has been collecting data about workplace violence for several years and provides a count of crime across England and Wales, including crimes not reported to or recorded by the police. It is conducted on a yearly basis.

The findings from the 2017/18 CSEW show:

- 374,000 adults of working age in employment experienced violence at work, including threats and physical assault;
- there were an estimated 694,000 incidents of violence at work: 330,000 assaults and 364,000 threats. This increased from an estimated 642,000 incidents in 2016/17;
- 1.4% of women and 1.5% of men were victims of violence at work once or more during the year prior to their interview;
- just over half (54%) of workplace violence offenders were strangers. Among the 46% of incidents where the offender was known, they were most likely to be clients, or a member of the public known through work;
- victims reported the offender was under the influence of alcohol in an estimated 32% of assaults and 27% of threats at work; and
- the survey found 41% of violence at work resulted in physical injury.

The CSEW also shows that those in protective service occupations (such as police officers) faced by far the highest risk of assaults and threats while working, at 11.4% – eight times the average risk of 1.4%. Health and social care specialists and health professionals had higher than average risk at 5.1% and 3.3% respectively. Other professions at higher risk include other managers and proprietors at 2.3%.

Figures from the Labour Force Survey (LFS) also provide an estimate of the number of workers injured at work as a result of physical acts of violence. The latest estimates show that around 43,000 workers in Great Britain sustained non-fatal injuries as a result of acts of physical violence at work each year in the three years up to 2017/18. These accounted for around 7% of all non-fatal workplace injuries. Around 9,000 were over seven-day absence injuries.

The Reporting of Injuries, Diseases and Dangerous Occurrences Regulations (RIDDOR) require employers to report certain non-fatal workplace injuries. These are generally the more serious types, resulting in more than seven days absence from work or as a result of a specified list of injuries.

The HSE cautions that RIDDOR is under-reported, so does not fully capture the full scale of such cases. In 2017/18, there were 5,055 such reported cases to employees in Great Britain, accounting for 7% of all RIDDOR reported non-fatal injuries.

The LFS shows that around nine out of every ten workers who sustain an injury resulting from violence at work are employed in public services (including human health and social work activities, education and public administration and defence). RIDDOR shows a similar overall picture for reported injuries, although the figures show a higher proportion occurring in the human health and social care sector and a lower proportion in education.

Work-related violence – the legislation

[V8004] In common with most countries, there is little UK legislation or regulation designed to deal specifically with work-related violence. Instead several more general laws provide for regulation, enforcement and remedy in such cases.

In simple terms, the law provides:

- Regulation to ensure that employers provide a working environment in which the risks of being subjected to violence are minimised and the fears of employees are taken into account. Employees also owe a duty of care to themselves and to others who may be affected by their acts or omissions;
- Enforcement to provide sanctions if employers fail to comply with the regulation and punishment of individuals who act violently towards another person in a workplace context; and
- Remedy for employees who become victims of workplace violence. This will usually be in the form of compensation through an employment tribunal, civil court, or Criminal Injuries Compensation Authority (CICA).

Changes to the Criminal Injuries Compensation Scheme (CICS) came into effect in November 2012 in England, Wales and Scotland and reduced or withdrew compensation payments to many people injured in violent crimes, including those injured as a result of work-related attacks. The scheme previously awarded compensation to between 30,000 and 40,000 people a year who were seriously injured in a crime of violence but could not obtain recompense from any other source, such as their assailant. Those who suffer minor injuries no longer get compensation. The CICA received 32,280 new applications in 2017/18.

The legal responsibilities and rights relating to workplace violence are contained in a range of provisions that include:

- Health and safety legislation.
- Employment law.
- Criminal law.
- Civil law.

Statutory health and safety requirements

Health and Safety at Work etc Act 1974

[V8005] Employers have a legal duty under the *Health and Safety at Work etc Act 1974 (HSWA 1974)* to ensure, so far as is reasonably practicable, the health, safety and welfare at work of their employees. [*HSWA 1974, s 2(1)*].

Besides a common law 'duty of care' to others, employers have a statutory duty to do everything 'reasonably practicable' to prevent or control the risk of harm from hazards to health – including violence.

Employers are also required to conduct their undertaking in such a way as to ensure, so far as reasonably practicable, the safety of other people who are not their employees, and to whom the premises have been made available.

The 1974 Act is considered as the most important piece of legislation dealing with health and safety at work. Although it was implemented to cater for the risks associated with industries such as chemicals, mining and construction work, the basic principles are translated relatively easily within the context of the protection required against the risks of workplace violence. Lord Skelmersdale chaired the DHSS Advisory Committee on Violence to Staff, set up in response to the death of a social worker, and wrote in the 1988 report:

> Where violent incidents are foreseeable employers have a duty under section 2 [of the Health and Safety at Work etc Act 1974] to identify the nature and the extent of the risk and to devise measures which provide a safe workplace and a safe system of work.

In *Cook v Bradford Community Health NHS Trust* [2002] EWCA Civ 1616, [2002] All ER (D) 329 (Oct) a healthcare assistant in a psychiatric hospital was attacked by a patient in the seclusion suite of a unit for violent patients. She had entered in order to take cups of coffee to colleagues. The Court of Appeal said that her employer had a duty not to place her unnecessarily in a position where there was a risk of foreseeable danger and she was successful in her compensation claim.

The duty of care also embraces employees. Employees are required under *section 7* of the *HSWA 1974* to 'take reasonable care for the health and safety of himself and of other persons who may be affected by his acts or omissions at work'. The duties placed on the employee do not reduce the responsibility of the employer to comply with his or her health and safety duties under the 1974 Act.

Management of Health and Safety at Work Regulations 1999

[V8006] The *Management of Health and Safety at Work Regulations 1999 (MHSWR) (SI 1999 No 3242)* require employers to undertake a 'suitable and sufficient' assessment of the risks to which employees are exposed while they are at work and put in place measures to eliminate, minimise or control the risks. In many workplaces, the potential for violence is a significant hazard and therefore the duty on employers extends to the risk of violence.

Employers must:

- Establish how significant these risks are;
- Identify what can be done to prevent or control the risks; and
- Produce a clear management plan to achieve this.

To satisfy the requirements where there is a significant risk, employers must put into place a comprehensive policy for dealing with work-related violence. This will include effective measures for identifying and assessing risk, clear guidelines and procedures for dealing with violent incidents and their aftermath, and the provision of appropriate training and equipment.

Several cases have looked at risk assessments and measures to reduce the risk of violence, including the following:

In *Collins v First Quench Retailing Ltd* [2003] SLT 1220, an employer was ordered to pay compensation to an off-licence manager who was attacked in an armed robbery. The woman had been working alone in the off licence,

Work-related violence – the legislation [V8008A]

which had a history of incidents including previous armed robberies. In addition, she had requested that the employer provide better security by fitting screens and ensuring that there were always two members of staff present. Although the employer had fitted panic buttons and CCTV and arranged for two staff to work on the evening shifts, the judge in the case said that lone workers were generally easier to attack, and that having two employees in the shop at all times would have materially reduced the risk.

In *Smith v Welsh Ambulance Service (Chester County Court) (22 March 2007, unreported)*, a paramedic went into a derelict building alone to treat an unconscious drug addict and was threatened by other drug addicts. The trial judge accepted that members of the emergency services can face risky situations, but said that control room staff should have been trained to do risk assessments to help them make 'an intelligent decision' about whether to send lone workers into a particular situation. In addition, the paramedic had not been given adequate training as to whether he had a choice about giving assistance in risky circumstances. The employer was found to have failed in their duty of care.

Reporting of Injuries, Diseases and Dangerous Occurrences Regulations 2013

[V8007] Employers must notify their enforcing authority in the event of an accident at work to any employee resulting in death, major injury or incapacity to work for seven or more days. This includes any act of non-consensual physical violence inflicted on a person at work [*Reporting of Injuries, Diseases and Dangerous Occurrences Regulations 2013 (RIDDOR) (SI 2013 No 1471)*]. In addition, employers must record over three-day injuries.

Safety Representatives and Safety Committees Regulations 1977 and the Health and Safety (Consultation with Employees) Regulations 1996

[V8008] Employers must inform and consult with employees in good time on matters relating to their health and safety. Employee representatives either:

- appointed by recognised trade unions under the *Safety Representatives and Safety Committees Regulations 1977 (SI 1977 No 500)*; or
- elected under the *Health and Safety (Consultation with Employees) Regulations 1996 (SI 1996 No 1513)*,

may make representations to their employer on matters affecting the health and safety of those they represent.

This places a responsibility on the employer to consult and inform on issues of violence at work.

Emergency Workers (Scotland) Act 2005, Emergency Workers (Obstruction) Act 2006 and the Emergency Workers (Offences) Act 2018

[V8008A] These pieces of legislation were introduced to give additional legal protection from assault for emergency workers. The *Emergency Workers (Scotland) Act 2005* makes it a specific offence to assault, obstruct, or hinder someone providing an emergency service or someone assisting an emergency worker in an emergency situation.

[V8008A] Violence in the Workplace

The *Emergency Workers (Obstruction) Act 2006* came into force in February 2007 and applies to England, Wales and Northern Ireland. Under the Act, it is an offence to 'obstruct or hinder' particular groups of emergency workers responding to 'blue light' situations. The Act defines emergency workers as firefighters, ambulance workers and those transporting blood, organs or equipment on behalf of the NHS, coastguards and lifeboat crews. (The police already have their own obstruction offence in the Police Act 1996.)

In November 2018, the *Emergency Workers (Offences) Act 2018* came into effect making it a specific offence to assault emergency workers including police and prison officers, firefighters and NHS staff providing services to the public. The Act doubled the maximum prison sentence for the offence of assault or battery against an emergency worker from six months to a year and judges must now also consider higher sentences for a range of other offences including grievous bodily harm (GBH) and sexual harassment.

The Corporate Manslaughter and Corporate Homicide Act 2007

[V8008B] The Act introduced a new offence of corporate manslaughter (corporate homicide in Scotland), so that in the case of a work-related death, companies and organisations can be found guilty of corporate manslaughter as a result of serious management failures resulting in a gross breach of a duty of care (see **CORPORATE MANSLAUGHTER**).

Employment law

The Employment Rights Act 1996

[V8009] *Section 44* of the *Employment Rights Act 1996 (ERA 1996)* relates to 'Health and safety cases' and reminds us that responsibility for safety at work is something shared between employers and employees. The section covers:

- Employees' statutory entitlement to a safe way of working, and to be able to fulfil their responsibility under the *HSWA 1974, s 7* to take care of themselves and others, without fear of recriminations ('any detriment') from their employer for doing so.
- The circumstances in which an employee should withdraw from danger (see **V8014** below).
- Employees' duty and right to be able to draw attention to safety deficiencies that may exist and to take 'appropriate action' to withdraw/remove themselves from 'serious and imminent' danger that they would be unable to avert.
- Warning employees that if they continue to knowingly and recklessly undertake work that is unsafe and get injured, they may not be able to make a claim against their employer for liability.

Contract law

[V8010] A contract of employment imposes two obligations on the part of the employer:

- The employer must provide a workplace in which employees are subjected to minimal exposure to risk.

- The employer must provide 'trust and support' to the employee in carrying out their role.

It is well established that workplace violence should be regarded as a risk in this context. The employer is expected to provide 'trust and support' to ensure that the employee is working under minimum stress and that the employer responds to issues of perceived risk as well as real risk.

Failure to provide appropriate support might reach a point where an individual can no longer tolerate the working conditions and decides to leave. This could be construed as constructive dismissal where the situation is treated as if the employee had been dismissed. The individual will usually seek a remedy through an employment tribunal or a county court.

The obligations of an employer are unambiguous. There is a clear requirement to assess the possible risk of an employee being subjected to violence and to provide appropriate measures which will minimise that risk. There is also a requirement to recognise and respond to the perceived fears of employees by providing appropriate support in relation to those fears.

Criminal law

[V8011] The laws in relation to violence, disorder, assault and threats are well established and cover almost every incident of work-related violence. Recent legislation offers greater protection against harassment and more serious penalties for racist incidents. Thankfully, the incidence of serious assault is comparatively rare but this brings with it a difficulty in relation to prosecuting the offender. In general, it requires the police to pursue the matter to prosecute the offender and many different issues have to be addressed when making the decision.

Quite often the nature of the assault is not serious enough for a criminal charge to be pursued or, for a variety of reasons, the circumstances do not warrant it. It is unlikely that an assault that results in only minor bruising, for example, would find its way into a court. While these decisions often make economic, procedural or legal sense, they leave the victims feeling frustrated and unsupported. A common cause of dissatisfaction felt by employees who have been assaulted is related to the apparent lack of action that follows the incident. Organisations need to be clear as to their stance on prosecution and their preparedness to support a private/civil action.

Civil Law

Negligence

[V8012] The tort of negligence imposes a duty to take reasonable care to avoid acts or omissions that may cause reasonably foreseeable harm to others. This common law duty of care runs in parallel to statutory health and safety requirements.

In cases where the victim has suffered serious injury, either mental or physical, they will undoubtedly look for substantial compensation for things like their loss of earnings, pain, suffering, and inconvenience. Although it is clear that the person who actually committed the act of violence is directly at fault, it is unlikely that the individual will have the funds to provide for any substantial compensation award. Consequently, the victim will be more likely to seek redress through his or her employer by claiming that the employer was negligent in providing appropriate measures to prevent the incident from happening, and/or in failing to provide adequate support post-incident.

In such cases, key questions will be asked:

- Was violence in the workplace something that ought to have been assessed?
- If so, what were the risks of violence that would have been identified in the risk assessment?
- What reasonably practicable control measures would the risk assessment have identified should it have been put in place?

An employer who has failed to assess the risk of violence will have great difficulty in defending a legal action for injury caused through employer negligence (see for example the cases highlighted at **V8006** above).

In *Selwood v Durham County Council* [2012] EWCA Civ 979 the Court of Appeal ruled that a social worker who was stabbed by a psychiatric patient for whose child she was responsible, had the right to sue two health trusts responsible for her attacker's care as well as her employer.

Vicarious liability

[V8013] In recent years, the boundaries have been extended even further to cover a company's responsibility for its employees' actions. 'Vicarious liability' is the legal concept whereby a company can be liable for the negligent conduct of its employees. A claim for compensation for personal injury will be successful if the conduct occurred during the course of employment. A number of cases have developed the law in this area, including the following:

In *Fennelly v Connex South Eastern Ltd* [2001] IRLR 390, a passenger was assaulted by a ticket inspector. The Court of Appeal held that the rail company was responsible for the conduct of the inspector and was therefore liable.

Two co-joined cases, *Weddall v Barchester Healthcare Ltd* [2012] EWCA Civ 25, [2012] IRLR 307 and *Wallbank v Wallbank Fox Designs Ltd* [2012] EWCA Civ 25, [2012] IRLR 307, involved managers who were assaulted by employees in the workplace. In Weddall, a care home manager phoned a member of staff at his home in the evening to ask if he could work an extra shift to replace a nightshift employee who had called in sick. The employee concerned was very drunk and thought that the manager was mocking him. He went to the care home and violently assaulted the manager. In this case the claimant was unsuccessful, and it was held that the employee was acting personally for his own reasons.

In Wallbank, the manager reprimanded an employee for the way he was working. The employee responded by violently assaulting him. In this case the court said that the risk of an over-robust reaction to an instruction was a risk created by the employment.

In *Vaickuviene v J Sainsbury plc* [2013] IRLR 792 a Lithuanian national working stacking shelves at a supermarket was murdered at work by a colleague who was a member of the British National Party. The incident followed frequent racist comments and aggression towards the victim. The victim's relatives sued the employer on the grounds that they were vicariously liable under the *Protection from Harassment Act* (see below) for the actions of their employee. The claim was rejected on the basis that there was not a sufficiently close connection between his actions and the work he was employed to carry out. "In the course of employment" does not have the same meaning as "at work".

In *Mohamud v W M Morrison Supermarkets plc* [2016] IRLR 362, the Supreme Court found that the supermarket was liable for the actions of an employee working in one of its petrol station kiosks when he violently attacked a man. Although he had grossly abused his position, the employee's conduct was sufficiently connected to the job for which he had been employed for his employer to be held responsible for his abuse of it.

At the same time as it ruled on *Mohamud*, the Court also considered *Cox v Ministry of Justice* [2016] UKSC 10, this time considering the liability of an organisation for a non-employee.

The Court ruled that an organisation can be liable for someone who is not their employee, but is carrying out activities as an integral part of the organisation's business and for its benefit, where the risk of the wrongdoing results from the organisation's decision to give responsibility to the wrongdoer.

The case arose after a prison catering manager was seriously injured when a prisoner dropped a 25kg bag of rice on her back, causing severe spinal injuries. The Ministry of Justice was liable, ruled the Supreme Court. Although not an employee, the prisoner was carrying out kitchen work after being selected for training and was being supervised by prison staff. He was placed in a position that enabled him to carry out the work, and he was the MoJ's responsibility.

In *Bellman v Northampton Recruitment Ltd* [2018] EWCA Civ 2214, [2019] 1 All ER 1133, [2019] IRLR 66, the Court of ruled that a recruitment agency was vicariously liable for the actions of its managing director after he punched one of his punched one of his employees at an event following a Christmas party, causing him brain damage. The incident happened after the main event when some members of staff continued to drink at a hotel bar in an unplanned extension of the party. The ruling clarifies that businesses can be held liable in some circumstances for the actions of an employee even where they take place outside the workplace.

Withdrawing services

[V8014] At what point is it appropriate for an employee to say that they are not prepared to continue because they feel they are in danger? This is a question which is faced from time to time where an employee feels that they are in danger of being physically hurt and they are not prepared to continue to deal with a situation. Legislation makes clear the duties upon employers to protect staff, and *s* 7 of the *HSWA 1974*, *Reg 8* of the *MHSWR 1999 (SI 1999 No 3242)* and *s 44(1)(d)* of the *ERA 1996* also place clear requirements upon employees to take action if their safety is compromised (see EMPLOYMENT PROTECTION).

The law relating to the use of force and self defence

[V8015] The use of force is considered under both common law and statute law and it is important that those employees likely to need to use force are aware of the law relevant to their circumstances. Although the powers are drawn from various sources, most specify that force can only be used where it is absolutely necessary, and that the amount of force used must be reasonable in the circumstances. The *Human Rights Act 1998* adds the term 'proportionate'.

In May 2019, the Office of the Chief Constable for Devon and Cornwall Police was handed down a fine of £234,500 after earlier pleading guilty to charges under the Health and Safety at Work etc Act 1974 in relation to the force's use of an Emergency Response Belt (ERB). In October 2012, police detained Thomas Orchard and he died in custody after being restrained.

Prosecutions arising from work-related violence

[V8015A] The HSE and local authorities have prosecuted several employers for health and safety failings following attacks on their employees. For example, in December 2018, NHS Oxleas Foundation Trust was fined £300,000 after two nurses suffered life-changing injuries when they were repeatedly stabbed by a service user in a medium-secure forensic unit in July 2016.

An investigation by the HSE found that although the centre routinely received high-risk patients, at the time of the incident there was no patient-specific risk assessment identifying the risks posed by a patient and the measures required to control those risks prior to admission to the ward. The investigation also found that the use of knives on an acute ward was fundamentally unsafe. Staff were entering and exiting the kitchen area several times while knives were in use, but there were no instructions or control measures in place regarding kitchen knives. Following the incident, all knives were removed from the acute wards.

Protection from Harassment Act 1997

[V8015B] The *Protection from Harassment Act 1997* can also be used in an employment context (see **H1726**). Action can be taken either through criminal prosecution or a private action for damages in the Civil Courts on behalf of an individual or a group. Sanctions include fines, imprisonment, or community sentences as well as a restraining order which prohibits the offender from continuing their offending behaviour.

Developing and implementing policy

[V8016] Effective policies translate priorities into action and set a clear direction for the organisation to follow with defined accountabilities. In any organisation that has workers at risk from violence, a comprehensive and workable policy is the cornerstone for building an effective response. Effective policy development requires the active commitment of senior management and the involvement of workers affected by violent incidents. Unless an implementation plan is put in place, supported by adequate resources to drive and embed change, the policy is likely to remain as words on a page. Once implemented, the processes, procedures, equipment, training and incidence of violence need to be monitored, analysed and evaluated to ensure a proactive and problem-solving approach to the ever-changing risks associated with work-related violence and aggression.

Problem recognition

[V8017] The most important first step in developing policy is to define what 'workplace violence' means within the organisation. Most organisations use the definition offered by the HSE:

> Any incident in which a person is abused, threatened, or assaulted in circumstances relating to their work.

This definition encompasses a range of unacceptable behaviour likely to cause physical or psychological harm. Importantly, it includes incidents where someone uses abusive language or behaviour that does not amount to a threat of or actual physical assault. It also covers incidents that might occur when people are not actually at work, for example, being followed home or confronted when arriving at or leaving work.

Conducting a review

[V8018] Having defined violence, the next step in developing an effective policy is to understand the nature and extent of the problem facing the organisation and the effectiveness, or otherwise, of existing policy, procedures and guidance. This is best achieved through a systematic review of existing arrangements. The review should include 'top down' and 'bottom up' elements of assessment to ensure that problems facing front line workers are understood and a 'reality check' made on the effectiveness of existing policy and procedures.

The purpose of such a review is to establish:

- the nature and extent of the problem;
- the specific roles that are at risk;
- the nature of the risks;
- the solutions to minimise the risks identified; and
- the effectiveness or otherwise of existing policy and procedures.

Collaboration

[V8019] An important element in the eventual success of such a review will be the approach adopted from the outset. It is relatively easy to secure employees' collaboration because the intention of the review is primarily to ensure they operate in an environment where risks from work-related violence are minimised.

Sources of information

[V8020] To carry out an effective review, there are a variety of sources from which data and information can be gathered. These include:

- Generic violence risk assessment of specific roles;
- Interviews, observation, focus groups and questionnaires; and
- Analysis of data.

Generic violence risk assessment of specific roles

[V8021] The people who face customers, clients, patients and other members of the public every day are best placed to identify and understand the risks they face from violence and aggression, and risk assessment of roles is a statutory requirement under health and safety legislation.

Interviews, observation, focus groups and questionnaires

[V8022] It is important to gather views from front line staff about the effects that the risks of workplace violence have upon them.

The information can be gathered through informal interviews with affected staff, focus groups of staff from a particular role or through a questionnaire which is designed to extract the appropriate information. Often, a combination of these techniques is used.

There are issues around confidentiality and individuals often fear their views will 'get back to management'. Confidentiality must be guaranteed, be evident in the procedure and be completely respected in practice.

Analysis of data

[V8023] The data available will vary according to the nature of the organisation and the sophistication of the reporting and recording procedures available. In the best case, a dedicated reporting system will provide information about the number of incidents of violence and aggression, as well as the times of day, locations, nature of injuries, frequency and types of injury, roles most at risk, and so on.

Designing the policy

Consultation and collaboration

[V8024] Once the nature and extent of the problem of violence is fully understood consideration can be given to designing policy that will provide direction and focus for action.

It is important that a policy development group is formed and made up of people who understand the issues and practical implications of the decisions that are made. In practice, such a group will probably exist for the general health and safety issues in the form of a health and safety committee or group (see figure 1 below).

Figure 1: Model for effective policy development group

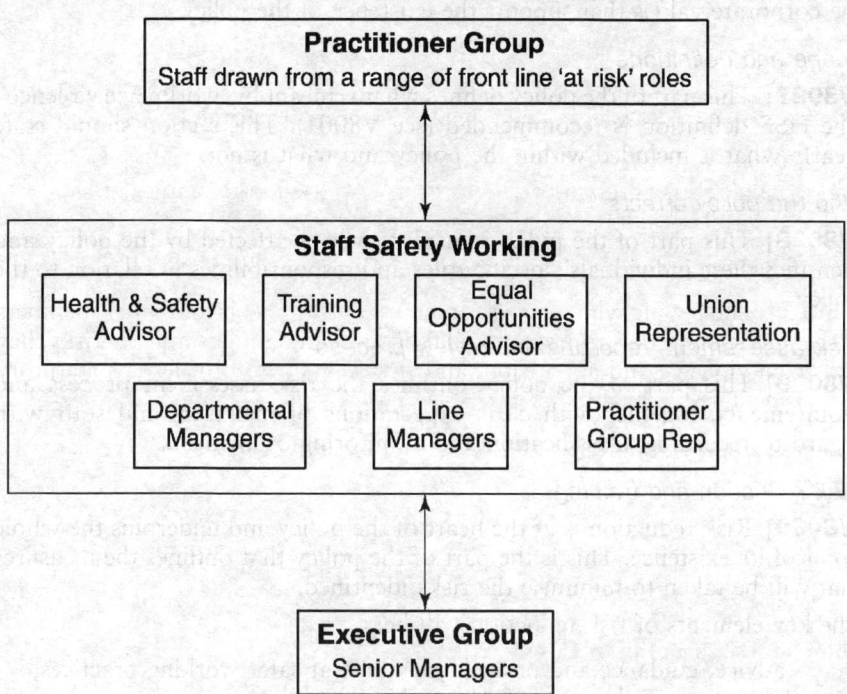

In this model, the practitioner group drives the policy formation, and front-line staff have ownership of the issues from the beginning and have been fully involved in the creation of the solution.

Policy content

[V8025] These are the basic elements of an effective policy:

- Purpose.
- Scope and definitions.
- Who the policy affects and their specific responsibilities.
- Risk assessment, reporting and review process.
- Risk reduction and training.
- Response to incidents:
  - Incident management.
  - Post-incident management.
- Reporting.
- Involvement of other agencies and sanctions.
- Maintenance.
- Communication strategy.
- Policy review and revision.

Purpose

[V8026] This part of the policy sets out the context and aims and expresses the corporate values that support the existence of the policy.

Scope and definitions

[V8027] This part of the policy defines what is meant by 'workplace violence'. The HSE definition is recommended (see **V8001**). This section should state clearly what is included within the policy and what is not.

Who the policy affects

[V8028] This part of the policy considers who is affected by the policy and identifies these individuals' specific roles and responsibilities in relation to the policy.

Risk assessment, reporting and review process

[V8029] This part of the policy outlines the risk assessment process and requirements, together with clear expectations of managers and staff with regard to recording, classification and monitoring of incidents.

Risk reduction and training

[V8030] Risk reduction is at the heart of the policy and underpins the whole point of its existence. This is the part of the policy that outlines the measures that will be taken to minimise the risks identified.

The key elements of risk reduction will be:

- advice, guidance and procedures aimed at safer working practices;
- equipment and design of the working environment;
- improved service delivery and reduction in triggers of violence;
- training and development in handling conflict and potentially violent situations.

Response to incidents

[V8031] The response will typically cover two key areas:

(1) Incident management
    This part of the policy outlines the procedures that managers and staff are required to follow should a violent incident occur, and provides clarity on roles, responsibilities and communications. It also includes communications with other agencies such as police, security and media. It should state the position with regard to the use of physical intervention and the provision of appropriate guidance and training.
(2) Post-incident management
    This part of the policy clearly defines post-incident management procedures including reporting requirements, operational support, and immediate and longer-term staff support. It should also state the responsibilities of line managers and their health and safety and occupational health functions.

Involvement of other agencies and sanctions

[V8032] This part of the organisation policy will outline guidance about the involvement of other agencies such as the police and local authorities. It should also outline the stance on prosecution and the options available including legal/financial support for employees and, where appropriate, the policy and guidance concerning exclusion and withdrawal of services.

Maintenance

[V8033] Performance indicators need to be established and managers held accountable for ensuring that breaches of policy and guidance are addressed.

Communication strategy

[V8034] This part of the policy states the corporate position with regard to internal and external publicity and identifies key messages/statements to be promoted. It should also include a strategy for managing media attention following an incident.

Policy review and revision

[V8035] This part of the policy establishes a monitoring and review process that allows periodic re-evaluation of the strategy and policy.

Implementation plan

[V8036] Implementation is about making the policy happen. It includes educating people about new processes and procedures, ensuring managers understand and monitor new practices, introducing and using new equipment, and setting up and running a programme of training and development. It is important that the implementation plan is adequately resourced to ensure that change is driven forward and embedded in the organisation.

Communication

[V8037] There are two aspects to this, internal communication to the people who work within the organisation, and external communication to the people who make up the clients, service users and the general public.

Internal communication

[V8038] This will depend on the structure, size and nature of the particular organisation. If it has a department that deals with communication and the media, its help will be invaluable. There are some factors that will help to make the communication process successful:

- Plan
  Thought needs to be given to the target audience, the communication structures in the organisation and how to make the information accessible.
- Policy launch
  It is much more positive if people recognise the start of a new process.
- Workplace briefings
  Front line managers should be given a full understanding of the policy first, and then should communicate this to their staff.
- Easy access 'help desk'
  A system is needed where people can quickly have their queries, worries and doubts dealt with.

- Reinforcement

  Reinforcement of the messages with a second wave of communication needs to be made some weeks later by checking out how well things are going.

External communication

[V8039] The media will naturally look for a story or slant and care must be taken that the original good intentions are not skewed to provide an interesting angle – at the expense of the launch.

New systems, procedures, practices and equipment

[V8040] The implementation of a policy will involve the introduction of new systems of working, new procedures, practices, and perhaps equipment, to learn about and to follow or use.

Implementing the policy

[V8041] If the organisation has developed an effective strategy and plan, the actual implementation should be relatively straightforward.

To safeguard the organisation, it is necessary to show a training and development plan that clearly outlines the phases of training being undertaken and the thinking that has gone into the development of those phases and the sequence of training for different roles.

Evaluating the policy

[V8042] It is very likely that policies and practices which made perfect sense in theory will run into trouble when applied in the workplace. It is important therefore to monitor and evaluate the implementation of the policy (see figure 2 opposite).

Figure 2: A model for monitoring and evaluating policy

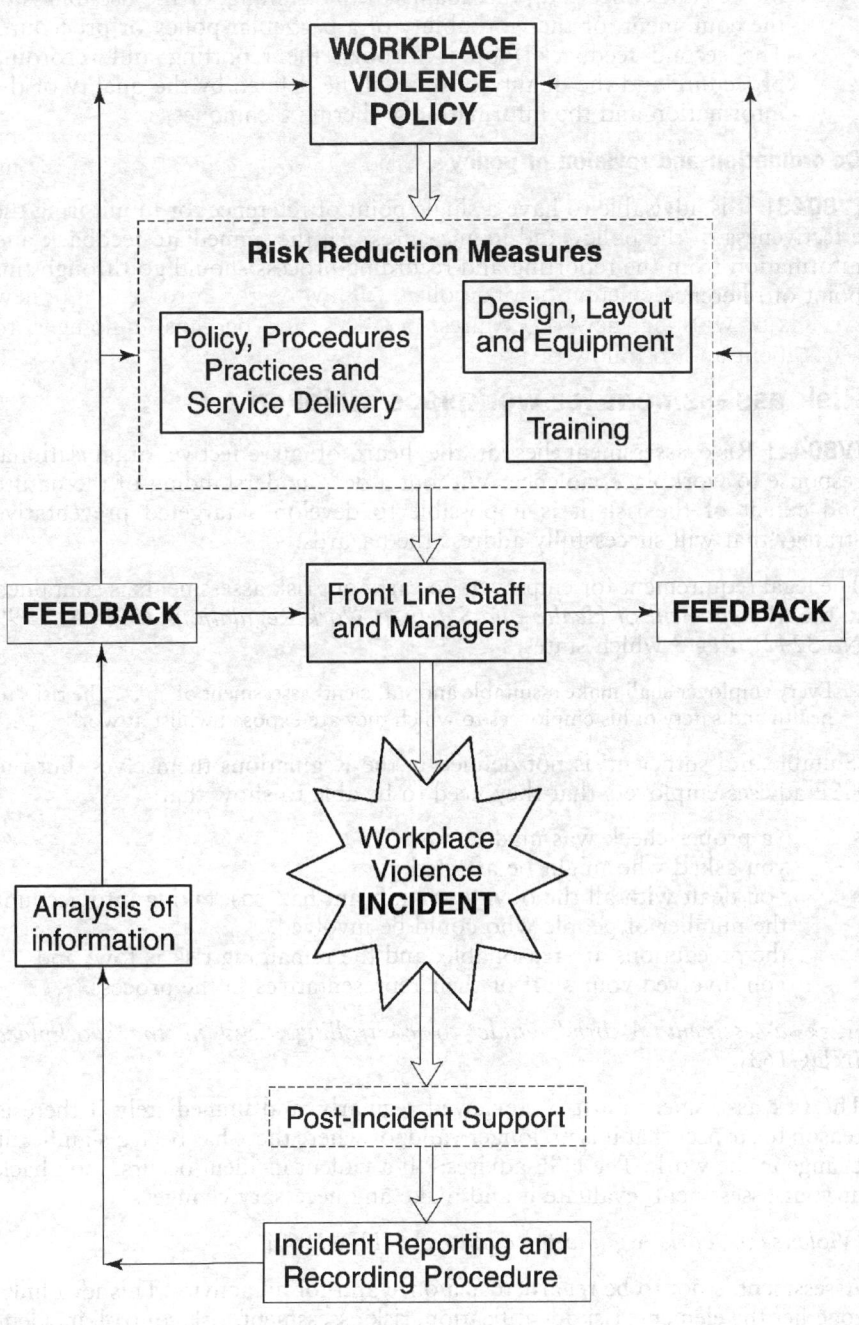

In this model there are two feedback loops.

- An 'immediate loop' where the quality of the measure is evaluated when first received. This will include information about the initial perceptions of the staff concerning the quality of the training, or the usefulness of the equipment, or the workability of a particular policy or procedure.
- The second feedback loop is through the reporting and recording procedure and the quality of this will be defined by the quality of the information and the information gathering techniques.

Co-ordination and revision of policy

[V8043] It is advisable to have a single point of reference for monitoring the effectiveness of the policy and its measures. All the immediate feedback and information from the reporting and recording process should go through this point of reference.

Risk assessment for workplace violence

[V8044] Risk assessment lies at the heart of an effective organisational response to workplace violence. Without a deep understanding of the nature and extent of the risk it is impossible to develop a targeted preventative strategy that will successfully address the hazards.

The legal requirement for employers to carry out risk assessments is contained in the *Management of Health and Safety at Work Regulations 1999 (SI 1999 No 3242), Reg 3* which states:

> Every employer shall make a suitable and sufficient' assessment of . . . the risks to health and safety of his employees to which they are exposed whilst at work . . . .

'Suitable and sufficient' is not defined in the Regulations themselves, but the HSE advises employers that they need to be able to show that:

- "a proper check was made;
- you asked who might be affected;
- you dealt with all the obvious significant hazards, taking into account the number of people who could be involved;
- the precautions are reasonable, and the remaining risk is low; and
- you involved your staff or their representatives in the process."

[*Risk assessment A brief guide to controlling risks in the workplace INDG163*]

The risk assessment must be reviewed regularly, and immediately if there is reason to suspect that it is no longer valid, or where there has been a significant change in the work. The HSE advises: "If a violent incident occurs, look back at your assessment, evaluate it and make any necessary changes."

[*Violence at work A guide for employers INDG169*]

Assessment is not to be regarded as a once-and-for-all activity. This idea links together the elements risk identification, risk assessment, risk control, incident report, incident investigation and risk review into a continuous improvement cycle.

There are three basic ways in which the risk of violence can be assessed:
- Generic violence risk assessment of role.
- Risk assessment of pre-planned event.
- Dynamic Risk Assessment.

Generic violence risk assessment of role

[V8045] This is perhaps the most recognised form of risk assessment. It involves an analysis of all the activities commonly associated with each job role that carry a level of risk from work-related violence. Each activity is examined, the possible risks are identified, and a judgement is made about the likelihood of the perceived risk and the severity of the consequences.

A good example to illustrate the point is any role that has an element of working alone and visiting clients or service users in their homes.

Some of the risks inherent in such a role are:
- Being targeted by criminals when travelling and in car parks, isolated locations, entrances to flats and apartments.
- Violence from a service user or other individual inside the person's home.
- Being held against will in premises.
- Being injured in an isolated location without access to help.

Measures to eliminate, minimise or control could include:
- Procedures in relation to callout, providing details of location, client visits, times of appointments to a central source and call in when complete.
- Personal safety guidance to reduce vulnerability when travelling and visiting.
- Training in Dynamic Risk Assessment (see **V8047** below), conflict resolution, communication skills etc.
- Provision of equipment where appropriate such as a mobile phone or attack alarm.

Each role should be examined in this way to determine all the risks and measures appropriate for that role.

Once this assessment and the resulting control measures have been established for a role, then everyone performing that role should be required to have the appropriate equipment, conform to the procedures and undertake the training required.

Management should ensure that the generic violence risk assessment is carried out on all roles where there is a potential danger and that the control measures are clearly stated and understood by everyone in each role.

Violence risk assessment – pre-planned event

[V8046] The generic violence risk assessment caters well for the main activities of a role. However, in every job there are events and activities that are out of the ordinary or, perhaps, done only rarely. It is quite possible that the generic risk assessment will not cover every circumstance, and in such cases it is important to assess the risk involved in a special event or activity as part of the planning process. The risk assessment will identify if there is an increased risk of potential for violence and a need for extra measures in these circumstances.

The risk assessment should be recorded and included in the planning documentation for the event itself.

Dynamic Risk Assessment

[V8047] There are many occasions when front line staff are confronted with situations which are unique and not catered for in a generic risk assessment or an assessment of a planned event. These are often the situations where people get hurt because they do not have a way of assessing the risks in the situation they are confronted with, and do not respond appropriately.

Dynamic Risk Assessment is a process which helps an individual to effectively assess a situation from a personal safety perspective, as it is unfolding. The person can continuously assess the circumstances and adjust his or her response to meet the risk presented moment by moment. Training which develops understanding and skills in Dynamic Risk Assessment can help an employee to assess situations more effectively and respond in the most appropriate way.

Risk assessment – who should do it?

[V8048] There is little doubt that the individual who is working at the customer interface is best placed to identify the risks faced. For generic violence risk assessment of roles and pre-planned events, the process needs to include someone who is trained and experienced in identifying risks and who can spot the hidden dangers in workplace environments, situations and incidents. The most effective solution is to combine the two – a person who is trained and competent in risk assessment and has experience in the management of violence should examine the roles alongside the people who carry out those roles on a day-to-day basis.

Risk assessment – how should it be approached?

[V8049] It is recommended that risk assessment be approached on a role-specific basis rather than purely on a geographical area or departmental basis. By definition, the risk of workplace violence will always involve some sort of interaction with another person. This unique feature means the risks of work-related violence are much more difficult to predict and to control because it is hard to predict the range of responses that someone might use in a situation involving conflict.

For each role, the following main areas need to be assessed:

- What contact is made with the clients, patients, service users or members of the public? This will usually be described in the job description. For example, enforcing role, caring profession, advising or dealing with complaints.
- What sort of people generally make up this contact? For example, travelling public, offenders, patients, clients or relatives.
- Where does the contact take place? For example, interview rooms, reception, platform, home address, classroom or on the streets.
- What sort of 'state' will the client/patient etc be in? For example, will they be influenced by drink or drugs, anxious, frustrated or hostile?
- What specific tasks are performed? For example, handling complaints, delivering 'bad news', denying benefits, evicting or arresting, or even treating an individual.

Content of the risk assessment

[V8050] The first question that should be asked is: 'Is there any likelihood that someone doing this role will be abused, threatened or assaulted in circumstances relating to their work?' If the answer is 'yes', then a risk assessment for workplace violence needs to be carried out.

Figure 3 below outlines five steps that provide a basic risk assessment model for both the generic risk assessment and for the risk assessment of a pre-planned event.

Figure 3: Risk assessment model

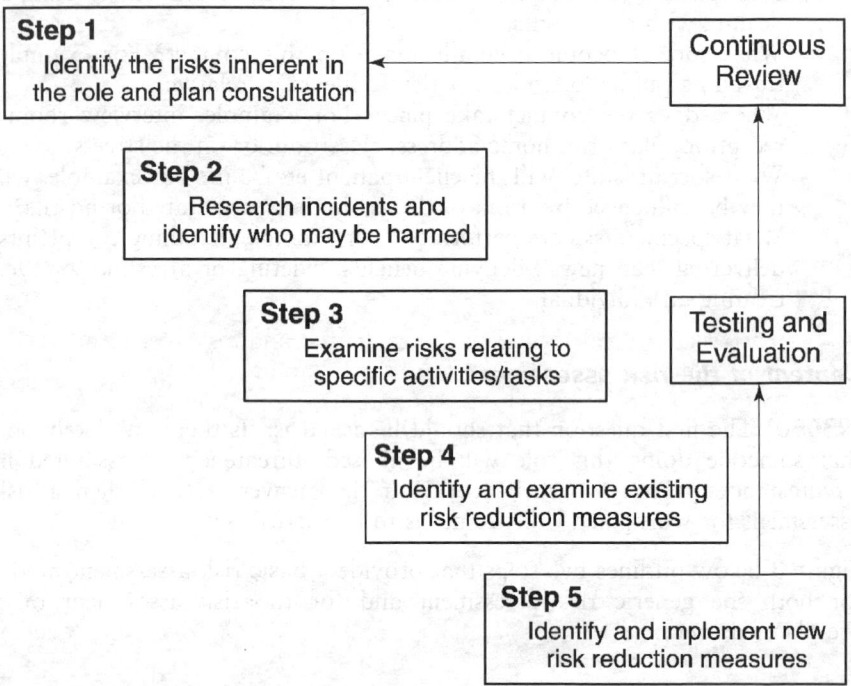

Step 1 – Identify the risks inherent in the role and plan consultation

[V8051] An examination of the job description should show the potential situations where the jobholder will be at risk. However, few job descriptions will capture all the possible scenarios that might arise and other sources of information should be used.

It is important to research other potentially useful sources of information such as HSE and sector-specific guidance and research.

Consultation with workers needs to be planned to ensure this covers a cross section of staff performing the roles being risk assessed. The consultation can take the form of a combination of the following:

- Structured and informal interviews.
- Small focus groups (typically 3–6 staff).
- Questionnaires.
- Workplace observation.

Step 2 – Research incidents and identify who may be harmed

[V8052] An examination and analysis of past-incident reports, local log books etc associated with workplace violence will provide invaluable data in a risk assessment. Unfortunately, violence is under-reported in most organisations, so it is important (in addition to putting measures in place to encourage

Risk assessment for workplace violence **[V8053]**

staff to report incidents and responding to their reports – see **V8063** and **V8070**) to try to establish the true extent of the problem by consulting with managers and staff and gathering anecdotal evidence of occurrences. Interviews, workplace observations, discussions and surveys should be carried out.

Step 3 – Examine risks relating to the specific activities and tasks performed

[V8053] Assessing the level of risk has its difficulties. By reviewing the actual activities and tasks performed it is possible to get a realistic measure of the risks involved and to ensure controls are both relevant and proportionate.

This stage involves an analysis of the data obtained in Step 2 and further consultation to establish:

- The risks associated with specific activities and tasks performed
  For example, delivering bad news, cash handling, or dealing with fare evaders. The risks may be compounded at certain times and locations, or by factors relating to drink or drugs.
- Practical risk reduction and support measures
  Those performing the job will often have the best ideas for reducing risk.

When considering the risks present in job activities and tasks, it is helpful to identify the risk level based on two elements:

(1) the *likelihood* of harm actually being caused, and
(2) the likely impact or *severity* of the harm caused.

The *likelihood* can be assessed on a sliding scale, such as:

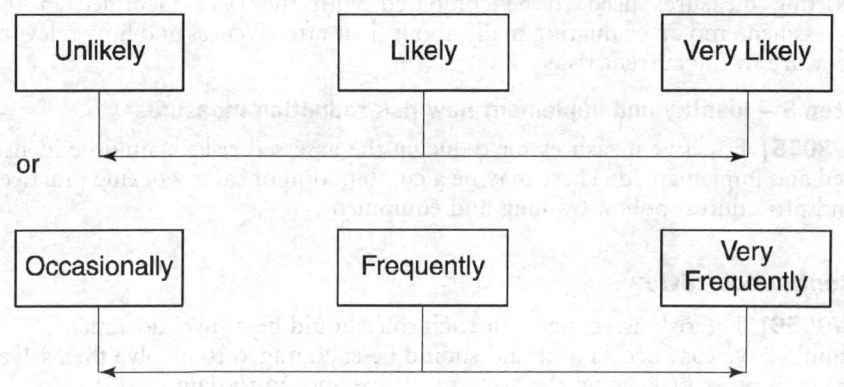

The *severity* of harm can be assessed on a similar scale such as:

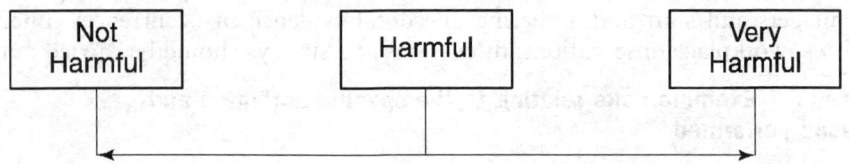

From this, a 'risk rating' can be established which identifies the risk as being low, medium or high. When considering the severity of an incident there are number of considerations including:

- Physical harm to those involved.
- Emotional impact on staff.
- Impact on the perpetrator and other service users.
- Numbers of people involved.
- Duration of the incident.
- Effect on operations and service delivery.
- Financial impact on both individuals and the organisation.

The likelihood/severity approach to risk assessment is long established and is helpful in providing a broad assessment of risk. However, it does not take into account the fear of violence experienced by the person doing the job. An employer has a duty to include this in the risk assessment and respond to this fear, for example through rationalising fears in training and introducing guidance to further minimise the risk.

Step 4 – Identify and examine existing risk reduction measures

[V8054] There will be risk reduction measures already in place. These existing measures need to be compared with the risks identified in the assessment and an evaluation made about their effectiveness and how relevant they are to the current risks.

Step 5 – Identify and implement new risk reduction measures

[V8055] Effective measures for reducing the assessed risks should be identified and implemented. These may be a combination of safer working practices and procedures, policy, training and equipment.

Continuous review

[V8056] The risk assessment for each role should be a 'live' document. Staff should have easy access to it and should be encouraged to involve themselves in the process of keeping the assessment live and up-to-date.

Reporting, recording and monitoring system for workplace violence

[V8057] Central to an effective strategy for combating workplace violence is an understanding of the nature and extent of the problem. Integral to this is a system that monitors and provides data and information about the trends surrounding violence in the workplace for a particular organisation.

Figure 4: A model for reporting and monitoring incidents of workplace violence

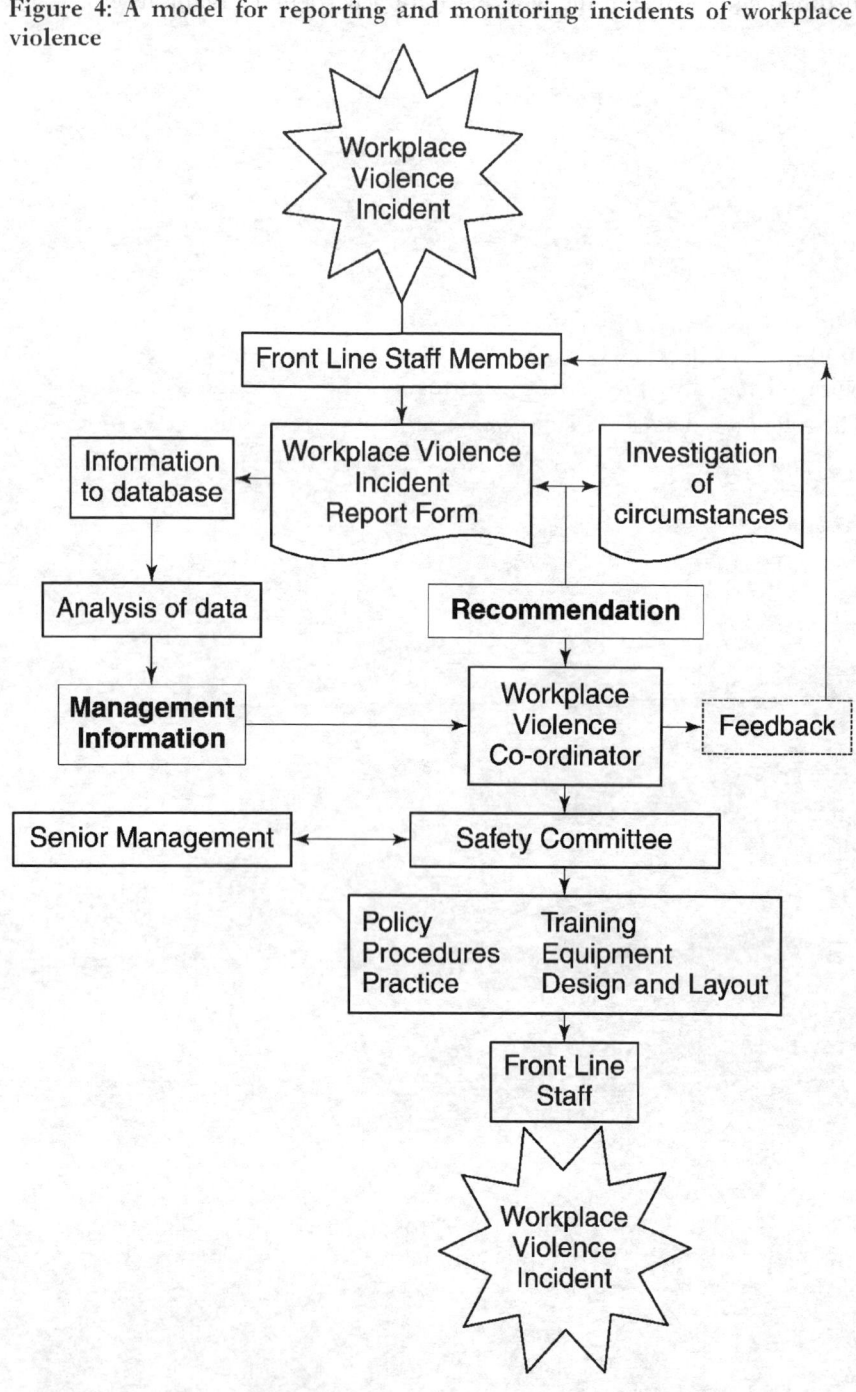

Designing the system

[V8058] The outcomes required of the system will be specific to the needs of a particular organisation but the essential elements of a system are:

- Information gathered through a dedicated workplace violence incident report form.
- An investigation process that will examine the circumstances of an incident and provide recommendations for improving risk reduction.
- An information database and analysis process.
- A workplace violence co-ordinator.
- A communication process.

Workplace violence incident report form

[V8059] The design of the form will depend on many factors, some of which will be dictated by the way the information will be recorded and analysed.

The following details are suggested as a minimum whether using a dedicated violence report or a combined use report form:

- Name and contact details of person making the report.
- Date, time and location of incident.
- Description of the incident and any injuries sustained.
- Details of any other staff involved.
- Details of the assailant(s).
- Details of any possible witnesses.

Further information will be required from the incident form, but this will differ between organisations based on the management data required.

The report form can also provide the basis for recording post-incident action by the line manager, including staff support, the investigation process, and subsequent recommendations and actions.

The investigation process

[V8060] The investigation process should aim to establish the facts of the incident and make recommendations that will help to reduce the risk of this type of incident happening again.

The recommendations from this investigation will be fed directly to the person responsible for co-ordinating issues of workplace violence for a decision as to how they should be taken forward.

Information database and analysis

[V8061] The information gathered from all the incidents of workplace violence can provide an invaluable insight into the trends, hotspots and risks that are not obvious in the day-to-day management of such incidents.

The database also provides a powerful tool for monitoring the effectiveness of risk reduction measures by monitoring over a similar period to see if the measures have had an impact on the risks identified.

Workplace violence co-ordinator

[V8062] This person (or department) will be responsible for:

- Maintaining the reporting and recording procedure.

- Monitoring the quality of the content and the investigation process.
- Monitoring the recommendations from the reporting and investigation process.
- Searching and monitoring the database for useful management information, trends, and hotspots that need further investigation.
- Informing the Safety Committee of recommendations and work-related violence issues for discussion and further action.
- Monitoring the effectiveness of risk reduction measures.
- Liaising with the police and other relevant agencies.
- Administering or assisting in the process of litigation or action against the attacker.

A communication process

[V8063] A clear communication process should be built into the system so that front line staff are kept fully aware of the most up-to-date policy, procedures and practice for their role. A common complaint is that individuals complete incident forms and then hear nothing more about the outcomes.

Implementation

[V8064] The following areas are essential for the successful implementation of a reporting process:

- Management involvement and training.
- Staff training.
- Monitoring the completion of reports.
- Providing feedback to those affected and information to the wider organisation.

Manager involvement and training

[V8065] Managers at every level need to understand and appreciate the complete reporting, recording and monitoring process.

Staff training

[V8066] Training helps to establish best practice and a sound consistent approach. It ensures that individuals understand the importance of the process and that completion is part of their duty of care.

Monitoring the completion of reports

[V8067] Four fundamental questions need to be asked on a routine basis:

- Are all relevant incidents being recorded?
- Are reports correctly entered?
- Are accounts of incidents adequate and professional?
- Is anything actually being done as a consequence?

Feedback and regular provision of information across the organisation

[V8068] Following the introduction of a reporting process, particular effort must be made to provide immediate and ongoing feedback in relation to reported incidents. This applies not only to the affected individual, but also across the whole organisation.

A *formal review* of the reporting system should be undertaken some months after initial implementation to establish whether the system is being used, whether it can be improved and what is being done with the information generated.

Measurable outcomes

[V8069] Measurable outcomes using information from the reporting process may include:

- The number of injuries to staff.
- The number of injuries to service users.
- The extent and type of injuries suffered.
- The number of working days lost due to assaults on staff.
- The number of prosecutions taken against those assaulting staff.
- Extent to which safety procedures have been followed.
- Which techniques taught in training have been used and their effectiveness.
- Incident and assault patterns.
- The groups at most risk.
- The highest risk tasks and activities.
- Profiles of perpetrators of violence.

The more sophisticated the database, the more comparisons and measures available.

Under-reporting

[V8070] It has been mentioned several times that most organisations suffer from under-reporting of workplace violence-related incidents. There are many reasons for this including, for example, over complicated and lengthy reporting forms and processes or fear of criticism after having brought an incident to light.

The organisation should work to eliminate as many barriers to reporting as possible.

Disclosure rules

[V8071] Written records of any incident that results in a court case may be called upon as evidence. This would include any internal incident reports, notes, debriefing notes, statements or correspondence.

The purpose of such 'disclosure' is to ensure that all potential sources of evidence are made available to defendants and the courts.

From a practical point of view, the purpose of the reporting process is to ensure that reports are essentially factual and as free as possible from conjecture and opinion.

Risk reduction measures for workplace violence

[V8072] The *Management of Health and Safety at Work Regulations 1999 (SI 1999 No 3242), Reg 4 and Sch 1*, require employers to design and implement protective and preventive measures on the basis of hierarchy of general preventative principles. The principles may be summarised as follows:

- Eliminate risk by avoiding or removing the hazard
- Substitute the process etc by one that carries less risk
- Control the risk at source
- Devise safe systems of work
- Provide adequate training and instructions
- Ensure adequate supervision is provided
- Provide personal protective equipment as a last resort

Most risk reduction measures fall under one of the following broad categories:

- Training and Information
- Work environment
- Job design

It may not be possible to completely eradicate the risk of violence in a workplace, but it is possible to minimise the risk and its impact by introducing appropriate control measures based on an understanding of the particular problems. Effective risk reduction is a blend of measures that embrace all the areas of policy, effective risk assessment, practice and procedures, training, equipment, design and layout.

Risk reduction – achieving a balance

[V8073] Appropriate risk reduction measures are a balance between understanding the identified risk and providing a balanced response in terms of acceptability, cost and relative effectiveness. This is not an easy equation to balance and it requires some thought before a reduction measure is introduced. The appropriate balance is, in the end, a matter of judgement, which can only be made by the organisation itself. The cost of a glass screen may be relatively modest and the simple balance may suggest it should be installed. However, other aspects can be added to the equation; is a glass screen acceptable in terms of the organisation's ethos and approach? Does it, in fact, reduce the level of violence – or does it provide a trigger for the level of aggression to increase? On the other hand, if there is a high risk of serious violence involving weapons, this may outweigh other considerations and it may be imperative that bullet-proof glass screen is in place.

Service delivery

[V8074] Workplace conflict can occur for many different reasons. Sometimes organisations and staff can create or exacerbate the environment within which a conflict develops and increase the risk of violence by the way they deliver services or approach their work.

Working practice

Staffing levels

[V8075] Queuing and apparent 'queue jumping', long waiting times, cancellations, missed connections and flights, poorly informed staff and rude or abrupt service are all common sparks which can ignite violence. Organisations should ensure that ratios of staff to customers, clients or service users are realistic and allow for the cost-effective provision of the best service.

Safe practice guidance

[V8076] Some roles have inherent risks associated with them and require specific guidance. For example, safe practice and communication systems need to be put in place for workers working alone, and for those working out in the community who can be isolated and vulnerable to crime.

Staff are more likely to adopt guidance they have been involved in developing.

Systems and procedures

[V8077] Effective risk assessment will anticipate and deal proactively with situations likely to cause conflict. High demands on service, running out of stock, maintenance work likely to cause delays and particularly unpalatable news are some of the situations that can be anticipated and contingencies for minimising and resolving conflict put in place.

Those working in certain urban areas may be more likely to be exposed to theft, robbery, aggressive begging, prostitution, public disorder, drug dealing and drug abuse. Rural areas present different challenges, as staff can be a long way from help.

Some roles, particularly in health care, necessarily involve close contact with a person to carry out checks, tests and medical procedures. Some workers deal with aggressive people in confined spaces where there are risks to other people and little prospect of immediate help from the police or security staff.

Each risk should be examined and appropriate guidance and practice designed which will minimise the risks involved in the working practice or situation.

Communications

[V8078] Poor information and communication often causes, or contributes to, conflict. Effective communication can quickly calm and resolve potential conflict. Many people will accept the inevitability of a delay or cancellation – as long as they know how long it will be or what alternatives are available.

Design and layout

[V8079] Designers are becoming more aware of the issue of work-related violence and significant developments have been made in this area as it becomes clear that personal safety and good service delivery go hand in hand. Some pub chains, for example, use clever layouts to reduce potential frustrations and conflict flash points, thereby 'enhancing the customer experience', selling more drinks and reducing the risk of violence and aggression.

'Queue jumping' is a very common cause of conflict and effective signage and queue management can reduce this.

Risk can be further reduced with the introduction of wider desks, raised floors, and areas offering access to those with special needs or requiring greater privacy. Staff need to have access to a secure place should a serious incident occur, and private areas need to be clearly identified and secured.

Proactive service delivery

[V8080] The quality of service delivery has a major bearing on the incidence of violence in the workplace. A climate for trouble is created when people are frustrated, their expectations are not met, and staff respond unprofessionally to their concerns.

Sometimes, staff cannot meet the customers' expectations because of circumstances beyond their control. The response should be to influence and adjust the customer expectation. The development of basic interpersonal skills and a service-oriented culture lays the foundations for a safer environment. This area is probably the most under-rated as a control measure.

Simple inexpensive gestures like providing refreshments, upgrades, free use of a telephone or concessions can help to reduce tension, and it is important that staff are empowered to offer these, with appropriate support guidance.

Warning 'flags'

[V8081] A contentious issue for many organisations is the placing of warning 'flags' against an individual service user or address. Many organisations cite client/patient confidentiality, data protection and human rights as reasons for not placing warnings. Some also argue that if staff know someone could be a problem then this will affect the way they deal with them and risk a self-fulfilling prophecy. Although each of these issues is valid and needs to be considered, employees also have rights and a balance must be sought in order to ensure staff safety.

Security responses

[V8082] Security responses available to organisations break down into three areas:

- Security personnel.
- Security procedures.
- Physical measures.

Security personnel

[V8083] Some organisations use external security contractors, others choose to employ an in-house team or mix the two.

If considering an external provider, it is important to establish the level of investment the provider makes in staff training and support. In a traditionally low paid area of work with a high turnover of staff, contractors are reluctant to invest in more than the bare minimum required for training.

More control over selection, training and performance is achieved through developing an in-house security team.

Whichever option is chosen it is important to remember that security staff need to be actively involved in the initiatives, so that they can win the trust and support of other staff and be made to feel part of the team.

Selection

[V8084] Security staff, by the very nature of their role, will deal with conflict on a regular basis. It is important to select staff who demonstrate the attitudes and behaviours that help to deal positively with situations where conflict is inevitable.

Training

[V8085] Security staff play a key role in service delivery and are often the first and last point of contact for a customer or service user. It is therefore a worthwhile investment to develop their communication and interpersonal skills.

Security staff are often expected to confront potential criminals, respond to violent situations, and to protect staff and service users. They need to be equipped with the necessary knowledge and skills to manage conflict and violent behaviour.

Security procedures

[V8086] Security staff face very difficult and sometimes dangerous situations where their actions may be subjected to detailed scrutiny. It is important therefore that staff are clear as to the expectations upon them for preventing and responding to violent incidents. Clear procedures need to be put in place that outline the action to be taken in areas such as incident management, refusing entry, searching and evicting people and making an arrest.

Other employees will also need guidance as to what they can expect of their security colleagues, and how they can help security staff when the need arises.

Physical measures

[V8087] There is a vast range of physical measures that can be introduced to reduce the risk of physical injury through a workplace violence incident.

Some of these measures can be extremely expensive and it is worth emphasising here that a thorough risk assessment needs to be undertaken before committing to any expenditure on specialised equipment.

Access control

[V8088] There are many variations of access control, ranging from barriers to open plan receptions. However, funding and existing building design are major influences on this. A well-designed and managed reception can contribute greatly to access control, yet be seen as offering a friendly and helpful service rather than looking intimidating and overtly tight on security.

Closed circuit television (CCTV)

[V8089] Although the outlay on CCTV can be high, it is one of the most popular security responses available for deterring and managing crime and violence. However, organisations need to be realistic about its purpose and use. One Home Office commissioned report suggested that CCTV is more effective as a detection tool than as a deterrent. CCTV needs to be monitored and used proactively to provide effective prevention.

Alarms

[V8090] There are many types of alarm system, each designed for a different purpose. The main alarm systems for enhancing personal safety are:

- Intruder alarms – commonly used to protect buildings after hours, or to protect restricted areas.
- Panic alarms – increasingly used in reception areas, interview rooms and isolated areas.
- Personal alarms – carried on the person as a means of attracting attention and temporarily distracting an assailant.

Before purchasing an alarm, be sure of the organisation's needs and consider them in the overall strategy for tackling violence.

Radios, paging and public address systems

[V8091] These communication systems have advanced greatly in recent years and can be an asset to staff safety. Some personal radios and pagers have built in panic alarms and sophisticated tracking facilities that pinpoint location.

Mobile phones

[V8092] Mobile phones can provide a means of communication and comfort to lone workers. They are useful for keeping colleagues informed of movements and of unforeseen travel/schedule difficulties.

Protective vests

[V8093] Often referred to as 'body armour' these vests can be designed to protect against the threats presented by firearms and edged weapons.

Any organisation can purchase protective vests as no specific authority or licence is required, and they can be a necessary risk reduction measure in some areas of work. It is however important that the decision to adopt protective vests is not taken lightly, as it can be a sensitive issue for employees, service providers, and the public generally. Great care needs to be taken when considering purchasing this expensive equipment and effective risk assessment, incident reporting and monitoring processes should be carefully scrutinised to help make any decision.

Restraints

[V8094] Restraints are common in policing, custodial services, airlines and some areas of healthcare. Although the concept of using restraints will understandably concern most employers, they may be a necessary measure to prevent a disruptive passenger putting an aircraft at risk, or to prevent a patient harming themselves or others. Use of such equipment needs to be justified by the role requirements and risk assessments, and staff need comprehensive training and guidance in its use.

Advice should be sought before adopting restraints to ensure an appropriate and lawful choice is made (see **V8015**).

Batons

[V8095] Batons are deemed offensive weapons and can only be used by the police.

Training

[V8096] Providing effective training is an essential element of an employer's duty of care to employees. Through training, employees can identify ways in which they can reduce risk and develop the skills they need to deal with potentially violent situations. Appropriate training for workplace violence can take many forms and can be sourced internally or externally. For training to be effective it is essential that it is not viewed as a 'stand-alone' solution but integrated into a wider learning programme that is actively supported by the organisation.

A training development model

[V8097] A well-designed training or development activity is developed through a three-stage process (see figure 5 below):

(1) Training Needs Analysis.
(2) Training design, development and testing.
(3) Implementing, monitoring and evaluation.

Each stage of this model is essential to the development of an effective training solution and should be evident in any training programme being offered, whether internal or external.

Figure 5: General training and development model

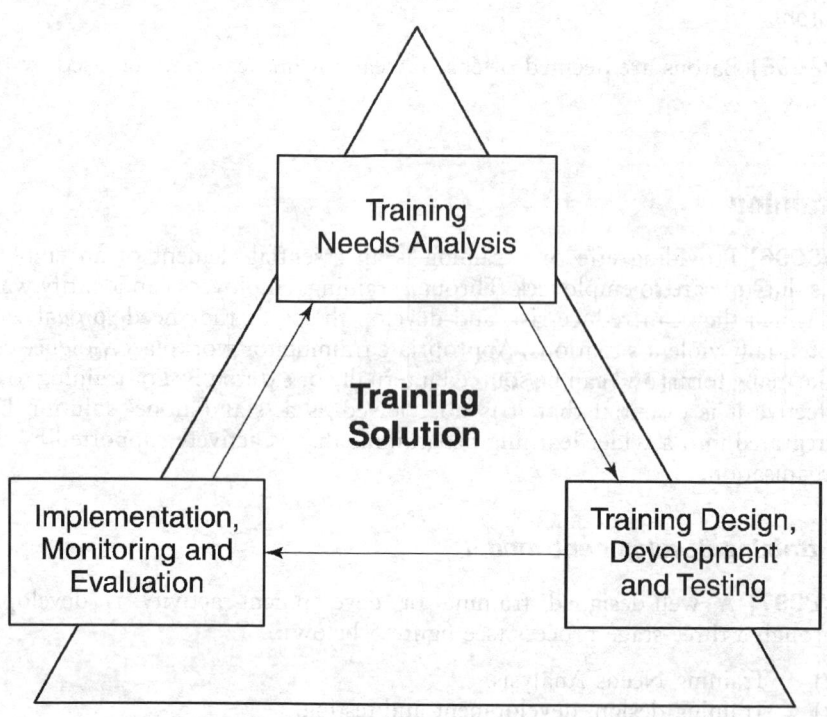

Training Needs Analysis

[V8098] The first step in a Training Needs Analysis is to identify the gap between how things are now and how things ought to be.

Any training or development solution should begin by asking two questions:

(1) How has the need for the proposed training been identified?
(2) What are the intended learning outcomes of the proposed training?

Identifying the need

[V8099] The variety of ways through which a training need can be identified are shown in figure 6 below.

A complete review of the risk of conflict and workplace violence across an organisation will identify a range of needs. Among these will be specific training and development requirements for staff performing specific roles and for line managers who are responsible for reducing risk and managing the issues that result from workplace incidents.

Figure 6: Identifying the need

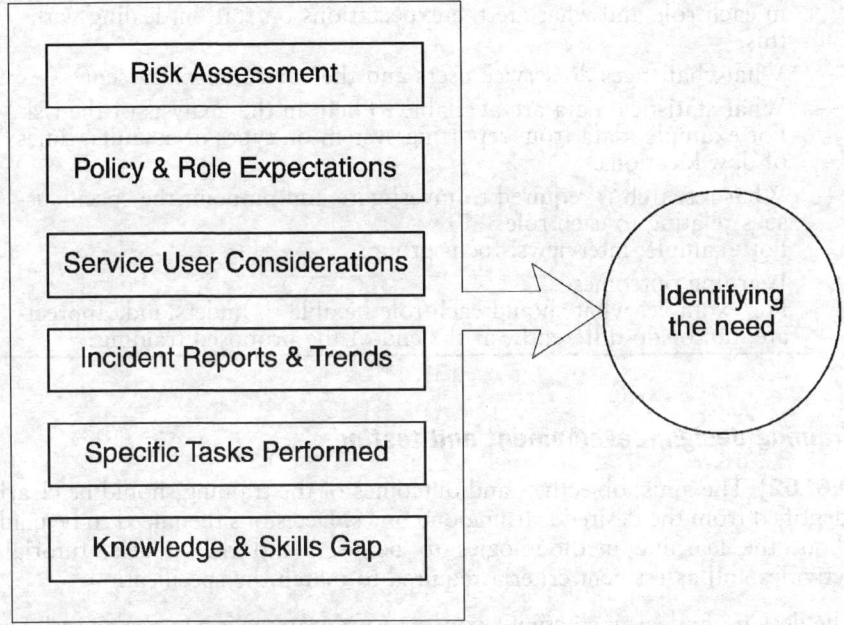

Defining the learning outcomes

[V8100] The outcomes should link directly to the identified needs. A good way to start is by saying: 'At the end of the training the delegates will . . . ' – and then describe the outcome. The outcome will describe what individuals will be able to do after they have attended the programme.

Here is a simple example. A review has shown the workplace violence reporting form is not being filled in correctly in 80% of cases. A training need has been identified. The training outcome will be: 'At the end of the training the delegates will be able to correctly complete the reporting form.'

The above example also highlights the importance of ensuring the need is correctly identified – in this case it may well be the form that is ambiguous, unclear or difficult to follow.

Checklist

[V8101] *Training Needs Analysis*

- How has the need for the training been identified?
  For example, through risk assessment; staff consultation; statistical data showing an increase in assaults.
- Who needs the training and why do they need it?
  For example, specific individuals, specific roles, specific departments, lone workers.

> - What are the specific roles – when, where and how are they exposed to the risk of conflict or violence?
> - What is the level of risk in terms of frequency and seriousness faced in each role and what are the expectations on staff in dealing with this?
> - What challenges do service users and the environment present?
> - What statistical data are available to help in the analysis of the risk? For example, data from reporting system on types of assaults, times of day, locations.
> - What research is required to investigate and pinpoint the specific issues relating to each role? For example, interviews, focus groups.
> - Learning outcomes For example, what should each role be able to understand, appreciate, do, or do differently, at the end of the proposed training?

Training design, development and testing

[V8102] The aims, objectives and outcomes of the training should be clearly identified from the desired learning outcomes. Decisions then need to be made about the learning methodologies to be used and the models, tutorials, activities and assessment criteria required to match the specification.

The design of a learning activity, course or programme is a specialist task and should be undertaken by the chosen training provider. It is essential to work closely with them to ensure the design meets all the requirements.

Generic and 'off-the-shelf' courses are unlikely to meet the needs of staff working in different roles with different problems and concerns.

Modern technology provides a great deal of choice in methods of delivering training including distance learning, video, audio and computer-based learning, virtual classrooms as well as traditional classrooms, workshops, and workplace-based learning such as coaching and mentoring. This choice can be somewhat bewildering and it must be recognised that not all these methods suit all occasions. Distance learning, in particular, can be attractive as a cheap alternative to other methods, but it will fail to deliver some important areas. Generally, distance learning is effective as a support element to a training programme, particularly for the 'knowledge' aspects. The classroom is good for practising skills, discussing issues and experiences, and problem solving. The workplace is good for coaching and applying the knowledge and skills.

Soft skills and physical intervention training

[V8103] Managing conflict demands a unique set of skills which fall into two distinct areas:

(1) *Soft skills* – communication and interpersonal skills that can be used to calm and control situations.
(2) *Physical intervention skills* – used when it is necessary to engage in physical contact with another person.

Core content – soft skills

[V8104] The Training Needs Analysis should identify the learning needs of people who perform a particular role. In general, the following constitute the core ingredients of any training solution which is intended to meet soft skills needs.

- *Organisational policy and values* – core values which underpin the organisation's approach and the policy on workplace violence.
- *Definitions of workplace violence* – what the organisation defines as 'violence' and what type of incidents should be considered for reporting.
- *Risk assessment* – dynamic assessment of a situation to assess the risk and consequent appropriate action.
- *Risk reduction and safety systems* – safe working practice, security procedures, the location, testing and use of alarms, panic buttons and CCTV equipment.
- *Theoretical models of aggression, violence and conflict management* – understanding the physiological and psychological processes that occur when people become angry, frightened or aggressive.
- *Triggers and escalation* – recognising the triggers and signs of increasing aggression so employees can anticipate and defuse situations before they become more serious.
- *Verbal and non-verbal communication skills* – understanding and practising the specific skills of communication and controlling personal space.
- *Special groups of people* – eg with drugs or alcohol issues, mental health issues, learning disabilities, cultural differences and the elderly.
- *De-escalation and calming skills and exit strategies* – de-escalation and calming through empathy, problem solving, win-win thinking and strategies for getting out of difficult situations.
- *Legal issues, self defence and use of force* – employees' understanding of their rights and responsibilities when confronted with aggressive and violent behaviour.
- *Support for staff* – post-incident – the support available through the organisation from line manger support to specialist counselling.
- *Post-incident reporting and debriefing* – understanding how to properly report an incident and being aware of the reasons why it this is important.

Core content – physical intervention skills

[V8105] It is more difficult to identify the core content for physical skills, as it will vary a great deal across different sectors. The Training Needs Analysis and risk assessment will identify whether there is a specific need for physical intervention in a particular role. Some self-protection systems include strikes and locks to joints in order to gain compliance through pain. Others are based on methods that do not employ aggressive techniques or use pain as a way of gaining compliance. Most organisations requiring physical intervention training are well advised to opt for an effective non-aggressive system, as this will reduce the likelihood of an escalation and the risk of injury to both staff and the assailant.

The following are the key pre-requisites to physical skills training:

- *Core soft skills* – core content described for soft skills should be regarded as an essential pre-requisite of any course involving physical intervention skills.
- *Skills for protecting oneself or another from unlawful assault* – commonly referred to as 'breakaway' or 'disengagement' techniques; skills to protect against strikes and skills to release holds and remove oneself or another from danger.
- *Interventions that are used to hold and restrain another person* – often referred to as 'control and restraint' or 'holding skills'; used to prevent someone from escaping lawful arrest or detention, or preventing them harming themselves or others.
- *Legal issues* – employees must be critically aware of their powers in relation to detaining someone.

Additional physical skills can be added to these core areas, including safer approaches to day-to-day tasks such as escorting or guiding.

Specialist skills can also be added to the highest-level training, such as the use of restraint equipment.

Reviewing the training content

[V8106] It is important at this stage to ensure the solution will satisfy three basic criteria:

(1) Relevance
Is the training content based upon a thorough Training Needs Analysis that considers the role and tasks performed and the risks associated with these?
(2) Legal review
Is the content of the course legally correct? Will it stand up to examination in legal proceedings if tested?
(3) Medical review
Are the learning methods and content safe? Will this stand up to examination in legal proceedings if tested?

The legal and medical basis for the training provided may be challenged in any legal proceedings and it is important that the organisation is confident that the training being provided will stand up to such scrutiny.

It is the responsibility of the training provider to show the appropriate legal and medical basis that underpins the learning solution being designed.

Testing the solution

[V8107] Before implementing the course or programme, it is a good idea to test it out using a pilot course or courses. The pilot course should be delivered under the same conditions that the fully implemented course will be delivered and feedback about the programme should be sought from delegates.

Implementation

[V8108] The rollout of the programme needs consideration, with some thought about who should receive training first. There are no hard and fast rules but it is important to develop a cohesive, prioritised strategy for implementation.

Checklist

[V8109] *Design and Implementation*

> — What are the aims and objectives of the training programme?
> — How does the content of the programme achieve the aims and objectives?
> — What areas of knowledge and understanding are required? For example, theory, models, legislation, policies and procedures.
> — What skills are needed by staff? For example, communication, problem solving, disengagement, escorting, restraint, detaining.
> — What are the appropriate attitudes and behaviours required? For example, avoiding physical contact, assertiveness.
> — What learning methodologies are being used to achieve the development and how appropriate are they to these circumstances?
> For example, distance learning, group discussion, practical role play, practice physical skills.
> — Has the training been tailored to the role and tasks performed?
> — Have staff been consulted in the design process?
> — Can any of the training be delivered in the workplace for realism and problem-solving opportunities?
> — How is the content reviewed to ensure it is tactically effective, legally correct and medically safe?
> — How will the programme be piloted and who should take part in the pilot?
> — How will the pilot be evaluated?
> — Who needs to know about the programme and how will it be communicated?
> — What backing and support (and from whom) is available at senior level?
> — How will the programme be rolled out? Who will receive it first – or will it be a mixed rollout? Can some groups, roles, individuals be trained together?

Monitoring and evaluating the solution

[V8110] The monitoring and evaluation of the programme needs to fit with the existing infrastructure of the organisation.

Figure 7 below shows a typical process that combines the individual development needs of the delegate with the feedback required to monitor and evaluate the programme. From an individual point of view, the delegate should define his or her personal learning objectives before the course, preferably involving his or her line manager. A short time after the course they should review the objectives and establish the learning outcomes achieved.

From an organisational point of view, delegates can complete an immediate post-course evaluation, which will establish reactions to, and satisfaction with, the programme. About four to six weeks later, delegates should provide further feedback about the impact of the programme on the way they perform in the workplace. This is then fed into the process of Training Needs Analysis and design for continuous improvements to the programme.

Figure 7: A typical delegate feedback process combining individual and organisational needs

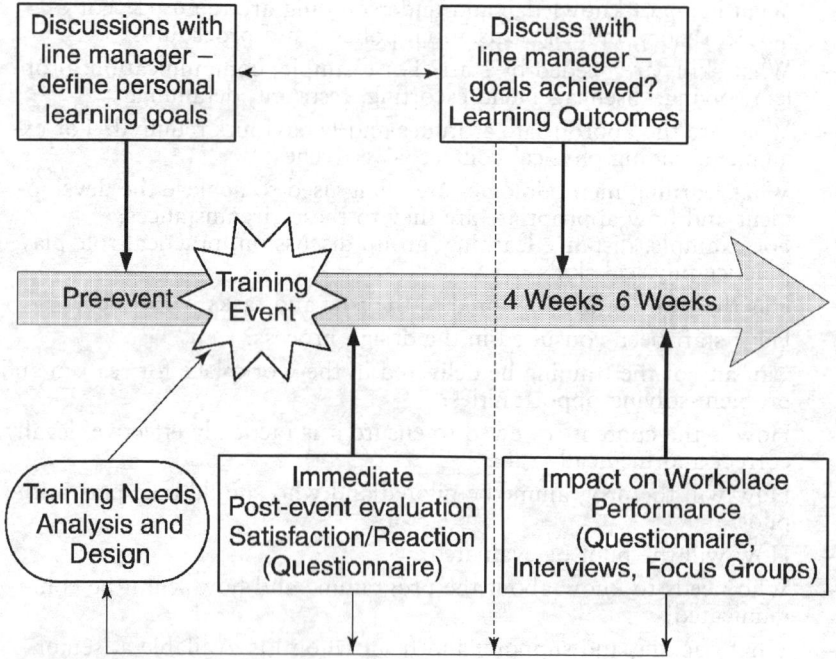

Internal or external training providers?

[V8111] The decision about whether to use an internal training department or an external training provider is quite a difficult one and needs some careful consideration.

Internal provider

[V8112] The most compelling reason for choosing an internal training department is usually that it costs less. However, there are hidden costs involved which include specialist research into the subject of violence in the workplace, design of the programme, and specific training for trainers necessary to deliver an effective conflict management training programme.

If an organisation's workplace violence issues require physical intervention training, then it must be remembered that the risks to the organisation increase with the level of physical intervention required. The internal provider must be able to satisfy managers that the interventions are both appropriate and medically and legally defensible.

Most internal training departments attract 'can do' people who are willing and eager to help and enthusiastic about new challenges. Care is needed to ensure that they can provide training specific to requirements – not what they may think is needed. The following checklist will help in this regard.

> *Internal provider*
> — How well is the department equipped to do a Training Needs Analysis on the requirements for dealing with workplace violence?
> — Can they design a properly researched solution that will provide effective development for the employees?
> — Are the trainers suitably experienced, qualified and credible to provide specialised training in managing conflict and workplace violence?
> — Does the trainer development programme include coaching and ongoing support?
> — Does the Training Needs Analysis indicate a requirement for skills in physical intervention?
> — How has the training solution been reviewed to ensure that it is tactically appropriate, legally correct and medically safe?
> — How will the training be monitored and evaluated to ensure it is effective and meeting the current needs of the workplace?

External provider

[V8113] The extra investment of engaging an external provider needs to be justified and the most compelling justification is that the provider has specialist knowledge, experience and expertise in the field of managing conflict and workplace violence. Of course, most external providers will claim to have such expertise in abundance and the difficulty lies in making the right choice from the plethora of providers who claim to have just the solution being sought.

In simple terms, the client will need to find out from the provider if:

- Their solution will equip the individuals at risk with the knowledge, skills, attitudes and behaviour to deal effectively with the incidents of workplace violence they may face.
- Their organisation has processes which ensure the training need is correctly identified, the solution is properly designed, and there is an evaluation process that provides feedback to validate and improve the product.
- Their training is properly designed to meet specific needs and is sufficiently robust to be medically and legally defensible if required.
- The people used to deliver the training are properly trained and experienced in delivering this type of training and understand the particular problems of the client.

The following checklist provides the basic questions that need to be asked.

> *External providers*
> — What evidence can they provide of their expertise in the field of managing conflict and workplace violence?
> — What experience do they have in the organisation's specific sector and work?
> — What do the other organisations they have worked with say about them?
> — What evidence can they provide as to the effectiveness of their training in the workplace?
> — What methodology will they use to identify the training needs of the organisation in the area of conflict management and workplace violence?
> — How will they develop their understanding of the business and organisation?
> — What learning methods are they proposing to use? Are they practically-based and in line with the training needs identified?
> — How do they ensure the content of their programmes is appropriate and legally correct?
> — How do they ensure the physical interventions used are medically safe?
> — How do they conduct their training to ensure the delegates are trained in a safe environment?
> — How do they develop their trainers?
> — What level of resources will they commit to the programme and how will they deal with short-term issues such as a trainer becoming ill?
> — What level of programme evaluation do they offer and how will it be fed back to the organisation?
> — Are they likely to provide credible expert witness support if required?

Combination of internal and external provision

[V8114] Combining the two options can provide a compromise and solution to this problem. It can be achieved by engaging an external provider to undertake a Training Needs Analysis and design a solution. The provider could be asked to deliver training directly to the higher risk and more demanding staff groups, and they could train internal trainers to deliver the bulk of the remaining programmes.

This solution is attractive because the visible costs are kept relatively low and the benefits of using an external provider are achieved to some degree.

> *Combine internal and external providers*
> — What experience can the external provider evidence to show they are capable of trainer training?
> — Will the training enable internal trainers to provide the programme in a way that is professional and credible?

> — What level of ongoing trainer coaching and support is offered?
> — Does the training require physical interventions? If so, review the decision and be sure the trainers can teach these safely and effectively.

Training the managers

[V8115] Managers are often forgotten when it comes to training in managing violence. The needs of managers, particularly line managers, are two-fold.

Firstly, they are often called into disputes involving customers or members of the public which have escalated and become too difficult for the member of staff to deal with. They may have already made the situation worse. Additionally, they are often dealing with situations when things have gone wrong and the potential for conflict is high. Consequently, there is a clear need for line managers in these situations to be personally skilled in conflict resolution as part of their role. They also need to be aware of the techniques and skills in this area provided to the people they manage.

Secondly, they have a vital role to play in the management of incidents of workplace violence and the provision of front-line support both during and after the process. They are key players in ensuring such incidents are properly recorded and reported and that staff carry out their roles in a way that achieves the organisation's goals in a safe and customer-friendly way. Staff can be affected by incidents in many different ways and need an individual response from their line manager.

The second point above is particularly important. Line managers have a great deal of influence in relation to the approach to dealing with conflict and violence in the workplace. Their individual attitude towards the whole issue will greatly influence how seriously the people who work for them approach the subject. This influence will affect attitudes towards risk assessment, approach to conflict situations, reporting and aftercare. Some people are badly affected by incidents that other members of staff might take in their stride. Sometimes the after effects of an incident do not set in until several hours or days later, and managers need to be aware of and look for the signs which indicate that the incident remains unresolved. The way in which the manager deals with the individual will have an influence on the speed that he or she will recover from the incident.

Incident management

[V8116] Until this point the focus has been on preventing violence or at least reducing it. This section focuses on the effective management of an incident should one occur, and how its impact can be reduced. It will then consider the steps that should be taken following the incident to support staff and learn from what has happened.

Although every eventuality cannot be planned for, it is important to plan and practise the response to both the most common and most serious scenarios that are likely to be experienced. Even though events will not always 'go to plan', the rigour of the planning process will help ensure that systems are in place and staff and managers are better prepared to deal with the situations that arise.

It is particularly important to prepare and practise a response to violent incidents, as during these highly emotive times employees will find it difficult to think clearly and objectively.

Figure 8: Incident management – planning and practice

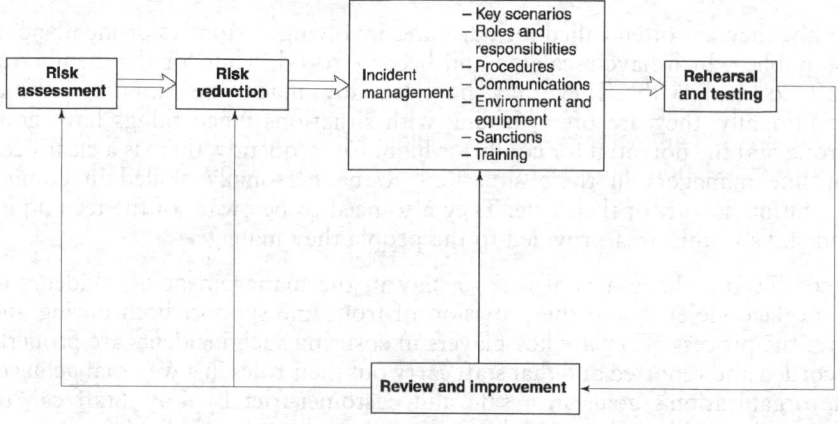

Figure 8 reinforces the importance of undertaking a violence risk assessment to establish key risk areas, and putting in place measures to reduce these risks. This process will help with the identification of key incident scenarios the organisation can then plan and train for.

Key scenarios

[V8117] The risk assessment process will have highlighted key areas of risk and control measures will be put in place to tackle these. In preparing for the management of incidents, it is important to use the risk assessment findings as the basis for the scenarios to be planned for and practised.

Roles and responsibilities

[V8118] It is important to be clear about roles and responsibilities prior to an incident occurring to prevent a breakdown in communication when under pressure. This extends further than a specific job role but to the wider team working in an area, for example, what part will the reception, security, domestic and office staff play, when an incident occurs in their area? Who takes control, how are they supported, who calls for help, who takes down descriptions? – these are all questions that need to be addressed. Everyone needs to be involved and has a part to play – even if it is in the background. In lone working environments, communications and expectations need to be clear between lone workers, office colleagues, control centre or switchboard colleagues.

Response teams

[V8119] Some organisations operate incident response teams. These multi-disciplinary teams follow a similar concept to 'crash teams' that respond to cardiac arrest. There are pro's and con's to this approach. Some organisations feel the response team is an intrusion and can damage relationships between local staff and patients/service users. If approached well, staff can however benefit from the knowledge that they will get competent help quickly. Careful selection of response teams and a high level of training are vital if they are to operate safely and earn the respect of staff.

Leadership

[V8120] The most critical role in incident management, as with any crisis, is leadership. Supervisors and managers need to have the confidence and skills to take control of an incident and its immediate aftermath, and staff will look to them for direction and example. Managers often miss out on staff conflict management training, yet they need these valuable skills and the additional knowledge in how to manage an incident and aftermath. Some managers will find difficulty in adjusting from their preferred leadership style to a more directive style required in a crisis. The concept of situational leadership is useful in training to help managers understand they need to adapt their style to the situation and staff will want clear direction and decisions. Senior management needs to provide clarity as to the level of authority given to managers and staff during an incident and how and when decisions should be escalated to a senior level.

Procedures

[V8121] The importance of clear role expectations and communications has already been discussed. These should be written down in the form of clear guidance and procedures. Flow charts will provide a useful means of simplifying and communicating responsibilities. Most organisations will have clear fire procedures in place and the same approach should be taken with violent incidents.

One key area requiring guidance and procedure is that of *preserving evidence*.

Violent incidents often result in the prosecution of the perpetrator. The success of the prosecution will depend on the quality of the evidence that supports it, and unfortunately many cases collapse at court due to simple errors made at the time of the incident in gathering and protecting evidence. Things that may seem trivial can have a dramatic impact in court as the defence seeks to discredit prosecution evidence.

The rules of disclosure highlight the duty of the prosecution to disclose all evidence, whether it will stand for or against the defendant. Organisations must therefore realise that anything that could be deemed as evidence relating to the case, such as any reports or records, will form part of the legal process. There are limited exceptions such as medical records.

Some basic tips and common sense will help to strengthen a prosecution and ensure justice is done. Guidance should cover:

- Importance of securing witnesses.
- Identification evidence.
- Use of Closed Circuit Television (CCTV) as evidence.
- Preserving a scene.

Communications

[V8122] Another important consideration in incident management is communications with other agencies such as the police, specialist consultants and the media. It is important to be proactive in responding to media interest following a violent incident. Considerable harm can be done to both the individuals involved and the reputation of the organisation if this is not handled well.

Environment and equipment

[V8123] Design, layout, security and equipment will be put to the test should an incident occur, and it is wise to ensure these are considered at the earliest opportunity and tested regularly. Panic alarms, for example, are all too often found to be disconnected or out of order.

Sanctions

[V8124] Many airlines and hospital trusts operate warning systems, for example they issue a letter to a disruptive individual or a yellow or red warning card. Clear sanctions play an important part in preventing and responding to violent behaviour. However, it must be recognised that confronting an individual or individuals could trigger an assault. It is essential staff are taught how to accurately assess people and situations, and are provided with training in when, where and how to confront a violent person. Some organisations are quick to launch these schemes but do not consider the position of the manager or member of staff who has to carry out these difficult and high-risk tasks.

Training

[V8125] Training will focus firstly on raising safety awareness and reducing risk. This is largely *knowledge* based and staff also need to develop the *skills* for dealing with the incident that does occur. If staff are to develop confidence in managing violence, the training will need to be dynamic and realistic, and cover the key scenarios identified.

Teamwork will be tested during stressful incidents and this should also be a key consideration in staff development to ensure staff communicate effectively, understand their responsibilities and respect each other. This should also extend to the wider team operating in a certain area, as communication often breaks down across roles or functions.

Physical fitness is particularly important and vital in some roles where staff are expected to respond to violence. Looking and feeling fit and behaving professionally will help earn respect and deter assault. Fitness will also play a vital part when restraining a violent person or even running away from one.

Rehearsal and testing

[V8126] Without doubt, rehearsal and testing is one of the most under-rated and least performed aspects of incident management. With the exception of some areas of the emergency services, psychiatric care and the armed forces, organisations rarely practise their response to violent incidents. This is somewhat strange as fire drills to test communications and procedures are regularly conducted within organisations.

In sport, coaches are often heard telling athletes 'you play as you train' and the same applies to incident management. Realistic practise and scenario-based training will make a big difference to the effectiveness of the incident response. Unforeseen problems with communications, equipment and procedures will be highlighted through a rehearsal and staff can be actively involved in solving these.

In some areas of work, it will be beneficial to set up multi-agency exercises involving police and other agencies. This helps to clarify expectations and iron out any problems.

Review and improvement

[V8127]–[V8128] Training, rehearsal and real incidents provide valuable feedback, and it is important that procedures are continuously reviewed to respond to this.

Checklist

— Assess the risk – does the organisation face a foreseeable risk of serious incidents of violence towards staff?

— Are roles, responsibilities and communication lines clearly identified for the management of an incident of workplace violence?

— Are the individuals and departments concerned aware of the roles, responsibilities and communication lines?

— Is the risk to staff and other service users great enough to warrant the development of specially trained response teams?

— Are the leadership roles and decision-making levels clear and have the appropriate people received training for taking a lead role in such incidents?

— Are there clear procedures outlined for managing the different aspects of an incident? In particular, for serious incidents are there clear procedures for securing and preserving evidence?

— Is there a clear strategy for managing communications and the media in the event of an incident?

— Has specialist equipment been identified and have staff been trained in its use in the event of an incident?

— Have all the people who might be called upon to respond to an incident received appropriate training?

— Is there a robust and regular method in place for rehearsing and testing the complete response to an incident?

> — Is there a review and evaluation process in place to provide feedback about the effectiveness of the incident management either after a test or a real incident?

Post-incident management

Investigation

[V8129] The post incident management process includes investigating the circumstances of the incident, attending to the needs of the victim and, where appropriate, dealing with the perpetrator.

Investigating incidents is an integral element of the risk management process. By establishing the factual circumstances surrounding an incident of workplace violence and gaining an understanding of why and how the incident took place it may be possible to take remedial action to prevent a similar one happening in the future. As with all other phases of the risk management process, it is important that employees are consulted and their suggestions taken into consideration.

An internal investigation should:

- collect facts on who, what, when, where and how the incident occurred;
- record information;
- identify contributing causes;
- encourage appropriate follow-up; and
- consider changes in controls, procedures and policy.

Incident investigation forms part of the wider legal requirement under the *Management of Health and Safety at Work Regulations (SI 1999 No 3242), Reg 5* for an employer to monitor and review preventive and protective measures. The HSE publication *Managing for health and safety* (HSG 65) provides further guidance.

HSE research clearly demonstrates the link between workplace violence and abuse and stress. Insensitive questioning of the victim during the early stages of an investigation may compound the traumatic effects of the incident itself, leading to longer-term psychological damage. Those responsible for conducting the investigation should be aware of this dynamic and carefully balance the priority of providing support to the victim against the requirement to investigate, question and learn from what happened to prevent recurrence.

How people are affected by workplace violence

[V8130] Perhaps the most important thing to recognise is that everyone has a different way of responding to and dealing with the aftermath of a violent or aggressive incident. There is no 'right' or 'wrong' way to react and people must be allowed to deal with it in their own way.

Being the victim of violence is particularly traumatic because it involves an interaction with another person at a very personal level and this can produce

Post-incident management [V8132]

some difficult and complex emotional reactions. These reactions will vary over the short, mid and longer term and it is important that support is provided at each stage, and where problems persist, professional support should be provided (see figure 9 below). It is important that those involved recognise that these reactions are quite 'normal' following an abnormal event, but equally no one says they have to experience them.

Figure 9: Timescale of reactions to workplace violence

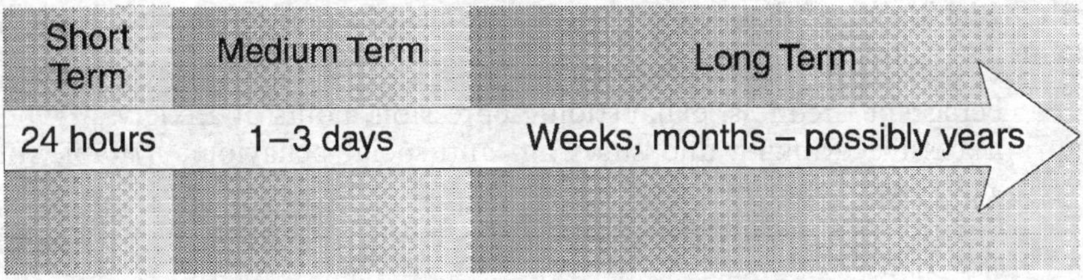

Short Term	Medium Term	Long Term
24 hours	1–3 days	Weeks, months – possibly years

Common reactions to an abnormal event such as an assault

Short-term reactions

[V8131] In the first few hours following the incident, the victim will have some initial reactions to the aggression and violence inflicted upon them. These reactions are predominantly emotional and are a direct response to the incident. Many factors will influence the severity of the reaction, not least of which is the individual's level of resilience towards traumatic situations.

The level of aggression, suddenness of the confrontation and physical injury sustained are also some of the factors which will influence how the victim will react. The following are the most likely reactions:

- Shock, confusion, disbelief, fear, helplessness.
- Anger, embarrassment, feeling of violation.

Many of these initial reactions will begin to lessen as the victim moves into the next phase.

Medium-term reactions

[V8132] The short-term reactions are characterised by their 'immediate' nature, formed before the victim has had time to think about, and begin to rationalise, what has happened. The medium-term reactions begin to appear when the victim has had a chance to consider the incident, work through what happened and think about the consequences, near misses and alternatives. This will be around 24 hours after the incident. Reactions can include:

- Feelings of loss, guilt, shame, embarrassment, humiliation.
- Exhaustion and tiredness, lack of sleep.
- Denial of effects, ready to get back to work.
- Anger, frustration and resentment.
- Lack of confidence, anxiety about similar situations or meeting the aggressor.

Moving successfully through this medium-term phase is often the key to recovery. Once the victim has acknowledged what has happened and come to terms with it, then he or she can move back towards a normal life. Line managers can provide vital support in this phase and are pivotal in the successful recovery of most of the victims.

Long-term reactions

[V8133] Generally, reactions that persist beyond a couple of weeks after the incident are indicative that the victim is finding difficulty in coming to terms with the incident and that he or she probably needs professional specialist help. Examples include:

- Persistent tiredness, exhaustion, depression, bouts of anxiety.
- Excessive drinking and smoking, anti-social behaviour, irritable and aggressive behaviour.
- Nightmares, flashbacks, headaches, nausea, difficulty in eating and sleeping.

A victim who displays these long-term reactions clearly needs specialised help, and an organisation that wants to provide a complete response to the range of issues that result from workplace violence will need to set up a procedure to facilitate this.

Returning to normal

[V8134] Many factors will influence the speed of recovery including the victim's life circumstances and their emotional resilience at the time of the incident. The turning point for most victims is the acceptance of what has happened to them. Once they accept the incident as a reality, they stop going through the 'if only' scenarios and stop blaming themselves for what took place.

Most people reach a point where they can move on from the event and get back to their normal daily lives. They achieve this when they regain confidence and self-esteem and recognise that, although life has changed in some aspects as a result of what occurred, they can become positive about being back in their working environment.

Although these various reactions have been described in discrete stages it will be rare for anyone to pass smoothly through them all. In reality, many things will cause a victim to progress and regress in the move towards normality. The return to work, for example, can be quite difficult and bring back feelings of insecurity and fear. Having to appear in court as a witness or learning of a colleague who has been involved in a similar incident may well trigger a recurrence of one or more of the reactions previously experienced.

For a few people, a return to 'normal' is virtually impossible, particularly if they have been permanently disabled by physical or mental injury. In such cases, the victim will need the most specialised care and support in trying to come to terms with their circumstances.

Thankfully, for most people the support of family, friends, colleagues and managers will be enough to help them recover from the trauma of the incident and return to a normal working life.

Post-incident management [V8136]

Providing post-incident support

[V8135] An appropriate post-incident support system should be an integral part of an organisation's overall response to workplace violence (see figure 10 below). The sophistication of the support system will depend upon the level of risk to which the staff are being exposed.

Staff working in the emergency services can often experience violent and other traumatic events, and therefore require a high level of post-incident support. This will range from line manager support, formal debriefing processes, to occupational health services, which will include access to specialist help for the most serious consequences of work-related violence. Some organisations adopt formal psychological debriefing processes, sometimes referred to as 'critical incident stress debriefing'. Although these approaches have considerable support in some areas, there is ongoing debate about their value, and concern at the potential risks associated with victims 're-living' the traumatic experience in the debriefing process. Whether or not the organisation adopts a formal debriefing process, it is essential that support mechanisms and procedures are put in place for managers and staff.

Even where the frequency of serious incidents is low, the organisation should have a post-incident support procedure and line mangers should be trained in the skills appropriate to helping a victim through the first stages of coping with an incident of work-related violence. It should also be possible to access specialist help if necessary.

Figure 10: Supporting the victim

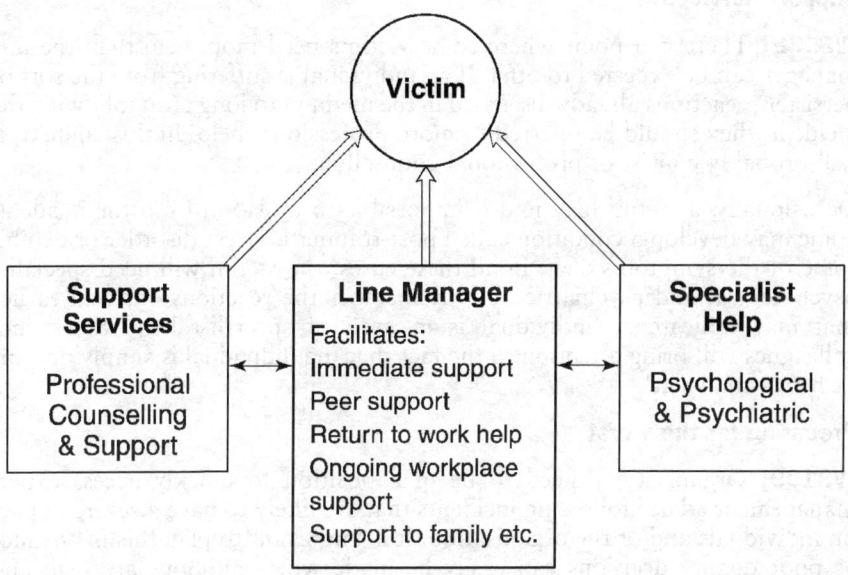

The role of the line manager

[V8136] In the vast majority of cases, support for the victim will be provided through his or her line manager. The short and medium-term reactions following an incident have been described earlier in this chapter and the focus of the line manager's support will be concentrated on helping the victim work though those reactions and in facilitating further support where necessary. Successfully dealing with these phases is vital in providing the optimum conditions for the individual to recover and return to a normal working life.

There are three points at which this support is crucial:

- Immediately after the incident has happened.
- During any absence from work.
- Preparing for and returning to work.

Organisations should provide guidance and training for managers to help them perform their key role sensitively and effectively.

Other immediate post-incident considerations

[V8137] Completing the workplace violence report form will require a formal narrative of the incident. For evidential reasons, the victim should personally complete this as soon as practicable after the incident. Every effort should be made to get the victim to complete at least the basic requirements while it is fresh in his or her memory.

It is tempting to leave this task but it can have far reaching consequences if the incident becomes the subject of criminal or civil proceedings. An accurate and early account of the incident can make all the difference to the outcome of a case.

Support services

[V8138] There is a point where some victims need more help than the line manager can be expected to offer. If an individual is suffering from the sort of persistent reactions already discussed in the medium to long term following the incident, they should be referred to more professional help. In this context, it will probably consist of professional counselling.

Occasionally, a victim may find it impossible to move on from the incident. Some may develop a condition called post-traumatic stress disorder or exhibit some of the symptoms of it. In all these cases, the victim will need specialist psychological and psychiatric help. Remember the reactions outlined earlier that may indicate an individual is in need of specialist help. Sometimes colleagues will bring attention to the fact that the individual is simply not 'his or her normal self'.

Preparing for the worst

[V8139] Organisations need to be in a position to quickly access expert management advice following incidents that are likely to have a severe impact on individuals and/or the organisation. It is important to plan this in advance as poor quality decisions can easily be made when emotions are high and everyone is under pressure. Plans should include access to advice and support concerning:

- Professional help for employees, service users, families and partners.
- Advice on legal issues.

- Advice on managing media attention.

The police investigation and court case

[V8140] Most victims want to see their aggressor brought to justice. However, it may prove to be quite a daunting process and it can help to be aware of how the process works and what might be expected. There are several prosecution routes including:

- Prosecution through the police.
- Private prosecution (a similar process but instigated through a solicitor, not police).
- Civil prosecution (instigated through a solicitor in a civil court).

Prosecution through the police is the most common route. However, employers should have clear policy and guidance on other avenues should the police or prosecution service decide not to progress the matter. This should include the degree of support they can provide to the victim, such as funding and providing time off for the legal process in preparing and giving evidence.

Criminal and civil courts provide further protection from harassment and violence in a number of ways, including:

- Injunctions.
- Restraining orders (eg preventing convicted persons contacting the victim/s).
- Anti-social behaviour orders.

When a court order is in place, the police are in a strong position to take action should it be breached.

Note: Offences can carry extra penalties if proven to be racially aggravated, and such offences will be recorded as crimes by the police and can constitute a more serious offence.

Giving evidence in court will understandably cause anxiety and it is especially difficult for the victim of violence 're-living' the trauma of what happened. It is vital that support is provided through this process by the organisation.

Victim Support (www.victimsupport.org.uk) is a national organisation with a great deal of experience in helping people to cope with being the victim of a crime, regardless of whether it has been reported to the police. The website for Victim Support Scotland is www.victimsupportsco.org.uk.

In addition, the Citizens Advice Witness Service provides free and independent support for witnesses in criminal courts across England and Wales (www.citizensadvice.org.uk/law-and-courts/legal-system/going-to-court-as-a-witness1/).

Figure 11: The process of prosecution

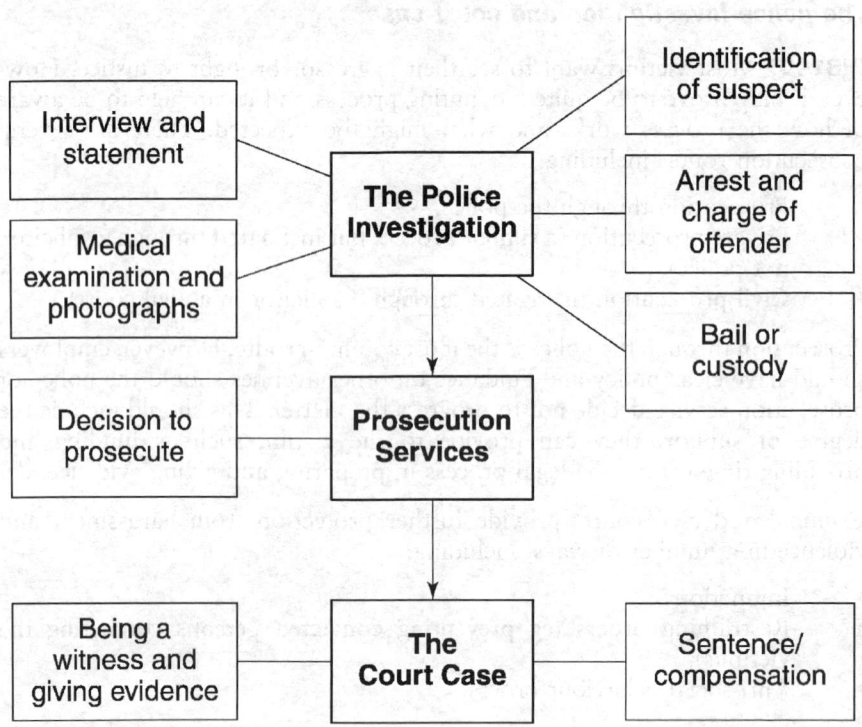

Support during the process of the investigation and court case

[V8141] For anyone who has to undergo the process of a court case it can prove to be a very difficult time.

As mentioned earlier, Victim Support is an organisation with many years of experience in helping individuals to cope with all the different aspects of surviving the trauma of being the victim of a crime. They have developed a Witness Support service, which provides practical help and support for anyone who has to attend a Crown Court as a witness.

Sources of further information

[V8142]–[V8143] The HSE has a work-related violence area on its website with links to a range of information, advice and general and sector-specific guidance at (www.hse.gov.uk/violence).

Acknowledgement

[V8144] This chapter was co-written by Maybo Limited, a specialist consultancy in managing work related conflict and violence (www.maybo.co.uk).

Figure 12: Maybo Risk Management Model

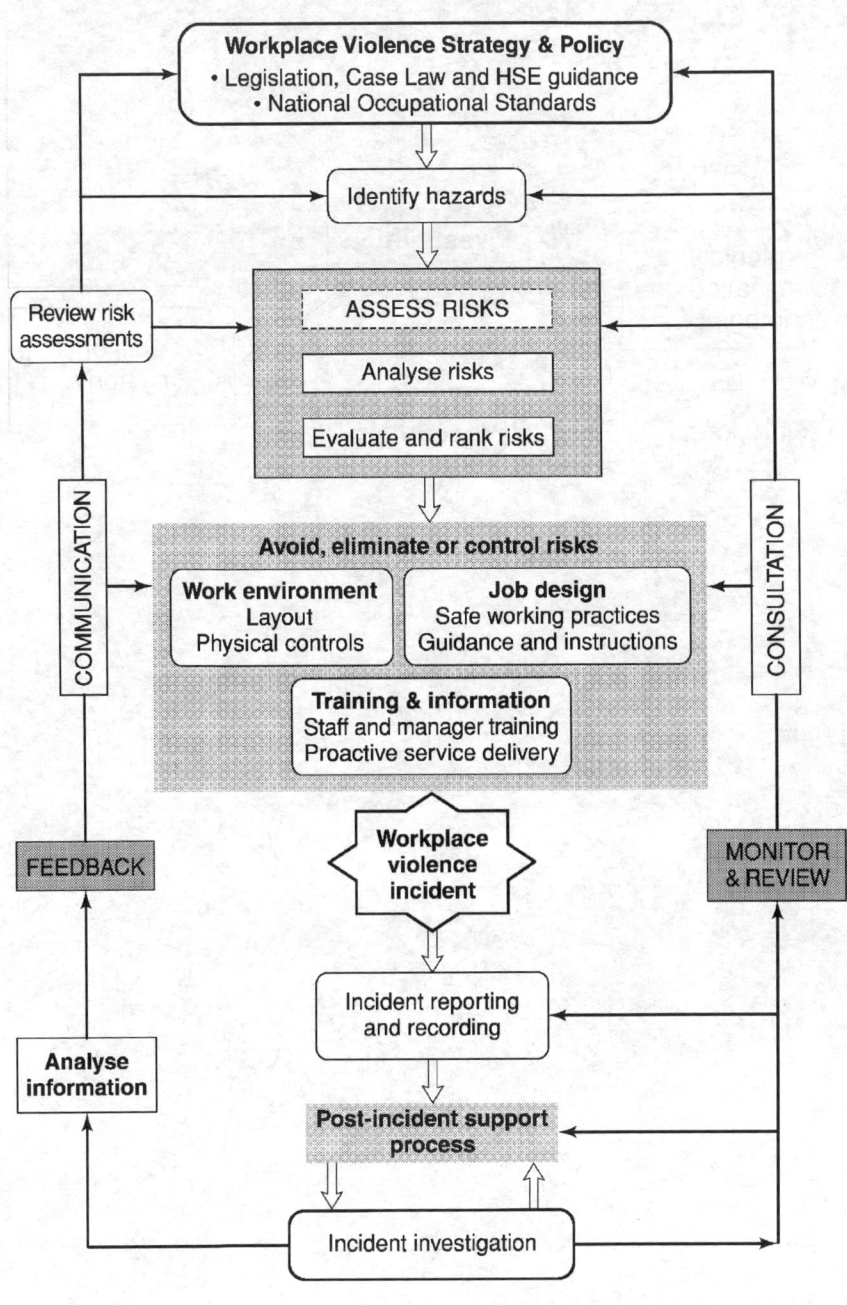

Vulnerable Persons

Andrea Oates

Introduction to vulnerable persons

[V12001] This chapter deals with groups of people who, for one reason or another, can be more vulnerable to health and safety risks. They include:

- children and young persons;
- new or expectant mothers;
- lone workers;
- people with disabilities; and
- inexperienced workers.

The groups above are subject to the protection of the full range of health and safety at work legislation – the *Health and Safety at Work etc Act 1974* ('*HSWA 1974*') and the various regulations made under it. Of particular relevance are the *Management of Health and Safety at Work Regulations 1999 (SI 1999 No 3242)* ('the *Management Regulations*') and their requirements for risk assessments. The potential vulnerability of some members of the workforce must be taken into account during the risk assessment process. In addition, the *Management Regulations* contain specific requirements in respect of children and young persons and also new or expectant mothers. These will be explained later in the chapter.

HSWA 1974, s 3 also places duties on employers (and relevant self-employed persons) in relation to people not in their employment such as visitors, occupants of premises, service users, neighbours or passers-by. Risks to these people must also be considered during risk assessments with appropriate attention given to those who are particularly vulnerable. The general principles of risk assessment are described in the chapter dealing with that subject while some specific aspects are covered in the chapter entitled COMMUNITY HEALTH AND SAFETY.

This chapter concentrates on health and safety legislation of particular relevance to vulnerable persons and the practical considerations to be taken into account when carrying out risk assessments relating to them.

Relevant legislation

[V12002] The *Management Regulations 1999 (SI 1999 No 3242)* impose specific requirements in relation to children and young persons and new or expectant mothers. Other health and safety regulations either directly recognise the vulnerability of particular workers or require this to be taken into account indirectly through some form of risk assessment. For example, individual capabilities must be considered when carrying out assessments of manual handling operations (see MANUAL HANDLING). The *Control of Electromagnetic Fields at Work (CEMFAW) Regulations 2016* require employers to consider the risks from exposure to electromagnetic fields (EMFs) and ensure that workers at particular risk, such as expectant mothers and workers with active or passive implanted or body-worn medical devices, are taken into account (see RADIATION).

Definitions and requirements in the Management Regulations requirements

[V12003] The terms 'child' and 'young person' are defined in *Regulation 1(2)* of the *Management Regulations 1999 (SI 1999 No 3242)*, while *subsections (4)* and *(5)* of *Regulation 3* contain requirements relating to risk assessments and young persons. *Regulation 19* of the Regulations deals with precautions, which must be taken to ensure the 'Protection of young persons'.

Health and safety law defines a *young person* as anyone under eighteen years of age.

A child is anyone who is not over compulsory school age – they have not yet reached the official age at which they may leave school, referred to as the minimum school leaving age (MSLA).

'New or expectant mother' is also defined in *Regulation 1(2)* of the *Management Regulations 1999 (SI 1999 No 3242)*:

'New or expectant mother' means an employee who is pregnant; who has given birth within the previous six months; or who is breastfeeding.

'Given birth' means 'delivered a living child or, after twenty-four weeks of pregnancy, a stillborn child'.

There are several regulations relating directly to them:

- *Regulation 16* – deals with risk assessments (and *Regulation 16A* deals with the alteration of working conditions in respect of new or expectant mothers who are agency workers);
- *Regulation 17* – covers possible suspension from night work (and *Regulation 17A* deals with certificates from registered medical practitioner in respect of new or expectant mothers who are agency workers);
- *Regulation 18* – contains requirements in respect of notification to the employer of pregnancy or having given birth (and *Regulation 18A* sets out general provisions in relation to agency workers).

These requirements are explained in full later in the chapter.

Other regulations directly affecting vulnerable persons

[V12004] Some regulations contain requirements for employees engaged in certain activities to be free from medical or physical conditions that make them unfit for that activity. Typical of these are the *Work in Compressed Air Regulations 1996 (SI 1996 No 1656)* and the Diving at Work Regulations 1997 (SI 1997 No 2776).

- **Children and young persons**
    Several sets of regulations contain age-related prohibitions or restrictions on specified activities.
    For example, under the *Ionising Radiations Regulations 2017 (SI 2017 No 1075)* no young person under the age of 18 can be employed to work with ionising radiation where they would need to be designated as classified (see RADIATION for more information). There are also specific dose limits for young people who may be exposed to ionising radiations while undertaking training or studying.
    Under the *Control of Lead Regulations 2002 (SI 2002 No 2676)* there are lower occupational exposure levels (action and suspension levels) for young people under 18 and they are prohibited from working in lead smelting and refining and in most jobs in the manufacture of lead-acid batteries. The *Working Time Regulations 1998 (SI 1998 No 1833)* define a young worker as being below 18 years of age and above the MSLA. Children are not covered by the Working Time Regulations, as established by *Addison t/a Brayton News v Ashby UKEAT, 17 January 2003*. Instead, the *Children (Protection at Work) Regulations 1998 (SI 1998 No 276)* control the working time of children aged 14 and 15 (see also V12006–V12007 below and WORKING TIME).

The Working Time Regulations specify a maximum working day of eight hours and a maximum working week of 40 hours for young workers – shorter than the maximum working hours specified for adult workers. These hours cannot be averaged out over a longer period. Young workers may not normally work at night between 10pm and 6am, or between 11pm and 7am where the contract of employment allows for work after 10pm. Young workers are also entitled to a rest break of at least 30 minutes, which should be consecutive if possible, where daily working time is more than four and a half hours. For more information on the limits on young workers' working time and their entitlements under the regulations, see WORKING TIME.

Also of relevance are the *Health and Safety (Training for Employment) Regulations 1990 (SI 1990 No 1380)* which give students on work experience programmes and trainees on training for employment programmes the status of 'employees'. The immediate provider of the training is treated as their 'employer'. While this particularly affects children and young persons, it also applies to adult participants in training programmes. Courses at schools, colleges and universities etc are not covered by these requirements.

HSE advice on work experience can be found online at: www.hse.gov.uk/youngpeople/workexperience/index.htm.

Guidance on the provision of post-16 work experience from the Education Funding Agency (EFA), Ofsted and the Health and Safety Executive (HSE) can also be found online at: www.gov.uk/government/publications/post-16-work-experience-as-a-part-of-16-to-19-study-programmes.

- **New or expectant mothers**
    Under the *Ionising Radiations Regulations 2017 (SI 2017 No 1075)* an employer who undertakes work with ionising radiation must ensure that, where they have been notified in writing that an employee is pregnant or breastfeeding:

>   '(a) in relation to an employee who is pregnant, the conditions of exposure are such that, after the employee's employer has been notified of the pregnancy, the equivalent dose to the foetus is as low as is reasonably practicable and is unlikely to exceed 1 mSv during the remainder of the pregnancy; and
>   (b) in relation to an employee who is breastfeeding, that employee must not be engaged in any work involving a significant risk of intake of radionuclides or of bodily contamination.'

For more information, see RADIATION.

Under the *Control of Lead Regulations 2002 (SI 2002 No 2676)*, employers should advise women of childbearing age on the special need to protect any developing foetus. When a woman declares she is pregnant, the employer should take appropriate action to remove her from work where her exposure to lead is significant. In addition, the action and suspension levels for women of reproductive capacity, set out in the Regulations, are lower than those for other employees.

The *Control of Electromagnetic Fields at Work (CEMFAW) Regulations 2016* require employers to consider the risks from their employee's exposure to electromagnetic fields (EMFs) and ensure that workers at particular risk, including expectant mothers, are taken into account (see RADIATION).

Example of regulations with indirect implications relating to vulnerable persons
[V12005]

- *Construction (Design and Management) Regulations 2015 (SI 2015 No 51) (CDM)*
  When construction work is to be carried out in or close to premises occupied by children (or those with special needs) the health and safety planning required by the CDM regulations must take particular account of their lack of awareness of risks and general curiosity. If normal access routes in premises are affected, this may create risks or problems for pregnant women, disabled people or the elderly.

- *Manual Handling Operations Regulations 1992 (SI 1992 No 2793)*
  The limited manual handling capabilities of, and potential risks to, both young persons and new or expectant mothers must be taken into account during risk assessments of manual handling operations. Similarly, risks to lone workers (with no source of assistance), people with temporary or permanent physical disabilities and inexperienced workers (who may be unaware of relevant handling techniques) must also be considered.

- *Control of Substances Hazardous to Health Regulations 2002 (SI 2002 No 2677) (COSHH)*
  Children and people with special needs may be at extra risk from hazardous substances as they may not be able to read or understand warnings or instructions. COSHH assessments should take account of the potential presence of such vulnerable persons and ensure the secure storage of hazardous substances where this is appropriate, eg by locking cleaners' cupboards.

- *Personal Protective Equipment at Work Regulations 1992 (SI 1992 No 2966)*
  The use of certain types of personal protective equipment ('PPE') may present difficulties for pregnant women or workers with physical disabilities. Young and inexperienced workers are likely to need more detailed explanations of the purposes of different types of PPE and how and when they are to be used, while those with special needs may need even closer attention.

- *Health and Safety (Display Screen Equipment) Regulations 1992 (SI 1992 No 2792)*
  The suitability of a display screen equipment ('DSE') workstation for a pregnant woman will need to be assessed, particularly as her size increases. Disabled workers or other staff with physical problems (eg past injuries) are likely to require special attention during workstation assessments, especially in relation to the suitability of chairs and other ergonomic aspects.

Children and young persons

[V12006]–[V12007] For many years, work done by children and young people was subject to an array of prohibitions, including on equipment that children and young persons were not allowed to use and processes or activities they must not be involved in. However, since 1997 the emphasis has switched from such prohibitions towards restrictions, based upon a process of risk assessment.

The Health and Safety Executive (HSE) advises that children below the minimum school leaving age (MSLA) must not be employed in industrial workplaces, such as factories or construction sites, except when on work experience.

It further advises: 'Children under 13 years of age are generally prohibited from any form of employment and local authorities have powers to make bylaws on the types of work, and hours of work, children aged between 13 years and the MSLA can do.'

An exception to this is where children are involved in areas like television, theatre and modelling. Children working in these areas will need a performance licence.

Risk assessment requirements

[V12008] *Regulation 3 of the Management Regulations 1999 (SI 1999 No 3242)* contains important requirements in relation to risk assessments and young persons. These are:

(4) An employer shall not employ a young person unless he has, in relation to risks to the health and safety of young persons, made or reviewed an assessment in accordance with paragraphs (1) and (5).

(5) In making or reviewing the assessment, an employer who employs or is to employ a young person shall take particular account of—

(a) the inexperience, lack of awareness of risks and immaturity of young persons;

(b) the fitting-out and layout of the workplace and the workstation;
(c) the nature, degree and duration of exposure to physical, biological and chemical agents;
(d) the form, range and use of work equipment and the way in which it is handled;
(e) the organisation of processes and activities;
(f) the extent of the health and safety training provided or to be provided to young persons; and
[(g) risks from agents, processes and work listed in the Annex to Council Directive 94/33/EC(b) on the protection of young persons at work as amended by Directive 2014/27/EU].'

Paragraph (1) of the Regulation contains the basic requirement for employers to carry out suitable and sufficient risk assessments. Directive 2014/27/EU contains requirements regarding the classification, labelling and packaging of substances and mixtures.

Protection of young persons

[V12009] In assessing the risks to young persons, the employer must take particular note of *Regulation 19* of the *Management Regulations 1999 (SI 1999 No 3242)* which states:

(1) Every employer shall ensure that young persons employed by him are protected at work from any risks to their health or safety which are a consequence of their lack of experience, of absence of awareness of existing or potential risks or the fact that young persons have not yet fully matured.
(2) Subject to paragraph (3), no employer shall employ a young person for work—
  (a) which is beyond his physical or psychological capacity;
  (b) involving harmful exposure to agents which are toxic or carcinogenic, cause heritable genetic damage or harm to the unborn child or which in any way chronically affect human health;
  (c) involving harmful exposure to radiation;
  (d) involving the risk of accidents which it may reasonably be assumed cannot be recognised or avoided by young persons owing to their insufficient attention to safety or lack of experience or training; or
  (e) in which there is a risk to health from—
    (i) extreme cold or heat;
    (ii) noise; or
    (iii) vibration,
  and in determining whether work will involve harm or risk for the purposes of this paragraph, regard shall be had to the results of the assessment.
(3) Nothing in paragraph (2) shall prevent the employment of a young person who is no longer a child for work—
  (a) where it is necessary for his training;
  (b) where the young person will be supervised by a competent person; and
  (c) where any risk will be reduced to the lowest level that is reasonably practicable.

Some of the practical implications of these provisions are considered later in the chapter.

Information on risks to children

[V12010] *Regulation 10* of the *Management Regulations 1999 (SI 1999 No 3242)* deals with 'information for employees' and contains two paragraphs relating to the employment of children. These state:

(2) Every employer shall, before employing a child, provide a parent of the child with comprehensible and relevant information on—

    (a) the risks to his health and safety identified by the assessment;
    (b) the preventive and protective measures; and
    (c) the risks notified to him in accordance with Regulation 11(1)(c).

(3) The reference in paragraph (2) to a parent of the child includes—

    (a) in England and Wales, a person who has parental responsibility, within the meaning of section 3 of the Children Act 1989, for him; and
    (b) in Scotland, a person who has parental responsibility, within the meaning of section 8 of the Law Reform (Parent and Child) (Scotland) Act 1986, for him.

This requirement to provide information to parents includes situations where children are on work experience programmes (where they have the status of employees by virtue of the *Health and Safety (Training for Employment) Regulations 1990 (SI 1990 No 1380)*) and also includes part-time or temporary work carried out by children who have not reached the minimum school leaving age.

Some of the practical implications of this requirement are considered later in this chapter.

Disapplication of requirements

[V12011] The wide-ranging impact of these provisions is lessened to a limited extent by *paragraph (2)* in R*egulation 2* of the *Management Regulations 1999 (SI 1999 No 3242)* – 'Disapplication of these Regulations'.

(2) Regulations 3(4), (5), and 10(2) and 19 shall not apply to occasional work or short-term work involving—

    (a) domestic service in a private household; or
    (b) work regarded as not being harmful, damaging or dangerous to young persons in a family undertaking.

However, the term 'family undertaking' is not defined in the Regulations.

Capabilities and training

[V12012] *Regulation 13* of the *Management Regulations 1999 (SI 1999 No 3242)* states in *paragraph (1)* that 'Every employer shall, in entrusting tasks to his employees, take into account their capabilities as regards health and safety'. Quite clearly the capabilities of children and young persons will be somewhat different from those of more experienced and mature employees. This requirement is also of relevance in considering other vulnerable groups including disabled and inexperienced workers.

Risk assessment in practice

[V12013] The requirements of *Regulations 3(4), (5)* and *19* of the *Management Regulations 1999 (SI 1999 No 3242)* are based on the requirements of European Directive (94/33/EC).

A good starting point for employers to identify exactly what he or she must do is to consider the three characteristics associated with young persons, which are mentioned in both *Regulations 3(5)* and *19(1)*:

- lack of experience;
- lack of awareness of existing or potential risks; and
- immaturity (in both the physical and psychological sense).

However, there will be different expectations of a school leaver who has already been playing a prominent role in a family business to those of a work experience student with no previous exposure to the world of work. Employers must also be aware that the physical and psychological maturity of young persons will vary hugely.

These three characteristics must then be considered in respect of the risks involved in the employer's work activities and particularly those identified in *Regulation 3(5)(b)–(g)* and *Regulation 19(2)* (as set out in **V12008** and **V12009** above).

The requirements set out in those Regulations (including the risks listed in the Annex to Council Directive 94/33/EC have been consolidated into a single checklist for employers in **TABLE B**.

The purpose of all risk assessments is to identify the measures the employer needs to take in order to comply with the law. Additional measures which the employer must consider in order to provide adequate protection for young persons are:

- preventing exposure of the young person to the risk;
- reducing the risk to the lowest level reasonably practicable by applying the principles of prevention outlined in a Schedule to the Management Regulations. These are:
- avoiding risks;
- evaluating the risks which cannot be avoided;
- combating the risks at source;
- adapting the work to the individual, especially as regards the design of workplaces, the choice of work equipment and the choice of working and production methods, with a view, in particular, to alleviating monotonous work and work at a predetermined work-rate and to reducing their effect on health;
- adapting to technical progress;
- replacing the dangerous by the non-dangerous or the less dangerous;
- developing a coherent overall prevention policy which covers technology, organisation of work, working conditions, social relationships and the influence of factors relating to the working environment;
- giving collective protective measures priority over individual protective measures; and

- giving appropriate instructions to employees.

Measures could include:

- providing additional training;
- providing close supervision by a competent person;
- carrying out additional health surveillance (as required by *Regulation 6* of the *Management Regulations 1999 (SI 1999 No 3242)* and regulations including the *Control of Substances Hazardous to Health (COSHH) Regulations 2002 (SI 2002 No 2677)*);
- taking other additional precautions.

Regulation 19(3) sets out that "Nothing in paragraph (2) shall prevent the employment of a young person who is no longer a child for work—

(a) where it is necessary for his training;
(b) where the young person will be supervised by a competent person; and
(c) where any risk will be reduced to the lowest level that is reasonably practicable."

In deciding what precautions are required, the employer must consider both young persons generally and the characteristics of individual young persons. Additional training and/or supervision may be necessary in respect of young persons with special needs.

Young people should gradually acquire more experience, awareness of risks and maturity, particularly as they pass through formal training programmes. As this occurs, restrictions on their activities may be progressively removed.

Provision of information

[V12014]–[V12015] *Regulation 10(1)* of the *Management Regulations 1999 (SI 1999 No 3242)* requires employers to provide all employees with comprehensible and relevant information on:

- risks to their health and safety identified by risk assessments (or notified by other employers);
- preventive and protective measures (ie appropriate precautions);
- emergency procedures and arrangements;
- the identity of those nominated to put these procedures and arrangements into place; and
- arrangements for co-operation and co-ordination in shared workplaces.

This requirement is of particular importance in relation to young persons. *Regulation 10(2)* requires employers *also* to provide information to parents of children (under the MSLA) on the risks the children will be exposed to and the precautions, which are in place.

Means of providing this information to young workers includes:

- induction training programmes;
- employee handbooks or rulebooks;
- job descriptions;
- formal operating procedures;

- information forms for parents of work experience students.

Further guidance on work experience is provided in the HSE publication (INDG364) *Young people and work experience – A brief guide to health and safety for employers* (www.hse.gov.uk/pubns/indg364.htm).

The information must be comprehensible – special arrangements may be necessary for young persons whose command of English is poor or for those with special needs. The type of information required to be provided will depend on the work activities and the risks involved. The content must be relevant – both to the workplace and the young person. It might include the following types of information:

- general risks present in the workplace
  eg fork lift trucks are widely used in the warehouse.
- general precautions taken in respect of those risks
  eg all fork lift drivers are trained to the standard required by the Approved Code of Practice ('ACoP').
- specific precautions in respect of the young person
  eg the induction tour includes identification of areas where fork lift trucks operate and indication of warning signs.
- restrictions or prohibitions on the young person
  eg X will not be allowed to drive fork lift trucks or any other vehicles. HSE guidance (L117) *Rider-operated lift trucks. Operator training and safe use. Approved Code of Practice and guidance* says that lift-truck operators must be over 16, except in ports, where they must be at least 18 years old, unless they are undergoing a suitable course of training, properly supervised by a competent person. Children under 16 should never operate lift trucks.
- supervision arrangements
  eg X will be supervised by the warehouse foreman (or other people designated by him or her).
- PPE requirements
  eg safety footwear must be worn by all employees working in the warehouse. This is supplied by the company.

Where restrictions or prohibitions are removed (eg after successful completion of training programmes) an appropriate record should be made, either on the original restriction/prohibition or within the individual's training record.

- Further guidance on organising work experience is available on the HSE website at: www.hse.gov.uk/youngpeople/workexperience/organiser.htm.

Work presenting increased risks for children and young persons

[V12016] The checklist below (TABLE B) is based on the requirements set out in *Regulations 3(5)* and *19(2)* of the *Management Regulations 1999 (SI 1999 No 3242)* and the Annex to the European Council Directive 94/33/EC as amended by Directive 2014/27/EU. It is intended to assist employers conducting risk assessments in respect of work by children and young persons.

These types of work or situations of exposure to risk are not necessarily prohibited, although the requirements of *Regulation 19(2)* must be taken into account. However, restrictions could be required for young persons (particularly children) and additional precautions may be required to provide them with adequate protection from risk. Many situations present no greater risk to young persons than adults and restrictions (eg close supervision) may only be necessary until the employer is sure that the young person is fully aware of the risks and is capable of taking the necessary precautions.

Table B
Excessively physically demanding work
• Manual handling operations where the force required or the repetitive nature could injure someone whose body is still developing (including production line work);
• Certain types of piece work (particularly if peer pressure may result in them tackling tasks or working at speeds that are too much for them).
Excessively psychologically demanding work
• Work with difficult clients or situations where there is a possibility of violence or aggression;
• Difficult emotional situations eg dealing with death, serious illness or injury;
• Decision-making under stress.
Harmful exposure to physical agents
• Ionising radiation (separate exposure limits apply to young persons);
• Non-ionising electromagnetic radiation, eg lasers, UV from electric arc welding or lengthy exposure to sunlight, infra-red from furnaces or burning/welding;
• Risks to health from extreme cold or heat;
• Excessive noise;
• Hand-arm vibration, eg from portable tools;
• Whole-body vibration, eg from off-road vehicles;
• Work in pressurised atmospheres and diving work.
Harmful exposure to biological or chemical agents
• Hazardous substances that are (for example) acutely toxic, corrosive to the skin, flammable, explosive, self-reacting, respiratory sensitisers, skin sensitisers and/or reproductively toxic;
• Carcinogenic and mutagenic substances;
• Asbestos (including asbestos-containing materials);
• Lead and lead compounds;
• Biological risks, eg legionella, leptospirosis (Weil's disease), zoonoses.
Work equipment
Where there is an increased risk of injury due to the complexity of precautions required or the level of skill required for safe operation eg:
• Woodworking machines (particularly saws, surface planing and vertical spindle moulding machines);
• Food slicers and other food processing machinery;
• Certain types of portable tools such as chainsaws;
• Power presses;
• Vehicles such as fork lift trucks, mobile cranes, construction vehicles;
• Other cranes and lifting hoists;
• Mobile elevating work platforms;
• Firearms.
Some young people may not have the physical size or strength necessary to operate equipment which has been designed for adults.

Dangerous materials or activities
• Work with explosives, including fireworks;
• Work with fierce or poisonous animals, eg on farms, in zoos or veterinary work;
• Certain types of electrical work, eg exposure to high voltage or live electrical equipment;
• Handling of flammable liquids or gases, eg petrol, acetylene, butane, propane;
• Work with pressurised gases;
• Work in large slaughterhouses (risks from animal handling, use of stunning equipment);
• Holding large quantities of cash or valuables.
Dangerous workplaces or workstations
• Work at heights, eg on high ladders or other unprotected forms of access;
• Work close to deep or fast-flowing water;
• Work in confined spaces, particularly where the risks specified in the *Confined Spaces Regulations 1997 (SI 1997 No 1713)* are present;
• Work where there is a risk of structural collapse, eg in construction or demolition activities or inside old buildings.

Further guidance on risks to young persons and appropriate precautions is provided on the HSE website at www.hse.gov.uk/youngpeople/index.htm.

New or expectant mothers

[V12017] Under the *Management Regulations 1999 (SI 1999 No 3242) Regulation 1*:

'New or expectant mother' means an employee who is pregnant; who has given birth within the previous six months; or who is breastfeeding.

'Given birth' means 'delivered a living child or, after twenty-four weeks of pregnancy, a stillborn child'.

Requirements of the Management Regulations

[V12018] The requirements for 'Risk assessment in respect of new and expectant mothers' are contained in *Regulation 16* of the *Management Regulations 1999 (SI 1999 No 3242)* which states in *paragraph (1)*:

Where—

(a) the people working in an undertaking include women of child-bearing age; and

(b) the work is of a kind which could involve risk, by reason of her condition, to the health and safety of a new or expectant mothers, or to that of her baby, from any processes or working conditions, or physical, biological or chemical agents, including those specified in Annexes I and II of Council Directive 92/85/EEC on the introduction of measures to encourage improvements in the safety and health at work of pregnant workers and workers who have recently given birth or are breastfeeding, [as amended by Directive 2014/27/EU]

the assessment required by Regulation 3(1) shall also include an assessment of such risk.

Directive 2014/27/EU deals with the classification, labelling and packaging of substances and mixtures.

References to risk from any infectious or contagious disease are references to a level of risk at work which is over and above the level to which a new or expectant mother may be expected to be exposed outside the workplace. The types of risk that are more likely to affect new or expectant mothers are described later in the chapter.

There have been several legal cases further interpreting employers' risk assessment requirements relating to new and expectant mothers.

In a recent European Court of Justice (CJEU) case, Elda Otera Ramos v Servicio Galego de Saúde, Instituto Nacional de la Seguridad Social Case C-531/15, a nurse employed in the accident and emergency unit of a Spanish hospital informed her employer she was breastfeeding when she returned to work after giving birth.

She was concerned that aspects of her job could expose her to health and safety risks that would affect her breast milk. These included a complex shift rotation pattern and exposure to ionising radiation, healthcare-associated infections and stress.

She requested a number of changes to her work but the hospital said her job did not involve risks to breastfeeding and rejected these. It also said her job was on an agreed list of 'risk-free' jobs and refused to issue a medical certificate confirming there was a risk to breastfeeding.

She argued she had been discriminated against when an inadequate risk assessment concluded her work was risk free without suitable investigation. The CJEU confirmed that where a breastfeeding mother can show a risk assessment is inadequate (or is not carried out) this can give rise to a potential discrimination claim.

Paragraphs (2) and (3) of Regulation 16 set out the actions employers are required to take if these risks cannot be avoided. *Paragraph (2)* states:

> Where, in the case of an individual employee, the taking of any other action the employer is required to take under the relevant statutory provisions would not avoid the risk referred to in paragraph (1) the employer shall, if it is reasonable to do so, and would avoid such risks, alter her working conditions or hours of work.

Consequently, where the risk assessment required under *Regulation 16(1)* shows that control measures would not sufficiently avoid the risks to new or expectant mothers or their babies, the employer must make reasonable alterations to their working conditions or hours of work.

In some cases, restrictions may still allow the employee to substantially continue with her normal work but in others it may be more appropriate to offer her suitable alternative work.

Any alternative work must be:

- suitable and appropriate for the employee to do in the circumstances;
- on terms and conditions which are no less favourable.

Paragraph (3) of Regulation 16 states:

If it is not reasonable to alter the working conditions or hours of work, or if it would not avoid such risk, the employer shall, subject to section 67 of the 1996 Act, suspend the employee from work for so long as is necessary to avoid such risk.

The 1996 Act referred to here is the *Employment Rights Act 1996* which provides that any such suspension from work on the above grounds is on full pay. However, payment might not be made if the employee has unreasonably refused an offer of suitable alternative work. Employment continues during such a suspension, counting as continuous employment in respect of seniority, pension rights and so on. Contractual benefits other than pay do not necessarily continue during the suspension. These are a matter for negotiation and agreement between the employer and employee, although employers should not act unlawfully under the *Equality Act 2010*. Enforcement of these rights is through the employment tribunals.

Regulation 16A contains requirements concerning the alteration of working conditions in respect of new or expectant mothers who are agency workers.

Regulation 17 of the *Management Regulations 1999 (SI 1999 No 3242)* deals specifically with night work by new or expectant mothers and states:

Where—
(a) a new or expectant mother works at night; and
(b) a certificate from a registered medical practitioner or a registered midwife shows that it is necessary for her health or safety that she should not be at work for any period of such work identified in the certificate,

the employer shall, subject to section 67 of the 1996 Act, suspend her from work for so long as is necessary for her health or safety.

Such suspension (on the same basis as described above) is only necessary if there are risks arising from the work.

HSE advice to new and expectant mothers in this area is that they can still work nights, but that 'if your GP or midwife has provided a medical certificate stating that you must not continue to work nights, then your employer must offer you suitable alternative day work on the same terms and conditions. If that is not possible, then your employer should suspend you from work on paid leave for as long as necessary to protect your health and safety and/or that of your child'.

Regulation 17A contains requirements concerning agency workers and sets out that where a new or expectant mother works at night; and a certificate from a registered medical practitioner or a registered midwife shows that it is necessary for her health or safety that she should not be at work for any period of such work identified in the certificate, the hirer should inform the temporary work agency straight away. The agency should then end the supply of that agency worker to the hirer.

The requirements placed on employers in respect of altered working conditions or hours of work and suspensions from work only take effect when the employee has formally notified the employer of her condition. *Regulation 18* of the *Management Regulations 1999 (SI 1999 No 3242)* states:

(1) Nothing in paragraph (2) or (3) of regulation 16 shall require the employer to take any action in relation to an employee until she has notified the employer in writing that she is pregnant, has given birth within the previous six months, or is breastfeeding.
(2) Nothing in paragraph (2) or (3) of regulation 16 or in regulation 17 shall require the employer to maintain action taken in relation to an employee—
   (a) in a case—
      (i) to which regulation 16(2) or (3) relates; and
      (ii) where the employee has notified her employer that she is pregnant, where she has failed, within a reasonable time of being requested to do so in writing by her employer, to produce for the employer's inspection a certificate from a registered medical practitioner or a registered midwife showing that she is pregnant;
   (b) once the employer knows that she is no longer a new or expectant mother; or
   (c) if the employer cannot establish whether she remains a new or expectant mother.

Regulation 18A contains similar provisions regarding agency workers who are new or expectant mothers.

Risks to new or expectant mothers

[V12019] HSE guidance on the risks which may be of particular relevance to new or expectant mothers is set out on its website (www.hse.gov.uk/mothers/faqs.htm) and are listed in the European Directive on Pregnant Workers (92/85/EEC).

They include the following:

Physical agents

[V12020]

Manual handling
Pregnant women are particularly susceptible to risk from manual handling activities as also are those who have recently given birth, especially after a caesarean section.
The HSE advises that special consideration should be given to new and expectant mothers whose capabilities may be affected by hormonal and physical changes. It explains that there is some evidence linking physically demanding work to premature birth, particularly where the work includes long periods of standing and/or walking. During the last three months of pregnancy there is an increased risk of musculoskeletal symptoms when heavy or repeated lifting is undertaken due to hormonal changes affecting the ligaments that support the joints. As pregnancy progresses it may become more difficult to achieve and maintain good postures and a shift of the centre of gravity can increase the risk of back pain for pregnant women. Manual handling assessments (as required by the *Manual Handling Operations Regulations 1992 (SI 1992 No 2793)*) are dealt with in more detail in the chapter MANUAL HANDLING in this publication.

Noise
According to the Institution for Occupational Safety and Health (IOSH) exposure of pregnant workers to high noise levels can affect the unborn child. It says research suggests that prolonged exposure of the unborn child to high noise levels during pregnancy may have an effect on a child's later hearing and that low frequencies have a greater potential for causing harm.

Radiation (ionising and non-ionising)
The foetus may be harmed by exposure to ionising radiation, including that from radioactive materials inhaled or ingested by the mother. As set out in **V12004** above, there are specific provisions in the *Ionising Radiations Regulations 2017 (SI 2017 No 1075)* concerning pregnant and breastfeeding women who are working with radiation.

Systems of work should be such as to keep exposure of pregnant women to radiation from all sources as low as reasonably practicable. Contamination of a nursing mother's skin with radioactive substances can create risks for the child and special precautions may be necessary to avoid such a possibility.

Several HSE publications provide detailed guidance on work involving ionising radiation. See L121 *Work with ionising radiation: Ionising Radiations Regulations 2017 Approved Code of Practice and Guidance*; and INDG334 'Working safely with radiation: Guidelines for expectant and breast-feeding mothers'.

Electromagnetic radiation
Health and Safety Executive (HSE) guidance to the *Control of Electromagnetic Fields at Work Regulations 2016* (HSG281) sets out that employers must give special consideration to the safety of workers at particular risk and assess if there are specific additional risks. These include expectant mothers who have advised employers of their condition. It advises employers to encourage workers to advise them in writing if they become pregnant as working with certain levels of electromagnetic fields (EMFs) could result in a greater risk to an expectant mother. It says employers may wish to take a practical approach and limit the exposure of expectant mothers to the public exposure limits set out in Council Recommendation 1999/519/EC (ec.europa.eu/health/electromagnetic_fields/docs/emf_rec519_en.pdf).

Sources of EMF which may pose specific risks to expectant mothers include work where workers need to be in close proximity to cables carrying high currents, automated induction heating systems: fault-finding and repair involving close proximity to the EMF source, automated welding systems: fault-finding, repair and teaching involving close proximity to the EMF source, and MRI equipment.

(For more information on both ionising and non-ionising, see RADIATION.)

Work in compressed air and diving work
Although pregnant women may not be at greater risk of developing the 'bends', potentially gas bubbles in the circulation could seriously harm the foetus should this condition arise. There is also evidence that women who have recently given birth have an increased risk of the bends.

The *Work in Compressed Air Regulations 1996 (SI 1996 No 1656), Reg 16(2)* states:

> ... the compressed air contractor shall ensure that no person works in compressed air where the compressed air contractor has reason to believe that person to be subject to any medical of physical condition which is likely to render that person unfit or unsuitable for such work.

This would appear to prohibit such work by pregnant women or those who have recently given birth. The HSE booklet (L96) *A guide to the Work in Compressed Air Regulations 1996* (1996) provides detailed guidance on the requirements of the *Work in Compressed Air Regulations 1996*.
The HSE advises that a diver who is pregnant or who suspects she may be pregnant should not dive.

Shocks and vibration
Major physical shocks or regular exposure to lesser shocks or low frequency vibration may increase the risk of a miscarriage. Activities involving such risks (eg the use of vehicles off road) should be avoided by pregnant women. The *Control of Vibration at Work Regulations 2005 (SI 2005 No 1093)* contain specific requirements for the assessment of vibration risks — see the chapter entitled VIBRATION.

Movements and postures
Fatigue from standing and other physical work has been associated with miscarriage, premature birth and low birth weight. Ergonomic considerations will increase as the pregnancy advances and these could affect display screen equipment workstations, work in restricted spaces (eg for some maintenance or cleaning activities) or work on ladders or platforms. Underground mining work is likely to involve movement and posture problems and will also be subject to some of the other 'physical agents' described in this section. Driving for extended periods or travel by air may also present postural problems.
Pregnant women should be allowed to pace their work appropriately, taking longer and more frequent breaks. They may need to be restricted from carrying out certain tasks. Seating may need to be provided for work that is normally done standing and adjustments to display screen equipment and other workstations may be necessary.

Biological agents

[V12021] Many biological agents in risk groups 2, 3 and 4 can affect the unborn child should the mother be infected during pregnancy (as set out in Council Directive 92/85/EEC of 19 October 1992 on the introduction of measures to encourage improvements in the safety and health at work of pregnant workers and workers who have recently given birth or are breastfeeding). Some agents can cause abortion of the foetus and others can cause physical or neurological damage. Infections may also be passed on to the child during or after birth, eg while breastfeeding.

Agents presenting risks to children include hepatitis B, HIV, herpes, TB, syphilis, chickenpox, typhoid, rubella (German measles), cytomegalovirus and chlamydia in sheep. Most women will be at no more risk from these agents at

[V12021] Vulnerable Persons

work than living within the community but the risks are likely to be higher in some work sectors, eg laboratories, health care, the emergency services and those working with animals or animal products.

Details of appropriate control measures for biological agents are contained in Schedule 3 of the ACoP booklet to the *COSHH Regulations 2002 (SI 2002 No 2677)* (L5). There is a separate HSE publication on *'Infection risks to new and expectant mothers in the workplace'*. When carrying out a risk assessment in respect of new or expectant mothers, normal containment or hygiene measures may be considered sufficient, but there may be the need for special precautions such as use of vaccines. Where there is a high risk of exposure to a highly infectious biological agent it may be necessary to remove the worker entirely from the high-risk environment.

Chemical agents

[V12022] The HSE lists the following chemical agents as possible risks to new and expectant mothers: toxic chemicals; mercury; antimitotic (cytotoxic) drugs; pesticides; carbon monoxide and lead.

> Control of these hazardous substances at work is required by the *COSHH Regulations 2002 (SI 2002 No 2677)* and separate regulations governing lead. Risk assessments in relation to pregnant women or those who have recently given birth may indicate that normal control measures are adequate to protect them also. However, additional precautions may be necessary, eg improved hygiene procedures, additional PPE or even restriction from work involving certain substances.
>
> The ACoP relating to the *COSHH Regulations 2002 (SI 2002 No 2677)* (contained in HSE booklet L5) and the HSE's online advice, COSHH Essentials (www.hse.gov.uk/coshh/essentials/index.htm) provide further details of the types of precautions which may be appropriate.
>
> *Mercury and mercury derivatives*
> Exposure to organic mercury compounds can slow the growth of the unborn baby, disrupt the nervous system and cause the mother to be poisoned. Mercury and its derivatives are subject to the *COSHH Regulations 2002 (SI 2002 No 2677)* and normal control measures (see above) may be adequate to protect new or expectant mothers.
> (See HSE publication L5.)
>
> *Antimitotic (cytotoxic) drugs*
> These drugs (which may be inhaled or absorbed through the skin) can cause genetic damage to sperm and eggs and some can cause cancer. Those workers involved in preparation or administration of such drugs and disposal of chemical or human waste are at greatest risk, eg pharmacists, nurses and other health care workers. Antimitotic drugs can present a significant risk to those of either gender who are trying to conceive a child as well as to new or expectant mothers, and all those working with them should be made aware of the hazards.
> Since there is no known safe threshold limit for them, exposure must be reduced to as low a level as is reasonably practicable.

Agents absorbed through the skin

Various chemicals including some pesticides may be absorbed through the skin causing adverse effects. These substances are identified in the tables of occupational exposure limits in HSE booklet EH40 '*Workplace exposure limits*' accompanied by an annotation 'Sk'. Many such agents (particularly pesticides which are subject to the *Control of Pesticides Regulations 1986 (SI 1986 No 1510)*) will also be identified by product labels.

Effective control of these chemicals is obviously important in respect of all employees (the COSHH Regulations 2002 (SI 2002 No 2677) applying once again) although risk assessments may reveal the need for additional precautions in respect of new or expectant mothers, eg modified handling methods, additional PPE, or even their restriction from activities involving exposure to such substances.

Carbon monoxide

Exposure of pregnant women to carbon monoxide can cause the foetus to be starved of oxygen with the level and duration of exposure both being important factors.

Once again it is important to protect all members of the workforce from high levels of carbon monoxide (the *COSHH Regulations 2002 (SI 2002 No 2677)* require it) but it may also be necessary to ensure that pregnant women are not regularly exposed to carbon monoxide at lower levels, eg from use of gas-fired equipment or other processes or activities.

Lead and lead derivatives

High occupational exposure to lead has historically been linked with high incidence of spontaneous abortion, stillbirth and infertility. Decreases in the intellectual performance of children have been attributed to exposure of their mothers to lead. Since lead can enter breast milk there are potential risks to the child if breastfeeding mothers are exposed to lead.

All women of reproductive capacity are prohibited from working in many lead processing activities. All work involving exposure to lead is subject to the *Control of Lead at Work Regulations 2002 (SI 2002 No 2676)*. Even where women of reproductive capacity are allowed to work with lead or its compounds the blood-lead concentrations contained within the Regulations as an 'action level' and a 'suspension level' are set at half those for adult males. This is intended to ensure that women who may become pregnant already have low blood lead levels.

Once pregnancy is confirmed, the doctor carrying out medical surveillance (as required by *Regulation 10* of *SI 2002 No 2676*) would normally be expected to suspend the woman from work involving significant exposure to lead. Detailed guidance on the Regulations is contained in HSE booklet L132 'Control of lead at work' (2002).

Working conditions

[V12023]–[V12024] The HSE lists the following as areas that workplace risk assessments must specifically consider:

- facilities (including rest rooms);
- mental and physical fatigue and working hours;
- stress (including post-natal depression);

- passive smoking;
- temperature;
- working with visual display units (VDUs);
- working alone;
- working at height;
- travelling;
- violence;
- personal protective equipment;
- nutrition.

Excessive physical or mental pressure could cause stress and lead to anxiety and raised blood pressure. Workplace stress is a complex issue and has received increased attention in recent years.

Fatigue and other possible causes of stress may need to be considered. These may be associated with the workload of individual pregnant employees (eg for those in management or administrative roles), the pressure of decision-making (eg in the health care or financial sectors) or the trauma of potential work situations (eg serious accidents dealt with by the emergency services).

Pregnant women are less tolerant of heat and may be more prone to fainting or heat stress. Although the risk is likely to reduce after birth, dehydration may impair breastfeeding. Exposure to prolonged heat at work, eg at furnaces or ovens, should be avoided. Maintenance or cleaning work in hot situations should also be avoided, particularly if this involves use of less secure forms of access such as ladders, where fainting could result in a serious fall.

The HSE advises that there is no need for pregnant women to stop working with Display Screen Equipment (DSE or VDUs). However, it recommends that, to avoid anxiety, women should talk to their doctor or someone who is well informed about current scientific information and advice. It also advises them to read HSE guidance aimed at new and expectant mothers.

Table C provides a summary of some of the issues that may affect new and expectant mothers at work.

Table C	
Issue	*Factors in work*
Morning sickness	Early shift work Exposure to nauseating smells Availability for early meetings Difficulty in leaving job
Frequent visits to toilet	Difficulty in leaving job/site of work
Breastfeeding	Access to private area Use of secure, clean refrigerators to store milk
Tiredness	Overtime/long working hours Evening work More frequent breaks Private area to sit or lie down
Increasing size	Difficulty in using protective clothing Work in confined areas Manual handling difficulties Posture at DSE workstations (Dexterity, agility, co-ordination, reach and speed of movement may also be impaired)

Table C	
Issue	*Factors in work*
Comfort	Problems working in confined or congested workspaces
Backache	Standing for extended periods Posture for some activities Manual handling
Travel (including fatigue, stress, static posture)	Reduce need for lengthy journeys Change travel methods
Varicose veins	Standing/sitting for long periods
Haemorrhoids	Working in hot conditions/sitting
Vulnerability to passive smoking	An effective smoking policy and ensuring compliance with legislative requirements on smoking
Vulnerability to stress	Modified duties Adjustments to working conditions or hours
Vulnerability to falls	Restrictions on work at heights or on potentially slippery surfaces
Vulnerability to violence (eg from contact with the public)	Changes to workplace layout or staffing Modified duties
Lone working	Effective communication and supervision Possible need for improved emergency arrangements

A practical approach to assessing risks

[V12025] Essentially risk assessment in respect of new or expectant mothers must be carried out by an employer if:

- there are women of childbearing age; and
- their work could involve risks to new or expectant mothers or their babies.

Most organisations employ women of childbearing age and, particularly in the case of large businesses, there may be a number of work activities that could create relevant risks. In some cases (eg those involving exposure to hazardous substances or radiation) it may be appropriate to stipulate that women are not allowed to work in certain activities, processes or departments once their pregnancy has been notified and/or for a finite period after they return to work after giving birth.

However, in many workplaces the issues will be far from clear cut. A practical approach is for the employer to develop a checklist similar to the sample provided at the end of this section. Such a checklist can be prepared by carrying out a review of the organisation's activities in order to identify risks which are present and may be of relevance in relation to new or expectant mothers. The risks described in the previous part of this chapter provide a good starting point from which a workplace-specific checklist can be developed. In some workplaces it may be appropriate to develop more than one such checklist (eg for production, maintenance and administration) because the profiles of risks are different.

Such a checklist should be completed by a suitable employer's representative (a personnel or health and safety specialist, or the pregnant woman's manager for example) together with the pregnant woman herself. The checklist is intended to provoke discussion about the employee's possible exposure to the risks that the employer has identified, so that any necessary additional precautions (or changes to work practices) can be agreed. There is also the opportunity to identify whether any further review of the situation is necessary. The process would be repeated once the mother returned to work after giving birth.

The sample checklist provided below is for a bakery. Female staff are likely to be involved in production work where there is significant potential for risk to new or expectant mothers, and they may also be employed in laboratory or maintenance work. Risks for staff working in administrative activities must also be considered, eg those associated with DSE workstations. However, the greatest problems could be for female staff delivering to retail outlets. As well as risks from manual handling, falls and possibly excessive driving, the difficulties for such staff in accessing suitable rest and toilet facilities must also be taken into account.

BENN THE BAKERS	NEW OR EXPECTANT MOTHERS	RISK ASSESSMENT CHECKLIST
Name of Employee:		Work location:
	Possible risks to consider	Precautions/changes agreed
Bakery and delivery work	Manual handling (production, maintenance, deliveries) Excessive standing (production lines) Awkward sitting (production lines) Difficult access (cleaning, maintenance) Excessive heat (near to ovens) Hazardous substances (eg cleaning materials, maintenance, laboratory) Significant risk of falls (eg slippery floors, maintenance work, trips on delivery work) Excessive driving (deliveries) Other	
Administrative activities	DSE workstation layout Unsuitable meeting times Manual handling (eg stationery, records) Significant travel (especially by car or air) Other	
All activities	Ability to leave workplace temporarily Availability of rest area/toilets Major pressure or stress situations Possible effects of fatigue Other	
Signature (for Benn the Bakers): Date:		Signature (new/expectant mother): Date for further review (if any):

This form should be completed when the pregnancy is first notified *and* when the new mother returns to work.

Risks to non-employees

[V12026] As part of their general risk assessments, employers must also consider risks to new or expectant mothers who are not part of their own workforce. Such people may be employed by another organisation or be self-employed but they may also be service users or members of the public. Even though other employers have duties to assess risks to their staff, 'host' employers still have a statutory duty (and a 'duty of care') to other people's employees, particularly in respect of risks of which they may be aware, but which might not be immediately apparent to others. In this context, risks to pregnant women especially could include:

- slips, trips and falls, eg due to slippery or uneven surfaces;
- accidental impact from people, eg running children or accidental contacts related to sports activities;
- accidental collision with mobile equipment;
- use of off-road transport;
- assault, eg on police or prison premises;
- ionising radiation;
- hazardous substances (see the earlier reference to risk phrases at V12022 above).

Employers are failing to protect the health and safety of new and expectant mothers

[V12026.1] Despite the legal requirements described above, a Women and Equalities parliamentary committee inquiry into pregnancy and maternity discrimination heard that as many as 21,000 women leave their jobs each year because employers are failing to tackle the pregnancy and maternity risks they face at work.

The committee carried out its inquiry in 2016 and found pregnant women and mothers were reporting more discrimination and poor treatment at work than they did a decade before.

A 2016 survey of 3,000 mothers carried out by the Equality and Human Rights Commission (EHRC) found two in five (41%) respondents felt there was a risk to or impact on their health or welfare. However, 38% said that their employer did not initiate a conversation about risks when they informed them of their pregnancy; almost one in five said they had identified risks that their employer had not identified; and 10% said that their employer had identified risks and had not tackled them.

The Women and Equalities Committee report, Pregnancy and maternity discrimination, can be found online at: www.parliament.uk/business/committees/committees-a-z/commons-select/women-and-equalities-committee/inquiries/parliament-2015/pregnancy-and-maternity-discrimination-15-16/.

Following publication of the report, the HSE revised its guidance on new and expectant mothers. This can be found online at: www.hse.gov.uk/mothers/.

Lone workers

[V12027] There are no specific legal restrictions on working alone – indeed many work activities would be extremely impractical if there were. However, the activities of lone workers must be subject to the same sort of risk assessment process as any other type of work. Even those regulations imposing the greatest restrictions on lone working (the *Electricity at Work Regulations 1989 (SI 1989 No 635)* and *Confined Spaces Regulations 1997 (SI 1997 No 1713)* for example) still require the application of risk assessment techniques in identifying suitable safe systems of work.

The various types of lone workers

[V12028] There are many different types of lone working. These can be divided into three categories:

People working alone in fixed workplaces
This category includes:
- workers in small shops and kiosks;
- people working in more remote parts of larger premises, eg isolated reception or enquiry desks, private interview rooms, security gatehouses;
- laboratories, stores etc;
- staff working late at night, early in the morning or at weekends (such as cleaners);
- staff responding to alarm calls etc, eg 'keyholders';
- staff working at home.

Staff working on a variety of premises belonging to other employers
They will be working independently of colleagues from their own employer, but may have others working around them for at least part of the time. They include:
- people carrying out construction or plant installation work;
- maintenance and repair workers;
- cleaners, painters, decorators etc;
- other types of contracting staff.

People working within the wider community
They may be working in urban, rural or remote locations, sometimes in places which are accessible to the public but also on private property (often in domestic premises). They include:
- mobile security staff;
- domestic installation, maintenance and repair staff;
- delivery staff (including postal workers);
- sales and technical representatives;
- architects, property surveyors and estate agents;
- rent collectors and meter readers;
- social workers, home carers and district nurses;
- police, traffic wardens and probation staff;
- agricultural and forestry workers;

- utility workers (electricity, gas, water, sewage, telecommunications etc);
- vehicle recovery staff;

and many other similar types of work.

Risks for lone workers

[V12029] In some situations, lone workers may be at risk because they are not physically capable of carrying out certain activities safely on their own. These may relate to:

- **Manual handling activities**
  – where team handling is necessary.
- **Use of certain types of equipment**
  – particularly heavy and bulky equipment, or where another person needs to be available to carry out adjustments.
- **Operation of controls etc**
  – controls may be physically separate from each other or from related instrumentation, requiring a second person to simultaneously activate controls or relay information (eg where a banksman may be required for lifting operations).
- **Achieving safe access**
  – a second person may be necessary to foot a ladder or ensure safe use of other forms of temporary access equipment.

Some work activities may involve such a high degree of risk that another person needs to be present to assist in the identification of risks or act as a deterrent to such risks. These situations include:

- **Some types of maintenance and repair work**
  – ensuring equipment has been correctly isolated and other precautions taken (two heads are better than one!).
- **Live electrical work**
  – ensuring the person carrying out the work is fully aware of what is live and what is not.
- **Work involving potential risks of aggression or even violence**
  – handling complaints, passing on unwelcome news, dealing with people with aggressive or violent histories, working in high risk areas of the community, handling cash or valuables.
- **Activities where staff feel highly vulnerable**
  – one-to-one encounters (possibly with unknown people), particularly in remote locations, empty premises or domestic property (female staff may feel more vulnerable but male workers should also be made aware of any risks).

There may also be risks to lone workers where a second person is needed to take appropriate emergency actions such as:

- **Isolating electrical supplies**
  – particularly where live work is being carried out.

- Rescuing a worker from inside a confined space
  – eg by use of a rescue line or wearing suitable respiratory protective equipment ('RPE') and PPE.
- Giving immediate first aid attention
  – particularly important for electrical accidents or people seriously affected by hazardous substances, or asphyxiated.
- Safely recovering a worker who has fallen while wearing a safety harness

The general potential for accidents and illness to lone workers from less foreseeable causes must also be considered, even though these may not justify the immediate presence of another person.

The examples above are of the more common types of risks that lone workers may encounter. Employers will no doubt be able to add to the list, particularly after consulting those who carry out lone work and their representatives.

Controlling risks to lone workers

[V12030] The previous section identified the types of risks that lone workers may be subject to. Set out below is a variety of measures which may be necessary to control risks. While some measures will be relevant to most, if not all risk situations, others may have only limited application.

- *General precautions*

  Lone workers must be made aware of the types of risks they may encounter and the precautions which must be taken to control those risks. This will require information, instruction and training, supported by effective ongoing supervision.

  Some types of lone working may justify the creation of formal written procedures or rules while in others the lone workers themselves may be expected to carry out a degree of dynamic risk assessment (see R3017 in RISK ASSESSMENT). Lone workers must be familiar with relevant procedures and rules and have received appropriate training if they need to make dynamic risk assessments.

  As a result of the employer's risk assessment, some activities may be identified as being ones which workers must never carry out alone. These must be communicated effectively to the lone worker, who must also be given guidance on where the necessary assistance can be obtained. This may be achieved by the temporary presence of work colleagues or assistance may be forthcoming from a customer's personnel or others working in the vicinity.

- *Communication*

  Good communications are essential in ensuring the safety of lone workers, including those who may work from a home base. This is likely to involve:
  – provision of suitable communication equipment, eg mobile phones, portable radios, direct radio links, monitored CCTV surveillance;

- availability of emergency alarms, eg fixed alarm buttons, portable attack alarms (some types of alarms detect an absence of movement by the alarm user or require a regular response from the user);
- availability of information on lone workers' whereabouts to others, eg work colleagues, control centres, security staff, partners, friends etc (this may involve the use of location boards, computerised or desk diaries, itinerary sheets or be verbal);
- ongoing communication from the lone worker, eg about delays, problems, changes in plans;
- ongoing checks on the lone worker's wellbeing;
- pre-emptive contacts from lone workers encountering unexpected risks (explaining where they are, who they are with or what they are doing and requesting a return call in an agreed period of time);
- checks to ensure the lone worker has safely returned to base (or home);
- arrangements in place for the repair, inspection, maintenance or replacement of equipment they use (including communication and alarm equipment).

- *Specific precautions*

  There are many specific precautions which can be taken to reduce the risks to lone workers. The nature of these will depend on the location of the lone worker and the type of activity involved. Examples include:
    - physical separation of the lone worker from people they must deal with, eg by use of screens or wide desks;
    - ensuring assistance is readily available to the lone worker when required, eg for certain manual handling operations or live electrical work;
    - placing lone workers where assistance from members of the public is at hand, eg arranging encounters, particularly with higher risk contacts, in public buildings or areas including hotels, cafes etc.

- *Emergency arrangements*

  Here too, the nature of the arrangements will vary according to the location and activity of the lone worker. Arrangements are likely to include:
    - response to alarms – who is expected to respond and what actions they should take;
    - follow-up of lone workers who do not return when expected or do not respond to calls;
    - provision of basic first aid equipment for lone workers, so they can carry out simple first aid for themselves or receive treatment from others coming to their aid.

- *Homeworkers*

  There are several risks likely to affect homeworkers which need to be properly controlled through:
    - suitable choice of equipment for use in the home environment;
    - training of staff in the correct use of equipment;

- arrangements for inspection and maintenance of equipment (particularly in respect of electrical risks);
- assessments of display screen equipment (DSE) workstations (the type of self-assessment checklist contained in the **DISPLAY SCREEN EQUIPMENT** chapter is useful for homeworkers;
- ensuring that loads can be manually handled safely (eg by using suitably sized containers);
- suitable arrangements for safe storage and use of any hazardous or dangerous substances kept by the homeworker;
- effective communication arrangements when working away from the home base (see above).

HSE leaflet INDG 226 provides further guidance on '*Homeworking*'.

Monitoring of lone workers

[V12031] Those responsible for managing and supervising lone workers must ensure their activities are monitored effectively. Since only a limited amount of observation of their work in the field is likely to be practicable, regular consultation with the staff involved will be important. Aspects to monitor for are:

- new risks which have become apparent;
- existing risks which may have increased;
- the effectiveness of existing precautions;
- workers' views on new precautions which may be appropriate.

It will also be necessary to actively monitor the effectiveness of precautions through actions such as:

- checking the effectiveness of communication equipment;
- testing alarms;
- enquiring as to the current whereabouts of lone workers;
- testing the awareness of staff on how they should react to emergency situations.

The HSE leaflet INDG73, *Working alone – Health and safety guidance on the risks of lone working* provides guidance in this area.

Disabled people

Types of disability

[V12032] Disabilities come in many different forms. Under the *Equality Act 2010* a person has a disability if they have a physical or mental impairment, and the impairment has a substantial and long-term adverse effect on their ability to carry out normal day-to-day activities. This includes sensory impairments which affect vision or hearing, cancer and HIV.

In *Banaszczyk v Booker Ltd* [2016] KEAT/0132/15/RN, the Employment Appeal Tribunal (EAT) involved a man who was dismissed after a spine injury he developed following a car accident affected his ability to do his job.

Mr Banaszczyk was employed as a picker at a Booker distribution centre. After the accident, he became unable to meet the hourly 'pick rate' of 210 25kg cases per hour and was dismissed. An employment tribunal found that because his spinal injuries were not severe enough to have an adverse effect on day-to-day activities such as shopping, eating and driving, he could not be considered legally disabled.

However, the EAT upheld Mr Banaszczyk's appeal against the decision. It said that for many people, in the context of work, moving and lifting cases of up to 25kg is a normal day-to-day activity and ruled that under the *Equality Act 2010*, he could be classified as disabled. His dismissal was therefore unfair and discriminatory. The case highlights a willingness to consider a variety of work activities, as well as those outside of professional life, when considering the *Equality Act 2010* and the definition of day-to-day activities.

The following impairments are among those capable of amounting to a disability:

- Asthma;
- Bipolar affective disorder;
- Cerebral palsy;
- Colitis;
- Congenital myotonic dystrophy;
- Depression;
- Dyslexia;
- Emphysema;
- ME or chronic fatigue syndrome;
- Migraine;
- Multiple sclerosis;
- Paranoid schizophrenia;
- Post-traumatic stress disorder; and
- Photo sensitive epilepsy.

Disabled people at risk

[V12033] Employers must take account of all disabled people who may be affected by their activities in their risk assessments. These include:

- permanent and temporary employees;
- people on training or work experience placements (who have the status of employees);
- visitors;
- customers or service users;
- neighbours and passers-by.

Access for disabled people

[V12034] *Regulation 25 of the Workplace (Health, Safety and Welfare) Regulations 1992 (SI 1992 No 3004)* requires that:

Where necessary, those parts of a workplace (including in particular doors, passageways, stairs, showers, washbasins, lavatories and workstations) used or occupied directly by disabled people at work shall be organised to take account of such people.

Factors which must be taken into account in complying with the above requirement and in preventing injury to disabled people are:

- the locations where disabled people work or must have access to;
- the width of access routes;
- maintaining access routes free from obstructions and slipping or tripping hazards;
- standards of guardrails and handrails;
- heights of door handles, light switches, alarms, controls etc.

It is not the intention of this chapter to consider the general accessibility of work premises to disabled members of the public.

Other issues of relevance

[V12035] There are a number of other issues which must be considered when carrying out risk assessments in respect of disabled people, whether members of the workforce or not, including:

- hazardous substances:
  - including their potential effects on those with relevant health conditions or allergies;
  - the abilities of exposed people to understand the risks or implement precautions.
- work equipment:
  - physical ability to handle the equipment;
  - capacity to understand risks and implement necessary precautions.
- hot (or very cold) items or surfaces:
  - the ability to identify such items or surfaces;
  - a perception of the risks involved;
  - the ability to keep clear of such risks.
- workstation layout:
  - ergonomically suited to the needs of the disabled worker;
  - provision of suitable seating is of particular importance.
- fire and other emergency arrangements:
  - the ability to hear or see alarm signals;
  - the physical ability to react to the alarm with the necessary speed;
  - the potential for panic amongst those with some types of disability.

Controlling risks to disabled people

[V12036] As a result of the risk assessment process employers must control risks to the standards required by legislation – either the general standards required by the *HSWA 1974* or the more specific requirements of individual sets of regulations. In respect of employees they must also comply with the requirements of *Regulation 13(1)* of the *Management Regulations 1999 (SI 1999 No 3242)* which states:

> Every employer shall, in entrusting tasks to his employees, take into account their capabilities as regards health and safety.

This will require employers to:

- Identify the disabilities of individual workers and others via:
  - questionnaires;
  - interviews and discussions;
  - medical examinations;
  - registration forms eg at hotel receptions.
- Identify situations where people with such disabilities may be at risk:
  - the types of issues referred to above.

In order to control risks to the standards required, it may be necessary for the employer to:

- modify existing equipment;
- provide alternative equipment;
- make modifications to the workplace;
- provide a work location suited to the individual's disability;
- provide additional or modified training (making special arrangements for those with learning difficulties);
- ensure emergency arrangements take account of disabled people;
- restrict the activities and/or locations of disabled people.

Employers have a duty under the *Equality Act 2010* ('DDA') to make 'reasonable adjustments' to the workplace or working arrangements of disabled people.

Avoiding discrimination

[V12037] Employers must find a balance between avoiding exposing disabled people to risk and discriminating against them unfairly.

Guidance for employers produced by the Equality and Human Rights Commission (EHRC) on the Equality Act 2010, *What Equality Law Means for You as an Employer: When You Recruit Someone to Work For You* (www.equalityhumanrights.com/en/publication-download/what-equality-law-means-you-employer-when-you-recruit-someone-work-you), sets out that:

> 'Except in very restricted circumstances or for very restricted purposes, you are not allowed to ask any job applicant about their health or any disability until the person has been:
>
> - offered a job either outright or on a conditional basis, or
> - included in a pool of successful candidates to be offered a job when a position becomes available (for example, if an employer is opening a new workplace or expects to have multiple vacancies for the same role but doesn't want to recruit separately for each one).'

The purpose of this is to ensure that job applicants are not ruled out just because of issues related to or arising from their health or disability, which may well say nothing about whether they can do the job. It is only once the job offer has been made, or an applicant is included in a group of successful candidates, that the employer can make sure that someone's health or disability would not prevent them from doing the job and they must also consider whether reasonable adjustments to enable them to do the job can be made.

Employers should not use health and safety reasons as an excuse for discriminating against disabled people. Research carried out on behalf of the HSE, *Extent of use of health and safety requirements as a false excuse for not employing sick or disabled people*, found health and safety was frequently being used as the rationale for non-recruitment or dismissal. However, it found considerably more evidence of employers overcoming health and safety difficulties to retain workers with a disability, health condition or injury.

There are many practical actions that can be taken to accommodate disabled workers safely in the workplace, as described in the section above. Even where restrictions must be placed on disabled workers, these can often relate to a very limited range of activities or locations, rather than preventing their employment altogether.

HSE guidance, health and safety for disabled people and their employers, can be found on its website at: www.hse.gov.uk/disability/largeprint.pdf.

Inexperienced workers

[V12038] Workers of all ages may (like young persons) be lacking in experience of some types of work or lack awareness of the risks associated with particular work activities. *Regulation 13* of the *Management Regulations 1999 (SI 1999 No 3242)*, as well as requiring employers to take into account the capabilities of employees in entrusting tasks to them (see above), contains several requirements about training.

Paragraph (2) of the *Regulation 13* states:

> Every employer shall ensure that his employees are provided with adequate health and safety training – on their being recruited into the employer's undertaking; and on their being exposed to new or increased risks . . .

Such new or increased risks are specified as:

- transfer or change of responsibilities;
- due to new or changed work equipment;
- because of new or changed systems of work.

Training and supervision of inexperienced workers

[V12039] General health and safety induction training and job specific training is necessary in all workplaces but particular account of how to cope with these training needs must be taken in:

- work activities with rapid staff turnover;
- employment utilising large numbers of temporary workers, eg to meet seasonal needs or special orders (special arrangements will be necessary for workers without a good command of English);

- major events utilising temporary staff (possibly including volunteers);
- workplaces providing placements for those on government-funded training and re-training schemes;
- work activities involving workers with special educational needs.

Even when inexperienced workers have received appropriate training they will still require ongoing supervision, in many respects similar to that required for young persons. The imposition of restrictions as described earlier in this section (in relation to children and young persons) may also be appropriate.

Work at Height

Andrea Oates

Introduction to work at height

[W9001]–[W9002] Ten million people in the UK work at height, according to the All-Party Parliamentary Group (APPG) on working at height. Falls from height are one of the biggest causes of immediate workplace fatalities and major injuries, with falls from ladders and through fragile roofs being common causes of death and injury. Annual health and safety statistics published by the Health and Safety Executive (HSE) show that falls from height account for around a quarter of the officially-recorded fatal injuries to workers each year. HSE statistics published in June 2019 show that of the 147 workers killed in 2018/19, 40 were killed in falls from height.

The *Work at Height Regulations 2005* aim to prevent these deaths as well as injuries due to falls from height. This Chapter summarises the main requirements of these regulations and highlights sources of more detailed advice and guidance.

Commencement and scope

[W9003] The Regulations came into force on 6 April 2005. They apply in Great Britain. Outside Great Britain they apply only as *HSWA 1974, ss 1–59* and *80–82* apply to certain premises and activities offshore by virtue of the *Health and Safety at Work etc Act 1974 (Application outside Great Britain) Order 2013 (SI 2013 No 240)*.

Equivalent Regulations, the *Work at Height Regulations (Northern Ireland) 2005, (SR 2005 No 279)*, apply in Northern Ireland and came into operation on 11 July 2005.

Who do the regulations apply to?

[W9004] The Regulations impose requirements on employers in relation to employees and other people working under the control of the employer.

They also apply to 'relevant' self-employed people at work and people working under their control. As a result of a change in the law on 1 October 2015 brought in by the *Deregulation Act 2015*, health and safety law now applies to self-employed people whose work activity is specifically mentioned in the *Health and Safety at Work etc Act 1974 (General Duties of Self-Employed*

[W9004] Work at Height

Persons) (Prescribed Undertakings) Regulations 2015, or if their work activity poses a risk to the health and safety of others. The activities specifically mentioned in these regulations are agriculture (including forestry), work with asbestos, work on a construction site, an activity to which the *Gas Safety (Installation and Use) Regulations 1998* applies, work with genetically modified organisms and work on the railways.

The Work at Height Regulations do not apply to shipping, unloading fishing vessels or offshore (where other legislation applies).

What is work at height?

[W9004.1] *Regulation 2* sets out that 'work at height' means—

(a)   work in any place, including a place at or below ground level;
(b)   obtaining access to or egress from such place while at work, except by a staircase in a permanent workplace,

where, if measures required by these Regulations were not taken, a person could fall a distance liable to cause personal injury.

The so-called 'two-metre' rule, contained in the revoked *Construction (Health Safety and Welfare) Regulations 1996 (SI 1996 No 1592)*, is no longer part of the wording of these regulations. Despite some strong lobbying for a two-metre threshold to be retained, the argument for goal setting regulations prevailed, supported by data showing that the majority of serious injuries have been caused by falls from a height of less than two metres. Height or level difference can be anywhere, below, above or at ground level. For example, standing on a chair, or indeed using it to climb onto a table or desk is work at height.

The HSE publication, *Working at height: A brief guide (INDG 401)*, provides the following examples:

- working on a ladder or a flat roof;
- where someone could fall through a fragile surface; and
- where someone could fall into an opening in a floor or a hole in the ground.

Access and egress, the route for getting to and from the place of work, also has to be considered, unless 'by a staircase in a permanent workplace'. Staircases are also covered by the *Workplace (Health, Safety and Welfare) Regulations 1992 (SI 1992 No 3004)*, *Reg 12*. The *Manual Handling Operations Regulations 1992 (SI 1992 No 2793)*, *Reg 4* and *Sch 1*, set out that factors to which the employer must have regard and questions they must consider when making an assessment of manual handling operations includes variations in level of floors or work surfaces. (See also **A1007 – ACCESS, TRAFFIC ROUTES AND VEHICLES**.)

The duty to cover or fence tanks, pits and other structures where there is a risk of a person falling into a dangerous substance is covered by *SI 1999 No 3004*, *Reg 13(5)–(7)*. In the approved code of practice (ACOP) to the *Workplace (Health, Safety and Welfare) Regulations 1992*, *Workplace health, safety and welfare* (L24) the tanks, pits and structures mentioned in *Reg 13(5)* are referred to as 'vessels' and include sumps, silos, and vats which people could fall into.

The HSE guidance to the Workplace Regulations sets out that:

- Every vessel containing a dangerous substance should be adequately fenced or covered to prevent a person from falling into it.
- Barriers should be sufficiently high, and filled in sufficiently, to prevent falls over or through the barrier. They should be of adequate strength and stability to restrain any person or object liable to fall on to or against them. Untensioned chains, ropes and other non-rigid materials should not be used.
- As a minimum, barriers should consist of two guardrails (a top rail and a mid-rail) at suitable heights. The top of the barrier should be at least 1,100 mm above the surface from which a person might fall.
- Covers should be capable of supporting the loads liable to be imposed on them, and any traffic which is liable to pass over them. They should be of a type which cannot be readily detached and removed, and they should not be capable of being easily displaced.
- Covers should be kept securely in place, except when they have to be removed for inspection or access purposes, and should be replaced as soon as possible.
- When barriers or covers cannot be provided, or have to be removed, effective measures should be taken to prevent falls. Access should be limited to specified people and others should be kept out by, for example, barriers. In high-risk situations suitable formal written permit-to-work systems should be adopted. A safe system of work should be operated which may include the provision and use of a personal fall-protection system. Adequate information, instruction, training and supervision should also be given.

In addition, *Reg 26* of the *Construction (Design and Management) Regulations 2015 (SI 2015 No 51)* is concerned with the prevention of drowning. Where, in the course of construction work, a person is at risk of falling into water or other liquid with a risk of drowning, suitable and sufficient steps must be taken to:

- prevent, so far as is reasonably practicable, the person falling;
- minimise the risk of drowning in the event of a fall; and
- ensure that suitable rescue equipment is provided, maintained and, when necessary, used so the person may be promptly rescued in the event of a fall.

Caving and climbing activities

[W9004.2] Recreational caving and climbing (including for sport, recreation, team building and similar activities, indoor or outdoor) are covered by the *Work at Height Regulations 2005*. The *Work at Height (Amendment) Regulations 2007 (SI 2007 No 114)* brought workers paid to lead or train others in climbing and caving in these activities within the scope of the *Work at Height Regulations 2005*.

The regulations introduced a new Regulation 14A setting out special provision for caving and climbing. These apply to work involving instruction or leadership in caving or climbing for sport, recreation, team building or similar activities.

The *Adventure Activities Licensing Regulations 2004 (SI 2004 No 1309)* also apply to caving and climbing activities for young people. The Adventure Activities Licensing Authority (AALA) licences centres or organisations that provide instruction and leadership of activities including: rock climbing, abseiling, ice climbing, gorge walking, ghyll scrambling, sea level traversing, caving, pot-holing and mine exploration.

The HSE sets out that caving (sometimes known as pot-holing) is the exploration of underground passages (other than those principally used as show-places open to the public) in parts of mines which are no longer worked; or in natural caves where the exploration of those passages requires, in order to be carried out safely, the use of rock climbing or diving equipment or the application of special skills or techniques.

Climbing is climbing, traversing, abseiling or scrambling over natural terrain or outdoor man-made structures (other than structures designed for such activities) which requires, in order to be carried out safely, the use of equipment for, or the application of special skills or techniques in, rock climbing or ice climbing. Climbing walls are exempt from licensing, as are abseiling towers and ropes courses.

The HSE has been reviewing the AALA for some time. The last update, published in November 2018, reported that the HSE Board had approved a proposal to remove the Adventure Activity Licensing Regulations and move to an "industry-led; non-statutory; not-for-profit scheme underpinned by the Health and Safety at Work etc. Act 1974, to provide assurance to users of outdoor activities". The HSE is pursuing this proposal, but says the final decision lies with Parliament and it will take a number of years before reaching a conclusion. There are no immediate changes to the licensing scheme and license holders will be notified well in advance of any changes.

Organisation and planning, competence and avoidance of risks

[W9005] The order of the first three Regulations of substance is significant. The first, *SI 2005 No 735, Reg 4*, covers 'Organisation and Planning', the second, *SI 2005 No 735, Reg 5*, covers 'Competence'. Only then, in *SI 2005 No 735, Reg 6*, 'Avoidance of risks from work at height', is the subject of risk assessment and the hierarchy of measures (or controls) addressed.

The HSE advises: 'Employers and those in control of any work at height activity must make sure work is properly planned, supervised and carried out by competent people. This includes using the right type of equipment for working at height.' (See www.hse.gov.uk/work-at-height/the-law.htm.)

Organisation and planning

[W9006] *SI 2005 No 735, Reg 4* states that employers must ensure that work at height is:

(a) properly planned (including the selection of work equipment, as expanded on in *SI 2005 No 735, Reg 7*);
(b) appropriately supervised; and,
(c) as so far as is reasonably practicable carried out in a safe manner.

[*Note*: (a) and (b) above are 'must do' absolute duties.]

Emergencies and rescue must be planned for. Other than the emergency services acting in an emergency, work at height must not be carried out if weather conditions jeopardise the health and safety of persons involved.

The HSE advises that employers should not consider working on a roof in poor weather conditions such as rain, ice, frost or strong winds (particularly gusting) or if slippery conditions exist on the roof. It advises that winds in excess of 23mph (Force 5) will affect a person's balance (www.hse.gov.uk/construction/faq-height.htm).

Competence

[W9007] *SI 2005 No 735, Reg 5* provides for people to gain competence to carry out their work by being trained under the supervision of a competent person. It requires employers to ensure that no person engages in any activity involving work at height or work equipment for use in work at height, including organisation, planning and supervision, unless they are competent to do so or are being supervised by a competent person if they are being trained.

No specific interpretation of competence is given in *SI 2005 No 735*, but the *Management of Health and Safety at Work Regulations 1999 (SI 1999 No 3242), Reg 13*, sets out employers' duties concerning capabilities and training.

In entrusting tasks to their employees, employers must take into account their capabilities as regards health and safety. They must ensure that employees are provided with health and safety training when they are recruited into the employer's undertaking and when they are exposed to new or increased risks because they are being transferred or given a change of responsibilities, new – or changes to – work equipment, new technology or new – or changes to – systems of work are being introduced. This training should be repeated periodically where appropriate; be adapted to take account of any new or changed risks to the health and safety; and take place during working hours.

The HSE advises that in the case of low-risk, short-duration tasks that take less than 30 minutes and involve ladders, 'competence requirements may be no more than making sure employees receive instruction on how to use the equipment safely (eg how to tie a ladder properly) and appropriate training', which could be on the job (www.hse.gov.uk/pubns/indg401.pdf).

However, it further advises: 'When a more technical level of competence is required, for example drawing up a plan for assembling a complex scaffold, existing training and certification schemes drawn up by trade associations and industry is one way to help demonstrate competence.'

Only having addressed the issues of organisation, planning, supervision and competence does *SI 2005 No 735, Reg 6(1)* require that the measures for the work to be carried out in safe manner must be identified with reference to the risk assessment carried out under *Management of Health and Safety at Work Regulations 1999 (SI 1999 No 3242), Reg 3.*

Avoidance of risks from work at height

[W9008] The HSE advises employers that before working at height they must work through the following simple steps:

- avoid work at height where it is reasonably practicable to do so;
- where work at height cannot be avoided, prevent falls using either an existing place of work that is already safe or the right type of equipment;
- minimise the distance and consequences of a fall, by using the right type of equipment where the risk cannot be eliminated (www.hse.gov.uk/pubns/indg401.pdf).

SI 2005 No 735, Reg 6(2) introduces this hierarchy with the requirement to avoid work at height where it is reasonably practicable to do so. A simple example is to have lighting units in premises that can be lowered to ground level remotely, for routine maintenance and lamp changes.

Where work at height cannot be avoided, *SI 2005 No 735, Reg 6(3)* stipulates that suitable and sufficient measures to prevent any person falling 'a distance liable to cause personal injury' be taken, so far as is reasonably practicable. In addition, *Regulation 7(1)* sets out that collective protection measures must be given priority over personal protection measures.

There is no reference to the so-called 'two-metre' rule (see **W9004.1** above) in the wording of *Regulation 6(3)*. It simply states:

'Where work is carried out at height, every employer shall take suitable and sufficient measures to prevent, so far as is reasonably practicable, any person falling a distance liable to cause personal injury.'

The HSE advises employers that they should:

- do as much work as possible from the ground;
- ensure workers can get safely to and from where they work at height;
- ensure equipment is suitable, stable and strong enough for the job, maintained and checked regularly;
- make sure they don't overload or overreach when working at height;
- take precautions when working on or near fragile surfaces;
- provide protection from falling objects; and
- consider emergency evacuation and rescue procedures (see www.hse.gov.uk/pubns/indg401.pdf).

SI 2005 No 735, Reg 6(4)(a) introduces *SI 2005 No 735, Sch 1* which stipulates the requirements for work areas from which people could fall and which are built or incorporated into premises, including the access to and egress from them. Examples of work areas covered by this Regulation include plant rooms and roof areas with roof-mounted equipment, and platforms and gantries provided for the cleaning and maintenance of glazing and atrium roofs. The schedule sets out that every existing place of work or means of access or egress at height shall:

(a) be stable and of sufficient strength and rigidity for the purpose for which they are intended to be or are being used;
(b) where applicable, rest on a stable, sufficiently strong surface;
(c) be of sufficient dimensions to permit the safe passage of persons and the safe use of any plant or materials required to be used and provide a safe working area, having regard to the work to be carried out there;
No absolute figures or dimensions are given here as it is recognised that there may be some localised obstruction that does not preclude an otherwise safe route or platform being used effectively. The ergonomics of how work is carried out must be considered.
(d) possess suitable and sufficient means for preventing a fall—including for example barriers, guardrails, and toe boards;
(e) possess a surface which has no – unguarded – gap:
  (i) through which a person could fall;
  (ii) through which any material or object could fall and injure a person; or
  (iii) giving rise to other risk of injury to any person, unless measures have been taken to protect persons against such risk;
(f) be so constructed and used, and maintained in such condition, as to prevent, so far as is reasonably practicable:
  (i) the risk of slipping or tripping; or
  (ii) any person being caught between it and any adjacent structure;
(g) where it has moving parts, be prevented by appropriate devices from moving inadvertently during work at height, such as a parking brake or brakes for example.

SI 2005 No 735, Reg 6(4)(b) covers the requirement for temporary work platforms and access, where they are not built into premises. Guardrails, toe boards, barriers and scaffolds are typical examples, covered in more detail in *SI 2005 No 735, Reg 8* and *SI 2005 No 735, Schs 2 and 3.*

SI 2005 No 735, Sch 2 sets out the requirements for guard rails, toe-boards, barriers and similar collective means of protection, and is linked to *SI 2005 No 735, Reg 8(a)*. It specifies that:

- the top guard-rail or other similar means of protection shall be at least 950 mm or, in the case of such means of protection already fixed when the regulations came into force, at least 910 mm above the edge from which any person is liable to fall;
- toe-boards shall be suitable and sufficient to prevent the fall of any person, or any material or object, from any place of work; and
- any intermediate guard-rail or similar means of protection shall be positioned so that any gap between it and other means of protection does not exceed 470 mm.

HSE investigations have indicated the importance of toe boards in preventing people from falling (not just materials or tools).

Schedule 2, paragraph 5(1) states that a lateral opening may be made at a point of access to a ladder 'where an opening is necessary'. However as proprietary self-closing/push-to-open scaffold gates are readily available and in common use, it is hard to envisage a situation where an ungated opening is 'necessary'.

Paragraphs 5(2) and *(3)* cover the requirements for removing or opening the outer barriers of loading bays, but specify that 'compensatory safety measures' must be in place to protect operatives on or adjacent to the loading bay. Typically this could be achieved by a gate mechanism that protects the back of the loading bay if the front is open, or by the use of harness and tether (a work restraint system as per *SI 2005 No 735, Sch 5, Pt 5*, so designed that it prevents the user from getting into a position from which a fall can occur) should a banksman need to operate on the platform when the outer gate is open.

SI 2005 No 735, Sch 3, Pt 1 specifies the requirements for all working platforms, and covers plant and machinery – mobile elevating work platforms including scissor lifts and boom-type cherry pickers as well as mobile aluminium towers, scaffolds and similar fabricated platforms.

The floor, ground or surface on which the equipment relies for support must be adequate for any load that may be imposed. Beware of inspection covers and the like that a wheel or outrigger, or a scaffold standard, may rest on.

SI 2005 No 735, Sch 3, Pt 2 specifies the additional requirements for scaffolds and requires:

- calculations, erection and dismantling plans (depending on the complexity of the scaffold) to be available and held on site. At its simplest, this means instructions showing the correct method of assembly and dismantling be delivered with hired mobile alloy towers, and for more significant scaffolds, the number of ties required, including the specification for proof load testing of any drilled anchors;
- the assembly, modification and dismantling by trained persons under competent supervision; and
- the marking by signs and 'delineation by physical means preventing access' when not available for use, typically during erection, modification or dismantling.

Descent down the hierarchy continues. *SI 2005 No 735, Reg 6(5)(a)(i)* stipulates the requirement to reduce the distance and consequences of a fall, if it is not reasonably practicable to provide a guarded work platform.

This is followed by *paragraph (ii)*, which states that the consequences of a fall must be minimised where is not reasonably practicable to minimise the distance fallen. Fall arrest nets, air bags and crash decks are examples of this type of measure.

Reliance on individual fall arrest harnesses, which provide only individual not collective protection, would have to have good justification.

The HSE says that employers should 'Always consider measures that protect everyone who is at risk (collective protection) before measures that protect only the individual (personal protection)' (www.hse.gov.uk/pubns/indg401.pdf).

It explains that:

collective protection is equipment that does not require the person working at height to act to be effective, for example a permanent or temporary guard rail;

personal protection is equipment that requires the individual to act to be effective, for example putting on a safety harness correctly and connecting it, via an energy-absorbing lanyard, to a suitable anchor point.

Finally, in *SI 2005 No 735, Reg 6(5)(b)*, the requirement for *additional* training and instruction is emphasised, so that the user fully understands the limitations and correct method of use of the chosen work equipment. Choosing inherently safer equipment alone is not enough. The number of incidents involving mobile elevating platforms overturning, or causing injury, illustrates this. The operator may have been shown how the controls work, but not trained to assess the factors that affect stability or the safety of others.

In *Bhatt v Fontain Motors Ltd* [2010] EWCA Civ 863, the judge commented: 'If an employee falls while working at height when he should not have been required to work at height at all, it is difficult to maintain that he was wholly to blame for the fall on the basis that the fall would not have occurred if he had followed the system prescribed by the employer.'

In this case, a garage employee had climbed a ladder in order to retrieve bumpers stored in a loft and had fallen and been injured. The employer argued that if the injured man had ensured the ladder was footed, the accident would not have happened. However, the court looked instead at the aim of the Regulations – to minimise the risk of injury involved in working at height and at the requirement for employers to remove the need to work at height, if reasonably practicable.

Work equipment

[W9009] *SI 2005 No 735, Reg 7* outlines the issues to be considered when selecting work equipment for work at height:

- measures that will provide protection for all are to be given priority over measures that protect only one person (similar to the concept in *Control of Substances Hazardous to Health Regulations 2002 (COSHH) (SI 2002 No 2677)* that dust suppression at source is better than a dust mask for an individual);
- the working conditions and the risks to the safety of people at the place where the work equipment is to be used;
- the duration and frequency of use;
- the length and/or height of means of access;
- the distance and consequences of a potential fall;
- the need for easy and timely evacuation and rescue in an emergency; and
- any additional risk posed by the use, installation or removal of the work equipment or by evacuation and rescue from it.

[W9009.1] Work at Height

Ladders

[W9009.1] *SI 2005 No 735, Sch 6* contains the requirements for ladders. The importance of *SI 2005 No 735, Sch 6* (and the wide range of guidance available) is for it to lead to users understanding *the correct methods of use of ladders*.

A ladder (or stepladder) is to be used only if it is for (a) low-risk work of short duration; or (b) where there are existing features on site which cannot be altered. This must be demonstrated by the risk assessment (carried out under *Reg 3* of the *Management Regulations*) showing that more suitable work equipment is not justified.

Specific requirements relating to how the chosen ladder is used are listed in *paragraph 10 of Sch 6*. A secure handhold and secure support must always be available to the user. The work must be such that the user can maintain a safe handhold when carrying a load unless, in the case of a stepladder, the maintenance of a handhold is not practicable when a load is carried and a risk assessment under *Reg 3* of the *Management Regulations* has demonstrated that the use of a stepladder is justified because of the low risk and the short duration of use.

Having decided on the use of a ladder, a pre-use check as required by *SI 2005 No 735, Reg 13* should be carried out. Guidance on pre-use checks is available on the HSE website (see www.hse.gov.uk/work-at-height/using-ladders-safely.htm).

The suitability of the ground must be assessed, with particular attention to the stability of the ladder. *Paragraph 5 of Sch 6* requires that a portable ladder be prevented from slipping. This can be achieved by securing, strapping or tying both stiles to the wall or fixed object, or by the use of an effective anti-slip or stability device.

HSE Research Report 205 (www.hse.gov.uk/research/rrpdf/rr205.pdf) sets out the results of an evaluation of the performance and effectiveness of ladder stability devices by Loughborough University.

The traditional footing of a ladder by a work colleague is not a solution encouraged by the regulations. Training and supervision is required. Assessment of the ground should also check for contaminants including mud, oil and grease that could reduce the grip of footwear on the ladder rungs during use.

The limit of nine metres vertical distance previously contained in the now revoked *Construction (Health, Safety and Welfare) Regulations 1996 (SI 1996 No 1592)* is retained in *Sch 6(9)*: 'Where a ladder or run of ladders rises a vertical distance of 9 metres or more above its base, there shall, where reasonably practicable, be provided at suitable intervals sufficient safe landing areas or rest platforms'.

As with all work equipment, users need adequate information and training to be able to use ladders and stepladders safely. Adequate supervision is also needed so that safe practices continue to be used.

HSE guidance, *Safe Use of Ladders and Step Ladders – A brief guide* (INDG455), contains practical guidance and is available on the HSE website as a downloadable PDF file at www.hse.gov.uk/pubns/indg455.pdf. The leaflet gives advice to employers, employees and self-employed people on 'simple, sensible precautions' to take when using ladders and stepladders to work at height.

Fragile surfaces

[W9010] Fragile surfaces are given a broad definition in the *Work at Height Regulations 2005*. A 'fragile surface' is 'a surface which would be liable to fail if any reasonably foreseeable loading were to be applied to it'.

SI 2005 No 735, Reg 9 requires the same level of consideration for a fall through a fragile surface as a fall into an open void. In addition, prominent warning signs are called for, or equivalent or better means of warning people, which may for example, include site inductions for visitors. With the proliferation of buildings with near flat roofs, deterioration of fixings or poor workmanship can result in a roof failing to support a person even if the material itself is not inherently fragile. Asbestos cement roofs and sky lights subject to ultraviolet degradation, 'traditional' fragile roof materials, will continue to require control of access for many years to come.

In detail, *Reg 9(1)* requires employers to ensure that workers do not pass across or near, or work on, from or near a fragile surface 'where it is reasonably practicable to carry out work safely and under appropriate ergonomic conditions without his doing so'.

Where this is not reasonably practicable, they must (so far as is reasonably practicable) provide suitable and sufficient platforms, coverings, guard rails or similar means of support or protection, and ensure they are used, 'so that any foreseeable loading is supported by such supports or borne by such protection'.

Where there is still a risk of a fall, they must 'take suitable and sufficient measures to minimise the distances and consequences of [the] fall' and provide prominent warning notices at the approach to any fragile surface or make people aware by other means. This requirement does not apply in the case of members of the emergency services acting in an emergency.

In 2003, the HSE brought an unsuccessful prosecution against the Metropolitan Police Commissioner after a police officer died and another was injured falling through roofs as they pursued suspects.

Falling objects and danger areas

[W9011] *SI 2005 No 735, Reg 10* relates to protecting people from falling, thrown or tipped material. This includes the need for consideration of how and where materials and objects are stored in order to provide protection against risks arising from the collapse, overturning or unintended movement of such materials or objects.

SI 2005 No 735, Reg 11 places a duty to exclude persons who may be harmed by falling objects by the use of:

(a) barriers – '... devices preventing unauthorised persons entering', and,
(b) signs – 'ensure ... such area is clearly indicated'.

Inspection for work at height

[W9012] *SI 2005 No 735, Reg 12* describes the requirements for the inspection of work equipment to facilitate safe work at height. This regulation applies to work equipment covered by *Reg 8* and *Schs 2–6* apply. This is:

- guard-rails, toe-boards, barrier or similar collective means of protection (Schedule 2);
- working platforms (Schedule 3);
- scaffolding (Schedule 3);
- nets, airbags or other collective safeguards for arresting falls (Schedule 4);
- personal fall protection systems (Schedule 5); and
- ladders (Schedule 6).

Regulation 12(2) requires employers to ensure that: 'where the safety of work equipment depends on how it is installed or assembled, it is not used after installation or assembly in any position unless it has been inspected in that position'.

Where work equipment is 'exposed to conditions causing deterioration which is liable to result in dangerous situations' it must be inspected at suitable intervals and 'each time that exceptional circumstances which are liable to jeopardise the safety of the work equipment have occurred' in order to maintain health and safety conditions and detect and remedy in good time any deterioration detected *(Reg 12(3))*.

SI 2005 No 735, Sch 7 lists the particulars to be recorded in a report of inspection:

- the name and address of the person for whom the inspection was carried out;
- the location of the work equipment inspected;
- a description of the work equipment inspected;
- the date and time of the inspection;
- details of any matter identified that could give rise to a risk to the health or safety of any person;
- details of any action taken as a result of any matter identified in *paragraph 5*;
- details of any further action considered necessary; and
- the name and position of the person making the report.

SI 2005 No 735, Reg 12(5) includes the requirement for the 'physical evidence' of the last inspection to accompany equipment sent from one undertaking to another. Equipment delivered from a hire company is a simple example. This mirrors the requirement of the *Lifting Operations and Lifting Equipment Regulations 1998 (LOLER) (SI 1998 No 2307), Reg 9(4)*, which applies to equipment used for lifting and lowering people, including cranes, as well as conventional mobile elevating work platforms.

Records of inspection reports must be completed in the same shift as the inspection, and delivered to the client on whose behalf the inspection was carried out within 24 hours (*Reg 12 (7)*). Records must be kept for at least three months after the end of the project during which the work equipment was used (*Reg 12 (8)*).

SI 2005 No 735, Reg 13 requires inspection of 'the surface and every parapet, permanent rail or other such fall protection measure of every place of work at height' on each occasion before use.

In 2016, the HSE launched a review of the requirements for the inspection or thorough examination of work plant and equipment contained in regulations including the Work at Height Regulations 2005. In March 2017 the HSE Board considered the outcome of the review, which concluded that the current framework for equipment inspection remains proportionate and fit for purpose. The Board agreed no action be taken at this time and brought the review to a close.

Duties of persons at work

[W9013] *SI 2005 No 735, Reg 14* repeats the duties of individuals to be responsible for their own safety and the safety of others as contained in *HSWA 1974, s 7*. *SI 2005 No 735, Reg 14 (1)* imposes the duty to report an activity or defect that may endanger that person or others and *SI 2005 No 735, Reg 14(2)* imposes the duty to use equipment in accordance with the training and instructions given.

Protection from falls

[W9014] *SI 2005 No 735, Sch 4* relates to collective safeguards for arresting falls, such as nets, airbags and landing mats. The risk assessment must demonstrate that they can work effectively for the work activity taking place. For example, if the fall of a person is likely to occur with or following the fall of materials such as sheet materials, steel or pre-cast concrete components, a fall protection system linked to an anchor above the work area may prove to be more effective. Employers should ask questions such as: Is there enough clearance from objects beneath that could cause injury? If the system requires support from newly-built walls, do they have enough strength, allowing for temperature and weather conditions?

SI 2005 No 735, Sch 5, Pt 2 outlines the requirements for personal fall protection systems. These are divided into four categories:

(i) work positioning systems. *SI 2005 No 735, Sch 5, Pt 2* calls for the inclusion and use of a backup to prevent or arrest falls. Where this is not *reasonably* practicable, all practicable measures (*note*: not all *reasonably* practicable measures) must be taken 'to ensure that the system does not fail'.

(ii) Rope access and positioning techniques. *SI 2005 No 735, Sch 5, Pt 3* describes the use of two separately anchored lines, a working line and a safety line. This has been standard practice for many years, and competency has been overseen by the Industrial Rope Access Trade Association (IRATA) (see www.irata.org for further information about its training and certification scheme).

(iii) Fall arrest systems. *SI 2005 No 735, Sch 5, Pt 4* requires fall arrest systems to incorporate a suitable means of absorbing energy and limiting the forces applied to the user's body. It also sets out that a system should not be used in a manner involving the risk of a line being cut or where its safe use requires a clear zone and there is no such zone, or in another manner which inhibits its performance or renders its use unsafe. Personal fall arrest shall be used only if the risk assessment demonstrates that the use of other safer work equipment is not reasonably practicable. The anchor point must be suitable and sufficient and have sufficient strength and stability to support any foreseeable impact and static load. A shock absorber, usually incorporated at the harness connection end of any lanyard or attachment line, is required to limit the forces applied to the user's body through the harness in the event of a fall. Assessment of the site needs to ensure that sharp edges that could cut a lanyard are avoided and that there is sufficient clearance to swing without striking objects that could cause injury in the event of a fall. Suspension trauma, similar to the effect of fainting on the parade ground when insufficient blood reaches the brain, is now well documented. If the blood supply to the brain of a person suspended in a near vertical position is not restored, death is the inevitable consequence. Therefore, provision for rescue of any person suspended in a harness after a fall has to be planned beforehand, and carried out within minutes.

(iv) Work restraint systems. *SI 2005 No 735, Sch 5, Pt 5* sets out that a work restraint system shall be designed so that, if correctly used, it prevents the user from getting into a position in which a fall can occur, and be correctly used. The benefit of a well-designed work restraint system is that a fall should be prevented, eliminating shock loads on both the anchor point and the human body wearing the harness, the need for rescue and the like. Frequently lanyards used for work restraint do incorporate shock absorbers, so that there is no possibility of them causing injury if used incorrectly as part of a fall arrest system.

British Standard, BS 8437:2005+A1:2012 *Code of practice for selection, use and maintenance of personal fall protection systems and equipment for use in the workplace* sets out the types of fall protection systems and equipment available and gives guidance on their selection, use and maintenance, as well as the training of users. It also provides direction on the rescue of people working at a height, in the event of an accident.

It is intended for use by employers, employees and self-employed persons who use personal fall protection systems and equipment as well as designers, eg architects and structural engineers and those responsible for the design of safe access routes on buildings and structures and those who commission work at height, including building owners and contractors.

APPG report on working at height

[W9014.1] In February 2019, the all-party parliamentary group (APPG) on working at height published its inaugural report on how to improve the safety environment for work at height, *Staying Alive: Preventing Serious Injuries and Fatalities While Working at Height*.

The report was the result of a year-long inquiry and the APPG called on the government and industry to undertake a major review of work at height culture, expand enhanced reporting and introduce reporting on near misses.

The APPG reported that the majority of the respondents to the inquiry believed that the principle of the Work at Height Regulations 2005 was broadly fit-for-purpose, although the interpretation and application of the legislation was varied.

It found that lack of planning during all stages of a project is a significant cause of falls from height. It identified lack of training, communication and supervision as a problem, and said: "Clients wield significant power yet do not fully understand the crucial role they play in ensuring contractors are working in the safest environments possible. Tight and changing deadlines, along with even tighter budgets can lead to the need to 'cut corners', thus increasing the likelihood of a fall from height."

The report makes a number of recommendations and highlighted areas for further consultation. It recommends:

- the introduction of enhanced reporting, without an additional burden, through the Reporting of Injuries, Diseases and Dangerous Occurrences Regulations (RIDDOR) 2013, which at a minimum, records the scale of a fall, the method used and the circumstances of the fall;
- the appointment of an independent body that allows confidential, enhanced and digital reporting of all near misses and accidents that do not qualify for RIDDOR reporting. The data collected by this independent body will be shared with government and industry to inform health and safety policy;
- the extension of the *Working Well Together – Working Well at Height* safety campaigns to industries outside of the construction sector; and
- an equivalent system to Scotland's Fatal Accident Inquiry process to be extended to the rest of the UK.

When the RIDDOR regulations were revised in 2013 there was a reduction in reporting requirements that brought to an end detailed data collection including the heights from which people fell.

In Scotland, ministers are required under s 29 of the Inquiries into Fatal Accidents and Sudden Deaths (Scotland) Act 2016 to report on fatalities. Fatal Accident Inquiries (FAIs) are the legal mechanism through which deaths in the workplace are investigated. They are mandatory for deaths occurring in the workplace, as well as those in custody or in circumstances deemed to be in the public interest, and are usually held in the sheriff court. The outcome of all FAIs since 1999 are publicly available and can be accessed online via the Scottish Courts and Tribunals Service.

The group recommends further consultation on:

- the creation of a digital technology strategy, to include a new tax relief for small, micro and sole traders, to enable them to invest in new technology; and
- a major review of work at height culture, including an investigation into the suitability of legally-binding financial penalties in health and safety which could be used towards raising awareness and training, particularly in hard to reach sectors.

The report also outlines a "huge potential for ever-evolving digital technology to make work at height safer". This includes the use of drones before or instead of climbing; the introduction of virtual and augmented reality to allow individuals to experience the work at height process without the associated risks; and the use of an app – an application downloaded onto a users' mobile device – that could help with enhanced inspection and reporting.

The report can be found at: https://workingatheight.info/wp-content/uploads/2019/02/Staying-Alive-APPG-REPORT.pdf.

Further information concerning work at height

[W9015] HSE guidance on work at height can be found on its website at www.hse.gov.uk/work-at-height/index.htm.

This includes *Working at height – A brief guide (INDG 401)* (www.hse.gov.uk/pubns/indg401.pdf).

Working Time

Mark Greaves, Pupil Barrister at Old Square Chambers and Nicola Coote, Personnel Health & Safety Consultants Ltd

Introduction to working time

[W10001] The *Working Time Regulations 1998 (SI 1998 No 1833)* ('the Regulations') came into force on 1 October 1998 and were the first piece of English law to set out specific rules governing working hours, rest breaks and holiday entitlement for the majority of workers (as opposed to those in specialised industries). The Regulations were introduced in order to implement the provisions of the Working Time Directive (93/104/EC) and the Young Workers Directive (94/33/EC). The Working Time Directive (93/104/EC) was amended in 2000 to cover certain excluded sectors and activities. It was ultimately consolidated by the Working Time Directive (2003/88/EC), which came into force on 2 August 2004 ('the Directive').

This chapter deals only with the Regulations as they govern adult workers, ie those aged 18 or over.

The Working Time Directive was adopted on 23 November 1993 as a health and safety measure under what was then Article 118a of the Treaty of Rome 1957 (now Article 155 of the Treaty on the Functioning of the European Union ('TFEU')), thereby requiring only a qualified majority, rather than unanimous, vote. As such, the UK Government, which was opposed to the Directive, was unable to avoid its impact. The Government challenged the Directive on the basis that it was not truly a health and safety measure and should, in fact, have been introduced under then Article 100 (now Article 115 of the TFEU) as a social measure which would have required a unanimous vote. The Court of Justice of the European Union ('CJEU') has nevertheless upheld the status of the Directive as a health and safety measure. The stated purpose of the Directive, acknowledged by the CJEU, is to lay down a minimum requirement for health and safety as regards the organisation of working time. This is, of course, significant to the extent that the English courts and tribunals will interpret the Regulations in accordance with the Directive by adopting the 'purposive' approach, which is now accepted under English law.

When they were first introduced, the Regulations (and accompanying guidance produced by the former Department of Trade and Industry (DTI), now the Department for Business Innovation and Skills (BIS)) were met with vociferous criticism from the business world. Just a year later the Government sought to address these initial concerns; following a limited consultation in July of 1999, amendments were tabled to two key areas of the Regulations (see **W10013** and **W10033** below).

Despite all the controversy regarding the Regulations and the implementation of the Directive it would appear that the Regulations continue to have little impact on the long-hours culture within the UK. Following the UK's decision to leave the EU, it will remain to be seen whether this is one of the regulations that the UK government will seek to review. It is likely to be some time, however, before any significant changes to EU-led legislation will appear and therefore these regulations will continue to be applicable in the UK for the foreseeable future.

The structure of the Regulations is to prescribe various limits and entitlements for workers and then to set out a sequence of exceptions and derogations from these provisions. The structure of this chapter will roughly follow that format.

Definitions relating to working time

[W10002] The *Working Time Regulations 1998 (SI 1998 No 1833), reg 2* sets out a number of basic concepts (some more familiar than others), which recur consistently and which underpin the legislation. These are as follows.

Working time

[W10003] Working time is defined, in relation to a worker, as:

(a) any period during which he is working, *at his employer's disposal* and carrying out his activity or duties;
(b) any period during which he is receiving relevant training; and
(c) any *additional* period which is to be treated as working time for the purposes of the Regulations under a relevant agreement (see **W10006** below).

Paragraph (a) of this definition is to be interpreted broadly, following the recent guidance of the Employment Appeal Tribunal ('EAT') in *Edwards and Morgan v Encirc Ltd* [2015] IRLR 528. In *Edwards*, the EAT confirmed that all three limbs of paragraph (a) must be satisfied for the period in question to fall under the definition, but emphasised the need for a purposive approach to be adopted in construing each limb (in accordance with the aims of the Directive). The test is not a narrow contractual one. Applying a broad approach, the EAT in *Edwards* thus held that time spent by a trade union representative at meetings with his employer at his workplace constituted 'working time'.

The breadth of the definition has led to numerous uncertainties, in particular, in relation to workers who are 'on call' or who operate under flexible working arrangements.

Despite this initial confusion, the CJEU decision in the cases of *SIMAP v Conselleria de Sandidad y Consumo de la Generalitat Valencia*: C303/98 [2001] IRLR 845 and *Landeshauptstadt Kiel v Jaeger*: C-151/02 [2003] IRLR 804 confirmed that time spent on call by doctors at health centres is 'working time' provided the doctors are available and required to be physically present at the workplace, even if they are allowed to sleep while on call. By contrast, time spent on call away from the health centre does not count as 'working time' unless the doctor is actually working. This reasoning was followed by the CJEU in *Vorel v Nemocnice Český Krumlov*: C-437/05. While these cases concerned doctors, a similar approach has been taken in other areas. In *MacCartney v Oversley House Management* [2006] ICR 210, the EAT ruled that the manager of sheltered accommodation, on call 24 hours a day for residents, was 'working' throughout, even though she was permitted to sleep and rest at a flat provided by her employer at the workplace. A similar approach was taken in relation to a residential home care assistant in *Hughes v Jones and anor t/a Graylyns Residential Home* (UKEAT/0159/08) and in *Anderson v Jarvis Hotels plc* (EAT/0062/05), where a hotel manager who was required to sleep overnight at the hotel and remain on call to cover emergencies was also held to be 'working' although he was free to sleep. In both cases, the EAT ruled that the likelihood of being called out is irrelevant. A more recent case *Whittlestone v BJP Home Support Ltd* (UKEAT/0128/13/1907) (19 July 2013, unreported), EAT in July 2013 reinforces this view, and further goes on to clarify that the national minimum wage must be paid where the worker is able to sleep whilst working overnight. The most recent case to date, *Truslove v Scottish Ambulance Service* [2014] ICR 1232, illustrates that time spent on call at work or at another place specified by the employer where the worker must be available to work at short notice will generally amount to working time. The decision of the Northern Ireland Court of Appeal in *Blakley v South Eastern Health and Social Services Trust* [2009] NICA 62 (14 December 2009, unreported) suggests that time spent on call at a place of the worker's choosing will probably not.

Because of the difficulty this has caused, in 2003, the European Commission proposed amending the Directive so that 'inactive time' on call, that is, time where a worker is available for work at his place of work but does not carry out any duties, would not count as working time. However, the proposals lapsed in 2009 without agreement being reached. In terms of whether work-related travel counts as working time, the GOV.UK website guidance states that working time 'includes travelling if part of the job' but does not 'include travelling between home and work'. In addition, the guidance states that 'travelling outside of normal working hours' is not working time. This has left open the difficult question of non-routine travel. For example, it has never been clear whether travel from London to Edinburgh on a Sunday, to be at a meeting in Edinburgh at 9am on Monday morning, counts as working time.

Further uncertainty arises from the classification of the travel time of peripatetic workers, such as domiciliary carers, or sales personnel. In *Federación de Servicios Privados del sindicato Comisiones obreras (CCOO) v Tyco Integrated Security SL, Tyco Integrated Fire & Security Corporation Servicios SA*, C-266/14, the CJEU held that, for workers who have no fixed place of work, travel time from home to the first and last customers counts as working time. The Court took into account that during this time these workers are at their employer's disposal, act under their employer's instructions and cannot use that time freely to pursue their own interests. The Advisory, Conciliation and Arbitration Service (ACAS) which provides free and impartial information and advice to employers and employees on workplace relations and employment law, is currently assessing the impact of this judgment on workers and employers in Great Britain. Employers and employees should be aware that break periods may need to be altered if this broader inclusion of travel time results in an extension of the working day.

It should be noted that the definition of working time can be clarified in a relevant agreement (see **W10009**) between the parties, but only by specifying any *additional* period which can be treated as working time. The relevant agreement cannot alter, and in particular cannot narrow, the statutory definition of working time.

Worker

[W10004] A worker is any individual who has entered into or works under (or where the employment has ceased), worked under:

(a) a contract of employment; or
(b) any other contract, whether express or implied and (if express) whether oral or in writing, whereby the individual undertakes to do or perform personally any work or services for another party to the contract whose status is not by virtue of the contract that of a client or customer of any profession or business undertaking carried on by the individual.

This definition is wider than just employees and covers any individuals who are carrying out work for an employer, unless they are genuinely self-employed, in that the work amounts to a business activity carried out on their own account (see also the reference to agency workers at **W10005**). When the Regulations were first introduced, the definition of 'worker' was a novel concept. It has now become a familiar one and appears in several other contexts such as the *Part Time Workers (Prevention of Less Favourable Treatment) Regulations 2000 (SI 2000 No 1551)* and the *National Minimum Wage Act 1998*.

A requirement to perform services "personally" is essential to qualify as a worker. Whether or not a contract includes an obligation to do work personally is determined by construing the contract in the light of the circumstances in which it was made and the common intention of the parties at that time: see *Wright v Redrow Homes (Yorkshire) Ltd* [2004] IRLR 720. Where the contract allows the individual to provide a substitute, this can cause difficulties. In *Byrne Bros (Formwork) Ltd v Baird & Others* [2002] IRLR 96 the EAT concluded, in relation to carpenters and labourers who had been offered work on a building site, that a limited power to appoint substitutes was not inconsistent with an obligation to provide work on a personal basis. Subsequent cases suggest that the circumstances in which substitution is permitted by the contract will be the relevant consideration. For example, in *James v Redcats (Brands) Ltd* [2007] IRLR 296 the EAT ruled that there was an obligation to perform work personally where a courier's contract allowed her to provide a substitute when she was "unable" to work, but not when she was simply unwilling. By contrast, in *Premier Groundworks Ltd v Jozsa* (EAT/0494/08) the EAT concluded that a groundworker was not a 'worker' in circumstances where he had an unfettered right to delegate work to someone who was "at least as capable experienced and qualified". The groundworker had no obligation to provide the services personally.

In *Bacica v Muir* [2006] IRLR 35 the EAT confirmed that the mere rendering of a service personally does not make an individual a worker. If an individual performs work for a person who is a customer of his profession, business or undertaking, then he is not to be regarded as a worker. In that case, a painter who provided personal services was held not to be a worker, as he was taxed at a special rate under the Construction Industry Scheme whereby he paid his own national insurance contributions, had business accounts prepared by an accountant, was free to work for others and had in fact done so, and was paid a rate that included an allowance for overheads. In *Cotswold Developments Construction Ltd v Williams* [2006] IRLR 181, the EAT suggested that the focus is on whether the individual markets his services to the public generally (in which case he is more likely to be self-employed) or whether he was recruited to work as an integral part of the principal's business (in which case he is more likely a worker). In *James v Redcats (Brands) Ltd* [2007] IRLR 296 the EAT stressed that no one factor is determinative and each case will depend on all the elements of the particular relationship. Elias J suggested that it may be helpful to consider whether the obligation to perform work personally is the dominant feature of the contract, as opposed to being merely incidental or secondary. If the dominant purpose is not personal service, but a particular outcome or objective, the individual will not be a worker.

Mobile Workers

[W10004A] Further rules apply to mobile workers: those working in haulage or on public service vehicles. As a general rule, drivers and crew of vehicles over 3.5 tonnes or passenger vehicles suitable for carrying more than nine people are subject to EU legislation (*European Directive 2002/15/EC and EC Regulation 561/2006*), implemented by the *Road Transport (Working Time) Regulations 2005*. Self-employed drivers are also bound by these rules subsequent to an amendment in 2012. Drivers and crew not falling within the EU rules are subject to the GB domestic drivers' hours rules. Both sets of rules set out maximum driving hours and minimum rest periods, and while there are statutory exceptions for certain vehicle types, no opt-out is possible. Guidance as to the interplay of these rules can be found at www.gov.uk/drivers-hours.

Agency workers

[W10005] The *Working Time Regulations 1998 (SI 1998 No 1833), Regulation 36* makes specific provision in relation to agency workers who do not otherwise fall under the general definition of 'workers'. This provides that where:

(i) any individual is engaged to work for a principal under an arrangement made between an agent and that principal, and
(ii) the individual is not a worker because of the absence of a contract between the individual and the agent or the individual and the principal, then the Regulations will apply as if that individual were a worker employed by whichever of the agent or principal is responsible for paying or actually pays the worker in respect of the work.

Again, individuals who are genuinely self-employed are excluded from this definition.

Collective, workforce and relevant agreements

[W10006] These play a significant role in the Regulations as employers and employees can, by entering into such agreements (where the Regulations so allow), effectively supplement or derogate from the strict application of the Regulations. Employers and employees who need a certain amount of flexibility in their working arrangements may well find one or other of these agreements will facilitate compliance with the Regulations.

Collective agreement

[W10007] This is an agreement with an independent trade union within the meaning of the *Trade Union and Labour Relations (Consolidation) Act 1992, s 178*. The definition in *s 178* is wide enough to cover agreements reached between trade unions and employers under the statutory recognition process.

Workforce agreement

[W10008] This is a new concept under English law which, since its introduction by the Regulations, has not been widely utilised. It is an agreement between an employer and the duly elected representatives of its employees or, in the case of small employers (ie those with 20 or fewer employees), potentially the employees themselves. In order for a workforce agreement to be valid, it must comply with the conditions set out in *Schedule 1* to the Regulations, which provides that a workforce agreement must:

(a) be in writing;
(b) have effect for a specified period not exceeding five years;
(c) apply either to
  (i) all of the relevant members of the workforce, or
  (ii) all of the relevant members of the workforce who belong to a particular group;
(d) be signed by
  (i) the representatives of the workforce or of the particular group of workers, or
  (ii) where an employer employs 20 or fewer workers on the date on which the agreement is first made available for signature, either appropriate representatives or by a majority of the workers employed by him;
(e) before the agreement was made available for signature, the employer must have provided all the workers to whom it was intended to apply with copies of the text of the agreement and such guidance as they might reasonably require in order to understand it fully.

'Relevant members of the workforce' are defined as all of the workers employed by a particular employer (excluding any worker whose terms and conditions of employment are provided for wholly or in part in a collective agreement). Therefore, as soon as a collective agreement is in force in respect of any worker, the provisions of any workforce agreement in respect of that worker would cease to apply.

Paragraph 3 of *Schedule 1* to the Regulations sets out the requirements relating to the election of workforce representatives. These are as follows:

(a) the number of representatives to be elected shall be determined by the employer;
(b) candidates for election as representatives for the workforce must be relevant members of the workforce, and the candidates for election as representatives of a particular group must be members of that group;
(c) no worker who is eligible to be a candidate can be unreasonably excluded from standing for election;
(d) all the relevant members of the workforce must be entitled to vote for representatives of the workforce and all the members of a particular group must be entitled to vote for representatives of that group;
(e) the workers must be entitled to vote for as many candidates as there are representatives to be elected;
(f) the election must be conducted so as to secure that (i) so far as is reasonably practicable those voting do so in secret, and (ii) the votes given at the election are fairly and accurately counted.

Relevant agreement

[W10009] This is an 'umbrella provision' and means a workforce agreement which applies to a worker, any provision of a collective agreement which forms part of a contract between a worker and his employer, or any other agreement in writing which is legally enforceable as between the worker and his employer. A relevant agreement could, of course, include the written terms of a contract of employment. It would only include the provisions of staff handbooks, policies etc where it could be shown that these were 'legally enforceable' as between the parties.

Maximum weekly working time

48-hour working week

[W10010] It is, of course, the 48-hour working week which initially caused so much controversy. The *Working Time Regulations 1998 (SI 1998 No 1833)*, *reg 4* provides that an employer shall take all reasonable steps, in keeping with the need to protect the health and safety of workers, to ensure that a worker's average working time (including overtime) shall not exceed 48 hours for each 7-day period.

Reg 4(6) provides a specific formula for calculating a worker's average working time over a reference period, which the Regulations have determined as 17 weeks (for exceptions, see **W10012**). This is as follows:

$$\frac{a+b}{c}$$

where
  a is the total number of hours worked during the reference period;
  b is the total number of hours worked during the period which
    (i) begins immediately after the reference period, and

(ii) consists of the number of working days equivalent to the number of 'excluded days' during the reference period; and

c is the number of weeks in the reference period.

'Excluded days' are days comprised of annual leave, sick leave or maternity leave and any period in respect of which an individual has opted out of the 48-hour limit (see **W10013** below).

When the case of *Barber v RJB Mining (UK) Ltd* [1999] IRLR 308 was first reported it was initially thought that this decision would add considerably to the importance of this provision. In *Barber*, Mr Justice Gage in the High Court held that *reg 4(1)* gives rise to a free-standing legal contractual right not to work more than 48 hours a week on average which a worker could enforce by a common law action. However, in practice this contractual right has proved of limited significance as, in the absence of a direct remedy for infringement, a worker must first resign and then claim constructive unfair dismissal or breach of contract.

It has also been held that a worker cannot claim tortious damages for breach of statutory duty based on the employer's failure to enforce the 48-hour limit, as the provision was not intended to confer private law rights of action: *Sayers v Cambridgeshire County Council* [2006] EWHC 2029 (QB), [2007] IRLR 29.

An employee who claims to have suffered psychiatric illness as a result of excessive working hours may seek to rely on a breach of *reg 4(1)* to bolster a claim in negligence. In *Hone v Six Continents Retail Ltd* [2005] EWCA Civ 922, [2006] IRLR 49, an employee successfully argued that his psychiatric injury was reasonably foreseeable in part because he had been working in excess of the 48-hour limit. However, the Court of Appeal confirmed in *Pakenham-Walsh v Connell Residential* [2006] EWCA Civ 90, [2006] All ER (D) 275 (Feb) that, while the employer's failure to comply with the 48-hour limit will be relevant, it cannot automatically lead to a finding that the employer has been negligent.

The issue of an employer's obligations under the Regulations where they are not the only employer of the worker is problematic. *Reg 4(2)* states 'an employer shall take all reasonable steps, in keeping with the need to protect the health and safety of workers, to ensure that the limit specified in paragraph (1) is complied with in the case of each worker employed by him in relation to whom it applies'. There is little guidance available regarding rights of workers and opting out of the Working Time Regulations. The initial guidance from the Department of Trade and Industry (DTI) suggested that workers with two jobs can either sign an opt-out, if the total time worked exceeds 48 hours a week, or reduce their hours to meet the 48-hour limit. If a worker does not tell their employer that they are concurrently employed elsewhere, and the primary employer has no reason to suspect this is the case, then it is extremely unlikely that this primary employer would be found in breach. However, the guidance on the GOV.UK website does not address this.

In *Brown v Controlled Packaging Services Ltd* (1999), unreported, a worker took up primary employment of 40 hours per week, with the understanding that on occasion he would be required to work overtime. It transpired that the overtime conflicted with bar work that he undertook two nights a week. After the Regulations came into force, the worker was asked to sign an opt-out agreement which was viewed by his manager as a formality to ensure that existing overtime arrangements complied with the Regulations. Brown refused to sign the agreement because long hours would be incompatible with his bar work. Brown left and claimed constructive dismissal. The Employment Tribunal found that the dismissal was automatically unfair under the *Employment Rights Act 1996, s 101A* – ie for a reason connected with the employee exercising his rights under the Regulations.

Young workers

[W10010A] The Regulations also provide for a maximum working time for young workers of 8 hours a day or 40 hours per week [*Working Time Regulations 1998 (SI 1998 No 1833), reg 5A*]. A young worker is defined as a person who is of above compulsory school leaving age but under 18.

Reference period

[W10011] The crucial issue in calculating the average number of hours worked is the question of when the reference period starts. The reference period is any period of 17 weeks in the course of a worker's employment, unless a relevant agreement provides for the application of successive 17-week periods. [*Working Time Regulations 1998 (SI 1998 No 1833), reg 4(3)*]. This means that unless the parties specify that the reference period is a defined period of 17 weeks, followed by a successive period of 17 weeks, then the reference period will become a 'rolling' 17-week period. This could be significant where there is a marked variation in the hours that an individual works in a particular period, from week to week. In such a situation an employer can agree with the workforce to amend the 17-week reference period, eg to 26 weeks. The reference period can be extended:

- up to a maximum of 52 weeks if there is a valid collective or workforce agreement in place;
- up to 26 weeks if the workers live far from their workplace (eg off-shore workers); if they work in security or surveillance that requires a permanent presence; or they do work that involves the need for continuity of service or production eg Press/Film/TV, hospital and care workers, farm workers, utility workers, dock and airport workers.

For the first 17 weeks of employment, the average is calculated by reference to the number of weeks actually worked. [*Working Time Regulations 1998 (SI 1998 No 1833), reg 4(4)*].

Exceptions and derogations to Regulation 4

[W10012] There are currently various exceptions to the maximum working week.

(a) reg 5 provides that an individual can agree with his employer to opt out of the 48-hour week (see **W10013** below).
(b) The 48-hour limit does not apply at all in the case of certain workers whose working time is unmeasured; in respect of other workers, part of whose working time is unmeasured, it only applies to that part of their working time which is measured. [*Working Time Regulations 1998 (SI 1998 No 1833), reg 20* – see **W10033** below].

(c) In special cases (as described in *reg 21* – see **W10034** below), the 17-week reference period over which the 48 hours are averaged will be automatically extended to 26 weeks. [*Working Time Regulations 1998 (SI 1998 No 1833), reg 4(5)*].

(d) A collective or workforce agreement can extend the 17-week reference period to a period not exceeding 52 weeks, if there are objective or technical reasons concerning the organisation of work. [*Working Time Regulations 1998 (SI 1998 No 1833), reg 23*].

Special rules apply to doctors in training, for whom the working time limit was phased in gradually. The maximum working week for doctors in training was initially 58 hours per week from 1 August 2004, and this reduced to 56 hours per week on 1 August 2007. For most doctors in training, the 48-hour limit now applies as of 1 August 2009. However, a further limited derogation applies from 1 August 2009 until 31 July 2011 to some doctors in training employed in specific areas at named hospitals, whose maximum working week is 52-hours per week [*Working Time Regulations 1998 (SI 1998 No 1833), reg 25A* and *Schedule 2A*]. The reference period for calculating average weekly working time for all doctors in training is any 26-week period in the course of employment, unless a relevant agreement provides for the application of successive 26-week periods.

Agreement to exclude 48-hour working week

[**W10013**] The 48-hour limit on weekly working time will not apply where a worker agrees with his employer in writing that the maximum 48-hour working week should not apply in the individual's case, provided that certain requirements are satisfied. These are:

(a) the agreement must be in writing;
(b) the agreement may specify its duration or be of an indefinite length – however, it is always open to the worker to terminate the agreement by giving notice, which will be the length specified in the agreement (subject to a maximum of 3 months), or, if not specified, 7 days; and
(c) the employer must keep up-to-date records of all workers who have signed such an agreement.

Until December 1999, employers were under controversial and onerous record keeping requirements in relation to workers who had signed opt out agreements. Since December 1999, employers only need to keep records of who has opted out.

A worker cannot be forced to sign an opt-out agreement. Moreover, the Regulations provide protection where an employee has suffered any detriment on the grounds that he/she has refused to waive any benefit conferred on him/her by the Regulations. [*Employment Rights Act 1996, s 45A(1)(b)*, inserted by the *Working Time Regulations 1998 (SI 1998 No 1833), reg 31(1)(b)* – see **W10041** below]. Furthermore, a worker will always have the right to terminate the opt-out agreement so that the 48-hour limit applies, on giving, at most, three months' notice.

On a practical note, the agreement by a worker to exclude the 48-hour working week must be a written agreement between the employer and the individual worker. It cannot take the form of or be incorporated in a collective or workforce agreement. It could, of course, be part of the contract of employment, (although it would be advisable for an employer to clearly delineate the agreement to work more than 48 hours from the rest of the contract of employment). It seems that employers' record keeping obligations in this regard would be satisfied by keeping an up-to-date list of workers who have signed opt-out agreements.

The ability of employers to seek the agreement of workers to 'opt out' of the 48-hour week (which is specifically permitted by the Directive) remains controversial as it arguably runs a 'coach and horses' through the spirit of the legislation. In 2003, the use of the opt-out was reviewed by the European Council, which recommended that further restrictions be placed on its use. Following years of negotiations, a Common Position was adopted by the European Council of Ministers that would have seen further safeguards for workers – for example, opt-outs could not be signed at the same time as an employment contract or within 4 weeks of starting work, they would have to be renewed in writing after a year, workers could withdraw with immediate effect in the first six months of employment or with two months' notice thereafter and any worker who opted-out would be subject to a maximum average working week of 60 hours per week or 65 hours per week if inactive on-call time was included (see **W10003** above). However, the European Parliament wanted to see the opt-out phased out altogether. Following negotiations, in which the UK Government in particular lobbied to retain the opt-out, no agreement was reached. As a result, the Directive and the Regulations remain unchanged and workers are still able to opt-out of the weekly working time limit.

Rest periods and breaks

Daily rest period

[W10014] Adult workers are entitled to an uninterrupted rest period of not less than 11 consecutive hours in each 24-hour period. [*Working Time Regulations 1998 (SI 1998 No 1833), reg 10*].

Employers must ensure that workers are allowed to take their rest breaks and that work arrangements provide a proper opportunity for workers to take their breaks. This was confirmed by the CJEU in a successful challenge brought by the European Commission to the Government's original guidance on rest breaks: see *Commission of the European Communities v United Kingdom*: C-484/04 [2006] ECR I-7471. The guidance originally stated that '*employers must make sure that workers can take their rest, but are not required to make sure they do take their rest*'. The CJEU ruled that this statement failed to properly implement the Directive. While employers do not need to force workers to take their breaks, according to the CJEU, the UK is bound to ensure that the minimum entitlements under the Directive are observed. In the CJEU's view, the words '*but are not required to make sure they do take their rest*' failed to do this, by suggesting that employers were not required to ensure workers actually exercised their rights. Guidance now available on the GOV.UK website is simple:

'Employers can say when employees take rest breaks during work time as long as:

- the break is taken in one go somewhere in the middle of the day (shift) not at the beginning or end)
- workers are allowed to spend it away from their desk or workstation (ie away from where they actually work)

It doesn't count as a rest break if an employer says an employee should go back to work before their break is finished.

Unless a worker's employment contract says so, they don't have the right to:

- take smoking breaks
- get paid for rest breaks.'

Exceptions and derogations

[W10015] The provisions regarding daily rest periods do not apply where a worker's working time is unmeasured. [*Working Time Regulations 1998 (SI 1998 No 1833), reg 20(1)*].

Derogations from the rule may be made with regard to shift work [*reg 22*], or by means of collective or workforce agreements [*reg 23*] or where there are special categories of workers [*reg 21*]. However, in all cases (except for workers with unmeasured working time under *reg 20(1)* – see **W10033** below) compensatory rest must be provided [*reg 24*].

Weekly rest period

[W10016] Adult workers are entitled to an uninterrupted rest period of not less than 24 hours in each seven-day period. [*Working Time Regulations 1998 (SI 1998 No 1833), reg 11*]. However, if his employer so determines, an adult worker will be entitled to either two uninterrupted rest periods (each of not less than 24 hours) in each 14-day period or one uninterrupted rest period of not less than 48 hours in each 14-day period.

The entitlement to weekly rest is in addition to the 11-hour daily rest entitlement which must be provided by virtue of *reg 10*, except where objective or technical reasons concerning the organisation of work would justify incorporating all or part of that daily rest into the weekly rest period.

For the purposes of calculating the entitlement, the 7 or 14-day period will start immediately after midnight between Sunday and Monday, unless a relevant agreement provides otherwise. The Regulations do not require Sunday to be included in the minimum weekly rest period.

Derogations from the entitlement to weekly rest periods are the same as those for daily rest (see **W10015** above).

Rest breaks

[W10017] By virtue of *reg 12* of the Regulations, adult workers are entitled to a rest break where their daily working time is more than six hours. Details of this rest break, including duration and the terms on which it is granted, can be regulated by a collective or workforce agreement. If no such agreement is in place, the rest break will be for an uninterrupted period of not less than 20 minutes and the worker will be entitled to spend that break away from his/her workstation, if he/she has one. The current guidance is that the employer can determine the timing of the break, but it must be offered during the shift and not at the beginning or end of it.

The essence of a rest break is that a worker knows at the start of it that he or she has 20 minutes free from work to do with as he or she pleases. A period of 'downtime' during working time therefore cannot constitute a rest break, nor could, for example, requiring workers to eat their lunch at their desks. In *Gallagher v Alpha Catering Services Ltd* [2005] IRLR 102 the Court of Appeal ruled that drivers and loaders of a catering business had not been given adequate breaks, despite the fact that they had several periods of uninterrupted downtime during the day, as they were required to be on duty and await instructions during that time. ACAS is monitoring this issue (see **W10003** – Mobile Workers).

Derogations from the entitlement to rest breaks include cases specified in *reg 21* of the Regulations (see above), where working time is unmeasured and cannot be predetermined [*reg 20(1)*], and also by means of a collective or workforce agreement [*reg 23*]. As before, compensatory rest must be provided other than where *reg 20* applies [*reg 24*].

In another case, Hughes v The Corps of Commissionaires Management Ltd (UKEAT/0173/10/SM) (22 November 2010, unreported), EAT, the claimant was a security guard who worked 12 hour shifts on his own. He could take his 20 minute rest breaks whenever he wanted but it could not be guaranteed that they would be uninterrupted breaks, which was the subject of his original complaint. He claimed that the respondent was in breach of reg 24, a claim which was rejected by the Tribunal, and when it was appealed, the matter was referred to a second Tribunal, which agreed with the original ruling. The present case is a second appeal by the claimant. The second Tribunal looked at both reg 24(a) and 24(b), but considered reg 24(b) as the most relevant issue:

> "(b) in exceptional cases in which it is not possible, for objective reasons, to grant such a period of rest, his employer shall afford him such protection as may be appropriate in order to safeguard the worker's health and safety."

The Tribunal came to the conclusion that, on the facts, the respondent had afforded the claimant with appropriate protection in order to safeguard his health and safety. They took account in particular of the fact that he was afforded breaks, and that although he was on call during them and could be called, he was also allowed to start his break again if that happened. In short, the Tribunal found that the claimant was afforded rest during each shift but rejected the notion that it was 'compensatory rest', regarding the definition of compensatory rest as requiring to have all the features of a 'Gallagher' rest break.

The EAT agreed with the Tribunal, although their approach to the interpretation of the Regulations was different in that they considered that the case fell within *reg 24(a)*, not *24(b)*:

> "(a) his employer shall wherever possible allow him to take an equivalent period of compensatory rest, and..."

The Court of Appeal dismissed the employee's further appeal ([2011] IRLR 915). The Court agreed with the approach of the EAT and emphasised that compensatory rest need not be, in every sense, a 'Gallagher' rest break. What is required was a break as near as possible to a 'Gallagher' break, applying the language of equivalence and compensation.

In this context, the decision is helpful to employers because it:

- confirmed that a period of compensatory rest does not need to have exactly the same characteristics as a rest break under the Working Time Regulations, provided that it gives a break from work and, so far as possible, lasts for at least 20 minutes; and
- it acknowledged that cost was an inevitable influence in an employer's assessment of possible alterations to working arrangements so as to provide a rest break. In this case, Mr Hughes worked on his own and it could not be guaranteed that his break would be uninterrupted.

The Court also confirmed that when considering if a worker is a "special case" worker (and therefore excepted from the entitlement to daily and weekly rest), it is the specific worker's activities that have to be considered, not those of the employer. Therefore, if a worker is engaged in continuous surveillance, but working arrangements can be made to enable the worker to carry out their duties without a permanent presence, they would not fall under the category of a "special case".

Monotonous work

[W10018] The *Working Time Regulations 1998 (SI 1998 No 1833)*, *reg 8* provides that where the pattern according to which an employer organises work is such as to put the health and safety of a worker employed by him at risk, in particular because the work is monotonous or the work rate is predetermined, the employer shall ensure that the employee is given adequate rest breaks. This Regulation is phrased in virtually identical terms to Article 13 of the Working Time Directive and unfortunately, its incorporation into the Regulations has not clarified its meaning in any way!

It is not clear how rest breaks in *reg 8* would differ from those under *reg 12*. It may be that in the case of monotonous work, an employer might have to consider giving employees shorter breaks more frequently as opposed to one longer continuous break. This was the suggestion in the Government's consultative document.

No derogations from these provisions apply except in the case of domestic workers.

Night work

Definitions

[W10019] Before considering the detailed provisions in relation to night work contained in *reg 6*, an understanding of the definitions of 'night time' and 'night worker' contained in *reg 2* of the Regulations is necessary. These are as follows.

Night time in relation to a worker means a period:

(a) the duration of which is not less than 7 hours; and
(b) which includes the period between midnight and 5am,

which is determined for the purposes of the Regulations by a relevant agreement or, in the absence of such an agreement, the period between 11pm and 6am.

Night worker means a worker:

(i) who *as a normal course* works at least three hours of his daily working time during night time (for the purpose of this definition, it is stated in the Regulations that (without prejudice to the generality of that expression) a person works hours 'as a normal course' if he works such hours on a majority of days on which he works – a person who performs night work as part of a rotating shift pattern may also be covered); or
(ii) who is likely during night time to work at least such proportion of his annual working time as may be specified for the purposes of the Regulations in a collective or workforce agreement.

The question of when someone works at least 3 hours of his daily working time 'as a normal course' was addressed in the case of *R v A-G for Northern Ireland, ex p Burns* [1999] IRLR 315. In this case the Northern Ireland High Court held that a worker who spent one week in each 3-week cycle working at least 3 hours during the night was a 'night worker'. The High Court said that 'as a normal course' meant nothing more than as a regular feature. The guidance on the GOV.UK website is not very helpful in this respect as it is very loose and general, merely stating that night work will apply if its likely that they will work a proportion of their annual working time during the night.

Length of night work

[W10020] If a worker falls within these definitions, then he or she is a night worker for the purposes of the Regulations. The *Working Time Regulations 1998 (SI 1998 No 1833), reg 6* then goes on to provide that an employer shall take all reasonable steps to ensure that the normal working hours of a night worker do not exceed an average of 8 hours in any 24-hour period. This is averaged over a 17-week reference period which is calculated in the same way as in *reg 4* (see **W10011** above).

As with the maximum working week, there is a formula for calculating a night worker's average normal hours for each 24-hour period as follows:

$$\frac{a}{b-c}$$

where:

a is the normal (not actual) working hours during the reference period;

b is the number of 24-hour periods during the applicable reference period; and

c is the number of hours during that period which comprise or are included in weekly rest periods under *reg 11*, which is then divided by 24.

Although the Regulations originally excluded night shift overtime hours from those which count towards 'normal' hours, this provision has been repealed. It is therefore possible for overtime to count towards 'normal hours' where it is regularly worked and forms part of a night worker's normal hours of work.

If the night work involves 'special hazards or heavy physical or mental strain', then a strict eight-hour time limit is imposed on working time in each 24-hour period and no averaging is allowed over a reference period. The identification of night work with such characteristics is by means of either a collective or workforce agreement which takes account of the specific effects and hazards of night work, or by the risk assessment which all employers are required to carry out under the *Management of Health and Safety at Work Regulations 1999 (SI 1999 No 3242)*. (see Risk Assessment).

The derogations in *reg 21* of the Regulations (special categories of workers) apply to the provisions on length of night work. Workers whose working time is unmeasured or cannot be predetermined (and where it is only partly unmeasured or predetermined, to the extent that it is unmeasured) are also excluded. [*Working Time Regulations 1998 (SI 1998/1833), reg 20*]. Other exemptions may be made by means of collective or workforce agreement. [*Working Time Regulations 1998 (SI 1998/1833), reg 23*].

Health assessment and transfer of night workers to day work

[W10021] An employer must, before assigning a worker to night work, provide him with the opportunity to have a free health assessment. [*Working Time Regulations 1998 (SI 1998 No 1833), reg 7*]. The purpose of the assessment is to determine whether the worker is fit to undertake the night work. While there is no reliable evidence as to any specific health factor which rules out night work, a number of medical conditions could arise or could be made worse by working at night, such as diabetes, cardiovascular conditions or gastric intestinal disorders.

Employers are under a further duty to ensure that each night worker has the opportunity to have such health assessments 'at regular intervals of whatever duration may be appropriate in his case'. [*Working Time Regulations 1998 (SI 1998 No 1833), reg 7(1)(b)*].

The Regulations do not specify the way in which the health assessment must be carried out, nor is there specific reference to medical assessments, so that strictly speaking, such assessments could be carried out by qualified health professionals rather than by a medical practitioner.

Contrast this with the position as regards the transfer from night to day work. If a night worker is found to be suffering from health problems that are recognised as being connected with night work, he or she is entitled to be transferred to suitable day work 'where it is possible'. In such a situation, the Regulations provide that a 'registered medical practitioner' must have advised the employer that the worker is suffering from health problems which the practitioner considers to be connected with the performance by that night worker of night work.

Annual leave

Entitlement to annual leave

[W10022] The right under the Regulations to paid holiday was the first time under English law that workers were given a statutory right to paid holiday. The Regulations originally provided for an entitlement to 4 weeks' paid holiday each year, which included leave taken on bank and public holidays. Although many employers developed a practice of giving workers paid leave on bank and public holidays in addition to the 4-week minimum, this disadvantaged those workers who received only the basic statutory entitlement.

The entitlement has now been increased to give all workers (subject to some exceptions) the right to 5.6 weeks' paid holiday each year, which equates to 28 days for individuals working five days a week. This comprises a 4-week basic entitlement [*Working Time Regulations 1998 (SI 1998 No 1833), reg 13*] and an additional entitlement of 1.6 weeks [*Working Time Regulations 1998 (SI 1998 No 1833), reg 13A*]. The statutory entitlement is subject to a statutory cap of 28 days for all workers [*Working Time Regulations 1998 (SI 1998 No 1833), reg 13A(3)*].

The purpose behind the increase was to give all workers the equivalent of eight bank and public holidays in addition to the existing entitlement of 4 weeks' paid holiday each year. However, the additional entitlement does not have to be given or taken on bank or public holidays and, as such, there is no statutory right to time off on bank or public holidays.

To ease the cost burden on employers, the increased entitlement was introduced in phases. On 1 October 2007, the statutory minimum entitlement increased from 4 to 4.8 weeks each year and, from 1 April 2009, this further increased to 5.6 weeks per year.

The right to paid annual leave begins on the first day of employment. Although the Regulations originally provided for a 13-week qualifying period for annual leave, this was ruled unlawful by the CJEU in *R (on the application of the Broadcasting, Entertainment, Cinematographic and Theatre Union) v Secretary of State for Trade and Industry*: C-173/99 [2001] IRLR 559 (ECJ) and was therefore removed.

Where a worker began work after 25 October 2001, the employer has the option to use an accrual system during the first year of employment. The amount of leave which the worker may take builds up monthly in advance at the rate of one twelfth of the annual entitlement each month. If the calculation does not produce a whole number the accrued leave entitlement is rounded up either to a half day (if the calculation is less than a half day) or a whole day (if the calculation exceeds a half day). Any rounded up element of leave taken would, of course, be deducted from the worker's annual entitlement to paid leave.

For the purposes of *regs 13* and *13A* of the Regulations, a worker's leave year begins on any day provided for in a relevant agreement, or, where there is no such provision in a relevant agreement, 1 October 1998 or the anniversary of the date on which the worker began employment (whichever is the later). In the majority of cases, written statements of terms and conditions contain a reference to the holiday or leave year. If they do not, employers are recommended to ensure that they do so. Failure in this regard could result in administrative confusion as each employee could have a different holiday year for the purposes of calculations under the Regulations.

The statutory leave entitlement may be taken in instalments, but the 4-week entitlement under *reg 13* can only be taken in the leave year to which it relates and a payment in lieu cannot be made except where the worker's employment is terminated (but see comments at **W10024** below). In relation to the additional entitlement under *reg 13A*, a relevant agreement may provide for this portion of the leave entitlement to be carried forward but only to the next leave year. There was a limited right to make payment in lieu of the additional entitlement between 1 October 2007 and 1 April 2009, but this is now gone and no payment in lieu is permitted except where the worker's employment is terminated.

These rules relate only to the statutory entitlement under the Regulations, so that if an employer provides for annual leave over and above the statutory entitlement, the enhanced element of the holiday can be carried forward or paid for in lieu as agreed between the parties. In *Miah v La Gondola Ltd* (1999), unreported, the Employment Tribunal found that a worker who was paid an additional week's pay instead of taking a week's holiday (due under the Regulations) during his employment, was entitled, on the termination of that contract, to payment for that proportion of annual leave which was due to him in accordance with *reg 14(3)(b)* and the payment paid to him during his employment in lieu of the week's statutory holiday had to be disregarded.

For many workers (and employers), the introduction of rights to paid annual leave was significant. It is understood that before the Regulations came into force some 2.5 million workers had no right to a holiday at all. Employment Tribunals have had to resolve many disputes arising from this right.

Compensation related to entitlement to annual leave

[W10023] Where an employee has outstanding leave due to him when the employment relationship ends, *reg 14* of the Regulations specifically provides that an allowance is payable in lieu. It states that in the absence of any relevant agreement to the contrary, the amount of such allowance will be determined by the formula

$$(a \times b) - c$$

where:

(a) is the period of leave to which the worker is entitled under the Regulations;

(b) is the proportion of the worker's leave year which expired before the effective date of termination;

(c) is the period of leave taken by the worker between the start of the leave year and the effective date of termination [*reg 14(3)*].

Regulation 14(4) provides that where there is a relevant agreement which so provides, an employee shall compensate his employer – whether by way of payment, additional work or otherwise – in relation to any holiday entitlement taken in excess of the statutory entitlement.

Interestingly in the case of *Witley & District Mens Club v Mackay* [2001] IRLR 595 the EAT held that where a relevant agreement (in this case a collective agreement) purported to allow an employer to dismiss an employee for gross misconduct without payment for any accrued holiday pay this provision would be void as *reg 14* specifically provides for a sum to be paid. It is open to debate whether a provision of a relevant agreement which permitted the payment of a notional sum (say, £1) on termination in the event of gross misconduct would be valid.

Payment, and notice requirements, for annual leave

[W10024] The *Working Time Regulations 1998 (SI 1998/1833), reg 16* specifies the way in which a worker is paid in respect of any period of annual leave: this is at the rate of a week's pay in respect of each week of leave.

In order to calculate the amount of a 'week's pay', employers are referred to *ss 221–224* of the *Employment Rights Act 1996* (although the calculation date is to be treated as the first day of the period of leave in question, and the relevant references to *ss 227* and *228* do not apply). The application of these provisions can lead to a disproportionately high level of holiday pay for workers paid on a commission only basis. This is because *s 223(2)* stipulates that if, during any of the 12 weeks preceding the worker's holiday, no remuneration was payable by the employer to the worker concerned, account should be taken of remuneration in earlier weeks so as to bring the number of weeks of which account is taken up to 12. In other words, it seems that the calculation of a 'commission only' worker's holiday entitlement will be based solely on the worker's earnings for any weeks (up to a maximum of 12) in which they received commission payments-any slow, unproductive weeks will be disregarded. Where commission is paid in addition to a base salary, the commission payments are normally excluded from the calculation of holiday pay: see *Evans v Malley Organisation Ltd (t/a First Business Support)* [2003] IRLR 156. However, in *Herring v Co-operative Insurance Society Ltd* [2006] All ER (D) 171 (Oct), the EAT held that it was at least arguable that insurance brokers who received a small base salary plus commission should have been paid projected commission during annual leave.

Whether bonus payments should be included as part of the holiday pay will depend on the nature of the bonus and whether it can properly be said that pay varies with the amount of work done: see *Adshead v May Gurney Ltd* [2006] All ER (D) 388 (Jul), EAT (UKEAT/0150/06). Overtime payments will not usually be counted towards a week's pay, unless the overtime is fixed under the employment contract: *Bamsey v Albon Engineering and Manufacturing plc* [2004] EWCA Civ 359, [2004] IRLR 457. In the case of *Smith v Chubb Security Personnel Ltd* (1999), unreported, employees who were obliged to work such hours as were detailed in their duty roster were paid holiday pay on the basis of a 40-hour week when in fact they worked a 3-week cycle of fourteen 12 hour days which averaged out at 56 hours per week. The Tribunal decided that their pay should be calculated on the basis of a 56-hour week.

The issue of commission payments in holiday pay calculations will hopefully soon be clarified by the Court of Appeal. British Gas's appeal from the EAT decision in *British Gas Trading Ltd v Lock* [2016] IRLR 316 was heard on 11 July 2016, and the much-anticipated judgment is expected before the end of the year. If it is held that commission should be included in holiday pay, this would have far-reaching effects. British Gas's representatives have stated that it has 1,000 potential claims hinging on the outcome of this case. Indeed, the 2004 Findings from the Workplace Employment Relations Survey found that 40% of all British workplaces had incentive pay schemes, payments which have often not been included in holiday pay.

The *Working Time Regulations 1998 (SI 1998 No 1833), reg 15* deals with the dates on which the leave entitlement can be taken, and sets out detailed requirements as to notice. This Regulation is extremely complicated and before going into the details, it should be noted that alternative provisions concerning the notice requirements can be contained in a relevant agreement. Bearing in mind the detailed nature of the provisions, it is recommended that employers consider specifying such notice requirements in their contracts of employment, thereby avoiding the need for confusion at a later stage.

Regulation 15 provides that a worker must give his or her employer notice equivalent to twice the amount of leave he or she is proposing to take. An employer can then prevent the worker from taking the leave on a particular date by giving notice equivalent to the number of days' leave which the employer wishes to prohibit. An employer can also, by giving notice equivalent to twice the number of days' leave in question, require an employee to take all or part of his or her leave on certain dates. This would, of course, be useful with regard to seasonal shutdowns over summer and Christmas holidays etc. An employee has no right to serve a counter notice and so, provided the employer has given proper notice, the employee must take the holiday as required. The Regulations do not require any notice to be given in writing. However, given the level of detail that must be included in any notice, employers would be advised to give the notice in writing and require their employees to do the same. This would also ensure that employers have accurate records of required holiday periods.

It is open for an employer to require a worker to take leave in single days rather than in a block. In addition, any day upon which a worker could be required to work can be designated by the employer as leave, even if the worker would not necessarily be required to work that day. In *Sumsion v BBC Scotland* [2007] IRLR 678, the EAT held that an employer could require a worker to take annual leave every second Saturday where the worker's contract required him to be available for work for up to 6 days per week. This was notwithstanding that work was rarely done on Saturdays, as the worker was required to be available. See also *Russell and Others v Transocean International Resources Ltd and Others* [2010] CSIH 82 where the Scottish Court of Session ruled that time off spent onshore by offshore workers as part of a rotating shift could constitute annual leave.

The employer's right to require a worker to take leave at a particular time is restricted during periods of maternity leave. In *Gomez v Continental Industrias Del Caucho S*: C-342/01, [2004] IRLR 407, the CJEU ruled that an employer was required to allow an employee to take her statutory leave entitlement at the end of her maternity leave, even though this was outside the agreed time for taking annual leave contained in a collective agreement.

Employers who give more than the statutory minimum entitlement and who are contractually obliged to pay their workers more for the statutory entitlement than they pay for the remaining leave will be interested in *Barton v SCC Ltd* (1999), unreported. In this case, an employer sought to argue that monies paid by way of holiday pay for leave in excess of the statutory minimum leave under the Regulations can go towards the employer's liability to pay a week's pay for each week of leave taken under the Regulations (calculated under the *Working Time Regulations 1998 (SI 1998 No 1833), reg 16*). In addition, the employer argued that they had the right to choose which days off would be paid at the statutory rate and which at the contractual rate. The Employment Tribunal found in favour of the workers on both issues. *Regulations 16(1)* and *(5)* make it clear that the Employment Tribunal has to focus on the individual holiday week in question and the payment made for that individual week. With regard to which holiday is paid at which rate, the Employment Tribunal said that the right to holiday pay is an entitlement and it is a matter for the worker to choose which holiday weeks he or she seeks to take pursuant to his or her statutory entitlement.

Rolled-up holiday pay

[W10024A] Historically, there were conflicting decisions in the UK as to whether it is permissible to roll-up holiday pay into the worker's hourly rate (see for example, *MPB Structures Ltd v Munro* [2003] IRLR 350 and *Caulfield v Marshalls Clay Products Ltd* [2004] EWCA Civ 422, [2004] IRLR 564). However, the CJEU in *Robinson-Steele v R D Retail Services Ltd*: C-131/104[2006] IRLR 386 ruled that rolled-up holiday pay is incompatible with the Directive. The CJEU reasoned that the purpose of holiday pay is to put the worker in the same position, as regards pay, as he or she would have been if he or she were at work, so the worker should receive his/her pay at the time he/she takes the leave. Accordingly, it is now clear that the practice of rolling up holiday pay is unlawful and employers must pay statutory holiday pay at the time leave is actually taken.

The CJEU in *Robinson-Steele* did go on to state that payments already made to a worker under genuine rolled up holiday arrangements could be offset against the worker's entitlement to pay during leave, provided the payments had been paid transparently and comprehensibly as holiday pay. Workers who historically benefited from rolled-up holiday pay could not, therefore, gain a windfall by claiming additional holiday pay when they took their leave. For the amounts to be off-set, there would have to have been a clear agreement that part of the hourly rate is attributable to holiday pay which is set out in the employment contract or the worker's payslips: see for example *Lyddon v Englefield Brickwork Limited* [2008] IRLR 198 and *Smith v J Morrisroes and Sons Limited* [2005] IRLR 72.

However, there is some uncertainty as to whether employers who continue to operate rolled-up holiday pay arrangements can benefit from this protection. Whilst the CJEU allowed genuine, historical rolled-up holiday payments to be offset against the employer's obligation to pay holiday pay, it also ruled that member states must take appropriate measures to ensure that practices such as rolled-up holiday pay are not continued. Following the ruling, the UK Government indicated that what was intended by the CJEU was to allow a transitional period for employers who had previously operated genuine rolled-up holiday pay arrangements to make alternative arrangements. Unfortunately no timeframe was provided, but given the time that has elapsed since the CJEU's judgment, operating a rolled-up holiday pay arrangement is fraught with risk and employers should therefore avoid it. From a practical point of view, this is likely to be very difficult for employers who engage casual workers or operate in industries with irregular or unpredictable working patterns, where rolled up holiday pay has become commonplace.

Holidays and sickness absence

[W10024B] An area which has caused great difficulty under the Regulations is the question of a worker's entitlement to annual leave while on sick leave. Does a worker continue to accrue statutory holiday entitlement whilst on sick leave? Can the worker take annual leave during a period that he or she is already on sick leave? What happens to annual leave which has not been taken due to periods of sickness absence on termination of employment?

In *Kigass Aero Components Ltd v Brown* ([2002] ICR 697 (EAT/481/00)) the EAT held that workers accrue, and are entitled to take, their paid annual leave under the Regulations even if they are on sick leave during that period provided that they comply with the notification requirements under the Regulations. The EAT also ruled that where employment has been terminated while the worker is on sick leave, the worker is entitled to claim a payment in lieu of unused statutory leave. The *Kigass* decision was initially overruled by the Court of Appeal in *Ainsworth v IRC* [2005] EWCA Civ 441, [2005] IRLR 465 but later reinstated by the then House of Lords in the same case which became known as *Revenue and Customs v Stringer* [2009] UKHL 31, [2009] 4 All ER 1205, [2009] ICR 985.

The claimants in *Ainsworth/Stringer* were all on long term sickness absence and were no longer receiving pay as they had exhausted their contractual and statutory sick pay entitlements. One of the claimants sought to designate part of her sickness absence as holiday and claimed holiday pay whilst still employed. Four other claimants had been dismissed whilst absent on unpaid sick leave and sought payment in lieu of their accrued annual leave entitlements. The employment tribunals and the EAT allowed the claims. However, on appeal, the Court of Appeal overturned the EAT, ruling that a worker on long-term sick leave is not entitled under *reg 13* to 4 weeks' paid annual leave in a year when he has not been able to attend for work. According to the Court, it followed from this that a worker whose employment was terminated during a year when he is absent from work on long-term sick leave was not entitled to compensation under *reg 14* for unused leave. The case was further appealed to the House of Lords, which referred various questions to the CJEU for determination.

The CJEU confirmed in Stringer and the combined case of *Schultz-Hoff v Deutsche Rentenversicherung Bund*: C-350/06 [2009] All ER (EC) 906, [2009] IRLR 214 that, under the Directive, a worker continues to accrue annual leave during a period of sickness absence. A requirement that the worker must actually have performed some work to qualify for annual leave would be contrary to the Directive. The CJEU went on to say that workers on long-term sick leave should not lose the opportunity to take their annual leave under the Directive. Accordingly, member states must either allow them to take it during their sick leave or at a later time when they are well enough, even if this means carrying it forward to a subsequent holiday year. When Stringer returned to the House of Lords, the parties agreed that the decision of the EAT should be reinstated. As a result, their Lordships did not address the issues relating to sickness absence and annual leave in any detail.

The case makes it clear that workers accrue their statutory annual leave entitlement during any period of sickness absence, even if this lasts for a full holiday year or more. Workers must also be allowed to take holiday, at full pay, even if they are already on sick leave and any entitlement to sick pay has run out. In addition, if a worker's employment ends during a period of sickness absence, they should be paid for any unused statutory annual leave for that leave year, even if they were absent for all or part of the leave year.

In practice, the ruling creates complications for workers in receipt of long-term sickness benefits, such as permanent health insurance (PHI), which typically pay part of the worker's salary. It is not clear, for example, whether workers who take annual leave while in receipt of PHI are entitled to full holiday pay on top of the PHI payments or whether it is possible to "top up" the difference between full salary and benefits received. Depending on the terms of the PHI scheme, taking holiday whilst in receipt of PHI payments may also cause the benefits to cease if the worker is no longer deemed under the scheme to be incapacitated.

It is also remains unclear whether workers in the UK who are ill during all or part of a holiday year are entitled to carry forward statutory annual leave. The *Working Time Regulations 1998 (SI 1998 No 1833), reg 13(9)* does not permit statutory annual leave to be carried forward to a subsequent leave year (apart from the additional 1.6 weeks' leave under *reg 13A*). However, in *Stringer* and *Schultz-Hoff* the CJEU said that it would be contrary to the Directive for a worker's annual leave entitlement to be extinguished in circumstances where they have not had the opportunity to take it due to illness. It is arguable that the Regulations may not comply with the Directive in this regard and it remains to be seen whether *reg 13(9)* can be interpreted purposively to allow for carry over of annual leave by workers who are prevented from taking it due to illness (as has happened in the case of *Shah v First West Yorkshire Limited* (ET/1809311/2009) – see below). In the meantime, the safest course for employers would appear to be to allow workers to take annual leave during a period of sickness absence.

The CJEU has also considered the question of whether workers who fall sick during their annual leave are entitled to reschedule their annual leave for another time. In the Spanish case of *Pereda v Madrid Movilidad SA*: C-227/08 [2010] 1 CMLR 103, [2009] IRLR 959, the claimant suffered an injury at work which mean he was unable to take prearranged holiday. The CJEU ruled that a worker in this situation who falls ill prior to the start of previously scheduled annual leave has the right to take his or her annual leave at a later period when he or she has recovered. This is so even if it means allowing the rescheduled leave to be taken in a subsequent leave year. In *Asociacion Nacional de Grandes Empresas de Distribucion (ANGED) v Federacion de Asociaciones Sindicales (FASGA)*: C-78/11, [2012] IRLR 779, the CJEU confirmed that Pereda applies equally to a situation where the worker falls ill during a period of annual leave. Accordingly, an employee who falls ill during annual leave is entitled to ask for that period to be treated instead as sick leave and to have 'replacement' annual leave at another time. The *Pereda* ruling was applied in the UK in *Shah v First West Yorkshire Limited* (ET/1809311/2009). In that case, an Employment Tribunal implied into the Regulations an exception to the 'use it or lose it' principle in *reg 13(9)* for workers who are prevented by illness from taking their holiday within the relevant leave year; in which case, they must be given the opportunity of taking that holiday in the following leave year.

Following *Pereda*, it is not clear whether the notice provisions in *reg 15(2)* can be used to require workers on sick leave to take their annual leave before the end of the holiday year. Arguably an employer can seek to rely on *reg 15(2)* but this will be subject to the worker's right to request to take the annual leave at a later time.

In terms of remedies, the House of Lords ruled in *Stringer* that non-payment of holiday pay gives a worker a right to claim unlawful deduction of wages under *section 13* of the *Employment Rights Act 1996*. In this regard, their Lordships overruled the Court of Appeal which had held previously that such a claim was not possible. A worker who is denied their holiday pay can therefore choose between a claim under the *Working Time Regulations 1998* (SI 1998 No 1833), *reg 30* or *section 13* of the *Employment Rights Act 1996*.

One consequence of this is that a worker on long-term sick leave, who has been denied their annual leave over several years, can potentially claim back-pay for previous years' annual leave. The failure to pay holiday pay is likely to be seen as a series of deductions and any claim would only need to be brought within 3 months of the last in the series of deductions (*section 23* of the *Employment Rights Act 1996*). By contrast, a claim for holiday pay under the Regulations must be brought within 3 months of the end of the holiday year and can only relate to the most recent holiday year (see **W10040** below).

There was initially conflicting authority on the issue of whether the worker must have either taken or requested annual leave before a claim for unlawful deductions can be made. In *Kigass Aero Components Ltd v Brown* [2002] IRLR 312 the EAT said that a worker must ask to take their leave before the entitlement to holiday pay arises. However, in *List Design Group Ltd v Douglas* [2003] IRLR 14 the EAT took a different view, holding that a worker can bring an unlawful deductions claim for unpaid statutory holiday regardless of whether they have actually taken or requested leave.

It is now clear, following the decision of the Court of Appeal in *NHS Leeds v Larner* [2012] EWCA Civ 1034, [2012] IRLR 825 that *Kigass* is no longer correct and that it is not necessary for a worker to seek to exercise the right to take annual leave in order to preserve it (an interpretation subsequently applied by the EAT in *Plumb v Duncan Print Group Ltd* [2015] IRLR 711, which also introduced an 18-month limit on holiday pay carry over). The practical effect of the decisions in *Larner* and *Plumb* is that if a worker, whether employed by a public or private employer, is unable or unwilling to take the four weeks' annual leave conferred by *reg 13* owing to sickness, the worker must be allowed to take it for a period of 18 months from the end of the leave year in which the leave arose.

Working time and flexible working, including home working

[W10024C] There is a growing trend for workers to either work from home for part (or all) of their working time, or to agree a more flexible approach to meeting their contractual working hours that also fits in with personal obligations or preferences. Whilst there are many mutually advantageous benefits to an effective flexible working agreement it can cause complications in monitoring working hours, rest breaks and ensuring the agreed hours are indeed being worked. Employers need to ensure that any agreements are formally agreed and understood, and that mechanisms are in place to monitor that arrangements are within the Regulations, unless the worker has formally opted out. Most individuals with employee status who have at least 26 weeks' continuous service with their employer may make one flexible working application to their employer in any given 12-month period to be allowed to vary their terms and conditions of employment. Requests can be made to:

- change the hours the employee is required to work;
- change the times when the employee is required to work;
- change where the employee is required to work.

Other types of requested contractual change, eg an application to change duties, do not fall within the statutory scheme.

Once a valid application has been submitted, the employer:

- is obliged to deal with it in a 'reasonable manner';
- is obliged to notify the employer of its decision on the application within a period called the 'decision period';
- is only entitled to refuse the application if it considers that one or more of certain defined grounds for refusal applies.

The employee may bring a claim to the employment tribunal if his employer fails in those obligations when dealing with the request. Note however that no employee has the right to insist that the employer accept his request to alter his terms and conditions.

Any such agreement should include arrangements for ensuring that flexible working arrangements do not exceed the regulations, and how this is monitored.

In December 2014 the Department for Business Innovation and Skills (BIS) produced a research paper (No 5) on the effects of working time titled 'The Impact of the Working Time Regulations on the UK labour market: A review of evidence' (BIS/14/1287). Relating to flexible working the conclusion was that technology changes since the 1990s have significantly changed the way people work. Increasingly UK employers are offering remote working arrangements which allow employees to work all or some part of their working time from home or at different times. In addition, the increasing availability of mobile internet means many employees work on public transport during commutes to work or meetings. Flexible working patterns mean that the traditional concept of a continuous working day does not apply to many workers and employees are increasingly fitting work around their private commitments. These developments give employees more control over their own working time patterns and enable them to use their time more efficiently. However, it does mean the concept of 'working time' has changed. Employers need to be cognisant of such changes that are developing when considering their policies and practices, to ensure continued compliance with the legislation.

Working time: records

[W10025] The publicity and controversy surrounding the introduction of the Regulations related primarily to the maximum working week and the annual leave entitlement. However, in practice the provisions relating to record keeping proved very significant for employers, in particular those with workers who had signed opt-outs, where the rigorous record keeping requirements arguably undermined the benefit to an employer of the freedom given by the opt-out. Accordingly, the UK Government significantly watered down the requirements for employers *vis a vis* opted out workers, and the current position is outlined below.

The *Working Time Regulations 1998 (SI 1998 No 1833), reg 9* introduces an obligation on employers to keep specific records of working hours. All employers (except in relation to workers serving in the armed forces – *reg 25(1)*) are under a duty to keep records which are adequate to show that certain specified limits are being complied with in the case of each entitled worker employed by him. These limits are:

(a) the maximum working week (see **W10010** above);
(b) the length of night work (including night work which is hazardous or subject to heavy mental strain) (see **W10020** above); and

(c) the requirement to provide health assessments for night workers (see **W10021** above).

These records are required to be maintained for two years from the date on which they were made.

In addition, where a worker has agreed to exclude the maximum working week under *reg 5* of the Regulations (see **W10013** above), an employer is required to keep up-to-date records of all workers who have opted out, which must also be kept for two years from the date on which they were made.

Excluded sectors

[W10026] Prior to 1 August 2003, the following sectors of activities (listed below) were excluded from the terms of the Regulations by *reg 18*:

(a) air, rail, road, sea, inland waterway and lake transport;
(b) sea fishing;
(c) other work at sea;
(d) the activities of doctors in training.

Following the adoption of five Directives aimed at extending the working time provisions to previously excluded sectors, this is no longer the case and most of the above sectors have gained some protection under the Regulations. The Directives are:

- Horizontal Amending Directive.
- Road Transport Directive.
- Seafarers' Directive (outside the scope of this chapter).
- Seafarers' Enforcement Directive (outside the scope of this chapter).
- Aviation Directive (outside the scope of this chapter).

The Horizontal Amending Directive was implemented by the UK with effect from 1 August 2003 *Working Time (Amendment) Regulations 2003 (SI 2003 No 1684)*.

The Road Transport Directive was implemented in the UK with effect from 4 April 2005 (see the *Road Transport (Working Time) Regulations 2005 (SI 2005 No 639)*.

Application of the new directives to previously excluded sectors

Mobile and non-mobile workers

[W10027] A mobile worker is defined in the *Working Time Regulations 1998 (SI 1998 No 1833)*, reg 2 as: 'any worker employed as a member of travelling or flying personnel by an undertaking which operates transport services for passengers or goods by road or air'. Non-mobile workers are undefined and are therefore any worker who is not a mobile worker, for example, all office based staff, warehouse staff etc. Unfortunately this does not address the situation where a worker carries out a combination of mobile and non-mobile activities (for example, a worker who spends part of his time based in a depot loading goods onto lorries and the remainder of his time driving an HGV delivering the goods.)

Road transport sector

Non-mobile workers

[W10028] Following the implementation of the *Working Time (Amendment) Regulations 2003 (SI 2003 No 1684)* on 1 August 2003, the Regulations were extended in full to non-mobile workers in the road transport sector.

Mobile workers

[W10029] Mobile workers who are not covered by the Road Transport Directive:

From 1 August 2003, mobile workers who are *not* covered by the Road Transport Directive (such as drivers of vehicles of less than 3.5 tonnes) are entitled to the following provisions of the Regulations:

(a) the average 48-hour working week (under *reg 4* of the Regulations – see **W10010** above);
(b) paid holiday (under *regs 13* and *13A* of the Regulations – see – see **W10022** above);
(c) appropriate health checks if they are night workers (under *reg 7* of the Regulations – see **W10021** above); and
(d) provision for adequate rest (unless there are exceptional circumstances as provided for in *reg 21(e)* of the Regulations *Working Time Regulations 1998) (SI 1998 No 1833), Regulation 24A(2)*].

Regulation 24A of the Regulations provides that they are not entitled to the daily rest requirements set out in *reg 10(1)* or the weekly rest periods set out in *reg 11(1)* and *(2)*. In addition, they are not entitled to the rest breaks set out in *reg 12(1)* or the limit on hours of work for night workers' in *reg 6(1)*. Further, the exemptions for special categories of workers may apply (see **W10034** below).

Mobile workers who are covered by the Road Transport Directive:

From 1 August 2003, mobile workers who are covered by the Road Transport Directive (mainly drivers of vehicles of over 3.5 tonnes) are entitled to the following provisions under the Regulations:

(a) paid holiday (under *regs 13* and *13A* of the Regulations – see **W10022** above); and
(b) appropriate health checks if they are night workers (under *reg 7* of the Regulations – see **W10021** above).

In addition, mobile workers who are covered by the Road Transport Directive are entitled under the *Road Transport (Working Time) Regulations 2005 (SI 2005 No 639)* to a maximum working week of 60 hours (provided always that the average working week over a 17-week period does not exceed 48 hours). Mobile worker's should not work for more than 6 hours without a break and are entitled to a 30 minute break where the worker's working time exceeds 6 hours but less than 9 hours and a 45 minute break where the worker's working time exceeds 9 hours. Night work is also restricted to 10 hours in any 24 hour period. Self-employed drivers are excluded from the ambit of the Road Transport Directive.

In the case of *R (oao United Road Transport Union) v Secretary of State for Transport* [2013] IRLR 890, the Union unsuccessfully challenged the refusal by the Secretary of State to extend the right of appeal to the Employment Tribunal to commercial road transport workers. The Union sought to rely on the EU law principles of equivalence and/or effectiveness to establish that commercial road transport workers were entitled to apply to the Tribunal for a remedy if required to work in contravention of EU regulations relating to breaks and rest periods during working hours. In rejecting these submissions, the crux of the Court of Appeal's reasoning was that rest breaks are not a right conferred upon the drivers but an obligation imposed for their health and safety.

Rail sector

[W10030] Following the implementation of the *Working Time (Amendment) Regulations 2003 (SI 2003 No 1684)* on 1 August 2003, the Regulations are, in principle, extended in full to all workers in the rail transport sector.

It is, however, possible to derogate from the Regulations in respect of railway workers whose activities are intermittent, whose working time is spent on board trains or whose activities 'are linked to transport timetables and to ensuring the continuity and regularity of traffic'. For more information on these partial exemptions see **W10034** below.

Special provisions are made for workers on cross-border rail services by the *Cross-Border Railway Services (Working Time) Regulations 2008 (SI 2008 No 1660)*, which came into force on 27 July 2008. In practice, these regulations are only relevant for workers on rail services operating through the Channel Tunnel. The regulations provide more favourable rest breaks, daily and weekly rest periods and limits on drivers' hours, and can be enforced by individual claims to an Employment Tribunal. The *Working Time Regulations 1998 (SI 1998 No 1833)* still apply subject to these more favourable provisions, apart from *reg 24* of the Regulations which deals with compensatory rest.

Other sectors

[W10031] It should be noted that the Horizontal Amending Directive (implemented in the UK in the form of the *Working Time (Amendment) Regulations 2003 (SI 2003 No 1684)*) together with the other sector-specific directives detailed above provide different entitlements for mobile workers in the sea-fishing sector, aviation sector, inland waterway and lake transport sectors and also in relation to offshore working. In relation to offshore workers, reg 25B of the Regulations provides specific exclusions for workers employed in offshore work.

In addition, in relation to doctors in training the Horizontal Amending Directive provided for a gradual phasing in of the Regulations. From 1 August 2004, the Regulations have applied in full to doctors in training except that the limit on average weekly hours of work was implemented in stages (see **W10012** above).

Partial exemptions

Domestic service

[W10032] The following provisions of the Regulations do not apply in relation to workers employed as domestic servants in private households:

(a) the 48-hour working week (see **W10010** above);
(b) length of night work (see **W10020** above);
(c) health assessments for night workers (see **W10021** above);
(d) monotonous work (see **W10018** above).

The provisions regarding minimum daily and weekly rest periods and rest breaks, the requirement to keep records and annual leave do apply.

Unmeasured working time

[W10033] The requirements of the Regulations listed below are, by virtue of *reg 20(1)* of the Regulations, not applicable to workers where, on account of the specific characteristics of the activity in which they are engaged, the duration of their working time is not measured or pre-determined or can be determined by the workers themselves. The provisions excluded are:

(a) the 48-hour working week (see **W10010** above);
(b) minimum daily and weekly rest periods and rest breaks (see **W10014** above);
(c) length of night work (see **W10020** above).

The provisions regarding monotonous work, the requirement for health assessments for night workers, the requirement to keep records and annual leave do apply. It should also be noted that there is no requirement to provide such workers with compensatory rest where the requirements of the Regulations are not complied with.

This provision is often thought to exclude the Regulations in relation to 'managing executives or other persons with autonomous decision-making powers'. However, it should be noted that this is simply one of the three examples given in *reg 20(1)*, as regards the type of worker with unmeasured working time. The application of the regulation could be much wider than this, and will depend on whether the worker in question is genuinely able to control his work, to the extent of being able to determine how many hours he works.

In December 1999, the UK Government introduced *reg 20(2)* to give a partial exemption from the 48-hour maximum working week for those workers who had some part of their working time that was unmeasured. *Regulation 20(2)* provided that the working time of workers did not effectively count towards their weekly working time to the extent that it was unmeasured, even if their time was not wholly unmeasured. However, *reg 20(2)* was later repealed with effect from 6 April 2006, following a complaint by the European Commission that it went beyond the permitted derogation in the Directive. Workers are therefore either within the *reg 20* 'unmeasured' exemption or they are outside it.

Special categories of workers

[W10034] The *Working Time Regulations 1998 (SI 1998 No 1833), Regulation 21* provides that in the case of certain categories of employees, the following Regulations do not apply:

(a) length of night work (see **W10020** above);
(b) minimum daily rest and weekly rest periods and rest breaks (see **W10014** above).

This leaves the 48-hour working week, requirements relating to monotonous work, the requirement for health assessments for night workers, the requirement to keep records and annual leave. However, it should be noted that where any of the allowable derogations are utilised, an employer will have to provide compensatory rest periods (see **W10037** below).

Further, for these categories of workers, the reference period over which the 48-hour working week is averaged is 26 weeks and not 17 [*Working Time Regulations 1998 (SI 1998 No 1833), Regulation 4(5)*].

The special categories of worker are as follows:

(i) where the worker's activities mean that his place of work and place of residence are distant from one another, or the worker has different places of work which are distant from one another (including cases where the worker carries out offshore work);
(ii) workers engaged in security or surveillance activities, which require a permanent presence in order to protect property and persons – examples given are security guards and caretakers;
(iii) where the worker's activities involve the need for continuity of service or production – specific examples are:
    (A) services relating to the reception, treatment or care provided by hospitals or similar establishments (including the activities of doctors in training), residential institutions and prisons;
    (B) workers at docks or airports;
    (C) press, radio, television, cinematographic production, postal and telecommunications services, and civil protection services;
    (D) gas, water and electricity production, transmission and distribution, household refuse collection and incineration;
    (E) industries in which work cannot be interrupted on technical grounds;
    (F) research and development activities;
    (G) agriculture;
    (H) the carriage of passengers on regular urban transport services;
(iv) any industry where there is a foreseeable surge of activity – specific cases suggested are:
    (A) agriculture;
    (B) tourism;
    (C) postal services;
(v) where the worker's activities are affected by:
    (A) unusual and unforeseeable circumstances beyond the control of the employer;

（B） exceptional events which could not be avoided even with the exercise of all due care by the employer;
（C） an accident or the imminent risk of an accident;
(vi) where the worker works in railway transport and:
（A） his activities are intermittent; or
（B） he works on board trains; or
（C） his activities are linked to transport timetables and to ensuring the continuity and regularity of traffic.

Shift workers

[W10035] Shift workers are defined by the *Working Time Regulations 1998 (SI 1998/1833), reg 22* as workers who work in a system whereby they succeed each other at the same work station according to a certain pattern, including a rotating pattern which may be continuous or discontinuous, entailing the need for workers to work at different times over a given period of days or weeks.

In the case of shift workers the provisions for daily rest periods (see **W10014** above) and weekly rest periods (see **W10016** above) can be excluded in order to facilitate the changing of shifts, on the understanding that compensatory rest must be provided (see **W10037** below). In addition, those provisions also do not apply to workers engaged in activities involving periods of work split up over the day – the example given is that of cleaning staff.

Collective and workforce agreements

[W10036] By virtue of the *Working Time Regulations 1998 (SI 1998 No 1833), reg 23(a)*, employers and employees have the power to exclude or modify the following provisions:

(a) length of night work (see **W10020** above);
(b) minimum daily rest and weekly rest periods and rest breaks (see **W10014** above),

by way of collective or workforce agreements. Compensatory rest must be provided (see **W10037** below).

In addition, *reg 23(b)* of the Regulations allows the reference period for calculating the maximum working week to be extended to up to 52 weeks, if there are objective or technical reasons concerning the organisation of work to justify this.

This Regulation gives the parties a good deal of flexibility (on the understanding that they are prepared to consent with each other) to opt out of significant provisions contained in the Regulations.

Compensatory rest

[W10037] The *Working Time Regulations 1998 (SI 1998 No 1833), reg 24* provides that where a worker is not strictly governed by the working time rules because of:

(a) a derogation under *Regulation 21* (see **W10034** above); or

(b) the application of a collective or workforce agreement under *Regulation 23(a)* (see **W10036** above);or
(c) the special shift work rules under *Regulation 22* (see **W10035** above);

and as a result, the worker is required by his employer to work during what would otherwise be a rest period or rest break, then the employer must wherever possible allow him to take an equivalent period of compensatory rest. In exceptional cases, in which it is not possible for objective reasons to grant such a rest period, the employer must afford the employee such protection as may be appropriate in order to safeguard his health and safety.

Enforcement of working time

[W10038] The Regulations divide the enforcement responsibilities between the Health and Safety Executive and the Employment Tribunals. In essence, the working time limits are enforced by the Health and Safety Executive and local authority environmental health departments. The entitlements to rest and holiday are enforced through Employment Tribunals.

Interestingly, despite the assertions that the Directive is a health and safety measure, there appear to be only two successful prosecutions by the Health and Safety Executive since the Regulations were introduced. In contrast, there have been numerous cases lodged at the Employment Tribunal or High Court since the implementation of the Regulations.

Health and safety offences

[W10039] The *Working Time Regulations 1998 (SI 1998/1833), reg 28* provides that certain provisions of the Regulations (referred to as 'the relevant requirements') will be enforced by the Health and Safety Executive (except to the extent that a local authority may be responsible for their enforcement by virtue of *reg 28(3)*). The relevant requirements are:

(a) the 48-hour working week;
(b) length of night work;
(c) health assessment and transfers from night work;
(d) monotonous work;
(e) record keeping;
(f) failure to provide compensatory rest, where the provision concerning the length of night work is modified or excluded; and
(g) failure to provide adequate rest to mobile workers, where the provision concerning the length of night work is excluded.

Any employer who fails to comply with any of the relevant requirements will be guilty of an offence and shall be liable on summary conviction (in the magistrates' court) to a fine not exceeding the statutory maximum and on conviction on indictment (in the Crown Court) to a fine.

In addition, the Health and Safety Executive can take enforcement proceedings utilising certain provisions of the *Health and Safety at Work etc Act 1974*, as set out in *regs 28* and *29*. Employers may face criminal liability under these provisions, the sanctions for which range, according to the offence, from a fine to two years' imprisonment. (For enforcement of health and safety legislation generally, see ENFORCEMENT.)

Employment tribunals

Enforcement of the Regulations

[W10040] By virtue of the *Working Time Regulations 1998 (SI 1998/1833), reg 30*, certain provisions of the Regulations may be enforced by a worker presenting a claim to an Employment Tribunal where an employer has refused to permit him to exercise such rights. These are:

(a) daily rest period;
(b) weekly rest period;
(c) rest break;
(d) annual leave;
(e) failure to provide compensatory rest, insofar as it relates to situations where daily or weekly rest periods or rest breaks are modified or excluded;
(f) failure to provide adequate rest, insofar as it relates to situations where daily or weekly rest breaks are excluded;
(g) the failure to pay the whole or any part of the amount relating to paid annual leave, or payment on termination in lieu of accrued but untaken holiday.

A complaint must be made within (i) three months (other than in the case of members of the armed forces – see below) of the act or omission complained of (or in the case of a rest period or leave extending over more than one day, of the date on which it should have been permitted to begin) or (ii) such further period as the tribunal considers reasonable, where it is satisfied that it was not reasonably practicable for the complaint to be presented within that time. (The time limit in respect of members of the armed forces is six months.)

Where an Employment Tribunal decides that a complaint is well-founded, it must make a declaration to that effect and can award compensation to be paid by the employer to the worker. This shall be such amount as the Employment Tribunal considers just and equitable in all the circumstances, having regard to (i) the employer's default in refusing to permit the worker to exercise his right and (ii) any loss sustained by the worker which is attributable to the matters complained of. With regard to complaints relating to holiday pay or payment in lieu of accrued holiday on termination, the tribunal can also order the employer to pay the worker the amount which it finds properly due.

There is no qualifying service period with regard to such complaints being presented to a tribunal.

In addition to a claim under the *Working Time Regulations 1998 (SI 1998/1833), reg 30*, failure to pay holiday pay also gives the worker the right to claim unlawful deduction of wages under *Section 13* of the *Employment Rights Act 1996* (see **W10024** above). Claims must be brought within 3 months of the deduction or the last in a series of deductions, which may allow a worker to claim back-pay in respect of previous holiday years.

Protection against detriment, and against unfair dismissal

[W10041] The *Working Time Regulations 1998 (SI 1998/1833), reg 31* inserts a new *section 45A* into the *Employment Rights Act 1996*, to protect a worker who is subjected to any detriment, where the worker has:

(a) refused (or proposed to refuse) to comply with a requirement which the employer imposed (or proposed to impose) in contravention of the Regulations;

(b) refused (or proposed to refuse) to forgo a right conferred on him by the Regulations;

(c) failed to sign a workforce agreement or make any other agreement provided for under the Regulations, such as an individual opt-out from the maximum weekly working time limit;

(d) been a candidate in an election of work place representatives or, having been elected, carries out any activities as such a representative or candidate; or

(e) in good faith, (i) made an allegation that the employer has contravened a right under the Regulations, or (ii) brought proceedings under the Regulations.

The right to bring these claims would (as with discrimination claims) allow an individual to pursue a claim relating to a breach of the Regulations whilst continuing in employment.

By virtue of the inserted *section 101A of the Employment Rights Act 1996*, the dismissal of an employee on all but ground (e) above is automatically unfair, although this will only apply to employees and not to the wider definition of worker (for the position of workers who are not employees, see below). The compensation available would be subject to any cap on unfair dismissal compensation (from 6 April 2016 the cap is £78,962, but it is revised in April each year). Employees whose contracts were terminated on ground (e) above would be protected from dismissal for assertion of a statutory right, by virtue of section 104 of the 1996 Act.

An employee would also be protected if selected for redundancy on any of the grounds listed above. This would be automatically unfair selection, by virtue of section 105 of the 1996 Act.

Workers (ie those who are not employees) whose contracts are terminated for any of the grounds set out above can claim that they have suffered a detriment, and they may claim compensation, which would be capped in the same way as an award for unfair dismissal.

Breach of contract claims

[W10042] Following the *RJB Mining case* (see **W10010**), there is an argument that workers may also be able to claim for breach of contract, if their employer breaches the obligations in *reg 4(1)* of the Regulations. (In practice, however, this type of claim has not arisen since the determination of this case.) It has also been held that a worker cannot claim tortious damages for breach of statutory duty based on the employer's failure to enforce the 48-hour limit, as the provision was not intended to confer private law rights of action: *Sayers v Cambridgeshire County Council* [2006] EWHC 2029 (QB), [2007] IRLR 29.

If the claim is for damages and is outstanding at the termination of the employment, then the worker could in principle take their case to the Employment Tribunal (within 3 months of the effective date of termination of the contract giving rise to the claim). If the claim is for a declaration or an injunction, then the worker will have to apply to the High Court, where special rules apply.

Contracting out of the Regulations

[W10043] Under the *Working Time Regulations 1998 (SI 1998/1833), reg 35*, an agreement to contract out of the provisions of the Regulations can be made via a conciliation officer or by means of a compromise agreement, and the provisions are similar and consistent with the current provisions in the *Employment Rights Act 1996, s 203* (as amended by the *Employment Rights (Dispute Resolution) Act 1998*).

Conclusion

[W10044] The *Working Time Regulations 1998 (SI 1998/1833)* broke new ground in English law, and continue to attract adverse publicity. Despite this, the most recent review, published by the by the Department for Business, Innovation and Skills in December 2014 (titled 'The Impact of the Working Time Regulations on the UK labour market: A review of evidence') (BIS/14/1287) makes some interesting conclusions. These include:

- There has been a reduction in the number of UK employees working long hours since the introduction of the legislation.
- The legislation did not lead to large numbers of workers' hours being capped at 48 hours – either because use of the opt-out was widespread amongst long-hours workers, or employers found other ways to adjust. Longer working hours may be offset by people working a shorter working week.
- The concept of 'working time' is different now compared to the 1990s due to the introduction of new technology enabling mobile work and the increasing trend of flexible working.
- The UK continues to have an excellent workplace health and safety record. The UK has one of the best workplace health and safety records and it is difficult, therefore, to make a direct link in the data between regulation of working hours and workplace health and safety outcomes.
- Retaining the opt-out is very important both to UK business and to UK employees. The evidence suggests that taking away the ability to opt-out would be harmful both to business and to the welfare of workers who currently opt-out.

Following the UK Referendum result to leave the European Union, the impact of these regulations and any subsequent changes to them agreed by the EU remain uncertain. There is likely to be little or no change in the foreseeable future to legislation on working time, at least whilst the UK Government negotiates the terms of Brexit.

Workplaces – Health, Safety and Welfare

Andrea Oates

Introduction to workplaces – health, safety and welfare

[W11001] The *Workplace (Health, Safety and Welfare) Regulations 1992 (SI 1992 No 3004)* (Workplace Regulations), as amended, set out a wide range of basic health, safety and welfare standards applying to most places of work. The Regulations are wider in scope than previous legislation and cover not only offices, shops and factories, but also apply to workplaces including schools, hospitals, theatres, cinemas and hotels.

The Workplace Regulations set down detailed standards for premises used as places of work in the following areas:

(a) stability and solidity [*SI 1992 No 3004, Reg 4a*] (see **W11003**);
(b) maintenance [*SI 1992 No 3004, Reg 5*] (see **W11003**);
(c) ventilation [*SI 1992 No 3004, Reg 6*] (see VENTILATION);
(d) temperature [*SI 1992 No 3004, Reg 7*] (see **W11005**);
(e) lighting [*SI 1992 No 3004, Reg 8*] (see LIGHTING);
(f) cleanliness and waste storage [*SI 1992 No 3004, Reg 9*] (see **W11007**);
(g) room dimensions and space [*SI 1992 No 3004, Reg 10*] (see **W11008**);
(h) workstations and seating [*SI 1992 No 3004, Reg 11*] (see **W11009**);
(i) conditions of floors and traffic routes [*SI 1992 No 3004, Reg 12*] (see **W11010**);
(j) freedom from falls and falling objects [*SI 1992 No 3004, Reg 13*] (see WORK AT HEIGHTS);
(k) windows and transparent or translucent doors, gates and walls (see **W11012**);
(l) windows, skylights and ventilators [*SI 1992 No 3004, Reg 18*] (see **W11013**);
(m) window cleaning [*SI 1992 No 3004, Regs 14–16*] (see also WORK AT HEIGHTS);
(n) organisation of traffic routes [*SI 1992 No 3004, Reg 17*] (see **W11015**);
(o) doors and gates [*SI 1992 No 3004, Reg 18*] (see **W11016**);
(p) escalators and moving walkways [*SI 1992 No 3004, Reg 19*] (see **W11017**);
(q) toilets and washing facilities [*SI 1992 No 3004, Regs 20, 21*] (see **W11019–W11023**);
(r) drinking water [*SI 1992 No 3004, Reg 22*] (see **W11024**);
(s) clothing accommodation and facilities for changing clothing [*SI 1992 No 3004, Regs 23, 24*] (see **W11026–W11027**); and
(t) resting and eating facilities [*SI 1992 No 3004, Reg 25*] (see **W11027**).

In addition, Reg 25A provides that: 'Where necessary, those parts of the workplace (including in particular doors, passageways, stairs, showers, wash-basins, lavatories and workstations) used or occupied directly by disabled persons at work shall be organised to take account of such persons.'

As well as the Workplace Regulations, other legislation contains health, safety and welfare standards for particularly high-risk industries and workplaces, including construction sites, mines, quarries and the railway industry. This legislation is not examined in this chapter, which looks at the provisions of the Workplace (Health, Safety and Welfare) Regulations 1992, together with guidance provided by the Health and Safety Executive (HSE) and the associated approved code of practice (ACoP). These are all set out in the HSE publication L24, *Workplace health, safety and welfare – Workplace (Health, Safety and Welfare) Regulations 1992 – Approved Code of Practice and guidance*. The chapter also sets out the law banning smoking in enclosed workplaces and public places and makes reference to the requirements of the Building Regulations 2010. These set out standards for building work in new and altered buildings to make them safe and accessible and limit waste and environmental damage. They apply where workplaces are being built, extended or modified. The fire safety aspects of the Building Regulations are under scrutiny following the Grenfell Tower fire in June 2017, which killed around 70 people when it spread rapidly throughout the 24-storey residential building.

It briefly looks at the *Equality Act 2010* which replaced a raft of equalities legislation including the *Disability Discrimination Act 1995* and requires employers to make reasonable adjustments to workplaces where necessary to ensure that disabled workers are not put at a substantial disadvantage (see the EQUALITY ACT 2010 for more information).

Finally, it sets out the provisions of the *Health and Safety (Safety Signs and Signals) Regulations 1996 (SI 1996 No 341)*.

Workplace (Health, Safety and Welfare) Regulations 1992

Definitions

[W11002] A 'workplace' is any non-domestic premises (or part of premises) available to any person as a place of work. Regulation 2(1) sets out that it includes:

(a) any place within the premises to which such person has access while at work; and
(b) any room, lobby, corridor, staircase, road or other place used as a means of access to or egress from that place of work or where facilities are provided for use in connection with the place of work other than a public road.

Regulation 3 sets out the exceptions and HSE guidance explains that:

- all operational ships, boats, hovercraft, aircraft, trains and road vehicles are excluded from the Regulations, although *Reg 13* (see **W11011**) applies where aircraft, trains and road vehicles are stationary in a workplace. In addition, the Regulations do apply to forms of transport that are fixed in position and are no longer being used as originally intended. The HSE gives the examples of those being used as restaurants or tourist attractions;
- they do apply to workplaces or parts of workplaces located at a quarry or above ground at a mine. However, they do not apply where the actual extraction of, or exploration for, minerals is being undertaken underground at mines, at quarries or other mineral extraction sites. Instead, separate sector-specific legislation applies;
- construction sites (including site offices) are excluded from the Regulations. Again, sector-specific legislation applies – the *Construction (Design and Management) Regulations 2015* (see Construction, Design and Management). Where construction work is in progress within a workplace and it is fenced off, it can be treated as a construction site and is excluded from the Workplace Regulations. If it is not fenced off, the Workplace Regulations and the *Construction (Design and Management) Regulations 2015* both apply;
- at temporary work sites the requirements for toilets, washing facilities, drinking water, clothing accommodation, changing facilities and facilities for rest and eating meals (*Regs 20–25*) apply so far as is reasonably practicable. The term 'reasonably practicable' means that the risk must be weighed against the costs necessary to avert it, including time and trouble as well as financial cost; and
- outdoor agricultural or forestry workplaces that are located away from the undertaking's main buildings are excluded from the Regulations, with the exception of the requirements regarding toilets, washing facilities and drinking water. These requirements apply so far as is reasonably practicable.

[*Workplace (Health, Safety and Welfare) Regulations 1992 (SI 1992 No 3004), Regs 2, 3*].

Employers must ensure that every workplace, modification, extension or conversion under their control, where any of their employees works, complies with the requirements of the Regulations. This duty also extends to others who have any control over workplaces, but not to self-employed people in respect of their own work or the work of any partner in their undertaking (Regulation 4).

The Regulations expand on employers' general duty under s 2 of the *Health and Safety at Work etc Act 1974* to ensure, so far as reasonably practicable, the health, safety and welfare of their employees at work, and the duty people in control of non-domestic premises have under s 4 of the Act towards people who are not their employees but use their premises.

The Regulations do not apply to private dwellings, but they do apply to hotels, nursing homes and to parts of premises where domestic staff are employed, for example in the kitchens of hostels.

Stability and solidity

[W11002.1] Where a workplace is in a building, the Regulations require the building to have a stability and solidity appropriate to the nature of the use of the workplace (Regulation 4A).

HSE guidance in L24 says that any building being used as a workplace should be capable of supporting all foreseeable loads imposed on it. There should be an inspection and maintenance regime, appropriate to the building's type and use, to ensure that any defect which may cause an unacceptable safety risk is detected in good time, so appropriate remedial action can be taken.

General maintenance of the workplace

[W11003] All workplaces, equipment and devices should be maintained

(a) in an efficient state,
(b) in an efficient working order, and
(c) in a good state of repair.

[Workplace (Health, Safety and Welfare) Regulations 1992 (SI 1992 No 3004), Reg 5].

Dangerous defects should be reported and acted on as a matter of good housekeeping (and to avoid possible subsequent civil liability). Defects resulting in equipment/plant becoming unsuitable for use, though not necessarily dangerous, should lead to decommissioning of plant until repaired – or, if this might lead to the number of facilities being less than required by statute, repaired forthwith (eg a defective toilet).

To this end, a suitable maintenance programme must be instituted, including:

(a) regular maintenance (inspection, testing, adjustment, lubrication, cleaning);
(b) rectification of potentially dangerous defects and the prevention of access to defective equipment;
(c) record of maintenance/servicing.

There are more detailed regulations dealing with plant and equipment used at work, including the *Provision and Use of Work Equipment Regulations 1998 (SI 1998 No 2306* as amended, which are examined in detail in MACHINERY SAFETY.

Ventilation

[W11004] *Workplace (Health, Safety and Welfare) Regulations 1992 (SI 1992 No 3004), Reg 6* deals with the provision of sufficient ventilation in enclosed workplaces. There must be effective and suitable ventilation in order to supply a sufficient quantity of fresh or purified air. If ventilation plant is necessary for health and safety reasons, it must give warning of failure.

The ACoP sets out that enclosed workplaces should be sufficiently well ventilated so that stale air, and air which is hot or humid because of the processes or equipment in the workplace, is replaced at a reasonable rate. Fresh air should, as far as possible, be free of any impurity which is likely to be offensive or cause ill health. Air which is taken from the outside can normally be considered to be 'fresh'; although air inlets should not be sited, for example, close to a flue, an exhaust ventilation system outlet, or an area in which vehicles manoeuvre. Where necessary, the inlet air should be filtered to remove particulates. In addition, workers should not be subject to uncomfortable draughts.

HSE guidance on the Regulations sets out that the fresh air supply rate should not normally fall below five to eight litres per second, per occupant and that when establishing a fresh air supply rate, employers should consider the following factors:

- the floor area per person;
- the processes and equipment involved; and
- whether the work is strenuous.

This area is dealt with in greater detail in VENTILATION.

Temperature

[W11005] The temperature in all workplaces inside buildings should be reasonable during working hours [*Workplace (Health, Safety and Welfare) Regulations 1992 (SI 1992 No 3004), Reg 7*]. In addition, workplaces should have adequate thermal insulation where necessary, taking into account the type of work being carried out and the amount of physical activity involved and the 'excessive effects of sunlight on temperature' should be avoided (*Reg 7(1A)*).

Workroom temperatures should enable people to work (and visit the toilets) in reasonable comfort, without the need for extra or special clothing. Although the Regulations themselves do not specify a maximum or minimum indoor workplace temperature, the ACoP sets out that workrooms should normally be at least 16°C for most types of work, and at least 13°C for work involving 'rigorous physical effort'.

For most kinds of work the acceptable range of thermal comfort lies between 16°C and 24°C.

The Chartered Institute of Building Services Engineers (CIBSE) recommends:

(a) heavy work in factories 13°C;
(b) light work in factories 16°C;
(c) hospitals 18°C; and
(d) offices 20°C.

Very sedentary work may still be performed with reasonable comfort around 24°C.

In rooms where food or other products/processes have to be kept at low temperatures and it is impractical to maintain the temperatures set out in the ACoP, the following measures should be applied as appropriate:

- enclosing or insulating the product;

- pre-chilling the product;
- keeping chilled areas as small as possible;
- exposing the product to workroom temperatures as briefly as possible;
- providing insulated duckboards or other floor coverings where workers have to stand for long periods on cold floors, unless special footwear is provided which prevents discomfort;
- excluding draughts from workstations, eg by using baffles;
- installing self-closing doors where such measures are practical and would reduce discomfort.

It also sets out that if the temperature in a workroom is uncomfortably high, for example because of hot processes or building design, reasonable steps to achieve a reasonably comfortable temperature include:

- insulating hot plants or pipes;
- providing air-cooling plant;
- shading windows;
- siting workstations away from places subject to radiant heat.

If a reasonably comfortable temperature cannot be achieved throughout a workroom, local heating or cooling should be provided. In extremely hot weather, fans and increased ventilation can be used instead of local cooling.

HSE guidance on the Regulations, *Workplace health, safety and welfare: A short guide for managers* (INDG244), explains how to carry out an assessment of the risk to workers' health from working in either a hot or cold environment. Employers should look at personal factors, such as body activity, the amount and type of clothing and duration of exposure, together with environmental factors, including the ambient temperature and radiant heat. If the work is outdoors, they should consider sunlight, wind velocity and the presence of rain or snow.

Any assessment should cover:

- Measures to control the workplace environment, particularly heat sources;
- Restriction of exposure by, for example, reorganising tasks to build in rest periods or other breaks from work;
- Medical pre-selection of employees to ensure that they are fit to work in these environments;
- Use of suitable personal protective clothing;
- Acclimatisation of workers to the working environment;
- Training in the precautions to be taken; and
- Supervision to ensure that the precautions the assessment identifies are taken.

HSE produces further guidance on temperature at work on its website www.hse.gov.uk/temperature.

Lighting

[W11006] Every workplace must be provided with suitable and sufficient lighting [*Workplace (Health, Safety and Welfare) Regulations 1992 (SI 1992*

No 3004), Reg 8]. This should be natural lighting, so far as is reasonably practicable. There should also be suitable and sufficient emergency lighting where necessary. The regulations do not define 'suitable and sufficient', but the ACoP says that the lighting should be sufficient to enable people to work, use facilities without experiencing eye-strain and move from place to place safely. Where necessary, artificial lighting should be provided at individual workstations, and at places of particular risk such as crossing points on traffic routes.

The HSE publication, *Lighting at work* (HSG38), gives detailed guidance in this area and is available to download free from the HSE website at www.hse.gov.uk/pubns/books/hsg38.htm.

The chapter on LIGHTING also deals with this area in greater detail.

General cleanliness

[W11007] All furniture and fittings of every workplace must be kept sufficiently clean. Surfaces of floors, walls and ceilings must be capable of being kept sufficiently clean and waste materials must not accumulate other than in waste bins *Workplace (Health, Safety and Welfare) Regulations 1992 (SI 1992 No 3004), Reg 9*.

The level and frequency of cleanliness will vary according to workplace use and purpose. Obviously, a factory canteen should be cleaner than a factory floor. Floors and indoor traffic routes should be cleaned at least once a week, though dirt and refuse not in suitable bins should be removed at least daily, particularly in hot atmospheres or hot weather. Interior walls, ceilings and work surfaces should be cleaned at suitable intervals and ceilings and interior walls painted and/or tiled so that they can be kept clean. Surface treatment should be renewed when it can no longer be cleaned properly. Spillages and waste matter from drains or toilets should be removed. Methods of cleaning, however, should not expose anyone to substantial amounts of dust, and absorbent floors likely to be contaminated by oil or other substances difficult to remove, should be sealed or coated, with non-slip floor paint for example (not covered with carpet).

Workroom dimensions/space

[W11008] Every room in which people work should have sufficient
(a) floor area,
(b) height, and
(c) unoccupied space

for health, safety and welfare purposes *Workplace (Health, Safety and Welfare) Regulations 1992 (SI 1992 No 3004), Reg 10*.

Workrooms should have enough uncluttered space to allow people to go to and from workstations with relative ease. The number of people who may work in any particular room at any time will depend not only on the size of the room but also on the space given over to furniture, fittings, equipment and general room layout. Workrooms should be of sufficient height to afford staff safe access to workstations. If, however, the workroom is in an old building with low beams or other possible obstructions, these should be clearly marked, eg 'Low beams, mind your head'.

The ACoP says that the total volume of the room (when empty), divided by the number of people normally working there, should be 11 cubic metres (minimum) per person, although this does not apply to:

(i) small structures where space is necessarily restricted, such as retail sales kiosks, attendants' shelters and machine control cabs; and
(ii) room used for lectures, meetings and similar purposes.

Workplace (Health, Safety and Welfare) Regulations 1992 (SI 1992 No 3004), Sch 1

In making this calculation, any part of a room which is higher than 3 metres is counted as being 3 metres high.

Where furniture occupies a considerable part of the room, 11 cubic metres may not be sufficient space per person. Here more careful planning and general room layout is required. Similarly, rooms may need to be larger or have fewer people working in them depending on the contents and layout of the room and the nature of the work.

Workstations and seating

[W11009] The Regulations stipulate the following.

Workstations

Every workstation must be so arranged that it is suitable for any person at work who is likely to work at the workstation, and any work likely to be done there. So far as reasonably practicable, outdoor workstations should provide protection from adverse weather, enable a person to leave it swiftly or to be assisted in an emergency, and ensure any person is not likely to slip or fall.
[*Workplace (Health, Safety and Welfare) Regulations 1992 (SI 1992 No 3004), Reg 11(1), (2)*].

Workstation seating

A suitable seat must be provided for each person at work whose work (or a substantial part of it) can or must be done seated. The seat should be suitable for the person doing the work and the work to be done.
Where necessary, a suitable footrest should be provided.
[*Workplace (Health, Safety and Welfare) Regulations 1992 (SI 1992 No 3004), Reg 11(3), (4)*].

It should be possible to carry out work safely and comfortably. Work materials and equipment in frequent use (or controls) should always be within easy reach, so that people do not have to bend or stretch unduly, and the worker should be at a suitable height in relation to the work surface. Workstations, including seating and access, should be suitable for special needs, including workers with disabilities. The workstation should allow people likely to have to do work there adequate freedom of movement and ability to stand upright, thereby avoiding the need to work in cramped conditions. More particularly, seating should be suitable, providing adequate support for the lower back and a footrest provided, if feet cannot be put comfortably flat on the floor.

Workstations with display screen equipment (DSE) are subject to the *Health and Safety (Display Screen Equipment) Regulations 1992 (SI 1992 No 2792)*, as amended (see DISPLAY SCREEN EQUIPMENT for more information).

The Health and Safety Executive (HSE) guidance on seating advises employers to use risk assessments in order to ensure that safe seating is provided. *Seating at Work*, (HSG57) is available, to download free from the HSE website www.hse.gov.uk/pubns/books/hsg57.htm.

Condition of floors and traffic routes

[W11010] The principal dangers connected with industrial and commercial floors are slipping, tripping and falling. Slip, trip and fall resistance are a combination of the right floor surface and the appropriate type of footwear. Employers should ensure that level changes, multiple changes of floor surfaces, steps and ramps etc are clearly indicated. Safety underfoot is at bottom a trade-off between slip resistance and ease of cleaning. Floors with rough surfaces tend to be more slip-resistant than floors with smooth surfaces, especially when wet. By contrast, smooth surfaces are much easier to clean but less slip-resistant. Slip-resistant vinyl floors should be periodically stripped, degreased and resealed with slip-resistant finish; linoleum floors similarly.

Apart from being safe, floors must also be hygienically clean and there has been a gradual transition to seamless floors in hygiene-critical areas. Whichever floor surface is appropriate and whichever treatment is suitable, underfoot safety depends on workplace activity (office or factory), variety of spillages (food, water, oil, chemicals), nature of traffic (pedestrian, cars, trucks).

Floors in workplaces must:

(a) be constructed so as to be suitable for use [*Workplace (Health, Safety and Welfare) Regulations 1992 (SI 1992 No 3004), Reg 12(1)*].
They should be of sound construction and adequate strength and stability to sustain loads and passing internal traffic and never overloaded (see *Greaves v Baynham Meikle* [1975] 3 All ER 99 for possible consequences in civil law).

(b) not have holes or slopes, or be uneven or slippery so as to expose a person to risk of injury [*Workplace (Health, Safety and Welfare) Regulations 1992 (SI 1992 No 3004), Reg 12(2)(a)*].
The surfaces of floors and traffic routes should be even and free from holes, bumps and slipping hazards that could cause a person to slip, trip or fall, drop or lose control of something being lifted or carried, or cause instability or loss of control of a vehicle.
Holes, bumps or uneven surfaces or areas resulting from damage or wear and tear should be made good, and pending this barriers should be erected or locations conspicuously marked. Temporary holes, result-

ing from the removal of floorboards for example, should be adequately guarded. Employers should consider the needs of disabled people, such as those with a visual disability or mobility issues.

Deep holes are governed by *Workplace (Health, Safety and Welfare) Regulations 1992 (SI 1992 No 3004), Reg 13* (see **W9013** WORK AT HEIGHTS).)

Where possible, steep slopes should be avoided, and if they cannot be avoided they should be provided with a secure handrail. Ramps used by disabled persons should also have handrails.

(c) be kept free from obstructions, and articles or substances likely to cause persons to slip, trip or fall, so far as reasonably practicable [*Workplace (Health, Safety and Welfare) Regulations 1992 (SI 1992 No 3004), Reg 12(3)*].

Floors should be kept free of obstructions impeding access or presenting hazards, particularly near or on steps, stairs, escalators and moving walkways; on emergency routes or outlets; in or near doorways or gangways; or by corners or junctions. Where temporary obstructions are unavoidable, access should be prevented and people warned of the possible hazard. When furniture is being moved, it should not be left in a place where it can cause a hazard.

In *Lowles v Home Office* [2004] EWCA Civ 985, [2004] All ER (D) 538 (Jul) the judge ruled that a two-inch step on a ramp leading into Armley prison, on which Ms Lowles slipped and injured herself, could be considered an obstruction.

(d) have effective drainage [*Workplace (Health, Safety and Welfare) Regulations 1992 (SI 1992 No 3004), Reg 12(2)(b)*].

Where floors are likely to get wet, in laundries, potteries and food processing plants for example, effective drainage should drain it away. Drains and channels should be situated so as to reduce the area of wet floor. The floor should slope slightly towards the drain and ideally have covers flush with the floor surface. Processes and plant which cause discharges or leaks of liquids should be enclosed, and leaks from taps caught and drained away. In food processing and preparation plants, work surfaces should be arranged so as to minimise the likelihood of spillage. Where a leak or spillage occurs, it should be fenced off or mopped up immediately.

Staircases should be provided with a handrail. Any open side of a staircase should have minimum fencing of an upper rail at 900mm or higher, and a lower rail.

It is also important to consider the dangers posed by snow and ice on, for example, external fire escapes.

Lewis v Avidan [2005] EWCA Civ 670, [2005] All ER (D) 136 (Apr) involved a claim from a care assistant who was injured slipping on a patch of water which had resulted from a burst water pipe. In this case, the Court of Appeal dismissed the claim as the judge ruled that an unexpected flood did not mean that the floor was inadequately maintained.

Falls and falling objects

[W11011] *Workplace (Health, Safety and Welfare) Regulations 1992 (SI 1992 No 3004), Reg 13* deals with falls and falling objects, which are covered in detail in WORK AT HEIGHTS.

It requires that tanks, pits and other structures containing dangerous substances are securely covered or fenced where there is a risk of a person falling in. Traffic routes over such open structures should also be securely fenced.

In this regulation, dangerous substances are those that are poisonous, corrosive or likely to scald or burn as well as fumes, gases and vapours likely to overcome a person, or any granular or free-flowing solid substance, or any viscous substance which is of a nature or quantity likely to cause danger to any person.

Windows and transparent or translucent doors, gates and walls

[W11012] Transparent or translucent surfaces in windows, doors, gates, walls and partitions should be constructed of safety material or be adequately protected against breakage, where necessary for health and safety reasons. The ACoP explains that this is where:

(a) any part is at shoulder level or below in doors and gates; or
(b) any part is at waist level or below in windows, walls and partitions, with the exception of glass houses.

Screens or barriers can be used as an alternative to the use of safety materials. Narrow panels of up to 250mm width are excluded from the requirement.

Transparent or translucent surfaces should be marked to make them apparent where this is necessary for health and safety reasons.

Workplace (Health, Safety and Welfare) Regulations 1992 (SI 1992 No 3004), Reg 14

Windows, skylights and ventilators

[W11013] Openable windows, skylights and ventilators must be capable of being opened, closed and adjusted safely. They must not be positioned so as to pose a risk when open.

The ACoP sets out that they should be capable of being reached and operated safely, with window poles or similar equipment, or stable platforms, made available where necessary. Where there is the danger of falling from a height, devices should be provided to prevent this by ensuring the window cannot open too far. They should not cause a hazard by projecting into an area where people are likely to collide with them when open. The bottom edge of opening windows should normally be at least 800mm above floor level, unless there is a barrier to prevent falls.

[*Workplace (Health, Safety and Welfare) Regulations 1992 (SI 1992 No 3004), Reg 15*].

Ability to clean windows etc safely

[W11014] *Workplace (Health, Safety and Welfare) Regulations 1992 (SI 1992 No 3004), Reg 16* requires windows and skylights to be designed and constructed so that they can be safely cleaned.

Organisation of traffic routes

[W11015] Traffic routes in workplaces should allow pedestrians and vehicles to circulate safely.

They should be suitable for the people or vehicles using them. They should be sufficient in number, in suitable positions and of sufficient size. This requirement applies so far as is reasonably practicable to a workplace which is not a new workplace, a modification, an extension or a conversion.

Where necessary for health and safety reasons, they should be suitably indicated.

They should be planned to give the safest route, wide enough for the safe movement of the largest vehicle permitted to use them, and they should avoid vulnerable items like fuel or chemical plants or pipes, and open and unprotected edges. Employers should give special consideration to the safety of people in wheelchairs.

There should be safe areas for loading and unloading. Sharp or blind bends should be avoided where possible, and if they cannot be avoided, one-way systems or mirrors to improve visibility should be used. Sensible speed limits should be set and enforced. There should be prominent warning of any limited headroom or potentially dangerous obstructions such as overhead electric cables. Routes should be marked where necessary and there should be suitable and sufficient parking areas in safe locations.

Traffic routes should keep vehicles and pedestrians apart and there should be pedestrian crossing points on vehicle routes. Traffic routes and parking and loading areas should be soundly constructed on level ground. Health and Safety Executive (HSE) guidance in this area can be found in the publication, *Workplace transport safety – guidance for employers*, HSG136 is available to download free from the HSE website at www.hse.gov.uk/pubns/books/hsg136.htm.

[*Workplace (Health, Safety and Welfare) Regulations 1992 (SI 1992 No 3004), Reg 17*].

This area is dealt with in detail in ACCESS, TRAFFIC ROUTES AND VEHICLES.

Doors and gates

[W11016] Doors and gates must be suitably constructed and fitted with safety devices. In particular,

(i)   a sliding door/gate must have a device to prevent it coming off its track during use;

(ii) an upward opening door/gate must have a device to prevent its falling back;
(iii) a powered door/gate must
    (a) have features preventing it from causing injury by trapping a person (eg accessible emergency stop controls),
    (b) be able to be operated manually unless it opens automatically if the power fails (where necessary for health and safety reasons);
(iv) a door/gate capable of opening, by being pushed from either side, must provide a clear view of the space close to both sides.

[*Workplace (Health, Safety and Welfare) Regulations 1992 (SI 1992 No 3004), Reg 18*].

Escalators and moving walkways

[**W11017**] Escalators and moving walkways must:

(a) function safely;
(b) be equipped with any necessary safety devices;
(c) be fitted with easily identifiable and readily accessible emergency stop controls.

[*Workplace (Health, Safety and Welfare) Regulations 1992 (SI 1992 No 3004), Reg 19*].

Welfare facilities

[**W11018**] 'Welfare facilities' is a wide term, embracing both toilets and washing accommodation at workplaces, provision of drinking water, clothing accommodation (including facilities for changing clothes) and facilities for rest and eating meals (see **W11027** below).

The need for sufficient suitable hygienic toilet and washing facilities in all workplaces is obvious. Sufficient facilities must be provided to enable everyone at work to use them without undue delay. They do not have to be in the actual workplace but ideally should be situated in the building(s) containing them and they should provide protection from the weather, be well-ventilated, well-lit and enjoy a reasonable temperature. Wash basins should allow washing of hands, face and forearms and, where work is particularly strenuous, dirty, or results in skin contamination (eg molten metal work), showers or baths should be provided. In the case of showers, they should be fed by hot and cold water and fitted with a thermostatic mixer valve. Washing facilities should ensure privacy for the user and be separate from the water closet, with a door that can be secured from the inside. It should not be possible to see urinals or the communal shower from outside the facilities when the entrance/exit door opens. Entrance/exit doors should be fitted to both washing and sanitary facilities (unless there are other means of ensuring privacy). Windows to toilets, showers/bathrooms should be obscured either by being frosted, or by blinds or curtains (unless it is impossible to see into them from outside).

[W11018] Workplaces – Health, Safety and Welfare

This section examines current statutory requirements in all workplaces. For requirements relating to sanitary conveniences and washing facilities on construction sites see C8056 CONSTRUCTION AND BUILDING OPERATIONS.

Sanitary conveniences in all workplaces

[W11019] Suitable and sufficient sanitary conveniences must be provided at readily accessible places. In particular,

(a) the rooms containing them must be adequately ventilated and lit;
(b) they (and the rooms in which they are situated) must be kept clean and in an orderly condition;
(c) separate rooms containing conveniences must be provided for men and women except where the convenience is in a separate room which can be locked from the inside.

[*Workplace (Health, Safety and Welfare) Regulations 1992 (SI 1992 No 3004), Reg 20*].

Washing facilities in all workplaces

[W11020] Suitable and sufficient washing facilities (including showers where necessary (see **W11018** above)), must be provided at readily accessible places or points. In particular, facilities must:

(a) be provided in the immediate vicinity of every sanitary convenience (whether or not provided elsewhere);
(b) be provided in the vicinity of any changing rooms – whether or not provided elsewhere;
(c) include a supply of clean hot and cold or warm water (running water so far as is reasonably practicable);
(d) include soap (or something similar);
(e) include towels (or the equivalent);
(f) be in rooms sufficiently well-ventilated and well-lit;
(g) be kept clean and in an orderly condition (including rooms in which they are situate);
(h) be separate for men and women, except where they are provided in a lockable room intended to be used by one person at a time, or where they are provided for the purposes of washing hands, forearms and face only, where separate provision is not necessary.

[*Workplace (Health, Safety and Welfare) Regulations 1992 (SI 1992 No 3004), Reg 21*].

In November 2017, the HSE announced that it had reviewed its approach concerning access to welfare facilities for visiting delivery drivers, including guidance to duty holders. It has re-examined the *Workplace (Health, Safety and Welfare) Regulations 1992*, in particular *Regs 20 and 21*, and will begin to update its guidance to make clear that drivers must have access to welfare facilities in the premises they visit as part of their work.

Minimum number of facilities – sanitary conveniences and washing facilities

[W11021] The ACoP sets out the minimum number of facilities required as follows:

(a) Where both women and men work:

Number of people at work	Number of WCs	Number of wash stations
1 to 5	1	1
6 to 25	2	2
26 to 50	3	3
51 to 75	4	4
76 to 100	5	5

(b) The alternative table can be used where the toilets are used only by men:

Number of men at work	Number of WCs	Number of urinals
1 to 15	1	1
16 to 30	2	1
31 to 45	2	2
46 to 60	3	2
61 to 75	3	3
76 to 90	4	3
91 to 100	4	4

For every 25 people above 100 (or fraction of 25) an additional WC and wash station should be provided; in the case of WCs used only by *men*, an additional WC per every 50 men (or fraction of 50) above 100 is sufficient (provided that at least an equal number of additional urinals is provided).

[*Workplace (Health, Safety and Welfare) Regulations 1992 (SI 1992 No 3004), Sch 1, Part II*].

Particularly dirty work etc.

[W11022] Where the work is particularly dirty, there should be one wash station for every ten people (or fraction of ten) at work up to 50 people; and one extra for every additional 20 people (or fraction of 20). Where sanitary and wash facilities are also used by members of the public, the number of conveniences and facilities should be increased so that workers can use them without undue delay.

Temporary work sites

[W11023] At temporary work sites suitable and sufficient sanitary conveniences and washing facilities should be provided so far as is reasonably practicable (see **E15039 ENFORCEMENT** for meaning). In other cases, mobile facilities should be provided wherever possible. If possible, these should incorporate flushing sanitary conveniences and washing facilities with running water for washing and meet the other requirements of the ACoP.

Remote work sites

[W11023.1] The ACoP sets out that for remote workplaces without running water or a nearby sewer, employers should provide enough water in containers for washing, or other means of maintaining personal hygiene, and enough chemical toilets. As far as possible, they should avoid chemical toilets that have to be emptied manually. If chemical toilets must be used, a suitable deodorising agent should be provided and they should be emptied and recharged at suitable intervals.

Drinking water

[W11024] An adequate supply of wholesome drinking water must be provided for all persons at work in the workplace. It must be readily accessible at suitable places and conspicuously marked, unless non-drinkable cold water supplies are clearly marked. In addition, there must be provided a sufficient number of suitable cups (or other drinking vessels), unless the water supply is in a jet.

[*Workplace (Health, Safety and Welfare) Regulations 1992 (SI 1992 No 3004), Reg 22*].

Where water cannot be obtained from the mains supply, it should only be provided in refillable containers. The containers should be enclosed to prevent contamination and refilled at least daily. So far as reasonably practicable, drinking water taps should not be installed in sanitary accommodation, or in places where contamination is likely, for instance, in a workshop containing lead processes.

Clothing accommodation

[W11025] Suitable and sufficient accommodation must be provided for:

(a) any person at work's own clothing which is not worn during working hours; and
(b) special clothing which is worn by any person at work but which is not taken home, for example, overalls, uniforms and thermal clothing.

[*Workplace (Health, Safety and Welfare) Regulations 1992 (SI 1992 No 3004), Reg 23(1)*].

Accommodation is not suitable unless it:

(i) is suitably secure for the person's own clothing where changing facilities are required;
(ii) includes separate accommodation for clothing worn at work and for other clothing, where necessary to avoid risks to health or damage to clothing;
(iii) allows or includes facilities for drying clothing (so far as is reasonably practicable); and
(iv) is in a suitable location.

[*Workplace (Health, Safety and Welfare) Regulations 1992 (SI 1992 No 3004), Reg 23(2)*].

The ACoP sets out that special work clothing includes all clothing that is only worn at work, such as overalls, uniforms, thermal clothing and hats worn for food hygiene purposes.

Workers' own clothing should be able to hang in a clean, warm, dry, well-ventilated place. If this is not possible in the workroom, then it should be put elsewhere. Accommodation should take the form of a separate hook or peg. Clothing which is dirty, damp or contaminated owing to work should be accommodated separately from the worker's own clothes.

Facilities for changing clothing

[W11026] Suitable and sufficient facilities must be provided for any person at work in the workplace to change clothing where:

(a) the person has to wear special clothing for work, and
(b) the person cannot be expected to change in another room for reasons of health or propriety.

Facilities are not suitable unless they include:

(i) separate facilities for men and women, or
(ii) separate use of facilities by men and women.

In addition, they must be easily accessible, of sufficient capacity and provided with seating.

[*Workplace (Health, Safety and Welfare) Regulations 1992 (SI 1992 No 3004), Reg 24*].

Changing rooms (or room) should be provided for workers who change into special work clothing and remove more than outer clothing and where necessary to prevent workers' own clothes being contaminated by a harmful substance. Changing facilities should be easily accessible from workrooms and eating places. They should contain adequate seating and contain, or connect with clothing accommodation, and showers or baths (see **W11018** above). Privacy of user should be ensured. The facilities should be large enough to cater for the maximum number of people at work expected to use them at any one time without overcrowding or undue delay.

Post Office v Footitt [2000] IRLR 243, involved an employers' appeal against an improvement notice requiring the construction of a separate changing room for women postal workers to change into and out of their uniforms. It examined the definition of 'special clothing' and the concept of propriety.

An environmental health officer had served an improvement notice under the *Workplace (Health, Safety and Welfare) Regulations 1992 (SI 1992 No 3004), Reg 24* as she had found that any female employees wishing to change their clothing could only do so in the general area of the women's toilet facilities.

In the High Court, the judge held that the uniform worn by postal workers was 'special clothing' for the purposes of *Regulation 24*. It was held that 'special clothing' is not merely limited to clothing that is worn only at work. Therefore, the fact that postal workers wear their uniform to and from work does not prevent it from being 'special clothing'. The changing facilities for women provided by the Post Office were therefore not 'suitable and sufficient' within the meaning of the Regulation.

The court also held that the fact that the changing facilities for men and women were separated was not in itself enough to satisfy the concept of propriety referred to in *Regulation 24(2)*. There is no reason why requiring one female to undress in the presence of another cannot be said to offend against the principles of propriety. The fact that many people would have no objection to changing in the company of others of the same sex does not absolve the employer from providing facilities for those who may prefer privacy.

Rest and eating facilities

[W11027] Suitable and sufficient rest facilities must be provided at readily accessible places (*Workplace (Health, Safety and Welfare) Regulations 1992 (SI 1992 No 3004), Reg 25(1)*).

(a) Rest facilities

A rest facility is:

(i) in the case of a new workplace, extension or conversion – a rest room (or rooms);
(ii) in other cases, a rest room (or rooms) or rest area; including
(iii) (in both cases):
- appropriate facilities for eating meals where food eaten in the workplace would otherwise be likely to become contaminated;
- suitable arrangements for protecting non-smokers from tobacco smoke. The regulations came into force before legislation banning smoking in enclosed workplaces and public places – see **W11028** below);
- suitable facilities for a pregnant or nursing mother to rest in. These should be near toilets and, where necessary, include the facility to lie down.

Canteens or restaurants may be used as rest rooms provided that there is no obligation to buy food there (ACoP) (*Workplace (Health, Safety and Welfare) Regulations 1992 (SI 1992 No 3004), Regs 25(2)–(4)*).

(b) Eating facilities

Where workers regularly eat meals at work, facilities must be provided for them to do so (*Workplace (Health, Safety and Welfare) Regulations 1992 (SI 1992 No 3004), Reg 25(5)*).

In offices and other workplaces where there is no risk of contamination, seats in the work area are sufficient, although workers should not be interrupted excessively during breaks, for example, by the public. In other cases, rest areas or rooms should be provided and in the case of new workplaces, this should be a separate rest room. Rest facilities should be large enough, and have enough seats with backrests and tables, for the number of workers likely to use them at one time.

Where workers regularly eat meals at work, there should be suitable and sufficient facilities. These should be provided where food would otherwise be contaminated, by dust or water for example. Seats in work areas can be suitable eating facilities, provided the work area is clean. There should be a means to prepare or obtain a hot drink, and where persons work during hours or at places where hot food cannot be readily obtained, there should be the means for heating their own food. Eating facilities should be kept clean.

The ACoP sets out that: 'Any area where smoking is permitted should be sited, where possible, far enough from work areas and non-smoking rest areas to prevent tobacco smoke getting into them – taking into account doors and windows that may open'.

Smoking

[W11028] Legislation banning smoking has been brought into force in England, Wales, Scotland and Northern Ireland.

The *Health Act 2006* (England and Wales) sets out a number of broad provisions for smoke-free legislation.

Five sets of smoke-free regulations set out the detail of the legislation:

- The Smoke-free (Premises and Enforcement) Regulations 2006 (SI 2006 No 3368);
- The Smoke-free (Signs) Regulations 2012 (SI 2012 No 1536);
- The Smoke-free (Exemptions and Vehicles) Regulations 2007 (SI 2007 No 765);
- The Smoke-free (Penalties and Discounted Amounts) Regulations 2007 (SI 2007 No 764); and
- The Smoke-free (Vehicle Operators and Penalty Notices) Regulations 2007 (SI 2007 No 760).

The Smoke-free (Premises and Enforcement) Regulations 2006 (SI 2006 No 3368)

The Smoke-free (Premises and Enforcement) Regulations apply only in England and set out the definition of 'enclosed' and 'substantially enclosed' premises, and outline the enforcement provisions.

Section 2 of the *Health Act 2006* sets out that premises that are:

- open to the public, or are used as a place of work by more than one person; or
- where members of the public might attend to receive or provide goods or services,

are to be smoke-free in areas that are enclosed or substantially enclosed.

The Regulations set out that premises are enclosed if they have a ceiling or roof and, except for doors, windows or passageways, are wholly enclosed, whether permanently or on a temporary basis.

Premises are substantially enclosed if they have a ceiling or roof, but have openings in the walls which are less than half of the total areas of walls, including other structures which serve the purpose of walls and constitute the perimeter of premises. No account can be taken of openings in which doors, windows or other fittings that can be open or shut. This is known as the 50% rule.

A roof includes any fixed or movable structures, such as canvas awnings. Tents, marquees or similar will also be classified as enclosed premises if they fall within the definition.

The Regulations are enforced by unitary and lower-tier local authorities and port health authorities within the areas for which they have responsibilities.

In *R (on the application of Black) (Appellant) v Secretary of State for Justice (Respondent)* [2017] UKSC 81, the Supreme Court held that Parliament must have intended that the Crown should not be bound by the smoking ban, otherwise it would have made express provision for it in the Act. The case was brought by a prisoner who argued that prisons should be smoke-free under the Health Act 2006.

The Smoke-free (Exemptions and Vehicles) Regulations 2007 (SI 2007 No 765)

These Regulations, which apply only in England:

- set out the limited exemptions from the smoke-free requirements of section 2 of the Health Act 2006; and
- specify that most public and work vehicles are to be smoke-free under section 5 of the Health Act 2006.

The Regulations also set out a number of exemptions from smoke-free legislation for premises that would otherwise be required to be entirely smoke-free in enclosed parts under smoke-free legislation. These concern:

- private dwellings;
- accommodation for guests and members;
- other residential accommodation;
- offshore installations;
- research and testing facilities; and
- specialist tobacconists.

They also set out provisions regarding performers: Where the artistic integrity of a performance makes it appropriate for a person who is taking part in that performance to smoke, the regulations allow for parts of premises in which a person performs to be not smoke-free in relation to that person only during the time of the performance.

The Regulations require enclosed vehicles to be smoke-free at all times, if they are used:

- by members of the public or a section of the public (whether or not for reward or hire); or
- in the course of paid or voluntary work by more than one person, even if those people use the vehicle at different times, or only intermittently.

An amendment introduced by the *Smoke-free (Private Vehicles) Regulations 2015 (SI 2015 No 286)* means that private vehicles must be smoke-free when:

- they are enclosed;
- there is more than one person in the vehicle; and
- a person under the age of 18 is present in the vehicle.

Vehicles are not required to be smoke-free when they are conveying people if they have a removable or stowable roof during the time the roof is completely removed or stowed.

The Regulations apply to all vehicles, except aircraft, and there are exceptions regarding ships and hovercrafts. There is also an exemption regarding a caravan or motor caravan that is stationary and not on a road; and a caravan or motor caravan that is stationary, is on a road and is being used as living accommodation (*Reg 11 (7)*).

The Smoke-free (Penalties and Discounted Amounts) Regulations 2007 (SI 2007 No 764)

These Regulations apply to England and Wales. There are different levels of fines in Scotland (see www.gov.uk/smoking-at-work-the-law).

The penalties are as follows:

- Workers can be fined up to £200 (up to £50 in Scotland) for smoking in the workplace.
- Businesses can be fined up to £2,500 if they do not stop people smoking in the workplace or up to £1,000 if they do not display 'no smoking' signs.
- In Scotland, there is a fixed penalty fine of £200, which can increase to £2,500 in the event of non-payment.

The Smoke-free (Vehicle Operators and Penalty Notices) Regulations 2007 (SI 2007 No 760)

These Regulations apply only in England and:

- set out who has legal duties corresponding to that in *section 8(1)* of the *Health Act 2006* to cause any person who is smoking in a smoke-free vehicle to stop smoking; and
- specifies the form of the fixed penalty notice for use by enforcement authorities.

The following have legal duties to cause any person who is smoking in a smoke-free vehicle to stop smoking:

- the driver;
- any person with management responsibilities for the vehicle; and
- any person in a vehicle who is responsible for order or safety on it.

These Regulations specify the form of fixed penalty notices to be used by enforcement authorities and how these can be adapted by enforcement authorities.

The Smoke-free (Signs) Regulations 2012 (SI 2012 No 1536)

The *Smoke-free (Signs) Regulations 2012 (SI 2012 No 1536)* replaced the detailed requirements for no smoking signs prescribed by the 2007 Regulations with a requirement that at least one legible no-smoking sign must be displayed. They require that at least one legible no-smoking sign to be displayed in smoke-free vehicles and in smoke-free premises.

People with disabilities

[W11029] *Regulation 25A of the Workplace (Health, Safety and Welfare) Regulations 1992* sets out that where necessary, employers must organise parts of the workplace (including doors, passageways, stairs, showers, washbasins, lavatories and workstations) used or occupied directly by people with disabilities at work to take account of their needs.

HSE guidance to the Regulations sets out that regardless of their disability, people should be able to gain access to buildings and be able to use the facilities. Employers may need to make some changes to a building or premises to take account of the needs of the disabled person. It says that this could include:

- taking into account the structure of a building, for example steps, changes of level, emergency exits or narrow doorways;
- handrails for support on staircases (essential);
- looking at the way the building has been fitted out, for example avoiding heavy doors, inaccessible toilets or inappropriate lighting;
- suitable toilets designed for wheelchair users and disabled people who can walk;
- specially designed cubicles in separate-sex toilet washrooms or a self-contained unisex toilet;
- outward opening doors to compartments; and
- workstation access widened for wheelchairs and changes to the height of workstations.

Further information is available from the Equality and Human Rights Commission (www.equalityhumanrights.com) and in the Building Regulations.

Safety signs at work – Health and Safety (Safety Signs and Signals) Regulations 1996 (SI 1996 No 341)

[W11030] Traditionally, safety signs, communications and warnings have played a residual role in reducing the risk of injury or damage at work, the need for them generally having been engineered out or accommodated in the system of work – a situation unaffected by these Regulations.

Types of signs

[W11031] Safety signs and signals can be of the following types:

(a) permanent (eg signboards);
(b) occasional (eg acoustic signals or verbal communications – acoustic signals should be avoided where there is considerable ambient noise).

Interchanging and combining signs

[W11032] Examples of interchanging and combining signs are:

(a) a safety colour (see **W11033** below) or signboard to mark places where there is an obstacle;
(b) illuminated signs, acoustic signals or verbal communication; and
(c) hand signals or verbal communication.

[*Health and Safety (Safety Signs and Signals) Regulations 1996 (SI 1996 No 341), Sch 1, Part I, para 3*].

Safety colours

[W11033]

Colour	Meaning or purpose	Instructions and information
Red	Prohibition sign	Dangerous behaviour
	Danger alarm	Stop, shutdown, emergency cut-out services
		Evacuate
	Fire-fighting equipment	Identification and location
Yellow or Amber	Warning sign	Be careful, take precautions
		Examine
Blue	Mandatory sign	Specific behaviour or action
		Wear personal protective equipment
Green	Emergency escape, first-aid sign	Doors, exits, routes, equipment and facilities
	No danger	Return to normal

[*Health and Safety (Safety Signs and Signals) Regulations 1996 (SI 1996 No 341), Sch 1, Part I, para 4.*]

Varieties of safety signs and signals

[W11034] Safety signs and signals include, comprehensively:

(a) safety signs – providing information about health and safety at work by means of a signboard, safety colour, illuminated sign, acoustic signal, hand signal or verbal communication;

(b) signboard – a sign providing information or instructions by a combination of geometric shape, colour and a symbol or pictogram, rendered visible by lighting of sufficient intensity;

(c) mandatory signs – signs prescribing behaviour (eg safety boots must be worn);

(d) prohibition signs – signs prohibiting behaviour likely to cause a health and safety risk (eg no smoking);

(e) hand signals – movement or position of arms/hands for guiding persons carrying out operations that could endanger employees;

(f) verbal communications – predetermined spoken messages communicated by human or artificial voice, preferably short, simple and as clear as possible.

[*Health and Safety (Safety Signs and Signals) Regulations 1996 (SI 1996 No 341), Reg 2*].

Duty of employer

[W11035] It is only where a risk assessment carried out under the *Management of Health and Safety at Work Regulations 1999 (SI 1999 No 3242)* indicates that a risk cannot be avoided, engineered out or reduced significantly by way of a system of work that resort to signs and signals becomes necessary. In these circumstances, all employers must:

(a) provide and maintain any appropriate safety sign(s) (see W11037–W11041 below) (including fire safety signals) but not a hand signal or verbal communication;
(b) so far as is reasonably practicable, ensure that correct hand signals or verbal communications are used;
(c) provide and maintain any necessary road traffic sign (where there is a risk to employees in connection with traffic); and
(d) provide employees with comprehensible and relevant information, training and instruction and measures to be taken in connection with safety signs.
[*Health and Safety (Safety Signs and Signals) Regulations 1996 (SI 1996 No 341), Regs 4, 5*]

Schedule 1 to the Regulations sets out the minimum requirements concerning safety signs and signals with regard to the type of signs to be used in particular circumstances, interchanging and combining signs, signboards, signs on containers and pipes, the identification and location of fire-fighting equipment, signs for obstacles and dangerous locations, for marking traffic routes, illuminated signs, acoustic signals, verbal communication and hand signals.

Schedule 1 was amended by the *Classification, Labelling and Packaging of Chemicals (Amendments to Secondary Legislation) Regulations 2015 (SI 2015 No 21)* and now requires that areas, rooms or enclosures used for the storage of significant quantities of hazardous substances or mixtures are indicated by a suitable warning sign, unless the labelling of individual packages or containers is adequate for this purpose.

Exclusions

[W11036] Excluded from the operation of these Regulations are:

(a) signs used in connection with the supply of any hazardous substance, mixture, product or equipment (except to the extent that any enactment which requires such signs makes reference to these Regulations);
(b) the transportation of dangerous goods;
(c) to signs used for regulating road, rail, inland waterway, sea or air traffic. Where there is a risk arising from the movement of traffic and the risk is addressed by a sign stipulated in the *Road Traffic Regulations Act 1984* (eg speed restriction sign), these signs must be used, whether or not the Act applies to that place of work. In effect this means that where road speed and other signs are needed on a company's road, these must replicate the signs used for the purpose on public roads; and
(d) activities on board ship.
[*Health and Safety (Safety Signs and Signals) Regulations 1996 (SI 1996 No 341), Reg 3(1), 4(6)*].

Examples of safety signs

Prohibitory signs

[W11037] Intrinsic features are a round shape and a black pictogram on white background, red edging and diagonal line (the red part to take up at least 35% of the sign area).

fig. 1 Safety signs (prohibitory)

No smoking

Smoking and naked flame forbidden

No access for pedestrians

Do not extinguish with water

Not drinkable

No access for unauthorised persons

No access for industrial vehicles

Do not touch

Warning signs

[W11038] Intrinsic features are a triangular shape; and a black pictogram on a yellow background with black edging (the yellow part to take up at least 50% of the area of the sign).

fig. 2 Safety signs (warning)

Flammable material or high temperature*

*in the absence of a specific sign for high temperature

Explosive material

Toxic material

Corrosive material

Radioactive material

Overhead load

Industrial vehicles

Danger: electricity

General danger

SI 1996/341 **[W11039]**

    Laser beam   Oxidant material   Non-ionizing radiation

    Strong magnetic field   Obstacles   Drop

This sign has been deleted from the list by the UK CLP Regulations and should not be used.

    Biological risk   Low temperature   Harmful or irritant material

Mandatory signs

[W11039] Intrinsic features:

- round shape
- white pictogram on a blue background (the blue part to take up at least 50% of the area of the sign).

fig. 3 Safety signs (mandatory)

Eye protection
must be worn

Safety helmet
must be worn

Ear protection
must be worn

Respiratory equipment
must be worn

Safety boots
must be worn

Safety gloves
must be worn

Safety overalls
must be worn

Face protection
must be worn

Safety harness
must be worn

Pedestrians must use this route

General mandatory sign (to be accompanied where necessary by another sign)

Emergency escape or first-aid signs

[W11040] Intrinsic features:

- rectangular or square shape
- white pictogram on a green background (the green part to take up at least 50% of the area of the sign).

fig. 4 Safety signs (emergency escape or first-aid)

Emergency exit/escape route

This way
(supplementary information sign)

 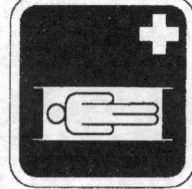  

First-aid post   Stretcher   Safety shower   Eyewash

Emergency telephone for first-aid or escape

Fire-fighting signs

[W11041] Intrinsic features:

- rectangular or square shape
- white pictogram on a red background (the red part to take up at least 50% of the area of the sign).

fig. 5 Safety signs (fire-fighting)

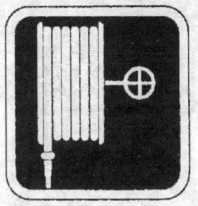
Fire hose

Ladder

Fire extinguisher

Emergency fire telephone

This way
(supplementary information sign)

Examples of hand signals

[W11042]
Meaning   *Description*   *Illustration*

A. General signals

fig 6 Hand signals

START Attention Start of Command	both arms are extended horizontally with the palms facing forwards.	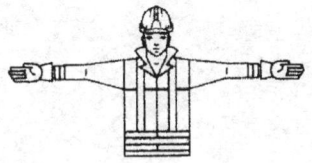

[W11042] Workplaces – Health, Safety and Welfare

| STOP
Interruption
End of movement | the right arm points upwards with the palm facing forwards. |  |

| END
of the operation | both hands are clasped at chest height. |  |

B. Vertical movements

| RAISE | the right arm points upwards with the palm facing forward and slowly makes a circle. |  |

| LOWER | the right arm points downwards with the palm facing inwards and slowly makes a circle. |  |

| VERTICAL DISTANCE | the hands indicate the relevant distance. |  |

SI 1996/341 [**W11042**]

C. Horizontal movements

MOVE FORWARDS	both arms are bent with the palms facing upwards, and the forearms make slow movements towards the body.	
MOVE BACKWARDS	both arms are bent with the palms facing downwards, and the forearms make slow movements away from the body.	
RIGHT to the signalman's	the right arm is extended more or less horizontally with the palm facing downwards and slowly makes small movements to the right.	
LEFT to the signalman's	the left arm is extended more or less horizontally with the palm facing downwards and slowly makes small movements to the left.	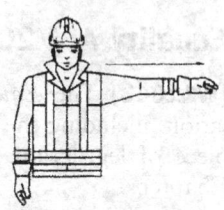
HORIZONTAL DISTANCE	the hands indicate the relevant distance.	

D. Danger

W110-33

DANGER both arms point upwards with
Emergency stop the palms facing forwards.

QUICK   all movements faster.

SLOW   all movements slower.

The HSE publication, *Safety signs and signals – The Health and Safety (Safety Signs and Signals) Regulations 1996 – Guidance on Regulations*, can be downloaded from its website at: www.hse.gov.uk/pubns/priced/l64.pdf. This revised edition was published in 2015 and brings the document up to date with regulatory and other changes, including those relating to the *Classification, Labelling and Packaging of Chemicals (Amendments to Secondary Legislation) Regulations 2015* (see above).

Equality Act 2010

[W11043] The *Equality Act 2010* brought previous discrimination legislation, including the *Disability Discrimination Act (DDA) 1995* into a single piece of legislation. The Act is set out in detail in the Equality Act 2010 Chapter.

It requires employers to make reasonable adjustments to working conditions or to the workplace to avoid putting disabled workers at a substantial disadvantage, and could involve making adjustments to the physical features of workplace premises.

Employers must make reasonable adjustments for staff to help them overcome disadvantage resulting from an impairment (for example by providing assistive technologies to help visually impaired staff use computers effectively).

More detailed guidance for employers on complying with disability discrimination legislation is provided by the Government Equalities Office at www.gov.uk/government/organisations/government-equalities-office and the Equality and Human Rights Commission at www.equalityhumanrights.com/.

Building Regulations

[W11044] Building regulations contain standards for building work in new and altered buildings (including workplaces) to make them safe and accessible and limit waste and environmental damage. In most cases, building work must be checked by an independent third party to make sure that the work meets the required standards.

The Regulations currently in force are the *Building Regulations 2010 (SI 2010 No 2214)*.

The Regulations impose requirements on people carrying out 'building work' which is defined in *Reg 3* as:

- the erection or extension of a building;
- the provision or extension of a controlled service or fitting in or in connection with a building;
- the material alteration of a building or controlled service or fitting;
- work required in relation to a material change of use;
- insertion of insulating material into a cavity wall;
- work involving underpinning of a building; and
- work to improve the energy performance of a building.

Building work must be carried out so that it complies with requirements set out in Parts A to P of Schedule 1 of the Regulations (Regulation 4). These relate to structure (Part A), fire safety (Part B), site preparation and resistance to contaminants and moisture (Part C), toxic substances (Part D), sound resistance (Part E), ventilation (Part F), sanitation, hot water safety and water efficiency (Part G), drainage and waste disposal (Part H), combustion appliances and fuel storage systems (Part J), protection from falling, collision and impact (Part K), conservation of fuel and power (Part L), access to and use of buildings (Part M), glazing – safety in relation to impact, opening and cleaning (Part N) and electrical safety (Part P).

Not all the provisions of Schedule 1 apply to all building work.

For more information on amendments to the regulations and where to find further guidance, see www.gov.uk/government/policies/providing-effective-building-regulations-so-that-new-and-altered-buildings-are-safe-accessible-and-efficient.

Following the Grenfell Tower fire in June 2017, in which around 70 people died, the Government appointed Dame Judith Hackitt to lead an independent review of building regulations and fire safety.

She published an interim report in December 2017. This sets out that the current system for ensuring fire safety in high-rise buildings is not fit for purpose. She said that 'the regulatory system for safely designing, constructing and managing buildings is not fit for purpose'. She found that the current system is highly complex and there is confusion about the roles and responsibilities at each stage. In many areas she said there was a lack of competence and accreditation.

The interim report sets out six broad areas for change:

- ensuring that regulation and guidance is risk-based, proportionate and unambiguous;
- clarifying roles and responsibilities for ensuring that buildings are safe;
- improving levels of competence within the industry;
- improving the process, compliance and enforcement of regulations;
- creating a clear, quick and effective route for residents' voices to be heard and listened to; and
- improving testing, marketing and quality assurance of products used in construction.

The final report is due to be published in Spring 2018. A copy of the interim report can be found online at: www.gov.uk/government/publications/independent-review-of-building-regulations-and-fire-safety-interim-report.

Factories Act 1961

[W11045] Although large parts of the *Factories Act 1961* have now been repealed and replaced by more recent legislation, a few provisions are still in force. Additionally, civil actions for injury, relating to breach of health and safety provisions of the *Factories Act* (though not welfare) may well continue for some time, since actions for personal injury can be initiated for up to three years after the injury or disease has occurred [*Limitation Act 1980, s 11*].

Residual application of the Factories Act 1961

[W11046] The *Factories Act 1961* applies to factories, as defined in *s 175*, including 'factories belonging to or in the occupation of the Crown, to building operations and works of engineering construction undertaken by or on behalf of the Crown, and to employment by or under the Crown of persons in painting buildings', eg hospital painters. [*Factories Act 1961, s 173(1)*].

Enforcement of the Factories Act 1961 and regulations

[W11047] Offences under the Act are normally committed by occupiers rather than owners of factories. Unless they happen to occupy a factory as well, the owners of a factory would not normally be charged. Offences therefore relate to physical occupation or control of a factory (for an extended meaning of 'occupier', see OCCUPIERS' LIABILITY). Hence the person or persons or body corporate having managerial responsibility in respect of a factory are those who commit an offence under *s 155(1)*. This will generally be the managing director and board of directors and/or individual executive directors. Moreover, if a company is in liquidation and the receiver is in control, he is the person who will be prosecuted and this has in fact happened (*Meigh v Wickenden* [1942] 2 KB 160; *Lord Advocate v Aero Technologies* 1991 SLT 134 where the receiver was 'in occupation' and so under a duty to prevent 'accidents by fire or explosion', for the purposes of the *Explosives Act 1875, s 23*).

Defence of factory occupier

[W11048] The main defence open to a factory occupier charged with breach of the *Factories Act 1961* is that the Act itself, or more likely regulations made under it, placed the statutory duty on some person other than the occupier. Thus, where there is a contravention by any person of any regulation or order under the *Factories Act 1961*, 'that person shall be guilty of an offence and the occupier or owner . . . shall not be guilty of an offence, by reason only of the contravention of the provision . . . unless it is proved that he failed to take all reasonable steps to prevent the contravention . . . '. [*Factories Act 1961, s 155(2)*].

Before this defence can be invoked by a factory occupier or company, it is necessary to show that:

(a) a statutory duty had been laid on someone other than the factory occupier by a regulation or order passed under the Act;
(b) the factory occupier took all reasonable steps to prevent the contravention (a difficult test to satisfy).

NB. This statutory defence is not open to a building contractor (in his capacity as a notional factory occupier).

Table of Cases

A

A v National Blood Authority [2001] 3 All ER 289, [2001] Lloyd's Rep Med 187, 60 BMLR 1, [2001] All ER (D) 298 (Mar) P9032
AB v Ministry of Defence [2012] UKSC 9, [2013] 1 AC 78, [2012] 3 All ER 673, [2012] 2 WLR 643, 125 BMLR 69, [2012] 22 LS Gaz R 19, [2012] NLJR 426, (2012) Times, 27 March, 156 Sol Jo (no 11) 31, [2013] 1 LRC 1, [2012] All ER (D) 108 (Mar) ... P9039
AM v WC [1999] ICR 1218, [1999] IRLR 410, EAT H1714
AMF International Ltd v Magnet Bowling Ltd [1968] 2 All ER 789, [1968] 1 WLR 1028, 66 LGR 706, 112 Sol Jo 522 .... O3015
Adams v Southern Electricity Board (1993) Times, 21 October, CA O3017
Addie (Robert) & Sons (Collieries) Ltd v Dumbreck [1929] AC 358, 98 LJPC 119, 34 Com Cas 214, [1929] All ER Rep 1, 140 LT 650, 45 TLR 267, HL O3008
Addison t/a Brayton News v Ashby UKEAT, (17 January 2003, unreported) V12004
Adsett v K and L Steelfounders and Engineers Ltd [1953] 2 All ER 320, [1953] 1 WLR 773, 51 LGR 418, 97 Sol Jo 419, CA R3004
Adshead v May Gurney Ltd [2006] All ER (D) 388 (Jul), EAT W10024
Advocate (Lord) v Aero Technologies Ltd (in receivership) 1991 SLT 134, Ct of Sess
.... W11047
Ahmed v The Cardinal Hume Acadamies (UKEAT/0196/18/RN) unreported H1708
Ainsworth v IRC [2005] EWCA Civ 441, [2005] IRLR 465, [2005] NLJR 744, (2005) Times, 16 May, [2005] All ER (D) 328 (Apr), sub nom IRC v Ainsworth [2005] ICR 1149; revsd sub nom Revenue and Customs v Stringer (sub nom Ainsworth v IRC) [2009] UKHL 31, [2009] ICR 985, [2009] IRLR 677, (2009) Times, 15 June, [2009] All ER (D) 168 (Jun) W10024B
Alcock v Chief Constable of South Yorkshire Police [1992] 1 AC 310, [1991] 4 All ER 907, [1991] 3 WLR 1057, 8 BMLR 37, [1992] 3 LS Gaz R 34, 136 Sol Jo LB 9, HL C6033
Allison v London Underground Ltd [2008] EWCA Civ 71, [2008] ICR 719, [2008] IRLR 440, [2008] PIQR P185, 152 Sol Jo (no 8) 34, [2008] All ER (D) 185 (Feb)
.... M1007
Aluminium Wire and Cable Co Ltd v Allstate Insurance Co Ltd [1985] 2 Lloyd's Rep 280 .... E13020
Andreas Kainz v Pantherwerke AG: C-45/13 (2014) ECLI:EU:C:2014:7, [2015] QB 34, [2014] 3 WLR 1292, [2014] 1 All ER (Comm) 433, [2014] All ER (D) 13 (Feb), ECJ P9046
Anderson v Jarvis Hotels plc (EAT/0062/05) (30 May 2006, unreported) W10003
Anderson v Newham College of Further Education [2002] EWCA Civ 505, [2003] ICR 212, [2002] All ER (D) 381 (Mar) O3011
Anslow v Norton Aluminium Ltd [2012] EWHC 2610 (QB), [2013] All ER (D) 03 (Jan) E5062
Anyanwu v South Bank Student Union (Commission for Racial Equality, interveners) [2001] UKHL 14, [2001] 2 All ER 353, [2001] 1 WLR 638, [2001] ICR 391, [2001] IRLR 305, [2001] ELR 511, [2001] 21 LS Gaz R 39, 151 NLJ 501, [2001] All ER (D) 272 (Mar) H1714
Aparau v Iceland Frozen Foods plc [1996] IRLR 119, EAT; revsd [2000] 1 All ER 228, [2000] IRLR 196, [1999] 45 LS Gaz R 31, CA E14004
Arafa v Potter. See Potter v Arafa
Archibald v Fife Council [2004] UKHL 32, [2004] 4 All ER 303, [2004] ICR 954, [2004] IRLR 651, 82 BMLR 185, [2004] 31 LS Gaz R 25, (2004) Times, 5 July, 2004 SC 942, 2004 SCLR 971, 148 Sol Jo LB 826, [2004] All ER (D) 32 (Jul) E16511
Armour v Skeen [1977] IRLR 310, 1977 JC 15, 1977 SLT 71, HC of Justiciary (Sc)
.... E15040, S7002

Table of Cases

Arnold v Central Electricity Generating Board [1988] AC 228, [1986] 3 WLR 171, 130 Sol Jo 484, [1986] LS Gaz R 2090; revsd [1988] AC 228, [1987] 2 WLR 245, 131 Sol Jo 167, [1987] LS Gaz R 743, CA; affd [1988] AC 228, [1987] 3 All ER 694, [1987] 3 WLR 1009, 131 Sol Jo 1487, [1987] LS Gaz R 3416, [1987] NLJ Rep 1014, HL N3022
Arthur v Anker [1997] QB 564, [1996] 3 All ER 783, [1996] 2 WLR 602, 72 P & CR 309, [1996] RTR 308, [1996] NLJR 86, CA O3017
Aspden v Webbs Poultry & Meat Group (Holdings) Ltd [1996] IRLR 521 .... E14004
Associated Dairies Ltd v Hartley [1979] IRLR 171, Ind Trib E15024
Asociacion Nacional de Grandes Empresas de Distribucion v Federacion de Asociaciones Sindicales: C-78/11 (2012) C-78/11, [2012] IRLR 779, [2012] ICR 1211, [2012] NLJR 912, [2012] All ER (D) 228 (Jun) W10024B
Attia v British Gas plc [1988] QB 304, [1987] 3 All ER 455, [1987] 3 WLR 1101, [1987] BTLC 394, 131 Sol Jo 1248, [1987] LS Gaz R 2360, [1987] NLJ Rep 661, CA C6033
Austin Rover Group Ltd v HM Inspector of Factories [1990] 1 AC 619, [1989] 3 WLR 520, [1990] ICR 133, [1989] IRLR 404, sub nom Mailer v Austin Rover Group plc [1989] 2 All ER 1087, HL O3019
Awan v ICTS UK Ltd [2018] UKEAT/0087/18/RN (15 November 2018, unreported) E14004

B

BP plc v Elstone (UKEAT/0141/09/DM) [2011] 1 All ER 718, [2010] ICR 879, [2010] IRLR 558, [2010] All ER (D) 140 (May) E14011.1
BS v Dundee City Council [2013] CSIH 91, [2014] IRLR 131, 2014 SC 254 .... M1517
Bacica v Muir [2006] IRLR 35, EAT W10004
Baker v Quantum Clothing Group [2009] EWCA Civ 499, [2009] PIQR P332, (2009) Times, 18 June, 153 Sol Jo (no 21) 29, [2009] All ER (D) 205 (May); revsd [2011] UKSC 17, [2011] 4 All ER 223, [2011] 1 WLR 1003, [2011] ICR 523, [2011] 17 LS Gaz R 13, (2011) Times, 14 April, 155 Sol Jo (no 15) 38, [2011] All ER (D) 137 (Apr) E15031, E15039, N3022
Baker v T E Hopkins & Son Ltd [1959] 3 All ER 225, [1959] 1 WLR 966, 103 Sol Jo 812, CA A1022
Balding v Lew-Ways Ltd (1995) 159 JP 541, [1995] Crim LR 878, DC P9545
Balgobin v Tower Hamlets London Borough Council [1987] ICR 829, [1987] IRLR 401, [1987] LS Gaz R 2530, EAT H1709, H1713, H1720, H1738
Bamsey v Albon Engineering & Manufacturing plc [2004] EWCA Civ 359, [2004] 2 CMLR 1353, [2004] IRLR 457, [2004] 17 LS Gaz R 31, (2004) Times, 15 April, 148 Sol Jo LB 389, [2004] All ER (D) 482 (Mar) W10024
Banaszczyk v Booker Ltd (2016) UKEAT/0132/15, [2016] IRLR 273, [2016] All ER (D) 173 (Feb), EAT M1507, M3007, M3007.1, R2005, V12032
Barber v RJB Mining (UK) Ltd [1999] 2 CMLR 833, [1999] ICR 679, [1999] IRLR 308, 143 Sol Jo LB 141, [1999] All ER (D) 244 .... E14002, W10010, W10042
Barber v Somerset County Council [2002] EWCA Civ 76, [2002] 2 All ER 1, [2002] ICR 613, [2002] IRLR 263, 68 BMLR 115, (2002) Times, 11 February, [2002] All ER (D) 53 (Feb); revsd [2004] UKHL 13, [2004] 2 All ER 385, [2004] 1 WLR 1089, [2004] ICR 457, [2004] IRLR 475, 77 BMLR 219, (2004) Times, 5 April, 148 Sol Jo LB 419, [2004] All ER (D) 07 (Apr) E14003, R2006, S11009
Barclays Bank plc v Kapur (No 2) [1995] IRLR 87, CA H1708
Barker v Corus UK Ltd [2006] UKHL 20, [2006] 2 AC 572, [2006] 3 All ER 785, [2006] 2 WLR 1027, [2006] ICR 809, 89 BMLR 1, [2006] NLJR 796, [2006] PIQR P390, (2006) Times, 4 May, [2006] 5 LRC 271, [2006] All ER (D) 23 (May) P9039
Barr v Biffa Waste Services Ltd [2012] EWCA Civ 312, [2013] QB 455, [2012] 3 All ER 380, [2012] 3 WLR 795, [2012] 2 P & CR 99, [2012] PTSR 1527, [2012] HLR 415, 141 ConLR 1, [2012] 2 EGLR 157, [2012] 14 LS Gaz R 21, [2012] BLR 275, [2012] 13 EG 90 (CS), 156 Sol Jo (no 12) 31, [2012] All ER (D) 141 (Mar) E5062
Bartoline v Heath Lambert and RSA [2006] EWHC 3598 (QB) P9526
Barton v SCC Ltd (1999), unreported W10024
Barton v Wandsworth Council (1995) IDS Brief 549 .... E14009

Baxter v Harland & Wolff plc [1990] IRLR 516, NI CA N3022
Beatt v Croydon Health Services NHS Trust [2017] EWCA Civ 401, [2017] IRLR 748, [2017] ICR 1240, [2017] All ER (D) 06 (Jun) E14011.1
Beattie v Secretary of State for Social Security [2001] EWCA Civ 498, [2001] 1 WLR 1404, 145 Sol Jo LB 120, [2001] All ER (D) 93 (Apr) C6041
Belhaven Brewery Co v McLean [1975] IRLR 370, Ind Trib E15024
Bellman v Northampton Recruitment Ltd [2018] EWCA Civ 2214, [2019] 1 All ER 1133, [2019] IRLR 66, 168 NLJ 7813, [2018] All ER (D) 54 (Oct) V8013
Bermingham v Sher Bros 1980 SC 67, 1980 SLT 122, HL O3014
Berry v Stone Manganese & Marine Ltd [1972] 1 Lloyd's Rep 182, 12 KIR 13, 115 Sol Jo 966 .... N3022
Berryman v Hounslow London Borough Council (1996) 30 HLR 567, [1997] PIQR P 83, CA O3004, O3006
Bett v Dalmey Oil Co (1905) 7F (Ct of Sess) 787 .... S3003
Betts v Tokley (t/a PDQ Cars Minibuses Limousines & Couriers) [2002] All ER (D) 99 (Jan), CA O3011
Bews v Scottish Hydro Electric plc 1992 SLT 749, Ct of Sess C6053
Bhatt v Fontain Motors Ltd [2010] EWCA Civ 863, [2010] All ER (D) 275 (Jul), CSRC vol 34 iss 9/2 .... W9008
Biesheuval v Birrell [1999] PIQR Q 40 .... C6023
Bishop v Baker Refractories Ltd [2002] EWCA Civ 76, [2002] 2 All ER 1, [2002] ICR 613, [2002] IRLR 263, 68 BMLR 115, (2002) Times, 11 February, [2002] All ER (D) 53 (Feb); revsd [2004] UKHL 13, [2004] 2 All ER 385, [2004] 1 WLR 1089, [2004] ICR 457, [2004] IRLR 475, 77 BMLR 219, (2004) Times, 5 April, 148 Sol Jo LB 419, [2004] All ER (D) 07 (Apr) R2006
Blakley v South Eastern Health and Social Services Trust [2009] NICA 62 .... W10003
Bixby, Case, Fry and Elliott v Ford Motor Group (1990), unreported N3022
Bottomley v Todmorden Cricket Club [2003] EWCA Civ 1575, [2003] 48 LS Gaz R 18, (2003) Times, 13 November, 147 Sol Jo LB 1309, [2003] All ER (D) 102 (Nov) O3014
Boston Scientific Medizintechnik GmbH v AOK Sachsen-Anhalt - Die Gesundheitskasse: Joined Cases C-503/13 and C-504/13 (2015) ECLI:EU:C:2015:148, 144 BMLR 225, [2015] All ER (D) 88 (Mar), ECJ P9032
Bowe v Mersey Rewinds Engineering Ltd [2018] EWCA Civ 72, 168 NLJ 7768, [2018] All ER (D) 11 (Feb) V5003, V5012
Bracebridge Engineering Ltd v Darby [1990] IRLR 3, EAT H1706, H1720, H1723
Bradburn v Great Western Rly Co (1874) LR 10 Exch 1, 44 LJ Ex 9, 23 WR 48, [1874–80] All ER Rep 195, 31 LT 464 .... C6053, E13007, P9509
Bradley v Eagle Star Insurance Co Ltd [1989] AC 957, [1989] 1 All ER 961, [1989] 2 WLR 568, [1989] 1 Lloyd's Rep 465, [1989] ICR 301, [1989] BCLC 469, 133 Sol Jo 359, [1989] 17 LS Gaz R 38, [1989] NLJR 330, HL E13024
Brice v Brown [1984] 1 All ER 997, 134 NLJ 204 .... C6033
Brioland Ltd v Searson [2005] EWCA Civ 55, sub nom Searson v Brioland Ltd [2005] 05 EG 202 (CS), [2005] All ER (D) 197 (Jan) O3002
British Airways Ltd v Moore [2000] IRLR 296 .... E14020, E14030
British Home Stores Ltd v Burchell [1980] ICR 303n, [1978] IRLR 379, 13 ITR 560, EAT E14013
British Gas Trading Ltd v Lock (2016) UKEAT/0189/15, [2016] 2 CMLR 1189, [2016] IRLR 316, [2016] ICR 503, [2016] All ER (D) 273 (Feb), EAT W10024
British Railways Board v Herrington [1972] AC 877, [1972] 1 All ER 749, [1972] 2 WLR 537, 116 Sol Jo 178, 223 Estates Gazette 939, HL F5073, O3008
British Transport Commission v Gourley [1956] AC 185, [1955] 3 All ER 796, [1956] 2 WLR 41, [1955] 2 Lloyd's Rep 475, 34 ATC 305, 49 R & IT 11, [1955] TR 303, 100 Sol Jo 12, HL C6022, C6026
Brookes v South Yorkshire Passenger Transport Executive [2005] EWCA Civ 452, [2005] All ER (D) 405 (Apr) V5012
Brown v Controlled Packaging Services Ltd (1999), unreported W10010
Brumfitt v Ministry of Defence [2005] IRLR 4, 148 Sol Jo LB 1028, [2004] All ER (D) 479 (Jul), EAT H1707
Buchanan v The Comr of Police of the Metropolis (2016) UKEAT/0112/16, [2016] IRLR 918, [2017] ICR 184, [2016] All ER (D) 29 (Oct), EAT M1510

Bunker v Charles Brand & Son Ltd [1969] 2 QB 480, [1969] 2 All ER 59, [1969] 2 WLR 1392, 113 Sol Jo 487 .... O3005
Bunning v GT Bunning & Sons Ltd [2005] EWCA Civ 983, [2005] All ER (D) 390 (Jul), CSRC vol 29 iss 1/1 .... R3026
Burnett v British Waterways Board [1973] 2 All ER 631, [1973] 1 WLR 700, [1973] 2 Lloyd's Rep 137, 117 Sol Jo 203, CA O3017
Burton v De Vere Hotels Ltd [1997] ICR 1, [1996] IRLR 596, EAT H1706
Butcher v Southend-on-Sea [2014] EWCA Civ 1556, [2014] All ER (D) 87 (Dec) ... O3003
Bux v Slough Metals Ltd [1974] 1 All ER 262, [1973] 1 WLR 1358, [1974] 1 Lloyd's Rep 155, 117 Sol Jo 615, CA P3001, P3026
Byrne Brothers (Formwork) Ltd v Baird [2002] ICR 667, [2002] IRLR 96, [2001] All ER (D) 321 (Nov), EAT W10004

C

Calder v Secretary of State for Work and Pensions (UKEAT/0512/08/LA) [2009] All ER (D) 106 (Aug) J3011
Campbell v Mylchreest [1999] PIQR Q 17, CA C6041
Campion v Hughes [1975] IRLR 291, Ind Trib E15024
Canada Life Ltd v Gray [2004] ICR 673, sub nom Gray v Canada Life Ltd [2004] All ER (D) 36 (Jan), EAT W10024B
Canterbury City Council v Howletts & Port Lympne Estates Ltd. See Langridge v Howletts & Port Lympne Estates Ltd
Cape Intermediate Holdings Ltd v Graham Dring [2018] EWCA Civ 1795, [2018] All ER (D) 16 (Aug) A5042.2
Capitano v Leeds Eastern Health Authority (1989) Current Law, October O3006
Cappoci v Bloomsbury Health Authority (21 January 2000, unreported) C6023
Carragher v Singer Manufacturing Ltd 1974 SLT (Notes) 28, Ct of Sess N3022
Castellain v Preston (1883) 11 QBD 380, 52 LJQB 366, 31 WR 557, [1881–5] All ER Rep 493, 49 LT 29, CA E13007, P9509
Caulfield v Marshalls Clay Products Ltd [2004] EWCA Civ 422, [2004] 2 CMLR 1040, [2004] ICR 1502, [2004] IRLR 564, 148 Sol Jo LB 539, [2004] All ER (D) 292 (Apr) W10024A
Century Insurance Co Ltd v Northern Ireland Road Tranport Board [1942] AC 509, [1942] 1 All ER 491, 111 LJPC 138, 167 LT 404, HL A1020
Chadwick v British Transport Commission (or British Railways Board) [1967] 2 All ER 945, [1967] 1 WLR 912, 111 Sol Jo 562 .... C6033
Chandler v Cape plc [2012] EWCA Civ 525, [2012] 3 All ER 640, [2012] 1 WLR 3111, [2012] ICR 1293, [2012] NLJR 621, [2012] All ER (D) 123 (Apr) A5042.2
Chesterton Global Ltd v Nurmohamed (Public Concern at Work intervening) [2017] EWCA Civ 979 .... E14011.1
Chessington World of Adventures Ltd v Reed. See A v B, ex p News Group Newspapers Ltd
Chief Adjudication Officer v Faulds [2000] 2 All ER 961, [2000] 1 WLR 1035, [2000] ICR 1297, (2000) Times, 16 May, 2000 SC (HL) 116, 2000 SLT 712, HL C6004
Chief Adjudication Officer v Rhodes [1999] ICR 178, [1999] IRLR 103, [1998] 36 LS Gaz R 32, 142 Sol Jo LB 228, CA C6006
Chief Constable of Lincolnshire Police v Stubbs [1999] ICR 547, [1999] IRLR 81, EAT H1706, H1712
Chrysler (UK) Ltd v McCarthy [1978] ICR 939 .... E15016
City of York Council v Grosset [2018] EWCA Civ 1105, [2018] 4 All ER 77, [2018] IRLR 746, [2018] ICR 1492, [2018] ELR 445, [2018] All ER (D) 88 (May) E14020
Clare v Perry (t/a Widemouth Manor Hotel) [2005] EWCA Civ 39, [2005] All ER (D) 67 (Jan) O3002
Clark v Hosier & Dickinson Ltd (in liq) [2003] EWCA Civ 1467, [2003] All ER (D) 225 (Oct) O3015
Clark v Nomura International plc [2000] IRLR 766 .... M1513
Coates v Jaguar Cars Ltd [2004] EWCA Civ 337, [2004] All ER (D) 87 (Mar) O3018
Coia v Portavadie Estates Ltd [2015] CSIH 3, 2015 SC 419, 2015 SCLR 479 ...... M1005

Coleman v EBR Attridge Law LLP [2010] 1 CMLR 846, [2010] ICR 242, [2010] IRLR 10, (2009) Times, 5 November, 153 Sol Jo (no 42) 28, [2009] All ER (D) 14 (Nov), EAT H1703
College of Ripon and York St John v Hobbs [2002] EWCA Civ 1074, EAT E16510
Collins v First Quench Retailing Ltd [2003] SLT 1220, 2003 SCLR 205, OH V8006
Collins v Wilcock [1984] 3 All ER 374, [1984] 1 WLR 1172, 79 Cr App Rep 229, 148 JP 692, [1984] Crim LR 481, 128 Sol Jo 660, [1984] LS Gaz R 2140, DC . H1723, H1730
Condogianis v Guardian Assurance Co [1921] 2 AC 125, 3 Ll L Rep 40, 90 LJPC 168, 125 LT 610, 37 TLR 685, PC E13005, P9505
Conn v Sunderland City Council [2007] EWCA Civ 1492, [2008] IRLR 324, [2007] All ER (D) 99 (Nov) H1726
Cook v Bradford Community Health NHS Trust [2002] EWCA Civ 1616, [2002] All ER (D) 329 (Oct), CA V8005
Corr (administratrix of Corr dec'd) v IBC Vehicles Ltd [2006] EWCA Civ 331, [2007] QB 46, [2006] 2 All ER 929, [2006] 3 WLR 395, [2006] ICR 1138, (2006) Times, 21 April, [2006] All ER (D) 466 (Mar); affd sub nom Corr (Administratrix of Corr dec'd) v IBC Vehicles Ltd [2008] UKHL 13, [2008] 1 AC 884, [2008] 2 All ER 943, [2008] 2 WLR 499, [2008] ICR 372, [2008] PIQR P207, (2008) Times, 28 February, 152 Sol Jo (no 9) 30, [2008] 5 LRC 124, [2008] All ER (D) 386 (Feb) C6033
Cotswold Developments Construction Ltd v Williams [2006] IRLR 181, [2005] All ER (D) 355 (Dec), EAT W10004
Cotton v Derbyshire Dales District Council (1994) Times, 20 June, CA O3017
Courtaulds Northern Spinning Ltd v Sibson [1988] ICR 451, [1988] IRLR 305, 132 Sol Jo 1033, CA E14004
Coventry City Council v Ackerman Group plc [1995] Crim LR 140, DC P9008
Cox v H C B Angus Ltd [1981] ICR 683 .... A1021
Cox v Hockenhull [1999] 3 All ER 577, [2000] 1 WLR 750, [1999] RTR 399, CA C6037
Cox v Ministry of Justice [2016] UKSC 10, [2016] AC 660, [2016] 2 WLR 806, [2016] IRLR 370, [2016] ICR 470, [2016] PIQR P139, 166 NLJ 7690, (2016) Times, 10 March, [2016] 3 LRC 468, [2016] All ER (D) 25 (Mar) V8013
Coxall v Goodyear Great Britain Ltd [2002] EWCA Civ 1010, [2003] 1 WLR 536, [2003] ICR 152, [2002] IRLR 742, (2002) Times, 5 August, [2002] All ER (D) 303 (Jul) R2006
Crawley v Mercer (1984) Times, 9 March C6052
Cresswell v Eaton [1991] 1 All ER 484, [1991] 1 WLR 1113 .... C6039
Crouch v British Rail Engineering Ltd [1988] IRLR 404, CA P3001
Cruickshank v VAW Motorcast [2002] ICR 729, [2002] IRLR 24, [2001] All ER (D) 372 (Oct), EAT E16510
Cunningham v Harrison [1973] QB 942, [1973] 3 All ER 463, [1973] 3 WLR 97, 117 Sol Jo 547, CA C6053

D

Da'Bell v National Society for Prevention of Cruelty to Children [2010] IRLR 19, [2009] All ER (D) 219 (Nov), EAT H1715
Dacas v Brook Street Bureau (UK) Ltd [2004] EWCA Civ 217, [2004] ICR 1437, [2004] IRLR 358, (2004) Times, 19 March, [2004] All ER (D) 125 (Mar) E13016
Daniel v Secretary of State for the Department of Health [2014] EWHC 2578 (QB), [2014] All ER (D) 290 (Jul) S11009
Darby v National Trust [2001] EWCA Civ 189, [2001] All ER (D) 216 (Jan) O3017
Dashiell v Luttitt [2000] 3 QR 4 .... C6023
Davie v New Merton Board Mills Ltd [1959] AC 604, [1959] 1 All ER 346, [1959] 2 WLR 331, 103 Sol Jo 177, HL P9025
Daw v Intel Corpn (UK) Ltd [2007] EWCA Civ 70, [2007] 2 All ER 126, [2007] ICR 1318, [2007] NLJR 259, [2007] All ER (D) 96 (Feb), sub nom Intel Corpn (UK) Ltd v Daw [2007] IRLR 355 .... M1524
Day v T Pickles Farms Ltd [1999] IRLR 217, EAT E14019, E14025

Table of Cases

Deary v Mansion Hide Upholstery Ltd [1983] ICR 610, [1983] IRLR 195, 147 JP 311 .... E15021
Deeley v Effer (1977) COIT No 1/72, Case No 25354/77, unreported E15023
Dhaliwal v Richmond Pharmacology [2009] ICR 724, [2009] IRLR 336, [2009] All ER (D) 158 (Feb), EAT H1702, H1706, H1708
Dickins v O2 plc [2008] EWCA Civ 1144, [2009] IRLR 58, [2008] All ER (D) 154 (Oct) E14003, M1524
Dietrich v Westdeutscher Rundfunk: C-11/99 [2000] ECR I-5589, ECJ D8202
Director of the Serious Fraud Office v Eurasian Natural Resources Corpn Ltd [2017] EWHC 1017 (QB), 167 NLJ 7746, [2017] All ER (D) 50 (May) E15026
Doherty v Rugby Joinery (UK) Ltd [2004] EWCA Civ 147, [2004] ICR 1272, (2004) Times, 3 March, 148 Sol Jo LB 235, [2004] All ER (D) 292 (Feb) V5012
Doleman v Deakin (1990) Times, 30 January, CA C6035
Donaldson v Hays Distribution Services Ltd [2005] CSIH 48, 2005 SLT 733, 2005 SCLR 717 .... O3018
Donoghue (or McAlister) v Stevenson. See M'Alister (or Donoghue) v Stevenson
Dornan v Department of the Environment for Northern Ireland [1993] NI 1, NI CA O3010
Dr Day v Health Education England [2017] EWCA Civ 329, [2017] IRLR 623, [2017] ICR 917, 158 BMLR 76, 167 NLJ 7746, [2017] All ER (D) 71 (May) E14011.1
Driskel v Peninsula Business Services Ltd [2000] IRLR 151, EAT H1707
Dunham v Ashford Windows [2005] IRLR 608, [2005] ICR 1584, [2005] All ER (D) 104 (Jun), EAT E16510
Durham v BAI (Run Off) Ltd (in scheme of arrangement) [2012] UKSC 14, [2012] 3 All ER 1161, [2012] 1 WLR 867, [2012] 2 All ER (Comm) 1187, [2012] Lloyd's Rep IR 371, [2012] ICR 574, 125 BMLR 137, [2012] NLJR 502, 156 Sol Jo (no 13) 31, [2012] All ER (D) 201 (Mar) A5042.2
Dunn v British Coal Corpn [1993] ICR 591, [1993] IRLR 396, [1993] 15 LS Gaz R 37, [1993] PIQR P 275, 137 Sol Jo LB 81, CA C6023
Dunn v Ovalcode Ltd (t/a UKR) [2003] All ER (D) 241 (Apr), EAT J3030

E

EC Commission v United Kingdom: C-484/04 [2006] ECR I-7471, [2006] 3 CMLR 1322, [2007] ICR 592, [2006] IRLR 888, (2006) Times, 21 September, [2006] All ER (D) 32 (Sep), ECJ W10014
Edwards v Encirc Ltd (2015) UKEAT/0367/14, [2015] IRLR 528, [2015] All ER (D) 344 (Feb), EAT W10003
Edwards v National Coal Board [1949] 1 KB 704, [1949] 1 All ER 743, 93 Sol Jo 337, 65 TLR 430, CA E15039, R3003, R3004
English v Thomas Sanderson Blinds Ltd [2008] ICR 607, [2008] IRLR 342, [2008] All ER (D) 282 (Feb), EAT; revsd [2008] EWCA Civ 1421, [2009] 2 All ER 468, [2009] 2 CMLR 437, [2009] ICR 543, [2009] IRLR 206, (2009) Times, 5 January, 153 Sol Jo (no 1) 31, [2008] All ER (D) 219 (Dec) H1707
Equal Opportunities Commission v Secretary of State for Trade and Industry [2007] EWHC 483 (Admin), [2007] 2 CMLR 1351, [2007] ICR 1234, [2007] IRLR 327, [2007] All ER (D) 183 (Mar) H1702, H1706
Esdale v Dover District Council [2010] EWCA Civ 409, [2010] All ER (D) 73 (Oct) O3002
Evans v Kosmar Villa Holidays plc [2007] EWCA Civ 1003, [2008] 1 All ER 530, [2008] 1 All ER (Comm) 721, [2008] 1 WLR 297, [2008] PIQR P126, [2007] All ER (D) 330 (Oct) O3017
Evans v Malley Organisation Ltd (t/a First Business Support) [2002] EWCA Civ 1834, [2003] ICR 432, [2003] IRLR 156, (2003) Times, 23 January, [2002] All ER (D) 397 (Nov) W10024

F

Fag og Arbejde (FOA), acting on behalf of Karsten Kaltoft v Kommunernes Landsforening (KL), acting on behalf of Billund Kommune: C-354/13 (2014) ECLI:EU:C:2014:2463, [2015] All ER (EC) 265, [2015] 2 CMLR 500, [2015] ICR 322, [2015] IRLR 146, (2014) Times, 26 December, [2014] All ER (D) 220 (Dec), ECJ M1509
Fairchild v Glenhaven Funeral Services Ltd (1978) Ltd [2002] UKHL 22, [2003] 1 AC 32, [2002] 3 All ER 305, [2002] ICR 798, [2002] IRLR 533, 67 BMLR 90, [2002] NLJR 998, (2002) Times, 21 June, [2003] 1 LRC 674, [2002] All ER (D) 139 (Jun) P9039
Fairhurst v St Helens and Knowsley Health Authority [1994] 5 Med LR 422, [1995] PIQR Q 1 .... C6028
Farrant v Thanet Disctrict Council (11 June 1996, unreported) O3006
Faulkner v Chief Adjudication Officer [1994] PIQR P 244, CA C6006
Fearon v Chief Constable of Derbyshire, UKEAT/0445/02/RN, [2004] All ER (D) 101 (Jan), EAT H1705
Federación de Servicios Privados del sindicato Comisiones obreras (CC.OO.) v Tyco Integrated Security SL: C-266/14 (2015) C-266/14, ECLI:EU:C:2015:578, [2016] 1 CMLR 795, [2015] IRLR 935, [2015] ICR 1159, (2015) Times, 07 October, [2015] All ER (D) 55 (Sep) W10003
Fennelly v Connex South Eastern Ltd [2001] IRLR 390, [2000] All ER (D) 2233, CA V8013
Fildes v International Computers (1984), unreported A1001
Fish v British Tissues (1994), unreported C6058
Fisher v CHT Ltd (No 2) [1966] 2 QB 475, [1966] 1 All ER 88, [1966] 2 WLR 391, 109 Sol Jo 933, CA O3004
Flanagan v Watts Bearne & Co plc [1992] PIQR P 144, CA C6052
Fletcher v Ministry of Defence [2010] IRLR 25, [2009] All ER (D) 187 (Nov), EAT H1706
Fletcher v Rylands and Horrocks (1865) 3 H & C 774, 34 LJ Ex 177; revsd (1866) LR 1 Exch 265, 30 JP 436, 4 H & C 263, 35 LJ Ex 154, 12 Jur NS 603, 14 LT 523; affd sub nom Rylands v Fletcher (1868) LR 3 HL 330, 33 JP 70, 37 LJ Ex 161, 14 WR 799, [1861–73] All ER Rep 1, 19 LT 220, HL P9522, P9528
Fox v British Airways plc [2013] EWCA Civ 972, [2013] ICR 1257, [2013] IRLR 812, (2013) Times, 17 October, [2013] All ER (D) 397 (Jul) M1517
Fraser v State Hospitals Board for Scotland (2000) Times, 12 September, 2001 SLT 1051, 2001 SCLR 357, Ct of Sess E14003
Frost v Aylesbury Dairy Co [1905] 1 KB 608, 74 LJKB 386, 53 WR 354, [1904–7] All ER Rep 132, 49 Sol Jo 312, 92 LT 527, 21 TLR 300, CA P9047, P9053
Frost v Chief Constable of South Yorkshire Police. See White v Chief Constable of South Yorkshire Police
Furmedge v Chester-Le-Street District Council [2011] EWHC 1226 (QB), [2011] All ER (D) 162 (May) O3004, O3005

G

Gallagher v Alpha Catering Services Ltd (t/a Alpha Flight Services) [2004] EWCA Civ 1559, [2005] ICR 673, [2005] IRLR 102, [2004] All ER (D) 121 (Nov) W10017
Gallagher v The Drum Engineering Co Ltd, COIT 1330/89, (1989), unreported J3011
Gallop v Newport City Council [2013] EWCA Civ 1583, [2014] IRLR 211, [2013] All ER (D) 145 (Dec) M1507
Gallop v Newport City Council (2016) UKEAT/0118/15, [2016] IRLR 395, [2016] All ER (D) 218 (Mar), EAT M1511
Geary v JD Wetherspoon plc [2011] EWHC 1506 (QB), [2011] All ER (D) 97 (Jun) O3017
Geddling v Marsh [1920] 1 KB 668, 89 LJKB 526, [1920] All ER Rep 631, 122 LT 775, 36 TLR 337 .... P9052
General Building and Maintenance plc v Greenwich Borough Council (1993) 92 LGR 21, [1993] IRLR 535, 65 BLR 57 .... S7010

Table of Cases

General Cleaning Contractors Ltd v Christmas [1953] AC 180, [1952] 2 All ER 1110, [1953] 2 WLR 6, 51 LGR 109, 97 Sol Jo 7, HL O3014
Gentle v Perkins Engines Peterborough Ltd (formerly Perkins Group Ltd) [2001] All ER (D) 360 (Jul), EAT E14022
Gerrard v Staffordshire Potteries Ltd [1994] EWCA Civ 31, [1995] ICR 502, [1995] PIQR P 169 .... P3002
Ghaith v Indesit Co UK Ltd [2012] EWCA Civ 642 .... M3002
Gibson v Skibs AS Marina (or Marena) and Orkla Grobe (or Grube) A/B and Smith Coggins Ltd [1966] 2 All ER 476, [1966] 2 Lloyd's Rep 39 .... T7002
Gitsham v C H Pearce & Sons plc [1992] PIQR P57, CA O3018
Glaister v Appleby-in-Westmorland Town Council [2009] EWCA Civ 1325, [2010] 1 LS Gaz R 14, [2009] All ER (D) 79 (Dec) P9504
Goldscheider v Royal Opera House Covent Garden Foundation [2018] EWHC 687 (QB), [2018] All ER (D) 09 (Apr) N3022, N3027.1, N3028
Gómez v Continental Industrias del Caucho SA: C-342/01 [2004] ECR I-2605, [2004] 2 CMLR 38, [2005] ICR 1040, [2004] IRLR 407, [2004] All ER (D) 350 (Mar), ECJ W10024
González Castro v Mutua Umivale, ProsegurEspaña SL, Instituto Nacional de la Seguridad Social (INSS) (2018) C-41/17, ECLI:EU:C:2018:736, [2018] IRLR 1142, [2019] ICR 339, EUCJ E14026; R3026
Goodchild v Vaclight [1965] CLY 2669 .... P9038
Goodwin v Bennetts UK Ltd [2008] EWCA Civ 1374, [2008] All ER (D) 220 (Dec) D8209
Goodwin v Patent Office [1999] ICR 302, [1999] IRLR 4, EAT E16510
Goold (W A) (Pearmak) Ltd v McConnell [1995] IRLR 516, EAT E14008
Gough v Thorne [1966] 3 All ER 398, [1966] 1 WLR 1387, 110 Sol Jo 529, CA ... O3011
Graney v Liverpool County Council (1995), unreported O3002
Grant v Australian Knitting Mills Ltd [1936] AC 85, 105 LJPC 6, [1935] All ER Rep 209, 79 Sol Jo 815, 154 LT 18, 52 TLR 38, PC P9052
Gray v Canada Life Ltd. See Canada Life Ltd v Gray
Greaves & Co (Contractors) Ltd v Baynham, Meikle & Partners [1975] 3 All ER 99, [1975] 1 WLR 1095, [1975] 2 Lloyd's Rep 325, 119 Sol Jo 372, 4 BLR 56, CA .. W11010
Greene v Chelsea Borough Council [1954] 2 QB 127, [1954] 2 All ER 318, [1954] 3 WLR 12, 52 LGR 352, 118 JP 346, 98 Sol Jo 389, CA O3004
Grieves v FT Everard & Sons Ltd. See Johnston v NEI International Combustion Ltd
Griffiths v Peter Conway Ltd [1939] 1 All ER 685, CA P9052
Grovehurst Energy Ltd v Strawson, HM Inspector, COIT No 5035/90, (1990), unreported E15024
Gwilliam v West Hertfordshire Hospital NHS Trust [2002] EWCA Civ 1041, [2003] QB 443, [2003] 3 WLR 1425, [2003] PIQR P99, (2002) Times, 7 August, [2002] All ER (D) 345 (Jul) O3014, P9504

H

Hadley v Hancox (1986) 85 LGR 402, 151 JP 227, DC E15011
Hague (Sarah Jane) (One of Her Majesty's Inspectors of Health And Safety) v Rotary Yorkshire Ltd [2015] EWCA Civ 696 .... E15021
Hainsworth v Ministry of Defence [2014] EWCA Civ 763, [2014] 3 CMLR 1053, [2014] IRLR 728, [2014] EqLR 553 .... M1511
Hairsine v Kingston upon Hull City Council [1992] ICR 212, [1992] IRLR 211, EAT J3011
Hampstead Heath Winter Swimming Club v Corpn of London [2005] EWHC 713 (Admin), [2005] LGR 481, [2005] NLJ 827, (2005) Times, 19 May, [2005] All ER (D) 353 (Apr), sub nom R (on the application of Hampstead Heath Winter Swimming Club) v Corpn of London [2005] 1 WLR 2930 .... O3019
Hardman v Mallon (t/a Orchard Lodge Nursing Home) [2002] IRLR 516, [2002] All ER (D) 439 (May), EAT M2019
Harris v Evans [1998] 3 All ER 522, [1998] 1 WLR 1285, [1998] NLJR 745, [1998] All ER (D) 148, CA E15028
Harrison (TC) (Newcastle-under-Lyme) Ltd v Ramsey [1976] IRLR 135, Ind Trib .. E15024

Hartley v British Railways Board (1981) 125 Sol Jo 169, (1981) Times, 2 February, CA O3014
Hartley v Sandholme Iron Co Ltd [1975] QB 600, [1974] 3 All ER 475, [1974] 3 WLR 445, [1974] STC 434, 17 KIR 205, 118 Sol Jo 702 ..... C6052
Hartley (R S) Ltd v Provincial Insurance Co Ltd [1957] 1 Lloyd's Rep 121 ..... E13020
Hartman v South Essex Mental Health and Community Care NHS Trust [2005] EWCA Civ 6, [2005] ICR 782, [2005] IRLR 293, 85 BMLR 136, (2005) Times, 21 January, [2005] All ER (D) 141 (Jan) S11009
Harvest Press Ltd v McCaffrey [1999] IRLR 778, EAT E14009
Harvey v Plymouth City Council [2010] EWCA Civ 860, 154 Sol Jo (no 31) 30, [2010] All ER (D) 328 (Jul), CA O3004, O3006
Hatton v Sutherland [2002] EWCA Civ 76, [2002] 2 All ER 1, [2002] ICR 613, [2002] IRLR 263, 68 BMLR 115, (2002) Times, 11 February, [2002] All ER (D) 53 (Feb); revsd [2004] UKHL 13, [2004] 2 All ER 385, [2004] 1 WLR 1089, [2004] ICR 457, [2004] IRLR 475, 77 BMLR 219, (2004) Times, 5 April, 148 Sol Jo LB 419, [2004] All ER (D) 07 (Apr) E14003, R2006, S11009
Hay (or Bourhill) v Young [1943] AC 92, [1942] 2 All ER 396, 111 LJPC 97, 86 Sol Jo 349, 167 LT 261, HL C6033
Health and Safety Executive v George Tancocks Garage (Exeter) Ltd [1993] Crim LR 605, DC E15014
Healy v Corpn of London (24 June 1999, unreported) E14004
Heil v Rankin [2001] QB 272, [2000] 3 All ER 138, [2000] 2 WLR 1173, [2000] Lloyd's Rep Med 203, [2000] IRLR 334, [2000] NLJR 464, [2000] PIQR Q 187, 144 Sol Jo LB 157, CA C6032
Herring v Co-operative Insurance Society Ltd [2006] All ER (D) 171 (Oct), EAT .. W10024
Heslop v Metalock (Great Britain) Ltd (1981) Observer, 29 November N3022
Hewett v Alf Brown's Transport Ltd [1992] ICR 530, [1992] 15 LS Gaz R 33, CA P3001
Hewson v Downs [1970] 1 QB 73, [1969] 3 All ER 193, [1969] 2 WLR 1169, 6 KIR 343, 113 Sol Jo 309 ..... C6053
Hinds v Keppel Seghers UK Ltd (2014) UKEAT/0019/14, [2014] IRLR 754, [2014] ICR 1105, [2014] All ER (D) 197 (Jun), EAT E14011.1
Hinz v Berry [1970] 2 QB 40, [1970] 1 All ER 1074, [1970] 2 WLR 684, 114 Sol Jo 111, CA C6033
Hone v Six Continents Retail Ltd [2005] EWCA Civ 922, [2006] IRLR 49 ..... S11009, W10010
Hood v Mitie Property Services (Midlands) Ltd [2005] All ER (D) 11 (Jul) O3014
Horkulak v Cantor Fitzgerald International [2003] EWHC 1918 (QB), [2004] ICR 697, [2003] IRLR 756, [2003] All ER (D) 542 (Jul); revsd in part [2004] EWCA Civ 1287, [2005] ICR 402, [2004] IRLR 942, 148 Sol Jo LB 1218, [2004] All ER (D) 170 (Oct) H1707, H1720
Horne-Roberts v SmithKline Beecham plc [2001] EWCA Civ 2006, [2002] 1 WLR 1662, 65 BMLR 79, (2002) Times, 10 January, 146 Sol Jo LB 19, [2001] All ER (D) 269 (Dec) P9044
Houweling v Wesseler [1963] 40 DLR (2d) 956 ..... O3005
Howard and Peet v Volex plc (HSIB 181) J3011
Hufford v Samsung Electronics (UK) Ltd [2014] EWHC 2956 (TCC), [2014] All ER (D) 60 (Sep) P9031, P9032
Hufton v Somerset County Council [2011] EWCA Civ 789, [2011] All ER (D) 72 (Jul) O3010
Hughes v The Corps of Commissionaires Management Ltd (UKEAT/0173/10/SM) (22 November 2010, unreported), EAT W10017
Hughes v Corps of Commissionaires Management Ltd [2011] EWCA Civ 1061, [2012] 1 CMLR 649, [2011] IRLR 915, [2011] 39 LS Gaz R 19, [2011] All ER (D) 123 (Sep) W10017
Hunt v Severs [1994] 2 AC 350, [1994] 2 All ER 385, [1994] 2 WLR 602, [1994] 2 Lloyd's Rep 129, [1994] 32 LS Gaz R 41, [1994] NLJR 603, 138 Sol Jo LB 104, HL C6028
Hunter v British Coal Corpn [1999] QB 140, [1998] 2 All ER 97, [1998] 3 WLR 685, [1999] ICR 72, [1998] 12 LS Gaz R 27, 142 Sol Jo LB 85, CA C6033

Table of Cases

Hussain v New Taplow Paper Mills Ltd [1988] AC 514, [1988] 1 All ER 541, [1988] 2 WLR 266, [1988] ICR 259, [1988] IRLR 167, 132 Sol Jo 226, [1988] 10 LS Gaz R 45, [1988] NLJR 45, HL C6052

I

IPC Magazines Ltd v Clements (EAT/456/99) E14022
Ilkiw v Samuels [1963] 2 All ER 879, [1963] 1 WLR 991, 107 Sol Jo 680, CA ... A1020.1
Indermaur v Dames (1866) LR 1 CP 274, 35 LJCP 184, Har & Ruth 243, 12 Jur NS 432, 14 WR 586, [1861–73] All ER Rep 15, 14 LT 484; affd (1867) LR 2 CP 311, 31 JP 390, 36 LJCP 181, 15 WR 434, 16 LT 293, Ex Ch O3014
IRC v Ainsworth. See Ainsworth v IRC
Insitu Cleaning Co Ltd v Heads [1995] IRLR 4, EAT H1706
Intel Corpn (UK) Ltd v Daw. See Daw v Intel Corpn (UK) Ltd
Interfoto Picture Library Ltd v Stiletto Visual Programmes Ltd [1989] QB 433, [1988] 1 All ER 348, [1988] 2 WLR 615, [1988] BTLC 39, 132 Sol Jo 460, [1988] 9 LS Gaz R 45, [1987] NLJ Rep 1159, CA O3016
Irvine v Metropolitan Police Comr [2004] EWHC 1536 (QB), [2004] All ER (D) 79 (May) O3018

J

J v DLA Piper UK LLP (UKEAT/0263/09/RN) [2010] ICR 1052, [2010] IRLR 936, 115 BMLR 107, 154 Sol Jo (no 24) 37, [2010] All ER (D) 112 (Aug), EAT E16510
James v London Borough of Greenwich [2008] EWCA Civ 35, [2008] ICR 545, [2008] IRLR 302, [2008] All ER (D) 54 (Feb) E13016
James v Redcats (Brands) Ltd [2007] ICR 1006, [2007] IRLR 296, [2007] All ER (D) 270 (Feb), EAT W10004
Jayes v IMI (Kynoch) Ltd [1985] ICR 155, [1984] LS Gaz R 3180, CA O3011
Joao v Jurys Hotel Management UK Ltd (2011) UKEAT0210/11 (11 October 2011, unreported) E14010
Joel v Law Union and Crown Insurance Co [1908] 2 KB 863, 77 LJKB 1108, 52 Sol Jo 740, 99 LT 712, 24 TLR 898, CA E13004, P9505
Johnson v Coventry Churchill International Ltd [1992] 3 All ER 14 .... E13010
Johnson v Unisys Ltd [2001] UKHL 13, [2003] 1 AC 518, [2001] 2 All ER 801, [2001] 2 WLR 1076, [2001] ICR 480, [2001] IRLR 279, [2001] All ER (D) 274 (Mar) H1728
Johnson v University of Bristol (17 October 2017, unreported), CA M1005.1
Johnston v NEI International Combustion Ltd [2007] UKHL 39, [2008] AC 281, [2007] 4 All ER 1047, [2007] 3 WLR 876, [2007] ICR 1745, 99 BMLR 139, [2007] NLJR 1542, (2007) Times, 24 October, 151 Sol Jo LB 1366, [2007] All ER (D) 224 (Oct), sub nom Grieves v FT Everard & Sons Ltd [2008] PIQR P95 .... C6033
Johnstone v Bloomsbury Health Authority [1992] QB 333, [1991] 2 All ER 293, [1991] 2 WLR 1362, [1991] ICR 269, [1991] IRLR 118, [1991] 2 Med LR 38, CA E14002, E14003
Jolley v Sutton London BC [2000] 3 All ER 409, [2000] 1 WLR 1082, [2000] LGR 399, [2000] 2 Lloyd's Rep 65, [2000] 2 FCR 392, [2000] All ER (D) 689, HL O3010
Jones v Sainsbury's Supermarkets Ltd (2000), unreported E14013
Jones v Sandwell Metropolitan Borough Council [2002] EWCA Civ 76, [2002] 2 All ER 1, [2002] ICR 613, [2002] IRLR 263, 68 BMLR 115, (2002) Times, 11 February, [2002] All ER (D) 53 (Feb); revsd [2004] UKHL 13, [2004] 2 All ER 385, [2004] 1 WLR 1089, [2004] ICR 457, [2004] IRLR 475, 77 BMLR 219, (2004) Times, 5 April, 148 Sol Jo LB 419, [2004] All ER (D) 07 (Apr) R2006
Jones v Tower Boot Co Ltd [1995] IRLR 529, EAT; revsd [1997] 2 All ER 406, [1997] ICR 254, [1997] IRLR 168, [1997] NLJR 60, CA H1706, H1711, H1712, H1715A, H1723
Jordan v Achara (1988) 20 HLR 607, CA O3004

K

Kelly v Covance Laboratories Ltd (2015) UKEAT/0186/15, [2016] IRLR 338, [2015] All ER (D) 316 (Oct), EAT H1708
Kemeh v Ministry of Defence [2014] EWCA Civ 91, (2014) Times, 11 March, [2014] All ER (D) 90 (Feb) H1715A
Kendall (Henry) & Sons (a firm) v William Lillico & Sons Ltd [1969] 2 AC 31, [1968] 2 All ER 444, [1968] 3 WLR 110, [1968] 1 Lloyd's Rep 547, 112 Sol Jo 562, HL P9053
Kennedy v Cordia (Services) LLP [2016] UKSC 6, [2016] 1 WLR 597, [2016] ICR 325, 149 BMLR 17, [2016] PIQR P154, 2016 SLT 209, 2016 SCLR 203, [2016] All ER (D) 99 (Feb), 2016 Scot (D) 8/2 .... R3021
Keown v Coventry Healthcare NHS Trust [2006] EWCA Civ 39, [2006] 1 WLR 953, [2006] 08 LS Gaz R 23, 150 Sol Jo LB 164, [2006] All ER (D) 27 (Feb) O3009
Kerr v Nathan's Wastesavers Ltd (1995) IDS Brief 548 .... E14009
Kigass Aero Components Ltd v Brown [2002] ICR 697, [2002] IRLR 312, [2002] All ER (D) 341 (Feb), EAT W10024B
Kilraine v London Borough of Wandsworth [2018] EWCA Civ 1436, [2018] IRLR 846, [2018] ICR 1850, [2018] All ER (D) 44 (Jul) E14011.1
King v Smith [1995] ICR 339, CA O3014
King v Sussex Ambulance NHS Trust [2002] EWCA Civ 953, (2002) Times, 25 July, [2002] All ER (D) 95 (Jul) M3002
Knott v Newham Healthcare NHS Trust [2002] EWHC 2091 (QB); affd [2003] All ER (D) 164 (May), CA M3002
Knowles v Liverpool City Council [1993] 4 All ER 321, [1993] 1 WLR 1428, 91 LGR 629, [1994] 1 Lloyd's Rep 11, [1994] ICR 243, [1993] IRLR 588, [1993] NLJR 1479n, [1994] PIQR P 8, HL M1005
Kolasa v Ealing Hospital NHS Trust [2015] EWHC 289 (QB), [2015] All ER (D) 262 (Feb) O3009

L

LMS International Ltd v Styrene Packaging and Insulation Ltd [2005] EWHC 2065 (TCC), [2006] BLR 50, [2005] All ER (D) 171 (Sep) P9528
Lambert v Lewis. See Lexmead v Lewis
Landeshauptstadt Kiel v Jaeger: C-151/02 [2003] ECR I-8389, [2004] All ER (EC) 604, [2003] 3 CMLR 493, [2004] ICR 1528, [2003] IRLR 804, 75 BMLR 201, (2003) Times, 26 September, [2003] All ER (D) 72 (Sep), ECJ W10003
Lane Group plc v Farmiloe (UKEAT/0352/03/2201) [2004] All ER (D) 08 (Mar) P3027
Langridge v Howletts & Port Lympne Estates Ltd (1996) 95 LGR 798, sub nom Canterbury City Council v Howletts & Port Lympne Estates Ltd [1997] ICR 925, [1997] JPIL 51 .... E15022
Larby v Thurgood [1993] ICR 66, [1992] 39 LS Gaz R 35, [1993] PIQR P 218, 136 Sol Jo LB 275, HC C6026
Larner v British Steel plc [1993] 4 All ER 102, [1993] ICR 551, [1993] IRLR 278, CA E15031
Larner v NHS Leeds [2012] EWCA Civ 1034, [2012] 4 All ER 1006, [2012] 3 CMLR 1101, [2012] IRLR 825, [2012] ICR 1389, 156 Sol Jo (30) 31, [2012] All ER (D) 273 (Jul) W10024B
Lask v Gloucester Area Health Authority (1995) Times, 13 December, CA J3015
Latham v R Johnson and Nephew Ltd [1913] 1 KB 398, 77 JP 137, 82 LJKB 258, [1911–13] All ER Rep 117, 57 Sol Jo 127, 108 LT 4, 29 TLR 124, CA O3010
Lau v DPP [2000] All ER (D) 224, DC H1726
Law Hospital NHS Trust v Rush [2001] IRLR 611, Ct of Sess E16510
Lea v British Aerospace plc [1991] 1 AC 362, [1991] 1 All ER 193, [1991] 2 WLR 873, 135 Sol Jo 16, HL C6043
Leonard v Southern Derbyshire Chamber of Commerce [2001] IRLR 19, [2000] All ER (D) 1327, EAT E16510
Lewis v Avidan Ltd [2005] EWCA Civ 670, 149 Sol Jo LB 479, [2005] All ER (D) 136 (Apr) O3018, W11010

Table of Cases

Lexmead (Basingstoke) Ltd v Lewis [1982] AC 225, sub nom Lambert v Lewis [1980] 1 All ER 978, [1980] 2 WLR 299, [1980] RTR 152, [1980] 1 Lloyd's Rep 311, 124 Sol Jo 50, CA; revsd sub nom Lexmead (Basingstoke) Ltd v Lewis [1982] AC 225, 268, [1981] 2 WLR 713, [1981] RTR 346, [1981] 2 Lloyd's Rep 17, 125 Sol Jo 310, sub nom Lambert v Lewis [1981] 1 All ER 1185, HL P9054

Lidl Italia Srl v Comune di Arcole (VR): C-315/05 [2006] ECR I-11181, [2006] All ER (D) 331 (Nov), ECJ F9023

Liffen v Watson [1940] 1 KB 556, [1940] 2 All ER 213, 109 LJKB 367, 84 Sol Jo 368, 162 LT 398, 56 TLR 442, CA C6053

Lim Poh Choo v Camden and Islington Area Health Authority [1980] AC 174, [1979] 2 All ER 910, [1979] 3 WLR 44, 123 Sol Jo 457, HL C6027, C6052

Lindop v Goodwin Steel Castings Ltd (1990) Times, 19 June C6043

List Design Group Ltd v Catley [2002] ICR 686, [2003] IRLR 14, [2002] All ER (D) 215 (Mar), EAT W10024B

Lister v Romford Ice and Cold Storage Co Ltd [1957] AC 555, [1957] 1 All ER 125, [1957] 2 WLR 158, [1956] 2 Lloyd's Rep 505, 121 JP 98, 101 Sol Jo 106, HL E13027

Livingstone v Rawyards Coal Co (1880) 5 App Cas 25, 44 JP 392, 28 WR 357, 42 LT 334, HL H1725

Longden v British Coal Corpn [1998] AC 653, [1998], [1998] 1 All ER 289, [1998] 3 WLR 1336, [1998] ICR 26, [1998] IRLR 29, [1998] 01 LS Gaz R 25, [1997] NLJR 1774, [1998] PIQR Q 11, 142 Sol Jo LB 28, HL C6053

Lowles v Home Office [2004] EWCA Civ 985, [2004] All ER (D) 538 (Jul) W11010

Lyddon v Englefield Brickwork Ltd [2008] IRLR 198, [2007] All ER (D) 198 (Nov), EAT W10024A

M

MWH UK Ltd v Victoria Susan Wise (H.M. Inspector of Health & Safety) [2014] EWHC 427 (Admin), [2014] All ER (D) 104 (Mar) E15015

McAdie v Royal Bank of Scotland [2007] EWCA Civ 806, [2007] IRLR 895, [2008] ICR 1087, [2007] NLJR 1355, [2007] All ER (D) 477 (Jul) R2005

McCafferty v Metropolitan Police District Receiver [1977] 2 All ER 756, [1977] 1 WLR 1073, [1977] ICR 799, 121 Sol Jo 678, CA N3022

McCamley v Cammell Laird Shipbuilders Ltd [1990] 1 All ER 854, [1990] 1 WLR 963, CA C6053

MacCartney v Oversley House Management [2006] ICR 510, [2006] IRLR 514, [2006] All ER (D) 246 (Jan), EAT W10003

McDermid v Nash Dredging and Reclamation Co Ltd [1987] AC 906, [1987] 2 All ER 878, [1987] 3 WLR 212, [1987] 2 Lloyd's Rep 201, [1987] ICR 917, [1987] IRLR 334, 131 Sol Jo 973, [1987] LS Gaz R 2458, HL O7029

Macdonald v Advocate General for Scotland [2003] UKHL 34, [2004] 1 All ER 339, [2003] ICR 937, [2003] IRLR 512, [2003] ELR 655, [2003] 29 LS Gaz R 36, (2003) Times, 20 June, 2003 SLT 1158, 2003 SCLR 814, 147 Sol Jo LB 782, [2004] 2 LRC 111, [2003] All ER (D) 259 (Jun) H1702, H1706

McDonald (dec'd) v National Grid Electricity Transmission plc [2014] UKSC 53, [2015] AC 1128, [2015] 1 All ER 371, [2014] 3 WLR 1197, [2014] ICR 1172, 141 BMLR 1, [2015] PIQR P98, 164 NLJ 7633, [2014] All ER (D) 257 (Oct) A5042.2

McEwan v Lothian Buses plc [2006] CSOH 56, 2006 SCLR 592, OH O3011

McFarlane v EE Caledonia Ltd [1994] 2 All ER 1, [1994] 1 Lloyd's Rep 16, [1993] NLJR 1367, [1994] PIQR P 154, CA C6033

McGarvey v EVE NCI Ltd [2002] EWCA Civ 374, [2002] All ER (D) 345 (Feb) ... O3019

McGinlay (or Titchener) v British Railways Board. See Titchener v British Railways Board

McGivney v Golderslea Ltd (1997), unreported O3003

McGuinness v Kirkstall Forge Engineering Ltd (1979), unreported N3022

McIlgrew v Devon County Council [1995] PIQR Q 66, CA C6027

McKay v Unwin Pyrotechnics Ltd [1991] Crim LR 547, DC P9012

MacLennan v Hartford Europe Ltd [2012] EWHC 346 (QB), [2012] 11 LS Gaz R 23, [2012] NLJR 363 .... S11009

McLoughlin v O'Brian [1983] 1 AC 410, [1982] 2 All ER 298, [1982] 2 WLR 982, [1982] RTR 209, 126 Sol Jo 347, [1982] LS Gaz R 922, HL C6033
McNulty v Marshalls Food Group Ltd 1999 SC 195, OH C6027
McTigue v University Hospital Bristol NHS Foundation Trust (2016) UKEAT/0354/15, [2016] IRLR 742, [2016] All ER (D) 211 (Jul), EAT E14011.1
Mailer v Austin Rover Group plc. See Austin Rover Group Ltd v HM Inspector of Factories
Majrowski v Guy's and St Thomas's NHS Trust [2005] EWCA Civ 251, [2005] QB 848, [2005] 2 WLR 1503, [2005] ICR 977, [2005] IRLR 340, (2005) Times, March 21, 149 Sol Jo LB 358, [2005] All ER (D) 273 (Mar); affd sub nom Majrowski v Guy's and St Thomas' NHS Trust [2006] UKHL 34, [2007] 1 AC 224, [2006] 4 All ER 395, [2006] 3 WLR 125, [2006] ICR 1199, [2006] IRLR 695, 91 BMLR 85, [2006] NLJR 1173, (2006) Times, 13 July, 150 Sol Jo LB 986, [2006] All ER (D) 146 (Jul) H1726
Makepeace v Evans Brothers (Reading) (a firm) [2001] ICR 241, [2000] BLR 737, (2000) Times, 13 June, [2000] All ER (D) 720, CA O3019
Malcolm v Broadhurst [1970] 3 All ER 508 .... H1724
Malik v BCCI SA (in liq) [1998] AC 20, [1997] 3 All ER 1, [1997] 3 WLR 95, [1997] ICR 606, [1997] IRLR 462, [1997] 94 LS Gaz R 33, [1997] NLJR 917, HL H1728
M'Alister (or Donoghue) v Stevenson [1932] AC 562, 101 LJPC 119, 37 Com Cas 350, 48 TLR 494, 1932 SC (HL) 31, sub nom Donoghue (or McAlister) v Stevenson [1932] All ER Rep 1, 1932 SLT 317, sub nom McAlister (or Donoghue) v Stevenson 76 Sol Jo 396, 147 LT 281, HL H1724, O3014, P9024, P9037, P9530
Marinello v City of Edinburgh Council [2010] CSOH 17, 2010 SLT 349, OH H1726
Masiak v City Restaurants (UK) Ltd [1999] IRLR 780, EAT E14009
Meigh v Wickenden [1942] 2 KB 160, [1942] 2 All ER 68, 40 LGR 191, 106 JP 207, 112 LJKB 76, 86 Sol Jo 218, 167 LT 135, 58 TLR 260, DC W11047
Miah v La Gondola Ltd (1999), unreported W10022
Miles Recruitment Ltd v Gangmasters Licensing Authority (2007), unreported P3027
Mills v Hassall [1983] ICR 330 .... C6053
Mohamud v WM Morrison Supermarkets plc [2014] EWCA Civ 116, [2014] 2 All ER 990, [2014] IRLR 386, [2014] ICR D19, [2014] All ER (D) 130 (Feb) V8013
Moloney v Lambeth London Borough Council (1966) 64 LGR 440, 110 Sol Jo 406, 198 Estates Gazette 895 .... O3003, O3004
Monk v PC Harrington Ltd [2008] EWHC 1879 (QB), [2009] PIQR P52, [2008] All ER (D) 20 (Aug C6033
Moonsar v Fiveways Express Transport Ltd [2005] IRLR 9, [2004] All ER (D) 110 (Nov), EAT H1706
Moorcock, The (1889) 14 PD 64, 58 LJP 73, 6 Asp MLC 373, 37 WR 439, [1886–90] All ER Rep 530, 60 LT 654, 5 TLR 316, CA E14004
Morris v Ford Motor Co Ltd [1973] QB 792, [1973] 2 All ER 1084, [1973] 2 WLR 843, [1973] 2 Lloyd's Rep 27, 117 Sol Jo 393, CA E13027
Moualem v Carlisle City Council (1994) 158 JP 1110, [1995] ELR 22, DC O3019
Muir (William) (Bond 9) Ltd v Lamb [1985] IRLR 95, EAT E14013
Munro v MPB Structures Ltd [2004] 2 CMLR 1032, [2004] ICR 430, [2003] IRLR 350, (2003) Times, 24 April, 2003 SLT 551, 2003 SCLR 542 .... W10024A
Murphy v Bradford Metropolitan Borough Council [1992] 2 All ER 908, [1992] PIQR P 68, CA O3002
Muzi-Mabaso v Revenue & Customs Comrs (2014) UKEAT/0353/14/DA, unreported M1511

N

Nagarajan v London Regional Transport [2000] 1 AC 501, [1999] 4 All ER 65, [1999] 3 WLR 425, [1999] ICR 877, [1999] IRLR 572, [1999] 31 LS Gaz R 36, 143 Sol Jo LB 219, HL H1736
Naylor (t/a Mainstreet) v Payling [2004] EWCA Civ 560, (2004) Times, 2 June, 148 Sol Jo LB 573, [2004] All ER (D) 83 (May), sub nom Payling v Naylor (t/a Mainstreet) [2004] 23 LS Gaz R 33 .... O3014

Neidel v Stadt Frankfurt am Main: C-337/10 [2012] 3 CMLR 92, [2012] ICR 1201, [2012] IRLR 607, ECJ M1514
New Southern Railway Ltd v Quinn (2005) UKEAT/0313/05, [2006] IRLR 266, [2006] ICR 761, [2005] Lexis Citation 1111, [2005] All ER (D) 367 (Nov), EAT E14024, E14025; R3026
Newsholme Bros v Road Transport and General Insurance Co Ltd [1929] 2 KB 356, 24 Ll L Rep 247, 98 LJKB 751, 34 Com Cas 330, [1929] All ER Rep 442, 73 Sol Jo 465, 141 LT 570, 45 TLR 573, CA E13003, P9502
Nico Manufacturing Co Ltd v Hendry [1975] IRLR 225, Ind Trib E15024
Nixon v Ross Coates Solicitors, UKEAT/0108/10ZT (6 August 2010,unreported) ... H1703
Norouzi v Sheffield City Council (2011) UKEAT/0497/10/RN, [2011] All ER (D) 77 (Aug), EAT H1706

O

O'Brien v MGN Ltd [2001] EWCA Civ 1279, (2001) Times, 8 August, [2001] All ER (D) 01 (Aug) O3016
O'Byrne v Sanofi Pasteur MSD Ltd: C-127/04 [2006] ECR I-1313, [2006] All ER (EC) 674, [2006] 1 WLR 1606, [2006] 2 CMLR 656, 91 BMLR 175, (2006) Times, 15 February, [2006] All ER (D) 117 (Feb), ECJ P9044
Ogwo v Taylor [1988] AC 431, [1987] 3 All ER 961, [1987] 3 WLR 1145, 131 Sol Jo 1628, [1988] 4 LS Gaz R 35, [1987] NLJ Rep 1110, HL O3014
O'Hanlon v Revenue and Customs Comrs [2006] ICR 1579, [2006] IRLR 840, [2006] All ER (D) 53 (Aug), EAT; affd [2007] EWCA Civ 283, [2007] ICR 1359, [2007] IRLR 404, (2007) Times, 20 April, [2007] All ER (D) 516 (Mar) M1511
Okerago v May & Baker Ltd (t/a Sanofi-Aventis Pharma) (2010) UKEAT/0278/09/ZT, [2010] IRLR 394, [2010] All ER (D) 79 (Mar), EAT H1706, H1714
O'Neill v DSG Retail Ltd [2002] EWCA Civ 1139, [2002] All ER (D) 500 (Jul) M3002
Osborne v Bill Taylor of Huyton Ltd [1982] ICR 168, [1982] IRLR 17, DC M2031, S7002
O'Neill v Buckinghamshire County Council (2010) UKEAT/0020/09, [2010] All ER (D) 15 (Feb), EAT R3026
O'Shea v Kimberley-Clark Ltd (1982) Guardian, 8 October N3022
Otero Ramos v Servicio Galego de Saúde, Instituto Nacional de la Seguridad Social (2017) C-531/15, ECLI:EU:C:2017:789, [2018] IRLR 159, [2018] ICR 965, [2017] All ER (D) 149 (Oct), EUCJ E14025; R3026
Oudahar v Esporta Group Ltd (UKEAT/0566/10/DA) [2011] ICR 1406, [2011] IRLR 730, [2011] All ER (D) 137 (Aug) E14010

P

PCW Syndicates v PCW Reinsurers [1996] 1 All ER 774, [1996] 1 WLR 1136, [1996] 1 Lloyd's Rep 241, CA E13004, P9505
Page v Freight Hire (Tank Haulage) Ltd [1981] 1 All ER 394, [1981] ICR 299, [1981] IRLR 13, EAT E14024
Page v Sheerness Steel plc [1995] PIQR Q 26, HC C6052
Page v Smith [1996] AC 155, [1995] 2 All ER 736, [1995] 2 WLR 644, [1995] RTR 210, [1995] 2 Lloyd's Rep 95, 28 BMLR 133, [1995] 23 LS Gaz R 33, [1995] NLJR 723, [1995] PIQR P 329, HL C6033
Pakenham-Walsh v Connell Residential [2006] EWCA Civ 90, [2006] 11 LS Gaz R 25, [2006] All ER (D) 275 (Feb) W10010
Palfrey v Greater London Council [1985] ICR 437 .... C6052
Palmer v R [1971] AC 814, [1971] 1 All ER 1077, [1971] 2 WLR 831, 55 Cr App Rep 223, 115 Sol Jo 264, 16 WIR 499, 511, PC O3009
Pape v Cumbria County Council [1992] 3 All ER 211, [1992] ICR 132, [1991] IRLR 463 .... P3026
Parker v South Eastern Rly Co (1877) 2 CPD 416, 41 JP 644, 46 LJQB 768, 25 WR 564, [1874–80] All ER Rep 166, 36 LT 540, CA O3016

Parkins v Sodexho Ltd (2001) EAT/1239/00, [2002] IRLR 109, [2001] Lexis Citation
  2218, [2001] All ER (D) 377 (Jun), EAT E14009, E14011.1
Parry v Cleaver [1970] AC 1, [1969] 1 All ER 555, [1969] 2 WLR 821, [1969] 1
  Lloyd's Rep 183, 113 Sol Jo 147, HL C6053
Parsons v Airplus International Ltd [2018] UKEAT/0111/17 (unreported) E14011.1
Paterson v Surrey Police Authority [2008] EWHC 2693 (QB), [2008] All ER (D) 85
  (Nov) M2009
Payling v Naylor (t/a Mainstreet). See Naylor (t/a Mainstreet) v Payling
Pearce v Governing Body of Mayfield School [2003] UKHL 34, [2004] 1 All ER 339,
  [2003] ICR 937, [2003] IRLR 512, [2003] ELR 655, [2003] 29 LS Gaz R 36, (2003)
  Times, 20 June, 2003 SLT 1158, 2003 SCLR 814, 147 Sol Jo LB 782, [2004] 2 LRC
  111, [2003] All ER (D) 259 (Jun) H1702
Pereda v Madrid Movilidad SA: C-277/08 (2009) C-277/08, [2009] ECR I-8405,
  [2010] 1 CMLR 103, [2009] IRLR 959, [2009] NLJR 1323, (2009) Times, 8 October,
  [2009] All ER (D) 88 (Sep) W10024B
Perry v Raleys Solicitors [2017] EWCA Civ 314, [2017] All ER (D) 162 (Apr); revsd
  [2019] UKSC 5, [2019] 2 All ER 937, [2019] 2 WLR 636, 169 NLJ 7829,
  [2019] All ER (D) 59 (Feb), SC E14011.1
Perth & Kinross Council v Gauld [2014] UKEATS 0046/13/JW M1516
Phipps v RochesterCorpn [1955] 1 QB 450, [1955] 1 All ER 129, [1955] 2 WLR 23, 53
  LGR 80, 119 JP 92, 99 Sol Jo 45 .... O3010
Pickford v Imperial Chemical Industries plc [1998] 3 All ER 462, [1998] 1 WLR 1189,
  [1998] ICR 673, [1998] IRLR 435, [1998] 31 LS Gaz R 36, [1998] NLJR 978, 142
  Sol Jo LB 198, [1998] All ER (D) 302, HL C6058
Pierce (a child by his litigation friend Annette Pierce) v West Sussex County Council
  [2013] EWCA Civ 1230, [2013] All ER (D) 166 (Oct) O3010
Pimlico Plumbers Ltd v Smith [2018] UKSC 29, [2018] 4 All ER 641, [2018] IRLR 872,
  [2018] ICR 1511, [2018] All ER (D) 65 (Jun), SC E14001
Ping v Esselte-Letraset [1992] PIQR P 74 .... C6058
Piper (Terence) v JRI (Manufacturing) Ltd [2006] EWCA Civ 1344, 92 BMLR 141, 150
  Sol Jo LB 1391, [2006] All ER (D) 181 (Oct) P9033, P9036
Plumb v Duncan Print Group Ltd (2015) UKEAT/0071/15, [2015] IRLR 711,
  [2015] All ER (D) 387 (Jul), EAT M1514A, W10024B
Polkey v A E Dauton (or Dayton) Services Ltd [1988] AC 344, [1987] 3 All ER 974,
  [1987] 3 WLR 1153, [1988] ICR 142, [1987] IRLR 503, 131 Sol Jo 1624, [1988]
  1 LS Gaz R 36, [1987] NLJ Rep 1109, HL E14013, M1517
Pollard v Tesco Stores Ltd [2006] EWCA Civ 393, [2006] All ER (D) 186 (Apr) P9032
Pollock v Cahill [2015] EWHC 2260 (QB), [2015] All ER (D) 330 (Jul) O3006
Polyflor Ltd v Health and Safety Executive [2014] EWCA Crim 1522, [2014] ICR 1142,
  [2014] All ER (D) 195 (Jul) E15037
Porcelli v Strathclyde Regional Council [1986] ICR 564, sub nom Strathclyde
  Regional Council v Porcelli [1986] IRLR 134, Ct of Sess .. H1702, H1706, H1707, H1723
Post Office v Footitt [2000] IRLR 243 .... W11026
Potter v Arafa [1995] IRLR 316, sub nom Arafa v Potter [1994] PIQR Q 73, CA ... C6027
Pratley v Surrey County Council [2003] EWCA Civ 1067, [2004] ICR 159, [2003] IRLR
  794, [2003] All ER (D) 438 (Jul) S11009
Preist v Last [1903] 2 KB 148, 72 LJKB 657, 51 WR 678, 47 Sol Jo 566, 89 LT 33, 19
  TLR 527, [1900–3] All ER Rep Ext 1033, CA P9052

Q

Quality Solicitors CMHT v Tunstall (UKEAT/0105/14), [2014] EqLR 679 .... H1706
Quinn v McGinty 1999 SLT 27, Sh Ct E13008

R

R v Adomako [1995] 1 AC 171, [1994] 3 All ER 79, [1994] 3 WLR 288, 158 JP 653, [1994] Crim LR 757, [1994] 5 Med LR 277, 19 BMLR 56, [1994] NLJR 936, [1994] 2 LRC 800, HL C9021
R v Associated Octel Co Ltd [1996] 4 All ER 846, [1996] 1 WLR 1543, [1996] ICR 972, [1997] IRLR 123, [1997] Crim LR 355, [1996] NLJR 1685, HL Intro9, IN15
R v A-G for Northern Ireland, ex p Burns [1999] IRLR 315 .... W10019
R v Barker (1996), unreported E15017
R v Boal [1992] QB 591, [1992] 3 All ER 177, [1992] 2 WLR 890, 95 Cr App Rep 272, [1992] ICR 495, [1992] IRLR 420, 156 JP 617, [1992] BCLC 872, [1992] 21 LS Gaz R 26, 136 Sol Jo LB 100, CA E15040
R (on the application of Aineto) v Brighton and Hove District Coroner [2003] EWHC 1896 (Admin), [2003] All ER (D) 353 (Jul) A3009
R (on the application of Junttan Oy) v Bristol Magistrates' Court [2003] UKHL 55, [2004] 2 All ER 555, [2003] ICR 1475, [2003] 44 LS Gaz R 30, (2003) Times, 24 October, 147 Sol Jo LB 1243, [2003] All ER (D) 386 (Oct) P9009
R v British Steel plc [1995] 1 WLR 1356, [1995] ICR 586, [1995] IRLR 310, [1995] Crim LR 654, CA E15039
R v Chapman (1992), unreported E15040
R v Chargot Ltd (t/a Contract Services) [2007] EWCA Crim 3032, [2008] 2 All ER 1077, [2008] ICR 517, [2007] All ER (D) 198 (Dec); affd [2008] UKHL 73, [2009] 2 All ER 645, [2009] 1 WLR 1, (2008) Times, 16 December, 153 Sol Jo (no 1) 32, [2008] All ER (D) 106 (Dec) E15031, O3019
R (on the application of Hampstead Heath Winter Swimming Club) v Corpn of London. See Hampstead Heath Winter Swimming Club v Corpn of London
R v D'Albuquerque, ex p Bresnahan [1966] 1 Lloyd's Rep 69 .... C6007
R v F Howe & Son (Engineers) Ltd [1999] 2 All ER 249, [1999] 2 Cr App Rep (S) 37, [1999] IRLR 434, 163 JP 359, [1999] Crim LR 238, [1998] 46 LS Gaz R 34, [1998] All ER (D) 552, CA E15043
R v Gateway Foodmarkets Ltd [1997] 3 All ER 78, [1997] 2 Cr App Rep 40, [1997] ICR 382, [1997] IRLR 189, [1997] Crim LR 512, [1997] 03 LS Gaz R 28, 141 Sol Jo LB 28, CA E15039
R v Goodman [1993] 2 All ER 789, 97 Cr App Rep 210, 14 Cr App Rep (S) 147, [1994] 1 BCLC 349, [1992] BCC 625 .... E15040
R v HIP (Bodycote) [2010] EWCA Crim 802, [2011] 1 Cr App Rep (S) 38 .... C7001
R v HTM Ltd [2006] EWCA Crim 1156, [2007] 2 All ER 665, [2006] ICR 1383, [2006] All ER (D) 299 (May) E15039, M2021, O3019
R v Hanson Quarry Products Europe Ltd (2011), unreported, Taunton Crown Court C7001
R (Health and Safety Executive) v Havering Borough Council [2017] EWCA Crim 242, [2017] 2 Cr App Rep (S) 54, [2017] All ER (D) 86 (Mar) E15036
R v Kerr (1996), unreported E15017
R (on the application of Unison) v Lord Chancellor [2017] UKSC 51, [2017] 4 All ER 903, [2017] 3 WLR 409, [2018] 1 CMLR 1047, [2017] IRLR 911, [2017] ICR 1037, 167 NLJ 7758, [2017] 4 Costs LR 721, [2017] All ER (D) 174 (Jul), SC E14001
R v MJ Allen Holdings Ltd (2016) [2016] EWCA Crim 2142, [2017] All ER (D) 58 (Jan) E15036
R v National Insurance Comr, ex p Michael [1977] 2 All ER 420, [1977] 1 WLR 109, [1977] ICR 121, 120 Sol Jo 856, CA C6006
R v Nelson Group Services (Maintenance) Ltd [1999] IRLR 646 .... E15039
R v Network Rail Infrastructure Ltd [2014] EWCA Crim 49, 178 CL&J 46, [2014] All ER (D) 111 (Jan) M2001
R v Owino [1996] 2 Cr App Rep 128, [1995] Crim LR 743, CA O3009
R (Black) v Secretary of State for Justice [2016] 1 WLR 3623, SC W11028
R (on the application of the Broadcasting, Entertainment, Cinematographic and Theatre Union) v Secretary of State for Trade and Industry: C-173/99 [2001] ECR I-4881, [2001] All ER (EC) 647, [2001] 1 WLR 2313, [2001] 3 CMLR 109, [2001] ICR 1152, [2001] IRLR 559, [2001] All ER (D) 272 (Jun), ECJ W10022

Table of Cases

R (on the application of United Road Transport Union) v Secretary of State for Transport [2013] EWCA Civ 962, [2014] 1 CMLR 723, [2013] IRLR 890, [2013] All ER (D) 359 (Jul) W10029
R v Sellafield Ltd [2014] EWCA Crim 49, 178 CL&J 46, [2014] All ER (D) 111 (Jan) M2001
R v Swan Hunter Shipbuilders Ltd [1982] 1 All ER 264, [1981] ICR 831, [1981] IRLR 403, [1981] Crim LR 833, sub nom R v Swan Hunter Shipbuilders Ltd and Telemeter Installations Ltd, HMS Glasgow [1981] 2 Lloyd's Rep 605, CA A1022
R v Tangerine Confectionary Ltd and Veolia ES (UK) Ltd [2011] EWCA Crim 2015 .... E15031
Rae v Mars (UK) Ltd [1990] 1 EGLR 161, [1990] 03 EG 80 .... O3006
Railtrack plc v Smallwood [2001] EWHC Admin 78, [2001] ICR 714, [2001] 09 LS Gaz R 40, 145 Sol Jo LB 52, [2001] All ER (D) 103 (Jan) E15015
Railways Comr (NSW) v Cardy (1961) 104 CLR 274, Aus HC O3007
Ratcliff v McConnell [1999] 1 WLR 670, [1999] 03 LS Gaz R 32, 143 Sol Jo LB 53, CA O3017
Readmans v Leeds City Council [1993] COD 419 .... E15015
Redfearn v Serco Ltd (t/a West Yorkshire Transport Service) [2006] EWCA Civ 659, [2006] ICR 1367, [2006] IRLR 623, (2006) Times, 27 June, 150 Sol Jo LB 703, [2006] All ER (D) 366 (May) H1703
Reed v Stedman [1999] IRLR 299, EAT E14019, H1702, H1706, H1707, H1708
Reid v Rush & Tompkins Group plc [1989] 3 All ER 228, [1990] 1 WLR 212, [1990] RTR 144, [1989] 2 Lloyd's Rep 167, [1990] ICR 61, [1989] IRLR 265, CA E13008
Reilly v Merseyside Health Authority [1995] 6 Med LR 246, 23 BMLR 26, CA C6032
Revenue and Customs v Stringer (sub nom Ainsworth v IRC). See Ainsworth v IRC
Reverend Canon Pemberton v Right Reverend Inwood [2018] EWCA Civ 564, [2018] IRLR 542, [2018] ICR 1291, [2018] All ER (D) 183 (Mar) H1708
Revill v Newbery [1996] QB 567, [1996] 1 All ER 291, [1996] 2 WLR 239, [1996] NLJR 50, 139 Sol Jo LB 244, CA O3009
Rhind v Astbury Water Park Ltd [2003] EWHC 1029 (QB), [2003] All ER (D) 217 (May); affd [2004] EWCA Civ 756, 148 Sol Jo LB 759, [2004] All ER (D) 129 (Jun) O3004
Ribee v Norrie (2000) 81 P & CR D37, 33 HLR 777, [2000] NPC 116, [2001] PIQR P 8, (2000) Times, 22 November, [2000] All ER (D) 1644, CA O3004
Richardson v Pitt-Stanley [1995] QB 123, [1995] 1 All ER 460, [1995] 2 WLR 26, [1995] ICR 303, [1994] PIQR P 496, [1994] JPIL 315, CA E13008
Ricketts v Torbay [2003] EWCA Civ 613 .... O3018
Ring v Dansk almennyttigt Boligselskab C-335/11 (2013) ECLI:EU:C:2013:222, [2014] All ER (EC) 1161, [2013] 3 CMLR 527, [2013] IRLR 571, [2013] ICR 851, 132 BMLR 58, [2013] EqLR 528, ECJ V12032
Roberts v Day (1977) COIT No 1/133, Case No 3053/77, unreported E15022
Roberts v Dorothea Slate Quarries Co Ltd [1948] 2 All ER 201, [1948] LJR 1409, 92 Sol Jo 513, 41 BWCC 154, HL C6004
Robertson and Rough v Forth Road Bridge Joint Board [1995] IRLR 251, Ct of Sess C6033
Robinson v Harman (1848) 18 LJ Ex 202, 1 Exch 850, 154 ER 363, [1843–60] All ER Rep 383, 13 LTOS 141, Exch Ct H1728
Robinson-Steele v RD Retail Services Ltd: C-257/04 [2006] ECR I-2531, [2006] All ER (EC) 749, [2006] ICR 932, [2006] IRLR 386, (2006) Times, 22 March, [2006] All ER (D) 238 (Mar), ECJ W10024A
Rose v Plenty [1976] 1 All ER 97, [1976] 1 WLR 141, [1976] 1 Lloyd's Rep 263, [1975] ICR 430, [1976] IRLR 60, 119 Sol Jo 592, CA A1020
Rotary Yorkshire Ltd v Hague [2014] EWHC 2126 (Admin), [2014] All ER (D) 74 (Jul) E15021
Royal Mail Ltd v Jhuti [2017] EWCA Civ 1632, [2018] IRLR 251, [2018] ICR 982, [2017] All ER (D) 126 (Oct) E14011.1
Rylands v Fletcher. See Fletcher v Rylands and Horrocks

S

Sabloff v Yamaha Motor Co 113 NJ Super 279, 273 A 2d 606 (1971) P9036
Saini v All Saints Haque Centre [2009] 1 CMLR 1060, [2009] IRLR 74, [2008] All ER (D) 250 (Oct), EAT H1703
Santos v Disotto Food Ltd (2014) UKEAT/0623/12, [2014] All ER (D) 139 (Apr), EAT S11009
Sarkar v West London Mental Health NHS Trust [2010] EWCA Civ 289, [2010] IRLR 508, [2010] All ER (D) 211 (Mar) E14012
Sayers v Cambridgeshire County Council [2006] EWHC 2029 (QB), [2007] IRLR 29 .... W10010, W10042
Scarth v East Hertfordshire District Council (HSIB 181) J3013
Scott and Swainger v Associated British Ports and British Railways Board [2000] All ER (D) 1937 .... O3017
Searson v Brioland Ltd. See Brioland Ltd v Searson
Secretary of State for Employment v Associated Society of Locomotive Engineers and Firemen (No 2) [1972] 2 QB 455, [1972] 2 All ER 949, [1972] 2 WLR 1370, [1972] ICR 19, 13 KIR 1, 116 Sol Jo 467, CA E14007
Secretary of State for the Department for Work and Pensions v Alam [2010] ICR 665, [2010] IRLR 283, [2009] All ER (D) 174 (Nov), EAT M1511
Select Managements Ltd v Westminster City Council. See Westminster City Council v Select Management Ltd
Selwood v Durham County Council [2012] EWCA Civ 979, [2012] All ER (D) 177 (Jul) V8012
Sheldrake v DPP [2004] UKHL 43, [2005] 1 AC 264, [2005] 1 All ER 237, [2004] 3 WLR 976, [2005] RTR 13, [2005] 1 Cr App Rep 450, 168 JP 669, [2004] 43 LS Gaz R 33, (2004) Times, 15 October, 148 Sol Jo LB 1216, 17 BHRC 339, [2005] 3 LRC 463, [2004] All ER (D) 169 (Oct) E15031
Shipham & Co Ltd v Skinner [2001] All ER (D) 201 (Dec), EAT J3030
Shirlaw v Southern Foundries (1926) Ltd [1939] 2 KB 206, [1939] 2 All ER 113, CA; affd sub nom Southern Foundries (1926) Ltd v Shirlaw [1940] AC 701, [1940] 2 All ER 445, 109 LJKB 461, 84 Sol Jo 464, 164 LT 251, 56 TLR 637, HL E14004
Showboat Entertainment Centre Ltd v Owens [1984] 1 All ER 836, [1984] 1 WLR 384, [1984] ICR 65, [1984] IRLR 7, 128 Sol Jo 152, [1983] LS Gaz R 3002, 134 NLJ 37, EAT H1703
Sibbald v Sher Bros (1980) Times, 1 February O3014
Siddorn v Patel [2007] EWHC 1248 (QB), [2007] All ER (D) 453 (Mar) O3009
Sienkiewicz v Greif (UK) Ltd [2011] UKSC 10, [2011] 2 AC 229, [2011] 2 All ER 857, [2011] 2 WLR 523, [2011] ICR 391, 119 BMLR 54, (2011) Times, 10 March, 155 Sol Jo (no 10) 30, [2011] All ER (D) 107 (Mar) A5042.2, P9039
Simpson v AI Dairies Farms Ltd [2001] EWCA Civ 13, [2001] All ER (D) 31 (Jan) O3014
Sindicato de Médicos de Asistencia Pública (Simap) v Conselleria de Sanidad y Consumo de la Generalidad Valenciana: C-303/98 [2000] ECR I-7963, [2001] All ER (EC) 609, [2001] 3 CMLR 932, [2001] ICR 1116, [2000] IRLR 845, [2000] All ER (D) 1236, ECJ W10003
Sloan v Governors of Rastrick High School [2014] EWCA Civ 1063, [2015] PIQR P1, [2014] All ER (D) 305 (Jul) M3002
Smith v A J Morrisroes & Sons Ltd [2005] ICR 596, [2005] IRLR 72, [2004] All ER (D) 291 (Dec), EAT W10024A
Smith v Chubb Security Personnel Ltd (1999), unreported W10024
Smith v Churchills Stairlifts plc [2005] EWCA Civ 1220, [2006] ICR 524, [2006] IRLR 41, [2005] All ER (D) 318 (Oct) E16511
Smith v Leech Brain & Co Ltd [1962] 2 QB 405, [1961] 3 All ER 1159, [1962] 2 WLR 148, 106 Sol Jo 77 .... C6022
Smith v Manchester City Council (or Manchester Corpn) (1974) 17 KIR 1, 118 Sol Jo 597, CA C6022
Smith v Northamptonshire County Council [2008] EWCA Civ 181, [2008] 3 All ER 1054, [2008] ICR 826, [2008] All ER (D) 132 (Mar); affd [2009] UKHL 27, [2009] 4 All ER 557, [2009] ICR 734, 110 BMLR 15, [2009] PIQR P292, (2009) Times, 21 May, 153 Sol Jo (no 20) 39, [2009] All ER (D) 170 (May) M1005

Smith v Welsh Ambulance Service (Chester County Court) (22 March 2007, unreported) V8006
Smoker v London Fire and Civil Defence Authority [1991] 2 AC 502; [1991] 2 All ER 449 .... C6053
Snowball v Gardner Merchant Ltd [1987] ICR 719, [1987] IRLR 397, [1987] LS Gaz R 1572, EAT H1715
South Surbiton Co-operative Society v Wilcox [1975] IRLR 292, Ind Trib E15025
South Yorkshire Fire & Rescue v Mansell [2018] UKEAT/0151/17/3001(30 January 2018, unreported) E14022
Southern Foundries (1926) Ltd v Shirlaw. See Shirlaw v Southern Foundries (1926) Ltd
Southern Portland Cement Ltd v Cooper [1974] AC 623, [1974] 1 All ER 87, [1974] 2 WLR 152, 118 Sol Jo 99, PC O3010
Speed v Thomas Swift & Co Ltd [1943] KB 557, [1943] 1 All ER 539, 112 LJKB 487, 87 Sol Jo 336, 169 LT 67, CA S3003
Spencer v Boots The Chemist Ltd [2002] EWCA Civ 1691, [2002] All ER (D) 465 (Oct) M2006
Spencer-Franks v Kellogg Brown and Root Ltd [2008] UKHL 46, [2009] 1 All ER 269, [2008] ICR 863, [2008] NLJR 1004, [2008] PIQR P389, (2008) Times, 3 July, 2008 SLT 675, 2008 SCLR 484, 152 Sol Jo (no 27) 30, [2008] All ER (D) 26 (Jul) M1005
Spurling (J) Ltd v Bradshaw [1956] 2 All ER 121, [1956] 1 WLR 461, [1956] 1 Lloyd's Rep 392, 7 LDAB 113, 100 Sol Jo 317, CA O3016
Stannard (t/a Wyvern Tyres) v Gore [2012] EWCA Civ 1248, [2013] 1 All ER 694, [2012] NLJR 1289, [2012] All ER (D) 44 (Oct) P9528
Staples v West Dorset District Council (1995) 93 LGR 536, [1995] 18 LS Gaz R 36, [1995] PIQR P 439, 139 Sol Jo LB 117, CA O3005, O3017
Stark v Post Office [2000] All ER (D) 276, CA P9010
Stevens v Blaenau Gwent County Borough [2004] EWCA Civ 715, [2004] HLR 1039, (2004) Times, 29 June, [2004] All ER (D) 116 (Jun) O3003
Stevenson v J M Skinner & Co (UKEAT/0584/07), unreported R3026
Strathclyde Regional Council v Porcelli. See Porcelli v Strathclyde Regional Council
Stringer v HM Revenue and Customs: C-520/06 [2009] All ER (EC) 906, [2009] 2 CMLR 657, [2009] ICR 932, [2009] IRLR 214, (2009) Times, 28 January, [2009] All ER (D) 147 (Jan), ECJ M1514, W10024B
Sumsion v BBC (Scotland) [2007] IRLR 678, EAT W10024
Swain v Denso Martin Ltd (2000) Times, 24 April M3002

T

TSB Bank plc v Harris [2000] IRLR 157, EAT E14019
Tacagni v Cornwall County Council [2013] EWCA Civ 702, [2013] All ER (D) 182 (Apr) O3002
Tawling v Wisdom Toothbrushes Ltd (1997), unreported W11043
Taylor v Furness, Withy & Co Ltd [1969] 1 Lloyd's Rep 324, 6 KIR 488 .... E14001
Tesco Stores Ltd v Constable [2008] EWCA Civ 362, [2008] Lloyd's Rep IR 636, [2008] All ER (D) 214 (Apr) P9523
Tesco Stores Ltd v Kippax COIT No 7605-6/90 (1990), unreported E15015
Tesco Supermarkets Ltd v Nattrass [1972] AC 153, [1971] 2 All ER 127, [1971] 2 WLR 1166, 69 LGR 403, 135 JP 289, 115 Sol Jo 285, HL E15039
Timis v Osipov (Protect intervening) [2018] EWCA Civ 2321, [2019] IRLR 52, [2018] All ER (D) 145 (Oct) E14011.1
Thomas v Brighton Health Authority [1999] 1 AC 345, [1998] 3 All ER 481, [1998] 3 WLR 329, [1998] IRLR 536, [1998] 2 FLR 507, [1998] Fam Law 593, 43 BMLR 99, [1998] NLJR 1087, [1998] PIQR Q 56, 142 Sol Jo LB 245, [1998] All ER (D) 352, HL C6027
Thomas v Bunn [1991] 1 AC 362, [1991] 1 All ER 193, [1991] 2 WLR 27, 135 Sol Jo 16, [1990] NLJR 1789, HL C6043
Thomas Sanderson Blinds Ltd v English, UKEAT/0316/10/JOJ (21 February 2011, unreported) H1705

Table of Cases

Thompson v Smith's Shiprepairers (North Shields) Ltd [1984] QB 405, [1984] 1 All ER 881, [1984] 2 WLR 522, [1984] ICR 236, 128 Sol Jo 225, [1984] LS Gaz R 741, sub nom Mitchell v Vickers Armstrong Ltd and Swan Hunter ShipbuidersLtd [1984] IRLR 93 .... IN15, N3022
Thompson v T Lohan (Plant Hire) Ltd (JW Hurdiss Ltd, third party) [1987] 2 All ER 631, [1987] 1 WLR 649, [1987] IRLR 148, [1987] BTLC 221, 131 Sol Jo 358, [1987] LS Gaz R 979, CA P9058
Thornett and Fehr v Beers & Son [1919] 1 KB 486, 88 LJKB 684, 24 Com Cas 133, 120 LT 570 .... P9051
Thornton v Shoe Lane Parking Ltd [1971] 2 QB 163, [1971] 1 All ER 686, [1971] 2 WLR 585, [1971] RTR 79, [1971] 1 Lloyd's Rep 289, 115 Sol Jo 75, CA O3016
Titchener v British Railways Board 1981 SLT 208, OH; on appeal 1983 SLT 269, Ct of Sess; affd [1983] 3 All ER 770, 127 Sol Jo 825, 1984 SLT 192, sub nom McGinlay (or Titchener) v British Railways Board [1983] 1 WLR 1427, HL O3010, O3017
Tomlinson v Congleton Borough Council [2003] UKHL 47, [2004] 1 AC 46, [2003] 3 All ER 1122, [2003] 3 WLR 705, [2003] 34 LS Gaz R 33, [2003] NLJR 1238, [2003] 32 EGCS 68, (2003) Times, 1 August, 147 Sol Jo LB 937, [2003] All ER (D) 554 (Jul) O3017, O3019
Trent Strategic Health Authority v Jain [2009] UKHL 4, [2009] 1 AC 853, [2009] 1 All ER 957, [2009] 2 WLR 248, [2009] PTSR 382, 106 BMLR 88, (2009) Times, 22 January, 153 Sol Jo (no 4) 27, [2009] All ER (D) 148 (Jan) E15028
Trim Joint District School Board of Management v Kelly [1914] AC 667, 83 LJPC 220, 58 Sol Jo 493, 111 LT 305, 30 TLR 452, [1914–15] All ER Rep Ext 1453, 7 BWCC 274, HL C6004
Tripp v Ministry of Defence [1982] CLY 1017 .... N3022
Truslove v Scottish Ambulance Service (2014) UKEATS/53/13, [2014] ICR 1232, [2014] All ER (D) 112 (Aug), EAT W10003
Twine v Bean's Express Ltd [1946] 1 All ER 202, 62 TLR 155; affd (1946) 202 LT 9, 62 TLR 458, CA A1020

U

*Uber BV and other companies v Aslam [2018] EWCA Civ 2748, [2019] 3 All ER 489, [2019] RTR 277, [2019] IRLR 257, [2019] ICR 845, [2018] All ER (D) 97 (Dec) . E14001
Unite the Union v Nailard [2018] EWCA Civ 1203, [2018] IRLR 730, [2019] ICR 28, [2018] All ER (D) 89 (Jun) H1715A

V

Vaickuviene v J Sainsbury plc [2013] CSIH 67, [2013] IRLR 792, 2014 SC 147, Ct of Sess V8013
Veakins v Kier Islington Ltd [2009] EWCA Civ 1288, [2010] IRLR 132, (2010) Times, 13 January, [2009] All ER (D) 34 (Dec) H1726
Vento v Chief Constable of West Yorkshire Police [2002] EWCA Civ 1871, [2003] ICR 318, [2003] IRLR 102, [2003] 10 LS Gaz R 28, (2002) Times, 27 December, 147 Sol Jo LB 181, [2002] All ER (D) 363 (Dec) H1715
Viasystems (Tyneside) Ltd v Thermal Transfer (Northern) Ltd [2005] EWCA Civ 1151, [2006] QB 510, [2005] 4 All ER 1181, [2006] 2 WLR 428, [2006] ICR 327, [2005] IRLR 983, [2005] 44 LS Gaz R 31, [2005] All ER (D) 93 (Oct) E13027

W

Walker v BHS Ltd [2005] All ER (D) 146 (May), EAT H1705
Walker v Northumberland County Council [1995] 1 All ER 737, [1995] ICR 702, [1995] IRLR 35, [1995] ELR 231, [1994] NLJR 1659 ... E14003, H1724, H1728, R2004, S11009

Wallbank v Wallbank Fox Designs Ltd [2012] EWCA Civ 25, [2012] IRLR 307,
  [2012] All ER (D) 01 (Feb) V8013
Waltons & Morse v Dorrington [1997] IRLR 488, EAT E14003, E14019
Waters v Metropolitan Police Comr [1997] ICR 1073, [1997] IRLR 589, CA; revsd
  [2000] 4 All ER 934, [2000] 1 WLR 1607, [2000] ICR 1064, HL H1712
Watson (administrators of) v Willmott [1991] 1 QB 140, [1991] 1 All ER 473, [1990] 3
  WLR 1103 .... C6039
Wattleworth v Goodwood Road Racing Co Ltd [2004] EWHC 140 (QB), [2004] All ER
  (D) 51 (Feb) O3011, O3017
Waugh v British Railways Board [1980] AC 521, [1979] 2 All ER 1169, [1979] 3 WLR
  150, [1979] IRLR 364, 123 Sol Jo 506, HL A3028, J3015
Waugh v London Borough of Sutton (1983) HSIB 86 .... J3011
Weathersfield Ltd (t/a Van & Truck Rentals) v Sargent [1999] ICR 425, [1999] IRLR 94,
  143 Sol Jo LB 39, CA H1703
Webster v Blue Ship Tea Room Inc 347 Mass 421, 198 NE 2d 309 (1964) P9031
Weddall v Barchester Healthcare Ltd [2012] EWCA Civ 25, [2012] IRLR 307,
  [2012] All ER (D) 01 (Feb) V8013
Wells v Wells [1999] 1 AC 345, [1998] 3 All ER 481, [1998] 3 WLR 329, [1998] IRLR
  536, [1998] 2 FLR 507, [1998] Fam Law 593, 43 BMLR 99, [1998] NLJR 1087,
  [1998] PIQR Q 56, 142 Sol Jo LB 245, [1998] All ER (D) 352, HL C6027
West (H) & Son Ltd v Shephard [1964] AC 326, [1963] 2 All ER 625, [1963] 2 WLR
  1359, 107 Sol Jo 454, HL C6030
West Cumberland By-Products Ltd v DPP [1988] RTR 391, DC E15042
West Midland Co-operative Society Ltd v Tipton [1986] AC 536, [1986] 1 All ER 513,
  [1986] 2 WLR 306, [1986] ICR 192, [1986] IRLR 112, 130 Sol Jo 143,
  [1986] LS Gaz R 780, [1986] NLJ Rep 163, HL E14013
Western Excavating (ECC) Ltd v Sharp [1978] QB 761, [1978] 1 All ER 713, [1978] 2
  WLR 344, [1978] ICR 221, [1978] IRLR 27, 121 Sol Jo 814, CA H1720
Westminster City Council v Select Management Ltd [1985] 1 All ER 897, [1985] 1 WLR
  576, 83 LGR 409, [1985] ICR 353, 129 Sol Jo 221, [1985] 1 EGLR 245,
  [1985] LS Gaz R 1091, sub nom Select Managements Ltd v Westminster City Council
  [1985] IRLR 344, CA O3001, O3019
Westwood v Post Office [1974] AC 1, [1973] 3 All ER 184, [1973] 3 WLR 287, 117 Sol
  Jo 600, HL A1001
Wheat v E Lacon & Co Ltd [1966] 1 QB 335, [1965] 2 All ER 700, [1965] 3 WLR 142,
  109 Sol Jo 334, CA; affd [1966] AC 552, [1966] 1 All ER 582, [1966] 2 WLR 581,
  [1966] RA 193, 110 Sol Jo 149, [1966] RVR 223, HL O3004
Whirlpool (UK) Ltd and Magnet Ltd v Gloucestershire County Council (1993) 159 JP
  123 .... P9008
White v Chief Constable of South Yorkshire Police [1999] 2 AC 455, [1999] 1 All ER 1,
  [1998] 3 WLR 1509, [1999] ICR 216, 45 BMLR 1, [1999] 02 LS Gaz R 28, [1998]
  NLJR 1844, [1999] 3 LRC 644, sub nom Frost v Chief Constable of South Yorkshire
  Police 143 Sol Jo LB 51, HL C6033
White v Pressed Steel Fisher [1980] IRLR 176, EAT J3011
White v St Albans City and District Council (1990) Times, 12 March, CA O3009
Whittlestone v BJP Home Support Ltd (UKEAT/0128/13) (19 July 2013, unreported),
  EAT W10003
Wieland v Cyril Lord Carpets Ltd [1969] 3 All ER 1006 .... C6022
Wileman v Minilec Engineering Ltd [1988] ICR 318, [1988] IRLR 144, EAT H1715
Williams v Gloucestershire County Council (10 September 1999, unreported) C6023
Willson v Ministry of Defence [1991] 1 All ER 638, [1991] ICR 595 .... C6040
Wilson v Graham [1991] 1 AC 362, [1991] 1 All ER 193, [1991] 2 WLR 27, 135 Sol Jo
  16, HL C6043
Wilson v National Coal Board 1978 SLT 129, OH; affd 1981 SLT 67, HL C6053
Wilsons and Clyde Coal Co Ltd v English [1938] AC 57, [1937] 3 All ER 628, 106 LJPC
  117, 81 Sol Jo 700, 157 LT 406, 53 TLR 944, HL S3003
Wilton v Timothy James Consulting Ltd (2015) UKEAT/82/14, [2015] ICR 765, [2015]
  IRLR 368, [2015] All ER (D) 70 (Apr), EAT H1706, H1718
Winterbottom v Wright (1842) 11 LJ Ex 415, 10 M & W 109, 152 ER 402 .... P9024
Withers v Perry Chain Co Ltd [1961] 3 All ER 676, [1961] 1 WLR 1314, 59 LGR 496,
  105 Sol Jo 648, CA R2006

Witley and District Men's Club v Mackay [2001] IRLR 595, [2001] All ER (D) 31 (Jun), EAT W10023
Woking Borough Council v BHS plc (1994) 93 LGR 396, 159 JP 427, 159 JPN 387 .... A3008
Wood v British Coal Corpn [1991] 2 AC 502, [1991] 2 All ER 449, [1991] 2 WLR 1052, [1991] ICR 449, [1991] IRLR 271, HL C6053
Woolfall and Rimmer Ltd v Moyle [1942] 1 KB 66, [1941] 3 All ER 304, 71 Ll L Rep 15, 111 LJKB 122, 86 Sol Jo 63, 166 LT 49, 58 TLR 28, CA E13020
Woolmington v DPP [1935] AC 462, 25 Cr App Rep 72, 104 LJKB 433, 30 Cox CC 234, [1935] All ER Rep 1, 79 Sol Jo 401, 153 LT 232, 51 TLR 446, HL E15031
Workvale Ltd (No 2), Re [1992] 2 All ER 627, [1992] 1 WLR 416, [1992] BCLC 544, [1992] BCC 349, CA E13024
Worsley v Tambrands Ltd (No 2) [2000] MCR 0280 .... P9035
Wren v Holt [1903] 1 KB 610, 67 JP 191, 72 LJKB 340, 51 WR 435, 88 LT 282, 19 TLR 292, [1900–3] All ER Rep Ext 1152, CA P9052
Wright v British Railways Board [1983] 2 AC 773, [1983] 2 All ER 698, [1983] 3 WLR 211, 127 Sol Jo 478, HL C6043
Wright v Redrow Homes (Yorkshire) Ltd [2004] EWCA Civ 469, [2004] 3 All ER 98, [2004] ICR 1126, [2004] IRLR 720, 148 Sol Jo LB 666, [2004] All ER (D) 221 (Apr) W10004

Y

Yearwood v Comr of the Police of the Metropolis (UKEAT 0310/03/2805) [2004] ICR 1660, [2004] All ER (D) 457 (May), EAT H1715A
Young v Charles Church (Southern) Ltd (1997) 39 BMLR 146 .... C6033
Young v Kent County Council [2005] EWHC 1342 (QB), [2005] All ER (D) 217 (Mar) O3010, O3011
Young v Post Office [2002] EWCA Civ 661, [2002] IRLR 660, [2002] All ER (D) 311 (Apr) R2004, R2009

Decisions of the European Court of Justice are listed below numerically. These decisions are also included in the preceding alphabetical list.

C-335/11: Ring v Dansk almennyttigt Boligselskab (2013) ECLI:EU:C:2013:222, [2014] All ER (EC) 1161, [2013] 3 CMLR 527, [2013] IRLR 571, [2013] ICR 851, 132 BMLR 58, [2013] EqLR 528, ECJ V12032
C-531/15: Otero Ramos v Servicio Galego de Saúde, Instituto Nacional de la Seguridad Social (2017), ECLI:EU:C:2017:789, [2018] IRLR 159, [2018] ICR 965, [2017] All ER (D) 149 (Oct), EUCJ E14025; R3026
C-41/17: González Castro v Mutua Umivale, ProsegurEspaña SL, Instituto Nacional de la Seguridad Social (INSS) (2018) ECLI:EU:C:2018:736, [2018] IRLR 1142, [2019] ICR 339, EUCJ E14026; R3026

Table of Statutes

A

Access to Medical Reports Act 1988
..... M1505, M1512, M1512A
    s 1 .... O1017
       68(2) O1017
Activity Centres (Young Persons' Safety) Act 1995 .... C6508.1
Administration of Justice Act 1982
    s 1(1) C6023, C6034
      5 .... C6052
     15 .... C6043
    Sch 1 .... C6043
Aviation and Maritime Security Act 1990
.... O7001

B

Boiler Explosions Act 1882
.... P7001
Building Act 1984 ... F5031, G1026

C

Celluloid and Cinematograph Film Act 1922 .... D0938, D0939
Charities Act 2006
.... C9002
Children Act 1989
    Pt 4 (ss 31–42) C9008
      5 (ss 43–52) C9008
Children (Scotland) Act 1995
    Pt 2 (ss 16–93) C9008
Children's Hearings (Scotland) Act 2011
.... C9008
Civil Aviation Act 1982
.... O7001
Civil Contingencies Act 2004
.... B8046, D6001, D6005, D6025, D6071, M1101
    Pt 1 (ss 1–18) B8043, B8044, B8045, D6003, D6005, D6006
    s 1(1)–(3) D6006
      2 .... D6006
        (1)(a)–(d) D6007
          (g) D6007
        (3) D6007
      4 .... D6006
        (1) D6007
     13 .... D6006
     17 .... D6006
    Pt 2 (ss 19–33) B8044, D6003, D6005, D6006
    s 19(1)–(6) D6006
     20 .... D6006
     23 .... D6006

Civil Contingencies Act 2004 – cont.
    s 24(3) D6006
     26 .... D6006
     28 .... D6006
    Pt 3 .... B8044
    Sch 1
      Pt 1
        para 1 .... D6007
      Pt 3 .... D6005
Civil Defence Act 1948
.... D6005
Civil Jurisdiction and Judgments Act 1982
.... P9046
Civil Liability (Contribution) Act 1978
.... E13027
Civil Protection in Peacetime Act 1986
.... D6005
Clean Air Act 1956
.... E5061
Clean Air Act 1968
.... E5061
Clean Air Act 1993
    Pt I (ss 1–3) E5061
      II (ss 4–17) E5061
      III (ss 18–29) E5061
Climate Change Act 2008
.... E5070
Communications Act 2003
.... F9012
Companies Act 1985
.... E16005
    s 141 .... E13024
     310 .... E15041
     651(5), (6) E13024
     722, 723 .... E13013
Companies Act 2006
..... A5042B, C9017, E16005, E16024a
    s 232 .... E15041
Company Directors' Disqualification Act 1986
    s 2 .... E15043.2
      (1) E15040
     13 .... E15040
Compensation Act 2006
.... A5042.2, E16024
    Pt 1 (ss 1–3) E13033
      2 (ss 4–15) E13033
      3 (ss 16–18) E13033
Consumer Protection Act 1987
...... Intro 19, A5030, P9027, P9038, P9543
    Pt I (ss 1–9) P9044
    s 1(2)(a)–(c) P9029
      2(1) P9028
      (2) P9028
        (b), (c) P9030

Table of Statutes

Consumer Protection Act 1987 – *cont.*
  s 2(3) P9030
    (5) P9039
    3(1) P9031
    (2) P9032
    4(1)(a)–(f) P9041
    5(1)–(4) P9040
    6(4) P9034
    7 .... P9043
  Pt II (ss 10–19) P9533
  s 41(1) P9026
    46(1)–(3) P9042
  Sch 1 .... P9044
    3 .... P9009, P9021, P9022, P9023

Consumer Rights Act 2015
  .... P9052
  Pt 1 (ss 1–60) P9048
  s 9(2) **P9049**
    (4)–(6) P9049
    10 .... P9052
    11 .... P9056
    13 .... P9050
    20, 21 .... P9048
    23, 24 .... P9048
    31 .... P9057
  Pt 2 (ss 61–76) P9047, P9059
  s 62(2) O3016
    (4) P9059
    (6) O3016
    (7)(a) O3016
    65 .... P9059
    (1) O3016
    68, 69 .... P9059
  Sch 2 .... P9060

Consumer Safety Act 1978
  .... E3029

Continental Shelf Act 1964
  .... O7001

Control of Pollution Act 1974
  .... E5061

Control of Pollution (Amendment) Act 1989 .... E5061

Coroners and Justice Act 2009
  s 125(1) E15043

Corporate Manslaughter and Corporate Homicide Act 2007
  . Intro 4, Intro 8, Intro 9, Intro 15, C6504, C6504.1, C8519, C 9001, C9024, D6029, D6030, E13001, E13032, E15001, E15038, IN13, M2001, M2103, P9524, V8008B
  s 1 .... C9023
    (1)–(5) C9002
    (6), (7) C9011
    2 .... C9002
    (1) C9003, E15038.1
    (d) C9001
    (2) ... C9001, C9003, E15038.1
    (4)–(6) C9003

Corporate Manslaughter and Corporate Homicide Act 2007 – *cont.*
  s 3(1)–(4) C9003
    4 .... C9005, E15038.1
    5 .... E15038.1
    (1)–(3) C9006
    6 .... E15038.1
    (1) C9007
    (3), (4) C9007
    (7), (8) C9007
    8 .... C8520, E15038.1
    (1), (2) C9009
    (4) C9009
    9 .... C9011
    (1) C9008, C9012
    (2), (3) C9012
    (4), (5) C9012
    10 .. C9011, E15038.1, E15043.2
    (1) C9013
    (4), (5) C9013
    11 .... C9002
    (1) C9014, E15045
    (2)–(5) C9014
    12 .... C9015
    13 .... C9016
    14(1), (2) C9017
    15(4) C9018
    16(1)–(3) C9019
    17 .... C9020
    18 .... C9021
    19 .... C9022
    20 .... C9023
    21 .... C9002, C9017
    22 .... C9024
    25 .... C9002
    27(4) C9023
    28 .... C9024
  Sch 1 ..... C9002, C9014, C9018, C9024

Countryside and Rights of Way Act 2000
  .... O3007

Crime and Disorder Act 1998
  .... H1729
  s 28 .... H1733
    32(1)(a), (b) H1733

Criminal Justice Act 1982
  s 37(2) E13014, E13015

Criminal Justice Act 2003
  s 144 .... E15043.1
    154(1) IN05
    162 .... E15043.2
    174 .... E15043.1
    240A E15043.2

Criminal Justice and Court Services Act 2000
  Pt 1 Ch 1 (ss 1–10)
  .... C9008

Criminal Justice and Courts Act 2014
  .... C 9003

Table of Statutes

Criminal Justice and Public Order Act 1994
s 34 .... E15026.1

D

Damages Act 1996
.... C6001, C6025
s 1(1) C6027
2 .... C6021
Dangerous Wild Animals Act 1976
.... P9516
Data Protection Act 1998
.... A3024, F7020, M1528,
O1017, R2007
s 1(1)(a)–(d) D6007
Defective Premises Act 1972
.... C9003, P9524, P9527
Deregulation Act 2015
.... C8018, IN01, IN19
s 108 .... E15012.2
Disability Discrimination Act 1995
..... A1001.1, H1701, V12036
s 6 .... W11001
18B E16511
Disability Discrimination Act 2005
.... R2005

E

Emergency Powers Act 1920
.... D6005
Emergency Workers (Obstruction) Act 2006 .... V8008A
Emergency Workers (Offences) Act 2018
.... V8008A
Emergency Workers (Scotland) Act 2005
.... V8008A
Employers' Liability (Compulsory Insurance) Act 1969
..... Intro 19, C6001, E13011,
E13017, E13020, E13021, O7030,
P9025, P9037, P9523
s 1(1) E13008, P9515
(3) E13008
2(2)(a) E13010
3 .... E13012
4(1) E13013
5 .... E13008
Employers' Liability (Defective Equipment) Act 1969 .... C6001, E13001
Employment Act 1989
.... C8018
Employment Act 2002
Pt 3 (ss 29–40) M1505
Employment Act 2008
.... M1505, M1517
Employment Relations Act 1999
.... E14013, E14019, J3005
s 10 .... E14008

Employment Rights Act 1996
... E14001, E14002, E14011.1,
E14017, M1530, O7001, O7056,
R2007
s 1 ...... E14006, M1505, M1513
3(2) E14006
13 . E14014, W10024, W10024B
23 .... W10024B
43B(1)(d) J3030
44 .... E14011, V8009
(1) J3030
(d) V8014
(e) E14010
(2), (3) E14009
45A E14022, W10041
(1)(b) W10013
47B J3030
48 ...... E14009, E14022, J3030
49 .... J3030
(2) E14009
64 .... E14020, E14030
67 .... E14025, E14029, R3026,
V12018
68 .... E14020, E14030
70(1) E14020, E14030
(4) E14029
72(1) E14028
86 .... E14002, E14006, E14011,
E14018
95(1) E14019
(c) H1719, H1720
98 ..... E14010, E14020, M1516
(4) E14021
100 .. E14009, E14011, E14022,
J3030
(1)(e) E14010
101A W10010, W10041
103A J3030
104 .... E14011, W10041
105 .... W10041
108 .... J3030
(3)(c) E14010
109 .... J3030
113–118 .... H1721
124(1A) J3030
203 .... W10043
221, 222 .... W10024
223(2) W10024
224 .... W10024
227, 228 .... W10024
Energy Act 1976 .... O7001
Energy Act 2013
Sch 8
para 3, 4 .... E15029
Enterprise and Regulatory Reform Act
2013 ...... C6058, E14011.1, IN01,
P9010
Environment Act 1995
.... A5030, D6001, D6035
s 14–18 .... D6025

Table of Statutes

Environment Act 1995 – *cont.*
  s 95(2)–(4) E15040
    108, 109 .... E15027
Environment and Safety Information Act 1988
  s 1 .... E15029
    2(2), (3) E15029
    3 .... E15029
  Sch E15029
Environmental Protection Act 1990
  .... A5030, E15043
  Pt 2 (ss 29–78) E5027.2
    3 (ss 79–85) E5062
  s 79 .... C8019
    157 .... E15040
Equality Act 2010 . Intro 18, A1001, A1001.1, E14002, E14020, E14021, E14022, E14023, E14024, E16501, E16502, E16509, E16512, E16515, E17070, M1505, M1506, M1507, M1508, M1509, M1511, M1523, M2115, M3007, O1003, O1007, O1020.1, O1020.4, R2005, R2006, R2009, R3028, V12018, V12032, V12037, W11001, W11043
  s 4 .... H1701
    5(2) E16513
    6 .... E16510
    9 .... H1703
      (1) E16517
      (2)(b) E16517
      (3) E16517
  Pt 2, Ch 2 (ss 13–27)
  .... H1709
  s 13 .... E16503
    (2) E16513
    (6) E16520
    15(1)(b) E16505
      (2) E16506
    19 .... E16504
    20(3)–(5) E16511
      (9) E16511
      (11) E16511
    21, 22 .... E16511
    24 .... H1705
    26 ...... E16507, H1701, H1702
      (1) H1703
        (b) H1707
      (2), (3) H1703
      (5) H1703
    27(3) E16508
    40(3) H1706
    109(1) H1711, H1714
      (2) H1715A
      (4) H1713, H1714

Equality Act 2010 – *cont.*
  s 110 .... H1714
    112 .... H1714
    120 .... E16522
    136 .... H1707
  Sch 1
    para 3 .... E16510
  Sch 8 .... E16511
    Pt 2 .... E16511
      3
    para 20 .... E16511
Explosives Acts 1875
  s 23 .... W11047
European Communities Act 1972
  .... N3002A
European Union (Withdrawal) Act 2018
  .... E5001

F

Factories Act 1937
  .... A5042.2
  s 14(3) E13020
Factories Act 1961
  ..... Intro 10, D0939, E15015, E15045, F5031, P7020, P9009, W11045
  s 14(1) E13008
    29 .... O3019
      (1) E15031, N3022
      (2) E15031
    30 .... A1023
    123–126 .... G1010
    155(1) W11047
      (2) W110487
    173(1) W11046
    175 .... G1010, W11046
Factory and Workshop Act 1901
  .... P7001
Fatal Accidents Act 1976
  .... C6033, C6040, E15001
  s 1(3) C6037
    1A C6035
    4 .... C6036
    5 .... C6037
Financial Services and Markets Act 2000
  .... E13024
Fire and Rescue Services Act 2004
  .... E15009
Fire Precautions Act 1971
  .... E15029, F5031, F5032, F5033, R3012
  s 23 .... E15040
Fire Safety and Safety of Places of Sport Act 1987 .... F5031, F5033
Fire (Scotland) Act 2005
  .... C8002
Fire Services Act 1947
  .... F5031

Fires Prevention (Metropolis) Act 1774			
			F5031
Food and Environment Protection Act 1985			F9007, O7001
s 1, 2			F9020
19			E15029
Food Safety Act 1990			
			F9001, F9002, F9022, F9072, F9078, O7001, P9524, P9543
s 1–3			F9007
5, 6			F9007.1
Pt II (ss 7–26)			P9533
s 7			F9006, F9008, F9009, F9010, F9023
8			F9006, F9008, F9010, F9011
9			F9008, F9010, F9013, F9014
10			F9008, F9010, F9013, F9015
11			F9008, F9010, F9013, F9016, F9017, F9018
12			F9008, F9010, F9013, F9018
13			F9008, F9010, F9013, F9019, F9020
14			F9008, F9011, F9012, F9024
15			F9008, F9012
16–18			F9008, F9020
19			F9008, F9021
20			F9008
21			F9008, F9070
(3), (4)			F9023
22			F9008
26			F9020
29–33			F9012
35			F9024
37			F9015
39			F9015
40			F9007, F9025
54(2)			E15045
Food Standards Act 1999			
			F9005
Fraud Act 2006			E15040
Freedom of Information Act 2000			
			E5041, E16009

G

Gas Act 1986			G1001, O7001
s 4A			G1002
7			G1002, G1005
8, 8A			G1002
11			G1002
18, 18A			G1002
21–22A			G1002
33A			G1002
Sch 2B			G1002
Gas Act 1995			G1001, G1002

H

Housing Act 2004			C6507
Health Act 2006			E13031, E14003
s 2			W11028

Health Act 2006 – *cont.*			
s 5			W11028
8(1)			W11028
Health and Safety at Work etc Act 1974			
			Intro 5, Intro 9, Intro 10, Intro 18, A1001, A5030, C6501, C6508, C6527, C7004, C7005, C7009, C8515, C9029, D0401, D0948, D8203, E3003, E13015, E13020, E14003, E14027, E15001, E15013, E17045, E17061, E18004, E18009, E18011, F3034, F5031, I2010, IN01, IN14, IN22, J3023, L5002, M1114, M2006, M2015, M2021, M2103, M3002, N3002A, O1056, O3003, O7001, O7003, O7009, O7010, O7033, O7046, O7055, P3027, P7031, P9010, P9011, P9524, R3001, S7004, V3001, V5015A, V8015, V12036, W9004.2, W10039
Pt 1 (ss 1–53)			E15012, O7004, O7009
s 1			E15005, W9003
(1)(a)–(c)			D6009
2			Intro 15, A1009, C9029, E13032, E14001, E14006, E14012, E14023, E14024, E15022E15034, E15035, E15036, E15037, E15039, E15043, E16509, M1002, M2002, M2014, M2016, M3015, O3019, P3004, P9009, R3005, V3015, W11002, W9003
(1)			Intro 11, C6503, D0951, E15031, G1026A, M2003, P3005, P7022, R3002S3001, S3004, S11006, V8004
(2)			Intro 11, E15031, M2003, R3002, T7003
(a)			C7008, L3001, R3017, S3004
(b)			L3001
(c)			S11040
(d)			A1001
(3)			Intro 11, Intro 23, M2003, M2031, S3004, S7001, S7002, S7003
(4)			J3004
(6)			J3002
(7)			J3017
3			C6512, C9024, P9527, E15034, E15035, E15037, E15039M2002, M2014, P9009, R3005, V3015, V12001, W9003
(1)			Intro 8, Intro 11, C6502, C6503, C6508, C9001, C9029, E15031F3027, G1026A, L3020, M2004, O3019, P7022, P9504, R3002, S3004
(a)			S3001
(2)			M2004, M3001, R3002

Table of Statutes

Health and Safety at Work etc Act 1974 – cont.

s 4	....	Intro 11, E15034, E15035, E15036, E15039, M2014, O3001, R3002, R3005, W11002, W9003
(1)–(3)	....	O3019
5	....	E15034, E15035, E15036, E15039, W9003
6	....	Intro 12, E15034, E15035, E15036, E15039, M1014, P1013, P3002, P3012, P9001A.1P9026, R3002, R3005, V3015, W9003
(1)	....	M1002, M1016
(a)	...	P1012, P9014, P9020, P9022
(b)–(d)	....	P9014
(2)	....	P9016
(3)	....	P9017
(4)	....	P9015
(a)	....	P9022
(5)	....	P9018
(6)	....	P9019
(8)	....	P9020, P9021
(8A)	....	P9009, P9021
(10)	....	P9022
7	....	Intro 12, A1020, E15034, E15035, E15036, E15043.2, J3009, M2017, S3004, V8004, V8009, V8014, W9003, W9013
(1)	....	A1020.1
8	....	Intro 12, E15034, E15036, W9003
9	....	Intro 12, E15034, E15036, P3004, W9003
11	....	E15005, W9003
12, 13	....	W9003
14	....	E15005, E15033, W9003
(1)	....	D6009
15	....	T7003, W9003
16	....	W9003
17	....	W9003
18	....	W9003
(1)	....	E15005, E15008
(7)(a)	....	E15008
19	....	W9003
(1)	....	E15007, T7003
20	....	W9003
(2)(e)	....	E15021
(j)	....	E15026.1
(7)	....	E15026.1
21	....	E15014, G1026A, J3003, W9003
22	....	W9003
(1)–(3)	....	E15015
(4)	....	E15016
23	....	W9003
(5)	....	E15014
24	....	E15024, W9003
(2)	....	E15019
(3)(a)	....	E15019
25	....	E15034, W9003

Health and Safety at Work etc Act 1974 – cont.

s 25(1)–(3)	....	E15027
25A	....	E15033, P9023, W9003
26	...	E15028, E15043.2, W9003
27	....	E15043.2, W9003
27A	....	P9023, W9003
28	....	W9003
(2)	....	D6007
(7)	....	D6007
(8)	....	J3016, O7050
29–32	....	W9003
33	....	Intro 15, W9003
(1)	....	E15034
(b)	....	E15036
(c)	....	E15036
(d)–(f)	....	E15036
(g)	....	E15017, E15031, E15036
(h)–(k)	....	E15036
(n), (o)	....	E15036
34	....	W9003
(1)	....	E15034
35	....	W9003
36	....	V3015, W9003
(1)	....	E15042
37	....	Intro 15, C9022, C9029, E13032, E15031, G1026A, O7002, S3004
(1)	....	C9029, E15040
38, 39	....	W9003
40	....	E15031, E15031, P9522, R3003, W9003
41	....	W9003
42	....	C9012, E15034, E15036, W9003
43–46	....	W9003
47	......	M2005, M2019, P9026, W9003
(2)	....	O3018
48	....	W9003
(1), (2)	....	E15045
49	....	W9003
50	....	W9003
51, 52	....	W9003
53	....	E15015, W9003
(1)	....	E15027, P9012
54–59	....	O7009, W9003
80, 81	....	O7009, W9003
82	....	O7009, W9003
(1)(c)	....	E15019
Sch 1	....	E15012
3A	....	E15017, E15032

Health and Safety (Offences) Act 2008
..... Intro 4, Intro 8, Intro 10, E15032, E15036

Highways Act 1980
s 1	....	D6006
168, 169	....	C8019

Hotel Proprietors Act 1956
.... P9527

Table of Statutes

Human Rights Act 1998 .... D6006, V8015

I

Immigration Act 2014 .... C9003

Income Tax (Earnings and Pensions) Act 2003
s 406 .... H1718

Industrial Diseases (Notification) Act 1981 .... A3010

Inquiries into Fatal Accidents and Sudden Deaths etc (Scotland) Act 2016
s 29 .... W9014.1

International Headquarters and Defence Organisations Act 1964 .... E15009, F5056

Interpretation Act 1978
Sch 1 .... E15034

J

Judgments Act 1838
s 17 .... C6043

L

Law Reform (Contributory Negligence) Act 1945
s 1(1) O3011

Law Reform (Miscellaneous Provisions) Act 1934 .... C6036, C6038

Law Reform (Personal Injuries) Act 1948 .... C6001

Legal Aid, Sentencing and Punishment of Offenders Act 2012 .... C6021, C6058
s 85 .... IN05, E15001, E15035

Licensing Act 2003 .... F5046

Limitation Act 1980 .... E13025, P9511
s 11 .... W11045
11A(3)–(5) P9044
33 .... E13024

Limited Partnerships Act 1907 .... C9002

Local Government Act 1972
s 138 .... D6005

Local Government (Miscellaneous Provisions) Act 1982 .... F5046

M

Magistrates' Courts Act 1980
s 127(1) E15034

Marine Safety Act 2003 .... O7005

Medical Act 1983 .... O7058

Medicines Act 1968 .... P9001A.1

Merchant Shipping Act 1988 .... C9001

Merchant Shipping Act 1995 .... O7008
s 100A O7008, O7033.2
193 .... O7033.1

Mineral Workings (Offshore Installations) Act 1971 .. O7001, O7003, O7004, O7004, O7039, O7046
s 5 .... O7011

Mines and Quarries Act 1954 .... G1010
s 180 .... F5034

N

National Health Service and Community Care Act 1990
s 60 .... E15045

National Minimum Wage Act 1998 .... W10004

New Roads and Street Works Act 1991
s 50 .... C8019

Nuclear Installations Act 1965 .... R1021.1

O

Occupational Safety and Health Act 1970 .... I2003

Occupiers' Liability Act 1957 ...... Intro 19, C6508, C8019, C9003, O3001, O3011, O3016, O3019, P9504, P9527
s 1(2) O3004, O3005, O3006
(3) O3003
(a) O3005, O3006
(4) O3007
2(1) O3002
(2) O3002, O3003, O3006, O3017
(3) O3006, O3010
(a) O3010
(b) O3014
(4)(b) O3015
(5) O3017
(6) O3006

Occupiers' Liability Act 1984 ...... Intro 19, C6508, C8019, C9003, F5073, O3001, O3008, O3010, O3016, P9527
s 1 .... O3007, O3013
(1)(a) O3009
(3) A1001, O3009
(4), (5) O3009

Table of Statutes

Occupiers' Liability Act 1984 – *cont.*
s 1(6) O3017
(9) O3009
Occupiers' Liability (Scotland) Act 1960
.... O3001
Offences Against the Person Act 1861
s 5 .... C9021
Offender Management Act 2007
s 13 .... C9008
Offices, Shops and Railway Premises Act 1963 .... Intro 10, E15025
s 16(1) E15023
Offshore Petroleum Development (Scotland) Act 1975
.... O7001
Offshore Safety Act 1992
.... O7001, O7033
s 1(1) O7004
Oil and Pipelines Act 1985
.... O7001

P

Partnership Act 1890
.... C9002, F5031
Petroleum Act 1879
.... D0402, F5031
Petroleum Act 1987
.... O7004
s 23 .... O7001
(1) O7033.1
Petroleum Act 1998
.... O7001, O7002, O7014
s 16 .... O7034
Pt IV (ss 29–45) O7015
Petroleum and Submarine Pipe-lines Act 1975 .... O7004
Petroleum (Consolidation) Act 1928
.... D0402, D0938, D0948, E15005, F5031, M1103
Petroleum (Production) Act 1934
.... O7001
Petroleum (Transfer of Licences) Act 1936
.... E15005, F5031
Pipe-lines Act 1962
.... O7004
Planning (Hazardous Substances) Act 1990
.... M1136
Police Act 1996 .... V8008
Police and Criminal Evidence Act 1984
.... E15026.1
s 78 .... E15026
Pollution Prevention and Control Act 1999
.... E5027.2
Prevention of Oil Pollution Act 1971
.... O7001
Protection from Harassment Act 1997
.... H1702, H1707, H1729, V8013, V8015B
s 1(2) H1731

Protection from Harassment Act 1997 – *cont.*
s 3 .... H1726
7(4) H1726
Public Health Act 1936
.... F5031
s 205 .... E14028
Public Health Act 1961
.... F5031
Public Health (Control of Disease) Act 1984
s 2(4) D6006
Public Interest Disclosure Act 1998
.... E14001, E14011.1
Public Order Act 1986
.... H1729
s 4A H1731

R

Race Relations Act 1976
.... H1701, H1702, H1706, O7001
s 32(1) H1711
Radioactive Substances Act 1993
.... E15029, O7066, R1018
Riding Establishments Act 1964
.... P9527
Riding Establishments Act 1970
.... P9516, P9527
Road Traffic Act 1960
.... F7012
Road Traffic Act 1988
s 145 .... E13001
185 .... P7013
Road Traffic (NHS Charges) Act 1999
.... E13031
Road Traffic Regulation Act 1984
.... A1005, C8002, C8019

S

Safety of Sports Grounds Act 1975
.... F5033
s 10 .... E15029
Sale and Supply of Goods Act 1979
s 13 .... P9056
55(i) P9057
Sale and Supply of Goods Act 1994
.... P9047, P9048, P9543
Sale of Goods Act 1893
.... P9048
Sale of Goods Act 1979
.... P9047, P9048
s 14 .... P9053, P9055
(2) P9049, P9051, P9052, P9054
(3) P9052, P9054
15(2) P9050
Serious Crime Act 2007
Pt 2 (ss 44–67) C9021

Serious Organised Crime and Police Act 2005
  s 73, 74 .... E15043.1
Sex Discrimination Act 1975
  .... H1701, H1707, O7001
  s 1(1)(a) H1702
    4A H1702
    6(2)(b) H1706
    41(1) H1714
      (3) H1714
    42 .... H1714
    51(1) E14024
Social Security Act 1989
  s 22(4)(c) C6036
Social Security Administration Act 1992
  .... C6001
Social Security (Contributions and Benefits) Act 1992 .... C6001, O7001
  s 94 .... C6003
    97–101 .... C6007
    103 .... C6011
  Pt XI (ss 151–163)
  .... M1505
  Sch 6 .... C6011
Social Security (Incapacity for Work) Act 1994 .... C6001
Social Security (Recoupment of Benefits) Act 1997 .... C6001
Social Security (Recovery of Benefits) Act 1997
  s 4 .... C6045
    6 .... C6045
    8 .... C6044
    9 .... C6045
    23 .... C6049
Social Work (Scotland) Act 1968
  s 27 .... C9008
Statutory Sick Pay Act 1994
  .... C6001
Supply of Goods and Services Act 1982
  .... P9047, P9543
  s 4 .... P9048
Supply of Goods (Implied Terms) Act 1973
  .... P9047, P9048
Supreme Court Act 1981
  s 35A C6043

T

Territorial Sea Act 1987
  .... O7001
  s 1 .... O7009
Third Parties (Rights Against Insurers) Act 1930
  s 1(1)(b) E13024
Trade Descriptions Act 1968
  .... F9003
Trade Disputes Act 1906
  .... J3009
Trade Union Act 2016
  .... E14001
Trade Union and Labour Relations (Consolidation) Act 1992
  .... E14011, O7001
  s 5, 6 .... J3005
    8 .... J3005
    20 .... J3009
    168 .... J3029
    178 .... J3032, W1007
      (3) J3005
    179(1) E14005
    199 .... E14008
  Sch A1 .... J3005

U

Unfair Contract Terms Act 1977
  .... P9047, P9048, P9543
  s 2 .... E14002, P9058
    (2) O3016
    (3) O3017
    11 .... O3016

V

Vehicle Excise and Registration Act 1994
  .... F5034
Visiting Forces Act 1952
  s 12 .... E15009

W

Welfare Reform Act 2012
  .... C6001

Table of Statutory Instruments

A

Abstract of Special Regulations (Highly Flammable Liquids and Liquefied Petroleum Gases) Order 1974 (SI 1974 No 1587) D0939
Adventure Activities Licensing Regulations 2004 (SI 2004 No 1309)
...... C6508, C6508.1, W9004.2
Agency Workers Regulations 2010 (SI 2010 No 93)
    reg 5 .... M2010
        8(a), (b) M2010
Agriculture (Tractor Cabs) Regulations 1974 (SI 1974 No 2034)
    .... E15005
    reg 3(3) N3015
Air Quality Standards Regulations 2007 (SI 2007 No 64) E5037, E5066
Air Quality Standards Regulations 2010 (SI 2010 No 1001)
    Sch 2 .... E5066
Asbestos Industry Regulations 1931 (SI 1931 No 1140) A5042.2
Asbestos (Licensing) Regulations 1983 (SI 1983 No 1649) .. A5005, A5006, A5007, H2138
Asbestos (Prohibitions) Regulations 1992 (SI 1992 No 3067)
    .... A5004, A5005, A5007

B

Borehole Sites and Operations Regulations 1995 (SI 1995 No 2038)
    .... E15005, F5034
Building Regulations 1991 (SI 1991 No 2768) W11001, W11029
Building Regulations 2000 (SI 2000 No 2531) F5034
    Sch 1 .... W11044
    Pt L
    L2 .... F3029, F3034
Building Regulations 2010 (SI 2010 No 2214) .. A1001, F5030.1, G1026, W11001
    reg 3, 4 .... W11044
    Sch 1
      Pt A E3033
      B E3033
      C E3033
      D E3033
      E E3033
      F E3033

Building Regulations 2010 (SI 2010 No 2214) – *cont.*
    Sch 1 – *cont.*
      Pt G E3033
      H E3033
      I E3033
      J E3033
      L E3033
      K E3033
      M E3033
      N E3033
      O E3033
      P E3033

C

Carriage of Dangerous Goods and Use of Transportable Pressure Equipment Regulations 2004 (SI 2004 No 568)
    .. D0402, D0444, O7036, P7012, P7013
Carriage of Dangerous Goods and Use of Transportable Pressure Equipment Regulations 2007 (SI 2007 No 1573)
    .... D0401, D0402, D0448
Carriage of Dangerous Goods and Use of Transportable Pressure Equipment Regulations 2009 (SI 2009 No 1348)
    .... Intro 13, A5030, D0401, D0443, D0448, D0909, D6001, D6025, D6071, H2148, P7031, R1018
    reg 5 .... D0402, D0418
        6 .... D0418, D0422
        12 .... D0444
    Pt 4 (regs 19–23) D0402
    reg 24(6) D0435
        31 .... D0438
Carriage of Dangerous Goods by Rail Regulations 1996 (SI 1996 No 2089)
    .... A5030, D0936, P7013
Carriage of Dangerous Goods by Road Regulations 1996 (SI 1996 No 2095)
    . D0402, D0936, D0938, D0939, D6001, P7013, V12015
Carriage of Dangerous Goods (Classification, Packaging and Labelling) and Use of Transportable Pressure Receptacles Regulations 1996 (SI 1996 No 2092)
    .... D0936, D0939
Carriage of Explosives by Road Regulations 1996 (SI 1996 No 2093)
    .... D0936, V12015
Celluloid etc Factories and Workshops Regulations 1921 (SR&O 1921 No 1825) D0939

TSI-1

Table of Statutory Instruments

Chemicals (Hazard Information and Packaging for Supply) Regulations 2002 (SI 2002 No 1689) H2101, H2117, H2141, P9035, R2510

Chemicals (Hazard Information and Packaging for Supply) (Amendment) Regulations 2008 (SI 2008 No 2337) .... Intro 9

Chemicals (Hazard Information and Packaging for Supply) Regulations 2009 (SI 2009 No 716) D0909, D0935, H2101, H2137, P9001A.1

Children (Northern Ireland) Order 1995 (SI 1995 No 755 (NI 2))
  Pt 5 (arts 49–61) C9008
    6 (arts 62–71) C9008

Cinematograph Film Stripping Regulations 1939 (SR&O 1939 No 571) D0939

Civil Contingencies Act 2004 (Contingency Planning) Regulations 2005 (SI 2005 No 2042) D6001, D6006
  Pt 2 (regs 4–12) D6007
    3 (regs 13–18) D6007
    4 (regs 19–26) D6007
  reg 20–26 .... D6007
  Pt 5 (reg 27) D6007
    6 (regs 28–35) D6007
  reg 29 .... D6007
    31–35 .... D6007
  Pt 7 (regs 36–44) D6007
  reg 36(a) D6007
    39–44 .... D6007
  Pt 8 (regs 44A–54)
    .... D6007
  reg 44A–54 .... D6007
  Pt 9 (regs 55, 56) D6007
    10 (regs 57, 58)
    .... D6007
    11 (reg 59) D6007

Civil Defence (Grant) Regulations 1953 (SI 1953 No 1777) D6005

Civil Jurisdiction and Judgments Order 2001 (SI 2001 No 3929) P9046

Civil Jurisdiction (Offshore Activities) Order 1987 (SI 1987 No 2197) O7009

Civil Procedure Rules 1998 (SI 1998 No 3132) .. Intro 19, E13029, P9546
  Pt 25 .... C6041
    36
  r 11 .... C6042
  Pt 41
  r 2(2) C6040

Civil Procedure Rules 1998 (SI 1998 No 3132) – *cont.*
  Pre-Action Protocols
  .... A3028

Classification and Labelling of Explosives Regulations 1983 (SI 1983 No 1140) D0936

Classification, Labelling and Packaging of Chemicals (Amendments to Secondary Legislation) Regulations 2015 (SI 2015 No 21) H2145, W11042

Companies Act 2006 (Strategic Report and Directors' Report) Regulations 2013 (SI 2013 No 1970) .. E5070, E16005, E16024A

Confined Spaces Regulations 1997 (SI 1997 No 1713) C7001, C7002, C7008, C7009, C7030, C8018, M2027, O7065, P7016, R3017, S3004, S3018, V12016, V12027
  reg 1 .... C7003
    (2) A1023, O7065
    3 .... A1023, C7004
    4 .... A1023
    (1) C7005
    (2) C7006
    5 .... C7007
    (1)–(3) D6024
    6 .... A1023

Construction (Design and Management) Regulations 1994 (CDM) (SI 1994 No 3140) .... Intro 9, Intro 13, IN21, M2022, V12003
  reg 15 .... S3004

Construction (Design and Management) Regulations 2007 (SI 2007 No 320) .... Intro 9, Intro 13, A5030, C8501, M1007, M2002, M2015, M2021, M2037, V12003
  Pt 1 (regs 1–3) C8001, M2022
    2 (regs 4–13) .. C8001, M2022, M2023
  reg 4 .... M2024
    (1) M2023
    5 .... M2023, M2024
    6 .... M2023, M2024
    7 .... M2023, M2024
    8 .... M2023
    9 .... C6506, M2024
    10 .... M2024
    11 .... M2025
    12 .... M2025
    13 .... M2026
  Pt 3 (regs 14–24) . C8001, M2022, M2027
  reg 14 .... M2027
    15–17 .... M2027
    18, 19 .... M2027

Table of Statutory Instruments

Construction (Design and Management) Regulations 2007 (SI 2007 No 320) – cont.
reg 20	M2027
21	M2027
22	M2027
23, 24	M2027
Pt 4 (regs 25–44)	C8001, M2022
5 (regs 46–48)	C8001, M2022
Sch 1	M2023

Construction (Design and Management) Regulations 2015 (SI 2015 No 51) ... C6506, C7002, C8001C8501, C8503, C8520, F3021, F3028, F3010.2, F3032, F3033, F3034F5062, G1010, G1026, IN15, IN16, IN21, W11002
Pt 1 (reg 1–3)	C8504
reg 2(1)	C8505
(2)	C8505
Pt 2 (reg 4–7)	C8504
reg 4	C8506, C8509, C8512
(2)(b)	C8002
5	C8506, C8507
(1)(a), (b)	C8505
(3), (4)	C8509
6	C8508, C8509
(1)	E15009
7	C8506, C8509
Pt 3 (reg 8–15)	C8504, C8510
reg 8	C8002, C8510
9, 10	C8511
11	C8505, C8506, C8512
12	C8505, C8506, C8513, C8518, R3017
(2)	C7002, C8516
13	C8505, C8506, C8514
(1)	C8521
(4)(c)	C8002
14	C8505, C8506, C8514, C8515
15	C8515
(2)	C8521
(8)	C8521
(11)	C8002
Pt 4 (reg 16–35)	C8504, C8516
reg 16(2)	C8002
17	C8002
18	C8002, C8018
19(2)	C8002
20	C8002
21	C8002
22(1)	C8002
(3)–(5)	C8002
23	C8002
24(2)–(4)	C8002
25	C8002
26	C8002, W9004.1
27(5)	C8002
28–34	C8002
35	C8002, L5005
Pt 5 (reg 36–39)	C8504, C8516

Construction (Design and Management) Regulations 2015 (SI 2015 No 51) – cont.
Sch 1	C8508, C8518
2	C8002, C8506, C8516, C8518, L5005
3	C7002
, C8513, C8518	
Sch 4	C8518
5	C8518
Appendix 1	C8519
2	C8519
3	C8519
4	C8519
5	C8519
6	C8519

Construction (Head Protection) Regulations 1989 (SI 1989 No 2209) C8018

Construction (Health, Safety and Welfare) Regulations 1996 (SI 1996 No 1592) .... Intro 13, IN21, W9004.1

Consumer Protection Act 1987 (Commencement No 1) Order 1987 (SI 1987 No 1680) A5030

Consumer Protection from Unfair Trading Regulations 2008 (SI 2008 No 1277) F9012

Control of Artificial Optical Radiation at Work Regulations 2010 (SI 2010 No 1140) R1010.11, R3016

Control of Asbestos at Work Regulations 1987 (SI 1987 No 2115) R3001

Control of Asbestos at Work Regulations 2002 (SI 2002 No 2675)
. A5007, H2101, H2138, R3013, S3018
reg 4	A5006

Control of Asbestos Regulations 2006 (SI 2006 No 2739) .. A5003, A5005, A5006, A5009, A5020, A5022, A5024, A5030, E15005, F3024, F3025, O1046, R3023, S3018, V3001
reg 4	Intro 13, A5017
8	A5012, A5027
9	A5027
21	A5011

Control of Asbestos Regulations 2012 (SI 2012 No 632) A5003, A5043, C8018, D6071, F3024, H2101, H2138, O1004, O1018, P3023, P9527, R3001, R3023, S3024, V3016
reg 2(1), (2)	A5007.1
3(1)	**A5007.1**

TSI-3

Control of Asbestos Regulations 2012 (SI 2012 No 632) – cont.
  reg 3(2) A5040
    (a), (b) **A5007.1**
    (c)(1)–(4) **A5007.1**
  4 .... A5010, P9528, R3013
    (2) A5020
  6 .... R3013
    (1)(a) V3015
    (2)(c) V3015
  9, 10 .... A5007.1
  11(1) A5033
    (2)(b) V3015
  12(1), (2) V3015
  13(1)(a), (b) V3015
    (2), (3) V3015
  14 .... P3008
  15 .... D6016
  17 .... A5007.1
  18 .... A5036
    (1)(a) A5007.1
  22 .... A5007.1
  25 .... A5042.1
  Sch 2 .... A5032

Control of Asbestos in the Air Regulations 1990 (SI 1990 No 556)
  .... A5030

Control of Electromagnetic Fields at Work Regulations 2016 (SI 2016 No 588)
  ... E14032, IN07, IN16, O1056, R1008, R1010.2, R1010.3, R1010.10, R3016
  reg 2 .... R1010.1
  3 .... R1010
  4 .... R1010.2
    (1) R1010.4, R1010.5
  5 .... R1010.4
  7 .... R1010.5
  8 .... R1010.6, R1010.8
  9 .... R1010.6
  10 .... R1010.8
  11 .... R1010.9
  Schedule R1010.4, R1010.5, R1010.9

Control of Explosives Regulations 1991 (SI 1991 No 1531) O7009

Control of Industrial Major Accident Hazards Regulations 1984 (SI 1984 No 1902) ... D6001, M1101, M1103, O7003

Control of Lead at Work Regulations 1980 (SI 1980 No 1248)
  .... R3001

Control of Lead at Work Regulations 2002 (SI 2002 No 2676)
  .. C8018, E14023, H2101H2140, O1004, R3016, V3001, V3016, V12003, V12004
  reg 2(1) V3015

Control of Lead at Work Regulations 2002 (SI 2002 No 2676) – cont.
  reg 5 .... V3015
  6 .... P3008, V3015
    (3)(b) V3015
    (4)(b) V3015
    (d) V3015
    (6) V3015
  8 .... P3008
    (1) V3015
    (2)(a) V3015
    (4) V3015
  10 .... V12021

Control of Major Accident Hazards Regulations 1999 (SI 1999 No 743)
  ... C6506, D6001D6003, D6007, D6035, D6039, D6045, M1101, M1109, M1115, M1118, M1131, O7003
  reg 2(1) .... Intro 14
  3(1), (2) .... Intro 14
  7 .... Intro 14, M1121
    (6), (7) M1117
  8 .... Intro 14, M1121
  9 .... Intro 14, M2039
    (3)(a) M1122
  10 .... Intro 14
  11 .... Intro 14, D6067
  12 .... Intro 14
  13 .... Intro 14
  14 .... Intro 14
  15(4) A3015
  Sch 1
  Pt 2
  column 2 .... Intro 14
    3 .... Intro 14
  Pt 3
  column 2 .... Intro 14
    3 .... Intro 14

Control of Major Accident Hazards (Amendment) Regulations 2005 (SI 2005 No 1088) M1109
  Sch 1
  Pt 3 .... M1104
  note 3 .... M1105

Control of Major Accident Hazards Regulations 2015 (SI 2015 No 483)
  .... A3015, C6506, D6003, D6054D6066, D6067, D6071, H2101, H2139, IN07, IN16, IN21, M1101, M1113AM1135, M1141, P7031, R3016
  reg 2(1) IN21, M1103, M1104
    (2) M1103
  3(1) M1104
  4(a), (b) M1102
  5 .... D6011
    (1) M1103
    (2) M1103, M1115
    (3), (4) M1103
  6(1) M1103

TSI-4

Control of Major Accident Hazards Regulations 2015 (SI 2015 No 483) – cont.

reg 6(1)(a)–(c)	M1116
(2)	M1103, M1116
(4)	M1103, M1116
(5)	M1103
(6)	M1103, M1116
7	M1103
(1)–(4)	M1117
(6), (7)	M1117
8	D6011, M1103
(1)	M1118
9	D6011, M1103
(1)	M1120
(2)	M1118
(6)	M1121
10	D6011, M1103
(1)	M1121
(2)(a)–(e)	M1121
(6)	M1121
11	D6011, M1103, M1122
12	D6011, M1103, M1118
(1)	M1123
(2)	M1123, M1124
(3)	M1123
(6)	M1123
13	D6011, M1103
(5)	M1124
(7)	M1124, M1126
14	D6011, M1103
(4)	M1126
15	M1103, M1123, M1124
16	D6011, M1103, M1126
17	D6011, M1103
(1)	M1111
(2)	M1111
(5)	M1111
18	D6011, M1103
19	M1103, M1111
20, 21	M1103, M1111
Pt 6 (ss 22–25)	M1103
reg 22	M1113
23(1)–(3)	M1113
24	D6011
25	M1113B
26	D6011, M1103
(1)–(5)	M1112
27	M1103
28–30	M1103
45	M1115
Sch 1	
Pt 1	
column 1	M1104
2	M1104
3	M1104
Pt 2	
column 1	M1104
2	M1104
3	M1104
Pt 3	M1104

Control of Major Accident Hazards Regulations 2015 (SI 2015 No 483) – cont.

Sch 2	M1103, M1117, M1120
3	M1103, M1118, M1120
4	M1103
Pt 1	M1123
2	M1124
Sch 5	M1112
6	M1103

Control of Noise at Work Regulations 2005 (SI 2005 No 1643)
...... Intro 10, Intro 14, E17032, N3002, N3002A, N3009, N3012, N3013, N3022, N3023, N3027.1, N3028, N3030, O1025, P3017, V12019

reg 5	R3008
7	P3008
9	O1026
(1)	C8018
10	N3029, P9014
(4)	N3029
Schedule	N3007

Control of Pesticides Regulations 1986 (SI 1986 No 1510) V12021

Control of Substances Hazardous to Health Regulations 1988 (SI 1988 No 1657)
.... Intro 8, R3001

Control of Substances Hazardous to Health Regulations 1999 (SI 1999 No 437)
.... D0401, H2140

Control of Substances Hazardous to Health Regulations 2002 (SI 2002 No 2677)
.... Intro 2, Intro 14, C7008, C8018, D0902, D0904, D0909, D0912, D6016, D6017, D6071H2101, H2117, H2118, H2120, H2122, H2123, H2126, H2138, IN14, M1024, M2009, M3002, O1005, O1018, O1041, O1045, O1046, P3023, R3001,R3023, R3028, S3018, V3001, V3011, V3016, V12003, V12013, V12021, W9009

reg 2	V3015
(1)	H2104, H2146
3(1)	C6505, H2105, H2106
4, 5	H2105A
6	H2105A, R3007, V3015
(1)(a)	H2107
(2)	H2106
(3)	H2125
7	H2105A, H2114, P3008
(1), (2)	H2107
(3)	H2107, H2109, H2113
(b)	V3015
(4)	H2107, H2109
(b)	V3015
(d)	V3015
(5), (6)	H2109, H2113
(7)	**H2111**
(a), (b)	V3015
(c)	H2146, V3015

Table of Statutory Instruments

Control of Substances Hazardous to Health Regulations 2002 (SI 2002 No 2677) – cont.
reg 7(9)	H2110
(10)	H2113
(11)	**H2111**
8	H2105, H2105A, H2114, H2127, P3008, V3015
9	H2105A, H2114, H2125, P3008
(1)	H2128, H2131
(a), (b)	V3015
(2)	H2128, H2129
(a)	V3015
(3)	H2128, H2131
(4)	H2128, V3015
(5)–(7)	H2128, H2131
10	H2105A, H2114, H2125
(1)–(7)	H2132
11	H2105A, H2114, H2125
(1)–(11)	H2133
12	H2105A, H2114
(1)–(3)	H2134
(4)	H2115, H2129, H2134, V3015
(5)	H2134
13	D6015, H2105A, H2114
(4), (5)	H2135
21	E15037
Sch 2	H2105
2A	H2111, V3015
3	H2113
4	H2130, V3015
5	H2132
6	H2133, H2146
7	H2134

Control of Substances Hazardous to Health (Amendment) Regulations 2004 (SI 2004 No 3386) .. O1041, O1046

Control of Vibration at Work Regulations 2005 (SI 2005 No 1093) .... Intro 14, E17033, O1031, R3016, V5001, V5004.1, V5009, V5012, V5015, V5016, V5024, V5029, V12019

reg 5	V5003
6	V5003
(1)	C8018
7	C8018, V5003
8	V5003

Control of Vibration at Work Regulations (Northern Ireland) 2005 (SI 2005 No 397) V5015

Controlled Waste Regulations 1992 (SI 1992 No 588) A5030

Controlled Waste (Registration of Carriers and Seizure of Vehicles) (Amendment) Regulations 1998 (SI 1998 No 605) A5030

Corporate Manslaughter and Corporate Homicide Act 2007 (Amendment) Order 2011 (SI 2011 No 1868) C9001, C9003

Corporate Manslaughter and Corporate Homicide Act 2007 (Commencement No 3) Order 2011 (SI 2011 No 1867) C9001, C9003

Criminal Jurisdiction (Offshore Activities) Order 1987 (SI 1987 No 2198) O7009

D

Dangerous Substances and Explosive Atmospheres Regulations 2002 (SI 2002 No 2776) . Intro 15, D0401, D0901, D0902, D0903, D0904, D0905, D0906, D0916, D0921, D0928, D0931, D0940, D0941, D0949, D0950, D0951, D0952, E3027, F5001, F5031, F5062, G1001, G1026, H2101, H2146, S3024, V3001, V3015

reg 1	D0911
2	D0908, D0909
3	D0908
4(1)(a), (b)	D0907
5	D0925, D0943, D0948, H2141, R3014
(1), (2)	D0912
(3)	D0914
(4)	D0913
(5)	D0913, D0914
6	D0943, D0948, H2141, R3014
(1)	D0915
(3)	D0917, D0918
(4)	D0917
(5)	D0918
(6)	D0919
(8)	D0919, G1016
7	D0943, G1026, H2141, R3014
(2)	D0921, D0924
(3)	D0922
(4)	D0923
(5)	D0929
8	D0943, D6017, H2141, R3014
(1)	D0934
(2)(a), (b)	D0934
(3)(a)	D0934
(4)	D0934
9	D0935, D0943, H2141, R3014
10	D0919, D0936, D0943
11	D0907, D0943, G1026

Table of Statutory Instruments

Dangerous Substances and Explosive Atmospheres Regulations 2002 (SI 2002 No 2776) – *cont.*
reg 12	D0908
13, 14	D0937
15	D0938
16	D0939
17	D0911, D0943
Sch 1	D0919, D0943
2	
para 1	D0920
2	D0921
Sch 3	D0924
4	D0922
5	D0936
6	D0943
Pt 1	D0938
2	D0938
Sch 7	D0943
Pt 1	D0939
2	D0939

Dangerous Substances in Harbour Areas Regulations 1987 (SI 1987 No 37) . D0404, D0938, D0939, E15005

Dangerous Substances (Notification and Marking of Sites) Regulations 1990 (SI 1990 No 304) H2101
Sch 1	D0909
3	H2141

Deregulation Act 2015 (Health and Safety at Work) (General Duties of Self-Employed Persons) (Consequential Amendments) Order 2015 (SI 2015 No 1637) M1003, M3002

Detention of Food (Prescribed Forms) Regulations 1990 (SI 1990 No 2614) F9014

Disability Discrimination (Employment Relations) Regulations 1996 (SI 1996 No 1456) W11043

Diving at Work Regulations 1997 (SI 1997 No 2776) O7004, O7057, V12003
reg 3(1), (2)	O7036
4	O7036
5(1)	O7037
(2)	O7038
6	O7037
(3)(b)	P7013
7	O7037
8	O7037, O7038
9	O7037
10, 11	O7037.1
12(1)	O7037.2
(3)	O7037.2
13(1)	O7037.2, V12019
14(1)	O7036
15	V12019
(6)	O7036

Diving Operations at Work Regulations 1981 (SI 1981 No 399) O7036

Docks Regulations 1988 (SI 1988 No 1655) V12015

Dry Cleaning (Metrication) Regulations 1983 (SI 1983 No 977) D0939

Dry Cleaning Special Regulations 1949 (SI 1949 No 2224) D0939

E

Electrical Equipment (Safety) Regulations 1994 (SI 1994 No 3260) E3029, E3031, M1017

Electrical Equipment (Safety) Regulations 2016 (SI 2016 No 1101) E3031, L3002

Electricity at Work Regulations 1989 (SI 1989 No 635) .. Intro 15, C7008, C8018, E3009A, E3032, O7063, V12027
Pt I (regs 1–3)	E3003
reg 2	E3003, E3004, E3021
Pt II (regs 4–16)	E3003, E3010
reg 4l	E3010
4(2)	E3021
(3)	E3011, E3022, E3028
(4)	E3024
5	E3010, E3012
6	E3010, E3013, E3025, E3026
(d)	F5062
7	E3005, E3010, E3014, E3024
8	E3005, E3010, E3015
9	E3010, E3016
10	E3010, E3017
11	E3010, E3018
12	E3010, E3019, E3023
13	E3010, E3019, E3022, S3004
14	E3010, E3019, E3022, E3024, S3004
15	E3010, E3020, L5006
16	E3010, E3011
Pt IV (regs 29–33)	E3003

Electricity at Work Regulations (Northern Ireland) 1991 (SI 1991 No 13) E3003

Electricity (Factories Act) Special Regulations 1944 (SI 1944 No 739) E3020

Electricity Safety, Quality and Continuity Regulations 2002 (SI 2002 No 2665) E3032

TSI-7

Table of Statutory Instruments

Electromagnetic Compatibility Regulations 1992 (SI 1992 No 2372) M1017

Electromagnetic Compatibility Regulations 2006 (SI 2006 No 3418) M1017

Electromagnetic Compatibility Regulations 2016 (SI 2016 No 1091) L3002

Employers' Health and Safety Policy Statements (Exception) Regulations 1975 (SI 1975 No 1584) M2031, S7002

Employer's Liability (Compulsory Insurance) Regulations 1998 (SI 1998 No 2573) .... Intro 15
- reg 1(2) E13009, E13018
- 2 .... E13017, E13021
- (1)(a)–(d) E13020
- (2) E13020
- 3(1) E13011
- 4(3)–(5) E13013
- 5(4) E13013
- 7, 8 .... E13013
- 9 .... E13001
- Sch 2 .... E13012
- para 14 .... E13001

Employers' Liability (Compulsory Insurance) (Amendment) Regulations 2004 (SI 2004 No 2882) E13012

Employers' Liability (Compulsory Insurance) (Amendment) Regulations 2008 (SI 2008 No 1765) .... Intro 15, E13013

Employment Equality (Age) Regulations 2006 (SI 2006 No 1031) H1701
- reg 6 .... H1702

Employment Equality (Religion and Belief) Regulations 2003 (SI 2003 No 1660) H1701
- reg 5 .... H1702

Employment Equality (Repeal of Retirement Age Provisions) Regulations 2011 (SI 2011 No 1069) M1505, M1511A

Employment Equality (Sex Discrimination) Regulations 2005 (SI 2005 No 2467) H1702

Employment Equality (Sexual Orientation) Regulations 2003 (SI 2003 No 1661) H1701, H1707
- reg 5 .... H1702

Employment Rights (Employment Particulars and Paid Annual Leave) (Amendment) Regulations 2018 (SI 2018 No 1378) E14006

Employment Rights (Miscellaneous Amendments) Regulations 2019 (SI 2019 No 731) E14006

Employment Tribunals (Constitution and Rules of Procedure) Regulations 2004 (SI 2004 No 1861)
- Sch 4 .... E15021

Employment Tribunals (Constitution and Rules of Procedure) Regulations 2013 (SI 2013 No 1237)
- Sch 1 .... E15019

Enterprise and Regulatory Reform Act 2013 (Commencement No 3, Transitional Provisions and Savings) Order 2013 (SI 2013 No 2227) H1706

Environmental Information Regulations 2004 (SI 2004 No 3391) E5041, M1111

Environmental Information (Scotland) Regulations 2004 (SI 2004 No 520) M1111

Environmental Permitting (England and Wales) Regulations 2007 (SI 2007 No 3538) A5030, E5027.2

Environmental Permitting (England and Wales) Regulations 2010 (SI 2010 No 675) E15043

Environmental Permitting (England and Wales) (Amendment) Regulations 2013 (SI 2013 No 390) E5027.2

Environmental Permitting (England and Wales) Regulations 2016 (SI 2016 No 1154) E5027.2, E5041
- reg 2(1) E5041
- 5 .... E5041
- 7, 8 .... E5041
- 12 .... E5041
- 17 .... E5041
- 20 .... E5041
- 22 .... E5041
- 34 .... E5041
- 36, 37 .... E5041
- 40 .... E5041
- Pt 5 (regs 45–56) E5041
- reg 57 .... E5041
- Pt 7 (regs 70–80) E5041
- Sch 1 .... E5041
- 5 .... E5041
- 7–Sch 26 .... E5041
- 27
- para 1 .... E5041

Environmental Protection (Duty of Care) Regulations 1991 (SI 1991 No 2839) A5030

Equality Act 2010 (Commencement No. 4, Savings, Consequential, Transitional, Transitory and Incidental Provisions and Revocation) Order 2010 (Amendment) Order 2010 (SI 2010 No 2337)
art 7 .... H1702

Equality Act 2010 (Disability) Regulations 2010 (SI 2010 No 2128)
.... E16510
reg 2, 3 .... E16510
4(1)(a)–(e) E16510
(2), (3) E16510
5 .... E16510

Equipment and Protective Systems Intended for Use in Potentially Explosive Atmospheres Regulations 1996 (SI 1996 No 192) ... D0902, D0924, D0928, E3027

F

Factories (Testing of Aircraft Engines and Accessories) (Metrication) Regulations 1983 (SI 1983 No 979) (Now revoked) D0939

Fire Certificates (Special Premises) Regulations 1976 (SI 1976 No 2003) D0938, F5031

Fire Precautions (Workplace) Regulations 1997 (SI 1997 No 1840) D0938, F5031

Fire Safety (Scotland) Regulations 2006 (SSI 2006 No 456) C8002

Firearms (Amendment) Regulations 2016, (SI 2016 No 425) A5029

Fixed-term Employees (Prevention of Less Favourable Treatment) Regulations 2002 (SI 2002 No 2034)
.... E14002

Food Hygiene (England) Regulations 2006 (SI 2006 No 14) .. F9002, F9034, F9060
reg 7 .... F9016
30 .... F9052

Food Hygiene (Scotland) Regulations 2006 (SI 2006 No 3) F9063
reg 30 .... F9064
Sch 4 .... F9064

Food Information Regulations 2014 (SI 2014 No 1855) F9003

Food Labelling Regulations 1996 (SI 1996 No 1499) .... Intro 15

Food Labelling (Declaration of Allergens) (England) Regulations 2008 (SI 2008 No 1188) .... Intro 15

Food Safety Act 1990 (Amendment) Regulations 2004 (SI 2004 No 2990) F9002, F9006

Food Safety and Hygiene (England) Regulations 2013 (SI 2013 No 2996) ... F9002, F9021, F9023, F9024, F9026, F9034, F9036
reg 6 .... F9015
8 .... F9018
9, 10 .... F9019.1
19(1) E15043
29 .... F9014
32 .... F9058
Sch 4 .... F9050
para 2 .... F9053
4 .... F9058

Food Safety (Improvement and Prohibition—Prescribed Forms) Regulations 1991 (SI 1991 No 100) F9015

Freight Containers (Safety Convention) Regulations 1984 (SI 1984 No 1890) E15005

G

Gas Appliances (Enforcement) and Miscellaneous Amendments Regulations 2018 (SI 2018 No 389) G1015

Gas Appliances (Safety) Regulations 1995 (SI 1995 No 1629)
.... G1015

Gas Safety (Installation and Use) Regulations 1998 (SI 1998 No 2451) G1001, G1010, G1025C, H2105, IN01, P9527
reg 2 .... G1010, G1020
3 ...... G1010, G1011, G1014
(1) G1013, G1026A
(2) R3002
(3) G1026A
(7) G1026A
4 .... G1013
5(1) G1013
(2) G1015
(3) G1015, G1026A
6(1)–(10) G1016
8(1) G1026A
9 .... G1020
12(6) G1022
13 .... G1022
14(1)(a)–(c) G1022
(2)–(4) G1022
15–17 .... G1022
18–23 .... G1023

Table of Statutory Instruments

Gas Safety (Installation and Use) Regulations 1998 (SI 1998 No 2451) – *cont.*
reg 26(1) G1026A
(2) G1017
(6)–(8) G1017
(9) G1017
27 .... G1017
(1) G1019
(2)–(5) G1019
28, 29 .... G1017
30(1) G1018
(2), (3) G1018, G1024
(4) G1018
31 .... G1017
32 .... G1019
33(1) G1024
(2) G1024
34 .... G1017
35 .... G1010, G1024
36 .... G1010
(2) G1024, G1026A
(3) G1024, G1025, G1026A
(4) G1026A
(6)(a), (b) G1025
(7), (8) G1025
(10) G1024
37 .... G1010
(1)–(3) G1012
(8) G1012
38 .... G1025A

Gas Safety (Management) Regulations 1996 (SI 1996 No 551) . G1001, G1009
reg 3 .... G1003
(1) G1004
5 .... G1005
6, 7 .... G1004, G1005
8 .... G1004
Sch 1 .... G1004

Gas Safety (Rights of Entry) Regulations 1996 (SI 1996 No 2535)
.... G1001
reg 4 .... G1006
5–8 .... G1007
9 .... G1008

General Food Regulations 2004 (SI 2004 No 3279) . F9002, F9006, F9008, F9034

General Product Safety Regulations 1994 (SI 1994 No 2328)
.... P9002

General Product Safety Regulations 2005 (SI 2005 No 1803)
.... P9001A.1, P9026, P9543
reg 2 .... P9003
3(1) P9002
5 .... P9006
(1) P9002
(3) P9002
6(1), (2) P9003

General Product Safety Regulations 2005 (SI 2005 No 1803) – *cont.*
reg 7(1) P9006
(a), (b) P9002
(3) P9002, P9006
(4) P9002
8(1)(a) P9002, P9006
(b) P9002
9(1) P9002, P9006
(2)(b) P9002
(4) P9002
11–15 .... P9002
20(2) P9006
29(1)–(4) P9008
21 .... P9007

Good Laboratory Practice Regulations 1999 (SI 1999 No 3106)
.... D0936

Greenhouse Gas Emissions Trading Scheme Regulations 2012 (SI 2012 No 3038)
.... E5070

H

Hazardous Waste (England and Wales) Regulations 2005 (SI 2005 No 894)
.... A5030

Health and Safety (Amendment) (EU Exit) Regulations 2018 (SI 2018 No 1370)
.... M1101, M1112

Health and Safety and Nuclear (Fees) Regulations 2015 (SI 2015 No 363)
reg 3–11 .... E15005
13–16 .... E15005
18 .... E15005
20, 21 .... E15005

Health and Safety and Nuclear (Fees) Regulations 2016 (SI 2016 No 253)
.... O7022

Health and Safety at Work etc. Act 1974 (Application Outside Great Britain) Order 1977 (SI 1977 No 1232)
.... O7003

Health and Safety at Work etc. Act 1974 (Application Outside Great Britain) Order 2001 (SI 2001 No 2127)
.... O7009, W9003

Health and Safety at Work etc. Act 1974 (Application Outside Great Britain) Order 2013 (SI 2013 No 240)
.... D0908, O7001
art 4–11 .... O7009

Health and Safety at Work etc. Act 1974 (Application Outside Great Britain) (Variation) Order 1995 (SI 1995 No 263)
reg 2(1) O7009

Table of Statutory Instruments

Health and Safety at Work etc Act 1974
(General Duties of Self-Employed
Persons) (Prescribed Undertakings)
Regulations 2015 (SI 2015 No 1583)
..... F7015, H2105, IN01, IN16,
M3002, R3002, W9004

Health and Safety (Consultation with
Employees) Regulations 1996 (SI 1996
No 1513) .... Intro 15, C8018,
C8515, D0948, E14009,
J3002, J3019, J3022, J3023,
J3030, J3031, V8008

reg 3	.... J3005, J3021
5(1)	.... A3027
(3)	.... J3027
6	.... A5025
7(1)	.... J3028
(b)	.... J3024
(2)	.... J3025
11(1)	.... A5025
Sch 1	.... J3026
2	.... J3029

Health and Safety (Display Screen
Equipment) Regulations 1992 (SI 1992
No 2792) ...... Intro 15, D8201,
D8203, D8210, D8213,
D8220, E17044, E17052,
E17064, L5002, O1005,
O1020.5, O1036, V12003,
W11009

reg 1	.... D8202, D8211
(4)(d)	.... D8221, D8221A
2	.... D8204, R3010
3	.... D8205
4	.... D8206
5	.... D8207
6	.... D8208
7	.... D8209
Schedule	.... D8205, S11015

Health and Safety (Enforcing Authority)
Regulations 1998 (SI 1998 No 494)
.... A3002, D0940, E15008

reg 2	.... E15009
4(4)(b)	.... E15009
3(1)	.... E15009, E15010
(3)	.... E15009
(4)(b)	.... E15009
(5)	.... E15009
5	.... E15011
6(1)	.... E15011
Sch 1	.... E15009, E15010
2	.... E15009

Health and Safety (Fees) Regulations 2005
(SI 2004 No 676) O7004

Health and Safety (Fees) Regulations 2009
(SI 2009 No 515) A5030

Health and Safety (First-Aid) Regulations
1981 (SI 1981 No 917)
.... Intro 16, D6021, D6071,
F7001, F7018, IN16

Health and Safety Information for
Employees (Amendment) Regulations
2009 (SI 2009 No 606)
.... Intro 10

Health and Safety (Miscellaneous
Amendments) Regulations 2002 (SI
2002 No 2174) . Intro 10, D8201,
W11001

reg 4(1) P3004

Health and Safety (Miscellaneous Repeals,
Revocations and Amendments)
Regulations 2013 (SI 2013 No 448)
.... D0909, L3029

Health and Safety (Safety Signs and Signals)
Regulations 1996 (SI 1996 No 341)
.... Intro 16, C8002, C8019,
D0936, N3013, W11001, W11030,
W11042

reg 2	.... W11034
3(1)	.... W11036
4, 5	.... W11035
25A	.... W11029
Sch 1	.... W11035
Pt I	
para 3	.... W11032
4	.... W11033

Health and Safety (Training for
Employment) Regulations 1990 (SI
1990 No 1380) . Intro 16, C6518,
R3025, V12003, V12010

Health and Safety (Young Persons)
Regulations 1997 (SI 1997 No 135)
.... R3025

High-activity Sealed Radioactive Sources
and Orphan Sources Regulations 2005
(SI 2005 No 2686)
.... R1018

Highly Flammable Liquids and Liquefied
Petroleum Gases Regulations 1972 (SI
1972 No 917) ... D0901, D0939,
H2141

I

Income Support (General) Regulations 1987
(SI 1987 No 1967)
.... C6041

Information and Consultation of Employees
Regulations 2004 (SI 2004 No 3426)
.... J3001, J3002

TSI-11

Table of Statutory Instruments

Ionising Radiation (Medical Exposure) Regulations 2017 (SI 2017 No 1322) R1018, R1019
Ionising Radiations Regulations 1999 (SI 1999 No 3232) E13020, E14024, E15005, E15012, H2101, IN16, IN21, O7066, R1019, V12003, V12015
   reg 7 .... R1019
   8(5) R1019
   9 .... P3008, R1019
   10 .... R1019
   13, 14 .... R1019
   15–20 .... R1019
   21(5) 20R1019
   24–26 .... R1019
   Sch 4 .... R1019
Ionising Radiations Regulations 2017 (SI 2017 No 1075) .. A3004, O1004, O1056, R1001, R1018, R3016, R3023, V3001, V12004, V12020
   reg 5–7 .... R1019
   8 ..... E14031, R1019, H2145
   9(1) E14031
   (6) **E14031**, R1019
   10–12 .... R1019
   13 .... R1019, R1021
   14 .... R1019
   15 .... E14031, R1019
   16–21 .... R1019
   22(5) R1019
   25–27 .... R1019
   Sch 3 .... R1019
   5 .... R1019

J

Justification of Practice Involving Ionising Radiation Regulations 2004 (SI 2004 No 1769) R1018

L

Lifting Operations and Lifting Equipment Regulations 1998 (SI 1998 No 2307) .. Intro 16, C8018, F3034, IN21, L3001, L3037, L3039, M1001, O7070, R3017
   reg 4 .... L3015
   5(1)(b) L3016
   6 .... L3017
   7 .... L3014, L3018
   8 .... L3019, L3020
   9 .... L3023, L3025
   (1), (2) L3021, L3027
   (3) L3021. L3026, L3027, S3004
   (a)(ii) L3024

Lifting Operations and Lifting Equipment Regulations 1998 (SI 1998 No 2307) – *cont.*
   reg 9(3)(b) L3028
   (4) L3021, W9012
   10 ..... L3023, L3025, L3026
   (1), (2) L3027
   11(1) L3027
   (2)(a), (b) L3027
   Sch 1 .... L3026
Lifts Regulations 1997 (SI 1997 No 831) L3021
   reg 4–7 .... L3004
   8 .... L3012
   (2) 3008
   9 .... L3012
   11 .... L3012
   13 .... L3011, L3012
   14 .... L3012
   Sch 3 .... L3004
   14 .... L3004
Lifts Regulations 2016 (SI 2016 No 1093) L3002, L3004
   reg 7(1) L3005
   (b) L3011
   8 .... L3007, L3010, L3011
   (1), (2) L3005
   9 .... L3007, L3008, L3010, L3011
   (2) L3005
   10 .... L3007, L3010
   12 ..... L3005, L3007, L3010
   13 .... L3013
   (2) L3007
   14 .... L3009
   18 .... L3007
   23 .... L3009
   (2) L3007
   33 .... L3006
   34(2) L3007
   41(2) L3007
   48 .... **L3006**
   49, 50 .... L3006
   51 .... L3013
   70 .... L3008
   Sch 2 .... L3004
Low Voltage Electrical Equipment (Safety) Regulations 1989 (SI 1989 No 728) P9008

M

Magnesium (Grinding of Castings and Other Articles) Special Regulations 1946 (SR&O 1946 No 2017) D0939
Management of Health and Safety at Work and Fire Precautions (Workplace) (Amendment) Regulations 2003 (SI 2003 No 2457) R2004

Management of Health and Safety at Work Regulations 1992 (SI 1992 No 2051)
. D6001, J3002, M2005, V12017

Management of Health and Safety at Work Regulations 1999 (SI 1999 No 3242)
.... Intro 16, Intro 18, A1010, A1013, A3023, A5030, C6501, C7002, C7008, C8015, C8516, D0904, D0948, D6015, D6067, E3003, E13006, E14024, E14025, E15012, E17044, E17062, F7004, H2127, H2135, J3002, J3019, L3019, M2001, M2002, M2103, O1005, O1020.5, O1020.9, O1036, O1041, O1045, O1056, P7031, R1001.1, R1008, R1018, R3021, R3047, V3001, V5016, V8006, V12002, W10020, W11035

reg 1	..... D6010, R3026, V12017
(2)	.... R3025
2(2)	.... V12011
3	...... Intro 8, A1023, R1019, M2005, M2008, M2009, M2020, R3005, S3004, S3016, S7006, S11007, V8044, W9007, W9009.1
(1)	... C6505, L3020, M1006, M1007, M2019, R3018, R3021, V12008, V12018
(a)	.... S3001
(2)	.... R3018
(3)	... A3029, R3018, R3022
(4)	. M2006, R3018, V12003, V12008, V12013
(5)	.... M2006, V12003, V12008, V12016
(b)–(g)	.... V12013
(6)	.... R3018, R3048
4	...... M2007, R3023, S3004, V8072
5	.... C6505, D6039, M1131, M2005, M2008, R3005, R3023R3024, R3048, S3004, S7006, S7007, V8129
6	..... M2009, O1009, R3023, V12013
7	... Intro 24, H2115, M1007, M2005, S3016, S7003
(8)	.... M2010, R3029
8	..... D0934, D6039, M1130, M1131, M2005, M2011, M2039, R3023, V8014
(1)	.... D6010
(2)	.... D6010
(3)	.... D6010
9	...... A1015, D6010, D6039, M1130, M1131, M2012
10	.... M3015
(1)	.... V12014
(2)	. J3015, M2013, V12010, V12014
(3)	.... V12010

Management of Health and Safety at Work Regulations 1999 (SI 1999 No 3242) – cont.

reg 11	... A3003, M2014, M2015, M3015, R1019
12	.... M2015, M3015
13	... M2016, M3015, W9007
(1)	.... V12012, V12036
(2)	.... T7004, V12038
14	.... M2017
15	.... M2013, M2018
16	... M2006, M2019, R3018, V5015
(1)	.... R3026, V12018
(2), (3)	.... E14025, E14027, R3026, V12018
(4)	.... R3026, V12018
16A	.... M2019, V12003, V12018
(1), (2)	.... **R3026**
17	... E14026, M2019, R3026, V12018
17A	.... M2019, V12003
(a), (b)	.... **R3026**
18	.. E14025, E14027, M2019, V12003, V12018
(2), (3)	.... R3026
18A	.. M2019, R3026, V12018
18AB	.... M2019
19	.... R3018, V5015, V12003
(1)	.... V12009, V12013
(2)	.... V12009, V12013, V12016
(3)	.... V12009, V12013
20	.... M2021
21	.... E15039
22	.... E15039, R2004
23–30	.... E15039
Sch 1	.... C85011, M2007, R3023, V8072

Management of Health and Safety at Work (Amendment) Regulations 2006 (SI 2006 No 438)

reg 6	.... R2004

Manual Handling Operations Regulations 1992 (SI 1992 No 2793)
.... Intro 10, Intro 17, C8018, E17044, E17063, IN18, M3001, M3007, M3008, M3017, O1005, O1020.5, V12003, V12019

reg 2	.... M3001
(2)	.... M3002
4	.... R3009, W9004.1
(1)	.... M3002
(b)(i)	.... M3003.2
(ii)	. M3003.2, M3015.1
(iii)	.... M3015.1
(2)	.... M3002, M3026
(3)	.... M3002, M3007

Manual Handling Operations Regulations
    1992 (SI 1992 No 2793) – cont.
  reg 4(3)(c) M3015.1
  Pt 1 .... M3003.2
  reg 4(3)(c) M3015
    5–8 .... M3002
  Sch 1 .. M3002, M3003, W9004.1
  para 1–3 .... M3003.2
Manufacture and Storage of Explosives
    Regulations 2005 (SI 2005 No 1082)
      .... E15005
  reg 9–11 .... A3035
Manufacture of Cinematograph Film
    Regulations 1928 (SR&O 1928 No 82)
      .... D0939
Maternity and Parental Leave etc
    Regulations 1999 (SI 1999 No 3312)
      .... E14002
Medical Devices Regulations 1994 (SI 1994
    No 3017) P7013
Medicines (Administration of Radioactive
    Substances) Regulations 1978 (SI 1978
    No 1006) R1018
Merchant Shipping (Accident Reporting and
    Investigation) Regulations 2012 (SI
    2012 No 1743) A3015
Merchant Shipping and Fishing Vessels
    (Control of Vibration at Work)
    Regulations 2007 (SI 2007 No 3077)
      .... V5015
Merchant Shipping (Dangerous Goods and
    Marine Pollutants) Regulations 1990
    (SI 1990 No 2605)
      .... D0404
Merchant Shipping (Dangerous Goods and
    Marine Pollutants) Regulations 1997
    (SI 1997 No 2367)
      .... O7068
Merchant Shipping (Oil Pollution
    Preparedness, Response and Co-
    operation Convention) Regulations
    1998 (SI 1998 No 1056) (OPRC 1998)
      .... O7009.10
Merchant Shipping (Oil Pollution
    Preparedness, Response and Co-
    operation Convention) (Amendment)
    Regulations 2015 (SI 2015 No 386)
      .... O7009.1
Mines Regulations 2014 (SI 2014 No 3248)
      .... M1003

N

Noise at Work Regulations 1989 (SI 1989
    No 1790) N3002, N3023,
                                V5009

Noise Emission in the Environment by
    Equipment for Use Outdoors
    Regulations 2001 (SI 2001 No 1701)
      .... N3016
Notification of Installations Handling
    Hazardous Substances Regulations
    1982 (SI 1982 No 1357)
      .... D0909, M1103
Notification of Installations Handling
    Hazardous Substances (Amendment)
    Regulations 2002 (SI 2002 No 2979)
  reg 7 .... D0909

O

Official Feed and Food Controls (England)
    Regulations 2009 (SI 2009 No 3255)
  reg 6(1) F9007.1
Offshore Electricity and Noise Regulations
    1997 (SI 1997 No 1993)
      .... E3003, O7063
Offshore Installations and Pipeline Works
    (First-Aid) Regulations 1989 (SI 1989
    No 1671) F7003, O7039,
                                O7056.1
  reg 5(1) O7057
    (a), (b) O7058
Offshore Installations and Pipeline Works
    (Management and Administration)
    Regulations 1995 (SI 1995 No 738)
      . O7004, O7005, O7011, O7024
  reg 2(1) O7009
    3 .... F5034
    (3) O7006
    5, 6 .... O7030
    7 .... O7031
    8 .... O7032
    9–20 .... O7030
    21A–21D O7030
    21F, 21G O7033
    21H O7033.1
Offshore Installations and Wells (Design
    and Construction, etc.) Regulations
    1996 (SI 1996 No 913)
      ...... O7001, O7004, O7009.9,
    O7010, O7011, O7016, O7017, O7022
  reg 5(1)(a)–(e) O7025
    (2) O7025
    6–11 .... O7025
    13–19 .... O7026
    20 .... O7027
    21 .... O7026
  Sch 1 .... O7025
Offshore Installations (Construction and
    Survey) Regulations 1974 (SI 1974 No
    289) O7012
Offshore Installations (Inspections and
    Casualties) Regulations 1973 (SI 1973
    No 1842) O7004

Table of Statutory Instruments

Offshore Installations (Offshore Safety Directive) (Safety Case etc) Regulations 2015 (SI 2015 No 398) (SCR 2015)
..... O7004.1, O7009, O7009.9, O7010, O7022, O7028, O7046
reg 2(10) O7009.10
6 .... O7009.12
21 .... O7009.11
30 .... O7009.10
Sch 6 .... O7009.1
7 .... O7009.1
14 .... O7009.13

Offshore Installations (Logbooks and Registration of Death) Regulations 1972 (SI 1972 No 1542)
.... O7004

Offshore Installations (Prevention of Fire and Explosion, and Emergency Response) Regulations 1995 (SI 1995 No 743) F5062, O7004, O7009.10, O7011
reg 4–10 .... O7028
11(1) O7028
12, 13 .... O7028
14(1)–(4) O7028
15–22 .... O7028

Offshore Installations (Safety Case) Regulations 1992 (SI 1992 No 2885)
. O7004, O7010, O7012, O7033
reg 4(1) O7022
(2) O7013.1
5 .... O7014, O7022
6 .... O7022
(1), (2) O7013.1
7 .... O7013.2, O7022
8 .... O7013.2, O7064
(1)(b) O7023
9 .... O7013.2
(2) O7022
10 .... O7019.1
11 .... O7014
12 .... O7016
17 .... O7020, O7022
19 .... O7017
20 .... O7018
Sch 1 .... O7013.1, O7013.2
2 .... O7013.2
4 .... O7019.1
5 .... O7014
6 .... O7020
7 .... O7017, O7018

Offshore Installations (Safety Case) Regulations 2005 (2005 No 3117)
.... O7004, O7009.1, O7012, O7022, O7028, O7046

Offshore Installations (Safety Representatives and Safety Committees) Regulations 1989

Offshore Installations (Safety Representatives and Safety Committees) Regulations 1989 – cont.
(SI 1989 No 971) . J3002, O7004, O7010, O7039, O7046, O7053
reg 5–10 .... O7047
16 .... O7048
17 .... O7049, O7050
(4)(a) O7051
18, 18A O7050
19, 20 .... O7052
22–24 .... O7054
26, 27 .... O7055
28(1), (2) O7056

Offshore Petroleum Licensing (Offshore Safety Directive) Regulations 2015 (SI 2015 No 385) (OPLR)
... O7009.2, O7009.3, O7009.4, O7009.5, O7010
reg 2(1) O7009.1, O7009.12
5, 6 .... O7009.1
7 .... O7009.1, O7009.6
8 .... O7009.1, O7009.7
9, 10 ...... O7009.1, O7009.8
11 .... O7009.1
(2)(a)(ii) . O7009.1, O7009.8
12 .... O7009.1
13 .... O7009.8
(1) O7009.7
14 .... O7009.1
15(1) O7009.1
(3) O7009.1
16–18 .... O7009.1
19(1) O7009.1
(7) O7009.1
20 .... O7009.1
21(1) O7009.1
(7) O7009.1
22–27 .... O7009.1
28 .... O7009.1, O7009.12
29(1)–(3) . O7009.1, O7009.12
30(1) O7009.1, O7009.12
(13), (14) O7009.1, O7009.12
31–34 .... O7009.1, O7009.12
35–37 .... O7009.1
40 .... O7009.1
Sch 1 .... O7009.1, O7009.6
2 .... O7009.1, O7009.6, O7009.7
3 .... O7009.1, O7009.7
4 .... O7008
Pt 1 .... O7009.1
2 .... O7009.1
Sch 5 .... O7009.1
6 .... O7009.1
para 5 .... O7009.7
Sch 7 .... O7009.1
para 5 .... O7009.7

TSI-15

Table of Statutory Instruments

Offshore Petroleum Licensing (Offshore Safety Directive) Regulations 2015 (SI 2015 No 385) (OPLR) – *cont.*
- Sch 8 .... O7009.1
- 9 .... O7009.1
- para 16 .... O7009.7
- Sch 10 .... O7009.1
- 12 .... O7009.1
- 20 .... O7009.1

Offshore Safety (Repeals and Modifications) Regulations 1993 (SI 1993 No 1823) O7004

Ozone-Depleting Substances Regulations 2015 (SI 2015 No 168) E5026.1

P

Packaging, Labelling and Carriage of Radioactive Material by Rail Regulations 1996 (SI 1996 No 2090) D0936

Part-time Workers (Prevention of Less Favourable Treatment) Regulations 2000 (SI 2000 No 1551) E14002, W10004

Part-time Workers (Prevention of Less Favourable Treatment) Regulations 2000 (Amendment) Regulations 2002 (SI 2002 No 2035) E14001

Personal Protective Equipment at Work Regulations 1992 (SI 1992 No 2966)Intro 17, A5030, . C7008, C8018, E3024, E17066, O1026, O1031, O1036, O1041, O1046, P3001, P3002, P3004, P3008, P3017, P3027, V12003
- reg 4(1) P3006
- (3) P3006
- (e) P3013
- 5 .... P3006
- 6 .... R3011
- (1)–(3) P3006
- 7–9 .... P3006
- 10 .... P3006
- (2) P3007
- (4) P3007
- 11 .... P3007

Personal Protective Equipment (EC Directive) Regulations 1992 (SI 1992 No 3139) .. P3001, P3002, P3009

Personal Protective Equipment (EC Directive) (Amendment) Regulations 1994 (SI 1994 No 2326) P3002, P3004

Personal Protective Equipment Regulations 2002 (SI 2002 No 1144) ... N3013, P3012, P3013, P3014, P3025, R1018, R1019

Petroleum (Carbide of Calcium) Order 1929 (SI 1929 No 992) D0939

Petroleum (Carbide of Calcium) Order 1947 (SI 1947 No 1442) D0939

Petroleum (Compressed Gases) Order 1930 (SI 1930 No 34) D0939

Petroleum (Consolidation) Act 1928 (Enforcement) Regulations 1979 (SI 1979 No 427) D0938

Petroleum (Liquid Methane) Order 1957 (SI 1957 No 859) D0938

Petroleum Spirit (Motor Vehicles etc) Regulations 1929 (SI 1929 No 952) D0938, M1103

Petroleum Spirit (Plastic Containers) Regulations 1982 (SI 1982 No 630) D0938, M1103

Pipelines Safety Regulations 1996 (SI 1996 No 825) .. G1001, O7024, P7031
- reg 3 .... E15009
- (4) G1009
- 5–10 .... G1009, O7035
- 11 ..... G1009, O7016, O7035
- 12–14 .... G1009, O7035
- 16 .... G1009, O7035
- 17, 18 .... G1009
- 19 .... O7035
- 20, 21 .... G1009, O7035
- 22 .... G1009
- (1), (2) O7035
- 23–26 .... G1009, O7035

Planning (Control of Major Accident Hazards) Regulations 1999 (SI 1999 No 981) M1102, M1136

Planning (Hazardous Substances) Regulations 1992 (SI 1992 No 656) M1136

Planning (Hazardous Substances) Regulations 2015 (SI 2015 No 627) M1141

Plugs and Sockets etc. (Safety) Regulations 1994 (SI 1994 No 1768) E3029, P9003

Police (Health and Safety) Regulations 1999 (SI 1999 No 860)
- reg 4 .... M1007

Pollution Prevention and Control (England and Wales) Regulations 2000 (SI 2000 No 1973) E5027.2, E5041

Table of Statutory Instruments

Pollution Prevention and Control (Industrial Emissions) Regulations (Northern Ireland) 2013 (SI 2013 No 160) E5027.2
Pollution Prevention and Control (Scotland) Regulations 2012 (SI 2012 No 360) E5027.2
Pressure Equipment Regulations 1999 (SI 1999 No 2001) ... P7006, P7013, P7016, P7017
Pressure Equipment (Safety) Regulations 2016 (SI 2016 No 1105) P7001, P7006, P7007
   Sch IV P7008
Pressure Systems and Transportable Gas Containers Regulations 1989 (SI 1989 No 2169) P7022
   reg 9(1)(a) P7022
Pressure Systems Safety Regulations 2000 (SI 2000 No 128) D0401, P7001, P7014, P7018, P7019, P7020, M1007
   reg 2 .... P7012, P7015
   3(5) P7011
   4 .... P7016
   5 .... P7017
   9 .... P7022, P7024
   10 .... P7025
   11, 12 .... P7028
   13 .... P7029
   14 .... P7030
   15 .... P7027
   16 .... P7011
   Sch 1
   Pt II P7013
      .... P7013
   Sch 2 .... P7011
   3 .... P7017
Prevention of Accidents to Children in Agriculture Regulations 1998 (SI 1998 No 3262) V12015
Probation Board (Northern Ireland) Order 1982 (SI 1982 No 713 (NI 10))
   art 4 .... C9008
Provision and Use of Work Equipment Regulations 1992 (SI 1992 No 2932) E17065, M1003
   reg 4 .... L3001
Provision and Use of Work Equipment Regulations 1998 (SI 1998 No 2306) .... Intro 17, A1009, A1010, A5030, C6510, C7008, C8002, E3003, F3032, G1026, L3051, M1001, M1006, M3002, O7024, P7031, S11015, T7004, V3015, V5015, W11003
   Pt I (regs 1–3) M1005
   reg 2(1) M1005
   Pt II (regs 4–24) .. M1005, M1007
   reg 4 .... L3109, M1007

Provision and Use of Work Equipment Regulations 1998 (SI 1998 No 2306) – *cont.*
   reg 5 ...... A1011, A1015, A1020, M1007
   (1) E15021, P9010
   6 .... A1012, M1006
   (2) M1007
   (5) M1007
   7 .... M1006, M1007, M1008, O1031
   8 .... M1007
   9 ...... M1006, M1007, P7026
   (1) S3001
   10 .... D0928, M1005, M1010
   (1), (2) M1007
   11 .... M1007.1
   (1) M1007, M1007.1
   (2) M1029, M1007.1
   (3) M1007, M1007.1
   12 .. E3026, M1007, M1007.1, M1008
   13–18 ...... M1007, M1007.1
   19 .... E3019, E3023, M1007, M1007.1
   20 .... M1007, M1007.1
   21 .. L5004, M1007, M1007.1
   22–24 ...... M1007, M1007.1
   Pt III (regs 25–30) M1005, M1008
   reg 25 .... M1008
   26 .... M1008
   27–30 .... M1008
   Pt IV (regs 31–35) M1005, M1008
   reg 32–35 .... M1008
   Pt V (regs 36–39) M1005
   Sch 1 .... M1007
   2 .... M1008
Public Information for Radiation Emergencies Regulations 1992 (SI 1992 No 2997) D6071
   reg 3 .... D6019
   Sch 2 .... D6019

Q

Quarries Regulations 1999 (SI 1999 No 2024) G1010
   reg 45(1) E15009

R

Race Relations Act 1976 (Amendment) Regulations 2003 (SI 2003 No 1626) H1702
Radiation (Emergency Preparedness and Public Information) Regulations 2001 (SI 2001 No 2975) . D6071, E15005, R1018, R1021
   reg 4–14 .... D6018

Table of Statutory Instruments

Radiation (Emergency Preparedness and Public Information) Regulations 2001 (SI 2001 No 2975) – *cont.*
  reg 16, 17 .... D6018
  Sch 5–Sch 10 .... D6018
Radioactive Material (Road Transport) (Great Britain) Regulations 1996 (SI 1996 No 1350) D0936
Registration of Births and Deaths Regulations 1987 (SI 1987 No 2088)
  Sch 2
  Form 14 .... A3014
Regulatory Reform (Fire Safety) Order 2005 (SI 2005 No 1541)
  . Intro 17, C8002, F5009, F5030, F5031, F5032, F5034, F5060.1, F5061.1, F5062, P9527, P9528, R3012
  Pt 2 .... F5046
  5 .... F5046
  art 3, 4 .... F5033
  8 .... C6507, F5033
  9 .... F5020, F5033, F5037, F5038
  10 .... F5038
  12 .... F5043, F5044
  13 .... F5042
  14 .... F5020, F5040
  15 ..... F5022, F5041, F5043
  16 .... F5043
  17 .... F5047
  18 .... F5048
    (5) F5049
  19 .... F5050
    (2) F5052
  20 .... F5051
  21 .... F5053
  22 .... F5054
  23 .... F5055, F5060
  25, 26 .... F5056
  29 .... E15029, F5057
  30 .... E15029, F5058
  31 .... E15029, F5059
  37 .... F5046, F5060
  38 .... F5046
  40 .... F5060
  49 .... F5061
  Sch 1
  Pt 2 .... F5038
Reporting of Injuries, Diseases and Dangerous Occurrences Regulations 1995 (SI 1995 No 3163)
  .... Intro 17, A5030, C6003, C6506, D6039, F7020, J3027, M2036, M2041

Reporting of Injuries, Diseases and Dangerous Occurrences Regulations 2013 (SI 2013 No 1471)
  . A3029, C8018, D6003, D6020, D6071, E3003, G1005, H2103, M1131, M2103, N3012, O1056, S7005, V5015, V8007
  reg 2(2) A3012
  3(1) A3001, A3003
  4 .... A3018
    (1) A3006
    (2) A3007
  5 ...... A3008, A3018, **C6506**
  6 .... A3018, C6506
    (3) A3005
  7 .... A3018
  8 ..... A3014, A3018, **E17067**, V5010
  9, 10 .... A3014, A3018
  11(1) A3010
    (2) A3011
  12(1)(b) O7050
    (c) A3017, A3018
    (2) A3017
  14(1) .. A3004, A3008, A3020
    (3), (4) A3009
    (5) A3004, A3020
    (6) A3004
  15(1), (2) A3020
  16 .... A3022
  Sch 1 ..... A3006, A3007, A3008, A3013, A3014
  para 4 .... A3015
  Pt 2 .... A3018
  Sch 2 .... A3013, A3035
  3 .... A3014
Road Transport (Working Time) Regulations 2005 (SI 2005 No 639)
  ...... M2103, W10026, W10029
Road Vehicles (Construction and Use) Regulations 1986 (SI 1986 No 1078)
  .... M2103

S

Safe work in confined spaces Confined Spaces Regulations 1997 (SI 1997 No)
Safety Representatives and Safety Committees Regulations 1977 (SI 1977 No 500) . Intro 15, Intro 17, A5025, C8015, D0948, J3002, J3006, J3010, J3011, J3017, J3021, J3030, T7017, V8008

Safety Representatives and Safety Committees Regulations 1977 (SI 1977 No 500) – cont.
reg 2	J3003
3	J3003
(2)	J3005
4(1)	J3007, J3009
(2)	J3011, J3013
4A(1)	J3007
(2)	J3004
5	J3008
(3)	J3014
6	J3007, J3008
7	J3007, J3027
(1)	J3015
(2)	A3027, J3015
8	J3005
9(1), (2)	J3018
11	J3013
Sch 2	J3012, J3013

Sale and Supply of Goods to Consumers Regulations 2002 (SI 2002 No 3045)
reg 3	P9049
5	P9048

Sex Discrimination Act 1975 (Amendment) Regulations 2008 (SI 2008 No 656)
.... H1702

Shipbuilding and Ship-repairing Regulations 1960 (SI 1960 No 1932)
.... D0939, V12015
reg 50, 51 .... P3008

Simple Pressure Vessel Regulations 1991 (SI 1991 No 2749) P7003

Simple Pressure Vessels (Safety) Regulations 1991 (SI 1991 No 2749)
.... P7003
Sch 1 .... P7005

Simple Pressure Vessels (Safety) Regulations 2016 (SI 2016 No 1092)
.... P7001, P7002

Site Waste Management Plans Regulations 2008 (SI 2008 No 314)
.... F3021A

Smoke-free (Exemptions and Vehicles) Regulations 2007 (SI 2007 No 765)
reg 11(7) W11028

Smoke-free (Penalties and Discounted Amounts) Regulations 2007 (SI 2007 No 764) W11028

Smoke-free (Premises and Enforcement) Regulations 2006 (SI 2006 No 3368)
.... W11028

Smoke-free (Signs) Regulations 2012 (SI 2012 No 1536) . F5018, W11028

Smoke-free (Vehicle Operators and Penalty Notices) Regulations 2007 (SI 2007 No 760) W11028

Social Security Benefits Up-rating Order 2013 (SI 2013 No 574)
.... C6017

Social Security (Claims and Payments) Regulations 1979 (SI 1979 No 628)
reg 24(1)	A3024
(3)	A3024
25	A3024
31	A3024

Social Security (Employed Earners' Employments for Industrial Injuries Purposes) Regulations 1975 (SI 1975 No 467)
reg 2–7	C6008
Sch 1, Sch 2	C6008
3	C6009

Social Security (General Benefit) Regulations 1982 (SI 1982 No 1408)
reg 11 .... C6011

Social Security (Industrial Injuries) (Prescribed Diseases) Regulations 1985 (SI 1985 No 967) N3012, N3018, N3019, V5010
reg 4(5)	N3020
25(2)	N3020
34	N3021
Sch 1	C6058, V5011
Pt 1	N3020
Sch 3	
Pt 2	N3021

Social Security (Industrial Injuries) (Prescribed Diseases) Amendment Regulations 1989 (SI 1989 No 1207)
reg 4(2) N3021

Social Security (Industrial Injuries) (Prescribed Diseases) Amendment Regulations 1990 (SI 1990 No 2269)
.... N3021

Social Security (Industrial Injuries) (Prescribed Diseases) Amendment Regulations 1993 (SI 1993 No 862)
reg 6(2) V5004

Social Security (Recovery of Benefits) (General) Regulations 1997 (SI 1997 No 2205)
reg 3	C6049
5	C6045, C6049
6	C6049
9	C6048
10	C6047

Submarine Pipe-lines (Inspectors etc.) Regulations 1977 (SI 1977 No 835)
.... O7004

Supply of Machinery (Safety) Regulations 1992 (SI 1992 No 3073)
.... M1001, M1018, V5015
reg 2(2)	M1009
4, 5	M1009

Table of Statutory Instruments

Supply of Machinery (Safety) Regulations 1992 (SI 1992 No 3073) – *cont.*
   reg 13–15 .... M1013
      29(a) P9009
   Sch 4 .... M1013

Supply of Machinery (Safety) (Amendment) Regulations 1994 (SI 1994 No 2063)
   reg 4 .... M1009
   Sch 2
      para 10 .... M1009

Supply of Machinery (Safety) (Amendment) Regulations 2011 (SI 2011 No 2157)
      .... L3002, M1009

Supply of Machinery (Safety) Regulations 2008 (SI 2008 No 1597)
   . E3003, L3002, M1007, M1009, M1010, M1014, N3014, N3035, R3016, V5020

T

Toy Safety Regulations 1995 (SI 1995 No 504) P9545

Transfer of Undertakings (Protection of Employment) Regulations 2006 (SI 2006 No 246) E14011

Transnational Information and Consultation of Employees Regulations 1999 (SI 1999 No 3323) J3001, J3002, J3031
   Pt III J3033
      IV J3033
   reg 17(1) J3033
      (3) J3033
      18(1) J3033
   Schedule
      para 6 .... J3033
           7(3) J3033

Transport of Dangerous Goods (Safety Advisers) Regulations 1999 (SI 1999 No 257) D0402

U

Unfair Terms in Consumer Contracts Regulations 1999 (SI 1999 No 2083) P9048
   reg 5(1), (2) E13004, P9505
      (4) E13004, P9505
      7 .... E13004, P9505
      12 .... E13004, P9505
   Sch 1 .... E13004, P9505
      2 .... E13004, P9505

W

Waste Management Licensing Regulations 1994 (SI 1994 No 1056) A5030

Work at Height Regulations 2005 (SI 2005 No 735) .... Intro 18, C6506, C7008, C8018, E13032, E15036, F3025, M1007, O7072, W11001, W9001, W9004.2
   reg 2 .... W9004.1
      3 .... R3015
      4 .... W9005, W9006
      5 .... W9005, W9007
      6 .... R3015, W9005
        (1) W9007
        (2), (3) W9008
        (4)(a), (b) W9008
        (5)(a)(i) W9008
          (b) W9008
      7 .... W9006, W9009
        (1) W9008
      8 .... W9012
        (a) W9008
      9(1) W9010
      10, 11 .... W9011
      12 .... M1006
        (2), (3) W9012
        (5) W9012
        (7), (8) W9012
      13 .... W9012
      14(1), (2) W9013
   Sch 1 .... W9008
      2 .... W9012
      para 5(1)–(3) W9008
   Sch 3 .... W9012
   Pt 1 .... W9008
      2 .... W9008
   Sch 4 .... W9012, W9014
      5 .... W9012
   Pt 2 .... W9014
      3 .... W9014
      4 .... W9014
      5 .... W9008, W9014
   Sch 6 .... W9012
      para 5 .... W9009.1
        9, 10 .... W9009.1
   Sch 7 .... W9012

Work at Height Regulations (Northern Ireland) 2005 (SR 2005 No 279) W9003

Work at Height (Amendment) Regulations 2007 (2007 No 114) W9004.2

Work in Compressed Air Regulations 1996
(SI 1996 No 1656)
.... C8018, P7013
reg 14 .... F5062
16(2) V12019
Working Time Regulations 1998 (SI 1998
No 1833) E14001, E14002,
E14011, J3001, J3031,
M1505, S11008, S11014,
S11019, V12004, W10001,
W10006, W10032, W10044
reg 2 ... J3032, W10002, W10019,
W10027
4 . W10012, W10020, W10029
(1) W10010, W10042
(2) W10010
(3), (4) W10011
(5) W10011, W10034
(6) W10010
5 .... W10012
6 .... J3002, J3032, W10019,
W10020
(1) W10029
7 .... W10029
(1)(b) W10021
8 .... W10018
9 .... W10025
10 ... J3002, J3032, W10014,
W10016
(1) W10029
11 .... J3002, J3032, W10016
(1), (2) W10029
12 ... J3002, J3032, W10017,
W10018
(1) W10029
13 .... W10022, W10024,
W10029
(9) W10024B
13A W10024B, W10029
(3) W10022
14 .... W10024, W10024B
(3) W10023
(b) W10022
(4) W10023
15 .... W10024
(2) W10024B
16(1) W10024
(5) W10024
18 .... W10026
(2) W10031
20 .... W10012, W10020
(1) W10015, W10017,
W10033
(2) W10033
21 .... W10015, W10020,
W10034, W10037
(e) W10029
22 .... W10015, W10035,
W10037
23 .... W10012, W10015,
W10017, W10020

Working Time Regulations 1998 (SI 1998 No
1833) – *cont.*
reg 23(a) J3032, W10036,
W10037
(b) J3032, W10036
24 .. J3032, W10015, W10037
(a), (b) **W10017**
24A(2) W10029
25(1) W10025
25A W10012, W10031
28(3) W10039
29 .... W10039
30 .... W10024B, W10040
31 .... W10041
35 .... W10043
36 .... W10005
Sch 1 .... J3002, J3032
para 3 .... W10008
Working Time (Amendment) Regulations
2003 (SI 2003 No 1684)
.... W10026, W10028, W10030,
W10031
Working Time (Amendment) Regulations
2007 (SI 2007 No 2079)
.... W10022
Workplace (Health, Safety and Welfare)
Regulations 1992 (SI 1992 No 3004)
.... Intro 18, A1001, A1002,
A5030, C7008, C8511, D0939, D8203,
E15012, E15023, E17029, E17034,
L5005, M2103, P7031, S11014, S11015,
V3001, V3016, W11001
reg 2(1) W11002
3 .... W11002
4 .... W11002
4A W11002
5 .... W11001, W11003
(1) O3018
(3) O3018
6 .... W11001, W11004
(1), (2) V3015
7 .... W11001
(1A) W11005
8 .... W11001, W11006
(3) L5002
9 .... W11001, W11007
10 .... W11001, W11008
11 .... W11001
(1)–(4) W11009
12 .... W9004.1, W11001
(1) O3018, W11010
(2)(a), (b) W11010
(3) O3018, W11010
13 . A1015, W11001, W11002,
W11010, W11011
(1)–(3) D0948
(5)–(7) W9004.1
14 .... W11001, W11012
15 .... W11001, W11013
16 .... W11001, W11014

Workplace (Health, Safety and Welfare) Regulations 1992 (SI 1992 No 3004) – *cont.*
- reg 17 .... W11001, W11015
  - (1) A1003
  - (2) A1004
  - (3), (4) A1004, A1005
- 18 .... W11001, W11016
- 19 .... W11001, W11017
- 20 .... W11001, W11019, W11020
- 21 .... W11001, W11020

Workplace (Health, Safety and Welfare) Regulations 1992 (SI 1992 No 3004) – *cont.*
- reg 22 .... W11001, W11024
- 23 .... W11001
  - (1), (2) W11025
- 24(2) W11026
- 25 .... V12034, W11001
  - (1)–(5) W11027
- 25A R3028, W11001
- Sch 1 .... W11008
- Pt 2 .... W11021

Index

A

ABATEMENT OF STATUTORY NUISANCE E5062
ABATTOIR
  controls as to F9005, F9007
ABRASION M1025
ABROAD
  see also European Union, International framework
  working E18002, E18022–E18023
ABSENCES *see* Managing Absence, Sickness Absence
ACAS
  code on disciplinary procedures in industry E14008, E14014, E14021
  dismissal
    Code of Practice on Disciplinary and Grievance Procedures M1517
    Early Conciliation notification E14008
    warnings E14014
ACCESS
  access/egress points
    confined spaces C7023
    employers to provide and maintain A1001
    hazards A1016–A1019
    safe workplace transport systems A1002, A1003
  business continuity planning B8008
  common law duty of care A1001, A1021
  community health and safety C6516, C6517
  confined spaces, *see* Confined Spaces
  definition A1001
  disabled persons, facilities for A1001, V12034, W11043
  disaster and emergency planning D6010
  electricity E3020
  employer's statutory duty A1001
  factories O3018
  hazards A1016–A1019
  heights, working at W9004.1, W9008
  non-English speakers, provision A1014
  occupiers' liability O3018
  parking spaces A1004

ACCESS – *cont.*
  pedestrians A1002, A1004, A1007
  safe workplace transport systems A1003
  speed limits regarding A1005
  traffic routes, *see* Traffic Routes
  trespassers C6513
  vehicles A1003–A1019
    access to A1015
    contractors A1014
    drivers, training of A1013, A1014
    hazardous operations with A1008, A1016–A1019
    maintenance A1012
    non-English speakers, provision A1014
    pedestrians and A1002, A1004, A1007
    provision of A1011
    statutory requirements concerning A1008, A1009
    subcontractors A1014
    unauthorised lifts in A1020
    work vehicles A1010–A1015
  violence in the workplace V8088
  workplace transport systems
    organisation of safe A1002, A1003
    parking areas A1004
    pedestrians A1004, A1007
    safe traffic routes A1006, A1007
    suitability of traffic routes A1004
    vehicles A1005
ACCIDENT HAZARDS
  see Major Accident Hazards Regulations
ACCIDENT INVESTIGATIONS A3029–A3033
  allocation of blame and liability A3030
  causes, analysis of A3032
  disciplinary procedures A3029
  elements of, main A3030–A3033
  evidence, collating A3031, A3032
  guidance A3029
  Health and Safety Executive A3030
  information A3029, A3032
  interviewing witnesses A3031
  lessons, learning A3029
  monitoring A3029

Index

ACCIDENT INVESTIGATIONS – *cont.*
performance A3029
rationale for A3029
reassurance and explanation A3029
reports A3030, A3032–A3033
risk assessment A3029
witnesses A3031

ACCIDENT REPORTING
accident book A3002
action to be taken by employers A3019
amputations A3006, C6506
axphyxia C6506
basic rules A3001
burns C6506
causes of accidents A3001
control of the premises,
  responsibility C6506
coverage A3005
crush injuries C6506
dangerous occurrences A3013
  list of A3035
death of employee A3018
defence for breach of Regulations A3022
disclosure of accident data
  generally A3026
  legal representatives A3028
  safety representatives A3026
disease
  duty to keep records of A3018
  reportable, list of A3002, A3018
employees, duties on A3023–A3025
ergonomics and E17067
Europe E18021
fatal accidents A3003, A3007
fines A3021
fractures A3006, C6506
further guidance A3034
gas fitters A3011
gas incidents A3002
  duty to report A3010
generally A3001
head injury C6506
hospital, at C6506
incident reporting centre A3003
injuries and dangerous occurrences C6506
  duty to keep records of A3015, A3030
  major A3005
legal representatives A3028
major accidents R3016, R3035
major injuries or conditions A3007
managing health and safety M2041
manual handling M3019
members of public A3001
non-workers A3008

ACCIDENT REPORTING – *cont.*
objectives A3025
occupational diseases A3014
offshore operations A3004
organ damage C6506
out of hours A3033
penalties A3021
procedures A3015–A3016
records, duty to keep A3017–A3018
  disease A3018
report, duty to A3001
requirements A3001
responsible persons A3002
RIDDOR A3002–A3022
  accident book A3002
  dangerous occurrences A3013, A3028
  defences A3022
  duties under A3002
  exemptions A3020
  gas incidents A3010
  major injuries or conditions A3007
  offshore installations O7069
  penalties A3021
  records, duty to keep, *see* records, duty
    to keep *above*
  responsible persons A3002, A3002road
    accidents
  scope of regulations A3002, A3005
  specified dangerous occurrences A3013
road accidents A3009
safety representatives A3027
scalping C6506
scope of duty A3001, A3002
scope of provisions A3002, A3005
sight injuries C6506
specified dangerous occurrences A3013
specified injuries to workers A3006
statistics A3001
unconsciousness C6506
visitors A3001

ACCIDENTS
asbestos A5033
causation M2028
community safety C6516
Control of Major Accident Hazards
  (COMAH) Regulations 2015 IN21
costs of H2139, IN24
fatal injuries, *see* Fatal Accidents
hazardous substances in workplace,
  where H2105, H2114, H2135,
  H2139
industrial injuries scheme C6001,
  C6005–C6006
major accident hazards, *see* Major
  Accident Hazards Regulations

Index

ACCIDENTS – *cont.*
 manual handling M3019
 reporting, *see* Accident Reporting
 Royal Society for the Prevention of Accidents IN26
ACCOUNTABILITY *see also* **Corporate Manslaughter**
 international comparison E18012
ACCOUNTS MODERNISATION DIRECTIVE
 Business Review E16005, E16024
 environmental management E16005, E16024, E16025
 Strategic Report E16005, E16024
ACCREDITATION E16005
 gas fitting operatives G1014
ACID RAIN E5001 E5014 E5015
ACOPs *see* **Approved codes of practice**
ACOUSTIC BOOTHS N3037
ACOUSTIC SHOCK
 call centres N3028
ACOUSTIC SIGNALS
 safety signs regulations W11030
ACTORS
 safety representatives J3005
ADJUSTMENTS
 disability discrimination E17070, M1511, R2005
 examples of R2005
 health and safety R2006
 justification R2006
 making R2004
 person associated with employee M1511
 reasonable E17070, R2005–R2006
 rehabilitation R2004–R2006
 sickness absence R2009
ADVENTURE ACTIVITIES
 regulations C6508, W9004.2
ADVERSARIAL
 proceedings E18007
ADVERSE ENVIRONMENT
 harassment H1702
ADVERTISEMENT
 food, as to F9020
ADVERTISING STANDARDS AUTHORITY (ASA) F9012
ADVICE ON HEALTH AND SAFETY LAW
 small businesses IN04
 sources of IN26
ADVISORY CONCILIATION & ARBITRATION SERVICE (ACAS)
 Code of Practice, disciplinary procedures E14008, E14021

AGE
 Employment Quality (Repeal of Retirement Age Provisions) Regulations 2011 M1505
 Equality Act 2010 E16513
 harassment H1702
 insurance exemption M1511
 manual handling M3007
 protected characteristic H1703
 risk assessment E16515
 violence in the workplace V8032
 workforce E16514
 young workers E16516
AGENCY WORKERS
 employers' liability insurance E13016
 equal treatment E14001
 mothers, new or expectant M2019
 visual display units D8202
 working time W10005
AGENTS
 liability of principal E16501
AGGRAVATING FACTORS
 fines C9011
AGRICULTURE
 air pollution and E5015
 climate change, effect of E5023
 enforcing authority E15009
 risk factors E16521
 tractor cabs E15005
 vibration V5001
AIR
 compressed air, *see* **Compressed Air, Work in**
 pollution, *see* **Air Pollution**
 quality, *see* **Air Quality**
AIR CONDITIONING L5007
AIRCRAFT
 food on board F9032–F9033
 occupiers' liability O3005
 vibration V5015
AIR MONITORING
 asbestos A5037
AIR PASSENGER DUTY (APD) E16006
AIR POLLUTION
 generally, *see* **Emissions into the Atmosphere**
AIR QUALITY
 assessment E5033–E5039
 emissions, *see* **Emissions into the Atmosphere**
AIR TESTING
 standards A5038
AIR TRANSPORT
 transport of dangerous goods D0404
 vibration V5015

Ind-3

Index

AIR TRANSPORT – *cont.*
  working time W10026, W10027, W10031
ALARMS
  fire F5017
  violence in the workplace V8090
ALLERGIC REACTION
  allergic rhinitis O1042
  extrinsic allergic alveolitis O1042
  food, to F9011, F9075
  respiratory diseases O1042, O1043
AMENITY LOSS OF
  kinds of C6031, C6034
AMMONIUM NITRATE D0908, M1104
ANIMAL PREMISES
  fire safety risk assessment F5061.1
ANIMAL PRODUCTS F9002 F9014 F9028
ANIMAL WORKERS
  respiratory diseases O1043
ANNUAL LEAVE
  accrual
    sickness absence, during M1514A
    system W10022
  claim for unlawful deduction of wages W10040
  compensation related to entitlement W10023
  enforcement of provisions W10040
  entitlement W10001, W10022
  instalments W10022
  notice requirements W10024
  payment W10024
    rolling up W10024A
  road transport W10029
  rolled up holiday pay W10024A
  sick leave and W10024B
ANNUAL REPORTS
  environmental data E5070
ANTHROPOMETRIC DATA E17007, E17010
ANTI-SOCIAL BEHAVIOUR ORDERS V8140
ANXIETY O1020.6–O1020.9
APC *see* Emissions into the Atmosphere
APPEAL
  compensation for work injuries C6018, C6050–C6058
  courts IN18
  enforcement E14014
  fire prevention and control
    enforcement notices, against F5060.1
    generally F5060
  improvement notice,
    against E15019–E15025
  prohibition notice,
    against E15019–E15025

APPLICANTS
  job, questions O1003
APPROACH
  'Plan, Do, Check, Act' IN05
APPROVED CODES OF PRACTICE
  asbestos A5005, A5012
  confined spaces C7008
  control of substances hazardous to health H2122
  dangerous substances D0902, D0933–D0933, D0934
  explosions D0933
  factories O3018
  ionising radiation R1019
  machinery safety M1007
  occupiers' liability O3018
  record assessments H2122
  training and competence T7010
  ventilation D0933
ARM
  vibration syndrome, *see* Hand Arm Vibration work-related Upper Limb Disorders, Work-Related Upper Limb Disorders
ARMED FORCES
  corporate manslaughter C9001, C9005, C9015
  manual handling M3002
ARRANGEMENTS
  statement of health and safety policy S7006
ARSON F5074
ARTS VENUES
  fire safety risk assessment F5061.1
ASBESTOS A5001–A5043
  2008 Mesothelioma Scheme C6058
  accidents A5033
  air
    monitoring A5037
    standards A5038
    testing A5038
  Approved Code of Practice A5005, A5012, A5043
    generally A5043B
    health A5043C
    legislation A5043A
    safety A5043C
  asbestos containing materials (ACMs) A5011–A5012, H2101, R3013
    management plan A5012–A5013
  asbestosis A5002
  assessment of work A5025
  buildings
    maintenance workers A5006

ASBESTOS – *cont.*
  buildings – *cont.*
    removal from A5005
    surveys A5023–A5026
  cleanliness A5035
  clearance indicators A5005
  compensation A5042B
  compliance strategy, development of A5016
  contractors A5012, A5043
  contractual agreements A5021
  Control of Asbestos Regulations 2012 IN21
  control of exposure A5024–A5042
    control measures A5030–A5031
    historical A5003
    seven step programme A5013–A5019
    statutory requirements governing H2101, H2138
  definition A5001
  designated areas A5036
  diseases A5002
  distribution A5042
  duty holders R3013
    identification of A5009
    occupiers A5006
    third parties A5020
  emergencies A5033
  employees A5025
    changing facilities A5041
    instruction and training A5028
    washing A5041
  employers' duties A5007–A5019
  employers' liability insurance E13001, E13026
  exception A5042A
  exemption A5042A
  exposure
    building maintenance workers A5006
    prevention A5029
    reduction A5029
  facilities management F3024
  guidance A5043
  health records A5040
  historical exposures A5003
  HSE publications A5043, R3013
  identification of presence A5024
  inspection A5010
  instruction A5028
  labelling A5042
  laboratories A5012
  leases A5021
  legislation A5043–A5043
  licensing of works A5027

ASBESTOS – *cont.*
  licensing regulations E15005, R3016, R3035
  long development period of E13001
  long tail disease P9553
  lung cancer A5002, A5003
  maintenance
    building maintenance workers A5006
    records A5011
  management
    asbestos-containing materials A5011
    leases A5021
    long-term plans A5018
    plans A5012, A5018–A5019, A5022
    third parties A5020
  medical surveillance A5040
  mesothelioma A5002, A5003
  minimum requirements A5022
  mixtures, control limit A5002
  notification of work with A5027
  other publication A5043D
  personal protective equipment P3004, P3008
  plans
    long-term A5018
    monitoring A5019
    review of A5019
  plant A5035
  premises
    affected A5008
    cleanliness A5035
    control of work on A5014
    removal A5005
  presence, identification of A5024
  prevention of exposure A5024–A5042
  priorities A5017
  prohibitions A5004, A5042A
  protective clothing A5032
  public liability insurance P9528
  raw asbestos A5042
  records A5011
  register A5011, A5022
  Regulations A5007–A5007A, D6016, H2138, IN21
  repairs
    leases A5021
  respiratory diseases O1042, O1046–O1055
  respirator zones A5036
  risk assessment A5007, A5010, A5017, IN14, R3013, R3016, R3035
  samples A5011, A5039
  self-employed A5007
  site clearance certification A5038
  spread of A5034

Index

ASBESTOS – *cont.*
standards
air testing A5038
analysis A5039
storage A5042
surveys A5023–A5026
guidance A5023
quality assurance A5023
risk assessments A5025
tenancy agreements A5021
third parties A5020
training A5028
types of A5001
UK Accreditation Service
laboratories A5012
ventilation regulations V3015
waste
labelling A5042
removal A5027
special A5027
storage A5042
work
assessment A5025
licence A5027
plan of A5026
Regulations H2138
workers A5006
works, licensing of A5027
ASPHYXIATION
confined spaces C7001
ASSAULT
harassment H1723, H1729–H1730
violence in the workplace V8002, V8011,
V8131–V8133
ASSEMBLY OR PRODUCTION
LINES M3004
fire safety risk assessment F5061.1
training and competence T7014–T7015
ASSOCIATION
discrimination by M1508
ASSURANCE
product liability P9544
ASTHMA
benefit claim C6057
employers' liability insurance E13026
hazardous substances in workplace,
where H2103
occupational disease C6057
respiratory diseases O1042, O1044
ATEX DIRECTIVE D0933 E3027
ATMOSPHERES
Dangerous Substances and Explosive
Atmospheres Regulations 2002 IN21
explosive
classification D0920–D0921

ATMOSPHERES – *cont.*
explosive – *cont.*
sources of D0921
ATMOSPHERIC HAZARDS
confined spaces A1020
ATTENDANCE MANAGEMENT R2007
AUDIOMETERS O1023
AUDITING
business continuity planning B8008,
B8034
disaster and emergency planning D6067
stress S11037
AUDIT REPORT
environmental management
system E16025
risk assessment, use for R3031
AUSTRIA
professional regulation E18017
AUTOMATION M3010
AUXILLIARY AID
reasonable adjustments M1511
AVIATION
managing health and safety M2042,
M2045
AWARDING ORGANISATIONS
training and competence T7009

B

BACK
manual handling M3004
pain O1020.1, V5005, V5023–V5024,
V5028
vibration V5005, V5023–V5024, V5028
whole-body vibration V5023–V5024
BAGS M3013
BANKS
environmental management E16007
Equator Principles E16007
BARRIERS
heights, working at W9008
Regulations W9004.1
mobile work equipment M1008
BATH HOISTS L3036
BATON
violence in the workplace V8095
BATTERY
electricity E3027
rooms E3027
BATTERY AND ASSAULT
common law H1730
criminal action H1729
harassment H1730

BATTERY AND ASSAULT – *cont.*
  trespass to the person H1723
BCF FIRE EXTINGUISHERS F5014
BEAT HAND AND BEAT ELBOW O1020.1
BEHAVIOURAL MEASURES
  machinery safety M1026
BELGIUM
  professional regulation E18017
BELIEF
  harassment, protected characteristic H1703
BENCHMARKING M2039
  consumer groups E16012
  small and medium-sized enterprises (SMEs) E13030
BENEFIT RATES *see* **Industrial Injuries**
BEREAVEMENT
  non-pecuniary losses, recoverability C6031, C6035
BIOLOGICAL AGENTS
  children and young people V12016
  new or expectant mothers V12021
  risk assessment/control of exposure R3027
  unborn child, hazardous to R3028
BITC (Business in the Community)
  CR Index E16002
  CSR E16002
BLASTING OPERATIONS
  personal protective equipment P3003, P3019
BLENDED LEARNING
  vocational qualifications T7007
'BODY ARMOUR' USE OF V8093
BORDER AGENCY OFFICES
  corporate manslaughter C9001
BOUNDARY
  construction operations C8019
BREACH OF CONTRACT
  harassment H1727
  remedies H1728
  working time, and W10042
BREACH OF STATUTORY DUTY IN22 E13002 E15021–E15035
BREAKDOWN MAINTENANCE F3015
BREAKS REST W10017
  stress S11008, S11014, S11019–S11020
  visual display screens D8206
BREASTFEEDING MOTHERS
  ionising radiation R1019
  risk assessment E14024–E14025, R3006
BRITISH SAFETY COUNCIL
  advice from IN26
BRITISH STANDARDS
  fire extinguishers F5016

BRITISH STANDARDS – *cont.*
  fire prevention and control F5017
  heights, working at W9014
  managing health and safety M2029
  vibration V5009
BROADCASTING
  enforcing authority E15009
BROKERS
  employers' liability insurance E13003
  professional indemnity insurance E13003
BSE F9020
BUILDING
  see also Facilities Management
  electricity E3033
  head protection, *see* Head Protection
  health and safety regulations W11042
  lifts L3009
  vibration, *see* Vibration
BUILDING AND CONSTRUCTION OPERATIONS
  see also Fire; Fire Precautions; Ladders; Work at Heights; Working Platforms
  asbestos
    prevention C8018
    surveys A5023–A5026
    workers A5006
  building work
    definition W11044
  CDM Regulations 2015. *see* Construction, Design and Management Regulations
  changing rooms C8002–C8017
  children and young people C8019, V12003–V12005
  cofferdams and caissons C8002–C8017
  compressed air, work in C8018
  confined spaces
    and see Confined Spaces
    generally C8018
  construction work
    definition IN21
  demolition or dismantling C8002–C8017
  drinking water C8002–C8017
  drowning, prevention of C8002–C8017
  egress C8002–C8017
  electricity supply, safety precautions C8018
  emergency procedures F5026
  energy distribution systems C8002–C8017
  ergonomic considerations C8002–C8017
  excavations C8002–C8017, C8018
  facilities management F3033
  fire prevention and control C8002–C8017
    building and construction operations C8002–C8017
    code of practice F5029

BUILDING AND CONSTRUCTION
  OPERATIONS – cont.
 fire prevention and control – cont.
  designing out fire F5027
  emergency procedures F5026
  emergency routes and
   exits C8002–C8017
  fire extinguishers F5028
  insurance F5029
  plant F5028
  precautions F5029
  site fire safety co-ordinator F5025
 first-aid areas F7025
 fresh air C8002–C8017
 good order C8002–C8017
 hazardous substances C8018
 head protection C8018
 health and safety file IN21
 health and safety plan IN21
 housekeeping C8018
 inspection, reports C8002–C8017
 insurance cover F5029
 introduction C8001
 ladders, *see* Ladders
 lifting operations A1024
 lighting C8002–C8017, L5004
 lockers C8002–C8017
 maintenance projects IN21
 mechanical vibration V5006
 new or expectant
  mothers V12003–V12005
 noise C8018
 non-English speakers A1014
 occupational diseases, *see* Occupational
  Diseases
 pedestrians C8002–C8017
 personal protective equipment P3003,
  P3004, P3008
 person in control C8002–C8017
 practical issues
  changing rooms C8002–C8017
  drinking water C8002–C8017
  fresh air C8002–C8017
  good order C8002–C8017
  lighting C8002–C8017
  lockers C8002–C8017
  rest facilities C8002–C8017
  safe place of work C8002–C8017
  sanitary conveniences C8002–C8017
  temperature C8002–C8017
  washing facilities C8002–C8017
  weather, protection from C8002–C8017
  welfare facilities C8002–C8017
 public liability exposure P9528

BUILDING AND CONSTRUCTION
  OPERATIONS – cont.
 Regulations. *See* Building Regulations
 rest facilities C8002–C8017
 risk assessment R3017
 safe place of work C8002–C8017
 sanitary conveniences C8002–C8017,
  W11018
 signing C8018
 site fire safety co-ordinator F5026
 site inductions and security C8019
 slips C8018
 temperature C8002–C8017
 traffic routes C8002–C8017, C8018
 training and competence C8018,
  T7018trips
 typical project issues C8018
 vehicles A1014, C8002–C8017, C8018
 vibration, risks relating to C8018
 visitor protection C8019
 vulnerable persons V12003–V12005
 washing facilities C8002–C8017
 waste management C8020
 weather, protection from C8002–C8017
 welfare facilities C8002–C8017
 working platforms, *see* Work at Heights;
  Working Platforms
BUILDING REGULATIONS IN21
 fire prevention and control F5030.1,
  F5030.2
 gas safety G1019
 requirements W11044
BUILDING WORK
 see also Building and Construction
  Operations
 definition W11044
 Factories Act, relating
  to W11045–W11048
 violence in the workplace, *see* Workplace
  violence
BULK CONTAINER
 intermediate D0416
BULLYING
 duty of care, breach E14003
 procedure E14006, E14012
BUNCEFIELD OIL DEPOT D0951
BURDEN OF PROOF
 criminal cases E15031
 employers' liability insurance E13002
 presumption of innocence E15031
 reverse E15031
 risk E15031
BURGOYNE REPORT O7003
BURNS
 electricity E3006

BURNS – *cont.*
  offshore operations O7071
BUSINESS CONTINUITY
    PLANNING B8001–B8042
  access, effects of denial of B8008
  after the crisis, communication B8042
  audits B8008, B8034
  British Standard B8006
  business case for B8010
  Business Continuity Institute B8001
  business impact analysis B8011, B8016, B8018
  buy-in, securing B8027
  Cadbury guidance B8006
  call deluge B8040
  command and control team B8024
  commercial drivers B8005
  communication B8038–B8042
  component approach B8014, B8022
  component testing B8036
  context of B8002
  creditworthiness, loss of B8017
  crisis communication B8038–B8042
  critical business approach B8015, B8018, B8022, B8037
  customer base, erosion of B8017
  damage assessment team B8024
  departmental teams B8024
  desk check/unit B8034
  development of B8011, B8020, B8025
  dilemma, the planning B8021
  disasters
    imported B8008
    natural, effects of B8008
    public relations B8009, B8038
    recovery B8001–B8003, B8023
  drivers B8004–B8009
  during crisis, communications B8041
  emergency response B8023
  fallback procedures B8023
  financial drivers B8007
  full simulation B8037
  hierarchical structure B8024
  implementation B8011, B8026–B8032
  importance of B8003
  individual responsibilities B8030
  insurance B8007
  key staff, loss of B8017
  maintenance B8011, B8033–B8037
  media training B8040
  model B8011, B8015
  objectives B8001
  outputs B8018
  phased roll-out B8031
  policy B8028

BUSINESS CONTINUITY PLANNING – *cont.*
  pre-crisis measures B8040
  press statements B8040
  project authorisation B8029
  public relations B8038–B8042
    disasters B8009, B8038
  quality drivers B8006
  quantification of business impact B8017
  regulatory and contracted impact B8017
  regulatory drivers B8006
  reputation, loss of B8017
  resumption of business B8023
  risk assessment B8012–B8018
  risk management B8007
  simulation B8037
  specialist services B8032
  standards B8006
  strategy B8011, B8019
  testing B8011, B8033–B8037
  Turnbull Committee B8001, B8006
  understanding your business B8011, B8012
  walkthrough test B8035
BUSINESS IN THE COMMUNITY (BITC)
  CR Index E16002
BUSINESS PLAN
  HSE IN12
BUSINESS REVIEW
  Companies Act 2006 E16005, E16024
  disclosure E16005, E16024
  environmental reporting E16005, E16024
  Strategic Report E16005, E16024
BYSSINOSIS
  occupational disease C6062

C

CABLES
  electricity E3028
  public liability P9523
  tripping over O3015
CADBURY REPORT IN05 B8006
CAISSON
  Building and construction sites C8002–C8017
CALL CENTRES
  acoustic shock N3028
  noise and N3028
  risk factors E16521
CALL DELUGE B8040
CANADA
  compensation systems I2011

Index

CANADA – *cont.*
  legal framework I2002, I2010
  professional development and recognition I2012
CANCER
  breast O1056
  carcinogens O1056
  employers' liability insurance E13026
  mobile phones R1008–R1009
  non-ionising radiation R1001, R1008–R1009
  occupational O1056
  respiratory H2103
  skin H2103, O1038
CANNED FOOD F9056
CAPABILITY
  definition M1516
  managing health and safety M2016
  manual handling M3007, M3015
CARBON ACTION SCHEME E16007
CARBON DIOXIDE FIRE EXTINGUISHERS F5015
CARBON DISCLOSURE PROJECT (CDP)
  reporting E16007–E16008
CARBON MONOXIDE
  confined spaces C7001
  gas safety G1001
  new or expectant mothers V12021–V12022
CARBON REDUCTION COMMITMENT
  sustainability F3018
CARBON TRUST
  carbon footprints E16008
CARCINOGENS M1104, O1043 O1056
CARCINOMA
  compensation claim C6055
CARPAL TUNNEL SYNDROME (CTS) O1020.1, V5004
  condition V5010
  reportable occupational disease V5011
CARRIAGE OF DANGEROUS GOODS. *See* Transportation of Dangerous Substances
CARRYING M3003
CASH AND CARRY WAREHOUSE
  food from F9060
CATARACTS O1033, O1034
CATERING AND HOSPITALITY
  See also Food
CAUSATION
  burden of proof E15031
'CAVEAT EMPTOR'
  public and products liability insurance P9505

CAVERS W9004.2
CE MARK
  equipment M1007
  lifts L3004, L3006, L3010
  machinery safety M1009, M1026
  personal protective equipment P3012, P3013
CENTRE FOR RADIATION CHEMICAL AND ENVIRONMENTAL HAZARDS R1017
CERTIFICATES
  emergencies V5015
  first aid training F7018
  Qualifications Credit Framework T7009
  vibration, emergency certificates for V5015
CHAINS L3043–L3044
  failure L3044
  types of L3043
CHANGES IN WORK IN06
CHANGING FACILITIES
  asbestos A5041
  construction and building operations, and C8002–C8017
  generally W11026
CHARITY
  corporate manslaughter C9002
  risk management IN05
CHARITY EVENTS
  food at F9007
CHARTERED INSTITUTE OF BUILDING SERVICES ENGINEERS (CIBSE) L5002
CHARTERED INSTITUTE OF ENVIRONMENTAL HEALTH
  advice from IN26
CHEMICAL HAZARDS
  Building and construction operations C8018
  confined spaces A1022
  major accident hazards, *see* Major Accident Hazards Regulations
  new or expectant mothers V12021–V12022
  supply chain E16008
CHEMICAL PLANT
  first-aid areas F7025
CHEMICAL SUBSTANCES
  children and young people V12016
  dangerous goods, carriage of D0402
  EU law D0901, D0908
  European Chemicals Agency (ECHA) R2502, R2511
  European Classification, Labelling and Packaging (CLP) Regulation H2137

Ind-10

CHEMICAL SUBSTANCES – *cont.*
  hazards, *see* Chemical Hazards; Major Accident Hazards Regulations
  information H2101, H2137
  labelling H2137
  packaging requirements H2101, H2137
  REACH *See* Registration, valuation and Authorisation of Chemicals in Europe
  skin diseases O1038, O1041
CHEWING GUM F9007
CHILDREN AND YOUNG PEOPLE V12001–V120026
  adventure activities C6508
  age V12006–V12007
  biological agents, exposure to V12016
  characteristics of V12013
  checklist V12016
  chemical agents, exposure to V12016
  child, definition of V12003
  construction work C8019, V12003–V12005
  contributory negligence O3011
  control of substances hazardous to health V12003–V12005
  dangerous materials or activities V12016
  dangerous workplaces or workstations V12016
  definition V12003
  employers' liability insurance E13016
  equipment V12003–V12005, V12016
  family undertakings V12011
  increased risks, works presenting V12016
  information V12010, V12014–V12015
  lead processes E14023
  legislation V12001–V12005
  Management of Health and Safety at Work Regulations V12001–V12003, V12009–V12016
  managing health and safety M2006, M2013, M2020
  manual handling V12003–V12005, V12016
  maturity V12013
  occupiers' liability O3010, O3011, O3017
  parents, information to V12010, V12014–V12015
  personal protective equipment V12003–V12005
  physical agents, harmful exposure to V12016
  physically demanding work, excessively V12016
  precautions V12013–V12015, V12016
  prohibitions, age related V12004
  psychologically demanding work, excessively V12016

CHILDREN AND YOUNG PEOPLE – *cont.*
  risk assessment M2013, R3025, V12002, V12006–V12007, V12008, V12013
  trainees in relevant training R3025
  training and competence V12003–V12005, V12012
  train surfing O3017
  trespass C6512
  volenti non fit injuria O3017
  warnings O3010
  work experience V12003–V12005, V12010
  young person, definition of V12003, V12006–V12007
CHLORACNE O1038
CHRONIC BRONCHITIS
  compensation claim C6055
CHRONIC UPPER LIMB PAIN O1020.1
CHUTES M3010
CINEMAS
  fire safety risk assessment F5061.1
CIVIL LAW
  civil action IN22
CIVIL LIABILITY IN22, vicarious liability
  see also Negligence
  breach of statutory duty IN22
  commencement of actions IN22
  contributory negligence IN22
  damages, assessment of IN22
  defences
    contributory negligence IN22
    factory occupier W11048
    foreseeability of injuries IN22
    generally IN22
    voluntary assumption of risk IN22
  director E15041
  duty of care IN22
  employers' liability insurance IN22
  evidence IN22
  expert evidence IN22
  fast track system IN22
  foreseeability of injuries IN22
  generally IN22
  initiating action IN22
  Löfstedt Review P9551
  manual handling F5073
  occupiers' liability IN22
  pre-action protocol IN22
  procedure IN22
  product liability insurance P9530
  product safety IN22, P9024–P9044
  proof IN22
  public liability insurance P9517
  settlement of cases IN22

Ind-11

Index

CIVIL LIABILITY – *cont.*
  stress S11009–S11009.2
  time limits IN22
  violence in the workplace V8012, V8140
  volenti non fit injuria IN22
  voluntary assumption of risk IN22
  Woolf Report
CIVIL PARTNERSHIP
  Equality Act 2010 E16501
  harassment H1703
CIVIL PROCEDURE RULES
  employers' liability insurance E13029
  pre-action protocols E13029
CLEANER
  fire prevention and control F5018
  risk assessment R3032
  risk factors E16521
CLEANING
  dangerous substances D0933
  facilities management F3030
  windows, *see* Window Cleaning
CLEANLINESS
  asbestos exposure A5032, A5035
  workplaces, of W11001, W11007
CLIMATE CHANGE
  acidifying substances E5015
  air passenger duty E16006
  Carbon Disclosure Project (CDP) E16007
  clean air strategy, new E5071
  consequences of E5023
  disclosure, Global Framework for climate risk E16007
  emissions trading E5027.1
  financial consequences of E16007
  Global Reporting Initiative E16007, E16024A
  greenhouse effect E5022
  international protocols E5026
  levy E16006
  reporting, greenhouse gas emissions E5070
  risk assessment in UK E5023.1
  Treaty E5026–E5027
  UK environmental policy, future strategy and legislation E5071
CLIMBERS W9004.2
CLOSED CIRCUIT TELEVISION
  violence in the workplace V8089
CLOTHING
  accommodation for workplaces S11014, W11001, W11025, W11026
  asbestos exposure A5032
  dangerous substances D0923
  explosions D0923

CLOTHING – *cont.*
  manual handling M3007
  workplaces S11014, W11001, W11025, W11026
CLUBS
  noise, and N3027
COAL
  see also Coal Miners
  burning, air pollution and E5002, E5061
COAL MINERS
  compensation, vibration white finger for V5003, V5012
  vibration V5003, V5012
CODES OF PRACTICE
  See also Approved Codes of Practice
  disability discrimination A1001.1
  explosions D0919
  food, as to F9006, F9025, F9035, F9038
  hazardous substances in the workplace H2105
  noise N3017
  pressure systems, and P7014
  Safety Case Regulations O7016
  vibration N3017
  violence in the workplace V8044, V8072, V8129
CODIFIED LEGISLATION
  international framework I2002–I2003
COFFERDAM
  building and construction sites C8002–C8017
COLLECTIVE AGREEMENTS E14005
COLOUR BLINDNESS O1033 O1034 O1035
COLOUR DISCRIMINATION
  Equality Act 2010 H1703
COMBINED CODE ON CORPORATE GOVERNANCE
  see also UK Corporate Governance Code replacement E16005
COMMERCIAL PREMISES
  health and safety law, enforcement of E15007
COMMON LAW
  gross negligence manslaughter C9021, C9023, E13032
COMMUNITY HEALTH AND SAFETY C6502–C6528
  access C6516, C6517
  accidents and emergencies C6516
  Adventure Activities Licensing Regulations 2004 C6508
  awareness, importance of worker C6511
  capabilities of customers C6516
  communication C6523

Index

COMMUNITY HEALTH AND SAFETY – *cont.*
community groups C6518
construction projects C6506
contractors, use of C6527
control of substances hazardous to health C6506
customers and users of facilities and services C6515–C6517
dangerous activities C6510
  planning of C6511
dangerous substances C6510
electrical equipment C6510
emergencies C6511, C6517
  customers and users, arrangements for C6517
  equipment C6514
  volunteers C6522
equipment C6510
  emergency C6514
  personal protective C6520
  selection and maintenance of C6517
  volunteers C6522
events
  communities, members of C6524
  contractors C6527
  HSE safety guide C6525
  information C6527, C6528
  risk assessment C6526
fencing C6514
fines C6503, C6504
fire safety C6507, C6522
groups, taking into account C6518–C6520
hazardous substances C6510
health surveillance C6506
height, working at C6506, C6510
home, entry into clients' C6522
individuals, taking into account C6518–C6520
informal visits C6518
information C6523, C6527, C6528
legal requirements C6502–C6508
lifting C6510
  planning of operations C6511
major accident hazards, *see* Major Accident Hazards Regulations
major public events C6524–C6526, C6527
Management Regulations C6505
manual handling C6522
monitoring of customers C6517
neighbours, duties to C6509–C6511
notices C6511, C6514
occupiers' liability C6508
passers-by, duties to C6509–C6511
personal protective equipment C6520

COMMUNITY HEALTH AND SAFETY – *cont.*
precautions
  contractors C6527
  customers and users of facilities and services C6517
  neighbours C6511
  passers-by C6511
  trespassers C6514
  visiting individuals and groups C6520
products, risks from C6516
prosecutions C6503, C6504
public events, major C6524–C6526
restricted areas, activities and groups C6520, C6523
risk assessment C6505, C6509–C6511, C6526, C6527
Safety Advisory Groups (SAGs) C6524
schoolchildren C6518
security of the workplace C6511
services, customers and users of C6515–C6517
signs C6511, C6514
standards C6527
storage locations and facilities C6511, C6514
students C6518
supervision and control C6517, C6520, C6523
traffic management C6511
training C6523
travel risks C6522
trespassers on work premises C6512–C6514
users of facilities and services C6515–C6517
vehicle traffic C6510
visiting individuals and groups C6518–C6520
volunteers C6521–C6523
worker awareness C6511
work experience students C6518

COMPANIES ACT 2006
Business Review E16005, E16024
corporate social responsibility E16005, E16024
environmental reporting E16005, E16024

COMPANY
director, *see* Director
health and safety offences E15039–E15042
junior staff, delegation to E15039
mergers and acquisitions E16007
occupiers' liability O3001
parent company C9002
Strategic Report E16005

Ind-13

Index

COMPENSATION
- acting in the course of employment C6006, C6007
- appeals C6018, C6050–C6058
- asbestos A5042B
- asthma C6057
- capitalisation of future losses C6027
- chronic bronchitis C6055
- coal miners, vibration white finger and V5003
- Compensation Act E13001, E13033, P9549
- compensation culture E13001, E13035, P9549
- compensator, duties of C6045
- complications C6046–C6049
- constant attendance allowance C6014
- contributory negligence C6049
- damages, *see* Damages
- deductible social security benefits, *see* Damages
- discrimination claims E16522
- emphysema C6055
- employers
  - persons treated as C6009
  - work-related violence V8015
- employers' liability insurance C6001, E13001
  - database E13026
- Employment and Support Allowance (ESA) C6015
- exceptionally severe disablement allowance C6014
- fatal injuries, *see* Fatal Injuries
- Financial Services Compensation Scheme E13024
- future earnings, loss of C6026
- home care, expenses of C6028
- industrial injuries benefits
  - benefits overlap C6020
  - weekly benefit rates payable C6017
- industrial injuries disablement benefit
  - claims in accident cases C6011
  - entitlement and assessment C6011
  - two rates of benefit C6013
- industrial injuries scheme C6001
- information provisions C6049
- institutional care, expenses of medical treatment C6028
- interest on damages C6043
- interim awards C6042
- Lord Young report E13035
- lost years, damages for C6029
- main types of injury to qualify C6023
- medical examination procedures C6001

COMPENSATION – *cont.*
- non-pecuniary losses
  - bereavement C6022, C6031, C6035
  - loss of amenity C6030, C6031, C6034
  - nervous shock C6031, C6033
  - pain and suffering C6031, C6032
  - quantification of C6030
  - recoverable C6031
- occupational deafness C6055
- osteoarthritis of the knee C6055
- pain and suffering C6032
- payments into court C6042
- pecuniary losses, assessment of C6022, C6042
- personal injuries provisions
  - acting in course of employment duties C6006, C6007
  - basis of claim for damages C6022
  - caused by accident C6005
  - non-pecuniary losses C6023
  - pecuniary losses C6023
  - relevant accident C6005
- pneumoconiosis, *see* Pneumoconiosis
- pre-existing disabilities C6011
- prescribed diseases and C6055
- primary carcinoma of the lung C6055
- provisional awards C6040
- public liability insurance P9517
- receipt of benefit, financial effects of C6019
- recovery of state benefit, *see* Damages
- relevant employment C6008
- repetitive strain injury C6058
- social security benefits C6015–C6020, C6044, C6045
- specified occupational diseases, claims for C6043
- statutory sick pay C6015
- structured settlements C6025, C6047
- unfair dismissal, for
  - additional award E14022
  - basic award E14022
  - compensatory award E14022
  - formula for E14017
  - special award E14022
- unlawful discrimination E16522
- vibration white finger V5003
- workplace violence V8004, V8012, V8013, V8015

COMPETENCE. *See also* **Training and competence**
- cycle of T7020
- definition T7002
- maintaining T7016

Index

COMPETENT PERSONS
appointment of M2010
definition T7002
lifting equipment L3019–L3022
noise N3029
permit to work systems S3016
pressure systems, and P7018
training T7001

COMPETITION
EC law M1009
machinery safety M1009

COMPLIANCE
large companies IN05

COMPRESSED AIR WORK IN
generally C8018
new or expectant
    mothers V12003–V12005,
    V12019–V12020
protective devices P7020

CONCERT HALL
noise exposure levels N3027.1

CONDITIONAL SALE AGREEMENT
product liability P9051

CONDUCT UNWANTED
purpose or effect H1703

CONFIDENTIALITY
harassment in the workplace H1736

CONFINED SPACES C7001–C7030
access and egress A1001, A1022, C7001, C7024
asphyxiation C7001
atmosphere, testing and monitoring the C7016
building and construction operations C8018
Code of Practice C7009
communications A1023, C7015
competence A1023, C7012
Confined Spaces Regulations 1997 C7003
    persons with duties under C7004
contents and residues, removal of C7019
contractors C7004
control measures C7010
definition of A1023, C7001, C7002
deliberately created C7003
designing safe system of work A1023
duty holders C7004
electrical equipment C7021
electricity A1022, A1023, C7026
emergencies C7007, C7029
employers' duties C7004
entry C7005, C7006, C7010, C7028
exemption certificates A1023
fire precautions A1023, C7025
formalised permit to work system A1022

CONFINED SPACES – *cont.*
free flowing solid, definition C7003
gas A1022, A1023
    equipment C7022
    flammable C7001
    purging C7017
    toxic C7001
hazards C7001
    atmospheric A1022
    chemical A1022
    physical A1022
hot work C7027
HSE guidance C7030
identification of risks C7010
independent contractors C7004
internal combustion engines C7022
introduction C8018
isolations C7020
Job Safe Procedure C7011
legal requirements, relevant C7002
lighting C7026
offshore operations O7065
oxygen concentration C7016
permit-to-work C7010–C7012
personal protective equipment C7023
precautions C7014–C7028
prohibition A1023
publications C7030
regulations as to C7001–C7007
respiratory protective equipment C7023
risk assessment A1023, C7005, C7006, C7008, R3017
roof space C7003
safe system of work A1023, C7007–C7013
self-employed C7004
smoking C7027
    exclusion area C7027
specified risk A1023, C7003
Standard Practice Instruction C7011
static electricity C7027
statutory requirements relating to work in A1001, A1023
supervision and training A1023, C7013
system of work C7003
task procedures C7011
training C7013
ventilation A1023, C7018
welding C7016
working time A1023, C7028

CONFORMITY ASSESSMENT
generally M1015–M1017
pressure systems, and P7008

CONSERVATION
Building Regulations W11044

Ind-15

CONSERVATION – *cont.*
  energy W11044
CONSTANT ATTENDANCE ALLOWANCE
  industrial injuries disablement benefit C6014
CONSTITUTION OF THE EU IN09
CONSTRUCTION AND BUILDING OPERATIONS
  see also Fire; Fire Precautions; Ladders; Work at Heights; Working Platforms
  accidents, prevention C8018
  asbestos
    surveys A5023–A5026
    workers A5006
  boundary matters C8019
  CDM Regulations 2015. *see* Construction, Design and Management Regulations
  changing rooms C8002–C8017
  children and young people C8019, V12003–V12005
  community health and safety C6506
  competence T7018
  compressed air, work in C8018
  confined spaces C8018
  construction work
    definition IN21, M2002
  drinking water C8002–C8017
  egress C8002–C8017
  emergency procedures F5026
  ergonomic considerations C8002–C8017
  facilities management F3033
  fire prevention and control
    code of practice F5029
    designing out fire F5027
    emergency procedures F5026
    enforcement C6507
    fire extinguishers F5028
    insurance F5029
    plant F5028
    precautions F5029
    site fire safety co-ordinator F5025
  first-aid areas F7025
  fresh air C8002–C8017
  good order C8002–C8017
  head protection C8018, P3019
  health and safety file IN21
  health and safety plan IN21
  housekeeping C8018
  insurance cover F5029
  introduction C8001
  ladders, *see* Ladders
  lifting operations A1024
  lighting
    generally C8002–C8017

CONSTRUCTION AND BUILDING OPERATIONS – *cont.*
  lighting – *cont.*
    statutory requirements L5004
  lockers C8002–C8017
  maintenance projects IN21
  management duties M2023
  new or expectant mothers V12003–V12005
  occupational diseases, *see* Occupational Diseases
  pedestrians C8002–C8017
  perimeter C8019
  personal protective equipment P3003, P3004, P3008
  person in control C8002–C8017
  practical issues
    changing rooms C8002–C8017
    drinking water C8002–C8017
    fresh air C8002–C8017
    good order C8002–C8017
    lighting C8002–C8017
    lockers C8002–C8017
    rest facilities C8002–C8017
    safe place of work C8002–C8017
    sanitary conveniences C8002–C8017
    temperature C8002–C8017
    washing facilities C8002–C8017
    weather, protection from C8002–C8017
    welfare facilities C8002–C8017
  Regulations IN21
  renewable energy structures C8505
  responsible person C8002–C8017
  rest facilities C8002–C8017
  risk assessment R3017
  safe place of work C8002–C8017
  sanitary conveniences C8002–C8017
  signing C8019
  site fire safety co-ordinator F5026
  site inductions and security C8019
  slips C8018
  temperature C8002–C8017
  training C8018, T7018trips
  typical project issues C8018
  vehicles A1014, C8002–C8017, C8018
  visitor protection C8019
  vulnerable persons V12003–V12005
  washing facilities C8002–C8017
  waste management C8020
  weather, protection from C8002–C8017
  welfare facilities C8002–C8017
  working platforms, *see* Work at Heights; Working Platforms

Index

CONSTRUCTION DESIGN
  AND MANAGEMENT
  REGULATIONS (CDM)
client
  definition C8505
  duties C8506
construction phase plan C8513
construction work
  definition C8505
contractor
  appointment C8507
  duties C8516
  meaning C8505
contractor, principal
  appointment C8607
  duties C8514
  engaging with workers C8515
  meaning C8505
corporate manslaughter C8520
Designer, principal
  appointment C8507
  duties C8512
  meaning C8505
designers
  duties C8511
  meaning C8505
domestic clients C8509
duty holders C8505
gas safety G1026
generally C8501, M2022
Health and Safety Executive (HSE)
  prosecutions C8521
health and safety file C8513A
information C8505
meaning of construction C8501
notifiable projects C8501
  conditions C8508
  employers C8501
  notifiable threshold C8518
  self-employed C8501
notification
  particulars C8518
pre-construction information C8505
principal contractor C8505
principal designer C8505
principles of prevention C8501
regulation, key IN21
self-employed C8514
welfare facilities C8518
CONSTRUCTION SITES *see* **Construction and Building Operations**
CONSTRUCTIVE DISMISSAL E14014
  E14019 V8010
  harassment, and H1720–H1721

CONSULTANTS
  health and safety IN26
CONSULTATION
  environmental reporting E16005
  food safety F9006, F9020
  generally J3001
  joint J3002
  regulatory framework J3002
  safety committees, *see* Safety Committees
  safety representatives, *see* Safety Representatives
  violence in the workplace V8002
CONSUMER AWARENESS E16012
  benchmarking E16012
CONSUMER PROTECTION
  food legislation and F9008, F9077
  product safety, *see* Product Liability; Product Safety
CONTAMINATED PREMISES
  cleaning of F9074
CONTAMINATION
  food, of F9011, F9036, F9038, F9047, F9074
  ionising radiation R1016
CONTINGENCY PLANNING M2037
CONTINUING PROFESSIONAL
  DEVELOPMENT T7021CONTRACT
  employment, of, *see* Contract of Employment
  facilities management F3022
  sale of goods legislation, *see* Sale of Goods
  unfair contract terms P9058–P9060
CONTRACT MEDICAL ADVISER
CONTRACT OF EMPLOYMENT
  additional duties, as to E14001
  collective agreements E14005
  consideration E14001
  disciplinary procedures E14006, E14008
  disciplinary procedures, *see* Disciplinary Procedures
  express terms E14002
  grievance procedures E14006, E14008
  implied duties E14003, E14006, E14008
  implied terms E14004
  other contracts contrasted E14001
  pension terms E14006
  sickness absence M1513, M1518
  sources of E14002–E14008
  statement of terms and conditions E14006
  statutory minimum notice E14002, E14018
  violence in the workplace and V8010
  working hours E14002, E14006
  works rules E14007, E14008

Ind-17

CONTRACTORS
asbestos A5012, A5043
community health and safety C6527
competence T7018
construction, design and management C8505–C8516
enforcing authority E15009
generally, *see* Construction and Building Operations
health and safety policy statements, approval systems under S7010
health and safety standards IN17
licensing A5027
major public events C6527
non-English speakers A1014
occupiers' liability O3015
permit to work systems S3015
public liability insurance P9528
risk assessment R3017
training and competence T7018
vicarious liability E13027

CONTRACTUAL LIABILITY
public liability insurance P9523

CONTRIBUTORY NEGLIGENCE IN22 O3011 P9028
children O3011
employers' liability insurance E13017
occupier's liability O3011
product liability, in cases of P9034

CONTROL DEVICES M1036

CONTROL MEASURES
asbestos A5030–A5031

CONTROL OF ELECTROMAGNETIC FIELDS REGULATIONS 2016
action level R1010.1
action plans R1010.5
employment protection E14032
equipment/work activities which may exceed ELVs R1010.3
EU directive on EMFs, implementation of requirements of R1008–R1009
exemptions R1010.2
exposure assessment R1010.4
exposure limit value R1010.1, R1010.3
further information R1010.10
health surveillance R1010.9
implementation R1010
information and training, employer to provide R1010.8
legal requirements R1010
low exposure equipment, list of R1010.3
risk assessment R1010.6
safety of employees R1010.7
training and information, employer to provide R1010.8

CONTROL OF ELECTROMAGNETIC FIELDS REGULATIONS 2016 – *cont.*
work activities/equipment which may exceed ELVs R1010.3

CONTROL OF LEAD AT WORK *see* Lead at Work

CONTROL OF MAJOR ACCIDENT HAZARDS REGULATIONS
see Major Accident Hazards Regulations

CONTROL OF SUBSTANCES HAZARDOUS TO HEALTH
see also Hazardous Substances
amendments H2145, H2146
biological agents, *see* Biological Agents
children and young people V12003–V12005
community health and safety C6506
dangerous substances D0901, D0903, D0910
lead at work, *see* Lead at Work
respiratory diseases O1046–O1055
vulnerable persons V12003–V12005

CONVICTION
health and safety offence, for publication of E15044

COOLING F5005–F5006, F9036

CORPORATE GOVERNANCE IN01–IN05
Cadbury Committee IN05
Combined Code E16005
compensation culture IN01
Financial Reporting Council (FRC) E16005
Greenbury Report IN05
Hampel Report IN05
large companies IN05
risk management IN05
Turnbull Report IN05

CORPORATE HOMICIDE (SCOTLAND).
see also Corporate Manslaughter
enforcement E15001, E15038.1
sentencing E15043
violence in the workplace V8008B

CORPORATE KILLING E15001 E15038 E15038 E13032 M2001
see also Corporate Manslaughter
Act, effect and requirements C9001
prosecutions E13001
small firms E13001

CORPORATE MANSLAUGHTER
Act, effect and requirements C9001, M2001
aggravating factors C9011
armed forces C9001, C9005, C9015
assessing seriousness C9011
Border Agency offices C9001, C9003

CORPORATE MANSLAUGHTER – *cont.*
causation C9002
charge E13032
charities C9002
child protection, functions relating to C9008
common law C9021, C9023
construction, design and management C8520
convictions C6504, C9029
corporations C9002, E15038
Crown bodies C9001, C9014
Crown immunity, none E15045
'culpability' C9011
custody
  Border Agency offices C9001, C9003
  Service premises C9001, C9003
directing mind C9001
directors, and E13032
disaster and emergency management systems D6029
DPP consent for proceedings C9001, C9020
duty of care C9001, C9002, E15038
  people owed, application of Act C9003
  relevant, meaning C9003
emergency services and emergencies C9007
employers' organisations C9002
enforcement E15001, E15038.1
evidence C9018
exclusively public functions C9004
exemptions C9001
fines C9011, E13032
government departments C9006, C9025, E15038
gross breach of duty C9001, C9002, C9009
gross negligence prosecutions C9021, C9023
guidance
  Crown Prosecution Service (CPS) C9028
  Health and Safety Executive (HSE) C9028
guidelines
  sentencing E13032
'harm' C9011
health and safety legislation, conviction under C9022
identification principle C9001
individual liability C9021
investigation C9026–C9027
law enforcement bodies C9006
management, senior C9002
management systems and practices C9002
mitigating factors C9011

CORPORATE MANSLAUGHTER – *cont.*
negligence C9001
no fault clause, where C9003
offence, generally C9002, E15038
organisation C9002
partnerships C9002, C9017, E15038
penalties C9001, C9010, E13032, E15043
  fines C9001, C9010, E13032, E15043
  publicity orders C9010, C9011
  remedial orders C9010, C9011
police forces C9001, C9002, C9006, C9016, E15038
probationary service C9008
procedure C9018
prosecutions
  generally C9026–C9027, P9548
  transfer of functions C9019
publicity orders C9013
public policy decisions C9001, C9004
remedial orders C9012
sentencing C9010, C9018, E13032, E15043
Sentencing Council C9010, E15001, E15043
sentencing guidelines IN13
seriousness of offence C9011
Service custody C9001, C9003
statutory inspections C9004
subcontractors C9002
territorial application C9024
trade unions C9002, E15038
transfer of functions C9019
violence in the workplace V8008B
voluntary organizations C9002
CORPORATE SOCIAL RESPONSIBILITY (CSR) IN05 E16002
BITC (Business in the Community) E16002
Companies Act 2006 E16005
CR Academy E16002
CR index E16002, E16007
international standard on E16002
International Strategic Framework for E16002
ISO international standard E16002
Memorandum of Understanding E16002
OSH practitioners and E18012
training and qualifications E16002
UN Global Compact Office E16002
CORROSIVE SUBSTANCES
manual handling M3002
COSTS
director's liability E15041
employers' liability insurance E13017
environmental management E16002

COSTS – *cont.*
  public and products liability
    insurance P9552, P9554
COSTS OF ACCIDENTS AND ILL-
    HEALTH IN24
COUNSELLING D6060–D6062 S11009
COURT OF APPEAL
  criminal jurisdiction IN18
CR ACADEMY
  services E16002
CRAMP O1020.1
CRANES L3029–L3031
  appointed persons L3030
  collapse L3029
  drivers L3030
  failure of L3031
  mobile
    offshore operations O7070
    safe lifting by L3030
  overturning L3029
  registration scheme L3029
  risk assessment L3018
  signallers L3030
  slingers L3030
  supervisors L3030
  types of L3029
    tower crane L3029
  uses of L3029
CRC ENERGY EFFICIENCY SCHEME
  reporting E16024
CREDIT SALE AGREEMENT
  breach of statutory duty,
    for E15021–E15035
  product liability P9051
CRIMINAL INJURIES COMPENSATION
    AUTHORITY (CICA)
  work-related violence V8004
CRIMINAL OFFENCES
  breach of statutory duty,
    for E15021–E15035
  burden of proof E15031
  corporate manslaughter E13032, IN05
    see also Corporate Manslaughter
  custodial sentences IN05
  employers' liability insurance E13008
  fire prevention and control F5060
  food, see Food
  harassment H1729
    battery H1730
    Crime and Disorder Act 1998 H1733
    Protection from Harassment Act
      1997 H1731
    Public Order Act 1986 H1730
  human rights E15031
  innocence, presumption of E15031

CRIMINAL OFFENCES – *cont.*
  lifting L3017
  lifts L3008
  machinery safety M1018
  penalties IN05, IN13, IN18, IN22
  personal protective equipment P3027
  process IN18
  public service cases E15036
  workplace violence V8011
CR INDEX
  reporting CSR E16002, E16007
CRISIS
  communication B8038–B8042
CROWN COURT
  jurisdiction IN18
CROWN THE
  enforcement E15009, E15013
  fire prevention and control F5061
  health and safety law and E15045
  immunity E15045
CRUSHING M1025
CSR see Corporate Social Responsibility
CULLEN REPORT
  see also Offshore Installations O7003,
    O7046
CUSTODIAL SENTENCES IN05
CUSTOM BUILT ARTICLES P9020
CUSTOMERS
  benchmarking E16012
  community health and
    safety C6515–C6517
  consumer awareness E16012
  customer base, erosion of B8017
  facilities management F3002
  precautions C65016
  small businesses IN04
CUTTING AND SEVERING M1025
CYBER SECURITY
  business continuity B8050
  risk guidance B8051
CYPRUS
  professional regulation E18017

D

DAIRY
  registration F9028
DAMAGES
  assessment C6039, IN22, R2004
  awards C6024, C6044
  bereavement, for C6031
  categories of injuries C6043
  coal miners V5003, V5012

DAMAGES – *cont.*
  contributory negligence C6046
  cost of care C6054
  damage-based agreements C6021
  deductible payments C6051, C6052
  director's liability E15041
  employers' liability insurance E13002
  fatal injuries C6036–C6039
  general C6024
  income tax status H1718
  interest on C6043
  interim awards C6041
  loss of amenity C6034
  loss of future earnings C6026
  lost years C6029
  noise N3012
  non-pecuniary losses, *see* Non-pecuniary losses
  occupational health and safety O1005
  occupational injuries and diseases
    basis of an award C6022
    capitalisation of future losses C6027
    hazardous substances, as to H2103
    loss of future earnings C6026
    non-pecuniary losses C6023, C6030
    pecuniary losses C6023, C6030
    prescribed diseases, for C6055–C6058
    structured settlements C6047
  pain and suffering C6031–C6033
  payments into court C6042
  pecuniary losses, assessment of C6026
  personal injuries at work
    accident, caused by C6005
    basis of award C6022
    fireman C6005
    non-pecuniary losses C6023
    pecuniary losses C6023
  personal protective equipment P3027
  prescribed diseases C6055–C6058
  product liability, *see* Product Liability
  psychiatric injury S11009
  public and products liability insurance P9501, P9552
  recovery of state benefits
    appeals against certificates of recoverable benefits C6050
    compensator's duties C6045
    complex cases C6046
    contributory negligence C6046
    exempt payments C6044
    information provisions C6045, C6046
    new regulations C6044
    rationale C6044
    retrospective effect of provisions C6044

DAMAGES – *cont.*
  recovery of state benefits – *cont.*
    structured settlements C6025, C6046
  rehabilitation R2004, R2006
  set-off
    deductible items C6042
    non-deductible items C6042
  special C6024
  stress S11009–S11009.2
  structured settlements C6025, C6047
  vibration white finger V5003, V5012, V5013
  wages, loss of C6053
  wrongful dismissal, for E14017
DANGER
  areas M2011
  serious and imminent danger, procedure for M2011
DANGEROUS ACTIVITIES
  community health and safety C6510, C6511
  neighbours and passers-by C6510
  planning C6511
DANGEROUS GOODS *see also* **DANGEROUS SUBSTANCE**
  carriage of IN21
DANGEROUS MACHINERY *see* Machinery Safety
DANGEROUS PRODUCTS
  liability for P9054
DANGEROUS SUBSTANCES
  see also Hazardous Substances; Major Accident Hazards Regulations; Transportation of Dangerous Substances
  Approved Codes of Practice D0902, D0934
  case histories D0901
  chemical agents D0901, D0908
  children and young people V12016
  cleaning D0933
  clothing D0923
  communication systems D0933
  community health and safety C6510
  containers D0933
    identification of hazardous contents of D0927
  control measures D0915, D0933
  Control of Lead at Work Regulations 2002 H2140
  control of substances hazardous to health D0901, D0903, D0910, D6015
  Dangerous Substances and Explosive Atmospheres Regulations 2002 D0901–D0917

DANGEROUS SUBSTANCES – *cont.*
  Dangerous Substances and Explosive
    Atmospheres Regulations 2002 –
    cont.
    application, general conditions
      for D0905
    commencement D0909, D0924
    exemptions D0928
    fire prevention and
      control F5062–F5072
    modernisation of legislation,
      amendments to D0929
    repeals and revocations of existing
      legislation D0930
    scope of D0903
  definition D0908
  design of plant equipment and
    workplaces D0933
  dusts, safe handling of combustible D0933
  duty holders D0906
  elimination of risk D0914
  emergencies, dealing with D0925, D0933
  employees
    information D0933
    training D0933
  enforcement of legislation D0931
  equipment D0920–D0924
  EU laws D0901
  explosive atmosphere D0920–D0921,
    H2141–H2143
  fire F5031
    risk of D0910
    safety legislation H2101,
      H2141–H2143
  garages D0901
  gas safety G1026
  general safety measures D0917
  guidance D0933–D0933, D0934
  Health and Safety Executive
    guidance D0914, D0933
  heat, sparks and flame, generation
    of D0933
  hot work D0933
  ignition sources D0933
  incidents, dealing with D0925
  information provision D0926, D0933
  inspection D0933
  instructions D0926
  labelling D0933
  leaflet D0933
  liquefied petroleum gas, unloading
    of D0933
  maintenance D0933
  mitigation measures D0916, D0933
  motor vehicle repair garages D0901

DANGEROUS SUBSTANCES – *cont.*
  NAMOS Regulations D0909
  national security D0928
  Permit-to-Work system D0933
  personal protective equipment D0906,
    P3017
  petrol D0901
    road tankers, unloading D0933
  petroleum D0929
  pipes, identification of hazardous contents
    of D0927, D0933
  plant equipment, design of D0933
  precautions D0920
  protection systems D0920–D0924
  radiation emergencies, *see* Radiation
    Emergencies
  radioactive substances, *see* Radioactive
    Substances
  recording findings D0911
  reduction of risk D0914
  regulatory impact assessment D0932
  repairs D0933
  risk assessment D0902, D0910–D0911,
    D0933, R3007, R3014, R3035
    equipment D0920
    guidance D0933
    reviews of D0912
  risk management D0902, D0910,
    D0913–D0917
  road tankers, unloading D0933
  ships D0904
  sites, notification and marking of H2144
  sources D0934
  storage D0933
  training D0926, D0933, T7005
  typical activities D0904
  ventilation
    Approved Code of Practice D0933
  verification of safety prior to first
    use D0922
  warning signs D0921, D0933
  waste D0933
  workplaces
    definition of D0907
    design of D0933
DATA
  ergonomic design and E17057, E17058
  workplace violence V8023
DATABASE
  Employers' Liability Tracing
    Office E13026
DATA PROTECTION
  occupational health and safety O1017
  rehabilitation R2007
  sickness absences R2007

DATA PROTECTION – *cont.*
  violence in the workplace V8081
DEAFNESS
  occupational
    compensation for C6056
    generally, *see* Noise-induced Hearing Loss
DEALER
  product safety P9052
DEATH
  see also Fatal Injuries
  asbestos related A5001
  corporate manslaughter E13032
    see also Corporate Manslaughter
  employers' liability insurance E13025
  reporting A3018
  work-related
    single greatest cause A5001
DECLARATIONS
  Declarations of Conformity
    European Union M1015–M1017
    example of M1015–M1017
    incorporation into other machinery M1015–M1017
  machinery safety M1015–M1017
DECOMPRESSION PROCEDURES
  work in compressed air, and C8018
DEDUCTIONS FROM WAGES E14014
DEEPWATER HORIZON
  oil rig explosion E16010, O7004.1
DEFECTIVE PRODUCTS
  liability for, *see* Product Liability
DEFENCES
  food legislation, under F9022, F9023, F9070, F9082
  HSWA, under E15029, E15031
  occupier's liability O3017
  offshore operations O7056
  public and products liability P9511, P9522
DELIVERY VEHICLES
  traffic routes for A1001, A1005, A1006
DEMOLITION
  building and construction sites C8002–C8017
DEMOTION E14014
DEMS *see* Disaster and Emergency Management Systems
DEPARTMENT FOR ENERGY AND CLIMATE CHANGE (DECC)
  environmental management E16024
DEPARTMENT FOR ENVIRONMENT FOOD AND RURAL AFFAIRS (DEFRA)
  environmental management E16024

DEPARTMENT OF HEALTH
  food, role as to F9005
DEPRESSION O1020.6–O1020.9
DE QUERVAIN'S DISEASE O1020.1
DEREGULATION
  self-employed IN01
  strategy IN19, P3008
DERMATITIS H2103 O1037 O1038 O1040 O1041 P3026
DESIGN
  dangerous substances D0933
  designers M1002, M1009
  electricity E3012–E3020, E3025, E3031
  ergonomic, *see* Ergonomic Design
  facilities management F3028, F3031–F3032
  fire prevention and control F5027
  floor M3014
  liability risk exposure P9544
  lifts L3009, L3011
  lighting E17031, L5014
  machinery safety M1002, M1003, M1009, M1026
  manual handling M3009, M3012, M3014
  plant equipment D0933
  pressure systems, and P7016
  processes or activities M3009
  role of E17001, E17002
  stress S11013
  task layout M3012
  violence in the workplace V8058–V8064, V8079, V8109
  workplaces D0933
DESIGNER
  product safety, as to, *see* Product Safety
DETERRING/IMPEDING DEVICES M1036
DEVELOPMENT
  competence T7020
DEVELOPMENT RISK
  defence, as P9045
DIFFUSE MESOTHELIOMA
  2008 Diffuse Mesothelioma Scheme C6058
DIGGING
  public liability P9523
DIGNITY
  harassment, violation H1701, H1708
  purpose or effect H1707
DILUTION VENTILATION V3014
DIPLOMAS
  Qualifications Credit Framework T7009
DIRECTIVES EU *see* European Union (EU Laws)
DIRECTORS
  civil liability E13008, E15041

DIRECTORS – *cont.*
  criminal liability E15040
  disclosure E16005
  employers' liability insurance E13008
  environmental management E16005
  Gross Negligence Manslaughter E13032
  health and safety responsibilities E15040, M2030
  indemnities E15041
  insurance E15041
  large companies IN05
  liability insurance E13023
  offences committed by E15032
DIRECT SALES E13003
DISABILITY DISCRIMINATION
  adjustments, reasonable A1001.1, R2005
  association, discrimination by M1508
  auxiliary aid M1511
  avoiding V12037
  codes of practice A1001.1
  disabled person, meaning M1506
  'discrimination arising from disability' R2005
  duty to make reasonable adjustments E16511, R2005
  Equality Act 2010 A1001, A1001.1, E16509–E16512
    disability, definition E16510, M1506
    discrimination by association M1508
    rehabilitation R2005
    sickness absence M1505
  ergonomics E17070
  European Law compatibility R2005
  harassment H1703
  health and safety excuse R2006
  indirect discrimination M1508, M1510
  job applicants V12037
  legal rights W11043
  long-term R2005
  musculoskeletal disorders O1020.4
  protected characteristic H1703, R2005
  reasonable adjustments A1001.1, E17070, M1511, R2005
  sickness absence M1506–M1511A
  unfair dismissal, and E14020–E14022
  workplace, in W11043
DISABLED PERSONS V12032–V12037
  access, facilities for A1001, R3028, V12034
  'associated with' M1508
  conditions M1509
    sight impairment exception M1509
  definition of A1001.1, E16510, M1506
  disablement benefit, *see* Disablement benefit

DISABLED PERSONS – *cont.*
  discrimination, *see* Disability discrimination
  disfigurement M1506
  duty to make reasonable adjustments E16511, R2005
  Equality Act 2010 E16501, R2005–R2006
  fire safety, means of escape F5061.1
  HSE guidance E16511
  impairment E16510
  long term E16510
  normal day to day activities R2005–R2006, V12032
  person associated with employee M1511
  rehabilitation R2005
  risk assessment E16512, R3028, R3032, V12033, V12035–V12036
  sickness absence M1506–M1511A
  types of disability V12032
  working tax credit C6016
DISABLEMENT BENEFIT
  occupational deafness, for
    assessment of benefit N3021
    conditions under which payable N3020
    entitlement to N3002, N3018
    statutory requirements N3012
DISASTER AND EMERGENCY MANAGEMENT SYSTEMS D6001–D6071
  access D6010
  active policy and pre-planning D6056
  audits D6067
  building contractors D6060–D6062
  business continuity D6055, D6060–D6062
  business management D6002
  CIMAH, pre D6001
  civil contingencies D6001
  Civil Contingencies Act 2004 D6005–D6008
  codes D6056
  command and control chart of arrangements D6041
  communication, establishing effective D6044
  competence D6046
  competent authority D6011–D6014
  conclusion D6070
  confined spaces D6024
  construction industry D6022–D6023
  contingency planning D6006
  control D6047
  co-operation D6007–D6008, D6045
  corporate culture and practice D6037
  corporate manslaughter D6029, E13032
    see also Corporate Manslaughter

Index

DISASTER AND EMERGENCY MANAGEMENT SYSTEMS – *cont.*
corporate reasons D6031
counselling support D6060–D6062
criminal offences D6029
dangerous substances D6009, D6017
definitions D6003
design and architecture D6037
effective DEM, need for D6004–D6032
   legal reasons D6005–D6029
emergency, meaning D6006
employers' duties D6010
enforcement D6032
environment D6033
   Environment Act 1995 D6028
establishing policy on D6041–D6042
European Union D6001, D6011–D6014
   radiation D6018
exclusions D6016
exercises D6054
explosives D6009, D6017
external factors D6035, D6036
Federal Emergency Management Agency (FEMA) D6001
first aid D6021
hazardous substances D6015
historical development D6001
holistic phase D6001
human error D6037
humanitarian reasons D6032
information
   public D6018, D6019
   radiation D6019
   sources of D6036, D6071
inquiries D6060–D6062
insurance D6002, D6030, D6060–D6062
internal factors D6035, D6037
investigations D6059, D6060–D6062
lead categories D6007–D6008
liberalisation D6001
local authorities D6011–D6014
loss adjusters D6060–D6062
lower tier duties D6017–D6020
Major Accident Prevention Policy (MAPP) D6011–D6014, M1117
major disasters E17002
   arrangements for D6041–D6042
management failure D6002
Management of Health and Safety at Work Regulations D6010
manslaughter D6029
maritime and coastal emergencies D6006
monitoring D6066

DISASTER AND EMERGENCY MANAGEMENT SYSTEMS – *cont.*
national security D6007–D6008
Northern Ireland D6007–D6008
notifications D6020
occupational health and safety D6002
organisation D6043–D6047
origins of D6002
phases D6001
Piper Alpha D6029
plans D6007–D6008
   after the event D6057–D6058
   before the event D6049–D6055
   decisional planning D6051
   during the event D6056
   generally D6048
   long-term D6063–D6064, D6065
   medium-term D6063–D6064
   monitoring D6066
   short-term D6060–D6062
   testing D6052
police D6060–D6062
policy
   establishing D6038–D6042
   statement of D6039
   summary D6042
procedures D6010
radiation D6018
   public information D6019
   regulations D6018
reporting duties D6020, D6035
responders D6007–D6008
responsiveness D6006
reviews D6067–D6069
risk assessment D6007–D6008, D6010, D6066
road transport D6026–D6027
safety reports D6021, D6022–D6023, D6024
   contents D6021
   preparation D6021
   purpose of D6022–D602
   review and revision of D6024
Scottish Environment Protection Agency D6011–D6014
security management D6002
sensitive information D6007–D6008
September 11, 2001, terrorist attacks D6001
Seveso III Directive D6011–D6014
   future developments M1141
testing D6052
top-tier duties D6021–D6028
training D6010, D6053

DISASTER AND EMERGENCY MANAGEMENT SYSTEMS – cont.
transport
   road D6025–6027
United States D6001
upper tier sites D6011–D6014
visitors D6060–D6062
warnings D6007–D6008

DISASTERS
see also Disaster and Emergency Management Systems
business continuity
   planning B8001–B8003, B8008, B8023
natural B8008
public relations B8009
recovery B8001

DISCIPLINARY PROCEDURES
ACAS procedures E14013, E14014
accident investigations A3029
appeal E14013
contract of employment, in E14006, E14008
essential elements E14013
hearing E14013
investigation E14013
rules, scope of E14012
sanctions
   demotion E14014
   dismissal, see Dismissal
   fines or deductions E14014
   suspension E14014
   warning E14014
sickness absence M1529

DISCRIMINATION
see also Harassment
age H1702–H1703
agents E16501
associative H1703, H1707
colour H1703
compensation E16522
direct E16501
   definition E16503
   test E16503
disability E16501, E16506, H1703
employees E16501
employers, liability E16501
Equality Act 2010 E16501–E16522
gender reassignment H1703
general claim H1703
history H1701–H1702
indirect E16501, E16504
marriage H1703

DISCRIMINATION – cont.
maternity H1703
nationality H1703
perceived H1703, H1707
pregnancy H1703
principals, liability E16501
racial H1703
remedies E16522
sexual H1703
training H1739
unlawful H1717, H1718

DISEASES
accident reporting A3002, A3016
asbestos-related, see Asbestos-related Diseases
cancer, see Cancer
compensation for workplace diseases, see Compensation for Work Injuries/Diseases
livestock H2103
new or expectant mothers V12018
non-ionising radiation R1008–R1009
pneumoconiosis, see Pneumoconiosis
prescribed industrial, benefits for C6010
record keeping duty A3016
specified, claims for C6048

DISFIGUREMENT
disability M1506

DISMISSAL
Code of Practice on Disciplinary and Grievance Procedures M1517
constructive E14014, E14019
health and safety grounds E14010
limited-term contract E14019
sickness absence M1516–M1517, M1530
unfair, see Unfair Dismissal
wrongful E14017, E14018

DISPLAY SCREEN EQUIPMENT see also Visual display units
ergonomics E17064
HSE guidance E17064
Regulations IN21

DIVERSITY
HSE stance E16501

DIVING OPERATIONS
equipment P3017
manual handling M3002
new or expectant
   mothers V12019–V12020
offshore operations
   diving contractor O7037
   general requirements O7036
   planning, dive O7038
   risk assessment O7038
safety devices for P3017

Index

DIVING OPERATIONS – *cont.*
  training and competence T7005
DOCTOR
  training W10026, W10031, W10034
  working hours W10003, W10012, W10031
DOMESTIC SERVICE
  working time W10032
DOORS AND GATES
  workplaces, safety regulations W11001, W11016
DOW JONES SUSTAINABILITY INDEX
  socially responsible investment E16007
DRAINS
  fire fighters O3014
  occupiers' liability O3014
DRILLING MACHINES M1030
DRINKING WATER
  construction and building operations, and C8002–C8017
  general provision W11024
DRIVERS
  access, when using vehicles, *see* Access
  training, generally A1013
  whole-body vibration V5023–V5024, V5026
DRIVE SHAFTS M1008
DROWNING
  prevention of C8002–C8017, P3017, P3024
DRUGS
DRY POWDER FIRE EXTINGUISHERS F5013
DUCTS F5007
DUST
  dangers of exposure to D0933
  explosions D0919, D0933
  fire D0919
  nuisance E5016
  safe handling D0933
  ventilation, *see* Ventilation
DUTY
  levels of IN20
DUTY OF CARE
  corporate manslaughter C9003
  occupational health and safety O1005
  rehabilitation R2004
  'relevant duty of care' C9003
  right to roam O3007
  trespassers O3007
  violence in the workplace V8004–V8005

E

EAR
  noise-induced hearing loss, *see* Noise-induced Hearing Loss
EARTHING E3015
EATING FACILITIES
  provision of W11027
ECO-LABEL SCHEME
  supply chain E16008
ECO-MANAGEMENT AND AUDIT SCHEME (EMAS) E16014 E16026
  differences with ISO 14001 E16016
ECZEMA P3026
EDUCATION
  see also Training and competence
  framework T7007–T7010
  higher T7010
  safety representatives, role J3002
EDUCATIONAL PREMISES
  fire safety risk assessment F5061.1
EFAW
  emergency first aid at work
    training F7001–F7001.1, F7018–F7018.3
EGGS
  production of F9030
EGRESS
  safe place of work C8002–C8017
ELECTRICAL HAZARDS
  confined work space, in A1022, A1023
  fire F5004, F5015
    extinguishers F5004, F5015
  public liability insurance P9528
  temporary installations C6516, C6517
ELECTRICITY
  access E3020
  adverse or hazardous environments E3013
  alternating current E3002
  arcing E3008
  assessing electrical risks E3009A
  ATEX Directives E3027
  battery rooms E3027
  Building Regulations W11044
  buildings E3033
  burns E3006
  cables E3028
  community health and safety C6510
  competence E3011, E3024
  conductors E3016
  confined spaces C7020, C7026
  connections E3017
  construction of system IN21
  current, measurement of E3002
  dangers E3004–E3009A

Index

ELECTRICITY – *cont.*
  design E3031
  design and construction E3012–E3020, E3025
  direct current E3002
  disturbances, damage E3002, E3032watts
  dwellings E3033
  earthing E3015
  Electricity at Work Regulations 1989 IN21
  electric shocks E3005, E3014–E3015, E3018
  electromagnetic damage E3032
  equipment E3003
    confined spaces C7020, C7022
    dangerous substances D0920
    Electrical Equipment (Safety) Regulations 2016 E3031
    explosions D0920
    isolated equipment, work on E3023
    live work E3024
    potentially explosive atmospheres E3027
    strength and capability E3012
  excess current protection E3018
  explosions D0920, E3009, E3027
  fatal incidents E3001
  fencing E3014
  fire E3007
  flashovers E3008, E3009
  guidance E3003, E3021, E3022, E3028
  injuries E3004
  insulation E3014, E3015, E3024
  isolation E3019, E3023
  legal background E3003
  lighting E3020
  lightning protection systems E3025
  live parts and work E3004, E3005, E3011, E3024
  maintenance E3021, IN21
  national vocational qualifications (NVQs) E3011
  offshore operations O7063
  Ohm's Law E3002
  overhead lines E3028
  permit-to-work E3023
  plugs E3029–E3030
  portable electrical appliances E3021
  potentially explosive atmospheres E3027
  precautions E3014–E3015
  public liability insurance P9528
  qualifications E3011
  quality E3032
  quarries E3003
  rectification process E3002
  referenced conductors E3016

ELECTRICITY – *cont.*
  residual current drive E3016
  resistance E3002
  risk assessment E3009A, R3016 App B
  risks, controlling E3010
  safe systems of work E3022–E3024
  shocks E3005, E3014–E3015, E3018
  sockets E3029–E3030
  standards E3003, E3024, E3027
  static electricity C7026, E3026
  trade associations E3011
  training E3011
  understanding E3002
  voltage E3004, E3014, E3031
  wiring regulations E3003
  working space E3020
ELECTROMAGNETIC FIELDS (EMF)
  European directive IN07–IN08, R1008–R1009
  non-ionising radiation R1002, R1004, R1006–R1010.7, R1010.9, R1010.10
ELECTROMAGNETIC RADIATION R1008–R1009 V12019–V12020
EMAS (ECO-MANAGEMENT AND AUDIT SCHEME) E16014 E16026
  differences with ISO 14001 E16016
EMERGENCIES
  corporate manslaughter C9007
  vibration, emergency certificates and V5015
EMERGENCY ARRANGEMENTS
  business continuity planning B8023
  community health and safety C6511, C6517
    customers and users, arrangements for C6517
    equipment C6514
    volunteers C6522
  confined spaces C7006, C7028
  customers and users, arrangements for C6517
  dangerous substances D0925, D0933
    carriage of *see* Transportation of Dangerous Substances
  equipment C6514
  escapes D0925
  explosions D0925
  heights, working at W9006
  managing health and safety M2039
  noise N3041
  sound systems N3041
  violence in the workplace V8094, V8135
  volunteers C6522
  warnings D0925

Ind-28

EMERGENCY EXITS
  fire prevention and control F5040
EMERGENCY PLANS
  see also Disaster and Emergency Management Systems; Emergency Procedures
  ionising radiation R1021
  major accident hazards regulations M1122–M1126
EMERGENCY PROCEDURES
  asbestos, release of A5033
  construction sites F5026
  disaster and emergency management systems, *see* Disasters and Emergency Management Systems
  fire prevention and control F5026
  food safety F9015, F9018, F9019
  hazardous substances, where H2139
  lone workers V12029, V12030
  offshore installations O7028
  radiation emergencies, *see* Radiation Emergencies
EMERGENCY PROHIBITION NOTICES AND ORDERS F9018, F9019
EMERGENCY SERVICES
  corporate manslaughter C9007
  employer's duties as to R3002, R3018, R3029
  managing health and safety M2012
EMERGENCY WORKERS
  legal protection from assault V8008A
EMISSIONS INTO THE ATMOSPHERE
  acidification E5001, E5015
  air quality assessment
    daughter directives E5033–E5039
    framework directive E5033
  clean air strategy, new E5071
  climate change, *see* Climate change
  CLRTAP, role of E5021, E5026
  coal burning E5002, E5061
  dark smoke E5062
  distance scales, range of E5001
  ecosystems, damage to E5015
  Environment Agency
    local authorities and E5046
    PPC
    role of E5043
  Europe
    air quality assessment, *see* air quality assessment, *above*
    BREF Notes E5028
    IPPC directive, *see* integrated pollution prevention control, *below*
    national emissions ceilings E5037
    policy makers, European E5026

EMISSIONS INTO THE ATMOSPHERE – *cont.*
  Europe – *cont.*
    protocols E5026
    solvent emissions E5026
    UNECE, role of E5021, E5026
    waste incineration directive E5026
  eutrophication E5015
  furnaces E5061
  ground-level ozone E5015
  human health
    effect on E5002
  integrated pollution control ('IPC')
    predecessor of IPPC, *see* integrated pollution prevention control E5061
  integrated pollution prevention control ('IPPC')
    application of E5027
    BAT concept E5027.3, E5028
    BREF Notes E5028
    European Bureau E5028
    Industrial Emissions Directive E5027.2
    role of E5027
    UK law, introduction into E5041
  local authorities
    Environment Agency and E5046
  National emission reduction ceilings directive (NECD) E5021
  nuisance E5016, E5062
  permits
    BAT concept, based on E5028
  pollutants
    ammonia E5002
    arsenic E5033
    benzene E5033, E5036
    cadium E5033
    carbon monoxide E5002, E5036
    dioxins E5026
    examples of E5001
    lead E5033
    medical effects of E5002
    nickel E5033
    nitrogen dioxide E5002, E5026, E5033
    non-methane volatile organic compounds E5002
    ozone E5002, E5033, E5037
    particulate matter E5002, E5033, E5034
    persistent organic E5002–E5006
    polycyclic aromatic hydrocarbons E5033
    sulphur dioxide E5002, E5033
  public liability insurance P9526, P9528
  regulation
    regulatory drivers E5002–E5016

Ind-29

EMISSIONS INTO THE ATMOSPHERE – cont.
  reporting E16024A
  smog E5002
  smoke E5061, E5062
  stratospheric ozone depletion E5024, E5025
  trans-boundary pollution E5021, E5064
  UNECE, role of E5021, E5026
  United Kingdom
    air quality standards E5064–E5067
    clean air legislation E5061
    Environment Agency, *see* Environment Agency, *above*
    local authorities, *see* local authority, *above*
    Northern Ireland E5045, E5069
    nuisance E5016, E5062
    regulators E5044–E5047
    Scotland E5044
    World Health Organisation guidelines E5064–E5065
EMISSIONS TRADING E16006
  EU Emissions Trading System E16006
EMPHYSEMA
  compensation claim C6055
EMPLOYEES
  see also Safety Committees; Safety Representatives
  accident reporting A3023–A3036
  competence T7001, T7006
  consultation with IN21
  contract, *see* Contract of Employment
  dangerous substances D0933
  defective equipment supplied to P9020
  disciplinary procedures, *see* Disciplinary Procedures
  duties of
    control of hazardous substances H2105, H2127, S7005
    manual handling M3002
    personal protective equipment, as to P3007
    statutory duties IN19
  employment protection rights E14009
  enforcement of health and safety laws against E15040–E15042
  Equality Act 2010 E16501
  heights, working at W9013
  information D0933
  managing health and safety M2017
  manual handling M3002
  negligence, *see* Negligence
  representatives T7017

EMPLOYEES – *cont.*
  safety rights
    not to be dismissed on certain grounds E14010, E14011, E14022
    not to suffer a detriment, *see also* Employment Protection E14009
  small businesses, focus on IN04
  suspension from work, *see* Suspension from Work
  training and competence D0933, IN21, T7001, T7006
  vibration V5015
  women workers, *see* Women Workers
  workers, distinction E14001
  workplace violence, *see* Workplace Violence
EMPLOYERS
  work-related road safety, duty to manage M2102
EMPLOYERS' LIABILITY INSURANCE E13001–E13037
  admission of liability E13029
  agency workers E13016
  apologies E13033
  apportionment E13026
  asbestos E13001, E13026
  assessment of premium E13022
  Association of British Insurers E13028
  asthma E13026
  benchmarking E13030
  book rate E13022
  breach of statutory duty E13002
  brokers E13003
    Financial Services Authority E13003
    proposal forms, completion of E13003
    regulation of E13003
  burden of proof, reversal of E13002
  business description, material fact E13021
  cancer E13026
  caveat emptor E13004
  certificates of insurance
    display E13013
    issue E13013
    retention E13013
  children E13016
  Civil Procedure Rules E13029
  claims handling E13017, E13033
  commercial insurance E13003
  compensation E13002, E13017
    Compensation Act E13033
    compensation culture E13001
  conditions of cover E13018–E13019
    prohibition of certain conditions E13020
  contributory negligence E13017

EMPLOYERS' LIABILITY INSURANCE – *cont.*
  corporate manslaughter E13001, E13032
    sentencing guidance E13032
  costs E13017
  cover provided by policy E13018–E13019
    business cover E13016
    compensation E13017
    conditions E13018–E13019
    costs E13017
    FSCS scheme. E13024
    geographical limits E13018
    interpretation of E13016
    legal expenses E13023
    necessary E13011
    persons E13016
    related covers E13023
    scope of cover E13017
  criminal offences
    corporate killing E13001
    failure to insure E13009
  damages E13017
  database E13026
  death of claimant E13025
  defence E13002
  degree of cover necessary E13011
  directors' personal liability E13008, E13023
  direct sales E13003
  disclosure, duty of E13004
  display of certificate E13013
    failure E13015
  dose, relevance E13026
  duty to take out E13008, IN22
  employees
    children and young people E13016
    consultation with IN21
    contributions, from E13017
    covered by Act E13009
    definition E13016
    duty of employers to E13002
    indemnities, from E13007
    not covered by Act E13010
  Employers' Liability (Compulsory Insurance) Regulations 1998 IN21
  Employers' Liability Tracing Office (ELTO) E13026, E13034
  evidence, asked to produce E13014
  failure
    criminal offences E13009
    display, to E13015
    insure, to E13001, E13014
    long development period, diseases with E13001

EMPLOYERS' LIABILITY INSURANCE – *cont.*
  failure – *cont.*
    maintain insurance, to E13014
  Fees for Intervention (FFI) E13037
  fire
    mitigation E13006
    sprinklers, installation of E13006
  first-aid F7012
  generally E13001, IN22
  geographical limits E13018
  good faith E13004
  hazardous work, conditions imposed for E13021
  HSE intervention E13037
  identification of insurer E13026
  indemnities E13007, E13024
  industrial diseases
    death of claimant E13025
    limitation periods E13025
  insolvent employers E13024
    indemnities E13024
    revival of dissolved companies E13024
  insurance contracts, law relating to E13003–E13006
    disclosure, duty of E13004
    loss mitigation E13006
    mitigation of loss E13006
    proposal form, filling in E13005
  issue of certificate E13013
  key principles E13035
  knowledge
    date of E13025
  legal expenses insurance E13023
  legal liability policy, as E13002
  limitation periods E13025
    accrual, date of E13025
    death of claimant E13025
    industrial diseases E13025
    knowledge, date of E13025
    mesothelioma E13025
  Löfstedt review E13036
  long tail diseases E13026
  Lord Young Report E13035
  loss mitigation E13006
  lung cancer E13026
  maintain, duty to E13008, E13014
  Making the Market Work Scheme
    no longer in operation E13028
  measure of risk E13022
  medical evidence E13029
  mesothelioma E13025, E13026, E13033
  motor claims, notification E13029
  NHS recoveries E13031

Index

EMPLOYERS' LIABILITY INSURANCE – *cont.*
noise E13026
notification of claims E13017
offshore operations E13001
penalties E13014, E13015
   display, failure to E13015
   failure to display E13015
   failure to insure E13014
   failure to maintain E13014
   insure, failure to E13014
   maintain, failure to E13014
performance indicators E13030
persons covered by policy E13016
Policyholders Protection Board E13024
Portal, notification E13029
potential liability E13033
pre-action protocol E13029
   voluntary nature E13029
premises, surveys E13022
premiums E13022
proposal form, filling in E13005
public liability policy E13002, E13016
purpose E13002
restrictive conditions E13020
retention of certificate E13013
risk, measure of E13022
scope of cover E13017
settlements E13017, E13029
small and medium-sized enterprises (SMEs) Performance Indicator E13030
Small Businesses and the Voluntary Sector document E13035
sole-employee incorporated companies (SEICs) E13012
statutory duty to take out E13001
subrogation E13007
tracing employers E13026
tracing insurers E13001
trade endorsements for certain types of work E13021
treatment, offer of E13033
uberrima fides E13004
vibration E13026
vicarious liability E13002, E13027
visits E13022
voluntary workers E13016
Woolf reforms E13017, E13029
young people E13016

EMPLOYERS' LIABILITY TRACING OFFICE (ELTO)
database E13026, E13034

EMPLOYERS *see also* **Employers' Liability Insurance**
asbestos A5007–A5019

EMPLOYERS *see also* **Employers' Liability Insurance** – *cont.*
breach of contract H1720
compensation *see* Compensation
competence, responsibility for T7006
confined spaces C7003
constructive dismissal H1720
contractors T7018
delegation to junior staff E15039
disaster and emergency planning D6010
discriminatory acts of employees E16501
duty of care E14003, IN22, M1524
employees' safety E14009
employees who drive road vehicles M2103
enforcement
   safety rules E14012, E14021
   unlawful discrimination E16522
Equality Act 2010 E16501
Europe E18011
fines IN18
   see also PENALTIES
harassment, liability
   causal link with employment H1712
   course of employment H1711
   defence H1713
   legal action H1716–H1733
   narrowing extent H1715
   putative agent H1710, H1715A
   remedies H1718, H1721, H1725, H1728
   third party, protection H1703
   tort H1717, H1722–H1725
   unfair dismissal H1719
   unlawful discrimination H1717, H1718
   vicarious liability H1710
harassment, policy H1734–H1740
   complaints procedure H1735
   confidentiality H1736
   formal policy H1738
   informal action H1737
   monitoring and review H1740
   training H1739
hazardous substances, where H2105
health and safety legislation, under R3001–R3006
health and safety policy statement, *see* Health and Safety Policy Statement
insurance, as to E13002
job applicants O1003
lighting, as to L5002
managing health and safety M2004
manual handling M3002
   duty of care IN22
   vicarious liability IN22

EMPLOYERS *see also* Employers' Liability Insurance – *cont.*
negligence
  workplace stress M1524
occupational diseases, as to, *see* Occupation Diseases
'permit to work' T7018
personal protective equipment, as to P3006, P3007
prison sentences IN18
  see also PENALTIES
risk assessment, *see* Risk Assessment
safe system of work S3003
statutory duty
  civil liability for breach IN22, W11045
  corporate manslaughter E13032
  criminal liability for breach IN18
  gross negligence manslaughter IN18
  health and safety legislation IN19, R3001–R3006
  written statement of health and safety policy, *see* Health and Safety Policy Statement
stress S11006
third parties and R3002, R3018
training T7001, T7003, T7006
unlawful discrimination E16522
violence in the workplace, *see also* Workplace Violence
  prosecutions V8015
volenti non fit injuria IN22
vulnerable persons V12003–V12005
EMPLOYMENT AND SUPPORT ALLOWANCE (ESA)
disability benefit C6015
EMPLOYMENT EQUALITY (REPEAL OF RETIREMENT AGE PROVISIONS) REGULATIONS 2011
insurance benefits M1511A
sickness absence M1505
EMPLOYMENT PROTECTION E14001–E14028
agency workers E14001
bullying E14006, E14012
civil action, right of IN22
constructive dismissal E14014, E14019
contractual terms of employment, *see* Contract of Employment
Control of Electromagnetic Radiation at Work Regulations 2016 E14032
employees' rights
  disclosures, as to E14009
  dismissal, in respect of, *see* Dismissal
  health and safety cases E14009, E14010
  safety representatives, for J3030

EMPLOYMENT PROTECTION – *cont.*
employees' rights – *cont.*
  scope of E14009–E14011
  enforcement of safety rules E14012–E14016
  Equality Act 2010 E14001
  medical adviser E15005
  parental leave E14001
  part-time E14001
  PIDA claim E14009
  psychiatric injury E14003
  suspension from work
    maternity grounds E14029–E14030
  unfair dismissal E14001, E14010
  whistle-blowers E14006, E14011.1
  women workers, *see* Women Workers
EMPLOYMENT TRIBUNALS
  complaints to E14009
  discrimination claims E16522
  fees E14001
  health and safety law E15014, E15016–E15025
  new or expectant mothers V12018
  working time, enforcement of provisions W10040–W10043
ENABLING DEVICES M1036
ENERGY CONSERVATION W11044
ENERGY EFFICIENCY
  Lighting L5001, L5002
ENERGY MANAGEMENT F3018 F3029 F3034
  Display Energy Certificate (DEC) F3034
  facilities management F3011
  operation F3018B
  strategy F3018A
  sustainability F3018C
ENERGY PERFORMANCE CERTIFICATE (EPC) F3034A
ENERGY SOURCES ISOLATION FROM M1007
ENFORCEMENT
  appeals E14013
  authority
    HSE IN18
  causation E15031
  corporate homicide (Scotland)
    offence E15001, E15038.1
    sentencing E15043
  corporate manslaughter (England and Wales)
    offence E15001, E15038.1
    sentencing E15043
  Crown, position of E15045
  dangerous substances legislation D0931
  deductions E14014

Ind-33

ENFORCEMENT – *cont.*
  demotion E14014–E14016
  disaster and emergency planning D6032
  employee
    against E15040–E15042
    careless E15037
  Enforcement Policy Statement E15001
  enforcing authority E15001
    appropriate E15008
    commercial premises, for E15007
    indemnification by E15028
    industrial premises, for E15007
    responsibilities of E15007
  environmental health officers, role of F9005
  explosions D0931
  fines E14014
  fire prevention and control, *see* Fire Prevention and Control
  food legislation, *see* Food
  forseeability E15031, E15039
  gross negligence manslaughter E15001
  Health and Safety Commission E15001
  Health and Safety Executive E15001, E15005, E15005.1
    enforcing authority E15007, E15008, E15009, IN18
    fees E15005, E15005.1
    publications E15006
    role of E15005
    transfer of responsibility by E15011
  health and safety policy statement E15001, S7002
  hearings E14013
  Improvement Notice, *see* Improvement Notice
  indictable offence
    definition E15032
    time limit E15032
  inspectors' powers
    improvement notice, as to, *see* Improvement Notice
    indemnification E15028
    interviews E15026.1
    investigation E15026
    prohibition notice, as to, *see* Prohibition Notice
    search and seizure E15027
  investigations E14013
  local authority
    enforcing authority, as E15007, E15008, E15010
    role of E15007
    transfer of responsibility by E15011
  manslaughter charges E15001

ENFORCEMENT – *cont.*
  National Enforcement Code E15012.1
  offences
    company officers E15040
    corporate offences E15039
    defences E15037, E15039
    directors E15040–E15041
    due to act of another E15042
    main offences E15032–E15035
    manslaughter E15038
    method of trial E15032
    penalties for E15036
    prosecution, *see* prosecution, *below*
    publication of convictions for E15044
    sentencing guidelines E15043
    statistics E15030
  penalties E15036
  procedures E14012
  product safety P9001
  Prohibition Notice, *see* Prohibition Notice
  prosecution E15001
    breach of statutory provisions E15030
    burden of proof E15031
    HSE name and shame website E15044
    offences, *see* offences, *above*
    publication of convictions E15044
    service of notice, coupled with E15018
  public register of notices E15029
  reasonable practicability E15031
  regulators'
    code, of E15012.2
    growth duty, of E15012.2
  'relevant statutory provisions'
    breach of, prosecution for E15030
    execution of E15028
    scope of E15012
  risk
    meaning E15031
    relationship with safety E15031
  risk assessment E15001
  rules E14012
  sanctions other than dismissal E14014
  sentencing guidelines E15043
  suspension without pay E14014
  violence in the workplace V8004
  warnings E14014

ENTANGLEMENT M1025

ENTERPRISE AND REGULATORY REFORM ACT 2013
  negligence IN22

ENTERTAINMENT INDUSTRY
  fire safety F5061.1
  noise control N3027–N3027.2

Index

ENVIRONMENT
dangerous goods, carriage of D0401
disaster and emergency planning D6028, D6033
ergonomic approach, *see* Ergonomics
ergonomic design and, *see* Ergonomic design
management of, *see* Environmental Management
thermal E17034

ENVIRONMENT AGENCY
environmental management and E16005
local authority associations and E5046
pollution control, *see also* Emissions into the Atmosphere E5043
Scottish Environment Protection Agency compared E5044
website E5047

ENVIRONMENTAL MANAGEMENT
accounts disclosure E16005
Acorn Scheme E16013
aggregates levy E16006
air passenger duty (APD) E16006
assurance E16024B
auditing of systems E16025
banks E16007
benefits of E16026
Business in the Community (BITC) E16002, E16008
Business in the Environment E16007
Business Review E16005, E16024
Carbon Action Scheme E16007
Carbon Disclosure Project (CDP) E16007
Carbon Trust E16008
chemicals E16008
climate change levy (CCL) E16006
communications E16022
Companies Act 2006 E16005
consumers E16012
corporate responsibility (CR) E16002
corporate social responsibility (CSR) E16002, E16005
costs E16002
CRC Energy Efficiency Scheme E16024
CR index E16002, E16007
customers E16012
Deepwater Horizon E16010, O7004.1
Department for Energy and Climate Change (DECC) E16024
Department for Environment, Food and Rural Affairs (DEFRA) E16024
directors E16005
disclosure E16005, E16024
Dow Jones Sustainability Index E16007
Eco-label Scheme E16008

ENVIRONMENTAL MANAGEMENT – *cont.*
effective management E16026
EMAS (Eco-Management and Audit Scheme) E16014
regulatory control, lessening of E16026
revision E16014
emissions trading E16006
Environment Agency, role of E16005
Equator Principles (EPs) E16007
Europe and E16002, E16005
external drivers
business community E16001, E16011
community pressure groups E16009
competitors' interests E16001
customers E16001, E16012
environmental crises E16010
environmental pressure groups E16001, E16009
financial community E16001, E16007
legislation E16001, E16005
market mechanisms E16001, E16006
role of E16001
facilities management F3018, F3029, F3031, F3034
financial performance, disclosure of E16005, E16024
Financial Reporting Council (FRC) E16005
future of E16027
green public procurement (GPP) E16008
guidelines
Eco-Management and Audit Scheme (EMAS) E16013, E16014, E16016, E16017
GRI Sustainability Reporting Guidelines E16024A
International Standards E16013, E16015, E16016
standards for environmental management E16013
WBCSD guidelines E16024
hazardous chemicals E16008
Hermes principle E16007
implementation E16016–E16024
information management E16023
insurers' interest E16007
internal controls E16005
internal drivers
corporate social responsibility E16001, E16002, E16005
employees E16001, E16004
management information needs E16001, E16003

Ind-35

ENVIRONMENTAL MANAGEMENT – *cont.*
international corporate social responsibility E16002
investors, and E16001, E16007, E16024
Landfill Communities Fund E16006
Landfill Tax E16006
loans for large projects E16007
meaning E16001
mergers and acquisitions E16007
Minister for Corporate Social Responsibility E16002
objectives E16020
operational controls E16023
Operator and Pollution Risk Appraisal (OPRA) regulation scheme E16026
policy as to E16019
practical requirements E16017
pressure groups E16001, E16009
Principles for Responsible Investment (PRI) E16007
proactive E16001
procurement E16008
project finance E16007
public procurement E16008
public reporting E16024
regulatory control, lessening of E16026
reporting guidelines E16024A
review E16018, E16025
shareholders, and E16001, E16007
small and medium-sized enterprises (SMEs) E16005
socially responsible investment (SRI) E16007
stakeholder dialogue E16001
Stakeholder Engagement Standard E16009
standards E16015
Strategic Report E16005, E16024
supply chain E16008
Supply Chain Leadership Collaboration (SCLC) E16007
sustainable development E16011
training E16022
UK Corporate Governance Code E16005
UK environmental policy, future strategy and legislation E5071
UN Environment Programme E16007
vehicle and fuel duty E16006
verification E16024B

ENVIRONMENT WORK
offensive H1707, H1708
violence V8123

EPICONDYLITIS O1020.1

EQUALITY
HSE stance E16501

EQUALITY ACT 2010
age E16513–E16516
 ageing workers E16514
 risk assessment E16515
 young workers E16516
belief E16501
civil partnership E16501
direct discrimination E16503
disability E16502
 discrimination E16506
 duty to make reasonable adjustments E16511, R2005
 example of protection E16506
 impairment E16510
 long term E16510
 meaning E16510, M1506
 risk assessment E16512
discrimination E16503–E16506
 association, by M1508
 direct E16503
 enforcement E16522
 remedies E16522
duty to make reasonable adjustments E16511
enforcement E16522
ergonomics E17070
gender E16520–E16521
 risk assessment E16521
gender reassignment E16501
harassment E16502
 definition H1703
 types of E16507
indirect discrimination E16502, E16504
marriage E16501
maternity E16501
migrant workers E16518
 risk assessment E16519
pregnancy E16501
prohibited conduct E16502
protected act E16508
protected characteristics E16501, H1703, R2005
race E16517–E16519
 migrant workers E16518
 risk assessment E16519
reasonable adjustments E16511, M1511, R2005
religion E16501
remedies E16522
risk assessment E16512, E16515, E16519
 gender issues E16521
sexual orientation E16501
sickness absence M1505
unlawful discrimination E16522
 compensation E16522

EQUALITY ACT 2010 – *cont.*
  victimisation E16502, E16508
  women E14023
  young workers E16516
EQUATOR PRINCIPLES E16007
EQUIPMENT
  see also Personal protective equipment, Visual display units
  Approved Code of Practice M1007
  buying M1007, N3035
  cartridge-operated tools, noise from N3003
  CE markings M1007
  checklist M1007
  children and young people V12016
  community health and safety C6510, C6520
    emergency C6514
    selection and maintenance of C6517
    volunteers C6522
  confined spaces C7018, C7022
  controls and control systems M1007
  dangerous parts of machinery M1007
  dangerous substances D0920, D0933
  design D0933
  directive on M1003
  Electrical Equipment (Safety) Regulations 2016 E3031
  electricity D0920, E3003, E3031
    dangerous substances D0920
    explosions D0920
    isolated equipment, work on E3023
    lighting requirements L5006, L5008
    live work E3024
    made dead E3019
    potentially explosive atmospheres E3027
    strength and capability E3012
  emergencies C6514
  energy sources, isolation from M1007
  ergonomic design, *see* Ergonomic Design
  European Union M1003, M1007
  explosions D0920
  fire F5001.1, F5019–F5027
  food F9038, F9039, F9043
  guards M1007
  guidance M1006, M1007
  hardware M1005
  HSE publications M1006, M1007
  information M1007
  inspection M1007
  instructions M1007
  lighting requirements L5006, L5008, M1007
  maintenance C6517, M1007

EQUIPMENT – *cont.*
  manufacturers M1007, N3013
  markings M1007
  mobile work M1005, M1007–M1008
  noise N3003, N3013, N3016, N3035
  outdoors, noise and equipment used N3016
  permit to work systems S3017
  plant equipment D0933
  power presses M1005
  precautions D0920
  Provision and Use of Work Equipment Regulations M1003, M1004–M1008
  structure if M1005
    summary of main requirements of M1007
  respiratory protective equipment C7022
  risk, definition of M1006
  risk assessment D0920, M1006, M1007
  second-hand M1007
  self-propelled C6510, M1008
  software M1005
  stability M1007
  standards M1007
  stress S11021
  suitability M1007
  suppliers M1007
  temperature M1007
  training M1007
  Use of Work Equipment Directive M1003
  vibration V5015
  violence in the workplace V8123
  volunteers C6522–C6523
  warnings M1007
  work, definition of IN21
ERGONOMIC DESIGN
  anthropometric data E17007, E17010
  building work C8002–C8017
  capabilities
    machinery, of E17009
    person, of E17009
  construction operations C8002–C8017
  disability discrimination E17070
  equipment
    controls, design of E17013, E17015
    design of, generally E17011
    displays, design of E17013–E17015
    display screen equipment (DSE) E17044, E17064
    hand tool design E17012
    information instruction E17016
    labelling E17016
    legislation as to E17060, E17064
    personal protective equipment E17066

Index

ERGONOMIC DESIGN – *cont.*
 equipment – *cont.*
  provision of E17042
  software design E17017
  suitability of E17065
  user trial E17056
 guidelines as to E17003, E17005–E17043, E17064–E17068
 job design
  enlargement of job E17036, E17038
  enrichment of job E17036, E17039
  organisational support and E17043
  rotation of work E17036, E17037
  scheduling of work E17036, E17040
  shift-work E17036, E17041
 legislation promoting E17003, E17060–E17070
 machinery, *see* equipment *above*
 musculoskeletal disorders and, *see* Musculoskeletal disorders
 people, for E17005
 physical work environment
  lighting and E17031
  noise and E17032
  physical hazards E17035
  role of E17030
  thermal environment E17034
  vibration and E17033
 population stereotypes E17008
 posture, role of E17006, E17010, E17021, E17031
 risk assessment E17059, E17062, E17064, E17068
 stress S11013
 task
  allocation of E17009
  designing E17010
 tools assisting E17003, E17051–E17059
 workstation
  arrangement of items on E17024
  characteristics of E17020
  factors to be considered E17018–E17029
  height of E17019
  layout of E17021–E17024
  lighting E17031, E17035
  normal working area E17023
  sitting versus standing E17029
  user trial E17056
  visual considerations E17025–E17028
  zone of convenient reach E17022

ERGONOMICS
 aim of E17001
 application of E17002
 core disciplines E17003

ERGONOMICS – *cont.*
 data
  collection of E17057
  existing, use of E17058
 design, *see* Ergonomic design
 display screen equipment (DSE) E17052, E17064
  checklist D8212
 employee involvement E17059
 Equality Act 2010 E17070
 ergonomic approach
  illustration of E17004
  overview of E17003, E17004
 frequency analysis E17055
 influencing behaviour E17059A
 legislation as to E17003, E17060–E17070
 manual handling E17044, E17052, E17063
 occupational health and safety O1014
 origin of term E17001
 reducing error E17059A
 risk assessment E17052, E17059, E17062, E17064, E17068
 role of E17001
 task analysis E17053
 tools of E17051–E17059
 upper limb disorders E17052, E17068
 use of term E17001
 workflow analysis E17054

ESCALATORS
 workplaces, safety regulations W11001, W11017

EU EMISSIONS TRADING SYSTEM E16006

EUROPEAN AGENCY FOR SAFTEY AND HEALTH AT WORK (EASHW)
 risk assessment E16521
 small and medium businesses IN04

EUROPEAN CHEMICALS AGENCY (ECHA)
 establishment R2511
 registration with R2502
 role R2511

EUROPEAN COMMISSION
 air pollution, and E5026
 CSR strategy E16002
 Directorate General Environment E5026
 website E5047

EUROPEAN ECONOMIC AREA (EEA)
 employers, advice for E18011
 state regulation E18017
 vocational training E18018
 working in E18002

Index

EUROPEAN ENVIRONMENT
   AGENCY(EEA)
  air pollution, environmental
     consequences E5015
EUROPEAN LAW
  application E18004
  civil law E18006
  common law, and E18006
  criminal law E18008
  European Court of Justice E18009
  harmonisation E18009
  inquisitorial approach E18007
  interpretation E18004
  legal structures E18005
  structure E18003
EUROPEAN STANDARDS D0919
EUROPEAN UNION (EU)
  accident reporting E18021
  air pollution, control of, see Emissions into
     the Atmosphere
  ATEX Directive D0933
  Brexit IN07–IN08
  Business Review E16005, E16024
  CE marking M1007, M1009,
     M1015–M1017, M1026, P3012,
     P3013
  chemical agents D0901, D0908
  competence E18016
  competition M1009
  constitution of the EU IN09
  corporate accountability E18012
  corporate social responsibility E16005
     strategy E16002
  dangerous substances D0901, D0908
     Declarations
        of Conformity M1015–M1017
  directives
     air quality and emissions, see Emissions
        into the Atmosphere
     ATEX Directives E3027
     chemical agents D0901
     dangerous substances D0901
     electricity E3027
     electromagnetic fields
        (EMF) IN07–IN08
     equipment M1003
     explosion D0901
     food, as to F9001, F9002, F9050
     Framework Directive E18011, IN09
     generally IN07–IN07
     health and safety J3002, R3005
     machinery safety M1001–M1003,
        M1015–M1017
     manual handling M3001

EUROPEAN UNION (EU) – cont.
  directives – cont.
     new or expectant
        mothers V12017–V12018, V12019
     Physical Agents (Vibration)
        Directive V5001, V5015
     training and competence T7003, T7022,
        V5015
     working time M1514, W10001,
        W10026
  disaster and emergency planning D6018
  Eco-label Scheme E16008
  electricity E3027
  Emissions Trading System E16006
  environmental management E16002,
     E16005
  equipment M1003, M1007
  European Commission, see
     European Commission
  EUROSTAT E18020
  explosion D0901, D0908, D0933
  fire extinguishers F5004, F5014
  Framework Directive IN09
  halon fire extinguishers F5004
  heights, working at W9001, W9009
  ill health reporting E18021
  Information and Consultation Procedures
     (ICPs) J3032
  judgments, transportability of P9046
  ladders W9009.1
  machinery safety M1001–M1003,
     M1015–M1017
     CE marking M1009, M1015–M1017,
        M1026
     competition M1009
     Declarations
        of Conformity M1015–M1017
     Transposed
        Machinery Standards M1001,
        M1003, M1010, M1015–M1017
  manual handling M3001
  new or expectant
     mothers V12017–V12018, V12019
  noise N3002A
  non-ionising radiation R1008–R1009
  Physical Agents (Noise) Directive N3002A
  Physical Hazards (Vibration)
     Directive O1031
  product safety, see Product Liability;
     Product Safety
  public procurement E16008
  qualifications E18016
  radiation D6018
  regulated profession E18014–E18017
  road transport W10029
  role of IN09

EUROPEAN UNION (EU) – *cont.*
　Social Charter declaration on health and safety IN07–IN08
　standards M1001, M1003, M1010, M1013–M1015
　training and competence T7003, T7022
　Transposed Machinery Standards M1001, M1003, M1010, M1013–M1015
　vibration V5001, V5015–V5016
　Vocational Occupational Standard (VOS) E18019
　worker participation E18013
　working abroad, resources E18022
　working time W10001, W10026, W10027–W10031
　Works Councils J3002, J3033
EVACUATION SIGNALS & SIGNS
　safety sign regulations W11033, W11040
EVENTS
　communities, members of C6524
　contractors C6527
　first aid F7012
　HSE safety guide C6525
　information C6527, C6528
　risk assessment C6526
　Safety Advisory Groups (SAGs) C6524
EVIDENCE
　accident investigations A3031, A3032
　violence in the workplace V8121
EXAMINATION
　lifting equipment L3021, L3023–L3026, L3028
　lifts L3011
　pressure systems, and
　　fitness for service P7023
　　frequency P7020
　　general P7022
　　imminent danger P7025
　　nature P7021
　　report P7024
　　scope P7019
EXCAVATION
　Building and construction regulations C8002–C8017
EXECUTIVES
　training and competence T7012
EXPECTANT MOTHERS
　see also New or Expectant Mothers
　display screen equipment R3027
　musculoskeletal disorders E17049
　risk assessment and E14024–E14026, R3018, R3026, R3027, R3030
　　biological agents R3027
　　chemical agents R3027
　　physical agents R3027

EXPECTANT MOTHERS – *cont.*
　risk assessment and – *cont.*
　　working conditions R3027
EXPLOSIONS
　approved codes of practice D0933
　ATEX Directive D0933
　Buncefield Oil Depot D0901
　case histories D0901
　classification of places with explosive atmospheres D0918, D0919
　clothing D0923
　codes D0919
　containers D0918, D0927
　Dangerous Substances and Explosive Atmospheres Regulations 2002 D0901, D0909
　　commencement D0924
　　enforcement D0931
　　exemptions D0928
　　modernisation of legislation D0929
　　repeals and revocations D0930
　drawings and plans D0919
　dust D0919
　　safe handling of combustible D0933
　duty holders D0906
　electricity D0920, E3009, E3027
　emergencies, dealing with D0925
　enforcement D0931
　equipment D0920–D0924
　EU law D0901, D0908
　European standards D0919
　explosive atmospheres D0918–D0919
　fire D0919
　fuel D0918
　gas safety G1001, G1026
　guidance D0933
　ignition sources D0933
　incidents, dealing with D0925
　information provision D0926
　instructions D0926
　national security D0928
　noise and hearing loss N3003
　offshore operations O7028
　petroleum D0929
　pipes D0918, D0927
　precautions D0918
　regulatory impact assessment D0932
　risk assessment and management D0902, D0919
　sources D0934
　spillages D0918
　storage D0918
　training D0926
　verification of safety prior to first use D0922

Ind-40

Index

EXPLOSIONS – *cont.*
warning signs on entry D0921
workplace, definition of D0907
zoning D0918, D0919

EXPLOSIVES
carriage by road D0407
Dangerous Substances and Explosive Atmospheres Regulations 2002 IN21
disaster and emergency planning D6009
fire prevention and control F5031
hazardous substances H2141–H2143, M1104
noise and E17032
risk assessment R3014, R3016 App B

EXPORTS
product safety P9002

EXTRINSIC ALLERGIC ALVEOLITIS O1042

EYEBOLTS L3046–L3047

EYES
binocular vision O1034
cataracts O1033, O1034
causes of diseases and disorders O1033
colour blindness O1033, O1034, O1035
colour vision O1033, O1034, O1035
corrective appliances D8207
detection O1034
diseases and disorders O1032–O1036
display screen equipment D8201, D8207, O1033, O1036
dryness and low humidity O1033
eye and eyesight test D8207
eye protection O1035, O1036
irritation O1033
legal requirements O1036
management O1035
offshore installations, injured on O7071
photokeratitis and photoconjunctivitis O1033
protection of P3021, P3026
risk assessment O1035, O1036
tests D8201, D8207, O1034
vision screening test D8207
visual acuity O1033, O1035

F

FACILITIES
see also Facilities management, Welfare Facilities
disabled person, for W11043

FACILITIES AND SERVICES CUSTOMERS AND USERS OF C6515–C6517

FACILITIES MANAGEMENT F3001–F3035
acquisition and disposal of buildings F3019
activities F3003
advantages F3009–F3010, F3028
asbestos F3024
brand strategy F3023A
breakdown maintenance F3015
change management F3023
cleaning F3030
communications F3010
company approach to F3006
Computer Aided Facilities Management (CAFM) software F3010
construction F3033
contract management F3022
contractors, management of F3027
core activities F3003
cost certainty F3010
coverage F3003
critical activities F3003
customer service F3017
data protection F3023B
definition F3002
design of buildings F3028, F3031–F3032
energy management F3018, F3029, F3034
  Display Energy Certificate (DEC) F3034
  Energy Performance Certificate (EPC) F3034A
environmental management F3018, F3029, F3031, F3034
  operation F3018B
  strategy F3018A
  sustainability F3018C
floor cleaning F3030
government departments, recognition from F3029
hard F3004
Health and Safety Executive F3029
height, working at F3025
help desks F3030
information technology F3030
in-house F3008
innovations F3030, F3035
Integrated Facilities Management (IFM) F3001, F3008, F3009
intelligent buildings F3030
key performance indicator (KPI) F3017
LED lighting F3016
legislation F3031–F3034
life cycle costing F3016
lifts F3032, F3034

Ind-41

FACILITIES MANAGEMENT – *cont.*
  maintenance
    breakdown F3015
    planned corrective F3014
    planned preventative F3013
    policy F3012
  management F3005
  non-core activities F3003
  operation F3034
  organisations F3001
  outsourcing F3003, F3008
  PESTLE analysis F3007
  planned corrective maintenance F3014
  Planned Maintenance (PM) F3013
  Planned Preventative Maintenance (PPM) F3013
  planning software F3030
  project management F3021
  qualification F3029
  recognition of F3029, F3035
  reduced cost F3010
  Reliability Centered maintenance (RCM) F3015A
  responsibilities of F3011–F3023
  security industry F3030
  service delivery F3010
  Site Waste Management Plans (SWMP) F3021A
  Six Sigma F3007, F3010
  soft F3004
  software F3030
  solutions, development and implementation of F3007
  space planning F3020
  stoddart review F3005A
  strategy F3007
  support services F3010
  SWOT analysis F3007
  telecommunications F3030
  Total Facilities Management (TFM) F3001, F3008–F3008A
  university degrees in F3029
  water quality F3026
  webinar F3030
  work equipment F3032
  workforce scheduling F3030
FACTORIES
  absence of overcrowding W11008
  access O3018
  changing clothes facilities W11026
  clothing accommodation W11025
  doors and gates W11016
  drinking water, supply of W11024
  eating facilities W11027
  enforcement of regulations W11047

FACTORIES – *cont.*
  fire prevention F5061.1
  floors, condition of, safety underfoot W11010
  general cleanliness W11007
  occupiers, actions against O3018
  occupiers, defences of W11048
  occupiers' liability O3018, W11047
  owner, liability of W11047–W11048
  pedestrian traffic routes W11015
  rest facilities W11027
  sanitary conveniences W11019–W11021
  sedentary comfort W11009
  slips, trips and falls O3018
  temperature control W11005
  traffic routes O3018
  washing facilities W11020
  work clothing W11025
  workroom dimensions/space W11008
  workstations W11009
FAIRGROUNDS
  accident reporting R3002
  enforcing authority E15009
FALLING
  arrest systems W9008, W9014
  factories O3018
  fatal accidents W9001
  harnesses W9008
  heights, working at W9001, W9004.1, W9008
    arrest systems W9008, W9014
    fatal W9001
    harnesses W9008
    objects W9011
    protection from W9014
    Regulations W9004.1
  mobile work equipment M1008
  objects W9011, W11011
FAMILY BUSINESSES
  children and young people V12011
FARM
  livestock on F9030
  protective equipment P3018
FAST TRACK CLAIMS E13029
FATAL INJURIES
  assessment of damages C6039
  corporate manslaughter E13012
  damages under fatal accident legislation C6036, C6037
  electricity E3001
  falls W9001
  heights, working at W9001
  interim awards C6041
  provisional awards C6040

FATAL INJURIES – *cont.*
  reporting A3005, A3007
  survival of actions C6036, C6038
  types of action for damages C6036
FAW
  first aid at work training F7001,
    F7018–F7018.3
FEDERAL EMERGENCY MANAGEMENT AGENCY (FEMA) D6001
FEE FOR INTERVENTION (FFI)
  HSE E13037, IN11, M1103
FENCING
  community health and safety C6514
  electricity E3014
FERTILISER D0908
FINANCIAL REPORTING COUNCIL
  environmental reporting E16005
  Strategic Report E16005
FINANCIAL SERVICES COMPENSATION SCHEME
  employers' liability insurance and E13024
FINES
  see also Enforcement
  accident reporting A3021
  aggravating factors C9011
  community health and safety C6503, C6504
  corporate manslaughter C6504, C9011, E15043, IN13
  machinery safety M1018
  mitigating factors C9011
  prosecutions IN13
  smoking W11028
FIREARMS
  noise and hearing loss N3003
FIRE EXTINGUISHERS
  BCF extinguishers F5014
  British Standard F5016
  carbon dioxide extinguishers F5015
  class 'F' extinguishers F5015.1
  colour coding F5016
  construction sites F5028
  distribution of portable
    extinguishers F5016
  dry powder extinguishers F5013
  electrical fires F5004, F5015
  European Union F5004, F5014
  fitting F5010
  foam extinguishers F5012
  halon, EU regulations on F5004, F5014
  portable, distribution of F5016, F5028
  siting of F5010, F5028
  suitability of F5003
  training F5015
  types of F5010–F5015.1

FIRE EXTINGUISHERS – *cont.*
  water extinguishers F5011
FIRE FIGHTERS
  calling F5008
  damages for psychological injury C6005
  drains O3014
  fire prevention and control F5073
  switches for luminous signs F5046
FIRE PRECAUTIONS
FIRE PREVENTION AND CONTROL
  active fire protection, passive and F5007
  active fire protection
    measures F5005–F5006
  alarms in certificated premises
    British Standards F5017
  alterations notice F5057
  appeals F5060
  arson F5074
  automatic detection systems
    British Standards F5017
    generally F5007
  breaching of fire resistant walls F5007
  building and construction
    operations C8002–C8017–C8017
  Building Regulations F5030.1, F5030.2, W11044
  Buncefield Oil Depot D0901
  child employees, and F5052
  civil liability F5073
  classification F5003
  cleaners F5018
  common causes of F5018
  common law, civil liability at F5073
  community safety C6507, C6522
    relevant persons C6507
  competent persons F5049
  confined spaces C7024
  construction sites, on
    code of practice F5029
    designing out fire F5027
    emergency procedures F5026
    insurance F5029
    plant F5028
    precautions F5029
    site fire safety co-ordinator F5025
  cooling F5005–F5006
  co-operation and co-ordination F5054
  costs of improvements F5001.1
  criminal offences F5060
  Crown
    RR(FS)O, and F5061
  dampers F5007
  dangerous substances F5043–F5045

Index

FIRE PREVENTION AND CONTROL – *cont.*
Dangerous Substances and Explosive Atmospheres Regulations (DSEAR) 2002 F5031, F5062–F5072
designated premises F5030, F5032, F5041
designing out fire F5027
detection C8002–C8017, F5007, F5017, F5042
ducts F5007
electrical E3007, F5004, F5015
elements of F5001.1
emergency procedures F5026
emergency routes and exits F5040
employees' duties. F5055
employers' liability insurance E13006
enforcement
  alterations notice F5057
  generally F5056
  notices
    alterations notice F5057
    appeals F5060.1
    generally F5058
    prohibition notice F5059
  nuclear installations F5056
  prohibition notice F5059
equipment F5001.1, F5009–F5017
evacuation M2011
external fire spread F5030.1
extinguishers, *see* Fire extinguishers
fire brigade
  calling the F5008
fire certificates
  change of conditions affecting exemptions F5042
  introduction F5032
fire detection F5042
fire doors F5007
fire drills
  generally F5022
  notice F5023–F5029
fire escapes F5007
  emergency routes and exits F5040
  means of F5030.1
  unsatisfactory means of F5021
fire fighters
  and see Fire fighters
fire fighting F5042
fire risk assessment
  areas, procedures for danger F5041
  carrying out F5038
  controlling residual risk F5039
  dangerous substances F5043–F5045
  emergency routes and exits F5040

FIRE PREVENTION AND CONTROL – *cont.*
fire risk assessment – *cont.*
  fire detection provision F5042
  fire fighting provision F5042
  introduction F5037
  serious and imminent danger to persons, procedures for F5041
fire statistics F5002
fuel
  generally F5001.1
  tanks F5028
gases F5003
generally F5001
good housekeeping F5018
guidance documents F5061
hazardous materials F5031
HSE guidance F5001.1
information provision F5051
insurance F5029, F5074
internal fire spread F5030.1
legislation
  Building Regulations F5030.1, F5030.2
  fire certification F5032
  fire safety F5030.2, F5031
  introduction F5030
  Regulatory Reform (Fire Safety) Order F5033–F5036
liquids and liquefiable solids F5003
maintenance F5047
managing health and safety
  evacuation procedure M2011
  risks, assessment of M2006
means of escape F5020–F5021
metals F5003
new buildings F5030
occupiers' liability of F5073
offences F5060
offshore operations O7028
oils and fats, cooking F5003
passive and active fire protection F5007
personal protective equipment P3017
petroleum D0929, F5031
plant F5028
precautions
  construction sites F5029
  criminal offences F5060
  enforcement F5056
  legislation F5030
  rationalisation of legislation on F5031
  regulations, specific F5032
premises, guidance F5061.1
pre-planning of fire prevention
  generally F5019

Ind-44

Index

FIRE PREVENTION AND CONTROL – *cont.*
pre-planning of fire prevention – *cont.*
  means of escape F5020–F5021
procedures F5008
prohibition notice F5059
public liability insurance P9528
Regulatory Reform (Fire Safety) Order
  application F5034–F5036
  generally F5033, !N21
responsible person C6507
risk assessment F5001.1, F5032, F5037, M2006, P9528, R3012, R3018, R3033, R3035
  guidance F5061.1
risk management F5001.1
rubbish F5018
safety assistance F5048
signs F5032, W11041, W11042
site fire safety co-ordinators
  emergency procedures F5026
  role of F5025
smoking F5018
smothering F5005–F5006
solid organic materials F5003
special risks F5001.1
sports, safety certificates and F5032
sprinkler systems E13006, F5007
starvation F5005–F5006
training F5053
triangle F5001.1
visitors, liability to F5073
volunteers C6522
walls, breach of fire resistant F5007
warnings, means of F5030.1
waste F5018
water F5005–F5006

FIRST AID
absences of first-aiders or appointed persons F7002, F7011
annual leave F7011
appointed persons
  absence F7002, F7011
  appointment of F7003
  role of F7019
arrangements
  employers F7001
assessment of first-aid needs
  absences of first-aiders or appointed persons F7002, F7011
  aim F7003
  annual leave F7011
  basic rule F7003
  checklist F7025 App B

FIRST AID – *cont.*
assessment of first-aid needs – *cont.*
  distance to emergency medical services F7008
  distant workers F7009
  doctors, qualified medical F7003
  factors affecting F7003
  hazards and risk in the workplace F7004
  history of accidents in organisation F7006
  HSE approval E15005
  insurance, employers' liability F7012
  lone workers F7009
  multi-occupied sites F7010
  nature and distribution of the workforce F7007
  no need to appoint first-aiders, where F7003
  nurses, registered F7003
  occupational health service advice F7003
  public, members of the F7012
  reassessment of needs F7013
  requirements F7003
  shared sites, employees on F7010
  size of organisation F7005
  trainees F7012
  travelling workers F7009
  workforce, nature and distribution of the F7007
  work pattern F7008.1
certificates F7018
  minimum information F7018.3
containers F7022
courses F7018
disaster and emergency planning D6021
distance to emergency medical services F7008
distant workers F7009
doctors, qualified medical F7003
emergency
  services, calling F7019
emergency first aid at work (EFAW) F7001, F7018–F7018.3
employer's duty
  arrangements F7001
  assessment of first-aid needs. *see* assessment of first-aid needs *above*
  first-aiders, provision of F7002
  informing employees of arrangements F7014
  requirements F7002
  scope of duty F7002
equipment, provision of IN21

Ind-45

FIRST AID – *cont.*
   events F7012
   eye irrigation F7023
   first aid at work (FAW) F7001,
       F7018–F7018.3
     subject areas F7025 App D
   first-aiders
     abilities F7017
     absence of first-aider F7002, F7011
     certificates F7018
     competencies F 7018
     courses
       e-learning F7018.2
       face to face F7018.2
     generally F7016
     no need to appoint first-aiders,
         where F7003
     numbers required F7016, F7017
     offshore operations O7058
     position in company F7017
     provision of F7002, F7016–F7018
     qualifications F7018
     refresher training F7018
     selection F7017
     special training F7018
     standard F7017
     training F7018
     unusual risks/hazards F7018
     updating skills F7014, F7018
   generally F7001–F7001.1
   hazards
     risk in the workplace, assessment
         of F7004
   history of accidents in organisation F7006
   insurance, employers' liability F7012
   internal memos as to F7014
   key employees, nomination of F7014
   Löfstedt review F7001
   lone workers F7009
   meaning F7001.1
   mental health first aid F7003.1
   multi-occupied sites F7010
   nature and distribution of the
       workforce F7007
   new employees, provision for F7014
   nurses, registered F7003
   offshore operations O7056.1–O7062
     equipment O7059–O7062
     facilities, duty to provide first-
         aid O7057
     medical personnel O7057, O7058
     offshore first-aider O7058
     offshore medic O7058
     sick bays O7059–O7062
     suitable persons O7057, O7058

FIRST AID – *cont.*
   public, members of the F7012
   qualifications of first-aid personnel F7018
   reassessment of needs F7013
   record keeping F7020
   records F7020
   refresher training F7018
   Regulations IN21
   resources, first-aid F7021–F7024
     additional resources F7023
     automated external defibrillators
         (AED) F7023.1
     containers F7022
     eye irrigation F7023
     materials F7022
     protective equipment F7023
     provision F7021
     stock of first-aid materials F7022
     travelling, kits for F7024
   rooms, first-aid F7025
   selection of first-aiders F7017
   self-employed F7015
   shared sites, employees on F7010
   signs, *see also* Safety Signs W11040,
       W11041
   size of organisation, assessment of F7005
   specially trained staff, action by F7001
   summary of requirements F7001
   trainees F7012
   training of first-aid personnel F7001,
       F7018, F7019, T7010
     additional training needs F7025 App F
     certificates F7018.3
     organisations F7018.1
     selecting provider F7018.2
   travelling, kits for F7024
   travelling workers F7009
   updating of skills and
       arrangements F7014, F7018
   workforce, nature and distribution of
       the F7007
FIT FOR WORK
   occupational health service O1015.1,
       R2011
   service IN06, M1504B
FITNESS FOR SERVICE
   pressure systems, and P7023
FITNESS FOR WORK M1504
FIT NOTE M1504
   occupational health O1015.2
   options for return to work M1504
   rehabilitation M1515, R2004, R2008
FIVE CAPITALS MODEL
   sustainability F3018

FIXED PENALTY NOTICES
  smoking W11028
FLAMMABLE SUBSTANCE
  explosive atmosphere D0921
  major accident hazards regulations M1104
FLEXIBLE WORK PATTERNS IN06
FLOORS
  condition of, in workplaces M3006, M3014, W11010
  design M3014
  facilities management F3030
  free from obstructions W11010
  hygienically clean W11010
  manual handling M3006, M3014
  variation in levels of M3006
FLUORESCENT TUBES
  stroboscopic effect L5008, L5020
FOAM FIRE EXTINGUISHERS F5012
FOCUS GROUPS V8022
FOOD
  adulteration of F9001
  advertising F9012, F9020
  Advertising Standards Authority (ASA) F9012
  aircraft, on F9032–F9033
  allergens, pre-packed foods, in F9075
  allergic reaction to F9011, F9075
  analysis of F9013
  animal products F9002, F9014, F9028
  business
    licensing F9021
    occasional F9041
    registration F9021
    scope of F9007
    temporary F9041
  canned foods F9056
  change in activities F9027
  charity events, at F9007
  chewing gum F9007
  chilled storage F9053–F9069
  choking, risk of F9010
  codes of practice F9007.1
    Codex Alimentarius F9035, F9038
    Food Safety Act 1990, penalties under F9025
    penalties under Food Safety Act 1990 F9025
    range of F9006, F9025
  complaints procedure F9080
  composite products, establishments producing F9028.1
  composition of F9020
  consultation F9006
  consumer concerns F9077
  consumer protection legislation F9008

FOOD – *cont.*
  contamination F9011, F9036, F9038, F9047, F9074
  cooking F9036
  cooling F9036, F9062–F9069
  criminal offences
    defences F9022, F9023, F9070, F9082
    enforcement officers, by F9013
    information requirements F9027
    prosecution of F9013, F9015
    strict liability of F9023
  definitions F9006, F9077
  Department of Health's role F9005
  descriptions F9012
  dietary preferences F9076
  display F9059, F9061
  domestic premises, exemption of F9031
  due diligence F9049
  emergency control orders F9019
  emergency prohibition notices and orders F9018
  enforcement procedures
    detention of food F9013, F9014
    emergency control order F9019
    emergency prohibition F9015, F9018
    enforcement officers' powers F9013
    entry to premises, powers of F9013
    government policy F9005
    improvement notice F9015
    licensing F9021
    offences, prosecution F9013, F9015
    personal prohibition F9017
    prohibition order F9016
    registration of businesses F9021
    scope of F9013
    seizure of food F9013, F9014
  equipment F9038, F9039, F9043, F9074
  EU
    directives F9001, F9002, F9050
    regulations F9006, F9034
  exempt premises F9030, F9032–F9033
  fair trading F9078
  falsely describing F9008, F9012
  falsely presenting F9008, F9012
  food business, definition F9026
  food rooms F9038, F9040
  Food Safety Act 1990, penalties under F9024, F9025
  Food Safety Manual
    contents of F9070–F9081
    value of F9082
  Food Standards Agency
    emergency control orders, issuing of F9019

Ind-47

Index

FOOD – *cont.*
  Food Standards Agency – *cont.*
    establishment of F9005
    role of F9005
    website F9019
  frozen storage F9047
  genetically modified F9020, F9077
  handler
    personal hygiene F9038, F9046, F9074
    policy as to F9074
    risk factors E16521
    supervision F9048, F9081
    training F9048, F9081
  hazard analysis F9035, F9036, F9073–F9074
  hazard warnings, use of F9019
  heat processing F9047
  hotels, in F9007
  hot food F9061
  human consumption, fitness for F9007, F9010, F9035
  hygiene
    definition F9035
    documentation and records F9037
    emergency prohibition notice F9018
    hazard analysis F9035, F9036
    improvement notices F9015
    industry guides F9049
    Meat Hygiene Service F9005, F9028
    'pre-requisites' F9038
    records and documentation F9037
    regulations, scope of F9034–F9049
    temperature control, *see* temperature control, *below*
  improvement notices F9015, F9016
  industry guides F9025, F9049
  information requirements F9027
  injurious to health F9009–F9010
  inspection F9027
  labelling
    non pre-packed foods F9004.1
    pre-packed foods F9004
    requirements F9003, F9020, F9023, F9075, F9078
  licensing F9027–F9028
  local authorities
    meat F9028
    monitoring of F9005
    role of F9005, F9007.1
  mail order foods F9054
  markets
    industry guides F9049
    registration of F9027
    stalls F9032–F9033

FOOD – *cont.*
  meat
    export cutting premises etc. for F9028
    Meat Hygiene Service F9005, F9028
    product plants F9028
  medicinal products F9007
  Medicines and Healthcare Products Regulatory Agency (MHRA) F9007
  microbiological food poisoning hazards F9036
  Ministry of Agriculture Fisheries and Food, and F9005
  motor vehicles F9032–F9033
  moveable premises F9027
  nature demanded, not of F9011
  new premises, notification of F9027
  nutrition labelling F9078
  packaging F9047
  parasites F9047
  pathogenic micro-organisms F9009, F9047
  penalties under Food Safety Act 1990 F9024, F9025
  pests F9047, F9074
  poisoning
    microbiological food poisoning hazards F9036
    risk of F9050
  premises
    cleaning F9074
    enforcement officers' powers F9013
    exemption from registration F9030, F9032–F9033
    facilities F9039
    mobile food premises F9027
    occasionally used only F9029
    pest control F9074
    regulations, scope of F9026–F9033
    ventilation F9039
  processed food F9057
  product recall F9079
  prohibition orders and notices F9016–F9018
  quality demanded, not of F9011
  raw products F9028, F9057
  registration F9026–F9033
  remedial action notices F9019.1
  safety
    definition of terms F9007
    enforcement procedures, *see* enforcement procedures, *above*
    legislation F9001, F9006, F9007.1
    offences under, *see* offences, *above*
    principal sections of F9006
    scope of F9001, F9007

FOOD – *cont.*
　sale of
　　ban on F9020
　　meaning of F9007
　　offences as to F9010, F9011
　　'use by' date F9007
　　vending machine,
　　　through F9032–F9033, F9041
　samples, taking of F9013
　sanitary conveniences F9039
　Scottish legislation, *see* Scotland
　short 'shelf life' F9055
　standards
　　Agency, *see* Food Standards Agency, *above*
　　control of F9001
　substance demanded, not of F9011
　supervision of staff F9048, F9081
　temperature control
　　four-hour rule F9059
　　regulations, scope of F9039, F9047
　　requirements F9036
　　time and temperature, relationship of F9051–F9052
　thawing frozen food F9047
　training of staff F9081
　transport F9038, F9042
　unfit for human consumption F9010
　vegetarian F9076
　vehicles F9032–F9033
　vending machines, use of F9032–F9033, F9041
　waste F9038, F9044
　water F9007, F9038, F9045
　wrapping F9047
FOOTWEAR
　manual handling M3004, M3007
FOREIGN DESIGNER
　product safety P9021
FOREIGN MANUFACTURER
　product safety P9021
FORESEEABILITY
　burden of proof E15031
　determination of reasonably practicable E15039
FORESTRY
　air pollution and E5015, E5021
　vibration V5001
FORK-LIFT TRUCKS L3034–L3035
　counterbalance trucks L3034
　failure L3035
　narrow aisle trucks L3034
　order pickers L3034
　overturning M1008
　pedestrian-operated stackers L3034

FORK-LIFT TRUCKS – *cont.*
　prosecution L3035
　reach trucks L3034
FOSSIL FUELS E16006
FRAGILE SURFACES W9010
FRAMEWORK FOR HIGHER EDUCATION QUALIFICATIONS (FHEQ) T7010 T7012
FRANCE
　professional regulation E18017
　reporting accidents E18021
FRESH AIR
　construction and building operations, and C8002–C8017
FRICTION M1025
FUEL
　see also Petroleum
　Building Regulations W11044
　Clean Air Act 1993 E5061
　emission controls, and E5035
　explosions D0918
　fire prevention and control F5001.1, F5028
FUMES
　ventilation, *see* Ventilation

G

GANGWAYS
　manual handling M3014
GARAGES D0901
GAS SAFE REGISTER G1013
GAS SAFETY
　accreditation of operatives G1014
　advice and assistance E15005
　appliances
　　installation and use, *see* installation and use, *below*
　　maintenance G1024
　　temporary installations C6516, C6517
　assessment of gas fitting operatives G1014
　Building Regulations G1019
　Bulk LPG installations, protective devices P7020
　carbon monoxide poisoning G1001
　chimney flues G1019
　competent person
　　qualifications G1014
　　quality control G1013
　　supervision G1014
　competition in domestic market G1002
　compressed natural gas D0933
　confined spaces C7016

Index

GAS SAFETY – *cont.*
  connection of service pipes G1002
  dampers G1019
  dangerous substances G1026
  definition of gas G1001
  disconnection G1007
    duty to notify G1002
  domestic market G1002
  duty holder, compliance etc by G1005
  entry, *see* rights of entry *below*
  equipment C7021
  escape incidents
    responsible persons' duties G1005
    suspecting leak G1005
  explosion G1001, G1026
  explosive atmosphere D0921
  fire F5003
  fitters A3011
  fittings G1020
  flues G1019
  gas fitting operatives G1014
    accreditation G1014
  Gas Safe Register G1013
  HSE guidance G1025C
  incidents, duty to report A3010–A3011
  Individual Gas Fitting Operatives G1014
  inspection G1007, G1025
  installation and use
    appliances G1010, G1012, G1017, G1018
    checks, safety G1024
    competent persons G1013, G1014
    dampers G1019
    duties G1024
    emergency controls G1021
    escape of gas G1010, G1012
    excluded supplies G1010
    flues G1019
    gas fittings G1020
    general safety precautions G1016
    landlords G1024
    maintenance requirements G1025
    materials G1015, G1022
    meters G1022
    pipework, installation G1023
    pressure fluctuations G1025A
    pressure systems, *see* Pressure Systems
    quality control G1013
    records G1025
    regulations
      interface with other legislation G1026
      scope of G1010
    regulators G1022

GAS SAFETY – *cont.*
  installation and use – *cont.*
    responsibilities G1011
    room-sealed appliances G1018
    supervision G1014
    temporary installations C6516
    tenants G1011
    testing gas appliances G1024
    unsafe gas appliances G1017
    use requirement G1010
    venting of waste gas G1010
    work G1010
    workmanship G1015
  landlords G1025C
  leaflets G1025C
  legislation G1001, G1026
  licensing arrangements G1002
  liquid petroleum gas (LPG) G1025B
  maintenance G1024, G1025
  management of supply
    compliance and co-operation G1005
    regulations as to G1003
    safe case document G1004
  materials, fittings G1015
  medical advice G1025C
  meters G1022
  network emergency co-ordinator G1003
  notification duty G1002
  pipelines G1001, G1009
  pipework installation G1023
  precautions G1016
  pressure fluctuations G1025A
  pressure systems, *see* Pressure Systems
  prosecutions G1026A
  protective device P7020
  public gas transporters G1002
  public liability insurance P9528
  purging C7016
  reconnection prohibition G1008
  regulations G1001, G1010, G1026
  reporting gas incidents A3002, A3010, G1005
  responsible persons G1005
  rights of entry
    disconnection G1007
    generally G1006
    inspection G1007
    reconnection prohibition G1008
    testing G1007
  room-sealed appliances G1018
  safety case G1003, G1004
  service pipes G1002
  suppliers A3010

Index

GAS SAFETY – *cont.*
supply
  domestic market G1002
  management G1003
suspecting leak G1005
testing G1007, G1024
training and competence T7005
transporters G1002
unloading compressed natural gas D0933
ventilation G1025C
workmanship of fitters G1015

GENDER
Equality Act 2010 E16520
manual handling M3007
risk assessment E16521
  table of relevant hazards E16521
sexual harassment H1702

GENDER REASSIGNMENT
harassment H1702
protected characteristic H1703

GENETICALLY MODIFIED ORGANISMS
applications as to E15005
food regulations and F9020, F9077
notifications as to E15005
risk assessment R3016, R3035

GERMANY
professional regulation E18017

GLASS
Building Regulations W11044

GLOBAL REPORTING INITIATIVE (GRI)
reporting guidelines E16024A

GLOVES V5019

GOOD ORDER
construction and building operations, and C8002–C8017

GREECE
professional regulation E18017

GREENBURY REPORT IN05

GREENHOUSE GASES
Carbon Disclosure Project (CDP) E16007–E16008
controls E16007
reporting E16024A

GREEN PUBLIC PROCUREMENT (GPP)
supply chain E16008

GRIEVANCE PROCEDURES
contract of employment, in E14006, E14008

GRIPS M3013

GROSS NEGLIGENCE MANSLAUGHTER
directors E13032
prosecutions C9021
Sentencing Council guideline IN18

GUARDS
adjustable M1030
automatic M1031
drilling machines M1030
equipment M1007
fixed M1029–M1030
heights, working at W9008
interlocking M1032
  mechanical M1035
  types of M1033
mobile work equipment M1008
position sensors M1032
power isolation M1034
risk assessment M1032
self-adjusting M1030
standards M1029, M1032
sweep away M1031
switches M1032, M1034–M1035
telescopic M1030
trapped key interlocking M1034

H

HALON FIRE EXTINGUISHERS F5004 F5014

HAMPEL REPORT IN05

HAND ARM VIBRATION (HAV)
advice V5017, V5018
carpal tunnel syndrome (CTS) V5004, V5010
causes O1028, V5010
daily exposure action value V5015
detection of O1029
diseases caused by V5010
effects V5010, V5014
emissions, testing V5020–V5021
ergonomic approach to E17012, E17033
E.U. law O1031
examples of causes of V5010
exposure action value V5015
frequency V5017
gloves V5019
guidance V5017, V5018
Health and Safety Executive (HSE) V5003
health surveillance O1029, O1030, V5015
legal requirements O1031
magnitude V5017
maintenance O1031
managing exposure O1030, V5018
occupational diseases O1027–O1031
occupations, examples of V5014
personal protection against V5019–V5022

Ind-51

HAND ARM VIBRATION (HAV) – *cont.*
  Physical Agents (Vibration)
    Directive V5015
  Physical Hazards (Vibration)
    Directive O1031
  prescription of V5014
  reportable occupational disease V5011
  risk assessment O1030, O1031, V5015, V5022
  symptoms O1027
  tools, vibration magnitudes of V5017
  vibration, meaning of O1027
  warnings V5020–V5021
HANDCUFFS V8094
HANDLES M3013
HAND SIGNALS
  safety signs W11032, W11042
HARASSMENT
  adverse environment H1702
  age, grounds of H1702–H1703
  assault H1723, H1730
  associative discrimination H1707
  battery
    criminal action H1729–H1730
    tort action H1723
  breach of contract H1727
    remedies H1728
  civil partnership, in respect of H1703
  claim
    elements of H1702
  commission of H1704
  compensation H1718, H1721
  concept H1702
  constructive dismissal H1720
  course of conduct cases H1706
  criminal action
    battery H1730
    Crime and Disorder Act 1998 H1733
    Protection from Harassment Act 1997 H1731, V8015B
    Public Order Act 1986 H1732
  definition H1702–H1703
    limitations H1709
  dignity H1701
  disability H1703
  discrimination claim, general H1703
  effect, purpose or H1702–H1703, H1707
  elements of claim H1703
  employers' liability H1710–H1715A
    causal link with employment H1712
    course of employment H1711
    defence H1713
    employees' liability H1714
    narrowing extent H1715

HARASSMENT – *cont.*
  employers' liability – *cont.*
    putative agent, as principal for H1715A
    vicarious liability H1710–H1712
  employers' policy H1734–H1740
    complaints procedure H1735
    confidentiality H1736
    formal policy H1738
    informal action H1737
    monitoring and review H1740
    training H1739
  environment, work H1707–H1708
  Equality Act 2010 E16507, H1701
    definition of harassment H1703
    employees' protection E16501
    protected characteristics H1703
  failures to investigate or protect H1706
  free standing claim H1701
  gender reassignment H1702
    protected characteristic H1703
  general discrimination claim H1703
  handling claims of H1734–H1740
  heads of claim H1702
  history H1702
  intimidating work environment H1707–H1708
  legal action H1716–H1733
    breach of contract H1720
    constructive dismissal H1720
    criminal action H1729
    remedies H1718, H1721, H1725, H1728
    tort H1717, H1722–H1725
    unfair dismissal H1719
    unlawful discrimination H1717, H1718
  marriage, in respect of H1703
  maternity, in respect of H1703
  monitoring H1740
  nature of conduct H1706
  negligence H1724
  perceived discrimination H1707
  policy H1734–H1740
  pregnancy, in respect of H1703
  preventing claims of H1734–H1740
  protected characteristics H1703
  protected statutory grounds H1702
  purpose or effect H1702–H1703, H1707
  racial H1702, H1703 H1733
  religion H1702–H1703
  remedies H1718, H1721
    breach of contract H1728
    tort H1725
  separate procedure E14006
  sexual H1701–H1703

HARASSMENT – *cont.*
  sexual – *cont.*
    definition H1702
    sexual orientation H1702
    single incident cases H1706
    third party, protection H1703
    torts
      assault and battery H1723
      negligence H1724
      remedies H1725
      unlawful discrimination H1717
    training H1739
    trespass to the person H1723
    types of E16507, H1701, H1703
    unfair dismissal
      claim H1719
      constructive dismissal H1720
      proof H1720
      remedies H1721
    unlawful discrimination
      elements of the tort H1717
      remedies H1718
    unwanted conduct H1702, H1704
    vicarious liability H1710–H1712
    violation of dignity H1701
    violence in the workplace V8011
    work environment H1707–H1708
HARNESSES M3013, W9008
HAZARDOUS SUBSTANCES
  accidents where H2105, H2114, H2135, H2139
  ammonium nitrate D0908
  Approved Code of Practice H2122
  assessment of risk
    discussion with staff H2119
    further investigations H2121
    guidance as to H2120
    information gathering for H2117
    ongoing review of H2125
    organisation of H2116
    records as to H2122, H2124
    requirements as to H2101, H2104–H2106, H2114–H2126
    who should carry out H2115
    workplace observations H2118
  biological problems H2103
  chemical substances, *see* Chemical hazards, Chemical Substances
  Codes of Practice H2105
  community health and safety C6510
  control of exposure
    adequacy of H2111, H2112
    biological agents, where H2113
    carcinogens, where H2113

HAZARDOUS SUBSTANCES – *cont.*
  control of exposure – *cont.*
    examination of H2128
    maintenance etc of H2114, H2126, H2128, H2129
    monitoring H2114
    other than PPE H2109
    PPE, using H2107, H2110, H2118, H2131
    steps required H2107
    suitability and sufficiency of H2114
    testing of H2128
    use of H2107, H2110, H2114, H2127
  Control of Substances Hazardous to Health Regulations 2002 IN21
  emergencies H2105, H2135
  employees' duties H2105
  employers' duties H2105
  employers' liability insurance, conditions on E13021
  fire prevention and control F5031
  first aid needs F7004
  guidance as to H2105
  harm to the body H2102
  health surveillance H2105, H2114, H2133
  incidents H2105, H2135
  instruction as to H2105, H2134
  legislation
    allied regulations H2101, H2137–H2146
    definitions H2101, H2104
    duties under. H2105
    main regulations H2101
    smoking ban H2102
    summary of H2101, H2104, H2105.1
  major accident hazards, *see* Major Accident Hazards Regulations
  maximum exposure limits, replacement with workplace exposure limits of H2112
  monitoring exposure H2111, H2112, H2132
  mutagens H2113
  notification requirements D0908
  personal protective equipment P3008, P3017
  poisoning H2103
  PPE, use of H2109, H2110, H2118
  prevention of exposure
    methods of achieving H2108
    steps requiring H2105, H2107
  REACH regulation H2137
  recommendations
    implementation of H2123
    review of H2123

Ind-53

HAZARDOUS SUBSTANCES – *cont.*
  records, preparation of assessment H2122
  risk assessments R3007, R3018, R3035
    examples. H2120
  self-employed, duties of H2105
  skin conditions H2103
  training as to H2105, H2114, H2134
  transportation of chemicals, *see*
    Transportation of Dangerous
    Substances
  ventilation H2130
  workplace exposure limits H2112
HAZARDS
  see also Hazardous Substances
  analysis M1024
  checklist M1024–M1025
  continuing M1024
  definition of M1023
  examples of M1025
  food, as to F9019
  identification of M1024–M1025
  machinery safety
    analysis M1024
    checklist M1024–M1025
    continuing M1024
    definition of M1023
    examples of M1025
    identification of M1024–M1025
  warnings F9019
HEAD PROTECTION
  British Standard P3025
  construction and building operations,
    and C8018
HEALTH
  air pollution and E5002–E5014
  checks M2115
  Health and Safety Commission, *see* Health
    and Safety Commission
  Health and Safety Executive, *see* Health
    and Safety Executive
  health and safety law, *see* Health and
    Safety Law
  manual handling M3001, M3002, M3007
  occupational health and safety O1007
  World Health Organisation E5064–E5065
HEALTH AND SAFETY COMMISSION
  agricultural equipment E15005
  guidance
    offshore operations O7028
  heights, working at W9001
HEALTH AND SAFETY EXECUTIVE
  (HSE)
  accident investigations A3030
  business plan IN12
  call centres N3028

HEALTH AND SAFETY EXECUTIVE
  (HSE) – *cont.*
  Construction Design and Management
    Regulation (CDM)
    prosecutions C8521
  dangerous substances D0933
  directors' duties IN05
  disabled persons V12032, V12037
    reasonable adjustments E16511
  diversity, stance E16501
  enforcement E15001
  enforcing authorities E15007–E15009,
    IN18
  equality, stance E16501
  equipment M1006, M1007
  events C6525
  facilities management F3029
  Fee for Intervention (FFI) E13037, IN11
  fees payable to E15005, E15005.1
  functions E15005
  funding IN11
  health and work strategy IN10
  heights, working at W9009
  inspectors E15005
  key messages V5004.1
  ladders W9009.1
  migrant workers E16518
  mines and quarries E15005, E15009
  named and shamed IN13
  new or expectant mothers V12019,
    V12023–V12024
  occupational health O1056,
    R2010–R2011
  prosecutions by IN13
  publications E15005, E15006, IN26
  rehabilitation R2003
  risk assessment, relating to
    age E16515
    disability E16512
    gender E16521
    migrant workers E16519
    young people E16516
  training and competence T7005
  transfer of responsibilities E15011
  vibration V5003, V5017, V5018, V5029
  violence in the workplace V8001–V8002,
    V8005
  volunteers C6523
  website E15044, R3054
  work experience students C6518
  working at heights W9001
HEALTH AND SAFETY LAW
  advice, sources of IN26
  Brexit IN07–IN08
  civil law, *see* Civil Liability

HEALTH AND SAFETY LAW – cont.
consultants IN26
criminal law IN18
enforcement, see Enforcement
European Union
  directives IN07–IN08
  role of IN07–IN08
Health and Safety at Work Act 1974
  application IN19
  duties under IN19
  employers IN19
  enforcement of, see Enforcement
  general public, protection for IN19
  levels of duty IN19
  premises, as to IN19
  reasonably practicable IN19
Health and Safety Commission, see Health and Safety Commission
Health and Safety Executive, see Health and Safety Executive
management of health and safety
  arrangements IN25
  organisational responsibilities IN25
  policies IN25
  risk assessment, see Risk Assessment
  sources of advice IN26
outside advice S7005
penalties and prosecutions IN13
regulations IN21
risk assessments, see Risk Assessment
structure IN17–IN22
supply-chain pressure IN15
training IN26

HEALTH AND SAFETY OFFENCES
penalties E15043
Sentencing Council E15001, E15043
sentencing guidelines E15043

HEALTH AND SAFETY OFFICER
responsibilities of S7003, S7005, S7009

HEALTH AND SAFETY POLICY STATEMENTS
arrangements for implementation of S7003, S7006
arrangements to be considered S7006
checklist for writing and reviewing S7011
communication of S7002.1, S7003, S7008
consultation mechanisms S7003
contents of S7003–S7006
contractors' approval systems S7010
development of S7002.1, S7011
effectively, ensuring arrangements work S7007
emergency procedures S7003
failure to implement E15040, S7002
generally S7001

HEALTH AND SAFETY POLICY STATEMENTS – cont.
guidance S7012
health surveillance S7006
inspections S7003, S7006
Intranet S7008
legal requirements S7002
management cycle S7007
managing health and safety M2032
organisation S7003, S7005
outside advice S7005
policy S7009
policy development S7002.1, S7011
review of S7002.1, S7003, S7009, S7011
risk assessments, recording of S7002.1, S7006
role of
  employees S7005
  health and safety officer S7003, S7005, S7009
  maintenance engineer S7005
  managers S7005
  managing director S7005, S7009
  occupational hygienist S7005
  personnel staff S7005
  supervisors S7005
statement of intent E14012, IN25, S7003, S7004
visitors and S7006
worker involvement S7005
young people and S7008

HEALTHCARE PREMISES
fire safety risk assessment F5061.1

HEALTH RECORDS O1016–O1018
asbestos employers A5040

HEALTH SCREENING
risk assessment R3006, R3016

HEALTH SURVEILLANCE
community health and safety C6506
hand-arm vibration syndrome O1029, O1030, V5015
hazardous substances in workplace, where H2105, H2114, H2133
ionising radiation R1019
managing health and safety M2009
manual handling M3007, M3019
night workers W10033, W10034
noise and hearing loss N3013, O1025
non-ionising radiation R1010.9
occupational health and safety O1004, O1009
respiratory diseases O1044, O1045
risk assessment R3006, R3016
skin diseases O1041
vibration V5015

HEALTH SURVEILLANCE – *cont.*
  vibration – *cont.*
    whole-body vibration O1029, O1030, V5028
  working time W10033, W10034
HEALTH WORKERS
  risk factors E16521
  violence in the workplace V8077
HEARING
  disciplinary E14013
HEARING LOSS *see* **Noise and occupational hearing loss**
HEAT
  processing, food safety F9047
  public liability insurance P9523, P9528
HEAVY METAL
  protocol as to E5026
HEIGHTS WORKING AT *see* **Working at Heights**
HERALD OF FREE ENTERPRISE D6029
HERMES PRINCIPLE E16007
HIGHER EDUCATION T7010
HIGH PRESSURE FLUID INJECTION M1025
HIRE CONTRACTS
  product liability P9048
HIRE PURCHASE
  product liability P9048
HOISTS L3032–L3033 L3036
  failure L3033
  patient/bath hoist L3036
  regulations L3032
HOLD-TO-RUN DEVICES M1036
HOLIDAYS *see* **Annual Leave**
HOME
  care workers C6028
  community health and safety C6522
  compensation for work injuries C6028
  entry into client's C6522
  volunteers C6522
HOME WORKERS
  display screen equipment D8202, D8213
  risk assessment D8213
HOMICIDE *see also* **Manslaughter**
  corporate E15038
  Scotland E15038
HONEY
  food premises regulations F9030
HOSPITAL
  occupiers' liability O3018
HOT DESKING D8219
HOTEL
  food premises regulations F9007

HOT WEATHER
  workplaces W11005
HOT WORKS C7026 D0933
HSE *see* **Health and Safety Executive**
HUMAN RIGHTS
  criminal cases and E15031
HUMIDITY
  manual handling M3006
HUNGARY
  professional regulation E18017
HYDROCARBON RELEASES
  offshore operations O7069
HYGIENE
  food legislation and, *see* **Food**
  skin diseases O1040
HYPERTHERMIA E17034

I

ICE AND SNOW SLIPPING ON M3014
IGNITION SOURCES D0933 D0933 D0933 D0933
ILL-HEALTH
  costs of IN24
  manual handling M3019
ILLUMINANCE
  average L5010
  lighting, *see* **Lighting**
  light measuring instruments L5010
  maximum ratios, adjacent areas L5011
  minimum measured L5011
  ratios L5012
  technical measurement of L5009
ILLUMINATED SIGNS
  safety signs regulations W11033
ILLUMINATING ENGINEERING SOCIETY
  limited glare index L5014
IMPORTER
  machinery safety M1013
  product safety, duties as to, *see* **Product Safety**
IMPROVEMENT NOTICE E15001
  appeal against E15019–E15025
  contents E15014
  contravention E15015, E15017, E15030
  Crown, position of E15045
  food safety F9015, F9016
  fundamentally flawed E15021
  Prohibition Notice compared E15016
  prosecution, coupled with E15018
  public register E15029, E15044
  service E15001, E15014
  use of E15013

Index

IMPROVEMENT NOTICE – *cont.*
  withdrawal E15014
INCAPACITY FOR WORK *see* Employment and Support Allowance (ESA)
  all work test C6002
INCIDENT MANAGEMENT POLICIES
  liability cover E18011
INDEMNITY
  directors, for E15041
  employers' liability insurance E13007, E13017, E13024
  inspectors, for E15028
  vicarious liability E13027
INDEPENDENT CONTRACTORS
  cables O3015
  occupiers' liability O3015, O3019
  safe system of work O3019
  supervision O3015, O3019
INDEX-LINKED GOVERNMENT STOCK
  inflation proof compensation C6028
INDICTABLE OFFENCE
  definition E15032
  time limit E15032
INDIRECT DISCRIMINATION
  Equality Act 2010 E16502, E16504
INDUCTION PROGRAMMES
  passport schemes, and T7019
INDUSTRIAL ATMOSPHERIC POLLUTION *see* Emissions into the Atmosphere
INDUSTRIAL DISEASES
  compensation for C6010
  employers' liability insurance E13001, E13025
  limitation periods E13025
  long tail diseases E13026
INDUSTRIAL INJURIES
  disablement pension C6011
INDUSTRIAL INJURIES DISABLEMENT BENEFIT (IIDB)
  accidents C6006, C6012
    defined C6005
  assessment C6011
  benefit overlaps C6020
  constant attendance allowance C6014
  disabled person's tax benefit C6016
  disablement benefit, *see* Industrial Injuries Disablement Benefit
  disablement pension C6011
  entitlement C6011
  exceptionally severe disablement allowance C6014
  noise at work N3001, N3018
  osteoarthritis of the knee C6055
  payment of C6017

INDUSTRIAL INJURIES DISABLEMENT BENEFIT (IIDB) – *cont.*
  personal injuries provisions
    acting in course of employment duties C6006, C6007
    caused by accident C6005
  pneumoconiosis C6055
  primary carcinoma of the lung C6055
  qualification for C6001
  rate of benefit C6013
  receipt of benefit, financial benefits of C6019
  statutory sick pay C6015
  weekly benefit rates payable C6017
INDUSTRIAL INJURIES SCHEME
  qualification for benefit C6001
INDUSTRIAL PREMISES
  health and safety law enforcement of E15007
INDUSTRIAL PRODUCTS
  product safety P9009
INDUSTRIAL RELATIONS
  policy and procedure E17043
INDUSTRY GUIDE
  food, as to F9025, F9049
INEXPERIENCED WORKERS
  training V12039
  vulnerable persons R3028, V12038–V12039
INFECTIOUS DISEASES O1043 O1045
INFORMATION PROVISION
  accident investigations A3029, A3032
  children and young people V12010, V12014–V12015
  community health and safety C6523, C6527
  dangerous substances D0926, D0933
  display screen equipment D8209
  environmental management E16024
  equipment M1007
  explosions D0926
  food safety F9027
  heights, working at W9009
  ladders W9009.1
  lifting equipment L3026–L3027
  managing health and safety M2013, M2036
  manual handling M3007, M3019
  permit to work systems S3014
  radiation emergencies, *see* Radiation Emergencies
  sensitive information D6007–D6008
  small businesses IN04
  stress S11040

Ind-57

INFORMATION PROVISION – *cont.*
  violence in the workplace V8008, V8020, V8061, V8068, V8141
INFORMATION TECHNOLOGY
  facilities management F3030
INJUNCTION
  violence in the workplace V8140
INJURED EMPLOYEES
  accident reporting, *see* Accident Reporting
INLAND REVENUE
  disclosures to E14009
INLAND WATERWAYS
  staff working hours W10026, W10031
INNOCENCE PRESUMPTION OF E15031
INQUIRIES
  disaster and emergency planning D6060–D6062
INQUISITORIAL
  proceedings E18007
INSECTS
  plagues of E5023
INSOLVENCY
  employers' liability insurance E13024
INSPECTIONS
  arrangements for S7003, S7006
  asbestos A5015
  building and construction operations C8002–C8017
  dangerous substances D0933
  enforcement of law, for, *see* Enforcement
  equipment M1007
  food safety F9027
  heights, working at W9012
  lifting equipment L3021
  managing health and safety M2041
INSTITUTIONAL CARE
  compensation for work injuries C6028
INSTITUTION OF OCCUPATIONAL SAFETY AND HEALTH (IOSH) IN23 IN26
  continuing professional development IN26, T7021
  qualification T7011
  recognition procedures T7022
INSTRUCTIONS
  equipment M1007
  product liability P9544
INSULATION E3014 E3015 E3024
  age discrimination, exemption M1511
  business continuity planning B8007
  construction and building operations, and F5029
  construction sites F5029
  directors E15041
  disaster and emergency planning D6002

INSULATION – *cont.*
  expenses, liability policies E13023
  fire prevention and control F5029, F5074
INSURANCE *See also* Employers' Liability Insurance, Public and Products Liability Insurance
  managing health and safety M1511, M2001
  occupiers' liability O3014
  professional indemnity insurance E13003
  public liability insurance E13016, O3003, O3014
  stress S11009
INTEGRATED FACILITIES MANAGEMENT (IFM)
  outsourced F3008
INTEREST
  damages, on C6041
INTERIM AWARDS
  damages, on C6041
INTERLOCKING GUARDS M1032–M1033 M1035
INTERMEDIARIES
  product liability and P9037
INTERNAL COMBUSTION ENGINE C7021
INTERNATIONAL COMMISSION ON NON IONISING RADIATION PROTECTION (ICNIRP) R1008–R1009 R1010.1
INTERNATIONAL FRAMEWORK
  Canada I2010–I2013
  Europe E18001–E18023
  legal E18003–E18010, I2002
  USA I2003–I2009
INTERNATIONAL STANDARDS see also International Framework
  environmental management E16015
  ISO 14000 series E16015
    EMAS differences E16016
INTRANET
  health and safety policy statements S7008
INTRUDER ALARMS V8090
INVESTIGATIONS
  See also Accident Investigations
  disaster and emergency planning D6059, D6060–D6062
  enforcement E14013
  risk assessment R3040
  violence in the workplace V8060, V8129, V8140–V8141
INVESTMENT
  environmental management E16024
  Principles for Responsible Investment (UN) E16007

IONISING RADIATION R1001,
    R1011–R1021.1
Approved Code of Practice R1019
approved dosimetry service R1019
breast-feeding E14031, R1019
Centre for Radiation, Chemical and
    Environmental Hazards R1017
classified persons, designation of R1019
consumer products R1014
contamination R1016
co-operation between employers R1019
cosmic radiation R1013
definition R1011
designated areas R1019
doses
    assessment and recording R1019
    exposure, limit for R1019
    Ionising Radiations Regulations
        2017 R1019
    lower radiation doses R1015
    units of measurement R1012
electromagnetic spectrum R1001.1
emergency planning R1021
enforcing authority E15009
female employees R1019
guidance R1019
harmful effects R1015
health surveillance R1019
Ionising Radiations Regulations
    2017 R1019
irradiation R1016
management R1019
man-made radiation R1014
medicine R1014, R1018
National Arrangements for Incidents
    involving Radioactivity
    (NAIR) R1021
natural radiation R1013
new or expectant mothers R1019,
    V12003–V12005, V12019–V12020
nuclear power R1014
occupational diseases O1056
outside workers R1019
overexposure R1019
passbooks R1019
personal protective equipment P3008,
    P3014, P3017, R1019
pregnant workers E14031, R1019,
    V12003–V12005, V12019–V12020
protection against radiation R1017
Radiation Protection Advisor R1019
radioactive substances
    assessment of risk R3016, R3027,
        R3035
    consumer products R1014

IONISING RADIATION – cont.
radioactive substances – cont.
    contamination R1016
    irradiation R1016
    man-made radiation R1014
    medicine R1014
    nuclear industry R1021.1
    nuclear power R1014
    radiography R1014
    regulations R1018, R1019
radioactivity R1012
radiography R1014
radon R1013
regulations IN21, R1018, R1019, R1020
risk assessment H2101, H2145, R1019,
    R3016, R3027, R3035
training R1019
types of R1011
units of measurement R1012, R1013

IOSH *See* **Institution of Occupational Safety
    and Health**
IRRADIATION R1016
ISO
    ISO 14000 series E16015
        EMAS differences E16016
    social responsibility E16002
ISOLATION PROCEDURES S3023
ITALY
    professional regulation E18017

J

JOB APPLICANTS
    discrimination O1003
JOB SAFETY ANALYSIS S3009
JOB SAFETY INSTRUCTIONS S3010
JOINT AND SEVERAL LIABILITY
    product liability P9039
JUDGMENT
    transportability of P9046

K

KEY PERFORMANCE INDICATOR
    (KPI) F3017
KEY POLLUTANTS AND ISSUES
    generally E5017–E5020
KITEMARK P3009

Ind-59

L

LABELLING
  dangerous goods, carriage of D0420
  dangerous substances D0933
  food, requirements as to F9003, F9020, F9023, F9075, F9078
  product safety P9002
  small businesses IN04
LADBROKE GROVE RAIL DISASTER D6029
LADDERS
  guidance W9009.1
  portable ladders W9009.1
  pre-check lists W9009.1
  step-ladders W9009.1
LAKE TRANSPORT
  working time W10026
LANDFILL COMMUNITIES FUND E16006
LANDFILL TAX E16006LANDLORD
LAND-USE PLANNING
  major accident hazards regulations M1135
LAPTOP COMPUTERS D8220
LARGE BUSINESSES
  compliance IN05
  corporate governance IN05
  corporate social responsibility IN05
  directors, duties of IN05
  Modern Slavery Act 2015 IN05
  performance, improving IN05
  plan-do-check-act management systems IN05
  public sector IN05
  risk management IN05
  voluntary sector IN05
LATVIA
  professional regulation E18017
LEAD
  children and young people V12002, V12016
  hazardous substances regulations and H2101, H2140
  new or expectant mothers V12003–V12005, V12021–V12022
  regulations H2140
  risk assessment R3016, R3027, R3035
  women, employment of E14023
LEASE
  asbestos A5021
LEAVE see **Annual Leave**
LEGAL ADVISER
  disclosures to E14009

LEGAL AID
  exceptional cases C6021
LEGAL EXPENSES INSURANCE E13023
LEGAL FRAMEWORK
  Brexit IN07–IN08
  civil law IN22
  criminal law IN18
  European Union E18003–E18010, IN07–IN08
  structure IN17
LEGAL REPRESENTATIVES
  accident reporting A3028
LEGIONNAIRES' DISEASE C6510, F3026, H2103, O1043
  outbreaks C6510
LEPTOSPIROSIS H2103
LEUCODERMA O1038
LIABILITY INSURANCE
LIABILITY see also **NEGLIGENCE**
  incident management policies E18011
  management regulations M2005
  strict P9522
LICENSING
  adventure activities C6508
  asbestos A5027
    contractors, asbestos removal A5027
    regulations as to E15005, R3016, R3034
  food business F9021
  training and competence T7020
LIFELONG LEARNING T7008
LIFTING EQUIPMENT
  accessories L3038–L3047
  bath hoist L3036
  chains L3043–L3044
  community health and safety C6510, C6511
  competent person L3019–L3022
  cranes L3018, L3029–L3031
  declarations of conformity L3027
  defects L3023–L3025
  duty holders L3014
  electrical equipment, controlled by L3002
  examination L3021, L3023–L3026, L3028
  eyebolts L3046–L3047
  failure of equipment L3028–L3031
  fork lift trucks L3034–L3035
  general requirements L3015–L3027
  hoists L3032–L3033, L3036
  individual operations, planning of L3020
  information, prescribed L3026–L3027
  inspection L3021
  installation L3017, L3021
  lifts, see Lifts

LIFTING EQUIPMENT – *cont.*
  manufacturers and owners/users, relationship between L3002
  markings L3018
  mobile equipment L3016, L3051
  once only use accessories L3050
  organisation L3019
  passers-by C6511
  patient/bath hoist L3036
  persons, lifting L3016
  planning C6511, L3019–L3020
  positioning L3017
  records L3027
  regulations L3014–L3027
  reports L3023–L3025
  risk assessment L3017
  ropes L3039–L3042
  shackles L3048–L3049
  slings L3045
  so far as is reasonably practicable (SFAIRP) L3016
  statutory requirements L3002
  strength and stability L3015
  vehicle lifting table L3037

LIFTING OPERATIONS L3001–L3052
  duty holders L3014
  equipment, *see* Lifting equipment
  lifts, *see* Lifts
  LOLER IN21, L3014–L3021
  manual handling M3010
  offshore operations O7070
  principals L3001
  Regulations IN21, L3014–L3027
  risk assessment R3017
  terminology L3001

LIFTS
  building or construction L3009
  CE markings L3004, L3010
  conformity assessment procedure L3005, L3006, L3011
  criminal offences L3008
  declarations of conformity L3005, L3006, L3009
  design L3009, L3011
  due diligence L3009
  examination L3011
  exhibitions L3010
  facilities management F3032, F3034
  failure L3033
  general requirements L3005–L3007
  goods L3003
  information, supply of L3008
  installations L3011–L3012
  maintenance L3032

LIFTS – *cont.*
  meaning L3003
  notified bodies L3013
  obstruction of lift shafts L3008
  penalties L3008, L3009
  persons L3003
  placing on the market L3006, L3010, L3012
  powered working platforms L3032
  quality assurance L3011
  regulations on L3004–L3013
  responsible persons L3005, L3006
  safety components L3005–L3007, L3010, L3012
  service safety components, putting into L3006, L3010
  standards L3005
  supply L3007
  technical documentation, retention of L3005, L3006, L3008
  testing L3011

LIGHTING L5001–L5020
  brightness L5014, L5017
  Chartered Institute of Building Services Engineers (CIBSE) L5002
  colour rendition L5014, L5019
  confined spaces C7025
  construction and building operations, and generally C8002–C8017
  statutory requirements L5004
  design L5014
  diffusion L5014, L5018
  disability glare L5015
  discomfort glare L5015
  distribution, British Zonal Method L5014, L5016
  economic lamp replacement policy L5013
  electricity E3020, L5009
  energy
    efficiency L5001, L5002
    saving bulbs L5009
  equipment M1007
  ergonomic design and E17004, E17031, E17035
  fitments, maintenance of L5013
  fluorescent tubes L5008, L5020
  glare L5015
  heights, working at W9008
  illuminance
    average L5011 (Table 26)
    light measuring instruments L5010
    minimum measured L5011
    ratios L5012
    technical measurement of L5009

LIGHTING – cont.
  incandescent light bulbs, phasing
    out L5009
  LED F3016, L5009
  lumen, meaning L5009
  maintenance of fitments L5013
  manual handling M3006, M3014
  natural L5007
  qualitative aspects L5014
  reflected glare L5015
  Society of Light and Lighting L5002
  solar L5008
  sources of light L5007, L5008
  specific processes L5003–L5006
  standards of L5009–L5013
  statutory requirements, general L5002
  stroboscopic effect L5020
  suitable and sufficient W11006
  traffic routes L5002
  VDUs L5002, L5003
  work equipment L5004
  workplace W11006
  workstations W11006
LIGHTNING PROTECTION
    SYSTEMS E3025
LIMITATION OF ACTIONS P9044
    W11045
  accrual, date of E13025
  death of claimant E13025
  employers' liability insurance E13001,
    E13025
    accrual, date of E13025
    death of claimant E13025
  industrial diseases E13025
  knowledge, date of E13025
  mesothelioma E13025
  public and products liability
    insurance P9511
LIMITING DEVICES M1036
LIQUEFIED PETROLEUM GASES
  dangerous substances D0933
  safety G1025B
  unloading D0933
LIQUIDS AND LIQUIFIABLE SOLIDS FIRE
    AND F5003
LISTED COMPANIES
  social and environmental data E5070
LOADING BAYS
  workplace access, generally A1004
LOADS
  see also Manual Handling
  compressed natural gas D0933
  dangerous substances
    carriage of D0439, D0440, D0443
    compressed natural gas D0933

LOADS – cont.
  regulated D0439
  segregation D0430–D0431
  vehicles A1006
LOCAL AUTHORITY
  air pollution, role as to, see Emissions into
    the Atmosphere
  disaster and emergency planning D6028
  Environment Agency and E5046
  food, role as to F9005, F9007.1, F9028
  land-use planning E5069
  National Enforcement Code IN01
  occupiers' liability O3001
  pollution control, role as to E5046
  statutory nuisance, abatement of E5062
  traffic management E5069
LOCKERS
  construction and building operations,
    and C8002–C8017
LÖFSTEDT REVIEW
  first aid F7001–F7001.1
  recommendations E13036, P9551,
    W11001
  result of IN16
  Work at Height Regulations W9014
LONE WORKERS V12027–V12031
  communication V12030
  community, persons working in
    the V12028
  emergencies V12029, V12030
  fixed workplaces V12028
  precautions V12030
  premises belonging to other employers,
    working at V12028
  risk assessment R3028, V12027
  risks for V12029
    control of V12030
  safe systems of work V12027
  types of V12028
  violence in the workplace V8092
  vulnerable workers R3028
LONG TAIL DISEASES E13026
  liability P9553
LONG-TERM SICKNESS
    ABSENCE M1502–M1505 O1002
    O1007 R2002
  definition M1502
  employment laws M1505
  statement of fitness for work M1504
LORD YOUNG REPORT
  compensation culture E13035
  insurance policyholders P9550
LOSS ADJUSTERS D6060–D6062
LOSS OF AMENITY
  compensation for C6030, C6031, C6034

Index

LUNG CANCER A5002–A5003 E13026 O1042
  benefit claim C6055
LUXEMBOURG
  professional regulation E18017

M

MACHINERY M1000–M1035
  abrasion M1025
  behavioural measures M1026
  CE marking M1009, M1015–M1017, M1026
  checklist M1024
  competition, EC law and M1009
  conformity assessment M1015–M1017
  control devices M1036
  criminal offences M1018
  crushing M1025
  cutting and severing M1025
  Declarations of Conformity
    European Union M1015–M1017
    example of M1015–M1017
    incorporation into other machinery M1015–M1017
  definition of a machine M1011
  design and, *see* Ergonomic Design M1002, M1003
    technical measures to be taken at M1026
  designers M1002, M1009
  deterring/impeding devices M1036
  directives on M1001, M1003, M1009–M1017
  drawing in M1025
  enabling devices M1036
  entanglement M1025
  equipment, *see* Equipment
  essential health and safety requirements M1015–M1017, M1026
  European Union M1001–M1003, M1015–M1017
    CE marking M1009, M1015–M1017, M1026
    competition M1009
    Declarations of Conformity M1015–M1017
    Transposed Machinery Standards M1001, M1003, M1010, M1015–M1017
  fines M1018
  foreseeable use M1002
  friction M1025

MACHINERY – *cont.*
  goal setting M1001
  guards M1029–M1035
  harm, definition of M1023
  hazards
    analysis M1024
    checklist M1024–M1025
    continuing M1024
    definition of M1023
    examples of M1025
    identification of M1024–M1025
  Health and Safety at Work etc. Act 1974 M1002
  high pressure fluid injection M1025
  hold-to-run devices M1036
  impact M1025
  importers M1013
  incorporation into other machinery M1015
  instruction manuals N3014
  legal requirements M1002
  limited movement control devices M1036
  limiting devices M1036
  manufacturers, duties on M1001, M1002, M1009–M1010, M1013
  markings M1009, M1015–M1017, M1026
  noise and hearing loss N3014, O1024
  power presses M1008
  procedural measures M1026
  product liability M1002, M1003
    directive on M1009
  puncture M1025
  records M1014
  reduction in risk, options for M1026
  risk
    analysis M1024
    definition of M1023
    evaluation M1024
    reduction, options for M1026
  risk assessment M1001, M1003, M1014, M1021–M1022
    definition M1022
    elements of M1024
    framework M1024
  safe, demonstration that the machinery is M1018
  safeguards and safety devices
    characteristics of M1028
    types of M1027–M1035
  shearing M1025
  sound level N3014
  stabbing M1025
  standards
    American National Institute M1019

Index

MACHINERY – *cont.*
  standards – *cont.*
    British M1019, M1021, M1024
    B type M1021
    C type M1021
    objectives M1020
    product safety compliance M1019
    risk assessment M1021
    status and scope of M1020
    structure of M1021
    Transposed Harmonised European Machinery M1001, M1003, M1010, M1013–M1015
  suppliers M1002, M1003, M1009–M1010, M1013, M1017–M1018
  Supply of Machinery (Safety) Regulations 1992 M1009–M1018
  technical file M1014
  trapping M1025
  trip devices M1036
  two-hand control devices M1036

MAGISTRATES' COURT
  jurisdiction IN18

MAINTENANCE
  asbestos
    building maintenance workers, exposure of A5006
    plans A5012
    records A5011
  business continuity planning B8011, B8033–B8037
  dangerous substances D0933
  electricity E3021
  equipment M1007
  facilities management F3012–F3015
  fire prevention and control F5047
  hand arm vibration O1031
  permit to work systems S3011
  pressure systems, and P7028
  records A5011
  vehicles A1012
  whole body vibration O1031
  workplace, of W11001, W11003

MAINTENANCE ENGINEER
  responsibilities of S7005

MAJOR ACCIDENT HAZARDS REGULATIONS
  action to be taken M1112
  'allmeasures necessary' M1115
  application M1104
  arrangements for major incident management M1131
  as low as is reasonably practicable (ALARP) M1115

MAJOR ACCIDENT HAZARDS REGULATIONS – *cont.*
  command and control chart of arrangements M1132
  community health and safety C6506, H2101, H2139
  Control of Major Accidents Hazards (COMAH) Regulations 2015 IN21
  dangerous substance, meaning M1104
  domino effects M1113A
  domino groups M1113A
  duties of operator M1103, M1112–M1113, M1115–M1118
  emergency plans M1122–M1126
  enforcement M1101
  environment, substances dangerous for M1104
  establishments M1104, M1109
  excluded sites and activities M1109
  exercises M1133–M1134
  explosive substances M1104
  Fee for Intervention (FFI) scheme, and M1103
  fees chargeable M1103, M1127
  flammable substances M1104
  future developments M1141
  history M1101–M1103
  industrial chemical process, meaning M1104
  information, sources of M1141
  inspections M1113B
  investigations M1113B
  land-use planning M1135
  list of substances covered by M1104
  loss of control M1104
  lower-tier establishments M1103–M1104, M1115
  major accident, what constitutes M1103
  notice of prohibition M1113
  notifications M1116
  off-site emergency plan, requirement for M1124–M1126
  on-site emergency plan, requirement for M1122–M1123, M1126
  operators M1103
  oxidising substances M1104
  places where applicable M1104, M1109
  processing of chemicals M1103
  prohibiting operation M1113
  provision of information to the public M1111
  quantity of substance 1103
  safety reports M1118–M1121
  Seveso Directives M1101–1102

Index

MAJOR ACCIDENT HAZARDS REGULATIONS – *cont.*
statement of policy for management M1130
storage of chemicals M1103
testing and validation of plans M1133–M1134
toxic substances M1104
training M1134
trigger events M1121
upper-tier establishments M1103–M1104, M1118, M1122

MAJOR ACCIDENT PREVENTION POLICY (MAPP) D6011–D6014, M1117

MAJOR PUBLIC EVENTS
communities, members of C6524
contractors C6527
HSE safety guide C6525
information C6527
risk assessment C6526
Safety Advisory Groups (SAGs) C6524

MANAGEMENT
see also Disaster and Emergency Management Systems, Environmental Management, Facilities management, Managing health and safety, Risk management
agents IN19
attendance R2007
community safety C6505
directors S7005, S7009
disaster and emergency planning D6002
eye diseases and disorders O1035
facilities management F3005
 Stoddart review F3005A
generally IN21
guidance IN23
hand arm vibration O1030
health and safety policy statements S7005
ionising radiation R1019
musculoskeletal disorder O1020.4
occupational health IN23, R2010–R2011
Occupational Safety and Health guidance on IN23
offshore workers O7029–O7033.2
organisation IN23
plan-do-check-act management systems IN05
policies IN25, S7005
Regulations IN21
respiratory diseases O1045
sickness absences, *see* Managing absence
small businesses, beliefs and attitudes of managers in IN03

MANAGEMENT – *cont.*
standards IN23
stress O1054, S11009–S11009.2, S11011–S11033, S11036–S11040.1
supply G1003, G1005, IN04
third parties
 asbestos A5020
 traffic E5069
training and competence T7004T7012
violence V8117–V8129
whole body vibration O1030
working time W10033

MANAGING ABSENCE M1501–M1531
causes M1523
certification M1521
deter, procedures to M1522
disability
 conditions that may amount to M1509
 definition M1507
 discrimination M1507–M1511
 reasonable adjustments M1511, R2005
 where condition amounts to M1506–M1511A
discrimination M1507–M1511
dismissal
 Code of Practice on Disciplinary and Grievance Procedures M1517
 fair M1516–M1517, M1530
 short-term absence M1530
 time limits M1517
elements R2007
employment contracts M1513, M1518
fit for work M1504B, O1015.1, R2007
GP medical statements M1504A, R2007
home/family responsibilities M1501
HSE guidance R2012
improvement, setting time limits for M1527
insurance benefits M1511A
keeping in touch with absent employees M1504
long-term absence M1502
 employment laws M1505
medical reports, access to M1512
 General Medical Council's Guidance M1512A
monitoring M1522
musculoskeletal injuries M1501
normal day to day activities M1507
notify, requirement to M1520
occupational medicine R2010–R2011
occupational physicians R2010–R2011
pandemics R2007
patterns, looking for M1525
popular sporting events R2007

Ind-65

Index

MANAGING ABSENCE – *cont.*
 presenteeism R2007
 progressive conditions M1509
 reasonable adjustments M1511, R2005
 records M1528
 rehabilitation M1515, R2005
 return to work interviews M1522
 reviews M1503, M1524
 short-term absences M1518–M1531
 sick pay M1513–M1514
 snow R2007
 statement of fitness for work M1504
 statutory sick pay M1514
 stress M1501, M1518
  negligence M1524
 substantial, effect M1507
 targets, setting M1527
 terms and conditions of employment M1513–M1517
 trigger points, defining M1526
 warnings, giving formal M1529
 workplace factors, reviewing M1524

MANAGING AGENT
 premises, duties at to IN19

MANAGING DIRECTOR
 responsibilities of S7005, S7009

MANAGING HEALTH AND SAFETY M2001–M2046
 accident reporting and investigation systems M2041
 agency workers M2019
 asbestos
  asbestos-containing materials A5013
  leases A5021
  long-term plans A5018
  plans A5012, A5018–A5019
  third parties A5020
 assistance with health and safety M2010
 audits M2042, M2045
 bench-marking M2039
 British Standards M2029
 capabilities M2016
 causes of accidents M2028
 checklist M2030
 children, risk assessment and M2013
 client, duties of M2024, M2027
 communication M2036, M2045
 competence M2023, M2037, M2045
 competent persons, appointment of M2010
 construction projects M2023
 contacts M2012
 contingency planning M2037

MANAGING HEALTH AND SAFETY – *cont.*
 contractor
  duties of M2024, M2027
  notifiable projects M2027
  principal M2027
 control M2034, M2045
 co-operation and co-ordination M2014, M2035, M20345
 Corporate Manslaughter and Corporate Homicide Act 2007 M2001
 danger
  areas M2011
  serious and imminent danger, procedure for M2011
 designer, duties of M2024, M2027
 emergencies
  plans and procedures M2039
  services M2012
 emergency services M2012
 employees
  duties M2017
  information for M2013
 employers' duties M2003–M2004
  employees, to M2004
  third parties M2004
 European Agency for Health and Safety at Work M2001
 external sources, contacts with M2012
 financial issues M2001
 fire
  evacuation procedure M2011
  risks, assessment of M2006
 health surveillance M2009
 host employers, persons working with M2015
 humanitarian issues M2001
 implementation M2040, M2045
 improvement, continual M2029, M2044
 information for employees M2013, M2036
 inspections M2041
 insurance M2001
 investigation systems M2041
 leases A5021
 legal issues M2001
 legal requirements M2002, M2039
 measurement performance M2045
 monitoring M2041
 new and expectant mothers and their babies, risk assessment and M2006, M2019
 objectives, setting M2039

Ind-66

MANAGING HEALTH AND SAFETY – cont.

occupational health and safety management systems M2029, M2038, M2040, M2044
organisation M2033
penalty M2001
performance
  measurement M2045
  reviews M2045
planning M2038–M2039, M2045
policy M2031–M2032
POPIMAR mnemonic M2029, M2032, M2042, M2045
prevention principles M2007
reasonably practicable, meaning of M2003
records M2008
recruitment and appointment M2037
regulations M2005–M2021
  civil liability M2005
reviews M2043
  checklist M2030
  performance M2045
risk assessment M2002, M2005–M2009, M2013, M2019
risk control systems M2039
self-employed M2015
small firms M2010
standards M2029, M2032, M2034
statements, contents of policy M2032
succession and contingency planning M2037
systems for M2029
temporary workers M2018
training M2016, M2037
visitors M2015
young people M2006, M2013, M2020

MANDATORY SIGNS W11038
see also Safety Signs

MANSLAUGHTER
corporate. *See* Corporate Manslaughter
corporate killing D6029, E13001, E13032, E15001, E15038, IN18, M2001
director E13032
disaster and emergency planning D6041–D6042
duty of care E15038
enforcement E15001
fines E15038
gross negligence C9001, C9021, C9023, E13032, E15038
Involuntary Homicide Bill D6041–D6042
reckless killing D6041–D6042
unlawful act manslaughter D6041–D6042
workplace, in E15038

MANUAL HANDLING OPERATIONS M3001–M3027
accident books M3019
accident investigation reports M3019
age M3007
armed forces M3002
assembly or production lines M3004
automation M3010
avoidance of risks M3008
back, lower M3004
bags M3013
body movements M3004
capability of the individual M3007, M3015
carrying M3003
charts M3003
children and young people V12003–V12005, V12016
chutes M3010
clothing 3007
community health and safety C6522
conveyors M3010
definitions M3002
deliveries, requiring direct M3009
design
  floor M3014
  processes or activities, of M3009
  task layout and M3012
directive on M3001
disabled workers M3007
discussions M3022
diving M3002
elimination of handling M3009
employees' duties M3002
employers' duties M3002
ergonomic approach E17044, E17052, E17063
excessive movement of loads M3004
exemptions M3002
fire procedures M3008
floors
  condition M3006, M3014
  design M3014
  uneven, slippery or unstable M3006
  variation in levels of M3006
footwear M3007
frequent or prolonged physical effort M3004
gangways M3014
gender M3007
grasp, making loads easier to M3013
gravity feed transfer systems M3010
grips M3013
handles M3013
harnesses M3013

Index

MANUAL HANDLING OPERATIONS – *cont.*
health sector M3001, M3002
health status M3007
health surveillance M3015, M3019
holding or manipulating loads at a distance from the trunk M3004
HSE guidance M3003, M3012, M3015, M3017, M3024, M3027
HSE sector guidance M3019
humidity M3006
ice and snow, slipping on M3014
ill-health records M3019
improving the work environment M3014
indents M3013
information and guidance M3007, M3019, M3028
  work sector M3019
levels, differing work M3014
lifting and lowering M3003
lifting tools M3010
lighting M3006, M3014
loads
  bulkiness or unwieldiness M3005, M3008
  grasp, difficult to M3005
  load related measures M3013
  meaning M3002
  related factors M3005
  sharp objects M3005
  size M3005, M3008, M3011
  stability M3005, M3013
  temperature M3005
  unwieldiness M3005, M3008
  weight M3005, M3008, M3011
Löfstedt Review M3002
manually-operated lifting devices M3010
mechanical handling aids M3001, M3010, M3012, M3013, M3025
mechanisation M3010
musculoskeletal problems, people with M3007
New and expectant mothers M3007
new or expectant mothers V12003–V12005, V12019–V12020
non-powered conveyors M3010
notes M3023
observations M3021
offshore activities M3002
operating procedures M3019
planning and preparation M3016–M3019
pneumatic transfer systems M3010
posture M3004
powered conveyors M3010

MANUAL HANDLING OPERATIONS – *cont.*
prevention of risks M3008–M3015
prolonged physical effort M3004
pushing and pulling M3003, M3004
  excessive M3004
rate of work imposed by a process M3004
'reasonably practicable' M3002, M3008, M3011
recommendations, review and implementation of M3025
records M3019, M3024
reduction of loads M3011
reduction of risks M3008, M3013
Regulations IN21, M3001–M3003
rest and recovery, insufficient M3004
risk assessment IN21, M3002
  after M3024–M3027
  arranging M3018
  checklist M3020, M3024
  consulting workers M3020–M3023
  filters M3003
  HSE guidance M3003
  information gathering M3019
  planning and preparation M3016–M3019
  recommendations, review of M3025
  reviews M3026
  team M3017
risk of injury M3002–M3015
  avoiding or preventing M3007–M3015
routines M3012
seated, handling when M3003
ship's crew M3002
size reduction M3013
slings M3013
slips, trips or falls M3006, M3014
social work sector M3001, M3002
space
  clear, providing M3014
  constraints M3006
stability M3005, M3013
stooping M3004
sudden movement, risk of M3004
surfaces, variation in levels of M3006
task related factors M3004
task related measures M3012
team handling M3012
temperature, extremes of M3006, M3013, M3014
toolkits M3003
toxic or corrosive substances, leaks of M3002
training M3001, M3007, M3015, M3027
transport M3009

MANUAL HANDLING OPERATIONS – *cont.*
trolleys and trucks M3010
twisting M3004
ventilation M3006, M3014
volunteers C6522
vulnerable persons V12003–V12005
weight
  limits M3003
  reduction M3013
whole-body vibration V5024
wind, gusts of M3006, M3014
working environment
  factors relating to the M3006
  improving the M3014
work sector information M3019

MANUFACTURER
equipment M1007, N3013
hand-arm vibration syndrome V5020–V5021
health and safety duties IN19
machinery safety M1001, M1002, M1009–M1010, M1013
noise and hearing loss N3013
product liability, *see* Product Liability; Product Safety

MANUFACTURING
asbestos exposure A5003
liability risk exposure P9544

MARITIME AND COASTAL EMERGENCIES D6006

MARKET
food premises regulations, *see* Food

MARKINGS
CE Mark, *see* CE Mark
lifting equipment L3018
pressure systems, and P7017

MARQUEE
food premises regulations F9032–F9033

MARRIAGE
harassment, in respect of H1703

MATERNITY. *See* New or expectant mothers

MATERNITY LEAVE
entitlement E14002, E14028

MEALS
facilities for W11027

MEANS OF ESCAPE
fire prevention and control, and F5020–F5021

MEAT
food safety legislation, *see* Food

MECHANICAL HANDLING AIDS M3001 M3010 M3012 M3013 M3025

MECHANICAL WORK
risk assessment R3016

MECHANISATION M3010

MEDIA
business continuity planning B8009, B8038–B8042
crisis communication B8038–B8042
disaster and emergency planning D6060–D6062

MEDICAL ADVISER
carbon monoxide poisoning G1025C
employment cases E15005
gas safety G1025C

MEDICAL CERTIFICATES M1504
statement of fitness for work M1504, M1515

MEDICAL DEVICES R1007

MEDICAL GROUNDS
suspension from work E14016

MEDICAL REPORTS
access to M1512
employer, obtaining own M1516

MEDICAL SERVICES
work injuries, for C6028

MEDICAL STATEMENTS M1504A, M1515

MEDICAL SURVEILLANCE
asbestos employers A5040
manual handling M3015

MEDICINAL PRODUCTS F9007 R1014 R1018

MEDICINES AND HEALTHCARE PRODUCTS REGULATORY AGENCY (MHRA) F9007

MENTAL HEALTH
first aid F7003.1
stress M1518

MERCURY POISONING
new or expectant mothers V12021–V12022

MESOTHELIOMA A5002–A5003 C6010 E13025 E13026
2008 Diffuse Mesothelioma Scheme C6058

METAL FUME FEVER H2103

METALS FIRE AND F5003

METER
gas G1022

MIGRANT WORKERS
race E16518
risk assessment E16519

MILK
distribution, of F9028

Ind-69

Index

MINES
  duties of managers E3003
  heights, working at W9001
  training and competence T7005
MINIMUM SPACE PER PERSON
  factories, offices and shops W11008
MITIGATING FACTORS
  fines C9011
MOBILE PHONE
  cancer risks R1008–R1009
  display screen equipment D8220
  non-ionising radiation R1008–R1009
  provision for travelling or otherwise absent employees F7009
  work place violence and V8092
MOBILE WORK EQUIPMENT
  definition L3051, M1008
  drive shafts M1008
  employees carried on M1008
  falling M1008
  falling object protective structures M1008
  forklift trucks, *see* Fork-lift trucks
  guards and barriers M1008
  lifting L3016, L3051
  mobile workers W10027–W10030
  overturning L3051
  power presses M1008
  Provision and Use of Work Equipment Regulations M1005, M1007–M1008
  remote controlled M1008
  restraining systems M1008
  rolling over M1008
  self-propelled C6510, M1008
  speed adjustment M1008
MONOTONOUS WORK
  working time W10019, W10033, W10034
MOTHERS *see* **New or Expectant Mothers**
MOTOR INSURERS BUREAU (MIB) E13026
MOTOR VEHICLES *see* **VEHICLES**
MOVING WALKWAYS
  workplaces, safety regulations W11017
MULTI-TRACK CLAIMS E13029
MUSCULOSKELETAL DISORDERS O1020.1–O1020.5
  back disorders E17045, O1020.1
  causes O1020.2
  clinical effects E17044
  detection O1020.3
  disability discrimination O1020.4
  European proposal D8201
  heavy manual work O1020.4
  HSE priority area M3001
  individual differences E17049
  legislation as to E17044, O1020.5

MUSCULOSKELETAL DISORDERS – *cont.*
  litigation E17044
  lower limb discomfort E17047
  management O1020.4
  managing absence M1501
  manual handling M3007
  occupational health professional, reporting to O1020.4
  osteo-arthritis O1020.1
  prevention of E17050
  psychosocial risk factors E17048
  repetitive strain injury O1020.1
  reporting O1020.4
  risk assessment O1020.4, O1020.5
  scope of E17003, E17044–E17050
  sickness absence O1020.3
  statistics as to E17002
  symptoms O1020.3, O1020.4
  task design and E17010
  tendons O1020.1
  upper limb disorders E17046, E17052, O1020.1
  use of term E17044
  work-related upper limb disorders O1020.1
MUSIC AND ENTERTAINMENT
  concert halls and theatres N3027.1
  outdoor music venues N3027.2
  pubs and clubs N3027
MUSICIANS
  noise exposure N3027–N3027.2
  safety representatives J3005
MUSTER AREAS
  offshore operations O7028
MUTAGENS H2113

N

NAMED AND SHAMED IN13
NAMOS REGULATIONS D0909
NATIONAL ARRANGEMENTS FOR INCIDENTS INVOLVING RADIOACTIVITY (NAIR) R1021
NATIONAL ENFORCEMENT CODE
  local authorities IN01
NATIONAL EXAMINATION BOARD IN OCCUPATIONAL SAFETY AND HEALTH (NEBOSH) IN26
NATIONALITY
  discrimination H1703
NATIONAL OCCUPATIONAL STANDARDS
  qualifications T7008

NATIONAL OCCUPATIONAL
STANDARDS – *cont.*
work-related violence V8142
NATIONAL OPEN COLLEGE NETWORK
(NOCN) T7017
NATIONAL SECURITY D0928
disaster and emergency planning D6097
NATIONAL VOCATIONAL
STANDARDS E3011 T7002
NATURAL GAS D0933
NATURAL LIGHTING
source of lighting L5007
NECK STRAINS
managing absence M1501
NEGLIGENCE
contributory negligence
damages and C6046
employee's IN22
product liability, as to P9029
duty of care C9021, E13032, IN22
employer's, necessity to prove M2005
Enterprise and Regulatory Reform Act
2013 IN22, M2005
gross E15038
harassment in the workplace H1724
legal aid, exceptional cases C6021
manslaughter E15038
noise N3022
occupiers' liability and O3003
product liability
comparison with P9038
contributory negligence P9034
public liability claims P9529
vicarious liability IN22
workplace violence and V8012
NEIGHBOURS
community safety C6509–C6511
dangerous activities, planning of C6511
employer's duty to R3002, R3032
precautions C6511
signs and notices C6511
traffic management and
precautions C6511
NERVOUS SHOCK
recoverable non-pecuniary losses C6033
NETTLE RASH O1038
NEW OR EXPECTANT
MOTHERS V12017–V12026
antimitotic drugs V12021–V12022
biological agents V12021–V12022
carbon monoxide V12021–V12022
checklist V12025
chemical agents V12021–V12022
compressed air, work in V12003–V12005,
V12019–V12020

NEW OR EXPECTANT MOTHERS – *cont.*
construction work V12003–V12005
definition V12003, V12017
directive on V12017–V12018, V12019
discrimination H1703
diseases, infectious or contagious V12018
diving V12019–V12020
electromagnetic radiation V12019–V12020
employment tribunal V12018
Equality Act 2010 E16520
European Union V12017–V12018,
V12019
health and safety of, protecting V12026
HSE guidance V12019, V12023–V12024
ionising radiation R1019,
V12003–V12005, V12019–V12020
lead V12003–V12005, V12021–V12022
legislation V12003–V12005
Management of Health and Safety at Work
Regulations V12001–V12003,
V12017–V12018
managing health and safety M2006,
M2019
manual handling M3007,
V12003–V12005, V12019–V12020
maternity and parental leave E14002,
E14028
mercury and mercury
derivatives V12021–V12022
movement V12019–V12020
musculoskeletal disorders E17049
night work V12018
noise V12019–V12020
non-employees, risks to V12026
personal protective
equipment V12003–V12005
physical agents V12019–V12020
physical or mental
pressure V12023–VI2024
posture V12019–V12020
pressure, physical and
mental V12019–V12020
risk assessment M2006, M2019, R3018,
R3026, R3027, R3032, R3035,
V12017–V12018, V12021, V12025,
V12026
risks to V12019–V12024
biological agents R3027
chemical agents R3027
display screen equipment R3027
physical agents R3027
skin, agents absorbed through
the V12021–V12022
stress S11014
suitable alternative work E14029, V12018

Index

NEW OR EXPECTANT MOTHERS – *cont.*
suspension from work E14029–E14030, V12018
table of hazards E16521
temperature V12019–V12020
vibration V5016, V12019–V12020
visual display equipment V12003–V12005, V12023
working conditions R3027, V12018, V12019–V12020, V12023–V12024
working time V12018

NIGHT WORK
definitions W10019
health assessment W10021, W10033, W10034
length of W10020
limits on E14002
new or expectant mothers E14026, V12018
night time W10019
night worker W10019
road transport W10029
stress E17041
transfer to day work W10021
working time W10029, W10033, W10034

NOISE AND OCCUPATIONAL HEARING LOSS N3001–N3041, O1021–O1026
acoustic booths N3037
acoustic shock N3028
acoustic signals W11031
action against employer at common law N3022
agriculture, specific legal requirements N3015
assessments, formal N3013, N3029–N3034
audiometers O1023
background noise level N3010
barriers to noise N3039
British Standard N3003, N3036, N3041
building and construction operations C8018
buying policy for equipment N3035
call centres N3028
cartridge-operated tools N3003
causes O1022
clubs N3027
code of practice N3017
competent person N3029
concert halls N3027.1
conditions for which deafness prescribed N3019
construction and building operations N3016–N3017

NOISE AND OCCUPATIONAL HEARING LOSS – *cont.*
construction sites
specific legal requirements N3016, N3017
continuous N3005–N3006
contribution N3022
Control of Noise at Work Regulations 2005 IN21
C-weighting N3009
daily personal noise exposure N3007, N3013, N3031
damage occurs, how N3003
deafness
noise-induced hearing loss N3003
defensive mechanisms of the ear N3003
definition O1021
designers of machinery, duties of N3014
disablement benefit, payment of
assessment of benefit N3021
claimants, number of N3002
conditions under which payable N3020
entitlement to N3018
occupational deafness N3012
statutory requirements N3012
dosemeter N3032
ear muffs N3036
emergency purposes, sound systems for N3041
employers' duties N3013
employers' liability insurance E13026
equipment
buying policy N3035
manufacturers, information provided by N3013
ergonomic approach to E17004, E17032
E.U. law N3002A
evidence N3022
explosives N3003
exposure limit values N3013
firearms N3003
frequencies O1022
general legal requirements N3012–N3017
hair cells O1022
hazardous environments N3033
health surveillance N3002A, N3013, O1025
hearing disability through ageing N3003
hearing protection N3036–N3041
hearing protection zones IN21, N3013, N3031, N3034
HSE guidance N3023
human ear, effect of sound on N3004
human responses to N3005
importers of machinery, duties of N3014

NOISE AND OCCUPATIONAL HEARING
    LOSS – cont.
  indices N3005–N3011
  industrial injuries disablement
      benefit N3001
  legal requirements relating
      to N3012–N3017, O1025
  liability of employer at common
      law N3022
  machinery, use of O1024
  manufacturers' duties N3014
  maximum sound pressure level N3008
  measurement of sound N3002A,
      N3005–N3011, N3029–N3032,
      O1023
  meters N3030, N3033
  musicians N3002A, N3027–N3027.2
  music systems N3040
  negligence N3022
  new or expectant
      mothers V12019–V12020
  night clubs N3027
  Northern Ireland N3022
  nuisance E5062
  occupational deafness
    assessment of disablement
        benefit N3021
    compensation for N3018–N3021
    conditions for benefit to be
        payable N3020
    damages for N3022
    disablement benefit N3012
    hearing loss, degree of
        disablement N3021 (Table 29)
  organisational and technical measures,
      reduction by N3013
  outdoor music venues N3027.2
  outdoors, equipment used N3016
  PA and music systems N3040
  peak sound pressure N3009
  personal exposure N3007, N3007.1
  personal protective equipment N3013,
      P3004, P3008
  Physical Agents (Noise) Directive N3002A
  precautions N3002A
  reducing noise N3035
    specific environments, in N3024–N3028
  reduction measures IN21
  regulations IN21, N3013, N3023
  reports N3034
  risk assessment N3002A, N3013, R3008,
      R3018, R3032, R3035
    assessors N3002A, N3029
    sufficient information, instruction and
        training N3002A, N3029

NOISE AND OCCUPATIONAL HEARING
    LOSS – cont.
  rotation of work N3038
  scale of problem N3002
  ships N3002A
  sound, perception of N3004, O1021
  sound level meters N3030, N3033, O1023
  sound power level N3011
  sound systems N3041
  statistical evidence N3022
  statutory controls over N3012–N3013
  successive employers N3022
  survey of N3002
  theatres N3027.1
  tinnitus O1022
  tractor cabs N3015
  upper exposure action levels N3013
  vibration E17033, N3012, N3013, N3017
  weekly personal noise exposure
      level N3007.1, N3013
  weighted sound pressure level N3006,
      N3008

NON-ENGLISH SPEAKERS
  contractors A1014

NON-IONISING
    RADIATION R1001–R1010.11
  body, interaction with the R1003
  cancer R1008–R1009
  Control of Artificial Optical Radiation at
      Work Regulations 2010 R1010.11
  Control of Electromagnetic Fields at Work
      Regulations 2016. see Control of
      Electromagnetic Fields Regulations
      2016
  definition R1001.1
  direct effects R1005
  diseases R1008–R1009
  Electromagnetic Fields (EMF) R1002,
      R1004, R1006–R1010.7, R1010.9,
      R1010.10
  electromagnetic spectrum R1001.1
  frequencies R1001.1, R1002, R1005,
      R1008–R1009, R1010.1, R1010.3,
      R1010.5–R1010.7
  health surveillance R1010.9
  indirect effects R1006–R1010.11
  International Commission on Non Ionising
      Radiation Protection
      (ICNIRP) R1008–R1009, R1010.1
  known risks R1003
  medical devices, implanted R1007
  mobile phone cancer risks R1008–R1009
  optical radiation R1010.11
  precautions R1010.11
  risk assessment R1010.6

Index

NON-IONISING RADIATION – *cont.*
uncertain health risks R1003–R1004
NON-PECUNIARY LOSSES
bereavement C6031
loss of amenity C6031, C6034
pain and suffering
illness following post-accident surgery C6031, C6032
nervous shock C6033
principal types C6023, C6031
recoverable C6033
NORTH AMERICA
codified legislation I2003
contracts, validity I2008
legal framework I2002—I2003
Occupational Safety and Health Authority (OHSA) I2003
products supplied to, insurance restriction P9538
professional insurance I2009
professional recognition I2005
punitive damages I2003
NORTHERN IRELAND
air pollution, regulation of E5046, E5069
disaster and emergency planning D6007–D6008
Food Standards Agency F9005
heights, working at W9003
Industrial Pollution and Radiochemical Inspectorate E5069
vibration regulation V5015
NOTICES
community health and safety C6511, C6514
neighbours and passers-by C6511
NOTIFIABLE CONSTRUCTION PROJECTS C8501
employers, application to C8501
notifiable threshold C8501, C8518
NOTIFICATION
see also Accident Reporting
disaster and emergency planning D6020
hazardous substances D0908
NAMOS Regulations D0909
violence in the workplace V807
work in compressed air, and C8018
NUCLEAR
industry R1021, R1021.1
power R1014
NUCLEAR INSTALLATIONS
fire prevention, enforcement C6507, F5056
NUISANCE
air pollution E5016, E5062
private nuisance P9529

NUISANCE – *cont.*
public liability claims P9529
public nuisance P9529
statutory nuisance E5062
NURSERY WORKERS
risk factors E16521
NURSES
occupational health R2010
rehabilitation R2010
risk factors E16521
NVQ (NATIONAL VOCATIONAL QUALIFICATION) T7011
assessment T7016

O

OCCUPATIONAL DEAFNESS
disablement benefit C6056, N3012
OCCUPATIONAL DISEASES O1021–O1056
cancer O1056
carginogens O1056
eye, diseases and disorders of the O1032–O1036
hand arm vibration and whole body vibration O1027–O1031
ionising radiation O1056
musculoskeletal disorders O1020.1–O1020.5
noise and hearing loss O1021–O1026
psychological disorders O1020.6–O1020.9
radiation O1056
reportable A3002, A3016
respiratory diseases O1042–O1055
skin diseases O1037–O1041
stress, anxiety and depression O1020.6–O1020.9
vibration white finger V5011–V5014
whole body vibration O1027–O1031
work-related upper limb disorders (WRULDs), *see* Work-Related Upper Limb Disorders (WRULDs)
OCCUPATIONAL HEALTH *see* Occupational safety and health
OCCUPATIONAL HYGIENIST
responsibilities, of O1013, S7005
OCCUPATIONAL MEDICINE
rehabilitation R2010–R2011
OCCUPATIONAL PHYSICIANS
qualifications R2010
OCCUPATIONAL SAFETY AND HEALTH O1001–O1020 O1056
accreditation scheme R2010

Ind-74

Index

OCCUPATIONAL SAFETY AND HEALTH
– cont.
buying services O1019
confidentiality O1016, O1018
damages O1005
data protection O1017
definition O1001
disaster and emergency planning D6002
duty of care O1005
ergonomics and, see Ergonomic Design; Ergonomics
Fit for Work service M1504B, R2011
fit notes O1015.2, R2008
GP fit notes O1015.2
guidance O1009
Health and Safety Executive, reporting to O1056
health education and promotion O1007
health records O1016–O1018
  confidentiality O1016, O1018
  data protection O1017
  keeping and making use of O1018
  obtaining O1017
  subject access O1017
health surveillance O1004, O1009
importance of O1001.1–O1005
legal duties O1004
long-term sickness O1002, O1007
management systems M2029, M2038, M2040, M2044
medical reports, obtaining O1017
musculoskeletal disorders, see Musculoskeletal Disorders
Occupational Safety and Health Consultants Register (OSCHR) R3029, T7022
personnel support O1003
prevention rather than cure O1007–O1009
priorities O1019
proactive, being O1007
reactive, being O1007
records O1016–O1018
register of consultants R3029
rehabilitation R2003–2004, R2009
relationship between occupational health and safety O1006
reporting O1056
risk assessment O1005, O1008
short-term sickness O1007
sickness absence O1002, O1007
'team', occupational health
  ergonomists O1014
  generally O1010
  hygienists O1013

OCCUPATIONAL SAFETY AND HEALTH
– cont.
'team', occupational health – cont.
  nurses O1011, R2010–R2011
  physicians O1012, R2010–R2011
  technicians O1012.1
  treatment services O1015
training and competence T7013, T7022
vulnerable persons O1005, O1008

OCCUPIER
control O3004
dangers to guard against O3012
definition O3004
duty to provide
  sanitary conveniences W11019
  washing facilities W11020
factory, actions against O3018
fire prevention and control F5073
liability, see Occupiers' Liability
number of people O3004

OCCUPIERS' LIABILITY
access O3018
causation O3017
children O3010, O3011, O3017
  trespass O3010
  warning signs O3010
common law duty of care O3002
community health and safety C6508
contributory negligence O3011
dangers to guard against O3012
drains O3014
duties under
  HSWA O3019
  OLA 1957 O3002
  owed. to who O3006
factories O3018, W11047
  access O3018
  Approved Code of Conduct O3018
  slips, trips and falls O3018
  traffic routes O3018
fire, as to O3014
foolhardy and negligent conduct O3002, O3014
foreseeability O3010
free will O3017
heights, working at O3014
independent contractors O3015, O3019
intruders O3009
latent dangers O3010
maintenance and repair of premises O3004
nature of duty O3003
obvious risks O3017
occupier, definition O3004

OCCUPIERS' LIABILITY – cont.
Occupiers' Liability Act 1984 A1001
premises
definition O3005
maintenance and repair O3004
visitors to O3006
work being done, dangers of O3015
public liability insurance O3014
public right of way O3006
recklessness O3014
risks willingly accepted O3017
safe system of work O3019
scope of duty O3004
Scotland, in O3001
slips, trips and falls O3018
standard of care C6508
structural defects O3013
supervision O3006
swimming O3017
traffic routes O3018
train surfing O3017
trespassers
children O3010
common law definition O3007
common law duty to O3008
duty to, under OLA O3009
latent dangers O3010
unfair contract terms O3016
visitors
community health and safety C6508
invitees O3003, O3006
licensees O3003, O3006
volenti non fit injuria O3017
waiver of duty O3016
warning notices O3006, O3009, O3016, O3017
children O3010
window cleaners, to O3014
workmen on premises O3014, O3015
ODOUR NUISANCE E5016 E5062
OFFENCES
see also Civil Liability, Criminal offences, Penalties, Sentencing
corporate manslaughter C9002
indictable E15032
offshore operations O7056
penalties E15036
OFFENSIVE WORK ENVIRONMENT
harassment H1707, H1708
OFFICE OF QUALIFICATIONS AND EXAMINATIONS REGULATION (OFQUAL) T7008
OFFICES
cleanliness, level and frequency of W11006

OFFICES – cont.
fire safety risk assessment F5061.1
window cleaning, *see* Window Cleaning
OFFSHORE OIL INSTALLATIONS
IPPC Directive E5027.2
OFFSHORE OPERATIONS O7001–O7075
abandonment of offshore installations O7001
access hazards O7072
administration. *see* management and administration *below*
applicable law O7009
application of criminal and civil law O7002
application outside Great Britain clause O7009
activities within territorial waters O7009
associated structure O7009
fixed installation O7009
jurisdiction O7009
mobile installation O7009
relevant waters O7009
territorial sea O7009
territorial waters O7009
vessel O7009
within territorial waters O7009
Approved Code of Practice O7028
Burgoyne Report O7003
burns O7071
carriage of dangerous substances O7068
civil law, application of O7002
confined spaces O7065
Corporate Major Accident Prevention Policies (CMAPPs) O7009.6
cranes O7070
criminal law, application of O7002
Cullen Report O7003, O7046
Deepwater Horizon disaster O7004.1
defences O7056
design and construction O7024–O7027
duty holders, duties of O7025
generally O7024
installations, integrity of O7025
personnel, duties of O7027
well-operators, duties of O7026
wells, integrity of O7026
diving
diving contractor O7037
general requirements O7036
planning, dive O7038
risk assessment O7038
drilling installations O7005
duty holders
defences O7056

Ind-76

Index

OFFSHORE OPERATIONS – *cont.*
  duty holders – *cont.*
    management and administration O7030
    offences O7056
  electricity O7063
  emergency response O7028
  employers
    defences O7056
    employers' liability insurance E13001
    offences O7056
  escapes O7028
  evacuation O7028
  exploration/exploitation licences O7001
  explosions O7028
  eye injuries O7071
  fire O7028
  first-aid O7056.1–O7062
    equipment O7059–O7062
    facilities, duty to provide first-aid O7057
    medical personnel O7057, O7058
    offshore first-aider O7058
    offshore medic O7058
    sick bays O7059–O7062
    suitable persons O7057, O7058
  fixed installation O7009
    abandonment O7015
    decommissioning O7014, O7015
    management and administration O7030
  hazards
    access hazards O7072
    burns O7071
    cranes O7070
    eye injuries O7071
    gassing incidents O7073
    lifting equipment O7070
    working at height O7072
  health and safety law applicable to O7009
  HSE Guidance O7028
  hydrocarbon releases O7069
  information provision O7028
  legislative background O7001
  lifting equipment O7070
  management and administration O7029–O7033.2
    defences O7056
    delegation of duties O7029
    duty holders O7030
    exclusion zones, temporary O7033.2
    general requirements O7029
    inspectors, powers of O7033
    installation manager's duties O7031
    offences O7056
    operators of fixed installation O7030

OFFSHORE OPERATIONS – *cont.*
  management and administration – *cont.*
    owners of mobile installations O7030
    persons on installation, duties of O7032
    safety zones O7033.1
  manual handling M3002
  meaning O7001
  medics O7058
  mobile installation O7009
    duties on owners O7030
    management and administration O7030
  model clauses O7001
  muster areas O7028
  noise at work O7063
  non-production installation O7005
  notification of hazardous activities
    combined operations O7019.1
    generally O7019
    well operations O7020–O7021
  offences O7056
  offshore installation
    abandonment O7001
    applicable law O7009
    application outside Great Britain clause. *see* application outside Great Britain clause *above*
    definition O7005
    design and construction. *see* design and construction *above*
    health and safety law applicable to O7009
    management and administration. *see* management and administration *above*
    parts of O7006–O7008
    Safety Case Regulations. *see* Safety Case Regulations
  oil pollution O7067
  permit to work O7074
  petroleum
    definition O7001
    provisions O7001
  Pipelines Safety Regulations O7035
  Piper Alpha disaster O7046
  production installations
    before commencing operations O7013.1, O7013.2
    type of offshore installations O7005
  protective equipment, emergency O7028
  radiation O7066
  repeal and revocation of outdated legislation O7004
  RIDDOR regulations O7069
  Robens Report O7003

OFFSHORE OPERATIONS – *cont.*
  Safety and Environmental Management Systems (SEMS) O7009.7
  safety committees O7052–O7054
    co-opted members O7052
    establishment of committee O7052
    functions O7054
    meetings O7053
    qualifications O7052
    training O7055
  Safety Directive (Safety Case etc) Regulations 2015 O7009.1–O7009.13
    Corporate Major Accident Prevention Policies (CMAPPs) O7009.6
    Safety and Environmental Management Systems (SEMS) O7009.7
    verification of environmental and safety-critical elements O7009.8
  safety representatives
    constituencies O7047
    establishment of constituencies O7047
    functions O7048
    generally O7046
    HSE Guidance O7046
    imminent risk of serious personal injury O7051
    inspection of equipment O7049
    risk of serious personal injury O7051
    safety case, inspection of O7050
    safety documents, inspection of O7050
    training O7055
  scope of provisions O7001
  Sea Gem incident O7001
  sick bays O7059–O7062
  specific legislation O7039–O7045, O7064
  stages O7001
  statistics report O7003.1
  Step Change O7003.1
  submarine pipelines
    construction O7034
    provisions O7001
    use O7034
  training O7069
  training of safety representatives O7055
  types O7005
  verification of safety-critical elements O7012, O7017, O7018
    environmental elements and O7009.8
  vibration V5015
  warnings O7028
  wells O7020–O7021, O7024, O7026
  working at height O7072
  working time W10031

OFFSHORE OPERATIONS – *cont.*
  written procedures for dealing with hazards O7069
OIL FOLLICULITIS O1038
OIL POLLUTION
  offshore operations O7067
OIL RIG EXPLOSION
  Deepwater Horizon E16010, O7004.1
OILS AND FATS FIRE AND F5003
OPEN AIR EVENTS
  fire safety risk assessment F5061.1
OPERATOR AND POLLUTION RISK APPRAISAL (OPRA) REGULATION SCHEME E16026
OPTICAL RADIATION
  regulations R1010.11
ORANGE BOOK D0404, D0407
OSTEOARTHRITIS O1020.1
  compensation claim C6055
OUTDOORS EQUIPMENT USED N3016
OUTDOOR WORKPLACE
  requirements W11002
OUTSOURCING
  facilities management F3008
  non-core activities F3003
OVERCROWDING W11008
OVERHEAD ELECTRIC LINES
  electricity E3028
OVERTURNING
  cranes L3029
  mobile equipment L3051
  platform lifts L3032
OZONE
  stratospheric ozone depletion E5024, E5025

P

PACKAGING F9047
  Dangerous goods, transport of D0413–D0416
PAGING V8091
PAIN AND SUFFERING
  damages for C6031–C6033
PANIC ALARMS V8090
PARENTAL LEAVE
  right to E14002
PARENT COMPANY
  liability C9002
PARKING AREAS
  workplace access, statutory duties A1009
PART-TIME WORKERS
  employment protection E14001–E14002

PASSERS-BY
  community health and
    safety C6509–C6511
  dangerous activities, planning of C6511
  duties to C6509–C6511
  employer's duties to R3002, R3032
  lifting operations, planning of C6511
  signs and notices C6511
PASSPORTS
  schemes T7019
  training and competence T7019
PA SYSTEMS N3040
PATIENT
  detained, corporate manslaughter E15038
PATIENT/BATH HOIST L3036
PAY
  definition J3026
  representatives of employee safety J3026
PAYMENT INTO COURT
  damages and C6042
PECUNIARY LOSSES
  principal types C6023
PEDESTRIANS
  workplace access, statutory duties A1007
PEDESTRIAN TRAFFIC ROUTES
  workplaces, safety regulations W11015
PENALTIES
  aggravating factors C9011
  corporate manslaughter C9010, E15043,
    P9548
    fines C9011
  examples IN05
  health and safety E15036, E15043
  imprisonment E15036
  mitigating factors C9011
  prosecutions IN13
  sentencing E15043
    individuals E15043.2
    organisations E15043.1
  Sentencing Council guidelines C9011,
    IN05, IN13
PENSION
  employee, for E14006
  terms and conditions of
    employment E14006
PERFORMANCE
  accident investigations A3029
  drivers for IN05
  improvements for IN03–IN05
  large businesses IN05
  managing health and safety
    measurement M2045
    reviews M2045
  public sector IN05
  small businesses IN03–IN04

PERFORMANCE – *cont.*
  voluntary sector IN05
PERFORMANCE INDICATORS
  small and medium-sized enterprises
    (SMEs) E13030
PERIMETER
  construction operations C8019
PERITENDINITIS O1020.1
PERMIT-TO-WORK SYSTEM S3011–S3022
  application, general S3020
  competent persons S3016, T7018
  confined spaces C7009–C7011
  contractors S3015
  dangerous substances D0933
  definition S3011
  description of work S3017
  documentation S3017
  electricity E3023
  equipment S3017
  essential features of S3013–S3019
  hazards S3018
  HSE guidance S3013
  information, provision of S3014
  isolation procedures S3023
  maintenance S3011
  offshore operations O7074
  operating principles S3022
  outline of S3021
  plant and equipment S3017
  precautions S3018
  procedures S3019
  risk assessments S3016
  selection S3015
  time for S3012
  training S3015, T7018
  written S3007
PERSONAL ALARMS V8090
PERSONAL HYGIENE
  food handlers, and F9038, F9046, F9074
PERSONAL INJURIES
  limitation of action W11045
  pre-action protocols E13029
**PERSONAL PROTECTIVE EQUIPMENT
  (PPE)**
  accommodation for P3002, P3006
  asbestos work P3004, P3008
  assessment of IN21
  breach of common law
    requirements P3027
  British Standards P3002, P3009, P3025
    (Table 1)
  CE mark of conformity P3012, P3013
  CEN standard P3009, P3011, P3013
  certification procedures P3009–P3014

PERSONAL PROTECTIVE EQUIPMENT (PPE) – *cont.*
children and young people V12003–V12005
clothing P3001, P3025 (Table 1)
common law requirements P3001, P3026, P3027
community health and safety C6520
confined spaces C7022
construction and building work P3003, P3004, P3008
criminal offences P3027
damages P3027
dangerous substances, for D0906, P3017
eczema P3026
employee
  contribution P3004
  duty to wear P3007
employers' duties P3006
ergonomic requirements E17035, E17066
European standards P3009–P3013
eye protection P3002, P3020, P3025, P3026
face protection P3020, P3025
farm workers P3018
floor protection P3022
foot protection P3022, P3025
general purpose P3016
general requirements P3015
hand protection P3021, P3025, P3026
hazardous substances H2109, H2110, H2118, P3008
head protection P3008, P3019, P3025
health and safety requirements P3015, P3025
hearing protection N3013, P3025
ionising radiation R1019
kitemark P3009
lead P3008
maintenance of H2131, P3002, P3005, P3006
mechanical impact risks P3017
mortal/serious dangers, protection against P3014
new or expectant mothers V12003–V12005
noise, for P3004, P3008, P3017
particular risks, additional requirements P3017
provision of IN21, P3002, P3006
quality control system P3015
radiation protection P3008, P3014, P3017, P3025
refusal to wear P3027
Regulations IN21

PERSONAL PROTECTIVE EQUIPMENT (PPE) – *cont.*
religious grounds P3019, P3027
respiratory protection P3008, P3014, P3023, P3025
risk assessment P3002, P3006, R3011, R3018, R3033, R3035
safe system of work P3026
safety training P3002, P3006
selection of P3002
self-employed persons P3006
shipbuilding operations P3008
sports-related occupations P3018
statutory requirements P3004, P3008
testing of P3011
uniform European standards P3009–P3013
use of P3002, P3006, P3007
vehicles P3018
visitors C6520
vulnerable persons V12003–V12005
whole body protection P3024
work activities requiring P3003
work regulations P3001–P3008

PERSONNEL
occupational health and safety O1003, R2010

PESTLE ANALYSIS
management tool F3007

PESTS F9047

PETROLEUM
dangerous substances D0933, M1104
fire F5031
legislation D0938
road tankers, unloading D0933

PHOTOKERATITIS AND PHOTOCONJUNCTIVITIS O1033

PHYSICAL AGENTS
children and young people V12016
new or expectant mothers V12019–V12020
skin diseases O1038

PHYSICAL HAZARDS
confined spaces A1020

PIPELINES
accident reporting A3005
dangerous substances D0927, D0933
explosives D0918, D0927
gas safety G1001, G1009
iron pipelines G1001
offshore operations safety regulations O7035

PIPER ALPHA DISASTER D6029

PITS
heights, working at W9004.1

PLACARDING D0421, D0422
PLAN DO CHECK ACT
  plan-do-check-act management systems IN05
  work-related road safety M2105
PLANNED MAINTENANCE (PM)
  facilities management F3013
PLANNED PREVENTATIVE MAINTENANCE (PPM)
  facilities management F3013
PLANS EMERGENCY see Emergency Plans
PLANT
  asbestos exposure A5035
  dangerous substances D0933
  design D0933
  fire prevention and control F5028
  permit to work systems S3017
  work in compressed air, and C8018
PLANT DISEASES E5023
PLATFORM LIFTS
  location L3032
  overturning L3032
  powered working platforms L3032
PLUGS E3029–E3030
PNEUMOCONIOSIS O1042 O1043
  compensation claim C6055
  hazardous substances and H2103
POINTING DEVICES D8221
POISONING
  carbon monoxide G1025C
  hazardous substances and H2103
POLICE
  corporate manslaughter C9001, C9002, C9006, C9016, E15038
  death of detainee C9001
  disaster and emergency planning D6060–D6062
  violence S11001, V8140–V8141
POLICE AUTHORITY
  Enforcing authority E15009
POLICIES HEALTH AND SAFETY see Health and Safety Policy Statement
POLICYHOLDERS PROTECTION BOARD
  claims handled by E13024
POLLUTION CONTROL
  air, see Emissions into the Atmosphere
  Environment Agency, see Environment Agency
  environmental management, see Environmental Management
  industrial, see Emissions into the Atmosphere
  integrated, see Emissions into the Atmosphere

POLLUTION CONTROL – cont.
  public liability insurance P9526, P9528
POPIMAR MNEMONIC M2029 M2032 M2042 M2045
PORTABLE ELECTRICAL APPLIANCES
  portable appliance testing (PAT) E3021
PORTAL
  motor claims notification E13029
PORTUGAL
  professional regulation E18017
POSITION SENSORS M1038
POST TRAUMATIC STRESS DISORDER (PTSD) C6032V8139
POSTURE D8214
POTASSIUM NITRATE D0704
POWER ISOLATION M1034
POWER PRESSES M1005, M1008
PRE-ACTION PROTOCOLS
  Civil Procedure Rules E13029
  employers' liability insurance E13029
  fast track claims E13029
  multi-track claims E13029
  personal injury E13029
  settlement E13029
  small claims E13029
PRECAUTIONS
  community health and safety contractors C6527
  customers and users of facilities and services C6517
  neighbours C6511
  passers-by C6511
  trespassers C6514
  visiting individuals and groups C6520
  confined spaces C7009, C7013–C7027
  contractors C6527
  customers and users of facilities and services C6517
  electricity E3014–E3015
  neighbours C6511
  noise and hearing loss N3002A
  non-ionising radiation R1010.11
  passers by C6511
  permit to work systems S3018
  trespassers C6514
  volunteers C6523
PRE-EMPLOYMENT
  health checks M2115
PREGNANT WORKERS see New or Expectant Mothers
PREMISES
  asbestos exposure A5035
  occupiers' liability O3005
  public liability insurance P9528
  work away P9528

Index

PRESENTEEISM R2007
PRESSURE EQUIPMENT
 Regulations IN21
PRESSURE FLUCTUATIONS
 gas safety, G1025A
PRESSURE GROUPS
 environmental management,
  and E16001–E16009
 tactics E16009
PRESSURE SYSTEMS
 certificate of adequacy P7004
 code of practice P7014
 competent persons P7018
 conformity assessment procedure P7008
 construction P7016
 design P7016
 documentation P7017
 duty holders P7011
  excepted systems and equipment P7013
  guidance P7014
 EC Directive P7006
 examination scheme
  fitness for service P7023
  frequency P7020
  general P7022
  imminent danger P7025
  nature P7021
  report P7024
  scope P7019
 excepted systems and equipment P7013
 fitness for service P7023
 guidance P7014
 health and safety legislation, and P7031
 imminent danger P7025
 information P7017
 installation P7016
 introduction P7001
 maintenance P7028
 marking P7017
 mobile plant P7018
 modification P7029
 notified bodies P7009–P7010
 operation P7026
 overpressure prevention precautions P7027
 Pressure Equipment Regulations
  conformity assessment procedure P7008
  coverage P7001
  generally P7006
  notified bodies P7009–P7010
  safety requirements P7007
 Pressure Systems Safety Regulations
  code of practice P7014
  coverage P7001, P7012
  definitions P7015

PRESSURE SYSTEMS – *cont.*
 protective device, meaning P6020
 record keeping P7030
 relevant systems and equipment P7012
 repair P7029
 safe operating limits P7018A
 safe operation P7026
 safety requirements P7007
 simple pressure vessels
  generally P7002
  regulations P7001, P7003
 Sound Engineering Practice (SEP) P7008
 steam generating boilers P7001
 storage vessels P7001
 type-examination certificate P7005
PRESUMPTION OF INNOCENCE E15031
PRIMARY CARCINOMA *see also* **Lung Cancer**
 compensation claim C6055
PRINCIPALS
 liability for discriminatory acts E16501
PRINCIPLE C
 UK Corporate Governance Code E16005
PRINCIPLES FOR RESPONSIBLE INVESTMENT (PRI) E16007
PRISONER
 corporate manslaughter E15038
PROCUREMENT
 environmental considerations E16008
 European Union E16008
 sustainable E16008
PRODUCTION LINES M3004
PRODUCTION METHODS
 risk assessment R3016
PRODUCT LIABILITY
 see also Product Safety, Public and Products Liability Insurance
 action, time limit for bringing P9044
 civil liability P9024–P9026
 community health and safety C6516
 contractual liability P9047
 credit sale P9055
 criminal liability P9011–P9023
 damages
  contributory negligence P9034
  persons liable for P9030
 dangerous products P9054
 defect in product
  definition of P9031
  key to liability, as P9031
  material time for existence of P9036
 development risk P9045
 EC Directive, *see* Product Safety
 exclusion of liability P9057

Index

PRODUCT LIABILITY – *cont.*
 hire contract P9048, P9055
 hire purchase P9048, P9055
 industrial products P9009
 intermediaries, role of P9037
 joint and several liability P9039
 judgments, transportability of P9046
 liability P9053
 machinery safety M1002, M1003, M1009
 negligence
  comparison with P9038
  contributory negligence P9034
 parameters of P9040
 producers
  damages, liability for P9028
  defences open to P9045
  definition of P9029
 sale of goods legislation, *see* Sale of Goods Acts
 strict liability
  introduction of P9028
  no contracting out of P9043
 substandard products, contractual liability for P9047
 suppliers' liability P9030
 supply
  definition of P9042
  time of P9033
 unfair contract terms P9058–P9060
 warnings, absence or inadequacy of P9035

PRODUCT SAFETY
 see also, Product Liability, Product Safety, Public and Products Liability Insurance
 articles for use at work
  definition of P9012
  designers' duties P9013, P9014, P9016
  importers' duties P9013, P9014
  installers' duties P9017
  manufacturers' duties P9013, P9015, P9016
  suppliers' duties P9013–P9016
 civil liability
  breach of statutory duty P9026
  generally IN22
 consumer expectation of P9032
 consumer products P9002–P9003
 Consumer Rights Act P9048
 criminal liability for breach of statutory duties P9011
 custom built articles P9020
 dangerous products P9054
 defective equipment supplied to employees P9025

PRODUCT SAFETY – *cont.*
 designers
  duties of P9013, P9014, P9016
  foreign, offences by P9021
 distributors, duties of P9002
 documentation P9002
 EC Directives
  scope of P9001, P9005
 enforcement P9011–P9012
 European Regulation
  Consumer Product Safety P9006
  Market Surveillance of Products P9006
 exports P9002, P9009
 general regulations
  conformity with, presumption of P9003
  defences under P9006
  distributors' duties P9002
  offences and penalties under P9006
  producers' duties P9002
  scope of P9002
 importers
  duties of P9013–P9015
  liability of, for
   foreigner's offences P9021
 industrial products P9009, P9019
 labelling P9002
 liability of suppliers P9030
 machinery M1019
 machinery safety M1009
 manufacturers
  duties of P9013–P9016, P9018
  foreign, offences by P9021
 market surveillance P9011
 negligence, comparison with P9038
 producers
  duties of P9002, P9005
  liability of, *see* Product Liability
 Product Safety and Market Surveillance Package P9001
 risk assessment P9002
 Sale of Goods Acts P9048
 substance for use at work
  definition of P9010
  designers' duties P9013
  duties of P9013–P9015
  importers' duties P9013, P9015
  industrial products, of P9019
  manufacturers' duties P9013, P9015
  sale of goods legislation, *see* Sale of Goods Acts
  suppliers' duties P9013, P9015
 supply chain P9002
 traceability P9002

Ind-83

Index

PRODUCT SAFETY – *cont.*
  unsafe imported products, powers to deal with P9023
  unsafe products, civil liability for P9024–P9026
  voluntary national standards P9003
  work equipment P9010

PRODUCTS LIABILITY INSURANCE *see* **PUBLIC AND PRODUCTS LIABILITY INSURANCE**

PROFESSIONAL ADVICE
  rehabilitation R2010–R2011

PROFESSIONAL INDEMNITY
  insurance E13003, P9513

PROFESSIONAL LIABILITY
  personal E18011

PROFESSIONAL REGULATION
  Canada I2012
  EU E18014–E18017
  USA I2005

PROGRAMMES
  training T7005

PROHIBITED CONDUCT
  Equality Act 2010 E16502

PROHIBITION NOTICE E15001
  appeals against E15019–E15025
  contents of E15015
  contravention of E15015, E15017, E15030
  Crown, position of E15045
  improvement notice compared E15016
  prosecution, coupled with E15018
  public register E15029, E15044
  service E15001
  temporary cessation of an activity E15015
  uses of E15013, E15015

PROHIBITORY SIGNS W11037
  see also Safety Signs

PROSECUTION
  see also PENALTIES
  community health and safety C6503, C6504
  corporate killing E13001
  enforcement of legislation by, *see* Enforcement
  gas safety G1026A
  penalties and IN13
  small firms E13001
  work-related violence V8015A

PROTECTED ACTS
  victimisation E16508

PROTECTED CHARACTERISTICS
  Equality Act 2010 E16501, H1703
  rehabilitation R2005

PROTECTIVE CLOTHING
  see also Personal protective equipment
  asbestos A5032
  British Standards P3025
  protection
    eye P3020
    foot P3022
    hand/arm P3021
    head P3019
    whole body P3024

PROTECTIVE DEVICES
  meaning P7020

PROTECTIVE VESTS
  violence in the workplace V8002

PSYCHOLOGICAL INJURY
  damages for C6005, S11009
  employer liability E14003
  stress S11009

PUBLIC ADDRESS SYSTEMS V8091

PUBLIC AND PRODUCTS LIABILITY INSURANCE P9501–P9554
  asbestos P9553
  claims procedures P9506
  Compensation Act P9549
  corporate manslaughter P9548
  costs, recovery P9554
  discharge of liability P9507
  duty of disclosure P9505
  duty to take out insurance P9504
  fees for intervention P9552
  Financial Conduct Authority P9502
  general law P9501–P9513
  Limitation Act P9511
  Löfstedt review P9551
  long tail disease P9553
  Lord Young report P9550
  policy exceptions P9510
  policyholders, requirements from P9550
  products liability insurance
    assurance P9544
    claims P9545
    compulsory P9532
    cover P9533
    defective warnings P9544
    design P9544
    efficacy P9539
    endorsements P9538
    examples of claims P9545
    extensions P9538
    financial loss P9540
    instructions P9544
    legislation P9543
    limits of indemnity P9536
    manufacture P9544

PUBLIC AND PRODUCTS LIABILITY INSURANCE – *cont.*
products liability insurance – *cont.*
  North America, products supplied to P9538
  persons P9534
  production P9544
  products guarantee P9541
  products recall P9542
  purpose P9531
  quality control P9544
  rating P9535
  risk exposures P9544
  territorial limits P9537
professional indemnity P9513
proposal forms P9503
public liability insurance
  asbestos P9528
  breach of statutory duty P9529
  civil action, specifically allowed P9522
  claims P9529
  claims notification P9523
  compulsory public liability insurance P9516
  contractors P9528
  contractual liability P9523
  cover provided P9517
  discharge of liability P9523
  duty to take out P9515
  electrical installations P9528
  endorsements P9524
  examples of claims P9529
  exclusions P9525
  extensions to cover P9524
  fire P9528
  gas safety P9528
  heat, use of P9523, P9528
  legislation P9527
  limits of indemnity P9520
  negligence P9529
  nuisance P9529
  persons P9518
  policy conditions P9523
  pollution P9523, P9526, P9528
  premises, ownership and control P9528
  purpose P9514
  rating P9519
  reasonable precautions P9523
  risk assessment P9528
  risk exposures P9528
  strict liability P9522
  suspension of cover P9523
  territorial limits P9521
  third party property P9528

PUBLIC AND PRODUCTS LIABILITY INSURANCE – *cont.*
public liability insurance – *cont.*
  underground services P9523
  work at height P9523
  work away P9528
  reasonable precautions P9508
  recovery of costs and benefits P9554
  Scotland P9547
  subrogation P9509
  vicarious liability P9512
  Woolf Reforms P9546
PUBLIC EVENTS MAJOR
  contractors C6527
  Event Safety Guide C6525
  health and safety C6524–C6526
  risk assessment C6526
  Safety Advisory Groups (SAGs) C0024
PUBLIC INTEREST ENTITIES
  social and environmental reporting E5070
PUBLICITY ORDER
  corporate manslaughter C9013
PUBLIC LIABILITY INSURANCE O3003
PUBLIC PROCUREMENT
  environmental considerations E16008
  European Union E16008
  sustainable E16008
PUBLIC SECTOR
  large companies IN05
  performance IN05
PUBS AND CLUBS
  noise at work N3027
PUNCTURE M1025
PUNITIVE DAMAGES
  USA i2003

Q

QUALIFICATIONS
  occupational health nurses R2010
  occupational physicians R2010
  other professions T7012
  outputs T7013
  'recognised qualification' T7010
  rehabilitation R2010
  stress S11034
  Vocationally Related Qualifications (VRQ) T7008
QUALIFICATIONS AND CREDIT FRAMEWORK (QCF) T7008
  Awarding Organisations (AO) T7009

QUALITY ASSURANCE AGENCY (QAA) T7010
QUALITY CONTROL
product liability P9544
QUARRIES
risk assessment C6510
QUESTIONS
job applicants O1003
QUEUING AND QUEUE JUMPING V8075 V8079

R

RACE
Equality Act 2010 E16517–E16519
harassment H1702, H1733
migrant workers E16518
protected characteristic E16517, H1703
risk assessment E16519
RACIST INCIDENTS V8011 V8140
RADIATION
Control of Electromagnetic Radiation at Work Regulations 2016 E14032
disaster and emergency planning D6018, D6019
emergencies, see Radiation Emergencies
European Union D6018
exposure
new or expectant mothers V12019–V12020
pregnant and breastfeeding mothers E14031
protection from P3008, P3015, P3017, P3025
guidance D6018
ionising radiation, see Ionising radiation
legal requirements R1010
non-ionising, see Non-ionising radiation
nuclear industry R1021, R1021.1
occupational diseases O1056
offshore operations O7066
radioactive substances, see Radioactive Substances
risk assessment R3016, R3027, R3035
training and competence T7005
RADIO
workplace violence, use in case of V8091
RADIOACTIVE SUBSTANCES
carriage of D0445–D0447
exposure to, see Radiation Exposure
ionising radiation. see Ionising Radiation
regulations on D0445–D0447

RADIOFREQUENCY ELECTROMAGNETIC FIELDS
uncertain health effects R1008–R1009
RADON R1013
RAILWAYS
cross-border services W10030
dangerous goods, carriage of D0404
Ladbroke Grove rail disaster D6029
staff working hours W10026, W10030, W10034
RAMPS W11010
RATE OF WORK IMPOSED BY A PROCESS M3004
RATIONALISATION
red tape challenge E18004
REASONABLE PRACTICABILITY
safety E15031
RECALL
product liability P9542
RECKLESS KILLING D6041–D6042
RECORDS
accidents, as to, see Accident Reporting
acoustic shock N3028
asbestos A5011
dangerous substances D0911
display screen equipment D8216A
first-aid F7020
lifting L3027
machinery safety M1014
managing health and safety M2008
manual handling M3019, M3024
occupational health and safety O1016–O1018
pressure systems, and P7030
risk assessments D8216A, R3048–R3051
Safety Case Regulations O7023
sickness absence M1528
violence in the workplace V8071
whole-body vibration V5025
working time W10033, W10034
RED TAPE CHALLENGE
government rationalisation E18004
REDUNDANCY
selection for E14010
REGISTRATION
food business F9021
Occupational Safety and Health Consultants Register (OSCHR) T7022
REGISTRATION EVALUATION AND AUTHORISATION OF CHEMICALS IN EUROPE (REACH)
authorisation list R2513
candidate substances R2513
chemicals covered by R2504

Index

REGISTRATION EVALUATION AND AUTHORISATION OF CHEMICALS IN EUROPE (REACH) – *cont.*
 compliance checking R2512
 customs supervision, substances under R2504
 dangerous substances H2137
 distributors R2505
 dossier evaluation R2512
 downstream users R2505, R2510
 duty holders R2505
 enforcement of provisions R2516
 evaluation of registration dossiers R2512
 exempt chemicals R2504
 food and foodstuff additives R2504
 free movement within EU R2502
 helpdesk and further information R2515
 import limits R2504
 information, exchange of R2516
 introduction R2501
 medicines, human and veterinary R2504
 preparation for registration 2018 R2506
 prioritised substances R2512
 purpose R2502, R2503
 radioactive substances R2504
 registrants R2505
 registration
  enforcement R2516
  information required R2509
  process R2507
  timescale for R2508
 restrictions, power to impose R2514
 Safety Data Sheets R2510
 substance evaluation R2512
 suppliers R2505
 transport, substances undergoing R2504
 UK competent authority R2515
 very high concern, substances of R2513
 waste R2504
 websites and further information R2517

REGULATED PROFESSION
 Europe E18014–E18019

REGULATIONS
 key IN20.1

REGULATORY IMPACT ASSESSMENTS
 dangerous substances D0932
 explosions D0932
 heights, working at W9009

REHABILITATION R2001–2009
 adjustments
  disability discrimination R2005
  examples of R2005
  health and safety R2006
  justification R2006

REHABILITATION – *cont.*
 adjustments – *cont.*
  making R2004
  reasonable adjustments R2005–R2006
  sickness absence R2009
 advice, using professional R2010–R2011
 apportionment of damages R2004
 attendance management R2007
 breach of duty R2004
 business argument R2002
 causation R2004
 common law R2004
 damages R2004, R2006
 data protection R2007
 definition R2001
 disability discrimination R2005
  definition R2005
  long-term disability R2005
  normal day-to-day activities R2005–R2006
  reasonable adjustments R2005
 'discrimination arising from disability' R2005
 duty of care R2004
 European Law compatibility R2005
 exacerbation of illness R2004
 foreseeability R2004
 guidance R2003
 Health and Safety Executive (HSE) R2003
 importance of R2002–R2006
 legal argument R2003
 long-term absence R2002, R2009
 occupational health R2003–R2004, R2010–R2011
 occupational health nurses R2010–R2011
 occupational physicians R2010–R2011
 presenteeism R2007
  conditions resulting from R2007
 professional advice R2010–R2011
 quantification of damages R2004
 reading and information R2012
 reasonable adjustments R2005–R2006
 return to work
  adjustments R2009
  alteration of hours R2009
  arrangements for R2004
  changing jobs R2009
  interviews R2007
  phased return R2009
  plans R2009
  preventing R2006
 sickness absences R2002
  attendance management R2007
  data protection R2007

Ind-87

REHABILITATION – *cont.*
  sickness absences – *cont.*
    long-term R2002, R2007, R2009
    lost time rate R2007
    managing M1515, R2007
    policy, contents of R2007
    recording R2007
    short-term R2002, R2007
  stress R2004, S11009–S11009.2
  vocational R2001
  working hours, alteration of R2009
RELIABILITY-
    CENTERED MAINTENANCE (RCM)
  facilities management F3015A
RELIGION
  harassment H1702
  head protection P3019
  personal protective equipment P3019, P3027
  protected characteristic H1703
REMEDIAL ORDERS
  corporate manslaughter C9012
REMOTE-CONTROLLED MOBILE WORK EQUIPMENT M1008
REMOTE WORK SITES
  water W11023
RENEWABLE ENERGY STRUCTURE
  construction work C8505
REPAIR
  dangerous substances D0933
  pressure systems, and P7029
REPETITIVE STRAIN INJURY (RSI)
  damages for C6058
  database C6058
  managing absence M1501
  visual display units, *see* Visual Display Units
REPORTABLE DANGEROUS OCCURRENCES *see* Accident Reporting
REPORTABLE OCCUPATIONAL DISEASES
  see also Accident Reporting
  accident reporting A3028
  carpal tunnel syndrome (CTS) V5011
  vibration white finger (VWF) V5011, V5012
REPORTING, ENVIRONMENTAL *see also* Accident Reporting
  greenhouse gas emissions E16024A
  guidelines E16024A
REPUTATION LOSS OF B8017
RESCUE
  heights, working at W9006
RESEARCH
  heights, working at W9001

RESEARCH – *cont.*
  product safety P9016, P9018, P9019, P9034
  violence in the workplace V8112
RESIDENTIAL CARE PREMISES
  fire safety risk assessment F5061.1
RESPIRATORY DISEASES O1042–O1055
  allergic responses O1043
  allergic rhinitis O1042
  animal workers O1043
  asbestos O1042, O1046–O1055
  asthma O1042, O1044
  carcinogens O1043
  causes O1043
  control of substances hazardous to health O1046–O1055
  detection of O1044
  engineering solutions O1045
  extrinsic allergic alveolitis O1042
  health surveillance O1044, O1045
  industries prone to causing O1042
  infectious diseases O1043, O1045
  legal implications O1046–O1055
  Legionnaire's disease O1043
  liability insurance P9553
  lung cancer O1042
  management O1045
  meaning O1042
  pneumoconiosis O1042, O1043
  respiratory protection O1045
  risk assessment O1045, O1046–O1055
  sensitisation O1043
  silicosis O1043
  spirometers O1044
  symptoms O1044, O1045
  tuberculosis O1042, O1043
  x-rays O1044
  zoonoses O1043
RESPIRATORY PROTECTION
  confined spaces C7022
  personal protective equipment/clothing P3023, P3024
RESPONSE TEAMS V8119
REST FACILITIES
  construction and building operations, and C8002–C8017
  general provision W11027
REST PERIODS
  breaks W10017
  daily W10014, W10015, W10034
  manual handling M3004
  monotonous work W10018
  stress S11008, S11014, S11019–S11020
  weekly W10016, W10034

Index

RESTRAINING ORDERS V8140
RESTRAINTS M1008 V8094 W9014
RETAIL PREMISES
  public liability insurance P9518
RETURN TO WORK
  adjustments R2009
  alteration of hours R2009
  arrangements for R2004
  changing jobs R2009
  interviews R2007
  phased return R2009
  plans IN006, R2009
  preventing R2006
REVERSING A1017
RHINITIS O1043
RIDDOR *see* **Accident Reporting**
RIDING SCHOOLS
  public liability insurance P9516
RIGHT TO ROAM O3007
RISK
  See also Risk Assessment
  analysis M1024
  burden of proof E15031
  definition of M1006, M1023
  evaluation M1024
  exposure
    products liability P9544
    public liability P9528
  foreseeability E15031
  liability insurance P9528, P9544
  lone workers V12029, V12030
  machinery safety M1023–M1024, M1026
  reduction, options for M1026
  safety and, relationship E15031
RISK ASSESSMENT
  absolute requirements R3004
  accident investigations A3029
  age E16515
  asbestos A5007, A5010, A5017, R3013, R3016, R3035
  assessment unit R3030
  biological agents E14025
  breastfeeding mothers E14023–E14024
  business continuity planning B8012–B8018
  carrying out R3036–R3041
  CDM health and safety plans R3017
  check list M3020, M3024, R3035
  chemical agents E14025
  children and young people M2013, R3025, R3032, V12002, V12006–V12007, V12008, V12013
  cleaners R3002, R3032
  community health and safety C6505, C6509–C6511, C6526, C6527
  confined spaces C7005, C7006, C7008

RISK ASSESSMENT – *cont.*
  contractors R3002, R3032
  control of R3024
  cranes L3018
  customers R3002
  danger areas R3023
  dangerous substances D0902, D0910–D0911, D0933, R3007, R3014, R3035
  disability discrimination, reasonable adjustments and E16511, W11043
  disabled persons E16512, R3028, V12033, V12035–V12036
  disaster and emergency planning D6007–D6008, D6010, D6066
  discussions R3038
  diving at work O7038
  dynamic risk assessment R3017
  emergency services R3002
  employees R3002, R3032
  enforcement E15001
  equipment M1006, M1007
  equipment, second-hand M1007
  ergonomic tool, as E17052, E17062, E17064, E17068
  examples H2120
  expectant mothers E14023, E14024, R3026, R3027
  explosives D0902, D0919, R3014, R3016
  eye diseases and disorders O1035, O1036
  fire precautions F5001.1, F5037, R3012, R3033–R3035
    public liability insurance P9528
  gender E16521
  genetically modified organisms R3016
  guards M1032
  guidance D0933
  hand-arm vibration syndrome O1030, O1031, V5015, V5022
  hazardous substances R3007
  health and safety at work requirements R3002–R3004
  health and safety policy statement S7002.1, S7006
  health surveillance R3023
  heights, working at R3015, W9004.1, W9007, W9008
  HSE publications and website R3031, R3053
  identification of issues R3033
  inexperienced workers R3028
  information required M3019, R3031
  investigations, further R3040
  ionising radiation H2101, H2145, R1019, R3016, R3027, R3035

Index

RISK ASSESSMENT – *cont.*
  lead R3016
  lifting equipment L3017
  lone workers R3028, V12027
  machinery safety M1001, M1003, M1014, M1021–M1022, M1024
  major incident potential R3016
  major public events C6526
  management regulations
    ACOP to R3042
    application of management cycle R3024
    children and young people R3025
    evaluation of precautions R3020
    hazards and risks R3019
    new or expectant mothers R3026, R3027
    scope of R3005, R3018, R3023
    'suitable and sufficient' R3021
  managing health and safety M2002, M2005–M2009, M2013, M2019
  manual handling operations IN21, M3002, R3009
    after M3024–M3027
    arranging M3018
    checklist M3020, M3024
    consulting workers M3020–M3023
    filters M3003
    HSE guidance M3003
    information gathering M3019
    planning and preparation M3016–M3019
    recommendations, review of M3025
    reviews M3026
    team M3017
  method statements R3017
  migrant workers E16519
  monitoring R3024
  musculoskeletal disorders O1020.4, O1020.5
  new mothers E14023, E14024, M2006, M2019, R3026–R3027, V12017–V12018, V12021, V12025, V12026
  noise N3013, R3008, R3018, R3032, R3035
    assessor N3002A, N3029
  non-ionising radiation R1010.6
  notes R3041
  observation R3037
  occupational health and safety O1005, O1008
  Occupational Safety Consultants Register (OSCHR) R3029
  organisation R3024
  permit to work R3017

RISK ASSESSMENT – *cont.*
  personal protective equipment R3011, R3018, R3033, R3035
  personnel, as to R3002, R3032
  physical agents E14025
  planning M3016–M3019, R3024, R3036
  practicable requirements R3004
  pregnant workers E14025, M2019
  preventative measures R3023
  public at large R3002
  race E16519
  reasonably practicable R3003
  recommendations M3025, R3045, R3046
  records, content of R3048–R3051
  regulations
    common regulations R3006–R3012
    specialist regulations R3016
  rehabilitation R2004
  respiratory diseases O1045, O1046–O1055
  review M3026, R3022, R3024, R3045, R3047
  safe systems of work R3017, S3001
  sales representatives R3032
  self-employment R3002
  serious and imminent danger, where R3023
  skin diseases O1039–O1041
  source of risk R3035
  staff absences R3034
  stress O1020.8, S11007, S11009–S11009.2
  substances hazardous to health R3007
  sufficient information, instruction and training N3002A, N3029
  team M3017
  temporary workers R3032
  test R3039
  training R3025, R3031, T7005
  variations of work practices R3034
  vibration V5015, V5027
  violence in the workplace V8001, V8006, V8021, V8044–V8055, V8057, V8077, V8094
  visitors R3002, R3032
  visual display units D8204, D8210–D8216A, R3010, R3018, R3027, R3036
    after the D8222, D8223
    approach to D8212
    checklist D8211–D8212, D8216A
    discussions with users and operators D8215
    homeworkers D8213
    observations at the D8214

Index

RISK ASSESSMENT – *cont.*
  visual display units – *cont.*
    persons carrying out the D8210
    re-assessments D8223
    records D8216A
    reduction of risk D8216
    reviews D8222, D8223
    self-assessment, guidance on D8217
    teleworkers D8213
    users, identification of D8211
  volunteer workers R3032
  vulnerable persons R3028, V12001
  whole-body vibration O1030, O1031, V5015, V5024–V5025
  who might be at risk R3032
  who should carry out R3032
  work activities examined under R3032
  working conditions E14025
  workplace violence, *see* Workplace Violence
  young people R3025, R3032
RISK MANAGEMENT
  asbestos A5007, A5010, A5017
  business continuity planning B8007
  corporate governance IN05
  cyber, guidance B8051
  dangerous substances D0902, D0913–D0917
  fire precautions F5001.1
  generally R3052
  HSE publications and website R3031, R3053
  large companies IN05
  reports IN05
  sensible R3052
  violence in the workplace V8001, V8129, V8141
ROAD ACCIDENTS
  reporting A3009
ROAD TRANSPORT
  annual leave W10029
  dangerous substances, transportation, *see* Transportation of Dangerous Substances
  directive on W10029
  disaster and emergency planning D6026–D6027
  night workers W10029
  staff working hours W10026, W10027, W10028–W10029
ROAM RIGHT TO O3007
ROBENS COMMITTEE
  enforcement powers of inspectors E15013
  offshore installations O7003
  report O7003

ROLLING OVER M1008
ROOFS
  heights, working at W9008
ROPES L3039–L3042
  breakage L3039
  failure L3040, L3042
  fibre rope failure L3040
  heights, working at W9014
  wire rope L304–L3042
ROYAL SOCIETY FOR THE PREVENTION OF ACCIDENTS IN26
RUBBISH FIRE AND F5018

S

SAFEGUARDS AND SAFETY DEVICES
  characteristics of M1028
  types of M1027–M1035
SAFE OPERATING PROCEDURES S3010
SAFE PLACE OF WORK
  construction and building operations, and C8002–C8017
SAFE SYSTEMS OF WORK S3001–S3024
  approach S3001
  common law S3003, S3005–S3006
  communication S3010
  components of S3002–S3006
  confined spaces C7008
  development of S3008–S3010
  documentation S3001
  electricity E3022–E2024
  employers' duty S3003
  hardware S3006
  heights, working at W9001
  independent contractors O3019
  isolation procedures S3023
  job safety analysis S3009
  job safety instructions, preparation of S3010
  legal components S3003–S3006
  lone workers V12027
  matrix S3007
  occupational health and safety practitioners S3001
  occupiers' liability O3019
  permit to work systems, *see* Permit to work systems
  pitfalls of not utilising S3001
  "Plan, Do, Check, Act" approach S3001
  policy statements S3004
  publications S3024
  review S3007
  risk assessment S3001

Index

SAFE SYSTEMS OF WORK – *cont.*
  safe operating procedures, preparation of S3010
  software S3006
  statute law S3004–S3006
  written arrangements S3004, S3007
SAFETY ADVISERS R3008
SAFETY ADVISORY GROUPS (SAGS) C6524
  major events C6501, C6524
SAFETY CASE REGULATIONS
  abandonment programme O7015
  audit O7010
  background O7010
  before commencement of operations
    generally O7013
    non-production installation O7013.3
    production installations
      establishment of O7013.1
      generally O7013.2
      relocation notification O7013.1
  compliance with health and safety requirements O7010
  copies O7023
  Cullen Report O7010
  decommissioning fixed installations O7014, O7015
  documents required to be kept on installation O7023
  duty holder O7011
  fixed installations
    abandonment programme O7015
    decommissioning O7014, O7015
  generally O7010
  major accident O7010
  recording of verification schemes O7018
  record keeping O7023
  regulations to underpin O7011
  reports O7023
  review of verification schemes O7018
  revision of verification schemes O7018
  safety-critical elements, verification of O7012, O7017, O7018
  safety representative inspection O7050
  specifications O7016
  summary of contents O7022
  2015 regulations O7009.1–O7009.13
  verification of safety-critical elements of installations O7012, O7017, O7018
SAFETY COLOURS W11032
  see also Safety Signs
SAFETY COMMITTEES
  'Brown Book' J3019
  consultation with IN21, J3018
  duty to establish J3017

SAFETY COMMITTEES – *cont.*
  establishment IN21, J3017, J3018
  function IN21, J3019
  offshore operations O7052–O7054
    co-opted members O7052
    establishment of committee O7052
    functions O7054
    meetings O7053
    qualifications O7052
    training O7055
  risk assessment, involvement in R3007
  role J3019
SAFETY CONCEPT
  relativity E15031
SAFETY HELMETS
  personal protective equipment P3020, P3025
SAFETY OF FOOD *see* Food
SAFETY PASSPORTS T7019
SAFETY POLICY *see* Health and Safety Policy Statement
SAFETY REPORTS D6021–D6024
SAFETY REPRESENTATIVES
  accident reporting A3027
  actors J3005
  appointment
    generally IN21
    unionised workforce J3005
  armed forces J3002
  'Brown Book' J3019
  Charter for J3002
  disclosure of information J3005, J3015, J3030
  duties of J3007
  education sector, in J3002
  employment protection legislation, overlap with J3002
  European developments J3001, J3002, J3033
  European Works Councils J3002, J3033
  extension of duty to consult J3002
  facilities to be provided by employer J3014
  general duty
    non-unionised workforce J3021
    unionised workforce J3004
  hazardous substances, discussions as to H2119
  independent J3005
  information disclosure
    non-unionised workforce J3027
    unionised workforce J3015
  musicians J3005
  non-unionised workforce
    general duty J3021

SAFETY REPRESENTATIVES – *cont.*
  non-unionised workforce – *cont.*
    generally J3020
    information provision J3002, J3027
    number of representatives for
      workforce J3022
    recourse for representative J3030
    rights and duties of
      representatives J3024–J3029
    role of safety representatives J3023
    time off with pay J3024–J3026
    training, relevant J3028
  number of representatives for workforce
    non-unionised workforce J3022
    unionised workforce J3006
  obligation to consult J3001, J3002
  offshore operations
    functions of representative O7048
    generally O7046
    HSE Guidance O7046
    imminent risk of serious personal
      injury O7051
    inspection of equipment O7049
    risk of serious personal injury O7051
    Safety Case, inspection of O7050
    safety documents, inspection of O7050
    training O7055
  prosecutions for non-compliance J3002
  protected disclosures J3030
  regulatory framework IN21, J3002
  rights and duties of safety representatives
    non-unionised workforce J3024–J3030
    unionised workforce J3010–J3013
  risk assessment, involvement in R3032
  role
    non-unionised workforce J3023
    unionised workforce J3007
  statutory protection for J3030
  technical information J3016
  time off with pay J3010–J3013, J3029
    non-unionised workforce J3024–J3026
    unionised workforce J3010–J3013
  training and competence E14001, J3028,
    T7003, T7017
  unionised workforce
    appointment of safety
      representatives J3005
    disclosure of information J3015
    duty to consult J3003, J3004
    facilities to be provided by
      employer J3014
    functions of safety representatives J3007
    general duty J3004
    generally J3003
    immunity, legal J3009

SAFETY REPRESENTATIVES – *cont.*
  unionised workforce – *cont.*
    information disclosure J3015
    'in good time', consultation J3007
    number of representatives for
      workforce J3006
    rights and duties of safety
      representatives J3001,
      J3010–J3013
    role of safety representatives J3007
    technical information J3016
    terms of reference J3007
    time off with pay J3010–J3013
    workplace inspections J3008
    workplace inspections J3008
SAFETY RULES
  enforcement by employer, *see* Disciplinary
    Procedures
SAFETY SIGNS
  see also Labelling of Hazardous Substances
  colour IN21
  community health and safety C6511,
    C6514
  design IN21
  emergency escape W11040
  employers' duties W11035
  exclusions W11036
  fire F5032, W11041, W11042
  hand signals W11032, W11042
  mandatory signs W11039
  neighbours and passers-by C6511
  prohibitory signs W11037
  Regulations IN21
  varieties of W11034
  warning signs W11038
SAFETY TRAINING AND COMPETENCE.
  See Training and competence
SALE OF GOODS
  dangerous products P9054
  dealer, sales by P9036
  fitness for purpose P9048–P9051
  food, *see* Food
  implied terms P9051
  quality of product P9051–P9053
  sample, sale by P9050, P9057
  strict liability P9053
SALES REPRESENTATIVE
  risk assessment R3032
SANITARY CONVENIENCES
  Building Regulations W11044
  construction and building operations,
    and C8002–C8017
  factories W11018–W11021
  food premises, for F9039
  minimum number of facilities W11021

Index

SANITARY CONVENIENCES – *cont.*
  occupier's duty to provide W11019
  temporary work sites W11023
SATNAV
  road safety M2113
SCAFFOLDING
  see also Working at Heights
  accident reporting A3028
  construction operations C8002–C8017
  guidance W9008
  heights, working at W9008
SCHOOLS
  community health and safety C6518
  occupiers' liability O3018
  visiting schoolchildren C6518
SCOTLAND
  corporate homicide E15038
  disaster and emergency
       planning D6011–D6014
     enforcement agencies F9005
  gas safety G1011
  legal system E18010
  liability claims P9547
  local air quality management E5069
  occupiers' liability O3001
  Scottish Credit and Qualifications
       Framework (SCQF) T7008
  Scottish Environment Protection
       Agency D6011–D6014, E5044
  smoking, prohibition of W11028
SCREENS COMPUTER *see* **Visual Display Units**
SEA FISHING
  working time W10026, W10031
SEATING WW11001 11008
  see also Workstation
SEA TRANSPORT
  manual handling M3002
  radiation emergencies, *see* Radiation emergencies
  staff working hours W10026
  vibration V5015
SECONDARY VICTIMS
  damages C6033
SECOND-HAND EQUIPMENT M1007
SECTOR SKILLS COUNCILS (SSCs) T7005
  standard setting T7008
SECURITY
  community health and safety C6511
  risk assessment R3016
  violence in the workplace V8082–V8095
  working time W10034
  workplace C6511

SECURITY PERSONNEL F3030
  V8083–V8086
see Employers' Liability Insurance, Public and Products Liability Insurance
SELF-ADJUSTING GUARDS M1030
SELF-DEFENCE
  use of force V8015
  violence in the workplace V8015
SELF-EMPLOYMENT
  asbestos A5007
  confined spaces C7003
  construction regulations C8501
  deregulation IN01
  first-aid F7015
  hazardous substances, duties as to H2105
  managing health and safety M2015
  no potential risk of harm to others,
       exemption R3002
  personal protective equipment P3006
  risk assessment R3002
  self-employed, meaning IN01
  vulnerable persons V12001
  working at height W9004
SELF-PROPELLED MOBILE EQUIPMENT M1008
SENTENCING
  guidelines IN05, IN13
  health and safety offences E15043
  mitigating factors C9011, E15043
  Sentencing Council C9001
  seriousness of offence C9011
SENTENCING COUNCIL E15001 E15043
  corporate manslaughter C9001, C9010, C9011
  fines C9011
  guidelines C9001, IN05, IN13
  publicity orders C9013
SERVICE CUSTODY PREMISES
  corporate manslaughter C9001, C9003
SERVICE INDUSTRIES IN06
SERVICES CUSTOMERS AND USERS OF C6515–C6517
SETTLEMENT
  employers' liability insurance E13029
  pre-action protocols E13029
SEVESO DISASTER D6011–D6014
SEX DISCRIMINATION
  Equality Act 2010 E14023
SEXUAL HARASSMENT
  see also Harassment
  concept H1702
  definition H1702
  detriment H1702
  nature of H1701
  protected characteristic H1703

SEXUAL HARASSMENT – *cont.*
  third party, by, employer's liability H1710
SEXUAL ORIENTATION
  harassment H1702
SHACKLES L3048–L3049
SHARED WORKSTATIONS D8219
SHAREHOLDERS
  environmental management and E16001, E16007
  pressure from E16007
SHEARING M1025
SHIFT WORKERS
  ergonomics and E17036, E17041
  stress S11019
  working time W10035
SHIPBUILDING
  first aid areas F7025
  personal protective equipment P3008
SHIPS D0907
  enforcing authority E15009
  fire safety enforcement C6507
  noise and hearing loss N3002A
  transportation of dangerous goods D0404
    UN Numbers and Proper Shipping Names D0408
SHOCK V12019–V12020
SHOPS
  doors and gates W11016
  drinking water, supply of W11024
  eating facilities W11027
  escalators W11017
  fire safety risk assessment F5061.1
  first-aid F7017
  floors, safety regulations W11010
  pedestrian traffic routes W11015
  rest facilities W11027
  sanitary conveniences W11018–W11021
  temperature control W11005
  washing facilities W11020
SICK BUILDING SYNDROME S11015
SICKNESS ABSENCES R2002
  see also Managing absence
  adjustments R2009
  attendance management R2007
  cost IN06
  data protection R2007
  Fit for Work service M1504B, O1015.1
  fit note R2008
  GP medical statements M1504A, R2007
  holidays and W10024B
  HSE guidance R2012
  insurance benefits M1511A
  long-term O1002, O1007, R2002
  lost time rate R2007

SICKNESS ABSENCES – *cont.*
  managing, *see* Managing absence
  musculoskeletal disorders O1020.3
  occupational health and safety O1002, O1007
  occupational medicine R2010–R2011
  occupational physicians R2010–R2011
  pandemics R2007
  policy, contents of R2007
  presenteeism R2007
  recording R2007
  short-term R2002
  stress O1020.6, S11017
SICK PAY E14006 M1513–M1514
SIGNALLERS
  cranes and L3030
SIGNALS *see* Safety Signs
SIGNS *see* Safety Signs
SIKH
  head protection P3008
SILICOSIS O1043
SITE WASTE MANAGEMENT PLANS (SWMP)
  facilities management F3021A
SIX SIGMA
  facilities management F3007, F3010
SKI ACTIVITY
  Enforcing authority E15009
SKIN CONDITIONS AND DISEASES H2103 O1037–O1041
  causes O1038
  chemical substances O1038, O1041
  chloracne O1038
  dermatitis O1037, O1038, O1040, O1041
    allergic contact O1037, O1038, O1040
    irritant contact O1037, O1038
  detection O1039
  engineering O1040
  experts O1039
  health surveillance O1041
  hygiene O1040
  legal requirements O1041
  leucoderma O1038
  management O1040
  oil folliculitis O1038
  personal protection O1041
  physical agents O1038
  risk assessment O1039, O1040, O1041
  skin cancers O1038
  skin ulcers O1038
SLAUGHTERHOUSE
  urticaria O1038
SLINGERS
  cranes and L3030

SLINGS L3045 M3013
  failure L3045
  safe working loads L3045
SLOVAKIA
  professional regulation E18017
SLOVENIA
  professional regulation E18017
SMALL AND MEDIUM-SIZED ENTERPRISES (SMEs)
  see also Small businesses
  benchmarking E13030
  Employers' Liability Insurance E13030
  environmental legislation E16005
  microbusinesses IN04
  occupational health R2011
  performance indicators E13030
  Work at Height Regulations W9014
SMALL BUSINESSES
  beliefs and attitudes of owners/managers in IN04
  control of substances hazardous to health IN04
  corporate killing E13001
  customers, focus on IN04
  Eco-Management Audit Scheme (EMAS) E16014
  employees, focus on IN04
  environmental legislation E16005
  influences over IN03
  insurers key principles document E13035
  labels IN04
  managing health and safety M2010
  microbusinesses IN04
  occupational health R2011
  operating conditions IN04
  performance IN03–IN04
  prosecutions E13001
  Small Business Performance Indicator E13030
  sole-employee incorporated companies (SEICs) E13012
  suppliers IN04
  supply chain management IN04
SMALL CLAIMS E13029
SMEs (SMALL AND MEDIUM-SIZED ENTERPRISES)
  see also Small businesses
  benchmarking E13030
  Employers' Liability Insurance E13030
  environmental legislation E16005
  microbusinesses IN04
  occupational health R2011
  performance indicators E13030
  Work at Height Regulations W9014

SMOG E5002 E5019
SMOKE
  statutory nuisance E5062
SMOKING
  confined spaces C7027
  exclusion area C7027
  fire F5018
  public places, prohibition in C8002–C8017, F5018, H2109, W11028
  smoking shelters or areas, provision C8002–C8017
  stress S11014
  workplaces, prohibition in C8002–C8017, F5018, H2102, W11028
SOCIAL CHARTER DECLARATION ON HEALTH AND SAFETY IN09
SOCIALLY RESPONSIBLE INVESTMENT (SRI)
  indices E16007
SOCIAL RESPONSIBILITY see Corporate Social Responsibility
SOCIAL SECURITY BENEFITS see Welfare Benefits
  compensation for work injuries C6044
SOCIAL WORK SECTOR
  manual handling M3001, M3002
SOCIETY OF LIGHT AND LIGHTING L5002
SOCKETS E3014, E3029–E3030
SOFTWARE D8206 F3030
  facilities management F3030
SOLAR LIGHTING L5008
  government grants L5008
SOLE-EMPLOYEE INCORPORATED COMPANIES (SEICs) E13012
SOLID ORGANIC MATERIALS FIRE AND F5003
SOLVENTS
  control of E5041
SOUND
  Building Regulations W11044
SOUND ENGINEERING PRACTICE (SEP) P7008
SOUND SYSTEMS N3041
SPAIN
  professional regulation E18017
SPECIFIED DISEASES
  claims for C6045–C6050
SPEED
  mobile work equipment M1008

SPILLAGES D0918
SPIROMETERS O1044
SPORT FIRE SAFETY AND F5032
SPORTS-RELATED ACTIVITY
  protective equipment P3018
SRI (SOCIALLY RESPONSIBLE INVESTMENT) E16007
STABBING M1025
STABILITY M1007
  Structural, building and construction operations C8002–C8017
STAFF FACILITIES S11014
STAIRS
  workplace, safety regulations on W11010
STAKEHOLDER
  reporting guidelines E16024A
  Stakeholder Engagement Standard E16009
STANDARDS
  American National Institute M1019
  British standards, *see* British Standards
  business continuity planning B8006
  community health and safety C6527
  corporate social responsibility. E16002
  dangerous substances, carriage of D0421, D0422
  electricity E3003, E3024, E3027
  environmental management E16015
  equipment M1007
  European D0919
    machinery safety M1001, M1003, M1010, M1013–M1015
  explosions D0919
  fire extinguishers F5016
  guards M1029, M1032
  Institution of Occupational Safety and Health (IOSH) T7021–T7022
  lead bodies T7008
  lifts L3005
  machinery safety
    American National Institute M1019
    British M1019, M1021, M1024
    B type M1021
    C type M1021
    objectives M1020
    product safety compliance M1019
    risk assessment M1021
    status and scope of M1020
    structure of M1021
    Transposed Harmonised European Machinery M1001, M1003, M1010, M1013–M1015
  managing health and safety M2029, M2032, M2034
  National Occupational Standards (NOS) T7008, T7013, T7016

STANDARDS – *cont*.
  National Vocational Standards T7002, T7021
  noise N3003, N3036
  packaging D0421
  Qualifications and Credit Framework (QCF) T7008
  setting national T7008
  stakeholder environment standard E16009
  training and competence T7008, T7022
  Transposed Harmonised European Machinery M1001, M1003, M1010, M1013–M1015
  violence in the workplace V8044
STATEMENT OF FITNESS TO WORK M1504
  options for return to work M1504
  rehabilitation M1515
STATEMENT OF TERMS AND CONDITIONS OF EMPLOYMENT
  contents E14006
STATEMENTS OF HEALTH AND SAFETY POLICY *see* Health and Safety Policy Statements
STATUTORY DUTY
  meaning IN18
STATUTORY NUISANCE E5062
STATUTORY SICK PAY M1514
STEAM
  pressure systems, *see* Pressure systems
STEP CHANGE
  offshore operations O7003.1
STORAGE
  asbestos waste A5042
  community health and safety C6511, C6514
  dangerous substances D0933
  explosions D0918
STRATEGIC REPORT
  environmental disclosure E16024
  greenhouse gas emissions E16005
STRATOSPHERIC OZONE DEPLETION E5024 E5025
STRESS ANXIETY AND DEPRESSION O1020.6–O1020.9
  causes O1020.7
  definition O1020.6
  ergonomics, role of E17002
  guidance O1020.9
  legal requirements O1020.9
  lost working days O1020.6
  management O1020.8
  managing absence M1501
  modifying factors O1020.7
  policies O1020.8

Index

STRESS ANXIETY AND DEPRESSION – *cont.*
  pressure, adverse reaction to O1020.6
  rehabilitation R2004
  rise in incidence of IN06
  risk assessment O1020.8
  support systems O1020.8
  symptoms O1020.6
  work, *see* STRESS AT WORK

STRESS AT WORK.
  administrative support and S11023
  ambiguity, role of S11022
  auditing of stress S11037
  Beacons of Excellence
    report S11009–S11009.2
  breaks S11008, S11014, S11019–S11020
  British Occupational Health Research
    Foundation (BOHRF)
    report S11009–S11009.2
  changes and arrangements,
    organisational S11027
  civil liability S11009–S11009.2
  clerical support and S11023, S11026
  communication S11027
  company structure, relevance
    of S11025–S11031
  competencies S11034
  conditions of work S11011–S11015
  conflict, role of S11022
  conflicting roles and
    functions S11017–S11018
  control
    hierarchy of S11026
    work, over S11033
  counselling, offers of S11009
  cover during staff absence S11024
  criminal liability S11009–S11009.2
  damages S11009–S11009.2
  deadlines, role of S11002, S11017
  decoration S11012
  definition S11002
  design S11013
  employers' duties S11006
  Employment Tribunal cases S11009
  ergonomics, role of S11013
  excessive pressure S11002
  extent of problem S11001
  factors to be taken into account, guidance
    on S11009
  foreseeability S11009
  guidance S11009
  hierarchy of control S11026
  individual perception of company S11029
  information and training S11040
  insurance S11009

STRESS AT WORK. – *cont.*
  interventions
    primary S11009.1, S11040.1
    secondary S11009.1, S11040.1
    tertiary S11009.1, S11040.1
  job organisation, relevance
    of S11025–S11031
  layout of workplace S11013
  maintenance S11012
  management standards S11009–S11009.2,
    S11011–S11033
  management techniques S11036–S11040.1
    auditing of stress S11037
    information and training S11040.1
    management policy S11038
    on-going monitoring S11039
    role of S11004–S11009
    surveys and S11037
  on-going monitoring S11039
  organisation of job S11025–S11031
  perceptions of role and
    company S11029–S11030
  personal factors S11032–S11035
  personal issues S11035
  physical environment and S11011
  pregnant workers S11014
  previous experience S11031
  psychiatric injury, damages for E14003,
    S11009
  qualifications S11034
  rehabilitation S11009–S11009.2
  research S11009.1
  response to, understanding S11003
  responsibility S11028, S11033
  rest times and facilities S11008, S11014,
    S11019–S11020
  retention strategies S11009.1
  risk assessment S11007
  risk factors S11010–S11015
  self-esteem, perception of one's S11030
  shift patterns S11019
  sick building syndrome S11015
  sickness absences S11017
  smoking S11014
  staff absences, cover during S11024
  staff facilities S11014
  standards S11009–S11009.2,
    S11011–S11033
  surveys and S11037
  targets, role of S11002, S11017
  technology, role of S11021
  time demands S11018
  training S11040
  travel and S11020

STRESS AT WORK. – *cont.*
  work environment O1020.6, S11009.1, S11011–S11015
  work equipment, role of S11021
  working time S11009, S11014, S11019
  work itself S11016–S11024
STRICT LIABILITY
  product safety, as to P9028, P9033
  public liability insurance P9522
STRUCTURED SETTLEMENTS
  method of paying damages C6025
STUDENTS
  visitors C6518
SUBCONTRACTOR
  corporate manslaughter C9002
SUBROGATION
  employer's liability insurance E13007
  legal principle P9509
SUBSTANCES
  see also Hazardous Substances
  Control of Substances Hazardous to Health Regulations 2002 IN21
  Dangerous Substances and Explosive Atmospheres Regulations 2002 IN21
SUBSTANCES, ESCAPE OF
  accident reporting A3028
SUBSTANDARD PRODUCTS
  contractual liability for P9047, P9055
SUITABLE ALTERNATIVE WORK
  suspension on maternity grounds E14029
SUPERVISION
  community health and safety C6517, C6520, C6523
SUPPLIER
  equipment M1007
  health and safety duties IN19
  machinery safety M1002, M1003, M1009–M1010, M1013, M1017–M1018
  product liability, *see* Product Liability; Product Safety
  small businesses IN04
SUPPLY CHAIN
  Carbon Trust, Carbon Footprints E16008
  chain management IN04
  hazardous chemicals E16008
  product safety P9002
  regulation IN17
SUPPORT SERVICES
  facilities management F3010
SUPREME COURT
  criminal law IN18
SURVEILLANCE
  health, policy statement S7006
  night workers W10034

SURVEYS
  asbestos A5023–A5026
SURVIVAL OF ACTIONS C6038
SUSPENSION FROM WORK
  maternity grounds V12018
    night work E14026
    notification E14027
    risk assessment E14025
    suitable alternative work E14029
  medical grounds E14016
  remuneration E14016
  without pay E14014
SUSTAINABILITY REPORTING GUIDELINES
  Global Reporting Initiative (GRI) E16024A
SUSTAINABLE DEVELOPMENT F3018
  business community E16011
  Carbon Reduction Commitment F3018
  corporate social responsibility (CSR) E16002
  definition F3018
  Five Capitals model F3018
  supply chain E16008
SUZY LAMPLUGH TRUST R3016
SWIMMING
  occupiers' liability O3017
  warnings O3017
SWITCHES M1032 M1034–M1035
SWITZERLAND
  professional regulation E18017
SWOT ANALYSIS
  management tool F3007

T

TANKS
  dangerous goods, carriage of D0416, D0421, D0422
  heights, working at W9001
TAX CREDIT
  disability element of working tax credit C6016
TELECOMMUNICATIONS F3030
TELEWORKERS
  risk assessment D8213
  visual display screens D8202, D8213
TEMPERATURE CONTROL
  construction and building operations, and C8002–C8017
  equipment M1007
  food, as to, *see* Food
  hot weather W11005

Ind-99

TEMPERATURE CONTROL – *cont.*
  manual handling M3006, M3013, M3014
  new or expectant
      mothers V12019–V12020
  sun protection W11005
  vibration V5015
  workplaces W11001, W11005
TEMPORARY WORK AND WORKERS
  managing health and safety M2018
TEMPORARY WORK SITES
  inclusion as 'workplace' W11002
TENANCY AGREEMENTS
  asbestos A5021
TENDINITIS O1020.1
TENDONS O1020.1
TENOSYNOVITIS O1020.1
TERRORISM D6001
TESTING
  disaster and emergency planning D6066
TEXTILES
  workers, risk factors E16521
THALIDOMIDE
  product safety and product liability P9002
THEATRES
  fire safety risk assessment F5061.1
  noise exposure levels N3027.1
THIRD PARTY
  harassment H1703
  public liability insurance P9528
THREATS
  violence in the workplace, as V8002
TINNITUS O1022
TOE-BOARDS W9008
TORTS
  harassment, and H1722
      assault and battery H1723
      negligence H1724
      remedies H1725
TOTAL FACILITIES MANAGEMENT (TFM)
  generally F3008A
  outsourced F3008
TOXICOLOGY
  air pollution E5002
TOXIC SUBSTANCES
  Building Regulations W11044
  major accident hazards M1104
  manual handling M3002
TRACEABILITY
  product safety P9002
TRACING
  Employers' Liability Tracing Office (ELTO) E13026

TRACTORS
  noise N3015
TRADE UNIONS
  collective agreements E14005
  recognition of E14001
  safety representatives IN21
  TUC Certificate in Occupational Health and Safety T7017
TRAFFIC MANAGEMENT C6511
TRAFFIC ROUTES
  above ground level, access routes to A1006
  building and construction sites C8002–C8017
  checklist
      pedestrians A1007
      vehicles A1006
  deliveries, for A1001, A1005, A1006
  heights, working at W9004.1
  internal traffic A1001, A1005
  lighting L5002
  markings A1005
  number of A1004
  obstructions A1005
  pedestrians A1002, A1004, A1007
  positions A1004
  safe access A1004
  sheeting operations, access to load tops A1008
  speed limits A1005
  speed ramps A1005
  statutory duties A1001, A1004
  sufficiency A1004, A1005
  suitability A1002, A1004
  vehicles, for A1005, A1006
TRAFFIC SYSTEMS
  see also Access
  common law duty of care A1001, A1021
  factories O3018
  occupiers' liability O3018
  organisation of safe A1002, A1003
  safe traffic routes A1006, A1007
  traffic routes, suitability of A1004
  vehicles allowed to circulate safely A1005
TRAINING AND COMPETENCE IN26, T7001–T7023
  adequacy T7004
  advice T7001, T7002, T7007
  Approved Codes T7010
  assessment T7014–T7015
  attitudes T7002
  Awarding Organisations (AO) T7009
  awards T7009
  blended learning T7007
  business continuity planning B8040

Index

TRAINING AND COMPETENCE – *cont.*
- capability T7006
- certificates T7009
- children and young people V12003–V12005, V12012
- community health and safety C6523
- competence
  - cycle of T7020
  - definition T7002
  - maintaining T7016
- competent, definition T7002
- competent person, definition T7002
- confined spaces C7011
- construction industry T7018
- Continuing Professional Development (CPD) T7021
- contractors T7018
- cycle of competence T7020
- dangerous substances D0926, D0933
  - transport of D0434, D0435, T7005
- definition of competence T7002
- development T7020
- diplomas T7009
- Directive T7003, T7022
- disaster and emergency planning D6010, D6046, D6053
- diving T7005
- doctors W10031
- education framework T7007–T7010
- electricity E3011, E3024
- employee representatives T7017
- employees T7001, T7003, T7006
- employer, responsibility T7003, T7006
  - contractors T7018
  - 'permit to work' T7018
- employment, for IN21
- environmental management E16022
- equipment M1007
- experience T7002
- explosions D0926
- fire extinguishers F5015
- first aid at work T7010
- food handlers, of F9036
- food legislation, as to F9081
- Framework for Higher Education Qualifications (FHEQ) T7010, T7012
- frameworks T7008
- gas fitters T7005
- global market, T7022
- harassment, and H1739
- hazardous substances, as to H2105, H2114, H2134
- Health and Safety at Work etc. Act 1974 T7003
- Health and Safety Executive (HSE) T7005

TRAINING AND COMPETENCE – *cont.*
- health and safety law
  - food legislation, as to F9081
  - relevant training IN21
- heights, working at W9007, W9008, W9009
- higher education T7010
- induction programmes M2016, T7019
- inexperienced workers V12039
- Institution of Occupational Safety and Health (IOSH) T7021–T7022
  - qualification T7011
- knowledge E17002, T7002 lack of
- ladders W9009.1
- liability T7003
- lifelong learning T7008
- maintenance of, T7016
- management M2016, M2037, M2045
- manual handling M3001, M3007, M3015, M3033
- mining T7005
- named training programmes T7005
- National Occupational Standards (NOS) T7008, T7013, T7016
- National Vocational Standards T7002, T7021
- need for, assessment of E17010
- new risks T7004
- non employees T7018
- Occupational Safety and Health Consultants Register (OSHCR) T7022
- occupational safety and health standards T7013
- Office of Qualifications and Examinations Regulation (Ofqual) T7008
- outputs T7013
- passport schemes T7019
- permit to work systems S3015, T7018
- personal safety T7005
- professions
  - developing programmes T7012
- programmes T7005
- qualifications T7007–T7013
- Qualifications and Credit Framework (QCF) T7008–T7009, T7011
- Quality Assurance Agency (QAA) T7010
- radiation protection advisers T7005
- recognised qualification T7010
- recognition, schemes of T7008
- registration T7022
- Regulations T7003–T7004
- repeated T7004
- resources E17042
- risk assessment R3025, R3031, T7005

Ind-101

Index

TRAINING AND COMPETENCE – *cont.*
safety passport schemes T7019
safety registration schemes T7005
safety representatives J3028, T7003, T7017
Scottish Credit and Qualifications Framework (SCQF) T7008
Sector Qualification Strategy (SQS) T7005, T7008
Sector Skills Councils (SSCs) T7005, T7008
skills T7002
societal risk T7005
stages of T7020
standards T7008, T7022
stress S11040
students R3025
taxonomy T7013
TUC Certificate in Occupational Health and Safety T7017
UK Commission for Employment and Skills (UKCES) T7008
universities T7010
validation processes T7008
visual display units D8208
Vocationally Related Qualifications (VRQ) T7008
Vocational Qualifications (VQ) T7007
work experience V12003–V12005, V12010
work in compressed air, and C8018
workplace violence, as to, *see* Workplace Violence

TRAINS
children and young people O3017
occupiers' liability O3017
surfing O3017

TRANSPORT
disaster and emergency planning D6026–D6027
enforcing authority E15009
Herald of Free Enterprise D6029
manual handling M3009
premises, fire safety F5061.1

TRANSPORTATION OF DANGEROUS SUBSTANCES
ADR Agreement D0402, D0403
  carriage D0426–D0427
  consignment procedures D0419
  documentation D0423–D0425
  exemptions D0440
  filling D0428
  handling D0426–D0427
  loading D0426–D0428
  marking and labelling D0420

TRANSPORTATION OF DANGEROUS SUBSTANCES – *cont.*
ADR Agreement – *cont.*
  packaging and containment D0413–D0416
  segregation D0430–D0431
  stowage D0429
  unloading D0426–D0428
  vehicle requirements D0432, D0433
air, carriage by D0404
bulk container and tank systems
  intermediate D0416
  placarding and marking D0421, D0422
carriage D0426–D0427
categories D0402, D0407
classification system D0406–D04012
consignment procedures D0417–D0422
  container, marking D0421, D0422
crew and vehicle D0432–D0438
dangerous goods list D0410, D0411
Dangerous Goods Safety Adviser D0405
documentation D0423–D0425
emergency procedures and information D0438
excepted quantities exemption D0441.1
exemptions D0439–D0444
filling D0428
generally D0401, D0402
handling D0426–D0427
identification D0406–D0412
labelling D0420
legal framework A1009–A1014, D0403, W11001
limited quantity packages
  exemptions D0441
loading D0426–D0428
marking and labelling
  packages D0420
  vehicles, tanks and containers D0421, D0422
operational procedures D0434
Orange Book D0404, D0407
package marking and labelling D0420
package selection D0408
packaging and containment D0413–D0416
packing groups D0412
placarding D0421, D0422
radioactive materials D0445–D0447
rail, carriage by D0402, D0404, D0419
road, carriage by D0402, D0403, D0409
  training D0434, D0435
  vehicle marking D0421, D0422
  vehicle requirements D0432, D0433
sea, carriage by D0404
segregation D0430–D0431

Index

TRANSPORTATION OF DANGEROUS SUBSTANCES – *cont.*
 small load exemptions D0442
 stowage D0429
 tank systems D0416
 training and competence D0434, D0435, T7005
 transport documents D0411
 United Nations
  classification/identification system D0407
  Orange Book D0404, D0407
  packaging and containment D0413–D0416
  Proper Shipping Names (PSNs) D0408
  UN Numbers D0408
 unloading D0426–D0427
 vehicle requirements D0432, D0433
  emergency information and documentation D0438
  equipment D0437
  filling and loading D0428
  placarding and marking D0421, D0422
  stowage D0429

TRAPPING M1025

TRAVEL
 community health and safety C6522
 stress S11020
 travelling time W10003
 volunteers C6522

TRAVELATORS *see* **MOVING WALKWAYS**

TREE FELLING
 personal protective equipment P3003, P3024

TRESPASSERS
 access C6513
 children C6512
 dangerous equipment or materials C6513
 duty to care O3007
 employer's duties to C6512–C6514, R3002, R3032
 hazardous or dangerous substances C6513
 occupier's liability, *see* Occupiers' Liability
 precautions C6514
 railway lines C6513
 right to roam O3007
 visitors O3007
 water C6513
 workplaces C6512–C6514

TRESPASS TO THE PERSON
 harassment H1723

TRIBUNAL *see* **Employment Tribunal**
TRIP DEVICES M1036
TROLLEYS AND TRUCKS M3010
TROPOSPHERE
 ozone E5002
TUBERCULOSIS O1042 O1043
TURNBULL REPORT IN05 B8001 B8006 E16005
TWO-HAND CONTROL DEVICES M1036
TYPE-EXAMINATION
 pressure systems, and P7005

U

'UBERRIMA FIDES'
 public and products liability insurance P9505
UK ATOMIC ENERGY AUTHORITY
 premises of E15005
UK COMMISSION FOR EMPLOYMENT AND SKILLS (UKCES) T7008
UK CORPORATE GOVERNANCE CODE
 principles and standards E16005
UNDERGROUND SERVICES
 public liability insurance P9523
UN ENVIRONMENT PROGRAMME E16007
UNFAIR CONTRACT TERMS
 occupiers' liability O3016
 product liability and P9047–P9048, P9059–P9060
UNFAIR DISMISSAL
 see also Enforcement
 assertion of statutory right, for E14011
 compensation for
  additional award E14022
  basic award E14022
  compensatory award E14022
  formula for E14017
  health and safety cases E14010
  special award E14022
 constructive dismissal E14014, E14019
 harassment, and H1719
  compensation E14022
  constructive dismissal H1720
  proof H1720
  remedies H1721
 health and safety grounds, on E14010, E14011, E14020
 procedural fairness E14012, E14013
 protection against E14017, E14019, J3030
 qualifying period for E14019

UNFAIR DISMISSAL – *cont.*
  reasonableness of
    employer's action E14021
  reasons for dismissal E14020
  remedies for
    compensation E14022
    re-engagement E14022
    reinstatement E14022
  statutory cap H1721
  wrongful dismissal and E14017
UNITED NATIONS
  Committee of Experts D0401, D0407
  dangerous goods, carriage of D0404,
    D0407, D0415
    Proper Shipping Names (PSNs) D0408
    UN Numbers D0408
  Economic Commission for Europe E5021,
    E5026
  Montreal protocol E5026.1
  Orange Book D0404, D0407
UNLAWFUL ACT MANSLAUGHTER
  D6041–D6042
UNSAFE PRODUCTS *see* **Public and**
  Products Liability Insurance
  liability for P9023, P9024
UNWANTED CONDUCT
  element of claim H1704
  harassment H1702
  purpose or effect H1703
  sexual nature H1703
UPPER LIMB DISORDERS E17052,
  O1020.1
  HSE guidance E17068
URTICARIA O1038
USA
  codified legislation I2003
  contracts, validity I2008
  legal framework I2002—I2003
  Occupational Safety and Health Authority
    (OHSA) I2003
  products supplied to, insurance
    restriction P9538
  professional insurance I2009
  professional recognition I2005
  punitive damages I2003
UTMOST GOOD FAITH E13004

V

VAPOUR
  explosive atmospheres D0921
VDU
  see Visual Display Units

VEGETARIAN FOOD F9076
VEHICLE EXCISE DUTY (VED) E16006
VEHICLES
  access to A1015
  authorisation to drive A1020.1
  building and construction
    operations C8002–C8017
  community health and safety C6510
  contractors A1014
  co-operation requirements A1015
  dangerous products
    carriage of D0432–D0438
    filling and loading D0428
    petrol D0901
    placarding and marking D0421, D0422
    stowage D0429
  deaths and injuries caused by A1008
  fall prevention A1008, A1015
  food safety. F9032
  garages, dangerous substances in D0901
  giving unauthorised lifts A1020
  hazardous operations, *see* potentially
    hazardous operations *below*
  loading A1008, A1018
  maintenance A1012
  Motor Insurers Bureau (MIB) E13026
  non-English speakers, provision A1014
  off-road, whole-body vibration and V5024
  parking areas A1009
  petrol D0901
  potentially hazardous
    operations A1016–A1019
    generally A1016
    loading A1008, A1018
    reversing A1008, A1017
    tipping A1008, A1019
    unloading A1008, A1018
  private vehicles A1009
  provision A1011
  repairs D0901
  reversing A1008, A1017
  safety of A1008–A1010
  speed limits A1009
  statutory requirements A1009
  subcontractors A1014
  tipping A1008, A1019
  traffic routes, *see* Traffic Routes
  training of drivers A1013
  unauthorised lifts A1020
  unloading A1008, A1018
  vehicle and fuel duty E16006
  vehicle lifting table L3037
  work equipment A1009
  work vehicles A1010–A1015

Index

VEHICLES – *cont.*
  work vehicles – *cont.*
    access to A1015
    contractors A1014
    co-operation requirements A1015
    fall prevention A1015
    generally A1010
    maintenance A1012
    provision A1011
    subcontractors A1014
    training of drivers A1013

VENDING MACHINE
  food, sale of F9032–F9033, F9041

VENTILATION
  adequate control V3015
  aerosol contamination V3002
  air cleaning devices V3009
  Approved Code of Practice D0922
  asbestos V3025
  Building Regulations W11044
  confined work spaces A1023, C7017
  contamination
    aerosols V3002
    gases V3002
    generally V3001
    sources of V3002
    toxicity V3004, V3005, V3014
  control options V3001
  dangerous substances D0933
  dilution (general) V3001, V3014
    fresh air V3014
  dilution vent V3014
  discharge V3010
  ducting V3003, V3007
    pressure V3013
  electrostatic precipitator V3009
  enclosures, partial V3004
  explosive atmospheres D0921
  fans V3008
  fresh air
    incoming V3014
    supply rate V3014
  fume cupboard V3004
  gases V3002
  generally V3001, V3016
  hoods V3003, V3005, V3006
  incoming air V3014
  lead processes, *see* Lead Processes
  legal requirements V3001, V3015
  local exhaust ventilation (LEV) H2127, H2130, V3001, V3003
    alternative to V3014
    components V3003
    design process V3011

VENTILATION – *cont.*
  local exhaust ventilation (LEV) – *cont.*
    enclosures, partial V3004
    examination and testing V3012, V3015
    HSE guidance V3015
    maintenance V3012, V3015
  low-volume high-velocity (LVHV) V3005
  maintenance V3012, V3015
  manual handling M3006, M3014
  natural V3014
  personal protective equipment V3001
  re-circulation systems V3014
  risk assessment V3015
  slots V3003, V3006
  system pressure V3013
    monitoring equipment V3013
    total volume flow V3013
  temperature control W11005
  testing V3012, V3015
  venturi scrubber V3009
  workplace W11005, W11013

VESSELS
  food, sale of F9032–F9033, F9041

VIBRATION
  aircraft in flight V5015
  back pain V5005
  blast-induced vibration V5009
  body E17033
  buildings and structures V5006, V5009
  code of practice, noise and N3017
  compensation scheme for miners, government V5003
  construction work C8018, N3017
  Control of Vibration at Work Regulations 2005 IN21
  controls V5027
  daily personal exposure values V5015
  damages V5003
  direction and frequency V5008
  dose value V5009
  effects on people of V5002–V5006
  emergency certificates V5015
  employers, impact on V5015
  employers' liability insurance E13026
  equipment, transitional period for V5015
  exposure limit values V5009, V5015
  fishing vessels V5015
  fixed structures V5006
  forestry sector V5001
  frequency of V5008
  gloves V5019
  Hand Arm Vibration (HAV), *see* Hand arm vibration (HAV)
  health surveillance V5015

VIBRATION – *cont.*
  measurement V5007–V5009
    direction and frequency V5008
    dose value V5009
    units V5007
  mechanical V5006
  new or expectant
    mothers V12019–V12020
  noise and E17032, N3012, N3013
  offshore workplaces V5015
  organisation and technical means,
    reduction of exposure by V5015
  Physical Agents (Vibration)
    Directive V5001, V5015
  regulations relating to N3012, V5001,
    V5015
  risk assessment V5015
  sea transport V5015
  shipping vessels V5015
  temperature V5015
  Vibration White Finger (VWF). *see*
    Vibration White Finger
  Whole-body vibration, *see* Whole-body
    vibration
  young workers V5016

VIBRATION WHITE FINGER (VRF)
  coal miners V5003, V5012
  common law, action at V5012
  damages V5003, V5012, V5013
  definition V5011
  dosage V5012
  guidance V5012
  occupational disease, as V5011–V5014
  reporting V5011
  stages of V5013
  Taylor-Pelmear Scale V5013
  warnings, wording of V5012

VICARIOUS LIABILITY IN22
  contractors E13027
  definition E13027
  harassment H1710–H1712
    course of employment H1711
    employer's defence H1713
    employer's liability, narrowing H1715
    liability of employees H1714
  indemnities E13027
  insurance E13002
  public liability insurance P9512
  workplace violence V8013

VICTIMISATION
  employees' protection E16501
  Equality Act 2010 E16508
  protected acts E16508

VICTIM SUPPORT
  SERVICES V8140–V8141
VIOLATION OF DIGNITY
  harassment H1701–H1702, H1708
  purpose or effect H1707
VIOLENCE IN THE
  WORKPLACE V8001–V8144
  access control V8088
  age groups V8002
  agencies, involvement of other V8032
  alarms V8090
  anti-social behaviour orders V8140
  assaults V8002, V8011, V8131–V8133
  batons V8095
  body armour V8093
  CCTV V8089
  checklist V8109, V8128
  civil law proceedings V8012, V8140
  codes of practice V8044, V8072, V8129
  collaboration V8019, V8024
  communication V8078, V8121, V8122
    external V8039
    internal V8038
    process of V8063
    strategy V8034, V8037
  compensation V8004, V8012, V8013,
    V8015
  constructive dismissal V8010
  consultation V8008, V8024, V8051
  contract law V8010
  co-ordinator, role of workplace
    violence V8062
  corporate homicide V8008B
  corporate manslaughter V8008B
  court case V8140–V8141
  Crime Survey V8002
  criminal law V8011
  data, analysis of V8023
  data protection, warning flags and V8081
  definition of V8001, V8017
  delegation V8110
  designing the system V8058–V8064,
    V8079, V8109
  disclosure V8071, V8121
  distance learning V8102
  duty of care V8004–V8005
  emergencies
    post-incident support V8135
    response belts V8094
  employee representatives V8008
  employees visiting individuals in their
    homes V8015
  enforcement V8004
  environment V8123
  equipment V8123

Index

VIOLENCE IN THE WORKPLACE – *cont.*
evidence, preserving V8121
experts V8139
external providers of training V8113, V8114
feedback V8068
focus groups V8022
forms
   design of report V8059
   post-incident considerations V8137
handcuffs V8094
harassment V8011
health and safety V8001–V8002, V8005–V8008
Health and Safety Executive V8001–V8002, V8005
health care workers V8077
identification of persons who may be harmed V8052
identification of risks V8051
implementation V8036–V8041, V8064–V8068, V8108–V8109
improvements V8127
incidents
   key scenarios V8117
   leadership V8120
   management V8116–V8128
   response teams V8119
   response to V8031
   roles and responsibilities V8118–V8120
information V8008, V8020, V8061, V8068, V8141
informing and consulting employees V8008
injunctions V8140
internal training providers V8111–V8112, V8114
interviews V8022
intruder alarms V8090
investigation process V8060, V8129, V8140–V8141
job description V8051
layout V8079
legislation V8004–V8016
lone workers V8092
long-term reactions V8133
maintenance V8033
managers
   involvement V8067
   post-incident support V8135
   training V8115
Maybo Risk Management Model V8144
measurable outcomes V8069
medium-term reactions V8132
mobile phones V8092

VIOLENCE IN THE WORKPLACE – *cont.*
monitoring and evaluating the solution V8110–V8114
nature and extent of problem V8002
negligence V8012
notification V8007
observation V8022
paging V8091
panic alarms V8090
personal alarms V8090
physical measures V8087–V8095
police investigation process V8140–V8141
policy
   affected by, persons V8028
   content of V8025
   co-ordination of V8043
   designing the V8024–V8035
   development of V8016–V8043
   evaluation of V8042
   group, policy development V8024
   implementation of V8016–V8043
   monitoring and evaluating V8042
   new systems, procedures, practices and equipment V8040
   purpose V8026
   review and revision V8035, V8043
   scope and definitions V8027
post-incident management V8129–V8141
   common reactions V8131–V8133
   effect on people V8130
   managers, role of V8136
   normal, return to V8134
   report forms V8137
   support, providing V8135–V8139
   support services V8138
   worst, preparing for the V8139
prevention programmes V8001
proactive service delivery V8080
problem recognition V8017
procedures V8121
prosecution of employers V8015
protective vests V8093
public, contact with the V8001
public address systems V8091
questionnaires V8022
queuing and queue jumping V8075, V8079
racist incidents V8011, V8140
radios, paging and public address systems V8091
receptions V8088
records, disclosure of V8071
rehearsals V8126
remedies V8004
reporting A3008, V8029, V8059, V8137

Ind-107

VIOLENCE IN THE WORKPLACE – *cont.*
  research V8112
  response teams V8119
  restraining orders V8140
  restraints V8094
  reviews V8127
    conducting V8018–V8023
    process V8029
  risk assessment V8001, V8006, V8021, V8029, V8044–V8056
    contents V8050–V8055
    continuous review V8057
    dynamic V8047
    generic V8045
    methods of V8049
    persons who should carry out V8048
    pre-planned events V8046
    rating of risks V8053
    reduction measures V8054–V8055
    restraints V8094
    specific activities and tasks performed V8053
    suitable and sufficient V8044
    systems and procedures V8077
  risk management V8001, V8129, V8141
  risk reduction V8030, V8072–V8073
  safe practice guidance V8076
  sanctions V8004, V8032, V8124
  security personnel V8083–V8086
    procedures V8086
    selection V8084
    training V8085
  security procedures V8086
  security responses V8082–V8095
  self-defence V8015
  service delivery V8074–V8081
  short-term reactions V8131
  soft skills V8103
  staffing levels V8075
  standards V8044
  statistics V8002
  support
    court case, during V8140–V8141
    providing V8135
    services V8138, V8140
    sources of V8142
  systems and procedures V8077
  testing V8102–V8107, V8107, V8126
  threats V8002
  training V8030, V8096–V8115, V8125
    checklist V8101
    core content V8104–V8105
    design V8102–V8107
    development model V8097

VIOLENCE IN THE WORKPLACE – *cont.*
  training – *cont.*
    external providers V8113, V8114
    identification of needs V8099
    internal providers V8111–V8112, V8114
    learning outcomes, defining V8100
    managers V8115
    physical intervention training V8103, V8105
    reviewing the content V8106
    soft skills V8103–V8104
    technology. V8102
    testing V8102–V8107, V8107
    Training Needs Analysis V8098–V8101
  under-reporting V8070
  use of force, self-defence and V8015
  vests V8093
  vicarious liability V8013
  victim support services V8140–V8141
  warnings V8009, V8081, V8124
  withdrawing services V8014
  working practice V8075
VIRAL HEPATITIS H2103
VISITORS
  accident reporting A3001
  activities to be carried out C6519
  areas to be visited C6519
  capabilities of visitors C6519
  community health and safety C6518–C6520
  construction and building operations, and C8018
  disaster and emergency planning D6060–D6062
  employers' duties to C0016–C0018, R3002, R3032
  fire prevention and control F5073
  health and safety policy statement S7006
  informal individual visits C6519
  information and instruction C6520
  managing health and safety M2015
  occupiers' liability C6508
  personal protective equipment C6520
  precautions C6520
  restricted activities C6520
  restricted areas and routes C6520
  schoolchildren and students C6519
VISUAL DISPLAY UNITS D8201–D8223
  agency workers D8202
  assessment approach D8212
  assessment review D8223
  breaks D8206
  causes of problems D8215
  daily work routine of users D8206

Index

VISUAL DISPLAY UNITS – *cont.*
definitions D8202
discomfort glare L5014
discussions with users and operators D8215
display screen equipment HSE guidance E17064
ergonomic approach E17044, E17052, E17064
ergonomic checklist D8212
exclusions D8203
eye diseases and disorders O1033, O1036
eyesight tests D8201, D8207, IN21
handheld devices D8221A
homeworkers D8202, D8213
hot desking D8219
information, provision of D8209
laptops D8220
lighting L5002, L5003
mobile phones D8220
new or expectant mothers V12003–V12005, V12023
pointing devices D8221
portable DSE D8220, D8221A
posture D8214
radiation exposure from R3027
recommendations, review and implementation of D8222
records D8216A
reflected glare L5014
Regulations D8201–D8213, IN21
risk assessment D8204, D8210–D8216A, R3010, R3018, R3027, R3036
  after the D8222, D8223
  approach to D8212
  checklist D8211–D8212, D8216A
  discussions with users and operators D8215
  homeworkers D8213
  observations at the D8214
  persons carrying out the D8210
  re-assessments D8223
  records D8216A
  reduction of risk D8216
  reviews D8222, D8223
  self-assessment, guidance on D8217
  teleworkers D8213
  users, identification of D8211
shared workstations D8219
smart phones D8221A
software tools D8206
special situations D8218–D8221A
teleworkers D8202, D8213
training D8208
users, definition of D8208

VISUAL DISPLAY UNITS – *cont.*
vision D8214
vulnerable persons V12003–V12005
workstations, requirements for D8205

VOCATIONAL QUALIFICATIONS (VQ) T7007
Europe E18018–E18019
Vocationally Related Qualifications (VRQ) T7008

VOLATILE ORGANIC COMPOUNDS
Protocol, as to E5026

VOLENTI NON FIT INJURIA
children O3017
defence to negligence, employer's IN22
occupiers' liability claims O3017

VOLUNTARY SECTOR
activities that volunteers can be involved in C6521
communication C6523
community health and safety C6521–C6523
corporate manslaughter C9002
emergencies C6522
employers' liability insurance E13016, E13035
entry into client's homes C6522
equipment C6522
fire C6522
information C6523
insurers key principles document E13035
large companies IN05
manual handling C6522
monitoring C6523
one-off volunteers C6521
performance IN05
precautions C6522
regular volunteers C6521
restrictions C6523
risk assessment C6522
supervision and support C6523
training C6523
travel C6522
work environment C6522

VULNERABLE WORKERS
see also Children and young people, Disabled persons, Lone workers
construction V12003–V12005
control of substances hazardous to health V12003–V12005
display screen equipment V12003–V12005
employers' duties V12001
Health and Safety at Work etc. Act 1974 V12001
inexperienced workers R3028, V12038–V12039

VULNERABLE WORKERS – *cont.*
legislation V12002–V12005
Management of Health and Safety at Work Regulations 1999 V12001, V12003
manual handling V12003–V12005
new or expectant mothers V12001–V12039
occupational health and safety O1005, O1008
personal protective equipment V12003–V12005
risk assessment R3028, V12001
self-employed V12001

W

WAGES
deductions from E14014
suspension from work E14016
WALES
Food Standards Agency F9005
Natural Resources Wales E5044, E16005
WAREHOUSES
fire protection F5061.1
WARNINGS
see also Warning Signs
ACAS Code E14014
children and young people O3010
disaster and emergency planning D6007–D6008
emergencies D0925
fire prevention and control F5030.1
formal M1529
hand-arm vibration syndrome V5020–V5021
occupiers' liability O3010, O3017
offshore operations O7028
product liability P9034, P9544
sickness absence M1529
swimming O3017
vibration white finger V5012
violence in the workplace V8009, V8081, V8124
WARNING SIGNS W11038
see also Safety Signs
dangerous substances D0921, D0933
entry, on D0921
explosions D0921
occupiers' liability and O3006, O3009, O3010, O3016
WASHING FACILITIES
asbestos A5041

WASHING FACILITIES – *cont.*
construction and building operations, and C8002–C8017
generally W11021
WASTE DISPOSAL
asbestos A5027
dangerous substances D0933
food, in respect of F9038, F9045
incineration, EC Directive E5026
WASTE MANAGEMENT
construction operations C8020
WATER
see also Legionnaires' disease
atmospheric pollution and E5015
facilities management F3026
fire extinguishers F5011
fire prevention and control F5005–F5006
food safety F9007
quality F3026
staff welfare L3035
trespassers C6513
WATER EFFICIENCY
Building Regulations W11044
WATER SUPPLY
food hygiene F9038, F9045
WATER SYSTEM
cleaning C6510
WEATHER
construction and building operations, and C8002–C8017
risk assessment and R3034
WEBINAR
facilities management industry F3030
WEBSITE
Department for Environment Food and Rural Affairs (DEFRA) E5047
Environment Agency E5047
European Commission E5047
Scottish Executive E5047
Sustainable Workplace E16011
WEFARE BENEFITS
system C6001
Welfare Reform Act 2012 C6001
WEIGHTS
manual handling M3003, M3013
WELFARE FACILITIES
clothing accommodation W11025
construction and building operations, and C8002–C8017
current specific statutory requirements W11018–W11029
drinking water, provision of adequate supply W11024
facilities for changing clothes W11026
minimum number of facilities W11021

WELFARE FACILITIES – *cont.*
  occupier's duties
    sanitary conveniences W11019
    washing facilities W11020
  particularly dirty work, wash stations required W11022
  rest and eating facilities W11027
  temporary work sites W11023
WELLS
  accident reporting A3005
  offshore operations O7026
WHISTLE-BLOWING
  compensation E14010
  employment protection E14011.1
  separate procedure E14006
WHITE FINGER
  vibration induced IN22
WHOLE-BODY VIBRATION
  advice V5023–V5028
  back pain V5023–V5024, V5028
  causes O1028
  daily exposure action value V5015
  detection of O1029
  driving V5023–V5026
  effects of V5005
  exposure assessments V5015
  guidance V5029
  health surveillance O1029, O1030, V5028
  HSE guidance V5029
  incidence of V5023
  maintenance O1031
  management O1030
  manual handling V5024
  occupational diseases O1027–O1031
  off-road vehicles V5024
  Physical Agents (Vibration) Directive V5015
  Physical Hazards (Vibration) Directive O1031
  record-keeping V5025
  risk assessment O1030, O1031, V5015, V5024–V5025, V5027
  risk reduction V5026
  symptoms O1027
WIND GUSTS OF M3006 M3014
WINDOWS AND WINDOW CLEANING
  general safety requirements W11001, W11014
  occupiers' liability O3014
  position and use of W11012–W11013
WIRING REGULATIONS E3003
WITNESSES
  accident investigations A3031
  interviewing A3031

WOMEN WORKERS
  breast feeding mothers E14024–E14025, E14031, R3017, R3026, R3032
  Control of Electromagnetic Radiation at Work Regulations 2016 E14032
  dismissal on medical grounds IN22
  generally E14023
  general statutory requirements E14023
  HGV driving E14023
  hours of work E14023
  ionising radiation R1019
  lead processes E14002
  maternity leave E14028
  new mothers E14024–E14026
  night work E14002, E14026
  notification of pregnancy, birth or breastfeeding E14027
  pregnant workers E14024, E14027, E14031
  risk assessment E14025, R3018
  sex discrimination legislation, effect of E14023
  statutory requirements E14002
  suspension from work
    remuneration E14030
    suitable alternative work E14029
  unfair dismissal, qualifying period for E14009
  unlawful sex discrimination E14023
WOOLF REFORMS
  protocols E13029
  rules E13029
WORK ENVIRONMENT
  see also Harassment
  intimidating H1708
  violence V8123
WORK EQUIPMENT
  definition M1005
  ergonomic design, *see* Ergonomic Design
  facilities management F3032
  heights, working at W9009
  machine safety, *see* Machinery Safety Regulations IN01
  risk assessment M1006
  self-propelled M1007
  training M1007
WORKERS
  Canada I2011
  definition of E14001, W11001
  employees, distinction E14001
  participation, EU E18013
  part-time E14002
  training, *see* Training and Competence
  USA I2007
  women, *see* Women Workers

WORK EXPERIENCE V12003–V12005, V12010, V12014–V12015
  HSE guidance C6518
WORKING ABROAD
  European Economic Area E18002
  resources E18022
WORKING AT HEIGHTS W9001–W9015
  access and egress W9004.1, W9008
  barriers W9004.1, W9008
  British Standards W9008, W9014
  causation W9001
  caving W9004.2
  climbers W9001, W9004.2
  community health and safety C6506, C6510
  competence W9005–W9008
  covers W9004.1
  danger areas W9011
  definition of W9004.1
  duties of persons at work W9013
  EC law W9001, W9009
  emergencies W9006
  employees' duties W9013
  equipment W9009
  facilities management F3025
  falls W9001, W9008
    arrest systems W9008, W9014
    fatal W9001
    harnesses W9008
    objects W9011
    protection from W9014
  fatal accidents W9001
  fragile surfaces W9010
  guard rails W9008
  guidance W9004.1, W9014
  Health and Safety Commission W9001
    Priority Programme W9001
  Health and Safety Executive
    ladders W9009.1
  inspection W9012
  instruction W9008
  ladders W9009.1
    guidance W9009.1
    portable W9009.1
    pre-check lists W9009.1
    step-ladders W9009.1
  level differences W9004.1, W9009
  lighting W9008
  Löfstedt review W9014
  mines W9001
  National Access and Scaffolding Confederation (NASC) W9008
  Northern Ireland W9003
  occupiers' liability O3014

WORKING AT HEIGHTS – *cont.*
  offshore operations O7072
  organisation W9005–W9008
  planning W9005–W9008
  public liability insurance P9523
  Regulations W9003–W9015
    application W9004
    commencement and scope W9003
    review W9014
  Regulatory Impact Assessment W9009
  rescue W9006
  research W9001
  restraints W9014
  risk assessment R3015, W9007–W9008
  risks, avoidance of W9005–W9008
  roofs W9008
  ropes W9014
  safe systems of work W9001
  scaffolding W9008
  sectors, list of relevant W9001
  self-employed W9004
  standards W9008, W9014
  toe boards W9008
  training W9007, W9008
  two metre rule W9008
  Work at Height Regulations 2005 IN21, W9014
    review of W9014
  work equipment W9009
  work platforms W9008
  work positioning systems W9014
WORKING AWAY
  public liability insurance P9528
WORKING TAX CREDIT
  disability element C6016
WORKING TIME
  48 hour working week W10010–W10013
  agency workers W10005
  air transport W10027, W10031
  annual leave, *see* Annual Leave
  breach of contract claims W10042
  breaks W10017
  case law W10001
  collective agreements
    definition W10007
    excluded sectors W10036
    role of W10006
  commencement of regulations W10001
  compensatory rest W10029, W10033, W10034, W10037, W10039
  continuity of service or production, need for W10034
  contracting out of regulations W10043
  daily rest periods W10014, W10015, W10029, W10034

Index

WORKING TIME – *cont.*
decision-making powers, persons with autonomous W10033
definitions W10001–W10009
  agency workers W10005
  collective agreements W10007
  mobile worker W10004A
  relevant agreement W10009
  worker W10004
  workforce agreement W10008
  working time W10003
detriment, protection against W10041
directives M1514, W10001, W10026, W10027–W10031
distant from each other, work and home W10034
doctors W10001, W10026
domestic workers W10032
DTI booklet W10001, W10017
EC Directives W10001, W10026, W10027–W10031
employment tribunals enforcement of provisions W10040–W10044
enforcement W10038–W10043
  detriment, protection against W10041
  employment
    tribunals W10040–W10044
  generally W10038
  health and safety offences W10039
  unfair dismissal W10041
exclusions
  collective agreements, by W10036
  directives to previously excluded sectors, application of W10027–W10031
  partial W10032–W10036
  shift workers W10035
  special categories of workers W10029, W10034
  workforce agreements W10036
flexible working W10024C
generally W10001
health assessments W10021, W10033, W10034
holidays, *see* Annual Leave
home working W10024C
impact of regulations W10044
inland waterways, workers on W10031
lake transport, workers on W10031
leave, *see* Annual Leave
  48 hour working
    week W10010–W10013
  agreement to exclude 48 hour working week W10001, W10013, W10038
  derogations W10012
  exceptions W10012

WORKING TIME – *cont.*
leave, *see* Annual Leave – *cont.*
  excluded days W10010
  reference period W10011
management W10033
mobile workers W10004A, W10027–W10030
  non-mobile activities and, combination of W10027
monotonous work W10018, W10033, W10034, W10039
new or expectant mothers V12018
night work W10034
  definitions W10019
  health assessment W10021, W10033, W10034
  length of W10020, W10029, W10039
  night time W10019
  night worker W10019
  offences W10039
  overtime hours W10020
  road transport W10029
  transfer to day work W10021
non-mobile workers W10027
notice W10024
offshore workers W10031
on-call workers W10001
purpose of provisions W10001
rail sector W10030, W10034
records W10025, W10033, W10034, W10039
Regulations IN21
rehabilitation R2009
relevant agreements
  definition W10009
  role of W10006
rest periods
  breaks J3002, J3032, W10017, W10034
  compensatory rest W10029, W10037, W10039
  daily J3002, J3032, W10014, W10015, W10029
  monotonous work W10018
  weekly J3002, J3032, W10016, W10029, W10034
return to work R2009
road transport workers W10026, W10027–W10029, W10038
  daily rest W10029
  directive on W10029
  maximum working week W10029
  night workers W10029
  non-mobile workers and W10028
scope of provisions W10001
seafarers W10026

Ind-113

Index

WORKING TIME – *cont.*
  sea fishing W10031
  security workers W10034
  shift workers W10035, W10037
  special categories of workers W10034
  stress and S11014, S11019
  structure of regulations W10001
  surges of activity, foreseeable W10034
  surveillance W10034
  transport workers W10026–W10031, W10034, W10038
  travelling time W10003
  unfair dismissal, protection against W10041
  unmeasured time W10033
  weekly rest periods W10016
  worker W10004
  workforce agreements
    definition W10008
    excluded sectors W10036
    role of W10006
  'working time' W10003
WORK PATTERNS F7008.1, IN06
WORKPLACE IN21
  buildings, *see* Building Work
  civil liability W11051
  cleanliness W11001, W11007
  clothing accommodation S11014, W11001, W11025, W11026
  condition of floor S11012, W11010
  construction site, exclusion W11002
  dangerous substances D0907
  definition of D0907, W11002
  dimensions/space W11008
  disabilities, people with W11029
  doors and gates W11001, W11016
  drinking water, provision of W11001, W11024
  eating facilities W11001, W11027
  enforcement of legislation as to, *see* Enforcement
  Equality Act 2010 W11043
  escalators and moving walkways W11001, W11017
  explosions D0907
  factories, *see* Factories
  falls from a height, safety measures W11001, W11011
  fire precautions, *see* Fire Precautions
  floor coverings W11001, W11010
  general cleanliness W11001, W11007
  general maintenance of W11003
  general requirements W11001
  hazardous substances in, *see* Hazardous Substances

WORKPLACE – *cont.*
  hot weather W11005
  inspections by safety representatives J3007
  layout S11013
  lighting, *see also* Lighting W11001, W11006
  loading bays, statutory duties A1009
  maintenance requirements W11001, W11003
  minimal standards W11001
  organisation of safe A1004
  overcrowding W11008
  parking areas, statutory duties A1009
  pedestrian access A1007
  pedestrian traffic routes W11001, W11015
  ramps W11010
  Regulations IN21
  remote work sites W11023
  rest facilities W11027
  safety signs, *see* Safety Signs
  sanitary conveniences S11014, W11001, W11019, W11021, W11023
  security C6511
  smoking ban C8002–C8017, F5018, H2102, W11028
  stability and solidity W11002
  staff facilities S11014
  steep slopes provided with handrail W11010
  stress, negligence claim M1524
  temperature controls W11001, W11005
  temporary work sites W11002
  traffic routes, *see* Traffic Routes
  vehicle traffic routes, *see also* Traffic Routes W11001
  ventilation of, *see also* Ventilation W11001
  washing facilities W11020
  welfare facilities W11018
  window cleaning, *see* Window Cleaning
  workroom dimensions W11001, W11008
  workstations W11001, W11009
WORKPLACE VIOLENCE *see* **Violence in the workplace**
WORK PLATFORMS W9008
WORK RELATED ROAD SAFETY
  assessing risks M2111
  business case for M2104
  checklist, HSE M2115
  employer's duties M2102
  evaluation, of M2117
  getting help M2124
  health and safety management systems M2105
  HSE guidance M2111, M2115

WORK RELATED ROAD SAFETY – *cont.*
legal responsibilities M2103
managing M2101
monitoring and evaluation M2117
  active monitoring M2118
  reactive monitoring M2119
    evaluation M2122
    gathering data M2120
    investigation M2121
Plan, Do, Check, Act approach M2105
road risk, control measures M2112
  control measures
  driver M2115
  implementing measures M2116
  journey task M2113
useful publications M2125
vehicle M2114

WORK-RELATED UPPER LIMB DISORDERS O1020.1
database C6058

WORKS COUNCILS J3002 J3033

WORKS MANAGER
responsibilities of S7005

WORKS RULES
incorporation into contract of employment E14007, E14008

WORKSTATIONS
adequate freedom of movement W11008
children and people V12016
design of, *see* Ergonomic Design
musculoskeletal disorders, *see* Musculoskeletal Disorders
rest breaks E17040
sedentary comfort W11008
suitable seating W11008
user trial E17056

WORK VEHICLES *see also* **Vehicles**
access to A1015
drivers of A1013

WORK VEHICLES *see also* **Vehicles** – *cont.*
loading A1018
maintenance A1012
provision of A1011
reversing A1017
statutory requirements relating to A1009
tipping A1019
training of drivers A1013
unauthorised lifts, giving of A1020
unloading A1018

WORLD HEALTH ORGANISATION
air quality guidelines E5064–E5065

WRITTEN POLICY STATEMENT *see* Health and Safety Policy Statement

WRONGFUL DISMISSAL E14017; E14018

Y

YOUNG PERSON
adventure activities C6508
Adventure Activities Licensing Regulations 2004 C6508
children and, *see* Children and Young People
definition V12003
health and safety policy and S7008
risk assessment E16516
training needs M2016
vibration V5016
work experience C6518, V12014–V12015

YOUNG WORKER
definition V12003

Z

ZONING D0918 D0921 D0919